THE RETINA AND ITS DISORDERS

THE RETINA AND ITS DISORDERS

EDITORS

JOSEPH C. BESHARSE
Department of Cell Biology,
Neurobiology and Anatomy
Medical College of Wisconsin
Milwaukee, WI
USA

DEAN BOK
Department of Neurobiology and Jules Stein Eye Institute
David Geffen School of Medicine at UCLA
University of California
Los Angeles, CA
USA

AMSTERDAM • BOSTON • HEIDELBERG • LONDON • NEW YORK • OXFORD
PARIS • SAN DIEGO • SAN FRANCISCO • SINGAPORE • SYDNEY • TOKYO
Academic Press is an imprint of Elsevier

ACADEMIC
PRESS

Academic Press is an imprint of Elsevier
525 B Street, Suite 1900, San Diego, CA 92101-4495, USA
The Boulevard, Langford Lane, Kidlington, Oxford OX5 1GB, UK

Notice
No responsibility is assumed by the publisher for any injury and/or damage to persons or property as a matter of products liability, negligence or otherwise, or from any use or operation of any methods, products, instructions or ideas contained in the material herein. Because of rapid advances in the medical sciences, in particular, independent verification of diagnoses and drug dosages should be made

British Library Cataloguing in Publication Data
A catalogue record for this book is available from the British Library

Library of Congress Cataloging-in-Publication Data
A catalog record for this book is available from the Library of Congress

ISBN: 978-0-12-382198-0

For information on all Academic Press publications
visit our website at www.elsevierdirect.com

CONTRIBUTORS

A P Adamis
University of Illinois, Chicago, IL, USA

J Adijanto
National Eye Institute, Bethesda, MD, USA

K R Alexander
University of Illinois at Chicago, Chicago, IL, USA

J Ambati
University of Kentucky, Lexington, KY, USA

B Anand-Apte
Cleveland Clinic, Cleveland, OH, USA

D H Anderson
University of California, Santa Barbara, CA, USA

R E Anderson
University of Oklahoma Health Sciences Center,
Oklahoma City, OK, USA

A C Arman
University of Southern California, Los Angeles,
CA, USA

V Y Arshavsky
Duke University, Durham, NC, USA

J D Ash
University of Oklahoma Health Sciences Center,
Oklahoma City, OK, USA

T J Bailey
University of Notre Dame, Notre Dame, IN, USA

J L Barbur
City University, London, UK

S Barnes
Dalhousie University, Halifax, NS, Canada

B-A Battelle
University of Florida, St. Augustine, FL, USA

M Bähr
University of Göttingen, Göttingen, Germany

J C Besharse
Medical College of Wisconsin, Milwaukee, WI, USA

P Bex
Schepens Eye Research Institute, Boston, MA, USA

P Bovolenta
Instituto Cajal (CSIC) and CIBER de Enfermedades Raras
(CIBERER), Madrid, Spain

N C Brecha
UCLA School of Medicine, Los Angeles, CA, USA;
VAGLAHS, Los Angeles, CA, USA

R Bremner
University of Toronto, Toronto, ON, Canada

S E Brockerhoff
University of Washington, Seattle, WA, USA

N L Brown
Cincinnati Children's Research Foundation, Cincinnati,
OH, USA

J A Brzezinski, IV
University of Washington, Seattle, WA, USA

B Burnside
University of California, Berkeley, Berkeley, CA, USA

P D Calvert
SUNY Upstate Medical University, Syracuse, NY, USA

J Carroll
Medical College of Wisconsin, Milwaukee, WI, USA

G J Chader
USC School of Medicine, Los Angeles, CA, USA

T Chan-Ling
University of Sydney, Sydney, NSW, Australia

D G Charteris
Moorfields Eye Hospital, London, UK

C F Chicani
University of Southern California-Keck School of
Medicine, Los Angeles, CA, USA

T Cogliati
National Institutes of Health, Bethesda, MD, USA

N J Colley
University of Wisconsin, Madison, WI, USA

I Conte
Telethon Institute of Genetics and Medicine (TIGEM),
Naples, Italy

M F Cordeiro
UCL Institute of Ophthalmology, London, UK

S W Cousins
Duke Eye Center, Durham, NC, USA

K M Coxon
UCL Institute of Ophthalmology, London, UK

T W Cronin
University of Maryland Baltimore County, Baltimore, MD, USA

M D Crossland
UCL Institute of Ophthalmology/Moorfields Eye Hospital, London, UK

J Cunha-Vaz
AIBILI, Coimbra, Portugal

M Cwinn
Ottawa Hospital Research Institute, Ottawa, ON, Canada

P A D'Amore
Schepens Eye Research Institute, Boston, MA, USA

M del Pilar Gomez
Universidad Nacional de Colombia, Bogotá, Colombia

J B Demb
University of Michigan, Ann Arbor, MI, USA

D Deretic
University of New Mexico, Albuquerque, NM, USA

W Drexler
Medical University Vienna, Vienna, Austria

J Duggan
UCL Institute of Ophthalmology, London, UK

R Ehrlich
Indiana University, Indianapolis, IN, USA

K Ford
Schepens Eye Research Institute, Boston, MA, USA

D H Foster
University of Manchester, Manchester, UK

P J Francis
Oregon Health and Sciences University, Portland, OR, USA

R N Frank
Kresge Eye Institute, Wayne State University School of Medicine, Detroit, MI, USA

F D Frentiu
University of Queensland, St. Lucia, QLD, Australia

Y Fu
Department of Ophthalmology and Visual Sciences, University of Utah, Salt Lake City, UT, USA

I Glybina
Kresge Eye Institute, Wayne State University School of Medicine, Detroit, MI, USA

M B Gorin
Jules Stein Eye Institute, Los Angeles, CA, USA

M S Gregory
Schepens Eye Research Institute, Harvard Medical School, Boston, MA, USA

L Guo
UCL Institute of Ophthalmology, London, UK

E V Gurevich
Vanderbilt University, Nashville, TN, USA

V V Gurevich
Vanderbilt University, Nashville, TN, USA

A Ha
Ottawa Hospital Research Institute, Ottawa, ON, Canada

A Harris
Indiana University, Indianapolis, IN, USA

M E Hartnett
Moran Eye Center, University of Utah, Salt Lake City, UT, USA

S Haverkamp
Max-Planck-Institute for Brain Research, Frankfurt/Main, Germany

S S Hayreh
University of Iowa, Iowa City, IA, USA

C C Heikaus
University of Washington, Seattle, WA, USA

K Hein
University of Göttingen, Göttingen, Germany

D B Henson
University of Manchester, Manchester, UK

A A Hirano
UCLA School of Medicine, Los Angeles, CA, USA

P Hiscott
University of Liverpool, Liverpool, UK; Royal Liverpool University Hospital, Liverpool, UK

J G Hollyfield
Cleveland Clinic, Cleveland, OH, USA

A Horsager
USC School of Medicine, Los Angeles, CA, USA

M S Humayun
USC School of Medicine, Los Angeles, CA, USA

D R Hyde
University of Notre Dame, Notre Dame, IN, USA

C Insinna
Medical College of Wisconsin, Milwaukee, WI, USA

P M Iuvone
Emory University School of Medicine, Atlanta, GA, USA

K Jian
Texas A&M University, College Station, TX, USA

L V Johnson
University of California, Santa Barbara, CA, USA

B Katz
Hebrew University, Jerusalem, Israel

V J Kefalov
Washington University School of Medicine, Saint Louis, MO, USA

M R Kesen
Duke Eye Center, Durham, NC, USA

C King-Smith
Saint Joseph's University, Philadelphia, PA, USA

H J Klassen
University of California, Irvine, Orange, CA, USA

M E Kleinman
University of Kentucky, Lexington, KY, USA

G Y-P Ko
Texas A&M University, College Station, TX, USA

M L Ko
Texas A&M University, College Station, TX, USA

T D Lamb
The Australian National University, Canberra, ACT, Australia

D C Lee
University of British Columbia, Vancouver, BC, Canada

S Lee
Yale University School of Medicine, New Haven, CT, USA

A A Lewis
University of Washington, Seattle, WA, USA

R Li
National Eye Institute, Bethesda, MD, USA

A Maminishkis
National Eye Institute, Bethesda, MD, USA

N A Mandal
University of Oklahoma Health Sciences Center, Oklahoma City, OK, USA

S C Mangel
The Ohio State University College of Medicine, Columbus, OH, USA

M B Manookin
University of Michigan, Ann Arbor, MI, USA

R E Marc
University of Utah, Salt Lake City, UT, USA

R Marco-Ferreres
Instituto Cajal (CSIC) and CIBER de Enfermedades Raras (CIBERER), Madrid, Spain

D G McMahon
Vanderbilt University, Nashville, TN, USA

B McNeill
Ottawa Hospital Research Institute, Ottawa, ON, Canada

S Meredith
Cambridge University Hospitals NHS Foundation Trust, Cambridge, UK

S S Miller
National Eye Institute, Bethesda, MD, USA

B Minke
Hebrew University, Jerusalem, Israel

J Mitchell
University of Toronto, Toronto, ON, Canada

T A Münch
University of Tübingen, Tübingen, Germany

J L Morgan
University of Washington, Seattle, WA, USA

O L Moritz
University of British Columbia, Vancouver, BC, Canada

A M Moss
Indiana University, Indianapolis, IN, USA

E Nasi
Universidad Nacional de Colombia, Bogotá, Colombia

I Nasonkin
National Institutes of Health, Bethesda, MD, USA

R W Nickells
University of Wisconsin, Madison, WI, USA

S Nusinowitz
UCLA School of Medicine, Los Angeles, CA, USA

T H Oakley
University of California, Santa Barbara, Santa Barbara, CA, USA

M Pacal
University of Toronto, Toronto, ON, Canada

K Palczewski
Case Western Reserve University, Cleveland, OH, USA

R Payne
University of Maryland, College Park, MD, USA

L Peichl
Max Planck Institute for Brain Research, Frankfurt am Main, Germany

M E Pennesi
Oregon Health and Sciences University, Portland, OR, USA

D C Plachetzki
University of California, Santa Barbara, Santa Barbara, CA, USA

I Provencio
University of Virginia, Charlottesville, VA, USA

L P Pulagam
Case Western Reserve University, Cleveland, OH, USA

E Pyza
Jagiellonian University, Kraków, Poland

Y D Ramkissoon
Royal Hallamshire Hospital, Sheffield, UK

I D Raymond
UCLA School of Medicine, Los Angeles, CA, USA

T A Reh
University of Washington, Seattle, WA, USA

C P Ribelayga
The Ohio State University College of Medicine, Columbus, OH, USA

L J Rizzolo
Yale University School of Medicine, New Haven, CT, USA

A A Sadun
University of Southern California-Keck School of Medicine, Los Angeles, CA, USA

A P Sampath
University of Southern California, Los Angeles, CA, USA

L Shi
Texas A&M University, College Station, TX, USA

W E Smiddy
Bascom Palmer Eye Institute, Miami, FL, USA

R G Smith
University of Pennsylvania, Philadelphia, PA, USA

M Snead
Cambridge University Hospitals NHS Foundation Trust, Cambridge, UK

N C Steinle
University of Kentucky, Lexington, KY, USA

S L Stella, Jr.
UCLA School of Medicine, Los Angeles, CA, USA

A Stockman
UCL Institute of Ophthalmology, London, UK

E Strettoi
Istituto di Neuroscienze CNR, Pisa, Italy

E E Sutter
The Smith-Kettlewell Eye Research Institute, San Francisco, CA, USA

W Swardfager
University of Toronto, Toronto, ON, Canada

A Swaroop
National Institutes of Health, Bethesda, MD, USA

D M Tait
Medical College of Wisconsin, Milwaukee, WI, USA

W B Thoreson
University of Nebraska Medical Center, Omaha, NE, USA

S I Tomarev
National Institutes of Health, Bethesda, MD, USA

G H Travis
UCLA School of Medicine, Los Angeles, CA, USA

B A Tucker
Schepens Eye Research Institute, Harvard Medical School, Boston, MA, USA

R L Ufret-Vincenty
University of Texas Southwestern Medical Center at Dallas, Dallas, TX, USA

S A Vinores
Johns Hopkins University School of Medicine, Baltimore, MD, USA

V A Wallace
Ottawa Hospital Research Institute, Ottawa, ON, Canada

Y Wang
University of Maryland, College Park, MD, USA

J Weiland
USC School of Medicine, Los Angeles, CA, USA

R G Weleber
Oregon Health and Sciences University, Portland, OR, USA

F S Werblin
UC Berkeley, Berkeley, CA, USA

A F Wiechmann
University of Oklahoma College of Medicine, Oklahoma City, OK, USA

D R Williams
University of Rochester, Rochester, NY, USA

D S Williams
UCLA School of Medicine, Los Angeles, CA, USA

P R Williams
University of Washington, Seattle, WA, USA

D Wong
University of Hong Kong, Hong Kong, People's Republic of China

R O L Wong
University of Washington, Seattle, WA, USA

S C Wong
Moorfields Eye Hospital, London, UK

S M Wu
Baylor College of Medicine, Houston, TX, USA

S Yazulla
Stony Brook University, Stony Brook, NY, USA

L Yin
University of Rochester, Rochester, NY, USA

M J Young
Schepens Eye Research Institute, Harvard Medical
School, Boston, MA, USA

D-Q Zhang
Vanderbilt University, Nashville, TN, USA

Z J Zhou
Yale University School of Medicine, New Haven,
CT, USA

M E Zuber
SUNY Upstate Medical University, Syracuse, NY, USA

PREFACE

During our recent efforts (2008–2009) at identifying authors and subsequently editing the retina section for Elsevier's *Encyclopedia of the Eye*, we were greatly impressed by the overall quality and depth of the chapters produced by our colleagues. What seemed at first to be a very large undertaking in fact became a pleasure when time had come to read and edit the chapters our colleagues had written. They had little incentive in this effort other than to expose their own research area to the broader community. In retrospect, we cannot thank them enough for what we now regard as a major service to students, postdoctoral fellows, residents, optometrists, and ophthalmologists. It was with this sentiment that we did not hesitate when we were asked to organize and reassemble their effort as a separate derivative volume for the retina community.

The *Retina and its Disorders* provides a readily accessible and comprehensive compendium on the retina in health and disease. Coverage extends from embryology and early patterning to age-related macular degeneration, a complex trait disease that now affects about 30% of individuals over the age of 75 in industrialized countries. Included are lucid descriptions of the anatomy, physiology, cell biology, neural pathways, and pharmacology of the retina. In addition, key experts cover its vasculature as well as state-of-the-art noninvasive testing of structure and function. Comprised of 111 chapters selected from Elsevier's *Encyclopedia of the Eye*, this volume provides a valuable desk reference for biomedical scientists, ophthalmologists, optometrists, and psychologists.

Joseph C. Besharse
Dean Bok
Editors

CONTENTS

Acuity

M D Crossland, UCL Institute of Ophthalmology/Moorfields Eye Hospital, London, UK

Glossary

Cycles per degree – The number of complete phases of a grating (e.g., the distance between the center of a white bar and the center of the next bright bar in a square-wave grating; or the distance between two adjacent areas of maximum brightness on a sine-wave grating) contained in 1° of visual angle.
Minimum angle of resolution – The size of the angle subtended at the eye of the smallest feature which can be reliably identified on an optotype.
Minute of arc – One-sixtieth of a degree.
Optotype – A letter, symbol, or other figure presented at a controlled size to measure vision.
Visual angle – The angle, which a viewed object subtends at the eye.

Detection and Resolution Acuity

Visual acuity can be defined in two broad ways. Detection acuity is measured by determining the size of the smallest object which can be reliably seen (is there a circle on the first or second screen?). Detection can be elicited reliably with targets, which subtend an angle at the eye as small as 1 s of arc ($1/3600°$). Even a small point of light will stimulate several photoreceptors due to the point-spread function of the eye: that is, the way in which light is diffracted through the eye's optics (**Figure 1(a)**).

Tests that require the identification of a target are a measurement of resolution acuity. These tests frequently involve identifying a letter or reporting an object's orientation (what direction is this letter C facing?). Acuity for these tests depends on the separation of the target features: if they are too close, the point-spread function from each element will overlap and they will not be identified (**Figure 1(b)**). The smallest separation of the elements required for identification of the target (**Figure 1(c)**) is known as the minimum angle of resolution (MAR). For an adult observer with good vision, a typical MAR for a centrally presented, high-contrast target can be as good as 30 s of arc ($1/120°$). **Figure 2** shows the feature critical for the MAR for some commonly used tests of visual acuity.

Measurement of Visual Acuity

Visual acuity tests have been used for millennia: the ancient Egyptians are reported to have used discrimination of the twin stars of Mizar and Alcor as a measurement of vision. The most familiar clinical test of visual acuity, the Snellen chart, was introduced in 1862, and is still widely used today.

Detection acuity is often measured psychophysically by means of a temporal two-alternative forced-choice experiment (did the light appear in the first or the second interval?). Detection acuity is rarely measured clinically.

In psychophysical experiments of the visual system, resolution acuity is commonly measured by asking observers to report the orientation of a grating with variable separation between each dark and light bar (**Figure 2(b)**). In clinical practice, gratings are rarely used, with the exception of forced-choice preferential looking tests in preverbal children. These tests consist of a uniform gray field with an isoluminant grating toward one side of the chart (**Figure 3(a)**). In a featureless room, the test is presented to the child and the clinician observes whether the child looks toward the grating. The finest grating toward which the child repeatedly looks is recorded as the visual acuity.

For cooperative patients, optotypes are more often used to measure clinical resolution acuity. The Landolt C (**Figure 2(c)**) is the standard to which letter visual acuity tests are compared. This target consists of a ring of fixed width with a gap, of height equal to the stroke width, at the top, left, right, or bottom of the circle. The observer is asked to report the position of this gap. The smallest gap whose position can be reliably reported is equivalent to the MAR.

The National Academy of Sciences standard for visual acuity measurement advocates the presentation of 10 optotypes, of equivalent difficulty to the Landolt C, at each acuity size. The horizontal spacing between each optotype should be at least one character width, and vertical spacing between lines should be 1–2 times the height of the larger optotypes. It suggests that the number of characters on each line should be equal, and that the size difference between consecutive lines is 0.1 log units: in other words, for each target size, the next line should be approximately 1.26 times smaller.

The Snellen chart (**Figure 3(b)**) does not meet these recommendations: the number of letters per line and step

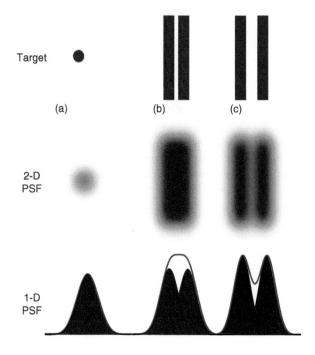

Figure 1 Schematic illustration of the point-spread function of three visual targets: (a) a point target; (b) two adjacent lines, too close to be resolved; and (c) two adjacent lines, with sufficient separation to be resolved. Middle row: two-dimensional representation of the target point-spread function; bottom row: one-dimensional representation of the point-spread function; and red line indicates the sum of energy incident on the retina. PSF, point-spread function.

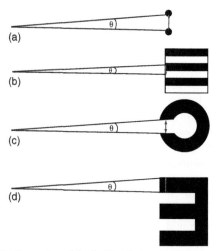

Figure 2 Examples of the limiting feature for four commonly used resolution tasks: (a) two-point discrimination task; (b) grating; (c) Landolt C; and (d) Sloan letter E (note that white gap size is equal in width to black bar elements).

size between the lines are variable, as is the horizontal and vertical spacing on the chart. There is also a marked difference in the legibility of different letters on the Snellen chart: a W, for example, has far less separation

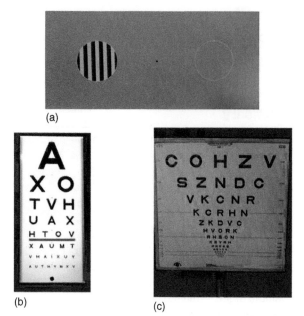

Figure 3 (a) A forced-choice preferential looking test consisting of a grating against an isoluminant background. Note the peephole in the center for the clinician to observe the child's visual behavior; (b) the Snellen chart; and (c) The ETDRS chart. ETDRS, Early treatment of diabetic retinopathy study.

between the elements of the letter and is more difficult to identify than a letter L. In the 1950s, Sloan suggested the use of 10 letters with a selection of vertical, horizontal, oblique, and round strokes which are each about as legible as a Landolt C. These Sloan letters are C, D, H, K, N, O, R, S, V, and Z. Each of the Sloan letters has a stroke width of the MAR and has a total height and width of five times the MAR.

The Bailey–Lovie chart, introduced before the recommendations of the National Academy of Sciences, conforms to most of these requirements, although it only has five letters per line. Further, the letters on the Bailey–Lovie chart are taller than they are wide: their height-to-width ratio is 5:4 and they are selected from the British Standards set of letters (D, E, F, H, N, P, U, V, R, and Z). The ETDRS chart (**Figure 3(c)**), developed for the early treatment of diabetic retinopathy study (ETDRS), is similar in design but does use the recommended 5 × 5 Sloan letters.

A criterion of 7/10 letters being read correctly for a line to be marked as seen was suggested by the National Academy of Sciences. This threshold reduces the chance of the line being scored correctly by chance (by a blind observer) to around 1 in 9 000 000. On a chart with five letters per line, recording a visual acuity where four of the five letters are read correctly equates to a chance success rate of 1 in 46 000. There is a theoretical advantage if the observer knows there are only 10 letters which can be presented on the chart: if an observer guesses from all 26 letters rather than the ten Sloan letters, the probability of the observer getting four out of five letters correct reduces to about 1 in 100 000.

Test–retest variability of the Snellen chart is around ±0.3 logMAR, while the ETDRS chart has far better repeatability (test–retest variability ~0.1–0.2 logMAR). Despite the many limitations of the Snellen chart, it is still widely used in clinical practice. While this is likely to be largely due to clinicians' familiarity with the Snellen chart, there is also a perception that Snellen acuity measurement is quicker than that on the Bailey–Lovie or ETDRS charts.

Various modified versions of the ETDRS chart exist: for example, a version with an altered letter set (A, B, E, H, N, O, P, T, X, and Y) has been developed for use by readers of most European languages, including those based on Cyrillic or Hellenic alphabets.

For observers unable to report letters on a sight chart, other frequently used optotypes include the tumbling E chart (formerly and less politically correctly known as the illiterate E chart), where a letter E is shown in each of four rotations; the HOTV chart, where only these four letters are used; symbols such as the Lea or Kay pictures; and simple shapes, such as the Cardiff card.

Reporting Visual Acuity

Clinicians have traditionally used Snellen fractions to record visual acuity, where the numerator is the test distance and the denominator the target size. The target size is expressed, counterintuitively, as the distance from which the target has an MAR of 1 min of arc. Therefore, a visual acuity of 6/6 indicates that from 6 m, letters with MAR 1-min arc are correctly identified, while a visual acuity of 3/36 indicates that from 3 m, the targets identified have a MAR of 1 min of arc when viewed from 36 m. The reciprocal of the Snellen fraction gives the visual acuity in MAR: so a visual acuity of 3/36 indicates a MAR of 12 min of arc.

In much of Europe, the Snellen fraction is reduced into a decimal fraction.

A further confusion with the Snellen system is that in countries not using the metric system, distances are expressed in feet rather than meters, with 20/20 being exactly equivalent to 6/6 but with a test distance of 20 ft rather than 6 m. Although Snellen recommended adoption of the metric system in 1875 and, in 1980, the US National Academy of Sciences favored adoption of a standard defined in meters, given the imminent adoption of the metric system, the feet system is still widely used in the USA, and among lay people in the UK.

The accepted standard for expressing visual acuity in clinical research, and increasingly in clinical practice, is to use the base 10 logarithm of the MAR (logMAR), such that 0.0 logMAR is equivalent to 6/6 or 20/20, and 1.0 logMAR is the same as 6/60 or 20/200. **Table 1** gives approximately equivalent values in MAR, cycles per degree, Snellen fractions in meters and feet, decimal acuity, and logMAR for a range of visual acuities.

Optical and Neural Limits on Visual Acuity

Visual acuity is limited by many factors: the optics and refraction of the eye; the clarity of the optical media; the spacing and function of the retinal photoreceptors; the ratio of retinal ganglion cells to photoreceptors; and the resolution of the primary visual cortex and higher areas of visual processing.

Each diopter of myopia reduces visual acuity: a −1.00DS myope will typically have uncorrected visual acuity of around 0.5 logMAR (6/18; 20/60) and a two-diopter myope will have vision of around 0.8 logMAR on a distance test. Hypermetropia can often be relieved by accommodation in young people, but each diopter of hypermetropia

Table 1 Visual acuity conversion table[a]

MAR (min)	Cycles/ degree	Snellen (metric)	Snellen (feet)	Decimal	Log MAR
60	0.5	1/60	20/1200	0.017	1.8
20	1.5	3/60	20/400	0.05	1.3
10	3	6/60	20/200	0.1	1
6.3	4.7	6/36	20/120	0.17	0.8
4	7.5	6/24	20/80	0.25	0.6
3.2	9.4	6/18	20/60	0.33	0.5
2	15	6/12	20/40	0.5	0.3
1.6	18.8	6/9	20/30	0.67	0.2
1.3	23	6/7.5	20/25	0.8	0.1
1	30	6/6	20/20	1	0
0.83	36	6/5	20/17[b]	1.2	−0.1
0.67	44	6/4	20/13[b]	1.5	−0.2
0.5	60	6/3	20/10	2	−0.3
0.33	91	6/2	20/7	3	−0.4

[a]Each row contains approximately equivalent values of visual acuity. Log MAR values have been rounded to 1 decimal place.
[b]On US Snellen charts, these lines are 20/16 and 20/12 respectively.

beyond the accommodative ability of the eye will reduce visual acuity by a similar amount to an equivalent degree of myopia. Astigmatism, particularly where the meridia of astigmatism are oblique, will also reduce uncorrected vision significantly.

Other aberrations of the eye beyond defocus and astigmatism further limit visual acuity. Retinal image quality can be improved by viewing monochromatic stimuli (to reduce chromatic aberration) and by using a deformable mirror to correct coma, trefoil, and other higher-order aberrations of the eye. Under these ideal conditions, Williams and colleagues have shown that subjects are able to resolve gratings of up to 55 cycles per degree, equivalent to a visual acuity of approximately −0.30 logMAR (6/3; 20/10).

Assuming that an image is perfectly focused on the retina, the next limit on visual resolution is the spacing of the retinal photoreceptors. In order to detect a grating, alternate black and white bars must fall on adjacent photoreceptors. This theoretical limit of vision, known as the Nyquist limit, is equivalent to a grating with light to dark separation of $1/\sqrt{D}$, where D is the center-to-center separation of two photoreceptors. In the fovea, D is approximately 3 μm, equivalent to a visual angle of approximately 55 cycles per degree – almost identical to the value found by Williams. This confirms that in people with good vision, all of the limits on visual acuity are precortical. Amblyopia, where vision is reduced despite the absence of any eye disease, is dealt with elsewhere in the encyclopedia.

Visual Acuity across the Retina

Nonfoveal vision is limited by many elements. First, the eye's optics are not optimized for viewing off the visual axis, and peripheral vision is subject to greater aberration than central vision. Second, the size of photoreceptors increases and their density falls with increasing eccentricity. The number of photoreceptors per retinal ganglion cell also increases, from less than one photoreceptor per ganglion cell in the fovea to more than 20 photoreceptors per ganglion cell in the far periphery. The volume of visual cortex devoted to noncentral retina is also proportionally lower. It is unsurprising, therefore, that visual acuity falls quickly with increasing distance from the fovea (**Figure 4**). This is one reason for the severely reduced visual acuity of people with central vision loss from diseases such as age-related macular disease.

Visual Acuity over Life

Over the first year of life, visual acuity assessed by a preferential looking test appears to be reasonably stable

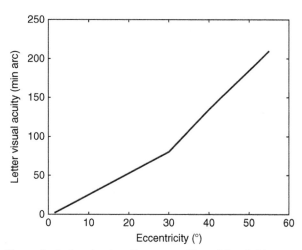

Figure 4 Letter visual acuity measured in peripheral vision as a function of degrees of eccentricity. Data from Anstis, S. M. (1974). Letter: A chart demonstrating variations in acuity with retinal position. *Vision Research* 14(7): 589–592.

at around 6 min of arc. Between a child's first and third birthday, visual acuity improves exponentially to reach 1 min of arc. A further small improvement in resolution ability to approximately 0.75 min of arc is achieved by age 5 years. In the absence of eye disease, this value remains relatively constant until the sixth decade. In a population-based study of nearly 5000 older adults, Klein found a decrease in visual acuity to a mean value of approximately 2 min of arc in those aged over 75 years. Of course, this reflects the age-related nature of many diseases which affect visual acuity, such as cataract, glaucoma, diabetic retinopathy, and age-related macular degeneration. **Figure 5** plots data from the studies of Mayer and Klein.

Visual Standards

In most countries, there is a visual-acuity requirement for car drivers. While the level and measurement technique varies between countries, the acuity limit is usually approximately 0.3 logMAR. Commercial airline pilots are required to have a binocular visual acuity of 0.0 logMAR.

Best corrected binocular visual acuity of 1.0 logMAR or poorer is used as a definition of low vision or partial sight in many countries, with acuity of worse than 1.3 logMAR being described as severe sight impairment.

Hyperacuity

Some visual tasks can be performed with a far greater degree of precision than would be suggested by the MAR. Alignment tasks such as Vernier discrimination (where the offset of one line with respect to another is detected, **Figure 6(a)**) can be performed with misalignment of less

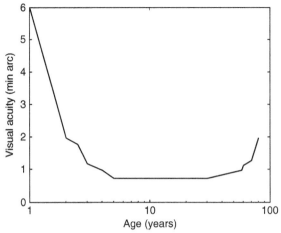

Figure 5 Variation in visual acuity over life. From Mayer, D. L. and Dobson, V. (1982). Visual acuity development in infants and young children, as assessed by operant preferential looking. Data from *Vision Research* 22(9): 1141–1151 and Klein, R., Klein, B. E., Linton, K. L., and De Mets, D. L. (1991). The beaver dam eye study: Visual acuity. *Ophthalmology* 98(8): 1310–1315.

(b)

(a)

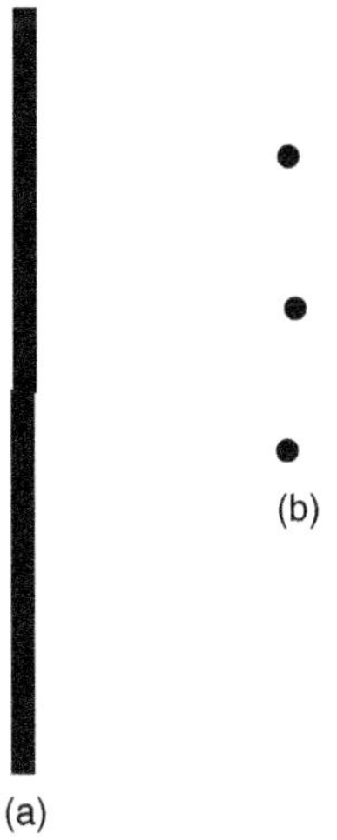

Figure 6 Examples of hyperacuity tasks. Misalignment of the lower element is easily visible. (a) Vernier alignment; (b) dot alignment: the offset of the middle dot with respect to the upper and lower dot is easily discerned.

than 5 s of arc – considerably less than the center-to-center spacing of a foveal photoreceptor. This is thought to be due to interpolation of the inputs of two or more adjacent neural elements.

Dynamic Visual Acuity

Throughout this article, visual acuity has been discussed for static targets. If the target is moved, central visual acuity decreases: the faster the target moves, the larger it

must be for it to be seen. If a target moves with velocity of $40°\,s^{-1}$, the MAR is increased to about 2 min of arc, while at $80°\,s^{-1}$, acuity is about 3 min of arc.

In peripheral vision, slow image motion (less than $10°\,s^{-1}$) slightly improves visual acuity for peripherally presented targets, perhaps because it breaks the phenomenon of Troxler fading.

Target motion at the retina can be induced by target movement, by eye motion, or by head motion. Many eye diseases, particularly those of the macula, are associated with poor fixation stability of the eye. This poor eye stability increases retinal image motion, and is significantly associated with poorer visual function. Small degrees of head motion do not significantly decrease visual acuity under normal conditions, but have a marked deleterious effect for subjects viewing through telescopic spectacles. Therefore, subjects with macular disease who have poor fixation stability and who view through telescopic low-vision aids have a marked impairment in their dynamic visual acuity.

See also: Chromatic Function of the Cones; Contrast Sensitivity; Photopic, Mesopic and Scotopic Vision and Changes in Visual Performance.

Further Reading

Anstis, S. M. (1974). Letter: A chart demonstrating variations in acuity with retinal position. *Vision Research* 14(7): 589–592.
Bailey, I. L. and Lovie, J. E. (1976). New design principles for visual acuity letter charts. *American Journal of Optometry and Physiological Optics* 53: 740–745.
Bennett, A. G. and Rabbetts, R. B. (eds.) (1989). Visual acuity and contrast sensitivity. In: *Clinical Visual Optics*, pp. 23–72. Oxford: Butterworth-Heinemann.
Brown, B. (1972). Resolution thresholds for moving targets at the fovea and in the peripheral retina. *Vision Research* 12(2): 293–304.
Committee on vision. (1980). Recommended standard procedures for the clinical measurement and specification of visual acuity. Report of working group 39. *Advances in Ophthalmology = Fortschritte der Augenheilkunde = Progres en Ophtalmologie* 41: 103–148. Assembly of Behavioral and Social Sciences, National Research Council, National Academy of Sciences, Washington, DC
Crossland, M. D., Culham, L. E., and Rubin, G. S. (2004). Fixation stability and reading speed in patients with newly developed macular disease. *Ophthalmic and Physiological Optics* 24: 327–333.
Demer, J. L. and Amjadi, F. (1993). Dynamic visual acuity of normal subjects during vertical optotype and head motion. *Investigative Ophthalmology and Visual Science* 34(6): 1894–1906.
Klein, R., Klein, B. E., Linton, K. L., and De Mets, D. L. (1991). The beaver dam eye study: Visual acuity. *Ophthalmology* 98(8): 1310–1315.
Liang, J., Williams, D. R., and Miller, D. T. (1997). Supernormal vision and high-resolution retinal imaging through adaptive optics. *Journal of the Optical Society of America. A, Optics, Image Science, and Vision* 14: 2884–2892.
Mayer, D. L. and Dobson, V. (1982). Visual acuity development in infants and young children, as assessed by operant preferential looking. *Vision Research* 22(9): 1141–1151.

Plainis, S., Tzatzala, P., Orphanos, Y., and Tsilimbaris, M. K. (2007). A modified ETDRS visual acuity chart for European-wide use. *Optometry and Vision Science* 84(7): 647–653.

Rosser, D. A., Cousens, S. N., Murdoch, I. E., Fitzke, F. W., and Laidlaw, D. A. (2003). How sensitive to clinical change are ETDRS logMAR visual acuity measurements? *Investigative Ophthalmology and Visual Science* 44: 3278–3281.

Thibos, L. N., Cheney, F. E., and Walsh, D. J. (1987). Retinal limits to the detection and resolution of gratings. *Journal of the Optical Society of America. A, Optics, Image Science, and Vision* 4: 1524–1529.

Westheimer, G. (1987). Visual acuity. In: Moses, R. A. and Hart, W. M. (eds.) *Adler's Physiology of the Eye: Clinical Application*, pp. 415–428. St Louis, MO: Mosby.

Adaptive Optics

L Yin and D R Williams, University of Rochester, Rochester, NY, USA

Glossary

Deformable mirror – A mirror equipped with an array of actuators on the back surface that can warp the mirror surface by small amounts into arbitrary shapes, allowing the correction of the eye's aberrations.

Diffraction-limited resolution – The resolution of an optical system that has no aberrations, the image quality of which is reduced only by the diffraction of the light in the pupil of the system.

Lipofuscin autofluorescence – Lipofuscin is composed of many molecules that are by-products of the visual or retinoid cycle. These accumulate in the retinal pigment epithelium with aging and can be visualized in the living eye because they fluoresce when exposed to short wavelength light.

Point-spread function (PSF) – The light distribution in the image plane of an optical system such as the eye, formed from light from a point source outside the eye, such as a very distant star.

Retinal densitometry – A method to measure the density of photopigment in the living eye.

Shack–Hartmann wavefront sensor – A device capable of measuring the optical defects of an optical system such as the human eye. The output of a wavefront sensor can be used to control the shape of a deformable mirror in an adaptive optics system.

Stiles and Crawford effect – The fact that the eye is far more sensitive to light entering through a point near the center of the pupil than the pupil margin even though the irradiance at the retina is very little dependent on entry point in the pupil. This effect is caused by the waveguide properties of cone photoreceptors, which are more sensitive to light falling on their optical axes (which point near the center of pupil) than obliquely incident light.

The Benefit of Adaptive Optics in Vision Science

The history of ophthalmoscopy after its invention by Helmholtz until today is marked by efforts to extract the most information possible from the light reflected from the retina. Over the last two decades, there has been a concerted effort to improve the resolution of the imaging process in all three spatial dimensions. The development of optical coherence tomography (OCT) improved the resolution in the axial dimension, and has allowed the routine imaging of individual layers of cells in the retina. More recently, the introduction of adaptive optics (AO) has improved the resolution of fundus cameras in both transverse dimensions. The transverse resolution of the conventional fundus camera is limited not by the camera itself but by the optics of the human eye. The sources of image blur in the eye's optics include diffraction, aberrations, and scatter. Diffraction is the image blur that results from the wave nature of light as it passes through the eye's pupil. Blurring by diffraction is not inevitable; there are exciting techniques on the horizon that may eventually overcome the fundamental resolution limit set by diffraction, but no one has yet demonstrated this in the eye. A hypothetical eye that suffered only from diffraction would allow a resolution no smaller than about 1.4 μm when imaging wavelengths of light in the middle of the visible spectrum (550 nm). This is smaller than the smallest cells in the retina, so that if one could make the natural human eye limited only by diffraction, cellular and even subcellular features that are invisible in conventional fundus imaging could be seen. As shown in **Figure 1** (upper panels), in a diffraction-limited eye, the larger the pupil, the smaller the image of a single point of light and therefore the better the resolution.

The monochromatic and chromatic aberrations of all eyes further blur the retinal image. The monochromatic aberrations of the human eye alone are greater than those of even a mediocre man-made optical system. As shown in **Figure 1** (lower panels), increasing the pupil reduces the effect of diffraction, but exacerbates the effect of aberrations, with the best trade-off between the two occurring for pupil sizes around 3 mm. Light scatter, the third cause of loss in retinal image quality, is relatively unimportant in young eyes but it can greatly reduce retinal image contrast in older eyes, especially those with cataract. Not only is the retinal image blurred by the optical factors mentioned above, it is also exceedingly dim. The amount of light that can be delivered to the retina is limited for safety reasons and the retinal reflectance is low: 10^{-3}–10^{-5} across the spectrum. This would be less of a problem if it were easy to integrate light over long exposures, but the eye is always moving; even an eye with excellent fixation moves about ~20 μm root mean square (rms) velocity. Eye motion artifacts are all the more troublesome in instruments with high magnification that are designed to look at cellular

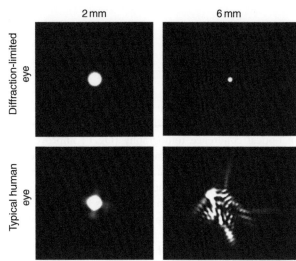

Figure 1 The point-spread function (PSF) for a diffraction-limited eye and a normal eye at two different pupil diameters. The PSF corresponds to the light distribution on the retina produced by a point source of light infinitely distant from the eye. For the hypothetical diffraction-limited eye, the PSF diameter decreases in inverse proportion to the pupil diameter such that large pupils produce the best image quality. However, in the typical human eye, aberrations increase with increasing pupil size, eliminating the benefit of escaping diffraction at the largest pupils. The goal of AO is to correct the aberrations to produce the PSF of a diffraction-limited eye with a large pupil. Adapted from Roorda, A. Garcia, C. A., Martin, J. A., et al. (2006). What can adaptive optics do for a scanning laser ophthalmoscope? *Bulletin de la Société Belge d'Ophthalmologie* 302: 231–244, Figure 1, with permission from Bulletin of the Belgian Societies of Ophthalmology (Copyright 2006).

structures that are often far smaller than 20 μm. Despite all these formidable limitations, it is possible to design fundus cameras that address all of them with varying degrees of success, making microscopic resolution of the living retina possible as described below.

Correcting the Eye's Monochromatic Aberration

It is possible to overcome the eye's monochromatic aberrations with AO, a two-step process in which the eye's wave aberration is measured and corrected, usually in real time. **Figure 2** describes the principle of AO for imaging the eye. The monochromatic aberration of the eye is measured with a wavefront sensor. The measured aberration data are used to control a wavefront compensation device, usually a deformable mirror that corrects the wave aberration. Ideally, it would completely remove all the monochromatic aberrations, leaving diffraction and scatter as the only remaining sources of image blur. It usually takes several iterations of the measurement and correction loop to achieve the best correction, at which point it is possible to obtain a retinal image that is almost

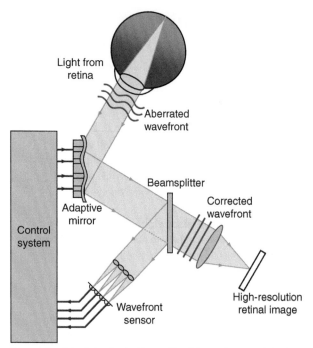

Figure 2 Principle of adaptive optics. The system contains two key parts: the wavefront sensor and wavefront corrector. The wavefront sensor, usually of the Shack–Hartmann type, measures the monochromatic aberration of the eye. It uses a 2-D array of lenslets conjugate with the eye's pupil to break the light from an infrared point source imaged on the retina into several hundred individual beams. Each beam is imaged on a CCD array. Its displacement on the CCD from where it would have landed had the eye been aberration-free indicates the slope of the wave aberration at that lenslet's location in the pupil. Information from all the lenslets is combined to compute the overall wave aberrations of the eye. These data are used to control a wavefront compensation device that corrects the wave aberration. The most commonly used device is a continuous surface deformable mirror. This mirror has a flexible surface overlying an array of actuators that can push or pull on the mirror surface locally. If the mirror surface is shaped so as to mimic the shape of the wave aberration but with half the amplitude of the wave aberration, the wavefront reflecting from the surface will be perfectly flat, and the monochromatic aberrations of the eye will have been corrected. Adapted from Carroll, J. Gray, D. C., Roorda, A., Williams, D. R. (2005). Recent advances in retinal imaging with adaptive optics. *Optics and Photonics News* 16: 36–42, Figure 1, with permission from Optical Society of America (Copyright 2005).

completely aberration free in eyes with normal amounts of aberrations. The rate of measurement and correction required to keep up with the temporal variations in the eye's wave aberration is relatively slow, at least compared with applications of AO in astronomy. Heidi Hofer has shown that measuring and correcting the wave aberration at 30 Hz or so is adequate to track the most important changes in the wave aberration, which are caused by microfluctuations in accommodation that have a temporal bandwidth of only a few Hertz. With complete wavefront correction, the point-spread function (PSF) is very

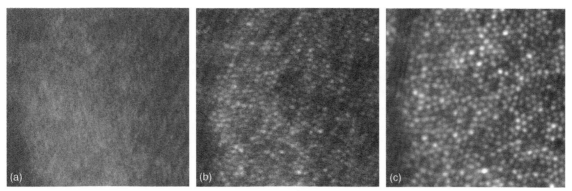

Figure 3 Adaptive optics and motion correction greatly improve the resolution of images of human cone mosaic. (a) Single frame of the reflectance image of the cone mosaic of a typical human eye at 1 degree of eccentricity imaged with all the aberrations of the eye, or (b) After the monochromatic aberrations of the eye were corrected with AO. (c) The summed frames of many images of the same cone mosaic with aberration and eye motion corrected with AO. The frames were registered before summing to correct for eye motion between frames, which increases the SNR over that obtained with single frames. The individual cones at this eccentricity are approximately 5 μm in diameter.

compact, approaching the light distribution produced by diffraction alone. Typically, one can achieve as much as an order of magnitude reduction in the rms wavefront error of a normal eye with this method. This provides a substantial improvement in image quality as shown in **Figure 3**.

Vision Correction with AO

One of the convenient features of AO correction is that the correction required to focus light onto the retina is the same as the correction required to image the retina at high resolution outside the eye. Many investigators have capitalized on the advantages of AO for vision correction as well as for retinal imaging. Correcting the higher order monochromatic aberrations (i.e., those other than defocus and astigmatism corrected by spectacles) produces a modest improvement in visual acuity and contrast sensitivity in the normal eye. The improvements can be dramatic in eyes with large amounts of higher-order aberrations such as those that suffer from keratoconus. The demonstration of these improvements in spatial vision with a deformable mirror has stimulated improvements in the control of laser ablation in refractive surgery as well as the fabrication of customized contact lenses that can correct higher-order aberrations. AO continues to be a valuable tool not only for correcting aberrations but also for generating specific patterns of aberrations so that their effects on vision can be studied conveniently. For example, it is possible to explore the design of contact lenses for presbyopes that increase the depth of field of the eye without the need to fabricate optical elements for each pattern one wishes to evaluate.

Retinal Imaging with AO

In retinal imaging, AO can be combined with almost any other imaging technology. David Williams' laboratory at

the University of Rochester first demonstrated the value of a closed-loop AO system for retinal imaging, incorporating AO into a flood-illuminated system that acquired single snapshots of the retina with a resolution adequate to resolve cone photoreceptors near the fovea. Austin Roorda, then at the University of Houston, demonstrated that AO could also improve the resolution of the scanning laser ophthalmoscope (SLO). SLOs are potentially confocal devices and AO offers improvements in both axial and transverse resolutions. Moreover, AO improves the focus of the light on the confocal pinhole in front of detector, which increases the available signal. The AOSLO has a high lateral resolution of less than about 2 μm. The axial resolution of better than ~60 μm, though poor by OCT standards, is nonetheless adequate for some optical sectioning of the retina. AO has also improved the transverse resolution of OCT, allowing a resolution of less than 3 μm in all three spatial dimensions in the retina. AO systems can also be combined with other imaging modalities such as phase contrast microscopy, polarization imaging, or fluorescence microscopy.

Compensating for Eye Motion

In many cases, the signals acquired in a single video frame of an AOSLO are weak enough to warrant frame averaging to increase the signal-to-noise ratio (SNR) of the image (**Figure 4(a)**). Eye motion between frames, which is substantial in the high-magnification images of AO systems, must be corrected before frame averaging can be achieved. One correction method is to register successive frames with normalized cross-correlation, the benefit of which is shown in **Figure 3(c)**. However, it is not uncommon for single frames, such as the one shown in **Figure 4(a)**, to have inadequate SNR for this method. In the case, illustrated here of autofluorescence imaging by Jessica Morgan at the University of Rochester, there was typically less than one photon for every 5 pixels in each frame. As shown in **Figure 4(b)**, this problem can be solved by simultaneously recording infrared reflectance images of the cone mosaic at the same retinal location. Eye motion between frames can be reliably recovered from reflectance images of cone mosaic with an accuracy of one-fifth of the diameter of a foveal cone, and this information can be used to register the low SNR images as shown in **Figure 4(b)**.

Eye movements can also produce warping artifacts within each frame. Austin Roorda and Scott Stevenson have developed methods in which the relative locations of local retinal features are compared across frames to compute and correct the eye movement warping within each frame as well as translation between frames. **Figure 5** shows

an AOSLO image motion before and after removing distortions from eye movements. Image motion after correction is reduced to a standard deviation of only 7 arcsec. David Arathorn at Montana State University, working in collaboration with Austin Roorda, has a fast software algorithm that can stabilize the retinal image in real time. This has very exciting applications for delivering light stimuli to single photoreceptors in both psychophysical and electrophysiological studies. Software registration approaches alone cannot address all the problems created by eye movements, particularly in AO instruments designed for routine clinical use where eye movements are much larger than the size of a single frame. In that case, successive frames do not share common features and cross-correlation cannot be

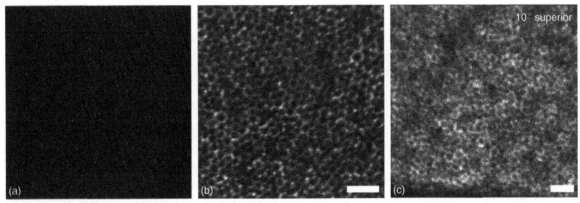

Figure 4 Dual registration improves the transverse resolution of autofluorescence images of the RPE mosaic. (a) Single frame of autofluorescence image of primate RPE mosaic. Because of the low SNR and photon density, the image does not have any apparent spatial structure. (b) 1000 frames of autofluorescence images of the same RPE mosaic shown in (a) summed with eye motion corrected using the dual-registration technique. The eye motion was calculated from reflectance images of the cone mosaic obtained simultaneously with the dim autofluorescence images, providing the translations necessary to register the autofluorescence images. The summed image reveals single cells in the RPE mosaic. (c) Autofluorescence image of human RPE mosaic at retinal eccentricity of 10 degrees. Bright regions in the images correspond to the accumulation of lipofuscin within the RPE cells. Dark regions correspond to the nuclei of RPE cells. Scale = 50 μm. Adapted from Morgan, J. I. Dubra, A., Wolfe, R., et al. (2009). In vivo autofluorescence imaging of the human and macaque retinal pigment epithelial cell mosaic. *Investigative Ophthalmology and Visual Science* 50: 1350–1359, Figure 5, with permission from Association for Research in Vision and Ophthalmology (Copyright 2009).

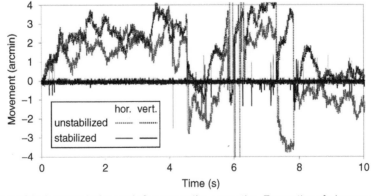

Figure 5 Eye motion in the AOSLO system before and after eye motion correction. Eye motion of a human subject within 10 s duration is captured at 480 Hz through AOSLO imaging (dotted red and blue lines: horizontal and vertical eye movements). After offline correction, the eye motion in the AOSLO images is reduced to flat lines (in red and blue: horizontal and vertical eye movement), with a standard deviation of 7 arcsec. This compares favorably with the most accurate methods to track the eye, having an accuracy that is roughly one-fourth of the diameter of the smallest foveal cones. Courtesy of A Roorda.

used. Dan Ferguson and Dan Hammer at Physical Sciences Incorporated have developed a hardware eye tracking system specifically for AO retinal imaging that complements the software approaches described above.

Imaging Cones

The ability to image cones at high resolution with AO opened a crucial window to examine both normal and abnormal processes in the retina. AO has made possible the first measurements of the antenna properties of single cones in the living human eye. Cone photoreceptors concentrate the image-forming light that passes through the pupil in the photopigment of the outer segment while simultaneously excluding stray light from sources that do not contribute to a sharp retinal image. This beneficial effect of the waveguide nature of cones gives rise to the psychophysical effect known as the Stiles and Crawford effect, in which the sensitivity of the eye declines dramatically for light beams that enter the margin instead of the center of the pupil. Austin Roorda and David Williams at the University of Rochester showed that nearby cones are remarkably well aligned with each other optically so that the angular tuning of a large group of cones is similar to the tuning function for a single cone.

Single cones imaged with AO in the living human eye show a striking variability in reflectance on a time scale of hours or days. The cause of this variability is unknown but could be related to the process of disc shedding in the cone outer segments. Don Miller at Indiana University has shown that, especially when the incoming light is highly coherent, there can be dramatic changes on very short time scales of 5–10 ms. These changes depend on the history of the cone's light exposure and provide an optical method to monitor to the response of cones to light. Kate Grieve and Austin Roorda have also reported increases in infrared reflectance following exposure to light that may provide an alternative method to monitor functional activity in single cones in the retina.

The Cone Mosaic and Color Vision

The first major scientific application of AO in the eye, undertaken at the University of Rochester, was to determine the organization of the human trichromatic cone mosaic. By combining AO retinal imaging with retinal densitometry, individual cones can be characterized by their sensitivity to long (L), middle (M), or short (S) wavelength light, according to the cone opsin it contains, as the example shown in **Figure 6**. Experiments using this method have shown that mosaic of L and M cones in the human are essentially randomly organized. It has been known for some time from indirect methods that the relative numbers of L and M cones varies greatly from eye to eye. AO revealed just how large this variation can be even across normal subjects where a variation in the order of 40-fold in L to M cone

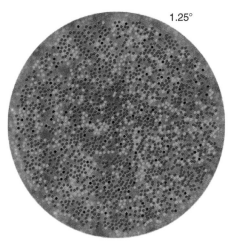

1.25°

Figure 6 Cone mosaic of a human subject at a retinal eccentricity of 1.25 degree with normal color vision. Individual cones in each mosaic were categorized as L, M, or S cone types, using retinal densitometry, and false colored, respectively in red, green, and blue. Courtesy of O. Masuda.

ratio has been observed. The human S cones are arranged randomly near the fovea but with slight tendency toward regular distribution, a tendency that is more pronounced in the macaque monkey.

The development of AO for the eye has also made it possible to study color vision in living human eyes in novel ways because it is now possible to deliver tiny flashes of light that are smaller than single cones. Heidi Hofer showed that near-threshold AO-delivered flashes of monochromatic light at a single wavelength produce a rich variety of color percepts. Indeed, for every subject she studied, the range of color experiences was too large to be explained by a simple model in which all cones of the same class produced the same color experience upon stimulation. David Brainard at the University of Pennsylvania has successfully described the range of color experiences in Heidi Hofer's data with a Bayesian model in which each cone feeds a specific circuit that provides the best estimate of the external stimulus given the local distribution of photon catches in the stimulated cone and its surrounding neighbors.

One of the most powerful applications of AO to date involves its use to characterize the topography of the cone mosaic in eyes in which the genotype is known. Until recently, it has been difficult to perform these studies with standard histological methods because of the difficulty in obtaining post-mortem tissue from eyes with specific and often rare genetic anomalies. Joe Carroll has shown that the cone mosaics of single gene dichromats appear completely normal despite the fact that they have two instead of the usual three cone photopigments (**Figure 7(b)**). On the contrary, dichromats with a specific polymorphism in one of the genes coding for a particular cone photopigment have a cone mosaic with numerous gaps corresponding to a spatially random loss of one class of cones (**Figure 7(c)**; also see **Figure 8** for another example).

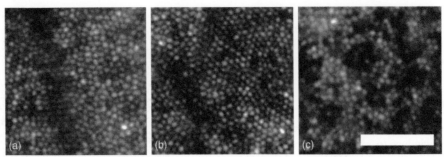

Figure 7   Pathological change of human cone mosaic in red or green dichromat. (a) Cone mosaic from a normal subject at approximately 1 degree of retinal eccentricity. (b) and (c) Cone mosaics at a similar eccentricity of two human subjects having red–green color-vision deficiency. In contrast to the normal cone mosaic in (a), the cone mosaic in (b) did not contain L cones, but appeared to have normal cone density. The cone mosaic in (c), did not contain M cones, and had reduced cone density with patches of functional loss of M cones. Scale = 50 μm for all panels. Reproduced from Carroll, J. Gray, D. C., Roorda, A., Williams, D. R. (2005). Recent advances in retinal imaging with adaptive optics. *Optics and Photonics News* 16: 36–42, Figure 3, with permission from Optical Society of America (Copyright 2005).

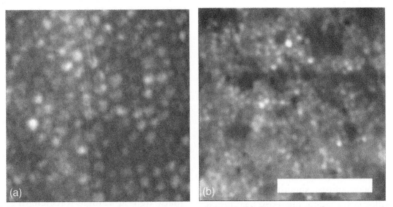

Figure 8   Pathological change of human cone mosaic in rod monochromat. (a) Cone mosaic from a normal subject at approximately 4 degrees of retinal eccentricity. (b) Possible rod mosaic of one subject who was a congenital achromat. The reason why the photoreceptors in (b) were believed to be rods is that their density and size match with the anatomical characteristic of the rod mosaic at this eccentricity, and differ greatly from the normal cones in (a). Scale = 50 μm for all panels. Reproduced from Carroll, J., Gray, D. C., Roorda, A., Williams, D. R. (2005). Recent advances in retinal imaging with adaptive optics. *Optics and Photonics News* 16: 36–42, Figure 3, with permission from Optical Society of America (Copyright 2005).

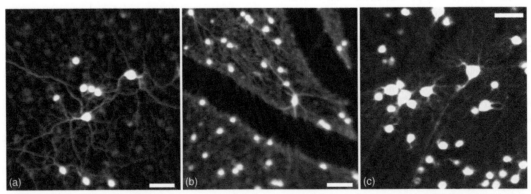

Figure 9   Fluorescence AOSLO images of primate retinal ganglion cells *in vivo*. (a–c) Fluorescence AOSLO imaging revealed the morphology of retinal ganglion cells labeled with fluorophore (rhodamine dextran) in living monkey eye. The transverse resolution of the images is fine enough to resolve the individual dendrites. The fluorophore was introduced into the ganglion cells through retrograde labeling through injections in the lateral geniculate nucleus (LGN). Scale = 50 μm for all panels. (a,c) Reproduced from Gray, D. C., Wolfe, R., Gee, B. P., et al. (2008). In vivo imaging of the fine structure of rhodamine-labeled macaque retinal ganglion cells. *Investigative Ophthalmology and Visual Science* 49: 467–473, Figures 1 and 5, with permission from Association for Research in Vision and Ophthalmology (Copyright 2008).

Imaging Retinal Pigment Epithelium

The retinal pigment epithelium (RPE) lies immediately behind the photoreceptors and plays several critical roles in maintaining their function. RPE cell damage is implicated in many retinal degenerative diseases such as age-related macular degeneration, retinitis pigmentosa, and Stargardt's disease. The ability to image these cells in the living retina and to track changes in them over time may prove valuable for understanding both normal RPE function and retinal disease. In AOSLO reflectance imaging, the more reflective photoreceptor mosaic normally obscures RPE cells. Occasionally, RPE cells can be seen in patients with retinal degenerative diseases, such as cone–rod dystrophy, where the overlying photoreceptors are absent. However, it has recently become possible to image the RPE mosaic in living human eyes in which the photoreceptor layer is intact, by taking advantage of the autofluorescence properties of lipofuscin in the RPE as shown in **Figure 4(c)**. Statistical characterization of the RPE mosaic, for example, packaging arrangement and cell density across eccentricity, in both normal subject and patients may eventually prove to be valuable for the clinical diagnosis of earlier stages of retinal degenerative disease. It may ultimately prove possible to use AO to image subcellular structures in RPE cells

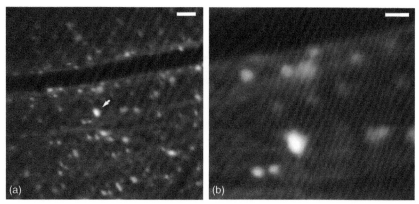

Figure 10 Fluorescence AOSLO images of rat retinal ganglion cells *in vivo*. (a) Fluorescence AOSLO image of rat retina with ganglion cell expressing EGFP. Scale = 50 µm. Image was taken at a large field of view (FOV). Gene encoding EGFP was delivered to ganglion cells through an AAV2 viral vector administrated intravitreally. The ganglion cell indicated by the white arrow was shown in (a) at higher magnification. Image at this view reveals the dendritic morphology of the cell. Such images could provide basis for morphological classification *in vivo*. Scale = 20 µm.

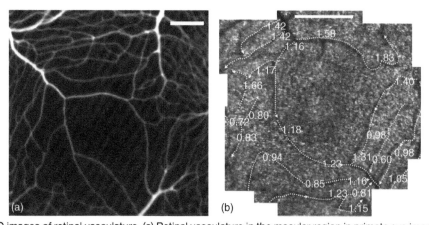

Figure 11 AOSLO images of retinal vasculature. (a) Retinal vasculature in the macular region in primate eye imaged by fluorescence AOSLO in combination with fluorescence angiography. Scale = 150 µm. (b) Direction of blood flow and leukocyte velocity within the capillaries of the macular region in human eye calculated from reflectance AOSLO images. Movement of discrete leukocytes was detected through an image-processing algorithm. The velocities of the leukocytes labeled on the images were in mm s^{-1}. Scale = 1 degree. (a) Reproduced from Gray, D. C., Merigan, W., Wolfing, J. I. et al. (2006). In vivo fluorescence imaging of primate retinal ganglion cells and retinal pigment epithelial cells. *Optics Express* 14(16): 7144–7158, Figure 6, with permission from Optical Society of America (Copyright 2006). (b) Adapted from Roorda, A., Garcia, C. A., Martin, J. A., et al. (2006). What can adaptive optics do for a scanning laser ophthalmoscope? *Bulletin de la Société Belge d'Ophthalmologie* 302: 231–244, Figure 6, with permission from Bulletin of the Belgian Societies of Ophthalmology (Copyright 2006).

and/or to monitor changes in RPE cells over time. Jessica Morgan has already shown that it is possible to use AO to track the time course of lipofuscin autofluoresence bleaching and RPE damage following exposure to bright light.

Imaging Retinal Ganglion Cells

The retinal image is conveyed to the brain through an array of 17 or more parallel ganglion cell pathways in the primate. We know remarkably little about the functional significance of many of these morphologically distinct ganglion cell classes. One of the major impediments to learning more about them is that the less numerous cells are difficult to target with a microelectrode that is capable of recording from only one single cell at a time. The development of optical methods to record from many ganglion cells

simultaneously in the intact living eye could greatly accelerate our understanding of ganglion cell function. There are numerous technical hurdles to overcome before this is feasible, but AO retinal imaging has now solved an important one: it is now possible to image ganglion cells along with some of their subcellular structure *in vivo* in nonhuman primate retina. This capability may also eventually prove useful in the study of diseases such as glaucoma, in which blindness results from the death of ganglion cells. New ways to image ganglion cells at a microscopic spatial scale may lead to earlier diagnosis and delivery of therapy.

Dan Gray, working in the AO group at the University of Rochester, imaged the dendritic morphology of macaque ganglion cells retrogradely labeled with fluorescent dye (rhodamine dextran) *in vivo*, with a transverse resolution of 1.6 μm (characterized as full width at half maximum;

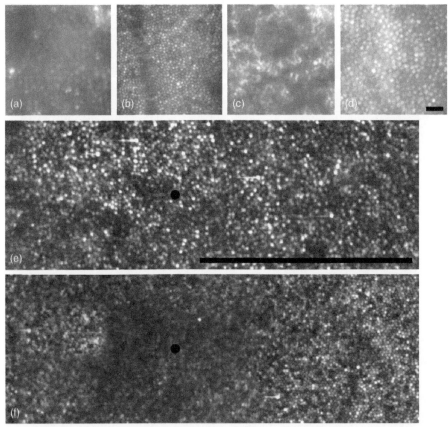

Figure 12 Structural change of cone mosaic of patients having retinal degenerative diseases. (a) and (c) Cone mosaics of a patient having cone–rod dystrophy compared, respectively with (b) and (d) cone mosaics of normal subjects at matched retinal eccentricity. For the patient, severe cone loss was apparent in (c), while in (a), cones still form a continuous mosaic, but the cone density is much decreased and cone size is much larger. Scale = 25 μm for panels (a–d). (e) and (f) Foveal cone mosaics of patients having disease caused by mitochondrial DNA (mtDNA) mutation. The two patients were affected by the disease at different severity. Cone mosaic of the less-affected patient in (f) showed normal cone spacing, while the cone mosaic of the more-affected patient in (e) showed increased cone size and cone spacing. The preferred fixation point of each subject is visualized as dark dots in (e) and (f). Scale = 1 degree for both panels. (a–d), Reproduced from Wolfing J.I., Chung, M., Carroll, J., et al. (2006). High-resolution retinal imaging of cone-rod dystrophy. *Ophthalmology* 113: 1014–1019, Figure 4, with permission from the American Academy of Ophthalmology, Elsevier Ltd (2006). (e,f) Adapted from Yoon, M.K., Roorda, A., Zhang Y., et al. (2009). Adaptive optics scanning laser ophthalmoscopy images in a family with the mitochondrial DNA T8993C mutation. *Investigative Ophthalmology and Visual Science* 50: 1838–1847, Figure 4, with permission from Association for Research in Vision and Ophthalmology (Copyright 2009).

Figure 9). Although in some cases, they were able to distinguish ganglion cell bodies at different depths in the retina, which is a valuable way to distinguish different classes, the depth resolution is modest, about 115 μm. Viral vector (e.g., AAV2)-mediated-gene delivery is another promising approach for delivering fluorophores (e.g., EGFP) to single cells. It requires a simpler intravitreal rather than an intracranial injection and can produce fluorescence that is essentially permanent in each cell that is transduced by the virus. Melissa Geng and Jason Porter at University of Rochester in collaboration with John Flannery at UC Berkeley combined this method with fluorescence AOSLO in the living rat eye (**Figure 10**). Another possibility is to use transgenics to create an animal whose cells express a fluorophore. Charles Lin and colleagues used this method to image microglia cells in living mouse eye. Work is underway to extend the viral vector method to nonhuman primate and to achieve expression of activity-dependent fluorophores such as G-CaMP (a green flurescent protein calcium probe) in single ganglion cells. These methods may ultimately allow AO retinal imaging to monitor the functional responses of ganglion cells.

Imaging Retinal Vasculature

Many retinal diseases, such as diabetic retinopathy or wet macular degeneration, have important effects on the retinal vascular structure and blood flow. AO combined with fluorescein angiography allows imaging of even the fine structure of the retinal capillary bed (**Figure 11(a)**). Austin Roorda, then at the University of Houston, showed that AO retinal imaging could image leukocyte motion in the smallest retinal capillaries without the need for fluorescein (**Figure 11(b)**). Steve Burns at University of Indiana has even succeeded in imaging the flow of erythrocytes by tracking movement of single erythrocytes in successive images of a single scan line superimposed on an oriented parallel to a small vessel.

Imaging Retinal Disease

The use of an AO to image retinal disease is one of the most exciting and also one of the most challenging applications. AO is at its best when the pupil is completely dilated and the ocular media are clear and scatter free. The challenge of imaging retinal disease lies in the fact that patients often have small pupils, poor fixation, and/or large amounts of light scatter. Despite these limitations, progress has been made in the deployment of AO in the clinical arena.

For many retinal degenerative diseases, the structural changes of the cone mosaic and the RPE mosaic of patients have been impressively documented *in vivo*. The diseases affecting cones include not only those relevant

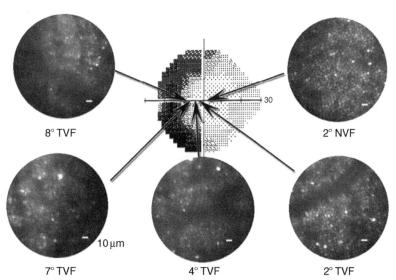

Figure 13 Correspondence between structural changes in the cone mosaic and functional measurement of visual sensitivity. Measured visual field sensitivity of a patient having rod–cone dystrophy is shown in the upper middle panel. The rest of the panels show the cone mosaics from retinal regions corresponding to locations in the visual field map. In those cone mosaic images, the darker regions indicate areas where cones are not reflecting light as in normal retina. From these images, it is not possible to determine if these dark regions represent loss of cones or structurally altered cones, hence became less effective in waveguiding and reflecting light. The extent of these dark regions strongly correlated with the level of visual sensitivity. The more extensive the dark regions (i.e., reduced cone density) were the lower the visual function (e.g., the comparison between the left and right panels). A strong positive relationship was found between the cone density and the level of visual sensitivity in patients with various types of retinal dystrophy. Adapted from Choi, S.S., Doble, N., Hardy, J.L., et al. (2006). In vivo imaging of the photoreceptor mosaic in retinal dystrophies and correlations with visual function. *Investigative Ophthalmology and Visual Science* 47: 2080–2092, Figure 5, with permission from Association for Research in Vision and Ophthalmology (Copyright 2006).

to color blindness mentioned previously, but also more common ones such as retinitis pigmentosa and photoreceptor dystrophy. As per the examples shown in **Figure 12**, cone loss can occur as a patchy, random dropout of cones that otherwise appear normal. In other cases, presumably when the mosaic has had the opportunity to remodel, the mosaic completely tiles the retina but the cones are larger and their density is reduced. Because the structure of the cone mosaic can be documented in living eyes, the specific structure in individual eyes can be correlated with the genotype of the patient, and also with functional measurements such as Humphrey visual field sensitivity, multifocal ERG, and contrast sensitivity (**Figure 13**).

Retinal degenerative diseases affect many other retinal structures, besides photoreceptors and RPE. As described in this article, AO imaging has enabled the accumulation of normative data of many retinal structures in both human and nonhuman primates, as well as data from animal disease models. These data will advance the clinical investigation of retinal degenerative disease. For example, the structure of lamina cribrosa was imaged in both normal primate eye and eye with experimentally induced glaucoma. Ultimately, AO imaging may play a role in the diagnosis of glaucoma and other retinal diseases in clinical practice.

See also: Color Blindness: Acquired; Color Blindness: Inherited; Optical Coherence Tomography.

Further Reading

Arathorn, D. W., Yang, Q., Vogel, C. R., et al. (2007). Retinally stabilized cone-targeted stimulus delivery. *Optics Express* 15: 13731–13744.

Artal, P., Chen, L., Fernandez, E. J., et al. (2004). Neural compensation for the eye's optical aberrations. *Journal of Vision* 4: 281–287.

Biss, D. P., Sumorok, D., Burns, S. A., et al. (2007). *In vivo* fluorescent imaging of the mouse retina using adaptive optics. *Optics Letters* 32: 659–661.

Carroll, J., Neitz, M., Hofer, H., Neitz, J., and Williams, D. R. (2004). Functional photoreceptor loss revealed with adaptive optics: An alternate cause of color blindness. *Proceedings of the National Academy of Sciences of the United States of America* 101: 8461–8466.

Carroll, J., Gray, D. C., Roorda, A., and Williams, D. R. (2005). Recent advances in retinal imaging with adaptive optics. *Optics and Photonics News* 16: 36–42.

Chen, L., Artal, P., Gutierrez, D., and Williams, D. R. (2007). Neural compensation for the best aberration correction. *Journal of Vision* 7(9): 1–9.

Geng Y., Greenberg K. P., Wolfe R., Gray D. C., et al. (2009). In vivo imaging of microscopic structures in the rat retina. *Investigative Ophthalmology and Visual Science* 50: 5872–5879.

Gray, D. C., Wolfe, R., Gee, B. P., et al. (2008). *In vivo* imaging of the fine structure of rhodamine-labeled macaque retinal ganglion cells. *Investigative Ophthalmology and Visual Science* 49: 467–473.

Hofer, H., Artal, P., Singer, B., Aragon, J. L., and Williams, D. R. (2001). Dynamics of the eye's wave aberration. *Journal of Optical Society of America A, Optics, Image Science, and Vision* 18: 497–506.

Hofer, H., Carroll, J., Neitz, J., Neitz, M., and Williams, D. R. (2005). Organization of the human trichromatic cone mosaic. *Journal of Neuroscience* 25: 9669–9679.

Hofer, H., Singer, B., and Williams, D. R. (2005). Different sensations from cones with the same photopigment. *Journal of Vision* 5: 444–454.

Liang, J., Williams, D. R., and Miller, D. T. (1997). Supernormal vision and high-resolution retinal imaging through adaptive optics. *Journal of Optical Society of America A, Optics, Image Science, and Vision* 14: 2884–2892.

Martin, J. A. and Roorda, A. (2005). Direct and noninvasive assessment of parafoveal capillary leukocyte velocity. *Ophthalmology* 112: 2219–2224.

Morgan, J. I., Dubra, A., Wolfe, R., Merigan, W. H., and Williams, D. R. (2009). *In vivo* autofluorescence imaging of the human and macaque retinal pigment epithelial cell mosaic. *Investigative Ophthalmology and Visual Science* 50: 1350–1359.

Morgan, J. I., Hunter, J. J., Masella, B., et al. (2008). Light-induced retinal changes observed with high-resolution autofluorescence imaging of the retinal pigment epithelium. *Investigative Ophthalmology and Visual Science* 49: 3715–3729.

Porter, J., Guirao, A., Cox, I. G., and Williams, D. R. (2001). Monochromatic aberrations of the human eye in a large population. *Journal of Optical Society of America A, Optics, Image Science, and Vision* 18: 1793–1803.

Porter, J., Queener, H., Lin, J., Thorn, K., and Awwal, A. A. S. (2006). Adaptive Optics for Vision Science: *Principles, Practices, Design and Applications*. Hoboken, NJ: Wiley-Interscience.

Roorda, A. and Williams, D. R. (1999). The arrangement of the three cone classes in the living human eye. *Nature* 397: 520–522.

Thibos, L. N., Hong, X., Bradley, A., and Cheng, X. (2002). Statistical variation of aberration structure and image quality in a normal population of healthy eyes. *Journal of Optical Society of America A, Optics, Image Science, and Vision* 19: 2329–2348.

Yoon, G. Y. and Williams, D. R. (2002). Visual performance after correcting the monochromatic and chromatic aberrations of the eye. *Journal of Optical Society of America A, Optics, Image Science, and Vision* 19: 266–275.

Zhong, Z., Petrig, B. L., Qi, X., and Burns, S. A. (2008). *In vivo* measurement of erythrocyte velocity and retinal blood flow using adaptive optics scanning laser ophthalmoscopy. *Optics Express* 16: 12746–12756.

Alternative Visual Cycle in Müller Cells

G H Travis, UCLA School of Medicine, Los Angeles, CA, USA

Glossary

Müller cell – A glial cell in the vertebrate retina that spans from the vitreal surface to the external limiting membrane, with apical processes that extend into the interphotoreceptor matrix. These cells perform multiple functions in the retina including the processing of visual retinoids.

Opsin visual pigment – A member of the G-protein-coupled receptor superfamily that functions as the light receptor in rods and cones. These pigments represent a complex between an opsin protein and a visual chromophore (see below).

Outer segment – An elongated light-sensitive structure attached to the connecting cilium of rod and cone photoreceptors. The rod outer segment in humans comprises a stack of approximately 1000 membranous disks. These disks are loaded with rhodopsin or cone opsin visual pigments.

Rpe65 – An abundant, membrane-associated protein in retinal pigment epithelium cells that functions as a retinoid isomerase. In particular, Rpe65 catalyzes the conversion of an all-*trans*-retinyl ester to 11-*cis*-retinol and a free fatty acid.

Visual chromophore – The light-absorbing molecular species in an opsin protein. The most common visual chromophore in vertebrates is 11-*cis*-retinaldehyde, which isomerizes to all-*trans*-retinaldehyde upon absorption of a photon.

Vision in vertebrates is provided by two types of photoreceptor cells, rods and cones. Rods mediate vision in dim light while cones mediate high-resolution color vision in bright light. Approximately 95% of photoreceptors in the human retina are rods. Nonetheless, cones are far more important for vision in humans. With the advent of artificial lighting, humans spend much of the time under conditions where the rod photoresponse is saturated and vision is mediated by cones. Both rods and cones contain a light-sensitive structure called the outer segment (OS) comprising a stack of densely packed membranous disks. These disks are packed with rhodopsin or cone-opsin visual pigment. The light-absorbing chromophore in most opsin pigments is 11-*cis*-retinaldehyde (11-*cis*-RAL). Absorption of a photon induces its isomerization to all-*trans*-retinaldehyde (all-*trans*-RAL), which activates the opsin pigment and stimulates the visual transduction cascade. After a brief period of activation,

the pigment dissociates to yield free all-*trans*-RAL and apo-opsin, which is no longer light sensitive. The pigment is regenerated by recombination of apo-opsin with another 11-*cis*-RAL. To sustain light sensitivity, all-*trans*-RAL released by bleached opsin pigments is converted back to 11-*cis*-RAL by a multi-step enzyme pathway called the visual cycle (**Figure 1**).

Visual Cycle in Retinal Pigment Epithelium Cells

All but the first step of the visual cycle takes place within cells of the retinal pigment epithelium (RPE), an epithelial monolayer adjacent to photoreceptor OS. In brief, all-*trans*-RAL is reduced by all-*trans*-retinol dehydrogenase (all-*trans*-RDH) in rod and cone OS to all-*trans*-retinol (all-*trans*-ROL) or vitamin A. The all-*trans*-ROL is released by the OS into the extracellular space or interphotoreceptor matrix (IPM), where it is bound to interphotoreceptor retinoid-binding protein (IRBP). The all-*trans*-ROL is taken up by apical processes of RPE cells, where it binds to cellular retinol binding protein type-1 (CRBP1). Holo-CRBP1 is the substrate for lecithin:retinol acyl transferase (LRAT), which transfers a fatty acid from phosphatidylcholine in internal membranes to the all-*trans*-ROL. The resulting fatty-acyl ester of all-*trans*-ROL (all-*trans*-RE) is the substrate for Rpe65-isomerase. Rpe65 catalyzes hydrolysis of the carboxylate ester and utilizes the energy released for isomerization of all-*trans*-ROL to 11-*cis*-retinol (11-*cis*-ROL). The final catalytic step in the visual cycle is oxidation of 11-*cis*-ROL, by one of several 11-*cis*-ROL dehydrogenases (11-*cis*-RDH) to 11-*cis*-RAL. Both 11-*cis*-ROL and 11-*cis*-RAL are bound to cellular retinaldehyde binding protein (CRALBP) in the RPE. The 11-*cis*-RAL is released by the apical RPE to the IPM, where it binds IRBP. IRBP carries the 11-*cis*-RAL to the photoreceptor OS, where it is taken up and recombines with apo-opsin to reform a rhodopsin or cone-opsin pigment.

A Role for Müller Cells in Visual Pigment Regeneration

Evidence suggests that Müller glial cells in the retina also participate in the processing of visual retinoids. Müller cells span nearly the full thickness of the retina. The apical microvilli of Müller cells extend into the IPM and hence are well situated to exchange visual retinoids

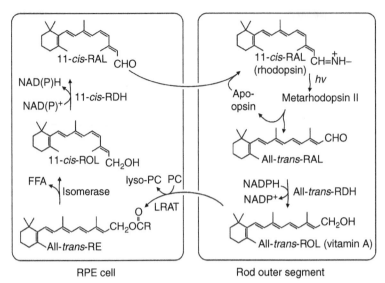

Figure 1 Visual cycle for rhodopsin regeneration. Absorption of a photon (*hv*) by a rhodopsin pigment molecule induces isomerization of 11-*cis*-RAL to all-*trans*-RAL, which converts the pigment to its active metarhodopsin II state. After deactivation, the pigment dissociates to yield apo-opsin and free all-*trans*-RAL. The all-*trans*-RAL is reduced to all-*trans*-ROL by one or more all-*trans*-RDHs that use NADPH as a co-factor. The all-*trans*-ROL is released by the OS to the IPM and is taken up by the apical RPE where it is esterified by LRAT to yield an all-*trans*-RE. The all-*trans*-RE is isomerized and hydrolyzed by Rpe65 to yield 11-*cis*-ROL. The 11-*cis*-ROL is oxidized by one or more 11-*cis*-RDHs to yield 11-*cis*-RAL chromophore. The 11-*cis*-RAL is released by the RPE into the IPM where it binds to IRBP. Finally, the 11-*cis*-RAL is delivered to the OS where it re-combines with apo-opsin to form a new visual pigment.

with rod and cone OS, similar to the apical microvilli of RPE cells. Importantly, Müller cells contain several proteins involved in the processing of visual retinoids. These include (1) CRALBP, (2) RPE-retinal guanine-nucleotide-binding protein (G-protein) coupled receptor (RGR-opsin), a non-photoreceptor opsin that effects light-dependent regulation of the visual cycle in RPE cells, (3) CRBP1, and (4) retinol dehydrogenases types 10 and 11. Müller cells neither express Rpe65 nor LRAT, which are both present in RPE cells. Therefore, Müller cells do not simply duplicate the function of RPE cells in the regeneration of visual chromophore.

A Second Source of Chromophore for Cones

Several lines of evidence suggest that an alternate source of chromophore precursor is available to cones but not rods. When frog retinas were separated from the RPE, cone opsins, but not rhodopsin, regenerated spontaneously after a photobleach. After photobleaching, isolated salamander cones recovered sensitivity with the addition of either 11-*cis*-ROL or 11-*cis*-RAL, while isolated rods only recovered sensitivity with the addition of 11-*cis*-RAL. Müller cells in primary culture were shown to take up all-*trans*-ROL and synthesize 11-*cis*-ROL, which they secreted into the medium. Salamander cones were shown to dark-adapt and regenerate visual chromophore in isolated retinas separate from the RPE. The intrinsic capacity of cones to recover sensitivity and regenerate

visual chromophore in salamander retinas was lost after exposure to a selective Müller-cell toxin (α-aminoadipic acid), suggesting that Müller cells play a role in these processes.

Functional Differences between Rods and Cones

Although morphologically similar, rods and cones differ greatly in sensitivity, dynamic range, and speed of the photoresponse. Rods are single-photon detectors and show saturation of the photoresponse under relatively dim background illumination, producing 500 or more photoisomerizations per second. At saturation, all cyclic guanosine monophosphate (cGMP)-gated cation channels are closed and the rod no longer responds to light. Cones, on the other hand, are 100-fold less sensitive than rods but still exhibit a photoresponse to a light flash under bright background illuminations producing up to 10^6 photoisomerizations per second. The response kinetics of cones is also several-fold faster than rods. These disparities reflect differences in the properties of rhodopsin and the cone-opsin pigments. Rhodopsin is exceedingly quiet, with a spontaneous activation rate in the dark of one thermal isomerization every 2000 years. Cone opsins are much noisier than rhodopsin for two reasons. First, the spontaneous isomerization rate of cone opsin is 10 000-fold higher than of rhodopsin. Second, 11-*cis*-RAL spontaneously dissociates from cone-opsins. A dark-adapted red cone contains approximately 10% apo-cone-opsin due to spontaneous dissociation of chromophore. In contrast, 11-*cis*-RAL combines almost irreversibly with apo-rhodopsin in rods. Apo-opsin, which results from dissociation of 11-*cis*-RAL, activates the transduction cascade, producing the phenomenon of dark light. These effects contribute to the noisy background of cones and explain their much lower sensitivity than rods.

Although the rod response is saturated in bright light, photon capture by rhodopsin continues unabated. Thus, in a rod-dominant retina under daylight conditions, the vast majority of photoisomerization events contributes nothing to useful vision. Under these circumstances, cones must compete with rods for the limited supply of 11-*cis*-RAL. The reversible binding of 11-*cis*-RAL to apo-cone-opsins and its irreversible binding to apo-rhodopsin confers a tendency of rods to steal chromophore from cones. This tendency aggravates the competition between rods and cones for limited chromophore.

An Alternate Retinoid Isomerase Activity in Cone-Dominant Retinas

Studies on the biochemistry of visual pigment regeneration in cone-dominant retinas suggest a mechanism whereby cones may escape competition from rods for visual chromophore. Experiments were performed on separated retinas and RPE from mice and cattle, as model rod-dominant species, and from chickens and ground squirrels, as model cone-dominant species. Measurement of endogenous retinoids showed that the rod-dominant species contain high levels of all-*trans*-retinyl esters (REs) in the RPE but no detectable REs in the retina. Cone-dominant species contain REs in both the RPE and retina, with predominantly 11-*cis*-REs in the retina. The activities of retinoid-processing enzymes were measured by incubating total homogenates or microsomes prepared from RPE or retinas with all-*trans*-ROL, then assaying for synthesis of new retinoids by high-performance liquid chromatography (HPLC). Incubation of RPE homogenates with all-*trans*-ROL resulted in rapid synthesis of all-*trans*-REs, due to the activity of LRAT. Only after accumulation of all-*trans*-REs was 11-*cis*-ROL synthesis detected, due to the activity of Rpe65. This profile of retinoid synthesis was observed with RPE homogenates from both rod- and cone-dominant species.

A strikingly different profile was observed when retina homogenates from cone-dominant chicken or ground squirrels were incubated with all-*trans*-ROL under similar conditions. Addition of all-*trans*-ROL resulted in rapid synthesis of 11-*cis*-REs and 11-*cis*-ROL, with negligible initial synthesis of all-*trans*-REs. Addition of palmitoyl coenzyme A (palm CoA) to the assay mixture increased synthesis of 11-*cis*-REs, again with minimal initial synthesis

of all-*trans*-REs. Synthesis of 11-*cis*-ROL and 11-*cis*-REs from all-*trans*-ROL without prior formation of all-*trans*-REs suggests that the mechanism of retinoid isomerization is different in retina versus RPE. Instead of converting an all-*trans*-RE to 11-*cis*-ROL and a free fatty acid, as catalyzed by Rpe65 in RPE cells, cone-dominant retinas appear to catalyze the direct conversion of all-*trans*-ROL to 11-*cis*-ROL. This energetically unfavorable reaction appears to be driven by mass action through secondary esterification of the 11-*cis*-ROL product. Here, the energy of isomerization is indirectly supplied by hydrolysis of the thio-ester in palm CoA versus hydrolysis of the carboxyl ester in phosphatidylcholine, as catalyzed by Rpe65. The isomerase machinery in retinas may therefore involve two catalytic activities, as depicted in **Figure 2**. One is an isomerase that catalyzes the passive interconversion of all-*trans*-ROL and 11-*cis*-ROL. This enzyme has not yet been identified. The second is a palm-CoA-dependent RE synthase. Several proteins with acyl CoA:retinol acyltransferase (ARAT) activity have been described. At least one of these, diacylglycerol acyltransferase type-1 (DGAT1), is expressed in the retina.

11-*cis*-ROL Dehydrogenase Activity in Cones but Not in Rods

As discussed above, cones can regenerate visual pigments and restore light sensitivity with the addition of either 11-*cis*-RAL or 11-*cis*-ROL, while rods can only regenerate rhodopsin pigment and recover sensitivity with addition of 11-*cis*-RAL. Since 11-*cis*-RAL is the visual chromophore for both cone and rod pigments, these observations suggest that cones, but not rods, express an 11-*cis*-RDH activity that oxidizes 11-*cis*-ROL to 11-*cis*-RAL. Robust nicotinamide adenine dinucleotide phosphate (NADPH)-dependent 11-*cis*-RDH activity has been detected in cone

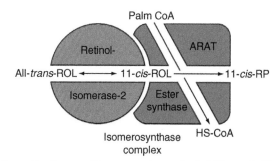

Figure 2 Proposed isomerase-2 complex in Müller cells. The isomerase in Müller cells appears to catalyze direct conversion of all-*trans*-ROL to 11-*cis*-ROL. This energetically unfavorable reaction appears to be driven by mass action through secondary esterification of 11-*cis*-ROL to an 11-*cis*-RE. This synthase uses palm CoA as an acyl donor, and hence is a type of ARAT. The isomerase-2/ARAT complex has been named isomerosynthase to denote its activities. Neither isomerase-2 nor its ARAT partner has been identified to date.

but not rod in photoreceptors. Reduction of all-*trans*-RAL to all-*trans*-ROL in bleached photoreceptors requires NADPH, and hence is an energy-consuming process. Oxidation of an 11-*cis*-ROL to 11-*cis*-RAL generates an NADPH reducing-equivalent. Since reduction of all-*trans*-RAL and oxidation of 11-*cis*-ROL occurs with 1:1 stoichiometry in a cone that is utilizing 11-*cis*-ROL as a chromophore precursor, the presence of reciprocal NADP$^+$/NADPH-specific dehydrogenases in cones affords a self-renewing supply of dinucleotide substrate at no energy cost to the cell. Eliminating the need for energy-dependent synthesis of NADPH may remove the metabolic bottleneck of all-*trans*-RAL-reduction in cones at high rates of photoisomerization.

An Alternate Visual Cycle that Mediates Pigment Regeneration in Cones

The observations outlined above can be explained by the hypothesized alternate visual cycle shown in **Figure 3**. According to this model, all-*trans*-ROL-isomerase (isomerase-2) and the 11-*cis*-ROL-specific ARAT are in Müller cells, which also express the 11-*cis*-ROL and 11-*cis*-RAL binding-protein, CRALBP. The major source of substrate for isomerase-2 is all-*trans*-ROL released by rods and cones during light exposure. Isomerization of all-*trans*-ROL to 11-*cis*-ROL and subsequent esterification of 11-*cis*-ROL to 11-*cis*-REs are probably limited by substrate

availability. It has been shown that apo-CRALBP stimulates 11-*cis*-RE-hydrolase (11-*cis*-REH) in RPE cells. These observations suggest a regulatory mechanism for the proposed visual cycle: in the dark-adapted state, cones contain fully regenerated opsin pigment and hence do not use 11-*cis*-ROL. CRALBP is saturated with 11-*cis*-ROL and the level of 11-*cis*-REs are high in Müller cells. Both isomerase-2 and 11-*cis*-RE-synthase are inactive due to the absence of available substrate. With the onset of light and the bleaching of photopigments, cones begin to take up 11-*cis*-ROL. This results in desaturation of CRALBP, activation of 11-*cis*-REH, and mobilization of 11-*cis*-RE-stores in Müller cells. Rods and cones begin to release all-*trans*-ROL, which permits replenishment of 11-*cis*-REs through activation of isomerase-2 and 11-*cis*-RE-synthase. This pathway becomes progressively more active with increasing light intensity, availability of all-*trans*-ROL substrate, and consumption of the 11-*cis*-ROL product by cones.

A New Role for IRBP

IRBP is the major extracellular retinoid-binding protein in retinas. It has been suggested that the primary function of IRBP is to bind all-*trans*-ROL and 11-*cis*-RAL during translocation of these retinoids between OS and the RPE. However, loss of IRBP in *irbp*$^{-/-}$ mice has only a mild effect on rhodopsin regeneration. Might IRBP play another role? The endogenous ligands of IRBP have been studied.

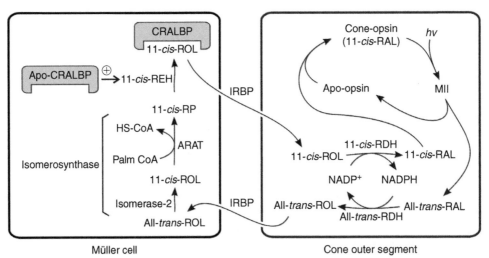

Müller cell   Cone outer segment

Figure 3 Proposed alternate visual cycle in Müller cells. Absorption of a photon (*hv*) induces 11-*cis* to all-*trans* isomerization of the retinaldehyde chromophore, resulting in activated opsin (MII) inside a cone outer segment. Decay of MII releases all-*trans*-RAL, which is reduced to all-*trans*-ROL by one or more NADPH-dependent all-*trans*-RDHs. The all-*trans*-ROL is released to the IPM where it binds IRBP and is carried to a Müller cell apical process. After uptake into the Müller cells, the all-*trans*-ROL is isomerized by isomerase-2 and esterified by ARAT to yield an 11-*cis*-RE (11-*cis*-RP). Müller cells also contain CRALBP, which binds 11-*cis*-ROL but not all-*trans*-retinol. 11-*cis*-REH is activated by apo-CRALBP to hydrolyze the 11-*cis*-RE, yielding 11-*cis*-ROL. Binding of 11-*cis*-ROL by CRALBP prevents its back-isomerization to all-*trans*-ROL. The 11-*cis*-ROL is released by CRALBP to IRBP in the IPM, where it is carried back to the cone OS. Within the cone, an NADP$^+$-dependent 11-*cis*-RDH oxidizes the 11-*cis*-ROL to 11-*cis*-RAL, which re-combines with apo-opsin to regenerate a new, light-sensitive cone pigment. The reciprocal reduction of all-*trans*-RAL and oxidation of 11-*cis*-ROL results in a self-renewing supply of NADP-co-factor at all rates of photoisomerization and no energy cost to the cell.

In light-adapted frog and bovine retinas, IRBP contains higher levels of 11-*cis*-ROL than 11-*cis*-RAL, in addition to all-*trans*-ROL. Moreover, IRBP immunoreactivity was found in association with cone but not rod OS. These observations suggest that the critical function of IRBP may not be exchange of all-*trans*-ROL and 11-*cis*-RAL between rods and RPE cells, as previously assumed, but rather exchange of all-*trans*-ROL and 11-*cis*-ROL between cones and Müller cells.

Alternate Visual Cycle in Rod-Dominant Species

Although clearly present in cone-dominant retinas, the evidence for an alternate visual cycle in Müller cells of rod-dominant species is less compelling. Endogenous 11-*cis*-REs and isomerase-2 catalytic activity were undetectable in homogenates from rod-dominant bovine and mouse retinas. The absence of 11-*cis*-retinoids in $rpe65^{-/-}$ mice has also been put forth as further evidence against the Müller-cell pathway in this species. However, this observation is difficult to interpret since $rpe65^{-/-}$ photoreceptors contain no visual chromophore, hence all-*trans*-ROL, the substrate for the alternative visual cycle, is not released following light exposure. The effective Michaelis constant (K_m) of all-*trans*-ROL for 11-*cis*-RE-synthesis by chicken retina homogenates is 13.5 μM. Thus, even in chickens, the alternate visual cycle is only active under conditions that yield high concentrations of all-*trans*-ROL. Recently, isolated mouse retinas were photobleached and analyzed physiologically for recovery of rod and cone light sensitivity. Cone sensitivity and the cone photoresponse recovered in these retinas despite the absence of RPE cells, while rod sensitivity and the rod response were dramatically attenuated. These physiological observations argue for existence of the alternate visual cycle in mouse retinas. The role of this pathway in rod-dominant retinas should be resolved upon identification of the proteins responsible for isomerase-2 activity, and measurement of their expression levels in rod- and cone-dominant retinas.

See also: Phototransduction: Phototransduction in Cones; Phototransduction: Phototransduction in Rods; Phototransduction: Rhodopsin; Phototransduction: The Visual Cycle.

Further Reading

Chen, Y. and Noy, N. (1994). Retinoid specificity of interphotoreceptor retinoid-binding protein. *Biochemistry* 33: 10658–10665.

Das, S. R., Bhardwaj, N., Kjeldbye, H., and Gouras, P. (1992). Muller cells of chicken retina synthesize 11-*cis*-retinol. *Biochemical Journal* 285: 907–913.

Fain, G. L., Matthews, H. R., Cornwall, M. C., and Koutalos, Y. (2001). Adaptation in vertebrate photoreceptors. *Physiological Reviews* 81: 117–151.

Jin, M., Li, S., Nusinowitz, S., et al. (2009). The role of interphotoreceptor retinoid-binding protein on the translocation of visual retinoids and function of cone photoreceptors. *Journal of Neuroscience* 29: 1486–1495.

Jones, G. J., Crouch, R. K., Wiggert, B., Cornwall, M. C., and Chader, G. J. (1989). Retinoid requirements for recovery of sensitivity after visual-pigment bleaching in isolated photoreceptors. *Proceedings of the National Academy of Sciences of the United States of America* 86: 9606–9610.

Kefalov, V. J., Estevez, M. E., Kono, M., et al. (2005). Breaking the covalent bond – a pigment property that contributes to desensitization in cones. *Neuron* 46: 879–890.

Mata, N. L., Radu, R. A., Clemmons, R., and Travis, G. H. (2002). Isomerization and oxidation of vitamin A in cone-dominant retinas. A novel pathway for visual-pigment regeneration in daylight. *Neuron* 36: 69–80.

Mata, N. L., Ruiz, A., Radu, R. A., Bui, T. V., and Travis, G. H. (2005). Chicken retinas contain a retinoid isomerase activity that catalyzes the direct conversion of all-*trans*-retinol to 11-*cis*-retinol. *Biochemistry* 44: 11715–11721.

Miyazono, S., Shimauchi-Matsukawa, Y., Tachibanaki, S., and Kawamura, S. (2008). Highly efficient retinal metabolism in cones. *Proceedings of the National Academy of Sciences of the United States of America* 105: 16051–16056.

Pugh, E. N. and Lamb, T. D. (2000). Phototransduction in vertebrate rods and cones: Molecular mechanisms of amplification, recovery and light adaptation. In: Stavenga, D. G., DeGrip, W. J., and Pugh, E. N., Jr (eds.) *Handbook of Biological Physics*, pp. 184–255. Amsterdam: Elsevier Science B.V.

Radu, R. A., Hu, J., Peng, J., et al. (2008). Retinal pigment epithelium-retinal G protein receptor-opsin mediates light-dependent translocation of all-*trans*-retinyl esters for synthesis of visual chromophore in retinal pigment epithelial cells. *Journal of Biological Chemistry* 283: 19730–19738.

Saari, J. C. and Bredberg, D. L. (1987). Photochemistry and stereoselectivity of cellular retinaldehyde-binding protein from bovine retina. *Journal of Biological Chemistry* 262: 7618–7622.

Stecher, H., Gelb, M. H., Saari, J. C., and Palczewski, K. (1999). Preferential release of 11-*cis*-retinol from retinal pigment epithelial cells in the presence of cellular retinaldehyde-binding protein. *Journal of Biological Chemistry* 274: 8577–8585.

Wang, J. S., Estevez, M. E., Cornwall, M. C., and Kefalov, V. J. (2009). Intra-retinal visual cycle required for rapid and complete cone dark adaptation. *Nature Neuroscience* 12: 295–302.

Yen, C. L., Monetti, M., Burri, B. J., and Farese, R. V., Jr. (2005). The triacylglycerol synthesis enzyme DGAT1 also catalyzes the synthesis of diacylglycerols, waxes, and retinyl esters. *Journal of Lipid Research* 46: 1502–1511.

Anatomically Separate Rod and Cone Signaling Pathways

S Nusinowitz, UCLA School of Medicine, Los Angeles, CA, USA

Glossary

Neurotransmitter – A chemical substance that is released at synaptic connections that are used to relay, amplify, and modulate signals between cells and neurons.

Photopic vision – The scientific term for color vision mediated by multiple cone photoreceptor types in bright light. In the human retina, photopic vision is tri-chromatic.

Phototransduction – A process by which light is converted into electrical signals in rod and cone photoreceptor cells in the retina of the eye.

Retinal circuitry – It refers to the neuronal pathways in the retina that carry information from photoreceptor cells to ganglion cells.

Scotopic vision – Vision mediated by rod photoreceptors in dim light. Scotopic vision is color blind.

Spatio-temporal vision – A term commonly used to describe the spatial and temporal properties of human visual processing, and the neural mechanisms which underpin them.

Visual adaptation – A process by which vision adjusts to or gets used to a change in overall brightness, color, and other spatio-temporal properties, in order to maximize visual sensitivity.

Rod and Cone Photoreceptors

The process of vision is initiated by the absorption of light by a photoreceptor pigment molecule. The mammalian retina contains two classes of photoreceptors, referred to as rods and cones, each distinct in their structural morphology and in their behavior in response to light. Rod photoreceptors are highly sensitive to light and are capable of responding to a single photon. In contrast, cone photoreceptors are less sensitive to light, but are exquisitely tuned to mediate color and spatio-temporal vision. There are three types of cone photoreceptors in the human retina, each with different, but overlapping, spectral sensitivities, and the interaction of the output from these photoreceptor types mediates the ability to discriminate small color differences.

Based on studies of the evolution of visual pigments, it is hypothesized that rods evolved from cones. A single amino acid substitution has been shown to exchange the molecular properties of rod and cone visual pigments. This suggests that a single spontaneous amino acid substitution could have been one of the key steps in the divergence of rod- and cone-mediated vision. The first step in the formation of rods themselves may have been the spontaneous formation of disk membranes that were separated from the plasma membrane of the outer segment. This step is thought to have increased the light sensitivity of the photoresponse, a physiological property that is typical of rod photoreceptors. The post-receptoral retinal circuitry for rods is superimposed on pre-existing retinal circuitry mediating cone function. This feature of retinal circuitry predates the evolution of the rod photoreceptor itself and likely represents the pathway of a redundant cone photoreceptor that evolved into a rod. As described below, there are multiple sites in the retinal circuitry where rods have the opportunity to penetrate the cone system circuitry.

The absorption of a photon of light by a photoreceptor pigment molecule initiates a sequence of events referred to as the phototransduction cascade. Much of what we know about the specifics of the phototransduction cascade comes from the study of the rod photoresponse. In mammalian rods, the absorption of a photon of light by rhodopsin, the visual chromophore in rods, initiates a sequence of biochemical events that ultimately leads to a decrease in the intracellular level of cyclic guanosine 3,5′-monophosphate (cGMP) and the closure of the cGMP-gated ion channels in the outer segment. The resultant closure of the cGMP-gated channels in the outer-segment membrane decreases the circulating current by blocking inward current flow. This results in an hyperpolarization of the photoreceptor cell membrane and a decrease in the release of neurotransmitter at the photoreceptor synapse with second-order neurons. The biochemical steps in cone outer-segment phototransduction are thought to be similar to that which occurs in rods. In fact, there are corresponding components in rod and cone phototransduction at each step in the cascade, including rod and cone versions of visual pigment, tranducin, phosphodiesterase, arrestin, kinase, cGMP-gated channels, and Na^+/K^+, Ca^{2+} exchangers.

Cone Postreceptoral Circuitry

Consider first the retinal circuitry mediating cone function. Cone photoreceptor cells synapse with bipolar and horizontal cells in the outer plexiform layer of the retina (see **Figure 1**). Cones release glutamate at a steady rate in darkness but this rate is slowed in a graded response that is correlated to the change in the outer-segment membrane potential after stimulation by light. The modulation of synaptic transmitter release drives signaling to second-order bipolar cells.

There are at least nine different types of cone bipolar cells in the mammalian retina that are distinguished on the basis of a number of features, including the number of

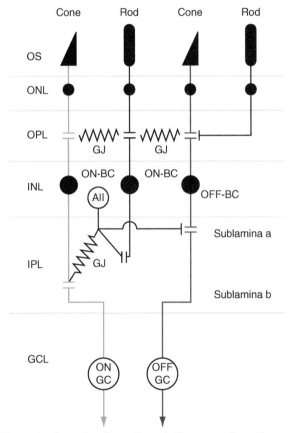

Figure 1 Cone signaling pathway in the mammalian retina. Cone photoreceptors respond to light with a graded hperpolarization and synapse with bipolar and horizontal (not shown) cells in the outer plexiform layer (OPL) of the retina. The release of glutamate at the synaptic terminal is modulated by the change in the outer-segment membrane potential. Cone signals are delivered to ON and OFF bipolar cells (BC) which carry signals to ON and OFF ganglion cells (GC) in the inner plexiform layer (green and red channels, respectively), thereby maintaining segregated cone pathways through the retina. ON and OFF bipolar cells provide excitatory and inhibitory inputs to amacrine cells and ganglion cells and respond with a graded depolarization (sign-inverting) or hyperpolarizing (sign-conserving) response, respectively. Cones communicate laterally with other cones (and rods) via electrical couplings called gap junctions (GJ). Lateral communication is also afforded by horizontal cells in the OPL and by amacrine cells in the IPL. Abbreviations: OS – outer segment, ONL – outer nuclear layer, OPL – outer plexiform layer, INL – inner nuclear layer, IPL – inner plexiform layer, GCL – ganglion cell layer, GJ – gap junction.

cone photoreceptors contacting the bipolar cell, the spread of the dendritic terminals, and the stratification of their axon terminals in the inner plexiform layer. The precise function of each of these types of bipolar cells is not well understood but they are presumed to be tuned to process different aspects of the visual stimulus. Morphologically, bipolar cells can be subdivided into classes commonly referred to as diffuse and midget bipolar cells. Diffuse bipolar cells (of which there are at least six types) have broad dendritic spreads and make contact with 5–20 long (L-) and middle (M-) wavelength cones; the number of connections depends on where they are located in the retina. These cells are likely involved in tasks requiring broad spatial averaging and color processing but have poor spatial resolution. In contrast, midget bipolar cells have a narrow dendritic tree and can make contact with a single cone and with a single ganglion cell. This cone-mediated circuitry provides a mechanism for high spatial resolution of a scene and is thought to mediate an acuity channel. The midget bipolar cell also transmits color information by virtue of its one-on-one connections with L- and M-sensitive cones. In addition, bipolar cells that make specialized contacts with short (S-) wavelength sensitive cones have been described in the primate, rat, and mouse retina.

Functionally, there are two broad classes of bipolar cells in the mammalian retina mediating the cone signaling pathway. They are commonly referred to as ON and OFF bipolar cells, because of their expression of either excitatory glutamate receptors (mGluRs) or inhibitory glutamate receptors (iGluRs). ON cone bipolar cells make contact with photoreceptors through invaginations in the cone pedicle and are flanked by a pair of horizontal cell dendrites, an arrangement commonly referred to as a triad. The triads are apposed to a presynaptic ribbon, which is the release site for neurotransmitter mediating signal transfer to second-order neurons. There are typically about 25 invaginations in a cone pedicle, but as many as 50 invaginations have been reported. A recent study has identified the protein pikachurin, a previously unknown dystroglycan-binding protein, as critical for the apposition of photoreceptor and bipolar cells dendrites at the ribbon synapse. OFF bipolar cells make contact directly at the cone pedicle base where up to 500 contacts are made with postsynaptic cells.

ON bipolar cells that express mGluRs are depolarized (sign-inverting) by the light response in cone photoreceptor cells. Their dendritic processes terminate in sublamina B of the inner plexiform layer (see **Figure 1**). In contrast, OFF bipolar cells express iGluRs. In these cells glutamate expression is linked to Na^+ influx. OFF cone bipolar cells are hyperpolarized (sign-conserving) by the light response in cones and have processes that end in sublamina A of the inner plexiform layer. The two types of bipolar cells provide excitatory and inhibitory inputs to downstream cells and tertiary neurons in response to light stimulation. In general, ON cone bipolar cells are excited by light that is brighter than its surround, whereas OFF cone bipolar cells are excited by light that is dimmer. This segregation of visual input is maintained in parallel signaling pathways to ON and OFF ganglion cells (see **Figure 1**).

Lateral Communication Networks

Signal transfer through the retinal layers, particularly for the rod system, depends on lateral communication between retinal cells. There are several cell types and neural connections within retinal layers that mediate lateral transfer of signal. The major cell types include amacrine and horizontal cells but lateral transfer is also accomplished by a class of low-resistance electrical gap junctions.

Gap Junctions

Low-resistance electrical gap junctions are ubiquitous in the mammalian retina. They enable the intercellular, bidirectional transport of ions, metabolites, and second-order messengers. These gap junctions are mediated by connexins of which several different types have been reported in the mammalian retina. The most abundant of these is neuronal Connexin 36. Connexin 36 is associated with processes in the outer and inner plexiform layers, consistent with expression in multiple cell types. In the outer plexiform layer, cone photoreceptors communicate laterally via Connexin 36-mediated gap junctions between cone pedicles (see **Figure 1**). The gap junctions also permit signal transfer between cone pedicles and rod spherules. In the inner plexiform layer, the expression of Connexin 36 co-localizes with the dendritic processes of AII-type amacrine cells. The latter gap junction mediates transfer of signal from AII amacrine cells to ON cone bipolar cells, a signaling pathway that is crucial for rod signals to infiltrate the cone postreceptoral signaling pathway. In addition, the gap junctions in the outer plexiform layer (between cone pedicles and rod spherules) allow rod signals to infiltrate the cone signaling pathway very early in visual processing. These pathways are described later in more detail.

Horizontal Cells

Horizontal cells are second-order, mainly inhibitory, neurons located in the outer plexiform layer. While these cells make synaptic connections with photoreceptors, they are also extensively coupled by either Connexin 50 or Connexin 57 gap junctions. Morphologically, three types of horizontal cells have been identified in the primate and human retina, referred to as HI, HII, and HIII. The HI-type horizontal cells have small dendritic fields (75–150 μm), but with long axons (300 μm) ending in a broad dendritic tree. HIII horizontal cells are similar to the HI horizontal cell

but have larger dendritic trees at all retinal locations. In addition, HIII cells contact many more cones than HI. HII horizontal cells are more spidery and intricate in dendritic field characteristics than either of the other types. The three types of horizontal cells in the human retina demonstrate evidence of color-specific coding. HI horizontal cells contact primarily with green- and red-sensitive cones, with a smaller number of contacts with blue-sensitive cones. HII horizontal cells contact blue-sensitive cones primarily, and HIII horizontal cells contact only green- and red-sensitive cones.

Most, if not all, of the input to horizontal cells is derived from the response of cone photoreceptors to light and, depending on the type of horizontal cell, can make contact with many cones over broad retinal areas. In addition, because horizontal cells are extensively coupled via electrical gap junctions, their receptive fields can be much larger than their dendritic spreads. The connection to rod photoreceptors has traditionally been thought to occur at the axon terminal process, implying that signal transfer from cones to rods is unidirectional. However, rod inputs have been recorded in horizontal cell somata in the cat retina, presumed to be delivered via cone-rod gap junctions or direct dendritic connections. Like cone OFF bipolar cells, horizontal cells express iGluRs at their dendrites and are hyperpolarized in response to light.

In the primate retina, the transmitter release from cones that drives horizontal cells and bipolar cells is also regulated by feedback from horizontal cells. This feedback loop provides a mechanism for the inhibition of signals from adjacent cone photoreceptors. The precise mechanism by which horizontal cells produce this lateral inhibition is not well understood but may occur as a result of the modulated release of the inhibitory transmitter γ-aminobutyric acid (GABA) at the synaptic terminal of cones or via the modulation of the Ca^{2+} channels that regulate the release of glutamate. Regardless of the precise mechanism of lateral inhibition, the lateral interconnections provided by horizontal cells contribute to the formation of the antagonistic surrounds of bipolar cell receptive fields. The antagonistic center-surround interaction is thought to enhance the detection of edges but have also been implicated in the processing of color information where the center-surround configuration modulates antagonistic (or opponent) color information, and in illusory surface filling effects. In addition, the observation of rod and cone inputs at horizontal cell somata provides a mechanism for integrating light signals over broad retinal areas to ensure optimal retinal sensitivity over the entire intensity range.

Amacrine Cells

There are up to 30 types of amacrine cells located in the inner retina of the mammalian retina that have been distinguished on the basis of morphological characteristics, physiological properties, and pharmacological criteria. While cells upstream from amacrine cells generate graded potentials in response to stimulation by light, the amacrine cell is the first site in the retina where action potentials are generated. Amacrine cells receive their input from bipolar cells, mediated by iGluRs at the synaptic terminals, and from ganglion cells and other types of amacrine cells. The main job of the amacrine cells is to provide a mechanism for transfer of signals from bipolar cells within and between sublamina of the inner plexiform layer, and with ganglion cells. However, amacrine cells, like horizontal cells, provide a mechanism for lateral signal communication between retinal cells, including providing a feedback loop to bipolar cells. They are assumed to play an important role in modulating activity in the antagonistic surrounds of ganglion cell receptive fields that shape higher visual functions, such as object segregation and spatio-temporal adaptation. The feedback from amacrine cells has also been implicated in switching the site of light adaptation between receptor and postreceptoral sites.

The extent of lateral transfer depends on the morphology of the amacrine cell. Wide-field amacrine cells transmit lateral information across a broad expanse of the inner plexiform layer and are present in many species, including the mouse, rat, cat, rabbit, salamander, and monkey. Small-field amacrine cells mediate local interactions between different sublaminae of the IPL. The best characterized of these is the AII amacrine cell which plays an important role in mediating signal transfer through the rod-mediated neural circuitry. Unlike cone bipolar cells, rod ON polar cells do not make direct contact with ON ganglion cells. Rather rod signals are transmitted to ON ganglion cells by AII electrical coupling (gap junctions) with ON cone bipolar cells and by synaptic connections with OFF-cone bipolar cells (see below). Another amacrine cell type, the A17 amacrine cell, has also been implicated in the rod-signaling pathway of the mammalian retina. Up to 11 other small-field amacrine cells have been identified in the cat, rabbit, and mouse, but their precise function is not well understood.

Ganglion Cells

Signals carried through the retinal layers converge on ganglion cells, the latter responsible for carrying information to higher-order visual centers. Up to 25 different types of ganglion cells have been identified in the mammalian retina, dependent on species. These retinal ganglion cells are broadly grouped into classes based on morphological characteristics and physiological properties. In the primate retina, for example, there are at least 18 different types of retinal ganglion cells that are classified morphologically into Pα (parasol), Pβ (midget), and Pγ types, and physiologically into two major types:

parasol or magnocellular (M), and midget or parvocellular (P) cells. The parvocellular cells project exclusively to the parvocellular layers of the lateral geniculate nucleus (LGN) and play a key role in central acuity. Parasol ganglion cells are motion-sensitive cells and primarily project to the magnocellular layers of the LGN. Intrinsically photosensitive retinal ganglion cells (ipRGCs) containing melanopsin as the photosenstitve pigment have also been described recently. These neurons, which receive input from both rod and cone photoreceptors, have been implicated in nonimage-forming responses to environmental light such as the pupillary light reflex and circadian entrainment.

Multiple Rod Signaling Pathways

Cone photoreceptors synapse directly with both ON and OFF bipolar cells, which then transmit signals in parallel pathways to ON and OFF ganglion cells, respectively (**Figure 1**). In contrast, rod photoreceptors, which synapse with rod ON bipolar cells, do not make direct contact with ganglion cells, but rather transmit their signals to these cells through several alternate pathways. Consider first the evidence for multiple rod pathways. The alternative signaling pathways are described later and shown in **Figure 2**.

It is generally accepted that rod-mediated vision, like cone vision, involves multiple signaling pathways. The evidence in support of this hypothesis derives from early psychophysical experiments in humans in which critical fusion frequency (CFF; the intensity at which flicker can just be detected) was measured under conditions mediated by rods. These experiments revealed two distinct branches in the function that relates CFF and stimulus intensity. In the lower branch, over dim flash intensities, the CFF was no better than 15 Hz, and remained at this level over a broad range of stimulus intensities. However, at higher intensities, covering mesopic light levels, CFFs increased rapidly and could be as high as 28 Hz. The double-branched CFF versus intensity response function implied the existence of at least two signaling pathways mediating rod function in the mammalian retina. This hypothesis was supported by the observation that patients with achromatopsia, a retinal abnormality in which cone function is absent, also display the same response properties.

Additional support for at least dual rod signaling pathways comes from psychophysical measurements of rod flicker perception in humans. These experiments demonstrated that for 15 Hz flickering stimuli (the optimal stimulus presentation frequency for demonstrating the interaction), there is an intensity region, well above flicker detection threshold, where the perception of flicker is minimized or nulled. The perceptual nulling of flicker has been assumed to result from the mutual cancellation of signals originating from at least two signaling pathways

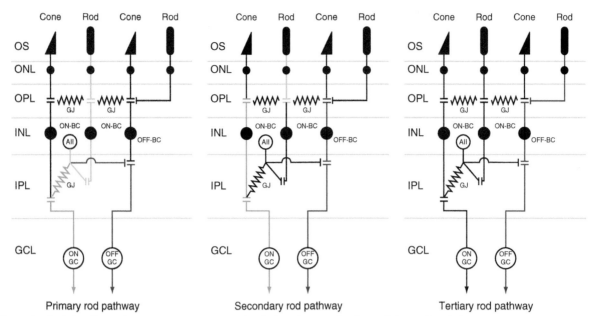

Figure 2 Rod signaling pathway in the mammalian retina. In the primary rod pathway (left panel), rod photoreceptors synapse with rod ON bipolar cells, which in turn make connections with amacrine AII cells in the inner plexiform layer (yellow pathway). Signals from the AII amacrine cells infiltrate the cone pathway by exciting ON cone bipolar cells via electrical gap junctions (green pathway) and via glycinergic (sign-inverting) synapses with OFF cone bipolar cells (red pathway). Rod signaling through the secondary rod pathway (middle panel) is mediated through rod-cone gap junctions between rod spherules and cone pedicles located in the outer plexiform layer. Through the rod-cone gap junction, rod signals have access to both ON and OFF cone bipolar cells and to ON and OFF ganglion cells (green and red pathways, respectively). A third pathway for rod signal transmission (right panel), in which rod photoreceptors bypass the rod ON bipolar cell and directly excite cone OFF bipolar cells, has also been hypothesized (red pathway). Abbreviations: OS – outer segment, ONL – outer nuclear layer, OPL – outer plexiform layer, INL – inner nuclear layer, IPL – inner plexiform layer, GCL – ganglion cell layer, GJ – gap junction.

having different speeds of signal transmission. The hypothesis argues that when signals are in phase, despite different speeds of transmission, they are mutually additive but, when out of phase, produce destructive interference, which contributes to the inhibition of signal strength and the flicker nulling perception. This mutual cancellation has also been demonstrated with the electroretinogram (ERG), which is a noninvasive measure of the massed response of the retina to light. It is assumed only to reflect activity of outer and middle retinal cells. As in the perceptual experiments described above, supportive evidence derives from a unique feature of the function that relates ERG signal amplitude and stimulus. A local response minimum is observed at an intensity that is well above ERG flicker detection threshold but still within the scotopic range of intensities and occurs at the same intensity where the perception of flicker in humans is also minimized.

Much of the electrophysiological evidence in support of the existence of multiple signaling pathways comes from experiments in which single unit extracellular recordings are made from ganglion cells in the mouse and rabbit retina. Using pharmacological agents to disrupt different cellular connections in the retinal circuitry, combined with animal models with known genetic defects affecting these connections, electrophysiological support for multiple rod

pathways has come from what signals remain detectable at the ganglion cell level. On the basis of these types of experiments, rod photoreceptor signals are presumed to be transmitted to ganglion cells via three alternate pathways.

The anatomical substrates mediating rod postreceptoral signaling in the retina seem to be well established and are assumed to be conserved across mammalian species. They are illustrated in **Figure 2**. In the primary rod pathway (left panel), rod photoreceptors synapse with rod ON bipolar cells, via sign-inverting glutamatergic synapses. The output from the rod ON bipolar cell is then transmitted to AII amacrine cells in the inner plexiform layer via sign-conserving glutamatergic synapses. Signals from the AII amacrine cells then converge onto the cone pathway by exciting ON cone bipolar cells via electrical gap junctions and inhibit cone OFF bipolar cells via sign-inverting glycinergic synapses (see **Figure 2** for color coding).

Rod signaling through the secondary rod pathway (middle panel) converges onto the cone circuitry at an even earlier stage. The secondary rod pathway is mediated through rod-cone gap junctions that exist between rod spherules and cone pedicles located in the outer plexiform layer. Through the rod-cone gap junction, rod photoreceptors can transmit signals directly to both ON and OFF cone bipolar cells and to ON and OFF ganglion cells. The circuitry for the primary and secondary rod pathways has been shown to exist in the cat, rabbit, primate, and more recently in the mouse.

A third pathway for rod signal transmission (right panel), in which rod photoreceptors bypass the rod ON bipolar cell and directly excite cone OFF bipolar cells, has also been hypothesized and supported by anatomical and physiological data. This alternative pathway has been demonstrated to exist using electrophysiological methods. In these experiments, a ganglion cell signal continues to be observed in animals without cones (thereby eliminating the rod–cone gap junction of the secondary pathway) and in which all signal transmission through the primary rod pathway is blocked with pharmacological agents.

Thus, the retinal circuitry comprising the rod system offers multiple signaling routes for carrying information from rod photoreceptors to inner retinal ganglion cells. While these signaling pathways provide the rod system with multiple opportunities for system redundancy, they also subserve specialized functions related to scotopic vision. It has been suggested that signal transfer from the primary to secondary rod signaling pathway affords the rod system the capability of enhanced temporal resolution at the expense of light sensitivity. However, further work is needed to better understand the precise role of each of the rod signaling pathways in the processing of visual information.

Concluding Statements

A major step in forming our perceptions of the visual world is accomplished in the retina, where information from rod and cone photoreceptors is filtered, processed, and channeled through multiple parallel signaling pathways. In addition to the different spectral sensitivities of rod and cone photoreceptors and the intensity range over which they operate, the different types of bipolar cells, amacrine cells, and horizontal cells are presumed to be tuned to capture or enhance specific attributes of a visual scene – color processing, brightness contrast, temporal processing, signal enhancement and integration, and adaptation mechanisms. Ultimately, these processed and filtered signals from the retina are transmitted to higher-order visual centers, such as the lateral geniculate nucleus and the primary visual cortex of the brain, where the information is optimized to form our perceptions of the visual environment.

Acknowledgments

The authors are grateful to Dr. William H. Ridder for reading the text and providing helpful comments and Bryan Chen for assistance with drawings.

See also: The Circadian Clock in the Retina Regulates Rod and Cone Pathways; Information Processing: Amacrine Cells; Information Processing: Bipolar Cells; Information Processing: Ganglion Cells; Information Processing in the Retina; Morphology of Interneurons: Amacrine Cells; Morphology of Interneurons: Bipolar Cells; Morphology of Interneurons: Horizontal Cells; Morphology of Interneurons: Interplexiform Cells; Phototransduction: Phototransduction in Cones; Phototransduction: Phototransduction in Rods.

Further Reading

Bloomfield, S. A. and Dacheux, R. F. (2001). Rod vision: Pathways and processing in the mammalian retina. *Progress in Retinal and Eye Research* 20: 351–384.

Dowling, J. E. (1999). Retinal processing of visual information. *Brain Research Bulletin* 50: 317.

Falk, G. and Shiells, G. (2006). Synaptic transmission: Sensitivity control mechanisms. In: Heckenlively, J. H., Arden, G. B., Nusinowitz, S., Holder, G., and Bach, M. (eds.) *Principles and Practice of Clinical Electrophysiology of Vision*, pp. 79–91. Cambridge, MA: MIT Press.

Fu, Y. and Yau, K. W. (2007). Phototransduction in mouse rods and cones. *Pflugers Archiv. European Journal of Physiology* 454: 805–819.

Kolb, H. (2006). Functional organization of the retina. In: Heckenlively, J. H., Arden, G. B., Nusinowitz, S., Holder, G., and Bach, M. (eds.) *Principles and Practice of Clinical Electrophysiology of Vision*, pp. 47–64. Cambridge, MA: MIT Press.

Kolb, H. and Famiglietti, E. V. (1974). Rod and cone pathways in the inner plexiform layer of cat retina. *Science* 186: 47–49.

Masland, R. H. (2001). The fundamental plan of the retina. *Nature Neuroscience* 4: 877–886.

Nickle, B. and Robinson, P. R. (2007). The opsins of the vertebrate retina: Insights from structural, biochemical, and evolutionary studies. *Cellular and Molecular Life Sciences* 64: 2917–2932.

Pugh, E. N., Jr. and Lamb, T. D. (1993). Amplification and kinetics of the activation steps in phototransduction. *Biochimica et Biophysica Acta* 1141: 111–149.

Schmidt, T. M., Taniguchi, K., and Kofuji, P. (2008). Intrinsic and extrinsic light responses in melanopsin-expressing ganglion cells during mouse development. *Journal of Neurophysiology* 100: 371–384.

Volgyi, B., Deans, M. R., Paul, D. L., and Bloomfield, S. A. (2004). Convergence and segregation of the multiple rod pathways in mammalian retina. *Journal of Neuroscience* 24: 11182–11192.

Wassle, H. (2004). Parallel processing in the mammalian retina. *Nature Reviews Neuroscience* 5: 747–757.

Anatomy and Regulation of the Optic Nerve Blood Flow

R Ehrlich, A Harris, and A M Moss, Indiana University, Indianapolis, IN, USA

Glossary

Anastamosis – A network of streams that both branch out and reconnect forming a communication between two blood vessels or other tubular structures.

Autoregulation – The intrinsic ability of a system to maintain constant blood flow despite changes in perfusion pressure and local vascular parameters to maintain homeostasis.

Choriocapillaris – A layer of capillaries in the choroid immediately adjacent to Bruch's membrane.

Central retinal artery – A branch of the ophthalmic artery which pierces the optic nerve close to the globe, sending branches to the internal surface of the retina.

Extraocular muscles – A group of six muscles that control the movements of the eye, including the superior, inferior, lateral, and medial recti, and the superior and inferior obliques.

Fenestration – From the Latin word for window (fenestra), a fenestration is an opening in a wall or membrane.

Hemodynamics – The study of the forces generated by the heart and the flow of blood through the cardiovascular system.

Ophthalmic artery – A branch of the internal carotid artery which enters the orbit through the optic canal, along with the optic nerve, to supply structures in the orbit.

Poiseuille's law – A physical law that describes slow, viscous, incompressible flow through a circular cross section. It states that for a laminar, nonpulsatile fluid flow through a uniform straight tube, the vascular resistance is inversely proportional to the fourth power of the radius of a vessel, and is directly proportional to the blood viscosity and length of the vessel.

Retrobulbar vessels – Blood vessels behind the eye.

Introduction

A thorough understanding of vascular anatomy is critical to appreciate the physiology of the optic nerve head (ONH). The study of blood flow and the metabolism of the eye are also important in understanding the role of the circulatory system in various eye diseases. The arterial supply to the optic nerve has been widely investigated; however, the precise anatomy of the anterior optic nerve microvasculature remains difficult to ascertain. Detailed assessment of this anatomy is limited by its small vessel caliber, its complex three-dimensional structure, and the relative inaccessibility of the microvascular bed. Although the evaluation of ocular hemodynamics continues to improve with the development of new imaging technologies, current techniques for measuring optic nerve blood flow do not directly evaluate the optic nerve. We summarize the vascular anatomy of the ONH, retina, and choroid, the regulation of blood flow in the ONH, and several imaging techniques used to measure blood flow in the eye.

Anatomy of the Vascular Supply

The ophthalmic artery (OA), which is the first branch of the internal carotid artery, provides the vast majority of the ocular blood supply (**Figure 1**). The OA enters the orbit through the optic canal and, in most individuals, runs inferolaterally to the optic nerve. After coursing nasally and anteriorly, the OA runs superior to the optic nerve, where it gives off most of its major branches. These branches include vessels to each of the extraocular muscles, the central retinal artery (CRA), and the posterior ciliary arteries (PCAs) (**Figure 2**). There are usually two to three PCA trunks, each dividing into approximately 10–20 short PCAs before, or, occasionally after, penetrating the sclera. The short PCAs supply the posterior choriocapillaris, peripapillary choroid, and the majority of the anterior optic nerve. The medial and temporal long PCAs pierce the sclera about 3–4 mm nasally and temporally from the optic nerve (**Figure 2**). They then travel anteriorly within the suprachoroidal space, along the horizontal meridians of the globe. Typically, the long PCAs divide in the vicinity of the ora serrata to supply the iris, ciliary body, and the anterior region of the choroid.

The CRA branches directly from the OA to pierce the medial aspect of the optic nerve sheath approximately 10–15 mm behind the globe. The CRA courses adjacent to the central retinal vein (CRV) through the center of the optic nerve. It emerges from the optic nerve within the globe, where it branches into four major vessels: the arteriola nasalis retinae superior, arteriola nasalis retinae inferior, arteriola temporalis retinae superior, and arteriola temporalis retinae inferior.

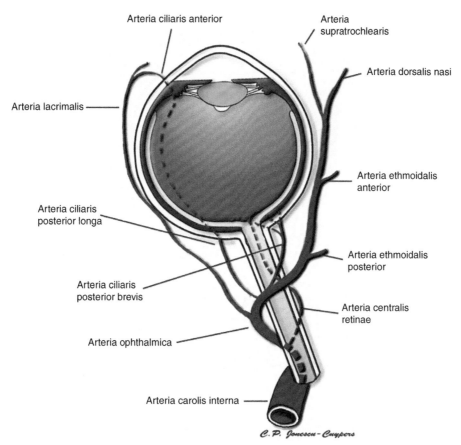

Arteria ciliaris anterior

Arteria supratrochlearis

Arteria dorsalis nasi

Arteria lacrimalis

Arteria ethmoidalis anterior

Arteria ciliaris posterior longa

Arteria ethmoidalis posterior

Arteria ciliaris posterior brevis

Arteria centralis retinae

Arteria ophthalmica

Arteria carolis interna

C. P. Jonescu - Cuypers

Figure 1 Schematic depicting the general vascular organization and the arterial feeds of the major and microciliary processes. The anterior and posterior arterioles that branch off the major arterial circle of the iris (MAC) supply the capillaries of the major and minor processes, respectively. Several smaller branches off the anterior arteriole feed into the marginal capillaries of the major process. Branches of the posterior arteriole feed the internal capillaries (ICs) of the major process. Blood supply to the capillaries of the minor processes is derived from more than one posterior arteriole. Anastomoses (green arrowheads) occur between the lateral branches, some marginal or central capillaries of the major processes and the basally located capillaries that extend posteriorly (white star). From Morrison, J. C. and van Buskirk, E. M. (1984). *American Journal of Ophthalmology* 97: 372–383 in Figure 11.25f in Bron, A. J., Tripathi, R. C., and Tripathi, B. J. (1997). The choroid and uveal vessels. In: *Wolff's Anatomy of the Eye and Orbit*, 8th edn., ch. 11. London: Chapman and Hall Medical.

The vasculature of the eye can be divided into two distinct systems: the retinal system and the uveal system. The retinal system provides blood flow to the inner two-thirds of the retina. The choroid and ciliary body are nourished by the uveal system. The retinal pigment epithelial layer, which is located between the retina and choroid, actively exchanges nutrients and metabolic waste products between the retina and the choroid. Thus, the outer layers of the retina receive their blood flow via the uveal system.

Retina

The retina receives its arterial blood supply from two distinct sources. The CRA provides blood flow to the inner two-thirds of the retina. The CRA branches on the surface of the optic disk, typically producing four main trunks which lie within the nerve fiber layer. Each trunk supplies its respective quadrant of the retina. The outer one-third of the retina, including the photoreceptors and bipolar cells, receives nourishment from the underlying choroid, specifically the choriocapillaris. Nutrients are actively transported between the choroid and retina via the retinal pigment epithelium. In approximately 30% of the people, a cilioretinal artery is present. Typically a branch of a ciliary artery, this vessel supplies a variably sized region of the retina temporal to the optic nerve. When present, the cilioretinal artery is an end artery, and therefore its territory receives no additional blood supply from any other vessels.

Retinal capillaries run parallel to the retinal nerve fiber layer, eventually coalescing into retinal veins, which empty into the CRV. The CRV exits the eye through the optic nerve, running parallel to the CRA. Once in the optic nerve, the CRV receives additional intraneural tributaries, and eventually empties into the superior ophthalmic vein. Although the CRV is normally the only outflow channel

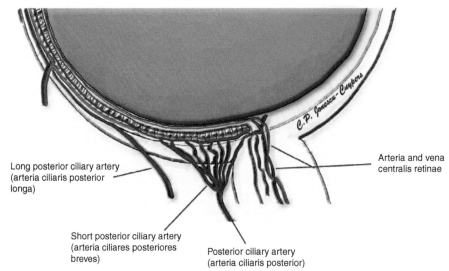

Long posterior ciliary artery
(arteria ciliaris posterior
longa)

Arteria and vena
centralis retinae

Short posterior ciliary artery
(arteria ciliares posteriores
breves)

Posterior ciliary artery
(arteria ciliaris posterior)

Figure 2 Drawing depicting the vascular territories of the ciliary processes. The first terrritory (outlined in the green box) includes the anterior arterioles, the lateral branches, and that feed the lateral branches, and the branches that drain into the basally located venules (white star). The second territory (indicated by the green arrows), includes the marginal capillaries and the capillary network (shown as short connections) that connect to these and the internal capillaries (IC) of the major process. The third territory (outlined in the purple box) includes the capillaries that branch off the posterior arterioles, and the vasculature of the minor processes. According to some authors the vessels in the posterior third of the major processes also fall within the third vascular territory. From Morrison, J. C. and van Buskirk, E. M. (1986). *Transactions of Ophthalmology Society* 105: 13 in Figure 10.28b in Bron, A. J., Tripathi, R. C., and Tripathi, B. J. (1997). The posterior chamber and ciliary body. In: *Wolff's Anatomy of the Eye and Orbit*, 8th edn., ch. 10. London: Chapman and Hall Medical.

for retinal circulation, potential anastamoses exist between the retinal and choroidal circulation. These alternate pathways are significant in the case of a CRV occlusion.

Choroid

The choroid supplies the outer retina with nutrients and maintains the temperature and volume of the eye. The choroidal circulation, which accounts for 85% of the total blood flow in the eye, is a high-flow system with relatively low oxygen content. The choroidal circulation is controlled mainly by sympathetic innervation and is considered not to be autoregulated. This lack of autoregulation makes the choroid more dependent on the ocular perfusion pressure.

The short PCAs supply the posterior choroid and the peripapillary region, while the anterior parts of the choroid are supplied by the long PCAs and the anterior ciliary artery. The anterior ciliary artery is a branch from the OA which accompanies the rectus muscle anteriorly to supply the iris and the anterior choriocapillaries.

The outer choroid, known as Haller's layer, is composed of large caliber, nonfenestrated, vessels. The inner choroid is referred to as Satler's layer, and is composed of significantly smaller vessels. The choriocapillaries of the innermost choroid are composed of richly anastomotic, fenestrated capillaries. The capillaries of the choriocapillaries are

separate and distinct from the capillary bed of the anterior optic nerve.

Venous drainage from the choriocapillaries is primarily through the four vortex veins. Minor drainage also occurs through the ciliary body and the anterior ciliary vein. Venous anastomoses are frequent in the choroid. The vortex veins drain into the inferior and superior ophthalmic veins, which then exit the orbit through the superior and inferior orbital fissures, respectively.

Optic Nerve

The anterior optic nerve is divided into four anatomical regions: the superficial nerve fiber layer, the prelaminar layer, the laminar region, and the postlaminar region (**Figure 3**). The arterial supply to the ONH is derived from branches of the OA. The short PCAs penetrate the perineural sclera at the posterior aspect of the globe to supply the peripapillary choroid and anterior ONH. The circle of Zinn-Haller is a noncontinuous arterial circle surrounding the ONH within the perineural sclera. Formed by a network of small branches of the short PCAs, the circle of Zinn-Haller provides multiple perforating branches to various regions of the anterior optic nerve, peripapillary choroid, and pial arterial system. The capillaries of the anterior ONH are nonfenestrated, contain tight junctions, and form a rich anastomotic plexus. Some investigators surmise that the division of the short PCAs into branches

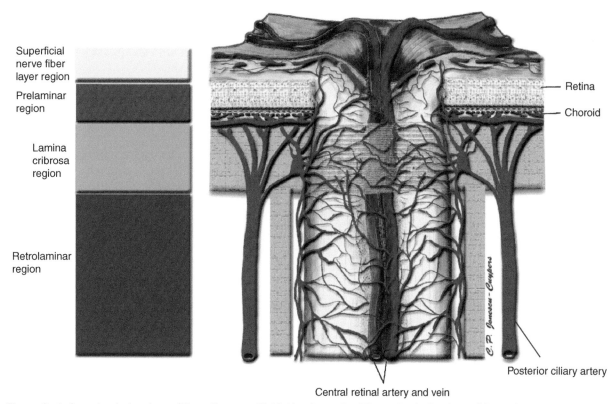

Superficial
nerve fiber
layer region

Prelaminar
region

Lamina
cribrosa
region

Retrolaminar
region

Retina

Choroid

Posterior ciliary artery

Central retinal artery and vein

Figure 3 Left: anatomical regions of the optic nerve. Right: blood vessels of the anatomical regions of the optic nerve.

that supply the choroid and those supplying the ONH form the watershed zone near the ONH.

The superficial nerve fiber layer, which is continuous with the nerve fiber layer of the retina, receives its blood supply from recurrent arterioles arising from branches of the retinal arteries (**Figure 4**). These vessels, referred to as epipapillary vessels, originate in the peripapillary nerve fiber layer and run toward the center of the ONH. The temporal nerve fiber layer may receive additional arterial contribution from the cilioretinal artery when present.

Immediately posterior to the nerve fiber layer is the prelaminar region, which lies adjacent to the peripapillary choroid. In this region, ganglionic axons are grouped into bundles, surrounded by glial tissue septa, as they prepare for passage posteriorly through the lamina cribrosa. The prelaminar region is supplied primarily by branches of the short PCAs and, when present, by branches of the circle of Zinn-Haller (**Figure 5**). The amount of choroidal contribution may be difficult to determine, as there are branches from both the circle of Zinn-Haller and the short PCAs which course through the choroid and ultimately supply the optic nerve in this region. These vessels do not originate in the choroid, but merely pass through it. The choroid contributes little, if any, blood supply to this area of the ONH.

The laminar region is continuous with the sclera and is composed of fenestrated connective tissue lamellae which allow the passage of neural fibers through the sclera. This region, called the lamina cribrosa, receives its blood supply either from centripetal branches of the short PCAs or from branches of the circle of Zinn-Haller (**Figure 6**). These branches pierce the outer aspect of the lamina cribrosa before branching centrally to form an intraseptal capillary network throughout connective tissue. The larger peripapillary choroidal vessels occasionally contribute small arterioles to the lamina cribrosa region.

The retrolaminar region lies posterior to the lamina cribrosa, and is discernible by the beginning of the axonal myelination. Surrounded by the meninges of the central nervous system (CNS), the retrolaminar region is supplied primarily by branches of the pial arteries and the short PCAs (**Figure 7**). The pial system is an abundant anastomotic network fed by the OA, the circle of Zinn-Haller, and recurrent branches of the short PCAs. The pial branches are located within the pia matter and extend centripetally to perfuse the axons of the optic nerve. In addition, the CRA occasionally contributes small branches within the retrolaminar optic nerve.

Like that of the retina, the venous drainage from the ONH is through the CRV. In the superficial nerve fiber

Superficial nerve fiber layer region

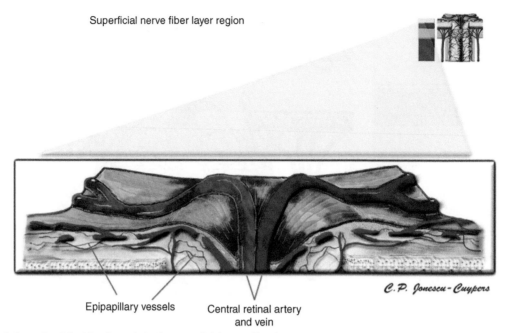

Epipapillary vessels   Central retinal artery
and vein

C. P. Jonescu-Cuypers

Figure 4 Schematic of the blood supply to the superficial nerve fiber layers.

Prelamina region

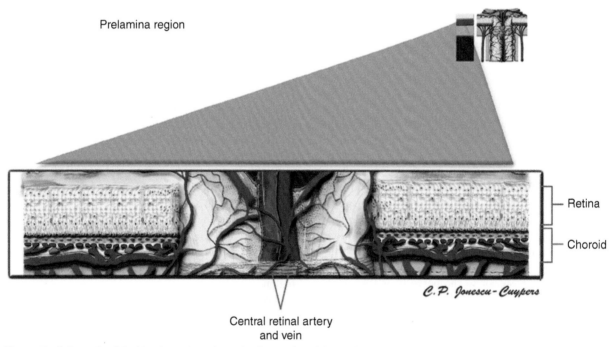

Retina

Choroid

C. P. Jonescu-Cuypers

Central retinal artery
and vein

Figure 5 Schematic of the blood supply to the prelaminar region of the optic nerve.

layer, blood is drained by small, converging veins that empty into the CRV. In the other layers of the ONH, centripetal veins serve as tributaries, eventually emptying into the CRV. In the prelaminar region, there is also a noteworthy contribution from the peripapillary choroidal veins. Small portions of the peripheral region of the ONH may partially drain into the pial venous network, which ultimately joins together with the CRV as well.

Histology of Blood Vessels in the Optic Nerve

The anterior optic nerve is composed of nerve axons, neuroglia, blood vessels, and connective tissue. Large-caliber arteries of the optic nerve contain a muscularis layer, composed of multiple layers of smooth muscle, surrounded by the adventitia. The latter consists of

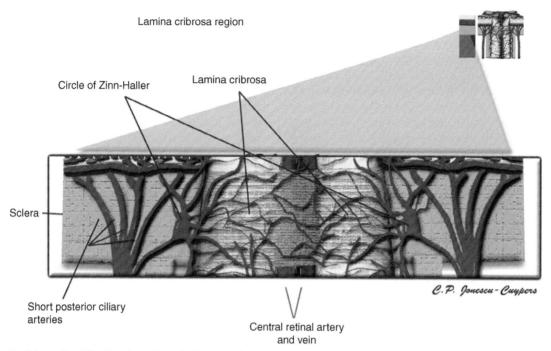

Figure 6 Schematic of the blood supply to the lamina cribrosa region.

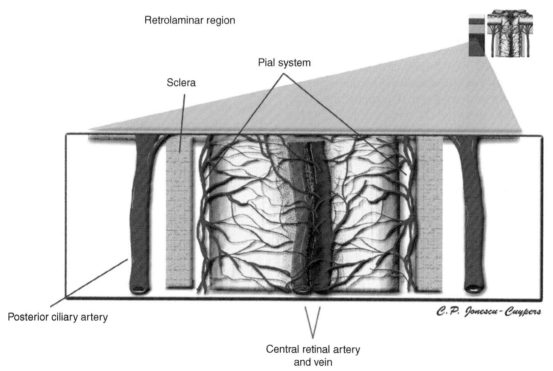

Figure 7 Schematic of the blood supply to the retrolaminar region of the optic nerve.

circumferential collagen fibers which blend with fibers from the perivascular space. On its luminal surface, the muscularis layer is separated from the endothelium by an inner elastic layer. The basement membrane lies between the endothelial cells and blends with the internal elastic lamina.

Arterioles of the optic nerve are much smaller in diameter, and have a single layer of smooth muscle.

They also possess minimal to no elastic lamina, but have a relatively dense reticular lining. The adventitia is continuous with collagen fibers of the intravascular space, and the endothelium is covered by a basement membrane.

Precapillaries and capillaries have a very thin wall and their basement membranes stain positively with the periodic acid-Schiff reaction. The endothelium, basement membrane, and mural cells of the venous structures are similar to that of the capillaries. The veins of the optic nerve are composed of an inner lining of endothelial cells, elastic fibers, thin adventitia, and intermittently, irregularly spaced, smooth muscle cells. When the venules enlarge, the basement membrane thickens accordingly.

Within the anterior ONH, the histology of the vasculature resembles that of the CNS, as the vessels contain a nonfenestrated endothelium with tight junctions. Capillaries predominate within the anterior ONH, and larger vessels are seldom visualized. Posteriorly, larger arterioles may be seen entering the lamina cribrosa. These capillaries and arterioles lack the internal elastic lamina and elastic tissue in the media that is characteristic of the larger vessels of the laminar and retrolaminar regions.

Regulation of Ocular Blood Flow

Autoregulation is the intrinsic ability of a system to maintain constant blood flow despite changes in perfusion pressure and local vascular parameters. The intrinsic control of blood flow involves chemical secretion by the cells in the immediate vicinity of the blood vessels. Within the eye, autoregulation is defined as local vascular constriction or dilation to alter vascular resistance, thereby maintaining a constant nutrient supply in response to perfusion pressure changes. Perfusion pressure is equal to the difference between the mean arterial pressure (MAP) and the venous pressure. The MAP is defined as the diastolic blood pressure plus one-third of the difference between systolic and diastolic blood pressure. Since the pressure in the central vein is normally slightly higher than the intraocular pressure (IOP), the IOP is used as an estimate of ocular venous pressure:

$$\text{Mean ocular perfusion pressure} = \frac{2}{3}\text{MAP} - \text{IOP}$$

$$\text{MAP} = \text{Diastolic BP} + \frac{1}{3}\left(\text{Systolic BP} - \text{Diastolic BP}\right)$$

The maintenance of appropriate ocular blood flow is challenged by physiological changes in IOP, blood pressure, ocular perfusion pressure, and local tissue metabolic demands. As per Poiseuille's law, vascular resistance is inversely proportional to the fourth power of the radius of a vessel, and is directly proportional to the blood viscosity and length of the vessel. In the eye, vascular resistance is therefore dependent on the regulation of vessel diameter. In normal subjects, autoregulation is usually maintained until the IOP reaches approximately 40–45 mmHg. Failure of stable blood flow regulation may lead to ischemic damage of the optic nerve or retinal ganglion cells, which likely contributes to further impairment in vascular regulation. Several mechanisms, including neurogenic-, metabolic-, myogenic-, humoral-, and endothelial-mediated factors, have been demonstrated to play a role in the vascular regulation of ocular blood flow.

Fed primarily by the CRA, the retinal system is generally a low-flow, constant rate system that supplies a highly metabolically active tissue. Although it may only account for as little as 15% of the total ocular circulation, the retinal circulation is capable of providing relatively constant blood flow over a substantial range of IOPs. The retinal and anterior optic nerve head do not possess direct autonomic innervations. Although the retinal and optic nerve head have adrenergic and cholinergic receptors their role remains unclear. Consequently, retinal blood flow is locally autoregulated.

Several vasoactive molecules mediate retinal vascular autoregulation. Endothelial tone is determined by the balance between the vasoconstricting and vasodilating effects of secreted factors. Nitric oxide (NO) is produced by the oxidation of L-arginine by endothelial-derived nitric oxide synthase, which is present in both a constitutively active, membrane-bound form, and an inducible, cytosolic form. NO diffuses to nearby pericytes and smooth muscle, where it activates guanylyl cyclase, leading to the increase of cGMP and subsequent vasodilation. There are numerous stimuli for the production of NO, including increased shear force, bradykinins, insulin-like growth factor 1, acetycholine, and thrombin. Additionally, NO also inhibits platelet aggregation, platelet granule secretion, and leukocyte adhesion.

Vasoconstriction of the retinal microvasculature is stimulated by several vasoactive molecules. Endothelins, the most potent vasoconstricting agents known, are molecules that bind to receptors on pericytes and smooth muscle cells. A second vasoconstrictive substance is angiotensin II. Angiotensinogen is an inactive molecule that is constitutively produced by the liver. In response to physiologic stimuli, the kidneys release renin, which converts angiotensinogen into angiotensin I. Angiotensin converting enzyme (ACE), which is present on the surface of luminal endothelial cells, converts angiotensin I to angiotensin II, and also inactivates bradykinin. Once in its active form, angiotensin II moderates retinal vasoconstriction through the activation of smooth muscle cells and pericytes.

The choroidal vasculature is controlled extrinsically through hormonal influence and stimulation from the autonomic nervous system. It is characterized as a high-flow, variable-rate system which is tightly regulated by the autonomic nervous system. The outer, larger vessels are composed of nonfenestrated endothelium, while the inner, smaller vessels form a richly anastomotic network of fenestrated capillaries. The vascular tone of the choroid is dominated by the sympathetic nervous system. Neurons course from the cranial cervical ganglion to the vascular bed, where vasoconstriction is mediated by the release of neuropeptide Y. The parasympathetic nervous system plays only a moderate and poorly defined role in the regulation of vascular tone. The presence of choroidal autoregulation is controversial and classically considered absent, though some autoregulatory response was reported during perfusion pressure changes.

Technology for Measuring Ocular Blood Flow

Color Doppler Imaging

Although originally developed for monitoring blood flow in the heart, carotid arteries, and peripheral vasculature, color doppler imaging (CDI) has proven to be useful in the study of retrobulbar vessels as well. By combining B-scan ultrasound images with velocity of measurements calculated from the Doppler shift of moving erythrocytes, CDI can be used to assess the velocity of blood flow through the retrobulbar vessels. The peak systolic velocity (PSV) and end diastolic velocity (EDV) can be measured and used to

calculate the mean flow velocity (MFV). An index of resistance (RI) can be calculated as $RI = (PSV - EDV)/PSV$ (**Figure 8**). Further research is necessary to determine the usefulness and application of the RI as it applies to the retrobulbar vasculature. Until recently, one critical limitation of this imaging technique had been that no quantitative information on vessel diameter was obtained, and therefore the calculation of total blood flow or flux was not possible. A recently developed analysis technique has made it possible to determine the diameter of the OA, allowing volumetric blood flow to be assessed. Further research is necessary to apply this technique to the assessment of blood flow through the other major vessels.

Angiography

Using sodium fluoroscein, angiography allows direct visualization of retinal blood flow. Most techniques measure the amount of time that it takes for the fluoroscein to pass through the retinal circulation. Using this technique, data can then be used to assess blood velocity through the retinal and optic disc circulation. Choroidal vessels are visualized in a similar fashion using indocyanine green, which is selected due to its increased binding to plasma proteins, preventing leakage from vessels to the surrounding tissue. Using videoangiography and scanning laser ophthalmoscopy, the arterio-venous passage time and retinal circulation time can be determined. This technology

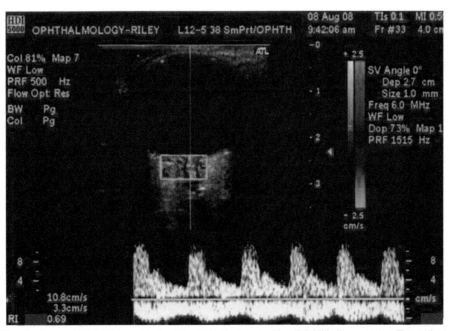

Figure 8 Color Doppler image of the central retinal artery and vein taken with a 7.5-MHz linear probe. The patient is placed comfortably in a half-supine position. An ultrasound probe is placed on closed eyelid and the optic nerve shadow of the optic nerve is identified. The vessels sampled include the ophthalmic artery, central retinal artery, and the nasal and temporal short posterior ciliary arteries. The Doppler-shifted spectrum (time–velocity curve) is displayed at the bottom of the image. Red and blue pixels represent blood movement toward and away from the transducer, respectively. The peak of the wave represents the peak systolic velocity (PSV) and the lower part of the wave the end diastolic velocity (EDV). The resistive index is calculated (PSV – EDV/PSV).

also has limitations, as it is based on the assumption that all of the blood of an area supplied by a specific artery is drained by a single corresponding vein.

Blue Field Entotpic Technique

The blue field entoptic phenomenon is produced by the different absorption of red and white blood cells when the retina is illuminated with blue light. Red blood cells absorb the short wavelength light, while passing white blood cells do not, thereby allowing the flux of the perimacular white blood cells to be estimated. This technique is limited by the assumption that leukocyte flux is proportional to retinal blood flow.

Laser Doppler Velocimetry

Laser Doppler velocimetry (LDV) is a technique that uses the optical Doppler shift of light to measure the blood flow velocities in retinal arterioles and venules. The Doppler shift of light is directly proportional to the blood velocity when the vessel is illuminated with a laser beam. The flow velocity in the vessel can be extrapolated from the range of frequency shifts of the power spectrum of the reflected laser light. The maximum frequency shift corresponds with the maximum velocity in the center of the vessel, assuming laminar flow.

Retinal Vessel Diameters

The aforementioned techniques can be utilized to provide information about ocular blood velocity, but lack the ability to calculate flux or flow rate. To determine blood flow, it is necessary to accurately measure the diameter of the vessel through which the blood is flowing. There are now commercially available systems that permit real-time assessment of retinal vessel diameter. The retinal vessel analyzer (RVA) is composed of a fundus camera and a sophisticated computer system which record vessel size in real time. One such system is the Canon Laser Doppler blood flowmeter, which combines the techniques of LDV and RVA. This approach is still limited to the study of larger vessels, and can only be performed on patients with clear ocular media.

Laser Speckle Technique

When the rough surface of the fundus is illuminated by coherent light, the backscatter of light produces a rapidly varying pattern. The rate of variation of this pattern produced by this laser speckle phenomenon can be measured to compute an estimate of the velocity of blood flowing through the retinal vessels. This laser speckle technique is limited by the fact that it provides only velocity information, as it cannot determine vessel

diameter and therefore cannot be used to measure volumetric flow.

Laser Doppler Flowmetry

The scattering theory for light in tissue, formulated by Bonner and Nossal, assumes a randomization of light directions impinging on the erythrocytes. By directing a laser light on vascularized tissue that contains no large vessels, relative mean velocity of erythrocytes and blood volume can be calculated. Through two-dimensional mapping of the optical Doppler shift, blood flow to the juxtapapillary retina and ONH can be accurately evaluated. There is, however, significant variation in scattering of light between test subjects, likely a result of varying vascular densities and orientations. Thus, this technique can be used to compare changes in a given subject, but has less use in comparison of values between subjects.

Laser Doppler flowmetry can be combined with scanning laser tomography to provide a two-dimensional map of blood flow to the optic nerve and surrounding retina. The Heidelberg retina flowmeter (HRF) is one such commercially available system (**Figure 9**). This technique, however, is most sensitive to blood flow changes in the superficial layers of the ONH, and therefore provides only limited information about the deeper regions. This limits the ability to account for the retinal blood flow that is supplied by the choriocapillaris from the uveal system.

Pulsatile Ocular Blood Flow

Based on the changes in ocular volume and pressure during the cardiac cycle, it is possible to estimate pulsatile ocular blood flow. The pulse amplitude, which is the maximum IOP change during a cardiac cycle, is measured using a modified pneumotonometer. Alternatively, the

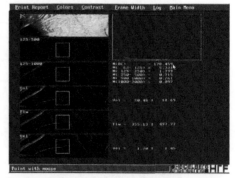

Figure 9 Confocal scanning laser Doppler flowmetry (Heidelberg retinal flowmeter) of optic nerve head and peripapillary retina. The patient is seated with the chin and forehead against the bar. The picture acquisition is performed without the need to dilate the patient's eye. The conventional 40 × 40 pixel measurement window collects flow values in arbitrary units from the entire retina except for large vessels.

ocular fundus pulsation amplitude can be determined by calculating the maximum change in distance between the retina and the cornea during a cardiac cycle. These two values have been shown to be useful in the calculation of pulsatile ocular blood flow, but lack information of the nonpulsatile component of ocular blood flow.

Optical Doppler Tomography

This technique combines the high-resolution cross-sectional imaging of optical coherence tomography with laser Doppler to measure velocity of blood flow in retinal arteries in real time.

Future Studies

Each of the previously discussed technologies quantifies some aspect of ocular blood flow. It is impossible, however, to interpret the impact of any single blood flow parameter measured within a single vascular bed on total retinal metabolism. The measurement of ocular blood flow is only a surrogate assessment of the metabolic status of the retina. Direct measurement of retinal tissue oxygenation would reveal the true impact of ischemia on retinal ganglion cell health and function.

New and emerging tools that assess metabolic parameters may help to reveal the relationship between reductions in ocular blood flow and tissue hypoxia. For example, Michelson and colleagues conducted a study in which they used imaging spectrometry to measure the oxygen saturation in retinal arterioles and venules in patients with glaucomatous optic neuropathy. In all examined eyes, the arteriolar oxygen saturation and the retinal arterio-venous differences in oxygenation were found to significantly correlate with the area of the patient's optic rim. Eyes with normal tension glaucoma, but not those with primary open angle glaucoma, showed significantly decreased arteriolar oxygen saturation. Although further advancements are still needed, these metabolic assessment tools may be very valuable in the evaluation of retinal hypoxia and in elucidating the effects of ocular ischemia. Obtaining accurate measurements of ocular tissue metabolism will greatly improve our understanding of disease pathophysiology, and will therefore lead to advancements in both diagnosis and treatment.

See also: Optic Nerve: Inherited Optic Neuropathies; IOP and Damage of ON Axons; Ischemic Optic Neuropathy; Optic Nerve: Optic Neuritis; Retinal Ganglion Cell Apoptosis and Neuroprotection.

Further Reading

Bron, A. J., Tripathi, R. C., and Tripathi, B. J. (1997). The choroid and uveal vessels. *Wolff's Anatomy of the Eye and Orbit,* 8th edn. London: Chapman and Hall Medical.

Drexler, W. and Fujimoto, J. G. (2008). State-of-the-art retinal optical coherence tomography. *Progress in Retinal and Eye Research* 27: 45–88.

Hardarson, S. H., Harris, A., Karlsson, R. A., et al. (2006). Automatic retinal oximetry. *Investigative Ophthalmology and Visual Science* 47: 5011–5016.

Harris, A., Jonescu-Cuypers, C. P., Kagemann, L., Ciulla, T. A., and Krieglstein, G. K. (2003). *Atlas of Ocular Blood Flow*. Philadelphia, PA: Elsevier.

Harris, A. and Rechtman, E. (2008). Optic nerve blood flow measurement. In: Yanoff, M. and Duker, J. (eds.) *Ophthalmology*, 3rd edn., ch. 10.8, section 2, pp. 52–55. Edinburgh: Elsevier.

Hayreh, S. S. (2001). Blood flow in the optic nerve head and factors that may influence it. *Progress in Retinal and Eye Research* 20: 595–624.

Hayreh, S. S. (2008). Patholophysiology of glaucomatous optic neuropathy: Role of optic nerve head vascular insufficiency. *Journal of Current Glaucoma Practice* 2: 6–17.

Mackenzie, P. J. and Cioffi, G. A. (2008). Vascular anatomy of the optic nerve head. *Canadian Journal Ophthalmology* 43: 308–312.

Michelson, G. and Scibor, M. (2006). Intravascular oxygen saturation in retinal vessels in normal subjects and open-angle glaucoma subjects. *Acta Ophthalmologica Scandinavica* 84: 289–295.

Morrison, J. C. and van Buskirk, E. M. (1984). *American Journal of Ophthalmology* 97: 372–383.

Orgül, S. and Cioffi, G. A. (1996). Embryology, anatomy, and histology of the optic nerve vasculature. *Journal of Glaucoma* 5: 285–294.

Orgül, S., Gugleta, K., and Flamer, J. (1999). Physiology of perfusion as it relates to the optic nerve head. *Survey of Ophthalmology* 43: S17–S26.

Schmetterer, L. and Garhofer, G. (2007). How can blood flow be measured? *Survey of Ophthalmology* 52: 134–138.

Simon, B., Moroz, I., Goldenfeld, M., and Melamed, S. (2004). Scanning laser Doppler flowmetry of nonperfused regions of the optic nerve head in patients with glaucoma. *Ophthalmic Lasers, Surgery, and Imaging* 34: 245–250.

Animal Models of Glaucoma

S I Tomarev, National Institutes of Health, Bethesda, MD, USA

Published by Elsevier Ltd., 2010.

Glossary

BAC – Bacterial artificial chromosome. It is a DNA construct based on a functional fertility plasmid, used for cloning in bacteria. The bacterial artificial chromosome's usual insert size is 150–350 kbp.

BAX – Proapoptotic BCL2-associated X protein. BCL2 is an integral outer mitochondrial membrane protein that blocks the apoptotic death of some cells.

Retrobulbar space – The area located behind the globe of the eye.

Synechia – An eye condition where the iris adheres to either the cornea (anterior *synechia*) or the lens (posterior *synechia*).

Tonometry – The procedure to determine the intraocular pressure.

TUNEL – Terminal deoxynucleotidyl transferase-mediated deoxyuridine triphosphate nick end labeling for detection of DNA fragmentation resulting from apoptotic programmed cell death.

Glaucoma is a complex disease, the initiation and progression of which involves interactions between different parts of the eye and brain. It is difficult to perform experiments directed toward elucidating pathogenic molecular mechanisms and potential treatments for glaucoma in human subjects and, as a rule, only postmortem material can be used for biochemical analysis. Experiments in cell culture or organ culture systems may only partially reproduce the complexity of the natural ocular environment. It is now well recognized that animal models may provide a very useful tool for understanding the underlying molecular mechanisms involved in glaucoma and for identifying new genetic components of the disease, including both causative and modifier genes. In addition, appropriate animal models are used to develop and test new regiments of glaucoma treatment as a prerequisite for clinical trials in humans. A number of animal models of glaucoma have been developed over the years. Since elevated intraocular pressure (IOP) is the most important risk factor in glaucoma, most of the animal models of glaucoma are based on elevation of IOP by surgical procedures or by genetic manipulations. Several models used to study death of the retinal ganglion cells (RGCs) include optic nerve crush or transaction, intravitreal injection of excitory amino acids (glutamate and *N*-methyl-D-aspartic acid (NMDA)), or retinal ischemia. Although these are not true glaucoma models, they allow the comparison of processes leading to RGC death induced by different initial insults. Such comparative analysis may lead to the identification of changes that are specific to glaucoma versus changes that are involved in more general RGC dysfunction. While none of the existing animal models is perfect, some of the existing models have been successfully used to uncover important features of glaucoma pathology in humans. Several factors should be considered in selecting a particular animal model of glaucoma for experimentation: (1) the similarity of the model visual system to the human eye; (2) the similarity in the time course of pathological changes in the model and human eyes; (3) ability to apply genetic manipulations; (4) training necessary to produce affected animals; (5) the size of the eye; (6) availability and difficulties of methods of analysis; (7) availability of animals; and (8) cost. This article briefly describes available animal models of glaucoma with emphasis on the strengths and weaknesses of each model.

Mammalian Models

Primate Models of Glaucoma

Monkey and human eyes are very similar both anatomically and functionally, making monkey models very attractive to study different eye pathologies including glaucoma. IOP in monkeys is measured using the same equipment that is used to measure IOP in humans. Moreover, tonometry and visual-field analysis can be performed in conscious, trained monkeys. This is an important factor since it is well documented that general anesthesia that is necessary to measure IOP in most other animal models results in rapid ocular hypotension. The main disadvantage of monkey models is that experiments with monkeys are expensive and require a highly skilled team of investigators. Moreover, large numbers of animals are required to assess effects of elevated IOP on the optic nerve head (ONH) and retina because of genetic variations between animals.

Several approaches have been used to develop pressure-induced glaucoma models in nonhuman primates. The most common method of IOP elevation in the monkey was originally developed more than 30 years ago and involves circumferential laser photocoagulation treatment of the trabecular meshwork. Several laser sessions are normally required to produce a sustained elevation of IOP. In the treated eyes, IOP rises several days after the laser treatment, normally to between 25 and 60 mmHg, and may last for more than a year. Other methods that have been used to produce elevated IOP elevation in monkeys are less consistent than laser coagulation. They include injection of ghost red cells, latex microspheres, cross-linked polyacrylamide gels, or enzymes into the anterior chamber or application of topical steroids. A non-IOP-related monkey model of glaucoma involves the delivery of endothelin-1 to the retrobulbar space through osmotic pump for 6–12 months; this induces ischemia and leads to the preferential loss of large RGC axons. Ischemia-induced focal axonal loss is similar to human glaucoma and this model may reproduce some aspects of normal tension glaucoma.

A number of important observations have been made using the monkey photocoagulation model. Apoptosis as the primary mechanism of glaucomatous RGC death was first demonstrated in this model before later being confirmed in other models and in human glaucoma. Multifocal electroretinogram (ERG) techniques were used in monkeys to demonstrate that not only RGCs but also cells in the inner and outer nuclear layers are damaged in advanced glaucoma. The monkey glaucoma model has been successfully used to study changes in retinal gene expression patterns after the induction of ocular hypertension. It is also being used to efficiently test new drugs and techniques to reduce IOP. For instance, recombinant adenoviral delivery of the human p21$^{WAF-1/cip-1}$ gene to cause cell cycle arrest before filtration surgery in ocular hypertensive monkey eyes has shown a beneficial effect in long-term control of IOP.

Rodent Models of Glaucoma

Several rodent models of glaucoma have been developed over the last 20 years and new models are at different stages of development in several laboratories. These models have proven useful because the drainage structures of the rodent eye are similar to those in humans. Their utility was enhanced further by the development of new methods to measure IOP and analyze glaucomatous changes in these small eyes. Rodent models, and especially mouse models, are relatively cheap and allow extensive genetic manipulations. Rodent models are preferred when a significant number of animals are required to conduct genetic screens or to test different drugs and agents for neuroprotective or IOP-lowering effects. One of the main disadvantages of rodent models is that there are anatomical differences between rodent and human eyes, including the arterial blood supply to the ONH and the absence of a well-developed, collagenous lamina cribrosa. These variations, as well as differences in general physiology, may explain why expression of certain genes in mouse and human eyes (e.g., mutated myocilin) have differential effects.

Rat Models

Rats are easy to handle. The relatively large size of their eyes allows multiple noninvasive IOP measurements in awake trained animals with commercially available equipment. The TonoPen was the instrument of choice for IOP measurements for many years but has recently been superseded by an induction/impact tonometer, marketed as the TonoLab rebound tonometer. This instrument is easy to operate and can be used in both rats and mice.

Several rat models of pressure-induced glaucoma have been developed over the last 15 years. IOP elevation in the rat eye may be achieved by injection of hypertonic saline solution into the episcleral vein that leads to sclerosis of the aqueous humor outflow pathway. Sustained IOP elevation occurs 7–10 days after injection in most but not all rats. The saline injection generally produces a range of IOP elevation in different animals from a very minimal rise to twofold increase over IOP in control eyes, which can remain elevated for up to several months. Cauterization of two or more of the four large episcleral veins is another method of IOP elevation. In this model, IOP elevation occurs very quickly and there are some indications that this procedure impedes blood outflow from the globe and leads to ischemia. Reports indicate that IOP elevation may last from several weeks to several months without requiring retreatment. IOP increase can be also achieved by laser photocoagulation of the trabecular meshwork with or without injection of Indian ink into anterior chamber. Intracameral injection of hyaluronic acid or latex microspheres is another method of IOP elevation in rats. However, the repeated weekly injections required by this method may produce undesirable effects and are labor consuming. Topical application of dexamethasone for 4 weeks may also be used to induce ocular hypertension. These methods of chronic IOP elevation in rats are accompanied by death of the RGCs by apoptosis, optic nerve degeneration, and ONH remodeling similar to those observed in glaucoma in humans. Acute ocular hypertension, on the other hand, may be produced in rats by cannulation of the anterior chamber with a needle attached to a saline reservoir. Although such treatment leads to retinal ischemic injury, it has been suggested that this model mimics acute angle-closure glaucoma in humans.

A mutant rat strain with a unilateral or bilateral globe enlargement and IOPs that range from 25 to 45 mmHg have been described. In this strain, cupping of the ONH as well as reduction in the number of RGCs progress with age. Unfortunately, this strain was obtained from the Royal College of Surgeons colony that has a mutation in the receptor tyrosine kinase gene, leading to degeneration of the photoreceptors. This drastically limits the utility of this strain to study phenomena that are specific to glaucoma and not confounded by other neurodegenerative processes.

Rat models of glaucoma have been used to study effects of elevated IOP on the ERG, changes in the gene expression patterns in the retina, RGCs and optic nerve, and changes in the protein spectrum of the retina. Rat models also are often used to study neuroprotection. For instance, the hypertonic saline model was used to demonstrate for the first time that agents targeting multiple phases of the amyloid-β pathway provide a therapeutic avenue in glaucoma management.

Mouse Models

Mouse models of glaucoma recently have become very popular. Although most mouse models of glaucoma are based on the elevation of IOP, information about IOP is essential even for the models that do not include experimental IOP manipulation. The mouse eye is much smaller than the human eye, and devices designed for tonometry in humans do not produce reliable data in the mouse. Thus, new methods to measure IOP in mice have been developed and, as a result, the development and acceptance of mouse models of glaucoma have been accelerated. Currently, several invasive and noninvasive methods of IOP measurements in mice exist. The oldest method remains as one of the most reliable and accurate methods and does not depend upon the mechanical properties of the cornea. It involves the insertion of a glass microneedle connected to a pressure transducer into anterior chamber of the eye. However, this procedure cannot be performed too frequently in the same eye, as adequate time is required for corneal wound healing. In addition, cannulation tonometry is technically difficult and training is required to develop sufficient expertise to obtain reliable IOP readings. Cannulation tonometry was used to demonstrate that common mouse strains exhibit different average IOPs in the range between 10 and 20 mmHg. Other methods of IOP measurements in mice were later developed including noninvasive techniques (TonoLab tonometer). Noninvasive techniques allow multiple IOP measurements within short periods of time without extensive training.

Pressure-induced mouse models

Surgical approaches similar to those that were used to produce elevated IOP in rats have also been developed in mice. Significant elevation of IOP in the C57BL/6J mouse eye is accomplished by combined injection of indocyanine green dye into the anterior chamber and diode laser treatment of the trabecular meshwork and episcleral vein region. IOP in operated eyes is significantly elevated 10 days after the surgery but returns back to normal 60 days after the procedure. Histological analysis of the treated eyes 65 days after the surgery revealed development of anterior synechia, loss of RGCs, thinning of all retinal layers, and damage to the optic nerve structures without evidence of prominent cupping. A reduction in the function of all retinal layers, as assessed by ERG studies, indicates that this model produces more dramatic

changes in the retina compared to glaucoma in humans. Elevation of IOP may also be induced by argon laser photocoagulation of the episcleral and limbal veins in C57BL/6J mouse eyes or by cauterization of three episcleral veins in CD1 mouse eyes. In one study, mean IOP in the eyes that underwent laser treatment was about 1.5 times higher than in control eyes for 4 weeks. RGC loss was $22.4 \pm 7.5\%$ at 4 weeks after treatment with the majority of terminal deoxynucleotidyl transferase mediated deoxyuridine triphosphate (dUTP) nick end labeling (TUNEL)-positive apoptotic cells detected in the peripheral areas of the retina. Episcleral vein cauterization produced a maximum IOP elevation within 2–9 days after the procedure, which decreased progressively after that to baseline values in the following 24–33 days. This was associated with a 20% decline in the number of RGCs 2 weeks after the surgery.

The DBA/2J strain has become a popular mouse model of secondary-angle-closure glaucoma and is one of the best-characterized mouse models of glaucoma in general. DBA/2J mice have mutations in two genes, *Tyrp1* and *Gpnmb*, which lead to pigment dispersion, iris transillumination, iris atrophy, and anterior synechia. IOP is elevated in most mice by the age of 9 months. IOP elevation was accompanied by the death of the RGCs, optic nerve atrophy, and optic nerve cupping. Although no group of the RGCs appears especially vulnerable or resistant to degeneration, fan-shaped sectors of cell death and survival radiating from the ONH have been detected. It has been suggested that axon damage at the ONH might be a primary lesion in this model. Several important observations have been made using DBA/2J model. It was shown that proapoptotic protein BAX is required for RGC death but not for RGC axon degeneration in this model of glaucoma, suggesting that BAX may be a candidate human glaucoma susceptibility gene. Unexpectedly, high dose of γ-irradiation accompanied with syngenic bone marrow transfer protected RGCs in DBA/2J. Similar to the results obtained with rat and monkey models, genes involved in the glial activation and immune response are activated in DBA/2J retina as shown by array hybridization. Complement component C1q is upregulated in the retina in several animal models of glaucoma and human glaucoma with timing, suggesting that complement activation plays a significant role in glaucoma pathogenesis. Recent data suggest that complement proteins opsonize central nervous system synapses during a distinct window of postnatal development and that the complement proteins C1q and C3 are required for synapse elimination in the developing retinogeniculate pathway. In DBA/2J mice, C1q relocalizes to adult retinal synapses at an early stage of glaucoma prior to obvious neurodegeneration. These data indicate that C1q in adult glaucomatous retina marks synapses for elimination at early stages of disease, suggesting that the complement cascade mediates synapse loss in glaucoma.

Another DBA/2 substrain, DBA/2NNia, also develops elevated IOP and demonstrates RGC loss and optic nerve degeneration when aged. However, depletion of cells in the inner and outer nuclear layers and significant damage of the photoreceptor cells in 15-month-old mice have also been observed.

Transgenic and knock-out approaches have been used to prospectively develop several mouse models of glaucoma. The main advantage of these approaches is that animals within a particular line produce more uniform responses in terms of IOP elevation and damage to the retina and optic nerve as compared to surgically induced models. A large number of animals may be obtained and no training is needed to produce affected mice. Several lines of transgenic mice have been developed that contain BAC DNAs with a Tyr423His point mutation in the mouse or Tyr437His point mutation in the human *MYOCILIN* (*MYOC*) genes. Tyr437His mutation in the *MYOC* gene leads to severe glaucoma cases in humans, and mouse Tyr423His mutation corresponds to this human mutation. However, expression of mutated mouse or human myocilin in the eye-drainage structures of mice leads to moderate (about 2 mmHg at daytime and 4 mmHg at nighttime) elevation of IOP which is much less dramatic than IOP elevation in humans carrying the same mutation in the *MYOC* gene. Since these mice demonstrate progressive degenerative changes in the peripheral RGC layer and optic nerve with normal organization of the drainage structures, it has been suggested that these mice represent a mouse model of primary open-angle glaucoma. Another model of primary open-angle glaucoma was developed by the expression of a mutated gene for the $\alpha 1$ subunit of collagen type I. This mutation blocks the cleavage of collagen by matrix metalloproteinase-1. Transgenic mice expressing mutated collagen demonstrate elevated IOP which increases to a maximum of 4.8 mmHg greater than controls at 36 weeks.

A transgenic model of acute angle-closure glaucoma was developed by expression of calcitonin-receptor-like receptor under the control of a smooth muscle α-actin promoter. Overexpression of this receptor in the papillary sphincter muscle results in enhanced adrenomedullin-induced sphincter muscle relaxation that leads to abrupt transient rises in IOP in some mice up to a mean level of about 50 mmHg between 30 and 70 days of age. Although the aberrant ocular functions of adrenomedullin and calcitonin-gene-related peptide have not been associated with the pathogenesis of human acute glaucoma, it has been suggested that adrenomedullin and its receptor in the iris sphincter may present novel targets for the treatment of angle-closure glaucoma.

Normal-tension mouse models

Mice deficient in the glutamate transporters GLAST or EAAC1 show RGC death and typical glaucomatous damage of the optic nerve without elevation of IOP. It has been shown that the glutathione levels are decreased in Müller cells of GLAST-deficient mice, while administration of glutamate receptor blocker prevents loss of RGCs. RGCs are more sensitive to oxidative stress in EAAC1-deficient mice. These mice represent a model of normal tension glaucoma and are currently being used to develop therapies directed at IOP-independent mechanisms of RGC loss.

Developmental mouse models

Defects in genes involved in the development of the anterior eye segment may lead to relatively rare developmental glaucomas, which account for less than 1% of all human glaucoma cases. Several genes have been implicated in congenital glaucoma and anterior segment dysgenesis. They include *CYP1B1*, *FOXC1*, *FOXC2*, *PITX2*, *LMX1b*, and *PAX6*. Although *Cyp1b1* knock-out mice do not develop elevated IOP, they have ocular abnormalities similar to defects in humans with primary congenital glaucoma: small or absent Schlemm's canal, defects in the trabecular meshwork, and attachment of the iris to the trabecular meshwork and peripheral cornea. *Foxc1*$^{-/-}$ mice die at birth, while *Foxc1*$^{+/-}$ animals are viable but have defects in the eye-drainage structures in the absence of IOP changes. Similar eye defects are observed in *Foxc2*$^{+/-}$ mice. It has been suggested that *Foxc1*$^{+/-}$ and *Foxc2*$^{+/-}$ mice are useful models for studying anterior segment development and its anomalies, and they may allow identification of genes that interact with *Foxc1* and *Foxc2* to produce a phenotype with elevated IOP and glaucoma.

Transgenic mice overexpressing the ocular development-associated gene (ODAG) in photoreceptors under the control of mouse Crx promoter exhibit gradual protrusion of the eyeballs with dramatically increased IOP that is not attributable to mechanical block of the aqueous humor outflow. These transgenic mice demonstrate optic nerve atrophy and impaired retinal development. All retinal layers of these transgenic mice are affected, thereby differentiating this model from a typical glaucomatous retina where morphological changes are detected only in the RGC layer.

Other Mammalian Models

Several other mammalian models of glaucoma have been developed. Pig eyes are relatively large and, although the drainage outflow system of the pig eye is slightly different from that of the human eye, the porcine retina is more similar to the human retina than that of other large mammals (i.e., dog, goat, and cow). Cauterization of three porcine episcleral veins leads to a 1.3-fold elevation of IOP that is apparent 3 weeks after the surgery and persists for at least 21 weeks. It has been shown that endothelium leukocyte adhesion molecule 1 (ELAM-1), a molecular marker for human glaucoma, is also elevated in the trabecular meshwork of pigs with elevated IOP.

Rabbits are a standard ophthalmic animal model for glaucoma filtration surgery and are often used for the development of new devices (e.g., drainage implants and degradable biopolymers) and medical therapies including gene therapy. At the same time, due to the unique anatomy of the rabbit eye, laser-induced elevation of IOP, like that in the monkey eye, is difficult to achieve. Alternatively, application of glucocorticoids has been successfully used to induce ocular hypertension in rabbit model. In addition, a line of rabbits with congenital glaucoma has been developed. Thick subcanalicular tissues and the deposition of extracellular matrix in the trabecular meshwork appear to contribute to the ocular hypertension exhibited by this model.

Several purebred dogs develop glaucoma with high frequency. Among North American breeds, the highest prevalence of primary glaucoma is observed in the American cocker spaniel (5.52%), basset hound (5.44%), and chow chow (4.70%), exceeding that in humans. Lens displacement resulting in secondary glaucoma is common in terrier breeds. The high prevalence of the glaucomas in these canine breeds suggests a genetic basis of pathophysiology.

It has been reported that topical application of corticosteroid induces reproducible elevation of IOP in the cow. The large amount of tissues available from the cow eye makes this model useful for biochemical studies.

Nonmammalian Models

Zebrafish

The zebrafish is an excellent model system to study complex diseases as it allows one to combine forward and reverse genetic approaches. The general organization of the zebrafish eye is similar to the human eye, although the fine details of individual ocular structures are rather different. In particular, there are significant differences in the organization of the iridocorneal angle between zebrafish and mammals. They include the trabecular meshwork and lack of iris muscles as well as ciliary folds in zebrafish as compared to mammals. Even with these limitations in mind, zebrafish have been used as a model organism for glaucoma studies. An accurate method exists to measure IOP in zebrafish which is based on servo-null electrophysiology. Using this method, baseline IOP differences have been demonstrated in genetically distinct zebrafish strains. Among tested strains, the long fin strain (LF) had the highest IOP (20.5 ± 1.2 mm Hg) while the Oregon AB strain (AB) has the lowest IOP (10.8 ± 0.3 mm Hg). At the same time, these differences in IOP do not lead to detectable defects of the retina or in visual function. Zebrafish have also been used to determine the function of several genes (*foxc1*, *lmx1b*, *wdr36*, *olfactomedin 1*, and *olfactomedin 2*)

implicated in glaucoma. It has been shown that wdr36 functions in ribosomal RNA processing and interacts with the p53 stress-response pathway, while olfactomedin 1 is essential for optic nerve growth and targeting of the optic tectum. Thus, zebrafish system may be very useful to complement studies with other model organisms, but by itself should be used with caution to study glaucoma.

Other Nonmammalian Models

Open-angle glaucoma characterized by elevated IOP can be induced in domestic chickens or in Japanese quails when they are reared under continuous light. Besides, an unknown autosomal dominant mutation in a Slate line of domestic turkeys has been identified that leads to secondary angle-closure glaucoma. Although these models might be useful to study certain aspects of glaucoma in humans, one should remember that structural and physiological differences between human and bird eyes complicate direct comparison.

Drosophila eyes have been suggested as a useful system for the discovery of genes that are associated with glaucoma. However, the general organization of human and *Drosophila* eyes are very different and data obtained with *Drosophila* may not always be relevant to glaucoma in humans.

Conclusion

Animal models have already provided interesting new information about potential mechanisms of glaucoma in humans. However, even in monkey models which most closely mimic the human form of the disease, the time course of changes in the glaucomatous eyes may be significantly accelerated as compared with human glaucomatous eyes. Indeed, all of the previously discussed systems are, after all, just models of human glaucoma. Reactions to the same insult (IOP, expression of the same mutated protein, etc.) may be somewhat different between various animal models and humans. Results obtained with these models should not automatically be applied to human condition and should be confirmed by testing in human subjects when possible. Nevertheless, information on molecular mechanisms of glaucoma obtained using animal models might be extremely valuable to develop new therapeutic approaches for glaucoma treatment and prevention in humans.

Further Reading

Anderson, M. G., Libby, R. T., Gould, D. B., et al. (2005). High-dose radiation with bone marrow transfer prevents neurodegeneration in an inherited glaucoma. *Proceedings of the National Academy of Sciences of the United States of America* 102: 4566–4571.

Baulmann, D. C., Ohlmann, A., Flügel-Koch, C., et al. (2002). Pax6 heterozygous eyes show defects in chamber angle differentiation that are associated with a wide spectrum of other anterior eye segment abnormalities. *Mechanisms of Development* 118: 3–17.

Harada, T., Harada, C., Nakamura, K., et al. (2007). The potential role of glutamate transporters in the pathogenesis of normal tension glaucoma. *European Journal of Clinical Investigation* 117: 1763–1770.

Iwata, T. and Tomarev, S. (2008). Animal models for eye diseases and therapeutics. In: Conn, P. M. (ed.) *Sourcebook of Models for Biomedical Research*, pp. 279–287. Totowa, NJ: Humana Press.

Levkovitch-Verbin, H., Quigley, H. A., Martin, K. R., et al. (2002). Translimbal laser photocoagulation to the trabecular meshwork as a model of glaucoma in rats. *Investigative Ophthalmology and Visual Science* 43: 402–410.

Libby, R. T., Anderson, M. G., Pang, I., et al. (2005). Inherited glaucoma in DBA/2J mice: Pertinent disease features for studying the neurodegeneration. *Visual Neuroscience* 22: 637–648.

McMahon, C., Semina, E. V., and Link, B. A. (2004). Using zebrafish to study the complex genetics of glaucoma. *Comparative Biochemistry and Physiology – Part C: Toxicology and Pharmacology* 138: 343–350.

Morrison, J. C., Johnson, E. C., Cepurna, W., and Jia, L. (2005). Understanding mechanisms of pressure-induced optic nerve damage. *Retinal Eye Research* 24: 217–240.

Pang, I.-H. and Clark, A. F. (2007). Rodent models for glaucoma retinopathy and optic neuropathy. *Glaucoma* 16: 483–505.

Rasmussen, C. A. and Kaufman, P. L. (2005). Primate glaucoma models. *Journal of Glaucoma* 14: 311–314.

Senatorov, V., Malyukova, I., Fariss, R., et al. (2006). Expression of mutated mouse myocilin induces open-angle glaucoma in transgenic mice. *Journal of Neuroscience* 26: 11903–11914.

Smith, R. S., John, S. W. M., Nishina, P. M., and Sundberg, J. P. (eds.) (2002). *Systematic Evaluation of the Mouse Eye*. Boca Raton, FL: CRC Press.

Weinreb, R. N. and Lindsey, J. D. (2005). The importance of models in glaucoma research Volume. *Journal of Glaucoma* 14: 302–304.

Blood–Retinal Barrier

J Cunha-Vaz, AIBILI, Coimbra, Portugal

Glossary

iBRB – The inner blood–retinal barrier is a situation of restricted permeability established at the level of the retinal vessels, between the blood and retinal tissue, by the tight junctions (zonula occludents) between neighboring retinal endothelial vessels and the retinal endothelial cells themselves.

oBRB – The outer blood–retinal barrier is a situation of restricted permeability established at the level of the retinal pigment epithelium, the blood, and the retinal tissue, by the tight junctions (zonula occludents) between neighboring retinal pigment epithelial cells and the retinal pigment epithelial cells themselves.

Retinal leakage analyzer – A method developed to perform localized measurements of blood–retinal barrier fluorescein leakage using a confocal scanning laser ophthalmoscope modified to obtain fluorescence measurements in the vitreous.

Tight junctions – Tight junctions are specialized junctions uniting neighboring cells by fusion of the outer leaflets of their cell membranes (zonulae occludentes) thus obliterating the intercellular space fusion and restricting paracellular diffusion.

Uveitis – Uveitis is inflammation of the uvea, which is the vascular layer of the eye sandwiched between the retina and the sclera. The uvea extends toward the front of the eye and consists of the iris, choroid layer, and ciliary body.

Vitreous fluorometry – A method developed to measure the fluorescence resulting from the presence of fluorescein in the vitreous after intravenous administration. It is a direct indicator of the permeability of the blood–retinal barrier to fluorescein.

The entire eye must function as the organ for vision and is organized with two major goals: normal function of the visual cell and the need to maintain ideal optical conditions for the light to access the visual cells, located in the back of the eye.

The blood–ocular barriers play a fundamental role in the preservation and maintenance of the appropriate environment for optimal visual cell function (**Figure 1**).

The blood–ocular barriers include two main barrier systems: the blood–aqueous barrier and the blood–retinal barrier (BRB) (**Figure 2**), which are fundamental to keep the eye as a privileged site in the body by regulating the contents of its inner fluids and preserving the internal ocular tissues from variations which occur constantly in the whole circulation. The blood–ocular barriers must not only provide a suitable, highly regulated, chemical environment for the avascular transparent tissues of the eye, but also serve as a drainage route for the waste products of the metabolic activity of the ocular tissues.

One of these barriers, the BRB, similar to the blood–brain barrier (BBB), is particularly tight and restrictive and is a physiologic barrier that regulates ion, protein, and water flux into and out of the retina.

It is also important to realize that once inside these barriers there are no major diffusional barriers between the extracellular fluid of the retina and adjacent vitreous; nor does the vitreous body itself significantly hinder the diffusional exchanges between the posterior chamber and the retinal extracellular fluid. This means that the functions of both barriers, blood–aqueous barrier and BRB, influence each other and must work in equilibrium.

Blood–Retinal Barrier

The presence of an intact BRB is essential for the structural and functional integrity of the retina and in clinical conditions where BRB breakdown occurs vision may be seriously affected.

The BRB consists of inner and outer components (inner BRB (iBRB) and outer BRB (oBRB)) and plays by itself a fundamental role in of the microenvironment of the retina and retinal neurons.

The BRB regulates fluids and molecular movement between the ocular vascular beds and retinal tissues and prevents leakage into the retina of macromolecules and other potentially harmful agents (**Figure 3**).

The iBRB is established by the tight junctions (TJs) (zonulae occludentes) between neighboring retinal endothelial cells. These specialized TJs restrict the diffusional permeability of the retinal endothelial layer to values in the order of $0.14 \times 10^{-5} \, \mathrm{cm \, s^{-1}}$ for sodium fluorescein. The retinal endothelial layer functions as an epithelium and in this way is directly associated with its differentiation and with the polarization of the BRB function. This continuous endothelial cell layer, which forms the main structure of the iBRB rests on a basal lamina that is covered by the processes of astrocytes and Müller cells. Pericytes are also present, encased in the basal lamina, in close contact with the endothelial cells but do not form a continuous layer and, therefore, do not contribute to the diffusional barrier. Astrocytes, Müller cells, and pericytes are considered to influence the activity of the retinal endothelial cells and of the iBRB by transmitting to endothelial cells regulatory signals indicating the changes in the microenvironment of the retinal neuronal circuitry.

The oBRB is established by the TJs (zonulae occludentes) between neighboring retinal pigment epithelial (RPE) cells. The RPE is composed of a single layer of RPE cells that are joined laterally toward their apices by

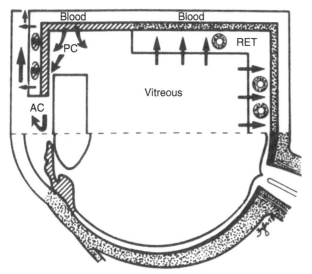

Figure 1   Schematic drawing of the blood–ocular barriers and main fluid movements. RET, retina; PC, posterior chamber; AC, anterior chamber.

TJs between adjacent lateral cell walls. The RPE resting upon the underlying Bruch's membrane separates the neural retina from the fenestrated choriocapillaries and plays a fundamental role in regulating access of nutrients from the blood to the photoreceptors as well as eliminating waste products and maintaining retinal adhesion. The metabolic relationship of the RPE apical villi and the photoreceptors is considered to be critical for the maintenance of visual function.

In both, iBRB and oBRB, the cell TJs restrict paracellular movement of fluids and molecules between blood and retina, and the endothelial cells and RPE cells actively regulate inward and outward movements. As a result, the levels in the blood plasma of aminoacids or fatty acids fluctuate over a wide range while their concentrations in the retina remain relatively stable.

Inner Blood–Retinal Barrier

Retinal endothelial cells

The endothelial cells of retinal capillaries are not fenestrated and have a paucity of vesicles. The function of these endothelial vesicles has been described as endocytosis or transcytosis that are receptor mediated. Pinocytotic residues are selectively decreased in the BRB endothelial cells. Receptor-facilitated transport mechanisms are used to move materials across the BRB (**Figure 4**). Channel-facilitated transport using transmembrane proteins is another mechanism for diffusion of specific substrates across the BRB. The glucose transporter Glut 1 is a good example, supplying the neuronal tissue with necessary glucose. Disruption of the iBRB in pathological conditions

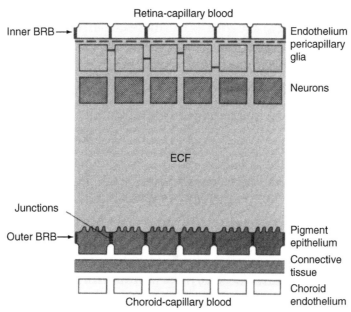

Figure 2   Schematic presentation of the inner and outer blood–retinal barriers (BRBs) and their relative location. ECF, extracellular fluid.

is associated with increased vesicle formation and disrupted endothelial membranes. These alterations may develop before opening of the TJs is detected on ultrastructural examination.

Retinal endothelial TJs

TJs or zonula accludentes of the retinal vascular endothelium are formed by fusion of the outer leaflets of adjacent endothelial cell membranes and were described for the first time in the retinal vessels in 1966.

The TJ obliterates the interendothelial space and confers highly selective barrier properties to the capillaries. Diffusion of molecules from the lumen to the tissue is significantly restricted by TJ.

These endothelial junctions have, like in the brain capillaries, extremely high electrical resistance, 1000–3000 ohm cm^2.

The TJ complex contains at least 40 proteins composing transmembrane and internal adapter proteins that regulated paracellular flux. Transmembrane proteins that make up the TJ are occludins, claudins, and junctional adhesion molecules (JAMs). Occludin is a 65-kDa protein and its changes correlate with permeability changes making it a likely candidate in regulating the opening and closing of the TJ. Claudins regulate small charged molecules and ion permeability. JAM is part of a family of proteins and is associated with adhesion molecules.

Numerous adapter proteins localize just below the membrane and act as TJ organizers and cytoskeleton anchors.

It is important to realize that TJs are dynamic structures that can be regulated by signal transduction through cyclic AMP levels, tyrosine kinases, etc.

Müller cells, astrocytes, and pericytes

A close spatial relationship exists between Müller cells and blood vessels in the retina suggesting a critical role for these cells in the formation and maintenance of the BRB, regulating the functions of barrier cells in the uptake of nutrients and in the disposal of metabolites under normal conditions. Barrier function is also impaired by matrix metaloproteinases (MMPs) from Müller cells as these MMPs lead to proteolytic degradation of the TJ protein occludin.

Astrocytes originate from the optic nerve and migrate to the retinal nerve fiber layer during retinal vascular development. They are associated closely with the retinal vessels and help to maintain their integrity. Astrocytes are known to increase the barrier properties of the retinal endothelium by enhancing the expression of TJ protein Z0.1 and may moderate TJ integrity. Astrocytes are considered to play an important regulatory role in the function of the BRB.

Finally, the pericytes have been shown to play a role in regulating vascular tone, secrete extracellular material, and being phagocytic. Pericytes are considered to play an accessory role in maintaining the integrity of the iBRB by inducing mRNA and protein expression of occludin and other protein junctions. There is evidence that pericytes interact with the endothelial cells contributing to their modulation.

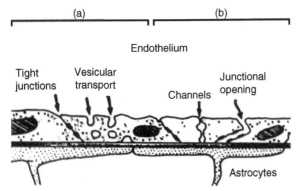

Figure 3 Pathways for solute movements across the inner blood–retinal barrier (retinal endothelial cells): (a) normal; (b) mechanisms of breakdown of iBRB.

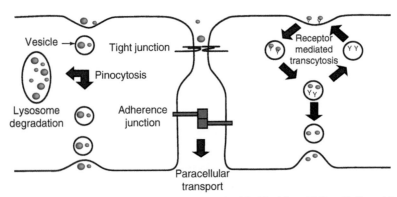

Figure 4 Transport mechanisms across retinal vascular endothelial cells. Modified from Philips, B. E. and Antonetti, D. A. (2007). Blood–retinal barrier. In: Joussen, A. M., Gardner, T. W., Kirchoff, B., and Ryan, S. J. (eds.) *Retinal Vascular Disease*, pp. 139–153. Berlin: Springer.

Outer Blood–Retinal Barrier

RPE cells

The RPE cells transport water from the subretinal space or apical side to the blood or basolateral side. Therefore, the RPE has the structural properties of an ion-transporting epithelium.

RPE cells regulate water content and lactic acid removal generated by the characteristic high metabolic rates in the retina. RPE cells transport water out of the retina and into the choroidal capillary plexus. The force generated by this water flux produces an adhesion force and helps to maintain retinal attachment. Water transport is linked with ion transport, organic anion transport, and other drainage mechanisms.

This outward molecular movement is largely dependent on active ionic transport associated with a relevant high oncotic pressure in the choroid.

RPE cells also have a fundamental role by transporting glucose and retinol, in the appropriate direction, from blood to the photoreceptors.

RPE tight junctions

Paracellular movement of larger molecules is restricted by the TJ between neighboring RPE cells. The paracellular resistance is 10 times higher than the transcellular resistance, classifying the RPE as a tight epithelium. Occludin, claudins, and adapter proteins have been detected at the RPE TJ as in TJ elsewhere. The TJs of the RPE are anchored to the actin cytoskeleton of RPE cells, interact with signaling molecules, and are important for the establishment of cell polarity.

In addition to TJ between RPE cells, the polarized distribution of RPE membrane proteins contributes to the function of the oBRB.

The outer retinal layers are nourished from the blood circulation through the fenestrated capillaries of the choriocapillaris and to subserve this function there is a necessity of a large baso-apical molecular movement from choroid to retina. Waste products of retinal metabolism are transported to the choroid through the oBRB.

Polarity of the outer and inner barriers: TJ modulation

Establishment of cell polarity is a characteristic of a tissue barrier. The endothelial cells of the retinal vessels and RPE develop distinct apical versus basal membrane surfaces. This cell polarity is associated with organization of the cytoskeleton, apical/basal cell membrane proteins, and organization of the junctional complexes between neighboring cells.

It is these TJ protein complexes that allow the establishment of the polarities of the BRB, restricting paracellular diffusion of blood–barrier compounds into the neuronal tissues.

Understanding the normal function of these TJs and the pathological changes that are induced by their alteration resulting in increased permeability is necessary to understand disease progression in retinal diseases such as diabetic retinopathy (iBRB primarily affected) and wet age-related macular degeneration (oBRB primarily affected).

Understanding the role of relevant proteins such as occludins, claudins, and JAMs in BRB physiology and in retinal pathology will certainly contribute to improved management of retinal disease.

Other factors regulating the molecular movement in the eye

Molecular movement from the retinal and choroidal vascular systems into, out, and across them is complex and limited by a variety of other ocular structures. There is a continuous molecular movement of small molecules (mainly water) from the vitreous cavity into the inner retina and through RPE to the choroid.

The major proportion of aqueous humor secreted by the ciliary body from its rich vascular supply provides a bulk flow of fluid through the anterior chamber of the eye but a smaller proportion enters the vitreous cavity where it is largely cleared across the retina and RPE to the choroidal circulation.

Molecular movement from vitreous to choroid is slowed by the cortical vitreous with its high concentration of hyaluronic acid stabilized in a relatively dense type II collagen matrix and the internal limiting membrane (ILM) of the retina. The ILM offers resistance to the diffusion of macromolecules of 148 kDa but allows the passage of smaller molecules. Movement through the retina of molecules that have crossed the iBRB into the retina is largely through the extracellular tissue spaces. Similary, Bruch's membrane that separates the basal RPE from the fenestrated capillaries appears, in health, to offer little resistance to molecular movement.

The presence of fenestrations in the choriocapillaris allows the passage of even large molecules such as albumin into the extravascular spaces of the choroid. The choriocapillaris, therefore, contributes little to the oBRB.

Molecular movement across the oBRB is, however, probably influenced by the very high rate of blood flow in the choroid. A possible explanation for the high blood flow in the choroid is that there is a need not only to supply oxygen and metabolites to the energy demanding retina and RPE, but also for the rapid removal of waste products of the retinal metabolism into the blood circulation.

Finally, the ciliary body may have a relevant regulatory role in the overall maintenance of the retinal microenvironment. The large surface covered by the ciliary processes, their location where the aqueous and vitreous meet, and the multiple transport functions of the ciliary epithelium are all factors suggesting an important role for the ciliary body in the regulation of the inner ocular fluids.

The microenvironment of the retina, which closely resembles brain extracellular fluid and is in equilibrium with the vitreous is, therefore, maintained by a variety of facilitated and active transport processes which are located mainly in the iBRB and oBRB with the retinal endothelial cells and RPE playing fundamental roles.

The Blood–Retinal Barrier and Ocular Immune Privilege

The immune response has developed and evolved to protect the organism from invasion and damage by a wide range of pathogens. With time, the immune system has developed destructive responses that are specific for pathogens as well as tissues. Such tissue injury might, however, have a devastating effect on the function of an organ such as the eye that needs to maintain optical stability.

The existence of ocular immune privilege is dependent upon multiple factors such as immunomodulatory factors and ligands, regulation of the complement system within the eye, tolerance promoting antigen-presenting cells (APCs), unconventional drainage pathways, and, with particular relevance, the existence of the blood–ocular barriers.

The blood–ocular barriers provide a relative sequestration of the anterior chamber, vitreous, and neurosensory retina from the immune system and create the necessary environment for the existence of ocular immune privilege. The evolution of immune privilege as a protective mechanism for preserving the function of vital and delicate

organs such as the eye has resulted in a complex system with multiple regulatory safeguards for the control of both innate and adaptative immurity. The consequences of inadvertent bystander tissue destruction by antigen-monspecific inflammation can be so catastrophic to the organ or host that a finely tuned regulatory system is needed to ensure the integrity of the ocular tissues and maintain optical relationships.

There are also several lines of evidence that points to immunosuppressive functions in the BRB cells, RPE, and retinal endothelial cells. These immunosuppressive effects are apparently due to the secretion of a variety of soluble factors, such as cytokines and growth factors.

Clinical evaluation of the blood–retinal barrier

Fluorescein angiography, an examination procedure performed routinely in the ophthalmologist's office, permits a dynamic evaluation of local circulatory disturbances and identifies the sites of BRB breakdown (**Figure 5**). It is, however, only semi-quantitative and its reproducibility depends on the variable quality of the angiograms.

Vitreous fluorometry was developed as a method capable of quantification of both inward and outward movements of fluorescein across the BRB system in the clinical setting. Protocols were devised, tested, and dedicated instrumentation developed.

With the development of vitreous fluorometry methodologies, a large number of clinical and experimental studies demonstrated convincingly the major role played by alterations of BRB in posterior segment disease.

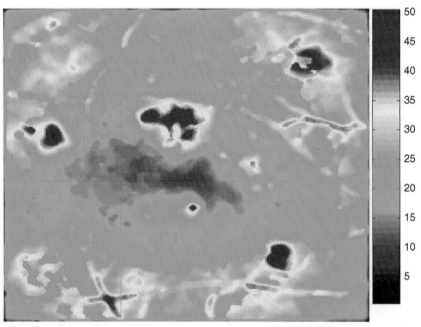

Figure 5 Sites of fluorescein leakage into the vitreous identified by the retinal leakage analyzer in an eye with nonproliferative retinopathy of a patient with diabetes type 2. Blue indicates minimal leakage; red indicates maximum leakage.

In clinical situations, alterations of the BRB have been measured in pathologies of the RPE, aged-related macular degeneration, and macular edema, as well as in hypertension and diabetes. The clinical use of vitreous fluorometry, however, has declined because it offers only an overall measurement over the posterior role and because at the time of its development there were no drugs available for stabilizing the BRB. Nowadays, vitreous fluorometry is mostly used in experimental research and in drug development.

More recently, retinal leakage mapping has been introduced to identify the sites of BRB breakdown. Further developments of this methodology based on confocal scanning laser ophthalmology (SLO-Retinal Leakage Analyzer) associated with improved optical coherence tomography imaging are expected to contribute to earlier diagnosis of BRB alterations in retinal disease as well as improved testing of the effect of new drugs that are now becoming available for treatment of retinal disease.

Blood–retinal barrier and macular edema

Macular edema is the result of an accumulation of fluid in the retinal layers around the fovea, contributing to vision loss by altering the functional cell relationship in the retina and promoting an inflammatory reparative response.

Macular edema is only a nonspecific sign of ocular disease and not a specific entity. It should be viewed as a special and clinically relevant type of macular response to an altered retinal environment, in most cases associated with an alteration of the BRB. It occurs in a wide variety of ocular situations such as uveitis, trauma, intraocular surgery, vascular retinopathies, hereditary dystrophies, diabetes, age-related macular degeneration, etc.

The increase in water content of the retinal tissue that characterizes macular edema may be initially intracellular or extracelullar. In the first case, also called cytotoxic edema, there is an alteration of the cellular ionic distribution.

In the second case, more frequent and clinically more relevant, the extracellular accumulation of fluid is directly associated with an alteration of the BRB. In this later situation, the protective effect of the BRB is lost and the Starling law applies. When there is breakdown of the BRB, any changes in the equilibrium between hydrostatic and oncotic pressure gradients across the BRB contribute to further water movements and progression of the macular edema.

It is also of great relevance to keep in mind that the BRB cells, retinal endothelial cells, and retinal pigment epithelial cells, are both the target and producer of ecosanoids, growth factors, and cytokines. Breakdown of the BRB leading to situations of macular edema may be mediated by locally released cytokines and induction of an inflammatory reparative response creating the conditions for further release of cytokines, growth factors, etc.

Macular edema is also one of the most serious consequences of inflammation in the retinal tissue. Inflammatory cells can alter the permeability of the TJs that maintain the iBRB and oBRB. Cell migration may occur primarily through splitting the junctional complex or through the formation of channels or pores across the junctional complex.

Macular edema has particular relevance for its frequency in diabetic retinopathy. Leukocyte adhesion to retinal vessels and breakdown of the BRB appear to be mediated by nitric oxide (NO). NO upregulates intercellular adhesion molecule-1 (ICAM-1) and promotes the downregulation of TJ protein expression.

Relevance of BRB to Treatment of Retinal Diseases

When administered systemically, drugs must pass the BRB to reach therapeutic levels in the retina. Drug entrance into the retina depends on a number of factors, including the plasma concentration profile of the drug, the volume of its distribution, plasma protein binding, and the relative permeability of the BRB. To obtain therapeutic concentrations within the retina, new strategies must be considered such as delivery of nanoparticles, chemical modification of drugs to enhance BRB transport, coupling of drugs to vectors, etc.

The BRB must be considered as a dynamic interface that has the physiological function of specific and selective membrane transport from blood to retina and active efflux from retina to blood for many compounds, as well as degradative enzymatic activities.

From the viewpoint of drug delivery, designing drugs (including peptides) with greater lipophilicity to enhance BRB permeability seems to be an easy approach. However, such a strategy would not only increase the permeation into tissues other than the retina, but also decrease the bioavailability due to the hepatic first pass metabolism in the case of oral administration. Accordingly, for the development of retina-specific drug delivery systems for neuroactive drugs the most effective approach is to utilize the specific transport mechanisms active at the BRB. That would mean designing drugs that mimic the substrates to be taken by particular transporters or receptors existing in the BRB.

Eye drops are generally considered to be of limited benefit in the treatment of posterior segment diseases. Newer pro-drug formulations that achieve high concentrations of the drug in the posterior segment may have a role in the future. Meanwhile, periocular injection is one modality that has offered mixed results.

Finally, the last years have seen a generalized and surprising safe utilization of intravitreal injections, a form of administration that circumvents the BRB. Steroids and a variety of anti-VEGF drugs have been administered through intravitreal injections to a large number of patients without significant side effects and demonstrating good acceptance by the patients. Intravitreal injections can achieve high drug concentrations in the vitreous and retina preserving the BRB function and its crucial protective function. At present the major challenge appears to be the need to decrease the number of intravitreal injections which in the case of anti-VEGF treatments are given every 6 weeks to maintain efficacy. The search for safe slow-delivery devices or implantable biomaterials is ongoing but the invasive approach to treat retinal diseases appears to be the only effective way of reaching rapidly therapeutic levels in the retina in the presence of a functioning BRB.

See also: Anatomy and Regulation of the Optic Nerve Blood Flow; Breakdown of the Blood–Retinal Barrier; Breakdown of the RPE Blood–Retinal Barrier; Developmental Anatomy of the Retinal and Choroidal Vasculature; Innate Immune System and the Eye; Macular Edema; Physiological Anatomy of the Retinal Vasculature; Retinal Pigment Epithelial–Choroid Interactions; RPE Barrier.

Further Reading

Cocaprados, M. and Escribano, J. (2007). New perspectives in aqueous humor and secretion and in glaucoma: The ciliary body as multifunctional neuroendocrine gland. *Progress in Retinal Eye Research* 26: 239–262.

Cunha-Vaz, J. G. (1979). The blood–ocular barriers. *Survey of Ophthalmology* 23: 279–296.

Cunha-Vaz, J. G., Faria de Abreu, J. R., Campos, A. J., and Figo, G. (1975). Early breakdown of the blood–retinal barrier in diabetes. *British Journal of Ophthalmology* 59: 649–656.

Cunha-Vaz, J. G. and Maurice, D. M. (1967). The active transport of fluorescein by retinal vessels and the retina. *Journal of Physiology* 191: 467–486.

Cunha-Vaz, J. G. and Maurice, D. M. (1969). Fluorescein dynamics in the eye. *Documenta Ophthalmologica* 26: 61–72.

Cunha-Vaz, J. G. and Travassos, A. (1984). Breakdown of the blood–retinal barriers and cystoid macular edema. *Survey of Ophthalmology* 28: 485–492.

Cunha-Vaz, J. G., Shakib, M., and Ashton, N. (1966). Studies on the permeability of the blood–ocular barrier. I. On the existence, development and site of a blood–retinal barrier. *British Journal of Ophthalmology* 50: 411–453.

Kaplan, H. J. and Niederkorn, J. Y. (2007). Regional immunity and immune privilege. In: Niederkorn, J. Y. and Kaplan, H. G. (eds.) *Immune Response and the Eye. Chemical Immubology Allergy* vol. 92, pp. 11–26. Basel: Karger.

Lobo, C., Bernardes, R., and Cunha-Vaz, J. G. (1999). Mapping retinal fluorescein leakage with confocal scanning laser fluorometry of the human vitreous. *Archives of Ophthalmology* 117: 631–637.

Partridge, W. M. (1998). Introduction to the blood–brain barrier: Methodology and pathology. In: Partridge, W. M. (ed.) *Introduction to the Blood–Brain Barrier: Methodology, Biology and Pathology*, pp. 1–10. New York: Cambridge University Press.

Peyman, G. A. and Bok, D. (1972). Peroxidase diffusion in the normal and laser-coagulated primate retina. *Investigative Ophthalmology* 11: 35–45.

Philips, B. E. and Antonetti, D. A. (2007). Blood–retinal barrier. In: Joussen, A. M., Gardner, T. W., Kirchhof, B., and Ryan, S. J. (eds.) *Retinal Vascular Disease*, pp. 139–153. Berlin: Springer.

Rapoport, S. I. (1976). *Blood–Brain Barrier in Physiology and Medicine.* New York: Raven Press.

Reese, T. S. and Karnovski, M. J. (1967). Fine structural localization of a blood–brain barrier to exogenous peroxidase. *Journal of Cellular Biology* 34: 207–217.

Shakib, M. and Cunha-Vaz, J. G. (1966). Studies on the permeability of the blood–retinal barrier. IV. Junctional complexes of the retinal vessels and their role on their permeability. *Experimental Eye Research* 5: 229–234.

Strauss, O. (2005). The retinal pigment epithelium in visual function. *Physiological Review* 85: 845–881.

Breakdown of the Blood–Retinal Barrier

S A Vinores, Johns Hopkins University School of Medicine, Baltimore, MD, USA

Glossary

Fenestrations – Spaces between vascular endothelial cells that allow free fluid exchange between vessel and tissue. Fenestrations are characteristic of vessels found in tissues that do not have a blood–tissue barrier.

Leukostasis – The adhesion of leukocytes to the vascular endothelium as part of an inflammatory reaction.

Macular edema – Fluid accumulation, due to blood–retinal barrier (BRB) breakdown, in the area of the human or primate retina of highest visual acuity.

Tight junctions – Also referred to as zonula occludens, tight junctions are complex arrangements of microfilaments and other proteins that connect retinal vascular endothelial (RVE) or retinal pigment epithelial (RPE) cells and restrict the flow between them. Tight junctions are an integral component of the blood–retinal, blood–brain, or blood–nerve barrier.

Vesicular transport – The nonspecific transcellular transport of fluid and high-molecular-weight molecules from the luminal to the abluminal surface of the vascular endothelium by means of pinocytotic vesicles.

Uveitis – An inflammation of the uvea, or the middle layer of the eye. The uvea consists of three structures: the iris, the ciliary body, and the choroid.

Introduction

The blood–retinal barrier (BRB), which is analogous to the blood–brain barrier, maintains homeostasis in the retina by restricting the entry of blood-borne proteins from the retina and by maintaining strict ionic and metabolic gradients. When this barrier breaks down, excess fluid accumulates in the retina and this can result in macular edema, which is associated with ischemic retinopathies, including diabetic retinopathy (DR) and retinopathy of prematurity (ROP), ocular inflammatory diseases, retinal degenerative diseases, and a variety of other ocular disorders, or following ocular surgery. BRB breakdown can occur at the inner BRB, which is established at the retinal vasculature, at the outer BRB, which consists of the retinal pigment epithelial (RPE) cells, or at both sites. The BRB is established by the formation of tight junctions between the retinal vascular endothelial (RVE) cells and the RPE cells and a paucity of endocytic vesicles within these cells. The establishment and maintenance of the BRB is regulated by the perivascular astrocytes and pericytes, but the mechanism for this regulation is not entirely clear. Some studies have shown that cell to cell contact is necessary to establish and maintain the BRB, while others provide evidence that a soluble mediator is sufficient. BRB breakdown can result from a disruption of the tight junctions, which are composed of a complex network of junctional proteins, an upregulation of vesicular transport across the RVE or RPE, or by degenerative changes to the barrier-forming cells or to the regulatory cells, the pericytes and glia. In some cases, BRB breakdown is related to identifiable structural defects, such as loss of pericytes, astrocytes, or RPE cells or changes to the vascular endothelial cells, as would be caused by microaneurysm formation. In other cases, where retinal vascular leakage is diffuse, such as in uveitis, or when the leakage is remote from a lesion, such as a surgical wound or tumor, it is clear that diffusible factors are involved. Blood–tissue barriers exist only in the retina, brain, and nerve. Vascular endothelial cells in the choroid and in other tissues are fenestrated (**Figure 1(a)**), allowing large molecular weight molecules to freely pass from the blood to the tissue, and thus do not have a barrier function.

Tight Junctions

Tight junctions or zonula occludens consist of complex arrangements of over 40 proteins in the peripheral cytoplasm and apical plasma membrane that connect RVE or RPE cells and restrict flow between them (**Figure 1(b)**). Occludin and the claudins (over 24 isoforms), which form the junctional strands and are believed to constitute the backbone of the tight junction, span the plasma membrane and bind junctional proteins in adjacent cells. Zonula occludens proteins 1, 2, and 3 (ZO-1, -2, and -3) are intracellular proteins that associate with the cytoplasmic surface of the tight junctions and organize the complex. The binding of ZO-1 to occludin establishes the tight junction. Other integral components of the junctional complex are the junctional adhesion molecules, tricellulin, cingulin, 7H6, and symplekin. A breach of the tight junctions (**Figure 1(c)**) can result from an alteration in the content of the junctional proteins, their redistribution, or their phosphorylation.

Adenosine, prostaglandin E1 (PGE1), interleukin-1β (IL-1β), tumor necrosis factor-α (TNFα), and vascular endothelial growth factor (VEGF) appear to be capable of causing a morphological and functional opening of the RVE tight junctions. A significant number of interendothelial cell tight junctions appeared open along their entire length within 6 h of intravitreal injection of each agent into rabbits with TNFα showing the greatest effect (35.6% of the interendothelial cell junctions appeared open, morphologically). The effect of PGE1 on tight junctions appeared to be transient, that of VEGF and IL-1β were partially reversible by 24 h, and the effect of the adenosine agonist, N-ethylcarboxamidoadenosine was not reversible by 48 h. The demonstration of immunoreactive albumin, which would normally be confined to the lumens of vessels with a blood–tissue barrier, along the entire length of these junctions, from the luminal to the abluminal surface, suggests that they are also functionally open (**Figure 2(c)**).

Vesicular Transport

Since the tight junctions restrict the flow of molecules across the BRB, a series of pumps, channels, and transporter molecules are necessary to transport specific essential molecules from the blood to the retina. The nonspecific transport of high molecular weight molecules and fluids across the RVE by way of pinocytotic vesicles (**Figure 2(a)**) or caveolae is referred to as vesicular transport (**Figure 2(b)**) and serves as a transcellular means of BRB breakdown. This mechanism appears to be the predominant means for BRB compromise associated with VEGF-A-induced hyperpermeability in monkeys and in DR in humans, rats, and rabbits.

In addition to causing the opening of interendothelial cell tight junctions in the retina, adenosine, PGE1, IL-1β, TNFα, and VEGF also promote the formation of pinocytotic vesicles in RVE cells and the distribution of albumin-containing intraendothelial vesicles across the entire RVE cell and at both the luminal and abluminal surfaces suggests that active vesicular transport is occurring. Although infrequently seen, the vesiculo-vacuolar organelle, which is associated with VEGF in the vascular endothelium of tumors, was also evident in the RVE of VEGF-treated rabbits, but not monkeys, and is likely to play a role in VEGF-mediated vascular permeability. The effect of these mediators on the outer BRB is less clear.

Role of Inflammation

Inflammation has been associated with BRB breakdown in DR, choroidal neovascularization (CNV) associated with age-related macular degeneration, aging, ocular inflammatory disease, and the administration of pro-inflammatory molecules. The increased adhesion of leukocytes to

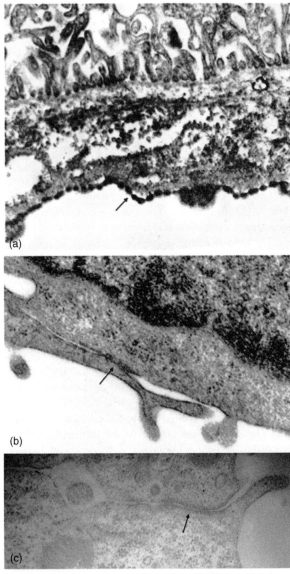

Figure 1 (a) Choroidal vessels are fenestrated (arrow) and, therefore, do not form a blood–tissue barrier. (b) A morphologically closed tight junction (arrow) in a normal retinal vessel. Note close apposition of vascular endothelial cells and an intact junctional complex. (c) A morphologically open tight junction (arrow) in a retinal vessel. The space between the vascular endothelial cells allows vascular leakage through the junction.

Occludin content at the tight junction is higher in cells that have a tighter barrier and decreased occludin correlates with increased BRB permeability, but occludin knockout mice appear to form functional tight junctions, so the association is complex and not simply regulated by occludin. Increased occludin phosphorylation is also associated with increased BRB permeability. Altered expression of claudins can lead to changes in selectivity of the junctions and claudin-5 appears to be particularly important for maintenance of a functional tight junction.

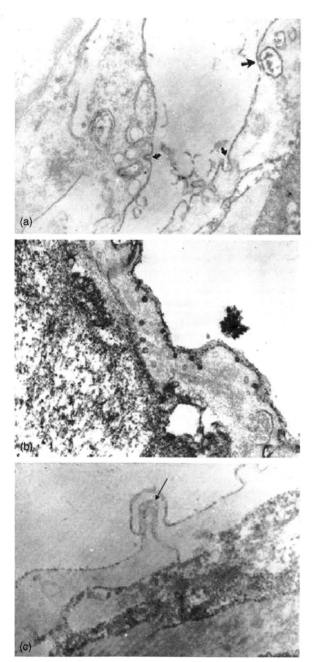

Figure 2 (a) Immunocytochemical staining for endogenous albumin shows the formation of pinocytotic vesicles (arrows) on the luminal surface of vascular endothelial cells in a 7 month galactosemic rat. Immunoreactive albumin is contained within the formed vesicles (top right). (b) Immunocytochemical staining for albumin demonstrates albumin filled vesicles throughout the cytoplasm of the vascular endothelial cells of a rabbit. The presence of these vesicles and the positive staining for albumin in the extracellular matrix (left) suggests that vesicles are actively transporting serum proteins across the endothelium and extruding their contents at the abluminal surface as a means of transcellular BRB breakdown. (c) Immunocytochemical staining for albumin along the entire length of the interendothelial cell junction (arrow) and in the basal lamina indicates that there is vascular leakage through the junction.

endothelial cells in the retina is associated with increased expression of intercellular adhesion molecule-1 (ICAM-1), vascular cell adhesion molecule-1 (VCAM-1), CD18, and other adhesion molecules, which are upregulated by VEGF and other pro-inflammatory molecules in DR and other ocular disorders, and appear to be regulated, at least in part, by protein kinase C (PKC). Diabetic CD18 and ICAM-1 knockout mice have significantly fewer adherent leukocytes than diabetic mice with normal CD18 and ICAM-1 and the decreased leukostasis is associated with fewer damaged endothelial cells and reduced BRB breakdown, supporting the role of adhesion molecules in increased inflammation and the correlation of an inflammatory response with endothelial cell damage and permeability. It is not clear whether the same molecules that facilitate leukostasis also mediate BRB breakdown or if this is attributable to molecules secreted by the recruited leukocytes, or both, but there appears to be a direct correlation between increased leukostasis and vascular permeability in the retina and pro-inflammatory molecules, such as TNFα and IL-1β, are among the most potent inducers of BRB breakdown. Leukocyte adhesion to the diabetic vascular endothelium can promote endothelial apoptosis and inhibition of leukocyte adhesion to the retinal vessels can not only prevent endothelial degeneration, but also reduce the diabetes-associated loss of pericytes, which support the vascular endothelium and help to confer BRB integrity. Inflammation can also alter the distribution of astrocytes and their ensheathment of retinal vessels, leading to alterations in BRB integrity. Leukocytes have also been shown to cause a downregulation and redistribution of tight junctional proteins, which leads to a disruption of tight junctions and a transient breakdown of the BRB during retinal inflammation.

Molecular Mechanisms

The induction of BRB breakdown is a complex process that is mediated, not by a single factor, but by the interaction of multiple factors operating through different receptors and signaling pathways. The list of molecules that have been identified as playing a role in BRB breakdown, which is by no means all-inclusive, includes VEGF, hypoxia-indicible factors-1 and -2 (HIF-1 and -2), placental growth factor (PlGF), TNFα, IL-1β, platelet-activating factor, adenosine, histamine, prostaglandins (PGE1, PGE2, and PGF2α), platelet-derived growth factors A and B (PDGF-A and -B), insulin-like growth factor-1 (IGF-1), ICAM-1, VCAM-1, P-selectin, and E-selectin. The key will be to determine what the initiating event is and which events are parts of the resulting cascade. By targeting the appropriate molecules, subsequent events leading to BRB failure may be blocked.

The various isoforms of VEGF and PlGF are members of the VEGF family. VEGF-A, a potent inducer of vascular permeability, binds to both VEGF type-1 (fms-like tyrosine kinase-1 or Flt-1) and type-2 (kinase insert domain-containing receptor, referred to as KDR in humans or Flk-1 in other species) receptors (VEGFR1 and VEGFR2), whereas PlGF binds only to VEGFR1, so comparing their activities may be a means of dissecting the respective roles of the VEGF receptor isoforms, since both receptors have been implicated in BRB breakdown. VEGFR1-mediated signaling appears to operate primarily through p38 MAPK (mitogen-activated protein kinase), while VEGFR2 signaling may be mediated through RAS, phosphoinositide 3-kinases (PI3K)/Akt, or phospholipase C (PLC)γ, but the interaction of VEGFR1 and VEGFR2 is complex and much remains to be learned about this interaction. Both receptors are associated with vascular permeability and angiogenesis, but in some circumstances, VEGFR1 can act as a negative regulator for VEGFR2.

VEGF is a key molecule in promoting increased retinal vascular permeability. This activity may be mediated, at least in part, by an upregulation of ICAM-1, E-selectin, and P-selectin as a means of facilitating its pro-inflammatory activity. VEGF-induced permeability showed a biphasic pattern with a rapid and transient phase followed by a delayed and sustained phase, the latter of which was blocked by antibodies to urokinase plasminogen activator or its receptor. VEGF receptor kinase inhibitors can suppress VEGF-mediated BRB breakdown, but this strategy shows that TNFα, IL-1β, and IGF-1 do not induce BRB leakage through an induction of VEGF, indicating that these mediators operate through distinct pathways that may also be targeted. Endothelial nitric oxide synthase activation and NO formation also appear to be implicated in VEGF-mediated vascular permeability, probably through activation of the serine/threonine protein kinase AKT/PKB, which can lead to an increase in nitric oxide production and ICAM-1 upregulation. Deletion of the hypoxia response element of the *Vegf* promoter also suppresses BRB breakdown in oxygen-induced ischemic retinopathy, demonstrating that HIF-induced VEGF is critical in this process.

TNFα and IL-1β are upregulated in DR and other ischemic retinopathies, as well as ocular inflammatory disease, and both molecules are associated with increased leukostasis and BRB breakdown. IL-1β has been shown to accelerate apoptosis of retinal capillary endothelial cells through activation of nuclear factor kappa light-chain enhancer of activated B cells and this is exacerbated in high glucose. IL-1β can also stimulate the production of reactive oxygen species, which in turn can induce the release of additional cytokines. Aspirin and etanercept are inhibitors of TNFα and each can reduce ICAM-1 levels, diabetes-related leukostasis, and BRB breakdown in diabetic rats without altering VEGF levels, showing that TNFα is involved in this process and that it operates through a distinct pathway from VEGF.

These data show that there are a number of potential target molecules for inhibitors to suppress BRB breakdown. The challenge will be to identify the best target or targets and develop the most effective therapeutic strategy.

Assessing BRB Breakdown

A variety of methods exist for the quantitative and qualitative assessment of the BRB, but each method has its particular limitations and sensitivity, so the choice of methods will largely depend on whether quantitative or qualitative data are desired and on the nature of the tissue being evaluated, whether it be fixed tissue, patients in a clinical setting, or experimental animal models. Since no single method can provide a quantitative assessment with precise localization of the site of BRB breakdown, multiple approaches may be necessary to provide an overall perspective. In addition, some methods can produce precise data in experimental models or on tissue specimens, but are not appropriate for use in the clinic.

To identify and compare factors that cause BRB breakdown and to evaluate the efficacy of new treatments designed to prevent or reduce macular edema, a reliable quantitative assay for assessing BRB function is essential. The most widely used protocols for the quantitative assessment of BRB breakdown utilize Evans blue or 3H-mannitol as tracers. With the Evans blue assay, the extracted dye is quantified in the retina following intravenous injection of the dye and subsequent perfusion with saline. A spectrophotometer, set at 620 nm, is used to quantify the leakage of dye into the retina. With the 3H-mannitol assay, a scintillation counter is used to determine the CPM/mg tissue, 1 h after an intraperitoneal injection of the tracer, and the data are expressed as a ratio of retina/lung or retina/kidney. Since the lung and kidney do not have a blood–tissue barrier, the ratio corrects for any variation in the amount of isotope injected or absorbed. These methods have been used to assess the BRB in several models of ocular disease and to determine the effect of various factors, agents, and genetic manipulations on the integrity of the BRB. Thus, these methods have been useful in identifying factors that initiate BRB compromise and for determining the relative efficacy of various agents at preventing or reducing BRB failure. Both methods produce highly reproducible results in an experimental setting, with the 3H-mannitol assay possibly being somewhat more sensitive due to the lower molecular weight of mannitol than Evans blue dye, but neither is applicable to the clinic. Vitreous fluorophotometry (VFP) is a more appropriate means of assessing BRB failure in a clinical setting. Although these methods can provide a quantitative assessment of the BRB, they cannot localize the site of leakage or provide any insight into the mechanism, so alternative techniques are required to provide this information.

Fluorescein angiography has been used extensively in the clinic to visualize BRB breakdown, but it does not allow resolution at the cellular level. Magnetic resonance imaging (MRI) enhanced by the paramagnetic contrast agent gandoliniumdiethylene-triaminetetraacetic acid has been used to localize and quantify BRB breakdown in living animals. MRI is not subject to the optical limitations of VFP and allows the investigator to distinguish between inner and outer BRB failure in the rabbit. Its resolution is not as great as that resulting from microscopic evaluation of fixed tissue, but MRI allows *in vivo* analysis, thus enabling the investigator to monitor progressive changes in BRB integrity within the same animal. The use of exogenous tracer substances can provide a higher resolution, but several limitations are associated with their use. The use of tracers is impractical for clinical studies, the introduction of exogenous material may alter BRB integrity, and retrospective studies cannot be done on archival tissues. The immunolocalization of endogenous albumin (**Figure 3**) or IgG can circumvent most of these limitations and offers the following advantages for BRB assessment. The technique can be used with fixed surgical, autopsy, or archival specimens, no exogenous substance is introduced, and it can be used at the light and electron microscopic (EM) levels. Although, by nature, this is not a quantitative method, it can show the location and extent of BRB breakdown and, if used at the EM level, it can demonstrate the means by which serum proteins are extravasated from the retinal vessels or may transverse the RPE layer. This technique has been used to assess the BRB in a variety of human and experimental ocular disorders, including DR, retinitis pigmentosa, vascular occlusive disease, neoplastic disease, ocular inflammation or infection, and other diseases that develop macular edema, but for which pathological defects do not reveal a cause for BRB breakdown. EM immunocytochemical staining for albumin reveals that BRB breakdown can occur by the opening of tight junctions between RVE or RPE cells (**Figure 2(c)**), by an upregulation of trans-endothelial vesicular transport (**Figures 2(a)** and **2(b)**), or by increased surface membrane permeability of RVE or RPE cells resulting from degenerative changes associated with the disease process. It has also provided insights into how various factors, such as VEGF, TNFα, IL-1β, prostaglandins, adenosine, and others promote BRB breakdown. VEGF transiently opens some tight junctions and some leakage through the interendothelial cell tight junctions is induced by VEGF, but electron microscopy has revealed that the predominant mechanism for VEGF-A-induced hyperpermeability of the RVE in monkeys and diabetes-related BRB breakdown in humans, rabbits, and rats is an upregulation of pinocytotic vesicular transport.

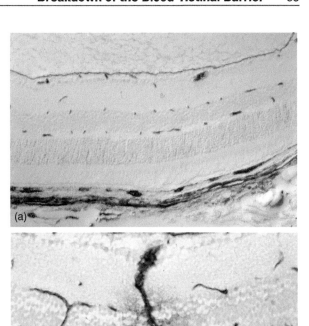

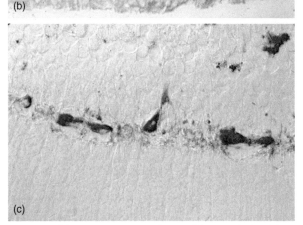

Figure 3   (a) In a normal mouse, immunohistochemical staining for albumin shows that, within the retina, albumin is confined to the vessels, indicating an intact BRB, but diffuse staining is demonstrated in the choroid (bottom) due to the fenestrated vessels and the absence of a blood–tissue barrier. (b) In a mouse infected with coronavirus, vascular leakage is demonstrated from a retinal vessel by immunohistochemical staining for albumin. (c) Immunohistochemical staining (red) shows that albumin has leaked from retinal vessels in a VEGF transgenic mouse. Vinores, S. A., et al. (2001). Blood–retinal barrier breakdown in experimental coronavirus retinopathy: Association with viral antigen, inflammation, and VEGF in sensitive and resistant strains. *Journal of Neuroimmunology* 119: 175–182, with permission from Elsevier.

Inhibiting BRB Breakdown

A variety of therapeutic approaches have shown success at inhibiting BRB breakdown, but generally not preventing it. Most of the agents currently in clinical trials target inflammatory processes or VEGF. Antibodies to key molecules, such as VEGF, PlGF, or TNFα, have been effective at suppressing BRB breakdown, as have inhibitors of these molecules. Drugs that block histamine receptors also reduce retinal vascular leakage in diabetic rats and humans. Bevacizumab (avastin), an anti-VEGF IgG1 antibody, Ranibizumab (lucentis), the Fab fragment of a humanized anti-VEGF antibody, Pegaptanib sodium (macugen), a VEGF aptamer, and VEGF trap, in which the binding domains of VEGFR1 and VEGFR2 are combined with the Fc portion of IgG to neutralize all VEGF family members, have all shown varying degrees of success in clinical trials for reducing macular edema by targeting VEGF. Corticosteroids inhibit BRB breakdown, but it is not clear whether this activity is mediated by their anti-inflammatory effect, which occurs, at least in part, through a downregulation of ICAM-1, their inhibition of VEGF expression, their induction of occludin and ZO-1 expression, their reversal of occludin phosphorylation, or a combination of these activities. Even though steroids may improve visual acuity, they carry a high risk of cataracts and glaucoma. The involvement of PKC in vascular permeability has been established and a PKC activator can promote BRB breakdown. PKC inhibitors can reduce retinal vascular permeability, particularly that mediated by VEGF or prostaglandins, but generalized inhibition of PKC is likely to have serious systemic consequences. A PKCβ inhibitor (LY333531) was also effective at suppressing retinal vascular permeability and may have fewer complications.

Prospects for the Future

As more studies are conducted, the complexity of BRB breakdown leading to macular edema becomes increasingly apparent. This process is not attributable to a single factor or event, but the interaction of an undetermined number of initiating events that generates a cascade of subsequent events, ultimately leading to BRB failure. Since this is a multifactorial process, a multifaceted or pleiotropic approach that restores the homeostatic balance is more likely to suppress BRB breakdown than targeting a single pathway with an inhibitor. That would explain why the currently used monotherapies may lead to a reduction of macular edema and improved visual acuity, but generally not a total resolution of the disorder. To develop more effective therapeutic strategies, a better understanding of the basic mechanisms in the pathogenesis of BRB breakdown is imperative. In addition, the frequent injections, high cost, and the occasional side effects associated with current therapeutic approaches emphasize the need for more effective treatment that is less invasive, less costly, and has little or no side effects.

Conclusions

BRB breakdown, leading to macular edema, occurs in a number of ocular disorders and can be due to structural changes or soluble mediators. Alteration in the content, distribution, or phosphorylation of junctional proteins can result in vascular leakage through the tight junctions. BRB breakdown can also result from an upregulation of transendothelial vesicular transport, which has been shown to be a major contributor to BRB failure caused by several mediators and in ocular disease models. Many of the same mediators can simultaneously promote opening of the tight junctions and upregulation of vesicular transport. Degenerative or structural changes to the RPE or RVE cells or to the pericytes and perivascular astrocytes that regulate the inner BRB can also lead to BRB breakdown. Inflammation promotes BRB breakdown, so the use of anti-inflammatory agents may be beneficial. BRB breakdown is not due to a single factor, but is a complex process involving multiple factors, receptors, and signaling pathways. Information on the molecular mechanisms is being revealed, but much remains to be learned. The complexity of the pathogenesis of BRB breakdown makes it likely that the greatest chance for success in preventing macular edema would be in targeting multiple molecules or pathways and a sensitive method for assessing the integrity of the BRB is necessary to monitor the efficacy of different therapeutic strategies.

Acknowledgment

Dr. Vinores is supported by grant R01EY017164 from the National Eye Institute, National Institutes of Health.

See also: Blood–Retinal Barrier; Breakdown of the RPE Blood–Retinal Barrier; Macular Edema; RPE Barrier; Retinal Vasculopathies: Diabetic Retinopathy; Retinopathy of Prematurity; Secondary Photoreceptor Degenerations: Age-Related Macular Degeneration.

Further Reading

Antonetti, D. A., Lieth, E., Barber, A. J., and Gardner, T. W. (1999). Molecular mechanisms of vascular permeability in diabetic retinopathy. *Seminars in Ophthalmology* 14: 240–248.

Erickson, K. K., Sundstrom, J. M., and Antonetti, D. A. (2007). Vascular permeability in ocular disease and the role of tight junctions. *Angiogenesis* 10: 103–117.

Gardner, T. W. and Antonetti, D. A. (2008). Novel potential mechanisms for diabetic macular edema: leveraging new investigational approaches. *Current Diabetes Reports* 8: 263–269.

Gardner, T. W., Antonetti, D. A., Barber, A. J., et al. (2002). Diabetic retinopathy: More than meets the eye. *Survey of Ophthalmology* 47(supplement 2): S253–S262.

Hofman, P., Blaauwgeers, H. G. T., Tolentino, M. J., et al. (2000). VEGF-A induced hyperpermeability of blood–retinal barrier endothelium *in vivo* is predominantly associated with pinocytic vesicular transport and not with formation of fenestrations. *Current Eye Research* 21: 637–645.

Joussen, A. M., Poulaki, V., Le, M. L., Koizumi, K., et al. (2004). A central role for inflammation in the pathogenesis of diabetic retinopathy. *FASEB Journal* 18: 1450–1452.

Leal, E. C., Santiago, A. R., and Ambrosio, A. F. (2005). Old and new drug targets in diabetic retinopathy: From biochemical changes to inflammation and neurodegeneration. *Current Drug Targets – CNS and Neurological Disorders* 4: 421–434.

Luna, J. D., Chan, C.-C., Derevjanik, N. L., et al. (1997). Blood–retinal barrier (BRB) breakdown in experimental autoimmune uveoretinitis: Comparison with vascular endothelial growth factor, tumor necrosis factor α, and interleukin-1β-mediated breakdown. *Journal of Neuroscience Research* 49: 268–280.

Rizzolo, L. J. (2003). Development and role of tight junctions in the retinal pigment epithelium. *International Review of Cytology* 258: 195–234.

Saishin, Y., Saishin, Y., Takahashi, K., et al. (2003). Inhibition of protein kinase C decreases prostaglandin-induced breakdown of the blood–retinal barrier. *Journal of Cellular Physiology* 195: 210–219.

Vinores, S. A. (1995). Assessment of blood–retinal barrier integrity. *Histology and Histopathology* 10: 141–154.

Vinores, S. A. (2007). Anti-VEGF therapy for ocular vascular diseases. In: Maragoudakis, M. E. and Papadimitriou, E. (eds.) *Angiogenesis: Basic Science and Clinical Applications*, pp. 467–482. Kerala, India: Transworld Research Network.

Vinores, S. A., Derevjanik, N. L., Mahlow, J., Berkowitz, B. A., and Wilson, C. A. (1998). Electron microscopic evidence for the mechanism of blood–retinal barrier breakdown in diabetic rabbits: Comparison with magnetic resonance imaging. *Pathology Research and Practice* 194: 497–505.

Vinores, S. A., Derevjanik, N. L., Ozaki, H., Okamoto, N., and Campochiaro, P. A. (1999). Cellular mechanisms of blood–retinal barrier dysfunction in macular edema. *Documenta Ophthalmologica* 97: 217–228.

Xu, Q., Quam, T., and Adamis, A. P. (2001). Sensitive blood–retinal barrier breakdown quantitation using Evans blue. *Investigative Ophthalmology and Visual Science* 42: 789–794.

Breakdown of the RPE Blood–Retinal Barrier

M E Hartnett, Moran Eye Center, University of Utah, Salt Lake City, UT, USA

Glossary

Adherens junction – Cell–cell junction having transmembrane proteins called cadherins that connect to cadherins of epithelial cells through extracellular domains and to anchor proteins, known as catenins, through intracellular tails. Anchor proteins also bind to a continuous belt of actin filaments along cytoplasmic side of the cell plasma membrane, ultimately holding neighboring cells together.

Focal adhesions – Cell–matrix junctions having transmembrane proteins called integrins that connect cells to extracellular matrix and also to the actin cytoskeleton. Transmembrane proteins can trigger signaling within and between adjacent cells.

Inner blood-retinal barrier – It regulates the transport of fluid, ions, and metabolites between the neurosensory retina and the retinal vasculature.

Na,K,ATPase – Enzyme providing active transport mechanism in the retinal pigment epithelial and other cells. By catalyzing an ATP-dependent transport of three Na^+ ions out and two K^+ ions into the cell, a transmembrane sodium gradient is created (necessary for other Na^+ coupled transport systems) and an osmotic gradient (drives water toward the choroidal or basal side of the RPE and away from the subretinal space).

Outer blood-retinal barrier – It regulates the transport of fluid, ions, and metabolites between the neurosensory retina and the choroidal vasculature.

Tight junction – A cell–cell adhesion usually at the apical lateral aspects of polarized endothelia but can be present in other cells consisting of a number of proteins, including ZO-1, -2, -3, transmembrane protein, occludin, and claudins. The tight junction regulates paracellular permeability between adjacent cells and protein and lipid movement between apical and basal compartments of the cell.

Introduction

The retinal pigment epithelium (RPE) forms the outer blood-retinal barrier (BRB), which regulates the transport of fluid, ions, and metabolites between the neurosensory retina and the choroidal vasculature. The inner BRB is formed by retinal endothelial cells (RECs), which regulate the transport of fluid, ions, and metabolites between the neurosensory retina and the retinal vasculature. Together, the RPE and RECs collectively form the BRB that serves to exclude blood-borne substances from the neurosensory retina and regulates the ionic and metabolic gradients required for normal retinal function. The RPE barrier structures and functions will be broadly described below and for further detail, the reader is directed to other articles in this encyclopedia.

RPE Barrier Structure

The RPE is a polarized monolayer located deep to the neurosensory retina with its apical processes adjacent to the photoreceptors and its basal aspect on Bruch's membrane, a semipermeable collagen and elastin sandwich that separates the choroid from the RPE and neurosensory retina (**Figure 1**). The RPE helps to maintain the outer BRB through tight junctions that are located at the apical–lateral junctions of adjacent cells. Much of what is known about tight junctions was learned from studies of epithelial cells other than the RPE.

Tight junction-associated proteins include ZO-1, -2, -3, and several transmembrane proteins, including junction adhesion molecules and occludin, which bind ZO-1 and -2 that then bind the cytoskeleton. Occludin also has extracellular domains, which bind to those of adjacent cells to form the tight junctions. Claudin proteins are also important in this structure and have tissue and regional specificity. There are a number of other proteins that are important in the development and function of the tight junction.

Besides tight junctions, there are other points of contact between cells or cells and extracellular matrix. Adherens junctions have transmembrane proteins called cadherins that connect to cadherins of adjacent RPE cells through their extracellular domains and to anchor proteins, known as catenins, through intracellular tails. The anchor proteins also bind to a continuous belt of actin filaments along the cytoplasmic side of the cell plasma membrane. The resulting structure holds neighboring cells together. Focal adhesions have transmembrane proteins called integrins that connect cells to extracellular matrix and also to the actin cytoskeleton. Besides providing structure and connections, transmembrane proteins can trigger signaling pathways within and between adjacent cells.

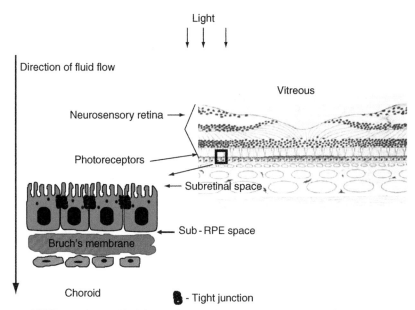

Figure 1 Artist diagram of RPE monolayer with tight junctions at the apical lateral junctions (see inset of RPE/Bruch's membrane/choroid in lower left corner) and in relation to the layers within the macula (shown as a diagram of cross-section of macula on right) and eye. Light is shown coming from above. Overall direction of fluid through the eye is represented by arrow on left.

In addition, a number of active processes regulate the flow of ions, glucose, and amino acids across the RPE. For example, ion fluxes and signaling processes are synchronized between cells through gap junctions, such as connexin 43.

RPE Barrier Functions

The RPE tight junctions regulate the passage of molecules across the paracellular pathway and this regulation depends on the selectivity and permeability of the tight junctions. The transcellular pathway requires pumps, such as Na,K-ATPase, channels, transporters, and metabolic modification to regulate transport across the cells of the monolayer. Breakdown of the RPE barrier may include disruption of the tight junction structure with reduced barrier properties of the RPE. To measure RPE barrier function in a living human, RPE barrier dysfunction is determined clinically by leakage of sodium fluorescein across the RPE seen on stereoscopic images of fluorescein angiograms following an intravenous injection of sodium fluorescein into the bloodstream. However, the distinction between dysfunction of the inner from the outer BRB is difficult in human diseases. This may be because broad BRB breakdown occurs from similar pathophysiologic mechanisms or because of limited resolution in optical imaging. Therefore, techniques are used *in vitro* to study the RPE barrier properties or *in vivo* to selectively poison the RPE and measure fluorescein leakage into the vitreous or eye in animal models (see below). When considering diseases associated with RPE barrier

breakdown, it is useful first to understand the fluid flow and forces within the eye.

Transport and Fluid Flow within the RPE and Eye

The vitreous gel is within the eye adjacent to the inner retina, the ciliary body, zonule, and posterior lens capsule. The neurosensory retina includes retinal layers extending from the inner limiting membrane adjacent to the vitreous to the photoreceptors above the RPE. The subretinal space is the potential space between the photoreceptor outer segments and the apical processes of the RPE. The sub-RPE space is a potential space beneath the RPE basal infoldings and Bruch's membrane. The choroid is deep to Bruch's membrane (**Figure 1**). Historically, fluid tends to accumulate more easily in the subretinal than the sub-RPE spaces.

Passive and active transport mechanisms within the RPE and ocular forces within the eye are important when considering the pathophysiology of disease (**Figure 1**). These mechanisms serve to maintain a directional flow from inside the eye, that is, vitreous, to the choroid. Passive transport mechanisms include intraocular pressure and the osmotic pressure from the choroid. However, studies have shown that the RPE reduces directional fluid movement from the subretinal space to the choroid. When the RPE barrier is damaged in animal models (such as through chemicals like sodium iodate), fluid egress from the subretinal space to the choroid is faster than in conditions in which the RPE has not been damaged. Therefore, in conditions in which there is increased fluid in the subretinal space and presumed RPE BRB breakdown, there is believed to be another pathophysiologic abnormality, such as increased perfusion pressure from the choroid from a vascular tumor, choroidal neovascular membrane, or hyperpermeable choroidal vessels, or reduced ability to accommodate extra fluid, such as with choroidal ischemia or inflammation.

RPE Na,K-ATPase pump

Since the RPE restricts fluid movement from within the eye toward the choroid, it is believed that one or more active transport mechanism is necessary to prevent fluid from accumulating in the subretinal space. The Na, K-ATPase pump is an important active transport mechanism in the RPE. This enzyme is located at the apical surface of the RPE and it is believed that the asymmetric positioning of this enzyme with other transporters, such as chloride channels, is important in creating directional fluid flow. The enzyme catalyzes an adenosine triphosphate (ATP)-dependent transport of three Na+ ions out and two K+ ions into the cell. This transmembrane sodium gradient is necessary for functions of other Na^+ coupled transport systems and creates an osmotic gradient, which helps drive water toward the choroidal or basal side of the RPE and away from the subretinal space. It has been shown *in vitro* using human RPE grown in polarized monolayers that Na,K-ATPase function is necessary for the structural integrity of tight junctions and their function. Inhibition of Na,K-ATPase using a K^+ free media or with ouabain led to increased ionic and nonionic permeability in association with a reduction in the number of contact points in the tight junctions as seen by transmission electron microscopy. Thus, Na,K-ATPase function is necessary to reduce paracellular permeability.

Methods to Assess RPE Barrier Structure and Function

Tight Junction Structure

ZO-1 staining is useful to define the hexagonal architecture of the RPE monolayer *in vitro*, but does not indicate that paracellular permeability is reduced and the barrier is tight. RPE is cultured in specific media to attain tight properties, and the tightness of the monolayer often correlates with the duration in culture and the immunolocalization of occludin to the cell junctions (**Figure 2**). However, lack of occludin can be associated with a functional tight junction and from this it has been shown that other proteins, including the claudins, are important. Therefore, ZO-1, occludin, and claudins, as well as other proteins, are

Actin (red)
Occludin (green)

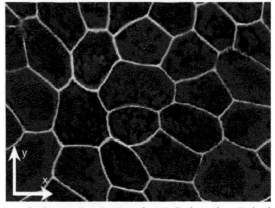

Figure 2 Immunolocalization of occludin (green), actin (red), and Hoechst nuclear stain (blue) in human fetal RPE cultured for >1 month.

important in the RPE tight junction function, and the reader is directed to other articles of this encyclopedia.

When viewed with electron microscopy, it is noted that functional tight junctions have multiple contact points between cells, and that when functional assays show reduced tightness of the barrier, the number of contact points was reduced. The protein components of the tight junction and phosphorylation or activation of certain component, such as occludin, can be determined by immunoprecipitation and/or Western blot analysis.

Barrier Function

Electrophysiologic methods are used to measure barrier function. Transepithial electrical potential (TEP) measures the ion gradient across the monolayer generated by energy-driven ion pumps that regulate passage across the cells. Transepithelial electrical resistance (TER) measures resistance of substances through the paracellular space mainly through the fine structure of the tight junction structure. Permeability to nonionic compounds such as inulin or mannitol can be measured.

Assays That Measure Barrier Properties of RPE

In vitro. Measures of barrier function can vary depending on the RPE cell type, the components making up the media, and the duration of time in culture. Although a TER of $100\,\Omega\,cm^2$ is sufficient to exclude movement from the apical and basal compartments of a monolayer of RPE, many investigators strive to obtain culture conditions in which the RPE monolayer develops a TER of $500\,\Omega\,cm^2$. However, in RPE-choroid explants, a TER of $200\,\Omega\,cm^2$ has been reported. Therefore, it remains unclear whether a higher TER than that *in vivo* indicates a physiologic condition or one in which the structure of the tight junction is altered by an abnormal microenvironment. It is important in studies that compare the effects of agents on a barrier properties that the cell culture conditions be standardized and replicable.

Ex vivo. Explants of the RPE and choroid have been used in modified Ussing chambers to study the movement of compounds and to determine the barrier properties.

In vivo. Assessing the BRB function is performed with several assays that measure the amount of a substance leaked into the retina. In animal studies, quantitative assays include comparing blood concentration of Evans' blue dye to that measured within the retina after administration of a known concentration, or measuring the extravasation of albumin or inulin into the retina. In addition, qualitative and semi-quantitative assessment (i.e., measuring the area of fluorescein leakage) of fluorescein angiograms can be performed. However, it is difficult to distinguish leakage from the inner versus the outer BRB using these methods.

Clinical assays. Since direct measure of specific functions of the outer BRB is not possible in humans *in vivo*, *in vitro* testing and fluorescein angiography or vitreous fluorophotometry are performed. Vitreous fluorophotometry measures fluorescein leakage into the vitreous by measuring fluorescence.

Fluorescein angiography. When sodium fluorescein (376 Da) dye is activated with blue light (490 nm), it emits fluorescence in the green wavelength (520–530 nm). In a fluorescein angiogram, the dye is injected as a bolus (usually 500 mg/5 mL over 7 s) into the antecubital vein, and images of the fundus are obtained digitally or on film over time. At ~10 s following injection, a choroidal flush from filling of the choroid that is partly blocked by the RPE melanin occurs. Nearly simultaneously, filling of the retinal arterioles occurs. Subsequently, the veins fill (**Figure 3(a)**). Areas of abnormal hyperfluorescence can occur from loss of the RPE melanin or cell (known as window defects because of permitted fluorescence from the underlying choroidal vasculature) or from filling of abnormal vessels or pooling or leakage into spaces where fluid is not naturally present. Hypofluorescence can occur from blockage by pigment or blood, or from nonperfusion within any of the vascular structures in the eye (choroid, retinal vasculature, vessels of the optic nerve). Stereoangiography is helpful to determine the relative position within the retina where abnormal fluorescence was seen. Currently, clinicians gain additional data from optical coherence tomography (OCT).

OCT. Light is directed through a dilated pupil and reflects off different optical interfaces corresponding to layers within the retina (**Figure 3(b)**). Disorganization of the retinal architecture, invasion of tissue, and fluid-like cysts can occur within layers and be used in conjunction with the fluorescein angiogram to diagnose and manage several diseases (see below). Spectral domain OCT has provided better image quality than time domain OCT.

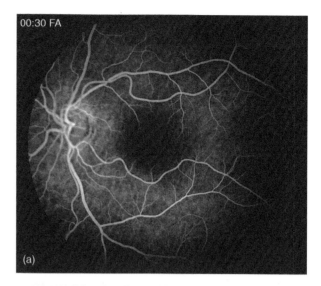

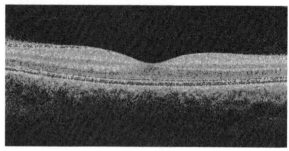

(b)

Figure 3   (a) Normal mid-phase fluorescein angiogram showing optic nerve at left (slightly out of focus) and filling of retinal arterioles and veins. (b) Spectral domain optical coherence tomogram (OCT) of adult macula showing reflective layers of retina (red – highly reflective) and natural foveal depression.

Clinical Conditions Associated with Breakdown of the RPE Barrier

Dysfunction of the BRB is believed to disrupt the normal health and function of the neurosensory retina in several diseases and lead to reduced visual function. RPE barrier dysfunction is diagnosed usually by the presence of hyperfluorescence in the deeper retina, unlike in normal, as determined by fluorescein angiography (cf. **Figures 3(a)** and **4(a)**). Occasionally, vitreous fluorophotometery is used, but this method does not distinguish leakage from the inner or outer BRB. Leakage of fluorescein from the RPE can appear as mild hyperfluorescence and staining, but may lead to intraretinal edema, seen as cystoid macular edema, shown by hyperfluorescence in early frames at the level of the RPE with petalloid hyperfluorescent within the retina in late frames (**Figure 4(b)**). However, most causes of CME are believed to involve several factors including leakage from the inner BRB and Müeller cell dysfunction.

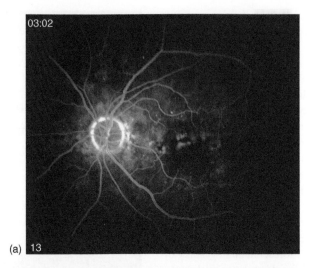

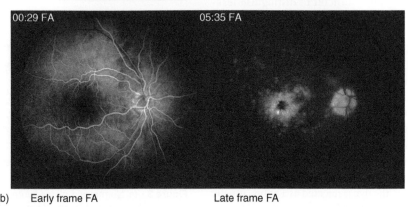

Figure 4   (a) Deep hyperfluorescence in macula of left eye seen on recirculation phase of fluorescein angiogram demonstrating subtle leakage at the level of the RPE barrier. (b) Early angiogram shows minimal hyperfluorescence surrounding the avascular zone of the macula, but in the recirculation frame of the angiogram, hyperfluorescence takes on a petalloid appearance in a patient with cystoid macular edema.

Inflammation/Infection

Inflammation reduces the BRB function and affects both the inner BRB of the RECs and the outer BRB of the RPE. Clinically, any disease that has an infectious or inflammatory component can compromise the BRB. Viral infections (such as human immunodeficiency virus (HIV), cytomegalovirus (CMV), Herpes) and bacteria (such as *Bacillus cereus*) have been shown to disrupt the tight junctions of the RPE *in vitro*. Factors such as tumor necrosis factor (TNF)-α, interleukins (IL)-6, -8, -1 disrupt tight junction barrier functions *in vitro* or *in vivo*. The RPE has been shown to have chemokine receptors, such as CXCR4, which can be activated by its ligand, stromal-derived factor (SDF-1), to release chemokines such as monocyte chemotactic protein-1 (MCP-1) and IL-8. Chemokines, like MCP-1, can attract leukocytes. Contact of RPE with monocytes results in additional secretion of IL-8 and MCP-1 through several pathways involving mitogen-activated protein kinases. Such pathways may amplify the effects of inflammation on BRB breakdown. As a model of human uveitis, experimental autoimmune uveoretinitis (EAU) is induced in animals by immunization with different proteins. For example, EAU can be induced in Lewis rats by immunization with S-antigen or in mice using peptides to interphotoreceptor retinoid-binding protein. A number of events occur in the EAU model and one is the movement of leukocytes into the retina. It has been shown that transient loss of ZO-1 occurs in RPE from IL-1 and MCP-1. (Transient loss of occludin-1 in RECs within venules was believed to play a role in permitting leukocyte diapedesis in the EAU model.) Inflammation in conditions such as Vogt Kayanagi Harada's disease, or choroidal inflammatory conditions can reduce RPE barrier function in humans.

Age-Related Macular Degeneration

Age-related macular degeneration (AMD) is a leading cause of nonreversible blindness worldwide. Clinical evidence of RPE BRB dysruption is seen in fluorescein angiograms showing hyperfluorescence from the choroidal vascular normally blocked by healthy RPE, but where loss of melanin or RPE cells has occurred, as in atrophic AMD (**Figure 5**). Hyperfluorescence can also occur from leaky blood vessels that appear to disrupt the RPE barrier and lead to fluid accumulation in neovascular AMD (**Figure 6**). OCT provides additional evidence in neovascular AMD as to the location of the fluid: beneath the RPE (sub-RPE), above it (subretinal), or within the retina itself (intraretinal) (**Figure 7**). Neovascular AMD occurs in about 10% of AMD cases but accounts for 90% of severe vision loss.

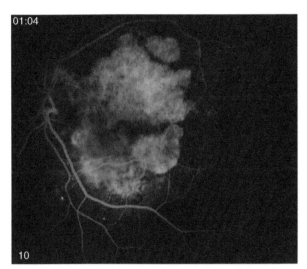

Figure 5 Presence of hyperfluorescence in late frame of fluorescein angiogram because of loss of RPE melanin, which naturally blocks fluorescence of underlying choroid.

Most of the neovascular AMD begins as ill-defined hyperfluorescence and late leakage noted on fluorescein angiography and had been initially called occult choroidal neovascularization (CNV). Later, occult CNV was characterized histopathologically as type 1 CNV, indicating that CNV remained beneath the RPE (**Figure 6**). This is in contrast to classic or well-defined CNV, characterized histopathologically as type II CNV in which the CNV has entered the neurosensory retinal space (compare **Figures 6** and **8**). (This distinction is useful as a model but in reality is not so clear-cut, since clinically there are often mixed types.) However, studies have provided support that when vision declines from neovascular AMD, more than half of the time it is from the transition from occult to classic CNV. This prompts the hypothesis that contact of the ECs with the RPE or its matrix can trigger signaling pathways leading to the release of angiogenic or chemotactic growth factors and the migration of choroidal ECs and movement of fluid into the neurosensory retina. These events would appear to involve a breakdown in the RPE BRB properties. Many stimuli that are believed to be involved in the pathophysiology of AMD also reduce BRB properties as determined in *in vitro* assays described earlier. These include, for example, oxidative stress (can disorganize tight junction proteins including occludin), inflammation (can reduce TER and disrupt tight junctions), activation of complement factors (RPE possess complement 5a receptors and can be stimulated by complement 5a to release several cytokines which can then regulate leukocyte function during inflammation), and contact of RPE or its extracellular matrix with endothelial cells (reduce TER and increase permeability of RPE through release of soluble forms of vascular

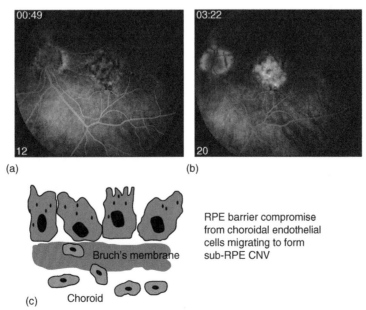

RPE barrier compromise from choroidal endothelial cells migrating to form sub-RPE CNV

Figure 6 (a) Ill-defined hyperfluorescence in early frame with (b) leakage in late frame of fluorescein angiogram secondary to RPE barrier disruption from occult or type 1 choroidal neovascularization beneath the RPE depicted in (c).

RPE layer →

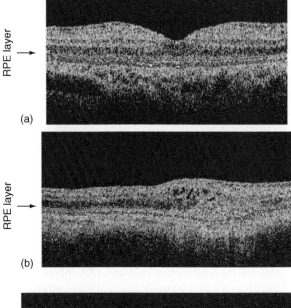

(a)

RPE layer →

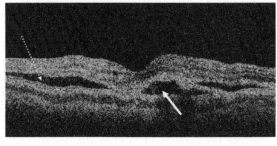

(b)

(c)

Figure 7   (a) Time domain optical coherence tomogram from patient with normal macula showing good definition of macular layers and normal foveal contour. (b) Optical coherence tomogram from patient with neurosensory retinal choroidal neovascularization, showing disruption of ordered architecture of the retinal layers, increased hyperreflectance (red) in layer near RPE/Bruch's membrane and intraretinal cysts, indicating intraretinal fluid (arrowhead). (c) Optical coherence tomogram showing subretinal fluid (dotted arrow – beneath the photoreceptors and above the RPE) and sub-RPE fluid (solid arrow – beneath the RPE and above Bruch's membrane).

endothelial growth factor (VEGF)). All these stimuli can lead to increased secretion of VEGF, which also increases permeability of blood vessels and RPE and choroidal endothelial migration, proliferation, and chemotaxis, all processes believed important in the development of neurosensory retinal choroidal neovascularization. Furthermore, RPE can release other cytokines that recruit leukocytes, including macrophages that can then release VEGF, or that interact with other processes important in the pathogenesis of AMD.

Diabetes Mellitus

In diabetic retinopathy, the inner BRB is impaired and is most easily appreciated clinically on fluorescein angiography, as leakage from microaneurysms, dilated capillaries, intraretinal microvascular abnormalities, and

neovascularization that grows above the inner limiting membrane into the vitreous (**Figure 9**). However, fluorescein staining at the level of the RPE is seen on fluorescein angiograms of patients with diabetes. Also, hyperglycemia has been shown to impair the function of tight junctions of the RPE *in vitro*. Diabetic retinopathy also impairs vision through the development of macular edema, which occurs when there is fluid and solutes that leak from the vasculature into the neurosensory retina (**Figure 10**). It can occur through a breakdown of the inner and potentially outer BRBs. Besides the finding that hyperglycemia can reduce TER in cultured RPE, animal models of diabetic retinopathy have found that there is reduced Na,K-ATPase activity in the RPE. So, once retinal blood vessels leak fluid, lipids, and protein into the neurosensory retina, mechanisms to transport fluid and compounds out of the retina also appear to be impaired in the diabetic state.

Inflammation has been shown to play a role in the pathophysiology of diabetic retinopathy. In animal models, leukostasis or adherence of white blood cells to retinal capillaries has been found and postulated to be a mechanism of later capillary nonperfusion and endothelial damage, which precede the development of proliferative diabetic retinopathy and macular edema. Furthermore, diabetes can also cause a choroidal vasculopathy associated with leukostasis, which may cause choroidal ischemia and later angiogenesis both of which can interfere with intrinsic ocular flow from the vitreous toward the choroid.

Proliferative Vitreoretinopathy

Proliferative vitreoretinopathy (PVR) occurs when cellular contractile membranes develop on the surface of the retina and contract it and pull open breaks in the retina, which can lead to complex retinal detachments. It is the most common cause of failed retinal detachment repair. Vitreous fluorophotometry readings in animal models of PVR show a breakdown in the BRB associated with released cells into the vitreous cavity. It is believed that RPE cells, serum, and other factors have access to the vitreous cavity and are responsible for further breakdown of the BRB and growth of preretinal membranes.

The treatment for PVR, currently, is surgical, requiring vitrectomy and stripping of the membranes from the retina, and then methods to reattach the retina and create a permanent chorioretinal adhesion. However, ongoing research may provide medical means to prevent the formation of preretinal membranes.

Drug Toxicity

Several drugs, including thioridazine (Mellaril), thorazine, hydroxychloroquine, and chloroquine, have been shown to cause vision loss and toxicity with pigmentary changes. The exact effects on the RPE and BRB are unclear.

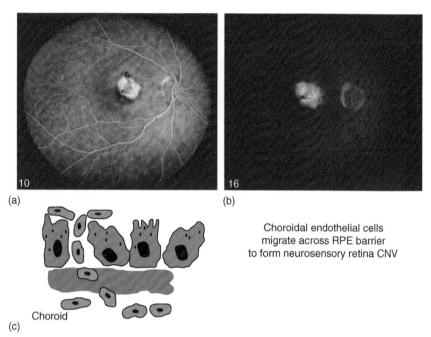

Choroidal endothelial cells
migrate across RPE barrier
to form neurosensory retina CNV

(c)

Figure 8 (a) Well-defined hyperfluorescence in early frame with (b) leakage in late frame of fluorescein angiogram from type II choroidal neovascularization (CNV) that has entered the neurosensory retina (depicted in (c)).

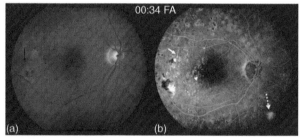

Figure 9 (a) Color images of right eye of patient with diabetic retinopathy previously treated with laser (examples of pigmented laser spots shown by arrows). (b) Fluorescein angiogram of same eye showing intraretinal microvascular abnormalities (arrow) and areas of avascular retina with dilated capillaries, an irregular foveal avascular zone, and microaneurysms (arrowhead). Dotted arrow shows area of hyperfluorescence from leaking neovascularization likely growing above the inner limiting membrane.

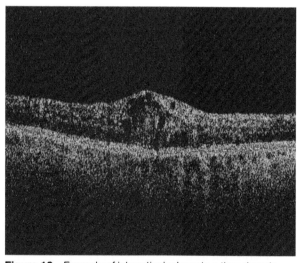

Figure 10 Example of intraretinal edema in a time domain optical coherence tomogram from a patient with diabetic macular edema. Note that the RPE reflective layer is intact in contrast to **Figure 7(b)** in which invasion of cells into the neurosensory retina has occurred in neovascular AMD.

Central Serous Retinopathy

Central serous retinopathy (CSR) is a clinical disease often occurring in young to middle-aged individuals, although it can also manifest or recur and become chronic in later life. Symptoms include reduced vision or inability to focus, and clinical examination shows the presence of a neurosensory retinal detachment within the macula. Fluorescein angiography shows focal areas of RPE leaks (**Figure 11**) or diffuse RPE disturbances. In chronic CSR, RPE decompensation occurs (**Figure 12**) and can lead to chronic subretinal leakage and accumulation of subretinal

and intraretinal fluid. Although cases of CSR usually resolve without permanent vision loss, recurrent or chronic CSR can lead to permanent loss of visual acuity. Even when a sole leak appears to be present, CSR is believed to be associated with broad RPE dysfunction, because if only one area of dysfunction were present, the Na,K-ATPases of surrounding healthy RPE would pump out fluid from the subretinal space.

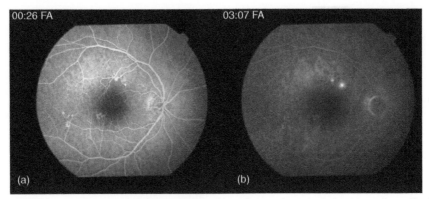

Figure 11 (a) Early hyperfluorescence at level of RPE in fluorescein angiogram from central serous retinopathy in 40-year-old male. (b) Late pooling of dye into the neurosensory retina from a leak in the RPE.

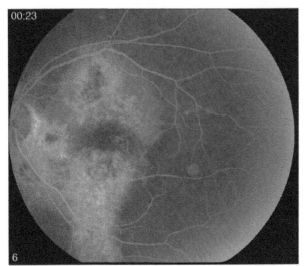

Figure 12 RPE decompensation showing broad area of hyperfluorescence in fluorescein angiogram from chronic long-standing central serous retinopathy in 60 year old male.

Angiograms using indocyanine green dye, which permits visualization of the choroidal vasculature, show that areas of choroidal hypoperfusion and later choroidal hyperpermeability are present in CSR. Although one might suspect inflammation to be a cause of the hyperpermeability, treatment with steroids can severely worsen CSR and should be avoided. Corticosteroids can affect the expression of adrenergic receptor genes and it is thought that this contributes to the overall effect of catecholamines on CSR. Some have postulated that the pathology may involve the adrenocorticotrophic hormone. Another unusual aspect of CSR is that treatment of a sole RPE leak on fluorescein angiography can hasten resolution of the serous detachment, even though it is believed that broad RPE dysfunction is present. Besides corticosteroids, hypertension also increases the risk.

CSR has long been believed to be associated with breakdown in the BRB particularly of the RPE. The cause remains unknown but it is associated with increased stress and a type A personality and is believed to be related to elevated cortisol and epinephrine, which affect the autoregulation of the choroidal circulation. In early studies, adult Japanese monkeys that received multiple daily (>30) injections of intravenous adrenalin developed serous retinal detachments and leaking RPE spots by fluorescein angiography similar in appearance to that seen in CSR. An intramuscular injection of prednisolone led to the same findings on fluorescein angiography but required fewer doses of adrenalin.

Retinitis Pigmentosa

Evidence of abnormalities in the localization of ZO-1, beta-catenin, and other associated adherens proteins in the $rho^{-/-}$ mouse, a model of autosomal dominant retinitis pigmentosa, provides support for tight junction and adherens junction-associated protein modifications in retinitis pigmentosa. Furthermore, in retinitis pigmentosa, there is cystoid macular edema often associated with hyperfluorescence of the RPE cells on fluorescein angiography, suggesting a breakdown of the RPE barrier. Toxins such as sodium iodate selectively poison the RPE and have been used to test the role of the RPE BRB in animal models in which fluid had been injected into the subretinal space or for studies in PVR. Although the toxin is selective for RPE, it poisons the RPE and therefore affects all functions of the RPE, not just the tight junctions.

Urethrane was used to test the BRB in earlier studies and was reported to lead to inhibition of intervesicular transport across endothelia and loss of RPE.

Growth Factors

Besides the role of VEGF in reducing RPE barrier properties, other growth factors play a role. Insulin-like growth factor-1 (IGF-1) can induce VEGF-related RPE barrier breakdown. Also, hepatocyte growth factor (HGF) can lead to disassembly of tight and adherens junctions in association with reduced barrier properties. HGF has also been shown to be increased in human PVR.

Studies to Increase the Function of the BRB

Carbonic anhydrase inhibitors lead to acidification of the subretinal space, which, in turn, leads to an increase in chloride ion transport into the choroid, thus eliminating water from the subretinal space and retina and increasing the adhesiveness of the RPE. Carbonic anhydrase inhibitors have been associated with improved BRB function, based on reduced fluorescein leakage into the retina, in small clinical studies. However, larger clinical trials have shown a smaller benefit.

See also: Breakdown of the Blood–Retinal Barrier; Photopic, Mesopic and Scotopic Vision and Changes in Visual Performance; Phototransduction in *Limulus* Photoreceptors.

Further Reading

Alberts, B., Johnson, A., Lewis, J., et al. (2002). Cell junctions, cell adhesion, and the extracellular matrix. In: Alberts, B., Johnson, A., Lewis, J., et al. (eds.) *Molecular Biology of the Cell*, pp. 949–1009. New York: Garland Science.

Bian, Z. M., Elner, S. G., Yoshida, A., and Elner, V. M. (2003). Human RPE-monocyte co-culture induces chemokine gene expression through activation of MAPK and NIK cascade. *Experimental Eye Research* 76: 573–583.

Campbell, M., Humphries, M., Kennan, A., et al. (2006). Aberrant retinal tight junction and adherens junction protein expression in an animal model of autosomal dominant retinitis pigmentosa: The Rho(–/–) mouse. *Experimental Eye Research* 83: 484–492.

Crane, I. J., Wallace, C. A., McKillop-Smith, S., and Forrester, J. V. (2000). CXCR4 receptor expression on human retinal pigment epithelial cells from the blood–retina barrier leads to chemokine secretion and migration in response to stromal cell-derived factor 1 alpha. *Journal of Immunology* 165: 4372–4378.

Dibas, A. and Yorio, T. (2008). Regulation of transport in the RPE. In: Tombran-Tink, J. and Barnstable, C. (eds.) *Ocular Transporters in Ophthalmic Diseases and Drug Delivery,* 1st edn., pp. 157–184. Berlin: Springer/Humana Press.

Fubuoka, Y., Strainic, M., and Medof, M. E. (2003). Differential cytokine expression of human retinal pigment epithelial cells in response to stimulation by C5a. *Clinical and Experimental Immunology* 131: 248–253.

Hartnett, M. E., Lappas, A., Darland, D., et al. (2003). Retinal pigment epithelium and endothelial cell interaction causes retinal pigment epithelial barrier dysfunction via a soluble VEGF-dependent mechanism. *Experimental Eye Research* 77: 593–599.

Hu, J. and Bok, D. (2001). A cell culture medium that supports the differentiation of human retinal pigment epipthelium into functionally polarized monolayers. *Molecular Vision* 7: 14–19.

Jin, M., Barron, E., He, S., Ryan, S. J., and Hinton, D. R. (2002). Regulation of RPE intercellular junction integrity and function by hepatocyte growth factor. *Investigative Ophthalmology and Visual Science* 43: 2782–2790.

Lodish, H., Berk, A., Zipursky, S. L., et al. (2000). Transport across cell membranes. In: *Molecular Cell Biology*. Basingstokes: WH Freeman.

Marmor, M. F. and Wolfensberger, T. J. (eds.) (1998). *The Retinal Pigment Epithelium*. New York: Oxford University Press.

Penn, J. S., Madan, A., Caldwell, R. B., et al. (2008). Vascular endothelial growth factor in eye disease. *Progress in Retina and Eye Research* 27(4): 331–371.

Rajasekaran, S. A., Hu, J., Gopal, J., et al. (2003). Na,K-ATPase inhibition alters tight junction structure and permeability in human retinal pigment epithelial cells. *American Journal of Physiology – Cell Physiology* 284: C1497–C1507.

Rizzolo, L. J. (2007). Development and role of tight junctions in the retinal pigment epithelium. In: Jeon, K. W. (ed.) *International Review of Cytology, a Survey of Cell Biology* vol. 258, pp. 195–234. San Diego, CA: Elsevier.

Sen, H. A., Robertson, T. J., Conway, B. P., and Campochiaro, P. A. (1988). The role of breakdown of the blood–retinal barrier in cell-injection models of proliferative vitreoretinopathy. *Archives of Ophthalmology* 106: 1291–1294.

Xu, H., Dawson, R., Crane, I. J., and Liversidge, J. (2005). Leukocyte diapedesis *in vivo* induces transient loss of tight junction protein at the blood–retina barrier. *Investigative Ophthalmology and Visual Science* 46: 2487–2494.

Yoshioka, H., Katsume, Y., and Akune, H. (1982). Experimental central serous chorioretinopathy in monkey eyes: Fluorescein angiographic findings. *Ophthalmologica* 185(3): 168–178.

Circadian Metabolism in the Chick Retina

P M Iuvone, Emory University School of Medicine, Atlanta, GA, USA

Glossary

Basic helix–loop–helix-Per-ARNT-Sim (bHLH-PAS) domain transcription factors – A family of transcription factors that heterodimerize and bind to E box enhancer elements in gene promoters. The family includes the clock gene products CLOCK, neuronal PAS domain protein 2 (NPAS2), and brain and muscle aryl hydrocarbon receptor nuclear translocator 1 (BMAL1), as well as the arylhydrocarbon nuclear receptor and the hypoxia-inducible factors.

Circadian rhythms – Changes in biological processes that occur on a daily basis; they are driven by autonomous circadian clocks; these rhythms provide selective advantage to organisms by allowing them to anticipate temporal changes in their environment.

Clock-controlled genes – Genes that are regulated by circadian oscillators via clock gene transcription factors; the proteins encoded by clock-controlled genes are rhythmically expressed and generate rhythms of physiology.

Clock genes – Genes that encode proteins that form the molecular basis of circadian oscillators; most clock gene proteins are transcription factors.

Scotopic vision – Vision in dim light that is mediated by rod photoreceptors and rod bipolar cell pathways.

Zeitgeber time – Time of day relative to the light–dark cycle; light onset corresponds to ZT0.

Introduction

The chick retina has been used extensively to study ocular circadian rhythms. It displays particularly robust rhythms of clock gene expression, clock-controlled gene expression, and several biochemical and physiological clock outputs. In addition, embryonic chick retinal cells can be cultured, and they maintain light responsiveness and circadian rhythm generation *in vitro*. These cell cultures facilitate pharmacological and molecular studies of clock signaling pathways.

Clock Gene Expression

The current model for the molecular basis of circadian clocks involves transcriptional–translational feedback loops that are comprised of highly conserved clock genes and the proteins that they encode (**Figure 1**). The clock gene proteins are characterized as positive and negative elements, based on their transcriptional activity. The positive elements include basic helix–loop–helix–PAS-domain (bHLH-PAS) transcription factors that heterodimerize and bind to circadian E box enhancer elements in promoters of other clock genes and clock-controlled genes. These positive elements include brain and muscle aryl hydrocarbon receptor nuclear translocator 1 (BMAL1; also called ARNTL, MOP3) and CLOCK. The BMAL1 heterodimerizes with CLOCK and the dimerized transcription factors stimulate the transcription of the genes encoding the negative elements, the cryptochromes (CRY1 and CRY2) and periods (PER1, PER2, and PER3). The CRY and PER proteins are imported into the nucleus and inhibit transactivation of their own promoters by BMAL1:CLOCK. In some clocks, neuronal PAS domain protein 2 (NPAS2, also called MOP4) can substitute for CLOCK and form active heterodimers with BMAL1. A second feedback loop involves E-box-mediated transcriptional activation of the orphan retinoic-acid-related receptor family genes, *Rev-erbα* and *Rorα*; the protein products of these genes contribute to the rhythmic regulation of *Bmal1* gene transcription. These feedback loops, coupled with a variety of post translational modifications, generate daily rhythms of gene expression that ultimately generate physiological circadian rhythms.

Most of the clock gene transcripts identified in mammalian circadian clocks have been identified in the chick retina and in cultured chick photoreceptors, and the regulation of their expression has been analyzed (**Figures 2 and 3**). Of the genes encoding the positive elements of the oscillator, *Bmal1* and *Npas2* transcripts show robust daily rhythms under light–dark cycles or in constant (24 h day^{-1}) darkness (**Figure 2**). These circadian rhythms peak near the time of subjective dusk (~ZT12) *in vivo* and *in vitro*. In contrast, *Clock* mRNA appears to be constitutively expressed in dark–dark (DD). Transcripts encoding the negative elements, the cryptochrome and period proteins, also show rhythmic expression (**Figure 3(a) and 3(b)**). The *Cry1* mRNA peaks in the middle of the day (~ZT8) *in vivo* and in photoreceptor cultures. *In vivo*, the most robust rhythms of *Cry1* mRNA expression are observed in the ganglion cell and photoreceptor cell layers. The *Cry2* mRNA is also expressed in retina, but analyses of whole retina mRNA suggest that it is not rhythmically expressed. The *Per2* mRNA expression is maximal in the early morning, while *Per3* transcript level appears maximal late in the night.

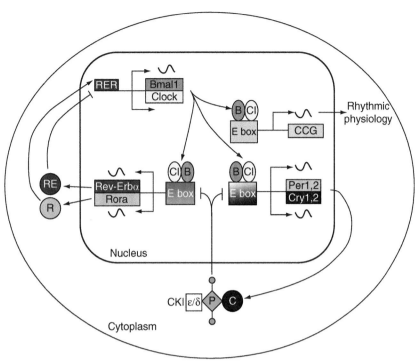

Figure 1 Circadian clockwork mechanism. Circadian clocks in a wide range of organisms are composed of two interdependent transcription–translation feedback loops that drive the periodic rhythms in the mRNA and protein levels of the clock components. In mammalian SCN, the first loop involves two bHLH-PAS-containing transcription factors, CLOCK (Cl) and BMAL1 (B). These transcription factors heterodimerize and activate the rhythmic transcription of three *period* genes (*Per1*-*Per3*, with *Per1* and *Per2* being critical to the circadian clock) and two *cryptochrome* genes (*Cry1* and *Cry2*). The PER (P) and CRY (C) proteins complex with casein kinase 1 δ and ε (CKIδ/ε), which phosphorylates PER. The resulting complex inhibits CLOCK/BMAL1-mediated transcription of *period* and *cryptochrome* genes, thus providing the negative feedback loop. The second loop involves CLOCK/BMAL1 driven rhythmic transcription of *Rev-erbα* and *Rora*, members of the retinoic acid-related orphan nuclear receptor family. The phase of *Rora* expression closely resembles those of *Per1* and *Per2*, and is opposite in phase with *Rev-erbα*. The resultant REV-ERBα and RORa proteins (RE and R, respectively) compete for the same promoter element, RRE (Rev-erb/Ror element) and drive the rhythm in *Bmal1* transcription. CLOCK/BMAL1 heterodimers also bind to circadian E-boxes in clock-controlled genes (CCGs), providing an output from the clock that drives rhythmic physiology. Adapted from Iuvone, P. M., Tosini, G., Pozdeyev, N., et al. (2005). Circadian clocks, clock networks, arylalkylamine N-acetyltransferase, and melatonin in the retina. *Progress in Retinal and Eye Research* 24: 433–456, with permission from Elsevier.

Thus far, *Per1* has not been identified in the chick retina, and the gene may be missing from the chicken genome. In addition to oscillating in a clock-dependent manner, *Per2* and *Cry1* are rapidly induced by light exposure (**Figure 3(c)** and **3(d)**). It is generally thought that this induction is a mechanism for circadian clock entrainment by light.

The rhythmic and light-regulated expression of circadian clock genes in the chick retina, particularly in cultured retinal cells, provides conclusive evidence that the retina contains autonomous circadian oscillators that can function independently of oscillators in the brain and pineal gland. Nevertheless, in the intact organism, retinal, pineal, and brain clocks are thought to interact to regulate physiology.

Circadian Regulation of Cyclic AMP in Retina

Cyclic AMP is a ubiquitous second messenger molecule that regulates multiple aspects of cellular metabolism and function. Effects of cyclic AMP are mediated by activation of cyclic AMP-dependent protein kinase (PKA), which phosphorylates proteins to regulate their function or activity. In so doing, cyclic AMP regulates intermediary metabolism, neurotransmission, and gene expression. Cyclic AMP can also affect cellular function by regulating cyclic nucleotide-gated channels or activating Epac, a cyclic AMP-dependent Rap GTP exchange factor. The Rap is a guanine nucleotide-binding protein of the Ras family.

In chick photoreceptor cell cultures, cyclic AMP levels are regulated by light and circadian clocks. Photoreceptor cyclic AMP levels are high in darkness and reduced by light exposure. Thus, when cells are exposed to a daily light–dark cycle, cyclic AMP fluctuates as a daily rhythm with high levels at night in darkness and low levels during the daytime in light (**Figure 4**). Exposure to light at night rapidly reduces cyclic AMP. The daily rhythm of cyclic AMP persists, albeit with reduced amplitude, when cells are transferred from a light–dark cycle to constant darkness. Thus, the combined effects of illumination and

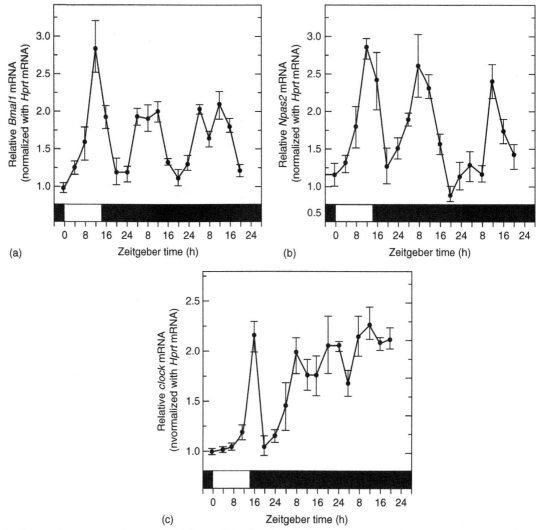

Figure 2 Temporal expression of positive modulators of the circadian clockwork system. Relative mRNA levels of *Bmal1* (a) and *Npas2* (b) in photoreceptor-enriched retinal cell cultures collected at the indicated Zeitgeber times (ZT) in light–dark (LD) and dark–dark (DD). Each data point represents clock gene transcripts normalized to hypoxanthine-guanine phosphoribosyl transferase (*Hprt*) mRNA and expressed relative to the lowest values in LD. The open horizontal bar at the X-axis represents times of light exposure; the black bars represent times of darkness. Analysis of variance (ANOVA) indicated significant rhythms of *Bmal1* and *Npas2* transcripts in LD and DD, with highest levels in the late day and early night. (c) *Clock* mRNA showed significantly higher values during the night (ZT 16) than during the day in LD; transcript levels increased on the first day of DD but there were no significant rhythms on DD1 or DD2. Reproduced from Chaurasia, S. S., Pozdeyev, N., Haque, R., et al. (2006). Circadian clockwork machinery in neural retina: Evidence for the presence of functional clock components in photoreceptor-enriched chick retinal cell cultures. *Molecular Vision* 12: 215–223, Copyright Molecular Vision 2006.

circadian influences interact to generate the daily rhythm of cyclic AMP.

The regulation of cyclic AMP formation in chick photoreceptor cells is Ca^{2+}-dependent, at least in part. Depolarization of the plasma membrane with high concentrations of extracellular K^+ stimulates cyclic AMP formation. This effect requires Ca^{2+} influx through L-type voltage-gated Ca^{2+} channels. The plasma membrane of photoreceptors is partially depolarized in darkness and is hyperpolarized by light. Thus, the dark–light difference in cyclic AMP formation in photoreceptors is likely due to high Ca^{2+} conductance in darkness and decreased Ca^{2+} conductance

following light exposure, due to closure of the voltage-gated channels. Accordingly, the circadian fluctuation in cyclic AMP levels is eliminated by nitrendipine, an L-type Ca^{2+} channel blocker.

Cyclic AMP is synthesized from ATP by adenylyl cyclase. There are 10 isoforms of adenylyl cyclase that are regulated by multiple mechanisms. The findings described above indicate that photoreceptor cyclic AMP levels may be regulated by a Ca^{2+}/calmodulin-stimulated cyclase. The type 1 and type 8 adenylyl cyclases are both stimulated by Ca^{2+}/calmodulin and are both expressed in the chick retina. There is a circadian rhythm in the expression of

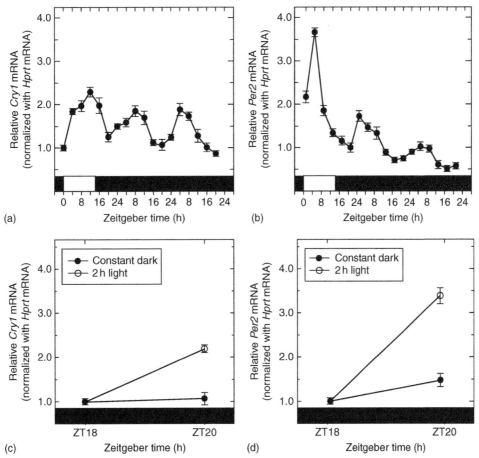

Figure 3 Temporal expression of negative modulators of the circadian clockwork system. Circadian profiles of *Cry1* (a) and *Per2* (b) transcripts in the photoreceptor-enriched retinal cell cultures collected at the indicated Zeitgeber time (ZT) in light–dark (LD) and dark–dark (DD). Each data point represents clock gene transcripts normalized to *Hprt* mRNA, expressed relative to the lowest values in LD. Acute light exposure at night induces *Cry1* (c) and *Per2* (d) mRNA expression. On day 9 *in vitro* (DIV9) cells were kept in constant darkness until ZT 18, when one group of cells was collected. Another group of cells remained in darkness for an additional 2 h (solid symbol), while a third group of cells was exposed to light for 2 h prior to cell harvesting (open symbol). Exposure to light significantly increased *Cry1* and *Per2* transcript levels. Reproduced from Chaurasia, S. S., Pozdeyev, N., Haque, R., et al. (2006). Circadian clockwork machinery in neural retina: Evidence for the presence of functional clock components in photoreceptor-enriched chick retinal cell cultures. *Molecular Vision* 12: 215–223, Copyright Molecular Vision 2006.

Adcy1, the transcript that encodes the type 1 adenylyl cyclase, in photoreceptor cell cultures and in the chick retina *in vivo*. In addition, there is a circadian rhythm in Ca^{2+}/calmodulin-stimulated adenylyl cyclase activity in membranes prepared from photoreceptor cell cultures, with high activity at night. Thus, the circadian regulation of cyclic AMP is due to clock-controlled expression of Ca^{2+}/calmodulin-stimulated adenylyl cyclase. The mammalian *Adcy1* gene contains an E-box in its promoter that can be activated by BMAL1:CLOCK heterodimers, and a similar mechanism may contribute to the circadian regulation of *Adcy1* expression in chick photoreceptor cells. However, this hypothesis has not yet been tested directly in the chick. The circadian rhythm of cyclic AMP formation may also be influenced by the clock-controlled expression and activity of the L-type Ca^{2+} channels in photoreceptors.

Circadian Regulation of Melatonin Biosynthesis

Melatonin is a neurohormone that is synthesized in the retinal photoreceptor cells and in the pineal gland. Melatonin synthesis in retinas of most vertebrate species, including chicken, is regulated in a circadian fashion, with high levels at night in darkness. Melatonin functions in the retina to optimize nighttime visual function. Like cyclic AMP formation, melatonin levels are regulated by both illumination and circadian clocks.

The key regulatory enzyme in melatonin biosynthesis is arylalkylamine N-acetyltransferase (AANAT). In chicken retina, AANAT activity undergoes a robust circadian rhythm with peak activity at night (**Figure 5**). Exposure to light at night causes a rapid decline in AANAT activity

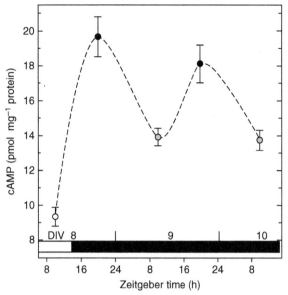

Figure 4 Circadian fluctuation of intracellular cAMP level. Cells were prepared from embryonic neural retinas and incubated for 8 days under LD. Illumination was switched from LD to DD before expected onset of light at the beginning of day 9 *in vitro* (DIV 9). White symbols represent cAMP level at Zeitgeber time (ZT) 10 in light; black symbols represent ZT 20 in darkness; gray symbols represent subjective day, ZT 10, in darkness. The horizontal white and black bars above the x-axis represent times of light and darkness, respectively. Level of cAMP was significantly higher at night than during the day in LD on DIV 8 and this fluctuation persisted in DD on DIV 9 and DIV 10. Reproduced from Ivanova, T. N. and Iuvone, P. M. (2003). Circadian rhythm and photic control of cAMP level in chick retinal cell cultures: A mechanism for coupling the circadian oscillator to the melatonin-synthesizing enzyme, arylalkylamine N-acetyltransferase, in photoreceptor cells. *Brain Research* 991: 96–103, with permission from Elsevier.

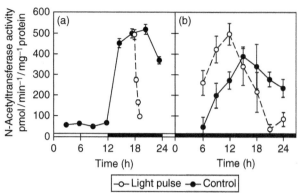

—o— Light pulse   —●— Control

Figure 5 Daily rhythm of retinal AANAT activity: effects of light. (a) AANAT activity fluctuates during the 12 h light–12 h dark cycle (filled circles). Unexpected light exposure at night (open circles) rapidly inhibits activity. (b) AANAT activity in constant (24 h day^{-1}) darkness. The activity rhythm persists on the second day of constant darkness (filled circles). The rhythm is phase advanced by a 6 h light pulse from 18 to 24 h 2 days prior to sampling in constant darkness (open circles). Filled bars on the x-axis represent darkness; open bars represent light. Activity was measured in retinal homogenates of 2-week old chickens. Reproduced from Iuvone, P. M. and Alonso-Gómez, A. L. (1998). Melatonin in the vertebrate retina. In: Christen Y., Doly, M., and Droy-Lefaix, M. -T. (eds.) *Retine, Luminiere, et Radiations*, vol. 9, pp. 49–62. Paris: Irvinn; and Iuvone, P. M., Tosini, G., Pozdeyev, N., et al. (2005). Circadian clocks, clock networks, arylalkylamine N-acetyltransferase, and melatonin in the retina. *Progress in Retinal and Eye Research* 24: 433–456, with permission from Elsevier.

($t^{1/2} \approx 20$ min), to insure that significant melatonin synthesis occurs in darkness only. Chicken AANAT activity is regulated by transcriptional and post-translational mechanisms (**Figure 6**). The retina displays daily rhythms of *Aanat* mRNA and AANAT protein in chickens exposed to a light–dark cycle or constant darkness. Levels of transcript, protein, enzyme activity, and melatonin all peak at night. The proximal promoter of the chicken *Aanat* gene contains a circadian E-box that can be activated by either BMAL1:CLOCK or BMAL1:NPAS2 heterodimers, and it is generally thought that this directly couples the circadian clock to the rhythmic expression of *Aanat*. In addition, the chicken *Aanat* 5′-flanking region contains cyclic AMP response elements. Thus, the circadian rhythm of cyclic AMP may also contribute to the rhythm of *Aanat* mRNA.

The AANAT protein is regulated by PKA-dependent phosphorylation and proteasomal degradation (**Figure 6**). The AANAT contains two consensus PKA phosphorylation sites. When cyclic AMP levels are high at night, AANAT is phosphorylated. Phospho-AANAT binds to 14-3-3 proteins, which are ubiquitous signaling proteins involved in

multiple cellular functions. The interaction of 14-3-3 with AANAT increases the affinity of the enzyme for its substrate, increasing catalytic activity. Exposure to light at night causes a very rapid decrease in AANAT activity and protein level ($t^{1/2} \approx 20$ min) without any initial change of *Aanat* mRNA (**Figure 6**). The decrease of AANAT is accompanied by a similarly rapid decline in melatonin levels, insuring that melatonin only functions in darkness. The decrease of protein and activity results from the light-induced decrease of cyclic AMP levels, resulting in dephosphorylation of AANAT, unbinding of 14-3-3, and rapid proteolytic degradation of the enzyme. The degradation can be blocked by proteasome inhibitors, indicating that this is a proteasome-dependent event.

During the daytime, multiple mechanisms cooperate to keep melatonin biosynthesis at a minimum (**Figure 6**). Low levels of cyclic AMP appear to play key roles in these mechanisms. The *Aanat* transcript levels are low, presumably due to reduced cyclic AMP-directed transcriptional activation and suppression by Cry1 of E-box-mediated transactivation (**Figure 6**). In addition, the low levels of cyclic AMP during the daytime, especially in light, favor the dephosphorylated state of the AANAT protein, resulting in its rapid degradation by proteasomes.

The majority of AANAT in the chick retina is expressed at night in photoreceptor cells. However, there is some

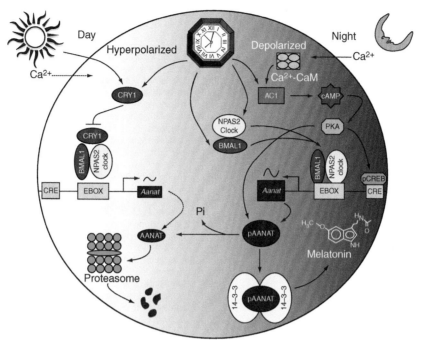

Figure 6 Model for circadian clock- and light-regulated melatonin biosynthesis in photoreceptor cells (see text for detailed description of the model). The left side of the figure depicts processes occurring in light while the right side shows processes occurring at night in darkness. Abbreviations: AANAT, arylalkylamine N-acetyltransferase; AC1, type 1 Ca^{2+}/calmodulin-stimulated adenylyl cyclase; CaM, calmodulin; cAMP, cyclic adenosine 3′,5′-monophosphate; CRE, cAMP-responsive element; pAANAT, phosphorylated AANAT; pCREB, phosphorylated cAMP response element-binding protein; PKA, cAMP-dependent protein kinase, with permission from Elsevier.

evidence that small amounts of AANAT may be expressed in ganglion cells during the daytime. The role of ganglion cell-derived AANAT is not known.

Other Rhythms of Gene Expression and Metabolism in the Chick Retina

Circadian clocks regulate phospholipid metabolism in the chick retina. Daily rhythms of phospholipid labeling with either ^{32}P or [^{3}H]glycerol have been observed in chicks kept on a light–dark cycle or in constant darkness. The ^{32}P labeling of total phospholipids in photoreceptors and ganglion cells peaks in the late night. The major phospholipid labeled under these conditions is phosphotidylinositol. In contrast, labeling with [^{3}H]glycerol peaks midday, with phosphatidylcholine as the major labeled species.

Melanopsin (Opn4) is a photopigment first discovered in *Xenopus* melanocytes, where it mediates photic regulation of melanin pigment aggregation. Subsequent cloning of orthologous melanopsin genes revealed that it is expressed in mammalian retina exclusively in a small subset of ganglion cells that are intrinsically photosensitive. These ganglion cells project to the suprachiasmatic nucleus (SCN) of the hypothalamus, the inferior olivary nucleus, and other brain regions involved in nonimage forming vision. The intrinsically photosensitive ganglion

cells mediate the pupillary light reflex and photic entrainment of the circadian oscillator in the SCN.

In contrast to mammals, chickens express two melanopsin genes, designated *Opn4x* and *Opn4m* because of their sequence homology to the *Xenopus* and mammalian melanopsin genes, respectively. The *Opn4x* and *Opn4m* transcripts are more widely distributed than their mammalian counterpart. Chicken melanopsin transcripts are found in a subpopulation of cells in the ganglion cell layer, and also in photoreceptors, retinal pigment epithelium (RPE) cells, inner nuclear layer (INL) cells, the pineal gland, and areas of the brain known to contain deep brain photoreceptors. The *Opn4x* is rhythmically expressed in the chicken retina in light–dark cycles and in constant darkness. Patterns of rhythmicity differ among retinal layers; *Opn4x* peaks in the morning in RPE and INL cells, but at night in photoreceptors. Similar to the retinal photoreceptors, *Opn4x* also peaks at night in chicken pinealocytes, which are directly photosensitive and contain a circadian clock that is entrained by light. The regulation and localization of *Opn4* is consistent with the hypothesis that this novel photopigment plays a role in circadian regulation in the retina and pineal gland.

Iodopsin is the photopigment of red-sensitive cone photoreceptors of avian species. Iodopsin mRNA levels fluctuate as a circadian rhythm in retinal photoreceptors *in vivo* and *in vitro*, in cultured photoreceptors, and in retinal explants. The rate of transcription of the iodopsin gene peaks late in the subjective day in constant darkness, ~3 h before the beginning of the subjective night, and mRNA levels peak early in the subjective night. The functional significance of the circadian rhythm of iodopsin mRNA is yet to be determined.

Ion channels in chicken cone photoreceptors are also subject to circadian regulation. There is a circadian rhythm in the affinity of the cone cyclic nucleotide-gated channel for cGMP, with highest affinity during the subjective night. In addition, L-type Ca^{2+} channels are regulated in a circadian fashion. The Ca^{2+} currents and immunoreactivity for $\alpha 1C$ and $\alpha 1D$ calcium channel subunits are greater at night than during the day. There is also a rhythm of $\alpha 1D$ mRNA level. More details on rhythms of ion channels can be found elsewhere in this encyclopedia.

Conclusion

The chick retina is a remarkably rhythmic tissue, with robust circadian control of gene expression, metabolism, physiology, and melatonin synthesis. Most attention has been paid thus far to photoreceptor rhythms, but inner retinal neurons also express clock genes and are likely to be subject to circadian control. The ability to generate retinal cell cultures, which maintain their circadian properties and can be manipulated pharmacologically and genetically, suggests that the chick retina will continue to be a valuable model system for exploring the circadian organization of the retina.

Acknowledgments

The author is grateful to the past and present members of his laboratory, especially Rashidul Haque, Nikita Pozdeyev, Shyam Chaurasia, and Tamara Ivanova, and to David Klein and the members of his laboratory, who contributed greatly to the body of knowledge contained within this article. The author also thanks Gianluca Tosini for his collaborative contributions and for critical comments and suggestions on the article. Research in the author's laboratory is funded by the National Institutes of Health EY004864 and EY06360, and by Research to Prevent Blindness.

See also: The Circadian Clock in the Retina Regulates Rod and Cone Pathways; Circadian Photoreception; Circadian Regulation of Ion Channels in Photoreceptors; Circadian Rhythms in the Fly's Visual System; Fish Retinomotor Movements; *Limulus* Eyes and Their Circadian Regulation; Neurotransmitters and Receptors: Dopamine Receptors; Neurotransmitters and Receptors: Melatonin Receptors.

Further Reading

Bailey, M. J., Beremand, P. D., Hammer, R., et al. (2004). Transcriptional profiling of circadian patterns of mRNA expression in the chick retina. *Journal of Biological Chemistry* 279: 52247–52254.

Bellingham, J., Chaurasia, S. S., Melyan, Z., et al. (2006). Evolution of melanopsin photoreceptors: Discovery and characterization of a new melanopsin in nonmammalian vertebrates. *PLoS Biology* 4: e254.

Bailey, M. J., Chong, N. W., Xiong, J., and Cassone, V. M. (2002). Chickens' Cry2: Molecular analysis of an avian cryptochrome in retinal and pineal photoreceptors. *FEBS Letters* 513: 169–174.

Bernard, M., Iuvone, P. M., Cassone, V. M., et al. (1997). Avian melatonin synthesis: Photic and circadian regulation of serotonin N-acetyltransferase mRNA in the chicken pineal gland and retina. *Journal of Neurochemistry* 68: 213–224.

Chaurasia, S. S., Haque, R., Pozdeyev, N., Jackson, C. R., and Iuvone, P. M. (2006). Temporal coupling of cyclic AMP and Ca/calmodulin-stimulated adenylyl cyclase to the circadian clock in chick retinal photoreceptor cells. *Journal of Neurochemistry* 99: 1142–1150.

Chaurasia, S. S., Pozdeyev, N., Haque, R., et al. (2006). Circadian clockwork machinery in neural retina: Evidence for the presence of functional clock components in photoreceptor-enriched chick retinal cell cultures. *Molecular Vision* 12: 215–223.

Chaurasia, S. S., Rollag, M. D., Jiang, G., et al. (2005). Molecular cloning, localization and circadian expression of chicken melanopsin (Opn4): Differential regulation of expression in pineal and retinal cell types. *Journal of Neurochemistry* 92: 158–170.

Chong, N. W., Bernard, M., and Klein, D. C. (2000). Characterization of the chicken serotonin N-acetyltransferase gene. Activation via clock gene heterodimer/E box interaction. *Journal of Biological Chemistry* 275: 32991–32998.

Chong, N. W., Chaurasia, S. S., Haque, R., Klein, D. C., and Iuvone, P. M. (2003). Temporal-spatial characterization of chicken clock genes: Circadian expression in retina, pineal gland, and peripheral tissues. *Journal of Neurochemistry* 85: 851–860.

Garbarino-Pico, E., Carpentieri, A. R., Contin, M. A., et al. (2004). Retinal ganglion cells are autonomous circadian oscillators synthesizing N-acetylserotonin during the day. *Journal of Biological Chemistry* 279: 51172–51181.

Guido, M. E., Pico, E. G., and Caputto, B. L. (2001). Circadian regulation of phospholipid metabolism in retinal photoreceptors and ganglion cells. *Journal of Neurochemistry* 76: 835–845.

Hamm, H. E. and Menaker, M. (1980). Retinal rhythms in chicks – circadian variation in melatonin and serotonin N-acetyltransferase. *Proceedings of the National Academy of Sciences of the United States of America* 77: 4998–5002.

Haque, R., Chaurasia, S. S., Wessel, J. H., III, and Iuvone, P. M. (2002). Dual regulation of cryptochrome 1 mRNA expression in chicken retina by light and circadian oscillators. *NeuroReport* 13: 2247–2251.

Iuvone, P. M. and Alonso-Gómez, A. L. (1998). Melatonin in the vertebrate retina. In: Christen, Y., Doly, M., and Droy-Lefaix, M.-T. (eds.) *Retine, Luminiere, et Radiations*, vol. 9, pp. 49–62. Paris: Irvinn.

Iuvone, P. M., Brown, A. D., Haque, R., et al. (2002). Retinal melatonin production: Role of proteasomal proteolysis in circadian and photic control of arylalkylamine N-acetyltransferase. *Investigative Ophthalmology and Visual Science* 43: 564–572.

Ivanova, T. N. and Iuvone, P. M. (2003). Circadian rhythm and photic control of cAMP level in chick retinal cell cultures: A mechanism for coupling the circadian oscillator to the melatonin-synthesizing enzyme, arylalkylamine N-acetyltransferase, in photoreceptor cells. *Brain Research* 991: 96–103.

Iuvone, P. M., Tosini, G., Pozdeyev, N., et al. (2005). Circadian clocks, clock networks, arylalkylamine N-acetyltransferase, and melatonin in the retina. *Progress in Retinal and Eye Research* 24: 433–456.

Pierce, M. E., Sheshberadaran, H., Zhang, Z., et al. (1993). Circadian regulation of iodopsin gene expression in embryonic photoreceptors in retinal cell culture. *Neuron* 10: 579–584.

Pozdeyev, N., Taylor, C., Haque, R., et al. (2006). Photic regulation of arylalkylamine N-acetyltransferase binding to 14-3-3 proteins in retinal photoreceptor cells. *Journal of Neuroscience* 26: 9153–9161.

Central Retinal Vein Occlusion

S S Hayreh, University of Iowa, Iowa City, IA, USA

Glossary

Demographic – The statistical study of a population, including geographical distribution, sex and age composition, and birth and death rates.

Electroretinography – The recording of the changes in electric potential in the retina by stimulating it by light.

Fluorescein fundus angiography – The visualization of blood vessels in the interior of the eye following intravenous injection of fluorescein.

Glaucoma – An eye disease caused by an increase in eye pressure, which causes changes in the optic nerve and loss of vision.

Hematological – Dealing with the blood and blood-forming tissues.

Histopathological – Dealing with the minute structure of diseased tissues.

Lamina cribrosa – The perforated portion of the back part of the white of the eye (sclera) through which nerve fibers from the retina exit.

Multifactorial – Related to, or arising through the action of many factors.

Neovascularization – The formation of abnormal new blood vessels.

Ophthalmoscopy – The examination of the interior of the eye with the instrument called ophthalmoscope.

Panretinal photocoagulation – The application of an intense beam of laser light to the entire retina.

Pathogenesis – The mechanisms of development of a disease.

Perimeter – An apparatus used to test the visual field.

Retinal vein occlusion is the most common retinal vascular occlusive disorder. In general, there is a tendency to regards this as one disease; that is not only incorrect but also causes much confusion. From the point of view of pathogenesis, clinical picture, prognosis, and management, retinal vein occlusion in fact consists of six distinct clinical entities that are categorized as follows:

1. Central retinal vein occlusion (CRVO), which comprises
   a. nonischemic CRVO, and
   b. ischemic CRVO.
2. Hemi-central retinal vein occlusion (HCRVO), which comprises
   a. nonischemic HCRVO, and
   b. ischemic HCRVO.
3. Branch retinal vein occlusion, which includes
   a. major BRVO, and
   b. macular BRVO.

It is beyond the scope of this article to discuss all the six types of retinal vein occlusion; hence, we restrict our discussion only to CRVO. Over the last 150 years, a large volume of literature has accumulated on the subject of CRVO. The objective of this article is to provide a brief review of the current state of our knowledge on the subject.

Pathogenesis

A good understanding of the pathogenesis of a disease is fundamental to a full grasp of the clinical features of the disease and its logical management. There is almost a universal tendency to blame one or two factors as causative factor(s) in the development of CRVO, but association does not necessarily mean there is a cause-and-effect relationship. Available evidence strongly suggests that the pathogenesis of CRVO, like many other ocular vascular occlusive disorders, is a multifactorial process. It seems that some risk factors predispose an individual or an eye to CRVO (predisposing risk factors), while others act as the final insult and produce clinically evident disease (precipitating risk factor(s)). Only when an eye and an individual have the critical number of risk factors required for the development of CRVO, does the CRVO develop. This must explain why bilateral CRVO is rare. Once this basic concept of multifactorial causation is understood, one can attach appropriate significance to the various risk factors. The various risk factors for CRVO may be divided into the following three categories:

Local. Two local factors are particularly important:
1. The central retinal vein and central retinal artery lie in the center of the optic nerve, surrounded by a fibrous tissue envelope (**Figure 1**). In elderly persons, sclerotic changes in the central retinal artery and the fibrous tissue envelope compress the thin-walled central retinal vein, resulting in narrowing of its lumen. This produces circulatory stasis. According to Virchow's triad, slowing down of the blood stream causes stagnation thrombosis.

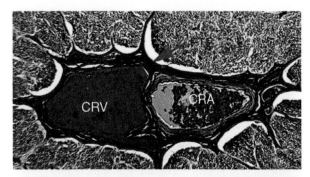

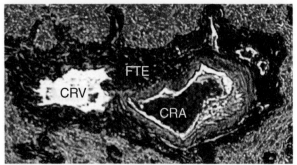

Figure 1   Histological sections (Masson's trichrome staining) showing the central retinal vessels and surrounding fibrous tissue envelope, as seen in a transverse section of the central part of the retrolaminar region of the optic nerve, in a normal rhesus monkey (above) and in a rhesus monkey with experimental arterial hypertension, atheroselerosis, and glaucoma (below). CRA, Central retinal artery; CRV, central retinal vein; FTE, fibrous tissue envelope.

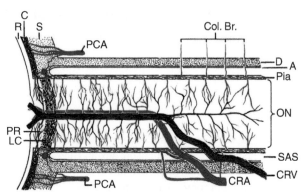

Figure 2   Schematic representation of blood supply of the optic nerve. A, arachnoid; C, choroid; CRA, central retinal artery; Col. Br., collateral branches from other orbital arteries to the optic nerve; CRV, central retinal vein; D, dura; LC, lamina cribrosa; ON, optic nerve; PCA, posterior ciliary artery; PR, prelaminar region; R, retina; S, sclera; SAS, subarachnoid space. Adapted from Hayreh, S. S. (1974). *Transactions – American Academy of Ophthalmology and Otolaryngology* 78: OP240–OP254, with permission from American Academy of Ophthalmology.

2. It is well established that CRVO is significantly more common in patients with raised intraocular pressure (IOP) and glaucoma.

Systemic. A significant association of CRVO has been reported with arterial hypertension, diabetes mellitus, cardiovascular disease, atherosclerosis, and thyroid disease.

Hematological. The literature is full of reports of hematological abnormalities in CRVO. The author recently critically reviewed the literature dealing with these and found no definite pattern – often the negative findings outweighed the positive ones. The idea of hematologic factors playing a role in CRVO is essentially based on the assumption that those hematological disorders, which play a role in development of systemic venous thrombosis (e.g., deep vein thrombosis), must also do so in CRVO. All the available evidence, however, indicates that the hematologic risk factor responsible for major systemic venous thrombosis occurs only sporadically in CRVO. Furthermore, CRVO is extremely rare in patients with systemic venous thrombosis. Moreover, the presence of a particular hematologic disorder in a patient does not necessarily mean it has a cause-and-effect relationship with CRVO. In view of this, there is no particular reason for conducting a detailed hematological investigation in all patients with CRVO.

Site of Occlusion in CRVO

Based on histopathological studies, there is a widespread misconception that the site of occlusion in CRVO is invariably at the lamina cribrosa. However, all the available anatomical, experimental, and clinical evidence (particularly fluorescein fundus angiography) shows that the actual site of occlusion in the central retinal vein is typically in the optic nerve, at a variable distance posterior to the lamina cribrosa, and not at the lamina cribrosa (**Figure 2**). The farther back the site of occlusion, the more collaterals are available, and the less severe is the retinal venous stasis. Thus, in nonischemic CRVO the site of occlusion most likely is farther back in the optic nerve, whereas in ischemic CRVO it is closer to the lamina cribrosa.

Demographic Characteristics

CRVO is more common in middle-aged and elderly persons, and patients with ischemic CRVO tend to be older than those with nonischemic CRVO. Contrary to the prevalent impression, CRVO is not at all rare in young persons, and the incidence in persons under the age of 45 years has been reported as high as 18%. Thus, no age is immune. In our series of 620 consecutive CRVO cases, 81% were nonischemic and 19% ischemic CRVO. The Kaplan–Meier estimate of the cumulative proportion of eyes that developed nonischemic CRVO in the fellow eye is about 6% within 1 year and 7% within 5 years from onset in the first eye; for ischemic CRVO it is 5.6% at 2.8 years.

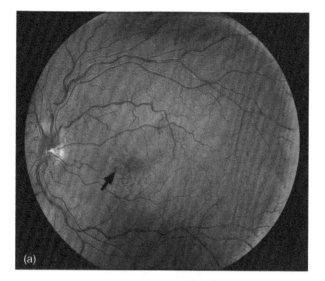

OCT image   Fundus image

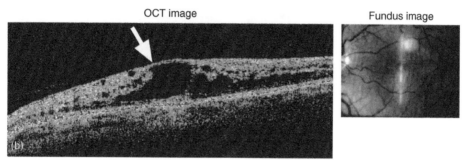

Figure 3 Fundus photograph (a) and OCT (b) of a nonischemic CRVO eye with resolution of retinopathy, except for the cystoid macular edema (arrow).

Clinical Features

With regard to symptoms, patients with nonischemic CRVO may have no symptoms and it may be detected as an incidental finding on a routine ophthalmic examination. Retinal venous stasis with mild retinal hemorrhages *per se* is asymptomatic. Occasionally there may be a history of episodes of transient visual blurring before constant visual deterioration. Almost invariably, it becomes symptomatic only when there is involvement of the foveal region by development of macular edema (**Figures 3** and **4**) and rarely by hemorrhages. Therefore, the most common complaint is gradual development of central visual blurring, usually more marked on waking up in the morning, improving to a variable extent after a few hours or in the afternoon. In ischemic CRVO, on the other hand, there is always marked deterioration of vision.

While the diagnosis of CRVO is not difficult because of its classical clinical features (**Figures 5** and **6**), the main problem is differentiation of nonischemic from ischemic CRVO, which is crucial for the correct management of CRVO. This is because nonischemic CRVO is a comparatively benign condition, with permanent central

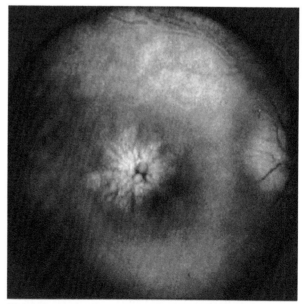

Figure 4 Late phase of fluorescein fundus angiogram of an eye with nonischemic CRVO showing classical petaloid pattern of cystoid macular edema.

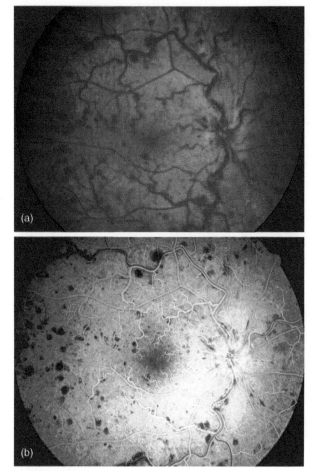

Figure 5 Fundus photograph (a) and fluorescein fundus angiogram showing intact retinal capillary network (b) of an eye with nonischemic CRVO. Reproduced from Hayreh, S. S. (1994). *Indian Journal Ophthalmology* 42: 109–132.

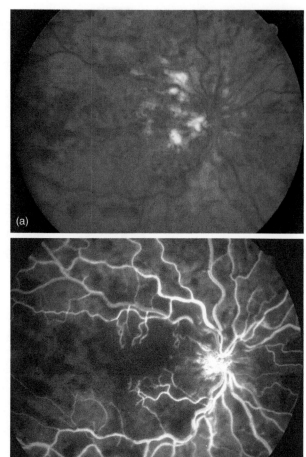

Figure 6 Fundus photograph (a) and fluorescein fundus angiogram showing complete nonperfusion of retinal capillary network (b) of an eye with ischemic CRVO. Reproduced from Hayreh et al. (1983). Ocular neovascularization with retinal vascular occlusion III. Incidence of ocular neovascularization with retinal vein occlusion. *Ophthalmology* 90: 488–506.

scotoma as the major complication in some eyes, but no ocular neovascularization (NV). In sharp contrast to this, ischemic CRVO is a blinding disease, with high risk of development of anterior segment NV, particularly neovascular glaucoma, which often results in blindness or even loss of the eye. Thus, the two types of CRVO can be compared to benign and malignant tumors.

Differentiation of Ischemic from Nonischemic CRVO

Ophthalmologists have almost universally used ophthalmoscopic and fluorescein angiographic appearances to evaluate and manage CRVO and to differentiate ischemic from nonischemic CRVO. However, these two morphological tests have much lower sensitivity and specificity to differentiate the two types of CRVO compared to the four functional tests – visual acuity, peripheral visual fields plotted with a Goldmann perimeter, relative afferent

pupillary defect, and electroretinography. **Table 1** gives sensitivity and specificity of various functional tests to differentiate ischemic from nonischemic CRVO.

On fluorescein fundus angiography, to differentiate ischemic from nonischemic CRVO, the presence of a 10 disk area or more retinal capillary obliteration has been regarded as the gold standard in practically all the reported studies, but there are several serious problems with this criterion, including the following:

1. During the early, acute stages of CRVO, to provide reliable information on retinal capillary obliteration, angiography has many serious limitations, including extensive retinal hemorrhages, poor-quality angiograms, inability to perform angiography for a variety of reasons, the time lag of several weeks after the onset of CRVO before retinal capillary obliteration is visible, and other limitations. A study showed that fluorescein angiography provided reliable information at best in only

50–60% of cases during the early, acute phase, which is clinically unsatisfactory for early management.

2. Most importantly, a criterion of a 10 disk area or more of retinal capillary obliteration has been widely advocated as the definitive yardstick for diagnosis of ischemic CRVO. However, this is an invalid criterion to differentiate nonischemic from ischemic CRVO. A multicenter study showed that eyes with <30 disk diameters of nonperfusion and no other risk factor are at low risk for iris/angle NV (i.e., ischemic CRVO), 'whereas eyes with 75 disk diameters or more are at highest risk'.

As regards ophthalmoscopy, there is a marked overlap between the two types of CRVO and virtually a continuous evolution of ophthalmoscopic lesions (i.e., retinal hemorrhages, venous dilatation, and cotton-wool spots, etc.; **Figures 5(a)** and **6(a)**), which makes it hard to use this to differentiate the two types of CRVO.

A study showed that to differentiate the two types of CRVO, the overall order of reliability of various tests is as follows:

1. *Relative afferent pupillary defect.* In unilateral CRVO, when the fellow eye is normal.
2. *Electroretinography.* This is the next best test and it can be done even when the fellow eye is not normal, as in bilateral CRVO.
3. *Relative afferent pupillary defect combined with electroretinography.* This proved to be the most reliable (in 97%).
4. *Peripheral visual fields plotted with a Goldmann perimeter.* This is next in order and better than visual acuity. Since central scotoma is present in all CRVO eyes, that does not help in differentiation.
5. *Visual acuity.* This is also helpful in many cases.
6. *Fluorescein angiography.* This proved to be much worse than any of the functional tests in early stages.
7. *Ophthalmoscopy.* This is the least reliable and most misleading parameter of all.

Table 1 Sensitive and specify of various functions tests to differentiate ischemic from nonischemic CRVO

Functional test		Sensitivity	Specificity
Visual acuity	20/400 or less	91%	88%
Peripheral visual fields	No I-2e	97%	73%
	Defective V-4e	100%	100%
Relative afferent pupillary defect	≥0.9 log units	80%	97%
Electroretinography	b-wave amplitude <60%	80%	80%

From Hayreh, S. S., Klugman, M. R., Beri, M., Kimura, A. E., and Podhajsky, P. (1990). Differentiation of ischemic from nonischemic central retinal vein occlusion during the early acute phase. *Graefe's Archive for Clinical and Experimental Ophthalmology* 228: 201–217.

Thus, we can conclude that no single test has 100% sensitivity and specificity to differentiate the two types of CRVO during the early, acute phase, such that no single test can be considered a gold standard; however, combined information from all the six tests is almost always reliable. The four functional tests overall are much superior to the two morphologic tests.

Course of CRVO

Both types of CRVO run a self-limited course, taking from a few weeks to many years for the retinopathy to resolve. In the meantime, some of these eyes can develop various complications, including those discussed below.

Complications

The main complications of the two types of CRVO are as follows:

1. *Macular edema.* This is the most common complication in both types of CRVO (**Figures 3** and **4**). However, it does not affect eyes with mild nonischemic CRVO. Chronic macular edema later on may produce cystoid macular degeneration (**Figure 3(a)**), macular pigmentary degeneration (very much resembling age-related macular degeneration; **Figure 7**), and/or epiretinal membrane – all resulting in central scotoma.

2. *Ocular NV.* This is the most dreaded complication of CRVO and a complication of ischemic CRVO alone. It is extremely important to note that ocular NV is almost never seen in nonischemic CRVO, unless it is associated with diabetic retinopathy or ocular ischemia.

The cumulative probability of developing various types of ocular NV in ischemic CRVO is shown

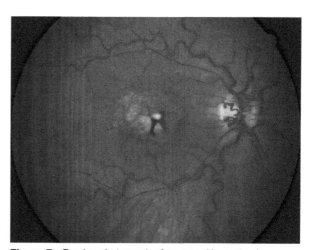

Figure 7 Fundus photograph of an eye with resolved nonischemic CRVO, showing macular pigmentary degeneration and retinociliary collaterals on the optic disk, as the permanent, residual changes.

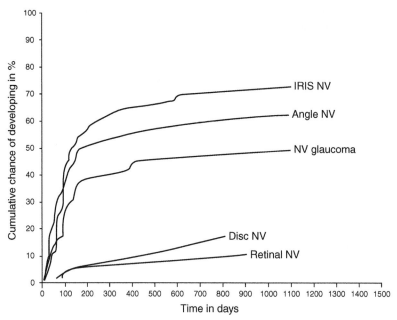

Figure 8   A graphic representation of cumulative chances (in %) of developing various types of ocular neovascularization in ischemic CRVO in relation to time from onset of the disease (in days).

graphically in **Figure 8**, which provides five very important pieces of information for management of CRVO:

a. Not every eye fulfilling the criteria given above for ischemic CRVO develops ocular NV.

b. When ocular NV does develop, the most common site is the anterior segment, much less frequently the posterior segment.

c. The greatest risk of developing anterior segment NV is during the first 7–8 months, after which the risk falls dramatically, to minimal. The old concept of 100-day glaucoma has no validity.

d. The maximum risk of developing neovascular glaucoma is about 50% – not 100%, as often stated.

e. About one-third of eyes with iris NV and about one-quarter of eyes with both iris and angle NV, contrary to the prevalent impression, never progress to develop neovascular glaucoma.

In order to place the overall incidence of ocular NV, particularly of neovascular glaucoma in CRVO, in true perspective, it is essential to point out two important facts:

i. ischemic CRVO constitutes only one-fifth of all CRVO cases (see above);

ii. neovascular glaucoma, the most dreaded complication of CRVO, is seen at the maximum in about 50% of ischemic CRVO cases only.

This means that the overall incidence of neovascular glaucoma in all CRVO cases is no more than 10% at the most – a key fact in any consideration of the management of CRVO.

3. *Vitreous hemorrhage.* In CRVO, this may be either secondary to retinal/optic disk NV or due to rupture of

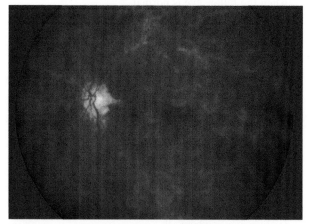

Figure 9   Fundus photograph of an eye with nonischemic CRVO in a patient on aspirin, showing extensive retinal hemorrhages.

the retinal blood through the internal limiting membrane, particularly in eyes with many subinternal limiting membrane hemorrhages (**Figures 9–11**). Therefore, it is important to be aware that the presence of vitreous hemorrhages in CRVO does not always mean that there is retinal/disk NV.

4. *Cilioretinal artery occlusion.* The major cause of serious visual loss in nonischemic CRVO is the development of associated transient occlusion of a cilioretinal artery due to hemodynamic block and not due to any thrombosis in the artery (**Figures 12** and **13**). This is particularly so when the artery supplies a large sector of the retina (**Figure 13**) or supplies the entire maculopapillar bundle (resulting in a large absolute centrocecal defect).

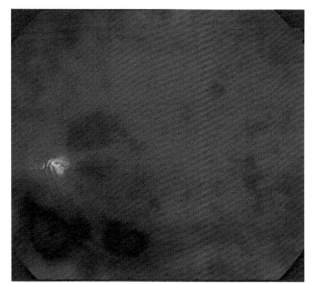

Figure 10 Fundus photograph of an eye with nonischemic CRVO in a patient on anticoagulant therapy, showing extensive retinal hemorrhages. Reproduced from Hayreh, S. S. (2006). *Retina* 26: S51–S62, with permission from Wolters Kluwer.

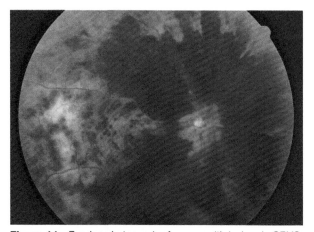

Figure 11 Fundus photograph of an eye with ischemic CRVO in a patient on anticoagulant therapy, showing extensive pre-retinal hemorrhages.

5. *Conversion of nonischemic CRVO to ischemic CRVO.* **Figure 14**, based on study of 620 consecutive CRVO eyes, shows Kaplan–Meier survival curves for cumulative probability of conversion of nonischemic CRVO to the ischemic type. This change can happen either overnight or gradually.

Management of CRVO

In the management of CRVO, the first, most crucial step is to determine whether one is dealing with nonischemic or ischemic CRVO because of their very different nature, prognosis, visual outcome, and management. Lack of such differentiation has resulted in major controversies on CRVO.

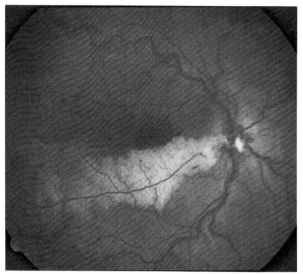

Figure 12 Fundus photograph of an eye with nonischemic CRVO and retinal infarct in a narrow strip below the foveola due to associated cilioretinal artery occlusion. Reproduced from Hayreh, S. S., Fraterrigo, L., and Jonas, J. (2008). Central retinal vein occlusion associated with cilioretinal artery occlusion. *Retina* 28: 581–594, with permission from Wolters Kluwer.

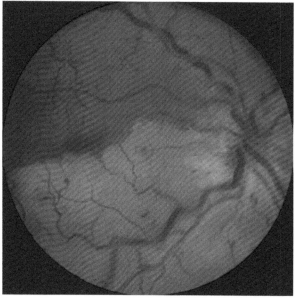

Figure 13 Fundus photograph of an eye with nonischemic CRVO and retinal infarct involving most of the lower half of the retina due to associated cilioretinal artery occlusion. Reproduced from Hayreh, S. S., Fraterrigo, L., and Jonas, J. (2008). Central retinal vein occlusion associated with cilioretinal artery occlusion. *Retina* 28: 581–594, with permission from Wolters Kluwer.

Over the years, many treatments have been advocated enthusiastically and success claimed. A review of the treatment options, which have been championed from time to time, reveals that they vary from the logical to the totally absurd. The most important consideration when evaluating any proposed therapy for any disease is

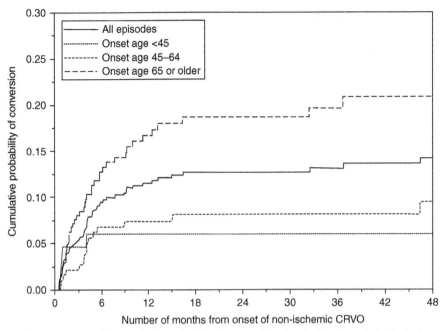

Figure 14 Kaplan–Meier survival curve for cumulative probability of conversion of nonischemic central retinal vein occlusion to ischemic type. Vertical axis gives cumulative probability of the conversion. Horizontal axis gives number of months from onset of nonischemic central retinal vein occlusion. Reproduced from Hayreh, S. S., et al. (1994). *American Journal of Ophthalmology* 117: 429–441.

to determine whether it is based on incontrovertible scientific facts and rationale. Treatments without such a logical foundation prove not only useless but also sometimes harmful. Most of the reported studies have been based on retrospective collection of information or on limited personal experience and, therefore, have a variety of limitations which make it hard to evaluate the claimed benefits; the limitations include lumping together of central and branch retinal vein occlusion, no differentiation of ischemic and nonischemic CRVO or use of invalid criteria to do so, therapies having no scientific validity, flawed study designs, and personal biases. All the proposed therapies must then be carefully scrutinized.

The main treatments advocated for CRVO can be divided into three categories, namely medical, surgical, and panretinal photocoagulation (PRP), which are discussed below.

Medical Treatments

These include anticoagulants, aspirin or other antiplatelet agents, ocular hypotensive therapy, hemodilution, systemic or intravitreal corticosteroids, intravitreal anti-VEGF (VEGF, vascular endothelial growth factor) drugs, systemic acetazolamide, and antihypertensive therapy. Of these, either aspirin or anticoagulants have been most widely used, based on the impression that a treatment which is beneficial for systemic venous thrombosis is also good for CRVO; however, the two conditions are totally different etiologically and pathogenetically. Neither anticoagulants nor aspirin has any scientific rationale in the management of CRVO. They increase retinal

hemorrhages (**Figures 9–11,** and **15**), thereby adversely influencing the visual outcome. Moreover, patients already on those drugs for other reasons do develop CRVO, indicating that they do not prevent an eye from developing CRVO. The so-called blood-and-thunder or tomato-ketchup fundus appearance described in CRVO is usually an iatrogenic phenomenon resulting from the use of aspirin or anticoagulants; it is not usually seen in regular CRVO. There is no scientific rationale for the commonly used ocular hypotensive therapy in CRVO eyes with normal IOP; however, if the fellow uninvolved eye has ocular hypertension or glaucoma (which is common), that eye must be treated to reduce the chances of its developing CRVO – thus, it is mostly the wrong eye (i.e., with CRVO) which is being treated. Recently, intravitreal corticosteroids and intravitreal anti-VEGF drugs have been widely advocated, primarily to manage macular edema; however, it is important to stress that the treatment with these agents is simply helping to reduce or eliminate macular edema transiently to prevent long-term permanent macular changes (**Figures 3** and **7**); it is not a cure for the CRVO, which has to take its own natural course. Moreover, both drugs require repeated intravitreal injections to maintain effectiveness and can have some side effects. As regards the rest of treatment modalities, there is not much scientifically valid evidence of beneficial effects.

Surgical or Invasive Treatments

These include (1) surgical decompression of the central retinal vein, (2) fibrinolytic therapy, and (3) laser-induced

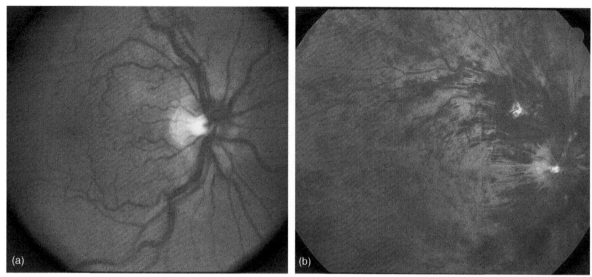

Figure 15 Fundus photographs of a 36-year-old man, who had nonischemic CRVO. When first seen by his ophthalmologist, he had a visual acuity of 20/20 and the fundus showed some peripheral retinal hemorrhages and a rare one posteriorly (a). The ophthalmologist started him on aspirin, and at next visit he had extensive hemorrhages all over (b). On follow-up, his visual acuity progressively deteriorated to finally 20/200. Reproduced from Hayreh, S. S. (2006). *Retina* 26: S51–S62, with permission from Wolters Kluwer.

chorioretinal venous anastomosis for treatment of non-ischemic CRVO, which are detailed below:

1. *Surgical decompression of central retinal vein.* It has been claimed that a procedure called radial optic neurotomy, in which a radial cut in the optic nerve head, from the vitreous side, extending all the way down to the lamina cribrosa and adjacent sclera, is beneficial. This procedure not only has no scientific rationale but also can actually be deleterious – it involves cutting thousands of optic nerve fibers, which results in visual loss in the distribution of the cut nerve fibers.
2. *Fibrinolytic therapy.* Currently, the most widely promoted procedure of this type is vitrectomy with branch retinal vein cannulation and infusion of tissue plasminogen activator (t-PA). This procedure lacks scientific rationale and can be associated with complications; beneficial claims made for it seem unwarranted.
3. *Laser-induced chorioretinal venous anastomosis for treatment of nonischemic CRVO.* This procedure has many immediate and late complications, which heavily outweigh any dubious benefits. Therefore, it is not a safe and effective mode of treatment for a condition which has a fairly good outcome if simply left alone (see below).

Panretinal Photocoagulation

Currently, PRP is almost universally considered the treatment of choice in CRVO to prevent development of ocular NV, and macular grid photocoagulation for the management of macular edema. The rationale for usefulness of PRP in CRVO is based primarily on its beneficial effect seen in proliferative diabetic retinopathy.

1. *Macular grid photocoagulation for macular edema.* A multicenter clinical trial showed no difference between treated and untreated eyes in visual acuity at any point during the follow-up period. This indicated that there is no role for this treatment in CRVO.
2. *PRP for prevention of ocular NV.* The theoretical justification for PRP in ischemic CRVO is to prevent development of ocular NV and associated blinding complications of neovascular glaucoma and/or vitreous hemorrhage. As discussed above, ocular NV and neovascular glaucoma are seen only in ischemic CRVO. Since nonischemic CRVO does not develop these complications, there is absolutely no indication or justification for PRP in nonischemic CRVO.

Two large prospective studies have been conducted to evaluate the role of PRP in ischemic CRVO. The first study, based on 123 eyes with ischemic CRVO (47 had PRP and 76 no PRP), showed no statistically significant difference between the two groups in the incidence of development of angle NV, neovascular glaucoma, retinal and/or optic disk NV, or vitreous hemorrhage, or in visual acuity. What it did show was that the PRP group suffered a statistically significant ($p \leq 0.03$) greater loss of peripheral visual fields than the nonlaser group (**Figure 16**).

The second study investigated the role of PRP in ischemic CRVO by a multicenter clinical trial in 181 eyes (90 had immediate prophylactic PRP, and 91 no PRP). The purpose of that study was twofold – first to determine whether prophylactic PRP prevents development of iris and angle NV, and second whether PRP prevents progression of iris/angle NV to neovascular glaucoma. In answer to the first question, the study

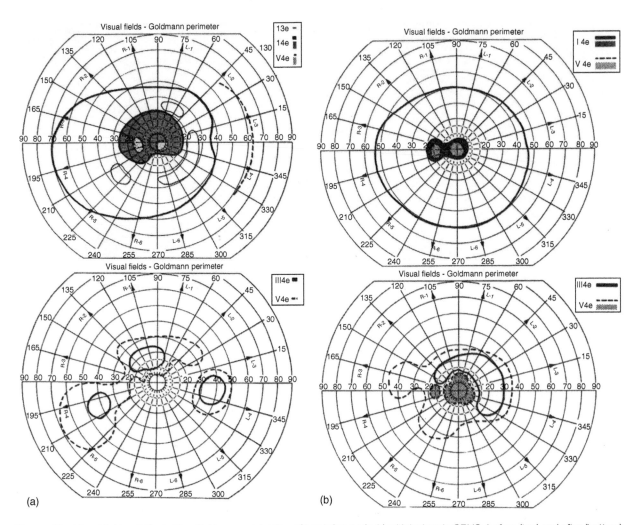

Figure 16 Visual fields, plotted with a Goldmann perimeter, of two left eyes (a, b) with ischemic CRVO, before (top) and after (bottom) panretinal photocoagulation. In both eyes, before photocoagulation (top) the peripheral fields were normal with I4e and V4e isopters, but after photocoagulation (bottom) the peripheral fields developed marked constriction and deterioration. Reproduced from Hayreh, S. S., Klugman, M. R., Beri, M., Kimura, A. E., and Podhajsky, P. (1990). Differentiation of ischemic from non-ischemic central retinal vein occlusion during the early acute phase. *Graefe's Archive for Clinical and Experimental Ophthalmology* 228: 201–217.

reported that "prophylactic PRP does not totally prevent" development of iris/angle NV; its recommendation was not to do that. As for the second aspect of the study, the authors recommended "careful observation with frequent follow-up examinations in the early months (including undilated slit-lamp examination of the iris and gonioscopy) and prompt PRP of eyes in which 2'clock iris/angle neovascularization develops." The most important feature of any study is its design, because that can determine its conclusions and their validity. This multicenter study unfortunately had flaws in its study design, which raised serious issues about the validity of its conclusions. There were three major flaws with the design of this study;

1. there was poor differentiation of ischemic and nonischemic CRVO, so that their treated and untreated groups had mixtures of both types of CRVO, instead

of only ischemic CRVO (PRP is indicated only for ischemic CRVO);

2. all eyes that developed 2'clock iris/angle NV were treated so that there was no control group to determine whether there was any difference in progression of iris/angle NV to neovascular glaucoma in treated versus the untreated eyes; and

3. no visual fields were plotted to evaluate the side effects and complications of PRP in CRVO.

The first study, as mentioned above, showed that following PRP, the large central scotoma combined with severe loss of peripheral visual fields might virtually blind the eye (**Figure 16**). It is unfortunate, that this multicenter study, in spite of all these serious flaws, has become the gold standard for treatment of CRVO.

In conclusion, as is quite evident from the brief discussion above, there is no scientifically valid proof so far that

PRP is safe and effective in prevention or management of neovascular glaucoma in ischemic CRVO.

Primary Factor Responsible for Blindness in Ischemic CRVO with Neovascular Glaucoma

The most important consideration in the management of neovascular glaucoma seen in ischemic CRVO is high IOP, which is the primary factor causing marked loss of vision or even blindness in the vast majority of these eyes, by producing glaucomatous optic neuropathy and anterior segment changes. Therefore, if neovascular glaucoma develops (the maximum risk of that is 50%; **Figure 8**) and the IOP is controlled satisfactorily by any means, the eye can maintain a reasonably good peripheral visual field once the retinopathy burns itself out in due course; the peripheral visual field constitutes a very important component of our visual function, because our navigational capacity depends essentially on that. As the retinopathy regresses, the stimulus for NV diminishes, resulting in spontaneous regression of NV – a fact too often overlooked.

Overall Conclusions About Advocated Treatments for CRVO

It is well known that 'a disease which has no treatment has many treatments' – each advocated enthusiastically for a while and then found to be not only no better than the natural history of the disease or sometimes actually harmful. CRVO has become a graveyard of such therapies, but, given that they were not backed by sound science, it is not surprising that there is so far no treatment for CRVO that has stood the test of time for safety, efficacy. and long-term beneficial and curative effect. It is unfortunate when enthusiastic claims made for different modes of treatment give desperate patients false hope, cost them so much, not only in money but also in trouble, unnecessary pain and disappointment, and, sometimes, do them more harm than good. Currently, these desperate patients shop for treatment via the Internet and are often misled because not all the information on the Internet is peer reviewed. In this review, we have challenged some of the conventional wisdom in the management of CRVO – that is never popular. To advocate watchful waiting rather than action is even more unpopular. We can only hope that science, logic, and common sense will prevail in the end.

Natural History of Visual Outcome in CRVO

The outcome of all advocated treatments has to be compared with the natural history of a disease, so that natural recovery is not attributed to the beneficial effect of a mode of treatment. Recently, the natural history of visual

Table 2 Visual acuity in 214 eyes with nonischemic CRVO – in 155 of them retinopathy had resolved completely and 59 eyes still had macular edema

	All eyes (214 eyes)	Resolved eyes (155 eyes)
Visual acuity		
20/20–20/25	50%	65%
20/30–20/40	14%	12%
20/50–20/70	11%	7%
20/80–20/100	4%	3%
20/200	12%	6%
20/400–CF[a]	10%	7%
Central scotoma		
None	56%	68%
Small size	34%	26%
Moderate size	5%	3%
Large size	1%	1%

[a]Mainly due to cataract, age-related macular degeneration, etc. but not due to CRVO.
From *Retina* 2007 27: 514–517.

outcome was reported in a prospective study of 214 untreated consecutive patients with nonischemic CRVO – in 155 of them the retinopathy and macular edema had resolved completely during follow-up and 59 eyes still had macular edema at the last visit. This study was based on evaluation of both visual acuity and visual fields plotted with a Goldmann perimeter – visual acuity provides information about the function of the fovea only and not of the entire retina, while visual fields provide information about the function of the entire retina involved by CRVO. **Table 2** gives the findings on visual acuity and permanent central scotoma in all the 214 eyes as well as in 155 eyes with completely resolved retinopathy. Peripheral visual fields were perfectly normal in both the groups all along. Among the 59 eyes that still had macular edema, resolution of macular edema would improve the visual acuity and central visual fields in due course. Thus, the natural history of visual outcome in nonischemic CRVO is much better than the results of any of the treatments advocated so far. By contrast, in ischemic CRVO, during the initial stages, the ganglion cells in the macular retina are irreversibly damaged by ischemia → a permanent central scotoma → little chance of improvement of visual acuity. Almost all eyes with ischemic CRVO initially have normal peripheral visual fields.

Investigations of CRVO Patients

Finally, a commonly asked question is: What systemic and hematologic investigations need to be done in patients with CRVO? A voluminous and confusing literature exists on systemic and hematologic disorders in CRVO. Our study of 612 patients with CRVO showed that, apart from a routine medical evaluation, the vast majority of patients with CRVO need not undergo an extensive and expensive workup for systemic diseases. As regards hematological evaluation, there is no good reason why all patients with CRVO should be subjected to the costly, exhaustive special hematologic and hypercoagulability investigations usually done in patients with spontaneous major systemic venous thrombosis – unless, of course, there is some clear indication. The routine, inexpensive hematologic evaluation is the one required by CRVO patients.

See also: Ischemic Optic Neuropathy; Physiological Anatomy of the Retinal Vasculature.

Further Reading

Hayreh, S. S. (1983). Classification of central retinal vein occlusion. *Ophthalmology* 90: 458–474.

Hayreh, S. S. (1996). The CVOS Group M and N Reports. *Ophthalmology* 103: 350–352.

Hayreh, S. S. (2002). *Radial optic neurotomy for central retinal vein occlusion. Retina* 22: 374–377; 827.

Hayreh, S. S. (2003). Management of central retinal vein occlusion. *Ophthalmologica* 217: 167–188.

Hayreh, S. S. (2005). Prevalent misconceptions about acute retinal vascular occlusive disorders. *Progress in Retinal and Eye Research* 24: 493–519.

Hayreh, S. S., Fraterrigo, L., and Jonas, J. (2008). Central retinal vein occlusion associated with cilioretinal artery occlusion. *Retina* 28: 581–594.

Hayreh, S. S., Klugman, M. R., Beri, M., Kimura, A. E., and Podhajsky, P. (1990). Differentiation of ischemic from non-ischemic central retinal vein occlusion during the early acute phase. *Graefe's Archive for Clinical and Experimental Ophthalmology* 228: 201–217.

Hayreh, S. S., Klugman, M. R., Podhajsky, P., Servais, G. E., and Perkins, E. S. (1990). Argon laser panretinal photocoagulation in ischemic central retinal vein occlusion – A 10-year prospective study. *Graefe's Archive for Clinical and Experimental Ophthalmology* 228: 281–296.

Hayreh, S. S., Rojas, P., Podhajsky, P., Montague, P., and Woolson, R. F. (1983). Ocular neovascularization with retinal vascular occlusion III. Incidence of ocular neovascularization with retinal vein occlusion. *Ophthalmology* 90: 488–506.

Hayreh, S. S., Zimmerman, M. B., Beri, M., and Podhajsky, P. A. (2004). Intraocular pressure abnormalities associated with central and hemi-central retinal vein occlusion. *Ophthalmology* 111: 133–141.

Hayreh, S. S., Zimmerman, M. B., McCarthy, M. J., and Podhajsky, P. (2001). Systemic diseases associated with various types of retinal vein occlusion. *American Journal of Ophthalmology* 131: 61–77.

Hayreh, S. S., Zimmerman, M. B., and Podhajsky, P. (1994). Incidence of various types of retinal vein occlusion and their recurrence and demographic characteristics. *American Journal of Ophthalmology* 117: 429–441.

Hayreh, S. S., Zimmerman, M. B., and Podhajsky, P. A. (2002). Hematologic abnormalities associated with various types of retinal vein occlusion. *Graefe's Archive for Clinical and Experimental Ophthalmology* 240: 180–196.

Ingerslev, J. (1999). Thrombophilia: A feature of importance in retinal vein thrombosis? *Acta Ophthalmologica* 77: 619–621.

The Central Retinal Vein Occlusion Group (1995). *A randomized clinical trial of early panretinal photocoagulation for ischemic central vein occlusion – the central vein occlusion study group N report. Ophthalmology* 102: 1434–1444.

The Central Retinal Vein Occlusion Group (1995). Evaluation of grid pattern photocoagulation for macular edema in central vein occlusion – the central vein occlusion study group M report. *Ophthalmology* 102: 1425–1433.

The Eye Disease Case-Control Study Group (1996). Risk factors for central retinal vein occlusion. *Archives of Ophthalmology* 114: 545–554.

Choroidal Neovascularization

M R Kesen and S W Cousins, Duke Eye Center, Durham, NC, USA

Glossary

Age-related macular degeneration (AMD) – An eye disease with its onset usually after age 60 that progressively destroys the macula, the central portion of the retina.

Angiomatous – Relating to a tumor consisting of a mass of blood vessels.

Brusch's membrane – The inner layer of the choroid, separating it from the pigmentary layer of the retina.

Choriocapillaris – The capillaries forming the inner vascular layer of the choroid.

Choriocapillaris pillars – Lateral walls of the choriocapillaris.

Choroid – The layer of the eye containing blood vessels that furnish nourishment to the other parts of the eye, especially the retina.

Leukostasis – Increased blood viscosity and tendency to clotting.

Subretinal space – An area below the retina.

Telangiectatic – Small dilated blood vessels.

Choroidal Neovascularization

Choroidal neovascularization (CNV) is the growth of pathologic new blood vessels originating from the choriocapillaris invading into the subretinal pigment epithelium (subRPE) and/or the subretinal space. In the early stages of CNV formation, these vessels are capillary-like, lacking the structural integrity of mature vessels causing them to leak plasma or blood. Components of CNV often mature into more organized vessels with features of arteries and veins, and CNV can ultimately evolve into fibrovascular scar tissue.

CNV is known to occur in many ophthalmic diseases. However, it is the most important sequella of age-related macular degeneration (AMD), causing the neovascular form of AMD (NVAMD). In developed countries, NVAMD is the leading cause of blindness in people older than 50 years of age.

Clinical Detection of CNV

Clinically, CNV is detected in patients who complain of blurred and distorted vision, often caused by leakage of plasma from the CNV into or under the overlying neurosensory retina. Because of the optical clarity of the ocular media, the retina can be directly observed by clinicians. Upon examination, eyes with CNV exhibit evidence of leakage such as retinal edema, subretinal fluid, or hemorrhage (**Figure 1(a)** and **1(b)**). If they are not leaking, CNV can sometimes be observed as circular gray or reddish structures under the retina (**Figure 2(a)**). However, in most cases, nonleaking CNV is not directly visible. In some cases, untreated CNV evolves into fibrovascular membranes, called disciform scars.

Clinical imaging technologies have greatly contributed to our understanding of CNV pathobiology and are utilized to enhance detection of CNV. Fluorescein angiography (intravenous injection of a fluorescent dye which can be photographed as it transits through the retina and CNV) is the standard diagnostic technology to confirm the presence of CNV. In classic CNV, fluorescein angiography often reveals CNV to manifest well-defined margins with homogeneous leakage of dye (**Figure 2(b)**). Yet, most cases of CNV are occult, demonstrating less well-defined margins with irregular leakage. Originally, an association was assumed between the fluorescein angiographic appearance and the anatomic location or the growth pattern of CNV. It was presumed that classic CNV was located subretinally, external to the retinal pigment epithelium (RPE) and occult CNV was located external to the RPE within Bruchs membrane. However, data from the Submacular Surgery Trials (SST) research group indicated a lack of correlation between fluorescein angiography pattern and anatomic localization of the CNV.

In the past decade, newer imaging technologies have revealed that the morphologic manifestation of pathologic new vessels in NVAMD (and presumably other diseases with CNV) is much more diverse than traditionally described using fluorescein angiography. Another angiographic imaging technology employs indocyanine green (ICG) dye, allowing visualization of the morphology and the flow of the choroidal vasculature. Dynamic (or video) ICG angiography demonstrates that most CNV is a complex of three components, including an intrachoroidal feeder artery, a subretinal or subRPE capillary network, and an intrachoroidal draining vein (**Figure 2(c)**).

By ICG angiography, CNV appears to manifest a spectrum of morphology ranging from capillary pattern to branching arteriolar pattern. Capillary pattern CNVs are composed mostly of capillaries with minimal arterioles and short, small-caliber feeder arteries. Branching arteriolar pattern CNVs are composed of long, large-caliber

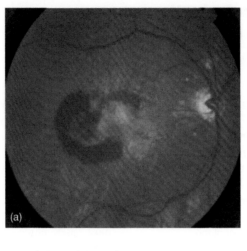

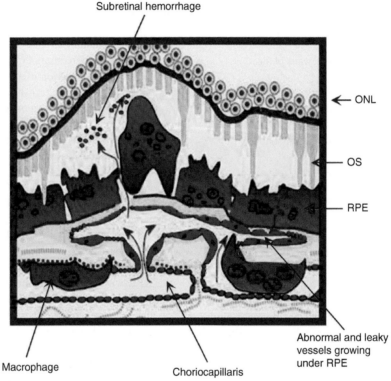

Subretinal hemorrhage

ONL

OS

RPE

Abnormal and leaky
vessels growing
under RPE

Macrophage   Choriocapillaris

ONL – outer nuclear layer
OS – outer segment
(b)   RPE – retinal pigmented epithelium

Figure 1   (a) The color fundus photograph demonstrates subfoveal choroidal neovascular membrane with surrounding subretinal hemorrhage. (b) Schematic diagram demonstrates the choroidal neovascular membrane located beneath the retinal pigment epithelium causing leakage of fluid or blood. Arrow indicate direction of fluid and blood leakage.

feeder arteries, many branching arterioles, and minimal capillary morphology (**Figure 3(a)–3(d)**). Mixed pattern CNV demonstrates feature of both extremes. Presumably, all CNVs initially begin as capillary pattern, then progress through a process of remodeling whereby small capillaries obtain a sheath of pericytes which stabilize the incipient new vessel. In some cases, the new vessels become surrounded by vascular smooth muscle and collagen, thus becoming arteriolarized. Hypothetically, the variability among CNV morphology is due to differences in the degree and extent of remodeling. One emerging hypothesis is that capillary-dominated CNVs are exquisitely responsive to anti-vascular endothelial growth factor (anti-VEGF) treatment, whereas branching arteriolar CNV with large feeder vessels are less responsive or even resistant to anti-VEGF therapy.

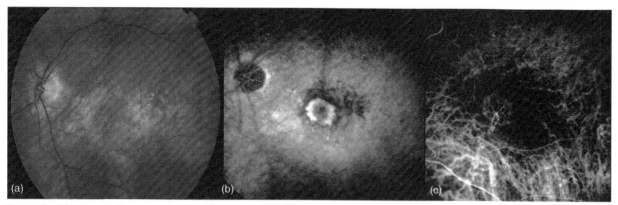

Figure 2   (a) The color fundus photograph demonstrates subtle gray discoloration of the retina due to underlying choroidal neovascular membrane. (b) Late phase fluorescein angiogram shows area of hyperfluorescence corresponding to classic choroidal neovascular membrane with progressive leakage of dye resulting in blurring of the margins. (c) Dynamic indocyanine green angiogram reveals branching vascular network composed of afferent arteriole, capillaries, and draining venule.

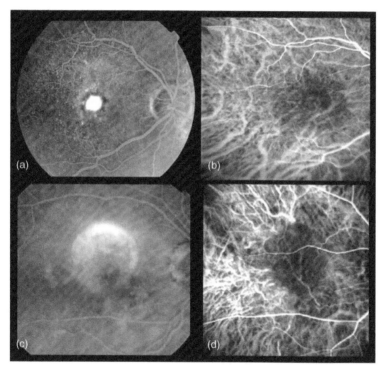

Figure 3   (a) Late arteriovenous phase angiogram shows bright hyperfluorescence corresponding to choroidal neovascular membrane. (b) Dynamic indocyanine angiogram reveals neovascular membrane composed of capillaries with minimal branching arterioles or venules. (c) Venous phase angiogram shows a well-demarcated area of hyperfluorescence corresponding to pigment epithelium detachment with pooling of dye. (d) Dynamic indocyanine angiogram reveals long, large-caliber choroidal feeder vessel perfusing many branching arterioles extending into the pigment epithelium detachment. Minimal capillary structure is observed.

Another imaging technology, called optical coherence tomography (OCT), is a high-resolution, cross-sectional imaging technique that enables visualization of tissue structures up to 2 mm below the surface of the retina. The projected light is reflected back from intraocular structures based upon their distance, thickness, and reflectivity. OCT has become the most commonly used imaging modality in diagnosis and management of patients with AMD since it can demonstrate retinal edema, subretinal fluid, pigment epithelial detachments, and, in some cases, the CNV itself (**Figure 4**). Being the preferred technique to assess treatment response, OCT plays a critical role in guiding the management of patients with AMD.

Histopathology of CNV

Histopathology of CNV has been performed by two types of analyses: evaluation of postmortem specimens within

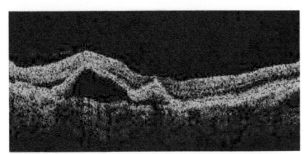

Figure 4 Optical coherence tomography (OCT) image demonstrates pigment epithelial detachments with overlying subretinal fluid.

intact eyes or after removal of CNV as part of a surgical procedure. Postmortem analysis demonstrates that CNV exhibits a wide morphological spectrum. Some cases demonstrate that endothelial-lined capillary tubes are within Bruch's membrane or under the retina. Most cases show that CNV demonstrates more complex morphology, with additional cellular components including RPE cells, pericytes, myofibroblasts, and inflammatory cells. Many cases also demonstrate extensive extracellular matrix deposition and fibrosis. Unfortunately, minimal information is available for the histopathology of the initial stages of CNV formation.

Analysis of the choroid in postmortem specimens has also revealed significant changes, including loss of choriocapillaris surface area, changes in large vascular architecture, increased extracellular matrix, and thickening of the choriocapillaris pillars. Additionally, intrachoroidal vascular changes with polypoidal-like outpouchings of some of the choroidal vessels have been observed.

Analyses of surgically excised CNV during submacular surgery have allowed examination of CNV at an earlier stage in the course of the disease than in postmortem eyes. Histopathology of many early CNV is dominated by macrophage infiltration in addition to the other cellular components (endothelial-lined channels, pericytes, RPE cells) and extracellular matrix. Some examples of surgically excised CNV demonstrate more fibrosis and myofibroblasts with minimal inflammation. In general, Grossniklaus and colleagues have suggested that CNVs in postmortem eyes appear to be more fibrotic than surgically excised CNV, implying that cellular morphology evolves from inflammatory active to inflammatory inactive stages.

Animal Models of CNV

No ideal animal exists for CNV or NVAMD. In current experimental models, CNV is induced by disruption of the Bruch's membrane/RPE complex using laser (argon, krypton, diode), by injections of growth factors, and by genetic manipulations (transgenic or gene transfer) to overexpress angiogenic growth factors. At present, the

few models of spontaneous CNV in animals are not reproducible enough to lend themselves to robust scientific analysis.

Laser-induced CNV

Placement of a small laser injury to the outer retina and choroid reliably induces CNV in rodents and subhuman primates. Some believe that a physical hole in Bruch's membrane is necessary. Others believe that thermal injury to the choroid and Bruch's membrane without physical disruption is adequate to trigger CNV, and the model represents a response to injury. Nevertheless, laser injury does not mimic the (unknown) natural stimulus for spontaneous neovascularization in humans. In spite of the artificial trigger, the model does accurately mimic many of the subsequent mechanisms for neovascularization and has predicted therapeutic response to drugs that have been validated in human studies.

The model is triggered by local inflammation at the laser site. Within 3–5 days, authentic pathologic neovascularization occurs. The initial lesion appears to be an invasion of fibrocytes and endothelial cells. Macrophages are prominent at all stages of development. Within a week of development, cell types within the growing vessels include endothelial cells, mesenchymal subtypes (pericytes, myofibroblasts, and smooth muscle cells), macrophages, RPE cells, CD34 cells, and bone-marrow-derived progenitor cells. Numerous growth factors are expressed within the growing CNV, including angiogenic factors, inflammatory cytokines, integrins, and connective tissue growth factor. Among agents that prevent the development of CNV include VEGF inhibitors, anti-inflammatory agents, and integrin antagonists. Recently, agents that block platelet-derived growth factor (PDGF) – important growth factor for mesencyhmal cells – have shown to cause regression of established laser-induced CNV.

Growth factor models

Overexpression of angiogenic growth factors in the outer retina can also induce CNV, but the interpretation of the pathobiology is surprisingly complex. A detailed review is beyond the scope of this article. Briefly, subretinal injection using adenoviral vector containing VEGF-164 causing RPE transduction and overexpression of VEGF induces robust CNV in rodents. Interestingly, transgenic overexpression of VEGF using choroid- or photoreceptor-specific promoters failed to induce CNV without additional injury by subretinal injection, suggesting another stimulus (inflammation or injury) is necessary to trigger CNV. Subretinal injection of pellets or spheres impregnated with basic fibroblast growth factor or other angiogenic factors could also induce CNV in rabbits and primates.

Pathobiology of CNV

Several theories regarding the pathogenesis of CNV have been proposed including RPE dysfunction, changes in choroid, changes in Bruch's membrane, oxidative stress, genetics, ocular perfusion abnormalities, inflammation, and ischemia.

Angiogenesis

Angiogenesis is the growth of new blood vessels originating from preexisting network of blood vessels. Angiogenesis plays a major role in normal vascular development and in wound healing (during which new vessels form, then regress). Angiogenesis has been proven to contribute significantly to the formation of pathologic new vessels in CNV and NVAMD.

Vascular endothelial growth factor

It is now well established that VEGF plays a key role as a stimulator in the neovascular forms of AMD. VEGF, a homodimeric glycoprotein of approximately 45 kDa, is a very potent vascular endothelial cell mitogen that stimulates proliferation, migration, and proteolytic activity of endothelial cells. VEGF is also a potent mediator of increased vascular permeability. Its additional functions include retinal leukostasis and neuroprotection. Three receptor tyrosine kinases have been identified for VEGF: VEGF receptor 1 (VEGFR1) has both positive and negative angiogenic effects; VEGFR2 is the primary mediator of the mitogenic, angiogenic, and vascular permeability effects of VEGF-A, and VEGFR3 mediates the angiogenic effects on lymphatic vessels.

The VEGF gene family is comprised of VEGF-A, -B, -C, -D, and placental growth factor (PlGF). VEGF-A has the strongest association with angiogenesis and is the target of most current anti-VEGF treatments. Nine major VEGF-A isoforms with varying heparin-binding capabilities and tissue distribution have been identified to date: 121, 145, 148, 162, 165, 165b, 183, 189, and 206. These isoforms are produced by alternative exon splicing of the VEGF-A gene. VEGF-165 has been shown to be the most abundant isoform and has a vital role in angiogenesis.

VEGF is essential for normal and abnormal angiogenesis. Deficiency of VEGF expression in the RPE results in the absence of choroid. VEGF is induced by hypoxia in RPE cells and is secreted from the basal side of these cells toward the choroid. High levels of VEGF receptors KDR and flt-4 are found on the choriocapillaris endothelium facing the RPE layer.

The strong evidence supporting its critical role in CNV has made VEGF an optimal target for the treatment of NVAMD. Ranibizumab, a humanized antibody fragment against all isoforms of VEGF, has been developed and proven to be efficacious. Data from two pivotal phase III clinical trials – MARINA (Minimally Classic/Occult Trial of the Anti-VEGF Antibody Ranibizumab in the Treatment of Neovascular AMD) and ANCHOR (Anti-VEGF Antibody for the Treatment of Predominantly Classic CNV in AMD) – led to the approval of Food and Drug Administration (FDA) of Ranibizumab in June 2006 for the treatment of all forms of NVAMD.

Other angiogenic factors

Although VEGF has the primary role in the development of CNV, it likely does so in concert with other growth factors including transforming growth factor-β (TGF-β), aFGF, and bFGF. Other cytokines including the integrins, angiopoetins, and matrix metalloproteinase (MMP) inhibitors may also play an important role. Activation of endothelial cells from the subjacent choriocapillaris results in the production of various degradatory enzymes, such as MMP-2 and -9, allowing them to invade Bruch's membrane and the subretinal space. Both MMP-2 and -9 are activated by VEGF and facilitate the spread of proliferating endothelial cells by digesting the extracellular matrix.

Natural inhibitors of angiogenesis

Endogenous inhibitors include endostatin, and pigment epithelium-derived factor (PEDF), both of which were demonstrated to be significantly reduced in Bruch's membrane in eyes with AMD. Thrombospondin 1, a 450-kDa glycoprotein, contributes to the regulation by functioning both as a stimulator and an inhibitor with its ability to bind to different matrix proteins and cell-surface receptors.

Inflammation Inflammation is typically divided into antigen-specific immunity (T cells, B cells, and antibody-mediated reactions) and innate immunity (neutrophils, macrophages, and amplification cascades, such as cytokines and complement). Although a role for specific immunity has not been excluded, a major role for innate immunity is emerging in the pathogenesis of CNV. Space does not permit a thorough review, but brief mention will be made of two key inflammatory components: monocytes/macrophages and complement.

Monocytes and macrophages Monocytes (the circulating cells) and macrophages (tissue infiltrating equivalent) represent 5% of the circulating white blood cell population. Macrophages are recruited into tissues for the purposes of scavenging, repair, and defense. In the process, they can impart either beneficial effects (like removing deposits and repairing extracellular matrices) or harmful ones (like bystander damage to healthy tissue or neovascularization). As mentioned above, macrophages are numerous in many examples of human CNV. In most experimental models of CNV, macrophages are positive regulators of increased severity of pathologic neovascularization, although some investigators have identified an anti-angiogenic subtype of macrophages. Macrophages can be induced to synthesize and release pro-angiogenic factors, including VEGF, PDGF, tumor necrosis factor-alpha, prostaglandin E2, and others. Conversely, VEGF is a moderately potent chemotactic factor to recruit and stimulate macrophages into tissues. In addition to their role in angiogenesis, macrophages may contribute to bystander damage in the retina. Numerous macrophage-like cells can be observed in the neurosensory retina overlying CNV in animal models and in human specimens. Presumably, these cells may be releasing factors that promote neuronal injury and decreased vision. Patients with circulating monocytes in the blood that express high levels of pro-angiogenic factor mRNA are more likely to have CNV. Macrophages are attractive targets for therapy. Antimacrophage drugs that are under investigation include those that block macrophage-derived pro-angiogenic cytokines, block cell adhesion molecules (especially integrin alpha 5, beta 1, necessary for monocytes to adhere to vessel surfaces and enter into tissues), block cyclo-oxygenase and block macrophage retention within the CNV or retina.

Complement   Complement is a series of over 30 proteins, produced by the liver and secreted into the blood, that serve as an amplification cascade for host defense. Complement activation is triggered by one of three distinct processes (immune complexes, lectin pathway, or alternative pathway) to activate a sequential cascade of interrelated proenzyme intermediates, whose main goal is to produce three types of effector molecules: C3b (coats structures to promote phagocytosis), anaphylatoxins (C3a, C5a, promotes chemotaxis of macrophages and neutrophils), and membrane attack complex (creates pores in cell membranes promoting activation or death). Recently, genetic variations in various complement molecules have been associated with AMD, especially polymorphism in CFH (a negative regulator of activation of the alternative pathway). C3, C3b, and other complement fragments have been demonstrated in specimens of human CNV. In experimental CNV, genetic knock-out or drug inhibition of complement, C3aR, and C5aR significantly diminish CNV formation. It is assumed that C5a may directly activate endothelial cells, or may serve to recruit macrophages into CNV to potentiate neovascularization. Accordingly, complement inhibitors are in clinical trial for the treatment of CNV, including inhibitors of C3, C5, and C5aR.

Vasculogenesis   Postnatal or adult vasculogenesis is the contribution of circulating bone-marrow-derived vascular progenitor cells to the repair of normal vessels or the formation of pathologic new vessels. Adult animals maintain a reservoir of stem cells in the bone marrow that may enter the circulation, become recruited into tissues, and differentiate into mature cell types that contribute to healthy repair in regenerating organs. A subset of these marrow cells has been shown to exist in humans and animals that specifically contribute to vascular repair, including both endothelial cells and mesenchymal subtypes (pericytes, myofibroblasts, vascular smooth muscle). Although this field is in its infancy, postnatal vasculogenesis has been shown to contribute to both endothelial and mesenchymal subsets in experimental CNV in several animal models. These cells can be positive or negative regulators of CNV severity, depending on prior environmental exposure of the bone marrow stem cells. Vascular progenitor cells can be identified in the blood of patients with AMD, and certain subsets may identify those with more severe pathology or worse therapeutic outcomes. If confirmed, identification of these cells may be useful as a biomarker for worse outcome and blocking their recruitment may be a therapeutic target.

Therapy of CNV

Prior to 1970, no effective therapy for CNV was available. Upon the recognition of the vascular character of CNV, thermal laser ablation of the CNV lesion as defined by the area of classic leakage on fluorescein angiography was introduced as therapy. The effectiveness of thermal laser as superior to the natural history was validated by several randomized controlled trials, and laser remained the standard of care until the twenty-first century. Photodynamic therapy (PDT) with the photosensitizing agent verteporfin was introduced in early 2000. After intravenous infusion of verteporfin, low energy red-laser light directed at the CNV was used to photoactivate verteporfin into a thrombotic agent, which caused occlusion of the CNV. PDT with verteporfin was proven superior for the treatment of classic CNV, but remained controversial for other subtypes.

Upon the realization that VEGF was the major factor driving angiogenesis and vascular permeability, several companies developed inhibitors of VEGF that could be administered by direct injection into the vitreous cavity of the eye. The first of these to become approved and clinically available was pegaptanib, a synthetic RNA aptamer which specifically bound extracellular VEGF-165 isoform. Although superior to PDT and standard of care, the therapeutic effect on vision and leakage control was modest. Subsequently, two pan-VEGF (i.e., blocking all isoforms) antibody-based inhibitors were introduced. Bevacizumab, a full-length antibody, was developed for intravenous inhibition of VEGF in cancer treatment. Ranibizumab, closely related to the Fab fragment of bevacizumab, was developed for intraocular injection. In 2005, ranibizumab's effectiveness was validated in two landmark clinical trials which showed that monthly injections resulted in outstanding leakage control, stabilization of CNV growth, and improved vision in 30–40% of patients with classic or occult CNV. However, effectivity requires once a month injection for at least 2 years, and less frequent regimens have not produced the same magnitude of positive outcomes. Subsequently, intravitreal injection of the full-length antibody, bevacizumab, has also been adopted by clinicians but without validation by a randomized controlled trial.

Other interesting agents are currently in clinical trial. Several target VEGF, including VEGF trap (a fusion protein between Fc heavy chain and VEGF receptor that blocks extracellular VEGF), bevasirinib (RNA interference, selectively targeting VEGF mRNA), and several small molecule drugs that inhibit VEGF receptor signaling (the so-called receptor tyrosine kinase inhibitors). Drugs targeting other molecular targets are also in trial, including inibitors of complement, integrins, and PDGF.

Variants of Pathologic Neovascularization in NVAMD

Over the past decade, the spectrum of neovascular AMD has been expanded beyond classic and occult CNV. Identification of additional variants, especially polypoidal choroidal vasculopathy (PCV) and retinal

angiomatous proliferation (RAP), has suggested that the pathobiology and treatment of neovascular AMD is a more complex than suggested by the previous review.

Polypoidal choroidal vasculopathy

PCV is a form of intrachoroidal neovascularization that accounts for up to 50% of NVAMD in Asians and may be underdiagnosed in Caucasians. PCV is characterized by large-caliber feeder vessels perfusing intrachoroidal branching arteriolar networks which terminate in one or more intrachoroidal polypoidal structures rather than typical subretinal capillaries of CNV (**Figure 5(a)–5(d)**). Clinical features include orange–red subretinal nodules corresponding to dilated, polyp-like lesions associated with hemorrhagic or serous pigment epithelial detachments, retinal edema, or subretinal hemorrhage. PCV most commonly occurs in the peripapillary and macular regions. ICG angiography is required for diagnosis and visualization of PCV.

The pathobiology of PCV is debated, especially whether it is a distinct disorder from CNV or a disease of similar biology but with different vascular morphology. Anecdotal cases illustrate examples of eyes with PCV

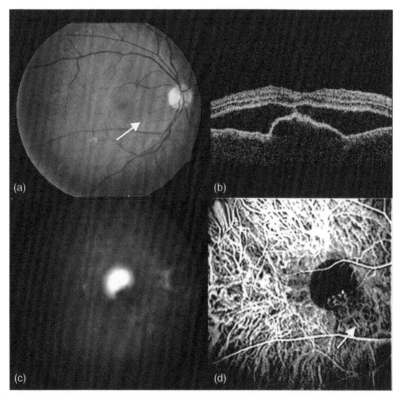

Figure 5 (a) A color fundus photograph shows pigment epithelial detachment and faintly visible feeder vessel extending into the detachment as pointed by the arrow. (b) Optical coherence tomography image shows pigment epithelial detachment with overlying subretinal fluid. (c) Late venous phase fluorescein angiogram shows an area of hyperfluorescence due to pooling of dye into the pigment epithelial detachment. The angiogram fails to show any details of the choroidal neovascular membrane. (d) Dynamic ICG angiogram demonstrates a long afferent arteriole perfusing terminal polypoidal structures consistent with classification as polypoidal choroidal vasculopathy as indicated by the arrow.

that later develop CNV, and vice versa. Histopathology demonstrates thin-walled, saccular polypoidal vessels that were initially regarded as being of venular origin. However, some excised specimens demonstrate the presence of both large choroidal arterioles and venules. Light microscopy demonstrated disruption of inner elastic layer and arteriolosclerotic changes of the blood vessels, and in some cases, the branching arterioles are associated with pericyte-like cells.

The natural history and treatment of PCV is not well established. Anecdotal reports suggest that about half of cases demonstrate deterioration of vision in spite of treatment and that anti-VEGF therapy is less effective than for traditional CNV. The efficacy and safety of PDT seems promising in small studies, but this treatment combined with anti-VEGF therapy must be investigated in future randomized clinical trials.

Retinal angiomatous proliferation

RAP is another recently recognized but common variant of NVAMD that is characterized by the formation of pathologic new vessels within the retina that secondarily invade into the subretinal space or into the choroid. The incidence of RAP variant of NVAMD is unknown, since it requires the use of ICG angiography for definitive diagnosis. Anecdotal reports suggest that at least 15% of the newly diagnosed patients with NVAMD manifest RAP. This variant may be less common in Asians.

High speed ICG angiography offers best visualization of the RAP lesion, particularly for localizing the retinal feeding arterioles and draining venules (**Figure 6(a)–6(e)**). Clinically, stage I RAP lesions present with intraretinal neovascularization with telangiectatic retinal capillaries and small angiomatous structures perfused by the retinal circulation. Stage II RAP lesions extend the new vessel complex into the subretinal space to form subretinal neovascularization often associated with a serous PED. Most stage II RAP lesions are also perfused by retinal arterioles and drained by retinal venules. In these cases, focal closure of the retinal feeding arteriole by thermal laser leads to loss of flow into the RAP lesion and disappearance of hemorrhage, retinal edema, and PED. In stage III RAP,

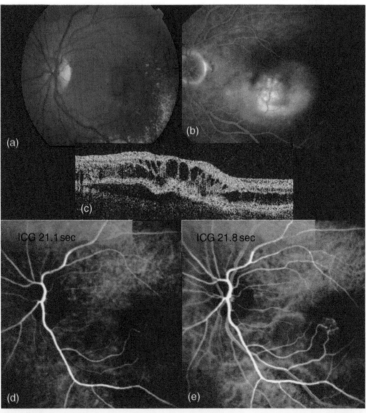

Figure 6   (a) A color fundus photograph shows pigment epithelial detachment associated with intraretinal hemorrhages and extensive exudates temporal to the detachment. (b) Late arteriovenous phase angiogram demonstrates an area of hyperfluorescence corresponding to the pigment epithelial detachment, however, fails to reveal details of the choroidal neovascular membrane to allow for proper classification. (c) Optical coherence tomography image demonstrates serous pigment epithelial detachment with cystic thickening of the overlying retina. Areas of intraretinal hemorrhage appear as hyperreflective spots. (d, e) Dynamic angiogram clearly demonstrates the details of the subretinal neovascular membrane including the afferent arteriole derived from the retinal artery, capillary network, and the draining venule merging with the retinal vein, consistent with classification of the lesion as stage II RAP lesion.

retinal–choroidal anastomosis (RCA) is formed through a break in RPE, often associated with an underlying choroidal neovascular membrane with both choroidal and retinal perfusion. Late stage III RAP lesions develop large fibrotic scars. The transition between purely retinal perfusion (stage I–II RAP) into choroidal perfusion (early stage III RAP) can be difficult to determine clinically.

As with PCV, the pathobiology of RAP is unknown. The histopathological findings in stage II cases demonstrate intraretinal neovascularization without any evidence of CNV. Stage III cases demonstrate both CNV and chorio–retinal anastomosis. RAP responds well to anti-VEGF therapy, suggesting that the underlying pathobiology is strongly mediated by VEGF. Identification of the cellular source of VEGF in RAP (i.e., retinal, RPE, or choroidal) to explain the origination of new vessel growth in the retina as differentiated from choroidal origin in typical CNV remains speculative.

Prior to the advent of anti-VEGF therapy, many treatments for RAP were tried, however, with variable results. Fortunately, anti-VEGF agents effectively control leakage, but rapid recurrences necessitate frequent injections. Focal thermal laser treatment of the afferent arteriole and extrafoveal angioma has been demonstrated to be effective as salvage therapy for recurrences. Advanced forms of the disease involving a vascularized RPE detachment and RCA are unlikely to respond well to any form of current treatment.

Future Directions

Our knowledge base regarding the pathobiology of CNV is expanding as new generation imaging modalities become available and provide us with further anatomical details allowing recognition of the variety in morphology of these lesions. A vast amount of research is focused on learning various aspects of AMD. Although we have come a long way in understanding and treating AMD, considerable amount of work remains to fully understand the pathogenesis, recognize different types, and tailor the therapy based upon special characteristics of each of these lesions to optimize the prognosis in these patients. Areas that especially require additional study include the role of VEGF isoform contribution to healthy choroid and CNV, role of complement and macrophage inflammation, role of endothelial and mesenchymal subtypes in vasculogenesis, predictive biomarker development, and clarification of pathobiology of NVAMD variants, especially PCV and RAP. Combination of comparison of imaging technologies, genotyping studies, and biomarker studies in the context of therapeutic trials will clarify some of these issues.

See also: Injury and Repair: Neovascularization; Retinal Pigment Epithelial–Choroid Interactions.

Further Reading

Ciardella, A. P., Donsoff, I. M., Huang, S. J., Costa, D. L., and Yanuzzi, L. A. (2004). Polypoidal choroidal vasculopathy. *Survey of Ophthalmology* 49: 25–37.

Cousins, S. W., Csaky, K. G., and Espinosa-Heidmann, D. G. (2005). Macular degeneration. In: *Clinical Strategies for Diagnosis and Treatment of AMD: Implications from Research*, pp. 167–191. Berlin: Springer.

Csaky, K. G. and Cousins, S. W. (2007). Immunology of age-related macular degeneration. In: Jennifer, I. and Lim, M. D. (eds.) *Age-Related Macular Degeneration*, 2nd edn., pp. 11–33. Boca Raton, FL: CRC Press.

Fukushima, I., McLeod, D. S., and Lutty, G. A. (1997). Intrachoroidal microvascular abnormality: A previously unrecognized form of choroidal neovascularization. *American Journal of Ophthalmology* 124: 473–487.

Green, W. R. (1999). Histopathology of age-related macular degeneration. *Molecular Vision* 5: 27.

Grossniklaus, H. E. and Green, W. R. (2004). Choroidal neovascularization. *American Journal of Ophthalmology* 137: 496–503.

Grossniklaus, H. E., Ling, J. X., Wallace, T. M., et al. (2002). Macrophage and retinal pigment epithelium expression of angiogenic cytokines in choroidal neovascularization. *Molecular Vision* 8: 119–126.

Lutty, G., Grunwald, J., Maiji, A. B., Uyama, M., and Yoneya, S. (1999). Changes in choriocapillaris and retinal pigment epithelium in age-related macular degeneration. *Molecular Vision* 5: 35.

Chromatic Function of the Cones

D H Foster, University of Manchester, Manchester, UK

Glossary

CIE, Commission Internationale de l'Eclairage –
The CIE is an independent, nonprofit organization responsible for the international coordination of lighting-related technical standards, including colorimetry standards.

Color-matching functions – Functions of wavelength λ that describe the amounts of three fixed primary lights which, when mixed, match a monochromatic light of wavelength λ of constant radiant power. The amounts may be negative. The color-matching functions obtained with any two different sets of primaries are related by a linear transformation. Particular sets of color-matching functions have been standardized by the CIE.

Fundamental spectral sensitivities – The color-matching functions corresponding to the spectral sensitivities of the three cone types, measured at the cornea. The spectral sensitivities may be normalized so that the maximum is unity or according to the nominal population densities of the cone types.

Heterochromatic flicker photometry – The adjustment of the radiant power of one of two spatially coextensive lights presented in alternating sequence at a temporal frequency such that there is a unique value of the radiant power where the sensation of flicker is minimum; that is, with a higher or lower radiant power, the sensation of flicker becomes greater.

Luminous efficiency function V_λ – The inverse of the radiant power of a monochromatic stimulus of wavelength λ that produces a luminous sensation equivalent to that of a monochromatic stimulus of fixed wavelength λ_0. The units and λ_0 may be chosen so that the maximum of this function is unity. It is also known as the relative luminous efficiency or relative luminosity function.

Optical density – Absorbance; see spectral absorbance.

Radiant power and quantum units – Radiant power is measured in watts but it is sometimes more appropriate to measure it in quanta s^{-1}. Sensitivities may be expressed as the logarithm to the base 10 of the reciprocal of the radiant power required to reach a criterion level of performance.

Spectral absorbance $A(\lambda)$ – Logarithm to the base 10 of the reciprocal of the spectral transmittance, $\tau(\lambda)$; that is, $A(\lambda) = -\log_{10} \tau(\lambda)$. It depends on the path length. If l is the path length and $a(\lambda)$ is the spectral absorptivity, then, for a homogeneous isotropic absorbing medium, $A(\lambda) = la(\lambda)$ (Lambert's law).

Spectral absorptance $\alpha(\lambda)$ – Ratio of the spectral radiant flux absorbed by a layer to the spectral radiant flux entering the layer. If $\tau(\lambda)$ is the spectral transmittance, then $\alpha(\lambda) = 1-\tau(\lambda)$. The value of $\alpha(\lambda)$ depends on the length or thickness of the layer. For a homogeneous isotropic absorbing medium, $\alpha(\lambda) = 1-\tau(\lambda) = 1-10^{-la(\lambda)}$, where l is the path length and $a(\lambda)$ is the spectral absorptivity. Changes in the concentration of a photopigment have the same effect as changes in path length.

Spectral absorptivity $a(\lambda)$ – Spectral absorbance of a layer of unit thickness. Absorptivity is a characteristic of the medium, that is, the photopigment. Its numerical value depends on the unit of length.

Spectral sensitivity – The inverse of the radiant power of a monochromatic stimulus of wavelength λ that produces a criterion response equal to that of a monochromatic stimulus of fixed wavelength λ_0. The units and λ_0 may be chosen so that the maximum of this function is unity. In specifications by the CIE, the lower limit of the wavelength range is generally 360–400 nm and the upper limit is generally 760–830 nm, but smaller ranges may be used. The spectral sensitivity of a cone is essentially the normalized spectral absorptance of its photopigment (discounting any geometrical factors, e.g., cone waveguide properties).

Spectral transmittance $\tau(\lambda)$ – Ratio of the spectral radiant flux leaving a layer to the spectral radiant flux entering the layer. The value of $\tau(\lambda)$ depends on the path length.

Unique hues – Hues that are judged to be unmixed. They form mutually exclusive pairs, so that no light can appear to contain both red and green or both blue and yellow.

Wavenumber – The reciprocal of wavelength, usually in cm^{-1}.

Introduction

Normal human color vision is governed by three types of retinal cone photoreceptors sensitive to light over different, but overlapping, regions of the spectrum and conventionally designated long-, medium-, and short-wavelength-sensitive (L, M, and S). Because the intensity and spectral properties of light cannot be distinguished once it has been absorbed (the principle of univariance), chromatic information is obtained by comparing the outputs of these different cone types. Signals from L, M, and S cones converge via excitatory and inhibitory pathways onto subsets of anatomically distinct ganglion-cell populations, which then project to the lateral geniculate nucleus, and thence to the higher cortical areas. The aim of this article is to provide an overview of how the cones contribute to chromatic function and how that contribution is modified by prereceptoral filtering and postreceptoral processing within the eye.

Photopigments and Phototransduction

The photopigment in the outer segment of the cone consists of two covalently linked parts, a protein called opsin and a chromophore based on retinal, an aldehyde of vitamin A. It is the latter that provides light sensitivity by isomerizing from 11-*cis* to all-*trans* forms. This leads to the activation of a guanine nucleotide-binding protein (G-protein), transducin, and a cascade of molecular events that result in a change in the rate of neurotransmitter release from the receptor to other neurons in the retina. In the course of the cascade, the signal provided by photon absorption is greatly amplified.

The wavelength λ_{max} of maximum absorption of the pigment (see the section entitled 'Cone Spectral Sensitivities') depends on the particular amino acid sequence of the opsin and its relationship to the chromophore. The L and M pigments are members of an ancestral medium/long-wavelength-sensitive class of vertebrate pigments. These two pigments have a high degree of homology, and the majority of the difference in λ_{max} arises at just a few amino acid sites. A common polymorphism of the L and M pigments, in which alanine is substituted for serine at codon 180, can create shifts in λ_{max} of several nanometers.

The S pigment is a member of a distinct ancestral ultraviolet (UV)-sensitive class of vertebrate pigments. Its spectral tuning is different from that of the L and M pigments, and the shift in λ_{max} from UV to the violet of normal human S cones arises from amino acid replacements at multiple tuning sites.

Fundamental Spectral Sensitivities

The spectral sensitivities of the cones, *in vivo*, have been estimated by many different methods. Historically, the most important method is based on psychophysical

color-matching, in which a mixture of three fixed primary lights is matched against a monochromatic light of variable wavelength and constant radiant power. These data form the basis of the system of colorimetry standardized by the Commission Internationale de l'Eclairage (CIE). In the normal trichromatic eye, such color-matching functions yield a family of spectral sensitivities that are a linear transformation of the unknown cone spectral sensitivities. This transformation is not unique in that it varies with the choice of primaries, but data obtained from dichromatic and sometimes monochromatic observers, who have, respectively, only two and one of the three normal cone types, provide good estimates of these so-called fundamental spectral sensitivities.

Figure 1 shows one set of normalized L-, M-, and S-cone fundamental spectra, with wavelengths λ_{max} of maximum sensitivity of *c.* 570, 543, and 442 nm. These estimates represent corneal cone spectral sensitivities, as color matching necessarily involves the spectral transmission properties of the ocular media (see the section entitled 'Prereceptoral Attenuation'). Several such sets of fundamentals have been derived, based on different sets of color-matching functions and different assumptions about cone sensitivity and prereceptoral absorption.

Other psychophysical methods of estimation involve selectively chromatically adapting the three cone types by a background light and measuring the spectral sensitivity of the most sensitive color mechanism at or near its threshold level of response, a technique associated primarily with W. S. Stiles (see the subsection entitled 'Isolating Cone Responses: Selective Chromatic Adaptation'). Psychophysical methods have the advantage that sensitivity can be determined over a range of two to five decades

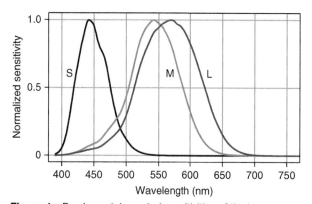

Figure 1 Fundamental spectral sensitivities of the long-, medium-, and short-wavelength-sensitive (L, M, and S) cones. Data are normalized to a maximum of unity on a radiance scale and are based on the 2° data of Table 2 of Stockman, A. and Sharpe, L. T. (2000). The spectral sensitivities of the middle- and long-wavelength-sensitive cones derived from measurements in observers of known genotype. *Vision Research* 40: 1711–1737; modified from the mean 10° color-matching functions of Stiles, W. S. and Burch, J. M. (1959). N.P.L. colour-matching investigation: Final Report (1958). *Optica Acta* 6: 1–26.

(i.e. 2–5 log units), although with normal trichromats, chromatic adaptation cannot generally isolate the response of a single cone type throughout the spectrum.

Nonpsychophysical but noninvasive methods of *in vivo* estimation include retinal densitometry (i.e., fundus spectral reflectometry) and electroretinography. In retinal densitometry, differences in reflectance of the fundus are measured, at various wavelengths, before and after selectively bleaching the cone pigments. The method is limited by its low signal-to-noise ratio and the need to distinguish between signals from L and M cones. It is also difficult to obtain signals reliably from S cones. In electroretinography, the summed electrical response of the retina to a flickering light is measured with a corneal electrode. This method provides a much larger dynamic range, of several log units, along with a good signal-to-noise ratio, but, as with retinal densitometry, signals from L and M cones need to be separated. Both techniques are best applied to well-characterized dichromats.

Prereceptoral Attenuation

The spectrum of the incident light reaching the cones is modified by two prereceptoral filters. The first is the lens (and cornea). In the young eye, the lens appears colorless, but, with increasing age, it becomes yellow, partly as a result of light scatter and absorption, which produce large transmission losses at short wavelengths (**Figure 2**). Even within the same age group, however, the size of the loss at 400 nm may vary over 1 log unit. Another important factor in the changing appearance of the lens is the accumulation of a particular fluorogen, which leads to increased lens fluorescence.

The second prereceptoral filter is a nonphotosensitive yellow pigment, usually referred to as the macular pigment. Its maximum absorption is at *c.* 460 nm (**Figure 2**). It is a carotenoid consisting of two xanthophylls, lutein and zeaxanthin, whose densities decline with increasing eccentricity, with zeaxanthin concentrated only in the fovea.

The optical density of macular pigment can be quantified, *in vivo*, by heterochromatic flicker photometry and by Raman spectroscopy, in which argon laser light is used to excite the xanthophylls. Values of the optical density at the fovea vary markedly across individuals, from >1.0 to almost zero, with a mean of *c.* 0.6. Some of the hypothesized benefits of the macular pigment include protection from the damaging effects of short-wavelength light, reducing the effects of longitudinal chromatic aberration and glare, and protection against reactive oxygen species.

In addition to the lens and macular pigment, there is another important factor modifying the effective spectral sensitivity of the cones, namely, self-screening by the photopigment, which acts to broaden its absorption spectrum. If at wavelength λ the spectral absorptivity is $a(\lambda)$ and the path length is l, then the spectral absorptance $\alpha(\lambda)$ is given by $1 - 10^{-la(\lambda)}$. Therefore, increasing l (or equivalently the concentration of the pigment) has a proportionally larger effect at values of λ where $\alpha(\lambda)$ is small than at values where it is large.

Cone Spectral Sensitivities

Estimates of cone spectral sensitivities can be obtained from corneal spectral sensitivities, that is, the fundamental spectra of **Figure 1**, by eliminating the estimated effects of prereceptoral losses (see the section entitled 'Prereceptoral Attenuation'). **Figure 3** shows the results. The normalized L-, M-, and S-cone spectral sensitivities have wavelengths λ_{max} of maximum sensitivity of *c.* 559, 530, and 421 nm, respectively. Comparing these spectra with corneal spectra (**Figure 1**) reveals the influence of prereceptoral filtering in shifting λ_{max} toward longer wavelengths, especially for S cones, where the increase in λ_{max} is *c.* 20 nm.

In vitro methods of estimating cone spectral sensitivities include recording from single cone outer segments with a suction microelectrode, direct microspectrophotometry of intact receptor outer segments, and differential spectrophotometry of recombinant cone pigments produced in tissue-culture cells. Microelectrode recordings take advantage of the amplification of the signal provided by the phototransduction cascade and so can provide sensitivities over 5–6 log units, but spectrophotometry, as with *in vivo* reflectance measurements, is limited by

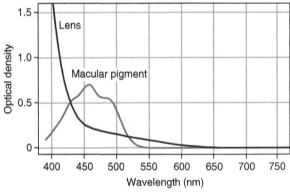

Figure 2 Variation of the optical density of the lens and of the macular pigment with wavelength. Data are from Table 2 of Stockman, A. and Sharpe, L. T. (2000). The spectral sensitivities of the middle- and long-wavelength-sensitive cones derived from measurements in observers of known genotype. *Vision Research* 40: 1711–1737; based on van Norren, D. and Vos, J. J. (1974). Spectral transmission of human ocular media. *Vision Research* 14: 1237–1244; and on Bone, R. A., Landrum, J. T., and Cains, A. (1992). Optical density spectra of the macular pigment *in vivo* and *in vitro*. *Vision Research* 32: 105–110. Plotted values are means over observers. Macular-pigment densities have been doubled for clarity.

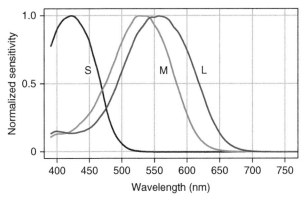

Figure 3 Spectral sensitivities of the long-, medium-, and short-wavelength-sensitive (L, M, and S) cones derived from mean L-, M-, and S-cone fundamental spectra (**Figure 1**). Data are normalized to a maximum of unity on a radiance scale and are based on Table 2 of Stockman, A. and Sharpe, L. T. (2000). The spectral sensitivities of the middle- and long-wavelength-sensitive cones derived from measurements in observers of known genotype. *Vision Research* 40: 1711–1737.

Table 1 Estimated wavelengths λ_{max} of maximum sensitivity of long-, medium-, and short-wavelength-sensitive cones and cone pigments from *in vivo* and *in vitro* measurements in humans and primates

Method	λ_{max} (nm)		
	L	M	S
Psychophysics[a]	559	530	421
Suction microelectrode recording from primate cones[b]	561	531	430
Microspectrophotometry of excised human cones[c]	558	531	419
Recombinant human cone pigments[d]	558	530	425

[a]Stockman, A. and Sharpe, L. T. (2000). The spectral sensitivities of the middle- and long-wavelength-sensitive cones derived from measurements in observers of known genotype. *Vision Research* 40: 1711–1737.
[b]Baylor, D. A., Nunn, B. J., and Schnapf, J. L. (1987). Spectral sensitivity of cones of the monkey *Macaca fascicularis*. *Journal of Physiology – London* 390: 145–160.
[c]Dartnall, H. J. A., Bowmaker, J. K., and Mollon, J. D. (1983). Human visual pigments: Microspectrophotometric results from the eyes of seven persons. *Proceedings of the Royal Society of London, Series B: Biological Sciences* 220: 115–130.
[d]Merbs, S. L. and Nathans, J. (1992). Absorption spectra of human cone pigments. *Nature* 356: 433–435; Oprian, D. D., Asenjo, A. B., Lee, N., and Pelletier, S. L. (1991). Design, chemical synthesis, and expression of genes for the three human color vision pigments. *Biochemistry* 30: 11367–11372.

its low signal-to-noise ratio, providing sensitivities over only 1 or 1–2 log units. All *in vitro* methods are constrained by the nonphysiological nature of the illumination (and sometimes by the method of specimen preparation and recording). Notwithstanding the differences in experimental methodology and analysis, there is broad agreement in the values of λ_{max} (**Table 1**).

Template for Cone Spectral Sensitivity

The response of a cone is determined by the number of photons it absorbs: the spectral sensitivity describes the probability of that absorption at each frequency v, which is invariant of the medium, unlike wavelength λ. The main absorption band of mammalian visual pigments is the α band (the β and other bands at progressively shorter wavelengths have less influence because of absorption in the ocular media). Despite the differences in λ_{max} of the L, M, and S pigments, plotting cone spectral sensitivity or spectral absorbance against the normalized frequency, that is, v/v_{max}, where v_{max} is the frequency of maximum sensitivity, yields closely similar dependencies, as illustrated by the data in **Figure 4** for suction microelectrode recordings from monkey cones (symbols).

These dependencies may be fitted empirically by a polynomial or other smooth function of normalized frequency (**Figure 4**, curve). This function may then be used as a template or nomogram and applied to sample data that are noisy or of limited range, allowing, for example, more accurate estimates of λ_{max}, as was done with the entries in **Table 1**. Templates are best obtained separately for sensitivities on logarithmic and linear ordinates, for the weighting of the fits is different and the templates are therefore not exact logarithmic transforms of each other (the data in **Table 1** were based on both kinds of templates).

Cone Gains and von Kries Scaling

The chromatic function of the cones is determined by not only their spectral sensitivities but also their relative gains. The effect of the spectrum of the prevailing light on those gains is summarized by the coefficient rule of von Kries, which refers to the idea that the sensitivity of each cone type depends only on activity in that cone type. Thus, if the radiant power in the light is biased toward long wavelengths, then the gain of the L cones is reduced in proportion to that bias. In fact, the evidence from microelectrode recordings in the retina of primates suggests that although substantial and independent chromatic adaptation does takes place among the three cone types, it is not complete. Adaptation also takes place at postreceptoral levels (see the section entitled 'Postreceptoral Spectra'). Even so, to a first-order approximation, von Kries scaling achieves a normalization of cone responses over a wide range of illumination conditions before bleaching dominates.

Isolating Cone Responses

The experimental investigation of chromatic function sometimes requires the selective stimulation of a particular cone type. There are two general methods for achieving this *in vivo*: one employs selective adaptation either by

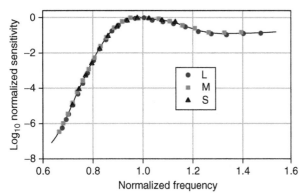

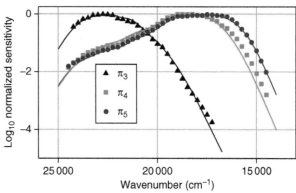

Figure 4 Spectral sensitivities of long-, medium-, and short-wavelength-sensitive (L, M, and S) monkey cones measured by suction microelectrode. Log mean absorbance, normalized to a maximum of zero (i.e., unity on a linear scale), is plotted against normalized frequency v/v_{max} for each cone type. Data are from Table 1 of Baylor, D. A., Nunn, B. J., and Schnapf, J. L. (1987). Spectral sensitivity of cones of the monkey *Macaca fascicularis*. *Journal of Physiology – London* 390: 145–160. The smooth curve is a locally weighted polynomial regression.

Figure 5 Stiles' background-field spectral sensitivities of the mechanisms π_3, π_4, and π_5 (symbols). Log mean sensitivity, normalized to a maximum of zero on a log quantum scale (i.e., unity on a linear scale), is plotted against wavenumber (proportional to frequency). The corresponding mean S-, M-, and L-cone fundamental spectral sensitivities, in the same units, are also shown (curves). Data are based on Table 2 (7.4.3) of Wyszecki, G. and Stiles, W. S. (1982). Color science: Concepts and methods, quantitative data and formulae. New York: John Wiley & Sons and on Table 2 of Stockman, A. and Sharpe, L. T. (2000). The spectral sensitivities of the middle- and long-wavelength-sensitive cones derived from measurements in observers of known genotype. *Vision Research* 40: 1711–1737.

a steady or temporally modulated background field so that only the chosen cone type mediates detection; the other constrains changes in the stimulus so that they are invisible to all but the chosen cone type.

Selective Chromatic Adaptation

The two-color threshold method of Stiles is based on the properties of the increment-threshold function. The minimum detectable radiance (the increment threshold) of a small test flash of wavelength λ is measured as a function of the radiance of a large steady background field of wavelength μ. The resulting monotonically increasing, threshold-versus-radiance (t.v.r.) curve may have more than one branch, but, over a limited range, each branch typically retains its shape while undergoing a vertical displacement with a change in λ (yielding a test spectral sensitivity) or a horizontal displacement with a change in μ (yielding a background-field spectral sensitivity). Provided certain assumptions hold, the test spectral sensitivity should coincide with the background-field spectral sensitivity, where both are defined. The assumptions are that the mechanisms underlying the t.v.r. curves – Stiles' π mechanisms – act independently of each other; that the observed t.v.r. curve depends only on the smallest increment threshold of the mechanisms available; and that each mechanism has a well-defined spectral sensitivity.

Figure 5 shows background-field spectral sensitivities for three mechanisms, π_3, π_4, and π_5 (symbols) superimposed on the corresponding S-, M-, and L-cone fundamental spectra (curves) (see the section entitled 'Fundamental Spectral Sensitivities').

The π mechanisms cannot represent individual cone activity, for there are too many of them: three S mechanisms (π_1, π_2, and π_3), two M mechanisms (π_4 and π_4'), and two L mechanisms (π_5 and π_5'), depending on the conditions of measurement. In addition to their multiplicity, there is evidence of nonadditivity, that is, the effect of a background field consisting of two monochromatic lights is different from the sum of the effects of the two lights alone. Yet, as **Figure 5** shows, π_3, π_4, and π_5 approach quite closely the corresponding cone fundamentals.

The mechanisms defined by the t.v.r. curves depend, of course, on where in the visual pathway the threshold for detection is determined. With an appropriate choice of the spatial and temporal properties of the test flash and the background field, it is possible to obtain both test and background-field spectra that appear to mainly represent not cone responses but opponent combinations of cone responses (see the subsection entitled 'Postreceptoral Spectra: Cone Opponency').

Silent Substitution

The principle of silent substitution is illustrated in **Figure 6**. At a particular criterion response level, here 0.9 of maximum (other levels might be chosen), there are two wavelengths, 543 nm and 594 nm, at which the L-cone sensitivities are equal (red crosses). The alternation of two monochromatic, equal-radiance lights at these two wavelengths is therefore invisible to L cones but visible to M cones, for the M-cone sensitivity at 543 nm is more

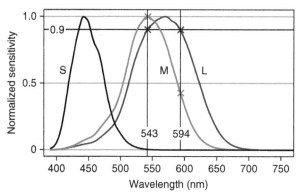

Figure 6 Silent substitution. Responses at 543 and 594 nm are the same for L cones (red crosses) but different for M cones (green crosses). Data are normalized cone fundamental spectra (**Fig. 1**).

than twice that at 594 nm (green crosses). At these two wavelengths, the S-cone sensitivity is vanishingly small and can be ignored. The same technique may be applied to M and S cones.

Unlike selective chromatic adaptation, the isolation achieved by silent substitution is a second-order one: it is not the stimulus but its alternation that is silent. Nevertheless, the method may be used to determine the spectral sensitivity of the isolated cone class (here the M cones); and even when the spectral sensitivity of the silenced class (the L cones) is known only approximately, it may still, in principle, be applied in an iterative way.

Rod Intrusion

The rod photoreceptors of the retina are not normally associated with chromatic function. But with large stimulus fields of low-to-moderate luminance, color vision can be influenced by rod activity, which biases the apparent hue and diminishes the saturation of extrafoveal stimuli. Even for foveal stimuli of just 1° visual angle, some rod involvement may be detected. Under daylight conditions, however, rod intrusion in color matches is small and is likely to be negligible above 100 cd m^{-2}.

Spatial Densities

The numbers of the three cone types are neither constant over individuals nor uniform across the retina. These variations have different implications for chromatic and luminance function.

Individual Variations

The relative frequencies of the different cone types, as proportions of the total number of cones, have been estimated across individuals by indirect psychophysical methods and by more direct methods, both *in vivo* and *in vitro*. The most important psychophysical method models the photopic luminous efficiency function obtained by heterochromatic flicker photometry as a linear sum of signals from L and M cones, under the working assumptions that the contribution of each is weighted by its numerosity and that the spectral sensitivity of each is known. Direct *in vivo* methods include retinal densitometry, electroretinographic flicker photometry, and high-resolution adaptive-optics imaging combined with retinal densitometry. Results from flicker photometry correlate with those from electroretinographic methods. *In vitro* methods include microspectrophotometry with single cones, immunocytochemical labeling of small retinal patches, and analysis of L and M opsin messenger RNA in homogenized retinal patches.

It is clear that the relative frequency of S cones is fairly constant over individuals, *c.* 5–6% for a patch of retina at 1° eccentricity. For L and M cones, it is their ratio that is normally specified, and, although a working value of 1.5:1 or 2:1 is often assumed in modeling their contributions to the luminous efficiency function, the actual ratio varies greatly from individual to individual, from *c.* 0.7:1 to 12:1, by direct measurement, and over a somewhat larger range by indirect measurement. This variation may reflect the different evolutionary and developmental histories of the S-cone and L- and M-cone pathways.

Retinal Distributions

That the three cone types have different distributions across the retina has been long known from psychophysical measurements of, for example, increment threshold and grating acuity with cone-specific stimuli. These findings have been confirmed and extended by some of the techniques mentioned in the preceding subsection.

The spatial density of L and M cones is greatest in the central fovea and it declines steadily with increasing retinal eccentricity. By contrast, the spatial density of S cones is zero in the central *c.* 20 arcmin of the fovea (0.1 mm); it increases to a maximum at *c.* 1° eccentricity, and then declines again. Because the decrease in S-cone density is less rapid than that of the other cone types, the relative frequency of S cones, as a proportion of the total number of cones at each eccentricity, increases slowly with eccentricity and levels off at *c.* 5° or a little more.

In the fovea, there seems to be little evidence for other than a random distribution of S cones and L and M cones, although occasional departures in the direction of a more regular S-cone distribution and some clumping of L and M cones have been observed. In the periphery, there are more systematic departures from randomness in the distributions of both S cones and L and M cones. The purpose of the cone-rich rim of the ora serrata is unknown.

Eccentricity and Chromatic Function

The absence of S cones in the center of the fovea accounts for the perceptual phenomenon of small-field tritanopia: for sufficiently small fields, color vision is dichromatic, so that just two fixed primary lights are needed to match an arbitrary test light (see the section entitled 'Fundamental Spectral Sensitivities'). The size of the tritanopic zone varies across individuals, and it may be absent in some.

The consequences of individual variation in L:M cone ratios are largely confined to luminance function. For chromatic function, the variation may be counterbalanced by changes in gain associated with excitatory and inhibitory inputs. In judgments of a perceptually unique yellow, in which the wavelength of a monochromatic light is adjusted so that it appears neither red nor green and which might be expected to reveal an imbalance between L and M cone inputs, the variation in the selected wavelength seems not to be attributable to ratio differences. One factor contributing to the stability of unique hues may be a form of chromatic adaptation to the spectrum of the environment.

For the individual eye, there are marked variations in chromatic function with retinal eccentricity. Thus, in general terms, color-matching performance with a small field, of the order of 1° extent, diminishes with increasing distance from the fovea, tending to dichromacy at 25–30° and to monochromacy at 40–50°. More specific changes are revealed by psychophysical measurements of hue and saturation and of chromatic contrast sensitivity with eccentricity. The latter is quantified by the detectability of a stimulus modulated along a red–green or blue–yellow color axis, defined with respect to stimuli specific for L, M, or S cones (i.e., an L vs. M axis and an S vs. L + M axis). The size of the stimulus is usually increased with eccentricity for optimum response. For red–green modulation, contrast sensitivity is high at the fovea and then declines rapidly with eccentricity, falling to zero at 25–30°, whereas for blue–yellow modulation, contrast sensitivity is much flatter and declines only slowly with eccentricity, and at a rate similar to that for luminance modulation. Since luminance contrast sensitivity is determined by the same cone signals as red–green contrast sensitivity, the decline in the latter with eccentricity is presumably due to changes in chromatic coding rather than in L and M cone densities (see the subsection entitled 'Postreceptoral Spectra: Cone Opponency').

Postreceptoral Spectra

Cone Opponency

Spectral sensitivity at postreceptoral levels is modified by antagonistic interactions between signals from the different cone types. This cone opponency – also occasionally called color opponency – is not the same as the red–green and blue–yellow color opponency identified in some perceptual experiments where the poles of the reference axes correspond to the unique hues, red, green, blue, and yellow.

The detailed organization of cone opponency has been difficult to characterize unambiguously, and retinal connectivity is not as specific as has sometimes been assumed. Thus, there is anatomical evidence of random, nonselective connectivity of L and M cones to ganglion-cell receptive fields, and physiological evidence of both randomness and selectivity. Nevertheless, because the response of a ganglion cell is determined by a weighted sum of its excitatory and inhibitory inputs, the nature of the connectivity has less effect on the spectral characteristics of cone-opponent responses than the weights associated with the inputs. **Figure 7** shows theoretical cone-opponent spectra (curves) from simple linear combinations of cone signals of the form $S - k(M + L)$ in the left panel, $M - kL$ in the middle panel, and $L - kM$ in the right panel, where the amount of inhibition is indicated by the weight constant k (the assumption of linearity is not essential).

With $k = 0.4$, the wavelength λ_{max} of maximum sensitivity for the $L - kM$ spectrum shifts from 570 to 589 nm, a difference of 19 nm, whereas for the $M - kL$ spectrum, λ_{max} shifts from 543 to 539 nm, that is, just 4 nm, and for the $S - k(M + L)$ spectrum, there is no shift at all. With larger k, both $L - kM$ and $M - kL$ spectra become more narrowed. Combined with a shift in λ_{max}, the effect is sometimes described as spectral sharpening.

These changes in spectral shape are similar to those observed in microelectrode recordings in primate retina and psychophysically in both test and background-field spectral sensitivities (see the subsection entitled 'Isolating Cone Responses: Selective Chromatic Adaptation') obtained under conditions designed to isolate cone-opponent activity. Examples of background-field cone-opponent spectra are also shown in **Figure 7** (symbols). Notice that the asymmetries in the shifts of λ_{max} do not themselves require asymmetric weighting of cone signals nor do they imply asymmetric $L - kM$ and $M - kL$ responses.

As already observed (see the subsection entitled 'Spatial Densities: Eccentricity and Chromatic Function'), red–green and blue–yellow contrast sensitivity varies with retinal eccentricity, but there is evidence that the opponent weights revealed in those measurements remain constant with retinal location up to $c.$ 10° eccentricity. Beyond that, differences emerge in the L versus M and M versus L responses.

It is at present unclear how these kinds of cone-opponent activity relate to the perceptual color opponency of unique hues.

Advantages of Cone Opponency

Along with providing information about the spectral content of a stimulus, cone opponency has two other

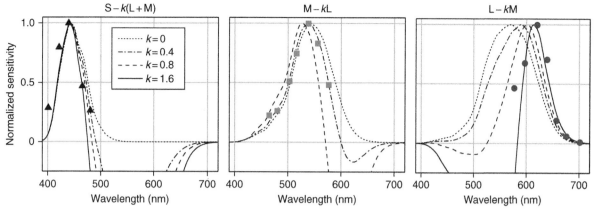

Figure 7 Theoretical and observed cone-opponent spectral sensitivities. Each set of curves represents spectra of the form S − k(M + L) (left panel), M − kL (middle panel), and L − kM (right panel), normalized to a maximum of unity on a radiance scale, with values of the weight constant k indicated (for k = 1.6, the function M − kL is mainly negative). Normalized L-, M-, and S-cone fundamental spectra were taken from Table 2 of Stockman, A. and Sharpe, L. T. (2000). The spectral sensitivities of the middle- and long-wavelength-sensitive cones derived from measurements in observers of known genotype. *Vision Research* 40: 1711–1737. The symbols show psychophysical background-field spectra obtained under conditions intended to isolate cone-opponent function, and are means calculated from Figure 1(a)–(c) of Foster, D. H. and Snelgar, R. S. (1983). Test and field spectral sensitivities of colour mechanisms obtained on small white backgrounds: Action of unitary opponent-colour processes? *Vision Research* 23: 787–797.

advantages. First, it reduces information redundancy. Because of the close overlap in the L- and M-cone spectral sensitivities, their responses are highly correlated. But spectrally sharpened L − kM and M − kL combinations provide greater independence and more efficient coding for neural transmission. Second, with chromatic adaptation at a cone-opponent level and not just at a receptoral level (see the subsection entitled 'Cone Spectral Sensitivities: Cone Gains and von Kries Scaling'), it is possible to compensate more completely for any bias in the spectrum of the prevailing light. This may be achieved by von Kries scaling, but of cone-opponent responses rather than cone responses.

See also: Color Blindness: Inherited; Photopic, Mesopic and Scotopic Vision and Changes in Visual Performance; Phototransduction: Phototransduction in Cones; Rod and Cone Photoreceptor Cells: Inner and Outer Segments.

Further Reading

Baylor, D. A., Nunn, B. J., and Schnapf, J. L. (1987). Spectral sensitivity of cones of the monkey *Macaca fascicularis*. *Journal of Physiology – London* 390: 145–160. The smooth curve is a locally weighted polynomial regression.

Bone, R. A., Landrum, J. T., and Cains, A. (1992). Optical density spectra of the macular pigment *in vivo* and *in vitro*. *Vision Research* 32: 105–110.

Bowmaker, J. K. (2008). Evolution of vertebrate visual pigments. *Vision Research* 48: 2022–2041.

Buzás, P., Blessing, E. M., Szmajda, B. A., and Martin, P. R. (2006). Specificity of M and L cone inputs to receptive fields in the parvocellular pathway: Random wiring with functional bias. *Journal of Neuroscience* 26: 11148–11161.

Crook, J. M., Lee, B. B., Tigwell, D. A., and Valberg, A. (1987). Thresholds to chromatic spots of cells in the macaque geniculate nucleus as compared to detection sensitivity in man. *Journal of Physiology – London* 392: 193–211.

Curcio, C. A., Allen, K. A., Sloan, K. R., et al. (1991). Distribution and morphology of human cone photoreceptors stained with anti-blue opsin. *Journal of Comparative Neurology* 312: 610–624.

Dacey, D. M. and Packer, O. S. (2003). Colour coding in the primate retina: Diverse cell types and cone-specific circuitry. *Current Opinion in Neurobiology* 13: 421–427.

Dartnall, H. J. A., Bowmaker, J. K., and Mollon, J. D. (1983). Human visual pigments: Microspectrophotometric results from the eyes of seven persons. *Proceedings of the Royal Society of London, Series B: Biological Sciences* 220: 115–130.

Foster, D. H. and Snelgar, R. S. (1983). Test and field spectral sensitivities of colour mechanisms obtained on small white backgrounds: Action of unitary opponent-colour processes? *Vision Research* 23: 787–797.

Hofer, H., Carroll, J., Neitz, J., Neitz, M., and Williams, D. R. (2005). Organization of the human trichromatic cone mosaic. *Journal of Neuroscience* 25: 9669–9679.

Lee, B. B., Dacey, D. M., Smith, V. C., and Pokorny, J. (1999). Horizontal cells reveal cone type-specific adaptation in primate retina. *Proceedings of the National Academy of Sciences of the United States of America* 96: 14611–14616.

Mansfield, R. J. W. (1985). Primate photopigments and cone mechanisms. In: Fein, A. and Levine, J. S. (eds.) *The Visual System*, pp. 89–106. New York: Liss.

Merbs, S. L. and Nathans, J. (1992). Absorption spectra of human cone pigments. *Nature* 356: 433–435.

Nickle, B. and Robinson, P. R. (2007). The opsins of the vertebrate retina: Insights from structural, biochemical, and evolutionary studies. *Cellular and Molecular Life Sciences* 64: 2917–2932.

Oprian, D. D., Asenjo, A. B., Lee, N., and Pelletier, S. L. (1991). Design, chemical synthesis, and expression of genes for the three human color vision pigments. *Biochemistry* 30: 11367–11372.

Sakurai, M. and Mullen, K. T. (2006). Cone weights for the two cone-opponent systems in peripheral vision and asymmetries of cone contrast sensitivity. *Vision Research* 46: 4346–4354.

Sperling, H. G. and Harwerth, R. S. (1971). Red–green cone interactions in the increment-threshold spectral sensitivity of primates. *Science* 172: 180–184.

Stiles, W. S. and Burch, J. M. (1959). N.P.L. colour-matching investigation: Final Report (1958). *Optica Acta* 6: 1–26.

Stockman, A. and Sharpe, L. T. (2000). The spectral sensitivities of the middle- and long-wavelength-sensitive cones derived from measurements in observers of known genotype. *Vision Research* 40: 1711–1737.

Stockman, A., Sharpe, L. T., Merbs, S., and Nathans, J. (2000). Spectral sensitivities of human cone visual pigments determined *in vivo* and *in vitro*. *Methods in Enzymology.*

Part B: Vertebrate Phototransduction and the Visual Cycle 316: 626–650.

Stromeyer, C. F., III, Lee, J., and Eskew, R. T., Jr (1992). Peripheral chromatic sensitivity for flashes: A post-receptoral red–green asymmetry. *Vision Research* 32: 1865–1873.

van Norren, D. and Vos, J. J. (1974). Spectral transmission of human ocular media. *Vision Research* 14: 1237–1244.

The Circadian Clock in the Retina Regulates Rod and Cone Pathways

S C Mangel and C P Ribelayga, The Ohio State University College of Medicine, Columbus, OH, USA

Glossary

Circadian clock – A type of self-sustained molecular oscillator with a period of approximately 24 h.

Dopamine – The main retinal catecholamine produced by a type of amacrine or interplexiform cell. Dopamine activates D_1 and D_2 receptors and plays key roles in light-adaptive processes and in the effects of the retinal clock. Most of the effects of retinal dopamine are through volume transmission; thus, dopamine acts as a neurohormone in the retina.

Electrical conductance – A measure of how easily electrical current flows along a certain path that has a difference in voltage or potential.

Entrainment of a circadian clock – Altering the phase of a circadian oscillator due to the presence of environmental input (e.g., light/dark, temperature) generally in the early or late night.

Ganglion cell – The output neuron of the retina that sends its axon to other parts of the brain.

Gap junction – A type of electrical synapse that is comprised of intercellular channels that directly connect the cytoplasm of two cells and facilitate the cell-to-cell passage of ions and small molecules.

Horizontal cell – The second-order interneuron in the outer retina that regulates photoreceptor–bipolar cell synaptic activity.

Melatonin – The neurohormone produced by the retina in the photoreceptor cells and in the pineal gland and whose production is increased at night under the control of circadian clocks.

Mesopic – An adjective that describes the dim ambient light levels between the scotopic and photopic ranges, such as those observed at dawn and dusk, to which both rods and cones can respond.

Photopic – An adjective that describes the bright ambient light levels, such as those observed during a sunny day, to which cones, but not rods, can respond.

Rod (cone) pathway – The ensemble of retinal cells (circuit) that sequentially relay the electrical signals through the retina from the rods (cones) to the ganglion cells.

Scotopic – An adjective that describes the very dim ambient light levels, such as those observed during a moonless night, to which rods, but not cones, which have been separated from the retina, can respond.

Synapse – A zone of contact between two cells that allows the transmission of an electrical signal from one cell to the other. Although a synapse may be chemical or electrical (gap junction), when used alone, the term usually refers to a chemical synapse.

Introduction

Most vertebrate retinas contain both rods and cones, photoreceptor cells that detect and transduce visual images into neural signals. These neural signals are then transmitted to bipolar cells, second-order neurons that then signal ganglion cells, the output neurons of the retina (**Figure 1**). As we live on a planet that rotates, the ambient or background illumination during a bright sunlit day, compared to a moonless night, changes by approximately 20-billion-fold on a daily basis. The retina must be able to function effectively during this dramatic daily environmental change. At least three adaptive mechanisms are thought to have evolved in the vertebrate retina to facilitate its day/night operation. First, because isolated rods and cones, which have been dissociated from the retina, display high and low light sensitivity, respectively, it has been accepted that rods and cones function under different illumination conditions, that is, that rods mediate dim light (scotopic) vision at night and cones mediate bright light (photopic) vision during the day. Second, external environmental factors, such as the level of ambient illumination itself, can modulate the light responses of retinal neurons and alter the operating characteristics of neural networks within the retina. For example, although ganglion cells exhibit a center-surround receptive field under light-adapted conditions, such as occurs in a bright sunlit day, their receptive fields exhibit only center responses if the retina is subsequently maintained in the dark for 30–40 min, such as occurs if an animal moves from a brightly illuminated area to the shade for a prolonged period. Third, intrinsic retinal processes, such as the circadian (24-h) clock or oscillator in the retina, can also modulate the light responses of retinal neurons and alter the operating characteristics of neural networks within the retina. Although it was originally thought that the transition between day vision and night vision is a passive process driven by the intensity of ambient illumination, it is

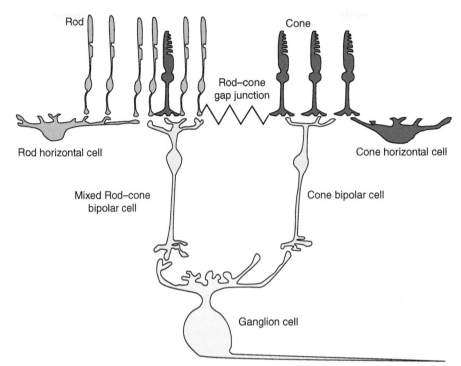

Figure 1   Rod and cone pathways in the goldfish retina. This schematic drawing shows retinal cell types and rod and cone pathways in the goldfish retina. In goldfish retina, as in all vertebrate retinas, rods and cones are anatomically connected or coupled by gap junctions, a type of electrical synapse. Both rods and cones synapse onto second-order neurons, the bipolar cells and horizontal cells, but with some degree of segregation. Mixed rod–cone bipolar cells in goldfish receive direct synaptic input from rods and cones. In the mammalian retina, a single class of rod bipolar cell makes synaptic contact exclusively with rods (not shown). In the goldfish, rod horizontal cells make synaptic contact only with rods and cone horizontal cells and cone bipolar cells make synaptic contact exclusively with cones. In goldfish, bipolar cells directly relay the photoreceptor light responses to ganglion cells, the output neurons of the retina. Individual ganglion cells in both fish and mammals can be driven by signals from both the rod and cone pathways. In mammals, however, rod bipolar cells do not directly synapse onto ganglion cells, but instead communicate rod signals to cone bipolar cells and then to ganglion cells through an interneuron, the AII amacrine cell (not shown).

now known that the retinal clock plays a key role in this mechanism.

A circadian clock is a biological oscillator that has persistent rhythmicity (i.e., a molecular process that rewinds itself) with a period of approximately 24 h under constant environmental conditions (e.g., constant darkness and temperature). Thus, day/night differences in biochemical, morphological, and physiological processes, which are observed under constant environmental conditions, can be attributed to the action of a circadian clock. In the vertebrate retina, many such day/night differences have been observed, including differences in neuronal light responses, dopamine and melatonin content and release, visual sensitivity, retinomotor movements, extracellular pH, photoreceptor disk shedding, and gene expression. A number of these day/night differences have been demonstrated to be under the control of the circadian clock in the retina, and not the circadian clocks located elsewhere in the brain (i.e., suprachiasmatic nucleus of the hypothalamus, pineal gland) or in peripheral organs (i.e., liver).

This article focuses on how the circadian clock in the retina regulates rod and cone pathways, especially concerning circadian regulation of rod–cone coupling.

Our review focuses primarily on findings that have been obtained from the fish retina, because this aspect of circadian function has been most thoroughly studied in the fish. However, the available evidence suggests that the circadian mechanisms, discussed in this article, occur in the vast majority of vertebrate species, including both mammals and nonmammals. These similarities will be noted when evidence from mammalian and other nonfish species is available. Information on how the retinal clock regulates other retinal processes is dealt with elsewhere in this encyclopedia and in other recent reviews and papers, which are listed under the section, titled 'Further reading'.

Rod and Cone Pathways in the Fish Retina

As shown in **Figure 1**, rod signals can reach ganglion cells through at least two separate pathways in all vertebrate species that have both rods and cones. First, rod input can reach ganglion cells through cones. In both mammalian and nonmammalian retinas, rods and cones are anatomically connected or coupled by gap junctions, a type of electrical synapse at which rod input can enter the cone circuit and thereby reach ganglion cells. Although it was thought that rod–cone electrical coupling is relatively weak, recent evidence has demonstrated that rod–cone coupling in both fish and mice is strong at night, but weak during the day due to the action of the retinal clock.

Second, rod input can reach ganglion cells through bipolar cell pathways that do not involve cones. Rods signal bipolar cells at chemical synapses in all vertebrates. In fish, these bipolar cells also receive synaptic contact from cones and, thus, are called mixed rod–cone bipolar cells (**Figure 1**). In contrast, in mammals, bipolar cells that receive rod input do not receive cone input. Individual ganglion cells in both fish and mammals can be driven by signals from both the rod and cone systems. In fish, mixed rod–cone bipolar cells synapse directly onto ganglion cells, whereas in mammals, rod bipolar cells do not directly synapse onto ganglion cells but instead provide indirect rod input to ganglion cells through AII amacrine cells, which then signal cone bipolar cells.

Day/Night Differences in the Light Responses of Neurons in the Fish Outer Retina

The first evidence that supported the idea that the retinal clock regulates rod and cone pathways by modulating the electrical synapses between rods and cones was obtained from electrical recordings of goldfish cone horizontal cells, second-order cells that receive synaptic contact from cones, but not from rods (**Figure 1**). Under dark-adapted conditions during the day, the hyperpolarizing light responses of these cells are cone driven (**Figure 2**). However, under dark-adapted conditions at night, the light responses of the cells are dominated by rod input. Specifically, the cells respond to light that is $100\times$ dimmer at night, than in the day. In other words, the threshold stimulus that evokes a response is in the low scotopic range (i.e., intensities to which isolated rods, but not isolated cones, respond) at night, but in the low mesopic range (i.e., the threshold intensity of isolated cones) in the day. In addition, the light responses of dark-adapted cone horizontal cells at night, but not in the day, have a time course that is similar to that of rod horizontal cells; they are slower and of longer duration, especially to brighter light stimuli. Additional support that the cells receive rod input at night, but not in the day, was the demonstration that the L-type (or H1) cone horizontal cells, which make synaptic contact with red (long-wavelength)-sensitive cones, are most sensitive to long-wavelength stimuli during the day, but are most sensitive to middle-wavelength stimuli at night, which is typical of rods. Interestingly, the light responses of rod horizontal cells, which make synaptic contact with rods, but not with cones (**Figure 1**), are similar in the day and night under dark-adapted conditions (**Figure 2**).

The day/night differences in the light responses of cone horizontal cells described above are due to the action

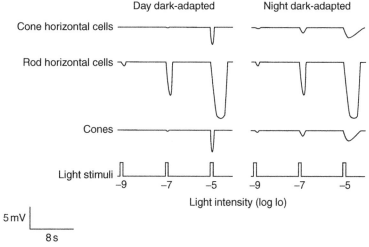

Figure 2 The light responses of goldfish cones and cone horizontal cells depend on the time of day. Schematic intensity-response series are shown for a cone horizontal cell (top trace), a rod horizontal cell (middle trace), and a cone (bottom trace) in the goldfish retina during the day under dark-adapted conditions (left column) and during the night under dark-adapted conditions (right column). At night, the rising and falling portions of the light responses of cones and cone horizontal cells are slower, response duration is longer than stimulus duration, and response threshold is approximately 2 log units lower, compared to daytime. These light-response characteristics are similar to those of rod horizontal cells (although the light responses of rod horizontal cells are larger at night), indicating the presence of significant rod input to cones and cone horizontal cells at night. In contrast, the low-threshold, slow rod-driven light responses of rod horizontal cells do not change between day and night. The light response traces were generated from the averaged response latency, time-to-peak, response duration and amplitude of the cells. Stimuli were full-field white light. Light intensity is relative to a standard (Io), with Io $= 2.0$ mW cm^{-2}.

of a circadian clock, and not the result of acute dark-adaptive or other environmental effects, because similar day/night differences are observed when the fish and/or their *in vitro* retinas are maintained in constant darkness and temperature for 24–72 h. In addition, prior reversal of the 12-h light/12-h dark cycle in which the fish are maintained, reverses the day/night differences in the light responses of the cone horizontal cells. As circadian clocks can be entrained or phase-shifted by the light/dark cycle, this finding demonstrates that the observed day/night differences in the light responses of L-type cone horizontal cells are circadian in nature, and not due to day/night differences in environmental factors.

More recently, electrical recordings of the light responses of goldfish cones in the day and night, following 24–72 h of constant darkness, revealed that they are also under circadian control. Specifically, the cone light response threshold is $100\times$ lower at night (i.e., low scotopic) than in the day (**Figure 2**). In addition, the light responses of dark-adapted cones at night, but not in the day, have a time course that is similar to that of rod horizontal cells; they are slower and of longer duration, especially to brighter light stimuli. Moreover, although recorded cones can be distinguished as either blue (short-wavelength)-sensitive, green (middle-wavelength)-sensitive, or red (long-wavelength)-sensitive during the day, all recorded cones at night are most sensitive to middle-wavelength light, and their sensitivity closely matches that of rods.

The circadian clock regulates the light responses of cones and cone horizontal cells in part through activation of the D_2 family of dopamine receptors. Dopamine receptors have been divided into the D_1 and D_2 receptor families based on their opposite effects on the level of intracellular cyclic adenosine monophosphate (cAMP) and the difference in their affinity for dopamine. D_1 receptors are approximately $500\times$ less sensitive to dopamine than D_2 receptors, and D_1 receptor activation increases cAMP, whereas D_2 receptor activation decreases it. Substantial evidence indicates that both the D_1 and D_2 families of receptors are found in the retina and that rods and cones express D_2 receptors, whereas rod and cone horizontal cells express D_1 receptors. Moreover, the retinal clock produces a circadian rhythm in dopamine release through melatonin. Specifically, the clock increases melatonin synthesis and release at night, compared to the day. As melatonin inhibits dopamine release, a circadian rhythm in dopamine release is generated, but one in which extracellular dopamine levels are higher in the day, than at night (i.e., the dopamine and melatonin rhythms are in antiphase).

The retinal clock produces a variety of circadian rhythms in the retina, including daily rhythms in retinomotor movements and cyclic guanosine monophosphate (cGMP)-gated cationic channels in cones, by increasing dopamine levels in the day so that D_2 receptors are activated. The clock does not appear to utilize D_1 receptors, apparently because it does not increase dopamine levels sufficiently to activate the low-affinity D_1 receptors in the retina. Similarly, the retinal clock modulates the strength of rod input to cones and cone horizontal cells by increasing the activation of the D_2 receptors on rods and cones during the day. When D_2 receptor activation is minimal at night, as evidenced by the lack of effect of D_2 receptor antagonists at night, rod input dominates the light responses of cones and cone horizontal cells. In contrast, when the retinal clock activates D_2 receptors in the day, rod signals do not reach cones and cone horizontal cells, as evidenced by the finding that application of D_2 receptor antagonists during the day increases rod input to cones and cone horizontal cells. Indirect evidence based on data obtained from cone horizontal cells further suggests that the decrease in intracellular cAMP in rods and cones that is evoked by D_2 receptor activation in the day eliminates rod input to cones and cone horizontal cells. Although this idea requires further direct testing on cones, a circadian rhythm of cAMP content in retinal photoreceptors, with high levels at night, has been described in several vertebrate species.

The Circadian Clock in the Retina, and Not the Retinal Response to the Ambient Illumination, Controls Rod–Cone Coupling

The day/night differences in the strength of the rod input to cones and cone horizontal cells originate in part from changes in the conductance of rod–cone gap junctions. Following the injection of a membrane-impermeant, gap-junction-permeant, biotinylated tracer (biocytin) during the day, the tracer accumulates in the injected goldfish cone. However, when the experiment is conducted at night, the tracer diffuses to many cones and rods in the vicinity of the injected cell, consistent with an increase in the conductance of the rod–cone gap junctions. The tracer experiments demonstrate that the modulation of the rod–cone gap-junctional conductance is the primary means through which the retinal clock controls the strength of the rod signal that flows into cones and cone horizontal cells (**Figure 3**).

Consistent with the effects of the clock on the light responses of cones and cone horizontal cells described above, the clock uses dopamine to control the extent of rod–cone tracer coupling. Specifically, the clock-controlled daytime increase in dopamine decreases rod–cone tracer coupling, whereas the nighttime drop in dopamine levels is required to increase the coupling. The day/night difference in rod–cone coupling can be pharmacologically manipulated and the results are consistent with the involvement of D_2, and not D_1, dopamine receptors. Thus, by increasing dopamine release and D_2 receptor activation during the day, the retinal clock decreases rod–cone coupling and

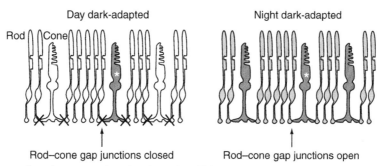

Figure 3 Day/night changes in rod–cone gap-junctional coupling. This drawing illustrates that following injection of the biotinylated tracer biocytin into a single goldfish cone (indicated by an asterisk) during the day (left panel), the tracer remains restricted to the injected cone (filled cone). In contrast, injection of biocytin into a single cone at night (right panel) leads to tracer staining of numerous rods and cones (filled cells) in the vicinity of the injected cell. Since biocytin is a membrane-impermeant molecule that can diffuse through gap junctions, these observations demonstrate that the rod–cone gap junctions are closed during the day and open at night. A similar day/night difference in rod–cone tracer coupling occurs in the mouse.

thereby decreases rod input to cones and cone horizontal cells. Conversely, the nighttime decrease in dopamine release enhances rod–cone coupling and increases rod input to cones and cone horizontal cells.

Interestingly, the effects of the clock on rod–cone tracer coupling and on cone light responses are not altered when dim background lights are present. More specifically, similar results are obtained in the day and night, with and without dopamine D_2 receptors blocked, when the intensity of the ambient illumination is very dim (i.e., low scotopic), as occurs on a moonless night, or is brighter, but still dim (i.e., mesopic), as occurs on a moonlit night. Most significantly, at night, the extent of rod–cone tracer coupling and the strength of rod input to cones and cone horizontal cells are not decreased even when the intensity of the ambient illumination is in the mesopic range. As the background light level at night normally varies between very dim starlight and dim moonlight conditions, these results indicate that the retinal clock, and not the retinal response to the normal visual environment at night, regulates rod–cone coupling. The clock decreases rod–cone coupling at dawn by increasing D_2 receptor activation and increases rod–cone coupling at dusk by reducing D_2 receptor activation.

The Circadian Clock in the Mammalian Retina Controls Rod–Cone Coupling

Although it may be thought that circadian regulation of retinal function occurs in nonmammals, and not in mammals, substantial evidence indicates that the circadian clock in the mammalian retina regulates a variety of cellular phenomena, including melatonin and dopamine production and release, neuronal activity, rod disk shedding, extracellular pH, and rod–cone coupling. Recent findings, using bath application of the gap-junction-permeant, tracer molecule neurobiotin onto mouse retinas that had been cut in

pieces with a razor, indicated that neurobiotin diffused extensively through photoreceptor cell gap junctions under dark-adapted conditions at night and in the day when D_2 receptors were blocked, but diffused significantly less under dark-adapted conditions in the day. Thus, it seems likely that the retinal clock controls the strength of rod–cone coupling in most, if not all, mammalian and nonmammalian retinas that have both rods and cones. Moreover, in most vertebrate species, including mammals, the fact that (1) rods and cones are connected by gap junctions, (2) D_2 receptors are expressed by rods and cones, but not by horizontal cells, and (3) the retina contains a circadian clock supports this view. Although the clock increases the conductance of rod–cone gap junctions at night, evidence to date does not address whether at night the clock also increases the conductance of cone–cone and/or rod–rod gap junctions, which are also found in vertebrates.

A Circadian Clock Pathway in the Retina

The circadian clock in the fish retina regulates the light responses of cones and cone horizontal cells in part through a melatonin/dopamine pathway that controls rod–cone coupling (**Figure 4**). Specifically, the clock regulates melatonin synthesis, so that melatonin synthesis and release are kept low during the day and dramatically increased at night. Melatonin inhibits dopamine release from dopaminergic interplexiform cells, and consequently, the extracellular levels of dopamine are the lowest at night. The relief in the inhibition of dopamine release during the day generates an increase in the extracellular levels of dopamine in the day, so that the D_2 receptors on photoreceptor cells are activated, which then lowers intracellular cAMP and protein kinase A (PKA) levels in the photoreceptors, decreasing the conductance of rod–cone gap junctions. As a consequence, rod input to cones and cone horizontal cells is decreased. At night, because the clock decreases

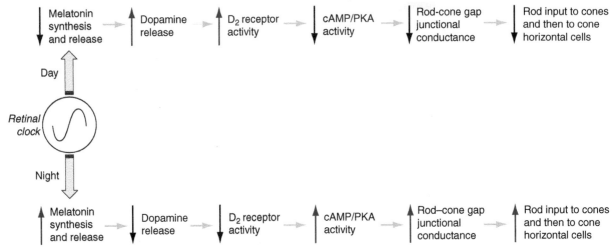

Figure 4 A retinal circadian clock pathway that controls rod–cone coupling. The circadian clock in the retina utilizes melatonin and dopamine to control rod–cone coupling. Specifically, the clock exerts transcriptional and post-translational control of melatonin synthesis, so that melatonin synthesis and release are kept low during the day and dramatically increased at night. Melatonin inhibits dopamine release and consequently extracellular levels of dopamine are the lowest at night. The relief in the inhibition of dopamine release during the day generates an increase in the extracellular levels of dopamine in the day so that the D_2 receptors on photoreceptor cells are activated, which then lowers intracellular cAMP and protein kinase A (PKA) levels in photoreceptors, decreasing the conductance of rod–cone gap junctions. As a consequence, rod input to cones and cone horizontal cells is decreased. At night, because the clock decreases dopamine levels below the threshold of D_2 receptor activation, the intracellular cAMP level in photoreceptor cells increases, raising the conductance of rod–cone gap junctions and increasing rod input to cones. As a result, rod signals are transmitted to cone horizontal cells, even though these cells do not make direct synaptic contact with rods. In the daytime and nighttime sequences of events that are illustrated, upward pointing red arrows and downward pointing black arrows indicate increases and decreases, respectively, in the events with which they are associated. Although the pathway illustrated here has been established from data collected in goldfish, accumulative evidence indicates that this pathway is likely conserved among vertebrates.

dopamine levels below the threshold of D_2 receptor activation, the intracellular cAMP level in photoreceptor cells increases, raising the conductance of rod–cone gap junctions and increasing rod input to cones. As a result, rod signals are transmitted to cone horizontal cells, even though these cells do not make direct synaptic contact with rods. Although the circadian pathway illustrated here has been established from data collected in goldfish, accumulative evidence indicates that this pathway is likely conserved among vertebrates, including mammals. However, it is still unresolved as to whether the clock in the mammalian retina increases dopamine release in the day by generating a melatonin rhythm, as occurs in nonmammals, or by direct control of dopamine metabolism in dopaminergic cells.

It is interesting to note that circadian clock pathways in the retina may be different from light-responsive and light/dark-adaptive pathways. As described above, the retinal clock, and not the retinal response to the level of ambient illumination, uses specific neurotransmitters (i.e., melatonin), neurotransmitter receptors (i.e., melatonin receptors, dopamine D_2 receptors), and synaptic processes (i.e., rod–cone gap junctions) to control the strength of rod–cone electrical coupling, These physiological processes, which participate primarily in circadian clock pathways within the retina, may be distinct from other processes that primarily respond to light stimuli, such as those that mediate bright light adaptation (i.e., D_1 receptors). In the case of dopamine

receptors, this segregation of pathways may be explained by the difference in the affinity of D_1 and D_2 receptors for endogenous dopamine. Although the retinal clock increases extracellular dopamine levels sufficiently to activate the high-affinity D_2 receptors on rods and cones, the low-affinity D_1 receptors on horizontal cells are not activated, as evidenced by the absence of a day/night difference in horizontal-cell gap-junctional coupling under dark-adapted conditions. Instead, horizontal cell coupling, which is modulated by D_1 receptor activation, is decreased by bright light stimulation during the day, which increases extracellular dopamine levels sufficiently to activate the D_1 receptors on horizontal cells. Viewed from this perspective, the D_1 and D_2 receptor systems in the retina function in a complementary manner; the retinal clock activates D_2 receptors at dawn and reduces their activation at dusk, whereas bright lights activate D_1 receptors during the day. Retinal function (neurotransmitters, synapses, pathways, adaptation, etc.) may therefore arise from the interplay of both light-responsive and circadian clock pathways.

Functional Implications of Circadian Clock Control of Rod–Cone Coupling

As the retinal clock increases the strength of the electrical synapses between rods and cones at night, very dim light signals from rods can reach cones and then cone horizontal cells at night. Thus, although individual cones that have been separated from the intact retina cannot respond to very dim light (i.e., low scotopic) stimuli, dark-adapted cones in the intact retina at night can do so because the clock opens the electrical synapse between rods and cones at night. Circadian control of rod–cone electrical coupling therefore serves as a synaptic switch for the direct introduction of rod signals to cones and cone pathways at night, but not in the day.

In addition to enabling rod signals to reach cones at night, the circadian increase in rod–cone electrical coupling may also enhance the detection of very dim, large light stimuli by the rod to bipolar cell to ganglion cell circuit. Many rods converge onto each bipolar cell, summing visual signals over a large spatial area at synapses that are highly nonlinear. Although noise in one photoreceptor cell is independent of the noise in nearby photoreceptor cells, dim, large visual objects will produce similar or correlated responses from nearby photoreceptor cells. As a result, an increase in electrical coupling between nearby photoreceptor cells will decrease photoreceptor noise more than it reduces their light responses to dim, large objects. Increased photoreceptor coupling at night will therefore augment the signal-to-noise ratio and the reliability of rod responses to dim, large stimuli before the nonlinear rod to bipolar cell synapse distorts the signal and the noise. Circadian control of rod–cone electrical coupling thus enhances the detection of very dim, large objects at night, and by decreasing rod–cone coupling at dawn, improves the detection of small objects in the day. The absence of rod signals in the cone pathways during the day facilitates the processing of high acuity and color information by the cone pathways during the day. In contrast, the increase in rod–cone coupling at night may maximize nighttime vision and tune the retina to detect large, dim objects. Moreover, circadian control of rod–cone coupling may also mediate in part the circadian rhythm in visual sensitivity that occurs in many vertebrates, including fish and human.

Finally, the nighttime increase in rod–cone coupling may influence photoreceptor survival. Specifically, the metabolic exchange of small signaling molecules and nutrients will likely occur between rods and cones each night because open gap-junctional channels are large enough to allow the diffusion of such small molecules between coupled cells. Healthy rods might improve cone survival by providing coupled cones with nutrients and protective factors at night and/or dying rods might facilitate the death of coupled cones through the diffusion of pro-apoptotic factors each night.

See also: Anatomically Separate Rod and Cone Signaling Pathways; Circadian Metabolism in the Chick Retina; Circadian Regulation of Ion Channels in Photoreceptors; Fish Retinomotor Movements; *Limulus* Eyes and Their Circadian Regulation; Morphology of Interneurons: Amacrine Cells; Morphology of Interneurons: Bipolar Cells; Morphology of Interneurons: Horizontal Cells; Morphology of Interneurons: Interplexiform Cells; Neurotransmitters and Receptors: Dopamine Receptors; Neurotransmitters and Receptors: Melatonin Receptors; The Physiology of Photoreceptor Synapses and Other Ribbon Synapses.

Further Reading

Barlow, R. B. (2001). Circadian and efferent modulation of visual sensitivity. *Progress in Brain Research* 131: 487–503.

Bloomfield, S. A. and Dacheux, R. F. (2001). Rod vision: Pathways and processing in the mammalian retina. *Progress in Retinal and Eye Research* 20: 351–384.

Copenhagen, D. R. (2004). Excitation in the retina: The flow, filtering, and molecules of visual signaling in the glutamatergic pathways from photoreceptors to ganglion cells. In: Chalupa, L. M. and Werner, J. S. (eds.) *The Visual Neurosciences*, pp. 320–333. Cambridge, MA: MIT Press.

Dowling, J. E. (1987). *The Retina, an Approachable Part of the Brain.* Cambridge, MA: Harvard University Press.

Green, C. B. and Besharse, J. C. (2004). Retinal circadian clocks and control of retinal physiology. *Journal of Biological Rhythms* 19: 91–102.

Iuvone, P. M., Tosini, G., Pozdeyev, N., et al. (2005). Circadian clocks, clock networks, arylalkylamine *N*-acetyltransferase, and melatonin in the retina. *Progress in Retinal and Eye Research* 24: 433–456.

Raviola, E. and Gilula, N. B. (1973). Gap junctions between photoreceptor cells in the vertebrate retina. *Proceedings of the National Academy of Sciences of the United States of America* 70: 1677–1681.

Ribelayga, C. and Mangel, S. C. (2003). Absence of circadian clock regulation of horizontal cell gap junctional coupling reveals two dopamine systems in the goldfish retina. *Journal of Comparative Neurology* 467: 243–253.

Ribelayga, C., Cao, Y., and Mangel, S. C. (2008). The circadian clock in the retina controls rod–cone coupling. *Neuron* 59: 790–801.

Ribelayga, C., Wang, Y., and Mangel, S. C. (2002). Dopamine mediates circadian clock regulation of rod and cone input to fish retinal horizontal cells. *Journal of Physiology (London)* 544: 801–816.

Ribelayga, C., Wang, Y., and Mangel, S. C. (2004). A circadian clock in the fish retina regulates dopamine release via activation of melatonin receptors. *Journal of Physiology (London)* 554: 467–482.

Tessier-Lavigne, M. and Attwell, D. (1988). The effect of photoreceptor coupling and synapse nonlinearity on signal:noise ratio in early visual processing. *Proceedings of the Royal Society of London, Series B* 234: 171–197.

Wang, Y. and Mangel, S. C. (1996). A circadian clock regulates rod and cone input to fish retinal cone horizontal cells. *Proceedings of the National Academy of Sciences of the United States of America* 93: 4655–4660.

Warrant, E. J. (1999). Seeing better at night: Life style, eye design and the optimum strategy of spatial and temporal summation. *Vision Research* 39: 1611–1630.

Witkovsky, P. (2004). Dopamine and retinal function. *Documenta Ophthalmologica* 108: 17–40.

Circadian Photoreception

I Provencio, University of Virginia, Charlottesville, VA, USA

Glossary

Circadian rhythm – An endogenously generated biological rhythm with a period of about 24 h.
ipRGC – The term stands for intrinsically photosensitive retinal ganglion cell, which is a small subset of retinal ganglion cells that are rendered light sensitive because they express melanopsin, implicated in nonvisual photoresponses.
Melanopsin – The opsin-based photopigment of intrinsically photosensitive RGCs.
Nonvisual photoresponses – Physiological or behavioral responses to light that do not require the formation of images. Circadian photoentrainment is an example of a nonvisual photoresponse.
Photoentrainment – The synchronization of circadian rhythms to the daily light:dark cycle.

Introduction

Circadian rhythms are biological rhythms that exhibit a period of about 24 h (Latin, *circa* around + *dies* day). They persist in an environment devoid of time cues (*Zeitgebers*; German, *Zeit* time + *Geber* giver). This persistence of rhythmicity in constant conditions indicates the presence of an internal circadian clock. Circadian clocks are ubiquitous, existing in organisms ranging from cyanobacteria to humans. In metazoans, many tissues are capable of autonomous circadian rhythmicity, however, the phases of such rhythms tend to be orchestrated by a master pacemaker. In mammals, this master circadian pacemaker resides in the hypothalamic suprachiasmatic nuclei (SCN), two bilateral structures that straddle the midline immediately dorsal to the optic chiasm and are separated from each other by the third ventricle. The SCN coordinate the phases of multiple oscillators located in peripheral tissues, including heart, liver, and lung. The intergeniculate leaflet (IGL) of the lateral geniculate complex is also considered a component of the circadian system and receives input from both of the eyes. The IGL has been proposed to function in assessing ambient illumination levels.

While the periods of circadian rhythms are about 24 h, rarely are they exactly 24 h. Similar to a timepiece that runs too slowly or too quickly, the utility of a circadian clock is dependent on occasional resetting of its phase. The primary resetting agent for most circadian systems is light. Daily fluctuation in ambient irradiance resulting from the Earth's rotation about its axis is the most predictable diurnally variable feature of the environment. Organisms have evolved photoreceptive mechanisms to communicate this most reliable of *Zeitgebers*, the day:night cycle, to the circadian machinery so that circadian phase can be synchronized (entrained) with the astronomical day.

The SCN receives retinal input via the retinohypothalamic tract which is comprised of the myelinated axons of retinal ganglion cells (RGCs) (**Figure 1**). The response of the SCN to phase-shifting stimuli varies throughout the course of the circadian day and is phase-dependent (**Figure 2**). For photic stimuli, illumination during an animal's early subjective night results in a phase delay of the circadian clock as measured through the recording of circadian locomotor activity rhythms. By contrast, light pulses administered during the late subjective night cause activity rhythms to be phase advanced. Illumination during the subjective day has a negligible effect on circadian phase. In addition to this phase dependence, the magnitude of the light-induced phase shifts also varies as a function of the duration, intensity, and wavelength composition (color) of the light pulse. Conclusive identification of the photoreceptors that mediate circadian phase shifting has been wrought with problems but significant strides have been made recently.

Nonvisual Photoreception

Nonmammalian vertebrates possess numerous extraocular photoreceptors, some of which are necessary to shift the phase of the circadian system, so it is in proper phase alignment with the day:night cycle. Amphibians, for example, have photoreceptors in the pineal gland, the frontal organ, paraventricular zones of the brain, iris, and skin, in addition to the classical rod and cone photoreceptors of the retina. By contrast, all mammalian photoreception is restricted to the eyes.

One notable exception was a report in humans claiming that blue-light illumination of the popliteal region behind the knee caused shifts in the phase of salivary melatonin and temperature rhythms. This surprising result, which suggested the presence of extraocular photoreception in the mammals, inspired others to investigate the possibility of extraocular photoreception in other animal models. One group found no effect of direct sunlight on bilaterally

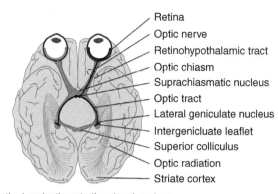

Retina
Optic nerve
Retinohypothalamic tract
Optic chiasm
Suprachiasmatic nucleus
Optic tract
Lateral geniculate nucleus
Intergenicluate leaflet
Superior colliculus
Optic radiation
Striate cortex

Retinal projections to the visual system
Retinal projections to the circadian system

Figure 1 Retinal projections to the visual and circadian systems. Projections to the visual system are shown in green and those to the circadian system are shown in blue.

enucleated golden hamsters whose backs were shaved to maximize skin exposure. In these animals, locomotor circadian rhythms and levels of pineal melatonin were unperturbed by sunlight exposure during the animals' subjective night. Intact control animals, however, showed dramatic phase shifts in activity rhythms and suppression of pineal melatonin levels. Several other groups using human subjects failed to replicate the initial observation in humans indicating that some uncontrolled, nonphotic aspect of the experimental paradigm in the original study caused the observed effects on the circadian system. It is now widely accepted that the anatomical site of mammalian photoreception is exclusively ocular.

Physiological responses to light may be classified as visual or nonvisual. The responses that require the construction of images are considered visual, while those responses that simply require detection of changes in irradiance, such as circadian photoentrainment, may be considered nonvisual. The mammalian retina, the primary focus of this article, subserves both visual and nonvisual photoreception.

Due to the anatomical accessibility of the eye and the profound quality-of-life deficits experienced by the blind, the retina has been one of the most intensely studied tissues. Additionally, the laminar organization of the retina has made it an exquisite model for understanding basic concepts of development and neural communication. Retinal cytoarchitecture has been extensively described for more than 150 years. In fact, the rod and cone cells were postulated to be the light-sensing elements of the retina, as early as the mid-1850s by the German anatomist and physiologist Heinrich Müller.

In view of this long history of study, it was reasonable to suspect that rods and cones mediated both visual and nonvisual photoreception because no other classes of photoreceptors were known to exist in the mammalian eye. However, several lines of evidence suggested the existence of a class of ocular photoreceptor that was neither rod nor cone. In 1927, Clyde Keeler recognized that blind mice lacking photoreceptor cells continued to exhibit a pupillary light reflex, although this response was slower and less sensitive than that of fully sighted mice. Seventy-four years later, it was confirmed that the pupillary light reflex does indeed persist in blind mice and is maximally sensitive to blue wavelengths. Phase shifting of the circadian clock in response to light pulses is also maintained in mice homozygous for the retinal degeneration 1 allele ($Pde6b^{rd1/rd1}$), resulting in the loss of rods and cones. Similar to the pupillary response, light-induced circadian phase shifting in these visually blind mice exhibits a peak spectral sensitivity in the blue wavelengths. Finally, elevated nighttime levels of serum melatonin can be acutely suppressed by blue light presented to the eyes.

All of these blue-sensitive responses exhibit a peak spectral sensitivity in the 460–480 nm wavelength range, a domain of the electromagnetic spectrum that does not coincide with the peak sensitivity of known rod and cone photopigments of the human or murine retina. Since photoreceptors are defined by their photopigments, which in turn dictate their spectral sensitivity, a common strategy for identifying new photoreceptors is to identify new photopigments which should eventually point to the identification of a new photoreceptor class. The accumulating evidence for novel photoreceptors that mediate nonvisual responses to light led to the discovery of many new opsin-based photopigments. First among these was pinopsin, which was originally identified in the pineal gland of chicken and whose function remains unknown, but most likely plays a role in the photoregulation of melatonin synthesis. Very ancient opsin (VA-opsin), parapinopsin, and parietopsin were subsequently identified in the eye, parapineal, and parietal eye, respectively, of nonmammalian vertebrates. Based on an inspection of completed genomes, it is unlikely that mammalian homologs of these opsins exist.

However, within the last decade new opsins were indeed revealed in the mammals, none of which were localized to rods or cones. Peropsin and retinal G-protein-coupled receptor (RGR) opsin are expressed in the retinal pigment epithelium and are likely to function as accessory photoisomerases which use photic energy to convert all-*trans*-retinoids to their respective 11-*cis* isomers. 11-*cis*-Retinal is the requisite chromophore of all vertebrate signaling opsins. Encephalopsin was originally identified only in the brain and spinal cord of mouse, sites not known to be inherently photoreceptive. A subsequent study claimed the presence of encephalopsin message in the eye, however, the role of encephalopsin in the central nervous system remains unknown. Opn5 is another opsin predicted to exist within the mammalian eye based on the presence of messenger ribonucleic acid (mRNA), although a translated gene product remains to be discovered (**Table 1**).

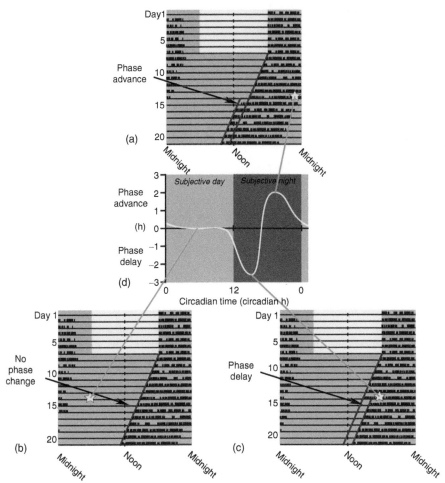

Figure 2 Phase response curve. Actograms (a, b, c) of daily activity obtained in running wheel cages containing a single mouse. Each horizontal line represents one day. Line thickenings indicate bouts of activity. Animals are maintained in a 12-h light:12-h dark cycle for the first week (gray background, dark; yellow background, light) and then maintained in constant darkness for the remainder of the experiment (gray background). After approximately 1 week in constant darkness, a standardized light pulse is administered (star symbol) and the phase shift of the onsets of activity on the days following the pulse are determined. The magnitude and direction (delay or advance) of the phase shift in response to a standardized light pulse varies as a function of the circadian phase of the circadian cycle during which the pulse is given. This relationship can be plotted as a phase:response curve (d). In general, light exposure during the early subjective night elicits phase delays, while illumination during the late subjective night causes phase advances. Pulses administered during the subjective day do not shift the clock.

Melanopsin and the Mammalian ipRGC

Among the nonvisual opsins, melanopsin has been the most extensively studied photopigment. It is expressed in a small subset of RGCs that project to sites in the brain not involved in the formation of images. Originally cloned from the photosensitive dermal melanophores of *Xenopus laevis*, homologs were subsequently identified in chicken and other nonmammalian vertebrates. A second melanopsin gene was recognized in the chicken genome and is also present in other vertebrate classes. However, based on chromosome synteny, it is clear that one of these two genes has been lost in mammals.

In the nonmammalian vertebrates, melanopsin is expressed in the retina and extraretinal tissues, including the iris, brain, pineal, and skin. Many of these sites had been shown previously to be inherently light sensitive, although the photopigments mediating the sensitivity were unidentified. Unlike the broad anatomical distribution observed in nonmammalian vertebrates, in mammals, melanopsin expression is restricted to less than 2% of the RGCs. Notably, extraretinal expression has not been observed in any mammal.

Pituitary adenylyl cyclase-activating peptide and glutamate are coexpressed within the melanopsin-containing cells. Melanopsin is what confers photosensitivity upon these intrinsically photosensitive retinal ganglion cells (ipRGCs). Loss of melanopsin renders these cells nonresponsive to light.

Murine ipRGCs elaborate two or three primary dendrites which bifurcate within 50 μm of the perikaryon. The

Table 1 Nonvisual opsins of mammals

Opsin	Gene name	Tissue distribution	Putative function
Peropsin	*rrh*	rpe	Photoisomerase
Retinal G protein-coupled receptor	*rgr*	rpe	Photoisomerase
Melanopsin	*opn4*	Retina	Nonvisual photoreception
gpr136/ neuropsin	*opn5*	Eye, brain, testis, and spinal cord	Unknown
Encephalopsin/ panopsin	*opn3*	Brain, testis	Unknown

dendrites ramify within the S1 (OFF layer) or S5 (ON layer) sublaminae of the inner plexiform layer. A small fraction of ipRGCs may have dendrites arborizing in both S1 and S5. Dendrites are studded with varicosities giving a rosary bead appearance. Melanopsin protein is found throughout the plasma membrane of the cell body, the segment of the axon contained within the eye, and the dendritic arbor. Arbors are sparse, regularly tiled across the retina with substantial overlap, and rather large, having mean field diameters of around 450 μm in mice and 600 μm in rats. The receptive field of ipRGCs corresponds to the dimension of the dendritic field, indicating that the entire arbor is capable of initiating phototransduction. Membrane density of melanopsin is about 10 000-fold lower than that of rods and cones. This is largely attributable to the lack of a specialized light harvesting organelle in ipRGCs comparable to the photopigment-dense outer segments of rods and cones. This relative paucity of photopigment in ipRGCs results in a very low photon-capture efficiency, thereby rendering this system effective only in very bright irradiances. Nevertheless, ipRGCs, similar to rods, can signal the absorption of single photons. The spectral sensitivity of dark-adapted melanopsin peaks in the blue wavelengths and coincides with the action spectra of many of the nonvisual responses previously described, including circadian photoentrainment and the pupillary light reflex of visually blind mice that lack rods and cones.

A comparison of the melanopsin peptide sequence against known opsins indicates that it more closely resembles the rhabdomeric opsins of invertebrates rather than the ciliary opsins of vertebrates. Not surprisingly, melanopsin employs a phototransduction cascade more similar to that of insect rhabdomeres rather than vertebrate outer segments. Illuminated amphibian melanophores darken and exhibit a light-stimulated increase in inositol trisphosphate and protein kinase C (PKC)-dependent phosphorylation. Presumably, these effects are mediated by a melanopsin-initiated cascade because overexpression of melanopsin

hypersensitizes melanophores to light. Inhibition of phospholipase C (PLC), PKC, or chelation of intracellular calcium blocks melanophore darkening suggesting that this light-initiated response must activate a phosphoinositide signaling cascade. Heterologous expression of melanopsin in HEK293 cells renders them light sensitive and leads to a light-dependent increase in intracellular calcium and depolarization of membrane potential. Depolarization can be blocked by pharmacologic inhibitors of the $G\alpha_{(q/11)}$ subunit of guanine nucleotide-binding protein or PLC. These lines of evidence again suggest that light-activated melanopsin triggers a phosphoinositide signaling cascade similar to that of rhabdomeric photoreceptors. These same transduction components are found within ipRGCs and are critical for eliciting photoresponses. Furthermore, responses to light can be elicited in excised, inside-out patches of ipRGC membrane indicating that the critical signaling molecules are closely associated or are within the plasma membrane. A hallmark of rhabdomeric opsins is the formation of a stable red-shifted metastate in response to illumination. Whether melanopsin forms such a state remains equivocal, although some dark-adapted responses to blue light appear to be primed by preexposure to red light, which presumably photoconverts a red-sensitive metastate back to the blue-sensitive dark-adapted state. The general similarity to rhabdomeric signaling indicates that ipRGCs and invertebrate photoreceptors may have shared a common evolutionary ancestor.

ipRGCs project to central sites not associated with vision but implicated in regulating different forms of nonvisual photophysiology. For example, the SCN of the hypothalamus, the site of the primary circadian pacemaker, receives a robust input from ipRGCs. By contrast, the classical, nonphotosensitive RGCs provide minimal innervation to the SCN. Photoreceptor-mediated effects on the circadian axis occur exclusively through ipRGCs, suggesting that classical retinal rod and cone circuits impinge upon ipRGCs at the inner plexiform layer. Other principle central targets of ipRGCs include the IGL, the olivary pretectal nucleus, and the lateral habenula. The subparaventricular zone, the ventrolateral preoptic nucleus, and the ventral lateral geniculate nucleus are secondary, less intensely innervated targets.

Functions of ipRGCs

The role of ipRGCs in mediating nonvisual responses to light has been largely elucidated through the use of genetic mouse models. Mice null for melanopsin exhibit rather modest phenotypes. The magnitude of light-induced phase shifts of circadian locomotor activity rhythms is attenuated in melanopsin knock-out mice to about 40% of that observed in wild-type sibling controls (**Figure 3(a)**). The lengthening of circadian period observed in wild-type

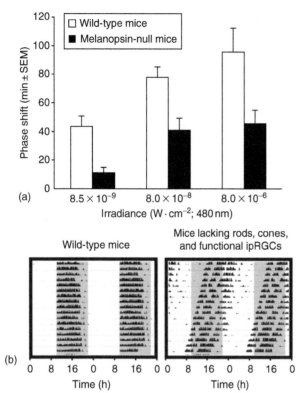

(a)

(b)

Figure 3 Circadian responses to light in retinal mutants. (a) Irradiance-dependent circadian phase shifting in wild-type and melanopsin-null mice. Mice lacking melanopsin were deficient in light-induced circadian phase shifting at all irradiances tested. (b) Melanopsin-null mice also lacking rods and cones fail to show synchronization (entrainment) of circadian locomotor activity to the light:dark cycle; white is the time that lights are on. Actograms are double plotted (i.e., horizontal lines correspond to 48 h) and darkness is indicated by a gray background. By contrast, the wild-type mouse is well entrained.

mice that are maintained in constant light is also diminished in melanopsin knockout mice. However, entrainment to standard light:dark cycles appears to be unaffected in mice lacking melanopsin. Other nonvisual responses such as the pupillary light response, the acute inhibition of activity by light (masking), and the suppression of the melatonin biosynthetic pathway are unaffected in melanopsin knockout mice. Interestingly, many of these same responses show no deficits in murine genetic models lacking rods and cones. Combined, these results suggest some level of redundancy between the rod/cone system and ipRGCs. To test this possibility, melanopsin-null mice have been crossed with mice either lacking functional rods and cones or lacking rods and cones all together. These animals exhibit no responses to light (**Figure 3(b)**). Essentially, they behave as though they have been bilaterally enucleated.

Many of the retinorecipient structures such as the SCN that mediate nonvisual responses to light receive negligible, if any, input from rod and cone pathways.

However, melanopsin-knockout mice with light-insensitive ipRGCs continue to exhibit some degree of light-driven nonvisual responses. This raises the possibility that in addition to their inherent photoreceptive capacity, ipRGCs may also serve to convey light signals initially detected by rods and/or cones to brain sites such as the SCN. Indeed, specific ablation of ipRGCs mimics the previously mentioned results obtained in animals lacking functional rods, cones, and ipRGCs. These results indicate that ipRGCs, while intrinsically light sensitive, also transmit information generated by the classical photoreceptors.

Apart from its role in entraining circadian rhythms, the eye is a circadian oscillator itself. This was first demonstrated in the isolated eyes of gastropods, which show a circadian rhythm in the firing frequency of compound action potentials. Shedding of disks from the distal tips of rods is also circadian and persists in isolated amphibian and mammalian eyes whose optic nerves have been severed. Teleost eyes exhibit circadian rhythms in retinomotor movements of photoreceptors and the migration of melanosomes within cells of the retinal pigment epithelium. Syntheses of melatonin and dopamine are perhaps the best-studied circadian outputs of the eye. These amines are synthesized in antiphase; melatonin levels peak in the night and dopamine levels peak during the day. This phase relationship is maintained in constant darkness indicating that these rhythms are indeed circadian. However, in the absence of melatonin, dopamine is only acutely synthesized in response to light and is no longer produced in a circadian fashion.

Determining which ocular cell types harbor the circadian clock has proved to be difficult. Isolated amphibian eyecups continue to produce melatonin rhythmically, even when the vast majority of inner retinal neurons are lesioned. These findings suggest that the photoreceptors contain a competent clock and the biosynthetic machinery to make melatonin. Additionally, action spectral analyses in the amphibian retina implicate the principal green-sensitive rods as critical to the acute light-mediated regulation of ocular melatonin synthesis. Whether these cells are the actual site of the amphibian retinal clock remains to be determined.

The mammalian eye also contains a circadian clock. Cultured hamster retinas maintained in constant darkness produce melatonin rhythmically with a period near 24 h. Retinas from *tau* hamsters whose period of locomotor rhythms are significantly shorter than those of wild-types also exhibit rhythms in retinal melatonin with periods that parallel the shortened activity cycles. Importantly, these rhythms can be reset by pulses of light demonstrating that similar to the amphibian eye, the mammalian eye is an autonomous circadian system with a circadian pacemaker that drives a measurable output, which can be reset if it comes out of phase with the light:dark cycle.

In summary, the mammalian eye has a dual function. Similar to the ear which supplies auditory and vestibular input to the brain, the eye provides visual input for the formation and interpretation of images and nonvisual input for the regulation of a myriad of light-regulated physiology such as the entrainment of circadian rhythms. The primary central circadian pacemaker within the SCN is entrained by light, the most reliable of external, daily time cues. The photoreceptors mediating photoentrainment include the classical photoreceptors (rods and/or cones), known primarily for their role in vision, and unique RGCs, which are inherently light sensitive because they express the photopigment melanopsin. The relative insensitivity of ipRGCs suggests that the classical photoreceptors mediate photoentrainment at low light levels, while ipRGCs are responsible for entraining the SCN at higher irradiances. However, even at low light levels, photoreceptive signals transmitted through rod or cone pathways reach sites of the brain involved in non-visual photoresponses via the ipRGCs.

The presence of an ocular photoreceptive system that is anatomically distinct from the rods and cones raises the possibility that individuals suffering from blindness due to photoreceptor loss may retain a full healthy complement of ipRGCs. Some blind individuals do indeed lack cognitive vision but maintain an ability to regulate pineal melatonin and remain synchronized to the prevailing day:night cycle. The high incidence of ocular infection in the blind drives many clinicians to replace the eyes with prosthetics. Care should be taken to consider the anatomical source of blindness. Those suffering from photoreceptor-derived or cortical blindness may develop circadian-based maladies upon removal of the eyes, thereby exacerbating an already diminished quality of life.

See also: The Circadian Clock in the Retina Regulates Rod and Cone Pathways; Circadian Regulation of Ion Channels in Photoreceptors; Evolution of Opsins; Microvillar and Ciliary Photoreceptors in Molluskan Eyes; The Photoreceptor Outer Segment as a Sensory Cilium; Phototransduction: Inactivation in Rods; Phototransduction: Phototransduction in Rods; Rod and Cone Photoreceptor Cells: Inner and Outer Segments.

Further Reading

Berson, D. M. (2007). Phototransduction in ganglion-cell photoreceptors. *Pflugers Archiv* 454: 849–855.

Berson, D. M., Dunn, F. A., and Takao, M. (2002). Phototransduction by retinal ganglion cells that set the circadian clock. *Science* 295: 1070–1073.

Cahill, G. M. and Besharse, J. C. (1993). Circadian clock functions localized in xenopus retinal photoreceptors. *Neuron* 10: 573–577.

Campbell, S. S. and Murphy, P. J. (1998). Extraocular circadian phototransduction in humans. *Science* 279: 396–399.

Czeisler, C. A., Shanahan, T. L., Klerman, E. B., et al. (1995). Suppression of melatonin secretion in some blind patients by exposure to bright light. *New England Journal of Medicine* 332: 6–11.

Do, M. T., Kang, S. H., Xue, T., et al. (2009). Photon capture and signalling by melanopsin retinal ganglion cells. *Nature* 457: 281–287.

Graham, D. M., Wong, K. Y., Shapiro, P., et al. (2008). Melanopsin ganglion cells use a membrane-associated rhabdomeric phototransduction cascade. *Journal of Neurophysiology* 99: 2522–2532.

Guler, A. D., Ecker, J. L., Lall, G. S., et al. (2008). Melanopsin cells are the principal conduits for rod–cone input to non-image-forming vision. *Nature* 453: 102–105.

Hattar, S., Liao, H. W., Takao, M., Berson, D. M., and Yau, K. W. (2002). Melanopsin-containing retinal ganglion cells: Architecture, projections, and intrinsic photosensitivity. *Science* 295: 1065–1070.

Keeler, C. E. (1927). Iris movements in blind mice. *American Journal of Physiology* 81: 107–112.

Kumbalasiri, T. and Provencio, I. (2005). Melanopsin and other novel mammalian opsins. *Experimental Eye Research* 81: 368–375.

Provencio, I. (2007). Melanopsin cells. In: Hoy, R. R., Shepherd, G. M., Basbaum, A. I., Kaneko, A., and Westheimer, G. (eds.) *The Senses: A Comprehensive Reference.* Oxford: Elsevier.

Tosini, G. and Menaker, M. (1996). Circadian rhythms in cultured mammalian retina. *Science* 272: 419–421.

Wright, K. P. Jr. and Czeisler, C. A. (2002). Absence of circadian phase resetting in response to bright light behind the knees. *Science* 297: 571.

Circadian Regulation of Ion Channels in Photoreceptors

G Y-P Ko, K Jian, L Shi, and M L Ko, Texas A&M University, College Station, TX, USA

Glossary

Circadian oscillator – A system that generates self-sustained oscillations or rhythms of about 24 h. Current models include self-sustained transcription and translation feedback loops that operate at the cellular level to provide outputs that are circadian in nature.

Cyclic GMP-gated cation channels (CNGCs) – The nonselective cation channels that are activated through direct binding of cyclic nucleotides onto the channel proteins. In general, cyclic-nucleotide gated channels are heterotetrameric complexes consisting of two or three different subunits (α, β, and γ), with important channel properties determined by the subunit composition.

ENSLI amacrine cells – A class of retinal amacrine cells that release the peptide, somatostatin. These cells get their name because they are immunoreactive for enkephalin, neurotensin, and somatostatin (ENSI, enkephalin-, neurotensin-, and somatostatin-like immunoreactive).

L-type voltage-gated calcium channels (L-VGCCs) – The membrane channels that mediate a voltage-dependent and depolarization-induced calcium influx. They regulate diverse biological processes such as contraction, secretion, neurotransmission, differentiation, and gene expression in many different cell types. The L-VGCCs are composed of a pore-forming $\alpha 1$-subunit and the auxiliary β-, $\alpha 2\delta$-, and γ-subunits. They can be blocked by divalent cations (e.g., cobalt) and organic L-VGCC antagonists, including dihydropyridines, phenylalkylamines, and benzothiazepines.

Retinoschisis – An X-linked retinal dystrophy that features disorganization of retinal cell layers, disruption of the synaptic structures and neurotransmission between photoreceptors and bipolar cells, and progressive degeneration of rod and cone photoreceptor cells. Retinoschisis results from mutations in retinoschisin.

Circadian Oscillators Regulate the Functions of the Visual System

Circadian oscillators are biological clocks that exist in almost all living organisms on the earth from bacteria to humans, with persistent rhythmic periods close to 24 h (*circa dian*) even in the absence of external timing cues. The circadian oscillators coordinate rhythmic changes in biochemistry, physiology, and behavior of living organisms, so that organisms can be synchronized with the 24-h oscillations of the external environment. The molecular nature of circadian oscillators varies from species to species. A generalized model for the generation of circadian rhythm involves the transcription, translation, and feedback of clock genes and their own transcriptional products. It is composed of two interlocking transcription–translation feedback loops as well as post-translational modulations. Collectively, these components comprise the core oscillator or the circadian oscillator. Visual systems have to detect images despite large daily changes in ambient illumination between day and night, and intrinsic circadian oscillators in the retina provide such a mechanism for visual systems to initiate more sustained adaptive changes throughout the course of the day. The retina is a heterogeneous tissue with multiple cell types organized in several cell layers. Early studies of circadian regulation in *Xenopus* and chicken retinas indicated that retinal circadian clocks are mainly located in the photoreceptors. Later research revealed that there are multiple oscillators present in various retinal cells in a species-dependent manner. The overall circadian regulation of the retina relies on the synaptic circuitry and feedback modulation among different retinal oscillators, so that the retina is able to anticipate and adapt to sustained daily illumination changes as well as acute light/dark adaptation.

Circadian Regulation of Photoreceptors

The circadian oscillators in photoreceptors are endogenous and able to function independently in the absence of other retinal inputs. These photoreceptor oscillators lead to morphological, physiological, biochemical, and molecular changes that ultimately regulate photoreceptor function and physiology in a circadian fashion. In vertebrate rod photoreceptors, outer segment shedding and renewal is a continuous process, but its rate is under circadian control. In some teleosts and anurans, the inner segments of rod and cone photoreceptors undergo contraction and elongation (i.e., retinomotor movement) in response to changes in ambient illumination as well as in the circadian cycles. While cones remain in a contracted state during the day, rods contract at night. Photoreceptors form specialized ribbon synapses with secondary neurons such as horizontal or bipolar cells. The numbers and ultrastructures of synaptic ribbons in both photoreceptors and bipolar cells undertake changes in relation to the time of day and light intensities. In avians, amphibians, reptiles, and other lower vertebrate species, melatonin is synthesized and secreted from photoreceptors at night, while its synthesis is inhibited by light. The transcription of arylalkylamine N-acetyltransferase (AANAT), the melatonin synthesis enzyme, is under circadian control. While the synthesis and release of dopamine from retinal amacrine cells shows light-driven and circadian fluctuations, the circadian nature of dopamine is dependent on the melatonin rhythm. In addition to the circadian oscillator genes, several photoreceptor-specific genes, ion channels, and enzyme activities are also under circadian control that ultimately contribute to the circadian regulation of photoreceptor physiology and function.

Ion Channels in Photoreceptors

Ion channels are macromolecular pores that allow charged ions to move across cell membranes and contribute to the excitability of neurons and muscles. Electrical signals are generated and modulated as different types of channels open and close in response to neurotransmitters, hormones, membrane potential changes, mechanical forces, and other agents. There are two major classes of ion channels: voltage-gated and ligand-gated ion channels. Voltage-gated ion channels open or close their pores in response to membrane potential changes. Ligand-gated ion channels gate ion movements and generate electrical signals in response to specific chemicals such as neurotransmitters or cyclic nucleotides. Vertebrate photoreceptors are highly polarized, so the distribution of ion channels on the plasma membrane is spatially compartmentalized. The outer segment membrane contains cyclic guanosine monophosphate (cGMP)-gated ion channels that are closed by light, whereas the inner segment and the synaptic terminal are furnished with at least six different ion channels. These six ion channels include:

1. hyperpolarization-activated and cyclic-nucleotide-modulated cation channels;
2. L-type voltage-gated calcium channels (L-VGCCs);
3. calcium-activated potassium (K_{Ca}) channels;
4. noninactivating voltage-gated potassium channels;

5. calcium-activated chloride channels; and
6. cGMP-gated ion channels.

Thus far, cGMP-gated ion channels and L-VGCCs have been studied the most rigorously, especially with regard to their gating properties and physiological functions in photoreceptors. Both ion channels are under circadian control. In the following sections, the circadian regulation of cGMP-gated ion channels and L-VGCCs are reviewed in detail.

Circadian Regulation of cGMP-gated Cation Channels

Cyclic GMP-gated cation channels (CNGCs) are non-selective cation channels that belong to the family of cyclic-nucleotide-gated channels. In general, cyclic-nucleotide-gated channels are heterotetrameric complexes consisting of two or three different subunits (α, β, and γ), with important channel properties determined by the subunit composition. Activation of these channels is through direct binding of cyclic nucleotides onto the channel proteins. The CNGCs of rods and cones have structurally similar but distinct α- and β-subunits. Rods contain CNGCα1-subunits that are more sensitive to calcium-induced inhibitions, while cones contain CNGCα3-subunits that are not as sensitive to calcium. In the vertebrate retina, phototransduction is mediated by a G-protein-coupled cascade that results in changes in the gating of CNGCs in the outer segments of rods and cones. Light initiates a fall in intracellular cGMP as the end point of a G-protein-mediated phototransduction cascade, resulting in closure of CNGCs, reduced cation influx, and membrane hyperpolarization. In the dark, the intracellular cGMP concentration is relatively high. This causes tonic activation of these channels and a steady transmembrane influx of sodium (Na^+) and calcium (Ca^{2+}) ions. Therefore, CNGCs carry the photoreceptor dark current and serve essential roles in the light-dependent changes in photoreceptor membrane potential and subsequent neural processing.

In chick cone photoreceptors, the apparent affinity of CNGCs for their activating ligand is under circadian regulation. There is roughly a twofold change in the apparent affinity of CNGCs for cGMP throughout the course of a day, with the affinity substantially higher at night than during the day even in constant darkness. Such changes in channel affinities can be expected to occur especially at the lower range of cGMP concentrations ($\leq 7 \mu M$ in the dark) that are within the photoreceptor physiological range. Other biophysical features of the gating of CNGCs, such as unitary conductance, numbers of cGMP-binding sites on the channels, density of channels in the plasma membrane, and the maximum current amplitudes do not vary as a function of the time of day.

The ion channel gating properties of photoreceptor CNGCs can be modulated by multiple processes, including direct phosphorylation or dephosphorylation of the channel subunits and the binding of Ca^{2+}/calmodulin or related molecules. Dephosphorylation of CNGCs on serine/threonine residues or tyrosine residues causes an increase in the apparent affinity of CNGCs for their activating ligand. Binding of Ca^{2+}/calmodulin or phosphorylation of CNGCs causes these channels to shift to a lower affinity state for cGMP. The clock regulation of cone CNGCs entails a post-translational modification of the channel molecules. More specifically, it is the circadian rhythmicity of tyrosine phosphorylation that underlies the circadian modulation of cone CNGC affinity to its ligand. Inhibition of tyrosine kinases during the day increases the apparent affinity of CNGCs for cGMP, whereas inhibition of tyrosine phosphatases at night produces the opposite effect. While the protein expression and phosphorylation of the channel pore-forming CNGCα3-subunits remain constant throughout the course of the day, tyrosine phosphorylation of an auxiliary subunit (probably the β-subunit, ∼85 kDa) displays a circadian rhythm. During the daytime, tyrosine phosphorylation on this 85-kDa protein is twice as high as at night. Therefore, circadian rhythmicity of tyrosine phosphorylation on the 85-kDa auxiliary subunit of cone CNGCs provides one of the final steps in the circadian regulation of CNGCs.

The CNGCs in photoreceptors are essential components of visual phototransduction cascades. As such, it is possible that they also play a role in the light entrainment of the circadian oscillators in photoreceptors. As the gating of CNGCs is under circadian control in cone photoreceptors, these channels represent a potential example of an entity that is both an input to and an output from the circadian oscillator. Roenneberg and Merrow have presented models of circadian oscillator systems in which pathways that lead to entrainment of the core oscillators (i.e., the circadian inputs) can themselves be regulated by the oscillators (i.e., they are also components of circadian outputs). One feature of these models is that they contain additional feedback loops that can markedly enhance the stability of the overall oscillator system. Therefore, the circadian regulation of CNGCs in retinal photoreceptors represents an adaptation to enhance the stability of the circadian oscillators in photoreceptors.

Signaling Pathways Leading to the Circadian Regulation of CNGCs

Even though the connection from the molecular oscillator to the regulation of CNGC affinity rhythm is still not completely understood, the small guanosine triphosphate (GTP)ase Ras and mitogen-activated protein kinase (MAPK) signaling pathway and calcium-calmodulin kinase II (CaMKII) are involved as circadian outputs to regulate CNGC rhythms. The activities of Ras and MAPK are themselves under circadian control and oscillate concurrently with the CNGC affinity rhythm. The activity of CaMKII also displays a circadian rhythm, but it runs antiphase to the MAPK rhythm, and CaMKII is a downstream target of MAPK. Perturbation of the activities of Ras, MAPK, or CaMKII causes phase-dependent changes in the gating properties of CNGCs. Furthermore, this Ras–MAPK–CaMKII pathway serves as part of a common circadian output pathway to regulate other molecules in photoreceptors (see **Figure 1**). However, neither MAPK nor CaMKII directly phosphorylates CNGCs in a circadian fashion. It is their downstream targets leading to tyrosine phosphorylation on the auxiliary subunit of CNGCs that ultimately govern the circadian regulation of CNGC gating properties.

The expression and activation of adenylate cyclase in the retina are under circadian control, and the cyclic adenosine monophosphate (cAMP) content in retinal photoreceptors is maximal at night. In avian retinas, this cAMP rhythm not only drives the rhythm of melatonin synthesis and secretion, but it also serves as part of the circadian output and leads to MAPK activation and modulation of the CNGC affinity rhythm. Besides serving as a circadian output to regulate ion channels, cAMP can stimulate retinomotor movements. In *Xenopus* retina, cAMP signaling resets circadian oscillators within photoreceptors. Hence, the second messenger, cAMP, may well serve as part of both input and output pathways in photoreceptors.

Circadian Phase-Dependent Modulation of Cone CNGCs by Dopamine

While circadian rhythms can occur in retinal photoreceptors cultured in the absence of other functional cell types, multiple cell types contribute to the overall circadian control of the intact retina. In avian retina, melatonin inhibits the release of dopamine from a subpopulation of amacrine cells. Consequently, retinal melatonin and dopamine are in antiphasic circadian cycles. As a result, dopamine functions as a feedback signal from the inner retina that refines and modulates circadian control mechanisms within the photoreceptors. Dopamine evokes a phase-dependent modulation of CNGC affinity rhythm in chick cone photoreceptors that is through the activation of the D2 family of dopamine receptors. Exposure to dopamine or D2 agonists causes a significant decrease in the apparent affinity of CNGCs at night but has no effect on CNGCs during the day. It is well established that D2 receptors are pertussis-toxin-sensitive G-protein-coupled receptors that mediate inhibition of adenylate cyclase and cause a decrease of cAMP formation in the

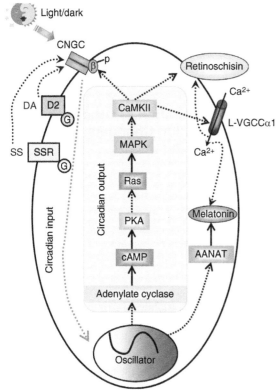

Figure 1 This model illustrates the circadian rhythm in retina cone photoreceptors. Light and dark (day and night) signals from the environment enter the photoreceptor through cGMP-gated cation channels, since cGMP-gated cation channels are essential in phototransduction for light detection. The light/dark signals through the circadian input pathway entrain the photoreceptor core oscillator that cause the photoreceptor to be in sync with the environment. From the core oscillator through the circadian output, the activities of different molecules, including cGMP-gated ion channels, L-type voltage-gated ion channels, and retinoschisin are under circadian regulation. This circadian output is composed of a series of signaling pathways. The circadian regulation of CNGCs in retinal photoreceptors represents an adaptation to enhance the stability of retinal circadian oscillators. It demonstrates a model that Roenneberg and Merrow have presented, in which the elements that lead to entrainment of the core oscillators (e.g., cGMP-gated cation channels) can themselves be regulated by the oscillators. One feature of this model is that it contains additional feedback loops that can markedly enhance the stability of the overall oscillator system at the cellular level (photoreceptors only) and maybe at the retinal network level (the retinal oscillators). The circadian phase-dependent regulation of cGMP-gated cation channels by dopamine and somatostatin is represented by brown and purple dotted arrows, respectively. The black dotted arrows indicate that there are multiple steps currently not known, while the solid arrows represent known signaling. Blue dotted arrows represent that the secretion of melatonin and retinoschisin is under the control of L-type voltage-gated calcium channels. AANAT, arylalkylamine N-acetyltransferase; Ca^{2+}, calcium ion; CaMKII, calcium-calmodulin kinase II; CNGC, cGMP-gated cation channel; DA, dopamine; D2, dopamine D2 receptor; G, G protein; L-VGCCα1, L-type voltage-gated calcium channel α1-subunit; SS, somatostatin; SSR, somatostatin receptor.

retina. However, the phase-dependent modulation of CNGCs in cone photoreceptors by dopamine does not involve pertussis-toxin-sensitive G protein or cAMP signaling, since exposure to cAMP or pertussis toxin does not reverse the phase-dependent modulation of dopamine on CNGCs. This circadian modulation of cone CNGCs by dopamine is partially through the MAPK–CaMKII signaling pathway.

Modulation of Cone CNGCs by Somatostatin

Somatostatin is released from a class of amacrine cells that are immunoreactive for enkephalin, neurotensin, and somatostatin, and thus are called ENSLI (enkephalin-, neurotensin-, and somatostatin-like immunoreactive) amacrine cells. There are two forms of somatostatin: somatostatin-14 (SS-14) with 14 amino acids and somatostatin-28 (SS-28) with 28 amino acids, and both forms are released from ENSLI cells. The release of somatostatin from the ENSLI cells is under circadian control, which is high at night and low during the day in mammalian retinas, and this rhythm is concurrent with the activities of these cells with a high sustained rate of activity in the dark and a low sustained rate of activity in the light. The five somatostatin receptors (referred to as Sst1–Sst5) are guanine nucleotide binding protein (G-protein)-coupled receptors. While all five subtypes of somatostatin receptors are expressed in mammalian retina, only Sst2–Sst5 are present in the chick retina. Both SS-14 and SS-28 modulate cone CNGC sensitivity to cGMP that depend on circadian phase and immediate history of illumination. Both SS-14 and SS-28 decrease CNGC sensitivity to cGMP at night, which are similar to the action of dopamine via the D2 receptor and resistant to pertussis toxin. In addition, SS-28, but not SS-14, evokes a transient increase in CNGC affinity to cGMP only during the early part of the day in cone photoreceptors that have been exposed to light for 1–2 h. This transient effect by SS-28 is mediated by pertussis-toxin-sensitive G-protein-coupled receptors and activation of phospholipase C and protein kinase C signaling cascades. Therefore, somatostatin modulation of retinal cones may serve to reinforce circadian processes intrinsic to the photoreceptors, and may also contribute to more rapid adaptive responses to changes in ambient illumination.

Circadian Regulation of L-VGCCs

L-VGCCs mediate a voltage-dependent and depolarization-induced calcium influx and regulate diverse biological processes, such as contraction, secretion, neurotransmission, differentiation, and gene expression in many different cell types. The L-VGCCs are composed of a pore-forming α1-subunit and the auxiliary β-, α2δ-, and γ-subunits, and they can be blocked by divalent cations (e.g., cobalt) and organic L-VGCC antagonists, including dihydropyridines, phenylalkylamines, and benzothiazepines. The α1-subunit serves as a voltage sensor to detect voltage changes across the plasma membrane, and it also controls the pore size to allow selective divalent cations to pass through. Mammalian α1-subunits are encoded by at least 10 distinct genes, and α1C, α1D, and α1F (also known as $Ca_v1.2$, $Ca_v1.3$, and $Ca_v1.4$, respectively) are expressed in the retina photoreceptors. Photoreceptors are nonspiking neurons, and they release glutamate continuously in the dark as a result of depolarization-evoked activation of L-VGCCs.

Circadian regulation of L-VGCCs has been observed in goldfish retinal bipolar cells, chick cone photoreceptors, and other nonretinal neurons. In both retinal cases, the average maximum current amplitudes of L-VGCCs are significantly larger at midnight than at midday. The activation voltages that elicit L-VGCC currents (i.e., current–voltage relationship) and the channel gating kinetics do not change throughout the course of a day. In chick retinas, the main factor contributing to the circadian regulation of L-VGCC current amplitudes is the expression of functional L-VGCCα1-subunits, and both messenger RNA (mRNA) levels and protein expression of VGCCα1D are rhythmic. The Ras–MAPK–CaMKII signaling pathway (**Figure 1**) serves as part of the circadian output to regulate L-VGCCs as well as CNGCs as described above. However, the varying maximum amplitudes of the L-VGCC currents is in stark contrast to the CNGC maximum currents, which remain constant throughout the day, and which instead exhibit changes in gating properties.

One major functional significance of the L-VGCC rhythm is the circadian control of melatonin release. In nonmammalian vertebrates, melatonin is synthesized and secreted from retinal photoreceptors and is under circadian control. Inhibition of L-VGCCs with dihydropyridines blocks the synthesis and release of melatonin. Another physiological aspect of the L-VGCC rhythm is the circadian control of retinoschisin secretion. Retinoschisin is a 224-amino-acid protein secreted mainly by retinal photoreceptors and bipolar cells. Retinoschisin is distributed throughout the retina but is mainly concentrated around outer and inner segments of photoreceptors and both retinal plexiform layers. Mutations in the retinoschisin gene (*RS1*) cause X-linked retinoschisis, a retinal dystrophy that features disorganization of retinal cell layers, disruption of the synaptic structures and neurotransmission between photoreceptors and bipolar cells, and progressive degeneration of rod and cone photoreceptor cells. Hence, retinoschisin is believed to play an important role in the development and maintenance of retinal cytoarchitecture. In chick retinas, the mRNA level and protein expression of retinoschisin are under circadian control, and the secretion of retinoschisin is higher at night than during the day. Inhibition of L-VGCCs with

dihydropyridines dampens the circadian rhythm of retinoschisin secretion where only nighttime secretion is affected. Therefore, the circadian control of L-VGCCs has a profound impact in regulation of photoreceptor physiology and synaptic transmission.

Circadian Regulation of Other Photoreceptor Ion Channels: Potassium Channels (K⁺ Channels)

Potassium (K⁺) channels are the most diverse ion channels with more than 100 genes of the pore-forming α-subunits identified to date. They can dampen membrane excitability and set the resting membrane potentials in neurons. According to their genetic homology and functional characteristics, there are four major families of K⁺ channels: voltage-gated K⁺ channels, Ca^{2+}-activated K⁺ channels, inward rectifier K⁺ channels, and leak K⁺ channels. In photoreceptors, the dark inward current through CNGCs in the outer segment is balanced by a K⁺ outward current in the inner segment, and this K⁺ outward current is mainly carried out by the voltage-gated K⁺ channels. The major subtypes of voltage-gated K⁺ channels present in photoreceptors are delayed-rectifier K⁺ channels (Kv1.2, 1.3, and 2.1) and A-type transient K⁺ channels (Kv4.2). The Ca^{2+}-activated K⁺ channels and outward-rectifying noninactivating K⁺ channels (Kv10.2; *eag2*) are also found in photoreceptors.

Thus far, the reports on circadian regulation of K⁺ channels have been limited to invertebrate studies. In *Aplysia* retinal pacemaker neurons (not photoreceptors), there is a robust circadian rhythm of potassium currents carried by voltage-gated K⁺ channels (I_{KV}), while A-type K⁺ currents and Ca^{2+}-activated K⁺ currents remain constant throughout the day. When I_{KV} peaks during the late night (predawn), the compound action potential firing frequencies reaches its nadir. The circadian rhythm of I_{KV} contributes to the circadian control of the frequency of compound action potentials in pacemaker neurons. The basal retinal neurons in the eye of the mollusk *Bulla gouldiana* express a circadian rhythm in optic nerve impulses. It is the circadian regulation of I_{KV}, but not other outward K⁺ channels, that drive the daily fluctuations in membrane conductance and membrane potential of these neurons.

Conclusion

The circadian oscillators in the retina play important roles in the regulation of retinal physiology and function, including sensitivity to light, neurotransmitter release, gene expression, morphological changes at the synaptic terminals, metabolism (as pH changes), and rod–cone dominance. The photoreceptor components of electroretinograms (ERGs), the electrophysiological recordings of retinal physiology, recorded from humans as well as animals, show daily rhythms. The circadian oscillators in photoreceptors are endogenous, and they can function independently without other retinal inputs. These circadian oscillators regulate retinomotor movement, outer segment disk shedding and membrane renewal, morphological changes at synaptic ribbons, gene expression, a delayed rectifier potassium current, the affinity of cGMP-gated ion channels, the L-type voltage-gated Ca^{2+} channel currents, and the activities of MAPK and CaMKII, among other photoreceptor activities. Photoreceptors are more sensitive to intense light damage during the subjective night than during the subjective day, even in animals that have been maintained in constant darkness for several days after circadian light–dark cycle entrainment. Several retinal degenerative diseases are associated with dampened or abnormal circadian rhythms in ERGs. In the Royal College of Surgeons rats, the first known animal model with inherited retinal degeneration, the diurnal changes in the c-wave of the ERGs are dampened prior to the beginning of retinal degeneration from postnatal day 17 to 24. Mutations in human cone–rod homeobox (CRX) are associated with retinal diseases, including cone–rod dystrophy-2, retinitis pigmentosa, and Leber congenital amaurosis, which all lead to blindness. In a mouse model of CRX mutation, the circadian rhythmicity of the ERG is abolished, and the mutant mice are not able to be entrained by light–dark cycles. The synthesis and release of dopamine from retinal amacrine cells is also under circadian control, and disruption of the circadian rhythmicity of dopamine synthesis causes a glaucoma-like disorder in quails. Fain and Lisman have proposed that circadian oscillators may provide a common pathway mediating several forms of retinal degeneration. Therefore, dysfunctions of circadian oscillators in the retina contribute to some forms of retinal degeneration, which in many instances may lead to blindness.

See also: Circadian Metabolism in the Chick Retina; The Circadian Clock in the Retina Regulates Rod and Cone Pathways; Circadian Rhythms in the Fly's Visual System; Fish Retinomotor Movements; *Limulus* Eyes and Their Circadian Regulation; Neurotransmitters and Receptors: Dopamine Receptors; The Photoreceptor Outer Segment as a Sensory Cilium.

Further Reading

Barnes, S. and Jacklet, J. W. (1997). Ionic currents of isolated retinal pacemaker neurons: Projected daily phase differences and selective enhancement by a phase-shifting neurotransmitter. *Journal of Neurophysiology* 77: 3075–3084.

Barnes, S. and Kelly, M. E. (2002). Calcium channels at the photoreceptor synapse. *Advances in Experimental Medicine and Biology* 514: 465–476.

Chae, K. S., Ko, G. Y., and Dryer, S. E. (2007). Tyrosine phosphorylation of cGMP-gated ion channels is under circadian control in chick retina photoreceptors. *Investigative Ophthalmology and Visual Science* 48: 901–906.

Chen, S. K., Ko, G. Y., and Dryer, S. E. (2007). Somatostatin peptides produce multiple effects on gating properties of native cone photoreceptor cGMP-gated channels that depend on circadian phase and previous illumination. *Journal of Neuroscience* 27: 12168–12175.

Green, C. B. and Besharse, J. C. (2004). Retinal circadian clocks and control of retinal physiology. *Journal of Biological Rhythms* 19: 91–102.

Kaupp, U. B. and Seifert, R. (2002). Cyclic nucleotide-gated ion channels. *Physiological Reviews* 82: 769–824.

Ko, G. Y., Ko, M. L., and Dryer, S. E. (2001). Circadian regulation of cGMP-gated cationic channels of chick retinal cones. Erk MAP Kinase and Ca2+/calmodulin-dependent protein kinase II. *Neuron* 29: 255–266.

Ko, G. Y., Ko, M. L., and Dryer, S. E. (2003). Circadian phase-dependent modulation of cGMP-gated channels of cone photoreceptors by dopamine and D2 agonist. *Journal of Neuroscience* 23: 3145–3153.

Ko, G. Y., Ko, M. L., and Dryer, S. E. (2004). Circadian regulation of cGMP-gated channels of vertebrate cone photoreceptors: Role of cAMP and Ras. *Journal of Neuroscience* 24: 1296–1304.

Ko, M. L., Liu, Y., Dryer, S. E., and Ko, G. Y. (2007). The expression of L-type voltage-gated calcium channels in retinal photoreceptors is under circadian control. *Journal of Neurochemistry* 103: 784–792.

Ko, M. L., Liu, Y., Shi, L., Trump, D., and Ko, G. Y. (2008). Circadian regulation of retinoschisin in the chick retina. *Investigative Ophthalmology and Visual Science* 49: 1615–1621.

Michel, S., Manivannan, K., Zaritsky, J. J., and Block, G. D. (1999). A delayed rectifier current is modulated by the circadian pacemaker in Bulla. *Journal of Biological Rhythms* 14: 141–150.

Molday, R. S. and Kaupp, U. B. (2000). Ion channels of vertebrate photoreceptors. In: Stravenga, D. G., Degrip, W. J., and Pugh, E. N., Jr. (eds.) *Molecular Mechanisms in Visual Transduction,* 1st edn., pp. 143–182. Amsterdam: Elsevier Science.

Pugh, E. N., Jr. and Lamb, T. D. (2000). Phototransduction in vertebrate rods and cones: Molecular mechanisms of amplification, recovery and light adaptation. In: Stravenga, D. G., Degrip, W. J., and Pugh, E. N., Jr. (eds.) *Molecular Mechanisms in Visual Transduction,* 1st edn., pp. 183–255. Amsterdam: Elsevier Science.

Tosini, G., Pozdeyev, N., Sakamoto, K., and Iuvone, P. M. (2008). The circadian clock system in the mammalian retina. *BioEssays* 30: 624–633.

Circadian Rhythms in the Fly's Visual System

E Pyza, Jagiellonian University, Kraków, Poland

Glossary

Circadian clock – A cell-autonomous, cyclical autoregulatory molecular mechanism, involving genes – clock genes and their proteins that generate circadian molecular oscillations.

Clock genes – The cyclically, with a cycle about a day, expressed genes engaged in molecular mechanism of circadian clock.

Cryptochrome – A blue-light absorbing protein, which contains a flavin-binding domain and functions as a circadian photoreceptor and an element of the molecular clock in the central and peripheral clocks of *Drosophila*, respectively. In the mammalian circadian clock, cryptochromes are the core clock elements.

Electroretinogram – A record of electric activity of cells, measured by inserting an electrode into the eye.

Entrainment – The synchronization of circadian oscillations of the clock to the external daily changes of day and night or to other cyclically changing environmental cues.

Green fluorescent protein (GFP) – A green-light-emitting protein originally found in *Aquorea victoria* and used, after insertion of its gene in transgenic animals, as a reporter of selected proteins and cell types.

Pigment-dispersing hormones – A family of peptides that regulates migration of pigment granules in the crustaceans' eye and in chromatofores of their body surface and also plays the role of neurotransmitters in circadian systems of crustaceans and insects. In insects, these peptides are called pigment-dispersing factors.

Rhabdomere – A light-absorbing part of the compound eye photoreceptor.

Subjective day – A part of the circadian cycle under constant darkness (DD) or in continuous light (LL) in which previously, in the day/night cycle (LD), was the day.

Subjective night – A part of the circadian cycle under constant darkness (DD) or in continuous light (LL) in which previously, in the day/night cycle (LD), was the night.

Zeitgeber time – A time of the 24-h day/night cycle in the environment measured in hours. Usually ZT0 means the beginning of the day.

The Fly's Visual System

The visual system is a major sensory system for flies and includes the large compound eyes, the ocelli – simple eyes located on the vertex of the head and the optic lobes of the brain. The compound eyes are image-forming eyes, while the ocelli play a role in orientation. In addition to the photoreceptors in the compound eyes and ocelli, light stimuli can be received by extraocular photoreceptors and deep brain photoreceptors.

The most external layer of the compound eye is the retina, which overlies three optic neuropils of the optic lobe: lamina, medulla, and lobula (**Figures 1(a)** and **1(b)**). In flies, the lobula is divided into the lobula and the lobula plate. The retina of the compound eye is composed of many single units called ommatidia whose number depends on the size of the eye. In the housefly *Musca domestica*, the compound eye is composed of 3000 ommatidia, while the fruit fly, *Drosophila melanogaster*, has about 800 ommatidia in each eye. Each ommatidium is composed of eight photoreceptors and six of them, R1–R6, terminate in the first optic neuropil or lamina. The other two photoreceptors, R7 and R8, have axons that bypass the lamina and terminate in the medulla. The light-sensitive pigment in the R1–R6 is the blue-sensitive rhodopsin Rh1, while the visual pigments in R7 and R8 are ultraviolet (UV)-sensitive: Rh3, Rh4, or Rh5.

The lamina also has a modular structure and is composed of cylindrical modules called cartridges (**Figures 1(c), 1(d)** and **2**). Each cartridge contains 12 cell types and is surrounded by three epithelial glial cells. Elements of the cartridge include the terminals of photoreceptors R1–R6, the axons and dendrites of the lamina monopolar cells L1–L5, amacrine cell processes, processes of cells located in the medulla, and processes of tangential neurons with cell bodies located in other parts of the brain. The structure and synaptic contacts of the medulla have not been as extensively studied as in the lamina; however, the medulla is known to consist of columns each of which contains 35 cells. The lobula and the lobula plate are centers for processing motion information, and they contain, among others, 10 individually identifiable vertical system neurons that respond to visual wide-field motions of arbitrary patterns.

The function of the retina is to receive photic and visual information, transduce this information into receptor potentials, then transmit the information to the lamina by regulating the release of the neurotransmitter

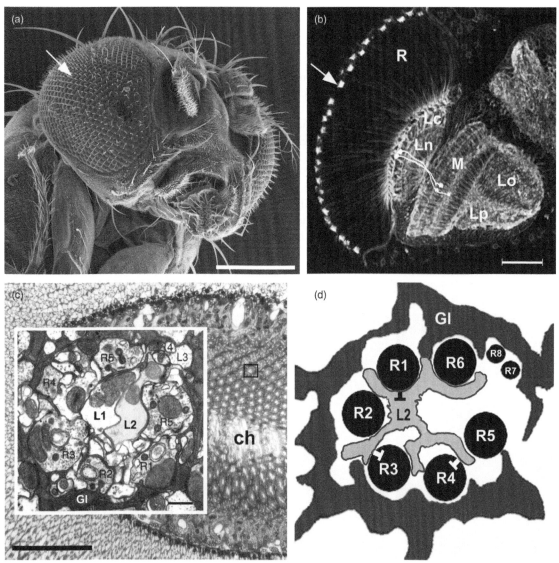

Figure 1 The visual system of *Drosophila melanogaster*. (a) Scanning electron micrograph of the head and compound eye of *Drosophila* viewed from the right side. The fly's compound eye is composed of hexagonally arranged units/facets called ommatidia. Arrow denotes the surface of the compound eye. Scale bar = 200 μm. (b) The structure of the compound eye and optic lobe in horizontal section as revealed by immunostaining with an antibody against α-tubulin. The ommatidial array of photoreceptors in the retina (R) innervates the first of a series of neuropils, the lamina (Ln), where the first-order interneurons, L1 and L2 monopolar cells (marked in white) have their cell bodies and axons. L1 and L2 axons terminate in the second optic neuropil or medulla (M). Lc, lamina cortex; Ln, lamina neuropil; M, medulla; Lo, lobula; Lp, lobula plate. Arrow indicates the surface of the eye. Scale bar = 50 μm. (c) The structure of the lamina in cross section. The lamina is composed of cylindrical modules called cartridges (the region enclosed in a small box), which comprise the same cell types and constitute synaptic units of this neuropil. ch, chiasma. Scale bar = 20 μm. Inset: EM micrograph showing the magnification of a single cartridge with the profiles of L1 and L2 monopolar cells at its axis and the surrounding photoreceptors (R1–R6) and glial cells (Gl). L3: monopolar cell L3; 4: monopolar cell L4. Scale bar = 1 μm. (d) Schematic representation of the lamina cartridge in cross section with the positions of photoreceptors axons (R1–R-8), and the axon of L2 monopolar cell (L2), as well as presynaptic elements (T-bars) of synaptic contacts that are formed between these cell types; tetrad synapses (white T-bars) and feedback synapses (black T-bar). Gl, epithelial glia that surround each cartridge. Reproduced from Pyza, E. and Górska-Andrzejak, J. (2008). External and internal inputs affecting plasticity of dendrites and axons of the fly's neurons. *Acta Neurobiologiae Experimentalis* 68: 322–333.

histamine at synaptic contacts called tetrad synapses. At these tetrad synapses, photoreceptors R1–R6 contact two large monopolar cells (LMCs), L1 and L2, the major output neurons of lamellar cartridges, and two other post-synaptic elements which may include processes of amacrine cells, epithelial glial cells, and L3 monopolar cell (**Figure 2**). L2 also forms feedback synapses back onto the photoreceptor terminals.

The Fly's Circadian System

Circadian rhythms are self-sustaining oscillations with a period of about 24 h that are maintained in constant conditions, such as constant darkness (DD). These rhythms are not only entrained to the external day/night cycle primarily by light, but they also can be entrained by other environmental cues. The eyes of flies and of other insect species provide light input to the circadian system which controls circadian rhythms in behavior and in physiological and biochemical processes, including those within the visual system. The circadian system in flies is composed of a central clock (pacemaker) located in the brain as well as peripheral clocks located in many tissues throughout the body. All circadian rhythms in the body are synchronized to each other and maintain the organism's homeostasis in time. As will be described in greater detail below, the circadian system controls structural changes within the lamina of the visual system and the numbers of specific synapse, and these changes are correlated with the animal's locomotor activity rhythm.

In *D. melanogaster*, clock genes are expressed in approximately 100 neurons in the brain belonging to seven groups on each side of the brain: small and large ventral lateral neurons (s-LN$_v$s and l-LN$_v$s), dorsal lateral neurons (LN$_d$s), lateral posterior neurons (LPNs), and three groups of dorsal neurons (DN$_1$, DN$_2$, and DN$_3$). These cell groups constitute the main circadian pacemaker. Four of five s-LN$_v$s and four l-LN$_v$s express the neuropeptide pigment-dispersing factor (PDF), the only output neurotransmitter of the circadian system known so far. Peripheral clocks, distributed throughout the animal,

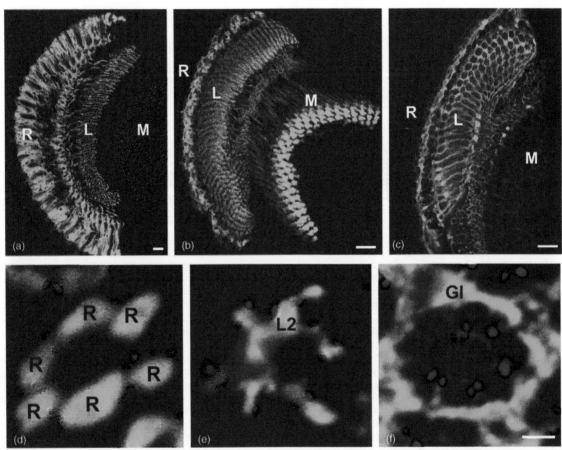

Figure 2   (a–c) Confocal images of frontal cryostat sections through the retina and optic lobes of lines of *Drosophila* showing targeted expression of green fluorescent protein (GFP) (green) in photoreceptors, L2 monopolar cell, and epithelial glial cells, respectively. The section shown in (a) was also immunolabeled with a monoclonal antibody specific for the presynaptic protein Bruchpilot (red). R, retina; L, lamina; M, medulla. Scale bar = 20 μm. (d–f) Cross sections of the lamina cartridges of the GFP (green)-expressing transgenic flies described above. Sections were also immunostained for the presynaptic protein Bruchpilot (red). R: axons of photoreceptors; L2: axon of L2 monopolar cell; Gl: processes of epithelial glial cells. Scale bar = 1 μm. Reproduced from Pyza, E. and Górska-Andrzejak, J. (2008). External and internal inputs affecting plasticity of dendrites and axons of the fly's neurons. *Acta Neurobiologiae Experimentalis* 68: 322–333.

generate circadian rhythms in peripheral organs, including antennae, Malpighian tubules, testes, and the retina photoreceptor cells. In contrast to the pacemaker neurons, in which the circadian oscillations of clock gene expression are self-sustained for many days in DD, oscillations in the peripheral clocks decline after several cycles in DD. Thus, the molecular mechanisms underlying central and peripheral clocks may be different.

Circadian clocks are cell autonomous, and a cell is said to contain a circadian clock if it expresses clock genes, those that show circadian expression and control the expression of a large population of other genes (so-called clock-controlled genes). In clock cells, clock genes and their proteins form transcriptional/translational negative- and positive-feedback loops that are the molecular basis of circadian clocks. Much of our understanding of molecular clocks comes from studies of *D. melanogaster*, and our ideas about how these molecular clocks are regulated and entrained continue to be refined. A basic model is presented below and in **Figure 3**.

In central pacemaker clock cells, two cyclically expressed core clock genes are *period* (*per*) and *timeless* (*tim*). When their proteins PERIOD (PER) and TIMELESS (TIM) accumulate in cytoplasm, they form heterodimers and enter the nucleus. In the nucleus, they bind to heterodimers of the transcription factors CLOCK(CLK)/CYCLE(CYC) encoded by *Clock* (*Clk*) and *cycle* (*cyc*) genes, respectively, and PER, and possibly TIM, repress their own transcription. This is the negative-feedback loop of the molecular clock. CLK/CYC also control the cyclical expression of two other core clock genes, *vrille* (*vri*) and *Par domain protein* 1ε (*Pdp1ε*) and their protein products, VRI and Pdp1ε, respectively, regulate the transcription of *clk*. VRI represses the transcription of *Clk* while Pdp1ε enhances it. The increase in *clk* transcription is a positive-feedback loop of the molecular clock. The level of PER and the transition of PER and TIM from the cytoplasm to the nucleus is controlled by phosphorylation so that the transcriptional feedback occurs at a particular time of the day.

Light entrainment of the molecular clock in central pacemaker neurons depends on an intracellular blue-light-photosensitive protein CRYPTOCHROME (CRY). CRY is involved in the light-dependent degradation of TIM. When the light is on, TIM does not accumulate in the cytoplasm, the negative-feedback loop that represses *per* and *tim* transcription is delayed and, as a result, the phase of the clock will be delayed. However, if TIM is degraded, while it is in the nucleus, the repression of CLK/CYC will be removed early, the transcription of *per* and *tim* will be advanced and the phase of the clock will be advanced.

The molecular mechanism in the peripheral clocks is also based on *per* and *tim* cyclical expression; however, in peripheral clocks, CRY may be a central element of the molecular mechanism in addition to its role as a circadian photoreceptor involved in the light entrainment of the molecular clock. It has been suggested that in peripheral clocks of *D. melanogaster*, CRY is a transcriptional repressor playing a role similar to that of CRY proteins (mCRY1 and mCRY2) in mammalian molecular clocks.

Similar circadian systems – consisting of many circadian clocks that are based on a transcriptional/translational feedback loop model – are probably present in other insect species. However, some aspects of the molecular mechanism of the clock may differ in even closely related dipteran species. For example, in *Lucilia cuprina*, the PER expression pattern is similar to that of *D. melanogaster*, but in the housefly, *M. domestica*, PER isolated from the whole head does not cycle.

As was described above, light is the major environmental cue that entrains the central pacemaker to the day/night cycle. The light input to the central pacemaker is complex, and the fly's compound eyes are not the only source. Ocelli photoreceptors, the extraocular photoreceptors known as the Hofbauer–Buchner (H–B) eyelet, and the intracellular blue-light-photosensitive protein CRYPTOCHROME (CRY), which is present in most cells harboring molecular circadian clocks, also provide light input. The H–B eyelet consists of four photoreceptors located between the retina and medulla. These photoreceptors contact the small LN$_v$s pacemaker neurons and synchronize the molecular rhythms in these neurons. The small LN$_v$s, in turn, synchronize the animal's behavioral rhythms by the rhythmic release of PDF. Since the pacemaker comprises several groups of neurons, each group may be important for producing circadian rhythms in different cells, tissues, and systems, and receive light inputs from different photoreceptors.

The visual system receives circadian input from pacemakers; however, it is still unknown which group or groups of pacemaker neurons send the circadian input to the visual system. In addition, the visual system possesses peripheral clocks in the retina photoreceptor cells.

Circadian Rhythms in the Retina of the Compound Eye

Most arthropods, including insects, show daily changes in the structure of photoreceptors manifested by changes in the size of the rhabdomere, the photosensitive part of photoreceptors. Moreover, daily changes in screening pigment movements have been documented in many insect and crustaceans species. The most pronounced structural changes in photoreceptors, regulated by a circadian clock, have been described in the horseshoe crab *Limulus polyphemus*. In contrast to crustaceans, locusts, praying mantis, and mosquitoes, the cross-sectional area

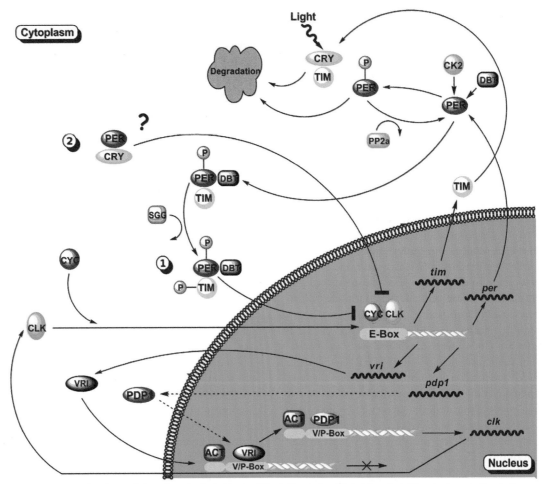

Figure 3 Molecular mechanisms of circadian clocks in *Drosophila*. 1. Current model of the circadian clock operating in the pacemaker neurons in the brain. In this model *period* (*per*) and *timeless* (*tim*) clock genes are cyclically expressed at the end of day and at the beginning of night. Later during the night, PER and TIM proteins accumulate in the cytoplasm and PER is phosphorylated by DOUBLETIME (DBT) and possibly by casein kinase CK2. The phosphorylation leads to PER degradation if TIM does not bind to PER forming PER/TIM heterodimers. PER is also stabilized by PP2a, which dephosphorylates PER. In the heterodimers PER/TIM, TIM is phosphorylated by SHAGGY (SGG) and PER is bound to DBT which enables PER and TIM to be transported to the nucleus. In the nucleus, PER and, possibly, also TIM inhibit transcription factors CYCLE (CYC) and CLOCK (CLK), which as heterodimers repress transcription of *per* and *tim*. Repression of *per* and *tim* transcription by their own proteins constitutes a negative-feedback loop of the molecular clock. CYC/CLK heterodimers also control cyclical expression of *vrille* (*vri*) and *pdp*1ε which proteins affect transcription of *clk* by competing for binding to *clk* promotor. PDP1ε stimulates, while VRI inhibits *clk* expression. The increase in *clk* transcription is a positive loop of the molecular clock. Transcription of *clk* can also be activated by a clock-independent activator, ACT. In this model, CRYPTOCHROM (CRY) protein is a clock photoreceptor. During the day, CRY is activated by light and binds TIM leading to TIM degradation. After TIM degradation PER cannot form heterodimers with TIM in the cytoplasm and be transported to the nucleus. In result, the clock is delayed in phase. When light stimulates the degradation of TIM while TIM is in the nucleus, the repression of *per* and *tim* transcription by PER/TIM is removed and the molecular clock advances in phase. 2. A possible mechanism of circadian clock functioning in peripheral clocks and in a clock controlling circadian rhythms in the visual system. In this model, PER forms heterodimers with CRY which are transported to the nucleus where CRY represses the transcription of *per* and *tim*.

of rhabdomeres in the retina photoreceptors of flies shows no (*L. cuprina*) or only little (*D. melanogaster*) change during the day and night cycle.

However, in many species, including flies, diurnal rhythms have been detected in the amplitude of the electrical activity of the eye recorded with the electroretinogram (ERG). The ERG originates from both the retina and the lamina L1 and L2 monopolar cells and measures

changes in eye sensitivity. In *D. melanogaster*, cyclical changes are observed in both the amplitude of the ERG and the level of visual pigment in photoreceptors as measured by microspectrophotometry. When *D. melanogaster* is maintained under normal LD conditions, the ERG-measured sensitivity and the concentration of visual pigment in photoreceptors are highest during the night. Then about 8 h before lights-on, the sensitivity begins to decline and, about 2 h after lights-on, the level of visual pigment in the eye starts to decrease. The sensitivity and the level of visual pigment continue to decrease until 4 h after the onset of the light. In white-eye *D. melanogaster* mutants, the level of visual pigment recovers 2–4 h later. The observations that changes in ERG sensitivity and visual pigment levels anticipate changes in illumination, and that these cyclic changes are maintained in DD conditions indicate both processes are under circadian control.

In the blowfly *Calliphora vicina*, a diurnal insect, three ERG components have been studied under LD and DD conditions; one component called the sustained negative potential originates at least partly from the retina photoreceptors, and two others, ON and OFF transients, originate from the lamina monopolar cells. The lamina transients exhibit circadian oscillations and increase during the subjective night in DD. The negative potential that originates from the retina changes in antiphase with the transients and becomes smaller during the subjective night, indicating that the retina and the lamina components of ERG seem to be regulated by two different circadian clocks. In the cockroach *Leucophaea maderae*, a nocturnal insect, the opposite changes to those recorded in the blowfly have been observed, and the negative potential of ERG in this species exhibits the highest amplitude during the subjective night. The changes in the ERG observed in these two different species indicate that the photoreceptor activity is correlated with the animal's pattern of locomotor activity. This activity is higher during the day than at night in diurnal insects and higher at night in nocturnal insects.

In addition to regulating the content of visual pigment in insect photoreceptors, circadian clocks may regulate phototransduction by controlling the expression of some genes such as the one that encodes a transient receptor potential channel involved in this process.

Much evidence indicates that retinal photoreceptors are themselves peripheral clocks. They show daily changes in the abundance of the clock protein PER that are maintained in DD conditions and they express other clock genes, including *cry*. As in the pacemaker neurons in the brain, the abundance of PER in the photoreceptors is highest at the end of the night. In photoreceptors, however, PER decays faster during the day than in the *per*-expressing neurons in the brain. The rhythmic changes in PER immunostaining in the photoreceptors are also affected by *per* mutations. For example, in flies expressing the *per^s* mutation, in which the period of circadian locomotor activity is shorter (19 h) than 24 h, PER in photoreceptors appears earlier than in wild-type flies. Evidence that the circadian clocks in retinal photoreceptors are cell autonomous comes from experiments that used transgenic *D. melanogaster* in which PER was expressed only in photoreceptors. In these animals, cyclic PER expression was observed in photoreceptors in DD condition, even

though no PER expression was detected elsewhere in the animal and the animals were otherwise arrhythmic. Although the retinal photoreceptors possess autonomous circadian clocks, it is unknown to what extent these clocks control the rhythms observed in photoreceptors. Again, it is unknown if the clocks in the fly's retina impact the rhythms in cells that are postsynaptic to the photoreceptors.

Circadian Rhythms in the First Visual Neuropil (Lamina)

Although structural changes in the photoreceptor cell bodies of flies are not dramatic, striking circadian rhythms have been detected in the lamina of the fly's visual system. Here, the number of synaptic contacts, the morphology of two visual interneurons, and the number of some organelles in the photoreceptor terminals change during the LD and DD conditions. These rhythms are examples of plasticity in the nervous system regulated by a circadian clock, and they may serve to gate the flow of information through the visual system to correlate with the animal's locomotor activity.

Since clock gene expression has not been detected in the lamina neurons, circadian rhythms observed in the lamina must be driven by circadian pacemaker neurons located in the brain and/or clocks present in the retina photoreceptors. These rhythms may also be maintained by clock-gene-expressing glial cells distributed between neurons and other glial cells in the optic neuropils.

Circadian Plasticity of Synaptic Contacts

In the lamina of the housefly *M. domestica*, cyclic changes have been found in the frequency of two types of synaptic contacts: tetrad presynaptic profiles or the so-called T-bars, and feedback synapses (**Figure 1(d)**). In the housefly, which is a diurnal species, the number of tetrad presynaptic profiles is highest at the beginning of the day when the flies also show the highest locomotor activity. In contrast to the tetrads, the number of feedback synapses peak at the beginning of the night. The oscillations in the number of feedback synapses are maintained in DD and are not affected by light. Therefore, they are clearly controlled by circadian clocks. By contrast, the number of tetrad synapses does not significantly change in DD, but increases after a 1-h light pulse presented during the subjective day or subjective night in DD. This could be interpreted to mean that the number of tetrad synapses is controlled by light, not by circadian clocks. However, studies with *D. melanogaster* suggest that circadian clocks are important regulators of the number of tetrad synapses.

In *D. melanogaster*, daily oscillations have been found in the abundance of a scaffolding protein of the T-bar ribbon in presynaptic tetrad synapses, the BRP protein encoded by the *Bruchpilot* gene (**Figure 2**). As in the housefly, the number of presynaptic sites immunoreactive to BRP and the abundance of the protein in the lamina are highest at the beginning of the day. This rhythm was not detected in DD, but it was also absent in cyclic light in the arrhythmic *per* null mutant. Taken together, these observations indicate that although the daily rhythm in the formation of T-bar presynaptic tetrads depends on light, signals from circadian clocks are also required.

Circadian Plasticity of Second-Order Neurons and Glial Cells

The daily oscillations in the frequency of photoreceptor tetrad synapses are correlated with morphological changes in the L1 and L2 monopolar cells of the lamina. In the housefly, both cells swell at the beginning of the day and shrink during the night, and this rhythm is maintained in DD and also in LL, a condition that disrupts circadian rhythms in behavior, causing animals to become arrhythmic. Examination of the morphology of the entire L2 cells in transgenic *D. melanogaster*, in which green fluorescent protein (GFP) was expressed specifically in L2 cells (**Figures 2(b)** and **2(e)**), has shown that circadian changes occur in the girth of axons and in the length of dendrites extending from these axons (**Figure 4**). The daily pattern of morphological changes of the L2 is slightly different in males than in females. Circadian changes also occur in the size of L2 nuclei but not in their cell bodies (**Figure 4**).

Changes in the size of monopolar cell axons are correlated with the pattern of the locomotor activity of fly species. In typically diurnal species such as *M. domestica* and *C. vicina*, L1 and L2 monopolar cell axons are largest at the beginning of the day when flies show highest locomotor activity. The sizes of these axons can be further increased, by 20–30%, when insects are additionally stimulated to fly intensively. In *D. melanogaster* the daily pattern of changes in sizes of L1 and L2 axons is different. Under LD, *D. melanogaster* has two peaks in locomotor activity, one in the morning and a second one in the evening, and in this species, the L1 and L2 cell axons enlarge twice during the day, in the morning and in the evening. Changes in size of L1 and L2 monopolar cell axons are offset by changes in size of the surrounding epithelial glial cells, which shrink during the day and swell during the night. Changes in the girth of L1 and L2 interneuron axons seem to be correlated with the perimeter of their dendritic trees. Studies of transgenic

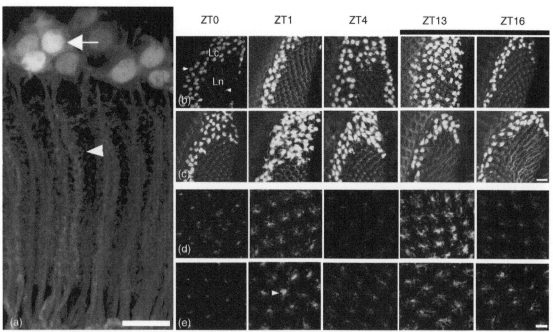

Figure 4   Daily changes in morphology of L2 monopolar cell in the lamina of transgenic *D. melanogaster* expressing GFP in the L2 cells (21D-GAL4/UAS-S65T-GFP). (a) Confocal image showing the structure of L2 monopolar cells in the lamina. GFP expression labels nuclei of the somata (arrow) in the lamina cortex, as well as axons and dendrites (arrowhead) in the neuropil. Scale bar = 10 μm. (b, c) Daily changes (ZT0–ZT16, ZT0: the beginning of the day. ZT 13: the beginning of the night, dark bar) in the size of L2 cell nuclei in males (b) and females (c). The cross-sectional area of the nuclei appears to be largest 1 h after lights-on (ZT1, females) and 4 h after lights-on (ZT4, males) and smallest in the middle of the night (ZT16, females and males). Scale bar = 20 μm. (d, e) Differences in morphology of the dendritic tree of L2 in males (d) and females (e). The cross-sectional area of L2 monopolar cell axons is larger at ZT1 and ZT13 than in other time points in both males and females. Scale bar = 5 μm. Reproduced from Pyza, E. and Górska-Andrzejak, J. (2008). External and internal inputs affecting plasticity of dendrites and axons of the fly's neurons. *Acta Neurobiologiae Experimentalis* 68: 322–333.

D. melanogaster, in which GFP is expressed specifically in L2 monopolar cells, revealed that the L2 dendrites postsynaptic to tetrad synapses are longest at the beginning of the day. This rhythm is absent in arrhythmic *per* null mutant flies (*per*[0]), and its pattern is changed from wild-type in mutant flies harboring a mutation in the CRY protein (*cry*[b]). These findings indicate that circadian plasticity of the L2 cells depends on cyclical expression of *per* and involves *cry* in maintaining the phase of the rhythm.

In contrast to L1 and L2 monopolar cells, photoreceptor terminals in the lamina do not change their girth. It is unknown, however, if other monopolar cells in the lamina show similar morphological circadian changes.

Circadian Rhythms in Organelles inside Photoreceptor Terminals

Although photoreceptor terminals in the lamina do not show circadian changes in their size, circadian structural changes have been detected in some organelles inside the terminals. For example, in the housefly, there is a vertical migration of pigment granules into and out of photoreceptor terminals. Radial movements of screening pigment within the cell bodies of photoreceptors R1–R6 are well documented. These movements are driven by light. Pigment granules move horizontally toward the rhabdomeres during light adaptation and away from the rhabdomeres during dark adaptation providing an important pupillary mechanism. However, the vertical migration of pigment granules in the housefly photoreceptors appears regulated by a circadian clock. Fewer pigment granules are present in photoreceptor terminals during the day compared to the night. This rhythm is maintained in DD, thereby indicating its endogenous, circadian regulation. These vertical movements of pigment granules may play a role in light adaptation providing more pigment granules to the photoreceptor cell bodies during the day.

Other organelles in the housefly photoreceptor cell terminals are formed from the invaginations of neighboring cells in the lamina: from other photoreceptor terminals and glial cells. Among these two types of invaginations, only the number of inter-receptor invaginations changes during the subjective day and subjective night in DD, and the number is greatest at the end of the subjective night. These organelles are rare in the photoreceptor terminals in flies reared in LD conditions, but increase in dark-reared flies. Their functions remain to be discovered.

Neurotransmitter Regulation of Circadian Rhythms in the Visual System

The lamina is invaded by several sets of wide-field tangential neurons with processes directed at right angles to the lamellar cartridges; thus, they are appropriately suited for the global regulation of visual processing. In the housefly's lamina, the most numerous of these processes are immunoreactive for serotonin (5-hydroxytryptamine, 5-HT). They originate from two giant ventral protocerebral neurons called LBO5-HT that project ipsi- and contralaterally to the optic lobes. Another group of processes is immunoreactive for PDF and originates from the LNvs. Still other processes are immunoreactive for the neuropeptide Phe-Met-Arg-Phe-NH_2 (FMRFamide). The neurotransmitter chemistry of remaining processes remains to be identified.

Processes of LBO5-HT neurons are also present in the lamina of *D. melanogaster*, but in this species, processes of other wide-field neurons, including PDH-immunoreactive processes, arborize in the distal medulla and only a few reach the lamina. Since the paracrine release of both biogenic amines and neuropeptides has been reported, neuromodulators released in the medulla may reach targets in the lamina of the fruit fly through volume transmission.

Among neurotransmitters detected in the lamina and the medulla, 5-HT has been found to influence both photoreceptors and L1 and L2 monopolar cells. 5HT modulates potassium channels in the photoreceptors of *D. melanogaster*, and in photoreceptors of locust, a diurnal modulation of potassium channel function can be mimicked by application of 5-HT. In the lamina of the housefly, 5-HT has robust effects on the structure of L1 and L2 cells, but the effects are not the same on both cells. Injections of 5-HT into the housefly's medulla increase the diameter of L1 but not L2 monopolar cell axons. However, when 5-HT and other biogenic amines are depleted from the lamina by injecting the drug reserpine, there is a decrease in the girth of the L2 axon but not in the axon of the L1 cell. This suggests that 5-HT influences the diameter of both L1 and L2 axons, but that the effects observed depend on its concentration.

Injections of other neurotransmitters into the medulla also affect L1 and L2 axon sizes, but their effects are less significant than that of 5-HT. PDF injections induce a small increase of both cells, while FMRFamide had the opposite effect. PDF applications also have an effect on the screening pigment granules in the photoreceptor terminals. Injections of histamine, a neurotransmitter released from photoreceptor terminals, increase the size of L1 but not L2 processes, while injections of either glutamate or gamma aminobutyric acid (GABA) decrease L1 and L2 axons. The effects of glutamate and GABA mimic the changes in L1 and L2 processes seen during the night under LD conditions and during the subjective night in flies held in constant darkness. Glutamate is a transmitter candidate for the L1 and L2 cells and amacrine cells of the lamina, and GABA is detected in cells that project from the medulla into the lamina. GABA injections also decreased the number of feedback synapses from L2 onto photoreceptors.

Larval Visual System

Circadian rhythms in the visual system exist not only in adult flies but also in larvae. The larval visual system, Bolwig's organ (BO) is located on both sides of the head and consists of 12 photoreceptors expressing either Rh-5 or Rh-6. All these photoreceptors project to the LNs of the larval circadian system, which includes four neurons on each side of the fruit fly's larval brain. The BO photoreceptors do not express clock genes, so they are not circadian oscillators like the retina photoreceptors; instead, the BO photoreceptors receive circadian information from the LNs as do other target neurons. The circadian input from the LNs seems to control sensitivity of the larval visual system since the *per* and *tim* null mutants are relatively insensitive to light, while mutations in the positive clock elements, *Clk* and *cyc*, increase the light sensitivity of larvae. In *D. melanogaster* larvae, the circadian clock regulates the light-avoidance behavior and visual sensitivity. Both are high at the end of the subjective night and low at the end of subjective day in DD conditions.

The BO photoreceptors survive metamorphosis, and in the adult fly, Rh6-expressing photoreceptors correspond to the extraocular photoreceptors of the H–B eyelet. In larvae and in adults these structures play an active, if subsidiary, role in the entrainment of circadian rhythms. In larvae, light is transmitted from the BO to the LNs where light-induced degradation of TIM is essential for the larval circadian clock entrainment. Moreover, when BO is eliminated, the LN dendrite structure is altered indicating direct interactions between the larval visual system and the larval circadian system. The larval optic nerve is also required for the development of a serotonergic arborization originating in the central brain and for the development of the dendritic tree of the circadian pacemaker neurons – the small LN$_v$s. The larval optic nerve and adult extraocular photoreceptors sequentially associate with the small LN$_v$s during *D. melanogaster* brain development.

Circadian Circuits in the Fly's Visual System

In flies, circadian rhythms in the visual system are correlated to the circadian rhythm in locomotor activity. This indicates that circadian oscillators, both central and peripheral, must communicate with each other, but the underlying circuitry is largely unknown. The circadian rhythms in the retina seem to depend mostly on the peripheral clocks inside the photoreceptors, although pacemaker neurons could impact rhythms in the retina by controlling the number of feedback synapses from L2 onto photoreceptor terminals. As it was described above, these synapses show robust circadian changes. However,

it is unclear if the rhythms in the lamina, and probably in the other optic neuropils, are regulated only by the LNs or by both the photoreceptor's clocks and the LNs. It is also unknown if the dorsal neurons, DN$_1$–DN$_3$, have any effects on the circadian rhythms in the visual system.

In the housefly, severing the optic lobe from the rest of the brain abolishes circadian changes in the sizes of the L1 and L2 monopolar cell axons. This suggests that circadian information is transmitted to lamina monopolar cells from central pacemaker neurons by neurotransmitters. So far, the only neurotransmitter identified in the pacemaker neurons is PDF, which is present in the LNvs of *D. melanogaster* and in a similar set of neurons in *M. domestica*. In the housefly, this peptide is cyclically released from PDF varicosities in a paracrine fashion. In this species, the circadian changes of PDF varicosity sizes have been observed in LD and DD, and the daily release of PDF from the dense-core vesicles of PDF-immunoreactive processes have been detected. If PDF, which has been detected in dense arborizations in the medulla of all fly species studied and clearly in the lamina of larger flies, transmits circadian information from the LNvs, then PDF receptors should be present in neurons in this part of the visual system. However, PDF receptors have not been detected in the distal part of the medulla or in the lamina, but in the proximal medulla and in the lobula. In *D. melanogaster*, however, the processes of PDF-immunoreactive l-LN$_v$s terminate in the distal medulla next to *per*-expressing glial cells, which may intermediate between the pacemaker neurons and neurons in the lamina cartridges showing circadian plasticity. Thus, glial cells in the optic lobe might be the third player in the regulation of circadian rhythms in the visual system. In the lamina, the epithelial glial cells change their sizes, decreasing during the day and increasing during the night, completely opposite to the neurons. Disrupting glial metabolism and closing gap-junction channels in the lamina have an effect on L1 and L2 monopolar cell sizes and their circadian morphological plasticity.

The Role of Circadian Clocks in the Visual System

The autonomous circadian oscillators located in retina photoreceptors, and clocks in the pacemaker neurons and in glial cells of the optic lobe, provide a circadian gating of visual information. This gating changes during the day and night and differs between diurnal and nocturnal animals. The activity of the visual system is the highest during the day in diurnal species and during the night in nocturnal species. Furthermore, this pattern correlates with the pattern of locomotor activity in each species. Because of the circadian gating of visual information, light stimuli received during the night in diurnal species do not disrupt its daily rhythm of activity and the daily pattern of animal behavior, but light stimuli can shift the phase of these rhythms. On the other hand, stressful conditions in the environment and a temporary increase in locomotor activity can result in an increase of the flow of information through the visual system, but only when this stimulation is correlated with the active period of the daily rest (sleep)/activity rhythm. Locomotor stimulation, however, may also shift the phase of the rhythm.

In addition to the gating visual information, the circadian system allows an organism to predict daily changes in the environment, and to prepare the visual and other sensory systems for a proper reception of sensory stimuli and efficient transmission of sensory information. The circadian clocks in the photoreceptors control the reception of light stimuli, while the transmission of sensory information is controlled by clocks in several types of cells, including the retina photoreceptors, the pacemaker neurons, and glial cells. These properties of the circadian and visual systems are typical not only for flies but also for other animals.

See also: The Circadian Clock in the Retina Regulates Rod and Cone Pathways; Circadian Metabolism in the Chick Retina; Circadian Regulation of Ion Channels in Photoreceptors; Fish Retinomotor Movements; Genetic Dissection of Invertebrate Phototransduction; *Limulus* Eyes and Their Circadian Regulation; Retinal Degeneration through the Eye of the Fly.

Further Reading

Battelle, B. A. (2002). Circadian efferent input to *Limulus* eyes: Anatomy, circuitry, and impact. *Microscopy Research and Techniques* 15: 345–355.

Collins, B. and Blau, J. (2007). Even a stopped clock tells the right time twice a day: Circadian timekeeping in *Drosophila*. *Pflügers Archives – European Journal of Physiology* 454: 857–867.

Hall, J. (2003). Genetics and molecular biology of rhythms in *Drosophila* and other insects. *Advances in Genetics* 48: 1–280.

Hardie, R. C. (2001). Phototranduction in *Drosophila melanogaster*. *Journal of Experimental Biology* 204: 3403–3409.

Hardin, P. E., Krishnan, B., Houl, J. H., et al. (2003). Central and peripheral circadian oscillators in *Drosophila*. In: Chadwick, D. J. and Goode, J. A. (eds.) *Novartis Foundation Symposia 253: Molecular Clock and Light Signalling*, pp. 140–150. Hoboken, NJ: Wiley.

Marrone, D. F., Boutillier, J. C., and Petit, T. L. (2005). Morphological plasticity of the synapse. Interaction of structure and function. In: Stanton, P. K., Bramham, C., and Scharfman, H. E. (eds.) *Synaptic Plasticity of Transsynaptic Signaling*, pp. 495–518. New York: Springer.

Meinertzhagen, I. A. and Pyza, E. (1999). Neurotransmitter regulation of circadian structural changes in the fly's visual system. *Microscopy Research and Techniques* 45: 96–105.

Meinertzhagen, I. A. and Sorra, K. E. (2001). Synaptic organization in the fly's optic lamina: Few cells, many synapses and divergent microcircuits. In: Kolb, H., Ripps, H., and Wu, S. (eds.) *Progress in Brain Research*, vol. 131, pp. 53–69. Amsterdam: Elsevier Science.

Pyza, E. (2001). Cellular circadian rhythms in the fly's visual system. In: Denlinger, D. L., Giebultowicz, J., and Saunders, D. S. (eds.) *Insect Timing: Circadian Rhythmicity to Seasonality*, pp. 55–68. Amsterdam: Elsevier Science.

Pyza, E. (2002). Dynamic structural changes of synaptic contacts in the visual system of insects. *Microscopy Research and Techniques* 58: 335–344.

Pyza, E. and Górska-Andrzejak, J. (2008). External and internal inputs affecting plasticity of dendrites and axons of the fly's neurons. *Acta Neurobiologiae Experimentalis* 68: 322–333.

Sehgal, A. (ed.) (2004). *Molecular Biology of Circadian Rhythms*. Hoboken, NJ: Wiley.

Tosini, G., Pozdeyev, N., Sakamoto, K., and Iuvone, P. M. (2008). The circadian clock system in the mammalian retina. *BioEssays* 30: 624–633.

Townes-Anderson, E. and Zhang, N. (2006). Synaptic plasticity and structural remodeling of rod and cone cells. In: Pinaud, R., Tremere, L. A., and De Weerd, P. (eds.) *Plasticity in the Visual System: From Genes to Circuits*, pp. 13–32. New York: Springer.

Relevant Websites

http://flybase.org – FlyBase: A Database of Drosophila and Genomes.
http://flybrain.neurobio.arizona.edu – Flybrain: An Online Atlas and Database of the *Drosophila* Nervous System.

Color Blindness: Acquired

D M Tait and J Carroll, Medical College of Wisconsin, Milwaukee, WI, USA

Glossary

Blue–yellow (tritan) defect – A color-vision deficiency in which colors along the blue–yellow axis are difficult to distinguish from one another and sensitivity to blue light is decreased. This is typically caused by a defect in the short-wavelength-sensitive cone photoreceptor pathway. The word tritanopia derives from the Greek *tritos*, meaning third, alluding to the defect being associated with the third of the three primary colors (blue) + *anopia*, meaning blindness.

Deuteranopia – A type of red–green defect in which the middle-wavelength-sensitive photopigment is nonfunctional, leading to decreased hue discrimination and a tendency to confuse reds and greens. Derives from the Greek *deuteros*, meaning second, alluding to the defect being associated with the second of the three primary colors (green) and *anopia*, meaning blindness.

Monochromacy – A complete lack of ability to distinguish colors.

Photopigment – A light-sensitive protein in the photoreceptors (rods and cones) that undergoes a conformational change when absorbing light and initiates the process of visual transduction.

Protanopia – A type of red–green defect in which the long-wavelength-sensitive photopigment is nonfunctional, leading to a loss of sensitivity to red light and a tendency to confuse reds and greens. Derives from the Greek *protos*, meaning first, alluding to the defect being associated with the first of the three primary colors (red) and *anopia*, meaning blindness.

Trichromacy – The fundamental ability to see in full color, based on the possession of three types of cone photoreceptor. The formal proposal of this theory to explain human color vision is credited to Thomas Young.

Introduction

"He *knew* the colors of everything, with an extraordinary exactness (he could give not only the names but the numbers of colors as these were listed in a Pantone chart of hues he had used for many years). He could identify the green of Van Gogh's billiard table in this way unhesitatingly. He knew all the colors in his favorite paintings, but could no longer see them, either when he looked or in his mind's eye. Perhaps he knew them, now, only by verbal memory." Oliver W. Sacks describes the 'Case of the colorblind painter' in his book *An Anthropologist on Mars* where after a car accident a painter tragically lost all ability to experience color.

While many color-vision deficiencies are inherited and present from birth (congenital), they may also be acquired, as in the case of the colorblind painter described above. Acquired defects, as the name implies, indicate a disruption in color discrimination as a consequence of a traumatic event, such as exposure to toxins, a cortical injury, or the defect may accompany an ocular disease. In most cases, the color-vision defect can be thought of as secondary because the causative event often has more serious consequences for the individual.

Classifying Acquired Color Blindness

Normal color vision relies on the presence of three spectrally distinct cone photoreceptor types in the retina, where the spectral sensitivity of the cone cell is determined by the photopigment it contains (short-, middle-, or long-wavelength sensitive; S, M, or L, respectively). Inherited disruptions in color discrimination are normally classified according to which of the cone photopigment classes is functionally disrupted or absent altogether. A defect in the S photopigment produces a tritan defect, corresponding to a loss along the blue–yellow axis, whereas defects in the M and L pigments produce a deutan or protan defect, respectively, though both correspond to a loss along the red–green axis of color vision. Consequently, in inherited defects there is a 1:1 relationship between the affected cone type and the perceptual deficit. However, acquired defects are far more complex and thus must be categorized not by the type of photoreceptor affected, but rather as a variable loss along a particular color axis: blue–yellow axis, red–green axis, or a nonspecific loss of color vision. The dimensionality of these losses can be appreciated by looking at **Figure 1** which illustrates the red–green and blue–yellow axes of color vision.

In addition to classifying the perceptual problems associated with acquired defects, they can be categorized based on the mechanism of the defect: absorption, alteration, and reduction. Absorption defects affect the absorption of a

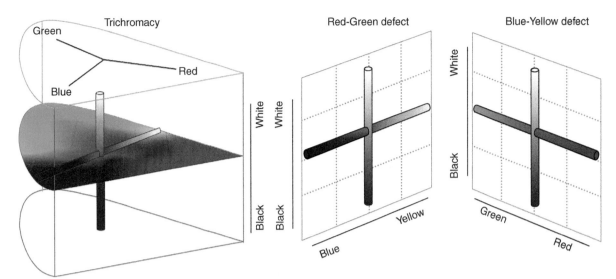

Figure 1 The geometric reduction of color discrimination that accompanies color-vision defects. Possession of three types of cone photopigment makes trichromatic color vision possible. In this case, there is good color discrimination along three distinct axes (black–white, red–green, and blue–yellow) and each discriminable hue (of which there are over 2 million) has a unique combination of activity along each of these axes (left). Acquired color-vision defects can typically be characterized by a loss along the blue–yellow or red–green axis. In red–green defects (middle), color vision is reduced to only the black–white and blue–yellow axes, reducing the dimensionality of color vision (c. 10 000 discriminable hues). Blue–yellow defects (right) show the same reduction in dimensionality. However, the residual color discrimination exists along the red–green and black–white axes. Modified from Neitz, J., Carroll, J., and Neitz, M. (2001). Color vision: Almost reason enough for having eyes. *Optics and Photonics News*, January 2001, pp. 26–33.

particular hue due to deposits affecting the lens or cornea. For example, the normal aging process introduces a yellowing of the lens of the eye, causing a selective absorption defect in discrimination in the blue–yellow axis of color vision. An alteration defect is one that shifts normal color perception and normally reflects a destruction of the macular cones. A reduction defect occurs typically as a result of diseases of the optic nerve and causes perceptive reduction of saturation within colors along the red–green axis as well as a more mild blue–yellow reduction.

Discriminating Acquired from Inherited Color Blindness

There are a number of key differences between inherited and acquired defects. Acquired defects are often difficult to classify because combined, and sometimes subtle, defects often occur. While congenital color-vision deficiencies are typically stable throughout life and both eyes are equally affected, acquired deficits may change in severity over time, and they are more likely to be asymmetric. Furthermore, acquired defects are predominantly tritan, have an equal incidence in males and females, and are typically accompanied by a reduction in visual acuity. Erroneously, color-vision defects are often thought of as being exclusively inherited. However, acquired defects are even more common, with 5% of the population being

Table 1 General differences between acquired and inherited color-vision defects

Acquired color-vision defects	Inherited color-vision defects
Acquired after birth	Present at birth
May progress with age	Stable throughout life span
Combined defects may occur making diagnosis difficult	Precise diagnosis of a given type (protan, deutan, tritan)
Monocular deficiencies can occur	Both eyes are equally affected
Visual acuity is often reduced	Visual acuity is unaffected (except in achromatopsia and blue-cone monochromacy)
Predominantly tritan deficiencies	Predominantly red–green deficiencies
Equal incidence in males and females	Higher incidence in males
Frequency (1 in 20 individuals)	Frequency (1 in 12 males affected with red–green defect)

affected. **Table 1** provides a complete comparison of acquired and inherited defects.

Clinically, there are a number of tests used to diagnose color-vision defects (acquired or inherited). While genetic tests are obviously the definitive approach for characterizing inherited disorders, the diagnosis of acquired color-vision defects relies on the patients' performance on psychophysical tests of color vision. Perhaps the most widely used are pseudoisochromatic plate tests. These plates contain a field of dots that vary in luminance, and embedded within the dots is a number or other shapes differing in chromaticity from the background. For normal trichromats, this shape will be distinguishable from the background. However, depending on the design of the plate, the shape may blend into the background or a different shape may appear altogether for the color-deficient observer. Different plates are used to test for defects along the primary color axes (red–green and blue–yellow). Another type of test is an arrangement test in which the patient orders materials in a specific hue or saturation order. The stimulus is prearranged in a random order and subjects are instructed to arrange the stimuli according to color. Correct arrangement of the various color samples requires not only normal color vision, but also good discriminative ability. Finally, the gold standard in assessing color vision is color matching. By asking the patient to match either a monochromatic yellow test light to a mixture of red and green primaries (Rayleigh match) or a cyan/yellow test mixture to a blue/green primary mixture (Moreland match), one can detect even the most mild red/green or blue/yellow defect, respectively.

Conditions Resulting in an Acquired Color-Vision Defect

Ocular Diseases

The most common acquired color-vision deficiencies arise from various diseases which can impact the eye in a number of ways. Such diseases include progressive cone dystrophies, retinal pigment epithelium (RPE) dystrophies, ocular media opacities, and impairments that affect the retinal neurons (ganglion cells). As a group, these deficiencies can be characterized by an abnormal cone electroretinogram (ERG), reduced visual acuity, central visual-field loss, and/or light sensitivity (photophobia). The age of onset, severity of color-vision defect, and rate of progression can vary widely among different diseases. A few of the most common diseases are discussed below. As an aside, the color-vision defect is often the least of these patients' worries; nonetheless, it is affiliated with the disease (which itself may be an inherited condition) and can sometimes contribute to the diagnosis.

Age-related macular degeneration

Age-related macular degeneration (AMD) is the leading cause of blindness in the developed world. As such, it also represents the most prevalent form of acquired color-vision deficiency. Although this disease has two forms (wet and dry), both involve a degeneration of the photoreceptors in the central retina (macula). It is in this area where acuity and color vision are at their best, so it comes as no surprise that individuals with AMD suffer deficits in

both. In the slower-progressing dry form of AMD, yellow deposits called drusen develop underneath the macula. As these drusen negatively affect the health of the RPE, which is required for normal function of the photoreceptors, a loss of vision slowly results. In the fast-progressing wet AMD, tiny blood vessels begin to grow behind the retina toward the macula. Bleeding, leaking, and scarring occur from these fragile blood vessels, eventually causing irreversible damage to the retina and rapid vision loss. Color vision becomes progressively distorted as the disease progresses and more photoreceptors are functionally compromised, but this does not usually contribute to the differential diagnosis.

Glaucoma

The term glaucoma represents a group of diseases of the optic nerve characterized by a loss of retinal ganglion cells with a corresponding loss of some or most of the fibers of the optic nerve. In this disease, constriction of the visual field occurs slowly over time, and may not be recognized until the impairment is quite severe, with as many as 60% of the retinal ganglion cells degenerating before a perceptual consequence is detected. Worldwide, glaucoma is the second leading cause of blindness, and interestingly the deterioration in color vision of these patients has been proposed as a predictor of significant visual-field loss. Field loss arises from diffuse changes in the neural retina, supporting the idea that acquired color deficiency in glaucoma is caused by damage to the neural pathway rather than the retinal photoreceptors. Despite these patients having a normal S-cone ERG, acquired tritanopia (with severity paralleling the visual-field loss) is often found. This tritan deficiency is consistent with other optic nerve diseases, which suggests that the S-cones themselves remain intact, but that postreceptoral mechanisms involved in the transmission of blue–yellow color discrimination must be selectively disrupted.

Retinitis pigmentosa

Retinitis pigmentosa (RP) is a hereditary retinal dystrophy in which abnormalities of the rods, or the RPE of the retina, lead to progressive visual loss. Affected individuals first experience night blindness and defective dark adaptation, followed by reduction in peripheral vision (tunnel vision), and often a loss of central vision late in the course of the disease. There is good evidence that the progressive degeneration of the rods leads to the secondary loss of cones, likely as a result of reduced rod-derived cone viability factors. Disruptions in color discrimination can occur as the cones degenerate, and these individuals can exhibit a severe tritan defect or occasionally pseudoprotanomaly (a mild red–green defect). Most RP sufferers eventually go blind, therefore again the color-vision defect is of relatively minor importance to the individual.

Cataract

An age-related cataract is an impairment of the crystalline lens of the eye in which the lens hardens, becomes opaque, and yellows. A normal lens selectively absorbs short-wavelength light, however, both normal aging and the development of cataracts enhance this effect and causes problems discriminating in the blue–green portion of the spectrum. This condition is known as xanthopsia and refers to an abundance of yellow which dominates the visual scene. Patients who have cataract surgery in which their natural lens is removed and replaced with a clear plastic lens often experience cyanopsia – or a blue-dominated visual scene. This cyanopsia results because the brain compensates the perceptual yellow tint caused by the cataract by adding a blue tint to the visual scene, and when the yellow lens is removed it takes the brain time to renormalize to the new chromatic input. The mild-blue perception after cataract removal may persist for weeks or even months but normal color vision gradually returns, highlighting the significant chromatic plasticity of the adult visual system.

Diabetic retinopathy

Diabetic retinopathy is an ocular manifestation of the systemic disease, diabetes, occurring when the disease is left untreated. Elevated blood sugar levels present in diabetics can induce changes in the retinal blood vessels severe enough to impact patients' vision. The swelling and growth of new vessels in the eye can cause these vessels to bleed, cloud vision, and destroy the retina by failing to supply adequate oxygen and meet the metabolic needs of the photoreceptors. Visual disturbances begin with a blur, but the pattern of vision loss in these patients is often diffuse, irregular, and achromatic, reducing visual perception to shades of gray interrupted by large blank patches.

Optic neuritis

Optic neuritis is an inflammation of the optic nerve which can cause partial or complete vision loss. These patients report sensitivity to light (photophobia), pain with eye movements, and a degradation of color vision. Direct axonal damage may also play a role in nerve destruction in many cases. The loss of color vision starts as an abnormal dimming, and apparent increased noise in the visual images, which then progresses to a general dimming or loss of color discrimination and blurring across the entire visual field. In many cases, only one eye is affected and patients are not aware of the loss of color vision until the doctor asks them to close or cover the healthy eye. Typically, these defects are similar to a deutan defect (a form of red–green defect), but with additional reduction in sensitivity to the shorter wavelengths. Nearly half of all patients with multiple sclerosis (MS) will develop an episode of optic neuritis, and up to one-third of the time optic neuritis is the presenting sign of MS. This stems from inflammation of the optic nerve, as a result of the degeneration of the myelin sheath surrounding it. MS patients may suffer periodic attacks of optic neuritis, which increase in severity, leading to a permanent loss of color vision.

Cortical Defects

Another way in which one can acquire a color-vision deficiency is through a cortical insult. Cortical defects result from a head injury, stroke, or neurodegenerative disease. Unlike other forms of acquired color-vision deficiencies, those arising from cortical defects are often unilateral, impacting one-half of the brain, and in turn only one part of the visual field, although full-field defects can occur. Injuries impacting color vision typically result from damage to either V2 (prestriate cortex), or V4 – the areas of the brain responsible for processing color information; however, uncertainty remains about precisely which cortical loci (if any) are responsible. As cortical damage is rarely localized to a small section of the brain, there are often other behavioral deficits resulting from the injury, making it difficult to interpret individual cases and generalize about an individual's symptoms.

Cerebral achromatopsia

One famous case of cerebral achromatopsia is that of the colorblind painter discussed in the introduction of this article. It is presumed that the painter's acquired monochromacy was due to cortical injury caused either by damage to the visual cerebral cortex during an automobile accident, or by a stroke affecting one of the visual areas. As V2 and V4 are among the most metabolically active areas of the cerebral cortex, they are among the first to suffer from the effects of reduced oxygen delivery, and may therefore be affected in the case of a stroke. In this case, color disturbances occur very rapidly and may precede other symptoms. The resulting loss of color vision may occur as a total loss (complete achromatopsia), or as a partial loss affecting only one-half of the visual field (hemiachromatopsia). Unlike degenerative disorders with the same perceptual appearance, in central achromatopsia, the retina is not damaged so acuity typically remains unchanged. This finding illustrates how different cortical areas can serve discrete aspects of vision.

Neurodegenerative diseases

Parkinson's disease (PD) is a disease that impairs cognitive, motor, and sensory function through the progressive loss of dopaminergic neurons, and a few studies have reported abnormal color vision in these patients. Dopamine deficiency is believed to alter retinal visual processing primarily by changing the receptive field properties of ganglion cells. The S-cone photoreceptors are sparse in the retina and are believed to have elevated susceptibility to retinal damage caused by progressive loss of the dopaminergic cells in the retina. This is certainly the case in retinal detachments where S-cone defects predominate. Consequently, those who have cited color-discrimination problems in patients with PD identified a tritan axis of confusion. There is debate on this, however, because most standard methods of testing color vision require the patient to make motor movements, which can provide an additional disadvantage and confounding variable for PD patients.

Another neurodegenerative disease, Alzheimer's disease, also has implications for color vision. The cortical atrophy which causes the hallmark dementia associated with this disease causes not only degeneration of the visual areas of the brain (gray matter), but may also induce optic nerve degeneration. As the disease progresses, patients' vision tends to decline from a mild tritan color-vision deficiency into complete achromatopsia.

Toxin-Induced Defects

In addition to the internal defects that cause color-vision abnormalities, there are numerous environmental factors that can alter color perception. Multiple therapeutic drugs have been reported to produce color-vision deficiencies in patients. While most of these occurrences have been documented as a result of toxic levels of the prescribed drugs, some can occur even at recommended dosages, though most side effects are transient. Many studies have found altered color perception in workers occupationally exposed to chemicals as well. A possible mechanism for these toxin-induced defects may be related to the direct effect of the chemicals on cone function, and/or an interference with neurotransmitter signaling.

Digitalis

The digitalis family of drugs (digoxin and digitoxin) is typically prescribed to control congestive heart failure as well as certain cardiac arrhythmias. Several retinal cells (i.e., photoreceptors, Müller's cells, and retinal pigment epithelial cells) express digitalis-sensitive isoforms of sodium–potassium adenosine triphosphate (ATP)ase, and inhibition of these pumps by digitalis is associated with alterations in electrical-response properties of the retina. Clinical ERG and *in vitro* cell studies have shown that toxic levels of digoxin lead to rod and cone dysfunction, with cones being affected to a greater extent. Patients experiencing disturbances of color vision while taking these drugs usually exhibit a red–green defect in which objects are covered with a reddish haze and sometimes exhibit symptoms similar to optic neuritis. Fortunately, color vision tends to recover after the drug is withdrawn. It has been speculated that the prevalent use of yellow in Vincent Van Gogh's work may be due to the use of digitalis, used at the time for its euphoric effects and as a treatment for epilepsy.

PDE5 inhibitors

Phosphodiesterase (PDE) inhibitors, including Viagra (sildenafil), Levitra (vardenafil), and Cialis (tadalafil), are used to treat erectile dysfunction by selectively inhibiting cyclic guanosine monophosphate (cGMP)-phosphodiesterase type 5 (PDE5), present in all vascular tissue, which leads to vasodilation. It also exerts an inhibitory action against PDE6, the phosphodiesterase isoform present in rod and cone photoreceptors, though the inhibition efficacy is only about 1/10 of that for PDE5. PDE6 is a key player in the phototransduction cascade, catalyzing the hydrolysis of cGMP in response to absorption of light by the photopigment molecule. This results in a reduction in cGMP, which results in a closing of the cyclic-nucleotide-gated ion channels and thus a hyperpolarization of the cell. As such, PDE5 inhibitors can interfere with this process and lead to transient changes in rod and cone outer segment function, and the retinal effects may include impaired blue–green color discrimination and decreased rod- and cone-driven ERG amplitudes. At high doses of PDE5 inhibitors, patients have reported a blue tinge to vision, increased apparent brightness of lights, and blurred vision, although some studies report no effects on vision. Color-vision disruptions that were experienced were transient, and no consistent pattern has emerged to suggest any long-term effect of PDE5 inhibitors on the retina or other ocular structures.

Chloroquine

This drug is used in the treatment and prevention of malaria and more recently has gained use as an immune system suppressant prescribed in autoimmune disorders such as rheumatoid arthritis and lupus erythematosus. The color-vision impairment from chloroquine is a consequence of the fact that it is a melanotropic compound with an affinity for pigmented, melanin-containing structures; this naturally results in accumulation of the drug in the highly pigmented RPE. Both the dose and duration of chloroquine treatment impact the level of toxicity because the compound remains in the retina for several years, even after treatment has been discontinued. A tritan defect is most common, but as toxicity increases, a protan-like red–green defect can also manifest. Unfortunately, this is one of the few cases in which withdrawal of the drug is typically insufficient to prevent further damage.

Ethambutol

Ethambutol is a drug prescribed in combination as a treatment for tuberculosis. It has been found to induce a secondary color-discrimination disturbance which shifts the threshold for wavelength discrimination without a change in absolute sensitivity, resulting in color-discrimination errors along the red–green axis. There is evidence that it may disrupt the morphology of the cone pedicle – disturbing the transmission of signals from the cone photoreceptors to their postreceptoral contacts (horizontal and bipolar cells). This disturbance is most profound for low-intensity stimuli. These effects can occur after either a lengthy therapy at a low dosage, or a short therapy at high doses. The color defect is similar to a deutan-like red–green defect, but with a reduction in sensitivity to the shorter wavelengths as well. Unlike many other acquired defects, in this case, impairments in color discrimination can be useful for detection of toxicity, thus color vision should be monitored in patients undergoing regular treatment for tuberculosis. Generally, color vision improves gradually after withdrawal of the drug; yet, in some cases color vision is permanently impaired.

Vitamin A deficiency

Common in developing countries, and associated with alcoholism and metabolic storage diseases in developed countries, vitamin A deficiency is a very preventable cause of acquired color-vision deficiency. This essential vitamin is required for regeneration of the visual pigments of both rods and cones in the retina. A deficit in vitamin A initially causes night blindness and, if left untreated, will contribute to complete blindness by making the cornea very dry and damaging the retina and cornea. In the more mild forms, tritan deficiencies occur, and transition to a general loss of hue discrimination. Approximately 400 000 children in developing countries go blind each year due to a deficiency in vitamin A. If reached before the onset of blindness, recovery can be shown following oral or injectable doses of the vitamin.

Exposure to metals and chemicals

Multiple studies have examined the color perception in groups of workers exposed to high levels of metallic mercury. Surprisingly, a dose-related loss of color vision is demonstrated even at presumed safe levels of mercury exposure. Although the mechanism of mercury's impact on the visual system is not fully understood, it seems to be localized to the retina because traces of mercury have been reported in the optic nerve, RPE, inner plexiform layer, vessel walls, and ganglion cells. Color-vision deficiencies revolve mainly around the blue–yellow axis, but may show some nonspecific losses. As with most of the other toxins, this effect seems to be reversible if individuals are removed from toxic exposure.

An additional collection of chemicals used in plastic, rubber, and viscose rayon manufacturing plants, rotogravure printing industries, as well as dry-cleaning facilities have similarly been implicated in reversible color-vision deficiencies. These chemicals include styrene, perchloroethylene (PCE), toluene, carbon disulfide, and n-hexane.

Conclusion

The range of acquired color-vision deficiencies is broad in both type and severity; some affect ocular structure directly while others perturb the neural pathways responsible for color vision, some are preventable and/or reversible while others are merely an indicator of a more serious condition. Despite their diverse etiology and manifestation, these defects can have a significant impact on the individual, for unlike those born with a color-vision deficiency, acquired sufferers know, and in most cases can remember, what they are lacking perceptually. Returning to the case of the colorblind painter gives those of us who are color normal a small taste of the devastating psychological effect of an acquired color-vision deficiency: "The wrongness of everything was disturbing, even disgusting, and applied to every circumstance of daily life."

See also: Color Blindness: Inherited; The Colorful Visual World of Butterflies; Cone Photoreceptor Cells: Soma and Synapse; Phototransduction: Adaptation in Cones; Phototransduction: Inactivation in Cones; Phototransduction: Phototransduction in Cones; Polarized-Light Vision in Land and Aquatic Animals; Primary Photoreceptor Degenerations: Retinitis Pigmentosa; Primary Photoreceptor Degenerations: Terminology; Rod and Cone Photoreceptor Cells: Outer Segment Membrane Renewal; Secondary Photoreceptor Degenerations: Age-Related Macular Degeneration.

Further Reading

Birch, J. (1993). *Diagnosis of Defective Colour Vision.* New York: Oxford University Press.

Cowey, A. and Heywood, C. A. (1995). There's more to colour than meets the eye. *Behavioural Brain Research* 71: 89–100.

Gegenfurtner, K. R. and Sharpe, L. T. (1999). *Color Vision: From Genes to Perception.* New York: Cambridge University Press.

Heywood, C. A. and Kentridge, R. W. (2003). Achromatopsia, color vision, and cortex. *Neurological Clinics of North America* 21: 483–500.

Iregren, A., Andersson, M., and Nylen, P. (2002). Color vision and occupational chemical exposures: I. An overview of tests and effects. *NeuroToxicology* 23: 719–733.

Jackson, G. R. and Owsley, C. (2003). Visual dysfunction, neurodegenerative diseases, and aging. *Neurologic Clinics of North America* 21: 709–728.

Lamb, T. D. and Pugh, E. N., Jr. (2006). Phototransduction, dark adaptation, and rhodopsin regeneration: The proctor lecture. *Investigative Ophthalmology and Visual Science* 47(12): 5137–5152.

Pokorny, J. (1979). *Congenital and Acquired Color Vision Defects.* New York: Grune and Stratton.

Pokorny, J. and Smith, V. C. (1986). Eye disease and color vision defects. *Vision Research* 26: 1573–1584.

Sacks, O. W. (1995). *An Anthropologist on Mars.* New York: Random House.

Zeki, S. (1990). A century of cerebral achromatopsia. *Brain* 113: 1721–1777.

Color Blindness: Inherited

J Carroll and D M Tait, Medical College of Wisconsin, Milwaukee, WI, USA

Glossary

Blue–yellow defect – A color-vision deficiency in which blue and yellow are difficult to distinguish from one another. In inherited defects, this is typically caused by a defect in the short-wavelength-sensitive photoreceptor.

Dichromacy – A color-vision deficiency in which only two primaries are required to perfectly match a monochromatic light. In this deficiency, one of the three types of photopigments is functionally absent so color vision is reduced to only two dimensions.

Monochromacy – A complete lack of ability to distinguish colors caused by defects in the morphology or function of the cones.

Photopigment – A light-sensitive protein in the photoreceptors (rods and cones) that undergoes a conformational change when absorbing light and initiates the process of visual transduction.

Red–green defect – A color-vision deficiency in which red and green are difficult to distinguish from one another. There are two inherited forms of this defect: Deutan, which is caused by a defect in the medium-wavelength-sensitive photoreceptor, and protan, which is caused by a defect in the long-wavelength-sensitive photoreceptor.

Trichromacy – The fundamental ability to see in full color, based on the ability to match a monochromatic light by using a mixture of any three primaries. This ability requires the possession of three distinct classes of photoreceptors.

Introduction

Beyond the occasional spousal argument over the color of a shirt or tie, a significant number of people are born with a defect in their ability to perceive and discriminate colors. While the existence of deficiencies in color discrimination had been appreciated for some time, it was the famous chemist John Dalton who gave the most well-known analytical description of the condition in 1798. This was due to the fact that he himself suffered from a color-vision defect: "Reflecting on these facts, I was led to conjecture that one of the humors of my eye must be a transparent, but colored, medium, so constituted as to absorb red and green rays principally, because I obtain no proper ideas of these in the solar spectrum; and to transmit blue and other colours more perfectly." While Dalton did not accurately identify the biological basis of his color-vision deficiency, he articulately described it, and thus the term Daltonism is now synonymous with color-vision deficiencies.

Photoreceptor Basis of Human Color Vision

Humans with normal color vision can discriminate some 2 million hues. The retinal substrate for this exquisite color vision is the cone photoreceptor mosaic. While the highly sensitive rods outnumber cones nearly 20:1 and subserve vision at low (scotopic) light levels, it is the cones that underlie the majority of our visual experience at high light levels (photopic), including high spatial acuity and color vision.

There are three subclasses of cone photoreceptor, each having a distinct spectral sensitivity (**Figure 1**). The spectral sensitivity of the photopigment simply reflects the probability of it absorbing a photon of light, and is determined by the particular photopigment present in the cone cell. Photopigments (and their associated cones) are classified according to the region of the visible spectrum they are most sensitive to – either short-, middle- or long-wavelength sensitive (abbreviated S, M, and L). All humans with normal color vision have the same S-cone pigment, with peak absorption around 417 nm. The M-cone pigment varies slightly among individuals, with an average peak of about 530 nm. There is widespread variation in the peak sensitivity of the L-cone pigment among humans with normal color vision, though there are two main variants that peak at 555 and 559 nm, respectively. Trichromatic color vision is afforded by the presence of one cone type from each of these three spectral classes, and these cone types appear to be randomly arranged within the cone mosaic (**Figure 1**). While the human retina contains nearly 100 million photoreceptors, only about 5 million of them are cones, and of the cones, about 95% are of the L/M type. If the function of one or more of the cone classes is disrupted or absent, the result is a compromised ability to make chromatic discriminations, that is, a color-vision deficiency.

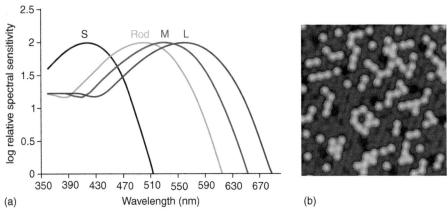

Figure 1 Photopigment basis for trichromatic color vision. (a) Photopigment absorption spectra. The human photopigments have different, but overlapping, spectral sensitivities. The cone photopigments, which dominate our photopic (daytime) vision, are named based on the region of maximal absorption in the visible spectrum – short-(S), middle-(M), or long-wavelength sensitive (L). Rods serve scotopic vision and are maximally sensitive at about 500 nm. (b) Simulation of the organization of the L (red), M (green), and S (blue) cones within the cone photoreceptor mosaic. Note the relative paucity of S cones in the mosaic, and the random arrangement of the L and M cones, with L cones outnumbering M cones by about 2:1 in individuals with normal color vision, though this is variable.

Genetic Basis of Human Color Vision

All inherited color-vision defects are associated with disruptions in the expression of normal cone photopigments. Human photopigments differ only in their opsin (protein) component, though they share the same chromophore, 11-*cis*-retinal. At the protein level the L and M pigments are about 96% homologous, though they show only 43% identity with the S pigment, while human rhodopsin (the opsin found in rods) is about 41% homologous with any of the cone opsins. Each of the human cone opsins is encoded by a different gene. The gene encoding rhodopsin is 5.0-kb long and found on chromosome 3, while the gene encoding the S opsin is located on chromosome 7 and is 3.2-kb long. The rhodopsin and S-opsin genes each have four introns and five exons, and as a result of their autosomal location, humans normally have two copies of each of these genes. In contrast, the L/M gene(s) are located on the X chromosome. While most mammals have only a single L/M gene on the X chromosome, most Old World primates (including humans) possess both L and M genes. In primates that have both the L and M genes on a single X chromosome, they are located in a tandem array near the end of the long arm of the X chromosome (Xq28), and a locus-control region (LCR) enables exclusive expression of one gene from the array in a given cone photoreceptor cell.

The location of the L/M gene array and their high homology allow for frequent unequal homologous crossovers (intermixing of the genes). During meiosis, this can occur between the genes (intergenic crossover) or within the genes (intragenic crossover) between two parental X chromosomes. Such recombination events are responsible for the observed variation in L/M gene number on the X chromosome (humans with normal color vision can have between two and nine genes in this array). In addition, genes with hybrid L and M sequences are produced from intragenic crossover events. These hybrid genes encode photopigments that have spectral sensitivities intermediate of the normal L- and normal M-cone photopigments. The combined variability in gene number and gene sequence means that in a group of 100 males, the probability that any two will have identical L/M arrays is less than 2%. Interestingly, there is no normal variation in the S-pigment gene.

As mentioned previously, there is a great deal of variability in the L and M photopigments. In fact, of the 364 amino acids there are 18 residues where differences have been identified between and among the L and M photopigments (positions 65, 111, 116, 153, 171, 174, 178, 180, 230, 233, 236, 274, 275, 277, 279, 285, 298, and 309; see **Figure 2(a)**). Positions 274, 275, 277, 279, 285, 298, and 309 co-segregate and can be used to identify a pigment as either L or M. This is because these amino acids, in particular the substitutions at positions 277 and 285, are responsible for generating the majority of the spectral separation between the L and M pigment classes. L genes specify tyrosine and threonine at positions 277 and 285, respectively, whereas M genes specify phenylalanine and alanine at these same positions, resulting in a shift in peak sensitivity toward shorter wavelengths of approximately 24 nm. Substitutions at the other amino acid positions can produce more subtle spectral shifts, driving the spectral variations within the L and M classes. The most common, a serine/alanine dimorphism at position 180, shifts the maximum absorption (λ_{max}) by about 3–7 nm, though the amino acid identity at other polymorphic positions can influence the magnitude of this shift.

This genetic variation can have a measurable impact on color vision. For example, among normal trichromats, the

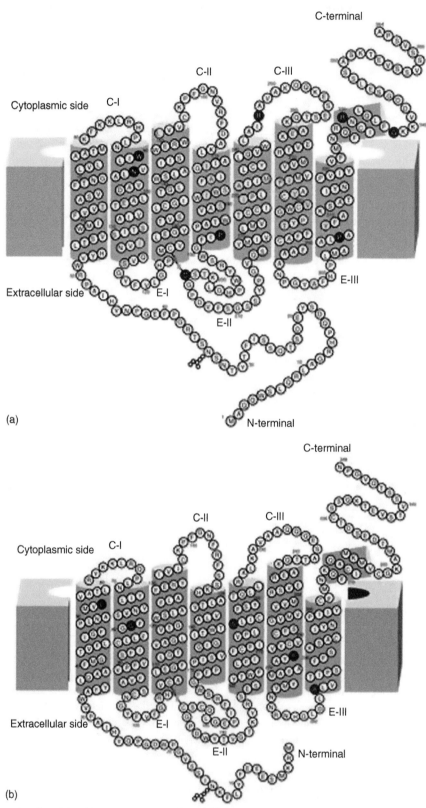

Figure 2 Two-dimensional models of the human cone opsins. (a) L/M opsin. Each circle represents a single amino acid, with mutations associated with the loss of L- or M-pigment function shown as filled black circles. Filled yellow circles represent the dimorphic sites that can differ between the L and M pigments. A number of these sites have been shown to be involved in the spectral tuning of the pigments (see text). The amino acid identities at all 18 dimorphic sites are those of the presumed primordial L opsin. (b) S opsin. Each circle represents a single amino acid. The five mutation sites associated with inherited tritan color-vision deficiency are indicated by black circles.

variation in spectral sensitivity within the L and M pigment classes can be readily observed on color-matching performance. In the Rayleigh match, an individual is asked to determine the mixture of a red and green primary needed to exactly match a monochromatic yellow light. Variability in color-matching behavior had long been recognized, but only in the last 20 years has it been directly linked to the variability in the L and M photopigments, and thus the L/M gene array.

Red–Green Color-Vision Deficiencies

Besides inducing subtle alterations in performance on laboratory color-vision tests, the genetic mechanisms that give rise to variability in the L/M-gene array can induce significant defects in color discrimination. Disruptions in normal L/M-photopigment expression result in an inherited form of color-vision deficiency that affects the red–green (L/M cone) system. Among individuals of Western European ancestry, about 8% of males have a red–green color-vision defect. The incidence is significantly lower among Africans and Asians, as well as smaller isolated populations such as Fijian Islanders and the Inuit. These defects are associated with the L/M array on the X chromosome and are inherited as X-linked recessive traits, so the incidence in females is much lower (approximately 0.4%). Nonetheless, approximately 15% of females are heterozygous carriers of a red–green color-vision defect. It has been suggested that there may be a heterozygous advantage for female carriers, as given the variability with the L and M spectral classes, these women can have four spectrally distinct cone types, providing the photoreceptor basis for tetrachromatic color vision.

The genetic causes of inherited red–green color-vision deficiency fall into two main categories. The most common cause is a rearrangement of the L/M genes resulting either in the deletion of all but one L or M pigment gene, or in the production of a gene array in which the first two genes encode a pigment of the same spectral class (i.e., L/L or M/M). Even though the L/M array can contain more than two genes, it is believed that only the first two genes in the array are expressed. The second general cause involves the introduction of a mutation in either the first or the second gene in the array, rendering the expressed pigment nonfunctional. The most prevalent inactivating mutation, accounting for nearly 10% of red–green dichromacy, results in the substitution of arginine for cysteine at position 203 (C203R) in the L/M opsin molecule. Mutating the corresponding cysteine residue in human rhodopsin (position 187) causes autosomal dominant retinitis pigmentosa (RP). This cysteine residue forms an essential disulfide bond and is highly conserved among all G-protein-coupled receptors. Mutant photopigments do not fold properly, and are retained in the endoplasmic reticulum and not targeted to the cone outer segment membrane. As such, L/M cones expressing this mutant pigment are nonfunctional, and individuals harboring this mutation base their color vision on an S cone and a single L- or M-cone type (rendering them dichromatic).

In the case where the first two genes encode pigments from the same spectral class, they can have either the same spectral sensitivity or slightly different spectral sensitivity (owing to the fact that there is variation within the L and M pigment classes). If the first two genes encode pigments with identical spectral sensitivities, the individual again bases their color vision on an S cone and a single L- or M-cone type and will be dichromatic. However, if the first two genes encode pigments with different spectral sensitivities (though from the same spectral class, L or M), the individual is technically trichromatic, in that they have three different cone types. However, since the spectral separation between their two types of L (or M) pigment is not as great as the separation between L and M, their discrimination is not normal. Thus, these individuals are known as anomalous trichromats. The standard L and M photopigments differ in their peak spectral sensitivity by nearly 30 nm. In anomalous trichromats with multiple L (or M) pigments, the λ_{max} may be separated by as few as 2 nm or as many as 10–12 nm, where the degree of spectral separation depends on the identity of the amino acids at sites 65, 111, 116, 153, 171, 174, 178, 180, 230, 233, and 236 within the L/M pigment (**Figure 2(a)**). In general, the further apart the pigments, the better the discrimination; in some cases, the deficiency is so mild that the individual is unaware of it until genetic and/or behavioral testing. Likewise, as the λ_{max} of the pigments gets closer together, the discrimination worsens, with the most severe individuals behaving nearly like dichromats.

The red/green defects can be separated based on the (1) dimensionality of the residual color vision (dichromat or anomalous trichromat) and (2) spectral subtype of the remaining cone (protan or deutan). **Figure 3** shows the different spectral sensitivity curves that underlie the different forms of red/green color-vision deficiency.

Individuals with an absence of L-cone function are said to have a protan defect. Protanopes are dichromats who possess an S pigment and an M pigment. Protanomalous trichromats possess a normal S pigment and two spectrally distinct M pigments. Perceptually, the absence of a cone type can have differing effects. Individuals with a protan defect are less sensitive to light in the long-wavelength (red) portion of the spectrum. Therefore, the brightness of red, orange, and yellow are reduced compared with a normal observer. Furthermore, they may have problems in distinguishing red from green, as well as difficulties differentiating a red hue from black.

Individuals with an absence of M-cone function are said to have a deutan defect. A deuteranope possesses an

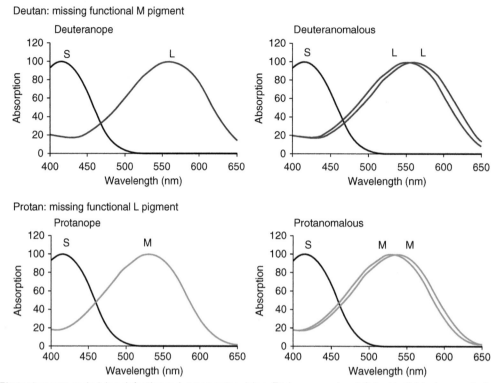

Figure 3 Photopigments underlying defective red–green color vision. Red–green color-deficient individuals are missing either all members of the M class or all members of the L class of pigment. Dichromats have only one pigment in the L/M region of the spectrum, whereas trichromats have two pigments in the L/M region of the spectrum. The degree of color-vision deficiency in persons with anomalous trichromacy depends on the magnitude of the spectral difference between their pigment subtypes. Deuteranopes and deuteranomalous trichromats have no functional M pigment, though deuteranomals have two slightly different L pigments. Likewise, protanopes and protanomalous trichromats have no functional L pigment, though protanomals have two slightly different M pigments.

S pigment and an L pigment, whereas deuteranomalous trichromats possess an S pigment and two spectrally distinct L pigments. Deuteranomalous defects are by far the most prevalent of any of the congenital color-vision defects, affecting nearly 5% of men. Individuals with deutan defects exhibit a reduction of sensitivity to colors in the green region of the spectrum, though the decrease in sensitivity is less pronounced than the long-wavelength depression in protans due to the manner in which the spectral sensitivities overlap (**Figure 1**). Deutans suffer similar hue-discrimination problems as the protans, but without the long-wavelength dimming. These discrimination errors are exploited in the design of color-vision tests; however, the real-world consequences of having certain red–green deficiency can be minimal. The perceptual consequences of having various color-vision defects are simulated in **Figure 4**. Interestingly, there is evidence that individuals with red–green defects may actually have an advantage when viewing camouflage, which is designed to blend into the environment, but this is done assuming a trichromatic visual system.

While the red–green color-vision defects can be behaviorally classified according to the cone subtype that is nonfunctional, the structural basis of the defects has been unclear. Recent work using high-resolution, *in vivo* imaging of the human retina has shown that while conceptually one can think of a deutan retina containing only S and L cones, and a protan retina containing only S and M cones, the residual cone mosaic in red–green-deficient individuals can vary dramatically, depending on the genetic cause of the defect. Shown in **Figure 5** are images of the cone mosaic from a normal trichromat and an individual with a red–green defect caused by an inactivating mutation in his M gene. Each circle is a single cone photoreceptor, and the dark spaces in the red–green-deficient retina represent cones that have degenerated or that are morphologically compromised. Thus, some individuals will have normal numbers of cones (though only two, instead of three types), while some will be structurally missing an entire cone class. The impact on vision in general is not yet known, and the genetic heterogeneity within the red–green defects remains to be reconciled with the mosaic phenotype.

Tritan Color-Vision Deficiencies

Tritan defects are an inherited autosomal dominant abnormality of S-cone function. These defects occur in

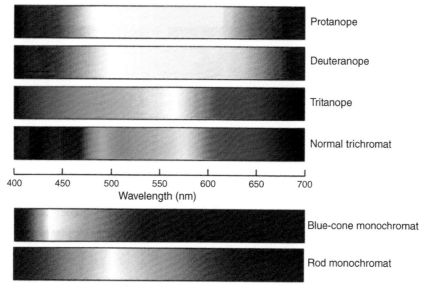

Figure 4 Perceptual consequences of inherited color-vision defects. Shown is a computer simulation of the color spectrum as it would appear to a protanope, deuteranope, tritanope, and normal trichromat. Each color-vision deficiency shows greatly reduced chromatic discrimination compared to that of a normal trichromat. The bottom two panels reveal the perceptual consequences of monochromacy on the appearance of the spectrum for a blue-cone monochromat and rod monochromat. Reproduced from Gegenfurtner, K. R. and Sharpe, L. T. (1999). *Color Vision: From Genes to Perception*. New York: Cambridge University Press, with permission of Cambridge University Press.

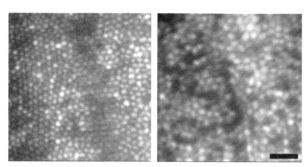

Figure 5 High-resolution images of the living retina obtained with adaptive optics. On the left is a retinal image from a patient with normal vision, at an eccentricity of approximately 1° temporal to fixation. Each circular structure is an individual cone photoreceptor. On the right is an image from a patient who is red–green color blind. Visible are numerous holes where cones have either died or degenerated, due to a mutation in one of the genes responsible for normal color vision. Despite the loss of nearly 33% of his functioning cones, this person has normal visual acuity. Scale bar = 20 μm.

both males and females with equal frequency and are believed to be rare, affecting as few as 1 in 10 000 people. As the S-pigment gene is on chromosome 7 and humans are diploid, each S-cone photoreceptor expresses the S-cone pigment genes from both copies of chromosome 7. Consequently, a defect in one S-pigment gene can be sufficient to cause tritanopia, just as a mutation in one rhodopsin gene is sufficient to cause RP, a retinal degeneration that involves the degeneration of the rod photoreceptors.

Unlike the red–green defects, there is no genetic/pigment basis for a tritanomalous defect, as there is no spectral variation among functional S-cone pigments.

This disorder exhibits incomplete penetrance, meaning that individuals with the same underlying mutation manifest different degrees of color-vision impairment, even within a sibship. Mutations in the S-cone-opsin gene, which encodes the protein component of the S-cone photopigment, have been identified and they give rise to five different single amino acid substitutions that have only been found in affected individuals but not in unaffected control subjects (**Figure 2(b)**). Each substitution occurs at an amino acid position that lies in one of the transmembrane alpha helices, and is predicted to interfere with folding, processing, or stability of the encoded opsin. For example, one of the identified mutations corresponds to an amino acid position at which a substitution in the rod pigment, rhodopsin, causes autosomal dominant RP.

A fundamental difference between S cones and L/M cones is the potential for dominant negative interactions between normal and mutant opsins. This is because each S cone expresses both autosomal copies of the S-opsin gene, whereas L and M cones each only express one gene from the L/M array on the X chromosome. Rod photoreceptors also express a rhodopsin gene from two autosomes, and for rhodopsin mutations underlying autosomal dominant RP, dominant negative interactions lead to the death of the photoreceptors and ultimately the degeneration of the retina. Curiously, tritan defects have not been reported to cause progressive retinal degeneration and are only slightly more rare than adRP; however, it has been suggested that the relative paucity of S cones compared to rods in the normal retina may be responsible for the absence of more general retinal degeneration. Nevertheless, the structural homology between the S-cone opsin and rhodopsin, taken together with the similar molecular mechanisms underlying the defects, suggest that S cones themselves degenerate in autosomal dominant tritan defects.

Blue-Cone Monochromacy

Blue-cone monochromacy (BCM) is a condition where both L- and M-cone function is absent. This disorder affects approximately 1 in 100 000 individuals. Since L and M cones comprise about 95% of the total cone population, these individuals have a rather severe visual impairment, including photophobia (light sensitivity), poor acuity, minimally detectable electroretinogram (ERG) responses, and diminished color discrimination. Any residual color vision in these individuals must be based solely on the S cones and rods. This means that under most conditions, they are monochromatic, though under mesopic (dim light; between photopic and scotopic levels) conditions they may gain some additional discrimination capacity.

As with the red–green defects, there are two main genetic causes of BCM, sometimes referred to as one-step or two-step mutations, though both lead to the absence of functional L and M cones. One-step mutations involve a deletion of essential *cis*-regulatory DNA elements needed for normal expression of the pigment genes. Two-step mutations involve a deletion of all but one of the X-chromosome visual pigment genes. This would normally confer red–green dichromacy (see above), but the one remaining gene contains a missense mutation. Due to the X-linked nature of the condition, female carriers are spared from a full manifestation of the associated defects, but they can show abnormal cone function on the ERG. One-step and two-step conditions may have important phenotypic differences in terms of the architecture of the cone mosaic in carriers. For the one-step mutations, the absence of an essential enhancer means that the cone photoreceptors cannot transcribe an opsin gene from the affected X chromosome. In contrast, for the two-step condition, there is a completely functional gene but the encoded opsin has a deleterious amino acid substitution, and the photoreceptor is expected to produce the mutant opsin. Depending on the nature of the mutation, it may either reduce or abolish proper folding of the encoded protein, or the gene may be transcribed but the message may be immediately targeted for degradation, and these may in turn ultimately affect the viability of the cone or its neighboring cells. While both conditions will likely compromise the viability of the cone, they may do so over different timescales.

Achromatopsia

Complete achromatopsia (i.e., rod monochromacy) is a congenital vision disorder in which all cone function is absent or severely diminished. It is typically characterized by an absolute lack of color discrimination, photophobia, reduced acuity, visual nystagmus, and a nondetectable cone ERG. Previously, exploration of this disease was limited to findings from histological and anatomical review and there was a debate surrounding whether the cones were absent, malformed, or merely minimally present. Recent work using *in vivo* cellular imaging, together with improved measures of cone function, suggest that not all achromats have the same cellular deficit, though in all cases their perception is dominated by the rod system and vision at normal light levels can be difficult if not impossible.

Rod monochromacy affects up to 1 in 30 000 people, and results from mutations in one of two components of the cone phototransduction cascade – transducin or the cyclic-nucleotide-gated (CNG) channel. Mutations in CNG alpha 3 (CNGA3) and CNG beta 3 (CNGB3), which encode the α- and β-subunit of the CNG channel, respectively, are by far the most common genetic cause of rod monochromacy, accounting for about 60–80% of the cases. A relatively small number of patients (~5%) have a mutation in the GNAT2 gene, which encodes the α-subunit of the cone G-protein transducin. The current theory is that the genetic heterogeneity underlies the observed phenotypic variability in patients with this disease, and this likely explains the previous discrepancies in histological reports. However, a systematic linkage of genotype and phenotype has not been done. Recently, a gene therapy approach has been used to restore cone function in dog and mouse models of the disease; however, one would predict that the degree of remaining cone structure would be a predictor of whether a particular human achromat could benefit from similar therapeutic intervention.

Conclusion

Inherited color-vision deficiencies range from mild difficulties discerning pale hues from gray to quite severe issues differentiating even the most saturated hues, and while most of these deficiencies are often categorized by the general misnomer color blindness, the defects are in fact quite different from one another and rarely indicate a condition in which the patient is truly blind to color. There is enormous genetic variation in all inherited color-vision defects, and it is becoming clear that this variation has consequences for the visual system that reach far beyond subtle differences in color perception/discrimination.

See also: The Colorful Visual World of Butterflies; Cone Photoreceptor Cells: Soma and Synapse; Phototransduction: Adaptation in Cones; Phototransduction: Inactivation in Cones; Phototransduction: Phototransduction in Cones; Polarized-Light Vision in Land and Aquatic Animals; Primary Photoreceptor Degenerations: Retinitis Pigmentosa; Primary Photoreceptor Degenerations: Terminology; Rod and Cone Photoreceptor Cells: Outer Segment Membrane Renewal.

Further Reading

Birch, J. (1993). *Diagnosis of Defective Colour Vision*. New York: Oxford University Press.

Carroll, J., Choi, S. S., and Williams, D. R. (2008). *In vivo* imaging of the photoreceptor mosaic of a rod monochromat. *Vision Research* 48: 2564–2568.

Cruz-Coke, R. (1970). *Color Blindness: An Evolutionary Approach*. Springfield, IL: Charles C Thomas.

Dalton, J. (1798). Extraordinary facts relating to the vision of colours: With observations. *Memoirs of the Literary and Philosophical Society of Manchester* 5: 28–45.

Gegenfurtner, K. R. and Sharpe, L. T. (1999). *Color Vision: From Genes to Perception*. New York: Cambridge University Press.

Hess, R. F., Sharpe, L. T., and Nordby, K. (1990). *Night Vision: Basic, Clinical, and Applied Aspects*. New York: Cambridge University Press.

Mollon, J. D. (2003). The origins of modern color science. In: Shevell, S. K. (eds.) *The Science of Color*, 2nd edn., pp. 1–39. New York: Elsevier.

Nathans, J., Piantanida, T. P., Eddy, R. L., Shows, T. B., and Hogness, D. S. (1986). Molecular genetics of inherited variation in human color vision. *Science* 232(4747): 203–210.

Pokorny, J. (1979). *Congenital and Acquired Color Vision Defects*. New York: Grune and Stratton.

Sharpe, L. T., Stockman, A., Jägle, H., et al. (1998). Red, green, and red–green hybrid pigments in the human retina: Correlations between deduced protein sequences and psychophysically measured spectral sensitivities. *Journal of Neuroscience* 18(23): 10053–10069.

The Colorful Visual World of Butterflies

F D Frentiu, University of Queensland, St. Lucia, QLD, Australia

Glossary

Gene duplication – Duplication of a region of DNA containing a gene, thought to be one of the most powerful evolutionary mechanisms for diversification.

Filtering pigment – Molecules present in photoreceptor cells that act as spectral filters by absorbing particular wavelengths of light.

λ_{max} – Wavelength of peak spectral absorbance of a visual pigment, measured in nm.

Neofunctionalization – The origin of a new gene function by mutation.

Phylogeny – A description of the evolutionary relationships among species inferred to have descended from a common ancestor.

Positive selection – Process through which advantageous mutations increase in frequency in a population.

Visual pigment – An opsin protein bound to a light-sensitive molecule, the chromophore, which is derived from retinal.

Introduction

Butterflies are some of the most colorful animals on the planet. The diversity of their wing colors and patterns has not only inspired human artistic expression but it has also made these animals prominent study systems in biology. For example, butterflies have become leading model animals for the study of evolution and development, particularly of how wings are patterned. Equally, our understanding of defensive signaling is based largely on butterflies, for example, mimicry in *Heliconius* species where two butterflies unpalatable to predators resemble one another in color and pattern. Butterflies also play an important ecologic and economic role by pollinating flowers and orchard crops, where they utilize vision among other senses to find sources of nectar. Understanding vision in butterflies therefore gives us a fascinating insight into how these animals might see their surroundings, their own wing colors, and may also inform us on their patterns of habitat use and pollination.

So what do butterflies see? Can they see in color and if so, how do they use color vision? Here, what we currently know about color vision in the butterflies within an evolutionary context has been reviewed. First, this article provides an overview of the butterfly group and its evolutionary history. Subsequently, a mechanistic account of how the structure of the butterfly eye makes color vision possible is provided, paying particular attention to the molecular basis of the visual pigments that enable this type of vision to occur. Next, is discussed the evolutionary processes that have led to a diversification of the visual systems present in this group that is unparalleled in other insects. Finally, the ecological significance of color vision in the butterflies is explored.

The Butterflies and Their Evolutionary History

The butterflies (Rhopalocera) are a charismatic group of insects that includes an estimated 15 000 species. This is a recently evolved group within the order Lepidoptera, which includes the moths. The Rhopalocera comprises the true day-flying butterflies (Papilionoidea), the skippers (Hesperioidea), and a newly identified group of nocturnal butterflies (Hedyloidea) (**Figure 1**). Within the true butterflies (Papilionoidea), which are the subject of most studies, five families are recognized. The families are: the Papilionidae (including the swallowtails and birdwings), the Pieridae (including the whites and sulfurs), the Lycaenidae (the blues and coppers), the Nymphalidae (the brush-footed butterflies, including the famous monarch, the morphos, and frittilaries), and the Riodinidae (the metalmarks) (**Figure 1**).

The butterflies are thought to have originated on the ancient continent of Gondwana during the Cretaceous period, sometime prior to the extinction of dinosaurs at the Cretaceous–Tertiary (K/T) boundary and possibly concurrent with the radiation of the flowering plants. However, the paucity of fossil butterflies has led to scientific disagreement as to the exact date of when butterflies first appeared, with estimates of their age ranging from 150 to 70 Ma. New molecular approaches are helping resolve some of these controversies. Molecular phylogenies (i.e., reconstructions of the lineages of species using DNA sequence data) have indicated that the Nymphalidae and the Pieridae families were probably present at the K/T boundary, around 65 Ma, and in the case of the Pieriae, may date to 100 Ma. Definitive resolution of the age of the butterflies and timing of divergence of the major families, however, awaits the development of additional molecular markers.

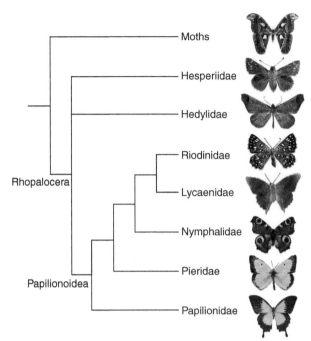

Figure 1 Evolutionary relationships of the major lineages in the Rhopalocera.

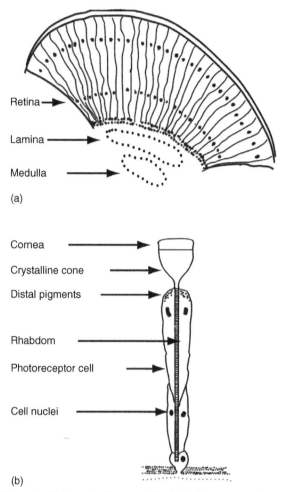

(a)

(b)

Figure 2 Schematic diagram of the butterfly compound eye. (a) Diagram of a longitudinal section through the eye showing the numerous repeated ommatidia comprising the butterfly retina as well as non photoreceptor regions (lamina and medulla). (b) Diagram of a longitudinal cross-section of an individual ommatidium. Reproduced from Frentiu, F. D., et al. (2007). © National Academy of Sciences USA.

Color Vision and the Butterfly Eye

Most organisms detect light through the presence of photosensitive molecules located in specialized organs, such as eyes. However, only arthropods (the phylum that includes insects, spiders, and crustaceans) and vertebrates have developed eyes capable of true color vision. Color vision is the ability to discriminate between two visual stimuli based on their wavelength regardless of their relative intensities. Color vision also requires the presence of at least two photoreceptors with overlapping spectral ranges so that the same point in space can be compared.

Despite the observations that butterflies use colored flowers as food sources and that they employ color-based signaling via their wings, it was only relatively recently that color vision was demonstrated in butterflies using behavioral tests. *Papilio* butterflies were trained to associate rewards containing sucrose with particular colors. When butterflies were behaviorally tested, a majority of animals chose colors previously associated with food rewards among an array of colors regardless of their intensity.

Butterflies, like other insects, have compound eyes that contain thousands of units called ommatidia (**Figure 2(a)**). In butterflies, each ommatidium contains nine photoreceptor cells (cells R1 to R9) (**Figure 2(b)**), that is, cells that possess light-sensitive visual pigments that make color vision possible. The cell membranes of the photoreceptors are folded into microvilli (cell membrane projections) that form the rhabdomeres. The rhabdomeres of one

ommatidium form a cylinder (the rhabdom) that acts as an optical waveguide for light passing through (**Figure 2(b)**). Light passes through the lens and is focused by the crystalline cone onto the rhabdom. When light propagates down the rhabdom, the majority of it is absorbed by the visual pigments in the photoreceptor cell membranes. Light that is not absorbed by the visual pigments is reflected back by the tapetum, a structure that sits at the base of the ommatidium and is found in most butterflies except papilionids and the pierid genus *Anthocharis*. In addition to visual pigments, butterflies also possess filtering pigments that surround the rhabdom and act as spectral filters by absorbing wavelengths of light and shifting the peak spectral sensitivity of their photoreceptors to longer wavelengths.

Key to color vision in both arthropods and vertebrates are the visual pigments and, most importantly, the presence of at least two spectrally distinct types of visual pigments. Neural inputs from at least two types of photoreceptor cells that bear different visual pigments are required for color discrimination. A visual pigment comprises of an opsin protein bound to a light-sensitive molecule, the chromophore (**Figure 3**) that, in butterflies, is 11-*cis*-3-hydroxyretinal. Photons reaching the chromophore cause its photoisomerization and induce a conformational change in the opsin protein. In turn this activates a guanine nucleotide-binding protein (G-protein) that initiates the phototransduction cascade that converts light into signals to the brain through a series of biochemical reactions. The eyes of invertebrates employ fundamentally different phototransduction cascades than those of vertebrates.

The absorbance spectrum of a visual pigment depends on the interaction of the chromophore with critical amino acids in the opsin protein. By itself, the chromophore has a wavelength of maximum absorption (the λ_{max} value) in the ultraviolet (UV) part of the light spectrum at approximately 380 nm. However, through the chromophore interacting with key amino acids in the binding pocket of the opsin protein, a diversity of visual pigment λ_{max} values can be achieved, a phenomenon called spectral tuning. The sensitivities of different photoreceptors to light (**Figure 4**) are determined by the opsins that they express and any associated filtering pigments they may contain. The visual pigments of butterfly species sampled to date range in λ_{max} from 340 to 600 nm (**Table 1**), although the exact opsin amino acids involved in producing this diversity remain to be elucidated.

The opsins that form the basis of visual pigments are ancient molecules that belong to the large G-protein-coupled receptor family. They have a seven transmembrane domain structure, with a diagnostic lysine residue in the seventh helix that binds to the chromophore. The opsin family predates the emergence of the major groups of animals present today. Phylogenetic reconstructions suggest that the ancestral arthropod may have been able to see in color, including in the ultraviolet. Insect ultraviolet/blue (UV/B) and blue-green opsins originated early in arthropod evolution. Distinct insect UV and B opsins then evolved via duplication of the ancestral UV/B opsin and the long wavelength (LW) opsin evolved from a duplication of the ancestral blue-green opsin. The blue-green opsin was however lost in most insects, with the exception of some flies (e.g., *Drosophila*). Insects such as bees, butterflies, and moths now possess at least three classes of visual pigment in their photoreceptors that enable them to see in the UV (UV, 300–400 nm), blue (B, 400–500 nm), and long wavelength (LW, 500–600 nm) parts of the light spectrum. However, compared to moths and bees, butterflies display an unusual diversification of their visual systems. In the following, I explore the diversification of butterfly eyes and the evolutionary processes that have produced it.

Spectral Heterogeneity of Butterfly Eyes

Although ommatidia are anatomically identical in structure, they are spectrally very heterogeneous: that is, they hold different complements of photoreceptors that express different visual pigments; different photoreceptors within an ommatidium are designated as R1-R9 and may express different pigments. Using molecular genetic techniques such as *in situ* hybridization to visualize patterns of opsin mRNA has allowed us to map the types of ommatidia present in the butterfly eye. Three types of ommatidia exist in the main retinas of butterflies, with all types of ommatidia expressing LW visual pigments but differing in the expression of the UV and B visual pigments.

The extent of spectral heterogeneity of butterfly eyes differs among the major butterfly families. For example,

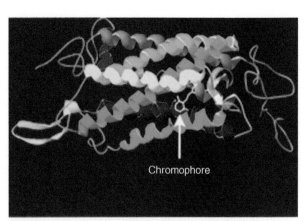

Figure 3 Three-dimensional model of a visual pigment comprising a seven-transmembrane opsin protein and a chromophore (indicated by the arrow). The seven transmembrane domains are shown in different colors. Reproduced from Frentiu, F. D., et al. (2007). © National Academy of Sciences USA.

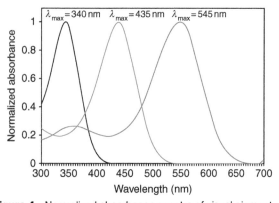

Figure 4 Normalized absorbance spectra of visual pigments in the eye of the monarch, *Danaus plexippus*. Wavelengths of peak absorbance (λ_{max}) for the three visual pigments are estimated to be 340, 435, and 545 nm. Peak absorbance values from Stalleicken et al. (2006) *Journal of Comparative Physiology A* 192: 321–331.

Table 1　Approximate λ_{max} of visual pigments in different butterfly species.

Family	Subfamily	Species	UV	B	LW
Lycaenidae	Lycaeninae	*Lycaena rubidus*	360	437, 500	568
		Lycaena heteronea	360	437, 500	568
		Lycaena dorcas	360	437, 500	568
		Lycaena nivalis	360	437, 500	568
Nymphalidae	Apaturinae	*Asterocampa leilia*	–	–	530
		Sasakia charonda	345	425, 440	540
	Charaxinae	*Archeoprepona demophon*	–	–	565
	Danainae	*Danaus plexippus*	340	435	545
	Heliconiinae	*Agraulis vanillae*	–	–	555
		Heliconius charitonia	–	–	550
		Heliconius erato	370	470	555
		Heliconius hecale	–	–	560
		Heliconius sara	–	–	550
	Limenitidinae	*Limenitis archippus archippus*	–	–	514
		Limenitis archippus floridensis	–	–	514
		Limenitis arthemis astyanax	–	–	545
		Limenitis lorquini	–	–	530
		Limenitis weidemeyerii	–	–	530
	Nymphalinae	*Aglais urticae*	380	460	530
		Anartia jatrophae	–	–	530, 565
		Euphydryas chalcedona	–	–	565
		Inachis io	–	–	530
		Junonia coenia	–	–	510
		Nymphalis antiopa	–	–	534
		Polygonia c-album	–	–	532
		Polygonia c-aureum	350	450	540, 565
		Siproeta stelenes	–	–	522
		Vanessa cardui	360	470	530
	Satyrinae	*Hermeuptychia hermes*	–	–	530
		Neominois ridingsii	–	–	515
		Oeneis chryxus	–	–	530
		Pararge aegeria	360	460	530
Papilionidae	Papilioninae	*Papilio xuthus*	360	460	530, 515, 575
Pieridae	Pierinae	*Pieris rapae*	360	425, 453	563
Riodinidae	Riodininae	*Apodemia mormo*	340	450	505, 600

the eyes of Nymphalidae species are quite simple in terms of ommatidial heterogeneity. Studies of Nymphalidae species to date show that they have only three types of ommatidia in the main retina and one type in the dorsal rim area (DRA) of the eye, which is an eye region specialized for the detection of polarized skylight. All ommatidial types in the main retina express long wavelength visual pigments in their R3-R8 photoreceptor cells, but the expression of the UV and B opsins in the R1 and R2 cells is variable (the R9 cell expresses the LW opsin in the main retina and may express the UV opsin in the DRA). One ommatidial type contains one UV and one blue receptor, the second has two blue receptors, and the third has two UV receptors. In the DRA ommatidia, the R1-R8 cells express the UV opsin. This type of eye employs a straightforward, one-to-one relationship between the type of visual pigment expressed and spectral phenotype of the photoreceptor cell. It may also best represent the ancestral butterfly eye and it resembles the eyes of bees and moths.

By contrast, the eyes of other families of butterflies are much more diverse in their spectral complements.

For example, the butterfly *Pieris rapae* (family Pieridae) expresses four opsins but photoreceptors with seven different peak sensitivities have been identified, due to the presence of filtering pigments. Perhaps the most spectrally diverse butterfly eye studied so far belongs to the papilionid butterfly, the Japanese swallowtail, *Papilio xuthus*. *Papilio xuthus* expresses five different opsins in the eye, has eight different types of photoreceptors, and employs tetrachromatic color vision. In these animals, both opsin gene duplications and pigments acting as spectral filters have led to spectral diversification of visual systems. Below, I consider the evolutionary mechanisms that have led to butterfly visual system diversity.

Diversification of Visual Pigments via Gene Duplication and Positive Selection

The eyes of some butterflies express a larger number of visual pigments than the three UV, B, and LW known to be present in bees and moths. The diversity of visual pigments is primarily due to opsin gene duplications that have occurred independently in different butterfly families (**Figure 5**). Duplications of the B opsin gene have occurred independently in two of the five butterfly families: the Pieridae and the Lycanenidae, giving rise to four different visual pigments. In the Pieridae, as exemplified by the well-studied *Pieris rapae* (the cabbage white), duplicate B opsins have diversified into violet- ($\lambda_{max} = 425$ nm) and blue-absorbing ($\lambda_{max} = 453$ nm) visual pigments. In the Lycaenidae, duplication of the B opsin has also facilitated the emergence of two visual pigments. However, the peak wavelength sensitivities in species such as *Polyommatus icarus* are different from those found in the Pieridae, with one visual pigment absorbing in the blue ($\lambda_{max} = 437$ nm) and the other in the blue-green ($\lambda_{max} = 500$ nm).

Duplications of the LW opsin gene have occurred independently in three butterfly families, also leading to diversification of visual pigments. In the Papilionidae, three LW opsins are now expressed in the eye, each the result of a round of gene duplication. In *Papilio xuthus*, these duplicated opsins now encode three different visual pigments with λ_{max} values ranging from 515 to 575 nm. The LW opsin gene duplications have also occurred in two species of butterflies in the Nymphalidae and one in the Riodinidae families. In total, of more than 50 butterfly species studied to date, 7 LW opsin duplicates have been identified. The most red-shifted visual pigment known to date in the riodinid butterfly *Apodemia mormo*, with a λ_{max} of 600 nm, has resulted from a gene duplication specific to the family Riodinidae. Molecular evidence has also

suggested that duplication was followed by elevated rates of amino acid evolution in one of the LW duplicates. However, not all opsin duplications have resulted in visual pigments expressed in the eye, with some expressed in the optic lobes and brain.

In addition to gene duplication, a mechanism that generates spectral diversity in butterflies is positive selection on single opsin genes, whereby novel mutations that enhance an organism's fitness spread through a population and may replace other genetic variants. The most striking example of this process to date comes from the North American butterfly genus *Limenitis*. Butterflies in this group have radiated across the North American continent from a European ancestor during the past 3–4 My. The genus is best known for species that display wing color pattern mimicry such as the viceroy *Limenitis archippus*, which mimics the monarch *Danaus plexippus*, and *L. arthemis astyanax*, which mimics the more toxic pipevine swallowtail *Battus philenor*. An unusual diversity in λ_{max} values of the LW visual pigment has been found in this group of butterflies, suggesting that their visual systems had diversified in tandem with wing color patterns (**Figure 6**). Reconstruction of phylogenetic relationships within the genus indicated that a shift in spectral sensitivity towards the blue part of the light spectrum had occurred (**Figure 6**). Wavelength sensitivities of LW visual pigments had diversified from an ancestral λ_{max} of 545 nm in *L. arthemis astyanax* to a λ_{max} of 515 nm in *L. archippus archippus*.

Using several molecular evolutionary analyses, the signature of positive selection was found at several amino acid sites in *Limenitis* LW opsins. The *Limenitis* opsin protein was modeled against the bovine rhodopsin crystal structure (homology modeling) in order to visualize where the amino acid sites were located. The results indicated that some of the amino acids found to be under positive selection were located in the chromophore-binding pocket, strongly suggesting that they interact with the chromophore to determine spectral sensitivities. The same amino acids were found to change in parallel in butterfly species that were distantly evolutionarily related to the *Limenitis* genus but that also showed a shift in λ_{max} in the same direction (**Figure 6**). Interestingly, one of the amino acid sites under positive selection in butterflies is evolutionarily homologous to an amino acid site in the cone opsin of humans that is responsible for a 5–7 nm shift to the blue part of the light spectrum and which is under balancing selection in New World monkeys. These findings suggest a common molecular basis for spectral shifts in insects and vertebrates, a feature that has been retained across more than 540 My of evolution. However, opsin amino acid sites suggested by molecular evolutionary analyses to be involved in the spectral tuning of butterfly visual pigments need to be tested and functionally characterized experimentally.

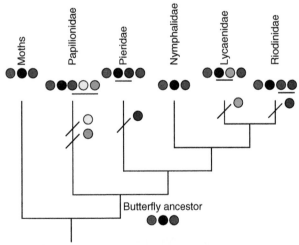

Figure 5 Evolution of visual pigment diversity in the butterflies and hypothesized complement of pigments in the ancestral butterfly eye. Circles indicate UV (gray), blue (blue), and long wavelength (orange) visual pigments. Diagonal lines along the branches of the phylogeny denote opsin duplications.

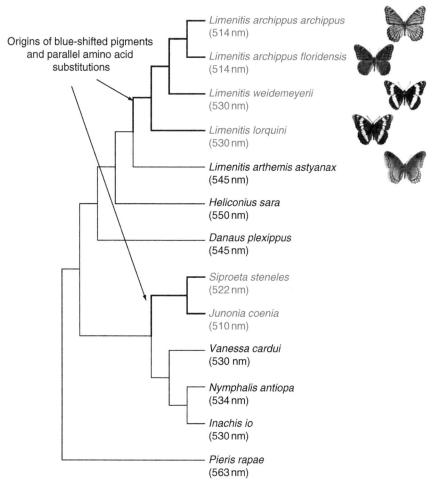

Origins of blue-shifted pigments
and parallel amino acid
substitutions

Limenitis archippus archippus
(514 nm)

Limenitis archippus floridensis
(514 nm)

Limenitis weidemeyerii
(530 nm)

Limenitis lorquini
(530 nm)

Limenitis arthemis astyanax
(545 nm)

Heliconius sara
(550 nm)

Danaus plexippus
(545 nm)

Siproeta steneles
(522 nm)

Junonia coenia
(510 nm)

Vanessa cardui
(530 nm)

Nymphalis antiopa
(534 nm)

Inachis io
(530 nm)

Pieris rapae
(563 nm)

Figure 6 Diversification of wing colors and LW visual pigment spectral sensitivities in the genus *Limenitis* following colonization of North America. Blue-shifted visual pigments ($\lambda_{max} \leq 530$ nm) in evolutionarily independent lineages of butterflies in the family Nymphalidae display parallel substitutions in the opsin protein at key amino acid sites that interact with the chromophore. Reproduced from Frentiu, F. D., et al. (2007). © National Academy of Sciences USA.

Diversification of Photoreceptor Types via Filtering Pigments

Some butterflies possess photoreceptor cells containing filtering pigments that coat the rhabdom (**Figure 1**), which are red, orange, and yellow in color. Filtering pigments act as spectral filters by absorbing particular wavelengths of light traveling down the rhabdom. Whilst the breadth of wavelength sensitivity of a photoreceptor is determined by the opsin it expresses, by absorbing shorter wavelengths of light, filtering pigments shift the λ_{max} of some photoreceptors towards longer wavelengths. The color discrimination abilities of butterflies in a particular wavelength range are thus enhanced by comparing neural inputs from two types of photoreceptors that express the same opsin but have differing peak spectral sensitivities (λ_{max}) due to the presence or absence of filtering pigments.

Filtering pigments play a role in the diversification of the peak spectral sensitivity of photoreceptors in the ommatidia of some butterflies. The same visual pigment, in conjunction with filtering pigments, is used to expand photoreceptor sensitivities in *Heliconius* butterflies. For example, in *Heliconius erato*, two types of LW photoreceptors express the same LW visual pigment but differ in peak spectral sensitivity as a result of the presence of a red-filtering pigment. This expanded wavelength discrimination in the LW range is used in foraging behavior. In *Pieris rapae*, filtering pigments expressed in conjunction with the same LW opsin are used to produce two additional photoreceptor types that enhance wavelength discrimination in the red part of the light spectrum, although the behavioral and ecological reasons for this pattern are unclear.

Currently very little is known about the genetic basis, molecular identity, evolution, and ecological significance of butterfly filtering pigments. Both *Pieris* and *Papilio* butterflies have lateral filtering pigments suggesting that filtering pigments are an evolutionarily old feature of the butterfly eye. Some of the pigments used in butterfly wing colors are molecularly similar to eye pigments found in *Drosophila*, although butterflies may employ a much larger repertoire of pigments than flies. Tantalizing clues from a handful of studies suggest that, aside from diversifying photoreceptor sensitivities, filtering pigments may play a role in the co-evolution of photoreceptors and color-based mating signals. The same genes that mediate both spectral sensitivities in photoreceptors via filtering pigments may also influence the expression of wing pigments expressed in butterfly wings. However, this hypothesis remains to be tested empirically by identifying and molecularly characterizing the genes involved. Filtering pigments and their expression in the butterfly eye offer a fruitful avenue for research into the co-evolution of animal color-based signals and the ability to perceive them.

Evolution of Sexually Dimorphic Eyes: Opsin Duplications and Sex-Specific Filtering Pigment Expression

One of the most interesting aspects of butterfly vision is that some species display sexually dimorphic eyes: males and females detect different colors in the same areas of the eye. This feature appears to have evolved independently in two butterfly families, the Lycaenidae and the Pieridae, via two different mechanisms. In the North American butterfly *Lycaena rubidus* (Lycaenidae), duplication of the B opsin has facilitated the evolution of a sexually dimorphic eye. The eyes of this butterfly express four opsin genes, one UV, two B (B1 and B2), and one LW. In the dorsal part of the eye, *L. rubidus* males exclusively express the B1 opsin in the R3-R8 photoreceptors, whereas in the female both B1 and LW are expressed. The R1 and R2 cells primarily express UV opsins in this dorsal part of the eye. Therefore, duplication and neofunctionalization of the B1 domain of expression have potentially allowed the evolution of dichromatic and trichromatic color vision in this part of the eye in males and females respectively. The male *L. rubidus* may use dichromatic color vision in the dorsal part of the eye in behaviors associated with defense of its mating territory. If opsin duplication has facilitated the evolution of sexually dimorphic eyes, we may yet discover these eyes to be more prevalent in the butterflies than previously thought given how frequently such duplications have occurred in this group of insects.

In *Pieris rapae* subspecies *crucivora* (Pieridae), sexually dimorphic eyes have also been documented, however in this case mediated by the sexually differentiated expression of filtering pigments in some photoreceptors. The sexual dimorphism occurs in the violet photoreceptors of this species present in some ommatidia. In females, violet photoreceptors match their expected peak spectral sensitivity ($\lambda_{max} = 425$ nm). However, in males, the expression of a violet-absorbing filtering pigment modifies violet photoreceptors into double-peaked blue photoreceptors that have a much narrower spectral sensitivity, with a λ_{max} of 416 nm. This sexual dimorphism may aid males in discriminating the different wing colors of males and females that differ in the short wavelength part of the spectrum, but this hypothesis requires empirical testing through behavioral tests. Not all subspecies of *Pieris rapae* show sexually dimorphic filtering pigment expression, suggesting that pigment-mediated changes in spectral sensitivities may rapidly facilitate fine-tuning of visual systems according to the ecology of a particular species or subspecies.

Ecological Significance of Butterfly Visual System Diversity

Why are butterfly eyes so diverse? What are the ecological reasons for such a diversity of color vision systems? To date, however, the ecological and life history factors driving the diversity observed have been far less thoroughly investigated than the molecular basis of this pattern. No clear adaptive matching of the visual pigments to ambient light, as demonstrated in fish, has ever been found in butterflies. For example, the coelacanth lives at a depth of 200 m and only receives light at around 480 nm in wavelength. Its two visual pigment sensitivities peak at around 475 and 485 nm, suggesting adaptive matching to the light environment in this species. In cichlids, visual system spectral sensitivities also seem to match the ambient light environments of these species. In butterflies, the limited evidence available to date suggests that visual pigment diversity may be driven by the social signaling and foraging needs of the animals rather than their particular light environment.

Like many other insects, butterflies utilize vision for a whole range of tasks. They use vision to locate food sources, select suitable oviposition sites, and in intraspecific communication. *Papilio* butterflies use two of their duplicated LW opsins to see in the green part of the light spectrum when ovipositing and foraging. The lycaenid butterfly *Polyommatus icarus* uses one of the duplicate B opsins together with the LW opsin-based visual pigment to see green up to 560 nm when foraging. Importantly, opsin gene duplications resulting in the diversification of photoreceptor types have clearly impacted on the foraging behavior of butterflies.

Mate choice in many butterfly species also occurs via color-based signaling. For example, *Heliconius* butterflies appear to use both colored and polarized light for finding mates. The size of UV-reflecting eye spots on the wings seems to correlate with mating success in *Bicyclus anynana*. These observations suggest that there might be selective pressure for visual systems to evolve to detect the specific spectral signals of wings. Correlative evidence suggests that photoreceptor sensitivities have evolved in tandem with wing coloration but empirical evidence is needed.

Conclusions

It is clear that the visual systems enabling butterflies to see in color are functionally very diverse compared to other insects, particularly their moth ancestors. Visual system diversity has been achieved primarily through opsin gene duplication, positive selection at single opsin loci, and heterogeneous expression of filtering pigments in photoreceptors. The evolutionary basis of this pattern of diversification has recently received significant attention, however the exact molecular basis of spectral tuning of butterfly visual pigments remains unknown. Butterfly visual systems offer an excellent opportunity to study the evolutionary and functional genetic links from genotype to phenotype through to behavioral and ecological consequences. To this end, further work is required to explicitly link the pattern of photoreceptor diversification in different butterflies to their ecology, behavior, and life history traits. Only then will we gain a complete understanding of the evolutionary significance of visual system diversification in these insects.

See also: Circadian Rhythms in the Fly's Visual System; Color Blindness: Acquired; Color Blindness: Inherited; Microvillar and Ciliary Photoreceptors in Molluskan Eyes; The Photoresponse in Squid; Phototransduction: Rhodopsin; Phototransduction: The Visual Cycle.

Further Reading

Arikawa, K., Wakakuwa, M., Qiu, X., Kurasawa, M., and Stavenga, D. G. (2005). Sexual dimorphism of short-wavelength photoreceptors in the small white butterfly, *Pieris rapae crucivora*. *Journal of Neuroscience* 25: 5935–5942.

Bernard, G. D. (1979). Red-absorbing visual pigment of butterflies. *Science* 203: 1125–1127.

Bernard, G. D. and Remington, C. L. (1991). Color vision in *Lycaena* butterflies: Spectral tuning of receptor arrays in relation to behavioral ecology. *Proceedings of the National Academy of Sciences of the United States of America* 88: 2783–2787.

Briscoe, A. D. (2008). Reconstructing the ancestral butterfly eye: Focus on the opsins. *Journal of Experimental Biology* 211: 1805–1813.

Frentiu, F. D., Bernard, G. D., Sison-Mangus, M. P., Brower, A. V. Z., and Briscoe, A. D. (2007). Gene duplication is an evolutionary mechanism for expanding spectral diversity in the long-wavelength photopigments of butterflies. *Molecular Biology and Evolution* 24: 2016–2028.

Frentiu, F. D., Bernard, G. D., Cuevas, C. I., et al. (2007). Adaptive evolution of color vision as seen through the eyes of butterflies. *Proceedings of the National Academy of Sciences of the United States of America* 104: 8634–8640.

Frentiu, F. D. and Briscoe, A. D. (2008). A butterfly eye's view of birds. *BioEssays* 30: 1151–1162.

Kelber, A. (1999). Ovipositing butterflies use a red receptor to see green. *Journal of Experimental Biology* 202: 2619–2630.

Kelber, A. and Pfaff, M. (1999). True color vision in the orchard butterfly, *Papilio aegeus*. *Naturwissenschaften* 86: 221–224.

Kinoshita, M., Shimada, N., and Arikawa, K. (1998). Colour vision of the foraging swallowtail butterfly *Papilio xuthus*. *Journal of Experimental Biology* 202: 95–102.

Sison-Mangus, M. P., Bernard, G. D., Lampel, J., and Briscoe, A. D. (2006). Beauty in the eye of the beholder: The two blue opsins of lycaenid butterflies and the opsin gene-driven evolution of sexually dimorphic eyes. *Journal of Experimental Biology* 209: 3079–3090.

Stalleicken, J., Labhart, T., and Mouritsen, H. (2006). Physiological characterization of the compound eye in monarch butterflies with focus on the dorsal rim area. *Journal of Comparative Physiology A - Neuroethology Sensory Neural and Behavioral Physiology* 192: 321–331.

Stavenga, D. G. and Arikawa, K. (2006). Evolution of color and vision of butterflies. *Arthropod Structure and Development* 35: 307–318.

Terakita, A. (2005). The opsins. *Genome Biology* 6: Article No 213.

Wahlberg, N., Braby, M. F., Brower, A. V. Z., et al. (2005). Synergistic effects of combining morphological and molecular data in resolving the phylogeny of butterflies and skippers. *Proceedings of the Royal Society B: Biological Sciences* 272: 1577–1586.

Wakakuwa, M., Stavenga, D. G., Kurasawa, M., and Arikawa, K. (2004). A unique visual pigment expressed in green, red and deep-red receptors in the eye of the small white butterfly, *Pieris rapae crucivora*. *Journal of Experimental Biology* 207: 2803–2810.

Cone Photoreceptor Cells: Soma and Synapse

R G Smith, University of Pennsylvania, Philadelphia, PA, USA

Glossary

Cascade – A sequence of biochemical reactions that process a neural signal.

Connexin – A molecule comprising a gap junction that sits in a neuron's membrane and can join with a connexin from another neuron to create a large nonspecific pore that conducts ions and small molecules.

Ephaptic – Electrical conduction across a synapse between neurons without the mediation of a neurotransmitter.

Henle fiber – The long axon of a foveal cone that extends laterally to terminate outside of the fovea.

Invagination – A permanent infolding of a cell's external membrane, associated in photoreceptors with their synaptic ribbons, and containing fine dendritic processes of bipolar and horizontal cells.

Invert – In a postsynaptic cascade, to convert a rising signal into a falling signal.

Low-pass filter – A filter that removes high frequencies and passes low frequencies.

Mesopic – The 3-log unit range of background illuminance in which rod and cone signals temporally sum in cones and the cone bipolar pathway.

Microtubule – A small tubule composed of the protein tubulin, often found as arrays in neural axons, involved in transport of cellular components.

Pedicle – The cone terminal, with a foot-like flat base, which contains synaptic ribbons.

Ribbon – A presynaptic structure that collects vesicles of neurotransmitter for release.

Telodendria – Fine axonal processes extending laterally from the base of the cone terminal, which contact neighboring rod and cone terminals.

Triad – A synapse in the cone terminal that contacts dendritic processes of two horizontal cells and a bipolar cell.

Introduction

The cone photoreceptor is responsible for vision in daylight and provides color vision for most vertebrates. The cone is specialized to compress the large range of environmental illumination into a neural signal that can be processed by bipolar and ganglion cells in the retina and passed to the brain. The cone soma and synapse support this compression by several mechanisms located in its biophysical properties and at its ribbon synapse. In addition, cones perform spatial filtering because their terminals are electrically coupled to their neighbors and to the surrounding rods through gap junctions.

Structure

Morphology and Topology

The cone is a specialized neuron consisting of an outer segment, inner segment and soma, axon, and axon terminal (**Figure 1**). The biochemical pathways responsible for transducing light into an electrical signal are contained in the outer segment. The electrical signal passes to the inner segment where it is processed by voltage-gated channels. The inner segment is contiguous with the soma but they lie in distinct retinal layers separated by the external limiting membrane. The cone soma lies in the upper row of the outer nuclear layer. For most mammalian cones, the soma is larger in diameter ($\sim$5 μm) than the outer segment. It contains the nucleus which holds the cell's DNA, necessary for development and to maintain the cell's biochemical machinery. Many mammalian species have $\sim$20-fold more rods than cones, so the remainder of the outer nuclear layer consists of several layers of rod somas. Cones of most species (and outside the fovea in primate) are spaced semirandomly with a nearest-neighbor distance that varies from $\sim$5 μm in central retina or the visual streak (e.g., cat, rabbit, and guinea pig) to $\sim$15 μm in peripheral retina, and are surrounded by rods.

Axon and Terminal

The cone axon carries the electrical signal from the soma to the axon terminal. The axon varies in length depending on the species and eccentricity (distance from central retina) of the photoreceptor. In foveal cones near the center of the primate eye, cone axons extend up to 400–600 μm laterally to allow the outer segments to be packed tightly together in the foveal pit and the axon terminals to be given adequate space outside the fovea to make their synaptic connections with other neurons. These long axons are termed Henle fibers, and the layer containing the horizontally projecting axon fibers is termed Henle's layer. In other mammals, the cone axon

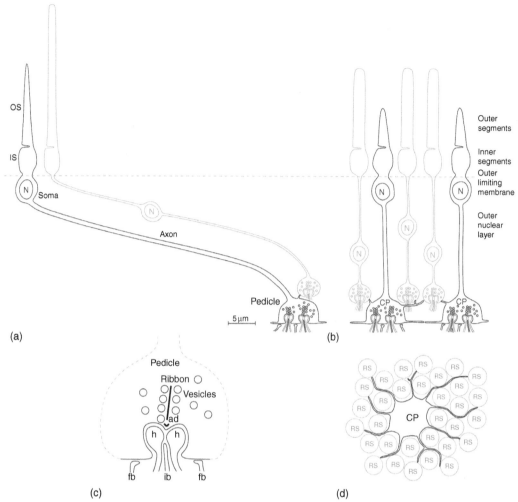

Figure 1 Morphology and topology of the cone. (a) Primate cone with its outer segment in parafovea, showing the lateral extension of the axon which allows the axon terminal sufficient space to make its synaptic connections. A few rods (gray) exist in the parafovea and their axons also extend laterally. Primate foveal cones extend farther, up to 600 μm. (b) In most mammals and in distal primate retina, cones extend vertically interspersed between rods (gray). The cone pedicles are interconnected by gap junctions between basal processes. (c) The cone ribbon synapse sits at the apex of an invagination into the basal surface of the cone pedicle. Attached to the ribbon are two rows of vesicles that contain glutamate. The vesicles are released at the active zone near the arciform density (ad). Two horizontal cell processes (h) and one or more invaginating ON bipolar cell dendrites (ib) extend into the pedicle to receive glutamate from vesicles released from the ribbon. Flat OFF bipolar cell dendrites (fb) contact the cone pedicle outside the invagination, and are thought to receive glutamate that diffuses from several nearby ribbons. (d) Tangential view of a cone pedicle (CP) surrounded by rod spherules (RS, gray). Telodendria extend from each cone pedicle to contact all the surrounding rods and cones with gap junctions.

extends vertically: in cat, the cone axon typically extends ~50 μm in the outer nuclear layer traversing 6–10 layers of rod somas, and in guinea pig, the axon is shorter, typically 10–25 μm. Typically, cone axons are 1–2 μm in diameter, interspersed between rod somas (35 μm dia.) and rod axons (0.4 μm dia.). The fourfold greater diameter of cone axons suggests a functional interpretation, because space in the outer nuclear layer is limited. One possibility is that the faster cone response might require a larger diameter in a long axon to transmit the cone's higher frequency response. This was tested with a computational model, and a thin diameter in a long axon was found

sufficient, also consistent with the maintenance of the difference in diameter between cone and rod axons even where they are shorter. Another possibility is related to the axon terminal: the cone terminal has ~20-fold more synaptic contacts than the rod, and its cross-sectional area is ~16-fold greater. Both rod and cone axons are filled with microtubules, which are essential for transporting synaptic proteins from the nucleus, so the larger diameter of the cone axon may be a consequence of its need to perform more synaptic maintenance. The cone axon terminal is typically ~5 μm in diameter, and is flat on its basal surface (toward the ganglion cells); hence, it looks like a foot and is commonly called a pedicle. The pedicles of adjacent cones align to form a flat plane which lies near the upper edge of the outer plexiform layer.

Synapse

The cone makes chemical synaptic contacts onto $\sim$10 types of bipolar cell and two types of horizontal cell (HA and HB in cat and rabbit; H1and H2 in primate) at its axon terminal. The synapse releases glutamate via packets of membrane called synaptic vesicles, which are small ($\sim$30–50 nm) organelles created by endocytosis from the cell's external membrane. The vesicles diffuse freely around the cytoplasm while they are being filled with glutamate by transporters. The presynaptic machinery in the terminal contains several dense structures known as ribbons because in cross section they are thin and extend vertically away from the cell's membrane. The active zone, where vesicles are released from the ribbon onto the external membrane, is defined by a trough-shaped band called the arciform density, which contains structural proteins for anchoring the ribbon to the cell membrane and the calcium channels responsible for triggering release. The ribbon allows high continuous release rates, for it collects several rows of vesicles and tethers them for release, to provide a larger readily releasable pool. The mechanism for vesicle release is complex, containing several dozen proteins, and its details are not yet understood, but it is initiated by calcium ions binding to a receptor protein which causes a vesicle to fuse with the external membrane and release its contents into extracellular space. To fuse with the external membrane, the vesicle must be docked near membrane calcium channels at the base of the ribbon. However, there is some evidence for compound fusion, where vesicles fuse with and release into nearby docked vesicles. After release, the vesicle membrane is recycled with a specialized mechanism involving the protein clathrin which coats an invaginating bit of external membrane with a buckyball-type structure and pinches it off as a cytoplasmic vesicle.

Invagination and Triad

The cone's ribbon synapse is located at the apex of an extension of extracellular space into the bottom surface of the cone terminal, called an invagination, where fine processes of horizontal and bipolar cells extend to form postsynaptic specializations. Each invagination holds two horizontal cell dendritic processes which emanate from different horizontal cells, and one fine dendritic process of an ON bipolar cell. The horizontal cell processes swell up inside the invagination on either side of the bipolar dendrite. Together, the ribbon and postsynaptic processes are called 'a triad'. The function of the invagination is unknown but it has been suggested to limit diffusion of

the neurotransmitter released by the cone or by horizontal cells for negative feedback. The cone terminal contains 20–50 ribbons, each in a separate invagination, so there are several dozen triads. Off-bipolar cells also contact the cone terminal. Some with alpha-amino-3-hydroxy-5-methyl-4-isoxazolepropionic acid (AMPA) receptors invaginate the cone near the ON bipolar process, and others with kainate receptors contact the base of the pedicle at the so-called flat or basal contacts which lack associated presynaptic vesicles. The flat contacts are located in the space between the invaginations, allowing them to sense the release from several ribbons.

Biophysical Properties

The inner segment and axon terminal of photoreceptors contain several membrane-bound ion channels, including K_v, K_{Cn}, BK, and L-type Ca^{2+}. The K_v channels of the inner segment are activated by depolarization, providing an outward current to balance the inward dark current through the light-modulated cyclic guanosine monophosphate (cGMP)-gated channels of the outer segment. These K^+ channels provide an adaptational influence, opposing the dark signal, and indirectly opposing the light signal by deactivating with hyperpolarization. The K_{Cn} channels, underlying the I_h current, provide a delayed depolarization when activated by hyperpolarization, and calcium-activated potassium channels (BK) help to maintain the resting potential near -40 mV. These are adaptational effects that tend to limit the hyperpolarization from bright light and depolarization in the dark. The cone terminal's L-type Ca^{2+} channels ($Ca_v1.4$) uniquely include alpha 1D subunits and may have special gating properties. In addition, the terminal contains a calcium-activated chloride current ($I_{Cl(Ca)}$) and calcium-induced calcium release (CICR). Calcium is highly regulated in the terminal, with internal buffering and plasma membrane transporters (protein misfolding cyclic amplifications (PMCAs)), because calcium is the trigger for vesicle release. Calcium is highly compartmentalized at the terminal because calcium in the outer and inner segments and soma performs other cellular functions. In the terminal, ryanodine receptors and IP_3 receptors modulate calcium release from internal stores. Voltage-gated sodium channels have been identified in primate cones, possibly to help amplify the cone signal along its axon. The cone terminal contains several transporters with special functions. A proton pump in the vesicles acidifies them so that protons are released along with the glutamate. The protons bind to the local calcium channels to block calcium entry, and thus regulate release. Several transporter proteins carry glutamate, including at least two isoforms that are inserted into the membrane of vesicles to load them with glutamate. A glutamate transporter sitting in the external membrane of the pedicle is important for uptake of glutamate from the extracellular space. The light response of postsynaptic cells depends on a reduction in glutamate, but when the transporter is blocked pharmacologically the postsynaptic light response is greatly reduced. This implies that the triad synapse is protected to some extent against diffusion by the cone pedicle's invagination.

Gap Junctions

Cone axon terminals are coupled by gap junctions which carry electrical signals and small molecules directly from each cone to its neighbors. Each gap junction comprises connexin molecules which can form a conductive pore when they are aligned in the membrane of both coupled neurons. The gap junctions are located on telodendria which are basal processes that emanate from the cone terminals to make contact with the neighbors. The cone gap junctions are punctate, that is, they comprise just a few connexin molecules. The connexin has been identified as Cx36. In addition, a cone's telodendria contact all the surrounding rods with punctate gap junctions. The cone side of the rod-cone gap junction also comprises Cx36, but the identity of the rod-specific connexin molecule is not known. In foveal cones, because their terminals are all located outside the fovea, some of the telodendria extend several hundred micrometers around the perimeter of the fovea to make contact with the neighboring cones, whose axons have extended in the opposite direction from central fovea.

Morphological Implications of Spectral Sensitivity

The cones of most vertebrates exist in several types that differ in their spectral sensitivity. Until recently, it was difficult to distinguish cone spectral type based on morphology, but antibodies are now available to specifically stain the outer segments by spectral type. Direct imaging is now able to routinely determine the spectral type. The inner segments of birds and some species of reptiles and mammals contain oil droplets which filter out some wavelengths, to narrow the outer segment's sensitivity and improve the ability for color discrimination. The primate S-cone, sensitive to short wavelengths, is smaller and its terminal makes fewer contacts, so it is distinguishable from middle (M)- and long (L)-wavelength-sensitive cones. In addition, the M- and L-type cones can be distinguished by the number and location of their synaptic connections with bipolar cells. The M- and L-cones are electrically coupled by gap junctions, which blur their spectral sensitivity. The coupling is a reasonable compromise because the coupling increases their contrast sensitivity and the spectral blurring effect is moderate due to the similarity of the M- and L-type spectral sensitivities.

The M- and L-type cones are located randomly in bunches of similar type, which along with the indiscriminate coupling may improve contrast sensitivity without causing much blurring of color. The S-type cone is not coupled strongly to its M- and L-type neighbors, apparently because its spectral sensitivity is farther from the M- and L-type curves.

Function

Adaptation and Predictive Coding

The role of the cone photoreceptor in daylight is crucial because it is the first step in the visual pathway. The cone's task is to transduce and relay information in the visual environment relevant to an organism's survival. A major problem faced by the cone is that daylight varies over 5 log units, but a synaptic signal is only able to vary only over 1–2 log units. The problem is one of dynamic range and also of noise because dynamic range limits the maximum signal, and noise limits the minimum discriminable signal. The amount of information available is related to the maximum signal divided by the minimum, that is, the number of distinguishable levels. Since cone transduction is noisy and the ribbon synapse is also noisy, to maximize the amount of information the cone must remove all irrelevant information and selectively transmit the most important signal components with high sensitivity. One way the cone deals with this problem is by removing information about the background level by adaptation. This is accomplished first in the outer segment. In addition, the cone has several more adaptational mechanisms: modulation of the light-evoked signal by voltage-gated channels in the inner segment and terminal, gain reduction at the ribbon synapse, and regulation of its neurotransmitter release by negative feedback. To preserve information about fine temporal detail, the adaptational mechanisms must be slow, and to preserve spatial detail, the adaptational mechanisms must be wide. The theory for such adaptational mechanisms is called predictive coding, in which a temporally slow or spatially wide average signal is subtracted from a photoreceptor. Thus, the cone ribbon synapse carries a signal that is optimally adapted in time and also in space.

Synaptic Transfer Function

Upon depolarization, the ribbon synapse releases glutamate with a nonlinear transfer function, which is controlled by the voltage sensitivity of the L-type calcium channel. The mechanism for release is linearly modulated by calcium because glutamate release is proportional to the terminal's local calcium concentration. Release is maximal in the dark when the cone is depolarized, and minimal in bright light when the cone is hyperpolarized

(**Figure 2**). Due to the nonlinear release function, the synaptic gain for small light increments is also maximal in the dark, and minimal in the light. Thus, the synaptic gain is adaptive, that is, it tends to oppose the light-evoked signal and reduce contrast gain in bright light. Negative feedback tends to oppose this effect, that is, it reduces the change in synaptic gain due to a change in background level. Glutamate released by the cone binds to the post-synaptic metabotropic glutamate receptor 6 (mGluR6) receptor to activate a second-messenger cascade which inverts and low-pass filters the signal in the bipolar cell. Glutamate also diffuses out of the invagination to bind to kainate receptors at the tips of off-bipolar cells.

Rate of Vesicle Release

A major source of noise for the cone signal is the random release of vesicles by the ribbon. Although the details of the release mechanism are not known, it is thought to be stochastic (noisy), similar to a modulated Poisson distribution, for which the standard deviation is equal to the square root of the mean. There is also some evidence for more regular release. Protons released along with the vesicle's packet of glutamate bind to the local calcium channels which may generate a short refractory period, allowing release to be more regular. The cone synapse is thought to release vesicles at $100–200\,s^{-1}$ per active zone in the dark, but in bright sunlight the rate is reduced to $<10\,s^{-1}$ per active zone. The ability of a ribbon synapse to transmit information about contrast, even with a high release rate, is extremely limited: at a Poisson rate of $200\,ves\,s^{-1}$, contrast threshold for a 100 ms response would be ~0.2. Therefore to increase sensitivity, several ribbons transmit the same cone signal in parallel. With 20 ribbons, the total release rate is $~2000\,s^{-1}$, for which the contrast threshold would be 0.07. Each bipolar cell contacts several (4–8) ribbons per cone, and bipolar cells (except foveal midget bipolars) normally contact several (4–6) cones, so a bipolar cell integrates signals from one to several dozen cone ribbons.

Electrical Coupling

The gap junctions between cone terminals function as a spatial filter, averaging cone signals to remove noise, and removing high spatial frequencies. The effect is to enlarge the cone receptive field beyond the effect of optical blur. Adjacent L- and M-cones in primate are electrically coupled by gap junctions, which blur to some extent their ability to support color discrimination, implying that their action to reduce noise provides a benefit to luminance contrast discrimination. S-cones are uncoupled to the other types, probably because their spectral sensitivity is so different that it would suffer out of proportion to the potential increase in contrast sensitivity. Cone coupling is paradoxical to some extent because in a linear system which collects cone signals, for example, convergence to a large ganglion cell's dendritic tree, gap junctions would provide no benefit. However, the cones apparently benefit from the reduction of noise before their signal passes

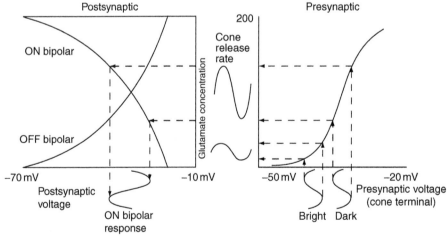

Figure 2 Transfer function of cone ribbon synapse. At right, presynaptic ribbon release function, showing two signals in the cone terminal, one more hyperpolarized (bright), and the other more depolarized (dark). The brighter signal releases less glutamate and has lower release gain because the release function is nonlinear. The fraction of modulated release is higher for the hyperpolarized (bright) signal than for the depolarized (dark) signal because the maintained release rate is lower in bright light. However, the system must compromise to set a relatively high release rate. At left, the postsynaptic transfer function from glutamate binding to voltage in the ON bipolar response is inverted. For the relatively high release of the dark signal, the ON and OFF bipolar responses are similar in amplitude and degree of saturation. The presynaptic release rate must be maintained relatively high to keep the postsynaptic signals symmetrical; if the release rate were lower, the OFF bipolar signal would decrease and the ON bipolar signal would saturate. Negative feedback maintains the cone ribbons' relatively high maintained release rate to optimize the signal/noise ratio in both ON and OFF bipolar cells.

through the ribbon synapse which is to some extent non-linear. The gap junctions coupling a cone to its neighboring rods allow rod signals at mesopic backgrounds, where a rod temporally sums several photon signals, to pass to cones to be transmitted through the cone pathway to ganglion cells. Since cones are 30–100-fold less sensitive than rods, this rod–cone coupling represents a form of adaptation for the cone pathway.

Negative Feedback

Adaptation from spatially pooled horizontal cell feedback at the cone triad is important for retinal signal processing. Glutamate released by the cone ribbon synapse binds to AMPA receptors on the horizontal cell dendrite to depolarize it, and in turn the horizontal cell provides negative feedback to control the release of glutamate by the cone. The negative feedback contributes to an antagonistic surround in bipolar cells and in all other postsynaptic neurons. In some species (e.g., turtles, fish, and putatively in some mammals), horizontal cells receive spectrally specific signals from cones and therefore their negative feedback contributes to cone color opponency. The mechanism for the negative feedback from horizontal cells is not known but several candidates are currently being studied. Horizontal cells synthesize γ-aminobutyric acid (GABA) and release it through a transporter working in reverse, and in some species (e.g., turtle, fish, salamander, and frog), GABAergic feedback has been found in the cone terminal. Horizontal cells also generate a feed-forward surround signal in both on- and off-bipolar cells, which have $GABA_A$ receptors in their dendrites. The effect is to generate a stronger surround signal in the bipolar cell, allowing it to more tightly control its release of neurotransmitter. However, standard GABA-receptor blockers do not affect the surround in most species. Some evidence suggests that feedback may be provided through release of protons in the invagination, because when pH buffers, such as 4-(2-hydroxyethyl)-1-piperazineethanesulfonic acid (HEPES) are applied via the perfusion bath *in vitro*, they block horizontal cell feedback proportionate to their buffering capacity. A third possibility is that the feedback is ephaptic (electrical). The pH and ephaptic theories describe a feedback signal that directly modulates the cone terminal's calcium current. One benefit of all these putative feedback mechanisms is their freedom from noise that would occur with vesicular release. Feedback is necessary to control neurotransmitter release, but it has one problem, that if too strong it can cause instability from delays in the feedback loop. Oscillations can be evoked by flashed full-field stimuli with a dark mask over the horizontal cell recording site. The dark depolarizes the cone, increasing gain in the feedback loop, which tends to generate oscillations.

Other adaptational mechanisms such as K^+ channels can generate similar resonances related to the delay in their response.

Theory of Ephaptic Feedback

The proposed ephaptic feedback modulates the cone pedicle's calcium channels by controlling the polarization of the external surface of the membrane inside the invagination. When a horizontal cell is hyperpolarized by a remote light (i.e., without modulating local glutamate release), the current that flows from extracellular space into its dendritic tips is increased. The increase in current causes a voltage drop inside the invagination, which shifts the voltage at the external membrane surface more negative. As both surfaces of the membrane have been shifted negative, the calcium channel senses a depolarization (lack of polarization) and thus opens further, which opposes the original light signal. The opposite effect occurs when a cone terminal is hyperpolarized by light but the horizontal cell is not. Light reduces glutamate release, which closes the horizontal cell's glutamate-gated (AMPA) channels, decreasing the current flow into the dendritic tip, which shifts the external cone membrane more positive, causing the calcium channel to sense a greater hyperpolarization, which comprises net positive feedback. However, the horizontal cell dendritic tips contain hemichannel gap junctions (connexins) which are not gated by glutamate but shunt the glutamate-gated channels. Their effect is to reduce positive feedback and to enhance negative feedback. One potential problem for the theory is that the hemichannels reduce the light response, and because their conductance is less than the AMPA receptor channels one imagines they should not cause a robust effect. However, the theory is supported by strong evidence, for example, the hemichannels can be blocked by carbenoxolone which reduces the negative feedback, and the same result occurs in knockout animals in which the hemichannels are removed. The relative magnitude of the positive and negative ephaptic feedback effects depend on several factors, including the conductances and the location of the underlying channels. The amount of voltage drop at the external membrane surface depends on the geometry of the current path through extracellular space in the invagination. A robust feedback signal, for example, the typical 5–10 mV shift seen in the calcium current in the cone pedicle, would require a relatively high resistance, and therefore the mechanism is currently controversial. Although the ephaptic theory has not generally been accepted by the field, it can explain some of the phenomena given as evidence for pH feedback. HEPES buffer is known to acidify neurons, and hemichannels are known to be modulated by internal pH, which is consistent with the well-known effect of

HEPES to block the cone surround. An advantage of ephaptic feedback is its short delay, reducing the tendency for instability. A further merit to the ephaptic theory is that it provides a functional role for the invagination, to generate the extracellular voltage shift. Putatively, then, the invagination creates a specialized local environment for the triad synapse, to limit chemical and/or electrical diffusion.

Bandpass Adaptation Filter

The cone terminal comprises a spatiotemporal filter consisting of a low-pass filter from electrical coupling between cones, and a high-pass filter from horizontal cell negative feedback. The signal in horizontal cells is low pass in space because they are large and well coupled: their receptive fields can be ~10-fold larger than their dendritic field. The feedback signal in some species has two components, the faster (pH or ephaptic) modulation of the calcium current, and a slower GABAergic component. Therefore, the horizontal cell signal, when it modulates the cone synaptic release, can be characterized as a subtraction of a low-pass filter, effectively making the cone synapse a high-pass filter. The overall effect of electrical coupling and negative feedback is to generate a band-pass filter at the cone terminal. The filter removes spatial and temporal frequencies that transmit less information, improving the performance of the visual pathway which utilizes noisy synapses for transmission.

Conclusion

The cone photoreceptor plays a crucial role in vertebrate vision because it is responsible for transducing fast changes in contrast while ignoring the background light level. To improve its sensitivity over 5 log units of background illumination, the cone contains several mechanisms for adaptation, originating in the transduction cascade, in its biophysical properties, and in the ribbon synapse. The ribbon is part of a complex local circuit called the triad that combines adaptation with spatial filtering. The triad maximizes the amount of information that can be transmitted through noisy synapses over a wide range of environmental light levels.

Acknowledgment

This work was supported by NEI grant EY016607.

See also: Phototransduction: Adaptation in Cones.

Further Reading

Copenhagen, D. R. (2004). Excitation in retina: The flow, filtering, and molecules of visual signaling in the glutamatergic pathways from photoreceptors to ganglion cells. In: Chalupa, L. M. and Werner, J. S. (eds.) *The Visual Neurosciences,* 2 vols, pp. 234–259. Cambridge, MA: MIT Press.

DeVries, S. H., Li, W., and Saszik, S. (2006). Parallel processing in two transmitter microenvironments at the cone photoreceptor synapse. *Neuron* 50: 735–748.

Gaal, L., Roska, B., Picaud, S. A., et al. (1998). Postsynaptic response kinetics are controlled by a glutamate transporter at cone photoreceptors. *Journal of Neurophysiology* 79: 190–196.

Heidelberger, R., Thoreson, W. B., and Witkovsky, P. (2005). Synaptic transmission at retinal ribbon synapses. *Progress in Retinal and Eye Research* 24: 682–720.

Hsu, A., Smith, R. G., Buchsbaum, G., and Sterling, P. (2000). Cost of cone coupling to trichromacy in primate fovea. *Journal of the Optical Society of America A: Optics, Image Science and Vision* 17: 635–640.

Hsu, A., Tsukamoto, Y., Smith, R. G., and Sterling, P. (1998). Functional architecture of primate cone and rod axons. *Vision Research* 38: 2539–2549.

Massey, S. (2008). Circuit functions of gap junction coupling in the mammalian retina. In: Basbaum, A. I., Kaneko, A., Shepherd, G., and Westheimer, G. (eds.) *The Senses: A Comprehensive Reference.* Vol. 1, pp. 457–471. San Diego: Elsevier.

Rodieck, R. W. (1998). *The First Steps in Seeing.* Sunderland, MA: Sinauer Associates.

Smith, R. G. (2003). Retina. In: Arbib, M. A. (ed.) *The Handbook of Brain Theory and Neural Networks,* 2nd edn., pp. 11–23. Cambridge, MA: MIT Press.

Smith, R. G. (2008). Contribution of horizontal cells. In: Basbaum, A. I., Kaneko, A., Shepherd, G., and Westheimer, G. (eds.) *The Senses: A Comprehensive Reference.* Vol. 1, pp. 341–349. San Diego: Elsevier.

Smith, R. G., Freed, M. A., and Sterling, P. (1986). Microcircuitry of the dark-adapted cat retina: Functional architecture of the rod–cone network. *Journal of Neuroscience* 6: 3505–3517.

Sterling, P. and Matthews, G. (2005). Structure and function of ribbon synapses. *Trends in Neuroscience* 28: 20–29.

Thoreson, W. B. (2007). Kinetics of synaptic transmission at ribbon synapses of rods and cones. *Molecular Neurobiology* 36: 205–223.

Tsukamoto, Y., Masarachia, P., Schein, S. J., and Sterling, P. (1992). Gap junctions between the pedicles of macaque foveal cones. *Vision Research* 32: 1809–1815.

van Hateren, J. H. (2007). A model of spatiotemporal signal processing by primate cones and horizontal cells. *Journal of Vision* 7: 3.

Contrast Sensitivity

P Bex, Schepens Eye Research Institute, Boston, MA, USA

Glossary

Channels – The groups of visual sensors that are selective for a narrow range of image spatial or temporal structure.

Contrast constancy – At high contrasts, apparent contrast is relatively independent of the parameters that strongly influence contrast-detection threshold.

Contrast-detection threshold – The statistical contrast boundary below which contrast is too low for an image to be detected reliably and above which contrast is high enough for frequent image detection. Often defined as the contrast that produces 75% correct target identifications in forced-choice paradigms.

Contrast sensitivity – The reciprocal of contrast-detection threshold that also represents the transition between visible and invisible images.

Critical flicker frequency – The highest flicker rate of a full contrast image that can be detected reliably.

Forced-choice paradigms – Robust behavioral method used to measure detection or discrimination thresholds. Observers are forced to select between two or more intervals, of which only one contains a target.

Fourier analysis – Analytical method that calculates the simple sine-wave components whose linear sum forms a given complex image.

Resolution limit – The highest spatial frequency of a full contrast image that can be detected reliably.

Spatial frequency – The number of image cycles that fall within a given spatial distance, typically 1° of visual angle.

Temporal frequency – The number of image cycles that fall within 1 s.

Wavelets/gabors – A local filter that is the point-wise product of a two-dimensional (2D) spatial sine wave and a 2D Gaussian envelope.

Most people are familiar with image brightness and contrast from their controls on computer and television displays. The brightness control adjusts the mean luminance of the display uniformly, in order that the intensity of every point in the image increases when brightness is increased or decreases when brightness is reduced. The contrast control adjusts the difference between the lightest and darkest areas of the image. Increasing contrast makes areas that are below mean luminance darker and areas that are above mean luminance lighter, without changing the mean value. Decreasing contrast draws all values toward the mean, thus making the whole image fainter, similar to viewing the image through fog.

Figure 1 illustrates the effect of changing the contrast of a sine-wave striped pattern (the reasons for using a sine-wave pattern are described below). The top panel shows images of gratings whose contrast increases from 12.5% on the left to 100% on the right. The mean luminance of each image is the same. The traces in the bottom row plot luminance versus position for a horizontal slice through each image.

Contrast-Detection Threshold

A powerful measure of visual sensitivity can be obtained by finding the minimum contrast that is necessary for an image to be detected. This minimum contrast is referred to as contrast-detection threshold (C_{thresh}) and it is important because it defines the transition at which an image moves from invisible to visible. One method to estimate C_{thresh} might be to allow a subject to adjust the contrast until an image is just visible. However, this method is highly subjective and large differences in individual criteria for just visible make this measure unreliable.

Psychophysical Assessment of Vision

To overcome these problems, most researchers employ forced-choice procedures that require an observer to identify which of two or more intervals (the more the better) contain the target. An example of a four-alternative forced-choice (4AFC) detection task is shown in **Figure 2(a)**. In this case, a computer presents a target in one of four positions at random around a central fixation point. The observer's task is to fixate the central dot and to indicate the location of the target, usually by pressing a computer button. Targets that are below C_{thresh} (sub-threshold) are rarely detected, whereas targets that are above C_{thresh} (supra-threshold) are usually detected. Contrast-detection thresholds are therefore probabilistic and are defined as the contrast at which they are correctly detected midway between chance and perfect performance.

It is difficult to cheat on forced-choice methods or to change criteria – the target is either seen, in which case its position is correctly identified, or it is not seen, in which

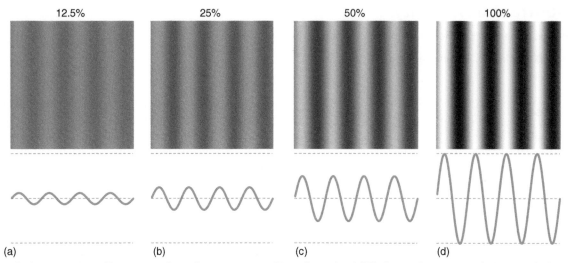

Figure 1 Image contrast. The top row shows the appearance of two-dimensional (2D) sine-grating patterns that are routinely used in vision research. The contrast of the sine grating increases from left (12.5%) to right (100%) as shown by the caption. The bottom row plots a horizontal section through each image and shows that contrast changes the luminance range separately from mean luminance.

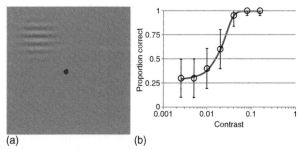

Figure 2 Contrast detection. (a) Example of a four-alternative forced-choice (4 AFC) task. The observer is required to fixate the central dot and to indicate whether the target appeared top left, top right, bottom left, or bottom right. The target contrast is adjusted by computer to a level that produces 75% correct detection. (b) A typical psychometric function. Circles show the proportion of trials the target was detected (ordinate) as a function of the target contrast (abscissa). Error bars show ± 1 standard deviation. The curve shows the best-fitting cumulative normal function, from which the interpolated 75% correct point is taken as contrast-detection threshold.

case the subject is forced to guess. Notice that when guessing, the subject is still correct sometimes (25% if there are four alternatives, 33% if there are three, or 50% if there are two, etc.), as shown in the frequency of seeing curve in **Figure 2(b)**, where, at low contrasts, performance is 25% correct. The data have been fit with a curve known as a psychometric function, in this case a cumulative Gaussian:

$$Y = g + (1 - g)^* \mathrm{erf}(z/\mathrm{sqrt}(2))/2$$

where $z = (X - \mu)/\sigma$; g is the guess rate (0.25 in a 4AFC experiment). The mid-point (μ) of the psychometric function is often taken as C_{thresh} – for a 2AFC task, this is 75% correct. In the example shown, the observer achieved

75% correct at approximately 2.5% contrast. The slope (σ) can be used to infer how easily nearby contrasts can be discriminated from one another – a shallow slope means that a large contrast difference is required to achieve a given change in performance, whereas a steep slope means that a small change in the stimulus produces a large change in performance.

Spatial Frequency Channels

Based on behavioral observations in humans and single unit recordings in mammalian visual systems, researchers discovered around half a century ago that the visual system analyses images at a series of relatively narrow spatial scales and orientations known as channels. Thus, fine and coarse image details are encoded separately and Fourier analysis can be used to study the image structure that is encoded by different visual processing channels. Fourier analysis computes the sum of basic sine waves whose linear sum produces the image. To illustrate the representations of an image that are available at different spatial scales, **Figure 3(a)** shows a typical image, together with its coarse (**Figure 3(b)**) and fine (**Figure 3(c)**) spatial structure.

Visually responsive neurons in primary visual cortex, the first cortical projection from the retina through the lateral geniculate nucleus of the thalamus, respond to images only within a limited area of the visual field, known as the classical receptive field, and are selective for a limited range of spatial frequencies and orientations. These receptive fields are now routinely modeled as Gabor or wavelet functions, defined as:

Figure 3 Spatial frequency in real images. (a) An image of Albert Einstein's face is encoded at a range of spatial scales, from (b) coarse – low spatial frequency to (c) fine – high spatial frequency.

$$G(x, y, \lambda, \varphi, \sigma, \gamma) = \exp\left(-\frac{x'^2 + \gamma^2 y'^2}{2\sigma^2}\right) \sin\left(2\pi \frac{x'}{\lambda} + \varphi\right)$$

where $x' = x \cos\theta + y \sin\theta$ and $y' = -x \sin\theta + y \cos\theta$, λ represents the wavelength, θ the orientation, and ψ the phase of the sine-wave component. For the Gaussian window, σ is the standard deviation and γ is the spatial aspect ratio. Examples of Gabors are illustrated in **Figure 4**. On the top row, spatial frequency increases from left to right and all Gabors are of the same orientation 0° and contrast. On the bottom row, spatial frequency is fixed, but orientation is 45°, 90°, or 135° (from left to right). The visual system encodes image structure with a bank of such wavelet filters that represent the retinal image through patchwise local analysis.

Figure 5 provides compelling demonstrations that our visual system employs a set of spatial frequency and orientation-selective channels. These demonstrations show that after prolonged viewing of a particular pattern (termed adaptation) the appearance of other patterns can be altered (termed an aftereffect). In these demonstrations, adapting to a pattern of one spatial frequency or orientation produces a loss in sensitivity in the channel that responds most to that pattern, but little change in channels tuned to other spatial frequencies or orientations. This localized loss in sensitivity produces a relative shift in the responses of our visual channels that cause us to experience changes in the appearance of the image.

These observations have led to the widespread use of sine-wave grating patterns in basic and clinical vision research. In order to derive a measure of vision that reflects the sensitivity across our set of visual channels and to reflect the fact that functional vision requires us to detect and interact with objects of various sizes, contrast-detection thresholds are measured for gratings of a range of bar widths, expressed as spatial frequency or the number of grating cycles per unit distance. **Figure 4** illustrates Gabors of differing spatial frequency; however, the size of one grating cycle on the retina depends on the distance

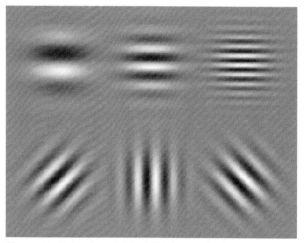

Figure 4 Gabor (wavelets) of differing spatial frequency and orientation. Top row: spatial frequency increases from left to right, orientation is fixed at 0°. Bottom row: Orientation increases from left to right: 45°, 90°, and 135°, spatial frequency is fixed.

from which it is viewed. Therefore, image sizes are usually calculated in terms of visual angle, which specifies the retinal image size. **Figure 6** shows how visual angle is calculated and its relationship to image size and viewing distance. A convenient rule is that 1 cm viewed from 57 cm subtends a visual angle of 1° and roughly corresponds to a finger nail viewed at arm's length.

Contrast Sensitivity Function

Many researchers have shown that for sine-grating patterns, C_{thresh} strongly depends on spatial frequency. This fundamental observation is demonstrated in the classic image shown in **Figure 7**. Spatial frequency increases from left to right and contrast increases from top to bottom, so that contrast is constant across any horizontal line. Contrast-detection thresholds can be visualized on this figure as the imaginary curve along

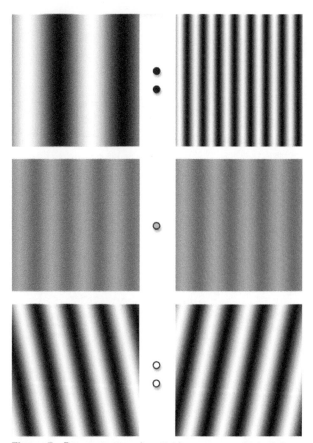

Figure 5 Demonstration of spatial frequency- and orientation-selective aftereffects. First note that when you fixate the centre gray dot, the gratings in the middle row are of the same spatial frequency and orientation. Next, look back and forth between the black dots in the top row for around 10 s. Now, when you look at the center gray dot, the grating on the left appears to be of higher spatial frequency than the grating on the right. Next, look back and forth between the white dots in the bottom row for around 10 s. Now, when you look at the center gray dot, the grating on the left appears tilted counterclockwise, while the grating on the right appears tilted clockwise. These aftereffects are robust even though you know that the gratings in the middle row are the same. These demonstrations provide compelling evidence that visual processing involves channels that are narrowly tuned for spatial frequency and orientation.

which the grating changes from invisible (toward the top of the figure) to visible (toward the bottom of the figure). Most people report that the function peaks somewhere near the middle of the figure. Notice that the peak shifts as you move the figure closer or further away. This demonstrates the importance of visual angle rather than physical image size. Note that the highest spatial frequency that can be detected at maximum contrast is given by the rightmost point on a contrast sensitivity function (CSF). This is referred to as the resolution limit and is a quick and convenient method of assessing visual sensitivity than measuring the entire CSF.

When measured with forced-choice procedures, (**Figure 2**) contrast-detection thresholds are lowest for

gratings around 2–5 cycles per degree of visual angle (c deg^{-1}). By convention, the inverse of C_{thresh} ($1/C_{thresh}$) is usually reported and is termed contrast sensitivity. The rationale for the use of contrast sensitivity over contrast-detection threshold is most likely because the shape of the CSF is the same as that of the underlying modulation transfer function of the system. The circles in **Figure 8** show the author's contrast sensitivity as a function of spatial frequency measured with a forced-choice procedure. Error bars show 95% confidence intervals. The data have been fit (green curve) with the outputs of a set of spatial frequency channels shown by the colored curves. The channels are log spaced in spatial frequency (with peaks at 0.5, 1, 2, 4, 8, 16, or 32 c deg^{-1}) and have the same bandwidth (1.4 octaves). The summed outputs of the set of filters provide a good fit to the data and this channel-based system is now a widely accepted model of early visual processing.

The spatial frequency aftereffect shown in **Figure 5** is easily explained with this channel-based model. Adapting to one spatial frequency reduces the responses of the channel that is most sensitive to that spatial frequency, but has little effect on the responses of other channels. When a different spatial frequency is subsequently viewed, the overall activity across the channels is shifted away from the adapted channel. This shift in the population response produces a shift in apparent spatial frequency away from the adapting frequency. An analogous model explains the shifts in orientation in the lower row of **Figure 5**, except that orientation-selective channels are adapted rather than spatial-frequency-selective channels.

The CSF is highly dependent on the mean luminance of the display on which it is measured. This can easily be experienced by viewing **Figure 7** with a pair of dark sunglasses (possibly two pairs), which moves the curve down (reducing sensitivity) and shifts the peak to lower spatial frequencies. The data in **Figure 8** were collected on a standard computer monitor that has a mean luminance of 50 cd m^{-2} (candelas per square meter). Photopic, mesopic, and scotopic vision and changes in visual performance show that sensitivity to high spatial frequencies increases with mean luminance. This property is important because CSFs are routinely measured on relatively dim displays (e.g., 50–100 cd m^{-2}) in the laboratory and in the clinic; however, the luminance of the real world is typically much greater. For example, the luminance of a cloudy sky is around 35 000 cd m^{-2}, suggesting that standard experimental conditions may underestimate sensitivity to fine spatial structure.

Temporal Contrast Sensitivity

In addition to a dependence on spatial frequency, contrast sensitivity also depends strongly on temporal frequency. **Figure 1** illustrates spatial variation in luminance, but

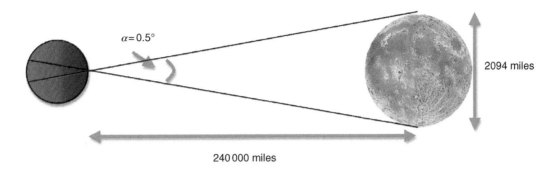

Figure 6 Visual angle and viewing distance. The angular size of an object is calculated as $2*\tan((0.5*h)/d)$, where h is the height of the object and d is the distance from which it is viewed. The example, which is not to scale, shows the angular size subtended by the moon is 0.5°. For comparison, the nail of the average index finger viewed at arm's length subtends 1°.

Figure 7 Illustration of the contrast sensitivity function (CSF). Spatial frequency increases from left to right, contrast increases from top to bottom. The contrast along any horizontal line is fixed. Different spatial frequencies become visible at different contrasts and define an imaginary curve that separates seen from unseen structure. Notice that if you move the image closer to your eye, the peak moves to the right and if you move it further away, the peak moves to the left. This demonstrates that contrast sensitivity depends on retinal not physical image size. If you wear one or two pairs of dark sunglasses, the curve shifts down and the peaks moves left, which demonstrates the dependence of the CSF on mean luminance.

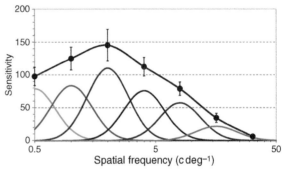

Figure 8 Spatial contrast sensitivity. Circles show contrast sensitivity (the reciprocal of contrast-detection threshold) for sine gratings of a range of spatial frequencies. Sensitivity peaks at around 2 c deg^{-1} under the conditions employed here and decreases at lower or higher spatial frequencies. The black curve is the summed sensitivity of the set of log-scaled channels shown by the colored curves and provides a good fit to the data.

used here (50 cd m^{-2}) and decreases at lower or higher temporal frequencies. These data are well fit (black curve) by a model with only two temporal channels, compared with the multiple channels that support spatial contrast sensitivity. One channel (red curve) is low-pass or sustained and is most sensitive to structure that is stationary or slowly changing over time. The second channel (blue curve) is band-pass or transient and is most sensitive to structure that changes at around 5 Hz.

The spatial resolution limit falls steadily with distance from the fovea, an effect that can be experienced by viewing **Figure 7** while fixating away from the center of the image. As you fixate further away, the threshold curve moves further down the figure and its peak shifts further to the left. Unlike spatial resolution, temporal resolution (the highest flicker rate that can be detected at any contrast, often called critical flicker fusion frequency)

imagine instead that the x-axis represents time, rather than space. Now the figure illustrates flicker. Flicker frequency can be varied in the same way as spatial frequency is varied in **Figures 3** and **7**. The circles in **Figure 9** show how the author's contrast sensitivity varies as a function of temporal frequency for a 2 c deg^{-1} grating pattern. Sensitivity peaks around 5 Hz, at the mean luminance

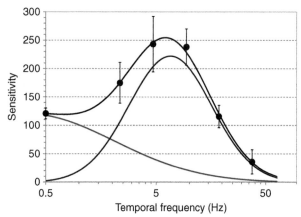

Figure 9 Temporal contrast sensitivity. Circles show contrast sensitivity (the reciprocal of contrast-detection threshold) for sine gratings of a range of temporal frequencies. The black curve is the summed sensitivity of the two log-scaled channels shown by the red and blue curves. The red curve has peak sensitivity at low temporal frequencies – that is, static images – and is termed a sustained channel. The blue curve has peak sensitivity around 6 Hz and is termed a transient channel.

increases moderately with distance from the fovea. This explains why older, 60-Hz computer displays can sometimes be seen to flicker when seen in the peripheral visual field, but not when viewed directly. Just as spatial contrast sensitivity depends on luminance, so does temporal contrast sensitivity. A 35-mm film is generally recorded at 24 frames per second, a refresh rate that could be easily detected at moderate light levels, as can be seen from **Figure 9**. For this reason, movie theaters are generally dark because sensitivity to high flicker rates is poor under those conditions. In addition, the visible 24-Hz image update rate is masked by flashing the illuminant at 48 Hz, so each frame is flashed twice.

At supra-threshold contrasts, apparent contrast is relatively independent of spatial or temporal frequency, a phenomenon termed contrast constancy. Contrast constancy can be experienced in **Figure 7** – while the transition between visible and invisible gratings has a curved shape, toward the bottom of the figure, the gratings appear to have similar contrast regardless of spatial frequency. This has important implications for image enhancement, which should therefore target only image components that are below their C_{thresh}.

See also: Acuity; Anatomically Separate Rod and Cone Signaling Pathways; Chromatic Function of the Cones; Information Processing: Contrast Sensitivity; Information Processing: Direction Sensitivity; Information Processing in the Retina; Information Processing: Retinal Adaptation; Photopic, Mesopic and Scotopic Vision and Changes in Visual Performance.

Further Reading

Bracewell, R. (1999). *The Fourier Transform and Its Applications*, 3rd edn. London: McGraw-Hill.

Campbell, F. W. and Robson, J. G. (1968). Application of Fourier analysis to the visibility of gratings. *Journal of Physiology* 197: 551–566.

Field, D. J. and Tolhurst, D. J. (1986). The structure and symmetry of simple-cell receptive-field profiles in the cat's visual cortex. *Proceedings of the Royal Society of London. Series B. Biological Sciences* 228(1253): 379–400.

Georgeson, M. A. (1990). Over the limit: Encoding contrast above threshold in human vision. In: Kulikowski, J. J. (ed.) *Limits of Vision*, pp. 106–119. London: Erlbaum.

Hubel, D. H. and Wiesel, T. N. (1959). Receptive fields of single neurones in the cat's striate cortex. *Journal of Physiology* 148: 574–591.

Kelly, D. H. (1961). Visual responses to time-dependent stimuli. 1. Amplitude sensitivity measurements. *Journal of the Optical Society of America. A, Optics, Image Science, and Vision* 51: 422–429.

Kulikowski, J. J. and Tolhurst, D. J. (1973). Psychophysical evidence for sustained and transient detectors in human vision. *Journal of Physiology* 232(1): 149–162.

Landis, C. (1954). Determinants of the critical flicker-fusion threshold. *Physiological Reviews* 34(2): 259–286.

O'Shea, R. P. (1991). Thumb's rule tested: Visual angle of thumb's width is about 2 deg. *Perception* 20(3): 415–418.

Rovamo, J., Virsu, V., Laurinen, P., and Hyvarinen, L. (1982). Resolution of gratings oriented along and across meridians in peripheral vision. *Investigative Ophthalmology and Visual Science* 23: 666–670.

Coordinating Division and Differentiation in Retinal Development

R Bremner and M Pacal, University of Toronto, Toronto, ON, Canada

Glossary

Cell birth – For the purposes of this article, cell birth is the activation of a new transcriptional program that is never seen in retinal progenitor cells (RPCs), but is limited to differentiating retinal transitional cell (RTC). This definition excludes cell-cycle exit because, as discussed in this article, differentiating RTCs can be generated independent of exit. In normal cells, the timing of cell birth is usually measured by briefly labeling replicating DNA with bromodeoxyuridine (BrdU) or tritiated thymidine (^{3}H-thymidine) and then, days/weeks later, counting intensely labeled cells which, by definition, exited the cell cycle soon after the label was introduced. This tool is useless in mutants where cell birth occurs in the absence of cell-cycle exit. In that case, birth must be noted when it occurs using markers induced in differentiating RTCs that are never present in RPCs (e.g., Nrl in newborn rods).

Cell-cycle rate – It reflects the rate of cell expansion, which correlates with cell-cycle length.

Competence – This activity influences the ability of RPCs to generate certain cell types. It is permissive, but not instructive in that it creates potential, but it is not sufficient to force cell birth on its own. The competence of RPCs changes throughout development. Early RPCs are competent to produce ganglion, horizontal, amacrine, and cone cells, while late RPCs produce bipolar and Müller cells. Rods are generated throughout, but predominantly are a mid–late-born cell type.

Exit – Normally, cell birth is closely coupled to cell-cycle exit and ensures that a newborn RTC remains post mitotic until and after it becomes a terminally differentiated cell.

G0 – Cells can be driven out of G1 into reversible G0 by serum starvation or contact inhibition, or into irreversible G0 by terminal differentiation or senescence.

G1, S, G2, and M – Most cell cycles have four phases: S phase is when DNA is synthesized; M phase is when cells undergo mitosis to generate two offspring (often termed daughters); G1 and G2 are gaps between M and S or S and M, respectively.

Interkinetic nuclear migration (INM) – Cellular processes connect the RPC to the apical (also called outer or ventricular) and basal (also called inner or vitreal) surfaces of the retina. RPC nuclei move as they traverse the cell cycle. M-phase nuclei are located at the apical surface, then move more basally to go through G1 and start S, then finish S migrating back to the apical side, and go through G2 just before they reach the end of that journey.

Mitogen – Extracellular factor that induces cell division.

R – The restriction point in G1, beyond which a cell will continue into S phase even if mitogens are withdrawn.

Retinal progenitor cell (RPC) – A dividing cell in the normal developing retina.

Retinal transition cell (RTC) – Also termed precursor. These are newborn differentiating cells. They are distinguished from RPCs by a novel transcriptome that includes many mRNAs/proteins that are never detected in RPCs. Normally, RTCs are post mitotic. However, failure to exit does not prevent induction of the RTC transcriptome. Birth (induction of the differentiation program) and exit can be uncoupled and are, therefore, viewed as separate in this article.

Introduction

Cell division is driven by extrinsic mitogens that influence the intrinsic core cell-cycle machinery. Extrinsic factors can also influence the decision to exit and differentiate during development. In the retina, the earliest born cells are ganglion neurons (**Figure 1**) and these cells secrete factors that have crucial roles in influencing division and fate, such as Sonic Hedgehog. Beyond this stage, however, although many factors can influence division and fate, it is not clear whether they actually do so *in vivo*. Dissociated, well-separated, individual retinal progenitor cells (RPCs) can generate clones of cells that resemble clones *in vivo* in both size and composition, except for ganglion cells which do not survive *in vitro*. These data suggest that apart from ganglion cell births, much of retinal development is intrinsically programmed and thus, apart from a general requirement for mitogens, extrinsic factors may not play a major role in determining whether RPCs divide or exit. Further, we have introduced a few of the basic issues around cell-cycle regulation, discuss the role of some

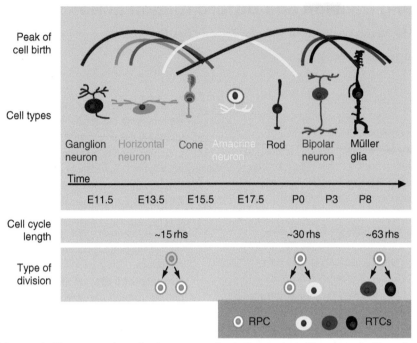

Figure 1 Retinal histogenesis. The upper schematic shows approximate time periods (E, embryonic; P, postnatal) of genesis for each of the seven major murine retinal cell types. The lower-half depicts when symmetric production of two RPCs gradually switches to symmetric production of two differentiating RTCs. This switch is matched by an increase in cell-cycle length. Dividing RPCs are depicted with white cytoplasm and green nuclei. Postmitotic differentiating RTCs are depicted with red nuclei and colored cytoplasm.

key cell-cycle regulators in retinal development, and finish with a discussion as to how RPC expansion versus neurogenesis is coordinated through the Notch pathway.

A Few Basics of Cell-Cycle Regulation

Extracellular positive (mitogenic) and inhibitory cues are sensed in cells through fluctuating levels of cyclins. These proteins are required to activate cyclin-dependent kinases (Cdks). Cyclin D-Cdk4/6 complexes form in G1, cyclin E-Cdk2 complexes act at G1-S, cyclin A-Cdk2 complexes drive S, and cyclin B-Cdk1 complexes drive M. However, these complexes may act in other phases and show considerable functional redundancy. Treatment of resting cells with mitogens induces *Cyclin D1* transcription through Ras–Raf–Erk-mediated activation of the immediate early gene Ap1 (Ap1/Ets). Glycogen synthase kinase 3 beta (Gsk3β)-mediated Thr phosphorylation of cyclin D reduces its stability and nuclear localization. Mitogens also activate the Ras–Pi3k–Akt pathway which phosphorylates and inactivates Gsk3β, doubling cyclin D1 half-life and promoting nuclear translocation. Cyclin D-Cdk4/6 complexes phosphorylate retinoblastoma (Rb) proteins on multiple sites. The cyclin D-Cdk4/6 complex also titrates the Cip/Kip family of Cdk2 inhibitors (CKI), discussed further below.

Rb inhibits division in two major ways (**Figure 2**). First, it binds the activating E2f transcription factors (E2f1, 2, and 3a). These are inducer genes that positively regulate the cell cycle (e.g., cyclins E and A) and other genes that are necessary for the nuts and bolts of DNA replication (e.g., PCNA, RRM1). Rb binds and quenches E2f activity, and recruits silencing cofactors to permanently shut down genes in terminally differentiating or senescent cells. Rb also binds E2f3b and E2f4, whereas its relatives p107 (Rbl1) and p130 (Rbl2) preferentially bind E2f4 and 5 – which are thought to be repressive E2fs that primarily mediate inactivation of target genes (although E2f3b can mimic some functions of E2f3a). However, p107/p130 can partner with activating E2fs if E2f4 is missing. E2fs 6–8 do not bind the Rb family of pocket proteins, but inhibit transcription by recruiting other co-repressors. The extent to which E2f target-gene induction involves direct activation (mediated by activator E2fs) versus derepression (loss of Rb pocket proteins from repressor E2fs) is not completely resolved.

Second, Rb – but not p107/p130 – binds the Cdh1 subunit of anaphase-promoting complex or cyclosome (APC/C) (**Figure 2**). APC/C ubiquitin ligase degrades securin and cyclins to permit passage through and escape from M phase, but in G1 it degrades Skp2 – part of another E3 ubiquitin ligase (SCF^{Skp2}) that degrades Cip/Kip CKIs to promote Cdk2 activity. Rb binds Skp2,

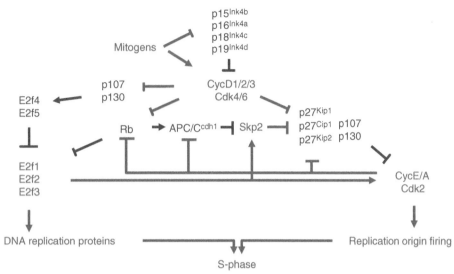

Figure 2 Some Key Regulators of G1/S transition. The two major axes of G1/S inhibition are the Rb–E2f and the Cip/Kip–Cdk2 axes. Mitogens stimulate division by inhibiting INK4a CKIs (e.g., p19^{Ink4d}) and increasing CycD levels. Activated Cdk4/6 phosphorylates and inhibits the Rb family. D-Cdk4/6 complexes also sequester Cip/Kip CKIs (e.g., p27). Active Rb binds and inhibits gene transactivation by E2f1-3. p107/130 form repressor complexes with E2f4/5 that target the same genes as E2f1-3. On the other axis, Cip/Kip CKIs bind and inhibit Cdk2 complexes. P107 and p130 – unlike Rb – can bind and inhibit Cdk2 complexes. There are also feed-forward and feedback links between the Rb–E2f and Cip/Kip–Cdk2 axes. Feed-forward effects include blockade of Skp2-mediated degradation of p27 by Rb-Cdh1, and E2f-mediated induction of CycE/A. Feedback effects include inhibition of Rb by Cdk2 (thus further activating both E2fs and Skp2). Positive regulators are in red, negative are in green. The figure does not, by any means, include all regulators and links.

presenting it for destruction to APC/C, thus preventing degradation of Cip/Kip CKIs and blocking Cdk2 action.

Rb phosphorylation weakens binding to E2f, resulting in induction of cyclin E, Cdk2 activation, further Rb phosphorylation, and induction of cyclin A – a sequential process that is necessary for cell-cycle progression. Cyclin D function is dispensable in Rb-deficient cells, but cyclin E is required for division even in the absence of Rb, indicating it has other targets. Indeed, independent of Cdk2 activation, cyclin E promotes loading of MCM proteins onto origins of replication, and E-Cdk2 phosphorylates this complex to trigger DNA replication. Cyclin E overexpression can drive the cell cycle even when E2f activity is blocked. Apart from unleasing E2f, Rb phosphorylation also releases it from both Skp2 and Cdh1 – thus activating Cdk2 by a second route (**Figure 2**). This positive-feedback loop allows cells to pass R and enter S; indeed, E2f1 behaves as a bistable switch to drive this irreversible transition.

Distinct CKI families bind and inhibit cyclin-Cdks (**Figure 2**). The Ink4 family, which includes p16^{Ink4a}, p15^{Ink4b}, p18^{Ink4c}, and p19^{Ink4d} (encoded by *Cdkn2a/b/c/d*, respectively), inhibits Cdk4/6. Ink4 CKIs act upstream of Rb-E2f and need Rb plus p107 or p130, and E2f4 or E2f5 to block division. The Cip/Kip family of CKIs, which includes p21^{Cip1}, p27^{Kip1}, and p57^{Kip2} (encoded by *Cdkn1a/b/c*, respectively), inhibit cyclin A/E-Cdk2 and cyclin B-Cdk1 and can act downstream of Rb. Notably, p107

and p130, but not Rb, also function as CKIs that inhibit Cdk2 activity as potently as p21^{Cip1} (**Figure 2**). In the complete absence of the Rb family, mouse embryo fibroblasts (MEFs) lack a G1 restriction point, and progress even if mitogens are withdrawn, but these cells arrest at G2 due to the combined action of Cip/Kip CKIs and p53 tumor suppressor.

Inhibitory mitogens – such as transforming growth factor-beta (TGFβ), block cyclin D induction or inhibit its activity by inducing the expression of Ink4 CKIs. TGFβ also inhibits cell-cycle progression by triggering nuclear translocation of an E2F4/E2f5–p107–Smad3 complex that associates with Smad4 protein, and then binds and silences the c-*Myc* promoter through a Smad-E2f element.

In summary (**Figure 2**), Rb and Cip/Kip CKIs cooperatively inhibit division by constraining E2f- and Cdk2-mediated induction of S-phase gene transcription and replication origin firing, respectively. Rb and p27^{Kip1} cross-talk positively by promoting Skp2 degradation and blocking Rb phosphorylation, respectively. Mitogen-activated cyclin-Cdks sequentially phosphorylate Rb, cyclin D-Cdk4/6 sequesters Cip/Kip CKIs, and Skp2, freed of Rb, stimulates CKI degradation, all of which leads to activation of E2f and Cdk2. E2f and Cdk2 cross-talk positively by inducing cyclins and phosphorylating Rb, respectively, This dual axis triggers the production and/or activation of the components needed for DNA replication.

Cyclins and Cdks in Retinal Development

RPCs express higher levels of cyclin D1 than any other embryonic tissue (**Figure 3**). D1 absence causes severe retinal hypocellularity. Downregulation/inhibition of D1 during differentiation is important since ectopic expression in differentiating photoreceptors prevents normal cell-cycle exit, mimicking the effect of pocket protein loss. The large induction in p27^{Kip1} protein translation in newborn retinal neurons likely dwarfs any remaining D1-Cdk4/6 in these cells. D1 loss reduces RPC division beyond E16.5 but not earlier, suggesting that the earliest phase of RPC expansion is D1-independent. This delayed

requirement for D1 may reflect the gradual increase in Rb and p107 expression in RPCs during development. Consistent with the role of D cyclins in inactivating Rb and sequestering CKIs (see above), the *D1*-null retina has hypophosphorylated Rb, no cyclin E–Cdk2 activity, and hypocellularity is rescued when p27^{Kip1} is also missing. A cyclin D1^{KE} point mutant binds but fails to activate Cdk4/6 and partially rescues the *D1*-null retina. Like normal D1-Cdk, D1^{KE}-Cdk complexes sequester p27^{Kip1}, and consequently both Cdk2 activity and Rb phosphorylation are increased in *D1*^{KE} versus *D1*-null retinas.

The D family has two other members – D2 and D3. The defect in *D1*-null mice is rescued in mice expressing

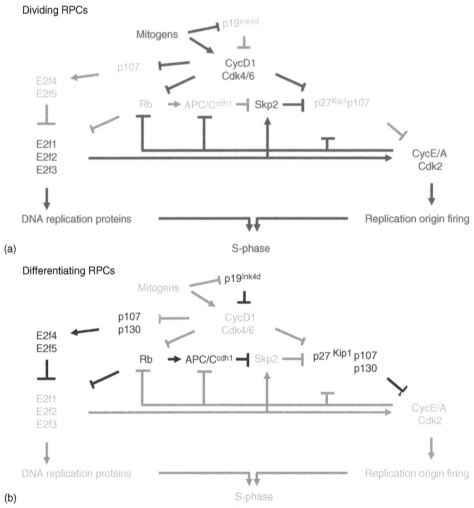

Figure 3 Cell-cycle regulation in RPCs versus differentiating RTCs. (a) In dividing RPCs, extremely high levels of CycD1 activate Cdk4/6 and maintain Rb and p107 in an inactive phosphorylated state and sequester p27^{Kip1}. The latter is also efficiently degraded, presumably due to high levels of Skp2. Inactive pocket proteins leave the activating E2fs free to induce transcription of genes required for DNA replication and CycE/A that activate Cdk2. p27^{Kip1} absence leaves CycE/-Cdk2 free to fire DNA origins and further inhibit negative cell-cycle regulators (e.g., Rb, p27^{Kip1}). (b) In differentiating RTCs, both p19^{Ink4d} and p27^{Kip1} proteins are induced and inhibit Cyc–Cdk complex activity. Cyclin expression is downregulated. Pocket proteins are activated by dephosphorylation and quench E2f activity. p107/p130 act in other cells to form repressor complexes with E2f4/5, although whether this is the case in the retina has not been shown explicitly. Rb may facilitate p27^{Kip1} stability by bringing Skp2 into contact with APC/C. However, in human retina, there is only brief overlap in Rb and p27^{Kip1} expression in differentiating neurons. The picture is more complex than depicted.

D2 from the *D1* locus, so they are interchangeable in this regard. Normally, however, D2 is not expressed in the retina and while D3 is expressed in Müller glia, it is present at very low levels in RPCs. There is no induction of D2 or D3 in *D1*-null retinas and *D1/D2*- or *D1/D3*-null retinas do not show a more severe phenotype than the *D1* knockout (KO). The loss of p27^{Kip1} in *D1*-null retinas also does not affect *D2* and *D3* expression. Triple-null mice survive to ~E15 and, consistent with the idea that D cyclins are not required for early RPC division the early triple-null retina also appears wild-type (WT).

The cyclin E family consists of E1 (formerly E) and E2, and both are expressed in retina. Apart from a spermatogenesis defect in *E2*-null males, *E1*- and *E2*-null mice are normal. $E1^{-/-};E2^{-/-}$ mice die around mid-gestation (~E10) due to defective endoreduplication (repeated S with no M-phase) in trophoblasts, and thus a defective placenta. By using tetraploid WT blastocysts, which form a normal placenta but do not contribute to the embryo proper – $E1^{-/-};E2^{-/-}$ embryonic stem (ES) cells injected into these blastocysts could, in about half the cases, generate a normal embryo that, like WT ES cells in this approach, survive to birth. Half of $E1^{-/-};E2^{-/-}$ embryos die with cardiac defects, all megakaryocytes (like trophoblasts) show defective endoreduplication, and while MEFs could divide normally, they were unable to exit quiescence. Importantly, no neural phenotypes were reported in $E1^{-/-};E2^{-/-}$ embryos, suggesting that cyclin E functions are redundant in RPCs, perhaps due to the combined actions of cyclins D1 and A.

Remarkably, knocking in cyclin *E1* to the *D1* locus partially rescues the cyclin *D1*-null retinal phenotype with retinal thickness reaching ~75% WT. Rb phosphorylation is only slightly higher than in the *D1* KO, suggesting that low levels of Rb phosphorylation are sufficient to permit RPC division. Indeed, cyclin E can override a cell-cycle block induced by overexpressing a mutated version of Rb lacking most of its phosphorylation sites. The Cdk-independent role of cyclin E in loading DNA-replication origins may be important in this context.

Like cyclins, there is also considerable redundancy among Cdks. Although $Cdk4^{-/-}$ mice are 20% smaller at birth, show pancreatic islet cell hypotrophy, and null MEFs divide more slowly due to elevated p27^{Kip1}, there are no obvious retinal defects. Apart from a mild hematopoietic defect $Cdk6^{-/-}$ mice are normal, and *Cdk4/6* redundancy in RPCs remains to be addressed since double-null mice die after E14.5. Cdk3 is rarely mentioned as it is defective in many mice and is thus dispensable. *Cdk2*-null mice appear normal except for a defect in meiosis, and deleting *Cdk2* does not affect fibroblast division. *Cdk2/4/6* triple-null mice die around E12.5 and retinal explant studies have not been attempted. The latter study also shows that Cdk1 (also called Cdc2) – the most ancient cell-cycle kinase best known for its role at G2/M in

mammals – can substitute for other kinases to mediate much of the proliferation needed for embryogenesis. Moreover, it is required for the first division following fertilization. A conditional allele will be needed to study its role in later stages, including RPC expansion.

In summary, cyclin D1 promotes RPC division both by phosphorylating Rb proteins and sequestering p27^{Kip1}, although early RPC expansion appears cyclin D-independent. E cyclins, Cdk4/6, or Cdk2 each seem redundant, but further studies are required to tease apart overlapping cyclin and Cdk roles and to determine if Cdk1 is essential for RPC division.

The Rb Family in Retinal Development

Rb is the tumor suppressor mutated in the familial cancer retinoblastoma. Rb and p107 pocket proteins are present in mouse and human RPCs as well as in postmitotic retinal transition cells (RTCs), whereas p130 seems to be confined to the latter. As expected, Rb loss triggers extra division. However, Rb and p107 seem to be inactive in RPCs as removing Rb or both Rb and p107 in mouse retina does not affect the number of M-phase or visual system homeobox 2 (Vsx2$^+$, also called Chx10) RPCs. Inactivation of Rb in RPCs may be due to the extremely high levels of cyclin D1 and, as noted above, Rb is hypophosphorylated in *D1*-null retinas. The logical explanation for Rb expression in RPCs may be that it is poised to act rapidly in newborn RTCs once D1 levels drop (**Figure 3**). Rb is also important for arresting division following DNA damage, which could also be employed in RPCs. Nevertheless, the irrelevance of Rb in controlling normal RPC division is a hard pill to swallow in view of multiple *in vitro* overexpression experiments implying that Rb tempers the expansion of all dividing cells. Yet Rb-null ES cells divide normally, and Rb-null embryos are not enlarged despite minimal apoptosis, and the same is true of p107/p130-null embryos, and there is not much compensation in either case. Moreover, Rb loss does not impact much of *Xenopus* embryonic development, again implying that it is already inactive.

Instead of tempering expansion, Rb is employed mainly to promote permanent cell-cycle exit, such as in terminally differentiating (**Figures 3** and **4**) or senescing cells, or to execute arrest in DNA-damaged cells. Thus, in contrast to the undetectable effect of *Rb* loss in RPCs, there is a dramatic effect in differentiating RTCs (**Figure 4**). RTCs missing Rb, or Rb and one or more of its relatives, divide ectopically. Some cell types choose apoptosis to defend the tissue from cancer, which is E2f1-dependent, but p53-independent. Indeed, most defects in the Rb-null retina are rescued in the *Rb/E2f1*-null retina, except for a notable cell-cycle and cell-death-independent differentiation defect in a subset of amacrine interneurons that is caused by E2f3a. Intriguingly, Rb repression of E2f3 is also

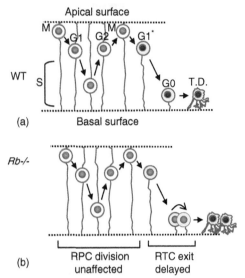

(a)

(b)

RPC division   RTC exit
unaffected   delayed

Figure 4 Uncoupling cell birth and cell-cycle exit. (a) In the WT retina, the RPC nucleus/cell body changes position as the cell progresses through the cell cycle. This process is termed interkinetic nuclear migration (INM). Following cell birth, the apical process begins to retract as the newborn RTC moves to its final destination and undergoes terminal differentiation (TD) Green and red nuclei depict whether the cell is dividing or post mitotic, respectively. Yellow cytoplasm indicates induction of a distinct transcriptome in the RTC. (b) In the absence of Rb, RPC division is unaffected, but differentiating RTCs divide ectopically (note green, instead of red, nucleus). However, the birth transcriptome (yellow cytoplasm) is activated. The response of different RTCs to Rb loss is cell-type specific. Some (as depicted here) survive, terminally differentiate and exit division via Rb-independent means. During ectopic division, these cells are at risk for neoplastic transformation (not shown). Other cell types (e.g., ganglion cells) escape transformation by undergoing apoptosis (not shown).

important to ensure proper migration of forebrain neurons against independent of cell cycle/apoptosis activities. Amacrine cells are one of a subset that survive loss of Rb proteins, and most escape tumorigenesis not by death, but by Rb-independent cell-cycle escape (**Figure 4**). However, rare *Rb/p107*-null or *Rb/p130*-null amacrine cells – likely through post-pocket protein events that override this arrest – can form sporadic retinoblastoma. Human retinoblastoma does not require p107 or p130 loss, likely reflecting broader expression of these proteins in other species. Intriguingly, long-term ectopic division of differentiating cells in fly tissues also requires disruption of multiple cell-cycle regulators. Ectopically dividing murine horizontal cells are also resistant to apoptosis, but they are better protected against transformation and require loss of *Rb*, *p130*, and one allele of *p107* to form tumors. Natural resistance to apoptosis is an attractive feature for a cancer cell-of-origin, especially one like retinoblastoma that requires fewer rate-limiting events than adult cancers. Several post-Rb events have been identified in human retinoblastoma and likely facilitate progression past a benign retinoma state, which ends in senescence.

In summary, pocket proteins apparently do not regulate the cell cycle in normal RPCs, but are poised to act in differentiating RTCs where Rb is required to quench E2f1 activity (**Figures 3** and **4**). p107 and p130, although nonessential, act as backups when Rb is removed. The CKIs $p19^{Ink4d}$ and $p27^{Kip1}$ may play an important role in activating Rb proteins in differentiating RTCs (**Figure 3**). Pocket protein loss creates a dangerous state where ectopically dividing RTCs risk neoplastic transformation. This risk is countered in some cells by apoptosis and others by Rb-independent means of cell-cycle exit.

Ink4 CKIs and p19^{Arf} in Retinal Development

Cdkn1c encoding $p18^{Ink4c}$ is expressed weakly in the embryonic NBL, but is dispensable for retinal development. *Cdkn2b*, which encodes $p15^{Ink4b}$, lies adjacent to *Cdkn2a*, which encodes two transcripts that have distinct first, but shared downstream, exons and encode $p16^{Ink4a}$ and the p53-activating protein $p19^{Arf}$ ($p14^{ARF}$ in humans) from different reading frames. $p15^{Ink4b}$ expression has not been reported in the retina and, while deleting both *Cdkn2b/2a* loci (which removes all three proteins) renders mice extremely susceptible to tumorigenesis, eye defects in addition to those seen in *Cdkn2a* or *$p19^{Arf}$*-null mice were not described. $p19^{Arf}$ is expressed in embryonic vitreal pericytes where it represses platelet-derived growth factor receptor beta (PDGFRB) expression independent of double minute 2 (MDM2) or p53, limiting expansion of these endothelial support cells, thus its absence triggers abnormal expansion and severe defects in the adult eye. As expected, *Cdkn2a*-null mice, lacking both $p16^{Ink4a}$ and $p19^{Arf}$, also have this defect, but no obvious retinal defects – consistent with the absence of these proteins in RPCs. The fourth Ink4 protein – $p19^{Ink4d}$ – is encoded by *Cdkn2d*. Its expression pattern is consistent with a role in facilitating cell-cycle exit (**Figure 3**) and, indeed, null mice show abnormal division followed by elevated apoptosis. The defects may be a milder version of those seen in the absence of Rb or following overexpression of E2f1 or cyclin D1 in differentiating retinal neurons.

Cip/Kip CKIs in Retinal Development

Some $p21^{Cip}$ – encoded by *Cdkn1a* – is expressed in the WT retina, which increases in the absence of *Rb*, implying a context-specific role in retinal cell-cycle control. $p21^{Cip}$ absence alone does not affect retinal development, so it would be interesting to know its effect when combined with Rb loss.

In the embryonic retina, a few cells express $p57^{Kip2}$ (e.g., 3%, E14.5) and its loss triggers extra division, which may reflect extra RPCs and/or ectopically dividing

RTCs; the latter fits the observation that p57^{Kip2} is expressed in late G1 or G0. Expression ceases around P0, and is reactivated in a subset of postmitotic amacrine cells, consistent with a role for p57^{Kip2} in differentiation.

The most influential Cip/Kip CKI in the retina is p27^{Kip1} (**Figure 3**), as suggested by its broader expression pattern in mouse and human retina (e.g., ~50%, E14.5). p27^{Kip1} mRNA is high in mouse RPCs and low in differentiating RTCs but vice versa for protein, implying that rapid translation and mRNA degradation parallels differentiation. In human RTCs, p27^{Kip1} expression seems, in general, to precede Rb protein expression, suggesting that a dual wave of inhibitors is employed to shut division down. p27^{Kip1} is detected at G2 in RPCs, earlier than p57^{Kip2}, consistent with an early role in promoting cell-cycle exit in newborn RTCs. The *Cdkn1b*-null retina has excess dividing cells at least until P10, which could reflect extra RPCs and/or ectopically dividing differentiating RTCs.

Some adult p27^{Kip1}-deficient retinas contain focal hyperplastic lesions, possibly due to reactive gliosis. These lesions are more severe when the *Cdkn1b* gene is replaced by an altered protein (p27^{CK-}) that cannot bind cyclin–Cdk complexes. *p27$^{+/CK-}$* mice do not get retinoblastoma, but do develop lung tumors. The increased phenotypic severity in *p27$^{+/CK-}$* mice relative to *p27^{Kip1}*-null mice reveals that p27^{Kip1}, when freed from cyclin–Cdks, has a dominant disruptive function. This activity may relate to the cytoplasmic role of p27^{Kip1} in regulating the Rho-Rock signaling pathway which, intriguingly, is also targeted by p21^{Cip} and p57^{Kip2}. p27^{Kip1} has other cytoplasmic activities such as binding the microtubule regulator stathmin. p21^{Cip} and 27^{Kip} are distributed between the cytoplasm and nucleus in mouse retina.

Cell-cycle defects in the p19^{Ink4d}- or p27^{Kip1}-deficient retinas are enhanced when both are missing, consistent with cooperative Rb activation and Cdk inactivation to promote cell-cycle exit. Apoptosis and dysplasia are also more severe. As in the $Rb^{-/-}$ retina, deleting p53 does not rescue apoptosis, yet surprisingly it does reverse the dysplasia.

E2Fs in Retinal Development

Multiple E2f mRNAs have been detected in the retina, and protein expression has been confirmed for E2f1 as well as both a and b isoforms of E2f3. Deleting *E2f1* slows RPC division approximately twofold, whereas *E2f3* loss has no effect, suggesting redundancy. The superior role for E2f1 in RPCs differs from that in MEFs where E2f3 is more important. As noted above, E2f1 drives ectopic division in differentiating $Rb^{-/-}$ RTCs and transgenic E2f1 expression in photoreceptors also impairs cell-cycle exit. E2f3a perturbs differentiation in some Rb$^{-/-}$ amacrine

cells, which is similar to its effect in the $Rb^{-/-}$ forebrain. E2f4 loss affects Shh expression in the telencephalon, but its role in the retina, similar to other E2fs, awaits further study.

Separating Rate, Birth, and Exit

The above discussion summarizes how cell-cycle regulators affect RPC expansion and RTC cell-cycle exit. We now turn to how the factors that promote cell birth influence the cell-cycle machinery. At this point, we encourage the reader to review the glossary definition of the terms competence, interkinetic nuclear migration (INM), cell birth, and exit. To facilitate discussion of the model below, we emphasize that birth and exit are viewed as separate activities that are temporally coupled. Birth is the new transcriptional program activated in a differentiating cell that defines its identity, and exit is the cessation of division.

Distance from Notch Predicts Birth

As already described, RPC nuclei undergo INM. Live imaging studies in zebrafish have indicated that nuclei that migrate deeper into the basal layer are more likely to generate a differentiating RTC when they return to the apical surface and undergo mitosis. These data raise the possibility that to achieve birth, nuclei must escape from a diffusable inhibitory signal. An attractive candidate is the antineurogenic factor Notch (**Figure 5**). Activated Notch is cleaved, releasing its intracellular domain (Nicd), which binds recombination signal-binding protein (Rbpj, also called Cbf1, Suh, Igkrb, Kbf2). Rbpj activates expression of hairy and enhancer of split (Hes) proteins, which are basic helix-loop-helix (bHLH) transcriptional repressors that block expression of proneurogenic genes. In line with the speculative model in **Figure 5**, Notch is more abundant at the apical side, but there is conflicting literature on whether the diffusable cleaved Nicd fragment or its gene targets (e.g., the Hes proteins) are more abundant/active in apically located nuclei. The Baier/Link groups found that antibodies to Nicd labeled apically located fish RPCs. Using live imaging, this group also found that a fluorescent protein expressed from the Nicd-inducible *her4* promoter was more abundant in RPCs with more apically located nuclei. However, the Reh group finds that Nicd is lowest in the apically located nuclei in the mouse retina, and both the Reh group and the Kageyama group report higher levels of Hes1 in progenitors when their nuclei are located at the basal side of the neuroepithelium in mouse and chick retina and mouse cortex (using highly destabilized Hes1 reporters). Given the highly transient nature of Notch signaling, it will be important to use highly destabilized reporters, like those used in cortex

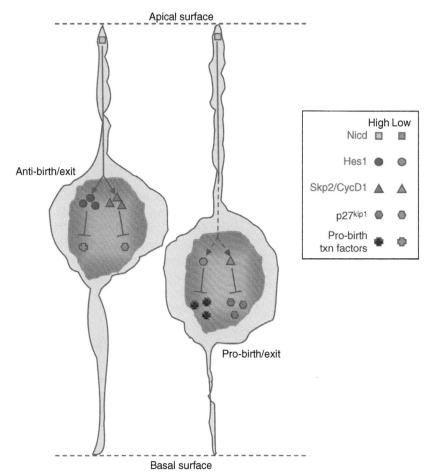

Figure 5 Notch and INM control birth/exit pathways. This Nicd gradient model is based on data from zebrafish studies, but is still controversial (see text for details). In addition, the link between Nicd and Skp2/cyclin D1 induction has been shown in other cell types but not, as yet, RPCs. In this speculative model, the RPC nucleus on the left migrates less basally than the RPC nucleus on the right. As a result, the former is under control of the apical gradient of Notch signaling mediated by the cleaved diffusable Nicd, which drives transcription of Hes1 (green) as well as Skp2 and Cyclin D1 (blue) which counter-expression of neurogenic bHLH factors and inhibit $p27^{Kip1}$, respectively. On the contrary, the nucleus that migrates deeper is less affected by Nicd because it has to diffuse further (dotted gray arrows), and the reduced levels of Hes1 and Skp2/CycD1 permit induction of pro-birth bHLH factors (red) and stabilization of $p27^{Kip1}$ (orange).

by Kageyama's group, for future studies in retina, to better characterize cell-cycle-dependent changes in Notch signaling. Genetic studies prove that Notch is antineurogenic, so it will be important to deduce exactly how it achieves this end.

Genetic analysis in fish shows that, whether or not Nicd/Hes gradients are involved, INM plays a key role in determining birth. In dynactin-1 (mok) mutants, it was found that RPC nuclei migrated deeper (thus appearing to escape Notch signaling) and generated more early-born ganglion cells. The mechanisms that control INM distance are unclear, but both actin- and microtubule-based motors have been implicated. Notably, RPCs are in S-phase when they approach the basal side and neurogenic bHLH genes are induced in S-phase. Moreover, transplant experiments indicate that both cortical and retinal progenitors can change fate during this period.

Together, these data build a model in which nuclei destined to undergo birth are moved a sufficient distance from the apical surface to escape influence of the Notch pathway, thus permitting induction of neurogenic factors required to activate the birth program.

Birth Does Not Require Exit

Birth and cell-cycle exit are intimately linked, so it is tempting to conclude they must be interdependent (birth needs exit). There is also a model proposing that slowing cell-cycle rate in the last division might also be critical to facilitating birth. Below, we summarize evidence for and against these issues.

In support of the notion that rate reduction favors birth, progenitor cell-cycle length increases with the gradual transition to symmetric production of two RTCs. In zebra fish, the average cell-cycle length does not predict birth, but, of two sibling RPCs, the one with the longer cycle is more likely to differentiate following M-phase. Moreover, amounts of the Cdk-inhibitor drug olomoucine that slow, but do not stop, division are sufficient to trigger premature neurogenesis in the telencephalon. In support of the idea that birth requires exit, the two are temporally coupled and p27^{Kip1} induction occurs in the last G2 just prior to RTC birth. Moreover, overexpression of this CKI induces early birth.

The above data are not totally conclusive. Correlations do not distinguish consequence from cause. A longer final cycle in sibling RPCs might reflect deeper migration of the neurogenic partner trying to escape Notch. Olomoucine has targets other than Cdks and importantly, mutations in *E2f1*, *cyclin D1*, or *Vsx2* that lengthen cell cycle by specific genetic means do not cause a switch to early-born cell types. Thus, whether lengthening the cell cycle is necessary for birth remains moot.

What about exit, is it required for birth? Notch-pathway defects trigger both birth and exit, but this is also a correlation that does not distinguish whether they are interdependent or can be uncoupled. Genetic evidence supports the latter since neurogenesis in the retina, forebrain, cerebellum, and inner ear goes ahead in the absence of Rb and differentiating neurons divide ectopically. As discussed earlier, a subset of ectopically dividing neurons eventually undergo apoptosis, while others survive and exit independent of Rb, and these are the source of sporadic retinoblastoma. However, clearly, birth occurs despite these downstream defects. As with Rb loss, there is also not a shift to late-born cells in retinas lacking p27^{Kip1} and/or p19^{Ink4}. The interphotoreceptor retinoid-binding protein (IRBP) promoter is activated just before photoreceptors are born, yet its use to overexpress cyclin D1 or E2f1 does not prevent initiation of the birth program, and the resulting photoreceptors divide ectopically. CKI overexpression assays suggest that exit can drive birth, yet do not prove that this is the physiologically relevant or necessary route. Indeed, arresting division with hydroxyurea or aphidicolin in *Xenopus* embryos does not disrupt most central nervous system (CNS) differentiation, indicating that arrest *per se* is not neurogenic. Perhaps, CKIs induce early birth by affecting a process other than exit, and, indeed, a mutated version of p27^{Kip1} incapable of binding and inhibiting Cdks induces neurogenesis through interaction and stabilization of neurogenin 2 (Neurog2 or Ngn2). Alternatively, active cell-cycle components may maintain expression and/or activity of Notch-pathway components and downregulation of the cell cycle would thus block Notch signaling, an intriguing possibility given that E2f regulates the expression of Hes family members.

In summary, while birth and exit are closely coupled, exit is not necessary for birth. They are both induced following escape from Notch-pathway signals, but in an apparently parallel rather than interdependent fashion.

Mechanisms Linking Birth to Exit

Despite functional independence, birth and exit are coupled temporally, which makes sense given that a single pathway, Notch, is so central in coordinating them both. It is well established that Notch downregulation leads to induction of neurogenic transcription factors through alleviation of Hes1-mediated repression. But how is Notch connected to cell-cycle exit? Hes1 also represses expression of p27^{Kip1} mRNA, but in mouse retina p27^{Kip1} protein induction during cell birth is regulated at a posttranscriptional level. There are many ways in which p27^{Kip1} translation and stability are regulated, yet the specific approach used in the retina is unclear. Activated Notch can induce Skp2 through the same mechanism it uses to activate Hes1 expression and, as noted earlier, Skp2 promotes p27^{kip1} ubiquitin-mediated degradation (**Figure 2**). Forkhead transcription factors can promote p27^{Kip1} stability by downregulating a proteosome subunit. Notch loss downregulates Vsx2 and Tlx4, and both are required for high cyclin D1 and low p27^{Kip1} levels, although the details are unclear. The intracellular cleaved domain of Notch can directly activate the cyclin D1 promoter, so this mechanism might be employed in the retina. Cyclin D1 downregulation likely ensures activation of Rb protein in newborn neurons. p19^{Ink4d} induction would also facilitate this event, but – although it is regulated at both transcriptional and posttranscriptional levels – the mechanisms used in retinal cells are unclear. Once Rb is activated, it shuts down E2f, and the most critical target is E2f1 since removing the latter rescues all ectopic division and apoptosis in the Rb-null retina.

Conclusion

Birth and exit are coordinately regulated through escape from Notch signaling, which may involve a distance-dependent mechanism that relies on polarity signals and INM. It is unclear exactly how the decision to evade Notch is made. Despite their temporal proximity birth is not dependent on exit. The factors that activate the new transcriptional program can be induced independent of signals necessary to shut off division and neurons attempt to develop as they divide ectopically. Failure to couple birth to exit can lead to retinoblastoma, underscoring the importance of maintaining tight coupling. To avoid this catastrophe, some neurons undergo apoptosis, but others adopt Rb-independent means of exiting, which provides a window of time during which neoplastic clones may evolve. Well-known cell-cycle regulators are downregulated and activated to ensure exit parallels birth, and many of these events are posttranscriptional. Exactly how evading Notch triggers these events in the retina remains to be resolved, although there are clues from work in other cell types.

Acknowledgments

We are grateful to Valerie Wallace, Tom Reh, and Brian Link for comments. Research on retinal development in the Bremner lab is funded by the Canadian Institute for Health Research (CIHR) and the Foundation Fighting Blindness. M. Pacal is supported by a Vision Science Research Program fellowship from the University of Toronto.

See also: Embryology and Early Patterning; Eye Field Transcription Factors; Histogenesis: Cell Fate: Signaling Factors; Intraretinal Circuit Formation; Photoreceptor Development: Early Steps/Fate; Retinal Histogenesis.

Further Reading

Baye, L. M. and Link, B. A. (2008). Nuclear migration during retinal development. *Brain Research* 1192: 29–36.

Besson, A., Dowdy, S. F., and Roberts, J. M. (2008). CDK inhibitors: Cell cycle regulators and beyond. *Developmental Cell* 14: 159–169.

Burkhart, D. L. and Sage, J. (2008). Cellular mechanisms of tumour suppression by the retinoblastoma gene. *Nature Reviews. Cancer* 8: 671–682.

Buttitta, L. A. and Edgar, B. A. (2007). Mechanisms controlling cell cycle exit upon terminal differentiation. *Current Opinion in Cell Biology* 19: 697–704.

Cayouette, M., Barres, B. A., and Raff, M. (2003). Importance of intrinsic mechanisms in cell fate decisions in the developing rat retina. *Neuron* 40: 897–904.

Cayouette, M., Poggi, L., and Harris, W. A. (2006). Lineage in the vertebrate retina. *Trends in Neuroscience* 29: 563–570.

Chen, D., Opavsky, R., Pacal, M., et al. (2007). Rb-mediated neuronal differentiation through cell-cycle-independent regulation of E2f3a. *PLoS Biology* 5: e179.

Del Bene, F., Wehman, A. M., Link, B. A., and Baier, H. (2008). Regulation of neurogenesis by interkinetic nuclear migration through an apical-basal notch gradient. *Cell* 134: 1055–1065.

Farkas, L. M. and Huttner, W. B. (2008). The cell biology of neural stem and progenitor cells and its significance for their proliferation versus differentiation during mammalian brain development. *Current Opinion in Cell Biology* 20: 707–715.

Gotz, M. and Huttner, W. B. (2005). The cell biology of neurogenesis. *Nature Reviews. Molecular Cell Biology* 6: 777–788.

Kageyama, R., Ohtsuka, T., Shimojo, H., and Imayoshi, I. (2008b). Dynamic Notch signaling in neural progenitor cells and a revised view of lateral inhibition. *Nature Neuroscience* 11: 1247–1251.

Levine, E. M. and Green, E. S. (2004). Cell-intrinsic regulators of proliferation in vertebrate retinal progenitors. *Seminars in Cell and Developmental Biology* 15: 63–74.

Malumbres, M. and Barbacid, M. (2009). Cell cycle, CDKs and cancer: A changing paradigm. *Nature Reviews. Cancer* 9: 153–166.

Nelson, B. R., Hartman, B. H., Georgi, S. A., Lan, M. S., and Reh, T. A. (2007). Transient inactivation of Notch signaling synchronizes differentiation of neural progenitor cells. *Developmental Biology* 304: 479–498.

Pacal, M. and Bremner, R. (2006). Insights from animal models on the origins and progression of retinoblastoma. *Current Molecular Medicine* 6: 759–781.

Shimojo, H., Ohtsuka, T., and Kageyama, R. (2008). Oscillations in notch signaling regulate aintenance of neural progenitors. *Neuron* 58: 52–64.

van den Heuvel, S. and Dyson, N. J. (2008). Conserved functions of the pRB and E2F families. *Nature Reviews. Molecular Cell Biology* 9: 713–724.

Wallace, V. A. (2008). Proliferative and cell fate effects of Hedgehog signaling in the vertebrate retina. *Brain Research* 1192: 61–75.

Developmental Anatomy of the Retinal and Choroidal Vasculature

B Anand-Apte and J G Hollyfield, Cleveland Clinic, Cleveland, OH, USA

Glossary

Angiogenesis – The formation of new blood vessels from preexisting ones, generally by sprouting.

Central retinal artery – A branch of the ophthalmic artery that enters the eye via the optic nerve.

Choriocapillaris – An exceptionally dense capillary bed that nourishes the posterior choroid up to the level of the equator of the eye.

Circle of Zinn – An annular artery surrounding the optic nerve. Its branches contribute to the pial circulation, the optic nerve at the level of the lamina cribrosa and to the nerve fiber layer of the optic disk.

Fenestrae – The circular openings in the choriocapillaris facing Bruch's membrane that measure approximately 800 Å. The fenestrae of the choriocapillaries have a diaphragm covering.

Posterior ciliary arteries – The branches of the ophthalmic artery that form the blood supply to the choroid.

Vasculogenesis – The formation of new blood vessels through *de novo* formation of new endothelial cells.

Vortex veins – The venous collecting vessels draining the choroid, ciliary body, and iris.

Choroidal Vascular Network

Embryology

At the fourth week of gestation in humans (5-mm stage), the undifferentiated mesoderm surrounding the optic cup begins to differentiate and form endothelial cells adjacent to the retinal pigment epithelium (RPE). These early vessels are the precursors of the choriocapillaris, which ultimately envelop the exterior surface of the optic cup. Concurrently, the hyaloid artery branches from the primitive dorsal ophthalmic artery and passes along the embryonic (choroidal) fissure to enter the optic cup. Shortly thereafter, the condensation of the mesoderm occurs at the site of future choroidal and scleral stroma with the gradual onset of pigmentation in the outer neuroepithelial layer of the optic cup, the RPE. By the fifth week of development (8–10-mm stage), the RPE becomes more melanized and the vascular plexus extends along the entire exterior surface of the cup from the posterior pole to the optic cup rim.

By the sixth week (12–17-mm stage), the choriocapillary network begins to develop a basal lamina. This, together with the basement membrane of the RPE, forms the initial boundaries of Bruch's membrane separating the neural retina from the choroid. Concomitantly, rudimentary vortex veins develop in all four quadrants of the eye. Except for the two basal lamina (of the RPE and choriocapillaris), the only other component of Bruch's membrane at this stage is a collagenous central core. The choroidal capillary network becomes almost completely organized by the eighth week (25–30-mm stage) with connections to the short posterior ciliary arteries. By the eleventh week (50–60-mm stage), the posterior ciliary arteries show extensive branching throughout the choroid. It is only following the third and fourth months of gestation that most of the choroidal vasculature matures.

While the molecular mechanisms regulating choroidal development have not been fully defined, it is largely accepted that the presence of differentiated RPE and its secretion of growth factors, such as vascular endothelial growth factor (VEGF) and fibroblast growth factor 9 (FGF-9), are critical for the physiological development and differentiation of the choroidal vascular network.

Gross Anatomy

Almost the entire blood supply of the eye comes from the choroidal vessels, which originate from the ophthalmic arteries. The left and right ophthalmic arteries arise as the first major branch of the internal carotid, usually where the latter break through the dura to exit the cavernous sinus. In some individuals (around 10%), the ophthalmic artery arises within the cavernous sinus, while in others (around 4%), it arises from the middle meningeal artery, a branch of the external carotid.

The ophthalmic artery shows a wide variation in the branching pattern as it approaches the eye. The posterior ciliary arteries, which form the blood supply to the choroid, and the central retinal artery, which enters the eye via the optic nerve, are branches of the ophthalmic artery (**Figure 1**). Other branches of the ophthalmic artery supply the lacrimal gland, extraocular muscles, and lids.

The choroid (also referred to as the posterior uveal tract) is vascularized by two separate arterial systems: (1) the short posterior ciliary arteries, which supply the posterior choroid and (2) the long posterior ciliary arteries, which supply the anterior portion of the choroid (as well as the iris and ciliary body).

Approximately 16–20 short posterior ciliary arteries penetrate the sclera in a circular pattern surrounding the optic nerve, with the distance between these vessels and the nasal side of the nerve being closer than that on the temporal side. These arteries anastomose within the sclera to form the circle of Zinn, an annular artery surrounding the optic nerve (**Figures 1** and **2**). The branches from the circle of Zinn contribute to the pial circulation, the optic nerve at the level of the lamina cribrosa, and the nerve fiber layer of the optic disk. Other branches from the circle of Zinn along with direct branches from the short posterior ciliary arteries enter the choroid to provide the arterial blood supply to the posterior uveal track. These arteries divide rapidly to terminate in the choriocapillaris, an exceptionally dense capillary bed that nourishes the posterior choroid up to the level of the equator of the eye.

The two long posterior ciliary arteries penetrate the sclera on either side of the optic nerve near the level of the horizontal meridian of the eye. The temporal long posterior ciliary artery enters the sclera approximately 3.9 mm from the temporal border of the optic nerve while the nasal long posterior ciliary artery enters approximately 3.4 mm from the nasal border of the optic nerve. Additional long posterior ciliary arteries are present that course farther toward the anterior segment before penetrating the sclera, usually more anterior than the entry of the temporal and nasal long posterior ciliary arteries. The long posterior ciliary arteries course through the suprachoroid, begin to branch just anterior to the equator, and contribute to the circulation of the iris and ciliary body. Just anterior to the equator, some branches of these vessels course down into the choroid and branch to terminate in the choriocapillaris from the ora serrata back to the equator of the eye.

In general, the larger diameter arteries of the choroid are found most proximal to the sclera, in an area referred to as the lamina fusca. These arteries continue to branch and ultimately form the extensive choriocapillaris adjacent to the acellular Bruch's membrane located on the basal side of the RPE (**Figure 3**). The capillary network of the choriocapillaris is approximately 3–18 μm in diameter and oval shaped in the posterior eye, becoming gradually wider and longer as it moves toward the equatorial region (approximately 6–36 μm wide by 36–400 μm long). The network becomes irregular in the peripheral third of the choroid owing to the entrance and exit of arterioles and venules. Compared to capillaries in other organs, the choriocapillary lumen are significantly larger, with a diameter of nearly 20 μm in the macular region and 18–50 μm in the periphery. A network of collagen fibrils surrounds the choriocapillaris and provides a structural supportive framework.

From the choriocapillaris, venous collecting vessels emerge that ultimately exit the eye through the vortex veins (**Figure 1**). In addition to the choroid, the vortex veins also drain the ciliary body and iris circulation. The number of vortex veins is variable, with at least one per

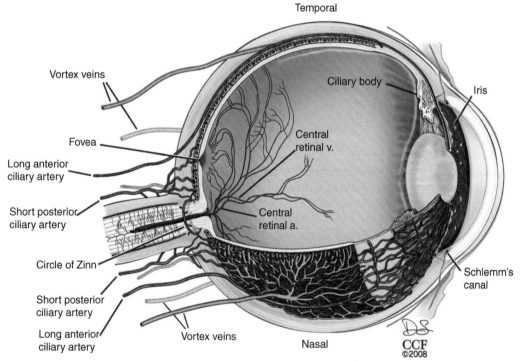

Figure 1 A cutaway drawing of the human eye showing the major blood vessels supplying the retina choroid and anterior segment. The view is from a superior position over the left eye and the horizontal section passes through both the optic nerve and the fovea. Drawing by Dave Schumick.

eye quadrant; however, the total number per eye is usually seven, with more found on the nasal side than are found temporally. The vortex veins usually exit the sclera at the equator or up to 6 mm posterior to this location after forming an ampulla near the internal sclera. The venous branches that open into the anterior and posterior regions of the vortex venular network are oriented along the meridian and are mostly straight, while those joining on the lateral and medial sides have a circular orientation. The vortex veins, in turn, empty into the superior and

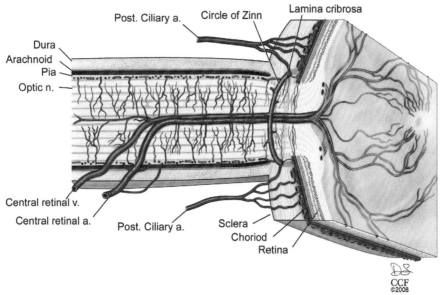

Figure 2   A cutaway drawing along the superior–inferior axis of a left human eye through the optic nerve, showing details of the vascular supply in this location. The fovea is present on the right side of the drawing at the center of the termination of the central retinal vessels. Drawing by Dave Schumick.

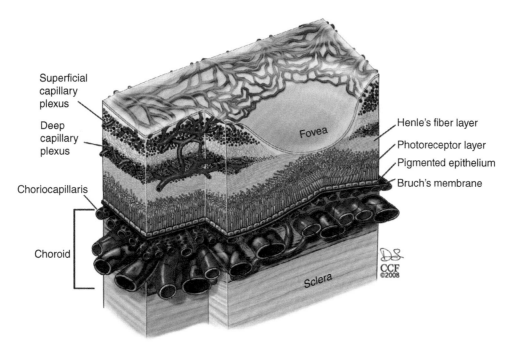

Figure 3   A diagram showing details of the retinal and choroidal vasculature and changes that occur at the level of the human fovea. The branches from the central retinal circulation form two distinct capillary plexi within the ganglion cell layer (the superficial capillary plexus) and in the inner nuclear layer (the deep capillary plexus). These two vascular plexi end as the ganglion cell and inner nuclear layer disappear in the foveal slope. The choroid contains a dense vascular network terminating with the fenestrated choriocapillaris adjacent to Bruch's membrane. Drawing by Dave Schumick.

inferior ophthalmic veins, which leave the orbit and enter the cavernous sinus.

Physiology

The primary function of the choroid vasculature is to provide nutrients and oxygen to the outer retina. The loss of the choriocapillaris, as occurs in central areolar choroidal sclerosis, results in atrophy of the overlying retina. The choriocapillary network is unique in that it lies in a single plane below Bruch's membrane. Normal choroidal circulation occurs when both choroidal arterial and venous pressures are above 15–20 mm of Hg, which is the normal physiological intraocular pressure. In addition, since the blood flow rate through the choroid is relatively high compared to other tissues and is regulated by the arterial vessel diameter, it causes relatively lower amounts of oxygen to be extracted from each milliliter of blood.

The wall of the choriocapillaris facing Bruch's membrane is fenestrated with circular openings (fenestrae) measuring approximately 800 Å. The fenestrae of the choriocapillaries are unique in that they have a diaphragm covering them, unlike those seen in the renal glomerulus. These fenestrae allow easy movement of large macromolecules into the extracapillary compartment. Fluid and macromolecules escaping from these leaky vessels percolate through Bruch's membrane and have access to the basal side of the RPE. The passive movement of fluid and macromolecules from this leaky circulation is blocked from reaching the subretinal space (interphotoreceptor matrix) by zonula occludens junctions that form a continuous belt-like barrier near the apical border of the RPE. Thus, the RPE is able to block the passive movement of the large molecules and fluid from the choroid, allowing the RPE to function as the outer portion of the blood retinal barrier.

Pathology

The exudative (wet form) of age-related macular degeneration (AMD) is characterized by an abnormal proliferation of the choriocapillaris with occasional invasion through Bruch's membrane into the subretinal space. Leakage of the fluid and/or blood often leads to retinal or RPE detachment and loss of central vision. VEGF, basic fibroblast growth factor (bFGF), and pigment epithelial-derived growth factor (PEDF) have been postulated to play a role in the development of choroidal neovascularization (CNV). Some studies have suggested that the thickening of Bruch's membrane that occurs with age and in AMD could result in decreased permeability. The survival factors for the choriocapillaris, such as VEGF, which are secreted basally from the RPE, could remain sequestered in Bruch's membrane and be unable to reach the endothelial cells of the choriocapillaris.

This could potentially lead to atrophy of the choriocapillaris that is seen in dry AMD and may be the initiating event for local hypoxia, and upregulation of VEGF and CNV. Myopia is the second most common cause of CNV. Presumed ocular histoplasmosis syndrome, multiple white dot syndrome, multifocal choroiditis, punctate inner choroidopathy, birdshot chorioretinopathy, and the healing phase of choroidal ruptures are the other causes of CNV.

Hyaloid Vascular Network

Embryology

During the first gestational month, the posterior compartment of the globe contains the primary vitreous comprised of a fibrillary meshwork of ectodermal origin and vascular structures of mesodermal origin. At the 5-mm stage, the primitive dorsal ophthalmic artery sprouts off the hyaloid artery, which passes through the embryonic choroidal fissure and branches within the cavity of the primary optic vesicle. Behind the lens vesicle, some of the branches make contact with the posterior side of the developing lens (capsula perilenticularis fibrosa), while others follow the margin of the cup and form anastomoses with confluent sinuses to form an annular vessel. The arborization of the hyaloid artery forms a dense capillary network around the posterior lens capsule (tunica vasculosa lentis, TVL) and surrounding the lens equator. In addition, capillary branches are given off that course throughout the vitreous (vasa hyaloidea propria). The capsulopupillary vessels anastomose with the annular vessel around the rim of the optic cup and connect to the choroidal vasculature through which venous drainage occurs. By the sixth week (17–18-mm embryo), the annular vessel sends loops forward and centrally over the anterior lens surface. By the end of the third month, the anterior portion of the TVL is replaced by the pupillary membrane, which is supplied via loops from the branches of the long posterior ciliary arteries and the major arterial circle. The development of the hyaloid vasculature is almost complete at the ninth week (40 mm) stage. The venous drainage from the vessels of the anterior lens capsule and, subsequently, from the pupillary membrane and from the capsulopupillary vessels occurs through vessels that assemble into a network in the region where the ciliary body will eventually form and will anastomose and feed into the venules of the choroid.

Normally, the hyaloid vascular system begins to regress during the second month of gestation. This process begins with atrophy of the vasa hyaloidea propria, followed by the capillaries of the TVL, and, finally, the hyaloid artery (by the end of the third month). The occlusion of the regressing capillaries by macrophages appears to be a critical step for atrophy to occur. As the vascular structures regress, the primary vitreous retracts, and collagen fibers and a ground substance of hyaluronic acid are produced, forming a secondary vitreous. By the fifth to the sixth month of gestation, the posterior compartment is primarily composed of the secondary vitreous, and the primary vitreous is reduced to a small, central structure, the Cloquet canal, which is a thin, S-shaped structure that extends from the disk to the posterior surface of the lens.

Physiology

The hyaloid vascular system apparently supplies the nutritive requirements of both the lens and the developing retina before the acquisition of the retinal vasculature. The localization of VEGF in the TVL and papillary membrane may be responsible for fenestrae being present only in the hyaloid capillaries facing the lens. The hypoxia-inducible factor and VEGF have been postulated to play a role in the development and regression of the hyaloid. Other factors that have a part to play in regression are Wnt receptor (Lrp5), Frizzled-4, collagen-18, Arf, Ang2, and BMP-4.

Pathology

Persistent hyperplastic primary vitreous occurs owing to the failure of the hyaloid vasculature to completely regress. The TVL, the anterior extension of the primary vitreous, is comprised of a layer of vascular channels originating from the hyaloid artery, the vasa hyaloidea propria, and the anterior ciliary vessels. The anterior part of this system is supplied by the ciliary system and the posterior part by the hyaloid artery and its branches. The posterior system usually regresses completely by the seventh month of gestation, while the anterior part follows by the eighth month. Two clinical forms of persistent hyperplastic primary vitreous have been identified based on the vascular system that fails to regress: (1) persistent TVL that mainly affects the anterior segment and is now termed anterior hyperplastic primary vitreous and (2) posterior hyperplastic primary vitreous. The hallmark of posterior hyperplastic primary vitreous is the presence of a retinal fold. The lens is usually clear, but may form a cataract over time if the membranous vessels grow forward to enter the lens through the posterior capsule. Other complications include secondary angle-closure glaucoma, microophthalmia, vitreous membranes, tractional retinal detachment at the posterior pole, and a hypoplastic (underdeveloped) or dysplastic (disorganized) optic nerve head. Patients with the anterior hyperplastic primary vitreous often have the best prognosis for visual recovery. The posterior pole is usually normal, with no evidence of a retinal fold or abnormalities in the optic nerve or macula. Patients characteristically present with the appearance of a whitish mass behind the lens (leukokoria) early in life. Occasionally, elongated ciliary processes are present and the eye is microphthalmic. Intralenticular hemorrhage can occur if the fibrovascular membrane invades the lens. Other complications include secondary angle-closure glaucoma, strabismus, and coloboma iridis.

Retinal Vascular Network

Embryology

Until the fourth month of gestation, the retina remains avascular as the hyaloid vasculature provides the nutrients to the developing retina. At the fourth month (100-mm stage), the primitive vascular mesenchyme cells near the hyaloid artery invade the nerve fiber layer. At later stages in development, the hyaloid artery regresses back to this point, which marks the posterior origin of the retinal circulation. In the fourth month of gestation, the first retinal vessels appear when solid endothelial cords sprout from the optic nerve head to form a primitive central retinal arterial system. By the sixth month, these vessels start developing a faint lumen that occasionally contains a red blood cell. At this time, the vessels extend 1–2 mm from the optic disk and continue to migrate outward extending to the ora serrata nasally and to the equator temporally by the seventh to eighth month. Pericytes or mural cells are conspicuously absent from the vessels at this time and do not appear until 2 months after birth. The retinal vasculature achieves the adult pattern by the fifth month after birth. The relative roles of angiogenesis versus vasculogenesis in the development of the retinal vascular network are still controversial. However, it is generally accepted that VEGF secreted in a temporal and spatial pattern by microglia and astrocytes plays a critical role in the development of the superficial and deep layers of the retinal vasculature. Platelet-derived growth factor (PDGF) produced by the neuronal cells induces the proliferation and differentiation of astrocytes, which respond to the hypoxic environment by secreting VEGF. These cells establish a gradient of VEGF and a track for endothelial cells to follow. The hypoxia-inducible factor has been shown to be critical in the hypoxia-induced regulation of retinal vascular development. VEGF isoforms have been shown to perform highly specific functions during developmental retinal neovascularization.

Gross Anatomy

Usually, the only arterial blood supply to the inner retina is from the central retinal artery that runs along the inferior margin of the optic nerve sheath and enters the eye at the level of the optic nerve head (**Figure 2**). Within the optic nerve, the artery divides to form two major trunks and each of these divides again to form the superior nasal and temporal and the inferior nasal and temporal arteries that supply the four quadrants of the retina. The retinal venous branches are distributed in a relatively similar pattern. The major arterial and venous branches and the successive divisions of the retinal vasculature are present in the nerve fiber layer close to the internal limiting membrane. The retinal arterial circulation in the human eye is a terminal system with no arteriovenous anastomoses or communication with other arterial systems. Thus, the blood supply to a specific retinal quadrant comes exclusively from the specific retinal artery and vein that supply that quadrant. Any blockage in blood supply therefore results in infarction. As the large arteries extend within the retina toward the periphery, they divide to form arteries with progressively smaller diameters until they reach the ora serrata where they return and are continuous as a venous drainage system. The retinal arteries branch either dichotomously or at right angles to the original vessel. The arteries and venules generated from the retinal arteries and veins form an extensive capillary network in the inner retina as far as the external border of the inner nuclear layer. Arteriovenous crossings occur more often in the upper temporal quadrants with the vein usually lying deeper than the artery at these crossings.

Branches from the central retinal vessels dive deep into the retina forming two distinct capillary beds, one in the ganglion cell layer (superficial capillary plexus) and the other in the inner nuclear layer (deep capillary plexus). Normally, no blood vessels from the central retinal arteries extend into the outer plexiform layer (**Figure 3**). Thus, the photoreceptor layer of the retina is free of the blood vessels supplied by the central retinal artery. The choriocapillaris provides the blood supply to photoreceptors. Since the fovea contains only photoreceptors, this cone-rich area is free of any branches from the central retinal vessels (**Figure 3**).

In some individuals (18%), a cilioretinal artery derived from the short posterior ciliary artery (from the choroid vasculature) enters the retina around the termination of Bruch's membrane, usually on the temporal side of the optic nerve, and courses toward the fovea, where it ends in a capillary bed and contributes to the retinal vasculature. In approximately 15% of eyes with a cilioretinal artery, the branches supply the macula exclusively, whereas in other individuals, they can nourish the macular region and regions of the upper or lower temporal retina.

Physiology

The rate of blood flow through the retinal circulation is approximately 1.6–1.7 ml g^{-1} of retina with a mean circulation time of approximately 4.7 s. The flow rate through the retinal vessels is significantly slower than that through the choroidal vasculature.

The arteries around the optic nerve are approximately 100 μm in diameter with 18-μm-thick walls. These decrease in diameter in the branched arteries located in the deeper retina to around 15 μm. The walls of the retinal arteries have the characteristics of other small muscular arteries and are composed of a single layer of endothelial cells: a subendothelial elastica, a media of smooth muscle cells, a poorly demarcated external elastic lamina, and an adventitia comprised of collagen fibrils. Near the optic disk, the arterial wall has five to seven layers of smooth muscle cells, which gradually decrease to two or three layers at the equator and to one or two layers at the periphery. These vessels continue to have the characteristics of arteries, and not arterioles, up to the periphery. The retinal arteries lose their internal elastic lamina soon after they bifurcate at the optic disk. This renders them immune from developing temporal arteritis and distinguishes them from muscular arteries of the same size in other tissues. As a compensatory mechanism, the retinal arteries have a thicker muscularis, which allows increased constriction in response to pressure and or chemical stimuli.

The major branches of the central vein close to the optic disk have a lumen of nearly 200 μm with a thin wall made up of a single layer of endothelial cells having a thin basement membrane (0.1 μm), which is continuous with the adjacent media comprised predominantly of elastic fibers, a few muscle cells, and a thin adventitia. The larger veins in the posterior wall have three to four layers of muscle cells in the media. As the retinal veins move peripherally from the optic disk, they lose the muscle cells, which are replaced by pericytes. The lack of smooth muscle cells in the venular vessel wall results in a loss of a rigid structural framework for the vessels, resulting in shape changes under conditions of sluggish blood flow (e.g., diabetes mellitus) or increased blood viscosity (polycythemia), or with increased venous pressure (papilledema or orbital compressive syndromes).

The retinal capillary network is spread throughout the retina, diffusely distributed between the arterial and venous systems. There are three specific areas of the retina that are devoid of capillaries. The capillary network extends as far peripherally as the retinal arteries and veins up to 1.5 mm from the posterior edge of the troughs of the ora serrata (ora bays), leaving the ora serrata ridges (ora teeth) without any retinal circulation. The 400-μm-wide capillary-free region centered around the fovea is another area lacking retinal capillaries. Finally, the retina adjacent to the major arteries and some veins lacks a capillary bed.

The retinal capillary wall is comprised of the endothelial cells, basement membrane, and intramural pericytes. The retinal capillary lumen is extremely small (3.5–6 μm in diameter), requiring the circulating red blood cells to undergo contortions to pass through. Unlike the choriocapillaris, the endothelial cells of retinal capillaries are not fenestrated. The edges of endothelial cells show interdigitation and are joined by the zona occludens at their lumenal surface. The endothelial cells of the retinal arteries are linked by tight junctions, which prevent the movement of large molecules in or out of the retinal vessels. These tight junctions, in concert with the Muller cells and astrocytes, establish the blood–retinal barrier that prevents the passage of plasma proteins and other macromolecules in or out of the capillary system. The pericytes (also called mural cells) are embedded within the basal lamina of the retinal capillary endothelium. The pericytes are believed to play an important role in the stabilization of the retinal capillary vasculature.

Pathology

Diabetic retinopathy (maculopathy) is the leading cause of vision loss in patients with type 2 diabetes and is characterized by the hyperpermeability of retinal blood vessels, with subsequent formation of macular edema and hard exudates. Early changes in the retinal vasculature include thickening of the basement membrane, loss of pericytes, formation of microaneurysms, and increased permeability. It is generally hypothesized that these changes lead to the dysfunction of the retinal vessels, loss of vessel perfusion, hypoxia, induction of retinal VEGF expression, and pathological neovascularization as seen in proliferative diabetic retinopathy. In severe cases, retinal detachment can occur as a result of traction caused by fibrous membrane formation. Laser photocoagulation and VEGF-blocking agents are currently being used as therapeutic approaches for this disease.

Retinopathy of prematurity is a potentially blinding disorder affecting premature infants weighing 1250 g or less and with a gestational age of less than 31 weeks. This disease is characterized by abnormal retinal vascularization and can be classified into several stages ranging from mild (stage I), with mildly abnormal vessel growth, to severe (stage V), with severely abnormal vessel growth and a completely detached retina. The pathophysiology of this disease has been extensively studied and several factors have been implicated. The premature birth of an infant before the retinal vasculature has extended to the periphery results in stoppage of the normal blood vessel growth that is driven by the hypoxia-mediated expression of VEGF. This is compounded by the oxygen therapy given to alleviate the respiratory distress, which reduces VEGF expression leading to vaso-obliteration. Once the infant is taken off oxygen, there is a relative hypoxia, upregulation of VEGF, and florid abnormal neovascularization.

See also: Blood–Retinal Barrier; Development of the Retinal Vasculature.

Further Reading

Hogan, M. J., Alvarado, J. A., and Weddell, J. (1971). *Histology of the human eye.* Philadelphia, PA: W.B. Saunders.
Jakobiec, F. A. (1982). *Ocular anatomy, embryology and teratology.* Philadelphia, PA: Harper and Row.
Saint-Geniez, M. and D'Amore, P. A. (2004). Development and pathology of the hyaloid, choroidal and retinal vasculature. *International Journal of Developmental Biology* 48: 1045–1058.
Wolff, E. (1933). *The anatomy of the eye and orbit.* Philadelphia, PA: P. Blakiston's and Co.

Development of the Retinal Vasculature

T Chan-Ling, University of Sydney, Sydney, NSW, Australia

Glossary

Angiogenesis – The growth of new blood vessels through the process of budding from existing vessels. Angiogenesis occurs in response to a stimulus such as hypoxia.

Inner plexus – The network of superficial blood vessels that lies on the inner surface of the retina.

Mural cells – Includes pericytes and smooth muscle cells. These cells surround the endothelial cell tubes and contribute to the formation of stable vessels.

Outer plexus – The deeper network of blood vessels that lies at the junction of the inner nuclear layer and the outer plexiform layer of the retina. This layer of vessels forms by budding from the inner plexus, and consists entirely of capillaries.

Pericytes – Mesenchymal cells that are associated with small vessels. These cells are critical for vessel stability and for the formation of the blood–retinal barrier.

Vasculogenesis – The process of vessel formation through the organization of vascular precursor cells into chords and vessels. Vasculogenesis proceeds in the absence of stimuli such as VEGF.

Introduction

The development of the vascular network of the human retina follows a very specific topography and series of events, producing a network of vessels that precisely meets the metabolic demands of the healthy adult retina. Disruption of this process can lead to the under- or overproduction of vessels and/or the formation of vessels with pathological characteristics, including a breakdown of the blood–retina barrier (BRB) and inappropriate pericyte ensheathment leading to vessel instability. In this article, we describe the process through which the vasculature develops, and the intrinsic and extrinsic signals that control its formation.

An Overview of Human Adult Retinal Vasculature

The retina has the highest metabolic demand of any tissue in the body. The conflicting requirements of sufficient blood supply and minimal interference with the light path to the photoreceptors are met by two vascular supplies: inherent intraretinal vessels supply the inner two-thirds of the retina, and the choroidal vasculature supplies the outer third of the retina.

The vasculature of the adult retina enters and exits through the optic disk (**Figure 1(a)**). As shown in the fundus photo (**Figure 1(b)**), four main artery/vein pairs supply the human retina. These vessels are termed the superior nasal, inferior nasal, superior temporal, and inferior temporal branches of the central retinal artery and vein. Smaller arterioles branch off these four main arteries, whereas terminal branches bifurcate toward the peripheral retina. The fovea, a region of the retina that contains the highest density of photoreceptors, remains avascular throughout development and in the adult retina. The inner plexus ramifies in the ganglion cell and nerve fiber layers. Some of these vessels dive down through the inner plexiform and inner nuclear layers to form the outer, or deep, plexus at the junction of the inner nuclear and outer plexiform layers of the retina (**Figure 1(c)**). Vessels in the inner plexus cover the spectrum from arterioles and venules to capillaries and postcapillary venules, whereas vessels in the outer plexus are mainly capillary in size. Both vascular layers reach almost to the edge of the adult retina, leaving an avascular rim that is thin enough to be adequately oxygenated through diffusion from the choroid. A third intraretinal vascular network, the radial peri-papillary capillaries (RPCs), forms a limited plexus in the nerve fiber layer around the optic disk. The RPCs radiate out from the optic nerve head (ONH) and supply the thick nerve fiber bundles where they exit the retina.

Figure 1(d) is a schematic representation of central nervous system (CNS)/microvascular interface, as would be found typically in the human retina. CNS vessels are characterized by tight junctions between adjacent vascular endothelial cells; the tight junctions are responsible for creating the BRB. Outside the endothelial cell layer lays the basal lamina, which is laid down by endothelial cells and pericytes. This layer is composed predominantly of collagen IV, fibronectin, and laminin. Pericytes are located within the basal lamina, between the vascular endothelial cells and the astrocytic endfeet that form the glia limitans, or glial limiting membrane. In addition to astrocytes, ultrastructural evidence suggests that the glia limitans also includes perivascular cells and microglia.

Vasculature of the Primordial Retina

During embryonic development, the retina forms as an extension of the diencephalon. The rudimentary structures

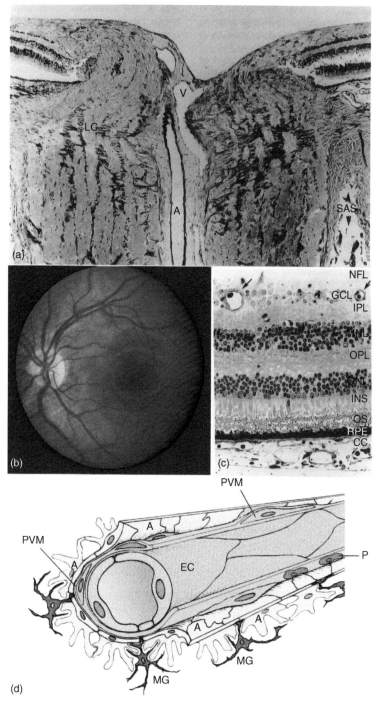

Figure 1 Structure of the adult retina. (a) Cross-section of the adult retina through the optic nerve. Shown are a vein (V), artery (A), lamina cribrosa (LC), and subarachnoid space (SAS). Forrester, J. V., Dick, A. D., McMenamin, P. G., and Lee, W. R. (2002). *The Eye: Basic Sciences in Practice*, 2nd edn. London: Elsevier. (b) Wide-field photograph of the normal human fundus. Visible are the four main artery/vein pairs, extending out from the optic nerve in a four-lobed pattern. Reproduced from Ms. Christine Craigie, Sydney Australia. (c) Low magnification micrograph of the human retina. Starting with the inner retina at the top of the image, visible are the nerve fiber layer (NFL), the ganglion cell layer (GCL), the inner plexiform, and inner nuclear layers (IPL/INL), the outer plexiform and outer nuclear layers (OPL/ONL), the photoreceptor inner segments (INS) and outer segments (OS), the retinal pigmented epithelium (RPE), and the choroid (CC). The arrows point to vessels of the inner (superficial) plexus. From Forrester, J. V., Dick, A. D., McMenamin, P. G., and Lee, W. R. (2002). *The Eye: Basic Sciences in Practice*. London: Elsevier. (d) Schematic diagram illustrating the components of a retinal (brain) vessel wall. From the vascular lumen outwards, shown are endothelial cells (EC; in pale orange), basal lamina (including collagen IV, fibronectin and laminin; in green), pericytes (P; in red), astrocytes (A; in lavender), perivascular cells (PVM; light blue), and perivascular microglia (MG; brown). McMenamin, P. G. and Forrester, J. V. (1999). In: Lotze, M. T. and Thompson, A. W. (eds.) *Dendritic Cells in the Eye*. London: Academic Press.

of the eye are distinguishable by 4–5 weeks gestation (WG). At the earliest stages in development, the primordial lens and retinal tissues are oxygenated through the hyaloid vasculature, a vascular network that begins as an artery entering the eye through the optic nerve (the central hyaloid artery), splits into hyaloid vessels as it continues forward through the vitreous and around the developing lens, and exits at the front of the eye. This vascular network is present early in human embryonic development and regresses as the retinal vasculature forms and is able to meet the increasing metabolic demands of the eye. Typically, the hyaloid vasculature has totally regressed by the ninth month of gestation.

Formation of the Human Retinal Vasculature Takes Place Through Vasculogenesis and Angiogenesis

Blood vessels can be formed by one of two distinct mechanisms. Vasculogenesis is the *de novo* formation of vessels by the aggregation of endothelial precursor cells. Vessels develop from vascular precursor cells (VPCs) that aggregate into solid vascular cords, which then become patent and differentiate to form primitive endothelial tubes. Formation of vessels by angiogenesis occurs through budding from existing vessels; this process takes place through proliferation of vascular endothelial cells and serves to vascularize neighboring tissues. The process of angiogenesis is driven by signals that are produced in response to hypoxia, including vascular endothelial growth factor$_{165}$ (VEGF$_{165}$), whereas the process of vasculogenesis is independent of hypoxia.

Vasculogenesis: Vascular formation through transformation from VPCs

The first event in the development of the retinal vasculature is the *de novo* formation of vessels by vasculogenesis. This stage, detectable in the human retina before 12 WG, initiates with the migration of spindle-shaped cells of mesenchymal origin from the ONH. The individual cells can be Nissl stained and express CD39, vascular endothelial growth factor receptor 2 (VEGFR2), and ADPase, an ecto-enzyme found on the luminal surface of endothelial cells in the adult retinal vasculature. The VPCs migrate outward from the optic disk, as shown in **Figure 2(a)**.

Patent vessels form through the transformation of solid vascular chords. The VPCs localize to the inner surface of the retina, between nerve fiber bundles, and orient their longitudinal axes along the direction of migration. The population of VPCs proliferates and differentiates to form a primordial vascular bed centered on the optic disk (**Figure 2(b)**). As early as 14–15 WG, vascular chords begin to coalesce on the surface of the retina behind the wave of spindle cells, beginning in the region proximal to the ONH and

progressing outward (**Figure 2(c)** and **2(e)**). **Figure 2(d)** shows the transition from vascular chords and the discrete spindle cells located more peripherally.

Vascular chords begin to establish themselves as vessels that express CD34 and support blood flow as early as 18 WG (**Figure 2(f)**). These first primordial vessels formed through vasculogenesis are typically radial, have uniform diameter and have low capillary density (**Figure 3(a)**). The formation of primitive vessels lags behind the leading edge of VPCs by a distance of at least 1 mm. Vasculogenesis and the formation of the primordial vessel architecture is complete by 21 WG, at which point spindle cells are no longer detectable in the retina (**Figure 3(b)**).

Spindle cells (and the formation of cords and vessels behind the leading edge of spindle cells) migrate outward from the optic disk in a four-lobed pattern, as seen in **Figure 3(b)**. The lobes extend the farthest in the temporal and superior directions, and they correspond to the future location of the four major artery–vein pairs in the adult retina (**Figure 1(b)**). The lobes curve around the region of the incipient fovea, leaving this area free of spindle cells and vascular cords. The fovea and perifoveal region remains avascular through 25 WG.

Although vasculogenesis is responsible for the primordial vessels that form in the inner two-thirds of the developing retina, this primitive network with its low capillary density is very inefficient in meeting the metabolic demands of the underlying retinal tissue. As a result, the remaining vessels in the retina form through the process of angiogenesis, which is driven by physiological levels of hypoxia. Angiogenesis is responsible for the increasing vascular density in the central retina (**Figure 3(c)**), vessel formation on the inner surface (the inner plexus) of the peripheral retina, and the formation of the outer, deep plexus and the radial RPCs near the ONH.

Angiogenesis. Angiogenesis in the retina is driven by physiological hypoxia, which occurs as a result of the increasing synaptic activity in the retinal tissues. As the retina thickens and the cell layers start to differentiate and become active, the tissue becomes hypoxic. In response to hypoxia, the astrocytes in the nerve fiber and ganglion cell layers (GCLs) express VEGF$_{165}$, which in turn stimulates endothelial cell growth. Once these vessels become patent and can direct blood flow, the demand for oxygen is met and local VEGF expression decreases. It is important to note that the physiologic levels of hypoxia are too low to damage the surrounding tissue, but are sufficient to signal the need for increased oxygen supply to the tissue.

Inner plexus: Angiogenic filopodial extension is mediated by an astrocytic template and basal lamina components. As with vasculogenesis of the retina, formation of the superficial vascular plexus through angiogenesis begins in the region around the optic disk and spreads outward toward the periphery. As early as 17–18 WG, an exuberant network of broad capillaries starts to grow out from the primary

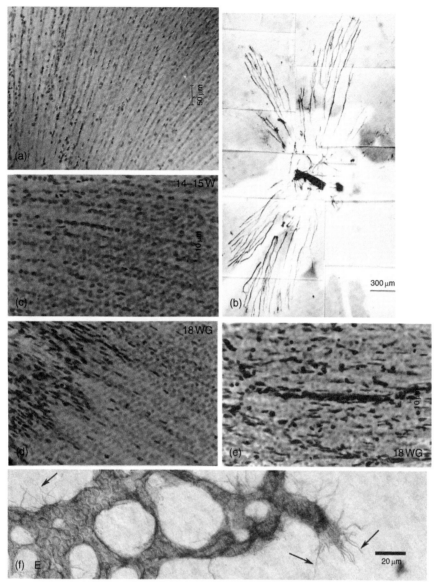

Figure 2 Formation of the early retinal vasculature. (a) Nissl-stained retinal whole mount of a 14–15 WG human fetal retina, immediately peripheral to the optic nerve head. Large numbers of spindle cells are visible, streaming from the optic nerve head. These cells were concentrated between nerve fiber bundles. (b) Immunohistochemical analysis of CD34⁺ vessels in a human fetal retina whole mount, 15 WG. (c) Nissl-stained retinal whole mount of a 14–15 WG human fetal retina, showing the alignment of vascular precursor cells, the first evidence of chord formation prior to the development of patent vessels. (d) Nissl-stained retinal whole mount of an 18 WG human fetal retina. Visible is the leading edge of migration of the spindle-shaped vascular precursor cells. Vascular chords are visible behind the leading edge. (e) A newly-formed solid chord of cells forming a vessel in a human fetal retina at 18 WG. From Hughes, S., Yang, H., and Chan-Ling, T. (2000). Vascularization of the human fetal retina: Roles of vasculogenesis and angiogenesis. *Investigative Ophthalmology and Visual Science* 41: 1217–1228. Copyright Association for Research in Vision and Ophthalmology. (f) CD34+ vessels in a human retina, 18 WG. Note the red blood cells in the patent vessels and the filopodial extensions (arrows) in the region of new vessel growth. From Chan-Ling, T., McLeod, D. S., Hughes, S., et al. (2004). Astrocyte-endothelial cell relationships during retinal vascular development. *Investigative Ophthalmology and Visual Science* 45: 2020–2030. Copyright Association for Research in Vision and Ophthalmology.

vasculature, filling the area that lies between the primordial vessels (**Figure 3(d)**). Between 18 and 30 WG, the sprouting of new vessels is led by the filopodia of endothelial tip cells (**Figures 3(e)** and **4(b)**). These filopodia orient along the net-like framework that has been set up by astrocytes in the wake of the leading edge of APCs, initially producing a capillary bed that closely follows the structure and organization of the astrocyte scaffold (**Figure 4(a)** and **4(c)**). Although endothelial tip cells serve to orient the growth of the capillary network, the tip cells themselves do not divide. Instead, rapid cell growth occurs at the level of the trailing stalk cell (**Figure 4(b)**).

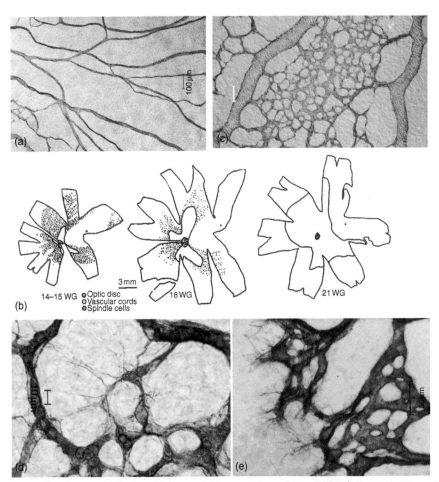

Figure 3 Vascularization of the human fetal retina. (a) Immunohistochemical analysis of CD34⁺ vessels in the central retina at 18 WG. These near-radial vessels all extend predominantly radially from the optic disk. (b) Topographical maps of the outer limits of spindle cells and vascular chords in the human fetal retina at 14–15, 18, and 21 WG. (c) CD34⁺ vessels in a human fetal retina at 21 WG. Note the extensive capillary network that has formed in the space between the major vessels. (d) CD34⁺ vasculature in a human fetal retina at 26 WG. Vessel development follows the extension of numerous filopodia. Capillary sprouting (angiogenesis) is responsible for increasing capillary density in central retina. (e) CD34⁺ vasculature in a region of filopodial extension at the leading edge of patent vessel formation in a human fetal retina at 25 WG. Red blood cells evident in the vessel lumen in the vessels just central to the region of filopodial extension demonstrate that these vessels are patent. From Hughes, S., Yang, H., and Chan-Ling, T. (2000). Vascularization of the human fetal retina: Roles of vasculogenesis and angiogenesis. *Investigative Ophthalmology and Visual Science* 41: 1217–1228. Copyright Association for Research in Vision and Ophthalmology.

In the human fetal retina, Pax-2 expression is limited to cells of the astrocyte lineage. Glial fibrillary acidic protein (GFAP) is expressed in mature astrocytes. Pax-2⁺/GFAP⁻ astrocyte precursor cells (APCs) can be detected at the ONH around 12 WG. This population of cells appears at a slightly later time than the VPCs, and can readily be distinguished from VPCs by their rounded shape and by the expression of Pax-2. Like the VPCs, the APCs migrate outward from the optic disk. The leading edge of APCs is immediately peripheral to the leading edge of vessel formation, preceding the region of vessel formation by no more than 120 μm (**Figure 4(c)**). Beginning at 18 WG, astrocytes loosely ensheath the newly formed vessels, where they play a role in the induction of the BRB. APCs migrate outward from the optic disk behind the spindle cells. Similar to the spindle cells, they also remain

excluded from the incipient fovea as they migrate outward. VPCs (and the primordial vasculature formed from these cells) are limited to the inner two-thirds of the retina (**Figures 2(b)** and **3(b)**) but APCs continue to migrate toward the peripheral retina. As they mature, APCs begin to express GFAP. The Pax-2⁺/GFAP⁺ astrocyte cells reach the edge of the retina around 26 WG.

In addition to driving the formation of capillaries in the space between the primitive vascular network, angiogenesis also controls the peripheral spread of vessels beyond the inner two-thirds of the retina. The growth of vessels in this region closely follows the central-to-peripheral migration of astrocytes. **Figure 4(c)** and **4(d)** show the relationship between patent vessel formation, APCs, and differentiated astrocytes. Again, the increasing metabolic demands of the maturing tissues lead to hypoxia and the

Development of the Retinal Vasculature

Figure 4 Angiogenesis and association of the developing vasculature with an astrocytic scaffold. (a) Low magnification (3 panels on the left) and high magnification (right panel) of an early postnatal mouse retinal whole mount, showing the overlap between the vascular plexus (isolectin, green) and astrocytic scaffold (GFAP; red) during development. (b) An illustration of endothelial tip cells, stalk cells and lumen. From Gerhardt, H., Gloding, M., Fruttinger, M., et al. (2003). VEGF guides angiogenic sprouting utilizing endothelial tip cell filopodia. *Journal of Cell Biology* 161: 1163–1177. (c) Developmental map of a 14-WG retina. Red: Pax2$^+$/GFAP$^-$ APCs; yellow: Pax2$^+$/GFAP$^+$ immature astrocytes; green: CD34$^+$ blood vessels. At this stage, Pax2$^+$/GFAP$^-$ APCs extend in advance of the leading edge of CD34$^+$ blood vessels by a small margin. At the midretina and near the optic nerve head, most cells are Pax2$^+$/GFAP$^+$ immature astrocytes, and CD34$^+$ vessels are clearly evident. (d) ADPase vasculature in a retinal wholemount from a human fetus at 16 WG. Note the four well-defined vascular arcades. From Chan-Ling, T., McLeod, D. S., Hughes, S., et al. (2004). Astrocyte-endothelial cell relationships during retinal vascular development. *Investigative Ophthalmology and Visual Science* 45: 2020–2030. Copyright Association for Research in Vision and Ophthalmology.

production of VEGF by the network of astrocytes that has formed on the surface of the retina; the dense capillary mesh forms along the astrocyte scaffold. Conversely, the endothelium may also influence astrocytic differentiation, as vascular endothelial cells have been shown to induce the expression of GFAP in APCs.

Outer plexus: Angiogenic growth is driven by neuronal maturation. The outer or deep layer of vessels also forms by the process of angiogenesis. Beginning around 25–26 WG, the superficial vessels start to bud and grow radially

down from the inner plexus of the retina (**Figure 5**). At the junction of the inner nuclear layer and the outer plexiform layer, the vessels start to ramify among the two layers. This stage in fetal development coincides with the peak period of eye opening, when the visually evoked potential, indicative of a functional visual pathway and photoreceptor activity, is first detectable in the human infant. The radial growth of these descenders is not preceded by either VPCs or APCs. VEGF expression in this region correlates with the soma of Müller cells, suggesting

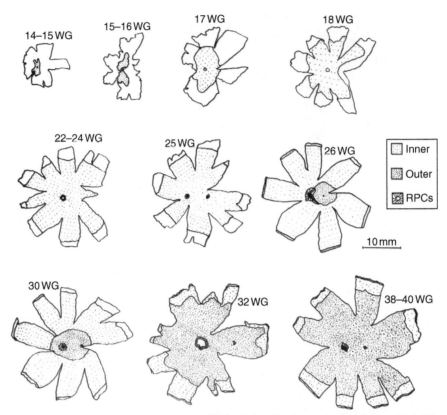

Figure 5 Outer limits of the inner and outer plexuses and the RPCs during development of the human fetal retina. Shown are representative maps of human vascular development from 14–40 weeks gestation. From Hughes, S., Yang, H., and Chan-Ling, T. (2000). Vascularization of the human fetal retina: Roles of vasculogenesis and angiogenesis. *Investigative Ophthalmology and Visual Science* 41: 1217–1228. Copyright Association for Research in Vision and Ophthalmology.

that these cells produce the VEGF that stimulates and guides endothelial cell growth in the outer vascular plexus.

Unlike the inner plexus, the formation of the outer plexus is centered around the fovea. Formation of the outer plexus exactly matches the pattern of maturation of the neuronal retina, suggesting that increased metabolic demands from active neurons produce local, physiological hypoxia and drive the growth of vessels to these tissues (**Figure 5**). In this region, the vasculature is limited to capillary-sized vessels; larger vessels are not found in the outer plexus.

Foveal and perifoveal region. The vasculature of the inner plexus surrounds, but does not enter, the fovea; the fovea is supported by the outer plexus alone. The area of greatest cone density in the retina lies in the center of the avascular area, in a region called the fovea centralis (**Figure 6(a)–6(c)**). The absence of surface vessels in this region allows for maximal light penetration with minimal shadowing of the densely packed cones. The fovea itself, an oval of $500 - 600 \,\mu m$ in diameter, remains completely avascular through 25 WG (**Figure 6(d)**). The developmental cues that direct this aspect of retinal vascularization are not well understood, but recent evidence suggests that the developing fovea expresses antiangiogenic factors

including pigment epithelium derived factor (PEDF), natriuretic peptide precursor B, and collagen type IVα2.

More recently, it has been shown that expression of Eph-A6 directs vascularization in the fovea and perifoveal region. Immunohistochemistry from fetal primate retinas suggests that a gradient of Eph-A6 is centered at the fovea. This gradient, which is highest at the inner GCL and lowest near the inner plexiform layer, serves to repel astrocytes and angiogenesis from the inner layer of the retina toward the inner plexiform layer, where the vessels are found in the primate fovea. Because the incipient fovea lacks an astrocytic scaffold to set up the framework for vascularization, the pattern of formation of vessels in the outer plexus of the fovea and perifoveal region is different from that of the capillary networks in the inner plexus. As shown in **Figure 6(e)**, from the time they are formed, the vessels are much more regular in their size and spacing than are newly formed capillary networks elsewhere in the retina (**Figure 6(f)**).

Radial peri-papillary capillaries. The third vascular plexus of the retina, the RPCs, begins to form around 21 WG (**Figure 5**). These capillary-sized vessels typically extend radially from the ONH (**Figures 5** and **7(a)**) and cover a small rim surrounding the optic papilla (ONH); for this

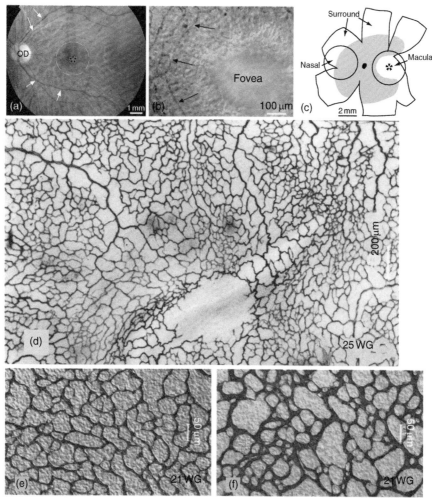

Figure 6 Vascularization in the foveal and perifoveal region. (a) Fundus of the adult human retina. Shown are the optic disk (OD), the macula (broken circle), and the fovea centralis (asterisk). (b) High magnification image of a flatmount of a human fovea. Note the vessels near the fovea (arrows); no vessels are seen on the surface of the fovea. (c) Diagram of a human fetal retina flatmount at 19 WG, showing the vascularized region (gray shading), the macula (circle), and the fovea (asterisk). Note how the region of vascularization curves around the fovea and does not enter. Reproduced from Kozulin, P., Natoli, R., O'Brien, K. M. B., Madigan, M. C., and Provis, J. M. (2009). Differential expression of anti-angiogenic factors and guidance genes in the developing macula. *Molecular Vision* 15: 45–59. (d) Immunohistochemical analysis of CD34[+] vessels in a human fetal retina at 25 WG. Visible is the foveal region, showing the absence of blood vessels. (e) and (f) Immunohistochemical analysis of CD34[+] vessels in a human fetal retina at 25 WG. Region adjacent to (e) ir distant from (f) the fovea. Note the finer caliber and more regular meshwork of the vasculature in the perifoveal region. From Hughes, S., Yang, H., and Chan-Ling, T. (2000). Vascularization of the human fetal retina: Roles of vasculogenesis and angiogenesis. *Investigative Ophthalmology and Visual Science* 41: 1217–1228. Copyright Association for Research in Vision and Ophthalmology.

reason, they have been given the name RPCs. These vessels extend from the inner vasculature to supply the nerve fiber layer near the ONH (**Figures 5** and **7(b)**). The RPCs form a very limited plexus, extending no further than 1 mm from the ONH throughout human fetal development (**Figure 5**). As with other retinal vessels that are formed through angiogenesis, the timing and extent of RPC formation suggests that it is driven by hypoxia-mediated VEGF resulting from increased metabolic demands of the underlying tissues. Their superficial location and lack of mural cell ensheathment, leading to a thin vascular wall, makes RPCs particularly prone to hypoxia resulting from raised

intraocular pressure in glaucoma (the ischemic model of glaucoma susceptibility).

Lack of Involvement of VEGF in Early Stages of Vascularization

Although retinal angiogenesis is driven by hypoxia-induced VEGF expression, vasculogenesis in the human retina is independent of metabolic demand and hypoxia-induced VEGF expression. Most of the vessels formed by vascularization develop prior to the expression of VEGF. By 18 WG, the inner plexus covers more than half of the

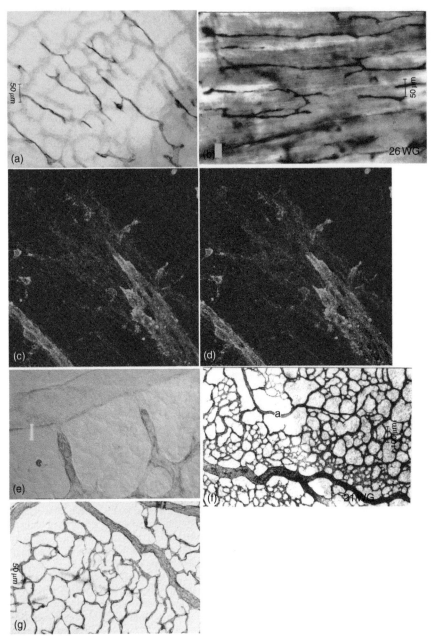

Figure 7   Later stages in the development of human retinal vasculature. (a) and (b) Radial peri-papillary capillaries in the region of the optic nerve head at 25 (a) and 26 (b) WG. These vessels were located superficially, were of fine caliber and extended radially from the optic nerve head. From Hughes, S., Yang, H., and Chan-Ling, T. (2000). Vascularization of the human fetal retina: Roles of vasculogenesis and angiogenesis. *Investigative Ophthalmology and Visual Science* 41: 1217–1228. Copyright Association for Research in Vision and Ophthalmology. (c) and (d) Newly formed capillaries at the leading edge of patent vessel formation in an 18 WG human retina are already closely ensheathed by NG2+ pericytes. Personal observations. (c) Shows NG2 immunolabeling whereas (d) shows co-localization of both NG2 and Cd34 in the same field of view. (e)–(g) Vascular remodeling of capillary beds in the inner plexus. Vessels retract, leaving behind a fine meshwork of capillaries. Formation of the capillary-free zone around arteries (f and g) is visible by 21 WG. (e) From Hughes, S. and Chan-Ling, T. (2000). Roles of endothelial cell migration and apoptosis in vascular remodeling during development of the central nervous system. *Microcirculation* 7: 317–333. (f) From Hughes, S., Yang, H., and Chan-Ling, T. (2000). Vascularization of the human fetal retina: Roles of vasculogenesis and angiogenesis. *Investigative Ophthalmology and Visual Science* 41: 1217–1228. Copyright Association for Research in Vision and Ophthalmology. (g) Personal observations.

retinal area. However, VEGF mRNA is not detectable in the human retina until 20 WG. Formation of the inner plexus is well underway by 14–15 WG, prior to the differentiation of most retinal neurons. Finally, the development of the inner plexus is centered at the optic disk, whereas neuronal maturation is centered at the fovea. In contrast, development of the outer plexus by angiogenesis is centered on the fovea.

Angiogenic and anti-angiogenic factors

Retinal angiogenesis is controlled by a balance of pro-angiogenic and antiangiogenic factors. VEGF is a pro-angiogenic factor that has several forms: VEGFA (there are four isoforms, including $VEGFA_{165}$), VEGFB, VEGFC, and VEGFD. There are at least three VEGF receptors: VEGFR1, 2, and 3. VEGFA is not only a pro-angiogenic factor, but it also has protective functions. Further, it prevents vessel regression and appears to stabilize the mature vasculature, rendering mature vessels resistant to fluctuations in tissue oxygen levels. VEGFA also has a protective effect in the maintenance and function of the neuronal and Muller cells of the retina. VEGFR3 is required for normal retinal vascular development and is upregulated during VEGFA-induced angiogenesis. Its ligand, VEGFC, can cause blood vessel enlargement and tortuosity, and *in vitro* has a synergistic effect on angiogenesis with VEGFA.

Anti-angiogenic factors in the retina include PEDF and vascular endothelial growth inhibitor (VEGI). PEDF is synthesized by cells of the retinal pigment epithelium (RPE), and is present in the developing embryonic and adult human RPE, choroid, and retina. In the human, it is unclear as to how the expression of PEDF relates to the developing vasculature. VEGF stimulates PEDF expression by RPE cells but suppresses PEDF expression by Müller cells. VEGI inhibits endothelial cell proliferation and causes the apoptosis of proliferating endothelial cells. It is expressed by endothelial cells and is thought to play a role in the maintenance of a quiescent adult vasculature. Little is known about VEGI expression in ocular tissues. $VEGI_{192}$ may not be expressed during retinal and choroidal vessel formation but instead may be expressed by the endothelial cells of the quiescent adult vasculature, and may be associated with mural cell maturation. The relative expression of pro- and antiangiogenic growth factors ultimately determines whether neovascularization takes place once a stable quiescent vascular plexus is reached in young adulthood.

Cell–Cell Interactions in the Formation of the Human Retinal Vasculature

Müller cell–endothelial cell interactions

Along with astrocytes, Müller cells are found surrounding the vessels of the inner plexus of the retina, and are thought to be involved in the establishment of the BRB. However, astrocytes are not found in the deeper plexus of the retina; these vessels are ensheathed solely by Müller cells. The existence of an intact BRB and the presence of tight junction proteins in the deep plexus suggest that in this layer of vasculature, Müller cells are capable of inducing the formation and maintenance of barrier properties.

Pericyte–endothelial cell interactions: Vessel stability

Pericytes are another integral component of the mature retinal vasculature. These cells are found within the basal lamina that ensheaths capillary-sized vessels, and play a role in vessel stabilization, control of blood flow, and possibly in modulating vascular permeability and maintenance of the BRB. Pericyte density in the retina is the highest of any tissue, and these cells are also the first to be lost in diabetic retinopathy. Although immature pericytes associate with newly formed retinal vessels (**Figure 7(c)** and **7(d)**), the presence of these immature mural cells does not prevent vessel regression. However, mature vessels that are ensheathed by mature pericytes are quite stable and resistant to $VEGF_{165}$ withdrawal, suggesting that pericytes must be mature and fully differentiated to contribute to vessel stability. A reciprocal relationship exists between endothelial cells and pericytes; whereas pericytes express $VEGF_{165}$ and contribute to vessel stability, retinal endothelial cells express PDGFB, which induces recruitment and proliferation of pericytes.

Pericytes and smooth muscle cells constitute the mural cells, an important component in the maturation and stabilization of vessels. Mural cells are recruited to ensheath immature endothelial tubes by the localized expression of platelet-derived growth factor (PDGF). The inappropriate expression of PDGF interferes with the recruitment of mural cells and leads to abnormal retinal vascular remodeling. More specifically, the endothelial tube must be ensheathed by three components for a new vessel to become patent and stable: mural cells, a basal lamina, and astrocytes. Without any one of these components, vessels are leaky and unstable. Although the presence of immature pericytes along vessels does not protect these vessels from hyperoxia-induced regression, a mature vasculature with mature mural cells, astrocytes, and basal lamina is stable and resistant to the effects of hyperoxia. It appears that the extent of desmin expression by pericytes ensheathing immature vessels imparts upon its associated vascular segment vascular stability.

Vascular Remodeling

With the formation of an exuberant capillary plexus (**Figure 3(c)**) coupled with decreased metabolic demand as endothelial cell growth slows, the newly vascularized region becomes hyperoxic. Hyperoxia in turn leads to vascular remodeling, a paring back of the new vascular network to more precisely meet the metabolic needs of the underlying retina. Remodeling of the vascular network of the retina includes two processes: coalescence of smaller vessels into larger vessels and the retraction of endothelial cells from vessels that are pruned back (**Figure 7(e)**), followed by redeployment of most of these cells to areas where the vascular network is actively forming. Most noticeably, capillaries retract near the arteries, forming a capillary-free space in this region (**Figures 6(d)** and **7(f)**). Although some apoptosis is evident in the retracting vessels, cell death is not the main mechanism for vessel retraction. As the vascular network is remodeled and the metabolic demands of the tissue stabilize, the capillary network becomes very well organized and regular in appearance (**Figure 7(g)**). The process of vascular remodeling occurs over an extended period of time during human embryonic development.

Vascularization and the Health of the Eye

Interruptions to the normal cellular and molecular mechanisms of retinal vascular formation/function can lead to blindness. In infants, retinopathy of prematurity (ROP) develops when premature delivery and supplemental oxygen exposes an infant's developing retina to elevated oxygen levels. This hyperoxygenation of the retina leads to a loss of the normal molecular cues that drives vessel formation such that when the infant is returned to room air, the rate of retinal vascularization is significantly delayed relative to neuronal maturation. A subsequent return to normoxic conditions generates retinal hypoxia that exceeds the normal physiologic levels, triggering the formation of leaky, tortuous vessels. This abnormal vasculature can lead to intraretinal hemorrhage and partial or complete retinal detachment.

In adults, the effects of tissue hypoxia and hyperglycemia can be seen in diabetic retinopathy. In this disease, the physiologic processes of diabetes prevent the retinal vasculature from meeting the metabolic demands of the active retina and generate locally high levels of hypoxia, in addition to the damaging effects of prolonged hyperglycemia. This retinal hypoxia stimulates angiogenesis in the context of a mature, fully developed retinal vasculature. Subsequent retinal neovascularization produces tortuous, fragile vessels over the surface of the retina. These vessels often leak fluid into the vitreous and can cause macular edema. In addition, the development of fibrous tissue in association with these vessels can lead to traction retinal detachments or vitreous hemorrhages.

Conclusions

The human retina is vascularized by a combination of two mechanisms: vasculogenesis and angiogenesis. Vasculogenesis is responsible for the formation of the earliest vessels of the inner plexus, starting at the ONH and reaching the inner two-thirds of the retina. Between that boundary and the inner periphery of the retina, in the space between these early vessels, reaching down to the outer plexus, under the fovea and over the nerve bundles at the optic nerve,

the process of angiogenesis drives the formation of the remaining vasculature. The formation of these vessels is controlled by physiological hypoxia – a level of hypoxia that is well tolerated by the tissue but is insufficient to cause tissue damage. Physiological hypoxia is sufficient to stimulate VEGF production (mediated by hypoxia inducible factor-1α (HIF-1α)) by astrocytes and Müller cells and to direct endothelial growth where it is needed. Once this proliferation of new vessels becomes patent and directs blood flow to the developing retinal tissues, a modest level of hyperoxia then orchestrates vascular remodeling and retraction, a process that is still ongoing in a full-term neonate. The balance between physiological hypoxia and hyperoxia, vessel growth and vessel retraction, leads to the formation of a vasculature that precisely meets the metabolic demands of the human retina.

See also: Physiological Anatomy of the Retinal Vasculature.

Further Reading

Adamis, A. P., Aiello, L. P., and D'Amato, R. A. (1999). Angiogenesis and ophthalmic disease. *Angiogenesis* 3: 9–14.

Chan-Ling, T. (2006a). Glial, neuronal and vascular interactions in the mammalian retina. *Progress in Retinal and Eye Research* 13: 357–389.

Chan-Ling, T. (2006b). The blood retinal interface: Similarities and contrasts with the blood–brain interface. In: Dermietzel, R., Spray, D. C., and Nedergaard, M. (eds.) *Blood–Brain Barriers – From Ontogeny to Artificial Interfaces*, pp. 701–724. Weinheim: Wiley-VCH.

Chan-Ling, T., Gock, B., and Stone, J. (1995). The effect of oxygen on vasoformative cell division. *Investigative Ophthalmology and Visual Science* 36: 1201–1212.

Chan-Ling, T., Halasz, P., and Stone, J. (1990). Development of retinal vasculature in the cat: Processes and mechanisms. *Current Eye Research* 9: 459–476.

Chan-Ling, T., McLeod, D. S., Hughes, S., et al. (2004). Astrocyte-endothelial cell relationships during retinal vascular development. *Investigative Ophthalmology and Visual Science* 45: 2020–2030.

Dorrell, M. I., Aguilar, E., and Friedlander, M. (2002). Retinal vascular development is mediated by endothelial filopodia, a preexisting astrocytic template and specific R-cadherin adhesion. *Investigative Ophthalmology and Visual Science* 43: 3500–3510.

Dorrell, M. I. and Friedlander, M. (2006). Mechanisms of endothelial cell guidance and vascular patterning in the developing mouse retina. *Progress in Retinal and Eye Research* 25: 277–295.

Flynn, J. T. and Chan-Ling, T. (2006). Retinopathy of prematurity: Two distinct mechanisms that underlie Zone 1 and Zone 2 disease. *American Journal of Ophthalmology* 142: 46–59.

Fruttinger, M. (2007). Development of the retinal vasculature. *Angiogenesis* 10: 77–88.

Gerhardt, H., Gloding, M., Fruttinger, M., et al. (2003). VEGF guides angiogenic sprouting utilizing endothelial tip cell filopodia. *Journal of Cell Biology* 161: 1163–1177.

Hughes, S. and Chan-Ling, T. (2000a). Roles of endothelial cell migration and apoptosis in vascular remodeling during development of the central nervous system. *Microcirculation* 7: 317–333.

Hughes, S., Yang, H., and Chan-Ling, T. (2000b). Vascularization of the human fetal retina: Roles of vasculogenesis and angiogenesis. *Investigative Ophthalmology and Visual Science* 41: 1217–1228.

Kozulin, P., Natoli, R., Madigan, M. C., O'Brien, K. M. B., and Provis, J. M. (2009). Gradients of *Eph-A6* expression in primate retina suggest a role in definition of the foveal avascular area. *Molecular Vision* 15: 2649–2662.

Kozulin, P., Natoli, R., O'Brien, K. M. B., Madigan, M. C., and Provis, J. M. (2009). Differential expression of anti-angiogenic factors and guidance genes in the developing macula. *Molecular Vision* 15: 45–59.

Lutty, G. A., Chan-Ling, T., Phelps, D. L., et al. (2006). *Proceedings of the Third International Symposium on Retinopathy of Prematurity*: An update on ROP from the lab to the nursery (November 2003, Anaheim, California). *Molecular Vision* 12: 532–580.

Stone, J., Itin, A., Alon, T., et al. (1995). Development of retinal vasculature is mediated by hypoxia-induced vascular endothelial growth factor (VEGF) expression by neuroglia. *Journal of Neuroscience* 15: 4738–4747.

Embryology and Early Patterning

P Bovolenta and R Marco-Ferreres, Instituto Cajal (CSIC) and CIBER de Enfermedades Raras (CIBERER), Madrid, Spain
I Conte, Telethon Institute of Genetics and Medicine (TIGEM), Naples, Italy

Glossary

Agenesis – Failure of an organ to develop during embryonic growth and development.

Anophthalmia – Absence of the eye as a result of a congenital malformation during development.

Coloboma – Congenital malformation due to failure of fusion of optic fissure.

Hyperplasia – Abnormal increase in cell proliferation within an organ or tissue due to the disregulation of cell-cycle control mechanisms.

Induction – Change in cell fate resulting from an external signal (inducer). This signal secreted by a group of cells instructs target cells to acquire specific properties.

Invagination – Morphogenetic process leading to the coordinated inward folding of a cell layer forming a pocket on the surface.

Mesenchyme – Embryonic mesodermal connective tissue composed of loosely packed unspecialized cells characterized by the ability to migrate into other tissues.

Phenotype – Expression of a specific trait based on genetic and environmental influences.

Prosencephalon – Anterior-most division of the developing vertebrate brain (forebrain) that further subdivides into the telencephalon and the diencephalon.

Embryology

Eye development starts at late gastrula stages with the specification of retinal precursor cells within the eye field in the anterior neural plate. At this stage, retinal progenitor cells occupy a medial position and are surrounded rostrally and laterally by telencephalic precursors and caudally and medially by cells that will form the diencephalon. During neurulation, coordinated rearrangement of the cells that form each of these presumptive regions results in a new cell organization, whereby retinal progenitor cells now become laterally positioned. *In vivo* monitoring of retinal progenitor cell movement has shown that, at least in fish, dorsal diencephalic cells – positioned caudally to the eye field – move inward and forward, displacing prospective eye and optic stalk tissue forward and then sidewise. At the same time, laterally positioned telencephalic precursors fold toward the midline leaving retinal precursors cells to the side of the neural tube (**Figure 1**). As these movements happen, the cells of the left and right eye, which were initially intermingled, segregate in their respective domains forming two visible optic vesicles.

At the present time, it is not clear whether similar morphogenetic movements occur in other vertebrate species. In mammals, for instance, these early events might be slightly different as the first visible sign of eye formation are two small depressions (optic pits) in the anterior neural tube, which appear well before folding of the tube occurs.

Independent of these initial differences, in all vertebrates the optic vesicles are formed by a single pseudostratified neuroepithelium that is surrounded by cephalic mesenchyme, including migrating neural crest cells, and limited at the most distal tip by the surface ectoderm. The cells that compose the optic vesicle neuroepithelium are initially morphologically and molecularly indistinguishable but reach their final complexity through a series of inductive and morphogenetic events that, in part, depend on their interaction with the surrounding tissues. The extension of thick cytoplasmic processes emanating from the basal surface of both optic vesicle and ectodermal cells establishes cell-to-cell contact and initiates a series of temporally correlated structural changes in both structures. A fibrillar extracellular matrix (ECM) builds up between the neuroectoderm and the ectoderm, favoring strong adhesion, while the ectoderm immediately adjacent to the optic vesicle thickens forming the lens placode (**Figure 2**).

Once formed, the lens placode and the adjacent optic vesicle begin a coordinated invagination to form the lens pit and the optic cup. ECM-mediated adhesion between the two tissues is important for coordinating morphogenetic movements. In addition, some of its components, such as laminin and fibronectin, may participate in lens cell differentiation, which occurs concomitantly. As a result, the lens pit deepens and gradually assumes a spherical shape that finally breaks away from the ectoderm forming the lens vesicle. A spatio-temporally restricted massive apoptosis contributes to this separation. Thereafter, the lens vesicle undergoes an asymmetric differentiation acquiring a characteristic polarity: cells located in the most proximal region (facing the neuroepithelium)

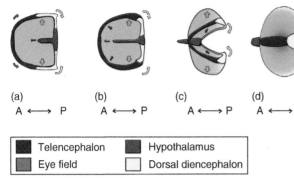

| ■ Telencephalon | ■ Hypothalamus |
| ■ Eye field | □ Dorsal diencephalon |

Figure 1 Optic vesicle morphogenesis in zebrafish. (a) The eye field (in green) in the anterior neural plate is surrounded anteriorly and peripherally by telencephalic precursors (blue), posteriorly by the dorsal diencephalon (light yellow) and medially by cells that will form the hypothalamus (red). (b, c) The hypothalamic and dorsal diencephalic precursors move forward beneath the eye field displacing it to the front and laterally. The lateral edges of the neural plate fold toward the midline. (c, d) The lateral telencephalon cells meet at the dorsal midline, leaving two optic vesicles at the side. Adapted from Egland, S. J., Blanchard, G. B., Mahadevan, L., and Adams, R. J. (2006). A dynamic fate map of the forebrain shows how vertebrate eyes form and explains two causes of cyclopia. *Development* 133: 4613–4617.

elongate to form the primary lens fibers, whereas those in the most distal portion (facing the ectoderm) form a cuboidal monolayer known as the anterior lens epithelium.

Concomitant with lens formation, the optic neuroepithelium undergoes a series of rearrangements that culminates with the formation of a bi-layered optic cup (**Figure 2(c)** and **2(c′)**). The distal inner layer will give rise to the future neural retina (NR), while the distal outer layer to the retinal pigmented epithelium (RPE), which acquires specialized epithelial characteristics. The proximal ventral region forms the optic stalk, from which the optic nerve derives. The transition zone between the future retina and the RPE forms the ciliary margin (CM), or periphery of the retina. Despite its neural origin, the CM differentiates into non-neural structures: the proximal part into the ciliary epithelium while the distal part becomes the iris. In fish and amphibians, the CM harbors bona fide stem cells, which continue to proliferate throughout life to increase the eye size. Cells with stem-like properties also appear to be present in the CM of birds and mammalian eyes. These cells, however, are quiescent in physiological conditions.

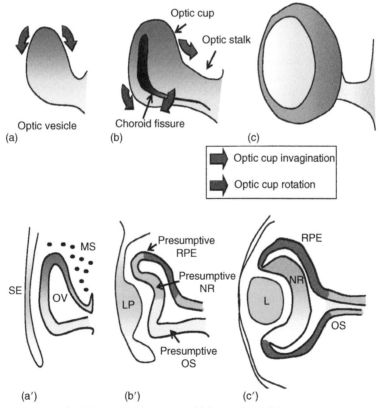

Figure 2 Schematic representation of initial eye morphogenesis. (a) The optic vesicle appears as a protrusion of the anterior neural tube. (b) Folding of the distal and ventral neuroepithelium generates the optic cup and the optic (choroids) fissure. (c) The optic fissure seals and a spherical eye cup forms. (a′) The optic vesicle neuroepithelium is composed of the cells that are morphologically and molecularly indistinguishable. (b′) As the vesicle folds, the dorsal neuroepithelium specifies as presumptive RPE (dark green), the distal region as presumptive neural retina (light green), while the ventral portion as optic stalk (light yellow). The surface ectoderm thickens forming the lens placode (blue). (c′) Complete folding of the vesicle results in an optic cup, where the RPE completely surrounds the neural retina. L, lens; LV, lens vesicle; MS, mesenchyme; NR, neural retina; OS, optic stalk; OV, optic vesicle; RPE, retinal pigment epithelium; SE, surface ectoderm.

Although there are common features in the way the optic cup is formed among vertebrates, differences do exist. In fish and amphibians, the optic vesicles, which develop as flat wing-like protrusions, undergo a process of cavitation, rotation, and invagination that shifts cells that were originally in a ventral position to the medial layer, from which the RPE progenitors derive. These progenitors stretch over the prospective NR, occupying the dorsal aspect of the optic cup. In mammals and birds instead, the optic vesicle neuroepithelium undergoes two simultaneous foldings (**Figure 2(b)**): (1) the spherical vesicle becomes concave outward, resembling the bowl of a spoon where the optic stalk is the handle and (2) at the same time, the ventral portion of the entire vesicle folds inward giving rise to the groove of the optic (choroid) fissure. Differential growth rates are associated with these events. In the future RPE as well as in the CM, the proliferation rate is quite slow in comparison to that of the perspective NR. There are also differences along the dorso-ventral axis, with an initial higher proliferation rate in the ventral part, which enables the formation of the optic fissure.

The optic fissure can be divided into two adjoining parts – the retinal fissure and the optic groove – which derive from the progressive invagination of the ventral surface of the optic vesicle and stalk, respectively. The transition between the retinal fissure and the optic groove dictates the position where the optic disk, or blind spot of the retina, will form. This structure, a real interface between the optic stalk and the retina, enables the entrance of surrounding mesenchymal cells into the developing eye chamber, which will form the hyaloid artery, the main blood supply for the eye. At the same time, it allows the egression of retinal axons from the eye cup.

As a last step in the formation of a complete optic cup, its dorsal portion begins to proliferate more rapidly enabling the shallow bowl to become deeper, while its lateral edges meet ventrally across the optic fissure. The eye rudiment now has the appearance of a hemispheric cup where the RPE completely surrounds the NR. The sealing of the optic fissure completes these morphogenetic events (**Figure 2(c)** and **2(c′)**). Although this is a poorly understood process, in humans the fusion usually begins in the middle portion of the fissure and extends anteriorly and posteriorly.

Early Patterning

The morphogenetic movements and the tissue interactions described above are orchestrated by specific genetic programs coordinated by cell-signaling mechanisms – inductive signals – and regulated by the activity of a relatively small number of transcription factors (see **Table 1**). Many of the genes involved in the specification of the eye

Table 1 Summary of extracellular molecules and transcription factors implicated in early eye development

Events in eye development	Secreted molecules	Transcription factors
Proximo-distal patterning of OV	Shh, nodal	Pax2, Pax6, Vax1
Specification of NR and RPE	FGF, TGFβ; Activin/BMPs	Chx10, Six3, Six6, Lhx2, Pax6, Mitf, Otx1, Otx2
Dorso-ventral patterning of the optic cup	BMP4, BMP7, Shh, RA, ventroptin, noggin	Pax6, Pax2, Vax, Tbx5

Shh, Sonic hedgehog; FGFs, Fibroblast grow factors; BMPs, bone morphogenetic proteins; RA, retinoic acid.

field, including retinal homeobox (Rx), paired box 6 (Pax6), hairy enhancer of split 1 (Hes1), orthodenticle homolog 2 (Otx2), LIM homeobox 2 (Lhx2), and SIX homeobox 3 (Six3), are also involved in the early patterning events that convert eye-field cells into the structures that compose the optic cup (**Figure 3(a′)**). However, changes in the expression levels, as well as combinatorial interaction with other factors, diversify the activity of these genes enabling the progressive acquisition of different fates and functions. As mentioned above, cells that compose the optic vesicle are initially morphologically and molecularly undistinguishable and, therefore, are all potentially competent to originate the NR, the optic stalk, or RPE. Information, in the form of signaling molecules, derived from the surrounding tissues and the neuroepithelium itself modulates and restricts the expression of different transcription factors to a specific domain of the vesicle, initiating its patterning along the proximo-distal and dorso-ventral axis.

Proximo-Distal Patterning of the Optic Vesicle

The optic vesicles are bulges of the neural tube and thus their subdivision is often described according to a proximal–distal axis. In this view, future optic stalk precursors occupy a proximal location, while distal cells constitute the NR and RPE precursors (**Figure 3(b)**).

Proximo-ventral cells of the optic stalk are generated under the influence of signaling molecules, emanating from the axial ventral midline, including Sonic hedgehog (Shh) and nodal, a member of the transforming growth factor-beta (TGFβ) family (**Figure 3(b)**). Mutations in these genes cause severe anterior neural tube defects that include the formation of a single midline (cyclopic) eye, lacking optic stalk tissue. The phenotype is the result of Shh-mediated regulation of the spatial expression of Pax6 and Pax2 – two homeobox transcription factors of the paired type – which normally

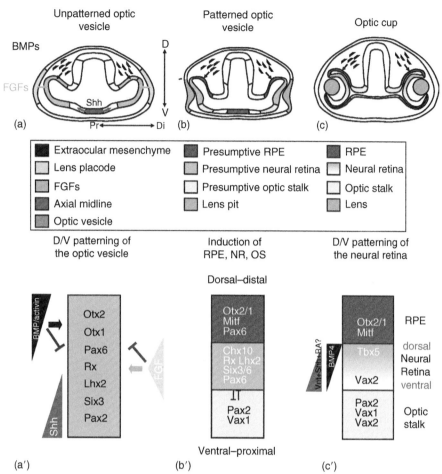

Figure 3 Distribution of inductive signals and transcription factors involved in early patterning of the eye. Progressive tissue specification during the transition from unpatterned (a) to patterned optic vesicle (b) and optic cup (c) where the different colors represent the distinct territories. In (a')–(c') color bar graphs represent the distribution of molecular components implicated in the patterning of the optic vesicle into optic cup. In (a) and (a') all the neuroepithelial cells are indistinguishable (light green) and express a common set of transcription factors. Shh signaling from the axial midline is required for the specification of the proximo-ventral optic stalk (orange in (b) and (b')). TGFβ-like signals from the extraocular mesenchyme promote RPE character (dark green in (b) and (b')), whereas FGF signals from the lens placode repress RPE and activate neural retina identity (light green in (b) and (b')). During optic cup formation the graded distribution of BMP4 dorsally and ventroptin, possibly together with RA and Shh, in the ventral side establish the dorso-ventral polarity of the neural retina (shaded green in (c) and (c')). Shh, Sonic hedgehog; FGFs, fibroblast growth factors; BMPs, bone morphogenetic proteins; RA, retinoic acid; Vnt, ventroptin.

demarcate the distal and proximal optic primordium, respectively (**Figure 3(b')**). Indeed, overexpression of Shh causes the expansion of the Pax2 domain at the expense of the Pax6-positive tissue, while the opposite occurs in absence of Shh. A similar dependence on Shh signaling has also been observed for the expression of Vax genes, other homeobox transcription factors involved in optic stalk generation.

Targeted inactivation of Pax2 or Vax1 in mice results in severe optic nerve abnormalities, consistent with a role for these genes in proximal eye development. The phenotype of these mice is characterized by the presence of coloboma of the ventral optic fissure, and, in the case of Pax2, an extension of the RPE into the optic nerve region. In contraposition, the Pax6 gene is necessary for the

formation of the distal vesicle, including the lens tissue. Mutations in this gene impair both lens and optic vesicle formation, ultimately causing anophthalmia. In these mice, the only remnant of the eye primordium has optic stalk characteristics. Conditional inactivation of Pax6 at different time points of eye development, however, has highlighted additional functions for this gene both in NR and RPE development (see below).

In summary, acquisition of the ventral optic stalk phenotype depends on the Shh-dependent expression of transcription factors that impose ventral character to optic vesicle precursors. The establishment of a precise boundary between the presumptive optic stalk and the distal optic vesicle may result from the reciprocal transcriptional repression between Pax6 and Pax2 (**Figure 3(b')**).

Specification of the RPE and NR

Extracellular signals, derived from the surface ectoderm in contact with the prospective NR or from the periocular mesenchyme surrounding the presumptive RPE, pattern the distal optic vesicle (**Figure 3(a)**).

Removal of the surface ectoderm induces prospective NR cells to acquire the characteristics of RPE cells. This is because the ectoderm secretes high levels of two members of the fibroblast growth factor (FGF) family of signaling molecules – FGF1 and FGF2 – while the prospective retina expresses the FGF receptor 1. The addition of either one of the two factors is sufficient to rescue NR formation after ectoderm removal. On the contrary, exposure of the presumptive RPE to FGF confers to the cells NR properties. Therefore, FGF signaling normally activates NR specification but inhibits RPE formation. The future retina itself expresses other FGFs, such as FGF3, FGF8, FGF9, and FGF15, which may contribute to maintain the properties of the retina, but are not sufficient to induce them, since addition of these FGFs cannot rescue the phenotype caused by surface ectoderm removal.

The activity of the FGF is complemented by other signals, originating from the dorsal extraocular mesenchyme, which have opposite effects on the specification of these territories. Members of the TGFβ superfamily of signaling molecules, such as activins or the related bone morphogenetic proteins (BMP) – Bmp4 and Bmp7 – are expressed in the surrounding mesenchyme and/or the presumptive RPE itself. Interference with the activity of these molecules prevents RPE development and induces NR-specific genes, whereas their addition activates RPE characteristics in the entire vesicle.

Thus, TGFβ/BMP and FGF signaling act antagonistically on the specification of RPE and NR precursors. Directly or indirectly, their activity seems to impinge upon the expression of specific transcriptional regulators, such as *Caenorhabditis elegans* homeo domain 10 (Chx10) and Otx2 or microphthalmia-associated transcription factor (Mitf), the function of which is instrumental to the acquisition of either RPE or NR identities. Indeed, FGF signaling activates the expression of Chx10 in the retina but represses that of Otx2 and Mitf in the RPE, while opposite results have been observed with manipulations of TGFβ/BMP signaling.

Besides those mentioned above, a number of additional transcription factors are also specifically expressed in the presumptive neural retina and RPE (**Figure 3(b′)**). Currently, it is not clear how many of them are really needed to impose tissue specificity. The paired-like homeobox gene Chx10 (also known as Vsx1) is possibly the best candidate to impose a neural retina character to the naive optic vesicle cells (**Figure 3(b′)**). The onset of Chx10 expression is restricted to the presumptive NR domain and mutations in the human or mouse gene

lead to microphthalmia, cataracts, and abnormal iris development. The expression of two members of the Six family of homeobox transcription factors, Six3 and Six6, initially expressed in the entire vesicle, is also soon restricted to the prospective NR. The overexpression of these factors in fish embryos causes retinal hyperplasia and ectopic formation of retinal-like tissue. In mice, genetic inactivation of Six3 leads to complete loss of the forebrain, while that of Six6 impairs retinal proliferation and differentiation. It is, therefore, still unclear whether the two genes are essential pieces of the genetic program that establishes NR identity. Nevertheless, overexpression of either one of them in the RPE cells leads to the acquisition of a NR phenotype, suggesting that their function is at least incompatible with the acquisition of the RPE character.

Similarly to Six3, the expression of Lhx2 becomes restricted to the prospective NR. Mice deficient in Lhx2 function are characterized by anophthalmia, like that observed in mice mutants for the Pax6 gene. Although the two genes seem to act independently, both are required during the period of close contact between the surface ectoderm and optic vesicle. Whether this implies that their activity participates in the specification of the NR domain still needs to be established.

The function of at least four transcription factors – Otx1, Otx2, Mitf, and Pax6 – is at the core of the transcriptional network required to establish the RPE character. Functional inactivation of either Mitf or Otx1 and Otx2 genes (bHLH and homeo-domain-containing transcription factors, respectively) impairs RPE differentiation. On the contrary, retinal cells transfected with either Otx or Mitf genes acquire a RPE phenotype and accumulate granules of pigment. Although it is unclear whether Otx and Mitf act in a feedback loop or in parallel, they synergize in the activation of melanogenic gene expression, including tyrosinase or the tyrosinase-related protein 2. Although Pax6 expression in the prospective RPE is transient, its function – possibly together with that of Pax2 – seems necessary to initiate Mitf expression. Indeed, Pax6 binds directly to the Mitf promoter and in embryos deficient in both Pax2 and Pax6 optic vesicles are small and composed of Mitf-negative cells that instead co-express Otx2 and the NR marker, Chx10.

As in the case of the proximo-distal patterning of the optic vesicle, subdivision of the distal vesicle in the NR and RPE is reinforced by reciprocal repression of transcription factors. Mitf is expressed ectopically in the neuroretina of Chx10-deficient mice, driving cells toward an RPE-like identity. Conversely, misexpression of Chx10 in the developing RPE caused downregulation of Mitf with a consequent loss of pigment, strongly suggesting an antagonistic interaction between Chx10 and Mitf.

Chx10 function also appears to be required to establish a boundary between central and peripheral retina, the CM. In Chx10 null mutants, in fact, peripheral structures, including the ciliary body, are abnormally expanded. Similar expansions are also observed upon stabilization of β-catenin, a key effector of canonical wingless pathway (Wnt) signaling. In addition, ectopic expression of Wnt ligand induces abnormal expression of CM markers, such as Otx1 and msh homeobox 1 (Msx1), in the central retina. This suggests that canonical Wnt signaling plays a crucial role in establishing a difference between central and peripheral retina, imposing identity to the ciliary body and iris.

Dorso-Ventral Patterning of the Optic Cup

As mentioned above, the transition between the optic vesicle and the optic cup establishes a new position for most of the optic vesicle precursors. Once this rearrangement is accomplished, the optic cup undergoes a new wave of polarization along the dorso-ventral axis.

A gradient of BMP signaling is primarily responsible for this new pattern (**Figure 3(c)** and (c′)). Bmp4 is expressed in the dorsal retina of most vertebrate species analyzed. Mis-expression of Bmp4 in the ventral portion of the optic cup suppresses the expression of Pax2 and Vax and activates that of Tbx5, a transcription factor of the T-box class normally expressed only in the dorsal portion of the optic cup. The counterbalance for this dorsalizing activity is ventroptin, a BMP signaling antagonist expressed in the ventral optic cup. Forced expression of ventroptin in the dorsal cup decreases Bmp4 expression and expands that of Vax genes. Thus, the ventral character of the optic cup seems to depend on the inhibition of Bmp4-mediated dorsalizing activity.

Structures that are typical of the ventral optic cup – such as the optic fissure, the hyaloid artery, and the pecten (a vascular-like structure in birds) – also depend on BMP signaling, although this may be due to actions of other members of the family. Overexpression of the BMP antagonist noggin in the chick ventral optic cup alters its development and induces ectopic expression of optic stalk markers in the region of the ventral retina and RPE. Furthermore, in Bmp7 null mice the optic fissure does not form. Retinoic acid (RA) is an additional signal that may be involved in establishing the ventral optic cup identity, although its precise functions are controversial. Enzymes involved in the synthesis and degradation of RA as well as RA receptors are expressed in the eye with complex and polarized patterns. RA treatment upregulates Pax2 expression in the eye of zebrafish embryos, while deprivation of vitamin A during embryogenesis, inactivation of RA receptors, or inhibition of RA synthesis all lead to embryos with defects in the ventral retina. Despite this, embryos deficient in Raldh1, Raldh2, and Raldh3, genes coding for RA-synthesis enzymes, develop an optic cup with a normal dorso-ventral patterning, questioning the relevance of RA in this process.

Equally unclear is whether Shh signaling contributes to impose a ventral character to the optic cup. Although Shh is needed to control the initial expression of Pax2 and Vax genes in the optic vesicle, its forced dorsal expression at slightly later stages of development does not ventralize the dorsal optic cup, although BMP4 expression is downregulated. Defects are, instead, observed in the optic disk and the NR, where Shh causes activation of Otx2 expression and pigmentation. Conversely, interference with Shh signaling perturbs Otx2 expression and pigment formation in the ventral RPE, which loses its characteristics. Thus, Shh signaling may have a more specific function in the specification of the ventral RPE cells as well as of those that compose the optic fissure.

Independent of the signaling mechanism, the transcriptional control of dorso-ventral polarization of the optic cup depends largely on the activity of Tbx5 in the dorsal portion of the cup and Vax genes in the ventral part. Overexpression of Tbx5 causes dorsalization of the optic cup while overexpression of Vax leads to its ventralization. Pax6 seems to regulate the expression of both genes but in opposite directions establishing an intermediate zone that separate the two domains.

The significance of a dorso-ventral polarity in the NR at this stage and, thus, of the function of Tbx5 and Vax genes might be tightly linked to the establishment of the proper spatial order of the retino-tectal projections through the regulation of the expression of members of the ephrin (Eph) family of receptor tyrosine kinases and their ephrin ligands, which are axon-guidance cues essential for target recognition of RGC axons.

Conclusions

Early eye patterning in vertebrates is controlled by an evolutionary conserved genetic program, although initial morphogenetic events may vary from species to species. This program is regulated by the interplay among a number of signaling pathways and transcription factors. The information available at present represents an important backbone that needs to be extended for full understanding of eye development. *In vivo* imaging of eye morphogenesis as well as additional characterization of the involved genetic networks may help to understand the causes of inborn eye malformation such as anophthalmia, microphthalmia, and coloboma.

See also: Developmental Anatomy of the Retinal and Choroidal Vasculature; Eye Field Transcription Factors;

Histogenesis: Cell Fate: Signaling Factors; Photoreceptor Development: Early Steps/Fate; Retinal Histogenesis; Zebra Fish–Retinal Development and Regeneration.

Further Reading

Adler, R. and Canto-Soler, M. V. (2007). Molecular mechanisms of optic vesicle development: Complexities, ambiguities and controversies. *Developmental Biology* 305: 1–13.

Barishak, Y. R. (2001). *Embryology of the Eye and Its Adnexa.* Basel: Karger.

Chow, R. L. and Lang, R. A. (2001). Early eye development in vertebrates. *Annual Review of Cell and Developmental Biology* 17: 255–296.

Egland, S. J., Blanchard, G. B., Mahadevan, L., and Adams, R. J. (2006). A dynamic fate map of the forebrain shows how vertebrate eyes form and explains two causes of cyclopia. *Development* 133: 4613–4617.

Fitzpatrick, D. R. and van Heyningen, V. (2005). Developmental eye disorders. *Current Opinion in Genetics and Development* 15: 348–353.

Lupo, G., Harris, W. A., and Lewis, K. E. (2006). Mechanisms of ventral patterning in the vertebrate nervous system. *Nature Reviews Neuroscience* 7: 103–114.

Martinez-Morales, J. R., Rodrigo, I., and Bovolenta, P. (2004). Eye development: A view from the retina pigmented epithelium. *BioEssays* 26: 766–777.

Rembold, M., Loosli, F., Adams, R. J., and Wittbrodt, J. (2006). Individual cell migration serves as the driving force for optic vesicle evagination. *Science* 313: 1130–1134.

Relevant Website

http://www.med.unc.edu – University of North Carolina at Chapel Hill, School of Medicine: Eye Development.

Evolution of Opsins

T H Oakley and D C Plachetzki, University of California, Santa Barbara, Santa Barbara, CA, USA

Glossary

Bilaterian – A major family of animals characterized by bilateral symmetry, including insects, nematode worms, mollusks, vertebrates, echinoderms, and other animals.

Chromophore – A light-reactive chemical that binds to a protein, such as opsin, and which allows biological sensitivity to light.

Ciliary – Relating to cilia, which are subdivided into motile and sensory types. Ciliary photoreceptors are those in which phototransduction occurs in a sensory cilium.

CNG (cyclic nucleotide-gated ion channel) – A family of proteins that acts (along with other functions) as the ion channel in ciliary photoreceptor cells. CNG proteins open in the presence of cyclic nucleotides to allow positively charged ions such as calcium to flow into the cell.

Cnidaria – A group of animals that includes jellyfish, anemones, hydras, and corals.

Co-duplication – The simultaneous duplication during evolution of interacting components (e.g., genes or proteins).

Co-option – The pattern or process whereby existing components (e.g., genes or proteins) are used in the evolution of a new structure or function.

Eumetazoa – A major group of animals characterized by the presence of numerous cell types, tissues, and organs and that includes bilaterians and cnidarians.

GPCRs – Guanine-nucleotide-binding protein (G-protein)-coupled receptors form a class of proteins that functions by embedding in cell membranes. These proteins function to detect a signal outside the cell and then signal to proteins inside through a G protein to the cell to change the cell's state in some way.

G protein (Gt, Gq, Go, Gs) – A family of multi-unit proteins that are activated by GPCR receptor proteins to transmit signals into cells. Gt, Gq, Go, and Gs are subfamilies of G-alpha proteins activated by opsins.

Photopic vision – The vision in high light levels, which may include color vision.

Phototransduction – The conversion of a light signal received by an animal opsin to an electrical signal that is transmitted to the nervous system.

Rhabdomeric – A type of photoreceptor cell with numerous microvilli. It is also called microvillar photoreceptors.

Rhodopsin – A name originally given to isolated visual pigments that contained both opsin protein and nonprotein chromophore. Today the term still applies to visual pigments, and is also used commonly to describe the opsin protein expressed in vertebrate rod (dim-light) photoreceptors, and the opsins of certain organismal groups, such as bacteria.

Scotopic vision – A monochromatic vision in low light levels.

Spectral tuning – The changes in specific amino acids of opsin proteins that mediate the maximum sensitivity to different wavelengths of light of different visual pigments.

Tetrachromatic – A visual system based on four classes of visual pigments with four different λ-max values.

Trichromatic – A visual system based on three classes of visual pigments with three different λ-max values.

TRP (transient receptor potential ion channel) – A family of proteins that acts (along with other functions) as the ion channel in rhabdomeric photoreceptor cells. TRP proteins open in the presence of products derived from the lipid diacylglycerol to allow positively charged ions to flow into the cell.

Type I opsin – The retinylidene proteins present in bacteria and algae, which are referred to by various names, including bacteriorhodopsin, bacterial sensory rhodopsins, channelrhodopsin, halorhodopsin, and proteorhodopsin. These proteins have varied functions, including bacterial photosynthesis (bacteriorhodopsin), which is mediated by pumping protons into the cell, and phototaxis (channelrhodopsin), which is mediated by depolarizing the cell membrane.

Type II opsin – The retinylidene proteins present for approximately 600 million years in eumetazoans (animals not including sponges), and unknown from sponges or any nonanimals. The proteins of this family have varied functions, including phototransduction and vision, circadian rhythm entrainment, mediating papillary light reflex (pupil constriction), and photoisomerization (recycling the chromophore).

λ-max – The wavelength of light to which a visual pigment (opsin plus chromophore) is maximally sensitive.

Introduction

Opsins are a group of proteins that underlie the molecular basis of various light-sensing systems, including phototaxis, circadian (daily) rhythms, eye sight, and a type of photosynthesis. They are sometimes called retinylidene proteins because they bind to a light-activated, nonprotein chromophore called retinal (retinaldehyde). Opsins are also, in some cases, called rhodopsins, a name originally given to isolated visual pigments that contained both opsin protein and nonprotein chromophore in a time before the two separate components were known. Today, the term rhodopsin is used commonly to describe the opsin expressed in vertebrate rod (dim-light) photoreceptors, and the opsins of certain organismal groups, such as bacteria.

General Opsin Structure and Function

Opsins are light-sensitive proteins that snake in and out of a cell membrane 7 times. Crystal structures of opsins have been generated, which allows a detailed understanding of this protein structure. Opsin light sensitivity is mediated by light-sensitive chemicals called chromophores. Opsin proteins covalently bind a chromophore through a Schiff base linkage to a lysine amino acid in the seventh membrane-spanning region of the protein. The absorption of a photon of light results in a photoisomerization, or shape change, of the chromophore. Photoisomerization then causes a shape change in the liked opsin protein.

Type I and Type II Opsins

Two major classes of opsins are defined and differentiated based on primary protein sequence, chromophore chemistry, and signal transduction mechanisms. Several lines of evidence indicate that the two opsin classes evolved separately, illustrating an amazing case of convergent evolution.

Type I opsins are present in bacteria and algae and are referred to by various names, including bacteriorhodopsin, bacterial sensory rhodopsins, channelrhodopsin, halorhodopsin, and proteorhodopsin. These opsins have varied function, including bacterial photosynthesis (bacteriorhodopsin),

which is mediated by pumping protons into the cell, and phototaxis (channelrhodopsin), which is mediated by depolarizing the cell membrane. Type II opsins are present in eumetazoans (animals not including sponges), but are unknown from sponges or any nonanimals. As opsins are known from cnidarians and bilaterian animals (animals with bilateral symmetry, including humans, flies, and earthworms), type II opsins are inferred to have been present in their common ancestor, which lived about 600 million years ago. These opsins have varied function, including phototransduction and vision, circadian rhythm entrainment, mediating pupillary light reflex (pupil constriction), and photoisomerization (recycling the chromophore).

Despite their functional similarity and despite both being seven-transmembrane proteins, multiple lines of evidence indicate that type I and type II opsins evolved independently. First, the primary amino acid sequences of type I and type II opsins are no more similar than expected by chance. Second, the orientation of the transmembrane domains differs between the major groups. Third, the major opsin groups differ in chromophore chemistry. Prior to light activation, the chromophore of type I opsins is an all-*trans* isomer. Light activation then involves isomerization of the chromophore to 13-*cis* retinal. In contrast, prior to light activation, the chromophore of type II opsins is 11-*cis* retinal. Light activation of type II opsins involves isomerization to all-*trans* retinal (**Figure 1**). Fourth, type II opsins belong to the larger protein family called G-protein-coupled receptors (GPCRs), which transmit varied signals from outside to inside cells by activating guanosine triphosphate (GTP)ase proteins, which in turn signal to second messengers that affect the state of the cell in various ways. Type I opsins do not activate G proteins. Furthermore, type II opsins are more closely related to nonopsin, light-insensitve GPCRs than they are to type I opsins.

Major Type II Opsin Classes

Based on phylogenetic analyses, there are four major classes of type II opsins in animals: cnidops, retinal G-protein receptor (RGR)/Go, rhabdomeric (Gq), and ciliary (Gt) (**Figure 2**).

Cnidops

In 2007, a new major class of opsins was described based on analyses of whole genome sequences of the phylum Cnidaria, which includes sea anemones and jellyfish. This cnidarian-specific class of opsins is called cnidops. The molecular details of cnidops-mediated phototransduction are not yet fully elucidated, but box jellyfish cnidops appears to activate Gs-class G proteins, which in turn activate adenylyl cyclase. For what might cnidarians be using opsins? Some jellyfish are known to use light cues to

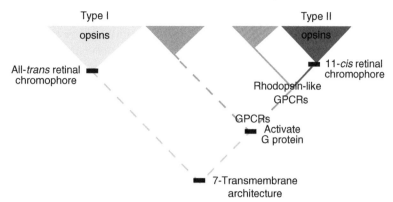

Figure 1 Evolutionary relationships of type I and type II opsins, and G-protein-coupled receptors (GPCRs). Triangles represent major groups with many genes from many species. Solid lines indicate good support for evolutionary relationship (common descent) based on sequence similarity and other evidence. Dashed lines indicate there is no evidence for common ancestry beyond random similarity of sequences and similar function. Type I opsins are well known from bacteria and function in a type of photosynthesis and, in some cases, to mediate movement toward or away from light. Type II opsins are found only in animals (see **Figure 2** and **Table 1**), and are likely related to nonopsin GPCRs, which include chemoreceptor and other proteins that detect a signal outside a cell, and mediate a response inside the cell. A chromophore is a light-sensitive chemical that is bound to opsin. Type I and type II opsins use different chromophores.

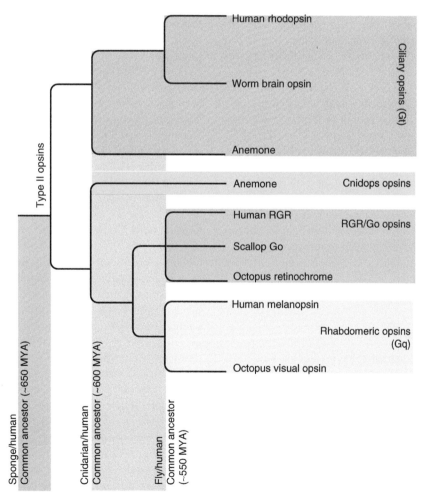

Figure 2 Illustrated are possible evolutionary relationships of the major groups of opsin genes (ciliary, cnidops, RGR/Go, and rhabdomeric). Here, each major group is represented by genes from one to three species, but each major opsin group is found in many other species as well. **Table 1** shows names of other opsins, some not on this figure, and the major opsin group to which they belong. Based on presence in anemone (a cnidarian) and bilaterian animals (humans and annelid worms), ciliary opsins and therefore all type II opsins originated before the common ancestor of cnidaria and bilateria, which lived some 600 million years ago.

regulate their depth in the water column and their swimming speed, while others (box jellyfish) possess sophisticated camera eyes analogous those of humans. These complex cnidarian eyes may be capable of image-forming vision, and are used for obstacle avoidance and perhaps for hunting prey. At present, there is no evidence for cnidops-mediated color vision.

Retinal G-Protein Receptor/Go

Retinal G-protein receptor (RGR) and Go opsins are functionally distinct from each other, but sometimes group together in phylogenetic analyses, indicating they may share more recent common ancestry with each other than with the other major opsin classes. Opsins grouping in this clade include those from vertebrates, cephalochordates (invertebrate chordates represented by the lancelet *Branchiostoma*), and mollusks.

In humans, three opsins often fall within the RGR/Go group: peropsin, neuropsin, and RGR. Cephalopod mollusks also possess a related opsin called retinochrome. Interestingly, members of this opsin class have not been noted from any arthropod or any other ecdysozoan (a group of molting invertebrates, including arthropods and nematode worms).

The RGR opsins are not well understood, but some important clues to their function exist. These opsins preferentially bind an all-*trans* isomer of retinal, contrasting with rod and cone opsin, which require 11-*cis* retinal for function. When tested in specific wavelengths of light that ranged from blue to ultraviolet (UV), RGR was able to convert all-*trans* retinal to the 11-*cis* isomer. Thus, it would seem that RGRs function to provide fresh retinal of the functional 11-*cis* conformation that is required for proper rod and cone opsin function. Retinochrome opsins from mollusks also have this photoisomerase function.

In addition, in mollusks and cephalochordates are opsins that may interact with a Go-class G protein, and as such are called Go opsins.

Rhabdomeric (Gq)

Opsins belonging to the rhabdomeric (Gq) class are present in bilaterian animals (animals with left and right sides), but unknown from Cnidaria. Where known, these opsins activate Gq-class G proteins, which drive phospholipase C second messengers. The opsins are sometimes called rhabdomeric opsins, because they tend to be expressed in specific cell types (rhabdomeric photoreceptor cells) that increase cell surface area using microvilli. Gq opsins are well known for mediating vision of many invertebrate animals, and are particularly well studied in the compound eyes of the fly *Drosophila melanogaster*. In chordates, including humans, Gq opsins are not expressed in the primary visual photoreceptors; instead, a Gq opsin,

called melanopsin is known to be expressed in retinal ganglion cells, which are among the cells that relay signals from the primary photoreceptors to the optic nerve. Melanopsin is also involved in mediating the papillary reflex.

Ciliary (Gt)

Opsins belonging to the ciliary (Gt) class are present in eumetazoan animals, including vertebrates, annelid worms, and cnidarians (anemones and jellyfish). In vertebrates, these opsins activate Gt-class G proteins (transducins), which drive phosphodiesterase (PDE) enzymes. In invertebrates, the G protein that is activated by ciliary opsins is unknown. This class of opsins is often called ciliary opsins because they are known to be expressed in ciliary photoreceptor cells, which assemble a phototransduction organelle from a sensory cilium membrane. In annelids and bees, ciliary opsins are expressed in the brain. Although the organismal function of ciliary opsins is unknown in invertebrates, some have suggested a role in entraining circadian rhythms. In vertebrates, ciliary (Gt) opsins are expressed in the primary photoreceptors, the rods and cones, with different subclasses of opsin expressed in rods versus cones and further subclasses expressed in different cones to mediate color vision. Ciliary opsins with extraretinal expression are also known from vertebrates, and have been given various monikers, often named for their site of expression, such as pinopsin, parapinopsin, and tmt opsin. Details about the molecular and organismal functions of these extraretinal ciliary opsins are scarce (**Table 1**). One opsin in this class, parietopsin, is expressed in the parietal eye of a lizard and activates a Go-class G protein.

Opsin and Color Vision

Color vision, the ability to discriminate between light information of different wavelengths, is present in various groups of animals including insects and vertebrates, and involves opsin genes. Although color vision is a behavioral phenomenon that involves processes of both physical detection and neurobiological comparison, duplicated opsin proteins expressed in the retina directly mediate the physical detection of different wavelengths of light.

The specificity of a given opsin protein to a given wavelength of light is summarized by its λ-*max*, a measure of the peak absorbance across the light spectrum. The specific λ-*max* of a given opsin protein is largely a function of its amino acid sequence. Opsins representing the different spectral classes are specifically expressed in individual photoreceptor neurons called cone cells. When stimulated by light of the proper wavelength, opsins trigger a cellular response, leading to a change in the resting electric charge, or potential, of a photoreceptor cell. Color vision emerges from this system through the comparison

Table 1 Various names for type II opsins

Name	Major clade	Taxon	Expression	Function
Encephalopsin/ panopsin	Ciliary		Brain	Retinoid Receptor?
Exo-rhodopsin	Ciliary	Zebra fish	Pineal	Unknown
Parapinopsin	Ciliary	Vertebrates	Parapineal, pineal	
VA/VAL opsin	Ciliary	Vertebrates	Horizontal and amacrine cells, pineal, brain	Unknown
Cone L-opsin	Ciliary	Vertebrates	Cone cells	Photopic Vision
Cone M-opsin	Ciliary	Vertebrates	Cone cells	Photopic Vision
Cone S-opsin	Ciliary	Vertebrates	Cone cells	Photopic Vision
Cone V-opsin	Ciliary	Vertebrates	Cone cells	Photopic Vision
Pinopsin	Ciliary	Vertebrates	Brain, pineal organ	
Pteropsin	Ciliary	Bees, Mosquitoes	Brain	Unknown
Rhodopsin	Ciliary	Vertebrates	Rod cells	Scotopic Vision
TMT opsin	Ciliary	Teleost fishes	Many tissues	Unknown
Parietopsin	Ciliary	Zebra fish, fugu, Xenopus, lizard	Parietal eye	Photoreception
Cnidops	Cnidops	Cnidarians (hydra, anemone, and jellyfish)	Neurons	Light perception
Neuropsin	RGR/Go	Mammals	Testes, brain, spinal cord, eye	Photoisomerase?
Peropsin	RGR/Go	Mammals	RPE	Photoisomerase?
RGR	RGR/Go		RPE and Müller cells	Photoisomerase
Retinochrome	RGR/Go	Cephalopoda	Retina	Photoisomerase
Go	RGR/Go	Scallop, lancelet	Scallop eye	Photoreception
Arthopsin	Rhabdomeric	Daphnia	Unknown	Unknown
Melanopsin	Rhabdomeric	Vertebrates	Retinal ganglion cells	Circadian rhythms, papillary reflex
Rhabdomeric (Gq) opsin	Rhabdomeric	Invertebrates	Eye	Vision

of the resultant potentials of cone cells with those that express opsins of differing λ-max expressed in adjacent cone cells. For instance, humans possess three cone-opsins with λ-max values falling roughly into the categories of blue, blue–green, and yellow–green. Compared to other vertebrates, human opsins represent a single blue S opsin and two M opsins (blue–green and yellow–green) that resulted from a primate-specific gene duplication event. When red light falls upon a cone cell that is sensitive to yellow–green, it activates opsin signaling because light of this wavelength falls within the activity spectrum of this protein. However, red light does not fall within the sensitivity spectra of either the blue or blue–green opsins. Together, comparisons made between activated and non-activated cone cells lead to the human perception of red light. In most vertebrates, these comparisons are made by additional cells in the retina (i.e., the horizontal and ganglion cells) and also in a portion of the brain called the visual cortex.

In vertebrates, four distinct opsin classes, each with distinct sensitivities, accomplish the full spectral range. These include violet-(V) short-(S), middle-(M), and long-(L) wavelength varieties. Studies in the sea lamprey, a distant vertebrate relative of humans, have provided data that allow scientists to infer that the four major classes of vertebrate retinal opsins are the result of ancient gene duplication mutations that likely occurred before the

origin of all living vertebrates. This implies that major opsin clades must have been lost in various vertebrate lineages. For instance, placental mammals likely originated from nocturnal ancestors that lost the V and M spectral classes of opsin. In addition, some marine mammals are known to have further lost the S class of opsin and are cone monochromats. Similar losses have been reported in a deep-dwelling, Lake Baikal fish. In general, each of these examples of opsin loss can be correlated to photoecology. When spectral sensitivity of a given wavelength is not required, selection against mutations at these loci is relaxed, allowing the loss of functional copies. However, on numerous occasions, photoecology has provided a generative context for the origination of new opsin genes and classes in vertebrate and invertebrate lineages alike. Often, these new visual repertoires converge both in spectral sensitivity (i.e., λ-max) and in the specific mutations that have lead to the functional shift.

Despite major differences between insect-faceted eyes and vertebrate camera-like eyes, both eye types utilize opsins in a similar way to achieve color vision. Similar to vertebrate color vision, insect color vision is based on a process of physical detection of light by opsins of differing λ-max, followed by neurobiological comparison of these light data. In insects and other arthropods, the neurobiological basis for color vision is less well understood but likely involves a part of the insect brain called the medulla. While the neurobiological basis for insect color vision requires further clarification, it is clear that opsin gene duplication has been a driving force for the evolution of color vision in insects and other invertebrates. For example, the duplicated opsin genes of fruit flies are well studied. Two fly opsins are maximally sensitive to UV (345, 375 nm), one to blue (437 nm), two to blue–green (420, 480 nm), and one to green (508 nm) light.

Molecular Basis of Wavelength Sensitivity

Different λ-*max* values of different opsins are mediated by specific amino acid differences, which in some cases have been experimentally demonstrated. For example, in primates, three amino acid sites have a dominant role in differentiating red-sensitive and green-sensitive opsins. In birds, one opsin gene is blue sensitive in some species, and UV-sensitive in other species, a difference that is mediated by a single amino acid change. Changing the 84th amino acid of a zebra finch opsin allowed the protein's sensitivity to change from UV to blue when the protein was expressed in cultured cells. Conversely, pigeon and chicken blue-sensitive opsins became UV sensitive in cell culture after a change of the 84th amino acid. The combination of gene duplication and differential changes in amino acids of duplicated opsins has generated a diversity of proteins in different species.

Opsin and Modes of Phototransduction Evolution

As discussed above, opsin genes were very often duplicated and retained during animal evolution. Early opsin gene duplications led to the major opsin groups and more recent duplications mostly led to additional specializations, such as the ability for color vision. As members of highly coordinated protein networks, changes in opsin proteins are sometimes correlated with changes in partnering proteins. The interaction of two evolutionary processes has resulted in the diversity of opsin-based phototransduction pathways observed today that contains a combination of shared and distinct interactions. First, co-option refers to instances where an opsin recruited different intracellular signaling components than its ancestor during evolution. Second, co-duplication involved the simultaneous duplication of multiple genes of an ancestral network. Co-option and co-duplication are not discrete alternatives; instead, some genes of a network originated by co-duplication, whereas others joined the network by co-option. This becomes especially clear if we examine the evolution of particular networks at multiple timescales, or at increasing spatial scales by increasing the number of interactions considered.

An example of phototransduction pathways that evolved largely by co-duplication is the different pathways used in rod and cone cells of vertebrate retinas. Rods are specialized for dim light (scotopic) vision, and cones are specialized for bright light (photopic) vision. Rods and cones utilize different, duplicated opsins as well as different duplicated genes downstream of opsin. Multiple rod-specific and cone-specific genes originated through large-scale segmental duplication of an ancestral vertebrate genome before the origin of gnathostomes (jawed vertebrates) and after the split of vertebrates and their closest nonvertebrate relatives. As such, the duplication of multiple genes in these pathways maps to the same time interval and is consistent with co-duplication.

Examples of the origin of new opsin pathways involving co-option probably include the origin the major opsin groups, which signal to different G proteins. In particular, the rhabdomeric opsins are the only opsins to signal through Gq G proteins and transient receptor potential (TRP) ion channels. Since rhabdomeric opsins originated in the ancestor of living bilaterally symmetric animals, and are unknown from Cnidaria, they have a more recent evolutionary origin than ciliary opsins, which are present in Cnidaria. This pattern indicates co-option: at or near their origin, rhabdomeric opsins began signaling to Gq and TRP, proteins that predate bilaterian animals.

See also: Circadian Rhythms in the Fly's Visual System; Color Blindness: Inherited; The Colorful Visual World of Butterflies; Genetic Dissection of Invertebrate Phototransduction; *Limulus* Eyes and Their Circadian Regulation; Microvillar and Ciliary Photoreceptors in Molluskan Eyes; The Photoreceptor Outer Segment as a Sensory Cilium; Phototransduction in *Limulus* Photoreceptors; Phototransduction: Phototransduction in Cones; Phototransduction: Phototransduction in Rods; Phototransduction: Rhodopsin; Polarized-Light Vision in Land and Aquatic Animals; Rod and Cone Photoreceptor Cells: Inner and Outer Segments.

Further Reading

Arendt, D. (2003). Evolution of eyes and photoreceptor cell types. *International Journal of Developmental Biology* 47(7–8): 563–571.

Briscoe, A. D. and Chittka, L. (2001). The evolution of color vision in insects. *Annual Review of Entomology* 46: 471–510.

Fernald, R. D. (2006). Casting a genetic light on the evolution of eyes. *Science* 313(5795): 1914–1918.

Hardie, R. C. and Raghu, P. (2001). Visual transduction in *Drosophila*. *Nature* 413(6852): 186–193.

Lamb, T. D., Collin, S. P., and Pugh, E. N. (2007). Evolution of the vertebrate eye: Opsins, photoreceptors, retina and eye cup. *Nature Reviews. Neuroscience* 8(12): 960–975.

Land, M. F. and Nilsson, D-E. (2002). *Animal Eyes.* (Willmer, P. and Norman, D. (eds)). Oxford: Oxford University Press.

Nathans, J. (1999). The evolution and physiology of human color vision: Insights from molecular genetic studies of visual pigments. *Neuron* 24(2): 299–312.

Oakley, T. H. and Pankey, M. S. (2008). Opening the "black box": The genetic and biochemical basis of eye evolution. *Evolution Education and Outreach* 4: 390–402.

Plachetzki, D. C. and Oakley, T. H. (2007). Key transitions during animal eye evolution: Novelty, tree thinking, co-option and co-duplication. *Integrative and Comparative Biology* 47: 759–769.

Salvini-Plawen, L. V. and Mayr, E. (1977). *On the Evolution of Photoreceptors and Eyes.* New York: Plenum Press.

Spudich, J. L., Yang, C. S., Jung, K. H., and Spudich, E. N. (2000). Retinylidene proteins: Structures and functions from archaea to humans. *Annual Review of Cell and Developmental Biology* 16: 365–392.

Terakita, A. (2005). The opsins. *Genome Biology* 6(3): 213.

Yokoyama, S. and Yokoyama, R. (1996). Adaptive evolution of photoreceptors and visual pigments in vertebrates. *Annual Review of Ecology and Systematics* 27: 543–567.

Eye Field Transcription Factors

M E Zuber, SUNY Upstate Medical University, Syracuse, NY, USA

Glossary

Aniridia – Congenital condition resulting in the underdevelopment or lack of an iris.

Anophthalmia – Congenital absence of one or both eyes.

Ectopic eye – An eye that has formed in an abnormal location (an additional eye).

Eye field – A region of the anterior neural plate of vertebrate embryos that is fated to generate the eyes.

Gene homology – Two genes are homologous if they are derived from the same gene in a common ancestor.

Hypomorphic mutant – A mutation in which the altered gene product has reduced activity or in which the wild-type gene product is expressed at a reduced level.

Iris – Colored portion of the eye surrounding the pupil.

Microphthalmia – Congenital defect resulting in small eyes.

Retina – Light-sensing part inside the inner layer of the eye. The vertebrate retina consists of seven major cell classes.

Discovery and Structural Features of Eye-Field Transcription Factors

Pax6

Vertebrate Pax6 (paired box gene 6) was identified, in 1991, in mouse and humans as a member of a multigene family of paired-box-containing (Pax) genes. Small-eye (Sey) mutant mice were found to have mutations in the Pax6 gene predicted to interrupt gene function. The *Drosophila melanogaster* (fly) mutant eyeless (ey) was subsequently shown to be homologous to vertebrate Pax6. Remarkably, the mouse Pax6 was shown to functionally rescue the eyeless phenotype in flies, demonstrating functional conservation of Pax6 from flies to mammals. Pax transcription factors contain a paired domain and either a partial or complete (as in the case of Pax6) homeodomain (see **Figure 1** and **Table 1** for EFTF structure and characteristics). In addition, Pax6 contains a proline–serine–threonine-rich (PST-rich) C-terminus.

The paired domain is an approximately 126-residue DNA-binding domain originally described in the *Drosophila* gene of the same name. Two distinct DNA-binding domains are present in the paired-type domain in the form of N-terminal and C-terminal subdomains. Alternative splicing of the Pax6 gene results in alternate isoforms with distinct DNA-binding activities with functional consequences. The 60-residue homeodomain forms a helix-turn-helix structure at the carboxy-terminal end that also binds DNA. As with the paired domain, an altered homeodomain can result in developmental defects. The PST-rich region acts as a trans-activating domain with binding sites for other proteins such as transcriptional coregulators. Pax6 mutants that truncate the C-terminal-half of the protein remove the PST-rich region, retain their DNA-binding capacity, but act dominantly to repress normal Pax6 function.

Six3

The first Six (Sine oculis homeobox) family member was identified in 1994 as the gene altered in the *Drosophila* mutant *sine oculis* (*so*), the most striking phenotype of which is the lack of eyes. Vertebrate Six3 (*sine oculis homeobox homolog 3*) was first identified in mouse in 1995 and shown to share significant amino-acid homology with *so*. Members of the Six family of homeobox transcription factors are characterized by two conserved domains – the Six domain (SD) followed by a more carboxy-terminal homeodomain (HD). The Six domain is typically 115 residues and is required for protein–protein interactions between Six genes and their binding partners. The homeodomain is a 60-residue DNA-binding domain (as described above). Both the SD and HD are required for normal function of the protein. The more recent identification of additional *Drosophila* Six family members also demonstrated vertebrate Six3 (and section titled 'Six6') share greater sequence homology with fly *Optix* than its namesake *sine oculis*. Based on sequence homology, intron–exon boundries, and the proteins with which they form complexes, Six family members are divided into three subfamilies. Six3 and Six6 have been assigned to the Six3/6 subfamily.

Six6

A second Six gene structurally related to Six3 was identified, in 1998, in zebrafish, chicken, and mouse as Six6 (*sine oculis homeobox homolog 6*) and Optx2 (optic six gene 2). Among the family of Six genes, Six6 shares greatest sequence homology and expression pattern with Six3, suggesting they may have arisen via gene duplication. Despite their similarities, Six6 and Six3 can be easily distinguished at the amino-acid level. The most highly divergent regions lie at the amino- and carboxy-terminals to the Six and homeodomains. In particular, the region of Six6 amino-terminal to the Six domain is dramatically smaller than the corresponding region of Six3. For example, only nine residues are present at the amino-terminal to the Six domain in mouse Six6, while 88 residues are present in the Six3 protein.

Rax

Vertebrate Rax (retina and anterior neural fold homeobox) was identified, in 1997, by three independent groups in Xenopus, mouse, and zebrafish, and is shown to be expressed in the developing eyes and required for normal mouse eye formation. Rax – like Pax6 – is classified as a member of the paired homeobox class of transcription factors. In contrast to Pax6, however, Rax contains a paired-like homeodomain, but no paired domain. Within

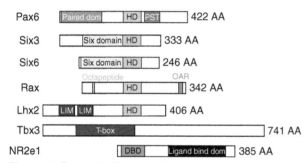

Figure 1 Schematic illustrating the location of structural domains in the mouse eye-field transcription factors. The EFTFs – with the exception of the Six genes Six3 and Six6 – have distinct domain structures and are members of different transcription-factor families. Five of the seven bind DNA targets via a homeobox domain (HD). While Tbx3 (via the T-box) and Nr2e1 bind DNA through DNA-binding domains (DBDs) specific to their transcription factor family. Each EFTF also contains protein motifs that allow it to complex with other proteins. These protein–protein interactions serve to facilitate (or sometimes block) the ability of the transcription factor to bind its DNA targets. Co-regulatory proteins that bind to these regions can serve as co-activators or co-repressors to increase or decrease, respectively, the rate of expression from a target gene. Courtesy of Andrea Viczian.

the amino-terminal, Rax contains an octapeptide motif consistent with this homeobox gene subfamily. In the carboxy-terminal, a 15-residue region present in other paired-like genes termed the paired tail or OAR (otp, aristaless, and rax) domain is also conserved. Finally, the sequence carboxy-terminal to the paired-like homeodomain is PST-rich. These regions (octapeptide, OAR, and PST-rich) may function as transactivation domains.

Lhx2

First identified in rat as LH-2 (LIM Hox gene 2) in 1993, vertebrate Lhx2 (LIM homeobox-2) is a member of a large LIM-domain (Lin 11, Isl-1, Mec-3)-containing family of genes with diverse structure and biological functions. Mouse Lhx2 was subsequently shown to be required for normal eye formation. The LIM domain (two of which are found in Lhx2) is a cysteine- and histidine-rich zinc-finger motif involved in protein–protein interactions. The LIM domains of Lhx2 protein are located in the amino-terminal region and followed by a carboxy-terminal, highly conserved, DNA-binding homeodomain. There is little evidence that LIM domains of the LHX class can bind DNA directly, but may regulate DNA binding of the protein via interactions with other proteins.

Tbx3

Vertebrate Tbx3 was identified, in 1994, in mouse as a member of a large gene family characterized by a shared T-box protein motif. Mutation of the founding member of this family caused truncated tails in mice and was short-hand named T for short-tail. Once the gene was identified, the DNA-binding region of the T protein (which is 180–190 residues and can be located at any position in the protein) was subsequently named the T-box. The more than 50 T-box family members identified have been classed into five subfamilies. Tbx3 is a member of the Tbx2 subfamily that includes Tbx2-5 and is most similar to Tbx2.

Nr2e1

Vertebrate NR2E1 (nuclear receptor subfamily 2, group E, member 1) was originally identified in 1994 as Tlx,

Table 1   Names and the null phenotypes of EFTF homologs

EFTF	TF family	Fly homolog	Vertebrate alias(es)	Eye phenotype of mouse null
Pax6	Paired homeobox	eyeless, twin of eyeless	AN2, WAGR	small or no eyes
Six3	Six homeobox	sine oculis/optix	Holoproencephaly 2 (HPE2)	lack of anterior brain (no eyes)
Six6	Six homeobox	sine oculis/optix	Optx2, Six9	small eyes
Rax	Paired-like homeobox	DRx	Rx	no eyes
Lhx2	LIM homeobox	apterous	LH-2	no eyes
Tbx3	T-box	optomotor blind	ET, XHL, UMS	embryonic lethal (eye normal)
NR2E1	Nuclear receptor type	tailless	XTll, Tlx	degenerated retina

a unique vertebrate nuclear receptor with structural homology to the previously described fly gene, *tailless* (*tll*). Tll was classified as a member of the steroid receptor superfamily of proteins. Nr2e1 (and tll) are orphan nuclear receptors, implying – although their protein sequence predicts a hormone-binding domain – the hormone ligand that binds to it has not been identified. The NR2E1 DNA and ligand-binding domains are located in the amino- and carboxy-terminal regions of the protein, respectively. NR2e1 shares the highest similarity with tll in these regions. *In vitro* DNA-binding assays, and misexpression of vertebrate Nr2e1 in the fly shows that the vertebrate and fly proteins share similar target-gene specificity, which is unique among the nuclear receptor superfamily.

The EFTFs are Expressed during and are Required for Normal Eye Formation

Transplantation experiments performed in the beginning of the last century demonstrated that the amphibian eye field (originally called the eye anlagen or, sometimes, eye primordia) is located in the neural plate. Removing the anterior portion of the neural plate resulted in the creation of eyeless animals. When this same region was transplanted to the belly-wall or grown in culture, a histologically normal eye would form. These remarkably simple experiments clearly demonstrated that the eye primordia is specified in embryos at the neural plate stage, and that all the genes required to initiate eye formation were being expressed in the tissue at that time. Subsequent, detailed mapping experiments using more modern tools precisely defined the location and shape of the eye field as a single crescent of cells spanning the breadth of the early anterior neural plate.

During embryonic development, eye and brain formation are tightly coordinated. Neural induction results in an early neural plate with a propensity to form anterior structures (including the eyes). This early neural plate is subsequently patterned in response to diffusible molecules that help to form the different brain regions (the forebrain, midbrain, and hindbrain). The signaling systems known to regulate this patterning include the wingless (Wnts), fibroblast growth factors (FGFs), bone morphogenetic proteins (BMPs), Nodals, and retinoic acid. Inactivation or modification of any one of these signaling systems can result in abnormal eye formation. These abnormalities are often coincident with changes in the expression of the EFTFs. These effects may sometimes be indirect, however, as some of these molecules are required for neural induction and not directly responsible for the formation of the eye field. Nevertheless, EFTF expression can be regulated by these signaling systems. For example, cultured embryonic stem cells treated with

BMP, Wnt, and/or Nodal antagonists express EFTFs and are directed to a retinal progenitor lineage. Precise control of the Wnt signaling system appears to be critical for patterning of the forebrain – the brain region in which the eye field and ultimately the eyes form. Disrupting Wnt activity by modifying the ligand or its signaling cascade can have dramatic effects on eye formation. Either too much or too little Wnt activity can lead to abnormal eye formation. The evidence suggests that a precise Wnt gradient is required to pattern and appropriately position the eye field within the forebrain.

The location and timing of eye-field specification in the anterior neural plate is synchronized with the expression of the EFTFs. The expression patterns of the individual genes are distinct and extend to other regions of the neural plate and beyond. However, all are to some extent expressed in the eye field in one or more species. For example, the expression pattern of the frog (*Xenopus laevis*) EFTFs overlap in the eye field during and immediately following its specification (**Figure 2**). The EFTFs of other species have similar patterns of coordinated expression. At neural plate stages, for example, Pax6, Rax, Six3, and Six6 are also observed in a single band of expression in chicken, zebrafish, and mouse embryos.

The single band of EFTFs expressing eye-field cells generates both eyes. Since all vertebrate embryos have a pair of eyes organized symmetrically across the body midline, the single eye field described above must be split into two lateral eye primordia. Infrequently, this separation does not take place due to environmental toxins or genetic mutations. This results in cyclopia, the formation of one large midline eye. One toxin long known to induce cyclopia is 11-deoxyjervine (also known as cyclopamine). A US Department of Agriculture study, in the 1950s, showed that pregnant ewes feeding on corn lily (a naturally occurring source of cyclopamine) bore one-eyed lambs (**Figure 3**). More recent studies have demonstrated that diffusible factors secreted by cells near the eye field regulate its splitting. Signaling pathways regulated by the ligands hedgehog and nodal converge and are required for eye-field separation. Hedgehog represses the EFTF Pax6 and since hedgehog is normally expressed underneath the eye field midline, it is likely that hedgehog expression is required for normal eye-field separation and the formation of two eyes (**Figure 3**). Consistent with this interpretation, humans, mice, and fish lacking functional nodal or hedgehog signaling are cyclopic. Interestingly, the action of cyclopamine also appears to be through the hedgehog signaling pathway. Cyclopamine regulates the activity of Smoothened, the receptor for the hedgehog ligand.

Evidence that the EFTFs are required for normal eye formation is abundant. The most extensively studied of the EFTFs, Pax6, was originally identified as the gene mutated in the small eye (Sey) mouse which, when

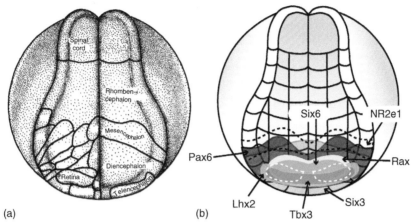

Figure 2 Schematic diagrams of stage 15 *Xenopus laevis* embryos. (a) Fate-map illustrating the areas of the neural plate that will eventually form the major brain regions. (b) Expression patterns of the frog eye-field transcription factors. The approximate location of the eye field is outlined in white. No two EFTFs have identical expression patterns, yet all (but Nr2e1) are expressed in the eye field. Nr2e1 is expressed posterior to the eye field proper at stage 15. However, only a few hours later, its expression overlaps that of the other EFTFs in the eye primordia. Courtesy of Andrea Viczian.

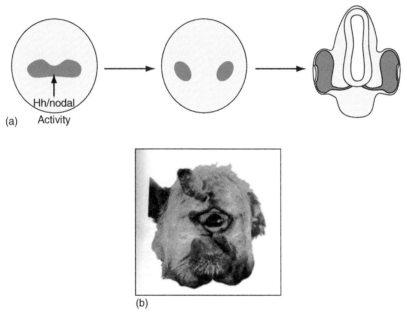

Figure 3 Generation of two eyes from a single eye field. (a) The single eye-field spans the entire anterior neural plate. For two eyes to form, this single EFTF-expressing domain must be split into the two (left and right) eye primordia. Cells posterior to and underlying the midline are required for this process. Signaling via the hedgehog (Hh) and nodal systems are thought to repress midline eye-field fate. (b) When these signaling systems are abnormal, eye-field separation does not take place and a single midline eye forms. The mother of this lamb fed on corn lily while pregnant. Cyclopamine made by corn lily blocks hedgehog signaling with dramatic effects. Courtesy of L. James, USDA Poison Plant Laboratory.

homozygous, lacks eyes. A large number of null and hypomorphic mutant alleles have been generated – with variable eye phenotypes including lens/cornea fusion, microphthalmia, cataracts, and anophthalmia. Flies with a mutant version of the *Drosophila* homolog of Pax6 (eyeless) also lack eyes, while frog eye development can be arrested using a function-blocking (dominant negative) version of Pax6. As with Pax6, functional inactivation of

Six3, Six6, Rax, Lhx2, and Nr2e1 results in flies, frogs, fish, and/or rodents with abnormal or no eyes.

Mutation of human EFTFs also results in abnormalities affecting the eyes. Pax6 mutations result in aniridia, microphthalmia, and Peters' anomaly – in which patients have cornea and iris abnormalities. Anophthalmia and microophthalmia have also been attributed to mutations in the human Six6 and Rax genes. Six3 mutation results in holoprosencephaly, a failure of the forebrain to divide properly into the two hemispheres. Holoprosencephaly is, in most cases, fatal. In less severe cases, facial deformities affecting the eyes are observed.

The EFTFs Form a Self-Sustaining Feedback Network

As described briefly above, many of the vertebrate EFTFs were originally identified as *Drosophila* genes required for fly eye formation or that have fly homologs. In some instances, the vertebrate versions can functionally restore the corresponding fly mutant or induce ectopic eyes in the fly. For example, mis-expression of mouse eyes absent (eya) and Pax6 in flies can induce ectopic eye development and rescue eye formation in loss-of-function mutants. Based on its remarkable evolutionary conservation, requirement for normal eye formation in multiple species, and the ability of mammalian Pax6 homologs to induce ectopic fly eyes, ey/Pax6 was originally hailed as a potential master regulator of eye formation. In initial models, eyeless was placed atop a hierarchy of genes that drove eye formation in the eye portion of the eye-antennal imaginal disk complex. However, a much more complex set of interactions has emerged. The genes twin-of-eyeless (toy), *sine oculis* (so), optix, eyes absent (eya), dachshund (dac), eye gone (eyg), and twin-of-eyegone (toe) are all required for fly eye formation. Individually or in combinations, these genes can also induce ectopic eyes, sometimes in the absence of eyeless. For instance, both toy and optix can induce ectopic eyes via an *ey*-independent mechanism. In the fly, a model has developed in which the retinal determination genes described above behave as a network with both hierarchical components and regulatory feedback loops.

In recent years, it has been discovered that an analogous – but by no means identical – transcription factor network may also be at work during vertebrate eye formation. As described above, fly retinal determination genes have the ability to induce ectopic eyes. Similarly, some EFTFs can induce ectopic retinal tissue and lenses in vertebrates. For example, in the frog, ectopic expression of Pax6 leads to the formation of ectopic eye-like structures. Lineage-tracing experiments show that the lens, retina, and retinal pigment epithelium of these eyes all arose in response to Pax6. Molecular markers confirmed the presence of some retinal cell types in the induced eyes. In a similar manner, ectopic expression of either fish or mouse Six3 promotes the formation of ectopic lens and retina cells. Remarkably, Six3 mis-expression was also reported to promote the formation of ectopic tissues with an optic vesicle-like structure in the hindbrain–midbrain region of developing mouse embryos. Experiments performed in fish and frogs also show that Rax homologs can induce

eye tissues. The mechanisms by which individual EFTFs induce the formation of eye tissue are still unclear. Dosage is important. In the case of frog Six6, relatively low doses stimulate the proliferation of eye-field cells, dramatically increasing the size of the eye field and eventually resulting in the growth of eyes nearly twice their normal size. At higher doses, ectopic eyes form in the rostral brain, suggesting that tissue which would normally form midbrain is respecified to form retina.

A likely mechanism is the ability of the vertebrate EFTFs to induce each other's expression. Pax6, Six3, Rax, and Six6 activate each other's expression, while inactivation of each, sometimes, reduces the expression of the others. When high doses of Six6 or Six3 sufficient to drive ectopic eye formation are injected into frog or fish embryos, the expression of genes normally present in the rostral brain is lost, while Pax6 and Rax are induced in their place. When eye-inducing levels of Pax6 are injected into frogs, ectopic expression of Rax and Six3 are detected. Some of these EFTF interactions are direct – that is, they involve one transcription factor binding to and activating the regulatory sequences of another. For instance, experiments using *in vitro* and transgenic approaches in mouse have demonstrated that Pax6 and Six3 bind directly to regulatory sequences in each other's gene. In addition to regulating each other's expression, both Pax6 and Six3 induce their own expression in developing embryos. Not all the inductive effects of the EFTFs on one another's expression are direct. For example, in Pax6$^{-/-}$ mice, early expression of Rax, Six3, and Lhx2 is unaffected. Similarly, Pax6, Six3, and Rax expression are normal in the small-eyed Six6$^{-/-}$ mouse. Intervening, yet-to-be-discovered signaling cascades remain to be identified. Nevertheless, these results demonstrated that EFTFs form a genetic network that has been partially conserved during evolution, which, in the anterior neural plate of vertebrate embryos, is required for eye primordia formation.

The Coordinated Expression of EFTFs is Sufficient for Eye Formation

As described above, the EFTFs are coordinately expressed in the eye of multiple species during its formation and they can regulate each other's expression. There is clear evidence from experiments with the retinal determination genes in flies that they can act coordinately, forming protein–protein interactions to activate downstream targets necessary for ectopic eye formation. For example, *Drosophila sine oculis* physically interacts with eyes absent through the Six domain, and, together, they can induce ectopic compound-eye formation. Vertebrate EFTFs have also been shown to act synergistically. In frogs, mis-expression of Six6 dramatically increases eye size by stimulating the proliferation of eye-field cells. In contrast, Pax6 alone has no effect on eye size. However, when co-expressed in the eye field of developing frogs, Pax6 potentiates the effect of Six6 on eye enlargement. Recent evidence also suggests that Pax6 and Lhx2 act synergistically by binding to and regulating Six6 expression. As in fly, functional interactions have also been detected among the vertebrate EFTFs. *In vitro* and *in vivo* assays show Rax physically associates with and can enhance the Pax6-mediated transactivation of a minimal promoter. Pax6 can also associate with Six3 and Lhx2 *in vitro*. Dach is the mouse homolog of the eye-inducing fly gene dachshund. Six6 physically associates with Dach and, together, they regulate mouse retinal cell proliferation by co-repressing the cyclin-dependent kinase inhibitor p27Kip1. Peculiarly, while fly dachshund mutants lack eyes, mouse Dach – which is expressed in developing eyes – does not appear to be required for mouse eye formation.

Together, these observations lead to the proposal that, in vertebrates – as in the *Drosophila* compound eye – specification might be driven by the coordinated expression of the EFTFs. This model was tested in vertebrates by co-expressing the EFTFs in developing *Xenopus* embryos. Coordinated expression of the EFTFs Tbx3, Rax, Pax6, Six3, Nr2e1, and Six6 with the anterior neural patterning gene Otx2 was found to be sufficient to induce ectopic eye fields and eye structures (**Figure 4**). In contrast to ectopic retinal tissues induced by individual EFTFs, cocktail-induced eyes were generated at a higher frequency, were larger, and develop outside the nervous system at various locations on the body including the belly. EFTF cocktail subsets and inductive analysis has also been used to begin to characterize the network of

interactions between vertebrate EFTFs in frogs and generate a working model for the interactions required for eye-field specification. In the model, coordinated expression of the EFTFs specifies the eye field within the presumptive forebrain (**Figure 5**). These results suggested the vertebrate EFTFs redirect noneye cells to an eye-field-like fate since their coordinate mis-expression induced eyes.

To some extent, the model outlined for frog eye-field specification is consistent with results generated using targeted knockout of mouse EFTFs (**Table 1**). In the frog model, Rax is predicted to be upstream of Pax6. Consistent with this order, Rax$^{-/-}$ mice lack normal Pax6 expression in the optic primordia, while Rax expression is unaffected in the Pax6$^{-/-}$ mice. Neither Six6 nor

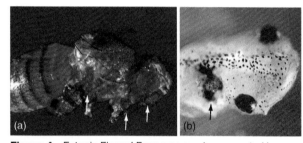

Figure 4 Ectopic Fly and Frog eyes can be generated by misexpression of the Eye Field Transcription Factors. (a) Ectopic eyes (red, white arrows) form on the wing, leg and antennae in response to the misexpression of the *Drosophila eyeless* (*ey*) gene. (b) Similarly, misexpression of a cocktail of vertebrate EFTFs including the frog *Pax6* (a homolog of *Drosophila ey*), *Tbx3, Rax, Six3, Nr2e1, Six6*, and *Otx2* induces the formation of extra frog eyes (white arrow). Fly and images courtesy of Donghui Yang-Zhou (F. Pignoni laboratory) and Andrea Viczian, respectively.

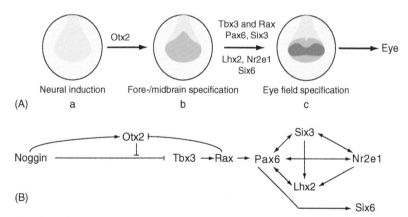

Figure 5 Simplistic model of eye field specification in the anterior neural plate of the frog embryo. (A) The initial neural plate (lightly shaded in a) is patterned to form fate-restricted areas. Otx2 expression is required for patterning of the early fore- and midbrain (b). Finally, the EFTFs form a self-sustaining feedback network that specifies the eye field (darkly shaded in c). (B) Neural induction (initiated by factors including noggin) results in an only partially patterned neural plate. Neural patterning genes like Otx2 then specify different neural regions. The early-expressed EFTFs Tbx3, Rax, Pax6, Six3, and Lhx2 act coordinately to specify the eye field. Although not required for the initial specification of the eye field, evidence suggests Nr2e1 and Six6 play later roles in eye formation. Arrows and bars indicate the induction and repression of target gene expression, respectively. Evidence for some, but not all, aspects of this predicted network are found in other species. Illustration courtesy of Andrea Viczian.

Nr2e1 are required for the initial specification of the eye field in frog. Six6$^{-/-}$ mice have normal (although small) eyes and Nr2e1$^{-/-}$ mice do not develop retinal defects until 3 weeks after birth. In mouse as well as frog, altering Six6 levels does not effect Pax6, Six3, or Rax expression.

Despite these similarities, inconsistencies and clear differences exist between the networks driving eye formation in vertebrates. In the frog model, Tbx3 is positioned upstream of Rax and is required for early eye formation. Although Tbx3$^{-/-}$ mice die as embryos, there are no obvious early eye abnormalities. Mutation of Rx3 (one of three Rax genes in fish) results in eyeless fish. Directly contradicting the frog model, Tbx3 expression is lost in the developing retina. These differences may be species specific or result from the use of different analytical methods. Additional work is needed to more clearly define the genetic interactions among these genes and their role in specification of the vertebrate eye field.

Conclusions

Experiments first performed over a century ago established that the eyes form from the most anterior region of the vertebrate neural plate. However, only recently have genes responsible for eye-field specification been identified. Using multiple techniques and model systems, researchers have identified a network of transcription factors that are required and, in some cases, sufficient for eye formation. In spite of recent work, the functional relationships among the EFTFs are still not clear. In addition to understanding how these gene products regulate each other's expression, future work is also needed to determine the upstream regulators and downstream targets of the EFTFs.

See also: Coordinating Division and Differentiation in Retinal Development; Embryology and Early Patterning; Histogenesis: Cell Fate: Signaling Factors; Retinal Histogenesis; Zebra Fish–Retinal Development and Regeneration.

Further Reading

Chow, R. L. and Lang, R. A. (2001). Early eye development in vertebrates. *Annual Review of Cell Developmental Biology* 17: 255–296.

Donner, A. L. and Maas, R. L. (2004). Conservation and non-conservation of genetic pathways in eye specification. *International Journal of Developmental Biology* 48: 743–753.

Hanson, I. M. (2001). Mammalian homologues of the Drosophila eye specification genes. *Seminars in Cell and Developmental Biology* 12: 475–484.

Kumar, J. P. (2001). Signalling pathways in Drosophila and vertebrate retinal development. *Nature Reviews Genetics* 2: 846–857.

Kumar, J. P. (2008). The molecular circuitry governing retinal determination. *Biochimica et Biophysica Acta* 1789(4): 306–314.

Pappu, K. S. and Mardon, G. (2004). Geneti control of retinal specification and determination in Drosophila. *International Journal of Developmental Biology* 48: 913–924.

Zaghloul, N. A., Yan, B., and Moody, S. A. (2005). Step-wise specification of retinal stem cells during normal embryogenesis. *Biology of the Cell* 97: 321–337.

Zuber, M. E., Gestri, G., Viczian, A. S., Barsacchi, G., and Harris, W. A. (2003). Specification of the vertebrate eye by a network of eye field transcription factors. *Development* 130: 5155–5167.

Zuber, M. E. and Harris, W. A. (2006). Formation of the eye field. In: Sernagor, E., Eglen, S., Harris, W., and Wong, R. (eds.) *Retinal Development*, pp. 8–29. Cambridge: Cambridge University Press.

Fish Retinomotor Movements*

B Burnside, University of California, Berkeley, Berkeley, CA, USA
C King-Smith, Saint Joseph's University, Philadelphia, PA, USA

Glossary

Calycal processes – The microvillus-like projections extending from the distal end of the photoreceptor inner segment ellipsoid to form a chalice-like support around the base of the outer segment.

Ciliary axoneme – The cytoskeletal scaffold of the photoreceptor outer segment, which is composed of nine microtubule doublets that extend from the basal body lying at the distal end of the photoreceptor ellipsoid.

Cyclic adenosine monophosphate (cAMP) – A nucleotide that is generated from adenosine triphosphate (ATP) in response to hormonal stimulation of cell surface receptors coupled to adenyl cyclase. cAMP acts as a signaling molecule by activating A-kinase; it is hydrolyzed to AMP by a phosphodiesterase.

Cytochalasin D – A drug that inhibits polymerization of actin monomers into actin filaments, and thus inhibits actin-dependent motility.

Dynein – A member of a family of large motor proteins that undergo ATP-dependent movement along microtubules toward the microtubule's slow-growing (minus) end.

Ellipsoid – The mitochondria-rich distal region of the photoreceptor inner segment between the myoid and the outer segment.

K_d – Dissociation constant, measure of the tendency of a complex to dissociate.

Kinesin – A member of a family of motor proteins that undergo ATP-dependent movement along microtubules toward the microtubule's fast-growing (plus) end.

Myoid – The neck-like region of the photoreceptor inner segment between the nucleus and the ellipsoid, so called because it elongates and contracts in lower vertebrates.

Polymerase chain reaction (PCR) – The technique for amplifying specific regions of DNA by multiple cycles of DNA polymerization, each followed by a brief heat treatment to separate complementary strands.

Sodium/potassium adenosine triphosphatase (Na^+/K^+ ATPase) – The sodium pump, transmembrane carrier protein found in the plasma membrane of most animal cells, which pumps Na^+ out of and K^+ into the cell, using the energy derived from ATP hydrolysis.

Nature and Occurrence

In fish, photoreceptors and retinal pigment epithelium (RPE) undergo retinomotor movements in response to changes in ambient light conditions and to circadian signals. Retinomotor movements include elongation and contraction of the myoid regions of rod and cone photoreceptors and migration of melanin pigment granules (melanosomes) within long apical projections of the RPE (**Figure 1**). In light, cone outer segments are positioned first in line to absorb incoming light and rod outer segments are buried in the dispersed shielding pigment granules of the RPE. In the dark, rods are positioned first in line and RPE pigment granules are aggregated, fully exposing the rod outer segments. Thus, the retina's entire light capture area is rod-dominated at night and cone-dominated in the day.

Cone length changes range from 15 to 80 μm; while rod length changes are generally more modest. Since rates of elongation and contraction are in the range of 1–4 μm min^{-1}, transitions between light-adapted and dark-adapted positions are slow, requiring as much as 60 min in some species. This slowness may explain why mammals evolved the more agile strategy of pupillary dilation for light adaptation.

As these dramatic morphological changes can be easily detected in light micrographs of sectioned retinas, many early studies of retinal morphology were focused on the properties and species distribution of retinomotor movements. After Kühne's initial paper in 1856, the late eighteenth and early nineteenth century produced a huge literature describing retinomotor movements in various vertebrate groups. Retinomotor movements are modest or nonexistent in mammals, urodeles, and reptiles, more robust and widespread in anurans and birds, and most extensive and widely distributed in fishes with duplex retinas. In the pure cone retinas of early fish larvae and in the pure rod retinas of benthic fishes, retinomotor movements are absent.

Retinomotor movements are regulated not only by light but also by an endogenous circadian rhythm. In animals maintained in constant darkness (DD), retinomotor movements persist at expected dawn and dusk. In constant illumination, the cycles of retinomotor movements are suppressed. Circadian cone retinomotor movements have been observed in most fish species studied;

*Adapted from Burnside, B. and King-Smith, C. (2009). Retinomotor movements. In: Squire, L. R. (ed.) *Encyclopedia of Neuroscience*. Oxford: Academic Press.

circadian rod and RPE retinomotor movements are considerably less common. In all species studied to date, cones have been found to contract to their light-adaptive position in anticipation of expected dawn. Thus, for fish in natural environments, a circadian signal, rather than light onset, is the natural trigger for light-adaptive retinomotor movements.

Mechanisms of Force Production for Retinomotor Movements

Many of the earliest studies of retinomotor movement grappled with whether retinomotor movements required energy or represented passive accommodations to other forces. More recent studies using isolated cells have demonstrated that rod, cone, and RPE retinomotor movements can each occur in the absence of other cell types; even in photoreceptor fragments consisting of inner/outer segments only. Thus, force production for photoreceptor elongation and contraction and for pigment migration appears to be locally autonomous for each cell type.

Force Production for Photoreceptor Elongation and Contraction

In photoreceptors, retinomotor movements are effected by cell shape change. Photoreceptors elongate and contract by changing the length of the photoreceptor inner segment, that is, the part of the photoreceptor between the photopigment-bearing outer segment and the junctional complexes of the outer limiting membrane, where photoreceptors are attached by adherens junctions to supportive Müller cells. The motile portions of photoreceptors protrude into the subretinal space between retina and RPE, where they interdigitate with the long apical projections of the RPE cells (**Figure 1**). The interphotoreceptor matrix filling the subretinal space thus lubricates, or at least permits, the shear produced by rods and cones moving in opposite directions. The photoreceptor inner segment is comprised of the distal ellipsoid, containing a mass of aggregated mitochondria, and the more proximal myoid, so named for its ability to undergo retinomotor contraction. Retinomotor shape change is primarily mediated by change in myoid length (**Figures 2** and **3**).

In most vertebrate cells, force production for cell shape change is generated by actin filaments or microtubles, or both. Rods and cones are highly asymmetric cells with highly polarized and paraxially aligned actin and microtubule cytoskeletons (**Figures 2** and **3**). From the distal end of the inner segment, microvillus-like calycal processes extend up around the base of the outer segment. Paraxial bundles of highly cross-linked actin filaments form the cores of the calycal processes, continue through the peripheral cytoplasm of the ellipsoid, and then splay

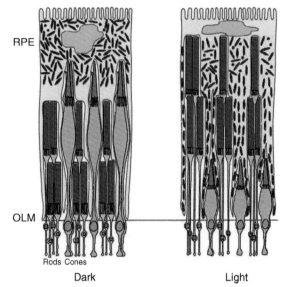

Figure 1   Retinomotor positions of rods, cones, and RPE pigment granules in light- and dark-adapted fish retinas. In dark-adapted retinas, cone myoids (purple) are long, rod myoids (pink) are short, and pigment granules (black) are aggregated to the base of the RPE cell. In light-adapted retinas, cone myoids (purple) are short, rod myoids (pink) are long, and pigment granules (black) have migrated into the long, interdigitating apical projections of the RPE. RPE, retinal pigment epithelium; OLM, outer limiting membrane.

out into a less tightly cross-linked, but still longitudinally oriented, actin filament array in the photoreceptor myoid. The calycal, ellipsoid, and myoid actin filaments are polarized and uniformly oriented with their fast-growing (plus) ends toward the outer segment. Since each type of myosin motor can walk in only one direction along an actin filament, the polarity of actin filaments determines the vector of force production for actin-based motility. Thus, a plus-end-directed myosin motor would walk along a photoreceptor inner segment actin bundle toward the outer segment, while a minus-end-directed motor would walk toward the nucleus. Of the more than 18 classes of myosins, only the class VI myosins have been found to be minus-end-directed motors. All other classes are plus-end directed. Myosins IIA, IIIA, IIIB, VI, VIIA, and IXB have all been shown to be present in fish photoreceptor inner segments. Thus, both plus- and minus-end-directed motors are present in or near the regions of elongation and contraction.

The microtubule cytoskeleton of the photoreceptor consists of two functionally distinct microtubule populations with different properties: the highly stable $(9+0)$ ciliary axoneme of the outer segment, and the more labile paraxially aligned cytoplasmic microtubules of the photoreceptor cell body. Relatively parallel microtubules populate the sleeve of cytoplasm surrounding the mitochondrial mass in the ellipsoid and extend throughout the myoid and axon. As with actin, microtubule polarity and the orientation of the microtubules within a cell play

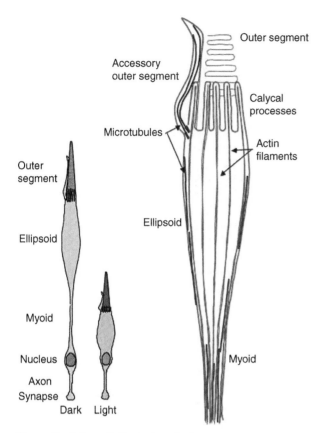

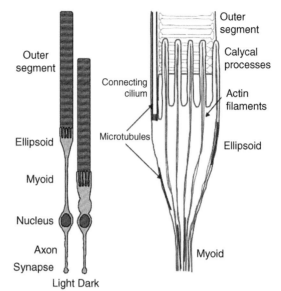

Figure 2 Schematic illustration of teleost cone retinomotor cell shape change and cytoskeleton. Retinomotor length change is mediated primarily by the cone's myoid region. Paraxially aligned actin filaments form the core bundles of cone calycal processes and course through the peripheral ellipsoid and the myoid. Cones possess two populations of microtubules: the axonemal microtubules of the accessory outer segment and the cytoplasmic microtubules of the ellipsoid, myoid, and axon.

Figure 3 Schematic illustration of teleost rod retinomotor cell shape change and cytoskeleton. Retinomotor length change is mediated primarily by the rod's myoid region. Paraxially aligned actin filaments form the core bundles of rod calycal processes, and course through the peripheral ellipsoid and the myoid. Rods possess two populations of microtubules: the axonemal microtubules of the connecting cilium and outer segment and the cytoplasmic microtubules of the ellipsoid, myoid, and axon.

critical roles in microtubule-based motility. Dynein motors move toward the microtubule minus end and most kinesins move toward the plus end. Both cytoplasmic dynein and kinesin II are found in photoreceptor inner segments. The polarity of the inner segment and axonal microtubules is uniform, with virtually all microtubules oriented with minus ends toward the outer segment. Thus, in a photoreceptor, a minus-end-directed microtubule motor, like dynein, would walk toward the outer segment, while a plus-end-directed microtubule motor, like kinesin, would walk toward the synapse.

Cytoskeletal inhibitors have been used to identify the roles of microtubules and actin filaments in teleost rod and cone retinomotor movements. In cones, dark-adaptive myoid elongation is microtubule dependent (and actin independent), while light-adaptive cone myoid contraction is actin dependent (and microtubule independent). The mechanisms of both microtubule-based elongation and actin-dependent cone myoid contraction are not established. Myosin IIA, a cytoplasmic myosin capable

of forming bipolar filaments, is concentrated in the myoids of trout and bass cones, suggesting that muscle-like sliding filaments of actin and myosin might be responsible for cone contraction. Since isolated cone inner/outer segment fragments undergo myoid contraction *in vitro*, it is clear that force production and regulatory machinery necessary for contraction both reside within the inner segment.

In rods, myoid elongation and contraction are both actin dependent and microtubule independent, although microtubules are present in the rod ellipsoid and myoid. Studies using the actin inhibitor cytochalasin D indicate that inhibition of elongation is produced at concentrations so low that actin assembly is blocked but preexisting actin filaments are not disrupted, suggesting that force production for rod elongation depends upon actin assembly. Since quantitative assessment of F-actin indicates that there is no net actin assembly in elongating rod inner outer segments, actin polymerization at the minus ends of filaments must be coupled with depolymerization at filament plus ends. The mechanism of rod contraction is less well studied. Since rod inner–outer segment fragments can both elongate and shorten in culture, it is clear that all necessary machinery is resident in the myoid/ellipsoid (**Figure 3**).

The RPE Cytoskeleton and Force Production for Pigment Granule Dispersion and Aggregation

Light-induced dispersion of melanin pigment granules (melanosomes) out into the apical projections of fish

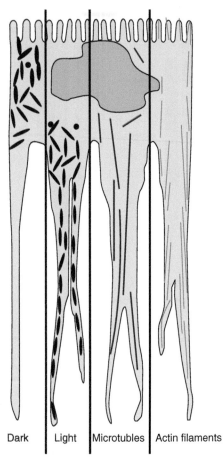

Dark | Light | Microtubles | Actin filaments

Figure 4 Schematic illustration of teleost RPE retinomotor pigment migration and cytoskeleton. Pigment granules aggregate into the cell body in the dark and disperse into apical RPE projections in the light. Both microtubules and actin filaments are found in the cell body and the apical projections.

RPE cells and dark-induced aggregation of granules back into the RPE cell body results from translocation of individual pigment granules within apical projections of the RPE cells, not from extension and retraction of the apical projections themselves. Apical projections extend from the RPE cell apical surface out into the subretinal space where they interdigitate with photoreceptor inner and outer segments. Thus, when the membrane-bound pigment granules are dispersed, they shield and protect rod outer segments from full bleach in bright light. RPE apical projections may be up to 100 μm long in some fish species and they remain extended throughout the light and dark phases of pigment granule migration.

Apical projections of fish RPE cells contain both actin filaments and microtubules. Microtubule disruption by intraocular injection of inhibitors *in vivo* blocks motility of pigment granules both in the cell body and in the apical projections; this finding would suggest that melanosome movement is microtubule dependent. However, when inhibitor studies are carried out using dissociated RPE

cells in culture, microtubule disruption has no effect on melanosome aggregation or dispersion. Presumably, this discrepancy results from the need for microtubules to provide structural support for apical projections, *in vivo*, or to maintain patent tracks or pathways through which pigment granules can move. In isolated cells, on the other hand, the attachment of the projections to the substrate likely keeps projections intact enough to permit pigment granule translocation. Isolated cell studies therefore suggest that microtubules are needed for the structural integrity of RPE apical projections, but the forces that actually power pigment granule translocation are microtubule independent.

In contrast, if the actin cytoskeleton is disrupted by cytochalasin D in dissociated RPE cells, both aggregation and dispersion of pigment granules are reversibly blocked, and maintenance of full aggregation or full dispersion is compromised. Thus, actin filaments are necessary and sufficient for both aggregation and dispersion. The actin dependence of pigment granule translocation in RPE cells suggests that myosin motor proteins are likely to play a role in force production. Polymerase chain reaction (PCR) screens of fish RPE have identified 11 distinct myosins expressed in RPE. Immunocytochemistry with antibodies to four of these myosins indicated that the fish RPE has abundant levels of myosins IIIA, VI, and IXb, and lower levels of myosin VIIa. Myosin VIIa has been immunolocalized on pigment granules in RPE of humans, rodents, and fish, and defective localization of pigment granules to the RPE apical projections in mice with a mutant myosin VIIA gene has been reported. *In vitro* motility assays have shown that isolated teleost RPE pigment granules do support plus-end-directed, actin-dependent motility.

A role for nonmuscle myosin II in RPE pigment granule movement has been implicated in studies using the myosin II inhibitor, blebbistatin, which partially blocked pigment granule aggregation in isolated sunfish RPE cells, but did not affect dispersion. Similarly, inhibitors of Rho guanosine triphosphate (GTP)-binding protein-activated kinase (ROCK), which is involved in myosin II activity, also blocked aggregation. These results suggest myosin II may contribute to pigment granule aggregation, possibly by contracting an actin network with which pigment granules are associated. Myosin IIa was identified in striped bass RPE by PCR amplification of a complementary DNA (cDNA) library.

Intracellular Regulation of Retinomotor Movements: The Role of Cyclic Adenosine Monophosphate

Numerous experimental observations demonstrate that cyclic adenosine monophosphate (cAMP) plays a critical role in the regulation of retinomotor movements in both photoreceptors and RPE. Studies using both intraocular injections *in vivo* and isolated retinas in culture have shown that a variety of treatments that elevate cAMP (but not cyclic guanosine monophosphate (cGMP)) produce retinomotor movements characteristic of darkness (night). Similar effects were obtained *in vitro* using isolated RPE cells, permeabilized cones, and isolated photoreceptor fragments (inner/outer segments), suggesting that increased cytoplasmic cAMP has a local and direct effect of triggering dark-adaptive retinomotor movements in each cell type. These findings are consistent with many other reports showing that elevated cAMP levels in the dark are a general feature of diurnal and circadian signaling cycles in vertebrate photoreceptors. For example, light and circadian control of photoreceptor melatonin release is closely correlated with changes in intracellular cAMP levels in photoreceptors.

Since inhibitors of cAMP-dependent kinase (protein kinase A (PKA)) block dark-adaptive retinomotor movements in both isolated cone and rod fragments and in intact retinas, it seems likely that cAMP regulates the photoreceptor cytoskeleton via protein phosphorylation. cAMP triggers elongation in detergent-permeabilized

cone models, further suggesting that PKA and its targets are detergent-insoluble, and thus likely to be physically associated with the cytoskeleton (or pigment granules). Pharmacological studies of rod inner–outer segment fragments suggest that PKA phosphorylation is enhanced in darkness and that dephosphorylation is required for light-induced elongation (**Figures 5** and **6**).

Regulation of Photoreceptor Retinomotor Movements by Paracrine Messenger

Dopamine

A role for paracrine signaling in light regulation of retinomotor movements is implicated by spectral sensitivity studies of cone retinomotor movements *in vivo*. The action spectrum for light-adaptive cone and RPE retinomotor movements most closely fits the absorption curve of the rod photopigment, suggesting that in cones and RPE, retinomotor movements are not directly triggered by light but are activated indirectly through a rod-mediated pathway. Two experimental observations demonstrate that the

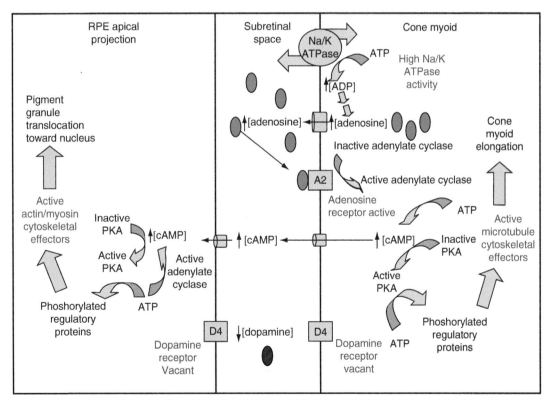

Figure 5 Model of first- and second-messenger signaling in teleost RPE, subretinal space, and cone myoid in darkness. The elevated dark activity of the cone Na/K ATPase elevates cytoplasmic adenosine, which exits through adenosine transporters to the subretinal space. Elevated extracellular adenosine enhances intracellular cAMP level via A2 receptors in cones but not in RPE. Extracellular dopamine levels are low. Elevated cone cytoplasmic cAMP triggers phosphorylation of regulatory proteins that activate microtubule-mediated cone elongation. Cone cytoplasmic cAMP also exits to the subretinal space through organic ion transporters in the cone plasma membrane. From the subretinal space cAMP enters the RPE cell through organic ion transporters, thereby elevating cAMP levels in the RPE cytoplasm and stimulating phosphorylation of regulatory proteins that activate actin-filament-mediated pigment aggregation.

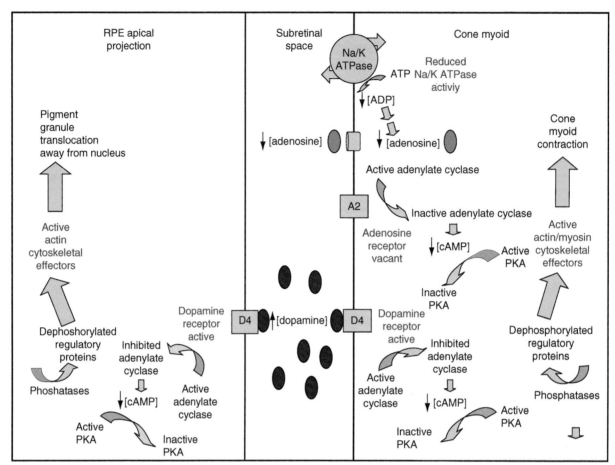

Figure 6   Model of first- and second-messenger signaling in teleost RPE, subretinal space, and cone myoid in the light. Reduced activity of the cone Na/K ATPase in the light reduces adenosine levels in the cone cytoplasm and subretinal space, thus decreasing cAMP production by cone A2 receptors. Increased dopamine release from interplexiform cells in the light produces elevated dopamine levels in the subretinal space. Dopamine activates D4 receptors on the cones and the RPE cells, thereby inhibiting adenylate cyclase and reducing intracellular cAMP in both cell types. Dephosphorylation of cytoskeletal regulatory proteins in cones inactivates microtubule-mediated cone elongation and triggers actin-filament-mediated contraction. Simultaneously, dephosphorylation of cytoskeletal regulatory proteins in RPE activates actin-filament-mediated pigment granule dispersion.

neuromodulator dopamine is a key player in this rod-mediated pathway:

1. intraocularly injected dopamine triggers light-adaptive cone contraction and RPE pigment granule dispersion in animals maintained in the dark, and
2. dopamine antagonists block light-induced cone contraction and RPE pigment granule dispersion in animals moved from darkness to light.

Further evidence that dopamine plays a central role in paracrine regulation of cone retinomotor movements is provided by pharmacological studies using intraocular injection *in vivo*, isolated retinas in culture, and isolated cone inner–outer segment fragments *in vitro*. In each of these preparations, light and/or dopamine promote cone contraction, whereas darkness and/or the adenylate cyclase stimulator, forskolin, promote cone elongation. Pharmacological profiles indicate that dopamine is acting

through D4-like or other D2 family dopamine receptors. This class of dopamine receptors causes inhibition of adenylate cyclase and, thus, would be expected to lower cone cytoplasmic cAMP levels, consistent with observations that experimentally increasing cAMP produces dark-adaptive movements.

The ability of light and dopamine to regulate retinomotor movement in isolated cone inner–outer segment fragments indicates that these agents can act directly on cones. Thus, light can trigger cone contraction in isolated cones, even though action spectra indicate that *in vivo* light acts through a rod-mediated pathway. These studies also indicate that D4 dopamine receptors are present on cone inner and/or outer segments. In teleost retinas, the source of dopamine is the interplexiform cell. Since projections of this inner retinal cell extend only into the outer plexiform layer, released dopamine would have to diffuse over tens of microns to reach receptors on the inner segments *in vivo*. The very high affinity of D4 receptors (dissociation constant (K_d) in the nM range) may facilitate this paracrine mode of signaling.

Light is much more effective than dopamine in triggering myoid elongation in rod fragments, and light and circadian regulation of rod movement was unaffected by treatment with 6-hydroxydopamine, which kills dopaminergic inter-plexiform cells. Light activation of myoid elongation in rod inner–outer segment fragments *in vitro* requires relatively high light intensities (20% bleach). The photoreceptive mechanism responsible for this activation can apparently count photons accurately for light pulse durations up to at least 10 min, suggesting that the critical factor in light acti-vation of rod elongation is quantum catch, rather than dura-tion of the light stimulus. Similar high-intensity thresholds and photon counting have been reported to mediate entrain-ment of circadian rhythms in several species.

The observation that dopamine plays a role in circa-dian as well as light regulation of cone retinomotor movement is suggested by results obtained by intraocular injection of dopamine agonists and antagonists in ani-mals maintained in DD. Cone myoid contraction can be induced at midnight by intraocular injection of dopamine or D4 receptor agonists. Partially contracted cones of DD animals at subjective midday can be induced to contract fully by intraocular injection of dopamine or D4 agonists, or to elongate by injection of the D4/D2 family antago-nists. The predawn cone contraction observed in DD animals in response to circadian signals can be completely eliminated by intraocular injection of the D4/D2 antago-nist shortly before the expected time of light onset. These observations suggest that circadian regulation of cone myoid length is mediated by endogenous dopamine, act-ing through D4/D2 family receptors.

Consistent with these findings, retinal dopamine release is higher in the light than in the dark, and dopamine release is higher in subjective day than in subjective night in DD in many species. These diurnal and circadian cycles of dopa-mine release are inversely correlated with diurnal and cir-cadian cycles of cAMP and protein phosphorylation levels in photoreceptors. Recent findings suggest that dopaminer-gic amacrine cells contain an autonomous circadian clock that drives dopamine release and metabolism. Although dopamine plays a critical role in regulating fish retinomotor movements, it is nonetheless unlikely to be the sole circa-dian regulator, since both light-induced and circadian cone movements persist (at reduced amplitude) in fish retinas after lesion of interplexiform cells by 6-hydroxydopamine.

Adenosine

This neuromodulator has an opposite effect to that of dopamine on cone retinomotor movements. In isolated cone inner–outer segment fragments, adenosine A2 recep-tor agonists activate and A2 antagonists inhibit cone myoid

elongation. Since A2 receptors are positively coupled to adenylate cyclase, adenosine effects on cone fragments are consistent with observations that elevating cAMP triggers dark-adaptive retinomotor movements. These findings indicate that A2 receptors must be present on cone inner (and/or outer) segments. Since cones elongate at subjective dusk in fish maintained in DD, and since adenosine stimu-lates cone elongation, adenosine could provide an endoge-nous circadian signal for expected dark onset in addition to the decrease in dopamine release occurring at that time.

Adenosine effects on cone movements could be a local effect, reflecting photoreceptor metabolic activity. Since an adenosine transporter is present in photoreceptors, increases in adenosine in photoreceptor inner segments are likely to be accompanied by an increase in adenosine level in the subretinal space. The increased sodium/potassium adenosine triphosphatase (Na^+/K^+-ATPase) activity in photoreceptor inner segments occurring in darkness is likely to be accompanied by an increase in intracellular adenosine levels, and consequently, increased release of adenosine into the subretinal space. The binding of released adenosine to photoreceptor A2 receptors might then enhance dark adaptation of the photoreceptors by stimulating adenylate cyclase, thereby reinforcing the dark signal. Other studies have shown that adenosine tri-phosphate (ATP) is released across the apical membrane of the RPE into the subretinal space in response to various triggers. This ATP is dephosphorylated into adenosine by extracellular enzymes (ectoenzymes) on the RPE apical membrane. Regulation of adenosine release and ectoenzme activity in response to light signals could also alter the balance of purines in subretinal space and, thus, influence retinomotor movements in cones.

Regulation of Retinomotor Movements in RPE Cells by Paracrine Messengers

RPE pigment migration in frogs has an action spectrum most closely resembling that of the rod photopigment. These observations suggest that for RPE, as for cones, a rod-mediated pathway triggers light-adaptive retinomo-tor movement. Pigment position is not affected by light in isolated sheets of teleost RPE, further demonstrating that light acts indirectly through a photoreceptor-mediated, paracrine pathway.

In isolated sheets of teleost RPE in culture, dopamine and D2 family agonists induce pigment granule disper-sion, while treatments that elevate cAMP induce pigment granule aggregation. In isolated RPE cells aggregated in cAMP, pigment dispersion is induced by microinjection of PKA inhibitors, suggesting that continuous phosphoryla-tion of PKA targets is required to maintain the aggregated state. Since aggregation can also be induced by the phos-phatase inhibitor okadaic acid, it seems likely that PKA and phosphatases are simultaneously active in RPE cells, and that their relative activities are altered by light and dark signals from the retina.

Surprisingly, underivatized cAMP is just as effective as membrane-permeant cAMP analogs at activating pigment granule aggregation in isolated RPE sheets. Washout of exogenous cAMP induces dispersion. This cAMP directly enters the RPE cell through organic ion transporters is suggested by observations that ATP and adenosine are ineffective at triggering aggregation and that organic ion transport inhibitors block cAMP – but not forskolin-induced pigment aggregation. Recently, it has been shown that isolated RPE takes up cAMP in a saturable manner. Thus, it seems clear that cAMP from the subretinal space can actually enter RPE cells through organic ion channels to activate pigment aggregation. These findings indicate that an increase in cAMP in the subretinal space in the dark would activate pigment granule aggregation in RPE cells. cAMP efflux has been shown to be associated with increased intracellular cAMP accumulation in many cell types, including pinealocytes. Thus, diurnal cycles of cAMP levels in retinal photoreceptors might be expected to produce increased cAMP efflux into the subretinal space at night. This extracellular cAMP in the subretinal space could then function as a retina-to-RPE signal for darkness.

cAMP and intracellular calcium often interact in cellular signaling pathways, either antagonistically or synergistically. Pigment granule migration is regulated by calcium in several types of dermal chromatophores. However, increasing extracellular or intracellular calcium has no effect on pigment granule position in either aggregated or dispersed isolated RPE cells. Furthermore, intracellular calcium levels are unaffected when RPE pigment granule motility is triggered by cAMP or cAMP washout. These findings suggest that RPE pigment granule movements are regulated directly by cAMP, that is, by phosphorylation and dephosphorylation of PKA targets, and that calcium does not act downstream of cAMP in the regulatory process.

Carbachol (an acetylcholine analog) triggers dispersion in isolated fish RPE. The carbachol receptor acts through a pathway commonly linked to calcium mobilization and carbachol-induced pigment granule dispersion is blocked by the calcium chelator 1,2-bis(*o*-aminophenoxy)ethane-*N,N,N',N'*-tetraacetic acid (BAPTA), suggesting that a rise in intracellular calcium plays a role upstream of cAMP in carbachol-induced dispersion.

Functions and Significance of Retinomotor Movements

What role retinomotor movements play in retinal function has been a topic of inquiry and speculation for more than a century. Direct experimental demonstration that retinomotor movements affect vision has been difficult to achieve, primarily because of the difficulty in interfering with retinomotor movements without compromising other aspects of retinal function. Nevertheless, several likely functions for retinomotor movements have been suggested. Occurring generally in species that lack pupillary movements, rod and RPE retinomotor movements provide an alternative mechanism to pupillary movements for shielding rods from full bleach in bright light while permitting optimal exposure for dim-light vision. Repositioning the rod and cone photoreceptors likewise provides an efficient mechanism for optimizing space by positioning the rods first in line to detect light at the focal plane across the entire retinal surface under dim-light conditions, and then moving cones to this position for bright-light vision. These movements permit the entire retinal surface to be used alternately for rod and cone vision. Some have noted that each photoreceptor type is elongated when it is expected to be silent, and suggested that the cable properties of the elongated myoid might contribute to attenuation of the signal from outer segment to synapse. The great morphological variation in retinomotor movements perhaps reflects the optimization of one or the other of these functions in different species.

Summary

Although retinomotor movements are most pronounced in lower vertebrates, their study contributes to a broader understanding of diurnal and circadian regulation of photoreceptor and RPE physiology. Retinomotor movements provide excellent models for investigating the roles of cytoskeletal elements in cell shape change and intracellular transport. Studies of retinomotor movements have called attention to the importance of cyclic changes in intracellular cAMP levels in the diurnal and circadian regulation of other aspects of photoreceptor physiology and metabolism. The recognition that light induction of cone and RPE movements depends on a rod-mediated paracrine pathway has contributed to recognition of the role that dopamine plays in light and circadian signaling. Since retinomotor movements exhibit properties of light and circadian regulation, characteristic of many other aspects of retinal physiology, they provide excellent experimental models for understanding light and circadian-regulatory processes.

See also: Circadian Metabolism in the Chick Retina; The Circadian Clock in the Retina Regulates Rod and Cone Pathways; Circadian Regulation of Ion Channels in Photoreceptors; *Limulus* Eyes and Their Circadian Regulation; Neurotransmitters and Receptors: Dopamine Receptors.

Further Reading

Ali, M. A. (1975). Retinomotor responses. In: Ali, M. A. (ed.) *Vision in Fishes*, pp. 313–355. New York: Plenum Press.

Arey, L. B. (1915). The occurrence and significance of photomechanical changes in the vertebrate retina – a historical survey. *Journal of Comparative Neurology* 25: 535–554.

Back, I., Donner, K. O., and Reuter, T. (1965). The screening effect of the pigment epithelium on the retinal rods in the frog. *Vision Research* 5: 101–111.

Blaxter, J. H. S. (1975). The eyes of larval fish. In: Ali, M. A. (ed.) *Vision in Fishes*, pp. 427–443. New York: Plenum Press.

Burnside, B. (1988). Photoreceptor contraction and extension: Calcium and cAMP regulation of microtubule- and actin-dependent changes in cell shape. In: Lasek, R. J. (ed.) *Intrinsic Determinants of Neuronal Cell Form*, pp. 323–359. New York: Alan R. Liss.

Burnside, B. (2001). Light and circadian regulation of retinomotor movement. *Progress in Brain Research* 131: 477–485.

Burnside, B. and Dearry, A. (1986). Cell motility in the retina. In: Alder, R. and Farber, D. B. (eds.) *The Retina*, Part 1, pp. 151–206. New York: Academic Press.

Burnside, B. and King-Smith, C. (2009). Retinomotor movements. In: Squire, L. R. (ed.) *Encyclopedia of Neuroscience*. Oxford: Academic Press. http://www.sciencedirect.com/science/referenceworks/9780080450469 (accessed July 2009).

Burnside, B. and Nagle, B. (1983). Retinomotor movements of photoreceptors and retinal pigment epithelium: Mechanisms and regulation. *Progress in Retinal Research* 2: 67–109.

Dearry, A. and Burnside, B. (1989). Regulation of cell motility in teleost retinal photoreceptors and pigment epithelium by dopaminergic D2 Receptors. In: Redburn, D. and Morales, H. P. (eds.) *Extracellular and Intracellular Messengers in the Vertebrate Retina*, pp. 229–256. New York: Alan R. Liss.

Douglas, R. H. (1982). The function of photomechanical movements in the retina of the rainbow trout (*Salmo gairdneri*). *Journal of Experimental Biology* 96: 389–403.

McNeil, E. L., Tacelosky, D., Basciano, P., et al. (2004). Actin-dependent motility of melanosomes from fish retinal pigment epithelial (RPE) cells investigated using *in vitro* motility assays. *Cell Motility and the Cytoskeleton* 58: 71–82.

Pozdeyev, N., Tosini, G., Li, L., et al. (2008). Dopamine modulates diurnal and circadian rhythms of protein phosphorylation in photoreceptor cells of mouse retina. *European Journal of Neuroscience* 27: 2691–2700.

Wagner, H. J., Kirsch, M., and Douglas, R. H. (1992). Light dependent and endogenous circadian control of adaptation in teleost retinae. In: Ali, M. A. (ed.) *Rhythms in Fishes*, pp. 255–292. New York: Plenum Press.

GABA Receptors in the Retina

S Yazulla, Stony Brook University, Stony Brook, NY, USA

Glossary

Disinhibition – A synaptic interaction in which the inhibitory input to a neuron is itself inhibited, thereby relieving the neuron of inhibitory control.

Endocannabinoids – Natural chemicals in the body whose actions on metabotropic cannabinoid receptors are mimicked by the active component of marijuana.

Ionotropic receptors – Membrane proteins that, when activated by specific ligands, directly alter membrane conductance.

IPSC – Inhibitory post synaptic current counteracts excitatory input to a neuron, for example by hyperpolarization or induction of a shunt current.

Metabotropic receptors – Membrane proteins that, when activated by specific ligands, indirectly alter a wide variety of cellular properties through G-protein-coupled enzyme cascades.

Introduction

Gamma-aminobutyric acid (GABA) is the major inhibitory amino acid transmitter in the retina. It is overwhelmingly represented in lateral inhibition, being most prominent in one class of horizontal cell and in numerous subtypes of amacrine cell. GABAergic transmission requires the synthesizing enzyme glutamic acid decarboxylase (GAD) and the degradative enzyme GABA transaminase (GABA-T) that may or may not be found in the same cells. The physiological actions of GABA are effected by receptors that may be broadly defined as proteins that bind to and respond to the presence of GABA. Under this scheme, there are three functional types of GABA receptor: (1) a vesicular transporter that concentrates cytoplasmic GABA into synaptic vesicles, (2) a membrane transporter that translocates GABA from the extracellular space into glia or neurons, and (3) plasma membrane receptors that mediate the cell's response to synaptically released GABA. This article highlights the properties and functions of these major types of the GABA receptor. Numerous sources on the history, physiology, and molecular biology of these receptor types are provided. **Table 1** illustrates the major types of the GABA receptor with representative agonists, antagonists, and most common

cellular locations in the outer plexiform layer (OPL) and inner plexiform layer (IPL). The pharmacology of GABA transporters (GATs) is less well developed than it is for the synaptic receptors and continues to be an area of intensive research.

Vesicular Transporters

Cytoplasmic GABA is concentrated into synaptic vesicles by a vesicular inhibitory amino acid transporter (VIAAT) that uses H^+-antiport activity to drive the uptake of GABA or glycine into synaptic vesicles. VIAAT is the only member of the solute carrier 32 (SLC32) family of H^+-coupled amino acid transporters; it is not related to the vesicular transporters for glutamate, monoamines, or acetylcholine. VIAAT, referred to as the vesicular GAT (VGAT), when applied to GABAergic neurons, is essential for the vesicular release of GABA and glycine. The existence of a common vesicular transporter for GABA and glycine is consistent with reports of a small percentage of amacrine cells that co-localize and likely release both GABA and glycine.

VIAAT was localized by immunohistochemistry (IHC) in zebrafish, mouse, rat, cat, and human retinas. In the inner retina, synaptic boutons throughout all layers of the IPL contain VIAAT immunoreactivity (IR), consistent with data showing that virtually all amacrine cells in the retina are either GABAergic or glycinergic. Data from salamander, cat, and human suggest the existence of subpopulations of bipolar cells that likely release both GABA and glutamate. For example, certain OFF cone bipolar cells in the cat retina contain not only VIAAT-IR and GAD65-IR, but also vesicular glutamate transporter-IR (VGLUT-IR). In the outer retina, VIAAT-IR is present in horizontal cell dendrites in zebrafish, mouse, rat, and human. The presence of $GABA_A$ receptors on the photoreceptor terminals of some mammals supports the notion that GABA is synaptically released in the OPL. VIAAT-IR in the OPL most likely includes not only the horizontal cell dendrites, but also the boutons of GABAergic and glycinergic interplexiform cells. These processes, however, have a relatively low density and are unlikely to be confused with the more numerous horizontal cell dendrites. Electron microscopy shows that VIAAT-immunoreactive horizontal cell dendrites innervate rods and cones in mouse and rat, suggesting that A-type and B-type horizontal cells of the mammalian retina are capable of vesicular GABA release. In fish horizontal cells, the subcellular

Table 1 GABA receptor types, agonists, antagonists, and localization in the retina

Synaptic receptors		Agonist	Antagonist	Location OPL	Location IPL
	GABA$_A$	Muscimol	Bicuculline	Cones	Amacrine cells
		Isoguavacine	Picrotoxin	Horizontal cells	Cone bipolar cells
		THIP	SR 95531	Bipolar cells	Ganglion cells
		TACA			Mueller's Cells
	α1–3,5βγ2 subunits	Benzodiazapines	Flumazenil	Horizontal cells	Amacrine and ganglion cells
	GABA$_B$	(R)-Baclofen	CGP 35348		Presynatic amacrine
		SKF 9751	SCH 50911		Postsynaptic ganglion
	GABA$_C$	TACA	Picrotoxin	Cones	Rod bipolar cells (mammals)
		CACA (partial)	Isoguavacine		Mixed bipolar cells (fish)
		Muscimol (partial)	TPMPA		
Transporter (plasma)		Uptake substrate	Uptake inhibitor		
	GAT-1	GABA	CI 966		Amacrine cells (major)
			NO 711		Mueller's cells (minor)
			(R)-tiagabine		
			SKF 89976A		
	GAT-2	GABA	EF1502	Horizontal cells (fish)	
		β-Alanine	Nipecotic Acid		
	GAT-3	GABA	Nipecotic Acid	Horizontal cells (fish)	Mueller's cells
		β-Alanine	(S)-SNAP 5114		
	BGT-1	GABA	(S)-SNAP 5114		
		Betaine			
Transporter (vesicular)	VIAAT (VGAT)	GABA	Vigabatrin Nipecotic Acid	Horizontal cells	Amacrine cells Bipolar cells (subset)

This listing is representative and not meant to be exhaustive regarding GABA receptor pharmacology. The most common sites for localization are listed to provide a general framework. Species-specific exceptions are listed in the text.

localization of VIAAT-IR has not been determined by electron microscopy; there are no data to indicate whether VIAAT-IR is restricted to the H1 GABAergic horizontal cells, innervating only cones, or distributed among all four types, innervating both rods and cones.

In the outer retina of mammals and nonmammals, the classification of neurons as GABAergic can be difficult. The presence of VIAAT alone does not necessarily indicate the presence of GABA for vesicular release. Other indicators, such as the presence of glutamate decarboxylase (GAD), the enzyme that catalyzes decarboxylation of glutamate to form GABA, or GABA uptake, are needed to support a GABAergic identity. For example, in most nonmammalian species, VIAAT-positive horizontal cell dendrites do not contain synaptic vesicles or synaptic specializations. Rather, a sodium-dependent GAT appears to facilitate GABA release from horizontal cells. To complicate matters further, the presence of GABA or GAD in an outer retinal neuron does not necessarily stipulate that GABA is synaptically released. For example, although GAD-IR and GABA-IR have been found in cone terminals in primate, lizard, and toad retinas, VIAAT-IR has

not been localized to photoreceptor terminals in any species. Lastly, although horizontal cells of some mammals contain VIAAT-IR, GABA-IR, and GAD-IR, neither GABA uptake nor GATs have been described in the horizontal cells of any mammal.

Plasma Membrane Transporters

GATs are members of the SLC6 family that belongs to the Na$^+$/Cl$^-$-dependent neurotransmitter transporter superfamily. GATs are arranged in 12 transmembrane domains. Molecular cloning studies identified three subtypes of high-affinity GAT (GAT-1, GAT-2, and GAT-3) and one lower-affinity betaine/glycine transporter that also transports GABA (BGT-1). GATs are responsible for the clearance of GABA from the extracellular space and, conceivably, could be present on the presynaptic neuron, postsynaptic neuron, or surrounding glia. In general, GAT-1 is present in neurons of the inner retina, GAT-3 in Müller's cells, and GAT-2 in the retinal pigmented epithelium (RPE) and ciliary epithelium of the

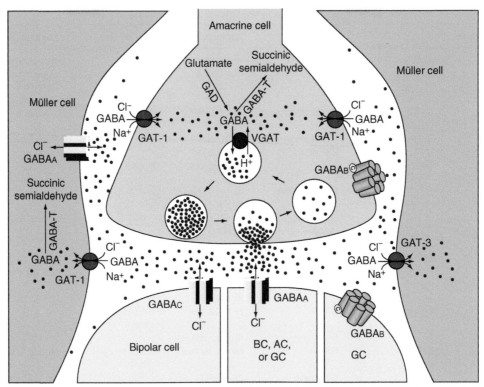

Figure 1 Schematic representation of GABA metabolism, receptor type, and localization in the inner retina. Receptor and transporter types have been placed in the most common locations. See text for details and listing of species-specific exceptions. GAD, glutamic acid decarboxylase; GABA-T, GABA transaminase; AC, amacrine cell; BC, bipolar cell; GC, ganglion cell.

mammalian retina (**Figure 1**). Overall, there is excellent agreement in the cellular distributions of ^{3}H-GABA uptake, GABA-T, which degrades GABA, and GATs in amacrine cells across species. However, in regard to horizontal cells and Müller's cells, there are notable species differences and inconsistencies.

Neuronal Localization

In a wide variety of mammals and nonmammals, from humans to teleost fish, GAT-1 is the predominant GAT/type in amacrine cells, displaced amacrine cells, and interplexiform cells. This distribution is consistent with the patterns of GAD-IR, GABA-IR, and ^{3}H-GABA uptake. Mammalian Müller's cells are labeled intensely with GAT-3 and, to a lesser extent, by GAT-1, consistent with observations that Müller's cells contain high levels of GABA-T and are the initial site for ^{3}H-GABA uptake. For the most part, nonmammalian Müller's cells neither take up ^{3}H-GABA nor display GAD-IR or GABA-IR. One exception is the skate retina, in which Müller's cells express GAT-1 and GAT-3, display a GAT current, and take up ^{3}H-GABA. Another exception is the salmon retina, in which Müller's cells have GABA-IR, but do not contain GAT-1. Cultured chick Müller's cells contain GAT-1 and GAT-3, while bullfrog Müller's cells contain GAT-1

and GAT-2 and display a GAT current. The implication of these findings in chick and bullfrog is not clear, given the absence of supporting evidence from ^{3}H-GABA uptake or GABA-T immunohistochemical studies. Given the high density of GAT on chick and bullfrog Müller's cells, including the end feet at the inner margin of the retina, it is surprising that ^{3}H-GABA uptake has never been demonstrated in Müller's cells of these species, regardless of how the ^{3}H-GABA was administered.

Although horizontal cells in many mammals contain GABA-IR and GAD-IR, they do not take up ^{3}H-GABA. As would be expected, GATs have not been localized to any mammalian horizontal cells. The evidence for a class of GABAergic horizontal cell is overwhelming for nonmammals. Horizontal cells of fish, amphibians, birds, and reptiles take up ^{3}H-GABA with great avidity. GAT-1, present in the inner retina, has not been localized to horizontal cells of any species. Studies on the localization of GAT-2 and GAT-3 in nonmammalian horizontal cells are sporadic, with studies usually focusing on one or the other. Exceptions are goldfish and zebrafish in which GAT-2 and GAT-3 were localized to a subtype of the cone horizontal cell, the H1 luminosity type. GAT-3 was not found in horizontal cells of skate, salamander, or bullfrog. GAT-2, absent in bullfrog horizontal cells, was not studied in skate, salamander, or salmon.

Except for goldfish and zebrafish, the lack of horizontal cell labeling by GAT in other nonmammals could be due to the general use of antibodies against the C-terminus of rat GAT-2 and GAT-3. There is a strong homology in the GAT amino acid sequence in the animal kingdom. Torpedo GAT-3 shows a 77% identical amino acid sequence with the mammalian GAT-3. Immunoblots of the goldfish brain with GAT-3 show a band between 60 and 75 kDa, consistent with the expected weight of 71 kDa. However, regardless of the overall homology or the sequence homology at the C-terminus, the site of antibody production is more critical in determining the validity of the labeling. Except for fish, the negative finding of GAT-2 and GAT-3 on horizontal cells in nonmammals has not been resolved. A curiosity is that ^{3}H-muscimol, a GABA$_A$ agonist, is transported avidly by amacrine cells in mammals and nonmammals, but not to any great extent by horizontal cells in any species, including fish and bird. This is a further indication of the difference in the type of GAT present on amacrine cells and nonmammalian horizontal cells.

A small percentage ($\sim$10%) of bipolar cells in some retinas may be GABAergic. In salamander, the vast majority of bipolar cells that contain GAT-1-IR also contain GABA-IR, as well as VGAT and GAD. Two types of OFF bipolar cell in zebrafish contain GABA and likely correspond to the OFF bipolar cells that display a GAT current. However, in primate retina, the variable reports of GABA-IR in bipolar cells are not supported by data localizing GAD, GAT-1, GAT-2, GAT-3, or GABA-T. These findings are significant because of the possibility that some bipolar cells release GABA and glutamate as neurotransmitters.

A GABA/taurine transporter was identified in bullfrog at the apical surface of the RPE. It was suggested that the RPE could take over the clearance of GABA released from horizontal cells in the distal retina of nonmammals because there are no reports that Müller's cells transport ^{3}H-GABA. Despite the reports of GATs in bullfrog Müller's cells, this seems to be an attractive idea that would be supported by the demonstration of GABA-T localization and GABA-T activity in the nonmammalian RPE.

Function

The major function of the GATs is to clear GABA from the extracellular space to either re-enter the vesicular pool or enter the tricarboxylic acid cycle after GABA-T action (**Figure 1**). There are at least three other consequences of the transporter activity. The first is that GABA transport is electrogenic with two Na^{2+}, one Cl$^-$, and one GABA per cycle, resulting in one-net inward positive charge. The depolarization that accompanies GABA uptake could modulate neuronal excitability by triggering Ca^{2+} influx and the subsequent release of Ca^{2+} from intracellular stores. For example, a nonsubstrate blocker of GABA transport

(NO-711) reduced cellular edema, presumably by blocking the ionic influx that accompanies GABA uptake. This strategy has proven useful in treating seizures and could apply to protecting retinal ganglion cells following ischemia or an excitotoxic release of glutamate from bipolar cells. As a second consequence of transporter activity, neuronal depolarization or a breakdown of the Na$^+$/Cl$^-$/GABA gradient can cause the GATs to operate in the reverse direction, thus releasing GABA. For example, an excitotoxic assault on the chick retina increases extracellular GABA by reversal of the GAT. This calcium-independent GABA release has been suggested to be the normal mode of release in horizontal cells, glia, and, perhaps, in starburst amacrine cells. This form of GABA release does not require adenosine triphosphate (ATP), except to maintain the ionic gradients. Third, the activity of GATs can limit the spillover of GABA in the extracellular space and regulate inhibition. This has been shown in the inner retina, in which inhibition of GATs by NO-711 enhances the light-evoked inhibitory postsynaptic currents (IPSCs) at GABA$_C$ receptors on bipolar cells, but has no effect on GABA$_A$ receptors on ganglion cells. The difference in response is due to the observation that GABA$_A$ receptors desensitize, whereas GABA$_C$ receptors do not. Thus, the GAT selectively modulated the transmission of signals from bipolar cells to ganglion cells. In all these cases, the effect of GAT activity is to regulate excitability.

Synaptic Receptors

There are three general classes of synaptic GABA receptors: two ionotropic and one metabotropic. These are differentiated by ligand-binding affinities and molecular-cloning techniques. GABA$_A$ receptors are heteropentameric structures that form a chloride channel using structurally related subunits α1–6, β1–4, γ1–3, δ, ε, and π. The most common pentamers are composed of $\alpha\beta\gamma$ subunits, with δ often substituting for γ. The subunit composition determines receptor pharmacology and kinetics. For example, GABA$_A$ receptors mediate the effects of benzodiazepines (BDZ), and the γ subunit is necessary for BZD sensitivity, while the type of α subunit determines BZD affinity. The β subunit affects channel properties and BZD efficacy. Substitution of an α4 or α6 subunit for an α1 subunit eliminates BZD sensitivity as does substitution of the δ subunit for a γ subunit. GABA$_A$ receptors are blocked by bicuculline, the nonselective chloride channel blocker, picrotoxin, and can be modulated allosterically by barbiturates, BZDs, ethanol, and steroids. The action of the GABA$_A$ receptor tends to be phasic. The ionotropic GABA$_C$ receptor also is a chloride channel that is composed mostly of homoligomeric ρ subunits (ρ_1, ρ_2, and ρ_3). GABA$_C$ receptors are insensitive to bicuculline, more sensitive to GABA than GABA$_A$ receptors, and can also be blocked by picrotoxin.

$GABA_C$ receptors do not desensitize; therefore, their effect is tonic. Metabotropic $GABA_B$ receptors are guanine nucleotide-binding protein-coupled receptors and regulate potassium or calcium channels. $GABA_B$ receptors are activated by baclofen, antagonized by phaclofen, insensitive to bicuculline and picrotoxin blockade, and less sensitive to GABA than $GABA_A$ receptors.

Photoreceptors

The source of GABA in the OPL of nonmammals is a type of horizontal cell, so-called H1 or luminosity cells. In mammals, GABA can be released from a type of interplexiform cell as well as from horizontal cells. $GABA_A$ receptors were localized to the OPL by *in vitro* autoradiography (ARG) of ^3H muscimol binding, $GABA_A$ receptor subunit-specific IHC, and single-cell electrophysiology. ^3H-BZD-binding sites were localized by *in vitro* ARG in the OPL of rat, monkey, and human, but were not present in the OPL of fish, salamander, or bird. There is evidence for $GABA_A$ receptors on rod terminals in goldfish and rat, and on the cone terminals in goldfish, tiger salamander, bullfrog, turtle, chicken, rat, mouse, pig, and cat. Most evidence shows $\beta2/3$ subunits on photoreceptor terminals. Data for α subunits are inconsistent, with negative results from *in situ* hybridization and positive results with IHC. In contrast to the localization of $\gamma1$ and $\gamma2$ subunits to salamander cone terminals by IHC, negative results with ^3H-BZD binding indicate an absence of γ subunits. This discrepancy could be due either to the nonspecific nature of IHC using antibodies that are not against salamander γ subunits or to low sensitivity of *in vitro* ARG using photoaffinity labeling of BZD ligands.

GABA-induced responses in cones of all species are reduced by bicuculline, indicating $GABA_A$ receptors. In addition, there is a component of GABA-induced responses in mammalian (pig and mouse) cones that is resistant to bicuculline, but sensitive to antagonists of $GABA_C$ receptors, indicating participation of $GABA_A$ and $GABA_C$ receptors. $GABA_C$ receptors have not been identified on photoreceptors in nonmammals as yet. In mouse, both $GABA_A$ and $GABA_C$ components of the response to GABA are potentiated by pentobarbital, suggesting heteromeric channels of $GABA_C$ $\rho1$ and $GABA_A$ receptor subunits. In bullfrog, $GABA_B$ receptors contribute to the GABA response along with $GABA_A$ receptors. GABAergic negative feedback onto cones is well established and involves a suppression of presynaptic Ca^{2+} currents. The function of GABA feedback onto photoreceptors is still controversial; it may have limited participation in the formation of the cone receptive field surround. More likely, the modulation of voltage-gated Ca^{2+} currents could set the gain at the cone synapse to maintain sensitivity to changing light conditions. The nondesensitizing nature of $GABA_C$ receptors on mammalian cones may compensate for the lower level of GABA release that is available from horizontal cells and interplexiform cells in the OPL of mammals compared to the high GABA content of H1 horizontal cells in nonmammals.

Horizontal Cells

The information regarding GABA receptors on horizontal cells is based primarily on data obtained with electrophysiology and, to a lesser extent, on immunohistochemical evidence. In addition, there is considerable species variability. $GABA_A$ and $GABA_B$ receptors have been localized by IHC on horizontal cells of some mammals and nonmammals, while $GABA_C$ receptors have been localized to horizontal cells of perch and goldfish. $GABA_B$ receptor-mediated responses have not been reported for horizontal cells in any species. $GABA_A$ responses have been recorded from horizontal cells in catfish, salamander, mouse, rabbit, and cat, while $GABA_C$ responses are prominent in horizontal cells in fish (goldfish, perch, and catfish) and salamander. Skate and zebrafish horizontal cells reportedly show neither $GABA_A$ nor $GABA_C$ responses. In goldfish, $GABA_A$ responses are present in isolated chromaticity horizontal cells, while $GABA_C$ responses are present in the cell body and axon terminal of the GABAergic H1 horizontal cells. In contrast, in perch, $GABA_C$ responses are restricted to the rod horizontal cells. The $GABA_A$ response in isolated mouse horizontal cells is enhanced by the BZD, diazepam, and pentobarbital, consistent with the localization of ^3H-BZD-binding sites in the mammalian OPL by ARG. $GABA_C$ receptor-mediated responses of GABAergic horizontal cells are complicated by the presence of the electrogenic GAT current. H1 horizontal cells release GABA at some steady rate under ambient illumination. They are depolarized by decrements in light intensity, resulting in increased GABA release by the transporter. During the hyperpolarizing response to light onset, the reuptake of GABA will depolarize the horizontal cells, reducing the light response and establishing a positive-feedback loop that allows for further GABA release from the H1 horizontal cells. In turn, the released GABA, via $GABA_A$ receptors, will depolarize the other cone and rod horizontal cells. In addition, the $GABA_C$ response shows a run-up with repetitive application of GABA, further depolarizing the horizontal cells. However, $GABA_A$ and $GABA_C$ receptors are suppressed by Zn^{2+} that is co-released with glutamate. This dampening effect by Zn^{2+} is especially important to downregulate the tonic $GABA_C$ response.

Bipolar Cells

Bipolar cells form functionally diverse groups that are differentiated on the bases of rod and cone input as well as ON or OFF response. On anatomical grounds, bipolar cells can receive GABAergic input from horizontal cells and interplexiform cells in the OPL and from a multitude of amacrine cell types in the IPL. Species differences among mammals and nonmammals and the effects of dissociation techniques on GABA receptors have resulted in considerable variability in the data. However, GABA$_A$ receptors are generally more prominent on dendrites in the OPL and the ratio of GABA$_A$ to GABA$_C$ receptors on axon terminals in the IPL is high for cone bipolar cells and low for rod bipolar cells. In mammals, IR to GABA$_A$ receptors (α1, β2/3, and γ2) and the GABA$_C$ ρ subunit is present at nonoverlapping, extrasynaptic sites on the dendrites of rod and cone bipolar cells. The contribution of GABA$_A$ and GABA$_C$ receptors to GABA-evoked IPSCs is balanced in dendrites of isolated mouse rod bipolar cells, whereas, in the ferret retinal slice, GABA$_A$ receptors overwhelmingly dominate the response to GABA puffed at dendrites in all types of the bipolar cell. Dendrites of bipolar cells are insensitive to GABA in salamander and display GABA$_A$ properties in goldfish and GABA$_C$ properties in zebrafish. In bullfrog, all bipolar cells show GABA$_A$ and GABA$_C$ sensitivity at dendrites, though GABA$_A$ receptors dominated, particularly for on bipolar cells.

In the IPL, GABA-evoked IPSCs are carried mostly by tonic GABA$_C$ receptors in rod bipolar cells and by phasic GABA$_A$ receptors in cone bipolar cells (**Figure 1**). This difference in GABA receptor type likely corresponds to the rapid kinetics of the photopic pathway and slower kinetics of the scotopic pathway. This is illustrated in the ferret retina in which activation of GABA$_C$ receptors more effectively inhibits the output to AII and A17 amacrine cells than GABA$_A$ receptors. When co-expressed ON bipolar cell axons, GABA$_A$ and GABA$_C$ receptors are differentially distributed. For example, in the ON mixed rod/cone bipolar cell of goldfish, GABA$_C$ receptors are clustered at the distal (vitreal) margin of the terminal, while GABA$_A$ receptors are more proximal to the cell body, near the connecting axon. Unlike the photoreceptors, there is a very high density of GABA receptors on glutamatergic bipolar cell terminals. Zn^{2+}, likely to be co-released with glutamate, suppresses the amplitude of responses mediated by GABA$_C$ receptors more so than GABA$_A$ receptors. The effects of zinc on the kinetics of GABA$_A$ receptors are variable and likely due to differences in GABA$_A$ subunit composition. The hyperpolarizing effect of GABA on all bipolar cells strongly inhibits glutamate release. This powerful inhibition is buffered by Zn^{2+} and other endogenous modulators including dopamine, neuropeptides, and endocannabinoids that control the transmission of signals to amacrine and ganglion cells.

Amacrine Cells and Ganglion Cells

At least two-thirds of the amacrine cells in all species utilize GABA as the primary neurotransmitter; GABAergic amacrine cells have additional secondary transmitters including acetylcholine, glycine, dopamine, and a wide selection of neuropeptides. Their processes are present in all layers of the IPL and they make extensive feedback contact with bipolar cells, serial contacts with other amacrine cells, and feed-forward contacts with ganglion cells. This makes for an inordinately complicated series of nested inhibitory circuits. In all species, GABA$_A$ receptors are expressed postsynaptically on amacrine cells and ganglion cells. GABA$_B$ receptors are located presynaptically on amacrine cells and postsynaptically on ganglion cells (**Figure 1**). GABA$_C$ receptors are found on ganglion cells in salamander, but apparently not in other species, and only rarely on amacrine cells. GABA$_A$ receptors on subtypes of GABA-receptive amacrine and ganglion cells are characterized by specific subunit compositions. Cholinergic amacrine cells are the only retinal neurons shown, as yet, to express the δ subunit, the presence of which should eliminate BZD sensitivity. The various α subunits aggregate at different synaptic sites. For example, the α2 subunit is found on cholinergic amacrine cells, while the α4 subunit is found on ganglion cells. Also, for the most part, GABA$_A$ receptors in the inner retina are enhanced by barbiturates and BZDs, indicating the presence of the γ subunit. In general, the effect of GABA$_A$ stimulation of amacrine cells is to suppress the transient component of the light response. The application of GABA tends to suppress the spontaneous activity and light responses of ganglion cells without having a great effect on the center-surround organization. However, GABA$_A$ antagonists abolish directional and orientation selectivity in ganglion cells, presumably by interacting with GABA$_A$ receptors on bipolar cells and other amacrine cells.

In general, the activation of GABA$_B$ receptors on amacrine cells and ganglion cells makes the light response more transient and reduces spike frequency in ganglion cells. In rabbit retina, baclofen facilitates the light-evoked release of ACh. GABA$_B$ receptors are found at limited but discrete locations on the dendrites of starburst and other types of amacrine cells. A feedback circuit involving the disinhibition of glycine onto starburst amacrine cells could account for the facilitatory effect of baclofen. GABAergic modulation of ganglion cell activity involves a complicated interaction of the subtypes of GABA$_A$, GABA$_C$, and GABA$_B$ receptors that are present ON bipolar and amacrine cells and act on chloride channels and a variety of calcium channels. While GABA$_A$ influence is rapid in onset and offset and likely participates in phasic inhibition, GABA$_B$ and GABA$_C$ influences are slower and likely participate in tonic inhibition. As GABA$_B$ receptors are less sensitive to GABA than GABA$_A$ or GABA$_C$ receptors, their influence may be more prominent during excessive GABA release that could occur with intense stimulation or excitotoxic release of glutamate from bipolar cells.

Müller's Cells

In addition to expressing GATs, Müller's cells in skate, baboon, and human express $GABA_A$ receptors (**Figure 1**). GABA-evoked currents in skate and human Müller's cells show $GABA_A$ pharmacology, including enhancement by Zn^{2+} and, in human, enhancement by barbiturates and BZDs. In human, the $GABA_A$ response is depolarizing and sensitivity is higher at the endfoot, soma, and sclerad margin. The efflux of Cl^- in response to $GABA_A$ activation on Müller's cells could have a variety of consequences. For example, it could regulate extracellular Cl^- homeostasis following Cl^- influx into neuronal $GABA_A$ and $GABA_C$ receptors. It could also accelerate the clearance of extracellular GABA by stimulating uptake, which requires cotransport of Na^+ and Cl^-. This latter function could be critical given data in mouse showing that $GABA_A$ activation could play a role in ganglion cell death following oxidative stress. In rabbit, Müller's cells synthesize and release the diazepam-binding inhibitor (acyl coenzyme A-binding protein, ACBP) in response to protein kinase A activation, mimicking intense neuronal activity. ACBP binds to an α subunit to reduce the Cl^- current, and hence, the level of inhibition. The reduction of Cl^- influx could reduce the osmotic entry of water into neurons, thereby preventing cell swelling. Thus, in mammals at least, Müller's cells that extend across the thickness of the retina play a major role in regulating neuronal GABAergic transmission, with the additional consequence of maintaining extracellular ionic and osmotic balance.

See also: Information Processing: Amacrine Cells; Information Processing: Bipolar Cells; Information Processing: Ganglion Cells; Information Processing: Horizontal Cells; Neurotransmitters and Receptors: Dopamine Receptors.

Further Reading

Brecha, N. C. (1992). Expression of $GABA_A$ receptors in the vertebrate retina. In: Mize, R. R., Marc, R. E., and Sillito, A. M. (eds.) *Progress in Brain Research*, vol. 90, pp. 3–28. New York: Elsevier Science.

Cueva, J. G., Haverkamp, S., Reimer, R. J., et al. (2002). Vesicular γ-aminobutyric acid transporter expression in amacrine and horizontal cells. *Journal of Comparative Neurology* 445: 227–237.

Gasnier, B. (2004). The SLC32 transporter, a key protein for the synaptic release of inhibitory amino acids. *Pflugers Archiv European Journal of Physiology* 447: 756–759.

Jellali, A., Stussi-Garaud, C., Gasnier, B., et al. (2002). Cellular localization of the vesicular inhibitory amino acid transporter in the mouse and human retina. *Journal of Comparative Neurology* 449: 76–87.

Nelson, H. (1998). The family of Na^{2+}/Cl^- neurotransmitter transporters. *Journal of Neurochemistry* 71: 1785–1803.

Olsen, R. W. and Sieghart, W. (2009). $GABA_A$ receptors: Subtypes provide diversity of function and pharmacology. *Neuropharmacology* 56: 141–148.

Raiteri, M. (2008). Presynaptic metabotropic glutamate and $GABA_B$ receptors. *Handbook of Experimental Pharmacology* 184: 373–407.

Schwartz, E. A. (2002). Transport-mediated synapses in the retina. *Physiological Reviews* 82: 875–891.

Wässle, H., Koulen, P., Brandstätter, J. H., Fletcher, E. L., and Becker, C. M. (1998). Glycine and GABA receptors in the mammalian retina. *Vision Research* 38: 1411–1430.

Yang, X-L. (2004). Characterization of receptors for glutamate and GABA in retinal neurons. *Progress in Neurobiology* 2004: 127–150.

Yazulla, S. (1986). GABAergic mechanisms in the retina. *Progress in Retinal Research* 5: 1–52.

Relevant Website

http://webvision.med.utah.edu – John Moran Eye Center, University of Utah, Webvision. The Organization of the Retina and Visual System.

Ganglion Cell Development: Early Steps/Fate

N L Brown, Cincinnati Children's Research Foundation, Cincinnati, OH, USA

Glossary

Atonal – Basic helix–loop–helix transcription factors involved in development. Atoh7, also called Ath5, is important in ganglion cell development.

Basic helix–loop–helix (bHLH) – A basic protein–DNA and helix–loop–helix protein–protein interaction domain found in some transcription factors.

Ganglion cell layer (GCL) – The innermost retinal cell layer, which contains retinal ganglion cell bodies and, in many vertebrates, displaced amacrine interneurons.

POU-domain transcription factors – A large group of transcription factors containing a bipartite DNA-binding domain called a POU domain.

Retinal progenitor cell (RPC) – A developing retinal cell that undergoes mitotic cell division to produce cells that can differentiate into retinal neurons or glia.

Transitional cell – A retinal progenitor cell that is newly postmitotic, but not yet differentiated as a particular retinal neuron or glial cell type.

Retinal Ganglion Cell Formation

Ganglion cells are the first neurons generated by retinal progenitor cells (RPCs) in the developing optic cup. Retinal ganglion cells (RGCs) differentiate first in all vertebrates. In birds and mammals, retinal neurogenesis begins in the center of the cup with the cell cycle exit of a group of RPCs near the optic stalk. Many, but not all, early neurons differentiate as RGCs. From this initiation zone, RGC genesis spreads outward to the periphery of the optic cup. However, in zebrafish, retinal neuron formation initiates in the ventral, nasal cup and propagates around the circumference of the eye as a wave. Ganglion cell layer (GCL) formation is the initial step of optic cup lamination, which results in a mature retina organized into three cell layers (outer nuclear layer, inner nuclear layer, and GCL) separated by two synaptic layers (outer and inner plexiform layers).

Over the past 15 years, considerable progress has been made toward understanding of how mitotically active RPCs choose the RGC fate, terminally differentiate as RGC neurons, extend their axons, and establish synaptic connections in the brain. Yet, this gene network remains incomplete, with more work needed to define RGC formation in molecular terms. Within the hierarchy, both intrinsic transcription factors and extrinsic signaling pathways provide integrated regulation of progenitor development into an RGC. While this article emphasizes intrinsic regulation of RGC fate specification, a short description of extrinsic signaling pathways known to regulate RGC development is also included.

Atoh7/Ath5 Function is Critical for RGC Development

The onset of retinal neurogenesis in a subset of optic cup RPCs is characterized by the activation of the basic helix–loop–helix (bHLH) transcription factor Atoh7/Ath5. The bHLH factors regulate many characteristics of retinal neuron formation. The encoded proteins contain both a basic DNA-binding and helix–loop–helix protein dimerization domains. These factors are expressed by RPCs, in which one or more cells ultimately becomes a neuron. Atoh7/Ath5 occupies a critical node in the RGC gene hierarchy, since zebrafish or mouse mutants lacking the gene have a profound to total loss of RGC differentiation and abnormal optic nerve formation. Adult mutant animals are viable, but have no optic nerves or chiasmata. Overexpression of frog and chick Ath5 during retinal development induces ectopic RGCs, but mammalian Atoh7 does not seem to have this ability. This difference correlates with the observation that, in the mouse retina, Atoh7-expressing cells (the Atoh7 retinal lineage) give rise to all seven major classes of retinal neurons and glia in the adult retina, and do not strictly produce RGCs. This indicates that Atoh7 function is needed to acquire the potential to become an RGC, but is insufficient to commit a cell to this fate. During embryonic retinal neurogenesis, the Atoh7 lineage is hypothesized to predominantly give rise to RGCs, but after the peak of RGC genesis, Atoh7-expressing cells are better able to adopt other retinal fates. This change in developmental ability is presumably due to input from signals secreted by differentiated RGCs and possibly other (non-Atoh7) retinal lineages. Such signals may either promote the adoption of non-RGC retinal neuron or glial fates, block RGC fates, or both.

One role for Atoh7 in the mouse retina is to facilitate cell cycle exit, since mutant cells appear to stall at a cell cycle checkpoint, unable to either resume mitosis or differentiate as RGCs. Eventually, these cells erroneously adopt the last retinal fate, that of Müller glia. Mouse retinal cells expressing Atoh7 fit the definition of transitional cells, because they are nonmitotic, migratory RPCs that commit to a particular fate, but remain undifferentiated when they express Atoh7. Atoh7+ cells co-express the cyclin kinase inhibitor Cdkn1b/p27^{Kip1}, which has multiple functions, including promoting exit from mitosis of cells at the G_1/G_0 checkpoint. Without Atoh7, the percentage of Cdkn1b+ cells fluctuates in both number and arrangement during the peak of RGC genesis. However, questions remain about the relationship between Atoh7 and Cdkn1b. Does mouse Atoh7 directly regulate the activation of Cdkn1b, or do Atoh7 and Cdkn1b act cooperatively to drive RPCs out of mitosis? The latter possibility has been proposed for frog Ath5 and cyclin-dependent kinase inhibitor xic1 (p27Xic1), although it is unclear whether this represents protein–protein interactions, coordinate regulation of shared downstream genes, or cross-regulation. Alternatively, the shifts in mammalian cyclin-dependent kinase inhibitor 1B (p27^{Kip1}) expression in Atoh7 mutant eyes may be correlative, but not indicative, of either type of interaction. Since cyclin-dependent kinase inhibitor 1B (Cdkn1b) transcription, translation, and protein stability are independently regulated during development, distinguishing among these possibilities will require further in-depth analyses.

Integrated Regulation of Atoh7/Ath5 Retinal Expression

Many proneural bHLH genes autoregulate their expression, including fruit fly atonal and mouse Atoh1, the most closely related bHLH factors to Atoh7. However, multiple lines of evidence show that Atoh7/Ath5 does not regulate its own expression. Atoh7/Ath5 appears in the optic cup slightly ahead of other proneural bHLH factors and has a very dynamic expression pattern first detectable in a subset of transitional cells that give rise to the earliest retinal neurons. Therefore, determining which transcription factors and signal transduction pathways directly influence Atoh7/Ath5 expression is essential for comprehending how broadly acting factors regulate the precise spatiotemporal patterns of retinal neurons. Among the early optic cup transcription factors, only Pax6 has been shown to directly activate retinal expression of Atoh7/Ath5 via binding to its highly conserved enhancer. Because Atoh7/Ath5 expression initiates in a subset of Pax6-expressing cells, other factors must simultaneously restrict Atoh7/Ath5 expression to newly postmitotic transition cells (**Figure 1**). Another equally important aspect of Atoh7/Ath5 regulation is its turnover, since Atoh7 mRNA disappears abruptly from transitional cells at their entry into the GCL. Therefore, repression is just as important as initial activation for the proper regulation of Atoh7/Ath5 (**Figure 1**).

There are several factors that genetically suppress Atoh7/Ath5 expression, including hairy and enhancer of split 1 (Hes1), neurogenic differentiation 1 (Neurod1), and neurogenic differentiation 4 (Neurod4/Ath3). Because Hes1 is a transcriptional repressor, it is not surprising that it blocks Atoh7 expression. Hes1 similarly suppresses the expression of other retinal proneural bHLH factors like neurogenin 2 (Neurog2) and achaete-scute complex homolog 1 (Ascl1). In the absence of Hes1, Atoh7, Neurog2, and Ascl1 are each prematurely activated in the mouse optic cup. Although Hes1 regulates the overall

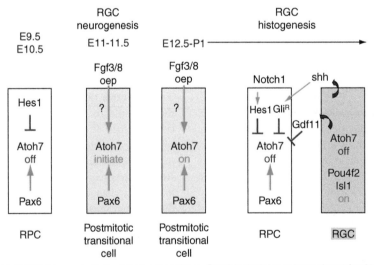

Figure 1 Regulation of Atoh7/Ath5 in postmitotic retinal progenitors. Combinatorial regulatory input for Atoh7/Ath5 transcription in the optic cup. Paired box gene 6 (Pax6) provides cells with the competence to express Atoh7, but Hes1 prevents activation from occurring prematurely. When Hes1 is downregulated, localized Fgf signaling triggers a subset of RPCs to express Atoh7. Retinal cells adopting the RGC fate express Atoh7 for a relatively short time, after which multiple signaling pathways (Notch, Shh, and/or Gdf11) abruptly shut off Atoh7, particularly as differentiating RGCs enter the GCL. Mature RGCs control their own number by secreting Shh and Gdf11, which block Atoh7 expression.

timing of neurogenesis, other regulatory inputs control the precise onset of each proneural bHLH gene. The timing function of Hes1 reflects its role as a downstream effector of Notch signaling, although not all Hes1 functions are Notch dependent. Other factors that suppress Atoh7 expression are Neurod1 and Neurod4, which do so synergistically. Embryonic mouse retinas that are mutant for both genes have expanded Atoh7 expression and a dramatic increase in RGCs, along with a profound loss of amacrine neurons. Thus, Neurod1 and Neurod4 repress Atoh7 expression (directly or indirectly) in RPCs that normally give rise to amacrine cells.

Multiple extrinsic signals influence vertebrate Atoh7/Ath5 expression, particularly for RGC formation. These include fibroblast growth factor 3 (Fgf3), fibroblast growth factor 8 (Fgf8), one eyed pinhead/nodal-like (oep/nodal-like), notch homolog 1 (Notch1), growth differentiation factor 11 (Gdf11), and sonic hedghog (Shh) signaling pathways (**Figure 1**). As shown in the zebrafish and chick retina, Fgf and oep signaling are required for Ath5 activation, while Notch (chick, frog, mouse), Gdf11 (mouse), and Shh (zebrafish, chick, mouse) each suppresses Atoh7/Ath5 expression. Intriguingly, chick Fgf3 and Fgf8 are both expressed in the central optic cup, just prior to the onset of retinal neurogenesis and Fgf protein-coated beads induce Ath5 expression. Although no role during early mouse retinal development has been reported for the orthologous genes, the secretion of stimulatory signals from the central optic cup (Fgf3 or 8) and/or optic stalk (Fgf8 and oep) would be sufficient to explain the onset of Atoh7/Ath5 expression in a restricted retinal domain. By contrast, the extrinsic pathways that negatively regulate Atoh7/Ath5 expression emanate from two distinct sources, RPCs and RGCs. The Notch pathway functions in multiple and complex ways during neurogenesis, but its classic role is to transduce a competitive signal between two adjacent progenitors, in which one cell (expressing more ligand) laterally inhibits the other (expressing more receptor) from differentiating. Therefore, Notch+ cells remain proliferative by suppressing neuron promoting factors like Atoh7. The other two signals, Gdf11 and Shh, are made and secreted by differentiated RGCs to provide negative feedback regulation to RPCs so that the number of neurons is not drastically overproduced. But, to date none of these pathways have been shown to directly activate or repress Atoh7/Ath5 transcription.

Atoh7/Ath5 Activates Pou4f/Brn3 Expression in RGCs

Atoh7/Ath5 activation of Pou4f/Brn3b expression is the principal molecular pathway directing RGC specification and differentiation. Studies in the chick, frog, and mouse eye all suggest that this activation is direct, although definitive biochemical assays have not yet been performed.

In the mouse optic cup, Atoh7 and Pou4f2 expression are 90% overlapping, with Atoh7 appearing slightly earlier than Pou4f2 in cells completing their final mitosis, migrating toward the GCL, and differentiating (**Figure 2**). Importantly, Atoh7 is abruptly turned off as nascent RGCs enter the GCL, while Pou4f2 expression is maintained by mature RGCs, even into adulthood.

The Pou4f/Brn subfamily of POU-domain transcription factors is comprised of Pou4f1/Brn3a, Pou4f2/Brn3b, and Pou4f3/Brn3c. Each of these genes acts as key regulators of sensory neuron development, particularly projection neurons. But, each POU-domain factor is required by different kinds of projection neurons. Although a particular class of projection neuron can express all three Pou4f factors, the one that is activated first performs the critical functions. For example, in the retina, all three factors are present in overlapping subsets of differentiated RGCs, but Pou4f2/Brn3b expression initiates expression

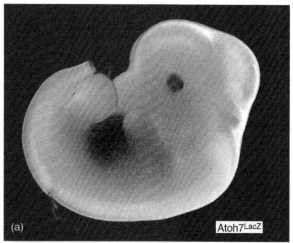
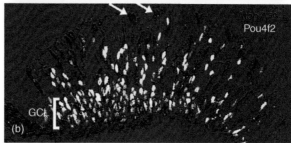

Figure 2 Atoh7 and Pou4f2 expression during RGC development. (a) E11.5 Atoh7^LacZ/+ embryo expressing β-galactosidase in the optic cup (detected through a chromagenic reaction that turns cells in the eye blue). (b) Anti-β-galactosidase (which detects cells with Atoh7^LacZ expression) and anti-Pou4f2 double immunofluorescent staining of E13.5 central optic cup section. Arrows point to Atoh7-expressing cells completing mitosis, which then migrate to the GCL. During migration, these cells initiate Pou4f2 expression (in yellow). Atoh7/Pou4f2 co-expressing cells are in white. Pou4f2 activates downstream target genes for RGC differentiation, axon outgrowth/guidance or survival. The bacterial β-galactosidase has a long half-life in mouse tissues, thus is present in mature RGCs, while endogenous Atoh7 is absent from the GCL (bracket).

about 2 days ahead of the others. The loss of either Pou4f1/Brn3a or Pou4f3/Brn3c has no effect on RGC development, although trigeminal and dorsal root ganglia critically require Pou4f1, and cochlear and vestibular neurons need Pou4f3. However, Pou4f2/Brn3b is a key regulator of RGC development, and mutant adult eyes have extremely thin optic nerves lacking 70% of the RGCs. In contrast to Atoh7 mutants, Pou4f2 mutants have differentiated RGCs that extend axons through the optic nerve, but subsequently fail to innervate the correct regions of the brain and die because these cells lack a critical survival factor. Although Pou4f1 and Pou4f3 seem dispensable for retinal development, they act redundantly with Pou4f2. Pou4f1−/−;Pou4f2−/− and Pou4f3−/−;Pou4f2−/− double mutants have a more profound loss of RGCs than Pou4f2 single mutants. Moreover, overexpression of individual Pou4f genes in chick RPCs demonstrated that each is fully capable of inducing RGCs. Finally, Pou4f3 rescues the loss of Pou4f2 in retinal explant cultures; and targeted replacement of Pou4f1 into the Pou4f2 locus restores RGC development and survival.

Pou4f2 regulates two distinct molecular pathways during RGC development. In the first, Pou4f2 turns on unknown downstream genes within the earliest differentiating RGCs. This subset of RGCs can develop as if Pou4f2 independent, because in Pou4f2 mutants Pou4f1 and Pou4f3 expression rescues them. The second, Pou4f2-dependent pathway produces the bulk of RGCs at later embryonic ages. Here, Pou4f2 regulates a distinct set of downstream genes from those found in the earliest RGCs, including the transcription factors distal-less homeobox 1 and 2 (Dlx1 and Dlx2). When Pou4f2 function is removed from the mouse optic cup, the early RGCs persist, due to Pou4f1 and Pou4f3 compensation. However, the larger, Pou4f2-dependent RGC population cannot complete neuronal differentiation, correctly innervate the brain or express survival factors. This results in mutant eyes with abnormally thin optic nerves. Interestingly, these circumstances point to the Pou4f2 mutant mouse as an excellent system in which to study optic nerve regeneration. For instance, could *ex vivo* cultured (and introduced) RPCs, or early differentiating RGCs, use the Pou4f2 mutant reduced optic nerve, as a substrate for projecting their axons to the correct regions of the brain? Or, could Pou4f1 and Pou4f3 expression be manipulated (activated or de-repressed) within the second, later group of developing RGCs to rescue their outgrowth and survival defects?

Pou4f2/Brn3b Controls Numerous RGC Processes

Multiple studies indicate that Pou4f2 is strictly a transcriptional activator, with numerous potential downstream target genes. The list of identified downstream genes includes Pou4f2 itself (autoregulation), Pou4f1 (cross-regulation), the transcription factor eomesodermin homolog (Eomes), the anti-apoptosis factor B-cell lymphoma-2 associated protein (Bcl2), the signaling molecule Shh and multiple cytoplasmic or cell surface proteins, including actin-binding LIM protein 1 (Ablim1), which participate in axon outgrowth, guidance, or vesicular trafficking at synapses. Among these Eomes, Ablim1, and Shh are described here as examples that highlight the diverse set of genes that Pou4f2 directly or indirectly regulates in nascent RGCs. First, Eomes is a transcription factor with a T-box DNA-binding protein motif. During the period of maximal Pou4f2 embryonic expression, Pou4f2 directly activates Eomes transcription in differentiated RGCs. Eomes is also expressed by a subset of inner nuclear retinal cells, potentially amacrine cells, but the role of this factor in the inner nuclear layer remains unexplored. When Eomes function was removed during retinal development, mutant mice have thinner than normal optic nerves, which reflect a postnatal loss of RGCs. Although initial RGC development does not require Eomes activity, myelination of RGC nerve fibers does, and this defect causes elevated RGC death in Eomes mutants. This myelination defect is very similar to the one that has been described for Pou4f2 mutant mice.

Pou4f2 function is also necessary for the proper expression of proteins that direct neurite outgrowth, axon guidance, and neuronal function at synapses. One such gene is Ablim1, encoding an actin-binding protein, with a LIM protein–protein interaction domain. Misexpression of a mutant form of Ablim1 during chick optic nerve development causes RGC axon guidance defects. But paradoxically, Ablim1 mutant mice have no retinal or optic nerve defects. This implies that Ablim1 acts redundantly with another molecule during RGC axon outgrowth. Because there are two other Ablim genes in mammals, future studies are needed to test the role of these genes, singly and in combination with each other and Ablim1, to understand if Pou4f2 regulation of Ablim expression is sufficient to explain the RGC axon guidance defects of Pou4f2 mutant mice. Finally, the search for additional genes regulated by Pou4f2 uncovered the secreted signaling molecule Shh.

Islet1 Acts Parallel to Pou4f2 During RGC Development

Although Atoh7 activation of Pou4f2 deploys an important molecular mechanism for generating RGCs, other factors also act at this same step in the genetic hierarchy of RGC. These genes, which largely encode transcription factors, require Atoh7/Ath5, but not Pou4f2, for their activation. There are several possibilities for how these additional factors act during RGC genesis. They may regulate a separate set of target genes, cross-regulate or interact with Pou4f2 and its downstream targets, or both. Several candidates that appear to act parallel with Pou4f2 are lim-homeodomain protein 1, Islet1 (Isl1), transducin-like enhancer of split 1 homolog (Tle1), and myelin transcription factor (Myt1). At present only Isl1 is known to function during RGC formation, and retinal-specific deletion of Isl1 results in adult mice lacking 95% of RGCs. This profound loss is more like the Atoh7 phenotype than it is to that of Pou4f2, with the main difference being the developmental age when RGCs are affected. The Isl1 transcription factor contains both a protein–protein interaction LIM-domain and a DNA-binding homeo-domain. Isl1 function is not only important during development in multiple tissues, like the spinal cord and pancreas, but also for cholinergic amacrine and bipolar neuron genesis in the retina. In optic cups devoid of Isl1 function, RGCs are specified and differentiate on schedule, including activating expression of Pou4f2 and other differentiation markers. But by birth, most RGCs downregulate Pou4f2 expression and then rapidly die. Closer examination of Isl1 mutant RGCs showed both aberrant axon projections and reduced Ablim1 expression, reminiscent of Pou4f2 mutants. Importantly, Pou4f2 and Isl1 act synergistically to control the activation of shared downstream genes, like Pou4f2, Pou4f1, and Shh. Thus, more work is needed to determine which genes Isl1 and Pou4f2 regulate in common versus those controlled by one factor or the other. Since presumably Tle1 and Myt1 also regulate similar processes during RGC formation, working out the combinatorial code of which transcription factors regulate each downstream gene is still needed.

Conclusion

As the genetic hierarchy of RGC development is deciphered, these studies draw us increasingly closer to understanding the sequence of steps for directing RPCs out of the cell cycle to differentiate as RGCs, and how these neurons make appropriate connections in the brain.

See also: Coordinating Division and Differentiation in Retinal Development; Embryology and Early Patterning; Eye Field Transcription Factors; Histogenesis: Cell Fate: Signaling Factors; Information Processing: Ganglion Cells; Intraretinal Circuit Formation; Photoreceptor Development: Early Steps/Fate; Retinal Histogenesis.

Further Reading

Brown, N. L., Patel, S., Brzezinski, J., and Glaser, T. (2001). Math5 is required for retinal ganglion cell and optic nerve formation. *Development* 128: 2497–2508.

Erkman, L., Yates, P. A., McLaughlin, T., et al. (2000). A POU domain transcription factor-dependent program regulates axon pathfinding in the vertebrate visual system. *Neuron* 28: 779–792.

Hatakeyama, J. and Kageyama, R. (2004). Retinal cell fate determination and bHLH factors. *Seminars in Cell and Developmental Biology* 15: 83–89.

Kanekar, S., Perron, M., Dorsky, R., et al. (1997). *Xath5* participates in a network of bHLH genes in the developing *Xenopus* retina. *Neuron* 19: 981–994.

Kay, J. N., Finger-Baier, K. C., Roeser, T., Staub, W., and Baier, H. (2001). Retinal ganglion cell genesis requires lakritz, a zebrafish atonal homolog. *Neuron* 30: 725–736.

Livesey, F. J. and Cepko, C. L. (2001). Vertebrate neural cell-fate determination: Lessons from the retina. *Nature Review Neuroscience* 2: 109–118.

Marquardt, T. (2003). Transcriptional control of neuronal diversification in the retina. *Progress in Retinal and Eye Research* 22: 567–577.

Mu, X., Fu, X., Beremand, P. D., Thomas, T. L., and Klein, W. H. (2008). Gene regulation logic in retinal ganglion cell development: Isl1 defines a critical branch distinct from but overlapping with Pou4f2. *Proceedings of the National Academy of Sciences of the United States of America* 105: 6942–6947.

Pan, L., Deng, M., Xie, X., and Gan, L. (2008). ISL1 and BRN3B co-regulate the differentiation of murine retinal ganglion cells. *Development* 135: 1981–1990.

Rapaport, D. H. (2006). Retinal neurogenesis. In: Sernagor, E., Eglen, S., Harris, B., and Wong, R. (eds.) *Retinal Development*, pp. 30–58. New York: Cambridge University Press.

Wang, S. W., Kim, B. S., Ding, K., et al. (2001). Requirement for Math5 in the development of retinal ganglion cells. *Genes and Development* 15: 24–29.

Genetic Dissection of Invertebrate Phototransduction

B Katz and B Minke, Hebrew University, Jerusalem, Israel

Glossary

Electroretinogram (ERG) – *In vivo* extracellular recording of the voltage response to light of the entire retina.

G protein – Guanine nucleotide-binding protein, a ubiquitous biological switch that is turned on by receptor-activation-induced exchange of bound guanosine diphosphate (GDP) with cytoplasmic guanosine triphosphate (GTP) and turned off by hydrolysis of the bound GTP.

Phosphoinositide (PI) signaling – Also designated as inositol-lipid signaling, it is a ubiquitous enzymatic cascade that uses phospholipids or their products for signaling.

Phospholipase C (PLC) – A superficial membrane-bound enzyme that hydrolyzes a minor plasma membrane phospholipid, phosphatidylinositol-4,5-bisphosphate (PIP_2), and produces two messengers: inositol 1,4,5-trisphosphate ($InsP_3$) and diacylglycerol (DAG).

Photochemical cycle – A multistep process that begins by absorption of a photon by the light-sensitive receptor, rhodopsin, which undergoes multiple modifications and ends by the regeneration of the original molecule.

Phototransduction – The process by which absorption of photons by light-sensitive receptors lead to production of electrical signal comprehensive to the nervous system.

Prolonged depolarizing afterpotential (PDA) – The light-induced electrical signal that continues in the dark long after light is turned off, produced in invertebrate photoreceptor cells. The PDA phenomenon has been widely exploited to screen for visual defective *Drosophila* mutants.

Rhabdomere – The photosensitive organelle of *Drosophila* photoreceptors which is composed of tightly packed microvilli.

Transient receptor potential (TRP) channels – A family of cation channel proteins that mediate a large variety of physical stimuli such as light, temperature, and chemical compounds.

Phototransduction is a process by which light is converted into electrical signals understood by the central nervous system. Initial studies of invertebrate phototransduction used *Balanus* and *Limulus* as the preparations of choice, due to their extremely large photoreceptor cells and the assumption that they constitute a simplified model system that represents phototransduction in general. However, later studies revealed that although both invertebrate and vertebrate photoreceptors use rhodopsin (R) as a receptor and a G (guanine nucleotide-binding)-protein-coupled signaling cascade to produce the electrical response to light; the rest of their enzymatic cascade of vision is entirely different. Phototransduction in vertebrate rods and cones uses cyclic guanosine monophosphate (cGMP) phosphodiesterase as an effector enzyme and cGMP-gated ion channels as its target. In contrast, phototransduction in the microvillar photoreceptors of invertebrates (photoreceptors in which the photosensitive organelle is composed of microvilli – the rhabdomere) uses a phosphoinositide (PI) signaling cascade, which is characterized by phospholipase C (PLC) as the effector enzyme and, at least in *Drosophila*, the TRP channels as its target. Nevertheless, phototransduction in invertebrate microvillar photoreceptors has become a model system for investigating inositol-lipid signaling and its role in TRP channel regulation and activation. The great advantage of using invertebrate microvillar photoreceptors is the accessibility of the preparation, the ease of light stimulation, the robust expression of key molecular components, and most importantly, the ability to apply the power of molecular genetics. The latter feature is mainly attributed to *Drosophila melanogaster* as a preferred preparation.

Although the phototransduction cascade of invertebrate microvillar photoreceptors is clearly different from that of vertebrate rods and cones, it may be similar to that in the intrinsically photosensitive retinal ganglion cells (RGCs) in vertebrates. The light-sensitive RGCs contain the visual pigment melanopsin and provide photic input to the circadian pacemaker of the master circadian clock. Both *Drosophila* photoreceptors and the light-sensitive RGCs are characterized by a bi-stable R that initiates a phototranduction cascade with TRP channels as the final target. It is now widely accepted that the detailed knowledge that has been obtained through studies of *Drosophila* photoreceptors constitute guidelines for research of the melanopsin-containing RGCs, and that a striking commonality has been found in both cell types.

Extensive genetic studies of the fruit fly, *D. melanogaster*, initiated at the turn of the twentieth century, by the Nobel prize laureate, Thomas Hunt Morgan, and greatly advanced by many other laboratories, has established the *Drosophila* as an extremely useful experimental model for genetic dissection of complex biological processes. The relatively small size of the *Drosophila* genome, ease of growth, rapid generation time, and high fecundity make this system ideally suited for screening large numbers of mutagenized individual flies for defects in virtually any phenotypically observable or measurable trait. The creation of balancer chromosomes – containing dominant markers and multiple inversions – which prevent recombination with the native chromosomes, allows any mutation – once recognized – to be rapidly isolated and maintained. Importantly, germ line transformation, using P-element transposition, combined with the availability of tissue-specific promoters, allows the introduction of cloned genes into specific cells of the organism. This provides a way to study *in vitro* modified gene products in their native cellular environment. These powerful molecular genetic tools, combined with the available genome sequence, allow screening for mutants defective in critical molecules, while devoid of *a priori* assumptions. Indeed, this methodology has produced large numbers of mutants defective in novel proteins the existence of which would have been difficult to otherwise predict. The following sections emphasize the unique contribution of the research in *Drosophila* photoreceptors, which has led to the discovery of novel molecules and processes important not only for phototransduction but also for many other biological mechanisms (**Figure 1**).

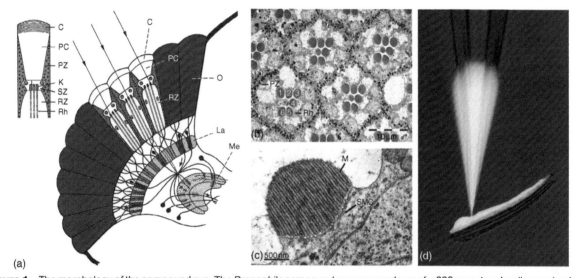

Figure 1 The morphology of the compound eye. The *Drosophila* compound eyes are made up of ~800 repeat and well-organized units termed ommatidia ((a) and (b)). The ommatidium consists of 20 cells, eight of which are the photoreceptor cells (RZ; (a)). (b) and (c) show an electron microscopic (EM) cross section of ommatidia and a rhabdomere (Rh) at the upper region of the photoreceptors, respectively. (d) shows an isolated ommatidium in which one photoreceptor cell is marked by a yellow fluorescent dye introduced by a patch pipette. Each ommatidium contains a dioptric apparatus composed of transparent chitinous cuticle, which forms the cornea (C;(a)) and an extracellular fluid-filled cavity, called the pseudocone (PC; (a)). The floor of the cavity is formed by four Semper cells (SZ; (a)) and the walls by primary pigment cells (PZ; (a) red and (b)), which together circle the pseudocone, shielding the photoreceptor from light coming from adjacent ommatidia. Below this rigid structure of the optical apparatus lie eight photoreceptor cells (RZ; (a), (b), and (d)). The photoreceptor cells are highly polarized epithelial cells, having a specialized compartment known as the rhabdomere (Rh; (a), (b), and (c)), consisting of a stack of ~30 000–50 000 microvilli (M; (c)) each ~2-µm long and ~60 nm in diameter. The transduction machinery is arrayed in these tightly dense structures, while the nucleus (N; (c)) and cellular organelles, such as submicrovillar cisternae (SMC; (c)), reside at the cell body. These highly ordered rhabdomeres form wave-guides that have been widely exploited for optical methods. The eight photoreceptors can be divided into two functional groups according to their position, spectral specificity and axonal projection. The R1–R6 cells (marked 1–6 in (b)) represent the major class of photoreceptors in the retina and are involved in image formation and motion detection. These cells have peripherally located rhabdomeres extending from the basal to the apical side of the retina where they terminate in a rhabdomere cap (K; (a)). They express a single opsin called Rh1, which when combined with 11-*cis* 3-hydroxy retinal, forms a blue-absorbing rhodopsin (R) or orange-absorbing metarhodopsin (M). The R1–6 cells (b) project their axons to the first optic lobe, the lamina (La; (a) green). The second group consists of two cells in the center of each ommatidium termed, R7 (marked 7 in (b)) and R8 (not shown) each spanning only half of the retina in length. The R7 rhabdomere is located distally in the retina and expresses one of two opsins, Rh3 or Rh4, characterized by a UV-absorbing R and blue-absorbing M. The R8 rhabdomere is located proximally in the retina, beneath the R7 rhabdomere (not shown) and expresses one of three opsins – Rh3, Rh5, or Rh6 – characterized by a UV, blue, or green-absorbing R, respectively. The R7 and R8 cells project their axons to the second optic lobe, the medulla (Me; (a) pink). On the basis of opsin expression in the R7 and R8 cells, three ommatidia subtypes can be distinguished. The R7 and R8 cells in ommatidia residing in the dorsal rim area both express Rh3 opsin. The 'pale' ommatidia subtype express Rh3 in R7 cells and Rh5 in R8 cells and constitute ~30% of the total ommatidia, while the 'yellow' ommatidia subtype express Rh4 in R7 cells and Rh6 in R8 cells and constitute ~70% of the total ommatidia. The central cells R7/8 function in color vision and detection of polarized light. This intriguing repeated structure has been a major scientific preparation for research of various aspects of retinal-cell differentiation and development. (a) Modified from Kirschfeld, 1967 and (b) modified from Minke and Selinger, 1996.

The *Drosophila* Phototransduction Cascade

Genetic Screens for Mutants Defective in Phototransduction Proteins

A genetic screen is a procedure to identify and select mutated individuals that possess a phenotype with a specific malfunction in a specific trait. This method requires a large number of mutated individuals, usually obtained by the use of mutagens or by random DNA insertions (transposons) and a simple, but powerful, isolation procedure. Additionally, this method requires genetic tools to isolate and maintain the stability of the mutation through generations. These requirements make *Drosophila* an ideal organism for genetic screens. Wide genetic screens targeting the two autosomes (II and III) and the X chromosome of *Drosophila*, have produced many mutants with identified phenotypes linked specifically to genes involved in visual behavior or photoreceptor cell function. These screens used defects in a number of visual functions such as optomotor response, phototaxis, and electrophysiological response to light. The isolation of mutants, specifically defective in the visual transduction pathway initially made use of the electroretinogam (ERG), which measures the electrical activity of the eye at the corneal level (**Figure 1(a)**). However, the ERG methodology failed to isolate the large numbers of mutants expected from a multistep and complex process such as the phototransduction cascade. The main reasons for this failure are that the phototransduction proteins are highly abundant and the upper limit in the depolarization signal is reached even when only a small fraction of the signaling molecules are excited. Therefore, mutations causing even a significant reduction in concentration or subtle malfunction of the phototransduction components could not be identified by this method. This necessitated the employment of a more sensitive, and yet simple, method for isolation of mutants.

In order to find mutants that are specifically affected in the phototransduction cascade and to identify those with subtle phenotypes, Pak and colleagues employed the electrical phenomenon devised by Hillman Hochstein and Minke designated prolonged depolarizing afterpotential (PDA; **Figure 2(a)**, upper trace). This method is based on a large net photoconversion of R to its dark stable photoproduct metarhodopsin (M) with a minimal conversion back to R (**Figure 3**), which brings the capacity of the phototransduction process to its upper limit. PDA is achieved in the fly by genetically removing the red screening pigment (**Figure 1(a)**) and by applying blue light, which is preferentially absorbed by the R state of the Rh1 photopigment. Rh1 is the photopigment expressed in photoreceptors R1–R6 in the *Drosophila* eye (**Figure 1(a)** and (**b**)). A large net photoconversion of R to M (**Figure 3**) prevents phototransduction termination at the photopigment level.

As a result, excitation is sustained long after the light is turned off, and the PDA-producing cells are unable to respond further (termed inactivation, **Figure 2(a)**, upper trace). Subsequently, application of orange light reconverts the activated M back to R and terminates the sustained excitation after the light is turned off. Since the PDA tests the maximal capacity of the photoreceptor cell to maintain excitation for an extended period and is strictly dependent on the presence of high concentrations of R and other signaling molecules, it detects even minor defects in R biogenesis, or exhaustion of critical signaling molecules. Thus, defects in the PDA easily reveal deficiencies in the concentration of phototransduction components. Indeed, the PDA screen yielded a plethora of novel and interesting visual mutants. One group of PDA mutants exhibited a loss in several features of the PDA. They were termed *nina* mutants, which stands for neither inactivation nor afterpotential (**Figure 2(a)**, lower trace). Most *nina* mutants were caused by reduced levels of R. The second group of PDA mutants lost the ability to produce the voltage response associated with the PDA, but were still inactivated by strong blue light and the inactivation could be relieved by orange light (**Figure 2(a)**, middle trace). These mutants, consisting of seven allelic groups, are termed *inaA–G*, which stand for inactivation but no afterpotential (no PDA-voltage response). The *ina* mutants were found to have normal R levels but are deficient in proteins associated with the function of the TRP channel. The *nina* and *ina* mutants have led to the identification of most of the crucial components of *Drosophila* phototransduction, many of which are novel proteins of general importance for many cells and tissues.

The Photochemical Cycle and the Mechanism Underlying Termination of M Activity

The *ninaE* mutant, having reduced Rh1 opsin levels, was isolated by Pak and colleagues using the PDA screen. R belongs to the super family of seven transmembrane G-protein-coupled receptors. It is activated by the absorption of a photon that isomerizes the chromophore (11-*cis* 3-hydroxy retinal), resulting in a conformational change of the opsin molecule and production of the physiologically active M state of the photopigment. To ensure high sensitivity, high temporal resolution, and low dark noise of the photoresponse, the active M has to be quickly inactivated and recycled (**Figure 3**). The latter requirement is achieved, in invertebrates, by two means: the absorption of an additional photon by the dark stable M – which photoconverts M back to R – or by a multistep photochemical cycle (**Figure 3**). The termination of vertebrate M activity is a two-step process initiated by M phosphorylation, followed by the binding of the protein arrestin (ARR) to phosphorylated M. Invertebrate M also undergoes

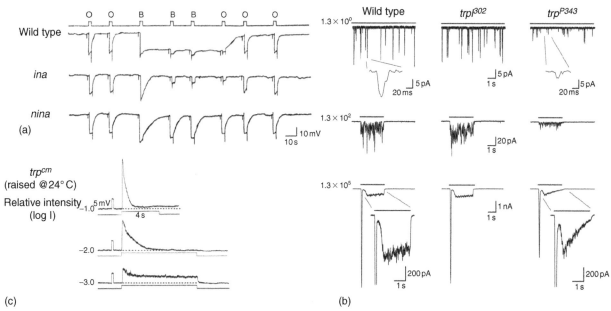

Figure 2 The electrophysiological responses to light of *Drosophila*. (a) The prolonged depolarization afterpotential (PDA) response of wild-type *Drosophila* and its modifications in the *nina* and *ina* mutants: Upper trace: ERG recordings from a wild-type fly in response to a series of intense blue (B, Schott, BG 28 broad-band filter) and orange (O, Schott OG 590 edge filter) light pulses used for induction and suppression of the PDA. This paradigm included two intense orange light pulses followed by an intense blue light pulse which converted ∼80% of the Rh1 photopigment from R to M and resulted in prolonged corneal negative response that continued in the dark. Two additional intense blue lights elicited small responses that originated from the central cells (R7 and 8) in which PDA was not induced, due to their UV-absorption spectra, while the R1–6 cells were nonresponsive (inactivated) due to maximal activation of the channels. The following orange light suppressed the PDA after light is turned off. Middle trace: The paradigm of the upper trace was repeated in an *ina* mutant. In contrast to wild type, the response to the intense blue light declined to baseline. However, the R1–6 cells remained inactivated and additional blue lights elicited responses only in R7 and 8 cells while the following orange light removed the inactivation and allowed recovery of the R1–6 cells response to light. Bottom trace: The paradigm of the upper traces was repeated in a *nina* mutant. The first blue light elicited a short PDA that quickly declined to baseline and additional blue lights elicited responses in all photoreceptor cells and allowed additional activation of R1–6 cells by orange lights. (b) (Top) Whole-cell patch-clamp recordings (clamped at −70 mV) from photoreceptor cells of wild-type (WT, left column), *trpl^{302}* mutant, expressing only the TRP channels (middle) and *trp^{P343}* mutant, expressing only the TRPL channels (right). Quantum bumps are elicited in response to very dim orange light (1.3 effective photons s$^{−1}$). The inset shows single bumps in faster time scale of WT fly and the *trp^{P343}* mutant. The bump amplitudes in the *trp^{P343}* mutant are significantly smaller than in WT or the *trpl^{302}* mutant. Middle: In response to 100-fold more intense light, the bumps sum up to produce a noisy light induced current (LIC, note the change in scale). Bottom: A further increase of 1000-fold in the light intensity induced a large response with an initial fast transient that declined to a small steady-state response, due to fast Ca^{2+}-dependent light adaptation in both WT and the *trpl^{302}* mutant. In the *trp^{P343}* mutant, the steady-state response declined to baseline during light. The inset shows amplified responses where the initial transient is off scale (note the change in scale). (c) Intracellular recordings from the *trp^{CM}* mutant, raised at 24 °C, in which the TRP channels are not functional. Responses to increasing intensities of orange light pulses are shown. The response to dim light (bottom trace) showed a sustained response during light. In contrast, the responses to medium and intense orange lights declined to baseline during illumination, showing the typical transient receptor potential of the *trp* mutant. (a) Modified from Pak, 1979.

light-dependent phosphorylation (**Figure 3**; Mpp), but this process is not required for response termination. This was demonstrated in experiments which showed that two *Drosophila* transgenic mutants, P[Rh1^{Δ356}] and P[Rh1$^{S\ to\ A}$], expressing Rs that lack the putative phosphorylation sites exhibit normal inactivation and ARR binding. However, as in the photoreceptors of vertebrates, the binding of ARR to M in *Drosophila* photoreceptors inactivates M (**Figure 3**). *Drosophila* has two protein homologs to the vertebrate ARR, both undergo light-dependent phosphorylation: the 49-kDa ARR (ARR2) and the 39-kDa ARR (ARR1). The phosphorylation of ARR2 is evident at dim light, while phosphorylation of ARR1 requires stronger light intensities. Both arrestins are phosphorylated by a

Ca^{2+}-calmodulin-dependent protein kinase II (CaMK II). ARR phosphorylation is required for the dissociation of ARR from the phosphorylated M (Mpp) upon photoconversion and for preventing endocytotic internalization of the ARR2–Mpp complex (**Figure 3**).

Based on an understanding of the photochemical cycle, Minke and Selinger explained the PDA phenomenon as follows: cellular ARR2 is present at a concentration which is insufficient to inactivate all the M generated by a large net photoconversion of R to M, leaving an excess of M persistently active in the dark. This explanation easily accounts for the elimination of the PDA response by mutations or by carotenoid deprivation which reduce the cellular level of R. It also accounts for the need to

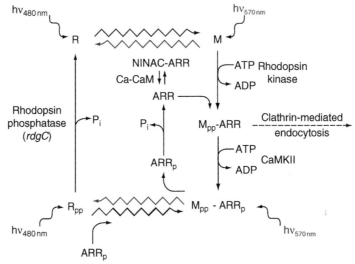

Figure 3 The photochemical cycle: the turn-on and turn-off of the photopigment. The arrestin (ARR) family of proteins play a key role in regulating the activity of G-protein-coupled receptors (GPCR). In *Drosophila,* two protein homologs of vertebrate arrestin exist. Both undergo light-dependent phosphorylation by Ca^{2+} calmodulin-dependent kinase (CaMK II), as shown by Matsumoto and colleagues. Their binding is essential for M inactivation. The study, which clarified the regulatory role of ARR2 was initiated by Selinger, Minke, and colleagues, using *in vitro* assays of ARR2 and M, in *Drosophila* and *Musca* eyes. They showed that upon photoconversion of R to M by illuminating with blue light (wavy blue arrow), the fly ARR2 was found predominantly in the membrane fraction, while photoconversion of phosphorylated M (Mpp) back to phosphorylated R (Rpp), by illuminating with orange light (wavy red arrow), resulted in the detection of ARR2 in the supernatant fraction (cytosol). ARR1 on the other hand always remains membrane bound (not shown). These studies indicated that the functional role of ARR2 binding to M is to terminate its activity. Binding of ARR2 also protects the phosphorylated metarhodopsin (Mpp) from phosphatase activity. Only upon photoregeneration of Mpp to Rpp, is ARR2 released and the phosphorylated rhodopsin (Rpp) is exposed to phosphatase activity by rhodopsin phosphatase (encoded by the *rdgC* gene discovered by O'Tousa and colleagues). These combined actions are crucial for preventing reinitiation of phototransduction in the dark as the reversible binding of ARR2 directs the protein phosphatase only toward the inactive Rpp. Subsequent studies by Dolph and colleagues and by Ranganathan and colleagues revealed that both CaMKII-dependent phosphorylation of ARR2 at Ser366 and the photoconversion of Mpp are required to release phosphorylated ARR2. They, furthermore, showed that the phosphorylation of ARR2 is required for its dissociation from Mpp upon photoconversion and that ARR2 phosphorylation prevents endocytotic internalization of the ARR2–Mpp complex. Montell and colleagues showed that ARR2 binds to myosin III kinase (NINAC). Recent studies by Hardie and colleagues suggest that, under low Ca^{2+} conditions, ARR2 is prevented from rapid binding to M because it is sequestered by NINAC. Ca^{2+} influx acting via CaM, during light (Ca-CaM), rapidly releases ARR2 and allows its binding to M and consequently, M inactivation.

photoconvert large net amounts of R to M to elicit the PDA response. Electrophysiological analysis performed by Zuker and colleagues on strong ARR2 (*arr2*) and ARR1 (*arr1*) mutant alleles revealed a set of phenotypes consistent with the stoichiometric requirement of ARR binding for M inactivation *in vivo*. In their study, they showed that a significant reduction in ARR2 levels leads to abnormally slow termination of light-activated currents and production of a PDA in mutants with low R levels.

Coupling of Photoexcited R to Inositol Phospholipid Hydrolysis

Light-Activated Gq Protein

It has been well established in photoreceptors of several invertebrate species that photoexcited R activates a heterotrimeric G protein. The first experiments, conducted on fly photoreceptors, were carried out by Minke and Stephenson who showed that when pharmacological agents known to activate G proteins were applied to

photoreceptors in the dark, they mimicked the light-dependent activation of the photoreceptor cells. Later studies used genetic screens, which resulted in the isolation of two genes encoding visual specific G-protein subunits. These genes – *dgq* and *gβe* – encode, respectively, the alpha and beta subunits of the guanine nucleotide-binding G protein ($G_q\alpha$ and $G_q\beta$). The *Drosophila* eye-specific $G_q\alpha$ ($DG_q\alpha$), cloned by Hyde and colleagues, showed ∼75% identity to mouse $G_q\alpha$ known to activate PLC. The most direct demonstration that $DG_q\alpha$ participates in the phototransduction cascade came from studies of mutants defective in $DG_q\alpha$ which showed highly reduced sensitivity to light. In the $G\alpha q^1$ mutant – isolated by Scott, Zuker, and colleagues – $DG_q\alpha$ protein levels are reduced to ∼1%, while $G_q\beta$, PLC, and R protein levels are virtually normal. The $G\alpha q^1$ mutant exhibits an ∼1000-fold reduced sensitivity to light and slow response termination, thus strongly suggesting that there is no parallel pathway mediated by another G protein. Manipulations of the $DG_q\alpha$ protein level by the inducible heat-shock promoter made it possible to show a strong correlation between the sensitivity to light and $DG_q\alpha$ protein levels, further establishing its major role in *Drosophila* phototransduction.

The eye-specific $G_q\beta$ ($G_q\beta_e$) shares 50% amino acid identity with other $G\beta$ homolog proteins. Analysis of mutants, isolated by Zuker and colleagues, which are defective in $G_q\beta_e$ ($G\beta_e^1$ and $G\beta_e^2$), showed a greatly (~100-fold) decreased sensitivity and slow response termination. Studies conducted by Selinger, Minke, and colleagues revealed that $DG_q\alpha$ is dependent on $G_q\beta\gamma$ for both membrane attachment and targeting to the rhabdomere, suggesting that the decreased light sensitivity of these mutants may result from the mislocalization of the $DG_q\alpha$ subunit. Previous studies showed that the attachment of $DG_q\alpha$ to $G_q\beta\gamma$ prevents spontaneous guanosine diphosphate and guanosine triphosphate (GDP–GTP) exchange. Analysis of Selinger, Minke, and colleagues of the stoichiometry between the $DG_q\alpha$ and $G_q\beta$ subunits revealed a twofold excess of $G_q\beta$ over $DG_q\alpha$. Genetic elimination of the $G_q\beta$ excess led to spontaneous activation of the visual cascade in the dark, demonstrating that $G_q\beta$ excess is essential for the suppression of dark electrical activity produced by spontaneous GDP–GTP exchange of $DG_q\alpha$.

Light-Activated PLC

Evidence for a light-dependent activation of PLC by $G_q\alpha$ in fly photoreceptors came from combined biochemical and electrophysiological experiments. These experiments were conducted by Selinger, Minke, and colleagues, in membrane preparations and intact *Musca* and *Drosophila* eyes. The eye membrane preparations responded to illumination with a $G_q\alpha$-dependent accumulation of inositol-1,4,5-trisphosphate ($InsP_3$) and inositol-bisphosphate ($InsP_2$), derived from phosphatidylinositol-4,5-bisphosphate (PIP_2) hydrolysis by PLC (**Figure 4**, lower panel). However, genetic elimination of the single $InsP_3$ receptor, performed by Zuker and colleagues and by Hardie and colleagues, had no effect on light excitation, thus putting in question the role of $InsP_3$ in phototransduction.

The key evidence for the participation of PLC in visual excitation of the fly was provided by Pak and colleagues, who isolated and analyzed the PLC gene of *Drosophila*, designated no receptor potential A (*norpA*). The *norpA* mutant has long been a strong candidate for a transduction defective mutant because of its drastically reduced receptor potential. The *norpA* gene encodes a β-class PLC that is predominately expressed in the rhabdomeres (**Figure 4**) and has extensive amino acid homology to a PLC extracted from bovine brain. Transgenic *Drosophila* carrying the *norpA* gene on a null *norpA* background rescued the transformant flies from all the physiological, biochemical, and morphological defects associated with the *norpA* mutants. The *norpA* mutant thus provides essential evidence for the critical role of inositol-lipid signaling in phototransduction, by showing that no excitation takes place in the absence of functional PLC. However, the events required for light excitation downstream of PLC activation remain unresolved.

The TRP Channels

The *trp* Mutant and the Discovery of the TRP Channel

A spontaneously occurring *Drosophila* mutant (isolated by Cosens and Manning), showing a decline in the receptor potential to baseline during prolonged illumination, was designated transient receptor potential (*trp*) by Minke and colleagues (**Figure 2(b)** and **(c)**). Minke and Selinger suggested that the *trp* gene encodes a Ca^{2+} channel/transporter mainly because application of the Ca^{2+} channel blocker La^{3+} to wild-type photoreceptors mimicked the *trp* phenotype. Subsequently, the cloning of the *trp* locus by Montell and Rubin revealed a novel membrane protein with a sequence of a channel. The available sequence of the *trp* gene led, several years later, to the discovery of mammalian TRPs and the TRP superfamily. However, the significance of the *trp* sequence – as a gene encoding a putative channel protein – was only first appreciated after a *trp* homolog, *trp-like* (*trpl*), was cloned by Kelly and colleagues. They used a screen for calmodulin-binding proteins and found a transmembrane (TM) protein. A comparison of its TM domain to that of voltage-gated Ca^{2+} channels and the TRP protein led to the conclusion that this protein is a putative channel protein with large identity to TRP (**Figure 5**). Strong evidence supporting the notion that TRP is the major light-activated channel came from a comparative patch-clamp study of isolated ommatidia (**Figure 1(d)**) of wild-type and the *trp* mutant conducted by Hardie and Minke. The use of Ca^{2+} indicator dyes and Ca^{2+}-selective microelectrodes, by Minke and colleagues, directly demonstrated that the TRP channel is the major route for Ca^{2+} entry into the photoreceptor cell. The final evidence showing that TRP and TRP-like protein (TRPL) are the light-activated channels came from the isolation of a null mutant of the *trpl* gene, by Zuker and colleagues, who demonstrated that the double mutant, *trpl;trp*, is blind. The notion that the *trp* gene encodes a Ca^{2+} channel has been recently confirmed by Hardie and colleagues by using a mutant with a point mutation at the suspected pore region of the channel, altering its Ca^{2+} permeability properties (**Figure 5**, D621 at the pore region of TRP channel).

Biophysical Properties of the TRP Channel

The *Drosophila* light-sensitive channels – TRP and TRPL – could be studied separately by utilizing the trp^{P302} and trp^{P343} null mutants, respectively. The channels are permeable to a

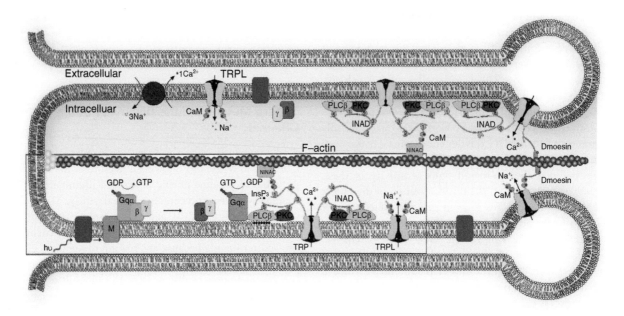

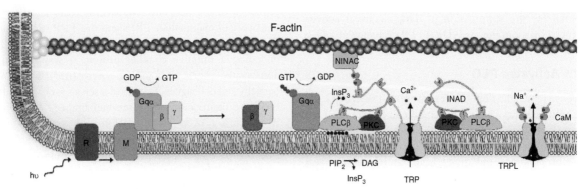

Figure 4   The phosphoinositide cascade of vision. The signaling proteins of the phototransduction cascade are tightly assembled in the microvillar structure (upper panel). The upper panel emphasizes the link of the signaling proteins to the actin cytoskeleton (F-actin) via two proteins: Dmoesin that binds the TRP and TRPL channels, at the base of the microvilli to F-actin and NINAC that associates INAD to F-actin. The only protein that diffuses during the phototransduction cascade is Gqα. The lower panel highlights the molecular components that directly participate in the phototransduction cascade. The numbers on the INAD indicate the various PDZ domains. Upon absorption of a photon, R (encoded by the *ninaE* gene, lower panel) is converted into the active state of the photopigment, M. This, leads to the activation of heterotrimeric G protein (G$_q$) by promoting the GDP to GTP exchange. In turn, this leads to activation of phospholipase Cβ (PLCβ, encoded by the *norpA* gene), which hydrolyzes the minor phospholipid, PIP$_2$ into the soluble InsP$_3$ and the membrane-bound diacylglycerol (DAG). Subsequently, two classes of light-sensitive channels, the TRP and TRPL, open by a still unknown mechanism. PLC also promotes hydrolysis of the bound GTP, resulting in G$_q$α bound to GDP and this ensures the termination of G$_q$α activity. The TRP and TRPL channel openings lead to elevation of calcium ions extruded by the Ca^{2+}/Na$^+$ exchanger CALX (violet circle, upper panel). Elevation of DAG and Ca^{2+} promote eye-specific protein kinase C (PKC, encoded by the *inaC* gene) activity, which regulates channel activity. PLC, PKC, NINAC (encoded by the *ninaC* gene), and the TRP ion channel form a supramolecular complex with the scaffolding protein INAD (encoded by the *inaD* gene). Recent studies by Minke and colleagues have shown a light- and phosphorylation-dependent dynamic association of Dmoesin to the *Drosophila* TRP and TRPL channels involved in rhabdomere maintenance. Dmoesin belongs to the ezrin–radixin–moesin (ERM) protein family, which is known to regulate actin–plasma membrane interactions in a signal-dependent manner. Dmoesin is localized to the base of the microvilli in adult *Drosophila* and its function is still unknown.

variety of monovalent and divalent ions including Na$^+$, K$^+$, Ca^{2+}, and Mg^{2+} and even to large organic cations such as tris(hydroxymethyl)aminomethane (TRIS) and tetraethylammonium (TEA). The reversal potential of the light-induced current (LIC) shows a marked dependence on extracellular Ca^{2+} indicating a high permeability for this ion. Hardie and colleagues measured the permeability

ratios for a variety of divalent and monovalent ions determined under bi-ionic conditions and confirmed a high Ca^{2+} permeability of $\sim$57:1 = Ca^{2+}:Cs$^+$ in the *trpl* mutant and $\sim$4.3:1 = Ca^{2+}:Cs$^+$ for the *trp* mutant.

The TRP and TRPL channels show voltage-dependent conductance during illumination. An early study by Hardie and Mojet revealed that the light response can

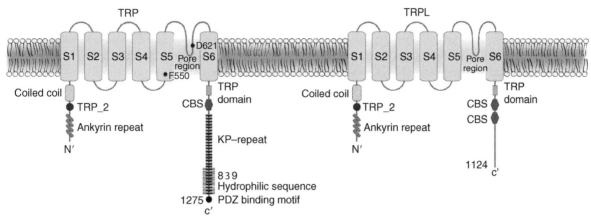

Figure 5 Structural features of the *Drosophila* TRP and TRPL channels. The *trp* and *trpl* genes encode a 1275- and 1124-amino acid membrane protein, respectively, which are presented schematically. The TRP channels share overall 40% amino acid sequence identity with TRPL and a greater similarity ($\sim$70%) in the putative transmembrane regions but little homology ($\sim$17%) in the C-terminal. The transmembrane domain (TM) of TRP and TRPL channels shows significant identity of $\sim$40% with the TM of vertebrate voltage-gated Ca^{2+} channels; all contain six TM helices and a pore loop between S5 and S6. The S5 pore loop and S6 region are likely the pore-forming region of TRP, as evidenced by a study of Hardie and colleagues. They replaced Asp^{621} (D621) with glycine or asparagine and the mutated TRP channel showed reduced Ca^{2+} permeability. Voltage-gated Ca^{2+} channels are formed by a single peptide and contain charged residues in the putative S4 helix, believed to form the voltage sensor. TRP and TRPL channels, on the other hand, assemble by four peptides as homo- or hetromultimeres and do not contain charged residues in S4. While the gating mechanism of the TRP and TRPL channels is unclear, a study conducted by Yoon, Pak, Minke, and colleagues showed that the substitution of Phe^{550} (F550) to isoleucine in S5 of the TRP channel (in the mutant fly trp^{P365}) forms a constitutively active channel leading to extremely fast light-independent retinal degeneration. The TRPL channel was originally identified as a calmodulin (CaM)-binding protein and contains two CaM-binding sites (CBS) in the C-terminal of the channel. One of these is unconventional in the sense that it can bind CaM in the absence of Ca^{2+}. The C-terminus fragment of the TRP sequence also has been reported to bind CaM in a Ca^{2+}-dependent manner. Both channels have a designated TRP domain located adjacent to the S6 with a EWKFAR motif found in many members of the TRP family. At the C-terminal region of the TRP there is a proline-rich sequence with 27 KP repeats, which overlap with a multiple repeat sequence, DKDKKP(G/A)D termed 8 × 9. Such proline-rich motifs occur widely and are predicted to form a structure involved in binding interactions with other proteins, including cytoskeletal elements such as actin. This region is unique to TRP and has not been found in any other member of the TRP family. The last 14 amino acids in the C-terminal of TRP are essential for the binding of INAD (see **Figures 4** and **7**) and form a PDZ binding domain. At the C-terminal, between the CBS and the KP-repeat domains, there is an unmarked sequence with low homology to a proline, glutamic acid, serine, and threonine (PEST) sequence commonly found in CaM-binding proteins and which is believed to be a target for rapid degradation by the Ca^{2+}-dependent protease calpain. The N-terminal regions of the TRP and TRPL proteins contain four ankyrin repeats and a coiled-coil domain. Both domains are believed to mediate protein–protein interactions. The N-terminal regions also contain a TRP_2 domain with unknown function, predicted recently as involved in lipid-binding and trafficking.

be blocked by physiological concentrations of Mg^{2+} ions. The block mainly influenced the TRP channel and affected its voltage dependence. Detailed analysis by Parnas, Katz, and Minke described the voltage dependence of heterologously expressed TRPL in S2 cells and of the native TRPL channel, using the *Drosophila trp*-null mutant. These studies indicated that the voltage dependence of the TRPL channel is not an intrinsic property, as is thought for other members of the TRP family, but arises from divalent open-channel block that can be removed by depolarization. The open-channel block by divalent cations is thought to play a role in improving the signal-to-noise ratio of the response to intense light and may function in light adaptation and response termination.

A comparison with voltage-gated K^+ channels and cyclic nucleotide-gated (CNG) channels, postulates that both TRP and TRPL are assembled as tetrameric channels, thus raising the question whether they assemble as homomultimers or as heteromultimers. Since both null *trp* and *trpl* mutants respond to light, each can clearly function without the other. However, a study by Montell and colleagues – based on heterologous coexpression studies and coimmunoprecipitation – led to the suggestion that the TRP and TRPL channels can assemble into heteromultimers. Detailed measurements of biophysical properties performed by Hardie and colleagues, questioned this conclusion since they found that the wild-type conductance could be quantitatively accounted for by the sum of the conductances determined in the *trp* and *trpl* mutants. In addition, Paulsen, Huber, Minke, and colleagues have demonstrated that the TRPL, but not the TRP, channel reversibly translocates from the rhabdomere to the cell body upon illumination, further implying that TRP and TRPL do not form heteromultimeres.

Lipids Activate the Light-Sensitive Channels in the Dark

The activation of PLC results in the hydrolysis of PIP$_2$ into DAG (**Figures 4** and **6**) and InsP$_3$. The pathway which recycles DAG back to PIP$_2$ – the PI pathway – has emerged to be most important for activation of the TRP and TRPL channels. The most familiar action of DAG is to activate the classical protein kinase C (PKC) synergistically with Ca^{2+}. However, mutations in the eye-specific PKC (PKC, *inaC*; **Figures 4** and **7**) lead to defects in response termination with no apparent effects on activation. A second role for DAG in *Drosophila* photoreceptors is to act as a precursor for regeneration of PIP$_2$ via conversion to phosphatidic acid (PA) by DAG kinase (DGK, *rdgA*; **Figure 6**). Mutations in DGK (*rdgA*) lead to a severe form of light-independent degeneration. Other mutations in the PI pathway – the CDP–diacylglycerol synthase (CDP–DAG Syntase, *cds*, **Figure 6**) and the PI transfer

protein (PITP, *rdgB*; **Figure 6**), also lead to severe forms of light-dependent degeneration. The degeneration phenotype – assumed to be caused by toxic increase in cellular Ca^{2+} due to constitutive activity of the TRP channels – suggested a possible involvement of the PI pathway in channel activation. Accordingly, a hypothesis has been put forward whereby DAG acts as an intracellular messenger leading directly to TRP and TRPL channel activation (see below). However, application of DAG to isolated *Drosophila* ommatidia did not activate the channels.

DAG is also a precursor for the formation of polyunsaturated fatty acids (PUFAs) via DAG lipase (**Figure 6**). Studies conducted by Hardie and colleagues showed that application of PUFAs to isolated *Drosophila* ommatidia, as well as to recombinant TRPL channels expressed in *Drosophila* S2 cells, reversibly activated the TRP and TRPL channels. Moreover, they showed that, in the *rdgA* mutant, TRP and TRPL channels are constitutively active, while elimination of TRP by a mutation (in the double-mutant

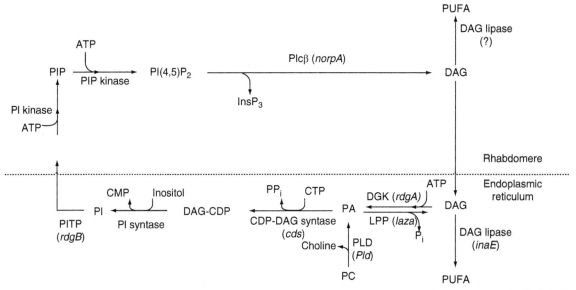

Figure 6 The inositol lipid cycle. In the phototransduction cascade, light triggers the activation of phospholipase Cβ (PLCβ). This catalyzes hydrolysis of the membrane phospholipid phosphatidylinositol-4,5-bisphosphate (PIP$_2$) into water-soluble inositol-1,4,5-trisphosphate (InsP$_3$) and membrane-bound diacylglycerol (DAG). The continuous functionality of the photoreceptors during illumination is maintained by rapid regeneration of PIP$_2$ in a cyclic enzymatic pathway (the PI pathway). DAG is transported by endocytosis to the endoplasmic reticulum (submicrovillar cisternae (SMC)) and inactivated by phosphorylation into phosphatidic acid (PA) via DAG kinase (DGK, encoded by the *rdgA* gene and cloned and localized by Hotta and colleagues) and to CDP–DAG via CDP–DAG syntase (encoded by the *cds* gene and cloned by Zuker and colleagues). PA can be converted back to DAG by lipid phosphate phosphohydrolase (LPP; also designated phosphatidic acid phosphatase (PAP)) encoded by *laza* (cloned by Raghu and colleagues and by Montell and colleagues). PA is also produced from phosphatidyl choline (PC) by phospholipase D (PLD, encoded by *Pld*). DAG is also hydrolyzed by DAG lipase (encoded by *inaE*) into polyunsaturated fatty acids (PUFA). This DAG lipase was predominantly localized outside the rhabdomeres, putting in question the participation of PUFA in channel activation *in vivo*. Subsequently, CDP–DAG is converted into phosphatidyl inositol (PI), which is transferred back to the microvillar membrane, by the PI transfer protein (PITP; encoded by the *rdgB* gene and cloned by Hyde, O'Tousa, and colleagues). Both RDGA and RDGB proteins have been immunolocalized to the SMC of the smooth endoplasmic reticulum at the base of the rhabdomere (**Figure 1(c)**). PIP and PIP$_2$ are produced at the microvillar membrane by PI kinase and PIP kinase, respectively. The SMC in *Drosophila* has been proposed to posses Ca^{2+} stores endowed with InsP$_3$ receptors (InsP$_3$R). Genetic elimination of the single InsP$_3$R did not produce any phenotype, suggesting that Ca^{2+} release from stores does not participate in the production of the response to light. It is important to realize that the PI-recycling enzymes – which are essential for continuous production of PIP$_2$ and hence for the maintained light response – are localized at the SMC. Therefore, the function of the SMC as a lipid-supplying organelle and its role in phototransduction should be further explored.

rdgA;trp), partially rescued the degeneration. The authors suggested a hypothesis in which the elimination of DAG kinase – by the *rdgA* mutation – leads to accumulation of DAG and hence of PUFAs. According to this model, a still undemonstrated accumulation of DAG or PUFAs – acting as messengers of excitation – constitutively activate the TRP and TRPL channels, which in turn leads to a toxic increase in cellular Ca^{2+} and thereby degeneration. Recently, Pak and colleagues identified the *inaE* gene as encoding a homolog of mammalian *sn-1* type DAG lipase and showed that it is expressed predominantly in the cell body of *Drosophila* photoreceptors. Mutant flies expressing low levels of the *inaE* gene product have an abnormal light response, while the activation of the light-sensitive channels was not prevented. Thus, the participation of DAG or PUFAs in TRP and TRPL activation *in vivo* needs further exploration.

It is important to realize that PLC activation, which converts PIP_2 – a charged molecule containing a large hydrophilic head-group – into DAG, which is devoid of the hydrophilic head-group, may cause major changes in lipid packing and lipid-channel interactions. It is, therefore, possible that neither PIP_2 hydrolysis nor DAG production affect the TRP and TRPL channel as second messengers but rather act as modifiers of membrane lipid-channel interactions. This may, in turn, act as a possible mechanism of channel activation.

Organization in a Supramolecular Signaling Complex via the Scaffold Protein Inactivation No Afterpotential D

An important step toward understanding *Drosophila* phototransduction has been achieved by the finding that some of the key elements of the phototransduction cascade are incorporated into supramolecular signaling complexes via a scaffold protein, inactivation no afterpotential D (INAD; **Figure 7**). The INAD protein was discovered by a PDA screen of defective *Drosophila* mutants (*inaD*). The first-discovered *inaD* mutant – the *InaD^P215* – was isolated by Pak and colleagues and subsequently was cloned and sequenced by Shieh and Zhu. Studies in *Calliphora* by Paulsen Huber and colleagues have shown that INAD binds not only TRP but also PLC (NORPA) and PKC (INAC). The interaction of INAD with TRP, NORPA, and INAC was later confirmed in *Drosophila*. It was further found by Zuker, Tsunoda, and colleagues that *inaD* is a scaffold protein, which consists of five ~90-amino acid protein-interaction motifs called postsynaptic density protein (PSD95), *Drosophila* disc large tumor suppressor (DlgA), and zonula occludens-1 protein (zo-1) (PDZ) domains. These domains are recognized as protein modules which bind to a diversity of signaling, cell-adhesion, and cytoskeletal proteins by specific binding to target sequences typically, though not always, in the final three residues of the C-terminal. The PDZ domains of INAD bind to the signaling molecules as follows: PDZ1 and PDZ5 bind PLC, PDZ2 or PDZ4 bind PKC, and PDZ3 binds TRP. This binding pattern is still in debate due to several contradictory reports. On the other hand, TRPL appears not to be a member of the complex, since unlike INAC, NORPA and TRP it remains strictly localized to the rhabdomeral microvilli in the *inaD^1*-null mutant isolated by Zuker, Tsunoda, and colleagues. Several studies by Montell and colleagues have suggested that, in addition to PLC, PKC, and TRP, other signaling molecules such as calmodulin, R, TRPL, and neither inactivation nor afterpotential C (NINAC) bind to the INAD-signaling complex. Such binding, however, must be dynamic. Biochemical studies by Huber, Paulsen, and colleagues conducted in *Calliphora* have revealed that both INAD and TRP are targets for phosphorylation by the nearby PKC. Accordingly, the association of TRP into transduction complexes may be related to increasing the speed and efficiency of transduction events as reflected by the proximity of TRP to its upstream activator, PLC, and its possible regulator, PKC.

TRP plays a major role in localizing the entire INAD multimolecular complex. Association between TRP and INAD is essential for correct localization of the complex in the rhabdomeres. This conclusion was derived by Tsunoda, Zuker, and colleague and by Montell and colleagues using *Drosophila* mutants in which the signaling proteins which constitute the INAD complex were removed genetically, and by deletions of the specific binding domains which bind TRP to INAD. These experiments show that INAD is correctly localized to the rhabdomeres in *inaC* mutants (where PKC is missing) and in *norpA* mutants (where PLC is missing), but severely mislocalized in null *trp* mutants. This indicates that TRP, but not PLC or PKC, is essential for localization of the signaling complex to the rhabdomere. To demonstrate that specific interaction of INAD with TRP is required for the rhabdomeric localization of the complex, the binding site at the C-terminal of TRP was removed or the three conserved residues in PDZ3 – which are expected to disrupt the interaction between PDZ domains and their targets – were modified. As predicted, both TRP and INAD were mislocalized in these mutants. The study of the above mutants was also used to show that TRP and INAD do not depend on each other to be targeted to the rhabdomeres; thus, INAD–TRP interaction is not required for targeting but for anchoring of the signaling complex. Additional experiments on TRP and INAD further show that INAD has other functions in addition to anchoring the signaling complex. One important function is to preassemble the proteins of the signaling complex. Another important function, at least for the case of PLC, is to prevent degradation of the unbound signaling protein.

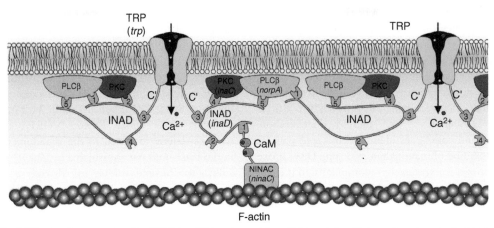

Figure 7 The INAD-signaling complex. The INAD is a 674-amino acid-scaffold protein with homology to cytoskeleton-associated proteins, which include PSD-95 (implicated in clustering of *N*-methyl-D-aspartate (NMDA) receptors and K$^+$ channels), *Drosophila discs large*, and the epithelial tight-junction protein ZO-1. These proteins all include a variable number of so-called PDZ domains – now recognized as protein modules – which anchor a variety of receptors, ion channels, and other signaling molecules, by specific binding to target sequences. The TRP protein anchors the INAD complex adjacent to the plasma membrane while the disruption of the interaction between these two proteins does not influence the target of either protein in pupae, while disrupting their retention at adulthood. The INAD sequence contained five consensus PDZ domains and identified specific interactions between PKC and PDZ2 or PDZ4, TRP and PDZ3, PLC with PDZ1 and PDZ5 and NINAC with PDZ1. This binding pattern is still in debate due to several contradictory reports. It was also reported that the INAD contains a Ca^{2+}-calmodulin-binding site which may be involved in its regulation. Several authors have commented that the close association of elements of the transduction cascade may be essential for the rapid response kinetics, which characterizes *Drosophila* phototransduction, by minimizing or eliminating diffusional delays. The slow response termination in *InaD^{P215}* mutant (disrupting INAD–TRP interaction) also leads to speculation that one function of the INAD complex is to target Ca^{2+} influx via the TRP channels directly to PKC, which appears to be required for Ca^{2+}-dependent inactivation of the light response. It is also speculated that PLC activity may be regulated by TRP-mediated Ca^{2+} influx. With regard to activation, any *norpA* or *inaD* mutation which prevents association of PLC and INAD leads to severe defects in excitation. This can be explained by a large reduction of PLC in the microvillar membranes, which would be expected to be essential for normal phototransduction in any model. Interestingly, the specific dissociation of TRP from the INAD complex in the *inaD^{P215}* mutant, leads to very minor, if any, defect in excitation, except in older flies where TRP levels begin to decline. Nevertheless, several authors have speculated that the INAD complex may provide a mechanism for direct gating of the light-sensitive channels by protein–protein interactions. At present, the only functional roles of INAD, as determined experimentally, are to protect its binding proteins from mislocalization and degradation. These functions are important for maintaining the specific stoichiometry among the signaling proteins (see the section titled 'The photoreceptor cells are sensitive to single photons').

A recent study by Ranganathan and colleagues has suggested that the binding of signaling proteins to INAD may be a dynamic process that allows additional levels of phototransduction regulation. Their study showed that two crystal structural states of the isolated INAD PDZ5 domain differ mainly by the formation of a disulfide bond. This conformational change has a light-dependent dynamic that was demonstrated by the use of transgenic *Drosophila* flies expressing the INAD with a point mutation that disrupts the formation of the disulfide bond. They proposed a model in which PKC phosphorylation of INAD at a still-unknown site promotes a light-dependent conformational change of PDZ5 which distorts its binding site for PLC and thus regulates phototransduction.

The Photoreceptor Cells Are Sensitive to Single Photons

Dim-light stimulation induces discrete voltage (or current) fluctuations in most invertebrate species, which are called quantum bumps (**Figure 2(b)**). Each bump is assumed to be evoked by the absorption of a single photon and these bumps sum up to produce the macroscopic response to more intense lights (**Figure 2(b)**). The bumps vary significantly in latency, time course, and amplitude, even when the stimulus conditions are identical. Bump generation is a stochastic process described by Poisson statistics whereby each effectively absorbed photon elicits only one bump. However, in at least three *Drosophila* mutants (*cam*, *ninaC*, and *arr2*) absorption of a single photon elicits a train of bumps with a rather fixed interval. This phenomenon is explained by a failure of M molecules to be inactivated in these mutants (**Figure 3**) and the existence of a positive and negative feedback and a refractory period in the bump-generating mechanism. The single-photon-single-bump relationship requires that each step in the cascade includes not only an efficient turn-on mechanism, but also an equally effective turn-off mechanism. The functional advantage is the production of a very sensitive photon counter with a fast transient response, very well suited for both the sensitivity and the temporal resolution required by the visual system. The requirement for efficient turn-off mechanisms is revealed when either M (see above) or PLC fail to inactivate and lead to continuous production of bumps long after light is turned off. Cook, Minke, and colleagues demonstrated that *Drosophila* PLC functions as a GTPase-activating protein (GAP). PLC thus terminates its own activity by accelerating the GTPase activity of the $G_q\alpha$ protein thereby causing the dissociation of $G_q\alpha$ from its target, the PLC (**Figure 4**). The inactivation of the G-protein by its target PLC ensures that each activated G-protein activates only one PLC and thus produces only one bump. It also ensures that every activated G-protein will eventually encounter a PLC molecule and thereby produce a bump. The accurate stoichiometry between PLC and G-protein molecules in the rhabdomere of wild-type flies is maintained by the binding of PLC to the scaffold protein INAD (**Figure 7**). This accurate stoichiometry ensures instantaneous inactivation of all activated G-proteins, when light is turned off, and fast response termination. However, in *norpA* or *inaD* mutants, in which the PLC level is reduced, excessive numbers of active $G_q\alpha$ proteins – relative to scarce PLC molecules – leads to continuous production of bumps in the dark until the last $G_q\alpha$-GTP encounters a PLC and is thereby inactivated by the GAP activity of PLC. This remarkable mechanism makes the photoreceptor cell a photon counter and ensures the high temporal and light-intensity resolutions of the response to light.

Conclusions

Drosophila photoreceptors use the ubiquitous inositol-lipid signaling for phototransduction with TRP channels as its target. While mammalian TRP channels are also modulated by the PI cascade and are activated by a large variety of stimuli, the physiological function and mechanism of activation is not entirely clear. In contrast, the physiological function of the *Drosophila* TRP channels as light-activated channels is well established, as well as the requirement of PLC for their activation. It is, therefore, important to direct future efforts toward investigating the activation mechanism of the various TRP channels in the native cells. Genetic tools seem to be especially suitable for such investigation. The similar structural features of the pore domain in all members of TRP channels and the involvement of PLC in their activation/modulation suggest that, in principle, they may share a common mechanism of activation. Thus, studies of *Drosophila* TRP channels is important for the understanding of this diverse and important family of channel proteins.

See also: Circadian Photoreception; Microvillar and Ciliary Photoreceptors in Molluskan Eyes; Phototransduction: Inactivation in Cones; Phototransduction in *Limulus* Photoreceptors; Phototransduction: Inactivation in Rods; Phototransduction: Phototransduction in Cones; Phototransduction: Phototransduction in Rods; Phototransduction: Rhodopsin.

Further Reading

Bloomquist, B. T., Shortridge, R. D., Schneuwly, S., et al. (1998). Isolation of a putative phospholipase C gene of *Drosophila*, *norpA*, and its role in phototransduction. *Cell* 54: 723–733.

Byk, T., Bar-Yaacov, M., Doza, Y. N., Minke, B., and Selinger, Z. (1993). Regulatory arrestin cycle secures the fidelity and maintenance of the fly photoreceptor cell. *Proceedings of the National Academy of Sciences of the United States of America* 90: 1907–1911.

Chyb, S., Raghu, P., and Hardie, R. C. (1999). Polyunsaturated fatty acids activate the *Drosophila* light-sensitive channels TRP and TRPL. *Nature* 397: 255–259.

Cook, B., Bar-Yaacov, M., Cohen Ben-Ami, H., et al. (2000). Phospholipase C and termination of G-protein-mediated signalling *in vivo*. *Nature Cell Biology* 2: 296–301.

Devary, O., Heichal, O., Blumenfeld, A., et al. (1987). Coupling of photoexcited rhodopsin to inositol phospholipid hydrolysis in fly photoreceptors. *Proceedings of the National Academy of Sciences of the United States of America* 84: 6939–6943.

Hardie, R. C. and Minke, B. (1992). The *trp* gene is essential for a light-activated Ca^{2+} channel in *Drosophila* photoreceptors. *Neuron* 8: 643–651.

Hardie, R. C. and Postma, M. (2008). *Phototransduction in Microvillar Photoreceptors of Drosophila and Other Invertebrates. The Senses: A Comprehensive Reference* vol. 1. Vision I, pp. 77–130. San Diego, CA: Academic Press.

Huber, A., Sander, P., Gobert, A., et al. (1996). The transient receptor potential protein (Trp), a putative store-operated Ca^{2+} channel essential for phosphoinositide-mediated photoreception, forms a signaling complex with NorpA, InaC and InaD. *EMBO Journal* 15: 7036–7045.

Minke, B. and Cook, B. (2002). TRP channel proteins and signal transduction. *Physiological Review* 82: 429–472.

Montell, C. and Rubin, G. M. (1989). Molecular characterization of the *Drosophila trp* locus: A putative integral membrane protein required for phototransduction. *Neuron* 2: 1313–1323.

Phillips, A. M., Bull, A., and Kelly, L. E. (1992). Identification of a *Drosophila* gene encoding a calmodulin-binding protein with homology to the *trp* phototransduction gene. *Neuron* 8: 631–642.

Scott, K., Becker, A., Sun, Y., Hardy, R., and Zuker, C. (1995). Gq alpha protein function *in vivo*: Genetic dissection of its role in photoreceptor cell physiology. *Neuron* 15: 919–927.

Tsunoda, S., Sierralta, J., Sun, Y., et al. (1997). A multivalent PDZ-domain protein assembles signalling complexes in a G-protein-coupled cascade. *Nature* 388: 243–249.

Vihtelic, T. S., Hyde, D. R., and O'Tousa, J. E. (1991). Isolation and characterization of the *Drosophila* retinal degeneration B (*rdgB*) gene. *Genetics* 127(4): 761–768.

Yamada, T., Takeuchi, Y., Komori, N., et al. (1990). A 49-kilodalton phosphoprotein in the *Drosophila* photoreceptor is an arrestin homolog. *Science* 248: 483–486.

Relevant Website

http://flybase.org – Flybase: Detailed information on all Drosophila Genes.

Hereditary Vitreoretinopathies

S Meredith and M Snead, Cambridge University Hospitals NHS Foundation Trust, Cambridge, UK

Glossary

Dominant negative effect – An altered gene product is produced that interacts with, and prevents the normal functioning of, the normal protein that is present.

Haploinsufficiency – A mutation resulting in a single copy of the normal gene is present so that only 50% of normal protein is produced, which is not enough for normal function.

OMIM – An online compendium of human genes and genetic phenotypes derived from the Mendelian Inheritance in Man (MIM) database.

Phenotype – The observable characteristics of an organism which are dependent on both genetic and environmental factors.

Pseudoexotropia – The appearance of a divergent squint when both eyes are fixing on the fovea measurable as a large angle kappa.

Introduction

The inherited vitreoretinopathies are collectively characterized by abnormalities in vitreous development and architecture. Since the development of the primary, secondary, and tertiary vitreous is largely complete by the first trimester, abnormalities of vitreous phenotype are present at birth, providing a useful and often pathognomonic diagnostic hallmark. In addition however, it is important to remember that the congenital anomalies of gel development and architecture will not be exempt from the more generalized age-related biochemical and structural changes.

Classification of the Hereditary Vitreoretinopathies

There have been significant, recent advances in both the clinical and molecular genetic analysis of the inherited vitreoretinopathies. Classification on clinical grounds (**Table 1**) is a helpful and convenient way to approach the investigation and diagnosis of the hereditary vitreoretinopathies. Four principal categories are currently recognized as follows: vitreoretinopathies that are associated with (1) skeletal abnormalities, (2) retinal dysfunction, (3) retinal vascular abnormalities, and (4) corneal guttata. Other varieties of inherited vitreoretinopathies do not fit easily into any of these subgroups and it is inevitable that a classification will be expanded as further clinical and molecular genetic heterogeneity is resolved.

Clinical Features of the Hereditary Vitreoretinopathies

The clinical features in each of the individual hereditary vitreoretinopathies are summarized in **Table 2**. Of the clinical features, the vitreous is the only category in which an abnormality has been universally described in the disorders collectively known as the hereditary vitreoretinopathies.

Vitreoretinopathies Associated with Skeletal Abnormalities

Stickler syndrome (OMIM #108300, #604841, and #184840)

The Stickler syndromes are a group of interrelated dominantly inherited disorders of collagen connective tissue which results in an abnormal vitreous embryology and a variable degree of orofacial abnormality, deafness, and arthropathy. Unlike some other hereditary arthro-ophthalmopathies, patients with Stickler syndrome have normal stature. In many patients with Stickler syndrome, systemic features may not be immediately obvious and the diagnosis of Stickler syndrome depends primarily on the recognition of characteristic congenital vitreous changes.

Stickler syndrome is the most common hereditary vitreoretinopathy. The majority of patients with Stickler syndrome have type 1 Stickler syndrome, which is part of the spectrum of type II collagen disorders. The type II collagenopathies, including type 1 Stickler syndrome, are characterized by a membranous vitreous appearance on biomicroscopy (**Figure 1**). Type 2 Stickler syndrome refers to Stickler syndrome associated with defects of type XI collagen and has a different vitreous appearance described as beaded or fibrillar (**Figure 2**). More recently, a recessive form of Stickler syndrome associated with mutations in a gene coding for type IX collagen has been described.

Table 1 Clinical classification of the hereditary vitreoretinopathies

Syndrome		OMIM classification
Vitreoretinopathies associated with skeletal abnormalities		
1. Stickler syndrome	Type 1	#108300
	Type 2	#604841
		#184840
2. Kniest dysplasia		#156550
3. Sponyloepiphyseal dysplasia congenital (SEDC)		#183900
4. Marshall syndrome		#154780
5. Knobloch syndrome		#267750
6. Marfan syndrome		#154700
Vitreoretinopathies associated with progressive retinal dysfunction		
7. Wagner syndrome		#143200
8. Goldmann–Favre syndrome/Enhanced S-cone dystrophy		#268100
Vitreoretinopathies associated with abnormal retinal vasculature		
9. Familial exudative vitreoretinopathy (FEVR)		#133780
		#601813
		#305390
10. Autosomal dominant vitreoretinochoroidopathy (ADVIRC)		#193220
Vitreoretinopathy associated with corneal changes		
11. Snowflake vitreoretinal degeneration		#193230

Weissenbacher–Zweymuller syndrome is sometimes included in the classification of the hereditary vitreoretinopathies. This syndrome is not associated with a vitreoretinopathy. The ocular features are hypertelorism, refractive errors, and strabismus. It results from mutations in the COL11A2 gene which is not expressed in the eye and is therefore not considered further in this article.

In contrast to the developmental myopia in the general population, myopia in Stickler syndrome is typically congenital, of high degree, and nonprogressive. It is important to remember that although the disorder is associated with congenital megalophthalmos, up to 20% of patients are not refractively myopic due to associated cornea plana.

Stickler syndrome is associated with a high risk of retinal detachment. The lifetime risk of retinal detachment is probably approaching 80% in type 1 Stickler syndrome, the majority of which is bilateral frequently as a result of a giant retinal tear at the pars plana (**Figures 3(a)** and **3(b)**). Other ocular findings are radial paravascular lattice, congenital quadrantic lamellar cataract (**Figure 4**), and early-onset nuclear sclerosis.

Orofacial abnormalities are common in Stickler syndrome and, even in the absence of a history of a cleft palate, it is important to examine the oral cavity for evidence of a submucous cleft or high-arch palate (**Figure 5**). Conductive hearing loss is an acknowledged association with any midline cleft via associated eustachian tube dysfunction, but the audiometry in Stickler syndrome will often reveal an additional high-tone (4000 Hz) sensorineural deficit as a

result of associated developmental abnormalities of the cochlea.

Both collagen type II alpha 1 (COL2A1) and collagen type XI alpha 1 (COL11A1) are expressed in cartilage accounting for the combined ocular and skeletal manifestations in Stickler syndrome. The arthropathy typically manifests as joint laxity and hypermobility from an early age and a premature degenerative osteoarthropathy by the third to fourth decade, characteristically affecting the lumbosacral spine, hips, and knees.

The majority of patients with Stickler syndrome will exhibit both vitreoretinal and systemic involvement. A smaller but important subgroup presents with a predominantly ocular, or ocular-only, phenotype as a result of a mutation which is preferentially expressed in ocular tissues. Ophthalmologists have a key role to play in recognition and diagnosis of this important subgroup who are otherwise frequently undiagnosed until retinal detachment occurs.

Other more severe chondrodysplasias result from dominant negative changes within the type II collagen triple helix and include kniest dysplasia, spondyloepiphyseal dysplasia congenita (SEDC), and dominant spondylo-epi-metaphyseal dysplasia (SEMD). Similar to Stickler syndrome, these disorders present with congenital vitreous abnormalities, retinal detachment, deafness, and midline clefting, but with more pronounced arthropathy than Stickler syndrome and disproportionate (rhizomelic) limb shortening.

Kniest dysplasia (OMIM #156550)

Kniest dysplasia is an autosomal dominant vitreoretinopathy associated with skeletal dysplasia presenting with shortening of the trunk and limbs (**Figure 6**). The vitreous is congenitally abnormal and associated with megalophthalmos, cleft palate, and midfacial hypoplasia. Both conductive and sensorineural hearing loss are common and the joints display characteristic enlargement, particularly of the proximal interphalangeal joints, with the fingers appearing long and knobbly (**Figure 7**).

Spondyloepiphyseal dysplasia congenital (OMIM #183900)

SEDC is an autosomal dominant vitreoretinopathy presenting at birth with shortening of the trunk and thereafter progressively disproportionate (proximal) limb shortening, with the hands and feet appearing relatively normal in contrast to Kniest dysplasia, and flattening of the midface (**Figure 8**). The chest is barrel shaped with associated kyphosis and increased lumbar lordosis. The ocular phenotype is very similar to Stickler syndrome with congenital megalophthalmos and a type 1 (membranous) vitreous anomaly.

Marshall syndrome (OMIM #154780)

There is some uncertainty as to whether or not the Stickler and Marshall syndromes are clinically separate entities.

Table 2   Clinical features of the hereditary vitreoretinopathies

Name of the syndrome	Description of the clinical features								
	Refractive error	Cornea	Lens	Vitreous	Retina	ERG	Optic disk	Other ocular features	Systemic features
Vitreoretinopathies associated with skeletal abnormalities									
Stickler syndrome	Congenital nonprogressive myopia		Quadrantic lamellar or early-onset nuclear sclerosis	Membranous or beaded vitreous anomaly	Giant retinal tears, retinal detachment, and radial perivascular lattice degeneration		Myopic/peri-papillary change	Megal-ophthalmos	Midfacial hypoplasia, cleft palate, joint hypermobility, arthropathy, and sensorineural deafness
Kniest dysplasia	Myopia			Membranous vitreous anomaly				Megal-ophthalmos	Shortening of trunk and limbs, midfacial hypoplasia, cleft palate, and hearing loss
SEDC	Myopia			Membranous vitreous anomaly				Megal-ophthalmos	Shortening of trunk and disproportionate proximal limb shortening, kyphosis, lumbar lordosis, and barrel-shaped chest
Marshall syndrome	Myopia		Congenital or juvenile cataracts	Syneresis				Ocular hypertelorism	Spondyloepiphyseal abnormalities, short stature, flat midface, cleft palate, sensorineural deafness, ectodermal dysplasia with hypotrichosis and hypohidrosis, dental structure abnormalities, calvarial thickening, and falx cerebri calcification
Knobloch	High myopia		Subluxated, cataract	Degenerative	Retinal detachment, macular degeneration				Occipital encephalocele, lung hypoplasia, cardiac dextroversion, flat nasal bridge, midface hypoplasia, joint hyperextensibility, unusual palmer crease, and unilateral duplicated renal collecting system
Marfan syndrome	Myopia	Cornea plana	Ectopia lentis, early nuclear sclerosis	Vitreous degeneration	Retinal detachment			Increased globe length, hypoplastic iris, and glaucoma	Increased height, scoliosis, lumbar lordosis, joint laxity, narrow high-arched palate, anterior chest deformity, mitral valve prolapse, mitral and aortic regurgitation, and dilatation of aortic root

Vitreoretinopathies associated with progressive retinal dysfunction

Wagner syndrome/ erosive vitreoretinopathy	Myopic	Early-onset cataract	Syneresis and condensation, membranes	Chorioretinal atrophy, tractional and rhegmatogenous retinal detachment, and retinal pigment epithelial change	Subnormal	Ectopic fovea with pseudoexotropia, and hemeralopia
Goldmann–Favre syndrome/ enhanced S-cone dystrophy		Cortical lens opacities	Liquefaction and fibrillar changes	Equatorial chorioretinal atrophy and pigment clumping, peripheral and macular schisis, and diffuse vascular leakage	Dysfunctional cone mechanism	

Vitreoretinopathies associated with abnormal retinal vascularization

FEVR			Peripheral 'snowflake' vitreous changes	Avascularity of peripheral retina, retinal exudates, retinal traction		Macular pucker, tractional and exudative detachment
ADVIRC	Micro cornea	Congenital cataract	Fibrillar condensation	Circumferential hyperpigmented band, punctuate white opacities, choroidal atrophy, retinal neovascularization, posterior staphyloma	EOG abnormal at level of RPE	Nano-ophthalmos, and closed-angle glaucoma

Vitreoretinopathy associated with corneal changes

Snowflake vitreoretinal degeneration	Corneal guttata	Cataract in young adulthood	Fibrillar degeneration with thickened cortical vitreous	Peripheral retinal abnormalities (small, shiny, snowflake like), sheathing and obliteration of retinal vessels		Waxy pallor

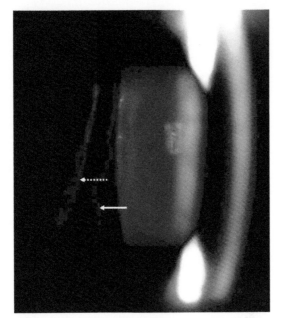

Figure 1   Membranous vitreous phenotype in type 1 Stickler syndrome. Dotted line indicates detached posterior hyaloid membrane; solid line indicates the type 1 membranous anomaly.

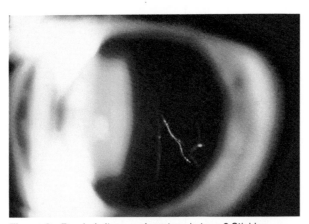

Figure 2   Beaded vitreous phenotype in type 2 Stickler.

Many features are shared such as midfacial hypoplasia, spondyloepiphyseal abnormalities, cleft palate, and sensorineural hearing loss, but patients with Marshall syndrome are reported to also have ectodermal dysplasia with hypertrichosis and hypohyidrosis, calvarial thickening, and ocular hypertelorism. The term Marshall–Stickler syndrome has been used but is potentially confusing and best avoided until Marshall syndrome is better characterized on clinical and molecular genetic grounds.

Knobloch syndrome (OMIM #267750)

Knobloch syndrome has been molecularly characterized and results from a mutation in the gene coding for type XVIII collagen. It has a more extensive systemic phenotype than Stickler or Marshall syndrome with developmental abnormalities described in several major systems of the body, including lung hypoplasia, cardiac dextroversion, and renal abnormalities as well as the ocular features of a degenerative vitreous appearance, high myopia, and retinal degeneration with retinal detachment. The distinguishing clinical feature of Knobloch syndrome is the presence of an occipital encephalocele.

Marfan syndrome (OMIM #154700)

Marfan syndrome is an autosomal dominant connective tissue disorder associated with an abnormal vitreous appearance, myopic astigmatism, and characteristic skeletal features of increased height with disproportionately long limbs and digits, scoliosis and lumbar lordosis, joint laxity, narrow, high-arched palate, and anterior chest deformity (**Figure 9**). Ectopia lentis (with the lens typically dislocated superior and temporally; **Figure 10**) is the ocular major Ghent criterion for the diagnosis of Marfan syndrome. Retinal detachment, myopia, increased globe length, cornea plana, hypoplastic iris, glaucoma, and early nuclear sclerotic cataract are all minor criteria. Marfan syndrome has cardiovascular manifestations of mitral valve prolapse,

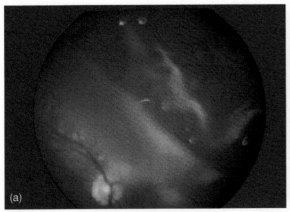
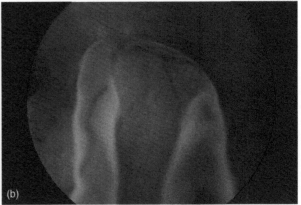

Figure 3   (a) Giant retinal tear characteristic of the retinal break seen in Stickler syndrome. (b) Retinal detachment caused by a giant retinal tear.

mitral regurgitation, dilation of the aortic root, and aortic regurgitation, with aneurysm of the aorta and aortic dissection being major life-threatening complications.

Vitreoretinopathies Associated with Progressive Retinal Dysfunction

The remaining hereditary vitreoretinopathies described in this article are purely ocular disorders. Those that are associated with retinal dysfunction – Wagner syndrome, erosive vitreoretinopathy, Goldmann–Favre syndrome, and enhanced S-cone syndrome (ESCS) – all have measurable electrophysiological changes of the retina demonstrated by recording a subnormal electroretinogram (ERG).

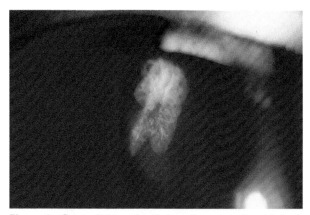

Figure 4   Congenital, quadrantic, lamellar cataract seen in both type 1 and type 2 Stickler syndrome.

Wagner syndrome (OMIM #143200)

The most striking finding in Wagner syndrome is the thickening and incomplete separation of the posterior hyaloid membrane, which tends to occur in a circular band and is variously described as a veil, sheets, or ropes. A large range of chorioretinal abnormalities have been described with the typical finding being chorioretinal atrophy with pigment migration into the retina (**Figure 11**). Electroretinographic responses are progressively subnormal (**Figure 12**) and visual-field testing demonstrates ring scotomas with eventual loss of central visual acuity. The chorioretinal pathology results in gradual progressive visual loss in the absence of retinal detachment, and has a progressive course. There is an association with early-onset cataract and mild myopia. Wagner syndrome is also associated with large angle kappas indicative of an ectopic fovea (**Figure 13**).

Prognosis in Wagner syndrome is poor with progressive visual loss. The risk of rhegmatogenous retinal detachment with Wagner syndrome is higher than that of the normal population, but the incidence does not appear to be as high as that seen in the Stickler syndromes.

The term Jansen syndrome has been used to describe a hereditary vitreoretinopathy but the clinical features reported are consistent with Wagner syndrome and linkage has been demonstrated in the original family to the same area as the causative gene in Wagner syndrome. The syndrome referred to as erosive vitreoretinopathy is reported to be associated with vitreous changes, hemeralopia with accompanying grossly reduced rod and cone responses, progressive chorioretinal atrophic changes, and combined traction-rhegmatogenous retinal detachments. Affected family members also had large angle kappas.

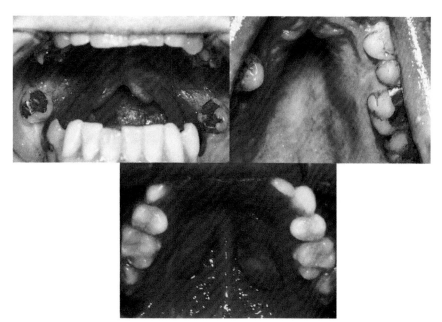

Figure 5   Bifid uvula and high-arch palate in Stickler syndrome.

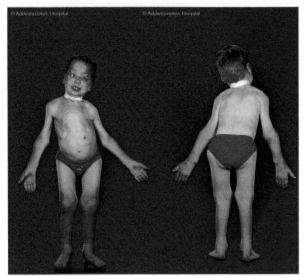

Figure 6   Kniest dysplasia demonstrating shortening of trunk and limbs.

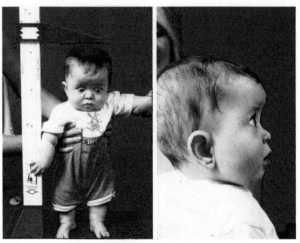

Figure 8   Spondyloepiphyseal dysplasia congenital demonstrating short stature and flattening of the midface.

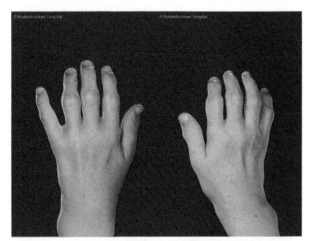

Figure 7   Interphalangeal dysplasia in Kniest dysplasia.

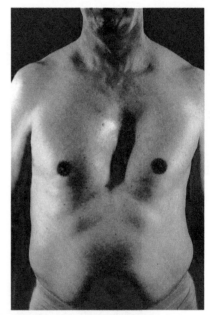

Figure 9   Pectus excavatum in Marfan syndrome.

Mutations in the same gene that cause Wagner syndrome have been implicated in erosive vitreoretinopathy. It is likely both Jansen syndrome and erosive vitreoretinopathy are phenotypic variants of Wagner syndrome.

Goldmann–Favre syndrome/enhanced S-cone dystrophy (OMIM #268100)

Patients with Goldmann–Favre syndrome have liquefaction and fibrillar changes of the vitreous, night blindness, equatorial chorioretinal atrophy and pigment clumping, peripheral and macular schisis, cortical lens opacities, rod–cone dysfunction, and diffuse vascular leakage on fluorescein angiography. Goldmann–Favre syndrome is inherited as an autosomal recessive disorder and is now known to be caused by a mutation in the same gene that causes ESCS (the NR2E3 gene). ESCS is the only inherited retinal disease

which exhibits a gain in photoreceptor function, with patients showing enhanced sensitivity to blue (short wavelength) light, with night blindness and loss of sensitivity to long and medium wavelengths. The ERG findings in this group of disorders demonstrate undetectable rod-isolated responses and reduced combined rod–cone responses.

Vitreoretinopathies Associated with Abnormal Retinal Vasculature

The clinical feature on ocular examination of the hereditary vitreoretinopathies in this category is an abnormal vascular pattern on the retina. This is demonstrated in more subtle cases using fluorescein angiographic studies.

Familial exudative vitreoretinopathy (OMIM #133780, #601813, and #305390)

The primary pathological process in familial exudative vitreoretinopathy (FEVR) is believed to be a premature arrest of retinal angiogenesis and vascular differentiation resulting in incomplete vascularisztion of the peripheral retina. The failure to vascularize the retina is the unifying feature seen in all affected individuals but, by itself, is usually asymptomatic. Secondary changes include neovascularization, retinal exudates, peripheral snowflake vitreous changes, and retinal traction. Epiretinal traction from vitreous and retinal surface membrane is the usual cause of

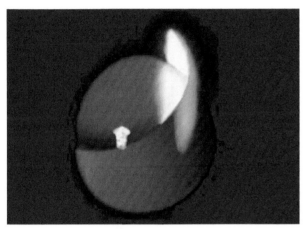

Figure 10 Ectopia lentis with dislocation of the lens superiorly in Marfan syndrome.

retinal detachment and severe retinal distortion is a common finding resulting in retinal folds and tractional or combined tractional and rhegmatogenous retinal detachment (**Figure 14**). The prognosis is highly variable with some individuals blind by age 10 and others (sometimes within the same family) asymptomatic throughout adult life. The eye with the milder phenotype may have only a small avascular area, visible only on fluorescein angiography.

Autosomal dominant vitreochoroidopathy (OMIM #193220)

Autosomal dominant vitreochoroidopathy (ADVIRC) is characterized by vitreous liquefaction with or without peripheral vitreal condensations. Peripheral pigmentary changes typically occur at the equatorial region with a discrete posterior boundary associated with diffuse retinal vascular leakage, cystoid macular edema, and early-onset cataract. The peripheral pigmented band extends from the ora serrata to the equator for 360° of the retina. Other ocular associations are vitreous cells and condensation, punctate opacities in the retina, choroidal atrophy, and early nuclear sclerosis. The ERG responses are normal although the electroculogram (EOG) has been shown to be abnormal in ADVIRC.

Vitreoretinopathy Associated with Corneal Changes

The remaining category contains only one hereditary vitreoretinopathy called snowflake vitreoretinal degeneration named after the retinal appearance.

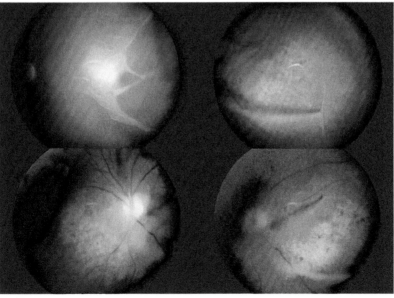

Figure 11 Fundal images showing vitreous condensations and chorioretinal atrophy characteristic of Wagner syndrome.

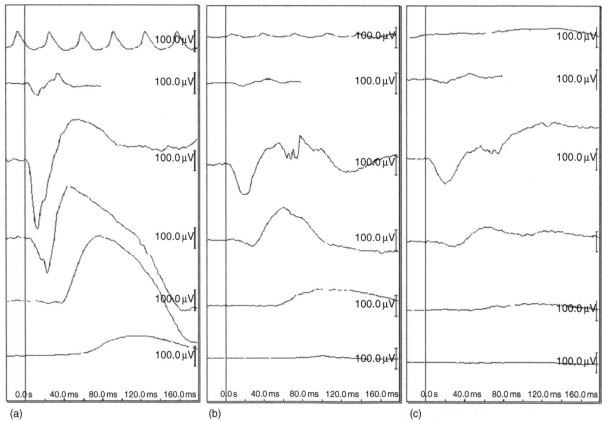

Figure 12 Rod and cone electroretinograms. (a) Representative normal responses, (b) responses from the left eye of a patient with Wagner syndrome aged 55, and (c) responses from the left eye of the daughter aged 19 who had inherited Wagner syndrome. The top two traces are light-adapted cone responses to 30-Hz flicker and to a bright flash; lower four traces are dark-adapted rod and mixed rod–cone responses to a range of stimulus intensities from dim (bottom) to bright (fourth from bottom).

Snowflake vitreoretinal degeneration (OMIM #193230)

The pathognomic feature of snowflake vitreoretinal degeneration is the association with corneal guttata. Disk pallor is also a key feature of snowflake vitreoretinal degeneration not described in any of the other hereditary vitreoretinopathies. The vitreous has a fibrillar appearance. The snowflake-like opacities in the retina are small and may not be immediately obvious. Minor vascular abnormalities of small retinal vessels have been described in snowflake vitreoretinal degeneration.

Molecular Genetics of the Hereditary Vitreoretinopathies

The classification in **Table 1** correlates with genotype. **Table 3** lists the causative genes that have been identified to date in the hereditary vitreoretinopathies with the protein affected (**Table 3**).

Mutations in genes coding for the structural components of vitreous result in a hereditary vitreoretinopathy. In common with connective tissues in other parts of the body, vitreous is composed of an extracellular matrix with relatively few cells and the arrangement of the collagenous proteins, along with the high water content, maintains transparency. Vitreous collagen fibrils are heterotypic and have a core of type II and V/XI collagen, with type IX collagen on the surface. Mutations in genes coding for chains which comprise these three types of collagen result in the most common hereditary vitreoretinopathy, Stickler syndrome.

The majority of patients with Stickler syndrome have type 1 Stickler syndrome which is part of the spectrum of type II collagen disorders along with Kniest dysplasia and SEDC. Most have premature termination mutations of the COL2A1 gene and are characterized by a membranous vitreous appearance on biomicroscopy. Other pedigrees exhibit a different beaded vitreous phenotype and are associated with mutations in one of the genes coding for type V/XI collagen (COL11A1). Type 1 Stickler syndrome, in the majority of cases, results from haploinsufficiency from nonsense-mediated decay through point mutations or frameshifts, whereas type 2 Stickler syndrome results from dominant negative mutations.

Collagen types II and IX are expressed in both vitreous and cartilage, so Stickler syndrome is characterized by an

Figure 13 Pseudoexotropia in Wagner syndrome.

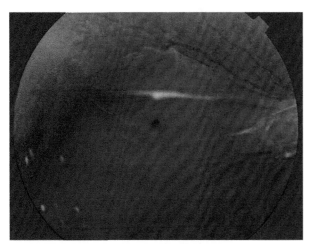

Figure 14 Retinal folds seen in familial exudative vitreoretinopathy.

ocular and skeletal phenotype. Cartilage also contains type XI collagen, which is similar to the type V/XI collagen of vitreous and they share the α1(XI) chain that is encoded by COL11A1; so mutations in this gene have an ocular and skeletal phenotype. A subgroup of Stickler syndrome, designated predominantly ocular, or ocular-only, has been shown to result from mutations of COL2A1 which are preferentially expressed in ocular tissues. Exon 2 of COL2A1 is spliced out of cartilage type II collagen and therefore mutations in exon 2 have a normal skeletal phenotype while displaying the characteristic ocular features of Stickler syndrome. Prior to detailed understanding of the molecular genetics of Stickler syndrome, the existence of patients demonstrating

ocular features without systemic findings was a source of confusion between Stickler syndrome and other vitreoretinal degenerative conditions without systemic involvement.

Fibrillins are one of the main constituents of extracellular microfibrils and these provide structural support in many tissues. Mutations in FBN1, encoding for fibrillin 1, are the major cause of Marfan syndrome. Mutations in a second gene, transforming growth factor beta receptor II (TGFBR2), have been found in patients fulfilling the clinical criteria for Marfan syndrome, but the mechanism for this remains to be clarified.

Wagner syndrome is a further example of a hereditary vitreoretinopathy which results from a mutation in a gene coding for a structural protein in the vitreous. CSPG2 encodes the core protein of a chondroitin sulfate proteoglycan called versican. Versican binds hyaluronan and therefore may contribute to the structure of the vitreous. Versican undergoes alternative splicing and several splice variants have been identified in the eye. All causative mutations to date affect the splicing of exon 7 of the CSPG2 gene which represents one of two glycosaminoglycan attachment sites. In Wagner syndrome there is an altered appearance to the vitreous, and progressive retinal dysfunction and hemeralopia even in the absence of retinal detachment, suggesting that versican has additional roles to being a structural component of the vitreous.

FEVR is also genetically heterogenous. All the causative genes to date in FEVR code for proteins involved in the Wnt signaling pathway. Frizzled homolog 4 (Drosophila) (FZD4) is a presumptive Wnt receptor, lipoprotein-receptor-related protein 5 (LRP5) can transduce Wnt signaling *in vitro*, and Norrie disease (Pseudoglioma) (NDP) encodes for norrin. Norrin is present in extracellular matrices and is thought to act as a ligand-receptor pair with FZD4. Although unrelated to Wnts, norrin may act on the Wnt signaling pathway through the frizzled receptor. Wnt receptors are implicated in retinal neovascularization. NDP mutations can also cause Norrie disease (characterized by incomplete retinal vascularization along with more extensive retinal degenerative changes, microphthalmia, and progressive mental disorder).

In general, the molecular genetics of the hereditary vitreoretinopathies demonstrate how mutations of key genes involved in ocular development and structure can result in different phenotypes. Goldmann–Favre syndrome is allelic with enhanced S-cone dystrophy (ESCD). The responsible gene encodes a retinal nuclear receptor involved in signaling pathways. ADVIRC is allelic with Best's macular dystrophy and results from mutations in the VMD2 gene encoding a transmembrane chloride channel. Snowflake vitreoretinal dystrophy is caused by a mutation in a gene encoding a different transmembrane channel.

Of the hereditary vitreoretinopathies associated with skeletal abnormalities, the majority are inherited as an autosomal dominant trait. The exceptions are Stickler

Table 3   Causative genes identified to date and the protein affected in the hereditary vitreoretinopathies

Vitreoretinopathy	OMIM number	Causative gene	Affected protein	First reference
Stickler syndrome	#108300	COL2A1	Type II collagen	Ahmad et al. (1991)
	#604841	COL11A1	Type XI collagen	Richards et al. (1996)
	#108300	COL9A1	Type IX collagen	van Camp et al. (2006)
Kniest dysplasia	#156550	COL2A1	Type II collagen	Winterpacht et al. (1993)
SEDC	#183900	COL2A1	Type II collagen	Tiller et al. (1995)
Knobloch syndrome	#267750	COL18A1	Type XVIII collagen	Sertie et al. (2000)
Marfan syndrome	#154700	FBN1	Fibrillin	Dietz et al. (1991)
		TGFBR2	Transmembrane receptor	Boileau et al. (1993)
Wagner syndrome	#143200	CSPG2	Chondroitin sulfate proteoglycan	Miyamoto et al. (2005)
Goldmann–Favre syndrome	#268100	NR2E3	Retinal nuclear receptor	Sharon et al. (2003)
FEVR	#133780	FZD4	Wnt receptor	Robitaille et al. (2002)
	#601813	LRP5	Wnt receptor	Toomes et al. (2004)
	#305390	NDP	Wnt receptor ligand	Fullwood et al. (1993)
ADVIRC	#193220	VMD2	Bestrophin	Yardley et al. (2004)
Snowflake vitreoretinopathy	#193230	KCNJ13	K^+ transporter	Hejtmancik et al. (2008)

syndrome specifically associated with COL9A1 mutations, and Knobloch syndrome, both of which have an autosomal recessive pattern of inheritance. Wagner syndrome/erosive vitreoretinopathy is inherited in an autosomal dominant pattern. Favre–Goldmann syndrome/ ESCD is autosomal recessive. Vitreoretinopathies associated with abnormal retinal vascularization are again inherited as autosomal dominant traits for the majority with the one exception being an X-linked form of FEVR. Snowflake vitreoretinal dystrophy is autosomal dominant.

See also: Rhegmatogenous Retinal Detachment; Secondary Photoreceptor Degenerations.

Further Reading

Ahmad, N. N., Ala-kokko, L. A., Knowlton, R. G., et al. (1991). Stop codon in the procollagen II (COL2A1) in a family with Stickler syndrome (arthro-ophthalmopathy). *Proceedings of the National Academy of Sciences of the United States of America* 88: 6624–6627.

Boileau, C., Jondeau, G., Babron, M-C., et al. (1993). Autosomal dominant Marfan-like connective-tissue disorder with aortic dilation and skeletal anomalies not linked to the fibrillin gene. *American Journal of Human Genetics* 53: 46–54.

Dietz, H. C., Cutting, G. R., Pyeritz, R. E., et al. (1991). Marfan syndrome caused by a recurrent *de novo* missense mutation in the fibrillin gene. *Nature* 352: 337–339.

Fullwood, P., Jones, J., Bundey, S., et al. (1993). X-linked exudative vitreoretinopathy: Clinical features and genetic linkage analysis. *British Journal of Ophthalmology* 77: 168–170.

Hejtmancik, J. F., Jiao, X., Li, A., et al. (2008). Mutations in KCNJ13 cause autosomal-dominant snowflake vitreoretinal degeneration. *American Journal of Human Genetics* 82(1): 174–180.

Miyamoto, T., Inoue, H., Sakamoto, Y., et al. (2005). Identification of a novel splice site mutation of the CSPG2 gene in a Japanese family with Wagner syndrome. *Investigative Ophthalmology and Visual Science* 46: 2726–2735.

Richards, A. J., Yates, J. R. W., Williams, R., et al. (1996). A family with stickler syndrome type 2 has a mutation in the COL11A1 gene

resulting in the substitution of glycine 97 by valine in alpha1(XI) collagen. *Human Molecular Genetics* 5(9): 1339–1343.

Robitaille, J., MacDonald, M. L., Kaykas, A., et al. (2002). Mutant frizzled-4 disrupts retinal angiogenesis in familial exudative vitreoretinopathy. *Nature Genetics* 32(2): 326–330.

Ryan, S. J., Hinton, D. R., Schachat, A. P., and Wilkensen, P. (eds.) (2006). *Retina,* 4th edn. Philadelphia, PA: Elsevier.

Sertie, A. L., Sossi, V., Camargo, A. A., et al. (2000). Collagen XVIII, containing an endogenous inhibitor of angiogenesis and tumor growth factor, plays a critical role in the maintenance of retinal structure and in neural tube closure (Knobloch syndrome). *Human Molecular Genetics* 9: 2051–2058.

Sharon, D., Sandberg, M. A., Caruso, R. C., Berson, E. L., and Dryja, T. P. (2003). Shared mutations in NR2E3 in enhanced S-cone syndrome, Goldmann–Favre syndrome, and many cases of clumped pigmentary retinal degeneration. *Archives of Ophthalmology* 121(9): 1316–1323.

Taylor, D. and Hoyt, C. S. (eds.) (2005). *Pediatric Ophthalmology and Strabismus,* 3rd edn. Philadelphia, PA: Elsevier.

Tiller, G. E., Weis, M. A., Polumbo, P. A., et al. (1995). An RNA-splicing mutation (G+5IVS20) in the type II collagen gene (COL2A1) in a family with spondyloepiphyseal dysplasia congenital. *American Journal of Human Genetics* 56(2): 388–395.

Toomes, C., Bottomley, H. M., Jackson, R. M., et al. (2004). Mutations in LRP5 or FZD4 underlie the common familial exudative vitreretinopathy locus on chromosome 11q. *American Journal of Human Genetics* 74(4): 721–730.

Van Camp, G., Snoeckx, R. L., Hilgert, N., et al. (2006). A new autosomal recessive form of stickler syndrome is caused by a mutation in the COL9A1 gene. *American Journal of Human Genetics* 79(3): 449–457.

Winterpacht, A., Hilbert, M., Schwarze, U., et al. (1993). Kniest and Stickler dysplasia phenotypes caused by collagen type II gene (COL2A1) defect. *Nature Genetics* 3(4): 323–326.

Yardley, J., Leroy, B. P., Hart-Holden, N., et al. (2004). Mutations of VMD2 splicing regulators cause nanophthalmos and autosomal dominant vitreoretinochoroidopathy (ADVIRC). *Investigative Ophthalmology and Visual Science* 45(10): 3683–3689.

Relevant Website

http://www.ncbi.nlm.nih.gov – National Center for Biotechnology Information, Online Mendelian Inheritance in Man.

Histogenesis: Cell Fate: Signaling Factors

M Cwinn, B McNeill, A Ha, and V A Wallace, Ottawa Hospital Research Institute, Ottawa, ON, Canada

2010 Elsevier Ltd. All rights reserved.

Glossary

Cell extrinsic signaling – It refers to the activation of a biological pathway within a particular cell by an extracellular factor such as a hormone and morphogen.

Cell intrinsic signaling – It refers to the activation of a biological pathway within a particular cell by a factor that is produced by that cell and is not secreted, such as a transcription factor.

Differentiation – The process whereby a less-specialized cell, such as a progenitor, becomes a specialized cell type. In general, a cell undergoing a differentiation program will exit the cell cycle and become restricted in the number of lineages it can adopt.

Lineage commitment/fate specification – The process whereby cell intrinsic and extrinsic cues will direct a progenitor cell toward differentiating into a specific cell lineage or cell type.

Multipotency – The ability of a particular progenitor cell to give rise to a limited number of different cell lineages.

Retinal progenitor cell (RPC) – An undifferentiated, proliferating, multipotent cell located in the neuroblast layer of the developing retina. RPCs are competent to give rise to the six neural and one glial cell types found in the mature retina; however, as retinal development proceeds, the competence of an RPC becomes restricted. The undifferentiated state, proliferative capacity, and multipotency of RPCs is controlled by both cell extrinsic and intrinsic signaling factors.

Introduction

The six different types of neurons and the one glial cell type in the adult vertebrate retina are generated in a conserved temporal sequence from a pool of multipotential retinal progenitors cells (RPCs). Retinal ganglion cells (GCs), cone photoreceptors, horizontal cells, and half of the amacrine neurons are born during the embryonic period, while the remaining amacrine neurons, bipolar neurons, and Müller glia are born in the postnatal period. Rod photoreceptors are generated throughout histogenesis.

To explain the temporal sequence of neuron generation from multipotential progenitors, a competence model has been proposed in which RPCs progress irreversibly through a series of stages where they are competent to generate a limited number of cell types. Because, depending upon the species, retinal histogenesis occurs over several days to weeks, factors that influence proliferation, timing of differentiation, as well as cell fate will impact this program. The role of transcription factors in lineage specification of RPCs is well established and a growing body of evidence also implicates a role for intercellular signaling in this process. Here, we review the impact of major developmental signaling pathways such as Hedgehog, Wnt, Notch, transforming growth factor-β (TGF-β), and retinoic acid (RA) on the histogenic program in the vertebrate retina.

Notch

The notch pathway is an evolutionarily conserved intercellular signaling mechanism that regulates numerous cellular programs. Activation of notch signaling is initiated through interaction of the notch extracellular domain on the surface of one cell with the DSL ligands (Delta, Serrate, and Lag2) on the surface of a neighboring cell (**Figure 1**). Upon binding to the DSL ligand, notch is cleaved by the γ-secretase complex releasing the notch intracellular domain (NICD). NICD translocates to the nucleus where it interacts with C-promoter binding factor 1/suppressor of hairless/longevity assurance gene 1 (CSL) complex and converts it from a transcriptional repressor to an activator. In the absence of notch signaling, CSL recruits corepressor proteins to form a transcriptional repressor complex, which inhibits target gene expression. The association of NICD with CSL displaces corepressors and recruits coactivators which results in transcriptional activation of basic helix–loop–helix (bHLH) transcription factors such as hairy and enhancer of split 1 and 5(Hes1 and Hes5).

Findings from early studies on the role of notch signaling in the retina suggest that one of its major functions is in controlling the timing of cell cycle exit and maintenance of the retinal progenitor pool. For example, in the frog, fish, rat, chick, and mouse retina the level of notch signaling is inversely proportional to the rate of cell cycle exit and neuronal differentiation. Similar to its actions in other regions of the central nervous system (CNS), the antidifferentiation effects of notch activity in the retina are associated with inhibition of proneural bHLH gene

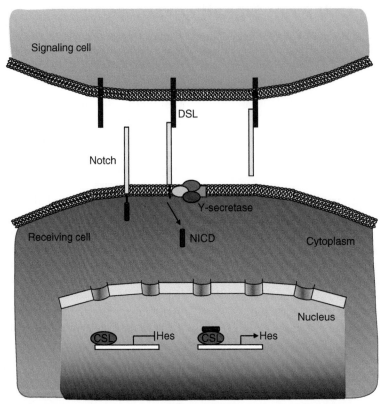

Figure 1   The notch signaling pathway. Ligand binding to the notch receptor leads to a γ-secretase-mediated proteolytic cleavage resulting in the release of the notch intracellular domain (NICD). NICD is able to enter the nucleus and associate with CSL (CBF1, Su(H), Lag1) resulting in transcription activation of target genes such as Hes genes.

expression. While notch activity does not appear to influence the temporal competence of retinal, it has been implicated in cell fate selection. Notch pathway inactivation at early stages of retinal development results in the preferential production of GC or cone photoreceptors and increased notch signaling or expression of its downstream effectors, Hes1 and Hes5, at later stages of retinal development promotes Müller glial cell development. If notch signaling maintains RPCs in cycle, it follows that notch signaling has to be reduced for RPCs to differentiation. Recent work in zebrafish suggests that regional differences in the localization of notch ligands contribute to the variability in notch signaling levels in the retinal neuroepithelium, thereby promoting the balance between cell cycle reentry and exit.

The Wnt Pathway

Wnt signaling has been shown to regulate a number of cellular processes, including differentiation, proliferation, polarity, and migration. Wnt molecules are lipid-modified secreted glycoproteins and 21 vertebrate Wnt homologs have been identified to date. Wnt binding to frizzled (Fzd) receptors and low-density lipoprotein-related receptor

5/6 (Lrp5/6) coreceptors induces the activation of the cytoplasmic phosphoprotein disheveled (Dsh). Three distinct signaling pathways have been described that function downstream of Dsh activation: canonical, planar cell polarity (PCP), and the Wnt/calcium pathway (**Figure 2**). The canonical Wnt pathway, the best characterized of the three, functions to regulate the stability and transcriptional activity of β-catenin. In the absence of Wnt signaling, β-catenin is phosphorylated and targeted for proteosome-dependent degradation by the axin/adenomatous polyposis coli (APC)/glycogen synthase kinase-3β (GSK-3β) destruction complex. Wnt signaling inhibits the activity of the degradation complex, which allows stabilized β-catenin to translocate to the nucleus where it interacts with T-cell factor (TCF)/lymphoid enhancing binding factor (LEF) transcription factors to regulate target gene expression. In the PCP pathway, which is involved in cell and tissue polarity, Rho GTPases are activated leading to alterations in the cytoskeleton. Finally, the Wnt/calcium pathway stimulates the release of intracellular calcium and activates protein kinase C and calcium/calmodulin-dependent protein kinase II to regulate cell fate and movement.

Gene expression profiling has revealed the expression of a number of Wnt pathway components, including Wnt ligands and Fzd receptors in the developing retina.

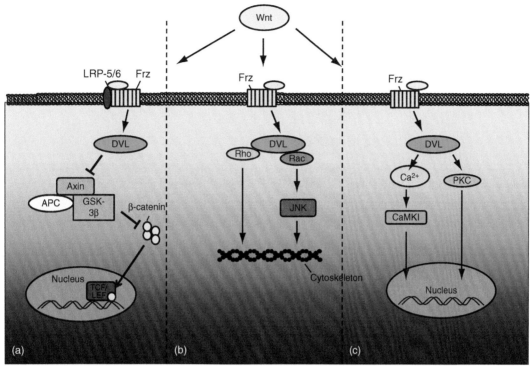

Figure 2 Wnt signaling pathway. Wnt signaling is initiated with the binding of a Wnt protein to its transmembrane receptor frizzled (Frz), leading to activation of one of three main branches of the Wnt signaling pathway. (a) Canonical signaling pathway, where the inhibition of the Axin/APC/GSK-3β destruction complex leads to an accumulation of β-catenin that is able to translocate into the nucleus and regulate transcription. (b) Planar cell polarity (PCP) pathway, where activation of Rho GTPases, Rho, and Rac results in changes to the cytoskeleton. (c) Wnt or calcium pathway, where an increase in intracellular calcium activates protein kinase C (PCK) and calcium or calmodulin-dependent protein kinase II (CaMKII) to regulate cell fate and movement.

The analysis of TCF/LEF transgenic Wnt reporter mice suggests that canonical Wnt signaling is active during eye development. Numerous studies published in recent years on chick, fish, frog, and mammals have implicated Wnt activity in retinal field establishment, retinal regeneration, and lens development. The role of Wnt signaling in retinal histogenesis is more controversial. In the frog, Fzd5 signaling through β-catenin is required for the induction of *Sex determining Y-box 2* (*Sox2*) expression and retinal neurogenesis. In the mouse and chick, however, canonical Wnt pathway activation suppresses neurogenesis and neuronal differentiation. The function of Fzd5 and β-catenin in retinal development is not conserved, as retinal cell specification and proliferation are normal in knockout mice for these genes, although β-catenin is required for lamination and cell adhesion in the retina. Moreover, the effect of Wnt/β-catenin on proliferation appears to be context and species dependent. Increased canonical Wnt signaling promotes proliferation in the frog and fish retina as well as at the marginal zone in the chick retina, but it inhibits proliferation in peripheral regions of the mouse and chick retina. One common theme that has emerged from these studies is the effect of Wnt/β-catenin signaling at the peripheral region of the developing eyecup. In mouse and chick, Wnt/β-catenin

promotes the acquisition of peripheral eyecup fates, including the ciliary marginal zone, ciliary body, and iris. Thus, the conservation of Wnt/β-catenin signaling in eye development might be at the level of the localization of the signal, rather than its developmental outcome.

The Hedgehog Signaling Pathway

Members of the Hedgehog (Hh) protein family are secreted signaling molecules that mediate patterning and growth in a number of tissues and organs during embryonic development. In mammals, there are three members of the hedgehog family: Sonic Hedgehog (Shh), Desert Hedgehog, and Indian Hedgehog. The biologically active amino-terminal fragment of Hh proteins is generated by autoproteolytic cleavage of the precursor protein, which is catalyzed by residues in the carboxy terminal domain of the precursor protein. Fully processed Hh-N is lipid modified with palmitate at the amino terminus and cholesterol at the carboxy terminus. For simplicity, we will refer to fully processed Shh-N as Shh. The Hh signaling cascade is initiated by Hh binding to its receptor, the 12-pass transmembrane protein, Patched (Ptc), and subsequent derepression of Smoothened (Smo), a seven-transmembrane domain

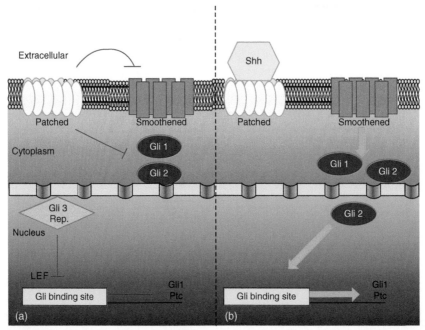

Figure 3 The hedgehog signaling pathway. (a) In the absence of Shh ligand, Patched (Ptc) represses Smoothened (Smo), leading to the inactivation of the Shh pathway. In this scenario, Gli3 repressor can also bind Gli consensus motifs to further repress hedgehog pathway activity. (b) The binding of Shh to Ptc derepresses Smo. The signaling cascade then converges on the Gli1 and Gli2 transcription factors that translocate into the nucleus to induce the transcription of target genes, such as Gli1 and Ptc.

receptor that mediates signal transduction leading to target gene activation (**Figure 3**). The signaling cascade converges on the activity of the *Gli* transcription factors, which regulate target gene expression, in part, through binding to conserved sequences in the promoters and enhancers of these genes. In vertebrates, there are three *Gli* proteins: *Gli1*, *Gli2*, and *Gli3*. *Gli1* and *Ptc* are direct targets of the Hh signaling pathway and the presence of mRNA for these genes indicates regions of active Hh signaling in tissues.

Retinal expression of Hh pathway components has been described in a number of vertebrate species. Several lines of evidence indicate that Shh expression in GCs is the primary mediator of Hh signaling during retinal development. The timing of Shh and Gli induction in GCs and RPCs, respectively, mirrors the central to peripheral wave of GC differentiation. GCs and Shh are required for Gli induction in RPCs, as GC ablation in explant culture or through conditional endotoxin expression abrogates *Gli1* expression, as does conditional inactivation of *Shh* in the retina. In the adult murine retina, *Shh* is also expressed in the inner nuclear layer most likely in amacrine neurons and *Ptc* is expressed in Müller glia; however, the functional significance of Hh pathway expression at this stage is not known.

Shh signaling has been shown to play a critical role in the histogenic program of RPC in a number of vertebrate species. Shh functions as a mitogen for RPCs, thereby regulating the total cell number in the retina. The mitogenic effect of Hh signaling in this context is mediated, in part, through its effects on the kinetics of the cell cycle and

the regulation of G1 and G2 cell cycle genes. Shh signaling also exerts species-dependent effects on the timing of cell cycle exit of RPCs. In the mouse, Hh is required to maintain RPC in the cell cycle, possibly by regulating expression of Hes1, an antidifferentiation bHLH transcription factor. Conversely, in zebrafish, Hh signaling is required for driving cell cycle exit, in part through the induction of p57, a cyclin-dependent kinase inhibitor. These Hh-dependent effects on cell cycle exit are associated with alterations in retinal histogenesis, as loss of Hh signaling in both species results in a deficit in late born retinal cell types. Shh signaling also exerts cell cycle-independent effects on GC development. Gain and loss of function studies in chick and mouse have demonstrated a requirement for Shh in negatively regulating GC development, thereby providing a mechanism by which GC control their number by feedback inhibition.

These studies on Hh signaling in the retina highlight the importance of GC-derived signals on the histogenic program in the retina and they also show how impacting the cell cycle and timing of differentiation contribute toward the establishment of the correct balance of retinal cell types.

Transforming Growth Factor-β

The TGF-β superfamily of signaling molecules is comprised of a number of subgroups, including the TGF-β, the bone morphogenic protein (BMP), and the growth and differentiation factor (GDF) subgroups. The TGF-β family members are involved in a variety of different developmental processes, including proliferation, migration, and cell death. Classical TGF-β signaling (**Figure 4**) involves ligand binding to a heteromeric receptor complex consisting of two type I and type II receptors that results in the activation and dimerization of homolog of mothers against Dpp (Smad) transcription factors, which then translocate into the nucleus where they interact with a number of cofactors to regulate target gene transcription. Signaling through different receptor combinations and Smad coactivators/corepressors allows members of the TGF-β superfamily to induce a multitude of cell-type and context-specific outputs. In the context of the retina, members of the TFG-β superfamily have been implicated in several aspects of retinal histogenesis including, the regulation of programmed cell death (PCD), progenitor proliferation, and patterning.

TGF-β signaling proteins (TGF-β1, -β2, and -β3) are important for the regulation of PCD, proliferation, and cell fate determination. Single TGF-β2 or double TGF-β2, and TGF-β3 null mice exhibit retinal hypercellularity likely as a consequence of reduced PCD during retinal development. TGF-β1 may also play a role in PCD as explants from early postnatal mice exhibit increased PCD in the presence of TGF-β1. In the early postnatal retina, TGF-β2 is the most abundantly expressed TGF-β family subtype, and it has been implicated in the induction of mitotic senescence of retinal progenitor cells and Müller glia. In addition to controlling PCD and proliferation, TGF-β2 plays a role in cell fate determination, as it negatively regulates amacrine cell genesis.

Studies to date have implicated BMP4 and BMP7 in early eye formation, retinal patterning, and retinal cell survival. The majority of BMP4 null embryos die prior to the stage of eye development; however, the cohort that survives to later stages fail to form a lens and some exhibit malformed optic vesicles. Consistent with a role for BMP4 signaling in eye development, mutations in BMP4 are associated with eye abnormalities in humans, the most severe being anopthalmia. Within the retina proper, BMP4 plays a role in the patterning of the retina by regulating transcription factors required for dorsal retina identity. Interestingly, BMP4 gene dosage is important for normal eye development, as haploinsufficiency for BMP4 in mice is associated with eye defects, such as a reduction in the number of ganglion and inner nuclear cells. BMP7 dosage is also important for eye development as gain and loss of function for BMP7 is associated with apoptosis in the retina and microopthalmia and anophthalmia. BMP signaling is also important for the specification of peripheral eye fates, such as the ciliary body. BMP4 and BMP7 are expressed in the ciliary margin, the distal part of the eye cup, and BMP signaling is required to maintain this structure, as it differentiates into neurons in the presence of ectopically expressed Noggin, a soluble BMP antagonist.

Studies employing mice that are conditionally null for type 1 BMP receptors have provided further evidence in support of a general role for BMP signaling in retinal

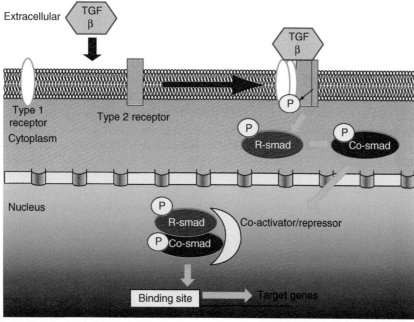

Figure 4 The TGF-β signaling pathway. The TGF-β signaling molecule binds to the heteromeric receptor complex consisting of type-1 and type-2 receptors, resulting in the phosphorylation of the type-1 receptor. This leads to the phosphorylation of a receptor Smad (R-Smad) which then dimerizes with a common-Smad (Co-Smad). The Smad complex then translocates to the nucleus where it can associate with a variety of coactivators or corepressors and modulate gene expression.

development. Two type I BMP receptors, BMPR-Ia and BMPR-Ib, are expressed in the retina. Deletion of either BMP-Ia or BMP-Ib alone does not have a drastic phenotype, likely due to compensatory functions of the two receptors. However, compound BMPR-1a$^{-/-}$BMPR-1b$^{+/-}$ mutant mice have dorsal retina patterning defects and mice null for both genes initially develop an eye structure but the retina undergoes extensive apoptosis and the mice are anopthalmic or microopthalmic at birth. Thus, the effects of BMPR-I deletions further demonstrate the roles of BMP's in eye development and patterning as well as cell survival and confirm the requirement for correct BMP dosage during eye development.

Although it is evident that TGF-β signaling is crucial for retinal development, the functions of other TGF-β family members in retinal development and histogenesis are still being uncovered. For example, GDF-11 has recently been shown negatively regulate ganglion cell genesis by controlling the time that RPCs are competent to produce this cell type. Thus, in GDF-11 null mice, an increase in RGCs occurs at the expense of amacrine cells and photoreceptors. Although it is not as defined as in the murine retina, some of the effects of TGF-β signaling in retinal development may also be conserved in other vertebrate species. For example, TGF-β and BMP signaling molecules are expressed in the chick and frog retina. In the chick, both TGF-β and BMP4 may play a role in mediating PCD.

Signaling Molecules and Photoreceptor Development

Cell extrinsic factors have also been implicated in rod photoreceptor development. Rods are highly specialized sensory neurons that have a complex morphology that is essential for their light-capturing function. Rod differentiation is a protracted process, in that there is a considerable lag between the cell cycle exit of rod precursors and the expression of end stage markers of the phototransduction cascade, including rhodopsin. Thus, differences in rhodopsin expression can be a consequence of altered cell specification or differentiation. To distinguish between these two possibilities, it is important to determine whether the signaling pathway under investigation exerts its effects on dividing or postmitotic cells and to assay early stage markers of rod development.

RA is a vitamin A derivative that regulates a number of aspects of development. RA enters the nucleus where it binds to RA receptor (RAR) heterodimers to regulate context-specific target gene expression (**Figure 5**). RA is synthesized by RPCs, differentiated neurons in the retina as well as the RPE in the adult eye. At early stages in ocular development, RA signaling plays an important role in regulating the growth of the ventral optic cup. RA signaling has also been shown to promote photoreceptor

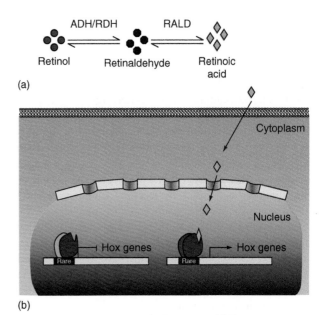

(a)

(b)

Figure 5 The retinoic acid (RA) pathway. (a) The metabolic pathway for the synthesis of RA from vitamin A. RA is synthesized through a series of oxidation reactions beginning with the oxidation of vitamin A (retinol) to retinal by aldehyde dehydrogenases (ADHs) and/or by retinol dehydrogenase (RDHs). Retinaldehyde dehydrogenase (RALD) converts retinal to RA. RA is able to enter the nucleus and bind to RAR/RXR heterodimers at RARE sites leading to transcriptional activation.

development in several vertebrate species. Whether RA mediates these effects by acting at the stages of rod specification or differentiation is more controversial. There is evidence that RA functions at the level of dividing progenitor cells to promote rod development. Conversely, RA has been shown in other studies to promote rhodopsin expression in postmitotic cells, and perturbation of RA signaling is associated with reduced or delayed rhodopsin expression rather than photoreceptor specification. Consistent with a role in differentiation, RA has been shown to regulate a number of photoreceptor-specific genes, including *arrestin*, *Cone rod homeobox* (*Crx*), and *Neural retina leucine zipper* (*Nrl*). Additional photoreceptor promoting molecules include taurine, an amino acid, that promotes the differentiation of postmitotic rod-precursor cells, b2-laminin and, depending on the context, fibroblast growth factor 2 (FGF2). Differential effects of Hh signaling on photoreceptor development have also been reported, with some studies finding positive effects of Hh signaling on rod development and others reporting an inhibitory effect.

In contrast, epidermal growth factor (EGF) has been reported to inhibit rod development. EGF might mediate this suppressive effect at the level of rod specification, as increased EGFR signaling in RPCs promotes Müller cell development at the expense of rods; however, a role for EGF signaling in rod development *in vivo* has not been established. Ciliary neurotropic factor (CNTF) exerts pleiotropic effects on neurons and glia in the developing and adult nervous system. In the retina, it exerts opposite effects on the development of postmitotic rod precursors in the rodent and chick. In rodents it is a powerful inhibitor of rod differentiation, whereas in chick it promotes rod development. The timing of CNTF receptor downregulation in rod precursors is coincident with the induction of rhodopsin expression. Thus, in conjunction with other extrinsic cues, CNTF signaling may coordinate the timing of final photoreceptor maturation with the development of other retinal components.

Conclusion

A primary theme that emerges from this overview of notch, Wnt, Hh, and TGF-β signaling in retinal histogenesis is the function of these pathways in controlling patterning, proliferation, and cell survival and to a more limited extent, cell fate. Key questions remain regarding the downstream events controlled by these pathways and how they converge on the cell intrinsic processes that regulate cell fate, competence, and the cell cycle machinery. Addressing these issues will lead to a deeper understanding of neurogenesis in general and will be beneficial in the development of techniques to manipulate stem cells for therapeutic applications.

See also: Coordinating Division and Differentiation in Retinal Development; Embryology and Early Patterning; Eye Field Transcription Factors; Injury and Repair: Stem Cells and Transplantation; Retinal Histogenesis.

Further Reading

Campo-Paysaa, F., Marletaz, F., Laudet, V., and Schubert, M. (2008). Retinoic acid signaling in development: Tissue-specific functions and evolutionary origins. *Genesis* 46: 640–656.

Ingham, P. W. and McMahon, A. P. (2001). Hedgehog signaling in animal development: Paradigms and principles. *Genes and Development* 15: 3059–3087.

Lad, E. M., Cheshier, S. H., and Kalani, M. Y. S. (2008). Wnt signaling in retinal development and disease. *Stem Cells and Development* 18(1): 7–16.

Levine, E. M., Fuhrmann, S., and Reh, T. A. (2000). Soluble factors and the development of rod photoreceptors. *Cellular and Molecular Life Sciences* 57: 224–234.

Lillien, L. (1995). Changes in retinal cell fate induced by overexpression of EGF receptor. *Nature* 377: 158–162.

Livesey, F. J. and Cepko, C. L. (2001). Vertebrate neural cell-fate determination: Lessons from the retina. *Nature Reviews Neuroscience* 2: 109–118.

Logan, C. Y. and Nusse, R. (2004). The Wnt signaling pathway in development and disease. *Annual Review of Cell and Developmental Biology* 20: 781–810.

Mann, R. K. and Beachy, P. A. (2004). Novel lipid modifications of secreted protein signals. *Annual Review of Biochemistry* 73: 891–923.

Massague, J. and Gomis, R. R. (2006). The logic of TGFbeta signaling. *FEBS Letters* 580: 2811–2820.

Ohsawa, R. and Kageyama, R. (2008). Regulation of retinal cell fate specification by multiple transcription factors. *Brain Research* 1192: 90–98.

Ross, S. and Hill, C. S. (2008). How the smads regulate transcription. *International Journal of Biochemistry and Cell Biology* 40: 383–408.

Schmierer, B. and Hill, C. S. (2007). TGFbeta-SMAD signal transduction: Molecular specificity and functional flexibility. *Nature Reviews Molecular Cell Biology* 8: 970–982.

Selkoe, D. and Kopan, R. (2003). Notch and Presenilin: Regulated intramembrane proteolysis links development and degeneration. *Annual Review of Neuroscience* 26: 565–597.

Wallace, V. A. (2008). Proliferative and cell fate effects of hedgehog signaling in the vertebrate retina. *Brain Research* 1192: 61–75.

Immunobiology of Age-Related Macular Degeneration

R L Ufret-Vincenty, University of Texas Southwestern Medical Center at Dallas, Dallas, TX, USA

Glossary

Adaptive immune system – An arm of the immune system that recognizes and responds to specific antigens. It is composed mainly of lymphocytes (T cells and B cells) and antigen-presenting cells that present antigen to T cells. B cells produce antibodies, which recognize antigens with high specificity. One of the properties of the adaptive immune system is immunological memory, which enables the body to respond in a faster and stronger manner to an antigen that is encountered a second time.

Drusen – The deposits of extracellular material lying between the basement membrane of retinal pigment epithelium (RPE) cells and the inner collagenous zone of Bruch's membrane. Drusen are readily visible as yellowish deposits lying deep to the retina. They vary in size and shape and may have a crystalline appearance due to calcification. They are often the earliest clinical sign of AMD.

Innate immune system – It comprises the cells and molecules that defend the host from infection by other organisms, in a nonspecific manner. The components of the innate (nonadaptive, non-antigen-specific) system recognize pathogens in a generic way. The innate immune system does not confer memory (long-lasting and enhanced immune responses). Components of the innate immune system include the complement system, and cells such as natural killer cells, mast cells, eosinophils, basophils, macrophages, and neutrophils.

M2 macrophages – Macrophages are cells that phagocytose (engulf and digest) cellular debris and pathogens. They can then stimulate or modulate lymphocytes and other immune cells. In a simplified scheme, macrophages can be sub-classified into M1 and M2. M1 macrophages are microbicidal and inflammatory, while M2 macrophages are immunomodulators. Among other things, M2 polarization is induced by interleukin-4, interleukin-10, interleukin-13, immune complexes, and glucocorticoid hormones. M2 macrophages express high levels of legumain (an endopeptidase) and this molecule can be used as a phenotypical marker for polarization. Macrophage polarization has been shown to be reversible.

Neoantigen – A newly acquired and expressed antigen that elicits an immune response. Examples include the antigens present in cells infected by viruses, those on tumor cells, or those induced by damage of normal molecules during tissue injury (e.g., oxidative damage to normal proteins).

Opsonization – A process by which a molecule (opsonin) enhances the binding of immune system cells to an antigen or cell surface, for example, by coating the negatively charged molecules on a cell membrane. This helps in the process of phagocytosis. Examples of opsonins are antibodies and complement molecules.

Single nucleotide polymorphism (SNP) – A type of mutation or allelic variant of a gene in which the DNA sequence differs only in a single nucleotide.

Vascular endothelial growth factor (VEGF) – A critical regulator of both physiological and pathological angiogenesis. It is involved in pathologic ocular neovascularization as seen in proliferative diabetic retinopathy, retinopathy of prematurity, and AMD. VEGF also promotes vascular permeability. It is strongly upregulated by hypoxia.

Epidemiology and Clinical Findings of Age-Related Macular Degeneration

Approximately 1.75 million Americans have advanced age-related macular degeneration (AMD). In fact, an estimated 30% of the population older than 75 years of age has some degree of macular degeneration. AMD is classified into a dry or non-neovascular form and a wet or neovascular form. About 12% of patients with AMD end up having advanced disease (neovascular AMD or central geographic atrophy (GA)). Presentations of dry AMD include drusen and a wide range of abnormalities of the retinal pigment epithelium (RPE), going from focal areas of hypopigmentation and hyperpigmentation, to the more advanced GA. Several studies have shown that self-reported quality-of-life scores for patients with choroidal neovascularization (CNV) secondary to AMD are worse than those reported by patients with acquired immune deficiency syndrome (AIDS; receiving treatment) and chronic obstructive pulmonary disease. Moreover, the levels of anxiety and depression were comparable. Genetic factors, smoking, and increased body mass index seem to have the highest impact on the likelihood of progression of AMD.

Despite recent advances in the treatment of CNV, clinical outcomes remain suboptimal. Furthermore, there is still no effective therapy for atrophic AMD. The Age-Related Eye Disease Study (AREDS) demonstrated that using high doses of certain vitamins and minerals (15 mg of β-carotene, 500 mg of vitamin C, 400 IU of vitamin E, 80 mg of zinc oxide, and 2 mg of cupric oxide) leads to a 25–30% reduction in the risk of development of advanced AMD. These vitamins and minerals mostly act as antioxidants. Their efficacy, combined with the observation that smoking is a significant risk factor for AMD, suggests an important role of oxidative damage in the early stages of macular degeneration. Current therapies for neovascular AMD include laser photocoagulation, photodynamic therapy, and anti-VEGF (VEGF, vascular endothelial growth factor) agents. All of these attempt to close or cause regression of the already-formed neovascular membranes. Although the newer anti-VEGF agents such as bevacizumab (Avastin) and ranibizumab (Lucentis) can lead to an average improvement in vision of up to two lines from the time of the detection of the neovascularization, this still means a significant loss in vision for many patients. The goal of newer therapeutic modalities should be to stop the process that leads to atrophy of the RPE and retina or to the formation of neovascularization before irreversible damage has occurred. This, of course, requires a better understanding of the pathogenetic mechanisms involved in the early stages of disease. Most experts now suspect that the immune system plays an important, although poorly understood, role in those early events.

The Immune System in AMD

Multiple lines of evidence imply a role of the immune system in AMD, including the presence of chronic inflammatory cells in retinal lesions and antiretinal antibodies in the serum of AMD patients. Macrophages, lymphocytes, and mast cells have been detected in CNV membranes and near ruptures of Bruch's membrane in eye specimens from AMD patients. In addition, AMD patients seem to have increased systemic levels of the inflammatory markers C-reactive protein (CRP) and interleukin-6 (IL-6). Yet, it is debated whether the immune system plays a role in initiating or exacerbating the disease, or if these findings are a response to the retinal damage.

The early stages of AMD are characterized by accumulation of debris (altered proteins and lipids) underneath the RPE. Recent literature suggests that an immune reaction to this debris may promote more tissue damage and/or a neovascular response. Both the innate and the antigen-specific arms of the immune system can cause local damage to RPE cells, Bruch's membrane, and the choriocapillaris, leading to either the atrophic or neovascular forms of AMD.

Histopathology of Drusen

For the last four decades there has been a debate surrounding the mechanism for the formation of drusen. However, whether drusen form from fragments of RPE cells, from abnormal material secreted by dysfunctional RPE cells, or whether some components come from the choriocapillaris, the composition of drusen gives us important information about the pathogenesis of AMD.

During the last decade Hageman, Anderson, Johnson, Penfold, and others have contributed a great deal to our knowledge of drusen composition by extensive histopathological analysis of human AMD specimens obtained within the first few hours of death. They have performed light microscopy, confocal microscopy, electron microscopy, and immunostaining in hundreds of specimens; and have also evaluated the levels of mRNA expression of inflammatory mediators in the RPE and choroid. From these studies they have identified a large number of molecules found in drusen, many of which are in some way related to inflammation:

1. acute-phase reactant proteins such as CRP;
2. inflammatory mediators such as serum amyloid P;
3. potential activators of the complement cascade, including nuclear fragments/nucleic acids, membrane-bound vesicles, mitochondria, modified lipids, cholesterol, and microfibrillar debris;
4. complement components such as C3 and its fragments, C5, and the membrane attack complex (C5b-9); and
5. complement regulatory proteins including vitronectin, clusterin, membrane cofactor protein (CD46), complement receptor 1, and complement factor H (CFH).

Role of Macrophages

Some studies have suggested that macrophages may play a protective role in AMD, while others suggest that they promote disease progression. Some investigators have suggested that macrophages promote CNV formation. They have shown in the mouse laser-injury model of CNV that depletion of macrophages diminishes the size of the CNV lesions. They have also shown that neovascular AMD patients tend to have higher levels of expression of tumor necrosis factor (TNF)-α in their circulating monocytes than controls. TNF-α is considered a marker for monocyte activation. Others propose that macrophages have a role in preventing AMD. Mice deficient in monocyte chemoattractact protein-1 (also known as CCL-2) or its receptor can develop CNV in old age, suggesting that impaired macrophage recruitment in these mice is responsible for decreased clearance of proinflammatory mediators, such as C5 and IgG, from the sub-RPE tissues, resulting in an increased risk for CNV. This idea is very attractive, but other groups have had difficulty reproducing the results. Mouse studies using the laser model of CNV have shown that IL-10 inhibits recruitment of macrophages into CNV lesions, but this leads to increased CNV size, suggesting that macrophages are antiangiogenic. Additional animal studies propose that senescence may cause a shift in the ocular microenvironment (including higher IL-10 levels) that makes it more proangiogenic; and that microenvironment, in turn, alters the polarization of macrophages toward an M2 phenotype which is unable to regulate angiogenesis. Thus, young macrophages appear to have an antiangiogenic role, while aging M2 macrophages change toward a proangiogenic phenotype. This could explain the different results obtained by different groups using the mouse laser model. Although it is still too early to determine if these conclusions apply to a noninjury model of CNV or to human AMD, the idea that "not all macrophages are made equal" and under different conditions may have opposing roles in AMD is very attractive.

Adaptive Immune System

The potential role of the innate arm of the immune system in AMD is further explored under the topic of genetics. The adaptive arm of the immune system is composed primarily of T cells and B cells. Although T and B cells have been identified in human AMD lesions, it is still not known whether they play causal roles in AMD versus being epiphenomena. Some investigators suggest that these T and B cells may be responding to subretinal neoantigens. Proteome analysis of the Bruch's membrane of AMD patients demonstrates higher levels of lipid–protein adducts derived from oxidative damage when compared to controls. These carboxyethyl pyrrole (CEP)–protein adducts are generated by the cross-linking of oxidized lipids and proteins. It has been proposed that the CEP–protein adducts may serve as neoantigens triggering an antigen-specific immune response in the subretinal/sub-RPE tissues that could promote the development of AMD. Moreover, immunization of mice with CEP-adducted mouse serum albumin led to an immune response and the formation of drusen and GA-like lesions in some animals.

Infectious Agents

Several mechanisms have been proposed by which infectious agents may promote the development of AMD. Molecular mimicry occurs when a microbial antigen resembles a self-antigen (in AMD it could be molecules on RPE cells or Bruch's membrane) so much that the immune system attacks the self-antigen while trying to eradicate the intended microbial antigen. By-stander injury refers to injury to self-tissues caused by the immune system as it tries to attack a microbial organism in close proximity. Finally, the infection can directly

injure the host tissues. It is not clear if in AMD any of these mechanisms can be invoked.

In 2005, the Nobel prize in medicine was awarded for the work associating *Helicobacter pylori* with gastritis and peptic ulcer disease. Several epidemiological studies have found that *Chlamydia pneumoniae* (serology) may be associated to atherosclerosis, a disease that has several risk factors in common with AMD. Moreover, recent findings indicate that patients with AMD may be more likely to have higher anti-*C. pneumoniae* antibody levels than matched controls. There was also an association with pathology specimens of surgically excised choroidal neovascular membranes. Two other studies in Japan and Australia confirmed the serological association. Other studies could not reproduce these results, but instead found an association between high anti-CMV antibody levels and neovascular AMD. Critics comment that these are mere statistical associations and not proof of causality. Furthermore, they may be the result of unintentionally biased statistical analyses. Finally, the numbers included in the studies were often low. In the case of *Chlamydia* and atherosclerosis, clinical studies found that antibiotics did not alter the risk for vascular events. This could mean that *Chlamydia* is merely a commensal organism that thrives on atheromatous plaques without causing the disease; or it could mean that *Chlamydia* triggers the disease, but is not needed for disease progression. The same reasoning applies to AMD. One potential hypothesis is that *Chlamydia* or other organisms could cause a chronic low-grade infection at the deep retina/RPE/choriocapillaris level, and that in susceptible individuals (e.g., those with the at-risk variant of complement factor H) this could lead to chronic low-grade inflammation and the formation of drusen or CNV. It is difficult to prove this hypothesis due to the lack of good experimental models. However, we expect further research may shed light on this.

Genetics of AMD and Evidence for a Role of Inflammation in the Pathogenesis of AMD

Mutations may account for up to 70% of the population-attributable risk for AMD. Specific single nucleotide polymorphisms (SNPs) that increase the risk for AMD have been identified in several genetic loci. Two of these loci are particularly interesting since they have been reproduced by different research groups and in different populations. The first locus is in chromosome 1q32 (regulator of complement activation or RCA locus), which includes the gene for complement factor H. Several SNPs that appear to be statistically associated with the development of AMD have been identified within this locus. The most consistent association has been with a polymorphism that causes a substitution of a tyrosine for a histidine at amino acid position 402 in complement factor H (CFH Y402H).

However, several other polymorphisms, including CFH I62V and some intronic SNPs, have also been found to be associated to AMD by different groups. Within the same locus 1q32, a deletion affecting factor-H-related genes (CFHR1 and CFHR3) may decrease the risk of AMD, perhaps by decreasing the competition with factor H for the binding of C3.

The second locus of interest involves a region of chromosome 10q26 containing three genes PLEKHA, LOC387715/ARMS2, and HTRA1. The strongest associations in this locus have been with SNPs in the hypothetical gene LOC387715 and in the promoter region for HTRA1.

A third, and perhaps less common association, involves a locus in chromosome 6p21. Mutations affecting complement factors B and C2 in this region have been shown to alter the risk for AMD.

It is difficult to determine if there is a causal relationship between all these SNPs and AMD phenotypes. The best rationale for causality is for the Y402H mutation. Complement factor H (CFH or Cfh) is a key regulator of the alternative complement pathway. A point mutation that leads to a histidine, instead of a tyrosine at amino acid position 402 of CFH (Y402H), increases the risk of AMD by sevenfold in homozygotes. Moreover, this histidine variant of CFH appears to make AMD patients resistant to therapy with anti-VEGF agents. Despite the eagerness of the pharmaceutical industry to dive into this field, our knowledge of the mechanisms explaining this increase in risk for disease and decrease in susceptibility to therapy is in its infancy. One hypothesis is that normally CRP opsonizes subretinal debris and helps macrophages clear this debris in a noninflammatory fashion. For this to work, CFH has to bind CRP, preventing CRP from fully activating the complement cascade. The at-risk variant of CFH (402H) has a reduced affinity for CRP, which may lead to reduced binding of CFH to CRP in the subretinal debris. Thus, the full complement cascade would be activated by CRP, resulting in a vicious cycle of accumulation of debris and CRP, activation of complement, and tissue damage. In support, one study demonstrated that individuals homozygous for the at-risk variant of CFH have elevated levels of CRP in the choroid. Moreover, 402 YY patients receiving a liver transplant from 402 HH donors quickly developed drusen. Nevertheless, some investigators suggest that the Y402H mutation may not be pathogenic; that it may simply be in linkage disequilibrium with a yet-unidentified pathogenic mutation. It is essential to determine if Y402H has causal relevance in AMD.

It is more difficult to explain causality for the 10q26 SNPs: LOC387715 is still considered a hypothetical gene since the protein it codes for has not been unequivocally identified. HTRA1 is a serine protease and it has been proposed that it may be involved in the facilitation of penetration of Bruch's membrane by neovascular fronds. Unfortunately, many of the identified SNPs are in regions of high linkage disequilibrium. In other words, it may be that the identified SNPs are not directly responsible for the disease, or that they are simply strongly associated with other mutations which are the direct culprits for the pathological events leading to AMD.

A large group of investigators from several institutions have recently published a single report showing an association between a mutation in the gene for toll-like receptor 3 (TLR3) and the development of GA using three separate case-control series. SNP rs3775291 in TLR3 seemed to offer mild-to-moderate protection against GA. It has no effect on drusen or CNV formation. Toll-like receptors are a class of membrane-spanning receptors that recognize structurally conserved molecules derived from microbes and play a critical role in innate immunity. TLR3, in particular, is expressed on dendritic cells and B lymphocytes and recognizes double-stranded RNA (dsRNA, usually from viruses). The hypothesis is that dsRNA from infecting viruses may trigger apoptosis of RPE cells through activation of TLR3. The SNP in question causes a reduction in TLR3-mediated cell death. If this hypothesis is correct, it not only implies a role for infectious agents in inducing AMD, but also raises concern for RPE damage with the new small interfering RNA (siRNA) investigational drugs (which are also small double-stranded RNA molecules). However, an earlier study was not able to find any association between the same SNP in TLR3 and AMD.

Mouse Models of AMD

Several groups have recently demonstrated that, despite the obvious anatomical differences with humans (lack of a macula), mice can develop drusen and CNV similar to those seen in clinical AMD. Some of the most interesting models will be discussed here.

One study examined mice deficient in either chemokine ligand 2 (CCL2) or chemokine receptor 2 (CCR2). The findings were discussed earlier. Investigators at the National Eye Institute have generated mice that are deficient in both chemokine receptors CCL2 and CX3CR1. These mice develop drusen-like and pigmentary changes by 6 weeks of age, and 15% of them develop CNV as early as 12 weeks of age. It should be noted that, at least in CX3CR1-deficient mice (not deficient in CCL2), another group in France has shown that the drusen-like lesions seen clinically are histopathologically different from drusen seen in AMD and consist of lipid-bloated subretinal microglial cells rather than subretinal pigment epithelium deposits. Both CCL2 and CX3CR1 are involved in the recruitment of microglia and monocytes into tissues. Monocytes can then further differentiate into macrophages and dendritic cells. Somewhat counter-intuitively, the CCL2/CX3CR1-deficient mice have increased numbers of macrophages and microglia in the subretinal space and retinal lesions.

Some investigators suggest that an alternative pathway for monocyte recruitment involving the upregulation of CCL5 is responsible for this finding. They also suggest that CX3CR1-deficient microglia may behave abnormally and produce local tissue injury. Furthermore, these mice have increased complement activation and retinal autoantibodies in their retinas. It is still not clear the mechanism by which these mice develop the AMD-like changes and if these mechanisms are relevant to human AMD. An association between two mutations in CX3CR1 and AMD has been reported. However, these mutations have not been identified in genome-wide scans and other studies. Further studies will be required to determine if mutations in CX3CR1 and/or CCL2 are truly involved in the pathogenesis of human AMD. In the meantime, it is clear from these studies that the migration and function of macrophages, microglia, and dendritic cells may lead to features similar to those seen in AMD and may, under certain circumstances or in certain subgroups of patients, have pathophysiologic relevance.

The effect of apolipoprotein (ApoE) variants on AMD pathogenesis has been examined in mice. It was shown that the eyes of aged, targeted replacement mice expressing human apoE4 and maintained on a high-fat and cholesterol diet develop diffuse sub-RPE deposits, drusenoid deposits, thickened Bruch's membrane, and RPE

pigmentary abnormalities. Close to 20% of the mice also developed CNV. Multiple genetic studies have given mixed results regarding the clinical relevance of ApoE variants and/or mutations.

Studies in superoxide dismutase 1 (SOD1)-deficient mice show that they develop drusen and CNV at an old age. This suggests that the increased oxidative stress in the retina leads to tissue injury and accumulation of debris. It should be noted that more recent work suggests that the yellow deposits seen clinically in these animals may not be drusen, but collections of lipid-laden macrophages. Moreover, there is no genetic evidence for an association between SOD1 and AMD.

Summary

The pathogenesis of AMD is complex and clearly involves genetic and environmental risk factors. At least three genes have been identified to be strongly correlated to the disease. However, it is likely that many others have an impact in the disease for some subpopulations of patients. A very simplified diagram on the pathogenesis of AMD is included in **Figure 1**. Note that the potential roles of neither the 10q26 locus and the TLR3 mutation nor the antibodies and T cells have been included.

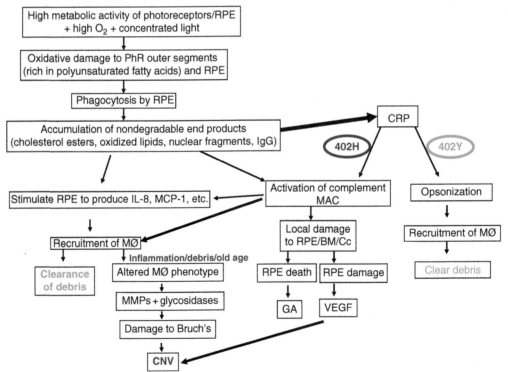

Figure 1 Proposed simplified pathogenesis of AMD. BM, Bruch's membrane; Cc, choriocapillaris; CNV, choroidal neovascularization; CRP, c-reactive protein; GA, geographic atrophy; IgG, immunoglobulin G; IL-8, interleukin 8; MAC, membrane attack complex; MCP-1, monocyte chemoattractant protein 1; MØ, macrophage; O_2, oxygen; PhR, photoreceptor; RPE, retinal pigment epithelium; VEGF, vascular endothelial growth factor; 402H, histidine-402 variant of complement factor H; 402Y, tyrosine-402 variant of complement factor H.

Based on the genetic, histopathologic, serologic, and animal model data, it is clear that the immune system has a significant role in the development of AMD. However, we are still far from determining the exact mechanisms by which it affects the clinical course of the disease. In fact, although the strongest incriminating data available point to the complement system and macrophages, it is not clear if the adaptive immune system (T cells, B cells, and antibodies) is also important.

The pharmaceutical industry, at present, has a great impetus to develop new therapeutic agents for AMD based on controlling inflammation. Several agents that block complement components 3 or 5 or factor B are in preclinical or early clinical trials. Furthermore, the effect of anti-inflammatory agents, in combination with anti-VEGF agents, is being studied. Yet, given our limited knowledge, the approach is far from ideal. Better animal models of AMD (based on genetic and environmental risk factors known to be important in AMD) would help us improve our understanding of the early steps in the pathogenesis of the disease, and perhaps allow us to identify new therapeutic targets and test new drugs.

Further Reading

Bressler, N. M., Bressler, S. B., and Fine, S. L. (2006). Neovascular (exudative) age-related macular degeneration. In Schachat, A. P. and Ryan, S. J. (eds.) *The Retina,* 4th edn., pp 1075–1114. Philadelphia, PA: Elsevier.

Bressler, S. B., Bressler, N. M., Sarks, S. H., and Sarks, J. P. (2006). Age-related macular degeneration: Nonneovascular early, AMD, intermediate AMD, and geographic atrophy. In Schachat, A. P. and Ryan, S. J. (eds.) *The Retina,* 4th edn., pp 1041–1074. Philadelphia, PA: Elsevier.

Edwards, A. O. and Malek, G. (2007). Molecular genetics of AMD and current animal models. *Angiogenesis* 10(2): 119–132.

Gehrs, K. M., Anderson, D. H., Johnson, L. V., and Hageman, G. S. (2006). Age-related macular degeneration – emerging pathogenetic and therapeutic concepts. *Annals of Medicine* 38(7): 450–471.

Haddad, S., Chen, C. A., Santangelo, S. L., and Seddon, J. M. (2006). The genetics of age-related macular degeneration: A review of progress to date. *Survey of Ophthalmology* 51(4): 316–363.

Nussenblatt, R. B. and Ferris, F. 3rd (2007). Age-related macular degeneration and the immune response: Implications for therapy. *American Journal of Ophthalmology* 144(4): 618–626.

Patel, M. and Chan, C. C. (2008). Immunopathological aspects of age-related macular degeneration. *Seminars in Immunopathology* 30(2): 97–110.

Information Processing: Amacrine Cells

R E Marc, University of Utah, Salt Lake City, UT, USA

Glossary

Buffer – A device that collects signals from a source and distributes them to targets without taxing the limited capacity of the source.

Dendrodendritic synapses – Unique synaptic motifs where both pre- and postsynaptic assemblies coexist on the same neuronal dendrite rather than segregating into dendrites and axons, respectively; they are the fundamental modes for biological feedback in sensory pathways.

Directional selectivity – The ability of a neuron to preferentially spike in response to stimuli originating from a given visual quadrant, but not others.

Feedback – The use of a portion of an output signal as an additional upstream input to modify the amplification in a network. Negative feedback reduces the gain and noise and improves the frequency response (spatial, temporal, or spectral). Positive feedback can be used to further tune the frequency response or generate resonance.

Feed forward – The use of a portion of an output signal as an additional parallel input to a downstream target. Negative feed forward selectively improves the frequency response (spatial, temporal, or spectral), but has little benefit on system noise or stability. Positive feed forward can be used to further amplify the effect of a signal.

Gain – The ratio of output and input amplitudes for a system, typically specified logarithmically as decibels of power [10 log (Pout/Pin)] or squared voltage [20 log (V²out/V²in)].

Nested feedback – The use of a portion of the feedback signal on the feedback process itself, allowing the effect of feedback to be more precisely tuned.

Sign-conserving synapses – Postsynaptic events mediated by receptors whose activation generates the same polarity of signal as the presynaptic terminal, that is, typically, a-amino-3-hydroxy-5-methyl-4-isoxazolepropionic acid, *N*-methyl D-aspartate, and nicotinic acetylcholine receptors.

Sign-inverting synapses – Postsynaptic events mediated by receptors whose activation generates the opposite polarity of signal as the presynaptic terminal, that is, typically, gamma aminobutyric acid and glycine.

Transfer function – The output of a given cell or device relative to its input, usually expressed as gain versus temporal or spatial frequency.

Amacrine Cells

Amacrine cells (ACs) and axonal cells (AxCs) are multipolar neurons that shape the ganglion cell (GC) function. While two to four classes of horizontal cells (HCs) provide lateral signal processing in the outer plexiform layer, ACs and AxCs are the most diverse neuronal group of any brain region, with over 30 classes in mammals and even more in highly visual nonmammalians, such as cyprinid fishes, where over 70 AC classes have been documented.

AC Classification, Form, and Patterns

The neural retina is composed of three superclasses of cells: (1) sensory neuroepithelial cells (photoreceptors and bipolar cells (BCs)); (2) multipolar neurons (ACs, AxCs, and GCs); and (3) gliaform neurons (HCs). Retinal multipolar neurons are further divisible into projection neurons (GCs and AxCs) and *bona fide* local circuit neurons, the ACs. This is an important distinction. The definition of the term amacrine refers to lacking an axon and ACs are neurons that mix presynaptic and postsynaptic specializations on their dendrites. These dendrodendritic specializations make them homologous to the anaxonic granule cells (GrCs) of the olfactory bulb (**Figure 1**). Both ACs and GrCs are inhibitory neurons that provide in-channel and cross-channel feedback. Conversely, GCs are classical projection neurons with exclusively postsynaptic retinal dendrites and presynaptic axon terminal ramifications in the central nervous system (CNS). The AxC group is a collection of diverse, nonamacrine retinal neurons that are similar to GCs in having distinct dendritic arbors and one or more long intraretinal axons. The AxC group includes polyaxonal cells in mammals and avians, certain dopaminergic neurons, and interplexiform cells. Interestingly, topologically similar AxCs in the olfactory bulb are the periglomerular cells and some of them are also dopaminergic.

Different kinds of ACs form two large superclasses with shared features: lateral and vertical ACs. Lateral ACs are precisely stratified in a single or a few layers of

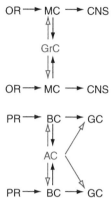

Figure 1 Signal processing in the olfactory bulb (top) and retina (bottom). Glutamatergic sensory neurons such as olfactory receptors (OR) and photoreceptors (PR) directly drive glutamatergic target neurons in the afferent chain, that is, olfactory mitral cells (MC) and retinal bipolar cells (BC), respectively. MCs and BCs then target the next glutamatergic elements as well as sets of inhibitory GABAergic local circuit neurons: olfactory granule cells (GrC) and retinal amacrine cells (AC). Both these neurons provide local inhibitory feedback onto their source neurons. ACs also provide inhibitory feed forward onto ganglion cells (GC) in the chain. Blue elements are glutamatergic. Red elements are GABAergic.

the inner plexiform layer, and spread their dendrites laterally either in a narrow to wide span as needed for their function; most of them release γ aminobutyrate (γ ACs). Vertical ACs distribute their dendrites across several levels of the inner plexiform layer, mostly in narrow and medium spans; most of them are glycinergic (gly ACs). **Figure 2** illustrates three very different structural patterns of vertebrate ACs. Radiate γ ACs (**Figure 2 (a)**) are cone BC-driven pyriform cells with a single, thick proximal dendrite that descends to a given level of the inner plexiform layer and then generates nearly 50 dendrites that each radiates unbranched through the retina at one level for up to 0.5 mm. It exists in both on and off varieties, each branching extensively in a narrow band of the inner plexiform layer as a sparse field of processes. Similar cells are present in most vertebrates, but are incompletely studied. Another lateral cell is the widely studied starburst γ AC (**Figures 2(a)** and **2(b)**). It exists in both on and off varieties, each branching extensively in a narrow band of the inner plexiform layer as a dense field of processes. In nonmammalians with thick inner plexiform layers, it is a pyriform cell; however, in most mammals, it is a multipolar cell. Mammalian rod gly ACs exemplify the vertical phenotype (**Figures 2(a)–2(c)**), with deeply branching arboreal dendrites that collect signals from rod BCs in the proximal inner plexiform layer, mid-inner plexiform layer gap junctions with on-cone BCs, and both inputs and outputs onto off BCs from lobular dendrites in the distal inner plexiform layer. As these cells illustrate, the inner plexiform layer is laminated according to cone BC type (**Figure 2(c)**), with off cells terminating in the distal

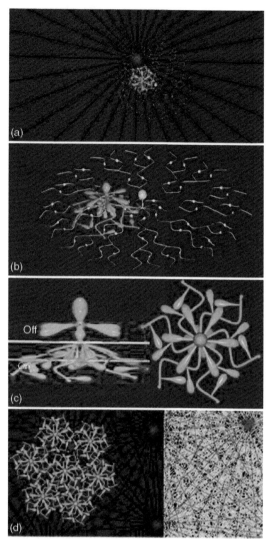

Figure 2 Schematic shapes and patterns of three canonical varieties of retinal ACs. (a) Top view of two lateral ACs and one vertical AC. Red: an off radiate γ AC from goldfish retina, with a wide-field, monostratified nonbranching dendritic arbor over 0.5 mm in diameter. These cells appear to be present in all vertebrates. Yellow: a goldfish off-starburst γ AC with a medium, monostratified, bifurcating, wavy arbor approximately 0.2 mm in diameter. Starburst cells are found in all vertebrates. In species with thick inner plexiform layers, they are pyriform with a single proximal dendrite; in those with thin inner plexiform layers (mammals), they are multipolar. Cyan: a mammalian vertical gly AC with a small, multistratified arbor nearly 0.05 mm in diameter. (b) Oblique view of the starburst γ AC and gly rod ACs showing that the starburst cell captures more synaptic space, but is laminated between the distal and proximal dendrites of the gly rod AC. (c) Vertical and top views of the gly rod AC. Immediately beneath the soma are several large lobular dendrites that contact off BCs. Deep in the inner plexiform layer are arboreal dendrites that capture rod BC inputs at their tips and, as they ascend through the inner plexiform layer, make gap junctions with cone BCs. The inner plexiform layer is divided into on and off layers exemplified by the yellow starburst AC dendritic bands. (d) Coverage patterns of ACs. The red radiate γ ACs are sparse and have coverage factors (CF) of approximately 4–6. The yellow starburst ACs are more frequent and have CF >100 and capture a great fraction of the synaptic space in a thin stratum of the inner plexiform layer. The cyan rod gly ACs are densely packed but have CF ≈ 1.

part and on cells in the proximal part. The AC processes fill the inner plexiform layer from the AC layer to the GC layer. The on- and off-starburst ACs indicate the positions of two fine on- and off-sublayers and exemplify the lateral motif of a large cohort of γ ACs. The gly rod ACs exemplify the vertical motif, bridging both on and off layers and transporting scotopic information in the on → off direction.

AC branching is quantified in many ways. The fractal box dimension (D) summarizes how effectively a dendritic pattern captures the synaptic space in a plane. $D = 1$ for a line and $D = 2$ for a complete plane. For radiate cells, $D \approx 1.2$ (low capture efficiency), while it approaches 1.5 for starburst cells (very high capture efficiency). In biological terms, radiate cells rigidly cover space, but not synapses, and contact inputs stochastically as they traverse wide swaths of the image. Conversely, starburst cells aggressively capture many synapses, rather than space, with curving dendrites. Thus, these cells will have very different signal-processing functions. Most ACs are distributed in coverings, which is one of the three types of two-dimensional (2D) cell patterns. The other two are packings (photoreceptors) and tilings (GCs). Patterns are partly defined by the extent of overlap that exists between members of a cell group (the coverage factor, CF) and how evenly the geometric centers of the cell fields are distributed over space (the conformity ratio, CR). Packings such as blue cones have CF < 1 and CRs ≈ 3 (weakly orderly) in mammals to >15 (crystalline) in fishes. GCs tile, with CFs ≈ 1 and CRs ≈ 3. Coverings shown by ACs have CF >1–100, indicating that for some ACs, dendritic fields of over 100 cells overlap a single point in the image plane, implying an intense synaptic control of signaling in that zone. Most CRs for ACs range from 1 to 4 in mammals. Using our examples from above (**Figure 2(d)**), radiate cells have CF ≈ 4–6 (center-to-center spacing) and form a sparse, but orderly, synaptic mesh over the retina. Starburst cells have CF ≈ 60 or more and densely pack the synaptic space. Narrow-field glycinergic ACs have CF ≈ 1 and act as nonspatial intercalary amplifiers or cross-channel controllers (discussed below).

AC Synapses and Neurochemistry

Almost all ACs are driven by BCs at ribbon synapses, decoding that glutamatergic input with a-amino-3-hydroxy-5-methyl-4-isoxazolepropionic acid (AMPA) receptors. ACs driven by rod BCs almost exclusively use AMPA-receptor-mediated currents, while most of those driven by cone BCs amplify and sustain the initial AMPA-driven currents with N-methyl D-aspartate (NMDA) receptors. Most native AMPA receptors are heteromers (some homomers may exist) of two or more of four known glutamate receptor (GluR) subunits: GluR1, GluR2, GluR3, and GluR4. There is no required stoichiometry; however, receptors with a mixture of GluR2 and GluR3 subunits are common in the brain and retina. There are post-translational modifications that change receptor conductances, kinetics, and glutamate sensitivity. The most important of these is the *gluR2* messenger ribonucleic acid (mRNA) pre-editing of a codon that converts a pore-lining neutral Q residue to a cationic R residue. Almost all mature GluR2 subunits are edited. The expressing of a GluR2 subunit has four important actions regardless of its partners: (1) it nearly abolishes the AMPA receptor Ca^{2+} permeability; (2) it decreases the channel conductance twofold or more; (3) it linearizes the I–V relation; and (4) it eliminates the GluR1 subunit phosphorylation-dependent increases in receptor conductance. Thus, ACs can express a range of sensitivities to glutamate. The most glutamate-sensitive neuron in the retina is the starburst AC population. Similarly, ACs express different levels and kinds of NMDA receptors, resulting in a broad spectrum of response attributes.

The outputs of ACs are conventional synapses targeting BCs, ACs, and GCs. Thus, ACs are positioned to powerfully shape all signal flow through the retina. However, there is a key difference in the high tonic vesicle fusion rate synapses of photoreceptor and BC ribbon synapses and the transient vesicle fusion of ACs. As a result, ACs act as temporal filters and signal through bursts of apparently precisely timed inhibitory transmitters. The presynaptic vesicle pools of ACs are generally very small, with tens to hundreds of vesicles, rather than the pools of thousands harbored by BCs. Thus, the actions of ACs are both swift and subtle. This is unlike the strong tonic influence of HCs as shown below.

Of the approximately 30 classes of ACs described for mammals, half to two-thirds of the classes are GABAergic and release GABA via small to large patches of vesicles at conventional synapses. Most of the remaining cells are narrow-field glycinergic cells. These two fast classical inhibitory transmitters act largely by opening anion-permeant channels that, in most cases, generate hyperpolarizing inward chloride currents. The purpose of the two kinds of anion channel transmitters is additivity: they do not occlude each other's receptors either by desensitization or by subadditivity, as in the cases of side-by-side GABA receptor patches.

The targets of GABAergic ACs decode GABA signals with either ionotropic GABAA or GABAC receptors (which modulate anion conductances) or metabotropic GABAB receptors that modulate K^+ conductances, or the operating range of voltage-gated Ca^{2+} channels. The difference between GABAA and GABAC receptors is similar to the difference between fast AMPA receptors and slower, more sustained NMDA receptors. GABAA receptors are heteromeric pentamers of GABA receptor subunits. Though they can have large anion conductances, they tend to desensitize quickly and act transiently. GCs and ACs predominantly express GABAA receptors. Ionotropic

GABAC receptors are homomeric assemblies of ρ subunits and are, arguably, a subset of the GABAA receptor group. When activated by GABA, GABAC receptor conductances are smaller, but more sustained, and they are more GABA sensitive. These receptors are primarily expressed by BCs. Since both ionotropic GABA and glycine receptor types activate chloride conductances, their efficacies can vary with the membrane potential of the target cell and the chloride gradient. Another mechanism that mediates the inhibition is shunting inhibition, where large conductance increases diminish the space and time constants of small dendrites. This inhibition is independent of the postsynaptic voltage and the chloride gradient and is most effective on small processes far from the major dendrites of a cell.

GABAB receptors are also highly effective regardless of membrane potential. They are guanine nucleotide binding protein coupled receptors (GPCRs) that either increase the voltage threshold of synaptic Ca^{2+} channels or the conductance of K^+ channels, strongly hyperpolarizing the cells (both slowing synaptic release). The former effect seems more prominent on BC axon terminals and the latter in GC dendrites or somas. Each mechanism can inhibit synaptic signaling in target processes with different timing and efficacy. However, the synaptic gain of such signaling is generally low and is often lesser than 1. Thus, individual GABAergic or glycinergic synapses are rather weak, and stimulus-dependent synchronicity is required for strong effects.

Several additional neuroactive signals associated with GABAergic ACs are co-transmitters: acetylcholine (ACh), certain catecholes, serotonin in nonmammalians, several peptides, and nitric oxide (NO). The differential regulation (if any) of GABA and ACh, catecholamine, serotonin, or peptide release is poorly understood. Glutamate may also be an excitatory AC co-transmitter for glycinergic ACs that express vGlut3, a vesicular glutamate transporter. Further, the roles of peptides and monoamines in information processing *per se* are also not completely clear, though some data suggest that peptides such as somatostatin may improve signal-to-noise ratios through an unknown mechanism.

ACs and Signal-Processing Fundamentals

ACs are components of small network submotifs (stereotyped aggregates of cell processes and synapses) that carry out analog signal-processing operations, including filtering (spatial, temporal, and spectral), gain control, signal–noise separation, signal buffering, and feature definition. Feedback is an essential control process in all amplification systems. Sign-inverting feedback (**Figure 3(a)**) is the primary mechanism by which most of these operations are effected in biology and electronics, and its essential nature is the addition of a scaled, inverted copy of the output of an

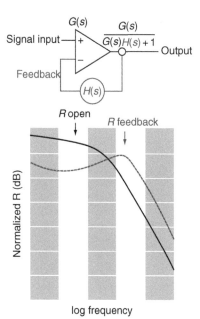

Figure 3 The basic principles of feedback. Top: an operational amplifier feedback circuit with an open loop sign-conserving transfer function $G(s)$ and part of the output fed back to the sign-inverting input, the transfer function $H(s)$, and net output transfer of $G(s)/[G(s)H(s)+1]$. Bottom: the effect of feedback on a given amplifier's performance. With no feedback, the response R as a function of frequency slowly rolls off. With feedback, the normalized peak response is at a higher frequency.

amplification stage back to that stage. Every system has an input–output response R that, for simple time-invariant linear systems, can be expressed as a transfer function of stimulus frequency $G(s)$, where G is the system gain and s is frequency. This is often referred to as the open-loop gain and has a shape similar to that in **Figure 3**. As inputs become faster (higher frequency), the gain begins to fall off. In a moving transient world, we would prefer to have the best gain at a higher frequency and negative feedback as the mechanism. The electronic negative-feedback amplifier was invented by Harold Stephen Black of Bell Laboratories in 1927, and the biological one by evolution some 3 billion years BP. By providing an inverted signal with transfer function $H(s)$ to the input, R changes from $G(s)$ to $G(s)/[G(s)H(s)+1]$. For $H(s) \geq 1$, the shape of the response function changes based on the shape of $H(s)$. If $H(s)$ is a slow, tonic response, then by depressing slow frequencies more than fast ones, R feedback has an improved frequency response. Similarly, feedback can reduce noise. Negative feedforward is also common biologically and can be even more potent in shaping outputs than feedback, although it is not as effective for frequency response or noise control.

We can consider the retina as having only two amplification stages: (1) photoreceptors → BCs and (2) BCs → GCs (**Figure 4**). Every stage of feed-forward gain also requires feedback or feed-forward control, whether made of silicon or cells. In the retina, an amplification stage is a

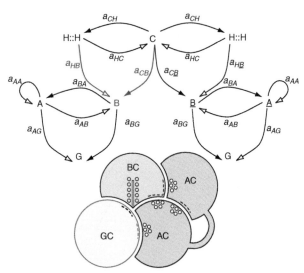

Figure 4   The basic connectivity of ACs. Top: the stages of amplification in the retina flowing from cones (C) to BCs (B) to GCs (G). Cones provide lateral excitatory signals to coupled sheets of HCs (H::H) which, in turn, provide lateral feedback to cones and lateral feed forward to BCs. Cone signals bifurcate, generating sign-conserving responses in off BCs (B) and sign-inverting responses in on BCs (B̲). Similarly, any feed-forward HC signals must also bifurcate to generate sign-inverting signals in off BCs and anomalous sign-conserving signals in on BCs. BC signals bifurcate and drive both ACs and GCs with sign-conserving mechanisms. ACs have the most complex topology, with feedback signals to BCs, feed-forward signals to GCs, and nested feed-forward/feedback by re-entrant loops. Every pairwise signal transfer has a transfer function, for example, B → A has transfer a_{BA}. Bottom: the topology of AC signaling at the BC → GC synapse. BCs provide a fast glutamatergic ribbon synapse input to ACs and GCs via AMPA receptors (blue dashes). ACs provide small conventional synapse GABA inhibition at predominantly ionotropic GABA receptors (red dashes) on ACs, BCs, and GCs.

small collection of properly connected excitatory and inhibitory synapses. Each synapse is itself a bicellular amplifier that encodes a presynaptic input voltage as a time-varying neurochemical signal, and postsynaptically decodes it as a direct current via ionotropic receptors or an indirect current or response modulation (e.g., intracellular Ca^{2+} release), typically through G-protein-coupled receptors (GPCRs). Excitatory synapses typically have amplifications or gains >>1, while inhibitory synapses typically have gains <1. ACs are the key feedback and feed-forward devices of the inner plexiform layer and their spatial distributions shape the nature of their interactions with BCs and GCs.

In detail, there are three primary submotifs that can strongly shape how ACs are involved in signal processing (**Figure 4**). The first is the classical reciprocal feedback synapse BCΔAC, wherein a signal from a BC ribbon onto an AC process is directly antagonized by an AC, typically a γ AC. This signaling pathway increases the frequency response and stabilizes the gain of the BC itself. The second is the parallel classical feed-forward synapse

chains BC1 G and BC1 AC G where the AC synapse antagonizes the BC excitation directly in the GC. However, as any BC noise has already been amplified by the BC1 G synapse, feed forward is ineffective in improving the signal-to-noise ratio. However, it is very effective in generating stimulus-specific lateral signals in GCs. The third submotif is the AC AC synapse. While extremely common, the kind of signal processing associated with this submotif is poorly understood. In its most concrete form, homotypic or even autotypic AC synapses appear to be a fundamental attribute of the reciprocal feedback system by adding a loop known as nested feedback: BC☐AC☐. This process is reminiscent of nested transconductance amplifiers where stages of looped self-inverting signals are used to tune up the amplifier, endowing it with better, high-frequency performance and giving it a wider bandwidth. Nested AC feedback may do the same.

Examples of AC Networks

There are almost no AC systems whose complete range of functions is completely understood, largely owing to their diversity and the difficulty of obtaining large physiological samples. Four generic mammalian AC networks and signal processing functions that are of particular interest are outlined here.

γ ACs of the Rod Pathway

γ ACs of the rod pathway mediate temporal processing, gain control, and noise suppression. Extensive anatomy, ultrastructure, physiology, and pharmacology reveal the serotonin-accumulating (S1/S2) γ ACs of the mammalian retina to be archetypal in-channel feedback devices. The mammalian rod pathway is unique among vertebrates (as far as is known) in using two cycles through BC amplification (**Figure 5**). All the outputs of the rod BC target ACs of two major classes: γ ACs that engage in feedback and gly ACs that engage in re-entrant feed forward to pass rod signals through the cone on and off channels. Every BC displays γ feedback synapses, which suggests that the narrowly stratified medium-to-wide field γ AC is a standard motif component. Rod BCs display clear GABAC-mediated inhibition and the S1/S2 γ ACs that provide feedback are strongly AMPA driven, on-center ACs. This simple feedback circuit will provide the BC with better temporal response, as scotopic light gets brighter, and improve signal-to-noise performance. Though the mammalian retina contains no serotonin, it is likely that the indoleamine-transporting S1/S2 cells are descendants of the nonmammalian serotoninergic/γ (S γ AC) system, which forms feedback synapses with both off and on mixed rod-cone BCs. The disappearance of serotonin signaling is clearly a recent evolutionary event, but the

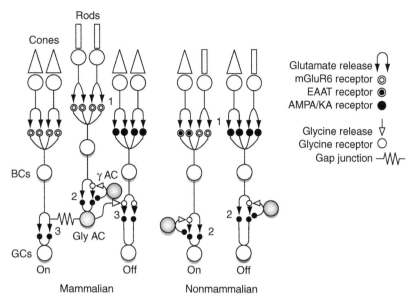

Figure 5 A comparison of rod and cone convergence onto GCs in mammalian and nonmammalian retinas. Nonmammalians use conventional mixed rod-cone BCs with direct input onto GCs and classical GABAergic feedback. Mammals predominantly exploit a rod BC that drives an intermediate stage of gain via the gly rod AC, which re-enters the cone pathways at the inner plexiform layer. The gly rod AC rives the off-BC pathway with sign-inverting glycine receptors and the on-BC pathway with gap junctions. Similar to nonmammalians, the γ rod AC provides BC feedback.

association of this simple BCΔAC network with rod BC pathways is clearly ancient.

Gly ACs of the Rod Pathway

Gly ACs of the rod pathway mediate cross-channel feedforward buffering. In electronics, a buffer is an element interposed between two device stages to provide improved transfer. Buffers are often used to fan out input signals to several different outputs while isolating the input stage from the output loads. The gly rod AC represents an evolutionary buffer to further amplify the rod signal and distribute it in a low-noise manner. All vertebrates send rod signals to the CNS via GCs that also carry cone signals. There is no private rod pathway to the brain, though almost all species have private pure cone photopic channels (e.g., the primate foveal midget pathway). Nonmammalians combine the rod and cone pathways at the first synapse in the outer plexiform layer by converging on mixed rod-cone BCs. This chain involves two high-gain ribbon synapses: rod → mixed BC → GC. Mammalians combine rod and cone pathways by fanning out from a pure rod BC pathway back into cone pathways via two scotopic motifs in the inner plexiform layer:

1. rod → rod BC → gly rod AC :: ON cone BC → GC and
2. rod → rod BC → gly rod AC → OFF cone BC → GC.

This circuit is more complex as there are two photopic motifs:

1. cone → OFF cone BC → gly rod AC and
2. cone → ON cone BC :: gly rod AC.

There are also γ AC → gly AC inputs of unknown provenance. Physiologically, dark-adapted rod ACs have strong transient AMPA-like depolarizations, while light-adapted cells have more sustained, small depolarizations. In what manner all of the inputs are regulated remains unknown.

Small γ ACs of the Primate Midget BC → GC Pathway

These cells may mediate the red–green (R–G) color opponent pathway. These ACs receive direct input from midget BCs of either the on or off pathways and provide reciprocal feedback as well as feed forward to a small patch of midget GC dendrites. Midget GCs apparently connect almost exclusively to cones that express the long-wave system (LWS) pigment gene: either LWS VP 560 nm (R) or LWS VP 530 nm (G) cones. By extension, midget BCs exist as R and G forms. Ultrastructural data suggest that each AC collects signals from several midget BCs and is presynaptic to each. This suggests that there is no R–G chromatic selectivity in the AC surround and that a midget-pathway AC has no mechanism to sort BC chromatic type. This sets the stage for one of the longest running controversies in primate vision. Are midget CG receptive fields truly center/surround (C/S) pure opponent [+R/–G, –R/+G, +G/–R, and –G/+R] or are they a collection of cells with a weighted distribution of surrounds that range from predominantly R or G or a mixture of R+G or yellow (Y) signals? A purely theoretical model has been proffered that involves glyc ACs as nulling tools to remove the inappropriate signal, making the surround appear pure. In any case, the appearance of small, spectrally pure antagonistic surrounds is inconsistent with large yellow-dominated HC surrounds observed in blue on-center GCs. How that signal disappears from R and G cone BCs remains a mystery in the face of strong evidence of coneΔHC feedback.

γ ACs of the Directionally Selective GC pathway

γ ACs mediate feature detection in the directionally selective (DS) GC pathway. The best-known feature detection event in the retina is the dependence of mammalian DS GC signaling on ionotropic GABA receptors. DS GCs exist in both on and on–off varieties; however, the qualitative feature of their signaling is similar. DS GCs have two response modes: (1) when a stimulus spot approaches the cell from the excitatory side, the cell fires, excited by both cone BC glutamate release activating AMPA receptors and starburst AC ACh activating nicotinic ACh receptors. Similar to the rod gly AC, starburst ACs provide the afferent flow with another burst of gain: cone → cone BC → starburst AC → GC. As starburst ACs are one of the most glutamate-sensitive cells in the retina, this makes DS cells exceptionally responsive. (2) When a stimulus spot approaches the cell from the null side, firing is inhibited by a GABAergic mechanism, possibly a γ AC or AxC

(see below) with an eccentric output field so that it receives excitation long before the DS GC does. The blockade of ionotropic GABAA receptors converts the DS GC into a highly excitable cell with no directional bias. It has been found that starburst ACs themselves have a directional bias in that stimuli starting at the tips of the dendrites tend to hyperpolarize and those starting dead center and moving away tend to depolarize; in addition, some authors argue that this is sufficient to build a DS network. However, given the CF of the starburst ACs (>100) compared to that of the DS GCs (≈1), the geometric requirements for this seem difficult to achieve. It is possible, even probable, that an additional γ AC or AxC is involved, as shown in **Figure 6**.

Axonal Cells

A subset of retinal neurons are the axonal or polyaxonal cells, which includes one of the dopaminergic neurons of the mammalian retina. Many are clearly GABAergic, while others, such as the dopaminergic neuron, may be dual glutamate/dopamine neurons. The essential feature of AxCs is the axon: a long–range output device that enables these neurons to send signals to regions substantially displaced from the soma and dendritic arbor. AxCs (including dopamine neurons) show either on or on–off responses. In nonmammalian retinas, object motion-selective (OMS) cells achieve their feature detection via polyaxonal γ AxCs. This raises the question of whether the mammalian DS GC uses the same strategy.

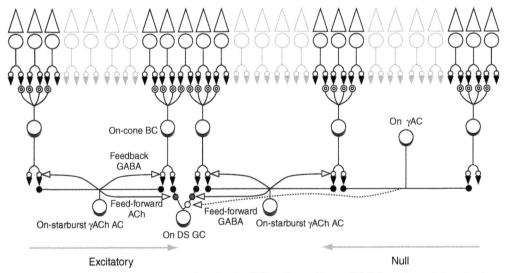

Figure 6   One topology for the generic on directionally selective (DS) pathway. The on DS GC receives glutamatergic excitation directly from on-cone BCs (black arrows and dots) and cholinergic feed-forward excitation (red dots) from local patches of starburst ACs. The starburst ACs also provide GABAergic feedback to BCs (open arrows and dots). While some argue that the starburst alone can generate directional surrounds for DS GCs, the symmetry of starburst AC fields makes this a challenging model. Similar to object motion sensitive (OMS) GCs, it is also possible that a γ polyaxonal AxC provides asymmetric inhibition for targets approaching from the null direction.

ACs and Disease

Over the past 5 years, it has become evident that, similar to the CNS, the retina undergoes extensive remodeling in response to sensory deafferentation effected by retinal degenerations, especially the retinitis pigmentosas. ACs are especially active components of remodeling, often sprouting new dendrites and participating in anomalous synaptic structures known as microneuromas. In addition, similar to other retinal neurons, they can be induced to relocate to ectopic somatic clusters loci (either to the distal retinal margin or GC layer) via anomalous migration columns. The most resilient cells of all, they are the last to die as the retina is slowly depleted of neurons. In fact, ACs may be a key component of neural survival in the remnant retina by providing a periodic source of neural activity in the absence of photoreceptors.

See also: GABA Receptors in the Retina; Information Processing: Bipolar Cells; Information Processing: Contrast Sensitivity; Information Processing: Direction Sensitivity; Information Processing: Ganglion Cells; Information Processing: Horizontal Cells; Information Processing in the Retina; Injury and Repair: Retinal Remodeling; Morphology of Interneurons: Amacrine Cells; Morphology of Interneurons: Bipolar Cells; Morphology of Interneurons: Interplexiform Cells; Neuropeptides: Function; Neuropeptides: Localization; Neurotransmitters and Receptors: Dopamine Receptors; Retinal Cannabinoids.

Further Reading

Baccus, S. A., Olveczky, B. P., Manu, M., and Meister, M. (2008). A retinal circuit that computes object motion. *Journal of Neuroscience* 28: 6807–6817.

Famiglietti, E. V., Jr. (1983). On and off pathways through amacrine cells in mammalian retina: The synaptic connections of "starburst" amacrine cells. *Vision Research* 23: 1265–1279.

Kittila, C. A. and Massey, S. C. (1997). Pharmacology of directionally selective ganglion cells in the rabbit retina. *Journal of Neurophysiology* 77: 675–689.

MacNeil, M. A., Heussy, J. K., Dacheux, R. F., Raviola, E., and Masland, R. H. (1999). The shapes and numbers of amacrine cells: Matching of photofilled with Golgi-stained cells in the rabbit retina and comparison with other mammalian species. *Journal of Comparative Neurology* 413: 305–326.

Marc, R. E. (1999). Kainate activation of horizontal, bipolar, amacrine, and ganglion cells in the rabbit retina. *Journal of Comparative Neurology* 407: 65–76.

Marc, R. E. (2004). Retinal neurotransmitters. In: Chalupa, L. and Werner, J. (eds.) *The Visual Neurosciences,* vol. 1, pp. 315–330. Cambridge, MA: MIT Press.

Marc, R. E. (2008). Functional neuroanatomy of the retina. In: Albert, D. M. and Miller, J. W (eds.) *Albert & Jakobiec's Principles and Practice of Ophthalmology,* 3rd edn, pp. 1565–1592. Philadelphia, PA: Saunders Elsevier.

Slaughter, M. M. (2004). Inhibition in the retina. In: Chalupa, L. and Werner, J. (eds.) *The Visual Neurosciences,* vol. 1, pp. 355–368. Cambridge, MA: MIT Press.

Tauchi, M. and Masland, R. H. (1984). The shape and arrangement of the cholinergic neurons in the rabbit retina. *Proceedings of the Royal Society of London B* 223: 101–119.

Vaney, D. I. (1990). The mosaic of amacrine cells in the mammalian retina. In: Chader, J. and Osborne, N. (eds.) *Progress in Retinal Research,* pp. 49–100. New York: Pergamon.

Vaney, D. I. (2004). Retinal amacrine cells. In: Chalupa, L. and Werner, J. (eds.) *The Visual Neurosciences,* vol. 1, pp. 395–409. Cambridge, MA: MIT Press.

Völgyi, B., Xin, D., Amarillo, Y., and Bloomfield, S. A. (2001). Morphology and physiology of the polyaxonal amacrine cells in the rabbit retina. *Journal of Comparative Neurology* 440: 109–125.

Wilson, M. and Vaney, D. I. (2008). Amacrine cells. In: Masland, R. H. and Albright, T. (eds.) *The Senses: A comprehensive reference,* pp. 361–367. Amsterdam: Elsevier.

Zhang, J., Jung, C. S., and Slaughter, M. M. (1984). Serial inhibitory synapses in retina. *Visual Neuroscience* 14: 553–563.

Zhang, J., Li, W., Trexler, E. B., and Massey, S. C. (2002). Confocal analysis of reciprocal feedback at rod bipolar terminals in the rabbit retina. *Journal of Neuroscience* 22: 10871–10882.

Information Processing: Bipolar Cells

S M Wu, Baylor College of Medicine, Houston, TX, USA

Glossary

Center-surround antagonistic receptive field (CSARF) – A receptive field of a visual neuron whose light response to light falling on the center region is of the opposite polarity (or spike increment/decrement) to the response to light falling on the surrounding regions of the cell's receptive field.

Depolarizing (ON-center) bipolar cell (DBC) – A bipolar cell exhibiting a depolarizing voltage response to light falling on the center region of its receptive field and a hyperpolarizing voltage response to light on its receptive field surround region.

Distal and proximal retina – In a retinal cross section, the distal retina refers to the photoreceptor side of the retina and the proximal retina refers to the ganglion cell side.

Feedback synapse – A synapse made from a higher-order neuron to a lower-order neuron, such as synapses made from horizontal cells to cones and from amacrine cells to bipolar cell axon terminals.

Hyperpolarizing (OFF-center) bipolar cell (HBC) – A bipolar cell exhibiting a hyperpolarizing voltage response to light falling on the center region of its receptive field and a depolarizing voltage response to light on its receptive field surround region.

Metabotropic glutamate receptor 6 (mGluR6) – A metabotropic glutamate receptor coupled with a second-messenger cascade found in DBC dendrites. It results in closure of cation channels in DBCs when bound with glutamate or its agonist L-2-amino-4-phosphonobutyrate (LAP4).

Rod bipolar cell – A bipolar cell whose light-evoked cation current is mediated primarily by rod synaptic inputs. In mammals, only rod DBCs have been identified.

Bipolar cells (BCs) are second-order neurons in the retina whose somas are located in the distal half of the inner nuclear layer (INL) (except for the displaced BCs found in some species whose somas are located in the outer nuclear layer (ONL)). BC dendrites branch in the outer plexiform layer (OPL) and make synaptic contacts with rod spherules, cone pedicles, and horizontal cell (HC) dendrites. Each BC bears an axon projecting to the inner plexiform layer (IPL) with terminals ramifying in different patterns at different sublaminae of the IPL. Based on their rod/cone contacts and axon terminal ramification patterns, BCs have been classified morphologically into several types. In mammals, one type of rod BC and 9 to 10 types of cone BCs have been identified (**Figure 1**). In fish, small BCs make synaptic contacts exclusively with cones and large BCs make contacts with both rods and cones, and in turtle, 11 morphological types of BCs have been found, with some contacting only cones and others contacting only rods. BCs in amphibian retinas contact both rods and cones, but some are clearly rod dominated where others are cone dominated.

BCs are the first neurons along the visual pathway that exhibit the center-surround antagonistic receptive field (CSARF) organization, the basic code for spatial information processing in the visual system. Light falling directly on the BC receptive field center region elicits voltage responses of the opposite polarity to the responses elicited by light falling on the surrounding regions of the BC's receptive field. BCs may be ON center with OFF surrounds (named ON-center BCs or depolarizing BCs (DBCs)), or OFF center with ON surrounds (named OFF-center BCs or hyperpolarizing BCs (HBCs)). The center input of BCs is mediated by rod and cone photoreceptors, which make chemical synapses on BC dendrites. BC receptive field centers are often found to be larger than their dendritic fields, suggesting that these cells may be electrically coupled. The surround input to BCs is mediated by interneurons that carry signals laterally from the surround region to the central region of the BCs' receptive field. These interneurons include HCs in the outer retina and amacrine cells (ACs) in the inner retina. HCs mediate BC surround responses through HC–cone–BC feedback synaptic pathways and/or HC–BC feed-forward (electrical or chemical) synapses, whereas ACs mediate BC surrounds through chemical synapses made on BC axon terminals in the IPL.

By using the whole-cell or microelectrode recording techniques in conjunction with fluorescent dye filling method, it has been shown that light response characteristics of various types of BCs are closely correlated with their axon terminal morphology. For example, it has been found in many species that axon terminals of DBCs ramify in the proximal half (sublamina B) of the IPL, whereas those of the HBCs ramify in the distal half (sublamina A) of the IPL. Axon terminals of rod BCs in mammalian retinas end near the proximal margin of the IPL, and axon terminals of cone BCs ramify in more central regions of the IPL. This agrees with the patterns of BC axon terminal stratification in the salamander: axon terminals of rod-dominated BCs ramify at the two margins of the IPL, whereas those of cone-dominated BCs ramify at central regions of the IPL. Additionally, despite the anatomical finding that mammalian rod and cone BCs make segregated synaptic contacts with rods and cones, it has been recently shown that some mammalian BCs exhibit mixed rod/cone responses similar to most BCs in lower vertebrates. Furthermore, although rod HBCs were not identified by earlier mammalian anatomical studies, recent physiological evidence suggests that they may exist at least in some mammals.

Based on studies of BC physiological properties in various species, it is reasonable to propose that vertebrate retinas have six major functional types of BCs: the rod (or rod dominated), cone (or cone dominated), and mixed (rod/cone) depolarizing and hyperpolarizing BCs (DBC_R,

DBC_C, DBC_M, HBC_R, HBC_C, and HBC_M). Each carries a characteristic set of light response attributes and projects them to the inner retina through axons that terminate at segregated regions (strata) of the IPL. Such stratum-by-stratum projection of light response attributes is exemplified by a large-scale voltage-clamp study of the salamander BC responses and morphology. This study reveals several rules for the function–morphology relationships of retinal BCs: (1) Cells with axon terminals in strata 1–5 (sublamina A) are HBCs (with outward light-evoked cation currents (ΔI_C)) and those in strata 6–10 (sublamina B) are DBCs (with inward ΔI_C). This agrees with the sublamina A/B rule observed in many vertebrate species (see, e.g., **Figure 1**, in which the IPL is divided into five sublaminae instead of 10). (2) Cells with axon terminals in strata 1, 2, and 10 are rod dominated, those in strata 4–8 are cone dominated, and

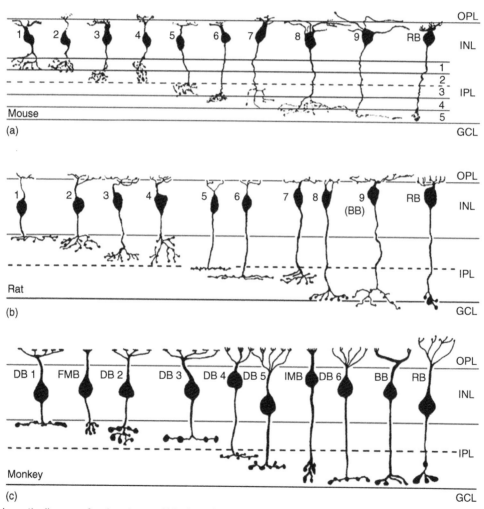

Figure 1 Schematic diagram of various types of bipolar cells in the mouse (a), rat (b), and monkey (c) retinas. Bipolar-cell images are derived from confocal images (a) or drawings (b) of Lucifer yellow/neurobiotin-injected cells, or drawings of Golgi-stained retinas (c). Inner plexiform layer (IPL) in this figure is divided into five sublaminae and IPL in the salamander retina (**Figures 2** and **3**) is divided into 10 sublaminae. OPL, outer plexiform layer; INL, inner nuclear layer; GCL, ganglion cell layer. From Ghosh, K. K., Bujan, S., Haverkamp, S., Feigenspan, A., and Wassle, H. (2004). Types of bipolar cells in the mouse retina. *Journal of Comparative Neurology* 469: 70–82.

those in strata 3 and 9 exhibit mixed rod/cone dominance. (3) Light-evoked ΔI_C at light onset in rod-dominated HBCs and DBCs are sustained, that of the cone-dominated HBCs exhibit a smaller sustained outward current followed by a transient inward current at the light offset, and that of the cone-dominated DBCs exhibit a sustained inward current followed by a small transient off outward current. (4) ΔI_{Cl} (light-evoked chloride currents) in rod-dominated BCs are sustained ON currents, whereas those in cone-dominated BCs are transient ON–OFF currents. ΔI_{Cl} in all BCs are outward, and thus they are synergistic to ΔI_C in HBCs and antagonistic to ΔI_C in DBCs. (5) BCs with axon terminals stratified in multiple strata exhibit *combined* light response properties of the narrowly monostratified cells in the same strata. (6) BCs with pyramidally branching or globular axons exhibit light response properties very similar to those of narrowly monostratified cells whose axon terminals stratified in the same stratum as the axon terminal endings of the pyramidally branching or globular cells.

In addition to projecting signals to various strata of the IPL, where ACs and ganglion cells (GCs) gather their inputs, BCs with various light response attributes have different CSARF organizations. **Figure 2** shows the morphology, patterns of dye coupling, light responses, CSARF properties, and membrane resistance changes associated with the center and surround voltage responses of the six functional types of BCs (HBC_R, HBC_M, HBC_C, DBC_C, DBC_M, and DBC_R) in the tiger salamander retina. These results suggest that the center and surround responses of various types of BCs in the retina are mediated by heterogeneous synaptic circuitry. The BC receptive field center diameters (RFCDs) vary with the relative rod/cone input: RFCD is larger in DBCs with stronger cone input, and it is larger in HBCs with stronger rod input. RFCD also correlates with the degree of dye coupling: BCs with larger RFCD are more strongly dye coupled with neighboring cells of the same type, suggesting that BC–BC coupling significantly contributes to the BC receptive field center.

BC center inputs are mediated by glutamatergic synapses. In darkness, glutamate is released from rod and cone photoreceptors, and it closes cation channels in DBCs and opens cation channels in HBCs. Postsynaptic responses of DBCs are mediated by metabotropic (metabotropic glutamate receptor 6 (mGluR6), L-2-amino-4-phosphonobutyrate (L-AP4)-sensitive) receptors coupled with a second-messenger cascade. In fish, the cone-to-DBC synaptic signal is mediated by a glutamate-activated chloride current that is suppressed by light and thus results in a membrane depolarization. Postsynaptic responses of HBCs are mediated by ionotropic receptors, and recent studies have shown that different subtypes of ionotropic glutamate receptors may be used by different types of HBCs in mediating rod/cone and transient/sustained signals.

Membrane resistance measurements in **Figure 2(e)** demonstrate that the center responses of all HBCs are associated with a resistance increase and the center responses of all DBCs are accompanied with a resistance decrease. This is consistent with the notion that glutamate released from rods and cones in darkness opens α-amino-3-hydroxyl-5-methyl-4-isoxazole-propionate (AMPA)/kainate receptor-mediated cation channels in HBCs and closes mGluR6-receptor-mediated cation channels in DBCs. Center light stimuli hyperpolarize rods and cones, suppress glutamate release, and result in a resistance increase (close ion channels) in HBCs and a resistance decrease (open ion channels) in DBCs.

Figure 3 is a schematic representation of the possible and unlikely synaptic pathways underlying surround inputs of various types of BCs, based on the surround response polarity and accompanying resistance changes shown in **Figure 2**. These results suggest that the HC–cone–BC feedback synapses may contribute to the surround responses of all six types of BCs. The negative HC–cone feedback synapses (pathway I) partially turn off the center responses by depolarizing the cones, as the membrane resistance changes associated with surround responses of all BCs are opposite to the resistance changes associated with center responses (**Figure 2(e)**).

Although all BCs share a common surround response pathway (the HC–cone–BC feedback pathway), various types of BCs use different HC and AC synaptic inputs to mediate their surround responses. It is unlikely, for example, that HBC surround responses are directly mediated by chemical synaptic inputs from hyperpolarizing lateral neurons, such as HCs and AC_{OFF}s, because of resistance change mismatch, and thus HBCs may only receive surround inputs from HC–cone–HBC and AC_{ON}–HBC synapses. On the other hand, resistance analysis suggests that DBC surround responses can be mediated by HC–cone–DBC, HC–DBC and AC_{OFF}–DBC chemical synapses, but not the AC_{ON}–DBC synapses. Moreover, dye coupling (**Figure 2(a)**) results indicate that DBC_Cs receive additional surround inputs from wide-field HCs through electrical synapses. Despite the heterogeneity, it is interesting to point out that an ON/OFF crossover inhibition rule applies here: cells with OFF (hyperpolarizing) responses (HCs and AC_{OFF}s) mediate surround inhibitory inputs to ON cells (DBCs); and cells with ON (depolarizing) responses (AC_{ON}s) mediate surround inhibitory inputs to OFF cells (HBCs). ON/OFF crossover inhibition from ACs to GCs have been reported in the salamander and mammalian retinas, and the data shown in **Figures 2** and **3** suggest that it may be a general rule for lateral inhibition in the visual system.

HCs mediate BC surround responses through the feedback synaptic pathway (HC→cone→BC) and/or the feed-forward synapses (HC→BC). Three synaptic mechanisms have been proposed for the HC feedback actions on cones. The first is that HCs release an inhibitory neurotransmitter (gamma aminobutyric acid (GABA) in several

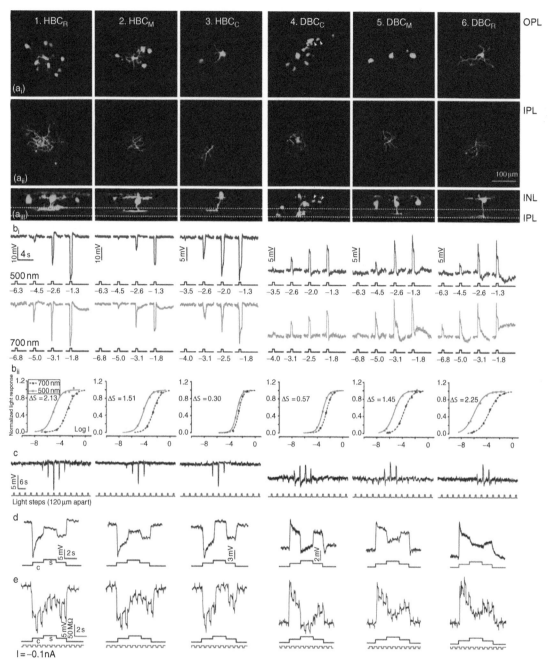

Figure 2 Morphology, light responses, and receptive fields of six types of bipolar cells in the tiger salamander retina. (a) Fluorescent micrographs of a neurobiotin-filled HBC$_R$ (column 1), an HBC$_M$ (column 2), HBC$_C$ (column 3), a DBC$_C$ (column 4), a DBC$_M$ (column 5), and a DBC$_R$ (column 6) viewed with a confocal microscope at the outer INL/OPL level (a$_i$), the IPL level (a$_{ii}$), and with z-axis rotation (a$_{iii}$). Scale bar = 100 μm. (b$_i$) BC voltage responses to 500 nm and 700 nm light steps of various intensities. (b$_{ii}$) Response-intensity (V-Log I) curves of the responses to 500-nm and 700-nm lights. ΔS (spectral difference, defined as $S_{700} - S_{500}$, where S_{700} and S_{500} are intensities of 700-nm and 500-nm light-eliciting responses of the same amplitude) of the six BCs are: 2.13, 1.51, 0.30, 0.57, 1.45, and 2.25. Since ΔS for the rods is about 3.4 and that for the cones is about 0.1 in the salamander retina, BCs with ΔS >2.0 are rod-dominated BCs, BCs with ΔS < 1.0 are cone-dominated BCs, and BCs with 1.0 < ΔS < 2.0 are mixed rod/cone BCs. (c) Measurements of BC receptive field center diameters (RFCDs) by recording voltage responses to a 100-μm-wide light bar moving stepwise (with 120-μm step increments) across the receptive field. (d) Voltage responses of the six types of BCs elicited by a center light spot (300 μm) and a surround light annulus (700 μm, inner diameter; 2000 μm, outer diameter). The surround light annulus was of the same intensity (700 nm, −2) for all six cells, whereas the intensity of the center light spot was adjusted so that it allowed the annulus to produce the maximum response. (e) Voltage responses of the six types of BCs elicited by a center light spot and a surround light annulus (same as in (d)), and by a train of −0.1-nA/200-ms current pulses passed into the cell by the recording microelectrode through a bridge circuit. From Zhang, A. J. and Wu, S. M. (2009). Receptive fields of retinal bipolar cells are mediated by heterogeneous synaptic circuitry. *Journal of Neuroscience* 29(3): 789–797, with permission from Society for Neuroscience.

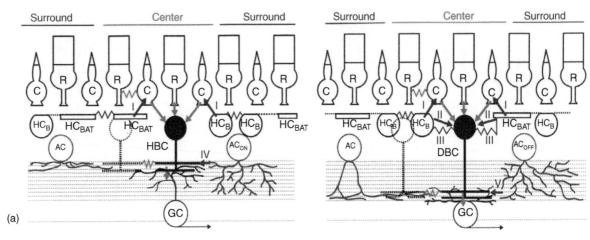

	Center synaptic organization		Surround synaptic pathways				
	Rod/cone inputs	BC–BC coupling	HC feedback (I) HC→cone→BC	HC feedforward HC→BC (II) chemical	(III) electrical	AC feedback (chemical) (IV) AC$_{ON}$→BC	(V) AC$_{OFF}$→BC
HBC$_R$	Rod: +++ Cone: +	+++	+ hyp $\xrightarrow{(-)}$ dep $\xrightarrow{(+)}$ dep	No	No	Yes (+) dep → dep	No
HBC$_M$	Rod: ++ Cone: ++	++	++ hyp $\xrightarrow{(-)}$ dep $\xrightarrow{(+)}$ dep	No	No	Yes (+) dep → dep	No
HBC$_C$	Rod: + Cone: +++	++	+++ hyp $\xrightarrow{(-)}$ dep $\xrightarrow{(+)}$ dep	No	No	Yes (+) dep → dep	No
DBC$_C$	Rod: + Cone: +++	+ +HC$_S$	+++ hyp $\xrightarrow{(-)}$ dep $\xrightarrow{(-)}$ dep	Yes (+) hyp → hyp	Yes (+) hyp → hyp	No	Yes (+) hyp → hyp
DBC$_M$	Rod: ++ Cone: ++	+	++ hyp $\xrightarrow{(-)}$ dep $\xrightarrow{(-)}$ dep	Yes (+) hyp → hyp	No	No	Yes (+) hyp → hyp
DBC$_R$	Rod: +++ Cone: +	0	+ hyp $\xrightarrow{(-)}$ dep $\xrightarrow{(-)}$ dep	Yes (+) hyp → hyp	No	No	Yes (+) hyp → hyp

(b)

Figure 3 Center-surround antagonistic receptive field organization of bipolar cells. (a) Schematic diagrams of center (green) and surround (red) synaptic pathways of HBCs (left) and DBCs (right). R, rod; C, cone; HC$_B$, B-type HC somas; HC$_{BAT}$, B-type HC axon terminals; HBC, hyperpolarizing bipolar cell. DBC, depolarizing bipolar cell; AC, amacrine cell; AC$_{ON}$, ON amacrine cell; AC$_{OFF}$, OFF amacrine cell; GC, ganglion cell. Arrows: chemical synapses; zigzags: electrical synapses; I–V: five surround synaptic pathways list in (b). (b) Variations in synaptic pathways mediating center (green) and surround (red) responses of the HBC$_R$, HBC$_M$, HBC$_C$, DBC$_C$, DBC$_M$, and DBC$_R$s. +++: strong; ++: intermediate; +: moderate; yes: possible; no: unlikely. For the possible pathways, response polarities (hyp, hyperpolarization; dep, depolarization) in each neuron and the synaptic sign ((+): sign preserving or (–): sign inverting) in each synapse in the pathways are indicated (e.g., in the HBC$_R$ HC–cone–BC pathway: (hyperpolarization in HC) through a sign-inverting synapse (–) (depolarization in cone) through a sign-preserving synapse (–) (depolarization in BC)). From Zhang, A. J. and Wu, S. M. (2009). Receptive fields of retinal bipolar cells are mediated by heterogeneous synaptic circuitry. *Journal of Neuroscience* 29(3): 789–797, with permission from Society for Neuroscience.

species) in darkness that opens chloride channels in cones, and surround light hyperpolarizes the HCs, suppresses feedback transmitter release, depolarizes the cones, depolarizes the HBCs, and hyperpolarizes the DBCs. The second mechanism is that surround light hyperpolarizes HCs, resulting in an outward current through hemichannels in their dendrites near the cones, charging the cone membrane and modulating calcium currents in cones, increasing their calcium-dependent glutamate release which depolarizes

the HBCs and hyperpolarizes the DBCs. The third mechanism is that surround-induced HC hyperpolarization elevates the pH in the HC–cone synaptic cleft, leading to an increase of calcium current in cones and a higher rate of glutamate release which depolarizes the HBCs and hyperpolarizes the DBCs. It is possible that different species under different conditions favor different feedback synaptic mechanisms, and/or different types of HC–cone synapses in the same animal may use one or more of these three HC–cone feedback mechanisms. This idea is supported by a recent study demonstrating that the responses of salamander GCs to dim surround stimuli are sensitive to GABA blockers and those to bright surround stimuli are sensitive to carbenoxolone, a gap-junction/hemichannel blocker.

The HC–BC feed-forward pathway requires a sign-preserving HC–DBC synapse and a sign-inverting HC–HBC synapse. Studies from salamander retina suggest that HC→DBC feed-forward synapses may be functional. However, application of GABA on BC dendrites in Co^{2+} Ringer's does not elicit any response, indicating that if feed-forward synapses are chemical, the neurotransmitter is not GABA. Histochemical evidence suggests that only about 50–60% of the HCs in the salamander retina are GABAergic, and the identity of neurotransmitter(s) in the rest of the HCs is unknown (but unlikely to be glycine). As illustrated in the salamander (**Figure 2(a)**) and the rabbit retinas, subpopulations of BCs are dye coupled with HCs, raising the possibility that HC–BC electrical coupling may be involved in mediating the HC–BC feed-forward synapses for DBC surround responses.

The AC–BC contribution to BC surround responses are mediated by GABAergic or glycinergic synapses. GABA receptors on ACs and GCs are largely $GABA_A$, and those on BC axon terminals are largely $GABA_C$. Glycine receptors have been localized in ACs, GCs, BC dendrites, and BC axon terminals. These receptors localized in the IPL are postsynaptic to glycinergic ACs, while those in the OPL are postsynaptic to glycinergic interplexiform cells. In the *Xenopus* retina, GABA suppresses the surround responses of the DBCs, but only slightly reduces the surround of the HBCs, and glycine suppresses the surround responses of both DBCs and HBCs. In the tiger salamander, one study shows that GABA reduces the surround responses of a subpopulation of HBCs, but another report reveals that application of picrotoxin and strychnine does not affect the surround responses of either DBCs or HBCs. Recent studies in the primate retina indicate that the HC feedback signal to cones as well as to the surround responses of GCs are not sensitive to GABAergic or glycinergic agents, but sensitive to carbenoxolone, suggesting that the surround responses in GCs are mainly mediated by HC actions on BCs in the outer retina, not by GABAergic or glycinergic AC actions in the inner retina. The reasons for these different GABA/glycine actions on surround responses are unclear. As illustrated in **Figures 2** and **3**, surround responses of different functional types of BCs in the salamander retina are mediated by different combinations of synaptic circuitries: HC→cone (GABA/hemichannel/proton) →BC, HC→BC (chemical/gap junction) and AC→BC (GABA/glycine). It is conceivable that the surround responses of different BCs/GCs from different animals under different conditions are mediated by different combinations of surround synaptic pathways, and thus they are sensitive to different synaptic blockers. The wide variation of synaptic circuitries underlying surround responses of various functional types of BCs allows for flexibility in function-specific modulation of BC/GC receptive fields. Hence different features of spatial and contrast information, such as rod/cone and ON/OFF signals, can be differentially modulated by different lighting and adaptation conditions.

Acknowledgements

This work was supported by grants from NIH (EY 04446), NIH Vision Core (EY 02520), the Retina Research Foundation (Houston), and the Research to Prevent Blindness, Inc.

Further Reading

Boycott, B. and Wassle, H. (1999). Parallel processing in the mammalian retina: The Proctor lecture. *Investigative Ophthalmology and Visual Science* 40: 1313–1327.

Dowling, J. E. (1987). *The Retina, an Approachable Part of the Brain.* Cambridge, MA: Harvard University Press.

Ghosh, K. K., Bujan, S., Haverkamp, S., Feigenspan, A., and Wassle, H. (2004). Types of bipolar cells in the mouse retina. *Journal of Comparative Neurology* 469: 70–82.

Kamermans, M. and Fahrenfort, I. (2004). Ephaptic interactions within a chemical synapse: Hemichannel-mediated ephaptic inhibition in the retina. *Current Opinion in Neurobiology* 14: 531–541.

Kolb, H., Ripps, H., and Wu, S. M. (2001). Concepts and challenges in retinal biology: A tribute to John E. Dowling. *Progress in Brain Research* 131: 23–29.

Pang, J. J., Gao, F., and Wu, S. M. (2004). Stratum-by-stratum projection of light response attributes by retinal bipolar cells of Ambystoma. *Journal of Physiology* 558: 249–262.

Sterling, P. and Demb, J. B. (2004). Retina. In: Sherpherd, G. M. (ed.) *The Synaptic Organization of the Brain*, 5th edn., pp. 217–269. Oxford: Oxford University Press.

Wu, S. M. (1992). Feedback connections and operation of outer plexiform layer of the retina. *Current Opinion in Neurobiology* 2(4): 462–468.

Wu, S. M. (1994). Synaptic transmission in the outer retina. *Annual Review of Physiology* 56: 141–168.

Wu, S. M. (2003). Intracellular light responses and synaptic organization of the vertebrate retina. In: Kaufman, P. L. and Mosby, A. A. (eds.) *Adler's Physiology of the Eye*, ch. 15, pp. 422–438. St. Louis, MO: Elsevier.

Wu, S. M. (2009). Bipolar cells. In: Squire, L., Albright, T., Bloom, F., Gage, F., and Spitzer, N. (eds.) *Encyclopedia of Neuroscience*, vol. 8, pp. 181–186.

Zhang, A. J. and Wu, S. M. (2009). Receptive fields of retinal bipolar cells are mediated by heterogeneous synaptic circuitry. *Journal of Neuroscience* 29(3): 789–797.

Information Processing: Contrast Sensitivity

M B Manookin and J B Demb, University of Michigan, Ann Arbor, MI, USA

Glossary

Adaptation – The change in response properties of a neuron, which enhance the ability to encode the immediate environment.

Contrast – The percent deviation in light intensity from the mean intensity (as defined over some period of time and region of visual angle).

Filter – The concept of a neuron's receptive field as a tuning function that is matched to certain spatial or temporal frequencies.

Receptive field – The area of space and period of recent time over which changes in light input can modulate the response of a neuron.

Threshold – The lowest contrast level of a spatiotemporal pattern where a neuron can respond to the stimulus reliably (physiology) or where an observer can perceive the stimulus reliably (psychophysics).

Contrast Processing and Adaptation

Humans can see and behave across a wide range of lighting conditions. For example, one can navigate through the woods on a starry night, where each rod photoreceptor absorbs a photon only about once per minute; and yet one can also navigate across the beach on a cloudless day, where cone photoreceptors absorb thousands of photons per second. The mean luminance between these extreme examples can differ by ~100-million-fold. This wide range of intensities poses a computational problem for the retina, because a ganglion cell can fire only about 20 action potentials (spikes) in the ~100 ms integration time of a postsynaptic neuron. Thus, the ganglion cell must continually adjust its sensitivity so that the wide range of light levels (~8 log units) can be encoded with the narrow range of output signals (~1–2 log units).

To deal with the mismatch between input and output, the retina adjusts its sensitivity depending on the mean intensity; and through mechanisms of light adaptation. These mechanisms are varied, and they include: the switch between rod photoreceptors (for night vision) and cone photoreceptors (for day vision); intrinsic properties of each receptor type that alter sensitivity depending on mean intensity; and postreceptoral mechanisms within the retinal circuitry. The apparent purpose of light adaptation is to adjust the ganglion cell's response to report, not the absolute intensity, but rather the contrast, or the percentage deviations from the mean intensity.

The contrast of a visual stimulus is a more robust property than the absolute intensity. To illustrate this point, consider a simple example, where an observer gazes at a bird on a background of leaves. Assume that the bird reflects 50% more light toward the observer's eye compared to the leaves (and ignore color in this example). Now imagine that the light reflected to the eye is reduced either by the observer's action (i.e., putting on a pair of sun glasses) or by a change in the light source reflecting off the objects (i.e., a cloud passes overhead, obscuring the sunlight). In either case, the light reflected into the eye is reduced 10-fold or more. However the relative reflectance is unchanged: the bird still reflects 50% more light than the leaves. Hence, it follows that the retina (and most of the visual system) is designed to encode contrast or the relative reflectance of objects within the same scene: the relative reflectance of objects represents a stable property of natural scenes, whereas absolute reflectance does not. Physiological measurements of retinal ganglion cells confirm this idea, showing that responses to a given contrast level are relatively constant over several orders of mean light level.

The Spatial Receptive Field

A ganglion cell calculates contrast over a specific retinal region known as its spatial receptive field. There are approximately 20 different types of ganglion cells whose axons form the optic nerve. These types encode different aspects of visual information, and some are highly selective for features such as wavelength of light or the direction of moving objects. Here, the focus is on the several types of ganglion cell that have a relatively conventional receptive field that can be described with an excitatory center region and an inhibitory surround region.

A ganglion cell's excitatory center corresponds to the retinal region aligned with its dendritic tree. Thus, the photoreceptors within the span of the ganglion cell's dendritic tree would all contribute to driving the excitatory center region. These photoreceptors synapse onto both ON- and OFF-types of bipolar cell, which express distinct glutamate receptors at their dendrites (metabotropic or ionotropic, respectively) and therefore have opposite responses to light: ON-type cells are excited by light increments, whereas OFF-type cells are excited by light decrements. Most ganglion cell types collect synapses from either ON-type bipolar cells or OFF-type bipolar cells and then inherit the ON- or OFF-center property from these presynaptic bipolar cells.

A ganglion cell's inhibitory surround corresponds to the retinal region that extends beyond the dendritic tree. Thus, an OFF-center cell that is excited by light decrements over the tree is inhibited by light decrements over the surround region, beyond the tree. The center and surround combine to report the relative contrast over space. The center is commonly stronger than the surround, so that a large object covering both the center and surround will drive a center response (e.g., a large bright object will provide some excitation to an ON-center cell). For some cell types (e.g., the X/beta-type ganglion cell of the cat or the midget/parvocellular-projecting ganglion cell of the monkey), the center and surround combine in an approximately linear fashion. Thus, the response to center plus surround stimulation can be predicted reasonably well by summing the separate responses to center and surround measured individually. For other cell types (e.g., the Y/alpha-type of the cat) there is a nonlinear combination of center and surround regions. For these nonlinear receptive fields, the presynaptic bipolar cells may be described by relatively linear receptive fields; the major nonlinearity of the ganglion cell receptive field may arise at the level of the synaptic output of the bipolar cells as they converge onto the ganglion cell. In general, a ganglion cell's excitatory

center is driven by the presynaptic bipolar cells, whereas the surround arises at two levels: the horizontal cells in the outer retina and the amacrine cells in the inner retina.

The Temporal Receptive Field

In addition to the two-dimensional spatial component, a ganglion cell's receptive field also has a temporal component. In many cases, ganglion cell responses can be described with a temporal filter that represents the best stimulus for driving a response. Thus, if the time course of the stimulus contrast over the past ~250 ms matches the filter, the response will be maximal. Ganglion cell firing rates can increase above the baseline rate much more than they can decrease below the baseline rate. In extreme cases, some ganglion cell types have essentially no background rate. Thus, a substantial nonlinearity in the ganglion cell's response is the threshold for firing action potentials (also known as rectification). The temporal filter concept must thus be combined with the concept of an output nonlinearity to predict a cell's response to a contrast modulation presented over time (**Figure 1**). A common method for modeling such responses is the linear–nonlinear model. The temporal filter can be measured by presenting a randomly flickering stimulus (commonly called white noise). For example, a spot over the receptive field center could have its intensity modulated over time. The response in the ganglion cell's firing rate could then be correlated with the stimulus to generate the filter. A separate step is used to relate the filtered output (a linear prediction of the temporal response) to

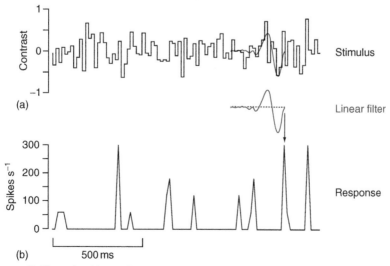

Figure 1 The retina temporally filters visual input. A temporal white-noise stimulus and firing rate response can be correlated to generate the linear filter. The filter represents a model of the temporal receptive field of the ganglion cell. The initial downward deflection (going backwards in time) indicates that the cell is an OFF-center type. The filter (red) is superimposed on the stimulus at a time when the stimulus closely matches the filter and the response output is large.

the actual firing rate; this step represents the nonlinearity associated with firing action potentials and typically includes a threshold for firing and also a point at which the firing rate saturates.

Receptive Field Properties Explain Contrast Sensitivity Functions

For cell types with linear receptive fields, the contrast sensitivity function can be explained by the receptive field properties described over space or time. For example, **Figure 2(a)** shows a temporal filter and its relationship to sinusoidal modulation of luminance of different temporal frequencies. The biphasic filter is best matched to the 6-Hz frequency and would thus predict a strong response to this frequency. At the lower frequency (1 Hz), the two phases of the filter are stimulated simultaneously and would therefore cancel to some extent; at the high frequency (12 Hz), the two phases of the filter are not stimulated strongly by the modulations and would thus, likewise, yield suboptimal responses. The response as a function of temporal frequency is plotted as the solid line in **Figure 2(b)**. This line represents the cell's temporal tuning function (or temporal contrast sensitivity function).

The same concept of filtering can be applied to the spatial domain (**Figure 2(c2)** shows a one-dimensional cut through a circular center–surround receptive field). In this case, sinusoidal modulation of light (over space; described in cycles per degree of visual angle) is presented to a center–surround receptive field. The center–surround filter best matches the 8 cycles deg^{-1} stimulus, whereas lower or higher frequencies are suboptimal. The tuning function (solid line in **Figure 2(c3)**) describes the spatial contrast sensitivity function.

Temporal and spatial contrast sensitivity functions can also be used to describe threshold measurements in human psychophysical experiments. In this case, the spatial or temporal filter is that of the entire visual system. A human is presented with brief stimuli of various spatial or temporal frequency contrast modulation and a threshold is determined (i.e., the lowest contrast at which the pattern can be reliably discriminated from the mean luminance). The perceptual contrast sensitivity function resembles the sensitivity functions of individual ganglion cells, at least superficially. It is difficult to determine how perceptual and neural sensitivity functions are related to one another, because perception depends on the output of many different types of ganglion cell as well as further processing at later stages in the brain.

Disrupting Specific Retinal Pathways Alters Perceptual Contrast Sensitivity in Selective Ways

The retinal ganglion cell axons exit the eye and travel to several different nuclei in the brain, including targets in the brainstem, midbrain, and thalamus. The pathway for

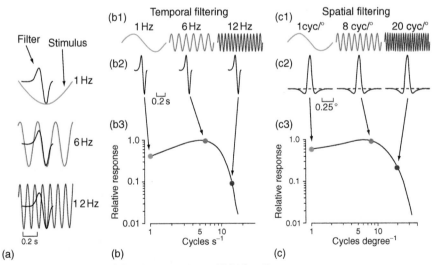

Figure 2 Spatial and temporal filtering explains the contrast sensitivity function.
(a) A temporal filter is shown with stimuli of three frequencies. The 6-Hz stimulus provides the best match to the filter.
(b) Temporal contrast sensitivity function. (b1) Three temporal frequencies (sine waves).
(b2) The filter is best matched to the 6-Hz sine wave. (b3) The sensitivity function represents the filter's normalized response to each frequency.
(c) Spatial contrast sensitivity function. (c1–c3) Same as for part (B) but for spatial frequency stimuli and the spatial filter. The filter shown is a one-dimensional cut through a circular center–surround receptive field.

conscious vision is believed to arise from the retinal projection to the thalamus, where relay cells then project retinal signals to the visual cortex. About a dozen or more different types of ganglion cell project to the lateral geniculate nucleus of the thalamus (LGN); and LGN relay cells provide a major input to primary visual cortex.

In primates, the LGN is organized into six prominent layers. The top four parvocellular (P) layers contain small cell bodies (ON and OFF midget/P cells), whereas the bottom two magnocellular (M) layers contain large cell bodies (ON and OFF parasol/M cells). Both P and M layers are separated by the eye where the ganglion cells originate; thus there are two P layers and one M layer that each receives input from the contralateral eye, whereas the other layers each receives input from the ipsilateral eye. The ON and OFF M and P cells thus account for four of the ~13 types that project to the LGN. The other nine ganglion cell types apparently project to relay cells that either reside in the layers between the M and P layers (the intercalated layers) or intermingle with the M and P cell bodies. These pathways are referred to collectively as koniocellular (K) cells.

To understand the role of the different ganglion cell types in perceptual contrast sensitivity, the M or P layers of the LGN were lesioned. We now understand that these lesion experiments were not entirely selective; the lesions to the P layers, for example, must have also affected some of the other (less numerous) K pathway ganglion cell types that project to the dorsal region of the LGN. Nevertheless, the M and P cells are the most numerous cell types that project to the LGN, and lesions to either the M or P layers yielded distinct deficits on contrast sensitivity measurements. Thus, these deficits are probably explained in large part by the lesions of either M or P cell types.

Lesions to the M layers reduced the monkey's sensitivity for high temporal frequency and low spatial frequency stimuli, whereas lesions to the P layers reduced sensitivity for low-temporal-frequency and high-spatial-frequency stimuli. Thus, the different ends of the monkey's contrast sensitivity function depended most heavily on distinct ganglion cell classes in the retina. There are two clear conclusions from these studies: there was no single cell type that could explain perceptual sensitivity for all possible patterns; and specific spatial or temporal domains of perceptual sensitivity depended most heavily on particular cell types.

Physical Limits to Contrast Sensitivity

Under optimal conditions, humans can detect small spots with contrasts of 1–3%. The most sensitive ganglion cell types can also detect small spots with contrasts of 1–3%. Thus, there may be certain conditions where perceptual thresholds are driven by a small number of ganglion cells and there may be relatively little information lost between the retina and the brain. However, there is a loss between the contrast threshold that could (theoretically) be computed at the level of photon absorptions by the photoreceptors and the threshold measured in the ganglion cell. Recent computational analysis suggests that, under certain conditions, this loss may be a factor of ~10–20.

The ability to detect contrast depends on the statistics of photon arrival and the statistical properties of various cellular processes. Photon arrival follows Poisson statistics, where the mean and the variance are equal. For example, consider a case where a ganglion cell integrates signals over 20 photoreceptors across the retina and over a 100-ms integration time and where the mean rate of photoisomerizations (i.e., absorbed photons) is 50 isomerizations (R^*) per photoreceptor per second (i.e., 5 R^*/photoreceptor/integration time). In this case, the mean R^* rate over the spatial/temporal integration ($20 \times 100 \times 5$) is 10 000 and the variance (across multiple trials) would be the same. Thus, the SD (or noise level) would be the square root or 100 R^*. The signal-to-noise ratio (mean/SD per integration time) would then be 10 000/100 = 100. Therefore, the cell in question would have difficulty detecting a difference of less than 1/100 (i.e., SD/mean) or 1% deviation from the mean level (i.e., 1% contrast). The contrast threshold would be worse (i.e., higher) when the mean luminance is lower, the number of integrated photoreceptors is fewer or the temporal integration time decreases.

Similar limitations on contrast sensitivity must arise within the neural circuit of the retina. For example, the release of neurotransmitter at retinal synapses probably obeys Poisson statistics similar to the case of photon arrival. Thus, the ability of the synapses to transfer information at low contrast depends on the release rates at these synapses and the number of synapses that are integrated by a given neuron. For example, the threshold of a cell would be best (i.e., lowest) in the presence of high release rates and a large number of integrated synapses (i.e., high degree of synaptic convergence within the circuitry). Thus, several physiological factors will place neural limits on the contrast threshold of retinal ganglion cells.

See also: Information Processing: Ganglion Cells; Morphology of Interneurons: Amacrine Cells; Morphology of Interneurons: Horizontal Cells; Phototransduction: Adaptation in Cones; Phototransduction: Adaptation in Rods; Phototransduction: Phototransduction in Cones; Phototransduction: Phototransduction in Rods; Retinal Pigment Epithelium: Cytokine Modulation of Epithelial Physiology.

Further Reading

Borghuis, B. G., Sterling, P., and Smith, R. G. (2009). Loss of sensitivity in an analog neural circuit. *Journal of Neuroscience* 29: 3045–3058.

Carandini, M., Demb, J. B., Mante, V., et al. (2005). Do we know what the early visual system does? *Journal of Neuroscience* 25: 10577–10597.

Chichilnisky, E. J. (2001). A simple white noise analysis of neuronal light responses. *Network* 12: 199–213.

Dacey, D. M., Peterson, B. B., Robinson, F. R., and Gamlin, P. D. (2003). Fireworks in the primate retina: *In vitro* photodynamics reveals diverse LGN-projecting ganglion cell types. *Neuron* 37: 15–27.

Dhingra, N. K., Kao, Y. H., Sterling, P., and Smith, R. G. (2003). Contrast threshold of a brisk-transient ganglion cell *in vitro*. *Journal of Neurophysiology* 89: 2360–2369.

Enroth-Cugell, C. and Robson, J. G. (1966). The contrast sensitivity of retinal ganglion cells of the cat. *Journal of Physiology* 187: 517–552.

Field, G. D. and Chichilnisky, E. J. (2007). Information processing in the primate retina: Circuitry and coding. *Annual Review of Neuroscience* 30: 1–30.

Geisler, W. S. (1989). Sequential ideal-observer analysis of visual discriminations. *Psychological Review* 96: 267–314.

Merigan, W. H. and Maunsell, J. H. R. (1993). How parallel are the primate visual pathways? *Annual Review of Neuroscience* 16: 369–402.

Rose, A. (1973). *Vision: Human and Electronic*. New York: Plenum Press.

Sterling, P. and Demb, J. B. (2004). Retina. In: Shephard, G. (ed.) *Synaptic Organization of the Brain,* 5th edn., pp. 217–269. New York: Oxford University Press.

Troy, J. B. and Enroth-Cugell, C. (1993). X and Y ganglion cells inform the cat's brain about contrast in the retinal image. *Experimental Brain Research* 93: 383–390.

Wandell, B. A. (1995). *Foundations of Vision*. Sunderland: MA: Sinauer.

Walraven, J., Enroth-Cugell, C., Hood, D. C., MacLeod, D. I. A., and Schnapf, J. L. (1990). The control of visual sensitivity: Receptoral and postreceptoral processes. In: Spillman, L. and Werner, J. (eds.) *The Neurophysiological Foundations of Visual Perception*, pp. 53–101. New York: Academic Press.

Wässle, H. (2004). Parallel processing in the mammalian retina. *Nature Reviews. Neuroscience* 5: 747–757.

Watson, A. B., Barlow, H. B., and Robson, J. G. (1983). What does the eye see best? *Nature* 302: 419–422.

Information Processing: Direction Sensitivity

Z J Zhou and S Lee, Yale University School of Medicine, New Haven, CT, USA

Glossary

Accessory optic nuclei – A series of small nuclei in the rostral midbrain, near where the optic tract enters the lateral geniculate nucleus. They receive inputs from motion-sensitive ganglion cells and project to the vestibular nuclei, triggering optokinetic movements in response to movement of the visual world across the retina. This system is very important for animals without foveae, but is poorly developed in humans.

Optokinetic nystagmus – The repeated reflexive responses of the eyes to ongoing large-scale movements of the visual scene.

Patch-clamp recording – A sensitive electrophysiological recording technique that permits the measurement of ionic currents flowing through individual ion channels. It uses a glass micropipette that has an open-tip diameter of about 1 μm. The micropipette tip is sealed onto the cell surface, enclosing a membrane surface area or patch that often contains only a few ion channels. Several variations of this technique are commonly applied.

Spatially offset inhibition – An inhibitory synaptic input from the receptive field surround that is asymmetrically offset to one direction.

Two-flash apparent motion stimulation – A visual stimulation paradigm, in which two nearby spots of light are flashed in rapid succession to simulate the motion of a light spot.

Detecting the direction of image movement is an essential task of the visual system. Neurons in many parts of the visual system show directional sensitivity to image motion. The initial computation of movement direction is accomplished in the retina. In the early 1960s, Barlow and co-workers discovered that a subset of output neurons in the rabbit retina responds vigorously to an image moving across its receptive fields in a particular (preferred) direction, but gives little or no response to the same image moving in the opposite (null) direction (**Figure 1**). These retinal neurons, termed direction-selective ganglion cells (DSGCs), display a directional preference that is independent of the nature of the image, such as contrast, size, and complexity. They are specialized in processing visual information regarding motion and motion direction.

Physiological Functions of DSGCs

In rabbit, a species for which retinal direction selectivity is best characterized, two types of DSGCs have been found: the ON type, which responds to the onset of a small spot of light flashed in the center of the cell's receptive field, and the ON–OFF type, which responds to both the onset and the offset of such a flash. ON DSGCs project their axons to the accessory optic system. They are believed to be involved in the feedback mechanism that stabilizes a moving image on the retina, for example, during the smooth tracking movements of the eyes. These cells respond best to slow image movements ($\sim 0.3^\circ\,s^{-1}$) and contribute significantly to the control of optokinetic nystagmus over the range of stimulus velocities that produce the most complete minimization of image motion on the retina. ON DSGCs are divided into three subtypes, distinguished by their preferred directions pointed roughly toward the anterior, superior, and inferior retina, respectively. Each of these three preferred directions is thought to correspond to a component of the head rotation that activates one of the three pairs of semicircular canals in the inner ears, suggesting that these cells may also detect the slippage of a visual image on the retina during head rotation and provide an error signal that tells the visual servo-system to compensate the difference between head rotation and eye rotation.

ON–OFF DSGCs, on the other hand, project to the superior colliculus and the lateral geniculate nucleus. They are further categorized into four subtypes, characterized by their preferred directions which, respectively, point roughly toward the four cardinal directions in the retina: superior, inferior, anterior, and posterior. It has been suggested that these four directions may correspond to the directions in which the eye is rotated by the four extraocular rectus muscles, indicating that ON–OFF DSGCs are also involved in the feedback mechanism that controls eye movement. However, unlike the ON type, ON–OFF DSGCs give largest responses to fast movements ($\sim 10^\circ\,s^{-1}$ or faster) and respond poorly to slow movements. Their contribution to image stabilization during the slow phase of optokinetic nystagmus may be significant only at stimulus velocities exceeding $\sim 3^\circ\,s^{-1}$. The fact that ON–OFF DSGCs project to the lateral geniculate nucleus in addition to the superior colliculus suggests that these cells may also process directional information of object movement for visual perception. However, the exact physiological functions of ON–OFF DSGCs remain to be understood.

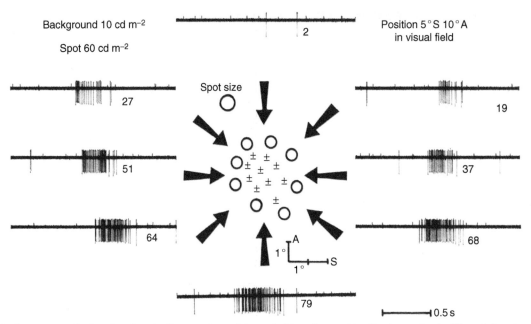

Background 10 cd m⁻²

Spot 60 cd m⁻²

Position 5°S 10°A
in visual field

Figure 1 Responses of a directionally selective ganglion cell (recorded from its axon) to stimulus motion in different directions. Map of receptive field in center. Traces show responses (spikes) elicited by the movement of a spot of light across the receptive field in the direction of adjacent arrow. This cell shows a preferred direction pointing upward. Anterior (A) and superior (S) meridians in the visual field are shown together with 1° calibration marks. The number of spikes is shown immediately after each response. Conventions are as follows: ±, response to stationary spot at both ON and OFF; O, no response; there are no responses outside the ring of Os. Adapted from Barlow, H. B., Hill, R. M., and Levick, W. R. (1964). Retinal ganglion cells responding selectively to direction and speed of image motion in the rabbit. *Journal of Physiology* 173: 377–407.

Synaptic Circuitry of DSGCs

As direction selectivity in the retina represents a fundamental form of information processing, it has been regarded as a model system for understanding neuronal computation in the brain. In their 1965 work, Barlow and Levick first suggested that the direction selectivity of ganglion cells is built from sequence-discriminating subunits. The primary mechanism for this discrimination was thought to result from a spatially offset lateral inhibition, which vetoes responses to sequences corresponding to stimulus movement in the null direction. Pharmacological experiments by Daw and colleagues subsequently found that gamma aminobutyric acid (GABA) receptor antagonists block direction selectivity by bringing out strong responses of DSGCs to stimulus movement in the null direction, suggesting that the spatially offset inhibition is mediated by GABA. Direct measurement of the inhibitory inputs to ON–OFF DSGCs became possible in the early 2000s, when whole-cell patch-clamp recordings were made successfully from these cells in the whole-mount rabbit retina by several labs. Fried and co-workers showed that the inhibitory input to ON–OFF DSGCs is larger during stimulus movement in the null direction than that in the preferred direction, and that the inhibitory input arrives ahead of the excitatory input when the stimulus moves in the null, but not in the preferred

direction. The spatial extent of the inhibitory input measured by whole-cell patch clamp is offset from the ON–OFF DSGC dendritic field toward the null direction. The excitatory current input to an ON–OFF DSGC is also directionally asymmetric: larger for preferred direction movement than for null direction movement. Similar results were also obtained from ON DSGCs in the mouse retina. These findings establish that the synaptic inputs to a DSGC are already directionally asymmetric.

An important clue to the origin of major synaptic inputs to DSGCs comes from the anatomical structure of these cells. DSGCs have a distinctive morphology, characterized by the looping dendritic branches (**Figure 2**). The dendritic stratification pattern of DSGCs follows the general rule of segregation between the ON and OFF channels in the inner plexiform layer (IPL) of the retina. Thus, ON DSGCs arborize only in the proximal half (ON sublamina) of the IPL, whereas ON–OFF DSGCs arborize in both the proximal and the distal half (OFF sublamina). Both the ON and OFF arborizations stratify narrowly in the IPL, juxtaposing the two narrow cholinergic strata, suggesting that DSGCs receive synaptic inputs from the relatively small number of bipolar and amacrine cell subtypes that terminate specifically in these strata. In particular, cholinergic amacrine cells, which are the only cholinergic cells in the mammalian retina, are expected to provide a significant amount of lateral input to DSGCs. Cholinergic amacrine

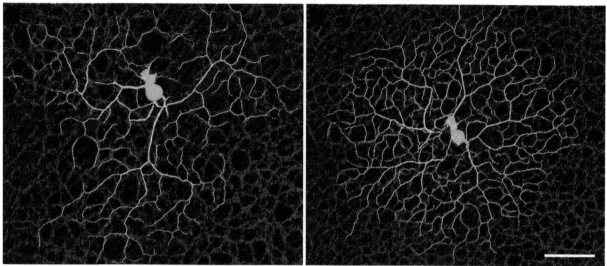

Figure 2 Dendritic morphology of ON–OFF direction-selective ganglion cell. Confocal micrographs of an intracellularly injected ON–OFF DSGC in the whole-mount rabbit retina, showing bistratified dendritic arbors in the ON (left panel) and OFF (right panel) sublamina of the inner plexiform layer. The characteristic looping dendrites co-stratify extensively with the cholinergic plexus (labeled with choline acetyltransferase in red). Scale bar = 100 μm. Adapted from Dong, W., Sun, W., Zhang, Y., Chen, X., and He, S. (2004). *Journal of Physiology* 556: 11–17.

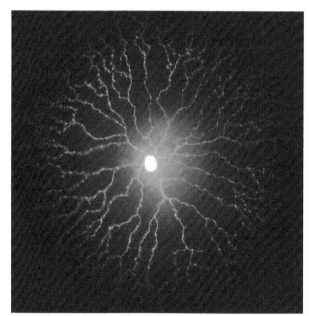

Figure 3 Dendritic morphology of starburst amacrine cell. Fluorescence micrograph of a SAC, with its soma located in the ganglion cell layer in a postnatal day 18 retina of a rabbit. The cell has a radially symmetric dendritic tree and a polar asymmetric synaptic structure. The input synapses from bipolar cell are distributed over the whole dendritic tree, but the output synapses are localized in the distal varicose zone, where synaptic vesicles are concentrated. Unpublished micrograph provided by Lee S and Zhou ZJ.

cells are also known as starburst amacrine cells (SACs; **Figure 3**). They exist as two mirror-symmetric populations across the IPL. They have four to five primary dendrites, each branching out regularly into secondary and tertiary processes, with numerous varicosities imbedded in the distal dendritic zone. The dendritic tree of a SAC emanates from the soma with a nearly perfect radial symmetry, rendering the cell a starburst appearance and hence its popular name. The processes of neighboring SACs overlap significantly and form honeycomb-shaped plexuses which co-fasciculate extensively with the looping dendrites of DSGCs, suggesting intimate synaptic interactions between the two cell types. Indeed, direct synapses from SACs to ON–OFF DSGCs have been observed at the electron microscopic level.

The neurochemical significance of the synapses between SACs and DSGCs became apparent when SACs were found to synthesize and release both acetylcholine (ACh) and GABA, a property that also rendered SACs the first known exception to Dale's principle of one neuron releases one fast neurotransmitter. The GABAergic nature of cholinergic amacrine cells suggests that these cells may provide the crucial, spatially offset GABAergic inhibition to DSGCs, an idea that has been tested and debated extensively in the literature. The co-release of ACh and GABA by SACs also raises the possibility that these two neurotransmitters may be used by SACs in a complimentary manner to enhance the sensitivity of DGSCs to motion and motion direction.

The first functional evidence for a key role of SACs in direction selectivity came from the finding that ablation of SACs by immunotoxins and neurotoxins results in an apparent elimination of the directional discrimination of DSGCs in the mouse retina, as well as a loss of optokinetic nystagmus. This finding demonstrates a critical link not only between SACs and direction selectivity, but also between direction selectivity and optokinetic

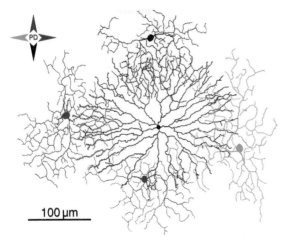

PD

100 μm

Figure 4 Synaptic circuit of ON–OFF direction selective ganglion cells. ON–OFF DSGCs receive GABAergic inhibitory inputs only from starburst amacrine cell dendrites that point in the null direction, but not in the preferred direction. A single SAC (shown in black) can provide null-direction inhibition to all four subtypes of ON–OFF DSGCs with preferred directions along four orthogonal directions (shown in different colors). For simplicity, only the circuitry in the ON sublamina is shown. Adapted from Taylor, W. R. and Vaney, D. I. (2003). New directions in retinal research. *Trends in Neurosciences* 26(7): 379–385.

nystagmus. Dual patch-clamp recordings from SAC–DSGC pairs in the rabbit retina subsequently demonstrated that DSGCs receive direct GABAergic synaptic inputs from SACs, and that such inputs come only from SACs located on the null side, but not the preferred side of the DSGC. This result establishes that SACs exert a spatially offset GABAergic inhibition on DSGCs through directionally asymmetric hardwiring between SACs and DSGCs, a landmark finding that forms the basis of the current model of the synaptic circuitry of DSGCs (**Figure 4**).

Direction-Selective Responses in SAC Processes

For the asymmetric wiring model of direction selectivity shown in **Figure 4** to work, the release of GABA from each branch of a SAC must also be directionally selective: larger during a null-direction movement and smaller during a preferred-direction movement. Otherwise, the SAC dendrites that synapse on a DSGC from the null direction would still inhibit the DSGC when an image moves to the DSGC receptive center along the preferred direction, even though such an inhibitory input may not precede the glutamatergic excitatory input from bipolar cells. Experiments using apparent motion stimulation also show that the GABAergic input to a DSGC (from SACs on the null side) is strongly suppressed by two-spot flashes that simulate a preferred-direction movement, indicating that GABA release from SAC dendrites is directionally selective. This

result suggests that a critical component of the direction-selective mechanism must reside upstream from DSGCs and in SACs.

How does a morphologically symmetric cell like the starburst produce the functional asymmetry required for direction-selective GABA release? A close anatomical examination of the SAC tells us that, despite the striking radial symmetry in the dendritic tree, the synaptic structure of the cell has a profound polar asymmetry. Input synapses on starburst dendrites distribute more or less uniformly along the entire dendritic length, whereas output synapses are localized in the distal varicose zones, where synaptic vesicles are concentrated (**Figure 3**). The distal varicose zones are believed to be electrotonically semi-isolated, due to the thin diameter of the dendrites and a heavy expression of K_{V3} voltage-gated potassium channels on the proximal dendrites. This polar asymmetry in input–output relation, combined with the short electrotonic length of the dendrites, would enable the distal dendrites of a SAC to process directional signals independently, with a preference to centrifugal stimulus movement, as predicted by computational models. In an elegant two-photon Ca^{2+}-imaging experiment, Euler and co-workers showed that distal starburst dendrites respond to spot illumination with intracellular Ca^{2+} transients that are restricted to distal dendrites. Importantly, these local Ca^{2+} responses tended to be directionally selective: stronger for centrifugal than for centripetal stimulus movement.

Two-flash apparent motion experiments further show that the direction selectivity in distal starburst dendrites involves both centrifugal excitation and centripetal inhibition. The underlying mechanism may involve both cell-autonomous properties and synaptic interactions at SAC dendrites. Cell-autonomous properties, such as nonlinear interactions between the activation of voltage-gated calcium channels and a gradient in membrane potential along the SAC dendrites, may contribute to an enhanced response of the distal dendrites to centrifugal image movement. On the other hand, synaptic interactions, particularly GABAergic inhibition from the receptive field surround, play a key role in centripetal inhibition. Dual patch-clamp recording experiments found that neighboring SACs inhibit each other through reciprocal GABAergic synapses. Zhou and Lee proposed that the synaptic mechanism for direction-selective release of GABA in distal starburst dendrites is built primarily from a classic center-surround receptive field structure. Synaptic inputs to the receptive field center are dominated by the glutamatergic input from bipolar cells, whereas synaptic inputs from the receptive field surround are dominated by the direct GABAergic input from neighboring SACs. Such a concentric receptive field structure would normally be directionally symmetric for a neuron that makes an integrative output decision at the soma or axon hillock. However, since the output decisions of a SAC are made independently in individual distal dendrites, such a concentric center-surround receptive field structure would produce a profound directional asymmetry at distal dendrites, where synaptic inputs received during centrifugal stimulus movement are dominated by excitation, but synaptic inputs received during centripetal stimulus movement are dominated by inhibition. This synaptic mechanism, together with an intrinsic mechanism that promotes centrifugal facilitation, would produce robust direction-selective GABA release from SACs.

Integration of Multiple Cooperative Mechanisms for Direction Selectivity

The direction-selective circuit so far identified consists primarily of DSGCs, SACs, and subsets of bipolar cells, although other amacrine cell types may also participate in the regulation of DSGC receptive fields. Distinct mechanisms of direction selectivity work in concert at three different levels. First, at the DSGC level, direction selectivity is shaped largely by a spatially offset inhibition from SACs, whose distal dendrites release GABA in a direction-selective manner and make GABAergic synapses onto DSGCs asymmetrically from the null direction only. Direction selectivity of DSGCs is also enhanced by the directionally asymmetric excitatory inputs, which may contain both glutamatergic and cholinergic components. In addition, a postsynaptic mechanism, involving local signal computation and spike generation in DSGC dendrites, may further sharpen direction selectivity. Second, at the SAC level, direction selectivity is formed in the distal starburst dendrites by a profound polar asymmetry between centrifugal excitation and centripetal inhibition. GABAergic inhibition, mediated largely by reciprocal inhibition between SACs, plays a key role in suppressing SAC responses to centripetal stimulus motion. However, the nature of lateral synaptic interaction during centrifugal motion is currently unclear. A model based on differential expression of two different chloride transporters proposes that the excitability of GABAergic input changes as the image moves centrifugally or centripetally along a SAC dendrite. Further experiments are required to test directly the predictions of this model. In addition to synaptic interactions, intrinsic cellular properties also play an important role in shaping direction selectivity in SAC dendrites, predominantly by contributing to centrifugal facilitation. Third, at the bipolar cell level, a direction-selective release of glutamate is strongly implicated by the finding of asymmetric excitatory inputs to DSGCs, but the underlying mechanism remains a major missing piece of the puzzle. A key issue is whether local terminals of a bipolar cell axon can function independently and make selective synapses with the dendrites of specific subtypes of DSGCs. An intriguing possibility in this regard is that SACs asymmetrically inhibit specific bipolar cell output synapses, which in turn synapse selectively on DSGCs in a spatially asymmetric manner.

While the critical GABAergic contribution of SACs to direction selectivity is well accepted, the cholinergic role of SACs remains elusive. Various models of cholinergic contributions to direction selectivity and motion sensitivity have been proposed over the years. However, the basic mode of nicotinic cholinergic action in the retina still remains obscure: Does ACh mediate fast neurotransmission at precise synaptic sites between SACs and DSGCs, or does it play a diffuse, paracrine role in modulating the activity of many ganglion cell types? More detailed experimental results are needed before it can be concluded as to how the release of ACh from SACs may enhance the sensitivity of DSGCs to image movement, whether cholinergic interactions enhance direction selectivity, how cholinergic synapses form a neural circuit, and how the cholinergic and GABAergic circuits of SACs interact with each other. Investigations into some of these questions are currently underway and may uncover additional levels of cooperation among synaptic mechanisms of motion and direction selectivity in the near future.

Concluding Remarks

Direction selectivity is a basic form of information processing produced by a relatively simple neuronal circuit in the inner retina. In order for this circuit to accomplish the robust computation required for detecting motion direction, multiple levels of cooperative mechanisms are integrated at each synapse. Such synaptic integration may represent an important feature of network computation in the central nervous system. From an anatomical point of view, the direction-selective circuit in the retina reveals a level of selectivity and precision in network organization that was previously unappreciated in most other parts of the central nervous system. Understanding the developmental mechanisms that control the establishment of the direction-selective circuit in the retina remains a challenging task of future investigation and will shed important light on the development of neuronal circuits in general.

Acknowledgments

This work is supported in part by grants from the National Eye Institute and Research to Prevent Blindness, Inc.

See also: GABA Receptors in the Retina; Information Processing: Amacrine Cells; Information Processing: Bipolar Cells; Information Processing: Ganglion Cells; Information Processing in the Retina; Morphology of Interneurons: Amacrine Cells.

Further Reading

Barlow, H. B. and Levick, W. R. (1965). The mechanism of directionally selective units in rabbit's retina. *Journal of Physiology* 178(3): 477–504.

Barlow, H. B., Hill, R. M., and Levick, W. R. (1964). Retinal ganglion cells responding selectively to direction and speed of image motion in the rabbit. *Journal of Physiology* 173: 377–407.

Dacheux, R. F., Chimento, M. F., and Amthor, F. R. (2003). Synaptic input to the on–off directionally selective ganglion cell in the rabbit retina. *Journal of Comparative Neurology* 456(3): 267–278.

Demb, J. B. (2007). Cellular mechanisms for direction selectivity in the retina. *Neuron* 55(2): 179–186.

Euler, T., Detwiler, P. B., and Denk, W. (2002). Directionally selective calcium signals in dendrites of starburst amacrine cells. *Nature* 418 (6900): 845–852.

Famiglietti, E. V. (1991). Synaptic organization of starburst amacrine cells in rabbit retina: Analysis of serial thin sections by electron microscopy and graphic reconstruction. *Journal of Comparative Neurology* 309(1): 40–70.

Fried, S. I., Munch, T. A., and Werblin, F. S. (2002). Mechanisms and circuitry underlying directional selectivity in the retina. *Nature* 420(6914): 411–414.

Lee, S. and Zhou, Z. J. (2006). The synaptic mechanism of direction selectivity in distal processes of starburst amacrine cells. *Neuron* 51(6): 787–799.

Oyster, C. W., Takahashi, E., and Collewijn, H. (1972). Direction-selective retinal ganglion cells and control of optokinetic nystagmus in the rabbit. *Vision Research* 12(2): 183–193.

Rodieck, R. W. (1998). *The First Steps in Seeing.* Sunderland, MA: Sinauer.

Tauchi, M. and Masland, R. H. (1984). The shape and arrangement of the cholinergic neurons in the rabbit retina. *Proceedings of the Royal Society of London. Series B. Biological Sciences* 223(1230): 101–119.

Taylor, W. R. and Vaney, D. I. (2003). New directions in retinal research. *Trends in Neurosciences* 26(7): 379–385.

Vaney, D. I. (1990). The mosaic of amacrine cells in the mammalian retina. *Progress in Retinal Research* 9: 49–100.

Wyatt, H. J. and Day, N. W. (1976). Specific effects of neurotransmitter antagonists on ganglion cells in rabbit retina. *Science* 191(4223): 204–205.

Yoshida, K., Watanabe, D., Ishikane, H., et al. (2001). A key role of starburst amacrine cells in originating retinal directional selectivity and optokinetic eye movement. *Neuron* 30(3): 771–780.

Information Processing: Ganglion Cells

T A Münch, University of Tübingen, Tübingen, Germany

Glossary

Excitation – The synaptic input to a cell that serves to depolarize and activate it.

Inhibition – The synaptic input to a cell that serves to hyperpolarize and suppress it.

Receptive field – For retinal cells, this refers to a region in the visual environment in which the presence of a visual stimulus will influence the activity of the neuron.

The eye is often compared to a camera, in which the retina plays the role of the film. However, this analogy does not do justice to the remarkable image- and information-processing capabilities of the retina. The roles that horizontal cells, bipolar cells, and amacrine cells play in the information processing in the retina have been discussed elsewhere in the encyclopedia. All their activity merges into the output cells of the retina, the ganglion cells. These cells form a sort of bottleneck through which all the information has to pass, which is destined to reach higher visual centers in the brain.

Ganglion Cell Types

The signals leaving the retina are transmitted by several channels, each channel carrying information about different features of the visual world. Each channel is embodied by a certain type of ganglion cell. It is still controversial how many different types of ganglion cells exist, but 20 should be a reasonable upper limit. Two ganglion cells are considered to belong to the same type if they share a defined set of properties, including morphological, physiological, and genetic properties. However, it is not trivial to define what such characterizing properties should be. Too fine a distinction would artificially separate ganglion cells into different types, even though they might belong to the same type, while a course classification will not do justice to the rich diversity of image-processing capabilities of the retina. For the purpose of this article, the term ganglion cell type will be loosely and pragmatically defined as the population of ganglion cells extracting the same features from the visual world and therefore performing the same function. Later in this article, we describe some specific examples of ganglion cell types and the visual features that activate them. First, however, we discuss some general principles of ganglion cell processing.

Principles of Ganglion Cell Processing

Neural activity in the retina is unusual in the central nervous system as most retinal neurons are analog: they do not fire action potentials; instead, transmitter release is a continuous function of membrane voltage. Ganglion cells are different; they use action potentials to transmit their signals along their axons which form the optic nerve. These cells integrate excitatory input from bipolar cells and inhibitory input from amacrine cells. It is the precise temporal and spatial integration of these inputs that leads to the final spiking output of the ganglion cells. It will become clear from the examples below that it is the inhibitory activity of the amacrine cells that creates most of the specificity in ganglion cell responses. Bipolar cells serve to set up a range of permissive stimuli, that is, the set of stimuli that a ganglion cell could, in principle, respond to. This stimulus space is then strongly restricted by the activity of the inhibitory amacrine cells, resulting in the specific responses of ganglion cell types.

Temporal Processing

Ganglion cells show diverse responses to step changes of light (**Figure 1(a)**). Generally, responses can be divided into ON (activated by increases of light intensity), OFF (activated by decreases of light intensity), and ON–OFF (activated by both increases and decreases of light intensity). The ganglion cells expressing these responses are accordingly termed ON cells, OFF cells, and ON–OFF cells. Within each of these classes, one can find cell types with transient or sustained responses. In addition, different cells have different latencies from stimulus onset until the time of the first spike. For some cell types, Gollisch and Meister demonstrated that the relative spike latency can code for the stimulus intensity (**Figure 1(b)**).

The different temporal response characteristics of ganglion cells are correlated with the level of stratification of ganglion cell dendrites. The dendrites of ON ganglion cells stratify in the inner half of the inner plexiform layer (sublamina b, close to the cell bodies of the ganglion cells), while the dendrites of the OFF ganglion cells stratify in outer half (sublamina a, toward the bipolar cells). The transient cells tend to stratify toward the center of the inner plexiform layer (IPL), while the sustained cells send their dendrites to the two borders of the IPL.

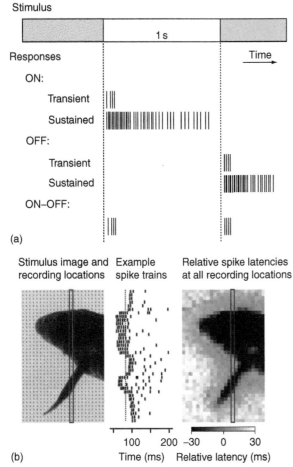

(a)

(b)

Figure 1 Temporal response characteristics of ganglion cells. (a) Different types of ganglion cells respond differently to the onset of a bright stimulus in the receptive field. ON cells respond at the onset of the stimulus; OFF cells respond at the offset; and ON–OFF cells respond at the both the on- and offset. Within these cell classes, one can find transient and sustained types of responses. (b) The latency of responses of some cell types indicates the intensity of the stimulus. Left: The stimulus image was presented to the retina multiple times at different locations, so that the recorded ganglion cell was located at the positions indicated by the dots. Middle: Spiking responses for the column of recording locations indicated by the red rectangle in the left panel. The first spike after stimulus onset is marked in red. The average time-to-first-spike is indicated by the dotted red line, and serves as time point 0 in the right panel. Right: Relative spike latency for each recording location coded as a grayscale value. Adapted from Gollisch and Meister (2007).

Spatial Processing

Center-Surround Receptive Fields

Many ganglion cell types express antagonistic center-surround receptive fields, first described by Kuffler in 1953. The terms ON and OFF cells defined in the previous paragraph are a shorthand for ON center cell and OFF center cell, describing the response characteristics of the cells when they are stimulated in their receptive field

center. When a stimulus spot falls in the surround region, an ON center cell might respond when a bright spot is removed, that is, with an OFF-type response (point 2 below). In his 2001 work, Ralph Nelson summarized the effects that the surround can have on ganglion cell responses. Point 4 of the list given below is equivalent to a size tuning of the response, that is, responses to larger spots can be weaker than responses to smaller spots.

1. Change from sustained to transient center response as stimuli are displaced from center.
2. Responses evoked to opposite stimulus phase with surround stimulation.
3. Active inhibition of the center response with surround stimulation.
4. A maximal response can be evoked only with an optimally sized spot.

Center-surround receptive fields can be successfully modeled as difference of Gaussians, describing two overlaying concentric receptive fields with opposite effects on the cell (**Figure 2**). For small stimuli, the effect of the strong center receptive field prevails. For larger stimuli, the weaker surround begins to show its effect due to its larger radius. Physically, center responses originate from the excitatory input of bipolar cells to the ganglion cell dendrites. Usually, the size of the receptive field center of a ganglion matches the size of its dendritic field quite well. Surround properties originate from the activity of laterally oriented inhibitory interneurons in the retina, namely horizontal cells and amacrine cells. The activity of these cells contributes to the surround at three levels. They suppress the activities of bipolar cells either in the outer retina (horizontal cells) or inner retina (amacrine cells). Amacrine cells also give direct input to ganglion cells, contributing to the inhibitory surround at a third level of interaction.

Tiling

One of the properties characterizing a ganglion cell type is that its members tile the retina. This means that, as a population, the ganglion cells of any type carry information about the whole scene; a single ganglion cell is contributing one pixel. Ganglion cells with small receptive fields (e.g., midget ganglion cells) can convey spatial details of the visual scene (**Figure 3(a)**, left column), while a ganglion cell with a large receptive field (e.g., parasol ganglion cells) will not be able to convey much detailed spatial information (**Figure 1(a)**, right column) – the cell's activity will signal that it was triggered from somewhere within its receptive field, but it is not possible to pinpoint the exact spatial location just from the activity of a single cell, or from the activity of the cell population. The situation is better when the ganglion cells of a type overlap (**Figure 3(b)**), a situation which is frequently

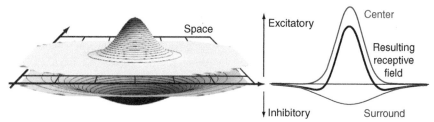

Figure 2 Difference of Gaussian model of center-surround receptive fields. According to this model, the receptive field of a ganglion cell consists of two bell-shaped components, a smaller excitatory component (center), and a larger inhibitory component (surround). The resulting overall receptive field has the shape of a Mexican hat (black curve, right).

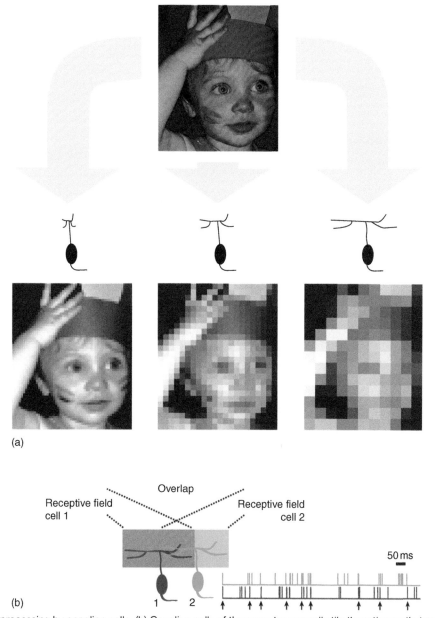

Figure 3 Spatial processing by ganglion cells. (b) Ganglion cells of the same type usually tile the retina so that each region in the visual environment is seen by at least one member of each type. One cell contributes one pixel of the image transmitted to the brain. Ganglion cells with small receptive and dendritic fields convey more detailed spatial information than larger cells. (b) Neighboring ganglion cells often fire action potentials synchronously. One possible mechanism is shared synaptic input. Then, the presence of synchronous spikes is an indication of the presence of an image feature in the region of overlap between two cells, which can increase the spatial resolution of a ganglion cell type. Spike trains adapted from Shlens, J., Rieke, F., and Chichilnisky, E. (2008). Synchronized firing in the retina. *Current Opinion in Neurobiology* 18: 396–402.

encountered (e.g., alpha ganglion cells). When a stimulus feature occurs in the region of overlap between neighboring cells, one can often observe a strongly increased probability of synchronous action potentials of the two cells. In many cells, the main reason for the synchronization is shared synaptic input. Consequently, if the two cells are triggered simultaneously by two separate stimuli occurring in the two nonoverlapping regions of their receptive fields, the firing is not synchronized. The precise relative timing of action potentials can therefore contribute to the spatial information conveyed by the ganglion cell population. More details on modes and mechanisms of synchronous firing in the retina can be found in the 2008 review by Shlens and co-workers.

Stimulus Features that Trigger Ganglion Cell Activity

As mentioned above, any ganglion cell type can be considered as a channel that conveys information about a certain image feature to higher visual centers of the brain. The activity of such a ganglion cell means that (one of) the trigger feature(s) has been present in the visual stimulus in the receptive field of the cell. What are the features that trigger ganglion cell activity?

Ganglion Cells as Spatiotemporal Filters

One way of looking at the processing of visual stimuli by ganglion cells is to interpret them as a bank of spatiotemporal filters (**Figure 4**). Each ganglion cell has certain frequency response characteristics in both space and time. For example, the size of the receptive field together with the surround properties may determine the spatial frequency to which a cell is tuned (small receptive field = high spatial-frequency response). The temporal frequency response is determined not only by cell-autonomous properties of the ganglion cells, but also by the properties of the presynaptic neurons. For example, if the presynaptic neurons are not able to follow a fast-flickering stimulus, neither will be the ganglion cell that is activated by those neurons. Interestingly, ganglion cells with a high spatial-frequency response tend to have a low temporal-frequency response and vice versa. In other words, ganglion cells tend to carry detailed information about space or time, but not both. In general, however, ganglion cells fill up the frequency range in space and time which is occupied by natural stimuli.

Ganglion Cells as Specific Feature Detectors

The following discussion is limited to one important category of visual events in the environment, namely

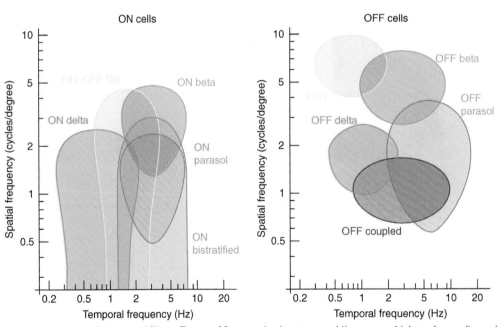

Figure 4   Ganglion cells as spatiotemporal filters. Range of frequencies in space and time over which each ganglion cell type in the rabbit retina responds. The colored blobs represent the full range of responses for each ganglion cell type when measured with a flashed square of light, 600 um on a side. Data have been generated by analyzing the response characteristics in space and time by Al Molnar and Frank Werblin based on data from Roska, B., Molnar, A., and Werblin, F. S. (2006). Parallel processing in retinal ganglion cells: How integration of space–time patterns of excitation and inhibition form the spiking output. *Journal of Neurophysiology* 95: 3810–3822; and Roska, B. and Werblin, F. (2003). Rapid global shifts in natural scenes block spiking in specific ganglion cell types. *Nature Neuroscience* 6: 600–608. Figure courtesy of Frank Werblin.

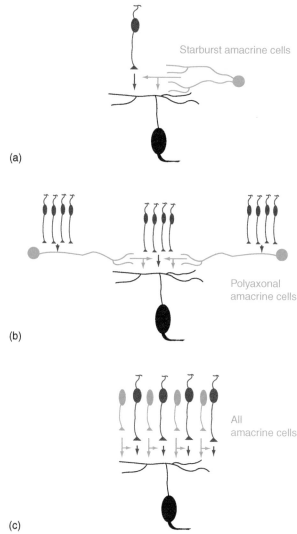

(a)

(b)

(c)

Starburst amacrine cells

Polyaxonal amacrine cells

All amacrine cells

Figure 5 Circuits of movement-encoding ganglion cells. These cells employ common computational principles. The inhibitory circuitry is wired such that unwanted responses are canceled. The targets of the inhibitory input are the dendrites of the ganglion cells and the terminals of the excitatory bipolar cells. The specific wiring diagram is different for each cell type to achieve the desired computation. (a) The inhibitory circuit of direction-selective (DS) ganglion cells is spatially directed and asymmetric. Inhibition is supplied by starburst amacrine cell processes that point in the null direction of the targeted DS cell. The spatial asymmetry creates a temporal difference in the synaptic inputs; inhibition arrives before a suppressed excitatory input for movement in the null direction. (b) Object-motion-sensitive (OMS) cells get long-range inhibitory input from polyaxonal amacrine cells. The OMS cell and the polyaxonal cells get activated by strongly rectifying bipolar cells with similar properties, reporting about the presence of movement. Different trajectories of movement create different temporal sequences of activity, and same trajectories create similar sequences. When an OMS cell sees the same trajectory as a majority of polyaxonal cells, its responses are canceled by the correlated inhibitory activity. When the OMS cell sees a different trajectory, for example, because an object moves within its receptive field relative to the background, it is allowed to respond. Polyaxonal cells have active properties so that the inhibition arrives

that of object motion. Evolution appears to have dedicated a significant amount of retinal hardware to the detection and decoding of moving stimuli. There are at least eight types of direction-selective (DS) ganglion cells (four ON–OFF DS cells, three ON DS cells, and one OFF DS cell) that encode information about the direction of object motion or image drift over the retina. In addition, several ganglion cell types seem to be specialized to recognize the movement of an object relative to global background motion (object-motion-sensitive (OMS) ganglion cells). Recently, a ganglion cell type has been described, which is sensitive to yet another important aspect of object motion, namely when an object is approaching the animal. In the remainder of this article, these different types of motion-sensitive ganglion cells are discussed in greater detail, with emphasis on the underlying mechanism of their response specificity. The basic principle is always the same: responses to unwanted visual stimuli are suppressed by specific inhibitory circuitry (**Figure 5**).

Direction-selective (DS) ganglion cells

The defining property of DS ganglion cells is that they respond well to a moving object when it moves in a so-called preferred direction. However, when the same object moves in the opposite so-called null direction, they remain silent. The best characterized are the ON–OFF DS cells. Their dendrites are bi-stratified in the ON and OFF sublaminae of the IPL, and the cells respond to both bright and dark objects. The neural circuitry computing the DS responses are present in both the ON and OFF systems, and seem to compute direction selectivity independently from each other. The importance of the inhibitory system can be appreciated by applying blockers of the inhibitory neurotransmitter gamma aminobutyric acid (GABA) to the retina. This results in equal responses to movement in all directions: responses to null-direction movement are uncovered, while responses to preferred-direction movement are hardly changed at all. The directionality of DS ganglion cell behavior is therefore obtained by actively suppressing responses to null-direction movement.

sufficiently fast at the OMS cell. (c) Approach-sensitive cells receive inhibitory input from AII amacrine cells, which are ON cells. They receive excitatory input from OFF bipolar cells. When dark and bright borders move concurrently within the receptive field, the inhibitory activity elicited by the bright borders cancels the excitatory activity elicited by the dark borders. This situation is encountered by laterally moving small objects or by image drift of a sufficiently detailed scene over the retina. An approaching dark object is expanding in size and has no bright moving image borders, allowing the cell to respond. AII cells get activated through electrical synapses with ON bipolar cells, so that the inhibition arrives sufficiently fast at the approach-sensitive ganglion cell.

The interneuron responsible for this suppression of null responses is the starburst amacrine cell. Starburst cells have peculiar properties that make them particularly suited for this task. For example, they have an asymmetric spatial distribution of incoming and outgoing synapses along their dendritic processes, which enables them to act at a distance: activation anywhere within the dendritic field can lead to release at the distal tips. In addition, their responses express DS properties themselves. Details of these are discussed elsewhere in the encyclopedia. The most important property responsible for the directional behavior of DS cells, however, is the geometrically specific connectivity between starburst cells and DS ganglion cells, and therefore a circuit property of the retina rather than a cell-autonomous property of either the DS cells or the starburst cells themselves. The geometry of the connectivity is such that only starburst cells located on the null side of the DS cell dendritic field inhibit the DS cell. As a consequence, a stimulus moving in the null direction will first encounter and activate this field of inhibitory starburst cells, which can then perform their action at a distance and inhibit the DS cell before the stimulus also enters the excitatory receptive field of the DS cell. This geometrically asymmetric connectivity can also be observed with static receptive field measurements: The inhibitory receptive field of DS cells is offset to the null side with respect to the excitatory receptive field, which matches the dendritic field quite well. DS cells are, therefore, an example of a cell type that does not have a classical center-surround receptive field.

Inhibitory activity acts at three levels in the DS circuitry to make the DS responses more robust. One level, already mentioned in the previous paragraph, is the geometrically asymmetric inhibition of DS ganglion cells by starburst amacrine cells. Geometric asymmetry is translated by a moving stimulus into a timing difference: during null movement, inhibition will reach the DS cell before excitation; however, during preferred movement, this will not be the case. A second level of inhibition targets the terminals of bipolar cells, and shows the same geometric properties as the inhibition of the DS cell dendrites. The properties of this presynaptic inhibition are consistent with the hypothesis that the same starburst cells that inhibit the DS cell also inhibit the presynaptic bipolar cells that make input to that DS cell. As a consequence, the excitatory input delivered by these bipolar cells is direction-selective because excitation during null-direction movement is strongly reduced. A third level of inhibition relates to the inhibitory input itself. During preferred-direction movement, the inhibitory input to DS cells is suppressed, as a consequence of the above-mentioned directional properties of starburst cell dendritic properties. These directional properties may be brought about, at least partly, by mutually inhibitory interactions between starburst cells.

In summary, because of inhibitory activity at many levels, DS ganglion cells receive directionally asymmetric synaptic inputs: during null-direction movement, they receive early-and-strong inhibitory input together with late-and-weak excitatory input; and during preferred-direction movement, they receive early-and-strong excitatory input with late-and-weak inhibitory input. Internal nonlinear properties of the DS cell itself emphasize the directional differences of the synaptic inputs: it appears that the DS cell contains local dendritic nonlinear properties. Calcium-driven spikes are generated locally in the dendrites when some threshold is crossed. These spikelets travel to the soma where they trigger standard sodium spikes.

Object motion sensitive (OMS) ganglion cells

The visual system constantly has to deal with global image jitter caused by eye, head, and observer movements. It poses a considerable challenge to neglect this global image movement and extract the more relevant movement of objects embedded in this scene. In other words, a ganglion cell with such properties would be active if an object moves within its receptive field (relative to the background), but it would be silent if there is only background movement within the receptive field, even if the local image trajectory is identical in both cases. Markus Meister and colleagues termed such ganglion cells as Object Motion Sensitive (OMS). The responsiveness of such a cell does not depend solely on the local image properties, but on a comparison of local and global properties. The next few paragraphs describe the computational principle of object motion detection.

The excitatory receptive field of OMS ganglion cells consists of small subunits, namely bipolar cells. Each of these subunits is strongly rectifying. This means that each subunit only gives input to the OMS cell if it is strongly activated. This results in relatively sparse excitatory input to the OMS cell.

Inhibitory input is provided by a class of amacrine cells called polyaxonal. Polyaxonal amacrine cells have a relatively small receptive field comparable in size to the OMS ganglion cell, but they have several axonal processes with active properties spreading across a considerable fraction of the whole retina. As a result, an OMS cell can receive inhibitory input from a polyaxonal amacrine cell that looks at a completely different area of the visual scene. Importantly, polyaxonal amacrine cells are electrically coupled with each other, so that they function as a coherent network through which activity can spread. Even though the axonal processes of a single polyaxonal cell are quite sparse, the network of these cells creates a dense mesh of axons. Polyaxonal cells receive similar excitatory input as OMS cells, namely temporally sparse input provided by strongly rectifying bipolar cells. These cells have a relatively high threshold, so that the polyaxonal network gets activated only if it receives near-simultaneous excitatory input from a large fraction of the visual scene. This condition is met during global image motion, because many of the subunits providing excitatory input to polyaxonal amacrine cells are activated simultaneously. Because of the coupling, all polyaxonal amacrine cells, as a network, fire a burst of action potentials. In time, this results in a relatively sparse sequence of activity in polyaxonal amacrine cells, and consequently in a sparse sequence of inhibitory events in OMS ganglion cells.

An OMS cell therefore receives sequences of excitatory and inhibitory inputs that are temporarily sparse. If there is an object in the receptive field of an OMS cell that moves uncorrelated relative to the global (background) movement, the temporal sequence of the sparse excitatory (local) and sparse inhibitory (global) events will also be uncorrelated. Therefore, the inhibitory events cannot cancel the excitatory events, and the OMS cell will be active. On the other hand, an OMS cell sitting in the background region of the image will receive correlated excitatory and inhibitory inputs, the inhibitory events suppress the excitatory events, and the cell remains silent. In fact, inhibitory input arrives at the OMS cell about 25 ms before the excitatory input, at least in salamander retina, presumably because of the fast active properties of polyaxonal amacrine cells. An OMS cell can, therefore, effectively report the existence of an object moving relative to background.

Recently, in 2008, Baccus and co-workers expanded on their original findings and reported that the main target of the inhibitory input by polyaxonal amacrine cells is not the OMS ganglion cell itself, but the bipolar cell terminals which provide input to the OMS cell.

Saccadic suppression

Roska and Werblin showed that in rabbit retina some ganglion cells are suppressed during saccades, sudden eye movements that serve to shift gaze direction. They suggest that this is due to the inhibitory activity of polyaxonal amacrine cells. This resets the activity of ganglion cells for each visual episode in-between saccades. It is possible that saccadic suppression is a consequence of the same circuitry and mechanism described for OMS ganglion cells in the paragraph above. Roska and Werblin provided a detailed list of ganglion cell types which receive saccadic suppression in rabbit retina, while Meister and colleagues have remained vague about the cell types in rabbit retina, which have the OMS property.

Approach-sensitive ganglion cells

The detection of approaching or looming optical stimuli is important for survival. Such stimuli trigger robust behavioral avoidance responses in basically all animals tested, from insects to humans. Babies as young as 2 weeks react

with widening of the eyes, turning of the head, lifting of the arms, and crying when they view a symmetrically expanding shadow on a screen directly in front of them. This suggests that hard-wired neural circuitry exists to detect such looming stimuli and to trigger these protective motor responses. Recently, we described a ganglion cell in the retina that is sensitive to dark approaching visual stimuli, such as widening shadows. Similar to the other cells described earlier, the function of these cells can only be appreciated when one considers the space of null stimuli, that is, the range of stimuli to which the cell does not respond because of active suppression by inhibitory circuitry. In the case of the approach-sensitive ganglion cells, these null stimuli comprise lateral movement of a small object within the receptive field, or lateral drift of the visual scene. In the case of a small, dark, moving object, for example, a hawk flying in sky, the cell will respond when the object is approaching the observer, but it will be suppressed when the object is moving on a lateral trajectory. Lateral drift of a visual scene is encountered during observer head or eye movements, similar to the OMS cells described in the above paragraph, but the mechanism for response suppression is substantially different.

The receptive field of approach-sensitive cells is comprised of small subunits for both the excitatory and inhibitory inputs. The excitatory input is mediated by transient OFF bipolar cells. Whenever a dark stimulus border is moving into a new subunit (i.e., into the receptive field of another bipolar cell), the approach-sensitive ganglion cell receives another episode of excitatory input, resulting in spiking output. This is true for any moving dark border within the receptive field of the ganglion cell, unless these events are canceled by simultaneously moving bright stimulus borders within the receptive field, as it would be encountered by the trailing edge of a laterally moving small dark object, but not by an approaching dark object, which does not create any trailing edges. The reason for the suppression by moving bright image borders is the structure of the inhibitory receptive field: it also consists of small subunits, but of the opposite polarity; the inhibitory input is mediated by ON amacrine cells with a small receptive field. The suppression of null responses is therefore computed on a local scale within the receptive field of ganglion cell.

The most interesting aspect of the approach-sensitive circuitry is the identity of the inhibitory amacrine cell and the specific synaptic connectivity employed. The amacrine cell responsible for the inhibition of the approach-sensitive ganglion cell is the well known AII (pronounced A-two) amacrine cell. AII amacrine cells are a main conduit for visual signals during nighttime vision. They are activated by rod bipolar cells through a glutamatergic chemical synapse, and then pass the signals on to both ON and OFF cone pathways. OFF pathways are activated through a glycinergic chemical synapse. The targets of these synaptic connections are both the terminals of OFF cone bipolar cells, and the dendrites of OFF ganglion cells. ON pathways are activated through electrical synapses (gap junctions) between the AII cells and the terminals of ON cone bipolar cells. During daytime vision, rods are saturated, and rod bipolar cells are not active. Under these conditions, AII cells are activated mostly by ON cone bipolar cells by virtue of the electrical synapse between these cells. Once activated, AII cells can release glycine and inhibit their targets in the OFF pathway. One such target is the approach-sensitive ganglion cell, and the bipolar cells that activate it.

This pathway for the inhibitory input of the approach-sensitive ganglion cell has properties ideally suited for the approach-sensitive computation. AII cells are small, with a receptive field not larger than that of bipolar cells. Therefore, the excitatory and inhibitory subunits are of compatible size. Another important aspect is the dynamics of the inhibition. As explained above, null stimuli for the approach-sensitive cell are those that have concurrently moving dark and bright borders. The inhibitory input, activated by the bright borders, suppresses responses that would be elicited by the dark borders. This requires the inhibition to arrive at its target fast enough, so that no excitatory signal slips through, similar to the fast inhibitory input arriving at DS cells and at OMS cells. In the case of the AII cells, the speed of inhibition seems to be ensured by the electrical synapse in the pathway. The electrical synapse effectively reduces the number of chemical synapses in the inhibitory pathway to the same number as those in the excitatory pathway.

See also: Information Processing: Direction Sensitivity.

Further Reading

Baccus, S. A., Olveczky, B. P., Manu, M., and Meister, M. (2008). A retinal circuit that computes object motion. *Journal of Neuroscience* 28: 6807–6817.

Demb, J. B. (2007). Cellular mechanisms for direction selectivity in the retina. *Neuron* 55: 179–186.

Fried, S. I., Münch, T. A., and Werblin, F. S. (2002). Mechanisms and circuitry underlying directional selectivity in the retina. *Nature* 420: 411–414.

Gollisch, T. and Meister, M. (2008). Rapid neural coding in the retina with relative spike latencies. *Science* 319: 1108–1111.

Kuffler, S. W. (1953). Discharge patterns and functional organization of mammalian retina. *Journal of Neurophysiology* 16: 37–68.

Nelson, R. (2001). Visual responses of ganglion cells. *From: Webvision – The Organization of the Retina and Visual System.* http://webvision.med.utah.edu/GCPHYS1.HTM (accessed May 2009).

Olveczky, B. P., Baccus, S. A., and Meister, M. (2003). Segregation of object and background motion in the retina. *Nature* 423: 401–408.

Rodiek, R. W. (1998). *The First Steps in Seeing.* Sunderland, MA: Sinauer.

Roska, B., Molnar, A., and Werblin, F. S. (2006). Parallel processing in retinal ganglion cells: How integration of space–time patterns of excitation and inhibition form the spiking output. *Journal of Neurophysiology* 95: 3810–3822.

Roska, B. and Werblin, F. (2003). Rapid global shifts in natural scenes block spiking in specific ganglion cell types. *Nature Neuroscience* 6: 600–608.

Shlens, J., Rieke, F., and Chichilnisky, E. (2008). Synchronized firing in the retina. *Current Opinion in Neurobiology* 18: 396–402.

Wässle, H. (2004). Parallel processing in the mammalian retina. *Nature Reviews* 5: 747–757.

Werblin, F. and Roska, B. (2007). The movies in our eyes. *Scientific American* 296: 72–79.

Information Processing: Horizontal Cells

A A Hirano, UCLA School of Medicine, Los Angeles, CA, USA
S Barnes, Dalhousie University, Halifax, NS, Canada
S L Stella, Jr., UCLA School of Medicine, Los Angeles, CA, USA
N C Brecha, UCLA School of Medicine, Los Angeles, CA, USA; VAGLAHS, Los Angeles, CA, USA

Glossary

Ephaptic transmission – A nonsynaptic electrical interaction mediating cellular communication. Closely apposed neural elements may share signals through an extracellular voltage change caused by the flow of current in a confined, resistive interstitial space between neurons. Changes of extracellular ion concentrations, such as high potassium levels caused by potassium extrusion from an active, depolarized neuron, also affect close-neighboring neurons.

Gain control – The gain or amplification of a graded potential synapse, such as that between photoreceptors and second-order neurons in the retina, is defined as the postsynaptic amplitude divided by presynaptic amplitude. To optimize signal-to-noise ratios, synaptic strength may be controlled by reciprocal signals that increase synaptic gain when signals are small, and reduce the gain when signals are large and saturating.

Gap junctions – The specialized electrical junctions between cells consisting of two connexons, each termed a hemichannel and made up of connexin proteins, which together form an intercellular pore. These mediate rapid electrical events between cells, allow diffusion of ions and small proteins, and are regulated by neuromodulators, including dopamine, nitric oxide, and retinoic acid, in the retina.

Photoreceptor synaptic triad – The synaptic arrangement consisting of two lateral horizontal cell endings or terminals and an ON bipolar cell dendrite that invaginate into a photoreceptor terminal.

Protons – The free hydrogen ions, as H^+ or hydronium ions (H_3O^+), in solution, usually bound (buffered) by other molecular constituents.

Receptive fields – The receptive fields of neurons in the visual system are the regions of retinal surface, receiving an optical projection of visual space, for which light stimulation causes a response in the cell. The classic center-surround antagonistic receptive field typically has a circular center and, forming a ring around the center, an annular surround. The sign of the cell's response to light in the center is the opposite of that in the surround. This receptive-field organization underlies edge detection by enhancing contrast between regions of varying brightness and/or color.

Roll back – A modest, slowly depolarizing trajectory seen in the horizontal cell voltage waveform, counteracting the rapid hyperpolarization of 30–50 mV during the response to a light stimulus. The roll back, whose amplitude is dependent on the intensity and diameter of the stimulus, is considered to be due to inhibitory feedback from horizontal cells to photoreceptors.

Introduction

Visual information is conveyed through the retina from photoreceptors via bipolar cells to ganglion cells, and the signal is modulated by inhibitory lateral interactions provided by horizontal cells in the outer plexiform layer and amacrine cells in the inner plexiform layer. Retinal neurons are characterized by having an antagonistic, concentric center-surround receptive-field organization.

Horizontal cells play a critical role in generating the inhibitory receptive-field surround of cone photoreceptors and bipolar cells in both mammalian and nonmammalian retinas. These cells are characterized by a broad, lateral spread of their processes as well as homologous coupling through gap junctions to other horizontal cells. In mammalian retinas, both A-type and B-type horizontal cells receive input from cone photoreceptors at their dendritic tips, and rod photoreceptors at their axon terminals. Horizontal cells transmit spatially broad visual signals back to photoreceptors and bipolar cells, to generate receptive-field surrounds (**Figure 1**). The antagonistic, center-surround receptive-field spatial organization contributes to visual processing that highlights changes in luminance and contrast to improve visual acuity. Horizontal cells also participate in setting the gain of the photoreceptor synapse through inhibitory feedback of information about local illumination levels, and in lower vertebrates, color-opponent receptive fields are generated through color-specific, antagonistic receptive-field surrounds.

Synaptic Interactions and Gap-Junction Coupling of Horizontal Cell Subclasses

There are two physiological classes of horizontal cells, which code for luminosity and chromaticity, in the vertebrate retina. The luminosity (L-type, or H1) horizontal cell responds to the whole range of visible light with a graded hyperpolarization, due to receiving inputs from all types of cone photoreceptors. The chromaticity (C-type, or H2, H3) horizontal cells respond to certain wavelengths of light with a hyperpolarization and a depolarization to others, in a biphasic or triphasic manner, due to selective connectivity with and feedback onto different spectral subtypes of cone photoreceptors. In this manner, the C-type horizontal cells begin the process of encoding color opponency. In addition, there are also horizontal cells that synapse selectively with rod photoreceptors in some species, including teleosts; however, most lower vertebrates exhibit mixed rod and cone inputs into horizontal cells. Mammals appear to have only the L-type horizontal cells, and these do not exhibit color-opponent properties.

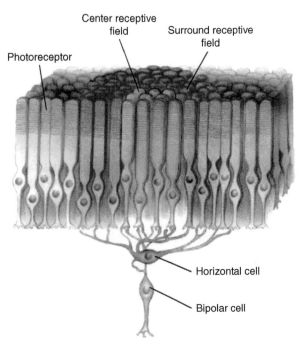

Figure 1 Formation of bipolar cell center-surround antagonistic receptive fields by integration of direct photoreceptor inputs at center of receptive field (shown in blue) and surround photoreceptor inputs (shown in red) carried through a large receptive-field horizontal cell. Horizontal cells are normally homologously coupled, except under the condition of extreme light adaptation. The horizontal cell signal is conducted to the bipolar cell through inhibitory feedback at the photoreceptor terminals and through direct feed-forward signaling to the bipolar cell. Reproduced from figure 9.22 of Bear, M. F., Connors, B. W., and Paradiso, M. A. (eds.) (2007). *Neuroscience: Exploring the Brain*, 3rd edn. Philadelphia, PA: Lippincott Williams & Wilkins.

In general, in mammals, A- and B-type horizontal cell dendrites form synapses with cone photoreceptors, whereas the axon terminal system of the B-type forms synapses with rod photoreceptors. Due to the high electrical resistance in the long axon connecting the axon terminals to the somatodendritic region, the two compartments of the B-type horizontal cell are thought to act as functionally independent units. Horizontal cells receive excitatory glutamatergic input from the photoreceptors and feedback onto the cone and rod terminals, presumably at the horizontal cell endings within the photoreceptor terminals. In darkness, photoreceptors release transmitter; thus, the membrane potential of horizontal cells is relatively depolarized. In response to light, all horizontal cells studied exhibit slow, graded hyperpolarizing potential changes due to the reduction in transmitter release from photoreceptors. These slow potentials were originally called S-potentials (in honor of Gunnar Svaetichin, who initially described them in 1953). With the advent of intracellular dyes, it was shown that S-potentials arose from horizontal cells in the outer retina.

A slow-adapting depolarization is seen in the horizontal cell voltage waveform, immediately following the hyperpolarization seen during the response to a light stimulus (**Figure 2**). This so-called roll back, whose amplitude is dependent on the diameter of the stimulus, as well as on the intensity and duration of the stimulus, is considered to be due to inhibitory feedback from horizontal cells to photoreceptors.

A key feature of horizontal cells is that they form a syncytium through gap-junction coupling between homologous horizontal cell subtypes, such that the networks of A-type horizontal cells are distinct from those formed by the B-type horizontal cells. As a result, the receptive fields of horizontal cells are considerably larger than the spread

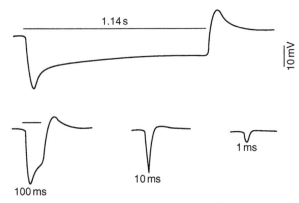

Figure 2 The S-potential in early recordings from the retina of cat, showing horizontal cell responses to bright light stimuli of various durations (1.14 sec, 100, 10, and 1 ms). Vertical bar on right corresponds to 10 mV. Reproduced from figure 8 of Brown, K. T. and Wiesel, T. N. (1959). Intraretinal recording with micropipette electrodes in the intact cat eye. *Journal of Physiology (London)* 159: 537–562, with permission from Wiley-Blackwell.

of their dendritic arborization. Furthermore, the number of coupled cells varies depending on the state of light adaptation of the retina. It is now known that several neuromodulators, including dopamine, nitric oxide, and retinoic acid, act through various signaling systems and distinct retinal circuitry to regulate properties of gap junctions within the coupled network of horizontal cells.

Ionic Conductances of Horizontal Cells and the Response to Light

The ionic conductances that shape the graded potentials of horizontal cells are attributable to the postsynaptic α-amino-3-hydroxyl-5-methyl-4-isoxazole-propionate (AMPA; GluR1, GluR2/3, and GluR4) and kainate (KA2, GluR6) ionotropic glutamate receptors that mediate photoreceptor input and an ensemble of voltage-gated ion channels, ion transport mechanisms, and voltage-insensitive ion channels. **Figure 3** shows a voltage-clamp recording from an isolated goldfish horizontal cell that illustrates several of the voltage-sensitive ion channel currents. In total, there are at least three types of voltage-gated K^+ channels (inward rectifier, delayed rectifier, and transient A-type), two types of voltage-gated Ca^{2+} channels (L- and N-type) and tetrodotoxin (TTX)-sensitive Na^+ channels. Horizontal cells also express hemichannels (formed from a variety of connexins in a species-dependent manner), amiloride-sensitive Na^+/H^+ channels (ENaCs), acid-sensitive $Na^+/$proton channels (ASICs), as well as a range of electrogenic exchangers for Na^+, Ca^{2+}, Cl^-, K^+, and

HCO_3^- (Na^+/Ca^{2+}, $Na^+/Ca^{2+}/K^+$, $Na^+/K^+/2Cl^-$, and AE3) and pumps (Na^+/K^+- and Ca^{2+}-adenosine triphosphate (ATP)ases, H^+-pumping plasma membrane V-ATPases) for maintaining ionic gradients. Finally, inositol 1,4,5-trisphosphate-sensitive and caffeine-sensitive intracellular Ca^{2+} stores contribute to calcium signaling and may influence synaptic transmission.

In the dark, horizontal cell membrane potential is driven by input from photoreceptors releasing L-glutamate acting at AMPA and kainate receptors to depolarize cells to a level near −20 mV (**Figure 4**). The balance of excitatory synaptic input and activation of delayed rectifier K^+ channels set this membrane potential. Reduction of glutamatergic input from photoreceptors by light terminates the depolarizing influence, and the membrane hyperpolarizes toward the potassium equilibrium potential (E_K) where inward rectifier K^+ channels (K_{IR}) activate and dominate the membrane potential. In fact, horizontal cell K_{IR} channels strongly clamp these cells at a potential near −70 mV in the absence of glutamate. The dihydropyridine-sensitive, L-type ($Ca_V1.2$ and 1.3, α1C and D) and N-type Ca^{2+} channels, which normally activate above −50 mV, may mediate vesicular fusion, and produce Ca^{2+} fluxes that activate Ca^{2+}-dependent enzymes, and trigger release of Ca^{2+} from internal stores. Na^+ channels are thought to have roles in accelerating the depolarizing phase of the horizontal cell response at the termination of a light step, and, in coordination with the A-type K^+ channels, participate in oscillatory potential production. Some of these channels are also modulated by neurotransmitters: inward

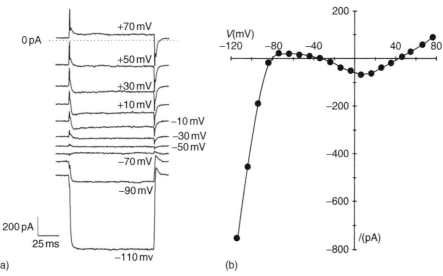

Figure 3 Current–voltage relation of isolated goldfish horizontal cell, recorded from a holding potential of −40 mV with steps in (a) to show temporal characteristics of the response and in (b) a depolarizing ramp voltage clamp protocol. The characteristic N-shape of the *I–V* relation illustrates the prominent inwardly rectifying current carried by K^+ at potentials more negative than −80 mV, an extended negative slope region due to L-type Ca^{2+} channels positive to −40 mV, and modest outward rectification positive to +40 attributable to delayed rectifier K^+ channels and hemichannels. Reproduced from figure 1 of Jonz, M. G. and Barnes, S. (2007). Proton modulation of ion channels in isolated horizontal cells of the goldfish retina. *Journal of Physiology (London)* 581(Pt 2): 529–541, with permission from Wiley-Blackwell.

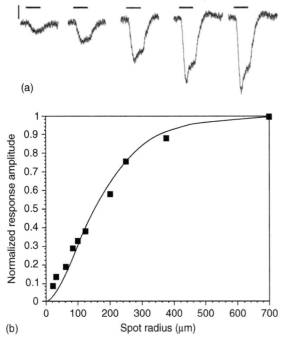

(a)

(b)

Figure 4 Responses of a mouse horizontal cell to (a) light spots of increasing radius (40, 90, 200, 375, and 700 μm) and constant intensity (log I/I_0 = −2.1) lasting 250 ms (horizontal bars). Note the prominent roll back of the hyperpolarization response with the largest spot stimulations, reflecting the engagement of the surround response. Vertical scale bar is 2 mV. (b) Plot of normalized response amplitudes against spot radius. Reproduced from figure 4 of Shelley, J., Dedek, K., Schubert, T., et al. (2006). Horizontal cell receptive fields are reduced in connexin57-deficient mice. *European Journal of Neuroscience* 23(12): 3176–3186, with permission from Wiley-Blackwell.

rectifier K^+ channels and L-type Ca^{2+} channels are suppressed by L-glutamate, and the L-type Ca^{2+} channels are modulated by dopamine in cone-driven horizontal cells.

Cellular Mechanisms of Horizontal Cell Neurotransmission

Lateral inhibition in the outer retina generates receptive-field surrounds of retinal neurons at photoreceptor terminals and at bipolar cell dendrites. In other words, the photoreceptor response can be modulated by light falling on neighboring photoreceptors through horizontal cell feedback. As the diameter of a spot of light increases and the horizontal cell feedback network is engaged, the hyperpolarizing membrane potential response of photoreceptors and horizontal cells exhibit a slow depolarizing sag or roll back in the membrane potential, which is attributed partly to this feedback (**Figure 4**). Although it is generally accepted that horizontal cells provide inhibitory feedback onto photoreceptor terminals, the synaptic and cellular mechanisms underlying this process are poorly understood.

There are several proposed mechanisms of synaptic transmission, involving gamma aminobutyric acid (GABA), protons, and an ephaptic mechanism through hemichannels, at the tips of horizontal cells. Discrepancy between findings from different groups may reflect technical issues and different biological strategies used by mammalian versus nonmammalian species, and perhaps even different synaptic mechanisms used by the different horizontal cell subtypes. Furthermore, it is also possible that different cellular mechanisms encode different aspects of visual information or may be used under different light conditions. However, it seems clear that the principal effect of horizontal cell feedback onto cone and rod photoreceptor terminals involves the modulation of their voltage-gated L-type Ca^{2+} channels.

Horizontal Cell Feedback and Feed-Forward

The inhibitory transmitter GABA has been proposed as the horizontal cell neurotransmitter, and the release of GABA is thought to occur through the voltage-modulated action of a plasma membrane GABA transporter (GAT) in nonmammalian retinas and by regulated vesicular release in mammalian retinas, as mammalian horizontal cells do not express GATs. An ephaptic model of feedback onto cone terminals was originally conceived by Byzov and colleagues, and later elaborated to involve hemichannels as current sinks. More recently, evidence has accumulated for protons as a key messenger. In any case, since the ON and OFF bipolar cells depolarize or hyperpolarize, respectively, in response to light-mediated reductions in glutamate release from photoreceptors, it seems simplest that for surround antagonism to affect the two types of bipolar cell responses in a coherent manner, the modulatory effect would be integrated at the presynaptic photoreceptor terminal by a feedback mechanism. To invoke feed-forward surround inhibition requires the horizontal cell transmitter to affect the postsynaptic ON and OFF bipolar cells through mechanisms of opposite polarity.

Several convergent findings indicate that GABA is a mammalian horizontal cell transmitter. One or both of the L-glutamate decarboxylase (GAD) isoforms are found in horizontal cells at the messenger RNA (mRNA) and protein levels. Many, but not all, studies have shown GABA immunoreactivity in horizontal cells of guinea pig, cat, rabbit, and primate retina. In contrast, there are studies reporting low or nondetectable levels of GAD_{67} or GABA immunoreactivity in horizontal cells of the adult mouse and rat retina, whereas GAD immunostaining is present at high levels in horizontal cells of the developing and juvenile mouse retina. The detection of GAD or GABA in the adult retina is influenced by several factors, including differential expression of GAD isoforms and levels of GAD and GABA in horizontal cells, as well as technical issues related to fixation protocols and antibody specificity. The presence of vesicular GABA transporter (VGAT) immunoreactivity in horizontal cells is consistent with a transmitter role for GABA in mammals, as VGAT mediates the accumulation of GABA into synaptic vesicles. The highest level of VGAT immunostaining is in horizontal cell processes underneath the photoreceptor terminals, and in the dendritic and axonal endings within the synaptic triad.

Feed-Forward onto Bipolar Cell Dendrites

Bipolar cells possess antagonistic receptive-field surrounds, which may be to some extent mediated by horizontal-cell feed-forward transmission onto bipolar cells. At the ultrastructural level, synapses between horizontal cell processes and bipolar cell dendrites have been reported. $GABA_A$ and $GABA_C$ receptor immunoreactivity is localized to bipolar cell dendrites adjacent to horizontal cell processes. In perforated-patch recordings, GABA elicits depolarizing inward currents when applied to dendrites of mouse rod bipolar cells and hyperpolarizing currents when applied to OFF-type cone bipolar cells, consistent with a GABAergic feed-forward input from horizontal cells in the creation of receptive-field surrounds. The inhibitory and excitatory actions of GABA at ON- and OFF-bipolar cell dendrites may be accounted for by the different Cl^- concentrations in bipolar cell dendrites maintained by chloride co-transporters. NKCC (Na-K-2Cl transporter), which accumulates Cl^- intracellularly, has been immunolocalized to dendrites of ON bipolar cells, whereas, KCC2, which extrudes Cl^-, has been localized to those of OFF bipolar cells. The localization of these chloride transporters would predict that chloride equilibrium potential (E_{Cl}) would be positive to the membrane potential (V_m) in the former situation, and E_{Cl} would be negative to V_m in the latter.

While horizontal cell inputs to bipolar cells could be demonstrated by direct current injection into or light stimulation of horizontal cells, it has been more difficult to sort out the contributions of the indirect pathway of horizontal cell to cone photoreceptor to ON bipolar cell from the direct horizontal cell to bipolar cell dendrite pathway. In experiments where L-2-amino-4-phosphonobytyric acid (L-AP4) was used to block photoreceptor input to ON bipolar cells, and by extension the indirect pathway, light stimulation produced a hyperpolarization response in these bipolar cells. From the relative magnitudes of the surround responses with and without L-AP4, about a quarter to a third of the surround response came through the feed-forward horizontal cell to bipolar cell synapse; the majority arose through the feedback to the cone photoreceptor terminals. Low cobalt concentrations (<500 μM) selectively block surround-evoked depolarization of cones (and receptive fields of downstream neurons), without affecting transmission from cones to horizontal cells. It has been proposed that low concentrations of cobalt may act by blocking GABA-induced current in turtle cones, but hemichannels and proton-permeable ENaCs have also been reported to be sensitive to low concentrations of cobalt. Finally, there is anatomical evidence from rabbit retina as well as electrophysiological and tracer evidence from salamander that indicate gap-junction coupling between horizontal cells and cone ON-type bipolar cells.

Feedback onto Photoreceptor Terminals

In turtle and salamander retina, there are extensive data supporting the GABA-mediated inhibitory feedback from horizontal cells to cones. A clear, nonzero reversal potential of this feedback synapse onto cones suggests that the underlying synaptic mechanism is likely chemical, rather than electrical. Moreover, bathing the retina in GABA blocks the feedback response of cone photoreceptors, as would be expected if a decrease in GABA release signals the horizontal cell feedback. In turtle cones, the effect of GABA was hyperpolarizing upon break-in in ruptured patch recordings, suggesting that the reversal potential for chloride in cones was negative to the resting membrane potential. The surround response would reflect a diminution of this hyperpolarizing current. However, the currents evoked by GABA in turtle cones in whole retina preparations were too small to account for feedback, unless enhanced by modulators of $GABA_A$ receptors (e.g., pentobarbital), suggesting that the contribution of the GABA-induced current to surround formation was minor. In salamander cones, the reversal potential for chloride in cones is close to the dark resting potential, suggesting GABAergic disinhibition near the dark potential should produce little membrane-potential change. This result is inconsistent with the postulated role for GABA in generating the feedback depolarization and supports other studies using GABA receptor antagonists, suggesting that GABA does not have a role in horizontal cell to cone photoreceptor feedback.

In contrast, the localization of GABA receptors to photoreceptor terminals is consistent with a feedback role for GABA. Morphological and molecular support for a feedback role includes the expression of $GABA_A$ receptor subunit messenger RNA (mRNA) by photoreceptors detected by *in situ* hybridization immunohistochemistry and single-cell RT-PCR. In addition, cone terminals of pig, mouse, and rat show $GABA_C$ receptor (ρ subunit) immunoreactivity. Similar to the GABA-induced currents in turtle cone pedicles, recordings from mouse and pig cones show the presence of functional $GABA_A$ and $GABA_C$ receptors, whereas rods in porcine and turtle retina exhibit little sensitivity to GABA. In addition to ionotropic GABA receptors, metabotropic $GABA_B$ receptors are expressed on horizontal cell processes, suggesting that GABA may also act presynaptically on horizontal cells. Finally, it is also likely that GABA acts on autoreceptors on horizontal cells as well as on neighboring horizontal cells, where it would likely be depolarizing as E_{Cl} is above V_m; this would provide positive feedback. Overall, these studies indicate multiple targets for GABA in the outer plexiform layer, which could mediate complex actions of GABA in the outer retina.

There are also findings that argue against the role of GABA in mediating horizontal cell feedback onto cones. Principally, GABA agonists and antagonists do not always appear to affect surround-evoked depolarization of cones. However, the presence of GABA receptors on horizontal cells, bipolar cells, as well as on cone photoreceptors complicates the interpretation of experiments using GABA antagonists. Furthermore, it is clear that small changes in the Ca^{2+} currents produced by feedback are not reflected in large changes in the membrane potential of the photoreceptor.

Ephaptic Transmission between Horizontal Cells and Photoreceptor Terminals

An ephaptic effect has also been hypothesized to underlie the feedback signal from horizontal cells to rod and cone photoreceptors. Originally, this electrical feedback mechanism was proposed to involve only glutamate-gated channels in the tips of horizontal cell processes that invaginate cone synaptic terminals. More recently, evidence for hemichannels at the tips of horizontal cell processes added this conductance mechanism as an additional current sink in the horizontal cell dendrites. Current flowing through the bulk resistance of the interstitial space and into hemichannels and glutamate receptor ion channels is proposed to produce a voltage drop in the synaptic cleft. This extracellular negative potential, in effect, shifts the activation of presynaptic Ca^{2+} channels in the positive direction, reducing the amount of glutamate released by the photoreceptor synapse.

The amplitude of the extracellular voltage drop is a critical function of the current density and the interstitial resistivity, and there is no consensus yet that the amplitude is sufficient to carry the 5–10-mV feedback signals that presynaptic Ca^{2+} channels apparently sense. Furthermore, the time course of an ephaptic response should be nearly instantaneous since it is electrical in nature, but it is well known that there is a slow time course of the roll back of the horizontal cell light response (**Figure 2**). The more rapid depolarization of the photoreceptor (or OFF bipolar cell) in response to strong surround inhibition seems better suited kinetically to the ephaptic effect. Carbenoxolone-induced block of hemichannels has been cited as evidence for ephaptic transmission. However this drug also blocks Ca^{2+} currents in photoreceptors at concentrations used to block ephaptic transmission.

Proton Mediation of Horizontal Cell Feedback

Recent work in turtle, salamander, ground squirrel, primate, zebra fish, and goldfish retina indicates that modulation of the extracellular pH in the synaptic cleft between horizontal cells and photoreceptors may be used to signal feedback to photoreceptors. Evidence that protons modulate voltage-gated Ca^{2+} channels at photoreceptor terminals, carrying the feedback signal, is substantial. The principal evidence is that 1–20-mM HEPES and other pH buffers reversibly block feedback onto photoreceptors by eliminating the shift in activation of the photoreceptor Ca^{2+} channel current. The exact cellular mechanism that controls the pH of the synaptic cleft between photoreceptors and horizontal cells remains undefined, but horizontal cell plasma membrane V-ATPases, ENaCs, ASICs, and hemichannels are possible membrane mechanisms by which horizontal membrane potential could alter cleft pH. Depolarized horizontal cells would produce or contribute to cleft acidification by extruding protons and/or by not taking them up, while hyperpolarized horizontal cells would facilitate cleft alkalinization by not releasing protons and/or by taking them up. The sensitivity of presynaptic Ca^{2+} channels to alterations in extracellular pH suggest that cleft pH is modulated between ~pH 7.0, during the peak of inhibitory feedback signaling, and ~pH 7.8, during maximal horizontal cell hyperpolarization and the absence of presynaptic inhibition. Physiological measurements and estimates of cleft pH changes support the possibility that pH changes of this magnitude occur.

Functional Roles of Horizontal Cells

The proposed functional roles of mammalian horizontal cells are derived from the notion that these cells provide inhibitory feedback onto photoreceptors. This idea is based largely on investigations of nonmammalian model systems that show: (1) a global contribution to retinal adaptation to different mean levels of illumination; (2) a local contribution to spatial processing and contrast enhancement by a spectrally broadband, but spatially restricted feedback to create the antagonistic receptive-field surrounds of photoreceptors, bipolar cells, and ganglion cells; and (3) a local contribution to chromatic processing by a chromatically selective feedback to create color-opponent receptive fields of cones, bipolar cells, and ganglion cells.

Retinal Adaptation

The syncytium of horizontal cells formed through electrical coupling of homologous cells shows characteristic changes with different states of light adaptation. Dark adaptation reduces coupling (from space constant determinations, tracer coupling), smaller receptive-field sizes (surround-to-center ratio), and decreased sensitivity of horizontal cells. The decreased electrical coupling would also increase input resistance of the cell, resulting in larger voltage changes to a given light stimulus. Several modulators of horizontal-cell gap-junctional coupling have been identified, including dopamine, nitric oxide, and retinoic acid, which appear to mediate the effects of adaptation.

The dopaminergic modulation of horizontal-cell gap-junction conductivity has been reported for both nonmammalian and mammalian retinas. During light stimulation, the levels of dopamine rise in the retina and, through activation of D1 dopamine receptors on horizontal cells in a cyclic adenosine monophosphate (cAMP)-dependent manner, the duration and frequency of gap-junction openings are reduced. This results in the uncoupling of horizontal cells, such that responses of a smaller pool of photoreceptors and, thus, input from a smaller visual area, influence the horizontal cell response, reflected in the reduced surround-to-center (annulus-to-spot) ratios. Findings in rabbit retina indicate that the network of coupled horizontal cells is greatest in dim, scotopic conditions (dim ambient light) and less extensive in darkness and in photopic conditions, that is, the degree of coupling reflects the adaptational state of the retina (**Figure 5**). The retinal circuit underlying the modulation of dopamine release appears to involve a pathway from photoreceptors to ON bipolar cells to dopaminergic amacrine cells to horizontal cells. In addition to differing adaptational states, retinal levels of dopamine vary with a circadian rhythm.

Similar to dopamine, nitric oxide appears to uncouple horizontal cells, modulating the electrical coupling through activation of soluble guanylate cyclase and a cyclic guanosine monophosphate (cGMP)-dependent cascade in horizontal cells. Interestingly, in rabbit retina, nitric oxide also appeared to increase the sensitivity of horizontal cells to light through possibly an indirect action on photoreceptor transduction or at the photoreceptor-horizontal cell synapse, in addition to the increased input resistance resulting from the reduced cellular coupling. This may occur to some extent through the modulation of the ionotropic glutamate receptors found on horizontal cells. It has been speculated that increased nitric oxide production by horizontal cells under dark-adapted conditions and by amacrine cells under light-adapted conditions may account for the biphasic modulation of horizontal cell coupling with adaptational state.

Illumination increases the levels of all *trans*-retinoic acid (at-RA), which is a byproduct of the phototransduction cycle, and thereby correlates with the amount of light illumination. at-RA can uncouple horizontal cells in mouse, rabbit, and carp in a stereospecific manner and can do so in the presence of D1 dopamine receptor antagonist, indicating that it is not acting through modulation of the dopaminergic pathway. In the presence of

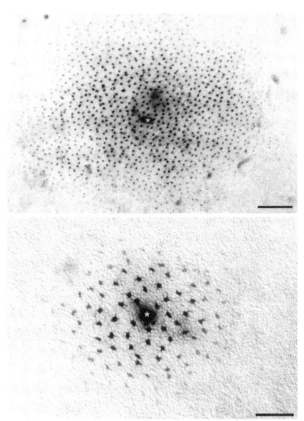

Figure 5 Regulation of horizontal cell tracer coupling by exposure of rabbit retina to different light intensities. (Top) Exposure to dim light intensity (log –6.0) produced coupling of over 1200 cells. (Bottom) Exposure to bright light intensity (log –1.0) produced coupling of 124 cells. Scale = 100 μm in top panel, 50 μm in bottom panel. Adapted from figures 7 and 8 of Xin, D. and Bloomfield, S. A. (1999). Dark- and light-induced changes in coupling between horizontal cells in mammalian retina. *Journal of Comparative Neurology* 405: 75–87. © 1999, John Wiley & Sons, Inc. Reprinted with permission of John Wiley & Sons, Inc.

at-RA, the receptive-field sizes of horizontal cells, as measured by annulus-to-spot ratios, were reduced. In addition to its effects on spatial response characteristics of horizontal cells, at-RA application in dark-adapted retinas could induce effects resembling light adaptation, such as reduced light responsiveness, changes in chromatic properties of H2 horizontal cells in teleosts, decreased gap-junctional permeability, as well as spinule formation in fish horizontal cells.

Gain Control of Synapses in the Outer Retina

The gain of the graded potential synapses between photoreceptors and horizontal cells is defined as the postsynaptic amplitude divided by presynaptic amplitude. High-gain synapses in an open-loop system (e.g., no inhibitory feedback) are inherently unstable, tending to saturate their postsynaptic targets. To maintain stability and to optimize signal-to-noise ratios, synaptic gain is controlled by reciprocal feedback signals so that gain is high and stable when signals are small, and gain is reduced when signals are large and potentially saturating. Having each photoreceptor synapse under closed-loop inhibitory feedback control imparts gain control to the system.

The ability of the horizontal cell network to adjust to different states of light adaptation permits in part the retina to operate optimally over 10 orders of magnitude of light intensities. The input-output relations of photoreceptor to second-order neurons are modulated by light and dark adaptation. For example, dim background illumination can increase the voltage gain of the rod output synapse by increasing the conductance of horizontal cell kainate receptors through a dopaminergic signaling pathway and enhancing rod Ca^{2+} channel activation. Tonic activation of the feedback synapse by the steady illumination of receptive-field surrounds shifts (or resets) the operating range of bipolar cells to the right along the intensity axis, such that a brighter light is now necessary to elicit a given response.

Spatial and Temporal Processing

The antagonistic center-surround receptive-field organization underlies our ability to detect edges, enhancing contrast between regions of varying brightness and color, and it is ultimately responsible for visual acuity. Baylor and colleagues demonstrated that the turtle cone photoreceptor light response was modulated by light intensity as well as by the pattern of light stimulation, where cone response to center illumination was modified by the stimulation of the surround. Moreover, hyperpolarization of horizontal cells could produce a depolarization in nearby cones. These data indicated that horizontal cells participate in generating antagonistic receptive-field surrounds of cone photoreceptors. The two morphological types of horizontal cells exhibit different spatial summation properties, owing to differing dendritic field sizes and degree of electrical coupling. The reduction of horizontal cell receptive-field sizes in darkness due to uncoupling would further improve spatial contrast ability. Because of their large dendritic fields and electrical coupling, in general, horizontal cells integrate light stimuli over a large area and thus respond well to large light stimuli (i.e., low spatial frequency), whereas small spot stimuli (i.e., high spatial frequency) evoke small responses. The knockout of connexin 57 (Cx57), a gap-junction protein specific to horizontal cells, in mice reduced the receptive-field sizes of horizontal cells, but did not eliminate the rollback in the horizontal cell light response, reflective of negative feedback to cones (**Figure 4**). The knockout diminished the tracer coupling of horizontal cells by 99%, which indicates that Cx57 is the principal connexin in horizontal

cells. Furthermore, the regulation of coupling by dopamine was lost, suggesting that gap junctions formed by Cx57 are the targets of this modulation.

In fish, D1 antagonists blocked the uncoupling of horizontal cells produced by flickering light, but not the uncoupling resulting from steady ambient light, suggesting that the temporal characteristics of the light stimulation may affect the pathway activated.

Chromatic Processing

The Stell model emerged from anatomical analysis of the connectivity of color-sensitive cones and three types of cone-driven horizontal cells in fish retina (**Figure 6**). The model accounted for findings that monophasic horizontal cells (H1) hyperpolarize irrespective of wavelength, but peak in the red or long wavelengths, biphasic horizontal cells (H2) hyperpolarize at short and medium wavelengths and depolarize at long wavelengths, and triphasic horizontal cells (H3) hyperpolarize at short and long wavelengths, while depolarizing at medium wavelengths. Hence, monophasic H1 cells signal luminance, whereas the other two types provide color-opponent signals. In the

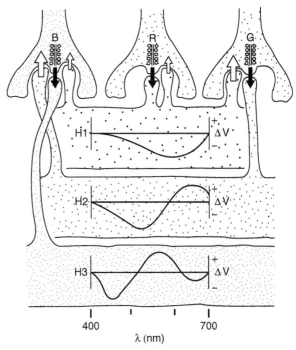

Figure 6 The Stell model of chromatic interactions in the goldfish retina. Cone types are labeled R, G, and B. The H1, H2, and H3 horizontal cells generate monophasic, biphasic, and triphasic spectral response functions by the pathways shown. Synaptic pathways from cones to horizontal cells are shown with filled arrows while feedback pathways are shown in open arrows. Reproduced from figure 2 of Stell, W. K., Lightfoot, D. O., Wheeler, T. G., and Leeper, H. F. (1975). Goldfish retina: Functional polarization of cone horizontal cell dendrites and synapses. *Science* 190(4218): 989–990. Reprinted with permission from AAAS.

case of the H2 red/green biphasic C-type cells, the depolarizing responses to long wavelengths appear to arise from feedback from red-sensitive L-type horizontal cells to green cones, which then synapse onto H2 cells, and, similarly, the depolarizations in the H3 come about through H2 feedback to short wavelength or blue cones, which then drive the H3 cells. Mammalian retinas do not appear to have color-opponent horizontal cells.

Conclusions

The visual system operates over a large range of stimulus intensities. That animals can see in the dark of a moonless night and in near-blinding sun-drenched scenes is a remarkable feat achieved by exploiting the dynamic range of numerous cascaded stages. Several adaptation stages have been described in rods and cones, beginning in the light-transducing outer segments. Modulation of the rod and cone output synapses is another critically important stage, and here adaptation results in large part from horizontal cell feedback and feed-forward, and may alter the form of some receptive fields. Many of the anatomical and biophysical features of the photoreceptor triad synapse are known in great detail, and regulation of the synapse by integrative feedback from horizontal cells is one area of critical importance in retinal neurobiology. Future studies will aid us in understanding the complex analysis of information in the visual system, and, in particular, how the visual system has evolved algorithms that optimize acuity under changing levels of ambient illumination.

Acknowledgment

This work was supported by NEI EY 15573 and a Veterans Administration Senior Career Scientist Award to NB, and CIHR grant MOP10968 to SB.

See also: The Circadian Clock in the Retina Regulates Rod and Cone Pathways; Cone Photoreceptor Cells: Soma and Synapse; GABA Receptors in the Retina; Information Processing in the Retina; Morphology of Interneurons: Bipolar Cells; Morphology of Interneurons: Horizontal Cells; Morphology of Interneurons: Interplexiform Cells; Neurotransmitters and Receptors: Dopamine Receptors; The Physiology of Photoreceptor Synapses and Other Ribbon Synapses; Rod Photoreceptor Cells: Soma and Synapse.

Further Reading

Baylor, D. A., Fuortes, M. G. F., and O'Bryan, P. M. (1971). Receptive fields of cones in the retina of the turtle. *Journal of Physiology (London)* 214: 265–294.

Bear, M. F., Connors, B. W., and Paradiso, M. A. (eds.) (2007). *Neuroscience: Exploring the Brain,* 3rd edn. Philadelphia, PA: Lippincott Williams & Wilkins.

Brown, K. T. and Wiesel, T. N. (1959). Intraretinal recording with micropipette electrodes in the intact cat eye. *Journal of Physiology (London)* 159: 537–562.

Burkhardt, D. A. (1993). Synaptic feedback, depolarization, and color opponency in cone photoreceptors. *Visual Neuroscience* 10: 981–989.

Davenport, C. M., Detwiler, P. B., and Dacey, D. M. (2008). Effects of pH buffering on horizontal and ganglion cell light responses in primate retina: Evidence for the proton hypothesis of surround formation. *Journal of Neuroscience* 28: 456–464.

Hirano, A. A., Brandstätter, J. H., and Brecha, N. C. (2005). Cellular distribution and subcellular localization of molecular components of vesicular transmitter release in horizontal cells of rabbit retina. *Journal of Comparative Neurology* 488: 70–81.

Hirasawa, H. and Kaneko, A. (2003). pH changes in the invaginating synaptic cleft mediate feedback from horizontal cells to cone photoreceptors by modulating Ca^{2+} channels. *Journal of General Physiology* 122: 657–671.

Jonz, M. G. and Barnes, S. (2007). Proton modulation of ion channels in isolated horizontal cells of the goldfish retina. *Journal of Physiology (London)* 581: 529–541.

Kamermans, M. and Fahrenfort, I. (2004). Ephaptic interactions within a chemical synapse: Hemichannel-mediated ephaptic inhibition in the retina. *Current Opinion in Neurobiology* 14: 531–541.

McMahon, D. G., Zhang, D. Q., Ponomareva, L., and Wagner, T. (2001). Synaptic mechanisms of network adaptation in horizontal cells. *Progress in Brain Research* 131: 419–436.

Perlman, I., Kolb, H., and Nelson, R. (2003). Anatomy, circuitry, and physiology of vertebrate horizontal cells. In: Chalupa, L. M. and Werner, J. S. (eds.) *The Visual Neurosciences* vol. 1, pp. 369–394. Cambridge, MA: MIT Press.

Shelley, J., Dedek, K., Schubert, T., et al. (2006). Horizontal cell receptive fields are reduced in connexin57-deficient mice. *European Journal of Neuroscience* 23: 3176–3186.

Stell, W. K., Lightfoot, D. O., Wheeler, T. G., and Leeper, H. F. (1975). Goldfish retina: Functional polarization of cone horizontal cell dendrites and synapses. *Science* 190: 989–990.

Thoreson, W. B., Babai, N., and Bartoletti, T. M. (2008). Feedback from horizontal cells to rod photoreceptors in vertebrate retina. *Journal of Neuroscience* 28: 5691–5695.

Verweij, J., Hornstein, E. P., and Schnapf, J. L. (2003). Surround antagonism in macaque cone photoreceptors. *Journal of Neuroscience* 23: 10249–10257.

Verweij, J., Kamermans, M., and Spekreijse, H. (1996). Horizontal cells feed back to cones by shifting the cone calcium-current activation range. *Vision Research* 36: 3943–3953.

Weiler, R., Pottek, M., He, S., and Vaney, D. I. (2000). Modulation of coupling between retinal horizontal cells by retinoic acid and endogenous dopamine. *Brain Research. Brain Research Reviews* 32: 121–129.

Wu, S. M. (1994). Synaptic transmission in the outer retina. *Annual Review of Physiology* 56: 141–168.

Xin, D. and Bloomfield, S. A. (1999). Dark- and light-induced changes in coupling between horizontal cells in mammalian retina. *Journal of Comparative Neurology* 405: 75–87.

Information Processing in the Retina

F S Werblin, UC Berkeley, Berkeley, CA, USA

Glossary

Directional sensitive ganglion cell – Ganglion cells that are particularly sensitive to movement of an image in a specific direction across the retina.
Local edge detector (LED) – A ganglion cell that appears to be unique in that it is activated by a local edge at the center of its receptive field, and suppressed by edges in the surround.
OFF pathway – Retinal circuitry within the retina responding at the offset of light.
ON pathway – Retinal circuitry within the retina responding at the onset of light.
Starburst amacrine cell – Retinal amacrine cells with a large and symmetrical dendritic arbor that uses both actylcholine and gamma aminobutyric acid. These cells play an important role in directional selectivity.

Introduction

The retina creates a dozen different representations of the visual world, each embodied at a separate sublayer of the inner plexiform layer, and carried by a separate class of ganglion cell. These representations are not just enhanced edges, but are often also complex space–time abstractions of the visual world. The early studies of Barlow; Maturana, Lettvin, McCulloch, and Pitts; and Campbell and Robson suggested that the optic nerve contains information carried by a number of different channels, each formed as a result of sophisticated neural computations. In the frog, for example, the retinal output contains the trigger features that guide the frog's behavior, including detectors for flys, edges, dimming, and contrast. As Barlow points out, the activity of individual neurons is not just a reflection of thought processes, but these activities are also thought processes. Therefore, what are these processes and how are they formed through neural interactions in the retina?

Over the last 40 years, visual neuroscientists have attempted to understand retinal function from a variety of perspectives. What is the nature of the representations of the visual world formed by the retina? Through what neural processes are these representations formed? Answering these questions requires a merging of information from a variety of different disciplines. Retinal anatomists have defined both the morphology and the conductivity between retinal neurons at the light electron microscope levels. Retinal electrophysiologists have defined the response properties of each class of retinal neurons. Retinal circuitry has been analyzed, aided in great measure by pharmacological studies where specific synaptic pathways have been defined for the use of receptor agonists and antagonists. A number of general principles of organization and function have been gleaned from these studies.

For example, as predicted by Horace Barlow and later verified through experiment, most retinal neurons operate in a spikeless mode, whereby membrane potential and synaptic transmission are continuous and graded. Furthermore, under most ambient conditions, most graded-potential retinal neurons operate at or near the midpoint of their response range, capable of signaling both increments and decrements. Embedded in a complex circuitry including feedback, most neurons are constantly active, talking to each other with a continuous stream of activity. The presentation of visual stimuli offsets this ambient state, generating at least a dozen different space–time patterns of neural activity at the retinal output that correspond to, and represent, the visual input.

The Outer Retina, Gain and Level Adjustment

One of the first steps in retinal processing performs the operations that match the response range of retinal neurons to the light input, setting the dynamic range. A significant amount of neural housekeeping is required to maintain the retina in a steady-state condition such that most neurons are operating near their mid-potential range. This set point control is achieved through a series of processes known as adaptation. Adaptation to luminance maintains the photoreceptors at an ambient potential level that varies little with the overall luminance state, over very many orders of magnitude. This is accomplished because cone gain decreases as ambient luminance increases. It is as though the cones put on sunglasses as the brightness of the environment increases. This adaptation is an inherent part of the transduction machinery, mediated at the outer segments of photoreceptors. Much more about the process of adaptation can be found in elsewhere in the encyclopedia.

Bifurcation of the Visual Pathways into ON and OFF streams

The pathway from cones to bipolar cells bifurcates at the bipolar dendrites into the ON and OFF streams of activity. Activity is initiated in OFF bipolars via ionotropic glutamate receptors in the bipolar dendrites; therefore, OFF bipolars polarize in phase with the photoreceptors. Activity is initiated in the ON bipolar dendrites via metabotropic receptors such that ON bipolar cells polarize out of phase with the photoreceptors. Both bipolar types drive ganglion cells via α-amino-3-hydroxyl-5-methyl-4-isoxazole-propionate (AMPA), kainate, and N-methyl-D-aspartic acid (NMDA) receptors; therefore, all ganglion cells respond in phase with their bipolar cell inputs. These differential streams are carried to different strata of the inner retina with most of the OFF streams in regions that are distal to the ON streams. These ON and OFF streams are then carried through many synapses in the visual system and can also be measured at the visual cortex. A sketch of the retinal pathways from cones to the ON and OFF bipolars, to the ON and OFF ganglion cells, is shown in **Figure 1**. More about bipolar cells can be found elsewhere in the encyclopedia.

Horizontal Cell Synaptic Interactions

The pathway from cones to bipolar cells is intersected by horizontal cells. Cones drive the horizontal cells, but horizontal cells are strongly electrically coupled; therefore, the neural image carried by horizontal cells is blurred. The coupled network of horizontal cells feeds forward to bipolar

cells via *gamma aminobutyric acid* (GABA)ergic synapse. ON and OFF bipolar cells respond with opposite polarities to light; therefore, horizontal cells must polarize the ON and OFF bipolar cells in opposite directions: GABA depolarizes ON bipolar cells and hyperpolarizes OFF bipolar cells because the chloride concentrations, and therefore the reversal potentials for GABA input, are different in the two bipolar cell types. Horizontal cells also feed back to cones, but the feedback mechanism from horizontal cells to cones remains controversial. Feedback may be mediated by an electrical synaptic feedback, by GABA, or it might be controlled by pH. Experiments in different animals under different conditions have led to these diverse theories. The circuitry is shown in **Figure 2**, added to the ON–OFF streams shown above. The role of GABA in horizontal cells is described elsewhere in the encyclopedia.

Horizontal Cells and Local Gain Control

Feedback from horizontal cells to cones provides a form of local adaptation or gain control so that local bright spots do not saturate neural activity the way they might in a conventional camera where the level of light arriving at the sensor is controlled by aperture size. Local gain control is achieved by the electrical coupled horizontal cells network whose blurred image is subtracted from the sharper image carried by the cones and bipolar cells. This subtraction normalizes activity across the retina with respect to the blurred activity of horizontal cells. The subtraction has two major results: the neural representation of edges is

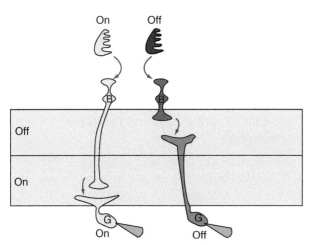

Figure 1 General glutamatergic pathways in the retina from photoreceptors (top) to bipolar cells (B) to ganglion cells. ON and OFF pathways are initiated at the dendrites of the bipolar cells (B, orange arrows). ON and OFF ganglion cells (G) are driven by their respective ON and OFF bipolar cells (black arrows). OFF activity is located in the distal half of the outer plexiform layer, while ON activity is located in the proximal half.

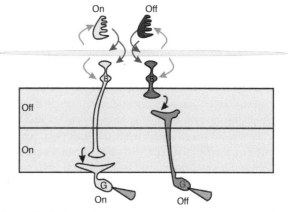

Figure 2 Horizontal cell feedback and feedforward added to the circuitry. Orange arrows represent glutamate pathways from cones to ON and OFF bipolar cells and to horizontal cells. Green arrows represent inhibitory pathways that mediate the horizontal-mediated antagonistic surround. The synaptic mechanism underlying this feedback pathway is still not fully understood. This feedback pathway is also implicated in local gain control. GABA is also thought to be fed forward to both bipolar cell types.

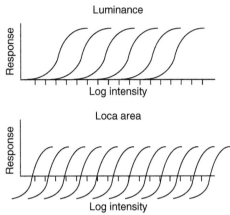

Figure 3 Luminance and local gain changes. Upper traces: gain changes at the photoreceptors shift the intensity–response curves along the log intensity axis; as ambient illumination increases, the transduction gain is reduced. Lower traces: local adaptation mediated by horizontal cell feedback subtracts the luminance level from the cone-to-bipolar signal, leaving mainly the contrast signal.

sharpened by the subtraction, resulting in the presence of Mach bands. Put differently, broadly distributed luminance changes are lost, subtracted from the cone to bipolar patterns by the blurred horizontal cell image. The neural image that remains after this subtraction is primarily related to local contrast. It is the local contrast image that is then brought to the inner retina at the synaptic terminals of the ON and OFF bipolar cells. A family of adaption curves is shown in **Figure 3**.

Interactions at the Inner Retina: Contrast Gain

The local gain-controlled neural image, brought to the inner retina, is subject to an additional adjustment by contrast gain control, that constrains signal magnitude so that postsynaptic neurons operate within their dynamic range. Contrast gain control is mediated by at least two mechanisms. Amacrine cell feedback to bipolar cells is thought to contribute to contrast gain control. In addition, a more significant mechanism, still not well understood, but likely located at the bipolar cell terminals, has been shown to increase contrast gain at low temporal contrast and to decrease gain at high temporal contrast. Contrast gain control is described elsewhere in the encyclopedia.

To summarize, there are three main gain control systems in the retina: luminance gain at the outer segments, local luminance gain at the horizontal cell system, and contrast gain control. These systems, taken together, allow the retina to detect dim objects at low contrast or bright objects at high contrast, with all neurons operating near their optimal gain and never saturating.

General Organizational Principles

Lateral Interactions are Concatenated

The results of an interaction that takes place early in visual processing are carried through and appear at later stages of processing. As an example, antagonistic lateral interactions at the outer retina, mediated by horizontal cell feedback and feedforward, form an initial antagonistic center-surround receptive field interaction. These concentric fields are first formed at the cone terminals, mediated by feedback from horizontal cells; however, this activity is read out by both the ON and OFF bipolar cell types. Therefore, bipolars, similar to cones, show concentric antagonistic receptive fields. Part of the antagonistic surround measured at ganglion cells derives from the interactions between cones and horizontal cells at the outer retina and carried to the ganglion cells via the bipolar cells. Additional antagonistic components are added through interactions at the inner retina, mediated by amacrine cells, and then measured in ganglion cells. Therefore, ganglion cells carry the results of lateral antagonism at the outer retina superimposed upon additional forms of lateral interaction formed at the inner retina. Ganglion cells can contain as many as five different lateral antagonistic components.

Mutual Antagonism is a Form of Amplification

Once again, lateral antagonism at the outer retina serves as an example of mutual antagonism because adjacent retinal regions, via horizontal cells antagonize, are mutually antagonistic. This mutual antagonism amplifies and generates the familiar Mach bands which enhance the neural image of edges in the visual world. As described later, the mutual antagonism between motion-detecting starburst amacrine cells amplifies directional differences. There are many other forms of mutual antagonism in the retina operating in different domains, and each of these serves to amplify the quality represented in that domain.

Redundant Feedforward and Feedback Interactions

As shown above, the lateral antagonistic signal that is fed back to cones is also fed forward to both ON and OFF bipolar cells. In many cases of retinal interaction, the GABAergic signal that is fed back to bipolar cells by amacrine cells is also fed forward to ganglion cells as shown in **Figure 4**.

Interaction between the Two Complementary Visual Streams in the Visual System

The ON and OFF pathways do not remain independent, but interact at every neural level in the retina and continue to interact at each of the higher visual centers. This interaction between the ON and OFF pathways, called

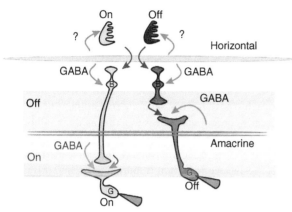

Figure 4 GABAergic amacrine cell feedback and feedforward can provide an additional lateral antagonistic interaction, generating an additional receptive field surround superimposed upon the feedback and feed-forward interactions mediated by horizontal cells at the outer retina.

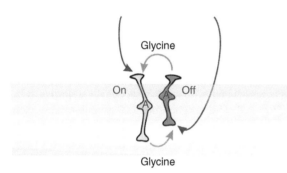

Figure 5 Crossover circuitry isolated. The ON system inhibits the OFF system, and the OFF system inhibits the ON system by way of glycinergic amacrine cells that span the ON–OFF sublaminae. This interaction compensates for the nonlinearities introduced by synaptic transmission. The interaction is found in most bipolar, amacrine, and ganglion cells.

crossover inhibition, is mediated by glycinergic amacrine cells, and serves to relinearize signals that have been distorted by nonlinear transmission at synapses. It is necessary to provide this circuitry compensation at each stage of processing to maintain a linear signal stream because if a nonlinear signal is filtered, linearity can never be reconstructed. In the transistor world, analog signal processing design requires interactions between the equivalent of ON and OFF circuitry to compensate for the rectifying nonlinearities inherent in transistors.

The circuitry underlying crossover inhibition involves ON amacrine cells feeding the OFF pathway, and OFF amacrine cells feeding the ON pathway. The circuit module representing this crossover activity is shown in **Figure 5**.

The crossover circuitry is superimposed upon and intersects the glutamate pathways and the GABAergic inhibitory pathways. The overall circuitry is summarized in **Figure 6**. More on amacrine cells can be found in elsewhere in the encyclopedia.

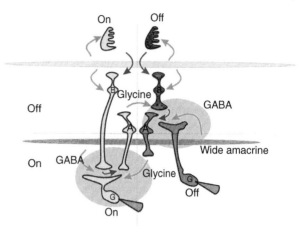

Figure 6 Generalized retinal circuit including horizontal cells, lateral amacrine cells, and vertical amacrine cells. This general circuit includes feedback and feed-forward horizontal cell interactions, feed-forward and feedback GABAergic amacrine cell interactions, and ON to OFF as well as OFF to ON crossover circuitry. Each of these five interactions can generate a different antagonistic receptive field surround for ganglion cells.

This is a summary of a basic retinal design. It accounts for most of the general activity measured in bipolar, amacrine, and ganglion cells. In almost all cases, the inhibition carried by the vertical amacrine cell elements is mediated by glycine, while inhibition carried by the laterally oriented amacrine cells is mediated by GABA. These general circuitry rules underlie a more specific circuitry that generates the physiological behavior of each ganglion cell type. This additional circuitry has been described for a few ganglion cells. In each case, special variations to the basic circuitry endows these cell types with their specific characteristics. More on neurotransmitters and receptors can be found elsewhere in the encyclopedia.

Inner Retinal Processing: Circuitry for Feature Extraction

Having solved both the housekeeping problems related to adaptation to luminance contrast and the nonlinearity problem introduced by synaptic transmission through appropriate crossover circuitry, the stage is set for the real work of the retina, which is creating the appropriate representations of the visual world. Most of the interesting interactions take place at the inner plexiform layer that is itself a multilayered and exquisitely organized structure. There appear to be about 10 strata in every retina species that has been studied. The strata are defined by the 10 levels at which the dendrites of different ganglion cell types ramify. Remarkably, there also appear to be about 10 different types of bipolar cells, and the axon terminals of each of these bipolar cell types also terminate roughly in the same 10 strata as a ganglion cell dendrites.

To a first rough approximation, it appears that each ganglion cell type receives input from a separate bipolar cell type as shown in **Figure 7**. However, a close look at the bipolar terminals shows that they are often more diffusely distributed, and there are ganglion cells with multistratified dendritic arborizations. Physiologically, the connections must be more diffuse because most ganglion cells ramify in either the ON or the OFF sublamina; however, most ganglion cells appear to receive both ON and OFF inputs (although activity tends to be dominated by either ON or OFF excitation).

Although the excitatory pathway from photoreceptors to bipolars to ganglion cells appears to be relatively straightforward, there exists a bewildering array of amacrine cells that generate a variety of interactions at the inner plexiform layer. Amacrine cells come in many diverse morphologies, and there appear to be approximately 30 different types. Functional properties of these amacrine cells can be thought of at three different levels of refinement: (1) there are two major morphological classes of amacrine cells identified by their vertical versus lateral orientation in the retina, (2) there are circuitries involving amacrine cells that mediate specific functions, and (3) there are amacrine cells that possess unique physiological properties. Each level of functionality is described below.

Amacrine Cell Morphological Types

The majority of amacrine cells falls into one of two general classes. One consists of narrowly ramifying vertically oriented amacrine cells that span the ON and OFF sublamina. This class contains and releases glycine as its

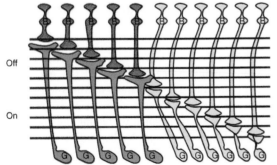

Off

On

Figure 7 Schematic of the layering of the inner plexiform layer (IPL). Each of the 10 bipolar cell types sends its axon terminal to a distinct region of the IPL, approximately a single substratum of the IPL. Each of the ganglion cell types sends its dendrites to a distinct substratum. Roughly speaking, each bipolar cell type is associated with a unique ganglion cell type. This description is only approximate. In fact, some ganglion cell dendritic fields are bistratified, such as the direction selective (DS) cells, and the axon terminals of some bipolar cells are more diffuse.

inhibitory transmitter. These amacrine cells appear to be strategically located to carry information from the ON to the OFF and the OFF to be ON sublamina, and are probably the key players in mediating crossover inhibition described earlier. The other major class of amacrine cell consists of widely ramifying, but monostratified, amacrine cells, each stratifying in a separate sublamina. These amacrine cells contain and release GABA as their inhibitory transmitter and may be responsible for various forms of lateral inhibition mediated at the inner retina. In some cases, this lateral inhibition is mutual and when this is the case, it serves as an amplifier of the visual quality carried by that specific form of inhibition.

Specific Ganglion Cell Circuitries

Directional selectivity

There are a few amacrine cells that have now been identified with very specific personalities. For example, starburst amacrine cells, named for the characteristic starburst pattern of their processes, span about 200 μm. Starburst cells contain and release both GABA and acetylcholine. They are the key elements in the organization of directional selectivity in the retina. A large population of starburst amacrine cells is associated with each directionally selective (DS) ganglion cell, and neighboring DS ganglion cells likely share many starburst amacrine cells. Starburst cells are inherently DS, generating more release for centrifugal movement. One likely mechanism involves calcium-initiated calcium release, but this remains an area of intense exploration. Release occurs along the outer one-third of the starburst processes. These processes not only release GABA, but are also GABA sensitive. This creates a mutual inhibition between starburst cells that acts to amplify directional motion sensitivity as shown in **Figure 8**. Starburst cells inhibit the DS cells asymmetrically, with stronger inhibition arriving from the null side than from the preferred side. These three mechanisms, inherent directional selectivity in the starburst cells themselves, mutual antagonistic interaction between neighboring starburst cells, and asymmetrical inhibition acting both pre- and postsynaptically at the ganglion and bipolar cells endow the DS cell with some of its directional properties as shown in **Figure 9**. More on DS circuitry can be found elsewhere in the encyclopedia.

Figure 8 Mutual inhibition between starburst amacrine cells amplifies the directional properties of the starburst network. Here, two starburst amacrine cells, themselves directionally selective, are mutually inhibitory (green arrows). In the circuit of this figure, the right starburst cell supplies inhibition to the ganglion cell for movement in the null direction from right to left. The right starburst cell is turned off by the left starburst cell for movement from left to right.

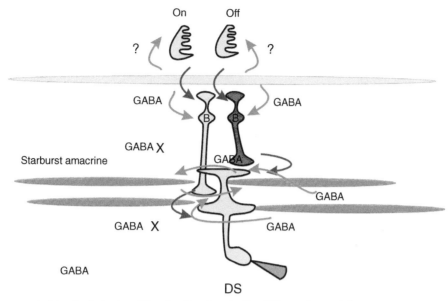

Figure 9 Pathways underlying the behavior of the directionally selective (DS) ganglion cell. Starburst amacrine cells (light blue), themselves directionally selective, are mutually inhibitory. They inhibit by feeding back to bipolar cells and forward to ganglion cells. Both feedback and feed-forward inhibition are asymmetric: they are stronger on the null side than on the preferred side, thereby endowing the DS ganglion cell with directional properties.

The circuitry puzzle regarding the DS cells is far from solved; however, the general organizational rules listed above still apply. The lateral inhibitory interneuron is GABAergic, following the GABA rule for laterally oriented cells.

Functions of AII Amacrine Cells

AII amacrine cells serve a very specific function to transcribe signals in the rod bipolar cells to both the ON and OFF cone pathways. The circuitry underlying this function is now well defined. AII amacrine cells are driven by rod bipolar cells that respond at ON. The AIIs are electrically coupled to cone bipolars, and make glycinergic synaptic contact with the OFF pathway. Recently, the AII amacrine cells have been shown to serve other functions. For example, the AII amacrine cells appear to be the key elements in providing glycinergic inhibition to the so-called looming detectors and in alpha cells, both described elsewhere in the encyclopedia (**Figure 10**).

There are numerous other examples of special purpose circuitry that utilize laterally oriented amacrine cell interneurons. The polyaxonal amacrine cells are thought to mediate saccadic suppression. In other cases, the same cell type has been implicated in mediating object motion sensitivity. It is likely that other amacrine cell types also serve specific functions, but their properties have not yet been identified.

At about the time that Levick was characterizing the DS ganglion cell, he also described another ganglion cell

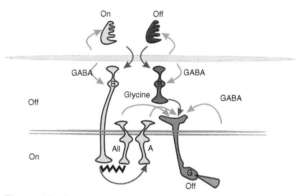

Figure 10 Pathways underlying the behavior of the alpha cell/looming detector. This ganglion cell appears to respond to a dark target at its receptive field center that increases in size, much like an approaching predator. It appears to receive glycinergic inhibition from two types of local amacrine cells: input from an AII amacrine cell that is electrically coupled to ON bipolar cells, and input from another amacrine cell class that receives excitatory glutamate input from ON bipolar cells.

that he termed the local edge detector (LED). This cell appears to be unique in that it was activated by a local edge at the center of its receptive field, and that activity was suppressed by edges in the surround. Van Wyk, Taylor, and Vaney have recently gone on to characterize some of the special temporal properties of this neuron. It receives excitation and also glycinergic inhibition at both ON and OFF, and is inhibited by edge stimuli presented at the surround via a GABAergic lateral pathway. Both inhibitory components follow the general rule of

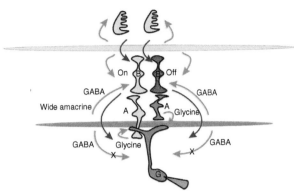

Figure 11 Pathways underlying the response properties of the local edge detector. This cell responds to subreceptive field detail at its receptive field center, most likely from the bipolar cells that drive it. It also receives local ON and OFF inhibition that is glycinergic (green arrows), as well as broad-field inhibition (green arrows) that is GABAergic and also responsive to fine detail. In this cell type, GABA is only fed back, not fed forward.

vertical glycinergic and lateral GABAergic activity as shown in **Figure 11**. The role of this neuron in the overall scheme of vision remains obscure, but it is likely involved in high-resolution, slow temporal response activity.

See also: GABA Receptors in the Retina; Information Processing: Amacrine Cells; Information Processing: Bipolar Cells; Information Processing: Contrast Sensitivity; Information Processing: Direction Sensitivity; Information Processing: Ganglion Cells; Information Processing: Retinal Adaptation.

Further Reading

Barlow, H. B. (1953). Summation and inhibition in the frog's retina. *Journal of Physiology* 119: 69–88.

Beaudoin, D. L., Borghuis, B. G., and Demb, J. B. (2007). Cellular basis for contrast gain control over the receptive field center of mammalian retinal ganglion cells. *Journal of Neuroscience* 27: 2636–2645.

Demb, J. B. (2008). Functional circuitry of visual adaptation in the retina. *Journal of Physiology* 586: 4377–4384.

Fried, S. I., Munch, T. A., and Werblin, F. S. (2002). Mechanisms and circuitry underlying directional selectivity in the retina. *Nature* 420: 411–414.

Fried, S. I., Munch, T. A., and Werblin, F. S. (2005). Directional selectivity is formed at multiple levels by laterally offset inhibition in the rabbit retina. *Neuron* 46: 117–127.

Hsueh, H. A., Molnar, A., and Werblin, F. S. (2008). Amacrine to amacrine cell inhibition in the rabbit retina. *Journal of Neurophysiology* 100(4): 2077–2088.

Lee, S. and Zhou, Z. J. (2006). The synaptic mechanism of direction selectivity in distal processes of starburst amacrine cells. *Neuron* 51: 787–799.

Levick, W. R. (1965). Receptive fields of rabbit retinal ganglion cells. *American Journal of Optometry and Archives of American Academy of Optometry* 42: 337–343.

Maturana, H. R., Lettvin, J. Y., McCulloch, W. S., and Pitts, W. H. (1960). Anatomy and physiology of vision in the frog (*Rana pipiens*). *Journal of General Physiology* 43(supplement 6): 129–175.

Molnar, A. and Werblin, F. (2007). Inhibitory feedback shapes bipolar cell responses in the rabbit retina. *Journal of Neurophysiology* 98: 3423–3435.

Roska, B., Molnar, A., and Werblin, F. S. (2006). Parallel processing in retinal ganglion cells: How integration of space-time patterns of excitation and inhibition form the spiking output. *Journal of Neurophysiology* 95: 3810–3822.

van Wyk, M., Taylor, W. R., and Vaney, D. I. (2006). Local edge detectors: A substrate for fine spatial vision at low temporal frequencies in rabbit retina. *Journal of Neuroscience* 26: 13250–13263.

Werblin, F. S. and Dowling, J. E. (1969). Organization of the retina of the mudpuppy, *Necturus maculosus*. II. Intracellular recording. *Journal of Neurophysiology* 32: 339–355.

Information Processing: Retinal Adaptation

K R Alexander, University of Illinois at Chicago, Chicago, IL, USA

Glossary

Contrast – Magnitude of luminance variation with respect to mean luminance, defined as Weber contrast (C_w) for discrete stimuli: $C_W = (I_T - I_B)/I_B$, where I_T and I_B refer to the retinal illuminance of a test probe and background, respectively; or as Michelson contrast (C_M) for periodic stimuli: $C_M = (I_{max} - I_{min})/(I_{max} + I_{min})$, where I_{max} is the maximum retinal illuminance and I_{min} is the minimum retinal illuminance.

Gain – Change in the neural response produced by either a change in luminance or a change in contrast over the range for which the stimulus–response function is reasonably linear, specified in units such as impulses per quantum or impulses per percent contrast.

Luminance – Amount of light given off by an extended source, either emitted or reflected, and usually specified in candelas per square meter ($cd\ m^{-2}$), although many alternative units exist, including apostilbs (asb), foot-lamberts (ftL), millilamberts (mL), and nits.

Retinal illuminance – Luminance in $cd\ m^{-2}$ multiplied by pupil area in square millimeters (mm^2) and specified in trolands (td).

Threshold – In psychophysics, the light level that marks the transition from invisibility to visibility, defined as either the absolute threshold ("yes, I see it") or the difference threshold ("yes, it is different"); in electrophysiology, the light level that elicits a criterion neural response amplitude or a criterion change in response amplitude.

The eye can potentially be exposed to an enormous range of light intensities, ranging from a few photons per second under extremely dim lighting conditions to light levels that can be more than 10-billion-fold higher (i.e., a factor of 10^{10} or 10 log units). Furthermore, there may be rapid temporal fluctuations in the light level due to eye movements. In addition, there can be marked changes in the chromatic properties of the visual environment, such as when viewing objects in incandescent room illumination versus outdoors at noon on a sunny day. Remarkably, the visual system is able to cope with the large range of illumination conditions through complex neural mechanisms that are collectively termed adaptation.

It should be noted, however, that there are actually a number of different uses of the term adaptation, ranging from an adjustment to the overall light level to more complex forms, such as spatial frequency adaptation, motion adaptation, and adaptation to artificially induced retinal image distortion or rotation. The emphasis of this article is on adaptation that is presumed to occur within the retina. How do we know which processes are retinal and which involve higher levels of the visual system? One method is to identify neurons within the retina that exhibit the physiological characteristics of the type of adaptation under investigation. This can be determined by recording from single neurons within the retina, by recording simultaneously from groups of neurons using a multi-electrode array, or by recording the electroretinogram (ERG), which is the massed electrical response of the retina. Further insight into the sites and mechanisms of adaptation can be gained through the study of transgenic animals that have a mutation in, or knockout of, putative components of the adaptation process, or of humans who have spontaneously occurring mutations in these components.

A complementary, behavioral method for investigating the site of adaptation is to employ dichoptic stimulation. In this technique, an adapting stimulus is presented to one eye and a test stimulus is presented to the other eye. The goal is to determine whether adaptation of the contralateral eye affects performance for targets presented to the tested eye. The first site at which there is a combination of information from the two eyes occurs at a cortical level. Therefore, if there is interocular transfer of the adaptation, then it is presumed that the primary site of the adaptation is cortical. On the other hand, if there is no evidence of interocular transfer, then the adaptation is presumed to occur at the retinal level.

Based on such considerations, the forms of adaptation that are thought to be predominantly or exclusively retinal in origin are: (1) light adaptation, which refers to the adjustment of the visual system to changes in the overall or mean illumination level; (2) contrast adaptation, which refers to the ability of the visual system to adjust to the variance of the illumination rather than to its mean; (3) chromatic adaptation, which refers to an adjustment to the spectral composition of light; and (4) dark adaptation, which refers to the time-dependent recovery of visual sensitivity in the dark following exposure to light. These forms of adaptation are the subject of this article, although the emphasis is on light adaptation.

In addition to adaptation, though, there are additional strategies that are employed by the visual system to cope with the broad range of illumination levels encountered by the visual system. One strategy is a change in pupil size, which is a mechanical way in which the visual system can partially adjust to varying light levels. As the overall light level increases, the pupil area decreases, which in turn decreases the retinal illuminance. However, the maximum change in pupil area is essentially 16-fold (i.e., a change in diameter from 2 to 8 mm); therefore a change in pupil size can compensate for only a small portion of the potential illumination range. Furthermore, owing to the directional sensitivity of the cone photoreceptors (the Stiles–Crawford effect), light entering the edge of the pupil is a less efficient stimulus for the cone system than light entering the pupil center. Therefore, the effective increase in retinal illuminance with increasing pupil size is actually less than would be predicted based on pupil area.

Another strategy used by the visual system to cope with the large range of environmental light levels is to split the load between the rod and cone systems. The rod system, which is extremely sensitive, covers the lower 3 log units of stimulation, termed the scotopic range. The cone system handles the highest 6 log units, termed the photopic range. Between these two ranges is the mesopic range, within which visual sensitivity can be rod-mediated or cone-mediated, depending on such factors as target wavelength, duration, size, and retinal eccentricity. Thus, the duplex nature of the retina provides a partial solution to the problem of dealing with the wide range of light levels impinging on the retina. However, adaptive processes within the rod and cone systems are also necessary in order to provide useful vision under all illumination conditions.

Light Adaptation

Light adaptation typically refers to the adjustment of the visual system to the overall illumination level. This adjustment allows for an amplification of neural signals relative to noise at low light levels, and prevents or minimizes saturation of the neural response at high light levels. Light adaptation also changes the way in which spatial and temporal information is processed. For example, spatial resolution is typically better at high illumination levels.

Characteristics of Light Adaptation

The typical light adaptation paradigm consists of the presentation of a brief test probe of retinal illuminance I_T against an adapting field of retinal illuminance I_B.

The dependent variable is the increment threshold ΔI, defined as:

$$\Delta I = (I_T - I_B) \qquad [1]$$

which is usually measured at different values of I_B. The typical threshold versus retinal illuminance or tvi function is illustrated in **Figure 1**. The data points in this figure represent psychophysical increment thresholds for a small achromatic test probe presented foveally against a large achromatic adapting field. The curve represents the log form of the equation:

$$\Delta I = K(I_B + I_0)^n \qquad [2]$$

where K and I_0 are fit parameters that represent the absolute threshold and the inflection point of the function, respectively, on log–log coordinates and n determines the slope of the function at high retinal illuminances.

At low adapting levels, the increment threshold is relatively independent of the retinal illuminance of the adapting field. Under these conditions, it is generally assumed that the threshold is governed by the internal noise within the visual system, also termed dark light or eigengrau. As the retinal illuminance of the adapting field increases, the increment threshold begins to rise as the internal noise becomes overwhelmed by the neural response to the adapting field. At high retinal illuminances, the increment threshold is proportional to the adapting field retinal illuminance, such that:

$$\Delta I / I_B = K \qquad [3]$$

This relationship is referred to as the Weber–Fechner relationship or Weber's law. On a log–log plot, Weber's law has a slope of 1. An important implication of Weber's

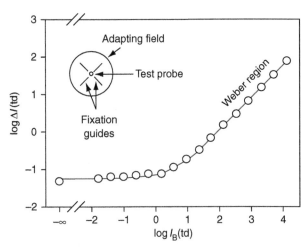

Figure 1 Increment threshold (circles) as a function of adapting field retinal illuminance for a foveally presented brief achromatic test probe in the center of an achromatic adapting field, as depicted in the inset. The curve represents the least-squares best fit of eqn [1].

law is that the visual system is organized to signal contrast rather than absolute luminance. In other words, because $\Delta I/I_B$ is constant within the Weber region, Weber contrast is also constant, regardless of the adapting level.

The data of **Figure 1** represent a tvi function for the foveal cone system, but an increment threshold function can also be obtained for the rod system if a short-wavelength test probe, to which the rods are sensitive, is presented in the visual field periphery against a long-wavelength adapting field that desensitizes the cone system. A schematic of a rod increment threshold function is shown in **Figure 2**. As is the case for the cone system, there is a range of low retinal illuminances over which the rod threshold remains constant. This is followed by a region over which the rod threshold is proportional to the square root of the retinal illuminance (the Rose–deVries region), and then there is a transition to Weber-law behavior. However, a primary difference between the rod and cone increment threshold functions is that the rod function shows saturation, or an upward turn at high retinal illuminances that is steeper than Weber's law. For adapting field retinal illuminances above rod saturation, the cone system mediates detection of the test probe.

Saturation can be observed in individual rod photoreceptors, in the sense that there can be a complete shutdown of the rod circulating current resulting from light exposure. However, there is considerable evidence that psychophysical rod saturation does not represent saturation at the level of the rod photoreceptors, but rather is a property of the pathway through which the

rod signals travel. For example, the value of I_B at which the onset of psychophysical rod saturation occurs depends on whether the adapting field is steady or flashed, on the size of the test probe and the wavelength of the adapting field, and on the level of cone stimulation, all of which indicate that postreceptoral factors are involved.

The tvi function of the cone system typically does not show saturation, owing to photopigment bleaching. When cone photopigment molecules become bleached as a result of light exposure, there is less total photopigment within a cone photoreceptor that is available to capture photons. As a result, a proportionally greater level of stimulation is needed to produce the same neural response. At equilibrium, the relationship between p, the fraction of unbleached cone photopigment, and retinal illuminance I in td is:

$$1 - p = I/(I + I_0) \qquad [4]$$

where I_0 is the half-bleaching constant of 4.3 log td. The effect of cone photopigment bleaching on vision has been likened to wearing sunglasses, which reduce the retinal illumination by a scaling factor. Photopigment bleaching is a way of avoiding saturation within the cone system but it is not a factor within the rod system, because the rod photoreceptors are saturated by illumination levels that bleach only a few percent of the rhodopsin molecules within an outer segment.

However, saturation within the cone system can be demonstrated if a probe-flash paradigm is used. In this paradigm, the test probe is presented simultaneously with a briefly flashed adapting field. A typical finding is that the threshold rises rapidly with increasing retinal illuminance of the flashed adapting field, such that at high retinal illuminances, the test probe itself is invisible and its presence can only be detected by virtue of an afterimage.

The probe-flash paradigm is a variant of Crawford masking or early light adaptation, in which the increment threshold for a test probe is measured with respect to the time of onset of a transient rather than a steady-state adapting field. Typically, the threshold begins to rise when the test probe is presented slightly before the masking flash. This curious result has been attributed to the differential latencies of the neural responses to the weak test probe and the stronger masking flash. The threshold is highest when the test probe and masking flash have simultaneous onsets. If the test probe is presented during the middle of a long-duration masking flash, then the threshold corresponds approximately to the steady-state level.

Traditionally, studies of light adaptation have used aperiodic test stimuli, such as the light pulses described above. However, there has been another experimental approach to light adaptation that has used periodic test stimuli, such as one whose retinal illuminance varies sinusoidally over time. In this approach, the dependent variable is contrast sensitivity, defined as the reciprocal of

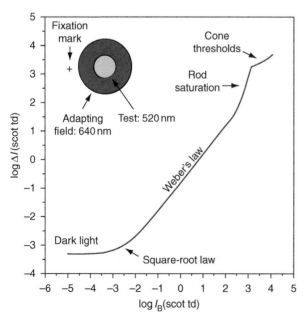

Figure 2 Schematic increment threshold function obtained under rod-isolating conditions. The inset depicts the stimulus configuration, which consists of a large short-wavelength (blue) test probe presented in the visual field periphery against a larger long-wavelength (red) adapting field.

the threshold contrast in Michelson units. A typical finding is that, at low temporal frequencies, contrast sensitivity is invariant with respect to mean retinal illuminance, which corresponds to Weber-law behavior. At high temporal frequencies, however, contrast sensitivity changes with mean retinal illuminance, such that the amplitude of the flicker rather than its contrast determines sensitivity. This finding indicates that there is a high-frequency linearity that discounts the mean retinal illuminance. In addition, the shape of the temporal contrast sensitivity function changes with adaptation level, becoming more band-pass at high retinal illuminances, so that sensitivity to intermediate frequencies becomes enhanced. This shape change has been attributed to a contrast gain-control mechanism.

The explanation for the high-frequency linearity, which is seen both psychophysically and in electrophysiological recordings, remains unclear, although several quantitative models have been proposed to account for it. In fact, a major theoretical challenge has been to develop a computational model that can encompass the adaptational features of data obtained with both periodic and aperiodic stimuli. To date, this attempt has not been entirely successful.

Mechanisms and Sites of Light Adaptation

As illustrated in **Figures 1** and **2**, the rod and cone systems can respond over a considerable range of retinal illuminances. However, individual neurons within the retina can only respond over an approximately 400-fold range of illumination levels. The typical response R of a retinal neuron as a function of retinal illuminance I is illustrated as the solid curve in **Figure 3**. This curve represents a plot of the Naka-Rushton equation:

$$R/R_{max} = I^n/(I^n + I_s^n) \qquad [5]$$

where R_{max} is the maximum neural response, I_s is the retinal illuminance that produces $R_{max}/2$, and n governs the steepness of the function, although n is usually set to 1. This response function is S-shaped when plotted on semilog coordinates, as in **Figure 3**. A major characteristic of the neural response function is that it shows saturation at high illuminance levels. The response function is considered to represent a static nonlinearity, or one which acts instantaneously with no change over time.

If a retinal neuron only operated according to the solid curve in **Figure 3**, then the presence of an adapting field would lead to response compression. For example, the adapting field indicated by the vertical line in **Figure 3** would produce a neural response that is toward the top of the response range, as indicated by the solid horizontal line. This would then leave little room for an additional response to an increment of light. In fact, if the retinal illuminance of the adapting field were high enough,

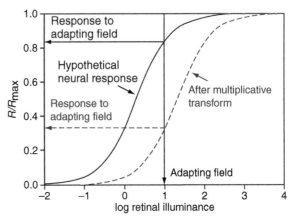

Figure 3 Normalized response amplitude vs. log retinal illuminance for a hypothetical retinal neuron before (solid curve) and after (dashed curve) a multiplicative transform. The vertical line with arrowhead indicates an arbitrary adapting field retinal illuminance, and the horizontal lines represent the hypothetical neural responses to the adapting field, based on the respective illuminance-response functions.

additional increments of light would be invisible due to response saturation.

One way in which a neuron can avoid saturation is by shifting its operating range in proportion to the mean level of illumination. This shift of the response function is termed multiplicative adaptation and is illustrated in **Figure 3** as the dashed curve. Multiplicative adaptation is also known as von Kries adaptation, dark glasses adaptation, and automatic gain control. After a multiplicative transform, the same adapting field retinal illuminance produces a smaller neural response, as indicated by the horizontal dashed line in **Figure 3**. This allows for the detection of light increments that would otherwise be invisible without the multiplicative response scaling. Multiplicative adaptation tends to be relatively fast acting, with a time constant on the order of a few tens of milliseconds, and is thought to be the result of a neural feedback circuit.

A second way in which a neuron can avoid saturation is through subtractive adaptation. This form of adaptation subtracts out the neural response to an adapting field, thus bringing the response down out of the saturating range, without affecting the response to a brief test probe. There are both fast and slow forms of subtractive adaptation, although both are typically much slower than multiplicative adaptation, with time constants on the order of seconds to tens of seconds. Fast subtractive adaptation is presumed to represent center-surround antagonism within neuronal receptive fields. Slow subtractive adaptation may be due to a change in the membrane hyperpolarization of retinal neurons.

The rod system is desensitized by dim backgrounds that produce a quantal absorption in only a very few rod photoreceptors. This observation has led to the concept of a rod adaptation pool, according to which signals from multiple rod photoreceptors are combined in controlling light adaptation. Pooling allows a retinal ganglion cell to respond to light when only a tiny fraction of the rods absorb photons, but it also increases the likelihood of neuronal saturation. Neural pooling can also be a factor with respect to light adaptation within the cone system. Signals from cone photoreceptors are processed by two major parallel pathways: magnocellular (M) and parvocellular (P), which are first organized at the retinal level and extend to the visual cortex. M retinal ganglion cells have a high contrast gain and saturate at relatively low levels of contrast. P ganglion cells have a low contrast gain and a more linear contrast-response function. There is generally a greater degree of spatial integration or pooling within M cells, due to their relatively larger receptive fields. Therefore, M cells are typically more light adapted by a given adapting field than are P cells.

The retinal site or sites of the various neural processes underlying light adaptation remain to be fully explicated. However, the following general conclusions can be drawn. With respect to the cone system, the site of adaptation appears to shift depending on the illumination level. At lower retinal illuminances, adaptation is dominated by postreceptoral processes, sometimes referred to as network adaptation. There is recent evidence that postreceptoral adaptation within the cone system occurs at the synapse between bipolar cells and ganglion cells. At higher retinal illuminances, adaptation occurs primarily within the cone photoreceptors themselves. Rod photoreceptors in the mammalian retina can show adaptation, in that there are changes in sensitivity and response kinetics as a function of illumination level. However, much of the light adaptation within the rod system appears to be postreceptoral in origin, occurring at the synapse between rod bipolar cells and AII amacrine cells.

A potentially powerful way to investigate the relationship between the phenomenology of light adaptation and retinal physiology is to study humans who have genetic mutations that can affect putative components of the adaptational process. An example is a visual condition termed bradyopsia (slow vision) that has been identified recently. Individuals with bradyopsia have mutations in the guanosine triphosphatase-activating protein RGS9 or its anchor protein R9AP, which impedes deactivation of the phototransduction cascade. Bradyopsia is characterized by a slow recovery of sensitivity following a sudden change in illumination and also a loss of motion sensitivity, particularly at low contrasts.

Contrast Adaptation

Contrast adaptation refers to an adjustment to the variance or contrast of the illumination, rather than to its mean level. There are two types of contrast adaptation:

spatial and temporal. Spatial contrast adaptation is selective for stimulus spatial frequency and orientation and is therefore predominantly cortical in origin. Temporal contrast adaptation, on the other hand, is observed in recordings from retinal ganglion cells in response to spatially uniform fields of light. Examples of stimuli used to study contrast adaptation are illustrated in **Figure 4**. Both involve the temporal modulation of a uniform field of light. One type of stimulus (**Figure 4**, top) is contrast-modulated sinusoidal flicker, whose mean luminance and temporal frequency remain constant but whose contrast is changed abruptly. The second (**Figure 4**, bottom) is contrast-modulated white noise.

Following a transition from a low-contrast to a high-contrast stimulus, there is an essentially instantaneous change in the gain and temporal response of retinal ganglion cells. This is followed by a slow change in the firing rate that may take several seconds to complete. There are similar fast and slow changes following a transition from a high-contrast to a low-contrast stimulus, including a temporary decrease in the maintained discharge rate, but these changes are typically more sluggish than for a transition to high contrast. Fast contrast adaptation represents the action of a gain-control mechanism, whereas slow contrast adaptation appears to involve membrane hyperpolarization. Temporal contrast adaptation may form the neural substrate for psychophysical flicker adaptation, in which exposure to high-contrast flicker reduces sensitivity for a subsequently viewed low-contrast flickering test stimulus.

Temporal contrast adaptation has been shown to occur at multiple sites within the retina, beginning in bipolar cells and including processes intrinsic to ganglion cells.

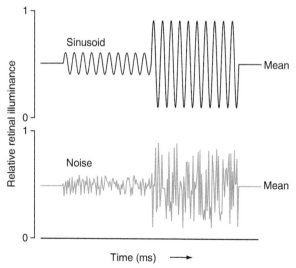

Figure 4 Examples of contrast-modulated stimuli used to study contrast adaptation. The top waveform is a contrast-modulated sinusoid; the bottom waveform represents contrast-modulated noise.

Temporal contrast adaptation is observed in M but not in P ganglion cells, owing in part to the faster temporal response and greater spatial pooling of M cells. Contrast adaptation and mean-luminance adaptation share certain similarities, but whether these forms of adaptation represent a common mechanism or distinct mechanisms remains to be determined.

Chromatic Adaptation

Chromatic adaptation refers to the effect of spectrally selective adapting fields on the detection and appearance of test stimuli of various wavelengths. When chromatic test probes and adapting fields are used to study light adaptation, the increment threshold function of the foveal cone system typically consists of more than one component, as illustrated in **Figure 5**. With the combination of test and adapting field wavelengths shown in **Figure 5**, the test probe is initially detected by a middle-wavelength-sensitive mechanism that becomes progressively more desensitized by the middle-wavelength adapting field. At high retinal illuminances, a short-wavelength-sensitive cone mechanism governs detection.

A change in test wavelength displaces a given tvi function vertically, and a change in adapting field wavelength displaces the tvi function horizontally. Initially, it was thought that the spectral sensitivities of the short-wavelength (S), middle-wavelength (M), and long-wavelength (L) cone photopigments could be derived by evaluating the relative displacements of the tvi functions as the test wavelength and

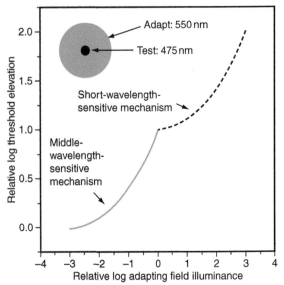

Figure 5 Schematic tvi functions for a 475-nm test probe (blue) presented foveally against a 550-nm adapting field (green), plotted in relative units. The curves represent plots of the tvi template of Stiles, and the lower (green) and upper (blue) branches represent detection by middle-wavelength-sensitive and short-wavelength-sensitive cone mechanisms, respectively.

adapting field wavelength were varied. This approach was used by W. S. Stiles to define what are known as π mechanisms. However, instead of identifying three π mechanisms corresponding to the three cone types, seven π mechanisms were derived from this method. The field sensitivities of three of the seven (π_1, π_4, and π_5) correspond approximately to the spectral sensitivities of the S, M, and L cones, respectively. Nevertheless, the shapes of the π_4 and π_5 mechanisms are broader than would be expected from the known spectra of the cone photopigments, and there are other failures of the π mechanisms to correspond to properties predicted by adaptation of the cone photoreceptors.

Partly to account for the mismatch between the properties of the π mechanisms and the cone spectral sensitivities, two-stage models of chromatic adaptation have been developed. The first stage consists of receptor adaptation, and the second stage combines signals from the S, M, and L cones in an opponent manner. The site of the opponent interactions among signals from different cone types has not yet been identified definitively. Candidates include horizontal cells (although these are not typically spectrally opponent), amacrine cells, and possibly gap-junctional connections between photoreceptors.

Chromatic adaptation also refers to a change in the color appearance of a test light as a result of a change in adapting field chromaticity. For example, a monochromatic light that appears yellow in isolation will appear greenish when superimposed on a long-wavelength adapting field. This change in color appearance has been attributed to a relative desensitization of the L-cone photoreceptors through von Kries or multiplicative adaptation. However, there are instances in which photoreceptor desensitization alone cannot account for the changes in color appearance, such as when the level of retinal illuminance is changed, or when a monochromatic light becomes desaturated during extended viewing. These changes in color appearance are presumably due to second-site adaptation in addition to cone photoreceptor adaptation. Two-stage models are also presumed to account for color constancy, in which colored surfaces maintain their appearance despite substantial changes in the spectral content of the illumination, such as the change from sunlight to incandescent lighting.

Dark Adaptation

Following the exposure of the eye to an adapting field that bleaches a significant fraction of photopigment, there is a systematic recovery of visual sensitivity that is referred to as dark adaptation. Whereas light adaptation occurs relatively quickly, dark adaptation can require a substantial period of time, on the order of tens of minutes. Strictly speaking, the term dark adaptation refers to the recovery of sensitivity to a briefly flashed test probe presented in complete darkness following the offset of a bleaching light. However, there are also variants of bleaching recovery in which the eye is not kept in darkness. These include the photostress recovery test, which measures the recovery of visual acuity following exposure to light from an ophthalmic instrument.

Characteristics of Dark Adaptation

The typical time course of dark adaptation following a bleach is illustrated in **Figure 6**. The data points in this figure represent thresholds for a test probe of 500 nm, a wavelength to which rod and cone systems are both sensitive. The test probe was presented at a retinal eccentricity of 20°, a retinal locus that contains both rod and cone photoreceptors. Thresholds are plotted relative to a baseline threshold that was measured in the fully dark-adapted state prior to a bleach. The recovery of sensitivity follows a two-branched course, each part of which is reasonably well fit by an exponential function. Immediately following the offset of the bleaching light, the recovery of sensitivity occurs relatively rapidly, and then sensitivity reaches a plateau region. This initial portion represents the recovery of cone system sensitivity. There is then a second region of rapid recovery followed by a slower phase, such that full recovery from a substantial bleach can take 45–50 min. This second region represents the return of rod system sensitivity. The transition point from cone-mediated to rod-mediated thresholds is termed the rod–cone break.

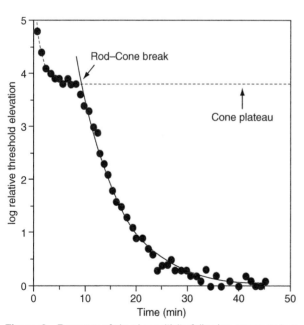

Figure 6 Recovery of visual sensitivity following exposure to a bleaching light, measured with a test probe of 500 nm presented to the peripheral retina in the dark. Thresholds are plotted with respect to the prebleach dark-adapted threshold. The dashed and solid curves are exponential functions fit to the cone-mediated (upper) and rod-mediated (lower) portions of the dark adaptation data, respectively.

Although the rod portion of the dark adaptation function in **Figure 6** has been fit with an exponential function, it is more accurately represented by several regions with linear slopes on log-linear coordinates, with each region representing a different physiological process.

The relative vertical placements of the rod and cone dark adaptation curves and the time course of the recovery of sensitivity depend on a number of factors, including the retinal location of testing, the wavelength of the test probe, and the retinal illuminance of the bleaching light. For example, dark adaptation testing at the rod-free fovea reflects only the cone portion of the curve. Long-wavelength test probes, to which the rod system is relatively insensitive, typically produce a less-pronounced and delayed rod–cone break. Weak bleaching lights produce a faster time course of recovery than strong bleaches.

The psychophysical threshold not only is elevated following the offset of a bleaching light, but it is also elevated by the presence of an adapting field. This correspondence has led to the concept of the equivalent background, which holds that the aftereffect of a bleach is equivalent to the presence of a background of real light. This equivalent background, which is sometimes termed dark light, is generally not visible because it is stabilized on the retina. However, it can sometimes be observed in the form of an afterimage. Real light and dark light may have similar properties under some test conditions, but they are not always identical.

Although dark adaptation typically follows the time course illustrated in **Figure 6**, it is not always the case that sensitivity improves following the offset of a bleaching light. Under certain conditions, sensitivity actually decreases as dark adaptation progresses. One example occurs when the task is to determine the hue threshold or Lie specific threshold, which refers to the retinal illuminance of a test probe at which it appears to have a color. The hue threshold begins to rise at the time of the rod–cone break, rather than remaining constant once the cones have recovered. This rise in the threshold for hue has been attributed to an influence of the recovering rod system on color appearance.

A second example occurs when the task is to determine whether a rapidly flickering test stimulus, presented at a frequency above the rod temporal resolution limit, appears to flicker. The threshold for flicker detection does not remain constant once the cone system has recovered, but instead rises as the rods recover their sensitivity. The rise in the flicker threshold has been attributed to a suppressive effect of the dark-adapting rod system that surrounds the test stimulus on the temporal sensitivity of the cone system.

A threshold elevation during dark adaptation also occurs when the task is to detect a short-wavelength test probe following the offset of a long-wavelength adapting field. In this case, there is a substantial threshold elevation immediately following the adapting field offset, which is followed by a gradual decrease in threshold. The brief threshold increase, termed transient tritanopia, is thought to represent a change in sensitivity within a postreceptoral opponent color mechanism. Transient tritanopia can also be observed in the ERG, indicating that it is retinal in origin.

Mechanisms of Dark Adaptation

The recovery of sensitivity in the dark, following light exposure, depends ultimately on the regeneration of bleached photopigment. For example, individuals with vitamin A deficiency, which limits the regeneration of rod photopigment, typically have a prolonged time course of rod dark adaptation, and rod thresholds may never reach a normal level. Yet, the recovery of rod sensitivity depends on more than the regeneration of a light absorber. This is demonstrated by the fact that rod thresholds remain elevated by 2 to 4 log units at a time when 90% of rhodopsin has been regenerated following a bleach. It is likely that the presence of various bleaching intermediates, such as metarhodopsin products and free opsin, contribute to the rod threshold elevation during dark adaptation.

In addition to physiological processes occurring within photoreceptors, it is apparent that postreceptoral factors are involved in the recovery of sensitivity during dark adaptation. For example, a change in the size of the test probe can influence the shape of the dark-adaptation curve, although test probe size should have no influence on the rate of photopigment regeneration. Furthermore, dim lights that bleach a trivial amount of photopigment and that have no effect on the receptor potential or on horizontal cell responses in the skate retina, nevertheless, result in an elevation of ERG b-wave and ganglion cell thresholds that requires several minutes to recover. Thus, dark adaptation appears to involve mechanisms at multiple levels within the retina.

Human gene mutations are continuing to provide important insights into the physiological processes underlying dark adaptation. For example, mutations in either the rhodopsin kinase gene or the arrestin gene produce Oguchi disease, in which there is a prolonged recovery of rod sensitivity following light exposure that is thought to be due to impaired deactivation of rhodopsin. Mutations in the RDH5 gene, which encodes the enzyme 11-*cis* retinol dehydrogenase, result in fundus albipunctatus, which is characterized by extremely prolonged rod dark adaptation, presumably due to an impairment in the conversion of 11-*cis* retinol to 11-*cis* retinal.

Conclusion

Retinal adaptation refers to diverse visual phenomena and neural processes, all of which represent the adjustment of the visual system to the prevailing conditions of light stimulation. Through the use of psychophysical and electrophysiological techniques, great progress has been made in understanding the fundamental mechanisms underlying retinal adaptation, but many of the details have yet to be clarified. Studies of transgenic animal models and of naturally occurring human mutations, in which there are alterations in the putative components of adaptation, show great promise in furthering our knowledge of this fundamental visual process.

See also: Anatomically Separate Rod and Cone Signaling Pathways; Information Processing: Contrast Sensitivity; Perimetry; Photopic, Mesopic and Scotopic Vision and Changes in Visual Performance; Phototransduction: Adaptation in Cones; Phototransduction: Adaptation in Rods; Unique Specializations – Functional: Dynamic Range of Vision Systems.

Further Reading

Demb, J. B. (2008). Functional circuitry of visual adaptation in the retina. *Journal of Physiology* 586: 4377–4384.

Dowling, J. E. (1987). *The Retina: An Approachable Part of the Brain.* Cambridge: Belknap Press.

Dunn, F. A., Doan, T., Sampath, A. P., and Rieke, F. (2006). Controlling the gain of rod-mediated signals in the mammalian retina. *The Journal of Neuroscience* 26: 3959–3970.

Dunn, F. A., Lankheet, M. J., and Rieke, F. (2007). Light adaptation in cone vision involves switching between receptor and post-receptor sites. *Nature* 449: 603–606.

Graham, N. and Hood, D. C. (1992). Modeling the dynamics of light adaptation: The merging of two traditions. *Vision Research* 32: 1373–1393.

Hess, R. F., Sharpe, L. T., and Nordby, K. (eds.) (1990). *Night Vision: Basic, Clinical and Applied Aspects.* New York: Cambridge University Press.

Hood, D. C. (1998). Lower-level visual processing and models of light adaptation. *Annual Review of Psychology* 49: 503–535.

Kaplan, E., Lee, B. B., and Shapley, R. M. (1990). New views of primate retinal function. In Osborne, N. N. and Chader, G. T. (eds.) *Progress in Retinal Research*, vol. 9, pp. 273–336. Oxford: Pergamon Press.

Lamb, T. D. and Pugh, E. N., Jr. (2006). Phototransduction, dark adaptation, and rhodopsin regeneration: The Proctor lecture. *Investigative Ophthalmology and Visual Sciences* 47: 5138–5152.

Nishiguchi, K. M., Sandberg, M. A., Kooijman, A. C., et al. (2004). Defects in RGS9 or its anchor protein R9AP in patients with slow photoreceptor deactivation. *Nature* 427: 75–78.

Reeves, A. (2003). Visual adaptation. In Chalupa, L. M. and Werner, J. S. (eds.) *The Visual Neurosciences*, pp. 851–862. Cambridge: MIT Press.

Shapley, R. and Enroth-Cugell, C. (1984). Visual adaptation and retinal gain controls. In Osborne, N. N. and Chader, G. T. (eds.) *Progress in Retinal Research*, vol. 3, pp. 263–346. London: Pergamon.

Shevell, S. K. (ed.) (2003). *The Science of Color*, 2nd edn. Oxford: Elsevier.

Stockman, A. and Sharpe, L. T. (2006). Into the twilight zone: The complexities of mesopic vision and luminous efficiency. *Ophthalmic and Physiological Optics* 26: 225–239.

Relevant Websites

http://webvision.med.utah.edu – Webvision: Light and Dark Adaptation.

http://cvision.ucsd.edu – CVRL Color and Vision database.

Optic Nerve: Inherited Optic Neuropathies

A A Sadun and C F Chicani, University of Southern California-Keck School of Medicine, Los Angeles, CA, USA

Glossary

Cecocentral scotoma – Visual field defect involving the optic disk area (blind spot) and papillomacular bundle (PMB) fibers.

Cybrid cells – These are a eukaryotic cell line produced by the fusion of a whole cell with a cytoplasmic mitochondria.

Donder's curve – An expected reduction of lenticular accommodation as a function of age.

Dyschromatopsia – A partial or complete loss of color vision.

Genetic penetrance – The degree to which individuals express a genetically determined condition.

Papillomacular fibers – Axons from the smaller retinal ganglion cells that carry the information from the macula.

Searching nystagmus – A condition where both eyes occasionally make a wide, comparatively slow, sweeping movement caused by poor vision.

Tapetoretinal – The region of the photoreceptors and retinal pigment epithelium in the retina.

Inherited optic neuropathies fall under the rubric of metabolic optic neuropathies. Metabolic optic neuropathies share many pathophysiologic and clinical characteristics. Most metabolic optic neuropathies involve derangements that affect mitochondria and oxidative phosphorylation. Acquired metabolic optic neuropathies are further divided into those of toxic and those of nutritional deficiency states. For example, both ethambutol toxicity and vitamin B12 deficiency produce bilateral symmetrical optic neuropathies that are very similar to inherited optic neuropathies.

This article discusses hereditary optic nerve diseases that affect the optic nerve in isolation. Common hereditary optic neuropathies include Leber's hereditary optic neuropathy (LHON), dominant optic atrophy (DOA), and congenital recessive optic atrophy (ROA). Like most acquired optic neuropathies, inherited optic neuropathies involve mitochondrial function. In all cases, mitochondria are impaired either by mutations of their own DNA (mtDNA), or mutations of nuclear DNA involved in the transcription of mitochondrial proteins or substrates for mitochondrial biochemistry. All three have a similar presentation inherited mitochondria optic neuropathy.

This makes the diagnosis both challenging and, at the same time, the clinical evaluation similar. The presentation is usually that of bilateral symmetric visual loss with dyschromatopsia, and central or cecocentral visual field defects, which involves the optic disk and papillomacular fibers. This archetypical set of clinical signs reflects the fact that in metabolic optic neuropathies there is a strong predilection for the papillomacular bundle (PMB). The distinctions between acquired and inherited optic neuropathies usually come down to issues of personal history, such as toxic exposure, decreased vitamin intake or malabsorption, and, especially for inherited optic neuropathies, family history. However, because of variable penetrance, the family history in inherited optic neuropathies may not be always positive. Prompt recognition of the characteristic elements from history and physical examinations often precludes unnecessary patient laboratory evaluation.

Leber's Hereditary Optic Neuropathy

LHON was first described in 1871 by Theodore Leber. Later von Hippel, Gowers, and Collins refined our understanding and introduced the term hereditary optic atrophy. As recently as the 1980s, LHON was considered to be a non-Mendelian inherited genetic disorder, since there was no male-to-male transmission. In 1988, Douglas Wallace demonstrated LHON as the first maternally inherited disease to be associated with point mutations in mitochondrial DNA, and it is now considered the most prevalent mitochondrial disorder.

LHON typically manifests as a subacute central loss of vision that predominantly affects young adult males. Age of onset is usually between 15 and 35 years; however, it has been reported to occur as young as 2 and as old as 80 years of age. Almost invariably the second eye is affected, within weeks to months. LHON is usually due to one of three pathogenic mitochondrial DNA (mtDNA) point mutations. These mutations affect nucleotide positions 11778, 3460, and 14484, respectively, in the ND4, ND1, and ND6 subunit genes of complex I which is integral for oxidative phosphorylation in mitochondria. These three primary mutations are responsible for about 95% of LHON cases; other rarer mutations continue to be described. In some pedigrees of LHON, associated systemic features have been reported; these include cardiac abnormalities such as pre-excitation syndromes and hypertrophic cardiomyopathy, reflex and sensory changes, Charcot-Marie-Tooth disease, and skeletal disorders. It is hypothesized that the respiratory chain dysfunction leads to energy depletion and reactive oxygen species (ROS) accumulation which in turn produce axoplasmic stasis and swelling, thereby blocking ganglion cell function and causing loss of vision. In some patients, this loss of function is reversible in a substantial number of ganglion cells, but in others, a cell-death pathway, probably apoptotic, is activated with subsequent extensive degeneration of the retinal ganglion cell layer and optic nerve. The retinal ganglion cell degeneration and axonal loss occur predominantly in the PMB of the optic nerve.

Biochemical and cellular studies in LHON point to a partial defect of respiratory chain function that may generate either an ATP synthesis defect and/or a chronic increase of oxidative stress through the accumulation of ROS. Histological evidence of myelin pathology in LHON also suggests a role for oxidative stress, possibly affecting oligodendrocytes of the optic nerves. In cell culture studies, LHON cybrid cells, constructed from the merging of heteroplasmic cytoplasm of a cell which had its nucleus removed and the cytoplasm from a second normal cell that had all its mitochondria removed, are forced by the reduced rate of glycolytic flux to utilize oxidative metabolism by a shift from glucose to galactose in the media. This causes these cells to be sensitized to a mitochondrial-mediated apoptosis. It has also been proposed that in cells carrying LHON mutations, there is a decrease in antioxidant defenses. Recent evidence shows that mitochondrial distribution reflects the different energy requirements of the unmyelinated prelaminar axons in comparison to the myelinated retrolaminar axons.

In LHON, the pathologic mutation may be either homoplasmic (involving all the mitochondria) or heteroplasmic (involving only a fraction of the mitochondria). Most heteroplasmic pedigrees have much lower penetrance but surprisingly, the disease is not milder in form.

Even with homoplasmic families, penetrance is highly variable. The rate of penetrance varies with the mutation and pedigree, although it is always greater in males. Hence, in a typical family with 11778 mtDNA, 8–10% of the women and 40–50% of the men may suffer devastating and sudden visual loss in young adulthood. This marked incomplete penetrance and gender bias imply that additional mitochondrial and/or nuclear genetic or epigenetic factors must be modulating the phenotypic expression of LHON. It is also likely that environmental factors contribute to the onset of visual failure. For example, there is increasing evidence that tobacco and alcohol consumption plays a role. There is an active search for nuclear genes that might modify the penetrance.

In LHON, fundus changes, such as microangiopathy and nerve fiber layer swelling, have been described to immediately precede or accompany the onset of visual loss (**Figure 1**). This process, although usually bilateral, occurs asynchronously over the course of several weeks to months and eventually evolves to severe optic atrophy and irreversible impairment of vision. The smaller caliber fibers of the PMB are selectively lost at a very early stage of the pathological process, which eventually extends to most of the rest of the nerve leading to optic atrophy.

The acute stage of LHON usually lasts a few weeks. The affected eye characteristically demonstrates an early dropout of the PMB; an edematous appearance of the rest of the nerve fiber layer, especially in the arcuate bundles; and enlarged or telangiectatic and tortuous peripapillary vessels (microangiopathy). There is absence of leakage from the optic disk or peripapillary region on fluorescein angiography These main features are seen on fundus examination, just before or subsequent to the onset of visual loss. In optical coherence tomography (OCT), LHON-affected patients showed extensive thinning of the retinal nerve fiber layer (RNFL), as would be expected in cases of optic atrophy. LHON carriers sometimes show significant RNFL thickening of the arcuate

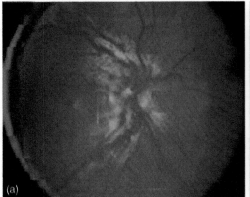 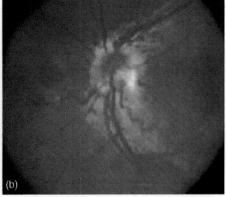

Figure 1 Right (a) and left (b) eyes of a patient with acute phase Leber's hereditary optic neuropathy (LHON). Fundus photographs of both eyes demonstrate hyperemic optic nerve heads, with dilated tortuous vessels, indistinct optic disk margins and swollen retinal nerve fiber layers (pseudoedema of the optic disk). Note also the telangectatic vessels.

bundles, in correlation with similar changes noted by funduscopy or the GDx nerve fiber analyzer, usually in the temporal sector, suggesting that the inferior temporal portion is affected first; this could be a subclinical sign. Several authors have described subclinical changes in the examination of asymptomatic carriers such as subtle optic disk findings, mild dyschromatopsia, OCT, and alterations in electrophysiology suggesting that LHON has some subtle chronic aspects.

In LHON-affected patients, the clinical examination reveals decreased visual acuity, dyschromatopsia, and cecocentral scotoma on visual field examination. There are a few reports of spontaneous recovery, especially with the 14484/ND6 mutation (up to 60% of cases) and in younger patients. Visual recovery may occur in one or both eyes and may happen as late as 10 years after the onset of visual loss. However, most often in LHON-affected individuals, the visual loss progresses and stabilizes within a year, with temporal optic atrophy.

There is currently no clinical evidence for an efficacious treatment to reverse vision loss in LHON. Theoretical considerations have led to the use of several agents involved with mitochondrial energy production and with anti-oxidant capabilities such as coenzyme Q10, succinate, L-carnitine, and vitamins K1, K3, C, B12, folate, and thiamine. Coenzyme Q10 is less likely to offer benefit since it is not transported to the mitochondria in sufficient concentration. A similar agent, Idebenone, has better drug delivery characteristics, and there have been a few encouraging reports and a prospective clinical trial with Idebenone in LHON. Case reports of successful recovery in LHON have to be considered in the light of spontaneous recovery. It remains unlikely that any of these agents alone or in combination will prove consistently useful in the treatment of acute visual loss in LHON or in the prophylactic therapy in asymptomatic family members at risk.

However, it is prudent to recommend the avoidance of agents that might induce oxidative stress or impair mitochondrial energy production. Most specifically, it is imperative to avoid exposure to smoke and alcohol as these environmental factors can trigger the loss of vision in susceptible individuals.

Dominant Optic Atrophy (DOA)

DOA, or Kjer's optic neuropathy, is one of the most common forms of hereditary optic atrophies, with estimated disease prevalence in the range of 1:10 000–1:50 000. Presentation usually occurs at latency age (7–10 years old). It often presents with imperceptible onset, a slowly progressive course, and leads to mild to moderate visual impairment (20/40–20/400). Insofar as onset is insidious and progression is slow, the young patient is often unaware of the visual loss. This usually comes to the parent's attention after a school-based visual screening.

The inheritance of DOA is autosomal dominant. Despite the variability in expression, the penetrance is actually very high. However, the vision impairment is often very mild; hence, the apparent absence of family history may not be accurate. We recommend the direct examination of the parents. DOA presents as mild, bilateral, sometimes asymmetric loss of visual acuity. On examination, there is a central, paracentral, or cecocentral visual field deficit; temporal optic disk pallor, often with a wedge shaped area of temporal excavation (**Figure 2**). There is mild generalized dyschromatopsia. In general, visual prognosis is good. However, patients with DOA often re-present around the age of 35 complaining of further visual loss. In fact, this represents premature presbyopia brought about by their lifelong habit of holding reading material much closer to their faces. Therefore, Donder's curve should not be applied to DOA patients. Instead, they should be offered plus lenses at about age 35 and graduated up to about plus 3.50 by age 50 to compensate for their closer near point.

In the year 2000 came the remarkable news that the genetic cause of dominant optic neuropathy had been identified. The gene OPA1 was, of course, nuclear and located on chromosome 3. However, the OPA1 protein encoded is imported to the mitochondria and serves structural roles in mitochondrial fission, fusion, and transport. Hence, though the genetics are somatic, the problem, like LHON, is in the mitochondria. Subsequently, other DOA genes have been found, as variations on the same theme. In addition to OPA1, there are OPA4 and other OPAs, which have been mapped to the 3q and 18q regions, respectively. All these genes are responsible for mitochondrial structural proteins.

Histological examination exhibited diffuse atrophy of the retinal ganglion cell layer, which is associated with atrophy and loss of myelin within the optic nerves. As in LHON, the retinal ganglion cells and axons lost are predominantly those of the PMB. However, the extent of axonal loss is significantly less in DOA.

In DOA, there is also an increased occurrence of associated sensorineural hearing loss so these patients should be advised to undergo audiology investigation. As in LHON, many agents have been tried as possible treatment options, and none were found to be effective. It remains unclear as to the role of environmental factors. These patients should be offered genetic counseling.

Recessive Optic Atrophy

This autosomal recessive somatic condition is the most uncommon form of inherited optic nerve disease.

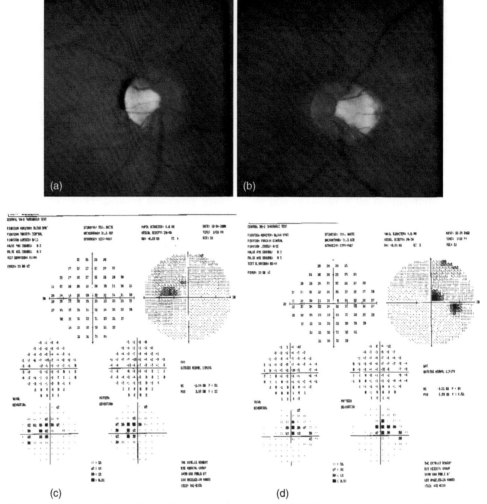

Figure 2 Right (a) and left (b) eyes in a patient with dominant optic atrophy (DOA): Fundus photographs show the characteristic wedge-shaped optic atrophy seen on the temporal side of both optic disks. 2(c) (Left) and 2(d) (Right): DOA. Humphrey Visual Fields demonstrate bilateral cecocentral visual field defects. Patient photographs are through the courtesies of Peter Quiros, MD.

Unlike LHON and DOA, ROA is usually discovered in the first 3–4 years of life. It often presents as severe visual impairment, frequently associated with searching nystagmus. Visual acuity ranges between no light perception (NLP) to 20/400. There is diffuse optic disk atrophy, sometimes with attenuation of the retinal arterioles, similar to that seen in tapetoretinal degenerations. Hence, the differential diagnosis of ROA is not so much a comparison to the other hereditary optic neuropathies as it is a comparison to retinal dystrophies. Electrophysiology using eletroretinography plays an important role in differentiating ROA from tapetoretinal degeneration, retinitis pigmentosa, or Leber's congenital amaurosis, this is normal in ROA and severely impaired in the retinal degenerations.

The gene/chromosome for ROA remains to be identified. Not surprising for an autosomal recessive condition, there are several reports of consanguinity related to ROA.

Other Inherited Conditions with Optic Atrophy

There are several well-characterized syndromes that also include optic atrophy. Wolfram's syndrome which is characterized by diabetes insipidus, diabetes mellitus, optic atrophy, and deafness is linked to the WFS1 gene located on chromosome 4p. Patients with Behr's syndrome, which consists of progressive encephalopathy, mental retardation, ataxia, nystagmus, and pes cavus, may also have optic atrophy. It is linked to the OPA3 gene located on chromosome 19q. Friederich's ataxia often includes vision loss and optic atrophy.

Conclusion

All of the inherited optic neuropathies serve to remind us that the optic nerve is highly dependent on mitochondrial function. Indeed, acquired optic neuropathies that affect mitochondrial metabolism are often in the differential diagnosis of genetic optic neuropathies. The genetics of LHON involve a mutation of mitochondrial DNA resulting in impairments of complex I, which in turn leads to decreased ATP production and increased ROS. This is in contradistinction to DOA in which the genetics involve nuclear DNA. However, despite the fact that these mutations are somatic, the OPA genes control structural mitochondrial proteins that also limit mitochondrial functions such that the final consequences are similar: decreased ATP production and increased ROS. Hence, inherited optic neuropathies and metabolic optic neuropathies, whether toxic or nutritional, all share the common pathophysiology of producing mitochondrial dysfunction and subsequent retinal ganglion cell loss. Not surprisingly, all mitochondrial optic neuropathies share a similar clinical presentation. If a treatment proves to be effective in one, it is likely to be of benefit in the other mitochondrial optic neuropathies.

Exciting new treatment options with agents that modulate mitochondrial function that have anti-apoptotic effects or are helpful for neuronal survival, are currently being studied. Inherited optic neuropathies may serve as the ideal model with which to test these purportive neuroprotective agents.

Further Reading

Barboni, P., Savini, G., Valentino, M. L., et al. (2005). Retinal nerve fiber layer evaluation by optical coherence tomography in Leber's hereditary optic neuropathy. *Ophthalmology* 112(1): 120–126.

Carelli, V., Ross-Cisneros, F. N., and Sadun, A. A. (2002). Optic nerve degeneration and mitochondrial dysfunction: Genetic and acquired optic neuropathies. *Neurochemistry International* 40(6): 573–584.

Carelli, V., Ross-Cisneros, F., and Sadun, A. (2004). Mitochondrial dysfunction as a cause of optic neuropathies. *Progress in Retinal and Eye Research* 23: 53–89.

Chalmers, R. M. and Schapira, A. H. V. (1999). Clinical, biochemical and molecular genetic features of Leber's hereditary optic neuropathy. *Biochimica et Biophysica Acta* 1410: 147–158.

Newman, N. J. (1998). Hereditary optic neuropathies. In Miller, N. R. and Newman, N. J. (eds.) *Walsh and Hoyt's Clinical Neuro-Ophthalmology*, pp 742–756. Baltimore: Williams and Wilkins.

Riordan-Eva, P., Sanders, M. D., Govan, G. G., et al. (1995). The clinical features of Leber's hereditary optic neuropathy defined by the presence of a pathogenic mitochondrial DNA mutation. *Brain* 118(2): 319–337.

Sadun, A. A. (2002). Metabolic optic neuropathies. *Seminars in Ophthalmology* 17(1): 29–32.

Sadun, A. A., Carelli, V., Salomao, S. R., et al. (2003). Extensive investigation of large Brazilian pedigree of Italian ancestry (SOA-BR) with 117788/Haplogroup J Leber's hereditary optic neuropathy (LHON). *American Journal of Ophthalmology* 136: 231–238.

Sadun, A., Salomao, S. R., Berezovsky, A., et al. (2006). Subclinical carriers and conversions in Leber's hereditary optic neuropathy: A prospective psychophysical study. *Transactions of the American Ophthalmological Society* 104: 51–61.

Sanchez, R. N., Smith, A. J., Carelli, V., et al. (2006). Leber's hereditary optic neuropathy possibly triggered by exposure to tire fire. *Journal of Neuroophthalmology* 26(4): 268–272.

Smith, J. L., Hoyt, W. F., and Susac, J. O. (1973). Ocular fundus in acute Leber optic neuropathy. *Archives of Ophthalmology* 90(5): 349–354.

Nikoskelainen, E. K. (1994). Clinical picture of LHON. *Clinical Neuroscience* 2: 115–120.

Wallace, D. C., Singh, G., Lott, M. T., et al. (1988). Mitochondrial DNA mutation associated with Leber's hereditary optic neuropathy. *Science* 242: 1427–1430.

Injury and Repair: Light Damage

N A Mandal, R E Anderson, and J D Ash, University of Oklahoma Health Sciences Center, Oklahoma City, OK, USA

Glossary

Acute light damage – Single exposure of intense light (1000–13 000 lux) for a brief period (20 min–24 h), which causes photoreceptor cell death, mainly in the central retina.

Apoptosis – A form of programmed cell death in multicellular organisms, which involves a series of biochemical events leading to a characteristic cell morphology and death. Death and the disposal of cellular debris do not damage the neighboring cells. Light-induced photoreceptor cell death occurs by apoptosis.

Chronic light damage – Exposure to sublethal doses of light (200–800 lux) in a cyclic manner (day–night) for a longer period of time (7–12 weeks). This causes overall thinning of the photoreceptor layer in the entire retina.

Inflammation – The complex biological response of vascular tissues to harmful stimuli such as pathogens, damaged cells, or irritants. It is a protective response by the organism to remove the injurious stimuli as well as initiate the healing process for the tissue.

Photoisomerization – The event in which a light-sensitive molecule absorbs a photon of light energy and undergoes a chemical change. In the current context, the 11-*cis*-retinal chromophore of rhodopsin is isomerized to all-*trans*-retinal, leading to activation of rhodopsin (R*) and initiation of the visual process.

Photostasis – Retinal adaptive response to different levels of cyclic illuminances. In bright cyclic light, rod outer segment (ROS) length is shortened, concentration of rhodopsin per unit of ROS is decreased, and ROS membranes become disorganized, compared to animals raised in dim cyclic light. This remarkable plasticity allows the retina of animals raised in dim and bright cyclic light to catch an equivalent number of photons each day.

Introduction

In 1966, Werner Noell discovered that the exposure of albino rats to bright visible light led to a rapid, specific, and irreversible loss of rod photoreceptor cells. Over the next four decades, this light-damage model has been used extensively to study mechanisms of retinal cell death and to test the efficacy of a variety of putative neuroprotective compounds. In his pioneering work, Noell discovered that photo-bleaching of rhodopsin was absolutely essential for light damage; rats raised on vitamin-A-deficient diets were not sensitive to light damage and the susceptibility to light damage followed the absorption spectrum of the visual pigment rhodopsin. Although the mechanism for initiation of the cell death signal in light damage is not known, recent studies have shown that while activation of rhodopsin is required, the visual transduction pathway downstream is not. Several outstanding reviews have been written on various aspects of light damage and therapeutic interventions. This article focuses on recent advances in identifying potential molecular mechanisms of cell injury and death, as well as the discovery of two independent mechanisms of endogenous self protection which are activated by light and stress in retinal photoreceptors.

There are several major advantages in using light damage to study retinal degeneration. Most inherited retinal degenerations occur over a long period of time with only a few photoreceptors dying each day. This makes it extremely difficult to biochemically study mechanisms of cell death or to study the effectiveness of therapeutic interventions. Light damage, on the other hand, has the advantage to more or less induce cell death in whole populations of photoreceptors at once. In addition, light damage can be tuned in terms of severity by varying the intensity of the light and duration of exposure. Hence, the intensity and duration required to cause a certain level of damage are correlated and a longer exposure can substitute for a higher intensity.

Over the years, two models of light damage have emerged: acute (short bright exposure) and chronic (less bright cyclic exposure). Both lead to retinal degeneration, but have quite different phenotypes. In the acute model, animals placed in light intensities ranging from 1000 to 13 000 lux in a time frame of 20 min to 24 h undergo a rapid and specific loss of rod photoreceptors. An example of acute light damage in a Sprague Dawley (SD) rat is shown in **Figure 1**. There is a clear demarcation of photoreceptor cell death, with normal-appearing retina (yellow arrow) adjacent to an area of moderate-to-massive cell death (white arrows). It is interesting that the subretinal space underlying the normal retina (yellow arrow) has been infiltrated with immune cells, suggesting that an inflammatory process is occurring prior to photoreceptor cell death. The role of inflammation in light damage is discussed subsequently in this article. A so-called spider graph of the thickness of the outer nuclear layer (ONL, rod nuclei) measured at defined distances from the optic nerve head to the inferior and superior ora serrata along the vertical meridian clearly demonstrates the specific loss of photoreceptor cells in the central superior region of the retina (**Figure 2**). The chronic light damage model involves raising albino rodents in sublethal levels of cyclic light (200–800 lux, depending upon the strain). Retinas of SD rats raised from birth to 7 weeks of age in 700-lux cyclic light show remarkably normal retinal morphology except for a thinning of the ONL (**Figure 3**). The loss of

photoreceptor cells is quite different from that which occurs in acute light damage, showing no regional differences (**Figure 4**). An interesting feature of the chronic light damage model is that, although the retina slowly loses photoreceptors, it is protected against the massive loss that occurs when these animals are placed in acute bright light. This point is discussed in detail later on in this article. These two models thus provide a means of testing putative neuroprotective compounds in two experimental paradigms. They are also useful to identify the molecular mechanisms that are responsible for light-induced damage, although the initial sequence of events leading to apoptosis may differ significantly from one protocol to another.

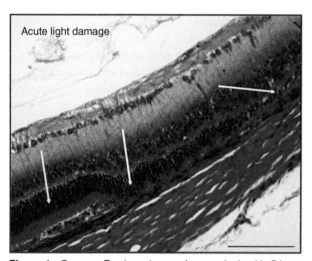

Figure 1 Sprague-Dawley rats were born and raised in 5 lux cyclic light. At 6–7 weeks of age, they were exposed to 2700-lux light for 6 h, after which they were returned to their cyclic light environment for 7 days. The white arrows point to areas of moderate to severe degeneration adjacent to an area of normal retina (yellow arrow).

Role of Rhodopsin Activation in Light Damage

While the precise mechanisms for cell damage are not fully understood, several genes have been identified that regulate photoreceptor sensitivity to light damage. When first identified, the action spectrum of light damage overlapped with the absorption spectrum of rhodopsin, suggesting a role for photoisomerization and phototransduction in cell injury. Indeed, a single, strong rhodopsin bleaching event is sufficient to induce light damage. The role of photoisomerization of 11-*cis*-retinal and rhodopsin activation as a requirement for light damage has been demonstrated pharmacologically and genetically. Pharmacological inhibition of rhodopsin regeneration with either the anesthetic halothane or by retinoic acid analogs, isotretinoin or 13-*cis*-retinoic acid, renders retinal photoreceptors less susceptible to light damage. Therefore, the severity of damage depends on the regeneration of rhodopsin during light exposure.

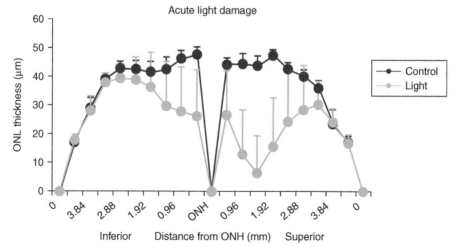

Figure 2 Sprague-Dawley rats were born and raised in 5-lux cyclic light. At 7 weeks of age, they were exposed to 2700-lux light for 6 h, after which they were returned to their cyclic light environment for 7 days. Plotted is the thickness of the outer nuclear layer (ONL) against distance from the optic nerve head (ONH) along the vertical meridian.

This sensitivity may be affected by polymorphisms or disease-causing mutations in opsin. The P23H opsin mutation is known to cause retinitis pigmentosa (RP) in humans. Work by Muna Naash and Daniel Organisciak, using transgenic mice or rats expressing human rod opsin with the P23H mutation, has shown that photoreceptors from these animals degenerate faster in bright cyclic light. Several recent studies have shown that mice with mutations in opsin, and phototransduction termination proteins arrestin and rhodopsin kinase are more sensitive to light damage. These results suggest that a prolonged R* or activated rhodopsin state is responsible for sensitivity to light damage. In more recent years, the Zurich group of Charlotte Reme, Andreas Wenzel, and Christian Grimm confirmed the

necessity for rhodopsin photobleaching by demonstrating that mice lacking the RPE65 gene, which encodes a protein necessary for photoisomerization of all-*trans*-retinyl esters, were protected from light-induced cell death. The same group has also shown that a slow rate of rhodopsin regeneration due to an L450M variation in the RPE65 protein also renders mice less susceptible to light damage. However, while activation of rhodopsin is required for acute light damage, activation of phototransduction through guanine nucleotide-binding protein α_t ($G\alpha_t$ or transducin) is not. This was demonstrated in an elegant study by Wenshan Hao using combinations of rhodopsin kinase, arrestin, and transducin knockout mice. The study clearly suggests that a transducin-independent but, rhodopsin-dependent, signaling is required for acute light damage.

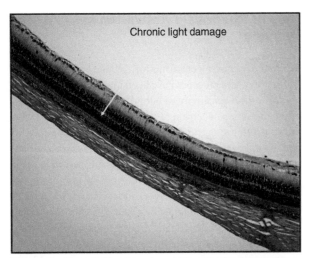

Figure 3 Sprague-Dawley rats were born and raised under 700-lux cyclic light until they were 7 weeks old. The retinas appear normal except for the thinning of the outer nuclear layer (ONL; yellow arrow). ONH, optic nerve head.

Mechanisms of Photic Injury: From Gene Expression to the Molecular Pathway

Gene expression analyses by DNA microarray studies have uncovered several possible molecular mechanisms of cell death in light damage. Our primary focus has been acute light damage (2700–3000 lux for 6 h) in albino mice or rats. This exposure paradigm induces oxidative stress that is sufficient to cause loss (>80%) of the photoreceptor cells in the central retina with no recoverable electroretinogram (ERG). However, cell death does not occur immediately following light exposure. It should be noted that by isolating RNA immediately after light exposure one can assess the stress response of photoreceptors before the onset of cell loss and detectable apoptotic cells in the retina. Therefore, by determining this early response one can decipher the molecular pathways that might regulate light-induced cell death. Retinas harvested

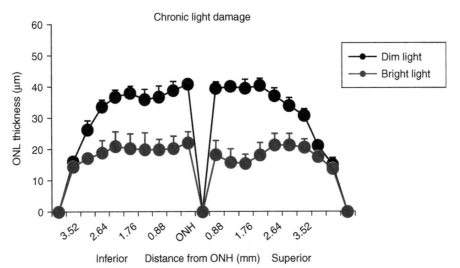

Figure 4 Sprague-Dawley rats were born and raised under 5 lux or 700 lux cyclic light until they were 7 weeks old. Plotted is the thickness of the outer nuclear layer (ONL) against distance from the optic nerve head along the vertical meridian.

immediately after the light exposure were used to generate complementary DNA (cDNA) to expressed messenger RNA (mRNA). The expression of the genes is studied by high-throughput DNA microarray analysis followed by quantitative (real-time) RT-PCR methods (RT-PCR, reverse transcriptase-polymerase chain reaction).

Apoptosis Genes

It has been well documented that apoptosis is the mechanism of photoreceptor cell death in light damage. Light damage induces expression of members of the activator protein 1 (AP-1) transcription factor family: c-fos, FosL1 (fra-1), junB, and c-Jun. The dimeric AP-1 transcription factor is formed by either c-fos or fra-1 binding to Jun proteins. The bright-light-induced apoptosis in retina is dependent upon activation of c-fos, and this c-fos upregulation plays a critical proapoptotic role, as evidenced from c-fos knockout mice, which showed marked resistance to light damage. Further, activation of the glucocorticoid receptor, which inhibits AP-1, also protects against light damage.

There are indications of involvement of other apoptosis genes in light damage. We have observed marked upregulation in the expression of tumor necrosis factor (TNF) receptor superfamily, member 1A (Tnfrsf1a), growth arrest and DNA-damage-inducible, beta (Gadd45b), and pleckstrin homology-like domain, family A, member 1 (Phlda1) genes. The protein encoded by Tnfrsf1a is one of the major receptors for the TNF-alpha (TNFα). This receptor can activate nuclear factor kappa-light-chain-enhancer of activated B cells (NFκB), mediate apoptosis, and function as a regulator of inflammation. In addition, the anti-apoptotic protein BCL2-associated athanogene 4 (BAG4/SODD) and adaptor proteins TRADD and TRAF2 have been shown to interact with this receptor. GADD45b can respond to environmental stresses by mediating activation of the p38/JNK pathway, which is involved in the regulation of growth and apoptosis. Phlda1 encodes an evolutionarily conserved proline–histidine-rich nuclear protein, which may play an important role in the anti-apoptotic effects of insulin-like growth factor-1. In summary, we observed a significant upregulation of many proapoptotic genes, which are mainly involved in the AP-1-mediated apoptotic pathway and did not detect any changes in the caspase-related genes in the initial phase of degeneration.

Role of Oxidant Stress in Light Damage

Oxidant stress has long been implicated in the pathogenesis of light damage. Intense light involves the generation of oxidants in the photoreceptors and the accumulation of oxidatively modified lipids, nucleic acids, and proteins. Oxidant stress is further supported by several reports describing protection from light damage induced by a variety of antioxidants, including ascorbate, dimethylthiourea, thioredoxin, and NG-nitro-L-arginine-methyl ester (L-NAME). We observed that pretreatment with the free radical trap phenyl-N-tert-butylnitrone (PBN) completely abrogates light-induced degeneration of the retina. The PBN protection not only involves the stabilization of the free radicals, but also suppresses the expression of many proinflammatory and apoptotic genes. Gene expression analysis has shown that light damage upregulates the expression of antioxidant genes heme oxygenase 1 (Ho-1), superoxide dismutase (Sod), thioredoxin, glutathione peroxidase, ceruloplasmin (Cp), and metallothioneins (Mt)-1 and-2. Mice expressing mutant SOD1 are highly susceptible to light damage and have accelerated age-dependent degeneration of the retina. CP is a ferroxidase that functions as an antioxidant by oxidizing iron from its ferrous to ferric form. Iron-derived hydroxyl radicals produced by the Fenton reaction may be important mediators of retinal photic injury as systemic administration of the iron chelating reagent desferrioxamine attenuates light-induced damage in rat retinas. Metallothioneins (MTs), another group of antioxidants, are copper- and zinc-binding proteins which can quench superoxide and hydroxyl radicals. Expression of oxidant defense genes further supports the hypothesis that light-induced retinal degeneration involves oxidative stress.

Role of Inflammation in Light Damage

We have noticed recurrent appearance of a host of inflammatory genes in the group of upregulated genes in many light-damage studies. In fact, invading immune cells are often observed in the retina near dying photoreceptors. These can be seen under the relatively normal photoreceptors (yellow arrow, **Figure 1**) adjacent to the injured photoreceptors (white arrows, **Figure 1**). While the connection between inflammation and light damage is not well known, two recent studies have suggested there is a strong link. Work from Barbel Rohrer has shown that genes related to complement activation are upregulated in response to light stress and that elimination of complement factor D reduces susceptibility to light damage. Work from Ying-qin Ni has shown that light stress leads to the upregulation of interlukin 1β (IL-β) and activation of resident microglia. Significantly, the study found that inhibiting microglia activation with naloxone protected photoreceptors from subsequent light damage. These studies suggest that an inflammatory response plays an active role in promoting photoreceptor death in light damage. Inflammation is also becoming recognized as a key factor in many retinopathies, including the epidemic forms of diabetic retinopathies and age-related macular degeneration (AMD). Therefore, a thorough investigation into the inflammatory process in the retina is essential for an understanding of the mechanism of light damage as well as other forms of retinal degeneration.

Light damage rodent models are slowly being recognized as models for retinal inflammation and are suitable models for human AMD. Here we summarize our observation of inflammatory gene expression changes. Photic injury upregulates the expression of many chemokine genes, Ccls (*Ccl2*, *Ccl3*, *Ccl4*, and *Ccl7*) and Cxcls (*Cxcl1*, *Cxcl11*, *Cxcl10*, and *Cxcl9*). Chemokines are a group of small (8–14 kDa), mostly basic, structurally related molecules that regulate cell trafficking of various types of leukocytes through interactions with a subset of guanine nucleotide-binding protein (G-protein-coupled) receptors. These molecules are divided into two major subfamilies, CXC and CC, based on the arrangement of the first two of four conserved cysteine residues; the two cysteines are separated by a single amino acid in CXC chemokines and are adjacent in CC chemokines. Chemokines also play fundamental roles in the development, homeostasis, and function of the immune system, and they have effects on cells of the central nervous system as well as on endothelial cells involved in angiogenesis or angiostasis. *Ccl2* or monocyte chemoattractant protein-1 (*Mcp-1*) upregulation is very robust and very early in the process of acute light damage. They may signal injury and recruit choroidal macrophages to scavenge retinal debris. Homozygous deletion of Ccl2 resulted in a mouse phenotype reported to be similar to human AMD. CCL3, also known as macrophage inflammatory protein-1, or monokine, is involved in the acute inflammatory state in the recruitment and activation of polymorphonuclear leukocytes. CCL4, a cytokine that is upregulated during the inflammatory response, is involved in the recruitment of neutrophils. CCL7 or monocyte chemotactic protein 3, a secreted chemokine, attracts macrophages during inflammation and metastasis. In addition, CXCL10 binds to its receptor CXCR3 and results in pleiotropic effects, including stimulation of monocytes, migration of natural killer and T-cells, and modulation of adhesion molecule expression. Upregulation of CXCL11 indicates probable activation of interferon gamma, which is a potent inducer of CXCL11 transcription.

Besides chemokines, we found significant upregulation of genes, which are either members of classic inflammatory proteins or involved in regulation of cellular inflammation. Expression of intercellular adhesion molecule 1 (*Icam1*), CCAAT/enhancer-binding protein (*Cebpb*/C/EBP, beta), cytokine-cardiotrophin-like cytokine factor 1 (*Clcf1*), lipopolysaccharide-induced TNF factor (*Litaf*), cyclooxygenase 2 (*Cox2*), ring finger protein 125 (*Rnf125*), and *Cd44* genes were significantly upregulated in acute light stress. The protein encoded by the *Cebpb* gene is a basic-leucine zipper (bZIP) transcription factor, which can bind as a homodimer to certain DNA regulatory regions of genes involved in immune and inflammatory responses and has been shown to bind to the IL-1 response element in the IL-6 gene, as well as to regulatory regions of several acute-phase and cytokine genes. The CLCF1 protein belongs to the IL-6 family of cytokines, which are involved in cell signaling through phosphorylation of gp130. Lipopolysaccharide is a potent stimulator of monocytes and macrophages, causing secretion of TNFα and other inflammatory mediators. COX2 is the key enzyme in prostaglandin biosynthesis and acts both as a dioxygenase and as a peroxidase and is involved in inflammation and mitogenesis. The *Rfn125* gene encodes a novel E3 ubiquitin ligase that contains an N-terminal RING finger domain, which may function as a positive regulator in the T-cell receptor signaling pathway. This is a small list of the inflammatory genes (~30% of all the upregulated genes) that are induced in acute light damage and clearly indicates a role in the pathogenesis of light damage.

Tissue Remodeling

Inflammatory signaling events are usually followed by tissue remodeling for the invasion of the macrophages and other cells, which then leads to the process of advanced cell death and removal of debris. We observed significant upregulation of matrix metallopeptidase 3 (*Mmp3*), tissue inhibitor of metallopeptidase 1 (*Timp1*), growth differentiation factor 15 (*Gdf15*), and plasminogen activator, tissue (*Plat*) genes. MMP3 is an endopeptidase that degrades extracellular matrix proteins and TIMP1 acts as an inhibitor of metalloprotease activity; PLAT is an enzyme that also plays a role in cell migration and tissue remodeling.

Transcription Factors

Many of the genes described under apoptosis, oxidative stress, and inflammation are transcription factors. However, there are other transcription factors upregulated in retinal light damage, which are involved in neuronal injury and various processes in cell death and survival. These include early growth response 1 (*Egr1*), zinc finger protein 36 (*Zif36*), activating transcription factor 3 (*Atf3*), and stress-signal transducer and activator of transcription 3 (*Stat3*).

Identification of Two Nonredundant Mechanisms of Endogenous Protection of Photoreceptors

Several recent studies have been designed to reveal pathways that protect photoreceptors from light damage. These studies have identified two independent protective pathways. One pathway is rapidly activated by phosphorylation following light stimulation, while the second pathway requires induction of new gene expression in response to chronic light stress.

Rhodopsin-Activated Endogenous Protection

Work from Raju Rajala has demonstrated that rhodopsin activation by light results in ligand-independent activation of the insulin receptor, which leads to activation of downstream signaling including increased PI3K and Akt kinase activity in photoreceptors. This activation was subsequently shown to be transducin independent. Genetic inactivation of the insulin receptor, insulin receptor substrate 2 (IRS2), AKT2, or BCL-XL have all led to increased photoreceptor susceptibility to acute light damage. These studies suggest that rhodopsin signaling, independent of transducin, is responsible for the activation of a defense mechanism through the insulin receptor and AKT2. Insulin receptor regulation of AKT2 is an ideal protective pathway for acute changes in light intensity since the entire pathway can be activated quickly through a series of phosphorylation events and does not require new gene expression. However, the AKT2 defense mechanism is overwhelmed as the duration of light exposure or intensity of light begins to induce cell death. It seems that the retina requires a secondary system of protection from chronic stresses such as prolonged bright light exposure and inherited genetic mutations.

Identification of Mechanisms for Chronic Light Stress-Induced Endogenous Protection

Noell's observation that animals raised in bright cyclic light were less susceptible to a subsequent light challenge than those raised in dim cyclic light or darkness inspired the discovery by John Penn and Ted Williams in the 1980s that albino rats born and raised in bright (but sublethal) cyclic light were protected from acute light damage. They discovered an enormous plasticity of the retina. Animals raised in dim cyclic light enhanced their chances of photon capture by lengthening their outer segments and increasing the packing density of rhodopsin in rod outer segment disk membranes. On the other hand, animals raised in relatively bright cyclic light reduced the length of outer segments, which also became somewhat disorganized, and decreased the packing density of rhodopsin in the disk membrane. The net result was to reduce the efficiency of photon capture by rhodopsin. Penn and Williams coined the term photostasis to describe the phenomenon of biochemical and morphological adaptation of the retina to modify efficiency of photon capture in animals exposed to different levels of cyclic light. It was suggested that these changes allowed photoreceptors to capture an equivalent number of photons each day regardless of their light environment. The morphological

changes associated with bright cyclic rearing were also accompanied by biochemical changes described by Penn and Anderson that included increased activity of glutathione enzymes (peroxidase, S-transferase, and reductase), elevation of retinal vitamins E and C, and decreased levels of polyunsaturated fatty acids (substrates for lipid peroxidation). These studies suggest that in response to chronic light stress, photoreceptors undergo photostasis to reduce activation of rhodopsin and induce an antioxidant defense. Similar findings have been shown in both mice and rats. Importantly, these stress-induced molecular and morphological adaptations were shown to protect photoreceptors almost completely from acute light damage.

The mechanism by which retinas of rodents raised in bright cyclic light are protected from acute light damage is an important area of research and significant progress has been made toward identifying the factors and receptors that are required, as well as identifying their relevant signal transduction pathways. Since the insulin receptor/PI3K/Akt2 pathway is involved in providing protection from acute light damage, this pathway was considered a likely candidate for chronic light stress-induced protection. However, recent findings suggest that chronic light stress-induced protection is independent of this pathway. Unlike the insulin receptor and AKT2-dependent protection, induced protection requires several days of preconditioning, suggesting that new gene expression is required. This indirectly suggests a different mechanism. In support of this, we have shown that induced protection is independent of AKT phosphorylation, and we have shown that *Akt2* knock-out mice have induced protection that is identical to wild-type mice. The insulin receptor and AKT2 appear to function as a rapid response to an acute injury, but is not necessary or perhaps is unable to protect from chronic injury. These studies demonstrate that there are two mechanisms of endogenous protection, one for acute injury that utilizes the insulin receptor and AKT2, and one for induced protection from chronic injury.

Role of Leukemia Inhibitory Factor

Several groups, including Steinberg, LaVail, Wen, and Stone, have shown that light preconditioning in rats induces the expression of several factors including basic fibroblast growth factor (FGF) (FGF2) and ciliary neurotrophic factor (CNTF). More recently in mice, the Grimm and Ash laboratories have shown that preconditioning induces the expression of leukemia inhibitory factor (LIF), cardiotrophin-like ligand (CLC), onchostatin M (OSM), FGF2, endothelin 2 (End2), and brain-derived neurotrophic factor (BDNF). While many factors are upregulated, it was not known which were responsible for induced protection. Early work from Steinberg and LaVail demonstrated that the intravitreal injection of FGF2, CNTF, LIF, or BDNF resulted in substantial protection of rat photoreceptors from a subsequent acute light damage independent of preconditioning. Additional work from Wen and Ash has similarly found that injection of LIF, CNTF, or BDNF protects mouse photoreceptors from a subsequent acute light damage. Of these, LIF and CNTF have tremendous therapeutic potential, given their ability to protect photoreceptors not only from acute light damage but also from inherited retinal degenerations.

Studies using intravitreal injections have demonstrated that multiple factors can induce protection, but did not demonstrate which factors were required for chronic light stress-induced protection. We have recently shown that LIF, CNTF, or CLC are the most likely candidates. This was demonstrated using an antagonist (LIF05) to the LIF receptor (LIFR), which blocks the activity of LIF, CNTF, and CLC. Intravitreal injection of LIF05 during light stress preconditioning greatly diminishes stress-induced protection from acute light damage. We also used conditional knock-out mice for the gp130 receptor and found that mice lacking gp130 expression in retinal photoreceptors also lose chronic light stress-induced protection. These recent studies demonstrate that the LIFR and gp130 expression in photoreceptors is essential for stress-induced protection and that the likely ligands are LIF, CLC, or CNTF, since these upregulated cytokines all utilize the LIFR and gp130. Our studies suggest that LIF may be the more important ligand since its expression is induced more than 100-fold following chronic light stress-induced preconditioning.

As described above, stress-induced protection was accompanied by upregulation of oxidant defense enzymes as well as photostasis. More recently, we have injected different doses of LIF into the vitreous of mice to determine whether LIF could induce both events. We found that at lower doses, LIF could induce protection from light damage without reducing photoreceptor sensitivity to light flashes. At higher concentrations, LIF not only induced protection, but also induced photostasis through decreased mRNA and protein expression of opsin, transducin (α and β subunits), and cyclic guanosine monophosphate (GMP) phosphodiesterase (PDE6A and PDE6B). Because of decreased expression of genes required for phototransduction, photoreceptors exhibited reduced efficiency in photon capture. These results demonstrate that LIF is upregulated by chronic bright-light stress and its induced expression is necessary for protection. LIF also has the ability to upregulate both photostasis and oxidant defense mechanisms. These studies clearly establish that in mice, LIF receptor, gp130, and perhaps LIF are essential players in chronic light stress-induced endogenous protection from acute light damage.

Many questions remain to be answered. In particular: Is LIF the essential factor? Which cells induce expression of LIF? Which signal transduction pathways are required for protection? What changes in gene expression are required for protection? In a series of studies, we have shown that intravitreal injection of LIF results in activation of ERK1/2 and STAT3. The pattern of activation changes over time. Within the first 30 min of injection, both phosphorylated ERK1/2 and STAT3 were detected only in Müller's cells and some ganglion cells. Within 4 h, however, all retinal cells were positive for STAT3, while ERK1/2 activation returned to basal levels. Detectable phosphorylated STAT3 remained up to 6 days following a single injection of LIF. Since phosphorylated STAT3 is a transcription factor, these results suggested that LIF could induce gene expression changes in all retinal cells including Müller's cells and photoreceptors.

Concluding Remarks

Light-damage models have led to major advances in our molecular understanding of retinal degeneration through oxidative injury mechanisms and, in the future, light-damage studies will be used to advance our understanding of injury induced by inflammation. Light-damage models have also led to the discovery of two independent mechanisms of endogenous neuroprotection, as well as to the discovery of promising new therapies to prevent or delay inherited retinal degenerations. Indeed, phase I clinical trials using encapsulated cells expressing CNTF have been completed with promising early results, and the study has since progressed into phase II trials. The protective effect of CNTF was initially described in the retina using the light-damage model. This alone should clearly establish the relevance for using light damage to identify potential therapeutic agents. By further defining the mechanisms of cell injury or protection, light-damage studies will likely lead to the development of new and more specific therapies.

See also: Injury and Repair: Retinal Remodeling; Phototransduction: Phototransduction in Rods; Phototransduction: Rhodopsin; Phototransduction: The Visual Cycle; Primary Photoreceptor Degenerations: Retinitis Pigmentosa; Primary Photoreceptor Degenerations: Terminology; Retinal Pigment Epithelium: Cytokine Modulation of Epithelial Physiology; Secondary Photoreceptor Degenerations: Age-Related Macular Degeneration.

Further Reading

Boulton, M., Rozanowska, M., and Rozanowski, B. (2001). Retinal photodamage. *Journal of Photochemistry and Photobiology* 64: 144–161.

Chen, L., Wu, W., Dentchev, T., et al. (2004). Light damage induced changes in mouse retinal gene expression. *Experimental Eye Research* 79: 239–247.

Marc, R. E., Jones, B. W., Watt, C. B., et al. (2008). Extreme retinal remodeling triggered by light damage: Implications for age related macular degeneration. *Molecular Vision* 14: 782–806.

Noell, W. K., Organisciak, D. T., Ando, H., Braniecki, M. A., and Durlin, C. (1987). Ascorbate and dietary protective mechanisms in retinal light damage of rats: Electrophysiological, histological and DNA measurements. *Progress in Clinical and Biological Research* 247: 469–483.

Penn, J. S. and Anderson, R. E. (1992). Effects of light history on the rat retina. In: Osborne, N. and Chader, G. (eds.) *Progress in Retinal Research,* vol. 11, pp. 75–98. New York: Pergamon Press.

Penn, J. S. and Thum, L. A. (1987). A comparison of the retinal effects of light damage and high illuminance light history. *Progress in Clinical and Biological Research* 247: 425–438.

Ranchon, I., Chen, S., Alvarez, K., and Anderson, R. E. (2001). Systemic administration of phenyl-*N*-tert-butylnitrone protects the retina from light damage. *Investigative Ophthalmology and Visual Science* 42: 1375–1379.

Reme, C. E., Grimm, C., Hafezi, F., Marti, A., and Wenzel, A. (1998). Apoptotic cell death in retinal degenerations. *Progress in Retinal and Eye Research* 17: 443–464.

Tanito, M., Agbaga, M. P., and Anderson, R. E. (2007). Upregulation of thioredoxin system via Nrf2-antioxidant responsive element pathway in adaptive-retinal neuroprotection *in vivo* and *in vitro*. *Free Radical Biology and Medicine* 42: 1838–1850.

Tanito, M., Kaidzu, S., Ohira, A., and Anderson, R. E. (2008). Topography of retinal damage in light-exposed albino rats. *Experimental Eye Research* 87: 292–295.

Wenzel, A., Grimm, C., Marti, A., et al. (2000). C-fos controls the ''private pathway'' of light-induced apoptosis of retinal photoreceptors. *Journal of Neuroscience* 20: 81–88.

Wenzel, A., Grimm, C., Samardzija, M., and Reme, C. E. (2005). Molecular mechanisms of light-induced photoreceptor apoptosis and neuroprotection for retinal degeneration. *Progress in Retinal and Eye Research* 24: 275–306.

Wu, J., Seregard, S., and Algvere, P. V. (2006). Photochemical damage of the retina. *Survey of Ophthalmology* 51: 461–481.

Injury and Repair: Neovascularization

M E Kleinman and J Ambati, University of Kentucky, Lexington, KY, USA

Glossary

Complement – A complex molecular cascade that directly removes pathogens and activates the host immune system.

Cytokines – The secreted proteins that mediate cellular communication through a specific set of cell-surface receptors.

Extracellular matrix – The extracellular material that supports cell structure and intercellular communication.

Laser-induced injury – A reproducible model of tissue injury with the application of a high-energy laser photocoagulation to the fundus.

Neovascularization – The formation of new blood vessels.

Vascular endothelial growth factor – A potent cytokine that is critical for the process of neovascularization.

Introduction

The retina is a highly organized multicellular system designed to efficiently convert light into electrical signals which can then be transmitted to the brain for cortical integration and, ultimately, visual perception. In an acute setting of retinal injury, the cellular components of the neural retina, retinal pigment epithelium (RPE), and choroid get activated in order to contain the wound, destroy any invading pathogens, and initiate the repair process. As a result of the subsequent proinflammatory surge, blood vessels invade the wound bed to revascularize injured and hypoxic tissue and provide a new source of nutrients to the remodeling tissues. Typically, as in several other peripheral tissues in the human body, such as skin, this vascular growth allows for tissue healing and scar remodeling; however, in the delicate architecture of the retina, these abnormal vessels grow in an unregulated fashion and are susceptible to leakage and rupture. These traits are hardly suitable for the retina's highly specialized function of vision and often lead to neovascularization (NV) as a cause of blindness in the context of retinal disease and injury. NV also contributes to other cellular responses in the injured retina, including gliosis and fibrosis. Through clinical observations and scientific investigation, critical insights into the

biology of NV in injury and repair in the retina, RPE, and choroid have modernized our current understanding of the cellular and molecular pathways that drive this pathological process.

Mechanisms of Injury

Retinal, RPE, and choroidal injury may be caused by a plethora of mechanical, cellular, physiologic, photochemical, and iatrogenic processes that are addressed in detail in other sections. Each of these instigating factors, either individually or in concert with one another, is capable of setting off the intricate cascade of factors leading to sub-retinal NV. A list of some of the common causes of injury-induced NV is provided next.

1. *Mechanical trauma.* Blunt trauma to the eye disorganizes the retinal architecture to varying degrees with displacement of reactive glial cells and RPE to the vitreo-retinal interface. These cells can form a fibrotic scar at the surface of the internal-limiting membrane (ILM), in the case of epiretinal membrane, and even grow into the vitreous as in the case of proliferative vitreo-retinopathy. More ominous is the presence of a choroidal rupture, a finding that is very likely to induce neovascular invasion of the retina through fractures in Bruch's membrane. Over time, these lesions eventually involute but do leave behind a residual fibrotic scar that significantly decreases visual acuity in the area.

2. *Age-related macular degeneration.* One of the most common forms of retinal injury is due to a combination of inflammation, oxidative stress, and photochemical damage secondary to age-related macular degeneration (AMD), a disease which accounts for the epidemic loss of vision in people over 60 years of age in the developed world. Between 10% and 20% of people with AMD will progress to the neovascular or wet form, which is responsible for about 90% of vision loss in patients with AMD. Recent evidence suggests that the gradual accumulation of inflammatory debris in drusen beneath the RPE and within the choroid ultimately leads to the production of proangiogenic cytokines and retinal invasion by abnormal and leaky vasculature. Some of these inflammatory components are comprised of photochemically altered proteins that are linked to intermediate or terminal products of the vitamin-A cycle or lipid-breakdown products released

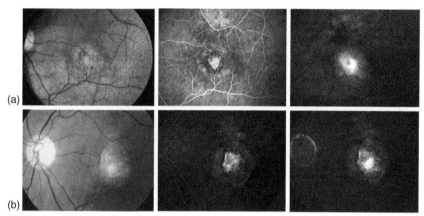

Figure 1 Retinal injury progression through inflammation, neovascularization, and fibrosis. (a) Left to right sequence of images representing CNV (far left, color fundus photography) after an inciting inflammatory event leading to leaky vasculature in the macula as evident on fluorescein angiography (middle, 45-s time point, far right, 10-min time point); (b) The same patient several years later showing subretinal fibrosis (far left) with fluorescein staining scar and RPE dropout (same sequence as above).

by dying photoreceptors and RPE. Neovascular AMD occurs either below the RPE in its occult form or invades the neural retina in its classic form. Choroidal NV (CNV) can wreak havoc through leakage into the intraretinal or vitreous spaces which acutely decreases visual acuity (**Figure 1**). Overtime, these NV membranes remodel into fibrotic scars that may still harbor abnormal vasculature and become sites of active NV recurrence.

3. *Diabetic retinopathy.* Diabetic retinopathy (DR) remains the primary cause of vision loss in patients aged less than 65 years. Hyperglycemia induces multiple microvascular abnormalities in the retina, eventually leading to retinal NV.

4. *Vaso-occlusive disease.* The retina receives dual blood supply with the central retinal artery feeding the inner layers and the choroidal vasculature feeding the outer layers, including the highly metabolic photoreceptor segments. Numerous pathologic processes can choke the retinal circulation, including atherosclerosis, vasculitis, thrombosis, and emboli. Retinal venous outflow is through the central retinal vein which is also at risk for collapse, stasis, or thrombosis. Choroidal venous flow exits through the vortex veins where thrombus may also form, albeit in a less-dramatic presentation than the central retinal vein occlusion. Another vascular disease that can trigger massive retinal ischemic injury is giant-cell arteritis, an inflammatory disorder that can occlude the central retinal artery. Ischemic damage to the retina is irreversible approximately 90–100 min after initial insult and can result in a widespread neovascular response in the retina, iris, and angle. These sequelae are the leading causes of enucleation after such catastrophic vascular events in the eye.

5. *Choroidal disease.* Histoplasma capsulatum is a fungal organism endemic to the Ohio and Mississippi river valleys that infects the choroidal tissue and causes localized areas of subretinal NV, which eventually scar over with RPE and glial cells. The organism is rarely isolated; however, patients often show a positive skin-allergy test thus earning this condition the title of presumed ocular histoplasmosis (POHS). Choroidopathies are a group of rare disorders that affect the tissues around the RPE due to either immune dysregulation or infectious disease, most often associated with nematode invasion. Although it is difficult to discern among the different forms of choroidopathy, several of them may lead to the formation of CNV. Multifocal choroiditis (MFC), multiple evanescent white-dot syndrome (MEWDS), bird-shot choroidopathy, serpiginous choroidopathy, and diffuse unilateral subacute neuroretinitis (DUSN) have all been associated with NV, hypothesized to be secondary to inflammation or immune dysregulation.

6. *Cryogenic injury.* Cryotherapy is a widely used treatment modality in vitreoretinal surgery to seal peripheral retinal tears in order to prevent rhegmatogenous retinal detachments. It is also used in combination with scleral buckling to destroy large areas of ischemic retina in order to prevent further NV. In some instances, cryotherapy itself can lead to subretinal NV.

7. *Laser-induced injury.* The use of laser photocoagulation for targeted retinal injury in the treatment of extrafoveal NV and proliferative DR has been widely used with success for several decades. In the treatment of CNV, focal laser injury is able to inhibit growth and decrease leakage from abnormal microvasculature. With panretinal photocoagulation, laser is used to ablate large swaths of the peripheral retina in DR that would otherwise be ischemic, producing proangiogenic factors. It is through the destruction of this pathologic tissue, that NV is spared in the precious visual real estate in the macula. Laser-injured tissues will commonly

remodel into small areas of RPE hyperpigmentation with fibrosis and only go on to develop CNV when high-energy laser photocoagulation is able to break through Bruch's membrane. It is critical to note that laser-induced injury which fractures Bruch's membrane serves as a very well-described animal model of CNV in many different laboratory animals, which is addressed in the following section.

Animal Models of NV after Laser-Induced Injury

In order to study the natural history of injury-induced NV *in vivo*, several experimental models have been developed to mimic the disease process which occurs in humans. Over 25 years ago, Stephen Ryan described a laser-induced injury model of subretinal NV in rhesus monkeys. The same technology has since been used to develop similar models in a variety of research animals, including the mouse. In this model, laser photocoagulation is used to fracture Bruch's membrane, which results in the formation of CNV (**Figure 2**). The laser-induced model captures many of the important features of the human condition including: migration of choroidal endothelial cells into the subretinal space via defects in Bruch's membrane (**Figure 3**), accumulation of subretinal fluid,

congregation of leukocytes adjacent to neovascular tufts, gliosis (**Figure 4**), leakage of fluorescein from immature new vessels into the subretinal space, and increased expression of angiogenic growth factors and their receptors in cells and ocular tissues. The cellular and molecular details of retinal injury and repair are discussed below.

Acute Responses to Retinal Injury

Blood–Retina Barrier Breakdown

In a normal uninjured retina, the blood–retina barrier (BRB), comprised of RPE and retinal endothelial cell tight junctions, serves to prevent the influx of circulating proinflammatory cells and proteins where their presence and actions would compromise vision; however, in animal models of high-energy laser injury, there is an immediate disruption in the neural retina, RPE tight junctions, and Bruch's membrane, leading to BRB breakdown.

Acute Release of Cytokines

With laser injury, thousands of retinal and RPE cells suffer instantaneous thermal damage or death and disperse their intracellular contents into the interstitium. Many pro- and antiangiogenic cytokines which are harbored in cytoplasmic granules are immediately released

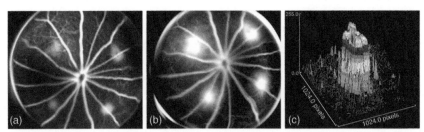

Figure 2 The mouse model of retina injury with laser photocoagulation. (a) An early time-point fundus fluorescein angiogram 7 days after laser injury of the retina, RPE, and Bruch's membrane showing hyperfluorescent hot spots. (b) Later time-point fluorescein angiogram of the same mouse eye exhibiting significant vascular leakage similar to the human form of CNV. (c) Vascular volumetrics and surface mapping reveal the size and shape of the CNV lesion in an important animal model that allows for scientific exploration into the molecular mediators of this process.

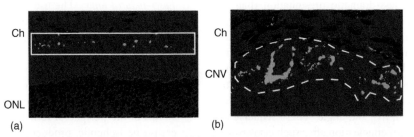

Figure 3 Choroidal vasculature before and after laser injury. (a) In the normal eye, the choroidal vasculature is organized into an extensive capillary network, the choriocapillaris, which exists in a single tissue plane below Bruch's membrane (white box encloses normal mouse choriocapillaris with vasculature appearing green). (b) After laser injury, the vasculature invades the retina where it grows into leaky neovascular membranes (dashed-line marks area of CNV formation 7 days after laser injury of the fundus).

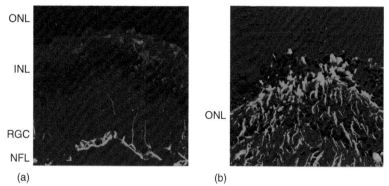

Figure 4 The participation of neural-retina derived cells to CNV pathogenesis. (a) Glial cells (retinal astrocytes shown here with glial-fibrillary acid protein (GFAP)) in green are located in well-demarcated areas of the un-injured neural retina. (b) Upon injury, glial cells migrate to areas of wound healing and proliferate along with endothelial cells while providing both structural and growth-factor support.

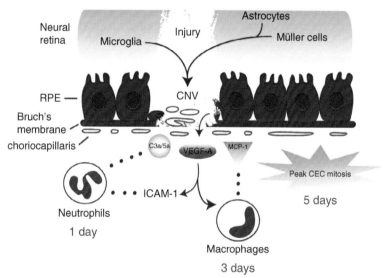

Figure 5 Critical mediators in CNV formation after acute and chronic injury. The initiating step of CNV is often an injury or disease process that induces local inflammation in the retina, RPE, and choroid. Locally produced leukocyte chemoattractants and provascular growth factors are secreted in the wound bed resulting in peak neutrophil infiltration at 1 day, maximal macrophage influx at 3 days, and a spike in choroidal endothelial cell proliferation at 5 days.

inducing the influx of circulating leukocytes into the retinal tissues (**Figure 5**). There is a paucity of evidence on which resident cells are the sources of these cytokines. RPE cells constitutively produce a wide array of cytokines that are usually secreted in a polarized fashion either toward choriocapillaris or the photoreceptor layer to maintain their relative functions. For example, pigment-epithelium-derived factor (PEDF), a potent antiangiogenic mediator, is secreted from the apical RPE surface toward the photoreceptor layer, while vascular endothelial growth factor-A (VEGF-A), a predominant proangiogenic cytokine, is secreted from the basal RPE toward the choriocapillaris. With injury, this directionality is lost allowing for the disinhibition of angiogenesis in normally avascular tissue planes.

VEGF-A belongs to the highly conserved platelet-derived growth factor (PDGF) family along with several other related proteins that constitute the VEGF subfamily, including placental growth factor (PlGF), VEGF-B, VEGF-C, and VEGF-D. For nomenclature purposes, the original VEGF protein was designated VEGF-A. Multiple isoforms of VEGF-A exist due to alternative splicing of the *Vegfa* gene; however, the 165-amino-acid variant in humans (VEGF-A$_{165}$) is the most potent and prevalent among the major isoforms expressed during pathologic NV. Many other important components within the VEGF family contribute to injury-related NV. Several genes important in VEGF regulation and activation, such as the coreceptors neuropilin-1 and -2 as well as PlGF, are acutely upregulated in the setting of laser injury and in some forms of human subretinal NV.

An acute elevation of VEGF-A released from various resident cells types, including RPE, microglia, Müller cells, and retinal astrocytes, transiently induces vascular permeability and upregulates endothelial cell expression of intercellular adhesion molecule (ICAM)-1. In response, there is a marked leukocytosis of injured tissues within 24 h. Resident cells also spew monocyte-chemoattractant protein (MCP)-1 into the wound matrix which acts as a homing device for circulating monocytes expressing its cognate receptor (CCR2). Interleukin-1β (IL-1β), tumor necrosis factor-α (TNF-α), and interleukin-6 (IL-6), all potent cytokines that sequester a wide range of inflammatory cells and induce their proliferation, are increased within 24 h of injury. IL-6 is instrumental in promoting continued expression of VEGF-A, ICAM-1, and MCP-1, which regulate the specific sequence of leukocyte chemotaxis that occurs over the initial 72 h after injury.

Injury-Induced Complement Activation

Complement factors are also a major participant in the robust and rapid retinal tissue response to injury. The complement pathways are comprised of multiple, complex, domino-like cascades engineered to generate massive proinflammatory induction through cytokine expression and leukocyte invasion in addition to promoting the self-assembly of molecular components that are capable of directly lysing bacteria and virus-infected cells. The two main active byproducts of complement activation, C3a and C5a, are predominantly responsible for the anaphylactic response. After laser injury, C3a and C5a levels are selectively and swiftly increased in the RPE and choroid within 4 h after photocoagulation with peak concentrations at 12 h. With blockade of C3a and C5a, as in mice that are deficient in C3a or C5a receptors, VEGF-A production is downregulated which may also inhibit leukocyte trafficking. Functional blockade of complement activation may be a valuable preventative therapy to arrest NV progression in patients with retinal disease and injury.

Other Angiogenic Mediators in the Retinal Response to Injury

Several other molecular factors have been implicated in the development of injury-induced NV through descriptive expression analyses in human specimens, including transforming growth factor-beta 1(TGF-β1), PDGF, fibroblast growth factor-2 (FGF-2), insulin-like growth factor-1 (IGF-1), and estrogen. Hepatocyte growth factor (HGF) is produced by RPE, upregulated as early as 6 h after laser injury, and hypothesized to mediate RPE proliferation during tissue repair.

Endogenous antiangiogenic molecules have also been identified such as PEDF and thrombospondin-1 that are decreased in CNV lesions and are capable of regulating NV responses in animal studies. These factors provide yet another level of control in this increasingly complex system of vascular regulation.

Modulation of the Extracellular Matrix

The biologic activities of VEGF-A and other cytokines, which acutely increased after injury, are influenced by the myriad of interactions between cells and the extracellular matrix (ECM). ECM proteins transmit cell–ECM communication and modulate tissue-remodeling events including NV. CNV membranes contain an abundance of ECM proteins similar to those found in granulation tissue during epithelial wound healing. Fibrin, fibronectin, and integrin receptors, all of which are important in ECM-guided angiogenesis, are present in surgically excised CNV membranes. The pathogenesis of injury-induced NV requires invasion of pre-existing ECM and the formation of new vascular conduits (**Figure 6**). Matrix metalloproteinases (MMP-2 and 9), which are capable of ECM dissolution, are upregulated in the acute-phase response. Tissue inhibitors of MMPs (TIMPs), which prevent excessive degradation of ECM by MMPs, are also expressed in CNV membranes thus offering a novel therapeutic target. Another proteinase mechanism, known to be involved in CNV progression, is the urokinase plasminogen activator pathway which is also being evaluated as an alternative treatment modality for pathologic NV.

More recently characterized ECM proteins may also be involved in VEGF-A signaling and NV response after injury. The matricellular protein SPARC (secreted protein, acidic, and rich in cysteine) is involved in tissue remodeling, cellular migration, and angiogenesis through its interaction with VEGF-A. After injury, SPARC expression is acutely decreased thus allowing VEGF-A to activate VEGFR-2 inducing endothelial cell proliferation and migration. In a paradoxical molecular mechanism, intravitreous administration of VEGF-A is able to suppress CNV when delivered after laser injury, whereas VEGF-A treatment prior to laser injury expectedly augments proangiogenic tissue response. These data may help improve our understanding of VEGF-A duality in NV induction and enhance the timing of administration of our targeted therapeutics to suppress this unwanted growth.

Cellular Response in Injury-Induced NV

Similar to epithelial wound healing, retinal injury is succeeded by an orchestrated arrival of various inflammatory cell types, a process that is determined by the acute response to injury discussed above. From other

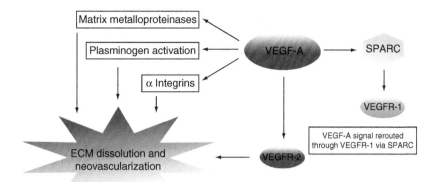

Figure 6 Modulation of the ECM with injury and VEGF-A upregulation. VEGF-A is a major factor in the pathogenesis of CNV through its direct effects on endothelial cell proliferation and modulation of the extracellular matrix. Integrins, plasminogen factors, and MMPs are all upregulated in response to VEGF-A contributing to ECM degradation that promotes neovascular invasion. VEGF-A is also able to curb the endothelial cell responses through its interaction with SPARC which reroutes signal transduction away from the proangiogenic pathways of VEGFR-2 to the nonproliferative cellular pathways of VEGFR-1.

wound-healing studies, it is known that inflammatory cell recruitment is directed by the expression of cytokines such as MCP-1, GROα (also known as CXCL1), and macrophage inflammatory protein-1α and-2 (MIP-1α, MIP-2). Many of these counterparts have been shown to be involved in NV progression in retinal injury.

Neutrophils

With breakdown of the BRB and acute VEGF-A-driven expression of ICAM-1 on the retinal and choroidal endothelia, there is an immediate and rampant neutrophil extravasation into the injured tissues which is maximal at 24 h. In addition to C5a, acute elevations in IL-1β and TNF-α may upregulate interleukin-8 (IL-8) which also acts as a potent neutrophil chemoattractant. These cells serve as a potent proinflammatory stimulus by secreting more VEGF-A along with numerous proangiogenic cytokines into the wound. Moreover, neutrophils are capable of engulfing invading pathogens and destroying them through the release of hypochlorous acid. Without neutrophil participation in wound healing, there is only a partial abrogation of the NV response to laser injury.

Macrophages

The next major influx of circulating proinflammatory leukocytes is the macrophage, which is driven by the enhanced expression of MCP-1 that peaks at 2 days after injury. These professional inflammatory cells home to sites of increased MCP-1 gradients through the expression of its cognate receptor, CCR2. This signaling axis is responsible for maximal macrophage infiltration into the choroid at 3 days post injury. Macrophages are able to respond to the proinflammatory milieu of vascular growth factors and secrete even more VEGF-A in order to aid revascularization

and tissue repair. After the arrival of macrophages, endothelial cells continue to increase their mitotic activity with peak proliferation at 5 days post injury. Approximately 1 week after injury, organized subretinal NV membranes are formed, which can be visualized *in vivo* with fluorescein angiography. Without macrophage influx, the ability of laser-injured areas to fully develop CNV is completely eliminated signifying the critical importance of this infiltrating cell in injury-induced NV.

Progenitor/Stem Cells

Pluripotent stem/progenitor cells are also recruited to sites of NV during injury and repair and incorporate into pathologic vasculature. Animal studies have now demonstrated that interfering with stem/progenitor cell homing to the sites of CNV through the modulation of the hypoxia responsive cytokine stromal-derived factor (SDF)-1α and its receptor, CXCR4, inhibits neovascular growth, a finding that offers another potential treatment for the human condition.

Other Infiltrating Cell Types

Although mast cells and eosinophils are able to generate significant proinflammatory stimuli, their contribution to injury-induced NV in the retina is believed to be negligible. B- and T-lymphocytes as well as natural killer cells are also able to infiltrate retina injuries, yet their particular effects and contribution to NV are still unclear but likely to be negligible.

Resident Tissue Cells

Among the resident cells of the retinal, RPE, and choroidal tissues that are involved in injury-induced NV, the effects of RPE cells, microglia, retinal astrocytes, and Müller cells are the most evident.

Microglia

Microglial cells are tissue macrophages dispersed throughout the normal retina, choroid, and central nervous system. Their primary functions are to respond to the local invasion of pathogens through activation of the innate immune system, promotion of inflammation, and phagocytosis of bacteria and virus-laden cells. Microglia express a unique set of surface markers, as well as the chemokine receptor CX3CR1 which binds to fractalkine (or CX3CL1), a cytokine that is secreted preferentially by inflamed retinal and endothelial cells. CX3CR1 is present on a number of differentiated cell types derived from myeloid progenitor cells, such as dendritic cells, infiltrating macrophages, neutrophils, and some endothelial cells, thus providing another signaling axis for the infiltration of professional inflammatory cells after laser injury.

As part of their immune-surveillance capabilities, microglia express a series of innate immune receptors, called toll-like receptors (TLRs), that recognize pathogen-associated molecular patterns and respond by alarming other immune system components, mediating cellular infiltration, and selectively inducing apoptosis to prevent infectious spread. Several TLRs, including TLR 2–9, all of which are expressed on microglia, are likely to be involved in modulating angiogenesis in the setting of injury, thus creating a significant link between the innate immune response and NV. This immunovascular phenomenon is an area of great interest at this time given the strong evidence that several neovascular-related diseases, including AMD, may be driven by immune-system activation.

VEGF-A, once thought to be a pure vascular mediator, has been discovered to impart significant cellular effects on neural cells including microglia. Similar to blood-derived macrophages, microglia express VEGF receptor-1 (VEGFR-1) enabling their migration toward areas of VEGF-A production. With injury, microglia switch from their quiescent state to become activated, demonstrated by increased cytokine secretion, immune receptor expression, and proliferation.

Retinal astrocytes

Activated microglia may also initiate the reparative process by inducing the chemotaxis and proliferation of retinal astrocytes through TGF-β1, IL-1β, and TNF-α. Unlike microglia, astrocytes may be far more robust in the promotion of immune-mediated inflammation and contribution to glial scarification after retinal injury.

Müller cells

Müller cells are specialized glia that provide structural and trophic support, among numerous other functions, in the retina. In the setting of injury, Müller cells are able to proliferate and dedifferentiate into progenitor-like cells thus providing a potential source for retinal regeneration. Whereas some vertebrates (avian, amphibian, and fish) have been found to demonstrate such reparative potential, there is still controversial evidence on whether Müller cells of the mammalian retina are capable of this phenomenon.

Cellular Responses to VEGF-A Receptor Binding

VEGF-A is a multifaceted cytokine that influences multiple endothelial cell pathways promoting growth, survival, and vascular permeability. Its signal transduction is mediated primarily through two receptor tyrosine kinases, VEGFR-1 and VEGFR-2; however, the latter is implicated as the principal proangiogenic transducer. Both VEGFR-1 and R-2 are expressed in normal human eyes and surgically excised CNV membranes. There is a continuing controversy on the role of VEGFR-1 as a decoy receptor designed to sequester free VEGF-A and prevent excessive VEGFR-2 activation. In multiple models of injury-induced NV, VEGFR-1 has been shown to function bilaterally as both a pro- and antivascular mediator. Importantly, it does not appear that VEGFR-1 does this by simply serving as a decoy, as VEGFR-1 can repress VEGFR-2 mediated endothelial cell proliferation through active signaling after retinal injury. Thus, VEGF-A may possess the ability to function dichotomously in both pro- and anti-angiogenic capacities.

Conclusion

NV in the context of injury and repair is designed to aid the healing process, but, in the fragile and transparent cellular layers of the retina, it wreaks havoc on neural function and visual acuity. Several decades of dedicated molecular science and centuries of clinical observations have yielded an exhaustive foundation of knowledge that has significantly improved our understanding of injury-induced blood vessel growth. These mechanisms are currently being elucidated with such resolution and speed that a detailed molecular map of this disease process may be within our reach in the near future. More importantly, the work has been a tremendous benefit to society, as several targeted therapeutics are now available to the millions of people around the world suffering from neovascular diseases. It is our hope that scientific investigation of the unique biologic responses to retinal injury will continue to reveal critical facets in vasomolecular medicine both in the eye and elsewhere that will aid in the design of advanced therapeutics.

See also: Breakdown of the Blood–Retinal Barrier; Breakdown of the RPE Blood–Retinal Barrier; Central Retinal Vein

Occlusion; Immunobiology of Age-Related Macular Degeneration; Injury and Repair: Retinal Remodeling; Pathological Retinal Angiogenesis; Primary Photoreceptor Degenerations: Terminology; Retinal Pigment Epithelium: Cytokine Modulation of Epithelial Physiology; Retinal Vasculopathies: Diabetic Retinopathy; Secondary Photoreceptor Degenerations: Age-Related Macular Degeneration; Secondary Photoreceptor Degenerations.

Further Reading

Ambati, J., Ambati, B. K., Yoo, S. H., et al. (2003). Age-related macular degeneration: Etiology, pathogenesis, and therapeutic strategies. *Survey Ophthalmology* 48(3): 257–293.

Carmeliet, P. (2005). Angiogenesis in life, disease and medicine. *Nature* 438(7070): 932–936.

Eter, N., Engel, D. R., Meyer, L., et al. (2008). *In vivo* visualization of dendritic cells, macrophages, and microglial cells responding to laser-induced damage in the fundus of the eye. *Investigative Ophthalmology and Visual Science* 49(8): 3649–4358.

Fisher, S. K., Lewis, G. P., Linberg, K. A., et al. (2005). Cellular remodeling in mammalian retina: Results from studies of experimental retinal detachment. *Progress in Retinal and Eye Research* 24(3): 395–431.

Friedlander, M. (2007). Fibrosis and diseases of the eye. *Journal of Clinical Investigation* 117(3): 576–586.

Holtkamp, G. M., Kijlstra, A., Peek, R., et al. (2001). Retinal pigment epithelium–immune system interactions: Cytokine production and cytokine-induced changes. *Progress in Retinal and Eye Research* 20(1): 29–48.

Nozaki, M., Raisler, B. J., Sakurai, E., et al. (2006). Drusen complement components C3a and C5a promote choroidal neovascularization. *Proceedings of the National Academy of Sciences of the United States of America* 103(7): 2328–2333.

Nozaki, M., Sakurai, E., Raisler, B. J., et al. (2006). Loss of sparc-mediated Vegfr-1 suppression after injury reveals a novel antiangiogenic activity of Vegf-A. *Journal of Clinical Investigation* 116(2): 422–429.

Osborne, N. N., Casson, R. J., Wood, J. P., et al. (2004). Retinal ischemia: Mechanisms of damage and potential therapeutic strategies. *Progress in Retinal and Eye Research* 23(1): 91–147.

Pournaras, C. J., Rungger-Brandle, E., Riva, C. E., et al. (2008). Regulation of retinal blood flow in health and disease. *Progress in Retinal and Eye Research* 27(3): 284–330.

Rattner, A. and Nathans, J. (2005). The genomic response to retinal disease and injury: Evidence for endothelin signaling from photoreceptors to glia. *Journal of Neuroscience* 25(18): 4540–4549.

Ryan, S. J. (1982). Subretinal neovascularization. Natural history of an experimental model. *Archives of Ophthalmology* 100(11): 1804–1809.

Ryan, S. J. (2006). *Retina*. Philadelphia, PA: Elsevier.

Vazquez-Chona, F. R., Khan, A. N., Chan, C. K., et al. (2005). Genetic networks controlling retinal injury. *Molecular Vision* 11: 958–970.

Wu, J., Seregard, S., and Algvere, P. V. (2006). Photochemical damage of the retina. *Survey Ophthalmology* 51(5): 461–481.

Injury and Repair: Prostheses

G J Chader, A Horsager, J Weiland, and M S Humayun*, USC School of Medicine, Los Angeles, CA, USA

Glossary

Age-related macular degeneration (AMD) – Retinal degeneration, primarily in the macula that is prevalent in the aging population.

Bare light perception (BLP) – A high level of visual loss but retaining light perception.

Lateral geniculate nucleus (LGN) – One of the central visual areas within the central nervous system.

No light perception (NLP) – A complete lack of visual sensation.

Phosphene – A phenomenon characterized by the experience of seeing light without light entering the eye.

Retinal degenerative diseases (RDDs) – A general term referring to many different types of diseases in which photoreception is lost mainly due to photoreceptor degeneration.

Retinotopy – The spatial organization of the neuronal responses to visual stimuli.

Retinitis pigmentosa (RP) – Retinal degeneration, primarily affecting rod photoreceptor cells starting in the retinal periphery.

Rationale for a Prosthetic Device

The idea of stimulating a portion of the central nervous system (CNS) to elicit formed vision is not new. The need to do this is particularly evident in the case of retinal degenerative diseases (RDDs) in which the retinal photoreceptor cells degenerate and die. The RDD family of diseases has two main branches: retinitis pigmentosa (RP) and macular degeneration (MD). RP primarily affects rod photoreceptor cells, leading initially to loss of dim-light and peripheral vision. It usually is early onset, leaving otherwise normal subjects blind or with severely affected vision for life. RP is thought of as a rare disease with a prevalence of approximately 1:3400 around the world. The MD branch primarily affects cone photoreceptors, degrading central and sharp vision. The most common form of MD, age-related macular degeneration (AMD), strikes later in life (usually >60 years of age), but affects millions around the world. In these diseases, although

---

*Dr. Humayun has a financial interest in Second Sight Medical Products.

photoreceptors cells degenerate and ultimately can be lost, other layers of the retina are not as severely affected and remain relatively intact for some time thereafter. Thus, there is a window of opportunity for replacing photoreceptor function with a prosthetic device that will capture visual images, electronically transfer these to remaining secondary neurons of the retina, with subsequent passage to the brain. However, much of the visual processing takes place in the brain so it may be possible to bypass the eye completely with a visual image transmitted directly to vision-processing areas of the brain such as the visual cortex. Both of these approaches have been studied for decades now with the ultimate goal of sight restoration not only in cases of RDD but even when the entire eye has been lost. In fact, each separate anatomical portion of the visual system, such as the retina, the optic nerve, the lateral geniculate nucleus (LGN) of the thalamus, and the primary visual cortex (V1), has been used as a target in attempts at producing visual images.

Brain Prosthetic Devices

In 1962, Brindley found that application of stimulating pulse trains to his own eye with a corneal electrode interfered with visual light stimulation. This established at least the possibility of direct electrical stimulation of the retina. In 1968, Brindley and Lewin first reported on the use of platinum disk electrodes to stimulate the occipital pole of a subject and elicit independent visual percepts. Dr. William Dobelle followed, starting in about 1974, implanting blind subjects with electrodes mainly over the surface of the visual cortex and evaluating phosphene maps, which represent vision without light actually entering the eye. Although neither the pioneering studies of Brindley nor Dobelle were able to yield a useful, commercial visual prosthetic device, they did establish that phosphenes could be evoked by current pulses from an electrode array and that smaller stimulating electrodes would probably be needed to allow for more focal stimulation of cortex neurons. This led to the fabrication of novel, small, high-density arrays made of biocompatible silicon capable of penetrating the cortex (e.g., the work by Wise et al.). About this time, the National Institutes of Health actively developed a group dedicated to stimulating the primary visual cortex (area VI) using fine-wire electrodes in the hope of sight restoration in blind subjects. Since then, a number of groups (Normann et al. Hambrecht et al., Troyk et al., etc.) have continued to develop a cortical prosthesis. In most of this work, a video camera is used to capture a visual image, which, through a processor, delivers electrical signals to intracortical electrodes that finally deliver the image to the appropriate brain area(s). To date, this body of work has established that electrical stimulation of the visual cortex can produce phosphenes and that penetrating electrodes allow for the use of low, relatively safe levels of stimulation. The LGN can also be electrically stimulated to produce visual percepts. For example, Pezaris and Reed in 2007 demonstrated that either electrical or visual stimulation of specific receptive fields of the LGN results in highly localized and repeatable eye movements in a nonhuman primate.

Unfortunately, the field of electrical stimulation of brain areas to produce formed vision has only slowly advanced in the last few years. There are currently only a few investigators doing basic research in the area, perhaps because of difficulties in continuing human testing and moving to *bona fide* clinical trials. In contrast, work on retinal prosthetic devices has proceeded briskly, with many groups around the world applying unique concepts and designs in human testing.

Retinal Prosthetic Devices

As the optic nerve is a bundle of ganglion cell axons, it rightfully can be considered as an extension of the neural retina. Optic nerve stimulation became feasible after the development of cuff electrodes for nerve fiber recordings by Hoffer and collaborators in the 1980s. Since then, several groups have demonstrated that such electrodes are effective in generating visual percepts in the blind. A problem with this approach though is that the optic nerve does not maintain spatial organization (retinotopy), making spatial mapping of the stimulus difficult. In the 1950s, a US patent was issued describing the insertion of a light-sensitive selenium cell behind the retina and the subsequent transient restoration of light perception. It was not until the 1990s though that the investigators demonstrated that there were enough inner neurons remaining in the retina of RP patients to warrant prosthesis implantation. Work by Milam, Humayun, and co-workers, for example, reported that about 30% of the ganglion cells remained in advanced RP retinas and an even larger proportion of cells in the inner nuclear layer. Similar studies on eyes procured from AMD patients revealed that both the ganglion cell layer and the inner nuclear layer were relatively well preserved. Thus, in both diseases, enough cells remain such that they can act as a platform for the array and could perhaps adequately transmit an electronic stimulus to the brain. In 1993, the first report was presented by Humayun and colleagues that direct electrical stimulation of the retinal surface could produce visual percepts in a blind human subject. In this case, a man with advanced RP, essentially no light perception (NLP), was able to perceive

a spot of light with the temporary electrical stimulation. At that time, many other groups were working in parallel to develop other versions of the retinal prosthetic device – some intraocular, some extraocular.

All these approaches have different pros and cons, both with respect to efficacy and safety. For example, in the extraocular approach, one major hurdle is the relatively large distance between the electrodes and the retina. Chowdhury and co-workers though have presented evidence in an animal model that electrodes placed on the exterior surface of the eye can, in fact, stimulate the retina, evoking responses from the visual cortex. Tano and co-workers have shown that suprachoroidal–transretinal stimulation (with the anode in the fenestrated sclera and the cathode in the vitreous) is able to elicit excitatory electrical potentials (EEPs) from the visual cortex. Within the eye, electronic arrays have been implanted either in the subretinal or epiretinal positions. Subretinal positioning, as used by Zrenner et al., Rizzo and Wyatt, Chow et al., etc., is logical since the array is in the natural position of the lost photoreceptors. This juxtapositioning of the array to the bipolar cells could theoretically maximize the computational processing of the retinal circuitry but could be made difficult or impossible by lack of fidelity of the bipolar–ganglion cell synapse. There is also growing evidence for a substantial gliotic wall in the RDD subretinal space that could effectively limit the signaling from the subretinal implant to the overlying neurons. As shown by Marc in cases of severe RDD, the outer retina (bipolar, horizontal, and, to some extent, amacrine cells) undergoes marked reorganization, altering the synapses, and hence making the approach of stimulating bipolar cells more challenging. In addition, surgical placement of the array in the subretinal space is technically more difficult than placement on the vitreal surface. In epiretinal placement, that is, on the vitreal side, surgery is easier as well as access of the array to other components of the device. In patients, it is not known in either the epiretinal or subretinal situations which cell type or types are stimulated and react to the electrical signal. The epiretinal approach, because it is closer to the ganglion cells, has the theoretical advantage of stimulating this layer with lower currents. This is a distinct advantage since this layer does not undergo as significant a reorganization process in end-stage RDD.

Along with position on the retina, power output is an important consideration. Too little and the device might not be effective; too much could induce pathological changes in addition to an already compromised retina. A possible example of the dangers of too small of a power output is the case of the artificial silicon retina (ASR). In this case, Chow and co-workers used passive photosensitive diode arrays powered by only ambient light within the eye. Specifically, the 2-mm diameter ASR contained 5000 microelectrode-tipped microphotodiodes for light-energy processing with no external power supply. In spite of initial positive reports from a clinical trial that the ASR could improve vision in a number of implanted RP patients, it was subsequently found by Chow, Pardue, and co-workers that the device itself was ineffective and that the improvement in vision probably resulted from a more generalized neurotrophic effect (i.e., release of neuron-survival factors) due to implantation of the device – active or nonactive. Failure of the ASR led others, such as Zrenner et al. and Palanker et al., to use microelectronics to power subretinal devices. In comparison, the epiretinal devices include radio-frequency (RF) power transmission for wireless operation of the implants.

Clinical Trials

The main clinical indication and usefulness of retinal prosthetic devices as discussed above will be to patients with inherited photoreceptor degenerations such as RP and, ultimately, AMD. Although other forms of therapy are now being considered for these diseases, the electronic prosthetic device should be useful in situations where other proposed therapies are not applicable. Current studies on gene therapy, for example, in specific types of RP such as Leber's congenital amaurosis, seem to be successful, but gene replacement therapy will only be useful if the specific gene mutation is known, and if a sufficient number of viable photoreceptor cells are present to allow for a visual response after therapy. Similarly, pharmaceutical therapy, that is, the use of neurotrophic agents and nutritional therapy, for example, the use of antioxidants, will only be possible if photoreceptors are yet present. In the case where few or no photoreceptors remain, photoreceptor cell transplantation or stem cell transplantation could certainly replenish the photoreceptor layer. However, studies on direct photoreceptor transplantation have yet to show significant functional efficacy and studies on stem cell development into mature photoreceptor neurons have yet to demonstrate safety and efficacy. Thus, to many RDD patients, particularly those with severe or advanced photoreceptor degeneration, the retinal electronic prosthesis could be the best hope for sight restoration.

Many groups around the world are working on prosthetic devices – including Japan, Belgium, Korea, Australia, China, Germany, and the USA – too many to cover in any detail in the current article. Only a few groups of the many excellent approaches have progressed to the stage of intensive human testing. To do this, most of the basic science groups have become associated with companies who provide critical functions such as manufacturing and quality control as well as regulatory expertise. The Zrenner group, for example, works with Retina Implant AG, a second group with Intelligent Medical Implants GmbH (IMI), and a third with Epi-Ret GmbH. In the USA, the Humayun consortium has worked with the company, Second Sight

Medical Products Inc. (SSMP), to establish a Food and Drug Administration (FDA)-approved clinical trial with retinal arrays chronically implanted in RP patients.

In the subretinal approach taken by Zrenner and Retina Implant AG, an external energy source is used along with a microphotodiode array (about 3 mm in diameter containing 1600 microphotodiodes) that is implanted in the subretinal space. Experiments on animal models of inherited retinal degeneration have shown that ganglion cells are indeed stimulated by this device. Cortical evoked potentials could also be recorded in some of the animal experiments. Based on these positive preclinical findings, Retinal Implants AG has proceeded to human testing, first studying the short-term effects of the implant in RP patients. More recently, a prospective, open pilot clinical trial has been initiated with the subretinal implant and functional placebo controls. Eligible participants must have inherited retinal degeneration, be blind in at least one eye, or have visual function not compatible with navigation and orientation. The primary outcome is a measure of activities of daily living and orientation as well as safety. Latest information indicates that Retina Implants AG has implanted eight subjects for 4 weeks to date in their trial (clinicaltrials.gov identifier: NCT00515814). IMI GmbH has reported on four subjects with an epiretinal array of 49 electrodes designed to be functional for 18 months (clinicaltrials.gov identifier: NCT00427180). A 25-electrode array, EPI RET3, was implanted for 4 weeks in six subjects (clinical trial identifier: DE/CA21/A07). Finally, SSMP has conducted a long-term trial with a 16-electrode array (clinicaltrials.gov identifier: NCT00279500) and a more recent study with a 60-electrode array (clinicaltrials.gov identifier: NCT000407602).

Due to space limitations, this article will mainly focus on the SSMP effort and its clinical trial as an example of progress with retinal prosthetic devices. In addition, many of the early results from the SSMP trial have been reported and are clearly documented in the scientific literature. Finally, testing of the SSMP device has been continuous since 2002 in specific patients who have not only had regular laboratory testing but also have been able to use the device at home for direct assessment of quality-of-life (QOL) improvements. The approach taken by Humayun and co-workers has been epiretinal with the electrode array tacked on to the vitreal surface of the retina. Work initially started with studies on animal models of inherited retinal degeneration. These studies demonstrated no major problems in terms of inflammation, neovascularization, or encapsulation of the device. These positive safety studies allowed for short-term human testing where electrodes were temporarily juxtaposed to the retina of subjects with advanced RP and activated to ascertain if phosphenes were elicited. Results were positive in that appropriate phosphene signals were reported by the subjects. Since safety was also apparent in these acute studies, more chronic implantation in the formal setting of a feasibility clinical trial was allowed by the FDA in 2002. The first-generation device used in the trial was called the Argus I; it had 16 electrodes affixed to the retina. Patients who had outer retinal degenerative diseases with a preoperative vision of bare light perception or worse and a nonrecordable electroretinogram (ERG) were chosen. The entire device is complex, with many interacting components that are the result of new technology modified from the cochlear implant. Functionally, the device has an external system that includes a tiny, lightweight video camera housed in the subject's glasses. This allows for image capture and initial processing. Image and power data are sent to the array using RF wireless transfer through a transmitter coil. The microelectronic implant is under the skin on the temporal bone behind the ear. A cable is threaded under the skin toward the orbit and around the eye under the rectus muscles. Once entering the eye through a sclerotomy, it is connected to the array, which is tacked to the retinal surface. The Argus 1 array has 16 platinum electrodes in a 4 × 4 arrangement within a silicone platform. The array is implanted in the macular region and is approximately 2.0 × 2.9 mm. It subtends an approximate 10° of visual angle with 14° on the diagonal.

Starting in 2002, six subjects were implanted with the Argus 1 device in the SSMP-sponsored feasibility clinical trial. The results were promising, in that safety was observed both for the subject and the device. No device failures were encountered, and no significant sequelae of the implantation were observed in the subjects. Because the device can be independently monitored and controlled (i.e., through the head-worn video camera or by custom software on a laptop computer), testing continues to date. Valuable longitudinal data have been obtained from this initial study. For example, although not surprising, it was found that the impedance and threshold values were inversely proportional to the distance of the electrode array from the retinal surface and that most of the electrodes had stimulatory thresholds below the safe charge injection limit. Importantly, subjects could spatially resolve the individual electrodes within the array with good resolution of the perceived locations of the elicited phosphenes, good two-point resolution, and good perception of motion direction. There was also a good correlation between percept brightness and stimulation level, demonstrating the ability of the implanted subjects to differentiate between different levels of illumination. Cortical responses were also evoked by the device. The electrically evoked responses (EERs) from the visual cortex are a primary measure of the possibility of functional vision. Most encouragingly, testing for object recognition showed that subjects with the use of the implant could differentiate between a cup, knife, and saucer. In mobility testing, patients could follow a line on the floor and identify and walk toward test doors. In the object recognition test, repeated testing gave scores significantly above chance, statistically at $P < 0.001$. Investigators at SSMP have also just concluded the first detailed study of spatial vision in an Argus I-implant patient definitively showing patterned visual perception with logMAR = 2.21, which is well matched to the sampling limit determined by the spacing of the electrodes. This provides strong evidence that it is possible for the brain to use the special information provided by individual electrodes. However, further testing is ongoing to understand the level of visual acuity obtainable with arrays that have more numerous or more closely spaced electrodes.

Testing of these subjects remains ongoing to determine if their QOL is enhanced, especially since they are able to use the devices at home while not under direct supervision. As might be expected though, differences between the performances of different subjects were observed. This could be based on differences in disease type and on the stage, that is, severity, of the degeneration in the individuals. Factors such as the site of implantation of the array and its proximity to the retinal surface must also be considered. All in all though, it is perhaps surprising that any functional efficacy is observed using the 16-pixel array. Theoretically, with such a simple array, only areas of light/dark should be discerned by the subjects, perhaps modified by shades of gray. That simple objects and letters can be identified though and mobility enhanced perhaps indicates that the brain can indeed make sense out of primitive signals and that, through a patient learning process, useful vision can indeed be restored.

Based on the positive results from both safety and efficacy in the feasibility portion of the trial, SSMP has initiated an international feasibility trial using a second-generation device, the Argus II. This new device has several improved designs, notably, an increased electrode count with up to 60+ electrodes on the retinal array. It also is much smaller than the first-generation device, allowing for an easier and shorter surgery procedure. Finally, as with the Argus I, the Argus II is designed to last the lifetime of the patient. Only severely affected RP patients are chosen to participate, with visual acuity and safety as the primary outcome measures. Secondary outcomes are improvements in orientation and mobility, activities of daily living, and general QOL. Early indications are that the Argus II device is safe (both to subject and device), although this certainly can only be properly assessed over a longer time period. Similarly, early data indicate that there is enhanced efficacy (i.e., improved vision). As alluded to above for the Argus I though, this does not come immediately and is a learning process over a period of months.

Challenges

Although the data presented above are encouraging, significant challenges yet remain in perfecting a prosthetic device that will restore a high level of functional vision, be it epiretinal, subretinal, or one of the other approaches. For example, it will be important to determine the actual mechanism of action of the array, in particular, which populations of neurons are stimulated. In this way, the efficacy and efficiency of the device might be enhanced, irrespective of future design improvements. Another important area of future investigation must be directed at gaining a better understanding of the different pathologies and how the usefulness of the device is affected by the state of each individual retina. Although significant numbers on inner retinal neurons remain alive in cases of advanced retinal degeneration, neural remodeling takes place such that morphology is altered and synaptic contacts rearranged. Investigators such as Milam, Marc, and Strattoi have presented compelling evidence that inappropriate neurite sprouting takes place in remaining rod, horizontal, and amacrine cells, with abnormal contacts with both neuronal and glial (Müeller) cells. These changes, along with the appearance of a dense fibrotic layer in the subretinal space, could severely affect proper inner retinal cell processing and signal transmission down the optic nerve. The effects of electrical stimulation on the normal retina have been well studied, but the effects on retinas affected with an inherited retinal degeneration are just beginning to be examined. These effects will have to be better defined in different genetic forms of RD and at different stages of disease such that the electrical properties of the device can be tuned and adjusted to individual situations. Similarly, the state of central connections could be a problem, particularly in patients with long-term degeneration (i.e., no or bare light perception) and with patients with very-early-onset degeneration such as Leber's congenital amaurosis in which central connections may not even have been initially formed. All of these factors will vary considerably from subject to subject and will have to be individually assessed in each case. That a level of functional vision has been restored though in a number of older human subjects with severe retinal degeneration indicates that, at least in many cases, enough functional inner retinal neurons and their synaptic connections do remain, and the visual cortex remains functional enough to process the electrically derived information into discernable forms.

Issues of safety also remain, particularly with regard to long-term implants. There is yet the possibility of low-level but cumulative damage from both electrical and mechanical effects of array implantation. Too high an electrical input could not only grossly burn the retina but, more subtly, also induce an altered state of conductivity and responsiveness within the remaining retinal cells. In addition, as preclinical and clinical studies continue, the properties of the implant itself may need to be modified as to materials, size, and shape. On the positive side, implantation of the Argus I began in 2002 with testing continuing to date, giving several years of evidence for not only patient and device safety but also continuation of efficacy.

Future Prospects

From an engineering standpoint, one of the greatest challenges but most exciting prospects for the future is in the enhancement of vision through improved and more sophisticated electronics, for example, an increased number of electrodes impinging on the retina. Theoretically, an electrode count of 1000+ should allow for face recognition and reading ability, leading to a marked improvement in QOL and mainstreaming of patients back into society. With this type of improved function, retinal electronic prosthetic devices should thus be helpful not only to the relatively limited number of patients with RP but also to the millions with AMD.

There are several other obvious but diverse factors that can be studied that should lead to improved action of a retinal prosthesis. One of these would be to assure close and uniform proximity of the array to the retinal surface in the specific cases of epiretinal and subretinal implantation. A bioadhesive glue that is biocompatible and reversible would be useful in this regard, particularly in an epiretinal device. Another area of improvement could be the use of an intraocular rather than extraocular camera. This would allow for more natural object tracking, as the eye rather than the head moves. A third area of interest would be to increase the relatively narrow visual field of the implants currently under study such that peripheral vision is restored. Finally, could it be possible to restore color vision? The implant subjects to date do describe different colors, including blue, yellow, orange, and green. Although not important to most of the visual tasks we perform, colors do abundantly increase our appreciation of our environment and certainly increase QOL.

See also: Injury and Repair: Neovascularization; Injury and Repair: Retinal Remodeling; Injury and Repair: Stem Cells and Transplantation; ; Primary Photoreceptor Degenerations: Retinitis Pigmentosa; Primary Photoreceptor Degenerations: Terminology; Retinal Vasculopathies: Diabetic Retinopathy; Secondary Photoreceptor Degenerations: Age-Related Macular Degeneration; Secondary Photoreceptor Degenerations.

Further Reading

Brindley, G. and Lewin, W. (1968). Short- and long-term stability of cortical electrical phosphenes. *Journal of Physiology (London)* 196: 479–493.

Caspi, A., Dorn, J., McClure, K., et al. (2009). Feasibility study of a retinal prosthesis. Spatial vision with a 16-electrode implant. *Archives of Ophthalmology* 127: 398–401.

Chader, G. (2007). Retina prosthetic devices: The needs, and future potential. In: Tombran-Tink, J., Barnstable, C., and Rizzo, J., III (eds.) *Visual Prosthetics and Ophthalmic Devices*, pp. 1–4. New York: Springer.

Dagniele, G. (2008). Psychophysical evaluation for visual prosthesis. *Annual Review of Biomedical Engineering* 10: 15.1–15.30.

Horsager, A., Greenwald, S., Weiland, J., et al. (2009). Predicting visual sensitivity in retinal prosthesis patients. *Investigative Ophthalmology and Visual Science* 50: 1483–1491.

Humayun, M., Weiland, J., Chader, G., and Greenbaum, E. (eds.) (2007). *Artificial Sight: Basic Research, Biomedical Engineering and Clinical Advances.* New York: Springer.

Javaheri, M., Hahn, D., Lakhanapal, R., Weiland, J., and Humayun, M. (2006). Retinal prostheses for the blind. *Annals of the Academy of Medicine, Singapore* 35: 137–144.

Lowenstein, J., Montezuma, S., and Rizzo, J., III. (2005). Outer retinal degeneration. An electronic retinal prosthesis as a treatment strategy. *Archives of Ophthalmology* 122: 587–596.

Normann, R., Maynard, E., Rousche, P., and Warren, D. (1999). A neural interface for a cortical vision prosthesis. *Vision Research* 39: 2577–2587.

Palanker, D., Vankov, A., Huie, P., and Baccus, S. (2005). Design of a high resolution optoelectronic retinal prosthesis. *Journal of Neural Engineering* 2: S105–S120.

Pezaris, J. and Reid, R. (2007). Demonstration of artificial percepts generated through thalamic microstimulation. *Proceedings of the National Academy of Sciences of the United States America* 104: 7670–7675.

Tombran-Tink, J., Barnstable, C., and Rizzo, J., III (eds.) (2007). *Visual Prosthesis and Ophthalmic Devices: New Hope in Sight.* New York: Springer.

Weiland, J. and Humayun, M. (2005). Retinal prosthesis. *Annual Review of Biomedical Engineering* 15: 361–401.

Yani, D., Weiland, J., Mahadevappa, M., et al. (2007). Visual performance using a retinal prosthesis in three subjects with retinitis pigmentosa. *Americal Journal of Ophthalmology* 143: 820–827.

Zrenner, E. (2002). Will retinal implants restore vision? Science 295: 1022–1025.

Injury and Repair: Retinal Remodeling

R E Marc, University of Utah, Salt Lake City, UT, USA

Glossary

Glial seal – A compaction of the distal microvillar processes of mature Müller cells and their stabilization by intermediate junction to form a barrier to diffusion and cellular movement.

Microneuromas – New, anomalous complexes of synapses by mature neurons.

Migration – The ability of a cell to relocate to a new position in a tissue. It is often thought that mature neurons cannot migrate without de-differentiating, but mature retinal neurons can migrate both intra- and extraretinally without losing their characteristic molecular profiles.

Neuritogenesis – The activation of new dendrites and axons from neurons; typically associated with developing cells, it is vigorously demonstrated by mature neurons in remodeling retinas.

Reprogramming – A major change in the gene expression profile of a cell, typically the expression of genes not characteristic of homeostasis.

Self-signaling – The ability of retinal networks to generate their own excitatory activity in the absence of a photoreceptor drive.

Overview

Retinal remodeling is a collection of molecular and cellular revisions triggered by primary inherited degenerative diseases such as retinitis pigmentosa (RP), Usher syndrome; secondary degenerative diseases with mixed environmental/genetic risks, such as age-related macular degeneration (AMD); and acquired retinal defects such as prolonged retinal detachment and light-induced retinal damage. These revisions include anomalous neuronal rewiring (targeting canonically inappropriate cells) and reprogramming (expressing canonically inappropriate genes or repressing characteristic genes); de novo neuritogenesis and synaptogenesis; spontaneous and corruptive self-signaling; bipolar cell (BC) dendrite truncation; supernumerary axon generation; neuronal migration along hypertrophic Müller cell (MC) columns; neuronal death; altered glial molecular profiles; vascular remodeling; and retinal pigmented epithelium

(RPE) invasion, vascular occlusion, and hyperpigmentation. Some revisions (e.g., reprogramming) begin as soon as photoreceptor stress is initiated, while others (e.g., synaptogenesis) are manifest only after complete local photoreceptor loss. Importantly, the local survival of even heavily altered cones can prevent much late-stage remodeling, apparently by stabilizing the dendritic compartment of BCs. Remodeling impacts the timing and potential outcomes of gene therapy, survival factor treatments, stem or progenitor cell implantation, retinal transplantation, and bionic implants.

It was not until the detailed imaging studies of Ann Milam (University of Pennsylvania) in the mid-1990s that the nature and scope of remodeling in human RP became evident. More recently, Enrica Strettoi (Centre National de la Recherche Scientifique, Pisa, Italy) and Bryan W. Jones (University of Utah) independently demonstrated that remodeling was also characteristic of animal models of human inherited retinal degenerations. Though remodeling was not initially given much credence despite strong homology to central nervous system (CNS) degenerative disorders, it is now gaining understanding as a serious denouement of retinal disease.

Progression

Remodeling kinetics are largely independent of the source of photoreceptor deafferentation and occur in distinct phases (**Figure 1**). In phase 1, neuronal and glial cells react to photoreceptor stress signals prior to photoreceptor death. In phase 2, neurons, glia, and microglia interact in the processes of photoreceptor death, outer nuclear layer (ONL) decimation, and formation of the glial seal, encapsulating the remnant retina. In phase 3, neural cells respond to deafferentation and non-neural cells form new cytoarchitectures in the remnant retina. The speed of these phases depends on the nature of the degeneration (**Figure 2**). Aggressive primary rod, cone–rod, or cone dystrophies, as well as RPE phagocytosis defects can rapidly transit phases 1 and 2 to extensive phase 3 remodeling. Slower photoreceptor degenerations (e.g., autosomal-dominant RP (adRP) models of rhodopsin mutations) can lead to extended periods of cone survival, delaying the onset of phase 3. In rodent models, the faster overall kinetics are partly due to the small eye and the likely constancy of cell–cell interaction areas. Greater loss of peripheral retina is tolerated in humans as long as the macula is spared.

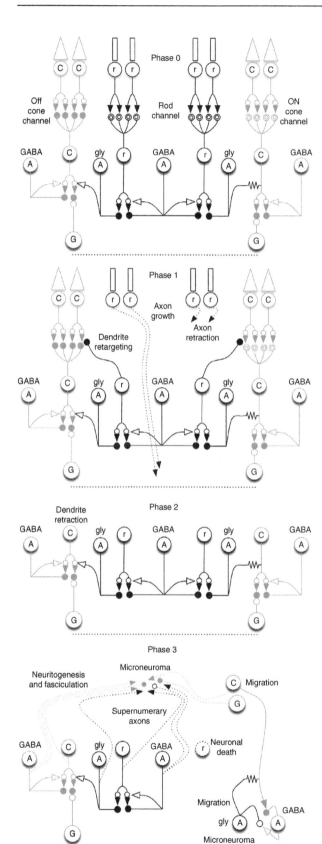

Figure 1 Remodeling phases in the mammalian retina. Phase 0 is the normal retina prior to the onset of acquired or inherited defect stress. The mammalian retina uses separate cone (C) and rod (r) channels prior to converging on retinal GCs (G).

Phase 1

The first evidence that remodeling precedes photoreceptor death was the demonstration that human rods harboring a rhodopsin defect were able to form new fascicles of axons, bypass their normal BC targets, and project into the ganglion cell (GC) layer prior to apoptotic stress and death. Some rodent model degenerations show the same ability. Furthermore, rodents with autosomal-recessive RP, such as the phosphodiesterase B 6 (*Pdeb6^{rd1}*) mouse (the rd1 mouse), show truncated BC dendritic arbors and horizontal cell (HC) axonal fields long before photoreceptor death. MC stress signals and protective alterations in neuronal glutamate receptor expression in light-induced retinal degeneration (LIRD) albino rodents are activated within hours of light-stress onset and long before photoreceptor death.

Phase 2

The degenerations transition to extensive cell-autonomous and/or bystander photoreceptor death. The ONL is dismantled with the involvement of activated microglia and hypertrophic MCs. The details are poorly understood and may vary according to gene defect. For example, dominant mutations that impair rhodopsin trafficking may activate endoplasmic reticulum (ER) stress in several ways such as anomalous protein multimerization, inhibition of proteasome cycling, and activation of the unfolded protein response. This results in a slow dismantling of the ONL by sporadic apoptosis. Conversely, mutations of transduction pathways that trigger calcium-dependent apoptosis are faster and more coherent. In either case, photoreceptor apoptosis and migroglial-initiated bystander cytolysis may create debris zones that must be cleared. The mechanisms of such clearance are unknown. Phase 2 ends with the entombment of the remnant neural retina by a thick seal of distal MC processes similar to the normal outer limiting membrane, with extensive intermediate junctions between the processes. Though often termed a glial scar, there is no evidence that the seal involves astrocyte proliferation as in

Cone BCs receive cone input with either ionotropic sign-conserving glutamate receptors (solid circle) or metabotropic sign-inverting glutamate receptors (double circle), and drive ACs (A) and GCs which decode signals with sign-conserving glutamate receptors. Rod BCs receive rod input with metabotropic sign-inverting glutamate receptors and then drive glycinergic rod ACs, which fan their signals out to ON cone channels through gap junctions (resistor symbol) and OFF cone channels through glycine release decoded by inhibitory receptors (open circles). In phase 1, rod photoreceptors reprogram to either bypass rod BCs with new axons or retract their synapses. Rod BCs both retract dendrites and retarget some to adjacent cone pedicles. In phase 2, photoreceptors are lost and all BCs lose their dendrites. In phase 3, extensive rewiring, neuritogenesis, microneuroma formation, cell migration, and neuronal death occur.

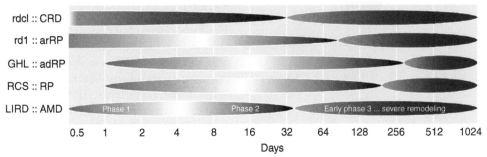

Figure 2 Different forms of retinal degeneration express different kinetics. The abscissa is exponential in time in either postnatal or postlight exposure days. Light-induced retinal degeneration (LIRD) is fast with phase 1 stress appearing almost immediately and phase 2 photoreceptor death extending for a couple of weeks before patches of phase 3 remodeling begin. In the the Royal College of Surgeons (RCS) rat, photoreceptor debris build up in the subretinal space causes stress, but photoreceptor death is not complete for 3 months or more, after which extensive remodeling occurs. The triple mutation valine 20 → glycine, proline 23 → histidine, proline 27 → leucine GHL mouse model of autosomal-dominant RP (adRP) does not begin to show photoreceptor stress until opsin synthesis begins and slowly traverses phase 2 for nearly a year before phase 3 remodeling begins. In contrast, the rd1 model of autosomal-recessive RP (arRP) is already stressed at birth with phase 2 photoreceptor death reaching completion close to 90 days. Similarly the rodless–coneless (rdcl) mouse is a model of cone–rod dystrophy (CRD) and cell death actually begins prenatally. This model spends most of its life in phase 3.

CNS glial scars. There are occasional breaks in the seal that are associated with phase 3 RPE and choroidal vascular invasion of the neural retina and neuronal escape into the choroid.

Phase 2+

In some degenerations, the death of rods is slow and seems to trigger variable bystander killing, leaving clusters of deconstructed cones with apparently functional synaptic contacts. This results in patches of retina suspended in late phase 2 (phase 2+) in a sea of phase 3 retina. The extent to which these preserve vision is unknown, but they definitely provide evidence that even marginal rescue of cones is a critical step if late-phase therapies are to be viable.

Phase 3

After loss of all photoreceptors, the stability of neuronal connectivity in the retina becomes progressively compromised through at least nine distinct processes (see below) including molecular reprogramming, individual cell rewiring, large-scale neuritogenesis, synaptogenesis and microneuroma formation, spontaneous self-signaling, neuronal migration to ectopic foci, progressive neuronal death, MC remodeling and altered gene expression, as well as RPE and vascular remodeling and migration. In severe degenerations, including human late-stage RP and geographic atrophy, the revision of the retina can be so severe that no visual function could ever be restored. In other cases, the neural retina survives but is likely to be so altered that upstream strategies such as subretinal implants, stem/progenitor cell transplants, or even fetal retinal transplants will likely not successfully deliver form

vision. Further, there is increasing evidence that bionic implants and transplants do not stabilize phase 3 remodeling and may accelerate it.

Remodeling Events

Reprogramming

Reprogramming is a shift, possibly reversible, in the gene expression of cells to anomalous molecular and anatomic states. This is best known for MCs in retinal stress where intermediate filament expression is elevated and is also becoming evident for retinal neurons, especially in terms of glutamate receptor expression. Receptor reprogramming can begin in phase 1–2 when rods lose the ability to directly signal rod BCs. In the $Pde6b^{rd1}$ mouse, the expression, localization, and function of the metabotropic glutamate receptor 6 (mGluR6) essential for proper ON pathway encoding decrease. In both mouse and human RP models, rod ON BCs appear to transiently upregulate expression of ionotropic glutamate receptors (iGluRs). In the LIRD mouse model, rapid upregulation of protective GluR2 subunits occurs within 24 h after initiation of photoreceptor stress. There is also evidence of gamma-aminobutyric acid (GABA) receptor redistribution in ON BCs of the $Pde6b^{rd1}$ mouse. Further, the expression of dendrites in all BCs appears to be completely suppressed after rod and cone loss, and supernumerary axons are formed. While the mechanisms initiating these changes are not yet known, they demonstrate that mature retinal neurons can change their architectures and receptor expressions, while glia can change their architectures and metabolic profiles. More reprogramming events are certain to be found.

Rewiring

Rewiring is the switching of dendritic or axonal targeting. The first known instance of this was the escape of rod synaptic terminals from BC dendrites to form long axonal fascicles (**Figure 1**), but with no established targets; cones behave similarly in some cases. In other forms of retinal degeneration, photoreceptors retract their synaptic terminals, which may be associated with alterations in the balance of cAMP/cGMP production (Camp, cyclic adenosine and guanosine monophosphate; cGMP, guanosine monophosphate). The numerous reports of BC sprouting in many model degenerations as well as aging mice likely reflect the simple extension of the dendrites still attached to retracting rod synaptic terminals and not true plasticity. More concretely, BCs bereft of rod input transiently retarget dendrites to cone terminals and, though they do not make structurally appropriate ribbon-associated synapses, they do express anomalous iGluRs. This is consistent with the disappearance of the electroretinogram b-wave, despite the fact that no rod BCs die early in phase 2, and with observations that ON center GC responses disappear long before those of OFF center GCs.

Neuritogenesis

Neuritogenesis is the large-scale evolution of new processes by amacrine cells (ACs), BCs, and GCs in phase 3. The mechanism of activation is unknown, but it encompasses all cell classes, suggesting a global signaling process and a pan-neural response. Many new fascicles course just distal to the glial seal and contain mixed neurites in patterns never observed in the normal retina. In other regions, especially near invading RPE processes, large tracts of tiny new neurites with only one or two microtubules form homogeneous fascicles. The tracts can traverse several hundred microns, suggesting significant alteration in the spatial patterning of circuits.

Synaptogenesis and Microneuromas

A logical extension of neuritogenesis is synaptogenesis. The extent to which new synapses are made in the inner plexiform layer (IPL) of phase 3 remodeling retina is not clear, but in regions of the MC's hypertrophy, surrounding migration columns (see below), in the GC layer, and especially in the remnant distal retina, numerous new AC, BC, and GC connections are made. The most distinctive zones are microneuromas, which range from 10 to 100 ìm in width and contain abundant conventional and ribbon synapses. The mechanisms that stimulate new synapse formation are also unknown but seem closely associated with RPE processes. As RPE cells are known to release several growth factors, it is plausible that they are the activators of synaptogenesis. The wiring within microneuromas seems chaotic

and, by serial section reconstruction and modeling, such networks appear to be resonant, which is incompatible with visual processing.

Self-Signaling

A number of observations suggest that retinal degenerations lead to spontaneous self-signaling in phase 3 retina. Photopsias (scintillating illusions common in RP) are initiated in the retina and the generator mechanisms can remain quiescent for decades in the absence of vision, only to be reactivated by experimental transocular currents. Spontaneous, erratic retinal waves of depolarization occur in rodent models of RP after light-driven responses are lost. Physiological measures show that rodent ACs and GCs are glutamate-activated in phase 3. GC activation in the phase 3 $Pdeb6^{rd1}$ mouse is clearly glutamatergic and not intrinsic. This means that BCs must be voltage modulated in some way. However, most remodeled BCs lack functional glutamate receptors and they must be depolarized by another mechanism. One plausible mechanism is periodic AC membrane potential fluctuations, leading to the modulation of BC anion currents and modulation in BC glutamate release. Some isolated ACs have shown endogenous oscillations in K^+ channel conductance, alternately hyperpolarizing and depolarizing them, presumably modulating GABA (or glycine) release. The isolation of ACs from the normal visual drive in retinal degenerations may unmask this intrinsic capacity. Once self-signaling is initiated, resonant networks created by rewiring or microneuroma formation can rapidly generate periodic activity. Such networks may be inimical to restoration.

Migration

Neuronal migration is a large-scale remodeling event. In most instances, migration is closely associated with both hypertrophy of MCs and anomalous vascular tangles (see below). All forms of cell mixing occur: collections of ACs and BCs can become displaced to the GC layer. Conversely, ACs and even GCs can migrate to the distal margin of the remnant retina. One fundamental question is whether these cells remain functional. Ultrastructurally, cells in migration columns such as GCs seem to have processes extending both distally and proximally much like neuroepithelial cells in development. However, after migration, the original orthotopic processes seem to be retracted. Migrated cells seem heavily connected to microneuromas. A more serious form of migration (emigration) occurs when RPE cells and the basement membrane are focally ablated and the distal seal can surge into the choroid. Large tracts of MCs and neurons can emigrate, decimating the remaining neural structures. This is especially evident in LIRD and suggests that it may play a role in loss of vision in severe nonvascular forms of AMD such as geographic atrophy.

Cell Death

There is little evidence of glial death in remodeling, but the proportions and numbers of survivor neurons change significantly. BCs form the largest cohort of neurons (other than photoreceptors) in normal retina; however, ACs always predominate in phase 3, suggesting that BC death is far more common than AC or GC death. This may be due to the fact that BCs are the only retinal cells that lose all of their glutamatergic input. Neurons require a basal level of Ca influx to maintain homeostatic gene expression and, through self-signaling, ACs and GCs clearly possess the critical glutamatergic input required to provide both transmitter-gated Ca flux and voltage-gated Ca-channel activation; BCs clearly lack that input. As subjects age, ACs and GCs also decrease in number. In any event, the loss of neurons has strong implications for all therapeutic interventions, including epiretinal implants.

MC Remodeling

MCs make up nearly 50% of the mass of the peripheral primate retina and are one of the major drivers of remodeling. While there has been little analysis of phase 1 MC function in inherited retinal degenerations, there is abundant evidence that they respond to rapid, coherent photoreceptor stress initiated by LIRD within hours by increasing intermediate filament expression, displaying distal process hypertrophy in the ONL, and increasing arginine expression (a marker of increased protein synthesis). In phase 2, MCs play a lead role in forming a seal between the remnant RPE and/or choroid. The transport characteristics of that seal are unknown, but its formation is paralleled by a massive increase in MC glutamine levels. MCs normally export glutamine from MCs to surrounding neurons. This transport is voltage sensitive and, similar to many Na-coupled transporters, allows increasing export with depolarization. Depolarization of MCs is closely coupled to light-activated events, and the loss of photoreceptors may play some role in progressive hyperpolarization of MCs and retention of gluamine. However, as the increase in MC glutamine is temporally linked to the formation of the distal seal, the mechanism of glutamine retention appears more complex than just constraining the MC voltage.

RPE Remodeling

In classical RP, invading RPE cells are one of the hallmarks of advanced disease. After formation of the MC seal, certain RPE cells and, sometimes, choriocapillaris endothelia are able to penetrate into the neural retina, forming large complexes of hypertrophic MCs, RPE with altered melanosomes encapsulating invading and remnant retinal capillaries, clusters of new neurites, and columns of migrating neurons. RPE cells seem to be the foci of large-scale morphologic derangements in the survivor retina, but whether they are initiators or responders is not clear. The RPE remains partially intact for long periods in many retinal degenerations, including the Royal College of Surgeons (RCS) rat defect; however, the RPE layer can become broken by patches of invading RPE cells and apical processes. In the RCS rat, apical RPE processes can extend to the GC layer.

Vascular Remodeling

New capillaries invade the neural retina in phase 3, emanating from both vitreal and choroidal sources. Little is known of the fundamental transport properties of these new vessels (e.g., whether they are fenestrated or not), but both molecular and genetic profiling show that neural retina in RP is metabolically deprived. New imaging data as well as electron microscopy suggest that the new vessels are too attenuated and too heavily invested by hypertrophic MCs and RPE to allow proper perfusion of the retina. These anomalous foci may trigger neuronal migration. The stimuli for vascular remodeling remain unknown, but VEGF secretion by invading RPE cells is one plausible source.

Impact of Remodeling on Therapeutics

The potential reversibility of remodeling events varies. For example, phase 1 and 2 changes in reprogramming of gene expression and neurite switching are reminiscent of normal plastic behavior and might respond to the appropriate therapeutic signals. However, phase 3 changes such as rewiring, neuritogenesis, and microneuroma formation, migration, and neuronal cell death are not reversible by any known means. Intermediate phenomena such as MC, RPE, and vascular remodeling are similarly challenging. Human RP patients show the full spectrum of remodeling defects; therefore, most therapies are impacted by a narrowing therapeutic window.

Primary Gene Therapy

All prospective primary gene therapies (those targeting known gene defects) depend on photoreceptor survival. However, it is clear that photoreceptor deconstruction starts long before photoreceptor death. The recent successes with gene therapy for replacing the deficient isomerase (RPE65) in that variant of Leber's congenital amaurosis (LCA) is not likely relevant to primary rod or cone gene defects, nor to defects associated with the accumulation of genotoxic and cytotoxic debris in the subretinal space such as those due to defects in RPE mer tyrosine kinase (MERTK). In human rod dystrophies, it is clear that mutations impacting rhodopsin trafficking also lead to changes in the inner segment function and photoreceptor architecture. Rod photoreceptor rewiring in human RP is unlikely to be reversible regardless of the success in replacing defective phototransduction genes. So far, gene therapy successes in animal models are largely restricted to prenatal or early postnatal genetic interventions. With the exception of soft diseases such as LCA and stationary night blindness that do not trigger photoreceptor deconstruction, most retinal degenerations clearly have only a tiny window for gene therapy. Primary gene therapies are currently restricted to phase 1.

Survival Factor Therapy

One strategy to retard retinal degeneration involves the use of survival factors such as neurotrophins (e.g., ciliary neurotrophic factor, CNTF) that slow photoreceptor apoptosis. These strategies were validated for slower models of adRP (the rat P23H and S344ter rhodopsin transgenic models). Some argue that the structural preservation afforded by CNTF in these models does not reflect a parallel functional rescue. There is evidence that early postnatal CNTF infusion in other rodent models of RP negatively alters photoreceptor gene expression profiles, activates MC stress signaling, and alters inner retinal organization. Simple survival factor therapy without a known cellular and molecular target is not likely to generalize well across human RP types. At present, survival factors offer little prospect of reversing or retarding remodeling. Survival factor therapies are restricted to phase 1 and early phase 2.

Stem/Neuroprogenitor Cell Therapy

Several groups have shown that isolated stem/ progenitor cells, particularly murine postnatal day 5 rod neuroprogenitor cells, have the potential to intercalate in the ONL and, possibly, repopulate the retina. The efficiency of such intercalation is low and it is not likely that these cells will survive in phase 3 retinas. Penetration of the MC seal is unlikely and several groups have failed to obtain significant numbers of exogenous photoreceptors to extend synapses through it. Direct injection into the retina is likely to activate microglial killing and, in any event, isolates any surviving photoreceptors from the remnant RPE. For photoreceptor progenitor cells to be successful in rescuing vision, they must also lead to cone survival and reconnect with existing neurons before morphologic remodeling begins. This excludes most current RP patients. Stem/neuroprogenitor cell therapies are currently restricted to phase 1 and early phase 2.

Retinal Transplantation

The team of Seiler and Aramant pioneered the effort to insert sheets of fetal retina into the subretinal space of degenerating retina. They have successfully demonstrated long-term photoreceptor survival and some visual driving in rats. Remodeling remains a major barrier in several ways. The extent of transplant-to-host neurite intermingling is small, and the glial seal predominates. Further, the transplanted retina begins to remodel and appears even more susceptible to alteration than the host. This suggests that remodeling signals emanate from the survivor retina. Certainly, after phase 3 BC dendrite truncation in the host retina, a semi-intact fetal retina cannot recapitulate normal circuitry by tandem connections. It is likely that most transplant-to-host connectivity is AC → AC driven and probably functionally random. Even so, such networks can generate light-driven behavior. Though lacking the spatiotemporal precision normal GCs, transplants have the potential for much higher sensitivity, range, and resolution than bionic mechanisms. Retinal transplantation therapies are largely restricted to phases 1–2, but may function if the connection can traverse the glial seal in phase 3 retinas.

Secondary Gene Therapies: Photosensitive Proteins

A recent development of import is the ability to induce expression of photosensitive proteins in survivor neurons by viral or molecular transfection, generating light responses directly in neurons. Several groups have shown that it is possible to express channelrhodopsin 2 (ChR2) in retinal BCs, and elicit ChR2-BC-driven photoresponses in GCs and light–dark preferences in behavioral search. This is an important advance and represents the potential to convert blind retinas into navigational systems without surgical intervention and ancillary equipment. The challenge is to demonstrate that this is possible in phase 3 profoundly blind patients, as they are the most obvious candidates for therapy. This technique has, once again, only been established in phase 2 or earlier models. Further, as with transplantation, there is no evidence that such secondary therapies will be resistant to remodeling, prevent cell death, or overcome signal corruption. A significant amount of basic research remains to be done but, despite promise, secondary gene therapies still seem limited to phases 1–2, although they may function in phase 3, even if randomly targeted.

Bionic Implants

The most successful schemes to restore vision to the profoundly blind from aggressive phase 3 retinal degenerations are epiretinal bionic implants. Surgically placed near the

GC layer, epiretinal implants provide direct current stimulation of retinal GCs or nearby circuitry to activate patches of visual sensation. The argument that such stimulation conflates ON and OFF responses seems irrelevant as several implanted patients can now successfully navigate with such devices. It is clear that the severe remodeling, especially neuronal death, limits candidacy. All implants, whether they are epiretinal, intraretinal, or subretinal, seem to induce further glial and neuronal remodeling. Severe remodeling, BC death, and microneuroma formation will more severely impact subretinal models.

The Importance of Cone Rescue

In every model, the survival of cones seems to delay the onset of phase 3, holding the retina in a phase 2+ state indefinitely. The effect is local and small patches of cones preserve connected BC dendrite structure even when surrounding BCs have lost all dendrites and glutamate receptors. The preservation persists even when cones are severely deconstructed, lacking outer segments and visual pigment expression, and significantly reduced in size. This suggests that cone contact alone, perhaps through synaptic integrins, may be sufficient to preserve BC function. Expression array studies suggest that cone deconstruction may be accelerated by metabolic deprivation. While preserving cones will not rescue vision, it will permit holding the survivor retina in suspended animation, making an array of interventions such as stem/progenitor cell transplantation, fetal retinal sheet transplantation, and secondary gene therapy viable for adults suffering from advanced retinal degenerations. Revisiting survival factor research with a focus on cones may be the critical advance needed for all intervention methods.

See also: Anatomically Separate Rod and Cone Signaling Pathways; Injury and Repair: Light Damage; Injury and Repair: Prostheses; Injury and Repair: Stem Cells and Transplantation; Primary Photoreceptor Degenerations: Retinitis Pigmentosa; Secondary Photoreceptor Degenerations: Age-Related Macular Degeneration; Secondary Photoreceptor Degenerations.

Further Reading

Gargini, C., Terzibasi, E., Mazzoni, F., and Strettoi, E. (2007). Retinal organization in the retinal degeneration 10 (rd10) mutant mouse: A morphological and ERG study. *Journal of Comparative Neurology* 500: 222–238.

Gupta, N., Brown, K. E., and Milam, A. H. (2003). Activated microglia in human retinitis pigmentosa, late-onset retinal degeneration, and age-related macular degeneration. *Experimental Eye Research* 76: 463–471.

Jones, B. W., Watt, C. B., Frederick, J. M., et al. (2003). Retinal remodeling triggered by photoreceptor degenerations. *Journal of Comparative Neurology* 464: 1–16.

Li, Z. Y., Kljavin, I. J., and Milam, A. H. (1995). Rod photoreceptor neurite sprouting in retinitis pigmentosa. *Journal of Neuroscience* 15: 5429–5438.

MacLaren, R. E., Pearson, R. A., MacNeil, A., et al. (2006). Retinal repair by transplantation of photoreceptor precursors. *Nature* 444: 203–207.

Marc, R. E., Jones, B. W., Anderson, J. R., et al. (2007). Neural reprogramming in retinal degenerations. *Investigative Ophthalmology and Visual Science* 48: 3364–3371.

Marc, R. E., Jones, B. W., Watt, C. B., et al. (2008). Extreme retinal remodeling triggered by light damage: Implications for AMD. *Molecular Vision* 14: 782–806.

Marc, R. E., Jones, B. W., Watt, C. B., and Strettoi, E. (2003). Neural remodeling in retinal regeneration. *Progress in Retinal and Eye Research* 22: 607–655.

Margolis, D. J., Newkirk, G., Euler, T., and Detwiler, P. B. (2008). Functional stability of retinal ganglion cells after degeneration-induced changes in synaptic input. *Journal of Neuroscience* 28: 6526–6536.

Peng, Y-W., Hao, Y., Petters, R. M., and Wong, F. (2000). Ectopic synaptogenesis in the mammalian retina caused by rod photoreceptor-specific mutations. *Nature Neuroscience* 3: 1121–1127.

Seiler, M. J., Thomas, B. B., Chen, Z., et al. (2008). Retinal transplants restore visual responses: Trans-synaptic tracing from visually responsive sites labels transplant neurons. *European Journal of Neuroscience* 28: 208–220.

Stasheff, S. F. (2008). Emergence of sustained spontaneous hyperactivity and temporary preservation of OFF responses in ganglion cells of the retinal degeneration (rd1) mouse. *Journal of Neurophysiology* 99: 1408–1421.

Strettoi, E., Pignatelli, V., Rossi, C., Porciatti, V., and Falsini, B. (2003). Remodeling of second-order neurons in the retina of rd/rd mutant mice. *Vision Research* 43: 867–877.

Sullivan, R., Penfold, P., and Pow, D. V. (2003). Neuronal migration and glial remodeling in degenerating retinas of aged rats and in nonneovascular AMD. *Investigative Ophthalmology and Visual Science* 44: 856–865.

Varela, C., Igartua, I., De la Rosa, E. J., and De la Villa, P. (2003). Functional modifications in rod bipolar cells in a mouse model of retinitis pigmentosa. *Vision Research* 43: 879–885.

Injury and Repair: Stem Cells and Transplantation

B A Tucker and M J Young, Schepens Eye Research Institute, Harvard Medical School, Boston, MA, USA
H J Klassen, University of California, Irvine, Orange, CA, USA

Glossary

Choriocapillaris – Vascular layer of the eye, located between the sclera and Bruch's membrane.
Ganglioside – A component of the cell membrane involved in signal transduction.
Neuroprotection – Halting apoptotic cell death of neurons through delivery of diffusible growth factors.
Proteoglycans – A unique class of heavily glycosylated glycoproteins.
Tetraspanins – A family of membrane proteins with four transmembrane domains.

Introduction

The use of surgical procedures for the treatment of blinding conditions has a long history, most notably for cataract. In the modern era, refinements in ophthalmologic techniques and pharmacology have greatly expanded the range of available interventions for a host of ocular conditions. Nevertheless, conditions involving the loss of nonregenerating cell types remain a persistent therapeutic challenge. In the case of diseases involving the loss of corneal endothelial cells, tissue transplantation has proven successful, even though the limited availability of donor corneas restricts the use of this approach.

More problematic are diseases involving the loss of retinal neurons and retinal pigment epithelial (RPE) cells where few, if any, options are currently available to the treating clinician. As a group, diseases of this type are common and examples include age-related macular degeneration, retinal detachment, retinitis pigmentosa, as well as various forms of optic neuropathy, including glaucoma. In addition, common retinal vascular diseases such as diabetic retinopathy and vascular occlusive disease frequently involve retinal cell loss. Given the prevalence of visual disability resulting from these conditions, novel methods of both preserving and replacing retinal neurons are sorely needed.

Early Transplant Work

Over the last several decades, experimental work in rodents has made considerable progress in neuroprotection as well as cell and tissue transplantation in the setting of retinal cell loss. Seminal work by LaVail and Steinberg demonstrated the power of growth factor-mediated neuroprotection for retinal degeneration, both hereditary and acquired. Work by Lund's group showed that transplanted retinal tissue can, to a certain extent, engraft and integrate with the recipient visual system. Work by the Turner and Gouras laboratories showed that RPE transplantation can confer survival benefits to host photoreceptors following placement in the subretinal space. This last observation was made in dystrophic Royal College of Surgeons (RCS) rats and was subsequently extended to a variety of manipulations in this model. Taken together, these early studies provided compelling evidence of efficacy in mammals.

Despite the above advances, however, significant limitations remained, thereby encumbering further development of these otherwise promising strategies. In the case of neuroprotection, a major challenge has been to find a method of sustained intraocular delivery that is not neutralized due to the rapid degradation of peptide growth factors by endogenous proteases. For neural transplantation, a perplexing challenge has been to find a method of promoting regeneration across injury-induced gliotic barriers. Furthermore, RPE cell transplantation was challenged by the apparent reluctance of grafted cells to reconstitute an orthotopic epithelial monolayer on Bruch's membrane. In all these cases, a compounding impediment to translation of these strategies was posed by the limited ability of the rodent eye to model an approach with tangible clinical relevance.

Within the past 10 years, new breakthroughs pertaining to the above challenges have emerged from the nascent field of stem cell transplantation, as will be discussed presently.

Neural Progenitor Cells

The first report of successful transplantation of stem-like cells to the retina was from Masayo Takahashi, then working in the laboratory of Fred Gage. The cells used were adult hippocampal progenitor cells (AHPCs), a clonally selected line of highly mitotic cells cultured from the brain of mature rats and genetically altered to express green fluorescent protein (GFP) as a reporter gene. This study showed a remarkable degree of apparently seamless integration of donor AHPCs into the immature retina of normal recipients. Although the cells fell short in terms of marker expression, their morphology, placement, and orientation were strikingly similar to resident retinal cell types.

In addition to working with Gage, we showed that AHPCs can also integrate into the diseased or injured retina of mature rats and that at least a degree of phenotypic marker expression was achievable, thereby underscoring the potential of using stem cells for cell replacement in the setting of common retinal disorders.

Together, these two studies with AHPCs stimulated considerable interest in stem cell research within the scientific community, particularly with respect to retinal applications. Initial work on cultured cells continued to focus on AHPCs as the first cell type with a demonstrated ability to integrate into the cytoarchitecture of the neural retina. These cells were shown to exhibit a number of nonimmunogenic properties and to express a number of important cytokines, as did similar brain-derived cells of human origin. Neural progenitor cells from the brains of human and GFP-mouse brain were compared in terms of surface marker expression. It was found that cultured neural progenitor populations consistently express 3-fucosyl-N-acetyl-lactosamine (CD15), GD(2) ganglioside, and the tetraspanin proteins CD9 and CD81 across species, whereas major histocompatibility complex (MHC) expression is more variable, specifically for class I antigens. Transplantation of the mouse cells showed that they exhibited characteristics of an immune-privileged cell type. Similar cells were later derived from the brain of the pig and cat.

Retinal Progenitor Cells

A significant limitation that emerged from the work with brain-derived progenitors was the difficulty in obtaining photoreceptor marker expression from these cells following engraftment in the retina. While it has been shown that recoverin expression can be induced under exceptional circumstances, it remained to be seen whether a stem-like cell could generate cells with outer segments that co-express multiple appropriate markers such as recoverin and rhodopsin. It was this challenge that led us to explore the possibility of deriving transplantable progenitor cell cultures from the mammalian neural retina.

Starting with newborn GFP-transgenic mice, we were able to grow cells from dissociated retinal tissue in the presence of mitogenic stimulation from epidermal growth factor (EGF). These cells were capable of expressing rhodopsin, either in culture under differentiation conditions or following engraftment in the host retina. We have obtained similar findings in work with porcine and human donor tissue of fetal origin. Following differentiation, either *in vitro* or *in vivo*, these cells have been shown to be capable of co-expressing rhodopsin and recoverin, consistent with rod photoreceptor phenotype. Retinal progenitor cells (RPCs) therefore represent a stem-like cell type that provides a potential source of new photoreceptors for

retinal repair in the setting of various forms of diseases and injuries.

As mentioned above, transplantation of mature RPE cells for cell replacement has been hampered by the apparent inability of these cells to recreate an epithelial sheet and adhere to Bruch's membrane. In contrast, retinal and various types of neural progenitor cells have also shown an ability to integrate into the RPE monolayer, although it is not yet known if such cells can replace native RPE cells at the functional level. Recent work with pluripotent stem cells suggests that these represent a source for new retinal cells with functional properties.

Induced Pluripotent Stem Cells

A new development in stem cell biology that has changed the way we think about commitment and differentiation is a process known as direct reprogramming (**Figure 1**). Yamanaka and colleagues were the first to demonstrate this technique, in which somatic cells are reprogrammed to become pluripotent stem cells through the addition of several stemness transcription factors. They began with a type of skin cell known as a fibroblast, and inserted four transcription factors (Oct3/4, Klf4, Sox2, and c-Myc) using retroviruses. A subpopulation of the fibroblasts began expressing other markers of embryonic stem (ES) cells, such as nanog, and formed teratomas when transplanted into immunocompromised mice. Careful evaluation of the resulting cells revealed that they indeed had all the properties of ES cells, and have been termed induced pluripotent stem (iPS) cells.

There are several important ramifications of this work. First, this technology offers the promise of creating patient-specific stem cells. One could harvest skin cells from a donor, turn these cells into iPS cells, and subsequently differentiate this population into the cells of interest, which would then be transplanted back into the donor. Importantly, this would obviate the need for chronic immunosuppression and eliminate the risk of immunorejection of the grafted cells. Many of the ethical issues associated with ES cells would also be circumvented by using this technique, although as with most such developments, new and unexpected ethical questions arise. Thus far, no one has created a viable embryo through this technology, and indeed there are many biological hurdles standing in the way of this. However, as the theoretical possibility of using iPS cells for reproductive cloning exists, this work raises novel moral issues.

A number of technical barriers must be overcome before iPS cells can be useful for the treatment of disease. Foremost among these is the use of oncogenes (cancer-associated transcription factors) to reprogram fibroblasts. These genes often lead to tumorogenesis, and indeed scientists have seen cancerous growths from the

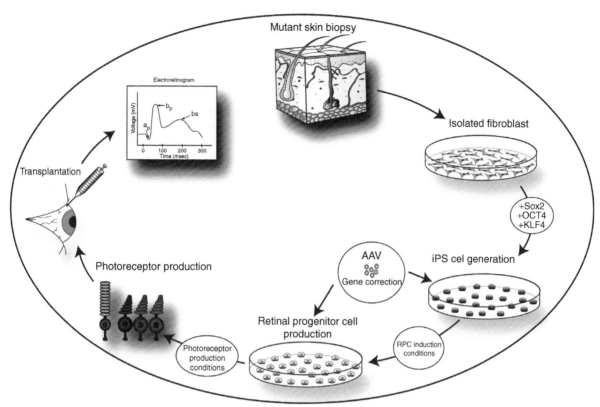

Figure 1   Schematic diagram illustrating the steps toward differentiation of reprogrammed fibroblasts into retinal neurons.

transplanted progeny of iPS cells. In addition, the use of retroviruses to deliver the four factors to the nuclei of the fibroblasts must be replaced by other means before using these cells in the clinic. It is worth noting, however, that in the months that followed the first publication of this technique, a flurry of scientific papers were published in which only two or three of the reprogramming factors were needed, and vectors other than retroviruses were used for gene insertion. This field is developing extremely rapidly, and it is likely only a matter of time until these methodological obstacles are overcome.

Stem Cells for Neuroprotection

In addition to cell replacement, stem-like cells have potential use in retinal neuroprotective strategies. We reported evidence that murine RPCs confer survival benefits to dysfunctional rods following transplantation to the retina of rhodopsin double knock-out mice. This result indicated that RPCs, or their more differentiated progeny, possess an inherent neuroprotective influence that is revealed in this setting. Interestingly, a similar finding was obtained in retinal dystrophic mice in work with a very different type of stem cell derived from the bone marrow.

Elucidating the mechanisms underlying such effects remains a subject of active interest; however, an alternative

method for inducing neuroprotection via stem cells is to modify them genetically so as to over-express the therapeutic gene of interest. Gamm and colleagues have shown that brain-derived progenitor cells modified to ever express a GDNF transgene were capable of rescuing photoreceptor cells in the RCS rat following transplantation.

Retinal Transplantation

As suggested above, to achieve functional repair in patients afflicted with retinal degenerative disorders, there is an urgent need to reconstruct, via cellular replacement, the damaged or lost layers of the retina. During the past 30 years attempts have been made in order to achieve retinal regeneration via transplantation of developing donor material. Although significant progress and invaluable information have been acquired, visual restoration using this technique has not been achieved. For instance, although embryonic rat retina was shown to survive without immune rejection, develop normally, and continue to respond to light for up to 3 months post-transplantation, extensive host–donor integration was not identified. Similarly, comparable experiments performed in pig models of retinal dystrophy showed that although healthy retinal tissue from fetal donors survived and maintained proper laminar structure and cellular organization for up 6 months post-transplantation, this tissue did not integrate with the host retina, and as such, did not aid in visual restoration. These findings collectively suggest that the eye is an amenable site for transplantation; however, the proper donor material or extracellular conditions have yet to be met.

Transplantation Strategies

Stem/progenitor cell transplantation as a means of inducing tissue reconstruction and functional regeneration has garnered extensive interest in the field of regenerative medicine. Unlike the solid tissue transplants mentioned above, stem cells have the ability to integrate within host tissue and develop into retinal specific cell types. For instance, in 2004, we were able to show that a subset of grafted RPCs gave rise to mature retinal ganglion and photoreceptor cells following transplantation. Since then, numerous studies reporting varying degrees of success have utilized an assortment of different cell types, including the fate-restricted photoreceptor precursor and the pluripotent ES cell. Regardless of the cell type used, efficient delivery methods are required. Traditionally, the two transplantation techniques most extensively utilized for cell delivery were vitreal cavity and subretinal space injections. Regardless of species, vitreal cavity injections generally involve insertion of a needle through the pars plana into the vitreal space, where in the case of mice, approximately 1 µl of cell suspension as a bolus is injected (**Figure 2(a)**). The subretinal space, unlike the vitreous, is actually a pseudo-space that is created upon detachment of the retina from the underlying RPE layer. Cellular injections into this space are performed by one of the two techniques. The first, which is very similar to that carried out during vitreal injections, involves entering the eye through the pars plana and inserting a needle beneath the retina via a precut hole, or retinotomy. Through this hole, fluid, most often saline, is injected, creating a retinal detachment and bleb into which the cell of choice can be delivered (**Figure 2(b)**). This approach generally requires a larger vitreal space in which to work and is often performed in larger models of retinal disease, such as the pig and human. The second technique, which is more often employed in smaller organisms such as rodents, is performed by making an incision through the sclera (sclerotomy), choriocapillaris, Bruch's membrane, and overlying RPE, through which a needle is inserted and saline is injected. This again creates a retinal detachment/bleb into which cells can be delivered (**Figure 2(c)**). The reasons for choosing one approach over the other are vast and may range from personal preference to surgical competence. However, more practical reasons for choosing vitreal or subretinal approaches for stem cell transplantation exist, including the type of injury and cell/retinal layer being targeted.

For instance, subretinal transplantation would be more beneficial for targeting the retinal photoreceptor layer in retinal degenerative diseases such as retinitis pigmentosa (RP) and age-related macular degeneration (AMD), while vitreal transplantation would be more beneficial for targeting the inner nerve fiber layer in diseases such as glaucoma and various other optic neuropathies. Tailoring the transplantation approach to the specific disease at hand allows for larger numbers of cells to be provided to the area required. For instance, performing vitreous cavity injections of retinal stem cells as a treatment regime for photoreceptor cell replacement requires that the cells migrate through the entire thickness of the remaining retina before arriving at the desired location. With an increased distance for cellular migration, there is a greater chance that cells will take up residence in unwanted retinal layers, thus decreasing the chance that they will actually reach the appropriate target.

Polymer Substrates

Despite choosing the correct transplantation technique, the use of bolus cell injection as a delivery method is actually quite inefficient and typically results in massive cellular efflux, low cellular survival, and poor integration. For example, it has previously been suggested that as little as 0.1% of transplanted retinal stem cells actually survive for an extended period of time following traditional bolus injection. Issues such as these, which hinder functional regeneration, are largely accounted for by the lack of cellular support and trauma associated with the injection processes. For instance, in a recent publication we were able to show that the shearing forces placed on retinal stem cells by passing them through the typical glass injection needle resulted in nearly 60% cell death at 3 days post-passage. When the glass injection needle was replaced with a large bore pipette, cell death was reduced to less than 2%. However, even if a large bore pipette could be used during the transplantation process, the fact remains that a large portion of the cells injected would still be lost to efflux via the implantation port. This is especially true when attempting to place cells into the subretinal space through an opening in either the retina or sclera. One approach that we have developed in an attempt to circumvent these issues is to apply tissue-engineering techniques focused on the use of biodegradable polymer scaffolds as cell delivery vehicles. In one study we were able to show that the transplantation of retinal stem cells on a biocompatible poly(lactic-*co*-glycolic acid) (PLGA) polymer resulted in greatly reduced cell death and prevented leakage and migration of retinal stem cells away from the transplantation site. This resulted in significantly improved cellular integration when compared to bolus cell injections.

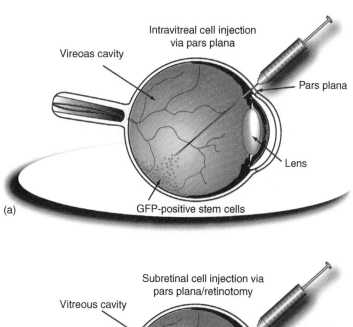

(a)

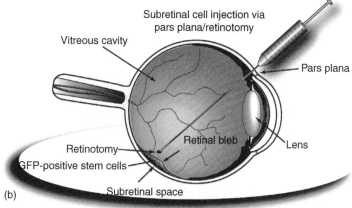

(b)

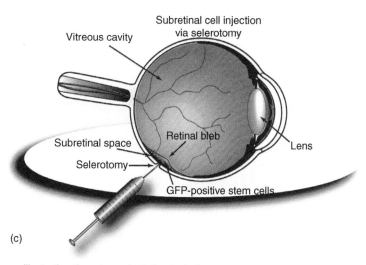

(c)

Figure 2   Schematic diagram illustrating three transplantation techniques.

Inhibitory Barriers

Although we have developed an efficient means of cell delivery, in order to achieve functional regeneration via retinal reconstruction/repopulation, a further increase in cellular integration and axonal extension/synapse formation is required. As suggested above, a major contributor to the lack of functional regeneration in the face of efficient cell delivery is the presence of an inhibitory extracellular environment. Unlike the peripheral nervous system (PNS), the central nervous system (CNS) is stricken with an abundance of myelin-associated extracellular matrix (ECM) proteins, namely MAG, Nogo, and Omgp, which are well known for their ability to inhibit process extension and functional regeneration. The optic nerve, like other CNS locations, is laden with these inhibitory proteins, and as such, their chemical and/or enzymatic neutralization has been shown to significantly enhance retinal ganglion cell axon extension and optic nerve regeneration. Thus, in order for stem cell therapies to be used as an effective treatment for diseases such as glaucoma, which affect the retinal ganglion cell layer and optic nerve, inhibition of these proteins will ultimately be required.

Unlike the optic nerve, the retina is void of myelin, oligodendrocytes, and their associated inhibitory ECM molecules. Thus, these factors are not a concern when attempting to stimulate regeneration in retinal degenerative diseases such as RP and AMD, both of which predominantly affect the retinal outer nuclear layer. Why then are attempts at stimulating retinal regeneration via stem cell transplantation still largely unsuccessful? The lack of functional integration following transplantation is in large part due to the presence of an inhibitory injury induced glial scar that forms during reactive gliosis. In the retina, reactive gliosis is predominantly characterized by Müller glial activation, as indicated by upregulation of intermediate filament proteins such as glial fibrillary acidic protein (GFAP) (**Figure 3(a)**, red) and the projection of processes from their original location (forming the outer limiting membrane) into the subretinal space. Here, these processes proceed to form a dense fibrotic barrier that contains a variety of growth inhibitory ECM/adhesion molecules,

which include the chondritin sulfate proteoglycan (CSPG) Neurocan (**Figure 3(a)**, blue) and the hyaluronan-binding glycoprotein CD44 (**Figure 3(b)**, red). Both Neurocan and CD44 have previously been shown to function as chemical inhibitors to axon growth and cellular migration, thus preventing regeneration and functional synapse formation. A variety of approaches have been taken in an attempt to remove these inhibitory ECM molecules, including enzymatic degradation of the proteins themselves. One such enzyme that we have utilized for this purpose is the matrix metalloproteinase 2 (MMP2). MMPs are well known for their ability to degrade a variety of ECM and cell adhesion molecules, including CD44, the CSPGs, and the aforementioned myelin-associated inhibitors. MMP2, in particular, has been shown to cleave both CD44 and Neurocan, thus releasing their negative hold on axonal extension and cellular migration. In a recent study, we discovered that endogenous MMP2 induction resulted in glial barrier-associated CD44 and Neurocan degradation at the outer limits of heavily scarred degenerating retinas, subsequently stimulating integration and synapse formation between the host and healthy transplanted tissue grafts. Conversely, chemical and/or genetic inhibition/removal of MMP2 was shown to completely abolish stem cell migration and retinal repopulation. In light of these findings, we have developed a biodegradable PLGA polymer scaffold that, as in previous permutations, is capable of efficient cell delivery and also possesses the ability to provide a sustained controlled release of active MMP2 following degradation (MMP2-PLGA). Subretinal transplantation of this polymer in conjunction with retinal stem cells as a composite graft has resulted in efficient glial barrier degradation and enhanced stem cell integration. This new polymer scaffold will undoubtedly act as an invaluable transplantation tool

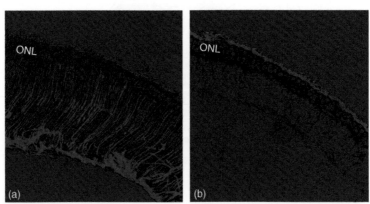

Figure 3 Injury-induced Müller cell activation stimulates inhibitory ECM molecule deposition and glial barrier formation. Eyes from adult retinal degenerative mice (Rho–/–) were enucleated, fixed, cryoprotected, sectioned, and immunostained for GFAP, CD44, and Neurocan. (a) Representative micrograph illustrating CD44 (blue) and GFAP (red) expression in the adult degenerative Rho–/– mouse retina. Activated Müller cells extend processes through the degenerating photoreceptor layer (ONL) into the subretinal space and deposit CD44 at the outer limits of the retina. (b) Representative micrograph illustrating neurocan (red) expression in the adult Rho–/– mouse retina. Inhibitory glial barrier-associated proteins, CD44 and Neurocan, deposited at the outer limits of the degenerating retina. Inhibitory CD44 and Neurocan molecules are intertwined within the degenerating photoreceptor layer, preventing cellular/axonal integration following subretinal transplantation.

for targeted repopulation of the retinal and restoration of vision.

Conclusions

Regardless of the source of the transplanted material, retinal transplants aim to rescue or replace injured photoreceptors. It is likely that each of these objectives will require different cell types as donor tissue. For example, rescue of diseased rods may be accomplished by delivering nonengrafting mesenchymal stem cells, or perhaps with engrafting neural stem cells. In these cases, rescue would be achieved by the delivery of growth factors such as glial cell-derived neurotrophic factor (GDNF) or brain-derived neurotrophic factor (BDNF) in a nonspecific fashion, possibly to cells other than photoreceptors (e.g., Müller glia). Engraftment and long-termed survival is a requirement if rescue of the diseased photoreceptors is to be prolonged. In contrast, repair through reconstruction involves a more complicated and indeed more difficult set of concerns. Cell replacement not only requires engraftment and long-termed survival, but also necessitates differentiation.

A cell type that can differentiate into fully mature, functional photoreceptors narrows the field of players somewhat. While a hematopoetic or stromal stem cell may by somewhat ill-suited to such a task, retinal stem cells or ES cells may be ideal candidates.

See also: Coordinating Division and Differentiation in Retinal Development; Injury and Repair: Retinal Remodeling; Primary Photoreceptor Degenerations: Retinitis Pigmentosa; Primary Photoreceptor Degenerations: Terminology; Retinal Ganglion Cell Apoptosis and Neuroprotection; Retinal Histogenesis; Zebra Fish–Retinal Development and Regeneration.

Further Reading

Caroni, P. and Schwab, M. E. (1988). Two membrane protein fractions from rat central myelin with inhibitory properties for neurite growth and fibroblast spreading. *Journal of Cell Biology* 106: 1281–1288.

Kajita, M., Itoh, Y., Chiba, T., et al. (2001). Membrane-type 1 matrix metalloproteinase cleaves CD44 and promotes cell migration. *Journal of Cell Biology* 153: 893–904.

Klassen, H. and Lund, R. D. (1987). Retinal transplants can drive a pupillary reflex in host rat brains. *Proceedings of the National Academy of Sciences of the United States of America* 84: 6958–6960.

Klassen, H. J., Ng, T. F., Kurimoto, Y., et al. (2004). Multipotent retinal progenitors express developmental markers, differentiate into retinal neurons, and preserve light-mediated behavior. *Investigative Ophthalmology and Visual Science* 45: 4167–4173.

Lamba, D. A., Karl, M. O., Ware, C. B., and Reh, T. A. (2006). Efficient generation of retinal progenitor cells from human embryonic stem cells. *Proceedings of the National Academy of Sciences of the United States of America* 103: 12769–12774.

LaVail, M. M., Unoki, K., Yasumura, D., et al. (1992). Multiple growth factors, cytokines, and neurotrophins rescue photoreceptors from the damaging effects of constant light. *Proceedings of the National Academy of Sciences of the United States of America* 89: 11249–11253.

MacLaren, R. E., Pearson, R. A., MacNeil, A., et al. (2006). Retinal repair by transplantation of photoreceptor precursors. *Nature* 444: 203–207.

Nakagawa, M., Koyanagi, M., Tanabe, K., et al. (2008). Generation of induced pluripotent stem cells without Myc from mouse and human fibroblasts. *Nature Biotechnology* 26: 101–106.

Osakada, F., Ikeda, H., Mandai, M., et al. (2008). Toward the generation of rod and cone photoreceptors from mouse, monkey and human embryonic stem cells. *Nature Biotechnology* 26: 215–224.

Park, I. H., Zhao, R., West, J. A., et al. (2008). Reprogramming of human somatic cells to pluripotency with defined factors. *Nature* 451: 141–146.

Sheedlo, H. J., Li, L., and Turner, J. E. (1989). Photoreceptor cell rescue in the RCS rat by RPE transplantation: A therapeutic approach in a model of inherited retinal dystrophy. *Progress in Clinical and Biological Research* 314: 645–658.

Takahashi, M., Palmer, T. D., Takahashi, J., and Gage, F. H. (1998). Widespread integration and survival of adult-derived neural progenitor cells in the developing optic retina. *Molecular and Cellular Neuroscience* 12: 340–348.

Tomita, M., Lavik, E., Klassen, H., et al. (2005). Biodegradable polymer composite grafts promote the survival and differentiation of retinal progenitor cells. *Stem Cells* 23: 1579–1588.

Zhang, Y., Klassen, H. J., Tucker, B. A., Perez, M. T., and Young, M. J. (2007). CNS progenitor cells promote a permissive environment for neurite outgrowth via a matrix metalloproteinase-2-dependent mechanism. *Journal of Neuroscience* 27: 4499–4506.

Relevant Websites

http://www.blindness.org – Foundation Fighting Blindness.

http://www.hsci.harvard.edu – Harvard Stem Cell Institute.

http://webvision.med.utah.edu – John Moran Eye Center, University of Utah, Webvision.

http://www.schepens.harvard.edu – Schepens Eye Research Institute, Michael Young.

Innate Immune System and the Eye

M S Gregory, Schepens Eye Research Institute, Harvard Medical School, Boston, MA, USA

Glossary

Endocytosis – The process by which cells absorb molecules, such as proteins, from outside the cell by engulfing them with their cell membrane to form an endosome.

Endophthalmitis – An infection of the posterior of the eye.

Opsinization – The process by which a pathogen or infected cell is marked for destruction by a phagocyte.

Phagocytosis – The engulfment of solid particles, such as bacteria, by the cell membrane to form an internal phagosome.

Introduction

Innate immunity comprises a large number of molecules and cells that recognize and respond rapidly to pathogens, providing immediate defense against infection. However, innate immunity also carries with it the potential of highly destructive inflammation that presents an important dilemma for the eye. Inflammation is necessary for successfully eradicating pathogens. An ideal response would eliminate the microorganisms before they are able to directly damage any ocular tissues. The innate immunity would be limited and produce little or no damage to the surrounding normal tissues. However, some types of ocular infections trigger inflammation that is either (1) insufficient to clear the microorganisms, resulting in direct destruction of ocular tissue by the pathogen, or (2) excessive inflammation that clears the microorganisms, but destroys a significant amount of normal tissue. Either of these two scenarios is undesirable and can lead to significant loss of vision. Therefore, a delicate balance must be achieved between the amount of inflammation required for pathogen clearance and the amount of nonspecific tissue damage.

The innate immune system of the eye is similar to other mucosal surfaces. The first tier is passive consisting of several anatomic, physical, and chemical barriers that work together to prevent infection without inducing inflammation. The second tier is active consisting of cellular and secretory components that together cause acute inflammation aimed at eradicating the pathogen. The delicate

tissues of the eye that make up the visual axis (cornea, lens, and retina) have a very low tolerance for inflammation, as a very small amount of damage can produce a significant loss of vision. The two-tiered system helps to prevent unnecessary inflammation and the active mechanisms of innate immunity are only turned on once the passive barriers have been breached. Both the passive and active arms of ocular innate immunity are the focus of this article.

Passive Innate Defense System

Anatomic and Physical Barriers

Several anatomic and physical barriers protect the anterior and posterior of the eye from invading pathogens (**Figure 1**). The active arm of innate immunity is only triggered when pathogens breach these barriers. The cornea is exposed to the external environment, making the anterior segment highly vulnerable to potential pathogen invasion. Therefore, the anterior segment possesses a multilayer barrier system that includes: eyelids and eyelashes, tear film, and the corneal epithelium. By contrast, the posterior segment is not exposed to the external environment, and is therefore less vulnerable to infection. The critical barriers of the posterior segment include (1) the retinal pigment epithelium (RPE), which. lies between the blood-rich choroid and the neural retina, and (2) the posterior lens capsule that forms the barrier between the anterior and posterior segments. Each component of the passive innate defense system is described briefly below.

Eyelids and eyelashes

The outermost barrier of the ocular surface consists of the eyelids and eyelashes. The eyelashes protect the ocular surface from dust and foreign debris. The regular blinking action of the eyelids moves the tears across the ocular surface, washing away potentially colonizing or infecting organisms.

Tear film

Tears form the second barrier, lubricating and protecting the ocular surface. Tears also posses a potent defense system that limits the growth, colonization, and survival of microorganisms. The tear film consists of three layers: the outermost lipid layer, an aqueous layer, and the inner mucus layer (**Figure 2**). The lipid layer lubricates the eyelid and slows evaporation of the aqueous tear film layer. The aqueous layer forms the major component of the tear film and contains numerous antimicrobial

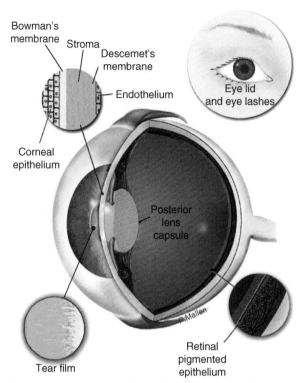

Figure 1 Anatomic and physical barriers of the eye. The eyelid, eye lashes, tear film, and corneal epithelium serve as barriers of the anterior segment of the eye. The posterior lens capsule and RPE serve as barriers of the posterior segment of the eye.

proteins including: lysozyme, lactoferrin, defensins, secretory IgA (sIgA), and complement. Many of these antimicrobial proteins are constitutively expressed and provide early, broad-spectrum protection against invading pathogens and also prevent the overgrowth of commensal bacteria. The innermost mucus layer of the tear film is made up of secreted and membrane-bound mucins that protect the epithelium from debris, pathogens, and desiccation. Mucins are high-molecular weight glycoproteins characterized by extensive O-glycosylation. Membrane-bound mucins expressed by the ocular surface epithelia include MUC1, MUC4, and MUC16. Secreted mucins are also found in the mucus layer and include MUC2, MUC5AC, and MUC19. The membrane-bound mucins anchor the ocular tear film to the corneal epithelium and are thought to act as a physical barrier against pathogen penetrance. Secreted mucins bind to pathogens in the tear film, facilitating their clearance from the ocular surface. Under normal conditions, mucin production and secretion by goblet cells and corneal epithelial cells are constitutive. However, mucin production can also be induced via Toll-like receptors (TLRs) expressed on the surface of corneal epithelial cells. Moreover, inflammatory cytokines, such as IL-1β, IL-6, and TNFα have also been shown to induce mucin production and secretion. Together, these data reveal that constitutively expressed mucins make up

a critical component of the passive defense system, while at the same time, upregulation of mucin production and secretion can also be a product of the active arm of innate immunity in the eye.

Corneal epithelium
The final barrier of the ocular surface consists of nonkeratinized stratified epithelial cells bound together by tight junctions. The corneal epithelium acts as a physical barrier to invasion of microorganisms due to the presence of epithelial intercellular tight junctions and the rapid renewal of epithelial cells with frequent shedding of the superficial layers of potentially infected epithelium. As mentioned in the previous section, the epithelium also expresses membrane-bound mucins that inhibit bacterial binding to the epithelial surface and produce several of the antimicrobial factors that are present in the ocular tear film.

Posterior lens capsule
The posterior lens capsule forms a physical barrier between the anterior and posterior segments of the eye after extracapsular cataract surgery and prevents the spread of microorganisms from the anterior chamber into the posterior chamber in the postsurgical eye. The best example of this is the fact that an intact posterior lens capsule is critical in preventing endophthalmitis following cataract surgery. Contamination of the aqueous humor can occur during cataract surgery. However, the pathogens are quickly cleared and endophthalmitis does not develop. By contrast, when the posterior capsule is breached, the rate of endophthalmitis increases significantly. This supports the finding that the anterior segment is much more efficient at clearing bacteria as compared to the posterior segment. Studies suggest the difference in ability to clear pathogens in the anterior versus posterior of the eye may be linked to expression of antimicrobial peptides. One major difference is that the AH is continuously secreted and drained, whereas the vitreous humor is not. The vitreous also offers greater opportunity for microbes to bind its fibrils. However, the exact molecular mechanisms involved remain unclear.

Retinal pigment epithelium
The RPE consists of a single layer of cells joined by tight junctions that lie between the photoreceptors of the neural retina and the blood-rich choroid. The RPE serves multiple functions aimed at protecting and maintaining the health of the neural retina. RPE cells (1) phagocytose shed disks from the photoreceptor outer segments and recycle their components; (2) transport nutrients from the choroid to the retina; (3) absorb light; (4) provide adhesive properties for the retina; and (5) serve as a rich source of cytokines, chemokines, and growth factors. More recently, RPE have also been linked to immunity and have been

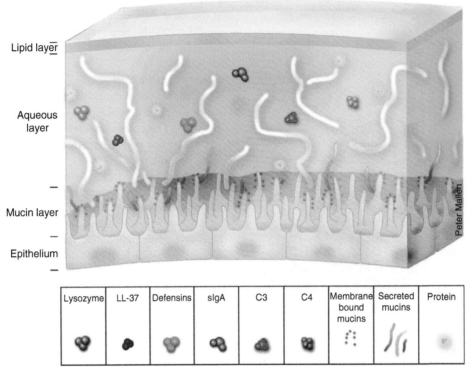

Figure 2 Ocular tear film. The ocular tear film is composed of three layers: the outermost lipid layer, the aqueous layer, and the innermost mucus or mucin layer. The lipid layer lubricates the eyelid and slows evaporation of the aqueous tear film layer. The aqueous layer forms the major component of the tear film and contains numerous antimicrobial proteins including: lysozyme, cathelicidin (LL-37), defensins, secretary IgA (sIgA), and complement components (C3 and C4). The mucin layer acts as a physical barrier against pathogen invasion and consists of both membrane bound and secreted mucins.

shown to behave as antigen-presenting cells that phagocytose pathogens, produce cytokines, and present pathogen-derived peptides to sensitized T cells. Therefore, while RPE acts as a physical barrier to the posterior segment, it is also important in shaping and regulating the adaptive immune response once this barrier has been breached.

Chemical Barriers

In addition to the anatomic and physical barriers designed to block invasive pathogens, there are also a number of soluble factors that inhibit bacterial growth, adherence, and survival. Several of these factors are constitutively expressed at low levels, providing a baseline of protection from foreign pathogens. However, many of these factors can also be upregulated in response to pathogens and inflammation; thus, these become an important product of the active arm of innate immunity. Some of the more important factors are discussed in more detail below.

Lysozyme

Lysozyme is bacteriacidal and makes up 20–40% of the total tear protein. Lysozyme kills bacteria by (1) binding to and creating pores in the bacterial cell wall, or (2) dissolving bacterial membranes by enzymatic digestion.

Secretory phospholipase A₂

Secretory phospholipase A_2 exhibits potent antibacterial activity against Gram-positive bacteria. Similar to lysozyme, secretory phospholipase A_2 dissolves bacterial membranes by hydrolyzing the principal phospholipid, phosphatidylglycerol.

Cathelicidin (LL-37)

LL-37 is a small cationic peptide with potent antimicrobial activity against Gram-positive and Gram-negative bacteria as well some viruses. The precise mechanism of action is incompletely understood, but it is widely believed that the antimicrobial activity is due to disruption of the microbial membrane or viral envelope.

Defensins

Beta defensins are expressed in epithelial cells that line mucosal surfaces such as the cornea. Similar to LL-37, defensins have a broad spectrum of antimicrobial activity and are effective against: Gram-positive and Gram-negative bacteria, fungi, and enveloped viruses. In addition to their antimicrobial activities, defensins modulate a variety of cellular activities including immune cell chemotaxis, epithelial proliferation, cytokine secretion, and stimulation of histamine release from mast cells. Human corneal epithelial cells constitutively express human beta defensin-1 (hBD-1) and hBD-3. By contrast, hBD-2 is induced in response to corneal injury, infection, or inflammation, and is approximately 10-fold more potent than hBD-1 with an even wider antibacterial spectrum. Therefore, while hBD1 and hBD-3 provide a baseline defense to protect the cornea from infection, upon injury or microbial invasion, hBD-2 is upregulated and displays increased antimicrobial activity.

Lactoferrin

Lactoferrin is bacteriostatic and binds to and depletes iron from the tear film, which is required for microbial metabolism and growth. Lactoferrin is also bactericidal and permeabilizes membranes of Gram-positive and Gram-negative bacteria. Additional functions of lactoferrin have also been described: (1) inhibits biofilm development, (2) inhibits bacterial adhesion to host cells, (3) inhibits intracellular invasion, (4) amplifies apoptotic signals in infected cells, and (5) enhances bactericidal activity of neutrophils.

Lipocalin-A

Lipocalin A prevents bacteria from obtaining iron, an essential nutrient for microorganism survival. However, unlike lactoferrin, lipocalin-A does not bind iron directly. Lipocalin-A inhibits the iron acquisition system of microbes by binding to and blocking microbial sidephores used to transport iron into bacteria.

Secretory IgA

sIgA protects the ocular surface against colonization and possible invasion by pathogenic microorganisms by binding to bacteria and facilitating clearance. In addition, sIgA can opsonize bacteria for phagocytosis.

Complement

Complement components (such as C3 and C4) are constitutively expressed in the tear film and are involved in phagocytic chemotaxis, opsonization, and lysis of bacteria. The eye is unusual in that there is a constitutive low level of activated complement that is present even in uninfected normal eyes. It is believed that this low level of activated complement provides innate immune surveillance of microbes and allows for a rapid activation of the full complement cascade upon pathogen invasion.

Active Innate Defense System

Pattern Recognition Receptors

Innate immunity develops rapidly and is described historically as nonspecific, while adaptive immunity develops slowly and is antigen-specific. However, the discovery of pattern recognition receptors (PRRs) that detect unique pathogen-associated molecular patterns revealed a level of specificity in innate immunity that not only allows discrimination between self and non-self, but also allows the development of innate immunity tailored to specific pathogens, such as bacteria, viruses, and fungi.

There are two classes of PRRs: (1) TLRs and (2) nucleotide-binding oligomerization domain (NOD)-like receptors (NLRs) (**Table 1**). TLRs are expressed on the cell surface and detect pathogens at the cell membrane. TLRs are also expressed within endosomes and detect pathogens that have been endocytosed. By contrast, NLRs are expressed within in the cytoplasm and detect the presence of microbial molecules inside the host cell. These two classes of PRRs are discussed in detail below.

Toll-like receptors

TLRs are type 1 transmembrane proteins with an extracellular domain for ligand binding composed of leucine rich repeats and a cytoplasmic domain for intracellular signaling which is known as the Toll/IL-1 receptor (TIR) domain. TLRs recognize bacteria, viruses, fungi, protozoa, and endogenous ligands, such as heat shock proteins and fibrinogen. Triggering of the TLR leads to activation of the transcription factor, nuclear factor-kappa B (NF-κB), and the expression of pro-inflammatory molecules, such as TNF-α, IL-1, and IL-2. To date, 10 human TLRs have been identified and each TLR has a unique ligand specificity (**Table 1**). TLRs were first identified on innate immune cells: neutrophils, macrophages, monocytes, and dendritic cells. More recently, TLRs were also identified on epithelial cells that lie at the host/environment interface: skin, gastrointestinal tract, respiratory tract, and urogenital tract. Furthermore, several TLRs have been identified on ocular tissue in both the anterior and posterior segment of the eye, including cornea, iris, ciliary body, choroid, and RPE.

Similar to other barrier epithelium, several TLRs have been identified on the RPE and corneal epithelial cells and are vital for sensing microbes and triggering a rapid response to eliminate the pathogen. The study of TLRs in ocular immunity is still relatively new and debate remains over which TLRs are expressed in the eye and whether TLRs are expressed on the cell surface or within endosomes. However, it is clear that TLRs are required for initiation of innate immunity in the eye. This is supported by studies using TLR knockout mice that demonstrate a significant decrease in inflammation characterized by decreased neutrophil infiltration and increased susceptibility to infection in the absence of TLRs.

NOD-like receptors

The NLRs comprise a large family of cytoplasmic PRRs that recognize bacteria and endogenous danger signals such as uric acid (**Table 1**). All members of the NLR family share a conserved NOD. However, the NLR family

Table 1   TLR and NLR expression within the eye

	Ligand	Location in the eye
TLR1	Triacyl lipopeptides	Cornea, conjunctiva, RPE[a]
TLR2	Glycolipds, lipopeptides, lipoproteins, PGN, LTA, HSP70, zymosan	Cornea, onjunctiva, RPE
TLR3	dsRNA (viruses)	Cornea, conjunctiva, RPE
TLR4	Lipid A (Gram-negative bateria), LPS, bacterial HSP60, RSV coat protein	Cornea, conjunctiva, iris, ciliary body, choroid, whole retina, RPE
TLR5	Flagellin	Cornea, RPE
TLR6	Diacyl lipopeptides	Conjunctiva, RPE
TLR7	ssRNA (viruses)	Cornea, conjunctiva, RPE
TLR8	ssRNA (virsues)	None detected
TLR9	Unmethylated CpG motifs of bacterial DNA, dsDNA (viruses and bacteria)	Cornea, conjunctiva, RPE
TLR10	N/d	Cornea, conjunctiva, RPE
TLR11	profillin	N/d
NOD1	iE-DAP	Anterior and posterior portions of the eye,[b] HCE-T and HCE[c]
NOD2	MDP	Anterior portion, HCE-T and HCE
NALP1	Cell rupture	HCE-T and HCE
NALP2	N/d	HCE-T and HCE
NALP3	Bacterial RNA, toxins and ATP, uric acid	HCE-T and HCE
NALP10	N/d	HCE-T

[a]RPE: primary cultures of human retinal pigment epithelial cells.
[b]Anterior portion contains corneal tissues; posterior portion contains all other ocular tissues.
[c]HCE-T, SV-40 immortalized human corneal epithelial cell line; HCE, human primary corneal epithelial cells.
TLR, Toll-like receptor; NLR, NOD-like receptors (caspase recruitment domain-containing NODs and pyrin domain-containing NALPs); PGN, peptidoglycan; LTA, lipoteichoic acid; iE-DAP, γ-D-glutamyl-meso-diaminopimelic acid; RSV, respiratory syncytial virus; MDP, muramyl-dipeptide; N/d, not determined.

can be subdivided into three groups based upon their N-terminal domains: caspase recruitment domain (NODs), pyrin domain (NALPs), or baculovirus inhibitor repeat (neuronal apoptosis inhibitor proteins, NAIPs). At present, the human NLR family comprises of 23 proteins, while at least 34 NLR genes have been identified in mice. NALPs are unique in that upon binding to their ligand, NALPs form a complex termed the inflammasome resulting in caspase-1 activation and the release of active IL-1β. Recently, NALP1, 3, and 10 were identified in corneal epithelial cells, but their potential function in regulating ocular innate immunity has not been determined. NAIPs inhibit caspase effectors and are mainly expressed in neurons where their primary role is to protect against apoptosis. NAIPs have not yet been found in ocular tissues. NODs are the most extensively studied members of the NLR family and are expressed in ocular tissue.

Both NOD1 and NOD2 are constitutively expressed within the eye; NOD1 in the anterior and posterior segments of the eye and NOD2 only in the anterior segment. However, which specific tissues express NODs is unknown. NOD1 and NOD2 recognize specific subcomponents of bacterial peptidoglycan. NOD1 recognizes D-γ-glutamyl-meso-diaminopimelic acid, while NOD2 recognizes muramyl dipeptide. Recently, a mutation in NOD2 was linked to the development of uveitis in patients with Blau syndrome, suggesting that cytoplasmic

NODs may be important in ocular inflammation. While the importance of NOD receptors in innate immunity and the pathogenesis of inflammatory disease are recognized outside the eye, the study of NODs within the eye is at an early stage. A complete understanding of ocular host defense will require a better understanding of where NODs and other NLRs are expressed and how they regulate innate immunity within the eye.

Complement
Similar to PRRs, the complement system acts as an innate immune surveillance system, detecting the first signs of pathogen invasion. The complement system was first identified as a biochemical cascade of serum proteins that help or complement antibodies to clear pathogens and mark them for opsinization by phagocytes. It is now known, however, that the complement system can also be activated directly by microbial products via the alternative pathway. Several studies demonstrate that complement is constitutively active at low levels in the eye. This is thought to be a primary defense mechanism of the eye against pathogenic infection. Upon pathogen invasion the complement system is further activated to clear the infection through (1) generation of inflammatory factors (C3 and C5a), (2) chemotaxis of phagocytes (C3 and C5a), (3) opsonization of Ab-coated cells (C3b), and (4) lysis of bacteria and virus-infected cells (C8 and C9).

Cytokines, Chemokines, and Effector Cells

Once pathogens breach the passive barriers of the eye and trigger the PRRs or the complement system, the primary function of innate immunity is to eliminate the invading pathogen as quickly as possible. To achieve this innate immunity must: (1) trigger an immediate immune response, (2) amplify the response, (3) clear the pathogens, and (4) activate the adaptive system if the pathogen cannot be cleared quickly. Cytokines, chemokines, and adhesion molecules play critical roles in regulating each of these stages.

Initiation and amplification

PRRs (TLRs and NLRs) recognize microbial products in the earliest phase of host defense and activate many immune and inflammatory genes, the products of which are important in initiating and amplifying antimicrobial immunity. Triggering TLRs on either the corneal epithelium or the RPE leads to activation of NF-κB via the (1) MyD88-dependent or (2) MyD88-independent pathway. The MyD88-dependent pathway is utilized via most TLRs (TLR1, 2, 4, 5, 6, 7, 8, and 9) and leads to the production of proinflammatory cytokines and chemokines (IL-6, IL-8, IL-18, MIP-1, and TNF-α). By contrast, only TLR3 and TLR4 utilize the MyD88-independent pathway and leads to the production of IFN-α and IFN-β. Recruitment of neutrophils into inflamed tissues is controlled predominantly by two chemokines (MIP-2 (= IL-8 in humans) and KC). In *Pseudomonas aeruginosa*-induced corneal keratitis, elevated MIP-2 and KC correspond with increased infiltration of neutrophils. In the posterior segment, elevated TNF-α, IL-1β, and CINC (rat homolog of IL-8) contribute to the breakdown of the blood–retinal barrier and the recruitment of neutrophils in response to *Staphylococcus aureus*. The adhesion molecules ICAM-1 and E-selectin are also upregulated early in iris, ciliary body, and retinal vessels, serving to enhance the infiltration of neutrophils to the site of infection.

Clearing the pathogen

In response to inflammatory cytokines and chemokines, neutrophils and macrophages are recruited to the site of infection. Bacterial clearance by neutrophils is accomplished by (1) phagocytosis, (2) generation of reactive oxygen species, and (3) the release of granule-associated enzymes: cathepsin G, myeloperoxidase, lactoferrin, and elastase. While the recruitment of neutrophils to the site of infection is essential for clearance of the pathogen, the persistence of neutrophils and the prolonged release of inflammatory mediators is also associated with nonspecific host tissue damage. Similar to neutrophils, macrophages also phagocytose and directly kill microbes as part of innate immunity. However, macrophages are also antigen-presenting cells, and as such participate in the development

of the adaptive immune response. The natural killer (NK) cell is another important innate effector cell in host defense against viral infections of the cornea such as herpes simplex virus type-1. NK cells respond in an antigen-independent manner and kill virus-infected host cells through the release of perforin and granzymes or through binding of the death receptors Fas and TRAIL-R on the target cell. If innate effector cells fail to clear the infection, adaptive immunity will take over and finish eradicating the pathogen. However, if the infection is successfully cleared, the final and most important step of innate immunity is preventing nonspecific host tissue damage. This is accomplished in the eye through several mechanisms that together make up innate immune privilege.

Innate Immune Privilege

The primary role of innate immunity is to rapidly eradicate invading pathogens through the induction of inflammation. As a general rule, the level of inflammation is proportional to the size and virulence of the infection. Small, nonvirulent infections are cleared by mild inflammation, while larger, more virulent infections induce intense inflammation. The potential danger of inflammation occurs when intense and/or prolonged inflammation threatens the surrounding normal tissue, resulting in nonspecific tissue damage and scarring. The potential danger of inflammation in the eye is magnified by the presence of irreplaceable and highly sensitive ocular tissues. Therefore, it is not surprising that immune privilege in the eye has multiple mechanisms to control innate immunity and limit nonspecific tissue damage. The aqueous humor contains multiple factors that directly inhibit innate immunity including: (1) TGF-β, soluble Fas ligand, and alpha-melanocyte stimulating hormone (inhibit neutrophil activation); (2) macrophage migration inhibitory factor (inhibits NK cell-dependent lysis of target cells); (3) calcitonin gene-related peptide (inhibits nitric oxide release from activated macrophages); and (4) complement regulatory factors: CD46, CD55, CD59, and Crry (inhibit complement activation). These factors work together to limit the damaging consequences of inflammation and to preserve the visual axis. Unfortunately, while these mechanisms evolved to limit local tissue destruction and preserve the visual axis, they may leave the eye more vulnerable to organisms whose virulence often requires a robust inflammatory response for eradication. Therefore, a delicate balance must be made between the amount of inflammation needed for eradicating the pathogen and the amount of nonspecific tissue damage.

Link between Innate and Adaptive Immunity

Innate and adaptive immunity have long been discussed as separate arms of the immune system. However, it is increasingly clear that they are indeed not separate but highly integrated. Several studies, both outside and inside the eye, identified the dendritic cell as a central link between innate and adaptive immunity. Immature dendritic cells reside in peripheral tissues and through triggering of TLRs, participate in the primary immune response against microbial infections. Immature dendritic cells encounter invading pathogens, capture bacterial antigens, and migrate to the draining lymph node. Once in the lymph node, only mature dendritic cells can efficiently prime naïve T cells and initiate adaptive immunity. Studies using TLR and MYD88 deficient mice, reveal that TLR signaling is required for bacteria-induced maturation of dendritic cells and induction of adaptive immunity. Therefore, TLRs on dendritic cells are essential for (1) sensing the microbe and initiating the immediate innate immune response, as well as (2) inducing the development of an adaptive immune response. While dendritic cells have been identified in the cornea and retina, additional studies are needed to completely understand their function in linking innate and adaptive immunity within the eye.

Conclusion

Innate immunity is a critical first line of defense against ocular infections. However, the regulation of innate immunity within the eye is just beginning to be unraveled. The identification of TLRs and NLRs has provided new insights into the mechanisms of host defense and the pathogenesis of inflammatory diseases. A better understanding of how microbial agents and endogenous host factors interact with TLRs and NLRs in the eye will be critical in advancing our knowledge of the pathogenesis of infectious and noninfectious eye diseases. Moreover, a better understanding of these mechanisms will lead to the identification of new therapeutic targets for treating and preventing sight-threatening infections.

See also: RPE Barrier.

Further Reading

Banchereau, J. and Steinman, R. M. (1998). Dendritic cells and the control of immunity. *Nature* 392: 245–252.

Creagh, E. M. and O'Neill, A. J. (2006). TLRs, NLRs and RLRs: A trinity of pathogen sensors that co-operate in innate immunity. *Trends in Immunology* 27: 352–357.

Gregory, M., Callegan, M. C., and Gilmore, M. S. (2007). Role of bacterial and host factors in infectious endophthalmitis. *Chemical Immunology and Allergy* 92: 266–275.

Haynes, R. J., McElveen, J. E., Dua, H. S., Tighe, P. J., and Liversidge, J. (2000). Expression of human beta-defensins in intraocular tissues. *Journal of Investigative Ophthalmology and Visual Science* 41: 3026–3031.

Holtkamp, G. M., Kijlstra, A., Peek, R., and de Vos, A. F. (2001). Retinal pigment epithelium–immune system interactions: Cytokine production and cytokine-induced changes. *Progress in Retinal and Eye Research* 20: 29–48.

Kaplan, H. J. and Niederkorn, J. Y. (2007). Regional immunity and immune privilege. *Chemical Immunology and Allergy* 92: 11–26.

Kawai, T. and Akira, S. (2007). TLR signaling. *Seminars in Immunology* 19: 24–32.

Kolls, J. K., McCray, P. B., and Chan, Y. R. (2008). Cytokine-mediated regulation of antimicrobial proteins. *Nature Reviews Immunology* 8: 829–835.

Pearlman, E., Johnson, A., Adhikary, G., et al. (2008). Toll-like receptors at the ocular surface. *Ocular Surface* 6: 108–116.

Rodriguez-Martinez, S., Cancion-Diaz, M. E., Jimenez-Zamudio, L., et al. (2005). TLRs and NODs mRNA expression pattern in healthy mouse eye. *British Journal of Ophthalmology* 89: 904–910.

Sack, R. A., Nunes, I., Beaton, A., and Morris, C. (2001). Host-defense mechanism of the ocular surfaces. *Bioscience Reports* 21: 463–480.

Sohn, J. H., Bora, P. S., Jha, P., et al. (2007). Complement, innate immunity and ocular disease. *Chemical Immunology and Allergy* 92: 105–114.

Streilein, W. J. and Stein-Streilein, J. (2000). Does innate immune privilege exist? *Journal of Leukocyte Biology* 67: 479–487.

Tosi, M. F. (2005). Innate immune responses to infection. *Journal Allergy and Clinical Immunology* 116: 241–249.

Van Vilet, S. J., den Dunnen, J., Gringhuis, S. I., Geijtenbeek, T. B., and Van Kooyk, Y. (2007). Innate signaling and regulation of dendritic cell immunity. *Current Opinion in Immunology* 19: 435–440.

IOP and Damage of ON Axons

R W Nickells, University of Wisconsin, Madison, WI, USA

Glossary

Apoptosis – The molecular and biochemical process by which a cell soma (cell body) is able to disassemble all of its organelles and break apart to be engulfed by neighboring cells. As the process is intrinsic, and tightly regulated by the cell itself, it has often been called a cell suicide program. Apoptosis is the mechanism of cell elimination during development (programmed cell death) and is often the end-stage result of cell loss in a variety of diseases, particularly chronic neurodegenerative disorders. Retinal ganglion cell soma apoptosis is the mechanism of cell loss in glaucoma.

Axons – The single extensions of neurons that act as the conduit for an electrical impulse or signal originating from the soma of the same neuron. They typically terminate in a synapse that contacts one of many dendrites of other neurons. In reference to glaucoma, axons of the retinal ganglion cells pass out of the eye into the optic nerve.

Glia – The abundant non-neuronal cell types in the nervous system. There are many different classifications of glia, but principally they are made up of astrocytes, Müller cells (in the retina), oligodendrocytes or Schwann cells (which synthesize myelin in the central and peripheral nervous systems, respectively), and microglia. Astrocytes and Müller cells are thought to function as neuronal support cells under normal conditions. Microglia are thought to play an important role in the innate immune response of neural tissue.

Lamina cribrosa – In higher primates, it is the connective tissue and neuronal structure comprising the scleral canal at the site where the optic nerve exits the eye. The connective tissue is formed as plates of collagen and basement membrane and the surface of these plates are occupied by astrocytes and a secondary cell type called lamina cribrosa cells. Pore structures exist between the plates through which retinal ganglion cell axons are bundled as they exit the eye. Rodents do not have a lamina cribrosa, but instead have columns of astrocytes that surround the bundles of exiting axons. Because of the lack of connective tissue, this structure has been termed the cellular lamina in these animals.

Optic nerve – It is anatomically described as the second of the 12 paired cranial nerves and is an extension of the central nervous system. The nerve begins at the lamina cribrosa and extends to the lateral geniculate nucleus in higher primates and the superior colliculus in rodents. Shortly after it exits the eye, the fibers of the nerve, which are axons of retinal ganglion cells, become myelinated.

Retinal ganglion cells – The projection neurons that reside in the innermost layer of the retina, and extend axons out of the eye and into the optic nerve. They receive electrical stimulation from photoreceptors by way of bipolar neurons and then convey that stimulus to optical centers in the brain.

Soma – Also known as the cell body, it is anatomically and functionally distinct from the dendrites, axon, and synapse, even though these compartments are all part of the same cell. The soma contains the basic cellular organelles including the nucleus and the majority of endoplasmic reticulum and Golgi bodies. Retinal ganglion cell somas reside in the ganglion cell layer of the retina.

Introduction – Intraocular Pressure as a Risk Factor for Glaucoma

Glaucoma is one of the world's leading causes of blindness, estimated to affect over 60 million people worldwide by the year 2010. It is typically a disease of the elderly and often goes undetected until later stages because progression is usually not associated with pain, or the devastating loss of central vision. Instead, glaucoma progresses slowly, creating a series of small peripheral defects in vision (called scotomas) that are compensated for by processing in the visual centers of an affected individual. The development of these pathological blind spots is principally the result of the regional degeneration of the retinal ganglion cells, involving the loss of both the axon in the optic nerve, and the cell body (soma) in the retina. Once a critical number of cells are lost in the retina, light information from that specific region is unable to be transmitted to the brain.

Because of the slow progressive nature of the disease, early detection is paramount to establish treatments that can attenuate further damage to the retina and optic nerve. A critical part of the early detection arsenal is the association between elevated intraocular pressure (IOP) and glaucoma. Nearly every person with glaucoma has either elevated IOP, or benefits from having existing levels of IOP lowered. In addition, experimentally induced ocular hypertension is a staple of virtually all animal models of the disease, while lowering elevated IOP in these models is associated with slowing the progression of glaucoma, thus providing experimental evidence for the causal relationship between eye pressure and optic nerve disease.

Currently, there is no clear mechanism that associates elevated IOP with the activation of optic nerve and retinal damage; however, studies over the last several decades have pointed to a viable model linking the two. At present, this model serves as a framework for both current and future hypothesis-driven studies that may help resolve this important question. Understanding the model requires a brief tutorial on the relevant anatomical structures of the eye. In essence, the mammalian eye is a closed hydrostatic system. The optics, which are designed to focus incoming light from the anterior segment of the eye onto the sensory retina lining the posterior segment of the eye, require that the globe be properly inflated. To do this, aqueous humor is secreted by cells of the ciliary body posterior to the iris and drained through the trabecular meshwork and Schlemm's canal, anterior to the iris. As light passes into the eye, and is focused onto the sensory retina, photons are captured by photoreceptors, where they are converted into chemical signals that are then processed by the neural network of the retina until finally reaching the retinal ganglion cells. Ganglion cells transmit these signals through axons that exit the globe, enter the optic nerve, and connect to neurons in visual centers in the brain. Anatomically, the axons from the ganglion cells typically form bundles and exit the globe through a posterior hole in the sclera called the lamina. In small mammals, such as rodents, the axon bundles in the laminar region are supported by columns of glial cells, principally astrocytes. This structure has been called the cellular lamina by some researchers to distinguish it from the lamina cribrosa (LC) found in the eyes of larger mammals such as primates. The LC accommodates larger numbers of ganglion cell axons (in the range of 1 million in higher primates versus 50 000–100 000 in rodents), which necessitates the presence of collagenous beams and plates that support the spaces through which the axon bundles pass (**Figure 1**). Similar to rodents, this region of the optic nerve is also populated with cells, including astrocytes, specialized cells referred to as LC cells, and microglia.

Biomechanical Engineering Studies

In a consideration of how elevated IOP causes the death of retinal ganglion cells, the lamina plays a central role. Several seminal studies on the pathophysiology of glaucoma have pointed to the laminar region of the optic nerve head as the first site of damage in glaucoma. The

evidence for this comes from a variety of observations, notably classical studies showing the disruption of both retrograde and anterograde axoplasmic transport in response to elevated IOP (**Figure 2**), and more recent electron microscopic studies showing very early axonal disruption in the nerve head of a mouse model of glaucoma.

Intuitively, the lamina causes something of a dilemma when considering that the eye must be under some kind of pressure in order to maintain normal function, since it is essentially a hole in an otherwise closed, spherical, hydrostatic system. Thus, one might expect that forces generated by hydrostatic pressure within this system would be focused onto the weakest point, such as this small hole. Given the requirement for some level of pressure in the eye, it is likely that the evolutionary development of the lamina would be sufficiently strong enough to support the axons passing through it, so that they are not adversely affected by the forces concentrated on this area. However, in the face of above-normal levels of IOP, which would increase the forces directed at the laminar structure, axons could be damaged if the forces focused on this region exceeded the structural capacity it was designed to withstand.

This rather speculative model has been tested by biomechanical engineers using finite element modeling. In this complicated field of applied engineering, real-life measurements of the tensile strength of biological tissues are taken and used to design small elements that are pieced together to create a three-dimensional model of the biological system in question. Early finite element models of the eye were created based on mathematics of a simple sphere under pressure, and these models helped confirm the basic intuitive idea that forces created by pressure inside the globe create stress and strain on the structural integrity of the lamina (**Figure 1**). Important advances in the field finite element modeling in the study of glaucoma have come from detailed element modeling of both the sclera of the eye and the different components of the primate LC.

Modulation of Glia Behavior at the Optic Nerve Head and Dysfunction of Ganglion Cell Axons

When IOP becomes elevated, resulting in an increase in the stress placed in the laminar region, the cells in this region respond. The precise nature of this response is not fully understood, but detailed gene expression profiles of the optic nerve head region of rats with experimental glaucoma clearly indicate that the glial cells in this region upregulate the expression of genes involved in proliferation and remodeling of the extracellular matrix (ECM). Similar changes in gene expression have been noted in optic nerve astroctyes and LC cells in higher primates

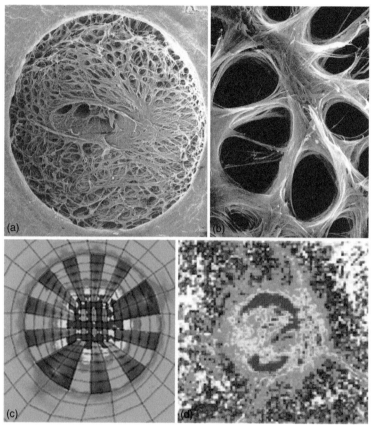

Figure 1 Increased strain associated with elevated IOP is directed onto the lamina cribrosa. (a) Scanning electron micrograph of a human lamina cribrosa after alkali treatment to remove soft tissue. The collagen-based laminar beams create an interacting network of plates and pores (higher magnification in (b)), through which bundles of retinal ganglion cell axons pass as they exit the eye and extend through the optic nerve. The diameter of the human scleral canal is approximately 1.7 mm, and the image shown is from the perspective of the vitreoretinal surface. (c) A contour stress plot of a finite element model of a human lamina. Models such as these predict that stresses generated by elevated IOP are concentrated on the laminar beams, with the highest stress (warm colors) occurring within the central beams. In addition to stress being applied to the beams, detailed histological studies also show an enlargement of the scleral canal. (d) A contour plot of tissue deformation of the lamina of a human donor eye. In this experiment, deformation was calculated as the topographical difference between the vitreoretinal interface when exposed to 15 and 50 mmHg. Experimentally, maximal deformation does not take place at the center of the lamina. (a and b) Courtesy of Dr. Harry A. Quigley, Wilmer Eye Institute, Johns Hopkins School of Medicine. (c) Reproduced from Bellezza, A. J., Hart, R. T., and Burgoyne, C. F. (2000). The optic nerve head as a biomechanical structure: initial finite element modeling. *Investigative Ophthalmology and Visual Science* 41: 2991–3000, with permission from the Association for Research in Vision and Ophthalmology. (d) Reproduced from Sigal, I. A, Flanagan, J. G., Tertinegg, I., and Ethier, C. R. (2004). Finite element modeling of optic nerve head biomechanics. *Investigative Ophthalmology and Visual Science* 45: 4378–4387, with permission from the Association for Research in Vision and Ophthalmology.

with experimental glaucoma and in the human disease. Early interpretations of these changes in gene expression generally assumed that cells in this region were responding to the loss of ganglion cell nerve fibers by laying down a glial scar. This notion has recently been challenged by biomechanical studies of the connective tissue composition of the laminas isolated from monkeys at the very early stages of experimental glaucomatous damage. In detailed three-dimensional reconstructions of serially sectioned nerve heads, in which each section was individually stained for ECM components, this group showed that connective tissue content increased in the laminar region well in advance of the loss of neuronal tissue. Surprisingly, the ratio of connective tissue to neural tissue remained relatively constant, indicative of an increase in

the thickness of the laminar region caused by the addition of new ECM material in the posterior region of the lamina and an actual widening of the scleral canal. Progressively, the connective tissue lamina begins to deform posteriorly partly as a function of the loss of ECM material in the anterior portion of the lamina, and, at later stages of glaucomatous progression, partly as a function of the loss of ganglion cell axons entering the optic nerve head. Eventually, the posterior deformation becomes permanent, resulting in the classic cupping of the optic nerve head observed clinically.

The interpretation of these changes in the optic nerve head in early glaucoma is that the glial cells in this region are responding to the tremendous increase in stress and strain induced by the elevation in IOP, by laying down

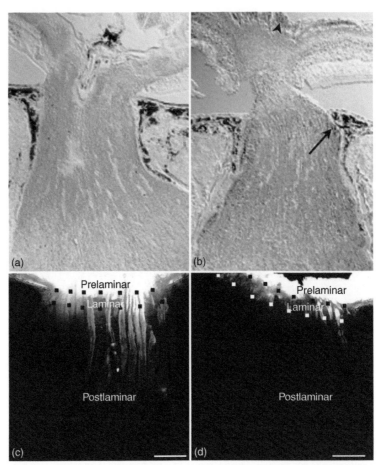

Figure 2 Ocular hypertension and glaucoma are associated with a blockage of axonal transport. (a and b) Immunohistochemical staining for the TrkB neurotrophin receptor in the optic nerve of rat eyes under normal IOP (a) or experimentally induced ocular hypertension (b). The TrkB receptor typically binds neurotrophic factors, such as brain-derived neurotrophic factor (BDNF), at ganglion cell synaptic connections in the brain. Once bound, TrkB is internalized and transported retrogradely to the ganglion cell somas. During periods of ocular hypertension, TrkB receptor accumulates at the level of the cellular lamina in the rat eye (arrow), while focal accumulations of this receptor can also be detected in the nerve fiber layer of the retina (arrowhead). These findings are consistent with a blockade of retrograde axonal transport during ocular hypertension. (c and d) Blockade of anterograde axonal transport in a porcine model of ocular hypertension. In this experiment, rhodamine β-isothiocyanate (RITC) was injected into the vitreous of pig eyes, which were then subjected to elevations in IOP for 6 h (d). Eyes under normal IOP (c) show uptake and transport of the RITC along the axons of the optic nerve. Ocular hypertensive eyes exhibit little to no transport of this dye past the laminar region. (a and b) Reproduced from Pease, M. E., McKinnon, S. J., Quigley, H. A., Kerrigan-Baumrind, L. A., and Zack, D. J. (2000) Obstructed axonal transport of BDNF and its receptor TrkB in experimental glaucoma. *Investigative Ophthalmology and Visual Science* 41: 764–774, with permission from the Association for Research in Vision and Ophthalmology. (c and d) Reproduced from Balaratnasingam, C., Morgan, W. H., Bass, L., et al. (2007) Axonal transport and cytoskeletal changes in the laminar regions after elevated intraocular pressure. *Investigative Ophthalmology and Visual Science* 48: 3632–3644, with permission from the Association for Research in Vision and Ophthalmology. Scale (in (c) and (d)) = 400 μm.

new connective tissue in an effort to increase the mechanical strength of the lamina. Addition of material to the posterior surface of the lamina ensures, at least in the short term, that axons will be minimally damaged in the process. Ultimately, this process appears to be a losing battle, as axons become progressively more damaged and begin to degenerate, the optic nerve head becomes a series of collapsed connective tissue plates. A critical question in the pathophysiology of glaucoma is what ultimately causes the damage to the axons in this region.

Several theories have been postulated for the source of the axonal damage. The earliest of these were contradictory models that postulated either (1) a decrease in blood flow

to the optic nerve head, creating an ischemic environment for the ganglion cell axons, or (2) direct mechanical damage to the axons as they passed through the connective tissue plates of the lamina such that they were being crimped as the plates underwent the process of posterior deformation. More recent models have predicted that the altered function of the astrocytes and microglia in this region leads to the release of toxic molecules. Several studies have suggested that these toxins could range from highly reactive nitric oxide radicals mediated by the activity of nitric oxide synthase-(NOS)-2 to tumor necrosis factor α (TNFα), which is produced by astrocytes under stress conditions and can activate retinal ganglion cell death. Regarding this hypothesis, although many questions remain unanswered, experimental evidence is clear that astrocytes, microglia, and possibly LC cells become reactive in the glaucomatous optic nerve head (**Figure 3**), leading to molecular and behavioral changes that have been associated with pathology in several other neurodegenerative disorders.

Others, however, have reasoned that the initial responses of glia in the nerve head are just as likely trying to protect the axons as they compensate for the increased mechanical strain. Protective effects of astrocytes can include becoming a source of trophic factors for damaged neurons. In addition, they also act as the principal source of glycogen stores in the central nervous system (CNS), and are typically involved in monitoring and balancing extracellular pH and ionic conditions, which both support neurons and prevent damaging effects. Thus, central questions in our understanding of glaucoma pathology

are whether the optic nerve glia simply become unable to continue their role in supporting the ganglion cell axons, or do they instead change to a molecular behavior that contributes to axonal pathology?

Another important consideration in our understanding of the effects of elevated IOP on the optic nerve head is the differential between IOP and arteriolar blood pressure. From a biomechanical standpoint, perfusion pressure to the small capillaries in the nerve head could meet much higher resistance in the face of elevated IOP. Thus, it is reasonable to predict that reduced blood flow would lead to micro-ischemic environments in this region. A similar consideration could also be made for the process of axoplasmic transport, particularly retrograde transport. Taken together, the pressure differential encountered at the lamina of a glaucomatous eye may consummate in a catastrophic stress where neurons are consuming greater amounts of adenosine triphosphate (ATP) to maintain

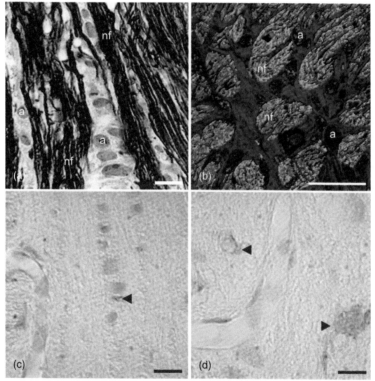

Figure 3 Ocular hypertension is associated with reactive astrocytes and microglia in the optic nerve head. (A) Columns of astrocytes (a) separate bundles of nerve fibers (nf) of ganglion cell axons in the cellular lamina of a young DBA/2J mouse without glaucoma. Longitudinal section, stained with silver impregnation. Scale = 10 μm. (B) In a mouse eye with glaucoma, astrocytes (a) become reactive and express the glial fibrillary acidic protein intermediate filament (stained red). This expression occurs early in disease, as evidenced by nerve fiber bundles (nf: stained green) still mostly intact. Transverse section through the lamina. Scale = 20 μm. (C, D) Images showing microglia expressing the marker gene IBA1 (brown reaction product) in a normal control rat lamina (C), or in the lamina of an eye with experimental glaucoma (D). Arrowheads point to immunopositive cells. Reactive microglia upregulate IBA1 expression. Scale (in (C) and (D)) = 3 μm. (A) Reproduced from Schlamp, C. L., Li, Y., Dietz, J. A., Janssen, K. T., and Nickells, R. W. (2006). Progressive ganglion cell loss and optic nerve degeneration in DBA/2J mice is variable and asymmetric. *BMC Neuroscience* 7: 66. (B) Reproduced from Howell, G. R., Libby, R. T., Jakobs, T. C., et al. (2007). Axons of retinal ganglion cells are insulted in the optic nerve early in DBA/2J glaucoma. *Journal of Cell Biology* 179: 1523–1537, with permission from the Rockefeller University Press. (C, D) Reproduced from Johnson, E. C., Jia, L., Cepurna, W. A., Doser, T. A., and Morrison, J. C. (2007). Global changes in optic nerve head gene expression after exposure to elevated intraocular pressure in a rat glaucoma model. *Investigative Ophthalmology and Visual Science* 48: 3161–3177, with permission from the Association for Research in Vision and Ophthalmology.

axoplasmic transport in the face of reduced access to nutrients and oxygen from both compromised astrocytes and reduced arteriolar blood flow. Ultimately, a scenario may develop where energy supplies are no longer sufficient to sustain transport against the pressure gradient and the axon is forced to execute a self-destruct pathway.

Compartmentalized Self-Destruct Pathways and the Pathology of Glaucoma

A major factor in understanding how elevated IOP leads to ganglion cell death is understanding how ganglion cells die in the first place. Ganglion cells are projection neurons of the CNS. The cell body, or soma, is located in the inner retina, where they take input from neurons in the outer retina through a complex network of dendrites located in the inner plexiform layer. The soma then projects an axon through the nerve fiber layer, which is adjacent to the vitreous, exiting the eye through the lamina into the optic nerve. In simple terms, a ganglion cell has four distinct compartments: the soma, the axon, a synapse at the end of the axon, and the dendritic tree.

Recently, the death of retinal ganglion cells has been evaluated in the context of effects on their individual compartments. Several studies, using genetically engineered mice, in which mutations in specific genes selectively affect the degenerative process of individual compartments, have shown that ganglion cell death can occur in an autonomous, compartment-specific, fashion. Two mutations, in particular, have contributed to the understanding of the process of ganglion cell death in glaucoma. The first of these is a knock-out mutation created in a gene called *Bax*, which is a member of a larger gene family that regulates the intrinsic apoptotic pathway by affecting permeability changes and dysfunction in mitochondria. Although apoptosis is generally considered in the same context as cell death, the term actually describes the self-destruct pathway executed by cell somas. In a defined, naturally occurring form of glaucoma in the DBA/2J inbred line of mouse, *Bax* deficiency was found to completely abrogate ganglion cell soma loss, but not the degeneration of the ganglion cell axons. A second mutant gene, which occurs naturally in mice, is called Wallerian degeneration slow (Wld^S). This mutant gene dramatically reduces the degenerative process called Wallerian degeneration, which is the self-destruct pathway executed by dying axons. In DBA/2J mice carrying the Wld^S mutation, ganglion cell axons are dramatically preserved in the face of ocular hypertension that causes glaucoma in wild-type mice. Importantly, in these mice, ganglion cell somas are also spared compared to wild-type animals. These data provide insight on the timing of events associated with ganglion cell loss. They indicate that axon loss is not prevented by blocking soma loss,

while soma loss can be attenuated if axon loss is prevented. Together, this strongly implicates glaucoma as an axogenic disease, distinct from other neurodegenerative disorders where the soma is the fist site of injury. Supporting this model, all studies examining the timing and progression of both optic nerve degeneration and retinal disease in the DBA/2J mouse (**Figure 4**), point to axonal damage preceding the loss of ganglion cell somas.

In a scheme that links elevated IOP to ganglion cell damage and death, the activation of an autonomous self-destruct pathway in the ganglion cell axon will, in turn, lead to the activation of the autonomous apoptotic pathway in the ganglion cell soma. The mechanism by which this occurs is also unknown, but several studies suggest that the loss of axonal transport will reduce the flow of neurotrophic growth factors to ganglion cell somas. This model for the activation of soma death in glaucoma has been coined the neurotrophin hypothesis. During development, the normal pruning of retinal ganglion cells, in which approximately 50% of the cells are eliminated, coincides with an increased dependence on neurotrophic factors for survival. Similarly, application of exogenous neurotrophins, after acute or chronic damage to the optic nerve, is able to prolong ganglion cell survival.

Although these studies are consistent with the neurotrophin hypothesis, there are several caveats that require further explanation and reveal the need for continued study. First, the level of protective effect of exogenous neurotrophins is transient, even if efforts are made to introduce a continuous supply of any given factor over a long period. Second, injury to the optic nerve is known to stimulate the expression of endogenous trophic factors by other cells, particularly macroglia, in the retina, leading to the question of why this increase is not an expression sufficient to sustain ganglion cell somas in the absence of retrograde transport? Part of the reason for this effect may lie within molecular changes occurring in the ganglion cell somas once they have become injured. An early component of the injury process is the silencing of normal gene expression and some of the genes that are downregulated appear to be for the receptors that neurotrophic factors bind to. An alternative explanation is that many of the studies showing the transient effects of exogenous trophic factors are usually examining the effects of only a single factor at a time. Compelling available data suggest that treatments that elicit the infiltration of cells, such as macrophages, provide a much greater and longer-term effect on the survival of ganglion cell somas. In this case, infiltrating cells are likely producing and secreting much more complex mixtures of trophic factors than could be applied by experimental manipulation. Lastly, the signaling pathways activated by neurotrophic factors are variable and highly regulated. One such level of regulation is by the localization of target receptors on neurons, such that receptors bound at the synapse are likely to activate distinct mitogen-activated protein kinase

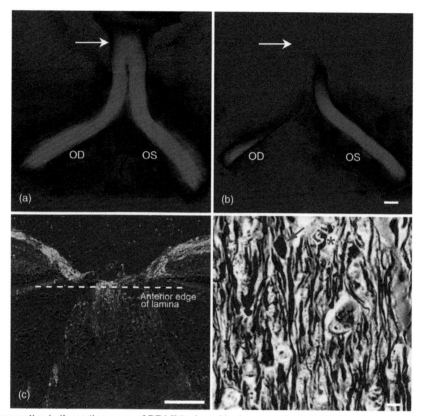

Figure 4   Axon degeneration in the optic nerves of DBA/2J mice with glaucoma. (a and b) Optic nerve tracts of a young mouse without disease (a) or an older mouse with bilateral disease (b). Axons are labeled using a postmortem technique in which DiI is allowed to diffuse along intact axonal tracts from the globe to the optic chiasm (denoted by arrow in both). Both nerves of young mice can be labeled all the way to the chiasm, while mice with glaucoma show partial or incomplete labeling that is typically asymmetric. Scale = 0.5 mm. (c) Axon loss in a DBA/2J mouse mutant for the proapoptotic *Bax* gene (*Bax$^{-/-}$*). Axons of retinal ganglion cells are labeled green. Mice lacking a functional *Bax* gene exhibit complete resistance of ganglion cell somas to glaucoma, but still undergo optic nerve degeneration. In this severe case, axons originating in the nerve fiber layer extend only to the laminar region. Past this point, axons have mostly degenerated, indicating that the point of axonal lesion is at the level of the lamina. Scale = 75 μm. (d) Silver-stained longitudinal section of the post-laminar region of a DBA/2J mouse with glaucoma. Axons, which are stained dark brown to black, are tortuous (asterisk) and often end in bulbous swellings (arrows) indicative of degeneration. Scale = 5 μm. (c) Reproduced from Howell, G. R., Libby, R. T., Jakobs, T.C., et al. (2007) Axons of retinal ganglion cells are insulted in the optic nerve early in DBA/2J glaucoma. *Journal of Cell Biology* 179: 1523–1537, with permission from the Rockefeller University Press.

(MAPK) pathways than the same receptors if they interact with a neurotrophin ligand at the plasma membrane of the cell soma. The activation of distinct MAPK pathways could certainly yield different effects on ganglion cell survival. This was demonstrated by recent studies showing increased survival and gain of function of ganglion cells in cat eyes after a partial optic nerve crush, when the neurotrophic factor, brain-derived neurotrophic factor (BDNF), was applied both as an intravitreal injection and simultaneously infused into the visual cortex.

Acknowledgments

The author would like to thank Dr. Harry A. Quigley for contributing scanning electron microscope images of the human lamina cribrosa and Dr. Cassandra L. Schlamp for preparing the figures. Some of this work was funded by NIH grant R01EY12223 to the author.

See also: Animal Models of Glaucoma; Retinal Ganglion Cell Apoptosis and Neuroprotection.

Further Reading

Hernandez, M. R. (2000). The optic nerve head in glaucoma: Role of astrocytes in tissue remodeling. *Progress in Retinal and Eye Research* 19: 297–321.

Hernandez, M. R., Andrzejewska, W., and Neufeld, A. (1990). Changes in the extracellular matrix of the human optic nerve head in primary open-angle glaucoma. *American Journal of Ophthalmology* 102: 180–188.

Howell, G. R., Libby, R. T., Jakobs, T. C., et al. (2007). Axons of retinal ganglion cells are insulted in the optic nerve early in DBA/2J glaucoma. *Journal of Cell Biology* 179: 1523–1537.

Johnson, E. C., Guo, Y., Cepurna, W. O., and Morrison, J. C. (2009). Neurotrophin roles in retinal ganglion cell survival: Lessons from rat glaucoma models. *Experimental Eye Research* 88: 808–815.

Johnson, E. C., Morrison, J. C., Farrell, S., et al. (1996). The effect of chronically elevated intraocular pressure on the rat optic nerve head extracellular matrix. *Experimental Eye Research* 62: 663–674.

Libby, R. T., Li, Y., Savinova, O. V., et al. (2005). Susceptibility to neurodegeneration in glaucoma is modified by Bax gene dosage. *PLoS Genetics* 1: 17–26.

Nickells, R. W., Semaan, S. J., and Schlamp, C. L. (2008). Involvement of the *Bcl2* gene family in the signaling and control of retinal ganglion cell death. *Progress in Brain Research* 173: 423–435.

Quigley, H. A., Addicks, E. M., Green, W. R., and Maumenee, A. E. (1981). Optic nerve damage in human glaucoma: II. The site of injury and susceptibility to damage. *Archives of Ophthalmology* 99: 635–649.

Roberts, M. D., Grau, V., Grimm, J., et al. (2009). Remodeling of the connective tissue microarchitecture of the lamina cribrosa in early experimental glaucoma. *Investigative Ophthalmology and Visual Science* 50: 681–690.

Schlamp, C. L., Li, Y., Dietz, J. A., Janssen, K. T., and Nickells, R. W. (2006). Progressive ganglion cell loss and optic nerve degeneration in DBA/2J mice is variable and asymmetric. *BMC Neuroscience* 7: 66.

Schwartz, M., Yoles, E., and Levin, L. A. (1999). 'Axogenic' and 'somagenic' neurodegenerative diseases: Definitions and therapeutic implications. *Molecular Medicine Today* 5: 470–473.

Soto, I., Oglesby, E., Buckingham, B. P., et al. (2008). Retinal ganglion cells downregulate gene expression and lose their axons within the optic nerve head in a mouse glaucoma model. *Journal of Neuroscience* 28: 548–561.

Tezel, G., Hernandez, M. R., and Wax, M. B. (2001). *In vitro* evaluation of reactive astrocyte migration, a component of tissue remodeling in glaucomatous optic nerve head. *Glia* 34: 178–189.

Whitmore, A. V., Libby, R. T., and John, S. W. M. (2005). Glaucoma: Thinking in new ways – a role for autonomous axonal self-destruction and compartmentalised processes? *Progress in Retinal and Eye Research* 24: 639–662.

Yang, Z., Quigley, H. A., Pease, M. E., et al. (2007). Changes in gene expression in experimental glaucoma and optic nerve transection: The equilibrium between protective and detrimental mechanisms. *Investigative Ophthalmology and Visual Science* 48: 5539–5548.

Intraretinal Circuit Formation

J L Morgan, P R Williams, and R O L Wong, University of Washington, Seattle, WA, USA

Introduction

The vertebrate retina is a laminated tissue in which the various component cell types and their synaptic connections are arranged in distinct layers or laminae. This laminar organization, together with a high correspondence between structure and function, makes the vertebrate retina an excellent model system for studying the development of neuronal circuits. The development of retinal circuits generally follows the same sequence of major events for all vertebrates studied thus far. In this article, we provide an overview of the assembly of circuits in the vertebrate retina, primarily focusing on the mouse and zebrafish that have in recent years become key model systems because of the availability of transgenic animals. Circuit assembly involving retinal neurons requires that each cell type forms connections with their appropriate synaptic partners and establishes the correct density of synaptic connections with these synaptic partners. We first briefly review the overall organization of the mature retinal circuitry and then provide the current views on how these circuits are assembled during the development.

Organization of Retinal Circuits of Vertebrates

Figure 1 shows the general morphological organization of the mature vertebrate retina. Cone photoreceptors transmit light-evoked signals to bipolar cells that relay these signals to the output neurons of the retina, the retinal ganglion cells. There are two major functional subclasses of bipolar cells: ON bipolar cells are depolarized by increased illumination, whereas OFF bipolar cells are hyperpolarized. Cone photoreceptors contact cone bipolar cells whereas in general, rod photoreceptors contact a single class of rod bipolar cells, which form a circuit dedicated to low light vision (the rod pathway). Unlike cone bipolar cells, rod bipolar cells do not contact retinal ganglion cells directly, and instead contact AII amacrine cells. Neurotransmission along the vertical pathway comprising photoreceptors–bipolar cells–retinal ganglion cells is mediated by glutamate. Signals transmitted along the vertical pathway are modulated by amacrine interneurons that provide feedforward inhibition directly onto retinal ganglion cells and feedback inhibition onto bipolar cells. In the inner plexiform layer (IPL), lateral inhibition is mediated by either gamma aminobutyric acid (GABA)-ergic or glycinergic transmission from the amacrine cells. In the outer plexiform layer (OPL), photoreceptor transmission is modulated by horizontal cells, although the exact mechanism(s) by which this modulation occurs remains highly debated.

Formation of Retinal Synaptic Laminae

The exquisite lamination of cell bodies and synaptic connections of the vertebrate retina have long attracted investigations into the mechanisms that organize this orderly arrangement. Lamination first emerges in the inner retina where retinal ganglion cells, the first-born neurons, migrate to the basal surface of the retina after cell division at the apical surface. As a ganglion cell layer forms, amacrine interneurons are generated and migrate to the inner retina where the majority of these cells accumulate and form a cell layer next to the ganglion cell layer. As amacrine cells and ganglion cells elaborate their processes, an IPL takes shape. An OPL becomes apparent later as the axons and dendrites of the photoreceptors, bipolar cells and horizontal cells differentiate and become confined to a narrow lamina. Interestingly, horizontal cells differentiate early, at around the time of amacrine and ganglion cell genesis, but they do not reach their laminar position until after the IPL emerges. This sequence of lamination, from inner to outer, thus does not depend strictly on the time course of cell generation and the subsequent migration of the various cell types (summarized in **Figure 2**). The lamination of the axons and dendrites of each cell type of the retina is described in detail in the following sections in the context of circuit assembly.

Assembly of the Vertical Pathway

The vertical pathway of the retina is the major excitatory pathway, conveying light information from the outer to inner retina through glutamatergic transmission. This pathway from photoreceptors to bipolar cells to retinal ganglion cells develops later than circuits involving lateral connections from amacrine cells and horizontal cells.

In order to understand how vertical circuits in the retina are assembled, many studies have focused on the mechanisms that organize the presynaptic axons and postsynaptic dendrites of cellular components of the vertical pathway into their appropriate laminae. How the axonal

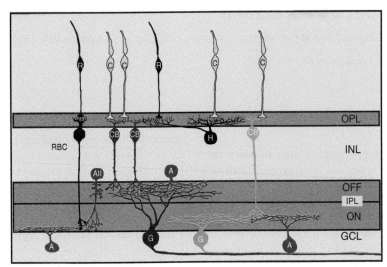

Figure 1   Organization of mature retinal circuits. Schematic of the laminar organization of the various cell types and their intraretinal axonal and dendritic projections is provided on the right. R, rods; C, cones, RBC, rod bipolar cell, CB, cone bipolar cell, H, horizontal cell, A, amacrine cell, AII, AII amacrine cell, G, ganglion cell, IPL, inner plexiform layer, OPL, outer plexiform layer. The IPL is approximately divided into OFF (outer two-fifths) and ON (inner three-fifths) sublaminae.

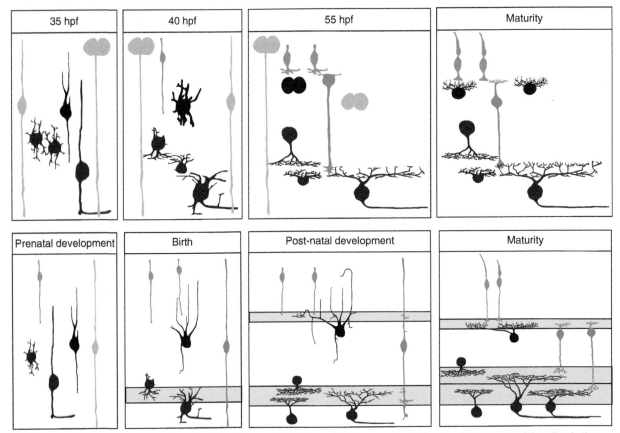

Figure 2   Developmental time-line in retinal circuit assembly. Schematic diagrams illustrating the developmental time-line in the structural assembly of the verteberate retina of mice (below) and zebrafish (above). Development in both species shares some common features, for example, the inner retinal circuits emerge before outer retinal circuits. However, there are some species differences in the developmental organization of the neuronal processes of some cells, especially the developmental remodeling of horizontal cell and ganglion cell dendrites. hpf: hours postfertilization; IPL: inner plexiform layer; OPL: outer plexiform layer.

and dendritic arbors of retinal ganglion cells, bipolar cells, and photoreceptors develop has largely been investigated by imaging labeled cells by light microscopy and by reconstructions from serial electron microscopy. Despite the static nature of the observations, these studies have collectively provided insightful views of how retinal ganglion cells, bipolar cells, and photoreceptors acquire the mature forms of their neuronal processes. More recently, live imaging approaches have captured the dynamic changes in axonal and dendritic architecture of developing retinal neurons, including the retinal ganglion cells and bipolar cells. These time-lapse studies confirmed previous observations and also revealed surprising cellular behaviors resulting in the stratified processes of some types of ganglion cells and bipolar cells.

Retinal Ganglion Cells

Dendritic stratification of retinal ganglion cells has been studied extensively in a variety of species. The stratification level of retinal ganglion cell dendrites is correlated with the type of functional input these neurons receive from cone bipolar cells. As indicated in **Figure 1**, retinal ganglion cells that depolarize to increased illumination have dendritic arbors stratifying in the inner three-fifths of the IPL, whereas ganglion cells stratifying in the outer two-fifths of the IPL hyperpolarize to illumination. In mammalian retinas including cat, rabbit, ferret, mouse, rat, and quokka, immature retinal ganglion cells have dendrites that course through the entire depth of the forming IPL (**Figure 3(a)**). This initially diffuse arrangement is altered with maturation, giving rise to laminated arbors that stratify at depths unique to each subtype of retinal ganglion cell. Such studies have largely focused on ganglion cells that later possess a single monostratified arbor. Retinal ganglion cells that have two stratified arbors

in mammalian retina (the bistratified ganglion cells) have not been followed throughout their early differentiation, mainly because of a lack of cell-specific markers for ganglion cell subtypes prior to when their dendrites stratify.

In zebrafish, a large proportion of retinal ganglion cells possess more than one stratified dendritic arbor. Because *in vivo* time-lapse imaging of individual retinal ganglion cells is possible in the transparent zebrafish, the sequence of events leading to the stratification of ganglion cell dendrites can be monitored. Such imaging studies demonstrated that ganglion cells with multiple stratified arbors appear to form each arbor, one stratum at a time (**Figure 3(b)**). These observations in zebrafish raise the possibility that dendrites in mammals and fish may adopt distinct mechanisms to achieve their stratification patterns. Alternatively, retinal ganglion cells that have multiple stratified arbors may develop differently from cells that become monostratified. Thus, future experiments in which bistratified ganglion cells in mammalian retinas can be identified early in their development are needed to distinguish between these possibilities.

What regulates retinal ganglion cell dendritic stratification? Studies in cat and ferret suggested that blockade of neurotransmission affects the emergence of stratified dendritic arbors of the retinal ganglion cells. Chronic application of the amino-phosphonobutyric acid (APB) agonist of the metabotropic glutamate receptor 6 (mGluR6) receptor found on the dendrites of mature ON bipolar cells led to a failure of ganglion cell dendrites to become stratified. However, APB has been shown to suppress the activity of all retinal ganglion cells in the immature ferret retina and thus its site of action is unclear. Examination of mice in which transmission along the vertical pathway is expected to be perturbed led to different observations. In mice lacking

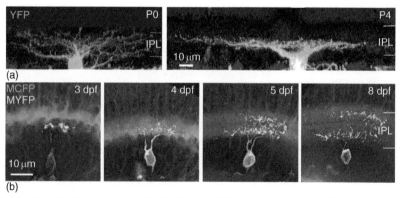

Figure 3 Dendritic stratification of retinal ganglion cells. (a) The dendritic arbors of neonatal retinal ganglion cells (blue) in the mouse retina become stratified with maturation; Red signal is reflected light. P: postnatal day; YFP: yellow fluorescent protein. (b) Dendritic stratification in many zebrafish retinal ganglion cells (yellow) emerges by progressive addition of new strata, one at a time. MYFP: membrane targeted yellow fluorescent protein. MCFP: membrane targeted cyan fluorescent protein (blue) in other retinal neurons. Reproduced with permission from **Figure 5** in Mumm, J. S., Williams, P. R., Godinho, L., et al. (2006). *In vivo* imaging reveals dendritic targeting of laminated afferents by zebrafish retinal ganglion cells. *Neuron* 52: 609–621, with permission from Elsevier.

the mGluR6 receptor, ganglion cell dendritic stratification is normal – in this case, however, one might expect that spontaneous release of neurotransmitter from the bipolar cells is unaffected even though visual stimulation along the ON pathway is abolished. Numerous studies have since been carried out to assess the role of visual experience in regulating dendritic stratification of retinal ganglion cells. Dark-rearing appears to decrease the developmental reduction in the proportion of ON–OFF retinal ganglion cells in mice. This developmental reduction is also less prevalent in mice lacking glycine receptors. In contrast, factors independent of activity have been found to guide dendritic lamination of ganglion cells. In particular, recent studies of the chick retina showed that members of the superfamily of immunoglobulin family of adhesion molecules (sidekicks and DsCam) influence the lamination patterns of retinal ganglion cells and amacrine cells. Although not yet completely worked out, it appears that molecular guidance cues involving adhesion molecules initially guide stratification of ganglion cell dendrites but later visually evoked activity may continue to shape the final lamination patterns of the ganglion cell dendrites.

Bipolar Cells

The development of the axonal terminals of bipolar cells has mostly been studied in a variety of species by single cell labeling methods using Golgi techniques and immunolabeling. Golgi-labeled cells that were presumed to be bipolar cells based on their radial morphology and terminal ending in the IPL appear to confine their axon terminals to specific depths within the IPL even before these cells form dendrites. Immunostaining for recoverin also suggested that ON and OFF bipolar cells stratify early in their differentiation. In transgenic mice, the mGluR6 promoter drives expression of green fluorescent protein in a subset of bipolar cells (ON bipolar cells). Expression is observed when these cells still possess apical and basal processes that are attached to the outer and inner limiting membranes, respectively. Time-lapse imaging of the retinas from these transgenic mice unequivocally demonstrated that axons and dendrites extend from the neuroepithelial-like processes as the bipolar cells differentiate (**Figure 4(a)**). Whether axons and dendrites in other species including zebrafish also develop in a similar manner is yet to be determined. Both axons and dendrites of the mouse ON bipolar cells explore retinal

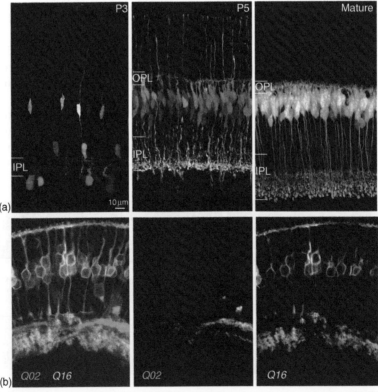

Figure 4 Lamination of bipolar cell axons and dendrites. (a) Green fluorescent protein driven by the mGluR6 promoter in bipolar cells at P3, P5 and mature retina. GFP expressing bipolar cells elaborate dendrites from apical and basal processes oriented vertically and neuroepithelial-like. IPL, inner plexiform layer. Note that upon maturation, dendrites are confined to the OPL (outer plexiform layer) and axons to the IPL. (b) Local perturbation in amacrine cell (blue) and bipolar cell (yellow) lamination visualized in a transgenic zebrafish in which a subset of amacrine cells express cyan fluorescent protein (Q02) and ON bipolar cells express yellow fluorescent protein (Q16) in the background of the *lakritz* mutant, which lacks ganglion cells.

depths outside their final lamination locations, but processes within the appropriate lamina are selectively stabilized with maturation. Process stabilization cues may be provided by other retinal neurons that have already grown stratified arbors in the forming plexiform layers. For example, the axonal terminals of the ON bipolar cells are largely distributed within the inner half of the IPL where they later contact ON ganglion cells. At this stage of axonal elaboration, as discussed later, amacrine cells that stratify in the ON sublamina have already contacted their target ganglion cells. Because amacrine cells can stratify independently of the presence of retinal ganglion cells, it is possible that these interneurons provide laminar cues for the bipolar axons. This hypothesis receives some support from observations in the zebrafish *lakritz* mutant that lacks retinal ganglion cells, in which there are local patches with lamination defects in the amacrine cell neurites although lamination largely occurs across the retina. In these locally perturbed regions, bipolar axon terminals are also abnormally placed (**Figure 4(b)**).

Photoreceptors

Photoreceptors connect to the bipolar cell dendrites once horizontal cells have already entered into synaptic partnerships with the photoreceptors (**Figure 2**). Apart

from early Golgi studies, relatively little is known about how individual photoreceptors terminate their axons to form a single lamina in the outer retina during development although several studies have described the maturation of their outer segments. Observations in the ferret retina showed that some photoreceptors even transiently project into the IPL during development, although the functional significance of these projections is not yet known.

Synaptogenesis in the Vertical Pathway

Synaptic connections in the vertical pathway have only recently been studied in detail by light microscopy. Early ultrastructural studies of the IPL and OPL of mammals and zebrafish suggest that ribbon synapses are formed between bipolar cells and retinal ganglion cells after conventional synapses are present (**Figures 5** and **6**). In rodents and zebrafish, contact between photoreceptors and bipolar cells occur about the same time as bipolar cells contact the retinal ganglion cells. However, ribbons appear late at bipolar synapses in the IPL. Detailed serial reconstruction of immature bipolar cell axon terminals in primate retina suggest that ribbons appear after clusters of vesicles are found at appositions between bipolar cell axonal processes and a neighboring process. Interestingly,

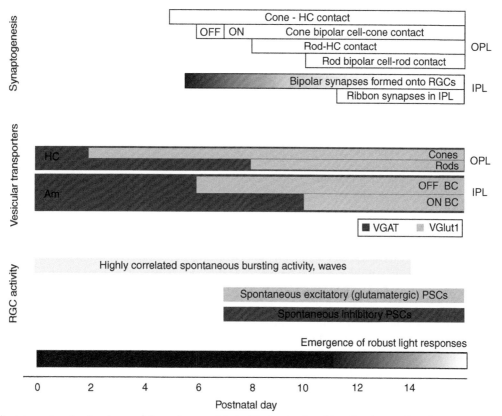

Figure 5 Summary of major developmental events associated with structural and functional assembly of mouse retinal circuits. VGAT, vesicular GABA/glycine transporter; VGlut1, vesicular glutamate transporter; HC, horizontal cells; BC, bipolar cells; PSCs, postsynaptic currents.

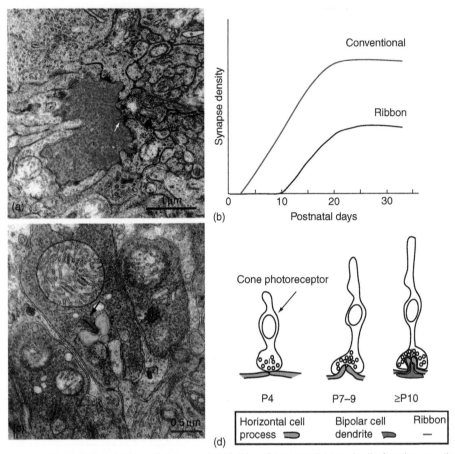

Figure 6 Synaptogenesis in the IPL and OPL. (a) Ultrastructure of ribbon (bipolar cell, blue shading) and conventional (amacrine, pink with asterisk and purple) synapses in a 2-week-old mouse retina. Shown here is a reciprocal synapse between a rod bipolar cell and amacrine cells. Arrow indicates ribbon. (b) Time-line of synaptogenesis in the IPL. (c) Example of a cone photoreceptor triad synapse in the OPL. Arrow indicates a ribbon. Green are horizontal cell processes and blue is bipolar cell process. Micrograph taken by Ed Parker, University of Washington. (d) Schematic illustrating time-line of triad formation in the mouse OPL. (b) Fisher, L. J. (1979). Development of synaptic arrays in the inner plexiform layer of neonatal mouse retina. *Journal of Comparative Neurology* 187: 359–372.

the earliest bipolar synapses appear to be monads, that is, there is only one postsynaptic process, whereas with maturation, bipolar cells synapses occur at junctions with two processes, forming a dyad synapse. The postsynaptic processes may comprise processes from two amacrine cells, an amacrine and a ganglion cell, or two ganglion cells. That bipolar synapses are already present on ganglion cell dendrites prior to the appearance of ribbons is supported by electrophysiological recordings from immature retinal ganglion cells. As discussed in more depth later, spontaneous excitatory postsynaptic currents (sEPSCs) can be recorded from ganglion cells in mammals prior to when ribbons appear. Electron microscope observations of developing mouse retina also suggest that bipolar synapses are present in the outer half (OFF) of the inner plexiform layer prior to their appearance in the inner half (ON) of the inner plexiform layer. There is support for the relatively earlier differentiation of synaptic connections, or at least their structures, in the OFF sublamina of the inner plexiform layer. For example, in mice, VGluT1 (vesicular

glutamate transporter) immunolableing is first observed in the OFF sublamina prior to its appearance in the ON sublamina (**Figure 5**). The significance of this sequence in maturation of the ON and OFF vertical pathways is unclear because light-evoked activity does not emerge until several days later.

Reconstructions of bipolar cell synapses at the ultrastructural level, however, cannot provide a view of the spatial distribution of such inputs on individual ganglion cell dendritic arbors unless extensive serial reconstructions are performed on immature ganglion cells. Using transient transfection methods, recent studies have successfully explored the spatial distribution of bipolar cell inputs on the dendritic arbors of retinal ganglion cells using light microscopy. Expression of fluorescently tagged postsynaptic density protein 95 (PSD95), a scaffolding protein found at glutamatergic postsynaptic sites, revealed the distribution of bipolar cell contacts across the dendritic arbors of ganglion cells in the mouse retina. Quantification of the spatial maps of these synaptic puncta indicated that in mice, there is a rapid acquisition of gluta-matergic postsynaptic sites from postnatal day 5 until eye-opening (**Figure** 7). Synaptogenesis between bipolar cells and retinal ganglion cells thus appears to proceed mostly prior to eye-opening (around 2 weeks after birth). Interestingly, the density of connections across the dendritic territory of the ganglion cells appears to be fairly invariant with age, even though there is significant structural remodeling (largely reduced dendritic branching density) of the dendritic arbors with maturation. This may reflect the developmental increase in the number of contact sites along the dendrites as dendritic density decreases with age.

Electron microscopy studies clearly indicate that cones form synaptic connections prior to rods. However, to date, there is little information concerning how bipolar cells and photoreceptors establish the specificity in their connectivity patterns during synaptogenesis. Retinas in which rod or cone populations are perturbed provide important insight into the specificity of wiring between rod and cone photoreceptors and their target bipolar cell types. In *nrl*-knockout mice in which rods fail to form and all photoreceptors become cones, rod bipolar cells are contacted by cones. Conversely, in in cyclic nucleotide gated channel A3 (CNGA3) knock-out mice in which cones are present but nonfunctional, cone bipolar cells form synaptic connections with rods. Thus, rod and cone bipolar cells can be targeted nonspecifically by cones and rods in the absence of proper interactions between these pre- and postsynaptic cell types. However, it is not yet clear whether specificity in the wiring between rods and cones and their target bipolar cells is obtained during development after a period of rewiring, or whether there is target selectivity during synaptogenesis. One way to address this question is to follow the development of photoreceptor-bipolar synaptogenesis over time *in vivo*. This can be readily achieved in zebrafish but is more difficult to perform for mammalian retinas. Furthermore, experiments that will determine how the different color cones wire up to the appropriate bipolar cells during development will be important in determining how color circuits are established in cone-dominated retinas.

Assembly of Lateral Circuits

Compared to the vertical pathway, the assembly of circuits that modulate transmission in the inner and outer retina are not as well understood. However, recent studies in which amacrine cells and horizontal cells can be identified early in development have shed new insight into the cellular behaviors of these cells as they form circuits.

Inner Retina – Amacrine Cells

Amacrine cells migrate freely toward the inner retina and appear to stratify shortly upon reaching the border with the forming ganglion cell layer (**Figure 2**). Serial electron microscopy suggested this pattern of amacrine cell migration and neurite development, which has

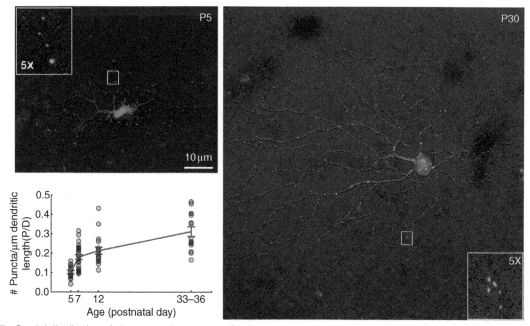

Figure 7 Spatial distribution of glutamatergic postsynaptic sites on developing ganglion cells. Examples of ganglion cells from an immature (postnatal day, P5) and a juvenile (postnatal day, P30) retina for which glutamatergic postsynaptic sites are labeled (blue). Dendrites are labeled by expression of the red fluorescent protein, td-Tomato, and gluamatergic postsynaptic sites by fluorescently tagged postsynaptic density 95 PSD95. Insets in upper left and lower right are 5× magnifications of the areas indicated by the boxes. Lower left: The density of PSD95-YFP (blue) puncta along the dendrites is plotted across ages studied.

subsequently been visualized in real time by *in vivo* imaging of these neurons in the zebrafish retina. Zebrafish amacrine cells very quickly stratify their neurites within the inner or outer half of the IPL as soon as their cell bodies reach their final locations. Thus, amacrine cell neurites do not appear to undergo a period of indiscriminate occupation of the IPL. Whether this pattern of neurite growth also occurs for mammalian amacrine cells, however, is not known although in rodents, cholinergic amacrine cells appear to form their mirror-symmetric laminations at two distinct depths in the IPL early in IPL formation.

The cues that influence the neuritic stratification of amacrine cells are not completely known, but it is evident that ganglion cells are not necessary for this process to occur. In mouse atonal 5 (*Math5*) knock-out mice and in the zebrafish mutant, *lakritz,* both lacking retinal ganglion cells, amacrine cells form stratified arbors. Because bipolar cells differentiate later than amacrine cells, these interneurons are unlikely to provide lamination cues for amacrine cells. The influence of adhesion molecules in the stratification of amacrine cell neurites has, however, been shown in the chick retina. The immunoglobulin superfamily of adhesion molecules, Down syndrome cell adhesion molecule (DsCam), DsCam-like, Sidekick-1 and Sidekick-2 are expressed in nonoverlapping strata in the chick inner plexiform layer. Manipulating the expression of these adhesion molecules in ovo suggests that they are involved in specifying the laminae within which amacrine cells and their target retinal ganglion cells stratify.

As yet, it is unknown what factors regulate synapse density between the amacrine cells and their target retinal ganglion cells. This issue can potentially be investigated by serial reconstructions of the IPL but because there are at least two dozen types of amacrine cells in the mammalian retina, and perhaps an even greater variety in the fish retina, mapping the inputs of specific subtypes of amacrine cells on the dendrites of their postsynaptic ganglion cells is extremely challenging. No doubt, future studies using transgenic approaches to visualize specific amacrine subtypes will be invaluable. Amacrine cells also provide feedback inhibition onto bipolar axon terminals. These feedback synapses have largely been studied in the rod bipolar cell pathway where A17 amacrine cells are known to contact the large axon terminals of rod bipolar cells (see **Figure 6 (a)**). The development of this highly localized reciprocal synapse has yet to be examined in detail but its well-characterized function makes this synapse a good model for studying the development of feedback circuits in general.

Outer Retina – Horizontal Cells

In the outer retina, modulation of transmission along the vertical pathway is provided by horizontal cells. In rodents, horizontal cell dendrites attain a laminated arbor after a period of reorganization (**Figure 2**). Cajal described immature horizontal cells as having a radial rather than lateral arbor. Since then, horizontal cells in rodents have been identified by immunolabeling for the GABA synthesizing enzyme, glutamic acid decarboxylase 67 (GAD67), and by calbindin immunoreactivity. Such labeling confirmed Cajal's early observations and also showed that horizontal cell dendritic stratification occurs as photoreceptors form synaptic connections onto the horizontal cells. Dendritic stratification of horizontal cells does not appear to rely on neurotransmission from photoreceptors because this process occurs even when cone photoreceptors are ablated during development. However, long-term loss of photoreceptor transmission does lead to elaboration of horizontal cell dendrites into the outer nuclear layer where they can receive ectopic contact.

Horizontal cells form contacts with cone photoreceptor terminals prior to the elaboration of bipolar cell dendrites that later invaginate into the cone terminal to form a synaptic triad (**Figure 6(c)** and **6(d)**). The triad comprises a single bipolar cell dendritic tip flanked by two horizontal cell processes at a location opposite to the ribbon in the cone terminal. The assembly of this triad structure has been described by elegant ultrastructural studies in the past but the cues that coordinate the assembly of each component of this synapse are still unknown. **Figure 5** summarizes the time-line of synaptogenesis between the various cell elements contributing to the OPL.

Emergence of Function – Spontaneous and Light-Evoked Activity

Visually evoked signals in the mammalian retina emerge shortly before eye-opening when photoreceptors have formed synaptic connections with the bipolar cells. However, the retina generates its own pattern of activity prior to photoreceptor transmission. In many mammalian species studied thus far, retinal ganglion cells have been found to exhibit bursts of action potentials that occur rhythmically and are synchronized among neighboring cells (**Figure 8**). This synchronized activity takes on the form of propagating waves that spread across the retina in different directions. The wave-like activity pattern has been linked to the refinement of the axonal projection patterns of the retinal ganglion cells to their subcortical targets, but as yet waves have not been found to influence the development of intraretinal circuits in mammals.

Changes in the properties of retinal waves, however, have been informative with regard to the organization and development of retinal circuits. For example, studies of retinal activity in chick, mice, ferret, and rabbit all suggest that early synchronized activity is mediated by gap junctions, then by cholinergic drive and later by glutamatergic transmission. Moreover, ON and OFF retinal

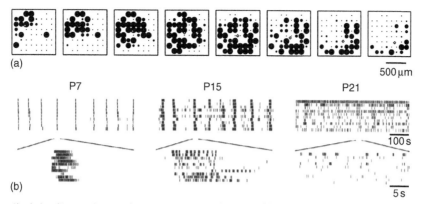

500 μm

P7   P15   P21

100 s

5 s

Figure 8 Immature retinal circuits spontaneously generate waves of activity. (a) Multielectrode array recordings (squares) for neonatal mouse retinas showing propagation of a retinal wave over time (left to right). Time between frames is 0.5 s. Each dot represents an electrode site and the size of the dot corresponds to spike rate. (b) Spike rasters from 10 representative cells recorded simultaneously on the array for three postnatal ages; P7, P15, and P21.

ganglion cells are synchronized in their bursting activity during the period when cholinergic drive is necessary for the wave activity. However, as glutamatergic drive takes over, the spontaneous bursting activity of ON and OFF mouse ganglion cells becomes desynchronized: ON ganglion cells burst before OFF ganglion cells. This asynchrony is generated by cross-over inhibition between these parallel pathways. Thus, prior to eye-opening, retinal circuits undergo changes in spontaneous activity patterns of the ganglion cells largely due to alterations in the type of excitatory drive (cholinergic then glutamatergic) and the maturation of inhibitory circuits. At the level of synapses, measurements of synaptic currents from individual retinal ganglion cells show that the amplitudes and frequency of sEPSCs increase with age, even after eye-opening.

Few studies have directly measured the early visual response properties of retinal ganglion cells. Such studies have been difficult to carry out *in vivo* because of the poor optics of the immature eye. As soon as it is possible to record light responses from the retina *ex vivo*, generally a few days before eye-opening, the characteristic center-surround organization of ganglion cell receptive fields can be identified in cat and rabbit. Although the antagonistic surround, mediated by inner and/or outer retinal inhibition, is present when the receptive field centers are first detected, the response properties of the early surrounds appear to be species dependent. The early light responses are weak and adapt rapidly but become robust several days later. However, ON and OFF responses are evident and there is no significant developmental change in the percentage of cells with ON- and OFF-center receptive fields in the cat. In contrast, in ferret and mice, there is a higher proportion of ganglion cells that receives converging ON and OFF input during development. The reduction in the incidence of ON–OFF cells in the ferret has been attributed to dendritic pruning, whereas in mice this reduction is considered to be the result of a decrease in the number of cells with dendritic arbors located at the

border of the ON and OFF sublaminae of the IPL. However, in mice, direct correlation between structure and function of individual ganglion cells that have converging ON and OFF input has yet to be obtained. Direction-selective ganglion cells can be found in rabbit retina before eye-opening and the development of the circuitry underlying this visual response property in mouse retinal ganglion cells does not depend on spontaneous retinal wave activity or visual experience. In contrast, the spatial receptive fields of turtle retinal ganglion cells appear to reorganize with maturation. Direction selectivity, orientation preference, and receptive field sizes of turtle ganglion cells are shaped by spontaneous and light-driven activity. Thus, the influence of neurotransmission, especially visual experience, in establishing the precision in wiring of intraretinal circuits appears to vary across species, perhaps due to different developmental constraints that are species specific. Alternatively, such differences may be due to how neurotransmission is altered *in vivo* and the consequence of each manipulation in the transmission of signals between specific cell types.

Conclusions

Much remains to be done in order to elucidate the molecular and cellular mechanisms that are responsible for establishing the many circuits within the vertebrate retina dedicated to the processing of light information. In recent years there have been significant technological advances that allow probing the structural and functional development of retinal circuits. However, cell-type specific markers that will enable tracking the same cell type throughout development are necessary for many future studies. With the increasing availability of transgenic models, it should be possible to directly address the mechanisms that are essential to circuit assembly *in vivo*. Investigating the development of intraretinal circuits is certainly well aided by the immense knowledge of the structure and function of the adult vertebrate retina, and the comparative anatomy and physiology of its circuits across many species.

See also: GABA Receptors in the Retina; Ganglion Cell Development: Early Steps/Fate; Histogenesis: Cell Fate: Signaling Factors; Information Processing: Amacrine Cells; Information Processing: Bipolar Cells; Information Processing: Ganglion Cells; Information Processing: Horizontal Cells; Information Processing in the Retina; Photoreceptor Development: Early Steps/Fate; Retinal Histogenesis.

Further Reading

Cajal, R. Y. (1960). *Studies on Vertebrate Neurogenesis.* Springfield, IL: Thomas.

Eglen, S. J., Sernagor, E., and Wong, R. O. (eds.) (2006). *Mechanisms of Retinal Development.* Cambridge: Cambridge University Press.

Elstrott, J., Anishchenko, A., Greschner, M., et al. (2008). Direction selectivity in the retina is established independent of visual experience and cholinergic retinal waves. *Neuron* 58: 499–506.

Fisher, L. J. (1979). Development of synaptic arrays in the inner plexiform layer of neonatal mouse retina. *Journal of Comparative Neurology* 187: 359–372.

Fuerst, P. G., Koizumi, A., Masland, R. H., and Burgess, R. W. (2008). Neurite arborization and mosaic spacing in the mouse retina require DSCAM. *Nature* 451: 470–474.

Galli-Resta, L., Leone, P., Bottari, D., et al. (2008). The genesis of retinal architecture: An emerging role for mechanical interactions? *Progress in Retinal and Eye Research* 27: 260–283.

Godinho, L., Mumm, J. S., Williams, P. R., et al. (2005). Targeting of amacrine cell neurites to appropriate synaptic laminae in the developing zebrafish retina. *Development* 132: 5069–5079.

Haverkamp, S., Michalakis, S., Claes, E., et al. (2006). Synaptic plasticity in CNGA3(−/−) mice: Cone bipolar cells react on the missing cone input and form ectopic synapses with rods. *Journal of Neuroscience* 26: 5248–5255.

Kay, J. N., Roeser, T., Mumm, J. S., et al. (2004). Transient requirement for ganglion cells during assembly of retinal synaptic layers. *Development* 131: 1331–1342.

Morest, D. K. (1970). The pattern of neurogenesis in the retina of the rat. *Zeitschrift für Anatomie und Entwicklungsgeschichte* 131: 45–67.

Morgan, J. L., Schubert, T., and Wong, R. O. (2008). Developmental patterning of glutamatergic synapses onto retinal ganglion cells. *Neural Development* 3: 8.

Mumm, J. S., Williams, P. R., Godinho, L., et al. (2006). *In vivo* imaging reveals dendritic targeting of laminated afferents by zebrafish retinal ganglion cells. *Neuron* 52: 609–621.

Nishimura, Y. and Rakic, P. (1987). Development of the rhesus monkey retina: II. A three-dimensional analysis of the sequences of synaptic combinations in the inner plexiform layer. *Journal of Comparative Neurology* 262: 290–313.

Schmitt, E. A. and Dowling, J. E. (1999). Early retinal development in the zebrafish, *Danio rerio*: Light and electron microscopic analyses. *Journal of Comparative Neurology* 404: 515–536.

Sernagor, E., Eglen, S. J., Harrris, W., and Wong, R. O. L. (eds.) (2006). *Retinal Development.* Cambridge: Cambridge University Press.

Sernagor, E., Eglen, S. J., and Wong, R. O. (2001). Development of retinal ganglion cell structure and function. *Progress in Retinal and Eye Research* 20: 139–174.

Strettoi, E., Mears, A. J., and Swaroop, A. (2004). Recruitment of the rod pathway by cones in the absence of rods. *Journal of Neuroscience* 24: 7576–7582.

Yamagata, M. and Sanes, J. R. (2008). Dscam and Sidekick proteins direct lamina-specific synaptic connections in vertebrate retina. *Nature* 451: 465–469.

Ischemic Optic Neuropathy

S S Hayreh, University of Iowa, Iowa City, IA, USA

Glossary

Afferent pupillary defect – Reduction in the response of the pupil to direct light.
Giant cell arteritis – An inflammation of the arterial wall, affecting medium and large size arteries.

Ischemic optic neuropathy constitutes one of the major causes of blindness or seriously impaired vision among the middle-aged and elderly population, although no age is immune. Its pathogenesis, clinical features, and management are discussed in this article.

Classification

Ischemic optic neuropathy is an acute ischemic disorder of the optic nerve. Based on the blood-supply pattern, the optic nerve can be divided into two very distinct parts: (1) the anterior part (called the optic nerve head) and (2) the rest of the optic nerve (**Figure 1**). Ischemic optic neuropathy is of two distinct types.

1. *Anterior ischemic optic neuropathy (AION).* This is due to ischemia of the anterior part of the optic nerve, which is supplied by the posterior ciliary artery (PCA) circulation. Etiologically and pathogenetically, AION is of the following two types: (a) arteritic AION (A-AION), which is due to giant cell arteritis (GCA), and (b) nonarteritic AION (NA-AION), which is due to all other causes and not GCA.
2. *Posterior ischemic optic neuropathy (PION).* This is due to ischemia of a segment of the rest of the optic nerve, which is supplied by multiple sources but not the PCA (**Figure 1**).

Nonarteritic Anterior Ischemic Optic Neuropathy

This is the most common type of ischemic optic neuropathy and has attracted the most controversy about its pathogenesis and management.

Pathogenesis

Nonarteritic anterior ischemic optic neuropathy (NA-AION) is caused by acute ischemia of the optic nerve head, whose main source of blood supply is from the PCA circulation (**Figure 1**). Marked interindividual variations in the blood supply of the optic nerve head and its blood-flow patterns profoundly influence the pathogenesis and clinical features of NA-AION.

Etiologically and pathogenetically, NA-AION is of two types:

1. *Transient nonperfusion or hypoperfusion of the optic nerve head circulation.* This is by far the most common cause of NA-AION. All the available evidence indicates that NA-AION is not a thromboembolic disorder. The mechanism causing transient nonperfusion or hypoperfusion of the optic nerve head circulation in NA-AION is multifactorial in nature. In the vast majority of cases, it is a transient fall of blood pressure, most commonly during sleep (nocturnal arterial hypotension – **Figure 2**) or a nap during the day. Any kind of shock also can cause a transient marked fall of blood pressure. A sharp rise in the intraocular pressure to high levels, as in neovascular glaucoma associated with ocular ischemia, or angle closure glaucoma, can also cause a transient fall in perfusion pressure in the optic nerve head, where perfusion pressure is equal to mean blood pressure minus the intraocular pressure.

A transient fall of perfusion pressure in the optic nerve head vessels below the critical autoregulatory range level in susceptible persons results in ischemia of the optic nerve head and development of NA-AION. The severity of the ischemia may vary from mild to marked, depending upon the severity and duration of the transient ischemia, and upon other factors influencing the blood flow in the optic nerve head.
2. *Embolic lesions of the arteries/arterioles feeding the optic nerve head.* This is a rare cause of NA-AION. Compared to the hypotensive type of NA-AION, the extent of optic nerve head damage in this type is usually massive, severe, and permanent.

Risk Factors for Development of NA-AION

All available evidence indicates that NA-AION is multifactorial in nature. The risk factors fall into two main categories:

1. Predisposing risk factors:
   (a) *Systemic.* These include arterial hypertension, nocturnal arterial hypotension, diabetes mellitus, ischemic heart disease, hyperlipidemia, atherosclerosis

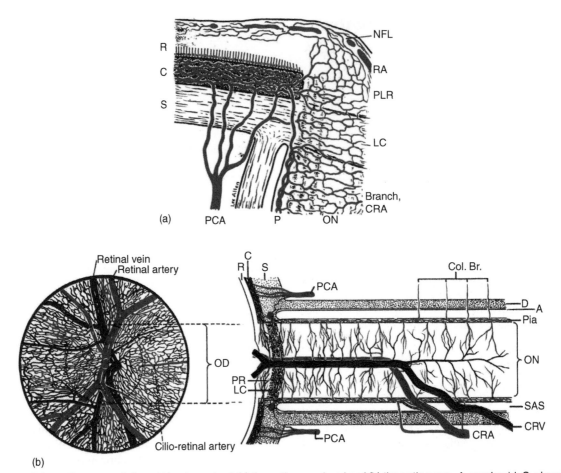

Figure 1 Schematic representation of blood supply of: (a) the optic nerve head and (b) the optic nerve. A, arachnoid; C, choroid; CRA, central retinal artery; Col. Br., collateral branches; CRV, central retinal vein; D, dura; LC, lamina cribrosa; NFL, surface nerve fiber layer of the disk; OD, optic disk; ON, optic nerve; P, pia; PCA, posterior ciliary artery; PR and PLR, prelaminar region; R, retina; RA, retinal arteriole; S, sclera; SAS, subarachnoid space. (a) Reproduced from Hayreh, S. S. (1978). In: Heilmann, K. and Richardson, K. T. (eds.) *Glaucoma: Conceptions of s Disease*, pp. 78–96. Stuttgart: Thieme. (b) Modified from Hayreh, S. S. (1974) *Transactions American Academy of Ophthalmology and Otolaryngology* 78: OP240–OP254.

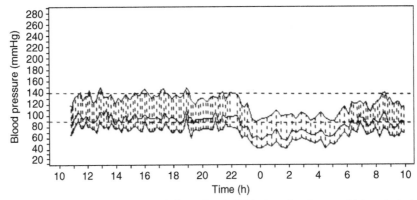

Figure 2 Ambulatory blood pressure monitoring records (based on individual readings) over a 24-h period, starting from about 11 a.m., in a 58-year-old woman with bilateral NA-AION, and on no medication. The blood pressure is perfectly normal during the waking hours but there is marked nocturnal arterial hypotension during sleep. Reproduced from Hayreh et al. (1999) *Ophthalmologica* 213: 76–96.

and arteriosclerosis, sleep apnea, arterial hypotension due to any cause, and migraine.

(b) *Risk factors in the eye and/or the optic nerve head.* These include absent or small cup in the optic disk, marked optic disk edema, raised intraocular pressure, and optic disk drusen. There is a misconception in the ophthalmic community that a small or absent cup is actually the primary factor in the development of the disease; this has resulted in catchy terms like "disk at risk". However, in the multifactorial scenario of the pathogenesis of NA-AION, an absent or small cup is simply a secondary contributing factor once the process of NA-AION has started.

2. Precipitating risk factor(s): in a person with predisposing risk factors present, a precipitating risk factor acts as the final insult (last straw), resulting in ischemia of the optic nerve head and NA-AION. Nocturnal arterial hypotension (**Figure 2**) is the most important factor in this category. Studies have shown that patients with NA-AION typically discover visual loss on waking in the morning, indicating that NA-AION developed during sleep when there is invariably a fall of blood pressure.

Conclusion

A host of systemic and local factors, acting in different combinations and to different extents, may derange the optic nerve head circulation, with some increasing optic nerve head susceptibility to ischemia and others acting as the final insult. Nocturnal arterial hypotension seems to be an important precipitating factor in the susceptible patient. Thus, the pathogenesis of NA-AION is complex but, not, as often stated, unknown.

There is a common perception among ophthalmologists and neurologists that NA-AION and cerebral stroke are similar in nature pathogenetically and in management. It is well established that stroke is a thromboembolic disorder. However, available evidence indicates that NA-AION is pathogenetically a hypotensive disorder, not a thromboembolic disorder. In NA-AION, unlike stroke, (1) there is no association between smoking and NA-AION; (2) aspirin has no beneficial effects in NA-AION; (3) no significant association has been found between NA-AION and thrombophilic risk factors; (4) fluorescein fundus angiography during the early stages of onset of visual loss invariably shows no evidence of complete occlusion of the vessels supplying the optic nerve head (**Figures 3(a)** and **3(b)**); and (5) 41% of NA-AION eyes show spontaneous visual improvement, which is rare in a thromboembolic disorder.

Clinical Features of NA-AION

NA-AION is the most common type of ischemic optic neuropathy. It usually has classical symptoms and signs. NA-AION is mostly a disease of the middle-aged and

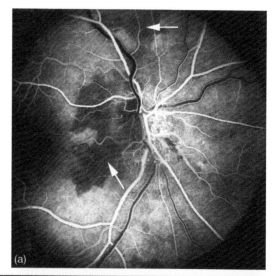

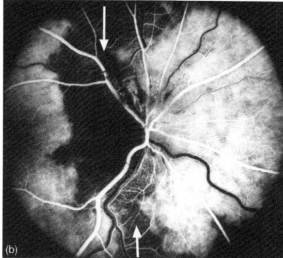

Figure 3 Fluorescein fundus angiograms of eyes with NA-AION. (a) Shows non-filling of temporal part of the peripapillary choroid (oblique arrow) and adjacent optic disk and the choroidal watershed zone (horizontal arrow). (b) Shows non-filling of the choroidal watershed zone (vertical dark band: arrows) between the lateral and medial PCAs and of the temporal part of optic disk. Reproduced from Hayreh, S. S. (1985). Inter-individual variation in blood supply of the optic nerve head. Its importance in various ischemic disorders of the optic nerve head, and glaucoma, low-tension glaucoma and allied disorders. *Documenta Ophthalmologica* 59: 217–246.

elderly population, although no age is immune. It has been reported in twenty-three percent of patients with NA-AION are under the age of 50 years. It is far more common among the white population than in other racial groups.

1. *Symptoms.* In the vast majority of patients, there is a sudden and painless deterioration of vision, usually discovered on waking in the morning. NA-AION patients often complain of loss of vision toward the nose and less often in the lower part. Later on,

photophobia is a common complaint. Simultaneous bilateral onset of NA-AION is extremely rare.

2. *Signs.* Initial visual acuity may vary from 20/20 (better than 20/40 in 33%) to marked loss. Therefore, a normal visual acuity does not rule out NA-AION. An inferior nasal visual field defect is the most common, followed by an inferior altitudinal defect; a combination of a relative inferior altitudinal defect with absolute inferior nasal defect is the most common pattern in NA-AION (**Figures 4(a)** and **4(b)**). The eye shows the presence of a relative afferent pupillary defect in unilateral NA-AION cases, and in some there may be raised intraocular pressure.

At the onset of visual loss, there is always optic disk edema (**Figures 5** and **6(a)**). There are several

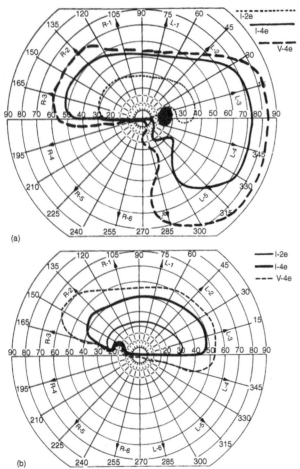

(a)

(b)

Figure 4 Visual-field defects in NA-AION, plotted with Goldmann perimeter using I-2e, I-4e, and V-4e targets, where the Roman numeral indicates target size, the Arabic numeral indicates relative intensity, and the lowercase letter indicates minor filter adjustment of the light intensity. (a) Shows an inferior altitudinal defect with I-2e and an inferior nasal defect with I-4e and V-4e. (b) Shows an absolute inferior altitudinal defect with I-2e, I-4e, and V-4e. The visual acuity in both eyes was 20/20. Reproduced from Hayreh et al. (2005) *Archives of Ophthalmology* 123:1554–1562.

misconceptions about optic disk edema in NA-AION. The most common one is that in NA-AION the optic disk edema is always pale, which is not at all true initially because the color of optic disk edema in NA-ION initially does not differ from that due to other causes; in some cases there may even be hyperemia of the optic disk (**Figures 5** and **6(a)**). A splinter hemorrhage at the disk margin is common (**Figure 7**). The optic disk edema resolves spontaneously in about 8 weeks, and the disk develops segmental or generalized pallor (**Figure 6(b)**). In the normal fellow eye, the optic disk usually shows either no cup or a small cup, which can be a helpful clue in the diagnosis of NA-AION in doubtful cases.

In diabetics, optic disk changes in NA-AION may have some characteristic diagnostic features. During the initial stages, the optic disk edema is usually, but not always, associated with characteristic prominent, dilated, and frequently telangiectatic vessels over the disk and many more peripapillary retinal hemorrhages than in nondiabetics (**Figures 8(a)** and **8(c)**). Because of these disk changes, NA-AION in diabetics has been mistakenly diagnosed as diabetic papillopathy. With the resolution of optic disk edema, vascular changes and hemorrhages also resolve spontaneously (**Figures 8(b)** and **8(d)**).

Fluorescein fundus angiography, at the onset of NA-AION, during the very early phase of dye filling in the fundus almost invariably shows delayed filling of the prelaminar region and the peripapillary choroid (**Figure 3(a)**) and/or choroidal watershed zone (**Figure 3(b)**).

Bilateral NA-AION

The cumulative probability of the fellow eye developing NA-AION has varied among different studies: 25%

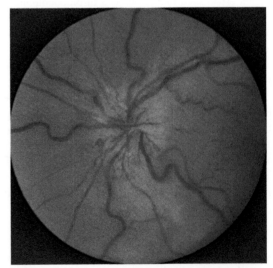

Figure 5 Left fundus photograph showing optic disk edema and hyperemia during the acute phase of NA-AION.

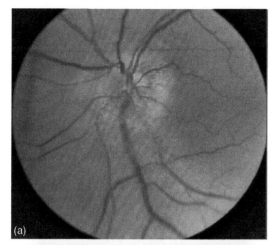

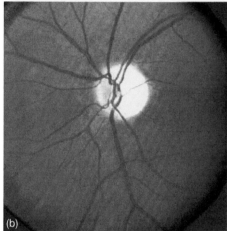

Figure 6 Fundus photographs of left eye of a 53-year-old man. (a) With optic disk edema during the active phase of NA-AION. (b) After resolution of optic disk edema and development of optic disk pallor – more marked in the temporal part than in the nasal part.

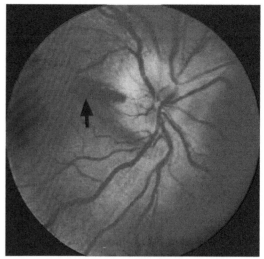

Figure 7 Right fundus photograph showing optic disk edema and hyperemia, with a splinter hemorrhage (arrow) during the acute phase of NA-AION.

within 3 years, 17% in 5 years, and 15% over 5 years. Diabetics have a significantly ($p = 0.003$) greater risk than non-diabetics of NA-AION in the second eye as well as earlier involvement. Simultaneous bilateral onset of NA-AION is extremely rare, except in patients who develop sudden, severe arterial hypotension, for example during hemodialysis or surgical shock.

Recurrence of NA-AION in the same eye
In a study of 829 NA-AION eyes, the overall cumulative percentage of recurrence of NA-AION in the same eye at 2 years was 5.8%. The only significant association for recurrence of NA-AION was with nocturnal arterial hypotension.

NA-AION and phosphodiesterase-5 inhibitors
These agents are currently popular for erectile dysfunction. A critical review of all the reported cases usually shows a good temporal relationship between the ingestion

of Viagra and other phosphodiesterase-5 inhibitors and the development of NA-AION, in persons who already have predisposing risk factors.

Amiodarone and NA-AION
There is a universal belief that amiodarone causes optic neuropathy, called "amiodarone-induced optic neuropathy". However, it is NA-AION produced in the multifactorial scenario by the systemic cardiovascular risk factors for which the drug is given rather than amiodarone *per se* causing it.

Familial NA-AION
There are five reports in the literature representing 10 unrelated families in which more than one member developed NA-AION. It is clinically similar to the classical non-familial NA-AION, with the exception that familial NA-AION occurs in younger patients and has much higher involvement of both eyes than the classical NA-AION. The role of genetic factors in familial NA-AION is not known.

Management of NA-AION

The outcome of all advocated treatments has to be compared with the natural history of a disease, so that natural recovery is not attributed to the beneficial effect of a mode of treatment. Therefore, it is essential to first consider the natural history of visual outcome in NA-AION. This has been investigated by two prospective studies in patients seen within 2 weeks of onset of visual loss and initial visual acuity of 20/70 or worse. Both showed spontaneous visual acuity improvement in 41–43% of the patients and worsening in 15–19% at 6 months. Evaluation of visual fields

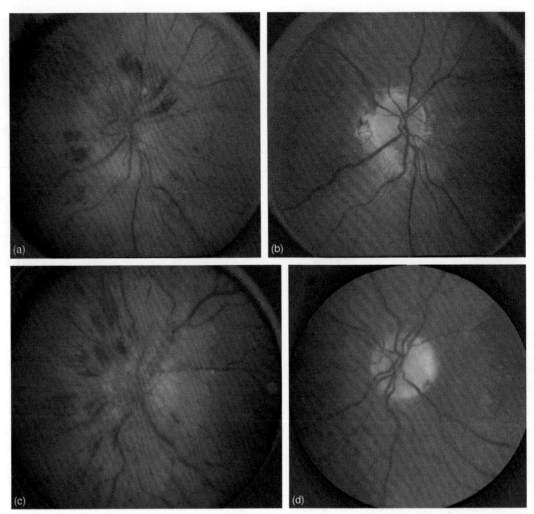

Figure 8 Fundus photographs of both eyes of a 51-year-old woman with adult-onset diabetes mellitus. She developed bilateral NA-AION, first in the right eye ((a) and (b)) and 8 months later in the left eye (c, d). (a, c) On first visit: Fundus photographs show massive optic disk edema with marked telangiectatic vessels on the optic disk, and many retinal hemorrhages. Visual acuity was 20/20 in the left and 20/15 in the right eye. Both eyes had an inferior nasal visual-field defect. (b, d) On resolution of optic disk edema: Fundus photographs show no optic disk edema but mild temporal pallor, no abnormal vessels on optic disk, and no retinal hemorrhages in the right eye (b), and a few resolving hemorrhages in the left eye (d). Reproduced from Hayreh et al. (1981) *Ophthalmologica* 182: 13–28.

with kinetic perimetry showed that of those with moderate-to-severe visual field defect, 26% showed improvement at 6 months. Visual acuity and visual fields showed improvement or further deterioration mainly up to 6 months, with no significant change thereafter.

Management of NA-AION has been a highly controversial subject. A number of treatments have been advocated. Following are the principal ones.

Optic nerve sheath decompression. A study in 1989 claimed that optic nerve sheath decompression improved visual function in "progressive" NA-AION. However, a recent multicenter clinical trial showed that this procedure is not effective and may be harmful, and thus is an inappropriate treatment for NA-AION, because 24% of the eyes with the optic nerve sheath decompression suffered further visual loss as compared to only 12% of eyes simply left alone.

Aspirin. Studies have shown that aspirin in NA-AION provides no long-term benefit in reducing the risk of NA-AION in the fellow eye.

Systemic corticosteroid therapy. A recent large, prospective study evaluated the role of steroid therapy in 696 NA-AION eyes, comparing the visual outcome in treated (364 eyes) versus untreated control (332 eyes) groups. In eyes with initial visual acuity of 20/70 or worse and seen within 2 weeks of onset, there was visual acuity improvement in 70% the treated group compared to 41% the untreated group (odds ratio of improvement: 3.39; 95% CI:1.62, 7.11; $p = 0.001$). Similarly, among those seen within 2 weeks of NA-AION onset and moderate to severe initial visual-field defect, there was improvement in 40% of the treated group and 25% of the untreated group (odds ratio: 2.06, 95% CI: 1.24, 3.40; $p = 0.005$).

In both treated and untreated groups, the visual acuity and visual fields kept improving for up to about 6 months after the onset of NA-AION, but very little thereafter.
Reduction of risk factors. As NA-AION is a multifactorial disease and many risk factors contribute to it, the correct strategy is to try reducing as many risk factors as possible to reduce the risk of developing NA-AION in the second eye or any further episode in the same eye. Nocturnal arterial hypotension is a precipitating risk factor in NA-AION patients with predisposing risk factors. In view of this, management of nocturnal arterial hypotension is an important step in both the management of NA-AION and the prevention of its development in the second eye. It seems NA-AION is emerging in some cases as an iatrogenic disease due to the use of currently available highly potent arterial hypotensive drugs.

Incipient NA-ION

This clinical entity initially presents with asymptomatic optic disk edema and no visual loss attributable to NA-AION. Available evidence indicates that it represents the earliest, asymptomatic clinical stage in the evolution of the NA-AION disease process; therefore, it shares most clinical features with classical NA-AION except for the visual loss initially.

Arteritic AION

The primary cause of A-AION is GCA, although other types of vasculitis can also cause it.

Pathogenesis

The primary cause is GCA, which has a special predilection to involve the PCA, resulting in its thrombotic occlusion. Since the PCA is the main source of blood supply to the optic nerve head (**Figure 1**), occlusion of the PCA results in infarction of a segment or the entire optic nerve head, depending upon the area of the optic nerve head supplied by the occluded PCA. This, in turn, results in development of A-AION.

Clinical Features of GCA and A-AION

GCA is a disease of late middle-aged and elderly persons and is almost 3 times more common in women than in men. There is evidence that GCA is far more common among Caucasians than other races.

Symptoms
GCA patients usually present with systemic symptoms, including anorexia, weight loss, jaw claudication, headache, scalp tenderness, abnormal temporal artery, neck pain, myalgia, malaise, and anemia. However, a study showed that 21% of patients with visual loss due to GCA have no systemic symptoms whatsoever (i.e., occult GCA).

One visual symptom of GCA is episodic transient visual loss, which is an important and an ominous sign of impending visual loss. In one series, it occurred in about one-third of the patients. Most patients with GCA develop visual loss due to A-AION suddenly without any warning. Occasionally there may be diplopia or ocular pain. A rare patient with GCA can suffer from euphoria and even deny any visual loss.

Signs
Visual acuity in eyes with A-AION varies between 20/20 and no light perception, but overall it is much worse than in NA-AION. The extent and severity of visual field defects depends upon the extent of optic nerve head damage caused by ischemia, usually much more extensive and severe than in NA-AION. When there is diplopia, there is extraocular motility abnormality. In uniocular A-AION, there is a relative afferent pupillary defect.

The optic disk always shows edema initially, which usually has a chalky white color (**Figure 9**) – a diagnostic characteristic of A-AION. When disk edema resolves, the optic disk in the vast majority shows cupping which is indistinguishable from that seen in glaucomatous optic neuropathy. In addition to optic disk edema, the fundus may show retinal cotton wool spots, central retinal artery occlusion, cilioretinal artery occlusion (**Figure 10(a)**), and/or choroidal ischemic lesions. Fluorescein fundus angiography done during the early acute stage of the disease shows evidence of occlusion of the PCAs (**Figures 10(b)** and **11**).

Laboratory investigations
Evaluation of erythrocyte sedimentation rate (ESR) and C-reactive protein (CRP) are the most important immediate diagnostic tests in the diagnosis of A-AION and its

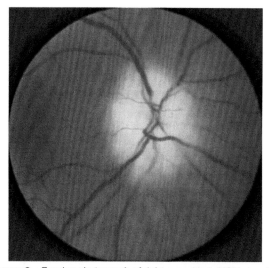

Figure 9 Fundus photograph of right eye with A-AION showing chalky white optic disk edema during the initial stages.

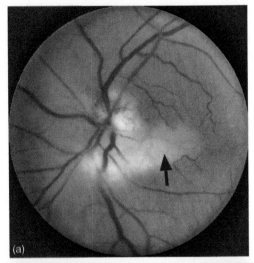

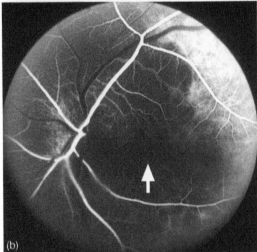

Figure 10 Fundus photograph (a) and fluorescein angiogram (b) of left eye with A-AION associated with cilioretinal artery occlusion. (a) Fundus photograph shows optic disk edema and retinal infarct (arrow) in the distribution of cilioretinal artery. (b) Fluorescein angiogram shows no filling of the choroid and entire optic disk supplied by the medial PCA and of the cilioretinal artery (arrow), but normal filling of the area supplied by the lateral PCA. (a) Reproduced from Hayreh, S. S. (1990). Anterior ischaemic optic neuropathy. Differentiation of arteritic from non-arteritic type and its management. *Eye* 4: 25–41; (b) reproduced from Hayreh S. S. (1978) *International Ophthalmology* 1: 9–l8.

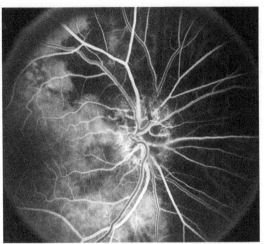

Figure 11 Fluorescein fundus angiogram of a right eye with A-AION, showing normal filling of the area supplied by the lateral PCA (including the temporal one-fourth of the optic disk) but no filling of the area supplied by the medial PCA (including the nasal three-fourth of the disk). Reproduced from Hayreh S. S. (1978) *International Ophthalmology* 1: 9–l8.

Management of A-AION

Management of A-AION is actually the management of GCA. Visual loss is preventable with (1) early diagnosis of GCA and (2) immediate and adequate steroid therapy. To establish a definite diagnosis of GCA without delay is the most critical step in the management of GCA. Classically the gold standard for diagnosis of GCA is the five criteria advocated by the American College of Rheumatologists: (1) age $\geq$ 50 years at onset, (2) new onset of localized headache, (3) temporal artery tenderness or decreased temporal artery pulse, (4) elevated ESR (Westergren) $\geq$ 50 mm/hour, and (5) positive temporal artery biopsy. The American College of Rheumatologists states: "A patient shall be classified as having GCA if at least 3 of these 5 criteria are met." However, studies indicate that these criteria are inadequate to prevent blindness in all GCA patients, particularly patients with occult GCA (21%) who never develop any systemic symptoms of GCA. Use of a positive temporal artery biopsy as the definite diagnostic criterion for GCA showed that the odds of a positive temporal artery biopsy were 9 times greater with jaw claudication ($p < 0.0001$), 3.4 times with neck pain ($p = 0.0085$), 2.0 times with ESR (Westergren) 47–107 mm/hour relative to those with ESR <47 mm/hour ($p = 0.0454$), and 3.2 times with CRP >2.45 mg dl compared to CRP <2.45 mg dl ($p = 0.0208$), and 2.0 times when the patients were aged $\geq$75 years as compared to those $\leq$75 years ($p = 0.0105$). Among the other systemic signs and symptoms, the only significant one was anorexia/weight loss ($p = 0.0005$); the rest showed no significant difference from those with negative temporal artery biopsy.

differentiation from NA-AION. Although high ESR is traditionally emphasized as a *sine qua non* for diagnosis of GCA, there are numerous reports of "normal" or "low" ESR in patients with positive temporal artery biopsy for GCA. Normal ESR does not rule out GCA. CRP, on the other hand, is a much more reliable test to diagnose GCA. Thus, both tests should be used for diagnosis of GCA and monitoring of steroid therapy in all patients. Other hematological tests which can help in the diagnosis of GCA include the presence of thrombocytosis, anemia, elevated white blood cell count, and low hemoglobin and hematocrit levels.

Differentiation of A-AION from NA-AION

When a patient is diagnosed as having AION, the first crucial step in patients aged 50 and over is to identify immediately whether it is arteritic or nonarteritic – missing A-AION can result in disastrous visual loss which is entirely preventable. Collective information provided by the following criteria helps to differentiate the two types of AION reliably.

1. *Systemic symptoms of GCA.* These are discussed above. However, 21% with occult GCA have no systemic symptoms of any kind at all. Patients with NA-AION have no systemic symptoms of GCA.
2. *Visual symptoms.* Amaurosis fugax is highly suggestive of A-AION and is extremely rare in NA-AION.
3. *Hematologic abnormalities.* Immediate evaluation of ESR and CRP is vital in all patients aged 50 and over. As discussed above, elevated ESR and CRP, particularly CRP, is helpful in the diagnosis of GCA. Patients with NA-AION do not show any of these abnormalities, except when a patient has some other concurrent systemic disease.
4. *Early massive visual loss.* There is a much more massive visual loss in A-AION than in NA-AION; however, the presence of perfectly normal visual acuity does not rule out A-AION.
5. *Chalky white optic disk edema* (**Figure 9**). This is almost always diagnostic of arteritic AION and is seen in the majority, but not all, of A-AION eyes.
6. *A-AION associated with cilioretinal artery occlusion* (**Figure 10**). This is almost always diagnostic of A-AION.
7. *Evidence of PCA occlusion on fluorescein fundus angiography* (**Figures 10(b)** and **11**). If angiography is performed during the first few days after the onset of A-AION, and the choroid supplied by one or more of the PCAs does not fill, this once again is almost diagnostic of A-AION.
8. *Temporal artery biopsy.* This finally establishes the diagnosis; however, the possibility of an occasional false negative biopsy has to be kept in mind.

Steroid Therapy to Prevent Blindness in GCA

This is a highly controversial subject because practically all the available information is from the rheumatologic literature. Rheumatologists and ophthalmologists have different perspectives which influence their recommendations on steroid therapy for GCA – the regimen advocated by the former primarily concerns managing benign rheumatologic symptoms and signs, whereas the latter confront the probability of blindness. In our 27-year prospective study on steroid therapy in GCA, marked differences were seen between the rheumatologic and ophthalmic steroid therapy regimens. In the light of information from that study, the following guidelines to prevent visual loss are suggested.

1. If there is a reasonable suspicion of GCA, as judged from systemic symptoms, high ESR and CRP (particularly high CRP) and sudden visual loss from A-AION or central retinal artery occlusion, high doses of systemic corticosteroid therapy must be started immediately as an emergency measure.
2. The physician should not wait for the result of the temporal artery biopsy because by the time it is available, the patient may have irreversibly lost further vision.
3. A high-dose steroid therapy (80–120 mg) must be maintained until both the ESR and CRP settle down to stable levels, which usually takes 2–3 weeks, with CRP reducing much earlier than ESR levels (**Figure 12**), followed by gradual tapering of steroid therapy.
4. A titration of the steroid dosage with the levels of ESR and CRP is the only safe and reliable method for tapering down and follow-up of steroid therapy; using clinical symptoms and signs of GCA as a guide, as often recommended by rheumatologists, is a dangerous practice for the prevention of blindness.
5. Patients with GCA show marked interindividual variation in the dosage of corticosteroids they require, their response to steroid therapy, and their therapeutic and tapering regimens of steroid therapy; therefore therapy must always be individualized. No generalization is possible; *NO* one size fits all.
6. The vast majority of GCA patients require a lifelong small dose of steroids to prevent blindness. As our study found no evidence that intravenous mega-dose steroid therapy was more effective than large-dose oral therapy in improving vision or preventing visual deterioration due to GCA, it is recommended that patients initially receive one intravenous mega dose (equivalent to 1 g of prednisone) followed by a high-dose (80–120 mg) of oral prednisone if that patient presents with: (a) a history of amaurosis fugax, (b) complete or marked loss of vision in one eye, or (c) early signs of involvement of the second eye.

Conclusion

If GCA patients are treated promptly and aggressively with an adequate dose of corticosteroids, and reduction of steroid therapy is regulated by using ESR and CRP as the only criteria, not a single patient will suffer any further visual loss 5 days after starting adequate steroid therapy. However, in our study, in spite of early high-dose steroid therapy, only 4% of GCA patients with visual loss showed any visual improvement, and during the first 5 days after the start of the therapy 4% developed further visual loss; but there was no further visual loss after that.

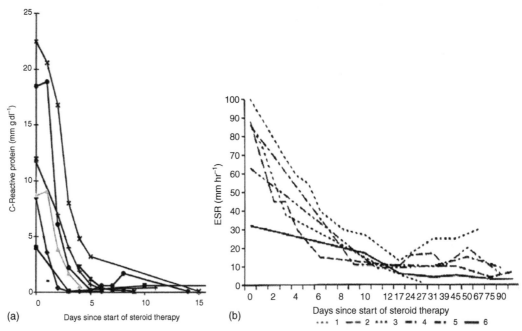

Figure 12 Graphs of (a) C-reactive protein levels and (b) erythrocyte sedimentation rates (ESRs) of six patients with giant cell arteritis, showing their initial responses to high-dose steroid therapy. Reproduced from Hayreh, S. S. and Zimmerman, B. (2003). Management of giant cell arteritis. Our 27-year clinical study: New light on old controversies. *Ophthalmologica* 217: 239–259.

Posterior Ischemic Optic Neuropathy

PION is much less common than AION and is due to ischemia of the optic nerve posterior to the optic nerve head, which is supplied by multiple sources but not the PCA (**Figure 1(b)**). It is a diagnosis of exclusion, and it should be made only after all other possibilities have been carefully ruled out, for example, macular and retinal lesions, NA-AION, retrobulbar optic neuritis, compressive optic neuropathy, other optic disk and optic nerve lesions, neurological lesions, hysteria, even malingering, and a host of other lesions.

Classification

Etiologically, PION can be classified into three types: (1) arteritic PION (A-PION), (2) nonarteritic PION (NA-PION), and (3) surgical PION.

Pathogenesis

Arteritic PION
This is due to GCA when arteritis involves the orbital arteries, which supply the posterior part of the optic nerve (**Figure 1(b)**). A-PION occurs much less commonly than A-AION.

Nonarteritic PION
An association between NA-PION and a variety of systemic diseases has been reported in the literature. In one series, comparison of NA-PION cases with a control population showed a significantly higher prevalence of arterial hypertension, diabetes mellitus, ischemic heart disease, cerebrovascular disease, carotid artery and peripheral vascular disease, and migraine than in NA-PION patients. Thus, the pathogenesis of NA-PION, similar to NA-AION, is multifactorial in nature, with a variety of systemic diseases, other vascular risk factors, and/or local risk factors predisposing an optic nerve to develop PION; defective autoregulation of the optic nerve may also play a role. Finally, some precipitating risk factor acts as the "last straw" to produce PION.

Surgical PION
This type of PION usually tends to cause bilateral, massive visual loss or even complete blindness, which is usually permanent; therefore, it has great medicolegal importance. It is almost invariably associated with prolonged systemic surgical procedures for a variety of conditions, including spinal and other orthopedic surgical procedures, radical neck dissection, coronary artery bypass, hip surgery, nasal surgery, and so on. The pathogenesis of surgical PION is multifactorial in nature. The main factors include severe and prolonged arterial hypotension, hemodilution from administration of a large amount of intravenous fluids to compensate for the blood loss, orbital and periorbital edema, chemosis and anemia, and even direct orbital compression due to the prone position.

Clinical Features of PION

NA-PION, similar to NA-AION, is seen mostly in the middle-aged and elderly population, but no age is immune.

Symptoms

Clinically, patients with A-PION and NA-PION typically present with acute, painless visual loss in one or both eyes, sometimes discovered upon waking up in the morning. In some eyes, it may initially be progressive. Patients with surgical PION discover visual loss as soon as they are alert postoperatively, which may be several days after surgery. Surgical PION usually tends to cause bilateral massive visual loss or even complete blindness, which is usually permanent.

Signs

Visual acuity may vary from 20/20 to no light perception. The most common visual-field defect is central scotoma, alone or in combination with other types of visual-field defects (**Figure 13**). A small number of PION eyes show the reverse pattern, that is, a normal central field with marked loss of peripheral fields (**Figure 14**).

Initially, apart from relative afferent pupillary defect in unilateral PION, the anterior segment, intraocular pressure, and optic disk and fundus are normal on ophthalmoscopy and fluorescein fundus angiography. The disk generally develops pallor within 6–8 weeks, usually more marked in the temporal part. The criteria to differentiate arteritic from nonarteritic PION are basically the same as those for arteritic and nonarteritic AION,

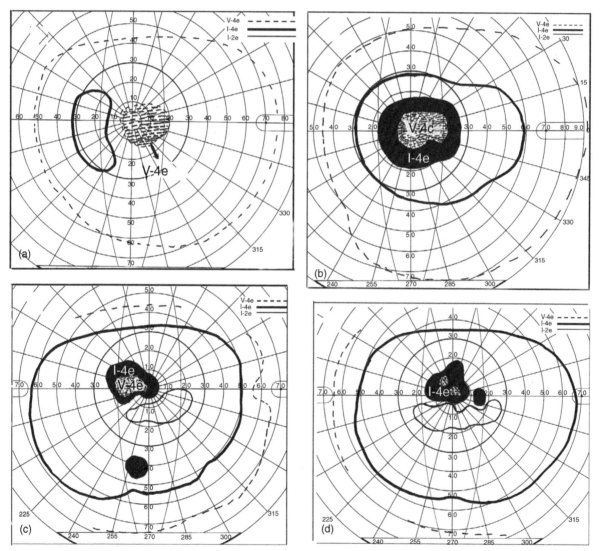

Figure 13 Four visual fields in nonarteritic PION showing varying sizes and densities of central scotoma and other field defects, with normal peripheral visual fields, where the Roman numeral indicates target size, the Arabic numeral indicates relative intensity, and the lowercase letter indicates minor filter adjustment of the light intensity. Reproduced from Hayreh, S. S. (2004). Posterior ischemic optic neuropathy: Clinical features, pathogenesis, and management. Eye 18: 1188–1206.

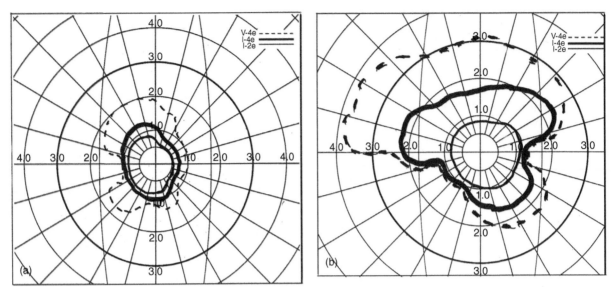

Figure 14   Visual fields of (a) right and (b) left eyes with arteritic PION, showing markedly constricted central visual fields, with complete loss of peripheral fields in both eyes. From Hayreh, S. S. (2004). Posterior ischaemic optic neuropathy: Clinical features, pathogenesis, and management. *Eye* 18: 1188–1206.

discussed above, except that the optic disk and fundus are initially normal in both types of PION.

Management of PION

The management of PION depends upon the type of PION. In all cases other than surgical PION, as in AION, the most important first step in persons aged 50 years or older is always to rule out GCA.

Arteritic PION

Management is similar to that of A-AION discussed above. However, there is usually no visual improvement with systemic steroid therapy.

Nonarteritic PION

The eyes of patients treated with high-dose systemic steroid therapy during the very early stages of the disease show significant improvement in visual acuity and visual field, compared to untreated eyes. However, spontaneous improvement in visual acuity and visual field may also occur to some extent in some eyes without steroid therapy. Since systemic risk factors may play a part in the development of NA-PION, in the management of these patients one should try to reduce as many risk factors as possible so as to reduce the risk of second eye involvement.

Surgical PION

Management amounts to prophylactic measures to prevent development, because once the visual loss occurs, it is usually bilateral, severe, and irreversible. No treatment has been found to be effective to recover or improve the lost vision. Prophylactic measures during surgery include

shortening the duration of surgery to a minimum and avoiding: arterial hypotension, excessive fluid replacement and hemodilution, pressure on the eyeball and orbit, and dependent position of the head. Since systemic cardiovascular risk factors may predispose a patient to a higher risk of developing surgical PION, it may be advisable to consider those factors in the decision to perform surgery.

Conclusions

Ischemic optic neuropathy is not a singular disease but a spectrum of several different types, each with its own etiology, pathogenesis, and management. Each must be considered a separate clinical entity. Overall, they constitute one of the major causes of blindness or seriously impaired vision, yet there is marked controversy on their pathogeneses, clinical features, and management. As the signs and symptoms can overlap, correct diagnosis is the key to producing the best visual outcome.

Further Reading

Hayreh, S. S. (1969). Blood supply of the optic nerve head and its role in optic atrophy, glaucoma, and oedema of the optic disc. *British Journal of Ophthalmology* 53: 721–748.

Hayreh, S. S. (1974). Anterior ischaemic optic neuropathy. I. Terminology and pathogenesis. *British Journal of Ophthalmology* 58: 955–963.

Hayreh, S. S. (1974). Anterior ischaemic optic neuropathy. II. Fundus on ophthalmoscopy and fluorescein angiography. *British Journal of Ophthalmology* 58: 964–980.

Hayreh, S. S. (1985). Inter-individual variation in blood supply of the optic nerve head. Its importance in various ischemic disorders of the optic nerve head, and glaucoma, low-tension glaucoma and allied disorders. *Documenta Ophthalmologica* 59: 217–246.

Hayreh, S. S. (1990). Anterior ischaemic optic neuropathy. Differentiation of arteritic from non-arteritic type and its management. *Eye* 4: 25–41.

Hayreh, S. S. (1996). Acute ischemic disorders of the optic nerve: Pathogenesis, clinical manifestations and management. *Ophthalmology Clinics of North America* 9: 407–442.

Hayreh, S. S. (2001). The blood supply of the optic nerve head and the evaluation of it – myth and reality. *Progress in Retinal and Eye Research* 20: 563–593.

Hayreh, S. S. (2004). Posterior ischaemic optic neuropathy: Clinical features, pathogenesis, and management. *Eye* 18: 1188–1206.

Hayreh, S. S. (2008). Non-arteritic arterior ischaemic optic neuropathy and phosphodiesterase-5 inhibitors. *British Journal of Ophthalmology* 92: 1577–1580.

Hayreh, S. S. (2009). Non-arteritic anterior ischemic optic neuropathy and thrombophila. *Graefe's Archive for Clinical and Experimental Ophthalmology* 247: 577–581.

Hayreh, S. S., Fingert, J. H., Stone, E., and Jacobson, D. M. (2008). Familial non-arteritic anterior ischemic optic neuropathy. *Graefe's Archive for Clinical and Experimental Ophthalmology* 246: 1295–1305.

Hayreh, S. S., Jonas, J. B., and Zimmerman, M. B. (2007). Non-arteritic anterior ischemic optic neuropathy and tobacco smoking. *Ophthalmology* 114: 804–809.

Hayreh, S. S., Podhajsky, P. A., Raman, R., and Zimmerman, B. (1997). Giant cell arteritis: validity and reliability of various diagnostic criteria. *American Journal of Ophthalmology* 123: 285–296.

Hayreh, S. S., Podhajsky, P. A., and Zimmerman, B. (1997). Nonarteritic anterior ischemic optic neuropathy: Time of onset of visual loss. *American Journal of Ophthalmology* 124: 641–647.

Hayreh, S. S., Podhajsky, P. A., and Zimmerman, B. (1998). Occult giant cell arteritis: Ocular manifestations. *American Journal of Ophthalmology* 125: 521–526, 893.

Hayreh, S. S., Podhajsky, P. A., and Zimmerman, B. (1998). Ocular manifestations of giant cell arteritis. *American Journal of Ophthalmology* 125: 509–520.

Hayreh, S. S. and Zimmerman, B. (2003). Management of giant cell arteritis. Our 27-year clinical study: New light on old controversies. *Ophthalmologica* 217: 239–259.

Hayreh, S. S. and Zimmerman, M. B. (2007). Incipient nonarteritic anterior ischemic optic neuropathy. *Ophthalmology* 114: 1763–1772.

Hayreh, S. S. and Zimmerman, M. B. (2007). Optic disc edema in non-arteritic anterior ischemic optic neuropathy. *Graefe's Archive for Clinical and Experimental Ophthalmology* 245: 1107–1121.

Hayreh, S. S. and Zimmerman, M. B. (2008). Nonarteritic anterior ischemic optic neuropathy: Clinical characteristics in diabetic patients versus nondiabetic patients. *Ophthalmology* 115: 1818–1825.

Hayreh, S. S. and Zimmerman, M. B. (2008). Nonarteritic anterior ischemic optic neuropathy: Natural history of visual outcome. *Ophthalmology* 115: 298–305.

Hayreh, S. S. and Zimmerman, M. B. (2008). Non-arteritic anterior ischemic optic neuropathy: Role of systemic corticosteroid therapy. *Graefe's Archive for Clinical and Experimental Ophthalmology* 246: 1029–1046.

Hayreh, S. S., Zimmerman, M. B., Podhajsky, P., and Alward, W. L. M. (1994). Nocturnal arterial hypotension and its role in optic nerve head and ocular ischemic disorders. *American Journal of Ophthalmology* 117: 603–624.

Hunder, G. G., Bloch, D. A., Michel, B. A., et al. (1990). The American College of Rheumatology 1990 criteria for the classification of giant cell arteritis. *Arthritis and Rheumatology* 33: 1122–1128.

Ischemic Optic Neuropathy Decompression Trial Research Group (1995). Optic nerve decompression surgery for nonarteritic anterior ischemic optic neuropathy (NAION) is not effective and may be harmful. *Journal of the American Medical Association* 273: 625–632.

Sadda, S. R., Nee, M., Miller, N. R., et al. (2001). Clinical spectrum of posterior ischemic optic neuropathy. *American Journal of Ophthalmology* 132: 743–750.

Light-Driven Translocation of Signaling Proteins in Vertebrate Photoreceptors

P D Calvert, SUNY Upstate Medical University, Syracuse, NY, USA
V Y Arshavsky, Duke University, Durham, NC, USA

Glossary

Arrestin – A protein which binds to and inactivates photoactivated rhodopsin. Arrestin binding results in termination of transducin activation.

Light adaptation – The ability of photoreceptors (and the visual system as a whole) to adapt the speed and sensitivity of light responses to ever-changing conditions of ambient illumination.

Phosducin – A protein which interacts with the $\beta\gamma$-subunit of transducin and reduces its membrane affinity.

Recoverin – A regulatory protein which is thought to regulate the speed at which arrestin can bind to photoactivated rhodopsin.

RGS9 – A protein responsible for returning activated transducin in its inactive form.

Transducin – A G protein that mediates the visual signal between the photoactivated visual pigment, rhodopsin, and the downstream effector enzyme, cGMP phosphodiesterase. Transducin consists of two functional subunits, α and $\beta\gamma$.

Introduction

Rod and cone photoreceptors are highly polarized cells which transduce information encoded by photons into electrical activity that can be processed by higher-order neurons. At the one end, photoreceptors have specialized ciliary organelles, outer segments, which are enriched in proteins directly involved in light detection and signal transduction. At the opposite end, synapses convey the information gathered by outer segments to downstream neurons. Vision begins when a molecule of rhodopsin in the outer segment becomes excited by light and activates a G protein, transducin. The transducin α-subunit stimulates its effector, cyclic guanosine monophosphate (cGMP) phosphodiesterase (PDE), which leads to the reduction in intracellular cGMP and to the electrical response mediated by the closure of the cGMP-gated cationic channels in the plasma membrane. The recovery of the photoresponse requires complete inactivation of all these molecular components. Photoexcited rhodopsin is inactivated through its phosphorylation by rhodopsin kinase and arrestin binding, which blocks transducin activation. The rate of rhodopsin phosphorylation, and thus its active lifetime, is regulated by the Ca^{2+}-binding protein, recoverin. Transducin (and accordingly PDE) activation is terminated upon the hydrolysis of GTP tightly bound to transducin α-subunit, a process markedly accelerated by the GTPase activating protein RGS9.

Importantly, three of the above-mentioned proteins (transducin, arrestin, and recoverin) undergo massive light-driven translocation between the major subcellular compartments of photoreceptors (**Figure 1**). In rods, transducin moves out of the outer segment and accumulates primarily in the inner segment, arrestin moves in the opposite direction, and recoverin shifts from the outer segment toward the synapse. In cones, in light arrestin moves in the same direction, whereas transducin moves very little, if at all. Recoverin translocation has not yet been analyzed in cones. A similar phenomenon involving the G protein (G_q), arrestin, and the transient receptor potential-like (TRPL) channel takes place in rhabdomeric invertebrate photoreceptors.

Light Dependency of Protein Translocation

Quantitative experiments revealed that the translocations of arrestin and transducin in rods take place in bright light. The outer segments of the rod contain very little arrestin (estimated under 7% of its total cellular content) in the dark and under moderate illumination. In mouse rods, arrestin begins to move to outer segments when the light intensity reaches a critical threshold, exciting over ~1000 rhodopsins per rod per second, which is within the upper limit at which mammalian rods can signal variations in light. Transducin translocation is also triggered at a threshold light intensity, although brighter, exciting ~5000 rhodopsins per rod per second, an intensity that completely saturates rods. The time required for the completion of protein translocation in saturating light in rods is on the order of tens of minutes. Although no such quantitative measurements are available for cones, available data indicate that cone arrestin translocation also requires fairly bright light. The existence of cone transducin translocation in intact cells remains somewhat

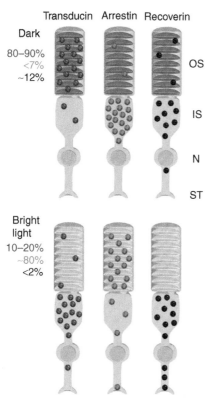

Figure 1 Schematic illustration of transducin, arrestin, and recoverin distribution in dark- and light-adapted rods. The numbers on the left, color-coded to the corresponding translocating proteins, represent the percentage of the proteins found in the outer segments in the dark or following bright light illumination. The subcellular rod compartments are abbreviated on the right: OS – outer segment; IS – inner segment; N – nucleus; ST – synaptic terminal. Reproduced from Calvert, P. D., Strissel, K. J., Schiesser, W. E., Pugh, E. N., Jr., and Arshavsky, V. Y. (2006). Light-driven translocation of signaling proteins in vertebrate photoreceptors. *Trends in Cell Biology* 16: 560–568.

controversial. Most investigators do not see it at any light intensity, whereas one group reports small degree of translocation observed in extremely bright light. The most recent report argues that the inability of cone transducin to efficiently translocate in light reflects specific physicochemical properties of its individual subunits.

Hypotheses on the Functional Roles of Protein Translocation

Photoreceptors adjust their sensitivity over a broad range of ambient light intensities, and protein translocation has been proposed to contribute to this process at the high end of adaptive light intensities. The reduction of transducin content in outer segments of rod in bright light correlates with a reduction in signal amplification in the rhodopsin–transducin–phosphodiesterase cascade, likely due to the reduction in transducin activation rate. This is likely to move the dynamic range over which rods

operate to higher light intensities. Although transducin translocation takes place in light that is saturating for rods, this range adjustment may be adaptive after the light is dimmed or extinguished. For example, such a mechanism could be useful as dusk approaches when vision is switching from being cone-dominant to rod-dominant. The fact that transducin translocation is triggered by light intensities that completely saturate rods makes it plausible to suggest that both phenomena, transducin translocation and response saturation, occur for essentially the same reason, the inability of rods to inactivate vast amounts of transducin beyond a certain light intensity. In this context, transducin translocation can be viewed as an elegant self-regulating mechanism triggered at the point where the rod exhausts other means of avoiding response saturation.

Although not yet tested experimentally, arrestin translocation is thought to be adaptive as well. Its increased concentration in the outer segment could reduce the response amplitude and/or accelerate recovery. It could also allow rods and cones to prepare for inactivation of large amounts of photoexcited rhodopsin and its bleach products produced in bright light. Similarly, recoverin translocation from outer segments may increase the amount of rhodopsin kinase available to phosphorylate rhodopsin, thus further reducing light sensitivity.

Light-driven protein translocation, particularly in rods, may also play a neuroprotective role. Rod saturation marks the transition from mesopic (mixed rod/cone) to cone-dominated photopic vision. Under these conditions, rods contribute little to vision and transducin translocation may prevent excessive energy consumption by rods by reducing the number of transducin molecules undergoing the cycle of activation/inactivation. This may, in turn, reduce the metabolic stress in the retina commonly believed to contribute to pathological processes. A reduced level of cellular signaling caused by transducin translocation may also reduce the chance of apoptotic death of the rod. At least in rodents, some forms of apoptosis are suggested to be caused by excessive signaling through the phototransduction cascade. The same argument may be applied to arrestin translocation, whose accumulation in the outer segment in bright light is likely to preamplify rhodopsin shutoff on a sunny day.

Finally, when vision is dominated by cones, rods may perform more of their housekeeping functions, such as checking the integrity of proteins by the ubiquitin proteasome system located in the inner segments.

What Is the Mode of Protein Translocation: Active Transport or Diffusion?

A major currently explored area in this field is whether phototransduction proteins translocate by diffusion or by active transport via molecular motors. In principle, the mode of translocation of each protein could be different between the light- and dark-induced directions. Most investigators argue against the involvement of molecular motors. One major argument is that all of these proteins (or at least individual subunits in the case of transducin) are fairly soluble and any of their relocation by molecular motors may be negated by the subsequent diffusion throughout the entire cellular volume. Another argument is that, although protein translocation by motors is rapid, it could be easily saturated by the very large number of translocation protein molecules. On the other hand, intracellular diffusion of soluble proteins in rods is also sufficiently rapid to explain the observed protein translocation rates and, unlike molecular motors, could not be saturated by the amount of protein molecules undergoing light-induced translocation. Yet, it should be noted that, based on the impediment of translocation by the cytoskeleton disrupting drugs, some believe that motor systems are involved, particularly in protein translocations in the dark-induced directions.

However, diffusion alone cannot account for the phenomenon because it does not explain any disequilibrium of protein distributions with the free cytoplasmic volume of the rod or cone. Thus, while diffusion serves as the mode of protein movement, the observed patterns of light-dependent protein redistribution may be explained by light-dependent appearance or disappearance of specific protein-binding sites in individual subcellular compartments. The next two sections illustrate these ideas in regard to transducin and arrestin.

Specific Mechanisms of Protein Translocation

Transducin

Recent reports suggest that transducin translocation could be explained by the differences in the membrane affinities of its αβγ heterotrimer compared to the individual α- and βγ-subunits. The heterotrimer is strongly membrane-associated due to the combined action of two lipid modifications: a farnesyl group on the γ-subunit and an acyl group on the α-subunit. Thus, in the dark-adapted rod, transducin heterotrimer is predicted to be concentrated on the disk membranes of the outer segment. When transducin is activated in light, α-subunits bind GTP and dissociate from the βγ-subunits (**Figure 2**). Both subunits become more soluble since each has only one lipid modification, allowing them to diffuse throughout the cytoplasm. Inevitably, GTP hydrolysis by the α-subunit would result in the restoration of the poorly soluble trimer. When this happens in the outer segment, transducin becomes re-attached to the disk membranes; but when it

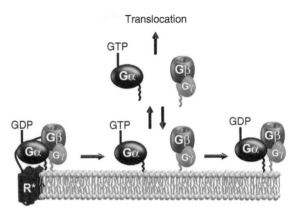

Figure 2 The putative role of transducin subunit dissociation in its translocation. When transducin is activated by photoexcited rhodopsin (R*), its α- and βγ-subunits become separated from one another and the membrane affinity of each is less than that of the heterotrimer. In this state, subunits may dissociate from photoreceptor disk membranes and translocate from the outer segment. The efficiency of translocation is dependent on the time during which transducin subunits stay apart before re-formation of the trimer. This time is determined by the rate of the GTP hydrolysis on the α-subunit and is dependent on the absolute amount of activated transducin. Reproduced from Calvert, P. D., Strissel, K. J., Schiesser, W. E., Pugh, E. N., Jr., and Arshavsky, V. Y. (2006). Light-driven translocation of signaling proteins in vertebrate photoreceptors. *Trends in Cell Biology* 16: 560–568.

happens in the inner segment, the trimer may adhere to the membranous structures there, causing its transient accumulation. The central assumption of this hypothesis, that transducin subunits move apart from one another, is supported by the difference in their translocation rates and by transducin translocation being facilitated by transgenic or pharmacological manipulations promoting transducin subunits to remain in the dissociated state. The translocation of transducin βγ-subunit is further enhanced by the protein called phosducin. Phosducin reduces membrane association of the βγ-subunit, which presumably allows it to more easily diffuse throughout the cytoplasm.

Another important mechanistic feature of transducin translocation is its light-intensity threshold. This threshold reflects the fact that the cellular content of transducin significantly exceeds that of RGS9, a protein responsible for rapid inactivation of transducin. Light of the threshold intensity produces activated transducin in the amount exceeding the capacity of RGS9 to inactivate it. Consequently, a fraction of activated transducin stays in the active, GTP-bound form (with α- and βγ-subunits dissociated) for a longer time, sufficient for dissociation from the disk membranes and diffusion through the photoreceptor cytoplasm. Accordingly, the knockout of RGS9 allows transducin translocation at a lower light intensity, whereas RGS9 overexpression shifts the threshold to brighter light.

Arrestin

It was first suggested that arrestin is equilibrated throughout the rod cytoplasm in the dark and is trapped in the outer segment in light upon binding to photoexcited rhodopsin. However, recent observations argue that the dark-adapted distribution of arrestin does not match the distribution of the free cytoplasm volume, indicating that most arrestin is bound to sites located in the inner segment. One hypothesis is that these sites are formed by microtubules. It was further proposed that arrestin translocation is explained by a simple competition between constitutive, low-affinity microtubule sites in the inner segment and transient, high-affinity sites in the outer segment formed upon rhodopsin photoexcitation. However, quantitative measurements indicated that, at the minimal light intensity sufficient to trigger arrestin translocation, the number of translocated arrestin molecules exceeds the number of photoactivated rhodopsin molecules by $\sim$30-fold. Even less rhodopsin activation is required to trigger arrestin translocation in knockout mice lacking RGS9 where transducin activation persists longer than normally. Therefore, triggering arrestin translocation is not dependent on the absolute amount of rhodopsin excited by light, but is rather dependent on arrestin release from the inner segment sites by a yet-to-be-identified signaling mechanism downstream from phototransduction.

Proteins' Return in the Dark

The rates of transducin and arrestin return to their dark-adapted locations upon switching from light to dark are much slower than their movement in the light-induced direction, with most measurements indicating that it takes at least an hour. This may be slow enough to be within the capacity of molecular motors, and indeed transducin and arrestin return can be blocked by cytoskeleton disrupting drugs. The specificity of these treatments remains unknown and artifacts such as clogging the connecting cilium could not be ruled out. On the other hand, arrestin and transducin return could also be explained by a combination of diffusion, removal of the light-induced binding sites, and restoration of the dark-adapted sites. Current evidence is insufficient to discriminate between the potential roles of molecular motors and diffusion in transducin and arrestin return to their dark-adapted cellular distributions.

See also: Phototransduction: Adaptation in Cones; Phototransduction: Adaptation in Rods; Phototransduction: Inactivation in Cones; Phototransduction: Inactivation in Rods; Phototransduction: Phototransduction in Cones; Phototransduction: Phototransduction in Rods.

Further Reading

Artemyev, N. O. (2008). Light-dependent compartmentalization of transducin in rod photoreceptors. *Molecular Neurobiology* 37: 44–51.

Brann, M. R. and Cohen, L. V. (1987). Diurnal expression of transducin mRNA and translocation of transducin in rods of rat retina. *Science* 235: 585–587.

Broekhuyse, R. M., Tolhuizen, E. F., Janssen, A. P., and Winkens, H. J. (1985). Light induced shift and binding of S-antigen in retinal rods. *Current Eye Research* 4: 613–618.

Calvert, P. D., Strissel, K. J., Schiesser, W. E., Pugh, E. N. Jr., and Arshavsky, V. Y. (2006). Light-driven translocation of signaling proteins in vertebrate photoreceptors. *Trends in Cell Biology* 16: 560–568.

Fain, G. L. (2006). Why photoreceptors die (and why they don't). *BioEssays* 28: 344–354.

Hanson, S. M., Francis, D. J., Vishnivetskiy, S. A., Klug, C. S., and Gurevich, V. V. (2006). Visual arrestin binding to microtubules involves a distinct conformational change. *Journal of Biological Chemistry* 281: 9765–9772.

Kerov, V., Chen, D. S., Moussaif, M., et al. (2005). Transducin activation state controls its light-dependent translocation in rod photoreceptors. *Journal of Biological Chemistry* 280: 41069–41076.

Nair, K. S., Hanson, S. M., Mendez, A., et al. (2005). Light-dependent redistribution of arrestin in vertebrate rods is an energy-independent process governed by protein–protein interactions. *Neuron* 46: 555–567.

Philp, N. J., Chang, W., and Long, K. (1987). Light-stimulated protein movement in rod photoreceptor cells of the rat retina. *FEBS Letters* 225: 127–132.

Reidel, B., Goldmann, T., Giessl, A., and Wolfrum, U. (2008). The translocation of signaling molecules in dark adapting mammalian rod photoreceptor cells is dependent on the cytoskeleton. *Cell Motility and the Cytoskeleton* 65: 785–800.

Slepak, V. Z. and Hurley, J. B. (2008). Mechanism of light-induced translocation of arrestin and transducin in photoreceptors: Interaction-restricted diffusion. *IUBMB Life* 60: 2–9.

Sokolov, M., Lyubarsky, A. L., Strissel, K. J., et al. (2002). Massive light-driven translocation of transducin between the two major compartments of rod cells: A novel mechanism of light adaptation. *Neuron* 34: 95–106.

Strissel, K. J., Lishko, P. V., Trieu, L. H., et al. (2005). Recoverin undergoes light-dependent intracellular translocation in rod photoreceptors. *Journal of Biological Chemistry* 280: 29250–29255.

Strissel, K. J., Sokolov, M., Trieu, L. H., and Arshavsky, V. Y. (2006). Arrestin translocation is induced at a critical threshold of visual signaling and is superstoichiometric to bleached rhodopsin. *Journal of Neuroscience* 26: 1146–1153.

Whelan, J. P. and McGinnis, J. F. (1988). Light-dependent subcellular movement of photoreceptor proteins. *Journal of Neuroscience Research* 20: 263–270.

Limulus Eyes and Their Circadian Regulation

B-A Battelle, University of Florida, St. Augustine, FL, USA

Glossary

Arhabdomeric cell – The secondary visual cell in *Limulus* median eyes that is not photoreceptive but is electrically coupled to photoreceptors and generates action potentials in response to photoreceptor depolarization.

Chelicerate – A type of arthropod belonging to the group Chelicerata that includes horseshoe crabs, scorpions, spiders, ticks, and mites.

Circadian clock – A biological clock that oscillates with a period of about 24 h even under constant conditions.

Compound eye – An eye that contains a few to many separate photoreceptive units.

Eccentric cell – A secondary visual cell in *Limulus* lateral eye that is not photoreceptive but is electrically coupled to photoreceptors and generates action potentials in response to photoreceptor depolarization.

Lateral inhibition – A mechanism of information processing in nervous systems that increases contrast and resolving power.

Ocellus – A type of eye with a single lens.

Octopamine – A major amine neurotransmitter in invertebrates that is the phenol analog of norepinephrine.

Ommatidium – The photoreceptive unit of compound eyes.

Quantum bump – The response of photoreceptors to a single photon of light.

Rhabdom – A photoreceptive membrane typically elaborated by arthropod photoreceptors that is composed of tightly packed microvilli.

The retinas of animals that live in cyclic light environments typically undergo rhythmic, daily changes in structure and function. These changes underlie the ability of animals to detect images over the large daily fluctuations in ambient illumination, and therefore they are critical for normal vision. Some of these changes, such as light and dark adaptation, are driven solely by fluctuations in illumination, while others are driven solely by signals from internal circadian clocks. Still other changes require interactions between light- and clock-driven biochemical cascades.

In the American horseshoe crab *Limulus polyphemus*, a chelicerate arthropod, circadian changes in vision are dramatic, and as a result, *Limulus* can see at night nearly as well as it can during the day. Since the circadian organization of the *Limulus* visual system is also advantageous for experimental manipulation, *Limulus* has been used to examine in detail the separate and combined effects of diurnal light and the clock on the retina.

The first part of this article describes the organization of the *Limulus* visual system and the circadian input to the eyes. The second part describes the impact of the circadian clock on *Limulus* retinal structure, the photoresponse, photosensitive membrane shedding, and gene expression. The synaptic mechanisms through which the clock influences the eyes are described as well as a clock-driven biochemical cascade in photoreceptors that may underlie some of the observed circadian changes in visual function.

Organization of the *Limulus* Visual System

Limulus has three major types of eyes: lateral, median, and rudimentary (**Figure 1**). The eyes that are obvious by examining the dorsal side of an adult animal are the lateral compound eyes (LE) on the dorsolataral carapace, and a pair of median eyes located on either side of the median dorsal spine near the front of the animal. Rudimentary eyes are found at three different locations, and in an adult animal they are largely hidden from view. Lateral rudimentary eyes are located below the cornea at the posterior edge of each LE, and a pair of fused median rudimentary eyes is located below the carapace between the two median ocelli. Ventral rudimentary eyes consist of a pair of optic nerves that extend anteriorly from the brain just below the cuticle on the ventral side of the animal. They terminate in an end organ located below a wart-like structure visible on the ventral cuticle in front of the mouth.

Lateral Compound Eyes

In an adult animal, each LE contains over 1000 conical lenses, and below each lens there is a single ommatidium, or mini-retina, that is aligned with the long axis of the overlying lens (**Figure 2(a)**) Each ommatidium is composed of 8–12 photoreceptors clustered like the sections of an orange around the dendrite of usually one secondary visual cell called the eccentric cell (**Figures 2(b)** and **2(c)**). The photoreceptors are electrically coupled to the dendrite of the eccentric cell. Two major non-neuronal cells types

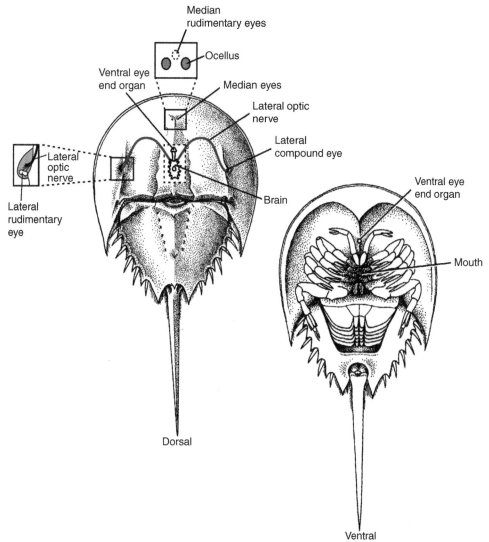

Figure 1 Schematics of dorsal and ventral views of an adult *Limulus* showing the locations of the eyes. The lateral compound eyes and median ocelli are visible on the dorsal carapace. The box at the left shows the location of a lateral rudimentary eye below the cornea at the posterior edge of the lateral compound eye. The box at the top shows the location of the fused median rudimentary eyes below the carapace between the median ocelli. The cutaway in the center shows the position of the brain and ventral rudimentary eyes. Ventral optic nerves extend anteriorly from the brain and terminate in an end organ located just beneath a wart-like structure that is visible on the ventral cuticle in front of the mouth. The long lateral optic nerves are shown in red. Modified from Calman, B. G. et al. (1991). *Journal of Comparative Neurology* 313: 553–562.

are present in ommatidia: cone cells and pigment cells. Cone cell processes occupy the aperture at the base of the lens and separate the lens from photoreceptors. Pigment cells surround each ommatidium and form partitions between photoreceptors.

LE photoreceptors respond with graded depolarizations to visible light, and the eccentric cells to which they are electrically coupled, encode these graded depolarizations as action potentials. Both photoreceptors and eccentric cells have axons that project to optic ganglia in the brain (lamina and medulla) through the lateral optic nerve (**Figure 3**(a)). However, because the lateral optic nerve in an adult animal is long, often over 10 cm, it is unlikely that the graded photoreceptor potentials reach the brain. Therefore, information from LEs is thought to reach the brain only through the activity of the eccentric cells.

In addition to projecting to the brain, eccentric cells extend axon collaterals laterally within the eye to form a neural plexus just below the ommatidia (**Figure 2**(a)). Within this plexus, eccentric cell collaterals from neighboring and even distant ommatidia make reciprocal inhibitory synapses which are the basis for lateral inhibition in the eye. Lateral inhibition, a fundamental mechanism of information processing in nervous systems that increases contrast

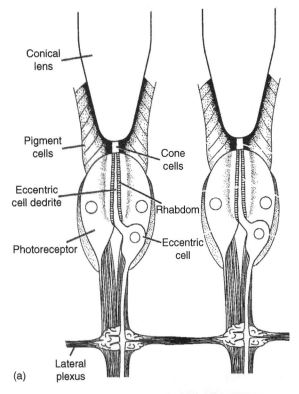

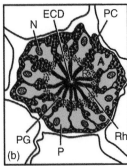

Figure 2 (a) Schematic of a longitudinal section through two ommatidia of the lateral eye, illustrating major cell types. The photoreceptors and eccentric cells project axons into the lateral optic nerve. Eccentric cell axons also project collaterals into a lateral plexus within the eye where they form the reciprocal inhibitory synapses that underlie lateral inhibition. (b) Schematic of a cross section through one ommatidium showing the arrangement of photoreceptors (P), shown in green, around a central eccentric cell dendrite (ECD). The nuclei (N) located in the arhabdomeral lobes (A) of photoreceptors is shown. Screening pigment granules within the photoreceptors (PG) define the junction between the arhabdomeral and rhabdomeral (R) lobes. The asterisk-like structure at the center of the ommatidium is the rhabdom (Rh) formed from the fused photosensitive membranes of neighboring photoreceptors. Pigment cells (PC) surround the ommatidium and form partitions between photoreceptors. (c) Confocal image of one ommatidium from a light-adapted daytime animal immunostained for LpMyo3. LpMyo3 distributes through photoreceptor cell bodies and concentrates at the rhabdom (Rh). Scale bar = 50 μm. (a) and (c) from Battelle, B-A. (2006) *Arthropod Structure and Development* 35: 261–274, with permission from Elsevier.

and resolving power, was first detected experimentally in *Limulus* LE. In vertebrate retinas, this same mechanism is known as center-surround inhibition.

Median Eyes

Median eyes are called ocelli because they have a single spherical lens (**Figure 4**). In median eye retinas, groups of 5–11 photoreceptors are typically associated with at least one secondary visual cell or arhabdomeric cell, but unlike LE ommatidia, these groups are not highly organized and difficult to discern anatomically. The photoreceptors and arhabdomeric cells are embedded in guanophores, cells that contain reflective guanine crystals, and the retina is surrounded by pigment cells.

An interesting feature of median eyes is that they contain two different types of photoreceptors: one that responds to visible light and another that is sensitive to ultraviolet (UV) light. These two photoreceptor types have not yet been distinguished anatomically. Median eye arhabdomeric cells are thought equivalent to LE eccentric cells because, similar to eccentric cells, they are electrically coupled to photoreceptors and generate action potentials in response to graded photoreceptor depolarizations. Median eye photoreceptors and arhabdomeric cells both project axons to ocellar ganglia in the brain through the median optic nerve (**Figure 3(a)**), but only the action potentials of the arhabdomeric cell are thought to reach the brain. Interestingly, arhabdomeric cell action potentials have been detected only in response to UV illumination.

Rudimentary Eyes

The lateral, median, and ventral rudimentary eyes differentiate in the embryo before the more complex LEs and median ocelli, and they presumably provide photic information to the developing embryo. Each rudimentary eye consists of a cluster of large photoreceptors ($100 \times 160 \,\mu m$) that are sensitive to visible light. Ventral eye photoreceptors are typically scattered along the length of the ventral optic nerves as well as clustered near the brain and in the end organ (**Figure 3(a)**).

Similar to the photoreceptors in lateral and median eyes, rudimentary eye photoreceptors produce graded depolarizations in response to illuminations, but they have no associated secondary visual cells. Rudimentary photoreceptors project axons to the brain through the lateral, median, or ventral optic nerves, and although their axon diameters are considerably larger than those of lateral and median eye photoreceptors, it still seems unlikely that their graded depolarizations reach the brain in an adult animal. An exception may be the graded

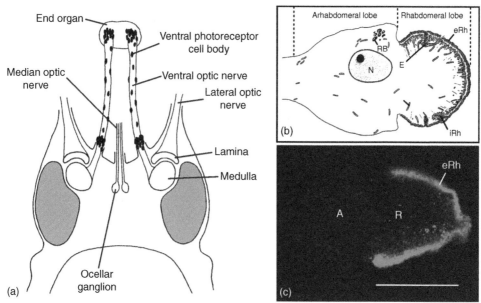

Figure 3   (a) Schematic of a dorsal view of the brain and ventral optic nerves. Ventral photoreceptor cell bodies (dark ovals) are scattered along the ventral optic nerves and clustered at their ends. Also shown are the median and lateral optic nerves and the optic ganglia. (b) Schematic of a ventral photoreceptor showing the rhabdomeral (R) and arhabdomeral (A) lobes, external rhabdom (eRh), internal rhabdom (iRh), efferent terminals near the rhabdom (E), mitochondria (M), ribosomes (RB), and nucleus (N). (c) Confocal image of a 1 µM optical section through a light-adapted ventral photoreceptor immunostained for visual arrestin. Visual arrestin concentrates at the rhabdom in the R lobe. Scale = 50 µm. (a) Modified from Calman, B. G. et al. (1991). *Journal of Comparative Neurology* 313: 553–562. (b) Modified from Calman, B. G. and Chamberlain, S. C. (1982). Distinct lobes of *Limulus* ventral photoreceptors. II. Structure and ultrastructure. *Journal of General Physiology* 80: 839–862, with permission form Rockefeller University Press. (c) from Battelle, B-A. (2006). *Arthropod Structure and Development* 35: 261–274, with permission from Elsevier.

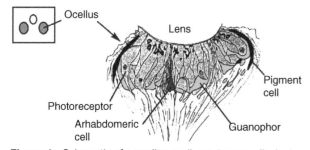

Figure 4   Schematic of a median ocellus cut perpendicular to the carapace. Most of the area proximal to the lens is filled with photoreceptor cells. The dark areas below the lens represent the rhabdoms, which are not highly organized in this eye. Arhabdomeric cells are interspersed among the photoreceptors. Pigment cells surround the structure and reflective guanophores lie nearer the base of the retina. Modified from Jones, J. N. and Brown, J. E. (1971). *Z. Zellforsch* 118: 297–309, with permission from Springer Verlag.

depolarizations produced by ventral photoreceptors located close to the brain.

The Photoreceptors

The photoreceptors are fundamentally similar in the three types of *Limulus* eyes. All are composed of two major compartments: a rhabdomeral lobe and an arhabdomeral lobe (**Figures 2(b)**, **3(b)**, and **3(c)**). The rhabdomeral lobe contains the photosensitive membrane or rhabdom, which in *Limulus*, as in other arthropods, is composed of tightly packed actin-rich microvilli. The arhabdomeral lobe is not photosensitive and contains the nucleus and metabolic machinery of the cell. As described above, all *Limulus* photoreceptors produce graded depolarizations in response to illumination, and with the exception of the median eye UV-sensitive photoreceptors, all respond to visible light with an absorption maximum of about 525 nm. Finally, all *Limulus* photoreceptors, as well as the eccentric cells of the LE and arhabdomeric cells of the median eye, release the inhibitory neurotransmitter histamine. Histamine is also the neurotransmitter used by the photoreceptors of insects and crustaceans.

The rhabdoms in the three types of eyes are organized differently, however, and they are highly organized only in LE ommatidia. In LEs, the rhabdoms of neighboring photoreceptors are fused so that in a cross section of an ommatidium, they appear like an asterisk with the eccentric cell dendrite at the center (**Figures 2(c)** and **2(c)**). The highly complex rhabdoms in median ocelli may project proximally into neighboring photoreceptors or infoldings of photoreceptor membranes may form internal rhabdoms (**Figure 4**). Rudimentary photoreceptors typically have both external and internal rhabdoms (**Figures 3(b)** and

3(c)) and often have more than one rhabdomeral lobe. In addition, rhabdoms of adjacent rudimentary photoreceptors frequently fuse forming a double layer of microvilli.

Among *Limulus* photoreceptors, those of the ventral eyes are the best studied, and their physiology is probably the most thoroughly characterized of any rhabdomeral photoreceptor. Ventral photoreceptors are also significant in the history of vision research. The first intracellular recordings of a photoresponse were made from ventral eye photoreceptors, and physiological studies of these photoreceptors provided the first evidence for the importance of Ca^{++} in light adaptation and a role for phospholipids in the photoresponse.

Circadian Organization of the *Limulus* Visual System

The circadian organization of the *Limulus* visual system is significantly different from that of vertebrates. In vertebrates, one or more circadian oscillators are present within the retina, and in some species, circadian oscillators are present in photoreceptors. In *Limulus*, the circadian oscillators that influence the eyes are located in the brain, not in the eyes. The *Limulus* brain contains bilateral circadian oscillators that remain synchronous through neuronal coupling, and they drive the activity of efferent neurons that project from the brain to all of the eyes.

The cell bodies of clock-driven efferent neurons that project to the eyes are clustered in the cheliceral ganglia located on either side of the base of the brain, and there are about 20 efferent neurons in each ganglion. The axon of each efferent neuron is thought to branch several times in the brain and its branches to project bilaterally out all of the optic nerves and innervate all of the eyes (**Figure 5(a)**). In the eyes, the efferent axons have neurosecretory-like terminals containing both clear vesicles and large, dense, crystalline granules. In LE ommatidia, efferent terminals innervate all cell types (**Figure 5(b)**). In median eyes, efferent terminals contact photoreceptors near the rhabdom, and in rudimentary eyes, efferent axons project to the rhabdomeral lobes of photoreceptors where they terminate directly adjacent to rhabdoms (**Figures 3(b)** and **5(c)**). Circadian efferent neurons, similar to those described in the *Limulus* visual system, also innervate the eyes of scorpions and spiders; therefore, the type of circadian regulation described for *Limulus* eyes appears to be a feature common to chelicerate arthropods, but not to crustaceans and insects.

In *Limulus*, efferent neurons innervating the eyes are active at night and silent during the day. They begin firing bursts of action potentials about 45 min before sunset. These bursts, which are synchronous in all optic nerves, reach a maximum frequency of about 2 Hz during the early evening. The burst rate slows after midnight, and then stops after dawn. This pattern of efferent nerve activity persists in constant darkness, and its phase (the time of day it starts and stops) can be shifted by changing the time of light onset. Thus, efferent nerve activity is clearly circadian.

Effects of the Clock on *Limulus* Eyes

Clock input increases the sensitivity of *Limulus* eyes to light at night, and this effect is most dramatic in LEs. In animals maintained in cyclic light, LEs become about 1 million times more sensitive to light at night compared to the day from the combined effects of dark adaptation and clock input. The effect of the clock is to roughly double the sensitivity obtained with dark adaptation alone.

In animals maintained in constant darkness, the onset of efferent nerve activity near dusk correlates with the increase in LE sensitivity. If the lateral optic nerve is cut during the day, severing the efferent axons and thus preventing clock signals from reaching the LE during the night, there is no nighttime increase in sensitivity. If the lateral optic nerve is cut during the night, interrupting clock input to the eyes, LE sensitivity falls rapidly toward the daytime level. However, if the distal end of a cut lateral optic nerve is stimulated electrically to activate the efferent axons, LE sensitivity increases even during the subjective day. Thus, increased LE sensitivity directly correlates with the activity of the efferent neurons.

In LEs, where the effects of the clock have been studied most thoroughly, the clock influences almost every aspect of retinal function, and studies with these eyes have revealed that the full range of some of the diurnal changes observed involves complex interactions between effects of clock and light (**Table 1**).

Structure

Much of the nighttime increase in LE sensitivity can be attributed to the changes in the structure of ommatidia (**Figure 5(b)**). As the LE transitions into the nighttime state, the aperture at the base of the lens, which is constricted during the day, widens and shortens permitting more photons to reach the underlying photoreceptors. The rhabdom also becomes shorter and wider, adjusting to the larger aperture. These structural changes, which increase photon catch during the night, persist with reduced amplitude in constant darkness in response to clock input alone. However, if clock input is eliminated, all of these rhythmic structural changes are abolished, even in cyclic light. Without clock input, ommatidia assume a structure that is never seen in an intact animal. Thus, these structural changes require clock input, and the effects of the clock are enhanced by cyclic light.

Another diurnal structural change observed in LE photoreceptors that probably contributes to increased nighttime sensitivity is the migration of photoreceptor screening pigment granules. However, unlike the

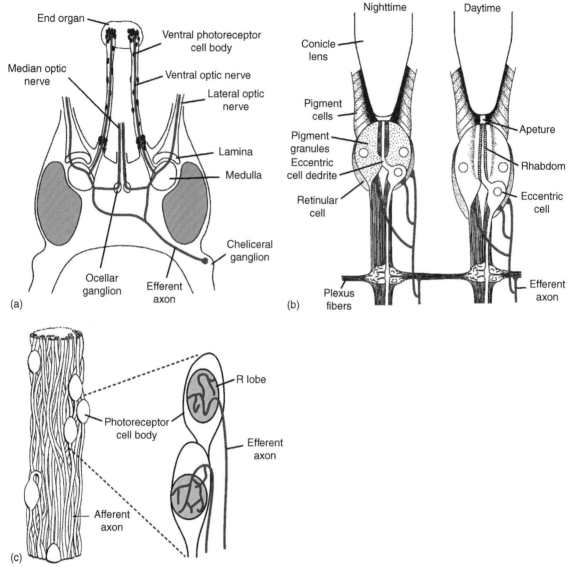

Figure 5 Schematics of circadian efferent neuron projections (shown in red). (a) The cell bodies of the circadian efferent neurons are located in the cheliceral ganglia at the base of the brain. A single idealized efferent neuron branches several times in the brain to project out all of the optic nerves. Modified from Calman, B. G. and Battelle, B-A. (1991). *Visual Neuroscience* 6: 481–495, with permission from Cambridge University Press. (b) Schematic of a longitudinal section through an ommatidium of the lateral eye illustrating projections and terminations of clock-driven efferent neurons, and the nighttime and daytime structure of ommatidia observed when the lateral eye is exposed to cyclic light and clock input. Modified from Chamberlain and Barlow (1979). *Science* 206: 316–363. (c) Schematic of efferent projections to ventral photoreceptors. On the left is an enlarged view of a portion of a ventral optic nerve showing ventral photoreceptor cell bodies and axons. On the right, a schematic of the two ventral photoreceptor cell bodies illustrates that the axons of the circadian efferent neurons project specifically to photoreceptor R lobes (shaded areas) and then ramify extensively. Modified from Evans et al. (1983). *Journal of Comparative Neurology* 219: 369–383.

structural changes in the aperture described above, pigment granule migration appears to be regulated entirely by the clock and is not significantly influenced by light. Photoreceptor pigment granules cluster near the junction of the rhabdomeral and arhabdomeral lobes during the day and disperse toward the periphery of the cell during the night (**Figure 5(b)**). This rhythm continues undiminished in eyes maintained in constant darkness and is

eliminated in eyes deprived of clock input even when the eyes are exposed to cyclic light.

Physiology

Clock-driven physiological changes in photoreceptors also contribute to increased LE sensitivity. Clock input at night increases the amplitude (gain) and duration of the

Table 1 Lateral eye responses to clock input are influenced by light and mimicked by octopamine and elevated cAMP.

Retinal response to clock input	Effect of light	Mimicked by octopamine and cAMP
Sensitivity increases	Enhances	Yes
Aperture widens and shortens	Enhances	Inferred[a]
Rhabdom widens and shortens	Enhances	Inferred[a]
Screening pigments disperse		Not tested
Quantum bump gain increases		Yes
Quantum bump duration increases		Yes
Noise decreases		Yes
Transient shedding primed	Triggers	Yes
Arrestin mRNA levels decrease		Yes

[a]Infusions of OA into the LE *in situ* increase LE sensitivity. Since increased sensitivity is largely due to a wider aperture and repositioned rhabdom, the effects of OA on these two parameters are inferred.

response to a single photon (the quantum bump) and decreases the frequency of spontaneous membrane depolarizations recorded in the dark (noise). Thus, the clock increases the signal-to-noise ratio of photoreceptors at night. The onset of these physiological changes correlates directly with the onset of efferent nerve activity and occurs well before the structural changes described above are detected.

Rhabdom Shedding

Similar to the photoreceptors of vertebrates, invertebrate photoreceptors shed some of their photosensitive membrane every day. *Limulus* photoreceptors exhibit two distinct mechanisms for rhabdom shedding: light-triggered transient membrane shedding and light-driven shedding (**Figure 6**). Both result in the internalization of photosensitive membrane into photoreceptors. Light-driven shedding is similar to the clathrin-mediated endocytosis of activated G-protein coupled receptors described in many other systems, including insect photoreceptors. It continues throughout the day in the light and is mediated by clathrin and arrestin. The clock does not influence this process. On the other hand, light-triggered transient shedding is a unique, synchronous process that rapidly removes membrane from the rhabdom at dawn. During light-triggered transient shedding, large whorls of rhabdomeral membrane are internalized from the base of rhabdomeral microvilli and trafficked into multivesicular bodies. The process is triggered by the dim light of dawn and is complete within an hour after first light. The mechanism for membrane internalization during transient shedding is not yet known. However, importantly from the perspective of this discussion,

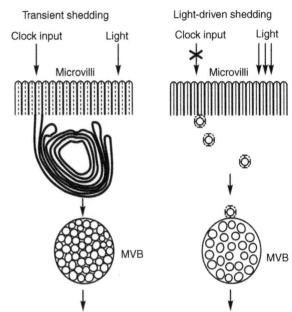

Figure 6 Schematic of transient rhabdom shedding and light-driven shedding. Transient rhabdom shedding must be primed by clock input during the night. It is triggered by the first light of dawn and characterized by the formation of large whorls of rhabdomeral membrane, which subsequently are processed in densely packed multivesicular bodies (MVBs). Light-driven shedding does not require clock input. This progressive process, which continues throughout the day, requires brighter light and involves the clathrin-mediated endocytosis of microvillar membrane from the base of the microvilli. The membranes endocytosed by this process aggregate and form loosely packed MVB. From Sacunas et al. (2002). *Journal of Comparative Neurology* 449: 26–42, with permission from John Wiley & Sons, Inc.

although transient shedding is triggered by light and probably involves the activation of protein kinase (PK) C, it occurs only after photoreceptors have received at least 3 h of clock input and the activation of cyclic adenosine monophosphate (cAMP)-dependent PK.

Gene Expression

In addition to driving structural and physiological changes in photoreceptors and priming transient rhabdom shedding, clock input influences the expression of at least one gene important for the photoresponse. Specifically, clock input to the LE controls an early step in the expression of the gene for visual arrestin (varr). Varr is the protein in photoreceptors responsible for quenching the phototransduction cascade, and as described above, it is also involved in the internalization of rhabdomeral membrane during light-driven shedding. In LEs maintained in constant darkness, clock input during the subjective night causes varr messenger RNA (mRNA) levels in photoreceptors to fall. Circadian fluctuations in varr mRNA levels may lead to a circadian fluctuation in varr protein levels and thus contribute to some of the circadian changes in the photoresponse described above.

The expression of other photoreceptor proteins probably is also under circadian control, but the clock does not regulate the expression of all proteins important for the photoresponse. For example, *Limulus* opsin mRNA levels are regulated by light, not by the clock. In LEs that receive normal clock input and are exposed to cyclic light, opsin mRNA levels rise during the late afternoon and fall during the night. This rhythm continues in LEs deprived of clock input if they are exposed to cyclic light, and it is eliminated in LEs maintained in the dark even when they receive normal clock input.

Biochemical Processes Mediating Clock Effects on *Limulus* Eyes

Octopamine and the Activation of a cAMP Cascade

When circadian efferent neurons are active during the night, they release the biogenic amine octopamine (OA) from their terminals, and OA, the phenol analog of norepinephrine, is probably most responsible for initiating the circadian changes observed in the eyes. The application of OA to *Limulus* eyes mimics many effects of the clock (**Table 1**). Other molecules are also released from efferent terminals. Specifically, γ-glutamyl conjugates of OA and tyramine, the precursor of OA, are released, but their physiological relevance is not yet clear. The presence of crystalline granules in efferent terminals suggests that one or more neuropeptides may be released. Crystalline granules are often associated with peptidergic synaptic terminals in invertebrates, but presumptive neuropeptides in the circadian efferent terminals in *Limulus* eyes have not been identified.

In *Limulus* eyes, OA activates membrane receptors that are coupled to adenylyl cyclase stimulating a rise in cAMP in photoreceptors, and any effects of the clock are mediated through the activation of this cAMP cascade (**Table 1**). There is no evidence for a direct effect of cAMP on *Limulus* photoreceptor physiology; however, there is good evidence that some of the clock-driven changes in the eye require the activation of cAMP-dependent PKA. Therefore, investigations of mechanisms underlying the circadian regulation of photoreceptor function have focused on identifying and characterizing clock-regulated photoreceptor phosphoproteins.

Clock-Driven Protein Phosphorylation

To date, one clock-regulated phosphoprotein in *Limulus* photoreceptors has been identified and partially characterized. It is an unconventional myosin III (LpMyo3), a homolog of the *Drosophila ninaC* gene product that is

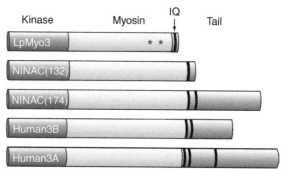

Figure 7 Schematic comparing the domain structure of LpMyo3 with that of the class III myosins expressed in *Drosophila* and humans. One isoform of myo3 has been found in *Limulus*. The two isoforms of myo3 expressed in *Drosophila* are splice variants of the same *ninaC* gene. Human class III myosins are products of two separate genes, *myo3A* and *myo3B*. Each protein contains an N-terminal kinase domain, a myosin domain, and one or more IQ calmodulin-binding motifs (black bars). The tail domains are most variable in length and sequence. LpMyo3 is a target of clock-stimulated phosphorylation within an actin-binding region near the C-terminus of its myosin-like domain (red asterisks).

required in *Drosophila* for normal photoreceptor function and survival. Since their discovery in *Drosophila* and *Limulus* photoreceptors, class III myosins also have been detected in photoreceptors of a variety of invertebrates and vertebrates, including octopus, fish, mice, and humans; thus, these proteins may be important for photoreceptor function in both vertebrates and invertebrates.

Class III unconventional myosins are characterized by having an N-terminal kinase domain, a myosin-motor-like domain, one or more IQ calmodulin-binding motifs, and a C-terminal tail of varying lengths (**Figure 7**). Thus, they are potential signaling molecules as well as potential molecular motors. LpMyo3 is photoreceptor specific, quantitatively a major protein in photoreceptors and distributed throughout the photoreceptor from its cell body to its terminals. During the day in the light, LpMyo3 concentrates over the rays of the photosensitive rhabdom (**Figure 2(c)**).

LpMyo3 becomes more highly phosphorylated in LEs *in vivo* in response to activation of the circadian efferent input. Its phosphorylation also becomes elevated when intact photoreceptors are incubated *in vitro* with OA, drugs that elevate intracellular cAMP levels and that activate PKA, respectively. Since LpMyo3 is also phosphorylated in response to light, this protein may be particularly important as a substrate upon which light- and clock-driven cascades converge.

Some biochemical properties of LpMyo3 are known. It is a kinase that phosphorylates its own myosin domain as well as other substrates, and its substrate specificity is similar, but not identical, to PKA. LpMyo3 also binds actin, but since its actin binding is insensitive to adenosine triphosphate (ATP) and the protein lacks ATPase activity, it probably does not function as a molecular motor. Clock-regulated phosphorylation sites in LpMyo3 have also been identified. This information provides clues to how the clock might regulate LpMyo3 and leads to speculations regarding how the phosphorylation of LpMyo3 might influence photoreceptor function.

Clock input to LEs increases the phosphorylation of two sites within the motor-like domain of LpMyo3 within and near an actin-binding interface called loop2 (**Figure** 7). Interestingly, LpMyo3 autophosphorylates these same sites through an intermolecular mechanism. The presence of phosphorylation sites within the actin-binding interface of a myosin is surprising. None of the other 22 classes of myosins that have been described to date becomes phosphorylated in this region. However, mutagenesis studies using other unconventional myosins indicate that changing the net charge in loop2 alters the affinity of myosin for actin. Phosphorylation of LpMyo3 within its actin-binding interface is predicted to reduce its actin affinity.

Actin is concentrated in the rhabdomeral microvilli. Functional consequences of a possible nighttime decrease in the affinity of LpMyo3 for microvillar actin are not yet known, but could be diverse. The Myo3 in *Drosophila* photoreceptors (neither inactivation nor afterpotential C (NINAC)) is thought to bind to and stabilize actin in rhabdomeral microvilli, and in *Drosophila* lacking NINAC, microvillar actin is fragmented or absent. These observations lead to speculations that LpMyo3 phosphorylation participates in circadian processes involving changes in the stability of rhabdomeral actin. One such process is light-triggered transient rhabdom shedding, which involves a transient breakdown of rhabdomeral actin. As described above, although this process is triggered by light, it must be primed by clock input and the activation of PKA. The clock-driven phosphorylation of LpMyo3 by PKA may contribute to the priming event.

LpMyo3 is also a kinase that may phosphorylate one or more proteins involved in the photoresponse. This idea is consistent with observations from *Drosophila*, which show that the photoresponse is abnormal in flies expressing NINAC without its kinase domain. If a change in the affinity of LpMyo3 for actin produces a nighttime decrease in the concentration of LpMyo3 at the rhabdom, the level of phosphorylation of LpMyo3 substrates may also fall and modify the photoresponse. Thus, multiple effects of the clock could be mediated through the modulation of LpMyo3.

Conclusion

The eyes of *Limulus*, with their diverse structures, large photoreceptors, and relatively simple organization, have long served as useful preparations for studying basic mechanisms of vision. Fundamental and broadly relevant aspects of photoreceptor function and sensory information processing were discovered through studies of the *Limulus* visual system. Moreover, studies using *Limulus* have revealed the fundamental importance of circadian rhythms for vision and the complex ways in which clock- and light-driven processes can interact to produce the full range of diurnal changes in retinas. The effects of circadian rhythms observed in *Limulus* LE have many parallels in vertebrate retinas, including those on photoreceptor structure, the photoresponse, and membrane shedding. In addition, circadian clocks in both *Limulus* and vertebrates exert their influence in retinas by regulating the release of biogenic amines and the activity of cyclic-nucleotide-signaling pathways. Mechanisms through which clock-driven signaling pathways influence specific retinal functions are just beginning to be identified, and their discovery remains an important challenge.

See also: Circadian Metabolism in the Chick Retina; The Circadian Clock in the Retina Regulates Rod and Cone Pathways; Circadian Regulation of Ion Channels in Photoreceptors; Fish Retinomotor Movements; Genetic Dissection of Invertebrate Phototransduction; Phototransduction in *Limulus* Photoreceptors; Rod and Cone Photoreceptor Cells: Outer Segment Membrane Renewal.

Further Reading

Barlow, R. B., Jr. (1983). Circadian rhythms in the *Limulus* visual system. *Journal of Neuroscience* 3: 856–870.

Barlow, R. B., Jr., Bolanowski, S. J., Jr., and Brachman, M. L. (1977). Efferent optic nerve fibers mediate circadian rhythms in the *Limulus* eye. *Science* 197: 86–89.

Barlow, R. B., Jr., Chamberlain, S. C., and Lehman, H. K. (1989). Circadian rhythms in the invertebrate retina. In: Stavenga, D. G. and Hardie, R. C. (eds.) *Facets of Vision*, pp. 257–280. Berlin: Springer.

Barlow, R. B., Jr., Chamberlain, S. C., and Levinson, J. Z. (1980). *Limulus* brain modulates the structure and function of the lateral eyes. *Science* 210: 1037–1039.

Battelle, B.-A. (2006). The eyes of *Limulus polyphemus* (Xiphosura, Chelicerata) and their afferent and efferent projections. *Arthropod Structure and Development* 35: 1–14.

Battelle, B.-A. (2008). Circadian rhythms in visual function in *Limulus*. In: Fanjul-Moles, M.-L. and Aguilar-Roblero, R. (eds.) *Comparative Aspects of Circadian Rhythms*, pp. 19–40. Kerala: Transworld Research Network.

Battelle, B.-A., Evans, J. A., and Chamberlain, S. C. (1982). Efferent fibers to *Limulus* eyes synthesize and release octopamine. *Science* 216: 1250–1252.

Calman, B. G. and Chamberlain, S. C. (1982). Distinct lobes of *Limulus* ventral photoreceptors. II. Structure and ultrastructure. *Journal of General Physiology* 80: 839–862.

Cardasis, H. L., Stevens, S. M., McClung, S., et al. (2007). The actin-binding interface of a myosin III is phosphorylated *in vivo* in response to signals from a circadian clock. *Biochemistry* 46: 13907–13919.

Chamberlain, S. C. and Barlow, R. B., Jr. (1984). Transient membrane shedding in *Limulus* photoreceptors: Control mechanisms under natural lighting. *Journal of Neuroscience* 4: 2792–2810.

Chamberlain, S. C. and Barlow, R. B., Jr. (1987). Control of structural rhythms in the lateral eye of *Limulus*: Interactions of natural lighting and circadian efferent activity. *Journal of Neuroscience* 4: 2794–2810.

Fahrenbach, W. H. (1975). The visual system of the horseshoe crab *Limulus polyphemus*. *International Review of Cytology* 41: 285–349.

Fein, A. and Payne, R. (1989). Phototransduction in *Limulus* ventral photoreceptors: Roles of calcium and inositol trisphosphate.

In: Stavenga, D. G. and Hardie, R. C. (eds.) *Facets of Vision*, pp. 173–185. Berlin: Springer.

Hartline, H. K., Wagner, H. G., and Ratliff, F. (1956). Inhibition in the eye of *Limulus*. *Journal of General Physiology* 39: 651–673.

Yeandle, S. (1958). Evidence of quantized slow potentials in the eyes of *Limulus*. *American Journal of Ophthalmology* 46: 82–87.

Macular Edema

R N Frank and I Glybina, Kresge Eye Institute, Wayne State University School of Medicine, Detroit, MI, USA

Glossary

Circinate lipid exudates – A round or oval assemblage of intraretinal lipid deposits that is often seen in macular edema, especially in diabetic macular edema. This presumably occurs because plasma exudes from a small retinal blood vessel, or a cluster of retina vessels, in a circular pattern around the abnormal vessel. The portion of this circle, farthest from the leaking vessels, is the thinnest, and as reabsorption of the fluid occurs at that location, the lipoproteins precipitate into the tissues, forming the visible exudates.

Cystoid macular edema – A form of macular edema in which much of the extracellular fluid collects in loculated spaces between the photoreceptor axons of Henle's fiber layer, forming round or oval, fluid-filled spaces. These are called cystoid because they are not lined with a layer of epithelial cells, which would make them true cysts; hence, these spaces are cystoid and not cystic.

Fovea – The central portion of the macula that contains only specialized cone photoreceptors, whose outer segments are narrow and elongated, rather than cone shaped as are extrafoveal cones.

Foveal avascular zone (FAZ) – A specialization of the capillary network in the foveal region of the retinal vasculature in humans and higher primates. There are no blood vessels in the FAZ, presumably as an evolutionary development to permit unimpeded access of light rays to the foveal cone photoreceptors. The FAZ can be readily seen as an approximately circular avascular zone about 1 mm in diameter in retinal vascular digest preparations from enucleated eyes or in good-quality fluorescein angiograms in living eyes.

Foveola – The central portion of the fovea, in which the axons of the cone photoreceptors are swept to the side, leaving only their inner and outer segments exposed to the incoming rays of light.

Henle's fiber layer – The obliquely directed axons of the foveal cone photoreceptors.

Irvine–Gass syndrome (aphakia, or pseudophakia, cystoid macular edema) – This entity was first described during the 1960s by the individuals whose eponymous names it bears. Presumably because of postoperative inflammation after cataract surgery, perifoveal capillaries begin to leak and edema fluid collects in a cystoid pattern in the central macula. The condition is often self-limited but may lead to considerably reduced vision. As demonstrated by fluorescein angiography, it may be accompanied by breakdown of the blood–retinal barrier in the capillary circulation of the optic nerve head.

Macula – The central region of the retina that is specialized for high-resolution vision. It is found in humans and higher primates and is notable histologically for the fact that it contains multiple layers of ganglion cells, while the extramacular retina contains only a single layer.

Macular edema – The thickening of the macular retina by extravasation of fluid, often containing lipid components (which may precipitate into the tissues) from breakdown of the normal blood–tissue barrier at the level of the endothelial cells of the retinal blood vessels in the macula. Fluid can also extravasate from the choroidal capillaries through the retinal pigment epithelium barrier, but this usually produces collections of subretinal fluid, which is distinguishable from the intraretinal fluid of macular edema.

Randomized controlled clinical trial (RCT) – A method of evaluating the efficacy of a new therapeutic procedure in which subjects with a disease under investigation volunteer to be allocated by random selection to a treatment group, or groups, which will receive the new therapy or therapies under investigation, or to a control group that receives either no treatment or treatment with a placebo or treatment by the method that has been accepted as the current standard of care. Evaluation of the results for each subject is done by observers who are masked, that is, who do not know to which of the treatment, or control, groups the various subjects belong. This method is considered the definitive way to determine the efficacy, or lack of efficacy, of a putative new therapy.

Macular edema, a swelling of the central portion of the human retina, or macula, is a common feature in many diseases. It is a widely recognized complication of diabetic retinopathy, where its prevalence (in the United States) may be of the order of 10% of all individuals with diabetes, varying somewhat by ethnic group but not by type of diabetes (i.e., type 1 or type 2). Macular edema also occurs commonly in retinal vein occlusions, in uveitis, following cataract surgery, as an accompanying feature of some ocular tumors, in hereditary retinal degenerations, such as the retinitis pigmentosa syndromes and, rarely, as an isolated genetic entity apparently in and of itself.

Diagnosis of Macular Edema

Recognition of macular edema as an important clinical entity has been increasing in recent years, and, in particular, since the 1960s with the advent of a variety of new diagnostic techniques. These have included high-magnification, stereoscopic ophthalmoscopy using the slit lamp and several types of contact and noncontact lenses that allow the observer to visualize the edematous macula in a better manner; stereoscopic retinal photography; fluorescein angiography to enable visualization of circulatory dynamics and the physiology/pathology of the macular circulation, even at the capillary level; and, most recently, optical coherence tomography (OCT), a rapid, noninvasive technique that permits accurate measurement of the thickness of the macula, and yields high-resolution tissue sections of the maculae of living subjects with almost histologic clarity.

Anatomy of the Macula

The macula is an anatomic specialization of the neural retina that is unique to humans and to higher primates (**Figure 1(a)–(d)**). It is defined histologically by the presence of multiple layers of retinal ganglion cells (there is only a single layer of ganglion cells in the extramacular retina), and ophthalmoscopically as the region extending from the temporal margin of the optic nerve head to the center of the macula (the foveola, the central depression that is evident clinically in the normal retina), and then for an equal distance temporal to the foveola. In the superior–inferior direction, the macula extends between the two major temporal retinal vascular arcades. It is not clear why, of the entire extent of the human retina, the macula is particularly susceptible to such a large number of disorders, for example, the macular degenerations, or to edema in a large variety of diseases. Speculations entail

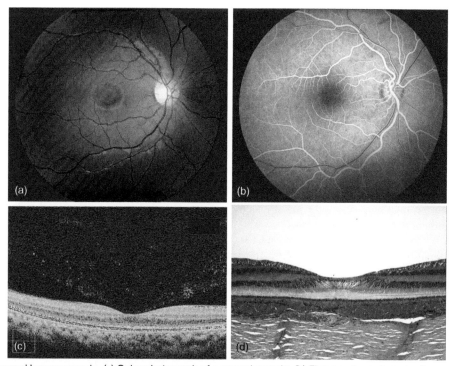

Figure 1 The normal human macula. (a) Color photograph of a normal macula. (b) Fluorescein angiogram of a normal macula. Note the striped pattern of blood flow in the retinal venule. This laminar flow, with the nonfluorescent column of blood cells in the center of the vessel, is typical of early flow in retinal venules. In later phases of the angiogram, the entire vessel will fill in. Note also the sharp margins of the vessels in this normal angiogram, in the normal retinal vasculature, the endothelial lining of the vessels is tight, even to the small (M.W. 327) fluorescein molecule, which is only about 65% bound to plasma proteins. (c) A spectral domain optical coherence tomogram (OCT) of a normal macula. In this high-resolution OCT scan, one can clearly see the cellular layers of the retina. (d) A histologic section of a normal human macula, stained with hematoxylin–eosin. This photomicrograph is courtesy of Ralph Eagle, M.D., Wills Eye Institute, Philadelphia, PA.

the likelihood that the macular retina, including both its neural elements and also the macular retinal pigment epithelium, has a higher level of metabolic activity because of its enhanced visual function, including a greater density of cone photoreceptors, a thicker ganglion cell layer, and a greater thickness of retinal pigment epithelial (RPE) cells; or that this enhanced activity necessitates greater blood flow with a greater likelihood of circulatory breakdown, especially in view of the very rich retinal capillary network within the macula and, particularly, in the perifoveal region. Finally, the anatomic specialization of the foveola, in which the foveolar cone axonal and synaptic processes are swept obliquely to all sides, permitting direct access of photic stimuli to the photoreceptor outer segments, may lead pathologically to the accumulation of extracellular edema fluid into ballooning spaces in this axonal region in the entity known as cystoid macular edema.

In studying macular edema in an experimental setting, investigators are hampered by the fact that the macula is an anatomic specialization of the retina that is limited to humans and some higher primates. While certain birds, for example, hawks and eagles, have a fovea, there is no macula in the sense that it is present in the human eye, and the blood circulation of the posterior avian globe is entirely different from that of humans.

Clinical Findings in Macular Edema

The thickening of the central retina is best appreciated by stereoscopic viewing methods, for example, slit-lamp ophthalmoscopy with a contact lens or a handheld high dioptric indirect ophthalmoscope lens. Until recently, multicenter clinical research studies, for example, the Early Treatment Diabetic Retinopathy Study (ETDRS) used stereoscopic retinal photography to evaluate macular thickening. However, relatively minimal central retinal thickening may not be apparent even to the experienced examiner and photographic methods are subject to variability. Therefore, the introduction of OCT, which is discussed in more detail below, has held recent very great interest, but it has also raised some unexpected questions.

In the absence of stereoscopic viewing methods, one can suspect the presence of macular edema by the presence of certain characteristic retinal lesions. Extensive intraretinal hemorrhages or microaneurysms (dilations of capillary-sized blood vessels, which appear ophthalmoscopically as tiny red dots) in the macular region are indicators of extravascular blood or plasma leakage with resultant macular edema (**Figure 2(a)** and **(b)**). An especially characteristic abnormality, particularly in individuals with diabetes, is the presence of intraretinal lipid deposits. These often have a circular (circinate) configuration (**Figure 2(a)**), which presumably is the result of

leakage of plasma from a central vascular lesion, or cluster of lesions (**Figure 2(b)**), with resorption of the edema fluid but not its lipid components around the periphery of the 360° circumference of the zone of leakage, where the thickness of the fluid layer is least and the less-soluble lipid therefore precipitates out of solution. Macular lipid, and edema, have characteristic appearances by OCT, and histologically, as demonstrated in **Figure 2(c)** and **(d)**. It has been known for a number of years that lower blood-lipid values are associated with lesser degrees of macular edema in diabetic subjects, although it is not clear that normalizing blood lipids can entirely prevent the development of macular edema in diabetic subjects.

There have been reports, at least one of which preceded by a number of years the development of OCT for objective measurements of macular thickness, that macular edema can vary over the course of a day. Some patients have noted that their vision, on arising in the morning, was much poorer than later in the day. Several investigators, using OCT, have found that macular thickness in at least some patients with diabetic macular edema is greater immediately after arising from a night's sleep than it is later. Although there are a number of possible causes, the most plausible is a simple gravitational effect, with edema fluid remaining in the upper body during recumbent posture and draining to the lower body when the individual is sitting or standing. However, although this diurnal effect does occur in some individuals, the Diabetic Retinopathy Clinical Research Network, in a study involving many institutions, found it to be infrequent.

Clinically Significant Diabetic Macular Edema

The ETDRS also introduced the term "clinically significant macular edema" to describe those cases in which the retinal thickening either involved the center of the macula (the fovea and foveola), or was sufficiently close to the center as to constitute a potential threat to the center with likely loss of central vision. The definition of clinically significant diabetic macular edema, as established by the ETDRS, is presented in **Table 1**.

Findings by Fluorescein Angiography

By intravenous fluorescein angiography, in which a solution of sodium fluorescein is injected intravenously through an antecubital vein and rapid sequence photographs are taken using fluorescence optics, it is possible to visualize some of the vascular dynamics of macular edema. Fluorescein is a small molecule (M.W. 327) that is only loosely bound to plasma proteins, for example, only about 65% to albumin, and this small size enables it to leak readily through patent interendothelial cell junctions. Hence, in successive frames of a fluorescein

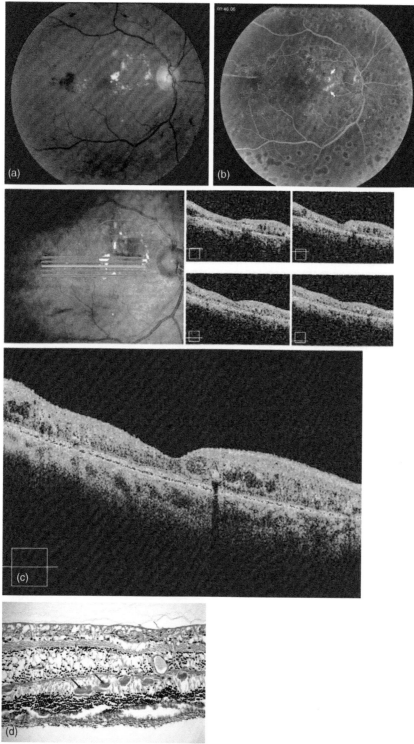

Figure 2   (a) Circinate lipid deposits form a ring in a case of diabetic macular edema. Multiple dot and blot hemorrhages and microaneurysms are also present. (b) A frame from a fluorescein angiogram from this patient, showing a cluster of leaking microaneurysms (arrows) at the center of the lipid ring. (c) Spectral domain OCT scans of this patient, showing intraretinal edema fluid, which appears as round, hollow spaces, and lipid plaques (red-orange densities). The vertical, empty spaces beneath the plaques occur because the plaques absorb the scanning laser beams, producing optical shadows. (d) Histologic section of the neurosensory retina, lacking retinal pigment epithelium and choroid, from an eye with diabetic macular edema (round or/and intraretinal lipid plaques, some of which are shown by arrows). The edema fluid appears as round or oval spaces with smooth borders, while the irregular spaces with ragged edges at the bottom of the micrograph are artifacts created by histologic sectioning. Hematoxylin–eosin stain. Photomicrograph courtesy of Ralph Eagle, M.D., Wills Eye Institute, Philadelphia, PA.

angiogram in patients with diabetic retinopathy or other edema-producing retinal vascular diseases, one can see an increasing fluorescent haze surrounding the microvascular lesions (**Figure 3**). These leaking lesions are often (though surprisingly, not always) associated with macular thickening by OCT examination.

Table 1   Clinically Significant Diabetic Macular Edema

1. Thickening of the retina at or within 500 μm (approximately one-half optic disk diameter) of the center of the macula.
2. Hard exudates at or within 500 μm of the center of the macula, if associated with thickening of adjacent retina (not residual hard exudates remaining after disappearance of retinal thickening).
3. A zone or zones of retinal thickening 1 optic disk area or larger, any part of which is within one disk diameter of the center of the macula.

From Early Treatment Diabetic Retinopathy Study Research Group (1985). Photocoagulation for diabetic macular edema. Early Treatment Diabetic Retinopathy Study Report No. 1. *Archives of Ophthalmology* 103: 1795–1805. © 1985 American Medical Association. All rights reserved.

Fluorescein angiography has also demonstrated a specific type of macular edema, so-called cystoid macular edema. The axons of foveal cone photoreceptors are displaced laterally in human and higher primate retinas, thereby permitting light to strike the foveal cone outer segments directly, without traversing the inner retinal layers. This lateral displacement produces a unique anatomic arrangement that is known as Henle's fiber layer. The center of the fovea itself is avascular; however, cyst-like extracellular fluid spaces collect within Henle's fiber layer. These can, although often with some difficulty, be visualized by white light ophthalmoscopy using a slit lamp and handheld or contact lens, or sometimes by ophthalmoscopy using a green filter to increase contrast. The characteristic appearance, using fluorescein angiography, is one or more oval, uniformly fluorescent spaces oriented radially around the foveal center similar to petals on a flower, giving rise to the term "petaloid appearance." A single such large cystoid cavity is shown in the fluorescein angiographic frame of **Figure 4(a)**. The cyst-like nature of this cavity is evident on OCT scans (**Figure 4(b)**). Although macular edema may produce severe loss of

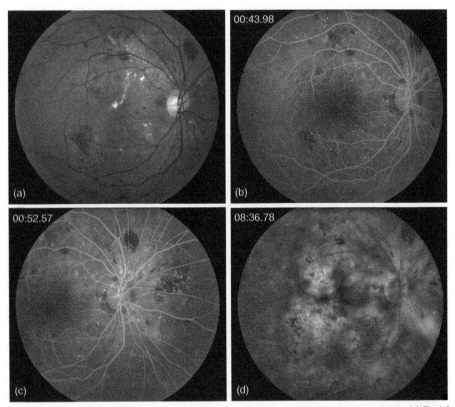

Figure 3   Sequence of angiographic frames showing increasing dye leakage in diabetic macular edema. (a) Red-free initial photograph of the right eye, taken before injection of the dye to show features of the posterior retina. (b) Frame from the fluorescein angiogram taken approximately 44 s from the time of dye injection (timing is indicated by numbers at the top left of the photograph). Arteries and veins have filled, and multiple microaneurysms, which appear as tiny white dots, are apparent. (c) Frame from the fluorescein angiogram taken at approximately 53 s after injection. Dye in the microaneurysms has intensified and has begun to leak from the vessel walls, creating a somewhat blurred appearance. (d) Frame from the angiogram taken approximately 8 min and 37 s from the time of injection. Dye and plasma have leaked profusely from the abnormal vessels into retinal tissue, contributing to the tissue edema.

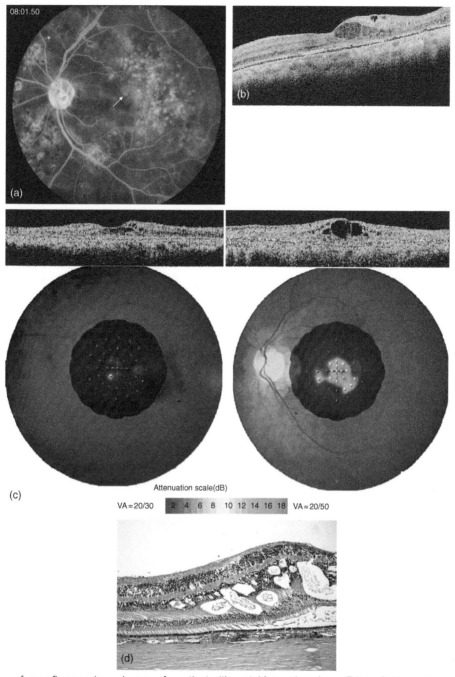

Figure 4 (a) Frame from a fluorescein angiogram of a patient with cystoid macular edema. Edema fluid containing the dye may fill many balloon-like spaces surrounding the center of the macula, producing an appearance similar to multiple radial petals on of flower, a so-called petaloid appearance. Only one such large space, indicated by the arrow, is seen in this photograph. The multiple small, round spots are previous laser treatment burns, which have been placed in a grid pattern. (b) Spectral domain OCT scan of this patient, showing the large cystoid spaces that appear in the angiographic frame in **Figure 4(a)**. (c) Microperimetric study of a patient of Dr. Tamer Mahmoud, who presented with bilateral cystoid macular edema. The microperimetry scans are here placed beneath OCT scans of the corresponding eyes. The color scale at the bottom of the illustration shows increasing loss of retinal sensitivity as one moves from left to right on the scale. (d) H-and-E-stained histopathological section of a retina with macular edema. Note the large fluid-filled vacuoles in the macula. This photomicrograph is courtesy of Ralph Eagle, M.D., Wills Eye Institute, Philadelphia, PA.

central visual acuity, vision is often surprisingly preserved in many instances. **Figure 4(c)** shows OCT scans of a patient with bilateral cystoid macular edema, with corresponding microperimetric data printed underneath.

This very sensitive visual sensitivity mapping device demonstrates only modest sensitivity loss in each eye, located almost directly corresponding to the position of the cystoid cavities. Visual acuity in each eye, also printed on the figure, is quite good. These cystoid cavities are, of course, not true cysts, since when histologic sections are available, they demonstrate that these fluid-filled spaces are not lined by layers of epithelium (**Figure 4(d)**). Cystoid edema may occur with any of the disorders that produce macular edema. One notable entity was first described during the 1960s, after fluorescein angiography had become widely used, when cystoid macular edema was observed angiographically in many patients who had just undergone seemingly uneventful intracapsular cataract surgery. Often, the postoperative visual acuity in these patients did not recover as fully as might have been expected, and fluorescein angiography revealed a pattern of cystoid macular edema, frequently accompanied by leaking of fluorescent dye from the optic nerve head. This pattern has been called the Irvine–Gass syndrome, after the individuals who initially reported it. With the advent of advanced extracapsular cataract surgical techniques, in which the posterior lens capsule is left in place and a posterior chamber intraocular lens is inserted, the incidence of postcataract surgical cystoid macular edema has been considerably reduced, but it still occurs. While this entity has been associated with various complications of the surgery, it often appears in cases where the surgical procedure was entirely uneventful.

Causes of Macular Edema

All investigators agree that the proximate cause of macular edema, associated with any systemic or ocular disease or drug, is a breakdown of the inner blood–retinal barrier, which is composed of junctional complexes of the endothelial cells of the retinal blood vessels, as opposed to the outer blood–retinal barrier, consisting of the retinal pigment epithelium, which serves as a boundary layer between the neural retina and the choroidal vasculature. Although contents of the vascular lumina can reach the extravascular space by transcellular mechanisms, directly through endothelial cell cytoplasm, most investigators believe that the most frequent pathologic mechanism is the breakdown of intercellular junctional complexes, which normally form a tight boundary between the vascular lumen and the extravascular milieu. This may be caused by specific molecular mechanisms, such as increased levels of vascular endothelial growth factor (VEGF, originally called vascular permeability factor because it was through that function that this molecular family was first discovered), or various inflammatory molecules or cytokines, or through the use of certain topical drugs in the treatment of glaucoma, such as epinephrine or, more recently, prostanoid compounds. Another molecular mechanism that has been of interest is the family of carbonic anhydrase enzymes. A recent paper, which describes proteomic analysis of the vitreous

fluid in a series of patients with proliferative diabetic retinopathy and vitreous hemorrhage, who underwent vitrectomy surgery, has reported that these vitreous samples demonstrated a substantial upregulation of the CA-1 isoform of carbonic anhydrase. Injection of this molecule into the vitreous cavity of rats produced generalized retinal edema. Another recent paper cast some doubt on the role of VEGF in diabetic macular edema. These investigators examined the relationship between various single nucleotide polymorphisms (SNPs) in the VEGF gene and its promoter in individuals with various levels of diabetic retinopathy severity in the very large and well-characterized Diabetes Control and Complications Trial/Epidemiology of Diabetes Interventions and Complications (DCCT/EDIC) cohort. They found that there were 18 SNPs in the VEGF gene and its promoter that were significantly associated with severe preproliferative and proliferative retinopathy, but no detectable significant associations of VEGF gene or promoter polymorphisms with diabetic macular edema.

Treatment of Macular Edema

Results from the ETDRS

The ETDRS, a randomized, controlled clinical trial (RCT) of laser treatment and aspirin therapy that involved over 3000 patients with moderate to severe nonproliferative and early proliferative diabetic retinopathy, and/or with macular edema, reported in 1985 that treatment with focal applications of small-diameter (50–100 μm) argon laser burns directly to macular microaneurysms (**Figure 5(a)–(c)**), or with grid laser applications could slow down or arrest the progress of diabetic macular edema with substantial preservation of vision by as much as 50% more in eyes that received treatment than in eyes that were randomly allocated to the no laser treatment group. The ETDRS results are frequently cited as showing that such laser treatment preserves vision, but does not improve it. However, in this study, a substantial number of the patients initially had good (i.e., better than 20/40) vision and hence were not likely to improve substantially. When one examines the results of the ETDRS patients whose initial vision was less than 20/40, it becomes apparent that laser treatment produced an actual improvement of visual acuity by more than one line on the ETDRS visual acuity chart in more than 40% of these subjects, compared to approximately 25% in the control group. In a more recent RCT, which compared two different techniques for applying focal argon laser treatment for diabetic macular edema, the Diabetic Retinopathy Clinical Research Network found that approximately 30% of patients with initial visual acuity less than 20/40 who received the modified ETDRS laser technique improved their visual acuity by more than three lines on

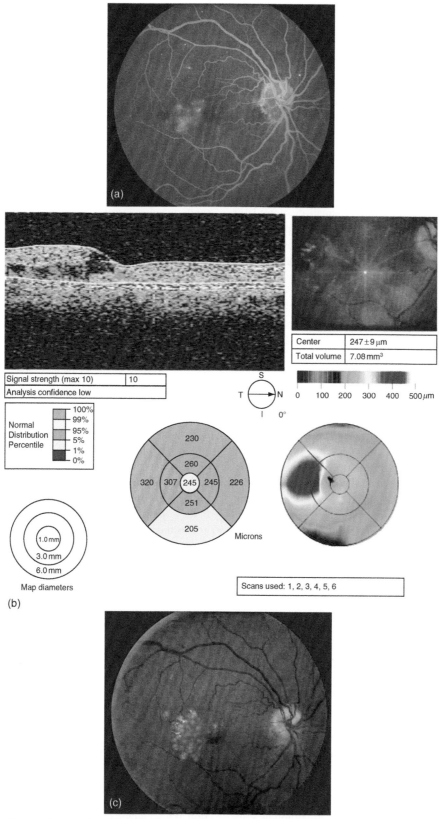

Figure 5 (a) Frame from a fluorescein angiogram showing intense leakage in the right eye of a patient with relatively focal macular edema; (b) Time domain OCT scan showing the region of edema (red-filled zone in the macular map). Color scales indicate retinal thickness. (c) Photograph taken immediately after application of focal argon laser burns, showing localization of the laser treatment to the edematous area.

the ETDRS vision chart, a halving of the visual angle, over a 1-year follow-up. At that time (fall, 2008), therefore, laser treatment, involving focal application of small, relatively light-intensity argon laser burns to areas of vascular leakage in the maculae of eyes with diabetic macular edema, and in particular clinically significant macular edema, was considered the treatment of choice. Application of this treatment, however, requires a degree of caution. Because eyes with clinically significant macular edema may not have suffered a loss of central vision (and most eyes with macular edema in the ETDRS had visual acuity at, or better than, 20/40), and because the laser treatment technique employed in this study involves the placement of moderately intense, small-diameter laser burns directly to microaneurysms identified ophthalmoscopically or by fluorescein angiography, placement of such burns to lesions very close to the center of the macula may result in damage to central vision. The use of focal laser photocoagulation for lesions very close to the macular center therefore does require an element of clinical judgment. Hence, the ETDRS recommended that laser treatment for clinically significant macular edema should be considered, but did not establish such treatment as an absolute standard of care.

Results of Other Clinical Trials

Another multicenter RCT, the Branch Vein Occlusion Study, found a significant beneficial effect of focal argon laser photocoagulation for macular edema, and for retinal neovascularization, in this disorder. However, eyes with central retinal vein occlusion, a much more severe condition, received no benefit from focal laser photocoagulation in the Central Vein Occlusion Study.

Focal laser treatment was also shown to be of clear benefit for the treatment of macular edema resulting from branch retinal vein occlusion in another RCT, the Branch Vein Occlusion Study. However, it was ineffective for central retinal vein occlusion in the Central Vein Occlusion Study, yet another RCT. There is no evidence for a beneficial effect of laser treatment for macular edema from any other cause.

As the pathogenesis of macular edema appears to have an inflammatory component, anti-inflammatory therapies have been widely used as treatment for some forms of this disorder. The ETDRS evaluated aspirin at a dose of 650 mg day^{-1}, compared to placebo for severe non-proliferative diabetic retinopathy and for diabetic macular edema and found no beneficial effect. In aphakia or pseudophakia cystoid macular edema, one RCT has demonstrated the efficacy of treatment with a topical nonsteroidal anti-inflammatory eyedrop, although some clinicians combine this with a topical steroid. More recently, a number of investigators have used intravitreal

injections of triamcinolone, a steroid molecule, for several forms of macular edema. Although several papers have reported good results of this procedure for diabetic macular edema, a very recent paper detailing the results of a large, multi-institutional RCT in the Diabetic Retinopathy Clinical Research Network found that, over a 2-year follow-up, intravitreal triamcinolone, at 1 and 4 mg doses, was significantly less effective in this disorder than focal laser treatment. These doses of intravitreal triamcinolone have been evaluated for their efficacy in treating macular edema resulting from branch or central retinal vein occlusion in the Standard Care versus COrticosteroid for REtinal vein occlusion (SCORE) study, two multi-institutional randomized, controlled clinical trials sponsored by the U.S. National Eye Institute. For central retinal vein occlusion, these studies found that triamcinolone injected intravitreally in either a 1 or a 4 mg dose produced a significantly higher number of eyes that gained three or more lines of vision over a 1-year follow-up interval than observation alone. One milligram was concluded to be the preferred dose for this indication because of its smaller number of adverse effects. This result is the first time that any treatment has been reported as significantly effective in improving vision in macular edema from a central retinal vein occlusion. However, the concurrent SCORE clinical trial showed that neither of these two steroid doses injected intravitreally produced gains in vision for eyes with macular edema secondary to branch retinal vein occlusion that were superior to those obtained by focal argon laser photocoagulation, the previously accepted standard of care for that entity. Reasons for the different efficacies of steroids and laser for macular edema in diabetic retinopathy and in central and branch retinal vein occlusions remain to be explained.

Another intravitreal steroid, fluocinolone, has been used successfully as a long-term, slow-release implant inserted surgically in the treatment of macular edema resulting from chronic uveitis. Like other steroids, but to a somewhat greater degree, the fluocinolone implant can result in substantial elevations of intraocular pressure, and in cataract formation. A full evaluation by RCT of the fluocinolone implant for macular edema in diabetes is underway, but clinical trials using this steroid implant involving other disorders in which macular edema is present have not yet been carried out.

Anti-VEGF Therapies and Macular Edema

Because the VEGFs were first recognized as a family of molecules that enhanced vascular permeability and the breakdown of blood–tissue barriers, before their recognition as angiogenic agents, the development of anti-VEGF agents with efficacy against neovascular retinal and choroidal diseases has been followed by clinical studies of these agents as potential treatments for macular edema. RCTs involving ranibizumab (Lucentis, Genentech), a humanized anti-VEGF monoclonal antibody injected intravitreally for diabetic macular edema and for macular edema in retinal vein occlusions are currently underway. A very similar monoclonal antibody, bevacizumab (Avastin, Genentech), developed initially as a cancer therapy, has also been employed in some preliminary clinical trials. One possible objection to the putative role of VEGF in diabetic macular edema, relating to the absence of significant VEGF gene polymorphisms in diabetic macular edema cases, compared to the presence of such changes in cases of severe preproliferative and proliferative diabetic retinopathy, has been discussed above. Another is the clinical observation that diabetic macular edema often occurs in the absence of retinal neovascularization and, conversely, neovascularization may occur without macular edema. If excessive VEGF secretion is essential to the cause of both types of diabetic retinal lesion, then both should occur together much more often than not. However, the issue may be much more complex, and its resolution will require, among other things, the completion of the ongoing clinical trials. There have been a number of reports describing encouraging results from much smaller trials. A report of a small, phase 2 trial of bevacizumab for diabetic macular edema from the Diabetic Retinopathy Clinical Research Network described suggestive evidence of a beneficial effect on macular thickness by OCT and visual acuity in patients injected with this agent, compared to focal argon laser photocoagulation. This trial, however, was quite small and complex in organization (a total of 100 patients, randomized into five treatment groups of 20 each, receiving various combinations of different doses of bevacizumab with or without laser therapy). A final determination of the efficacy of this mode of therapy must therefore await the completion of larger, longer controlled clinical trials.

Other agents that are currently being investigated for the treatment of diabetic macular edema include VEGF-TNF receptor-associated protein (TRAP; aflibercept, Regeneron), a VEGF receptor that is solubilized by being complexed to an immunoglobulin and that, upon intravitreal injection, acts by sponging VEGF molecules from the vitreous, and sirolimus (rapamycin), an antibiotic with immunomodulatory and anti-inflammatory properties, that is also capable of interfering with neovascularization.

A Puzzling Question

The use of OCT evaluations to measure macular thickness and to determine pathologic changes in macular anatomy has led to an unexpected finding. It has been generally assumed, that macular thickness beyond the

normal range, and visual acuity, were inversely, and fairly closely, correlated, and central macular thickness determined by OCT has been used as an endpoint (usually a secondary endpoint) of several clinical trials of new therapies for diabetic retinopathy and some other diseases. Although a rough correlation does exist, the Diabetic Retinopathy Clinical Research Network and others who have investigated this question have found that, for diabetic macular edema, the correlation is surprisingly poor (**Figure 6**). We, and others, have found that this poor correlation extends to macular edema in other disorders as well, and our own preliminary results suggest that the slope of the correlation curve of central macular thickness versus visual acuity differs for diabetic macular edema and for pseudophakia macular edema (Irvine–Gass syndrome).

The reasons for this unexpected observation are yet to be established. What is the effect of intracellular fluid accumulation versus large amounts of intercellular edema fluid, as in cystoid macular edema? Does edema of much longer duration have an adverse effect on visual acuity, by contrast with more acute occurrences? What kinds of anatomic alterations have an adverse effect on the visual acuity outcome? As a rule, patients whose fluorescein angiograms show extensive nonperfusion of the perifoveal capillary network, or who have a large lipid plaque in the center of the macula, will have poor visual acuity

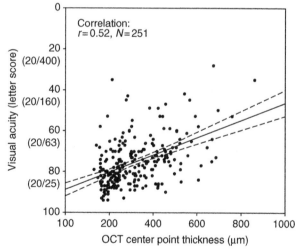

Figure 6 Best-corrected visual acuity, measured on the Early Treatment Diabetic Retinopathy Study (ETDRS) chart, of a large series of patients from the multicenter Diabetic Retinopathy Clinical Research Network, and correlated with macular central point thickness (measured by time domain OCT). The solid line indicates the best-fitting linear correlation curve, while the dashed lines indicate the 95% confidence interval. Although a correlation exists, it is surprisingly poor. Reprinted from DRCR. net Study Group (2007). Relationship between optical coherence tomography-measured central retinal thickness and visual acuity in diabetic macular edema. *Ophthalmology* 114: 525–536, with permission from Elsevier.

regardless of macular thickness. Are there more subtle anatomic changes that can be detected by newer, high-resolution OCT methods (**Figures 1(c)**, **2(c)**, and **5(b)** and (**c**); compare with **Figure 4(b)**)? Are there aberrations in photoreceptor structure or orientation that can be detected by high-resolution OCT, or by adaptive optics techniques? What prognostic information can be obtained by electrophysiologic methods such as the multifocal electroretinogram (**Figure 7**), that can detect functional alterations in very small regions of the macula? These and other questions remain subjects for investigation in the study of macular edema.

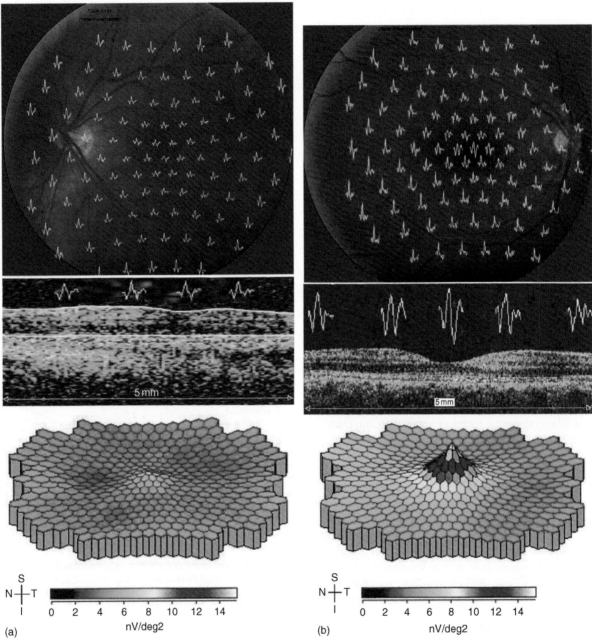

(a)   nV/deg2

(b)   nV/deg2

Figure 7   Multifocal electroretinogram (mfERG) traces of (a) A patient with diffuse macular edema and (b) a normal subject. In both figures, the mfERG traces are placed overlying a digital photograph of the individual's posterior retina, such that the site of the mfERG trace corresponds to its anatomic location. Note the diminution of the amplitudes and widening of the traces, indicating a prolongation of the latencies (implicit times) of the electrophysiologic responses to the light stimulus, in the retina with macular edema compared to the normal. Underneath the fundus images are placed time domain OCT scans of the normal and edematous retinas taken in the 180° meridian with overlying, corresponding mfERG responses. At the bottom of each figure is a three-dimensional pseudocolor map of mfERG P1 amplitudes (corresponding to the positive b-wave of a full-field ERG).

See also: Adaptive Optics; Blood–Retinal Barrier; Breakdown of the Blood–Retinal Barrier; Optical Coherence Tomography.

Further Reading

Aiello, L. P., Avery, R. L., Arrigg, P. G., et al. (1994). Vascular endothelial growth factor in ocular fluid of patients with diabetic retinopathy and other retinal disorders. *New England Journal of Medicine* 331: 1480–1487.

Al-Kateb, H., Mirea, L., Xie, X., et al. (2007). Multiple variants in vascular endothelial growth factor (VEGFA) are risk factors for time to severe retinopathy in type 1 diabetes: The DCCT/EDIC genetics study. *Diabetes* 56: 2161–2168.

Bearse, M. A., Jr., Adams, A. J., Han, Y., et al. (2006). A multifocal electroretinogram model predicting the development of diabetic retinopathy. *Progress in Retinal and Eye Research* 25: 425–448.

Diabetic Retinopathy Clinical Research Network (2008). A randomized trial comparing intravitreal triamcinolone acetonide and focal/grid photocoagulation for diabetic macular edema. *Ophthalmology* 115: 1447–1449.

Diabetic Retinopathy Clinical Research Network, Scott, I. U., Edwards, A. R., et al. (2007). A phase II randomized clinical trial of intravitreal bevacizumab for diabetic macular edema. *Ophthalmology* 114: 1860–1867.

Drexler, W. and Fujimoto, J. G. (2008). State-of-the-art retinal optical coherence tomography. *Progress in Retinal and Eye Research* 27(1): 45–88.

Early Treatment Diabetic Retinopathy Study Research Group (1985). Photocoagulation for diabetic macular edema, Early Treatment Diabetic Retinopathy Study Report No. 1. *Archives of Ophthalmology* 103: 1796–1806.

Frank, R. N. (2004). Medical progress: Diabetic retinopathy. *New England Journal of Medicine* 350: 48–58.

Frank, R. N. (2006). Etiologic mechanisms in diabetic retinopathy. In: Ryan, S. J., Schachat, A. P., Wilkinson, C. P., and Hinton, D. (eds.) *Retina,* 4th edn., ch. 66, pp. 1240–1270. London: Elsevier.

Gao, B. B., Clermont, A., Rook, S., et al. (2007). Extracellular carbonic anhydrase mediates hemorrhagic retinal and cerebral vascular permeability through prekallikrein activation. *Nature Medicine* 13: 181–188.

Gass, J. D. and Norton, E. W. (1966). Cystoid macular edema and papilledema following cataract extraction. A fluorescein fundoscopic and angiographic study. *Archives of Ophthalmology* 76: 646–661.

Huang, D., Swanson, E. A., Lin, C. P., et al. (1991). Optical coherence tomography. *Science* 254: 1178–1181.

Jampol, L. M., Sanders, D. R., and Kraff, M. C. (1984). Prophylaxis and therapy of aphakic cystoid macular edema. *Survey of Ophthalmology* 28(supplement): 535–539.

Schuman, J. S., Puliafito, C. A., and Fujimoto, J. G. (2004). *Optical Coherence Tomography of Ocular Diseases,* 2nd edn. Thorofare, NJ: Slack.

Tezel, T. H., Del Priore, L. V., Flowers, B. E., et al. (1996). Correlation between scanning laser ophthalmoscope microperimetry and anatomic abnormalities in patients with subfoveal neovascularization. *Ophthalmology* 103: 1829–1836.

The SCORE Study Research Group (2009). A randomized trial comparing the efficacy and safety of intravitreal triamcinolone with observation to treat vision loss associated with macular edema secondary to central retinal vein occlusion. The Standard Care vs Corticosteroid for Retinal Vein Occlusion (SCORE) Study Report 5. *Archives of Ophthalmology* 127: 1101–1114.

The SCORE Study Research Group (2009). A randomized trial comparing the efficacy and safety of intravitreal triamcinolone with standard care to treat vision loss associated with macular edema secondary to branch retinal vein occlusion. The Standard Care vs Corticosteroid for Retinal Vein Occlusion (SCORE) Study Report 6. *Archives of Ophthalmology* 127: 1115–1128.

Wojtkowski, M., Srinivasan, V., Fujimoto, J. G., et al. (2005). Three-dimensional retinal imaging with high-speed ultrahigh-resolution optical coherence tomography. *Ophthalmology* 112: 1734–1746.

Microvillar and Ciliary Photoreceptors in Molluskan Eyes

E Nasi and M del Pilar Gomez, Universidad Nacional de Colombia, Bogotá, Colombia

Glossary

Ciliary photoreceptors – Visual receptor cells in which the photosensitive region is derived from a cilium, a structure protruding from the cell body and characterized by the internal presence of nine longitudinal microtubules arranged in a circular fashion (and sometimes two additional microtubules in the center). Folding of the membrane of the cilium enhances its surface area to accommodate a great number of photopigment molecules.

Circumpallial nerve – Nerve that loops around the mantle of mollusks. It collects the axons from the eyes and projects to central neural ganglia of the animal.

Cyclic nucleotide-gated channels (CNG channels) – Family of ion channels that are ancestrally related to certain voltage-activated channels, but have evolved a different gating mechanism that responds to the binding of cGMP and cAMP to a site located in the intracellular side. They are responsible for generating the receptor potential in a variety of sensory cells, like olfactory neurons and many photoreceptors.

Diacylglycerol (DAG) – Product of the breakdown of phosphatidylinositol bisphosphate (PIP_2) by phospholipase C (PLC). This is the lipid moiety of the molecule and remains membrane bound. It is a prime activator of protein kinase C (PKC).

Guanylate cyclase (GC) – Enzyme responsible for generating the intracellular messenger cyclic guanosine monophosphate (cGMP), using guanosine trisphosphate (GTP) as substrate.

Inositol trisphosphate (IP_3) – Soluble, bioactive product of the breakdown of PIP_2, resulting from the cleavage of the polar head of this phospholipid. It can diffuse from the membrane to the endoplasmic reticulum (ER), where it binds to IP_3 receptors, causing them to expose a calcium-permeable pore and to release calcium contained within the endoplasmic reticulum (ER).

Microvillar photoreceptors – These are also called rhabdomeric photoreceptors. Photoreceptor cells in which the light-sensitive region presents evaginations of the cell membrane in the form of thin cylindrical protrusions, called microvilli. These are internally packed with longitudinal bundles of actin filaments that confer structural stability. The resulting increase in the surface area of the membrane makes it possible to accumulate a large population of photopigment molecules.

Phosphatidylinositol bisphosphate (PIP_2) – Minor phospholipid present in the inner leaflet of the cell membrane. Its enzymatic breakdown by phospholipase C (PLC) yields two bioactive products, namely inositol trisphosphate (IP_3) and diacylglycerol (DAG).

Phosphodiesterase (PDE) – Enzyme that cleaves a phosphodiester bond. In vertebrate photoreceptors, PDE breaks down cGMP, forming the inert compound 5′-GMP, thus leading to the closure of ion channels gated by cGMP.

Phospholipase C (PLC) – Enzyme that hydrolyzes inositol phospholipids in the membrane. A subclass, known as PLC-β, is activated by a guanine-binding protein (G-protein) of the G_q type.

Protein kinase C (PKC) – A class of enzymes capable of transferring a phosphate from an ATP molecule to a serine or threonine residue in a protein. Members of a subclass of these enzymes, known as conventional PKCs, are activated by calcium and/or diacylglycerol, two messenger molecules whose intracellular concentration increases in microvillar photoreceptors after stimulation with light.

pS, Pico-siemens (10^{-12} S) – A measurement unit of conductance. The conductance of a body is 1 S, such that upon applying 1 V across it, an electrical current of 1 A would flow.

Transient receptor potential (TRP), transient receptor potential like (TRPL) – Proteins first identified in the eyes of *Drosophila* as the ion channels subserving the late and the early phase, respectively, of the receptor potential. Genetic mutations leading to the failure to express TRP cause the receptor potential to consist of only the initial, transient phase, hence the name transient receptor potential (TRP).

Visual cells in the animal kingdom are usually partitioned into two distinct categories on the basis of the structure of the light-sensing organelle: in some photoreceptors, all the machinery necessary to absorb photons and generate a receptor potential is contained in a modified cilium; this is the case of the rods and cones in the vertebrate retina. In other cells, that function is subserved by microvilli, as it occurs, for example, in the photoreceptors of the compound eye of insects. In both designs, the folding of the membrane greatly increases its surface area to accommodate a large number of photopigment molecules – which are integral membrane proteins – thus conferring a high optical density to the cell, making it an efficient and compact light collector.

It is generally accepted that this dichotomy also marks a sharp boundary between the design of vertebrate versus invertebrate retinae. However, the eyes of several marine mollusks, such as those of *Pecten irradians* (**Figure 1**) and *Lima scabra* challenge such dogma, because they possess a double retina, comprised of microvillar cells in the proximal layer, and ciliary cells in the distal layer. Because each retinal layer gives rise to a separate branch of the optic nerve, which in turn produces either ON or OFF neuronal discharges in response to light, both cell types were thought to be visual receptors. This conjecture was corroborated by intracellular recordings of light-evoked responses in the intact retina. Surprisingly, these light responses have opposite polarities: proximal cells depolarize – as all other invertebrate photoreceptors were known to do – while distal cells hyperpolarize. Because those measurements were conducted under conditions designed to hamper synaptic transmission, it was unlikely that some of the light responses may have been evoked indirectly. Lingering doubts about the coexistence of functionally and structurally different primary visual receptors were dispelled by recordings of light-dependent changes in membrane voltage and membrane current in enzymatically isolated microvillar and ciliary cells.

Microvillar (Rhabdomeric) Photoreceptors

Excitation

The basic properties of the light response in microvillar visual receptors of mollusks are generally similar to those found in other invertebrate eyes. Light produces a depolarizing receptor potential, which is due to an increase in membrane conductance. Because there are no local second-order neurons, light information must be directly encoded by action potentials (**Figure 2**) propagating along the axons

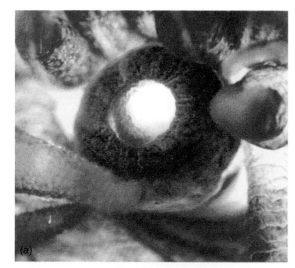

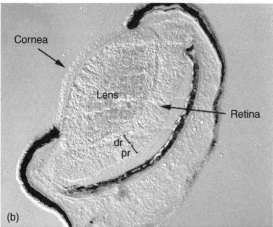

Figure 1 (a) The eye of the scallop, *Pecten irradian,* is of the simple type, possessing a single cornea and lens. The shiny appearance of the pupil is due to the presence of a reflector at the back of the eye, which helps focus light onto the retina.
(b) Cryosection of a fixed eye. The retina is comprised of two separate layers, proximal (pr) and distal (dr), each giving rise to a separate branch of the optic nerve.

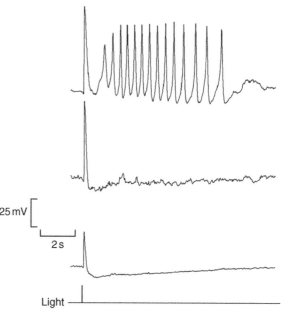

Figure 2 Current-clamp recording of the light response in an isolated microvillar photoreceptor of *Lima*. The membrane voltage was measured through a patch electrode in the whole-cell configuration. Flashes of 100-ms duration were presented, as indicated at the bottom, increasing their intensity at 0.6 log units (from bottom trace to top). A graded depolarization is evoked by light, eventually triggering one or more action potentials. The resting potential was ~–54 mV. Traces were offset vertically for clarity.

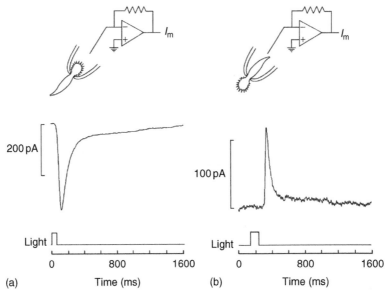

Figure 3 (a) Suction-electrode recording of light-evoked currents in isolated rhabdomeric photoreceptors of *Pecten* with the microvillar lobe inside the electrode. (b) Inverting the orientation of the cell in the pipette reverses the polarity of the recorded current. The active (inward) current is confined to the microvillar lobe of the cell.

that directly emanate from the photoreceptor cells and form the circumpallial nerve. The circumpallial nerve loops around the mantle of mollusks, collects the axons from the eyes, and projects to central neural ganglia of the animal. Voltage-dependent calcium currents contribute to the initiation of regenerative spikes, and, together with voltage- and calcium-dependent potassium currents, help shape the light response. The photocurrent is segregated to the photosensitive microvillar lobe, as clearly demonstrated by suction-electrode measurements (**Figure 3**). The light-sensitive conductance is cationic, permeable to sodium ions, and, to a lesser extent, to K (permeability ratio, $p_{Na}:p_K, \sim 1.8:1$). Additionally, there is a small contribution of calcium ions, which, however, is quantitatively minute (<3% of the light-evoked inward current). In this respect, these cells differ from *Drosophila*, where the Ca:Na selectivity ratio of light-dependent channels may be as high as 40:1. Such discrepancy may point to the evolution of different strategies for light-triggered calcium mobilization (a phenomenon universally observed in all microvillar receptors tested to date): in *Pecten* and *Lima*, like in some other species (e.g., *Limulus*), photo-induced Ca transients are impervious to the removal of extracellular calcium and reflect release from IP$_3$-sensitive intracellular stores; such a scheme would have little use for a Ca influx pathway. By contrast, other photoreceptors display minimal – if any – internal release and must therefore depend essentially on Ca-permeable channels at the plasma membrane.

The photo-induced calcium transients of *Pecten* and *Lima* are very large, best monitored with low-affinity fluorescent indicators like Fluo 5F and Calcium Green 5N ($K_D \sim 2.3$ and 14 μM, respectively; **Figure 4**); they are initiated

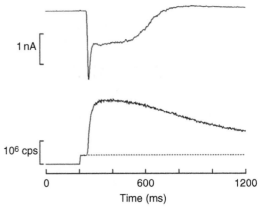

Figure 4 Simultaneous recording of membrane current under voltage clamp (top trace) and fluorescence of the indicator Calcium Green 5N, which was dialyzed through the patch pipette at a concentration of 75 μM (bottom trace). Emitted light was measured with a photomultiplier tube in photon-counting mode; the calibration bar at the bottom refers to counts per second (cps). A large increase in fluorescence above the initial level (indicated by the dashed line) occurred shortly after activating the epifluorescence beam, coincident with the beginning of the photocurrent.

in the light-transducing microvillar lobe, and precede the activation of the electrical response by ~1 ms. Blunting the light-triggered Ca changes with Ca buffers (internal 1,2-bis(o-aminophenoxy)ethane-N,N,N′,N′-tetraacetic acid (BAPTA) or high concentration of ethylene glycol-bis-(β-amino-ethyl ether) N,N,N′,N′-tetraacetic acid (EGTA)) interferes with the light response, dramatically reducing its amplitude and slowing down its kinetics. However, although cytosolic Ca levels are controlled by light and in turn exert a pivotal regulatory function, Ca does not, in all likelihood, directly activate light-dependent channels. Other events that follow phospholipase C (PLC) activation play a fundamental role; light-activated single-channel currents remain temporally viable in excised, perfused inside-out patches, pointing to membrane-bound signaling elements, rather than soluble messengers like inositol 1,4,5 triphosphate (IP_3) or Ca^{2+}. Diacylglycerol (DAG), the other, membrane-confined product of the breakdown of phosphatidylinositol-bisphosphate (PIP_2), is a plausible candidate. In *Lima*, a variety of DAG analogs, such as 2-dioctanoyl-sn-glycerol (DOG), phorbol 12-myristate 13-acetate (PMA), and (−)-indolactam, increase membrane conductance and evoke an inward current (**Figure 5**) with a similar ionic selectivity as the photoconductance; moreover, DAG analogs and light interact occlusively, suggesting that their effects converge onto a common target. Calcium elevation greatly potentiates and accelerates the effects of DAG analogs, an observation in line with the aforementioned effects of Ca manipulations. The generality of the role of DAG – or some metabolite thereof – in phototransduction remains to be systematically assessed in different species; moreover, even in *Lima*, the robust effects of DAG analogs fall quantitatively short of the speed and magnitude of the electrical response elicited by light. While the discrepancy could be ascribed to technical limitations in the application of chemicals to the cell, the participation of additional messenger molecules should not be ruled out. In other systems, phosphatidylinositol 4,5 bisphosphate

(PIP_2), long viewed simply as the substrate of phospholipase C-β (PLC-β) and the precursor of IP_3 and DAG, has emerged as a signaling molecule in its own right. Because PIP_2 levels in the rhabdomeric membrane are bound to drop with light-triggered activation of PLC, a direct participation of this phospholipid in light signaling would call for a negative messenger role, one that helps maintain some element(s) of the transduction cascade in the inactive state. In *Lima*, intracellular administration of PIP_2 specifically antagonizes the light-evoked current while sparing voltage-dependent currents. Moreover, in excised patches of *Pecten* rhabdomeric membrane screened for the exclusive presence of light-activated channels, functional depletion of PIP_2 by antibodies (to avoid confounding by the concomitant generation of its bioactive hydrolysis products) induces the appearance of single-channel currents, which can be silenced by exogenous replenishment of PIP_2. Thus, the visual excitation process may be complex, and involve the interplay of several signaling and modulator molecules, rather than constituting a linear cascade.

An additional complexity in the visual excitation scheme of molluskan microvillar receptors is the presence of separate populations of ion channels underlying the photocurrent, as subsequently corroborated in *Drosophila* where transient-receptor potential (TRP) and transient-receptor potential-like (TRPL) channels were molecularly identified. Two components of the macroscopic photocurrent of *Lima* can be distinguished by their time course and ion-conduction properties. In *Pecten*, cell-attached patch-clamp recordings in the exposed light-sensitive membrane reveal single-channel currents specifically activated by photostimulation (**Figure 6**) with a unitary conductance of ~48 pS and an additional population of smaller-amplitude currents ~18 pS. The identity of these single-channel currents as the constitutive elements of the macroscopic photoresponse was confirmed by the similarity of kinetics and light sensitivity, extending down to stimulation intensities that only evoke single-photon responses.

Light Adaptation

Modulation of sensitivity is a fundamental process in photoreceptor function, and enables the cell to maintain responsiveness over widely varying levels of ambient illumination. Once again, calcium had been singled out as a critical player as early as in the 1970s in *Limulus*, but its downstream effector(s) remained unclear. Protein kinase C (PKC) is a likely mediator, because some PKC subtypes are activated by calcium (and DAG), and also because in *Drosophila* it is associated with the macromolecular 'light-transduction complex'. Ca-dependent PKCα is selectively expressed in *Lima* eyes, as established by Western blot analysis using isoform-specific antibodies, and localizes in the light-sensing lobe of microvillar photoreceptors. Moreover, upon illumination it translocates from the cytosol to the membrane

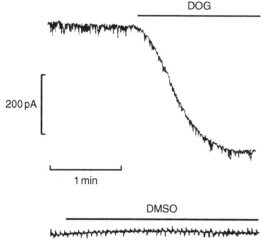

Figure 5 Diacylglycerol (DAG) analogs stimulate the light-sensitive conductance in *Lima* microvillar photoreceptors. Puffer-pipette application of the 2-dioctanoyl-sn-glycerol (DOG, 100 μM) activates an inward current, several hundreds of pA in amplitude, in a cell held under voltage clamp. Control application of dimethyl sulfoxide (DMSO) at the same concentration as that used to dissolve DOG is inert (bottom trace). The current activated by DOG and other diacylglycerol DAG analogs has the same ionic selectivity as the photocurrent; moreover, the analogs interact occlusively with light.

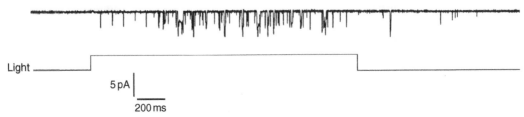

Figure 6 Cell-attached patch-electrode recording on the microvillar lobe of a *Pecten* rhabdomeric photoreceptor. Upon presenting a sustained, dim-light stimulus, single-channel inward currents are elicited. Two populations of light-activated currents are observed, one with a unitary conductance of $\sim\approx$48 pS and an additional population of smaller-amplitude currents $\sim$18 pS. No responses are produced by voltage stimulation in the dark (not shown).

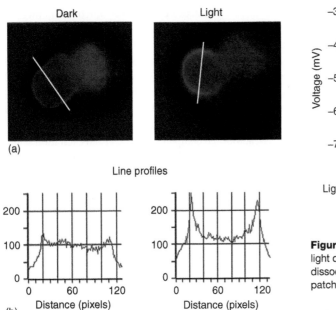

Figure 7 Light-induced translocation of PKCα.
(a) Photoreceptor dark adapted for 1 h and directly fixed in paraformaldehyde (left), or illuminated for 3 s just before fixation (right). Anti-PKCα antibodies revealed a different spatial pattern of immunofluorescence in the two conditions, whereas in control cells PKCα is largely distributed diffusely in the cytosol of the microvillar lobe, after light the fluorescence forms a much more pronounced ring around the edges and is nearly absent from the central portion. (b) Intensity profiles of the immunofluorescence, measured along a line cutting across the rhabdomeric lobe.

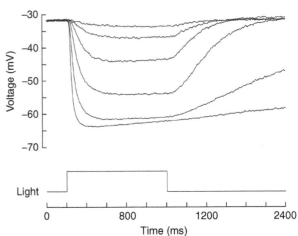

Figure 8 Hyperpolarizing receptor potentials elicited by a 1-s light of increasing intensity (top to bottom trace) in a dissociated ciliary photoreceptor of *Pecten*, measured with a patch electrode in whole-cell current-clamp mode.

(a functional assay of its activation), on a similar timescale as the onset of light adaptation (**Figure 7**). Chemical stimulation of PKC specifically depresses the light response, consistent with its role in desensitization, while pharmacological antagonists of PKC reduced light adaptation. These observations strongly support the involvement of PKC in the calcium-dependent regulation of response sensitivity.

Ciliary Photoreceptors

Photoreceptors of the distal retina function in a profoundly different way. Their resting potential is relatively depolarized ($\sim$−35 mV) owing to a high resting g_{Na}/g_K ratio, and illumination produces a hyperpolarizing receptor potential graded with light intensity (**Figure 8**). The light sensitivity is significantly lower than in proximal photoreceptors, but the purpose here is not range fractionation (unlike rods and cones of the vertebrate retinas). Instead, ciliary photoreceptors play a fundamentally different role from that of their microvillar counterparts: the information output ultimately entails action potentials, which, of course, could not be caused by light-induced hyperpolarization. It is a reduction of illumination that causes firing in these axons: the effect of light is to remove, in a time- and intensity-dependent way, the steady-state inactivation of voltage-dependent calcium channels, such that when illumination is decreased (e.g., when an approaching predator casts a shadow on the animal's visual field), the return to a depolarized membrane potential triggers a Ca spike. As such, these cells function as dark detectors and activation of the phototransduction cascade serves the function of priming them to respond to light dimming.

Excitation

The hyperpolarizing receptor potential of ciliary visual receptors also arises from an increase in membrane

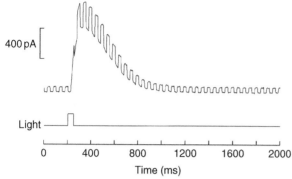

Figure 9 Light-induced increase in membrane conductance in a ciliary photoreceptor, assessed by superimposing a repetitive voltage step on the steady holding potential (10 mV, 100 Hz, and 0.5 duty cycle). When a light response was elicited, the size of membrane current perturbations grew several fold, indicating the opening of ion channels.

conductance (**Figure 9**). The reversal potential is ~-80 mV, close to the calculated value for E_K, the equilibrium potential for potassium, and exhibits a near-perfect Nernstian dependency on the concentration of extracellular potassium. Photostimulation, therefore, opens K-selective channels. Cell-attached patch recording on the ciliary appendages of the cell, presumably the light-sensitive organelles, reveals outwardly directed single-channel currents activated by light but not by voltage, with a unitary conductance ~26 pS.

Light-signaling in distal photoreceptors diverges sharply from that of their microvillar counterparts: the photoresponse is insensitive to both IP_3 and antagonists of the IP_3 receptor, and is also impervious to Ca elevation and to Ca buffering. Moreover, the guanine nucleotide-binding alpha Q protein ($G\alpha_q$) is not expressed in the distal photoreceptors of another member of the *Pectinidae*. These data strongly argue against the involvement of PLC – the canonical transduction cascade of invertebrate vision.

By contrast, substantial data have accumulated in support of a role for cGMP: intracellular application of cGMP or slowly hydrolyzing analogs (but not cAMP) elicits an outward current (**Figure 10(a)**) with similar ion-conduction properties as the light-evoked current: the reversal potential and its dependency on $[K]_o$ are the same (**Figure 10(a)**), and so is the characteristic outward rectification. Both the photocurrent and the current elicited by cGMP are inhibited by the same antagonists, such as l-*cis*-diltiazem. Furthermore, illumination and exogenous cGMP analogs interact occlusively. Activation of cGMP-dependent currents can be obtained in excised membrane patches, suggesting that cGMP operates directly on the channels. It can be concluded that cGMP is the internal final messenger for visual excitation, as it occurs in vertebrate photoreceptors.

Despite the structural and functional similarities of *Pecten* distal photoreceptors with rods and cones, a key difference places these receptors in a novel, distinct subcategory because the photoresponse is due to the opening, rather than the closing, of cGMP-gated channels. This implies that light must elevate cGMP levels, and this calls for an enzymatic machinery different from that of rods and cones. Several clues on the nature of this light-signaling pathway have emerged.

Photopigment, G Protein, and Arrestin

The molecular identity of the photopigment of ciliary receptors was elucidated in the Japanese scallop *Patinopecten yessoensis*, where a novel form of rhodopsin, dubbed SCOP2, was cloned and localized to the distal retinal layer by *in situ* hybridization. As in other invertebrates, the photopigment of scallop ciliary receptors is thermally stable upon illumination. The action spectrum, measured by both the late or the early receptor potential, peaks at 500 nm, and rhodopsin photoisomerization red-shifts its absorption curve by 75 nm; as a consequence, the fractional state of the pigment (rhodopsin (R)/metarhodopsin (M)) is a photoequilibrium that can be manipulated by varying the wavelength of illumination. A massive R to M conversion by blue-light illumination gives rise to prolonged hyperpolarizing after-potentials (or outward after-currents under voltage clamp) that can be reset by illumination with red light. Little is known about the detailed mechanisms that terminate the light response, but antibodies against bovine arrestin label a single band in Western blots of *Pecten* retinal homogenates and decorate ciliary photoreceptors both in cryosections of the eye and in dissociated retinas. Moreover, intracellular dialysis with the same antibodies slow down the falling phase of the photocurrent and allow prolonged after-currents to be elicited by spectrally neutral flashes, indicating that an arrestin-like molecule is implicated in visual excitation turn-off.

Seven transmembrane domain receptors (like rhodopsin), signal through a heterotrimeric G protein, and ciliary photoreceptors are no exception: GTP-γ-S, which interferes with G-protein deactivation, because it is resistant to the GTPase activity of $G\alpha$ and associated GTPase activating proteins (GAPs), causes the flash response to become sustained. Conversely, GDP-β-S inhibits phototransduction. However, the identity of the G-protein is unusual for visual cells: the only detectable $G\alpha$ form expressed in the distal retina was molecularly identified as a $G\alpha_o$, by Shichida and colleagues in *P. yessoensis*. A similar $G\alpha_o$, differing in a stretch of 22 amino acids but otherwise identical at the nucleotide level, has been cloned in *P. irradians*; this was also demonstrated to be confined to the layer of ciliary photoreceptors by *in situ* hybridization. Physiological and pharmacological data corroborate the participation of $G\alpha_o$ in light transduction: (1) mastoparan peptide activators of $G\alpha_o$ induce an outward current, which is suppressible by blockers of the light-sensitive conductance; (2) the light response is

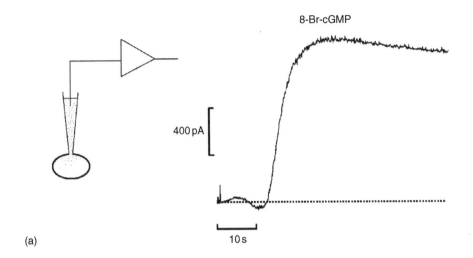

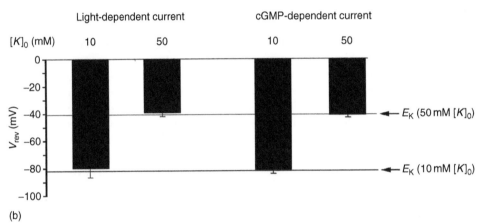

Figure 10 (a) Intracellular dialysis of an isolated ciliary photoreceptor cell with 20 μM of the cGMP analog, 8 bromo-cGMP (8Br-cGMP). Several seconds after rupturing the membrane patch to access the cell interior, a large outward current was evoked. (b) Similarity of the ion selectivity of the current elicited by light and by cGMP analogs. The reversal potential (V_{rev}) of the photocurrent was determined along with that of the current elicited by intracellular application of 8Br-cGMP. The measurements were conducted either in normal extracellular potassium (10 mM), or after elevating its concentration to 50 mM. In all cases, reversal potential V_{rev}, tracked exactly the predicted value of the K equilibrium potential, E_K.

inhibited by application of the A-protomer of pertussis toxin (the holotoxin moiety where the ribosyl-transferase activity resides), as predicted from the presence of a cysteine in the fourth position from the carboxy terminus of the $G\alpha_o$ sequence. This is the hallmark for susceptibility to ADP-ribosylation.

Guanylate Cyclase

The downstream mechanisms of visual excitation call for a light-induced elevation of cGMP, which could conceivably arise by either of two schemes: (1) inhibition of a phosphodiesterase (PDE) over a background of constitutive guanylate cyclase (GC) activity (i.e., the mirror image of the cGMP cascade of rods and cones) and (2) light-dependent stimulation of a cyclase (i.e., parallel to the cAMP cascade of olfactory neurons). Pharmacological antagonists of PDE fail

to mimic the effects of light (i.e., they do not directly activate the photoconductance, although some of them augment the amplitude of the light response). By contrast, GC antagonists reversibly inhibit the photoresponse, suggesting that light may control cGMP production, rather than degradation (i.e., scheme (2)). Such a notion, unprecedented for visual cells, parallels the well-established role of adenylate cyclase in ciliary neurons of the olfactory epithelium, where cAMP is the internal messenger. The putative light-regulated GC, however, must not be one of the canonical soluble (sGC) or membrane (mGC) forms. First, the light response is impervious to manipulations of the nitric oxide (NO) pathway, which suggests exclusion of a soluble GC. Second, changes in intracellular Ca concentration do not alter the photocurrent, which indicates an essential divergence with respect to the regulatory mechanisms that operate in vertebrate membrane GCs. Most importantly, like in all G-protein-coupled receptors, rhodopsin signals through a G-protein, and such a regulatory mechanism has not been reported by either class of GC. However, a new family of GCs has been uncovered in lower organisms (e.g., *Plasmodium*, *Paramecium*, and *Dictyostelium*), with the same topology as the G-protein-regulated adenylyl cyclase (class III) of olfactory neurons. In *Dictyostelium*, it has been reported that its activity is regulated by a heterotrimeric G-protein of the G_o subtype. G protein dependency for GC activity has recently been documented in Leydig tumor cells and may thus not be confined to protozoa. In *Pecten* retinal lysates, polyclonal antibodies against the multitransmembrane domain GC of *Paramecium* label a distinct band, with an apparent molecular mass of $\sim$240 kDa, similar to that of the multitransmembrane domain GCs, and greatly exceeding that of either soluble (70–82 kDa) or other membrane-bound GCs (up to 140 kDa).

Light-Dependent Ion Channels

The molecular identity of the ion channels underlying the receptor potential of ciliary photoreceptors has yet to be established. Interestingly, in Western blots of *Pecten* retinal lysates, antibodies raised against CNG-2 (the α-subunit of the transduction channel of olfactory neurons) label a single band of the appropriate apparent molecular mass, 73 kDa; by contrast anti-CNG-1, the vertebrate retina form, and anti-CNG-3 (which is expressed in mammalian heart, kidney, and sperm) produce no signals. In whole-eye cryosections, the same antibodies selectively decorate the distal layer of the retina, where ciliary photoreceptors are found. By confocal fluorescence microscopy in dissociated cells, the target was localized in the ciliary appendages, presumed to be the light-transducing organelles. The results suggest the presence of an olfactory-like CNG channel in distal photoreceptors, with a subcellular distribution compatible with a role in visual excitation.

The light-dependent channels of distal photoreceptors are uniquely interesting because of their gating and ion-selectivity properties. It has long been known that CNG channels are homologous to some voltage-gated K channels, and a common evolutionary origin has been proposed. Nonetheless, the two classes differ sharply in terms of ion permeation, which in CNG channels is characteristically cationic nonselective. The light-dependent channels of hyperpolarizing invertebrate photoreceptors are remarkably similar to both voltage-gated K channels and CNG channels: on the one hand, they are strongly selective for potassium and highly susceptible to blockage by certain K-channel antagonists like 4-aminopyridine. On the other hand, they are gated by cGMP and blocked by various antagonists of the light-sensitive conductance of rods, such as l-*cis*-diltiazem. As such, they seemingly constitute a missing link, bridging the gap between these two super-families of ion channels. This raises a question about the origin of their gating mechanism. In *Pecten*, the light-dependent K conductance exhibits a pronounced outward rectification due to voltage-dependent occlusion of the permeation pathway by Ca^{2+} and Mg^{2+} ions, which bind at a site located about half-way through the membrane (electrical distance $\delta \sim 0.6$ from the external surface). Blockage by Ca^{2+} and Mg^{2+} requires an open pore, and the channels can close with a divalent ion trapped inside. These observations suggest that the cGMP-controlled gate must reside near the extracellular side of the channel protein, in sharp contrast with the intracellularly located gate of its voltage-dependent K channel relatives.

Light Adaptation

Because the primary function of ciliary cells is to produce an OFF discharge upon the dimming of a continuous light, the photoresponse is bound to have a prominent sustained component. Nonetheless, during prolonged light stimulation the photocurrent decays to a plateau, and background illumination or conditioning flashes produce all the classical manifestations of light adaptation: shift in the sensitivity curve, compression of the response amplitude range, and acceleration of response kinetics. However, the underlying modulatory mechanisms operate in an unusual way: the lack of a detectable Ca permeability of the light-activated channels and of a functional IP_3 signaling pathway implies that light stimulation is not coupled to either influx or internal release of calcium. In fact, unlike all other known photoreceptors, fluorescent Ca indicators report no discernible light-induced changes in cytosolic calcium. As a consequence, this ion would not be in a position to play a significant role in light adaptation. Not surprisingly, direct manipulations of intracellular Ca, either buffering it with the rapid Ca chelator BAPTA, or, conversely, elevating it to μM levels fails to significantly change basal light sensitivity or to alter adaptation. The Ca-independent signaling pathway responsible for light adaptation appears to implicate cGMP, the same messenger that governs visual excitation: application of cGMP analogs not only activates the photoconductance, but, on a slower timescale, also depresses the light response to an extent that far exceeds what one would expect from the decreased pool of available channels (i.e., a simple competition for a common effector mechanism). This excess reduction of the photoresponse amplitude is accompanied by a shift in the sensitivity curve and acceleration of response kinetics, the hallmark signs of light adaptation (**Figure 11**). Tests with pharmacological antagonists indicate that the changes in sensitivity during light adaptation mediated by cGMP may be in part controlled by a cGMP-dependent protein kinase.

In summary, ciliary photoreceptors found in the retina of several bivalve mollusks diverge sharply from classical (microvillar) invertebrate photoreceptors, and partake

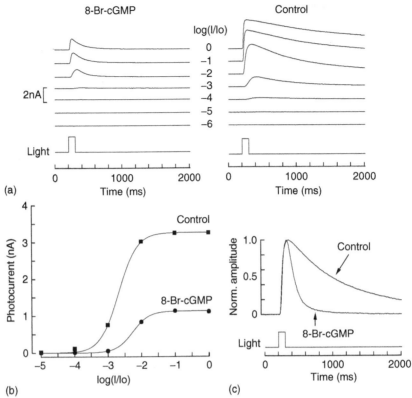

Figure 11 Photoresponse desensitization by cGMP analogs. (a) Current evoked by flashes of increasing intensity (bottom to top trace) delivered to ciliary photoreceptors internally dialyzed with 20 μM 8-Br-cGMP (left) vs. control solution (right). Log intensity for each trace indicated between two sets. (b) Intensity-response relation in the two cases in (a). In addition to the compression of response amplitude, 8-Br-cGMP shifts the curve to the right. (c) Normalized photocurrents (at –2 log) in control conditions vs. 8-Br-cGMP, highlighting the acceleration of the response decay.

instead of several morphological and structural features of vertebrate rods and cones; nonetheless, they utilize a fundamentally different cascade both for light transduction and for light adaptation, warranting their inclusion in a novel separate class of light-transducing cells. The parallelism with the odor-transduction cascade of olfactory neurons suggests a common lineage with an ancestral chemoreceptor cell.

See also: Genetic Dissection of Invertebrate Phototransduction; Phototransduction: Inactivation in Rods; Phototransduction in *Limulus* Photoreceptors; Phototransduction: Phototransduction in Cones; Phototransduction: Phototransduction in Rods.

Further Reading

Barber, V. C., Evans, E. M., and Land, M. F. (1967). The fine structure of the eye of the mollusk *Pecten maximus. Zeitschrift für Zellforschung und Mikroskopische Anatomie* 76: 295–312.

Cornwall, M. C. and Gorman, A. L. F. (1979). Contribution of calcium and potassium permeability changes to the off response of scallop hyperpolarizing photoreceptors. *Journal of Physiology* 291: 207–232.

Gorman, A. L. F. and McReynolds, J. S. (1969). Hyperpolarizing and depolarizing receptor potentials in the scallop eye. *Science* 165: 309–310.

Gomez, M. and Nasi, E. (1994). The light-sensitive conductance of hyperpolarizing invertebrate photoreceptors: A patch-clamp study. *Journal of General Physiology* 103: 939–956.

Gomez, M. and Nasi, E. (1995). Activation of light-dependent potassium channels in ciliary invertebrate photoreceptors involves cGMP but not the IP3/Ca cascade. *Neuron* 15: 607–618.

Gomez, M. and Nasi, E. (1998). Membrane current induced by protein kinase C activators in rhabdomeric photoreceptors: implications for visual excitation. *Journal of Neuroscience* 18: 5253–5263.

Gomez, M. and Nasi, E. (2000). Light transduction in invertebrate hyperpolarizing photoreceptors: Involvement of a G_o-regulated guanylate cyclase. *Journal of Neuroscience* 20: 5254–5263.

Gomez, M. and Nasi, E. (2005). A direct signalling role for PIP2 in the visual excitation process of microvillar receptors. *Journal of Biological Chemistry* 280: 16784–16789.

Gomez, M. and Nasi, E. (2005). Calcium-independent, cGMP-mediated light adaptation in ciliary photoreceptors. *Journal of Neuroscience* 25: 2042–2049.

Kojima, D., Terakita, A., Ishikawa, T., et al. (1997). A novel G_o-mediated phototransduction cascade in scallop visual cells. *Journal of Biological Chemistry* 272: 22979–22982.

Nasi, E. (1991). Two light-dependent conductances in the membrane of *Lima* photoreceptor cells. *Journal of General Physiology* 97: 55–72.

Nasi, E. and Gomez, M. (1992). Light-activated ion channels in solitary photoreceptors from the eye of the scallop *Pecten irradians. Journal of General Physiology* 99: 747–769.

Nasi, E. and Gomez, M. (1999). Divalent cation interactions with light-dependent K channels: Kinetics of voltage-dependent block and requirement for an open pore. *Journal of General Physiology* 114: 653–671.

Piccoli, G., Gomez, M., and Nasi, E. (2002). Role of protein kinase C in light adaptation of microvillar photoreceptors. *Journal of Physiology* 543: 481–494.

Morphology of Interneurons: Amacrine Cells

E Strettoi, Istituto di Neuroscienze CNR, Pisa, Italy

Glossary

Coverage factor – The product of mean dendritic field size and cell density. It gives a measure of the number of neurons of a particular type whose receptive fields overlap a particular point on the retina. Because of their great varieties in shapes and frequencies, amacrine cells of different types can have coverage factors anywhere between 1 and 500.

Retinal mosaic – The ordinate spatial pattern formed over the retinal surface by cells of a certain type, arranged regularly. Each retinal cell type has a unique mosaic that can be described formally in mathematical terms. Starburst amacrine cells tile the retinal surface very regularly, while dopaminergic amacrines have highly irregular mosaics.

Near the end of the nineteenth century, Santiago Ramon y Cajal named amacrine cells those retinal neurons whose cell bodies occupy the innermost tier of the inner nuclear layer (INL) and that ramify at various depths in the inner plexiform layer (IPL). Literally, their name means cells without an axon. From Cajal's initial observations and following early studies based on Golgi-impregnated neurons, the notion emerged that amacrine cells come in all shapes, sizes, and stratification patterns (**Figure 1**). In time, many morphological types were described thanks to the development of various techniques. These included intracellular recordings followed by dye injections, immunocytochemical staining, assorted anatomical tracing methods and, more recently, transgenic expression of fluorescent molecules. At present, it is accepted that the retina of mammals contains more than 26 varieties of amacrine cells; their catalog can be considered substantially complete. Classification schemes with a tendency to emphasize subtle differences among presumptive categories separate amacrine cells in as many as 40 different types. Surely enough, amacrine cells represent the most diverse cell types in the retina, as they comprise more varieties than bipolar and ganglion cells.

In the IPL, amacrine cells receive their excitatory input from bipolar cells and provide inhibitory output onto the dendrites of both other amacrine and ganglion cells. They also establish inhibitory synapses onto the axonal arborizations of bipolar cells. Functionally, amacrine cells are capable of modulating the activity of ganglion cells by direct inhibition, or by inhibiting the activity of bipolar cells that carry excitatory inputs to the inner retina. Many amacrine cells are also coupled by means of gap junctions.

Coupling can be with amacrine cells of the same (homologous) or different (heterologous) type. In addition, amacrine cell processes can be coupled to ganglion cell dendrites. Although many amacrine cells, such as the AIIs, generate graded signals in response to light, it has been shown that certain amacrine cells are capable of producing true action potentials that initiate locally and then propagate into every dendrite, exciting the entire cell. Thus, each amacrine cell can mediate both local as well as long-range lateral inhibition, regulating the spatial and temporal pattern of synaptic outputs from its dendrites.

Despite their name, certain wide-field amacrine cells have long, axon-like processes which probably function as true axons as they represent output fibers of the cell. These particular amacrines have been described as polyaxonal and their soma are often found in an interstitial position, that is, within the IPL. However, their long processes remain confined within the retina and do not contribute to the optic nerve as do the axons of ganglion cells.

Traditionally, amacrine cells have been divided into the broad categories of narrow-field (30–150 μm), small-field (150–300 μm), medium-field (300–500 μm), and wide-field (>500 μm) cells, based on the size of their dendritic field diameters. It is worth pointing out that the dendritic spread of the cells is correlated quite strongly with their degree of stratification. Hence, wide-field cells are highly stratified, medium-field cells are less so, and almost all narrow-field cells span radially across two or more levels of the IPL (**Figure 2**).

The mere size of their dendritic field, however, cannot account for the great variability in the morphology of these interneurons. Multiple classification schemes coexist: some amacrine cells have traditionally maintained numbers from old, partial nomenclatures (i.e., AII amacrines and A17 amacrines), while others are indicated with names reminiscent of their shapes, such as starburst cell, fountain amacrine, flag amacrine, or spider cell. As for bipolar and ganglion cells, a more functional, and therefore relevant criterion of identification involves knowing the stratification level of the cells. It is well known that the IPL can be subdivided into five equally thick strata or sublayers to which amacrine, bipolar, and ganglion cell processes can be assigned. The level of stratification of a given neuron in the IPL is predictive of the polarity of the cell response to light stimuli, so that neurons confined in the outermost two layers of the IPL, the so-called sublamina a, are excited when the light goes off, while neurons ramified in the innermost three tiers of the IPL, the sublamina b, are excited at the light onset. This strong correlation between anatomy and function makes it convenient to classify neuronal types primarily according to the strata of the IPL in which their processes are confined. Altogether, the combination of dendritic field size, pattern, and depth of

stratification within the IPL are sufficient criteria to separate cells into unique types. Supplementary information includes dendritic caliber, size and distribution of varicosities, and the course of dendrites.

The fact that branching of individual amacrine cells takes place at various levels in the IPL implicates that different amacrine cells contact different types of bipolar and ganglion cells and end up having different functional properties. Indeed, amacrine cells are major players in the retina's processing of visual information. They constitute at least 40% of all neurons in the INL of mammalian retinas and contribute to 64–87% of all synapses in the IPL according to the species. While it was once generally believed that primate retinas were more bipolar dominated as opposed to lower-mammal (i.e., rabbits) retinas, considered more amacrine dominated, quantitative morphology demonstrated that these cells represent some 40% of the INL cells in rabbits, mice, and primates.

Neurotransmitter immunocytochemistry (based on the localization of amino acids or rate-limiting enzymes) has shown that amacrine cells either contain gamma-aminobutyric acid (GABA) or glycine, two inhibitory amino acids used as neurotransmitters ubiquitously in the central nervous system. Remarkably, GABAergic amacrine cells are morphologically wide-field cells, while glycinergic amacrines are small- and medium-field neurons. Glycinergic amacrines are slightly more numerous, at least in the rabbit retina, in which they account for some 56% of all the amacrines. This fits well with the percentage of small-field amacrines (55%) revealed by the technique

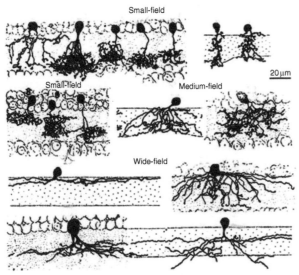

Figure 1 Small-, medium-, and wide-field amacrine cells from the human retina. Golgi silver impregnation. Modified from Poljak, S. (1941). The Retina.

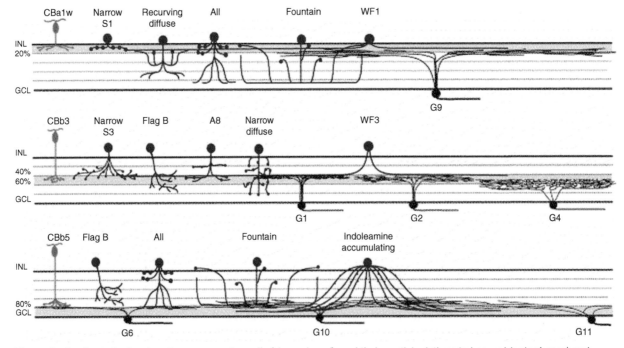

Figure 2 A schematic drawing of some amacrine cells (shown in red), and their spatial relations to known bipolar (green) and ganglion (blue) cell types. From Masland, R. H. (2004). Neuronal cell types. *Current Biology* 14, R497–R500.

of random photofilling. Glycinergic amacrines, with their narrow fields and little overlapping dendrites, mediate radial inhibition in the retina. These cells are meant for vertical transmission rather then for modulation of light signals.

GABAergic amacrines, instead, are more designated for later inhibition, a function traditionally ascribed to amacrine cells in general. In GABAergic amacrines, the fast neurotransmitter GABA usually coexists with another neurotransmitter or neuromodulator; this is typically a peptide, such as somatostatin, substance P, vasoactive intestinal peptide (VIP), and so on. Because they contain high levels of one amino acid, amacrine cells can be identified through their typical amino acidic signature in appropriately stained histological sections of the retina. Noticeably, cholinergic (starburst) amacrine cells are stained with methods revealing both GABA and acetylcholine and are known to corelease both transmitters.

Each particular point of the retina is covered by the dendritic trees of cells of the same type, overlapping to different extents. The parameter indicating the degree of overlapping is known as the coverage factor. This is obtained essentially by histological measurements and is defined as the product of cell density and area of the dendritic tree. A coverage factor of 10 means that a retinal point is covered by the dendritic trees of 10 cells of a given type. Amacrine cells have highly variable coverage factors. In general, small-field amacrine cells have lower coverage factors than wide-field amacrine cells. The well-known indoleamine-accumulating cells (IACs), which comprise two varieties of wide-field amacrines, narrowly stratified in the deepest stratum of the IPL, have coverage factors of 500–900. This means that in each point of the retina, a stack of processes from overlapping IAC cells are regularly piled in sublamina 5 of the IPL. Low-power electron microscopy shows bundles of IAC cells fasciculating in sublamina 5, running parallel to each other and interrupted by the large varicosities of rod bipolar axonal endings, also located at the border with ganglion cell bodies.

Individual types of retinal neuron exhibit regular spacing, such that cells of a certain type maintain a minimum distance from other neurons of the same type. Each cell is surrounded by an exclusion zone from which other cells of the same type are barred. On the contrary, spacing of neurons of different types is completely casual. Amacrine cells adhere to this rule. Because of the availability of cell-specific markers, the mosaics of certain amacrine cells have been described in detail. Starburst amacrine cells can be stained by antibodies against choline acetyl trasferase (ChAT), the rate-limiting enzyme in acetylcholine synthesis; their mosaic is highly characteristic. Similarly, AII amacrine cells can be labeled with antibodies against the protein disabled 1, or, in certain mammals including humans, by parvalbumin antibodies. Dopaminergic amacrines (historically, the first to be described, using enzyme histochemistry) can be revealed by antibodies against tyrosine hydroxylase. Each of these amacrine cells forms over the retinal surface a peculiar, well-recognizable mosaic that can be described formally in rigorous, mathematical terms. Whatever the mechanism leading to mosaic formation during retinal development might be, the regularity of a mosaic is a powerful means to assess that a given cell has been properly identified and correctly assigned to a type. Mutations in various genes controlling retinal development can alter mosaic formation and lead to retinal structural abnormalities.

Multidisciplinary studies on the currently well-known amacrine cell types have led to the notion that, as for other retinal neurons, diversity of shape means diversity of function. This concept is supported by a number of considerations: first, cells whose morphological features are very different have different membrane properties associated with the size of the dendritic tree, the caliber of individual dendrites, and the number and position of input and output synapses with respect to the cell soma. In addition, amacrine processes that occupy different sublayers of the IPL communicate with different sets of bipolar and ganglion cells. Since connectivity shapes function, the functional properties of amacrine cells with different patterns of connections must be different. Also, the area of the dendritic field (which varies greatly among various cells types) is strictly related to the retinal sampling capability (or area of visual space) of a given cell. A different coverage factor also reflects different sampling rates of cells.

The following sections describe two types of amacrine cells whose morphological and functional properties have been studied in detail and appear extremely different. They can be considered as paradigmatic of amacrine-cell-variegated structural and functional aspects.

AII Amacrine Cells

These small-field amacrines were first described in the cat retina by Kolb and co-workers in 1978 and can be considered as hallmark components of the retina of mammals. They bear a distinctive morphology as their dendrites are organized in two different layers: an ovoidal cell body gives rise to a thick, primary dendrite producing two orders of branches – a series of long, thin processes mostly restricted in sublamina 5 of the IPL, near the cell bodies of ganglion cells; and a second set of globose varicosities, called lobular appendages, clustered in a bushy and narrow ramification, spanning vertically through sublaminae 1–2 of the IPL (**Figure 3**). Electron microscopy has demonstrated that AII amacrine cells are placed in a pivotal position along the rod pathway. Scotopic signals transmitted from rods to rod bipolar cells (which belong to the category of ON-center, depolarizing neurons) are conveyed through sign-conserving, glutamatergic synapses to the vitreal

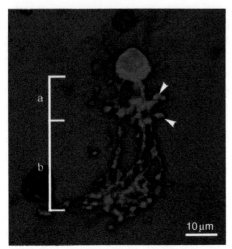

Figure 3 An individual AII amacrine cell from the mouse retina labeled after delivering the fluorescent probe DiI with a gene gun. Note the typical bistratified morphology, with lobular appendages (arrowheads) confined in the outermost part of the inner plexiform layer (IPL), the sublamina a. Long and thin processes of the same amacrine reach the innermost portion of the IPL, the sublamina b.

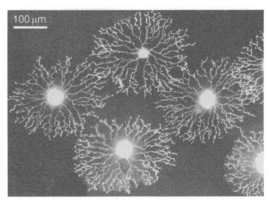

Figure 4 Example of three adjacent starburst amacrines stained by intracellular injection with Lucifer yellow. Reproduced from Vaney, D.I. (1999). Neuronal coupling in the central nervous system: Lessons from the retina. *Novartis Foundation Symposium* 219: 113–125.

dendrites of AII amacrine cells in sublamina 5 of the IPL. At this point, a dichotomy is generated in the rod pathway, as the signal is literally split into two channels by means of the connections of AII amacrines. These cells establish sign-conserving gap junctions with the axonal arbors of cone bipolar cells located in sublamina ON of the IPL, and glycinergic, sign-inverting synapses with axonal endings of cone bipolar cells terminating in sublamina OFF of the IPL. Finally, rod-initiated signals are transferred to the ganglion cell through sign-conserving chemical synapses established by axonal endings of cone bipolar cells in the ON and OFF halves of the IPL. Therefore, the rod pathway ultimately makes use of the axonal endings of cone bipolar cells to gain access to ganglion cells. The five-neuron chain in which the main rod pathway is organized is usually called piggyback arrangement, to indicate the fact that rod-generated signals use a common route, represented by cone bipolars, to exit the retina. AII amacrines are key elements of the piggyback pathway: because of their peculiar morphology and connectivity, they occupy a strategic position when illumination conditions switch from the scotopic range, in which rod photoreceptors are active, to the photopic range, in which cones become functional. AII amacrines, therefore, have to be informed about adaptation in the inner retina. This information is provided, among others, by dopaminergic amacrine cells, which play a relevant role in adaptation processes in the neural retina, and which provide a rich innervation of the primary dendrites of AII amacrine cells in sublamina 1 of the IPL.

Because of their synaptic circuitry, cells such as the AII do not conform to the general concept of amacrine cells as laterally placed, modulatory elements of the retinal connectivity, deputed to lateral transmission of electric

signals. Within the rod pathway, AII amacrines occupy a vertical position and rod-generated signals cannot enter ganglion cells without this strategically placed neuronal type. Hence, in very dim light, the AII amacrine is an obligatory connection in the retina's through-pathway. Generally speaking, small-field amacrine cells introduce little more lateral conduction than bipolar cells do; besides the AII, the function of other small-field amacrines is yet to be clarified.

As rod photoreceptors are numerous, AII amacrine cells are also abundant in the retina. Indeed, they are the largest population of amacrine cells accounted for, reaching 11–13% of all the amacrines in the retina of rabbits and mice. The remaining amacrine cell types come at a lower frequency, each reaching 3% maximum of the total population.

Starburst Amacrines

Starburst amacrine cells, so defined thanks to the shape of their circular and rich dendritic tree, are the second, most numerous amacrine cells in mammals, accounting for about 3% of the whole amacrine population. They are a recurrent finding of vertebrate retinas, from dogfish to primates. These neurons have a characteristic, radially symmetric morphology, with higher-order dendritic branches emerging from dichotomous ramification of thicker fibers. In the rabbit retina, their dendritic field size is approximately 400 μm and therefore they belong to the category of wide-field cells (**Figure 4**). It has been shown that starburst amacrines overlap extensively and their dendrites occupy a large percentage of the volume of the IPL. In the rabbit, calculations show that each millimeter of IPL is covered by 6 m of dendrites from starburst amacrine cells, so that the whole retina contains some 2 km of processes from this neuronal type.

Starburst amacrines occur in two, mirror-symmetric populations, composed of almost equal numbers of cells. In the first group, cell bodies reside in the innermost tier of the INL and dendrites form a narrow plexus in sublamina 2 of the IPL.

These are OFF-starburst cells. In the second population, cell bodies are located in the ganglion cell layer instead and therefore contribute to the heterogeneous group of displaced amacrine cells. Displaced starbursts can be easily identified in the ganglion cell layer even with simple nuclear staining because of the smaller nuclear size and visible regularity of their mosaic. Their dendrites form a second plexus in sublamina 4 of the IPL and are therefore ON-starbursts. Because of their early formation during retinal development and precise positions in the ON and OFF halves of the IPL, the two tiers of starburst dendrites (also known as cholinergic bands) are landmarks often used as references to define the stratification levels of other neurons in the IPL.

In the first days of postnatal development, when retinal neurons are being generated and assembled in regular tiers, waves of light-independent electrical activity traverse the inner retinal surface intermittently. Retinal waves can be recorded from the ganglion cell layer in the form of correlated activity of these neurons. Waves are thought to play an instructive role in the formation of topographic maps of projection neurons in the central-most visual areas. Pharmacological experiments indicate that developmental waves require the presence of starburst amacrines, releasing both GABA and acetylcholine, the latter acting on nicotinic receptors. The action of both neurotransmitters is excitatory, since GABA has an excitatory role during development. During retinal maturation, starburst cells communicate directly with each other, so that electrical activity generated at one retinal location can propagate to distant areas. However, during the subsequent developmental stages, excitation between starbursts becomes almost undetectable while GABAergic synapses among starburst cells switch from being excitatory to inhibitory.

Starburst cells participate in transretinal waves during early development but their role is totally different in the adult retina. The exact cofasciculation of starburst processes and the dendrites of a highly distinctive ganglion cell type, the ON-OFF directional selective (DS) ganglion cell, which senses stimuli moving in one direction, led to the hypothesis that, in the adult retina, starburst amacrines could provide a major source of synaptic input to this category of neurons. A wealth of data generated the notion that acetylcholine and starburst amacrines in particular, do contribute to the functional properties of directional selectivity. A current view of this complex problem that has fascinated physiologists for decades is that starburst amacrines are capable of releasing the neurotransmitter in a directional fashion, contributing to the tuning of DS ganglion cells for the preferred direction of motion.

Only for a few other types of amacrine cells has a clear role been elucidated. Among them, dopaminergic amacrines, also known as interplexiform cells, have a crucial function in neural adaptation and circadian rhythms; A17 amacrine cells, a long-time known type of wide-field amacrine, are typical local interneurons, providing feedback inhibition to the axonal ending of rod bipolar cells. For most of the amacrine cell types, however, the role is simply inferred on the basis of their costratification with other retinal neurons. Obviously, the fact that our inventory of amacrine cells is presumably complete represents only the first stage in the comprehension of the role of these interneurons. The next step toward a true understanding of retinal architecture will be to learn synaptic and functional interactions of the amacrine cell with individual types of bipolar and ganglion cells, constituting parallel pathways across the retina – a challenge for future years.

See also: Information Processing: Amacrine Cells; Morphology of Interneurons: Interplexiform Cells.

Further Reading

Baccus, S. A. (2007). Timing and computation in inner retinal circuitry. *Annual Review of Physiology* 69: 271–290.
Demb, J. B. (2007). Cellular mechanisms for direction selectivity in the retina. *Neuron* 55: 179–186.
Galli-Resta, L. (2002). Putting neurons in the right places: Local interactions in the genesis of retinal architecture. *Trends in Neuroscience* 25: 638–643.
Jeon, C. J., Strettoi, E., and Masland, R. H. (1998). The major cell populations of the mouse retina. *Journal of Neuroscience* 18: 8936–8946.
MacNeil, M. A., Heussy, J. K., Dacheux, R. F., Raviola, E., and Masland, R. H. (1999). The shapes and numbers of amacrine cells: Matching of photofilled with Golgi-stained cells in the rabbit retina and comparison with other mammalian species. *Journal of Comparative Neurology* 413: 305–326.
MacNeil, M. A. and Masland, R. H. (1998). Extreme diversity among amacrine cells: Implications for function. *Neuron* 20: 971–982.
Marc, R. E., Murry, R. F., Fisher, S. K., et al. (1998). Amino acid signatures in the normal cat retina. *Investigative Ophthalmology and Visual Science* 39: 1685–1693.
Masland, R. H. (2004). Neuronal cell types. *Current Biology* 14: R497–R500.
Masland, R. H. (2005). The many roles of starburst amacrine cells. *Trends in Neuroscience* 28: 395–396.
Masland, R. H. and Raviola, E. (2000). Confronting complexity: Strategies for understanding the microcircuitry of the retina. *Annual Review of Neurosciences* 23: 249–284.
O'Malley, D. M., Sandell, J. H., and Masland, R. H. (1992). Co-release of acetylcholine and GABA by the starburst amacrine cells. *Journal of Neuroscience* 12: 1394–1408.
Sharpe, L. T. and Stockman, A. (1999). Rod pathways: The importance of seeing nothing. *Trends in Neuroscience* 22: 497–504.
Strettoi, E. and Masland, R. H. (1995). The organization of the inner nuclear layer of the rabbit retina. *Journal of Neuroscience* 15: 875–888.
Taylor, W. R. and Vaney, D. I. (2003). New directions in retinal research. *Trends in Neuroscience* 26: 379–385.
Torborg, C. L. and Feller, M. B. (2005). Spontaneous patterned retinal activity and the refinement of retinal projections. *Progress in Neurobiology* 76: 213–235.
Vaney, D. I. (1999). Neuronal coupling in the central nervous system: Lessons from the retina. *Novartis Foundation Symposium* 219: 113–125.

Morphology of Interneurons: Bipolar Cells

S Haverkamp, Max-Planck-Institute for Brain Research, Frankfurt/Main, Germany

Glossary

Clomeleon – A ratiometric genetically encoded indicator comprising a fusion of cyan and yellow fluorescent protein that allows noninvasive chloride measurements in living tissue.

Dyad – Synaptic arrangement of the bipolar cell output synapses within the inner plexiform layer of the retina; the presynaptic ribbon of a dyad is surrounded by vesicles and two postsynaptic elements.

Genuine S-cones – In ventral mouse retina, the great majority of cones express both medium-wavelength (M) and short-wavelength (S) opsin. Blue cone bipolar cells contact only those cones, which express S-opsin only. These are the genuine S-cones.

Landolt club – Process that extends distally from many bipolar cells in cold-blooded vertebrates from the outer plexiform layer to the outer limiting membrane. The function of the Landolt club is not known in any retina.

Midget bipolar cell – The term midget refers to the small spread of their dendritic and axonal arbors. The one-to-one relationship between midget bipolar cells and cones on the dendritic end, and between midget bipolar cells and ganglion cells on the axonal end, is a distinctive property of midget bipolar cells around the fovea.

Type – Members of each cell type show very similar properties, that is, release the same transmitter, make the same connections to other cell types, and generally have the same morphology.

Introduction

The retina contains a large diversity of individual cell types, each carrying out a specific set of functions. They are mainly defined by their morphological appearance. Bipolar cells differ in their dendritic branching pattern, the number of photoreceptors contacted, and the shape and stratification level of their axons in the inner plexiform layer. The cells transfer the light signals from the photoreceptors to amacrine and ganglion cells. They can be subdivided, according to their light responses, into ON and OFF bipolar cells. This functional dichotomy is the result of the expression of different glutamate receptors at the synapses between photoreceptors and bipolar cell dendrites. OFF bipolar cells make flat or basal contacts with cone pedicles, ON bipolar cells make invaginating contacts. The axon terminals of OFF cone bipolar cells terminate in the outer half of the IPL and synapse with the dendrites of OFF ganglion cells, whereas those of ON bipolar cells terminate in the inner half of the IPL and contact the dendrites of ON ganglion cells.

Bipolar Cell Types of the Mammalian Retina

Bipolar cells of the mammalian retina can be subdivided into many different morphological types (**Figure 1**). There are at least nine types of cone bipolar cells and one type of rod bipolar cell. Most mammalian retinas are rod dominated. Therefore, rod bipolar cells form the numerically superior part of the bipolar cell population. Their dendrites make invaginating contacts with rod spherules (**Figure 3(b)**) and their axons terminate in the innermost part of the IPL (**Figure 1**: rod bipolar cell, RB). The number of rods converging on a single rod bipolar cell varies greatly between the species, and, within a species, with retinal eccentricity. In the peripheral human retina, each rod bipolar cell contacts 40–50 rods. Near to the fovea, the dendritic trees become smaller and 15–20 rods are connected. Regardless to retinal eccentricity, convergence in the rod pathway is usually higher than in the cone pathway.

Several types of cone bipolar cells have been recognized in different mammalian species (rabbit: 13, human: 10, cat: 8–10, rat: 9–11, ground squirrel: 6–8). The diagram in (**Figure 1**) compares the bipolar cells of the mouse retina with those of the peripheral macaque monkey retina. The nine putative cone bipolar cell types (labeled 1–9) and the rod bipolar cell of the mouse retina are arranged according to the stratification level of their axon terminals in the IPL. The cells were drawn from vertical sections following intracellular injections. Selective markers, which stain the whole population, are now available for most of the bipolar cell types of the mouse retina (see below). Types 1–4 are OFF cone bipolar cells; types 5–9 are ON cone bipolar cells. The cells contact on average between five and eight neighboring cone pedicles with one exception: type 9 has a wide dendritic tree that appears to be cone selective and it will be shown later that it contacts S-cones.

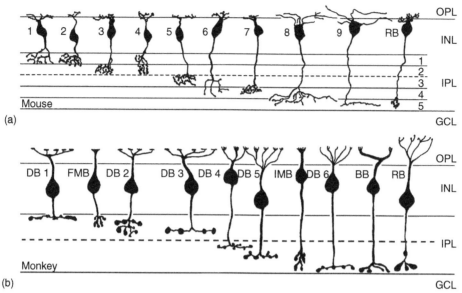

Figure 1 Schematic diagrams of bipolar cells of mouse (a) and primate retina (b). The retinal layers are indicated in (a) for the mouse retina. The inner plexiform layer (IPL) can be subdivided into five sublayers of equal width. The bipolar cell types of the mouse retina were named according to the level of stratification of their axon terminals in the IPL. The dashed horizontal lines dividing the IPL in (a) and (b) represent the border between the OFF-(upper) and the ON-(lower) sublayers. Bipolar cells with axons terminating above this line represent OFF bipolar cells, those with axons terminating below this line represent ON bipolar cells. DB, diffuse bipolar cells; FMB, flat midget bipolar cells; IMB, invaginating midget bipolar cells; BB, blue cone bipolar cells; RB, rod bipolar cells; OPL, outer plexiform layer; INL, inner nuclear layer; GCL, ganglion cell layer. Adapted from Ghosh, K. K., et al. (2004). See Erratum in *Journal of Comparative Neurology* 476: 202–203.

The mouse retina is considered to be rod dominated because only 3% of their photoreceptors are cones. However, the perspective changes if one examines the absolute number of cones; the cone density is about 13 000 cones/mm^2, similar to peripheral cat, rabbit, and macaque monkey retinas. Consequently, the types and retinal distributions of cone bipolar cells are closely similar between mammalian species. The bipolar cell types of the monkey retina (**Figure 1(b)**) were determined initially from Golgi-stained whole mounts. There is a striking similarity between mouse and monkey bipolar cells with respect to the shapes and stratification levels of their axons. However, there is also a clear difference; midget bipolar cells (flat midget bipolar cell (FMB); invaginating midget bipolar cell (IMB)) are only found in the monkey retina. FMB and IMB cells have dendritic trees, which contact a single cone. Bipolar cells contacting several neighboring cone pedicles were named diffuse bipolar cells (DB1–DB6).

Midget Bipolar Cells of the Primate Retina

Primates have trichromatic color vision based on three spectral types of cones: long-wavelength (red or L-), middle-wavelength (green or M-), and short-wavelength (blue or S-) sensitive cones. Midget bipolar cells receive inputs either from red or green cones. Thus, in terms of their input and the polarity of their response, there are four types of midget bipolar cells: red ON, red OFF, green ON, and green OFF. The term midget refers to the small spread of their dendritic and axonal arbors (**Figure 1b**:

FMB, IMB). In the central retina, a midget bipolar cell receives direct input from just one cone (**Figure 2(a)**). The bipolar cell axon terminal, which contacts just one ganglion cell, is correspondingly small. This one-to-one relationship between midget bipolar cells and cones on the dendritic end, and between midget bipolar cells and ganglion cells on the axonal end, is a distinctive property of midget bipolar cells around the fovea. Four to five millimeters beyond the fovea, in the near periphery, the midget bipolar cells become two and three headed connecting to two and three cones, respectively (**Figure 2(b)**).

Reconstructions of Golgi-impregnated midget bipolar cells of the primate retina by serial electron microscopy (EM) revealed a clear dichotomy of their dendritic contacts at the cone pedicle base: IMB cells made exclusively invaginating contacts, whereas FMB cells made only flat contacts (**Figure 3(a)**). In central retina, they are all in the vicinity of the ribbons (triad associated, TA), and at eccentricities beyond 3–4 mm, approximately 20% are nontriad associated (NTA). Individual IMB bipolar cells make up to 25 contacts with a cone pedicle, and an FMB cell makes approximately 2–3.5 times that number of basal synapses.

Blue Cone Bipolar Cells

Placental mammals other than primates have only two types of cones: M-cones, in which the visual pigment has an absorption maximum of >500 nm and S-cones with an absorption maximum at <500 nm. They are, therefore, dichromats. In an evolutionary comparison of color

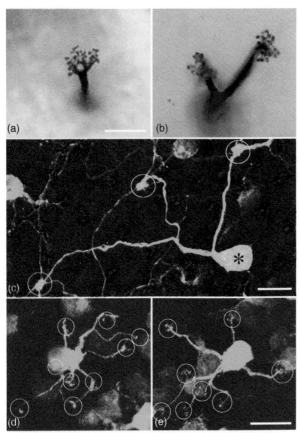

Figure 2 Bipolar cells and their cone contacts in the primate and mouse retina. (a, b) Horizontal view of Golgi-impregnated midget bipolar cells, with the plane of focus at their dendritic tips in the OPL. The micrographs show examples of a single-cone contacting (a) and a two-cone-contacting (b) midget bipolar cell. (c) Horizontal view of a Clomeleon mouse retina (Clm1 line) double labeled for GFP (green) and for GluR5 (red). The clusters of GluR5 puncta represent individual cone pedicles. The dendrites of the blue cone bipolar cell (asterisk) contact three-cone pedicles (circles) and avoid all other pedicles. The BB cell is cone-selective for S-opsin expressing cones. (d, e) Dendritic trees of two type 7 bipolar cells in the Gus-GFP mouse and their cone contacts (circles). Type 7 cells contact on average 8.4 cones. Scale bars = 10 μm. (a, b) Images: courtesy of H. Wässle. From Puller, C., Haverkamp, S., and Grünert, U. (2007). OFF midget bipolar cells in the retina of the marmoset, *Callithrix jacchus*, express AMPA receptors. *Journal of Comparative Neurology* 502: 442–454.

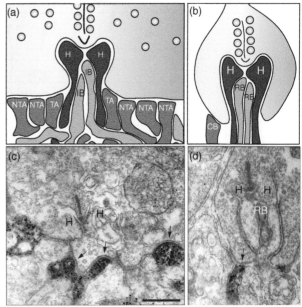

Figure 3 Bipolar cell contacts at photoreceptor terminals. (a, b) Schematic drawings showing the arrangement of contacts at a cone pedicle (a) and rod spherule (b). Horizontal cell processes (H) always end laterally and deeper in the invaginations of both rod and cone terminals. In cone pedicles, the central processes derive from invaginating ON bipolar cells (IB), whereas in rod spherules, the central elements derive from rod bipolar cells (RB). Flat contacts at the base of the cone pedicle are subdivided into triad associated (TA) or nontriad associated (NTA) contacts, depending on their relative distance from the triad. Some cone bipolar cells receive direct input from rods (CB in (b)). (c, d) Type 4 cone bipolar cells of the mouse retina express calsenilin and contact cones as well as rods. Preembedding electron micrographs showing several flat contacts of a calsenilin-positive dendrite at a cone pedicle base (arrows in (c)) and a calsenilin-positive dendrite making a flat contact on a rod spherule (arrow in (d)). Scale bar = 0.5 μm. (a, b) Schematic drawings: courtesy of C. Puller.

pigments, it has been estimated that the L/M separation in the old world primate lineage occurred ~35 million years ago. The separation of the M- and S-cone pigments occurred >500 million years ago and thus represents the phylogenetically ancient, primordial color system. The morphological substrate for the dichromatic color vision common to most placental mammals is the S-cone pathway. Bipolar cells selective for S-cones in the macaque monkey retina have long, smoothly curved dendrites and contact between one and three cone pedicles. Their axons terminate in rather large varicosities in the innermost part

of the IPL, close to the ganglion cell layer (BB cells in **Figure 1(b)**) and innervate the inner tier of the dendritic tree of the small bistratified ganglion cells: These cells are color-opponent and respond to increasing blue light (blue – ON) and decreasing yellow light (yellow – OFF).

Immunostaining with antisera specific for S-opsin has shown that S-cones constitute approximately 10% of the cones in most mammalian retinas. However, in some species S-cones have a very uneven topographical distribution across the retina and many cones express both M- and S-opsin. Recently, a transgenic mouse line could be studied, where Clomeleon, a genetically encoded fluorescence indicator, was expressed under the thy1 promoter. Clomeleon was expressed in ganglion cells, amacrine cells, and bipolar cells.

Clomeleon-labeled ganglion cells, amacrine cells and bipolar cells

Among the bipolar cells the S-cone-selective (blue cone) type could be identified, and the cone-selective contacts and the retinal distribution could be studied (**Figure 2(c)**). The morphological details of the blue cone bipolar cells match type 9 cells of mice (**Figure 1(a)**) and they are closely similar to the blue cone bipolar cell of the primate retina. It is interesting that in the ventral mouse retina, where most cones express both M- and S-opsin, blue cone bipolar cells contact only those cones, which express S-opsin only. They are the genuine S-cones of the mouse retina. Meanwhile, S-cone-selective bipolar cells have also been verified in ground squirrel and rabbit retina.

Diffuse Bipolar Cells

Most bipolar cell types of the mammalian retina contact between 5 and 10 neighboring cones (**Figure 2(d)** and **2(e)**). In mouse, the number of cone pedicles contacted by individual bipolar cells varied from an average of 5.6 for type 2 cells to an average of 8.4 for type 7 cells. Each and all cones are contacted by at least one member of any given type of bipolar cell (leaving aside the blue cone pathway). Consequently, each cone pedicle is connected to a minimum of eight different bipolar cells. They represent eight separate channels that transfer the light signal into the IPL. Parallel processing, therefore, starts at the first synapse of the retina, the cone pedicle.

Diffuse bipolar cells of the primate retina contact L- and M-cones in their dendritic field non-selectively. Whether all diffuse bipolar cell types also contact S-cones is still a matter of discussion, and it has been proposed that one type of diffuse bipolar cell avoids S-cones in the primate retina. This type would be a good candidate to transfer a yellow (L-plus M-cone) signal into the IPL, where it could contact the outer tier of the dendritic tree of the small bistratified ganglion cells. For the dichromatic ground squirrel retina, it has been shown that diffuse bipolar cells sample cone signals differently: some types receive mixed S/M-cone input and other types receive an almost pure M-cone signal. Bipolar cells that sum signals from S- to M-cones are therefore involved with the transfer of luminosity signals, whereas bipolar cells that carry M-cone signals can have, together with S-cone bipolar cells, a role in color discrimination. Alternatively, these bipolar cells may mediate – due to their relatively small dendritic fields – high acuity vision. This idea would correspond to the fact that the center of the human fovea, which mediates the highest acuity vision, also excludes S-cones.

Like the midget bipolar cells, diffuse bipolar cells also differ in their synaptic contacts with cone pedicles, making either flat or invaginating contacts. EM reconstructions of Golgi-impregnated diffuse bipolar cells of the macaque monkey retina revealed that DB1, DB2, and DB3, which have their axon terminals in the outer IPL and are putative OFF bipolar cells (**Figure 1(b)**), make exclusively basal junctions with the cone pedicle.

They always have triad associated (TA) and nontriad associated (NTA) contacts (**Figure 3(a)**). The proportion of TA and NTA contacts varies according to the cell type, as does the average number of contacts per cone, which is between 10 and 20. Bipolar cells DB4, DB5, and DB6 have their axon terminals in the inner part of the IPL and are putative ON bipolar cells. They have an average of between four and eight invaginating synapses per cone pedicle. In addition, they also form basal junctions, in a predominantly TA position. Thus, while the dichotomy 'invaginating = ON, flat = OFF' holds for midget bipolar cells, it does not conform so clearly for diffuse bipolar cells. Therefore, the type of synapse made by a bipolar cell at a cone pedicle, flat versus invaginating, is not the decisive feature; it is rather the glutamate receptor expressed there.

Cone Bipolar Cells with Rod Input

Recent results from rodent and rabbit retina have shown that some OFF cone bipolar cells make also basal contacts with rod spherules and thus receive a direct input from rods. This represents a third route for the rod signal, in addition to the rod bipolar cell circuit, and the gap junctions between rods and cones. In mouse, true-cone-selective OFF bipolar cells (types 1 and 2, **Figure 1(a)**) can be distinguished from types with mixed rod-cone input (types 3 and 4, **Figure 1(a)**). Type 4 bipolar cells make several basal contacts at the cone pedicle base (**Figure 3(c)**) and an individual cell contacts five to eight cones. In addition, some dendrites extend further out into the OPL and contact rod spherules as flat contacts (**Figure 3(d)**). On average, we counted 10 rod spherule contacts per type 4 bipolar cell, and approximately 10% of rods contacted by type 4 bipolar cells.

Immunocytochemical Markers and Transgenic Mouse Lines

The morphological classification of bipolar cells has been made more objective and more quantitative by immunocytochemical markers that selectively label specific cell types. Some of the markers label the same cell types across different species. For instance, rod bipolar cells of all mammals are immunoreactive for protein kinase C α (PKCα). However, other markers, such as calcium-binding proteins, label different cell types in different mammals. Calbindin antibodies label the DB3 OFF cone bipolar cell in the primate retina. However, in the rabbit retina, the calbindin-immunoreactive cell is an ON cone bipolar cell, and in the rat and mouse retina, no bipolar cell expresses calbindin immunoreactivity. Recoverin is another example of a marker that selects different types of bipolar cells in different species. While in rat and rabbit retinas two types of bipolar cell, an OFF and an ON cone bipolar cell, appear to be labeled, only the OFF midget bipolar cell is labeled in the macaque monkey retina. Even in closely related species, different bipolar cell types can be selected by the same marker. The antibody against the carbohydrate epitope CD15 labels a single population of ON bipolar cells (DB6) in macaque monkey, whereas DB6 cells and OFF midget bipolar cells are labeled in marmoset monkeys. In rabbit, CD15 antibodies label an ON cone bipolar cell, whereas in mouse CD15 is expressed in OFF cone bipolar cells. In contrast, antibodies against the calcium-binding protein CaB5 immunolabel at least three types of bipolar cells in a variety of mammalian species. In all species, rod bipolar cells, one ON cone bipolar cell and at least one OFF bipolar cell were labeled (**Figure 4**).

The IPL can be subdivided into five strata of equal thickness (**Figures 1** and **5**). In mouse, these strata can be easily defined by immunolabeling the retina for the calcium-binding protein calretinin (**Figure 5(b)**), which reveals three densely labeled horizontal bands of processes. The outer band (between stratum 1 and 2) contains the processes of the OFF cholinergic amacrine cells and the outer dendritic branches of direction-selective ganglion cells. The band in the inner IPL (between stratum 3 and

4) contains the processes of the ON cholinergic amacrine cells and the inner dendritic branches of direction-selective ganglion cells. The band in the middle of the IPL (between stratum 2 and 3) represents the level of stratification of a nitric oxide synthase (NOS) immunoreactive amacrine cell type and separates the OFF sublamina (outer) from the ON sublamina (inner). The calcium-binding protein 5 (CaB5)-immunoreactive bipolar cells stratify in three strata; in stratum 2 where the type 3 bipolar cells stratify, in stratum

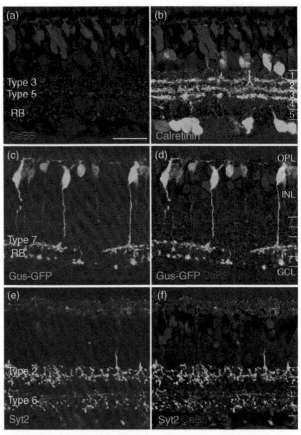

Figure 5 Immunocytochemical staining of mouse bipolar cells. (a, b) Vertical section through a mouse retina that was double immunostained for CaB5 (red) and calretinin (green). Three dendritic strata within the IPL express calretinin and subdivide the IPL into four sublaminae. Three bipolar cell types (type 3, type 5, and RB) express CaB5. Their axons terminate in the IPL in sublamina 2, sublamina 3, and sublamina 5, respectively. (c, d) Vertical section through the Gus–GFP mouse retina immunostained for GFP (green) and CaB5 (red). The retinal layers are indicated (OPL, INL, IPL, subdivided into five sublayers of equal thickness; GCL). Type 7 bipolar cells express high levels of GFP and their axons terminate at the border of sublaminae 3/4. Rod bipolar cells also express GFP, but weakly (double labeled with Cab5 in (d)). (e, f) Vertical section through the mouse retina immunostained for synaptotagmin 2 (Syt2 , green) and CaB5 (red). Type 2 and type 6 bipolar cells express Syt2. Type 2 axons terminate in sublamina 1/2, above the CaB5-labeled type 3 axons; type 6 axons terminate mainly in sublamina 4/5 and intermingle with CaB5-labeled RB axon terminals. Scale bar = 25 μm.

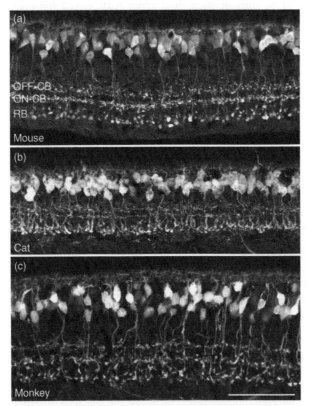

Figure 4 Comparison of CaB5 immunoreactivity in mouse (a), cat (b), and monkey (c) retina. In all three species, rod bipolar cells (RB) and at least one ON cone bipolar (ON-CB) and one OFF cone bipolar cell (OFF-CB) are labeled. Scale bar = 50 μm.

3 where type 5 cells stratify, and in stratum 4/5 where the rod bipolar cells terminate (**Figure 5(a)** and **5(b)**).

We have used several selective markers, either antibodies or the specific expression of fluorescent proteins in transgenic mouse lines, for analyzing the different types of bipolar cells in the mouse retina. Five putative OFF cone bipolar cells were analyzed by selective markers. Type 1 bipolar cells were found to be immunoreactive for the neurokinin 3 receptor (NK3R) and they could also be identified in Clm1 transgenic mice. Type 2 bipolar cells expressed NK3R and synaptotagmin II (Syt2) immunoreactivity (**Figure 5(e)** and **5(f)**). Type 3a and type 3b cells were immunostained for the hyperpolarization-activated cyclic nucleotide-gated potassium channel 4 (HCN4) and the protein kinase A regulatory subunit II β (PKA$_{RIIβ}$), respectively, and type 4 cells expressed the calcium-binding protein, calsenilin. In the case of ON cone bipolar cells, markers for four types were described. Type 5 bipolar cells were labeled in the 5-hydroxytryptamine 3 receptor-EGFP (5HT3R-EGFP) transgenic mouse; however, they represent two types (named 5a and 5b). Type 6 bipolar cells were partially identified because their axons express Syt2 (**Figure 5(e)** and **5(f)**). Type 7 bipolar cells were labeled in the

Gus-GFP mouse, a transgenic mouse line where GFP is expressed under the control of the gustducin promoter (**Figures 2(d)**, **2(e)**, **5(c)** and **5(d)**) Type 9, the blue cone bipolar cell, has been identified in Clm1 mice. Rod bipolar cells have been immunostained for PKCα. This list suggests that, with the exception of type 8 bipolar cells, selective markers, which stain the whole population, are available for all bipolar cell types of the mouse retina.

Synaptic Contacts of Bipolar Cells in the Inner Plexiform Layer

The axons of bipolar cells terminate in the IPL in lobular swellings (**Figure 6(a)**). Some bipolar cell types, such as DB3 and DB6 of the primate retina and type 7 of the mouse retina, keep their axon terminals within a narrow stratum (**Figure 5(c)** and **5(d)**). Hence, their output will be restricted to the amacrine and ganglion cell dendrites they meet within that stratum. Other bipolar cells such as type 4 and type 6 of the mouse retina occupy with their axon terminals the complete OFF or ON sublamina respectively (**Figure 5(e)** and **5(f)**). They are possibly engaged in contacts with a wider variety of postsynaptic

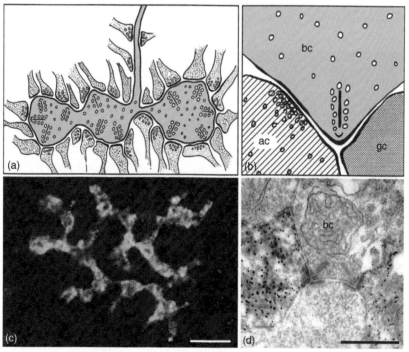

Figure 6 Synaptic output of bipolar cells in the IPL. (a) Schematic diagram of the axon terminal of a cone bipolar cell. It contains many presynaptic ribbons that are flanked by synaptic vesicles. (b) Magnified view of a cone bipolar cell ribbon synapse (dyad). The presynaptic bipolar cell (bc) releases glutamate and the two postsynaptic partners express different sets of glutamate receptors. The amacrine cell, in turn, makes a synapse back onto the bipolar cell terminal (reciprocal synapse). (c) Horizontal view of a GFP-labeled type 7 axon terminal (green) in a Gus-GFP retina. The output synapses are marked (red) by their expression of the ribbon associated C-terminal binding protein 2 (CtBP2/RIBEYE). Altogether, 128 output synapses have been counted at this axon terminal. (d) Electron micrograph of a bipolar cell axon terminal (bc) with two output synapses. Two of the postsynaptic elements are immunolabeled amacrine cell profiles (pre-embedding with glycogen phosphorylase in primate retina). Scale bar = 5 μm in (c), 0.5 μm in (d). ac, amacrine cell; gc, ganglion cell.

neurons. Midget bipolar cells of the primate retina represent a special case, because their axon terminals precisely match in width and depth the dendritic tops of midget ganglion cells, and they together form a densely interconnected glomerulus. The axon terminals of neighboring bipolar cells of a given type usually tile the retina without overlap in the horizontal direction. An interesting question is how the precisely layered and territorial arrangement of axon terminals is formed during embryonic development. Type-specific interactions have to be postulated because terminals of different types can overlap within the same sublamina. Proteins such as the Down's syndrome cell adhesion molecule (Dscam) and an immunoglobulin superfamily protein (Sidekick) are involved in lamina-specific segregations of neuronal processes within the IPL during embryonic development.

Bipolar cell axon terminals provide synaptic output through multiple ribbon synapses (**Figure 6(c)**). We have counted the number of output synapses of the type 7 bipolar cells in the Gus–GFP mouse. The numbers varied between 74 and 128 ($n = 8$) depending on the size of the axon terminals. The number of ribbon synapses made by rod bipolar cells of the rabbit retina was up to 30, compared to only 15 in the rat retina, which reflects the smaller size of rod bipolar axon terminals in rats. The fine structure of the bipolar cell output synapses is shown in **Figure 6(b)**. The presynaptic ribbon is surrounded by vesicles and two

postsynaptic elements. This synaptic arrangement is named a dyad. One of the postsynaptic partners at cone bipolar cell dyads is usually a ganglion cell dendrite, while the other one is an amacrine cell process. The amacrine cell process often makes within about 0.5–1.0 μm of the dyad a conventional synapse back onto the bipolar cell axon terminal. This arrangement appears to be a reciprocal synapse and because most amacrine cells are inhibitory, it is the structural correlate of negative feedback at the dyad. Bipolar cell axons receive, in addition to reciprocal synapses, input from amacrine cells not related to the dyads. In the case of rod bipolar cell dyads, both postsynaptic partners are amacrine cells (AI and AII); and AI cells provide the reciprocal synapses.

Costratification of Pre- and Postsynaptic Partners in the Inner Plexiform Layer

Bipolar axons terminate at distinct levels within the IPL, and different types of amacrine and ganglion cells also keep their processes at specific levels within the IPL, which leads to the prediction that they are also engaged in mutual synaptic contacts (**Figure 7**). However, this simple rule has only been verified in a few instances. Midget bipolar cells of the primate retina – both ON and OFF midget – contact midget ganglion cells. Parasol

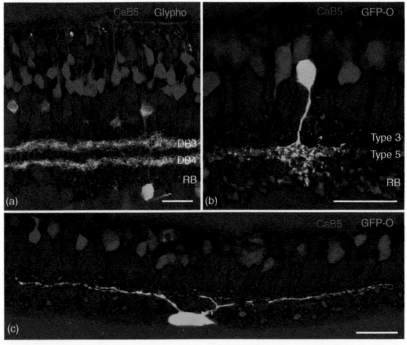

Figure 7 Costratification of pre- and postsynaptic partners in the inner plexiform layer. (a) Vertical section through a primate retina that was double immunostained for CaB5 (red) and glycogen phosphorylase (glypho, green). Axon terminals of CaB5-labeled DB3 and DB4 bipolar cells costratify with the glypho-immunoreactive amacrine cell processes. (b, c) Vertical sections through a transgenic mouse retina where a small set of small-field amacrine cells and ganglion cells express GFP under the control of the thy1 promotor (GFP-O line). Both the small-field amacrine cell with dendrites in the ON-sublamina (b) and the monostratified putative ON ganglion cell (c) costratify with the axon terminals of the type 5 ON bipolar cell. Scale bar = 20 μm.

ganglion cells of the primate retina also occur as OFF and ON pairs and their dendrites stratify in sublamina 2 and sublamina 4, respectively. The OFF parasol cells receive their major, excitatory input from DB3 bipolar cells, and ON-parasol cells most likely from DB5 bipolar cells. In rabbit, ON α ganglion cells stratify just below the ON cholinergic amacrine cells and cofasciculate (distribute together) with the axon terminals of the calbindin-immunoreactive bipolar cells.

The bistratified ON/OFF direction-selective (DS) ganglion cell coincides with the level of stratification of ON and OFF cholinergic amacrine cells. In the rabbit retina, DS ganglion cells receive the majority of their synaptic input from amacrine cells. The axon terminals of the CD15-immunoreactive bipolar cells stratify slightly more distally of the ON cholinergic band. In addition, they follow the pattern of the ON cholinergic dendrites, and are, therefore, good candidates for providing synaptic input to the DS circuitry.

Figure 7(a) shows a double labeling of glycogen phosphorylase (glypho) and CaB5 in the macaque monkey retina. The glypho-immunoreactive cells occur – like the cholinergic amacrine cells – as mirror-symmetrical populations of regular and displaced wide-field amacrine cells. The regular amacrine cells branch in sublamina

2 and co-stratify with DB3 cells; the displaced amacrine cells branch in sublamina 3 and co-stratify with DB4 cells.

Transgenic mouse lines are extremely helpful to study potential contacts between bipolar cells and their postsynaptic partners in the mouse retina. For instance, in the GFP-O line, each mouse expresses GFP in a small and variable set of ganglion cells and small-field amacrine cells. See **Figure 7(b)** for a small-field amacrine cell and **Figure 7(c)** for a monostratified ON ganglion cell, both of which coincide with the type 5 bipolar cell.

Bipolar Cells of Nonmammalian Vertebrates

It has been shown that bipolar cells of a particular cell are usually found in one-half of the IPL or the other, but not in both. In several nonmammalian vertebrates, however, the axon terminals of the bipolar cells are highly stratified, ending on one or several levels in the inner plexiform layer. Most cold-blooded vertebrate retinas contain large bipolar cells, long thought to be rod-related bipolar cells, and small bipolar cells, believed to be cone-related bipolar cells. In fish, it has been shown that many bipolar cells, particularly the larger ones, contact both rods and cones.

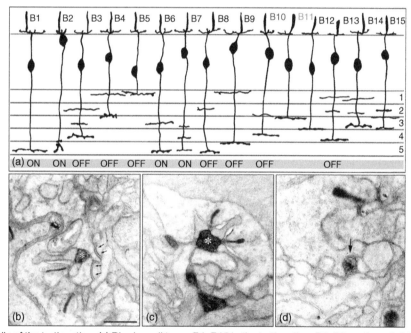

Figure 8 Bipolar cells of the turtle retina. (a) Bipolar cell types B1–B15 in the turtle retina. B1–B9 are from Golgi-impregnated retinae, B10, B11, and B13 from intracellular recordings with subsequent dye injection, and B12, B14, and B15 from Lucifer Yellow injections. The cells are drawn in vertical views with cell bodies, dendrites, and Landolt clubs in the outer retina and the stratification of their axons in the five strata (S1–S5). Several cells are bi- or tristratified. B1, B2, B6, and B7 are ON-center cells; B3, B4, B5, B8, B9, B10, and B13 are OFF-center cells. B10 and B11 are color-opponent cells (B10: red-ON, green/blue-OFF; B11: red-OFF, green/blue-ON). (b)–(d) Electron micrographs showing invaginating, ribbon-associated synapses of an HRP-stained B10 bipolar cell with L-cone pedicles (asterisks in (b) and (c)) and a noninvaginating, basal junction with an M-cone pedicle (arrow in (d)). Two unstained wide-scleft basal junctions in (b) are marked by small arrows. Scale bar = 0.5 μm.

It has also been shown that certain of these bipolar cells are likely to be color-coded, because they contact specific sets of cones. In turtle retina, at least 15 different morphological types of bipolar cells were found (**Figure 8(a)**). Some are monostratified with only a single axon terminal (B1, B2, B5, B11, and B12), several are bistratified (B4, B6, B8, B9, and B10), and some are tristratified (B3, B7, B13, B14, and B15). All seem to have Landolt clubs arising from their dendrites in the outer plexiform layer (OPL) to extend into the outer nuclear layer. A functional organization of the turtle IPL into OFF sublaminae (strata 1 and 2) and ON sublaminae (strata 3, 4, and 5), as has been described for other vertebrate retinas, is quite clear for two types of OFF bipolar cells, which stratify in the two distal strata (B4, B5) and for all four types of ON bipolar cells (B1, B2, B6, and B7). However, some OFF bipolar cells (B3, B9, B10, and B13) have axon terminals in strata 3–5 in addition to their terminations in stratum 1 or 2 (**Figure 8(a)**).

Color-Coded Bipolar Cells in the Turtle Retina

The turtle has excellent color vision, and is at least tetrachromatic. Three cones are sensitive to long wavelengths, one to medium wavelengths, one to short wavelengths, and a further cone to ultraviolet light. The chromatic types of cones can be morphologically identified by the presence and colors of their oil droplets and the shape of their pedicles. Two of the bipolar cell types in turtle are color-opponent: B10 is a red-ON, green/blue-OFF bipolar cell with axons in S2 and S4 and B11 is a red-OFF, green/blue-ON bipolar cell with an axon terminal in S3 (**Figure 8(a)**).

We have analyzed the cone contacts of a horse-radish peroxidase labeled B10 cell by serial EM reconstruction: 45 ribbon-associated synapses were found in single and double L-cones; basal junctions were found in M-cones (**Figure 8(b)**–**8(d)**). No contacts were found with rods, S-cones, and UV-cones. The results showed that invaginating synapses with L-cones and noninvaginating synapses with M-cones formed the basis of color opponency in an identified bipolar cell for red versus green light stimulation. We suggested sign-inverting transmission from L-cones at invaginating synapses mediated by G-protein-coupled metabotropic glutamate receptors, and sign-conserving transmission from M-cones at wide-cleft basal junctions mediated by ionotrophic (ion-gated) glutamate receptors. For B11 bipolar cells, we would predict that the dendrites express ionotropic glutamate receptors at their contacts with L-cones (red-OFF) and

metabotropic glutamate receptors at their contacts with M-cones (green-ON).

See also: Cone Photoreceptor Cells: Soma and Synapse; Information Processing: Bipolar Cells; Morphology of Interneurons: Amacrine Cells; Morphology of Interneurons: Interplexiform Cells; Rod and Cone Photoreceptor Cells: Inner and Outer Segments; Rod Photoreceptor Cells: Soma and Synapse.

Further Reading

Ammermüller, J. and Kolb, H. (1996). Functional architecture of the turtle retina. *Progress in Retinal Research* 15: 393–433.

Boycott, B. B. and Wässle, H. (1991). Morphological classification of bipolar cells in the macaque monkey retina. *European Journal of Neuroscience* 3: 1069–1088.

Boycott, B. B. and Wässle, H. (1999). Parallel processing in the mammalian retina. The Proctor Lecture. *Investigative Ophthalmology and Visual Science* 40: 1313–1327.

Chan, T. L., Martin, P. R., Clunas, N., and Grünert, U. (2001). Bipolar cell diversity in the primate retina: Morphologic and immunocytochemical analysis of a new world monkey, the marmoset *Callithrix jacchus*. *Journal of Comparative Neurology* 437: 219–239.

Euler, T., Schneider, H., and Wässle, H. (1996). Glutamate responses of bipolar cells in a slice preparation of the rat retina. *Journal of Neuroscience* 16: 2934–2944.

Famiglietti, E. V. (1981). Functional architecture of cone bipolar cells in mammalian retina. *Vision Research* 21: 1559–1563.

Ghosh, K. K., Bujan, S., Haverkamp, S., Feigenspan, A., and Wässle, H. (2004). Types of bipolar cells in the mouse retina. *Journal of Comparative Neurology* 469: 70–82.

Haverkamp, S., Möckel, W., and Ammermüller, J. (1999). Different types of synapses with different spectral types of cones underlie color opponency in a bipolar cell of the turtle retina. *Visual Neuroscience* 16: 801–809.

Haverkamp, S., Wässle, H., Dübel, J., et al. (2005). The primordial, blue cone color system of the mouse retina. *Journal of Neuroscience* 25: 5438–5445.

Haverkamp, S., Specht, D., Majumdar, S., et al. (2008). Type 4 OFF cone bipolar cells of the mouse retina express calsenilin and contact cones as well as rods. *Journal of Comparative Neurology* 507: 1087–1101.

Kolb, H. (1970). Organization of the outer plexiform layer of the primate retina: Electron microscopy of Golgi-impregnated cells. *Philosophical Transactions of the Royal Society, London, B* 258: 261–283.

Li, W. and DeVries, S. H. (2006). Bipolar cell pathways for color and luminance vision in a dichromatic mammalian retina. *Nature Neuroscience* 9: 669–675.

MacNeil, M. A., Heussy, J. K., Dacheux, R. F., Raviola, E., and Masland, R. H. (2004). The population of bipolar cells in the rabbit retina. *Journal of Comparative Neurology* 472: 73–86.

Sherry, D. M. and Yazulla, S. (1993). Goldfish bipolar cells and axon terminal patterns: A Golgi study. *Journal of Comparative Neurology* 329: 188–200.

Wässle, H., Puller, C., Müller, F., and Haverkamp, S. (2009). Cone contacts, mosaics and territories of bipolar cells in the mouse retina. *Journal of Neuroscience* 29: 106–117.

Morphology of Interneurons: Horizontal Cells

L Peichl, Max Planck Institute for Brain Research, Frankfurt am Main, Germany

Glossary

Cone pedicle – The axonal synaptic ending of the cone photoreceptor in the outer plexiform layer of the retina; it is commonly a relatively large conical structure that contains about 30 synaptic sites (triads), each having a presynaptic ribbon and three postsynaptic processes.

Connexin – A transmembrane protein; six connexins form a connexon or hemichannel and two connexons form a gap junction. Connexins are a diverse family and can combine into homomeric or heteromeric connexons and gap junctions.

Ephaptic – An action mediated by an electrical contact between nerve cells, without the mediation of a neurotransmitter. In the case of horizontal cells it is mediated by a hemichannel.

Gap junction – A specialized connection between neighboring cells, forming an electrical synapse in the neurons. A gap junction consists of two connexons (hemichannels) that form an intercellular pore across the touching cell membranes. The connexons are connexin hexamers. The pore allows various ions and molecules (e.g., neurobiotin) to pass between the cells. Gap junctions can be regulated (opened and closed) by neuromodulators and thus implement a flexible functional syncytium.

Rod spherule – The axonal synaptic ending of the rod photoreceptor in the outer plexiform layer of the retina; it is a small globular structure that contains two synaptic sites with a presynaptic ribbon and four postsynaptic processes.

Triad – A synaptic arrangement at the synaptic ending of a photoreceptor, located at an invagination of the cone pedicle or rod spherule. The presynaptic side of the triad contains an electron-dense synaptic ribbon for the docking of transmitter vesicles. The postsynaptic side has three invaginating processes, a central bipolar cell dendritic terminal and two lateral horizontal cell terminals. Rod triads receive two biploar cell terminals and two horizontal cell terminals.

General Morphology and Connectivity

Basic Morphology

Horizontal cells are interneurons of the outer retina and the largest neurons present in that region. Their somata are located in the outer part of the inner nuclear layer. Their processes ramify in the outer plexiform layer and form synaptic contacts with the photoreceptors. The horizontal cells of mammals – to which this article is limited – comprise two types, commonly termed A-type and B-type; the terms A-HC or HA and B-HC or HB are also in use. The B-type has a smaller, densely branched dendritic tree with relatively fine dendrites and also has an axon ending in a profusely branched axon terminal system. The A-type has a larger, more sparsely branched dendritic tree with fewer and stouter primary dendrites, and has no axon (**Figure 1(a)**). The dendrites of both types carry clusters of terminals (terminal aggregates) that synapse exclusively with cones. The axon terminal system of the B-type has unclustered terminals that exclusively contact rods (**Figures 1(b)–1(g)**).

The horizontal cell terminals, together with the dendrites of invaginating bipolar cells (ON bipolar cells, depolarizing to the onset of light), insert into invaginations at the base of the photoreceptor terminal (cone pedicle or rod spherule). The synaptic complexes thus formed are referred to as triads. Each cone pedicle possesses a substantial number of invaginations with triads (20–50 in primates), whereas a rod spherule commonly only has one invagination with two triads (**Figures 1(e)–1(g)**). Presynaptically, the triad is marked by a synaptic ribbon that is thought to play a crucial role in the continuous vesicular release of the photoreceptor transmitter glutamate. Postsynaptically, each mammalian triad has a bipolar cell dendritic terminal (two at rod spherules) as a central element, flanked by two horizontal cell terminals as lateral elements. The bipolar cells, via their axonal contacts, pass their signals on to the inner retina. The horizontal cells have no spatially segregated output synapses. The only synaptic contacts they form are with the photoreceptors. These are their input and output sites, representing a local feedback mechanism for the cones (by A-type and B-type dendrites) and the rods (by B-type axon terminals).

Horizontal cells modulate signal transmission from the photoreceptor to the invaginating bipolar cell processes.

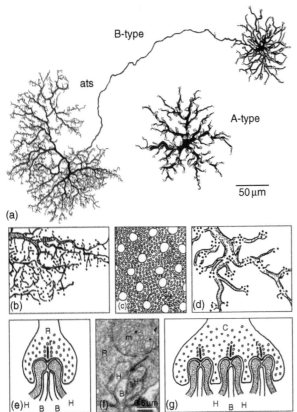

B-type

ats

A-type

50 µm

(a)

(b)

(c)

(d)

(e) H B B H

(f)

(g) H B H

Figure 1 Basic mammalian horizontal cell morphologies.
(a) Flat views of a Golgi-stained B-type horizontal cell with axon
terminal system (ats) and axonless A-type horizontal cell of cat.
Middle row: (b) enlarged part of the B-type ats with single
terminals that are the contacts with rods; (c) a schematic
photoreceptor pattern with the rather regularly spaced cones
surrounded by the more numerous and smaller rods; and (d) part
of an A-type dendrite with clustered terminals that are the
contacts with cones. Bottom row: (e) schematic triad
arrangements at a rod spherule; (f) electron micrograph of a triad
in a mouse rod spherule; and (g) schematic triad arrangements at
a cone pedicle. R, rod spherule; C, cone pedicle; H, horizontal
cell process; B, bipolar cell process; r, presynaptic ribbon; and
m, mitochondrion. For details, see text. (a) Adapted from
Boycott, B.B., Peichl, L., and Wässle, H. (1978). Morphological
types of horizontal cell in the retina of the domestic cat.
Proceedings of the Royal Society (*London*) *B* 203: 229–245.
(f) Image kindly provided by Silke Haverkamp. Adapted from
Figure 1 in Peichl, L., Sandmann, D., and Boycott, B. B. (1998).
Comparative anatomy and function of mammalian horizontal
cells. In: Chalupa, L. M. and Finlay, B. L. (eds.) *Development
and Organization of the Retina: From Molecules to Function*,
pp. 147–172. New York: Plenum Press. With kind permission of
Springer Science and Business Media.

The horizontal cell feedback is inhibitory and supposedly
creates the antagonistic receptive field surrounds of
bipolar cells, thus contributing to the receptive field
properties of the ganglion cells. However, the exact nature
of the feedback synapse is unknown. There is some evidence that mammalian horizontal cells use the inhibitory
transmitter gamma aminobutyric acid (GABA); however,

a feedback mediated by electrical conduction (ephaptic,
via hemichannels) and pH modulation (proton hypothesis) is also considered. In addition, the horizontal cells
probably have feed-forward connections to ON bipolar
cells. OFF bipolar cells, which make flat (noninvaginating) contacts at the photoreceptor base, are not in direct
apposition to the horizontal cells. It is assumed that the
horizontal cell modulation affects them indirectly via the
changed photoreceptor output.

The dendritic spread of a horizontal cell is much larger
than that of a cone bipolar cell, and activation by cones
anywhere in its dendritic field results in an output across
the entire dendritic field. Thus, information about the
illumination state of cones outside a bipolar cell's input
region is picked up by the overlying horizontal cells and
negatively fed back to the few cones providing the direct
bipolar cell input – the classical lateral inhibition that
sharpens contrast sensitivity and acuity. Most likely, the
same function is provided for the rod pathway by the
B-type axon terminal system.

Photoreceptor Contacts

In the mammalian retina, the rod and cone signals are
separately carried by cone and rod bipolar cells, respectively. These major rod and cone pathways only converge
in the inner retina. In addition, there is gap-junctional
coupling between rods and cones, and a few cone bipolar
cells also synapse with rods. Mammalian horizontal cells
separately serve the rod and cone pathways.

Contacts with cones

The dendritic fields of horizontal cells are commonly
shaped round to oval. Depending on the retinal location,
A-type dendritic trees in cat are 80–220 µm in diameter
and contact 120–170 cones, whereas B-type dendritic trees
are 70–120 µm in diameter and contact 60–90 cones.
A- and B-type cells in rabbit are slightly larger but contact
similar numbers of cones. In their cone connections, the
two horizontal cell types share a common input. Most
mammals are cone dichromats with an approximately 90%
majority of medium-to-long-wavelength-sensitive (L)
cones and a 10% minority of short-wavelength-sensitive
(S) cones. In most studied species, both A- and B-type cells
contact the vast majority of cones within their dendritic
fields; this was considered as evidence that each spectral
cone type contacted each horizontal cell type. Soma
recordings from cat and rabbit horizontal cells with spectral stimuli confirmed that both types hyperpolarize to all
wavelengths. Direct anatomical evidence for the nonselective contacts of both horizontal cell types with both spectral cone types has been obtained in rabbit and tree shrew
(**Figures 2** and **3**). In cone triads, the two lateral elements
are either two A-type terminals, two B-type terminals, or
one A-type and one B-type terminal; in rabbit, these three
combinations are found in equal numbers. The current

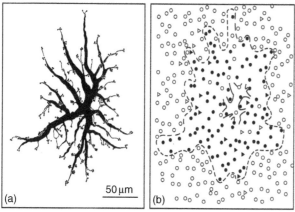

Figure 2 Cone contacts of an A-type horizontal cell in rabbit. (a) Drawing of the Lucifer yellow-injected cell. (b) Cone mosaic overlying the cell, revealed by labeling all cones with the marker peanut agglutinin and identifying the S-cones by an S-opsin antiserum. S-cones are shown as triangles and L-cones as circles. Cones contacted by the A-type cell (dendritic field outline given by broken line) are shown as filled symbols and noncontacted cones as open symbols. The cell contacts most of both spectral cone types in its reach. Reproduced from Figure 4 in Hack, I. and Peichl, L. (1999). Horizontal cells of the rabbit retina are non-selectively connected to the cones. *European Journal of Neuroscience* 11: 2261–2274. With kind permission of Wiley-Blackwell.

view is that mammalian horizontal cells do not participate in color opponency (i.e., the antagonistic interpretation of color opponent channels).

Contacts with rods

Mammalian rods have horizontal cell contacts only with B-type axonal terminals; conversely, B-type axonal terminals connect only to rods. The only reported exception is the cone-dominated gray squirrel retina, where the axon of an H1 cell (B-type equivalent) was described to contact cones and, perhaps, rods. Unlike some other vertebrates, mammals have no type of horizontal cell that is exclusively connected to rods. This role is taken over by the B-type axon terminal system. The B-type axon is not an axon in the functional sense; it does not conduct information from the dendrites and soma to the axon terminal system. The latter is electrically uncoupled from the dendritic part; apparently, the axon is too thin and long (several hundred microns) to conduct the graded potentials that horizontal cells use for signaling. The axonal synaptic connections with the rods resemble those of the dendrites with the cones (**Figures 1(e)–1(g)**). It is thought that the B-type axon terminal system is the independent rod horizontal cell of mammals. Thus, while being one metabolic entity, the B-type cell represents two functional units. Actually, a small rod input is found in soma recordings of both B- and A-type cells; however, this is attributed to direct rod/cone coupling and not to signal transfer through the B-type axon.

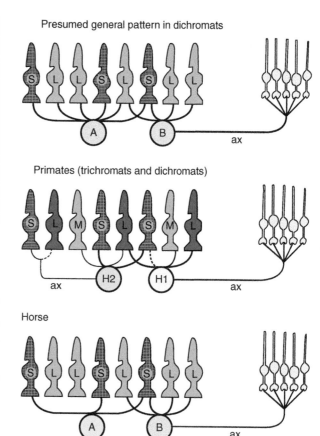

Figure 3 Scheme of the cone and rod contacts of horizontal cells in different mammals. (Top) The presumed general connectivity pattern in dichromats, confirmed in rabbit and tree shrew. A- and B-type cells indiscriminately contact S-cones and L-cones; the B-type axon terminal system contacts rods. (Middle) The connectivity pattern in primates. H1 cells contact M- and L-cones, but largely avoid S-cones (broken contact line); their axon contacts the rods. H2 cells contact S-cones strongly (thick contact line) and M/L-cones less strongly; their axon contacts S-cones, but may also contact M/L-cones. In dichromatic primates, M- and L-cones are only one spectral L type. (Bottom) The connectivity pattern in horse (and probably other equids). The A-type contacts S-cones exclusively, whereas the B-type makes indiscriminate contacts with S-cones and L-cones. The coloring of horizontal cell somata indicates their spectral tuning. Adapted from Figure 9 in Peichl, L., Sandmann, D., and Boycott, B. B. (1998). Comparative anatomy and function of mammalian horizontal cells. In: Chalupa, L. M. and Finlay, B. L. (eds.) *Development and Organization of the Retina: From Molecules to Function*, pp. 147–172. New York: Plenum Press. With kind permission of Springer Science and Business Media.

Why are the rods contacted by only one type of process (the B-type axonal terminals) and the cones by two types (A- and B-type dendrites) and what detailed functional difference does that make? An answer is currently not available. The receptive field surround of the ganglion cells becomes broader and shallower with decreasing light levels, and this is interpreted as a useful image-processing strategy. Nevertheless, it is unknown whether and how the rod/horizontal cell connections may be involved.

Population Properties and Gap-Junctional Coupling

The A- and B-type cells form populations that completely cover the retinal surface and thus provide their signals at all points of the retina. Within each population, the cells form a rather regular mosaic, that is, an even tiling of the retinal surface. The horizontal cells, similar to many other neurons, decrease their population density from the central to peripheral retina and conversely increase the size of the individual cells such that the dendritic overlap, or coverage factor, remains approximately constant across the retina. In cat, the density of A-type cells decreases from $800\,\text{mm}^{-2}$ to $100\,\text{mm}^{-2}$, and that of B-type cells from $2300\,\text{mm}^{-2}$ to $300\,\text{mm}^{-2}$. The coverage factor is approximately four for each type; therefore, each cone can, on average, contact eight horizontal cells. In rabbit, the central–peripheral density gradient is shallower ($550\,\text{mm}^{-2}$ to $250\,\text{mm}^{-2}$ for the A-type and $1375\,\text{mm}^{-2}$ to $400\,\text{mm}^{-2}$ for the B-type); the coverage factor is approximately six for the A-type and 8–10 for the B-type, somewhat larger than in cat. Several staining methods can be used to stain entire horizontal cell populations. In many mammals, neurofibrillar stains and immunocytochemical staining of the neurofilament proteins specifically reveal the A-type population (**Figure 4(a)**), but in horse, the B-type population is specifically stained. Antibodies against the calcium-binding protein calbindin (CaBP 28 kDa) stain both horizontal cell populations in various mammals (**Figure 4(b)**).

Mammalian horizontal cells are electrically coupled by gap junctions. The coupling is homotypical; there are no gap junctions between A- and B-type cells (**Figure 5**). The gap junctions of A-type cells in rabbit are composed from connexin 50 (Cx50) subunits, and are larger than those of the B-type cells where the connexin is Cx57. The gap junctions of B-type cells in mouse are also formed by Cx57; therefore, the cell-type-specific connexins may be conserved across species. The rod connectivity is similar through the coupling of B-type axon terminals; notably, this coupling is segregated from that of the B-type dendrites (**Figure 5**). Thus, with respect to electrical coupling, the horizontal cells form three separate networks, even though A-type and B-type dendrites largely share their synaptic input and output partners, the cones. The gap junctions are regulated by the ambient light level in a triphasic manner. At intermediate light levels, the horizontal cells are strongly coupled and their signals spread over large distances; in bright and in very low light, the gap junctions are closed and the horizontal cells act as smaller inhibitory units. Dopamine and other neuromodulators are involved in this regulation. Horizontal cell coupling is thought to play an important role in photoreceptor adaptation to different ambient light levels as it serves to collect light information over large areas of the retina.

Diversity of Morphology and Connectivity across Species

None of the basic features listed so far has provided compelling arguments to explain why there should be two horizontal cell types in mammals. Are there really

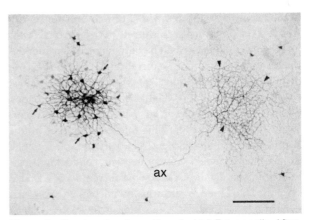

Figure 5 Gap-junctional coupling of rabbit B-type cells. After injection of neurobiotin into a B-type soma (asterisk), the dye spreads to neighboring B-type somata and dendritic trees (some arrowed) via gap junctions between dendrites, but not to neighboring B-type axon terminals or A-type cells. The dye that has diffused intracellularly through the axon (ax) to the axon terminal system of the injected cell then spreads to neighboring axon terminals via gap junctions (right side, arrowheads), and thereon diffuses to some B-type somata through their axons. This shows that gap-junctional coupling is type-specific (homotypic), and that coupling is segregated between B-type dendrites and axons. The scale bar = 100 μm. Image kindly provided by David I. Vaney.

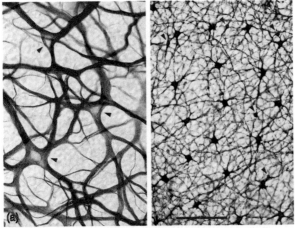

Figure 4 Flat views of population stains of horizontal cells. (a) Neurofibrillar staining of the A-type population in rabbit and (b) calbindin immunostaining of the A- and B-type populations in horse. In both images, the arrowheads indicate A-type somata. The scale bar = 50 μm for (a) and 100 μm for (b).

two types in every mammal? Which morphological differences are of functional significance? Are there differences in their specific connection with the cones and, thus, in chromatic processing? A- and B-type horizontal cells have been identified in a range of orders including primates, carnivores, lagomorphs, rodents, and ungulates; both types are present in rod-dominated and cone-dominated retinas. On this basic pattern, an unexpected diversity in morphology and connectivity is superimposed, which has led to modifications in the general definition of mammalian A- and B-type cells.

Variations of Shape

Old World primates including humans were perceived early on as deviating from the general mammalian pattern. Both the H1 cell (B-type equivalent) and the H2 cell have an axon (**Figures 6(a)** and **7**). Hence, H2 was regarded as unique to the primate retina. The axon of the H1 cell connects to rods, while that of the H2 cell connects to cones; the dendrites of both types only synapse with cones. As the H2 axon has no rod contacts, this cell can be equated to the A-type of other mammals. However, H2 cells have finer dendrites than H1 cells, the reverse of the presumed characteristic distinction between the A-type and B-type. Cells with the same morphologies also exist in New World monkeys. Moreover, the cone contacts of H1 and H2 cells in both Old World and New World primates differ from those of the B- and A-type cells of other mammals (see below).

In line with the overall more fine-grained processing in primate retina, particularly near the fovea, H1 and H2 cells have higher densities and smaller dendritic fields than the horizontal cells of other mammals. In macaque, H1 and H2 densities near the fovea are $18\,400\,\text{mm}^{-2}$ and $4{,}600\,\text{mm}^{-2}$, respectively, dropping to $1000\,\text{mm}^{-2}$ and $500\,\text{mm}^{-2}$, respectively, in the peripheral retina. A central H1 cell only contacts 6–7 cones and a peripheral one, 40–50 cones. H2 cells have larger dendritic fields than H1 cells at corresponding locations, conforming to the A-type/B-type differences in other mammals.

In artiodactyls (ox, sheep, pig, and deer), the B-type cells have a robust dendritic tree, while the A-type dendritic tree is delicate (**Figures 8(c)** and **7**); with regard to this, they resemble primates. The B-type has a single axon ending in an axon terminal system. The A-type has no axon; however, sometimes, one or a few dendritic processes extend beyond the perimeter of the dendritic field. Our evidence indicates that they are conventional dendrites connected to cones. The suggestion is that the artiodactyl A-type represents an intermediate shape between the roughly symmetric A-type dendritic fields of many mammals and the singular asymmetry of the primate H2 cell's axon, thus placing primate H2 cells at one end of a spectrum of A-type morphologies.

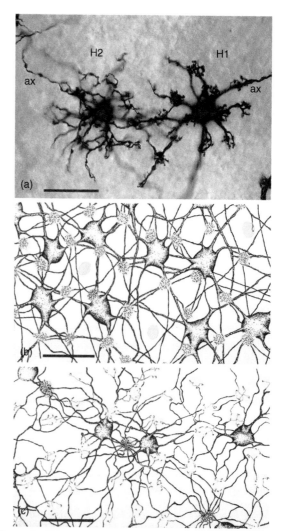

Figure 6 Primate horizontal cells and their cone contacts. (a) Golgi-stained macaque H1 and H2 cells in flat view, focused on the dendritic trees and terminal aggregates (ax, axon). (b) Plexus of macaque H1 cells, stained by a neurobiotin injection in one of the cells; tracer spread to the neighboring H1 cells occurred through the gap junctions. The H1 cells form dense terminal clusters at most cone pedicles (yellow, presumed M- and L-cones), but nearly completely miss the three presumed S-cones (blue). (c) Similarly neurobiotin-labeled macaque H2 cells. They not only strongly innervate the three presumed S-cones (blue), but also contact the other (M and L) cones. The scale bar = 25 μm. Images kindly provided by Dennis Dacey.

In perissodactyls (horse, ass, mule, and zebra), the B-type also has a robust dendritic tree. The single axon is very long, straight, and unusually thick (**Figures 8(d)** and **7**). The dendrites of the A-type are very fine and sparsely branched, and there is no indication of an axon (**Figure 7**). The most interesting feature of this A-type cell, however, is its selective connection to the S-cones (see below).

The cone-dominated retina of the tree shrew (*Tupaia belangeri*, Scandentia) has particularly unusual A-type cells (**Figure 9**). They are large and have stout radial

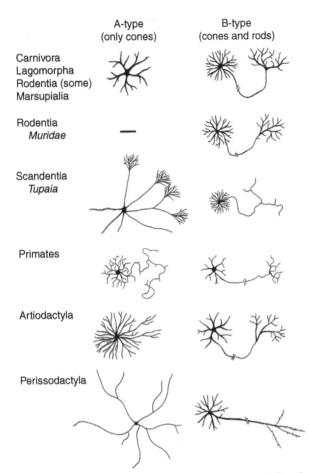

Figure 7 Schematic drawings to show interordinal variations in mammalian A- and B-type horizontal cell morphology. The diagram shows the basic branching patterns but not the synaptic terminals. Interruptions on B-type cells' axons indicate that the axons are longer than drawn. Adapted from Figure 5 in Peichl, L., Sandmann, D., and Boycott, B. B. (1998). Comparative anatomy and function of mammalian horizontal cells. In: Chalupa, L. M. and Finlay, B. L. (eds.) *Development and Organization of the Retina: From Molecules to Function*, pp. 147–172. New York: Plenum Press. With kind permission of Springer Science and Business Media.

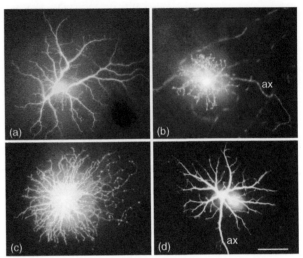

Figure 8 Species variations in horizontal cell morphology; Lucifer yellow-injected cells in flat view. (a) Guinea pig A-type; (b) gerbil B-type; (c) pig A-type; and (d) horse B-type. ax, axon. All cells shown at the same magnification and scale bar = 50 μm.

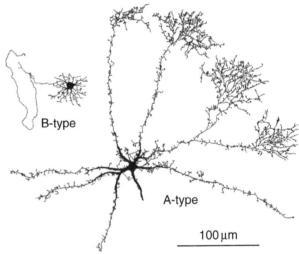

Figure 9 Drawings of Lucifer yellow-injected A- and B-type cells from the cone-dominated retina of the tree shrew. The B-type cell has a conventional dendritic tree and a sparsely branched axon terminal system, whereas the A-type cell has a unique branching pattern. Adapted from Figure 4 in Peichl, L., Sandmann, D., and Boycott, B. B. (1998). Comparative anatomy and function of mammalian horizontal cells. In: Chalupa, L. M. and Finlay, B. L. (eds.) *Development and Organization of the Retina: From Molecules to Function*, pp. 147–172. New York: Plenum Press. With kind permission of Springer Science and Business Media.

primary dendrites that rarely branch until the periphery of the dendritic field, where some ramify into unique bushy arborizations. The first description of these cells termed them multiaxonal because the arborizations were reminiscent of B-type axon terminal systems. We then showed that all connections of these cells, including those at the peripheral arborizations, are with cones. The cells thus conform to the basic mammalian A-type connectivity and can be interpreted as a further variety of A-type shape. This shape is not overtly associated with the high cone density in the tree shrew retina because the H2 cell (A-type equivalent) of the equally cone-dominated ground squirrel retina has a rather conventional shape. The B-type cells of the tree shrew are small and have a conventional dendritic tree; however, the axon is very sparsely branched

and has only a few terminals (**Figure 9**). This is to be expected because less than 10% of the tree shrew's photoreceptors are rods.

The rodents are the most diverse of the mammalian orders, and their horizontal cells have provided the biggest surprise. The muroid species rat, mouse, gerbil, and Syrian hamster possess only one type, the axon-bearing B-type (**Figures 8(b)** and **7**); intracellular injections and population analysis gave no evidence for a further type of horizontal cell. So far, the muroid rodents are the only known instance where the basic mammalian pattern of two horizontal cell types has not been found. In some other rodent groups, for example, sciurids and caviomorphs, both types are present (**Figure 8(a)**). The absence of A-type cells does not appear to be associated with nocturnality since the gerbil has active phases at both day and night and possesses a rather high cone proportion of 10–20%. Rat and mouse have a lower cone proportion (1% and 3%, respectively), which is, however, comparable to the cone proportions found in cat and rabbit (2–4%). Apparently, there is no correlation between a low cone/rod ratio and the absence of the A-type.

The morphology of the B-type axon terminal system also varies significantly across mammalian species. Most species have a thin B-type axon of a few hundred microns length that meanders randomly; in horse, the axon is thick, straight, and a few millimeters long. In cat, rabbit, and rat, the axon terminal system is rather densely branched and appears to innervate most of the rods in its field (as many as 3000 in cat). In other species such as primates or horse, the branching is less dense or even sparse. Here, only a minority of the rods present are innervated by any one terminal system. However, as a population, the overlapping axons of several cells ensure full coverage of the rods.

The existence of a third type of horizontal cell has been claimed for primates, the rabbit, and the South American opossum on the basis of individual cells that markedly differed from the other two types in morphology and presumed connectivity. However, other studies have concluded that such cells are encompassed in the normal variation of the standard two types; the most parsimonious interpretation is that these cells are extreme individuals. If horizontal cells with new or unusual morphological features are to be classified as a new type, it should also be demonstrated that this type exists as a population and adequately covers the retina.

The species variations in horizontal cell morphology add to the recognition that the basic blueprint of the mammalian retina is more flexible than is commonly assumed (**Figure 7**). Currently, there is little clarity regarding the meaning of these variations. The absence of the A-type in murids is of practical significance since genetically modified mouse strains are heavily used to elucidate the general principles of mammalian retinal wiring and function. Researchers have to be aware that mouse data on the contribution of horizontal cells to retinal processing may not be readily transferable to other species. On the other hand, mice may provide key insights into the effects that one versus two horizontal cell types have on the properties of ganglion cell receptive fields.

Species with Selective Cone Contacts

Most mammals are cone dichromats with a majority of L-cones and a minority of S-cones. In Old World primates and man, the mammalian L opsin gene has diverged into separate genes for the M (green) and L (red) cone opsins, making these species cone trichromats with refined color vision. Most New World monkeys are cone dichromats by genotype; however, some species were shown to have an L opsin polymorphism that results in a trichromatic phenotype in many females, while the males are dichromats. Hence, the retina of primates was an interesting place to investigate whether the horizontal cells are chromatically selective and might contribute to color processing.

Observations on Golgi-stained human and monkey horizontal cells had indicated that H1 dendrites specifically avoid or undersample the S-cones, whereas H2 dendrites contact all cones and H2 axons exclusively contact S-cones. In a seminal study, Dennis Dacey and his colleagues recorded the responses of H1 and H2 cells to chromatic stimuli in the isolated living macaque retina and demonstrated that H2 cells show the same hyperpolarization at all wavelengths, whereas H1 cells hyperpolarize to M- and L-cone stimuli but do not respond to S-cone stimuli. The recorded cells were injected with the tracer neurobiotin, labeling all members of patches of H1 or H2 cells around an injected cell through the homotypic gap junctions. The population of H1 cells innervates the majority of the cones, but hardly contacts the small fraction of presumed S-cones (**Figure 6(b)**). The H2 population, on the other hand, connects to all cones, but makes particularly numerous contacts with the small fraction of presumed S-cones (**Figure 6(c)**). Studies combining dye-labeling of individual horizontal cells and cone opsin labeling confirmed this connectivity pattern for macaque, orangutan, and chimpanzee. They showed that only close to 15% of the H1 cells contact S-cones, but then only sparsely; the H2 cells contact all cones within reach and have more synapses with each S-cone than with each of the M- and L-cones contacted. The axon of the H2 cell definitely contacts S-cones, but whether it does so exclusively has yet to be determined. Despite the differences in cone connectivity, the physiological recordings indicate that neither the H1 nor the H2 cell is involved in creating spectrally opponent receptive fields of bipolar and ganglion cells.

New World monkeys (where there are dichromatic and trichromatic individuals) have the same cone connectivity pattern of H1 and H2 cell dendrites as macaque and man. This suggests that the special horizontal cell connectivity with S-cones is not correlated with the evolution of trichromacy in the Old World primates. Notably, both horizontal cell types of trichromatic primates are nonselective in their M-cone and L-cone contacts. Apparently, when the red/green processing pathway evolved from a presumably dichromatic early primate retina, this did not involve alterations in horizontal cell connectivity.

A differential cone connectivity of horizontal cells was also found in the dichromatic horse (**Figure 3**). Its A-type cells have particularly large and sparsely branched dendritic fields with few and widely spaced terminal aggregates. In one Lucifer yellow-filled A-type cell, counterstained with an S-cone marker, all but one of its 45 contacts were with S-cones. Further cells need to be studied to confirm this finding. However, the A-type, similar to the B-type, is a consistently occurring cell population that covers the horse retina (**Figure 4(b)**). Therefore, the horse (and other equids) may possess an A-HC that is selectively connected to S-cones, while the B-type connects to both types of cone.

A similarly S–cone-selective horizontal cell may be present in the cone-dominated retinas of the dichromatic sciurids. The red squirrel and ground squirrel possess an axon-bearing H1 cell (B-type) and an axonless H2 cell (A-type). The density of dendritic terminal aggregates on the H1 cell is high enough to contact all cones present. In contrast, the terminal aggregates on H2 dendrites are spaced so far apart that they can only contact a small fraction of the cones, which presumably are S-cones. It would be worthwhile and now feasible to determine, experimentally, if an S-cone selective horizontal cell is present.

Conclusions and Open Questions

The principal horizontal cell dichotomy of an axon-bearing B-type, which serves rods and cones, and a commonly axonless A-type that only serves cones, holds for most mammals. Thus, the A-type and B-type cells represent basic components of the mammalian retinal blueprint, suggesting an indispensable function. The B-types are necessary in all mammals as they all have cones and rods. However, why is there an additional A-type in the cone pathway? What functional differences are there between the A-type and the B-type? Modeling suggests that the two types with their different dendritic field sizes and spatial summation properties can explain the assumed receptive field characteristics of the cones, and, hence, the receptive field organization of bipolar cells and ganglion cells. The requirements for temporal processing may be another reason for having more than one horizontal cell type. Indeed, horizontal cells in cat differ in their flicker-response properties. However, despite the absence of the A-type, rat and mouse have ganglion cells with center/surround receptive field organizations very similar to rabbit and cat.

Mammalian horizontal cells are rather diverse across species. This diversity encompasses morphological features as well as details of cone connectivity (**Figures 7** and **3**). Thus, some amendments to the textbook characterizations are necessary:

1. Dendritic thickness and dendritic branching pattern are not defining characteristics of either type, unless the order of mammal is specified. Within a given species, the two types differ in their dendritic morphology, which suggests some physiological difference. Does it matter whether the finer dendrites are on the B-type (as in cat and rabbit) or on the A-type (as in primates and artiodactyls)?

2. Across species and orders, the A-type is more variable than the B-type. A-type variability includes axon-like processes in primates, an S-cone preference in primates and horse, and a complete lack of the A-type in some rodents. B-type variability includes dendritic and axonal branching patterns and the near avoidance of S-cones by primate H1 cells.

There is no obvious correlation between these horizontal cell variations and the phylogenetic distance of the corresponding mammalian orders. For example, Carnivora and Lagomorpha are phylogenetically less close than Primates and Scandentia, but the former have very similar horizontal cell morphologies, while those of the latter differ significantly. The peculiarity of a B-type/H1 cell with fine dendrites and an A-type/H2 cell with stout dendrites is shared by artiodactyls, perissodactyls, and primates. Within any one taxon, horizontal cell features are commonly more conserved across species.

There is no obvious correlation between these horizontal cell variations and specific adaptations to different visual requirements and other retinal specializations. Different cone/rod ratios, indicative of a nocturnal or diurnal lifestyle, do not predict the presence of one or two horizontal cell types. Primates with highly developed trichromatic color vision have cone-selective horizontal cells, but so does the dichromatic horse. In addition, the cone selectivity of H1 and H2 cells is the same in New World and Old World primates, which have different levels of color vision.

See also: Cone Photoreceptor Cells: Soma and Synapse; Information Processing: Bipolar Cells; Information Processing: Horizontal Cells; Morphology of Interneurons: Bipolar Cells; The Physiology of Photoreceptor Synapses and Other Ribbon Synapses; Rod Photoreceptor Cells: Soma and Synapse.

Further Reading

Ahnelt, P. and Kolb, H. (1994). Horizontal cells and cone photoreceptors in human retina: A Golgi-electron microscopic study of spectral connectivity. *Journal of Comparative Neurology* 343: 406–427.
Boycott, B. B., Peichl, L., and Wässle, H. (1978). Morphological types of horizontal cell in the retina of the domestic cat. *Proceedings of the Royal Society (London) B* 203: 229–245.
Chan, T. L. and Grünert, U. (1998). Horizontal cell connections with short wavelength-sensitive cones in the retina: A comparison

between New World and Old World primates. *Journal of Comparative Neurology* 393: 196–209.

Dacey, D. M., Lee, B. B., Stafford, D. K., Pokorny, J., and Smith, V. C. (1996). Horizontal cells of the primate retina: Cone specificity without spectral opponency. *Science* 271: 656–659.

Hack, I. and Peichl, L. (1999). Horizontal cells of the rabbit retina are non-selectively connected to the cones. *European Journal of Neuroscience* 11: 2261–2274.

Mills, S. L. and Massey, S. C. (1994). Distribution and coverage of A- and B-type horizontal cells stained with neurobiotin in the rabbit retina. *Visual Neuroscience* 11: 549–560.

Peichl, L. and González-Soriano, J. (1994). Morphological types of horizontal cell in rodent retinae: A comparison of rat, mouse, gerbil and guinea pig. *Visual Neuroscience* 11: 501–517.

Peichl, L., Sandmann, D., and Boycott, B. B. (1998). Comparative anatomy and function of mammalian horizontal cells. In Chalupa, L. M. and Finlay, B. L. (eds.) *Development and Organization of the Retina: From Molecules to Function*, pp. 147–172. New York: Plenum Press.

Perlman, I., Kolb, H., and Nelson, R. (2003). Anatomy, circuitry, and physiology of vertebrate horizontal cells. In Chalupa, L. M and

Werner, J. S. (eds.) *The Visual Neurosciences* vol. I, pp. 369–394. Cambridge, MA: The MIT Press.

Smith, R. G. (2008). Contributions of horizontal cells. In Masland, R. H. and Albright, T. (eds.) *The Senses: A comprehensive reference* vol. 1, pp. 341–349. Amsterdam: Elsevier.

Wässle, H., Peichl, L., and Boycott, B. B. (1978). Topography of horizontal cells in the retina of the domestic cat. *Proceedings of the Royal Society* (*London*) *B* 203: 269–291.

Wässle, H., Dacey, D. M., Haun, T., et al. (2000). The mosaic of horizontal cells in the macaque monkey retina: With a comment on biplexiform ganglion cells. *Visual Neuroscience* 17: 591–608.

Relevant Website

http://webvision.med.utah.edu – Webvision: The Organization of the Retina and Visual System.

Morphology of Interneurons: Interplexiform Cells

D G McMahon and D-Q Zhang, Vanderbilt University, Nashville, TN, USA

Glossary

Retinitis pigmentosa – A group of inherited disorders characterized by progressive loss of photoreceptors.

Sublamina of inner plexiform layer – Inner plexiform layer is divided into sublamina a (the distal sublamina) and sublamina b (the proximal sublamina). Sublamina a is further divided into two strata, 1 and 2, whereas sublamina b is divided into three strata, 3–5.

Introduction

The canonical flow of visual information in the retina is for light stimuli to be transduced into neurochemical signals by the rod and cone photoreceptors in the outer retina and then passed through the synaptic layers (outer plexiform layer, OPL; inner plexiform layer, IPL) to ganglion cells which then transmit them to the rest of the brain via the optic nerve. Interplexiform neurons are a unique class of retinal amacrine cells that form an intraretinal feedback pathway transmitting adaptational visual signals in the opposite direction, from the inner retina back to the outer retina (**Figure 1**). They receive their synaptic input in the IPL and make their output via interplexiform processes terminating in the OPL. These intraretinal feedback neurons are present in most vertebrate retinas, and in the majority of species these neurons secrete the neurotransmitter dopamine.

Dopaminergic interplexiform cells (IPCs) exert widespread influence on the physiology and function of the retina, reconfiguring retinal circuits and altering the processing of visual signals in the retina, by initiating slow and sustained changes in the physiology of retinal neurons and synapses. In particular, retinal dopamine has been found to act on electrical synapses, or gap junctions, to restrict the flow of visual signals in retinal neural networks at the level of photoreceptors, horizontal cells, and amacrine cells. In addition to the direct synaptic contacts from interplexiform processes in the outer plexiform layer, dopaminergic retinal neurons exert influence throughout the retina by volume transmission, via perfusion of dopamine beyond synaptic zones. Dopamine has been found to affect all major classes of retinal neurons, from photoreceptors to ganglion cells. An overall effect on retinal function of these feedback signals from dopaminergic IPCs is to enhance signaling in cone pathways and to decrease the signaling on rod pathways, optimizing retinal circuitry for photopic visual processing during periods of relatively high light levels. However, roles for dopamine in dark-adapted retinal responses have also been shown. Thus, dopamine and IPCs play key roles in optimizing retinal function over the wide dynamic range of light intensities encountered in the visual environment.

Secretion of dopamine by IPCs is controlled by two factors: background illumination and an intrinsic daily clock in the retina, enhancing cone signals in conditions of bright light and during the day, and rod signals in conditions of dim light and during the night. In addition to this adaptational effect on retinal neural networks, dopamine secreted from IPCs has trophic effects on photoreceptor survival and eye growth that are associated with eye diseases. In the following, we discuss the anatomy, physiology, and significance for human visual health of dopaminergic IPCs.

Morphology of Dopaminergic Interplexiform Neurons

Dopaminergic amacrine (DA) cells are a subpopulation of amacrine cells whose cell bodies lie in the innermost cell row of the inner nuclear layer (INL). DA cell processes ramify extensively in the outermost layer of the IPL where they are both presynaptic to and postsynaptic to other neurons. The total number of DA cells averages 500 per retina, although it varies slightly in different species. IPCs are a class of DA cells in which additional fine processes arising either from the cell body or from one of the dendrites ascend through the INL to the OPL where they form an output plexus for the secretion of dopamine in the outer retina. The IPC was first described in the retina of cat by Gallego in 1971, and was extensively studied by Dowling and co-authors. The percentage of DA-IPCs among DA cells is species dependent. In goldfish, Cebus monkey, and mouse retinas, almost all DA cells are likely to be DA-IPCs, whereas in rabbit, turtle, salamander, and human retinas, interplexiform processes are rarely observed on DA neurons. Approximately 50% of DA cells are DA-IPCs in the rat, *Xenopus* and *Rhesus* monkey retinas.

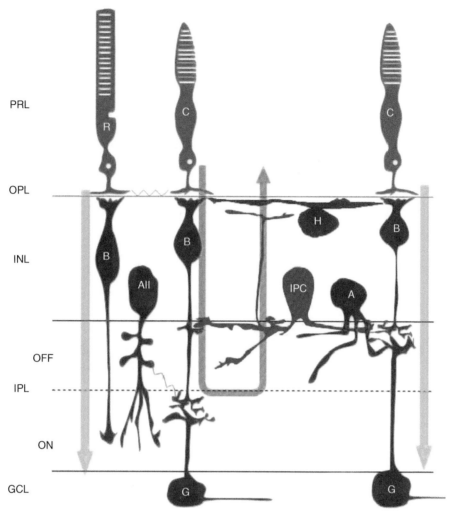

Figure 1 Interplexiform neurons mediate intraretinal feedback. An interplexiform cell (red) is located in the middle of the retina. Straight arrows indicate the canonical flow of visual signals from the outer retina to the inner retina. The U-shape arrow indicates an intraretinal feedback from the inner retina to the outer retina. R, rod; C, cone; H, horizontal cell; B, bipolar cell; AII, AII amacrine cell; IPC, interplexiform cell; A, inhibitory amacrine cell; G, ganglion cell, PRL, photoreceptor layer; OPL, outer plexiform layer; IPL, inner plexiform layer; GCL, ganglion cell layer; OFF, OFF sublamina of the IPL; ON, ON sublamina of the IPL.

Morphology and Distribution

Antibodies against tyrosine hydroxylase, the rate-limiting enzyme of catecholamine synthesis, are commonly used to detect DA cells for characterization of their localization, shape, size, dendritic arborization, and distribution (**Figure 2**). In order to enable targeting of living DA cells for morphological, functional, and molecular analysis, transgenic mouse lines have been created in which the DA cells are labeled with chemical or fluorescent protein reporters, driven by the tyrosine hydroxylase gene promoter. DA cells have three descriptive components to their morphology: soma, dendrites, and axon-like process. The morphology of dopaminergic IPCs is similar to that of other DA cells except for the presence of an additional axon-like process ascending to the OPL.

Soma

Somata of DA cells are either round or ovoid with diameters of 12–15 μm and areas of 100–150 μm² (**Figure 2**). In most species, DA cell somata are regularly and sparsely distributed in the innermost aspect of the INL across the entire extent of the retina. The distribution of the DA cells varies across the retina in the rat, with the peak density in the superior temporal quadrant and the lowest density in the inferior nasal quadrant. Occasionally, DA cells are displaced into the ganglion cell layer and, in this case, there is no evidence that these displaced amacrine cells have interplexiform processes.

Dendrites

Two to six primary dendrites arise from the cell body, and branch 4–6 times in the outermost aspect of the OFF

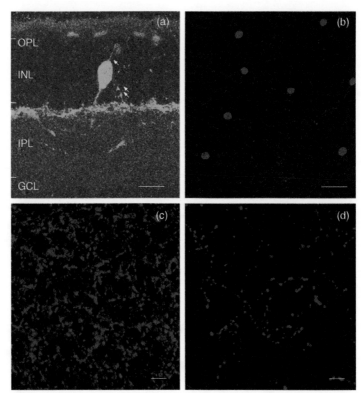

Figure 2 Immunohistochemical localization of TH in the mouse retina. (a) TH-positive soma situated in the INL stratifies at the border between the INL and the IPL in a vertical section. Single arrow and double arrows indicate ascending axon-like processes emerging from the DA cell body and from the plexus in the IPL, respectively. The axon-like processes traverse the INL toward the OPL. Scale bar: 20 μm. (b) Flat-mount view focused on the INL demonstrates the regular distribution of DA cells. Scale bar: 50 μm. (c) Flat-mount view focused on the border between the INL and the IPL shows a dense plexus of dopaminergic processes. Scale bar: 10 μm. (d) Flat-mount view focused on the OPL illustrates a loose plexus of fine dopaminergic processes. Scale bar: 10 μm.

sublamina (stratum 1) of the IPL. These branches radiate symmetrically from the soma and follow a straight soma-tofugal direction. The thickness of the primary dendrites is up to 4.5 μm and decreases at each bifurcation. The primary dendrites are usually smooth, whereas terminal branches exhibit varicosities (up to 0.5 μm in diameter). The long axis of the dendritic field of individual DA cells is approximately 800 μm. The dendritic fields of adjacent DA cells overlap extensively with a coverage factor of 2–4, forming a dense plexus in stratum 1 of the inner plexifom layer (**Figure 2**). Dendrites from teleost DA cells spread proximally throughout the IPL. In contrast, in mammalian retinas, only occasional processes from the plexus in stratum 1 run deeper into other stratums.

Axon-like fine process

Axon-like processes are quite distinct from the dendrites of DA cells. They are thin, straight, long, and sparsely branched at close to right angles. Teleost DA-IPC axon-like processes form a pronounced plexus in the OPL. DA-IPCs in Cebus monkey have dense plexuses of fine processes in both the outer and inner plexifom layer. In other species, DA cell processess form an extensive dense network in the IPL, whereas the ascending axon-like processes either do not branch or branch within clusters in the OPL. For instance, in the mouse retina, two to three axon-like processes arise either from the soma or from the primary dendrites. Each axon-like process, with multiple successive bifurcations, covers the extensive area of the retina: total length up to 10–25 mm. Thus, axon-like processes can overrun each other and, together with overlapping dendrite networks, form a particularly dense plexus of dopaminergic processes in stratum 1 of the IPL. This process network forms small rings around the origin of the primary dendrites of AII amacrine cells at the INL–IPL margin. Axon-like processes bear varicosities that are thought to be sites of dopamine release. Compared to the dense plexus in the IPL, the OPL plexus of fine dopaminergic processes is more loosely arranged. Axon-like processes also traverse the INL to the OPL, sometimes forming clusters of fine processes (**Figure 2**).

Synaptic Input to DA Neurons

Dopaminergic IPC processes extend widely into the outer and the IPLs. There is anatomical evidence that DA neurons receive synaptic input in the IPL from bipolar and amacrines cells, as well as centrifugal fibers arising from brain nuclei, and there is functional evidence for retrograde input to DA neurons from ganglion cells.

Input from bipolar cells and amacrine cells

DA cell interplexiform processes are presynaptic to horizontal and bipolar cells in the OPL, and there is no evidence that interplexiform processes are postsynaptic to any cells in the OPL. Electron microscopy of tyrosine hydroxylase immunoreactivity in rhesus monkey, cat, and rabbit has shown that DA cells receive synaptic input from both bipolar cells and inhibitory amacrine cells in mammalian retinas. The size, structure, and position of the bipolar to DA neuron synapses in the IPL suggest that they are from bistratified bipolar cells. Bistratified bipolar cells have also been found in the mouse, fish, turtle, and salamander retinas. In addition, close apposition of bipolar and DA neuron processes has also been observed in deeper layers of the IPL. Therefore, light information generated by photoreceptors may reach DA cells directly from bipolar cells or indirectly from bipolar cells through inhibitory amacrine cells (**Figure 3**).

Input from intrinsically photoreceptive ganglion cells

Intrinsically photoreceptive ganglion cells (ipRGCs) are endogenously photosensitive because they express melanopsin, a photopigment first described in photosensitive dermal melanophores of the frog. They comprise 1–2% of retinal ganglion cells forming a photosensitive network in the inner retina. The ipRGC dendritic processes ramify in the OFF layer of the IPL where they have close contact with DA neuron processes (**Figure 3**). Physiological experiments have suggested that DA cells receive synaptic input from ipRGCs (below). Whether DA cells are postsynaptic to the dendrites of ipRGCs is under investigation.

Input from centrifugal fibers

Many vertebrate retinas receive input from other parts of the brain via retinopetal axons. The efferent input to the retina originates from the ventral thalamus in reptiles, the olfactory bulb in fish, the isthmo-optic nucleus in birds, and the tuberomammillary nucleus of the posterior hypothalamus in mammals. In fish centrifugal fibers make synapses directly onto DA-IPCs. Histamine has been localized to retinopetal axons in the guinea pig, monkey, and rat, and evidence suggests that DA cells receive input from histaminergic centrifugal fibers in mammalian retinas. Thus, IPCs likely form a conduit for feedback to the retina from the rest of the brain as well as for intraretinal feedback signals.

Physiology of Dopaminergic Interplexiform Neurons

Dopamine Reconfigures Retinal Circuits

DA neurons comprise the central neuromodulatory system of the retina, forming an intraretinal feedback

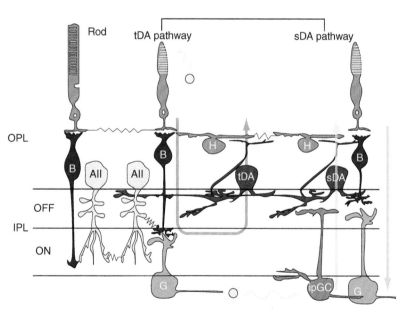

Figure 3 Dopamine neuron light response pathways. DA cells exhibit two classes of light responses: ON-transient and ON-sustained. ON-transient DA cells (tDA) are driven by rod or cone photoreceptors through ON-bipolar cells (tDA pathway), whereas ON-sustained DA neurons (sDA) are driven by melanopsin-expressing intrinsically photoreceptive ganglion cells (sDA pathway). Green straight arrow indicates the canonical flow of rod/cone visual signals from the outer retina to the inner retina. The U-shape red arrow indicates an intraretinal feedback of rod/cone initiated signals from the inner retina to the outer retina. The blue arrow indicates an intraretinal retrograde flow of ganglion cell photoreceptor signals from the inner retina to the outer retina. tDA, transient dopaminergic interplexiform cell; sDA, sustained dopaminergic interplexiform cell; ipGC, intrinsically photoreceptive ganglion cell.

pathway that reconfigures retinal circuits according to prevailing illumination conditions and signals from the retinal circadian clock. Through dopaminergic signaling, they restructure retinal function by modulation of chemical and electrical synapses, as well as by modification of the functional properties of retinal neurons. Both through direct synaptic contacts from interplexiform processes and through volume transmission, DA neurons influence all levels of retinal circuitry and all major classes of retinal neurons. In particular, dopamine regulates multiple neural circuits in the retina by modulating electrical synaptic transmission through gap junctions. This type of circuit modulation was first described in relation to dopaminergic uncoupling of gap junctions mediating electrical synaptic transmission between horizontal cells in teleost fish retinas, an action thought to reduce the size of retinal receptive fields. Subsequently, the regulatory action of dopamine on retinal gap junctions has been found to be a more general phenomenon, with coupling between rods and cones, AII amacrine cells and ganglion cells also being modulated by dopamine. Dopaminergic regulation of both rod–cone coupling and AII amacrine cell coupling serves to restrict the flow of visual signals from rods to retinal ganglion cells during light-adapted conditions in the day, while allowing flow of rod signals in the dark and at night.

In addition to regulating electrical synaptic transmission, retinal dopamine influences retinal circuit function by modifying chemical synaptic transmission, via modulation of glutamate and gamma aminobutyric acid (GABA) receptors. It also alters intrinsic functional properties of retinal neurons via modulation of membrane ion channels, such as voltage-gated sodium currents in bipolar and ganglion cells. Many of the synaptic and cellular changes induced by dopamine are consistent with a role for this transmitter in mediating light-adaptive changes in retinal function. For example, uncoupling of horizontal cells is consistent with a reduction in receptive field size and increased spatial resolution in light-adapted conditions. Restriction of rod signals from the cone pathways in the light and during the day is consistent with this notion as well. Modulation of sodium currents in bipolar cells and ganglion cells is consistent with accelerating light responses in cone pathways and shifting ganglion cell function from photon detection to contrast signaling, respectively. Finally, although IPCs do not directly synapse on photoreceptors, the outermost cells of the neural retina, they significantly influence their function, presumably through volume transmission of dopamine. In addition to regulating circadian rhythms in rod–cone coupling, dopamine influences regulation of the intracellular second messenger cyclic adenosine monophosphate and the timing of circadian secretion of melatonin by photoreceptors.

Light Drives Interplexiform Neurons via Both Conventional and Novel Pathways

One of the persistent puzzles regarding IPCs has been to understand the precise mechanisms by which they are affected by light and thus provide feedback signals within retinal circuits for background illumination. Studies of retinal dopamine release revealed a surprising heterogeneity in the lighting conditions that evoked dopamine release, including flickering light, steady light, and even prolonged darkness. Although studied intensively for more than three decades, the neural mechanisms by which light influences the activity of dopaminergic interplexiform neurons have remained incompletely understood, in part, because the sparse nature of retinal dopaminergic neurons had prevented analysis within intact retinal circuits with electrophysiological approaches that directly measure neuronal activity. Previous studies with anatomical approaches had suggested that retinal DA neurons are a homogeneous population of cells, that their primary input was from inhibitory GABA/glycine amacrine cells in the OFF sublamina of the IPL, and thus that their light responses could be due to disinhibition of OFF responses. Recently, the limitation on *in situ* electrophysiological recording of DA neurons was overcome by creating a transgenic mouse in which DA neurons were genetically marked with a fluorescent protein. Electrophysiological studies of dopaminergic interplexiform neuron light responses yielded the surprising finding that despite being considered a morphologically homogenous class, these neurons are functionally heterogeneous, with distinct neuronal subpopulations exhibiting transient and sustained light responses (**Figure 3**). This cellular heterogeneity is likely the substrate for the observed functional heterogeneity in dopamine regulation by light.

Light-responsive DA neurons fall into two distinct subpopulations (ON-transient and ON-sustained), which are driven by separate synaptic circuits. ON-transient cells are driven by rod or cone photoreceptors through ON-bipolar cells, in a conventional retinal ON-pathway circuit. ON-sustained DA neurons, however, are driven by melanopsin-expressing ipRGCs in a novel intraretinal retrograde light pathway in which ganglion cell photic signals are used for network adaptation within the retina. This novel intraretinal feedback pathway completely reverses the canonical direction of visual signaling, with photoreception being initiated in ganglion cells and dopaminergic signaling potentially modulating rod and cone photoreceptors (**Figure 3**). Thus, the dopaminergic interplexiform system is composed of two functional subpopulations of neurons tuned to distinct aspects of environmental light signals: transient DA neurons (t-DA, **Figure 3**), driven by rod and cone input through ON-bipolar cells that are tuned to rapidly changing background signals and which may mediate the observed dopamine release to flickering light, and sustained DA neurons (s-DA, **Figure 3**), driven by ipRGCs that are tuned to maintained background illumination and which may mediate the observed dopamine release to steady light.

While light stimulates dopamine secretion by dopaminergic retinal neurons, there is also basal release of dopamine in the dark. Studies by Raviola and colleagues showed that isolated DA neurons are spontaneously active, with sodium and calcium currents that support spontaneous spiking and dopamine secretion. Recordings from DA neurons in intact retinas have shown that DA neurons exhibit spontaneous bursts of spikes that are enhanced when input from inhibitory amacrine cells is blocked pharmacologically. In intact retinal circuits, the OFF-channel-driven GABA and glycine input from inhibitory amacrine cells partially suppresses the intrinsic bursting activity of DA neurons. As burst spiking is associated with increased dopamine secretion from midbrain dopaminergic neurons, the input from inhibitory amacrine neurons likely regulates basal secretion of dopamine by retinal DA neurons in the dark.

IPCs Signal Time of Day from the Retinal Clock

The retinal circadian clock shapes retinal function into 'day' and 'night' states timed to the local day–night cycle. This requires intraretinal signaling mechanisms for both the output of clock signals to the downstream retinal processes and the input of the local light cycle to set the phase of the retinal clock. Retinal dopamine, which is under both circadian and light-evoked control, appears to play a critical role in both of these processes. Among its manifold demonstrated effects on retinal neurons and circuits, dopamine regulates circadian rhythms in rod–cone coupling, and the resulting rhythms in spectral sensitivity and rod–cone balance of retinal responses. Several recent results also suggest that dopamine plays a key role in the signaling of light stimuli to the retinal circadian clock. Recent findings that retinal dopamine is a key component of light resetting of the retinal clock and that retinal dopaminergic neurons receive light input from ipRGCs suggest that retinal circadian photoreception is likely mediated by a novel intraretinal retrograde light-transmission circuit in which light signals originate in the ipRGCs and reset the circadian clock through their influence on sustained DA cells.

DA Neurons and Retinal Degenerative Disease

Parkinson's Disease

Parkinson's disease is a neurodegenerative disease caused by degeneration of doapminergic neurons in striatum and substantia nigra. Visual functions such as contrast sensitivity detected by pattern electroretinograms (PERGs) are also impaired in Parkinson's disease since there is concurrent loss of DA neurons in the retina. One possible cause is that dopamine deficiency increases receptive field size of horizontal cells in the outer retina, and of AII amacrine cells in the inner retina, through loss of dopaminergic control of gap junctions, which results in altering receptive field properties of ganglion cells (the primary origin of the PERG response).

Diabetic Retinopathy

Diabetic retinopathy, which is classically defined as a microvasculopathy, is now being viewed as a neurodegenerative disease of the retina. Functional visual deficits detected by electroretinogram (ERG) often occur before the appearance of overt retinal lesions in diabetic retinopathy. The most prominent change of the ERG in early diabetes is a distortion of oscillatory potentials that are thought to derive from the inner retinal neurons including DA cells. Indeed, DA cells initiate programmed cell death in early diabetes, and are lost during the later phases of the disease. Dysfunction of DA cells in diabetic retinas has also been suggested by studies examining retinal dopamine content and tyrosine hydroxylase activity. Together, these studies suggest that DA cells undergo degeneration in diabetes, but the cause remains unclear.

Human Health Implications of the Retinal Clock

Retinal circadian clock signaling modulates human vision, is associated with retinal degenerative diseases, and modifies photoreceptor survival in animal models of human ocular disease. Humans have daily psychophysical rhythms in absolute visual sensitivity, as well as the temporal resolution of vision, with peak scotopic sensitivity occurring at night and peak temporal resolution occurring during the day. These visual rhythms are mediated by corresponding rhythmic changes in retinal function with scotopic ERG b-wave sensitivity, and implicit time decreased during the day. Interestingly, these ERG rhythms are altered in patients with retinitis pigmentosa, indicating a link between circadian regulation of retinal function and this disease, which is a major cause of adult blindness. In addition, macular edema in diabetic retinopathy, intraocular pressure associated with glaucoma, and refractive errors associated with myopia in primates also exhibit circadian regulation. Finally, photoreceptors exhibit circadian-dependent vulnerability to light damage, being more susceptible in the night phase, and that dopamine and melatonin, two neurochemical messengers of the circadian clock, modulate photoreceptor survival in retinal degeneration animal models.

Summary

Retinal interplexiform neurons are a specialized subclass of amacrine cells that mediate intraretinal feedback of photic signals from photoreceptors, ganglion cells, and from brain nuclei to reconfigure retinal circuits according to prevailing illumination and circadian factors. They act primarily through the secretion of the neurotransmitter dopamine, both synaptically and via volume transmission. DA neurons and interplexiform neurons play critical roles in overall retinal function and in visual health by modulating retinal circuits, synchronizing the retinal clock and regulating eye and photoreceptor tropism.

See also: The Circadian Clock in the Retina Regulates Rod and Cone Pathways; Circadian Photoreception; Neurotransmitters and Receptors: Dopamine Receptors.

Further Reading

Besharse, J. C. and Luvone, P. M. (1992). Is dopamine a light-adaptive or a dark-adaptive modulator in retina? *Neurochemistry International* 20: 193–199.

Boelen, M. K., Boelen, M. G., and Marshak, D. W. (1998). Light-stimulated release of dopamine from the primate retina is blocked by 1-2-amino-4-phosphonobutyric acid (APB). *Visual Neuroscience* 15: 97–103.

Dacey, D. M. (1990). The dopaminergic amacrine cell. *The Journal of Comparative Neurology* 301: 461–489.

Djamgoz, M. B. A., Hankins, M. W., Hirano, J., and Archer, S. N. (1997). Neurobiology of retinal dopamine in relation to degenerative states of the tissue. *Vision Research* 37: 3509–3529.

Dowling, J. E. and Ehinger, B. (1975). Synaptic organization of the amine-containing interplexiform cells of the goldfish and Cebus monkey retinas. *Science* 188: 270–273.

Gustincich, S., Feigenspan, A., Wu, D. K., Koopman, L. J., and Raviola, E. (1997). Control of dopamine release in the retina: A transgenic approach to neural networks. *Neuron* 18: 723–736.

Hokoc, J. N. and Mariani, A. P. (1987). Tyrosine hydroxylase immunoreactivity in the rhesus monkey retina reveals synapses from bipolar cells to dopaminergic amacrine cells. *Journal of Neuroscience* 7: 2785–2793.

McMahon, D. G., Knapp, A. G., and Dowling, J. E. (1989). Horizontal cell gap junctions: Single-channel conductance and modulation by dopamine. *Proceedings of the National Academy of Sciences USA* 86: 7639–7643.

Mills, S. L. and Massey, S. C. (1995). Differential properties of two gap junctional pathways made by AII amacrine cells. *Nature* 377: 734–737.

Ribelayga, C., Cao, Y., and Mangel, S. C. (2008). The circadian clock in the retina controls rod–cone coupling. *Neuron* 59: 790–801.

Ruan, G. X., Allen, G. C., Yamazaki, S., and McMahon, D. G. (2008). An autonomous circadian clock in the inner mouse retina regulated by dopamine and GABA. *PLoS Biology* 6: e249.

Witkovsky, P. and Schütte, M. (1991). The organization of dopaminergic neurons in vertebrate retinas. *Visual Neuroscience* 7: 113–124.

Zhang, D. Q., Stone, J. F., Zhou, T., Ohta, H., and McMahon, D. G. (2004). Characterization of genetically labeled catecholamine neurons in the mouse retina. *Neuroreport* 15: 1761–1765.

Zhang, D. Q., Zhou, T. R., and McMahon, D. G. (2007). Functional heterogeneity of retinal dopaminergic neurons underlying their multiple roles in vision. *Journal of Neuroscience* 27: 692–699.

Zhang, D. Q., Wong, K. Y., Sollars, P. J., Berson, D. M., Pickard, G. E., and McMahon, D. G. (2008). Intraretinal signaling by ganglion cell photoreceptors to dopaminergic amacrine neurons. *Proceedings of the National Academy of Sciences USA* 105: 14181–14186.

Neuropeptides: Function

N C Brecha, UCLA School of Medicine, Los Angeles, CA, USA; VAGLAHS, Los Angeles, CA, USA
I D Raymond and A A Hirano, UCLA School of Medicine, Los Angeles, CA, USA

Glossary

Calcium imaging physiology – An experimental technique used for detecting and measuring calcium (Ca^{2+}) levels in cells or tissues. Calcium imaging techniques take advantage of calcium indicator dyes, which are molecules that respond to the binding of Ca^{2+} ions by changing their spectral properties.

G-protein-coupled receptor – A large protein family of seven-pass transmembrane receptors that bind molecules and activate intracellular signal transduction pathways to generate cellular responses by coupling to heterotrimeric guanine-triphosphate-binding proteins.

Ion channels – The transmembrane protein complexes that form a water-filled channel across the plasma membrane through which specific inorganic ions can diffuse across their electrochemical gradients.

Paracrine – Acting by volume transmission of transmitters or modulators rather than at chemical synapses in a point-to-point fashion.

Patch-clamp electrophysiology – The study of the electrical properties of biological cells and tissues using electrodes with a high-impedance (gigaohm) seal that permits measurements of small (pA) current and voltage changes.

Introduction

The vertebrate retina contains numerous transmitters and related signaling molecules, including peptides and growth factors, which have multiple roles in the retina, including cellular signaling, growth, and maintenance. The presence, expression, and distribution of peptides, and their receptors in the retina are discussed elsewhere in the encyclopedia. This article mainly focuses on the cellular actions of peptides in the retina. Peptides are characterized by their small size, ranging from 5 to 35 amino acids, and their actions are mediated through guanine-nucleotide-binding protein (G-protein)-coupled receptors (GPCRs) to influence multiple intracellular effectors, including cyclic adenosine monophosphate (cAMP), cyclic guanosine monophosphate (cGMP), Ca^{2+}, protein kinases, and phosphatases. These intracellular effectors modulate ion channels, ligand-gated channels, gap junctions, transporters, and receptors, and affect nuclear transcription factors.

The presence of peptides in the retina was firmly established in the 1970s on the basis of peptide bioactivity and immunoreactivity in retinal extracts, and peptide immunostaining, mainly of amacrine cells. Consistent with the action of peptides in the retina, peptide receptor binding sites have been reported in retinal extracts, and later the expression of their receptors by different retinal cell populations was demonstrated. There is a wide distribution of receptors, with bipolar, amacrine, and ganglion cell populations, expressing different complements of peptide receptors. Interestingly, there is often a difference between the cellular distribution of peptide-containing amacrine cells and their processes compared to the cellular distribution of their receptors, which suggests that peptides can act in a paracrine manner to influence multiple retinal circuits in both the outer and inner retina.

Peptides modulate the cellular activity of retinal neurons and circuits by influencing intracellular signaling pathways. For instance, several different peptides reported in the retina have an effect on cAMP and Ca^{2+} levels, and somatostatin (somatotropin-release-inhibiting factor, SRIF) modulates K^+ and Ca^{2+} currents. SRIF and vasoactive intestinal polypeptide (VIP) also modulate gamma aminobutyric acid-A ($GABA_A$) receptor currents. Through these actions, peptides can change the efficacy of synaptic transmission in the retina by regulating cellular excitability as well as by modulating the release of the fast-acting transmitters, GABA and glutamate, from presynaptic axonal terminals. The overall actions of peptides on retinal neurons are generally characterized as being slow in onset, long lasting and potent at low concentrations, suggestive of a role in adaptive mechanisms. Together, these findings provide support for multiple physiological roles for peptides in the retina.

Peptide Receptor Expression

In general, peptide actions are mediated by GPCRs that influence intracellular signaling pathways, which in turn modulate neuronal excitability and transmitter secretion. Furthermore, there are multiple peptide receptor subtypes for most peptides that are differentially coupled to these intracellular effectors, which markedly increase the diversity of peptide action. For instance, numerous receptor subtypes mediate the action of neuropeptide Y (NPY), somatostatin (SRIF), and VIP in the nervous system. In retina, there is a similar situation, with peptides expressed by a limited number of rarely occurring amacrine cell populations and multiple peptide receptor subtypes expressed by multiple and distinct retinal cell populations.

Numerous studies report the presence of numerous peptides in the vertebrate retina (**Table 1**). There are also multiple peptide receptors, including those for atrial natriuretic peptide, angiotensin II of the renin-angiotensin system, corticotropin-releasing hormone, NPY, opioid (enkephalin) and opioid-related peptides, SRIF, the takykinin (TK) peptides (substance P (SP), neurokinin A (NKA), neurokinin B (NKB)), and VIP in some vertebrate retinas. These observations are based on biochemical, molecular biological, and immunohistochemical findings. The best-documented peptides and peptide receptors in the mammalian retina are currently, NPY, SRIF, the TK peptides, and VIP.

Peptide-Binding Sites and Localization

In the late 1980s and early 1990s, the presence of high-affinity peptide-binding sites was reported for several peptides in the vertebrate retina. Biochemical studies showed peptide-binding sites in retinal homogenates, and autoradiographic techniques reported their localization to the plexiform layers. These findings were suggestive of peptide-mediated actions through specific receptors in the retina, although at that time very few peptide receptors had been identified and cloned.

Peptide Receptor messenger RNAs

The identification and cloning of the key peptide receptor genes in the early to mid-1990s provided the tools for determining the expression of specific peptide receptors in the retina. NPY, SRIF, and the TK peptide receptors

Table 1   Peptides and peptide receptors in the retina

Peptides	Preferred receptors
Angiotensin II	AT_{1A}, AT_2
Enkephalin (leu$_5$- and met$_5$-enkephalin)	delta-, kappa-, mu-opioid receptors
CRH	CRH-binding sites
Neuropeptide Y	Y1, 2, 4, 5
Pituitary adenylate-cyclase-activating polypeptide	PAC1, VAC1, VAC2
Somatostatin	sst$_{1-5}$
Tachykinin peptides	
• Substance P	NK-1
• Neurokinin A	NK-2
• Neurokinin B	NK-3
Vasoactive intestinal polypeptide	VAC1, VAC2

were reported in the mammalian retina. There are typically multiple subtypes of these receptors, based on both molecular and immunohistochemical findings, and these different subtypes vary in their abundance. These findings confirm that the mammalian retina, as indicated by binding and autoradiographic studies, synthesizes multiple peptide receptors.

Peptide Receptor Localization

Peptide receptors, including those that mediate the cellular actions of NPY, SRIF, and the TKs, have been principally studied in the mouse and rat retinas. Several receptor subtypes are expressed in the retina, often by one or more distinct cell types, including photoreceptors, bipolar, amacrine, and ganglion cells. For instance, in rat retina, the neurokinin-1 (NK-1) receptor, which is the preferred receptor for SP, is expressed by GABAergic amacrine cells, while, the NK-3 receptor, which is the preferred receptor for NKB, is expressed by OFF-type bipolar cells.

A frequent observation that has emerged from immunostaining studies is the mismatch between the distribution and number of peptide-containing processes and their receptors. For example, there is a differential distribution of SRIF processes and SRIF subtype (sst) receptors in the inner plexiform layer (IPL). There are examples of several peptides acting at receptors that are located away from the peptide-containing processes, which support the notion of a paracrine mode of action, although this does not rule out direct transmitter actions occurring at or near synaptic specializations where there is a close apposition of peptide- and peptide-receptor-expressing processes.

Intracellular Signaling

The activation of signal transduction cascades by peptides and related analogs in retinal extracts also supports the idea that endogenous peptides are present in the retina and their actions are mediated by specific receptors.

However, there are differences in observations concerning peptide actions on intracellular signaling pathways that may be due to the experimental approaches and sensitivity of the assays, as well as species differences. For instance, both NPY and SRIF inhibit forskolin-induced cAMP accumulation in some, but not all, vertebrate retinas. In contrast, VIP and VIP-related peptides potently stimulate adenylate cyclase activity in many vertebrate retinas. For example, SRIF is reported to stimulate cAMP accumulation in chick and ovine retinas, but SRIF does not affect cAMP accumulation in fish, pigeon, mouse, and rabbit retinas. Furthermore, SRIF inhibits VIP-stimulated adenylate cyclase activity in sheep retina. There is further complexity in evaluating SRIF's action in retinal homogenates, based on a recent report using both wild-type and sst receptor knock-out mouse lines. This study reported that interactions of sst receptor subtypes as well as levels of sst receptors influenced SRIF potency on adenylate cyclase activity. Finally, the sst_2 receptor is reported to mediate SRIF inhibition of adenylate cyclase activity through a G protein of the $G_{o\alpha}$ type in both mouse and rabbit retina.

In contrast, VIP's potent stimulation of adenylate cyclase activity is well established in many vertebrate retinas. Peptide histidine isoleucine, which is co-expressed with VIP, also stimulates adenylate cyclase activity. In addition, the VIP-related peptides, pituitary adenylate-cyclase-activating polypeptide-27 (PACAP-27) and PACAP-38, are positively coupled to adenylate cyclase activity, and their actions are more potent than VIP, indicating the presence of the PAC1 receptor in retinal homogenates. Furthermore, in rat retinal homogenates, PACAP-27 and PACAP-38, but not VIP, show a dose-dependent stimulation of inositol phosphate levels, suggesting multiple signal transduction pathways for PACAP peptides.

SRIF acting through the sst_2 receptor is also reported to induce nitric oxide production in the retina. Nitric oxide, in turn, would activate soluble guanylate cyclase and increase levels of cGMP. The presence of functional NK receptors has been shown by the dose-dependent stimulation of inositol phosphate accumulation and intracellular Ca^{2+} ($[Ca^{2+}]_i$) mobilization by the TK peptides, SP, NKA, and NKB.

Cellular Signaling

Ca^{2+} Imaging and Ion Channel Physiology

Several peptides, including met_5-enkephalin, NPY, SP, and SRIF, have an action at the cellular level in both nonmammalian and mammalian retinas. For instance, in goldfish retina, met_5-enkephalin, SP, and SRIF partially inhibit voltage-dependent Ca^{2+} currents in isolated mixed bipolar cell axon terminals. In the mammalian retina, low concentrations of NPY, SRIF, or VIP modulate both voltage- and ligand-gated channels in multiple cell types, as discussed below.

Pharmacological studies used isolated rat rod bipolar cells in acute preparations, coupled with Ca^{2+} imaging techniques to characterize the cellular actions of NPY and SRIF. Low concentrations of NPY did not result in detectable changes in $[Ca^{2+}]_i$ levels in rod bipolar cell axon terminals, suggesting that NPY alone does not influence $[Ca^{2+}]_i$ levels. In contrast, there is a dose-dependent inhibition of K^+-evoked increases of $[Ca^{2+}]_i$ with NPY; inhibition is maximal with 1 μM NPY and is also observed with 0.1 nM NPY. Maximal inhibition is also seen with 1 μM C2-NPY and NPY(13–36), selective Y2 receptor agonists.

In contrast, no inhibition is observed with Y1, Y4, and Y5 agonists. These findings indicate that NPY acts presynaptically through Y2 receptors to regulate glutamate release from rod bipolar cell axon terminals (**Figure 1**).

There have been several studies of SRIF action at the cellular level. In both isolated rat and rabbit rod bipolar cells, there is no detectable change in $[Ca^{2+}]_i$ levels following direct application of SRIF. However, SRIF strongly inhibits a K^+-stimulated increase of $[Ca^{2+}]_i$ through L-type Ca^{2+} channels in a dose-dependent manner in rat and rabbit rod bipolar cell axon terminals (**Figure 2**). SRIF also enhances GABA-evoked whole-cell currents in amacrine cells likely due to $GABA_A$ receptor phosphorylation, following activation of adenylate cyclase. In rabbit retina, SRIF and octreotide, a SRIF agonist, reduce a K^+-stimulated increase in $[Ca^{2+}]_i$ in rod bipolar cell axon terminals. In addition, SRIF inhibits Ca^+-activated K^+ currents (I_{BK}) in these cells. The octreotide effect is prevented by L-Tyr8Cyanamid 154806, an sst_2 receptor antagonist, indicating that these SRIF effects are likely to be mediated by sst_2 receptor activation.

In salamander photoreceptors, low concentrations of SRIF modulate both voltage-activated K^+ and L-type Ca^{2+} currents. SRIF enhances a delayed outwardly rectifying K^+ current in both rod and cone photoreceptors (**Figure 3**). It differentially modulates L-type Ca^{2+} channels currents: SRIF reduces the Ca^{2+} current in rods and increases the Ca^{2+} current in cones (**Figure 4**). Ca^{2+} imaging experiments of isolated rod and cone photoreceptors produce findings consistent with the electrophysiological findings. Together, these observations suggest that SRIF has a role in the regulation of glutamate release from photoreceptors based on its modulation of both voltage-gated K^+ and Ca^{2+} currents.

Membrane currents are not altered following direct application of VIP to isolated rat rod bipolar and ganglion cells. However, VIP does influence GABAs action at $GABA_A$ receptors. It potentiates GABA-evoked whole-cell currents in bipolar cells by 65% and ganglion cells by 54% (**Figure 5**). The onset of VIP action is slow with a long recovery period. VIP-induced potentiation of whole-cell GABA currents is mediated through phosphorylation of $GABA_A$ receptors, following adenylate cyclase activation. The activation of a cAMP-dependent pathway is consistent with biochemical findings that VIP increases cAMP levels in the mammalian retina. In contrast to these observations in rat, VIP reduces GABA-evoked whole-cell currents of rabbit bipolar cells by about 40%. Finally, in rat amacrine cells, VIP, SRIF, and enkephalin also enhance GABA-evoked whole-cell currents through $GABA_A$ receptor phosphorylation by protein kinase A to increase affinity of $GABA_A$ receptor for GABA. Thus, the peptides appear to fine-tune the inhibition by GABA.

These findings of NPY, SRIF, and VIP action at the cellular level are consistent with biochemical studies

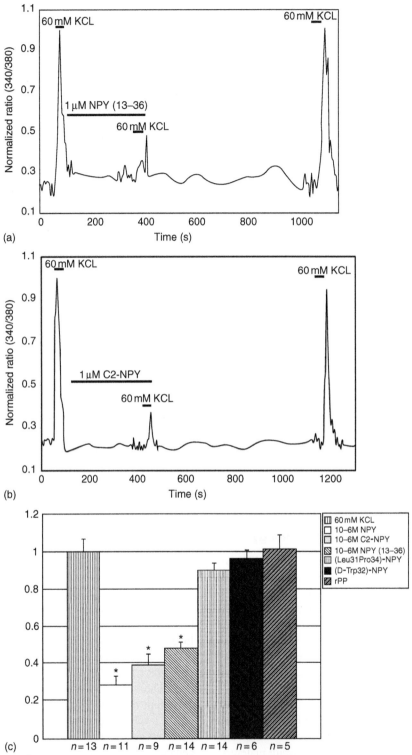

Figure 1 NPY inhibits Ca^{2+} influx into rod bipolar cells through Y2 receptor activation. (a) Micromolar concentrations of Y2 receptor selective agonists NPY (13–36) and (b) C2-NPY inhibits high K^+-induced increases in $[Ca^{2+}]_i$ in rod bipolar cell axonal terminals. The Ca^{2+} responses recover to baseline 10 min after the removal of the peptide. (c) Effect of NPY and selective Y-receptor agonists on depolarization-induced increases in intracellular calcium $[Ca^{2+}]_i$. Summary bar graph comparing the effects of NPY, Y2 receptor selective agonists NPY (13–36) and C2-NPY, Y1 receptor selective agonist [Leu^{31}Pro^{34}]-NPY, and Y5 receptor selective agonist [D-Trp^{32}]-NPY. (*) Significant inhibition when compared with high K^+ control ($P < 0.05$). From Figure 4 in D'Angelo, I. and Brecha, N. C. (2004). Y2 receptor expression and inhibition of voltage-dependent Ca^{2+} influx into rod bipolar cell terminals. *Neuroscience* 125: 1039–1049, with kind permission from Elsevier.

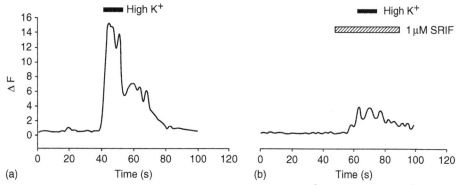

Figure 2   SRIF strongly inhibits a K⁺-stimulated increase of intracellular calcium [Ca²⁺]ᵢ through L-type Ca²⁺ channels in a rod bipolar axon cell terminal. (a) K⁺-stimulated increase of [Ca²⁺]ᵢ in an axonal terminal. (b) Inhibition of K⁺-stimulated increase of [Ca²⁺]ᵢ in an axonal terminal by SRIF treatment (striped bar). From Figure 3 in Johnson, J., Caravelli, M. L., and Brecha, N. C. (2001). Somatostatin inhibits calcium influx into rat rod bipolar cell axonal terminals. *Visual Neuroscience* 18: 101–108.

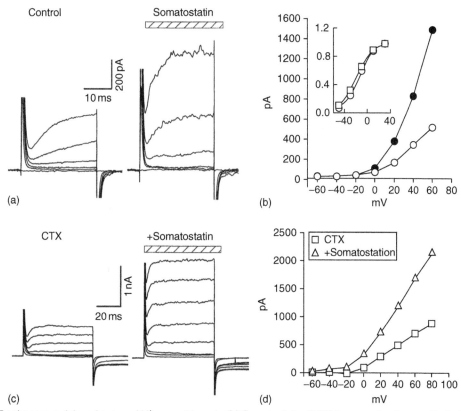

Figure 3   SRIF enhances a delayed outward K⁺ current in rods. (a) Somatostatin (SRIF) increased voltage-activated K⁺ current. (b) Current–voltage (I–V) relationship of the K⁺ current obtained by holding cells at –70 mV and stepping from –60 to +40 mV in 20 mV increments. (c) Charybdotoxin (CTX), which blocks calcium-activated K⁺ current, reduced outward current but did not prevent SRIF-induced increases in K⁺ currents. (d) I–V relationship of outward current in the presence of CTX. From Figure 2 in Akopian, A., Johnson, J., Gabriel, R., Brecha, N., and Witkovsky, P. (2000). Somatostatin modulates voltage-gated K(⁺) and Ca(2⁺) currents in rod and cone photoreceptors of the salamander retina. *Journal of Neuroscience* 20: 929–936.

reporting the presence of peptide receptors and the activation of intracellular effectors in retinal homogenates. They are also consistent with the localization of peptide receptors to different retinal cell types. Together, they show that peptides have multiple cellular actions including modulation of voltage- and ligand-gated ion channels, which would influence intrinsic cellular properties and transmitter release from retinal cells.

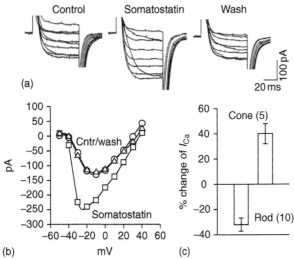

(a)

(b) mV

(c)

Figure 4 Excitatory effects of somatostatin (SRIF) on the Ca^{2+} current in cones. (a) Cones were held at − 70 mV, and depolarizing pulses were applied from − 40 to +40 mV in 10 mV steps. (b) Current–voltage plot of the data in (a) is shown. (c) A summary of the changes induced in the peak Ca current of rods and cones by SRIF is shown. From Figure 6 in Akopian, A., Johnson, J., Gabriel, R., Brecha, N., and Witkovsky, P. (2000) Somatostatin modulates voltage-gated $K^{(+)}$ and $Ca^{(2+)}$ currents in rod and cone photoreceptors of the salamander retina. *Journal of Neuroscience* 20: 929–936.

Functional Studies

Electroretinogram Recording

A limited action of peptides in the retina has been shown using electroretinogram (ERG), which reflects the contribution of electrical activity from the different retinal cell types. The initial a-wave appears to arise from photoreceptors; the b-wave, the activity of primarily second-order bipolar cells and Müller glial cells; the slower c-wave, the activity of the retinal pigment epithelium and Müller cells; and the d-waves, the activity of the OFF pathway. In rabbit retina eyecups, low concentrations of SRIF decreased the amplitude of the b-wave in rabbit eyecup preparations, but several other peptides, including cholecystokinin (CCK) and SP do not appear to affect the ERG.

Extracellular Recordings

In contrast, extracellular recordings have shown that SP, SRIF, and VIP influence multiple retinal cell types in both nonmammalian and mammalian retinal eyecup preparations. Typically, low concentrations of peptides increase the general excitability of cells. In the mudpuppy eyecup preparation, neurotensin and SP increased the firing activity of ganglion cells; however, met_5-enkephalin

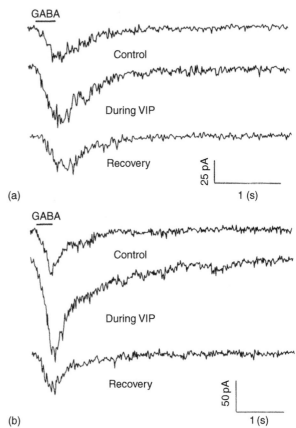

(a)

(b)

Figure 5 VIP potentiation of GABA-evoked Cl^- current in a bipolar cell (a) and a ganglion cell (b). The whole-cell GABA-evoked currents in both the bipolar and ganglion cells were potentiated after a 11-s application of VIP. From Figure 4 in Veruki, M. L. and Yeh, H. H. (1992). Vasoactive intestinal polypeptide modulates $GABA_A$ receptor function in bipolar cells and ganglion cells of the rat retina. *Journal of Neurophysiology* 67: 791–797.

inhibited them. However, the distribution of μ-opioid receptors is unknown in mudpuppy retina and the action of met_5-enkephalin could be an indirect effect on amacrine cells. These actions are characterized as slow and long lasting. There is also some specificity in peptide action. For example, using a rabbit eyecup preparation as described below, SRIF acts on all ganglion cell types; in contrast, SP acts on most brisk ganglion cells and VIP acts on ON- and OFF-center brisk ganglion cell types, respectively. Overall, peptide actions in this preparation are characterized as being modulatory and too slow to participate in fast, light-evoked responses.

Application of SP at low-to-moderate concentrations excites most brisk ganglion cells, including ON- and OFF-center and ON–OFF directionally selective ganglion cells in the rabbit retina (**Figure 6**). SP exerts these excitatory effects without affecting ganglion cell receptive field properties. The latency of the SP response

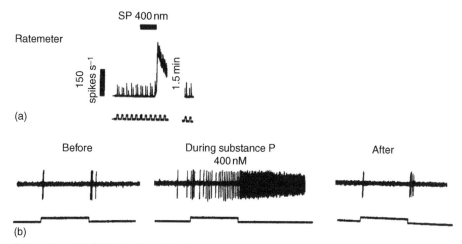

Figure 6 SP excitation of an ON–OFF direction selective ganglion cell responding to a stationary flashing spot. (a) Ratemeter records showing response to flashes (bottom trace) before, during and after application of 400 nM SP (upper bar); calibration bar on right is for 150 spikes s^{-1}. (b) Individual light responses before, during, and after application of SP. Bottom traces show duration of light response and upper traces show spikes. From Figure 1 in Zalutsky, R. A. and Miller, R. F. (1990). The physiology of substance P in the rabbit retina. *Journal of Neuroscience* 10: 394–402.

was shorter than that of SRIF. Furthermore, intracellular recordings show that SP depolarizes some amacrine cells, including GABA-containing amacrine cells. These experiments are consistent with the expression of the SP-specific NK-1 receptor by GABA-containing amacrine cells. SP did not affect horizontal cell activity, consistent with the lack of TK-binding sites and NK-1- or NK-3-receptor immunostaining of horizontal cells. Investigations also report that SP excites most ganglion cells in mudpuppy and fish retina. Together, these findings indicate that TK peptides act in the inner retina and affect the general excitability of ganglion cells and their level of spontaneous activity, rather than altering the characteristics of ganglion cell receptive field properties, as reported for SRIF and VIP in the rabbit retina.

In rabbit retina, low concentrations of SRIF excite all ganglion cell types, change their signal-to-noise ratio discharge activity and shift their center-surround balance toward a more dominant center (**Figure 7**). Similar to other peptides, SRIF actions are characterized as being slow in onset and having a long latency. SRIF is also reported to act on multiple cells in the inner and outer retina; it directly affects bipolar, amacrine, and ganglion cells and influences the horizontal cell network. Moreover, SRIF is reported to increase input resistance of amacrine and bipolar cells, suggesting an action on ion channels or gap junctions. This would be consistent with patch-clamp experiments showing that SRIF affects K$^+$ and Ca^{2+} currents in retinal neurons. In addition, modulation of gap junctions could be mediated by SRIF-induced dopamine release or nitric oxide production, as suggested by pharmacological studies. Together, these

observations indicate that SRIF acts on multiple retinal circuits to produce long-lasting changes in ganglion cell activity and receptive field organization, consistent with the idea that this peptide acts as a modulator to mediate the effects of light adaptation in the retina.

VIP excites ON- and OFF-center brisk ganglion cells in the rabbit retina. Similar to observations using SRIF and SP, VIP is potent at low concentrations and there is a delay in the onset of its action on ganglion cells. The maintained activity of ON- and OFF-center ganglion cells is increased by VIP and their excitatory responses to flashes of light are unaffected or slightly reduced in the presence of VIP. In contrast, VIP has little effect on the maintained activity of ON/OFF directionally selective ganglion cells, and the response of these cells to a moving stimulus. Finally, the action of VIP on ganglion cells in rabbit retina is congruent with other reports of the action of VIP on GABA currents on isolated ganglion cells in the rat retina.

Peptide Influence on Transmitter Release

Several investigators have evaluated the influence of peptides on transmitter release from the retina. The general experimental paradigm is to preload the retina with a radiolabeled transmitter, such as GABA, glycine, or dopamine, and, following a washout period, to add the peptide and measure changes of radiolabeled transmitter levels induced by the peptide. An alternative experimental design is to measure endogenously released transmitter

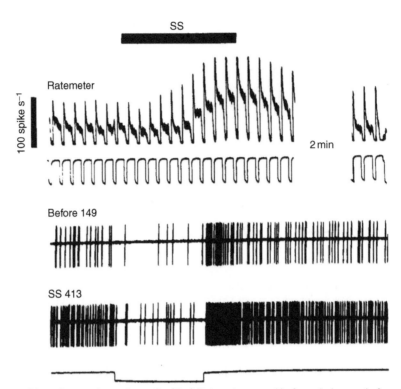

Figure 7 SRIF (somatostatin) excites most ganglion cells. (Top) Ratemeter record before, during, and after application of somatostatin indicated by the bar at top. (Middle and lower) An OFF ganglion cell response before and after 400-nM SRIF application. SS, somatostatin. From Figure 1 in Zalutsky, R. A. and Miller, R. F. (1990). The physiology of somatostatin in the rabbit retina. *Journal of Neuroscience* 10: 383–393.

levels, using radioimmunoassays or by high-pressure liquid chromatography.

Low concentrations of exogenously applied NPY stimulate the release of glycine, dopamine, acetylcholine, and 5-hydroxytryptamine from the vertebrate (frog and rabbit) retina. SRIF stimulates release of dopamine from the rat retina, and pharmacological studies indicate this action is mediated by sst receptors. SP also evoked the release of dopamine from rat retina, and the modulation of Ca^{2+} currents in fish bipolar cells by SP suggests that it may affect transmitter release.

A variant of this experimental approach involves the use of flashing light to stimulate transmitter release to determine if peptides influence light-evoked transmitter release. For example, the μ-opioid receptor agonist, (D-Ala2, MePhe4, Gly-ol5)-enkephalin (DAMGO), increases light-evoked release of acetylcholine from rabbit retina. These latter findings are consistent with-opioid-binding sites distributed homogenously over the IPL and GCL of the rat and monkey retina, and the localization of μ-opioid receptor immunoreactivity by bistratified ganglion cells of the rat retina. Furthermore in rabbit retina, nociceptin, the endogenous ligand for opioid-receptor-like 1, inhibits acetylcholine release evoked by flickering light. In contrast, SP and SRIF do not change the level of light-evoked release of acetylcholine from rabbit retina.

Peptide Function

Experimental findings are consistent with peptides acting as slow transmitters or modulators in the retina. For instance, NPY, PACAP, SP, SRIF, and VIP, acting through GPCRs, influence adenylate cyclase or phospholipase C activity in retinal homogenates and whole retina. Peptides also have potent modulatory effects on both Ca^{2+} and K^+ currents in retinal neurons. Furthermore, SRIF acting through sst_{2A} receptors inhibits the release of the excitatory transmitter glutamate from rod bipolar cells. In addition, the action of GABA at $GABA_A$ receptors is modulated by SRIF and VIP through phosphorylation of $GABA_A$ receptors by protein kinase A in rat rod bipolar, amacrine, and ganglion cells. These examples illustrate the functional role of peptides in regulating both transmitter release and the excitability of retinal neurons.

The concept that peptides act as modulators of retinal circuitry or networks is based on both anatomical and functional findings. Peptide-containing cells in most cases are wide-field amacrine cells that are characterized by a very-low–to-medium cell density. These cells ramify widely and have overlapping processes that cover the entire retinal surface. Therefore, they would have a broad influence on a large number of cells and are unlikely to mediate discrete point-to-point information processing. Connectivity studies show that most synaptic input onto peptide-containing amacrine cell processes are from amacrine cells, and these cells in turn terminate mainly on amacrine and ganglion cells. Peptides released from these cells could therefore influence multiple cells in a local fashion. In addition, a consistent finding for several peptides is a distinct distribution of peptide-containing and peptide-receptor-bearing processes and cells. These findings suggest that peptides are likely to diffuse from their release sites and act at a distance in a paracrine fashion. Together, these observations support the concept that peptides have a widespread effect on multiple types of retinal neurons.

A feature of many wide-field amacrine cells is the co-expression of GABA and a peptide. For example, in rat, cat, and monkey retina, GABA is co-localized with NPY, SP, and VIP in different amacrine cell types. In addition, glutamate, the predominant ganglion cell transmitter, is co-expressed with PACAP in ganglion cells that innervate the suprachiasmatic nucleus of the hypothalamus. These observations are consistent with findings elsewhere in the nervous system reporting the co-expression and co-release of classical transmitters and peptides from the same cell. Interestingly, a differential release of classic transmitters and peptides depending on the frequency and pattern of cell firing has been shown for several neuronal systems. GABA or glutamate and peptides can act together at the same site, or the peptides can diffuse through the tissue and act at more distant cellular sites. There is evidence for both modes of action in the retina. GABA and peptides released from wide-field amacrine cells may act locally at GABA$_A$ receptors; for instance, both VIP and GABA act on GABA$_A$ receptors expressed by bipolar cell axons and ganglion cells. Peptides can also diffuse from their release site and act in a paracrine manner, as suggested for example, by the different distributions of SRIF immunoreactive processes and sst receptors, which are expressed by multiple retinal cell types.

Functional studies also support the concept that peptides act as modulators of retinal circuitry. In general, the cellular effects of peptides are slow, and they occur at multiple locations in the retina. Both SRIF and VIP produce excitatory changes in the range of seconds to minutes in the spontaneous activity and neuronal discharge patterns of ganglion cells. Moreover, SRIF affects the center-surround balance of all types of ganglion cells. This action of SRIF suggests a role in light/dark adaptation. The role of other peptides in visual function is less well understood. However, as their wide-ranging actions are also likely to be too slow to mediate fast signaling, these peptides will probably also act in modulatory processes that occur on longer timescales and participate in adaptive mechanisms that globally affect the state of retinal circuits and networks.

Conclusion

The vertebrate retina is richly endowed with multiple peptides and peptide receptors; peptides are often localized to a single or at the most a few amacrine cell types, and their receptors are typically expressed by multiple cell types. Peptides act through GPCRs to modulate voltage- and ligand-gated ion channels, and influence neuronal excitability and transmitter release. The pattern of peptide and peptide receptor expression, and the cellular action of peptides being slow in onset, long lasting and potent at low concentrations is congruent with a modulatory role of peptides that would influence multiple cells and cellular networks. This broad modulatory role is consistent with peptides participating in slow signaling events in the retina and influencing adaptive mechanisms.

Acknowledgment

Support for this work was provided by NEI EY 04067 and a Veterans Administration Senior Career Scientist Award.

See also: GABA Receptors in the Retina; Neuropeptides: Localization; Neurotransmitters and Receptors: Dopamine Receptors; Neurotransmitters and Receptors: Melatonin Receptors.

Further Reading

Akopian, A., Johnson, J., Gabriel, R., Brecha, N., and Witkovsky, P. (2000). Somatostatin modulates voltage-gated K($^+$) and Ca(2$^+$) currents in rod and cone photoreceptors of the salamander retina. *Journal of Neuroscience* 20: 929–936.
Brecha, N. C. (1983). Retinal neurotransmitters: Histochemical and biochemical studies. In: Emson, P. C. (ed.) *Chemical Neuroanatomy*, pp. 85–129. New York: Raven.
Brecha, N. C. (2003). Peptide and peptide receptor expression and function in the vertebrate retina. In: Chalupa, L. and Werner, J. (eds.) *Visual System*, pp. 334–354. Boston, MA: MIT Press.
Casini, G., Catalani, E., Monte, M. D., and Bagnoli, P. (2005). Functional aspects of the somatostatinergic system in the retina and the potential therapeutic role of somatostatin in retinal disease. *Histology and Histopathology* 20: 615–632.
Cervia, D., Casini, G., and Bagnoli, P. (2008). Physiology and pathology of somatostatin in the mammalian retina: A current view. *Molecular and Cellular Endocrinology* 286: 112–122.
D'Angelo, I. and Brecha, N. C. (2004). Y2 receptor expression and inhibition of voltage-dependent Ca^{2+} influx into rod bipolar cell terminals. *Neuroscience* 125: 1039–1049.
Feigenspan, A. and Bormann, J. (1994). Facilitation of GABAergic signaling in the retina by receptors stimulating adenylate cyclase. *Proceedings of the National Academy of Sciences of the United States of America* 91: 10893–10897.
Hökfelt, T., Broberger, C., Xu, Z. Q., et al. (2000). Neuropeptides – an overview. *Neuropharmacology* 39: 1337–1356.
Jensen, R. J. (1993). Effects of vasoactive intestinal peptide on ganglion cells in the rabbit retina. *Visual Neuroscience* 10: 181–189.

Johnson, J., Caravelli, M. L., and Brecha, N. C. (2001). Somatostatin inhibits calcium influx into rat rod bipolar cell axonal terminals. *Visual Neuroscience* 18: 101–108.

Thermos, K. (2003). Functional mapping of somatostatin receptors in the retina. *A review.Vision Research* 43: 1805–1815.

Veruki, M. L. and Yeh, H. H. (1992). Vasoactive intestinal polypeptide modulates GABA$_A$ receptor function in bipolar cells and ganglion cells of the rat retina. *Journal of Neurophysiology* 67: 791–797.

Zalutsky, R. A. and Miller, R. F. (1990). The physiology of somatostatin in the rabbit retina. *Journal of Neuroscience* 10: 383–393.

Zalutsky, R. A. and Miller, R. F. (1990). The physiology of substance P in the rabbit retina. *Journal of Neuroscience* 10: 394–402.

Neuropeptides: Localization

N C Brecha, UCLA School of Medicine, Los Angeles, CA, USA; VAGLAHS, Los Angeles, CA, USA
A A Hirano and I D Raymond, UCLA School of Medicine, Los Angeles, CA, USA

Glossary

Autoradiography – A histochemical technique used to localize the binding site of a ligand in tissue using radiolabeled ligands (usually ^{131}I, ^{3}H, or ^{35}S) and photographic techniques to detect the isotope.
BAC – A bacterial artificial-chromosome-generated transgenic animals, which are a relatively potent way to generate knock-in transgenic mice that is thought to better recapitulate normal gene expression due to use of a large, chromosomal amount of regulatory genomic DNA surrounding the protein-coding region of the gene.
***In situ* hybridization histochemistry** – A histochemical technique used to localize the cellular localization of messenger RNA.
Isoforms – The different forms of a protein, derived from a single gene that result from alternative splicing or from a family of related genes.
Paracrine – Acting by volume transmission of diffuse messengers rather than at chemical synapses in a point-to-point fashion.

Introduction

The vertebrate retina contains numerous transmitters and related signaling molecules, including peptides and growth factors (**Table 1**). Peptides and growth factors have multiple roles in the retina, including cellular signaling, growth, and maintenance. This article focuses on the localization of peptides that are primarily involved in cellular signaling, including neurotransmission, and participate in retinal circuitry functions mediating visual information processing. Peptides are characterized by their small size ranging from 5 to 35 amino acids and slow actions that are mediated through multiple guanine-nucleotide-binding protein (G-protein)-coupled receptors (GPCRs), which influence intracellular signaling pathways.

The presence of peptides has been documented in both nonmammalian and mammalian retinas. The most studied peptides in the mammalian retina are neuropeptide Y (NPY), somatostatin (somatotropin-release-inhibiting factor, SRIF), the tachykinins (substance P (SP), neurokinin A (NKA) and neurokinin B (NKB)), and vasoactive intestinal polypeptide (VIP). This article primarily uses these peptides as exemplars of the pattern of peptide expression in the vertebrate retina.

Evidence for abundant peptide expression in the vertebrate retina began with multiple descriptions in the late 1970s of peptide activity in retinal extracts and peptide immunostaining of amacrine cells. These studies established that peptides are usually localized to low-density populations of wide-field amacrine cells; some ganglion cells also express peptides. Consistent with the presence of peptides in the retina is the expression of their receptors, which have a wide distribution, and have been reported in bipolar, amacrine, and ganglion cell populations. Interestingly, there is often a mismatch between the cellular distribution of peptide-containing amacrine cells and their processes compared to the cellular distribution of their receptors. A striking example is the distribution of SRIF and its receptor, SRIF subtype 2A (sst$_{2A}$, as discussed below), which suggests that peptides mainly act in a paracrine manner, and therefore influence multiple retinal circuits in both the outer and inner retina.

Peptides influence the cellular activity of retinal neurons and circuits by modulating multiple intracellular signaling pathways, which affect transmitter release and intrinsic neuronal properties. Peptide actions are characterized as being slow in onset, long-lasting, and potent at low concentrations, suggestive of a role in adaptive mechanisms. Together, these findings provide strong support for a functional role of peptides in the retina.

Peptide Expression

Bioassays and Radioimmunoassays

In the 1950s, Euler and his colleagues reported the presence of SP bioactivity in dog and bovine retinal extracts. However, it was not until the late 1970s with the development of additional bioassay systems and radioimmunoassays (RIAs) that a rich variety of peptides, including cholecystokinin, the enkephalin peptides, glucagon, SRIF, SP, thyrotropin-releasing hormone, and VIP, was described in vertebrate retinal extracts. At the present time, over 20 peptides have been reported in the vertebrate retina (**Table 2**) with a greater number of peptides and peptide families in nonmammalian compared to mammalian retinas.

In general, bioassay and RIA studies report low-to-moderate levels of peptides in retinal extracts compared to other tissues and brain regions. In most cases, findings

Table 1   Peptides and peptide receptors in the retina

Peptides	Preferred receptors
Neuropeptide Y	Y1, 2, 4, 5, 6
Pituitary adenylate-cyclase-activating polypeptide	PAC1, VAC1, VAC2
Somatostatin	sst$_{1-5}$
Tachykinin peptides	
• Substance P	NK-1
• Neurokinin A	NK-2
• Neurokinin B	NK-3
Vasoactive intestinal polypeptide	VAC1, VAC2

Table 2   Peptides reported in the vertebrate retina

Angiotensin II
Cholecystokinin
Corticotropin-releasing hormone
Enkephalin
    • Leu$_5$-enkephalin
    • Met$_5$-enkephalin
ß-Endorphin
FMRFamide
Glucagon
Luteinizing–hormone-releasing hormone
Natriuretic peptides
    • Atrial natriuretic peptide (α-ANP and γ-ANP)
    • Brain natriuretic peptide
    • C-type natriuretic peptide
Neurotensin
Neuropeptide Y
Nociceptin
Pituitary adenylate-cyclase-activating polypeptide
Peptide histidine isoleucine
Somatostatin
Tachykinin peptides
    • Substance P
    • Neurokinin A
    • Neurokinin B
Thyrotropin-releasing hormone
Vasoactive intestinal polypeptide

from these investigations are consistent with immuno-histochemical studies showing that these peptides are expressed in amacrine and ganglion cells. For instance, SRIF bioactivity and immunoreactivity are in retinal extracts from numerous species, and this peptide is localized to sparsely occurring amacrine and displaced amacrine cells in all species studied to date. Similarly, SP bioactivity and immunoreactivity are reported in the retina of numerous species, and SP immunoreactivity is localized to both amacrine and ganglion cells.

Peptides detected in the retina correspond to those detected in other tissues as to their molecular structure based on gel electrophoresis or high-pressure liquid chromatography and, in more limited cases, on peptide sequencing. These studies revealed that both forms of SRIF, SRIF-14 and SRIF-28, are differentially expressed in the mammalian retina; SRIF-14 is the predominant form in the rat and human retina and SRIF-28, in addition to SRIF-14, is in guinea pig and rabbit retina. Functionally, this finding is likely to be of importance, since SRIF-14 and SRIF-28 preferentially bind to different SRIF receptor subtypes.

Peptide Localization

Peptide Messenger RNA

There have been few investigations documenting peptide messenger RNAs (mRNAs) in the retina. Most studies evaluated mouse, rat, and human retinal extracts using Northern blots and reverse transcription polymerase chain reaction (RT-PCR), and rat retinal sections using *in situ* hybridization histochemistry. A combination of these approaches have been used to show that preprotachykinin (PPT) I mRNA, which generates SP and NKA, and PPT II mRNA, which generates NKB, are expressed in rat retinal extracts. *In situ* hybridization histochemical studies have extended these findings to establish a differential distribution of the tachykinin mRNAs with PPT I mRNA in cells distributed to the inner nuclear layer (INL), inner plexiform layer (IPL), and ganglion cell layer (GCL), and PPT II mRNA in cells distributed to the GCL. *In situ* hybridization histochemical studies have also established the presence of sparsely distributed SRIF mRNA-containing cells in the INL and GCL, while another study showed the co-expression of SRIF mRNA and immunoreactivity in amacrine and displaced amacrine cells. Finally, VIP mRNA and immunoreactivity are in sparsely distributed cell bodies in the inner retina. These findings extend earlier biochemical studies showing that the mammalian retina synthesizes multiple peptides, and it is reasonable to assume that other peptides located in the retina by RIA or immunohistochemistry are also synthesized by retinal neurons.

An alternative approach to evaluating the cellular localization of a peptide is the employment of a transgenic mouse line with the peptide promoter driving the expression of a reporter gene. Although this type of genetic approach has been commonly used in other regions of the nervous system, there have been limited findings reported to date in the retina. The best example is the detection of NPY in amacrine cells in the INL and GCL of a mouse line with ß-galactosidase (ß-gal) expression driven by the NPY promoter. ß-Gal-containing amacrine and displaced amacrine cells are distributed to all retinal regions and they have widely ramifying processes. The pattern of ß-gal expression matched the pattern of NPY immunostaining established using a highly characterized antibody to NPY; there is about 85% co-localization between β-gal expression and NPY immunoreactivity. Thus, independent experimental approaches confirm the pattern of NPY expression in the retina.

Peptide Immunostaining

Beginning in the late 1970s and early 1980s, numerous peptide immunolabelings were described in both non-mammalian and mammalian retinas. Peptide immunostaining was usually localized to distinct populations of amacrine cells, and in some cases, to ganglion cells and their central-nervous-system projections.

Peptide immunoreactivity is commonly localized to wide-field amacrine cell populations distributed to both the proximal INL and the GCL. These cells are characterized by processes that ramify in one or more laminae of the IPL. In addition, these processes arborize widely and overlap to form a network across the retinal surface. Some wide-field amacrine cell types that contain peptide immunoreactivity also have axon-like processes. These amacrine cell types are likely to be polyaxonal amacrine cells characterized by a more restricted dendritic field and multiple axons that extend beyond their field of dendrites. Finally, there are several examples of sparse peptide-containing processes crossing the INL and ramifying in the outer plexiform layer (OPL). These processes are likely to be derived from interplexiform cells, which are characterized by processes that ramify in both the IPL and OPL, and they are often categorized as an amacrine cell variant. These peptide-containing amacrine cell populations are characterized by a very-low-to-moderate cell density. In most cases, their cell bodies are distributed to all retinal regions. In all cases, peptide-containing processes cover the entire retinal surface.

Three examples that illustrate the general features described above are the NPY-, SRIF- and VIP-containing amacrine cells in the mammalian retina.

NPY immunoreactivity is localized to wide-field amacrine cells that are located in both the proximal INL and the GCL. NPY cells have a similar appearance in different mammalian retinas. In the INL, most immunoreactive cells were characterized by small cell bodies and fine processes that ramify primarily in lamina 1 of the IPL. A few cells also ramified in lamina 3 of the IPL. In the GCL, small-to-medium immunoreactive cells ramify primarily in lamina 5 of the IPL. A few immunoreactive processes, originating from somata in the INL and processes in the IPL, ramified in the OPL (**Figure 1**). In rat retina, immunoreactive cells had a regular distribution across the retina and an overall cell density of 280 cells mm^{-2} in INL and 90 cells mm^{-2} in GCL.

SRIF immunoreactivity is localized to sparsely distributed, wide-field amacrine and displaced amacrine cells (**Figure 2**). Immunoreactive somata give rise to thin varicose fibers that form a narrow and continuous plexus in lamina 1 of the IPL. In many species, there is also a narrow plexus of varicose fibers in laminae 3 and 5 of the IPL. Frequently, a smooth, thin-caliber, axon-like process is observed to arise from a cell body or a primary process. A few immunoreactive fibers also cross the INL to ramify in the OPL in both ventral and dorsal retina. These fibers can be traced to the plexus in lamina 1 of the IPL. Finally, in rabbit and cat retinas, there is a dense accumulation of SRIF-immunoreactive fibers along the retinal margin, which form a circumferential band in all retinal regions.

A major feature of the SRIF expression in the mouse, rabbit, cat, and human retina is the predominant distribution of SRIF-containing cell bodies to the ventral retina. These cells form a very low-density cell population. For example, in rabbit retina, cell density ranges from 6 cells mm^{-2} in ventral retina to 11 cells mm^{-2} at the retinal margins. In addition, the total number of SRIF-immunoreactive amacrine cells in these retinas is correspondingly very low.

VIP immunoreactivity is localized to sparsely distributed wide-field amacrine cells, mainly located in the proximal INL. VIP cells have a similar appearance and

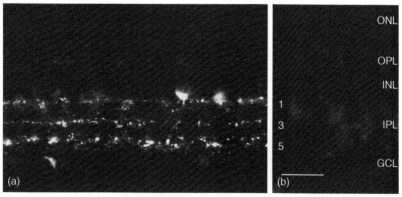

Figure 1 – NPY immunoreactivity in the rat retina. (a) NPY immunostaining of an amacrine cell body in the INL and varicose processes that ramify in laminae 1, 3, and 5 of the IPL. (b) Control experiment; NPY immunostaining was absent from a retinal section incubated with the NPY antibody that was preabsorbed with 10^{-6} M NPY. Vertical sections: scale bar = 30 μm. From Oh, S. J., D'Angelo, I., Lee, E. J., Chun, M. H., and Brecha, N. C. (2002). *Journal of Comparative Neurology* 446(3): 219–234. Copyright 2002, John Wiley & Sons, Inc. Reprinted with permission of John Wiley & Sons, Inc.

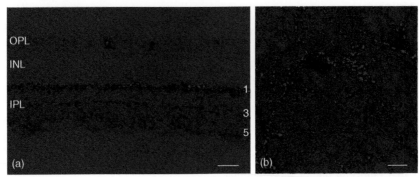

Figure 2 – SRIF immunoreactivity in the mouse retina. (a) SRIF immunoreactive processes are mainly in lamina 1 of the IPL. Sparse occurring processes also ramify in laminae 3 and 5 of the IPL, as well as crossing the INL and ramify in the OPL. (b) Plexus of SRIF immunoreactive processes in lamina 1 of the IPL. (a) Vertical section; scale = 50 μm; (b) Whole-mount preparation; scale = 50 μm.

distribution in different mammalian retinas. VIP immunoreactivity is distributed to multiple varicose processes and collaterals that ramify in laminae 1, 3, and 5 of the IPL in all retinal regions. VIP mRNA-containing and immunoreactive cells also form low-density cell populations in both the INL and GCL. For example, VIP-immunoreactive cells have an overall cell density of 25 cells mm^{-2} in rabbit retina, and there is an overall cell density of 50 cells mm^{-2} in the INL and 12 cells mm^{-2} in the GCL in the monkey retina.

Tachykinin and pituitary adenylate-cyclase-activating polypeptide (PACAP), a peptide in the VIP family, are reported in ganglion cells. These peptide-containing ganglion cells innervate multiple retinorecipient targets.

SP-immunoreactive ganglion cells are present in the frog, rat, hamster, rabbit, monkey, and human retina. These cells have been identified in co-staining experiments following either retrograde labeling of ganglion cells after fluorescent tracers are injected into retinorecipient nuclei or by the loss of ganglion cell immunostaining, following optic nerve section. In hamster, a small number of ganglion cells contain SP immunoreactivity; SP immunoreactivity in the lateral geniculate nucleus (LGN) is eliminated following optic nerve section. In rabbit, about 30% of the ganglion cells contain SP immunoreactivity. These cells have medium-to-large somata and dendrites that ramify extensively in the IPL. Their axons terminate in several retinorecipient nuclei, including the LGN, superior colliculus, and the accessory optic nuclei. In human retina, weakly staining SP-containing ganglion cells have also been identified on the basis of their morphology. Finally, in the *Macaca* monkey retina, the presence of SP-containing ganglion cells is suggested by the partial loss of SP immunostaining in the pregeniculate nucleus and the olivary pretectal nucleus, following bilateral eye enucleation.

There have been a modest number of ultrastructural investigations concerning the connectivity of peptide immunoreactive cells in rat, guinea pig, and primate retina. In general, the main input to peptide immunoreactive processes is from amacrine cells, whereas the major output formed by conventional synapses is onto amacrine and ganglion cells. There is a smaller percentage of input and output connections with bipolar cell axonal terminals. The large number of synaptic connections with amacrine cells indicates that peptide-containing cells are influenced principally by other amacrine cells and to a lesser degree by bipolar cells. There is more limited information about the connectivity of peptide-containing cells in other species, although overall the pattern of connectivity of these cells appears to be quite similar to that observed in monkey retina.

A feature of many wide-field amacrine cells is the co-expression of gamma aminobutyric acid (GABA) and a peptide. For example, in rat, cat, and monkey retina, GABA is co-localized with NPY, SRIF, and VIP. In addition, glutamate, the predominant ganglion cell transmitter, is reported to be co-expressed with PACAP in ganglion cells that innervate the suprachiasmatic nucleus of the hypothalamus. The co-expression of glutamate and SP in ganglion cells is also likely, although not formally demonstrated to date. These observations are consistent with findings elsewhere in the nervous system, reporting the co-expression and co-release of classical transmitters and peptides from the same cell. Interestingly, a differential release of classic transmitters and peptides, depending on the frequency and pattern of cell firing, has been shown in several systems. GABA or glutamate and peptides can act together at the same site, or the peptides can diffuse through the tissue and act at more distant cellular sites. There is evidence for both modes of action in the retina. GABA and peptides released from wide-field amacrine cells may act locally at GABA$_A$ receptors; for instance, both VIP and GABA act on GABA$_A$ receptors expressed by bipolar cell axons and ganglion cells. Peptides can also diffuse from their release site and act in a paracrine manner, as suggested, for example, by the differential distribution of SRIF-containing amacrine cell processes and SRIF receptors, which are expressed by multiple retinal cell types located away from the SRIF-containing processes.

Peptide Receptor Expression

As mentioned above, peptide actions are mediated by multiple GPCRs that influence intracellular signaling pathways, which regulate transmitter secretion and neuronal excitability. For instance, SRIF's cellular actions are mediated by five distinct GPCRs, sst_1–sst_5. There are also two sst_2 isoforms, sst_{2A} and sst_{2B}, from alternative splicing. The cellular actions of the tachykinin peptides are mediated by three receptors, known as NK-1, NK-2, and NK-3, whose preferred ligands are SP, NKA, and NKB, respectively.

Peptide-Binding Sites and Localization

The distribution of peptide receptors was initially evaluated using binding sites with autoradiographic approaches. In the late 1980s and early 1990s, the presence of high-affinity peptide-binding sites for several peptides, including SRIF, tachykinins, and VIP, was reported using radiolabeled peptides. Autoradiographic techniques were also used to define the regional distribution of peptide-binding sites. For example, SRIF-binding sites, identified with radiolabeled SRIF or a radiolabeled SRIF analog, were over the photoreceptor and both plexiform layers of the mouse and rat retina. High-affinity SP- and VIP-binding sites were homogeneously distributed over the IPL and GCL. In addition, NKA- and NKB-binding sites were evenly distributed over the IPL of the guinea pig retina. However, there are major difficulties in defining the cell types associated with these binding sites because of the low resolution of the autoradiographic technique. Furthermore, there are very few selective peptide agonists or antagonists available to distinguish among the different receptor isoforms. For example, octreotide binds to several SRIF receptor subtypes, and there is a high likelihood that multiple receptor subtypes were detected in these autoradiographic studies.

Peptide receptor mRNAs

The identification and cloning of peptide receptor genes in the 1990s have provided the tools for determining the expression of specific peptide receptors in the retina. NPY, SRIF, TK, and VIP receptors have been described most often in the mouse and rat retina. Multiple subtypes of these receptor mRNAs are found in the retina and they vary in their abundance; for example, in rat retina, the sst_2 and sst_4 mRNAs are the most abundant compared to the other sst mRNAs. Similarly, mRNAs of NPY and NK isoforms are present in retinal extracts with different levels of expression. Finally, mRNAs of the selective PACAP receptor, PAC1, and the VIP/PACAP-preferring receptors, VPAC1 and VPAC2, have been reported in retinal extracts. These findings illustrate that multiple peptide receptors are synthesized by the mammalian retina, in agreement with binding, autoradiographic, and immunohistochemical studies.

Pharmacological Studies

The activation of intracellular signaling pathways by peptides and related analogs in retinal extracts also supports the notion that endogenous peptides are present in the retina and they have a functional role that is mediated by specific receptors. For instance, both NPY and SRIF potently inhibit forskolin-induced cyclic adenosine monophosphate (AMP) accumulation, and VIP potently stimulates adenylate cyclase activity in retinal extracts. In addition, PACAP-27 and PACAP-38 are positively coupled to adenylate cyclase activity, and their actions are more potent than VIP, indicating the presence of PAC1. Furthermore, in rat retinal homogenates, PACAP-27 and PACAP-38, but not VIP, show a dose-dependent stimulation of inositol phosphate levels, suggesting multiple signal transduction pathways for PACAP peptides. Finally, the presence of functional NK receptors has been shown by the dose-dependent stimulation of inositol phosphate accumulation and $[Ca^{2+}]_i$ mobilization by the tachykinin peptides, SP, NKA, and NKB.

Pharmacological studies with isolated retinal cells and peptide receptor agonists and, in more limited cases, antagonists have also indicated the presence of pharmacological subtypes of peptide receptors. For instance, Y2, an NPY receptor, and sst_{2A} modulate L-type calcium channels expressed by rod bipolar cells. These investigations along with peptide binding and autoradiographic studies have been valuable in establishing the presence of functional peptide receptors in the retina.

Peptide Receptor Localization

Peptide Receptor mRNAs

In situ hybridization histochemical experiments have documented the cellular expression of NK-1 and NK-3 mRNAs in the rat retina. NK-1 mRNA is located in the INL and GCL, and NK-3 mRNA is mainly distributed to the middle and outer regions of the INL, corresponding to the location of bipolar cell bodies. This pattern of expression matches the NK-1- and NK-3-immunostaining patterns in the mouse and rat retina.

Peptide Receptor Localization

Peptide receptors, including those that mediate the actions of NPY, SRIF, and the TKs, have been mainly studied in the mouse and rat retinas. Several receptor subtypes are expressed in the retina, often by one or more distinct cell types, including bipolar, amacrine, and ganglion cells.

Sst and NK receptor expressions illustrate the general pattern of cellular localization of peptide receptors, although many details of individual receptor expression remain to be determined in future studies. In some cases, there are marked differences in the pattern of expression of these receptors, when studied using different, well-characterized antibodies. These disparate observations suggest that antibody specificity is a major factor influencing the understanding of peptide receptor expression in the retina. To date, animal models using genetic reporters driven by peptide receptor promoters to localize the cellular expression of peptide receptors have not been reported in the retina. The use of molecular approaches and knock-out lines will be needed to fully characterize the antibodies used for the immunostaining studies, and bacterial artificial chromosome (BAC) transgenics and knock-in mouse lines will be important for establishing the cellular localization of these receptors, independent of immunostaining approaches.

There are major differences in the reported pattern of SRIF receptor expression. For instance, some groups report that numerous amacrine cells in the rat and rabbit retina express sst_1 immunoreactivity. In contrast, other groups using a different antibody report sst_1 immunoreactivity in a distinct population of ganglion cells based on their size, and distribution of dendrites to lamina 3 of the IPL. Several independent studies, using different antibodies, agree that sst_{2A} immunoreactivity is the most abundant SRIF receptor subtype, and it is mainly localized to bipolar cells in mouse, rat, and rabbit retinas (**Figure 3**). Sst_{2A} immunoreactivity is also reported in some wide-field amacrine cells, photoreceptor terminals, and horizontal cells in some species. All investigators agree that ganglion cells express sst_4 immunoreactivity. Sst_4 immunoreactivity is in multiple ganglion cells and numerous multistratified processes in the IPL. Consistent with the immunostaining studies are *in situ* hybridization histochemical experiments describing sst_4 mRNA expression in cells distributed to the INL and GCL. Together, these findings provide strong evidence for the expression of SRIF receptors by ganglion cells. There are no reports of the cellular localization of sst_3 to retinal cells and there is a single report of sst_5 expression in nearly all amacrine cells.

NK-1 immunoreactivity is in numerous amacrine cell bodies in the INL and in some small and large cell bodies in the GCL in the rat retina (**Figure 4**). Immunoreactivity is prominent in all IPL laminae, and in addition there are fine caliber and varicose processes in the OPL, and a few processes in the ganglion cell axon layer. NK-1 immunoreactivity is observed in tyrosine-hydroxylase-containing amacrine cell bodies and their processes. Finally, most NK-1 immunoreactive amacrine cells contain GABA immunoreactivity. The large number of NK-1 immunoreactive

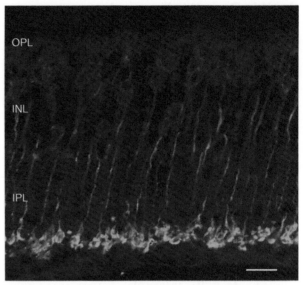

Figure 3 – Sst_{2A} immunoreactive rod bipolar cells in the mouse retina. Sst_{2A} immunostaining of rod bipolar cells, as well as some amacrine cells. Note very high level of expression (bright green) in rod bipolar cell terminals in laminae 5 of the IPL. Vertical section: scale = 15 μm.

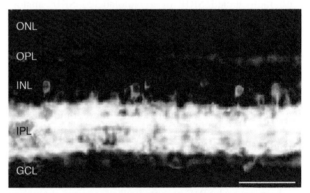

Figure 4 – NK-1 immunoreactivity in amacrine cells of the rat retina. NK-1 immunostaining of multiple amacrine cell bodies in the INL and heavy staining of processes in all laminae of the IPL. The pattern of immunostaining suggests that multiple amacrine cell types express this receptor. Vertical section: scale bar = 50 μm. Adapted from figure 1 in Casini, G., Rickman, D. W., Sternini, C., and Brecha, N. C. (1997) Neurokinin 1 receptor expression in the rat retina. *Journal of Comparative Neurology* 389(3): 496–507. Copyright 1997, John Wiley & Sons, Inc. Reprinted with permission of John Wiley & Sons, Inc.

cell bodies and the distribution of processes to the IPL suggest that this receptor is expressed by multiple amacrine cell populations. Some ganglion cells, based on cell body size, appearance, and position, are also likely to express NK-1 immunoreactivity.

In contrast, NK-3 immunoreactivity is in bipolar cell bodies distributed to the middle of the INL, and in their dendritic and axonal processes in the OPL and IPL, respectively (**Figure 5**). In the IPL, they ramify in

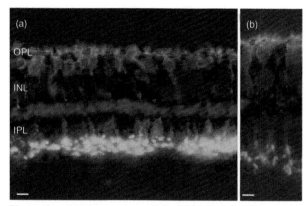

Figure 5 – NK-3 immunoreactivity in bipolar cells of the mouse retina. (a) NK-3 immunoreactive bipolar cells (red) are localized to the INL and labeled processes are in the OPL and laminae 1 and 2 of the IPL. PKC immunoreactivity in rod bipolar cells (green) that ramify in lamina 5 of the IPL. (b) The cell bodies of the NK-3 immunoreactive bipolar cells are clearly distinct from those of the rod bipolar cells. The pattern of NK-3 expression indicates OFF-type cone bipolar cells express this receptor. Vertical section: scale = 10 μm.

laminae 1 and 2, distal to where the type 3 cone bipolar cells stratify, suggesting NK-3 expression in OFF-type bipolar cells. A subset of the NK-3 immunoreactive bipolar cells expresses synaptotagmin 2. This same subgroup of NK-3 immunoreactive bipolar cells contains recoverin, a marker for type 2 bipolar cells. Together, these findings show that NK-3 immunoreactivity is localized to at least two types of OFF-type bipolar cells. In addition, NK-3 immunoreactivity is in tyrosine-hydroxylase-containing amacrine cells. There are no reports of specific NK-2 immunostaining in the retina, although NK-2 mRNA is detected in retinal extracts.

A frequent observation from these studies is the mismatch between the distribution of peptide-containing processes and their receptors. This mismatch supports the notion of a paracrine mode of action for peptides acting at receptors that are located away from the peptide-containing processes, although this does not rule out direct transmitter actions that could occur at synaptic specializations in areas of the IPL where there is apposition of peptide and peptide receptor processes.

Acknowledgment

Support for this work was provided by NEI EY 04067 and a Veterans Administration Senior Career Scientist Award.

See also: Information Processing: Amacrine Cells; Morphology of Interneurons: Amacrine Cells; Neuropeptides: Function.

Further Reading

Brecha, N. C. (1983). Retinal neurotransmitters: Histochemical and biochemical studies. In: Emson, P. C. (ed.) *Chemical Neuroanatomy*, pp. 85–129. New York: Raven.

Brecha, N. C. (2003). Peptide and peptide receptor expression and function in the vertebrate retina. In: Chalupa, L. and Werner, J. (eds.) *Visual System*, pp. 334–354. Boston, MA: MIT Press.

Brecha, N., Johnson, D., Bolz, J., et al. (1987). Substance P-immunoreactive retinal ganglion cells and their central axon terminals in the rabbit. *Nature* 32: 155–158.

Casini, G., Catalani, E., Dal Monte, M., and Bagnoli, P. (2005). Functional aspects of the somatostatinergic system in the retina and the potential therapeutic role of somatostatin in retinal disease. *Histology and Histopathology* 20: 615–632.

Casini, G., Rickman, D. W., Sternini, C., and Brecha, N. C. (1997). Neurokinin 1 receptor expression in the rat retina. *Journal of Comparative Neurology* 389: 496–507.

Cervia, D., Casini, G., and Bagnoli, P. (2008). Physiology and pathology of somatostatin in the mammalian retina: A current view. *Molecular and Cellular Endocrinology* 286: 112–122.

Marshak, D. W. (1989). Peptidergic neurons of the macaque monkey retina. *Neuroscience Research.Supplement* 10: S117–S130.

Thermos, K. (2003). Functional mapping of somatostatin receptors in the retina: A review. *Vision Research* 43: 1805–1815.

Zalutsky, R. A. and Miller, R. F. (1990). The physiology of somatostatin in the rabbit retina. *Journal of Neuroscience* 10: 383–393.

Neurotransmitters and Receptors: Dopamine Receptors

P M Iuvone, Emory University School of Medicine, Atlanta, GA, USA

Glossary

Amacrine cells – Neurons with cell bodies located in the proximal part of the inner nuclear layer, and neuronal processes ramifying in the inner plexiform layer.

Circadian rhythms – Changes in biological processes that occur on a daily basis, are driven by autonomous circadian clocks, and provide selective advantage to organisms by allowing them to anticipate temporal changes in their environment.

Interplexiform cells – Neurons with cell bodies in the amacrine cell layer; they are distinguished from amacrine cells by having neuronal processes that project in both the inner and the outer plexiform layers.

Neuromodulators – Chemical transmitters that mediate slow, long-lasting modulatory effects on neurons and neuronal circuits through synaptic or extrasynaptic receptors; they usually have no effect on neuronal membrane potential alone, but modulate the response to other neurotransmitters.

Photopic vision – The vision in bright light that is mediated by cone photoreceptors and cone bipolar cell pathways.

Localization of Dopamine Neurons in the Retina

Dopamine is a member of the catecholamine family of neurotransmitters, which also includes norepinephrine (also known as noradrenaline) and epinephrine (also known as adrenaline). It is the primary catecholamine in the retina; only trace amounts of epinephrine and norepinephrine are found in this tissue. It was first detected in the retina in the 1960s and has been the subject of intensive study ever since.

Dopamine is released from a unique population of neurons with cell bodies in the innermost region of the inner nuclear layer (INL). These neurons have been visualized by formaldehyde-induced histofluorescence and by immunohistochemistry, with antibodies against tyrosine hydroxylase (see **Figure 1**), the rate-limiting enzyme in dopamine biosynthesis. The dopamine neurons have long axons that ramify in sublamina 1 of the inner plexiform layer (IPL), at the border of the INL. The axons contain varicosities (presumptive sites of dopamine release) along their entire length. Although the number of dopamine neurons is small (approximately 500 in mouse retina), the extensive length and branching of the axons result in considerable overlap of processes throughout the retina. Some of the axons form rings around the perikarya of AII amacrine cells, which transmit rod pathway signals in the inner retina. Some processes, likely dendritic, descend further into the IPL. In many species, the processes of the dopamine neurons project toward the outer plexiform layer (OPL; see arrow in **Figure 1(a)**), designating them as interplexiform cells.

Regulation of Dopamine Neuronal Activity

Retinal dopamine neurons spontaneously fire action potentials, stimulating the release of dopamine throughout the cell. Light increases the firing of action potentials and stimulates dopamine synthesis, release, and metabolism. Both transient and sustained light-evoked firing patterns have been observed.

Although the processes of dopamine neurons ramify in the outer portion of the IPL, where OFF bipolar cells make synapses, most evidence suggests that the light responses of dopamine neurons are driven by the ON pathway. Light-evoked dopamine release from monkey and frog retinas is inhibited by L-(+)-2-amino-4-phosphonobutyric acid (L-AP4), which pharmacologically blocks synapses between photoreceptors and ON bipolar cells. Light-evoked firing of most, but not all, dopamine cells in mouse retina is also blocked by L-AP4. In addition, light-evoked dopamine metabolism, commonly observed in the vertebrate retina, is absent in the no b-wave (*nob*) mouse; this mouse has no functional ON pathway due to a mutation in the nyctalopin (*nyx*) gene, but has an intact OFF pathway. Retinal dopamine neurons are excited by glutamate, a bipolar cell transmitter, and inhibited by the amacrine cell transmitters gamma aminobutyric acid (GABA) and glycine. Once released, dopamine diffuses to act on extrasynaptic receptors found on multiple cells types throughout the retina.

Dopamine Neuronal Activity Is Coupled to Dopamine Synthesis and Metabolism

The light-evoked increase in dopamine neuronal activity is not associated with a depletion of the neuromodulator,

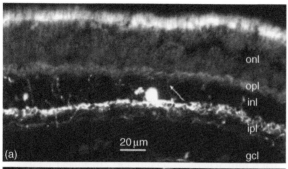

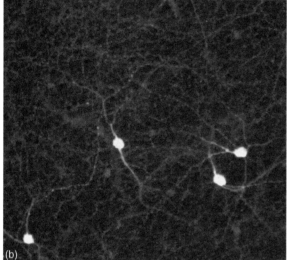

Figure 1 Dopamine amacrine/interplexiform cells of the rat retina. (a) A vertical section through the retina showing a dopamine cell body and processes labeled with an antibody to tyrosine hydroxylase. Note the extensive labeling of processes in sublamina 1 of the ipl and processes ascending to the opl (arrow). onl, outer nuclear layer; opl, outer plexiform layer; inl, inner nuclear layer; ipl, inner plexiform layer; gcl, ganglion cell layer. (b) Whole mount preparation of rat retina showing tyrosine hydroxylase immunoreactive cell bodies and processes from a horizontal view. Note the long, branching processes of the dopamine neurons and the extensive coverage of the ipl with these processes. Reproduced from Figure 1 in Witkovsky, P. (2004). Dopamine and retinal function. *Documenta Ophthalmologica* 108, 17–40. © Kluwer Academic Publishers. With kind permission of Springer Science and Business Media.

and is sometimes correlated with an increase in the steady-state levels of dopamine and it's primary metabolite 3,4-dihydroxyphenylacetic acid (DOPAC). Several mechanisms appear to account for these observations. The dopamine neurons contain plasma membrane dopamine transporters, which recapture a fraction of the dopamine released into the extracellular space. Following reuptake, the dopamine can be repackaged into vesicles for subsequent release or metabolized by monoamine oxidase to form DOPAC. A light-evoked increase of DOPAC levels is a common feature among many vertebrate retinas. More important in maintaining a readily releasable pool of dopamine is a light-evoked increase of dopamine biosynthesis.

This occurs through phosphorylation and activation of tyrosine hydroxylase, the rate-limiting enzyme in dopamine synthesis. While tyrosine hydroxylase can be phosphorylated by several protein kinases, the cyclic adenosine monophosphate (cAMP)-dependent kinase (protein kinase A (PKA)) appears to play a prominant role in the light-evoked regulation of tyrosine hydroxylase. The phosphorylation of the enzyme by PKA results in a conformational change that decreases the Km for tetrahydrobiopterin (BH4), the co-factor of tyrosine hydroxylase. This increases the activity of the enzyme at the subsaturating concentrations of BH4 found in retinal dopamine cells.

Circadian Control of Dopamine Release and Metabolism

In addition to being regulated by light, retinal dopamine release and metabolism, in many species, are controlled by circadian clocks that generate daily rhythms that persist in constant $(24 \, h \, d^{-1})$ darkness. In these instances, dopamine release and DOPAC levels increase during the subjective day and decrease during the subjective night. In mouse retina, dopamine neurons express circadian clock genes and, therefore, may possess an autonomous circadian clock. However, evidence obtained in several species, including mouse, suggests that the circadian rhythms of dopamine release and metabolism are dependent on another neuromodulator, melatonin, which is synthesized in photoreceptor cells and released in a circadian fashion during the subjective night. Retinal dopamine neurons express melatonin receptors, the activation of which inhibits dopamine release. In turn, dopamine released in the daytime suppresses melatonin formation. Since dopamine receptors are widely distributed in the retina, dopamine plays important roles in the circadian organization of the retina.

Dopamine Receptors

Dopamine receptors are guanine nucleotide-binding protein (G-protein)-coupled receptors. There are five subtypes, organized into two families based on structural and pharmacological similarities: the D1-like and D2-like families (**Table 1**). The individual subtypes are named based on the chronological order in which they were cloned. The D1 family includes the D1 and D5 receptors, while the D2 family consists of D2, D3, and D4 receptors. The D1 and D5 (also known as D1B) receptors are derived from genes without introns; thus, there are no splice variants of the D1-like receptors. They show very similar pharmacologies with respect to selective agonists and antagonists; however, dopamine has a much higher affinity for the D5 receptor than for the D1 subtype. The D2, D3, and

Table 1   Dopamine receptors in the retina

Receptor name	G protein	cAMP response	Cellular expression
D1-like family			
D1	G_s	$\Delta\uparrow$	Horizontal cells, cone bipolar cells, and amacrine cells
D5	G_s	$\Delta\uparrow$	Retinal pigment epithelial cells
D2-like family			
D2	G_i	$\Delta\downarrow$	Dopamine neurons and amacrine cells
D4	G_i	$\Delta\downarrow$	Photoreceptor cells

D4 receptor genes contain introns, and splice variants of the D2 receptor have been characterized. They show similar pharmacologies; however, some selective agonists and antagonists have been identified. A common feature of all dopamine receptors is that they couple to G proteins that regulate cAMP. The D1 family receptors stimulate cAMP formation, while the D2 family members inhibit it. Dopamine receptors have also been reported to affect phospholipase C, Ca^{2+} currents, K^+ currents, and the protein kinases AKT, glycogen synthase kinase-3β (GSK3β), and extracellular signal-regulated kinases (ERK1/2).

Nearly all retinal cell types appear to express dopamine receptors, consistent with dopamine's role as a paracrine neuromodulator that acts on extrasynaptic receptors. Of the D1 family members, dopamine D1 receptor immunoreactivity is found on processes in both the OPL and IPL. The receptors are expressed by horizontal cells, subtypes of cone bipolar cells, and some amacrine cells. There is evidence for the expression of functional D1-like receptors in ganglion cells of goldfish. D5 receptors are expressed by mammalian retinal pigment epithelial (RPE) cells. Of the D2 family members, dopamine D2 receptors are highly expressed by retinal dopamine neurons, and function as autoreceptors that inhibit dopamine release. D2 receptors are also expressed by other unidentified amacrine cells and possibly by some bipolar and ganglion cells. D2 receptors have also been shown to be expressed and functional in guinea pig Müller cells. Dopamine D4 receptors are highly expressed by photoreceptor cells in mammals. The photoreceptors of nonmammalian vertebrates also have D2-like receptors; in addition, these are likely to be D4 receptors based on pharmacological criteria, but no molecular proof of this has been provided yet. D4 messenger ribonucleic acid (mRNA) expression has also been observed in the IPL and ganglion cell layer; however, due to the lack of specific antibodies, the types of cells in these layers have not been identified. Interestingly, neither mRNA nor protein for the dopamine D3 receptor has been found in retina.

Functions of Dopamine in the Retina

The widespread distribution of dopamine receptors coupled with the ability of dopamine to diffuse throughout the retina suggests that this neuromodulator has many functions in regulating retinal physiology, particularly in serving as a chemical signal promoting light adaptive functions. This section describes the effects of dopamine on different cell types within the retina, beginning with the distal retina, and gives examples on how dopamine impacts retinal network adaptation, circadian rhythmicity, and ocular growth.

Retinal Pigment Epithelium

Dopamine D5 receptors stimulate cAMP formation, which inhibits rod outer segment phagocytosis by RPE cells. Cultured bovine RPE cells express D5 receptors and dopamine inhibits phagocytosis by these cells. Thus, dopamine may function as a component of the regulatory mechanism that controls rod outer segment turnover.

Photoreceptor Cells

Dopamine D2/D4 receptors on photoreceptor cells have numerous effects due to their ability to inhibit cAMP formation and decrease intracellular Ca^{2+} concentrations. Dopamine receptor activation inhibits the synthesis of melatonin in photoreceptor cells by inhibiting cAMP formation. It regulates the amplitude of diurnal rhythms of phosphorylation of phosducin by cAMP- and Ca^{2+}-dependent protein kinases. Dopamine has been reported to increase the rate of dephosphorylation of rhodopsin, and probably affects the phosphorylation state of many photoreceptor proteins. Dopamine has been reported to inhibit the hyperpolarization-activated current (I_H) in rods, as well as the activity of Na^+/K^+-adenosine triphosphate (ATP)ase, which balances the dark current. It decreases the intracellular Ca^{2+} concentration of photoreceptor cells and also affects rod–cone coupling, but the effects may vary by species. These effects on currents, Ca^{2+}, and coupling may contribute to a reduction of rod synaptic transfer to bipolar and horizontal cells during the daytime.

Horizontal Cells

Horizontal cells mediate lateral inhibition and synaptic feedback to photoreceptor cells. Different horizontal cell subtypes couple together through gap junctions to form networks. In retinas of both mammalian and nonmammalian vertebrates, the activation of D1-like receptors uncouples the horizontal cells, narrowing their receptive fields. The activation of dopamine D1 receptors in dark-adapted retinas depolarizes horizontal cells and reduces responses to flickering lights. The depolarization is due to the cAMP-dependent enhancement of glutamate-gated currents through α-amino-3-hydroxyl-5-methyl-4-isoxazole-propionate (AMPA)/kainate glutamate receptors in the horizontal cell membrane; other voltage-gated channels may also be involved.

Bipolar Cells

Although dopamine receptor immunoreactivity has been observed in subtypes of mammalian cone bipolar cells, its function is yet to be investigated. In salamander retina, glutamate responses in OFF cone bipolar cells are enhanced by dopamine, similar to the enhancement observed in horizontal cells. In addition, dopamine has been shown to modulate gamma-aminobutyric acid (GABA)-mediated inhibition of Ca^{2+} influx and neurotransmitter release from bipolar cell terminals. GABAergic amacrine cells synapse onto bipolar cell terminals to provide feedback inhibition of glutamate release via GABA-C receptors. Dopamine, acting on D1 receptors, reverses this inhibition and increases glutamate release.

Amacrine Cells

AII amacrine cells are critical mediators of the rod pathway. They receive excitatory input from rod depolarizing bipolar cells and, thus, depolarize in response to light. They transmit hyperpolarizing light signals to OFF cone bipolar cells through inhibitory glycinergic synapses and depolarizing light signals to ON cone bipolar cells through gap junctions. In addition, AII amacrine cells form homotypic gap junctions with neighboring AII amacrines. The axons of the dopamine neurons surround the perikarya of AII amacrine cells. Dopamine and D1 receptor agonists promote the uncoupling of gap junctions in AII amacrine cells by cAMP-dependent phosphorylation of connexin proteins. The gap junctions between amacrine cells are more sensitive to the uncoupling action of dopamine than those between AII amacrine cells and ON cone bipolar cells. In addition to secreting dopamine, the dopamine amacrine/interplexiform cells also synthesize and release the inhibitory neurotransmitter GABA. The dopamine amacrine/interplexiform cells synapse onto the AII amacrines, where their processes form rings around the perikarya of the AII cells. The active zones of the synapses have GABA-A receptors. Thus, activation of the dopamine amacrine/interplexiform cells is likely to decrease the light response of AII amacrine cells through GABA. Interestingly, D1 dopamine receptors have not been observed in these active zones, suggesting that dopamine diffuses to distal receptors to promote AII–AII cell uncoupling.

Dopamine stimulates acetylcholine release in the retina through D1-like receptors; most acetylcholine in mammalian retina is released from starburst amacrine cells, which function in the circuitry regulating directional selectivity to moving stimuli.

Most of the synapses made by dopamine amacrine/interplexiform cells are onto other amacrine cells. Thus, it is likely that dopamine affects many other functions within the inner retinal circuitry, but the details and functional consequences for visual processing are less clear.

Ganglion Cells

The effects of dopamine on ganglion cells are complex and it is difficult to discern which responses result from direct actions on ganglion cells and which responses reflect effects on retinal circuitry. In the mammalian retina, dopamine appears to decrease the sensitivity to light and strengthen the center surround of ganglion cells. These effects are mediated by dopamine D1-like receptors, and D1 receptor antagonists have opposite effects. A D1 receptor blocker also reduces the response of ON–OFF directionally sensitive ganglion cells to the leading edge of a moving light stimulus. In addition, dopamine strengthens cone pathway input to ganglion cells and reduces rod pathway input.

A small subpopulation of ganglion cells is referred to as intrinsically photosensitive retinal ganglion cells (ipRGCs). These ganglion cells project to the suprachiasmatic nucleus (SCN) of the hypothalamus, the site of the master circadian clock in mammals; to the olivary pretectal nucleus, a relay nucleus in the pupillary light reflex circuit; and to other nuclei involved in nonimage forming vision. These ipRGCs mediate photic entrainment of the circadian clock and are essential for the normal pupillary light reflex. They contain a photopigment, melanopsin, and are directly responsive to light, showing a depolarizing response to illumination. The ipRGCs also receive input from rods and cones through retinal circuitry. In mammals, dopamine amacrine/interplexiform cells are a component of this circuitry and synapse directly onto the ipRGCs. In addition, dopamine drives a circadian rhythm of melanopsin expression in the ipRGCs, at least in rats.

Müller Glial Cells

Müller cells are giant glia that span the width of the neural retina from the outer limiting membrane to the inner limiting membrane. They play important roles in retinal physiology by regulating the extracellular ionic milieu, especially the K^+ concentration. Müller cells are also an important source of trophic factors within the retina. Mammalian Müller cells express dopamine D2 receptor immunoreactivity. The application of dopamine or a D2 receptor agonist decreases K^+ conductance through an inwardly rectifying K^+ channel. These findings indicate that dopamine may influence the extracellular K^+ clearance, thereby affecting neuronal excitability and visual processing in the retinal circuitry.

Role of Dopamine in Photopic Visual Processing

Overall, the effects of dopamine on individual retinal cell types are consistent with an important role for this neuromodulator in visual processing in light-adapted retinas. Dopamine facilitates cone input to the inner retina and diminishes rod input. Extensive rod–cone coupling at night may shunt cone currents to rods, accounting for the apparent lack of cone input at night in the absence of dopamine. The uncoupling of the rod–cone network in the daytime in response to dopamine isolates the cones, strengthening their input signals to the inner retina. The effect of dopamine on horizontal cell coupling narrows receptive fields and makes the center surround organization more compact, enhancing spatial contrast sensitivity. The enhancement of glutamate-gated currents on OFF cone bipolar cells would also be expected to strengthen cone input. In addition, the uncoupling of the AII amacrine cell network by dopamine, together with the inhibitory effect of co-released GABA on AII amacrine cell light responses, should further suppress information flow through the rod pathway. Collectively, activation of the dopamine amacrine/interplexiform cell appears to fine-tune the retinal circuitry for high-resolution, high-contrast, low-sensitivity visual processing under photopic conditions.

Dopamine and Circadian Organization of the Retina

Many cells types in the vertebrate retina express circadian clock genes, including the dopamine amacrine/interplexiform cells. Dopamine can entrain or phase-shift the circadian clock in amphibian photoreceptor cells that regulates melatonin biosynthesis. Dopamine affects the circadian expression of a clock gene reporter (period 2:: luciferase) in the inner retina of mice. In addition, circadian-clock-driven dopamine release drives circadian rhythms of rod–cone coupling in mice and fish, and photoreceptor protein phosphorylation in mouse photoreceptors. With the widespread distribution of dopamine receptors in retina, the ability of dopamine to diffuse throughout the retina, the expression of clock genes in the dopamine amacrine/interplexiform cells, and the large increase in dopaminergic activity at dawn, dopamine may play important, conserved roles in the circadian organization of the retina by entraining and coordinating multiple circadian oscillators.

Dopamine, Retinal Development, Ocular growth, and Myopia

In chick embryo retinal cell cultures, dopamine has been reported to inhibit growth cone motility. In addition, the signaling mechanisms utilized by dopamine receptors appear to change with the developmental stage. Dopamine may play a role in stabilizing synapses during retinal development. However, many of these studies were done with *in vitro* systems, and may not necessarily reflect the development of the retina *in vivo*.

Dopamine may also be involved in ocular development and establishing emmetropia (perfect, focused vision). At birth, most animals are hyperopic, with images focused posterior to the photoreceptors. During the postnatal period, the eye elongates, increasing its own focal length. Under normal conditions, the eye will elongate only to the point that the image is focused on the photoreceptor cells, yielding the condition referred to as emmetropia. However, if the eye grows excessively long, the image will focus in front of the retina, causing myopia (nearsightedness). The growth of the eye in the axial dimension is regulated by the retina through visually guided feedback mechanisms that coordinate the growth of the eye with its optics. If the retinal image is degraded, as in the case of congenital infant cataract, the eye grows excessively long, resulting in myopia. This process is referred to as form-deprivation myopia and has been studied extensively in chicken hatchlings. When a diffuser goggle is placed over the eye of a newly hatched chick, the eye will grow excessively long, resulting in a large myopic refractive error within 1–2 weeks. Interestingly, allowing short periods of vision each day without the diffusers prevents the development of myopia.

Although some controversy exists, dopamine appears to contribute to the retinal circuitry involved in the feedback mechanism controlling emmetropization. Retinal dopamine levels and metabolism are reduced in eyes of chicks exposed to form deprivation. The administration of apomorphine, a dopamine receptor agonist, prevents the development of form-deprivation myopia. Moreover, dopamine receptor antagonists block the ability of short periods of unobstructed vision to prevent the development of myopia.

Summary

Dopamine release is driven by light and regulated by circadian clocks. The neuromodulator regulates multiple functions within the retina primarily by extrasynaptic mechanisms. Dopamine receptors are widely distributed in the retina, and dopamine has the capacity to diffuse from sites of release to distant receptors. A primary function of dopamine is to regulate network adaptation within the retina to optimize vision during the daytime.

Acknowledgements

The research in the author's laboratory is supported by NIH grants EY004864 and EY006360.

See also: Circadian Metabolism in the Chick Retina; Circadian Photoreception; Circadian Regulation of Ion

Channels in Photoreceptors; Information Processing: Amacrine Cells; Morphology of Interneurons: Amacrine Cells; Morphology of Interneurons: Interplexiform Cells; Neurotransmitters and Receptors: Melatonin Receptors; The Circadian Clock in the Retina Regulates Rod and Cone Pathways.

Further Reading

Djamgoz, M. B. A., Hankins, M. W., Hirano, J., and Archer, S. N. (1997). Neurobiology of retinal dopamine in relation to degenerative states of the tissue. *Vision Research* 37: 3509–3529.

Green, C. B. and Besharse, J. C. (2004). Retinal circadian clocks and control of retinal physiology. *Journal of Biological Rhythms* 19: 91–102.

Nguyen-Legros, J., Versaux-Botteri, C., and Vernier, P. (1999). Dopamine receptor localization in the mammalian retina. *Molecular Neurobiology* 19: 181–204.

Ruan, G. X., Allen, G. C., Yamazaki, S., and McMahon, D. G. (2008). An autonomous circadian clock in the inner mouse retina regulated by dopamine and GABA. *PLoS Biology* 6: e249.

Tosini, G., Pozdeyev, N., Sakamoto, K., and Iuvone, P. M. (2008). The circadian clock in the mammalian retina. *BioEssays* 30: 624–633.

Witkovsky, P. (2004). Dopamine and retinal function. *Documenta Ophthalmologica* 108: 17–40.

Zhang, D. Q., Zhou, T. R., and McMahon, D. Q. (2007). Functional heterogeneity of retinal dopaminergic neurons underlying their multiple roles in vision. *Journal of Neuroscience* 27: 692–699.

Neurotransmitters and Receptors: Melatonin Receptors

A F Wiechmann, University of Oklahoma College of Medicine, Oklahoma City, OK, USA

Glossary

Circadian rhythm – The term from the Latin *circa* which means around and *diem* which denotes day that is an approximate daily (24-h) periodicity in physiological processes of many organisms.

Diurnal – Activities that are repeated every 24 h, but are not necessarily under the control of a biological clock.

Orthologs – Homologous sequences which are similar to each other because they originated from a common ancestor.

Paracrine – The term from the Latin *para*, which means near, is a form of cell signaling in which the target cell is located within the same tissues as the cell that releases the chemical signal.

Pinealocytes – The main cells of the pineal gland that produce and secrete melatonin into the circulation.

Xenopus – The term from the Latin *strange foot*, which is a genus of frog native to Africa, and commonly used in research as a model organism.

Introduction

Melatonin (*N*-acetyl-5-methoxytryptamine) is an indolamine hormone synthesized by pinealocytes and retinal photoreceptors. The rate of melatonin synthesis, in most species studied, is highest at nighttime, and is considered to be a chemical signal of darkness that entrains circadian rhythms. This hormone synthesized in the pineal gland is secreted immediately into the circulation and acts as an endocrine hormone on distant target sites throughout the body. Melatonin produced in the retina, however, is thought to have a local, or paracrine role. It is thought that melatonin is synthesized and released by the photoreceptors at night, and diffuses throughout the retina to bind to melatonin receptors located on a variety of retinal cells. Since melatonin is a very lipophilic molecule, it diffuses freely through plasma membranes.

The three major subtypes of melatonin receptors are members of the superfamily of guanine nucleotide binding (G-protein)-coupled receptors. Most studies have shown that melatonin receptor activation is coupled to an inhibition of adenylate cyclase activity, although many reports demonstrate that other signaling mechanisms are conveyed by the melatonin signal. Melatonin receptors have been identified in many different retinal cells, including amacrine cells, horizontal cells, ganglion cells, photoreceptors, and the adjacent retinal pigment epithelium (RPE).

Sites of Retinal Melatonin Synthesis

Melatonin Synthesis by Photoreceptors

The photoreceptors appear to be the sites of melatonin synthesis in the retina. They express all of the enzymes involved in melatonin synthesis. Melatonin is synthesized from tryptophan in a series of four enzymatic steps: (1) tryptophan is converted into 5-hydroxytryptophan by tryptophan hydroxylase (TPH); (2) the 5-hydroxytryptophan is then converted into 5-hydroxytryptamine (serotonin) by aromatic amino acid decarboxylase; (3) serotonin is then converted into *N*-acetylserotonin by arylalkylamine *N*-acetyltransferase (AANAT); and (4) *N*-acetylserotonin is converted into melatonin (*N*-acetyl-5-methoxytryptamine) by hydroxyindole-O-methyltransferase (HIOMT). The enzyme activity and messenger RNA (mRNA) encoding TPH and AANAT exhibit circadian rhythms of expression, with highest levels occurring at night.

There is strong evidence that identifies the photoreceptors as the sites of retinal melatonin synthesis. Melatonin immunoreactivity is localized in the outer nuclear layer (ONL) of the retina which contains the cell soma of the photoreceptors. HIOMT and AANAT protein and mRNA are localized to photoreceptor cytoplasm, and a cyclic rhythm of AANAT activity persists following chemical lesion of the inner retina. The mRNA encoding TPH is localized to photoreceptors, and the photoreceptor layer of the amphibian retina continues to produce melatonin rhythmically in darkness after isolation from the inner retina.

In addition to the photoreceptors, some neurons of the inner retina may have the capacity to produce a small amount of melatonin. Melatonin immunoreactivity is observed in the inner retina, and a low level of AANAT mRNA has been detected in the inner nuclear layer (INL) and ganglion cell layer (GCL). The INL contains the cell soma of amacrine, horizontal, bipolar, and Müller cells. Since some cells in these layers also have melatonin receptors, the synthesis of melatonin by inner retinal neurons may be involved in the circadian activity of cells of the inner retina.

Phylogenetic Relationships between Photoreceptors and Pinealocytes

The ability of photoreceptors and pinealocytes to synthesize melatonin appears to be the consequence of an ancestral relationship between the retina and the pineal gland. Some primitive animals possessed three eyes which may have produced melatonin and were also capable of phototransduction. Pinealocytes of some lower vertebrates are morphologically very similar to retinal photoreceptors, and they synthesize melatonin as well as many proteins that are characteristic of retinal photoreceptors. Furthermore, the photoreceptors of the nonmammalian pineal gland are directly photosensitive, and during the embryologic development of the mammalian pineal gland, the pinealocytes undergo a transient photoreceptor-like differentiation.

It is suggested that the middle, or third eye, eventually evolved into an endocrine organ specialized for the secretion of melatonin into the circulation, in which the melatonin-producing pineal photoreceptors eventually lost their phototransduction capabilities, and the melatonin-producing cells of the lateral eyes evolved into photoreceptors specialized for phototransduction, but maintained their ability to synthesize melatonin. Genes expressing melatonin receptors in peripheral tissues may have also become expressed in ocular tissues, which enabled local paracrine signaling by melatonin in the retina.

Classification of Melatonin Receptors

Melatonin receptor expression has been identified in the retinas of several species. The major types of melatonin receptors that have been cloned are members of the superfamily of G-protein-coupled receptors. Melatonin receptors have been classified as Mel_{1a}, Mel_{1b}, and Mel_{1c} subtypes. These three subtypes are expressed in tissues, including the retinas, of lower vertebrates such as amphibians, fish, and birds. Mammalian melatonin receptors are classified according to their homology to the nonmammalian receptors, and according to their pharmacological properties. In mammals, the Mel_{1a} and Mel_{1b} receptor subtypes are designated as the MT1 and MT2 receptor subtypes, respectively. The mammalian ortholog of the Mel_{1c} subtype has been designated as GPR50, and does not bind melatonin. The G-alpha proteins coupled to melatonin receptors are inhibitory (G_i) to the activation of adenylate cyclase and cyclic adenosine monophosphate (cAMP) production in most tissues studied. However, receptor coupling to other G-alpha proteins ($G_{i\alpha2}$, $G_{i\alpha2}$, $G_{i\alpha q}$, $G_{i\alpha s}$, $G_{i\alpha z}$, and $G_{i\alpha 16}$), and hence other signaling pathways, have been reported. Nuclear melatonin receptors, which are members of the RAR-related RZR/ROR orphan nuclear receptor superfamily, have been reported to exist in some tissues, and a melatonin-binding site on the enzyme quinine reductase 2 has been identified and is referred to as the mammalian MT3 melatonin receptor.

Most G-protein-coupled receptors interact with each other to form homodimers or heterodimers. The mammalian MT1 and MT2 melatonin receptors can exist as homodimers and as heterodimers. Dimerization of G-protein-coupled receptors has important functional consequences in regard to receptor affinity, trafficking, and signaling. The relative expression levels of melatonin receptor subtypes in the retina may have a significant impact on the function of melatonin in the target cells.

Sites of Melatonin Receptors in the Retina

Melatonin receptors have been identified not only in several retinal neurons of the inner retina, but also in photoreceptors cells and RPE cells. The identification of melatonin receptors in photoreceptor cells was unanticipated, given that the photoreceptors are the site of retinal melatonin synthesis. The presence of melatonin receptors in photoreceptor cells suggests the possibility of what can be characterized as an intracrine (autocrine) signaling mechanism in response to melatonin, in which a cell synthesizes and releases a signaling molecule, and also has receptors to which the molecule binds and triggers an intracellular response. Another potential role of melatonin receptors in photoreceptor cells is that they may be involved in a negative-feedback mechanism which would enable melatonin to regulate the expression of the receptors to which it binds.

Melatonin Receptors in Photoreceptor Cells

Mel_{1a}, Mel_{1b}, and Mel_{1c} melatonin receptor subtype protein and mRNA have been identified in the photoreceptors of nonmammalian vertebrates, and MT1 receptors in photoreceptors of the mammalian retina. The receptor immunoreactivity has been identified primarily in the photoreceptor membranes of the inner segments, although some cytoplasmic immunoreactivity has been reported for some subtypes, and may represent newly synthesized receptors that have not yet been transported to the plasma membrane, or receptors that have been internalized after activation.

Mel_{1b} and Mel_{1c} melatonin receptor RNA and/or protein are expressed in *Xenopus* photoreceptors, and the MT1 (Mel_{1a}) receptor is localized to photoreceptors of the human retina. In the chicken retina, Mel_{1a} immunoreactivity and melatonin receptor mRNA expression (Mel_{1a}, Mel_{1b}, and Mel_{1c}) are localized to the photoreceptor layer. In the *Xenopus* retina, Mel_{1c} immunoreactivity is observed in the plasma membrane of photoreceptor inner segments, whereas Mel_{1b} receptor immunoreactivity appears in a punctate pattern in the proximal portion of photoreceptor inner segments. The differential pattern of Mel_{1b} and Mel_{1c} may reflect differential regulation of expression or trafficking of melatonin receptors in photoreceptor cells.

Melatonin Receptors in RPE

The RPE is a monolayer of cuboidal cells located between the vascular choroid layer and the neural retina. It is very closely associated with the retinal photoreceptor cells. This close association reflects the vital function of the RPE to provide physical and metabolic support to the photoreceptors. Circadian signals may play a role in influencing the coordinated interactions between the RPE and its adjacent tissues. The RPE, photoreceptors, retinal neurons, and choroidal cells interact in a coordinated manner for optimal function. Melatonin may play a role in the timing of the circadian phagocytosis of shed photoreceptor outer segments. The distal tips of rod photoreceptor outer segments are shed on a circadian rhythm as part of a renewal process, with peak shedding occurring early in the light period. The shed outer segment tips are phagocytized by the RPE, and melatonin is thought to be involved in this process. Melatonin secreted from photoreceptors at night may activate melatonin receptors on the RPE to regulate some circadian activities of the RPE that are important for optimal photoreceptor activity.

Melatonin inhibits forskolin-stimulated cAMP synthesis in RPE cell cultures, and melatonin affects the RPE membrane potentials and resistances at the apical or basal membrane. The mRNA encoding all three melatonin receptor subtypes is expressed in *Xenopus laevis* RPE but the Mel_{1b} receptor protein is localized only to the apical surface of the *Xenopus* RPE, and is not present on the basal surface. The Mel_{1c} receptor protein has also been localized to the *Xenopus* RPE. The presence of melatonin receptors on the apical microvilli, which directly contact the photoreceptors, but not on the basal membrane of the RPE, suggests that the photoreceptors are more likely to be the source of melatonin that activate melatonin receptors on the RPE rather than melatonin that is produced by the pineal gland and secreted into the general circulation.

Melatonin Receptors in Inner Retinal Neurons

Using autoradiography with ^{125}I-melatonin, it has been demonstrated that melatonin binding occurs in the inner plexiform layer (IPL) of many species. The IPL contains the synaptic terminals between bipolar cells, amacrine cells, horizontal cells, and ganglion cells. Since melatonin inhibits dopamine release from the retina, and high-affinity melatonin binding occurs in the IPL of the retina, the dopaminergic amacrine cell, which forms synaptic contacts in the IPL, has long been considered to be a candidate for the site of action of melatonin in the inner retina. Another candidate cell for melatonin receptor expression is the GABAergic amacrine cell (GABA, gamma aminobutyric acid) of the INL since $GABA_A$ receptor antagonists block melatonin-induced suppression of dopamine release. This suggests that the effect of melatonin on dopamine release may not be mediated only by direct action on dopaminergic cells, but that indirect action on GABAergic amacrine cells may also contribute to the inhibition of dopamine release via melatonin.

The autoradiographic localization of melatonin-binding sites in cells of the inner retina has been confirmed by both *in situ* hybridization and immunocytochemistry. In *Xenopus*, Mel_{1b} and Mel_{1c} RNA expression is localized to the INL, GCL, and photoreceptor inner segments. In the chicken retina, the mRNA encoding the Mel_{1a}, Mel_{1b}, and Mel_{1c} receptor subtypes is present in the INL, GCL, and photoreceptor inner segments. The INL contains the cell soma of bipolar, amacrine, horizontal, and Müller cells. In the human retina, Mel_{1b} receptor mRNA is much more highly expressed than is Mel_{1a} receptor RNA, suggesting that the Mel_{1b} receptor has a more significant role in human retinal physiology.

Using antibodies against specific melatonin receptor subtypes for immunocytochemistry, all three melatonin receptor subtypes (Mel_{1a}, Mel_{1b}, and Mel_{1c}) have been observed in the outer plexiform layer (OPL; the layer that contains the synaptic contacts between photoreceptors, bipolar cells, and horizontal cells) and the IPL. The MT1 (Mel_{1a}) receptor has been localized to horizontal cells in several mammalian species, including human. All three melatonin receptor subtypes appear to be present in horizontal cells of fish and *Xenopus* retina. The MT1 (Mel_{1a}) receptor has also been localized to AII amacrine and GABAergic amacrine cells of the mammalian retina.

Some Mel_{1a} and Mel_{1c} receptor immunoreactivity co-localizes with GABAergic and dopaminergic amacrine cells in the *Xenopus* retina. The presence of Mel_{1a} and Mel_{1c} receptors on dopaminergic and GABAergic amacrine cells is consistent with the observation that melatonin modulates the cyclic release of GABA and dopamine from retinal amacrine cells. In contrast, Mel_{1b} receptor immunoreactivity does not appear to co-localize with markers for dopaminergic and GABAergic neurons in the *Xenopus* retina, suggesting that melatonin does not act directly on GABAergic and dopaminergic amacrine cells through the Mel_{1b} receptor in this species.

In the *Xenopus* retina, the Mel_{1a}, Mel_{1b}, and Mel_{1c} receptor proteins are differentially distributed throughout the retina. In the OPL, for example, presumptive horizontal cell processes are immunoreactive for Mel_{1a} and Mel_{1b} receptor subtypes, but the immunoreactive labels appear to be in different cell processes. Cell somas in the INL are immunoreactive either for Mel_{1b} or Mel_{1a}, or for Mel_{1a} or Mel_{1c}, but not for both. All three melatonin receptor subtypes appear to be expressed in different populations of ganglion cells in the *Xenopus* retina, and the MT1 subtype is present in ganglion cells of the human and macaque retina.

Melatonin receptor mRNA and protein are rhythmically expressed in *Xenopus* and chicks, with peak levels of Mel_{1c} expression occurring in the day. In chicks, the rhythms of Mel_{1a} and Mel_{1b} receptor protein generally appear to be the opposite to that of Mel_{1c}, with lowest levels occurring in the early morning and higher levels in the evening. The patterns of cyclic rhythms appear to be distinctive for each receptor subtype in the retina. Circadian rhythms in melatonin receptor expression may perhaps be superimposed on the rhythm in retinal melatonin levels to provide an additional level of regulation of the responsiveness of retinal target to melatonin.

Effects of Melatonin on Retinal Function

Modulation of Neurotransmitter Release

Melatonin released from photoreceptors at night diffuses into the extracellular milieu and binds to melatonin receptors on dopaminergic amacrine cells. The activation of the melatonin receptors results in a decrease of dopamine release at nighttime. Thus, retinal dopamine levels are higher during the day and lowest during the night due to the circadian release of melatonin. Resultant lower dopamine levels at night cause a reduction in D_2 dopamine receptor activity on photoreceptors, causing an increase in photoreceptor intracellular cAMP levels that in turn causes an increase in coupling of gap junctions between rod and cone photoreceptors so that rod input dominates the cone horizontal cells. The increased sensitivity of horizontal cells to light at nighttime is therefore mediated at least in part by activation of D_2-like receptors by dopamine released from amacrine cells. A reduction in endogenous retinal dopamine levels causes hyperpolarization of horizontal cells and enhanced dark adaptation.

D_1 dopamine receptors, which are positively coupled to cAMP synthesis, are located on horizontal cells. Melatonin may therefore postsynaptically regulate horizontal cell activity by inhibiting the stimulation of cAMP synthesis in response to D_1 receptor activation. It may bind to receptors on GABAergic amacrine cells, stimulating them to inhibit dopamine release from nearby dopaminergic amacrine cells. In addition, since horizontal cells express melatonin receptors, and melatonin increases horizontal cell sensitivity to light, melatonin may act directly on horizontal cells to increase gap-junctional coupling of horizontal cells. Increased horizontal cell coupling would cause an increase in receptive field size, which would potentially increase the sensitivity of the retina to light during the dark period, since more second-order neurons would respond to a light stimulus.

Melatonin may therefore modulate dopaminergic transmission by a combination of directly reducing dopamine release from amacrine cells, and indirectly by stimulating GABAergic amacrine cells to inhibit dopamine release from dopaminergic amacrine cells, both of which would increase horizontal cell coupling. The resulting increased visual sensitivity at nighttime could be due to increased rod–cone coupling through dopamine binding to D_2 receptors on photoreceptors, or to horizontal cell coupling stimulated by the binding of melatonin to melatonin receptors on horizontal cells. A summary diagram of the known locations of melatonin receptors and the possible interactions with the various target cells is presented in **Figure 1**.

Melatonin increases horizontal cell sensitivity to light in salamander retina, and also potentiates glutamate-induced currents from isolated cone-driven horizontal cells in carp retina by increasing the efficacy and affinity of the glutamate receptor. These observations suggest that melatonin acts directly on melatonin receptors of horizontal cells. Melatonin modulates cyclic guanosine monophosphate (cGMP)-dependent glutaminergic transmission from cones to cone-driven horizontal cells by activation of the Mel_{1a} receptor, causes a depolarization of the H1 horizontal cell membrane potential, and reduces its light responses. These observations suggest that melatonin enhances the circadian sensitivity of rod photoreceptor signaling.

Melatonin has been shown in fish retina to potentiate responses of rod ON bipolar cells to simulated light flashes. This action of melatonin is mediated by the Mel_{1b} receptor, and increases cGMP levels by inhibiting phosphodiesterase activity. Melatonin may bind directly to Mel_{1b} receptors on rod ON bipolar cells to improve the signal/noise ratio for rod signals by enhancing signal transfer from rod photoreceptors to rod bipolar cells. The presence of melatonin receptors by immunocytochemistry has not yet been definitively established in bipolar cells.

Melatonin treatment of isolated rat retinal ganglion cells potentiates glycine-induced currents by increasing the efficacy and channel conductance of a glycine receptor. The inhibitory modulation of glycinergic inputs to ganglion cells may thus be strengthened by stimulation of melatonin receptor activation. This suggests that melatonin may regulate circadian changes in receptive field organization and light sensitivity by binding to melatonin receptors on ganglion cells.

Modulation of Photoreceptor Function

Several reports support the concept of a direct action of melatonin on retinal photoreceptor function. Melatonin induces membrane conductance changes in isolated frog rod photoreceptors, binds with low affinity to structures in the OPL in frog retina, enhances the rate of photoreceptor

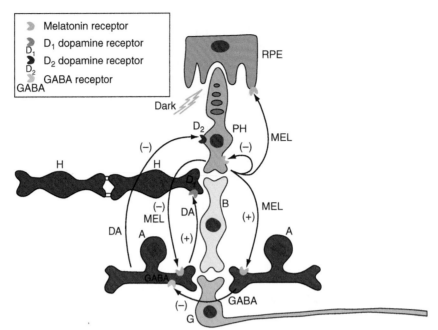

Figure 1 Summary diagram of locations of retinal melatonin receptors and potential interactions among target cells. Melatonin (MEL) is produced by photoreceptors (PH) at nighttime, and diffuses to target cells within the retina. Arrows represent the movement of melatonin to the target cells. The melatonin receptors are represented by a black crescent symbol. Dopamine and GABA receptors are represented by crescent symbols with D_1, D_2, or GABA as identifiers. Melatonin may bind to amacrine cells (A) that release GABA or dopamine (DA) as their neurotransmitter. Melatonin is thought to stimulate (+) GABA release and/or inhibit (−) dopamine release from these cells. GABA inhibits dopamine release from dopaminergic amacrine cells. A lower rate of dopamine release from amacrine cells at nighttime results in lower stimulation of D_1 receptors on horizontal cells (H), which leads to increased coupling of horizontal cells through gap junctions (=), which could cause an increase in receptive field size and increased retinal sensitivity to light. The decreased binding of dopamine at nighttime to D_2 receptors on photoreceptor cells results in an increase in melatonin synthesis, since dopamine inhibits melatonin synthesis in photoreceptors. Melatonin may potentially bind to horizontal cells to directly inhibit the cellular response to D_1 receptor binding. Melatonin may also bind to receptors located on the photoreceptor membrane, which could directly increase rod sensitivity to light, increase rod–cone coupling, and/or regulate synthesis of melatonin. Melatonin may bind to receptors on the apical membrane of retinal pigment epithelial (RPE) cells, to coordinate circadian interactions with photoreceptor outer segments. The possible expression of melatonin receptors on bipolar (B) cells, and the confirmed expression of melatonin receptors on ganglion cells (G) are not indicated by arrows.

outer segment disk shedding, and increases the degree of light-induced photoreceptor cell death. It causes a stimulation of the amplitude of the *a*-wave (rod photoreceptors) and the *b*-wave (inner retinal cells responding to the photoreceptor input) of electroretinogram (ERG) recordings of transgenic frogs that overexpress the Mel_{1c} receptor in rod photoreceptors. This suggests that melatonin acts directly on rod photoreceptors to increase retinal sensitivity to light as part of a dark-adaptation mechanism.

The role of melatonin in dark adaptation suggests a potential mechanism by which melatonin increases the degree of light-induced photoreceptor cell death. Since melatonin appears to increase the sensitivity of the retina to light as part of a dark-adaptation mechanism, an undesirable consequence of this may be an increased sensitivity to the deleterious effects of light. Although signals from the inner retina obviously play a significant role in the circadian activities of retinal photoreceptors, direct action of melatonin on receptors located on photoreceptors may contribute substantially to the functions of melatonin in circadian-regulated activities of the retina.

See also: Circadian Metabolism in the Chick Retina; Circadian Photoreception; The Circadian Clock in the Retina Regulates Rod and Cone Pathways; Circadian Regulation of Ion Channels in Photoreceptors; Neurotransmitters and Receptors: Dopamine Receptors.

Further Reading

Besharse, J. C. and Dunis, D. A. (1983). Methoxyindoles and photoreceptor metabolism: Activation of rod shedding. *Science* 219: 1341–1342.

Boatright, J. H., Rubim, N. M., and Iuvone, P. M. (1994). Regulation of endogenous dopamine release in amphibian retina by melatonin: The role of GABA. *Visual Neuroscience* 11: 1013–1018.

Dubocovich, M. L. (1983). Melatonin is a potent modulator of dopamine release in the retina. *Nature* 306: 782–784.

Dubocovich, M. L., Cardinali, D. P., Guardiola-Lemaitre, B., et al. (1998). Melatonin receptors. In: Girdlestone, D. (ed.) *The IUPHAR Compendium of Receptor Characterisation and Classification*, vol. I, pp. 188–193. London: UCPHAR Media.

Fujieda, H., Scher, J., Hamadanizadeh, S. A., et al. (2000). Dopaminergic and GABAergic amacrine cells are direct targets of melatonin: Immunocytochemical study of mt_1 melatonin receptor in guinea pig retina. *Visual Neuroscience* 17: 63–70.

Huang, H., Lee, S. C., and Yang, X. L. (2005). Modulation by melatonin of glutamatergic synaptic transmission in the carp retina. *Journal of Physiology* 569: 857–871.

Iuvone, P. M., Tosini, G., Pozdeyev, N., et al. (2005). Circadian clocks, clock networks, arylalkylamine N-acetyltransferase, and melatonin in the retina. *Progress in Retinal and Eye Research* 24: 433–456.

Lundmark, P. O., Pandi-Perumal, S. R., Srinivasan, V., and Cardinali, D. P. (2006). Role of melatonin in the eye and ocular dysfunctions. *Visual Neuroscience* 23: 853–862.

Reppert, S. M., Godson, C., Mahle, C. D., et al. (1995). Molecular characterization of a second melatonin receptor expressed in human retina and brain: The Mel$_{1b}$ melatonin receptor. *Proceedings of the National Academy of Sciences of the United States of America* 92: 8734–8738.

Scher, J., Wankiewicz, E., Brown, G. M., and Fujieda, H. (2002). MT$_1$ melatonin receptor in the human retina: Expression and localization. *Investigative Ophthalmology and Visual Science* 43: 889–897.

Wiechmann, A. F. and Smith, A. R. (2001). Melatonin receptor RNA is expressed in photoreceptors and displays a cyclic rhythm in *Xenopus* retina. *Molecular Brain Research* 91: 104–111.

Wiechmann, A. F. and Summers, J. A. (2008). Circadian rhythms in the eye: The physiological significance of melatonin receptors in ocular tissues. *Progress in Retinal and Eye Research* 27: 137–160.

Wiechmann, A. F., Vrieze, M. J., Dighe, R. K., and Hu, Y. (2003). Direct modulation of rod photoreceptor responsiveness through a Mel$_{1c}$ melatonin receptor in transgenic *Xenopus laevis* retina. *Investigative Ophthalmology and Visual Science* 44: 4522–4531.

Wiechmann, A. F., Udin, S. B., and Summers Rada, J. A. (2004). Localization of Mel$_{1b}$ melatonin receptor protein expression in ocular tissues of *Xenopus laevis*. *Experimental Eye Research* 79: 585–594.

Young, R. W. and Bok, D. (1969). Participation of the retinal pigment epithelium in the rod outer segment renewal process. *Journal of Cell Biology* 42: 392–403.

Non-Invasive Testing Methods: Multifocal Electrophysiology

E E Sutter, The Smith-Kettlewell Eye Research Institute, San Francisco, CA, USA

Glossary

Base interval of stimulation – Stimuli are presented at intervals that are integral multiples of a base interval.

Binary stimulation – The stimulation that switches between two states such as flash/no flash, pattern–contrast-reversing pattern.

First-order kernel – This response component is computed by adding the response signal following base intervals with a stimulus and subtracting the signal following intervals without stimulus. If the response is linear, response contributions from subsequent stimuli must cancel as they are added and subtracted the same number of times. If the response is nonlinear, the following responses depend on the presence of the preceding stimulus. The difference that appears in the first-order kernel is called an induced component.

Inion – The prominent projection of the occipital bone at the lower rear part of the skull.

Lamina cribrosa – A mesh-like structure through which the optic nerve exits the sclera.

m-Sequence stimulation – The binary stimulation controlled by a special class of pseudorandom binary sequences called m-sequences. These sequences have ideal properties for the analysis of nonlinear systems.

Myelination – An electrically insulating material that forms a layer, surrounding the axons of many neurons. Axons of retinal ganglion cells become myelinated at the lamina cribrosa in the optic nerve head. Myelination of axons greatly increases the propagation velocity of action potentials.

Nonlinear response – In the case of binary stimulation, nonlinear means that the contribution of a response to the signal may depend on preceding and immediately following responses or responses in the neighboring areas.

Saltatory nerve conduction – The nerve conduction in myelinated fibers whereby the action potential jumps from gap to gap between the myelinated section.

Second-order, first slice – This response component can roughly be thought of as the difference between the response to two consecutive stimuli and the two stimuli individually presented.

Why Multifocal?

Visual electrophysiology has enjoyed decades of useful applications in the clinic and research. It provides objective measures of function at different stages of visual processing. Until relatively recently, it has been restricted to testing of a single area in the visual field. In the clinic it was generally used as a full field test or a test of a single area or spot in the field. This has greatly limited its sensitivity in diagnosis and in measuring progression and recovery from disease. It was clear from the start that mapping responses across the retina recording focal responses from one small area at a time would have taken very long and would not be practical. Multifocal techniques have overcome this limitation by stimulating a large number of areas concurrently whereby the response of each area is encoded in its temporal stimulation pattern. The special encoding using binary m-sequences permits clean extraction of the focal response contributions from a single signal derived from the cornea or from the scalp over the visual cortex of the brain. The m-sequence encoding has the additional benefit that it provides information on nonlinear response properties such as fast adaptation and recovery from photo stress that can be important clinical indicators.

There are many multifocal protocols available to clinicians. Some of them test the central visual field with lower spatial resolution and short recording times useful for screening. Others offer high resolution for the detection of small scotomata and sensitive assessment of changes in the spatial extent of dysfunction. Yet, other protocols enhance inner retinal and specifically ganglion-cell contributions. These tools permit tailoring the test for a specific clinical purpose. In many laboratories, only one protocol is commonly used and ordered by the clinicians as a multifocal electroretinogram (mfERG). Judicious selection of one of many available tests can lead to faster and more accurate diagnosis and, in some cases, avoid errors. Careful data analysis by a knowledgeable user can tease out local-response abnormalities that might otherwise be missed. A few examples shown below are selected to illustrate this point.

The Basic Principle

The basic principle of mfERG and multifocal visual-evoked potential (mfVEP) recording is schematically illustrated in **Figure 1**. With both types of recordings, the size of the focal stimuli is scaled with eccentricity to

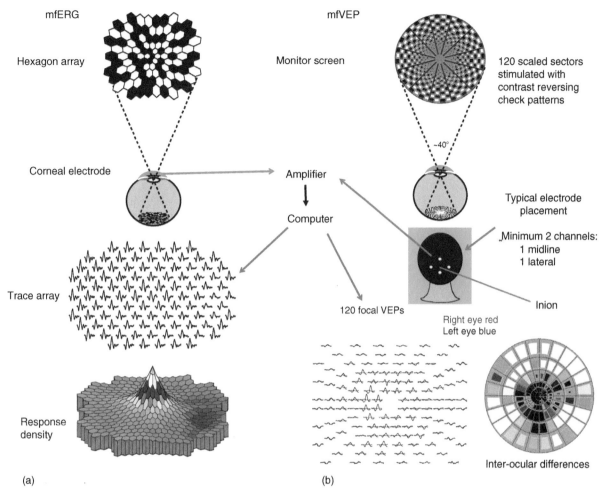

Figure 1 Schematic representation of the derivation of mfERG and mfVEP records. Commonly used stimulus arrays are shown on top. For the mfERG, hexagonal arrays are scaled with eccentricity to achieve similar signal-to-noise ratios across the stimulated field. The focal stimulation is m-sequence modulated flicker. The corneal signal can be derived with any of the available electrode techniques. For mfVEP, recording a dartboard pattern of stimulus elements is used to account for the steep eccentricity scaling of the representation of the visual field on the primary visual cortex. Each sector is stimulated with a contrast-reversing check pattern. Two perpendicular electrode pairs pasted to the scalp over the primary visual cortex cover the different directions of current flow. The data are presented as an expanded array of traces (bottom left) and as a pseudocolor field map (bottom right). The example is from an asymmetric patient. The left eye largely dominates (blue areas) except in the area of the right-eye blind spot (pink patch).

generate approximately the same signal amplitudes across the stimulated field. In the mfERG, we use scaled hexagonal arrays of the kind shown in **Figure 1(a)**. The steeper scaling of the cortical response requires the use of dartboard patterns **Figure 1(b)**. While the mfERG stimulus is usually focal flicker, the cortical responses of the multifocal visual-evoked cortical potential (mfVECP) are best elicited with focal contrast reversal of a check pattern.

Recording of Multifocal Data

All the data presented here have been recorded, analyzed, plotted, and exported with VERIS science 6.0 (Electro-Diagnostic Imaging, Inc, Redwood City, CA, USA). The mfERG data were recorded with a Burian–Allen bipolar

contact lens electrode. This electrode and others of similar construction provide the best signal-to-noise ratio. Disposable monopolar electrodes such as the DTL fiber or the HK loop are considered less invasive but provide noisier signals and require longer recording times to achieve the same-quality data. Contact lens electrodes offer the advantage that they correct for corneal astigmatism and prevent drooping eyelids. However, care must be taken that the corneal ring of the electrode is reasonably well centered on the dilated pupil. To be comfortable, the electrode should be selected to fit the size of the eye.

All mfERG records shown here were recorded with a stimulus screen calibrated to produce multifocal flash intensities of 2.7 cd s m^{-2}. In most cases, the patient's fixation stability was monitored with an eye camera. For more recent recordings, fixation was monitored with an infrared (IR) fundus camera that shows the position of the stimulus array on the fundus of the eye throughout the recording.

Multifocal Stimulators

Originally monochrome cathode ray tubes (CRTs) with a fast white phosphor were commonly used for stimulation. Few color CRTs could reach the required stimulus intensity. While, at this time, some suitable CRT monitors are still available, this technology is rapidly disappearing from the market and is being replaced by flat-panel liquid crystal display (LCD) monitors. The large panels are now bright enough, but achieve high brightness by leaving the pixels on during the entire display frame. This is ideal for pattern-reversal stimulation such as the mfVEPs, but it is not recommended for mfERGs. The switching speed of these panels is also marginal for ERG recording but adequate for mfVEPs. For multifocal flash ERG recording, we would like to have brief focal flash stimuli of no more than 2–3-ms duration at the beginning of each frame. This can be achieved with some of the available microdisplays. DLP projection displays can be used after substantial internal modifications.

Patient Positioning and Data Collection

In many laboratories chin rests are used for stabilization of the patient's head during recording. This is acceptable, but not the best solution. Patients are more comfortable in a reclining chair with a stable adjustable headrest. The reclined position helps to keep contact lens electrodes from dropping out. This arrangement works well in combination with a small stimulator mounted on an articulating arm. It allows adjustment of the angle of the stimulator so that the corneal ring of the contact lens electrode is centered on the patient's pupil.

The stimulus normally consists of a single cycle of a binary m-sequence. Using a longer m-sequence rather than averaging several shorter ones prevents contamination by higher-order kernels and, thus, provides cleaner separation of the local response contributions. For patient comfort, the record is collected in slightly overlapping segments that permit smooth splicing before data processing. Protocols for contact lens electrodes use a segment size of about 30 s. When recording with fiber electrodes, the patient must suppress blinks. To make this easier for the patient, the segment size is reduced to about 15 s.

Data Analysis and Presentation

Focal responses are extracted from the recorded signal using a cross-correlation executed by the fast m-transform. The focal responses may be contaminated with noise from blinks and small eye movements. A special artifact subtraction algorithm helps clean up the data. If the quality is not sufficient for this purpose, the operator can apply spatial filtering whereby each local waveform is averaged with a certain percentage of its nearest neighbors. This greatly improves the waveforms, but leads to some local smearing and is not recommended in cases where dysfunctional areas are suspected to be very small.

Clinically useful parameters of the local response waveforms can be extracted and presented as pseudo-color 2- or 3-dimensional topographic maps. The two most important ones are maps of response density and peak implicit time (time from the stimulus onset to the selected peak). Response densities are derived by dividing each focal-response amplitude by the solid visual angle of the corresponding stimulus patch. Estimating amplitudes of the often noisy focal responses using peak-to-trough measurements is very inaccurate. For this reason, a method of template matching is used. Each focal waveform is multiplied point by point with a normalized template of the underlying response. The template is derived as the average of waveforms in the same retinal region.

Implicit times of specific features of the response waveform, that is, the time from the stimulus onset to the feature are sensitive clinical measures. Of particular importance is the implicit time of the main positive peak $P1$. Estimating $P1$ implicit times using the highest point in the waveform is much too inaccurate when dealing with the noisy local responses. Much more accurate estimates are achieved by means of a template of the underlying waveform. On each focal waveform, the template is shifted into the position of best mean square fit. The location of the peak on the template waveform is then taken as the implicit time estimate. As with the amplitude estimation described above, the template waveform used at each location is the average of waveforms in the same neighborhood.

Dealing with Noisy Data

The main contaminations in mfERG responses are artifacts from blinks and small eye movements. It is possible to detect these artifacts and subtract them from the recorded signal. The process is capable of recovering some of the responses superimposed on the artifact, as long the artifact did not saturate data acquisition. This noise-reduction procedure works very well with mfERG records, but is less effective with mfVEPs where the contamination is usually noise from muscle tone required to maintain head posture rather than discrete artifacts. Here, it is more important to position the head in order to minimize tension.

How Long Does the Test Take?

The kernel extraction is an averaging procedure and, thus, follows the law of averages. Increasing the recording time by a factor k improves the signal-to-noise ratio by $k^{1/2}$. The duration of a test must be determined by the amount of information to be gained and the signal-to-noise ratio of the derived response.

A multitude of tests can be designed from very short tests to tests lasting perhaps 1 h or more. The longer tests are used when more precise information is required. In the selection of a test, one should consider all the information already known regarding the patient.

Figure 2(a) and 2(b) show the results from two very short recordings derived from the same subject. Sixty-one patches stimulated the central field. Interpolating hexagons were inserted at plotting time to permit a more accurate estimation of the topography. The signal quality was very good in this case and does not reflect what one might get from an average patient in such ultra-short recording times. However, recordings of 2-min duration can be quite adequate for patient screening.

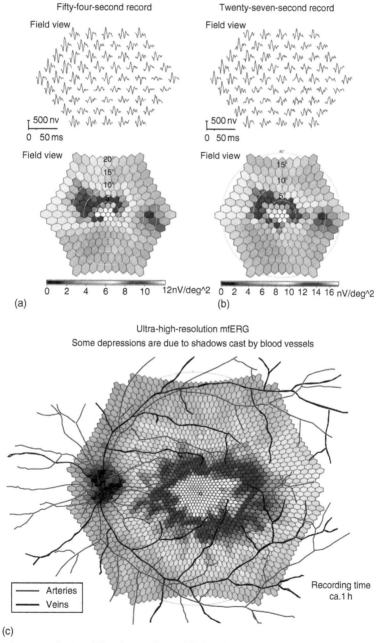

(a)

(b)

(c)

Figure 2 (a) Results from two very short multifocal recordings with 61 stimulus areas. The signals were derived with a bipolar Burian–Allen electrode. (b) High resolution recording using 509 stimulus areas within the central 45° of the visual field. The overlaid vasculature of the subject suggests that the depressed responses in some areas are caused by shadows of major blood vessels and their bifurcations.

Figure 2(c) illustrates a high-resolution recording with a stimulus array of 509 patches. The recording time of about 1 h necessary for this resolution is clearly not feasible in the clinic. The most popular array size uses 103 hexagons and records of 2–7 min length. Some clinics have used 241 hexagons in 7-min tests. Ideally, the spatial resolution of the test should be adjusted to meet the requirements of the clinical problem at hand.

Some Examples from the Retina Clinic

The samples presented below are selected to illustrate uses of different recording and analysis protocols. The clinical significance of the results cannot be discussed here due to space limitations.

Central Serous Retinopathy

The first example is from a male patient with central serous retinopathy (CSR). He was not diagnosed at the time of the recording but complained about a relative central scotoma in his left eye. We selected a 7-min test with 241 hexagonal patches because of the anticipated small spatial extent of the problem. The plots shown represent the average of two records. Individual records are slightly noisier, but would have been quite adequate in this case. **Figure 3(a)** shows the response density plots of the two eyes. The affected area in the left eye is clearly visible as an excavation in the central peak extending to the area surrounding the optic nerve head. The right eye is normal.

Some pathologies cause local delays in the main positive peak of the first-order waveform called the $P1$ peak indicated by the marks on the traces in **Figure 3(c)**. Such delays are mainly attributed to photoreceptor sensitivity loss. Mapping the peak implicit time can help distinguish between different pathologies. In this CSR case, such delays are found in areas with reduced response density (**Figure 3(b)**).

Plots of $P1$ implicit time generally show delays in the vicinity of the disk. This feature must be ignored. It is attributed to a contribution to the focal response from scattered light. Some light from the focal stimulation of the highly reflective optic disk is scattered onto the peripheral retina where it elicits a delayed response contribution.

Hydroxychloroquine Retinopathy

A small percentage of patients who take this drug for autoimmune disease develop a bull's eye retinopathy. This dysfunction is, at best, only partially reversible when the patient is taken off the drug. For disease prevention, the mfERG is now proposed as a test for patient screening.

The left column in **Figure 4(a)** shows a typical example of hydrohychloroquine toxicity. Only the right eye is

shown here as the presentation is usually bilaterally symmetric. The array of traces is shown at the top followed by the plots of response density and $P1$ implicit time below. Responses surrounding the center are substantially reduced to about 7–10° eccentricity with some relative sparing of the center.

The bottom plots shows that hydroxychloroquine toxicity does not cause significant changes in peak implicit times. This suggests that the changes are post-receptoral.

Amplitude ratios of ring averages around the fovea are used for diagnosis and assessment of severity of the bull's eye presentation. A plot from an automated screening protocol with statistical evaluation is shown in **Figure 5**. Ring averages are plotted in **Figure 5(a)**. In **Figure 5(b)**, response densities are plotted against eccentricity in degrees of visual angle and compared to data from a normal cohort. The data are interpolated with a cubic spline. For this analysis, ring averages have been normalized to the outermost ring that is usually minimally affected. The green bands around the normal mean (red line) represent the 2 standard deviation (SD) limit. The heavy black line is the patient's response amplitude. The area where the patient curve falls outside the 2-SD band is colored red. ICS is the index of central sparing used to evaluate the bull's eye configuration. It is computed as the ratio between the red area outside 3° eccentricity and the total red area below the 2-SD band. An ICS number of 1 indicates that all the losses in amplitude relative to the peripheral ring are outside 3° with complete sparing of the center. In a typical maculopathy patient, we find ICS values of around 0.2–0.4.

The importance of considering implicit times as well as response densities in standard clinical testing is illustrated with the example shown in the right-hand column of **Figure 4** showing a patient with unknown vision loss.

The patient's visual fields showed a bilateral bull's eye configuration reminiscent of hydroxychloroquine toxicity of **Figure 4(a)**. The perifoveal relative scotoma seen in the visual field is confirmed by the response density distribution of the mfERG (**Figure 4(a)**). However, in this case we see substantial delays in the areas with depressed response amplitudes. This finding suggests loss in receptor sensitivity. In this case, topographic mapping of $P1$ implicit times was needed to clearly distinguish this case from hydroxychloroquine toxicity. Latency plots are often necessary to distinguish different pathologies with similar response density topography.

Juvenile X-Linked Retinoschisis

The first-order trace array (**Figure 6(a)**) and the first-order response density plots (**Figure 6(b)**) show two areas where amplitudes are still within normal range. However, closer scrutiny reveals that the two areas are very different in their functional properties. Area 1 is normal in the

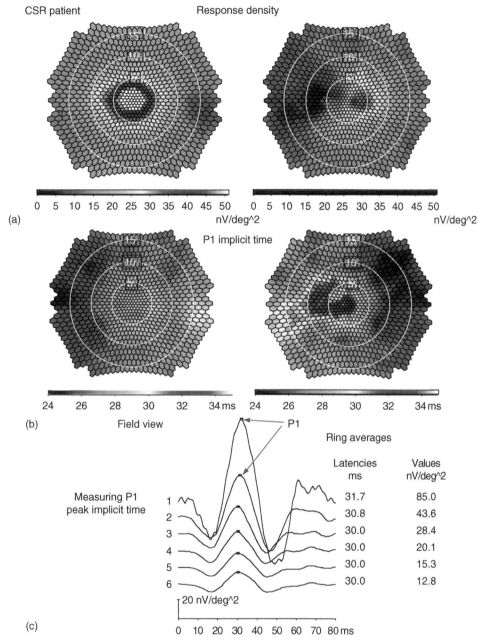

Figure 3 A patient complaining of a relative central scotoma diagnosed as central serous retinopathy with the help of this test. (a) A response density plot derived from two 7-min records. (b) The plot of *P*1 implicit time from the same record. *P*1 implicit times are measured from the time of the focal flash to the first positive peak as shown in (c).

first-order response amplitude as well as the *P*1 implicit time. The dominant higher-order component, the first slice of the second-order kernel, is also within normal range. This component reflects interactions between consecutive flash responses and is thought to originate predominantly in the inner retina. Area 2, on the other hand, has highly abnormal implicit times of close to 40 ms. The second-order component is almost completely absent in this area. This case illustrates the importance of looking not only at response densities but also at peak implicit times and the dominant higher-order kernel.

Detecting Small Central Dysfunction

The patient complained of a very small scotoma. It did not show in the visual field (**Figure 7(b)**) and there was no abnormal ophthalmoscopic finding. The patient drew the scotoma in the Amsler grid at 5° eccentricity (**Figure 7(a)**). Due to the small size of the suspected central dysfunction, we selected a protocol that places all 103 hexagons within the central 12°. The net recording time was 7 min. A distinct depression is seen at the location corresponding to the Amsler grid drawing (**Figure 7(a)**). In such cases, no spatial

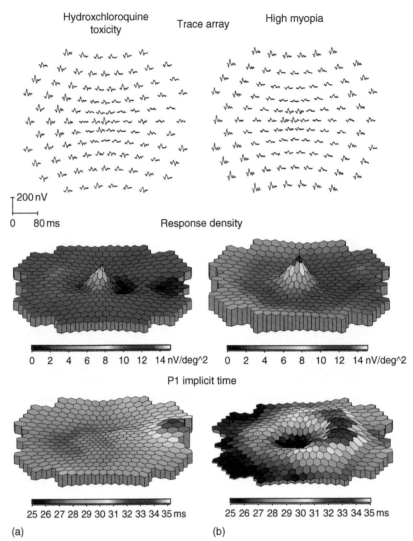

Figure 4   Column (a) shows a typical case of hydroxychloroquine toxicity. The characteristic peri-central amplitude loss is clearly seen in the trace array and the response density plot. The plot of *P*1 implicit time is normal suggesting inner retinal abnormalities. The enhancement in the area of the optic disc is due to scattered light and can be ignored. (b) a patient (high myopia) with a peri-central scotoma resembling hydroxychloroquine retinopathy. The increased implicit times in the depressed areas suggests a different pathogenesis, namely photoreceptor sensitivity loss. Plots of implicit time were needed to distinguish the two pathologies.

filtering could be applied to improve the quality of the focal waveforms as this would greatly reduce the small depression seen in the plot.

Another application of high-resolution central recording of this type is follow-up to macular hole surgery. OCT records document repair of structure, while the mfERG can be used to map recovery of function over time.

Applications to Neuro-Ophthalmology and Glaucoma

Some of the most important future clinical applications of multifocal electrophysiology are expected in the area of neuro-ophthalmology and glaucoma. The integrity of

the optic pathway can be tested using the visual-evoked cortical response derived by means of electrodes placed over the visual cortex as schematically illustrated on the right in **Figure 1**. The response to contrast reversal of a check pattern has been shown to be most sensitive for the detection of conduction losses to the cortex. The introduction of the multifocal pattern visual-evoked response (mfVEP) raised hopes for objective visual field mapping. The main obstacle to this aim is found in the convoluted cortical anatomy onto which the visual field is mapped. Different patches in the visual field generate dipole signal sources of different orientations. Adequate coverage of the visual field requires at least two electrode pairs at perpendicular orientation as well as their sum and difference signals. Uniform coverage is not possible. Comparison

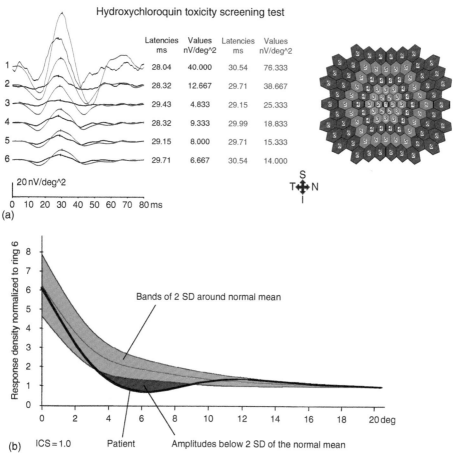

Hydroxychloroquin toxicity screening test

Figure 5 Automated statistical analysis for hydroxychloroquine toxicity screening. (a) responses in six rings for the patient (black) and the mean of 46 normal eyes (red). In (b), the amplitudes are plotted as a function of eccentricity. All data have been normalized to the outermost ring that is relatively unaffected in this condition. The ring averages of 5(a) are plotted as a function of eccentricity in degrees of visual angle with cubic spline interpolation. The green bands indicate 2-SD around the normal mean. The patient plot is in black.

with normative data is problematic due to the intersubject differences in the gross cortical anatomy. However, interocular comparison makes the mfVECP a powerful tool when the pathology is monocular or asymmetric between the eyes. Stimuli at corresponding locations in the visual field of the two eyes project to the same cortical patch and, in normal subjects, elicit virtually identical responses.

The cortical fold at the bottom of the calcarine fissure presents a problem that cannot be addressed with multiple electrodes. Sectors of the stimulus whose projections fall in this cortical area may wrap around the fold such that the same stimulus patch stimulates opposing surfaces of the sulcus generating mutually canceling dipole sources. This leads to characteristic signal dropout in the vicinity of the horizontal meridian. A substantial improvement was achieved by subdividing the sectors further and recording with 120 rather than 60 sectors. To compensate for resulting loss in signal to noise, the signal of each sector is averaged with a given percentage of the surrounding sectors after they have been brought to the same signal polarity. A typical plot of a trace array from

a normal subject is shown in **Figure 8(a)**. Traces from the right eye are red, left eye traces blue. In this presentation, the position of the traces is not topographic. They are approximately equally spaced to best utilize the rectangular plot surface. The approximate eccentricities of the waveforms are indicated with black contour lines. The plot was derived from two perpendicular electrode placements, one from electrodes on the midline 4 cm apart and the other from electrodes 4 cm lateral to the inion on both sides. At each stimulus location, the signal-to-noise ratios of the two recorded channels and their sum and difference signals were compared and the best combination was selected. The pseudo-color topographic map of **Figure 8(b)** was derived from the same record. It graphically represents the local differences in response amplitude between the two eyes. The saturation of the color in each stimulus patch indicates the interocular difference while the amount of gray in each patch is a measure of uncertainty. It is derived from the noise level in the record. While some data can also be displayed in numeric form, this graphic representation permits us to capture

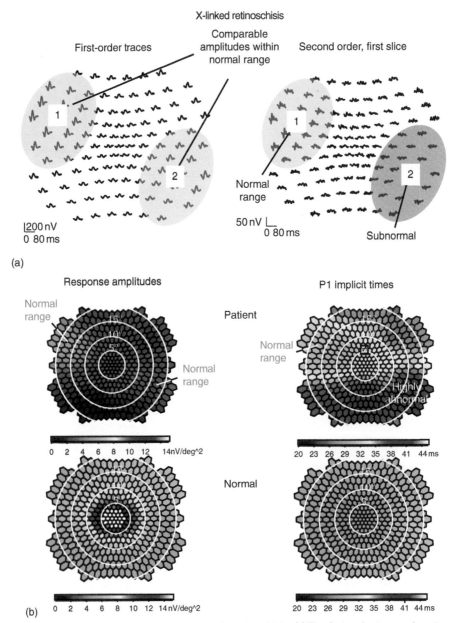

Figure 6   Record from right eye of a patient with x-linked juvenile retinoschisis. (a) The first-order traces show two areas with amplitudes within normal range. However, in area 2, the dominant second-order response contributions are almost completely extinct. (b) The response density plot shows two islands of relatively normal amplitudes. While the *P*1 implicit times are normal in area 1, they are highly abnormal in area 2.

the essence at a glance. Note that even in normal subjects, areas with small interocular differences in the peripheral areas are not unusual.

The mf VECP in Optic Neuritis

An obvious application of the mf VECP is in optic neuritis and multiple sclerosis. It is illustrated here with an example of acute optic neuritis. **Figure 9** shows mf VECP plots recorded in the acute phase and at 2-month intervals during recovery. In the acute phase, the responses within the central 3° of the affected eye are almost extinguished.

The first follow-up record shows substantial recovery of amplitudes in this area, while increased implicit times indicate areas of demyelination. In the second follow-up record, the implicit times in the lower part of the fovea are back to normal, while in the upper field the delays persist.

Comparison of the mfVEP and the mfERG in Optic Neuropathies and Glaucoma

It is well known that the ERG signal contains contributions from retinal ganglion cells and that losses of this cell population can, through a retrograde process, affect other

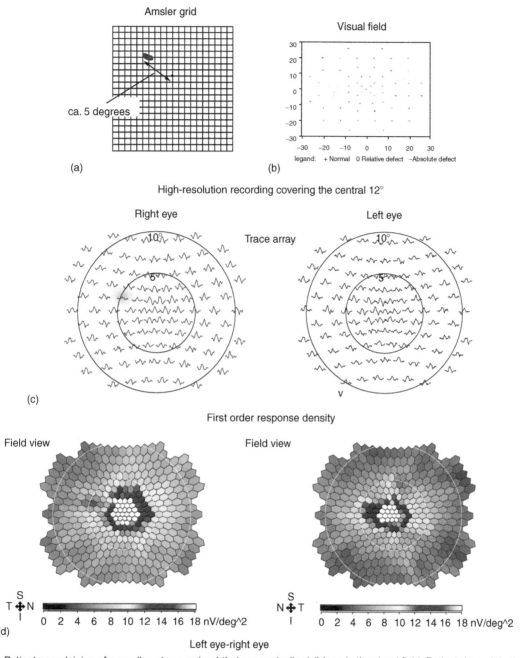

Figure 7 Patient complaining of a small scotoma not ophthalmoscopically visible or in the visual field. Patient draws it in the Amsler grid. A 7-min high resolution mfERG record reveals the dysfunctional patch objectively. (a) Amsler grid, (b) visual field, (c) trace arrays, (d) response density topography.

inner retinal response contributions. However, changes in the ERG signal are rather subtle and have proven difficult to detect in early stages of disease. The ganglion cell contributions to the ERG are small and overlap with signal contributions from other retinal sources, making their isolation and quantitative estimation extremely difficult. The discovery of the optic nerve head component (ONHC) of the ERG raised hope that isolation and mapping of ganglion cell function might become possible. The ONHC is a contribution from optic nerve fibers

near the optic nerve head. Its generation is schematically illustrated in **Figure 10**. The contribution of the ONHC is delayed by the amount of time it takes action potentials to travel from the stimulated retinal patch to the nerve head. It is recognized by its timing, which varies depending on the length of the nerve fibers connecting the stimulus site with the disc. Possible mechanisms for its generation are the sharp bend in the axons where they descend into the cup, the beginning of myelination near the lamina cribrosa and conductive changes in the extracellular

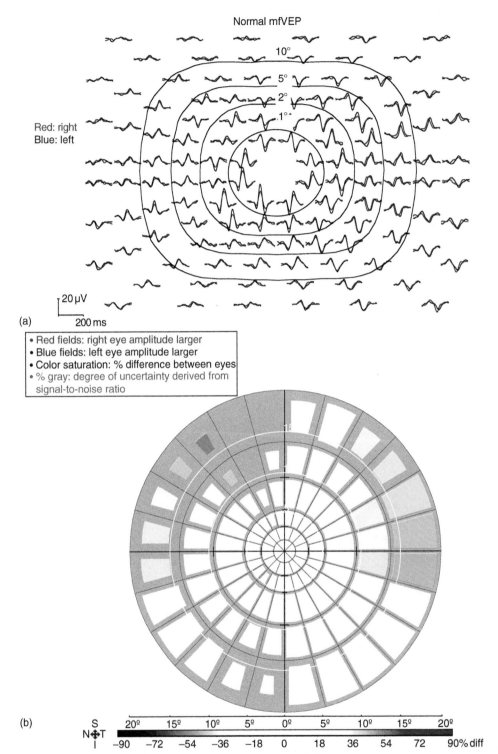

Figure 8 MfVEP of a normal subject. (a) interocular comparison of response traces: right eye red, left eye blue. The distribution of the traces is not topographic but has been arranged for best presentation of the traces in the rectangular plot area. Approximate eccentricities are indicated. (b) Pseudocolor topographic map of interocular differences.

medium. The observation that it can disappear in demyelinating disease while cortical responses are preserved, strongly points toward the transition from membrane conduction to saltatory conduction at the point where myelination begins as the main source of this signal. While mfVEP amplitudes are affected by the cortical anatomy, ONHC amplitudes directly reflect function. Their local evaluation does not require interocular response

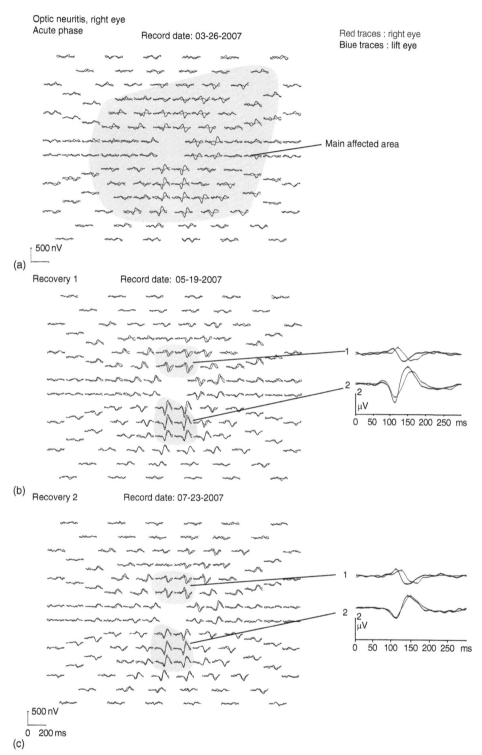

Figure 9 MfVEPs of a patient with optic neuritis. (a) during the acute phase: The red shading indicates the affected area. (b) After 2 months recovery: Amplitudes have largely recovered, but increased implicit times in the central area indicated demyelination. (c) After 4 months recovery: delays in the lower field have largely disappeared while those in the upper field remain.

comparison. In bilateral disease, the ONHC, thus, promises to be more reliable than the mf VECP.

In records collected with the commonly used mf ERG protocol, the ONHC is too small to be evaluated

(**Figure 10(b)** left). The traces are from a ring around the fovea starting from the area near the nerve head, proceeding through the upper field and returning through the lower field (insert on the right). The contributions

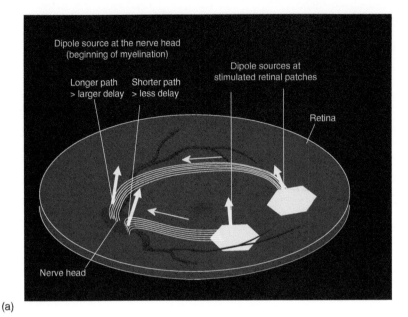

(a)

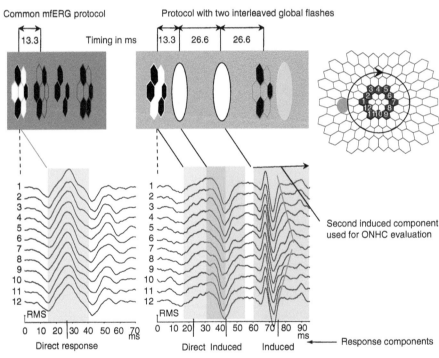

(b)

Figure 10   (a) This is a schematic illustration explaining the generation in the optic nerve head component (ONHC). Two signal sources are thought to contribute to the signal derived from the cornea, one from the stimulated retinal area and the second from the location in the nerve fibers where myelination begins. The latter is delayed relative to the retinal response by travel time of action potential from the stimulation site to the nerve head. It is recognized by this location-dependent delay. (b) Enhancement of the ONHC through the global flash paradigm. (Top) Schematic comparison of the stimulation used in the common m-sequence protocol on the left and the interleaved global flash protocol on the right. (Bottom) First-order traces derived from a normal subject: left, common protocol; and right, global flash protocol. The traces are from a ring around the fovea (insert on right) starting from the vicinity of the nerve head. The thin blue lines indicate the main feature of the ONHC contribution. The second induced component, epoch 60–90 ms, is normally used for visual evaluation.

from the ONHC should thus increase in implicit time to trace 7 and then decrease again. A stimulation protocol is now available that greatly enhances the ONHC as well as inner retinal response contributions. The protocol uses

global flashes interleaved at specific intervals between the multifocal stimuli. The principle is illustrated on the right in **Figure 10(b)**. The ONHC is now readily recognized. It consists of those features in the waveforms that shift to longer implicit times with increasing distance of the stimulus patch from the nerve head. Its main peak is indicated with a thin blue line.

The ONHC can be visually evaluated when traces on rings around the fovea are plotted in vertical columns. A plot of this kind from a normal subject is shown in **Figure 11(a)**. Traces in each ring are plotted starting from the patch nearest the disk, proceeding through the upper field and returning through the lower field.

The delays of the ONHC relative to the retinal contributions to the ERG are known. Propagation velocity of action potentials in the unmyelinated nerve fiber layer is determined by the fiber diameter and varies little between subjects. Aligning the traces to the ONHC before applying some spatial filtering further enhances its visibility and our ability to evaluate it by visual inspection. It is now seen as a vertical ridge in the columns of **Figure 11(b)**.

Until an algorithm for estimation of the ONHC that performs better than a visual evaluation becomes available, the ONHC is mapped and scored as follows: using the computer mouse, traces are marked pink for moderately deficient and red for severely deficient in the ONHC. The corresponding areas automatically appear pink and red in the topographic insert below together with a numeric score (see example of **Figure 12**).

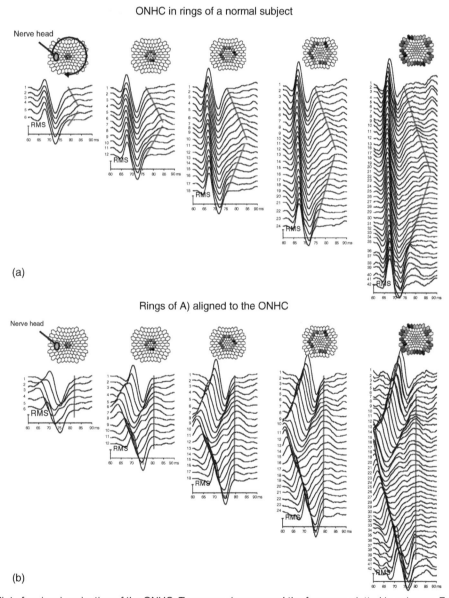

Figure 11   (a) Plots for visual evaluation of the ONHC. Traces on rings around the fovea are plotted in columns. Each ring begins on the side of the optic nerve head, proceeds through the upper field and returns through the lower field. (b) To enhance the appearance of the ONHC, the traces are aligned to ONHC implicit times.

Unilateral glaucoma patient

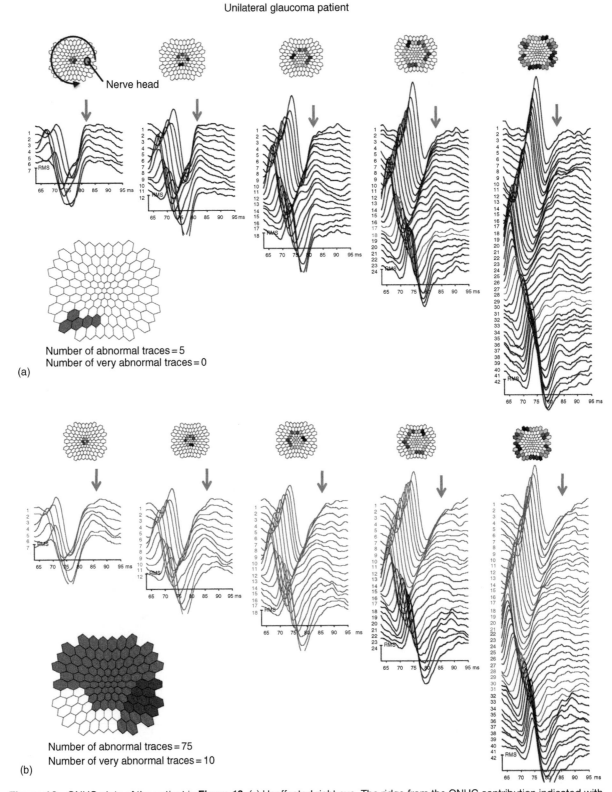

Figure 12   ONHC plots of the patient in **Figure 13**. (a) Unaffected right eye. The ridge from the ONHC contribution indicated with blue arrows is clearly visible and appears normal in most areas; (b) The ONHC ridge has largely disappeared. A residue is visible in the lower field.

Comparison of the mfVEP and the ONHC of the mf ERG in an Asymmetric Glaucoma Patient

To illustrate the performance of currently available electro-physiological function tests, a highly asymmetric glaucoma patient was selected. In **Figure 13(a)**, the mfVEP trace arrays of the two eyes are compared. **Figure 13(b)** shows an automated topographic analysis of the same data set. The color of the field indicates the eye with the larger response, red for right, and blue for left. The color saturation indicates the percent difference in response amplitude

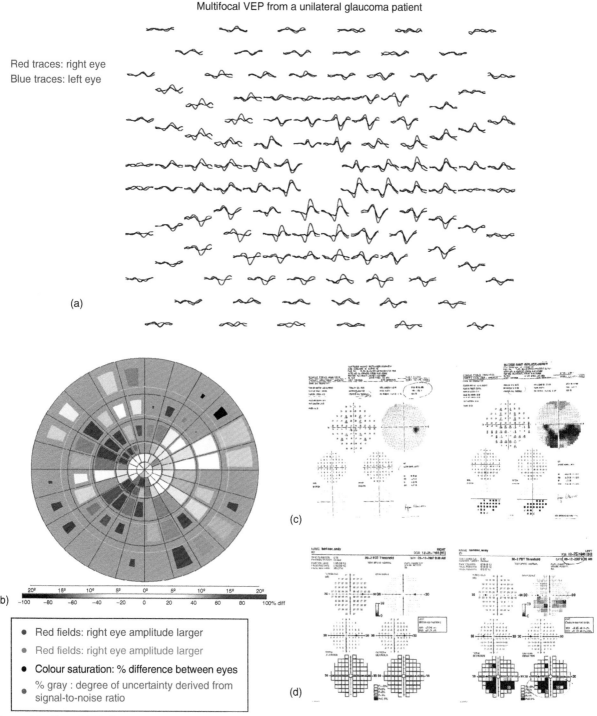

Multifocal VEP from a unilateral glaucoma patient

Red traces: right eye
Blue traces: left eye

(a)

(b)

20° 15° 10° 5° 0° 5° 10° 15° 20°
−100 −80 −60 −40 −20 0 20 40 60 80 100% diff

- Red fields: right eye amplitude larger
- Red fields: right eye amplitude larger
- Colour saturation: % difference between eyes
- % gray : degree of uncertainty derived from signal-to-noise ratio

(c)

(d)

Figure 13   The mfVEP of an asymmetric glaucoma patient: (a) trace arrays of the two eyes; (b) topographic representation of interocular differences shows areas with response deficit in the left eye; (c) visual fields; and (d) perimetry using frequency-doubling technology (FDT) maps for comparison.

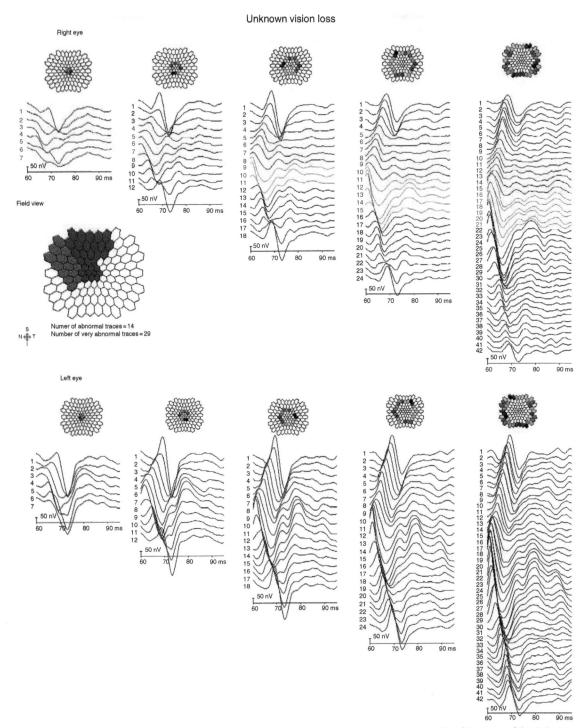

Figure 14   ONHC plots of a person with unknown vision loss. The complete disappearance of the ONHC in areas of the right eye is not consistent with the full visual field (not shown). Conclusion: The disappearance is due to loss of the transition from unmyelinated to myelinated fiber through retrolaminar demyelination rather than conduction loss.

between the eyes. The percentage of gray in each field represents the uncertainty in the estimate based on the signal-to-noise ratio in the record. The numeric version of the plot is not shown here. The inserts (c) and (d) show the visual field and the field mapped with the frequency-doubling technique (FDT) for comparison. **Figure 12(a)**

and **12(b)** show the ONHC analysis of the mfERG for the two eyes of the same subject. Traces judged deficient in the ONHC are marked pink or red by the operator depending on the degree of abnormality. The corresponding stimulus areas are marked in the same color in the inset below.

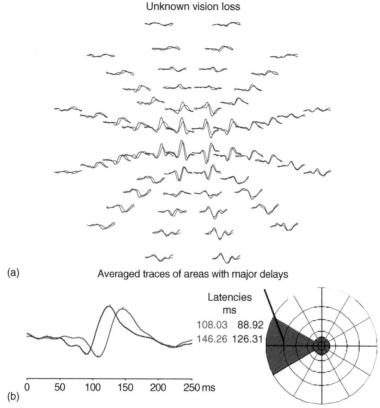

Figure 15 The conclusion from **Figure 14** is confirmed with an mfVEP. In the right eye, we find areas with substantial delays.

Both, the mfVECP and the ONHC data show more extensive losses than the two psychophysical function tests. In such highly asymmetric cases, the mfVECP can be a very sensitive test. It becomes more problematic when both eyes are affected. The ONHC, on the other hand, does not require interocular comparison.

Patient with Unknown Vision Loss

A 53-year-old male patient noticed blurring in his right eye. His visual acuity was V/A 20/30 on the right and 20/20 on the left. Visual fields were normal in both eyes. The common first-order response is somewhat lower in the right eye, but its topographic distribution is within normal range (not shown). The ONHC, on the other hand, is highly abnormal in the right eye, particularly in the center and in some portions of the upper field and completely absent in some areas (**Figure (14)**). The left eye is within normal range. The gross abnormalities in the ONHC are not consistent with the normal visual fields of the patient (not shown). This finding could, therefore, not be attributed to loss of nerve conduction. It could, however be explained by retrolaminar demyelination eliminating the source of the ONHC. This reasoning led to the mfVEP records shown in **Figure 15**. In the right eye, the responses are indeed substantially delayed in the areas with an abnormal ONHC confirming local demyelination. This case is interesting as electrophysiology led to the diagnosis of a patient not suspected of demyelinating disease.

Note that the ONHC is abnormal in some cases of optic neuritis even when substantial cortical delays are observed but not in others. In the above example, there are large cortical delays in areas where the ONHC is preserved. In the case of acute optic neuritis (**Figure 9**), the ONHC was judged within normal range in both eyes. In combination, the three tests (common mfERG, ONHC, and mfVEP protocols) allow us often to localize a dysfunction along the visual pathway.

Summary and Conclusion

Multifocal electrophysiology is not a single tool, but rather a collection of tools. Stimulation and analysis protocols can be optimized for specific clinical applications such as diabetes, AMD, optic neuropathies, etc. Much can still be done to shorten the tests and improve their efficiency. The protocols can be semiautomated to make clinical testing easy. However, the tools have to be paired with the expertise necessary to select the proper tool for each case and interpret the data.

There is a great deal of information contained in multifocal records that is only partially understood and still largely unexploited. Combining tests of function and structure promises to advance the understanding of both types of data and the pathogenesis of diseases. A case in point is the example of x-linked retinoschisis shown in **Figure 6**. The availability of OCT data on the different areas of the retina might have greatly helped our understanding of the connection between function and stricture in this and other similar cases.

Acknowledgment

The studies on which this article is based have been supported in part by NIH grant EY06961.

See also: Adaptive Optics; Optical Coherence Tomography; Primary Photoreceptor Degenerations: Retinitis Pigmentosa; Primary Photoreceptor Degenerations: Terminology; Secondary Photoreceptor Degenerations: Age-Related Macular Degeneration.

Further Reading

Hood, D. C. (2000). Assessing retinal function with the multifocal ERG technique. *Progress in Retinal and Eye Research* 19: 607–646.

Miyake, Y. (2008). *Electrodiagnosis of Retinal Diseases*. Tokyo: Springer.

Sutter, E. E. (2001). Imaging visual function with the multifocal m-sequence technique. *Vision Research* 41: 1241–1255.

Sutter, E. E. (1992). A deterministic approach to nonlinear systems analysis. In: Pinter, R. B. and Nabet, B. (eds.) *Nonlinear Vision*, pp. 171–220. Cleveland, OH: CRC Press.

Optical Coherence Tomography

W Drexler, Medical University Vienna, Vienna, Austria

Glossary

Adaptive optics (AO) – A technique originally developed for improving imaging performance in astronomy. Applied in vision sciences to correct wave front aberrations introduced by imperfect optics of the human eye to reduce the spot size in the retina and hence transverse resolution of retinal imaging.

Optical coherence tomography (OCT) – Optical analog to ultrasound, for noninvasive three-dimensional micrometer resolution visualization of superficial (up to 2 mm) tissue morphology.

Optophysiology – Optical analog to electrophysiology that enables noninvasive depth resolved optical probing of retinal physiology.

Optical coherence tomography (OCT) is a rapidly emerging noninvasive, optical diagnostic imaging modality enabling *in vivo* cross-sectional tomographic visualization of internal tissue microstructure in biological systems at resolution levels of a few micrometers. Novel high-speed detection techniques as well as development of ultrabroad bandwidth and tunable light sources have recently revolutionized imaging performance and clinical feasibility of OCT. In this view, OCT can now be considered as an optical analog to computed tomography (CT) and magnetic resonance imaging (MRI), not enabling full body imaging, but noninvasive optical biopsy, that is, micrometer/cellular resolution 3D visualization of tissue morphology.

The eye provides easy optical access to the anterior segment and the retina due to its essentially transparent nature. Axial and transverse resolutions are decoupled in OCT. While the axial one is mainly determined by the optical bandwidth of the employed light source, the transverse one is mainly given by the numerical aperture of the optics that is focusing the beam to the tissue. As a consequence, the axial OCT resolution for retinal imaging is not limited by the low-numerical aperture and large depth of focus of the human eye as it does in scanning laser ophthalmosocopy. For this reason, ophthalmic and especially retinal imaging has so far not only been the first, but also the most successful clinical application for OCT. Objectively this is evidenced by the fact that nearly 2500 (49%) of the 5000 OCT publications (including only peer-reviewed articles) have been published in ophthalmic journals. About 1300 (25%) have been published in optical or biomedical journals demonstrating the significant emphasis on OCT technology development since its invention. In addition, more than half a dozen companies offer this technology in its fourth generation for 3D retinal OCT. Considering the fact that OCT has been introduced only about two decades ago, retinal OCT therefore represents the fastest adopted imaging technology in the history of ophthalmology.

Figure 1 depicts an overview of OCT development with respect to axial resolution and data acquisition speed (measurement time) for morphologic (as opposed to functional) retinal imaging. After its first *in vitro* demonstration in 1991 and first *in vivo* imaging studies of the human retina in 1993 (cf. **Figure 1(a)–1(d)**), OCT has rapidly developed as a noninvasive, optical medical diagnostic imaging modality providing two-dimensional information of retinal structure with resolution about one order of magnitude better than ultrasound. Since its invention, the original idea of OCT was to enable noninvasive optical biopsy, that is, the *in situ* imaging of tissue microstructure with a resolution approaching that of histology, but without the need for tissue excision and postprocessing. As a consequence, two milestone developments that improved key technological OCT parameters – axial resolution and measurement time – significantly contributed to realize the optical biopsy idea of OCT. A first step toward this goal was the introduction of ultrahigh resolution OCT enabling a noticeably superior visualization of tissue microstructure, for example, all major intraretinal layers as well as cellular resolution OCT imaging in nontransparent tissue by improving axial OCT resolution by one order of magnitude from the 10–15 μm to the (sub)micrometer region, accomplishing 2–3 μm in the living human retina (cf. **Figure 1(e)–1(g)**). Advances in photonics technology including the development of ultrabroad bandwidth and high-speed tunable light sources as well as high-speed detection techniques have enabled a considerable improvement in data acquisition speed from several hundreds of A-scans/s to 30 000 A-scans/s and recently 300 000 A-scans/s (cf. **Figure 1(h)–1(j)**).

Despite numerous successful and clinically valuable OCT applications in the anterior eye segment (mainly in the cornea and anterior chamber angle), retinal OCT had a significantly higher clinical impact so far. This is mainly due to the lack of competing noninvasive techniques that can provide comparable wealth of information about the living human retina. Hence, the technological improvements that have been accomplished in the last decade not only enabled unprecedented OCT performance that is unique among noninvasive diagnostic

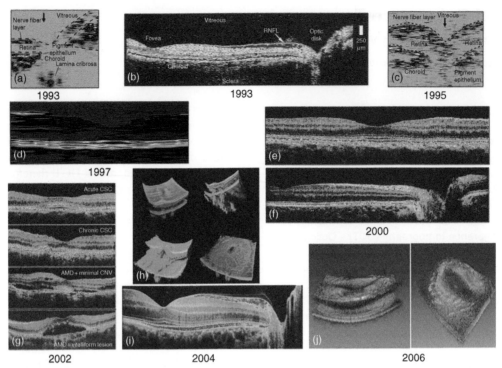

Figure 1 Development of retinal OCT regarding axial resolution and data acquisition. (a, c): First *in vivo* retinal OCT with 10–15 μm axial resolution and 2 A-scans/s; (b) 100 A-scans/s; (d): improved axial resolution (7–9 μm) with 2 A-scans/s; (e, f) first *in vivo* ultrahigh resolution (2–3 μm) with 160 A-scans/s of normal subjects; (g) first ultrahigh resolution OCT in patients; (h) three dimensional; (i) high definition ultrahigh resolution (2–3 μm) with 29 000 A-scans/s; (j) 300 000 A-scans/s and 10 μm axial resolution 3D retinal imaging at 1060 nm with enhanced penetration into the choroid.

techniques, but it also established retinal OCT as a state-of-the art noninvasive, complementary ophthalmic diagnostic methodology. For this reason, this article will mainly focus on retinal OCT.

3D Ultrahigh Resolution Retinal OCT

While third generation commercial retinal OCT systems (Stratus OCT) were based on the so-called time domain OCT, enabling up to 400 A-scans/s (one-dimensional (1D) measurements), fourth generation (spectral or Fourier-domain based) commercial retinal OCT systems, nowadays can perform up to 100 times more measurements enabling either highly sampled (high definition) 2D tomograms or 3D (volumetric) imaging of the retina. This is due to an alternative detection method that uses either a CCD camera-based spectrometer (spectral of Fourier domain), that can be read out quickly or a fast tuneable laser (swept source OCT or frequency domain OCT) as compared to a moving mirror (time domain) to perform a single A-scan. The first version is inherently more efficient and enables to raster scan the retina analog to a scanning laser ophthalmoscope (SLO) but thereby not only acquiring tissue information from a single 2D plane (in focus), but the full morphological depth information from an entire volume

with a depth resolution mainly given by the optical bandwidth of the employed light source.

A volumetric data set is acquired for the imaged area that can then be viewed and analyzed in ways similar to those used with CT or MRI scans (cf. **Figure 2(a)** and **2(b)**). Hence, the imaged volume can arbitrarily be cut according to the necessary diagnostic needs, for example, like an ultrahigh resolution SLO in en face (C-mode) tomograms (cf. **Figure 2(c)** and **2(d)**). **Figure 2** demonstrates 3D UHR OCT in the foveal region of a patient with retinal pigment epithelium (RPE) atrophy. Note that the axial dimension is twofold enlarged as compared to the other two dimensions for better visualization. The 3D representation of the macular region is presented at different-angled views (cf. **Figure 2(a)**) depicting the pathological change in the topography of the foveal depression as well as enabling unprecedented views in which the retina can be observed from any direction, including from below (cf. **Figure 2(b)**). **Figure 2(e)–2(h)** present virtual biopsy/surgery using 3D UHR OCT in combination with 3D data rendering which allows the user to excise and remove any given layer or part of the retinal volume to visualize intraretinal morphology.

The clinical benefit of these spectral of Fourier domain-based OCT instruments is demonstrated in highly sampled 3D visualization of the retina during a reasonable short data-acquisition time resulting in more

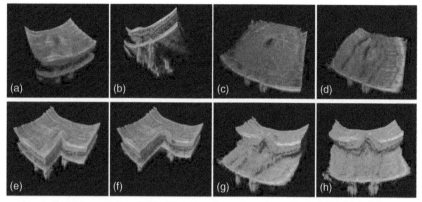

Figure 2 Three-dimensional ultrahigh resolution OCT of the macular region of a patient with retinal pigment epithelium atrophy; (a–d) at different-angled views; (e–h) virtual biopsy allows removal of any given layer or part of the retinal volume to visualize intraretinal morphology.

reliable and reproducible 2D thickness maps of (intra) retinal layers. Furthermore, retinal locations are less likely to be missed that are important for diagnosis since an entire volume is measured instead of deciding the location of B-scans during OCT acquisition. This might result in improved diagnosis of retinal pathologies, better understanding of retinal pathogenesis, as well as enhanced (objective) monitoring of novel therapy approaches.

3D Wide-Field Choroidal OCT

So far, commercially available retinal OCT has mainly been performed in the 800-nm wavelength region. This is mainly due to easy availability of broad bandwidth light sources and detectors (CCD cameras) in this wavelength region. Although OCT systems centered at 800 nm can resolve all major intraretinal layers, they only enable limited penetration beyond the retina, due to multiple scattering and absorption in the melanin-rich RPE. This results in limited visualization of the choriocapillaris and choroid. Moreover, in clinical OCT, turbid ocular media (e.g., cataract or corneal haze) represent a significant challenge when imaging the retina. Since scattering in biological tissues decreases monotonically with increasing wavelength, OCT imaging at 1060 nm can deliver deeper tissue penetration enabling delineation of choroidal structure. Being less sensitive to scattering eye media and enabling enhanced penetration into the choroid up to the choroidal–scleral interface at 1060 nm might therefore significantly improve the clinical feasibility of retinal OCT.

2D time domain-based OCT was initiated about 5 years ago demonstrating improved visualization of superficial choroidal structure. With the introduction of more efficient and hence more sensitive spectral of Fourier domain OCT, 3D OCT at 1060 nm demonstrated wide-field visualization of the entire choroid up to the sclera. In this approach, a cost-effective, easy-to-implement system

based on a high-speed InGaAs linear 1024 pixel array (SUI-Goodrich) enabling 47 000 A-scans/s, 5–8 μm axial resolution, and 2.6 mm scanning depth in tissue was developed. **Figure 3** depicts the comparison of 3D OCT at 800 nm (cf. **Figure 3(a)** and **3(b)**) versus 1060 nm (cf. **Figure 3(c)** and **3(d)**) in the same normal eye. Enhanced visualization of the choroid up to the choroidal sclera interface is accomplished using 3D 1060-nm OCT as compared to 3D 800-nm OCT (cf. yellow arrows in **Figure 3(d)**). In the subject with light fundus pigmentation, 3D 1060-nm OCT enables penetration beyond the choroid into the sclera (cf. red arrows in **Figure 3(g)**). High-speed 3D 1060-nm OCT also enables en face visualization of the choroidal vasculature without the use of any contrast agent (cf. **Figure 3(c), 3(e),** and **3(f)**).

High-speed 3D 1060-nm OCT therefore now enables unprecedented visualization of all three choroidal layers giving access to the entire choroidal vasculature. In addition, 2D choroidal thickness maps might have significant impact in the early diagnosis of retinal pathologies such as glaucoma, age-related macular degeneration and might contribute to a better understanding of myopigenesis.

Cellular Resolution Retinal OCT

For retinal OCT imaging, the cornea and the lens act as the imaging objective, thereby determining the numerical aperture and hence the beam diameter in the retina. This diameter specifies the transverse OCT resolution that is of the order of ~20 μm for a beam of ~1-mm diameter (at 800 nm) – approximately one order of magnitude worse than the best axial OCT resolutions accomplished so far. This can be improved by dilating the pupil and increasing the measurement beam diameter. In practice, however, for large pupil diameters, ocular aberrations limit the minimum focused spot size on the retina, even for monochromatic illumination. An alternative and promising

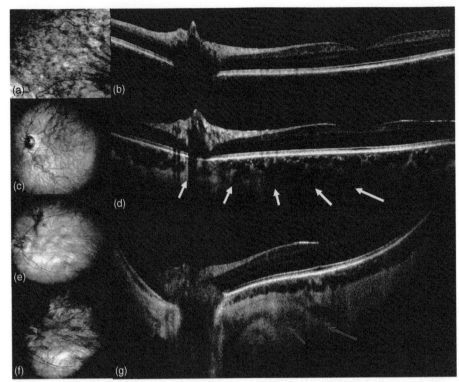

Figure 3   Wide-field three-dimensional choroidal OCT. (a, b) 3D-OCT at 800 nm and (c–g) 1060 nm of a normal retina: (b) high definition (4096 depth scans) 800-nm 3D-OCT scan over 35°; (d, g) high definition (2048 pixel) 1060-nm 3D-OCT scan over 35°; (a) en face view of the choroid using 3D OCT at 800 nm; (c, e, f) en face wide-field (35° × 35°) view using 3D OCT at 1060 nm; yellow arrows indicate enhanced choroidal visualization; red arrows indicate visualization of the sclera.

approach is to use adaptive optics (AO), which was originally developed to improve the resolution of astronomical imaging, to minimize ocular aberrations, reduce retinal spot size, and hence to improve transverse OCT resolution.

In ophthalmic AO a wave front sensor measures the individual ocular aberrations of the investigated eye, calculates an inverse wave front, and sends this information to a correcting device – a deformable mirror – for aberration correction. This procedure is performed in real time in a closed-loop configuration. In a couple of tenths of a second, ocular aberrations are compensated and continuously corrected during the entire measurement procedure. In the present approach, a deformable mirror (Mirao52, Imagine Eyes, France) with a unique performance in terms of amplitude (±50-μm stroke) and linearity was used, allowing for correcting highly aberrated normal or pathologic eyes. Furthermore a 140–160-nm Ti:sapphire laser (Femtolasers Integral, Femtolaser, Vienna, Austria) in combination with a CMOS Basler sprint spL4096-140k camera (Basler AG Germany) enabling 160 000 A-scans/s with 1536 pixels was used, resulting in ultrahigh speed cellular resolution retinal imaging with isotopic resolution of 2–3 μm. Furthermore, the chromatic aberrations of the eye in the 700–900 nm wavelength region of the employed light source have been compensated with a special lens.

The combination of high stroke deformable mirror-based AO, with ultrahigh speed, ultrahigh resolution OCT employing compensation of the eye's chromatic aberrations enabled isotropic OCT resolution of 2–3 μm. It is noteworthy that the extremely high measurement speed in combination with sufficient system sensitivity at this speed is essential to maintain cellular resolution morphology information despite motion artifacts. In analogy to AO, scanning laser ophthalmoscopy measurements, AO OCT raster scans a retinal area, but acquires full morphological information as a function of depth in the region of interest, without the need to scan the depth of focal plane. As a consequence, 3D morphology of single photoreceptor (PR) outer segments in addition to cellular microstructure at the level of the RPE and choriocapillaris can be visualized in the living human retina. **Figure 4** depicts *in vivo* cellular resolution retinal imaging in a normal human retina at about 4° parafoveal. A rendered volume of about 350 × 80 × 100 μm is presented. En face representations at different depths of the volume reveal cellular resolution intraretinal microstructure.

3D information of intraretinal morphology at cellular level might not only revolutionize ophthalmic diagnosis, but also significantly contribute to a more precise interpretation of OCT tomograms. Despite numerous clinical

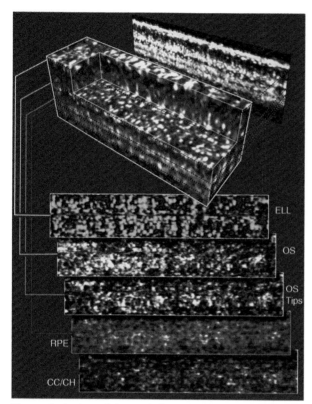

ELL

OS

OS
Tips

RPE

CC/CH

Figure 4 *In vivo* cellular resolution retinal adaptive optics OCT. 3D tomogram and en face cross sections of a normal human retina acquired at 4° parafoveal at 160 000 A-scans/s. Various retinal layers in photoreceptor layer may be distinguished with this technique. From top to bottom slice: ellipsoids (ELL), outer-segment tips (OS), outer segment tips (OS tips), retinal pigment epithelium (RPE), choriocapillaris (CC), and choriocapillaris/choroid (CC/CH). Width of slides corresponds to ~350 × 80 μm.

investigations using UHR OCT and systematic studies comparing histology with UHR OCT in *in vitro* animal models to correctly interpret OCT images, it is noteworthy that a comprehensive, reliable interpretation of all OCT intraretinal layers has not yet been accomplished. Although the state-of-the-art ophthalmic OCT technology enables significantly improved visualization of intraretinal layers, caution is therefore imperative regarding proper interpretation of OCT tomograms, especially of the distal part of the retina. **Figure 5** depicts state-of-the-art OCT interpretation of intraretinal layers in a high definition ultrahigh resolution OCT of the optic disk (a) and foveal (b) region of a normal subject. From the proximal to the distal part of the retina the nerve fiber layer (cf. **Figure 5(a)**, NFL) as well as plexiform layers (cf. **Figure 5(a)**, inner (IPL) and outer plexiform (OPL) layer) appear back-reflecting and therefore as a strong signal in the OCT tomogram. The ganglion cell layer (cf. **Figure 5(a)**, GCL) as well as nuclear layer (cf. **Figure 5(a)**, inner (INL) and outer nuclear (ONL) layer) appear less back-reflecting and therefore as a low signal in the OCT tomogram. The external limiting membrane (cf. **Figure 5(b)**, ELM), the inner

and outer PR segments (cf. **Figure 5(b)**, IS PR, OS PR), as well as the choroid (cf. **Figure 5(b)**) are also properly confirmed by literature. Red-labeled layers indicate that caution is imperative and the correct interpretation needs more conclusive studies to confirm correctness. This applies to the internal limiting membrane (cf. **Figure 5(a)**, ILM) and the distal part of the retina involving probably the distal tips of the PRs – sometimes also referred to Verhoeff's membrane (cf. **Figure 5(b)**), the RPE including Bruch's membrane (cf. **Figure 5(b)**, RPE/BM) as well as choriocapillaris (cf. **Figure 5(b)**).

To date, the most distal layer, that is, strongest continuous distal signal in OCT tomograms, has been interpreted as the RPE layer. Although literature describing the light–RPE interaction in the near-infrared region around 800 nm would confirm this interpretation, the relatively thick (up to 30 μm) appearance in OCT tomograms is not supported by histological findings which describe the RPE as a monocellular layer. In addition, the appearance of the RPE and distally adjacent layers as visualized by OCT varies in eyes with different pathologies. **Figure 6** depicts the region indicated with a red rectangle in **Figure 5** but imaged with AO OCT. 3D information at cellular resolution level significantly helps to correlate OCT findings with well-known anatomy from histology and therefore more precisely interpret 2D OCT tomograms. Hence, from the external limiting membrane (cf. ELM in **Figure 6**) toward the distal part of the retina, the first bright (white) signal indicates the junction between inner and outer PR segments. En face cellular resolution representation clearly reveals single ellipsoids. After a thicker, less bright (black) layer indicating a part of the PR outer segment, five signal bands with alternating bright (white) and weak (black) signal appearance is revealed by AO OCT. En face information of the first bright (white) band suggests that this layer corresponds to the outer part/outer tips of the outer PR segments. It is noteworthy that this structure as well as the junction between the inner and outer segment of the PR cannot be resolved within a circular zone of less than 2° parafoveal. Their too-tight spacing and the limited numerical aperture of the human eye in addition to weak contrast has not yet allowed the visualization of PRs in the center (±2° central region) of the fovea. Nevertheless, they are included in this figure for more precise interpretation. The next weak (black) signal band might correspond to the RPE somas, since the adjacent bright (white) signal band is clearly identified to correspond to the RPE layer due to the hexagonal structure of single RPE cells as visualized by AO OCT in the indicated en face view. The next weak (black) signal band might correspond to Bruch's membrane, although literature describes the thickness of this membrane close to the resolution limits of AO OCT. This interpretation is nevertheless supported by the fact that the cellular resolution en face image of the next bright (white) signal band indicates that it might correspond to

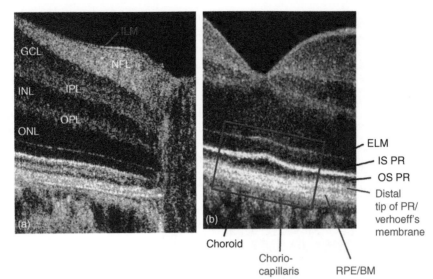

Figure 5 State-of-the-art interpretation of intraretinal layers in OCT. High definition ultrahigh resolution OCT of the optic disk (a) and foveal (b) region of a normal subject. Labeling of intraretinal layers that could properly confirmed by numerous previous studies are labeled white or black. Red-labeled layers indicate that caution is imperative and the correct interpretation needs more conclusive studies to confirm correctness. Nerve fiber layer (NFL), inner (IPL) and outer plexifrom layer (OPL), ganglion cell layer (GCL), inner (INL) and outer nuclear layer (ONL), external limiting membrane (ELM), inner (IS PR) and outer photoreceptor segment (OS PR), choroid, internal limiting membrane (ILM), and distal tips of the photoreceptors – sometimes also referred to Verhoeff's membrane, retinal pigment epithelium including Bruch's membrane (RPE/BM), choriocapillaris. Red square indicates region depicted in **Figure 5**.

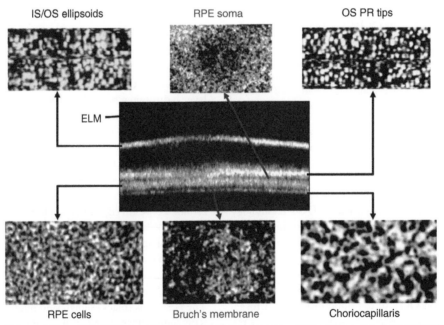

Figure 6 Revised interpretation of the RPE signal band in OCT. Three-dimensional information at cellular resolution level significantly helps to correlate OCT findings with well-known anatomy from histology and therefore more precisely interpret two-dimensional OCT tomograms; external limiting membrane (ELM); photoreceptors can only be resolved within a circular zone of less than 2° parafoveal. Nevertheless, they are included in this figure for more precise interpretation. Two layers are still labeled red indicating that more studies are needed to finally confirm their proper interpretation.

the choriocapillaris. It is noteworthy that despite the superb imaging performance of AO OCT, two layers are still labeled red in **Figure 6** indicating that more studies are needed to finally confirm their proper interpretation.

Figure 7 demonstrates the concept of clinical cellular resolution AO OCT in a patient with type 2 macular telangiectasia. A commercial 3D OCT is used to pre-screen a larger volume to identify suspicious locations.

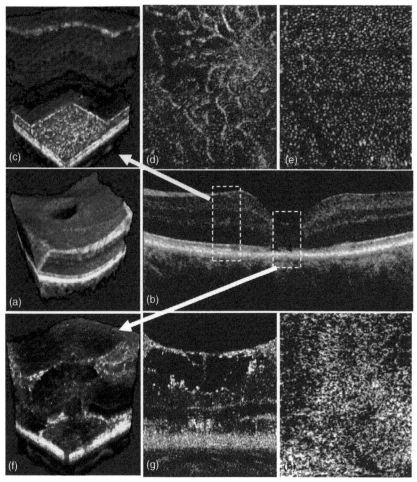

Figure 7 Possible strategy for clinical cellular resolution OCT in a patient with type 2 macular telangiectasia: (a) prescreening over 20°×20° (512 × 128 depth scans) using a commercially available 3D-OCT at 800 nm; (b) detection of impaired intraretinal morphology using a representative cross section from (a); zoom in at a normal (c–e, yellow-dashed square in (b)) and a pathologic ((f–h), white-dashed square in (b)) smaller volumes using cellular resolution OCT; volumetric rendering at 6° parafoveal (c) and 0° (f); en face images at the level of the capillaries in the inner nuclear layer at 6° (d) and at the level of the tips of the outer photoreceptors at 6° (e) extracted from (c); cross sections (g) and en face images at the level of the retinal pigment epithelium (h) extracted from (f).

These areas are then investigated (at the moment still with a separate system) with AO OCT at cellular resolution level revealing quite normal vasculature and PR appearance at 6° parafoveal whereas there is severe impairment at 0°.

Functional Retinal Imaging Using OCT

Numerous functional OCT extensions have been developed in the past of which Doppler OCT, measuring the blood flow velocity and polarization sensitive OCT, imaging depth resolved tissue birefringence have been the most developed and successfully applied in retinal imaging. Noncontact, depth-resolved optical probing of retinal responses to visual stimulation with 10 μm spatial resolution, achieved using functional ultrahigh resolution OCT, has recently been demonstrated. This method relies on the observation that physiological changes in dark-adapted retina caused by light stimulation can result in local variations in tissue reflectivity. This functional extension of OCT can be considered as an optical analog to electrophysiology and has therefore been called optophysiology.

To determine the sensitivity of optophysiology for the detection of changes in retinal reflectivity triggered by light stimulation, a dark-adapted living *in vitro* rabbit retina was exposed to a single flash of white light and optophysiology data were acquired synchronously with electroretinogram (ERG) recordings (cf. **Figure 8**). Throughout the functional experiments the isolated retinas were stimulated with single, 200-ms long white light flashes (cf. yellow rectangle in **Figure 8(c)**, **8(d)**, **8(f)–8(h)**, and **8(j)–8(l)**). A morphological B-scan was first taken from the measurement location (cf. **Figure 8(a)**). Multiple OCT depth reflectivity profiles (A-scans) were then acquired at one transverse location in the retina synchronously with ERG

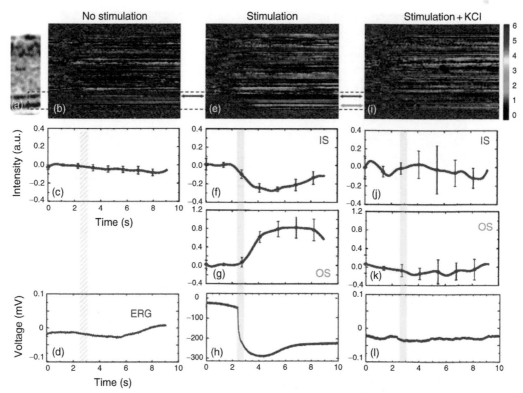

Figure 8 *In vitro* optophysiology in the rabbit retina – optical probing of depth resolved retinal physiology: (a) OCT retinal image of the rabbit retina; (b–d) response during no light stimulus; (e–h) representative single flash stimulus differential M-tomogram and extracted responses from the inner (f) and outer (g) photoreceptor layer; (i–l) case of KCL inhibited photoreceptor function; yellow boxes mark the time duration of the light stimulus; (d, h, l) simultaneous ERG recordings. OS: outer segment; IS: inner segment; ERG: electroretinogram; and KCL: potassium chloride.

recordings. The OCT A-scans were combined to form 2D raw data M-tomograms presenting the retina reflectivity profile as a function of time. The optical data were processed using a cross-correlation algorithm to account for any movement of the retina caused by the solution flow and for calculation of the optical background (average over the pre-stimulation A-scans of each M-tomogram) and generation of differential M-tomograms (cf. **Figure 8(b)**, **8(e)**, and **8(i)**). Optophysiological signals could be extracted from various retinal layers, so that depth-resolved optical back scattering changes that resulted from physiological processes induced by the optical stimulus could be detected. As expected, in the nonstimulated retina (cf. **Figure 8(b)**–**8(d)**) the optical reflectivity of the PR layer did not change significantly with time. When the retina is exposed to the light stimulus (marked by the yellow box), changes were seen in optical backscattering at locations corresponding to the inner (cf. **Figure 8(f)**) and outer (cf. **Figure 8(g)**) segment of the PR layer which correlated with changes in the corresponding ERG (cf. **Figure 8(h)**). When potassium chloride (KCL) was applied to the retinal sample to inhibit PR function (cf. **Figure 8(i)**–**8(l)**), the optical changes observed in the PR inner segment (IS) and outer segment (OS) of the PR layer were close to the optical background level and showed no correlation to

the onset of the light stimulus. Depolarization of the cell membranes can occur during conduction of an action potential which could be detected by UHR OCT, but also by detection of spatially resolved change in backscattering over time. The exact origin of the detected optophysiologic signals is unclear but might be related to the dipole reorientation (and therefore refractive index changes) at the PR membrane. Alternatively they could arise from light-induced isomerization of Rhodopsin in the outer PR segment or metabolic changes in the mitochondria of the inner PR segments.

These optophysiological findings in the *in vitro* rabbit model have recently been tried to transfer to the *in vivo* human retina. Due to limitations in coordinating stimulus and data acquisition and the very slow time course of PR recovery after photobleaching stimulus, a protocol that included imaging a normal subject at intervals of about 1 min was used. Dark-adapted human eyes were briefly subjected to localized photobleach. For 20 min prior to, and 30 min after stimulus, volumetric optical coherence tomograms were collected partially overlapping with the bleached region. A location at the peak rod density at 11° temporal was scanned because rod responses are interesting in regards to dark adaptation and rod responses had been successfully observed the rabbit *in vitro* experiments.

The scanned patch was sampled at 512 A-scans in the long axis of the rectangle (fast axis) and 256 samples in the narrow axis (slow axis). Given the 47-kHz A-scan rate of the camera, this volume could be collected in 2.8 s. Twenty double volumes were completed before the subject was exposed to a white, 48 000 cd m^{-2} stimulus, bleaching over 99% of both photopic (duration 6 s) and scoptopic photopigments. As the eye recovered dark accommodation, 35 more double volumes were recorded at 1 min intervals. Tomograms were segmented into retinal layers by a newly described algorithm exploiting information in adjacent B-scans. En face fundus images extracted from major intraretinal layers were laterally registered manually. Time series summarizing the observed backscatter in selected layers for the bleached and unbleached areas are shown with a variety of corrections and normalizations applied: tomograms were corrected for inherent sensitivity roll off; the ratio between other layers and an assumed unchanging layer (RPE) as well as the ratio of stimulated area to unstimulated area were calculated.

Figure 9 depicts results obtained from one normal subject. For each tissue layer in each measured volume, the result of the experiment is summarized by the mean voxel intensity underneath the retinal surface receiving the

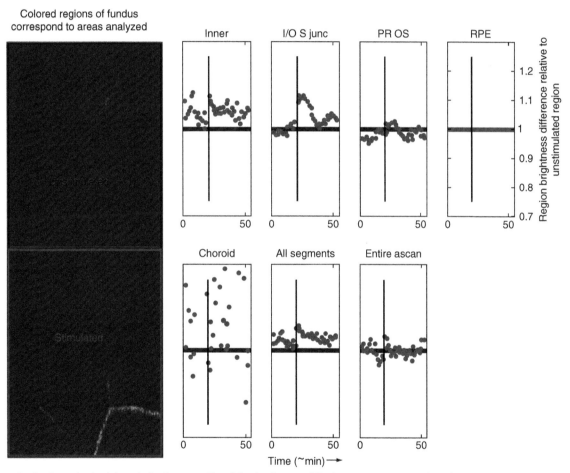

Figure 9 *In vivo* optophysiology in the human retina–A fundus image of the inner–outer segment junction layer is shown at the left to indicate the measured areas defined as unstimulated (blue) and stimulated (red). The volume has been segmented into five anatomical layers using only the most reliably segmented boundaries. The inner retina consists of the NFL, GCL, INL, IPL, and ONL. The choroid is defined from the lower RPE boundary to a thickness 20 pixels deeper (about the thickness of the RPE). All segments is defined from the inner boundary of the NFL to the outer boundary of the RPE. Entire A-scan ignores the segmented boundaries and finds an average pixel intensity over the entire thickness of the recorded A-scan. For each anatomical layer, the average intensity is computed in the stimulated and unstimulated areas and represented as a point in the time series. Twenty volumes acquired before the bleaching stimulus are represented by the points to the left of the vertical black line on each time series plot. Thirty-five volumes recorded after the bleaching stimulus are represented by the point on the right side of the vertical line. Time between acquisitions is about 1 min, so the time axis can be approximately read in units of minutes. Looking at the first four layers, the average pixel intensity appears to track roughly with the IOS distance from the zero delay indicated in the lower right plot. After RPE normalization and a ratio of the stimulated area compared to the unstimulated area, it again appears that there is a relative increase in average pixel intensity in the stimulated area after the stimulus that is most noticeable in the inner–outer segment junction layer that is decaying over a period of about 30 min after the stimulus.

stimulus, and a mean voxel intensity underneath the area without the stimulus. For any particular tissue layer, a time series of this plotted data is expected to show no change in the area without stimulus, and some change in the stimulated area, if there is a change to be observed. The hypothesis is that some layers will show change after stimulus while other layers will not. After the above mentioned normalizations, it appears that an increase larger than the noise level may occur in the inner outer segment region. The decay time for the change (20–30 min) is on the order of the time that the subject observed a persistent after image.

Although the experimental protocol differs from published protocols of studies performed *in vitro* in rabbit and *in vivo* in mouse, the key results are consistent in location and type of change. These, however, could not be convincingly repeated and are therefore presented as an illustration of the techniques used rather than an established observation. Furthermore, the results reported here were only observable in the data after several normalizations. These normalizations are logical, but should be applied with caution. Hence, measuring intrinsic optical signals with OCT in an awake human is challenging. Much needs to be done to deliver both clinically and neurologically significant results. Finally, and probably most importantly, because there is currently a lack of understanding of the physiology that produces intrinsic optical signals, it is difficult to search for the signal using the correct stimulus and measurement protocols. It seems likely that doing more work *in vitro* and with animal models where CCD camera techniques have been highly successful will help identify parameters that are most likely to yield a strong and relevant signal.

Conclusion

Currently, commercial fourth generation 3D ophthalmic OCT systems seem to have a significant diagnostic impact in daily clinical routine as well as novel therapeutic trials. After introducing 3D visualization of the healthy and pathologic human retina, the emphasis now shifts toward filtering out proper biomarkers from this significantly increased amount of information for improved diagnostic decisions. In this respect, it is at the moment not obvious where retinal OCT is heading. Recent developments in OCT technology significantly improved its potential for successful biomedical and clinical applications. These will also be important prerequisites for functional OCT extensions like Doppler OCT, polarization sensitive OCT, and optophysiology.

Despite several proof-of-principle demonstrations of significantly improved state-of-the-art retinal OCT imaging performance sound clinical studies on larger, properly selected patient cohorts are necessary to demonstrate the improved clinical impact and therefore verify the increased technological effort of these novel improvements in retinal OCT modalities.

Acknowledgments

The author wants to acknowledge all members of the biomedical imaging group at the School of Optometry and Vision Sciences, Cardiff University, Alan C. Bird and Catherine A. Egan, Medical Retina Service, Moorfields, London, United Kingdom. Financial and equipment support by the following institutions is also acknowledged: Cardiff University, FP6-IST-NMP-2 STREPT (017128), Action Medical Research (AP1110), DTI (1544C), European Union project FUN OCT (FP7 HEALTH, contract no. 201880); FEMTOLASERS GmbH, Carl Zeiss Meditec Inc., Maxon Computer GmbH, Multiwave Photonics.

See also: Adaptive Optics; Primary Photoreceptor Degenerations: Retinitis Pigmentosa; Primary Photoreceptor Degenerations: Terminology; Rod and Cone Photoreceptor Cells: Inner and Outer Segments; Secondary Photoreceptor Degenerations: Age-Related Macular Degeneration; Secondary Photoreceptor Degenerations.

Further Reading

Bizheva, K., Pflug, R., Hermann, B., et al. (2006). Optophysiology: Depth-resolved probing of retinal physiology with functional ultrahigh-resolution optical coherence tomography. *Proceedings of the National Academy of Sciences of the United States of America* 103: 5066–5071.

Choma, M. A., Sarunic, M. V., Yang, C. H., and Izatt, J. A. (2003). Sensitivity advantage of swept source and Fourier domain optical coherence tomography. *Optics Express* 11: 2183–2189.

de Boer, J. F., Cense, B., Park, B. H., et al. (2003). Improved signal-to-noise ratio in spectral-domain compared with time-domain optical coherence tomography. *Optics Letters* 28: 2067–2069.

Drexler, W. and Fujimoto, J. G. (2008). *Optical Coherence Tomography: Technology and Applications.* New York: Springer.

Drexler, W., Morgner, U., Ghanta, R. K., et al. (2001). Ultrahigh-resolution ophthalmic optical coherence tomography. *Nature Medicine* 7: 502–507.

Drexler, W., Morgner, U., Kartner, F. X., et al. (1999). In vivo ultrahigh-resolution optical coherence tomography. *Optics Letters* 24: 1221–1223.

Fercher, A. F., Hitzenberger, C. K., Drexler, W., Kamp, G., and Sattmann, H. (1993). *In-vivo* optical coherence tomography. *American Journal of Ophthalmology* 116: 113–115.

Fernandez, E. J., Hermann, B., Povazay, B., et al. (2008). Ultrahigh resolution optical coherence tomography and pancorrection for cellular imaging of the living human retina. *Optics Express* 16: 11083–11094.

Fujimoto, J. G., Brezinski, M. E., Tearney, G. J., et al. (1995). Optical biopsy and imaging using optical coherence tomography. *Nature Medicine* 1: 970–972.

Huang, D., Swanson, E. A., Lin, C. P., et al. (1991). Optical coherence tomography. *Science* 254: 1178–1181.

Huber, R., Wojtkowski, M., and Fujimoto, J. G. (2006). Fourier Domain Mode Locking (FDML): A new laser operating regime and applications for optical coherence tomography. *Optics Express* 14: 3225–3237.

Leitgeb, R., Hitzenberger, C. K., and Fercher, A. F. (2003). Performance of fourier domain vs. time domain optical coherence tomography. *Optics Express* 11: 889–894.

Považay, B., Hofer, B., Torti, C., et al. (2009). Impact of enhanced resolution, speed and penetration on three-dimensional retinal optical coherence tomography. *Optics Express* 17: 4134–4150.

Schuman, J. S., Puliafito, C. A., and Fujimoto, J. G. (2004). *Optical Coherence Tomography of Ocular Disease*. Thorofare, NJ: Slack.

Srinivasan, V. J., Adler, D. C., Chen, Y., et al. (2008). Ultrahigh-speed optical coherence tomography for three-dimensional and en face imaging of the retina and optic nerve head. *Investigative Ophthalmology and Visual Science* 49: 5103–5110.

Swanson, E. A., Izatt, J. A., and Hee, M. R. (1993). *In-vivo* retinal imaging by optical coherence tomography. *Optics Letters* 18: 1864–1866.

Tumlinson, A., Hermann, B., Hofer, B., et al. (2009). Techniques for extraction of depth resolved *in vivo* human retinal intrinsic optical signals with optical coherence tomography. *Japanese Journal of Ophthalmology* 53(4): 315–326.

Unterhuber, A., Považay, B., Hermann, B., et al. (2005). *In vivo* retinal optical coherence tomography at 1040 nm-enhanced penetration into the choroid. *Optics Express* 13: 3252–3258.

Optic Nerve: Optic Neuritis

K Hein and M Bähr, University of Göttingen, Göttingen, Germany

Glossary

Apoptosis – A genetically determined process of cell death that is marked by the fragmentation of nuclear DNA, occurs under both pathological and physiological conditions (known as programmed cell death).

Gadolinium – An element of the rare earth family of metals (atomic symbol Gd) used as a paramagnetic contrast agent in magnetic resonance imaging.

Isoelectric focusing – The most sensitive method for protein separation in serum and CSF using agarose gel electrophoresis.

Multiple sclerosis – A chronic autoimmune inflammatory demyelinating disease affecting the central nervous system (CNS) (synonym: encephalomyelitis disseminata).

P100 latency – The time it takes for the signal from the retina to reach the occipital cortex in the brain.

Plasmapheresis (plasma exchange) – A technology for removing pathologic plasma constituents from anticoagulated whole blood by a machine and returning the rest to the donor, usually with saline solution and albumin.

Snellen chart – Chart that contains rows of standardized letters, numbers, or symbols routinely used to assess visual acuity at a distance of 20 feet.

Visual evoked potential – A diagnostic test that measures alterations in the speed of responses to visual events. The patients watch a black-and-white checkerboard and an electroencephalogram (EEG) measures occipital cortex responses.

Definition

Optic neuritis (Latin – *neuritis nervi optici*) is defined as an inflammation of the optic nerve causing a relatively rapid onset of visual failure. Inflammation may occur in the portion of the nerve within the globe (neuropapillitis or anterior optic neuritis) or in the most commonly involved portion behind the globe (retrobulbar neuritis or posterior optic neuritis).

Optic Nerve Anatomy

The optic nerve, ~45 mm in length, can be divided into several segments: intraocular (1 mm), intraorbital (28 mm), canalicular (4–10 mm), and intracranial (3–16 mm). The diameter of the nerve increases from 1 mm at the intraocular point, 3–4 mm in the orbit, to 4–7 mm within the cranial cavity. The orbital section of the optic nerve has a slightly sinuous shape to allow for movements of the eyeball. As part of the brain, the optic nerve is surrounded by meninges. The outer dural sheath and inner arachnoid sheath merge with the sclera. Between the arachnoid sheath and pial sheath lies a cerebrospinal fluid (CSF)-filled space. The optic nerve consists mainly of fibers derived from the ganglion cells of the retina. The fibers emanating from the nasal part of each retina cross over to the other side in the optic chiasm. The nerve fibers originating in the temporal part of the retina do not cross over. From there, the mixed fibers from the two nerves become the optic tract passing through the thalamus and turning into the optic radiation until they reach the visual cortex located in the occipital cortex of the brain.

Immunopathogenesis

An early event in the development of inflammation is considered to be the activation of circulating autoreactive T lymphocytes by factors such as infection, superantigen stimulation, or effects of reactive metabolites or metabolic stress. Activated T lymphocytes interact with the venule endothelium surface to harm and break down the blood–brain barrier. Upregulation of vascular cell adhesion molecules (e.g., intercellular adhesion molecule-1 (ICAM-1)) increases vascular permeability and enables migration of T cells, B cells, and macrophages into the central nervous system (CNS). The activated antigen-specific T cells appear to develop a Th1-dominant profile with production of proinflammatory cytokines (e.g., IL-2, IFNγ, and TNFα). These cytokines exacerbate the inflammation resulting in further recruitment of pathogenic inflammatory cells, which play a direct role in demyelination. Other factors such as demyelinating antibodies, proinflammatory cytokines, and other soluble mediators contribute to myelin and axonal injury. Also the humoral immunity with autoreactive B-lymphocyte expansion with IgG production plays a relevant role in autoimmune inflammation. Despite significant research progress in this field, the antigen driving the putative cascade of autoimmune inflammation has not been identified, leaving a substantial gap in the understanding of pathogenesis.

Epidemiology

Optic neuritis typically affects young adults with a mean age of 30–35 years. The majority of patients are female. The annual incidence of optic neuritis ranges from 1.4 to 6.4 new cases per 100 000 population. The prevalence is estimated to be 115 per 100 000 population.

Etiology

Various diseases and conditions may cause optic neuritis. The first description of a relationship of optic neuritis to multiple sclerosis (MS) came from Buzzard in 1893. Nowadays, MS is recognized to be the most common etiology for optic neuritis. Up to 50% of MS patients will develop an episode of optic neuritis.

Some other causes of optic neuritis include inflammation of vessels supplying the optic nerve, infectious diseases (e.g., viral encephalitis, sinusitis, meningitis, HIV, tuberculosis, and syphilis) and autoimmune disorders (e.g., lupus erythematosus). Toxic agents can clearly influence optic nerve and retinal function. Among the common offenders are tobacco and alcohol. Drugs (e.g., salicylates, digitalis, amiodaron, and some antibiotics) can also damage the optic nerve. Rarely, tumor metastasis to the optic nerve, diabetes, pernicious anemia, Graves' disease, and trauma lead to optic neuritis.

Some people, especially children, develop optic neuritis following a viral illness such as mumps, measles, or a cold.

Symptoms

Optic neuritis typically presents with a triad of symptoms: loss of vision, dyschromatopsia, and pain during eye movements. The major symptom is vision loss, which occurs over a period of 3–5 days and generally improves within 2–8 weeks. The visual deficit varies from a small central or paracentral scotoma to complete blindness. The initial attack is unilateral in 70% of adult patients and bilateral in 30%. Both eyes are affected with equal regularity. The visual symptoms are usually described as blurring or fogging of vision. Interestingly, some patients become aware of a unilateral visual disturbance by accident when, for some reason, the unaffected eye is covered.

Many patients with optic neuritis may lose some of their color vision in the affected eye, called dyschromatopsia. The colors appear slightly washed out compared to the other eye. In the majority of cases, patients experience tenderness in the eye, which is associated with pain at rest or on eye movement. In some cases, pain on eye movement precedes the visual loss. Positive visual phenomena, such as a sensation of sparks and flashes of light induced by certain eye movements, occur infrequently.

Symptoms of optic neuritis include one or more of the following:

- reduced visual acuity (near or far)
- pain on eye movement
- central scotoma
- dyschromatopsia or impaired color vision
- impaired contrast sensitivity
- afferent pupillary defect, and
- headache

Diagnosis

Optic neuritis can usually be diagnosed from a patient's history, clinical examination, visual-evoked potentials (VEP), CSF investigation and observations from magnetic resonance imaging (MRI).

Clinical Examination

Despite the wide use of MRI and electrophysiological techniques, the diagnosis of optic neuritis depends on accurate clinical assessment. The medical history should also be assessed to determine if exposure to toxins such as lead or methanol may have caused the visual disturbance. Patients with optic neuritis typically contact an ophthalmologist first. A complete visual examination, including a visual acuity test, color vision test, and visualization of the retina and optic disc by indirect ophthalmoscopy, should be performed. However, frequently there is no abnormal appearance of the optic disk during ophthalmoscopy. Occasionally, there may be hyperemia of the optic disk and distension of the large retinal veins. At later stages the margins of the optic disk are blurred and papilledema can be found. Clinical signs such as impaired pupil response may be apparent during an eye examination, but in some cases the eye may appear normal. Central field defects predominate in patients with optic neuritis.

Visual Evoked Potential

VEP is a noninvasive technique to detect pathological changes of the visual system during optic neuritis and is used as a routine clinical test. Commonly used visual stimuli are flashing lights or patterns (e.g., checker boards) on a video screen. The main interest in electrophysiological findings for optic neuritis focus on the so-called P100 latency and VEP amplitude. During the acute phase of the

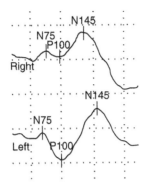

Figure 1 VEP from a 25-year-old woman with a 4-day history of acute right-sided optic neuritis. VEP shows a decrease in amplitude in the affected right eye. VEP amplitude consists of potentials originating in different parts of the brain (N75: area 17; P100: area 18, 19 and partially 17; N145: area 18 and 19).

disease, a decrease in the VEP amplitude, caused by a conduction block of inflamed optic nerve fibers, is a common pathological feature (**Figure 1**). The depression of the VEP amplitude correlates with the visual acuity and is associated with the degree of atrophy. The prolongation of P100 latency is the main characteristic in the chronic phase and persists over years in up to 70% of patients who suffer from optic neuritis.

Magnetic Resonance Imaging

The principal use of MRI is the evaluation of patients when optic neuritis is the first demyelinating disease event. In a typical case, if accurate clinical assessment has been done, MRI is not required to diagnose and distinguish optic neuritis from other common optic neuropathies. However, the patients with other pathologies such as ischemic or acute compressive optic neuropathy due to a cerebral aneurysm or pituitary tumor may not have the typical features that distinguish these disorders from optic neuritis. In these situations, MRI, when performed with the appropriate examination technique, can make a relevant contribution to the differential diagnosis. The small size and mobility of the optic nerve and artifacts caused by surrounding CSF and orbital fat are technically challenging in optic nerve imaging. Significant progress has been made in developing fat and CSF-suppressed high-resolution imaging. Use of these sequences increases the sensitivity in detecting inflammatory demyelination and optic nerve atrophy. Symptomatic lesions can be detected with sensitivities of 95% in fat-saturated fast spin-echo (FSE) imaging and 94% for the enhancing lesion on fat-saturated T1-weighted imaging following intravenous (IV) gadolinium administration (**Figure 2**). The abnormal gadolinium enhancement of the affected optic nerve indicates a blood–optic nerve barrier breakdown and is a consistent feature of acute optic neuritis.

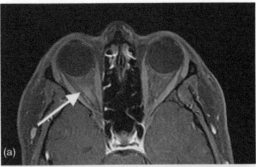
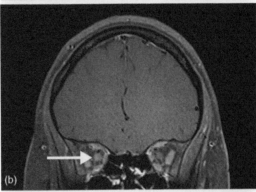
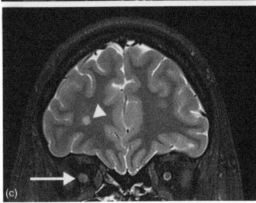

Figure 2 (a) Axial and (b) coronal T1-weighted magnetic resonance images after the IV application of gadolinium. (c) Coronal T2-weighted image from a 25-year-old woman with a 4-day history of acute right-sided optic neuritis demonstrating optic nerve swelling (arrow: diseased optic nerve; arrowhead: demyelinated cerebral lesion).

Cerebrospinal Fluid

CSF analysis is of great relevance to detect an inflammatory process in the brain. The diagnostic sensitivity of single parameters in CSF diagnostics depends on the quality of the applied techniques, especially with isoelectric focusing, and does not have a worldwide standard, such as MRI. In optic neuritis related to MS, typical changes are present in the CSF: slight pleocytosis (max. 50 leucocytes/µl), intrathecal (within the meningeal sheath) IgG production, and additional oligoclonal IgG bands, which are not present in the simultaneously examined serum. The so-called MRZ reaction (intrathecal synthesis of antibodies against measles, rubella, and varicella-zoster) has a higher diagnostic specificity and is suggested in a European consensus report. Before the use of brain MRI, the detection of CSF pleocytosis and oligoclonal IgG bands provided paraclinical evidence of the dissemination of lesions and met the criteria for the diagnosis of laboratory-supported MS in isolated optic neuritis.

In other autoimmune disorders (e.g., lupus erythematosus and neurosarcoidosis) the cell count is moderately elevated. Verification of oligoclonal IgG bands can be achieved in up to 70% of the cases.

CSF abnormalities in optic neuritis resulting from a viral infection are not specific concerning number of cells, differential cell count, and total protein content and depend particularly on the stage of the disease.

Optical Coherence Tomography

OCT is a relatively new, noninvasive and easy-to-use technology that quantifies the thickness of the peripapillary retinal nerve fiber layer (RNFL), fovea, and macula in real time with highly refined resolution. Originally developed to investigate retinal axonal loss in glaucoma patients, OCT demonstrated a significant reduction of mean RNFL thickness in patients with optic neuritis. Hence, OCT may be useful for observing occult neurodegeneration and monitoring the efficacy of potential neuroprotective therapies in optic neuritis. Serial clinical studies in optic neuritis correlating MRI, clinical, and electrophysiological data are required to establish the use of OCT as an *in vivo* biomarker for axonal loss.

Differential Diagnosis

The differential diagnosis can be divided into several pathophysiological categories: inflammatory/autoimmune diseases, infectious diseases, genetic/hereditary disorders, neoplastic disease, and other demyelinating diseases. Errors in the diagnosis can be minimized if strict clinical criteria are applied and the neurologist is alert to discrepancies in the clinical picture presented (e.g., absence of pain, age of onset, and failure to remit). Also the character of visual symptoms and the type of field defect give additional important information. A slowly progressive unilateral visual failure should always suggest the possibility of compressive etiology. A central field defect can occur with a tumor but, with time, will extend to the periphery. Interestingly, spontaneous improvement in vision may occur in this case. Pain on eye movement is not confined to optic neuritis and can occur in optic nerve glioma, meningioma, and aneurysm. Any of the fundoscopic changes found in optic neuritis may be reproduced by compressive or infiltrative lesions. In such cases, MRI will significantly contribute to the correct diagnosis. The various vascular optic neuropathies generally affect a different age group from optic neuritis and tend to have a poor prognosis.

When both eyes are involved in optic neuritis, simultaneously or sequentially, the disorder must be distinguished from Devic's disease, Leber's hereditary optic neuropathy, nutritional/toxic amblyopia, and functional blindness. Devic's disease, neuromyelitis optica, is characterized by a combination of transverse myelitis and optic neuritis occurring within a finite period of time from each other, typically within weeks. Clinically, the disorder usually has monophasic presentation. The identification of serum anti-aquaphorin-4 antibodies should be helpful for differential diagnosis. Of the hereditary disorders, Leber's optic neuropathy causes optic atrophy with acute or subacute visual loss. This disorder is maternally transmitted and can be diagnosed through genetic testing. In optic neuropathies caused by toxic agents, the overlap of the visual symptoms with those due to optic neuritis is minimal. It often presents as a painless, progressive, bilateral, and symmetrical visual disturbance. Optic nerve pallor is often found in the ophthalmologic examination. Vision loss can be presented as a symptom of Lyme borreliosis optic neuropathy, syphilis, HIV-associated optic neuropathies, and other infectious disorders. In these cases optic nerve involvement tends to be bilateral and MRI and CSF examinations should show a reliable distinction.

Treatment

Treatment of optic neuritis depends on the underlying cause. In a MS-related episode of optic neuritis a course of IV methylprednosolone followed by oral steroids has been found to be helpful. No treatment is also a viable option. The optic neuritis treatment trial (ONTT) has shown that high dose IV steroids (250 mg prednisone every 6 h for 3 days) followed by oral prednisone (1 mg kg^{-1} d^{-1} for 11 days) accelerated visual recovery but did not have any impact on the 6-month and 1-year visual outcome compared with a placebo. An interesting finding of this study was that patients who received IV corticosteroids followed by oral corticosteroids had a temporarily reduced risk of development of a second demyelinating event consistent with MS compared with those who received an oral placebo or treatment with oral corticosteroids only. Oral prednisone has been found to increase the likelihood of recurrent episodes of optic neuritis, and is not recommended for treating the disorder. Patients with severe optic neuritis, who do not respond sufficiently to corticosteroids, can undergo escalating immunotherapy with plasma exchange within 1 month of the first symptoms. In clinical trials up to 60% of patients with optic neuritis had functional improvement after 5 plasmapheresis sessions with an exchange volume of 3000 ml. Long term prophylactic immunomodulatory therapy with interferon-beta or glatiramer acetate is recommended in patients when optic neuritis is an event of demyelinating disease with a presence of further demyelinated brain lesions shown on the MRI.

Vision loss caused by viral or bacterial infection usually resolves itself once the virus/bacteria are treated. For herpes virus infection the IV treatment with aciclovir ($10\,mg\,kg^{-1}$ every 8 h for 14 days) is recommended. In cases of borrelia infection with neurological symptoms, patients should undergo a treatment session with doxycycline $200\,mg\,d^{-1}$ for 14–21 days. Optic neuritis resulting from toxin damage may improve once the source of the toxin is removed.

Neuroprotective Treatment Strategies

The increasing importance of axonal damage as a major substrate of clinical disability in autoimmune inflammation has led to ongoing research in developing treatment strategies that inhibit degeneration of axons and protect the neuronal cell body from apoptotic cell death. The hypothesis of achieving neuroprotective effects as a secondary phenomenon resulting from the treatment of inflammation and autoimmunity is supported by studies showing close association of axonal damage and inflammation. However, the elimination of the inflammatory component does not necessarily stop the disease progression. MRI studies showed that treatment with methylprednisolone did not limit ongoing lesion lengthening triggered by an episode of optic neuritis nor did it prevent optic nerve atrophy. A detrimental effect of corticosteroids on retinal ganglion cells (RGCs) has even been described in experimental optic neuritis. The effects of cytokines and trophic factors (e.g., erythropoietin, ciliary neurotrophic factor) on neuronal apoptosis appear to be limited; despite preventing neuronal apoptosis, these substances failed to improve visual acuity due to severe and ongoing degeneration of optic nerve fibers in autoimmune optic neuritis. For this reason, anti-inflammatory/immunomodulatory therapies should be combined with primary neuroprotective agents. The potential neuroprotective substances are still in the experimental stage of development, and controlled clinical trials are needed to prove the efficacy and tolerance of these agents.

Prognosis

The vision loss associated with optic neuritis is usually temporary. Spontaneous remission occurs within 2–8 weeks. The majority of patients (65–80%) recover visual acuity of 20/30 (on the Snellen chart) or better. Long-term prognosis depends on the underlying cause of the condition. If a viral infection has triggered the episode, it frequently resolves itself with no aftereffects. If optic neuritis is associated with MS, future episodes are common. Thirty-three percent of optic neuritis cases recur within five years. Each recurrence results in less recovery and worsening vision. There is a strong association between optic neuritis and MS. According to the latest literature, the probability of developing MS within 15 years after onset of optic neuritis is 50% and strongly related to the presence of lesions on a baseline non-contrast-enhanced magnetic resonance image of the brain.

See also: IOP and Damage of ON Axons; Non-Invasive Testing Methods: Multifocal Electrophysiology; Optical Coherence Tomography; Retinal Ganglion Cell Apoptosis and Neuroprotection.

Further Reading

Andersson, M., Alvarez-Cermeno, J., Bernardi, G., et al. (1994). Cerebrospinal fluid in the diagnosis of multiple sclerosis: A consensus report. *Journal of Neurology, Neurosurgery, and Psychiatry* 57: 897–902.

Beck, R. W. and Gal, R. L. (2008). Treatment of acute optic neuritis: A summary of findings from the optic neuritis treatment trial. *Archives of Ophthalmology* 126: 994–995.

Buzzard, T. (1893). Atrophy of the optic nerve as a symptom of chronic disease of the central nervous system. *British Medical Jornal* 2: 779–784.

Diem, R., Sättler, M. B., and Bähr, M. (2007). Neurodegeneration and protection in autoimmune CNS inflammation. *Journal of Neuroimmunology* 184: 27–36.

Hickman, S. J. (2007). Optic nerve imaging in multiple sclerosis. *Journal of Neuroimaging* 17: 42S–45S.

Hickman, S. J., Brierley, C. M. H., Brex, P. A., et al. (2002). Continuing optic nerve atrophy following optic neuritis: A serial MRI study. *Multiple Sclerosis* 8: 339–342.

Kahle, W., Leonhardt, H., and Platzer, W. (2002). *Color Atlas and Textbook of Human Anatomy,* 4th edn. Stuttgart: Thieme.

Maier, K., Kuhnert, A. V., Taheri, N., et al. (2006). Effects of glatiramer acetate and interferon-beta on neurodegeneration in a model of multiple sclerosis: A comparative study. *American Journal of Pathology* 169: 1353–1364.

Maurer, K. and Eckert, J. (1999). *Praxis der Evozierten Potentiale.* Stuttgart: Enke.

Perkin, G. D. and Rose, F. C. (1979). *Optic Neuritis and Its Differential Diagnosis.* Oxford: Oxford University Press.

Poser, C. M. and Brinar, V. V. (2007). The accuracy of prevalence rates of multiple sclerosis: A critical review. *Neuroepidemiology* 29: 150–155.

Sergott, R. C., Frohman, E., Glanzman, R., and Al-Sabbagh, A: OCT in MS Expert Panel (2007). The role of optical coherence tomography in multiple sclerosis: Expert panel consensus. *Journal of the Neurological Sciences* 263: 3–14.

The Optic Neuritis Study Group (2008). Multiple sclerosis risk after optic neuritis. *Archives of Neurology* 65: 727–732.

Trapp, B. D., Peterson, J., Ransohoff, R. M., et al. (1998). Axonal transection in the lesions of multiple sclerosis. *New England Journal of Medicine* 338(5): 278–285.

Wingerchuk, D. M. and Lucchinetti, C. F. (2007). Comparative immunopathogenesis of acute disseminated encephalomyelitis, neuromyelitis optica, and multiple sclerosis. *Current Opinion in Neurology* 20: 343–350.

Pathological Retinal Angiogenesis

A P Adamis, University of Illinois, Chicago, IL, USA

Glossary

Angiogenesis – The formation of new vessels from differentiated vasculature.

Chemoattractant – A chemical agent that induces cell migration (chemotaxis) toward the agent.

Embryogenesis – The phase of prenatal development involved in the establishment of the basic body form.

Hypoxia – A decrease below normal in the oxygen levels of a tissue.

Ischemia – Local loss of blood supply, with concomitant hypoxia, due to mechanical obstruction or degeneration of blood vessels.

Neovascularization – Proliferation of blood vessels in tissues not normally containing them or proliferation of blood vessels of a different kind than usual.

Uveitis – Inflammation of the uveal tract, which includes the iris, ciliary body, and choroid.

Vasculogenesis – Blood vessel formation occurring *de novo* from embryonic endothelial precursor cells (angioblasts).

Introduction

Research into the physiological development of the retinal vasculature and various ocular neovascular diseases has led to a deeper understanding of the molecular and cellular mechanisms underlying normal and pathological retinal vascularization. During embryogenesis, the developing retina is initially supplied by the hyaloid vasculature, which is then supplanted by a vascular plexus that originates at the optic nerve head and spreads peripherally. Subsequent remodeling and pruning leads to the formation of two additional parallel networks in the nerve fiber and inner plexiform layers, together with a circular avascular zone at the fovea.

Two basic mechanisms, vasculogenesis and angiogenesis, underlie the formation of new blood vessels; their relative contribution to the formation of the superficial retinal vasculature is a focus of some disagreement. Vasculogenesis, the differentiation of blood vessels from angioblasts, has been proposed as the mechanism for the formation of primary retinal vascular plexus, with deeper layers developing by angiogenesis; in contrast, it has been argued that all three vascular layers develop through angiogenesis, the formation of new vessels from differentiated vasculature. This process involves both sprouting from differentiated tip cells as well as intussusception, the splitting of existing vessels. The question as to the importance of vasculogenesis ultimately depends on the cellular identity of retinal precursors, and it is possible that some of the discrepancies may reflect species differences among mammals.

With respect to pathological retinal neovascularization, there is general agreement that angiogenesis is the driving mechanism. Many retinal neovascular conditions are associated with ischemia, with the resultant hypoxia leading to the upregulation of proangiogenic molecules. Depending on the particular disease, different pathophysiological mechanisms lead to a common endpoint of local ischemia and neovascularization. In central retinal vein occlusion, ischemia is a direct result of venous blockage. In proliferative diabetic retinopathy (PDR), the mechanism is more complex; the accumulation of polyols, reactive oxygen intermediates, and advanced glycation end products results in inflammation accompanied by retinal leukostasis leading to vascular injury, capillary blockage, and dropout. A third mechanism for ischemia-mediated neovascularization is seen in retinopathy of prematurity (ROP) in which premature birth interrupts the normal retinal vascular development; not only is postnatal tissue oxygen significantly higher than *in utero* but the effects are compounded by the use of oxygen therapy as well. Ultimately, these high oxygen levels lead to an increased rate of vascular pruning with ischemia-mediated increases in vascular endothelial growth factor (VEGF) levels and the resultant rebound aberrant neovascularization, which can be accompanied by retinal detachment. Finally, it should be noted that not all pathological retinal neovascularization is related to hypoxia since it is also observed in inflammatory conditions such as severe uveitis.

Promoters and Inhibitors of Angiogenesis

Over the past two decades, the identification and study of modes of action of proangiogenic and antiangiogenic factors have yielded an extensive list in both categories (see **Table 1**). While much of this work has been directed to developing treatments for cancer, considerable progress has also been made in elucidating the importance of these factors in ocular neovascular diseases. This article

Table 1   Proangiogenic and antiangiogenic factors

Proangiogenic and antiangiogenic factors	
Proangiogenic factors	*Antiangiogenic factors*
Angiogenin	Angioarrestin
Angiopoietin-1	Angiostatin (plasminogen fragment)
Complement factors C3 and C5	Antiangiogenic antithrombin III
Cryptic collagen IV fragment	Cartilage-derived inhibitor (CDI)
Developmentally regulated endothelial locus 1 (Del-1)	CD59 complement fragment
Ephrins/Ephs	Endostatin (collagen XVIII fragment)
Erythropoietin	Fibronectin fragment
Fibroblast growth factors; acidic (aFGF) and basic (bFGF)	Growth-related oncogene (Gro-β)
Follistatin	Heparinases
Granulocyte colony-stimulating factor (G-CSF)	Heparin hexasaccharide fragment
Hepatocyte growth factor (HGF)/ scatter factor (SF)	Human chorionic gonadotropin (hCG)
Interleukin-8 (IL-8)	Interferon α/β/γ
α5 integrins	Interferon-inducible protein (IP-10)
Leptin	Interleukin-12
Matrix metalloproteinases	Kringle 5 (plasminogen fragment)
Midkine	Metalloproteinase inhibitors (TIMPs)
Notch/D114	2-Methoxyestradiol
Pigment epithelium-derived growth factor	Pigment epithelium-derived growth factor
Placental growth factor	Placental ribonuclease inhibitor
Platelet-derived endothelial cell growth factor (PDECGF)	Plasminogen activator inhibitor
Platelet-derived growth factor-B (PDGF-B)	Platelet factor-4 (PF4)
Pleiotrophin (PTN)	Prolactin 16-kD fragment
Progranulin	Proliferin-related protein (PRP)
Proliferin	Retinoids
Transforming growth factor-α (TGF-α)	Slit/Robo4
Transforming growth factor-β (TGF-β)	Soluble VEGFR1
Tumor necrosis factor-α (TNF-α)	Tryptophanyl-tRNA synthase fragment
Vascular endothelial growth factor (VEGF)	VEGFxxxb
	Tetrahydrocortisol-S
	Thrombospondin-1 (TSP-1)
	Transforming growth factor-β (TGF-β)
	Vasculostatin
	Vasostatin (calreticulin fragment)

Reproduced from: Angiogenesis Foundation. Understanding angiogenesis. List of known angiogenic growth factors. Available at: http://www.angio.org/understanding/content_understanding.html.

Table 2   Actions of VEGF in promoting angiogenesis

- Endothelial cell mitogen
- Endothelial cell survival factor
- Chemoattractant for bone marrow-derived endothelial cells
- Chemoattractant for monocyte lineage cells
- Inducer of synthesis of endothelial nitric oxide synthase and consequent elevation of nitric oxide, itself a promoter of angiogenesis
- Inducer of synthesis of enzymes promoting blood vessel extravasation
  Matrix metalloproteinases
  Plasminogen activator

examines the evidence for those molecules whose roles are best understood in the context of pathological retinal angiogenesis.

Promoters of Angiogenesis

Vascular endothelial growth factor

VEGF (also known as VEGF-A) is a 45-kDa homodimeric glycoprotein that is the most potent known promoter of angiogenesis. It exists in a variety of isoforms and acts as a ligand for two receptor tyrosine kinases: VEGF receptor-1 (VEGFR1) and VEGF receptor-2 (VEGFR2). As a key regulator of angiogenesis, VEGF has a variety of proangiogenic actions (**Table 2**). Given the importance of ischemia in many retinal neovascular diseases, it is of particular relevance that retinal expression of VEGF is upregulated by hypoxia. In addition, it is an extremely potent known promoter of vascular permeability, 50 000 times stronger than histamine, thereby contributing to the edema that is often a central factor in vision loss.

The importance of VEGF in pathological retinal neovascularization has been established from both clinical and preclinical studies. The clinical work is correlative, demonstrating that ocular VEGF levels are increased in diseases such as diabetic retinopathy (DR), diabetic macular edema (DME), ROP, neovascular glaucoma, and retinal vein occlusion and in choroidal neovascular membranes from eyes of patients with age-related macular degeneration (AMD). With respect to the preclinical work, elevations of VEGF, whether by laser-induced retinal vein occlusion, direct injection, or transgenic approaches, all have proved capable of inducing ocular neovascularization.

Moreover, VEGF elevation has also been shown to be necessary for the neovascular response since blocking VEGF signaling prevents blood vessel growth. The approaches employed in these experiments have included the use of anti-VEGF antibodies or fragments thereof, an anti-VEGF aptamer, VEGFR fusion proteins, a nonangiogenic isoform of VEGF, antisense oligonucleotides, and small interfering RNAs. This work already has resulted in the clinical approval of two agents for the treatment of

AMD, pegaptanib and ranibizumab. Additionally, bevacizumab, a monoclonal antibody related to ranibizumab, increasingly is being used off-label for the treatment of AMD and other ocular neovascular diseases while VEGF-Trap, a fusion protein combining components of both VEGFR1 and VEGFR2, currently is being examined in a phase 3 trial as a treatment for AMD.

Studies have further demonstrated that ocular neovascular diseases such as AMD and DR bear many hallmarks of an inflammatory process, with VEGF acting as a proinflammatory cytokine. Genetic studies have revealed that the risk of AMD is strongly correlated with polymorphisms of several components of the complement cascade. In addition, retinal leukostasis is believed to be important in the capillary dropout characteristic of DR, leading to the development of ischemia. Furthermore,

while the molecular mechanisms involved in ischemia-mediated ocular neovascularization remain to be fully elucidated, the influx of inflammatory cells, including monocytes/macrophages (**Figure 1**) and neutrophils, is important for its development. Both of these cell types release VEGF. Since VEGF acts as a chemoattractant for macrophages while upregulating the expression of intercellular adhesion molecule-1 (ICAM-1), a molecule that promotes leukocyte adhesion, on influx of these inflammatory cells provides an inherent positive-feedback mechanism promoting neovascularization.

In detailed studies of VEGF action, the VEGF165 isoform proved to be especially active as an inflammatory cytokine, since it was highly expressed during ischemia-mediated neovascularization. VEGF165 was also more potent than VEGF121, another common isoform, both

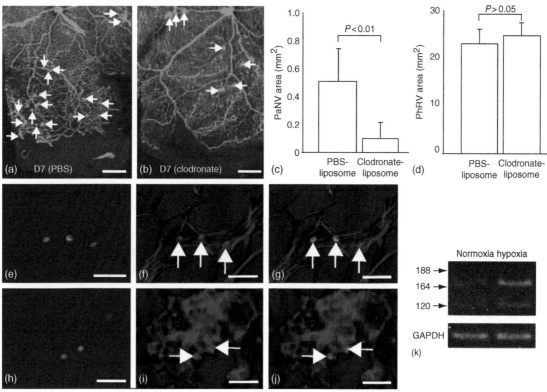

Figure 1 Monocytes contribute to pathological retinal neovascularization. In a ROP model, postnatal day zero (P0) rats were maintained for 10 days in 80% oxygen, interrupted daily by 30 min in room air followed by a progressive return to 80% oxygen. This treatment led to an avascular retina. On P10, corresponding to study day 0 (D0), retinal revascularization was induced by maintaining the rats in room air for an additional seven days (D7). (a–c) At D7, pathological neovascularization (PaNV; arrows in (a) and (b)) was significantly inhibited by treatment with clodronate liposomes compared to phosphate buffered saline (PBS) control liposomes (n = 8 for both treatments; mean ± standard deviation). (d) Physiological retinal vascular area (PhRV) was not significantly affected by treatment with clodronate liposomes (P > 0.05). (e–j) Influx of monocytes was observed just before and during pathological neovascularization (h–j). Monocytes were labeled with a fluorescein conjugated antibody to CD13 (e and h) while rhodamine-conjugated Concanavalin A was used to label the retinal vasculature and adherent leukocytes (f and i). As shown by superposition of these figures (panels (g) and (j)), the Concanavalin A and CD13 staining co-localized, indicating that the adherent leukocytes were monocytes. (k) In cultured peripheral blood monocytes obtained from retinopathologic rats at D7, exposure to hypoxia (1% oxygen) led to a marked increase in the expression of vascular endothelial growth factor mRNA compared to exposure to normoxia (21% oxygen). GAPDH; glyceraldehyde 3-phosphate dehydrogenase (used as a loading control). Scale bars: (a and b) 0.5 mm and (e–j) 50 μm. © Ishida et al., 2003. Originally published in *The Journal of Experimental Medicine*. doi:10.1084/jem.20022027.

as a macrophage chemoattractant and as a promoter of ICAM-1 upregulation.

Placental growth factor

Placental growth factor (PlGF) is a structurally related member of the same superfamily as VEGF, with all four PlGF isoforms acting as ligands for VEGFR1. PlGF has been implicated in a variety of physiological processes, including serving as a chemoattractant for monocytes and endothelial progenitor cells. Although ablation of the PlGF gene did not prevent embryonic angiogenesis, loss of PlGF led to impaired angiogenesis during conditions of ischemia and inflammation. In murine models, PlGF was found to be essential for the full expression of pathological neovascularization in laser-induced choroidal neovascularization (CNV) as well as for angiogenesis in tumors that were resistant to VEGFR inhibitors. Moreover, intravitreal injection of PlGF was shown to induce retinal edema in a rodent model, reflecting a disruptive effect on the retinal pigment epithelium. PlGF accordingly has attracted interest as a possible therapeutic target, and inhibition of its function is believed to contribute to the efficacy of VEGF-Trap, which, as discussed above, has been engineered to include the binding sites of both VEGFR1 and VEGFR2.

Platelet-derived growth factor

The platelet-derived growth factor (PDGF) family includes four members (PDGF-A through PDGF-D), with active forms being dimers and usually occurring in the homodimeric form. They serve as ligands for two receptor kinases, PDGFR-α and PDGFR-β. PDGF-B, acting through PDGFR-β, mediates most of the actions of PDGF in vascular development. Gene knock-out of either PDGF-B or PDGFR-β resulted in perinatal death owing to vascular deformities. In addition to promoting endothelial cell proliferation and capillary tube formation, PDGF-B signaling is especially important for the proliferation and recruitment of mural cells (pericytes and vascular smooth muscle cells) to the embryonic vasculature and for the maintenance of pericyte coverage (**Figure 2**). These actions are also essential for proper retinal vascular development. Endothelial cell-restricted ablation of PDGF-B produced defective pericyte coverage of retinal vessels and associated proliferative retinopathy; a similar phenotype resulted from the administration of an inhibitor to PDGFR signaling.

Investigations have demonstrated a complex interaction between PDGF and VEGF. During embryonic development, pericyte coverage was found to lag behind growth of retinal capillaries for several days, giving rise

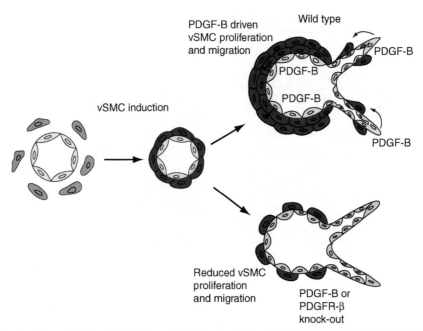

Figure 2   Platelet-Derived Growth Factor (PDGF)-B regulates the development of blood vessel walls. During blood vessel development, the nascent endothelial tube (yellow) is surrounded by undifferentiated mesenchymal cells (gray) that are induced to differentiate into vascular smooth muscle cells (vSMC) and to form a surrounding sheath (red). During further development of the vascular network, with concomitant growth and sprouting of blood vessels, PDGF-B derived from the endothelium further promotes vSMC proliferation and migration. These proliferative and migratory responses are reduced in mice in which PDGF-B or the corresponding receptor, PDGFR-β, have been genetically ablated, leading to defective coating of capillaries by pericytes as well as to vSMC hypoplasia in larger vessels. Reproduced with permission from Hellstrom, M., Kalen, M., Lindahl, P., Abramsson, A., and Betsholtz, C. (1999). Role of PDGF-B and PDGFR-beta in recruitment of vascular smooth muscle cells and pericytes during embryonic blood vessel formation in the mouse. *Development* 126: 3047–3055.

to an optimum window in which nascent blood vessels were especially vulnerable to VEGF depletion. Further investigations have revealed that the inhibition of both PDGF and VEGF signaling is quite effective in inhibiting ocular neovascularization. In experiments with three different murine models, VEGF inhibition was more effective for immature vessels while mature vessels were sensitive to PDGF-B. In every case, however, the most effective inhibition was achieved by interference with both signaling pathways (**Figure 3**).

Notch

The Notch family of cell surface proteins, consisting in mammals of four members, Notch1 through Notch4, is activated by ligands of the Jagged and Delta families and regulates pattern formation of numerous tissues, including those of the kidney, nervous system, and cardiovascular system. In its actions in the vascular system, the principal ligand for Notch is Dll4. Several recent studies have identified a key role for this signaling pathway as a negative regulator of retinal vessel patterning. Even heterozygous ablation of the dll4 gene led to dramatic increases in branching while other approaches to inhibiting the Notch pathway also caused excessive tip-cell (specialized endothelial cells which form the leading edge of vascular sprouts to initiate vessel branching) formation and endothelial cell proliferation. In a tumor model, blockade of the Notch pathway increased tumor vascularity but the vessels were nonproductive, resulting in decreased tumor growth. These findings are believed to reflect the importance of Dll4/Notch signaling in the negative regulation of VEGF-mediated angiogenic sprouting (**Figure 4**). Finally, the loss of Notch3 signaling mechanism has been found to interfere with mural cell differentiation, an action that may reflect Notch-mediated upregulation of PDGFR-β in vascular smooth muscle cells.

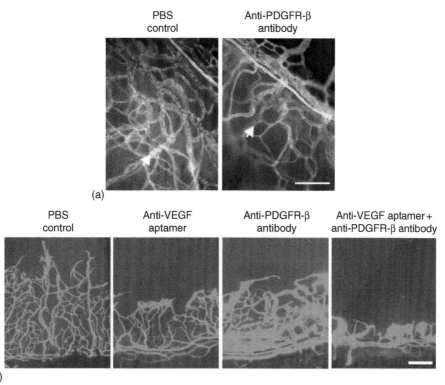

Figure 3 The role of platelet-derived growth factor (PDGF)-B in blood vessel growth and mural cell coverage in a corneal neovascularization model. (a) Endothelial cells were labeled by staining with lectin (green) and mural cells were stained with an antibody against smooth muscle actin (red). Starting at 10 days following corneal injury, mice received daily intraperitoneal injections of an anti-PDGFR-β antibody or phosphate-buffered saline (PBS) and were sacrificed at 20 days postinjury. Treatment with the anti-PDGFR-β antibody led to reduced mural cell coverage compared to controls (arrows). Scale bar = 20 μm. (b) Following induction of corneal injury, mice received daily intraperitoneal injections of one of the following: PBS; a polyethylene-glycolated antivascular endothelial growth factor (VEGF) aptamer; an anti-PDGFR-β antibody, or both the antiVEGF aptamer and the anti-PDGFR-β antibody. Neovasculature (green) was stained by fluorescein isothiocyanate–Concanavalin A. Neovascularization was significantly reduced by the antiVEGF aptamer compared with either PBS or the anti-PDGFR-β antibody ($P < 0.01$); inhibition of both VEGF and PDGF-B signaling led to a further significant reduction ($P < 0.05$), compared to inhibition of VEGF signaling alone. Scale bar = 100 μm. Adapted from Jo, N., Mailhos, C., Ju, M., et al. (2006). Inhibition of platelet-derived growth factor B signaling enhances the efficacy of anti-vascular endothelial growth factor therapy in multiple models of ocular neovascularization. *American Journal of Pathology* 168: 2036–2053. Copyright – The American Society for Investigative Pathology.

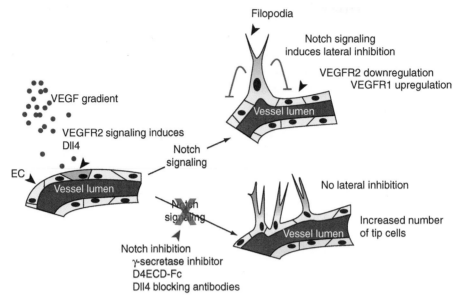

Figure 4 Role of Dll4/Notch in preventing tip-cell formation and branching. Schematic representation showing how Dll4/Notch signaling prevents excessive branching by inhibiting tip-cell phenotype. In response to a local gradient of vascular endothelial growth factor (VEGF), endothelial cells become tip cells and induce local Dll4 expression. This in turn activates Notch signaling in the neighboring cells and prevents these cells from becoming tip cells (lateral inhibition) through the induction of VEGFR1 and the downregulation of VEGFR2. Subsequently, these cells will become trunk/stalk cells and adopt the proliferative behavior necessary for vessel elongation. In the absence of Notch signaling, lateral inhibition is not induced and multiple branching occurs. Adapted from Sainson, R. C. and Harris, A. L. (2008). Regulation of angiogenesis by homotypic and heterotypic notch signalling in endothelial cells and pericytes: From basic research to potential therapies. *Angiogenesis* 11: 41–51, with permission from Springer.

Tumor necrosis factor-α

Tumor necrosis factor-α (TNF-α), a member of a large superfamily of cytokines and their receptors that are involved in regulating numerous physiological processes, is an important mediator of inflammation. Clinical studies have detected TNF-α in the fibrovascular membranes of patients with PDR and CNV. In two small case series, intravenous administration of infliximab, a monoclonal antibody directed against TNF-α, led to the regression of CNV in patients with AMD and to the alleviation of macular edema in patients with DME. In addition, a case report has described a similar regression of uveitis-induced neovascularization with this approach.

The mechanisms underlying the actions of TNF-α in ocular neovascularization have yet to be elucidated. TNF-α has been found to upregulate the synthesis of a number of genes that are important for angiogenesis, including VEGF, VEGFR2, angiopoietins 1 and 2 (Ang1 and Ang2), matrix metalloproteinases (MMPs) 2 and 9, PDGF-B, and the Notch ligand Jagged-1. The increase in Jagged-1 may account for the recent finding that TNF-α can induce the formation of tip cells.

Gene ablation studies in mice have been inconsistent, with some studies finding inhibition of retinal neovascularization in response to interference with TNF-α signaling, while others did not. Inhibition of laser-induced CNV by agents targeting TNF-α has been observed, though, both with intravitreal injection of infliximab and

intraperitoneal administration of etanercept, a fusion protein containing the TNF-α receptor. Also, TNF-α signaling was determined to be important for ischemia-mediated neovascularization in the hind limbs of mice. Taken together, these data suggest that TNF-α may serve as a potential molecular target for controlling ocular neovascularization.

Ephrins and Ephs

Ephrins are a family of ligands that bind to the Eph receptor tyrosine kinases. The ephrins fall into two broad classes, the ephrinAs, which are attached to the cell membrane by a glycosylphosphatidyl anchor, and the ephrinBs, which span the cellular membrane and possess a cytoplasmic signaling domain (**Figure 5**). There is a corresponding division among the Eph kinases, which fall into A and B subclasses, with ephrinAs binding primarily to EphAs and ephrinBs to EphBs. Because of the tethered nature of the ephrins, ephrin–Eph signaling requires cell–cell contact and can proceed in either the forward or reverse direction. In addition to angiogenesis, ephrin–Eph interactions are required for the proper development of the nervous and cardiovascular systems as well as such processes as insulin secretion and trafficking of immune cells.

Evidence has been adduced supporting the roles of both major Eph/ephrin classes in angiogenesis, including retinal neovascularization. A soluble EphA-containing fusion protein significantly inhibited VEGF- or ephrinA1-induced

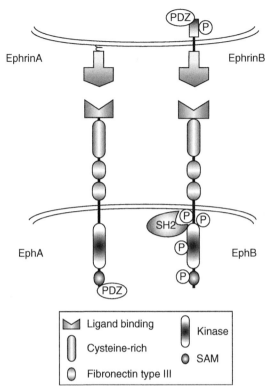

Ligand binding

Cysteine-rich

Fibronectin type III

Kinase

SAM

Figure 5 Ephrins and their Eph receptors. While both ephrins and Eph receptors are membrane-tethered proteins, ephrinBs traverse the membrane and possess a cytoplasmic signaling domain while ephrinAs do not. Ephrin–Eph binding results in receptor clustering followed by autophosphorylation of multiple tyrosine residues and docking of downstream effectors through *src*-homology domains. The presence of a sterile alpha motif (SAM) and a PDZ domain (shown here for the carboxy terminus of EphA but also present in EphB) promotes ligand-induced receptor clustering. Adapted from Dodelet, V. C. and Pasquale, E. B. (2000). Eph receptors and ephrin ligands: Embryogenesis to tumorigenesis. *Oncogene* 19: 5614–5619, with permission from Nature Publishing Group.

endothelial cell migration and assembly into capillary tubes *in vitro*, and when injected intravitreally the fusion protein also suppressed retinal neovascularization in a murine ROP model.

More extensive data are available implicating ephrinB/ Eph signaling. In clinical studies, both EphB2 and EphB3 were found to be expressed in fibroproliferative membranes of patients with ischemic ocular neovascularization. Genetic ablation of either ephrinB2 or EphB4 led to homozygous embryonic lethality owing to aberrant development of the vasculature. Analysis of expression patterns suggested that ephrinB2 is expressed mainly on arteries and EphB4 on veins, and that these molecules act in defining arterial or venous identity.

Preclinical studies of the role of ephrinB/EphB in ocular neovascularization have not been consistent, however. In corneal models, neovascularization was promoted by ephrinB2, or fusion proteins containing it, and by an EphB1 fusion construct. In contrast, administration of

soluble forms of EphB4 inhibited laser-induced CNV in rats as well as ischemia-induced retinal neovascularization in mice; the inhibition of retinal neovascularization also was seen when using soluble ephrinB2. Further work clearly is required if the ephrin/Eph systems can be considered as potential therapeutic targets.

Angiopoietins

Ang1 and Ang2 are members of a family of secreted ligands for Tie2, a receptor tyrosine kinase whose function is essential for vascular development and remodeling. Although Ang1 and Ang2 both activate Tie2, they usually act antagonistically. Similar lethal phenotypes involving gross vascular defects resulted from genetic ablation of Tie2 or Ang1 as well as from the overexpression of Ang2. While the effects of both Ang1 and Ang2 are complex, the overall action of Ang1 is to stabilize the quiescent vasculature while Ang2 is a destabilizing agent for the vascular endothelium (**Figure 6**).

Ang1, which is secreted by vascular smooth muscle cells, acts as a survival factor for endothelial cells *in vitro* and also can promote endothelial cell sprouting and tissue invasion of new blood vessels. In transgenic mice, co-expression of VEGF and Ang1 led to an additive effect on angiogenesis, but co-expression of Ang1 was found to suppress the leakiness of vessels induced by VEGF alone. Recently, this stabilizing effect has been shown to involve Ang1-mediated inhibition of the VEGF-induced internalization of vascular endothelial cadherin, a key component of adherent junctions. In addition, Ang1 inhibited other VEGF-mediated pro-inflammatory actions, including the upregulation of tissue factor, ICAM-1, and vascular cell adhesion molecule-1. Intravitreal injection of Ang1 also prevented many of the inflammatory changes characteristic of DR. In transgenic models, overexpression of Ang1 inhibited the development of laser-induced CNV and retinal neovascularization.

In contrast to the stabilizing effects of Ang1, Ang2 is primarily a destabilizing agent. In clinical studies, elevated levels of Ang2 and VEGF were detected in the vitreous of patients with DR as well as in choroidal neovascular membranes. Ang2 is synthesized by arterial smooth muscle cells and by endothelial cells where it is stored in Weibel-Palade bodies upon activation in response to stimuli such as thrombin or histamine, and sensitizes the endothelium to pro-inflammatory cytokines such as TNF-α. In addition to being upregulated by hypoxia and VEGF, Ang2 appears to act in concert with VEGF in many contexts, including enhancement of endothelial cell permeability. In several different rodent models, Ang2 was found to co-operate with VEGF in promoting neovascularization, but if VEGF levels were low, overexpression of Ang2 led to its regression.

Taken together, these findings suggest that Ang1 could have therapeutic potential as an antiangiogenic agent on

Quiescent/resting vasculature

Ang-1/Ang-2

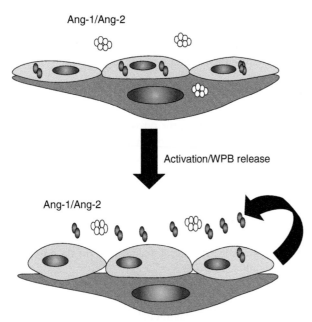

Activation/WPB release

Ang-1/Ang-2

Activated/responsive vasculature

Figure 6 Regulation of vascular responsiveness by Ang1 and Ang2. Angiopoietin 1 (Ang1) (multimeric, white) is secreted constitutively at a low level by mural (periendothelial) cells, and acts on the quiescent endothelium to sustain a low-level activation of Tie2, thereby helping to maintain the luminal cell surface in an antithrombotic and antiadhesive state. Ang2 (dimeric, gray) is stored in Weibel-Palade bodies (WPB) in the endothelium and during endothelial cell activation is released from them, along with other stored factors, leading to the Ang1/Ang2 ratio being altered more in favor of Ang2. As a result, the endothelial cell layer becomes destabilized and more responsive to pro-inflammatory stimuli. Reproduced with permission from Pfaff, D., Fiedler, U., and Augustin, H. G. (2006). Emerging roles of the Angiopoietin–Tie and the ephrin–Eph systems as regulators of cell trafficking. *Journal of Leukocyte Biology* 80: 719–726.

its own. For Ang2, the situation is more complex, although it may have promise if administered together with a VEGF-suppressive agent.

Erythropoietin

While primarily identified as a promoter of erythropoiesis, erythropoietin also acts as a neuroprotective agent and as a promoter of angiogenesis. It plays an essential role in establishing the vascular network following vasculogenesis in that homozygous deletion of erythropoietin or its receptor leads to embryonic lethality with extensive vascular anomalies. In ischemic conditions, erythropoietin (which is itself upregulated by hypoxia) upregulates VEGF and VEGFR2 and promotes the recruitment of bone marrow-derived endothelial progenitor cells.

Correlative clinical findings implicate erythropoietin in ocular neovascular disease. Elevated ocular levels of erythropoietin were detected in patients with DME and PDR; in addition, treatment of premature infants with erythropoietin has been correlated with an increased risk of ROP. Preclinical studies are limited to one report in which intravitreal administration of a soluble erythropoietin receptor inhibited ischemia-induced neovascularization in a murine model. While not extensive, the available data suggest that further studies are warranted.

Integrins

Integrins comprise a large family of transmembrane cell surface receptors that transduce signals from ligands in the extracellular matrix to the cytoplasm. Each integrin is composed of one α and one β subunit; each subunit class contains many representatives, and more than 24 α–β combinations have been observed. The principal focus of integrin research has been their relevance to cancer, and in this context they play important roles in angiogenesis.

Studies with small molecule antagonists have identified a role for the $\alpha_5\beta_1$ integrin in murine models of ocular neovascularization. These inhibitors include JSM5562, which inhibited corneal neovascularization, and a related molecule, JSM6427, which inhibited laser-induced CNV and ischemia-induced retinal neovascularization. Studies indicated that $\alpha_5\beta_1$ contributed to an angiogenic pathway that was distinct from VEGF-mediated angiogenesis, suggesting that combinatorial approaches may offer the promise of even greater efficacy.

Matrix metalloproteinases

The MMPs comprise a large family of enzymes that degrade the extracellular matrix, facilitating the tissue penetration of nascent blood vessels. While the primary focus of their physiological impact has been in the context of cancer, they also have been shown to be involved in ocular neovascularization. MMP action has been implicated in the exposure of a cryptic epitope of collagen IV that promoted laser-induced CNV in a murine model. MMPs also generated soluble fragments of VEGF from matrix-bound forms; as MMP expression by cultured retinal pigment epithelium cells is upregulated by VEGF, this may provide for a positive feedback amplification of the angiogenic response.

Components of the complement cascade

In addition to the genetic studies implicating specific mutations in the complement system as risk factors for AMD (see above), both clinical and preclinical studies have directly implicated factors C3a and C5a in promoting ocular neovascularization. Both of these factors have been found in the drusen of patients with AMD as well as in the eyes of mice with laser-induced CNV; genetic ablation of their corresponding receptors reduced VEGF expression, leukocyte recruitment, and induction of CNV. Moreover, these factors induced upregulation of VEGF expression when injected intravitreally in mice. Similar effects on CNV and VEGF expression were found in mice in which factor C3 was ablated. Finally, depletion of complement by administration of cobra venom factor has been shown to inhibit the development of the inflammation, which is often accompanied by neovascularization, in an animal model of uveitis. Taken together, these findings suggest that targeting certain components of the complement cascade could provide a therapeutic option for treating ocular neovascularization.

Inhibitors of Angiogenesis

Pigment epithelium-derived factor

Pigment epithelium-derived factor (PEDF) is a 50-kDa glycoprotein, originally found to be expressed in the retinal pigment epithelium that exhibits properties directly opposing VEGF: inhibition of VEGF signaling through VEGFR1, downregulation of VEGF expression in the context of ischemia, and inhibition of VEGF-induced increases in endothelial cell permeability. PEDF induces macrophage apoptosis, suggesting that it serves also as a modulator of inflammation. Systemic or intravitreal administration of PEDF, as well as its expression from a transgene, was found to inhibit ischemia-induced retinal neovascularization in mice. PEDF also induced apoptosis of cultured endothelial cells and suppression of VEGF-induced endothelial cell proliferation and migration. However, one study reported that the effects of PEDF were dose dependent, with both laser-induced CNV and endothelial cell differentiation *in vitro* being inhibited at low doses and promoted at high doses. Clinical data have been similarly inconsistent. While the expression of PEDF in Bruch's membrane was reduced in patients with AMD relative to controls, vitreous levels of PEDF were elevated in patients with PDR. The potential of PEDF as an antiangiogenic agent for ocular neovascularization thus remains to be established.

VEGFxxxb isoforms

VEGFxxxb denotes a parallel family of alternately spliced VEGF isoforms that are identical in length to the canonical VEGF family but differing in the last six amino acids. VEGFxxxb isoforms can bind VEGFR2 but have little or no ability to initiate downstream pathways. They therefore act as competitive inhibitors of VEGF. In clinical studies, VEGFxxxb was found to predominate in the vitreous of control eyes but to constitute only 12% in patients with PDR; similar downregulation of VEGFxxxb has been observed in several cancers. In preclinical studies, VEGFxxxb inhibited both corneal and retinal neovascularization. These findings suggest that downregulation of VEGFxxxb levels may contribute to pathological neovascularization and that members of this family may have potential as antiangiogenic therapies.

Soluble VEGF receptor 1

Soluble VEGFR1, also generated by alternative splicing, is a natural inhibitor that can bind VEGF but lacks the exons required for signaling. It is essential for maintaining the avascularity of the cornea. As an integral component of the engineered therapeutic agent VEGF-Trap (see above), the ligand-binding site currently is being evaluated in clinical trials as a VEGF-targeting agent.

Complementary regulatory protein CD59

In addition to containing components that promote pathological neovascularization, the complement cascade also includes regulatory components that are potential antiangiogenic agents. One such candidate protein is CD59, whose ablation facilitated the development of CNV in mice; in the converse experiment, laser-induced CNV was inhibited by intravitreal or intraperitoneal administration of a soluble fusion protein containing CD59.

Tryptophanyl-tRNA synthase fragment

Tryptophanyl-tRNA synthase fragment (T2-TrpRS), another naturally occurring molecule, was found to inhibit physiological retinal angiogenesis in mice as well as ischemia-induced preretinal neovascularization. Recently, it was shown that intravitreal injection of T2-TrpRS, together with an anti-VEGF aptamer, led to potent inhibition of ischemia-induced retinal neovascularization, suggesting yet another avenue for combinatorial therapy.

Slit/Roundabout4

Slit/Roundabout4 (Robo4) is a member of a family of transmembrane receptors for the Slit family of secreted ligands. While Slit/Robo signaling was initially shown to be important in neuronal patterning, it also has been implicated in angiogenesis. Recently, intravitreal injection of Slit2 was found to inhibit both ischemia-induced neovascularization and VEGF-retinal permeability; none of these effects were seen if Robo4 was genetically ablated. Slit2-mediated inhibition of VEGF-induced endothelial cell migration and tube formation also was dependent on the presence of Robo4. Taken together, these data suggest that Slit/Robo4 axis acts as a negative regulator of VEGF-induced proangiogenic signaling.

Other inhibitors

In addition to the inhibitors discussed herein, numerous other molecules have been identified that also possess antiangiogenic activity but whose involvement in ocular neovascularization is not as well characterized (**Table 1**).

New Directions in Antiangiogenic Therapy

While most research into pathological angiogenesis has been directed toward ablating the unwanted vasculature, recent studies into the control of malignant tumors suggest that antiangiogenic therapy may act to normalize, rather than to destroy, the tumor vasculature. This normalization facilitated tumor penetration of chemotherapeutic agents and increased their efficiency. Similar concepts have emerged from studies of Dll4/Notch signaling where inhibition led to more profuse tumor vasculature but slower tumor growth. The molecular mechanisms underlying vascular normalization are yet to be determined. Inhibition of the regulator of G protein signaling 5 (RGS5), a protein found in pericytes, has been shown to promote vascular normalization in a mouse tumor model; this allowed the influx of immune cells into the tumor parenchyma and led to increased survival. Upregulation of RGS5 resulted in a significant reduction in the susceptibility of nascent vessels to regression by VEGF inhibition, suggesting that RGS5 contributes to pericyte-endothelial cell interactions and vascular maturation.

Normalization of the vasculature also has been proposed as a contributor to the utility of antiangiogenic therapy in pathological ocular neovascular diseases. Retinal neovascularization may reflect an adaptive response to hypoxia that has gone awry. It remains to be seen whether inducing regression or normalization of aberrant retinal vasculature will be the most effective in providing favorable visual outcomes.

Conclusions

The systematic study of the mechanisms underlying physiological and pathological angiogenesis has yielded a wealth of knowledge as to underlying molecular and cellular mechanisms involved in retinal neovascular diseases, including the actions of proangiogenic and antiangiogenic factors that modulate these processes. Drugs targeting VEGF have already proved their clinical utility. It is clear, moreover, that a variety of other agents offer promise, either alone or as adjunctive therapy with agents targeting VEGF, and that there is reason for optimism that the range of treatments for retinal neovascular diseases soon will be expanded.

See also: Breakdown of the RPE Blood–Retinal Barrier; Development of the Retinal Vasculature; Retinal Vasculopathies: Diabetic Retinopathy; Retinopathy of Prematurity.

Further Reading

Adamis, A. P. (2002). Is diabetic retinopathy an inflammatory disease? *British Journal of Ophthalmology* 86: 363–365.

Andrae, J., Gallini, R., and Betsholtz, C. (2008). Role of platelet-derived growth factors in physiology and medicine. *Genes and Development* 22: 1276–1312.

Arcasoy, M. O. (2008). The non-haematopoietic biological effects of erythropoietin. *British Journal of Haematology* 141: 14–31.

Avraamides, C. J., Garmy-Susini, B., and Varner, J. A. (2008). Integrins in angiogenesis and lymphangiogenesis. *Nature Reviews Cancer* 8: 604–617.

Bora, N. S., Jha, P., and Bora, P. S. (2008). The role of complement in ocular pathology. *Seminars in Immunopathology* 30: 85–95.

Dodelet, V. C. and Pasquale, E. B. (2000). Eph receptors and ephrin ligands: Embryogenesis to tumorigenesis. *Oncogene* 19: 5614–5619.

Ferrara, N., Damico, L., Shams, N., Lowman, H., and Kim, R. (2006). Development of ranibizumab, an anti-vascular endothelial growth factor antigen binding fragment, as therapy for neovascular age-related macular degeneration. *Retina* 26: 859–870.

Fiedler, U. and Augustin, H. G. (2006). Angiopoietins: A link between angiogenesis and inflammation. *Trends in Immunology* 27: 552–558.

Gragoudas, E. S., Adamis, A. P., Cunningham, E. T., Jr., et al. (2004). Pegaptanib for neovascular age-related macular degeneration. *New England Journal of Medicine* 351: 2805–2816.

Hellstrom, M., Kalen, M., Lindahl, P., Abramsson, A., and Betsholtz, C. (1999). Role of PDGF-B and PDGFR-beta in recruitment of vascular smooth muscle cells and pericytes during embryonic blood vessel formation in the mouse. *Development* 126: 3047–3055.

Ishida, S., Usui, T., Yamashiro, K., et al. (2003). VEGF164-mediated inflammation is required for pathological, but not physiological, ischemia-induced retinal neovascularization. *Journal of Experimental Medicine* 198: 483–489.

Jo, N., Mailhos, C., Ju, M., et al. (2006). Inhibition of platelet-derived growth factor B signaling enhances the efficacy of anti-vascular endothelial growth factor therapy in multiple models of ocular neovascularization. *American Journal of Pathology* 168: 2036–2053.

Pfaff, D., Fiedler, U., and Augustin, H. G. (2006). Emerging roles of the Angiopoietin-Tie and the ephrin-Eph systems as regulators of cell trafficking. *Journal of Leukocyte Biology* 80: 719–726.

Ribatti, D. (2008). The discovery of the placental growth factor and its role in angiogenesis: A historical review. *Angiogenesis* 11: 215–221.

Rosenfeld, P. J., Brown, D. M., Heier, J. S., et al. (2006). Ranibizumab for neovascular age-related macular degeneration. *New England Journal of Medicine* 355: 1419–1431.

Sainson, R. C. and Harris, A. L. (2008). Regulation of angiogenesis by homotypic and heterotypic notch signalling in endothelial cells and pericytes: From basic research to potential therapies. *Angiogenesis* 11: 41–51.

Perimetry

D B Henson, University of Manchester, Manchester, UK

Glossary

Change probability – Change probability compares each test location with that of a baseline measure and establishes whether or not there has been any significant change. Results are presented in the form of a visual field plot where each location is classified according to a series of cut-off probability levels, for example, $p < 0.05$.

Kinetic perimetry – With the kinetic examination strategies, the perimetrist selects a stimulus of a given size and intensity and moves it from outside the visual field toward its center noting the position at which it first becomes visible.

Perimetry – Perimetry, or campimetry, is the technique used to measure the extent of the visual field or to assess the sensitivity of the visual system to stimuli presented within the visual field (IPS standards).

Standard achromatic perimetry (SAP) – SAP is the most widely used form of perimetry where a white stimulus is projected onto a white background, often called white-on-white perimetry.

Static threshold perimetry – In static threshold perimetry, an estimate of the patient's sensitivity is derived at a series of predetermined test locations.

Visual field – Tate and Lynn defined the visual field as "all the space that one eye can see at any given instant," it normally extends from the fixation axis: 60° up, 75° down, 100° temporally, and 60° nasally. The superior and nasal fields are limited by facial contours.

Perimetric Techniques

Kinetic Perimetry

With the kinetic examination strategies, the perimetrist selects a stimulus of a given size and intensity and moves it from outside the visual field toward its center noting the position at which it first becomes visible. This is repeated along a series of different meridians and the points at which the stimulus first became visible are then joined together by a line which is called an isopter. Scotomas within the area of an isopter are detected by continuing to move the stimulus toward the center of the visual field

after it has first been detected. The patient is asked to report, if at any time, it disappears.

The perimetrist can repeat the whole process with stimuli of different sizes and/or intensities, in order to build up a map of the patient's visual field, such as that given in **Figure 1**. Kinetic examination strategies were very popular in the early days of perimetry. They have, however, been largely replaced by static techniques although they are still retained for specific cases such as small residual islands of vision and where patients have difficulty in performing automated perimetry.

Static Threshold Perimetry

In static threshold perimetry, an estimate of the patient's sensitivity is derived at a series of predetermined test locations. There are many different algorithms that can be used to establish the threshold and a great deal of effort has gone into deriving ones that minimize test time and error (difference between the true and the measured threshold).

The first widely used threshold algorithm (Full Threshold) was developed by Spahr and Bebie and their colleagues. It is a two-reversal staircase strategy in which the step size reduced from 4 to 2 dB after the first reversal.

This algorithm suffers from four major drawbacks:

1. long test time,
2. demanding and exhausting for the patient,
3. poor repeatability, and
4. a significant learning effect.

In an attempt to shorten test times, the fast threshold staircase algorithm was introduced, in which there was a single step size of 3 dB with only one reversal. While being faster, this algorithm is more variable than the Full Threshold algorithm and thus has not been widely adopted in clinical practice.

Tendency-orientated perimetry was similarly developed in order to reduce test times. The algorithm combines data from several neighboring locations and reduces test times by effectively reducing spatial resolution.

The Swedish interactive threshold algorithm (SITA) is currently one of the most widely used algorithms. It is considerably faster than the full threshold algorithm (approximately 5 min per eye for a normal field) with similar repeatability. This has been achieved by:

1. using maximum-likelihood procedures in combination with a 4–2 dB staircase algorithm;
2. removing the need for false-positive catch trials; and

3. speeding up the rate of stimulus presentation in patients who respond quickly. The SITA algorithm monitors the patient's response rate and adjusts the presentation rate accordingly.

The SITA algorithm is currently available in two forms, standard and fast. SITA Fast has more liberal terminating criteria to further reduce overall test times.

Static Suprathreshold Perimetry

In a suprathreshold examination, the stimuli are initially presented at an intensity that is calculated to be above the patient's threshold. If the stimuli are seen then it is assumed that no significant defect exists. This strategy has largely been developed as a screening procedure for conditions such as glaucoma.

There are many different types of suprathreshold tests and **Table 1** highlights some of the major differences.

Selection of the test intensity is an important part of a suprathreshold test. If the intensity is set too high then the test will become insensitive to shallow defects. If the intensity is set too low then the test looses specificity. Most suprathreshold tests set the intensity at 5–6 dB above the threshold estimate.

Multiple stimulus tests are faster (approximately twice as fast in a person without a defect). The need to verbally respond to each presentation helps maintain patient attention, improves threshold estimates and reduces variability.

Test Targets

Most modern perimeters use stimuli whose sizes were defined by Goldmann (see **Table 2**).

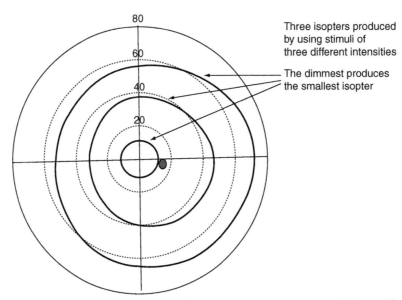

Three isopters produced
by using stimuli of
three different intensities

The dimmest produces
the smallest isopter

Figure 1 A map of a patient's visual field produced through kinetic perimetry using test stimuli of three different intensities. The dimmest test produces the smallest isopter.

Table 1 The different types of suprathreshold visual field test strategies

Test intensity	Fixed across the whole field	Used to test the visual field of drivers where the intensity is well above the estimated threshold
	Increases with eccentricity	Most common, with the increase matched to a database of threshold measures and age
Threshold setting	Age related	The test intensity is based upon the age of the patient.
	From measurement	The test intensity is based upon a series of threshold estimates at a few test locations, usually 4
Presentation	Single stimulus	Test locations are presented one at a time and the patient indicates when they see the stimulus by pressing a response button
	Multiple stimulus	Test locations are presented in patterns of 2–4 stimuli and the patient verbally reports the number seen
Algorithm	Standard	If stimulus is missed, it is presented a second time and only if missed twice is it marked as a miss. Pass criterion is one seen out of up to two presentations
	Multisampling	Pass criterion is raised, a typical one being three seen out of up to five presentations

Table 2   Goldmann stimulus sizes

Goldmann size	Nominal size (mm²)	Angular subtence (min of arc)
0	0.0625	3.78
I	0.25	7.68
II	1.0	15.36
III	4.0	30.71
IV	16	61.3
V	64	122.56

Standard Achromatic Perimetry

Standard achromatic perimetry (SAP) is the most widely used form of perimetry where a white stimulus is projected onto a white background, often called white-on-white perimetry. It normally uses a size III or V Goldmann stimulus on a background luminance of 31.5 apostilbs (10 cd m^{-2}). The intensity of the stimuli is given in decibels of attenuation from a fixed value of 10 000 apostilbs (this can vary from instrument to instrument). Thus, 0 dB corresponds to 10 000 abs, 10 dB to 1000 abs, and 20 dB to 100 abs.

Short-Wavelength Automated Perimetry

Patients with glaucoma often have an associated color vision defect in which their sensitivity to blue light is reduced. This observation led researchers to question whether or not the blue sensitive mechanism was more susceptible to glaucomatous damage and whether or not a perimeter that specifically targeted the blue mechanism might not be more sensitive than those that use the conventional SAP stimuli.

A problem encountered when trying to test the blue mechanism is the relatively low sensitivity of blue cones. Even at short wavelengths (440 nm) the blue sensitive cones are only marginally more sensitive than the red and green cones. Selective damage to the blue cones would have relatively little effect upon sensitivity as the red and green cones would simply step in once the blue cone sensitivity dropped below that of the other receptors. To overcome this problem, the red and green cones are desensitized by adapting the eye to a yellow light. **Figure 2** shows the sensitivity of the three receptors after adaptation. The blue cones are exposed at 440 nm wavelength. The extent of exposure is important. If we consider that any damage to the blue cones mechanism is likely to lower its sensitivity we can see from **Figure 2** just how much loss can be tolerated before the red and green receptors again become the most sensitive (approximately 1.5 log units).

There have been a number of studies that have demonstrated that blue-yellow defects precede those for SAP, however, the test has not been widely adopted because:

1. blue-yellow perimetry is particularly sensitive to lens opacities;

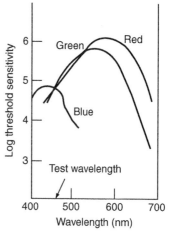

Figure 2   The log threshold sensitivity of the blue, green, and red cone systems after adaptation with a yellow light that desensitizes the red and green cones. Under these conditions the blue cones can be tested using 440 nm wavelength test light. Note that even after adaption, blue cones are approximately 1.5 log units less sensitive than green and red cones.

2. short-wavelength automated perimetry is more variable than SAP;
3. patients find the test difficult; and
4. most patients seen in a glaucoma clinic either have no visual field loss or have defects that can be detected with SAP.

High-Pass Resolution Perimetry (Ring Perimetry)

High-pass resolution perimetry uses ring-shaped targets of varying size (see **Figure 3**). The luminance inside each ring target is the same as the background while the core of the ring is brighter and its inner and outer edges darker. The overall intensity profile of the ring is such that when it cannot be resolved it cannot be detected. Patients press a response key when they see the stimulus (presentation time 165 ms) and a repetitive bracketing strategy is used to establish the minimum resolvable ring size.

High-pass resolution perimetry has good sensitivity and specificity when compared to SAP. It has also been shown to have threshold variability that is independent of sensitivity and can detect progressive loss earlier than conventional perimetry. High-pass resolution perimetry is also fairly fast taking on average only 5.5 min to test 50 locations.

On the negative side, the technique is sensitive to blur, either refractive or due to media changes, unable to measure defect depth within small circumscribed lesions, and is unable to detect scotomata whose size is less than the local liminal test target. With the current monitor technology, there is also a relatively small dynamic range of stimuli which limits the test ability to monitor loss in patients with significant loss.

Figure 3 High-pass resolution perimetry uses ring-shaped targets of varying size. The luminance inside each ring target is the same as the background while the core of the ring is brighter and its inner and outer edges darker. The overall intensity profile of the ring is such that when it cannot be resolved it cannot be detected. Patients press a response key when they see the stimulus (presentation time 165 ms) and a repetitive bracketing strategy is used to establish the minimum resolvable ring size.

Frequency-Doubling Technology Perimeter

The frequency-doubling technology (FDT) perimeter is based upon an illusion known as frequency doubling, in which an alternating sinusoidal grating of low spatial frequency (<4 cpd) appears, at certain temporal frequencies (>15 Hz), to have twice as many lines. This illusion is believed to be mediated by the magnocellular (M-cell) pathway and in particular by the large fiber diameter *My* ganglion cells which constitute 1.5–2.5% of all retinal ganglion cells.

Recent investigations of ganglion cell fiber loss in glaucoma have led to the development of a number of theories. One theory is that there is selective loss of M-cell fibers during the early stages of this condition. Another is that there is selective loss of large diameter axons and a third is that mechanisms with little redundancy (sparse representation) are likely to show losses earlier than those with greater redundancy. As FDT is believed to be based upon large-diameter M-cell fibers, which are sparsely represented, a screening test based upon this illusion should, according to all these theories, be particularly sensitive to early glaucomatous loss. However, while FDT perimeter uses the appropriate spatial and temporal frequencies for the illusion, the task presented to the patient is one of contrast sensitivity, that is, the patient is being asked when he/she can detect the appearance of a target in the peripheral field not when he/she sees the frequency-doubling illusion.

The FDT perimeter uses 0.25 cpd stimulus that subtend 10° at a temporal frequency of 25 Hz. A modified binary search strategy is used to test 17 locations. Threshold and suprathreshold test strategies have good sensitivities and specificities and the instrument has many attractive characteristics for glaucoma screening. It is a small, self-contained, portable instrument that is not sensitive to background illumination levels. It does not require a corrective lens for refractive errors (results are reported to be independent of refractive error up to ±7.00 D) and it is relatively fast.

A second-generation FDT perimeter, Humphrey Matrix, was produced in 2005. This instrument uses a range of stimulus sizes (10°, 5°, and 2°), two spatial frequencies (0.25 and 0.5 cpd) and three temporal frequencies (25, 18, and 12 Hz) to give a wider range of test programs including those that mimic those widely used in SAP (24-2, 10-2). These modifications overcome the major drawback of poor spatial resolution in the earlier instrument.

Reliability Estimates

Most modern perimeters incorporate methods for estimating the reliability of the patient's responses. Widely used measures are fixation accuracy, false-positive response rate, and false-negative response rate. The relationship between these reliability measures and test–retest variability is, however, poor.

Fixation Accuracy

The simplest technique for monitoring fixation is observation by the perimetrist either with the aid of a telescope or camera. Such techniques are totally dependent upon the perimetrist's judgment and continued vigilance.

Some perimeters incorporate a fixation monitor that indicates when fixation has been lost. Most fixation monitors cannot differentiate between rotations of the eye, which occur when the patient looks away from the fixation target, and translations of the eye. A translation of the eye, such as a slight sideways movement of the head, does not necessarily mean that fixation has been lost or that the angular subtence of the perimetric stimuli has been changed by a significant amount. Automatic fixation monitors are also generally insensitive to small, but significant, fixation errors (e.g., 1°).

Most modern perimeters incorporate the Heijl–Krakau technique for sampling fixation. With this technique stimuli are occasionally presented in the region of the patient's blind spot. If fixation is maintained during these presentations the stimulus will not be seen. If, on the other hand, fixation is lost then the stimulus is likely to fall outside of the blind spot and elicit a response from the patient. A frequently quoted cut off for reliable fixation errors is less than 33% of presentations. There is no evidence on which to base this cut-off value and the relatively small number of trials (on average there are only 10–12 fixation trials in a perimetric test) mean that the precision of the estimated rate is very low. As the position of the blind spot varies from one individual to another it is necessary, at the onset of the examination, to have a little routine which establishes the blind spot's location. The accuracy of this routine is important as errors may later manifest themselves as numerous fixation errors in a patient who has maintained good fixation.

The major advantages of the Heijl–Krakau technique are its simplicity and ease of implementation. Its disadvantages are:

1. It only samples fixation.
2. It increases the examination time.
3. It is unlikely to detect small fixation errors as the standard target subtends $\approx 0.5°$ and the blind spot an area of $5° \times 7°$.

At present, the best method is direct observation by the perimetrist. This technique is not only sensitive but also, via verbal feedback, can result in an improvement in the fixation accuracy for subsequent presentations.

False-Positive Responses

Single stimulus static test programs derive an estimate of the false-positive response rate with a number of catch trials (when the instrument goes through the motions of presenting a stimulus but does not actually present one). If the patient responds to one of these catch trials then this is classified as a false-positive response. In the SITA test algorithms this catch trial method has been replaced by one based upon response times (interval between the presentation of a stimulus and the patient pressing the response button). Olsson and colleagues were able to show a good relationship between the number of response times that fell outside a patient's normal response window and the false-positive response rate. They also demonstrated that estimates based upon response times were more repeatable than those based upon catch trials. An added advantage of using response times is that it reduces, slightly, the overall test time. Many patients do not give a single false-positive response during a visual field test and it is important to remember the false-positive response rate is a poor predictor of test–retest variability. A less than 33% false-positive rate is often used to differentiate reliable from unreliable results, again there is no evidence on which to base this cut-off value and the relatively small number trials means that the precision of the estimate is poor. More stringent cut-offs are often used in research studies.

False-Negative Responses

An estimate of the false-negative response rate is obtained by retesting a location with a stimulus whose intensity is above the already established threshold. If the patient fails to respond positively to this presentation then this is classified as a false-negative. False-negative responses are often found to increase with the extent of visual field damage and this is believed to be due to the relationship between variability and sensitivity. Most patients have a low false-positive rate and those with a rate above 33% are

often classified as unreliable although more stringent cut-offs are often used in research trials. Again there is no evidence to support this cut-off value and the number of trials means that the precision of the estimate is poor.

Clearly, if the patient makes a high proportion of errors then his/her results must be viewed with a certain amount of suspicion. The judgment of an attending perimetrist is, however, often of greater value than these reliability estimates.

Analytical Techniques

Total Deviation and Pattern Deviation Plots

Total deviation and pattern deviation maps take the threshold data from a visual field examination and calculate, for each test location, whether or not the threshold values are significantly different from those of a normal eye of the same age. They rely upon the perimeter having a database of threshold values from normal eyes in order to perform these calculations. Pattern deviation values differ from total deviation values in that they have been adjusted for overall shifts in sensitivity. For example, we might get a patient whose overall sensitivity is below that of a normal patient of the same age. In this case, some locations may be highlighted as being abnormal on the total deviation map but not on the pattern deviation map. This adjustment is based upon the findings from some of the most sensitive test locations, that is, if these are above or below normal then all values are shifted down or up.

Total and pattern deviation probability maps use the distribution of total and pattern deviation values within a normal population, to calculate the probability of each threshold estimate coming from a normal eye. The results are expressed in the standard statistical way, that is, as being beyond the 5%, 2%, 1%, or 0.5% probability level.

Global Indices

Flammer and colleagues proposed that the visual field defects associated with glaucoma could be divided into three different categories: (1) those that cause an overall depression in the sensitivity of the eye, (2) those that cause local defects, and (3) those that cause an increase in the variability of results; and that each category could be represented by an index. They called these three indices mean defect, loss variation, and short-term fluctuation.

The concept of using three separate indices to represent a glaucomatous visual field has found very wide acceptance within the ophthalmological professions and most visual field instruments now incorporate a set of algorithms which compute these or very similar indices. The name given to these indices varies from one instrument to another, as does the form of calculation. In the

Humphrey instrument they are called mean deviation, pattern standard deviation, and short-term fluctuation and the values are weighted according to the normal variance seen at each location.

Mean defect (deviation) (MD) gives the average difference between the threshold values and those from an age-matched control. Loss variance is a measure of the spread of defect values which increases when there is a local defect. An alternative measure of spread is the standard deviation (SD). This statistic has the advantage of giving a measure of spread in decibels and is now used in preference to variance measures. In the Humphrey perimeter this is called pattern standard deviation (PSD) while in the Octopus it is called sLV, which stands for the square root of the loss variance, that is, the SD. Short-term fluctuation is a measure of variability in threshold estimates derived from repeat testing a subset of locations. However, a global estimate of variability is of limited value as it is now recognized that variability is dependent upon sensitivity and varies across the visual field.

MD and sLV or PSD can be plotted over time to aid the detection of progressive loss. They can also be plotted against each other to give a measure of the extent of loss, Brusini staging system. The sensitivity of MD to media opacities, such as cataract, has led to the development of a new global index for the Humphrey perimeter called the visual field index (VFI). This index gives a percentage measure of the residual field and uses pattern deviation values and a weighting system based upon cortical magnification to give a more representative measure of a patient's visual function.

The Glaucoma Hemifield Test (GHT) (Humphrey) and cluster analysis (Octopus) are additional indices designed to help detect rather than quantify glaucomatous visual field loss. They are designed to detect vertical asymmetry between the superior and inferior hemifields. The GHT looks at the asymmetry in pattern deviation probability values between five areas in the superior field and their mirror images in the inferior field, while the Octopus cluster analysis uses the defect values within each of five slightly different clusters within the superior and inferior visual field. Both use a database of normal values to establish whether or not a response is outside normal limits.

Linear Regression

Linear regression can be applied to a longitudinal series of:

1. global indices (MD, PSD, VFI),
2. clusters of test locations (Octopus cluster analysis), or
3. individual test locations (point-wise analysis).

The significance of any change over time and the gradient of the regression line can be used for predicting long-term outcomes.

The Progressor software package presents the findings from a point-wise regression analysis in a particularly novel way that retains the information on both the depth of the defect and the significance of any change while the Peridata software package (Peridata) color codes each test location according to the significance of any change. From these charts the clinician can ascertain whether or not any progressive changes are close to fixation or at the edge of the visual field, where they may have been influenced by artifacts such as a droopy upper lid.

The accuracy of a regression analysis and any predictions is dependent upon the number of examinations. Several research groups have concluded that we really need about five visual field results before we can reliably calculate the gradient of the regression line.

Change Probability

Change probability compares each test location with that of a baseline measure and establishes whether or not there has been any significant change. Results are presented in the form of a visual field plot where each location is classified according to a series of cut-off probability levels, for example, $p < 0.05$. The baseline value is often the average of two visual field results (to give a better estimate) that need not be the first ones recorded, that is, we can establish whether there has been any significant change from intermediate results.

Change probability analysis takes into account the relationship between variability and sensitivity. One of the criticisms of change probability is that it does not use the information obtained in intermediate examinations, that is, it only compares the current finding with the baseline value. The recently introduced Progression Analysis Probability plot for the Humphrey Visual Field Analyzer uses different symbols to code whether or not the change has occurred in just the current, last two, or last three examinations.

See also: IOP and Damage of ON Axons; Retinal Ganglion Cell Apoptosis and Neuroprotection.

Further Reading

Anderson, A. J., Johnson, C. A., Fingeret, M., et al. (2005). Characteristics of the normal database for the Humphrey Matrix perimeter. *Investigative Ophthalmology and Visual Science* 46: 1540–1548.

Artes, P. H., Henson, D. B., Harper, R., and McLeod, D. (2003). Detection and quantification of visual field loss: A comparison of perimetric strategies by computer simulation. *Investigative Ophthalmology and Visual Science* 44: 2582–2587.

Bengtsson, B. (2000). Reliability of computerised perimetric threshold tests as assessed by reliability indices and threshold reproducibility in patients with suspect and manifest glaucoma. *Acta Ophthalmologica* 78: 519–522.

Bengtsson, B. and Heijl, A. (2008). A visual field index for calculation of glaucoma rate of progression. *American Journal of Ophthalmology* 49: 66–76.

Bengtsson, B., Olsson, J., Heijl, A., and Rootzen, H. (1997). A new generation of algorithms for computerised perimetry, SITA. *Acta Ophthalmologica* 75: 368–375.

Brusini, P. and Filacorda, S. (2006). Enhanced glaucoma staging system (GSS2) for classifying functional damage in glaucoma. *Journal of Glaucoma* 15: 40–46.

Cello, K. E., Nelson-Quigg, J. M., and Johnson, C. A. (2000). Frequency doubling technology perimetry for detection of glaucomatous visual field loss. *American Journal of Ophthalmology* 129: 314–322.

Chauhan, B. C., House, P. H., McCormick, T. A., and LeBlanc, R. P. (1999). Comparison of conventional and high-pass resolution perimetry in a prospective study of patients with glaucoma and healthy controls. *Archives of Ophthalmology* 117: 24–33.

Flammer, J., Drance, S. M., Augustiny, L., and Funkhouser, A. (1985). Quantification of glaucomatous visual field defects with automated perimetry. *Investigative Ophthalmology and Visual Science* 26: 176–181.

Frisen, L. (1987). A computer-graphics visual field screener using high-pass spatial frequency resolution and multiple feedback devices. *Documenta. Ophthalmologica Proceedings Series* 49: 441–446.

Henson, D. B. and Artes, P. H. (2002). New developments in supra-threshold perimetry. *Opthalmic and Physiological Optics* 22: 463–468.

Henson, D. B., Chaudry, S., Artes, P. H., Faragher, E. B., and Ansons, A. (2000). Response variability in the visual field: Comparison of optic neuritis, glaucoma, ocular hypertension and normal eyes. *Investigative Ophthalmology and Visual Science* 41: 417–421.

Johnson, C. A., Adams, A. J., Casson, E. J., and Brandt, J. D. (1993). Blue-on-yellow perimetry can predict the progression of glaucomatous damage. *Archives of Ophthalmology* 111: 645–650.

King, A. J. W., Taguri, A., Wadood, A. C., and Azuara-Blanco, A. (2003). Comparison of two fast strategies, SITA fast and TOP, for the assessment of visual fields in glaucoma patients. *Archives of Ophthalmology* 240: 481–487.

Miranda, M. and Henson, D. B. (2008). Perimetric sensitivity and response variability in glaucoma with single stimulus automated perimetry and multiple stimulus perimetry with verbal feedback. *Acta Ophthalmologica* 86: 202–206.

Morales, J., Weitzman, M. L., and Gonzalez de la Rosa, M. (2000). Comparison between Tendency-Oriented Perimetry (TOP) and octopus threshold perimetry. *Ophthalmology* 107: 134–142.

Olsson, J., Bengtsson, B., Heijl, A., and Rootzen, H. (1997). An improved method to estimate frequency of false-positive answers in computerized perimetry. *Acta Ophthalmologica* 75: 18–183.

Viswanathan, A. C., Fitzke, F., and Hitchins, R. A. (1997). Early detection of visual field progression in glaucoma: A comparison of PROGRESSOR and STATPAC 2. *British Journal of Ophthalmology* 81: 1037–1042.

Relevant Websites

http://webeye.ophth.uiowa.edu – University of Iowa Health Care; IPS Standards: Imaging and Perimetric Society.

http://www.peridata.org – Peridata; Peridata Software GmbH.

Photopic, Mesopic and Scotopic Vision and Changes in Visual Performance

J L Barbur, City University, London, UK
A Stockman, UCL Institute of Ophthalmology, London, UK

Glossary

Contrast sensitivity function (CSF) – The reciprocal of contrast threshold measured as a function of spatial frequency in a spatial CSF or as a function of temporal frequency in a temporal CSF.

Higher order aberrations (HOAs) – These aberrations in the eye describe imperfections in the optics that cannot, in general, be corrected for with conventional refraction (i.e., sphere and cylinder components).

Mesopic – The range of intermediate light levels between cone threshold and rod saturation where both rod and cone signals contribute to a visual response.

Photopic – The range of high light levels above rod saturation where vision is mediated by signals from cone photoreceptors.

Scotopic – The range of low light levels below cone threshold where visual responses rely entirely on rod signals.

Spectral luminous efficiency (SLE) – The SLE function represents an appropriate measure of detector spectral responsivity, which in the case of the human eye is used to convert radiant flux to an equivalent photometric flux.

Spectral power distribution (SPD) – This describes the wavelength distribution of radiant flux (power) emitted by a source or surface.

Spectral radiance – A quantity that describes the radiant flux (power) within a narrow wavelength interval emitted by a source or surface in a given direction per unit solid angle, per unit area of the source.

Spectral responsivity (SR) – The SR of a detector of radiation is a measure of the spectral sensitivity of the detector and represents the signal generated per unit incident radiant flux as a function of wavelength.

Introduction

Arguably, the most important feat of human vision is its ability to operate effectively over the enormous 10 000-million-fold range of illumination levels to which it can be exposed, from starlight to bright sunlight (see **Figure 1**). Moreover, it achieves this despite the limitations imposed by individual neurons in the visual system, many of which have dynamic ranges of no more than about 100-fold. Sensitivity regulation over such a massive range cannot be achieved without compromise. One compromise is to share the range between two different types of photoreceptors: the sensitive rods, functioning at lower scotopic levels of illumination, and the less-sensitive cones, functioning at higher photopic levels, the two working together at intermediate mesopic levels. Another important compromise is the trade-off between increased sensitivity at lower light levels, and improved temporal and spatial acuity at higher light levels. As the light level increases, high sensitivity is no longer needed and is traded for improvements in spatial and temporal acuity. These adjustments occur separately within the rod and the cone systems, but the rod system has a lower acuity and higher sensitivity compatible with its operation at lower light levels.

There are ~100 million rods, but only 5 million cones in the human retina. The 5 million cones are most densely packed in a small region of the retina at the center of our vision, where photopic vision is best. Cones are less sensitive to light but respond more quickly than rods. The cones and their postreceptoral pathways operate well at higher light levels in the upper mesopic and photopic range. Over most of this range, processes of light adaptation within the retina ensure that the operation of the cone photoreceptors adjusts to changes in light level to provide a useful dynamic range of about 100, which remains relatively independent of light level. When the light level is lower, the cone system increases its sensitivity by increasing the spatial and temporal extents over which light is summed, as a result of which spatial and temporal acuity is reduced. Even at very high light levels, bleaching of the light-sensitive pigment in the cone protects its output from saturating and thus failing to signal changes in light. By contrast, rods and their pathways operate well at lower light levels in the scotopic and lower mesopic ranges. The rods and rod pathways are optimized for these levels. The rods are more sensitive and much slower than the cones, so that they integrate light over time, and their postreceptoral pathways integrate light over space. Capable of single-photon detection at the lowest levels, the rod system ceases to operate effectively at the top of the mesopic range.

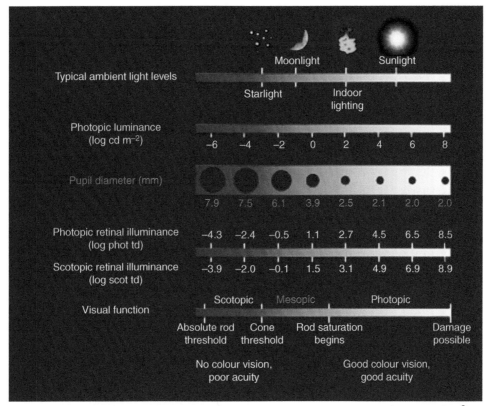

Figure 1 Illumination levels. Typical ambient light levels are compared with photopic luminance (log phot. cd m^{-2}), mean pupil diameter (mm), photopic and scotopic retinal illuminance (log photopic and scotopic trolands, respectively), and visual function. The scotopic, mesopic, and photopic regions are defined according to whether rods alone, rods and cones, or cones alone contribute to a visual response. The conversion from photopic to scotopic values assumes a white standard CIE D$_{65}$ illumination. Based on **Figure 1** of Stockman, A. and Sharpe, L. T. (2006). Into the twilight zone: The complexities of mesopic vision and luminous efficiency. *Ophthalmic and Physiological Optics* 26: 225–239.

Mesopic vision describes the range of light levels over which signals from both rods and cones contribute to the visual response. This range extends just over a 1000-fold from cone threshold to rod saturation, and encompasses light levels that are often found in occupational environments. At mesopic levels, the spectral composition of the illuminant plays an important role in determining the relative strengths of rod and cone signals (see **Figure 2**). Marked changes in visual performance that vary over the visual field are observed as the level of illumination is lowered and/or the spectral composition of the illuminant is modified. The gradual increase in rod signals as the light level decreases causes changes in the overall spectral sensitivity of the eye that has consequences for the visual effectiveness of the illuminant spectral power distribution (SPD) (i.e., the intensity of the light as a function of wavelength). Mesopic luminous efficiency is inherently complex, and therefore difficult to standardize or model, because it depends on the outputs of both the rod and the cone photoreceptors. Not only are there differences in the photoreceptor spectral sensitivities, but there are also differences in the properties of the postreceptoral pathways through which the rod and cone signals are transmitted.

The nature and balance of these pathways are constantly changing with changes in light level in the mesopic range.

Much effort has gone into establishing spectral responsivity functions that are appropriate for photopic, mesopic, and scotopic lighting conditions. This approach rests on the assumption that visual performance can be adequately predicted by weighting the intensity of the light reaching the eye (the radiant flux) with functions that reflect the spectral responsivity of the eye at high, intermediate, and low light levels. Some success has been achieved at low light levels in the scotopic range, but this success reflects the fact that the spectral responsivity is determined by a single photoreceptor type, the rods, which have a single type of photosensitive molecule. The approach has been less successful at photopic levels, where responsivity depends on up to three cone photoreceptors, and least successful at mesopic levels where it depends on different types of rod and cone photoreceptor systems with markedly different properties. In this article, we present data on how visual performance changes with light level. We argue that the changes in performance are much more profound than the changes in spectral sensitivity as captured by changes in luminous efficiency functions.

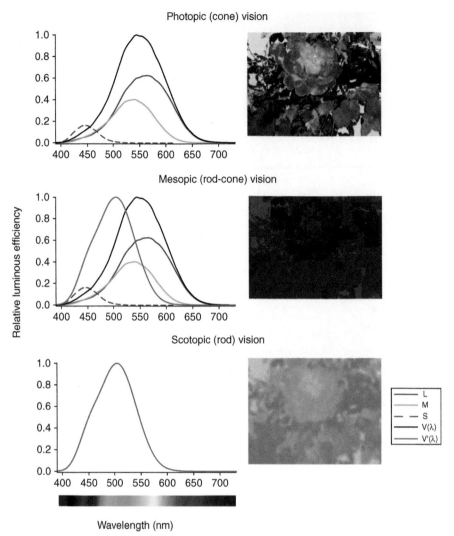

Figure 2 The useful range of light levels can be divided into three regions: the photopic range that yields best temporal and spatial contrast acuity and color discrimination, the mesopic range where color signals become less important, our temporal responses become more sluggish, and the world appears dark, and the scotopic range where the world appears brighter, but we can no longer discriminate color differences and can only see large, high-contrast objects. The spectral sensitivities of the photoreceptors are shown. The relative heights of the M- and L-cone spectral sensitivities reflect their assumed relative contributions to V(λ). The S-cone function is dashed to signify that its contribution to luminous efficiency is minimal. The photographs on the right illustrate how our vision changes as the eye adapts to mesopic and scotopic illumination.

The Concept of Luminous Efficiency Function

The traditional development of photometry has been strongly influenced by radiometry and the properties of an ideal detector of radiation, which is assumed to exhibit response linearity and additivity and has a well-defined spectral responsivity (in the case of the eye at photopic levels, $V(\lambda)$). Let the radiant light fluxes Θ_1 and Θ_2, respectively, produce signals S_1 and S_2 when the detector is exposed to each flux separately. Response linearity means that when the detector is exposed to $\Theta_1 + \Theta_2$ together, the signal generated must equal $S_1 + S_2$, irrespective of the intensity and the spectral composition of Θ_1 and Θ_2.

When this is the case, the spectral responsivity curve of the detector becomes a particularly useful quantity since it provides the means of computing the detector signal in response to any broadband, spectral distribution of light flux. This is achieved by simply integrating the radiant flux (weighted at each wavelength by the spectral luminous efficiency function or $V(\lambda)$) over the spectral range for which $V(\lambda)$ and $\Theta(\lambda)$ are both nonzero to give:

$$\Theta_V \sim \int V(\lambda)\Theta(\lambda)d\lambda, \qquad [1]$$

where Θ_V is usually described as the luminous flux. The simplicity of this approach continues to be important and relevant in the more applied areas of vision science, for example, in lighting engineering and photometry. Spectral luminous efficiency functions (i.e., curves that specify the spectral responsivity of the eye, and thus the effectiveness of lights) have been produced using a number of different experimental approaches for both photopic vision (when only cone photoreceptor signals are involved) and for scotopic vision (when vision relies entirely on rod signals). The response linearity of the eye is, however, very limited. Lights that are intense enough to adapt the visual response give rise to nonlinearities that break down the additivity required by eqn [1]. At photopic and mesopic levels, where up to four photoreceptors can contribute to the visual response, additivity is even more limited. As the wavelength or SPD of the illuminant changes, and the photoreceptors selectively adapt, so also do the relative contributions of the different photoreceptors to luminous efficiency. Mesopic and photopic luminous efficiency functions, in general, are not therefore fixed in spectral sensitivity.

Spectral luminous efficiency functions are currently used to provide a means of quantifying the total luminous flux associated with broad-band and narrow-band sources, and for computing the luminance contrast of an illuminated object with respect to that of the surrounding background. When the relative SPD of an object matches that of its surrounding background, the computed contrast is independent of the assumed spectral responsivity curve. However, for objects that differ in SPD from their surrounding background, the computed contrast is strongly dependent on the choice of spectral luminous efficiency function. Moreover, at photopic levels, perceived color differences can also contribute significantly to perceived object contrast, as a result of which any efforts to produce a single spectral luminous efficiency function to account for both the quantity of luminous flux and perceived contrast has severe limitations. In the mesopic range, when rod contrast signals also contribute to object conspicuity, prediction of effective contrast becomes

even more challenging. The relative rod and cone contributions to luminous efficiency vary continuously with the illumination level, the spectral composition of the adapting background, the location of the stimulus in the visual field, and the temporal and spatial characteristics of the stimulus. A number of studies have attempted, with limited success, to model these complex interactions. Yet, any model based only on changes in spectral sensitivity is unlikely to capture the effects of the large number of parameters that have been shown experimentally to affect object appearance.

In the remainder of this article, we will be less concerned with the need to produce fixed spectral luminous efficiency functions appropriate for either the photopic or the mesopic range. Instead, we focus on describing how key aspects of visual performance change with ambient light level.

Quality of Vision and Light Level

Spatial Acuity, Spatial Contrast Sensitivity, and Light Level

An important aspect of visual performance is the ability to see fine spatial detail. A simple functional test is to measure the threshold luminance contrast (i.e., $\Delta L/L_b$, where ΔL represents the increment in stimulus luminance with respect to the surrounding background (L_b)) needed to resolve letters or some other visual stimuli when the angular subtense of stimulus elements is larger than the high-contrast resolution limit of the eye. The latter is often taken to be $1'$ of arc and corresponds to a letter size of $\sim 5'$ of arc. In order to avoid eye strain and to minimize the effects of microfluctuations of accommodation, a stimulus size three times the spatial resolution limit of the eye or larger is often employed in various occupational environments. For these reasons, the data shown in **Figure 3** were measured using a Landolt ring stimulus of

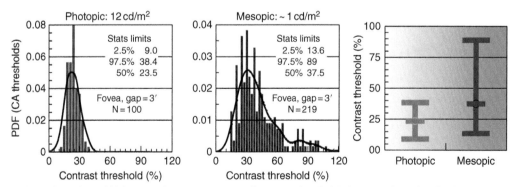

Figure 3 Low photopic and high mesopic measurements of contrast thresholds for gap orientation discrimination using Landolt ring stimuli of $3'$ gap size. One hundred subjects participated in the photopic study and 219 subjects carried out the mesopic test. All subjects had normal Snellen acuity and the age distribution was as follows: photopic: mean = 23 years, standard deviation (sd) = ± 7, mesopic: mean = 28 years, sd = ± 14. The histogram shows the probability distribution function (PDF) of the contrast acuity (CA) thresholds measured as % luminance contrast (i.e., $\Delta L/L_b$, where ΔL represents the increment in stimulus luminance with respect to the surrounding background (L_b)). The statistical limits from each graph are shown as vertical bars to illustrate the overlap in photopic and mesopic thresholds.

$3'$ gap size. The orientation of the gap was restricted to four possible locations (i.e., top right, top left, bottom right, and bottom left) and the subject's task was to press one of four buttons to indicate the position of the gap. Two different groups of normal subjects (with high-contrast acuity of $1'$ of arc or better) were involved in this study. The photopic thresholds follow a tight distribution, while the mesopic thresholds are larger and exhibit significantly increased variability. Interestingly, about 50% of the 219 subjects examined under mesopic conditions exhibit thresholds that fall within the normal 2σ (σ: standard deviation) limits of the photopic range.

The results show no significant correlation with either age or the quality of the observer's optics (in terms of the mean-wavefront, higher order aberrations (HOAs)). **Figure 4** shows similar data as a function of retinal illuminance for a single subject measured at the fovea and $\pm 2.5°$ away from fixation, along the horizontal meridian. In this experiment, the pupil size was measured every 20 ms and the luminance of the visual display was adjusted appropriately in order to maintain constant retinal illuminance. The results show a large ~ 2.2-log unit increase in foveal contrast acuity thresholds over the 1000-fold change in retinal illuminance. Contrast acuity at low light levels in the mesopic range is extremely poor and below 0.6 photopic trolands (phot. td, a measure of photopic retinal illuminance) the periphery becomes more sensitive than the foveal region because of the distribution of rods and cones across the retina.

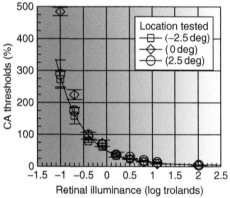

Figure 4 Contrast acuity thresholds ($\Delta L/L_b$) measured at the fovea and $\pm 2.5°$ in the periphery, as a function of retinal illuminance, for a $3'$-gap Landolt ring stimulus. Spectrally calibrated neutral density filters were employed to achieve the full range of retinal illuminance. Pupil size was monitored continuously and this measurement was used to adjust the luminance of the display to maintain constant retinal illuminance. The correlated color temperature of the background field was $\sim 6500K$ (CIE (x, y) – chromaticity coordinates: 0.305, 0.323). The three stimulus locations were interleaved randomly and the subject's task was to report the orientation of the gap (i.e., top right, top left, bottom right, or bottom left) by pressing one of four buttons located at the corners of a square, each button corresponding with one of the gap locations.

Contrast acuity measurements, although functionally important, only partly characterize how the spatial properties of the visual system change with retinal illuminance. A more complete characterization is provided by the spatial contrast sensitivity function or CSF, which defines the sensitivity (i.e., the reciprocal of the contrast threshold) for the detection of sinusoidal gratings (smoothly changing black-and-white stripes) as a function of their spatial frequency (i.e., as they change from coarse to increasingly fine stripes). Loosely speaking, the contrast describes local, spatial differences between the object of interest (the stripes, in this case) and its background (the mean background level), and often correlates with subjective measures such as object conspicuity. In a typical CSF experiment, the observer is presented with sinusoidal gratings of a given spatial frequency, and is asked to vary its contrast to find the contrast at which the grating is just visible. This threshold measurement is then repeated at a series of spatial frequencies to build up a complete CSF. Several spatial CSFs measured by van Nes and Bouman as a function of retinal illuminance are shown in **Figure 5**. As the illumination level increases, the CSFs change in shape from being low-pass (i.e., falling monotonically with increasing spatial frequency) to being much broader, slightly band-pass functions (i.e., peaking in sensitivity at some intermediate frequency and falling off in sensitivity at lower and higher frequencies).

There are two notable features of the data in **Figure 5**. First, at low spatial frequencies, the contrast sensitivity becomes roughly constant above about 0.09 phot. td. Thus, the contrast sensitivity becomes independent of the illumination level at higher photopic levels. Second, at higher spatial frequencies, the highest spatial frequency that can just be seen, which corresponds to the spatial acuity measurements just discussed, improves markedly, increasing from 5 cycles per degree at 0.0009 phot. td to 55 cycles per degree at 5900 phot. td. These two features reflect, in part, the decrease in the spatial extent of visual integration with increasing light level.

In general, the visual system is good at spatial-contrast detection to the extent that, under optimal conditions, the threshold contrast can be as low at 0.2% (see **Figure 5**). By contrast, the visual system is poor at estimating the amount of total luminous flux, a task which is presumably evolutionarily less important than the detection of faint edges and boundaries that reveal the presence of objects in the visual field.

Pupil Size, Higher order Aberrations, and Light Level

The size of the pupil affects retinal illuminance, depth of field, diffraction, and HOAs. In addition, it can also affect scattered light in the eye, when the light scattering is nonuniform over the pupil. The most important and

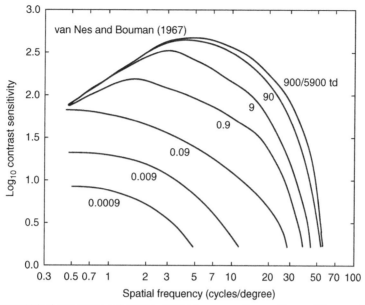

Figure 5 Spatial contrast sensitivity functions from 0.0009 to 5900 td measured by van Nes and Bouman (1967). The CSFs for 5900 td and 900 td are identical. A 2-mm diameter entrance pupil was used, so that these CSFs will be diffraction limited at highest illumination levels. Replotted from van Nes, F. L. and Bouman, M. A. (1967). Spatial modulation transfer in the human eye. *Journal of the Optical Society of America* 57: 401–406.

best-studied afferent visual signal that drives the pupil is generated by changes in the ambient illumination, and this pathway is associated entirely with subcortical projections. Other pathways are also involved, but the corresponding pupil changes are more transient and smaller in amplitude. Aging affects steady-state pupil size, but in general the pupil varies from 2 mm in bright sunlight to over 8 mm in the dark (see **Figure 1**). The HOAs of the eye, mostly spherical aberration, increase rapidly with pupil diameter. Under natural conditions, this corresponds to either mesopic or scotopic vision. **Figure 6(a)** shows how the average wavefront aberration of the eye varies with pupil diameter. At low light levels, the pupil of the eye is large and consequently the quality of the retinal image is affected by increased aberrations and scattered light. When the light level is high, HOAs become very small and diffraction becomes more important and can often limit most the quality of the retinal image that can be achieved.

The potential of improving spatial vision by reducing HOAs using customized, wavefront-guided corneal refractive surgery has been of considerable interest. Since the size of the pupil significantly affects the aberrations in the eye (**Figure 6(a)**), any experimental assessment of the benefit of HOA correction must compare performance with and without correction using the natural pupil size. The visual benefit for contrast acuity that results from HOA correction is shown in **Figure 6(b)** for ambient light levels in the range 140–0.01 cd m^{-2}. Although HOA correction undoubtedly improves retinal image quality for large pupil sizes in mesopic and scotopic vision, the visual

benefit for everyday visual performance as measured by contrast acuity is limited, largely as a result of the poor spatial resolution of mesopic and scotopic vision. The results of **Figure 6(b)** suggest that the spatial CSFs shown in **Figure 5** are well matched to the increases in the HOAs with increasing pupil size. As the light level decreases and the pupil size and HOAs increase, the retina is less able to resolve higher spatial frequencies, and thus the most deleterious effects of the increasing aberrations are invisible.

Flicker Perception

Similar to the changes in the spatial properties of the visual system that accompany light adaptation, the changes in its temporal properties can also be characterized by contrast sensitivity measurements. A temporal CSF defines the sensitivity for the detection of flicker as a function of temporal frequency. In a typical experiment, the observer is presented with a uniform disk that flickers sinusoidally at some temporal frequency, and is then asked to adjust its flicker contrast (sometimes called ripple ratio or modulation) until the flicker is just visible. This threshold measurement is subsequently repeated at a series of temporal frequencies to build up a complete CSF. **Figure 7** shows temporal CSFs measured by Kelly as a function of retinal illuminance. These functions have several characteristics in common with the spatial CSFs shown in **Figure 5**. As the illumination level is increased, the temporal CSFs, like the spatial CSFs, change from being low-pass to slightly band-pass.

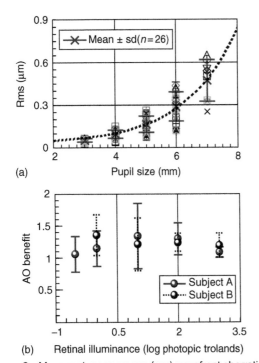

(a)

(b)

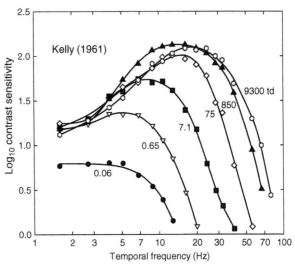

Figure 7 Temporal contrast sensitivity functions from 0.006 to 9300 td measured by Kelly (1962). Replotted from Kelly (1962). Visual responses to time-dependent stimuli I. Amplitude sensitivity measurements. *Journal of the Optical Society of America* 51: 422–429.

Figure 6 Mean root mean square (rms) wavefront aberration (a parameter that relates to the quality of the retinal image) was measured in 26 subjects and is plotted as a function of pupil diameter. The error bars show ±sd for this subject group. Section B plots the visual benefit (defined as the ratio of contrast sensitivities measured with and without correction of higher order aberrations in the eye), as a function of retinal illuminance. The visual benefit that follows correction of higher order aberrations is small when the retina is adapted to the corresponding ambient illumination. Replotted from Van Kvansakul, J., Rodriguez-Carmona, M., Edgar, D. F., et al. (2006). Supplementation with the carotenoids lutein or zeaxanthin improves human visual performance. *Ophthalmic and Physiological Optics* 26: 362–371; and Dalimier, E., Dainty, J. C., and Barbur, J. L. (2008). Effects of higher-order aberrations on contrast acuity as a function of light level. *Journal of Modern Optics* 55: 791–803.

Other features are also shared between the spatial and temporal CSFs. First, at low temporal frequencies, like low spatial frequencies, the contrast sensitivity becomes roughly constant and therefore independent of light level – in this case, at 0.65 phot. td and above. Second, as the retinal illuminance is increased, the highest temporal frequency that can just be seen (i.e., the temporal acuity limit, which is also known as the critical fusion frequency) increases from about 13 Hz at 0.06 phot. td to 80 Hz at 9300 phot. td.

The changes in the shapes of the temporal CSFs and the reduction in integration time both reflect a speeding up of the visual response with light adaptation. This speeding up can be assessed more directly by measuring the differences in visual delay between the two eyes when they are in different states of adaptation. **Figure 8(a)** shows how the cone response speeds up as the retinal illuminance in one eye is increased (relative to the cone

response in the other eye fixed at a retinal illuminance of 4.16 log phot. td). In terms of phase delay (where 360° is one cycle of flicker), the response speeds up, for example, at 10 Hz, by about 150° between 1.05 and 4.16 log phot. td. This is equal to 150/360 or 0.42 of a cycle, which, given that one 10-Hz cycle lasts 100 ms, represents a speeding up of roughly 42 ms. The visual response mediated by rods also speeds up over the scotopic range. **Figure 8(b)** shows the rod response in one eye with varying retinal illumination relative to the rod response in the other eye fixed at a retinal illuminance of 1.3 log scot. td. At 4 Hz, the response speeds up by 135° or 0.38 cycles between −3.3 and −0.8 log scot. td. This translates to a speeding up of roughly 94 ms. These results illustrate that changes in visual delay can be substantial. Visual performance, in terms of the speed of the visual response, improves markedly as the rod or the cone systems light adapt.

At mesopic levels, the situation is more complicated. Although the rod system is relatively light adapted and the cone system relatively unadapted at these levels, the rod system is still more sluggish than the cone system. These differences reflect intrinsic differences between the speeds of the responses of the rod and cone photoreceptors, as well as differences between the rod and cone postreceptoral pathways. The situation is further complicated by an abrupt transition that occurs within the rod system at mesopic levels from a slow, sensitive postreceptoral pathway to a faster, more insensitive one. Due to the complex differences between the temporal properties of rod- and cone-mediated vision, mesopic measures of luminous efficiency and visual performance will be strongly dependent not only on the relative sensitivities

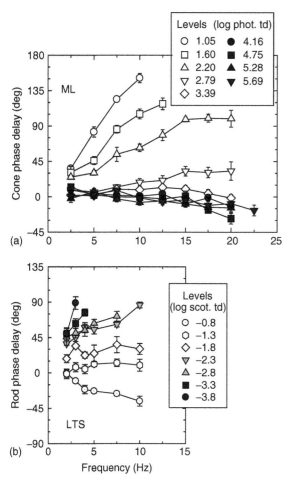

Figure 8 Binocular phase-delay measurements made between the two eyes in different states of adaptation. (a) M-cone phase delays in degrees between signals generated in the left eye and those generated in the right eye for observer ML. The adaptation level in the right eye was fixed at 4.16 log phot. td; that in the left eye was varied according to the key. Data replotted from Figure 5 of Stockman, A., Langendörfer, M., Smithson, H. E., and Sharpe, L. T. (2006). Human cone light adaptation: From behavioral measurements to molecular mechanisms. *Journal of Vision* 6: 1194–1213. (b) Rod phase delays degrees between signals generated in the left eye relative and those generated in the right eye for observer LTS. The adaptation level in the right eye was fixed at −1.30 log scot. td; that in the left eye was varied according to the key. Data replotted from Figure 9 of Stockman and Sharpe (2006). Into the twilight zone: The complexities of mesopic vision and luminous efficiency. *Ophthalmic and Physiological Optics* 26: 225–239.

of the rods and cones, but also on the temporal characteristics of the stimuli used to make those measurements. Thus, targets of different duration are likely to produce mesopic results with different rod to cone weightings.

Color Vision and Light Level

Color signals contribute significantly to the detection and appearance of objects, particularly when the photopic luminance contrast of the stimulus is low. In fact, in terms of cone-contrast signals under optimal conditions, the eye sees color better than it sees luminance. Other studies also show that even the pupil of the eye responds more vigorously to chromatic than to luminance signals when small stimuli are involved. Color can be perceived at or near the cone-detection threshold. For instance, the appearance of a spectrally green target is achromatic at scotopic levels, but takes on a clearly green tinge at mesopic levels. Color affects the conspicuity of stimuli. In general, the effective contrast of visual stimuli can be expressed as a complex function of photopic and scotopic luminance contrast with significant contributions from red/green and yellow/blue color signals in the photopic range. Importantly, chromatic signals also contribute to the perception of brightness (but not to luminance). In the mesopic range, cone signals become less effective, but the stronger rod signals do not appear to contribute significantly to color vision, particularly when threshold measurements are involved. When the stimulus is well above threshold, rod signals affect both the color appearance and the overall conspicuity of the stimulus.

Conclusions

The idea that the performance of the visual system can be usefully characterized over a 10 000-million-fold range of light levels by just three spectral luminous efficiency functions, corresponding to the scotopic, mesopic, and photopic ranges, is overly ambitious. Any model based solely on spectral sensitivity is unlikely to capture the effects of the large number of parameters that can influence object appearance. Here, we have focused on the way in which other key aspects of visual performance, such as spatial and temporal contrast sensitivity and acuity, visual delay, and color sensitivity, change with light level. We argue that only by linking such changes with changes in spectral luminous efficiency could we reasonably hope to predict visual performance, particularly at mesopic levels where the characteristics of the rod and cone systems are so different.

See also: Acuity; Chromatic Function of the Cones; Color Blindness: Acquired; Color Blindness: Inherited; Contrast Sensitivity; Information Processing: Contrast Sensitivity; Information Processing: Retinal Adaptation; Phototransduction: Adaptation in Cones; Phototransduction: Adaptation in Rods; Phototransduction: Inactivation in Cones; Phototransduction: Phototransduction in Cones; Phototransduction: Phototransduction in Rods; Phototransduction: Rhodopsin; Phototransduction: The Visual Cycle.

Further Reading

Chisholm, C. M., Evans, A. D., Harlow, J. A., and Barbur, J. L. (2003). New test to assess pilot's vision following refractive surgery. *Aviation, Space, and Environmental Medicine* 74(5): 551–559.

Dalimier, E., Dainty, J. C., and Barbur, J. L. (2008). Effects of higher-order aberrations on contrast acuity as a function of light level. *Journal of Modern Optics* 55: 791–803.

Kelly, D. H. (1961). Visual responses to time-dependent stimuli I. Amplitude sensitivity measurements. *Journal of the Optical Society of America* 51: 422–429.

Kvansakul, J., Rodriguez-Carmona, M., Edgar, D. F., et al. (2006). Supplementation with the carotenoids lutein or zeaxanthin improves human visual performance. *Ophthalmic and Physiological Optics* 26: 362–371.

Sharpe, L. T. and Stockman, A. (1999). Two rod pathways: The importance of seeing nothing. *Trends in Neurosciences* 22: 497–504.

Stockman, A., Jägle, H., Pirzer, M., and Sharpe, L. T. (2008). The dependence of luminous efficiency on chromatic adaptation. *Journal of Vision* 8(16): 1–26.

Stockman, A., Langendörfer, M., Smithson, H. E., and Sharpe, L. T. (2006). Human cone light adaptation: From behavioral measurements to molecular mechanisms. *Journal of Vision* 6: 1194–1213.

Stockman, A. and Sharpe, L. T. (2006). Into the twilight zone: The complexities of mesopic vision and luminous efficiency. *Ophthalmic and Physiological Optics* 26: 225–239.

van Nes, F. L. and Bouman, M. A. (1967). Spatial modulation transfer in the human eye. *Journal of the Optical Society of America* 57: 401–406.

Walkey, H. C. and Barbur, J. L. (guest editorial) (2006). Shedding new light on the twilight zone. *Ophthalmic and Physiological Optics* 26: 223–224.

Walkey, H. C., Barbur, J. L., Harlow, A., and Makous, W. (2001). Measurements of chromatic sensitivity in the mesopic range. *Color Research and Application* 26: 36–42.

Walkey, H. C., Barbur, J. L., Harlow, J. A., et al. (2005). Effective contrast of colored stimuli in the mesopic range: A metric for perceived contrast based on achromatic luminance contrast. *Journal of the Optical Society of America* 22: 17–28.

Photoreceptor Development: Early Steps/Fate

I Nasonkin, T Cogliati, and A Swaroop, National Institutes of Health, Bethesda, MD, USA

Published by Elsevier Ltd., 2010.

Glossary

Cell competence – In the context of retinal development, it refers to the characteristic ability of retinal progenitor cells (RPCs) to generate particular retinal cell types. It is determined by a combination of intrinsic and extrinsic factors.

Cell-fate determination – An early step (generally irreversible) in the developmental process leading to the acquisition of a differentiated cellular phenotype.

Cell-fate specification – Developmental process leading to the acquisition of a reversible cell fate, generally preceding determination.

Conditional knockout mouse – A genetically engineered mouse in which targeted deletion of gene(s) of interest can be temporally and/or spatially controlled.

Knockout mouse – A genetically engineered mouse carrying targeted deletion of one or more genes that is transmitted through the germ-line. Knockout mice in which gene function is completely ablated are referred to as *null* for that gene.

Phototaxis – Movement in response to light stimuli.

Precursor – For the purposes of this article, and in agreement with previously published material, the term precursor is used for a postmitotic cell committed to a specific cell fate but not having the differentiated cell function or phenotype. Other authors may use the term precursor interchangeably with the term progenitor.

Progenitor – For the purposes of this article and in agreement with previously presented material, a progenitor is a proliferating, undifferentiated, multipotent, yet developmentally restricted, cell that expresses a combination of genes biasing its fate toward one or multiple differentiated retinal phenotypes.

Rhabdomere – A visual pigment-containing organelle, characterized by numerous microvilli, normally found on the apical surface of photoreceptors, mostly in arthropods. Its counterpart in vertebrates is the outer segment of rods and cones.

Transgenic mouse – A mouse genetically engineered with random integration of one or more exogenous genes (transgenes) in its genome. Expression of the transgene can be cell/tissue specific and controlled over time.

Introduction

Rod and cone photoreceptors are highly specialized sensory neurons responsible for the detection of visual stimuli. They convert quanta of light into signals that are transmitted via interneurons (bipolar cells) to projection neurons (retinal ganglion cells (RGCs)) and then to the brain, where neuronal electrical stimuli are interpreted as visual sensation. In vertebrates, rod and cone photoreceptors are characterized by the expression of rhodopsin and cone opsins, respectively. Opsins belong to a family of membrane-bound guanine nucleotide binding protein-coupled receptors, which are covalently linked to a vitamin A-derived retinaldehyde chromophore and are responsible for initiating the phototransduction cascade.

The capacity to detect and respond to changes in environmental illumination is a fundamental survival skill present, to different degrees, in all living beings. Opsin-like molecules first appeared in prokaryotes (bacteria and archaea) and could function as proton-pumps to transfer energy within a unicellular organism. Activation of opsin-like molecules in response to circadian rhythms and dark–light cycles might sustain a primitive form of flagella and pili phototaxis. The capacity to perceive light stimuli further evolved in protozoa, where opsins allow for rudimental vision to sense the environment and guide adaptive movements. However, only metazoans (animals) display anatomical structures dedicated to detection and processing of light stimuli, with graded complexity from the pit eyes of worms and mollusks to the camera-type eyes of vertebrates, all of which contain light-responsive photoreceptors.

Evolution has resulted in two primary types of photoreceptor cells: rhabdomeric and ciliated. In both, the cell membrane has evolved to accommodate the maximum number of opsin molecules to increase the probability that one is activated by a single photon of light. Photoreceptors in most invertebrates are rhabdomeric with thousands of opsin-containing microvilli, whereas most vertebrates have a ciliated structure containing multiple invaginations of opsin-rich stack of membranous disks, known as the outer segment. Furthermore, as described elsewhere in this encyclopedia, other marine mollusks – such as the Pectin – contain rhabdomeric and ciliary photoreceptors in separate layers in the same retina. While phototransduction leads to membrane depolarization in rhabdomeric photoreceptors, it results in hyperpolarization in ciliated photoreceptors. Developmental processes also differ between invertebrates and vertebrates, even in those species displaying a camera-type structure. In the former (e.g., the octopus), the eye is of epidermal origin and mature photoreceptors face the vitreous cavity. In the latter, the eye develops from the neural plate (neuroectodermal origin) and mature photoreceptors are oriented away from the vitreous.

This article follows the developmental pathway that results in the generation of rods and cones from pools of retinal progenitor cells (RPCs) in the mammalian retina. We describe the factors involved in photoreceptor cell-fate specification, determination, and maturation. We have focused on human photoreceptors for their relevance in retinal degenerative diseases and on the mouse because of abundant available literature in this model pertaining to photoreceptor development.

Photoreceptor Development

Rod and Cone Pattern in Human and Mouse Retina

Photoreceptors are characterized by distinct morphology and synaptic connections that reflect their unique functions in light detection and visual process. Rod photoreceptors function in dim light, whereas cones are responsible for chromatic vision and visual acuity in bright light. These differences in light sensitivity are made possible by distinct visual pigments (opsins) and other proteins that mediate the phototransduction cascade.

All rod photoreceptors express rhodopsin from the *OPN2/RP4/RHO* gene and display peak sensitivity at a wavelength of approximately 500 nm. Cone photoreceptors can be further distinguished in different subtypes, based on the wavelength sensitivity conferred by their characteristic opsin photopigment. Humans and old world primates have three subtypes of cone photoreceptors. Cones sensitive to blue light express short-wavelength-sensitive (S)-opsin from the *OPN1SW* gene (S cones); cones sensitive to green light express medium-wavelength-sensitive (M)-opsin from the *OPN1MW2* gene (M cones); whereas those sensitive to red light express long wavelength-sensitive (L)-opsin from the *OPN1LW* gene (L cones). On the other hand, only S-opsin and M-opsin from *Opn1sw* and *Opn1mw* genes, respectively, are expressed in mouse cone photoreceptors.

In humans and mice, rods represent over 95% of the photoreceptors. The remaining 5% (human) and 3% (mouse) are cone photoreceptor subtypes. Rods and cones are arranged in a spatial mosaic in the human retina, where cones are concentrated in a region at the center of the macula – the fovea. The innermost 100-μm-wide central fovea – known as the pure cone area – contains a dense population of L and M cones and it is completely devoid of S cones, which are distributed in the peripheral fovea. Rods

appear only approximately 300 μm from the center of the fovea and their density peaks outside the fovea at the eccentricity of the optic disc. A smaller proportion of cones are found interspersed within the more numerous rods in the peripheral retina. The ratio of L-to-M cones in the pure cone area and between central and peripheral retina varies considerably among individuals, and no characteristic cone-subtype distribution pattern is evident.

In the mouse retina, rods are relatively homogeneously distributed, and there is no structure resembling the human fovea. However, S- and M-opsins are expressed in cones in a reciprocal dorsal-to-ventral gradient with S-opsin-expressing cones localized predominantly in the ventral portion of the retina and M-opsin cones predominantly in the dorsal region. Differences in photoreceptor subtypes and distribution are reflected in the developmental events that generate photoreceptors in the human and mouse retina.

Development of Cone and Rod Photoreceptors

The retinal neuroepithelium is populated by proliferating, undifferentiated, multipotent RPCs (neuroblasts), from which all neural retinal cell types – including rod and cone photoreceptors – are generated during retinogenesis. Cell birth refers to the time when a cell exits the final mitotic cycle and commits to a specific differentiated fate. Birth-dating studies in the murine retina have shown that RPCs destined to become cone photoreceptors exit the cell cycle between embryonic day (E) 11 and E18 starting from the central retina and proceeding toward the periphery (**Figure 1(a)**). In an overlapping wave, rod progenitors exit the cell cycle between E12 and postnatal day (P) 10. Initiation of opsin protein expression occurs later, with S-opsin protein first detected around E19, followed by rhodopsin around P2 and M-opsin around P7.

Immunohistochemical analysis of the developing human retina, together with data from Macaque monkeys, suggests that cone photoreceptors are generated in the human retina starting from the prospective fovea in a period of time from fetal week (Fw) 8 to Fw34 (**Figure 1(b)**). In humans, rod genesis begins after cone birth around Fw10 but overlaps with cone genesis and continues into the first 8 months of postnatal life. Unlike the mouse, S-opsin expression in humans initiates around Fw12, preceding rhodopsin and L-/M-ospin expression, which are evident by Fw15. In humans, synaptic and neurotransmitter proteins are expressed in association with the initiation of synaptogenesis during the lag period between cone genesis and production of cone opsin. However, synaptogenesis in rods occurs only after the onset of rhodopsin synthesis.

In the mouse, the above-mentioned events occur first in the center of the retina and proceed toward the periphery; this is similar to humans where, generally, new developmental events start in the pure cone area. In the early postnatal human retina, the fovea is still immature with a

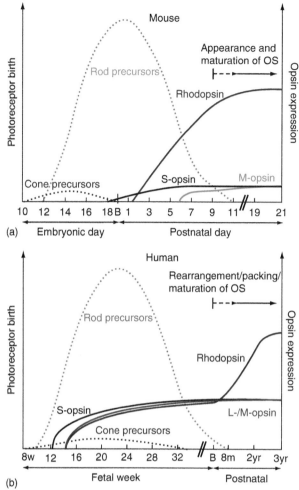

Figure 1 Photoreceptor development in mouse (a) and human (b) retina. In the first stage of photoreceptor development, prospective cone and rod precursors exit the cell cycle in sequential (cone first and rod after), yet overlapping waves (dotted lines). Postmitotic cells lose their apical process while retaining their connection to the outer limiting membrane and have elongated cuboidal morphology (cones) or a round shape (rods). The amplitude of the dotted curves in (a) and (b) represents the proportion of cells becoming postmitotic and fated to differentiate into cones (purple) and rods (orange), and the period of time in which all cells exit the cell cycle. Synthesis of specific opsins initiates the differentiation stage. Solid lines in (a) and (b) illustrate the temporal protein expression pattern of rhodopsin (brown), S-opsin (blue), M-opsin (green), and L-/M-opsin (red). In the mouse (a), S-opsin (blue) is expressed first, followed by rhodopsin (brown) and then M-opsin (green). The relative amount of rhodopsin expressed is considerably higher than S- and M-opsin, consistent with the higher proportion of rods (>95%) in the photoreceptor cell population. Photoreceptor maturation is completed between the second and the third postnatal week (arrow). In the human (b), S-opsin is synthesized first, followed shortly after by overlapping rhodopsin and L-/M-opsins. Rhodopsin production increases dramatically after birth, coincidental with outer segment formation and maturation. Cone and rod photoreceptors are rearranged after birth, with packing of cones in the fovea (dashed arrow). Outer segments continue to expand until they reach their mature size by the third year of age (solid arrow). B, birth; OS, outer segment.

greater proportion of rods than cones at the edge of the pure cone area. The mature fovea becomes evident by 1 year of age with the characteristic packaging of cones in the center. Rod outer segments continue to grow in length well into the first postnatal years until the human retina reaches full maturation by approximately 3–5 years of age.

Factors Affecting Photoreceptor Genesis

In the stereotypical progression of cell differentiation in the retina, cone genesis is initiated from common proliferating RPCs, in a sequential manner after RGCs and before horizontal and amacrine neurons. Rod birth is initiated later compared to cone genesis, but it precedes bipolar and Müller glia cells. However, the time intervals during which full complements of cone and rod photoreceptors are generated largely overlap (see **Figure 1**). Lineage and birth-dating analyses of mouse retina indicate that photoreceptor cell-fate decisions are made at the time of terminal mitosis; however, additional investigations are necessary. Two key elements control the commitment and differentiation of photoreceptors: gene-regulatory networks that confer competence to the RPC and dictate differentiation events (**Figures 2** and **3**), and extrinsic factors that modulate transcriptional cascades in differentiating RPCs and later modulate cell–cell communication (**Figure 2**). In addition, epigenetic mechanisms (such as chromatin modifications) appear to play a significant role in directing photoreceptor specification and differentiation.

Early Stages in Photoreceptor Development

From RPC to Photoreceptor Precursor

At the time of their last mitosis and prior to exiting the cell cycle, RPCs fated to become photoreceptors upregulate the expression of paired-class homeodomain transcription factor orthodenticle protein homolog 2 (OTX2) (**Figure 2**). In the mouse, OTX2 is first detected at E11.5–12.5 in the retinal neuroblastic layer (**Figure 3**). OTX2 expression increases thereafter, concomitantly with early cone and rod development, and persists in the postnatal mouse retina in bipolar and photoreceptor cells. *Otx2* plays an important role as an early factor that specifies photoreceptor cell fate, and probably as a late factor that may promote terminal differentiation by participating in upregulation of photoreceptor-specific genes. When *Otx2* is conditionally knocked out in the developing mouse eye, photoreceptors do not develop and are replaced by amacrine-like cells. Furthermore, when *Otx2* is ubiquitously expressed in RPCs, all cells follow the photoreceptor fate. In humans, *OTX2*

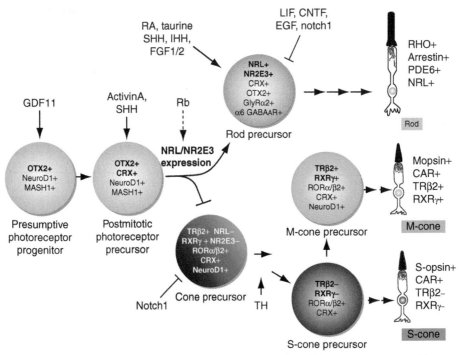

Figure 2 A model of photoreceptor development in the murine retina. Rods, S and M cones are generated from a common pool of retinal progenitor cells (RPCs), which exit the cell cycle at specific times during retinogenesis. A synergistic interaction between intrinsic and extrinsic factors progressively restricts cell fate, biasing individual cells toward differentiation into a unique mature cell type. Some of the known molecules involved are illustrated here. Proteins that appear to be major contributors to specification/determination of cell fate are shown in bold. Expression of OTX2 and CRX defines the postmitotic pool of cells fated to become photoreceptors (photoreceptor precursors). Expression of NRL and its target NR2E3 determines rod photoreceptor cell fate, whereas their absence leads to cone fate. NRL/NR2E3-expressing rod precursors progress through multiple transition stages and become functional rod photoreceptors when rhodopsin and phototransduction protein synthesis occurs and outer segments are mature. Photoreceptor precursors that do not express NRL/NR2E3 progress toward the cone lineage. Upon downregulation of TRβ2 and RXRγ, cone precursors become S cone precursors and synthesize S-opsin. For M cone precursors to develop, S-opsin expression must then be repressed by the heterodimer TRβ2/RXRγ, and M-opsin synthesis initiated by another TRβ2-containing complex. S- and M-opsin expression followed by outer segment maturation complete cone differentiation. The current model supports the existence of a default S-cone pathway that requires active inhibition by NRL/NR2E3 and by TRβ2/RXRγ to allow differentiation of rods and M cones, respectively. The roles of other cell intrinsic and extrinsic factors are further described in the text. Arrows indicate active promotion and truncated lines indicate inhibition of a developmental stage. Dotted lines indicate tentative roles. See the text for explanation of abbreviations.

mutations are associated with retinal diseases, such as anophthalmia, microphthalmia, and coloboma.

At the molecular level, OTX2 is shown to activate the promoter of cone–rod homeobox (CRX) transcription factor, a closely related homeodomain transcription factor. The two transcription factors (OTX2 and CRX) appear to promote completion of photoreceptor differentiation programs (**Figure 2**). CRX does not contribute to photoreceptor cell specification, but is essential for terminal differentiation and maintenance. In *Crx-null* mice, despite normal photoreceptor genesis, morphogenesis is incomplete as outer segments fail to elongate, and photoreceptors do not produce a full complement of phototransduction proteins. These abnormal photoreceptors undergo synaptogenesis, but their synaptic endings appear malformed and eventually degenerate. Mutations in the human *CRX* result in retinopathies, including Leber

congenital amaurosis (LCA), cone–rod dystrophy, and retinitis pigmentosa (RP).

Another protein suggested to participate in photoreceptor cell development is the basic helix-loop-helix (bHLH) transcription factor NeuroD1 (**Figure 2**). *NeuroD1* is expressed in developing and differentiated rod and cone photoreceptors. Its targeted deletion in the mouse, however, causes only a modest decrease in photoreceptor number. NeuroD1 acts in combination with another bHLH transcription factor, MASH1, which may contribute to regulating the timing of photoreceptor differentiation as its targeted deletion in mouse leads to delay in photoreceptor development.

Very early, at the time of photoreceptor specification, postmitotic precursor cells become committed toward rod or cone fates. The Maf family basic motif-leucine zipper transcription factor NRL and its target, the

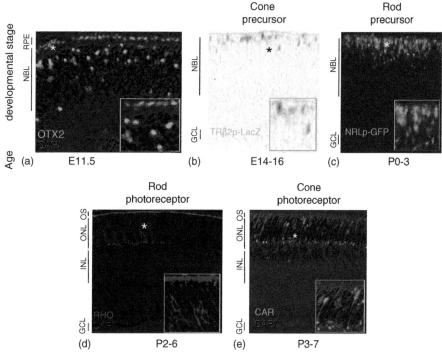

Figure 3 Stages of photoreceptor development in murine retina. Developmental stages are characterized by unique markers. Postmitotic photoreceptor precursors (a) can be detected by immunohistochemistry (IHC) with anti-OTX2 antibodies (green). Intermediate stages leading to cone or rod precursors are less delineated. However, cone precursors are characterized by the expression of TRβ2 and RXRγ, while rod precursors can be distinguished for NRL and NR2E3 expression. When the *lacZ* reporter gene driven by the *Trβ2* promoter is expressed in transgenic mice (b), cone precursors, normally expressing TRβ2, express LacZ (cyan). Similarly, rod precursors in a transgenic mouse expressing green fluorescent protein (GFP) under control of *Nrl* promoter (c) become fluorescent (green). Mature photoreceptors are identified by expression, at the RNA level first and protein level after, of their characteristic opsins and phototransduction proteins. Rod (d) and cone (e) photoreceptors are visualized by IHC with antirhodopsin (red) and anticone arrestin (green) antibodies, respectively. Areas labeled with an asterisk in (a)–(e) are enlarged in the inlets. Images are provided by Dustin Hambright ((a), (d), and (e)), Li Jia and Douglas Forrest (b), and Jerome Roger (c). RPE, retinal pigment epithelium; NBL, neuroblastic layer; GCL, ganglion cell layer; OS, outer segments; ONL, outer nuclear layer; INL, inner nuclear layer; E, embryonic day; P, postnatal day; DAPI, 4'-6-diamidino-2-phenylindole (nuclear staining); CAR, cone arrestin.

photoreceptor-specific orphan nuclear receptor NR2E3, are the primary determinants of rod versus cone cell-fate determination. It appears that S cone cell fate is specified as a default pathway and that inhibition of this pathway by NRL/NR2E3 is permissive to rod specification (**Figure 2**).

From Cone Precursor to Cone Photoreceptor

Cones are the first photoreceptor cell type to exit the cell cycle in all vertebrates (**Figure 1**). However, their differentiation takes a few days (mouse) to weeks (human) and is complete only after a majority of rods have become postmitotic. Cone precursors must go through an additional specification process that determines distinct cone subtypes (L, M, or S in humans, and M or S in mouse).

Studies of cone-photoreceptor-fate determination rely 8w?>on the detection of specific opsins at the mRNA or protein level, thus overlapping with studies on opsin gene regulation. In mouse and human cone precursors, S-opsin expression is detected first, followed by M-opsin expression (**Figure 1**). Variable numbers of photoreceptors go through a transitional state in which both opsins are expressed, before S-opsin is downregulated and M-opsin predominates. In the adult mouse retina, a prevalent population of photoreceptors coexpresses both pigments, whereas cone photoreceptors in the human retina express a single opsin.

In mice, onset of cone opsin expression is regulated by numerous factors; these include CRX, members of the nuclear receptor family, thyroid hormone nuclear receptor beta 2 isoform (TRβ2) and retinoid X nuclear receptor gamma (RXRγ), and of the retinoic acid (RA) receptor-related orphan receptor – RORα and RORβ2. RORα and RORβ2 are expressed in postmitotic cone-photoreceptor precursors and directly regulate S-opsin expression synergistically with CRX. Furthermore, RORα also appears to regulate the expression of M-opsin through direct binding to its promoter, highlighting a potentially more complex role in cone differentiation. Notably, *Rorβ2-null* mice lack outer segments, suggesting RORβ2 plays an additional role in promoting photoreceptor maturation.

TRβ2 and RXRγ are also expressed in postmitotic cone precursors in the neonatal mouse retina (**Figures 2** and **3**) and act as transcriptional repressors of S-opsin expression. In fact, both receptors are downregulated concomitant with the initial onset of S-opsin expression and upregulated at a later stage in M cones in the dorsal retina to suppress S-opsin. Further evidence comes from studies of *Rxrγ-null* and *Trβ2-null* mice. Both nuclear receptors are necessary to repress S-opsin expression and probably act in concert on the S-opsin promoter. However, only TRβ2 is required for M-opsin expression, making it a key regulator of M cone differentiation. *Trβ2* gene appears to be regulated by a transcriptional complex containing NeuroD1. Although NeuroD1 alone is not sufficient to initiate *Trβ2* expression, it is required for sustained gene activation, thus supporting its role in M cone differentiation.

To date, there is no evidence involving RA in cone development in the mouse or human. On the other hand, TH has been suggested to regulate the ratio and patterning of cone-photoreceptor subtypes through changes in geographical distribution in the developing retina.

From Rod Precursor to Rod Photoreceptor

NRL is the master regulator of rod cell fate and serves as the unique defining signature of newborn rod photoreceptors (see **Figures 2** and **3**). NRL induces the expression of NR2E3. Following this, NRL and NR2E3 – together with CRX – lead to the expression pattern typical of mature rod photoreceptors (**Figure 2**). In mice *null* for the *Nrl* gene, all rods are converted to cones. Targeted deletion of *Nr2e3* in the mouse leads to enhanced S-cones with hybrid photoreceptors. Transgenic expression of *Nr2e3* in *Nrl-null* mice suppresses cone differentiation and leads to the generation of rod-like photoreceptors, yet NR2E3 is not sufficient to produce functional rods. These data support a model in which active induction by NRL and NR2E3 in CRX-expressing photoreceptor precursors is required for rod cell-fate determination with simultaneous inhibition of the cone pathway (**Figure 2**).

Mutations in NRL are associated with retinal degenerative diseases, including autosomal dominant and recessive RP, and retinopathies with varying phenotypes. Loss of NR2E3 in humans leads to enhanced S-cone syndrome, Goldmann–Favre syndrome, and similar retinopathies with increased S-cone function.

The retinoblastoma (*Rb*) gene appears to be an important intrinsic regulator of rod development. In the postnatal mouse retina, *Rb* is expressed in mitotic RPCs, where it regulates timely exit from the cell cycle, and in differentiating rod photoreceptors. When *Rb* is deleted, RPCs continue to proliferate and rod photoreceptors do not develop. Unlike *Nrl*- and *Nr2e3-null* mice, rod photoreceptors in *Rb-null* mice do not change their fate to cones (cone number remains unmodified). Rather, their development is arrested at the progenitor stage. It remains unclear from the current literature whether Rb is instructive or permissive for rod-photoreceptor cell fate.

Several extrinsic factors contribute to signaling for rod development (**Figure 2**). Cell–cell interaction mediated by Notch1 is known to sustain the undifferentiated and proliferating state of RPCs, repressing neuronal fate in general. Recently, Notch1 signaling has been shown to specifically inhibit photoreceptor fate, that is, cone in the embryonic and rod in the postnatal mouse retina. This function of Notch1 may allow differentiation of other neuronal cell types as light detection evolved from uniquely photoreceptor-based, in lower species, to multiple cell-type-mediated, in higher species. Other extrinsic factors inhibiting rod fate are leukocyte inhibitory factor (LIF), ciliary neurotrophic factor (CNTF), and epidermal growth factor (EGF).

Taurine, secreted by RGCs, is suggested to stimulate rod photoreceptor production through glycine (Gly) and gamma aminobutyric acid (GABA(A)) receptors. The hypothesis that taurine and its receptor GlyRα2 have a role in promoting rod cell fate awaits further validation.

Proper maturation and/or migration and integration of young rods into the outer nuclear layer (ONL) may require gradients of Indian Hedgehog (IHH) secreted by the retinal pigment epithelium (RPE), and Sonic Hedgehog (SHH) secreted by RGCs. IHH may also have a role in the specification of photoreceptor fate, possibly inducing *Nrl* in photoreceptor precursors. Similarly, fibroblast growth factor (FGF) family members – acidic FGF (FGF1) and basic FGF (FGF2) – are implicated in rod maturation and may induce NRL expression.

The role of RA in promoting rod photoreceptor differentiation is still unclear, though it can induce *Nrl* expression and RA-responsive sites are present in the *Nrl* promoter. However, retinoic acid receptor (*RAR*)β2/*RAR*γ2 double *null* mutant mice contain a normal photoreceptor complement.

Activin A – a transforming growth factor beta (TGFβ)-like protein expressed by extraocular mesenchyme and RPE – promotes photoreceptor development *in vitro* and *in vivo*, and increases the number of photoreceptor cells in rat retinal cultures. *In vitro*, it causes RPCs to exit the cell cycle and biases them to become rods, but not cones. *In vivo*, mice with homozygous deletion of *activinA* show substantial decrease in the number of photoreceptors. Finally, growth differentiation factor 11 (GDF11) and related TGFβ family members control the competence of mitotic progenitors to acquire rod cell fate. *GDF11-null* retinas have more RGCs at the expense of photoreceptors.

Maturation of Photoreceptors

Cones are generated in mice during the prenatal period. Rod photoreceptor birth overlaps with the genesis of all retinal cell types, though most rods are born postnatally (**Figure 1**). Maturation of committed precursors to differentiated functional photoreceptors is a lengthy process and involves expression of cell-type-specific phototransduction genes, biogenesis of outer segments, and formation of synapses with specific interneurons. Expression of most photoreceptor-enriched genes depends on the synergistic or antagonistic actions of NRL, NR2E3, and CRX and their interaction with other regulatory proteins. In most instances, these proteins co-occupy the promoter/enhancer regions of their target genes. Mutations in the target genes of NRL and CRX are associated with retinal dysfunction.

Studies with transgenic and knockout mice, together with microarray and chromatin immunoprecipitation analysis, have yielded valuable information about gene-regulatory networks that guide photoreceptor differentiation and maturation. As NRL and its direct target nuclear receptor NR2E3 determine rod cell fate, these two transcription factors activate the expression of rod-specific genes and repress cone-gene expression. NRL, even in the absence of NR2E3, can activate all rod-specific genes (e.g., rhodopsin, PDE-α, PDE-β) but the repression of cone genes (e.g., S-opsin) is not efficient, leading to hybrid photoreceptors in mice expressing NRL but not NR2E3 (rd7 mice). NR2E3, on the contrary, can repress cone genes but it is unable to efficiently activate rod genes in the absence of NRL. Both NRL and NR2E3 interact with a multitude of regulatory proteins to accomplish transcriptional regulation. NRL is a highly phosphorylated protein and its function is modulated by several kinases. NRL also interacts with TATA-binding protein (TBP) and, presumably, brings the basal transcriptional machinery to target gene promoters. Recent studies have shown a key role for the protein inhibitor of activated STAT3 (PIAS-3) in the sumoylation of NR2E3, adding another level of control in gene expression and consequently rod differentiation. A combined action of NRL and NR2E3 is essential to generate functional rod photoreceptors.

CRX, on the other hand, acts as an enhancer of both rod and cone genes. While photoreceptors are produced even in the absence of CRX, these cells do not elaborate outer segments because of the low expression of most, if not all, phototransduction and structural proteins. CRX is, therefore, necessary to produce functional photoreceptors. CRX also interacts with coactivator proteins that possess histone acetyltransferase (HAT) activity and recruits HATs to promoter/enhancer regions to acetylate histone H3, thereby inducing and maintaining chromatin configurations that facilitate binding of NRL, NR2E3, and RNA polymerase II. These recently described molecular mechanisms underscore the importance of as yet poorly understood epigenetic factors in determining retinal photoreceptor cell fate and maturation.

Current Research in Photoreceptor Development

Cell interactions and intrinsic cellular mechanisms modulate gene expression and regulate photoreceptor differentiation. Among these, chromatin remodeling and small regulatory RNAs have attracted attention in recent years.

Histone methylation/acetylation are key epigenetic modifications that govern chromatin dynamics. The role of chromatin-modifying activities in directing tissue-specific development is an active area of investigation. Histone acetylation is emerging as an important mechanism in regulating photoreceptor development. Acetylation of histone H3 by HATs – recruited by CRX – appears important in maintaining chromatin configurations permissive to NRL/NR2E3 transcriptional activity in developing rods. Furthermore, histone deacetylase 4 (HDAC4) activity is shown to promote the survival of newly differentiated photoreceptors. More recently, the chromatin-remodeling complex Baf60c has been identified in differentiating, but not mature, retinal cells.

A family of three DNA methyltransferases, Dnmt1, Dnmt3a, and Dnmt3b, partially cooperate to establish and maintain genomic DNA methylation patterns. The presence of high levels of Dnmt3a and of Dnmt3b in the mouse rostral neural tube, including the optic grooves and cranial neural folds at E8.5, and in the area of evaginating optic vesicles at E9.5 is suggestive of the role Dnmts play in eye development. Epigenetic chromatin-remodeling mechanisms are also active in retinal neuroblasts undergoing cell-fate commitment, and are likely to contribute to retina-restricted patterns of gene expression. For example, hypomethylation is suggested to modulate interphotoreceptor retinoid-binding protein (IRBP) gene activation during photoreceptor genesis. Little is known about the influence of methylation on retinal cell fate, yet it is plausible that DNA methylation is actively involved in establishing RPC competence.

MicroRNAs (miRs) are short (18–24 nucleotides), noncoding, RNA sequences that modulate gene expression by binding the 3' (and 5') untranslated region (UTR) of their target RNAs, thus regulating their stability and translation. They originate as longer RNA transcripts that are processed and cleaved by the subsequent activity of two RNase III endonucleases – the Drosha-DGCR8 complex and Dicer. MiRs could play a gene-regulatory role in development (including retinal) that is comparable to that of transcription factors. For example, in the *Drosophila* eye, miR-7 is activated by EGF-receptor (EGFR) signaling in cells undergoing the initial steps in photoreceptor differentiation. Furthermore, conditional *Dicer-null* mutation in the developing retina results in apparently normal retinal structure – albeit interspersed with rosettes and fated to progressive degeneration – pointing to the role of miRs in normal photoreceptor genesis and function. Further studies are warranted to fully elucidate the role of miRs in photoreceptors in general and in their development in particular.

Conclusions

When investigating cellular events and molecular mechanisms, the accuracy of models is only as good as the methods used to collect the data and the relevance to human retinal development of the animal model used. Future studies in other vertebrate species (e.g., zebrafish) and in human embryonic stem cells, with more sophisticated tools to dissect cell-fate specification/determination mechanisms, and the analysis of the vast amount of *omic* data available will permit integration of current photoreceptor development models to more closely represent the relevance to human disease. It is the authors' auspice that this article be read as the starting point to stimulate the reader's curiosity to further investigate the field of retinal developmental neurobiology for more in-depth understanding and up-to-date breakthroughs.

See also: Coordinating Division and Differentiation in Retinal Development; Embryology and Early Patterning; Histogenesis: Cell Fate: Signaling Factors; Microvillar and Ciliary Photoreceptors in Molluskan Eyes; Phototransduction: Phototransduction in Cones; Phototransduction: Phototransduction in Rods; Retinal Histogenesis; Zebra Fish–Retinal Development and Regeneration.

Further Reading

Adler, R. and Raymond, P. A. (2008). Have we achieved a unified model of photoreceptor cell fate specification in vertebrates? *Brain Research* 4: 134–150.

Chen, J., Rattner, A., and Nathans, J. (2005). The photoreceptor-specific nulear receptor Nr2e3 represses transcription of multiple cone-specific genes. *Journal of Neuroscience* 25: 118–129.

Cheng, H., Aleman, T. S., Cideciyan, A. V., et al. (2006). *In vivo* function of the orphan nuclear receptor NR2E3 in establishing photoreceptor identity during mammalian retinal development. *Human Molecular Genetics* 15: 2588–2602.

Fishman, R. S. (2008). Evolution and the eye: The Darwin bicentennial and the sesquicentennial of the origin of species. *Archives of Ophthalmology* 126: 1586–1592.

Hatakeyama, J. and Kageyama, R. (2004). Retinal cell fate determination and bHLH factors. *Seminars in Cell and Developmental Biology* 15: 83–89.

Hendrickson, A., Bumsted-O'Brien, K., Natoli, R., et al. (2008). Rod photoreceptor differentiation in fetal and infant human retina. *Experimental Eye Research* 87: 415–426.

Lamba, D., Nelson, G., Kari, M. O., and Reh, T. A. (2008). Specification, histogenesis, and photoreceptor development in the mouse retina. In: Chalupa, L. M. and Williams, R. W. (eds.) *Eye, Retina, and Visual System of the Mouse*, pp. 299–310. Cambridge, MA: MIT Press.

Livesey, F. J. and Cepko, C. L. (2001). Vertebrate neural cell-fate determination: Lessons from the retina. *Nature Reviews Neuroscience* 2: 109–118.

Mears, A. J., Kondo, M., Swain, P. K., et al. (2001). Nrl is required for rod photoreceptor development. *Nature Genetics* 29: 447–452.

Ng, L., Hurley, J. B., Dierks, B., et al. (2001). A thyroid hormone receptor that is required for the development of green cone photoreceptors. *Nature Genetics* 27: 94–98.

Nishida, A., Furukawa, A., Koike, C., et al. (2003). Otx2 homeobox gene controls retinal photoreceptor cell fate and pineal gland development. *Nature Neuroscience* 6: 1255–1263.

Oh, E. C., Khan, N., Novelli, E., et al. (2007). Transformation of cone precursors to functional rod photoreceptors by bZIP transcription factor NRL. *Proceedings of the National Academy of Sciences of the United States of America* 104: 1679–1684.

Onishi, A., Peng, G.-H., Hsu, C., et al. (2009). Pias3-dependent SUMOylation directs rod photoreceptor development. *Neuron* 61: 234–246.

Roberts, M. R., Srinivas, M., Forrest, D., et al. (2006). Making the gradient: Thyroid hormone regulates cone opsin expression in the developing mouse retina. *Proceedings of the National Academy of Sciences of the United States of America* 103: 6218–6223.

Zhang, J., Gray, J., Wu, L., et al. (2004). Rb regulates proliferation and rod photoreceptor development in the mouse retina. *Nature Genetics* 36: 351–360.

The Photoreceptor Outer Segment as a Sensory Cilium

J C Besharse and C Insinna, Medical College of Wisconsin, Milwaukee, WI, USA

Glossary

Axoneme – The cytoskeletal backbone of cilia and flagella composed of nine microtubule doublets aligned in a cylindrical array. Motile axonemes, generally referred to as 9 + 2 axonemes, contain a pair of central singlets and dynein arms.

Basal body – A centriole that becomes associated with the cell membrane for the nucleation of a cilium. Doublet microtubules of cilia in most animals are extensions of triplets of the basal body.

Centriole – A barrel-shaped or cylindrical organelle composed in most animals of nine triplet microtubules. A pair of centrioles, surrounded by an amorphous zone containing many proteins, constitutes the centrosome.

Cilia – The extensions of the cell surface that have a plasma membrane and contain a cytoskeletal core of microtubules, called an axoneme, which extends out from the basal body. They may be motile, as in the airway epithelium in humans, or nonmotile and sensory in function. Cilia and flagella have the same organization, but flagella are generally motile and are longer.

Intraflagellar transport – A microtubule-based trafficking pathway required for the formation and maintenance of cilia and flagella. Intraflagellar transport (IFT) involves the bidirectional transport of a protein adaptor complex composed of highly conserved IFT proteins. The pathway requires kinesin 2 family motors in the outward (anterograde) direction and a cytoplasmic dynein in the return (retrograde) direction. The pathway is equally important in cilia and flagella.

Kinesins – The plus-end-directed motor proteins powered by adenosine triphosphate hydrolysis to walk along microtubules. These are involved in multiple cellular functions, including transport of cargo, mitosis, and meiosis. Many different genes encode kinesins with different properties and functions.

Microtubules – The cytoskeleton tubules assembled from protofilaments composed of linear polymers of α- and β-tubulin heterodimers. These are asymmetric hollow cylinders with plus and minus ends that differ in polymerization rate and binding proteins. They form a dynamic network within the cell and organize themselves into complex structures, such centrioles, basal bodies, and cilia.

Microvilli – The membrane protrusion from the cell surface composed of cytoplasm and dense bundles of actin filaments serving to increase the surface area of the cell.

Sensory cilia – The cilia that contain membrane receptors and signaling components. These often have a nonmotile 9 + 0 axoneme, but actively motile cilia may also serve a sensory function.

Introduction

There are at least two fundamentally different designs for visual photoreceptors, generally referred to as the rhabdomeric type and the ciliary type. Rhabdomeric photoreceptors are widely represented among invertebrates and have been intensively studied in the horseshoe crab, fruit fly, and squid. Rhabdomeric photoreceptors concentrate the visual pigment for photon capture in a highly replicated array of microvilli, each containing an actin cytoskeletal core. The array of microvilli is called the rhabdomere. The visual pigment of ciliary photoreceptors is also concentrated in a photon capture organelle called the outer segment (OS). However, the OS is derived from the plasma membrane of a cilium and retains the microtubule-based cytoskeletal core, called an axoneme, which is common to all cilia. Often, rhabdomeric and ciliary photoreceptors are referred to as invertebrate and vertebrate types, respectively, but this distinction can be misleading. Both photoreceptor types appeared early in animal evolution and are present in multiple invertebrate phyla. Furthermore, some animals have both photoreceptor types, and the retina in the compound eye of the scallop contains a layered juxtaposition of both rhabdomeric and ciliary photoreceptors. Although vertebrate photoreceptors with true rhabdomeres have not been identified, the recent discovery of intrinsically light-sensitive ganglion cells (ipGCs) has led to the realization that their melanopsin photopigment and transduction pathway have features in common with rhabdomeric rather than ciliary photoreceptors. This suggests that ipGCs may have originated from an evolutionary ancient rhabdomeric precursor.

Turnover of the OS and Phototransduction Machinery

The light-sensitive OS of vertebrate photoreceptors is an elegantly complex organelle that serves as the starting point for highly sensitive nighttime vision (rods) as well as high acuity, color vision (cones) in the daytime. The OS consists of a stack of membrane disks containing a photopigment (opsin) of the guanine-nucleotide-binding protein-coupled receptor (GPCR) family, and a large array of cytoplasmic, cytoskeletal, membrane, and membrane-associated proteins that are essential for phototransduction. The phototransduction cascade begins with photon absorption by the vitamin-A-derived chromophore of opsin and proceeds through activation of transducin, a guanine-nucleotide-binding protein (G-protein), which in turn activates a phosphodiesterase that reduces cytoplasmic cyclic guanosine monophosphate (cGMP) levels. The photoresponse is a hyperpolarization event that occurs when declining local cGMP levels result in closure of the cGMP-gated channel in the OS plasma membrane. The high sensitivity of the rod cells, which can respond to single-photon absorption events, and the rapid responses of cone cells in bright light, depend on the integrity of the disk stack and close juxtaposition of the protein components of the transduction cascade. Optimal function also depends on other OS proteins that support transduction such as membrane guanylyl cyclases (GC1 and GC2), which regenerate cGMP, a Na^+/Ca^{2+} exchanger which regulates Ca^{2+} levels, anaerobic glycolysis, which supplies a portion of the adenosine triphosphate (ATP), and the pentose phosphate shunt which supplies nicotinamide adenine dinucleotide phosphate (NADPH), essential for conversion of the all-*trans* retinaldehyde to retinol in the visual pigment cycle.

Two additional interdependent features of OS organization, essential for optimal visual function, are proper targeting of the phototransduction proteins to the OS and long-term maintenance through OS renewal. A striking feature of normal rods and cones is the high level to which phototransduction proteins are concentrated in the OS. Since those proteins are all synthesized at polyribosomes in the inner segment, a great deal of recent focus has been on those mechanisms that are essential for proper trafficking of OS proteins. This is a major problem throughout the life of the cell because OSs renew at a particularly high rate (see **Figure 1**). In the 1960s, Richard Young, then at the University of California at Los Angeles, thoroughly documented the fact that rod OSs are renewed through continuous new disk assembly adjacent to the inner segment. Furthermore, Young along with Dean Bok determined that OS length is maintained through a compensatory shedding of disks from the distal tip where they are phagocytized and degraded by the retinal pigment epithelium. In the 1970s, these same concepts were extended to include the cone OSs. The turnover of OSs is highly conserved, but occurs at very different rates in different species. In mice, rod OSs turnover once every 10 days.

The Photosensitive Organelle as a Sensory Cilium

With the advent of transmission electron microscopy in the 1950s, photoreceptors were an early subject of analysis, particularly by Eduardo de Robertis in Argentina and Kiyoteru Tokuyasu and Eichi Yamada in Japan. Those early studies showed that developing photoreceptors have the basic organization of cilia. During early differentiation, a centriole pair moves to the plasma membrane where one member serves as a template for assembly of a microtubule cytoskeletal structure called an axoneme (**Figure 2**); the other centriole is generally seen next to the basal body as an accessory centriole. Centrioles consist of an array of nine triplet microtubules and each triplet has an A-, B-, and C-tubule. The A-tubule is a complete microtubule comparable to cytoplasmic microtubules, but the B subtubule is a partial microtubule built on the wall of the A-tubule. Likewise, the C-tubule is built on the wall of B. The centriole that associates with the plasma membrane is called a basal body because it serves as a template for outgrowth of doublet microtubules that form the axoneme. As a consequence of basal body templating, the axoneme grows out as an array of nine doublet microtubules that are direct extensions of the A- and B-tubules of the basal body. The photoreceptor and other sensory cilia are generally said to have a 9 + 0 axoneme, in contrast to motile cilia, which have a 9 + 2 axoneme; the 2 in the latter designation refers to a central pair of single microtubules within the core of the axoneme. Although the 9 + 2 axoneme is found in a wide array of motile cilia, 9 + 0 cilia are sometimes motile. For example, the rotatory cilia of the embryonic Henson's node have a 9 + 0 axoneme.

As the axoneme elongates in rodents, the plasma membrane expands at the distal end of the cilium, and membrane vesicles and tubules accumulate (**Figure 2**). At this early stage, the photopigment apoprotein, opsin, is localized within the plasma membrane of the distal end of the cilium, and membrane vesicles and tubules accumulate. This region expands in the early stages of OS formation, quickly taking the form of an orderly stack of membrane disks (**Figure 2**). Frog OSs, as described by S. E. Nilsson in 1964, exhibit ordered disks from the very beginning of differentiation. At later stages, and, presumably during early development, new membrane disks are thought to form as evaginations of the ciliary membrane. The most proximal part of the cilium emanating from the basal body, called the connecting cilium by Eduardo de Robertis in 1956, connects the photosensitive OS with the cells synthetic machinery in the inner segment; components of the phototransduction machinery in the OS are synthesized in the inner segment and must be transported to the OS through the connecting cilium.

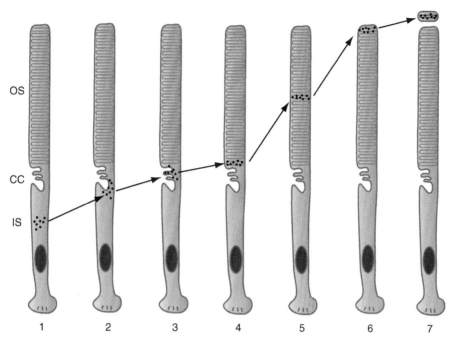

Figure 1 Conceptual diagram of a pulse chase autoradiography experiment revealing OS turnover in rod cells. Radioactive amino acids provided as a pulse at 1 (left) were incorporated into protein, mainly rhodopsin, in the inner segment (IS). Over the next few hours (2–3) radioactive protein was transported to the apical inner segment and disk-forming region at the connecting cilium (CC), and incorporated into newly formed disks creating a discrete radioactive band (4). Over the next ≈10 days (mammals), the band was gradually displaced toward the distal end of the OS (5–6) until discarded in a process called disk shedding (7). Discarded disks (7) are phagocytized by adjacent retinal pigment epithelium (not shown). From Young, R. W. (1967). The renewal of photoreceptor cell outer segments. *Journal of Cell Biology* 33: 61–72.

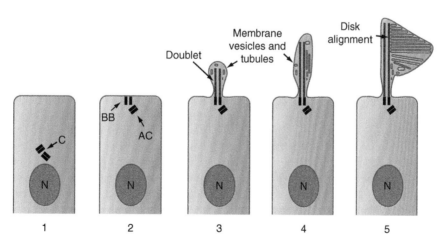

Figure 2 Early development of the outer segment. After the last mitotic division the centriole pair (1, left) moves to the cell surface (2). One member of the centriole pair associates with the plasma membrane where it is called a basal body (BB); the second centriole is often seen adjacent to the basal body in EM images where it is called the accessory centriole (AC). The basal body nucleates the extension of doublet microtubules at the cell surface to elongate the cilium (3). In the early stages of cilium elongation membrane vesicles and tubules are seen in the ciliary cytoplasm (3–4). Finally, disks align perpendicular to the axoneme composed of doublet microtubules (5). Based on early EM analysis by Tokuyasu, K. and Yamada, E. (1959). The fine structure of the retina studied with the electron microscope. IV. Morphogenesis of outer segments of retinal rods. *Journal of Biophysical and Biochemical Cytology* 6: 225–230; De Robertis, E. (1960). Some observations on the ultrastructure and morphogenesis of photoreceptors. *Journal of General Physiology* 43(6) supplement, 1–13; and Greiner, J. V., Weidman, T. A., Bodley, H. D., and Greiner, C. A. M. (1981). Ciliogenesis in photoreceptor cells of the retina. *Experimental Eye Research* 33: 433–446.

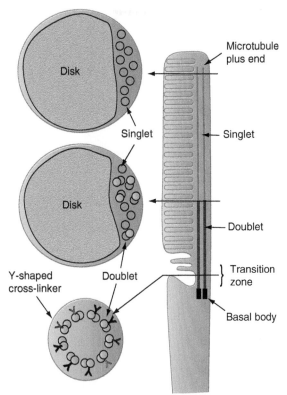

Figure 3 Structure of the ciliary axoneme of a mature photoreceptor. Longitudinal view is shown on the right and magnified cross sections at the position of the arrows on the left. The axoneme grows out of the basal body as an array of nine doublet microtubules. The microtubule plus ends extend distally and the minus ends are anchored in the basal body. The region adjacent to the basal body is a transition zone in which doublets are closely linked with the plasma membrane by Y-shaped cross-linkers (lower diagram on left); this region is often referred to as the connecting cilium. The region immediately distal to the transition zone is the site of new disc assembly. Within the OS the doublets loose their B-tubule to become singlets in the distal OS. This is shown as an abrupt transition on the right, but conversion to singlets appears to occur gradually leaving mixtures of singlets and doublets (middle diagram on left). The distal OS only has singlets.

The term connecting cilium refers to that portion of the cilium between the basal body and the disk-forming region of the OS. As originally demonstrated by Pal Röhlich in 1976, the connecting cilium is actually comparable to the transition zone, a structure common to all eukaryotic cilia and flagella (**Figure 3**). Here, the plasma membrane is closely linked to the doublet microtubules of the axoneme by cross-linkers that extend from the doublet microtubule to the plasma membrane. These cross-linkers are stable structures that remain bound to the axoneme after detergent extraction and link the axoneme to cell surface glycoconjugates through transmembrane connections. Although these microtubule-membrane cross-linkers exhibit a conserved Y-shaped structure across many cilium types, their molecular composition has not been determined. Recently, a number of cilium proteins relevant to human photoreceptor

degenerative disease, such as retinitis pigmentosa guanosine triphosphate (GTP)ase regulator (RPGR) and RPGR-interacting protein 1 (RPGRIP), have been localized to the transition zone and some may be components of the cross-linking structures. Further high-resolution analysis of the composition of cross-linkers may lead to a better understanding of the function of connecting cilium proteins that are relevant to human disease.

Distally, beyond the transition zone, the axoneme extends deep into the OS (**Figure 3**). This point requires emphasis because the term connecting cilium refers to the link between inner and OS and the term is often used with the implication that this is the entire cilium. However, both early electron microscopic studies and numerous immunocytochemistry studies have shown that photoreceptor axonemes extend through much of the length of the OS (**Figure 4**); in some cases, they extend all the way to the distal tip. The recent finding of extremely long axonemes in mouse, frog, and zebrafish OSs, along with earlier studies showing much shorter axonemes, suggests that they may vary significantly in their length. The reason for variability in axoneme length observed in various studies is not known. A possible explanation, however, is that the distal axoneme is dynamic and unstable, resulting in some cases in poor preservation for morphological studies. The principles governing these length variations in either rods or cones remain unknown, but are likely to be relevant to the finding that OSs maintain a relatively constant length through many cycles of OS turnover.

Evidence for Intraflagellar Transport in Photoreceptors

Recently our work has demonstrated that the assembly of photoreceptor OSs depends on a highly conserved microtubule-based trafficking pathway called intraflagellar transport (IFT). IFT was originally discovered in the motile flagella of the green alga *Chlamydomonas rheinhartii* and quickly extended to the sensory cilia *Caenorhabditis elegans*. The essential components of IFT are the kinesin and dynein molecular motors that drive movement along axonemal microtubules and multiprotein IFT particles that are thought to link cargo such as cytoskeletal and phototransduction proteins to the IFT motors (see **Figure 5**). For example, at least 16 different proteins assemble to form two large protein complexes referred to as IFT particles (**Figure 5**), and all of these proteins are highly conserved between *C. rheinhartii* and man. IFT transport is bidirectional. In anterograde IFT plus-end-directed motors of the kinesin 2 family move IFT particles with attached cargo toward the plus end of the axoneme, while in retrograde IFT a minus-end-directed dynein motor returns the IFT machinery for exchange with a pool in the cell body. Again, the pathway is highly conserved in that the same molecular

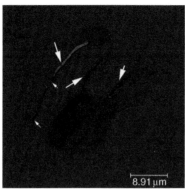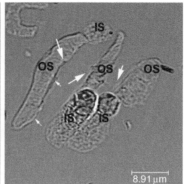

Figure 4   Immunocytochemical labeling of the axoneme of cone OSs from zebrafish. An antibody to α-tubulin was used to label (red) the microtubules of the axoneme (large arrows). An immunofluorescence image is on the left and for orientation this image is merged with a phase image of the same cells on the right. The image includes a long single cone (upper left) and a double cone (lower right). Note that the axoneme staining is attenuated distally (small arrows), particularly in the long single cone. Magnification bars in lower right equal 8.91 μm. IS, inner segment; OS, outer segment.

motors identified in *C. rheinhartii* and *C. elegans* perform similar functions in virtually all eukaryotic cilia including those in man. For example, heterotrimeric kinesin II, assembled from the kinesin family member 3A and 3B (KIF3A, KIF3B) along with kinesin-associated protein 3 (KAP3) proteins, is the canonical anterograde motor, while a dynein containing the cytoplasmic dynein 2 heavy chain (DNCH2) is the canonical retrograde motor.

A prominent role for IFT in photoreceptor OS formation is now well established. Four of the IFT proteins (IFT88, IFT57, IFT52, and IFT20) have been localized along photoreceptor axonemes and photoreceptor OS assembly defects have been fully characterized in rods of mice with a mutation in IFT88. This has been extended to cone cells with targeted, cone-specific deletion of IFT20, which results in disrupted OS assembly. A central feature of the IFT model is the multi-protein IFT particle (**Figure 5**). IFT particles containing IFT88, IFT57, IFT52, and IFT20 can be isolated from bovine photoreceptor outer segments. The photoreceptor IFT particle is large, fractionating in sucrose gradients at a peak size of ~500–750 kDa and has properties remarkably similar to those originally described in *C. rhein-hartii*. This implies that the photoreceptor particle contains additional IFT proteins for which antibodies have not yet been generated. Evidence for photoreceptor IFT is also based on analysis of its canonical motors. All three subunits of kinesin II as well as DNCH2 have been localized to photoreceptor axonemes. Furthermore, conditional deletion of the KIF3A subunit of kinesin II disrupts OS assembly, causes rhodopsin mislocalization, and results in photoreceptor cell death.

A Special Role for KIF17 in Photoreceptors

The foregoing description of photoreceptor IFT describes conditions and expectation of a canonical IFT model that has

applicability in virtually all ciliated cells. However, a novel feature of photoreceptor IFT is the critical involvement of an additional kinesin motor, the homodimeric kinesin family member 17 (KIF17). As illustrated in **Figure 5**, both kinesin II and KIF17 are associated with the IFT particle along doublet microtubules in the proximal OS, but KIF17 alone is associated with movement of the IFT particle along singlet microtubules in the distal OS. Our recent work has demonstrated that KIF17 is required to form outer segments, but does not simply replace kinesin II function; both kinesins are required. An interesting feature of this work is that while reduced kinesin II function disrupts both photoreceptor and kidney cilium elongation, knockdown of KIF17 results in failed OS assembly with no apparent effect in the kidney. The importance of KIF17 is likely related to the presence of singlet microtubule extensions in photoreceptors (see **Figure 3**). While singlet extensions are prominent in photoreceptor cilia, as originally illustrated in the older EM literature, singlet extensions in kidney sensory cilia are either very short or not present at all.

Work from the laboratory of Jonathan Scholey at the University of California at Davis has shown that the *C. elegans* KIF17 homolog, osmotic avoidance abnormal protein (OSM-3), is also required for sensory cilium elongation, specifically in cilia with singlet extensions. In *C. elegans* cilia OSM-3 serves as an accessory IFT motor along with kinesin II. Specifically, either kinesin motor can function and compensate for loss of the other motor in the proximal cilium, which contains doublet microtubules, but only OSM-3 can extend and move along the distal singlets. In fact, it was the existence of singlet extensions in photoreceptors that drew our attention to KIF17 and led to our finding of a prominent role of this accessory kinesin motor. However, the simple model for dual IFT kinesins in *C. elegans* does not fully explain findings in photoreceptors. The *C. elegans* model predicts that knockdown of KIF17 would result in short OSs that fail to elongate.

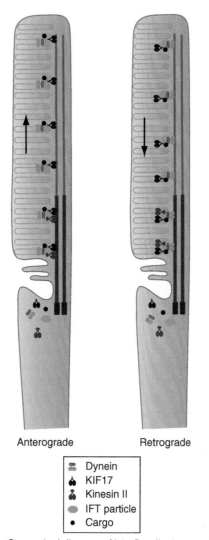

Anterograde   Retrograde

☷	Dynein
♠	KIF17
♟	Kinesin II
⬤	IFT particle
•	Cargo

Figure 5 Conceptual diagram of intraflagellar transport (IFT) in photoreceptors. (Left) Anterograde IFT uses the plus end directed kinesin motors, kinesin II and KIF17, to move IFT particles with attached 'cargo' toward the distal, plus end of the axoneme (arrow). Note that the minus-end-directed motor, dynein, is illustrated as cargo for anterograde IFT. Based largely on work in C. elegans, it is hypothesized that kinesin II and KIF17, move cooperatively on doublets in the proximal outer segment and that KIF17 operates alone on singlets in the distal outer segment. (Right) Retrograde IFT uses a cytoplasmic dynein motor to return IFT particles and cargo to the inner segment. Note that kinesin II and KIF17 are illustrated as cargo for retrograde IFT. IFT complexes along with cargo and molecular motors are thought to assemble into large complexes in the inner segment in the region adjacent to the basal body.

Results from direct comparison of mutated forms of the KIF3B subunit of kinesin II and KIF17 suggest that the motors carry out nonredundant functions and suggest that, in addition to a role for KIF17 in distal cilium extension, the motor may also be involved in transport of proteins that are essential for disk assembly in the proximal outer segment. Although knockdown of KIF17 results in ablation of OS formation, a recent mutagenesis study of a

consensus calcium, calmodulin kinase II (CaMKII) phosphorylation site in the KIF17 tail region, demonstrates that KIF17 plays a critical role in the distal OS in the control disk shedding (see **Figure 1**).

What Is the IFT Cargo?

The requirement for IFT in ciliogenesis has led to the general idea that the IFT particle serves as an adaptor to link IFT cargo to the requisite IFT motors. Some of the early evidence for cargo relates to the motors themselves. Since the axoneme plus end is at the distal end of the cilium, the minus-end-directed dynein motor required for retrograde IFT cannot move from the cell body to the distal tip on its own and available data suggest that it is cargo during anterograde IFT. Likewise, recycling of the plus-end-directed kinesins back to the inner segment appears to require retrograde IFT (see **Figure 5**). Another feature of axoneme organization is that its microtubules can assemble or disassemble only at the plus end, which is located distally. Since the axoneme can both elongate and shorten, its building blocks are likely cargo in both directions. It is important to point out that since the axoneme provides the principle cytoskeletal scaffold for the OS, IFT would be of great significance if its main purpose were to maintain axoneme structure. Recent evidence, however, suggests that IFT transports membrane components of the OS as well.

The photoreceptor OS provides a special challenge in that the components of phototransduction are present in high abundance and turn over rapidly. The question is: Which of these components is moved to the OS by IFT? One reasonably clear conclusion is that not all OS components require IFT. For example, some phototransduction proteins, such as arrestin and transducin, move between inner and OS in relationship to light and dark adaptation and a strong case can be made for free diffusion as the underlying mechanism for movement of abundant freely soluble proteins. Current evidence favors the idea that many membrane and membrane-associated proteins may require IFT. In C. rheinhartii and C. elegans, where real time imaging is feasible, movement of specific membrane proteins has been detected along the cilium. Rigorous proof that a particular protein is IFT cargo requires direct demonstration that it is linked to the IFT machinery and moves with IFT components, and there are no examples of this type of evidence outside of C. rheinhartii and C. elegans.

Nonetheless, mutation of the KIF3A subunit of the kinesin II motor, and mutation of the IFT88 subunit of the IFT particle both result in rhodopsin mislocalization, suggesting that rhodopsin is carried as IFT cargo. In support of the specificity of this effect, we recently demonstrated that expression of a dominant-negative form of the KIF3B subunit of kinesin II in zebrafish cones results in cone opsin mislocalization, but dominant-negative KIF17 blocks OS elongation without causing opsin mislocalization. Since studies of protein mislocalization in cells carrying mutations is an inherently indirect measure, we have recently used a variety of pull-down assays to isolate IFT protein complexes containing the two kinesin motors as well as rhodopsin and another OS membrane protein, retinal guanylyl cyclase 1 (RetGC1 also called GUCY2E). The work with GC1 is particularly interesting because we have identified a small chaperone protein (DNAJB6) that binds specifically to IFT88 in the IFT particle and to the kinase homology domain in the cytoplasmic tail region of GC1. We have proposed that this is the key linkage required for association of membrane GC1 with the IFT machinery and that the binding can be regulated directly through the ATPase activity of heat-shock cognate protein 70 (HSC70). This represents the first specific model for the molecular linkage of a cargo protein with the IFT machinery.

Summary and Perspective

The fact that photoreceptor OSs are sensory cilia has been known for many years, but recent advances in understanding the mechanisms underlying ciliogenesis, including the IFT pathway, are likely to have a large impact on our understanding of the cell biology of photoreceptors. Although mutations in genes involved in ciliogenesis often cause embryonic lethal phenotypes and will not frequently appear among the causes of photoreceptor-specific degenerative disease, our understanding of disease mechanisms are likely to improve through understanding which photoreceptor proteins are IFT cargo. For example, human GC1 contains three human disease causing mutations in the domain that we have recently shown to bind a linker chaperone that couples GC1 to the IFT machinery. This has led us to propose that those human mutations lead to abnormal ciliary trafficking. The importance of photoreceptor cilia also provides a basis for understanding syndromic diseases such a Bardet–Biedl syndrome, which includes RP among a complement of other abnormalities. Such diseases are now referred to as ciliopathies because cilia provide a common basis for understanding pathology in the different tissues that are affected. Finally, placement of photoreceptor OSs in their appropriate niche as a special type of sensory cilium provides evolutionary perspective on pathways common to all cilia that emerged early in eukaryote evolution.

See also: Circadian Photoreception; Genetic Dissection of Invertebrate Phototransduction; *Limulus* Eyes and Their Circadian Regulation; Light-Driven Translocation of Signaling Proteins in Vertebrate Photoreceptors; Microvillar and Ciliary Photoreceptors in Molluskan Eyes; The Photoresponse in Squid; Phototransduction: Adaptation in Rods; Phototransduction in *Limulus* Photoreceptors; Phototransduction: Inactivation in Cones; Phototransduction: Inactivation in Rods; Phototransduction: Phototransduction in Cones; Phototransduction: Phototransduction in Rods; Phototransduction: Rhodopsin; Phototransduction: The Visual Cycle; Retinal Degeneration through the Eye of the Fly; Rod and Cone Photoreceptor Cells: Outer Segment Membrane Renewal.

Further Reading

Baker, S. A., Freeman, K., Luby-Phelps, K., Pazour, G. J., and Besharse, J. C. (2003). IFT20 links kinesin II with a mammalian intraflagellar transport complex that is conserved in motile flagella and sensory cilia. *Journal of Biological Chemistry* 278: 34211–34218.

Besharse, J. C. and Horst, C. J. (1990). The photoreceptor connecting cilium. A model for the transition zone. In Bloodgood, R. A. (ed.) *Ciliary and Flagellar Membranes*, pp. 389–417. New York: Plenum.

Bhowmick, R., Li, M., Sun, J., et al. (2009). Photoreceptor IFT complexes containing chaperones, guanylyl cyclase 1, and rhodopsin. *Traffic* 10(6): 648–663.

De Robertis, E. (1960). Some observations on the ultrastructure and morphogenesis of photoreceptors. *Journal of General Physiology* 43(6): supplement, 1–13.

Eckmiller, M. S. (1996). Renewal of the ciliary axoneme in cone outer segments of the retina of *Xenopus laevis*. *Cell Tissue Research* 285: 165–169.

Hong, D. H., Yue, G. H., Adamian, M., and Li, T. S. (2001). Retinitis pigmentosa GTPase regulator (RPGR)-interacting protein is stably associated with the photoreceptor ciliary axoneme and anchors RPGR to the connecting cilium. *Journal of Biological Chemistry* 276: 12091–12099.

Inglis, P. N., Ou, G., Leroux, M. R., and Scholey, J. M. (2007). The sensory cilia of *Caenorhabditis elegans*. *WormBook* March 8: 1–22.

Insinna, C. and Besharse, J. C. (2008). Intraflagellar transport and the sensory outer segment of vertebrate photoreceptors. *Developmental Dynamics* 237: 1982–1992.

Insinna, C., Pathak, N., Perkins, B., Drummond, I., and Besharse, J. C. (2008). The homodimeric kinesin, Kif17, is essential for vertebrate photoreceptor sensory outer segment development. *Developmental Biology* 316: 160–170.

Maerker, T., van Wijk, E., Overlack, N., et al. (2008). A novel Usher protein network at the periciliary reloading point between molecular transport machineries in vertebrate photoreceptor cells. *Human Molecular Genetics* 17: 71–86.

Marszalek, J. R., Liu, X., Roberts, E. A., et al. (2000). Genetic evidence for selective transport of opsin and arrestin by kinesin-II in mammalian photoreceptors. *Cell* 102: 175–187.

Pazour, G. J., Baker, S. A., Deane, J. A., et al. (2002). The intraflagellar transport protein, IFT88, is essential for vertebrate photoreceptor assembly and maintenance. *Journal of Cell Biology* 157: 103–113.

Rohlich, P. (1975). The sensory cilium of retinal rods is analogous to the transitional zone of motile cilia. *Cell Tissue Research* 161: 421–430.

Tokuyasu, K. and Yamada, E. (1959). The fine structure of the retina studied with the electron microscope. IV. Morphogenesis of outer segments of retinal rods. *Journal of Biophysical and Biochemical Cytology* 6: 225–230.

Young, R. W. (1967). The renewal of photoreceptor cell outer segments. *Journal of Cell Biology* 33: 61–72.

Young, R. W. and Bok, D. (1969). Participation of the retinal pigment epithelium in rod outer segment renewal process. *Journal of Cell Biology* 42: 392–403.

The Photoresponse in Squid

J Mitchell and W Swardfager, University of Toronto, Toronto, ON, Canada

Glossary

Arhabdomeral lobe – The proximal segment of the photoreceptor cell containing the nucleus and other organelles.

Arrestin – A protein that binds to metarhodopsin and arrests phototransduction by inhibiting metarhodopsin activation of its target gaunine nucleotide-binding protein.

Calpain-like protease – An enzyme isolated from squid photoreceptors that, like calpain, requires millimolar concentrations of calcium for its proteolytic activity.

Metaretinochrome – The conformation of retinochrome when retinal has been photoisomerized to 11-*cis*-retinal.

Metarhodopsin – The light-activated conformation of opsin when bound to all-*trans* retinal.

Retinochrome – A photosensitive protein that binds all-*trans* retinal.

Rhabdomeral lobe – The distal segment of the photoreceptor cell containing the molecular machinery of phototransduction.

Rhabdomere – The collective surfaces of the distal segment composed of densely packed microvilli.

Rhodopsin kinase – An enzyme that adds phosphates onto serine or threonine residues in the carboxyl-terminus of metarhodopsin.

Introduction

Squid, like most invertebrates, have light-sensing organs. The visual systems of squid are composed of camera-type eyes. Incoming light passes through a single lens and an image is formed on the light-sensing cells of the retina in the anterior chamber of the eye. The retina consists of a single layer of photoreceptive neurons that are segmented in structure. The outer segment, also known as the rhabdomeral lobe, contains the protein machinery of phototransduction. The inner segment, or the arhabdomeral lobe, contains the cellular organelles involved in protein synthesis and the soma contains the cell nucleus. Axons arising from the soma of the photoreceptors comprise the optic nerve that projects to the squid brain (**Figure 1**).

The photoreceptor outer-segment membrane forms densely packed microvilli that greatly increase the surface area of the membrane available to capture incoming photons of light (see enlargement in **Figure 1**). Embedded in the membrane are the rhodopsin receptors containing light-sensitive chromophores. Light activation of rhodopsin sets off a cascade of molecular interactions that culminate in depolarization of the photoreceptor membrane (**Figure 2**). In recent years, many of the molecular components of the light-activated signaling system have been identified and a review of our current knowledge of these components is outlined here. Equally important to vision are the molecular mechanisms that inhibit signal transduction after transmission of the light signal (**Figure 3**). These mechanisms are not well understood; however, many of the components of the inactivation pathway have been revealed and these will also be discussed.

Molecular Components of Squid Visual Signal Transduction

The squid visual signal transduction system is composed of a light-sensitive receptor, rhodopsin, that transduces its activation signal via a heterotrimeric G protein (G_q) to a phospholipase C (PLC) enzyme. Activated phospholipases hydrolyze membrane phospholipids liberating soluble inositiol 1,4,5-trisphosphate (IP_3) and membrane-bound diacylglycerol (DAG). While it is still not clear how these second messengers stimulate membrane depolarization, it probably involves release of calcium from the submicrovillar tubules as a result of IP_3 stimulation of receptors on these organelles and perhaps direct stimulation of transient receptor potential (TRP)-like channels in the membrane by DAG. Together, these mechanisms raise intracellular calcium and increase membrane depolarization.

Squid Rhodopsin

Squid rhodopsin consists of a guanine nucleotide-binding protein-coupled receptor (GPCR) or opsin bound to a light-absorbing retinoid chromophore. Squid opsin genes have been cloned from *Loligo forbesi*, *Loligo pealei*, and *Todarodes pacificus*, and they share 45% sequence identity with other invertebrate and vertebrate opsins. However,

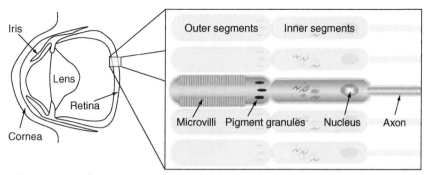

Figure 1 Structure of the squid eye and photoreceptors. Squid have camera-type eyes in which incident light enters the eye and an image is focused through a lens onto the retina in the anterior chamber of the eye. The enlargement on the right shows the orientation of the photoreceptors in the retina. The outer (rhabdomeral) segments contain the microvillar membranes in which rhodopsin and all of the photoreceptor proteins are embedded or associated. The inner (arhabdomeral) segments contain the cell organelles while the photoreceptor cell axons compose the optic nerve that extends to the brain.

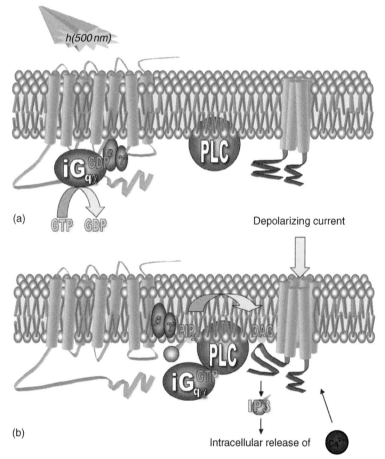

Figure 2 Activation of the squid visual system. (a) In the dark state, rhodopsin is likely coupled to an inactive (GDP-bound) invertebrate G_q (iG_q) protein. On absorption of a photon, 11-*cis*-retinal isomerizes to all-*trans*-retinal changing rhodopsin to metarhodopsin, which stimulates GDP–GTP exchange on the i$G_q\alpha$ subunit. (b) Activated i$G_q\alpha$-GTP changes conformation to interact with PLC stimulating the enzyme to hydrolyze phosphoinositol 4,5-bisphosphate (PIP$_2$) to inositol 1,4,5-trisphosphate (IP$_3$) and leaving DAG in the membrane. IP$_3$ stimulates the release of calcium (Ca^{2+}) from submicrovillar stores and, together with DAG, stimulates opening of an ion channel in the membrane. Membrane depolarization is transmitted to the optic lobe of the squid brain through the photoreceptor cell axons.

squid opsins are about 100 residues larger than those of vertebrates, primarily owing to the addition of a proline-rich carboxyl-terminal tail. Squid rhodopsins bear structural hallmarks of the GPCR superfamily, including

amino-terminal sites of N-linked glycosylation, carboxyl-terminal sites of palmitoylation, a disulfide bridge between two extracellular loops, proline residues in the α-helices of the transmembrane domains, and a (D/E)R(Y/W)

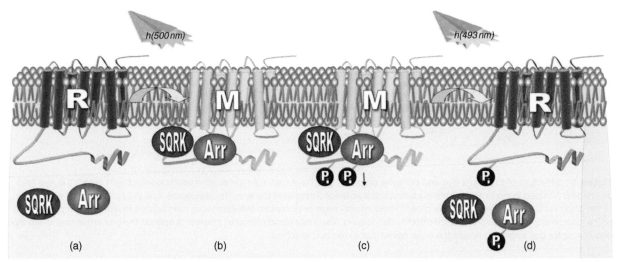

Figure 3 Inactivation of the squid visual system. (a) Rhodopsin (R) stimulation by light (*h* 500 nm) stimulates a conformational change to metarhodopsin (M). (b) SQRK and arrestin (Arr) both bind with high affinity to metarhodopsin. Arrestin binding obstructs further interaction of metarhodopsin with iG$_q$, uncoupling metarhodopsin from further stimulation of the signal transduction system. (c) SQRK phosphorylates metarhodopsin and following an increase in intracellular Ca^{2+} concentration, SQRK also phosphorylates arrestin. (d) A second light stimulus (*h* 493 nm) isomerizes all-*trans*-retinal back to 11-*cis*-retinal and converts metarhodopsin back to rhodopsin. Phosphorylation of arrestin and rhodopsin by SQRK may facilitate dissociation of the two proteins to return the system back to the dark state, which is primed to receive subsequent stimulation by light.

sequence in the third helix which is involved in G protein interactions.

The presence of a proline-rich carboxyl-terminal tail is unique to cephalopod rhodopsins. This motif consists of 9–10 repeats of the pentapeptide Pro–Pro–Gln–Gly–Tyr that may facilitate receptor trafficking and morphogenesis. Though unique among rhodopsins, tandem repeats of proline-rich sequences are found in other protein families, where they are often associated with protein–protein interactions. Rhodopsin–rhodopsin interactions may be of structural importance in the cephalopod rhabdomere, since rhodopsin networks form in the microvilli of the rhabdomeral lobe. In native membranes, electron microscopy has revealed both poorly ordered rhodopsin clusters of 4–10 molecules and ordered rhodopsin pentameres. Intermolecular interaction is mediated in part by the rhodopsin carboxyl-termini, which aggregate and extend intracellularly from the membrane surface. In contrast, when the carboxyl-terminal is cleaved, crystalline lattice formation is observed both in reconstituted membranes and in crystallized rhodopsin. These more highly ordered crystalline arrays may be favored by interactions between transmembrane domains of adjacent molecules. It has therefore been suggested that the proline-rich region may function, in part, to limit crystalline array formation in native membranes, contributing instead to the formation of less-ordered rhodopsin clusters that confer membrane superstructure. Unfortunately, technical considerations prevent the observation of the squid rhodopsin crystal structure with an intact proline-rich tail, precluding definitive structural analysis.

The majority of opsins are covalently linked to an 11-*cis*-retinal chromophore and squid rhodopsin employs the 4-hydroxy-retinal derivative. In the squid, retinal is attached via a protonated Schiff's base linkage to a lysine residue in transmembrane domain 7 (residue 303 in *Loligo* and 305 in *Todarodes*). In contrast to the mammalian opsins, however, where the Schiff's base is stabilized using a glutamic acid residue of transmembrane domain 3 as a counterion, the crystal structure of *Todarodes* reveals the counterion to be a glutamic acid at residue 180, which is located between transmembrane helices 3 and 4.

On absorption of a photon, 11-*cis*-retinal isomerizes to all-*trans*-retinal in a few hundred femtoseconds, which induces conformational changes in the opsin component. In invertebrates, the product of photoexcitation is an active metarhodopsin that is comparable in stability to inactivated rhodopsin. This contrasts the situation in vertebrates, where rhodopsin undergoes rapid sequential transitions between several unstable intermediate states. In vertebrates, all-*trans*-retinal is released during thermal relaxation of photoexcitation products and rhodopsin must be regenerated through subsequent recombination of opsin with another molecule of 11-*cis*-retinal. In the squid, however, the retinoid chromophore can remain attached to opsin throughout the rhodopsin activation cycle, and in a subsequent deactivating photoconversion event, invertebrate metarhodopsin can be converted to inactive rhodopsin by photon absorption. Thus, photobleaching may not be inevitable in the squid retina. The spectral sensitivities of squid rhodopsin and metarhodopsin differ by only a few nanometers (493 and 500 nm, respectively) suggesting that the same visual stimulus could be both activating and inactivating. Due to the comparable stabilities and interconverting wavelengths of rhodopsin and metarhodopsin, it has been suggested that substantial populations of each could exist at steady state in the squid eye. However, retinas obtained from freshly caught squid are found to contain almost exclusively rhodopsin, a finding that suggests a highly efficient inactivation pathway.

In addition to rhodopsin, cephalopod photoreceptor cells contain retinochrome, a second photosensitive retinal-binding protein implicated in rhodopsin regeneration. Squid retinochrome from *Todarodes pacificus* has been cloned, revealing a 301-amino-acid protein. The structure of retinochrome resembles that of rhodopsin, but its retinoid occupancy is reversed; whereas rhodopsin binds 11-*cis*-retinal and produces all-*trans*-retinal on photoisomerization to metarhodopsin, retinochrome binds and photoisomerises all-*trans*-retinal to 11-*cis*-retinal in its conversion to metaretinochrome. Metaretinochrome then releases 11-*cis*-retinal, providing it to rhodopsin via a shuttling protein known as retinal-binding protein (RALBP). In the dark, metaretinochrome is localized in the arhabdomeral lobe, where it releases 11-*cis*-retinal and, subsequently, binds all-*trans*-retinal released from metarhodopsin. Soluble RALBP then shuttles 11-*cis*-retinal to the rhabdomeral microvilli, where it binds retinal-free opsin, which may have arisen as a photoproduct or by synthesis *de novo*, in order to generate rhodopsin. Light-dependent translocation of both rhodopsin and retinochrome has been documented. In the dark, rhodopsin and retinochrome colocalize at the base of the microvilli, while in the light rhodopsin redistributes along the entire area of the microvillar membrane, and retinochrome becomes more plentiful in the rhabdomeral lobe. Dynamic control of the availability of rhodopsin (and other light-absorbing pigments and signaling proteins) in signaling compartments may modulate the cascade and rhodopsin regeneration. Thus, the eye can adjust to dim or bright ambient light conditions, accurately perceive objects in each, and regenerate rhodopsin when necessary.

Squid Visual Guanine Nucleotide-Binding Protein, G_q

Squid rhodopsin couples to its effector enzyme, a PLC, via a heterotrimeric G protein belonging to the G_q subfamily. Like all G proteins in this family, squid G_q is composed of three nonidentical subunits, α, β, and γ. These subunits are associated with each other in the inactive state with GDP bound to the α subunit. In this state, the G protein is tightly bound to the rhabdomeric membrane, likely in association with rhodopsin. Upon 11-*cis*-retinal isomerization by light, a conformational change in the receptor causes a conformational change in the G protein subunits, which opens up the gunanine nucleotide-binding site on the α subunit allowing GDP to be exchanged for GTP. In the GTP-bound state, $G_q\alpha$ has a lower affinity for rhodopsin and its $\beta\gamma$ partners, but gains a higher affinity for its effector, PLC. PLC is recruited to the rhabdomeric membrane to interact with activated $G_q\alpha$ and its substrate phosphotidylinositol 4,5-bisphosphate. $G_q\alpha$ binding activates the PLC enzyme and at the same time the PLC stimulates the GTPase activity of $G_q\alpha$ resulting in hydrolysis of the terminal phosphate group on the bound GTP, thus rendering the G protein once again inactive in the GDP-bound state. Reassociation of the G protein subunits completes the cycle of activation and inactivation back to the basal state. The G protein is then ready to receive the next signal from rhodopsin.

G_q protein subunits have been purified and cDNA sequences encoding the proteins have been reported from *L. forbesi* and *L. pealei*. *Loligo* $G_q\alpha$ is similar in amino acid sequence to $G_q\alpha$ proteins of other species with a conserved guanine nucleotide-binding domain and three switch regions that are the major sites of conformational change in the protein when GDP is exchanged for GTP on the protein. The sites of interaction between a G protein α subunit and its receptor are primarily in the two ends of the protein. The carboxyl-terminus of squid $G_q\alpha$ is identical to that found in G_q proteins from other species; however, the amino-terminus of the protein in all invertebrates lacks a six-amino-acid extension found in mammalian $G_q\alpha$. Studies using *L. pealei* $G_q\alpha$ expressed in mammalian cells suggest that the modified amino-terminus found in invertebrate proteins increases the efficacy of G protein activation by receptors.

The amino-terminus is also the site of posttranslational addition of palmitic acid on one or more of the two cysteine residues at position 3 and 4 of the $G_q\alpha$ subunit. This lipid modification helps maintain membrane association of the G_q protein and is particularly important for keeping $G_q\alpha$ attached to the membrane following activation by the receptor when $G_q\alpha$ dissociates from the receptor and $G_q\beta\gamma$ subunits.

The $G_q\beta\gamma$ subunits have not been examined extensively. They appear to have similar functions to other G protein $\beta\gamma$ subunits in that they associate with the $G_q\alpha$ subunit when it is bound to GDP and dissociate when $G_q\alpha$ is bound to GTP. The $\beta\gamma$ subunits are tightly associated with the retinal membranes at all times and lipid modifications of the γ subunit may contribute to this localization. *Loligo* $G\beta$ has a similar sequence to that of all other G protein β subunits, whereas *Loligo* $G\gamma$ is quite distinct from $G\gamma$ subunits of other species. $G_q\beta\gamma$ subunits do not activate purified PLC from squid eyes. Further studies will be required to determine if squid G protein $\beta\gamma$ subunits have any additional roles in visual signal transduction.

Squid Visual PLC

The protein stimulated by activated G_q in the squid visual system is PLC. The protein has been purified and the amino acid sequence determined from *L. pealei*. Immunoblot analysis of many squid tissues showed that the visual PLC is uniquely expressed in the photoreceptor membranes. Squid visual PLC is a 140 kDa protein that has significant sequence similarity and a domain structure that is common to phospholipase β enzymes; a PH domain that helps the protein bind to membranes, X and Y catalytic domains, and a C2 domain that is likely the site of calcium binding to the enzyme. In the absence of light stimulus to the photoreceptors, intracellular calcium concentrations are low and the PLC has very little catalytic activity. Upon activation of rhodopsin, the catalytic activity of the phospholipase is highly stimulated by the binding of activated $G_q\alpha$ to domains in the carboxyl-terminal end of the protein known as P and G boxes. The enzyme hydrolyzes membrane phospholipids with preference for phosphatidylinositol 4,5-bisphosphate, which is converted into inositol 1,4,5-trisphosphate (IP_3) and DAG. A network of submicrovillar tubules has been observed beneath the microvilli that may express IP_3 receptors and release stored calcium in response to IP_3 activation. A rapid rise in intracellular calcium may help to maintain PLC activity as these enzymes are stimulated in the presence of elevated calcium concentrations.

In addition to the role that calcium plays in visual signal transduction, high (millimolar) calcium concentrations can activate a calpain-like protease found in the squid retina. This protease can cleave several proteins in the visual signaling pathway. Calpain cleaves PLC near the carboxyl-terminus and renders the protein insensitive to $G_q\alpha$ activation. The protease can also cleave rhodopsin near its carboxyl-terminus removing the proline-rich repeat sequence from the rest of the molecule. Calcium activation of this protease may therefore play a role in freeing rhodopsin and G_q to allow for greater membrane turnover following prolonged light exposure.

Light-Activated Ion Channel

The final component in the visual signal transduction system is the channel regulated by PLC hydrolysis of membrane lipids. This component of the squid visual system has been characterized in only one report; an electrophysiological recording made from an isolated *L. pealei* photoreceptor. Light stimulation of the squid photoreceptor evoked an inward current with a peak amplitude greater than 1000 pA and channel activity was most frequently recorded when the patch electrode was placed near the apical tip of the cell where microvilli are present.

A squid photoreceptor channel has been identified and cloned from *L. forbesi* and it was found to be most homologous to that of the visual transient receptor potential (TRP) ion channels identified in *Drosophila*. TRP and an additional TRP-like channel have been shown to constitute the light-activated response in *Drosophila*. Analysis of the amino acid sequence of the squid channel suggests a protein with six or eight membrane spanning segments. The amino-terminus contains an ankyrin-like repeat sequence similar to that found in *Drosophila* TRP channels, and may account for the association of the protein with the cytoskeleton during purification. The carboxyl-terminus of the squid channel is considerably shorter than that found in the *Drosophila* TRP channel and lacks the proline-rich repeat sequence suggested to link TRP to intracellular Ca^{2+} stores. Expression of a peptide composed of the squid channel carboxyl-terminus bound to calcium-calmodulin *in vitro*. These studies suggest that the squid channel may be regulated differently from that of *Drosophila*, however, since the squid channel has not yet been characterized, its properties and the mechanisms by which it is regulated by the PLC signaling cascade remain speculative.

Desensitization of Visual Signal Transduction

Once rhodopsin has been activated, a series of protein–protein interactions occur within the photoreceptor cell to terminate signal transduction and restore the photoreceptor to its inactive state. Like most GPCRs, activated squid rhodopsin is phosphorylated by a G-protein-coupled receptor kinase (GRK) also called rhodopsin kinase. Light-activated rhodopsin also binds arrestin and biochemical studies using purified arrestin have demonstrated that this uncouples rhodopsin from activation of G_q.

Squid Rhodopsin Kinase

The squid visual system expresses a kinase that has sequence and functional similarity with other GRKs. The most extensively studied GRKs are the mammalian rhodopsin kinases (GRK1 and GRK7) and β-adrenergic receptor kinases (GRK2 and GRK3). Interestingly, molecular cloning of squid rhodopsin kinase (SQRK) revealed much higher sequence similarity to the mammalian β-adrenergic receptor kinase GRK2 (66%) than to GRK1 (33%), which terminates signaling in the mammalian visual system. This is a common theme among invertebrate rhodopsin kinases, as eye-specific GRK1 cloned from *Drosophila* and octopus rhodopsin kinase (ORK) also bear higher sequence identity to GRK2 than to GRK1. The structural similarities between SQRK and GRK2 include a central serine/threonine kinase catalytic domain, a structurally conserved amino-terminal domain bearing an RGS domain, a conserved carboxyl-terminal sequence and a PH domain in the carboxyl-terminal. In GRK2, it has been established that a carboxyl-terminal region partially overlapping the PH domain associates with the G protein $\beta\gamma$ subunits hence facilitating membrane localization in a stimulus-dependent manner. These structural similarities suggest that the phosphorylation of squid rhodopsin may more closely resemble the phosphorylation event of the mammalian β-adrenergic receptor than that of mammalian rhodopsin. SQRK structural motifs bearing high homology to the GRK2 Ca^{2+}/CaM-binding domain and the GRK2 clathrin box motif suggest that Ca^{2+} may have a role in regulating SQRK activity and that SQRK may also bind to the clathrin heavy chain and play a role in endocytosis. To date, functional characterization of SQRK has confirmed that purified SQRK is able to phosphorylate squid rhodopsin in rhabdomeric membranes in a light-dependent manner. SQRK phosphorylation of rhodopsin requires GTP and Mg^{2+} ion cofactors that may relate to the need to activate the G_q protein to allow SQRK access to rhodopsin.

Squid Visual Arrestin

Squid visual arrestin (sArr) has been cloned, purified, and characterized with respect to its functional interactions with rhabdomeric membranes. Squid arrestin from *L. pealei* is a 400-amino-acid protein with an estimated mass of 55 kDa that is expressed exclusively in eye tissue. Sequence identity between sArr and those from *Drosophila* and *Limulus* are 42% and 37%, respectively (sequence similarity including conservative substitutions is considerably higher, at 61% and 60%, respectively). Of the mammalian arrestins, visual arrestin from *L. pealei* shares highest identity with β-arrestins (44% identity and 64% similarity to β-arrestin1; 42% identity and 63% similarity to β-arrestin2). Squid visual arrestin is only 32% identical to mammalian visual arrestin, conservatively substituted to 49% similarity. This pattern of similarity parallels that between invertebrate rhodopsin kinase and mammalian β-adrenergic receptor kinases, and suggests that the functional interactions with invertebrate rhodopsin resemble those of the mammalian β-adrenergic receptor more closely than those of mammalian rhodopsin.

For many GPCRs, including mammalian rhodopsin, it has been demonstrated that receptor phosphorylation enhances high-affinity binding of arrestin. Accordingly, the primary structure of squid arrestin contains both conserved residues associated with both high- and low-affinity phosphate interactions. However, purified sArr does not seem to require rhodopsin phosphorylation to bind light-activated rhodopsin. This parallels biochemical studies in *Drosophila*, where purified Arr2 can bind to phosphorylated and unphosphorylated light-activated rhodopsin with comparable affinity. Further, light-dependent binding of Arr2 was equivocal in wild type and mutants where rhodopsin cannot be phosphorylated (both a truncation mutation lacking the phosphorylation site, and a serine to alanine point mutation that retains a similar structure but which cannot be phosphorylated). These findings are consistent with observations that the phosphorylation-deficient rhodopsin mutants display similar deactivation kinetics. Thus for invertebrates, the role of rhodopsin phosphorylation in arrestin binding is unclear, and light activation of rhodopsin may be sufficient.

Squid visual arrestin associates with the rhabdomeric membrane in a light-dependent manner and inhibits light-activated GTPase activity. This is consistent with the notion that recruitment and stoichiometric binding to the intracellular surface can uncouple G_q from the receptor and terminate signaling by a competitive mechanism. In the dark, squid arrestin also has appreciable affinity for the rhabdomeric membrane, an interaction which can be abrogated by inositol 1,2,3,4,5,6-hexakisphosphate (IP$_6$), a soluble analog of the membrane lipid phosphatidylinositol 3,4,5-triphosphate. This suggests that arrestin can also bind to membrane phospholipids, an interaction that studies in *Drosophila* suggest may mediate arrestin trafficking along an elaborate system of cytoplasmic structural components.

The primary structure of sArr includes five fingerprint motifs that correspond with domains of distinct functional importance conserved among the arrestin family; a region that recognizes receptor activation, a domain rich in hydrophobic interactions, a domain that recognizes receptor phosphorylation, a carboxyl-terminal regulatory domain, and a conserved amino-terminal domain. Overall, the arrestin molecule adopts a concave saddle-like conformation consisting of amino- and carboxyl-terminal domains rich in antiparallel β-sheets that hinge on a central polar core region of buried salt bridges that are disrupted when the molecule encounters the activated receptor. The primary sequence of sArr contains ^{26}Asp, ^{169}Arg, ^{293}Asn, ^{300}Asp, and ^{381}Arg, which are identical to bovine visual arrestin (except ^{293}Asn, which is a conservative substitution). These five essential conserved residues are contained within consensus sequences of five to eight identical or conservatively substituted amino acids that form the salt bridges buried in the polar core. Both the amino- and carboxyl-terminal domains mediate high-affinity binding to rhodopsin intracellular loops and these regions are located exclusively on the concave side of arrestin. Arrestin binding to rhodopsin thus occludes the binding of the G protein, preventing catalysis of GTP exchange, and uncoupling the receptor from its effector.

Interestingly, sArr can be phosphorylated by SQRK, a novel function among the GRK family. This contrasts the phosphorylation of *Drosophila* and *Limulus* visual arrestins, which are phosphorylated by Ca^{2+}/calmodulin-dependent kinase II. Phosphorylation of purified sArr requires SQRK, membranes, light-activated rhodopsin, and the presence of Ca^{2+}. Though the kinase is different, Ca^{2+} dependence is common to *Limulus* and *Drosophila*. Studies in *Drosophila* show that the kinetics of invertebrate visual arrestin phosphorylation are fast (seconds after illumination, 43% of *Drosophila* Arr2 is phosphorylated) and that arrestin phosphorylation can facilitate its release from rhodopsin once arrestin-bound metarhodopsin is photoconverted back to its inactive state. A similar role for squid arrestin phosphorylation has been found, as phosphorylated arrestin dissociates from dark-adapted membranes more readily than unphosphorylated arrestin.

Conclusion

Many of the molecular components of the squid visual system have been identified and characterized. Studies have suggested many similarities between the squid and other invertebrate visual systems. The strength of the squid system has been in the abundance of retinal tissue that makes protein purification and characterization feasible. These biochemical studies have complemented the power of genetic manipulations in the *Drosophila* system and the ease of electrophysiology in *Limulus*. Further studies are still required to determine the roles of rhodopsin and arrestin phosphorylations as well as identification of the many proteins involved in resetting the signaling components in the dark. We still know very little about how the activation of PLC results in membrane depolarization and the characteristics of the ion channels in the microvillar membranes. Future studies may eventually reveal all the molecular machinery of this fascinating visual system.

See also: Circadian Photoreception; Genetic Dissection of Invertebrate Phototransduction; Microvillar and Ciliary Photoreceptors in Molluskan Eyes; Phototransduction in *Limulus* Photoreceptors; Phototransduction: Inactivation in Cones; Phototransduction: Inactivation in Rods; Phototransduction: Phototransduction in Cones; Phototransduction: Phototransduction in Rods; Phototransduction: Rhodopsin; Evolution of Opsins; The Photoreceptor Outer Segment as a Sensory Cilium.

Further Reading

Go, L. and Mitchell, J. (2007). Receptor coupling properties of the invertebrate visual guanine nucleotide-binding protein iGqα. *Cellular Signalling* 19: 1919–1927.

Mayeenuddin, L. H., Bamsey, C., and Mitchell, J. (2001). Retinal phospholipase C from squid is a regulator of Gq alpha GTPase activity. *Journal of Neurochemistry* 78: 1350–1358.

Mayeenuddin, L. H. and Mitchell, J. (2001). cDNA cloning and characterization of a novel squid rhodopsin kinase encoding multiple modular domains. *Visual Neuroscience* 18: 907–915.

Monk, P. D., Carne, A., Liu, S.-H., et al. (1996). Isolation, cloning and characterization of a trp homologue from squid (*Loligo forbesi*) photoreceptor membranes. *Journal of Neurochemistry* 67: 2227–2235.

Murakami, M. and Kouyama, T. (2008). Crustal structure of squid rhodopsin. *Nature* 453: 363–367.

Nasi, E. and Gomez, M. (1992). Electrophysiological recordings in solitary photoreceptors from the retina of squid, *Loligo pealei*. *Visual Neuroscience* 8: 349–358.

Ryba, N. J., Findlay, J. B., and Reid, J. D. (1993). The molecular cloning of the squid (*Loligo forbesi*) visual Gq-alpha subunit and its expression in *Saccharomyces cerevisiae*. *Biochemical Journal* 292: 333–341.

Swardfager, W. and Mitchell, J. (2007). Purification of visual arrestin from squid photoreceptors and characterization of arrestin interaction with rhodopsin and rhodopsin kinase. *Journal of Neurochemistry* 101: 223–231.

Venien-Bryan, C., Davies, A., Langmack, K., et al. (1995). Effect of C-terminal proline repeats in ordered packing of squid rhodopsin and its mobility in membranes. *FEBS Letters* 359: 45–49.

Walrond, J. P. and Szuts, E. Z. (1992). Submicrovillar tubules in distal segments of squid photoreceptors detected by rapid freezing. *Journal of Neuroscience* 12: 1490–1501.

Phototransduction: Adaptation in Cones

T D Lamb, The Australian National University, Canberra, ACT, Australia

Glossary

Avoidance of saturation – The ability of cones (but not rods) to continue functioning in steady background illumination of arbitrarily high intensity.

Light adaptation – The rapid adjustment of sensitivity and kinetics (of the entire visual system, or of the photoreceptors) that occurs in response to altered ambient light intensity. The adjustment is rapid irrespective of whether the change is an increase or decrease in intensity, provided that the change is not too great.

Saturation – The failure of the photoreceptors (or of the visual system) to respond to incremental illumination in the presence of appropriately bright illumination. The rod photoreceptors saturate at a relatively low background intensity; in contrast, the cone photoreceptors avoid saturation by steady backgrounds, no matter how bright the light is.

Weber's law – The reduction in visual sensitivity that occurs in inverse proportion to the intensity of the ambient background illumination. This corresponds to a rise in visual threshold in direct proportion to the ambient illumination.

Performance of the Photopic (Cone) System

Workhorse of Vision

For human vision, the photopic cone system can be considered the workhorse of vision, because it is operational under almost all of the conditions that we (in a modern world) experience. Thus, it is our photopic system that provides our sense of vision under all lighting conditions, apart from exceptionally low levels such as starlight conditions. Under moonlight conditions, our scotopic and photopic systems are both functional, over an intensity range that is termed mesopic. If you are ever in doubt as to whether you are using your photopic system under twilight or nighttime conditions, there is a simple test: if you are able to detect any color in the scene, then your cones are active; your rods may also be active, but this is one test of whether there is any cone activity at that level of intensities. In addition, the photopic system remains functional at all higher intensities, up to the brightest sunlit conditions than we ever experience.

It is interesting to consider that, despite their enormous importance to our vision, cones make up perhaps only 5% of the population of photoreceptors over most of our peripheral retina. This relatively low proportion of cone photoreceptors in the peripheral retina is entirely adequate for our normal peripheral vision, which requires only relatively low spatial acuity. Even though the great majority of peripheral photoreceptors are rods, they are simply not used under most of the circumstances that we think of as vision – thus, they are only used at exceedingly low ambient lighting levels. The reason for the great numerical preponderance of rods is to be able to capture every available photon at those very low light intensities.

Rapid Response and Moderate Sensitivity

In survival terms, one of the greatest advantages of cones over rods is their much faster speed of response. The responses of our rods, even when they are light adapted, are much too slow to allow us to function visually at the speeds that are required to escape predators and to capture prey. Cones, instead, are specialized so as to permit extremely rapid signaling of visual stimuli to the brain.

Cones are often described as having much lower sensitivity than rods, but this view is misleading, especially when considered in terms of the rapidly changing visual stimuli that the cones are specialized for signaling. Although the peak sensitivity to a brief flash of light may be perhaps 30-fold lower in a cone than in a rod, the sensitivity to rapidly fluctuating stimuli is considerably higher in cones than in rods; thus, the slow response of the rods makes them quite insensitive to rapidly changing stimuli. When expressed in terms of the efficacy of activation within the G-protein cascade of phototransduction, the amplification in cones and rods appears to be essentially indistinguishable – the real difference is in the speed of inactivation.

Avoidance of Saturation

The second crucial feature of cones, in terms of survival advantage, is their amazing ability to avoid saturation no matter how intense the steady background illumination becomes. This property stands in stark contrast to the situation in rods, which are completely incapable of responding once the background exceeds a relatively low level (corresponding roughly to twilight illumination). One of the major challenges in photoreceptor research is to provide a clear understanding of how it is that cones are able to avoid saturation at arbitrarily high light intensities, whereas rods succumb to saturation at very low light intensities. As will be described below, considerable advances have recently been made toward providing this understanding.

Light Adaptation of the Cones

Cone photoreceptors undergo light adaptation over an enormously wide range of intensities, and it is likely that almost all of the adaptation that is observed in the overall photopic visual system is mediated by these changes occurring at the level of the receptor cells. This section describes the adaptational effects that occur in the cone photoreceptors.

Flashes on Backgrounds: Desensitization and Acceleration

Figure 1 illustrates the responses of a cone photoreceptor to the same set of flashes presented under three different conditions; the flashes *A–F* were of progressively greater intensity from left to right, but were exactly the same in each of the three panels. In (a), the flashes were presented in darkness, and represent a standard dark-adapted flash family; in (b), the same flashes were presented shortly after a dim steady background had been turned on; and in (c), the same flashes were presented after the onset of a brighter background. In the presence of the background illumination, the amplitude of the responses to dim flashes was smaller. For example, for the second flash intensity, *B*, the response amplitude becomes markedly smaller from (a) to (b) to (c). In other words, backgrounds of increasing intensity progressively desensitized the cone's incremental

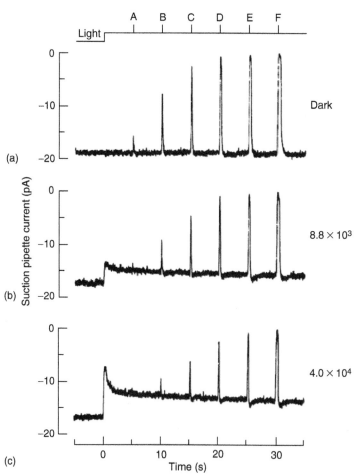

Figure 1 Circulating current of a salamander cone in response to flashes and steps of illumination. Timing of illumination is indicated by the marker trace at the top; flashes *A–F* increased in intensity by factors of ~4, and were the same intensity in each of (a)–(c). In (a), these flashes were presented in darkness; in (b) and (c) the same flashes were presented on steady backgrounds that had been switched on at time zero; the background in (c) was ~4 times brighter than in (b). Reproduced from Matthews, H. R., Fain, G. L., Murphy, R. L. W., and Lamb, T. D. (1990). Light adaptation in cones of the salamander: A role for cytoplasmic calcium concentration. *Journal of Physiology* 420: 447–469.

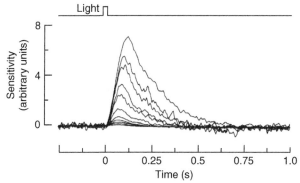

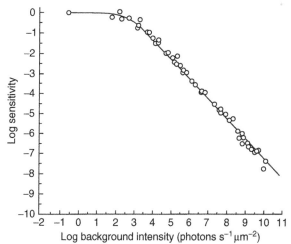

Figure 2 Incremental responses of a salamander cone to test flashes presented on backgrounds of increasing intensity. The largest trace is for a dim flash presented in darkness, while the other traces correspond to the same test flash presented on backgrounds. In fact, in the presence of brighter backgrounds, the test flash intensity was increased in order to obtain measurable responses, and the plotted traces have therefore been scaled as response divided by test flash intensity, so as to provide a direct measure of sensitivity. Reproduced with permission from Matthews, H. R., Fain, G. L., Murphy, R. L. W., and Lamb, T. D. (1990). Light adaptation in cones of the salamander: A role for cytoplasmic calcium concentration. *Journal of Physiology* 420: 447–469.

Figure 3 Cone sensitivity as a function of background intensity. Data are plotted in double logarithmic coordinates, and were obtained from intracellular measurements averaged from 15 cones in the turtle eyecup preparation. The experiments used laser illumination to achieve very high background intensities, and monitored step sensitivity rather than the more conventional flash sensitivity. The smooth curve plots Weber's law, given by eqn [1]. Reproduced from Burkhardt, D. A. (1994). Light adaptation and photopigment bleaching in cone photoreceptors *in situ* in the retina of the turtle. *Journal of Neuroscience* 14: 1091–1105.

response. Such behavior is very characteristic of photoreceptors, and these responses from cones are qualitatively similar to those obtained from rods.

The manner in which the response to a dim flash is modified by the presence of backgrounds of different intensity is illustrated in **Figure 2**. The largest trace is the response to a dim flash presented under fully dark-adapted conditions, while the other traces are for the same flash presented on backgrounds of progressively brighter intensity. (In fact, in order to maintain responses of measurable amplitude, the flash intensity was increased in the presence of backgrounds, and the traces actually plot response divided by flash intensity; i.e., the response sensitivity.)

The traces in **Figure 2** demonstrate that the effect of backgrounds of increasing intensity is to both desensitize and accelerate the response to an incremental dim flash. This behavior of cones is very similar to that exhibited by rods.

Dependence of Sensitivity on Background Intensity: Weber's Law

By plotting the peak amplitude of each of the traces in **Figure 2** as a function of the background intensity on which it was measured, one obtains a sensitivity versus background plot of the type illustrated in **Figure 3**.

The results plotted in **Figure 3** were obtained over an extremely wide range of background intensities by Dwight Burkhardt using a laser source of illumination. Importantly, the preparation was the intact eyecup (of the

turtle), so that the photoreceptors remained in contact with the retinal pigment epithelium (RPE) and thereby experienced normal regeneration of visual pigment, in order that meaningful results could be obtained even at very high background intensities. (One methodological difference between the results plotted in **Figure 3** and the results that are more usually plotted is that step sensitivities rather than flash sensitivities are plotted; however, this does not, in practice, make much of a difference.)

For background intensities from 10^3 to 10^{11} photons $\mu m^{-2} s^{-1}$, the relationship between log sensitivity and log background intensity is a straight line with a slope of -1; in other words, over roughly 8 log units of background, the turtle cone's sensitivity declines inversely with background intensity. The curve plotted near the points in **Figure 3** represents Weber's law, described by:

$$\frac{S}{S_D} = \frac{1}{1 + (I/I_0)} \qquad [1]$$

where S is flash sensitivity, S_D is its dark-adapted value, I is the background intensity, and I_0 is the half-desensitizing intensity, also known as the dark-adapted equivalent background intensity. The good fit of the Weber's law expression shows, very importantly, that cone photoreceptors in the intact eyecup are able to completely avoid saturation, even at enormously high intensities of steady illumination. This feature represents a crucial distinction between the properties of cones and rods. The circulating current of rods is shut off at quite low background intensities, so that the rods are unresponsive to superimposed stimuli, in the presence of background illumination of moderate intensity.

In the same set of experiments, Burkhardt measured the intensity at which 90% of the pigment was in the bleached state, and found this to be around 10^6 photons $\mu m^{-2} s^{-1}$. For all intensities above that level, the observed Weber's law behavior can be accounted for in terms of pigment bleaching. For each additional 10-fold increase in intensity, there will be a 10-fold reduction in the amount of pigment remaining available to absorb light, and hence there will necessarily be a 10-fold reduction in sensitivity in the absence of any other change of parameters of transduction in the outer segment. In other words, if the photoreceptor is able to avoid saturation up to intensities that cause substantial bleaches, then it will be able to exhibit Weber's law desensitization at higher intensities purely by means of pigment bleaching. Cones are able to function up to this critical intensity, whereas rods saturate at much lower intensities.

Extremely Rapid Recovery of Cone Photocurrent

In order to measure the performance of mammalian cones in the presence of extremely bright background illumination, it is necessary to use a preparation in which the cones are in contact with the RPE (as was the case in the experiments above with turtle cones). Accordingly, it is not appropriate to use suction-pipette experiments at very high intensities. On the other hand, experiments measuring the electroretinogram (ERG) in the intact eye are very suitable.

Results from an experiment designed to measure the kinetics of recovery of the circulating current of human cone photoreceptors, upon extinction of steady illumination that bleached 90% of the visual pigment, are illustrated in **Figure 4**. This Figure shows recordings of the a-wave of the human ERG, which monitors primarily the response of photoreceptors; and at these incredibly high intensities, only the cone photoreceptors are responding. The left panel shows the response to a bright flash superimposed on the intense steady background, while the right panel shows the response obtained at extinction of that background. As indicated by the right-hand pair of vertical scales, the bright flash responses established the zero level of circulating current, as well as the level of circulating current during the intense background (i.e., unity on the inner scale). Separate measurements (not shown) established the dark current (i.e., unity on the outer right-hand scale). Even during the presence of the intense steady background, the cone circulating current was roughly 50% of its original level in darkness.

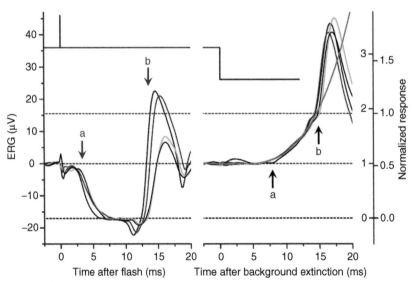

Figure 4 Extremely rapid recovery of human cone photocurrent upon extinction of intense illumination, measured with the ERG. The four colored traces plot ERG responses from two subjects, at two different flash intensities. The light stimulus is monitored by the black traces at the top. Left panel is for an intense flash presented on the intense steady background; right panel is for extinction of that background. The ON a-wave and b-wave elicited by the bright flash are indicated by the red arrows; the OFF a-wave and b-wave elicited by extinction of the background are indicated by the blue arrows. The b-wave is roughly similar in the two cases, and arises from postreceptoral activity. The OFF a-wave represents recovery of the cone circulating current, and begins around 7 ms after the intense background is turned off. Dashed horizontal lines represent the following levels of cone circulating current (from bottom): zero level, steady level during intense background, dark level, as indicated by the two normalized scales on the right. Reproduced from Kenkre, J. S., Moran, N. A., Lamb, T. D., and Mahroo, O. A. R. (2005). Extremely rapid recovery of human cone circulating current at the extinction of bleaching exposures. *Journal of Physiology* 567: 95–112.

The traces on the right show the ERG a-wave upon extinction of the intense background. Little change occurs for the first 7 ms, but thereafter a substantial upward response occurs, the OFF a-wave indicated by the blue arrow, until about 15 ms after extinction of the background; at this point, the a-wave is obscured by spike-like activity of the b-wave. There is compelling evidence that the a-wave traces for these subjects monitor the recovery of the cone circulating current. On this basis, the cone circulating current is essentially fully recovered within about 15 ms after extinction of illumination so intense that it bleaches 90% of the cone pigment. This is extremely rapid recovery, at least when compared with the time course of recovery following intense flashes delivered from darkness. The next section considers the speed that is required for the shut-off reactions of phototransduction, in order to be able to account for recovery of the circulating current as fast as is shown in **Figure 4**. The smooth gray curve near the measured traces in the right-hand panel of **Figure 4** was calculated from the model presented in the next section, using the short time constants listed in **Table 1**.

Extremely rapid recovery of cone circulating current, as inferred from the results of **Figure 4**, is also required in order to account for classical experiments on the flicker-fusion frequency of human subjects. Even at quite low photopic intensities, human subjects are able to detect square-wave flicker at a frequency of around 50 Hz using peripheral vision. However, at higher intensities, the flicker-fusion frequency increases to 100 Hz or more. At a frequency of 100 Hz, the illumination is being switched on and off at intervals of 5 ms each. Thus, in

order for the flicker to be detectable, some degree of recovery of cone circulating current must have occurred within 5 ms. Hence, this finding is broadly consistent with the time course inferred from **Figure 4**.

Molecular Basis of Cone Light Adaptation

Reaction Steps Underlying Rapid Recovery of the Cone's Light Response

Figure 5 presents a schematic of the reaction steps underlying phototransduction in cones, where shut-off reactions and the lifetimes (or turnover times) of important intermediates are indicated in red.

For mammalian cones, the speed of the various shut-off reactions shown in the schematic of **Figure 5** has been estimated in a number of recent studies using intact preparations. In the case of monkey cones, the parameters were extracted through theoretical modeling of results obtained from intracellular recordings of horizontal cells in the retina–RPE–choroid preparation. In the case of human cones, the parameters were extracted through theoretical modeling of ERG results, including those of the type illustrated in **Figure 4**. The shut-off reactions have been found to be extremely rapid, and the parameters that have been reported are summarized in **Table 1**.

The collected estimates in **Table 1** are consistent with the notion that all four of the shut-off time constants in human cones are extremely short, with values in the range of 3–18 ms; in fact, it appears that three of the time constants could be around 5 ms or less, and one around

Table 1 Shut-off time constants estimated for mammalian cones and rods

	τ_R ms	τ_E ms	τ_{Ca} ms	$\tau_{cG} = 1/\beta$ (intense) ms	Preparation	Ref.
Cone		18			Human ERG	1
	5	13		4	Human ERG	2
	3	9	3	4	Monkey retina	3
	3	10	3	6	Human ERG	4
Rod	70	200			Mouse suction pipette	5

Estimates for the four shut-off time constants (τ_R, τ_E, τ_{Ca}, and τ_{cG}) obtained from recent experiments are tabulated.
Note that it is not usually possible to determine which is which for the two time constants τ_R and τ_E; however, in the case of the rod results, this identification was made using other experiments.
The time constants for cones are around 20 times faster than for rods.
The studies from which these results were obtained are as follows:
1: Friedburg, C., Allen, C. P., Mason, P. J., and Lamb, T. D. (2004). Contribution of cone photoreceptors and post-receptoral mechanisms to the human photopic electroretinogram. *Journal of Physiology* 556: 819–834.
2: Kenkre, J. S., Moran, N. A., Lamb, T. D., and Mahroo, O. A. R. (2005). Extremely rapid recovery of human cone circulating current at the extinction of bleaching exposures. *Journal of Physiology* 567: 95–112.
3: van Hateren, J. H. (2005). A cellular and molecular model of response kinetics and adaptation in primate cones and horizontal cells. *Journal of Vision* 5, 331–347.
4: van Hateren, J. H. & Lamb, T. D. (2006). The photocurrent response of human cones is fast and monophasic. *BMC Neuroscience* 7, 34.
5: Krispel, C. M., Chen, D., Melling, N., Chen, Y.-J., Martemyanov, K. A., Quillinan, N., Arshavsky, V. Y., Wensel, T. G., Chen, C.-K. & Burns, M. E. (2008). RGS expression rate-limits recovery of rod photoresponses. *Neuron* 51, 409–416.

10–15 ms. (Note that the turnover time for cyclic guanosine monophosphate (cGMP), τ_{cG}, listed in **Table 1** represents the value applicable during very intense illumination; under dark-adapted conditions, when the light-induced activity of the phosphodiesterase (PDE) is relatively low, this time constant is likely to be much longer.)

The shut-off time constants for the activated visual pigment (R*) and the activated G-protein/PDE (E*), τ_R and τ_E, are around 20-fold shorter in human cones than in mouse rods, with values of ~80 and ~200 ms having been reported in rods.

Cone Avoidance of Saturation

Work done in collaboration with Edward N. Pugh, Jr. has shown that the ability of the cones to avoid saturation is explicable in terms of the combination of these two 20-fold shorter time constants and the bleaching of cone visual pigment.

In human rod photoreceptors *in vivo*, the circulating current is halved at a steady intensity of ~70 scotopic trolands (600 R* s^{-1}), with complete saturation occurring at ~1000 scotopic trolands (~10^4 R* s^{-1}). If the activation gain of transduction is the same in human cones as in human rods, then the two very short cone time constants would elevate the intensities required for half and full saturation by some ~400×, to levels of ~240 000 and ~4 × 10^6 R* s^{-1} in cones. An additional factor is that the cGMP-gated channels of mammalian cones (in contrast to those of rods) show increased cGMP-binding affinity when Ca^{2+} falls, thereby further increasing the R* rate required for saturation.

How do these estimated rates of isomerization compare with the maximum rate at which the cone visual pigment can be bleached during steady illumination? At steady state, the rate of photoisomerization equals the rate of pigment regeneration, which is set by the delivery of 11-*cis* retinal to opsin. For human L/M cones, the maximal rate of regeneration has been measured as ~45% min^{-1}, or 0.75% s^{-1}. If the outer segment contains ~40 million pigment molecules, then the maximal rate of photoisomerization during intense steady light will be ~300 000 R* s^{-1}. This rate cannot be exceeded in the steady state because a higher rate of isomerization would lead to such a low level of cone pigment available to absorb light that the rate could simply not be maintained.

Hence, from the numbers in the preceding paragraphs, the rate of photoisomerization required to saturate the human cone current exceeds the highest rate of isomerization of cone pigment molecules (~300 000 R* s^{-1}) that can be elicited by a steady light of arbitrarily high intensity. Therefore, the human cone photoreceptor cannot be saturated by steady lights, no matter how bright they are. This is not to say that the cone can never be saturated;

if an intense light is presented from dark-adapted conditions (when the cone initially has a full complement of visual pigment), then it will transiently be driven into saturation until bleaching reduces the amount of visual pigment to a suitably low level.

Modeling of Human Cone Light Adaptation

A computational model of human cone light adaptation has been developed by Hans van Hateren and Herman Snippe, which puts factors of the type described in the preceding section into a comprehensive theoretical/numerical model. They take a molecular description of the steps in cone phototransduction closely similar to that illustrated in **Figure 5**, including pigment bleaching, and

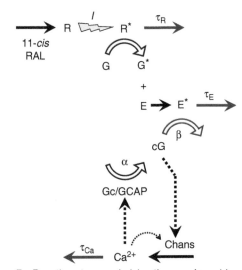

Figure 5 Reaction steps underlying the cone's rapid recovery. Visual pigment (R) is formed by delivery of 11-*cis* retinal (11-*cis* RAL). Light (of intensity *I*) activates the visual pigment, and the activated pigment (R*) is inactivated with a time constant τ_R. R* activates the guanine-nucleotide-binding G protein to G*, which then binds to phosphodiesterase (E) to form active G*–E* (E*). The active G*–E* complex (E*) has a lifetime τ_E. E* hydrolyzes cGMP (cG) with a rate constant β. The increase in β is proportional to steady intensity *I*, and to the product of the R* and E* lifetimes; that is, $\beta = \beta_{Dark} + (A/n_{cG}) \tau_R \tau_E I$ (see Nikonov, S., Lamb, T. D., and Pugh, E. N., Jr. (2000). The role of steady phosphodiesterase activity in the kinetics and sensitivity of the light-adapted salamander rod photoresponse. *Journal of General Physiology* 116: 795–824). cGMP is formed by guanylyl cyclase (GC), under regulation by the Ca^{2+}-sensitive guanylyl-cyclase-activating proteins (GCAPs). When present, cG causes opening of ion channels (chans) in the plasma membrane, admitting Ca^{2+}. Ca^{2+} has a powerful negative-feedback action through GCAPs onto the rate α of cGMP synthesis. Ca^{2+} is removed from the cytoplasm with a time constant τ_{Ca}. Reproduced from Lamb, T. D. and Pugh, E. N., Jr. (2006). Avoidance of saturation in human cones is explained by very rapid inactivation reactions and pigment bleaching. *Investigative Ophthalmology and Visual Science* 47, E-Abstract 3714, with permission from the Association for Research in Vision and Ophthalmology.

they express the system in terms of differential equations. Their simulations predict that human cones will indeed conform to Weber's law over a very wide range of background intensities, and that they will not saturate with steady intensities of any level. Thus, there is now a comprehensive description of the process of light adaptation in human cones and, in particular, of the ability of human cones to avoid saturation.

See also: Phototransduction: Adaptation in Rods; Phototransduction: Phototransduction in Cones.

Further Reading

Burkhardt, D. A. (1994). Light adaptation and photopigment bleaching in cone photoreceptors *in situ* in the retina of the turtle. *Journal of Neuroscience* 14: 1091–1105.

Friedburg, C., Allen, C. P., Mason, P. J., and Lamb, T. D. (2004). Contribution of cone photoreceptors and post-receptoral mechanisms to the human photopic electroretinogram. *Journal of Physiology* 556: 819–834.

Kenkre, J. S., Moran, N. A., Lamb, T. D., and Mahroo, O. A. R. (2005). Extremely rapid recovery of human cone circulating current at the extinction of bleaching exposures. *Journal of Physiology* 567: 95–112.

Lamb, T. D. and Pugh, E. N., Jr. (2006). Avoidance of saturation in human cones is explained by very rapid inactivation reactions and pigment bleaching. *Investigative Ophthalmology and Visual Science* 47, E-Abstract 3714.

Matthews, H. R., Fain, G. L., Murphy, R. L. W., and Lamb, T. D. (1990). Light adaptation in cones of the salamander: A role for cytoplasmic calcium concentration. *Journal of Physiology* 420: 447–469.

Nikonov, S., Lamb, T. D., and Pugh, E. N., Jr. (2000). The role of steady phosphodiesterase activity in the kinetics and sensitivity of the light-adapted salamander rod photoresponse. *Journal of General Physiology* 116: 795–824.

Pugh, E. N., Jr., Nikonov, S., and Lamb, T. D. (1999). Molecular mechanisms of vertebrate photoreceptor light adaptation. *Current Opinion in Neurobiology* 9: 410–418.

Pugh, E. N., Jr. and Lamb, T. D. (2000). Phototransduction in vertebrate rods and cones: Molecular mechanisms of amplification, recovery and light adaptation. In: Stavenga, D. G., de Grip, W. J., and Pugh, E. N., Jr. (eds.) *Handbook of Biological Physics. Molecular Mechanisms of Visual Transduction* vol. 3, ch. 5, pp. 183–255. Amsterdam: Elsevier.

van Hateren, J. H. and Lamb, T. D. (2006). The photocurrent response of human cones is fast and monophasic. *BMC Neuroscience* 7: 34.

van Hateren, J. H. and Snippe, H. P. (2007). Simulating human cones from mid-mesopic up to high-photopic luminances. *Journal of Vision* 7(4): 1.

Phototransduction: Adaptation in Rods

T D Lamb, The Australian National University, Canberra, ACT, Australia

Glossary

Dark adaptation – The very slow recovery of visual sensitivity that occurs upon the return to darkness following exposure of the eye to extremely intense (and possibly prolonged) illumination. Full recovery of the human visual system takes about 45 min, after a total bleach of all the rhodopsin.

Light adaptation – The rapid adjustment of sensitivity and kinetics (of the entire visual system or of the photoreceptors) that occurs in response to altered ambient light intensity. The adjustment is rapid irrespective of whether the change is an increase or decrease in intensity, provided that the change is not too great.

Saturation – The failure of the photoreceptors (or of the visual system) to respond to incremental illumination in the presence of appropriately bright illumination. The rod photoreceptors saturate at a relatively low background intensity; in contrast, the cone photoreceptors avoid saturation by steady backgrounds, no matter how bright the light is.

Weber's law – The reduction in visual sensitivity that occurs in inverse proportion to the intensity of the ambient background illumination. This corresponds to a rise in visual threshold in direct proportion to the ambient illumination.

Vision over a Billion-Fold Range of Light Intensities

Our visual system operates effectively over an enormously wide range of intensities, of at least a billion-fold, from around 10^{-4} cd m^{-2} under shaded starlight conditions to around 10^5 cd m^{-2} under intense sunlight. Changes in pupil area account for only about 1 log unit of this >9-log-unit range, since the pupil diameter changes from a maximum of 8 mm to a minimum of 2.5 mm, corresponding to a 10-fold reduction in area. Instead, the great bulk of the operational range is achieved by the combination of, first, a switch between the rod (scotopic) and cone (photopic) pathways in our duplex visual system and, second, the ability of each of these photoreceptor systems to operate over a range of 5 log units (100 000-fold) or more.

This ability of the visual system (or of any of its component parts, such as the photoreceptors) to adjust its performance to the ambient level of illumination is known as light adaptation; the adjustment typically occurs very rapidly (within seconds), whether the light intensity is increasing or decreasing. The term dark adaptation is reserved for the special case of recovery in darkness, following exposure of the eye to extremely bright and/or prolonged illumination that activates (bleaches) a substantial fraction of the visual pigment, rhodopsin. Dark adaptation occurs slowly, and the full recovery of the scotopic visual system after a very large bleach can take as much as an hour.

The changes that accompany light adaptation are beneficial to the possessor of the eye. At very low intensities, the sensitivity is increased to the utmost that is possible so that the rod photoreceptors reliably signal the arrival of individual photons and the scotopic visual system operates in a photon-counting mode. The ability of the scotopic system to operate at incredibly low intensities is enhanced by two deliberate trade-offs – of reduced spatial resolution (increased spatial summation) and reduced temporal resolution (increased temporal integration) – that permit more reliable detection of small signals in the presence of noise. Similar trade-offs are used in the photopic system so that as the ambient illumination decreases from daylight levels toward twilight levels, one's spatial and temporal resolution deteriorate; this is why, in cricket, bad light stops play.

In contrast, the changes that characterize dark adaptation are disadvantageous. To be essentially blind to dim stimuli, for some considerable time following intense light exposure, cannot in any way be useful to an organism. Indeed, for a caveman, entering a cave from bright sunshine, it may have been a serious handicap to have been unable to see well for tens of minutes. Why should such an apparently unsatisfactory situation have persisted? A possible reason could be because it represents an unfortunate downside that has somehow resulted from the enhancements that were needed in order to enable the scotopic system to detect individual photons, and thereby be able to function at incredibly low light levels.

Performance of the Scotopic (Rod) System

For the rod pathway, the dominant mechanisms of scotopic light adaptation result from alterations of signal processing at postsynaptic stages within the retina, and the rods themselves adapt over only a modest range of intensities before being driven into saturation. This is illustrated in **Figure 1**, which compares the changes in sensitivity of the rod photoreceptors and of the overall visual system during scotopic light adaptation.

The blue symbols and curve plot the relative sensitivity of the overall visual system, measured psychophysically, while the red symbols and curve plot the relative sensitivity of primate rod photoreceptors. Importantly, the overall scotopic visual system begins desensitizing at intensities around 1000 times lower than those required to begin desensitizing the rod photoreceptors. This occurs because the postreceptoral scotopic system is able to integrate photon signals from large numbers of rod photoreceptors, thereby gaining increased sensitivity, while introducing the need to begin desensitizing at much lower background intensities in order to avoid saturation. Hence, the rod photoreceptors maintain their maximal sensitivity over several log units of the lowest intensity regime (up to ~10 isomerizations per second) where the visual system needs to exhibit gradual desensitization.

When the background intensity is reduced from relatively high scotopic intensities (moving from right to the left along the x-axis in **Figure 1**), the sensitivity of rods, and of the scotopic visual system, steadily rises. However, below the intensities indicated by the blue and red arrows, the sensitivity of, first, the rods and, second, the visual system fails to continue increasing, as if the respective mechanism were experiencing a phenomenon equivalent to light. Accordingly, the arrowed intensities for the rods and for the scotopic visual system have been referred to as equivalent background intensities. Clearly, the equivalent background for the scotopic system (around 0.016 photoisomerizations per second) is several log units lower than the equivalent background intensity for the rods (around 50 isomerizations per second).

The curves in **Figure 1** plot desensitization according to the combination of Weber's law with saturation at high intensities, as described by

$$\frac{S}{S_{\mathrm{D}}} = \frac{1}{1 + (I/I_0)} \exp(-I/I_{\mathrm{sat}}) \qquad [1]$$

where S is flash sensitivity, S_{D} is its dark-adapted value, and I is the background intensity. The first term on the right-hand side expresses Weber's law, where I_0 is the equivalent background intensity mentioned above. This first term indicates that, at low background intensities (when $I \ll I_0$) the sensitivity S approaches a constant level (its dark-adapted value, S_{D}), while for brighter backgrounds (when $I \gg I_0$) the sensitivity declines inversely with background intensity.

At higher scotopic intensities, both the rods and the overall scotopic system exhibit saturation, characterized by a steep decline in sensitivity with increasing background intensity. This behavior is described by the second term on the right-hand side in eqn [1], where I_{sat} is termed the saturation intensity of around 2500 isomerizations per second. It is almost certain that saturation of the overall scotopic system results directly from saturation of the rods.

The span of intensities from I_0 (the equivalent background) to I_{sat} (the saturation intensity) is known as the Weber region and, in this range of background intensities, the sensitivity declines inversely with background intensity; that is, $S \propto 1/I$. Since the contrast in a visual stimulus is, likewise, inversely proportional to background intensity (i.e., contrast $= \Delta I/I$), this Weber region is characterized by a fixed level of contrast sensitivity; that is, a given level of contrast elicits a fixed size of response. Thus, an important feature of Weber's law light adaptation is that it provides automatic extraction of visual contrast.

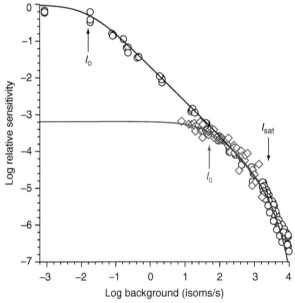

Figure 1 Sensitivity of the human scotopic visual system (blue) and of monkey rod photoreceptors (red) as functions of background intensity in double logarithmic coordinates. The blue symbols are from human psychophysical measurements and the blue curve plots Weber law desensitization in conjunction with saturation, as described by eqn [1], with parameters $I_0 = 0.016$ photoisomerizations per second (blue arrow) and $I_{\mathrm{sat}} = 2500$ photoisomerizations per second (black arrow). The red symbols are from suction pipette measurements from isolated rod photoreceptors of monkeys (*Macaca fascicularis*); these symbols have been shifted vertically to align with the blue symbols in the upper intensity range. The red curve also plots eqn [1] with the same value of I_{sat}, but with $I_0 = 50$ photoisomerizations per second (red arrow). Data for the blue symbols are from Figure 3 of Aguilar, M. and Stiles, W. S. (1954). Saturation of the rod mechanism of the retina at high levels of stimulation. *Optica Acta* 1: 59–65. Their troland values were converted using a factor of $K = 8.6$ photoisomerizations per second per troland. Data for the filled symbols are from Figure 9A and Table III of Tamura, T., Nakatani, K., and Yau, K.-W. (1991). Calcium feedback and sensitivity regulation in primate rods. *Journal of General Physiology* 98: 95–130.

For mammalian rods, the Weber region encompasses only 1–2 log units of intensity, though for the larger rods of lower vertebrates, it may encompass a slightly wider range of about 3 log units. On the other hand, for the overall scotopic system, the Weber region covers a much wider range of at least 5 log units (i.e., over 100 000-fold). In addition, for cone photoreceptors, it extends over an even wider range.

The Purpose of Light Adaptation: Optimization of Performance

The purpose of light adaptation is to permit the visual system (or any neuron within it) to provide the best performance possible at that particular level of illumination. However, it is not always clear what constitutes best. For example, for the rod photoreceptors, it is clear that at very low ambient levels of illumination, their sensitivity should be as high as possible. However, we cannot readily anticipate the time course of their response that will be optimal.

Avoidance of Saturation: Range Extension

As the ambient light intensity increases, it is important that the rod (or any other cell) should avoid saturating, or else it will be unable to signal. By preventing saturation, light adaptation permits a photoreceptor to extend the range of intensities over which it operates. Although the rods achieve light adaptation over a limited range of intensities, the cones excel, and are able to avoid saturation no matter how bright the steady illumination becomes. Why has evolution permitted the rods to be driven into saturation by relatively low intensities? In part, it is because the photopic (cone) system is functional at these intensities, so that there is no disadvantage if the rods saturate. Not only is there no disadvantage – in fact, there is a distinct advantage when the rods saturate, in conserving energy during daylight conditions. Maintenance of the rod circulating current, in darkness and at low light levels, represents an extremely high metabolic load on the cells, and the elimination of this load when the cones are functional provides a major benefit to retinal metabolism. From this perspective, the limited range of rod light adaptation is beneficial, whereas an extended range (as occurs in cones) would be detrimental.

Extraction of Contrast Information and Optimization of Response Kinetics

In addition to the very important function of extending the operating range of the photoreceptor, there are two other ways in which photoreceptor light adaptation optimizes the cell's response. First, as described above in relation to eqn [1], it permits the extraction of contrast in the visual scene, independent of the absolute level of illumination. Second, it provides real-time adjustment of the time course of the response to an incremental flash of light, in a manner that is presumed to be optimal for the visual system. Thus, at very low background intensities, the response is sluggish, and postreceptoral elements are able to integrate visual signals over relatively long times. At progressively higher background intensities, the response becomes progressively accelerated, thereby improving the time resolution of the system. However, we do not have sufficient information yet to be able to describe exactly how it is that kinetic changes of this kind are actually optimal for the visual system.

Light Adaptation of the Rod Photoreceptors: Range Extension, Desensitization, and Acceleration

In the presence of background illumination, it is not only the overall visual system that adapts, but also the rod photoreceptors themselves display light adaptation, characterized by an extension of their operating range and by desensitization and acceleration of the incremental flash response.

Prevention of Rod Photoreceptor Saturation: Range Extension

The response of a salamander rod to the onset of steady illumination at different intensities is illustrated in **Figure 2**. At the beginning of the step of light, the rod's response begins rising according to what is predicted from the time integral of the flash response, but very soon deviates, falling well below the linear prediction (upper panel). Characteristically, the response to such a step of light typically exhibits an early peak followed by a sag. This deviation from the simplest linear prediction is a crucial aspect of light adaptation – if this deviation did not occur, then the rod would be driven into saturation by lights of very low intensity. Such saturation can be induced by exposing the rod to a solution that clamps the cytoplasmic calcium concentration; in the presence of calcium-clamping solution (lower panel), the responses of the rod follow the predictions of the smooth theoretical curves, and a very low intensity (labeled 2) saturates the rod. This result shows that at least a part of the rod's ability to continue operating in backgrounds of moderate intensity (i.e., the extension of its operating range) is a consequence of changes in cytoplasmic calcium concentration; the molecular mechanisms that contribute to this will be discussed below.

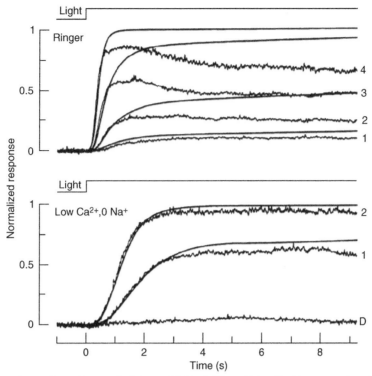

Figure 2 Responses of a salamander rod to onset of steps of light of different intensity. Upper panel: under control conditions (Ringer solution). Lower panel: in the presence of Ca^{2+}-clamping solution. The step intensities increased by factors of ~4 for traces labeled 1–4; D, darkness. The smooth curves are predictions obtained by integrating the measured dim flash response (not shown), and represent the step responses that are predicted in the absence of any adaptation. Reproduced from Fain, G. L., Lamb, T. D., Matthews, H. R., and Murphy, R. L. W. (1989). Cytoplasmic calcium as the messenger for light adaptation in salamander rods. *Journal of Physiology* 416: 215–243.

Desensitization and Acceleration

The manner in which background illumination modifies the rod's response to a dim test flash is illustrated in **Figure 3**. The uppermost trace is for a dim flash presented in darkness, while the remaining traces are for exactly the same flash presented on backgrounds of successively higher intensity. Characteristically, the flash response becomes progressively more desensitized and accelerated with backgrounds of higher intensity. Thus, the peak of the incremental flash response moves downward and leftward as the background intensity increases.

By plotting the peak amplitude of the flash response as a function of the background intensity upon which it was elicited, one obtains a plot of the kind indicated by the red symbols in **Figure 1**, where sensitivity declines according to Weber's law, given above in eqn [1].

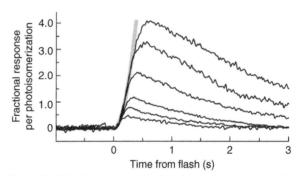

Figure 3 Dim flash responses of a salamander rod obtained in the dark (top trace) or in the presence of backgrounds of progressively higher intensity. Each trace was obtained by taking the raw response and dividing by the circulating current, and then dividing by the flash intensity. Reproduced from Pugh, E. N., Jr., Nikonov, S., and Lamb, T. D. (1999). Molecular mechanisms of vertebrate photoreceptor light adaptation. *Current Opinion in Neurobiology* 9: 410–418.

Unaltered Rising Phase, but Accelerated Recovery

For the incremental flash responses in **Figure 3**, the vertical scale has been adjusted to take account of changes

in the level of circulating current remaining in the presence of the different background intensities. Thus, rather than plotting raw sensitivity (response per photoisomerization), **Figure 3** instead plots the fractional response (i.e., the incremental response as a fraction of the circulating current at that background) per photoisomerization. This has been done in order to provide a direct measure of the level of activation of the guanine nucleotide-binding protein (G protein) cascade of phototransduction; thus, it can be shown that the level of cascade activation is best measured by the fractional channel opening, which in turn is measured by the incremental response expressed as a fraction of the existing circulating current. When plotted in this manner, the incremental responses in **Figure 3** demonstrate the remarkable property that the onset phase of the response is invariant; that is, the traces for different background intensities exhibit a common rise at early times, indicated by the smooth gray trace. This behavior indicates that the amplification parameter describing the activation steps in phototransduction is unaltered during light adaptation; in other words, light adaptation causes no change in the efficacy of the activation steps in phototransduction. Instead, it is clear that light adaptation causes a marked speeding up of the shut-off steps in the transduction cascade. The molecular identity of the steps that are accelerated is analyzed below.

Saturation of the Rod Photocurrent at Higher Background Intensities

At higher background intensities, the rod circulating current is completely suppressed. Thus, in the upper panel of **Figure 2**, intensities higher than those labeled 4 cause the response simply to rise to its maximum level (corresponding to the closure of all cyclic guanosine monophosphate (cGMP)-gated channels in the outer segment) and, as a result, incremental stimuli are unable to elicit any incremental response so that the cell's response is saturated. Typically, such saturation sets in exponentially with increasing background intensity, as described by the second term on the right of eqn [1].

Calcium-Dependent Mechanisms of Rapid Light Adaptation in Rod Photoreceptors

The mechanisms that contribute to light adaptation in photoreceptors (i.e., to the alteration in response properties of the photoreceptors upon exposure to background illumination) are closely associated with the mechanisms of response recovery. These mechanisms of adaptation can be classified broadly as (1) those that are calcium dependent and (2) those that do not involve calcium. Both categories are important; yet, the noncalcium-dependent mechanisms have frequently been overlooked.

Role of Calcium: Resensitization through Prevention of Saturation

When cGMP-gated ion channels in the outer segment are closed in response to light, the cytoplasmic concentration of calcium drops. This drop in Ca^{2+} concentration is vitally important to light adaptation, though it is crucial to emphasize that it does not cause the desensitization that characterizes photoreceptor light adaptation. Quite the contrary: the drop in Ca^{2+} concentration actually rescues the rod from the saturation that would otherwise be induced by light, and thereby prevents the onset of massive desensitization at relatively low intensities of background illumination. Thus, the light-induced drop in Ca^{2+} acts to increase the rod's sensitivity above the drastically reduced level that would occur either if the Ca^{2+} concentration did not alter or if the rod's calcium-dependent mechanisms were inoperative.

Powerful Negative-Feedback Loop Mediated by Calcium

Calcium is the cytoplasmic messenger for a very powerful negative-feedback loop that tends to stabilize the rod's circulating current. If ever the Ca^{2+} concentration drops (e.g., in response to light, or as a result of some other perturbation), then, as described below, a number of changes occur very rapidly. These changes are stimulated by the unbinding of Ca^{2+} from at least three classes of calcium-sensitive protein: (1) guanylyl cyclase activator proteins (GCAPs) 1 and 2, which activate guanylyl cyclase; (2) recoverin, which regulates the lifetime of activated rhodopsin; and (3) calmodulin, which modulates the opening of cGMP-gated channels. Calcium's action via each of these pathways leads to the opening of cGMP-gated channels, thereby increasing the circulating current and admitting Ca^{2+} ions from the extracellular medium. This influx of Ca^{2+} ions tends to counteract the initial reduction in Ca^{2+} concentration, thereby completing a negative-feedback loop.

Each of these molecular mechanisms contributing to the calcium negative-feedback loop contributes toward extending the rod's operational range of light intensities by helping prevent saturation of the circulating current. Thus, each of these three molecular mechanisms assists in rescuing the rod from saturation and hence increasing, rather than decreasing, the rod's sensitivity compared with the case that would exist if the mechanism were absent. Each of the three mechanisms is most effective over some range of calcium levels, and a corresponding range of light intensities. Overall, the most powerful of the three (at least in rods) is the GCAPs' activation of guanylyl cyclase.

Since the various components of the calcium negative-feedback loop act quite rapidly, they contribute to determining not only the photoreceptor's sensitivity in the presence of background illumination, but also the kinetics of its response to an incremental flash presented on the background. The importance of altered Ca^{2+} concentration in setting the incremental flash response kinetics can be demonstrated by incorporating a calcium buffer (such as 1,2-bis(o-aminophenoxy)ethane-N,N,N′,N′-tetraacetic acid (BAPTA)) into the outer segment. Although the flash response begins rising exactly as in control conditions, it does not begin recovering as soon and, instead, rises to a substantially larger and later peak with slower final recovery (see also **Figure 4**).

Three Calcium-Sensitive Molecular Pathways

Guanylyl cyclase activation

In response to a drop in calcium concentration, Ca^{2+} will unbind from the GCAP proteins (GCAP1 and GCAP2), thereby activating guanylyl cyclase and stimulating the production of cGMP at a greatly increased rate, leading to the opening of cGMP-gated channels.

The cyclase activity increases roughly as the fourth power of the drop in Ca^{2+} concentration, and furthermore (as also applies for the other two routes considered below), the number of channels open increases approximately as the cube of the cGMP concentration. Because of the cascading of two such steep dependencies, any small fractional change in Ca^{2+} concentration stimulates a large and opposite fractional change in channel opening; that is, the fractional change in channel opening is opposite in sign to, and up to 12× the magnitude of, the originating fractional change in Ca^{2+} concentration. As a result, this molecular mechanism is the most potent of the three that contribute to the calcium negative-feedback loop and, hence, to setting the adaptational state in rods. It is especially dominant at relatively bright background intensities, corresponding to low Ca^{2+} concentrations, and is

therefore the most important in extending the rod's operating range to high intensities.

The role of the GCAPs/guanylyl cyclase component of the Ca^{2+} feedback loop in setting the waveform of the incremental flash response is illustrated in **Figure 4**, where averaged responses are shown for two classes of rod: rods from wild-type (WT) mice and rods from GCAPs knock-out mice. In a manner very similar to that seen in rods containing the calcium buffer BAPTA, the response in the GCAPs knockout case begins rising exactly as for the control (WT) case, but it does not recover as soon; therefore, the response continues rising and reaches a larger and later peak.

Shortened R* lifetime

Activated rhodopsin (R^*) is inactivated by multiple phosphorylation steps mediated by rhodopsin kinase (GRK1) followed by binding of arrestin. It is generally assumed that the decline in R^* activity follows exponential kinetics, and can therefore be described by a characteristic lifetime, τ_R; however, it is worth bearing in mind that there is no direct evidence for this assumption. It was established by Satoru Kawamura that GRK1's phosphorylation of R^* is calcium dependent and that the effect is mediated by the calcium-binding protein recoverin. The molecular mechanism of this dependence is not entirely clear; however, some evidence suggests that the calcium-bound form of recoverin binds to GRK1, thereby preventing it from interacting with R^*. In any case, it is proposed that a reduction in Ca^{2+} concentration leads to a shortened R^* lifetime, τ_R.

The slowest time constant in the phototransduction cascade (the so-called dominant time constant, τ_{dom}) can be estimated from the steepness of the relationship between the duration that the rod is held in saturation by a bright flash and the flash intensity. Over the years, there has been considerable debate as to whether this dominant time constant is set by the R^* lifetime, τ_R, or, instead, by the lifetime τ_E of the transducin–phosphodiesterase (PDE) complex (the effector). The situation may be species dependent; however, in mouse rods, it has now been clearly established by Marie Burns' group that, under dark resting conditions, the dominant time constant is that of transducin–PDE, with $\tau_E \approx 200$ ms, while the R^* lifetime is shorter, with $\tau_R \leq 80$ ms. In the scenario where the R^* lifetime is shorter than the transducin–PDE lifetime, further light-induced shortening of τ_R is likely to have very little effect on the response kinetics, but will instead cause a reduction in sensitivity because fewer molecules of transducin will be activated during the R^* lifetime.

Although it remains difficult to establish the effectiveness of any individual mechanism in an intact rod with a functional calcium feedback loop, it appears that the recoverin-mediated reduction in R^* lifetime plays a moderate role, especially at relatively low background intensities.

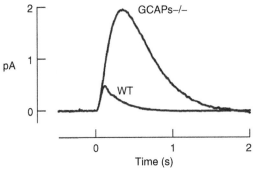

Figure 4 Single-photon responses from rods of wild-type (WT) and GCAPs knock-out (GCAPs−/−) mice. Suction pipette recordings from single rods, analyzed to extract the mean response to a single photoisomerization. Circulating current in rods of both strains averaged 12 pA. Reproduced from Burns, M. E., Mendez, A., Chen, J., and Baylor, D. A. (2002). Dynamics of cyclic GMP synthesis in retinal rods. *Neuron* 36: 81–91.

Channel reactivation

In response to a drop in calcium concentration, Ca^{2+} unbinds from calmodulin (in the case of the rods), leading to a lowered dissociation constant ($K_{1/2}$) for the binding of cGMP to the channels. The effect of the lowered $K_{1/2}$ is that any given concentration of cGMP will cause the opening of a larger fraction of the cGMP-gated channels, leading to an increase in circulating current and the influx of more calcium. However, the potency of this effect is low in rods, and the mechanism contributes only weakly to rod adaptation. In contrast, cones possess a much more powerful mechanism, mediated by a different calcium-sensitive protein.

Rod Photoreceptor Light Adaptation Independent of Calcium

There are at least three classes of noncalcium-dependent phenomena that represent mechanisms of light adaptation in rod photoreceptors, insofar as the properties of the response to light are altered in comparison with the dark-adapted state. First, there is response compression, whereby the reduced level of circulating current in the presence of steady background illumination reduces the size of the flash response. This phenomenon will not be discussed here, in part because it is both very well known and very simple and also because (in philosophical terms) it can be viewed as a failure of light adaptation; in comparison, cones cope much better and effectively avoid response compression by feedback mechanisms that maintain the circulating current. Second, there is pigment depletion. However, this is never relevant in rod light adaptation because the rods are driven into saturation even by very low levels of bleached pigment (see section titled 'Dark adaptation of the rods: Very slow recovery from bleaching'). Third, there is a direct effect of PDE activation, which is now considered.

Accelerated Turnover of cGMP

In a rod outer segment in darkness, the activity of the PDE is low; therefore, the turnover rate constant for cGMP (denoted β) is low, with a correspondingly long turnover time constant for cGMP, $\tau_{cGMP} = 1/\beta$, of around 1 s in amphibian rods and around 200 ms in mammalian rods. The magnitude of this parameter has a major effect on both the sensitivity and the kinetics of the rod's response to a flash. Thus, when the PDE activity increases in steady illumination, the shorter turnover time for cGMP contributes to both desensitization and acceleration of the photoresponse (compared with the case that would have applied, had the steady level of PDE activity not increased).

To provide an intuitive understanding of this mechanism, it is helpful to consider what we have referred to previously as the bathtub analogy. Imagine a container of water, such as a tall cylinder, and let the height of water in the cylinder represent the level (concentration) of cGMP in the outer segment. The rate at which water runs out of the cylinder, through a drain hole at the base, is proportional both to the height of water and to the size of the opening, representing the cGMP level and the PDE activity, β, respectively. Likewise, the rate at which water flows in to the cylinder through a tap at the top represents the activity of guanylyl cyclase, α. When a steady state is reached, the height of water will equal the rate of influx divided by the size of the drain hole; that is, $cGMP = \alpha/\beta$. Importantly, whenever the water level is perturbed from this steady-state level (e.g., upon a brief opening of an additional drain hole), the level will re-equilibrate with a time constant $\tau_{cGMP} = 1/\beta$ (provided that the rate of influx through the tap remains constant). Hence, if the drain hole is small (and the inflow via the tap correspondingly small), then any perturbation in water level elicited by a transient additional outflow will be corrected only slowly; if the drain hole is large (and the influx correspondingly large), then any perturbation will be rapidly corrected. Furthermore, though perhaps less intuitively, it can be shown that for a noninstantaneous perturbation, corresponding to the normal flash response, not only will the kinetics of recovery be faster, but the peak also will be smaller.

Hence, the effect of the increased PDE activity during steady illumination is both to accelerate the response kinetics and to reduce the peak amplitude (i.e., reduce the sensitivity) to an incremental flash. Calculations show that in rods the 20-fold increase in β during steady illumination provides the primary mechanism underlying the measured shortening of the time to peak and the decrease in flash sensitivity.

Slow Changes in Rods: Light Adaptation or Dark Adaptation?

In addition to the conventional features of rod photoreceptor light adaptation that occur extremely rapidly (on a subsecond time scale), other changes have been reported to occur over a time frame of minutes of exposure, in response to lights that saturate the cell's response. As the effects of these changes are very slow, and can only be observed in darkness when the adapting exposure is extinguished, there is a semantic issue as to whether these phenomena should be thought of as light adaptation or as dark adaptation.

Light-Induced Change in the Dominant Time Constant

It has recently been shown by Marie Burns' group that exposure of mouse rods to a just-saturating intensity of around 1000 photoisomerizations per second, for 1 min or more, leads to a persistent speeding of the bright-flash response upon extinction of the background. The change did not involve any reduction in the activation phase of transduction, but instead involved a reduction in the dominant time constant of response recovery; typically, the dominant time constant τ_{dom} dropped from around 200 ms under dark-adapted conditions to around 100 ms immediately after extinction of the saturating light. The adaptational effect developed relatively slowly, building up over 60 s or so, and it required a rhodopsin bleach level of around 2% for full effect. The effect was relatively long lasting, declining with a time constant of around 80 s.

The molecular mechanism giving rise to this adaptational effect is not known, though some evidence suggests that it corresponds to a reduction in lifetime of the activated transducin–PDE complex. If so, it represents a phenomenon distinct from the actions of dimmer adapting lights.

Light-Induced Translocation of Proteins

The light-induced translocation of transducin, recoverin, and arrestin in photoreceptors is dealt with in detail elsewhere in this encyclopedia and, therefore, mentioned only briefly here. Movements of protein are elicited only at quite bright intensities (generally in the saturating range) and occur over a time scale of many minutes. In mouse rods, intensities above 3000 photoisomerizations per second for 30 min (which bleach a substantial fraction of the rhodopsin) trigger the movement of transducin from the outer segment to the inner segment, while slightly lower intensities of 1000 photoisomerizations per second or more trigger the movement of arrestin in the opposite direction; recoverin also leaves the outer segment in bright light.

Protein movements of these kinds may well affect the adaptational state of the rod, though this is yet to be established clearly. Since the movements are triggered only by saturating light intensities, the electrical effects cannot readily be observed during the illumination because the circulating current is completely suppressed. One possibility is that the protein translocation contributes to some form of conservation – for example, lowering the guanosine 5′-triphosphate (GTP) consumption involved in the continual (and maximal) activation of transducin during daylight conditions. Alternatively, it may be that the changes help prepare the rod for its return to lower intensities, as occurs around dusk. Interestingly, in one attempt that was made to detect any change in the amplification constant of human rods (using the electroretinogram (ERG)) following exposures to intensities that elicit transducin translocation in mouse rods, no change in amplification was detectable.

Dark Adaptation of the Rods: Very Slow Recovery from Bleaching

Following exposure of our eye to very intense illumination, our visual threshold is greatly elevated and may take tens of minutes to recover fully. Closely comparable effects can be measured in the overall visual system and at the level of the rod photoreceptors or the rod bipolar cells. The slow recovery of sensitivity is referred to as dark adaptation or bleaching adaptation; however, it should be noted that this use of the term adaptation is something of a misnomer. Adaptation normally refers to beneficial adjustments; yet, the changes that accompany intense illumination are distinctly disadvantageous – thus, there can be no advantage in being almost blind following exposure to intense light.

The recovery of visual threshold for a human subject is plotted in **Figure 5**, following the cessation of nine light exposures that bleached from 0.5% to 98% of the rhodopsin. For a bleach of 20%, the visual threshold was initially elevated by 3.5 log units. This indicates that the elevation of threshold is out of all proportion to the fraction of pigment remaining unbleached; even though 80% of the rhodopsin remained functional, the threshold was raised 3000-fold. Instead, there is overwhelming evidence that the phenomenon arises from the presence within the outer segment of unregenerated opsin (i.e., the presence of the protein part of the visual pigment, prior to its recombination with the regenerated 11-*cis* retinal).

Remarkably, the recovery of scotopic (rod-mediated) threshold exhibits a region of common slope across all the bleach levels, as indicated by the parallel red lines in **Figure 5**. This region is termed the S2 component of recovery, and has a slope $\Psi_{S2} = -0.24$ log unit min^{-1} that is characteristic of dark adaptation recovery in normal (young adult) human eyes; also characteristic is the nature of the rightward shift of the recovery traces as a function of increasing bleach level – the form of this shift is as expected for a rate-limited (zero-order) recovery process, as distinct from an exponential (first-order) recovery process.

From a detailed analysis of results of this kind, in combination with knowledge of the retinoid cycle, Trevor Lamb and Edward Pugh developed a cellular model that can account for human dark adaptation behavior. They postulated that (1) the presence of opsin (without chromophore) gives rise to a phenomenon closely equivalent to light, through activation of the G protein cascade of transduction and (2) the elimination of opsin via its reconversion to rhodopsin follows rate-limited kinetics because of a limitation in the supply of 11-*cis* retinal that results from the movement of this substance from a pool in the retinal pigment epithelium.

Application of this cellular model has provided an accurate account of (1) the regeneration of visual pigment

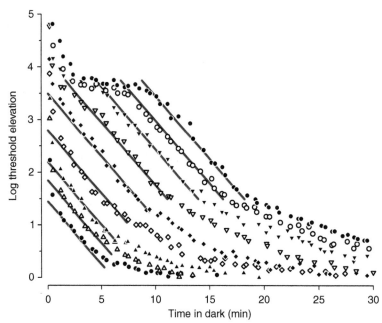

Figure 5 Human psychophysical dark adaptation. Recovery of log threshold elevation in a normal human observer is plotted as a function of time in darkness, after a wide range of bleaching exposures (from 0.5% to 98%). Parallel red lines represent component S2, with a slope of –0.24 decades min^{-1} (see text). The lateral shift between the lines is consistent with the rate-limited delivery of 11-*cis* retinal from the RPE to opsin in the outer segments. Reproduced from Lamb, T. D. and Pugh, E. N., Jr. (2006). Phototransduction, dark adaptation, and rhodopsin regeneration. The Proctor Lecture. *Investigative Ophthalmology and Visual Science* 47: 5138–5152, with permission of the Association for Research in Vision and Ophthalmology.

in humans and other mammals, measured by retinal densitometry; (2) normal human dark adaptation behavior (as in **Figure 5**); and (3) the slowed regeneration of pigment and the slowed dark adaptation that is characteristic of a number of diseases that affect the photoreceptors and/or retinal pigment epithelium.

See also: Light-Driven Translocation of Signaling Proteins in Vertebrate Photoreceptors; Phototransduction: Adaptation in Cones; Phototransduction: Inactivation in Rods; Phototransduction: Phototransduction in Rods; Phototransduction: The Visual Cycle.

Further Reading

Cameron, A. M., Mahroo, O. A. R., and Lamb, T. D. (2006). Dark adaptation of human rod bipolar cells measured from the *b*-wave of the scotopic electroretinogram. *Journal of Physiology* 575: 507–526.

Krispel, C. M., Chen, C-K., Simon, M. I., and Burns, M. E. (2003). Novel form of adaptation in mouse retinal rods speeds recovery of phototransduction. *Journal of General Physiology* 122: 703–712.

Lamb, T. D. and Pugh, E. N., Jr. (2004). Dark adaptation and the retinoid cycle of vision. *Progress in Retinal and Eye Research* 23: 307–380.

Lamb, T. D. and Pugh, E. N., Jr. (2006). Phototransduction, dark adaptation, and rhodopsin regeneration. The Proctor Lecture. *Investigative Ophthalmology and Visual Science* 47: 5138–5152.

Nikonov, S., Lamb, T. D., and Pugh, E. N., Jr. (2000). The role of steady phosphodiesterase activity in the kinetics and sensitivity of the light-adapted salamander rod photoresponse. *Journal of General Physiology* 116: 795–824.

Pugh, E. N., Jr. and Lamb, T. D. (2000). Phototransduction in vertebrate rods and cones: Molecular mechanisms of amplification, recovery and light adaptation. In: Stavenga, D. G., de Grip, W. J., and Pugh, E. N., Jr (eds.) *Handbook of Biological Physics, Vol. 3, Molecular Mechanisms of Visual Transduction, ch. 5*, pp. 183–255. Amsterdam: Elsevier.

Pugh, E. N., Jr., Nikonov, S., and Lamb, T. D. (1999). Molecular mechanisms of vertebrate photoreceptor light adaptation. *Current Opinion in Neurobiology* 9: 410–418.

Tamura, T., Nakatani, K., and Yau, K-W. (1991). Calcium feedback and sensitivity regulation in primate rods. *Journal of General Physiology* 98: 95–130.

Phototransduction: Inactivation in Cones

V V Gurevich and E V Gurevich, Vanderbilt University, Nashville, TN, USA

Glossary

Arrestin – A protein that selectively binds light-activated phosphorylated photopigment and blocks further signal transduction. Cones express two subtypes, arrestin1 and arrestin4 (often termed rod and cone arrestins, respectively).

Cone opsins – Light receptors, consisting of the protein part (opsin) and 11-*cis*-retinal covalently attached via Schiff base to a lysine in the seventh transmembrane domain. All opsins are members of superfamily of G-protein-coupled receptors (GPCRs), the largest family of signaling proteins in animals (mammals have ~1000 different GPCRs).

GCAP – Guanylyl cyclase activating protein is a member of the superfamily of EF-hand-containing calcium-binding proteins. Cones express two homologs, GCAP1 and GCAP2, which in the calcium-liganded form inhibit and in magnesium-liganded form enhance the activity of retinal guanylyl cyclase (retGC).

GRKs – G-protein-coupled receptor kinases that specifically phosphorylate active forms of their cognate receptors. Cones express rhodopsin kinase (systematic name: GRK1) and a cone-specific form GRK7. However, mice do not have GRK7; therefore, photopigments in mouse cones and rods are phosphorylated by a single isoform, GRK1.

Phosphodiesterase (PDE) – The photoreceptor-specific cyclic guanosine monophosphate (cGMP) PDE, PDE6. Cone PDE6 is a heterotetramer, consisting of two identical catalytic α'-subunits and two inhibitory γ-subunits. PDE rapidly hydrolyzes cGMP upon its activation by transducin, when its catalytic activity approaches the theoretical limit set by the rate of cGMP diffusion.

RetGC – Retinal guanylyl cyclase is structurally related to receptor guanylyl cyclases. Cones predominantly express RetGC1, in contrast to rods that express RetGC1 and RetGC2 at comparable levels.

RGS9-1 – Photoreceptor-specific short isoform of the regulator of G-protein signaling 9 expressed in both rods and cones. It interacts with the complex of the guanosine triphosphate (GTP)-liganded active α-subunit of transducin with PDEγ and facilitates its intrinsic GTPase activity, thereby directly inactivating

transducin and indirectly PDE. Cones express much more RGS9-1 than rods.

Transducin – Photoreceptor-specific heterotrimeric G protein that couples to light-activated opsins. Its α-subunit belongs to Gi/o family. All types of cones express the same α-subunit that is different from the rod variant.

Rod photoreceptors are often described as a marvel of molecular engineering, which creates an impression that cones are just noisier and less-sensitive rods. In fact, as light sensors, cones are just as amazing: their adaptability gives cones a much wider dynamic range covering more than seven orders of magnitude of light intensity without saturation. Cones begin to function in the light of the full moon reflected from objects in the night and are still adequate for a direct look at the sun. Mostly for technical reasons, the biochemistry of cone photoreceptors, particularly the molecular mechanisms underlying adaptation, is not as well studied as the signaling in rods. The assumption that the signaling and shutoff mechanisms in cones and rods are qualitatively similar is often used to fill the gaps in our knowledge of cone biochemistry. To avoid repetition, here we emphasize known differences between the cone and rod inactivation mechanisms.

Cone Signaling Cascade

Cone opsins are closely related to rhodopsin and belong to the same branch of the G-protein-coupled receptor superfamily. Gene duplication events in early vertebrate evolution produced five groups of light receptors: rhodopsins and four classes of cone opsins. Mammals lost half of cone opsin classes, retaining only two. Light activates cone opsins via induced isomerization of 11-*cis*-retinal covalently attached to a lysine in the seventh transmembrane domain. Cone opsins use the same 11-*cis*-retinal as rhodopsin, but have very different spectral sensitivity, with maxima ranging from 360 nm (ultraviolet) to 575 nm (red). Spectral tuning of covalently linked retinal is achieved by changing its environment in the retinal-binding pocket of opsin. Light-activated cone opsins couple to cone transducin, which, in turn, activates the cone subtype of phosphodiesterase 6 (PDE6). Subsequently, a decrease of cytoplasmic cyclic guanosine monophosphate (cGMP) reduces the influx of Na^+ and Ca^{2+} through the cone variant of cGMP-gated channels, resulting in cell hyperpolarization. Similar to rods, the decrease in Ca^{2+} concentration and consequent replacement of bound Ca^{2+} with Mg^{2+} converts guanylate cyclase activating proteins (GCAPs) from inhibitors to activators of retinal guanylate cyclase (RetGC). The latter replenishes cGMP lost during the light response, which opens the channels, thereby restoring cytoplasmic Ca^{2+} to the original levels. Thus, the activation and deactivation mechanisms employed by rods and cones are quite similar. However, subtle differences at every step of the pathway, including differences in the subtypes of signaling proteins involved, their expression levels, and the geometry of the cell, result in striking functional specialization of the two types of photoreceptors.

Shutoff of the Light-Activated Cone Opsins

As far as signaling is concerned, the key difference between rhodopsin and cone opsins is lower thermal stability of the latter. Because cone opsins spontaneously activate with much higher probability than rhodopsin and readily release retinal even in the dark, cones generate noise that is orders of magnitude higher than in rods. This rules out the detection of signals below the noise level (e.g., a few photons) and makes even completely dark-adapted cones pre-desensitized and ready to operate at light levels they can detect. Loose attachment of the retinal to opsin also results in a significantly faster spontaneous decay of the light-activated cone opsin. For example, mouse cone S-opsin transgenically expressed in rods lacking arrestin was estimated to decay ~40 times faster than rhodopsin coexpressed in the same cell. However, this spontaneous decay with a time constant of ~1.3 s is still much slower than the rate of recovery in cones.

Evolution equipped cones with a more elaborate (and presumably more efficient) photopigment shutoff machinery than that found in rods. The opsin inactivation is accelerated by pigment phosphorylation followed by arrestin binding. In most species, cones express two G-protein receptor kinase (GRK) subtypes: GRK1 (shared with rods) and cone-specific GRK7. It is likely that the co-expression of GRK7 with higher enzymatic activity accelerates opsin phosphorylation in cones. However, it should be noted that mice and rats are rare exceptions: these nocturnal rodents have only GRK1 in both types of photoreceptors. Cones also express two arrestin subtypes, arrestin1 and cone-specific arrestin4 (formerly known as rod and cone arrestins, respectively). Arrestin1 is present in cones at ~50-fold molar excess over arrestin4. A recent study in knock-out animals shows

that both arrestins contribute comparably to the shutoff of the photopigment in cones. It is not entirely clear why cones express two arrestin subtypes, rather than a higher level of one subtype, especially considering that the cone opsin transgenically expressed in mouse rods is rapidly and efficiently deactivated by rod arrestin1.

Two functional differences between these arrestins provide some clues. Arrestin1 has high propensity to self-associate, cooperatively forming dimers and tetramers at physiological concentrations. Even in the dark-adapted rod, where the outer segment contains a small fraction of the total arrestin1, most of arrestin1 is a tetramer. It has been unambiguously shown that only monomeric arrestin1 is an active rhodopsin-binding species; therefore, oligomers appear to be storage forms. In contrast, cone-specific arrestin4 does not self-associate at physiologically relevant concentrations; therefore, the whole complement of arrestin4 present in cones is an active monomer. Recent estimates of their expression and arrestin1 self-association constants suggest that dark-adapted cones have in the outer segment ~60 μM of arrestin4 and ~30 μM (5% of the total) of monomeric arrestin1 ready to bind phosphorylated opsin at any time, in addition to a huge backup supply of arrestin1 oligomers.

The second important difference lies in the stability of the arrestin complex with phosphorylated opsin. Arrestin1 forms very stable complexes that take the bound molecule of phosphopigment out of the game for a long time. This is important in the rod to ensure the fidelity of the shutoff. In order to release completely inactive rhodopsin upon dissociation, arrestin1 must stay bound until metarhodopsin II (Meta II) slowly decays and likely until it is regenerated with 11-*cis*-retinal. In contrast, arrestin4 forms fairly transient complexes with phosphorylated cone opsins, likely to ensure the quick return of the opsin back into the active pool. This is important for cone photoreceptors that function at a high rate of pigment bleaching. Although we do not know with certainty why cones express both arrestin subtypes, one scenario appears to provide a plausible explanation. Given the concentrations of the two arrestins in cones, in moderately bright light arrestin4 likely has an advantage, so that the majority of phosphorylated cone opsin would be rapidly recycled to the signaling-competent pool. Increasing levels of illumination inducing massive pigment bleaching would force the cell to draw increasingly on the virtually inexhaustible supply of arrestin1, which forms long-lived complexes with the opsin. The recent finding that arrestin1 plays a more prominent role in cone recovery after very bright flashes is consistent with this model. The formation of arrestin1-opsin complexes would take larger and larger fraction of the pigment out of action for a relatively long time, possibly serving as one of the mechanisms of light adaptation. Even though a cone-specific visual cycle involving

Müller glia provides 11-*cis*-retinal faster than the canonical retinal pigment epithelium-based visual cycle supplying rods, very bright light bleaches cone pigment faster than it can be regenerated. This loss of functional opsin was proposed to reduce light capture, acting as a mechanism of adaptation. It is entirely possible that in bright light, both incomplete regeneration of opsin and its binding by arrestin1 cooperate to limit the active pool, thereby reducing light sensitivity of cones.

Overall, cones combine less-stable photopigment with more sophisticated machinery of its inactivation (**Figure 1**). These factors apparently contribute to faster shutoff at the opsin level and likely provide cone-specific mechanisms for light adaptation.

Inactivation of Transducin and PDE

It is generally accepted that the activation of cone transducin by cone opsins and that of cone PDE6 by the guanosine triphosphate (GTP)-liganded α-subunit of cone transducin proceeds in similar ways to corresponding processes in rods. All three subunits of cone transducin differ from their rod counterparts, but the significance of this specialization is uncertain. In fact, cone S-opsin transgenically expressed in mouse rods efficiently activates the signaling cascade coupling to rod transducin. Cone PDE6 is an $\alpha'_2\gamma_2$ heterotetramer, in contrast to the $\alpha\beta\gamma_2$ version in rods, but the functional significance of the use of different catalytic subunits remains to be elucidated. There is one biochemical difference that undoubtedly contributes to the much faster inactivation of transducin-PDE6 complex in cones: $\sim$10-fold higher level of the regulator of G-protein signaling 9-1 (RGS9-1) expression. It has been convincingly shown that the deactivation at this step rate limits the recovery kinetics in rods, and that the level of RGS9-1, which accelerates self-inactivating GTPase of transducin α-subunit, sets the speed of transducin-PDE6 inactivation. Thus, the shutoff at the opsin and transducin-PDE6 level in cones is much faster than corresponding processes in rods; however, it is still not clear which step is rate limiting in cone recovery.

Restoration of cGMP and Intracellular Calcium Level

Similar to the situation in rods, cone activation results in a drop in the intracellular Ca^{2+} concentration due to the closure of the cGMP-gated channels mediating the bulk of Ca^{2+} entry. In order to return to the initial state after opsin and PDE6 are fully inactivated, cones need to restore cytoplasmic cGMP hydrolyzed by PDE6. Cones and rods use the same negative-feedback mechanism that translates the drop in Ca^{2+} resulting from the reduction in the cGMP level into a signal to make more cGMP. Ca^{2+} dissociates from GCAPs when its cytoplasmic concentration drops in the light. The replacement of lost Ca^{2+} by Mg^{2+} converts GCAPs from inhibitors to activators of RetGC. The generated cGMP opens the channels, and the consequent increase in cytoplasmic Ca^{2+} stops further cGMP synthesis. Cones apparently express the same combination of GCAP1 and GCAP2 (which differ in their Ca^{2+} sensitivity) as rods. The functional significance of the predominance of the RetGC1 isoform in cones (in contrast to similar levels of RetGC1 and RetGC2 in rods) is not clear.

Several important differences between rods and cones are known to be responsible for much faster cone recovery. First, the rate of recovery depends on the absolute amounts of cGMP and Ca^{2+} that need to be replenished. Here cones hold an obvious advantage due to the much smaller volume of their outer segments: the hydrolysis or synthesis of the same absolute amount of cGMP leads to a more significant change in its concentration. Similarly, the closure of the same fraction of cGMP-gated channels leads to a more profound drop in intracellular Ca^{2+} in cones. However, geometry is only part of the story. The channel expressed in cones has a different subunit composition and ion preference. About 35% of the inward current via the cone cGMP-gated channel is carried by Ca^{2+}, whereas in rods this fraction is only $\sim$20%. Thus, the closure of the same fraction of channels upon PDE6 activation results in a substantially greater change in the absolute number of Ca^{2+} ions entering the cell. Increased Ca^{2+} influx in cones is balanced by its accelerated extrusion via $Na^+/K^+–Ca^{2+}$-exchanger, so that the turnover of Ca^{2+} in cone outer segments is more rapid. The combination of faster constitutive extrusion, larger fraction of the current carried by Ca^{2+} through cGMP-gated channels, and much smaller outer segment volume greatly increases the rate of Ca^{2+} drop in response to light stimulus, speeding up RetGC activation and cGMP resynthesis. High intracellular Ca^{2+} reduces the sensitivity of the channels to cGMP, so that when the intracellular Ca^{2+} drops, the channels become more sensitive to the cytoplasmic cGMP and therefore reopen faster. This mechanism operates in both types of photoreceptors, but it is more powerful in cones, further contributing to accelerated recovery.

Conclusions

Cone photoreceptors use essentially the same molecular mechanisms of signal shutoff at the opsin level as rods. At this step, cones achieve much higher speed of inactivation by employing, in addition to GRK1 and arrestin1

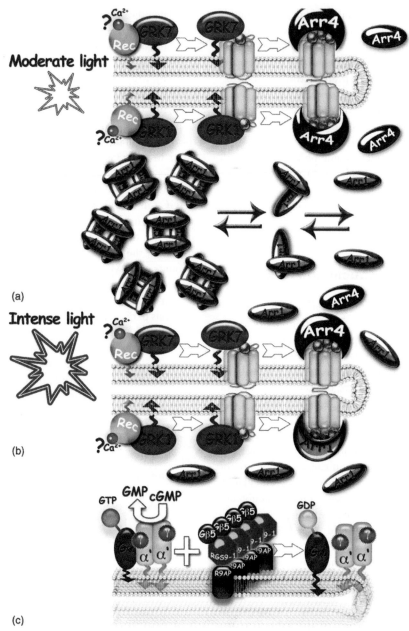

Figure 1 Biochemical mechanisms of rapid inactivation in cones. (a) Cone opsins are phosphorylated by both GRK1 and GRK7 coexpressed in cones of most vertebrates, including humans. At moderate light levels, the signaling by phosphorylated photopigment is largely quenched by constitutively monomeric arrestin4, which forms transient complexes with the receptor. (b) During massive opsin activation in very bright light, the amount of expressed arrestin4 becomes insufficient to quench all active opsins; therefore, cones increasingly use coexpressed arrestin1, which forms longer-lived complexes with phosphorylated cone opsins. The consumption of monomeric arrestin1 by the photopigment shifts its monomer–dimer–tetramer equilibrium toward dissociation of oligomers, which generates virtually inexhaustible supply of binding-competent monomer. (c) RGS9-1 is expressed at ∼10-fold-higher level in cones than in rods, ensuring much faster inactivation of transducin and PDE. Cone opsin is shown as a bundle of seven transmembrane domains; opsin-attached phosphates are shown as spheres; lipid modifications anchoring recoverin, GRK1, GRK7, α-subunit of transducin, and catalytic α′-subunits of PDE are shown as membrane-imbedded arrows. Rec: recoverin, Arr1: arrestin1, and arr4: arrestin4.

used by rods, cone-specific GRK7 and arrestin4. The presence of two GRKs speeds up the phosphorylation of light-activated opsin, whereas the expression of two arrestin subtypes with very different functional characteristics

likely results in a gradual switch from rapidly reversible arrestin4 interaction with phospho-opsin at moderate light levels to semi-irreversible binding of arrestin1 in very bright light. Inactivation at the transducin/PDE level is accelerated by a 10-fold higher expression of RGS9-1 in cones. Two key features ensure faster recovery in cones than in rods. Faster Ca^{2+} turnover due to higher influx through cone-specific cGMP-gated channels and efflux via $Na^+/K^+–Ca^{2+}$-exchanger generate greater net changes in the number of Ca^{2+} ions in the outer segment when the same fraction of the channels is closed. Due to much smaller outer-segment volume, the same net change in the number of cGMP molecules or Ca^{2+} ions produces greater changes in the concentration of these second messengers. Rapid response and recovery gives cones better temporal resolution than rods. The high speed of activation and inactivation in combination with more powerful adaptation mechanisms (many of which still need to be elucidated at the molecular level) allows cones to function in a broad range of light levels without saturation.

See also: Phototransduction: Adaptation in Cones; Phototransduction: Inactivation in Rods; Phototransduction: Phototransduction in Cones; Phototransduction: Phototransduction in Rods; Phototransduction: Rhodopsin.

Further Reading

Cote, R. H. (2006). Photoreceptor phosphodiesterase (PDE6): A G-protein-activated PDE regulating visual excitation in rod and cone photoreceptor cells. In: Beavo, J. A., Francis, S. H., and Houslay, M. D. (eds.) *Cyclic Nucleotide Phosphodiesterases in Health and Disease*, pp 165–193. Boca Raton, FL: CRC Press.

Dizhoor, A. M., Olshevskaya, E. V., and Peshenko, V I. (2006). Calcium sensitivity of photoreceptor guanylyl cyclase (RetGC) and congenital photoreceptor degeneration: Modeling *in vitro* and *in vivo*. In: Philippov, P. P. and Koch, K.-W. (eds.) *Neuronal Calcium Sensor Proteins*, pp 203–219. New York: Nova Science Publishers, Inc.

Gurevich, V. V., Hanson, S. M., Gurevich, E. V., and Vishnivetskiy, S. A. (2007). How rod arrestin achieved perfection: Regulation of its availability and binding selectivity. In: Kisselev, O. and Fliesler, S. J. (eds.) *Signal Transduction in the Retina. Methods in Signal Transduction Series*, pp 55–88. Boca Raton, FL: CRC Press.

Hanson, S. M., Van Eps, N., Francis, D. J., et al. (2007). Structure and function of the visual arrestin oligomer. *European Molecular Biology Organization Journal* 26: 1726–1736.

Knox, B. E. and Solessio, E. (2006). Shedding light on cones. *The Journal of General Physiology* 127: 355–358.

Korenbrot, J. I. and Rebrik, T. I. (2002). Tuning outer segment Ca^{2+} homeostasis to phototransduction in rods and cones. *Advances in Experimental Medicine and Biology* 514: 179–203.

Nikonov, S. S., Brown, B. M., Davis, J. A., et al. (2008). Mouse cones require an arrestin for normal inactivation of phototransduction. *Neuron* 59: 462–474.

Phototransduction: Inactivation in Rods

V V Gurevich and E V Gurevich, Vanderbilt University, Nashville, TN, USA

Glossary

Arrestin (also known as S-antigen, 48-kDa protein, and rod or visual arrestin; systematic name: arrestin1) – A protein that selectively binds light-activated phosphorylated rhodopsin and blocks further signal transduction.

Guanylyl cyclase activating protein (GCAP) – A member of the superfamily of EF-hand-containing calcium-binding proteins. Rods express two homologs, GCAP1 and GCAP2, which in calcium-liganded form inhibit and in magnesium-liganded form enhance the activity of retinal guanylyl cyclase (retGC).

Phosphodiesterase (PDE) – Photoreceptor-specific cGMP phosphodiesterase, PDE6. Rod PDE6 is a heterotetramer, consisting of two nonidentical catalytic subunits (α- and β-) and two inhibitory γ-subunits. PDE6 rapidly hydrolyzes cyclic guanosine monophosphate (cGMP) upon its activation by transducin. In fully activated state, its catalytic activity approaches the limit set by the rate of cGMP diffusion.

RetGC – It is structurally related to receptor guanylyl cyclases. Rods express comparable levels of two homologs, RetGC1 and RetGC2.

Regulator of G-protein signaling 9 (RGS9-1) – Photoreceptor-specific short isoform of the regulator of G-protein signaling 9 expressed in both rods and cones. It interacts with the complex of the guanosine triphosphate (GTP)-liganded active α-subunit of transducin with PDEγ and facilitates its intrinsic GTPase activity, thereby directly inactivating transducin and indirectly PDE.

Rhodopsin – Light receptor, consisting of the protein part (opsin) and 11-*cis*-retinal covalently attached via Schiff base to a lysine in the seventh transmembrane domain. A member of the superfamily of G-protein-coupled receptors (GPCRs), also known as seven transmembrane domain receptors (7TMRs), the largest family of signaling proteins in animals (mammals have $\sim$1000 different GPCRs).

Rhodopsin kinase (RK) (systematic name: GRK1) – It is a member of the G-protein-coupled receptor kinase (GRK) family expressed in both rods and cones.

Transducin – Photoreceptor-specific heterotrimeric G protein that couples to light-activated rhodopsin. Its α-subunit belongs to Gi/o family.

As light sensors, vertebrate rod photoreceptors are a remarkable evolutionary achievement: rods yield amazingly low noise despite the presence of 10^8–10^9 molecules of the light receptor rhodopsin, and demonstrate single-photon sensitivity and a dynamic range of seven orders of magnitude of light intensity. This level of perfection is achieved through several unique structural and biochemical adaptations. The rod outer segment (OS) is a specialized signaling compartment containing rhodopsin molecules tightly packed in disks. It is separated from the inner segment (IS), which is a mitochondria-rich power station providing huge amounts of energy. Several soluble signaling proteins move between the two compartments depending on the illumination, ensuring their on-demand delivery to the OS. The OS concentrations of transducin (Td) and arrestin, the proteins that transmit and shut down rhodopsin signaling, respectively, change by at least 10-fold. An important functional feature of the rod is that every biochemical step in the pathway between photon capture and the change in synaptic output has its own dedicated shutoff mechanism.

What Needs to Be Inactivated: Overview of the Signaling Cascade

Rhodopsin activation by a photon of light is the first step in visual signaling. Due to extremely high concentration of its cognate G protein, Td, and rapid diffusion of both active rhodopsin (Rh*) and Td in the plane of the disk membrane, Rh* activates a molecule of Td every few milliseconds, generating 50–100 active Td (Td*) during its lifetime. These events occur in the two-dimensional space on the cytoplasmic surface of disk membranes. Each Td* binds the inhibitory γ-subunit of cyclic guanosine monophosphate (cGMP) phosphodiesterase (PDE6), turning the enzyme on. Each molecule of active PDE6 hydrolyzes several cGMP molecules per millisecond, producing a rapid drop in the cGMP concentration in the three-dimensional cytoplasmic space. This results in closure of cGMP-gated Na^+/Ca^{2+} channels on the plasma membrane. In rods, light activation of a single rhodopsin translates into the hydrolysis of $\sim$100 000 cGMP molecules. The channels are heterotetramers, with each subunit carrying a cGMP-binding site in its C-terminal domain. Highly cooperative cGMP binding to the four sites in the channel greatly increases its response to the change in cGMP concentration. The decrease of the inward current hyperpolarizes the rod, reducing neurotransmitter release in its output synapse.

Several features of the rod signaling machinery bring its light sensitivity within the range of its physical limit: the detection of single photons. First, the concentration of signaling molecules in the rod OS is orders of magnitude higher than in normal cells: $\sim$3 mM rhodopsin (compared to low nanomolar concentrations of related receptors elsewhere), $\sim$0.3 mM Td, $\sim$60 μM PDE, and so on. Second, all three signaling proteins involved have much lower basal activity than their counterparts in other cells. This results in an incredibly low noise level, making signal-to-noise ratio favorable for the detection of even an extremely weak signal. Third, very efficient shutoff mechanisms at every step of the pathway rapidly terminate the signaling, allowing for an exquisite subsecond temporal resolution of mammalian rods.

Shutoff of the Light-Activated Rhodopsin

Rhodopsin is a prototypical G-protein-coupled receptor. In contrast to $\sim$1000 other members of this superfamily, it has virtually no basal activity, because it is effectively suppressed by the covalently attached inverse agonist, 11-*cis*-retinal. Retinal is converted by light into the all-*trans* form, which is a potent agonist of rhodopsin. The fact that it remains covalently attached to the receptor (in contrast to other GPCRs where bound and free agonists are in dynamic equilibrium) ensures a powerful burst of signaling. Through a series of short-lived photoproducts, light-activated rhodopsin reaches the Metarhodopsin II (Meta II) state, which is an active form (Rh*) that couples to Td. Meta II is in equilibrium with the two other states, Meta I and Meta III, which are believed to be inactive, or at least considerably less active than Meta II. Ultimately, all-*trans*-retinal dissociates, yielding empty protein opsin, which has orders of magnitude lower ability to activate Td than Meta II. However, this spontaneous deactivation of rhodopsin is inadequate as a shutoff mechanism for two reasons. At physiological temperatures, rhodopsin decay takes about a minute, which would greatly compromise temporal resolution. Moreover, the activity of opsin, which is much higher than that of dark rhodopsin, would generate considerable noise, compromising rod sensitivity. Therefore, rods use a sophisticated two-step mechanism to achieve rapid and complete rhodopsin deactivation.

First, light-activated rhodopsin is phosphorylated by rhodopsin kinase (RK). Similar to Td, RK is activated by binding to Rh*. Therefore, it selectively phosphorylates the active form of rhodopsin. It should be noted that at low light levels, RK was reported to phosphorylate multiple rhodopsin molecules for each light-activated one, likely by targeting neighboring inactive rhodopsins in the crowded disk membrane. In mammals, RK activity is believed to be held in check by its interaction with Ca^{2+}-loaded recoverin. As a result, RK is fully unleashed only after a brief delay, which allows Td activation to continue until the Ca^{2+} concentration in the rod actually drops. However, low affinity of recoverin for Ca^{2+} ($K_D \sim 5$ μM) is the weak point of this model. Current estimates of the Ca^{2+} concentrations in the dark- and light-adapted mouse OS are $\sim$250 nM and $\sim$25 nM, respectively, so that only a very small fraction of recoverin would be Ca^{2+}-loaded in either. In addition, unlike RK, recoverin predominantly localizes in the inner segment. Still, rods express 50 molecules of recoverin for each RK, so a relatively small change in the Ca^{2+} occupancy of a fraction of recoverin present in the OS could conceivably play a role in RK regulation.

Phosphorylation *per se* reduces, but does not abolish the ability of rhodopsin to activate Td. In the next step, arrestin binds active phosphorylated rhodopsin (P-Rh*), shielding its cytoplasmic tip and precluding further Td interaction. Arrestin apparently remains bound until rhodopsin decays to opsin, and very likely even longer, until opsin is regenerated with 11-*cis*-retinal to the truly inactive dark rhodopsin. Arrestin has several dedicated phosphate-binding residues and other elements that specifically interact with light-activated rhodopsin independently of its phosphorylation state. These partial interactions mediate relatively low-affinity binding to dark P-Rh and unphosphorylated Rh*, respectively. Arrestin elements participating in these interactions also serve as sensors, allowing arrestin to test the functional state of the rhodopsin molecule it encounters and then quickly dissociate from its low-affinity targets, dark Rh, dark P-Rh, or Rh*. In contrast to all other forms, P-Rh* simultaneously engages both sets of elements. This turns the two sensors on at the same time, allowing the arrestin transition into a high-affinity rhodopsin-binding state. This transition involves a global conformational change in arrestin, which mobilizes additional arrestin elements for the interaction. Thus, arrestin works as a molecular coincidence detector, swinging into action only when the rhodopsin molecule it encounters is both active and phosphorylated. The model of sequential multisite interaction readily explains exquisite arrestin selectivity, that is, manifold difference in arrestin binding to Rh* and dark P-Rh on the one hand, and to its preferred target P-Rh* on the other. The salt bridge between positively charged Arg175 and negatively charged Asp296, which is one of the intramolecular interactions holding arrestin in its basal state, was identified as the main phosphate sensor in arrestin. Rhodopsin-attached phosphates bind Arg175 and neutralize its charge, thereby breaking the salt bridge and facilitating arrestin transition into its active conformation. The reversal of either charge by targeted mutagenesis yields mutants with reduced need for rhodopsin-attached phosphates that bind Rh* with much higher affinity than wild-type protein.

Rhodopsin has multiple phosphorylation sites in its C-terminus. The issue of the number of rhodopsin-attached phosphates necessary for high-affinity arrestin binding was resolved only recently. Studies performed *in vitro* with rhodopsin carrying defined number of phosphates and *in vivo* with mice expressing rhodopsin mutants with different number of sites show that rhodopsin multi-phosphorylation is required. A single rhodopsin-attached phosphate does not appreciably increase arrestin affinity, two somewhat enhance the binding, and three phosphates are necessary for high-affinity interaction *in vitro* and for the rapid shutoff of photoresponse *in vivo*. Whereas arrestin binding does not further increase when Rh* has more than three phosphates, the presence of additional phosphorylation sites on Rh accelerates the shutoff of the photoresponse *in vivo*. This likely reflects the kinetic effect of the abundance of sites that remain available to RK on partially phosphorylated rhodopsin. For example, rhodopsin with three sites would have only one possible RK target left after the incorporation of two phosphates, whereas rhodopsin with six sites would still have four available targets at the same level of phosphorylation. Thus, supernumerary sites would ensure that the magic number of three phosphates per rhodopsin is achieved faster.

Inactivation of Td and PDE

Td is a prototypical heterotrimeric G protein consisting of α-, β-, and γ-subunits. In the inactive state, the αβγ-trimer has guanosine diphosphate (GDP) in the nucleotide-binding site of the α-subunit. In this state, lipid modifications of both α-(N-terminal myristoyl) and γ-(C-terminal farnesyl) subunits provide a fairly strong membrane anchor. This restricts the Td diffusion to the plane of the disk membrane and enforces the orientation favorable for Rh* interaction, thereby maximizing its chances of encountering active rhodopsin and being activated by it. The Td interaction with Rh* opens its nucleotide-binding pocket, whereupon GDP promptly falls out and is immediately replaced by GTP simply because the latter is much more abundant in the cytoplasm. The GTP-liganded α-subunit dissociates from Rh* and βγ-dimer. Tdα-GTP binds the inhibitory γ-subunit of cGMP PDE, greatly increasing PDE activity by relieving the inhibition. Importantly, the separation of the two parts of Td heterotrimer dramatically weakens their membrane anchoring, so that active Tdα-GTP can jump off the disk where it was generated by Rh* and activate PDE on neighboring discs, spreading the signaling in three dimensions. Due to its very high catalytic activity (kcat $\sim$2000 s^{-1} per subunit), active PDE rapidly reduces cGMP concentration in its vicinity, which leads to the closure of cGMP-gated channels and hyperpolarization of the rod within milliseconds of rhodopsin activation by light (**Figure 1**).

Similar to other heterotrimeric G proteins, Tdα has GTPase activity, which serves as a built-in self-inactivation mechanism. However, the intrinsic GTPase of free Tdα is very slow. Interaction of Tdα with PDE γ-subunit increases the activity of its GTPase. The interaction of Tdα–GTP–PDEγ complex with rod-specific GTPase activating protein (GAP) increases the GTPase activity even further. GAP consists of the short isoform of RGS9, Gβ5 (homolog of G-protein β-subunits), and another protein that provides membrane anchor for the complex, RGS9 anchoring protein (R9AP). Low basal GTPase of Tdα gives it time to diffuse around searching for PDE to activate without losing the signal in the transmission. The dramatic acceleration of the GTPase activity of Tdα by the PDE and GAP ensures that the signal is terminated quickly after it is received by PDE, improving the temporal resolution of the photoreceptor cell. The recent finding that the expression of the GAP complex in rods increases the rate of the response shutoff in a dose-dependent manner convincingly demonstrated that the inactivation of Tdα–GTP–PDEγ complex is the rate-limiting step in this process. These elegant experiments also revealed that when this step is maximally accelerated, the recovery kinetics becomes dominated by some other process with the time constant of $\sim$80 ms. This number sets the upper limit for the next slowest step, which could be one of the following: the average lifetime of active Rh*; the release of PDEγ from Tdα-GDP; reassociation of PDEγ with PDE catalytic subunits; or even the time free Tdα-GTP spends searching for PDE and/or docking to it.

Resynthesis of cGMP and Restoration of Calcium Level

Obviously, to return to its initial state and become ready to respond to the next photon with the same vigor, the rod photoreceptor needs to do more than just turn off Rh* and all Td and PDE molecules activated by it. The response leaves, in its wake, substantially reduced cytoplasmic cGMP concentration and very low intracellular calcium due to the closure of the cGMP-gated channels that are responsible for the bulk of Ca^{2+} entry into the OS. Photoreceptors are equipped with an

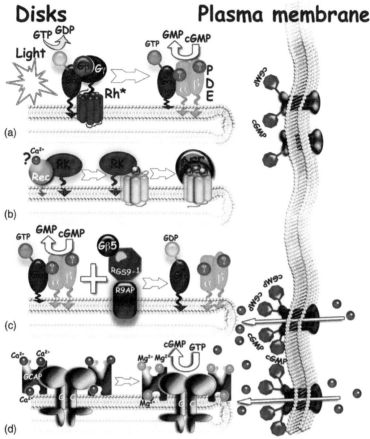

Figure 1 Biochemical mechanisms of signal inactivation in rods. (a) Visual amplification cascade. Light-activated rhodopsin (Rh*) catalyzes GDP/GTP exchange on visual G protein, Td, sequentially activating dozens of Td molecules. Inactive Td is an αβγ-heterotrimer, whereas upon activation the GTP-liganded α-subunit dissociates from the βγ-dimer and binds the inhibitory γ-subunit of rod PDE (which is an αβγ₂ heterotetramer). This activates PDE, which hydrolyzes massive amounts of cGMP (over 100000 molecules per one Rh*). The decrease in cytoplasmic cGMP closes cGMP-gated cation channels on the plasma membrane (right panel). Channel closure reduces the influx of Na$^+$ and Ca^{2+}, hyperpolarizing the cell up to 1 mV per one Rh*. (b) Rh* is phosphorylated by the rhodopsin kinase (RK, systematic name GRK1), which is expressed in rods and cones of all vertebrates. In the dark, RK may be kept away from rhodopsin via its interaction with Ca^{2+}-liganded recoverin (Rec). Multiphosphorylation prepares Rh* for arrestin (Arr) binding. Arrestin shields the cytoplasmic tip of rhodopsin, sterically blocking its interactions with transducin, thereby completing rhodopsin inactivation. (c) The intrinsic GTPase activity of the Td α-subunit serves as a built-in inactivation mechanism. Its interaction with PDEγ and RGS9-1 (which exists in constitutive complex with Gβ5 and membrane anchoring protein R9AP) greatly facilitates GTP hydrolysis, ensuring rapid inactivation of Td and PDE. (d) In the dark, retinal guanylyl cyclase (GC) is inhibited by Ca^{2+}-liganded GCAPs. Light-induced closure of the cGMP-gated channels results in the drop in cytoplasmic Ca^{2+}. Its replacement with Mg^{2+} on the metal-binding sites of GCAPs converts them into GC activators. GC replenishes the cytoplasmic cGMP and consequent opening of the channels restores cytoplasmic Ca^{2+}, thereby turning off GC. Rhodopsin is shown as a bundle of seven transmembrane domains; Rhodopsin-attached phosphates are shown as spheres; lipid modifications anchoring recoverin, GRK1, α- and γ-subunits of Td, and catalytic α- and β-subunits of rod PDE are shown as membrane-imbedded arrows. Arr, arrestin1; GC, gyanylyl cyclase; Rec, recoverin; Rh*, light-activated rhodopsin; RK, rhodopsin kinase.

ingenious negative-feedback mechanism that translates the drop in Ca^{2+} resulting from the reduction in cGMP level into a signal to replenish it. In photoreceptors, cGMP is synthesized by retinal guanylyl cyclases (retGCs). RetGCs are related to a family of hormone-regulated guanylyl cyclases, such as atrial natriuretic factor receptor, which have extracellular hormone-binding domain connected via a single transmembrane helix to the intracellular guanylyl cyclase domain. Similar to these receptors, retGCs are dimeric, with each monomer

equipped with a catalytic domain and an extracellular domain. Interestingly, Mg^{2+} and GTP are bound by two different subunits forming the active catalytic site. As far as we know, the extracellular domain of retGCs neither binds any ligands nor participates in the enzyme regulation. Instead, the activity of retGCs is tightly regulated by their interaction via intracellular elements with GCAPs. GCAPs, as well as recoverin, are members of the neuronal calcium sensor protein branch of the superfamily of calcium-binding proteins containing EF hands (that includes calmodulin).

Similar to other members of this family, GCAPs have four EF hands, three of which actually bind divalent cations. Strictly speaking, GCAPs are Ca^{2+}–Mg^{2+}-binding proteins. The word activating in their name is a bit misleading: Mg^{2+}-liganded GCAPs activate retGCs, whereas Ca^{2+}-liganded forms actually inhibit cGMP synthesis. Thus, in dark-adapted rods with high free-Ca^{2+} concentrations (estimates range from 250 to 600 nM in different species), GCAPs keep the retGC activity at low level. This makes perfect sense, because high Ca^{2+} indicates that there is enough free cGMP ($\sim$2–5 μM) to keep the channels open. Light-induced decrease of intracellular Ca^{2+} (to 5–50 nM, based on different estimates) is the direct result of channel closure, reflecting reduced cGMP in need of replenishing. The loss of bound Ca^{2+} and its replacement by Mg^{2+} (which is always $\sim$1 mM in the cytoplasm) switches GCAPs from the inhibitory to the activating mode exactly when rapid cGMP synthesis is necessary to restore its level. Increasing cGMP opens more channels, thereby gradually restoring Ca^{2+}. Rising Ca^{2+} displaces Mg^{2+} on GCAPs, progressively reducing retGC activity, so that the cell returns to the initial state. After a dim flash, this process often overshoots, leading to a transient increase in the cGMP and Ca^{2+} concentration, likely because PDE is inactivated faster than retGC. The absence of GCAPs slows down cGMP resynthesis, so that light-induced PDE activity results in a more profound decrease of cGMP than in the normal rod. This results in closure of more channels and greatly increases the amplitude of single-photon response. This compromises temporal resolution, prolonging the rising and falling phase of the light response, and limits the working range of rods to lower light levels.

Interestingly, vertebrate photoreceptors express two isoforms of retGC, retGC1 and retGC2, and at least two GCAPs, GCAP1 and GCAP2. The presence of two isoforms of each protein in rods of all vertebrates, including fish, clearly indicates that the different isoforms have nonredundant functions. RetGCs are membrane proteins, suggesting that retGC1 and retGC2 may be localized to different membranes within the OS. RetGC1 was reliably detected in disks, and the possibility that a fraction may also be present in the plasma membrane remains open. The localization of RetGC2 was not studied with sufficient spatial resolution. Since retGC is a dimer, two isoforms of RetGC could give rise to three types of dimers, two homo- and one heterodimer. The fact that each subunit interacts with either GCAP1 or GCAP2 further expands the number of combinatorial possibilities. Definitive experiments, such as knockouts of individual isoforms of either protein, singly and in different combinations, are needed to fully elucidate the biochemistry of the Ca^{2+} feedback mechanism. A recent study of GCAP2 knockout mice revealed that although each GCAP is responsible for about half of the total retGC activation, the functions of the two proteins are quite distinct. Due to lower affinity for Ca^{2+}, GCAP1 switches to the activation mode as soon as the concentration of Ca^{2+} begins to fall, whereas GCAP2 responds later, when Ca^{2+} levels drop further. Thus, together the two GCAPs ensure graded increase in retGC activity in a wider range of Ca^{2+} concentrations than either one could have covered alone.

Another issue in need of clarification is the physiological role of a remarkable buffering capacity of the OS cytoplasm for both second messengers. According to current estimates, total cGMP in the OS is as high as $\sim$50 μM, with only 2–5 μM of it being free and the rest bound to the noncatalytic sites on PDE α- and β-subunits. The polycationic region of PDEγ appears to stabilize the interaction of cGMP with these sites. Reciprocally, the presence of cGMP in noncatalytic sites enhances the interaction of the α- and β-subunits with PDEγ. This mechanism implies that PDE activation by Td would release cGMP from noncatalytic sites. The role of this event in the photoresponse remains unclear. Similarly, free Ca^{2+} represents only a small fraction of the total Ca^{2+} in the OS cytoplasm, the rest being bound to several abundant proteins, such as recoverin, GCAPs, and calmodulin. Ca^{2+} binding by these proteins is a two-way street: on the one hand, it critically regulates their function, on the other hand, by soaking up Ca^{2+}, they significantly change its concentration, modulating the input they respond to. Obviously, Ca^{2+} and cGMP buffering cannot be separated by purely experimental means from other functional modalities of the proteins involved. Therefore, rigorous experimentation must be supplemented with detailed biochemically realistic mathematical modeling to distinguish between the effects of binding on protein activity and on the concentration of free second messenger in the cytoplasm, which is necessary to elucidate the exact biological roles of both.

Light-Dependent Protein Translocation and Rod Signaling

Arrestin localization to the OS in the light and to the IS in the dark was first described in 1985, before the role of this protein in signal termination was established. The subsequent discovery that Td also translocates in a light-dependent fashion, moving in the opposite direction, suggested an idea that translocation may underlie well-known adaptation of rods to different light levels. Preferential localization in the dark-adapted rod of a signal transducer to the rhodopsin-rich OS and a signal terminator to the IS could increase light sensitivity by slowing down shutoff. Conversely, the removal of Td from the OS and accumulation of arrestin in this compartment in the light could significantly reduce it by decreasing the number of Td molecules activated by

Rh* and speeding up rhodopsin inactivation. Subsequent studies showed that the fraction of recoverin, which presumably slows down rhodopsin phosphorylation by keeping RK away from Rh*, in the OS decreases dramatically from 12% in the dark to less than 2% in the light (the bulk of recoverin localizes to the IS in both conditions). The amplification constant in light-adapted rods was found to decrease ~10-fold, in line with the reduction of Td concentration in the OS. While these results support the idea that Td translocation plays a role in rod adaptation, by fully explaining the changes in sensitivity by Td movement alone, they effectively rule out any significant contribution of the translocation of other proteins. The recent finding that phosducin, which interacts with free βγ-dimer released upon Td activation and demonstrates robust translocation from the OS in the light, does not contribute to rod adaptation, supports this notion. The movement of Td, arrestin, phosducin, and recoverin in both directions is a relatively slow process that takes many minutes, which does not seem adequate to explain much faster photoreceptor adaptation. The translocation of arrestin in both directions is energy independent. It is driven in the dark by its low-affinity interactions with microtubules, particularly abundant in IS, and in the light by its binding to light-activated forms of rhodopsin. These findings suggest that the translocation of arrestin and other proteins is more likely to play a role in rod survival during daytime than in relatively fast light/dark adaptation. However, the translocation of different proteins may have distinct functions, which to a large extent remain to be elucidated.

Why Rods Do Not Have an Action Potential

In most neurons, extracellular Ca^{2+} enters presynaptic terminals during an action potential. A brief increase in its local concentration triggers transient exocytosis of neurotrasmitter-containing vesicles. In contrast, vertebrate rod photoreceptors work backward. In the dark, rods are partially depolarized (OS membrane potential is about −35 mV) due to massive influx of Na^+ and Ca^{2+} ions through cGMP-gated channels. This results in continuous release of the neurotransmitter (L-glutamate) from ribbon synapses. By virtue of closing cGMP-gated channels, light of increasing intensity induces progressive hyperpolarization of rods up to −60 mV (a change of ~25 mV). The activation of a single rhodopsin changes the membrane potential by as much as 1 mV. Thus, light intensity is encoded in the extent of hyperpolarization, which determines the magnitude of the decrease of neurotransmitter release. This mechanism makes the signaling graded, in contrast to the all-or-nothing type in neurons with a conventional action potential. It

directly couples the change in membrane potential with synaptic activity, so that both the closure of the cGMP-gated channels upon light stimulation and their reopening upon signal termination described above immediately translate into corresponding changes in neurotransmitter release. The absence of thresholds ensures that the information is not lost in transmission, so that the brain can take full advantage of the single-photon sensitivity of the rod photoreceptors. In addition, this mechanism creates a natural ceiling: a full stop of neurotransmitter release is the maximum possible effect of the illumination of any intensity.

Conclusions

In many respects, rod photoreceptors are virtually perfect light sensors that combine single-photon sensitivity with a surprisingly wide dynamic range. Exquisitely timed and extremely efficient inactivation at every step of the signaling cascade between light absorption by rhodopsin and changes in the membrane potential plays an important role in their function. Not surprisingly, molecular errors in this complex multistep inactivation mechanism due to mutations in key proteins underlie a variety of congenital visual disorders in humans, ranging in severity from night blindness to retinal degeneration.

See also: Light-Driven Translocation of Signaling Proteins in Vertebrate Photoreceptors; Phototransduction: Adaptation in Cones; Phototransduction: Adaptation in Rods; Phototransduction: Inactivation in Cones; Phototransduction: Phototransduction in Cones; Phototransduction: Phototransduction in Rods; Phototransduction: Rhodopsin.

Further Reading

Burns, M. E. and Arshavsky, V. Y. (2005). Beyond counting photons: Trials and trends in vertebrate visual transduction. *Neuron* 48: 387–401.

Cote, R. H. (2006). Photoreceptor phosphodiesterase (PDE6): A G-protein-activated PDE regulating visual excitation in rod and cone photoreceptor cells. In: Beavo, J. A., Francis, S. H., and Houslay, M. D. (eds.) *Cyclic Nucleotide Phosphodiesterases in Health and Disease*, pp. 165–193. Boca Raton, FL: CRC Press.

Dizhoor, A. M., Olshevskaya, E. V., and Peshenko, I. V. (2006). Calcium sensitivity of photoreceptor guanylyl cyclase (RetGC) and congenital photoreceptor degeneration: Modeling *in vitro* and *in vivo*. In: Philippov, P. P. and Koch, K.-W. (eds.) *Neuronal Calcium Sensor Proteins*, pp. 203–219. New York: Nova Science Publishers, Inc.

Gurevich, V. V. and Gurevich, E. V. (2004). The molecular acrobatics of arrestin activation. *Trends in Pharmacological Science* 25: 105–111.

Gurevich, V. V., Hanson, S. M., Gurevich, E. V., and Vishnivetskiy, S. A. (2007). How rod arrestin achieved perfection: Regulation of its availability and binding selectivity. In: Kisselev, O. and Fliesler, S. J. (eds.) *Signal Transduction in the Retina, Methods in Signal Transduction Series*, pp. 55–88. Boca Raton, FL: CRC Press.

Phototransduction in *Limulus* Photoreceptors

R Payne and Y Wang, University of Maryland, College Park, MD, USA

Glossary

Confocal fluorescence microscopy – An optical microscopy method that uses a focused laser spot to measure fluorescence within an extremely small volume.

Cytosol – The liquid portion of a cell's contents, or cytoplasm.

Depolarization – A positive-going change in the membrane potential of a cell.

Diacylglycerol (DAG) – A glyceride consisting of two fatty acid chains covalently bound to a glycerol molecule through ester linkages.

d-*myo*-Inositol (1,4,5) trisphosphate (IP$_3$) – An inositol derivative having three phosphate groups covalently bound to the inositol ring.

Guanine-nucleotide-binding protein (G protein) – A family of signaling proteins that are activated by the exchange of guanine triphosphate for guanine diphosphate bound to the protein.

On-cell patch clamp – An electrophysiological recording technique that uses a polished glass micropipette to measure currents flowing through a small patch of cellular plasma membrane containing ion channels.

Phosphoinositide – A phospholipid containing a polar inositol headgroup.

Smooth endoplasmic reticulum (SER) – A network of membrane sacs found in animal cells that, among other things, functions as a store of calcium ions.

Transient receptor potential (TRP) channels – A widespread family of ion channels. The light-activated channels of *Drosophila* are founding members of the family.

Arrangement of Eyes in *Limulus*

American horseshoe crabs (*Limulus polyphemus*) have 10 eyes. They have two large lateral compound eyes, each containing about 1000 clusters of photoreceptors or ommatidia. A small lens within each ommatidium focuses light from a small patch of visual space onto each photoreceptor cluster, which transmits information about local changes in light intensity to the brain through a nerve fiber. There are five additional eyes on the top side of its first (anterior) major

body section, two median eyes, one endoparietal eye, and two rudimentary lateral eyes. Two ventral eyes are located on the underside of the animal above the mouth (**Figure 1**). Photoreceptors located on the telson (tail) constitute the 10th eye. Of these eyes, the lateral compound, median ocellar and ventral eyes have been extensively studied. All have microvillar photoreceptors, in which the visual pigment, rhodopsin, is embedded in the membrane of finger-like projections of the plasma membrane called microvilli (**Figure 2**). Each eye has allowed for a study of different aspects of invertebrate vision. The large compound eyes have been favorite preparations for the study of image processing by compound eyes, the biochemistry of visual transduction, and the circadian control of visual sensitivity. The medial eye has been studied for its sensitivity to ultraviolet (UV) light which has allowed the elucidation of the physiological consequences of the reversible photoisomerization of invertebrate rhodopsin. Lastly, the ventral eyes have played a role in the understanding of phototransduction in invertebrate photoreceptors.

The Microvillus is the Cellular Structure Mediating Visual Transduction

Central to the performance of each photoreceptor cell is the rhabdomere – an array of photoreceptive microvilli positioned so as to maximally absorb light entering the eye. Each microvillus is a cylindrical outgrowth of the plasma membrane, 50–80 nm in diameter and 0.5–2 μm in length (**Figure 2**). Electron micrographs often show an axial filament within each microvillus. The axial filament contains a bundle of actin filaments with their + ends directed toward the tip of the microvillus. The actin filaments appear to extend through the bottom of the microvillus into the cytoplasm, through fenestrations in submicrovillar cisternae (SMC) of smooth endoplasmic reticulum (SER). The actin cytoskeleton is responsible for the presence of an unconventional motor protein, myosin III, in the microvilli of *Limulus* lateral eye photoreceptors. The membrane of a typical microvillus contains 1000 or more particles, which are presumed to be mostly molecules of the visual pigment, rhodopsin, which absorbs light and initiates the physiological response of the photoreceptor. A photoreceptor might possess 10^5 microvilli, resulting in a total rhodopsin content of $\sim 10^8$ molecules, comparable to that of vertebrate retinal photoreceptors.

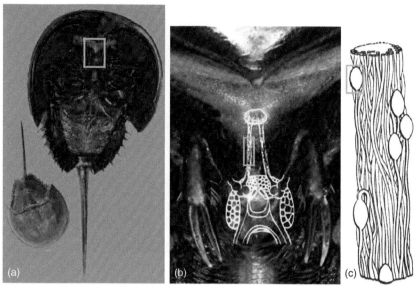

Figure 1   (a) Underside and top side (inset) of the Atlantic horseshoe crab (*Limulus polyphemus*). The green box shows the area enlarged in (b). (b) Enlarged view of the underside, just above the mouthparts. The drawn overlay shows the arrangement of the ventral nerves (green box, enlarged in (c)) and ventral eyes under the skin. The ventral nerves lead from the brain (just above the mouthparts at the bottom of the figure) to the ventral eyes above. (c) Diagram of ventral nerve axons and attached photoreceptor cell bodies (green box). Drawings in (b) and (c) are adapted from Calman, B. G. and Chamberlain, S. C. (1982). *Journal of General Physiology* 80: 839–862.

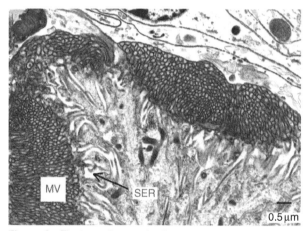

Figure 2   Electron micrograph of a section through the R-lobe of a ventral photoreceptor, showing the microvilli (MV) and sub-microvillar cisternae of smooth endoplasmic reticulum (SER; arrow). Adapted from Dabdoub, A., Payne, R., and Jinks, R. N. (2002). *Journal of Comparative Neurology* 442: 217–225. Copyright 2009 Wiley-Liss, Inc.

Studies of Visual Transduction Using *Limulus* Ventral Photoreceptors

The *Limulus* ventral photoreceptor (**Figure 3**) is a highly polarized cell divided into two lobes, analogous to the inner and outer segments of vertebrate retinal photoreceptors. The rhabdomeral (R) lobe bears microvilli on its plasma membrane and is therefore light sensitive. The light-insensitive arhabdomeral (A) lobe contains the cell's

nucleus. An axon projects from the A lobe toward the animal's central nervous system. Ventral photoreceptors were originally chosen as a model for invertebrate phototransduction because of their large size (>200 μm). This facilitates insertion of multiple electrodes and makes it possible to clamp the membrane potential of the cells, measure electrical current flow across the plasma membrane, and inject compounds of interest into the cytoplasm (**Figure 4**).

The essential electrical response to illumination is the activation of a very large flow of current into the cell, carried mostly by sodium ions (**Figure 4**). The result is a depolarization (positive-going change) of the cell membrane which is graded with light intensity, of up to 60 mV. The reversal potential of the light-sensitive current is between +10 mV and +20 mV and its dependence on extracellular ion concentrations indicates that this conductance is sodium- and potassium-, but not Ca^{2+}-permeable. This is in contrast to the light-activated transient receptor potential (TRP) channel conductance in the photoreceptors of the fruit fly, *Drosophila*, which is highly Ca^{2+} permeable.

The study of ventral photoreceptors has revealed that they have remarkable performance characteristics, most notably the very large amplification of the transduction process. Amplification refers to the amount of charge that is carried across the plasma membrane as a result of excitation by a single photon. In ventral photoreceptors, this gain can be directly measured because the single-photon response, termed a quantum bump, is easily recorded using glass micropipettes. In response to very dim illumination,

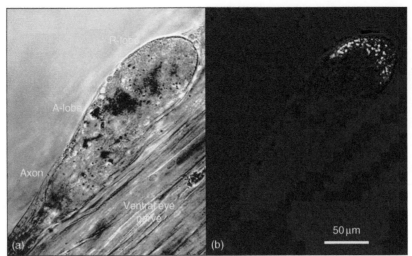

Figure 3   (a) Light micrograph of a ventral photoreceptor. R-lobe, rhabdomeral lobe; A-lobe, arhabdomeral lobe. (b) Immunolocalization (blue) of rhodopsin within the same photoreceptor cell. The ventral photoreceptor has two lobes, a light-sensitive rhabdomeral lobe (R-lobe), which bears rhodopsin-containing microvilli on its plasma membrane and is analogous to the outer segments of vertebrate photoreceptors, and a light-insensitive arhabdomeral lobe (A-lobe), which is analogous to the inner segment of vertebrate photoreceptors. From Battelle, B. A. et al. (2001). *Journal of Comparative Neurology* 435: 211–225. Copyright 2009 Wiley-Liss, Inc.

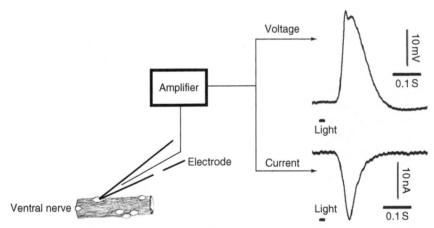

Figure 4   Experimental arrangement for recording electrical activity (membrane potential and voltage clamp currents, right) from ventral photoreceptors impaled with a glass micropipette, left.

quantum bumps over 10 mV in amplitude occur randomly as individual photons are effectively absorbed by rhodopsin molecules. Under voltage clamp, the peak current across the membrane generated by an effectively absorbed photon can exceed 1 nA and appears to be generated over several square microns of membrane surface, containing hundreds of microvilli. The comparatively large currents flowing indicate that quantum bumps are caused by the passage across the plasma membrane of hundreds of millions of cations through ion channels. By contrast, in the smaller *Drosophila* photoreceptors or amphibian rods, a single-photon event involves a maximum current of less than 10 pA. *Limulus* photoreceptors achieve this large amplification in only 100–200 ms, faster than amphibian rods. A further remarkable feature of *Limulus* photoreceptors is their broad dynamic range. Whereas most vertebrate rods

work over about a 4-log-unit range of light intensity before saturating, *Limulus* photoreceptors work over 8. They achieve this range through a strong adaptation process that reduces amplification at high light intensity. The large dynamic range of these cells obviates the need for the dual system of photoreceptors (rods and cones) used to achieve a large dynamic range in the vertebrate eye.

The Light-Sensitive Conductance Consists of the Summed Effect of Conventional Ion Channels

The glial cells that surround individual ventral photoreceptors can be removed, exposing the plasma membrane surface. This preparation allows On-cell patch-clamp recording of single ion channels in the microvillar membrane. As expected, channels whose opening was triggered by light were recorded by this method, with single-channel conductance ranging from 18 to 50 pS. Identification of these channels as the light-activated conductance requires establishing a close correspondence between the properties of individual channel currents and those of photocurrents recorded from the whole cell. Depolarizing potentials applied in the dark did not open the channels; therefore the light-activated opening was not just a secondary consequence of the depolarizing receptor potential. Light-evoked single-channel activity was graded with light intensity, reduced by light adaptation, and the single-channel currents reversed in the same membrane potential range as the macroscopic photocurrent. While these characteristics are consistent with the recorded channels being those that carry the light-activated current, definitive molecular or pharmacological proof is still needed.

The Response of the Ventral Photoreceptor is Mediated by the Phosphoinositide Cascade

A large body of evidence now demonstrates that the phosphoinositide (PI) pathway links the absorption of light to the activation of ion channels in the plasma membrane of invertebrate microvillar photoreceptors (**Figure 5**). The PI pathway is a mechanism for releasing intracellular messengers upon the activation of a receptor

protein, using inositol phospholipids as a substrate. In microvillar photoreceptor cells, the receptor protein is rhodopsin and the cascade is localized to the membrane of the microvilli that cover the plasma membrane of the light-sensitive R-lobe. Activated rhodopsin catalyzes the exchange of guanine triphosphate (GTP) for guanine diphosphate (GDP) bound to the alpha subunit of a heterotrimeric GTP-binding protein of the G_q subfamily ($G_{q\alpha}$), which in turn activates phospholipase C (PLC). PLC cleaves phosphatidylinositol 4,5 bisphosphate (PIP_2), a minor membrane phospholipid, into a lipid messenger, diacylglycerol (DAG), and the water-soluble messenger, d-*myo*-inositol 1,4,5 trisphosphate (IP_3). In support of this biochemical pathway, light-activated PIP_2 hydrolysis and/or IP_3 production has been reported in ventral photoreceptors of *Limulus*. $G_{q\alpha}$ has been amplified and sequenced from *Limulus* ventral eye tissue and immunolocalized to the rhabdomeral microvilli. Pharmacological agents that inhibit PLC, neomycin and U-73122, dramatically desensitize the light response of *Limulus* photoreceptors.

The PI Cascade Generates at Least Two Intracellular Messenger Molecules

The PI cascade generates two messenger molecules with very different properties (**Figure 5**) DAG is essentially confined to the plasma membrane, while IP_3 can diffuse into the surrounding cytoplasm. In addition, the decline of the precursor, PIP_2 may act as an additional signal within

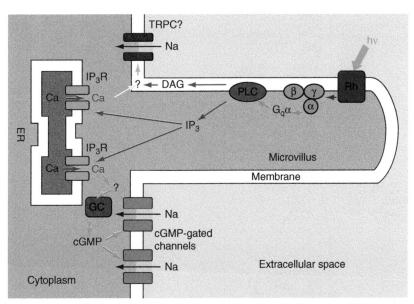

Figure 5 Diagram of mechanisms proposed to mediate phototransduction within a microvillus of the *Limulus* photoreceptor. cGMP, cyclic guanosine monophosphate; DAG, diacylglycerol; ER, endoplasmic reticulum; $G_q\alpha$, alpha subunit of a heterotrimeric GTP-binding protein of the G_q subfamily; β,γ, beta and gamma subunits of a heterotrimeric GTP binding protein of the G_q sub-family; GC, guanylate cyclase; IP_3, d-*myo*-inositol (1,4,5) trisphosphate; IP_3R, d-*myo*-inositol (1,4,5) trisphosphate receptor; PLC, phospholipase C; Rh, rhodopsin; TRPC, transient receptor channel C.

the membrane. The difficult task of assessing the relative roles of these messengers in activating the light-sensitive ion channels has dominated research on the invertebrate phototransduction cascade.

Roles of IP₃ and Intracellular Ca²⁺ Ions in Excitation of *Limulus* Ventral Photoreceptors

There is no evidence that IP_3 can itself activate the light-sensitive channels of ventral photoreceptors. The best-characterized alternative target of IP_3 is a Ca^{2+} channel, the IP_3 receptor protein (IP_3R), located in the membrane of endoplasmic reticulum (ER). IP_3 therefore releases Ca^{2+} from intracellular stores within the ER. In *Limulus* photoreceptors, suitable calcium stores are located close to the base of the rhodopsin-containing microvilli – the subrhabdomeral cisternae (SMC) of SER. Extensions of the SMC are juxtaposed less than 100 nm from the bases of the microvilli. The cytoplasm of both the R lobe containing the SER and, to a lesser extent, the A lobe are immunoreactive when probed with an anti-IP_3R antibody.

In darkness, ventral photoreceptors, like most cells, maintain a low free cytosolic Ca^{2+} concentration ($[Ca^{2+}]_i$) of 200–600 nM. Release of calcium from the SMC results in a very large elevation of $[Ca^{2+}]_i$ during the first few hundred milliseconds of the light response. Confocal fluorescent light microscopy has enabled the measurement of the light-induced elevation of $[Ca^{2+}]_i$ in ventral photoreceptors at spots within 4 μm of the microvillar membrane in the R lobe (**Figures 6** and **7**).

Following a very bright flash, $[Ca^{2+}]_i$ begins to rise after a latent period of approximately 20 ms. Thereafter, $[Ca^{2+}]_i$ rises at an initial rate of 1–$2\,mM\,s^{-1}$ to reach a peak of 100–200 μM within 200 ms. For less-intense flashes, the peak elevation of $[Ca^{2+}]_i$ is graded with flash intensity, increasing from 2 μM to more than 140 μM as light intensity increases from 10 effective photons to 10 000 effective photons. For the dimmest flashes so far investigated, these concentrations represent ∼600 free Ca^{2+} ions per effective photon generated within the confocal measurement volume.

Very high levels of $[Ca^{2+}]_i$ are reached only transiently upon illumination. Following a dim or moderate-intensity light flash, $[Ca^{2+}]_i$ falls close to its resting value within 1 s, although a small lingering elevation may persist for up to a minute afterward. Even during sustained intense illumination, $[Ca^{2+}]_i$ falls within 5 s to a sustained plateau elevation of less than 20 μM. As expected, if calcium release from stores occurs, the light-induced rise in $[Ca^{2+}]_i$ is unaltered by removal and chelation of extracellular Ca^{2+}, but is severely reduced by pharmacological agents expected to deplete Ca^{2+} stores.

IP₃ Can Release Ca²⁺ from the SER

Microinjections of IP_3, or photolysis of caged IP_3, rapidly release Ca^{2+} from the R lobe in darkness (**Figure 8**). Photolysis of caged IP_3 by UV light delivered to a spot beneath the microvillar membrane results in local elevations of Ca^{2+} that are comparable in magnitude and rate

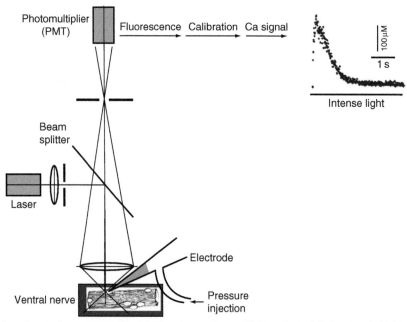

Figure 6 Diagram of confocal microscopy method (left) and recordings of light-activated Ca2+ signals (right) from a Ca²⁺-sensitive fluorescent indicator dye (indicated in green) injected into the photoreceptor and recorded close to the microvillar membrane of the R-lobe.

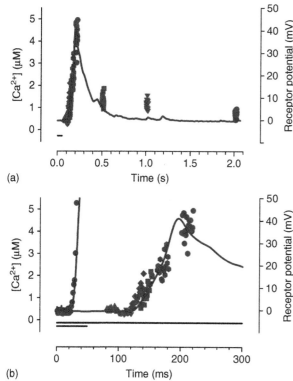

(a)

(b)

Figure 7 (a) Receptor potential (solid line) and reconstructed elevation of $[Ca^{2+}]_i$ (red symbols) recorded following a flash from an attenuated laser beam that delivered ∼50 effective photons to a photoreceptor. The bar beneath the trace indicates the timing of the flash. (b) The responses to the dim flash used in (b) are shown on an expanded timescale on the right. On the left is shown the rising edge of the responses to a much brighter step of light produced by the unattenuated laser beam, delivering ∼10^8 effective photons per second. The bars below the traces indicate the onset and duration of the stimuli. Adapted from Payne, R. and Demas, J. (2000). *Journal of General Physiology* 115: 735–748.

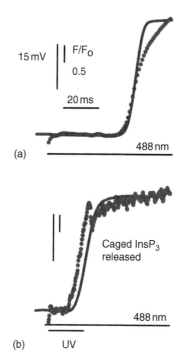

(a)

(b)   UV

Figure 8 Timing of Ca^{2+} release by caged InsP3. Photoreceptors were loaded with caged IP_3 and the Ca^{2+} indicator dye fluo-3; 488 nm and UV laser beams were focused onto the edge of the R-lobe. As an indicator of $[Ca^{2+}]_i$, uncalibrated dye fluorescence is shown, expressed as a fraction of the background fluorescence during the latent period of the response (F/F_0). (a) Membrane potential (solid line) and fluo-3 fluorescence (red dots) recorded during illumination by the 488-nm laser alone, stimulating Ca^{2+} release and depolarization through the photoisomerization of rhodopsin, that is, through the natural phototransduction pathway. Laser stimulation began at the beginning of the fluorescence trace. (b) Effect of superimposing a 20-ms duration UV flash and so releasing caged InsP3 into the cytoplasm of the cell, stimulating an earlier release of Ca^{2+} directly from the SER. Adapted from Ukhanov, K. and Payne, R. (1997). *Journal of Neuroscience* 17: 1701–1709.

of rise to those elicited by bright visible light. However, the latency of the Ca^{2+} signal that follows illumination by visible light is ∼30 ms longer than that of the response to the release of caged IP_3 (compare **Figures 8(a)** with **8(b)**), the difference being presumably the time required for light to activate the PI pathway. There is therefore convincing evidence that the light-induced release of Ca^{2+} from internal stores is mediated by the PI pathway acting on IP_3Rs in SER that are closely juxtaposed to the microvillar membrane.

Released Ca²⁺ Ions can Activate an Inward Current

Pulsed pressure injections of solutions containing 1–2 mM Ca^{2+} into the R lobe of ventral photoreceptors activates a current in the plasma membrane of up to 20 nA, with a similar reversal potential, sodium dependence, and outward rectification to that activated by light. Release of Ca^{2+}

ions through IP_3 activates the same conductance, indicating that the $[Ca^{2+}]_i$ generated through the endogenous Ca^{2+} release pathway is sufficient. The current, typically 5–20 nA in amplitude following a pulse of 100 μM IP_3, generates a transient depolarization of the photoreceptor lasting for less than 1 s. The coupling between the elevation of $[Ca^{2+}]_i$ and the depolarization of the photoreceptor is rapid. Depolarization follows caged IP_3-induced Ca^{2+} release after 2.5 ± 3.3 ms (**Figure 8**), while photolysis of caged Ca^{2+} (*O*-nitrophenyl ethylene glycol tetraacetic acid (EGTA)) at the edge of the R lobe activates current within 1.8 ± 0.7 ms.

Light-Induced Ca²⁺ Release can be Detected before the Electrical Response

The above experiments indicate that micromolar $[Ca^{2+}]_i$, released from internal stores by IP_3, can activate an inward current through the plasma membrane within a few milliseconds. It follows that if light-induced Ca^{2+} release is to act similarly, then a component of the photocurrent must be initiated a few milliseconds after $Ca2^+$ is released. Certainly, the onset of the two signals is highly correlated if measured confocally beneath the microvillar membrane. The time for $[Ca^{2+}]_i$ to exceed 2 μM is approximately equal to that for the receptor potential to exceed 8 mV (mean difference: 2.2 ± 6.4 ms). However, the question of which event occurs first is difficult to address, since two signals with different noise levels are compared (**Figure 7**). The detection of the Ca^{2+} signal lead the electrical response by up to 5 ms within about one-third of cells examined. The lag in other cells could indicate the presence of an early Ca^{2+}-independent component of the response present in some cells, but it is difficult to be sure of this because the placement of the confocal measuring spot relative to the microvillar membrane is critical. However, in any case, given the rapidity with which Ca^{2+} can elicit an inward current (1–3 ms; see above) and the fact that it takes the photocurrent 50 ms to rise to peak, this timing appears to be sufficient for released Ca^{2+} ions to contribute to the activation of the photocurrent during the rising edge of the response to light.

How Does IP₃-Induced Ca²⁺ Release Activate Inward Current and is this Current Flowing through the Light-Sensitive Conductance?

Given the detailed evidence above, it seems reasonable to propose that IP_3-induced Ca^{2+} release can activate the light-activated conductance in ventral photoreceptors (**Figure 5**). However, the molecular nature of the channel and the site of calcium's action are not yet known. An ideal preparation for electrophysiology, ventral photoreceptors do not have the advantage of molecular genetic approaches that have allowed the identification of the light-sensitive channels in *Drosophila* as members of the TRP family. Still, two hypotheses have been developed for the nature of the light-activated ion channels based on physiological and molecular evidence.

The first hypothesis is that the channels are not Ca^{2+}-gated, but are cyclic guanosine monophosphate (cGMP) gated. This seems a remote possibility at first, given the evidence that IP_3-induced Ca^{2+} release can rapidly activate a plasma membrane conductance in intact cells. However, the proposal that a further messenger exists downstream from Ca^{2+} is driven by the experimental inability of Ca^{2+} to directly activate ion channels when applied to the inside of patches of plasma membrane excised from the rhabdomeral lobe. Instead, application of cGMP activated channels in a minority of excised patches. The channel events activated during application

of cGMP had a similar conductance to light-activated channels, similar reversal potential when bathed in media mimicking intracellular and extracellular ion concentrations and a similar increase in open probability upon membrane depolarization. Since intracellular injections of cGMP or its analogs depolarize the photoreceptor, it was proposed that cGMP might be a terminal messenger in the visual cascade in ventral photoreceptors, as well as in vertebrate rods and cones. A putative cGMP-gated channel has been sequenced from ventral photoreceptors and localized to the microvillar photoreceptors. However, to reconcile this hypothesis with the large body of evidence for the initiation of the light response by the PI pathway, some coupling mechanism must be found to link IP_3-induced Ca^{2+} release to the production of cGMP. There is some pharmacological evidence that a Ca^{2+}-activated guanylate cyclase (GC) might provide this link (**Figure 5**), but no biochemical or molecular evidence has so far been obtained for this hypothesis.

The second hypothesis is that *Limulus* channels are members of the TRP family which, unlike *Drosophila* TRP and TRPL, do not display a high Ca^{2+}- permeability (**Figure 5**). TRP channel homologs have been cloned from ventral photoreceptor messenger RNA (mRNA), and an established activator of some TRP-family channels, the synthetic lipid, 1-oleoyl-2-acetyl-*sn*-glycerol (OAG), activates a conductance with reversal potential similar to that activated by light. Activation of this conductance by OAG apparently requires the presence of free Ca^{2+} ions in the injection pipette, which may explain why extracellularly applied OAG has no effect and why there is a complete dependence of the light response on light-induced Ca^{2+} release from intracellular stores. The hypothesis of a *Limulus* homolog of the TRP channel provides a testable alternative to the proposed role of a cyclic nucleotide channel and may resolve the differences in phototransduction between *Limulus* and *Drosophila* photoreceptors.

Adaptation, a Decrease in the Sensitivity of the Visual Cascade, is Mediated by Small, Lingering Elevations of Ca²⁺

The onset of prolonged illumination of ventral photoreceptors, or the huge elevations of $[Ca^{2+}]_i$ that occur following flashes of light are not sustained but fall back to the micromolar range within seconds. Concurrently, the photocurrent also falls from tens or hundreds of nA to a few nA. These declines are the result of a decrease in the photoreceptor's sensitivity that prevents saturation of the photoreceptors in bright light and so extends its dynamic range. The initial light-induced elevation of $[Ca^{2+}]_i$ therefore appears to function as a feedback signal that subsequently reduces the sensitivity of the visual cascade.

In support of this concept, slow injection of Ca^{2+} ions into the cytosol of ventral photoreceptors diminishes the sensitivity and latency of the light response, mimicking the effect of an adapting light. Injection of Ca^{2+} chelators, such as EGTA or 1,2-bis(o-aminophenoxy)ethane-N,N, N',N'-tetraacetic acid (BAPTA), not only slows down and diminishes the initial photocurrent, but also blocks the decline of the light-induced current during prolonged illumination. The site of this feedback inhibition of the light response by Ca^{2+} could be at several points in the phototransduction cascade. For example, IP_3-induced Ca^{2+} release is known to be inhibited by lingering elevations of $[Ca^{2+}]_i$ in ventral photoreceptors. In addition, pharmacological activation of protein kinase C greatly reduces the sensitivity of the light-induced current, apparently acting upstream of IP_3-induced Ca^{2+} release.

Drosophila and *Limulus* Photoreceptors Operate Differently and Illustrate Two General Mechanisms Coupling the PI Cascade to an Electrical Response

The two most extensively studied microvillar photoreceptors, those of *Limulus* and *Drosophila*, have many aspects of their phototransduction mechanism in common. Both are thought to utilize the PI cascade to open non-selective cation channels and both exhibit large elevations of $[Ca^{2+}]_i$, which are necessary for signal amplification and speed, as well as light adaptation. However, there are also clear differences: *Limulus* photoreceptors utilize IP_3-induced Ca^{2+} release to elevate intracellular free Ca^{2+} ion concentrations, while *Drosophila* photoreceptors are thought to utilize DAG or derived products to open Ca^{2+}-permeable TRP channels that allow Ca^{2+} entry from the extracellular space. The photoreceptors of the two species are therefore specific examples of two general mechanisms in cells for coupling the PI pathway to $[Ca^{2+}]_i$ elevation. Why is there this difference? One explanation is the need in *Limulus* for the extra amplification provided by IP_3-induced Ca^{2+} release. One IP_3 molecule can release hundreds of Ca^{2+} ions from the ER through the IP_3R channel. These Ca^{2+} ions can then diffuse along the inner surface of the plasma membrane to activate downstream targets, such as the hundreds of channels required to open in order to produce a significant quantal event in these giant photoreceptors. The trade-off for this amplification is slower speed. The extra amplification step (Ca^{2+} release) required in the visual cascade may explain why *Limulus* photoreceptors are slower than fly photoreceptors to respond to light. This speed difference is entirely reasonable, given the animals' differing demands on their visual systems. Horseshoe crabs do not fly; rather they use their lateral eyes to find mates near moonlit beaches.

Extension of Phototransduction Mechanisms to other Microvillar Photoreceptor Types

Detailed knowledge of the mechanisms of both *Limulus* and *Drosophila* photoreceptors may outline general approaches used by other cell types to the problem of tailoring the outcome of the PI cascade to the need for either speed or amplification. An outstanding question is the extent to which other photoreceptor types utilize these mechanisms. It is apparent that light-induced Ca^{2+} release from intracellular stores occurs in photoreceptors of the file clam, leech, and honeybee. Furthermore, the light-sensitive channels that mediate the immediate electrical response in the file clam are relatively impermeable to Ca^{2+} ions, like those of *Limulus*, but not *Drosophila*. Barnacle photoreceptors, on the other hand, like those of *Drosophila*, display prominent light-induced Ca^{2+} influx. Thus, elements of the two mechanisms for coupling phototransduction to an elevation of Ca^{2+} seem to be scattered across the invertebrate phyla. Whether there is a functional or evolutionary rationale for these differences remains to be seen. Indeed, it is not known yet whether photoreceptors in the other eight eyes of *Limulus* function similarly to the much-studied ventral eye.

See also: Circadian Rhythms in the Fly's Visual System; Evolution of Opsins; Genetic Dissection of Invertebrate Phototransduction; *Limulus* Eyes and Their Circadian Regulation; Microvillar and Ciliary Photoreceptors in Molluskan Eyes; The Photoresponse in Squid.

Further Reading

Bandyopadhyay, B. C. and Payne, R. (2004). Variants of TRP ion channel mRNA present in horseshoe crab ventral eye and brain. *Journal of Neurochemistry* 91: 825–835.

Battelle, B. A. (2006). The eyes of *Limulus polyphemus* (Xiphosura, Chelicerata) and their afferent and efferent projections. *Arthropod Structure and Development* 35: 261–274.

Brown, J. E. and Blinks, J. R. (1974). Changes in intracellular free calcium concentration during illumination of invertebrate photoreceptors. *Journal of General Physiology* 64: 643–665.

Brown, J. E., Rubin, L. J., Ghalayini, A. J., et al. (1984). *myo*-Inositol polyphosphate may be a messenger for visual excitation in *Limulus* photoreceptors. *Nature* 311: 160–162.

Chen, F. H., Baumann, A., Payne, R., and Lisman, J. E. (2001). A cGMP-gated channel subunit in *Limulus* photoreceptors. *Visual Neuroscience* 18: 517–526.

Fein, A., Payne, R., Corson, D. W., Berridge, M. J., and Irvine, R. F. (1984). Photoreceptor excitation and adaptation by inositol 1,4,5 trisphosphate. *Nature* 311: 157–160.

Hardie, R. C. and Minke, B. (1992). The *trp* gene is essential for a light-activated Ca^{2+} channel in *Drosophila* photoreceptors. *Neuron* 8: 643.

Nasi, E., Gomez, M., and Payne, R. (2000). Phototransduction mechanisms in microvillar and ciliary photoreceptors of invertebrates. In: Hoff, A. J., Stavenga, D. G., de Grip, W. J., and Pugh, E. N. (eds.) *Molecular Mechanisms in Visual Transduction – Handbook of Biological Physics*, vol. 3, pp. 389–448. Amsterdam: Elsevier Science.

Phototransduction: Phototransduction in Cones

V J Kefalov, Washington University School of Medicine, Saint Louis, MO, USA

Glossary

Dark adaptation – The mechanism that allows photoreceptors to recover their sensitivity to dark-adapted levels following exposure to bright light.

Light adaptation – The mechanism that allows photoreceptors to reduce their sensitivity in the presence of steady light.

Phototransduction cascade – A series of reactions in the outer segments of photoreceptors through which the energy of a photon is converted into a change in the membrane potential of the cell.

Visual cycle – A series of reactions initiated by the activation of the visual pigment by light and terminating in resetting the pigment to its inactive, ground state. It involves the decay of the photoactivated visual pigment to free opsin and all-*trans* retinal, the recycling of chromophore from all-*trans* to 11-*cis* outside of photoreceptors, and the regeneration of the visual pigment molecule.

Visual pigment – A G-protein-coupled receptor consisting of protein, opsin, covalently linked to a chromophore, 11-*cis* retinal. The absorption of a photon by the visual pigment is the initial step in activating the phototransduction cascade.

Introduction

Cone photoreceptors mediate our vision during the day and provide us with fine spatial and temporal resolution as well as color perception. In most species, cones are located mostly in the central area of the retina where the image directly in front of the eyes is projected. Unlike rods, where the signal from hundreds of photoreceptors is integrated for optimized photon detection in low light conditions, signals from individual cones are relayed to the brain. As a result, the spatial resolution of our central vision, driven primarily by the cones, is excellent, whereas that of our peripheral vision, driven by the rods, is significantly lower. Color discrimination is achieved as each cone typically expresses a single type of visual pigment which conveys different spectral sensitivity to different cone types. While single photoreceptors cannot discriminate colors as the degree of photoactivation depends not only on the wavelength of the stimulus but also on its intensity, the visual system extracts that information by comparing the signals coming from the different cone types. An interesting exception to the one cell–one pigment rule is the mouse retina where green and ultraviolet cone visual pigments are coexpressed in the same cells. The functional significance of that arrangement is not clear.

Functional Properties of Cones

Cones use a phototransduction cascade, similar to the one well characterized in rods, to convert the energy of light into an electrical signal. In addition, cone phototransduction proteins are homologous, or sometimes even identical, to the ones found in rods. Yet, cones have functional properties that are distinct from those of rods and that are suited for their role as bright-light detectors. First, cones are significantly less sensitive than rods. The rod phototransduction cascade is tuned for high amplification which allows rods to achieve the maximal physically possible sensitivity and generate a detectable single photon response. As such enormous gain requires buildup of the reactions of the phototransduction cascade, the trade-off is the slow kinetics of rod responses. Cones, on the other hand, are 30- to 100-fold less sensitive than rods (**Figure 1**) and require the simultaneous activation of tens to hundreds of visual pigment molecules to generate a detectable response. As a result of the low amplification of their phototransduction cascade, cones are not sensitive enough to function under low light conditions, depriving us of color vision in dim light. Instead, the low cone phototransduction gain shifts their dynamic range toward brighter light conditions and enables cones to function during the day. The low signal amplification in cones is made possible by the rapid inactivation of their phototransduction cascade. This results in the second notable difference from rods, namely, that cone responses are typically several fold faster than rod responses. The rapid activation and subsequent inactivation of the cone phototransduction cascade reactions provides the basis for the high temporal resolution of cone-mediated vision (**Figure 1**). The rapid activation of cones results in short latency of detection, whereas their rapid inactivation enables discrimination of stimuli spaced closely in time. In contrast, the slower rod responses limit the temporal resolution of rod-mediated vision. Third, following exposure to bright light, cones fully recover their sensitivity within a few minutes. Rods, in contrast, experience a long

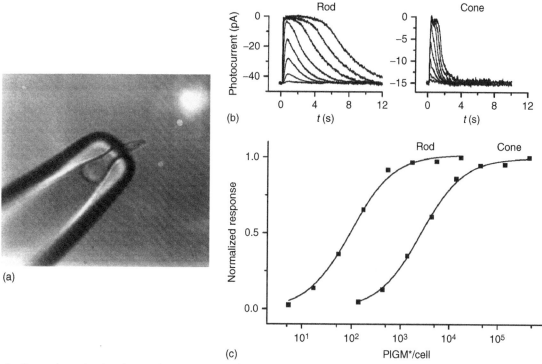

Figure 1 Comparison of rod and cone photoresponses. (a) Salamander red cone drawn in a suction pipet electrode with the outer segment protruding out. (b) Families of photoresponses from a salamander rod (left) and a red cone (right) to brief test flashes of increasing intensity delivered at $t = 0$. Note the significantly faster response kinetics of cone responses compared to rod responses. (c) Normalized intensity–response curves for the same two cells. Note the significantly lower cone sensitivity compared to the rod sensitivity.

refractory period following exposure to bright light and can take up to an hour for a complete recovery of their sensitivity. This process, known as dark adaptation, prevents cones from becoming refractory and allows us to retain visual perception in a quickly changing light environment. Finally, cones have a remarkable ability to adjust their sensitivity over a very wide range and remain photosensitive even in extremely bright light. Rods, in contrast, saturate in even moderately bright light and remain nonfunctional during most of the day. This process, known as light adaptation, prevents cones from saturating in bright light and allows us to see throughout the day. With rods saturated, cones are responsible for most of the visual information reaching our brain during the day. In fact, with the introduction of artificial lighting, humans rely almost exclusively on cones both during the day and at night. This is why cone disorders, such as macular degeneration, the most common cause of blindness in the elderly, have a devastating effect on vision.

Obstacles for Studying Cone Phototransduction

The last several decades have seen a tremendous advance in our understanding of the function of photoreceptors.

The development of electrophysiological tools for studying the function of single photoreceptors, together with biochemical and genetic tools have revealed the mechanism of phototransduction and provided quantitative description of the reactions involved in it. Unfortunately, these advances have been almost exclusively limited to rods. The great abundance of rods in most mammalian retinas (95% of all photoreceptors in human and 97% in mouse retinas) has facilitated the purification and biochemical study of rod phototransduction proteins. In contrast, the small fraction of cones and the homology between rod and cone phototransduction proteins have rendered comparable studies from cone proteins technically challenging. A further obstacle has been the fragility of mammalian cone photoreceptors, which has rendered physiological studies from cones also significantly more challenging than comparable rod studies. As a result, while mammalian rod phototransduction has been characterized in quantitative details, most of what we currently know about cone phototransduction is derived from studies of amphibian and fish photoreceptors. Based on the similarities in structure and transduction proteins between rods and cones, it has been assumed that phototransduction in cones follows the same set of reactions as phototransduction in rods. There exist, however, important quantitative phototransduction differences in rods and cones pertinent to their function in dim and bright light, respectively. The phototransduction cascade in cones will be discussed here in the context of the much better understood rod phototransduction cascade.

In both, rods and cones, phototransduction takes place in specialized compartments, called outer segments, which consist of stacks of membrane disks, similar to a stack of coins. Unlike in rods, where these disks are surrounded by, but not connected to, the plasma membrane, in cones these disks are formed from invaginations of the plasma membrane. As a result, the plasma membrane of cone outer segment has significantly higher area, a factor possibly important for the rapid flow of molecules in and out of the cell. The transduction channels are cGMP-gated non-selective cation channels held open in darkness by the binding of free cGMP in the outer segment. Cone cGMP channels are homologous to those found in rods and in olfactory neurons and consist of two cyclic nucleotide-gated alpha 3 (CNGA3) and two cyclic nucleotide-gated beta 3 (CNGB3) subunits. In darkness, the influx of Na^+ and Ca^{2+} through these channels depolarizes the cells to about $-40\,mV$, which results in the steady release of the neurotransmitter glutamate from the cone synaptic terminal. Photoactivation of the cell results in the hydrolysis of cGMP, closure of the transduction channels, hyperpolarization of the cell, and reduction in the release of neurotransmitter from the cone synaptic terminal.

Cone Visual Pigment and Phototransduction

Phototransduction in cones is initiated by the activation of cone visual pigments by the absorption of a photon. The cone visual pigments, similar to rod pigments, consist of protein, opsin, covalently attached to a chromophore, typically 11-*cis* retinal. Cone opsins have a moderate level ($\sim50\%$) of homology to rod opsins. The visual chromophore is a derivative of vitamin A (all-*trans* retinol), which is converted in the pigment epithelium into 11-*cis* retinal and then transported to the photoreceptor's outer segments where it combines with opsin to form the visual pigment. The visual pigment is expressed at very high levels in the disks of the outer segment (3.5 mM), so that a photon traveling along the outer segment has a $\sim40\%$ chance of activating a pigment molecule. Interestingly, the concentrations of rod and cone visual pigments in the outer segment as well as their extinction coefficients are similar. In addition, the probability that a pigment molecule will become activated once a photon has been absorbed (quantum efficiency) is also comparable between rod and cone pigments. Thus, with respect to the pigment distribution and optical properties, only the typically smaller size of the cone outer segment compared to that of the rod contributes to the lower sensitivity of cones.

Studies with amphibian photoreceptors indicate that the different stability of rod and cone pigments modulates their respective phototransduction cascades. First, studies of transgenic *Xenopus* rods expressing red cone opsin have allowed the direct observation of physiological responses to the activation of a single cone pigment molecule. This has made possible the determination of the rate of spontaneous thermal activation of red cone pigments, which produces a response identical to the activation by a photon. The molecular rate of thermal activation measured in this way is $\sim10\,000$ times higher for red cone pigment than for rod pigment. As a result, amphibian red cones experience ~200 pigment activations per second in darkness. This level of dark activity is comparable to the total dark noise measured from salamander red cones, indicating that most of the noise in these cells originates in the thermal activation of the pigment. This spontaneous activity acts as background light to induce adaptation and, therefore, desensitization and acceleration of the flash response. A second mechanism by which the stability of the visual pigment contributes to the differences between rods and cones is based on the covalent bond between opsin and retinal in their respective pigments. Both biochemical and physiological studies indicate that the formation of the covalent bond between opsin and chromophore is reversible in cones but not in rods. As a result, the visual pigment in cones, but not in rods, can spontaneously dissociate into free opsin and 11-*cis* retinal. The very low level of free 11-*cis* retinal in the outer segment (only $\sim0.1\%$ of the pigment content) shifts the equilibrium between free and chromophore-bound cone opsin so that even in dark-adapted cones, there is $\sim10\%$ free opsin. At this high level, the total catalytic activity of free opsin, though weak per single molecule, is sufficient to induce adaptation and further reduce the sensitivity and accelerate the kinetics of the cone flash responses.

The effects of cone pigment properties on mammalian photoreceptor function have not been well characterized. Interestingly, studies from transgenic mouse rods expressing cone pigments indicate that, though still significantly higher than that of rod pigment, the rate of thermal activation of cone pigment is not high enough to affect cone photosensitivity significantly. A possible explanation for the relatively low thermal activity of cone pigments in mammalian species compared to amphibians might be that they use a slightly different chromophore (11-*cis* retinal or A1) than most amphibian photoreceptors (11-*cis* 3-dehydroretinal or A2). The reversibility of cone pigment formation and its possible effect on cone function have not yet been examined in mammalian cones. Finally, differences in the properties of rod and cone visual pigment also contribute to the very different rates of dark adaptation in rods and cones.

Activation of Cone Phototransduction

Once activated, the visual pigment binds to and activates a heterotrimeric G protein, called transducin (G_t). This triggers the exchange of GDP for GTP on the α-subunit of transducin ($G_{t\alpha}$) and the dissociation of $G_{t\alpha}$·GTP from $G_{t\beta\gamma}$. This represents the initial amplification step in phototransduction as one visual pigment molecule can activate multiple G_t molecules. Rod and cone transducins are closely related and the primary structures of their α-subunits are ~80% identical, with even higher identity in the region of interaction with the visual pigment. Biochemical studies indicate that rod and cone pigments have comparable binding affinities for rod transducin and that they activate rod transducin with similar kinetics. Furthermore, studies with transgenic animals coexpressing rod and cone visual pigments in the same photoreceptor have shown that rod and cone pigments produce comparable responses. Thus, cone pigments expressed in rods produce a response with rod-like amplification and kinetics and, conversely, rod pigments expressed in cones produce a response with cone-like amplification and kinetics. These results indicate that the activation of the phototransduction cascade by the visual pigment and the inactivation of the visual pigment are not determined by its properties but rather, by the downstream transduction reactions, including the activation of transducin. Indeed, biochemical studies of fish photoreceptors have shown that the activation of transducin is ~25 times less effective in cones compared to rods. This lower activation efficiency would contribute to the lower amplification of the signal and, therefore, to the lower sensitivity of cones. It is not clear yet whether the lower activation of transducin in cones is due to the properties of the cone isoform of transducin or due to the faster inactivation of pigment in cones compared to rods. An interesting recent observation is that exposure to bright light in rods triggers translocation of the subunits of activated transducin from the outer to the inner segment. In contrast, in cones, such translocation does not occur, possibly because transducin subunits are inactivated and re-form a trimer faster than in rods. The mechanism of this light-dependent translocation is still not well understood and is an active area of research.

Once activated by the visual pigment, $G_{t\alpha}$·GTP in turn activates cGMP phosphodiesterase (PDE) by binding to its inhibitory subunit PDEγ and removing its inhibition on the catalytic PDEαβ. The resulting hydrolysis of cGMP by PDE leads to the closure of cGMP-gated channels in the cone outer segment and the hyperpolarization of the photoreceptor to produce the light response. While cone PDE has 60% identity to rod PDE, biochemical studies of fish photoreceptors indicate that the activation of PDE by transducin might also be ~10 times less effective in cones compared to rods, contributing further to the lower cone sensitivity.

Inactivation of Cone Phototransduction

Response termination is achieved as the visual pigment, transducin, and PDE are inactivated and the concentration of cGMP is restored to its dark, preflash level. Though these reactions in cones are not well characterized, it is clear that, similar to their activation, quantitative differences in the inactivation of phototransduction reactions in rods and cones contribute to the lower sensitivity and faster response kinetics of cones. The activity of the visual pigment is initially partially quenched when it is phosphorylated by a G-protein receptor kinase (GRK). Phosphorylation of activated visual pigment is ~50 times faster in cones compared to rods. It appears that this faster phosphorylation is the result of two factors – higher expression of GRK in cones and higher efficiency of cone GRK (GRK7) compared to rod GRK (GRK1). While most species, including human, express GRK1 in rods and GRK7 in cones, the mouse retina is unusual as its rods and cones share the same kinase, GRK1. In this case, the faster pigment inactivation in cones is most likely due to the higher concentration of GRK1 and possibly also to differential modulation of that reaction by the calcium-binding protein recoverin.

Following phosphorylation, complete inactivation of the phosphorylated visual pigment is achieved by the subsequent binding of a protein called arrestin. The cone isoform of arrestin (Arr4) has about 50% identity to rod arrestin (Arr1). The mouse retina again represents an unusual case, as in addition to Arr4, mouse cones also express Arr1. Interestingly, the ratio of arrestin to visual pigment is 7 times higher in cones compared to rods. In dark-adapted rods, most of arrestin is in the inner segment and does not, therefore, contribute to the inactivation of rod visual pigment. As a result, the quantity of arrestin in the outer segments of rods is only a few percent of their visual pigment. Exposure to bright light triggers the translocation of arrestin from the inner to the outer segment for more efficient pigment inactivation. While arrestin also transloactes in cones, the total quantity of arrestin in their outer segments in darkness is comparable to that of their visual pigment. Recent studies from mouse cones lacking both rod and cone arrestins reveal that either arrestin is capable of inactivating cone visual pigment though Arr1 is much more abundant than Arr4 in cones. Studies with transgenic rods expressing cone S-opsin and either rod or cone arrestin further demonstrate that rod arrestin is more efficient at inactivating cone pigment than cone arrestin. The relatively low expression of Arr4 in cones and its relative inefficiency suggest a possible additional role for this protein. The coexpression of two arrestins and their high concentration in cone outer segments would contribute to the rapid cone pigment inactivation and are consistent with the more rapid pigment inactivation and faster response termination in cones compared to rods.

$G_{t\alpha} \cdot GTP$ is inactivated as GTP is hydrolyzed into GDP. This reaction is catalyzed by PDEγ as part of a GTPase-activating protein (GAP) complex that consists, in addition, of regulator of G-protein signaling (RGS9), RGS9 anchoring protein (R9AP), and a G_β subunit ($G_{\beta 5}$). The rod and cone PDEγ have comparable potencies for inhibiting PDE and also for enhancing the hydrolysis of GTP by the GAP complex. In contrast, even though the identical RGS9 protein is present in rods and in cones, its concentration is more than 10 times higher in cones compared to rods. Deletion of RGS9 in the mouse greatly retards cone response inactivation, and mutations in RGS9 have been associated with slow cone deactivation in patients. Thus, while the extent to which the differences in GAP activity in rods and cones contribute to their functional differences is not well understood, RGS9 and the GAP complex clearly play an important role in the inactivation of cone phototransduction.

The final step in photoresponse termination involves the upregulation of synthesis of cGMP by guanylyl cyclase (GC) to restore the concentration of free cGMP in the outer segment and reopen the cGMP-gated channels. While rods express two isoforms of GC, that is, GC1 and GC2, cones appear to express predominantly, if not exclusively, GC1. The role of GC2 in rods is not clear as its deletion produces only a mild change in rod physiology. It is also not understood how modulation of GC by the pair of GC-activating proteins (GCAP1 and GCAP2) contributes to the unique functional properties of cones. Although the distribution of GCAPs between rods and cones in different species is ambiguous, it appears that GCAP2 is prevalent in rods, while GCAP1 is expressed at high levels in cones. The possible role of GCAPs in mediating light adaptation in cones is discussed below in the context of light adaptation.

Dark Adaptation of Cones

Quantitative differences between the phototransduction cascades of rods and cones not only contribute to the difference in sensitivity and kinetics of photoresponses as discussed above, but also play a role for the very different adaptation properties of rods and cones. The ability to recover their sensitivity rapidly following exposure to bright light, or dark-adapt, is critical for the function of cones as daytime photoreceptors. The absorption of a photon by the visual pigment not only triggers its activation, but also results in its eventual decay into free opsin and all-*trans* retinal. Dark adaptation of both, rods and cones, after exposure to bright light requires regeneration of the visual pigment from opsin and 11-*cis* retinal. However, the speed of pigment regeneration, and hence sensitivity recovery, is very different in rods and cones, with full recovery requiring less than 5 min in cones and up to an hour in rods (see **Figure 2**).

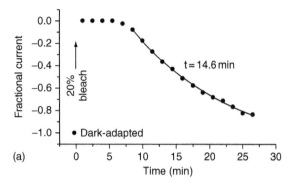

(a)

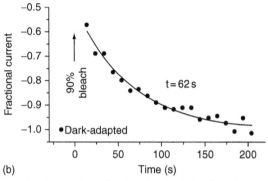

(b)

Figure 2 Comparison of rod and cone dark adaptation. Recovery of the circulating (dark) current in salamander rod (a) and red cone (b) measured with a suction electrode. Cells were exposed to bright light that activated (bleached) 20% of the rod pigment and 90% of the cone pigment. The recording was done in the presence of exogenous 11-*cis* retinal to enable pigment regeneration in the isolated cells. Current recovery is fit by a single exponential decay function (solid line). Note the significantly faster recovery of the current in cone compared to the current in the rod.

Several factors contribute to the rapid pigment regeneration in cones. First, the decay of the photoactivated pigment to free opsin and all-*trans* retinal occurs in seconds for cone pigments compared to minutes for rod pigments. Second, the reduction of all-*trans* retinal into all-*trans* retinol, which takes place in the outer segment and is catalyzed by retinol dehydrogenase (RDH), is also 10–40 times faster in cones compared to rods. The reduction reaction requires the cofactor nicotinamide adenine dinucleotide phosphate oxidase (NADPH). While it is possible that the faster reduction of all-*trans* retinal in cones is due to the different properties of rod and cone RDH enzymes, a more likely hypothesis is that the reduction reaction is limited by the supply of NADPH from the inner segment. Third, single-cell measurements from amphibian photoreceptors indicate that the clearance of all-*trans* retinol from the outer segment is ~25 times faster in cones compared to rods. However, the actual difference in these rates in the intact retina might be affected by factors such as the proximity to the pigment epithelium and the action of extracellular chromophore-binding proteins such as interphotoreceptor retinoid-binding protein (IRBP). Finally, the formation of the covalent bond between opsin and 11-*cis* retinal during pigment regeneration occurs in seconds in cones and minutes in rods. Together, these factors contribute to the faster turnover of cone visual pigment and the faster dark adaptation of cones compared to rods.

In addition to the effects of faster visual pigment decay and regeneration, cone dark adaptation is accelerated by the noncovalent interaction between opsin and 11-*cis* retinal. Pigment regeneration requires the initial binding of 11-*cis* retinal in the chromophore pocket of free opsin. While in rods, the noncovalent binding of retinal activates the opsin molecule and desensitizes the rods, in cones, this reaction has the opposite effect and inactivates cone opsin. As a result, the noncovalent binding of 11-*cis* retinal to opsin delays dark adaptation in rods but accelerates it in cones, as it allows cones to substantially recover their sensitivity even before the regeneration of their visual pigment.

Recent biochemical studies indicate that another mechanism contributing to the faster dark adaptation of cones compared to rods is based on the supply of recycled chromophore for pigment regeneration. The canonical visual cycle involves the pigment epithelium, where all-*trans* retinol is converted into 11-*cis* retinal via a series of enzymatic reactions and then transported back to the photoreceptors for incorporation into opsin. The rapid dark adaptation of cones and their ability to maintain adequate levels of pigment and remain light sensitive even in steady bright light require rapid pigment regeneration, hence rapid recycling of chromophore for cones. However, the slow rate of chromophore turnover in the pigment epithelium and the competition for recycled chromophore between cone opsin and overwhelming levels of rod opsin in most rod-dominant species indicate that the canonical pigment epithelium visual cycle might not be sufficient to meet the chromophore demand of cones. Indeed, recent biochemical studies from cone-dominant species have brought up the idea of a second, cone-specific pathway for recycling of chromophore located within the retina and possibly relying on the Müller cells. The role of this novel cycle in mammalian rod-dominant species is still controversial. However, recent physiological experiments with amphibian photoreceptors demonstrate the function of a retina visual cycle under physiological conditions in a rod-dominant retina. Importantly, the combined action of the pigment epithelium and the retina visual cycles is required for the rapid and complete dark adaptation of cones.

Light Adaptation in Cones

In contrast to rods, which saturate in moderate light and are not responsive during the day, cones have the ability to adapt their sensitivity and remain functional over a very wide range of light intensity. Studies with amphibian and fish photoreceptors indicate that, similar to the case of rods, cone adaptation is mediated by intracellular calcium, modulated by the activation of the phototransduction cascade. In the dark, the continuous current entering the outer segment through the cGMP-gated channels is carried in part by calcium, which is returned to the extracellular space via a $Na^+/(Ca^{2+}, K^+)$ exchanger. Following photoactivation and the closure of cGMP channels, calcium continues to be exported out of the cell through the $Na^+/(Ca^{2+}, K^+)$ exchanger until a new equilibrium is reached. As a result, activation by light causes a decline in the concentration of calcium in the outer segment of the cell. This triggers the calcium-mediated negative feedback on phototransduction, which in rods is required for eventually terminating the signal and for adapting the cell in response to light. Interestingly, calcium constitutes a larger fraction of the total ionic flux in and out of the outer segment of cones compared to rods. Thus, in cones of amphibians and fish, the fraction of photocurrent carried by calcium is about 35% compared to 20% in rods. As would be expected from the need to maintain a steady calcium concentration in darkness, the matching rates of extrusion of calcium via the $Na^+/(Ca^{2+}, K^+)$ exchanger are also higher in cones compared to rods. The combination of faster turnover of calcium in cones and their smaller volume compared to rods allows calcium in cones to decline several times faster upon light stimulation. In addition, their range of calcium concentrations from darkness to bright light is threefold wider than that in rods. These quantitative differences create the potential for more powerful modulation of phototransduction by calcium in cones compared to rods consistent with the ability of cones to adapt better and faster to various light conditions than rods.

The mechanisms by which calcium modulates the cone phototransduction cascade are not well understood. However, comparison between cone and rod phototransduction reveals several interesting points. One mechanism by which calcium modulates phototransduction in rods involves inactivation of the visual pigment via phosphorylation by rhodopsin kinase. This reaction is modulated by the calcium-binding protein recoverin (also known as S-modulin). Recoverin is a member of the EF-hand superfamily and exerts its effect by inhibiting phosphorylation of rhodopsin by rhodopsin kinase at high calcium levels. In rods, inhibition of rhodopsin kinase by recoverin regulates phototransduction in darkness, in high calcium conditions, but has little effect during light adaptation, in low calcium conditions. The role of recoverin in modulating cone phototransduction in darkness and during light adaptation is not known. However, rods and cones share the same isoforms of recoverin and rhodopsin kinase. In addition, calcium modulates the sites and extent of pigment phosphorylation in cones but not in rods. Finally, unlike in rods, in cones, the calcium-dependent inactivation of cone visual pigment could be the rate-limiting step for the shutoff of the cone photoresponse.

Another mechanism by which calcium modulates phototransduction in rods involves the synthesis of cGMP by GC. As discussed above, this reaction is modulated by GCAP1 and GCAP2. GCAPs modulate GC in rods up to 20-fold as they inhibit it at high intracellular calcium levels and activate it at low calcium levels. While the simultaneous deletion of GCAP1 and GCAP2 delays the recovery of cone light responses, the extent to which GCAPs modulate cone phototransduction in darkness and during light adaptation is not known.

Finally, calcium is also believed to directly modulate the cGMP-gated channels in cones. The Ca^{2+}-dependent modulation of cGMP current is minimal in amphibian and undetectable in mammalian rods. In contrast, cone cGMP channels are directly modulated by Ca^{2+} both in fish and in mammalian retina. The molecular mechanism of cone channel modulation remains to be discovered. While calmodulin binds to and modulates heterologously expressed cGMP-gated channels, its role in the intact cone photoreceptor has been questioned.

Epilog

These are exciting times for studying cone phototransduction. Until recently, technical issues such as the low abundance of cone photoreceptors in rod-dominant retinas and the fragility of mammalian cone photoreceptors have held back the biochemical and physiological studies of cones. As a result, despite the crucial role of cones for our daytime vision, mammalian cone phototransduction has been poorly understood. Recent development of several genetically modified mice has turned the tables. One example is the Nrl knockout mouse. Nrl is a transcription factor required for rod photoreceptor differentiation and its deletion produces a retina populated exclusively by cone-like photoreceptors. This makes possible the purification and biochemical characterization of mammalian cone phototransduction proteins. The Nrl knockout retina has also been used recently for physiological studies of cone photoreceptors. Other examples of useful genetically modified mice include those lacking the rod visual pigment (rhodopsin knockout) and the rod $G_{t\alpha}$ subunit (transducin α knockout). The lack of functional rods in both of these retinas makes possible the physiological identification and study of cone photoreceptors. This approach was most recently used to investigate the role of Arr1 and Arr4 in the inactivation of mouse cone pigments. The combination of new genetic models and improved physiological tools provides

promise for studies of mammalian cone photoreceptors using the full range of tools that have been so successful in characterizing the function of mammalian rods. This should allow not only quantitative characterization of the cone phototransduction cascade but also understanding the mechanisms for cone dark and light adaptation which make cones invaluable as our daytime photoreceptors.

See also: Light-Driven Translocation of Signaling Proteins in Vertebrate Photoreceptors; Phototransduction: Adaptation in Cones; Phototransduction: Adaptation in Rods; Phototransduction: Inactivation in Cones; Phototransduction: Inactivation in Rods; Phototransduction: Phototransduction in Rods; Phototransduction: Rhodopsin; Phototransduction: The Visual Cycle.

Further Reading

Donner, K. (1992). Noise and the absolute thresholds of cone and rod vision. *Vision Research* 32: 853–866.

Ebrey, T. and Koutalos, Y. (2001). Vertebrate photoreceptors. *Progress in Retinal and Eye Research* 20: 49–94.

Fu, Y. and Yau, K. W. (2007). Phototransduction in mouse rods and cones. *Pflugers Archive: European Journal of Physiology* 454: 805–819.

Hecht, S., Haig, C., and Chase, A. M. (1937). Rod and cone dark adaptation. *Journal of General Physiology* 20: 831–850.

Holcman, D. and Korenbrot, J. I. (2005). The limit of photoreceptor sensitivity: Molecular mechanisms of dark noise in retinal cones. *Journal of General Physiology* 125: 641–660.

Kawamura, S. and Tachibanaki, S. (2008). Rod and cone photoreceptors: Molecular basis of the difference in their physiology. *Comparative Biochemistry and Physiology Part A: Molecular and Integrative Physiology* 150: 369–377.

Kefalov, V., Fu, Y., Marsh-Armstrong, N., and Yau, K. W. (2003). Role of visual pigment properties in rod and cone phototransduction. *Nature* 425: 526–531.

Kefalov, V. J., Estevez, M. E., Kono, M., et al. (2005). Breaking the covalent bond – A pigment property that contributes to desensitization in cones. *Neuron* 46: 879–890.

Korenbrot, J. I. and Rebrik, T. I. (2002). Tuning outer segment Ca^{2+} homeostasis to phototransduction in rods and cones. *Advances in Experimental Medicine and Biology* 514: 179–203.

Mata, N. L., Radu, R. A., Clemmons, R. C., and Travis, G. H. (2002). Isomerization and oxidation of vitamin A in cone-dominant retinas: A novel pathway for visual-pigment regeneration in daylight. *Neuron* 36: 69–80.

Nikonov, S. S., Brown, B. M., Davis, J. A., et al. (2008). Mouse cones require an arrestin for normal activation of phototransduction. *Neuron* 59: 462–474.

Rebrik, T. I. and Korenbrot, J. I. (2004). In intact mammalian photoreceptors, Ca^{2+}-dependent modulation of cGMP-gated ion channels is detectable in cones but not in rods. *Journal of General Physiology* 123: 63–75.

Rieke, F. and Baylor, D. A. (2000). Origin and functional impact of dark noise in retinal cones. *Neuron* 26: 181–186.

Tachibanaki, S., Arinobu, D., Shimauchi-Matsukawa, Y., Tsushima, S., and Kawamura, S. (2005). Highly effective phosphorylation by G protein-coupled receptor kinase 7 of light-activated visual pigment in cones. *Proceedings of the National Academy of Sciences of the United Sates of America* 102: 9329–9334.

Wald, G., Brown, P. K., and Smith, P. H. (1955). Iodopsin. *Journal of General Physiology* 38: 623–681.

Yau, K. W. (1994). Phototransduction mechanism in retinal rods and cones. The Friedenwald Lecture. *Investigative Ophthalmology and Visual Science* 35: 9–32.

Phototransduction: Phototransduction in Rods

Y Fu, Department of Ophthalmology and Visual Sciences, University of Utah, Salt Lake City, UT, USA

Glossary

Dark current – Also called circulating current. Current generated by constant influx of Na^+/Ca^{2+} into the rod outer segment through cGMP-gated channels, which is balanced by an outward current flowing across the inner segment membrane that mainly carried by potassium channels.

Dark light – Signals produced by thermal activation of rhodopsin in the dark, which adapt the visual system like real background light.

Phototransduction – The conversion of a light signal to an electrical signal in a photoreceptor cell.

Quantum efficiency – The ratio between the number of photoactivated molecules and the number of molecules that absorbed a photon.

Single-photon response – Electrical signal triggered by a single photon in a rod cell.

Suction-electrode recording – The recording of light-sensitive current of a single rod (or cone) by drawing its outer segment (or inner segment) into a suction electrode.

Introduction

Image-forming vision in vertebrates is mediated by two types of photoreceptors: the rods and the cones. Rods are specialized for dim-light (scotopic) vision while cones mediate vision in bright light (photopic). Great progress has been made in understanding rod phototransduction since the introduction of the suction-electrode recording technique in the late 1970s. The light-sensitive current of individual amphibian and mammalian (including primate) photoreceptors can be recorded with this method. Bovine retina, on the other hand, has been a favorite preparation for studying phototransduction by biochemists because of the abundance of tissue available. The mouse, however, has become an increasingly popular animal model for study in the past decade through the advent of gene-targeting techniques. When combined with electrophysiology, mouse genetics provides unmatched power in elucidating the *in vivo* functions of key phototransduction proteins, most of which have been knocked out, overexpressed, or mutated in rods, yielding a rich body of information on the mechanisms underlying the amplification, recovery, and adaptation of rod photoresponses. The details of the activation phase of rod phototransduction are now established. A quantitative description, the Lamb–Pugh model, is achieved that reproduces the activation kinetics of the rod response under physiological conditions. In this article, the focus is on the activation phase of rod phototransduction with particular emphasis on the molecular mechanisms underlying its high signal amplification feature.

Vertebrate Rods Are Highly Efficient Photon Detectors

Psychophysical experiments performed by Hecht, Schlaer, and Pirenne in 1942 suggested that human retinal rods can detect single photons. Thirty-seven years later, suction-electrode recordings from isolated toad rods by Baylor, Lamb, and Yau confirmed this remarkable ability of vertebrate rods (**Figure 1**). The amazing ability of vertebrate rods to detect single photons can be attributed to at least three factors: high quantum efficiency of photoactivation, low intrinsic noise, and a powerful signal amplification cascade. Two other factors greatly increase the photon capture ability of vertebrate rods, numerical dominance of rods over cones, and a highly specialized outer segment structure. The dense stack of disks of the rod outer segment ensures that virtually every photon traveling axially will be captured. In a sense, vertebrate rods can be viewed as sophisticated three-dimensional photon capture devices.

Phototransduction in Rods: A G-Protein-Signaling Pathway

Rod phototransduction is one of the best-characterized G-protein-signaling pathways. The receptor is rhodopsin (R), the G protein is transducin (G), and the effector is cyclic guanosine monophosphate (cGMP) phosphodiesterase (PDE or PDE6). Upon photon absorption, the rhodopsin molecule becomes enzymatically active (R^*) and catalyzes the activation of the G-protein transducin to G^*. Transducin, in turn, activates the effector PDE to PDE^*. PDE^* hydrolyzes the diffusible messenger cGMP. The resulting decrease in the cytoplasmic-free cGMP concentration leads to the closure of the cGMP-gated channels on the plasma membrane. Channel closure leads to localized reduction on the influx of cations into the outer segment, which results in membrane hyperpolarization, that is, the intracellular voltage becoming more negative (**Figure 2**).

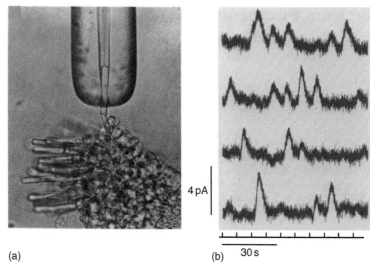

(a)   (b)   30 s

Figure 1 Suction-electrode recording on the membrane current of a single toad rod. (a) The outer segment of a rod projecting from a piece of retina was sucked in position in a suction electrode. Proximal end of cell remains attached to retina. Boundary between inner and outer segments is visible. (b) Response of rod outer segment to a series of 40 consecutive dim flashes, 20 ms flash delivering 0.029 photons μm^{-2} at 500 nm, flash timing monitored below. The rod showed no response to some flashes, or a small response of $\sim$1 pA to others, and occasionally a larger response. This suggests that the flash response is quantized, as might be expected when on average very few photons are absorbed. With further analysis, the authors demonstrated that each quantal electrical event resulted from a single photo-isomerization with mean amplitude of $\sim$1 pA – the single-photon response. Modified from Baylor et al. (1979) *Journal of Physiology*, 288: 589–611 (a) and 288: 613–634 (b), with permission from Blackwell Publishing.

This hyperpolarization decreases or terminates the dark glutamate release at the synaptic terminal. The signal is further processed by other neurons in the retina before being transmitted to higher centers in the brain.

Following light activation, a timely recovery of the photoreceptor is essential so that it can respond to subsequently absorbed photons, and signal rapid changes in illumination. This recovery from light requires the efficient inactivation of each of the activated components: R*, G*, and PDE*, as well as the efficient regeneration of rhodopsin (R) and the rapid restoration of the cGMP concentration. The termination rates of the activation steps set the time course of the photoresponse.

Although rod phototransduction is the best-characterized sensory transduction pathway, rods differ from other sensory cells in that light leads to hyperpolarization rather than depolarization. Rods respond to light with graded hyperpolarization whose amplitude increases monotonically as a function of flash intensity until saturation. One hallmark of rod phototransduction is the reproducibility of its single-photon response in both amplitude and kinetics. This is quite remarkable considering the fact that events generated by single molecules are stochastic in nature. The study on the underlying mechanisms has long been a hot topic in the vision field. Recent research pointed to two possible mechanisms: (1) Rhodopsin inactivation is averaged over multiple shutoff steps so that the integrated R* activity varies less than otherwise controlled by a single step. (2) Averaging over the deactivation of multiple G-protein molecules.

High Quantum Efficiency of Photoactivation

The quantum efficiency of photoactivation measures the probability that the adsorption of a photon initiates photoactivation. This probability is defined as the ratio between the number of photoactivated molecules and the number of molecules that absorbed a photon. Quantum efficiency of visual pigments is wavelength independent at $\sim$0.7 in the spectrum of visible light. This suggests that every absorbed photon in the visible range can activate rhodopsin equally well. The quantum efficiency of 0.7 is very similar across all visual pigments. This high efficiency seems to be a common feature of most vertebrate visual pigments.

The Great Thermal Stability of Rhodopsin

Unlike chemosensory systems, phototransduction is not triggered by the binding of a chemical ligand to the receptor, rhodopsin. Instead, the chemical, 11-*cis*-retinal in birds and land-based animals (or 11-*cis*-3,4-dehydro-retinal in aquatic animals), is prebound to rhodopsin. Photon absorption triggers the *cis*- to *trans*-isomerization of the retinoid. This isomerization rapidly converts the ligand from a powerful antagonist to a powerful agonist, leading to the formation of a series of spectrally distinct intermediates of rhodopsin in the order of bathorhodopsin, lumirhodopsin, metarhodopsin I (Meta I), and metarhodopsin II (Meta II) within a few milliseconds. Meta II is

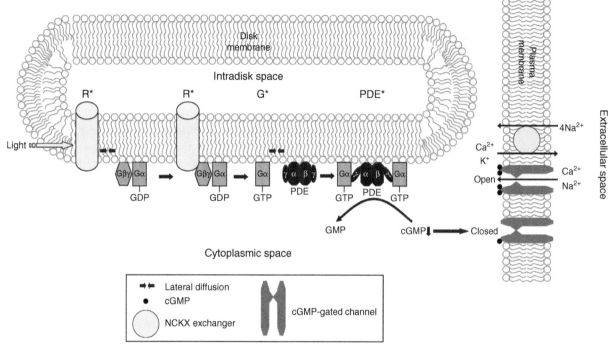

Figure 2 Schematic representation on the activation of vertebrate rod phototransduction. Following photon absorption, the activated rhodopsin (R*) activates the heterotrimeric G protein, catalyzing the exchange of GDP for GTP, producing the active Gα*-GTP. Two Gα*-GTPs bind to the two inhibitory γ-subunits of PDE, thereby releasing the inhibition on the catalytic α- and β-subunits, forming PDE*, which in turn catalyzes the hydrolysis of cGMP. The consequent decrease in the cytoplasmic-free cGMP concentration leads to the closure of the cGMP-gated channels on the plasma membrane and blockage of the influx of cations into the outer segment, which results in the reduction of the circulating dark current.

the active form of rhodopsin (R*), which in turn activates the downstream G protein, transducin. Because free opsin can weakly activate the transduction cascade, the antagonist role of 11-*cis*-retinal in the dark is important to keep the noise low in rods.

Even with 11-*cis*-retinal attached, rhodopsin occasionally undergoes spontaneous (thermal) activation in the dark, producing responses identical to those triggered by photons. This noise is often expressed as dark light because the noise adapts the visual system like real background light. This activity sets the limit on scotopic sensitivity, the visual sensitivity in darkness or dim light. To achieve the single-photon-detection sensitivity, rods not only need to have a high amplification system, but also need to have extremely low noise, or to be very quiet in the dark. This quietness can be partly attributed to the great thermal stability of rhodopsin. In a toad rod, the rate of thermal activation of rhodopsin was measured to be $\sim$0.03 event s^{-1} rod^{-1} at 22 °C, corresponding to an average wait of 2000 years for the spontaneous activation of a given rhodopsin molecule to occur, based on a total of 2×10^9 rhodopsin molecules per cell. This great stability makes it possible for rods to pack many rhodopsin molecules to the rod disks to increase its photon-capture ability while keeping the dark noise low.

It should be mentioned that the question of dark noise in vision has had a long intellectual history from the point of view of psychophysics and system neuroscience. As early as 1940s and 1950s, Hecht and Barlow have estimated the amount of dark light in human rods based on psychophysical experiments. More than 30 years later, Baylor and colleagues used suction-electrode recording technique on primate rods to demonstrate that the very low quantal noise from rhodopsin, corresponding to $\sim$0.01 event s^{-1} rod^{-1} in darkness, indeed matches the human psychophysical scotopic threshold. The quantitative agreement between the quantal noise measured from single rods and that measured in human psychophysics was considered a breakthrough in the vision field and a wonderful convergence between cell physiology and human psychophysics/system neuroscience – the goal of modern neuroscience after all.

The Activation of Transducin Constitutes the First Amplification Step

The second component of the rod phototransduction is the 81-kDa heterotrimeric G-protein, transducin (G_t, or $G\alpha_{t1}\beta_1\gamma_1$), which forms a subfamily of heterotrimeric G proteins. The molecular weight for α-, β-, and γ-subunits of rod transducin is approximately 39, 36, and 6 kDa, respectively. Transducin is present at $\sim$10% the amount of rhodopsin in the disk membrane. Although transducin subunits are soluble, the holo-transducin is firmly anchored to the disk membranes by a farnesyl lipid group that is post-translationally attached to the carboxy-terminal of the γ-subunit and an acyl group on the amino-terminal of the α-subunit. Therefore, photo-activated rhodopsin (R*) can only interact and activate transducin through lateral diffusion at the membrane surface.

Like many other G proteins, transducin exists in one of the two states: the GDP-bound inactive state and the GTP-bound active state. Binding of R* catalyzes the exchange of GTP for GDP on the α-subunit. The active Gα-GTP (G*) dissociates from R* as well as its native partner, Gβγ, and interacts with PDE to carry the signal forward. In the meantime, R* is able to activate additional molecules of transducin. Transducin activation by R* represents the first amplification step in the phototransduction cascade. The estimated rate of transducin activation by a single R* varied from 10 to over 3000 s^{-1} at room temperature. A rate of ~120 s^{-1} was later reported to be more consistent with biochemical, light-scattering, and electrophysiological measurements. The rate is roughly doubled in mammalian rods due to the higher body temperature. Until recently, it was believed that over a hundred transducins are activated during the lifetime of a single R* in mammalian rods. This number is now revised to be ~20 in mouse rods, based on the shorter life time of R* (80 ms) and activation rate of transducin by R* (240 s^{-1}).

The importance of transducin for conveying the signal from R* to PDE was manifested from animal models deficient in Gα and human patients carrying Gα mutations. It was found that rods of Gα$_{t1}$-null mice (Gnat1$^{-/-}$) lost all light sensitivity. In human, a mis-sense mutation in Gnat1 (encoding the rod Gα$_{t1}$) is implicated in autosomal-dominant congenital stationary night blindness of Nougaret, caused by constitutive activation of rod phototransduction. The gnat1$^{-/-}$ mouse line has proven to be a valuable tool for blocking rod phototransduction to study cone phototransduction and circadian photoreception. It was also used successfully to delineate two apoptotic pathways in light-induced retinal degeneration. Bright light triggers apoptosis of photoreceptors through a mechanism requiring the activation of rhodopsin but not transducin signaling. In contrast, low-intensity light induces apoptosis that is predominantly dependent on transducin signaling.

The High Catalytic Power of PDE Accounts For the Second Amplification Step

PDE is the third component of rod phototransduction. It is a hetero-tetrameric protein consisting of two catalytic subunits, α- and β-, and two identical γ-subunits. PDE is anchored to the disk membrane by a hydrophobic isoprenyl group (compounds that are derived from isoprene, 2-methylbuta-1,3-diene, linearly linked together) post-translationally attached to the C-termini of the two catalytic subunits. As for transducin activation, PDE activation by G is through lateral diffusion on the rod disk membrane. Each catalytic subunit has two high-affinity noncatalytic binding sites and one catalytic binding site for cGMP. The noncatalytic sites were suggested to modulate the binding affinity between PDEγ and PDEαβ. The amount of PDE is ~1–2% of rhodopsin. Thus, the first three components of phototransduction are present in the ratio of 100R:10G:1PDE. In the dark, the two γ-subunits act as inhibitory subunits by binding to the two catalytic subunits and significantly reducing the hydrolysis of cGMP. In the light, Gα-GTP encounters PDEγ and sterically displaces the latter, therefore relieving its inhibitory effect on the catalytic subunits and permitting the hydrolysis of cGMP to proceed (**Figure 2**). Since each G* can only activate one PDEγ, two G*'s are required to fully activate a holo PDE. This is likely the scenario in vivo during light activation due to the excess amount of G over PDE and the presence of many molecules of G* activated by rhodopsin.

In contrast to the amplification achieved during transducin activation by R*, the activation of PDE by G* constitutes no gain, that is, with an efficiency approaching 1 (one G*, one activated PDE catalytic subunit) or 0.5 in terms of PDE holoenzyme. It is the catalytic power of PDE* that provides the second amplification step. It was reported that PDE* hydrolyzes cGMP at a rate close to the limit set by aqueous diffusion, with a K_m of ~10 μM and a K_{cat} of 2200 s^{-1}, making it one of the most efficient enzymes in vivo.

In addition to the noise produced by spontaneous activation of rhodopsin, spontaneous activation of individual catalytic PDE subunits produces the continuous noise, which accounts ~30–80% (depending on the species) of the total dark noise variance in rods. The basal spontaneous PDE activity balances constitutive guanylate cyclase activity in the dark, therefore maintaining a steady-free cGMP level. It also has the function of increasing the rate of cGMP turnover and consequently speeding up the dim flash response.

One might have expected that the deletion of PDEγ from mouse rods would unleash the full catalytic power of PDEαβ. However, it was found, in the absence of PDEγ that the PDEαβ dimer actually lacked catalytic activity, and the photoreceptors of the mutant mouse rapidly degenerated. Thus, the inhibitory PDEγ subunit appears to be necessary for the integrity of the catalytic PDEαβ subunits. The degeneration might be caused by an abnormally high cGMP concentration due to the lack of hydrolysis. A related example is the rd mouse, which is the oldest and one of the best-known models for retinal degeneration. The rod cells in the rd mouse begin to degenerate at about postnatal day 8, followed by cones;

by 4 weeks, virtually no rod photoreceptors are left. Degeneration in this mouse model is preceded by the accumulation of cGMP in the retina, correlated with deficient activity of the rod PDE due to a mutation in the PDEβ subunit. It is worth noting that the *rd* mouse was instrumental in suggesting that inner retinal neurons could mediate non-image-forming vision.

cGMP Is the Second Messenger Mediating Rod Phototransduction

By 1970, scientists generally believed that a second messenger was required to mediate the rod photoresponse based on several lines of evidence. First, light absorption occurs on the rod disk membrane, whereas the light-sensitive conductance is in the plasma membrane. Since rod disks are separate from the plasma membrane, a second messenger is required to connect the two. Second, the dim-flash response of rods lasts a few seconds, which is too long to be accounted by the open time of known membrane conductance. However, it took more than a decade before the identity of the second messenger was finally determined to be cGMP. The fierce battle was fought on the validity between two competing candidates, Ca^{2+} and cGMP. According to the Ca^{2+} hypothesis, which was first proposed by Hagins, the concentration of intracellular free Ca^{2+} is low in the dark and rises in the light to block light-sensitive current. The main supporting evidence is that reducing the concentration of external Ca^{2+} dramatically increases the dark current, suggesting that internal Ca^{2+} inhibits the dark current. On the other hand, the cGMP hypothesis proposed that the concentration of cGMP was high in the dark to maintain a cGMP-dependent conductance. Light led to the hydrolysis of cGMP and the subsequent closing of the conductance. The supporting evidence is that intracellular injection of cGMP increases the amplititude and latency of the photoresponse. Adding to the complexity is the finding that the free cGMP concentration varies inversely with the free Ca^{2+} concentration in rods, making it difficult to separate the effect of the two.

This debate was finally settled with the discovery of cGMP-gated channels in rods by Fesenko and colleagues in 1985. By using the patch-clamp technique, they showed that cGMP increased a cation conductance of inside-out patches of outer-segment plasma membrane without the need of ATP. The direct channel gating by cGMP is surprising because cyclic nucleotides were generally believed to act through cyclic-nucleotide-dependent kinases and protein phosphorylation on target proteins at that time. This dogma partially explained scientists' reluctance to embrace the cGMP hypothesis because protein phosphorylation was too slow. Another monumental work by Yau and Nakatani was published at the same year that helped the anointment of cGMP as the

right candidate. An identical cGMP-gated cation conductance was found on a truncated rod outer segment with an intact plasma membrane. Most importantly, this conductance could be suppressed by light, suggesting that the long-sought light-sensitive conductance is the cGMP-gated conductance. The publications by Fesenko and Yau marked the end of the Ca^{2+} hypothesis.

The cGMP-Gated Channel Provides the Final Step of Signal Amplification

The cGMP-gated channel belongs to the family of cyclic-nucleotide-gated (CNG) channels, which are nonselective cation channels. The channel is located on the plasma membrane with a density of $400-1000 \, \mu m^{-2}$ and is the last component in the activation phase of phototransduction. Rod CNG channels consist of CNGA1 (or α1) and CNGB1 (or β1) subunits. CNGA1 subunit forms functional homomeric channels by themselves when heterologously expressed. Although CNGB1 does not form functional channels by themselves, it confers several properties typical of native channels when coexpressed with the CNGA1 subunit: flickery opening behavior, increased sensitivity to L-*cis*-diltiazem, (a CNG channel–specific inhibitor) and weaker block by extracellular calcium. For a long time, the rod channel was believed to be a hetero-tetramer consisting of two CNGA1 and two CNGB1 subunits. In 2002, a number of laboratories made the surprising discovery that the rod channel actually has a 3CNGA1:1CNGB1 subunit composition. In humans, mutations in CNGA1 cause retinitis pigmentosa. CNGB1 subunits were found to be crucial for the targeting of the native CNG channel in rods. Thus, only trace amounts of the CNGA1 subunit were found on the rod outer segments in CNGB1-null mice and the majority of rod photoreceptors failed to respond to light.

The gating of the rod channel by cGMP is cooperative with a Hill coefficient of ∼3; therefore, the light-triggered suppression of the dark current is 3 times larger than the decrease in the intracellular cGMP concentration. This is the last step of signal amplification in rod phototransduction. The combined amplification provided by rhodopsin, PDE, and CNG channels is very high ($\sim 10^5-10^6$), ensuring the high sensitivity of rods, including the ability of rods to detect single photons.

In the dark, the concentration of free cGMP in the rod outer segment was estimated to be several μM, which is lower than the $K_{1/2}$ ($\sim 10-40 \, \mu M$ depending on Ca^{2+} concentration), the concentration of cGMP necessary to half-maximally activate the channel. As a result, only ∼1% of the CNG channels are open! In other words, 99% of the channels are already closed in the dark and light can only suppress the remaining 1% channels. This explains why current induced by cGMP injection is more than 10 times larger than the dark current.

The inward current through the cGMP channel is composed of $\sim$85% Na^+ because Na^+ is the predominant external cation and the channel is nonselective to monovalent cations. The remaining current is mainly carried by Ca^{2+} with a minor contribution from Mg^{2+}. Extracellular Ca^{2+} actually partially blocks the channel to reduce its conductance under physiological conditions. The inward current is balanced by an outward current flowing across the inner-segment membrane, which is mainly carried by potassium channels. This circulating current is also called dark current in both rods and cones. Unlike other ligand-gated channels, the CNG channel does not desensitize to cGMP, which is important for rods to maintain a steady dark current ranging between 20 and 70 pA in vertebrate rods. The rod photoresponse is essentially a transient suppression of the circulating current. It was estimated that the dark current was carried by $\sim$10 000 channels. The participation of large numbers of micro-channels averages out the channels noise, that is, reduces an otherwise substantial stochastic channel noise if the dark current were carried out by a few macro-channels. This feature improves the sensitivity of rods.

Two extrusion mechanisms are critical in maintaining ionic balance in rods. An energy-dependent Na–K ATPase at the inner segment pumped Na^+ out and K^+ into the cells. A Na/Ca,K exchanger (NCKX) in the outer-segment plasma membrane extrudes one Ca^{2+} and one K^+ outward in exchange for four Na^+ inward producing the net entry of one positive charge. The exchanger and the CNG channel were found to form a stable complex on the plasma membrane, likely as a way to control the stoichiometry between the two, which is critical for regulating Ca^{2+} concentration in the rod outer segment. During the light response, the influx of Ca^{2+} is reduced due to the closure of some CNG channels while the efflux of Ca^{2+} through the exchanger is maintained. The resulting Ca^{2+} decline triggers negative feedback to produce light adaptation.

See also: Phototransduction: Adaptation in Cones; Phototransduction: Adaptation in Rods; Phototransduction: Inactivation in Cones; Phototransduction: Inactivation in Rods; Phototransduction: Phototransduction in Cones; Phototransduction: The Visual Cycle; Primary Photoreceptor Degenerations: Retinitis Pigmentosa; Primary Photoreceptor Degenerations: Terminology; Secondary Photoreceptor Degenerations: Age-Related Macular Degeneration; Secondary Photoreceptor Degenerations.

Further Reading

Arshavsky, V. Y., Lamb, T. D., and Pugh, E. N., Jr. (2002). G proteins and phototransduction. *Annual Review of Physiology* 64: 153–187.

Baylor, D. A., Lamb, T. D., and Yau, K. W. (1979a). The membrane current of single rod outer segments. *Journal of Physiology* 288: 589–611.

Baylor, D. A., Lamb, T. D., and Yau, K. W. (1979b). Responses of retinal rods to single photons. *Journal of Physiology* 288: 618–634.

Burns, M. E. and Arshavsky, V. Y. (2005). Beyond counting photons: Trials and trends in vertebrate visual transduction. *Neuron* 48: 387–401.

Burns, M. E. and Baylor, D. A. (2001). Activation, deactivation, and adaptation in vertebrate photoreceptor cells. *Annual Review of Neuroscience* 24: 779–805.

Fu, Y. and Yau, K. W. (2007). Phototransduction in mouse rods and cones. *Pflugers Archiv – European Journal of Physiology* 454: 805–819.

Luo, D. G., Xue, T., and Yau, K. W. (2008). How vision begins: An odyssey. *Proceedings of the National Academy of Sciences of the United States of America* 105: 9855–9862.

Pugh, E. N., Jr. and Lamb, T. D. (2000). Phototransduction in vertebrate rods and cones: Molecular mechanisms of amplification, recovery and light adaptation. In: Stavenga, D. G., de Grip, W. J., and Pugh, E. N., Jr. (eds.) *Handbook of Biological Physics, Vol. 3: Molecular Mechanisms of Visual Transduction*, pp. 183–255. Amsterdam: Elsevier.

Phototransduction: Rhodopsin

L P Pulagam and K Palczewski, Case Western Reserve University, Cleveland, OH, USA

Glossary

GPCRs – G protein-coupled receptors are membrane receptor proteins with a seven-transmembrane helical topology that are capable of activating G proteins.

G proteins – Heterotrimeric intracellular proteins so named because they bind to the guanine nucleotides, guanosine diphosphate in an inactive state and guanosine triphosphate in an active state.

GRK1 – G protein-coupled receptor kinase 1 (rhodopsin kinase) is a highly specific protein kinase that catalyzes phosphorylation of photoactivated rhodopsin thereby triggering its deactivation.

LCA – Leber's congenital amaurosis is an inherited degenerative disease of the retina that results in a severe loss of vision.

Photoisomerization – The structural change between chromophore geometric isomers (*cis* to *trans*) caused by photoexcitation. For rhodopsin, photoisomerization of its chromophore leads to activation of this receptor.

ROSs – Rod outer segments are the cylindrical outer portions of rod cells, each containing hundreds of membranous disks enveloped by the cellular (plasma) membrane.

RP – Retinitis pigmentosa is the name of a heterogeneous group of progressively blinding degenerations of the human retina caused by mutations in genes encoding photoreceptor proteins with an autosomal dominant (adRP), autosomal recessive (arRP), or X-linked pattern of inheritance.

Schiff base – A functional chemical group containing a carbon–nitrogen double bond. The retinylidene moiety is a chemical link between retinal and an amino group, for example, Lys[296] of opsin.

Vision is an important biological sensing mechanism that involves conversion of light signals received by the eye into electrical nerve impulses transmitted to the brain by a process called phototransduction, consisting of a cascade of biological processes that occur in photoreceptor cells (rod and cone cells) of the retina. In the absence of light, photoreceptors are depolarized to a membrane resting potential of −40 mV. In the presence of light, the plasma membrane of the photoreceptor cells becomes hyperpolarized to −70 mV, resulting in a reduced amount of neurotransmitter released to downstream neurons. This article focuses on rhodopsin structure that relates to its function as a G protein-coupled receptor (GPCR).

Rod Cells and Rhodopsin

The vertebrate rod cell, a highly differentiated postmitotic neuron, is characteristically long, cylindrical, and primarily consists of an outer segment connected to an inner segment via a cilium (**Figures 1(a)** and **1(b)**). The rod outer segment (ROS) contains a stack of disk membranes enclosed by the plasma (cell) membrane, whereas the rod inner segment (RIS) contains the metabolic machinery for this cell. Rhodopsin is processed in the endoplasmic reticulum and transferred to the Golgi membranes of the RIS for additional processing of its carbohydrate moieties. Then rhodopsin-containing Golgi vesicles fuse with the apical plasma membrane of the inner segment and the rhodopsin molecules are transported through the rod cell cilium to the ROS where they form disk membranes. Mutations in the C-terminal region of rhodopsin inhibit transport of rhodopsin to ROS, indicating that this region is essential for recognition by the transport machinery.

A mammalian ROS consists of a stack of 1000–2000 disks enclosed by the plasma membrane. A cryo-electron tomography image of murine ROS (**Figure 1(c)**) also reveals that the thickness of a single disk membrane is 8 nm. Rhodopsin comprises >90% of all proteins in disk membranes and occupies 50% of the disk membrane volume. It is also present at a lower density in the plasma membrane of rod cells, and its expression is essential for ROS formation, which is absent in knock-out Rho$^{-/-}$ mice. In wild-type mice, there are approximately 8×10^4 rhodopsin molecules per disk and 3.96×10^{14} per eye. The power spectra of negatively stained disk membranes from bovine ROS (**Figure 1(d)**) reveals a diffuse diffraction ring at $\sim(45\,\text{Å})^{-1}$, indicating paracrystallinity of rhodopsin. Organization of the seven helices of rhodopsin was first ascertained by using a low-resolution imaging method called electron crystallography. Rhodopsin is unequally distributed in disk membranes. Electron tomographs of the ROS reveal both high- and low-density regions (**Figure 1(e)**). An atomic force microscopic image of native disk membranes showed that the average packing density of rhodopsin monomers is $48\,300 \pm 8000\,\mu\text{m}^{-2}$. Recent atomic force microscopic studies disclosing the arrangement of rhodopsin in native mouse disk membranes revealed that rhodopsin and opsin form structural dimers arranged in

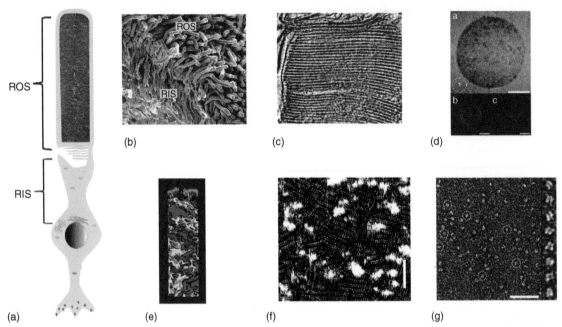

Figure 1   Vertebrate retina and rhodopsin. (A) Diagram of a rod cell. Vertebrate rod cells are postmitotic neurons with highly differentiated rod outer segments (ROSs) connected to rod inner segments (RISs) that generate proteins and energy to sustain phototransduction. ROSs consist of hundreds of stacked disk membranes enveloped by a plasma membrane. The main component of disk membranes is rhodopsin. Biochemical processes involving rhodopsin in the ROS allow rapid transduction of a light signal to graded hyperpolarization of the plasma membrane resulting from a decrease of light-sensitive conductance in ROS cGMP-gated cation channels. (B) Scanning electron micrograph of mouse retina. Rod cells comprise ~70% of all 6.4 million retinal cells, whereas cone cells represent <2%. (C) An x–y slice electron tomogram of vitrified mouse ROS. Disk membranes consist of a phospholipid bilayer studded with rhodopsin. (D) Transmission electron microscopy of negatively stained native disk membranes from mouse rod photoreceptors adsorbed on carbon film. (a) Morphology of a native bovine disk membrane. (b) Average of five power spectra calculated from broken circle outlined region shown inside the disk membrane in (a). A diffuse power diffraction signal is evident, indicating paracrystallinity of rhodopsin. (c) Average of five power spectra calculated from broken circle outlined region shown outside the disk membrane in (a). No power diffraction is evident. Scale bars: (a) = 2000 Å; (b and c) = 40 Å^{-1}. (E) The density of disk membranes is not uniform, indicating that rhodopsin is unevenly distributed in the mouse membrane. The distribution of high- (dark gray) and low- (light gray) density regions is shown in a top view of a single disk. Areas of low density (gray value < 0.33) represent 29% and areas of high density (gray value > 0.33) represent 71% of the disk volume. Gray values were obtained by computing the gray value for each voxel (three-dimensional pixel) in a disk membrane volume of 10 disks. Spacer proteins connecting two disks are colored red and the plasma membrane is colored blue. (F) Topograph obtained by using atomic-force microscopy shows the paracrystalline arrangement of rhodopsin dimers in the native disk membrane of mouse rod photoreceptors. Vertical brightness ranges: 1.6 nm. Scale bar = 50 nm. (G) Transmission electron microscopy of negatively stained disk membranes solubilized by n-dodecyl-β-D-maltoside. Rhodopsin dimers are clearly discerned on the carbon film. Magnified selected particles marked by broken circles are shown on the right. Scale bar = 500 Å. Frame size of the magnified particles in the gallery is 104 Å. (C and E) From Nickell et al. (2007). Originally published in *The Journal of Cell Biology*. (doi:10.1083/jcb.200612010). (F) Adapted from figure 2a in Fotiadis D. (2003). Atomic-force microscopy: Rhodopsin dimers in native disc membranes. *Nature* 421: 127–128. (D and G) Adapted from figure 3 in Suda K. (2004). The supramolecular structure of the GPCR rhodopsin in solution and native disc membranes. *Molecular Membrane Biology* 21: 435–446: Taylor and Francis.

paracrystalline arrays of rows (**Figure 1(f)**). Previous studies also support the importance of dimerization/oligomerization as necessary for the function of many, if not all GPCRs. Negative staining of detergent (*n*-dodecyl-β-D-maltoside)-solubilized disk membranes shows bi-lobed, roughly conical structures with lengths of ~65 Å, and these lobes are separated from each other by ~32 Å (**Figure 1(g)**).

Structure of Rhodopsin

Rhodopsin is a transmembrane protein consisting of an apoprotein, the 348 amino acid residue-long opsin

(**Figure 2**), linked to a chromophore, 11-*cis*-retinal. The chromophore is bound covalently via a protonated Schiff base to the Lys^{296}-containing side chain of the opsin. Bovine rhodopsin also is post-translationally modified. The N-terminal Met is acetylated, and Cys^{322} and Cys^{323} of the C-terminus are palmitoylated, which is very common in GPCRs. In addition, a disulfide bond exists between Cys^{110} $^{(H-III)}$ and Cys^{187} $^{(E-II)}$. Rhodopsin is glycosylated at Asn^{2} and Asn^{15} by the hexasaccharide sequence (Man)$_3$Glc(Nac)$_3$. The molecular mass of bovine opsin with its post-translational changes (palmitoylation, acetylation of the N terminus and glycosylation) is 42 002.

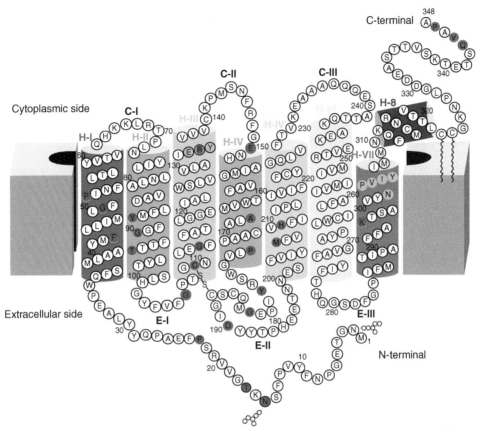

Figure 2 Two-dimensional representation of rhodopsin. Rhodopsin has seven transmembrane α-helices, C-I, C-II, and C-III depict its cytoplasmic loops, and E-I, E-II, and E-III represent the extracellular loops in this diagram. Stability of the helical segment is increased by the disulfide Cys[110]–Cys[187] bridge (shown as -S-S-), a highly conserved feature among many GPCRs. The chromophore, 11-*cis*-retinal, not shown here, is attached to Lys[296] via a protonated Schiff base. Asn[2] and Asn[15] are sites of glycosylation by conserved glycans, Met[1] is acetylated and Cys[322] and Cys[323] are palmitoylated. The predominant phosphorylation sites are Ser[334], Ser[338], and Ser[343]. The whole C-terminal region is highly mobile but, as shown by using a model peptide, it may become more rigid when bound to arrestin. The highly conserved domains among GPCRs, (D/E)R(Y/W) (colored in pink) in helix 3 and NPXXY in helix VII (colored in green), are important for transforming the receptor from an inactive to a G protein-coupled conformation. Nonsense/missense mutations in rhodopsin leading to retinitis pigmentosa are colored in dark gray.

Rhodopsin was crystallized from detergent solutions. The three-dimensional structure of bovine rhodopsin at 2.8 Å, the first high-resolution structure reported for a GPCR, reveals the internal organization of this receptor molecule (**Figure 3**). Rhodopsin folds into seven transmembrane helices (H-I–H-VII) that vary in length from 20 to 33 amino acid residues, and one cytoplasmic helix (H-8). The transmembrane residues are irregular and tilted at various angles due to their Gly and Pro residues. Further advances in rhodopsin crystallization were recently reviewed.

The N-terminal region of rhodopsin is located intradiscally (extracellular) and the C-terminal region is cytoplasmic (intracellular), each region possessing three interhelical loops. The extracellular region consists of four distorted β-strands, three interhelical loops and the doubly glycosylated N-terminal domain. The extracellular region from residues 173–198 acts as a plug for the chromophore-binding pocket (**Figures 4(a)** and **4(b)**). The cytoplasmic surface contains 14 positively charged residues, whereas the extracellular side contains only three positively charged residues, an arrangement that agrees with the positive inside rule for multispanning eukaryotic membrane proteins. The intracellular/cytoplasmic region of rhodopsin is essential for its vectorial transport from the site of synthesis to the ROS and it also plays important roles in G protein activation and photoactivated rhodopsin desensitization.

A highly conserved (D/E)R(Y/W) motif in the GPCR A family is formed by the tripeptide, Glu[134]-Arg[135]-Tyr[136], located in the cytoplasmic region of bovine rhodopsin (**Figure 4(a)**). The carboxylate group of Glu[134] forms a salt bridge with Arg[135], a highly conserved residue among GPCRs, and Arg[135] also interacts with Glu[247] and Thr[251] in H-VI. The ionization state of Glu[134] is sensitive to its environment, such that protonation of this residue causes rhodopsin activation (from meta IIa to meta IIb). This motif plays an important role in conformational changes in the structure of GPCRs that lead to their activation.

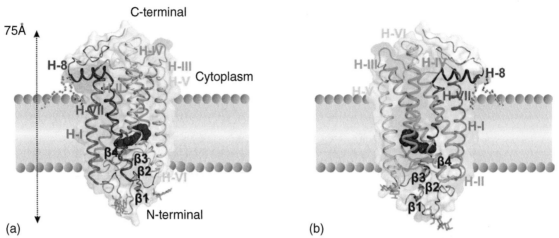

Figure 3 Structure of rhodopsin. (a) Rhodopsin has seven transmembrane helices (H-I to H-VII) and one peripheral helix (H-8). The transmembrane segments are α-helical, but these helices are highly distorted and tilted. Helices are displayed as ribbons and colored from blue (H-I) to red (H-8), as in the visible light spectrum. Helix labels are shown in the same color as their respective helices. N-terminal and C-terminal ends are colored gray. A palmitoyl group (orange-colored ball and stick representation) is attached to each of the two Cys residues at the end of helix H-8. Removal of this group has only a minor effect on phototransduction. $\beta_1-\beta_4$ are distorted β strands. The carbohydrate moieties (cyan-colored stick representation) are at Asn[2] and Asn[15]. The Gly[3] to Pro[12] region forms the first β-hairpin that runs parallel to the expected plane of the membrane. Arg[177] to Asp[190] leave helix H-4 to form a second twisted β-hairpin on the extracellular side. Rhodopsin is represented in a space-filled background and the plane of the lipid bilayer is shown. (b) Opposite side of rhodopsin shown in **Figure 2(a)**. (pdb code: 1U19.pdb)

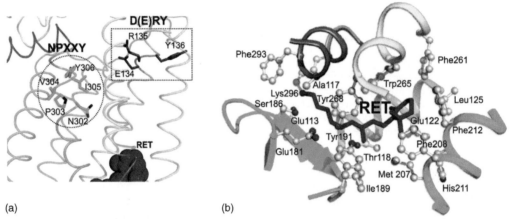

Figure 4 Functional regions of rhodopsin. (a) Conserved NPXXY and (D/E)R(Y/W) motifs in rhodopsin. Amino acid residues of these motifs are rendered as sticks, retinal (RET) is shown in the CPK color scheme as red, and helices are displayed as ribbons. (b) Retinal (RET)-binding site. Amino acid residues within about 5 Å are displayed to show the side chain environment surrounding the 11-*cis*-retinylidene group. Residues are represented as balls and sticks and retinal is shown in stick form (red). See text for explanation/discussion. Helices are colored as in **Figure 3**.

Another highly conserved NPXXY (Asn-Pro-Xaa-Xaa-Tyr) motif located at the end of helix VII and the beginning of H-8, is also close to the cytoplasmic region. Both the D (E)RY and NPXXY regions control the meta II (active) state of rhodopsin, and the NPXXY sequence is likely to be involved in G protein coupling (**Figure 4(a)**). The greatest distortion in H-VI is imposed by Pro[267] (a highly

conserved residue among GPCRs) and H-VII is elongated due the presence of Pro[291] and Pro[303] (parts of NPXXY) located in close proximity to Lys[296], the retinal-binding residue. Highly conserved Glu[122] and His[211] are located at the Zn^{2+}-binding site. The interaction between the β ionone ring and H-III occurs at Glu[122], one of the residues that determine the rate of meta II decay (**Figure 4(b)**).

Chromophore-Binding Site

Visual pigments of both rod and cone cells contain the chromophore, 11-*cis*-retinal, bound covalently to a Lys side chain (Lys^{296} in bovine rhodopsin) via a protonated Schiff base. The absorption maximum (λ_{max}) of free solubilized 11-*cis*-retinal is about 380 nm. When this chromophore binds to opsins, its λ_{max} shifts toward longer wave lengths (a red shift) ranging from 435 nm (frog rods) to 560 nm (human cones). The protonated Schiff base linkage is responsible for about 70 nm of this shift. A further red shift results from the retinal-binding-pocket environment, especially its counter ion, which is Glu113 in vertebrate rhodopsins. In bovine rhodopsin, the λ_{max} of mutant E113Q (Glu to Gln) is dramatically shifted from 498 nm to ~380 nm. The absorption maximum also varies according to the interaction sites of the opsin molecule with the chromophore, especially dipolar interactions near the β ionone ring. Therefore, the λ_{max} absorption of visual pigments varies from species to species that differ with respect to their opsin protein sequences.

The chromophore in rhodopsin is located in the core of the seven transmembrane helices, closer to the extracellular side of the disk membrane. 11-*cis*-Retinal helps maintain rhodopsin in an inactive state (**Figure 4(b)**). The retinal-binding pocket is formed by helices H-III, H-V, H-VI, and H-VII and the antiparallel β sheet of the N-terminal plug, part of extracellular loop II (E-II in **Figure 2**). Although the retinal-binding site is very hydrophobic (**Figure 4(b)**), four charged residues, specifically Glu^{113} $^{(H-III)}$, Glu^{122} $^{(H-III)}$, Glu^{181} $^{(β\ sheet)}$, and Lys ^{296} $^{(H-VII)}$, are located near the chromophore. Lys^{296} $^{(H-VII)}$ donates an amino group to form a protonated Schiff base and Glu^{113} $^{(H-III)}$, which is 3.6 Å away from the Schiff base, acts as a counter ion for this linkage. The positive charge on the protonated Schiff base is energetically unstable in the hydrophobic protein core but the Glu^{113} $^{(H-III)}$ counter ion stabilizes the base by shifting its pK_a from neutral to alkaline. Highly conserved among all known vertebrate visual pigments, Glu^{113} $^{(H-III)}$ plays three important roles: (1) It keeps rhodopsin in its resting state by participating in the salt bridge with the Schiff base. Disruption of this bridge allows the H-VI motion that occurs upon photoactivation. (2) It prevents spontaneous hydrolysis of the Schiff base by stabilizing the protonated Schiff base via increasing its K_a by as much as 10^7. (3) It causes a major bathochromic (longer wavelength) shift in the maximum wavelength absorption of visual pigments. Longer wavelength absorption is essential because the front of the eye in most animals does not allow ultraviolet (UV) light to reach the retina. Steric hindrance, resulting either from mutating Gly^{121} $^{(H-III)}$ or substituting larger R groups at the C9 position, causes transducin (G$_t$) activation in the dark whereas lack of the C9 methyl group impedes photoactivation.

Opsin reacts within minutes with 11-*cis*-retinal to form rhodopsin. Similarly, 7-*cis* and 9-*cis* retinal also form visual pigments. In contrast, all-*trans*-retinal and 13-*cis*-retinal cannot regenerate opsin. The crystal structure of rhodopsin reveals that the chromophore-binding pocket is well defined, suggesting that the binding pocket has high specificity for the Schiff base and the β ionone ring. The exact location of these two components restricts the length of the chromophore-binding site. Therefore, 7-*cis*, 9-*cis*, and double and triple *cis* retinal analogs that have similar lengths and structures can regenerate the opsin, whereas chromophore analogs that are either shorter or longer than 11-*cis*-retinal cannot.

Rhodopsin Cycle – Retinal Isomerization

Inactive rhodopsin is activated upon light absorption, which induces a *cis*–*trans* isomerization that converts 11-*cis*-retinal to all-*trans*-retinal. The activation process can be divided into three phases (**Figure 5**).

1. Light-induced *cis*–*trans* isomerization of the retinylidene.
2. Thermal relaxation of the retinylidene–protein complex.
3. Hydrolysis of the Schiff base linkage, leading to formation of rhodopsin's active meta II state.

Light absorption induces isomerization of 11-*cis*-retinylidene to all-*trans*-retinylidene, resulting in a transient intermediate called photorhodopsin, formed by the fastest chemical reaction (200 fs) in the rhodopsin cycle. Photorhodopsin is converted first into a thermally stable and high-energy intermediate product called bathorhodopsin. About 60% of the incident photon energy is stored in bathorhodopsin and then used to drive further conformational changes. In this state, the chromophore is in an 11-*trans*-15-*anti* conformation, a distorted all-*trans* conformation that results from steric restriction caused by the polyene chain of the retinal and the protein side chains. The β ionone ring and Schiff base are located in a conformation similar to that of rhodopsin in the dark, but Thr^{181} and Glu^{113} are slightly moved. Then the blue-shifted intermediate (BSI) is produced during the thermal relaxation of bathorhodopsin, but BSI can be observed only by time-resolved measurements during subsequent formation of lumirhodopsin. The distorted all-*trans*-retinal in bathorhodopsin relaxes by dislocation of the β ionone ring in lumirhodopsin. Displacement of this ring reflects the movement of helix III aided by interactions between other helices, that result in a slightly disordered structure during the transition from the dark state to lumirhodopsin. In particular, Thr^{181} and Glu^{122} (which are moved slightly in bathorhodopsin) become significantly moved due to the β ionone ring

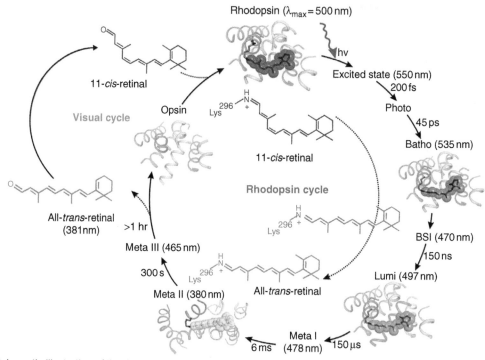

Figure 5 Schematic illustration of the rhodopsin cycle. Rhodopsin consists of an apoprotein called opsin and the bound chromophore, 11-*cis*-retinylidene, a geometric isomer of vitamin A in aldehyde form that imparts a red color to this protein. Upon light activation, rhodopsin transforms into opsin via many intermediate states. In the dark, rhodopsin contains the 11-*cis*-retinal chromophore attached to Lys[296] (rendered as sticks) of helix H-VII in a protonated Schiff base linkage. Upon absorption of a photon, the chromophore is isomerized with ~65% probability from a *cis* C^{11}–C^{12} double bond to a *trans* conformation. In addition, with light activation, rhodopsin transforms into the photointermediate, bathorhodopsin (Batho), which thermally relaxes to BSI followed by lumirhodopsin (Lumi), which then changes to Meta I. During the transition of meta I to meta II, the all-*trans*-retinylidene Schiff base becomes deprotonated. Meta II, the signaling state capable of G protein activation, ultimately decays to free all-*trans*-retinal and opsin. The released photoisomerized chromophore, all-*trans*-retinal, is reduced to an alcohol by short chain alcohol dehydrogenases, such as prRDH, retSDR, and RDH12. The all-*trans*-retinal then diffuses into the retinal pigmented epithelium. There it undergoes enzymatic transformation back to 11-*cis*-retinal in a metabolic pathway known as the visual cycle. The replenished 11-*cis*-retinal then combines with opsin to form rhodopsin, thereby completing the rhodopsin cycle. The λ_{max} in nm as well as the duration (fs-s) of the various components are shown.

displacement. Lumirhodopsin then relaxes further into meta I. Although the conformation of the chromophore in meta I closely resembles that in lumirhodopsin and the Schiff base proton is still hydrogen bonded, the overall structure is similar to rhodopsin. Meta I is further converted to meta II in two steps that can be separated by 20 ms in detergent-solubilized samples: (1) Conversion of meta I to meta IIa, accompanied by proton transfer from the Schiff base to the counterion Glu[113]. (2) Subsequent uptake of a proton from the cytoplasm leading to meta IIb formation. The proton acceptor here is Glu[134 (H-III)], thus meta IIa and meta IIb are in a pH-dependent equilibrium regulated by proton uptake at Glu[134] (part of E/DRY motif). However, only meta IIb can trigger G_t activation. Deprotonation of the Schiff base is characterized by a large UV shift of the absorption maximum from 478 nm in meta I to 380 nm in meta II. A low-resolution crystal structure of a bovine-deprotonated Schiff base meta II-like rhodopsin has been elucidated. Finally, meta II decays

into opsin and all-*trans*-retinal. Free opsin exhibits the largest conformational changes as compared to dark-state rhodopsin (**Figure 5**).

Visual Cycle – Rhodopsin Regeneration

The visual cycle consists of a series of reactions by which all-*trans*-retinal released from opsin isomerizes back into 11-*cis*-retinal that again binds to the opsin (**Figure 5**). This cyclic process does not require light. The released all-*trans*-retinal is transformed first to all-*trans*-retinol by a retinal dehydrogenase (RDH) in the ROS. The all-*trans*-retinol is transferred to the retinal pigment epithelium (RPE) where it is esterified by lecithin:retinol acyltransferase (LRAT), and later isomerized to 11-*cis*-retinol by a retinol isomerase. 11-*cis*-RDH then converts 11-*cis*-retinol back to 11-*cis*-retinal, which leaves the RPE to regenerate the opsin in the ROS.

Vertebrate versus Invertebrate Rhodopsins

Enzymatic regeneration of rhodopsin does not occur in invertebrate visual systems where rhodopsin and meta-rhodopsin are photoconvertible. Upon photon absorption, 11-*cis*-retinal (or its analogs) of rhodopsin is converted into all-*trans*-retinal of metarhodopsin, and then irradiation of metarhodopsin changes the all-*trans*-retinal back to 11-*cis*-retinal by a process called photoregeneration.

Notably, invertebrates differ from vertebrates in the photoactivation of rhodopsin. Absorption of a photon by invertebrate rhodopsin leads to a stable meta II, but retinal remains in the retinal-binding pocket. This contrasts to vertebrate rhodopsin where retinal leaves the meta II-binding pocket. Invertebrate phototransduction also involves an inositol-1,4,5-triphosphate signaling cascade (cyclic-guanosine monophosphate (GMP) in vertebrates) and a G_q-type G protein (G_t in vertebrates) that is stimulated by photoactivated rhodopsin.

Three-dimensional structures of bovine (vertebrate) rhodopsin and squid (invertebrate) rhodopsin show structural similarities in the arrangement of their transmembrane helices (**Figure 6**). Squid rhodopsin is a 50-kDa protein composed of 488 amino acid residues. A proline-rich 10-kDa C-terminal extension compared

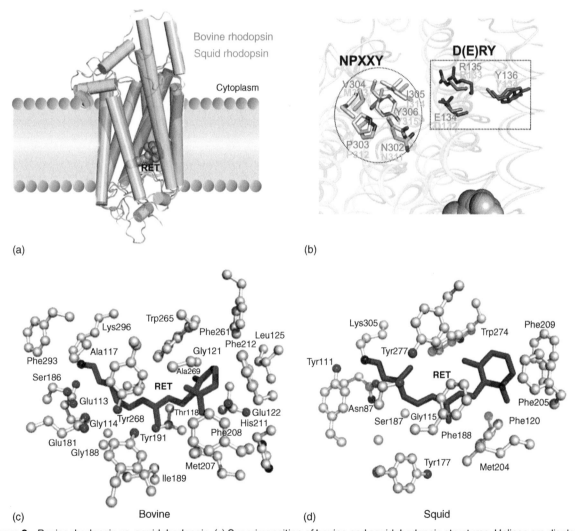

(a)

(b)

(c)   Bovine

(d)   Squid

Figure 6 Bovine rhodopsin vs. squid rhodopsin. (a) Superimposition of bovine and squid rhodopsin structures. Helices are displayed in cartoon style, retinal (RET) is rendered as CPK balls, and the planar lipid bilayer is shown. (b) Conserved NPXXY and (D/E)R(Y/W) motifs of both bovine (green) and squid (yellow) rhodopsins are compared. Amino acids of these motifs are rendered as sticks, retinal is rendered as CPK balls, and helices are displayed as ribbons. (c) Retinal-binding site of bovine rhodopsin. Amino acid residues within about 5 Å are displayed to show the side-chain environment surrounding the 11-*cis*-retinylidene group. Residues are rendered as balls and sticks and retinal is shown in stick form (red). (d) Retinal-binding site of squid rhodopsin. Amino acid residues within about 5 Å are displayed to reveal the side-chain environment surrounding the 11-*cis*-retinylidene group. Residues are rendered as balls and sticks and retinal is shown in stick form (red). See text for explanation/discussion.

with bovine rhodopsin is important for intracellular trafficking of this rhodopsin, and its deletion does not affect G protein activation. Like other GPCRs, squid rhodopsin forms a disulfide bridge between Cys^{108} and Cys^{186} for proper folding of the E-II loop. The NPXXY and (D/E)R (Y/W) regions of squid rhodopsin are similar to those of bovine rhodopsin. However, Val in the NPXXY motif, and Glu in the (D/E)R(Y/W) motif of bovine rhodopsin are replaced by Met and Asp, respectively, in squid rhodopsin (**Figure 6(b)**). Their structural similarity implies that the functional difference between vertebrate and invertebrate rhodopsins might be due to specific interactions between retinal and the amino acid sequence of the retinal-binding site. These binding sites are compared in **Figures 6(c)** and **6(d)**. In squid rhodopsin, the highly conserved Glu^{180} is too far away from the retinal-binding pocket and Asn^{185} is located between Glu^{180} and the Schiff base. Asn^{185} is assumed to move after photoisomerization of retinal, to mediate an indirect interaction between Glu^{180} and the retinal Schiff base. In the dark state, either Asn^{87} or Tyr^{111}, which is highly conserved among all invertebrates, might act as a hydrogen-binding partner (counter ion) for the Schiff base. These residues are replaced by Gly^{89} and Glu^{113} in bovine rhodopsin.

Signaling Cycle

After photoactivation, an active meta II state of rhodopsin triggers the activation of transducin (G_t protein). This form of rhodopsin is capable of activating G_t proteins during the relatively prolonged period of its activation. Accordingly, rhodopsin activity is regulated by its phosphorylation, a common feature among many GPCRs. This regulatory process is also important for rod cells to recover their responsiveness during dark adaptation. Desensitization of rhodopsin involves two steps: (1) phosphorylation of meta II, reducing the rate of transducin activation and (2) binding of arrestin to meta II, completely ending transducin activation by rhodopsin (**Figure 7**). This phosphorylation is carried out by rhodopsin kinase, a specific kinase also known as GRK1. Rhodopsin kinase phosphorylates specific serines in the rhodopsin C-terminal sequence. However, upon light illumination, GRK1 is released from recoverin and phosphorylates rhodopsin at multiple sites. Ser^{343}, Ser^{338}, and Ser^{334}, located in the C terminal domain on the cytoplasmic side of rhodopsin, are the main sites for this phosphorylation. Some threonine residues in this domain are also phosphorylated. Phosphorylation of this cytoplasmic domain reduces the ability of the G_t protein to bind to meta II, but it does not completely stop G_t activation. By binding to phosphorylated-meta II, arrestin prevents the interaction of G_t with meta II, and thus completely terminates G_t activation. At least three phosphorylated sites are required for high-affinity binding

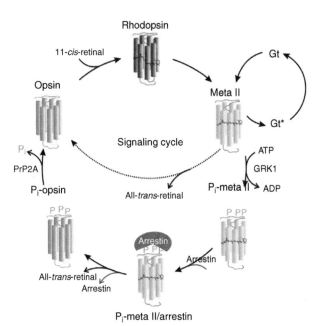

Figure 7 Interaction of rhodopsin with partner proteins. Phototransduction starts with the absorption of light by rhodopsin that causes photoisomerization of 11-*cis*-retinal to all-*trans*-retinal. Photoisomerization of this chromophore induces conformational changes in rhodopsin leading to formation of meta II, the signaling state of rhodopsin. Meta II binds and activates a large number of photoreceptor-specific G protein molecules, transducins (Gt), by catalyzing the exchange of guanosine triphosphate (GTP) for guanosine diphosphate (GDP) on transducin's α-subunit, Gtα. Deactivation of meta II and consequent Gt-mediated signaling starts with binding of the GPCR kinase called GRK1 (or rhodopsin kinase) that catalyzes subsequent phosphorylation of Ser and Thr residues in the C-terminus of rhodopsin. Phosphorylated rhodopsin then is capped by binding to arrestin, which prevents any residual Gt activation by meta II. The complex of phosphorylated rhodopsin–arrestin loses all-*trans*-retinal and then arrestin, after which the phosphorylated opsin is dephosphorylated by the action of protein phosphatase 2A (PrP2A). All-*trans*-retinal is transformed, through a series of steps, to 11-*cis*-retinal, which rebinds to opsin (as shown in **Figure 4**), thereby continuing rhodopsin signaling. A fraction of meta II loses all-*trans*-retinal (without phosphorylation) and directly transforms to opsin.

of the rhodopsin–arrestin complex. Arrestin dissociates from rhodopsin as meta II decays and loses all-*trans*-retinal. Then phosphorylated opsin is dephosphorylated by protein phosphatase 2A (PrP2A). The resulting free opsin is readily regenerated by 11-*cis*-retinal and continues recycling through the signaling cascade. A fraction of meta II directly dissociates into all-*trans*-retinal and opsin.

Rhodopsin Interaction with Other Proteins

According to X-ray crystallographic models and atomic force microscopic studies, H-IV–H-V of rhodopsin contact each other in a rhodopsin dimer. The sizes of G_t, GRK1, and arrestin proteins also favor the hypothesis of their interaction with the rhodopsin dimers. Previously reported models of rhodopsin with these proteins are discussed below. These complexes have yet to be resolved by crystallography.

Rhodopsin–G$_t$

In its inactive state, G$_t$ is a membrane-associated protein that consists of α, β, and γ subunits with one guanosine diphosphate (GDP) noncovalently bound to the α-subunit. Post-translational modification of the α-(myristoylation) and γ-(farnesylation) subunits help this protein to associate with the membrane. A model for the rhodopsin–G$_t$ complex has been reported (**Figure 8(a)**). Spectroscopic, biochemical, and peptide competition experiments reveal that cytoplasmic loops II, III, H-8, and the C-terminal tail of rhodopsin interact with transducin. The interacting sites of transducin are the C-terminal tail, N-terminal helix, the α4–β6–α5 region of the α-subunit, and the farnesylated C-terminal region of the γ-subunit.

Rhodopsin–GRK1

GRK1 phosphorylates multiple sites on the C-terminal tail that is freely accessible in both active and inactive states of rhodopsin (**Figure 8(b)**). A single activated rhodopsin (meta II) molecule can induce the phosphorylation of hundreds of other rhodopsins. Cytoplasmic loops II and III of rhodopsin are the most important sites for binding of GRK1, and the N-terminal 30 residues of GRK1 are important for this interaction. Inactivating mutations in GRK1 are found in human patients with Oguchi disease, a stationary form of night blindness characterized by a substantial delay in recovery of dark vision after photobleaching.

Rhodopsin–Arrestin

Arrestin binds to photoactivated-phosphorylated rhodopsin (**Figure 8(c)**). Biochemical analysis of the arrestin–rhodopsin complex reveals that several domains of arrestin are essential for this interaction. In particular, the region from residues 163 to 189 is essential for binding to activated-phosphorylated rhodopsin, but not to unphosphorylated rhodopsin. Lysine and arginine residues of arrestin are also very important for specific binding, but only to phosphorylated rhodopsin.

Mutations in Rhodopsin and Retinal Diseases

Mutations in the genes encoding many proteins involved in phototransduction and the visual and signaling cycles have been implicated in causing blinding diseases of humans such as Leber's congenital amaurosis (LCA), Stargardt

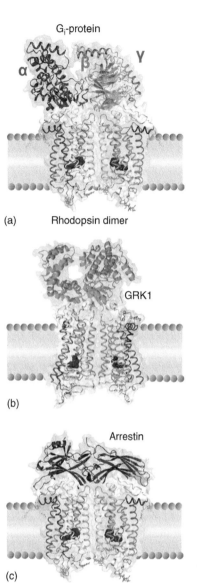

Figure 8 Conceptual models of rhodopsin dimers interacting with Gt, GRK1 and arrestin. (a) Rhodopsin dimer bound to one heterotrimeric Gt. Gtα is colored red, Gtβ is colored green, and Gtγ is colored blue. Gt occupies a single rhodopsin dimer, with only one rhodopsin monomer requiring activation. Helices of rhodopsin are colored as in **Figure 2**. (b) A rhodopsin monomer is modeled such that its third cytoplasmic loop (C-III) lies close to the proposed receptor-docking site for GRK1. This allows the GRK1 active site to have easy access to the C-tail of activated rhodopsin or of a neighboring unactivated rhodopsin in the same membrane plane, thereby allowing high gain phosphorylation of the ROS. (c) This theoretical model reflects the interaction of one arrestin molecule with a rhodopsin dimer. Molecules are represented in a space-filled background and the plane of the lipid bilayer is shown. No structural optimization was performed.

macular degeneration, congenital cone–rod dystrophy, and retinitis pigmentosa (RP). More than 100 rhodopsin mutants resulting in human eye diseases have been identified (**Figure 2**). Some mutations result in degeneration of

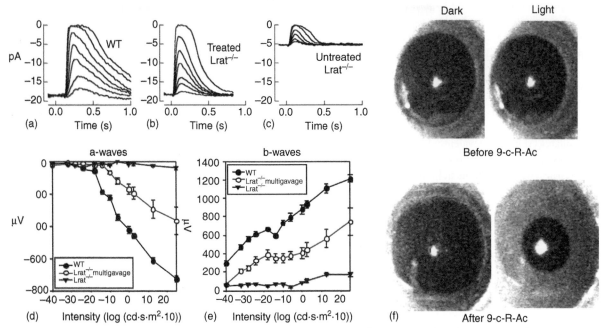

Figure 9   Rescue of visual responses measured by single-cell recording and ERG responses of single *WT* and *Lrat*$^{-/-}$ mouse rod cells. (a) Flash families measured for a *Lrat*$^{+/+}$ mouse (WT) rod, (b) a *Lrat*$^{-/-}$ rod from a mouse that had received a single gavage with 9-*cis*-R-Ac (9-*cis*-retinyl acetate), and (c) a control *Lrat*$^{-/-}$ mouse rod. Rods were obtained from 8-week-old mice. Each panel superimposes averaged responses to 5–20 repeats of a flash; flash strength was increased by a factor of 2 to produce each successively larger response. (d and e) Comparisons of scotopic single-flash ERG recordings from *Lrat*$^{+/+}$ (WT) control mice, 9-*cis*-R-Ac gavaged *Lrat*$^{-/-}$ mice and *Lrat*$^{-/-}$ untreated mice. *Lrat*$^{-/-}$ mice were gavaged 9 times with 5 μmol of 9-*cis*-R-Ac during a 1-month period. (f) Light-induced pupillary constriction of *Lrat*$^{-/-}$ mice before and after treatment with 9-*cis*-R-Ac. All together, these experiments show that 9-*cis*-retinyl acetate restored retinal function in this animal model of LCA. Adapted from figures 4 and 6 in Batten M. L. (2005). Pharmacological and rAAV gene therapy rescue of visual functions in a blind mouse model of Leber congenital amaurosis. *PLOS Medicine* 2(11): e333.

rod cells, while some affect the function of rhodopsin. Mutations at the C-terminal tail impair rhodopsin trafficking from RISs to the ROSs. Mutations, for example Pro[23] to His, which lead to rhodopsin misfolding will not allow the protein to reach disk membranes of ROS. Nonetheless, the ROSs degenerate and finally cause blindness. The Lys[296] mutant is unable to bind chromophore, thereby compromising rhodopsin function. Mutations, for example Ala[292] to Glu, which lead to human congenital night blindness do not involve ROS degeneration but rather compromise human vision under dim light. Mutations of proteins in the visual cycle also cause eye diseases. For example, inactivating mutations in the *LRAT* gene cause LCA. The *Lrat*$^{-/-}$ knock-out mouse with *LRAT*-mediated retinal dystrophy evidences only traces of retinoid compounds in ocular tissues, resulting in impaired vision from birth. The ROS are shortened in *Lrat*$^{-/-}$ mice, and photoreceptors degenerate very slowly. This disease can be treated by dietary intake of active chromophores or their 9-*cis*-precursors. Oral supplementation of *Lrat*$^{-/-}$ mice with 9-*cis*-retinyl acetate restored retinal function (**Figure 9**).

See also: Phototransduction: Phototransduction in Rods; Phototransduction: The Visual Cycle; Rod and Cone Photoreceptor Cells: Inner and Outer Segments; Rod Photoreceptor Cells: Soma and Synapse.

Further Reading

Arshavsky, V. Y., Lamb, T. D., and Pugh, E. N., Jr. (2002). G proteins and phototransduction. *Annual Review of Physiology* 64: 153–187.

Filipek, S., Stenkamp, R. E., Teller, D. C., and Palczewski, K. (2003). G protein-coupled receptor rhodopsin: A prospectus. *Annual Review of Physiology* 65: 851–879.

Fotiadis, D., Liang, Y., Filipek, S., et al. (2003). Atomic-force microscopy: Rhodopsin dimers in native disc membranes. *Nature* 421: 127–128.

Hargrave, P. A., McDowell, J. H., Curtis, D. R., et al. (1983). The structure of bovine rhodopsin. *Biophysics of Structure and Mechanism* 9: 235–244.

Menon, S. T., Han, M., and Sakmar, T. P. (2001). Rhodopsin: Structural basis of molecular physiology. *Physiological Reviews* 81: 1659–1688.

Muller, D. J., Wu, N., and Palczewski, K. (2008). Vertebrate membrane proteins: Structure, function, and insights from biophysical approaches. *Pharmacological Reviews* 60: 43–78.

Okada, T., Sugihara, M., Bondar, A. N., et al. (2004). The retinal conformation and its environment in rhodopsin in light of a new 2.2 Å crystal structure. *Journal of Molecular Biology* 342: 571–583.

Palczewski, K. (2006). G protein-coupled receptor rhodopsin. *Annual Review of Biochemistry* 75: 743–767.

Palczewski, K., Kumasaka, T., Hori, T., et al. (2000). Crystal structure of rhodopsin: A G protein-coupled receptor. *Science* 289: 739–745.

Park, J. H., Scheerer, P., Hofmann, K. P., Choe, H. W., and Ernst, O. P. (2008). Crystal structure of the ligand-free G-protein-coupled receptor opsin. *Nature* 454: 183–187.

Park, P. S., Lodowski, D. T., and Palczewski, K. (2008). Activation of G-protein-coupled receptors: Beyond two-state models and tertiary conformational changes. *Annual Review of Pharmacology and Toxicology* 48: 107–141.

Rao, V. R. and Oprian, D. D. (1996). Activating mutations of rhodopsin and other G protein-coupled receptors. *Annual Review of Biophysics and Biomolecular Structure* 25: 287–314.

Ridge, K. D. and Palczewski, K. (2007). Visual rhodopsin sees the light: Structure and mechanism of G protein signaling. *Journal of Biological Chemistry* 282: 9297–9301.

Salom, D., Lodowski, D. T., Stenkamp, R. E., et al. (2006). Crystal structure of a photoactivated deprotonated intermediate of rhodopsin. *Proceedings of the National Academy of Sciences of the United States of America* 103: 16123–16128.

Travis, G. H., Golczak, M., Moise, A. R., and Palczewski, K. (2007). Diseases caused by defects in the visual cycle: Retinoids as potential therapeutic agents. *Annual Review of Pharmacology and Toxicology* 47: 469–512.

Phototransduction: The Visual Cycle

G H Travis, UCLA School of Medicine, Los Angeles, CA, USA

Glossary

Lipofuscin – Fluorescent pigment granules found within cells of the RPE. Lipofuscin contains oxidized fatty acids and condensation of products of retinaldehyde with phosphatidylethanolamine, such as A2E. Lipofuscin is thought to arise from the incomplete digestion of phagocytosed outer segments. Components of lipofuscin, including A2E, are cytotoxic and thought to play a role in the etiology of macular degeneration.

Opsin visual pigment – Opsin pigments are light-sensitive complexes containing a protein and an 11-*cis*-retinaldehyde chromophore.

Outer segment – An elongated light-sensitive structure attached to the connecting cilium of rod and cone photoreceptors. The outer segment comprises a stack of approximately 1000 membranous disks. These disks are loaded with rhodopsin or cone opsin visual pigments.

Retinyl ester – A conjugate of vitamin A (retinol) with a fatty acid. Retinyl esters represent stable, nontoxic, and water-insoluble storage forms of retinol. Retinyl esters are also the substrate for Rpe65-isomerase in RPE cells.

Retinaldehyde – An oxidized form of retinol. Retinaldehydes are highly reactive and potentially cytotoxic. The 11-*cis* isomer of retinaldehyde (11-*cis*-RAL) is the light-sensitive chromophore in rhodopsin and cone-opsin visual pigments.

Schiff base – It is also called an imine. Results from the reaction of a primary amine (as in lysine or phosphatidylethanolamine) with a carbonyl group (as in retinaldehyde) to form a carbon–nitrogen double bond with loss of a water molecule. Formation of a Schiff base is reversible.

The vertebrate retina contains two classes of light-sensitive cells, rods and cones. Both cell types contain a membranous structure called the outer segment (OS), which are loaded with rhodopsin or cone-opsin visual pigments. These pigments are members of the G-protein-coupled receptor superfamily. Each rod OS contains approximately 10^8 rhodopsin pigments. The ligand for these pigments is 11-*cis*-retinaldehyde (11-*cis*-RAL), which is covalently coupled to a lysine in the opsin protein through a Schiff-base linkage. Absorption of a photon by an opsin pigment induces photoisomerization of the 11-*cis*-RAL chromophore to all-*trans*-retinaldehyde (all-*trans*-RAL). This isomerization converts the pigment to active metarhodopsin II, which stimulates the visual-transduction cascade. After a brief period, metarhodopsin II is inactivated by rhodopsin kinase-mediated phosphorylation and subsequent capping by arrestin. Next, all-*trans*-RAL dissociates from the inactivated opsin pigment. To restore light sensitivity, the bleached apo-opsin recombines with another 11-*cis*-RAL, forming a new rhodopsin or cone-opsin pigment. To maintain continuous vision in light, the all-*trans*-RAL released by bleached pigments must be converted back to 11-*cis*-RAL. This process is carried out by a multistep enzyme pathway called the visual cycle (**Figure 1**). The first two catalytic steps of this pathway occur in photoreceptors, while the remaining steps take place in cells of the retinal pigment epithelium (RPE). The RPE is an epithelial monolayer adjacent to the photoreceptors. Apical processes of RPE cells interdigitate with the photoreceptor OS. The regeneration of visual chromophore is one of the several collaborations between photoreceptors and RPEs; and cone opsins may have access to an alternative source of 11-*cis*-RAL chromophore. This alternative retinoid pathway is present in Müller glial cells.

Clearance of All-*trans*-RAL from OS Disks

Following photoactivation and subsequent deactivation of the opsin pigment, all-*trans*-RAL probably exits between transmembrane (TM) helices, TM1 and TM7, into the lipid bilayer. The all-*trans*-RAL diffuses within the bilayer until it encounters the amine headgroup of a phosphatidylethanolamine, which may condense with the all-*trans*-RAL to form the Schiff base, *N*-retinylidene-phosphatidylethanolamine (*N*-ret-PE). This condensation reaction is reversible. On the cytoplasmic surface of the OS disk-membrane, all-*trans*-RAL is reduced to all-*trans*-retinol (all-*trans*-ROL), driving dissociation of *N*-ret-PE (see below). However, all-*trans*-RAL can be temporarily trapped as *N*-ret-PE on the intradiscal surface. An adenosine triphosphate (ATP)-binding cassette transporter called ABCA4 (also ABCR or rim-protein) is present in the disks of rod and cone OS. Mice with a knockout mutation in the *abca4* gene show delayed clearance of all-*trans*-RAL and elevated *N*-ret-PE in the retina following exposure to light. *In vitro* studies suggest that ABCA4 is an outwardly directed flippase for

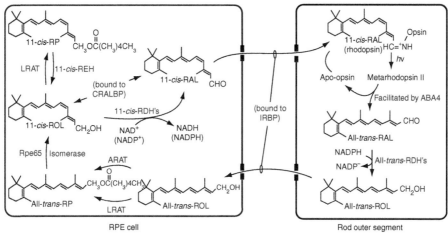

Figure 1 Visual cycle. Following absorption of a photon (*h*ν), 11-*cis*-RAL Schiff base in rhodopsin is isomerized to all-*trans*-RAL, converting the receptor to active metarhodopsin II. Subsequently, the all-*trans*-RAL dissociates from apo-opsin. ABCA4 transports all-*trans*-RAL (as *N*-ret-PE) across the disk bilayer from the interior to the cytoplasmic leaflet. The all-*trans*-RAL is reduced to all-*trans*-ROL by one or more all-*trans*-RDH's that use NADPH as a cofactor. The all-*trans*-ROL is released by the OS to IRBP in the IPM. The all-*trans*-ROL is carried by IRBP to the apical RPE, where it is taken up and esterified by LRAT or ARAT to yield an all-*trans*-RE such as all-*trans*-RP. The all-*trans*-RP is isomerized and hydrolyzed by Rpe65 to yield 11-*cis*-ROL. The 11-*cis*-ROL may be oxidized by one or more 11-*cis*-RDH to yield 11-*cis*-RAL chromophore. Alternatively, the 11-*cis*-ROL may be secondarily esterified by LRAT or ARAT to yield an 11-*cis*-RE, such as 11-*cis*-RP, representing a preisomerized storage form of chromophore precursor. When needed, the 11-*cis*-RP is hydrolyzed by 11-*cis*-REH to yield 11-*cis*-ROL. 11-*cis*-ROL and 11-*cis*-RAL are bound to CRALBP in RPE cells. The 11-*cis*-RAL is released by the RPE into the IPM where it binds to IRBP. Finally, the 11-*cis*-RAL is delivered to the OS where it recombines with apo-opsin to form a new visual pigment.

N-ret-PE, consistent with the biochemical phenotype in *abca4*$^{-/-}$ mice. Thus, ABCA4 appears to facilitate the removal of all-*trans*-RAL from disk membranes for subsequent reduction to all-*trans*-ROL. Mutations in the human *ABCA4* gene cause Stargardt macular degeneration and a subset of recessive cone–rod dystrophy in humans. Stargardt patients and *abca4*$^{-/-}$ mice accumulate toxic lipofuscin pigments in RPE cells. Buildup of these fluorescent pigments is important in the pathogenesis of photoreceptor degeneration in Stargardt's disease.

Reduction of All-*trans*-RAL to All-*trans*-ROL

This reaction is carried out in photoreceptor OS by a member of the short-chain dehydrogenase/reductase family called photoreceptor retinol dehydrogenase (prRDH) or RDH8. RDH8 uses nicotinamide adenine dinucleotide phosphate oxidase (NADPH) as a co-factor. In *rdh8*$^{-/-}$ knockout mice, reduction of all-*trans*-RAL to all-*trans*-ROL is slowed but not halted, suggesting that RDH8 function is complemented in photoreceptors by at least one other retinol dehydrogenase. Photoreceptors contain a second retinol dehydrogenase called RDH12 that also catalyzes NADPH-dependent reduction of all-*trans*-RAL to all-*trans*-ROL. Mice with a knockout mutation in the *rdh12* gene show mildly slowed reduction of all-*trans*-RAL to all-*trans*-ROL, and protection from light-induced

photoreceptor degeneration. Unlike RDH8, which is expressed in photoreceptor OS, RDH12 is expressed in photoreceptor inner segments. This distribution is unexpected given that all-*trans*-RAL is released following light exposure into the OS. RDH12 may play a detoxifying role in the inner segment by reducing all-*trans*-RAL that escaped reduction by RDH8 in the OS. Mutations in *RDH12* cause a severe recessive blinding disease called Leber congenital amaurosis (LCA). No mutations in *RDH8* have been associated with a retinal dystrophy in humans. Mice with a knockout mutation in the *rdh8* gene show normal kinetics of rhodopsin regeneration and delayed recovery of sensitivity following exposure to bright light. An identical pattern is seen in *abca4*$^{-/-}$ mice. ABCR and all-*trans*-ROL dehydrogenase act sequentially in the visual cycle to remove all-*trans*-RAL following a photobleach (**Figure 1**). Delayed dark adaptation in *rdh8*$^{-/-}$ and *abca4*$^{-/-}$ mice is probably due to noncovalent reassociation of all-*trans*-RAL with apo-opsin to form a noisy photoproduct that activates transducin.

Transfer of All-*trans*-ROL from Photoreceptors to the RPE

Interphotoreceptor retinoid-binding protein (IRBP) is secreted by photoreceptors and present at a high concentration in the extracellular space. Besides IRBP, this space is filled with extracellular matrix material and is called the interphotoreceptor matrix (IPM). IRBP contains binding sites for both 11-*cis*- and all-*trans*-retinoids. IRBP has been shown to accelerate the removal of all-*trans*-ROL from bleached photoreceptors. The uptake of all-*trans*-ROL by IRBP may involve a receptor on the OS plasma membrane. Retinoids bound to IRBP are protected from oxidation and isomerization during transit through the IPM. Mice with a knockout mutation in the *irbp* gene show accumulation of all-*trans*-ROL in the retina, and reduced all-*trans*-REs in the RPE following light exposure. These mice also show accumulation of 11-*cis*-RAL in the RPE and reduced 11-*cis*-RAL in the retina following light exposure. These results suggest that IRBP functions to extract all-*trans*-ROL from bleached photoreceptors, and 11-*cis*-RAL from RPE cells. Mutations in the *RBP3* gene for IRBP cause the inherited blinding disease, recessive retinitis pigmentosa in a small subset of cases.

Another all-*trans*-ROL-binding protein, cellular retinol–binding protein type-1 (CRBP1), is present in RPE cells. CRBP1 is a soluble protein that binds all-*trans*-ROL with 100-fold higher affinity than does IRBP. This difference in affinity drives the uptake of all-*trans*-ROL from the IPM into RPE cells. Compared with wild-type mice, *crbp1* $^{-/-}$ knockout mice contain reduced all-*trans*-REs in the RPE and higher all-*trans*-ROL in the retina following light exposure. This biochemical phenotype is similar to the phenotype in *irbp* $^{-/-}$ mice.

Synthesis of Retinyl Esters

The major retinyl-ester synthase in RPE cells is lecithin:retinol acyl transferase (LRAT), which catalyzes the transfer of a fatty-acyl group from the *sn*1 position in phosphatidylcholine to all-*trans*-ROL (see **Figure 1**). The resulting all-*trans*-retinyl esters (all-*trans*-REs) are water-insoluble, and represent a stable and nontoxic storage form of vitamin A. Mice with a knockout mutation in the *lrat* gene contain virtually no all-*trans*-REs or other visual retinoids in their ocular tissues. Accordingly, *lrat* $^{-/-}$ mice are totally blind. Mutations in the human *LRAT* gene are yet another cause of recessive LCA.

Another retinyl-ester synthase activity, called acyl-CoA:retinol acyltransferase (ARAT), is present in RPE cells. Unlike LRAT, ARAT uses palmitoyl coenzyme A (palm CoA) as an acyl donor. Two enzymes have been shown to posses ARAT activity. Diacylglycerol acyltransferase type-1 (DGAT1), which catalyzes palm CoA-dependent synthesis of triglycerides from diacylglycerol, also catalyzes palm CoA-dependent synthesis of all-*trans*-REs from all-*trans*-ROL. Multifunctional *O*-acyltransferase (MFAT) also possesses ARAT catalytic activity. The very low level of all-*trans*-REs in the RPE of *lrat* $^{-/-}$ mice despite the presence of ARAT activity is due to the 10-fold higher K_M for all-*trans*-ROL substrate of ARAT versus LRAT. ARAT preferentially uses free all-*trans*-ROL as a substrate in contrast to LRAT, which uses holo-CRBP1.

Retinoid Isomerization

Conversion of a planar all-*trans*-retinoid to the strained 11-*cis* configuration is energetically unfavorable. Rpe65-isomerase uses all-*trans*-REs as substrate and catalyzes two reactions: hydrolysis of the carboxylate ester, and *trans* to *cis* isomerization of the C11–C12 double bond in the retinoid. Accordingly, the energy released by ester hydrolysis (-5.0 kcal mole^{-1}) is used to drive isomerization ($+4.1$ kcal mole^{-1}). Rpe65 is homologous to β-carotene oxygenase in mammals and apocarotene oxygenase (ACO) in cyanobacteria. The X-ray diffraction analysis showed that ACO has a seven-bladed β-propeller structure, with a Fe^{2+}-4-His arrangement at its axis. The four His residues that define the Fe^{2+}-binding site are conserved in all members of the ACO family including Rpe65. Rpe65 was shown to bind Fe^{2+}, which is required for its catalytic activity. Rpe65 is strongly associated with membranes but contains no membrane-spanning segments. Mice with a knockout mutation in the *rpe65* gene contain high levels of all-*trans*-REs in the RPE and no detectable 11-*cis*-RAL. Accordingly, *rpe65* $^{-/-}$ photoreceptors contain only apo-opsin, and the mice have no detectable visual function. Despite blocked synthesis of visual chromophore, photoreceptor morphology is nearly normal in *rpe65* $^{-/-}$ mice. Visual function has been restored in *rpe65* $^{-/-}$ mice and dogs by administering exogenous visual chromophore. Injection of recombinant adeno-associated virus (AAV) containing a wild-type *rpe65* gene into the subretinal space (between RPE cells and photoreceptors) of *rpe65* $^{-/-}$ mice partially rescued the blindness phenotype. More recently, patients with *RPE65*-mediated LCA received subretinal injections of a similar *RPE65*-containing AAV. Encouragingly, these blind patients partially recovered visual function with expression of wild-type Rpe65 in their RPE.

The 11-*cis*-ROL synthesized by Rpe65 binds to cellular retinaldehyde-binding protein (CRALBP) in RPE cells. CRALBP also binds 11-*cis*-RAL. Mutations in the gene for CRALBP (*RLBP1*) cause several inherited retinal dystrophies including recessive retinitis pigmentosa. A newly synthesized molecule of 11-*cis*-ROL has two potential fates. As discussed below, it can be oxidized to 11-*cis*-RAL for use as visual chromophore. Alternatively, it can be esterified by LRAT to form an 11-*cis*-RE. 11-*cis*-REs represent a storage form of preisomerized chromophore precursor. Hydrolysis of 11-*cis*-REs is catalyzed by 11-*cis*-retinyl ester hydrolase (11-*cis*-REH) in the plasma membrane of RPE cells. The protein responsible for 11-*cis*-REH activity in RPE cells has not yet been identified.

Synthesis of 11-*cis*-RAL Chromophore

The final step in the visual cycle is oxidation of 11-*cis*-ROL to 11-*cis*-RAL. This reaction is catalyzed by 11-*cis*-ROL-dehydrogenase type-5 (RDH5), which uses NAD^+ as a cofactor. Mice with a knockout mutation in the *rdh5* gene show accumulation of 11-*cis*-ROL and 11-*cis*-REs in the RPE, and delayed recovery of rod sensitivity following light exposure. 11-*cis*-RAL is synthesized in *rdh5* $^{-/-}$ mice, albeit at a reduced rate, suggesting that RPE cells express at least one other 11-*cis*-ROL-dehydrogenase. RDH11 catalyzes $NADP^+$-dependent oxidation of 11-*cis*-ROL to 11-*cis*-RAL in the RPE. Surprisingly, *rdh5* $^{-/-}$, *rdh11* $^{-/-}$ double-knockout mice also synthesize 11-*cis*-RAL, although more slowly than in *rdh5* $^{-/-}$ or *rdh11* $^{-/-}$ single-knockout mice, and much more slowly than in wild-type mice. Thus, extensive functional redundancy exists for the oxidation of 11-*cis*-ROL in RPE cells, similar to the functional redundancy for reduction of all-*trans*-RAL in photoreceptors. 11-*cis*-RAL is strongly bound to CRALBP in RPE cells. CRALBP has been shown to interact with a protein complex on the cytoplasmic surface of the apical plasma membrane. From this position, 11-*cis*-RAL is transferred across the plasma membrane to bind IRBP in the IPM. This process may involve a receptor for IRBP on RPE cells.

Regeneration of Rhodopsin or Cone Opsin

The final step in the visual cycle is regeneration of a visual pigment from an apo-opsin and 11-*cis*-RAL. The mechanism whereby 11-*cis*-RAL is transferred from IRBP in the IPM to apo-opsin in the OS disk is unknown. It may involve an IRBP receptor on the OS plasma membrane, or simple diffusion of the 11-*cis*-RAL. No retinoid-binding protein has been identified in OS. The interaction of 11-*cis*-RAL with an apo-opsin involves a two-step process. First, a weak noncovalent complex is formed with 11-*cis*-RAL binding to a hypothesized entrance site on the opsin. Second, the 11-*cis*-RAL moves into the hydrophobic pocket and forms a Schiff base. This step is virtually irreversible in the case of rhodopsin. Once formed, rhodopsin is extremely quiet, with a spontaneous thermal-activation rate of one isomerization every 2000 years. In contrast to rhodopsin, recombination of 11-*cis*-RAL with the apo-cone-opsins is less favorable thermodynamically. Unlike rhodopsin, 11-*cis*-RAL freely dissociates from cone-opsins. For example, a dark-adapted red cone contains approximately 10% apo-cone-opsin due to spontaneous dissociation of chromophore. This effect contributes to the higher noise and much lower sensitivity of cones versus rods. It is also explains the tendency of rods to steal visual chromophore from cones when the availability of 11-*cis*-RAL is limited.

Regulation of the Visual Cycle

In the dark, photoreceptors stop releasing all-*trans*-ROL. Residual all-*trans*-ROL is esterified by LRAT. The major retinoids present in a dark-adapted eye are all-*trans*-REs in the RPE and 11-*cis*-RALs in photoreceptor visual pigments. How does the visual cycle know to stop converting all-*trans*-REs into 11-*cis*-RAL chromophore in the dark? One mechanism is the strong inhibition of Rpe65 by its product, 11-*cis*-ROL. When rhodopsin is fully regenerated and CRALBP is saturated, further synthesis of 11-*cis*-ROL by Rpe65 is inhibited.

A second mode of visual-cycle regulation involves an opsin protein called RPE-retinal G-protein receptor (RGR) opsin, expressed in RPE cells. Within RPE cells, all-*trans*-REs are stored in two compartments, internal membranes and oil droplets. Rpe65 associates with internal membrane but not lipid droplets. Hence, RPE internal membranes contain a pool of all-*trans*-REs available as substrate for isomerization, while lipid droplets contain a storage pool of all-*trans*-REs. This storage pool is potentially much larger than the isomerase pool in membranes. RGR opsin mediates light-dependent transfer of all-*trans*-REs from the storage compartment to the membrane compartment for isomerization. In light, where the requirement for visual chromophore is high, RGR opsin stimulates synthesis of 11-*cis*-ROL by increasing substrate availability to Rpe65. Consistently, mice with a knockout mutation in the *rgr* gene synthesize less 11-*cis*-RAL in the light and accumulate all-*trans*-REs. Mutations in the human *RGR* gene cause autosomal dominant retinitis pigmentosa.

See also: Phototransduction: Inactivation in Cones; Phototransduction: Inactivation in Rods; Phototransduction: Phototransduction in Cones; Phototransduction: Phototransduction in Rods; Phototransduction: Rhodopsin.

Further Reading

Batten, M. L., Imanishi, Y., Maeda, T., et al. (2004). Lecithin–retinol acyltransferase is essential for accumulation of all-*trans*-retinyl esters in the eye and in the liver. *Journal of Biological Chemistry* 279: 10422–10432.

Beharry, S., Zhong, M., and Molday, R. S. (2004). N-retinylidene-phosphatidylethanolamine is the preferred retinoid substrate for the photoreceptor-specific ABC transporter ABCA4 (ABCR). *Journal of Biological Chemistry* 279: 53972–53979.

Cideciyan, A. V., Aleman, T. S., Boye, S. L., et al. (2008). Human gene therapy for rpe65 isomerase deficiency activates the retinoid cycle of vision but with slow rod kinetics. *Proceedings of the National Academy of Sciences of the United States of America* 105: 15112–15117.

Gollapalli, D. R. and Rando, R. R. (2003). All-*trans*-retinyl esters are the substrates for isomerization in the vertebrate visual cycle. *Biochemistry* 42: 5809–5818.

Jin, M., Li, S., Moghrabi, W. N., Sun, H., and Travis, G. H. (2005). Rpe65 is the retinoid isomerase in bovine retinal pigment epithelium. *Cell* 122: 449–459.

Kaschula, C. H., Jin, M. H., Desmond-Smith, N. S., and Travis, G. H. (2006). Acyl coa:retinol acyltransferase (ARAT) activity is present in bovine retinal pigment epithelium. *Experimental Eye Research* 82: 111–121.

Kefalov, V. J., Estevez, M. E., Kono, M., et al. (2005). Breaking the covalent bond – a pigment property that contributes to desensitization in cones. *Neuron* 46: 879–890.

Lamb, T. D. and Pugh, E. N. (2004). Dark adaptation and the retinoid cycle of vision. *Progress in Retinal and Eye Research* 23: 307–380.

Maeda, A., Maeda, T., Imanishi, Y., et al. (2006). Retinol dehydrogenase (RDH12) protects photoreceptors from light-induced degeneration in mice. *Journal of Biological Chemistry* 281: 37697–37704.

Mata, N. L., Weng, J., and Travis, G. H. (2000). Biosynthesis of a major lipofuscin fluorophore in mice and humans with ABCR-mediated retinal and macular degeneration. *Proceedings of the National Academy of Sciences of the United States of America* 97: 7154–7159.

Radu, R. A., Hu, J., Peng, J., et al. (2008). Retinal pigment epithelium-retinal g protein receptor-opsin mediates light-dependent translocation of all-*trans*-retinyl esters for synthesis of visual chromophore in retinal pigment epithelial cells. *Journal of Biological Chemistry* 283: 19730–19738.

Redmond, T. M., Yu, S., Lee, E., et al. (1998). Rpe65 is necessary for production of 11-*cis*-vitamin A in the retinal visual cycle. *Nature Genetics* 20: 344–351.

Travis, G. H., Golczak, M., Moise, A. R., and Palczewski, K. (2007). Diseases caused by defects in the visual cycle: Retinoids as potential therapeutic agents. *Annual Review of Pharmacology and Toxicology* 47: 469–512.

Weng, J., Mata, N. L., Azarian, S. M., et al. (1999). Insights into the function of rim protein in photoreceptors and etiology of Stargardt's disease from the phenotype in ABCR knockout mice. *Cell* 98: 13–23.

Winston, A. and Rando, R. R. (1998). Regulation of isomerohydrolase activity in the visual cycle. *Biochemistry* 37: 2044–2050.

Physiological Anatomy of the Retinal Vasculature

S S Hayreh, University of Iowa, Iowa City, IA, USA

Glossary

Blood flow autoregulation – The property of a tissue or an organ (i.e., retina) to maintain constant blood flow during changes in perfusion pressure.

Blood–retinal barrier – The system of occluding cellular junctions in the retinal pigmented epithelium and retinal vascular endothelium that prevents free movement of fluid and macromolecules from the blood into the retina.

Cilioretinal artery – A retinal artery which arises from the chorioid or the posterior ciliary artery, and is found as a variant in some individuals.

Cotton wool spots – White patches on the retina observed on fundus examination and are caused by local obstruction of the tiny arteries supplying that area.

Fluorescein fundus angiography – Fundus photographs taken in rapid sequence following injection of the fluorescent dye, fluorescein. This provides important information about blood flow as the dye reaches the retinal and choroidal vasculature.

The retina has a dual blood supply; the retinal vasculature supplies only the inner retinal layers up to the inner part of the inner nuclear layer (**Figure 1**), while the choroidal vascular bed supplies the outer 130 μm – up to the outer part of the inner nuclear layer, so that retinal vessels supply only 20% of the retina while the choroid supplies 80%. In this article, the discussion is restricted to the retinal component.

Arterial Supply of the Retina

The main arterial supply of the retina is by the central retinal artery (CRA). In some eyes, another artery, called the cilioretinal artery, may supply a highly variable part of the retina.

Central Retinal Artery

The CRA is usually the first branch of the ophthalmic artery, arising as an independent branch or in common with one of the posterior ciliary arteries (**Figure 2**). Its course can be divided into three distinct parts: (1) intra-orbital (lying below the optic nerve (ON) – (**Figure 2**)), (2) intravaginal (lying in the space between the ON and its sheath), and (3) intraneural (lying in the ON) (**Figure 3**). It enters the ON about 10 mm posterior to the eyeball (**Figures 2** and **3**). A variable number of branches arise from each of its three parts, which anastomose with the surrounding branches from other arteries, mostly in the pial plexus of the ON (**Figure 3**). At the optic disk, the CRA usually first divides into two and then each of them further divides into its various branches (**Figure 4**). The lumen of the intraneural part of the CRA is approximately 200 μm.

Cilioretinal Artery

The cilioretinal artery is either a direct branch of one of the posterior ciliary arteries or arises from the peripapillary choroid and enters the retina by hooking around the Bruch's membrane at the disk margin – usually on the temporal side (**Figures 3** and **5**). Based on ophthalmoscopy, the incidence of the occurrence of the cilioretinal artery reported by different authors varies from 6% to 25%. However, fluorescein fundus angiography provides the most reliable data because the cilioretinal artery fills synchronously with the choroidal filling, which usually starts to fill before the retinal circulation (**Figure 6(a)**). An artery which, on ophthalmoscopy, may look similar to a cilioretinal artery may in fact be an intraneural branch of the CRA emerging at the optic disk – not a true cilioretinal artery. A fluorescein fundus angiographic study of 2000 eyes showed one or more cilioretinal arteries in 32% of the eyes and in both eyes in 15% of persons. There is great variability in size, number, and distribution of the cilioretinal arteries. The area of the retina supplied by the cilioretinal arteries varies markedly, from a tiny region to a large sector of the retina. Eyes where one-fourth to half of the retina is supplied by a cilioretinal artery have been observed (**Figure 6(a)**). Rarely, the CRA is missing and the entire retina is supplied by the cilioretinal artery (**Figure 6(b)**). The outer part of the entire retina is always supplied by the posterior ciliary artery. When a cilioretinal artery is present, in the part of the retina supplied by it, the entire thickness of the retina receives its blood supply from the posterior ciliary artery.

The blood flow in the retinal vascular bed depends upon the perfusion pressure, which is equal to the difference

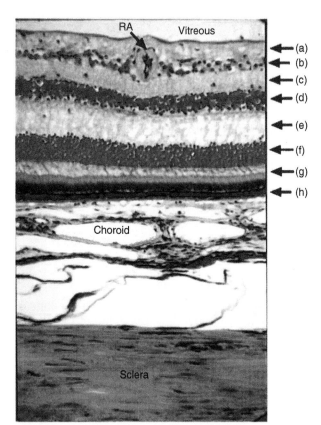

Figure 1 Light micrograph of the retina, choroid, and sclera. The retinal layers are identified as: (a) nerve fiber layer; (b) ganglion cell layer; (c) inner plexiform layer; (d) inner nuclear layer; (e) outer plexiform layer; (f) outer nuclear layer; (g) rod and cone layer; and (h) retinal pigment epithelial layer. RA, retinal arteriole.

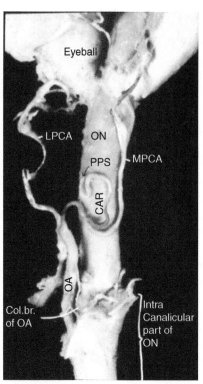

Figure 2 View from under surface of the human eyeball and optic nerve (ON) showing central artery of the retina (CAR) and its site of penetration into the optic nerve sheath (PPS), medial (MPCA) and lateral (LPCA) posterior ciliary arteries, and ophthalmic artery (OA). Reproduced with permission from Singh (Hayreh), S. and Dass, R. (1960). The central artery of the retina I. Origin and course. *British Journal of Ophthalmology* 44: 193–212.

between the retinal arterial and venous pressures. The CRA and cilioretinal artery belong to two arterial systems with different physiological properties. This raises an important physiological issue in eyes with a cilioretinal artery when that eye develops central retinal vein occlusion in which the retinal venous pressure rises suddenly to a high level. The CRA arises directly from the ophthalmic artery and the retinal vascular bed supplied by it has an efficient blood flow autoregulation (see below), so that when there is a fall in perfusion pressure in the retinal arterial bed, caused by a rise in the retinal venous pressure, the autoregulatory mechanism in the central retinal arterial vascular bed kicks in trying to maintain retinal circulation. By contrast, the cilioretinal artery belongs to the choroidal vascular system, which has no autoregulation, so that when the venous pressure rises, there is no corresponding compensatory autoregulatory mechanism. Moreover, the perfusion pressure in the choroidal vascular bed normally is lower than that in the CRA, and there is no corresponding rise of pressure in the choroidal venous bed. In view of all these factors, the following scenario occurs in an eye with cilioretinal artery developing central retinal vein occlusion: sudden occlusion of the central retinal vein results in a

marked rise of intraluminal pressure in the entire retinal capillary bed; when that intraluminal pressure rises above the pressure in the cilioretinal artery, the result is a hemodynamic block in the cilioretinal artery, producing cilioretinal artery occlusion (**Figure 7**).

Intraretinal Branches of the CRA

Each of the two main branches (superior and inferior) of the CRA at the optic disk usually divides into temporal and nasal branches, which supply the four quadrants of the retina (**Figure 4**); however, there is marked variation in their vascular pattern. In the retina, the arrangement of the branches and their subdivisions is highly variable, so much so that each eye has a different pattern (**Figures 4 and 7**). It has been suggested that the pattern could be used for personal identification like a finger print. Usually, there is a dichotomous or right-angle branching pattern. The various branches, by multiple divisions, finally end in terminal or precapillary arterioles, which are usually not visible on ophthalmoscopy. Terminal arterioles play an important role in the regulation of retinal blood flow by constriction or dilatation.

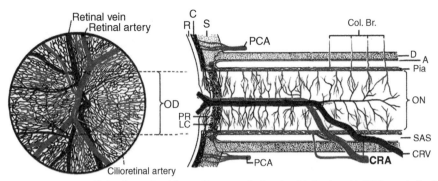

Figure 3 Schematic representation of blood supply of the optic nerve. A, Arachnoid; C, choroid; CRA, central retinal artery; Col. Br., collateral branches; CRV, central retinal vein; D, dura; LC, lamina cribrosa; OD, optic disc; ON, optic nerve; PCA, posterior ciliary artery; PR, prelaminar region; R, retina; S sclera; SAS, subarachnoid space. Modified with permission from Hayreh, S. S. (1974). Anatomy and physiology of the optic nerve head. *Transactions of the American Academy of Ophthalmology and Otolaryngology* 78: OP240–OP254.

There has been a controversy about the true nature of the arteries in the retina. According to some accounts these are small arteries. Others, however, consider them as arterioles after the first branching in the retina because they possess the following anatomic properties, typically seen in arterioles: (1) the widest part of the lumen of the retinal arterioles is near the optic disk and there its diameter is about 100 μm, which is typically the diameter of an arteriole; and (2) unlike arteries, they possess neither an internal elastic lamina nor a continuous muscular coat. This differentiation from the arteries is important in understanding their pathological involvement in some diseases, such as giant cell arteritis. In the retina, there are no interarterial or arteriovenous anastomoses, so that the retinal vascular bed is an end-arterial system.

These intraretinal arterial branches mainly lie in the nerve fiber and ganglion cell layer, usually under the internal limiting membrane (**Figure 1**); however, at the arteriovenous crossing they may extend down to the inner nuclear layer.

Retinal Capillary Bed

Each terminal arteriole gives out a plexus of 10–20 interconnected capillaries (**Figure 8**). Capillaries lie between the feeding arterioles and venules (**Figure 8**). Around the retinal arteries, there is a capillary-free zone (**Figure 8**). The retinal capillaries are arranged in two layers (**Figure 9**): (1) a superficial layer in the ganglion cell and nerve fiber layers, and (2) a deeper layer in the inner nuclear layer which is denser and more complex than the superficial layer. However, in the posterior retina, there may be three layers in the peripapillary region and there is only one layer in the perifoveal region. Furthermore, in the peripheral retina the deep layer disappears and only the superficial layer is left, with a wider network. At the extreme periphery of the retina, there is an avascular zone about 1.5 mm width.

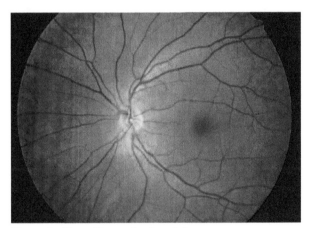

Figure 4 Normal human fundus – left eye.

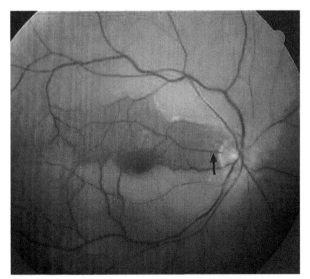

Figure 5 Ophthalmoscopic appearance of right eye with central retinal artery occlusion (pale retina) and a normal cilioretinal artery in the area of normal retina (arrow).

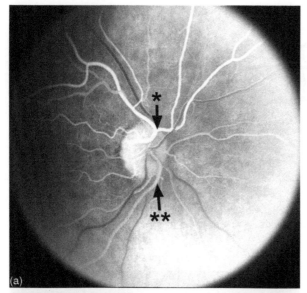

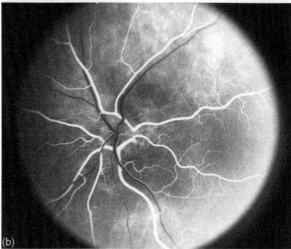

Figure 6 Fluorescein fundus angiograms of (a) right and (b) left eyes of a person. (a) The right eye has one large cilioretinal artery (one asterisk) supplying the superior one-third of the retina that starts to fill before the central retinal artery (two asterisks) which supplies the rest of the retina. (b) The left eye has two large branching cilioretinal arteries (white) – the one above branches in the upper half and the other below branches in the lower half. This eye lacks a central retinal artery.

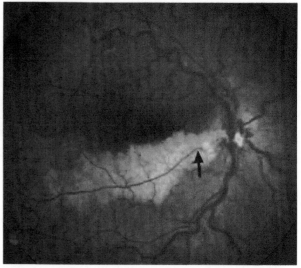

Figure 7 Fundus photograph of right eye with nonischemic central retinal vein occlusion associated with cilioretinal artery occlusion (arrow).

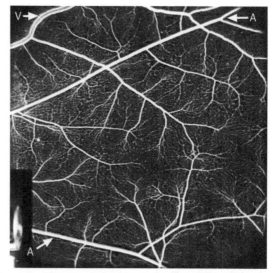

Figure 8 Fluorescein fundus angiogram showing retinal vessels and capillary network. A, retinal arteriole; V, retinal vein.

In addition to the retinal capillary bed described above, there is a distinct retinal capillary bed called the radial peripapillary capillaries, which were first described in 1940 (**Figures 9** and **10**). They have the following special characteristics compared to other retinal capillaries:

1. They are long, straight capillaries, measuring several hundred microns to several millimeters.
2. They form the most superficial layer (**Figure 9**) lying among the superficial nerve fibers, along the superior and inferior temporal arcades of retinal vessels and the peripapillary region (**Figure 10**).
3. They rarely anastomose with one another.

4. They arise from the peripapillary retinal arterioles lying deeper in the retina, and drain into retinal venules or veins on the optic disk (**Figure 3**).

Because of these characteristics, the radial peripapillary capillaries assume importance in the development of several lesions. For example, cotton wool spots are often located in the distribution of the radial peripapillary capillaries, which indicates that the latter may play a role in the pathogenesis of cotton wool spots. In addition, in chronic optic disk edema these capillaries become dilated and develop microaneurysms and hemorrhages.

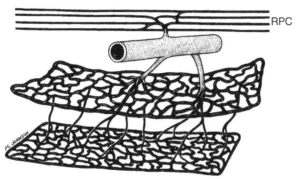

Figure 9 Schematic representation of two layers of the retinal capillaries and radial peripapillary capillaries (RPC). Reproduced with permission from Henkind, P. (1969). Microcirculation of peripapillary retina. *Transactions of the American Academy of Ophthalmology and Otolaryngology* 73: 890–897.

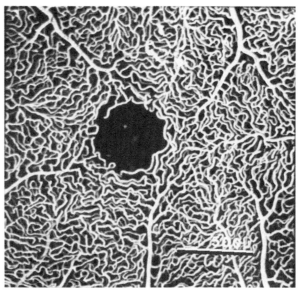

Figure 11 A cast of retinal capillaries in the macular region of a monkey showing the foveal avascular zone in the center. Courtesy of Professor Koichi Shimizu.

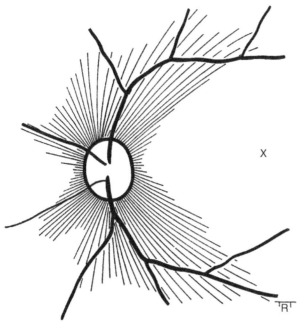

Figure 10 Schematic representation of radial peripapillary capillaries. Site of foveola (X). Reproduced with permission from Henkind, P. (1967). Radial peripapillary capillaries of the retina: I. Anatomy: Human and comparative. *British Journal of Ophthalmology* 51: 115–123.

In the macular region, the capillaries are supplied by arterioles arising from the superior and inferior temporal arteries (**Figure 4**). Their thickness decreases toward the center of the macula where they are arranged in a single layer. The capillaries are absent in the foveal region, with a capillary-free zone of about 400–500 μm in diameter (**Figure 11**).

The wall of the retinal capillaries consists of endothelial cells, pericytes, and basement membrane. Their diameter varies from 3.5 to 6 μm. The endothelial cells have tight-cell junctions, which constitute a blood–retinal barrier (see below). In addition to the endothelial cells, there are also pericytes which form a discontinuous layer within the basement membrane of the capillaries. They have a contractile property, by virtue of which they may play a role in regulating blood flow in the capillaries and autoregulation of blood flow (see below). Pericytes are lost preferentially in diabetes so that they may have a role in diabetic retinopathy; it is suggested that diabetic pericyte loss is the result of their migration. Migration of pericytes is also involved in the regulation of angiogenesis.

Retinal Venous Drainage

The postcapillary venules drain the blood from the capillaries but, occasionally, capillaries may join a major vein directly. The terminal arterioles and postcapillary venules are situated in an alternating pattern, with the capillary bed in between the two (**Figure 8**). The postcapillary venules drain into bigger venules and finally into the branch retinal veins. The lumen of the major branch retinal veins, just before they join to form the central retinal vein, is about 200 μm. In the central part of the retina, the branch retinal veins and arteries usually run in close association and at places cross one another (**Figures 4, 5,** and **7**). On the other hand, in the peripheral retina, the veins do not follow the course of the arteries. Various retinal arteries and veins in the retina cross each other at arteriovenous crossings (**Figures 4, 5,** and **7**). In a study of 189 normal eyes, at the sites of arteriovenous crossing, the artery crossed over the vein in 68% and was the reverse in the remainder. However, in eyes with branch retinal vein occlusion, the artery crossed over the vein at the site of occlusion in 98% of the cases, indicating that pattern of arteriovenous crossing plays a role in the development of branch retinal vein occlusion. At the site of arteriovenous crossing, the artery and vein share a common fibrous coat and are separated by only a thin endothelial lining and basement membrane.

The superior branch veins usually join to form a superior trunk and the inferior branch veins, an inferior trunk. These superior and inferior trunks join on the optic disk to form the central retinal vein (**Figure 3**). However, in 20% of eyes, the superior and inferior trunks do not join together at the disk but enter the optic disk as two separate trunks; this represents a congenital anomaly (**Figure 12**). During the third month of intrauterine life, there are always two trunks of the central retinal vein in the ON, one on either side of the CRA (**Figure 13**), and one of the two trunks usually disappears before birth; however, in 20% of eyes, a dual-trunked central retinal vein persists into adult life. In such eyes, only one of the two trunks may develop occlusion in the ON, resulting in development of the clinical entity called hemi-central retinal vein occlusion (**Figure 12**).

The central retinal vein travels in the ON temporal to the artery, where the central retinal vein and artery lie in the center of the ON, surrounded by a fibrous tissue envelope (**Figure 14**). During its intraneural course, the vein receives many tributaries (**Figure 3**). The central retinal vein exits the ON and its sheath (**Figure 15**), and finally drains into either the superior ophthalmic vein or directly into the cavernous sinus.

Nerve Supply

The intraorbital and intraneural portions of the CRA have an adrenergic nerve supply from a sympathetic nerve called the nerve of Tiedemann (**Figure 16**); however, the retinal branches of the CRA have no adrenergic nerve supply. Therefore, there is no autonomic innervation of the retinal vascular bed.

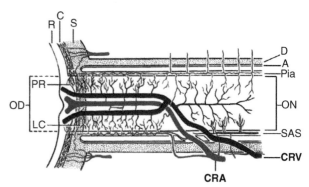

Figure 13 Schematic representation of two trunks of the central retinal vein in the anterior part of the optic nerve. A, Arachnoid; C, choroid; CRA, central retinal artery; Col. Br., collateral branches; CRV, central retinal vein; D, dura; LC, lamina cribrosa; OD, optic disc; ON, optic nerve; PCA, posterior ciliary artery; PR, prelaminar region; R, retina; S, sclera; SAS, subarachnoid space.

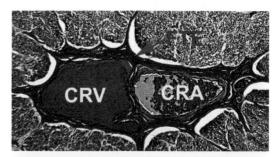

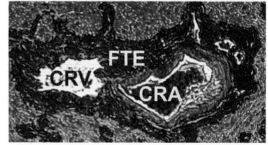

Figure 14 Histological sections (Masson's trichrome staining) showing the central retinal vessels and surrounding fibrous tissue envelope, as seen in a transverse section of the central part of the retrolaminar region of the optic nerve, in a normal rhesus monkey (above) and in a rhesus monkey with experimental arterial hypertension, atherosclerosis and glaucoma (below). CRA =Central retinal artery, CRV = central retinal vein; FTE = fibrous tissue envelope.

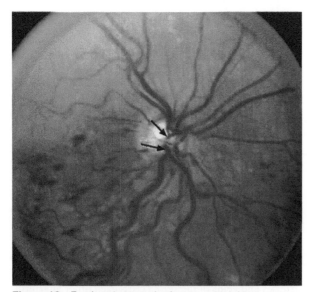

Figure 12 Fundus photograph of an eye with inferior hemicentral retinal vein occlusion, involving the lower trunk of the central retinal vein. Two trunks (arrows) of the central retinal vein enter the optic disk separately above and below.

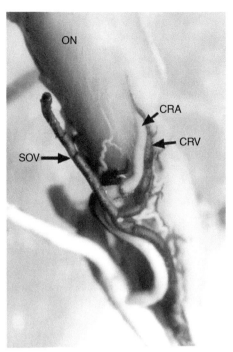

Figure 15 View from under surface of optic nerve in a rhesus monkey showing the intraorbital part of the central retinal vessels and their site of penetration into the sheath of the optic nerve. CRA, central retinal artery; CRV, central retinal vein; ON, optic nerve; SOV, superior ophthalmic vein. Reproduced with permission from Hayreh, S. S. (1965). Occlusion of the central retinal vessels. *British Journal of Ophthalmology* 49: 626–645.

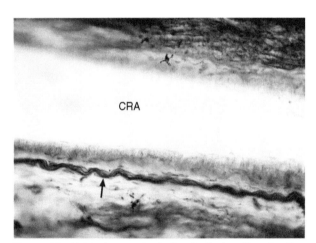

Figure 16 Light micrograph of a longitudinal section of optic nerve of a rhesus monkey, showing central retinal artery (CRA) in the center of the nerve and nerve of Tiedemann (arrow) running parallel with the wall of the central retinal artery. (Gros-Schultze's stain). Reproduced with permission from Hayreh, S. S., Vrabec, Fr. (1966). The structure of the head of the optic nerve in rhesus monkey. *American Journal of Ophthalmology* 62: 136–150.

Blood–Retinal Barrier

The retina has two types of blood–retinal barriers.

Inner blood–retinal barrier. This lies in the retinal vessels. It is produced by the tight cell junctions between the endothelial cells of the vessels (due to the presence of extensive zonulae occludentes). The tight interendothelial ·cell junctions block movement of macromolecules from the lumen toward the interstitial space. Pericytes, Müller cells, and astrocytes also contribute to the proper functioning of this barrier.

Outer blood–retinal barrier. Tight cell junctions between the retinal pigment epithelial cells (**Figure 1**) also produce a blood–retinal barrier, preventing the leakage of fluid from the choroid into the retina. This barrier breaks down when the retinal pigment epithelial cells are destroyed or subjected to ischemia, as in hypertensive choroidopathy.

The blood–retinal barrier plays an important role in the regulation of the microenvironment in the retina. An intact blood–retinal barrier is essential for maintaining retinal structure and function. Breakdown of this barrier results in increased vascular permeability of the capillaries, which causes retinal edema, as seen in a variety of retinopathies. Breakdown of the inner blood–retinal barrier may be caused by acute distension of the vessel walls, ischemia, chemical influences, defects in the endothelial cells, or failure of the active transport system.

The retinal tissue itself has no barrier in its stroma, therefore fluid may diffuse from one part to the adjacent areas.

Autoregulation of Retinal Blood Flow

The object of blood flow autoregulation in a tissue is to maintain relatively constant blood flow during changes in perfusion pressure. This is an important mechanism to regulate blood flow. The retinal circulation has efficient autoregulation. The exact mechanism and site of autoregulation are still unclear except that it most probably operates by altering the vascular resistance. It is generally considered as a feature of the terminal arterioles; so with the rise or fall of perfusion pressure beyond normal levels, the terminal arterioles constrict or dilate, respectively, to regulate the vascular resistance and thereby the blood flow. Recent studies have suggested that pericytes in the retinal capillaries play a role in autoregulation as well because of their contractile property. The metabolic needs of the tissue also regulate the autoregulation. Autoregulation works within a critical range of perfusion pressure, and it breaks down with any rise or fall of the perfusion pressure beyond the critical autoregulatory range.

The vascular endothelium plays an active role in the vasomotor function of both macro- and microvasculatures, including maintenance of vascular tone and regulation of blood flow. Recent studies suggest that vascular-endothelial-derived vasoactive agents (e.g., endothelin-1, thromboxane A_2, and prostaglandin H_2 – vasoconstrictors; and nitric oxide – a vasodilator) profoundly modulate local vascular tone and, thereby, may also play a role in autoregulation. Mechanical stretching and increases in arteriolar transmural pressure induce the endothelial cells to release contracting factors affecting the tone of arteriolar smooth muscle cells and pericytes. Therefore, damage to vascular endothelium (as in arteriosclerosis, atherosclerosis, hypercholesterolemia, aging, diabetes mellitus, ischemia, and possibly from other causes) may be associated with abnormalities in the production of endothelial vasoactive agents, and consequent autoregulation abnormalities.

See also: Blood–Retinal Barrier; Breakdown of the Blood–Retinal Barrier; Breakdown of the RPE Blood–Retinal Barrier; Central Retinal Vein Occlusion; Pathological Retinal Angiogenesis.

Further Reading

Anderson, D. R. (1996). Glaucoma, capillaries and pericytes 1. Blood flow regulation. *Ophthalmologica* 210: 257–262.

Cunha-Vaz, J. G. (1976). The blood–retinal barriers. *Documenta Ophthalmologica* 41: 287–327.

Duke-Elder, S. and Wybar, K. C. (1961). Anatomy of the visual system. In: Duke-Elder, S. (ed.) *System of Ophthalmology* vol. 2, pp. 363–382. London: Kimpton.

Haefliger, I. O., Meyer, P., Flammer, J., and Lüscher, T. F. (1994). The vascular endothelium as a regulator of the ocular circulation: A new concept in ophthalmology? *Survey of Ophthalmology* 39: 123–132.

Hayreh, S. S. (1963). The cilio-retinal arteries. *British Journal of Ophthalmology* 47: 71–89.

Hayreh, S. S. and Hayreh, M. S. (1980). Hemi-central retinal vein occlusion. Pathogenesis, clinical features, and natural history. *Archives of Ophthalmology* 98: 1600–1609.

Hayreh, S. S., Fraterrigo, L., and Jonas, J. (2008). Central retinal vein occlusion associated with cilioretinal artery occlusion. *Retina* 28: 581–594.

Henkind, P. (1967). Radial peripapillary capillaries of the retina: I. Anatomy: Human and comparative. *British Journal of Ophthalmology* 51: 115–123.

Henkind, P. (1969). Microcirculation of peripapillary retina. *Transactions of the American Academy of Ophthalmology and Otolaryngology* 73: 890–897.

Justice, J. Jr. and Lehmann, R. P. (1976). Cilioretinal arteries. A study based on review of stereo fundus photographs and fluorescein angiographic findings. *Archives of Ophthalmology* 94: 1355–1358.

Kaur, C., Foulds, W. S., and Ling, E. A. (2008). Blood–retinal barrier in hypoxic ischaemic conditions: Basic concepts, clinical features and management. *Progress in Retinal and Eye Research* 27(6): 622–647.

Pournaras, C. J., Rungger-Brändle, E., Riva, C. E., Hardarso, S. H., and Stefansson, E. (2008). Regulation of retinal blood flow in health and disease. *Progress in Retinal and Eye Research* 27: 284–330.

Singh (Hayreh), S. and Dass, R. (1960). The central artery of the retina I. Origin and course. *British Journal of Ophthalmology* 44: 193–212.

Singh (Hayreh), S. and Dass, R. (1960). The central artery of the retina II. Distribution and anastomoses. *British Journal of Ophthalmology* 44: 280–299.

Weinberg, D., Dodwell, D. G., and Fern, S. A. (1990). Anatomy of arteriovenous crossings in branch retinal vein occlusion. *American Journal of Ophthalmology* 109: 298–302.

Wise, G. N., Dollery, C. T., and Henkind, P. (1971). *The Retinal Circulation*, pp. 20–54. New York: Harper and Row.

The Physiology of Photoreceptor Synapses and Other Ribbon Synapses

W B Thoreson, University of Nebraska Medical Center, Omaha, NE, USA

Glossary

Ca²⁺ microdomains – Local submembrane regions of elevated intracellular Ca^{2+} caused by the influx of Ca^{2+} through nearby Ca^{2+} channels.

Calcium-induced calcium release (CICR) – Release into the cytoplasm of Ca^{2+} ions stored in the endoplasmic reticulum that is triggered by the Ca^{2+}-dependent activation of ryanodine receptors.

ERG b-wave – The electroretinogram (ERG) is a massed electrical response of the retina to light that can be recorded by electrodes on the surface of the cornea. ON bipolar cells provide the source of the b-wave of the ERG. A selective reduction in the b-wave indicates a reduction in signaling between photoreceptors and ON bipolar cells.

L-type calcium currents – Currents from high-voltage-activated CaV1.1 (alpha 1S), CaV1.2 (alpha 1C), CaV1.3 (alpha 1D), or CaV1.4 (alpha 1F) calcium channels which show sustained activation and can be selectively blocked by dihydropyridine antagonists.

OFF bipolar cells – Second-order retinal bipolar cells which exhibit a hyperpolarizing response to light mediated by non-NMDA ionotropic glutamate receptors.

ON bipolar cells – Second-order retinal bipolar cell subtypes which exhibit a depolarizing response to light mediated by mGluR6 metabotropic glutamate receptors.

Readily releasable pool – A pool of synaptic vesicles primed for rapid release by elevation of intracellular calcium.

SNARE proteins – SNAP and NSF attachment receptors are a family of proteins participating in vesicle fusion. They can be subdivided into vesicle SNAREs (v-SNAREs), which attach to vesicles, and target SNAREs (t-SNAREs), which associate with the plasma membrane.

Synaptic ribbon – An electron dense presynaptic structure that tethers synaptic vesicles in nerve terminals of sensory neurons including photoreceptors, retinal bipolar cells, hair cells, pinealocytes, and electroreceptors.

Total internal reflectance (TIRF) microscopy – By taking advantage of the subwavelength evanescent field of light created by reflections at the interface between a cell and coverslip, TIRF microscopy can be used to visualize subwavelength structures such as synaptic vesicles.

Anatomy of the Ribbon Synapse

Structures of the electron dense ribbons found at the synapses of retinal photoreceptors and bipolar cells differ depending on cell type. In cross section, photoreceptor ribbons appear as ~35 nm thick bars, but in three dimensions they form flat, ribbon-like structures. Mammalian rods have one to two ribbons that can be up to 2 μm in length and extend up to 1 μm into the cytoplasm. Each wraps around the synaptic ridge to form crescent or horseshoe shapes. The ribbons in mammalian cone are smaller, typically less than 1 μm long and extend only a few hundred nanometers into the cytoplasm. They are shaped like surfboards, and typically a dozen or more ribbons are in each cone terminal. Sitting just below the photoreceptor ribbon is a trough-like arciform density. Ribbons in bipolar cells are planar, like those in rods, but they are smaller than both rod and cone ribbons.

The focus of this review is on ribbon synapses in retina; however, ribbon synapses are not unique to retina. They are present in other sensory neurons, including pinealocytes, electroreceptors in the lateral line organ of fishes, and hair cells of the cochlea and vestibular apparatus. Again, the ribbons in the synapses of these cells have different structures depending on cell type. For example, hair cell ribbons are small spheres.

The ribbon synapses of photoreceptor terminals contain many more synaptic vesicles than conventional synapses. Conventional synapses have 10–100 vesicles near each presynaptic density compared to the synaptic terminal of a lizard cone, which contains ~170 000 vesicles or 7000 vesicles per ribbon. Furthermore, ~85% of the vesicles at ribbon synapses are freely mobile and readily participate in release compared to ~20% of vesicles at conventional synapses. The expanded mobility of vesicles at ribbon synapses may be due to the absence of synapsins which have been proposed to tether synaptic vesicles at conventional synapses. The small subset of vesicles in photoreceptor terminals that are tethered to synaptic ribbons are attached by fine filaments (**Figure 1**).

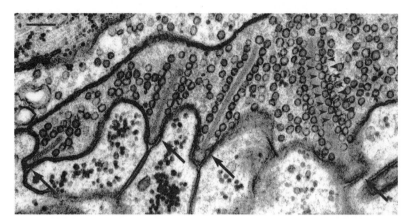

Figure 1 Ribbon-style active zones (arrows) in a salamander rod photoreceptor. A tangential section through the ribbon at right shows the hexagonal packing of vesicles on the ribbon (arrowheads). Scale = 200 nm. Adapted from Thoreson, W. B., Rabl, K., Townes-Anderson, E., and Heidelberger, R. (2004) A highly Ca^{2+}-sensitive pool of vesicles contributes to linearity at the rod photoreceptor ribbon synapse. *Neuron* 42: 595–605, with permission from Elsevier.

Vesicle Pools and Vesicular Release at Synaptic Ribbons

Newly tethered vesicles can be found throughout the ribbon, indicating that vesicles either freely enter the ribbon at any position or redistribute rapidly about the ribbon face after attachment. However, once attached to the ribbon, vesicles exit only from the base. Vesicles tethered on the bottom first to third rows of the ribbon contact the plasma membrane along the synaptic ridge. These vesicles constitute a pool that can be released rapidly in response to increased Ca^{2+} levels. In bipolar cells, the fastest component of vesicular release equals the number of vesicles tethered along the bottom row of the ribbon, and maintained depolarization stimulates release of the total number of vesicles lining the entire ribbon. Similarly, the total releasable pool in rod photoreceptors equals the number of vesicles tethered to the entire ribbon (~700 vesicles in amphibian rods), and there is an ultrafast component similar to the number of vesicles tethered at the ribbon base (~30 vesicles). Because of their smaller size, cone ribbons have a smaller releasable pool than rod ribbons.

In contrast with bipolar and photoreceptor cells, the ultrafast release component in hair cells exceeds the number of vesicles lining the bottom row by nearly 10-fold, and the total releasable pool exceeds the number of vesicles lining the ribbon six- to eightfold. The large ultrafast or rapidly releasable pool could be explained by compound fusion between adjacent vesicles as discussed later. The large size of the total pool suggests that vesicles released from the ribbon are replenished rapidly from the surrounding cytoplasm.

Vesicular transmitter release clearly occurs at the ribbons in ribbon synapses, but the ribbons may not be the only site of transmitter release. Photoreceptors also contact bipolar cell dendrites at flat or basal junctions identified by pre- and postsynaptic membrane densities. In mammalian retina, the basal junctions of cones only contact OFF-type bipolar cells and the dendrites of many OFF-type bipolar cells are not directly apposed to synaptic ribbons. In salamander retinas, the architecture of photoreceptor input to bipolar cells is different, and OFF bipolar cells receive ~80% of their contacts from ribbon synapses and only ~20% from basal junctions. On the other hand, salamander ON-type bipolar cells receive ~ 20% of their contacts from ribbons synapses and ~80% from basal junctions.

Basal junctions lack the vesicle clusters which typify conventional synapses; however, the absence in mammals of direct ribbon contacts onto many OFF bipolar cells led to the suggestion that they necessarily receive synaptic input from basal junctions. This idea was tested in experiments that compared synaptic events recorded simultaneously from two OFF bipolar cells – one that contacted photoreceptors only at basal junctions and a neighbor whose dendrites approach the ribbon. These studies showed that glutamate released at a synaptic ribbon can diffuse rapidly to bipolar cell processes at basal junctions. This finding shows that glutamate released at the ribbon can reach basal junctions, although it does not exclude the possibility of additional nonribbon release events.

The question of whether nonribbon release events occur in ribbon synapses has been addressed further by using total internal reflectance (TIRF) microscopy to visualize single-vesicle fusion events. Studies of release from bipolar cells indicate that up to 1/3 of fusion events occur at ectopic sites away from the ribbon including much of the release during sustained depolarization. Studies on hair cells from mice with disrupted ribbon anchoring also suggest a role for nonribbon sites in sustained release by showing that sustained release is unchanged although fast release is diminished. However, in photoreceptors, as discussed later, there is evidence suggesting that sustained release occurs predominately at the ribbon.

Role of the Ribbon in Release

The functions of the ribbon in release are not fully understood. It has been widely suggested that the ribbon may operate like a conveyor belt, acting as a molecular motor to accelerate delivery of vesicles to their release sites. Consistent with this possibility is the presence of a kinesin motor protein, KIF3A, at the ribbon. However, release of vesicles attached to the ribbons does not require ATP, although ATP is needed for subsequent replenishment and priming of vesicles. Ribbons are also not necessary for sustaining high-frequency release since it can be observed at conventional central nervous system (CNS) synapses without ribbons. In fact, during sustained release from photoreceptor ribbons, the delivery of vesicles to the base of the ribbon is actually slower than the rate predicted from delivery by simple diffusion. Thus, rather than accelerating release, synaptic ribbons in cones appear to slow the rates of sustained release by constraining the rate of vesicle delivery to the base. By making release more regular, such a mechanism could improve the ability to detect small intensity changes that produce small changes in release rate.

Another hypothesis of ribbon function is that it may operate as a vesicle trap. In this scenario, vesicles moving about the terminal by Brownian motion occasionally collide with the ribbon face where they are captured like flies to fly paper and then delivered to the base for release.

Ribbons may assist in vesicle priming. The match between releasable pool size and the number of vesicles tethered to rod or bipolar cell ribbons indicates that tethered vesicles are primed for release. Furthermore, members of the RIM family of proteins involved in vesicle priming localize to the ribbon. RIM proteins are synaptic proteins required for normal neurotransmitter release.

Another proposed role for the ribbon is to facilitate compound fusion during depolarization by promoting fusion between neighboring vesicles on adjacent rows of the ribbon. Compound fusion could explain the finding in hair cells that many more vesicles fuse rapidly after stimulation than are anchored at the base of the ribbon. The possibility of compound fusion in hair cells is supported by statistical properties of fluctuations in postsynaptic currents and presynaptic exocytotic capacitance changes. There is also ultrastructural and electrophysiological evidence for compound or coordinated fusion of multiple vesicles at bipolar cell synapses.

Ribbons may perform more than one of these functions. They may capture vesicles, prime them for release, regulate their delivery to release sites, and facilitate compound fusion. To sort out the unique functions of ribbons and ribbon synapse, the biochemistry and electrophysiological properties of ribbon synapse are being studied in detail.

Synaptic Proteins

Ribbon synapses contain many of the same proteins as conventional synapses. For example, conventional and ribbon synapses both employ SNARE proteins in vesicle fusion: synaptobrevin (VAMP1 and 2), SNAP-25, and syntaxin. Ribbon synapses also possess Munc 18-1, Munc 13-1, RIM, and rab3A proteins which assist with assembly of the SNARE complex and vesicle priming. RIM1 is distributed across the ribbon, whereas RIM2 localizes to the ribbon base, suggesting that they may have different roles.

In addition to many similarities, there are also differences among the proteins at ribbon and conventional synapses. In place of syntaxin-1 found at many conventional synapses, ribbon synapses utilize syntaxin-3b. In place of complexins 1 and 2 that interact with the SNARE complex at many synapses, ribbon synapses possess complexins 3 and 4. Ribbon synapses lack synapsins. The functional consequences of these differences remain to be explained.

As in conventional synapses, a rise in intracellular free Ca^{2+} is required for neurotransmitter release, but the identity of the calcium sensor(s) at ribbon synapses is unsettled. Synaptotagmin I is the principal calcium sensor at most conventional synapses. Antibodies to synaptotagmin I/II label photoreceptor and bipolar cell ribbon synapses in mouse and bovine retina but do not label these synapses in goldfish and salamander retina, which are instead labeled by antibodies to synaptotagmin III. In hair cells, it has been proposed that otoferlin, not synaptotagmin, is the principal calcium sensor for exocytosis.

The most conspicuous difference between ribbon and conventional synapses is the presence of the ribbon-specific protein, ribeye. Ribbons are constructed from interdomain interactions between adjoining ribeye molecules. Each bipolar cell ribbon is formed from ~4000 ribeye molecules and the 10-fold greater surface area of rod ribbons suggests that they are built from ~40 000 ribeye molecules. Ribeye is an alternative transcript of the gene for transcriptional repressor C-terminal-binding protein 2 (CtBP-2) with a unique ribbon-specific A domain and an enzymatic B domain. Interactions between ribeye molecules are regulated by NAD and NADH levels, suggesting a mechanism by which changes in the metabolic state of the terminal could contribute to observed circadian changes in ribbon structure.

As mentioned above, the kinesin KIF3A is located at the ribbon, but since ATP-driven molecular motors are not needed for vesicle release from the ribbon, the role of KIF3A at the ribbon is unclear. However, kinesin motor proteins interact with plant homologs of CtBP, indicating that the interaction of kinesins with the major structural protein in ribbon synapses is highly conserved. It has also been proposed that KIF3A may assist with circadian changes in ribbon structure.

Interactions among synaptic proteins help maintain the structure of the ribbon. Interactions between ribeye and the cytomatrix scaffold protein bassoon anchor ribbons to the active zone. Bassoon also tethers ribbons in hair cells and pinealocytes. Bipolar cells lack bassoon but possess a related protein, piccolo. In photoreceptors, piccolo is located further up the ribbon than bassoon.

Interactions between cytoskeletal proteins, membrane-spanning dystroglycan proteins, and extracellular matrix pikachurin proteins are important for maintaining contacts between photoreceptor ribbon synapses and their postsynaptic targets. Disrupting these interactions leads to reductions in the electroretinogram (ERG) b-wave (indicating a reduction in ON bipolar cell responses) in patients with muscular dystrophy.

Interactions between ribeye and Unc119, a protein that is highly expressed in photoreceptors, may help localize Ca^{2+} channels to the base of the ribbon. Unc 119 can bind to both ribeye and the calcium-binding protein CaBP4. CaBP4 binds in turn to Ca^{2+} channels in photoreceptor terminals. Mutations in Unc 119 lead to cone/rod degeneration.

Photoreceptor Calcium Channels

Glutamate release from photoreceptors requires an influx of Ca^{2+} through Ca^{2+} channels. However, unlike conventional synapses that use N- and P-type calcium channels, photoreceptors and other ribbon synapses rely on dihydropyridine-sensitive, L-type calcium channels. L-type calcium channels are classified by their α1 pore forming subunits into CaV1.1-CaV1.4 subtypes. Mutations in CaV1.4 (also known as α1F) cause incomplete congenital stationary night blindness and antibodies to CaV1.4 label mammalian rod terminals, suggesting that CaV1.4 is the principal subtype in rods. Rods and long-wavelength sensitive cones also appear to possess CaV1.3 channels but CaV1.3 antibodies do not label short-wavelength-sensitive cones, suggesting that they possess a different channel subtype.

Calcium channels cluster beneath the ribbon. Freeze fracture electron micrographs from mammalian cones show clusters of ~500 polyhedral transmembrane particles, each with a central dimple, beneath the arciform density of each ribbon. Sites of calcium influx co-localize with

ribeye-binding peptides and antibodies to L-type calcium channels co-localize with antibodies to bassoon and ribeye.

Properties of single-calcium channels recorded from amphibian rod photoreceptors are similar to those of L-type calcium channels in other tissues. Single-CaV1.4 channels expressed in HEK293 cells showed a tiny single-channel conductance and extremely low open probability. However, it is unlikely that these properties are retained *in vivo* since they imply an unrealistically large number of channels per ribbon (>15 000).

Photoreceptor calcium currents (I_{Ca}) exhibit a sigmoidal voltage dependence (**Figure 2**). When measured under the same experimental conditions, the voltage dependence of I_{Ca} (calcium current) in rods and cones of different species are remarkably similar. Typically, photoreceptor I_{Ca} activates above −60 mV and is fully activated around −20 mV. I_{Ca} attains about a third of its

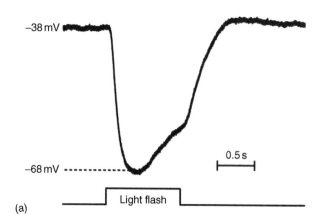

(a)

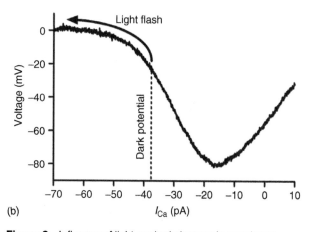

(b)

Figure 2 Influence of light-evoked changes in membrane potential on cone calcium currents. (a) Response of a salamander cone to a bright flash of light. (b) Calcium current averaged from eight salamander cones evoked by a ramp voltage protocol (0.5 mV ms^{-1}). By convention, the influx of positively charged Ca^{2+} ions into cones is shown as negative or inward current. The dark potential of the cone in (a) is denoted by the dashed line and the reduction in I_{Ca} caused by the hyperpolarizing light response is shown by the arrow.

peak amplitude at the dark resting potential (ca. −40 mV). The hyperpolarizing light response of a rod or cone photoreceptor diminishes I_{Ca} and thereby diminishes Ca^{2+}-dependent release. The sigmoidal voltage dependence of I_{Ca} contributes to response compression at higher intensities. As illustrated in **Figure 2**, the first 10 mV of membrane hyperpolarization during a cone light response causes a much greater decrease in I_{Ca} than the next 10 mV. This diminished responsiveness at more hyperpolarized potentials is sufficiently pronounced at rod synapses that it has been described as response clipping. There is evidence that Ca^{2+} influx through cGMP-gated cation channels can stimulate synaptic release from cones but the role that these channels play in release under normal physiological conditions is not clear.

Photoreceptor I_{Ca} shows limited and slowly developing inactivation involving both voltage- and calcium-dependent mechanisms. Limited inactivation is important for sustaining synaptic release in darkness when photoreceptors are continuously depolarized. Although the amplitude of I_{Ca} declines slowly in darkness, changes in I_{Ca} produced by brief changes in illumination mirror the sigmoidal voltage dependence of I_{Ca}.

Along with pore-forming α1 subunits, calcium channels possess accessory β and α2/δ subunits. Knockout of β2 subunits almost completely abolishes both the ERG b-wave and staining for CaV1.4 in the outer plexiform layer indicating that β2 subunits are the predominant subtype at photoreceptor synapses. Mutations in α2/δ type 4 subunits lead to disordered ribbons, reduced scotopic b-waves, absent photopic b-waves, and a human cone dystrophy, suggesting that this accessory subunit is associated with calcium channels in photoreceptors, particularly cones.

The calcium-binding protein, CaBP4, is closely associated with photoreceptor calcium channels. When heterologously expressed in the presence of CaBP4, the voltage dependence of CaV1.3 and CaV1.4 I_{Ca} is shifted to more negative potentials, similar to the voltage dependence of I_{Ca} in photoreceptors. By shifting activation to more positive potentials, mutations in CaBP4 reduce the amplitude of I_{Ca} in the normal physiological range and thereby reduce synaptic output from photoreceptors. The reduction in rod output accompanying CaBP4 mutations causes congenital stationary night blindness.

Role of Intracellular Ca^{2+} in Release

Free calcium levels in the photoreceptor synapse are tightly regulated by a variety of mechanisms to maintain synaptic output and prevent calcium overload during the continual influx of Ca^{2+} in darkness.

One way to remove Ca^{2+} from the cytoplasm is to pump it into the endoplasmic reticulum using sarco- and endoplasmic reticulum ATPases (SERCA). SERCA 2A predominates in photoreceptor terminals. The release of calcium from these sequestered stores by calcium-induced calcium release (CICR) can amplify calcium entry through L-type calcium channels. CICR in retina is mediated by a retina-specific variant of the type 2 ryanodine receptor. In rods, CICR amplifies synaptic release and increases the likelihood of the simultaneous fusion of multiple vesicles. CICR also contributes to synaptic release from ribbon synapses in vestibular hair cells. Calcium imaging studies show evidence for CICR in cone cell bodies but it does not appear to contribute to synaptic release. Immunohistochemical studies suggest IP3 receptors, which mediate the release of intracellular Ca^{2+} in many cell types, may be present in cone terminals. However, there is no physiological evidence for IP3-mediated release of calcium in photoreceptors.

Depletion of calcium from intracellular stores triggers the opening of calcium-permeable channels in the plasma membrane to facilitate store refilling. Store-operated calcium entry can influence synaptic release by regulating basal calcium levels in photoreceptor terminals.

Calcium can also be removed from the cytoplasm by pumping it out of the cell. Calcium is removed from outer segments by a Na/Ca exchanger whereas extrusion from inner segments and synaptic terminals rely more on plasma membrane calcium ATPases (PMCA). PMCA2 antibodies label photoreceptor terminals and PMCA2 knockout mice show significant reductions in rod-driven responses, suggesting that this subtype is particularly important in regulating calcium levels in rod terminals.

Calcium buffering by cytoplasmic proteins provides a much more rapid way to reduce free calcium levels than extrusion. The principle calcium buffers are calbindin, calretinin, and parvalbumin although many signaling proteins (e.g., calmodulin, synaptotagmin, and CaBP4) also bind calcium. There is considerable species variability, but cones typically possess the fast, low mobility buffer calbindin whereas the higher mobility buffers calretinin and parvalbumin are less common in photoreceptors.

Supplementing these mechanisms, large calcium increases in rods and cones can be buffered by mitochondrial uptake.

Physiology of Release at Photoreceptor Synapses

Photoreceptors hyperpolarize to light and decreases in light intensity cause photoreceptors to depolarize. Depolarization increases the open probability of Ca^{2+} channels clustered beneath the ribbon. The opening of Ca^{2+}

channels increases $[Ca^{2+}]_i$, stimulating fusion of vesicles at the base of the ribbon. The calcium sensors in bipolar and hair cells have a low affinity for calcium requiring $>10\,\mu M$ calcium to stimulate exocytosis. Release from bipolar cells and most other CNS neurons exhibits high cooperativity consistent with the binding of as many as five Ca^{2+} ions required for release. The release mechanism employed by photoreceptors differs from these synapses by showing a much higher affinity for Ca^{2+} whereby submicromolar calcium levels can stimulate release. This high affinity is consistent with the possible involvement of synaptotagmin III, which has a higher affinity for calcium than synaptotagmin I or II. In addition to higher Ca^{2+} affinity, release from photoreceptors shows lower cooperativity for Ca^{2+} binding ($N \leq 3$).

Ca^{2+} channels are close to the ribbon and thus opening of a channel will expose nearby release sites to high $[Ca^{2+}]$. Opening only a few channels is sufficient to stimulate fusion of a vesicle. Increasing the number of active Ca^{2+} channels increases the number of active Ca^{2+} microdomains and this in turn increases the number of active release sites. Because the number of active release sites increases linearly with I_{Ca} amplitude, there is a linear relationship between I_{Ca} and release at hair cell synapses. This mechanism may also contribute to linearity between I_{Ca} and release at photoreceptor synapses although linearity at this synapse is also promoted by use of a sensor with a low cooperativity for calcium binding ($N \leq 3$).

Synaptic release at ribbon synapses involves two components: a transient burst of release stimulated by abrupt membrane depolarization and the slower, sustained release that accompanies maintained depolarization. In bipolar cells, as with many other neurons, fast release requires very high levels of Ca^{2+} whereas slow sustained release is triggered by low Ca^{2+} levels. The separate control of fast release by low Ca^{2+} affinity sensors and slow release by high Ca^{2+} affinity sensors is consistent with the idea that fast and slow release in bipolar and hair cells may occur at ribbon and nonribbon sites, respectively.

Unlike bipolar cells, fast and slow release from photoreceptors exhibit the same high affinity for Ca^{2+}, suggesting that both components of release occur at the same site. Thus, sustained release from photoreceptors appears to be predominately due to continued release of vesicles from the ribbon, albeit at lower rates than those attained during fast transient release. Because of the high affinity for Ca^{2+} exhibited by the release apparatus in photoreceptors, micromolar levels of Ca^{2+} present at the base of the ribbon in darkness are sufficient to stimulate fusion of vesicles at the base of the ribbon almost immediately after docking. As a consequence, the base of the cone ribbon is largely devoid of vesicles in darkness. This means that in darkness, calcium channel openings often occur beneath empty release sites. The rate of sustained release in darkness is

therefore not directly controlled by the stochastic opening of individual Ca^{2+} channels, but by the rate at which vesicles are delivered and readied for release at the base of the ribbon.

Release rates decline when photoreceptors hyperpolarize, allowing vesicles to be replenished at release sites along the base of the ribbon. With a sufficiently long and bright flash of light, the entire readily releasable pool of vesicles can be replenished. When the cone depolarizes at light offset, the rapid release of this replenished pool of vesicles can evoke a large off response in second-order neurons.

Photoreceptors release vesicles continuously at a rate of ~10–20 vesicles per ribbon per second in darkness. Cones can respond to light intensities spanning a 10 000-fold range, but this sustained release rate can encode only 10–20 distinguishable levels of steady light if synaptic release exhibits Poisson release statistics. If sustained release is controlled by the rate of vesicle delivery down the ribbon rather than the stochastic openings of individual calcium channels, this will make the rate of sustained release more regular. Regularization allows discrimination of a greater number of light levels than predicted for a Poisson release process. The high rates of release from cones that can be attained at light offset allow for the encoding of up to 100 distinguishable light decrements. This may account for psychophysical results showing a greater sensitivity to decrements than increments of light.

Rods exhibit slower release kinetics than cones, roughly matched to the slow kinetics of rod light responses. Rod and cone synapses have similar ribbons, I_{Ca} with similar properties, and similarly rapid, high affinity calcium sensors, suggesting that differences in Ca^{2+} handling and buffering may be responsible for rod/cone differences in release kinetics.

The continuous release of vesicles from photoreceptor synapses in darkness is balanced by compensatory endocytosis of vesicles. Photoreceptors rely largely on clathrin-mediated endocytosis whereas bipolar cells and hair cells rely more on bulk retrieval of large endosomes. Visualization of single vesicles at bipolar cell terminals by TIRF microscopy show that the vast majority of vesicles undergo full collapse during fusion indicating that kiss-and-run retrieval of fully formed vesicles is minimal at this synapse.

Disease-Related Mutations in Synaptic Proteins at the Photoreceptor Synapse

Given that all visual information must pass through the photoreceptor synapse, it is not surprising that mutations in synaptic proteins of photoreceptors can produce visual deficits. For example, rod–cone dystrophies can be caused by mutations in Rab3 interacting protein (RIM1),

UNC-119, or Ca^{2+} channel $\alpha2/\delta$ subunits. Congenital stationary night blindness can be caused by mutations in the rod CaV1.4 Ca^{2+} channel or CaBP4. Misregulation of glutamate release by photoreceptors and bipolar cells may also contribute to excitotoxic damage in neurodegenerative diseases of the retina.

See also: Cone Photoreceptor Cells: Soma and Synapse; Rod Photoreceptor Cells: Soma and Synapse.

Further Reading

Choi, S. Y., Borghuis, B. G., Rea, R., et al. (2005). Encoding light intensity by the cone photoreceptor synapse. *Neuron* 48: 555–562.

Daiger, S. P., Sullivan, L. S., and Browne, S. J. (2009). *RetNet – Retinal Information Network*. http://www.sph.uth.tmc.edu/retnet (accessed July 2009).

DeVries, S. H., Li, W., and Saszik, S. (2006). Parallel processing in two transmitter microenvironments at the cone photoreceptor synapse. *Neuron* 50: 735–748.

Dowling, J. E. (1987). *The Retina: An Approachable Part of the Brain.* Cambridge, MA: Harvard University Press.

Heidelberger, R., Thoreson, W. B., and Witkovsky, P. (2005). Synaptic transmission at retinal ribbon synapses. *Progress in Retinal and Eye Research* 24: 682–720.

Jackman, S., Choi, S.-Y., Thoreson, W. B., et al. (2009). Role of the synaptic ribbon in transmitting the cone light response. *Nature Neuroscience* 12: 303–310.

Kolb, H., Fernandez, E., and Nelson, R. (2009). *Webvision: The Organization of the Retina and Visual System.* http://webvision.med.utah.edu (accessed July 2009).

Krizaj, D. and Copenhagen, D. R. (2002). Calcium regulation in photoreceptors. *Frontiers in Bioscience* 7: 2023–2044.

LoGiudice, L. and Matthews, G. (2007). Endocytosis at ribbon synapses. *Traffic* 8: 1123–1128.

Prescott, E. D. and Zenisek, D. (2005). Recent progress towards understanding the synaptic ribbon. *Current Opinions in Neurobiology* 15: 431–436.

Rodieck, R. (1998). *The First Steps in Seeing.* Sunderland, MA: Sinauer.

Sterling, P. and Matthews, G. (2005). Structure and function of ribbon synapses. *Trends in Neuroscience* 28: 20–29.

Thoreson, W. B., Rabl, K., Townes-Anderson, E., and Heidelberger, R. (2004). A highly Ca^{2+}-sensitive pool of vesicles contributes to linearity at the rod photoreceptor ribbon synapse. *Neuron* 42: 595–605.

tom Dieck, S. and Branstatter, J. H. (2006). Ribbon synapses of the retina. *Cell and Tissue Research* 326: 339–346.

Polarized-Light Vision in Land and Aquatic Animals

T W Cronin, University of Maryland Baltimore County, Baltimore, MD, USA

Glossary

Chromophore – As used here, a small molecule that when bound to a protein causes the complex to absorb light at visible or near-visible wavelengths. The chromophore for visual pigments, in which the protein component is opsin, is either 11-*cis* retinal or a very similar molecule.

Circular polarization – As used here, a type of polarization of light in which the electric vector, or *e*-vector, rotates one full circle for each wavelength traveled by the light, thus describing a circle as seen from the wave front or a helix as seen from the side. Circular polarization can be either right-handed or left-handed, depending on the direction of rotation.

Dichroism – The property of a substance to absorb polarized light of one *e*-vector orientation more strongly than of other orientations, thus transmitting linearly polarized light.

Dielectric – Refers to chemical compounds or substances that do not conduct electricity. Water and most biological molecules are dielectric.

e-Vector – The electrical vector of an electromagnetic wave. For polarized light, the *e*-vector orientation is usually taken to be the plane of polarization.

Linear polarization – Sometimes called plane polarization. Refers to light in which the *e*-vectors of the constituent photons are all oriented on the same axis, or in the same plane.

Microvillus – A membranous protrusion from a cell surface shaped like a tiny tube, typically only a few cell membrane thicknesses in radius.

Polarized light – The light in which the *e*-vector lies in a plane (for linearly polarized light) or rotates through a full circle once for each wavelength (circularly polarized light).

Rayleigh scattering – A type of scattering of electromagnetic energy caused by interactions of the energy with particles much smaller than the wavelength of the energy. Rayleigh scattering produces the blue color of the sky and also produces a celestial polarization pattern by scattering of sunlight.

Specular reflection – Reflection as from a mirror, where the reflected ray leaves the surface at the same angle that the incident ray arrived. Specular reflection is typical of shiny surfaces; examples in nature include shiny leaves, insect cuticle, wet skin, or the surface of smooth water.

Light is made up of streams of photons, the elementary particles that carry electromagnetic energy. Each of these photons can be thought of as a miniature electromagnetic wave, which has a single wavelength related to the energy it carries (the distance the photon travels from one energy maximum to the next, inversely proportional to the photon's frequency) and a single plane within which the electrical energy vibrates – the polarization angle, properly called the *e*-vector (for electrical vector) angle. Note that since the energy is electromagnetic, there are both electrical vectors and magnetic vectors present, normal to each other. For consistency, throughout this article reference is made only to the *e*-vector. Therefore, a beam of light, containing countless photons, is characterized by its intensity (the number of photons delivered per unit time), its spectrum (the distribution of wavelengths of all the photons in the beam), and its polarization (the distribution of the planes of vibration, or *e*-vector angles, of all the photons in the beam).

The most common form of polarization, linear (or plane) polarization, has two descriptors: the overall *e*-vector angle, which is the mean angle of all planes of vibration of the constituent photons, and the degree of polarization, which is the fraction of energy of all photons vibrating within the plane of the *e*-vector angle. Of course, in a typical beam consisting of photons of mixed wavelengths, these polarization parameters generally vary with wavelength, creating a polarization spectrum. In this article, only linearly polarized light is discussed unless otherwise noted.

Like many vertebrates, humans are not generally aware of light's polarization properties, but the visual systems of most animals perceive light's polarization and use this ability to regulate their behavior. To help us understand what visualizing polarization would be like, the polarization properties of light can be analogized to its color properties. The spectrum of light produces the sensation of color, with a perceived hue (the predominant wavelength of constituent photons) and purity or saturation (the overall distribution of wavelengths around that of the hue itself). Hue is therefore analogous to the *e*-vector (the predominant angle of polarization of constituent photons) and saturation to the degree of polarization (the distribution of angles around this). In fact, polarization fields are often portrayed as images using false colors where angle is coded into hue and degree of polarization into saturation. Such a display can also include the coding of overall brightness as the intensity at each point, to provide a complete description of the polarized-light field.

Polarized Light in Nature

There are no natural sources of polarized light of known biological significance. Nevertheless, linearly polarized

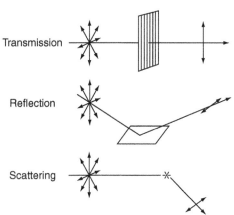

Figure 1 The three most common ways by which linearly polarized light is created either in nature or in the laboratory. At the top part of the figure, the polarization is produced by transmission through some dichroic material, with its preferential plane of transmission symbolized by the vertically parallel lines. Light emerging from a perfect dichroic material becomes fully linearly polarized. Dichroic polarizers are relatively rare in nature, and account for only a minor fraction of the polarized light observed in natural light fields. The middle part of the figure illustrates polarization by reflection from a smooth, dielectric surface. At a particular angle, known as Brewster's angle, the reflected light is fully polarized parallel to the surface. Most biological surfaces are dielectric, as is the surface of water, so much light reflected from shiny natural surfaces is highly polarized. The bottom section of the figure illustrates polarization induced by scattering. When the scattering angle is orthogonal to the axis of the ray being scattered, the scattered light is fully polarized at an e-vector angle perpendicular to the plane containing the original ray and the scattered ray.

light is abundant in natural scenery. Light can become polarized in many ways, but the most important processes in nature are through differential absorption, differential reflection, or differential scattering (**Figure 1**).

Some natural or artificial transparent materials preferentially transmit one e-vector plane while absorbing others, usually because of aligned molecules within the material. This property, known as dichroism, is not particularly common in biological systems, but there are important exceptions, including the inherent dichroism of visual pigment molecules described later. Reflection of light from dielectric surfaces produces polarization parallel to the surface (**Figure 1**). Therefore, bodies of water and many surfaces in natural scenery reflect horizontally polarized light. Rayleigh scattering from molecules and suspended particles in air produces a well-known pattern of polarization in the sky. Scattering-induced polarization varies with the scattering angle, being greatest (often near 100% polarization) for scattering perpendicular to the axis of the incoming ray (**Figure 1**). As a result, skylight polarization reaches its maximum in a band that stretches across the sky at 90° to the sun. The axis of the e-vector of the scattered ray is perpendicular to the plane defined by the incoming ray and the scattered ray, such that the band of maximum sky polarization has its e-vectors oriented tangentially to the great circle 90° from the sun's position. At dawn or dusk, this band stretches vertically across the celestial hemisphere (**Figure 2**). Since Rayleigh scattering is most effective at short wavelengths, skylight polarization is strongest in the ultraviolet. Scattering from water molecules and very small particles suspended in natural waters also produces polarization (**Figure 3**), although it rarely reaches the very high degrees of polarization seen in the sky. Light scattering in water is optically different from the processes operating in air, and polarization in

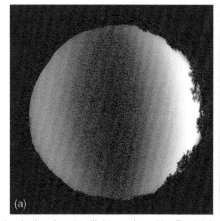 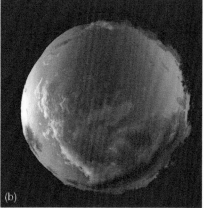

Figure 2 Polarization in the sky at twilight produced by Rayleigh scattering, imaged through a fisheye lens fitted with a linear polarizer with the transmission axis oriented to the right and left. Thus, vertically polarized light is not transmitted to the camera and shows as a dark band in the sky. In these photographs, taken at the same location and not enhanced or retouched in any way, North is to the top and West to the right. (a) Polarization in a clear sky at dusk. Note the clearly visible band of strong polarization passing from North to South through the zenith. (b) Polarization in a partly cloudy sky at dawn. The polarization is still clearly visible, but the presence of clouds depolarizes the skylight.

Figure 3   Polarization of light underwater produced by scattering from water molecules and suspended particles. (a) An unaltered image of an underwater scene at a depth of about 7 m, showing coral reef and rubble. (b) The same scene shown as a polarization image, with the degree of polarization encoded by brightness. The maximum degree of polarization in this scene is about 50%. Note that the parts of the scene that are fairly near the camera and that appear darkest in the normal photograph are most polarized. This occurs because the water between these dark regions and the camera scatters mostly horizontally polarized light.

water typically reaches its maximum value at blue-green wavelengths.

Thus, in the sky and underwater, scattering of incoming light produces partial polarization that varies with solar position and direction of view, and reflection of light from the air–water interface or from shiny surfaces (e.g., leaves, wet surfaces, animal skin, scales, or cuticle) produces strong polarization in geometrically favorable circumstances. If a terrestrial animal has polarized-light vision, the sky presents a reliable pattern useful for navigation, but in contrast, the chaotic and unpredictable pattern of polarized-light reflection produces false, pointillistic images that can mask or taint the true colors and locations of objects. Consequently, as described later, photoreceptors in animals that would normally be sensitive to the polarization of light are sometimes structurally modified to destroy polarization sensitivity.

The situation is almost always simpler in water than in air, particularly at depths greater than a few meters. Due to refraction at the air/water interface, illumination from the sun or moon is confined to within 46° of overhead position. The resulting polarization field, while variable to some extent, has horizontally oriented e-vectors much of the time, and the degree of polarization is almost always lower than in air. The pointillistic reflection of polarized light from objects is virtually gone underwater, as the refractive index gradient between water and most natural objects is much lower than in air, such that there is little specular reflection of light (required to produce polarization from dielectric surfaces). The predictable surround, typically low degree of polarization, and minimal polarized-light reflective noise would seem to make polarization vision in water of little utility, yet many aquatic animals have excellent polarization sensitivity. Currently, it is not

always clear what biological advantages are provided by such visual abilities.

Polarization Sensitivity and Polarization Vision

Polarization Responses of Photoreceptor Cells

Light is absorbed in visual photoreceptors of all animals by molecules of visual pigment, which consist of a chromophore (derived from vitamin A or a close chemical analog) linked to a protein, termed opsin. Just as each visual pigment molecule has its characteristic absorption spectrum, which ultimately determines the spectral sensitivity of the photoreceptor within which it resides, it also has an inherent polarization sensitivity. This exists because the chromophore itself is dichroic, absorbing preferentially when the e-vector of an incident photon is parallel to the long axis of the molecule. Since the chromophores of visual pigment lie nearly parallel to the membranes of photoreceptors, when light arrives perpendicular to these membranes the incident photons are likely to be polarized parallel to the absorption axes of some of the visual pigment chromophores. The way in which these chromophores are aligned within the photoreceptor cell as a whole determine whether or not the receptor responds differentially to polarized light, and thus whether it has inherent polarization sensitivity (**Figure 4**).

In the rod and cone photoreceptors of vertebrate retinas, photoreceptive membranes are arranged in a series of parallel layers, either in flattened disks (rods) or lamellae formed from folded membrane sheets (cones). Since light generally strikes these layers normal to their surfaces, it encounters visual pigments that are arrayed at

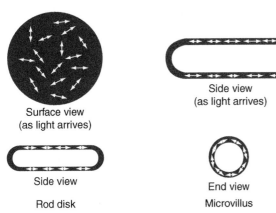

Surface view
(as light arrives)

Side view
(as light arrives)

Side view

End view

Rod disk

Microvillus

Figure 4 An illustration to show absorption of polarized light by vertebrate rod photoreceptors (left; cone photoreceptors would have similar properties) and by microvillar photoreceptors like those of arthropods or cephalopod mollusks (right). In life, light arrives normal to the flat surfaces of rod disks and encounters randomly oriented chromophores of visual pigment lying within the disk membrane (symbolized by double-headed arrows to indicate the preferred axis of polarization for best absorption). Since the orientation is fully random, there is no preferential absorption of any given e-vector angle. If light were to arrive from the side of the disk, it would encounter chromophores that are restrained to angles near that of the membranes themselves, favoring the absorption of horizontally polarized light. In microvillar photoreceptors (right), light arrives orthogonal to the long axis of each microvillus, and encounters visual pigment chromophores that are oriented roughly parallel to the axis of the microvillus. Thus, the microvillus as a whole preferentially absorbs light polarized parallel to its axis. If light were to impinge on the microvillus from the end, it would encounter chromophores arrayed at all possible angles around the circumference of the microvillus, and no preferred absorption orientation would exist.

all possible orientations (**Figure 4**, top-left). Consequently, rods and cones rarely have an overall polarization response, even though the individual molecules of visual pigment are dichroic. The situation would be very different if light impinged on rods or cones from the side (i.e., normal to the long axes of their outer segments). It would then meet chromophores lying in the planes of the membrane layers, and all chromophores would preferentially absorb light polarized nearly parallel to the membrane. Note that while the individual chromophores have random arrangements in the membrane's plane, and thus absorb light from this direction with varying effectiveness (suggested by the variable lengths of the double-headed arrows), they always absorb light polarized in the membrane's plane most effectively.

The photosensitive membranes of photoreceptor cells of arthropods (crustaceans, insects, etc.) and cephalopods (octopus, squid, and cuttlefish) are constructed from bundles of microvilli. In each microvillus, for reasons that are not yet fully understood, the molecules of visual pigments are arranged such that their chromophores are roughly

parallel to the axis of the microvillus. The microvilli typically extend out perpendicular to the axis of the receptor cell as a whole, such that light arrives perpendicular to each microvillus. In this orientation, each microvillus preferentially absorbs light polarized parallel to its axis, such that if microvilli are arranged parallel throughout the receptor as a whole, the cell will be polarization-sensitive. Note that this property requires no other cellular specializations, and as a result, almost all microvillar photoreceptor cells have some level of polarization sensitivity.

Polarization sensitivity

Having receptor cells that respond differentially to polarized light is only the first requirement for polarization sensitivity at higher levels of neural analysis. For the nervous system to be able to analyze light's polarization, sets of photoreceptors with different preferred polarization orientations must be compared, typically through opponent processing. This type of analysis is like that of color vision, where sets of photoreceptors with differential spectral sensitivity are compared for color processing. Recall that polarization of light has three attributes: intensity, degree of polarization, and polarization angle. Thus, for full awareness of light's polarization at a given point in the visual field, independent inputs from three receptor sets must be analyzed. Interestingly, few animals do this; in almost all cases, only two receptor sets with orthogonal microvilli are compared. This is reasonably effective in practice, because natural polarization tends to be predictable, such that if the receptor sets are appropriately oriented, the polarization is well analyzed. Two-channel polarization analysis can be extended to full polarization sensitivity if the receptors are rotated relative to the stimulus, although this has rarely been observed in practice.

The animal groups for which the mechanisms of polarization sensitivity are best understood are the insects, the crustaceans, and the cephalopod mollusks (octopus and squid). All of these have microvillar photoreceptors, and in most species the receptors are arranged orthogonally. Crustaceans and insects have compound eyes that are particularly well designed for analyzing polarized light. Each unit of the eye contains a group of photoreceptor cells, such that each unit of the compound eye can potentially serve as an independent polarization detector. For two-axis polarization sensitivity, subsets of receptors in each group have orthogonal microvilli. In insects, these subsets often extend into a central, fused photoreceptor from four sides, with microvilli entering from opposite sides abutting near the center. Viewed from the tip of the receptor group, the overall arrangement ends up having two cell sets with horizontal microvilli and two with vertical microvilli (**Figure 5**). Crustaceans have a similar system, but here the orthogonal sets of microvilli exist in successive layers, making the entire composite

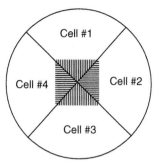

Figure 5 A schematic diagram of the structure of typical polarization-sensitive photoreceptors such as would be found in the compound eyes of insects. The receptor is viewed in a section as seen on the axis of the photoreceptor group. In life, these receptors would form a bundle of cells arranged in a circle and together forming a tall cylinder with the microvilli arranged in a smaller cylinder running down its middle. Each cell forms a section of the overall receptor (a single cell in this diagram can represent either one receptor cell or two cells lying side-by-side with parallel microvilli). Note that cells on opposite sides of the receptor extend parallel microvilli toward the junction in the center, and thus have parallel polarization sensitivity. Since two sets of receptors exist, with either horizontally oriented or vertically oriented microvilli, the receptor group as a whole can provide information for two-axis polarization analysis. Receptors of crustaceans are similar to this, but each layer of the joint photoreceptor contains microvilli from only one subset of cells, with either horizontally arrayed or vertically arrayed orientations. Successive layers of the receptor contain microvilli from the other subset of cells, and thus the receptor cylinder has stacks of mutually orthogonal microvillar layers which can contribute to two-axis polarization analysis. Photoreceptors of cephalopods are somewhat different from these, since they are arrayed continuously side-by-side throughout a retina, but each junction of four cells forms a set of microvilli organized like those in the center of the diagrammatic insect photoreceptor illustrated here. Cephalopod receptor cells also form parallel microvilli on the opposite side of each cell. Consequently, each cell contributes to two junctions of microvilli, one on each side of the cell. Again, with two primary axes of orientation of microvilli, separate cells can contribute to two-axis polarization sensitivity.

photoreceptor like a pile of a large number of circular segments, each having microvilli orthogonal to the segments immediately above and below. Finally, the cephalopods (which do not have compound eyes, but instead have a single lens eye structured much like a vertebrate camera eye) arrange their microvillar receptors such that each cell has microvilli on two opposing sides (like a two-sided toothbrush). The mosaic of cells forms junctions similar to what is pictured in the center of **Figure 5**, except that each of the four cells in this figure would form another junction on the opposite side with yet other cells. In all these cases, the cells with parallel microvilli viewing one point in space join to form one polarization channel, and those with microvilli orthogonal to these join for the opponent channel.

Vertebrate polarization sensitivity is more difficult to explain, and it is fair to say that we are still not able to account for it satisfactorily. Nevertheless, there is no doubt at all that some vertebrates sense light's polarization. Recall that end-on stimulation of rods and cones is unlikely to produce any differential sensitivity to the plane of polarization, because chromophores are randomly oriented for such light (**Figure 4**). If vertebrate photoreceptor cell outer segments were slanted relative to the axes of impinging rays of light, this would confer some polarization sensitivity. It appears that in at least some fishes, the outer segments of some classes of cones lie on their sides, tilting their lamellae vertically in the retina. If all cones of a given class lay parallel, or were organized into orthogonal classes, this could permit the retina as a whole to achieve an overall polarization sense. There is recent evidence that some rod or cone classes are measurably dichroic to end-on illumination. The origin of this dichroism is unclear, but it could be caused by parallel tilting of the rod disks or cone lamellae.

Polarization vision

If an animal has polarization sensitivity, it can obviously respond in some way to a polarization stimulus. As described later, these responses are frequently hard-wired and inflexible, and the polarization sense that drives them does not correspond to what is normally conceived of as vision, which implies a perception of space, form, and individual objects. The term polarization vision refers to a polarization sense analogous to color vision, whereby animals visualize polarization attributes of features within the overall field of view and use polarization variations to enhance the visibility, contrast, or features of particular objects. In principle, an animal that is capable of polarization vision perceives the visual world as a pattern varying in polarization features among receptive fields. While polarization sensitivity is most useful for orientation or for organizing simple responses, polarization vision offers the potential to direct complex behavior such as predation, camouflage generation or breaking, and signal detection. There is only weak evidence of this ability in some vertebrates, but many species of both arthropods and cephalopods probably use true polarization vision in ways that are discussed later.

Disentangling polarization and color sensitivity

Many − perhaps most − animals that are sensitive to polarized light also have color vision. This presents both perceptual and sensory-processing challenges, as it is generally not desirable to mix these visual modalities. For example, if a receptor cell that contributes to a perceptual color channel has some residual polarization sensitivity, color appearance will be altered by stimuli that contain polarized light. This is most often a problem for animals with microvillar photoreceptors due to their inherent polarization bias. The cephalopods have firmly dealt with this issue by discarding color vision entirely – the great majority of octopuses, squids, and cuttlefishes have only a single spectral receptor class in their retinas, restricting vision entirely to the intensity and polarizational domains. Many crustaceans have reached a similar solution, devoting nearly all of their receptors to polarized-light reception. Some crustacean species, however, separate color and polarization processing, using a single spectral class for polarization analysis while reserving a set of other polarization-insensitive classes for color vision. This solution is used, for example, by stomatopod crustaceans, also known as mantis shrimps. Insects also commonly separate color-sensitive from polarization-sensitive receptors, isolating their polarization receptors to just one part of the visual field and often using only ultraviolet receptors for polarization analysis. Some insect species destroy polarization sensitivity in photoreceptors by twisting the entire receptor group around its long axis. In a few cases, surprisingly, insects unify polarization and color perception in the same receptor cells, interpreting some stimuli by combining these two modalities into a single signal. Some butterfly species, for instance, examine potential oviposition sites in this way. In vertebrates, however, as with other aspects of polarized-light photoreception, it is unknown how (or even if) polarized-light processing is kept separate from color processing. This could be a difficult problem, as it is thought that some vertebrates use different spectral types of cones to sense different polarization planes, a technique that immediately must mix color and polarization information at the first level of light detection.

The Contributions of Polarized-Light Perception to Behavior

Sensing polarized light seems strange to us, but for most animals it is as fundamental to their visual perception as color vision is to humans. Indeed, as will be described shortly, in many animals polarized-light perception plays similar roles to those assumed by color vision, and it can even work together with color vision to improve visual interpretation of stimuli. However, there are many situations where polarized light is used for special purposes unique to this modality. Among these are water surface detection and skylight navigation.

Water surfaces reflect horizontally polarized light, as illustrated in **Figure 1**. This is why sunglasses with polarizing lenses make it easier for fisherman to see fish – the lenses are oriented to block horizontally polarized light, reducing the glare from the water's surface and clarifying the visibility of objects in the water itself. Many flying insects, including adult water beetles and mayflies, use the reflected polarization in the opposite way – their eyes are adapted to respond strongly to large expanses of horizontal polarization below the horizon, which in nature invariably correspond to water surfaces. The insects respond by diving into the water, or by alighting on its surface to oviposit. This simple response is extremely reliable in nature, but can lead to disastrous consequences for the insects today, when many shiny, horizontal surfaces are man-made. Parking lots, oil ponds, and even the painted surfaces of cars and other manufactured objects induce the same response, which in these cases is frequently lethal.

Scattering of sunlight in the clear sky produces a highly reliable pattern of polarization (**Figure 2**), recognized by the visual systems of many insects (including bees, ants, and crickets) as well as by other arthropods including some spiders and crustaceans. The pattern is used for navigation, as it is a perfect indicator of the current position of the sun, persisting even when the sun is not visible behind an obscuring object or landscape feature, or when the sun is hidden by clouds. Thus, navigation is possible even on quite cloudy (but not wholly overcast) days. Most insects that navigate using skylight polarization devote a small region of the compound eye, called the dorsal rim, to perceiving the pattern, and most require only a small patch of clear sky to orient. Navigation using the location of the sun or skylight polarization patterns is not simple, as the solar position drifts through the sky with changing dynamics throughout the seasons, and insects must be able to compensate for these changes each day as they manage their foraging excursions. Some insects, including dung-foraging scarab beetles, use skylight polarization created by moonlight to navigate during their nocturnal rambles. In an interesting vertebrate example, migrating birds are thought to use skylight patterns of polarization at twilight to calibrate their magnetic compasses.

The tasks described so far are analogous to map senses, simple reflexes, or other perceptual abilities that are not strictly visual in the sense that we humans understand it. In other words, these types of abilities do not examine features or objects in the outside world except in very general ways. However, there are animals that actually see patterns of polarization in a fashion that is quite analogous to the way that we perceive the external world – they use polarization vision to recognize objects, to enhance contrast of prey, or to see signals of conspecific animals. Some animals, in fact, can be trained to discriminate objects that we see as identical but that differ in the patterns of polarization that they reflect or transmit. Octopus and mantis shrimps learn such tasks.

Near the water's surface, the skylight polarization pattern penetrates and is therefore available as an orientation cue. Deeper than this, underwater polarization is only rarely usable for navigation (although it can be used to orient vertical migration) because it is frequently weak, often oriented near the horizontal plane, and quite variable depending on the background (**Figure 3**). Nevertheless, the background polarization can be used to enhance the contrast of the visual world and make midwater objects more visible. Squids take advantage of this, making them more effective predators on fishes and transparent planktonic prey against a polarized background than against a depolarized one. Presumably, they use their polarized-light sense to detect differences in polarization between the prey and the background, making objects of similar overall brightness appear distinguishable.

If squids and cuttlefishes can distinguish objects based on polarization, it should not be surprising that they also have body patterns that reflect polarized light, undetectable by many marine animals, and apparently employ these patterns to communicate with each other. The polarization patterns are actively controlled by the animal producing them and can appear and disappear in fractions of seconds. When displayed, the polarization reflections remain highly visible to a conspecific individual even as the signaler changes its posture or moves its arms about.

Two other groups of animals, one marine and one terrestrial, are currently known to recognize and respond to polarization signals. Mantis shrimps produce an abundance of signals based on patterns of polarized light reflected from their carapace (**Figure 6**), and use these signals during mating and aggressive displays. Like the cephalopods, of course, they are marine invertebrates and conduct their diplays underwater. One group of insects, however, uses polarized-light signals in the open air. Many species of tropical butterflies find mates in the diffuse light under the rainforest canopy. Here, the background polarization is relatively weak, and the strong polarization pattern of the sky is rarely visible. Thus, the polarization produced by reflection from scales on butterfly wings can act as an unusually strong, visible signal.

Sensitivity to Circularly Polarized Light

This discussion of polarized-light sensitivity would not be complete without mention of a recently discovered visual modality, sensitivity to circularly polarized light. Circular polarization differs from linear polarization, the type discussed exclusively until now, in that the *e*-vector does not remain within a single plane, but instead rotates around the axis of the beam of light. Circularly polarized light is not common in nature, and its presence cannot be detected with standard polarization-sensing systems. Despite this, one group of animals, the mantis shrimps, perceives circularly polarized light and produces circularly polarized signals by reflection. This ability is particularly unexpected because there is no known source of circular polarization underwater other than signals from other mantis shrimps, so it is difficult to explain how and why the ability originally arose. It is possible that circular polarization sensitivity in these animals first appeared as an accidental epiphenomenon related to the unusual way in which their linear polarization system is assembled, and that this led to the elaboration of signals based on circularly polarized light. See the suggested reading for a more detailed account of this unusual finding.

Summary

The ability to perceive and respond to linearly polarized light is widespread among animals, occurring in many vertebrates and invertebrates. Some of these species use polarization for general tasks that do not require precise imaging, such as finding water or navigating using patterns of scattered polarized light in the sky. Others truly see polarized objects and use this imaging ability to detect prey and recognize signals from conspecifics. Our poor understanding of the biology of polarized-light sensitivity

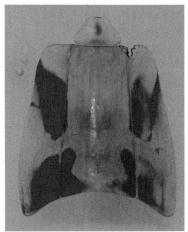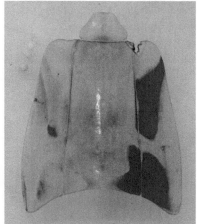

Figure 6   Polarization signals reflected from the shed carapace (or molt) of the stomatopod crustacean, or mantis shrimp, *Odontodactylus cultrifer*. These patterns of polarization are visualized through a linearly polarizing filter rotated to two orientations at 90° to each other. Signals like these are used during aggressive or mating displays of mantis shrimps. Photograph by T. H. Chiou.

in vertebrates, and the recent discovery of circular polarization sensitivity in mantis shrimps, suggest that there are other aspects to polarized-light sensitivity and to polarization vision that still remain to be revealed.

See also: Microvillar and Ciliary Photoreceptors in Molluskan Eyes; The Photoresponse in Squid; Phototransduction: Rhodopsin; Rod and Cone Photoreceptor Cells: Inner and Outer Segments; The Colorful Visual World of Butterflies; Evolution of Opsins.

Further Reading

Chiou, T. -H., Kleinlogel, S., Cronin, T. W., et al. (2008). Circular polarisation vision in a stomatopod crustacean. *Current Biology* 18: 429–434.

Cronin, T. W., Shashar, N., Caldwell, R. L., et al. (2003). Polarization vision and its role in biological signaling. *Integrative and Comparative Biology* 43: 549–558.

Dacke, M., Doan, T. A., and O'Carroll, D. C. (2001). Polarized light detection in spiders. *Journal of Experimental Biology* 204: 2481–2490.

Hawryshyn, C. W. (1992). Polarization vision in fish. *American Scientist* 80: 164–175.

Horváth, G. and Varjú, D. (2004). *Polarized Light in Animal Vision: Polarization Patterns in Nature*. Berlin: Springer.

Muheim, R., Phillips, J. B., and Akesson, S. (2006). Polarized light cues underlie compass calibration in migratory songbirds. *Science* 313: 837–839.

Shashar, N., Rutledge, P., and Cronin, T. W. (1996). Polarization vision in cuttlefish: A concealed communication channel? *Journal of Experimental Biology* 199: 2077–2084.

Sweeney, A., Jiggins, C., and Johnsen, S. (2003). Polarized light as a butterfly mating signal. *Nature* 423: 31–32.

Waterman, T. H. (1981). Polarization sensitivity. In: Autrum, H. (ed.) *Handbook of Sensory Physiology* VII/6B, pp. 281–469. Berlin: Springer.

Wehner, R. (2001). Polarization vision – a uniform sensory capacity? *Journal of Experimental Biology* 204: 2589–2596.

Post-Golgi Trafficking and Ciliary Targeting of Rhodopsin

D Deretic, University of New Mexico, Albuquerque, NM, USA

Glossary

Cell polarity – The asymmetry in cell shape, protein distributions, and cell functions.

Golgi complex – A biosynthetic organelle comprised of a stack of membranous cisternae that are involved in protein modifications, such as processing of N-linked sugars as well as sorting and transport of membrane proteins. Newly synthesized membrane and secretory proteins enter the stack through the cis-Golgi, progress through the medial cisternae and exit at the trans-Golgi.

Phosphatidylinositol-4,5-bisphosphate (PI(4,5)P$_2$) – An essential second messenger as well as lipid regulator of membrane trafficking. Along with its precursor phosphatidylinositol-4-phosphate (PI(4)P), it regulates Arfs and is regulated by them, providing a positive-feedback loop that regulates membrane trafficking.

Post-Golgi transport carriers (TCs) – The post-Golgi vesicles that carry cargo to the plasma membrane. It is now clear that these carriers are large pleiomorphic structures, rather than small vesicles, as previously believed, thus the term vesicles has been replaced with transport carriers.

Small GTPases – The members of the low-molecular-weight (20–25 kDa) Ras super family of guanosine-triphosphate (GTP)-binding proteins comprised of at least four large families, including the Arfs and the Rabs. Small GTPases function by providing directionality to membrane traffic through the molecular switch whose ON and OFF states are triggered by binding and hydrolysis of GTP. The nucleotide-bound state determines the affinity of interactions with regulatory proteins and the downstream effectors of small GTPases.

SNARE proteins – The soluble N-ethylmaleimide-sensitive factor attachment protein receptor (SNARE) proteins are major components of the intracellular machinery responsible for targeted membrane delivery. SNAREs were identified as membrane receptors for the soluble N-ethylmaleimide-sensitive factor (NSF) attachment protein (SNAP) in the cell-free system that reconstituted intra-Golgi trafficking. SNARE proteins form complexes, which are generally composed of a four helical bundle that bridges opposing membranes and brings them into close proximity to initiate fusion.

The trans-Golgi network (TGN) – A tubular network in the close proximity to the trans-Golgi cisternae that represents the central sorting station of the cell, where proteins and lipids destined for different subcellular domains are segregated from each other and sorted into post-Golgi TCs.

Introduction

Retinal rod photoreceptors are exquisitely complex polarized cells that carry photon detection and visual transduction that are essential as the first step in vision. Following the final cell division of their precursors, photoreceptors attain a level of polarity that is nearly unmatched in other cells of the body. Maintaining this organization throughout the lifetime of the organism is a prerequisite for vision. Proper targeting and retention of the macromolecular complexes involved in the visual transduction cascade are accomplished by the highly coordinated action of protein and lipid regulators that together constitute the membrane trafficking machinery. Specific components of this machinery involved in the directed delivery of rhodopsin and its associated proteins and lipids are only beginning to emerge.

Photoreceptor Polarity

Rod photoreceptors are modified neurons with specialized light-sensing organelle, the rod outer segment (ROS). The ROS is filled with membranous disks housing the phototransduction machinery that converts photon absorption by rhodopsin into changes in neurotransmitter release, thus transmitting photosensory information to the visual cortex. The light-sensing machinery is comprised of peripheral and integral membrane proteins of the ROS. It is continuously replenished through ROS disk membrane renewal, followed by its removal through daily shedding and phagocytosis by retinal pigment epithelial (RPE) cells.

The ROS disk membrane proteins are embedded in a low-viscosity lipid bilayer milieu comprised of unsaturated long-chain phospholipids highly enriched in omega-3 docosahexaenoic acid (DHA, 22:6(n-3)), which is essential for sensory membrane function and for cell survival. The exceptionally high content of polyunsaturated DHA phospholipids renders ROS membranes highly susceptible to light and oxidative damage.

ROSs initially form from primary cilia, and a short 9+0 (nonmotile) connecting cilium remains in the adult as the only path of communication between the ROS and the photoreceptor rod inner segment (RIS). The RIS houses mitochondria and biosynthetic membranes involved in oxidative metabolism and membrane protein and lipid biosynthesis, respectively. The photoreceptor-connecting cilium corresponds to the transition zone of primary cilia, which is considered a gateway for the admission of specific proteins to this privileged intracellular compartment.

The primary cilium is the site of assembly of large molecular complexes involved in intraflagellar transport (IFT). IFT protein 20 (IFT20) subunit links mammalian IFT complex with the microtubule motor, kinesin II. The base of the cilium is a region of particularly high lipid ordering, separating the ciliary membrane from the surrounding plasma membrane due to high cholesterol content and glycosphingolipid products of the phosphatidylinositol 4-phosphate (PI(4)P)- and Arf-dependent effector four-phosphate-adaptor protein 2 (FAPP2). Cholesterol rings have also been reported to surround the photoreceptor connecting cilium. Lipid ordering might be important for the docking of the basal body to the plasma membrane or the extension of the ciliary axoneme, both essential processes in ciliogenesis.

The RIS is separated from the nuclear and synaptic domains by adherens junctions (AJs) that comprise a continuous adhesion belt, the outer limiting membrane (OLM). Interestingly, this junctional region lacks tight junctions that normally confine plasma membrane proteins to their respective domains. Nonetheless, rod membrane proteins are strictly confined to their specialized domains and the maintenance of polarity is essential for the cell's function and survival. The photoreceptor synaptic terminal contains specialized ribbon synapses that are responsible for the tonic release of neurotransmitters, which is interrupted by photon capture.

Photoreceptor cytoskeletal networks and molecular motors play a major role in the cell polarity. Microfilaments provide structural support by encircling the RIS beneath the plasma membrane and are anchored at the AJs. Other polar dynamic actin networks and filaments are also dispersed through the cell, most notably in the distal portion of the cilium, at the sites of ROS disk formation where they regulate the growth of nascent disks. Actin-based motility through the cilium is thought to be mediated by myosin VIIa, the product of the Usher syndrome (USH) 1B (Usher1B) gene. An array of microtubules radiates into the RIS from the microtubule-organizing center nucleated by a pair of centrioles located below the cilium. Microtubules generate polar networks that generally determine the position of membrane organelles

and allow intracellular motility. Heterotrimeric molecular plus end-directed motor kinesin-II and homodimeric KIF17 mediate microtubule-dependent trafficking into the ROS. The absence of KIF3A subunit of kinesin-II causes membrane accumulation in the RIS and cell death. Cytoplasmic dyneins 1 and 2 mediate retrograde trafficking to the centrosome from the RIS and the ROS, respectively.

Photoreceptor Biosynthetic Membrane Trafficking: Endoplasmic Reticulum, Golgi, and Post-Golgi Transport Carriers

ROS integral membrane proteins are synthesized on the rough endoplasmic reticulum (ER), pass through the Golgi apparatus, and are then incorporated into post-Golgi vesicles or, in current terminology, rhodopsin transport carriers (RTCs) that bud from the *trans*-Golgi network (TGN), the central membrane sorting station of the cell, and are transported to a docking site near the base of the cilium. Along with rhodopsin, DHA phospholipids are co-transported on RTCs, which fuse with the specialized domain separating the ciliary membrane from the surrounding RIS plasma membrane, thereby regulating the replenishment of light-sensitive ROS membranes. Rhodopsin represents 90% of the newly synthesized protein in rod photoreceptors, and its biosynthetic pathway is best understood. In the 1980s, the laboratories of David S. Papermaster and Joseph C. Besharse combined electron microscope (EM) immunocytochemistry, autoradiography, and freeze-fracture analysis and demonstrated that newly synthesized rhodopsin is transported vectorially to the base of the cilium on membranous carriers. Membrane biosynthesis was also studied by pulse–chase experiments that established the kinetics of movement of newly synthesized rhodopsin through membrane compartments separated by subcellular fractionation. This methodology was subsequently refined by Deretic and Papermaster to incorporate high-resolution linear sucrose gradients, which separated the low-density post-Golgi carriers from other subcellular organelles, including the Golgi, TGN, plasma membrane, and synaptic vesicles.

Successful isolation of post-Golgi RTCs provided not only insight into their molecular composition, but was also the basis for the development of the retinal cell-free assay that reconstitutes RTC budding *in vitro*. The development of this assay in our laboratory led to the discovery that rhodopsin contains a sorting signal within its five C-terminal amino acids that regulates its incorporation into RTCs as they bud from the TGN. Abundant evidence points to the role of the amino acid sequence valine-x-proline-x (VxPx motif) in rhodopsin's C-terminal domain as a sorting signal into RTCs and delivery to the cilium and the ROS. In our retinal cell-free system a monoclonal antibody, whose antigenic site is within the five C-terminal amino acids of rhodopsin, and synthetic peptides corresponding to the C-terminus of rhodopsin inhibit RTC budding from the TGN. Studies in a number of laboratories employing transgenic animals expressing rhodopsin lacking the sorting signal, or rhodopsin–C-terminal fusion proteins carrying autosomal dominant retinitis pigmentosa (adRP) mutations, confirmed that the absence of the correct sorting information results in the targeting to the RIS plasma membrane and the synapse *in vivo*.

Because of the exceptionally high membrane turnover, photoreceptors are vulnerable to mutations that affect membrane trafficking. Among the rhodopsin mutations with the most severe phenotypes are those that alter the rhodopsin C-terminal VxPx targeting motif. Rhodopsin C-terminal mutations cause rapid photoreceptor cell death, retinal degeneration, and blindness in adRP. Within the VxPx targeting motif, V345 and P347 are the primary sites of C-terminal adRP mutations involving single amino acid substitutions.

To dissect the sorting machinery that regulates RTC budding, targeting, and fusion, it was essential to identify the resident proteins of this organelle. These studies indicated a succession of binding on RTCs of members of the small guanosine triphosphate (GTP)-binding protein families (G proteins or small GTPases) that are the known regulators of membrane trafficking.

Small GTPases of the Rab and Arf Families and Their Regulators in Rhodopsin Trafficking

Rabs

Directed delivery of membrane cargo is mediated through vesicular transport regulated by the small GTPases of the Rab and Arf families, which play a central role in organizing intracellular membrane trafficking. All GTPases function by providing a universal molecular switch whose ON and OFF states are triggered by binding and hydrolysis of GTP. Small GTPases are in a dynamic equilibrium between the cytosol and the membranes maintained by the interactions with a number of regulatory proteins, including nucleotide exchange factors (GEFs), GTPase-activating proteins (GAPs), and their downstream effectors, and the affinity of these interactions is determined by the nucleotide-bound state. The Rab family of small GTPases includes over 60 members that are generally designated numerically (i.e., Rab6, Rab8, and Rab11). Upon GTP binding, activated Rabs recruit a multitude of effectors that organize membrane domains involved in the tethering of membranes to other membranes and to cytoskeletal elements, thus conferring directionality to membrane traffic. Macromolecular complexes organized by Rabs provide a unique identity to membrane microdomains within cellular organelles. Consequently, a change in Rabs,

termed Rab conversion, changes the membrane identity and accompanies cargo progression through intracellular compartments.

Rab6 regulates retrograde transport between the Golgi and the ER, but in cells with hypertrophied synthesis of simple membranes it also associates with post-Golgi carriers. In photoreceptor cells, Rab6 is associated with the Golgi, TGN, and RTCs. It was demonstrated in *Drosophila* that Rab6 regulates rhodopsin trafficking, and that the expression of the GTPase-deficient mutant of Rab6 leads to retinal degeneration.

Rab11 has been localized to both the Golgi and the recycling endosomes, which are involved in return of plasma membrane receptors to the cell surface through endocytosis. However, the Golgi-associated function of Rab11 is less well understood. The specific functions of Rab11 in apparently divergent cellular processes are probably based on its ability to interact with different effector molecules that belong to the family of Rab11-interacting proteins (FIPs), which are localized in different trafficking pathways. Rab11 is associated with photoreceptor TGN and RTCs where it interacts with FIP3. At the TGN, Rab11 and FIP3 are incorporated into a ciliary targeting complex regulated by the small GTPase Arf4, described below. Sustained presence of Rab11 on RTCs suggests that Rab11 might also interact with the conserved octameric Sec6/8 complex, also known as the exocyst in yeast. The Sec6/8 complex tethers the Rab11/FIP3-positive membranes and is involved in tethering RTCs to the RIS plasma membrane. Exocyst complex localizes to the cilia in polarized epithelial cells, and we also find it at the base of the photoreceptor cilium. In *Drosophila*, interaction of the Sec6/8 complex with Rab11 plays a role in the tethering of membranes carrying rhodopsin. Rab11 may also cooperate with another RTC-associated Rab, Rab8. A handover from Rab11 to Rab8, or Rab conversion, may occur at the base of the cilium to couple the final stages of traffic along the ciliary pathway.

Rab8 regulates polarized trafficking in epithelial cells and neurons through its activity on cytoskeleton remodeling necessary for membrane outgrowth and the formation of cellular protrusions. It has recently emerged as a major player in ciliogenesis. In retinal photoreceptors, Rab8 regulates RTCs fusion and ROS biogenesis. It acts at the base of the cilium in conjunction with another small GTPase (Rac1), phosphatidylinositol-4,5-bisphosphate ($PI(4,5)P_2$), actin and the phosphoinositide- and actin-binding protein moesin. In addition, Sec6/8 complex is also likely to function as one of Rab8 effectors in ciliogenesis. Transgenic *Xenopus* with photoreceptor-specific expression of GFP-Rab8Q67L dominant-active mutants show normal photoreceptor cell morphology, but cause slow retinal degeneration, whereas GFP-Rab8T22N GTPase-deficient dominant-negative mutants show a defect in membrane tethering and accumulate RTCs in the vicinity of the cilium, leading to rod cell death and rapid retinal degeneration. Emerging evidence points to Rab8 as a central regulator of the biogenesis of primary cilia, suggesting that its regulation of rhodopsin trafficking may be a part of a broad and more general role for Rab8 in the regulation of ciliogenesis.

Mutations that affect regulatory proteins of small GTPases and their interacting proteins have been found to cause X-linked retinitis pigmentosa (RP), autosomal recessive Leber congenital amaurosis (LCA), and choroideremia. Rab escort protein 1 (Rep1), encoded by the choroideremia gene is a subunit of geranylgeranyl transferase, the enzyme that isoprenylates Rab proteins. Rab8-interacting proteins, Optineurin and Huntingtin, are also linked to the neurodegenerative diseases primary open-angle glaucoma (POAG) and Huntington's disease, respectively. Rab8, Optineurin, and Huntingtin are collectively involved in the linkage of membrane organelles to the cytoskeleton, suggesting that the breakdown of this linkage may be a common theme in retinal degeneration and in other neurodegenerative diseases.

Arfs

The Arf family of small GTPases includes three different groups of proteins: the Arfs, Arf-like proteins (Arls), and SARs. Arfs were originally discovered as ADP-ribosylation factors, but in 2006 the new nomenclature for the human Arf family of GTP-binding proteins, formerly known as ARF, ARL, and SAR proteins, has been established. Arfs are no longer called ADP-ribosylation factors (ARFs), since ADP-ribosylation appears unrelated to their physiological function. Arf family members regulate membrane trafficking, lipid metabolism, organelle morphology, and cytoskeleton dynamics. These functions were elucidated for the abundant class I Golgi Arfs, Arf1 and Arf3, and the plasma membrane-associated Arf6. The loss of a single class I or class II Arf has little effect on membrane trafficking, but the deletion of pairs of Arfs causes distinct defects. This suggests a pair-wise engagement of Arfs and a certain redundancy in their function. Arf function depends on GTP hydrolysis mediated by Arf GAPs, which are essential for coupling the proofreading of cargo incorporation to the budding of membrane carriers and are often incorporated into protein coats.

The selection and packaging of sensory receptors and membrane cargo targeted to the primary cilia and cilia-derived sensory organelles are critical to replenish the ciliary membrane, yet it remains poorly understood. Our recent studies have demonstrated that rhodopsin C-terminal VxPx targeting signal binds Arf4 to regulate incorporation of rhodopsin into RTCs at the TGN. The function of the class II Golgi-associated Arfs, Arf4 and Arf5, is least understood, yet the direct and specific binding of the VxPx-targeting motif to Arf4 suggests a distinct role for this particular Arf in the generation of RTCs. The targeting VxPx motif binds Arf4 and recruits it to the TGN, leading to assembly of a ciliary targeting complex. This complex is comprised of two small GTPases, Arf4 and Rab11, the Rab11/Arf effector FIP3, and an Arf-GAP/effector ASAP1. The localization of the ciliary targeting complex in photoreceptors is illustrated in **Figure 1**. ASAP1 catalyzes phosphatidylinositol 4,5-bisphosphate (PIP$_2$)-dependent GTP hydrolysis on Arf4. Transgenic frogs expressing an Arf4 mutant impaired in ASAP1-mediated GTP hydrolysis, display dysfunctional rhodopsin trafficking and cytoskeletal and morphological defects, resulting in retinal degeneration. FIP3, which binds Arf4, also forms a ternary complex with Rab11 and ASAP1 and stimulates Arf GAP activity of ASAP1. Emerging evidence points to the role of ASAP1 and FIP3 as a functional module that provides temporally and spatially restricted hydrolysis of GTP bound to Arf4 at the TGN. Since ASAP1 and FIP3 act as homodimers, they may oligomerize to form a protein coat that regulates ciliary targeting, a specialized form of the TGN-to-plasma membrane trafficking.

Rhodopsin provides the spatial control for the ciliary targeting module by recruiting Arf4 to the carrier budding sites at the TGN through its VxPx targeting signal. Surprisingly, the VxPx motif is not unique to rhodopsin, but is present in other membrane proteins targeted to primary cilia such as polycystins 1 and 2, and the cyclic nucleotide-gated channel CNGB1b subunit. The VxPx from polycystin-2 also binds Arf4, suggesting that the targeting complex recruited through Arf4 is a part of conserved machinery involved in the selection and packaging of the cargo destined for delivery to the cilium.

SNAREs and their Regulators in Rhodopsin Trafficking

In addition to GTPases and their effectors, the soluble N-ethylmaleimide-sensitive factor attachment protein receptor (SNARE) proteins are major components of the intracellular machinery responsible for targeted membrane delivery. SNAREs are considered to be directly involved in membrane fusion. After the tethering step, SNAREs are activated on the opposing donor and target membranes to form a complex that bridges the two membranes and brings them into close proximity to initiate fusion. Although SNARE pairing alone is not sufficient to determine the specificity of organelle fusion, cognate SNAREs are correctly paired in biological membranes, based on proofreading and polarized distribution leading to their relative enrichment at the appropriate fusion sites. Rabs also function by concentrating and activating SNAREs, accessory proteins and lipids, at the sites of membrane fusion and are thus required for carrier docking and fusion with the target organelle.

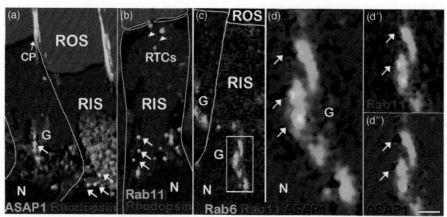

Figure 1 The Arf GAPASAP1 co-localizes with Rab11 on nascent RTCs at the *trans*-Golgi network (TGN). (a) A confocal optical section (0.7 µm) of frog retina labeled with anti-rhodopsin C-terminal mAb 11D5 (red) and anti-ASAP1 (green). Anti-rhodopsin antibody labels the rod outer segment (ROS) and the Golgi (G) in the rod inner segment (RIS), where ASAP1-positive puncta (yellow, arrows) line up with regular periodicity. ASAP1 is also detected in calycal processes (CP) that evaginate from the RIS and surround the base of the ROS. Nuclei (N) are stained with TO-PRO-3 (blue). (b) A confocal optical section labeled with anti-rhodopsin C-terminal mAb 11D5 (red) and anti-Rab11 (green). Rab11-positive puncta (yellow, arrows) aligned with rhodopsin-laden Golgi. Rab11 is also present on RTCs (arrowheads). Nuclei (N) are stained with TO-PRO-3 (blue). (c) ASAP1 (blue) and Rab11 (red) colocalize in the bud-like profiles at the tips of the *trans*-Golgi (Rab6, green) (boxed area magnified in (d). (d) Magnified *trans*-Golgi area from panel C, with ASAP1- and Rab11-positive buds (arrows), which likely represent the TGN. (d' and d''). Rab11 (red) and ASAP1 (blue) are shown separately. Scale bar = 3 µm in (a)–(c), 0.7 µm in (d), 1 µm in (d') and (d''). Modified from Mazelova, J., Astuto-Gribble, L., Inoue, H., Tam, B. M., Schonteich, E., Prekeris, R., Moritz, O. L., Randazzo, P. A., and Deretic, D. (2009). Ciliary targeting motif VxPx directs assembly of a trafficking module through Arf4. *EMBO J* 28: 183–192.

SNARE complexes are generally composed of a four-helical bundle bridging opposing membranes and bringing them into close proximity to initiate fusion. Fusion with the plasma membrane requires formation of a complex between syntaxins (Qa SNAREs) and VAMPs (R-SNAREs), each contributing one helix to the four-helix SNARE bundle, and Qbc SNAREs, either neuronal SNAP-25 or non-neuronal SNAP-23, which provide two helices to the central layer of the core complex. SNAREs are targeted to appropriate membrane domains based on specific sequences. The polarized distribution of Qa SNAREs is likely to contribute additional specificity of membrane targeting by promoting fusion with only certain target membranes. Recent evidence suggests that the local lipid environment, particularly phospholipids enriched in omega-3 and omega-6 fatty acids, also contributes to regulate SNARE function.

The membrane fusion event through which RTCs deliver rhodopsin to the cilium is mediated by a SNARE complex. Syntaxin 3 and SNAP-25 are the Q-SNAREs for the fusion of incoming RTCs with the RIS plasma membrane and, therefore, regulators of ROS biogenesis in photoreceptors. The distribution of these SNAREs in photoreceptors is illustrated in **Figure 2**. Remarkably, omega-3 DHA enhances syntaxin 3 incorporation into SNARE complexes at RTC fusion sites and promotes ciliary membrane expansion and ROS biogenesis. Microtubules direct the restricted distribution of syntaxin 3, consistent with the membrane cytoskeleton playing an essential role in

concentrating RTC fusion regulators around the cilium. Syntaxin 3 is the major partner for SNAP-25 in photoreceptor cells; however, SNAP-25 pairing with syntaxin 1A, or 1B, may regulate a distinct trafficking pathway in the RIS. Interestingly, Syntaxin 3 is also found at the base of mouse ROS, suggesting an additional role for this SNARE in rodent rods.

ROS is a Modified Primary Cilium

Almost all cells possess primary cilia that house an array of signal transduction modules. Long underappreciated, the cilium has recently received a great deal of attention due to the ciliary involvement in a wide range of human diseases, including retinal degeneration, polycystic kidney disease (PKD), Bardet–Biedl syndrome (BBS), and neural tube defects. Many cilium disease proteins were detected in the mouse photoreceptor ciliary proteome. Ciliary involvement in a wide range of retinal diseases has come into sharp focus in the past several years, with the molecular mechanism underlying these diseases being rapidly elucidated. Several human syndromes, including Senior–Loken syndrome, Jeune syndrome, and BBS, are also characterized by both cystic kidneys and retinal degeneration, which are often found in combination with skeletal defects or other abnormalities such as obesity, polydactyly, hypogenitalism, and developmental delay that might also be caused by defects in cilia. The organization of the small GTPases,

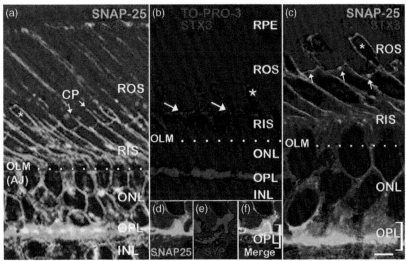

Figure 2 SNAP-25 is a photoreceptor RIS plasma membrane (PM) and synaptic SNARE, whereas syntaxin 3 is concentrated in the RIS PM. (a) A confocal optical section (0.7 μm) of frog retina labeled with anti-SNAP-25 N-terminal mAb (green) and anti-syntaxin 3 (STX3, red). Nuclei are stained with TO-PRO-3 (blue). Anti-SNAP-25 (green) outlines the photoreceptor PM. The calycal processes (CP), which are in continuum with the RIS PM, also contain SNAP-25. The ROS, which are visible by DIC, are completely devoid of this SNARE. AJ, adherens junctions that form the outer limiting membrane (OLM, dotted line). The retinal layers are: ONL-outer nuclear, OPL-outer plexiform, INL-inner nuclear. SNAP-25 co-localizes with synaptophysin (SYP, red) in the OPL, which encompasses the synapses of rods and cones with the rod bipolar, cone bipolar, and horizontal cells. Asterisks indicate the protruding RIS of green rods, a minor subpopulation that accounts for ~5% of total rods. (b) A confocal optical section labeled with anti-syntaxin 3 (STX3, red), which is highly concentrated in the RIS PM (arrows), but is absent from the ROS, and from the RPE. Syntaxin 3 is also abundant in the OPL. Nuclei are stained with TO-PRO-3 (blue). (c) SNAP-25 and syntaxin 3 co-localize (yellow) in the RIS at the RTC fusion sites in the vicinity of cilia (arrows). Syntaxin 3 is also abundant in the inner (lower) half of the OPL (large bracket), where bipolar and horizontal cells are localized, but not in the outer (upper) half where photoreceptor synapses are localized. (d–f). SNAP-25 (green in (d)) co-localizes with synaptophysin (red in (e)) in the photoreceptor synapses (yellow in (f)). However, SNAP-25 is abundant in the bipolar and horizontal cell processes in the OPL (green in (f)), where synaptophysin is not detected. Scale bar = 8 μm in (a), 10 μm in (b), 7 μm in (c)–(f). Modified from Mazelova, J., Ransom, N., Astuto-Gribble, L., Wilson, M. C., and Deretic, D. (2009). Syntaxin 3 and SNAP-25 pairing, regulated by omega-3 docosahexaenoic acid (DHA), controls the delivery of rhodopsin for the biogenesis of cilia-derived sensory organelles, the rod outer segments. *Journal of Cell Science* 122: 2003–2013.

SNAREs and their regulators involved in the ciliary targeting of rhodopsin is schematically illustrated in **Figure 3**.

Twelve BBS genes have been identified. A complex composed of seven BBS proteins, the BBSome, localizes to the base of the cilium and is required for ciliogenesis. BBS3, which is not a part of BBSome encodes the Arf family GTPase Arl6. Strikingly, Rabin8, the GDP/GTP exchange factor that activates Rab8, localizes to the basal body and contacts the BBSome. In cultured epithelial cells, activated Rab8 enters the primary cilium and promotes extension of the ciliary membrane. This explains the accumulation of RTCs below the cilium in photoreceptors expressing mutant Rab8. Strikingly, activated Rab8, in its GTP-bound form interacts with another centrosomal/ciliary protein CEP290/BBS14/NPHP6, which is not a part of the BBSome. Thus, BBS may be caused by defects in Rab8-mediated vesicular transport to the cilium.

An extraordinary array of retinopathy-associated ciliary proteins includes X-linked RP1, which is localized to the proximal cilium and involved in disk morphogenesis, and RP2, which was recently identified as a GAP for the Arf family GTPase Arl3 that is involved in kidney and photoreceptor development. A major player in ciliary morphogenesis is retinitis pigmentosa GTPase regulator (RPGR), which is homologous to RCC1, the nucleotide exchange factor for the small GTPase Ran. Mutations in the retina-specific ORF15 isoform of RPGR (RPGR (ORF15)) were found in X-linked RP3, which is associated with 10–20% of RP. RPGR appears to be a part of a ciliary and basal body protein network that, when disrupted, can result in Leber congenital amaurosis, Senior–Loken syndrome, nephronophthisis, or Joubert syndrome. RPGR (ORF15) co-localizes with RPGRIP1 at centrioles and basal bodies and interacts with nucleophosmin. RPGRIP1, which is affected in patients with LCA, anchors RPGR to the photoreceptor connecting cilium and participates in disk morphogenesis. RPGR also interacts with calmodulin and nephrocystin-5, a ciliary IQ domain protein, which is mutated in Senior–Loken syndrome, and with the centrosomal/ciliary protein CEP290/BBS14/NPHP6, which is truncated in early-onset retinal degeneration in the rd16 mouse. In addition,

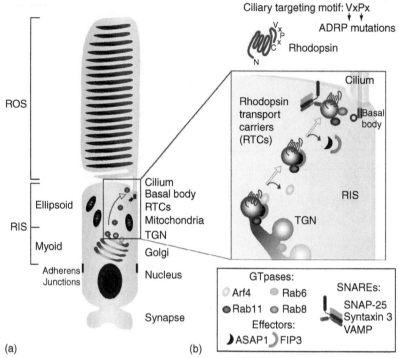

Figure 3 Post-Golgi trafficking and ciliary targeting of rhodopsin (a) Diagram of the rod photoreceptor cell. Cilium protrudes from the cell body (RIS) and elaborates the ROS filled with membranous disks containing photopigment rhodopsin and associated phototransduction machinery. Following synthesis in the RER, newly synthesized rhodopsin traverses the Golgi and the TGN, localized in the myoid region of the RIS, where it is incorporated into transport carriers (RTCs). RTCs travel from the TGN, though the mitochondria-laden ellipsoid region of the RIS, to the base of the cilium where they fuse with the RIS PM. Adherens junctions separate the RIS from the synapse. Little is known about membrane targeting to the RIS PM and the synapse. (b) Polarized trafficking of post-Golgi RTCs is dependent on the rhodopsin C-terminal VxPx ciliary targeting motif. ADRP mutations in the VxPx motif are indicated. Selected proteins involved in the recognition of the VxPx motif, sorting of rhodopsin into the RTCs and their targeting to the cilium are shown in the enlarged area of the RIS. Rhodopsin C-terminal binds to, and recruits Arf4 to the TGN membrane, leading to assembly of a ciliary targeting complex. This complex is comprised of two small GTPases Arf4 and Rab11, the Rab11/Arf effector FIP3, and an Arf-GAP/effector ASAP1. The small GTPases Rab6 regulates trafficking through the Golgi, whereas Rab8 regulates RTC fusion. Syntaxin 3 and SNAP-25 are a part of the SNARE complex that catalyzes RTC fusion at the base of the cilium.

mutations in the gene encoding the basal body protein RPGRIP1L (RPGRIP-like), a nephrocystin-4 interactor, cause Joubert syndrome.

Another ciliary and basal body protein network linked to myosin VIIa is disrupted in human USH, the most frequent cause of combined deafness–blindness. USH is genetically heterogeneous with three clinical types, USH1–3. The scaffold protein harmonin (USH1C) integrates USH1 and USH2 molecules into protein networks. The Usher protein network is organized by the scaffold proteins SANS (USH1G), which provides a linkage to the microtubule transport machinery, and whirlin (USH2D), which anchors USH2Ab and very large G-protein-coupled receptor 1b (VLGR1b). Remarkably, the USH protein network is also a part of the periciliary ridge complex (PRC), a specialized membrane domain for docking and fusion of RTCs in Xenopus photoreceptors. Finally, the Usher protein network is linked to the Crumbs polarity complex in the retina and mutations in Crumbs cause retinitis pigmentosa (RP12).

Conclusions and Summary

Numerous diseases arise from defects in proteins that participate in membrane protein trafficking, vectorial transport, and assembly of outer segment membranes. Thus, maintenance of photoreceptor cell polarity is of utmost importance for their health and survival, and ultimately for vision.

See also: Genetic Dissection of Invertebrate Phototransduction; The Photoreceptor Outer Segment as a Sensory Cilium; The Physiology of Photoreceptor Synapses and Other Ribbon Synapses; Primary Photoreceptor Degenerations: Retinitis Pigmentosa; Primary Photoreceptor Degenerations: Terminology; Retinal Degeneration through the Eye of the Fly; Rod and Cone Photoreceptor Cells: Inner and Outer Segments; Rod and Cone Photoreceptor Cells: Outer Segment Membrane Renewal; Secondary Photoreceptor Degenerations: Age-Related Macular Degeneration; *Xenopus laevis* as a Model for Understanding Retinal Diseases.

Further Reading

Cai, H., Reinisch, K., and Ferro-Novick, S. (2007). Coats, tethers, Rabs, and SNAREs work together to mediate the intracellular destination of a transport vesicle. *Developmental Cell* 12: 671–682.

Deretic, D., Schmerl, S., Hargrave, P. A., Arendt, A., and McDowell, J. H. (1998). Regulation of sorting and post-Golgi trafficking of rhodopsin by its C-terminal sequence QVS(A)PA. *Proceedings of the National Academy of Sciences of the United States of America* 95: 10620–10625.

Deretic, D., Williams, A. H., Ransom, N., et al. (2005). Rhodopsin C-terminus, the site of mutations causing retinal disease, regulates trafficking by binding to ARF4. *Proceedings of the National Academy of Sciences of the United States of America* 102: 3301–3306.

Gillingham, A. K. and Munro, S. (2007). The small G proteins of the Arf family and their regulators. *Annual Review of Cell and Developmental Biology* 23: 579–611.

Green, E. S., Menz, M. D., LaVail, M. M., and Flannery, J. G. (2000). Characterization of rhodopsin mis-sorting and constitutive activation in a transgenic rat model of retinitis pigmentosa. *Investigative Ophthalmology and Visual Science* 41: 1546–1553.

Leroux, M. R. (2007). Taking vesicular transport to the cilium. *Cell* 129: 1041–1043.

Malsam, J., Kreye, S., and Sollner, T. H. (2008). Membrane fusion: SNAREs and regulation. *Cellular and Molecular Life Sciences* 65: 2814–2832.

Mazelova, J., Astuto-Gribble, L., Inoue, H., et al. (2009). Ciliary targeting motif VxPx directs assembly of a trafficking module through Arf4. *EMBO Journal* 28: 183–192.

Mazelova, J., Ransom, N., Astuto-Gribble, L., Wilson, M. C., and Deretic, D. (2009). Syntaxin 3 and SNAP-25 pairing, regulated by omega-3 docosahexaenoic acid (DHA), controls the delivery of rhodopsin for the biogenesis of cilia-derived sensory organelles, the rod outer segments. *Journal of Cell Science* 122: 2003–2013.

Moritz, O. L., Tam, B. M., Hurd, L. L., et al. (2001). Mutant rab8 Impairs docking and fusion of rhodopsin-bearing post-Golgi membranes and causes cell death of transgenic *Xenopus* rods. *Molecular Biology of the Cell* 12: 2341–2351.

Nachury, M. V., Loktev, A. V., Zhang, Q., et al. (2007). A core complex of BBS proteins cooperates with the GTPase Rab8 to promote ciliary membrane biogenesis. *Cell* 129: 1201–1213.

Papermaster, D. S., Schneider, B. G., and Besharse, J. C. (1985). Vesicular transport of newly synthesized opsin from the Golgi apparatus toward the rod outer segment. Ultrastructural immunocytochemical and autoradiographic evidence in *Xenopus* retinas. *Investigative Ophthalmology and Visual Science* 26: 1386–1404.

Shi, G., Concepcion, F. A., and Chen, J. (2004). Targeting od visual pigments to rod outer segment in rhodopsin knockout mice. In: Williams, D. S. (ed.) *Photoreceptor Cell Biology and Inherited Retinal Degenerations*, pp. 93–109. Singapore: World Scientific Publishing.

Tam, B. M., Moritz, O. L., Hurd, L. B., and Papermaster, D. S. (2000). Identification of an outer segment targeting signal in the COOH terminus of rhodopsin using transgenic *Xenopus laevis*. *Journal of Cell Biology* 151: 1369–1380.

Wandinger-Ness, A. and Deretic, D. (2008). Rab8a. *UCSD-Nature Molecule Pages*. Nature Publishing Group. doi:10.1038/mp. a001997.001901.

Relevant Websites

http://www.retina-international.com – Retina International.

http://www.signaling-gateway.org – The UCSD-Nature Signaling Gateway.

http://www.sph.uth.tmc.edu/RetNet/ – Retinal Information Network.

http://webvision.med.utah.edu – WEBVISION: The organization of the retina and visual system.

Primary Photoreceptor Degenerations: Retinitis Pigmentosa

M E Pennesi, P J Francis, and R G Weleber, Oregon Health and Sciences University, Portland, OR, USA

Glossary

Allied disorders – Retinitis pigmentosa (RP) is often grouped with a class of more stable, inherited retinal disorders collectively referred to as RP and allied disorders. Some of these allied disorders cause similar clinic findings as RP, e.g., nyctalopia (night blindness), but usually do not show progression and deterioration with time. An example is congenital stationary night blindness (CSNB), which can present with nyctalopia and decreased rod and cone function on the electroretinogram (ERG). Unlike RP, most patients with CSNB have stable visual function. X-linked CSNB is caused by mutations in nyctalopin (NYX) and L-type voltage dependent calcium channel (CACNA1F). Although the majority of mutations of rhodopsin causes typical RP, rare mutations, such as G90D in rhodopsin, produce night blindness with such mild progression late in life that they have been called stationary night blindness. Another allied disorder is achromatopsia, which is caused by mutations in cyclic nucleotide-gated channel subunits (CNGA2, CNGB3) or guanine nucleotide alpha-binding protein 2 (GNAT2). Achromatopsia is associated with severely decreased central and color vision, photophobia, and nystagmus. These symptoms are similar to those that can be seen with some cone–rod dystrophies. Indeed, later in life some modest foveal atrophy can occur and cases of progressive cone–rod dystrophy have been associated with mutations of some of the achromatopsia genes. However, unlike cone–rod dystrophies, which invariably progress, achromatopsia is, in the vast majority of cases, stationary.

Cone dystrophy – Cone photoreceptors are affected and rod photoreceptors are minimally affected or spared in cone dystrophy. Many cases of early cone dystrophies with time will develop significant rod abnormalities.

Cone–rod dystrophy – Cone-rod dystrophy, as a group, involves both photoreceptors with cones affected more than rods. Certain forms of RP present with greater cone than rod involvement on ERG and these patients have been termed to have cone–rod RP. However, in cone–rod dystrophies as a group the primary defect lies in cones and secondary rod loss

occurs with time. Most investigators consider primary cone–rod dystrophy separate from RP.

Extrinsic factor – An agent external to the organism that contributes to or is causative of a disease state. This can include drugs, foods, normal nutrients (excess or deficiency), toxins, inhaled chemicals, infectious agents, and exposures to radiation such as light, sound, and high-energy particles.

Intrinsic factor – An agent that is inherent to the organism that contributes to or is causative of a disease state.

Mixed intrinsic and extrinsic etiology for a secondary photoreceptor degeneration – This occurs when a person has a genetic variant that creates a toxic metabolite in the presence of an extrinsic molecule that would normally not be encountered.

Mixed model of primary and secondary photoreceptor degeneration – This is considered when a genetic alteration within the photoreceptors is insufficient to cause photoreceptor degeneration by itself, but predisposes to degeneration in the presence of an extrinsic or intrinsic agent. A second mode of combined primary and secondary photoreceptor degeneration is when one group of photoreceptors, such as the rod photoreceptors, undergoes a primary degenerative process that is due to a mutation in a gene that is expressed in those photoreceptors and precipitates apoptosis, which leads to a secondary degenerative process, in this example cones, due to alterations in the cellular environment induced by death of neighboring cells.

Primary retinal degeneration – This occurs when cells in the retina, usually photoreceptors, die secondary to a process that originates within the retina itself. An example of a primary retinal degeneration is RP, which is caused by mutations in genes that encode proteins important for retinal function. A disease can be classified as a primary retinal degeneration if the genetic defect is such that correction of expression of the normal gene product in the photoreceptors is required to correct the abnormality and arrest the degeneration.

Primary retinal degeneration with secondary photoreceptor degeneration – This occurs when photoreceptor degeneration is the result of mutation(s) of a gene that exists in other retinal cells, for example, retinal pigment epithelial (RPE) cells. Correction of the genetic defect would require modification of the effects of those other retinal cells (e.g, RPE cells).

Retinal atrophy – A broad term encompassing not only processes that occur with retinal degenerations, but also abnormal retinal tissue or cellular loss due to developmental defects and malnutrition.

Retinal degeneration – A process whereby cells in the retina undergo cell death by apoptosis. Most retinal degenerations affect both rod and cone photoreceptors, but some disorders reflect damage that occurs principally in other cell types, e.g., the RPE in Stargardt's disease and other ABCA4-related retinopathies. Secondary degeneration of the RPE is also common. Transsynaptic degeneration of higher-order cells, bipolar and ganglion cells, can also occur. The general term retinal degeneration should be distinguished from the more specific term, photoreceptor degeneration.

Retinal dystrophy – A broad term that not only encompasses retinal degenerations, but also includes abnormal retinal function due to developmental defects and malnutrition.

Retinitis pigmentosa (RP) – A heterogeneous group of diseases that result in degeneration of the rod and cone photoreceptors and secondarily the RPE. This degeneration usually leads to a loss of night vision due to the early degeneration of rods, constricted visual fields, decreased responses on ERG, and ultimately a decrease in visual acuity once macular cones begin to degenerate. Typical fundus findings include midperipheral atrophy of the pigment epithelium, bone spicule pigments, retinal vessel attenuation, and waxy pallor of the optic nerve. The term RP usually refers to only rod-cone dystrophies; however, cone-rod dystrophies and cone dystrophies are sometimes grouped under this term.

Rod–cone dystrophy – A retinal dystrophy in which the rod photoreceptors are affected more than the cones. Most forms of RP manifest as rod–cone dystrophies.

Secondary photoreceptor degeneration of the extrinsic type – A secondary photoreceptor degeneration of the extrinsic type exists if, despite the underlying molecular defect, one could avoid the photoreceptor degeneration by preventing an individual's exposure to an extrinsic agent or condition (e.g., toxin, drug, infectious agent, light, and trauma).

Secondary photoreceptor degeneration of the intrinsic type – If one can prevent photoreceptor degeneration by correcting or reversing a systemic or ocular metabolic or immune process, then it is a secondary photoreceptor degeneration of the intrinsic type.

Background

Retinitis pigmentosa (RP) is caused by a large number of genetic defects that result in a characteristic pattern of degeneration of the rod and cone photoreceptors and the retinal pigment epithelium (RPE). This degeneration usually leads to a loss of night vision due to the early degeneration of rods, constricted visual fields, decreased responses on electroretinogram (ERG), and ultimately a decrease in visual acuity once macular cones begin to degenerate. The typical fundus exam in RP reveals midperipheral atrophy of the pigment epithelium, bone spicule pigmentation, retinal vessel attenuation, and waxy pallor of the optic nerve (**Figure 1**).

RP was first named by the Dutch ophthalmologist, Frans Cornelius Donders, in the mid-nineteenth century, although earlier clinical descriptions of the disease exist. The term retinitis pigmentosa is somewhat of a misnomer because inflammation is not thought to be the primary pathological mechanism. Rather, mutations in over 100 genes have been shown to cause RP and its allied disorders, and there still remain a significant number of genes yet to be identified. To keep track of the ever-growing list of genes implicated in this disease, a comprehensive online database, Retnet, has been established by Dr. Stephen Daiger.

One of the most fascinating aspects of RP is that mutations in genes that encode functionally distinct

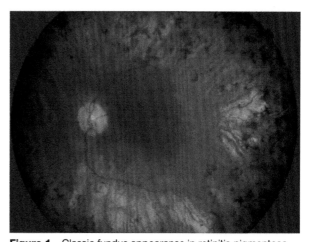

Figure 1   Classic fundus appearance in retinitis pigmentosa demonstrating bone spicule pigmentation, vascular atrophy, retinal pigment epithelium atrophy, and waxy pallor of the optic nerve. From Weleber, R. and Evan, K. G. (2006). Retinitis pigmentosa and allied disorders. In: Ryan, S. J. (ed.) *The Retina*, 4th edn., vol. 1, chap. 17, pp. 395–498. Philadelphia, PA: Elsevier.

proteins result in a common degenerative pathway. Some of the many examples include genes involved in the structural integrity of the photoreceptors and cilia, the retinoid cycle, the phototransduction cascade, the extracellular matrix, cellular metabolism, intracellular trafficking, and RNA processing.

Prevalence

The worldwide prevalence for all forms of RP has been reported to be approximately 1:4000. While most studies have focused on the prevalence in European/Caucasian populations, the occurrence of RP has been reported throughout the world.

Inheritance

All forms of Mendelian inheritance have been reported but autosomal dominant, recessive, and X-linked traits are most frequently seen. Rarely RP is inherited as a digenic disorder or through the maternal line as a mitochondrial disease.

Autosomal Recessive

RP patients with a family history of similarly affected relatives are called multiplex, whereas those with no family history are classified as simplex. Simplex individuals are usually assumed to represent autosomal recessive inheritance, although some of these cases may be *de novo* dominant mutations or unrecognized X-linked inheritance. When simplex cases are included, autosomal recessive cases of RP have been reported to account for approximately 50–60% of all cases, with the exact percentage varying from country to country. Some of the most commonly affected genes are usherin (*USH2A*), a gene that is involved in both Usher syndrome and autosomal recessive RP, and the phosphodiesterase beta subunit (*PDE6B*), a gene involved in phototransduction.

X-Linked RP

X-linked RP results from mutations of genes on the X chromosome and represents approximately 5–15% of patients with RP. To date, six genes that cause retinal degeneration have been linked to the X chromosome. Two genes, retinitis pigmentosa GTPase regulator (*RPGR*) and retinitis pigmentosa 2 (*RP2*), are known, and several genes remain to be identified.

Males with X-linked RP typically have more severe retinal degeneration compared to autosomal recessive and dominant forms of the disease. The actual rate of degeneration is likely similar to the other forms, but the age of onset appears to be earlier.

Female carriers are thought to have a mosaic retina in which some of the cells express the normal allele, while others express the mutant allele. The fundus findings in female carriers of X-linked RP can vary from very subtle changes, such as mottling of the RPE, to more severe disease with some patients showing the classic bone spicule pigmentation. Even in the cases of female carriers with a normal appearing fundus, changes are usually apparent on the ERG.

Autosomal Dominant

Patients with autosomal dominant RP often have a family history of the disease, although there are cases of incomplete penetrance and *de novo* mutations. Autosomal dominant mutations account for approximately 30–40% of patients with RP. In general, patients with autosomal dominant RP tend to be less affected than patients with X-linked or autosomal recessive RP. Some of the most commonly mutated genes include rhodopsin (*RHO*) and retinitis pigmentosa 1 (*RP1*).

Nonsyndromic versus Syndromic Retinal Degeneration

Most cases of RP are nonsyndromic and the pathology is limited to the eye. However, RP can also be associated with dysfunctions in other organ systems, with many of these cases comprising defined syndromes.

The most common syndromic association with RP is Usher syndrome, which is an autosomal recessive disorder and is divided into three subtypes based on clinical findings. Patients with type I Usher syndrome present with severe, but nonprogressive congenital hearing loss, balance problems, and RP. In type II Usher syndrome, patients have less severe hearing loss, RP, and normal balance. Patients with type III Usher syndrome, start with symptoms similar to type II but later progress to type I. The retinal findings in Usher syndrome are indistinguishable from those characteristic of nonsyndromic autosomal recessive RP. Eleven genes have been found to cause Usher syndrome. Considering the presumed shared evolutionary ancestry of photoreceptors and cochlear hair cells, it is likely that some of these genes share similar functions.

Bardet–Biedl syndrome (BBS) is an autosomal recessive disorder in which RP is a universal finding. Other commonly associated features include postaxial polydactyly, truncal obesity, abnormalities of cognition, and renal disease. Mutations in 12 genes have been implicated in BBS. Many of these genes encode proteins that are important for the formation or function of the cilia (**Figure 2**).

Some other syndromes that can present with RP include: abetalipoproteinemia (Bassen–ornzweig disease), Alström

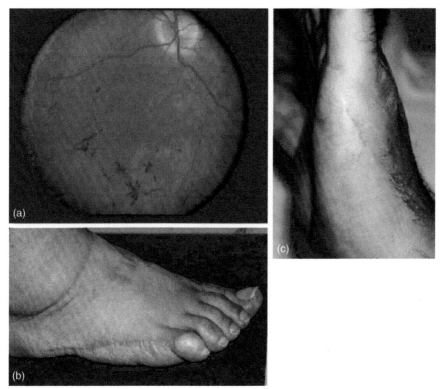

Figure 2 (a) Fundus photos of patient with Bardet–Biedl syndrome (BBS), demonstrating the classical changes of RP that include bone spicule pigmentation, vascular attenuation, and waxy pallor of the optic nerve. (b) Scars on the foot of a patient with BBS from removal of an extra digit. (c) Similar scars on the hand.

syndrome, chronic progressive external ophthalmoplegia (CPEO), Friedreich's ataxia, incontinentia pigmenti (Bloch–Schulzberg syndrome), Joubert syndrome, Kearns–Sayre syndrome, mucopolysaccharide disorders, neuronal ceroid lipofuscinoisis (Batten disease), Refsum disease (infantile and adult), Senior–Loken syndrome, and spinocerebellar ataxia type 7.

Classification of RP

One often confounding feature of RP is the many different ways in which the disease can be classified. RP can be classified by its mode of inheritance, age of onset, fundus appearance, pattern of functional vision loss, or by genetic mutation.

As mentioned previously, RP is often characterized by its pattern of inheritance. With the advent of genetic testing, patients are increasingly being tested and classified according to which genes are mutated (see **Table 1** for the most common mutations and **Table 2** for description of genes). The phenotype and course of the disease can show significant variation with different mutations in the same gene. Likewise, there can also exist significant phenotypic variations between two people who harbor the same mutation. Ultimately, classification by genetic

Table 1 Most common genes causing retinitis pigmentosa by inheritance

	Most common genes causing retinitis pigmentosa
Autosomal recessive RP (including Usher syndrome)	*USH2A, PDE6B, PDE6A, MYO7A, CRB1, RGR, CNGB1, RPE65*
Autosomal dominant RP	*Rho* (rhodopsin), *RP1, PRPF31, PRPF3, RDS/ROM*
X-linked RP	*RPGR, RP2*

defects will likely prove to be the most useful way to segregate and treat patients with RP. However, many genes remain to be discovered and genetic testing is not yet universally available to test for all mutations in known genes. For these reasons, it is useful to examine the ways in which RP has been categorized in the past.

Classification by Age of Onset

Severe forms of RP that manifest before the first year of life are referred to as Leber congenital amaurosis (LCA). The forms of RP occurring between 1 and 5 years have been termed juvenile RP or severe early childhood-onset retinal dystrophy (SECORD). LCA is characterized by

Table 2   Genes, protein, diseases, and function

Gene symbol	Protein	Diseases	Function
ALMS1	Alström syndrome protein 1	Alström syndrome	Exact function unknown, may play a role in ciliogenesis
CACNA1F	Calcuim-channel, voltage-dependent, alpha 1F subunit	Incomplete CSNB, AIED, and other X-linked CRD (CORDX3, Maori disease, CSNB with retinal and optic atrophy)	Acts as a subunit in the major voltage-sensitive calcium channel in rod and cone photoreceptor terminals. Required for the calcium flux into photoreceptors (rods and cones) that is needed for sustaining the tonic neuro-transmitter release from presynaptic terminals. Required for formation/maintenance of ribbon synapses
CEP290	Centrosomal protein CEP290	Joubert syndrome, LCA	Localizes to cilium, may mediate G-protein trafficking
CHM	Rab escort protein 1 (REP-1)	Choroideremia	Participates in post-translational lipid modifications of proteins to enable membrane attachments that are essential in membrane trafficking. Acts as a geranylgeranyl transferase and appears required for specific Rab pathways.
CNGA2	Cyclic nucleotide-gated channel – subunit alpha 2	Achromatopsia	Codes for the alpha subunit of the cyclic nucleotide-gated channels in cones
CNGB1	Cyclic nucleotide-gated channel – subunit B1	arRP	Codes for the beta subunit of the cyclic nucleotide gated channels in rods
CNGB3	Cyclic nucleotide gated channel – subunit beta 1	Achromatopsia	Codes for the beta subunit of the cyclic nucleotide-gated channels in cones
CRB1	Homolog of crumbs	arRP, LCA	Homologous to crumbs in the *Drosophila*, where it plays a role in cell–cell interactions and photoreceptor polarity
GNAT2	Transducin alpha 2	Achromatopsia	Plays a role in the photoreceptor phototransduction cascade. Forms a complex with the beta and gamma subunits and acts to convert cGMP to GMP
MYO7A	Myosin-VIIA	Usher syndrome type IA	May play a role in trafficking of ribbon-synaptic vesicle complexes and renewal of the outer photoreceptors disks
NYX	Nyctalopin	CSNB	Predicted secreted protein important for development of ON bipolar cell signaling pathways
OAT	Ornithine-∂-aminotransferase	Gyrate atrophy	Catalyzes the conversion of L-ornithine and a 2-oxo acid to L-glutamate 5-semialdehyde and an L-amino acid
PDE6A	Phospodiesterase alpha subunit	arRP	Plays a role in the photoreceptor phototransduction cascade. Forms a complex with the beta and gamma subunits and acts to convert cGMP to GMP
PDE6B	Phospodiesterase beta subunit	arRP	Plays a role in the photoreceptor phototransduction cascade. Forms a complex with the alpha and gamma subunits and acts to convert cGMP to GMP
PHYH	phytanoyl-CoA 2-hydroxylase	Refsum disease	Catalyzes the first step in the alpha-oxidation of phytanic acid
PRPF3	PRPF3	adRP	Forms part of a spliceosome complex for mRNA processing
PRPF31	PRPF31	adRP	Forms part of a spliceosome complex for mRNA processing
RDH5	11-*cis* retinol dehydrogenase	Fundus albipunctata	11-*cis* RDH is found in the RPE, where it catalyzes the final step in the biosynthesis of 11-*cis* retinaldehyde, the visual chromophore for both rods and cones
RDS	Peripherin	RP, pattern dystrophy	Interacts with Rom-1 as the major morphogen for disk formation and to stabilize photoreceptor disks
RGR	Retinal g-protein-coupled receptor	arRP, adRP	Expressed in the RPE and Müller cells and plays a role in retinoid recycling
RHO	Rhodopsin	adRP, arRP CSNB	Mediates the detection of photons through light-induced isomerization of 11-*cis* to all-trans retinal, which triggers a conformational change leading to G-protein activation and release of all-trans retinal

Continued

Table 2   Continued

Gene symbol	Protein	Diseases	Function
RLBP1	Cellular retinaldehyde-binding protein 1-like protein 1	arRP, retinitis punctata albescens, Bothnia dystrophy, Newfoundland rod–cone dystrophy	Plays a role in visual pigment regeneration: carrier for endogenous 11-*cis*-retinol and 11-*cis*-retinal, major 11-*cis*-retinol acceptor in the isomerization step of the rod visual cycle, stimulating isomerization of all-*trans*- to 11-*cis*-retinol, facilitates oxidation of 11-*cis*-retinol to 11-*cis*-retinal by 11-*cis*-retinol dehydrogenase (RDH5)
ROM	Rom-1	Digenic RP	Interacts with peripherin to stabilize photoreceptor disks
RP1	RP1	adRP	Localizes to photoreceptor cilium, may play a role in transport of proteins between inner and outer segments
RP2	RP2	xlRP	Exact function unknown. Stimulates GTPase activity of tubulin and may function to link cell membrane with cytoskeleton
RPE65	RPE65	arRP, LCA	Expressed in the RPE and acts in retinoid metabolism to isomerizes all-trans-retinal ester to 11-*cis* retinol
RPGR	Retinitis pigmentosa GTPase regulator	xlRP	Exact function unknown. Localizes to cilium and may act to maintain protein polarization across the cilium
USH2A	Usherin	Usher syndrome type II, arRP	Exact function unknown, Interacts with collagen IV and fibronectin and may be required for stable integration into the basement membrane

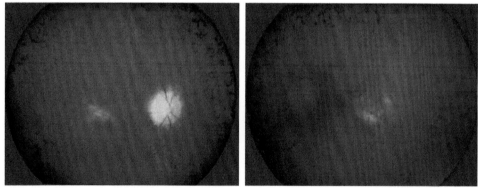

Figure 3   Fundus photographs of a 16-year-old patient with Leber congenital amaurosis. There is waxy pallor of the optic nerve, severe vascular attenuation, RPE atrophy most notable in the macula, and bone spicule pigmentations.

severe vision loss, nystagmus, unrecordable ERGs, and poorly responsive pupils (amaurosis). Mutations in at least 16 genes have been found to cause LCA and most of these are inherited in an autosomal recessive fashion. Juvenile RP/SECORD is thought to be caused by less severe mutations in the same set of genes as LCA, and more mild mutations in some of these genes have been implicated in recessive RP (**Figure 3**).

Classification by Fundus Appearance

The classic fundus appearance in RP is described below (see the subsection titled 'Fundus findings'). Deviations from this classic fundus appearance have given rise to several alternative terms for pigmentary retinopathies including: inverse RP, concentric RP, sector RP, retinitis punctata albescens, fundus albipunctatus, RP with preserved peri-arteriolar RPE, pigmented perivenous retinochoroidal atrophy, and retinitis sine pigmento. Some of these terms are falling out of usage as genetic characterization of the disease is becoming more common.

There have been many cases of unilateral RP described and it has been proposed that this could be caused by a somatic mutation. However, there is yet to be a histologically confirmed case of RP caused by a somatic mutation. Most of these cases likely represent other diseases that can mimic RP and cause a pigmentary retinopathy.

Classification by Functional Loss

Historically, adult RP was categorized as either type I (rod dysfunction) or type II (rod and cone dysfunction) based on psychophysical testing. These classifications were further refined on the basis of electrophysiological findings to include the categories: rod–cone dystrophy, cone–rod dystrophy, or cone dystrophy. Most forms of RP are rod–cone dystrophies in which rod photoreceptor death occurs first and is later followed by subsequent cone photoreceptor death. This article will use the term RP to describe the most common form, namely rod–cone dystrophy. The forms of RP that cause cone and cone–rod dystrophy will be denoted accordingly.

Mechanism of Disease

RP and its allied disorders are caused by mutations in over 100 genes and likely the same number remains to be elucidated. Ultimately, these mutations lead to photoreceptor death by apoptosis. It is still not fully understood how mutations in genes, which code for an array of functionally different proteins, result in a common pathway to photoreceptor death.

Mutations can result in decreased expression of a given protein, cause loss of function of that protein, or imbue a gain of function. In autosomal recessive forms of RP, there is a loss of expression or function when both copies of a given gene are mutated. In contrast, autosomal dominant forms of RP are thought to be caused by gain-of-function mutations, where the mutated protein becomes toxic or interferes with the function of the remaining normal forms of that protein (dominant negative effect). In autosomal dominant RP, the most common mutations are found in *RHO*. These dominant mutations can lead to forms of *RHO* that do not inactivate properly or are not transported to the outer segment.

An example of one well-studied mutation that causes autosomal dominant RP is the P23H mutation in the *RHO* gene. This mutation results in misfolding of the protein such that it is sequestered in the endoplasmic reticulum and is never transported to the outer segment. The misfolded proteins accumulate creating aggregations that activate an unfolded protein response. Dysregulation of these responses may lead to photoreceptor death although the exact mechanism has yet to be determined.

Clinical Presentation

Symptoms

Most frequently, the earliest symptom of RP is night blindness that precedes visual-field change and, in some, retinal pathology. In children, parents may comment that their child is afraid or becomes distressed in the dark. Older children often comment on poor vision compared with their fully sighted peers. In adults, difficulties with night driving are frequent. The second cardinal symptom of RP is progressive peripheral visual-field loss. Since central vision is spared early in the course of the disease, some patients do not notice this loss of visual field until the degeneration has become quite advanced.

Another common symptom of RP is problems with dark adaptation, such as difficulties adjusting to dim illumination when entering a movie theater. Additionally, as the disease progresses, patients can develop photophobia. Color vision is typically normal early in the disease but with progression, blue–yellow defects become apparent. Flashes of light, or photopsias, are experienced by most patients.

Patients with cone and cone–rod dystrophies can present with early photophobia, decreased central vision, and impaired color vision. These patients will typically have worse visual acuity than patients with rod–cone dystrophies due to earlier involvement of macular cones.

Refraction

Refractive errors in patients with RP have been studied and, on average, these patients are more myopic and have a greater degree of astigmatism than those in the general population. By contrast, patients with early-onset forms of RP, such as LCA or SECORD, tend to have hyperopic refractions.

Anterior Segment and Cataract

The external ocular exam and anterior segment are typically unremarkable in nonsyndromic forms of RP. However, there does appear to be a higher rate of keratoconus and glaucoma in patients with RP. Posterior subcapsular cataracts are common and often can become visually significant. Cataract extraction is beneficial in patients with RP if the cataract is thought to be the vision-limiting factor.

Fundus Findings

The characteristic signs on fundus examination include midperipheral atrophy of the pigment epithelium, intraretinal pigment accumulation (bone spicules; **Figure 1**), retinal vessel attenuation, and waxy pallor of the optic nerve. A yellowish ring of peripapillary atrophy is sometimes seen in patients with RP as well as with optic nerve head drusen. A minor number of vitreous cells are commonly observed in patients with RP. Cystoid macular edema is common in patients with RP and can often result in significantly decreased vision. Rarely, patients can develop a Coats-like retinopathy.

Diagnostic Tests for RP

Dark Adaptation

Dark adaptation can be a useful test in patients with RP. Patients who manifest with a rod–cone dystrophy will usually have a detectable increase in final dark-adapted thresholds and show delayed dark-adaptation curves. Prolonged dark adaptation is especially common among patients with *RHO* mutations. Elevations of the early cone segment of the dark-adaptation curve may be particularly noticed by patients, more so than elevations of the rod segment (**Figure 4**).

Visual Fields

Visual fields are not only useful for making the diagnosis of RP, but are also one of the most useful objective

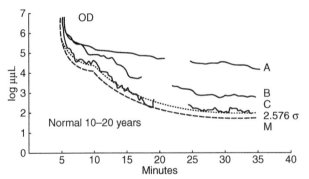

Figure 4 Example of dark-adaptation curves in a normal subject (dashed lines represent the mean normal response, dotted lines represent the upper limit of normal) and patients with retinitis pigmentosa (solid lines). From Weleber, R. and Evan, K. G. (2006). Retinitis pigmentosa and allied disorders. In: Ryan, S. J. (ed.) *The Retina*, 4th edn., vol. 1, chap. 17, pp. 395–498. Philadelphia, PA: Elsevier.

methods for monitoring progression of the disease. Decreased visual-field sensitivity results from photoreceptor loss (**Figure 5**).

The earliest change seen as measured by kinetic perimetry is concentric constriction or decreased sensitivity with static perimetry in diffuse disease and relative midperipheral scotomas seen in the in regional disease. As these midperipheral scotomas or regions of decreased sensitivity enlarge and deepen, severe tunnel vision results. Eventually, macular function fails and visual field becomes difficult or impossible to measure by conventional perimetry. Although visual function may be reduced to light perception only, it is rare for patients to become completely blind. With the exception of female carriers in X-linked RP, visual-field loss is usually symmetrical. Marked asymmetry should raise concern for diseases that mimic RP (**Figure 6**).

The rate of visual-field loss has been shown to be exponential. This rate is thought to be similar for the different forms of inheritance once correction has been made for the critical age of onset. Massof and Finkelstein found that patients lost about 50% of their visual field every 4.5 years. The superior visual field, which corresponds to the inferior retina, is often more affected than inferior visual fields. Based on this finding, it has been suggested that increased levels of light may play a role in accelerating retinal degeneration and this in turn may play a role in the forms of RP with greater damage in the inferior retina.

Electroretinograms

ERGs play a crucial role in the diagnosis of RP because these electrophysiological recordings are sensitive enough to detect decreased photoreceptor function early in the disease when fundus findings and visual fields may be minimally altered. In addition, ERGs are particularly

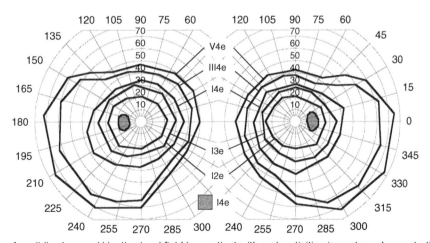

Figure 5 Example of a mildly abnormal kinetic visual field in a patient with early retinitis pigmentosa demonstrating the responses to different-sized targets. The gap between the size III4e and size I4e isopters is greater than normal, indicating loss of sensitivity in this region. The blind spot (region containing the optic nerve head and therefore no photoreceptor cells) is plotted in each eye just temporal to the fovea.

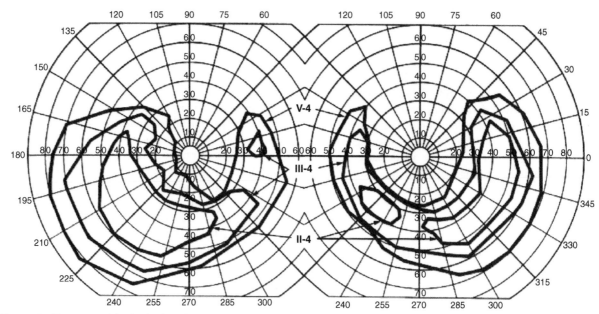

Figure 6 Kinetic visual fields obtained from patient with retinitis pigmentosa. Note the relative preservation of inferior fields, which correlated with preserved superior retina. From Weleber, R. and Evan, K. G. (2006). Retinitis pigmentosa and allied disorders. In: Ryan, S. J. (ed.) *The Retina*, 4th edn., vol. 1, chap. 17, pp. 395–498. Philadelphia, PA: Elsevier.

useful to assess visual function in preverbal infants and children. Almost all patients with symptomatic RP will have detectable changes on the ERG at the time of diagnosis. While the ERG is useful for the diagnosis of RP, visual fields are better for monitoring of the course of the disease. In severe cases of RP, such as LCA, the ERG may be not recordable.

Patients with RP can show decreased amplitude and timing of the major components of the ERG. Caution must be taken when interpreting decreases in the amplitude of an ERG because poor contact of the electrodes, deviations of the eye, and high myopia can affect the amplitude of the signal. When present, delayed timing tends to be a more robust indicator of dysfunction.

By analyzing the different components of the ERG, different forms of RP can be classified. Degeneration of the rod and cone photoreceptors leads to a decrease in the amplitude of different waveforms of the ERG and can also increase the timing or latency of the peaks of these waveforms. The most common forms of RP manifest as a rod–cone dystrophy and the first detectable changes will be apparent on the scotopic ERG. Decreases in the b-wave amplitude and timing of the peak of the b-wave are indicative of early rod photoreceptor death. Further loss of rod cells leads to further decreases in the b-wave amplitude and decreased amplitude of the a-wave responses at higher intensities. Patients with a cone–rod dystrophy have normal, or lesser defect of b-wave responses to dim scotopic stimuli, but typically have more markedly abnormal ERGs to 30-Hz flicker or single-flash stimuli measured under photopic conditions (**Figures 7–9**).

Fundus Photography/Fluorescein Angiography

Documentation by fundus photography can assist in monitoring changes in patients with RP. Fluorescein angiography in patients with RP will demonstrate hyperfluorescence in areas of RPE atrophy and can highlight areas of cystoid macular edema. However, fluorescein angiography has largely been supplanted by optical coherence tomography (OCT) for detecting cystoid maculopathy. In addition, concerns about light exposure accelerating certain forms of RP in animal models have prompted many ophthalmologists to exercise caution in obtaining excessive photographs.

Optical Coherence Tomography

OCT provides a noninvasive cross-sectional image of the retina. It is very useful in patients with RP when there is a question of cystoid macular edema. The ability to detect cystoid macular edema by OCT often obviates the need to get a fluorescein angiogram.

Differential Diagnosis

It is important to realize that RP is not the only cause of a pigmentary retinopathy but many other diseases can mimic RP. Significant asymmetry or the onset of symptoms in an elderly patient should raise suspicion for one of the diseases that mimics RP.

Trauma to the eye can disrupt the retina and result in pigment migration of the RPE into the retina with the formation of bone spicules. By a similar mechanism,

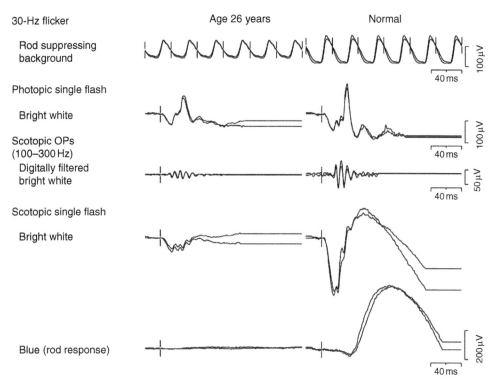

Figure 7 ERGs recorded from a patient with autosomal recessive RP (left column) compared to a control patient (right column). This patient is demonstrative of a rod–cone dystrophy. There is a flat response to the dim blue flash under scotopic conditions, which specifically stimulates rods. The bright flash under scotopic conditions normally elicits mixed responses from both rods and cones. In this case, the response is severely attenuated and the small amount of signal is likely coming from the cone system. Under light-adapted conditions (photopic single flash and 30-Hz flicker), which selectively stimulate the cones, the response is only slightly decreased consistent with the categorization of a rod–cone dystrophy.

ophthalmic artery occlusions and old retinal detachments can present with a pigmentary retinopathy.

Additionally, infections caused by syphilis, toxoplasmosis, and herpes viruses can lead to a pigmentary retinopathy. Congenital rubella infection can often be misdiagnosed as RP or Usher syndrome because these patients present with deafness and a fine, speckled pigmentary retinopathy. The key to differentiating patients with rubella from those with RP is that patients with rubella retinopathy will have normal or near normal responses by ERG (**Figure 10**).

Diffuse unilateral neuroretinitis (DUSN) is caused by a chronic infection with a nematode. In the early stages, this disease can be distinguished due to the appearance of crops of yellowish, deep choroidal infiltrates, neuroretinitis, and sometimes, visualization of the worm itself. However, late in the disease, with the exception of being unilateral, the fundus appearance is identical to RP with the fundus showing bone spicules, vascular attenuation, and optic atrophy (**Figure 11**).

Inflammatory diseases that cause a posterior uveitis can cause chronic changes that mimic the pigmentary changes of RP. Some examples include sarcoidosis, birdshot choroidoretinopathy, serpiginous retinopathy, Behcet disease, and acute zonal occult outer retinopathy (AZOOR).

Certain drugs can cause pigmentary retinopathies and their usage must be excluded prior to making a diagnosis of RP. One example is thioridazine (Mellaril), an antipsychotic drug, which mimics RP by causing decreased night vision, RPE atrophy, and a pigmentary retinopathy. Hydroxychloroquine (Plaquenil) is used to treat systemic lupus erythematous and rheumatoid arthritis and when taken for extended period of time or at higher doses can lead to central vision loss. Other drugs that have been found to cause pigmentary retinopathies include chlorpromazine, chloroquine, and quinine.

Autoimmune retinopathy is an incompletely understood disease resulting from antibodies to retinal antigens and can present with many of the same features of RP, such as decreased vision, visual-field loss, and decreased ERGs. Unlike the other diseases that mimic RP, autoimmune retinopathy does not present with a pigmentary retinopathy and, in many cases, the fundus appearance can be normal or only show vascular attenuation. A subset of cases of autoimmune retinopathy is associated with carcinomas in other parts of the body. Two examples of this entity are cancer-associated retinopathy (CAR), which often arises from small cell carcinoma of the lung and melanoma-associated retinopathy (MAR). CAR patients may test positive for antibodies directed against retinal

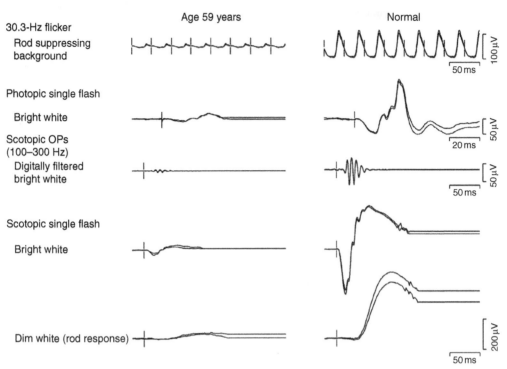

Figure 8 ERGs recorded from a patient with peripherin/*RDS* null mutation (left column) compared to a control patient (right column). This patient is demonstrative of a rod–cone dystrophy where the rods and cones are equally affected. It is of importance that peripherin is expressed in both rods and cones. There is a severely diminished response to the dim white flash under scotopic conditions, which specifically stimulates rods. The bright flash under scotopic conditions normally elicits a mixed response from both rods and cones. In this case, the response is severely attenuated. Under light-adapted conditions (photopic single flash and 30-Hz flicker), which selectively stimulate the cones, the response is also severely decreased consistent with the categorization of an equal rod–cone dystrophy.

antigens, such as anti-recoverin or anti-enolase, while MAR patients may test positive for antibodies directed against bipolar cells.

Prognosis in RP

It is uncommon for a patient with RP to lose all light perception. A study by Grover et al. in 1999 showed that only 0.5% of patients over the age of 45 had no light perception and 50% retained vision better than 20/40. In the typical rod–cone dystrophies, visual loss usually starts in the midperiphery with central visual acuity being spared for many years. Most patients will eventually qualify as being legally blind, often secondary to decreased visual fields prior to being disqualified on the account of decreased central visual acuity. RP is a slowly progressive disease and many patients will eventually experience decreased central acuity, most often from decrease in cone photoreceptor density, macular edema, epiretinal membranes, or retinal pigment defects. However, when there is an unexpected decrease of central acuity, the development of cataracts or cystoid macular edema should be suspected because these sequelae are more amenable to treatment.

Current Treatments

Currently, there is no known cure for most forms of RP, although some treatments have been shown to slow down the progression of the disease and future therapies are promising.

Treatable Forms of RP

A few rare forms of RP are amenable to specific treatments. It is important to rule out these treatable forms of RP because prompt therapy can prevent further damage. The treatable forms of RP include abetalipoproteinemia and adult Refsum disease.

Resources/Support for Patients with RP

The diagnosis of RP can be both frightening and confusing to patients. In addition to having their questions answered by the physician, patients can benefit by meeting with a genetic counselor who can take a detailed family history and answer questions about heritability. Patients should be referred to the many organizations or websites dedicated to providing support for this disease, such as the Foundation Fighting Blindness.

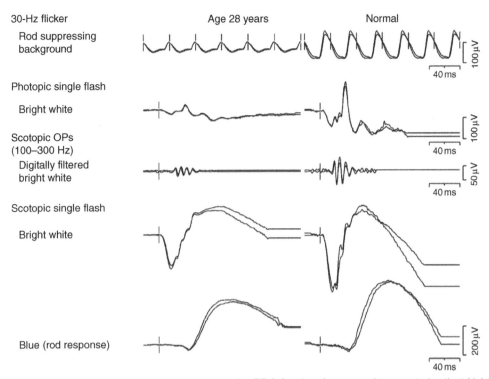

Figure 9 ERGs recorded from a patient with autosomal recessive RP (left column) compared to a control patient (right column). This patient is demonstrative of a cone–rod form of RP. There is a mildly diminished response to the dim blue flash under scotopic conditions, which specifically stimulates rods. The bright flash under scotopic conditions normally elicits a mixed response from both rods and cones. In this case, the response is only moderately attenuated. Under light-adapted conditions (photopic single flash and 30-Hz flicker), which selectively stimulate the cones, the response is severely decreased consistent with the categorization of a cone–rod dystrophy.

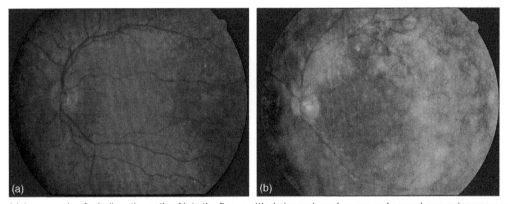

Figure 10 (a) An example of rubella retinopathy. Note the fine, mottled pigmentary changes and normal-appearing nerve and vessels. Unlike RP, ERG testing in this patient would be expected to be normal. (b) Example of a pigmentary retinopathy caused by syphilis.

Optimizing Remaining Vision

It is important that patients with RP have an up-to-date refraction to optimize remaining vision. Patients with significantly decreased visual acuity greatly benefit from a referral to low vision services. A variety of different magnifying devices are available to assist patients with RP. For example, night-vision devices can assist with navigation in dim conditions, although a bright, wide, beam flashlight may be a more cost-effective solution. Various

magnifiers and closed-circuit televisions (CCTVs) are available to enhance reading vision.

Cataractogenesis is frequent in patients with RP. Functional visual improvement can be achieved by cataract surgery in carefully selected individuals with suitable remaining retinal function. The risks of surgery are increased in these individuals who should be counseled specifically regarding the risks of postoperative cystoid macular edema.

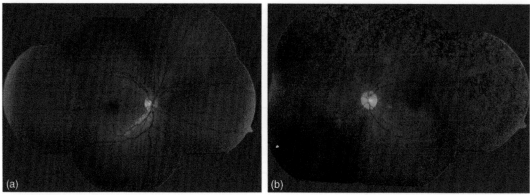

Figure 11 (a) Normal fundus from a patient with diffuse unilateral subacute neuroretinitis (DUSN). (b) The affected eye in the same patient. Note that appearance is identical to RP, demonstrating waxy pallor of the nerve, vascular attenuation, bone-spicule pigmentation, and RPE atrophy.

Cystoid macular edema in RP has been treated with carbonic anhydrase inhibitors such as oral acetazol-amide (Diamox) or topical dorzolamide (Trusopt). These medicines can improve vision in some RP patients with cystoid macular edema; however, their efficacy can decrease with time and many patients cannot tolerate the side effects induced by these medicines.

Vitamin A

Observational studies of patients with RP taking vitamin A and vitamin E supplementation demonstrated a slower decline in cone ERGs than expected and led to a rando-mized, controlled, double-masked trial, to assess if these supplements could slow down retinal degeneration. Addi-tional subgroup analyses in this study suggested that oral supplementation with 15 000 IU of vitamin A modestly slowed down the loss of ERG amplitude over a 5-year period in certain individuals with RP. Currently, vitamin A supplementation is frequently recommended but its use is not universal. High doses of vitamin A supplementation have also been associated with elevated liver enzymes, elevated triglycerides, and an increased risk of osteoporo-sis. It seems prudent to check annual liver function tests and triglyceride levels for all patients and bone density scans in older patients taking vitamin A supplementation. Vitamin A should be avoided in children, pregnant women, and those with decreased liver function.

Additionally, from a mechanistic disease perspective, vitamin A should be avoided in forms of RP caused by mutations in the gene *ABCA4* due to evidence in animal models of accelerated retinal degeneration. *ABCA4* codes for the ABCR protein, which is important for transport of vitamin A-derived all-trans-retinal from the disk to the photoreceptor cytoplasm. Mutations in this gene are responsible for Stargardt macular dystrophy and rarely can also cause autosomal recessive RP.

Docosahexanoic Acid

Docosahexanoic acid (DHA) is an important omega-3 fatty acid that comprises 30–40% of fatty acids in the retina. The exact role of DHA is not known, but it has been proposed to play a role maintaining membrane fluidity, mediating 11-*cis* retinal transport, and acting as a precursor for neuroprotec-tive factors. Studies in patients with X-linked RP suggested that decreased levels of DHA correlate with decreased ERG responses and have prompted studies to evaluate if supplementation with DHA might slow down retinal degeneration. Two prospective, randomized, double-masked studies (one in patients with X-linked, the other in patients with all forms of RP) failed to show a significant benefit of DHA. However, considering the low risk of adverse effects, many centers do recommend DHA supplementa-tion to patients with RP.

Neuroprotection/CNTF

A relatively new strategy for the treatment of RP is to prevent photoreceptor loss by the delivery of neuroprotec-tive factors. Numerous studies in animal models have docu-mented the successful rescue of photoreceptor degeneration by neurotrophic factors. A delivery system for one of these factors, human ciliary neurotrophic factor (CNTF), has been developed using encapsulated cell technology. These devices use RPE cells that have been transfected to express CNTF and are enclosed by a semipermeable membrane which allows nutrients to diffuse in, but prevents immune attack on the cells. Phase I studies, implanting this device in patients with RP, have been completed without any major adverse events. Phase II studies, which will be able to better assess any visual improvement, are underway.

Gene Therapy

Replacement of defective genes in autosomal recessive forms of RP holds much promise. Currently, three groups have used an adeno-associated viral vector to deliver normal copies of the all-trans-retinol isomerase (*RPE65*) in patients with LCA. The treatment has thus far been well tolerated and some patients have demonstrated improvement in subjective vision and to some more objective tests such as pupillary reflexes. Gene therapy holds much promise for treating RP, but several challenges remain. RP and its allied disorders are caused by mutations in over 100 genes. Designing a multitude of vectors for each of these genes poses an arduous challenge.

Autologous RPE Transplantation

Replacement of RPE cells has been attempted using suspensions or sheets of cultured RPE cells or autologous grafts. Engraftment of these injected cells has been demonstrated in animal models. Rescue of photoreceptor degeneration has been achieved suggesting that transplanted RPE cells may modulate photoreceptor death. However, in spite of early photoreceptor rescue in these animal models, long-term restoration of vision has been disappointing.

Stem-Cell-Based Therapies

Cell-based therapy using stem cells is currently being intensely explored for rescuing vision. There are two fundamentally different strategies: one is to limit the progress of photoreceptor loss by introducing cells before such loss has progressed too far; the other is to replace lost photoreceptors. Stem cells are multipotent cells capable of self-renewal and have the potential to develop into many specific cell types. Their capacity for proliferative expansion to a large scale and their ability to produce a number of growth factors make them attractive candidates to be used to replace or repair damaged cells in adult organisms.

Microelectrode Implants

One novel concept for treatment of RP is to bypass the loss of the photoreceptors by electronically stimulating the retina, optic nerve, or visual cortex using microelectrode implants.

Expression of Photosensitive Proteins

A very recent approach for treating RP has been the strategy to bypass photoreceptor loss by using viral vectors to express light-sensitive proteins, such as channel rhodopsin, into postreceptor ganglion cells. Early experiments in small animals have demonstrated successful responses from ganglion cells using this strategy. Much work remains to be done, including how to obtain high enough expression to provide adequate sensitivity for useful vision.

Conclusions

RP is a significant cause of vision loss in adults and children. Diagnosis is best made by careful history and clinical examination combined with retinal electrophysiology and psychophysical testing. For some individuals, genetic testing can identify causative mutations. While there is currently no cure for RP, many treatment options are emerging and future therapies are promising.

See also: Adaptive Optics; Injury and Repair: Prostheses; Primary Photoreceptor Degenerations: Terminology.

Further Reading

Berson, E. L., Rosner, B., Sandberg, M. A., et al. (1993). A randomized trial of vitamin A and vitamin E supplementation for retinitis pigmentosa. *Archives of Ophthalmology* 111: 761–772.

Berson, E. L., Rosner, B., Sandberg, M. A., et al. (2004). Clinical trial of docosahexaenoic acid in patients with retinitis pigmentosa receiving vitamin A treatment. *Archives of Ophthalmology* 122: 1297–1305.

Fishman, G. A., Farber, M. D., and Derlacki, D. J. (1988). X-linked retinitis pigmentosa. Profile of clinical findings. *Archives of Ophthalmology* 106: 369–375.

Grant, C. A. and Berson, E. L. (2001). Treatable forms of retinitis pigmentosa associated with systemic neurological disorders. *International Ophthalmology Clinics* 41: 103–110.

Grover, S., Fishman, G. A., Anderson, R. J., et al. (1999). Visual acuity impairment in patients with retinitis pigmentosa at age 45 years or older. *Ophthalmology* 106: 1780–1785.

Hamel, C. P. (2007). Cone rod dystrophies. *Orphanet Journal of Rare Diseases* 2: 7.

Hartong, D. T., Berson, E. L., and Dryja, T. P. (2006). Retinitis pigmentosa. *Lancet* 368: 1795–1809.

Heckenlively, J. R. (1988). *Retinitis Pigmentosa.* Philadelphia, PA: Lippincott.

Hoffman, D. R., Locke, K. G., Wheaton, D. H., et al. (2004). A randomized, placebo-controlled clinical trial of docosahexaenoic acid supplementation for X-linked retinitis pigmentosa. *American Journal of Ophthalmology* 137: 704–718.

Radu, R. A., Yuan, Q., Hu, J., et al. (2008). Accelerated accumulation of lipofuscin pigments in the RPE of a mouse model for ABCA4-mediated retinal dystrophies following vitamin A supplementation. *Investigative Ophthalmology and Visual Science* 49: 3821–3829.

Sieving, P. A., Caruso, R. C., Tao, W., et al. (2006). Ciliary neurotrophic factor (CNTF) for human retinal degeneration: Phase I trial of CNTF delivered by encapsulated cell intraocular implants. *Proceedings of the National Academy of Sciences of the United States of America* 103: 3896–3901.

Weleber, R. G. and Gregory-Evans, K. (2006). Retinitis pigmentosa and allied disorders. In: Ryan, S. J. (ed.) *The Retina* 395–498. Philadelphia, PA: Elsevier.

Relevant Websites

http://www.ncbi.nlm.nih.gov – Online Mendelian Inheritance in Man (OMIM).

http://www.sph.uth.tmc.edu – Retinal Information Network (Retnet).

Primary Photoreceptor Degenerations: Terminology

M E Pennesi, P J Francis, and **R G Weleber,** Oregon Health and Sciences University, Portland, OR, USA

Glossary

Allied disorders – Retinitis pigmentosa (RP) is often grouped with a class of more stable, inherited retinal disorders collectively referred to as RP and allied disorders. Some of these allied disorders cause similar clinic findings as RP, e.g., nyctalopia (night blindness), but usually do not show progression and deterioration with time. An example is congenital stationary night blindness (CSNB), which can present with nyctalopia and decreased rod and cone function on the electroretinogram (ERG). Unlike RP, most patients with CSNB have stable visual function. X-linked CSNB is caused by mutations in nyctalopin (NYX) and L-type voltage dependent calcium channel (CACNA1F). Although the majority of mutations of rhodopsin causes typical RP, rare mutations, such as G90D in rhodopsin, produce night blindness with such mild progression late in life that they have been called stationary night blindness. Another allied disorder is achromatopsia, which is caused by mutations in cyclic nucleotide-gated channel subunits (CNGA2, CNGB3) or guanine nucleotide alpha-binding protein 2 (GNAT2). Achromatopsia is associated with severely decreased central and color vision, photophobia, and nystagmus. These symptoms are similar to those that can be seen with some cone–rod dystrophies. Indeed, later in life some modest foveal atrophy can occur and cases of progressive cone–rod dystrophy have been associated with mutations of some of the achromatopsia genes. However, unlike cone–rod dystrophies, which invariably progress, achromatopsia is, in the vast majority of cases, stationary.

Cone dystrophy – Cone photoreceptors are affected and rod photoreceptors are minimally affected or spared in cone dystrophy. Many cases of early cone dystrophies with time will develop significant rod abnormalities.

Cone–rod dystrophy – Cone–rod dystrophy, as a group, involves both photoreceptors with cones affected more than rods. Certain forms of RP present with greater cone than rod involvement on ERG and these patients have been termed to have cone–rod RP. However, in cone–rod dystrophies as a group the primary defect lies in cones and secondaryrod loss occurs with time. Most investigators consider primary cone–rod dystrophy separate from RP.

Extrinsic factor – An agent external to the organism that contributes to or is causative of a disease state. This can include drugs, foods, normal nutrients (excess or deficiency), toxins, inhaled chemicals, infectious agents, and exposures to radiation such as light, sound, and high-energy particles.

Intrinsic factor – An agent that is inherent to the organism that contributes to or is causative of a disease state.

Mixed intrinsic and extrinsic etiology for a secondary photoreceptor degeneration – This occurs when a person has a genetic variant that creates a toxic metabolite in the presence of an extrinsic molecule that would normally not be encountered.

Mixed model of primary and secondary photoreceptor degeneration – This is considered when a genetic alteration within the photoreceptors is insufficient to cause photoreceptor degeneration by itself, but predisposes to degeneration in the presence of an extrinsic or intrinsic agent. A second mode of combined primary and secondary photoreceptor degeneration is when one group of photoreceptors, such as the rod photoreceptors, undergoes a primary degenerative process that is due to a mutation in a gene that is expressed in those photoreceptors and precipitates apoptosis, which leads to a secondary degenerative process, in this example cones, due to alterations in the cellular environment induced by death of neighboring cells.

Primary retinal degeneration – This occurs when cells in the retina, usually photoreceptors, die secondary to a process that originates within the retina itself. An example of a primary retinal degeneration is RP, which is caused by mutations in genes that encode proteins important for retinal function. A disease can be classified as a primary retinal degeneration if the genetic defect is such that correction of expression of the normal gene product in the photoreceptors is required to correct the abnormality and arrest the degeneration.

Primary retinal degeneration with secondary photoreceptor degeneration – This occurs when photoreceptor degeneration is the result of mutation(s) of a gene that exists in other retinal cells, for example, retinal pigment epithelial (RPE) cells. Correction of the genetic defect would require modification of the effects of those other retinal cells (e.g, RPE cells).

Retinal atrophy – A broad term encompassing not only processes that occur with retinal degenerations, but also abnormal retinal tissue or cellular loss due to developmental defects and malnutrition.

Retinal degeneration – A process whereby cells in the retina undergo cell death by apoptosis. Most retinal degenerations affect both rod and cone photoreceptors, but some disorders reflect damage that occurs principally in other cell types, e.g., the RPE in Stargardt's disease and other ABCA4-related retinopathies. Secondary degeneration of the RPE is also common. Transsynaptic degeneration of higher-order cells, bipolar and ganglion cells, can also occur. The general term retinal degeneration should be distinguished from the more specific term, photoreceptor degeneration.

Retinal dystrophy – A broad term that not only encompasses retinal degenerations, but also includes abnormal retinal function due to developmental defects and malnutrition.

Retinitis pigmentosa (RP) – A heterogeneous group of diseases that result in degeneration of the rod and cone photoreceptors and secondarily the RPE. This degeneration usually leads to a loss of night vision due to the early degeneration of rods, constricted visual fields, decreased responses on ERG, and ultimately a decrease in visual acuity once macular cones begin to degenerate. Typical fundus findings include midperipheral atrophy of the pigment epithelium, bone spicule pigments, retinal vessel attenuation, and waxy pallor of the optic nerve. The term RP usually refers to only rod–cone dystrophies; however, cone–rod dystrophies and cone dystrophies are sometimes grouped under this term.

Rod–cone dystrophy – A retinal dystrophy in which the rod photoreceptors are affected more than the cones. Most forms of RP manifest as rod–cone dystrophies.

Secondary photoreceptor degeneration of the extrinsic type – A secondary photoreceptor degeneration of the extrinsic type exists if, despite the underlying molecular defect, one could avoid the photoreceptor degeneration by preventing an individual's exposure to an extrinsic agent or condition (e.g., toxin, drug, infectious agent, light, and trauma).

Secondary photoreceptor degeneration of the intrinsic type – If one can prevent photoreceptor degeneration by correcting or reversing a systemic or ocular metabolic or immune process, then it is a secondary photoreceptor degeneration of the intrinsic type.

Histological and Fundus Features of Retinitis Pigmentosa

The fundus exam in retinitis pigmentosa (RP) reveals mid-peripheral atrophy of the pigment epithelium, bone spicule pigments greater in the periphery than centrally, retinal vessel attenuation, and waxy pallor of the optic nerve.

Bone Spicule Pigmentation

Bone spicules are intraretinal accumulations of melanin pigment that result from the migration of retinal pigmented epithelial (RPE) cells into the retina after photoreceptor death. These spicules are commonly found in a perivascular pattern and may encircle and occlude these vessels, and are a typical feature of RP. However, there are cases of RP that do not present with bone spicules as well as other disease processes, such as trauma and infection, can result in an appearance that mimics RP by presenting with bone spicules (**Figure 1**).

Waxy Pallor of the Optic Nerve

Much as the name implies, waxy pallor of the optic nerve refers to the funduscopic appearance of the optic nerve seen in many patients with RP. When retinal photoreceptors die, Müller cells and astrocytes in the retina undergo gliosis to form scar tissues. It is thought that this process may lead to the waxy pallor of the optic nerve (**Figure 2**).

Peripapillary/Optic Nerve Head Drusen

Peripapillary drusen/optic nerve drusen are found more commonly in patients with RP. Peripapilliary drusen are histologically different from the drusen found in macular degeneration and are found near or within the optic nerve head. They are thought to result from accumulations of materials in the axons of ganglion cells by axoplasmic stasis and can become calcified with time. Such drusen can cause isolated and asymmetrical visual-field defects, which can be slowly progressive (**Figure 3**).

Bull's Eye Maculopathy

A bull's eye maculopathy results from photoreceptor loss and retinal thinning in a parafoveal distribution. It is often apparent on color fundus photos, but can be visualized on

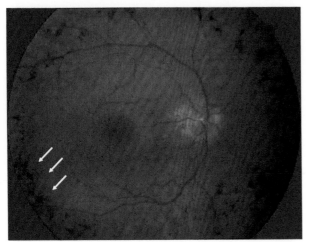

Figure 1 A fundus photo of a patient with retinitis pigmentosa demonstrating intraretinal pigment accumulations (white arrows), also known as bone spicules, which are a common finding in retinitis pigmentosa. They are caused by the migration of retinal pigment epithelial cells into the retina after photoreceptor degeneration. Modified from Weleber, R. and Evan, K. G. (2006). Retinitis pigmentosa and allied disorders. In: Ryan, S. J. (ed.) *The Retina*, 4th edn., vol. 1, chap. 17, pp. 395–498. Philadelphia, PA: Elsevier.

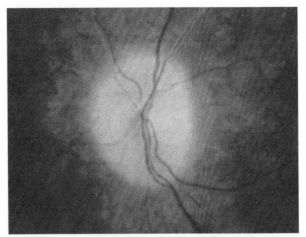

Figure 2 An example of the waxy nerve pallor seen in patients with RP. Also note the vascular attenuation and sheathing.

fluorescein angiography as well. Bull's eye maculopathies are most commonly not only seen in cone–rod dystrophies, but can also be seen in other diseases such as Stargardt's macular dystrophy, Batten's disease, and with hydroxychloroquine (Plaquenil) toxicity (**Figure 4**).

Coats-Like Response

Coats disease is characterized by peripheral retinal telangiectasias, which are dilations of retinal blood vessels, and is usually seen in young males. Rarely, patients with RP can develop localized areas of vascular telangiectasias with exudation similar to those seen in Coats disease. Extravasation of fluid from these vessels can lead to an exudative retinal detachment. This process is termed a Coats-like response (**Figure 5**).

Classification of RP by Fundus Pattern
Classic Pattern for RP

The classic pattern on fundus exam in RP reveals midperipheral atrophy of the pigment epithelium, bone spicule pigments in the periphery greater than centrally, retinal vessel attenuation, and waxy pallor of the optic nerve (**Figure 6**).

Inverse RP

The term inverse RP refers to a funduscopic pattern where the central retina exhibits more pigmentary changes and RPE atrophy than the periphery. This term is falling out of favor, as it is now understood that this pattern is often seen with advanced cone and cone–rod dystrophies, as well as some instances of secondary pigmentary retinopathies.

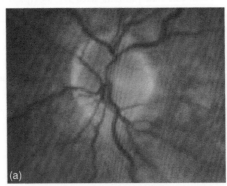
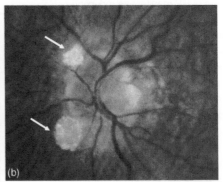

Figure 3 (a) Patient with retinitis pigmentosa – note the lack of a normal physiological cup, which raises the concern for buried drusen. (b) The same patient seen years later who demonstrates peripapillary drusen (arrows).

Concentric RP

Concentric RP is a defined by a subgroup of patients who show a consistent pattern of centripetal vision loss from the far periphery toward the center. This distinguishes it from the classic pattern of RP, which starts in the mid-periphery and progresses both outward and inward. Histological studies have shown an abrupt transition between diseased and normal areas of retina.

Sector RP

Sector RP is a term that refers to the fundus appearance where abnormal pigmentation and atrophy is confined to only one area of the retina, usually an arc inferior to the macula. Debate exists whether this term is specific for certain gene defects or is a stage in evolution of what will become a more generalized process with time. Autosomal recessive or autosomal dominant regional disease can be characterized by such a sectorial pattern at least in

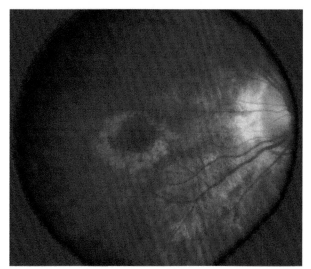

Figure 4 A fundus photo demonstrating a Bull's eye maculopathy (dark surrounded by light area).

early disease. The pattern is usually symmetrical between the two eyes. Mutations in rhodopsin (*RHO*) have been associated with sector RP. These patients are often less symptomatic and their electroretinograms (ERGs) are much less reduced than patients with diffuse or more widespread RP. Early generalized RP can mimic sector RP and, therefore, the diagnosis of sector RP must remain provisional for at least 10 years (**Figure 7**).

RP Sine Pigmento

The term RP sine pigmento is applied to the early stages of RP where the classical findings of visual-field loss, decreased ERGs, and vascular attenuation may be present, but there is a lack of bone spicules on the fundus exam. This is thought to represent an early manifestation of the disease and most patients will eventually develop bone spicules, although often not until later stages. Another term in use is pauci-pigmentary retinopathy. In the case of patients who develop symptoms later in life and have a normal, or minimally abnormal, fundus appearance, a work-up for autoimmune retinopathy should be considered.

Tapetal-Like Reflex/Sheen

Tapetal-like sheen is a yellowish-white metallic reflex that can be seen in young males with X-linked RP, women who are carriers of this disease, and patients with cone–rod dystrophy. As the disease advances and RPE atrophy progresses, this reflex can fade. This finding is not pathognomonic of RP because it can be seen in other allied disorders (**Figure 8**).

RP with Preserved Peri-Arteriolar RPE

RP with preserved peri-arteriolar RPE (PPARPE) has been found in families with mutations in the crumbs

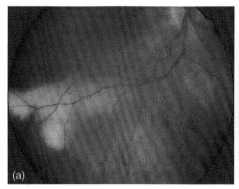
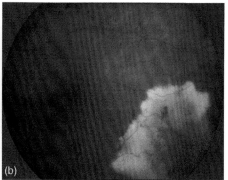

Figure 5 Both (a) and (b) show fundus photos from a patient with RP and a Coat's-like response. Note the yellowish subretinal exudates.

homolog 1 gene (*CRB1*). These patients present with the typical findings of RP, but for reasons not understood, have preservation of the RPE near arterioles.

Pigmented Paravenous Retinochoroidal Atrophy

In pigmented paravenous retinochoroidal atrophy (PPRCA), degeneration and accumulation of bone spicules are limited to the areas around the retinal veins. These patients are often asymptomatic, but with careful testing, scotomas corresponding to the areas of degeneration can be identified. Typically, the ERG is minimally reduced and disease is thought to be, at most, slowly progressive. Although PPRCA has been thought not to be inherited, mutations in the *CRB1* gene have been found in families demonstrating dominant inheritance (**Figure 9**).

Figure 6 Fundus photo of a typical patient with RP. Note the pallor of the optic nerve, the vascular attenuation, atrophy of the pigment epithelium, and bone-spicules. Reproduced from Weleber, R. G., Butler, N. S., Murphey, W. H., Sheffield, V. C., and Stone, E. M. (1997). X-linked retinitis pigmentosa associated with a two base-pair insertion in codon 99 of the RP3 gene RPGR. *Archives of Ophthalmology* 115: 1429–1435.

Fundus Albipunctata

Fundus albipunctata is an autosomal recessive disease that does not result in a retinal degeneration like RP, but can cause a congenital stationary night blindness. It is more accurately classified as an allied disorder. It is caused by mutations in the retinol dehydrogenase 5 gene (*RDH5*), which results in problems regenerating visual pigment. As a result, these patients have severely decreased rod recovery with prolonged dark adaptation. The fundus exam shows many discrete yellowish-white dots at the level of the RPE. Atrophic lesions in the macula can occur in many patients in later years (**Figure 10**).

Retinitis Punctata Albescens

Retinitis punctata albescens is an autosomal recessive disease caused by mutations in the retinaldehyde-binding protein 1 gene (*RLBP1*). Similar to fundus albipunctata, this disease presents with severe night blindness and small, discrete, yellowish-white lesions in the fundus, but can have a pigmentary retinopathy as well. Later stages of RPA may develop diffuse disease similar to advanced RP.

Gyrate Atrophy

Gyrate atrophy is an autosomal recessive form of diffuse choroidal atrophy caused by mutations of the gene (*OAT*) for ornithine-∂-aminotransferase (OAT). The deficiency of this enzyme results in elevated plasma and tissue levels of ornithine, which exert a cytotoxic effect on the RPE, possibly by endpoint inhibition of a common intermediate for proline synthesis, L-Δ^1-pyrroline-5-carboxylic acid (P5C), which is normally formed from ornithine by OAT and from glutamic acid by P5C synthase. The early stage of gyrate atrophy is associated with sharply demarcated areas of peripheral chorioretinal atrophy. Later stages develop more diffuse and generalized total vascular choroidal atrophy. Dietary restriction of arginine, the precursor for ornithine, can be beneficial. Additionally, a rare subset of patients has been shown to respond with lowered

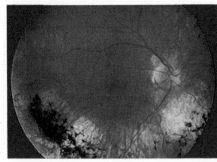
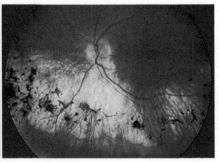

Figure 7 An example of sector RP. From Weleber, R. and Evan, K. G. (2006). Retinitis pigmentosa and allied disorders. In: Ryan, S. J. (ed.) *The Retina*, 4th edn., vol. 1, chap. 17, pp. 395–498. Philadelphia, PA: Elsevier.

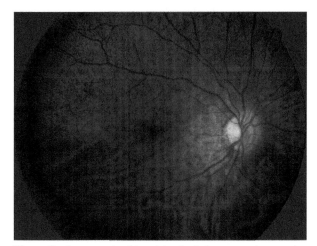

Figure 8 An example of tapetal-retinal sheen seen in a carrier of X-linked retinitis pigmentosa.

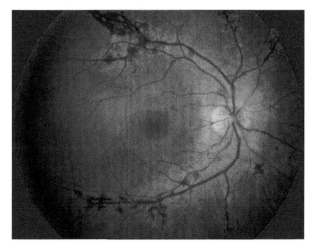

Figure 9 Example of pigmented paravenous retinal choroidal atrophy.

ornithine levels to treatment with vitamin B6 (pyridoxine HCL), which acts as a co-factor for the defective enzyme. Disorders such as gyrate atrophy blur the distinction between primary and secondary retinal degenerations leading to the realization that it is difficult to draw the line between RP and allied diseases (**Figure 11**).

Choroideremia

Choroideremia is an X-linked disease, which is caused by a mutation in the *CHM* gene and leads to progressive degeneration of the retina, RPE, and choroid. The *CHM* gene encodes the homolog of the Rab escort protein 1 (REP1) which is thought to be important in the function of a Rab geranylgeranyl transferase. Choroideremia can often be mistaken for X-linked RP, as the two diseases can share several features including: nyctalopia, retinal RPE atrophy, pigmentary changes, and decreased ERGs and X-linked inheritance. Unlike the waxy nerve pallor seen in RP, patients with choroideremia will often have a normal-appearing nerve and relative preservation of the macula and peripapillary retina (**Figure 12**).

Syndromic Forms of RP

Abetalipoproteinemia

Also known as Bassen–Kornzweig syndrome, abetalipoproteinemia results from a deficiency of beta lipoproteins, which are necessary for normal absorption of fat-soluble vitamins from the gut, leading to poor absorption of vitamins A, D, E, and K. The syndrome is characterized by low levels of fat-soluble vitamins, ataxia, acanthocytosis,

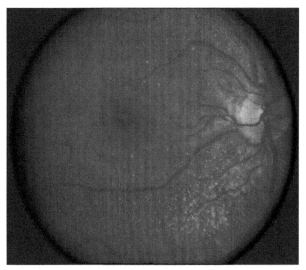

Figure 10 A fundus photograph of fundus albipunctata. Note the characteristic punctate white dots.

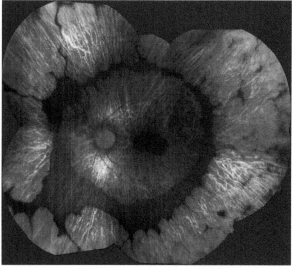

Figure 11 Gyrate atrophy. Note the scalloped peripheral areas of chorioretinal degeneration as well as central atrophy.

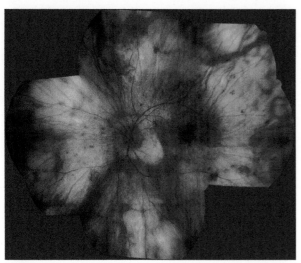

Figure 12 An example of choroideremia demonstrating large areas of chorioretinal atrophy with some macular sparing.

and RP. Treatment with combined vitamin A and vitamin E has been shown to prevent or slow retinal degeneration and, in rare patients, reverse the dark adaptation and ERG defects.

Alström Syndrome

Alström syndrome is an autosomal recessive disease characterized by deafness, obesity, diabetes, cardiomyopathy, and RP. This syndrome shares many overlapping features with Bardet–Biedl syndrome (BBS) and, indeed, much like the genes implicated in BBS, the gene involved in this syndrome, Alström syndrome 1 (*ALMS1*), is thought to play an important role in the structure and function of the cilium.

Bardet–Biedl Syndrome

The details on Bardet–Biedl syndrome are discussed elsewhere in this encyclopedia.

Chronic Progressive External Ophthalmoplegia/ Kearns–Sayre Syndrome

Chronic progressive external ophthalmoplegia (CPEO) and Kearns–Sayre syndrome are mitochondrial myopathies that cause progressive muscle paralysis and pigmentary retinal degeneration. Kearns–Sayre syndrome is associated with sudden cardiac death.

Friedreich's Ataxia

This is an autosomal recessive neurodegenerative disease caused by a trinucleotide repeat expansion in an intron of the frataxin gene (*FXN*), leading to silencing of the gene's

protein product, frataxin. The resulting disease causes muscle weakness, ataxia, cardiac hypertrophy, deafness, and retinal degeneration.

Vitamin E Deficiency

Mutations in the alpha-tocopherol transferase protein gene lead to vitamin E deficiency and cause a Friedreich-like ataxia associated with RP. Treatment with oral vitamin E has been shown to halt both the neurological and visual manifestations of this disease.

Incontinentia Pigmenti (Bloch–Schulzberg Syndrome)

Incontinentia pigmenti (Bloch–Schulzberg syndrome) is an X-linked dominant disorder caused by mutations in the nuclear factor kappa-light-chain-enhancer of activated B cells (NFκB) essential modulator gene, *NEMO*, and is usually lethal in males. Affected females demonstrate abnormal teeth and nails, hyperpigmentation of the skin, and central nervous system defects. Retinal findings in these patients include peripheral telangiectasias, hypopigmentation of the fundus, and a pigmentary retinopathy.

Joubert Syndrome

Also known as cerebellooculorenal syndrome, Joubert syndrome is an autosomal recessive disease caused by mutations in 11 different genes. One causative gene, the centrosomal protein 290 (*CEP290*), has also been associated with Senior–Loken syndrome and nonsyndromic Leber congenital amaurosis. Patients with Joubert syndrome have hypoplasia of the cerebellar vermis, and also renal problems and retinal dystrophy. The hypoplasia of the cerebellar peduncles of the midbrain creates a characteristic radiological finding on CT scans termed the molar tooth sign.

Mucopolysaccharide Disorders

Mucopolysaccharide (MPS) disorders can also present with a pigmented retinopathy, some examples of which are Hurler syndrome (MPS type IH), Scheie syndrome (MPS type IS), and Hunter syndrome (MPS type II).

Neuronal Ceroid Lipofuscinosis (Batten Disease)

Neuronal ceroid lipofuscinosis (Batten disease) is a group of autosomal recessive neurodegenerative diseases that result from the accumulation of lipofuscin. This disease is characterized by vision loss, seizures, progressive motor and cognitive dysfunction, and retinal degeneration. Eight genes (*CLN1–CLN8*) have been associated with this disease.

Infantile Refsum Disease

This disease is caused by defective peroxisomes and patients with this disease present with RP, hearing loss, hepatomegaly, and mental retardation. Unlike the adult form of the disease where specific peroxisomal enzymes are defective, the infantile form is characterized by a complete defect in perixosomal biogenesis. Levels of phytanic acid, very long chain fatty acids, and pipecolic acids are elevated. The disease is ultimately fatal.

Adult Refsum Disease

This disease is an autosomal recessive peroxisomal disorder characterized by elevated levels of phytanic acid, which lead to RP, anosmia, deafness, ataxia, cardiac arrhythmias, skeletal anomalies, and polyneuropathy. Mutations in the gene *PAHX*, which encodes phytanoyl-CoA 2-dydroxylases, have been shown to be responsible for some cases of adult Refsum disease. Treatment for adult Refsum disease includes plasmapheresis and dietary restrictions of phytanic acid and its precursors.

Senior–Loken Syndrome

This syndrome is an autosomal recessive disease that also falls under the umbrella of ciliopathies and can share some of the same features as Joubert syndrome (see the section titled 'Syndromic forms of RP'). This disease is characterized by Leber's congenital amaurosis and nephronophthisis (cystic kidneys).

Spinocerebellar Ataxia Type 7

This is an autosomal dominant neurodegenerative disease caused by a trinucleotide expansion in the ataxin 7 gene (*ATXN7*). The disease is characterized by cerebellar ataxias, dysphagia, dysarthria, and a cone–rod dystrophy.

Usher Syndrome

The details of Usher Syndrome are discussed elsewhere in this encyclopedia.

Visual Testing in RP

Terminology of Light Adaptation

The retina can respond to an astonishing 9 log units of light. Rods and cones differ in their sensitivity, temporal characteristics, and response to background lights. Depending on the intensity of a stimulus and background light present, different classes of cells respond. Scotopic refers to conditions under which only rods are functional. Mesopic refers to conditions under which both rods and cones are functional. Photopic refers to conditions under which only cones are functional.

Dark Adaptation

Dark adaption is measured with a Goldmann–Weekers dark adaptometer. Each eye is tested separately by first bleaching the rod and cone photopigments with an intense light and then measuring the brightness of a second light needed to achieve a threshold response. A dark adaptation curve can be drawn by repeating this threshold measurement over time after the bleaching light has been turned off. The normal recovery curve can be separated into two segments. The first segment occurs as cones recover from the bleaching light. The rods are much slower to recover and contribute to the second segment of the curve.

Visual Fields

The details on visual fields are discussed elsewhere in this encyclopedia.

ERG Terminology

The ERG is a fundamentally important test for studying RP and allied disorders. Different forms of RP can be classified based on the ERG response. Just as the electrocardiogram (EKG) measures the electrical activity of the heart through surface electrodes placed on the chest, the ERG measures the electrical activity generated by the retina to flashes of light through electrodes placed on the eye. The standard ERG is most commonly measured using an electrode embedded in a contact lens. Reference electrodes are placed on the forehead and ears. The International Society of Clinical Electrophysiology in Vision (ISCEV) has developed standardized methods for eliciting and recording the ERG.

Full-Field ERG

With the full-field ERG, flashes of light are delivered into a Ganzfeld diffuser, which ensures a uniform distribution of the light to the retina. It is important to realize that the full-field ERG measures a summed response from all of the cells in the retina. Therefore, the visual acuity and the ERG response may not correlate. For example, a patient could have a small scar in the center of the fovea that significantly decreases the vision, but the recordings of the ERG could be normal; the opposite can also be true. A patient could have an extinguished ERG due to peripheral degeneration, but still maintain 20/20 visual acuity due to preservation of the central fovea. For this reason, it is important to correlate visual acuity and ERG recordings along with visual-field results.

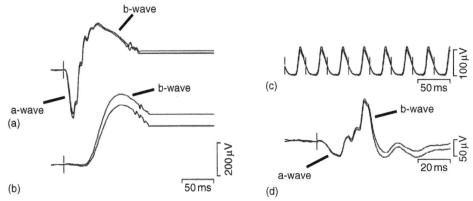

Figure 13 Components of a normal ERG. (a) Scotopic response to bright flash, which stimulates both rods and cones. Note the mixed a-wave, b-wave, and oscillatory potentials (superimposed on b-wave). (b) Scotopic response to a dim white flash, which stimulates only the rods. Note the rod-driven b-wave. (c) Photopic response to a 30-Hz flash, which isolates cones. (d) Response to a single flash under photopic conditions, which also isolates cones. Note the cone a-wave, b-wave, and oscillatory potentials (superimposed on b-wave).

Multifocal ERG

The full-field ERG can demonstrate a large disparity between the ERG signal and the visual acuity. To better detect localized retinal dysfunction in the macula, the multifocal ERG (mfERG) is very useful. The mfERG uses an m-sequence-derived check board pattern to selectively stimulate and isolate electrical activity from the retina. Specifically, the mfERG tests the macula under mesopic or photopic conditions, and therefore is primarily indicative of macular cone function.

Rod-Isolated ERG Response

The ERG can be recorded under dark-adapted (scotopic) conditions or light-adapted (photopic) conditions. When dim flashes are delivered under scotopic conditions, only the rod photoreceptors are stimulated. Such a flash elicits a positive electrical potential termed the b-wave. The scotopic b-wave arises from responses elicited by the rod bipolar cells. As the intensity of the flash increases, the b-wave grows in amplitude and, at higher intensities, a negative potential generated by the rod photoreceptors emerges and is termed the a-wave. Oscillations imposed on the b-wave are generated by higher-order retinal circuitry and are termed the oscillatory potentials.

Mixed Rod–Cone ERG Response

At even higher intensities, both the rod and cone photoreceptors are stimulated to generate a mixed scotopic response. Contributions from the cones contribute to the growing a-wave, while activity of the cone bipolar cells contributes to the b-wave. Once the intensity of the light has become intense enough to saturate the photoreceptors, the a-wave will cease to increase in amplitude.

The amplitude from baseline to the peak of the a-wave is termed the saturated a-wave amplitude and the time to the peak of the a-wave is the implicit time. The b-wave amplitude is measured from the trough of the a-wave to the peak of the b-wave.

Cone-Isolated ERG Response

The electrical activity of the cone photoreceptors can be isolated in two ways. The first is to measure the response to 30-Hz flicker flashes for which the cones have sufficient time to recover but the rods do not. The repetitive stimulus elicits a sinusoidal-like response. The peak-to-peak amplitude and the time to peak of this response relative to the stimulus can both be measured. The second method is to turn on a background light, which saturates the rod photoreceptors but minimally affects the cones (photopic conditions). Flashes under these conditions will elicit a cone-driven a-wave and b-wave as well as oscillatory potentials (**Figure 13**).

See also: Anatomically Separate Rod and Cone Signaling Pathways; Non-Invasive Testing Methods: Multifocal Electrophysiology; Primary Photoreceptor Degenerations: Retinitis Pigmentosa; Secondary Photoreceptor Degenerations: Age-Related Macular Degeneration.

Further Reading

Berson, E. L., Rosner, B., Sandberg, M. A., et al. (1993). A randomized trial of vitamin A and vitamin E supplementation for retinitis pigmentosa. *Archives of Ophthalmology* 111: 761–772.

Berson, E. L., Rosner, B., Sandberg, M. A., et al. (2004). Clinical trial of docosahexaenoic acid in patients with retinitis pigmentosa receiving vitamin A treatment. *Archives of Ophthalmology* 122: 1297–1305.

Fishman, G. A., Farber, M. D., and Derlacki, D. J. (1988). X-linked retinitis pigmentosa. Profile of clinical findings. *Archives of Ophthalmology* 106: 369–375.

Grant, C. A. and Berson, E. L. (2001). Treatable forms of retinitis pigmentosa associated with systemic neurological disorders. *International Ophthalmology Clinics* 41: 103–110.

Grover, S., Fishman, G. A., Anderson, R. J., et al. (1999). Visual acuity impairment in patients with retinitis pigmentosa at age 45 years or older. *Ophthalmology* 106: 1780–1785.

Hamel, C. P. (2007). Cone rod dystrophies. *Orphanet Journal of Rare Diseases* 2: 7.

Hartong, D. T., Berson, E. L., and Dryja, T. P. (2006). Retinitis pigmentosa. *Lancet* 368: 1795–1809.

Heckenlively, J. R. (1988). *Retinitis Pigmentosa*. Philadelphia, PA: Lippincott.

Hoffman, D. R., Locke, K. G., Wheaton, D. H., et al. (2004). A randomized, placebo-controlled clinical trial of docosahexaenoic acid supplementation for X-linked retinitis pigmentosa. *American Journal of Ophthalmology* 137: 704–718.

Radu, R. A., Yuan, Q., Hu, J., et al. (2008). Accelerated accumulation of lipofuscin pigments in the RPE of a mouse model for ABCA4-mediated retinal dystrophies following vitamin A supplementation. *Investigative Ophthalmology and Visual Science* 49: 3821–3829.

Sieving, P. A., Caruso, R. C., Tao, W., et al. (2006). Ciliary neurotrophic factor (CNTF) for human retinal degeneration: Phase I trial of CNTF delivered by encapsulated cell intraocular implants. *Proceedings of the National Academy of Sciences of the United States of America* 103: 3896–3901.

Weleber, R. G. and Gregory-Evans, K. (2006). Retinitis pigmentosa and allied disorders. In: Ryan, S. J. (ed.) *The Retina*, pp. 395–498. Philadelphia, PA: Elsevier.

Relevant Websites

http://www.ncbi.nlm.nih.gov – National Center for Biotechnology Information, OMIM.

http://www.sph.uth.tmc.edu – The University of Texas School of Public Health, Retinal Information Network (Retnet).

Proliferative Vitreoretinopathy

P Hiscott, University of Liverpool, Liverpool, UK; Royal Liverpool University Hospital, Liverpool, UK
D Wong, University of Hong Kong, Hong Kong, People's Republic of China

Glossary

Cytokeratins – A family of proteins found in the cytoskeleton (intermediate filaments) of epithelial cells.
Glial fibrillary acidic protein – A protein found in the cytoskeleton (intermediate filaments) of glial cells.
Immunohistochemistry – The localization of antigens, especially proteins, in tissue sections by antigen-antibody reactions. The reaction sites are visualized by a label such as a chromogen (typically a brown or red dye).
Myofibroblast – A fibroblast-like cell with some of the features of smooth muscle, including contractile properties.

Introduction

In 1983, the Retina Society Terminology Committee published a landmark paper in which the term proliferative vitreoretinopathy (PVR) was proposed for a condition that had been recognized as the major cause of failure of retinal detachment surgery. Previous names for this disease included preretinal organization, massive vitreous retraction (MVR), massive preretinal retraction (MPR), and massive periretinal proliferation (MPP) – terms that highlighted some of the main clinical features of the disorder.

In this condition, following a retinal detachment, cells leave their normal location in the retina and migrate to the retinal surfaces. Here, the cells proliferate to form membranes. Although many of these membranes consist of only a thin layer of (glial) cells and produce no clinical problems or symptoms, in about 10% of retinal detachment patients the membranes develop into thicker scar-like tissues that are able to contract. It is this ability of the membranes to contract that leads to the clinical picture of PVR.

Definition

PVR is strictly defined as a complication of rhegmatogenous retinal detachment that is characterized by the formation of membranes on both surfaces of the detached retina and on the posterior surface of the detached vitreous gel.

Location of PVR Membranes

The membranes on the vitreous surface of the retina are usually called epiretinal membranes and these are the most common membranes of PVR. Membranes that form beneath the detached retina (i.e., between the neuroretina and the retinal pigment epithelium) are termed sub- or retro-retinal membranes. Membranes on the posterior surface of the vitreous gel are known as posterior hyaloid membranes and typically are continuous with epiretinal membranes in PVR (hence they tend to have a similar composition). The expression periretinal membrane is sometimes used to encompass all membranes around the retina.

In addition, membranes can extend into the vitreous base and anteriorly over the pars plicata to the back of the iris: a condition known as anterior PVR. There is also evidence that PVR can have a distinct component within the neuroretina itself – a situation that may be called intraretinal PVR (iPVR).

It is clear that cellular proliferation can occur at several of the above sites in the same eye. It is also important to recognize that membrane formation can occur in a wide variety of diseases other than retinal detachment. For example, the ischemic retinopathies can also lead to epiretinal and posterior hyaloid membranes although the membranes in these conditions are usually heavily vascularized and thus differ from PVR membranes (see below).

The Significance of Membrane Formation in PVR

In PVR, the membranes may contract so that they exert traction on adjacent tissues. Epiretinal membranes tend to produce tangential traction on the retinal surface. The effect of this retinal traction is to cause retinal folding and/or (re)detachment of the sensory retina (**Figure 1**). Such an effect can be localized, as for example when an epiretinal membrane over the macular (epimacular membrane) causes folding of the macular (macular pucker) or in a peripheral membrane causes a star fold (**Figure 2**). On the other hand, epiretinal membranes may be diffuse, rather than localized, covering much of the retinal surface.

The traction from epiretinal membranes can be mild, causing no more than subtle surface wrinkling of the inner retina. Moderate traction can cause marked folds of the retina. Severe traction can be associated with displacement of retinal vessels and ectopia of the fovea. The effect of the traction is, however, not just two dimensional (2-D). The newer generation of Optical Coherence Tomography systems provides high-resolution cross-sectional views or reconstructed 3-D images of the retina (**Figure 2**). These images clearly show that the traction may be superficial but sometimes the whole thickness of the retina is involved. Thus, these signs can indicate the development of PVR.

Clinically, we often see the effect of traction rather than the epiretinal membranes themselves: on the attached retina, epiretinal membranes can be difficult to visualize (**Figure 1**). Nonetheless, the epiretinal membranes are present and exerting isometric' traction on the retina. The effect of this traction may only become apparent once the retina becomes detached, at which point one may observe star folds or retinal breaks with rolled edges when the traction in the tissue is focal. In addition,

when the epiretinal membrane is diffuse, the detached retina will appear to have reduced mobility. Anterior PVR membranes will draw the pre-equatorial retina anteriorly toward the ciliary processes, whereas epiretinal membranes in the post-equatorial retina tend to contract the retina into a cone configuration. The combination of the anterior and posterior traction systems on a totally detached retina will give rise to a so-called closed funnel (**Figure 3**).

Subretinal membranes, particularly in the form of bands, are apt to elevate the neuroretina like a sheet on a washing line or, in the case of a circular band, like a napkin in a napkin ring. Posterior hyaloid membranes can produce circumferential or, if the vitreous base is involved, anteroposterior traction.

Particularly in eyes that have undergone previous vitrectomy or trauma, proliferative tissue can extend into the vitreous base and anteriorly toward the pars plicata of the ciliary body, the iris and even as far as the pupil margin. Traction in this latter situation may lead to traction on the ciliary body (hypotony can result) and posterior displacement of the iris.

The Cells Involved in PVR

It is now accepted that PVR combines reaction to damage by astrocytes of the central nervous system (gliosis) with fibrosis. In this context, the gliosis involves Müller cells and retinal astrocytes, while the fibrosis includes metaplasia or transdifferentiation of the retinal pigment epithelial (RPE) cells. Although retinal glia and RPE cells are thus major players in PVR, it is also apparent that a variety of other cell types are involved in the disease.

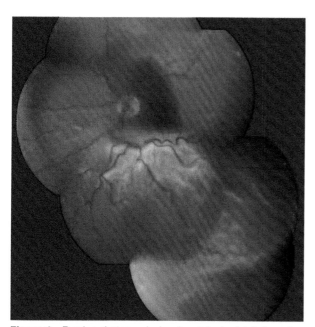

Figure 1 Fundus photograph showing detached, inferior retina with diffuse PVR.

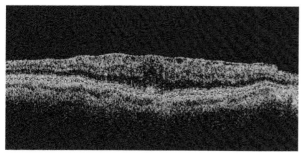

Figure 2 OCT image of retina showing a thin epiretinal membrane and subjacent retinal distortion.

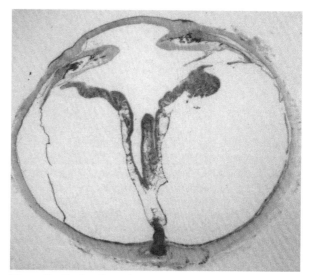

Figure 3 Section through an eye with PVR and a closed-funnel retinal detachment. Retinal folds can be seen. The eye is aphakic and there are anterior PVR membranes that also involve the iris.

Glial Cells

There is compelling evidence from a variety of morphological, immunohistochemical, *in vitro*, and experimental studies that astrocytes and Müller cells are involved in PVR membranes, particularly in epiretinal and, to a lesser extent, subretinal membranes.

Within PVR epiretinal membranes, glial cells often form layers in the tissue (**Figure 4**). These layers are adherent to fibrous components of the membranes and, in both human and experimental PVR, the cells can sometimes be traced through defects in the retinal inner limiting lamina into the retina itself. These observations

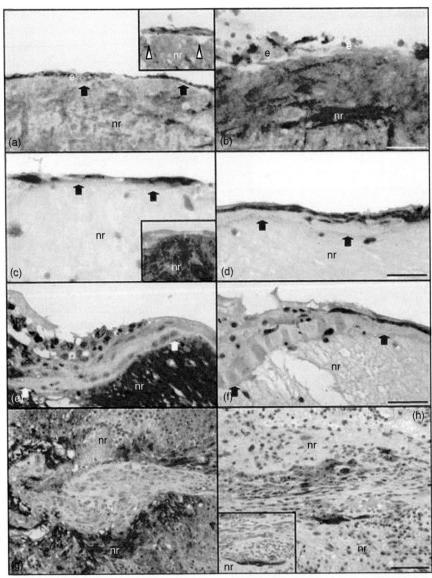

Figure 4   (a–h) Sections through neuroretina (nr) with PVR membranes in enucleated eyes. The sections have been stained with the immunohistochemical method for glial elements with an antibody to glial fibrillary acidic protein (see insets in (a–c) , and also (e) and (g)) or for RPE elements with antibody to cytokeratin 7 ((c), (d), (f), and (h)) or to a range of cytokeratins (inset in (h)): red-brown reaction product, hematoxylin counterstain. (a) The retina is gliotic and there is an epiretinal membrane 'e' composed of a glial monolayer. The internal limiting lamina is marked (arrows). Inset: this epiretinal membrane is composed of a double layer of glia and there is some distortion of the internal limiting membrane (arrowheads). (b) The epiretinal membrane 'e' is composed of glial and nonglial cells. Again, the underlying retina is gliotic. (c and d) Epiretinal RPE cells are seen. The subjacent retina is gliotic (inset in (c)). The internal limiting lamina is marked (arrows). Parts (e) and (f) show the same area of the same specimen: the epiretinal membrane contains both glial (e) and RPE (f) components. The internal limiting lamina is marked (arrows). Parts (g) and (h) show the same area of the same specimen: the retina is folded, disorganized, and gliotic. There is a PVR membrane containing fibroblastic RPE cells and a few glia. Some of the fibroblast-like cells do not label for glial or RPE markers, even with an antibody that reacts with a range of cytokeratins (inset in (h)): their origin is unclear. Scale bars: (a–f) 50 μm; (g and h) 100 μm.

have led to the concept that glial cells traversing the vitreoretinal interface serve to anchor the epiretinal membrane to the retina and that the epiretinal glia may form a substrate or scaffold that other cells might use to produce the rest of the epiretinal tissue. In support of this notion, experimental models have demonstrated that glial cells can break through the retinal inner limiting lamina early in the formation of epiretinal membranes and spread across the retinal surface to form sheets that other cells may adhere to. Failure of glial sheets to become populated by other cells may account for arrest of the disease process at an early stage and the production of asymptomatic or nonproblematic subtle membranes on either surface of the neuroretina.

RPE Cells

The same sorts of methods that were used to detect glia in PVR membranes have been employed to demonstrate RPE cells. Indeed, it is now widely accepted that RPE cells are major components of PVR membranes and the presence of large numbers of RPE cells in epiretinal membranes is one of the distinguishing factors between PVR membranes and epiretinal membranes caused by other diseases. Early in the development of PVR, RPE cells leave their normal location at the chorioretinal interface and either migrate or are swept with subretinal fluid movements to the surfaces of the detached retina. Here, the cells may attach to early membranes (**Figure 4**). In the case of epiretinal membranes, RPE cells may adhere to glia that have already arrived, while in subretinal membranes it has been suggested that the cells may in addition settle on fibrin deposits. Some RPE cells may also migrate into or through the neuroretina.

RPE cells have a remarkable propensity to undergo metaplasia or transdifferentiation. As a result, an RPE cell may change from a polarized, sedentary pigmented cuboidal epithelial cell to a nonpigmented fibroblastic cell, a migratory macrophage-like cell, a cell in a gland-like structure or even a bone-forming cell (**Figure 4**). In PVR membranes, RPE cells most often adopt fibroblastic or macrophagic phenotypes, though they may attempt to (re-)form a polarized monolayer as well.

Fibroblastic Cells

PVR membranes often have a substantial fibrous element that consists of fibroblastic cells in extracellular matrix. Many of these cells are of RPE origin (**Figure 4**). Others may be derived from perivascular sources such as adventitial cells of larger retinal vessels.

There has been much interest in the role of fibroblastic cells in PVR membranes for two reasons. First, they are believed to be responsible for generating most of the tractional forces within the tissue. Second, the cells are

thought to be responsible for the production of the bulk of the extracellular matrix in the membranes. There is still debate about how these cells may produce traction: theories include smooth muscle-like contraction of the (myofibroblastic) cells and cell–matrix interactions that cause shortening of matrix elements.

Macrophages

In addition to macrophagic RPE cells, macrophages of hematogenous origin are involved in PVR. Macrophages are especially abundant in membranes arising in the presence of some tamponade agents (see below).

Irrespective of their origin, macrophages are believed to have a number of important roles in PVR. In the early stages of the disease, they produce mitogens, chemotactic agents, and growth factors that probably have a role in cell recruitment to and proliferation in the membranes. Growth factors and enzymes produced by macrophages later in the disease process may be involved in matrix synthesis and remodeling.

Vascular Elements

Blood vessels are found in around 10–20% of PVR membranes but, in contrast to periretinal membranes of conditions such as central retinal vein occlusion and proliferative diabetic retinopathy, blood vessels usually do not form a major component of the tissue. However, vessels are probably more abundant in anterior PVR membranes (**Figure 5**).

Other Cells

T lymphocytes, including CD4 and CD8 positive cells, have been found in PVR membranes. Some of these cells express interleukin-2 receptor, suggesting that they are activated and capable of promoting the cellular events in the tissue. Ciliary body epithelial cells have been reported in the membranes of anterior PVR (**Figure 5**). Hyalocyte-like cells have been described in epiretinal membranes generally and it has been suggested that hyalocytes may give rise to some of the fibroblastic and macrophagic cells in the epiretinal tissues. Recently, ganglion cell neurites have been observed in experimental and human epi- and subretinal membranes. They are co-localized with glia, suggesting active outgrowth into the periretinal tissues.

The Extracellular Matrix in PVR Membranes

PVR membranes contain a matrix that, in terms of composition, is similar to the one seen in a healing skin wound (**Figure 6**). These components include structural proteins

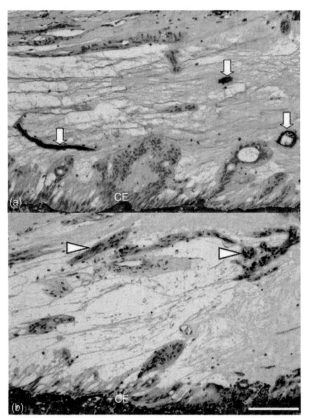

Figure 5 Anterior PVR membrane extending over distorted ciliary epithelium (CE) in an enucleated eye. (a) Stained with the immunohistochemical method for the endothelial marker CD34: note that there are blood vessels in the membranes (arrows). (b) Stained with the immunohistochemical method for cytokeratins: note that in addition to cells in the membranes (arrowheads), the CE normally expresses cytokeratins too. Thus, it is possible that at least some of the epithelial cells in anterior PVR membranes are of CB rather than RPE origin (hematoxylin counterstain, scale bar: 200 μm).

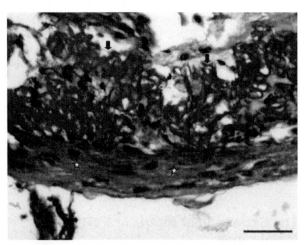

Figure 6 Section through a surgically excised PVR epiretinal membrane. The tissue contains a fibrous element (stars). Convoluted retinal internal limiting membrane is also present (arrows). Periodic acid Schiff reagent, hematoxylin counterstain. Scale bar: 50 μm.

such as collagens and elastic fiber precursors, adhesive glycoproteins like fibronectins and laminins, glycosaminoglycans, matricellular proteins such as tenascins and thrombospondins, and matrix enzymes like matrix metalloproteinases together with their inhibitors.

Much of the matrix in PVR is produced locally by the membrane cells themselves, though there is evidence that some components enter from elsewhere. For example, some blood derivatives like plasma fibronectin are found in PVR membranes. Irrespective of origin, the extracellular matrix in PVR membranes increases with time and there is a corresponding decrease in cellularity of the tissue. Indeed, this change in cellularity together with the contractile nature of the tissue and the presence of fibroblastic cells in matrix gave rise to comparisons between PVR membranes and healing wounds.

As in healing wounds generally, the matrix is more than a passive space filler: there is good evidence that PVR matrix is an important regulator of cell behavior (notably migration and proliferation) in the tissue. For example, there have been a number of reports demonstrating that cells in PVR membranes express a range of cell-surface receptors for the various matrix components around them (e.g., integrins) and experimental studies implicating the importance of such receptors in contraction in PVR models. Moreover, early matrix may have adhesive properties that aid cohesion of the developing tissue. There is also evidence of matrix remodeling in more established membranes. Longstanding PVR membranes are often densely fibrous. It is also worth noting that, when surgically excised, PVR membrane specimens often contain internal limiting membrane from the retina. This material tends to become convoluted in the tissue (**Figure 6**), presumably as a result of tractional forces upon it.

Pathogenesis and Natural History

Although the pathogenesis of PVR is not fully elucidated, our understanding of the disease is at the point where logical therapies can be designed and implemented.

Fundamental to the development of PVR is retinal detachment, itself dependent upon degenerative changes in the vitreous and retina. Vitreous degeneration (syneresis) is a normal aging process that can be accelerated by conditions such as myopia or trauma and results in the formation of fluid and formed components. Attachments between formed vitreous and retina may permit rotational tractional forces, such as dynamic traction from saccadic eye movements, to be transmitted between the two structures. In turn, a retinal tear may form and fluid vitreous pass through the hole in the retina to give rise to a (rhegmatogenous) retinal detachment.

There is experimental evidence that some changes associated with PVR can occur within a day of retinal detachment. In retinal detachment, together with entry of fluid vitreous to the subretinal space, there is breakdown of the blood–retina barrier with accumulation of plasma proteins in the vicinity of the retina and influx of blood-borne inflammatory cells. Thus, there is aggregation of hematogenous and locally derived proteins and cytokines, including plasma glycoproteins, and growth and differentiation factors. Many of the components of this collection have chemotactic and/or mitogenic properties for RPE and glial cells. In fact, RPE cells can be observed to detach from Bruch's membrane and may be transported to the surfaces of the detached retina while glia may breech the retinal inner limiting lamina (see above) in retinal detachment.

The above reaction to retinal detachment might set the scene for PVR formation, as we have seen only a minority of retinal detachment patients develop the condition. It thus appears that additional factors are required and, indeed, a number of clinical risk factors have been identified (see below). These include factors that would be expected to elevate the concentrations of chemotactic and mitogenic chemical mediators and increase the influx of inflammatory cells to the retinal surfaces, such as hemorrhage, multiple surgical interventions, and large surface area of RPE cells exposed to detachment.

Cells displaced to the surfaces of the detached retina can be seen to adhere to each other as well as the retinal surface. These very early membranes lack matrix but they do possess adhesive glycoproteins like fibronectins. Moreover, a number of matricellular proteins including thrombospondin 1 are also present. It has been hypothesized that fibronectin and thrombospondin 1 can form a provisional matrix in healing wounds. Thus, it is possible that they provide some integrity to developing PVR membranes as well as an adhesion substrate for the cells.

Once in the developing membranes, the cells proliferate. Proliferation is assumed to increase the amount of membrane tissue and it can be detected as long as a year after the onset of the disease. In this respect, PVR differs from skin wounds where proliferation is restricted to a short wave early in the process. It is thought that cells also migrate along the retinal surfaces toward developing membranes, perhaps in response to local chemotactic agent production in the new tissue.

In addition to cell recruitment, cell migration in PVR membranes might also be a mechanism by which tractional forces are generated (motile cells impart a force on their substrate). Thus, it may be that cell migration, myofibroblastic contraction, and cell–matrix interactions (see above) can all be involved in membrane contraction.

The buildup of extracellular matrix with time in PVR membranes is matched by a reduction in cellularity of the tissue (**Figure 6**). Cell loss in PVR membranes probably occurs through apoptosis and nonapoptotic pathways. Ultimately, untreated PVR membranes become paucicellular and fibrous in nature.

Intraretinal PVR

The concept that retina-shortening cellular changes may occur within the neuroretina itself in PVR is relatively recent. Nevertheless, it is now clear that several cell types, including Müller cells and astrocytes, not only become reactive but also do replicate in the retina in PVR and it is thought that this gliosis contributes to retinal shortening. RPE cells are also involved in this process, although their numbers appear to be small compared to the numbers in periretinal membranes. Moreover, gliosis, with or without epiretinal and/or subretinal membranes, can cause marked retinal distortion and localized retinal thickening that can lead to the formation of a focal mass (**Figure 7**). Indeed, similarities in the microscopic appearances of localized PVR masses and vasoproliferative tumors of the retina, including the presence of RPE cells in both lesions (**Figure 8**), have led to speculation that some vasoproliferative tumors may be part of the spectrum of PVR.

Incidence and Risk Factors

It is often stated that PVR afflicts around 10% of all patients with retinal detachment. However, it is clear that some patients are at much greater risk of developing the condition than others. The most important risk factor for postoperative PVR seems to be preoperative PVR. A variety of clinical risk factors have been identified or suggested, including size and number of retinal holes, extent and duration of the detachment, the presence of blood and/or intraocular inflammation, aphakia, preoperative choroidal detachment, early stages of the disease or poor visual acuity prior to initial surgery, and multiple previous attempts at re-attaching the detached

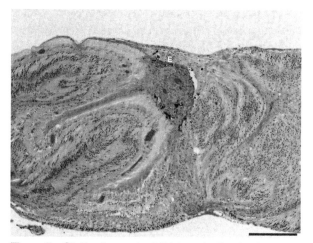

Figure 7 Gliotic, disorganized, thickened retina that has full-thickness folds in association with epiretinal membrane (E), in a section of an eye removed for complications of PVR. The retinal changes give rise to the formation of a localized mass (hematoxylin and eosin; scale bar: 500 μm).

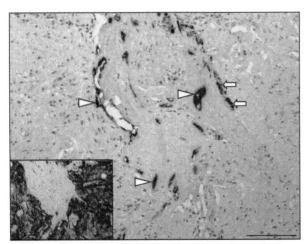

Figure 8 Section showing part of a vasoproliferative tumor (reactive retinal glioangiosis) in an eye removed for complications of retinal detachment. The lesion contains scattered RPE cells (arrowheads), revealed (red) by immunohistochemical staining for cytokeratin 7 (hematoxylin counterstain). A few of these cells contain melanin pigment (*arrows*). Staining of the lesion for glial cells (*inset*: immunohistochemical staining red, hematoxylin counterstain) confirms that the retinal tissue is gliotic and disorganized. Scale bar: 200μm

retina. Methods employed during the primary retinal detachment surgery may also increase the risk of PVR. Thus, there is evidence that cryopexy, choice of tamponade agent, and vitrectomy impact on the risk of developing PVR.

Clinical Classification of Proliferative Vitreoretinopathy

There are several attempts to classify PVR according to the clinical features of the retinal detachment. These classifications are descriptive and help surgeons communicate, especially with regard to surgical approaches in the management of the condition. Thus, they are useful for surgical planning and are not based on pathobiology. PVR classifications are also not prognostic. They do not correlate well with visual prognosis or anatomical success with treatment. The classification is also not related to the stages of the disease. It is just the clinical picture at one snapshot in time. Despite their many shortcomings, clinical PVR classifications are widely used in clinical trials and for clinicopathological correlates.

Several classifications have been suggested, based on the clinical manifestations of the disease. For the most part, there is commonality between these systems with regard to the earlier or milder stages of the disease, whereas the schemes tend to diverge in their classification of later or more advanced stages of PVR.

Within these systems, stage A (minimal) is usually regarded as the presence of vitreous haze and pigment

clumps, whereas stage B (moderate) is typically recognized as wrinkling of the retinal surface, decreased vitreous mobility, and increased retinal stiffness.

With respect to the more severe or later stages of PVR, some schemes separate the advanced stages by the number of quadrants of the retina involved. Thus, for example, the Retina Society Terminology Committee classification of 1983 associates stage C (marked) disease with fixed retinal folding and adds a number to reflect the number of quadrants involved (e.g., C-3 is fixed folds involving three quadrants of retina). In this system, stage D (massive) reflects the involvement of all four retinal quadrants. In addition, stage D is graded 1–3 depending on how extensive the folding of the retina is (D-1 being an open funnel of totally detached retina, D-3 being a closed funnel so that the optic nerve head cannot be seen: **Figure 3**).

Other schemes classify the more severe stages of PVR into anterior and posterior groups, according to the location of the disease with reference to the retinal equator. In these systems, the number of retinal quadrants involved by the disease is again used so that PVR involving fixed folds in, say, two quadrants of retina posterior to the equator would be classed P2 or CP2 (stage D is generally discarded in these schemes). The schemes employing anterior and posterior also add a contraction type, depending on the extent of epiretinal membrane (focal or diffuse), the presence of subretinal membrane, or posterior hyaloid/vitreous base proliferation.

Management

There is much interest in preventing proliferation of membranes after retinal detachment surgery by treating high-risk patients with combinations of agents. One such combination is the antiproliferative drug 5-fluorouracil and low-molecular-weight heparin (which binds growth factors). This combination has been shown to reduce the incidence of PVR in high-risk retinal detachment patients undergoing vitrectomy.

Once established, PVR membranes are removed by microsurgery so that the retina can be reattached. Again, there is much interest in the use of pharmacological measures to stop membrane recurrence after PVR surgery. For example, there is evidence that daunomycin used preoperatively can reduce the requirement for repeat surgery in these patients.

Tamponade agents to maintain retinal attachment are frequently employed during and after surgery for retinal detachment and PVR. These agents incorporate gases (e.g., air) and liquids (e.g., silicone oil). Liquids have a tendency to emulsify in the eye, particularly if they are of low viscosity. The result is the formation of droplets of various sizes in the vitreous cavity or even elsewhere in the eye (such as in the aqueous if tamponade gains access

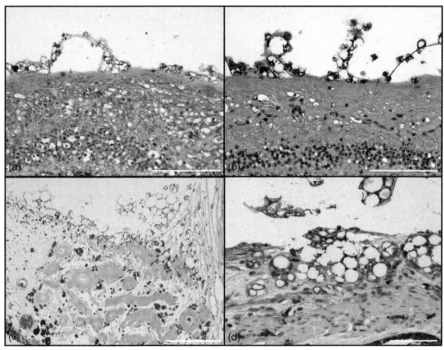

Figure 9 Photomicrographs from sections of an eye removed following liquid tamponade use in the treatment of PVR. Parts (a) and (b) show the vitreoretinal interface with numerous vacuolated cells that label for the macrophage marker CD68 (b), but not the glial marker glial fibrillary acidic protein (a). Similar cells are seen in the drainage angle (c) and in the anterior chamber and iris (d). The features are consistent with macrophage reaction to emulsified liquid tamponade agent (hematoxylin and eosin (c); immunohistochemical staining, red reaction product (a), (b), and (d); scale bars (a), (b), and (d): 100 μm; (c) 200 μm).

to the anterior chamber: **Figure 9**). Thus, emulsification of tamponade agent may impact on the pathology of PVR membranes. It appears that the droplets can stimulate a foreign body-type reaction and attract macrophages to the retinal surface (**Figure 9**). New membranes may develop and these tissues characteristically have the microscopic appearances of PVR membranes containing granulomata to emulsified oil.

Outcomes

Successful treatment in PVR has often been measured in terms of final retinal reattachment rates. The assumption is that effective anti-PVR therapy would lead to an increased rate of successful reattachment. This assumption may or may not hold true. It is often missed or untreated holes that lead to retinal redetachment and not necessarily PVR, which may or may not be controlled by the anti-PVR drugs. Indeed, a totally ineffective anti-PVR treatment may be compatible with anatomical success so long as the epiretinal membranes do not act on the retina to produce another retinal break or cause tractional retinal detachment. Hence, anatomical success rate is a poor proxy for PVR control. Another important point is that anatomical success does not guarantee visual recovery. The advancement of surgery in the last few

years has greatly increased the final anatomical success rate. Disappointingly, this success rate has not been translated into visual improvement. In retinal detachment patients, the fellow eye is also likely to be involved in sight-threatening pathology so that many PVR patients end up with visual impairment in both eyes.

Conclusions

Despite intense research over the last 25 years that has improved our understanding of the condition, PVR remains the major cause of failure after retinal detachment surgery. Nevertheless, continuing advances in both surgical and pharmacological manipulation of the disease, based on an expanding knowledge of PVR pathobiology, can be expected to reduce the impact of the disease in the future.

See also: Rhegmatogenous Retinal Detachment.

Further Reading

Asaria, R. H., Kon, C. H., Bunce, C., et al. (2001). Adjuvant 5-fluorouracil and heparin prevents proliferative vitreoretinopathy: Results from a randomized, double-blind, controlled clinical trial. *Ophthalmology* 108: 1179–1183.

Charteris, D. G. (1995). Proliferative vitreoretinopathy: Pathobiology, surgical management, and adjunctive treatment. *British Journal of Ophthalmology* 79: 953–960.

Colthurst, M., Williams, R. L., Hiscott, P., and Grierson, I. (2000). Biomaterials used in the posterior segment of the eye. *Biomaterials* 21: 649–665.

Fisher, S. K., Lewis, G. P., Linberg, K. A., and Verardo, M. R. (2005). Cellular remodeling in mammalian retina: Results from studies of experimental retinal detachment. *Progress in Retinal and Eye Research* 24: 395–431.

Heimann, K. and Wiedemann, P. (1989). *Proliferative Vitreoretinopathy.* Heidelberg: Kaden.

Hiscott, P. and Mudhar, H. (2008). Is vasoproliferative tumour (reactive retinal glioangiosis) part of the spectrum of proliferative vitreoretinopathy? *Eye* 23: 1851–1858.

Hiscott, P. and Sheridan, C. (1998). The retinal pigment epithelium, epiretinal membranes and proliferative vitreoretinopathy. In: Marmor, M. F. and Wolfensberger, T. J. (eds.) *Retinal Pigment Epithelium – Function and Disease*, pp. 478–491. New York: Oxford University Press.

Hiscott, P., Morino, I., Alexander, R., Grierson, I., and Gregor, Z. (1989). Cellular components of subretinal membranes in proliferative vitreoretinopathy. *Eye* 3: 606–610.

Hiscott, P., Sheridan, C., Magee, R., and Grierson, I. (1999). Matrix and the retinal pigment epithelium in proliferative retinal disease. *Progress in Retinal and Eye Research* 18: 167–190.

Kampik, A., Kenyon, K. R., Michels, R. G., Green, W. R., and de la Cruz, Z. C. (1981). Epiretinal and vitreous membranes. Comparative study of 56 cases. *Archives of Ophthalmology* 99: 1445–1454.

Kirchhof, B. and Wong, D. (eds.) (2005) *Vitreo-Retinal Surgery. Essentials in Ophthalmology.* Berlin: Springer.

Machemer, R. and Laqua, H. (1975). Pigment epithelium proliferation in retinal detachment (massive periretinal proliferation). *American Journal of Ophthalmology* 80: 1–23.

Pastor, J. C., de la Rúa, E. R., and Martín, F. (2002). Proliferative vitreoretinopathy: Risk factors and pathobiology. *Progress in Retinal and Eye Research* 21: 127–144.

The Retina Society Terminology Committee (1983). The classification of retinal detachment with proliferative vitreoretinopathy. *Ophthalmology* 90: 121–125.

Wiedemann, P., Hilgers, R. D., Bauer, P., and Heimann, K. (1998). Adjunctive daunorubicin in the treatment of proliferative vitreoretinopathy: Results of a multicenter clinical trial. Daunomycin study group. *American Journal of Ophthalmology* 126: 550–559.

Relevant Website

http://www.youtube.com – Number of videos concerning PVR and its management.

Retinal Cannabinoids

S Yazulla, Stony Brook University, Stony Brook, NY, USA

Glossary

Age-related macula degeneration (AMD) – It comprises a variety of diseases, mainly of the elderly, that involves loss of vision in the central region of the retina.

Endocannabinoids – Natural chemicals in the body that are mimicked by the active component of marijuana.

Monoamine oxidase (MAO) – An enzyme that degrades dopamine.

Snellen acuity – A test of visual acuity that uses a standard sized "E" in four orientations.

Transient receptor potential type vanilloid 1 receptor (TRPV1) – An ionotropic receptor that increases intracellular calcium either by entry through the plasma membrane or from intracellular stores. It is activated by noxious heat, capsaicin, and the endocannabinoid, anandamide.

Vernier acuity – A test of visual acuity that tests the ability to detect displacement of two lines end-to-end.

Marijuana and the Endocannabinoids

The active component of the marijuana plant *Cannabis sativa*, Δ^9-tetrahydrocannabinol (THC), mimics endogenous chemicals, endocannabinoids (eCBs) that activate membrane receptors. eCBs include a variety of amide, ester, and ether derivatives of arachidonic acid. The most widely studied of these are arachidonoyl ethanolamide (anandamide, AEA) and *sn*-2 arachidonoyl glycerol (2-AG) (**Figure 1**). Other eCBs have been identified with varying degrees, or no affinity for cannabinoid receptors, and also compete with AEA and 2-AG for metabolizing enzymes. In this way, they modulate activity by competition at the receptors or by affecting substrate availability for metabolism.

Synthesis and Release

Unlike water-soluble transmitters, AEA and 2-AG are lipophilic and not stored in synaptic vesicles. Rather, membrane phospholipids are metabolized on demand to liberate AEA and 2-AG by calcium-dependent phospholipases. The precursor of AEA is *N*-arachidonyolphosphatidyl ethanolamine (NAPE), formed by calcium-dependent transfer of arachidonic acid (AA) from arachidonoylphosphatidylcholine to phosphatidylethanolamine (PE). There are multiple pathways for AEA liberation from the membrane. First, NAPE is hydrolyzed by phospholipase D (PLD) to release AEA and phosphatidic acid. Second, NAPE is hydrolyzed to *N*-acyl-lyso-PE by phospholipase A_1/A_2; then, AEA is released by lysophospholipase D. Third, phospholipase C (PLC) cleaves NAPE to generate phosphoanandamide, which is dephosphorylated to liberate AEA. The PLC pathway may be involved in the on-demand synthesis of AEA rather than in maintaining basal tissue levels of AEA. The primary pathway for 2-AG synthesis involves hydrolysis of diacylglycerols (DAG) by DAG lipase isozymes, DAGLα and DAGLβ. DAGs may be produced by the PLC β-catalyzed hydrolysis of phophotidylinositol or hydrolysis of phosphatidic acid by a phosphohydrolase. AEA and 2-AG freely diffuse within the membrane where they interact with the active sites of degradative enzymes and receptors. AEA binds reversibly to serum albumin, and it is likely that such binding is critical for the movement of AEA and 2-AG in blood, the extracellular matrix, and the cytoplasm. The presence and localization of AEA and 2-AG are inferred from the distribution of receptors, synthesizing and inactivating enzymes as well as physiological effects on identified cells.

Inactivation

AEA and 2-AG are inactivated following intracellular accumulation by fatty acid amide hydrolase (FAAH), monoacylglycerol lipase (MGL), cyclooxygenase-2 (COX-2), and lipoxygenase (LOX). AEA and 2-AG are hydrolyzed by FAAH into AA and ethanolamine or glycerol, respectively. 2-AG, but not AEA, is hydrolyzed by MGL. Following hydrolysis of AEA or 2-AG, AA is incorporated into membrane phospholipids. COX-2 oxidizes arachidonic acid, AEA, and 2-AG to prostamides or prostaglandin glyceryl esters, leading to prostaglandins. In addition, oxidation of AA by LOX produces 12-(S)-hydroperoxyeicosatetraenoic acid (15-(S)-HPETE), 5-(S)-HETE, and leukotriene B_4, all of which are agonists of TRPV1 receptors (**Figure 2**). The effects of AEA and 2-AG are modulated by the balance of metabolic enzymes that is specific to each cell type.

Receptors

Effects of cannabinoids are mediated by metabotropic (G-protein-coupled receptors (GPCRs)) and ionotropic

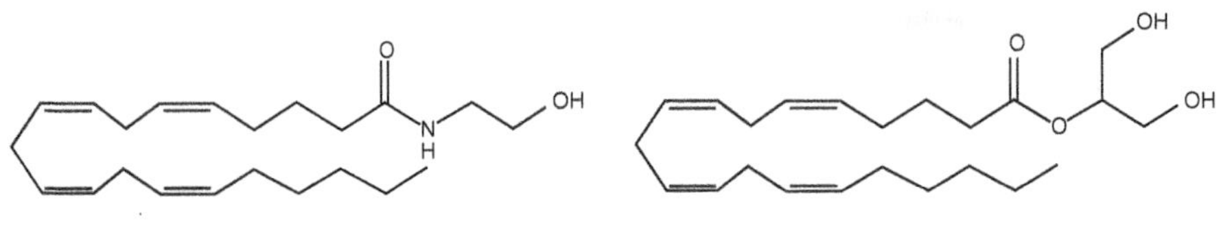

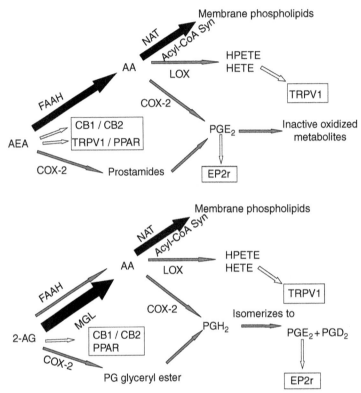

Anandamide

2-arachidonoyl-glycerol

Figure 1 Chemical structures of endocannabinoids: arachidonoyl-ethanolamide (anandamide, AEA) and 2-arachidonoyl-glycerol (2-AG).

Figure 2 This schematic illustrates some of the metabolic pathways for the degradation of AEA and 2-AG. In the dominant pathways (bold arrows), AEA and 2-AG are hydrolyzed to arachidonic acid (AA), and then rapidly incorporated into membrane phospholipids via N-acyltransferase (NAT) and acyl-Coenzyme A synthetase. Lesser pathways (shaded arrows) involve oxidation by cyclooxygenase-2 (COX-2) of AEA, 2-AG, and AA to prostaglandins (PGE$_2$ and PGD$_2$). Additionally, AA may be oxidized by lipoxygenase (LOX) to 12-(S)- and 15-(S)-HPETE and 5-(S)-HETE. Hollow arrows show that AEA and 2-AG are endoligands for CB1, CB2, and PPAR receptors, while AEA also activates TRPV1 receptors. Metabolites of COX-2 oxidation activate EP2 receptors, and metabolites of LOX oxidation activate TRPV1 receptors.

(ion channel) receptors (**Figure 2**). In general, activation of cannabinoid 1 receptors (CB1Rs), via heterotrimeric guanosine-5'-triphosphate (GTP)-binding proteins Gi/o (Gi/o), modulates voltage-gated K$^+$ and Ca^{2+} conductances, resulting in a reduction of neurotransmitter release, particularly γ-aminobutyric acid (GABA) and glutamate. CB2 receptors, which also signal through Gi/o, are expressed in cells of the immune system and the central nervous system (CNS), particularly in astrocytes. There is evidence for additional cannabinoid receptors, perhaps GPCR 55. AEA, but not 2-AG, activates the ionotropic transient receptor potential type vanilloid 1 receptor (TRPV1) that increases intracellular calcium

either by entry through the plasma membrane or from intracellular stores. Prostamides and prostaglandin glycerol esters, produced by eCB oxidation by COX-2, bind to a variety of prostaglandin receptors. eCBs are ligands for peroxisome proliferator-activated receptors (PPARs), members of the nuclear receptor superfamily that are involved in lipid metabolism, insulin sensitivity, regulation of inflammation, and cell proliferation.

Distribution and Function

CB1Rs are the most numerous GPCRs in the brain. eCBs, their receptors, and metabolizing enzymes are enriched in brain regions associated with the physiological and psychomotor effects of cannabis. AEA and 2-AG have short- and long-term effects on synaptic plasticity and neuroprotection. The effects depend largely on retrograde transmission in which postsynaptic dendrites release an eCB that binds to presynaptic CB1Rs to reduce transmitter release. Retrograde release of eCBs is evoked by two mechanisms. In a voltage-dependent mechanism, depolarization of postsynaptic dendrites by L-glutamate opens voltage-gated calcium channels. The increase in intracellular Ca^{2+} activates Ca-dependent PLD to release an eCB. A second mechanism involves activation of heterotrimeric GTP-binding protein $G_{q/11}$ ($G_{q/11}$) coupled metabotropic receptors, usually group I metabotropic glutamate receptors (mGluRs), mGluR1 and mGluR5, and muscarinic receptors (M1 and M3). By enzymatic cascades that may or may not release calcium from intracellular stores, eCBs are released from the plasma membrane. The eCB-induced reduction of presynaptic glutamate and GABA release contributes to synaptic plasticity, while the reduction of glutamate release inhibits excitotoxicity following ischemia. Evidence implicates 2-AG more so than AEA in plasticity, while both AEA and 2-AG are involved in neuroprotection.

Cannabinoids and Ocular Tissues

Marijuana induces conjunctival vasodilation and reduces intraocular pressure (IOP), but is not a mydriatic. These effects are mediated locally by eCBs as demonstrated in the ciliary body, iris, choroid, and trabecular meshwork in mammalian tissues. THC, as low as 10^{-12} M, increases monoamine oxidase (MAO) activity in the bovine trabecular meshwork, choroid, and ciliary processes but not in the iris. Hydrolysis of anandamide has been measured in the porcine iris, choroid, lacrimal gland, and optic nerve. CB1 mRNA and CB1R-immunoreactivity (IR) have been detected in the ciliary body, trabecular meshwork, and conjunctival epithelium of rat, mouse, bovine, and human. AEA and 2-AG have been measured by gas chromatography in human ocular tissues. The content of eCBs varies in certain disease states, suggesting the importance of eCBs in maintaining ocular homeostasis. For example, 2-AG levels are lower in the ciliary body of patients with glaucoma. However, in diabetic retinopathy there are higher levels of 2-AG only in the iris, and increased levels of AEA in the retina, ciliary body, and cornea. Eyes of patients with age-related macula degeneration (AMD) also show increases of AEA in the retina, choroid, ciliary body, and cornea. Topically applied AEA reduces IOP by activation of CB1R and activation of the prostaglandin E 2 receptor (EP2R) after conversion of AEA to prostamides (see **Figure 2**). Administration of either AEA or THC to human nonpigmented epithelium (NPE) cells

induces COX-2 expression, indicating a relationship among prostaglandins, COX-2, and eCBs in lowering IOP. In addition, EP2 receptors have been localized in the NPE of mouse, porcine, and human ciliary body.

Cannabinoids – Retinal Anatomy

Early studies of the effects of cannabis on vision were performed in concert with the effects of alcohol in order to examine the influence on visual motor behaviors as they related to driving. Anecdotal reports also came from studies citing side effects of cannabis when used as an analgesic. Effects on vision are subtle and include blurred and double vision, a reduction in vernier and Snellen acuity, alterations in color discrimination, an increase in photosensitivity and an increase in recovery from foveal glare. It is unlikely that all of these effects of marijuana are due to cortical or preretinal sites because processes of light–dark adaptation take place in the retina. Knockout mice that lack CB1Rs or FAAH are not blind, but the effects on vision have not been studied.

Biochemical Assay

The first evidence for cannabinoids in the retina was the demonstration that THC induced an increase in MAO activity, indicating a role in dopaminergic transmission. Later, FAAH-mediated hydrolysis of ^{3}H-AEA was shown in homogenates of porcine, bovine, and goldfish retinas. AEA and 2-AG were detected in mammalian retina by gas chromatography. Release of AEA from bovine retinal extracts in a physiological buffer demonstrated that the extracts contained the metabolic machinery necessary for eCB release, the precursor NAPE, and PLD.

Localization – Cannabinoid Receptors

CB1Rs have been localized by immunohistochemistry in the retinas of numerous species, including human, monkey, mouse, rat, chick, salamander, and goldfish. Despite differences in detail, there is a common theme. In general, the most prominent label is in cells of the through pathway: photoreceptors, bipolar cells, and ganglion cells. Cone pedicles in all species contain CB1Rs. Rod spherules appear to be labeled in all species except goldfish. Ultrastructural analysis has been performed exclusively on goldfish cones. CB1R-IR is on plasma membrane at the perimeter of the pedicle as well as within the invagination. CB1R-IR is not immediately apposed to the synaptic ribbon, but is at some distance from it. Regarding bipolar cells in mammals, CB1R-IR is restricted to rod bipolar cells as confirmed by double labeling with antisera against PKC. In goldfish, there is a higher proportion ($\sim$3:1) of CB1R-IR in ON bipolar cells compared to OFF bipolar cells. This difference holds for mixed rod–cone bipolar cells as well as for cone bipolar cells. CB1R-IR, on the bipolar cell synaptic terminal membrane, is not adjacent to the synaptic ribbons. Rather, the CB1R-IR is always some distance removed from the ribbon, the same as observed for the cone pedicles.

Regarding rat horizontal cells, CB1R-IR is confined to the cell bodies and is not present on the dendrites, unlike bipolar cells. CB1R-IR is also found on a population of large amacrine cells, identified in rat as a rare type that is immunoreactive for PKC and GABA. In goldfish, CB1R-IR is on presynaptic membrane of amacrine cell boutons. These boutons appear throughout the depth of the inner plexiform layer and are presynaptic to bipolar cell terminals and small processes derived from ganglion cells. It is likely that these CB1R-immunoreactive processes are from a single type of diffuse amacrine cell.

CB1R-IR is on Müller's cells in goldfish but not in any other preparation. There are inconsistent reports of CB1R-IR in mammalian astrocytes, microglia, and oligodendrocytes. Activation of CB1Rs inhibits excitatory amino acid transport and induces glutamate release from astrocytes in the mammalian brain. CB1R and CB2R are involved in gliotic responses to injury. The interaction of eCBs and glia has not been investigated in the retina. CB2 mRNA was described in all cellular layers of the rat retina; this could include glial labeling, particularly Müller's cells.

Localization – Metabolizing Enzymes

There is relatively little information regarding the distribution of eCB metabolizing enzymes in the retina. The distribution of FAAH-IR in the rat and mouse is quite different from that in the fish. FAAH-IR, in rat and mouse, is most prominent in medium size and large ganglion cells, while weaker FAAH-IR is observed in the soma of horizontal cells, large dopaminergic amacrine cells, dendrites of starburst amacrine cells, and Müller's cells. FAAH-immunoreactive bipolar cells in rat and mouse are exclusively cone bipolar cells, in contrast to CB1R-IR that is exclusively in rod bipolar cells. In goldfish, FAAH-IR is present over cone photoreceptors, Müller's cells, and some amacrine cells, not ganglion cells as in mouse and rat. The distribution of FAAH-IR as it relates to FAAH activity was studied in goldfish retina. ^{3}H-AEA is hydrolyzed by FAAH with ^{3}H-AA rapidly incorporated into membrane phospholipids. Silver-grain deposition represents the trapping of ^{3}H-arachidonic acid in the plasma membrane. FAAH-IR and specific ^{3}H-AEA uptake showed the same pattern over cone photoreceptors, Müller's cells, and some amacrine cells. The co-distribution of FAAH-IR and ^{3}H-AEA uptake indicates that the bulk clearance of AEA from the extracellular space in the retina occurs as a consequence of a concentration gradient across the plasma membrane created by FAAH activity.

AEA is a ligand for the TRPV1 receptor whose binding site is on an intracellular domain. As FAAH and TRPV1 are integral membrane proteins of the endoplasmic reticulum and plasma membrane, respectively, FAAH activity may regulate the levels of AEA for TRPV1 activation. Also, following the hydrolysis of AEA by FAAH, LOX metabolites of AA could activate TRPV1. AEA then could act as an intracellular mediator by being produced from and/or degraded by the same neurons that express TRPV1 receptors. Supporting anatomical evidence for this scheme was provided first in goldfish in which co-localization of TRPV1-IR with FAAH-IR occurs in three types of amacrine cells, two of which are GABAergic. These cells ramify in interplexiform layer sublaminae *a* and *b*, indicating a general function in the OFF, ON, and ON/OFF pathways. The role would depend on the downstream cascade following the increase in calcium concentration.

MGL-IR co-localizes with FAAH-IR over medium size and large ganglion cells and with CB1R-IR in all rod bipolar cells in rat and mouse. In rat retina, COX-2-IR is constitutive in horizontal, amacrine, and ganglion cells. Following transient ischemia, COX-2 is upregulated in these cell types and induced in Müller's cells. This pattern in rat differs from mouse in which COX-2 is restricted to bipolar cell bodies and their axons. COX-2 bipolar cells are a mixed group, with 65% rod bipolar cells and 35% cone bipolar cells. Also rod bipolar cells are of two types: 68% contain COX-2-IR and 32% do not.

The cannabinergic system in vertebrate retinas, as indicated by CB1R-IR, FAAH-IR, COX-2-IR, and MGL-IR, is concentrated in the through pathway of photoreceptors, bipolar cells, and ganglion cells (**Table 1**). These cells for the most part use L-glutamate as their neurotransmitter. Cells of the inhibitory lateral pathways, horizontal cells and amacrine cells, do not feature as prominently. Exceptions are some horizontal cells, dopaminergic amacrine cells, and cholinergic starburst amacrine cells that label weakly for FAAH-IR. Bipolar cells tend to be of the on type and differentiated by rod and cone input. CB1R-IR and MGL-IR are restricted to rod bipolar cells, FAAH-IR to cone bipolar cells, and COX-2-IR to subtypes of rod and cone bipolar cells.

Cannabinoids – Retinal Physiology

Effects on Transmitter Release

Stimulation of CB1Rs via $G_{i/o}$ reduces voltage- and Ca^{2+}-evoked release of [^{3}H]-noradrenaline and [^{3}H]-dopamine in guinea pig retina. Agonists of CB1Rs, but not CB2Rs, inhibit K^+- and ischemia-evoked [^{3}H]

Table 1 The general distribution of CB1 receptors, FAAH, MGL, and COX-2 immunoreactivities in the retina of a variety of species as determined by immunohistochemistry

Species	CB1	FAAH	MGL	COX-2
Fish	Cones	Cones		
	25% OFF BC / 100% ON BC			
	Diffuse AC	TRPV1 AC		
	Müller's cells	Müller's cells		
Rat/Mouse	Rods/cones			
		HC (weak)		
	Rod BC	Cone BC	Rod BC	Rod/cone BC
	PKC AC	DA AC/ACh AC (weak)		
	Ganglion cells	Ganglion cells	Ganglion cells	
Salamander	Rods/cones/ganglion cells			
Chick	Rods/cones/ganglion cells			
Monkey	Rods/cones/ganglion cells			

BC – bipolar cells, AC – amacrine cells, DA – dopamine. Specific details may be found in the citations indicated for each species.

D-aspartate release from isolated bovine retina. Uptake of [^{3}H] D-aspartate identifies high-affinity uptake sites for L-glutamate and L-aspartate in photoreceptors, a small percentage of ganglion cells and Müller's cells. The rank order of potency for the CB1 agonists differs for K^+- and ischemia-evoked release. As photoreceptors are more resistant to ischemia than ganglion cells, the difference in the rank order could reflect the relative potencies of the agonists on CB1Rs on these cell types.

Effects on Ganglion cells

CB1R-mediated activity was demonstrated in rat retinal ganglion cells by [^{35}S]GTPγS autoradiography and reverse transcription polymerase chain reaction (RT-PCR). Voltage-activated Ca^{2+} currents in cultured rat ganglion cells are suppressed by cannabinoid agonist, WIN 55,212-2, an effect that is blocked by CB1 antagonists, SR141716A and AM281. The presence of CB1R function on rat retinal ganglion cells appears unusual in that CB1Rs tend to be at presynaptic boutons. One possibility is that CB1Rs are present on associational ganglion cells, whose axons and axon collaterals do not leave the retina. Rat and mouse ganglion cells also contain FAAH and MGL, putting them in position to regulate AEA and 2-AG as potential retrograde transmitters for suppression of bipolar cell and amacrine cell activity.

Effects on Bipolar Cells

CB1-mediated inhibition of L-type calcium (I_{ca}) and delayed rectifier ($I_{K(V)}$) currents has been reported for ON-bipolar cells of salamander and goldfish. As yet there are no data on OFF-bipolar cells. The voltage-activation range of the currents is not altered, but simply scaled down over the entire activation range. Goldfish mixed rod–cone (Mb) bipolar cells also have D1 dopamine

receptors that enhance I_{Ca} and $I_{K(V)}$ via G protein G_s. CB1R agonists and dopamine oppose each other to modulate $I_{K(V)}$ of Mb bipolar cells. Co-application of WIN 55,212-2 (0.1–0.25 μM) reversibly blocks the enhancement induced by 10 μM dopamine even though low concentrations of WIN 55,212-2 have no effect when applied alone. The effects of dopamine and cannabinoid agonists on $I_{K(V)}$ occur within the physiological range of Mb bipolar cell function (~−25 to 0 mV). $I_{K(V)}$ would be activated during the on portion of the response and, as a counter current, would modulate the peak:plateau ratio of the response. CB1R activation should make the Mb bipolar cell on response more tonic by suppressing the hyperpolarizing effect of $I_{K(V)}$, whereas D1 receptor activation should make the on response more phasic by enhancing $I_{K(V)}$. The effect on ganglion cells should be relatively tonic responses in scotopic (dark-adapted) conditions and relatively phasic responses in photopic (light-adapted) conditions. CB1-induced suppression of calcium currents should reduce transmitter release and reset sensitivity to further increments.

Cannabinoids and Photoreceptors
Voltage-Gated Currents

CB1-mediated modulation of photoreceptor membrane currents has only been reported for tiger salamander and goldfish. The voltage-activation ranges of these currents are not affected. Salamander rods and cones responded differently to WIN 55,212-2. I_K is suppressed in single cones and rods, whereas I_{Ca} is suppressed in cones but enhanced in rods. The differential effect on I_{Ca} and I_K in rods would increase transmitter release, resulting in the reduction of sensitivity, which is an apparent counteradaptive effect. Goldfish cones show a biphasic response to WIN 55,212-2: an enhancement of I_K, I_{Cl}, and I_{Ca} via G_s at concentrations $<1\,\mu M$, and suppression via $G_{i/o}$ at concentrations $>1\,\mu M$. The data obtained with retrograde suppression, to be described below, suggest that the enhancement produced by WIN 55,212-2 may be due to agonist-specific trafficking in which binding of agonists to CB1Rs favors coupling to different G proteins. For example, WIN 55,212-2 increases intracellular calcium by $G_{q/11}$ coupling in human trabecular meshwork cells, while other CB1 agonists, including THC, 2-AG, CP55940, and methanandamide couple to $G_{i/o}$ but not to $G_{q/11}$.

The caution is that the data obtained with WIN 55,212-2 may or may not apply to other agonists or the eCBs.

An effect of evoked released of eCBs in the retina was demonstrated in goldfish. Retrograde suppression of membrane currents in goldfish cones in a retinal slice was achieved by applying a puff of saline with 70 mM KCl or an mGluR1 agonist, RS-3,5-dihydrophyenylglycine (DHPG), through a pipette at an Mb bipolar cell body while recording $I_{K(V)}$ from cone inner segments under whole-cell voltage clamp (**Figure 3(a)**). Retrograde inhibition of $I_{K(V)}$ was reversible and stable over several hours

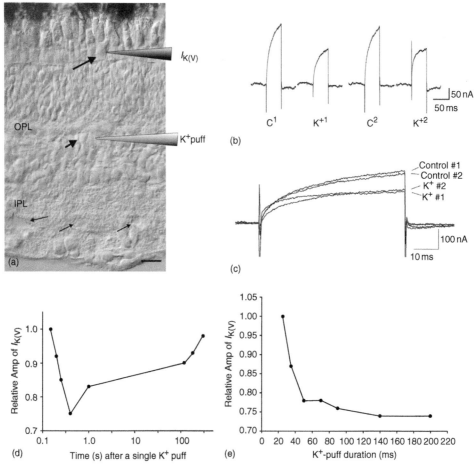

Figure 3 Properties of the retrograde responses of cones. (a) An illustration of the method used to detect retrograde responses in goldfish cones in a retinal slice. Whole cell recordings of $I_{K(V)}$ were obtained from long-single cones (long arrow). A puff pipette, containing 70 mM KCl, was positioned slightly upstream and at the cell body of an Mb bipolar cell (short arrow). Thin arrows indicate the synaptic terminals of Mb bipolar cells. OPL – outer plexiform layer, IPL – inner plexiform layer. Calibration bar = 20 μm. (b), (c) Sequential and overlay of raw records of $I_{K(V)}$ from a single cone evoked by a 50-ms depolarizing pulse to +54 mV from a holding potential of –70 mV. The records have not been normalized. A 50-ms K⁺ puff was delivered twice. $I_{K(V)}$ in response to K⁺ puff #1 was reduced compared to that evoked for the prepuff control #1. The cone was allowed to recover for 30 min after K⁺ puff #1. $I_{K(V)}$ returned to control amplitude (C2, control #2). The K⁺ puff #2 produced an equivalent reduction in $I_{K(V)}$. (d) Time course (log scale) of the reduction of $I_{K(V)}$ in response to a single 50-ms puff of K⁺ shows a latency of about 200 ms following the puff, a peak response at about 500 ms, and a gradual return to control level by 5 min. (e) Effect of K⁺ puff duration on $I_{K(V)}$. These data were obtained from a single cone over 4 h. After a prepuff control value of $I_{K(V)}$ was obtained, a 25-ms K⁺ puff was administered and the effect on $I_{K(V)}$ was determined. The cell was allowed to recover for 30 min and another prepuff control and a K⁺ puff of a longer duration was administered. This sequence was followed for all puff durations. Thus, the value plotted for each puff duration is relative to its own prepuff control. There was no effect with a puff of ≤25 ms duration. Near maximal suppression of $I_{K(V)}$ at about 25% was achieved with a puff of 50 ms and there was little additional effect with puffs as long as 200 ms. Reproduced from Fan, S. F. and Yazulla, S. (2007). Retrograde endocannabinoid inhibition of goldfish retinal cones in mediated by 2-arachidonoyl glycerol. *Visual Neuroscience* 24: 257–267.

(**Figures 3(b)** and **3(c)**). It had a latency of about 200 ms after a K^+ puff, was reduced on average by 25%, and had a halftime of 3.4 min to recover (**Figure 3(d)**). Retrograde suppression of $I_{K(V)}$ was unaffected by a combination of the GABA receptor antagonist, picrotoxin, and α-amino-3-hydroxyl-5-methyl-4-isoxazole-propionate (AMPA) glutamate receptor antagonist, 6-cyano-7-nitroquinoxaline-2,3-dione (CNQX), but blocked completely by the CB1 antagonist, SR141716A, indicating mediation by CB1 receptors. Experiments with the FAAH inhibitor (URB597), a COX-2 inhibitor (nimesulide), and a blocker of 2-AG synthesis (Orlistat) indicated that 2-AG, rather than AEA, is the retrograde eCB.

Two conditions evoke 2-AG release from Mb bipolar cells, strong depolarization and activation of mGluR1, corresponding to voltage-dependent and voltage-independent mechanisms. Rods and cones release glutamate at a steady rate under any ambient illumination; this rate is increased by decrements of light intensity and decreased by increments in light intensity. Voltage-dependent release of 2-AG would occur following depolarization of the Mb bipolar cell in response to a light flash. The retrograde suppression of glutamate release from cones would be a positive feedback that would amplify the reduction in cone transmitter release initially caused by increasing light intensity. This may not be physiologically relevant because the halftime to recover from a 50-ms stimulus is several minutes. The long half-life of the suppressive effect should make this mechanism insensitive to rapid changes in intensity.

The voltage-independent mechanism that provides negative feedback on glutamate release may be more functional. Hypothetically, glutamate, released during ambient illumination, stimulates mGluR1α on Mb bipolar cells tonically to maintain a steady release of 2-AG via a $G_{q/11}$ mechanism. The degree of feedback inhibition of glutamate release from cones varies inversely with ambient illumination; the

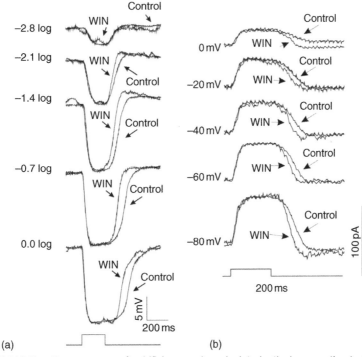

Figure 4 Effect of WIN 55,212-2 on the responses of goldfish cones in an isolated retinal preparation to flashes of light. (a) Voltage-light responses of an L-cone under current clamp to a 200-ms light stimulus of increasing intensities (log unit changes, top to bottom) for Control conditions and after 8 min in 10 μM WIN 55,212-2. Indicated at the left are relative stimulus intensities. The response amplitudes in the control and WIN conditions differed from each other by about 10%. To facilitate comparison, the traces were normalized and superimposed. Except for the dimmest intensity (-2.8 log), there was a speeding up of the response to light offset and an enhancement of the overshoot at two intermediate intensities. There was no effect on the response to light onset or on the plateau phase of the response. The 5 mV calibration refers to the control response. (b) Current-light responses of an L-cone at different holding potentials to a 200-ms light stimulus of approximately half-maximal intensity in control and 10 μM WIN 55,212-2. The timing of the light stimulus is indicated at the bottom of the figure. The amplitude of the light response decreased with decreasing holding potential because the holding potential approached the reversal potential of the photocurrent. The response amplitudes in the control and WIN conditions differed from each other by 5–20%. To facilitate comparison, the traces were normalized and superimposed. Speeding up of the response to light offset in response to WIN 55,212-2 is apparent at all holding potentials. There was no effect of WIN 55,212-2 on the response to light onset or plateau phases of the light response. The holding potential did not change the kinetics of the light responses. The 100 pA calibration refers to the control response. Modified from Struik, M., Yazulla, S., and Kamermans, M. (2006). Cannabinoid agonist WIN 55212-2 speeds up the cone light offset response in goldfish. *Visual Neuroscience* 23: 285–293.

dimmer the background, the stronger the negative feedback. As background is increased, feedback is reduced. Thus, ambient illumination produces an eCB tone that maintains transmitter release from cones within narrow limits. In this way, the ability of the cone to respond to increases and decreases of light intensity is maintained regardless of background. Retrograde transmission occurs even though Mb bipolar cells were hyperpolarized by glutamate acting on either the excitatory amino acid transporter (EAAT) or mGluR6 receptor. The retrograde effect was suppression of currents and not biphasic as expected from data obtained with low concentrations of WIN 55,212-2. It was this finding that led to the idea that the enhancing effect of WIN 55,212-2 on goldfish cones and salamander rods was due to agonist specific trafficking.

WIN 55,212-2 Affects the Cone Light Response

WIN 55,212-2 affects not only cone membrane currents, indicative of presynaptic modulation, but also the response of cones to light. Goldfish cones in an isolated retina preparation were stimulated by light in combination with voltage- and current-clamp protocols (**Figure 4**). WIN 55,212-2 (10 μM) has no effect on the absolute sensitivity of the cones or the kinetics of the onset response. However, the light offset response is faster and the depolarizing overshoot is enhanced. This effect is seen at all but dim intensities (**Figure 4(a)**) and is independent of holding potential (**Figure 4(b)**). This is found under current-clamp as well as under voltage-clamp conditions, indicating modulation of the cyclic guanosine monophosphate (cGMP)-gated channels in the cone outer segment rather than by voltage-dependent currents. The effects of WIN 55,212-2 are not blocked by SR141716A, indicating that CB1Rs are not involved. Given a train of flashes, the photocurrent recovers more quickly with WIN 55,212-2, such that the peak-to-peak response to succeeding flashes is increased. This effect, combined with the shortened recovery time to the offset of bright flashes, could increase contrast detection or critical flicker frequency. A concern is whether the effect of WIN 55,212-2 on the photoresponse would be observed with other CB1 agonists or eCBs because the effect of WIN 55,212-2 is not mediated by CB1 receptors.

In summary, cannabinoids presynaptically suppress the synaptic output of photoreceptors and on-bipolar cells. The effect is subtle as might be expected since smoking marijuana does not produce blindness. Evidence for effects of cannabinoids on amacrine cells is strongest for their suppression of dopamine release. Dopamine, a signal for light adaptation in the retina, antagonizes the action of cannabinoids in on-bipolar cells. eCBs are critically involved in neuronal plasticity. This also appears to include light and dark adaptation, processes of neuronal plasticity that occur in the retina.

Cannabinoids – Development and Neuroprotection

Studies regarding the effects of prenatal-marijuana use on children show deficits on visual habituation, tremors, and startle responses in neonates of 4–30 days old, but no effects on children of 1–6 years old. Problems with behavior, visual perceptual tasks, language comprehension, attention, and memory in 9-year-olds are attributed to effects on the prefrontal cortex, an area enriched in CB1Rs. Although CB1R localization and effects on GABA release have been studied in embryonic rat and chick retinas, no studies have investigated or commented on the effects of manipulating eCBs on retinal development.

The end point of glaucoma is ganglion cell death by apoptosis that may be caused by optic nerve injury following compression or ischemia. CB1 agonists (THC and cannabidiol) as well as inhibition of FAAH protect ganglion cells from glutamate excitotoxicity and ischemia caused by increased IOP. In contrast, COX-2 contributes to neuronal cell death following ischemia or NMDA-toxicity in glial cells, retinal pigment epithelium (RPE), and ganglion cells, while COX-2 blockers prevent ganglion cell apoptosis. Despite progress on the interaction of eCBs, COX-2 metabolites, and EP2 receptors in neuroprotection in the brain, such information is lacking in the retina.

Conclusion

The cannabinergic system is concentrated in the through pathway of the retina. Cannabinoids suppress dopamine release from amacrine cells and presynaptically inhibit potassium currents and glutamate release from cones and on-bipolar cells. How this relates to light and dark adaptation, receptive field formation, temporal properties of ganglion cell responses, and ultimately visual behavior needs to be addressed. eCBs are the most recently described neuromodulators to be studied extensively in neural and non-neural tissues. The existence of multiple eCBs, degradative enzymes, and receptors paints a picture of great complexity. They are important for their role in neuroplasticity and neuroprotection. Further study will verify the importance of eCBs in the retina as well.

See also: Information Processing: Amacrine Cells; Information Processing: Bipolar Cells; Information Processing: Ganglion Cells; Information Processing: Horizontal Cells; Neurotransmitters and Receptors: Dopamine Receptors; Phototransduction: Adaptation in Cones.

Further Reading

Fan, S. F. and Yazulla, S. (2007). Retrograde endocannabinoid inhibition of goldfish retinal cones is mediated by 2-arachidonoyl glycerol. *Visual Neuroscience* 24: 257–267.

Glaser, S. T., Deutsch, D. D., Studholme, K. M., Zimov, S., and Yazulla, S. (2005). Endocannabinoids in the intact retina: ^{3}H-anandamide uptake, fatty acid amide hydrolase immunoreactivity and hydrolysis of anandamide. *Visual Neuroscience* 22: 693–705.

Iversen, L. L. (2000). *The Science of Marijuana.* New York: Oxford University Press.

Nucci, C., Gasperi, V., Tartaglione, R., et al. (2007). Involvement of the endocannabinoid system in retinal damage after high intracellular pressure-induced ischemia in rats. *Investigative Ophthalmology Visual Science* 48: 2997–3004.

Onaivi, E. S., Sugiura, T., and Di Marzo, V. (eds.) (2006). *Endocannabinoids: The Brain and Body's Marijuana and Beyond.* Boca Raton: CRC Press.

Straiker, A. and Sullivan, J. M. (2003). Cannabinoid receptor activation differentially modulates ion channels in photoreceptors of the tiger salamander. *Journal of Neurophysiology* 89: 2647–2654.

Straiker, A., Stella, N., Piomelli, D., et al. (1999). Cannabinoid CB1 receptors and ligands in vertebrate retina: Localization and function of an endogenous signaling system. *Proceedings of the National Academy of Sciences of the United State of America* 96: 14565–14570.

Struik, M., Yazulla, S., and Kamermans, M. (2006). Cannabinoid agonist WIN 55212-2 speeds up the cone light offset response in goldfish. *Visual Neuroscience* 23: 285–293.

Tomida, I., Pertwee, R. G., and Azuara-Blanco, A. (2004). Cannabinoids and glaucoma. *British Journal of Ophthalmology* 88: 708–713.

Yazulla, S. (2008). Endocannabinoids in the retina: From marijuana to neuroprotection. *Progress in Retinal and Eye Research* 27(5): 501–526.

Yazulla, S., Studholme, K. M., McIntosh, H. H., and Deutsch, D. G. (1999). Immunocytochemical localization of cannabinoid CB1 receptor and fatty acid amide hydrolase in rat retina. *Journal of Comparative Neurology* 415: 80–90.

Yazulla, S., Studholme, K. M., McIntosh, H. H., and Fan, S. F. (2000). Cannabinoid receptors on goldfish retinal bipolar cells: Electron-microscope immunocytochemistry and whole-cell recordings. *Visual Neuroscience* 17: 391–401.

Relevant Websites

http://cannabinoidsociety.org/ – This is the official website of the International Cannabinoid Research Society. It provides updates and background information on all aspect of the endocannabinoid field.

http://webvision.med.utah.edu/ – This website from the University of Utah provides extensive coverage of retinal anatomy and physiology, particularly mammals.

Retinal Degeneration through the Eye of the Fly

N J Colley, University of Wisconsin, Madison, WI, USA

Glossary

ABC-type multidrug transporter – A family of adenosine-triphosphate-binding cassette (ABC) transmembrane proteins that transports various molecules, including proteins, ions, sugars, and lipids, across extracellular and intracellular membranes using energy derived from adenosine triphosphate (ATP).

Allele – One member of a pair of genes occupying a specific location on a chromosome (locus) that controls the same trait, for example, eye color.

cGMP phosphodiesterase (PDE) – This enzyme is found in several tissues, including the rod and cone photoreceptor cells, and it belongs to a large family of cyclic nucleotide PDEs that catalyze the hydrolysis of cyclic adenosine monophosphate (cAMP) and cyclic guanosine monophosphate (cGMP) into AMP and GMP, respectively.

Class B scavenger receptors – A family of proteins, which includes the scavenger receptor class B type I (SR-BI), and CD36, which are cell surface receptors that mediate lipid uptake. They are also thought to play an important role in vitamin A metabolism by mediating the uptake of carotenoids into cells.

Cyclophilin – A protein that binds the immunosuppressant drug, cyclosporin, which is often used to suppress tissue rejection following an organ transplantation. The protein displays peptidyl prolyl *cis–trans* isomerase activity, which catalyzes the *cis/trans* isomerization of peptide bonds on proline residues and is thought to play a role in protein folding.

Flippases – The enzymes located in the membrane that aid in the movement of phospholipid molecules between the two leaflets that comprise a cell's membrane. The process requires energy derived from ATP.

Homeodomain-containing transcription factors – A homeobox is a DNA sequence found within genes that regulates developmental processes in animals, fungi, and plants. A homeobox is about 183-DNA-bp long and it encodes a 61-amino-acid protein domain, called the homeodomain, which binds DNA and plays a key role in the regulation of gene expression.

Paired domain – A conserved domain found in a set of transcription factor proteins which are important in regulating gene expression during development.

Paired box (PAX) genes belong to the PAX family of transcription factors.

Rab-GTPase – These are small guanine-nucleotide-binding proteins (G proteins) conserved from yeast to humans and are members of the Ras superfamily of small GTPases. They function in distinct steps in membrane trafficking pathways, including vesicle formation, actin- and tubulin-dependent vesicle movement, and membrane fusion events.

Retinitis pigmentosa (RP) – A heterogeneous group of genetically inherited retinal degeneration disorders leading to progressive loss in vision. Many people with RP retain some sight all their life, others become legally blind in childhood, and some become legally blind in their 40s or 50s. The progression of RP is different in each case.

Rhabdomere – The light-sensing organelle of the *Drosophila* photoreceptor cell. It is the functional equivalent of the outer segment of vertebrate rod and cone cells. A rhabdomere is made up of 60 000 tightly packed microvilli, and each microvillus is 50 nm in diameter and about 1–2 μm in length.

Second-site modifier screens – The genetic screens that are designed to detect a mutation in a second locus (gene) that enhances or suppresses the effect of an existing mutation.

'Prized pest'
You hover over the soft, brown
bananas like a floater, in and out
of my vision, you whose eyes
are so much like my own.

Who am I, dear one,
to swat at you, send you
swirling toward the ceiling
when, really, we share
the same humble beginnings.
And you in your simplicity
hold the key to my complexity.

Let me set out a plate of the sweetest
peaches, invite you to rest
on the arm of my chair,
however late it is.

Marilyn Annucci

Each model organism used to study retinal degenerative diseases has the advantage that others lack. Frogs, fish, rats, and mice have all provided great insights, but it is the tiny fruit fly, *Drosophila melanogaster*, that has played a central role in elucidating the molecular genetics of eye development and the early identification of mutations that cause retinal degeneration. The first mutations in *Drosophila* known to cause retinal degeneration were identified in the 1960s by the pioneering studies of Bill Pak and co-workers. At that time, these findings were only of interest to a few investigators. It was thought that animals as different as flies and humans could not share a similar genetic makeup and, therefore, the amount of transferable knowledge would be limited.

The revolutionary finding that put flies into the spotlight was the one showing that genes controlling pattern formation and development in flies could also do so in humans. In the 1980s, homeodomain-containing transcription factors were found to be essential during development in *Drosophila* for directing the production of appendages, such as the legs and the antennae. Almost identical homeodomain-containing genes were found in the genomes of a wide range of organisms, including humans and mice. This knowledge led to the conclusion that organisms as different as flies and humans contain nearly identical genes. A few years later, working on eye development, Walter Gehring's lab cloned the *eyeless* gene in *Drosophila*. They discovered that *eyeless* is a transcription factor, containing a paired domain and a homeodomain, that directs eye formation. The *eyeless* gene is related to the mouse and human *Pax6* genes (paired box), and the *eyeless/Pax6* genes regulate a cascade of genetic processes involved in eye development. Mutations in these genes result in aniridia in humans, a Small eye (*Sey*) phenotype in mice and an *eyeless* phenotype in *Drosophila*. Aniridia is a congenital condition that is characterized by incomplete iris formation. Further, when expressed in flies, both the *Drosophila eyeless* gene and the mouse *Pax6* gene (*Small eye*, or *Sey*) were able to direct the production of ectopic compound eyes. That *Sey* induced the formation of compound eyes and not mouse structures revealed that mice and flies share signaling components that are interchangeable. The proteins encoded by these genes share 94% identity in the paired domain and 90% identity in the homeodomain. It is remarkable that *eyeless* is not only essential for eye formation, but also its ectopic expression can override other developmental processes in a variety of tissues. For example, *eyeless* is capable of directing leg, wing, and antennal tissues to form eyes. As a result of these findings, *eyeless/Pax6* was dubbed a master regulatory gene for eye formation during development. These landmark discoveries led to an explosion of exciting work in the 1990s that prompted a reassessment of the evolution of eyes. A complex network of eye determination genes direct eye formation, including another Pax6, *twin*

of *eyeless* (*toy*), *sine oculis* (*so*), *eyes absent* (*eya*), and *dacshund* (*dac*). Counterparts of these genes play a role in mammalian eye development and have been implicated in a variety of human diseases. These studies reveal that even though the compound eye of the fly looks very different from mammalian eyes, both share similar signaling pathways that are able to substitute for each other to form an eye.

Due to these elegant findings, it is now widely accepted that many genes are functionally equivalent between flies and humans. In addition, the same (or similar) mutations cause disease in both species. In fact, nearly three-fourths of all human disease genes have related sequences in *Drosophila*. Examples include gene mutations involved in retinal degeneration, deafness, skeletal malformations, cognitive impairment, cancer, immunity, alcoholism, cocaine and nicotine addiction, heart disease, metabolic and storage diseases, Alzheimer's disease, Parkinson's disease, and Huntington's disease.

At the turn of the twentieth century in the hands of Thomas Hunt Morgan, the founding father of *Drosophila* research, *Drosophila* emerged as a powerful genetic workhorse. In 1910, Morgan identified the first mutation in *Drosophila*, which was a spontaneous *white* eye-pigment mutation that caused a normally red-eyed fly to be white eyed (**Figure 1(a)**). This first allele transformed our understanding of genetics and heredity. The *white* gene encodes a membrane-associated, adenosine triphosphate (ATP)-binding, ATP-binding cassette (ABC)-type multidrug transporter required for the transport of pigment precursors involved in eye pigment biosynthesis. In humans, mutations in the *ABCA4* gene (also *ABCR*) account for approximately 3% of autosomal recessive retinitis pigmentosa (RP) and are linked to both recessive cone–rod dystrophy and recessive Stargardt macular dystrophy. Similar to the *Drosophila white* gene, the *ABCA4* gene encodes a membrane-bound, ATP-binding transporter that in humans localizes to the rims of rod and cone outer segment disks. The ABCA4 transporter serves as a flippase in the retinoid cycle. When the *ABC4* gene is mutated, toxic detergent-like by-products accumulate in the retinal pigment epithelium (RPE) leading to severe pathology. Therefore, the *white* gene, discovered at the turn of the century, was subsequently found to encode an ABC-type transporter required for eye pigment biosynthesis in *Drosophila*, and is related to another ABC-type transporter in the human eye that is involved in the retinoid cycle and several types of retinal diseases.

Not only do we share many genes in common with flies, but we also share a great deal of the same metabolic and signaling pathways. Flies are now being used as genetic models for the National Aeronautics and Space Administration (NASA) astronauts and are providing vital information on how space travel and gravitational changes alter gene expression. Work on flies continues to reveal

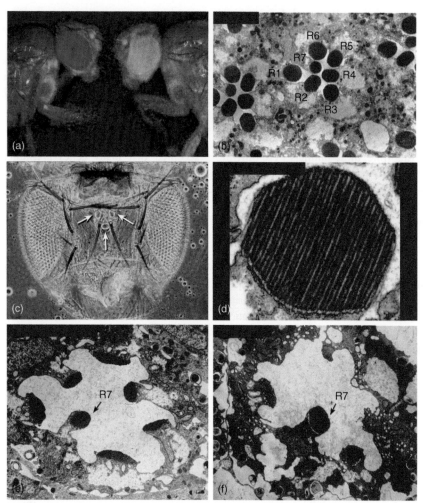

Figure 1   (a) Wild-type red-eyed fly, Canton S compared to a white-eyed mutant fly, w^{1118}. (b) Cross section through the compound eye showing the R1-7 photoreceptor cells and their photosensitive rhabdomeres (R). The R8 photoreceptor cell is located below the plane of the section. (c) The adult *Drosophila* visual system showing the two compound eyes and the three simple eyes (ocelli) located on the top of the head (arrows). (d) A higher magnification of a rhabdomere showing the microvilli. The rhabdomeres are made up of about 60 000 microvilli and are 50 nm in diameter and 1–2 μm in length. (e) A newly eclosed $ninaE^{I17}$ mutant fly, showing the reduced size of the rhabdomeres. $ninaE^{I17}$ is a null allele, so the flies completely lack Rh1 rhodopsin expressed in the R1-6 photoreceptor cells. (f) Six-day-old $ninaE^{I17}$ fly, showing that the rhabdomeres of the R1-6 photoreceptor cells are almost completely gone, but the R7 cell rhabdomere remains.

general principles that are fundamental to a wide spectrum of biological processes. Studies in *Drosophila* have led to conceptual and technical breakthroughs in the areas of development, gene expression, learning and memory, sleep, alcoholism, cocaine and nicotine addiction, ecology and evolution, olfaction, taste, mechanotransduction, vision, hearing, aging, pigmentation, biological clocks and circadian rhythms, courtship and mating behaviors, and human disease and the development of new pharmaceuticals.

In addition to sharing genes and signaling pathways with humans, flies are a powerful model for providing insights into human health and disease for other reasons. In spite of their small size, flies display complex rituals such as courtship behavior, so questions related to the genetic basis of complex behavior are tractable in the fly.

The fruit fly uses the same or similar genes to develop from a fertilized egg to an adult, but they do it in the short time of about 11 days. A female will lay hundreds of eggs, allowing large numbers of genetically identical offspring to be obtained. Flies have a short life span of about 2 months, so the onset and progression of age-related retinal degeneration disorders or any other age-related degenerative process can be studied quite rapidly.

The eye is not essential for viability or fertility of the flies, therefore genes encoding proteins that are uniquely required for visual function may be easily manipulated and studied. Large-scale mutagenesis screens have been carried out, producing hundreds of thousands of mutant flies whose phenotypes can be analyzed to identify genes required for vision. In addition, second-site modifier screens have been used to identify novel genes in signaling pathways. Fly models can be used to dissect the cell biological basis and physiological basis of retinal degeneration, and therefore can be used to obtain insights into mechanisms of degenerative disorders. Just like in humans, electroretinogram (ERG) recordings can be carried out, and they have proved to be an indispensable means for uncovering visual system defective phenotypes that would otherwise have remained unnoticed.

Transgenic flies can be easily produced and, as a result, mutant genes may be introduced and mutant phenotypes may be complemented with wild-type transgenes. Using specialized promoters, genes may be targeted to specific tissues and may also be overexpressed. The fly has a relatively small genome, made up of about 13 600 genes in four pairs of chromosomes. However, despite the dramatic differences in size and apparent complexity between humans and flies – we have less than twice as many genes as a fly – our genome is estimated to be made up of only 20 000–25 000 genes contained in 23 pairs of chromosomes. Therefore, despite the fly's perceived simplicity, or our perceived complexity, our genetic makeup may not be all that different. Its versatility for genetic manipulation and convenience for unraveling fundamental biological processes continue to guarantee the fly a place in the spotlight for unraveling the basis of and therapeutic treatments for human disease.

The Compound Eye and Phototransduction

The *Drosophila* compound eye is composed of approximately 800 individual eye units called ommatidia, each containing the outer, R1-6 photoreceptor cells that extend the full length of the retina and express the major rhodopsin in the eye, the blue-sensitive rhodopsin, Rh1 (**Figure 2**). Rh1 is encoded by the *ninaE* gene, and it displays 22% amino acid identity with human rhodopsin. The inner photoreceptor cells, R7 and R8, are arranged such that a subset of the R7 cells expresses the ultraviolet (UV)-opsin, Rh3. They pair with the R8 cells expressing the blue-sensitive opsin Rh5 (p-ommatidia), while the R7 cells expressing the UV-opsin Rh4 pair with the R8 cells expressing the green-sensitive opsin Rh6 (y-ommatidia). The R7 cells are located above their partner R8 cells (**Figures 1(b)** and **2**). Above the photoreceptor cells are four cone cells and two lens components; the pseudocone (also called the crystalline cone) and the corneal lens. Two primary pigment cells surround the cones and each ommatidium is optically isolated by a sheath of secondary and tertiary pigment cells (**Figure 2**). The adult visual system also contains three simple eyes, ocelli, located on the top of the head (**Figure 1(c)**). The ocelli express the violet-sensitive, Rh2 opsin. *Drosophila* photoreceptor cells contain specialized portions of the plasma membrane,

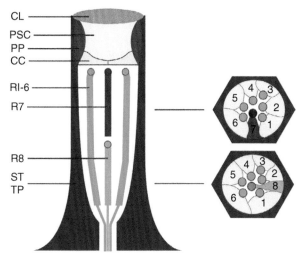

Figure 2 Schematic of an ommatidium. CL, corneal lens; PSC, pseudocone; PP, primary pigment cells; CC, cone cells; R1-6, R7, and R8 photoreceptor cells; SP and TP, secondary and tertiary pigment cells. Adapted from Tomlinson, A. and Ready, D. F. (1987). Cell fate in the *Drosophila ommatidium*. *Developmental Biology* 123: 264–275.

called rhabdomeres, which comprises approximately 60 000 tightly packed microvilli containing rhodopsin photopigments and other components of the phototransduction cascade (**Figure 1(b)** and **1(d)**). The microvillar processes of the rhabdomeres are functionally similar to the photo-transducing disk membranes present in the vertebrate photoreceptor outer segments (**Figure 3**).

Phototransduction in *Drosophila* utilizes a signaling cascade in which light stimulation of rhodopsin leads to the activation of the heterotrimeric guanine-nucleotide-binding G protein (Gq) and the stimulation of phospholipase C beta (PLC-β), leading to the opening of the cation-selective transient receptor potential (TRP) and TRP-like (TRPL) channels. The photoreceptors depolarize as intracellular calcium dramatically rises from about 100 nm to about 10 μM. In the rhabdomeres, calcium rises even higher, to about 1 mM, and it is required for amplification, rapid response kinetics, and light adaptation in *Drosophila*. Calcium is subsequently removed from the rhabdomeres by a combination of sodium/calcium exchange and diffusion into the cell body where calcium increases to about 10 μM. Calcium in the cell body is buffered by calcium-binding proteins and is removed by uptake into intracellular stores by the sarco/endoplasmic reticulum (ER) calcium ATPase. The *Drosophila* phototransduction cascade shares some similarities with the phototransduction cascade in mammalian rod and cone photoreceptor cells. Both cascades are initiated by light-activation of rhodopsin that in turn leads to the stimulation of heterotrimeric G proteins. Phototransduction in *Drosophila* as well as in humans is terminated when the protein arrestin binds to light-stimulated rhodopsin and blocks the binding of rhodopsin

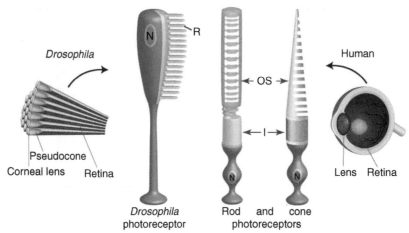

Figure 3 The *Drosophila* photoreceptor cell compared with human rod and cone photoreceptor cells. In *Drosophila*, the pseudocone cone (also called the crystalline cone) and the corneal lens are the lens elements, and they are secreted by the underlying cone cells. The *Drosophila* lens is comprised of droscrystallin, which is similar to insect cuticular proteins. R, rhabdomere; OS, outer segments; I, inner segments; N, nucleus. Photoreceptor cell drawings adapted from Chang, H. Y. and Ready, D.F. (2000). *Science* 290: 1978–1980.

to Gq. However, notable differences are that rod and cone channels are gated by cyclic nucleotides and they close in response to light, leading to a hyperpolarizing response.

Although certain features of phototransduction in *Drosophila* differ from rod and cone phototransduction, *Drosophila* phototransduction shares many common features with the cascade in intrinsically photosensitive retinal ganglion cells (ipRGCs). These cells function in circadian rhythm entrainment and pupil constriction. The light response in ipRGCs is initiated with absorption of light by melanopsin, which is more similar to *Drosophila* rhodopsins than to the photopigments in rods and cones. Light-stimulated melanopsin is thought to activate a phophoinositide cascade leading to the opening of channels that display similar properties to TRP channels. Therefore, *Drosophila* photoreceptor cells and ipRGCs share similar phototransduction cascades, and studies in flies will continue to provide insights into ipRGC function.

Genetic Screens Identify Retinal Degeneration Loci

Several forward genetic screens in *Drosophila* led to an explosion in the identification of many genes involved in retinal degeneration, and their counterparts in human disease. The approach has been to chemically mutagenize flies to disrupt photoreceptor cell function. The mutagenized flies are tested for function by ERG analysis, morphology by a deep pseudopupil (DPP), or Western blotting for the loss of candidate gene expression (such as arrestin). The ERG approach was pioneered by Bill Pak and co-workers in the 1960s and led to the isolation of over 200 ERG-defective mutants. In the 1980s, the development of gene-cloning techniques for *Drosophila* made it

possible to isolate the corresponding genes. For example, *neither inactivation nor after potential E* (*ninaE*), a mutant isolated in the original ERG screen was later found to harbor a mutation in the structural gene for the major rhodopsin in *Drosophila*, Rh1. In 1985, almost a decade prior to the cloning of eyeless, *Drosophila* rhodopsin, Rh1, was cloned and sequenced by two groups and found to display 22% amino acid identity with bovine rhodopsin. In addition, the first evidence that mutations in a rhodopsin gene led to retinal degeneration came from elegant studies in the 1980s in *Drosophila*. The use of the ERG in *Drosophila* was an effective strategy to identify mutants in phototransduction and also in retinal degeneration.

The DPP is a sensitive phenotype in the eye that can be easily assessed in live flies. It is based on the precise packing of the photoreceptor cells. Any mutation leading to, even subtle, structural alterations in photoreceptor cells will cause attenuation in the DPP. For example, a reduction in rhodopsin levels in the R1-6 photoreceptor cells in *ninaE* mutants, leads to structural alterations in the photoreceptor cells (**Figures 1(e)** and **1(f)**) and attenuation in the DPP. A variety of mutants were isolated by this method, including dominant alleles of *ninaE* (rhodopsin), and alleles of two chaperone proteins, *ninaA* (cyclophilin) and *calnexin*. Both the ERG and the DPP screens accelerated the pace of identifying mutations that cause retinal degeneration in *Drosophila*.

Retinal Degenerations in Flies and Humans

Mutations in rhodopsin are the leading cause of blinding disease in RP. RP is a heterogeneous group of inherited disorders that is characterized by progressive retinal degeneration and eventual blindness. RP may be inherited as an X-linked (about 5–15% of cases), autosomal recessive (50–60%), and autosomal dominant trait (30–40%). It affects one person in 4000 worldwide and is often restricted to the eye, but not always. In about 20–30% of the cases of RP, the genetic defects are not eye specific. There are approximately 30 syndromes that involve RP. One of the most common syndromes is Usher's syndrome. This syndrome is characterized by vision and hearing impairment, and mutations in myosin VIIA are responsible for one form, Usher 1B syndrome. Interestingly, loss of myosin VIIA function leads to deafness in *Drosophila*. In flies, like in humans, there are many examples of mutations in which the phenotype caused by the mutation is restricted to the eye, whereas there are others that are not. For example, mutations in the gene encoding rhodopsin (*ninaE*) and the *arrestin* gene cause defects that are restricted to the eye. Mutations in genes such as *retinal degeneration B* (*rdgB*, encoding a phosphatidylinositol transport protein), involve olfactory as well as visual defects.

Since the initial findings, in 1983, that mutations in *Drosophila* rhodopsin lead to retinal degeneration, over 100 mutations in human rhodopsin have been found to cause autosomal dominant RP (adRP). The first mutation identified in adRP patients, published by Dryja and co-workers in 1990, was a mutation that caused a proline residue located near the N-terminus of rhodopsin to be replaced by a histidine residue (Pro23His). A great majority of these mutants, including Pro23His, produce misfolded rhodopsin that is improperly transported through the secretory pathway. However, the mechanism by which the mutant rhodopsins cause dominant retinal degeneration was not known. In 1995, studies in *Drosophila* on rhodopsin mutations that act dominantly to cause retinal degeneration revealed that the retinal degeneration results from the interference in the maturation of normal rhodopsin by the mutant protein. These studies in *Drosophila* provided a mechanistic explanation for the cause of certain forms of adRP.

Mechanisms of Retinal Degenerations

Light-Dependent Retinal Degenerations

It is now widely appreciated that retinal defects and retinal degeneration can be triggered by mutations in almost every component of the photoreceptor cells. These mutations can be divided into two distinct classes. One class pertains to the unregulated activities of phototransduction and/or calcium toxicity. Mutations in this class lead to retinal degenerations that are dependent on or influenced by light stimulation of the cascade and the opening of the TRP and TRPL channels, and these are termed light dependent.

For example, some mutations in rhodopsin itself or mutations in the *arrestin* gene lead to light-dependent retinal degeneration. Arrestin is required for deactivating rhodopsin, and loss of arrestin causes unregulated rhodopsin and hence excessive activation of phototransduction. It is also thought that the loss of arrestin causes decreased endocytosis of Rh1 and all of these defects lead to retinal degeneration.

The precise spatial and temporal regulation of calcium is also essential for photoreceptor survival in flies and people. Prolonged elevation of cytosolic calcium or low levels of calcium can be toxic, leading to cell death and retinal degeneration. In *Drosophila*, mutations in *arrestin*, the $Na^+/Ca2^+$ exchanger (*calx*), the diacylglycerol kinase (*retinal degeneration A, rdgA*), and constitutively active TRP channels are all thought to trigger cell death by causing abnormally high levels of calcium. In humans, a lack of cyclic guanosine monophosphate (cGMP) phosphodiesterase (PDE), caused by mutations in *PDE6A* and *PDE6B*, leads to an elevation in the cGMP concentration in the outer segments, which in turn causes cGMP channels to be open, resulting in excessive levels of calcium. Defects in PDE6A and PDE6B cause recessive RP.

Light-Independent Retinal Degenerations

A second class of retinal degenerations involves defects in rhodopsin maturation and does not require activation of phototransduction by light. These are termed light independent. In *Drosophila*, as in humans, Rh1 is synthesized and glycosylated in the ER, binds its vitamin-A-derived chromophore (11-*cis* 3-hydroxyretinal), at a lysine residue in the seventh transmembrane domain, is transported through the various compartments of the Golgi, and is delivered to its final destination for phototransduction. The mechanisms that regulate rhodopsin maturation, such as its folding, glycosylation, chaperone interaction, chromophore attachment, and transport, are key to photoreceptor survival in flies and humans.

In flies, the transport of Rh1 from the ER to the rhabdomere requires the cyclophilin, NinaA. Cyclophilins are known to display peptidyl-prolyl *cis–trans* isomerase and are thought to play a role in protein folding during biosynthesis. Consistent with a role in protein folding, NinaA resides in the ER. In addition, NinaA is detected in secretory transport vesicles together with Rh1, and forms a specific and stable complex with Rh1, consistent with a broader role as a chaperone in the secretory pathway. Similarly, in mammals a cyclophilin-like protein (RanBP2/Nup358) modulates protein biogenesis. The *Drosophila* cyclophilin, NinaA, is a chaperone that is specifically required for Rh1 biosynthesis and maturation. Another chaperone required for Rh1 biosynthesis in *Drosophila* is calnexin and mutations in *ninaA* (cyclophilin), *ninaE* (Rh1), and *calnexin* all lead to severe retinal pathology in flies. In mammalian photoreceptors, calnexin is also expressed in the ER. Although calnexin is not required for the expression of rod rhodopsin, cone M-opsin, or melanopsin (in the ipRGCs) in the mouse, it is required for proper retinal morphology.

Rhodopsin in both mammals and flies undergoes N-linked glycosylation during biosynthesis, and in flies, elimination of the glycosylation site, asparagine 20 (N20I), results in the retention of rhodopsin in the secretory pathway. Moreover, in both mammals and flies, genes involved in rhodopsin chromophore biosynthesis and transport are critical to rhodopsin maturation and expression as well as photoreceptor function. Defects in chromophore production in the *Drosophila* mutants *ninaB*, *ninaD*, *ninaG*, and *santa maria*, cause a failure in Rh1 transport from the ER to the rhabdomere, resulting in a severe reduction in Rh1 and retinal pathology. The *ninaB* gene encodes an enzyme that catalyzes the conversion of carotenoids to retinal (β, β′ – carotene-15, 15′-monooxygenase) and *ninaG* encodes an enzyme that acts to convert retinal to 3-hydroxyretinal (oxidoreductase). Two additional *Drosophila* loci, *ninaD* and *santa maria*, are both similar to the mammalian class B scavenger receptors and play a role in transporting β-carotene to cells. In flies, β-carotene is required in the diet for the production of all-*trans* retinol, which is in turn converted to 11-*cis* 3-hydroxyretinal. Upon light stimulation 11-*cis* 3-hydroxyretinal is photoconverted to all-*trans* 3-hydroxyretinal. Mutations in chromophore biosynthesis result in defective Rh1 maturation, low levels of Rh1, and retinal pathology, establishing the importance of vitamin A in the fly.

Once rhodopsin exits the ER, it requires several Rab-guanosine triphosphate (GTP)ases for vesicular transport through the secretory pathway in flies and in mammals. Rab-GTPases are conserved from yeast to humans and are members of the Ras superfamily of small GTPases. They function in distinct steps in membrane trafficking pathways including vesicle formation, actin- and tubulin-dependent vesicle movement, and membrane fusion events. In *Drosophila*, Rab1, Rab6, and Rab11 mediate vesicular fusion between the ER and the Golgi (Rab1), intra-Golgi (Rab6), and post-Golgi (Rab11) transport of rhodopsin in *Drosophila*. Defects in Rab function cause inadequate Rh1 transport and retinal pathology. Therefore, the mechanisms that regulate Rh1 maturation, such as its folding, chaperone interaction, and chromophore binding and transport are essential for photoreceptor health in flies and humans.

Retinal Degenerations Caused by Mutations in Dual-Role Proteins

Although most retinal degenerations are classified as either light dependent or light independent, there is a growing list of retinal degenerations that fall into both

classes. In these cases, the corresponding mutant proteins play dual roles. For example, as was described above, calnexin is a chaperone required for rhodopsin maturation. In addition, it is a calcium-binding protein for regulating calcium in photoreceptor cells. Mutations in *calnexin* lead to defects in Rh1 maturation and retinal degeneration. The degeneration due to defects in rhodopsin maturation is light independent, but *calnexin* mutants also display prolonged and elevated levels of calcium, following light stimulation. In the *calnexin* mutants, the retinal degeneration is enhanced by the stimulation of phototransduction by light. Therefore, calnexin plays a dual role: one in rhodopsin maturation and another in calcium modulation.

Summary

In the 1980s, *Drosophila* took on a surprising new role, as an animal model for retinal disease, when the genetic similarities and fundamental processes between flies and humans became apparent. It became clear that information obtained in flies was transferable to human blinding diseases. As a result, and since then, there has been an explosion in the use of *Drosophila* as an animal model for unraveling the molecular genetic basis of retinal degeneration disorders. Despite its perceived simplicity, the fruit fly is, indeed, a remarkably complex creature with a genetic makeup that is surprisingly similar to our own. Investigators continue to capitalize on a whole host of versatile genetic techniques together with the accessibility of the fly to dissect fundamental photoreceptor cell mechanisms *in vivo*. The short life span of the fly, only 2 months, allows for monitoring the onset and progression of retinal degeneration in a short time. These advantageous features place *Drosophila* at the forefront of current research efforts, aimed at unraveling the basis of and therapeutic treatments for retinal degenerative disorders.

Acknowledgments

Our research, on retinal degeneration in *Drosophila*, is supported by funding from the National Eye Institute, the Retina Research Foundation, and the Retina Research Foundation/Walter H. Helmerich Research Chair. I gratefully acknowledge C. Vang, E. Rosenbaum and B. Larson for assistance with preparing the figures. For the poem, I thank M. Annucci, author of Luck (Parallel Press) and member of the Department of Languages and Literatures at the University of Wisconsin-Whitewater.

See also: Circadian Photoreception; Circadian Rhythms in the Fly's Visual System; Coordinating Division and Differentiation in Retinal Development; Embryology and Early Patterning; Evolution of Opsins; Ganglion Cell Development: Early Steps/Fate; Genetic Dissection of

Invertebrate Phototransduction; Histogenesis: Cell Fate: Signaling Factors; Photoreceptor Development: Early Steps/Fate; The Photoreceptor Outer Segment as a Sensory Cilium; Phototransduction: Phototransduction in Rods; Phototransduction: Rhodopsin; Primary Photoreceptor Degenerations: Retinitis Pigmentosa; Primary Photoreceptor Degenerations: Terminology; Retinal Histogenesis; Secondary Photoreceptor Degenerations: Age-Related Macular Degeneration; Secondary Photoreceptor Degenerations; *Xenopus laevis* as a Model for Understanding Retinal Diseases; Zebra Fish as a Model for Understanding Retinal Diseases; Zebra Fish–Retinal Development and Regeneration.

Further Reading

Bok, D. (2007). Contributions of genetics to our understanding of inherited monogenic retinal diseases and age-related macular degeneration. *Archives of Ophthalmology* 125: 160–164.

Colley, N. J., Baker, E. K., Stamnes, M. A., et al. (1991). The cyclophilin homolog *ninaA* is required in the secretory pathway. *Cell* 67: 255–263.

Colley, N. J., Cassill, J. A., Baker, E. K., et al. (1995). Defective intracellular transport is the molecular basis of rhodopsin-dependent dominant retinal degeneration. *Proceedings of the National Academy of Sciences of the United States of America* 92: 3070–3074.

Graham, D. M., Wong, K. Y., Shapiro, P., et al. (2008). Melanopsin ganglion cells use a membrane-associated rhabdomeric phototransduction cascade. *Journal of Neurophysiology* 99: 2522–2532.

Greenspan, R. J. and Dierick, H. A. (2004). 'Am not I a fly like thee?' From genes in fruit flies to behavior in humans. *Human Molecular Genetics* 13(2): R267–R273.

Halder, G., Callaerts, P., and Gehring, W. J. (1995). Induction of ectopic eyes by targeted expression of the eyeless gene in Drosophila. *Science* 267: 1788–1792.

Hardie, R. C. and Postma, M. (2008). Phototransduction in microvillar photoreceptors of *Drosophila* and other invertebrates. In: Allan, A. K., Basbaum, I., Shepherd, G. M., and Westheimer, G. (eds.) *The Senses: A Comprehensive Reference* vol. 1, pp. 77–130. San Diego, CA: Academic Press.

Hartong, D. T., Berson, E. L., and Dryja, T. P. (2006). Retinitis pigmentosa. *Lancet* 368: 1795–1809.

Pak, W. L. (1995). *Drosophila* in vision research. The Friedenwald lecture. *Investigative Ophthalmology and Visual Science* 36: 2340–2357.

Reiter, L. T., Potocki, L., Chien, S., et al. (2001). A systematic analysis of human disease-associated gene sequences in *Drosophila melanogaster*. *Genome Research* 11: 1114–1125.

Rosenbaum, E. E., Hardie, R. C., and Colley, N. J. (2006). Calnexin is essential for rhodopsin maturation, Ca^{2+} regulation, and photoreceptor cell survival. *Neuron* 49: 229–241.

Rubin, G. M. and Lewis, E. B. (2000). A brief history of Drosophila's contributions to genome research. *Science* 287: 2216–2218.

Tomlinson, A. and Ready, D. F. (1987). Cell fate in the *Drosophila ommatidium*. *Developmental Biology* 123: 264–275.

Wang, T. and Montell, C. (2007). Phototransduction and retinal degeneration in Drosophila. *Pflugers Archiv* 454: 821–847.

Wernet, M. F., Celik, A., Mikeladze-Dvali, T., et al. (2007). Generation of uniform fly retinas. *Current Biology* 17: R1002–R1003.

Relevant Website

http://www.sph.uth.tmc.edu – Genes and mapped loci causing retinal diseases: Homepage.

Retinal Ganglion Cell Apoptosis and Neuroprotection

K M Coxon, J Duggan, L Guo, and M F Cordeiro, UCL Institute of Ophthalmology, London, UK

Glossary

Apoptosis – The process of programmed cell death, whereby the cell proceeds through a highly regulated series of morphological changes resulting in the controlled disassembly of the affected cell.

Bax, Bak, Bad, and Bid – Proapoptotic proteins.

Bcl-2 – B-cell CLL/lymphoma 2, an antiapoptotic protein.

Excitotoxicity – The process by which raised levels of neurotransmitters trigger cell death.

Glaucoma – A major cause of blindness worldwide, resulting from the loss of retinal ganglion cells, with raised intraocular pressure as a major modifiable factor.

Neuroprotection – The use of therapeutic agents to prevent or reverse neuronal damage thereby retaining physiological function.

Neurotrophic factors – The growth factors that promote the growth, differentiation, and survival of neurons.

Retinal ganglion cells – The neurons that relay visual information from their cell soma located in the retina, through their axons which project along the optic nerve to the brain.

Introduction

Retinal ganglion cells (RGCs) relay visual information from their cell soma located in the retina, through their axons which project along the optic nerve to the brain. Their loss is associated with various optic neuropathies such as Leber hereditary optic neuropathy, optic neuritis, anterior ischemic optic neuropathy, and glaucoma. The most prominent of these is glaucoma, which is a major cause of irreversible blindness worldwide. In this article, glaucoma is used to highlight the challenges involved in unraveling the complexities of RGC apoptosis and neuronal cell death in order to facilitate more effective treatment strategies.

Apoptosis

Unlike necrosis, apoptosis is a regulated process of cell death leading to chromatin condensation, DNA fragmentation, oxidative damage, and autophagic degeneration, commonly proceeding through either extrinsic or intrinsic caspase-dependent pathways outlined in **Figure 1**. Identifying which aspect of apoptosis regulation is susceptible in glaucoma has important implications for the elucidation of pharmacological targets.

Apoptosis in Glaucoma

The exact mechanism triggering RGC apoptosis in glaucoma is still unidentified, although the major modifiable risk factor identified is that of raised intraocular pressure (IOP) and is commonly used to investigate apoptotic mechanisms. Mitochondria play a fundamental role in RGC apoptosis, with raised hydrostatic pressure inducing apoptosis that is at least partially dependent on mitochondria. More specifically, Bax, a regulator of membrane permeability, has been suggested to be essential in triggering apoptosis, with RGCs-expressing mutant Bax showing complete resistance to raised hydrostatic pressure, even after axonal loss.

While the primary site of damage in RGCs appears to be the axon, leading to axonal degeneration, and a positive correlation between raised IOP and axon loss has been established, it does not inevitably lead to cell soma death. The contribution of axonal degeneration and secondary challenges to the cell soma are evaluated below.

Diagnosis and Measuring Glaucoma Progression

Traditionally, glaucoma is screened by monitoring changes in IOP using tonometry, a technique lacking sensitivity in glaucoma detection, as damage to the RGCs can occur in the absence of raised IOP. This lack of sensitivity sparked the development and introduction of alternative screening methods. Examples of these are standard automated perimetry to measure visual-field loss, optical coherence tomography, allowing the quantification of the retinal nerve fiber layer (RNFL) thickness, and disk tomography to assess any structural damage to the optic nerve head (ONH). The major drawback common to these methods is the inability to detect the disease before considerable damage to the retina has occurred. It is estimated that death of 50% of the RGCs occurs before there is a significant-enough visual loss to diagnose glaucoma. The development of detection of apoptosing retinal cell (DARC), a method of detecting RGC apoptosis in glaucoma before the onset of visual loss, may be instrumental in early diagnosis and successful treatment

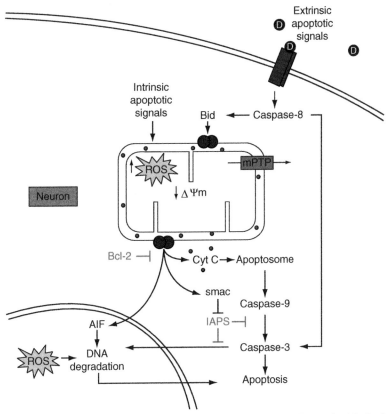

Figure 1 The intrinsic and extrinsic pathways of apoptotic cell death. Apoptosis can be triggered extrinsically through the binding of a death ligand to death receptors at the cell surface. The recruitment of procaspase-8 to the death receptors by adaptor proteins such as FADD follows allowing the conversion of procaspase-8 to its active form, caspase-8. Caspase-8 can cleave procaspase-3 triggering caspase cascade and the activation of the various effecter caspases which degrade DNA and various proteins. It can also act through a mitochondria-dependent manner, cleaving Bid, facilitating its translocation to the mitochondrial membrane and thereby initiating release of death mediators. Various intrinsic signals are also able to mediate apoptosis, through disruption of mitochondrial activity causing a subsequent decrease in the mitochondrial membrane potential ($\Delta\psi m$) and the activation of the mitochondrial permeability transition pore (mPTP), or through the actions of proapoptotic proteins such as Bax, Bak, Bad, and Bid. Following loss of mitochondrial stability various death mediators are released including cytochrome C (Cyt C), which in conjunction with APAF-1 and procaspase-9 forms the apoptosome which activates caspase-9 and triggers a caspase cascade. The release of Smac facilitates this process by inhibiting the actions of IAPS, an inhibitor of the caspase cascade. Various compounds, such as apoptosis-inducing factor (AIF), are also released from the mitochondria and are able to translocate to the nucleus where they trigger chromatin condensation and apoptosis. Mitochondrial dysfunction also leads to the increased generation of reactive oxygen species (ROS), which can trigger cell death through the modification of various molecules, including lipids, proteins, and DNA.

of the disease as well as a high-throughput screening method for new neuroprotectants.

Current Treatments for Glaucoma

Once diagnosed, the current treatments for glaucoma concentrate on lowering the raised IOP. First-line pharmacological therapies include prostaglandin analogs which increase aqueous humor outflow and beta-blockers to reduce aqueous humor formation. Alpha-agonists to increase the uveoscleral outflow of aqueous humor and carbonic anhydrase inhibitors suppressing enzymes involved in aqueous humor production are used as second-line treatments, while third- or fourth-line treatments rely on increasing trabecular outflow using

cholinergic agonists and miotic agents. The disadvantages all the drugs share are the potential side effects and the requirement for topical administration up to four times daily. Furthermore, the incidence of poor compliance and persistence with the topical application is high and so the success of glaucoma management is limited. In addition, glaucoma appears to progress in many sufferers despite the continued use of pharmacological treatments. In these cases, surgery such as trabeculoplasty to reduce resistance to the outflow of aqueous humor by modifying the trabecular meshwork or trabeculotomy to remove part of the trabecular meshwork, allowing enhanced drainage of the aqueous humor, may prove successful in reducing IOP. As the degeneration of visual field in glaucoma is known to progress through RGC apoptosis, research is currently underway to develop neuroprotective treatments for glaucoma.

Neuroprotection

Neuroprotection is defined as the use of therapeutic agents to prevent or reverse neuronal damage, thereby retaining physiological function. For example, neuroprotective treatments are being researched for diseases which progress through the death of neurons of the central nervous system such as Alzheimer's, Parkinson's, and Huntington's. The efficacy of these therapeutic agents in clinical trials has been somewhat controversial. A number of clinical trials performed on stroke patients testing different neuroprotective agents showed either little success or adverse side effects, while treating patients of spinal cord trauma with the neuroprotective agent methylprednisolone resulted in improved motor function. The variation in success of these agents is thought to be due to the mechanism of neuronal death in the different diseases. In strokes, neuronal injury occurs at the cell body, so irreversible cell death occurs instantly; therefore, treatment with a neuroprotective agent would be given too late. However, in spinal cord trauma, neuroprotective treatment may have the desired effect as injury occurs at the axon and death of the cell body results hours later. For this reason, the use of neuroprotection as a treatment for glaucoma, thought to be brought about by death of RGC axons, seems a promising option.

Research Models

Research into the eye disease uses models which are important tools allowing researchers to monitor the progression of the disease and test new therapies. In order for these models to provide informative and transferable data, similarity to the human disease and reproducibility is required. A number of different models are used in the study of the disease, all of which have their strengths and limitations.

In Vitro Glaucoma Models

Initial experiments for neuroprotection in glaucoma are likely to be carried out *in vitro*. A number of ocular cells have successfully been cultured and present a cost-effective alternative to animal models for studying the effects of apoptosis and neuroprotection. Cell cultures have the advantage of allowing rapid screening of potential therapies and observation of direct effects on the cultured cells, under strict environmental control. Cell culture also provides a greater understanding of how compounds function at a cellular level; however, the response of the isolated cells, maintained in an artificial environment, may differ from that *in situ*. For this reason and the contentious use of immortalized cells to study apoptosis, a number of *in vivo* glaucoma models have been developed.

In Vivo Glaucoma Models

Pioneered in rat and more recently developed in monkey, the optic nerve crush is a well-calibrated and reproducible model of glaucoma. Damage to the optic nerve results in cell-body death and subsequent secondary injury to adjacent neurons, as seen in glaucoma. Alternatively, RGC apoptosis can be induced by raising the IOP and causing retinal ischemia, typically achieved through blockage of the aqueous humor outflow, including injection of hypertonic saline into the episcleral veins, cauterization of the episcleral veins, and laser photocoagulation of translimbus. As an alternative, excessive exposure to excitotoxins, such as glutamate or *N*-methyl-D-aspartic acid (NMDA) by intravitreal injections can be used to induce RGC apoptosis. A further model can be generated by laser coagulation at the retina where RGC apoptosis is induced adjacent to the site of laser contact. The DBA/J2 mouse is a genetically determined glaucoma model showing increased IOP, RGC apoptosis, optic nerve atrophy, and ONH cupping.

Mechanisms of Apoptosis and Development of Neuroprotective Agents

Neurotrophic Factor Withdrawal

Deprivation of neurotrophic factors (NFs) induces apoptosis, and has been suggested to play a role in glaucoma as outlined in **Figure 2**. Raised IOP is proposed to lead to a blockade of anterograde and retrograde transports, preventing transport of NFs. The neuroprotective effects of neurotrophins (NTs), a family of NTFs, are thought to be mediated through the activation of phosphoinositide (PI)-(3)-kinase, the inhibition of which is sufficient to block the survival effects of NT. PI(3)K phosphorylates and activates Akt which is known to target several key apoptosis regulators, including Bcl-2 antagonist of cell death (Bad) and cyclic adenosine monophosphate (cAMP) response element binding protein (CREB). The phosphorylation of Bad promotes its sequestration by the chaperone protein 14-3-3, while CREB is activated by Akt (also known as protein kinase B), leading to the upregulation of B-cell CLL/lymphoma 2 (Bcl-2). The actions of Akt have downstream consequences for mitochondrial function and both caspase activation and increased ROS production have been attributed to loss of NT-stimulated pathways.

NFs as Neuroprotective Agents

Brain-derived neurotrophic factor (BDNF) has been showed in a number of studies to have neuroprotective effects. Intravitreal injections of the NT administered to rats, following optic nerve transection and IOP-induced ischemia, increased the survival of RGCs. In addition to BDNF, ciliary neurotrophic factor (CNTF), glial-cell-line-derived neurotrophic factor (GDNF), and pigment-epithelium-derived

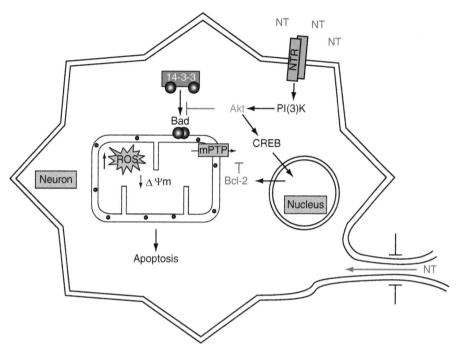

Figure 2 Apoptosis induced by neurotrophic factor withdrawal. Axotomy prevents the retrograde transport of both neurotrophins (NTs) and neurotrophic receptors (NTRs). NTs are able to suppress apoptotic signaling through the binding of NTR, the activation of phosphotidylinositol-3-kinase (PI(3)K), and subsequent activation of Akt. Akt promotes cell survival through multiple pathways, including the phosphorylation of Bad, which promotes its sequestration by the scaffold protein 14-3-3 and the activation of CREB, which promotes increased expression of Bcl-2. Bcl-2 is an essential antiapoptotic protein, which inhibits the release of death mediators, thereby preventing initiation of the caspase cascade.

growth factor (PEDF) have also demonstrated protection of RGCs in rat glaucoma models. The neuroprotective effects of the NFs are short-lived, with reduced survival of RGCs only weeks after a single injection. The longevity of the treatment has been addressed using a number of techniques. Injection of PEDF-peptide-loaded nanospheres into an ischemic rat model reduced RGC apoptosis over a longer period. The same technique was used to administer GDNF to the DBA/2J mouse glaucoma model and an ischemic rat model; again, increased survival of RGCs was observed over a longer period. An alternative method, meant to prolong the effects of the treatment, used an osmotic pump to administer NFs to an axotomized rat model. The technique was successful in reducing RGC apoptosis but most cells were dead within 1 month.

As an alternative to prolonged administration of NTs, researchers in this field have also looked toward gene transfer as a way of maintaining the required level of the NFs for long-term RGC survival. Transformation of a rat glaucoma model with the BDNF gene extended the life of the axotomized RGCs for a similar period. This work was furthered in the same model by injecting BDNF while simultaneously expressing the gene encoding the BDNF receptor, TrkB, known to show reduced expression in glaucoma models. The results showed increased survival of 76% of RGCs over a prolonged period.

Excitotoxicity

Glutamate is a neurotransmitter reported in raised concentrations in the vitreous of glaucoma patients and in animal models. RGCs are known to be highly susceptible to cell death through not only glutamate excitotoxicity and but also treatment with the glutamate analog, NMDA. This has lead to glutamate excitotoxicity being proposed as both a mechanism of primary insult upon RGCs and as a secondary insult following the death of RGCs and the release of further excitory amino acids and glutamate. At elevated levels, glutamate triggers the excessive activation of ionotrophic receptors, such as NMDA receptors, resulting in a subsequent influx of Ca^{2+} ions and an increase in oxidative stress leading to cell death as depicted in **Figure 3**.

Ca^{2+} is shown to increase mitochondrial permeability to ions and solutes and thought to regulate the mitochondrial permeability transition pore (mPTP), which is associated with the release of proapoptotic factors such as cytochrome C (Cyt C). Ca^{2+} has also been proposed to activate nitrous oxide synthase (NOS), leading to the generation of pathological quantities of nitric oxide (NO) and the subsequent production of reactive NO free radicals capable of triggering cell death, the full implications of which are discussed later.

Calpains are a family of cytoplasmic-calcium-activated cysteine proteases. Axotomy, ocular hypertension, and

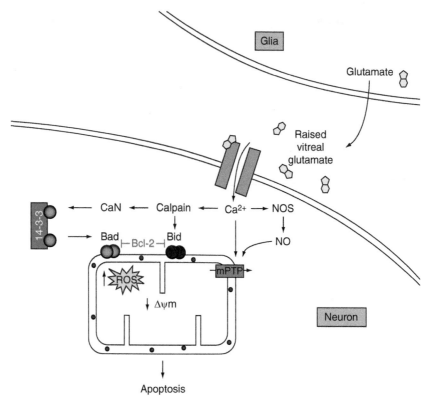

Figure 3 Excitotoxicity-induced apoptosis. Raised vitreal glutamate triggers the activation of ionotrophic calcium channels such as NMDA receptors. The resultant increase in intracellular calcium activates calpain and, in turn, calcineurin (CaN), which promote the translocation of Bad and Bid to the mitochondrial membrane, and the subsequent release of death mediators. Ca^{2+} also activates nitric acid synthase (NOS) which promotes mitochondrial dysfunction, ROS production, and activation of the mitochondrial permeability transition pore (mPTP). Increased cellular ROS and the release of death mediators, such as Cyt C and AIF, leads to apoptosis.

NMDA excitotoxicity all demonstrated calpain activity with inhibition showing reduced RGC loss. Increased Ca^{2+} in the retina correlates with increased calpain activity, which can activate the caspase cascade as well as lead to the phosphorylation of spectrin, Tau, and p35, and cleavage of known calpain substrates, including the auto-inhibition domain of calcineurin (CaN).

In addition to activation by calpains, CaN has also been shown to undergo activating cleavage by caspases and interestingly raised IOP, although the mechanism of the latter is unknown. Sustained Ca^{2+} has also been shown to lead to increased levels of CaN, which induces apoptosis through the dephosphorylation of Bad, releasing the pro-apoptotic protein from sequestration by the chaperon protein 14-3-3. Bad is then able to translocate from the cytosol to the mitochondria, allowing it to heterodimerize with Bcl-2 and B-cell leukemia X_L (Bcl-X_L), leading to the release of Cyt C and the triggering of apoptosis. A role for CaN is further supported by the observation that dosing of cells with glutamate led to the translocation of a green fluorescent protein (GFP) fusion protein of Bad from the cytosol to the mitochondria, and the prevention of this in cells containing an inactive mutant form of CaN.

As mentioned previously, Bad phosphorylation is regulated by NT and treatment with NT is sufficient to confer

resistance to excitotoxic insult. This suggests that excitotoxicity rather than working alone could act in conjunction with withdrawal of NFs to facilitate RGC apoptosis. Recent findings, however, have thrown doubt upon the hypothesis of glutamate excitotoxicity. Crucially, the raised glutamate levels observed in humans and animals have failed to be reproduced in follow-up studies. In addition, NMDA was able to induce apoptosis in the absence of Bax, a proapoptotic factor, suggested to be essential in apoptosis of RGCs in glaucoma. An alternative explanation for increased intracellular calcium and the subsequent triggering of apoptotic pathways is the activation of stress-activated channels by raised hydrostatic pressure.

NMDA-Antagonists and Neuroprotective Agents

This breakthrough of excess glutamate resulting in RGC apoptosis has led to considerable interest in the use of NMDA-receptor antagonists as neuroprotective therapies, many of which have been identified and characterized. MK801, a noncompetitive NMDA-receptor blocker, has been shown to protect RGCs from apoptosis in a number of experimental models including increased IOP and following intravitreal injections of NMDA in rats. Despite showing promise as a successful neuroprotectant in the animal models, MK801 was never used in clinical trials due to its adverse side effects of inducing vacuole formation in neurons and neuronal necrosis.

Other NMDA-receptor antagonists were also studied for use as potential neuroprotectants. For example, dextromethorphan was shown to aid the recovery of retinal activity in rabbits suffering from IOP-induced ischemia and flupirtine had a similar effect on the same model. Riluzole, an agent used for treatment of amyotrophic lateral sclerosis, reduced neuronal death following retinal ischemia brought about by raised IOP in rats. The most promising neuroprotective agent to date, memantine, is similar to MK801 but lacks the neurotoxic effects. Successfully used in the treatment of Alzheimer's disease (AD) and Parkinson's disease, memantine has also been shown to protect neurons from apoptosis in many different glaucoma models. Disappointingly, a recent second phase III clinical trial of memantine has revealed the compound to have no positive effect on visual-field deterioration in glaucoma patients.

Reactive Oxygen Species

Oxidative stress leading to neuronal apoptosis is an early event, occurring within hours of raised IOP, both *in vitro* and *in vivo*. Following atoxomy of RGCs, there is both an increase in reactive oxygen species (ROS) and cell death. Inhibition of ROS demonstrated a 50% reduction in RGC loss as did increasing expression of superoxide dismutase (SOD), known to be reduced in rat models with raised IOP.

The principal mechanism for ROS production is disruption of mitochondrial function, shown in **Figure 4**, leading to loss of mitochondrial membrane potential and subsequent release of death factors. In addition to raised IOP, light exposure has been suggested as a risk factor in glaucoma, through the increased generation of ROS. In addition, ROS can mediate apoptosis by reacting directly with various molecules including DNA, proteins, and lipids.

Mitochondrial Dysfunction and ROS Generation

ROS generation, particularly through the inhibition of complexes I and IV of the electron transport chain, has been implicated in various mechanisms of apoptosis, with treatment of antioxidants preventing apoptosis by inhibiting ROS generation and Cyt C release. A wide variety of antioxidants have been suggested as neuroprotectants in glaucoma, including vitamin E, 3-methyl-1,2-cyclopentanedione (MCP), and melatonin. Derivatives of catechin have been suggested to be potent antioxidants, and intravitreal co-addition of epigallocatechin was shown to attenuate the

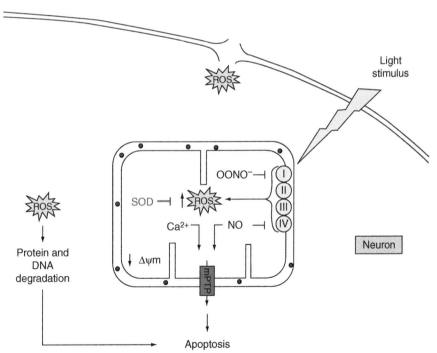

Figure 4 Reactive oxygen species (ROS) and mitochondrial dysfunction in apoptosis. Increased concentrations of ROS can lead to apoptosis through the modification of lipids, proteins, and DNA. Superoxide dismutase (SOD), a key enzyme involved in the removal of ROS, is reduced in glaucoma, potentiating the cells to increased damage by ROS. Inhibition of the electron transport chain by compounds, such as nitric oxide (NO) and peroxynitrite (OONO$^-$), leads to reduced mitochondrial membrane potential ($\Delta\psi$m) and increased generation of ROS, as has exposure to light. NO and Ca^{2+} can also promote mitochondrial dysfunction through activation of the mitochondrial membrane permeability transition pore (mPTP), and the release of death mediators.

retinal damage caused by treatment with the NO donor, sodium nitroprusside.

Ubiquinone (CoQ_{10}), a member of the electron transport chain, has been demonstrated to prevent lipid peroxidation and DNA damage. Furthermore, in recent tests in rats, the intraocular administration of CoQ_{10} afforded neuroprotective effects.

Antioxidants as Neuroprotective Agents

The most promising antioxidant treatment to date is that of orally administered *Gingko biloba* extract, which contains a number of substances shown to be effective in preventing mitochondrial damage through oxidative stress. Visual improvement has also been demonstrated in a double-masked long-term placebo-controlled study, with efficacy and safety reports suggesting a daily dose of 120 mg to be sufficient.

Protein Misfolding

The observation that AD patients demonstrated RGC loss typically associated with glaucomatous changes, including optic neuropathy and visual impairment, suggested possible mechanistic similarities. Supportive of mechanistic similarities between Alzheimer's and glaucoma is the observations of abnormal and phosphorylated Tau, a major plaque component, in the retina of glaucoma patients. β-Amyloid (Aβ), generated by the abnormal processing of amyloid precursor protein (APP), is another major component in Alzheimer's plaques. APP and Aβ are present in RGCs following elevation of IOP in rat models and in the DBA/2J mouse model. It has been shown *in vivo* that Aβ is neurodegenerative and that disruption of these APP processing pathways is sufficient to reduce RGC apoptosis in glaucoma models in rats.

In AD, mutations altering the function of APP processing proteins lead to increased Aβ production and, subsequently, increased apoptosis. A complementary mechanism, shown in **Figure 5**, has been implicated for glaucoma whereby normally rapid anterograde transport is blocked resulting in increased somal concentrations of APP. This has been proposed to trigger the abnormal processing of APP by caspase-3 in RGCs in glaucoma models generating increased Aβ in the retina and is supported by observations of decreased vitreal Aβ (suggesting increased deposition) in glaucoma patients.

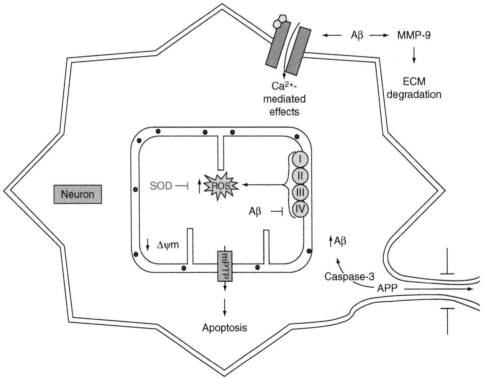

Figure 5 Aβ-induced apoptosis. The anterograde transport of amyloid precursor protein (APP) is blocked by axotomy, leading to increased somal concentrations of APP. At elevated concentrations, APP triggers the abnormal activation of caspase-3 which cleaves APP to yield Aβ, which has both intracellular and extracellular actions in promoting apoptosis. Aβ is shown to target the electron transport chain promoting increased reactive oxygen species (ROS) production and mitochondrial dysfunction. Additionally, Aβ upregulates the activity of ionotrophic calcium channels, increasing intracellular calcium and therefore promoting calcium-mediated apoptosis. Metalloproteinase-9 (MMP-9) also shows increased activity in the presence of Aβ, promoting increased extracellular matrix (ECM) degradation and apoptosis through anoikis.

Aβ is believed to induce apoptosis through elevated intracellular calcium and increased oxidative stress, as seen in AD, where oxidative damage is observed before significant plaque formation. Increased oxidative damage is indicative of mitochondrial dysfunction and both APP and Aβ have been shown to target mitochondria. At raised levels associated with glaucoma, APP has been shown to interact with mitochondria clogging them and preventing normal function. Aβ has been implicated in the inhibition of ketoglutarate dehydrogenase and complex IV of the electron transport chain, inhibition of both of which leads to increased ROS generation. This process could constitute a positive-feedback loop accelerating cell death, as the presence of ROS has been shown to facilitate Aβ production. Following mitochondrial dysfunction, apoptosis has ultimately been suggested to be mediated through the initiation of a caspase cascade.

Reduction of Misfolded Proteins in Neuroprotection

Recent *in vivo* studies have shown that a reduction in RGC apoptosis can be achieved by targeting the Aβ

pathway. Inhibition of β-secretase, responsible for Aβ production, reduced plaque formation and RGC apoptosis. Increased plaque removal and inhibition of plaque formation had a similar effect. All of these mechanisms were shown to be effective in reducing RGC apoptosis, with the greatest benefit seen through the use of combination therapy, which resulted in a maximal reduction in RGC apoptosis of greater than 80%.

Glial–Neuronal Interactions

Glial cells, comprised of astrocytes, microglia, and Müller cells, act as support cells for neurons, maintaining their regular function by providing both neurotrophins and sustenance while removing toxic neurotransmitters and ions. Glaucoma induces dramatic changes in the glia gene expression patterns, potentially converting them from supportive to neurotoxic, as summarized in **Figure 6**. In the scope of this article, the impact of distinct types of glial cells are not differentiated, but evaluated as a whole.

Glial cell activation has been demonstrated in the glaucomatous optic nerve, in the retina of glaucoma

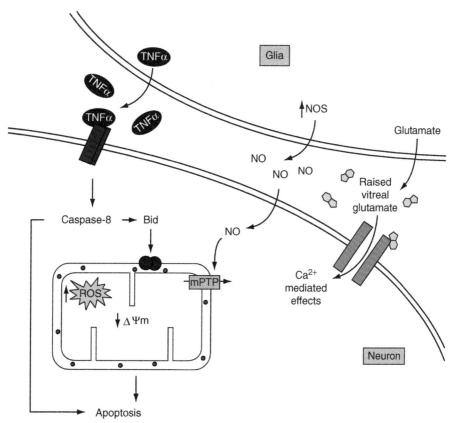

Figure 6 Apoptosis induced by glial–neuronal interaction. Glial cells release neurotoxic molecules such as TNF-α, glutamate, and nitric oxide (NO), which are known to induce apoptosis in RGCs. TNF-α is a death ligand and triggers caspase-8-mediated extrinsic apoptosis. Elevated intravitreal glutamate, due to increased secretion and decreased uptake by glia, leads to apoptosis through calcium-mediated mechanisms, while increased NO causes mitochondrial dysfunction and the subsequent release of death mediators as well as increased reactive oxygen species (ROS) and cytotoxic modifications to key cellular components.

patients and in glaucoma models showing altered levels of secretion. Increased secretion of NT appears to represent an attempted neuroprotective function, supporting damaged neuronal cells, while other changes, such as decreased secretion of interleukin (IL)-6 and increased secretion of NO and tumor necrosis factor (TNF)-α, which have been implicated in RGC apoptosis.

TNF-α, a trigger for extrinsic apoptosis, activates caspase-8 leading to activation of the caspase cascade, Bid activation, the loss of mitochondrial membrane potential, and the subsequent release of cell-death mediators such as apoptosis-inducing factor (AIF) and Cyt C. Increased secretion of TNF-α by glia has been shown to correlate with increased disease severity in glaucoma patients and is further accompanied by the upregulation of TNF receptor 1 (TNF-R1) in glial cells and in RGCs and their axons in glaucoma patients. TNF-α also represents a possible genetic component for glaucoma with the TNF-α-308 gene polymorphism identified in glaucoma patients suggesting a role for TNF-α signaling.

Glial cell secretion of NO, potentially stimulated by TNF-α, is raised in glaucoma with elevated NO concentrations shown to trigger axonal degradation and cell death in RGCs. There are three isoforms of NOS: neuronal (nNOS), inducible (iNOS), and endothelial (eNOS). Initially, iNOS received the most attention as it is induced under various stress conditions. It was suggested to be raised in the ONH of glaucoma patients and also shown to be raised in glaucoma rat model utilizing cauterization, where treatment with inhibitors was sufficient to abate apoptosis.

However, validation of this work in alternative models failed to produce correlating results, with both a murine model and a rat glaucoma model of IOP raised through intravitreal injections, failing to show the involvement of NOS or a reduction in RGC apoptosis by NOS inhibitors. It appears that iNOS upregulation could be due to secondary factors caused by the cauterization process and not raised IOP. While the role of iNOS in glaucoma has been called into doubt, it does not undermine the potential importance of NO signaling. nNOS and eNOS have been showed to be expressed at low levels in normal eyes in glial cells in the ONH, but shows dramatically raised levels in the ONH of glaucoma sufferers.

The main mechanism, through which NO triggers cell death, appears to be disruption of mitochondrial function. An interaction between NO and the mPTP directly facilitates the release of Cyt C and AIF, with mPTP inhibitors abating NO-induced neuronal apoptosis. Caspase-3 activation, however, has been shown without initial loss of mitochondrial membrane potential, suggesting activation may result from blockade of the electron transport chain and subsequent increased levels of ROS. Loss of mitochondrial membrane potential can be triggered as a downstream event, following prolonged inactivation of the electron transport chain.

NO binds complex IV of the electron transport chain, also known as cytochrome oxidase, reducing the enzyme's affinity for oxygen. Prolonged exposure to NO or peroxynitrite ($OONO^-$), a highly destructive molecule formed from ROS and NO, can lead to blockade of complex I of the electron transport chain. This facilitates further ROS production and $OONO^-$ synthesis, which are capable of extensively modifying cellular components leading to apoptosis, while antioxidant treatment in NO-induced apoptosis was shown to abate caspase activation.

Disruption of mitochondrial respiration has also been suggested to facilitate the disruption of Ca^{2+}, associating NO with excitotoxicity-induced apoptosis. This link was strengthened by observations that nNOS-deficient mice were resistant to NMDA-induced RGC death. NO induced modest increases in the amplitude of Ca^{2+} channels and the induction of the Ca^{2+}-mediated death effectors, the calpains.

Extracellular Matrix Degradation

Glaucomatous changes include extensive remodeling of the extracellular matrix (ECM), altering the levels of the various ECM components, including collagen I and IV, matrix metalloproteinases (MMPs), tissue inhibitors of metalloproteinases (TIMPs), transforming growth factor beta 2 (TGF-β2), and laminin.

MMP expression in particular has received attention, with expression in glaucoma patients being raised in comparison to normal patients. The condition of the ECM both regulates and is regulated by the expression and release of MMPs, with the subsequent reduced levels of laminin contributing to cell death, and significant RGC loss in the retina. Supportive evidence from the ability of TIMP-1 to abate neuronal apoptosis, as well as observations that MMP-9-deficient mice demonstrate reduced laminin degradation and increased resistance to neural trauma, further highlights the potential importance of MMPs in glaucoma.

It was recently demonstrated *in vivo* that raised IOP induced remodeling of the ECM within the retina. Raised IOP was shown to correlate to decreases in laminin and TGF-β2 and increased MMP-9, TIMP-1, and RGC apoptosis. Loss of survival signals from the ECM is thought to induce a specific form of apoptosis called anoikis.

Whether glaucomatous changes are initiated at the retina or ONH is still a matter of contention. Astrocytes in the ONH have been shown to be activated by raised IOP and produce MMPs that are able to remodel the ECM, possibly resulting in axonal compression and thereby facilitating apoptotic mechanisms associated with the blockade of anterograde and retrograde transport.

In addition to raised IOP, potential apoptotic mechanisms such as Aβ, NO, ROS, excitotoxicity, and TNF-α have all been implicated in triggering an increase in

MMP-9, suggesting that MMP-9 may represent an important downstream executor of apoptosis rather than a primary affecter.

Neuroprotective Vaccine

Studies using rat and mouse models of glaucoma have suggested a potential role for autoimmunity in the protection of neurons from secondary damage. Following the initial insult, nonspecific T-lymphocytes have been shown to accumulate at the primary lesion site. While nonspecific T-lymphocytes did not exhibit neuroprotection, effects were observed upon injection with myelin basic protein (MBP)-specific T-lymphocytes or immunization with MBP. Unfortunately, injections of anti-MBP-specific T-lymphocytes and MBP immunization induced the paralytic condition – experimental autoimmune encephalomyelitis (EAE).

Copolymer 1 (Cop 1), a synthetic peptide based on MBP, was discovered to suppress EAE and shown in clinical trials to be beneficial to patients with multiple sclerosis. Further studies have revealed that immunization with Cop 1 increases RGC survival following optic nerve crush, glutamate injections, and increased IOP in rats. This research shows that there is potential for the development of a glaucoma vaccine.

Summary

The primary mechanism of RGC loss in glaucoma is through apoptosis, which can be triggered through a variety of mechanisms both intrinsic and extrinsic. Apoptotic signaling shows a large level of redundancy with extensive cross talk. This redundancy poses a major problem in terms of neuroprotection with inhibition of one apoptotic pathway merely delaying apoptosis before mediation through an alternative pathway or the triggering of necrosis. It is still unclear as to the primary mechanism of apoptosis induction in glaucoma, although an increased understanding would aid a more effective development of neuroprotective strategies. Fundamental to the successful development of neuroprotective strategies is targeting the apoptotic pathway upstream and at multiple points to maximize effectiveness.

The studies of neuroprotective agents on glaucoma models carried out to date have shown great promise with many different agents demonstrating efficacy in a large number of models, summarized in **Table 1**. Unfortunately, translation of this research from animals through to clinical trials has exposed complications with side effects and inefficacy, hindering progression. However, due to the complicated nature and a number of different overlapping pathways inducing apoptosis, the identification of a multitude of potential neuroprotectants has been possible. One major problem still faced in the analysis of these compounds is the difficulties monitoring efficacy *in vivo*.

Table 1 Comparison of the models used to test the different neuroprotective agents

Neuroprotectant	Animal	Model
MK801	Rat	Increased IOP
		NMDA injection
Detromethorphan	Rabbit	Increased IOP
Flupirtine	Rabbit	Increased IOP
	Rat	Increased IOP
		NMDA injection
Riluzole	Rat	Increased IOP
Memantine	Monkey	Increased IOP
	Mouse	DBA/2J
	Rat	Glutamate injection
		Increased IOP
		Optic nerve crush
Brain-derived neurotrophic factor (BDNF)	Rat	Increased IOP
		Optic nerve transection
Ciliary neurotrophic factor (CNTF)	Rat	Increased IOP
		Optic nerve transection
Glial-cell-line-derived neurotrophic factor (GDNF)	Mouse	DBA/2J
	Rat	Increased IOP
Pigment-epithelium-derived growth factor (PEDF)	Rat	Increased IOP
		Optic nerve transection
Myelin basic protein (MBP) T-lymphocytes	Rat	Optic nerve transection
		Spinal cord contusion
Myelin basic protein (MBP)	Rat	Spinal cord contusion
Copolymer 1 (Cop 1)	Rat	Glutamate injection
		Increased IOP
		Optic nerve crush
Monoclonal anti-Aβ IgG	Rat	Increased IOP
Congo Red	Rat	Increased IOP
β-Secretase inhibitor	Rat	Increased IOP
Epigallocatechin (EGC)	Rat	NO donor
Ubiquinone (CoQ$_{10}$)	Rat	Increased IOP
Vitamin E	–	Primary cultures
	Human	Glaucoma patients
Gingko biloba	Rat	Increased IOP
	Human	Glaucoma patients

It is hoped that the development of the novel DARC technique for early diagnosis of glaucoma may also be a valuable tool in the analysis of potential neuroprotectants.

See also: Information Processing: Ganglion Cells; IOP and Damage of ON Axons.

Further Reading

Cheung, W., Guo, L., and Cordeiro, M. F. (2008). Neuroprotection in glaucoma: Drug-based approaches. *Optometry and Vision Science* 85(6): 406–416.

Cordeiro, M. F., Guo, L., Luong, V., et al. (2004). Real-time imaging of single nerve cell apoptosis in retinal neurodegeneration. *Proceedings of the National Academy of Sciences of the United States of America* 101(36): 13352–13356.

Grossmann, J. (2002). Molecular mechanisms of ''detachment-induced apoptosis – anoikis'' *Apoptosis* 7(3): 247–260.

Guo, L. and Cordeiro, M. F. (2008). Assessment of neuroprotection in the retina with DARC. *Progress in Brain Research* 173: 437–450.

Kuehn, M. H., Fingert, J. H., and Kwon, Y. H. (2005). Retinal ganglion cell death in glaucoma: Mechanisms and neuroprotective strategies. *Ophthalmology Clinics of North America* 18(3): 383–395, vi.

Lebrun-Julien, F. and Polo, A. D. (2008). Molecular and cell-based approaches for neuroprotection in glaucoma. *Optometry and Vision Science* 85(6): 417–424.

Lin, M. T. and Beal, M. F. (2006). Mitochondrial dysfunction and oxidative stress in neurodegenerative diseases. *Nature* 443(7113): 787–795.

Lipton, S. A. (2003). Possible role for memantine in protecting retinal ganglion cells from glaucomatous damage. *Survey of Ophthalmology* 48(supplement 1): S38–S46.

Wein, F. B. and Levin, L. A. (2002). Current understanding of neuroprotection in glaucoma. *Current Opinion in Ophthalmology* 13(2): 61–67.

Weinreb, R. N. and Lindsey, J. D. (2005). The importance of models in glaucoma research. *Journal of Glaucoma* 14(4): 302–304.

Yuan, J. and Yankner, B. A. (2000). Apoptosis in the nervous system. *Nature* 407(6805): 802–809.

Zhong, Y. S., Leung, C. K., and Pang, C. P. (2007). Glial cells and glaucomatous neuropathy. *Chinese Medical Journal (Engl)* 120(4): 326–335.

Retinal Histogenesis

J A Brzezinski, IV and T A Reh, University of Washington, Seattle, WA, USA

Glossary

Birthdate or born – The time when a progenitor permanently exits the cell cycle. Marking cells in their final division is referred to as birthdating.

Cell fate determination – The process by which a progenitor is programmed to become a specific cell type and its functional maturation. In the retina, a progenitor needs to acquire competence, exit the cell cycle, become specified, and differentiate.

Competence – The potential to adopt a certain cell fate(s).

Differentiation – The functional maturation of a specified cell.

Histogenesis – The development of a mature tissue from a naive progenitor population.

Lineage – A cohort of cells derived from division(s) of a common progenitor. This is also referred to as a clone.

Lineage tracing – The use of an indelible marker to label cells and all their descendants. Analysis at a later time point allows the inference of lineal relationships. This is also referred to as fate mapping.

Multipotent progenitor/progenitor – A cell that has competence to adopt several different fates. These cells may or may not be proliferative.

Progressive restriction model – A model of cell fate determination where multipotent progenitors lose the competence to adopt multiple cell fates over time. By this model, early progenitors have the competence to form all fates in a tissue.

Serial competence model – A variation on the progressive restriction model whereby multipotent progenitors transiently gain and subsequently lose competence for a subset of fates in a tissue over time. By this model, progenitors do not have the competence to form late fates at the earliest time points.

Specification or commitment – The point when a progenitor cell has irreversibly decided on a cell fate.

Birthdating

An interesting property of retinal cells is that they do not continue to divide after they differentiate. This means that at some point in development, a progenitor permanently exits the cell cycle, referred to as its birthdate. Investigators took advantage of this property and designed a clever pulse-chase experiment to investigate whether different cell types exited the cell cycle at characteristic times. Animals at various stages of development were given a pulse of ^{3}[H]-thymidine (3HdT). The 3HdT is incorporated into replicating DNA during synthesis (S)-phase and any excess is cleared from the body quickly. The incorporated 3HdT is maintained in the newly synthesized DNA permanently. If the cell continues to divide, it will dilute the 3HdT signal by one-half each division. Next, autoradiography was conducted after the retina was fully formed (the chase). Cells that retained maximum labeling are those that exited the cell cycle (born) on the day of 3HdT administration. More recently, birthdating studies have been conducted with synthetic nucleotides, such as 5-bromo-2-deoxyuridine (BrdU), which can be detected by antibodies instead of autoradiography.

About 50 years ago, Sidman used birthdating studies to test what order, if any, retinal cells were formed in the rodent retina. The observations of Sidman and future investigators revealed a stereotypical birth order that was broadly broken down into early and late groups (**Figure 1**). Retinal ganglion cells (RGCs) were born first, followed closely by horizontal cells, cones, and amacrines. The late cohort comprised rods, bipolar cells, and Müller glia. Birthdating has been conducted in several vertebrate species. While there are some small differences in the order, the first cells born in all species examined are RGCs. Although there is clearly an overall birth order, which is evident from the production onset of each cell type, there is also considerable overlap in the genesis of the cell types, such that multiple cell types are born on the same day of development (**Figure 1**). This overlap is also observed in species where retinal histogenesis is long, such as in monkey. These data implied that cell fate determination is not strictly regulated by measuring time (or cell cycles) during development. Nonetheless, the observance of a birth order indicated that there is a temporal input into cell fate determination.

These studies raised several questions about retinal progenitors. Are all retinal cell types derived from the progenitors in the optic cup? Are there different progenitors for each retinal cell type? Is fate choice predetermined or stochastic? These questions were addressed by tracing the fate of individual progenitors.

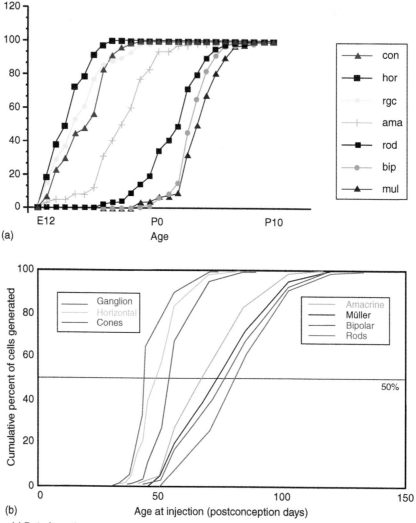

Figure 1 Birthdating. (a) Data from the rat retina showing the birthdates for each cell type as a percentage of its total. The progression from early to late cell-type generation that was first observed by Sidman can be clearly identified. CON, cone; HOR, horizontal cell; RGC, retinal ganglion cell; AMA, amacrine cell; BIP, bipolar cell; and MUL, Müller glia. (b) Similar thymidine labeling study in monkey shows a similar, though not identical, pattern of birthdates for the various types of retinal cells. Despite the fact that the patterns of generation are not identical, the early and late fates are largely segregated. (a) Data from Rapaport, D. H., Wong, L. L., Wood, E. D., Yasumura, D., and LaVail, M. M. (2004). Timing and topography of cell genesis in the rat retina. *Journal of Comparative Neurology* 474(2): 304–324. (b) Modified from La Vail, M. M., Rapaport, D. H., and Rakic, P. (1991). Cytogenesis in the monkey retina. *Journal of Comparative Neurology* 309(1): 86–114.

Lineage Tracing

To understand the behavior of progenitor cells, a way to trace the fate of individual progenitors was needed. Starting in the late 1980s, investigators designed elegant lineage-tracing (fate-mapping) experiments to study the retina. In rodents, Turner, Snyder, and Cepko used replication-incompetent retroviruses encoding a marker gene (e.g., *LacZ*) to infect retinal progenitors at various time points. These retroviruses can only infect dividing progenitors. Since the virus integrates into the genome, the progenitor and its descendents become permanently labeled. By adjusting the titer of the virus, individual progenitor lineages (clones) can be mapped from mature retinas (**Figure 2**).

The first set of retroviral lineage-tracing studies examined postnatal infections of rat retinas. In postnatal day 0 (P0) infections, majority of clones contained rods and were predominantly small (one to four cells). A large number of clones contained only rods, whereas others contained rods, amacrines, bipolars, and/or Müller glia (**Figure 2** and **Table 1**). There were few clones that did not contain any rods. Cell types born before P0 (e.g., cones, horizontals, and RGCs) were not observed in these clones, consistent with previous birthdating analyses. Clones generated from P2, P4, and P7 infections become progressively smaller. This reflects the progressive decrease in progenitor division that occurs in the first postnatal week. Clones were heterogeneous; some contained multiple cell types and others one cell type or just a single cell. From all these time points, there was no obvious lineage hierarchy that could be constructed from the clone composition data. This showed that mammalian retinal cell fate determination is stochastic, or nondeterministic. Importantly, they observed two-cell clones that had different fates. This implied that fate choice is decided during or after the last cell division.

To see if the progenitors in the optic cup can give rise to all the retinal cell types, the investigators generated clones from embryonic (E) time points when few cells have exited the cell cycle. For this reason, they injected their retroviruses into the subretinal space of E13 and E14

mice *in utero*, a difficult procedure. The mice were allowed to mature and clone composition was examined. The clone size and composition were highly heterogeneous (**Table 1**). All seven cell fates were represented in these clones. Moreover, these clones never contained any other cell types (i.e., astrocytes, vascular endothelial cells, pigment epithelium, etc.). These early lineage traces showed that all seven retinal cell types derive solely from a retina-restricted progenitor pool. The heterogeneity in clone composition (there were few multicell clones alike) reinforced that fate choice is apparently stochastic. Consistent with retinal cell frequency (~78% are rods in mice), nearly all clones contained one or more rods. Clone size varied from 1 to

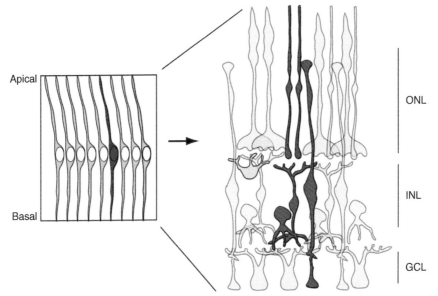

Figure 2 Lineage analysis. Diagram showing the basic strategy to track the lineages of the retinal progenitor cells. A progenitor cell is labeled (red) at an early stage of development, by using either a retroviral vector with a reporter gene, or a direct injection of a tracer. The cell undergoes multiple rounds of division in this case, and when the retina is examined in the adult animal, the different types of retinal cells can be identified by their laminar position and their morphology. In this case, two rods, a bipolar cell, an amacrine cell, and a Muller glial cell were derived from the progenitor. See **Table 1** for more examples of the types of clones found in these experiments.

Table 1 Retroviral lineage tracing in rodents

Age of infection	E13	E14	P0	P2	P4	P7
Ave. clone size	46.4	26.3	2.5	1.6	1.4	1.1
Max. clone size	217	234	22	7	7	2
Cell fates[a]	r, b, a, c, m, g, h	r, b, a, c, m, g, h	r, b, a, m	r, b, m, a	r, b, m, a	r, m
Clone examples	1g	2c	4r	2r	1r, 1b	1r
	1c	1c, 1g	1a	1b	1r, 1m	1m
	1c, 1h	8r, 1c	3r, 1b	1r, 1b	2r	1r, 1m
	17r, 1c, 3b, 1m, 1a	51r, 6b, 1m	2r, 1b, 1a	1r, 1m	3r, 1b	2r
	43r, 1c, 7b, 2a	164r, 2c, 1h, 12b	2r, 1b, 1m	1r, 1a	1r, 1a	
	111r, 16b, 1m, 5a, 1g	28r, 9b, 3a, 1g	10r	3r, 1m	1r, 1b, 1a	

[a]Cell fates represented in clones (r, rod; b, bipolar; a, amacrine; c, cone; m, Müller; g, ganglion; h, horizontal) listed in decreasing frequency observed. In mice, the cell frequency in decreasing order is r, a, b, m, c, g, and h.
Data from Turner, D. L. and Cepko, C. L. (1987). A common progenitor for neurons and glia persists in rat retina late in development. *Nature* 328: 131–136; and Turner, D. L., Snyder, E. Y., and Cepko, C. L. (1990). Lineage-independent determination of cell type in the embryonic mouse retina. *Neuron* 4: 833–845.

234 cells and had a bimodal distribution. There were abundant small clones (<5 cells) and abundant large clones (>20 cells). The largest clones are simply too big to have been generated by simple asymmetric (one progenitor, one neuron) divisions in the time allotted for development. Based on the distribution of clone sizes, it is possible that progenitors preferentially utilize symmetric divisions during retinal histogenesis. Lineage-tracing studies have been conducted in other vertebrates, yielding similar results.

These lineage-tracing studies answered several questions. First, retinal progenitors in the optic cup are multipotent and give rise to all seven cell types. Second, there are no separate progenitors for each cell type. Nonetheless, rod-only clones were observed, raising the possibility of a rod-only progenitor. Since rods make up 78% of the mouse retina, small- to medium-sized rod-only clones are expected from multipotent progenitors. Fourth, cell fate choice is stochastic. Fifth, cell fate specification occurs during the last cell cycle or later. These data raised several more questions about the mechanisms of retinal histogenesis. Is fate determination a cell autonomous process, a cell nonautonomous process, or a combination of both? Do retinal progenitors have broad competence that is gradually lost, or are progenitors more limited in their cell fate choices during development? These questions have been addressed by manipulating the local cellular environment.

Environmental Challenge

Several approaches have been undertaken to discriminate between cell autonomous and cell nonautonomous control of retinal cell fate determination. Most of these experiments involve challenging retinal progenitors with different environments. Unlike the previous birthdating and lineage-tracing studies, these environmental challenge experiments are harder to interpret and, at times, yield conflicting results.

One type of experiment was designed to answer the question: What is the default state of retinal progenitors? For these experiments, retinal progenitors were labeled with 3HdT and dissociated to single cell density to examine their potential for differentiation in isolation. Reh and Kljavin saw that when rat progenitors were isolated from early stages of development, the majority of the progenitors differentiated into RGCs, while progenitors isolated from late stages of development differentiated into cell types normally generated late in development, like rods. A somewhat different result was obtained from similar studies in the chick embryo by Adler and colleagues, suggesting that cones were the default cell fate. However, subsequent studies in chick revealed a ganglion cell bias for the progenitors isolated from the earliest stages of

retinal development. Together, these results led to the concept of a rolling default, or shifting competence in the progenitors; in other words, there is an intrinsic bias to the types of neurons generated by the progenitors that shifts progressively over developmental time. This idea has been recently supported by Notch-signaling studies. The Notch receptor is active in progenitor cells throughout development, and inhibition of signaling leads to premature progenitor differentiation. Inhibition of Notch in early progenitors leads to the overproduction of RGCs and cones, whereas inhibition of Notch function later in development results in overproduction of rods, but not of RGCs or cones. These results show that retinal progenitors change their competence intrinsically over time.

In the next type of experiment, investigators asked whether inductive interactions among the retinal cells played a role in their cell fate determination. Studies from *Drosophila* eye imaginal disk had shown an important role for cell–cell interactions in directing the ommatidial progenitors to their individual identities. In that tissue, the data were best fit by a sequential cell induction model in which the first cells, the R8 photoreceptors, induce the recruitment of the next type of photoreceptor, and so on. To test whether similar sequential inductions occurred in vertebrate retinas, investigators used heterochronic co-cultures, surrounding early embryonic progenitors with late-generated retinal cells. The groups of Raff and Reh co-cultured early embryonic rodent retinal cells with an excess of postnatal retinal cells. In both sets of experiments, the early progenitors were more likely to develop into rhodopsin expressing (rod photoreceptors) cells when compared to early progenitors cultured alone (which developed primarily early retinal fates). This increase was not seen when the challenging (older) cells were derived from the brain instead of the retina. Watanabe and Raff also found that the increase in rods was observed when the two cell populations were separated by a cell-impermeable membrane. Together, these data showed that a soluble factor(s) from late retina can promote rod fate in younger cells. In addition, very early mouse retinal progenitors (E11–E12) could form rods when cultured with an excess of older rat retinal cells; birthdating analyses indicate that rods are not normally generated by progenitors from this very early retina, which suggests that rod competence precedes rod genesis by at least 1 day. These experiments led to the idea that although progenitors may have an intrinsic bias in the types of cells they generate at any time in development, their fate can be influenced by factors in the microenvironment. Many subsequent studies have identified signaling factors in the developing retina that can influence the fate of the progenitors and, particularly, factors that can increase the percentage of these cells that differentiate into rods; however, since low-density cultures do not support robust rod differentiation (i.e., expression of rhodopsin and other markers), it is possible that most of the factors identified to date have more of an effect on the expression of these identifiers, rather than the choice of the rod fate *per se*.

Another concept that emerged from the early studies of retinal development was the idea that specific cell types use feedback regulation to control their density. Elimination of dopaminergic amacrine cells in the developing frog leads to an overproduction of these cells by progenitors at the ciliary marginal zone. This suggested that there is a nonautonomous feedback regulation to negatively control amacrine cell number. *In vitro* experiments found a similar effect in developing rat retina. When E16 retinal cells were co-cultured with an excess of P0 cells, fewer amacrines were generated. However, when the P0 cells were depleted of amacrine cells, there was an increase in the number of amacrine cells generated by the E16 cells. When the converse experiment was done, P0 cells co-cultured with an excess of E16 cells, an increase in amacrines and bipolars was seen along with a decrease in rods. This confirmed the presence of negative feedback on amacrine cells and implied feedback regulation of bipolar cell genesis. The concept of feedback regulation of retinal cell production was extended to ganglion cells by Waid and McLoon. They examined the influence of a late retinal environment on RGC fate determination using heterochronic co-cultures in chick. Early cells were inhibited from RGC fate when co-cultured with late retinal cells. This inhibition was not observed when RGCs were depleted from the challenging (late) cell population, which showed that RGCs can nonautonomously feed back to inhibit further RGC production. In a subsequent study from the McLoon lab, they blocked Notch signaling (an RGC inhibitor) at different times. Interestingly, when later time points were examined, they saw newborn RGCs in areas where RGC genesis had ceased in controls. This suggested that RGC competence extends beyond RGC genesis.

Although the studies described above have supported a role for cell–cell interactions in the regulation of cell fate in the vertebrate retina, some types of studies have failed to find nonautonomous effects. Rapaport and colleagues conducted heterochronic transplants in frogs. When younger retinal tissue was transplanted into older hosts, the donor tissue did not adopt later fates or differentiate early. This result was also seen in co-cultured cells *in vitro*. Importantly, the donor cells directly adjacent to the older host cells were not fate shifted. This suggested that cell fate determination is not regulated by changing environmental stimuli, rather, that cell competence is limiting in frogs. Cayouette and colleagues combined lineage tracing and single cell culturing of rodent progenitor cells. In this experiment, E16–E17 rat progenitors were plated at single cell density and the resulting clones were examined 7–10 days later. In parallel, they conducted a retroviral

lineage trace (similar to above) in E16–E17 retinal explants (intact tissue) cultured the same amount of time. The clones in both cases were screened for rod, bipolar, amacrine, and Müller glial cell fates. The clone composition and size in both experimental systems were similar in isolation and in explants. This suggested that fate choice is largely cell autonomous and that any given cell type is not required to induce (specify) another. In sum, while many studies have shown a role for cell–cell interactions in the control of retinal cell fate, the relative importance of intrinsic and extrinsic regulation is still not resolved.

These data have been used to assemble cell fate determination models (**Figure 3**). One model, progressive restriction, argues that early progenitors have competence to adopt all retinal cell fates and that this competence is gradually lost (restricted) over time. By this model, nonautonomous contributions are expected to specify cell fate in these multipotent progenitors over time. Early progenitors should be able to adopt late fates (if stimulated properly) and competence should extend beyond the normal genesis window to allow for feedback inhibition. Another model, serial competence, contends that progenitors have competence for only a few cell fates at any given time and that progenitors serially cycle through numerous restricted competence states. By this model, cell nonautonomous inputs are not required for fate specification. Presumably, the mechanism that controls competence dynamics would be reflected by the complement of transcription factors expressed in progenitors. In the next section, we discuss evidence that transcription factors regulate competence.

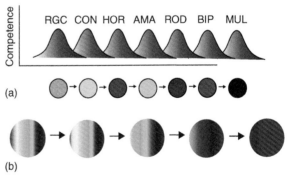

(a)

(b)

Figure 3 Cell fate determination models. (a) Serial competence model. The progenitor changes over time in their potential to generate different types of retinal cells. These changes in competence would be mediated by changes in the complement of transcription factors present in the cells. (b) Progressive restriction model. The progenitor can initially generate all types of retinal neurons, but over time loses one or another of the transcription factors needed for a specific fate. This is shown in the figure as what is initially a rainbow-colored cell, progressing into a cell that is only red. In principle, the progressive restriction could be due to a loss in key transcription factors, or alternatively by the addition of new repressors.

Transcription Factors and Competence

How transcription factors regulate retinal progenitor competence is beginning to be characterized (**Figure 4**). Although most of these transcription factors are expressed in the postmitotic cells, and therefore are unlikely to regulate the competence of the progenitor cells *per se*, there are members of several different transcription factor families that are expressed in the mitotically active cells. One transcription factor expressed by progenitors that appears to convey competence is *Pax6*. This eye-field transcription factor is expressed in all progenitors and directly promotes the expression of other transcription factors (*Ascl1, Ngn2, Atoh7*, etc.). Specific deletion of *Pax6* from progenitors leads to the apparent loss of competence to generate all retinal cell types except amacrine cells. A complementary result is obtained when *FoxN4*, a transcription factor expressed in a subpopulation of progenitors, is deleted in mice. In *FoxN4* null mice, amacrine cells fail to develop. This suggests that the combination of FoxN4 and Pax6 can convey progenitors with competence for all retinal cell fates. Ikaros is a transcription factor that is expressed primarily by early-staged progenitors, and mice deficient in this gene have fewer early-born neurons.

While these examples show that changes in transcription factors can affect the types of neurons produced by progenitors, and hence their competence to generate specific neuronal types in the retina, it has also become clear that a more general level of competence, to generate neurons versus glia, is also conveyed by these factors. Another transcription factor expressed in a subset of progenitors is *Ascl1* (*Mash1*), a member of the bHLH class. Prior to E15 in the rat retina, there is little *Ascl1* expression in the retina, though progenitors by this age are producing RGCs, cones, rods, and horizontal and amacrine cells. Even after E15, *Ascl1* is expressed in most, but not all, progenitors. Deletion of this gene in mice leads to an overproduction of Müller glia, apparently at the expense of rods and bipolar cells. It appears that *Ascl1* imparts retinal progenitors with the competence for late-generated retinal neurons, while also inhibiting Müller glial fate specification (**Figure 5**) in part by maintaining the expression of *Hes6*, but also by driving expression of key components of the Notch pathway, including *Hes5* and *Hes1*, to maintain the progenitors in an undifferentiated state while they generate additional neurons. When Notch signaling is reduced, even for times as short as 6 hours, the retinal progenitors are irreversibly committed to exit the cell cycle and differentiate. Since the Notch effector genes, *Hes1* and *Hes5*, are normally regulated during the cell cycle, such they are lowest during the G2 and M-phases, one model for the mechanism by which progenitors initiate differentiation is through a progressive slowing of the cell cycle as development proceeds (**Figure 5**). Immunolabeling for the active Notch intracellular domain in mouse retina has confirmed that Notch signaling is lowest in cells whose nuclei are at the apical surface. However, this is apparently not the case in the fish retina. Live imaging studies in cortex by Kageyama's group confirm that Notch signaling oscillates through the cell cycle, with the lowest levels in the progenitors with nuclei located at the apical surface, those cells that are in G2- or M-phases of the cell cycle. Given the highly transient nature of Notch signaling and the very fast cell cycles in fish, further studies using highly destabilized reporters will be needed in the fish to determine whether this difference is real.

Another member of the bHLH class of transcription factors, *Atoh7* (*Math5*), is transiently expressed by a small subset of postmitotic progenitors. Lineage-tracing and genetic-deletion studies have shown that it is necessary for RGC competence. In addition to these transcription factors expressed in progenitors, there are many that are expressed in subpopulations of nascent neurons. For example, *Ptf1a*, *NeuroD1*, and *Math3* are expressed in amacrine cells, and loss of one or more of these genes leads to defects in amacrine cell fate determination or survival. Moreover, overexpression of transcription factors such as *NeuroD1* can drive progenitor differentiation into specific cell types, though the types appear to vary depending on the species and the method of

	bHLH	Homeodomain		Other
Progenitor	Ascl1 Ngn2 Olig2 Hes5 Hes1	Pax6 Rax Prox1 Six3/6 Chx10 Lhx2		Sox9 Sox2 FoxN4
Precursor	Ath5 Ath3			
Differentiation cell type RGC		Brn3 Islet1		
CON	NeuroD1	Crx Otx2		TRbeta2 RXRg RORbeta
HOR		Pax6 Prox1		
AMA	NeuroD1 Ath3 Ptf1	Pax6 Prox1		
BIP	bHLHb4	Chx10 Otx2 Islet1		
ROD	NeuroD1	Crx Otx2		Nrl Nr2e3
MUL	Hes1	Rax		

Figure 4 Transcription factor code. Kageyama and others have proposed that a combination of bHLH and homeodomain transcription factors specify each retinal cell type. Although most of these factors are expressed primarily in postmitotic cells, and therefore might not be candidates for the changing competence models described above, several lines of evidence indicate that these factors are necessary for the full differentiation of these cell types.

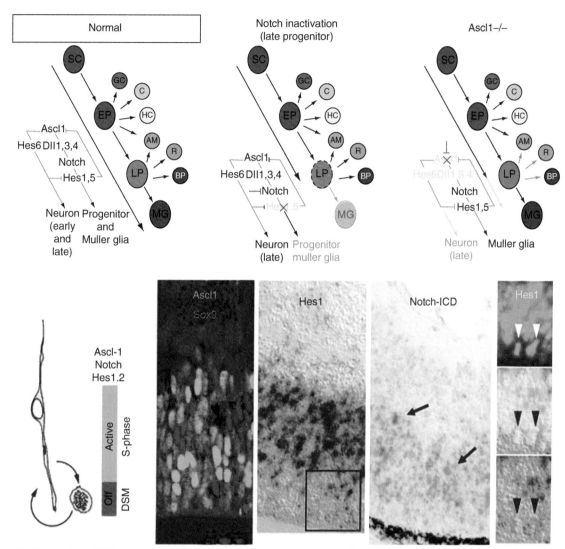

Figure 5 (Top row) *Ascl1/Mash1* functions to maintain neuronal competence in the progenitors. Deletion of this gene in mice leads to an overproduction of Müller glia, apparently at the expense of rods and bipolar cells. It appears that *Ascl1* imparts retinal progenitors with the competence for late-generated retinal neurons, while also inhibiting Müller glial fate specification by maintaining the expression of Hes6, and by driving expression of key components of the Notch pathway to maintain the progenitors in an undifferentiated state while they generate additional neurons. SC, stem cell; EP, early progenitor; LP, late progenitor; GC, ganglion cell; C, cone; HC, horizontal cell; AM, amacrine cell; R, rod; BP, bipolar cell; and MG, Müller glia. (Bottom row, left) Model of how Notch signaling changes with the cell cycle in mouse and chick retina. Notch signaling is high during the S-phase of the cell cycle and low to absent at the apical (ventricular) surface when cells are in G2 and M-phases of the cell cycle. This can be seen in the lower middle and right panels by the Notch ICD immunoreactivity (arrows) and the expression of Hes1, a downstream effector of Notch. By contrast, progenitor markers like Sox9 (red) and Ascl1-GFP (green), are expressed throughout the cell cycle in progenitor cells. Modified from Nelson, B. R., Hartman, B. H., Ray, C. A., et al. (2009). Acheate-scute like 1 (Ascl1) is required for normal delta-like (Dll) gene expression and notch signaling during retinal development. *Development Dynamics* 238: 2163–2178; and from Nelson, et al. (2007) Transient inactivation of notch signaling synchronizes differentiation of neural progenitor cells. *Developmental Biology* 304: 479–498.

overexpression. These studies show that few transcription factors fit cleanly in the simple models of progressive restriction or serial competence. Moreover, since few targets of these transcription factors have been identified, the nature of competence regulation remains unclear. Nevertheless, the transcription factor networks for two cell types, RGCs and rod photoreceptors, are beginning to be worked out.

Conclusions

A great deal has been done over the past 50 years to understand the developmental mechanisms of vertebrate retinal histogenesis. Birthdating studies have revealed a characteristic genesis order of the retinal cell types. Lineage-tracing experiments have shown that retinal progenitors are multipotent and that fate choice is stochastic.

Experiments that challenge the environment of retinal progenitors have revealed that both cell autonomous and cell nonautonomous factors contribute to fate determination. More recent techniques, such as expression fate mapping (lineage by gene expression) and live imaging of cell lineages, will further our understanding of retinal fate determination. In addition, the recent increase in early cell-type-specific markers will allow us to more precisely examine the effects of environmental challenges. While several factors have been identified, many more experiments are needed to elucidate the molecular mechanisms of retinal fate determination.

See also: Coordinating Division and Differentiation in Retinal Development; Ganglion Cell Development: Early Steps/Fate.

Further Reading

Cayouette, M., Barres, B. A., and Raff, M. (2003). Importance of intrinsic mechanisms in cell fate decisions in the developing rat retina. *Neuron* 40(5): 897–904.

Elliott, J., Jolicoeur, C., Ramamurthy, V., and Cayouette, M. (2008). Ikaros confers early temporal competence to mouse retinal progenitor cells. *Neuron* 60(1): 26–39.

Li, S., Mo, Z., Yang, X., et al. (2004). Foxn4 controls the genesis of amacrine and horizontal cells by retinal progenitors. *Neuron* 43(6): 795–807.

Marquardt, T., Ashery-Padan, R., Andrejewski, N., et al. (2001). Pax6 is required for the multipotent state of retinal progenitor cells. *Cell* 105(1): 43–55.

Nelson, B. R., Hartman, B. H., Ray, C. A., Hayashi, T., Bermingham-McDonogh, O., and Reh, T. A. (2009). Acheate-scute like 1 (Ascl1) is required for normal delta-like (Dll) gene expression and notch signaling during retinal development. *Development Dynamics* 238: 2163–2178.

Ohsawa, R. and Kageyama, R. (2008). Regulation of retinal cell fate specification by multiple transcription factors. *Brain Research* 1192: 90–98.

Rapaport, D. H., Wong, L. L., Wood, E. D., Yasumura, D., and LaVail, M. M. (2004). Timing and topography of cell genesis in the rat retina. *Journal of Comparative Neurology* 474: 304–324.

Reh, T. A. (1992). Cellular interactions determine neuronal phenotypes in rodent retinal cultures. *Journal of Neurobiology* 23: 1067–1083.

Reh, T. A. and Kljavin, I. J. (1989). Age of differentiation determines rat retinal germinal cell phenotype: Induction of differentiation by dissociation. *Journal of Neuroscience* 9(12): 4179–4189.

Shimojo, H., Ohtsuka, T., and Kageyama, R. (2008). Oscillations in notch signaling regulate maintenance of neural progenitors. *Neuron* 58: 52–64.

Turner, D. L. and Cepko, C. L. (1987). A common progenitor for neurons and glia persists in rat retina late in development. *Nature* 328: 131–136.

Turner, D. L., Snyder, E. Y., and Cepko, C. L. (1990). Lineage-independent determination of cell type in the embryonic mouse retina. *Neuron* 4: 833–845.

Waid, D. K. and McLoon, S. C. (1998). Ganglion cells influence the fate of dividing retinal cells in culture. *Development* 125: 1059–1066.

Watanabe, T. and Raff, M. C. (1992). Diffusible rod-promoting signals in the developing rat retina. *Development* 114: 899–906.

Retinal Pigment Epithelial–Choroid Interactions

K Ford and P A D'Amore, Schepens Eye Research Institute, Boston, MA, USA

Glossary

Angiogenesis – The formation of new blood vessels from preexisting ones.

Atrophy – To wither or deteriorate.

Blood–retinal barrier – Specialized, nonfenestrated, tightly joined endothelial cells that form a transport barrier for certain substances between the retinal capillaries and the retinal tissue.

Choroidal neovascularization – A condition whereby new blood vessels that originate from the choroid grow and break through Bruch's membrane into the subretinal pigment epithelium (sub-RPE) or subretinal space.

Dominant negative – A genetic mutation where the gene product adversely affects the normal, wild-type gene product within the same cell.

Fenestration – A small pore (60–80 nm in diameter) within the endothelium wall that allows the passage of small molecules and a limited amount of proteins.

Microphthalmia – An abnormal smallness of the eyes, occurring as the result of disease or of imperfect development.

Phagocytosis – The process in which phagocytes engulf and digest microorganisms and cellular debris.

Quiescence – The absence of proliferation.

Trophic factor – A molecule that promotes cellular growth and/or survival.

Introduction

Located at the back of the eye, the retinal pigment epithelium (RPE)–choroid complex is comprised of the RPE, a polarized, epithelial monolayer and the choroid, a highly fenestrated vascular bed. Separated by Bruch's membrane (BrM), an elastic lamina, the RPE and choroid each play a vital role in normal eye physiology. Considerable evidence indicates that there is not only a great deal of interaction between the RPE and choroid, but that the integrity of this interaction is critical to normal eye function. Examination of the morphology of the choroidal microvasculature reveals that the vessels are "polarized," with fenestrations preferentially localized to the capillary surface proximal to the RPE and BrM. Whereas the

cytoplasm of the endothelial cell is thinnest in this region, endothelial cell bodies and nuclei are more prominent distal to the RPE. These observations led to the speculation that the RPE exerts an inductive effect on the choriocapillaris by releasing a factor that diffuses across BrM to provide a trophic effect and to mediate the anatomic specializations in the capillary endothelial cells. Preservation of a normal interaction between the RPE and choroid is required for proper eye physiology. This review explores the intricacies of the RPE and choroid, and their interactions during development, in the adult and with aging.

RPE–Choroid Complex Development

RPE Development

The development of the RPE is dependent upon the coordination of transcription factor expression and inductive signals received from the tissues surrounding the developing eye. Eye development proceeds from two principal tissue components: the neural ectoderm, which buds from the wall of the forebrain to form the optic vesicle, and the surface ectoderm, which forms the lens (**Figure 1**). When the optic vesicle comes in contact with the surface ectoderm, it invaginates, forming the optic cup. The optic cup consists of an inner layer, which gives rise to the neural retina, and an outer layer, which forms the RPE. At the optic cup stage, the presumptive RPE and retina are separated by a thin remnant of lumen, which becomes filled by a material known as the interphotoreceptor matrix (IPM). Coincident with this is the onset of the expression of RPE65, which encodes a protein involved in the conversion of all-*trans*-retinal to 11-*cis*-retinal. Prior to this stage, the RPE is a ciliated and pseudo-stratified epithelium; however, following IPM formation, RPE maturation commences. The onset of RPE maturation is marked by melanogenesis, which requires the activation of the tyrosinase promoter.

As the RPE continues to differentiate, it displays a complete apical to basolateral polarity, with short apical microvilli and small basolateral membrane infoldings, and the formation of tight junctions between the RPE cells, which can be divided into three stages. The early stage of tight junction formation is characterized by the expression of key tight junction proteins, such as zona occludens 1 (ZO-1), occludin, and claudins. However, these tight junctional complexes are rudimentary so the RPE lacks complete barrier properties and is therefore leaky. As the

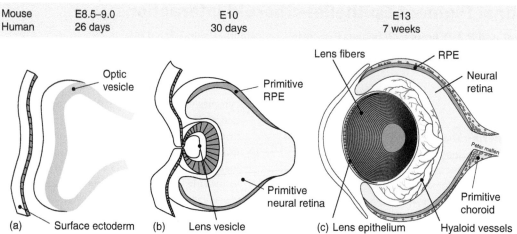

| Mouse | E8.5–9.0 | E10 | E13 |
| Human | 26 days | 30 days | 7 weeks |

Figure 1 Development of the RPE and choroid. (a) Eye development begins with the budding of the neural ectoderm from the wall of the forebrain to form the optic vesicle. (b) When the optic vesicle comes in contact with the overlying surface ectoderm, it invaginates, forming the optic cup, which consists of an inner layer and an outer layer. The inner layer gives rise to the neural retina, and the outer layer eventually forms the RPE. Choroidal development proceeds from two embryonic tissues: the mesoderm and cranial neural crest cells. The endothelial cells of the choroidal blood vessels are derived from the mesoderm, whereas neural crest cells give rise to the stromal cells, melanocytes, and pericytes. (c) Choroid development begins early in eye development, as the oxygenation of the retina is supplied solely by the choroid and the transient hyaloid vascular system. Initially, tubes and spaces form in the surrounding periocular region of the optic vesicle, and eventually expand to form a plexus. Primitive capillaries develop from this plexus adjacent to the RPE as the optic vesicle invaginates. In humans, the choroidal plexus fuses to form a singular vessel known as the annular vessel at the anterior region of the optic cup during the second and third months of gestation. This primitive plexus is then organized into a complex network, and a well-defined choriocapillaris layer also appears at this stage.

early phase ends, Na^+ K^+ ATPase (ATPase, adenosine triphosphatase) becomes concentrated at the apical surface, and the apical microvilli begin to elongate. In the second stage, the tight junctions become increasingly less permeable, presumably due to the alterations in the distribution of various tight-junction proteins. At this stage, the RPE now prevents the free diffusion of membrane proteins, and the basolateral membrane is remodeled. In the last stage of tight-junction formation, the composition of tight-junction protein isoforms stabilizes and the RPE begins to display barrier properties characteristic of a tight epithelium. Following completion of tight-junction formation, the RPE begins to express specialized proteins, such as the glucose transporter, that aid in the transport of essential nutrients from the RPE to the photoreceptors. As RPE maturation concludes, the RPE becomes fully functional and is able to interact with the photoreceptors.

Choroid Development

Blood vessels develop by vasculogenesis and angiogenesis. In vasculogenesis, primitive vascular cells assemble into a primitive capillary plexus, which is remodeled to define the pattern of the vascular architecture, whereas in angiogenesis, new vessels develop as sprouts from preexisting vessels. It is thought that the superficial vessels of the retina form by vasculogenesis at the optic nerve and expand along a gradient from the posterior to the anterior retina. However, new vessels then sprout via angiogenesis

and invade the retina to form intermediate and deep capillary beds. Nonetheless, the primitive vessels derived via vasculogenesis or angiogenesis must still be remodeled before they are considered mature. Remodeling involves the growth of new vessels and regression of others. In addition, alterations in lumen diameter and vessel wall thickness, which are dictated by the local needs of the tissue, must also occur.

The choroid develops from two principle embryonic tissues: the mesoderm and cranial neural crest cells. The mesoderm gives rise to the endothelial cells of the choroidal blood vessels, whereas the stromal cells, melanocytes, and pericytes are all derived from the neural crest cells. Choroid development begins early in eye development, which is not surprising considering that the differentiation of the ocular tissues relies upon the oxygen and nutrients supplied by the primitive vascular system. In early eye development, the oxygenation of the retina is supplied solely by the choroid and the transient hyaloid vascular system. The vascularization of the retina itself is actually a late event. Initially, tubes and spaces form in the surrounding periocular region of the optic vesicle. These tubes, which are lined by the mesodermal endothelium, expand to form a plexus as eye development proceeds. Concomitant with optic vesicle invagination, the primitive capillaries develop from this plexus adjacent to the RPE. In humans, choroidal capillaries completely encircle the optic cup by the 13-mm stage of development and remain separated from the retina by the basement membrane of the RPE. At the anterior region of the optic cup, the choroidal plexus fuses to form a singular vessel known as the annular vessel, and during the second and third months of gestation, this primitive plexus is organized into a complex network. A well-defined choriocapillaris layer also appears at this stage. Pigmentation does not appear in choroidal melanocytes until late gestation – between 6 and 7 months; however, it is complete at birth.

BrM Development

In mice, the formation of the choroidal vessels precedes the deposition of BrM, and maturation of the BrM layers is not complete until 6 weeks following birth. However, in humans and monkeys, BrM layers are completely formed *in utero*. BrM formation commences near the RPE, and is initially comprised of a single basement membrane derived from the RPE. The RPE is able to synthesize many of the extracellular matrix (ECM) components that comprise BrM; therefore, it is not surprising that BrM development begins near the RPE. On the 17th to 18th day of gestation in rats, collagenous fibrils gradually accumulate in the space between the RPE and choriocapillaris endothelium and begin to form a three-dimensional meshwork within BrM. The meshwork consists of three sublayers: the basement membranes of the RPE and choriocapillaris endothelium, and a collagenous layer. The final component of BrM, the central elastic layer, appears on the fifth postnatal day. Initially, dense elastic deposits appear within the collagenous layer, and subsequently accumulate to form a single continuous layer, which separates the collagenous layer into an inner and outer layer. Finally, an immature, but five-layered BrM is formed by the ninth day after birth.

RPE–Choroid Complex: Structure and Function

RPE Structure and Function

It is apparent that the RPE performs a variety of complex functions that are essential for proper visual function. If any aspect of one of the many roles that the RPE serves is disrupted or fails, retinal degeneration, loss of visual function, and ultimately blindness may result. Therefore, the RPE is an indispensable component of the RPE–choroid complex. Anatomically, the RPE is juxtaposed to the outer segments of the photoreceptors at its apical surface, extending long apical microvilli that surround the outer segments. These microvilli provide a means for a complex structural interaction between the RPE and photoreceptors. BrM, which is at the basolateral side of the RPE, separates the RPE from the fenestrated endothelium of the choriocapillaris.

The RPE serves many functions that are vital to normal eye function and physiology. These functions have been categorized based on the characteristics of three classes of cells: epithelium, macrophage, and glia. The major function of an epithelium is to control the passage of substances from one extracellular space to another; therefore, the epithelial aspect of the RPE is its role in actively transporting ions, water, and metabolic end products from the subretinal space to the blood. The RPE is also involved in the uptake of nutrients, such as glucose, retinol, and fatty acids from the blood to nourish the photoreceptors. One of its most critical functions is the exchange of retinal with the photoreceptors. The photoreceptors are unable to reisomerize all-*trans*-retinal; therefore, it is transported to the RPE where it is reisomerized to 11-*cis*-retinal, and then transported back to the photoreceptors, a process known as the visual cycle of retinal.

The RPE also functions as a macrophage in that it phagocytoses the photoreceptor outer segments, which are constantly being renewed. As new membrane is added to the base of the outer segment, the older membrane is advanced toward the tip, and subsequently shed. The phagocytosis of the shed outer segments by the RPE is critical to normal photoreceptor function because it maintains the excitability of the photoreceptors. The vital nutrients, such as retinal, which are recycled and returned to the photoreceptors following outer segment digestion help rebuild the light-sensitive outer segments from the base of the photoreceptors.

The final role that the RPE plays is that of a glial cell. RPE resembles glial cells in the manner in which they respond electrically to changes in the concentration of extracellular K^+, which is determined by the photoreceptors. The RPE possesses tight junctions at both its basolateral and apical membranes, which generate transepithelial resistance (TER); furthermore, the tight junctions at the basolateral surface form the RPE portion of the blood–retinal barrier.

Choroid Structure and Function

The choroid is a vascular bed that supplies nutrients and oxygen to the RPE and outer nuclear layer of the retina. It consists of larger vessels and a highly fenestrated choriocapillaris (**Figure 2**), whose endothelial basement membrane comprises one layer of BrM. Fenestrated endothelium is a characteristic of tissues that are involved in secretion and/or filtration. The photoreceptors are metabolically very active and the fenestrated choriocapillaris facilitates the transport of oxygen and nutrients to fulfill their metabolic needs. RPE tight junctions, in combination with sophisticated transport systems, determine which components of the choriocapillary secretions are transported to the photoreceptors.

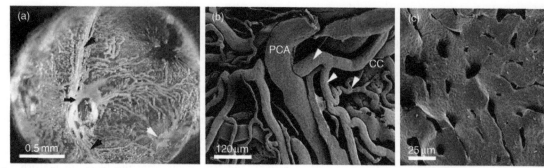

Figure 2 Corrosion cast of adult mouse choroidal vasculature. The choroid is a vascular bed that supplies nutrients and oxygen to the RPE and outer nuclear layer of the retina. It consists of larger vessels and a highly fenestrated capillary bed known as the choriocapillaris. (a) Scleral view of the entire adult choroid. Major vessels can be easily identified: posterior ciliary artery (black arrow), long posterior arteries (black arrowhead), and vortex vein (white arrowhead). (b, c) Scanning electron micrographs of vascular cast of the choroidal vasculature. (b) Posterior view showing one posterior ciliary artery (PCA) around the optic nerve. On each side, the artery divides regularly into smaller branches (arrows) and choriocapillaris (CC). (c) Anterior view showing the extremely dense choriocapillaris plexus.

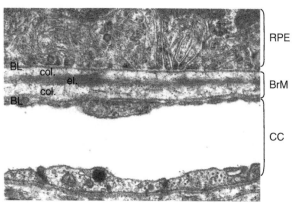

Figure 3 Ultrastructure of the RPE–BrM–choriocapillaris. The RPE and choriocapillaris are separated by a thick (1–4 μm) elastic lamina, BrM, which consists of five layers: the RPE basement membrane (BL), an inner collagenous layer (col), a central elastic layer (el), an outer collagenous layer (col), and the choriocapillaris (cc) endothelium basement membrane (BL), respectively, from top to bottom. BrM acts as a barrier for macromolecules and regulates the diffusion of small molecules between the RPE and the choriocapillaris.

BrM Structure and Function

Located between the RPE and the choriocapillaris, BrM is a thick (1–4 μm) pentalaminar ECM, which consists of the RPE basement membrane, an inner collagenous layer, a central elastic layer, an outer collagenous layer, and the choriocapillaris endothelium basement membrane (**Figure 3**). BrM is comprised of several types of extracellular components, including many types of collagen, laminin, fibronectin, and proteoglycans. The collagenous layers of BrM are primarily composed of type I collagen and collagenous-associated proteins. It is believed that these components fortify the framework of BrM and aid in the resistance to the force that intraocular pressure exerts on the back of the eye. The basement membranes of the RPE and choriocapillaris endothelium are composed of type IV collagen, which can indirectly interact with cells via laminin, and also binds to heparin and heparan sulfate proteoglycans. The central elastic layer consists of cross-linked elastin fibers that are associated with fibulin 5, an ECM protein that is thought to aid in elastin fiber assembly and serve as a link between the elastin fibers and cell surface receptors. BrM acts as a support and a barrier between the retina and the choroid, and is thought to support the many functions of the RPE. BrM is semipermeable, and therefore controls the transfer of molecules and cellular components between the RPE and the choroidal vasculature.

RPE–Choroid Interactions

Interactions During Development

The presence of the RPE is required for proper choroid development. Studies have shown that an intact, fully differentiated RPE is required for normal choroidal development; when the RPE is transdifferentiated into a neural retina by expressing fibroblast growth factor (FGF)-9 under the control of the tyrosine-related protein 2 (TRP-2) promoter, the choroid fails to develop. This is further illustrated in humans with colombas where RPE differentiation has failed and there are abnormalities in development of both the choroid and sclera. Basic fibroblast growth factor (bFGF) has also been implicated in choroidal development. Transgenic mice in which a dominant-negative bFGF receptor (FGFR1) was overexpressed in the RPE displayed choroidal abnormalities, such as incomplete and immature choroidal vessels.

The RPE also secretes a variety of growth factors and an increasing body of evidence indicates that RPE-derived vascular endothelial growth factor (VEGF) is essential to choroidal development. Studies in humans and rodents have illustrated that both VEGF and its receptor, VEGFR2, are highly expressed by the RPE and the underlying mesenchyme, respectively, at the time of choriocapillaris formation. Furthermore, mice with an RPE-specific deletion of VEGF presented a variety of defects, including microphthalmia, loss of visual function, and the complete absence of the choriocapillaris. In addition, the RPE itself was discontinuous, suggesting that RPE-derived VEGF is not only important for choroidal development, but also for RPE survival. Whether the RPE abnormality is secondary to the defects in choroidal development or due to a direct effect of VEGF on RPE has not been elucidated.

Interactions in the Adult

Normal RPE–choroid interactions are not only important during development, but also in the adult. Several studies have revealed the impact of RPE loss on choroidal structure and function. The presence of an intact RPE is critical to proper choroidal function, as surgical RPE removal causes several changes throughout the choroid. Both large choroidal vessels and the choriocapillaris display a reduction in circulation, and depending upon the extent of RPE removal, choroidal nonperfusion can be permanent due to fibroblast infiltration. A landmark study in which the RPE was selectively destroyed by sodium iodate treatment illustrated that within 1 week following iodate injection, the choriocapillaris had reduced fenestrations and displayed signs of atrophy, such as degenerating endothelial cells, and pericapillary basal laminae that had begun to separate from the apparently shrunken endothelium. Together, these data illustrate the critical role the RPE plays in both survival and in the maintenance of the choriocapillaris.

Growth factor secretion

RPE secretes a variety of growth factors, including VEGF, FGFs, transforming growth factor-β (TGF-β), and ciliary neutrophic factor (CNTF). The RPE also secretes pigment-epithelium-derived factor (PEDF), which functions not only to maintain the retina by acting as a neuroprotective factor, but also to provide antiangiogenic activity to inhibit endothelial cell proliferation, thereby stabilizing the choriocapillaris endothelium. VEGF, a well-characterized angiogenic factor, also plays a role modulating blood vessel permeability. Differential splicing of VEGF pre-messenger RNA (mRNA) gives rise to multiple isoforms, with the most notable being VEGF120, VEGF164, and VEGF188 in mice, and VEGF121, VEGF165, and VEGF189 in humans. VEGF120 does not bind heparin sulfate proteoglycans (HSPGs) and is readily diffusible, whereas VEGF164 is partially sequestered on the cell surface and in the ECM. VEGF188, with high affinity for heparan sulfate, is therefore primarily cell-surface- and matrix-associated. The RPE primarily expresses the diffusible isoforms, VEGF164 and VEGF120, while VEGF188 is virtually undetectable. RPE-derived VEGF is essential for both survival and maintenance of the underlying choriocapillaris endothelium. Interestingly, PEDF and VEGF are secreted by the RPE in a polarized fashion in opposing directions. PEDF is secreted to the apical surface of the RPE to support the neurons and photoreceptors of the retina, whereas VEGF is primarily secreted to the basolateral surface where it acts on the choroidal endothelium (**Figure 4**). Despite the fact that a small fraction of total VEGF, a potent angiogenic factor, is secreted to the apical surface by the RPE, the outer nuclear layer of retina remains completely avascular, presumably due to the balance between antiangiogenic (PEDF) and angiogenic (VEGF) factors.

Receptor expression

In parallel with the many factors secreted by the RPE, the choroid expresses a number of corresponding receptors, including FGF receptors, type I and II TGF-β receptors, and VEGFR2. Of particular interest is the fact that VEGFR2 is observed primarily in the choriocapillaris adjacent to the RPE, with substantially less VEGFR2 expression observed in the major vessels of the choroid. Furthermore, VEGFR2 is localized to the apical surface of the choriocapillaris endothelium and to the photoreceptors where it is constitutively activated, which is surprising given that adult choroidal vasculature is mature and quiescent.

Isoform-specific VEGF mouse model

In support of a critical role for RPE-derived VEGF in the maintenance of the adult choriocapillaris, we have shown that the absence of soluble VEGF isoforms in the RPE leads to changes that recapitulate the classical features of dry age-related macular degeneration (AMD). Mice expressing only VEGF188 (i.e., lacking the diffusible isoforms that they normally express) display signs of RPE dysfunction, such as increased autofluorescence, loss of barrier proprieties, and accumulation of basal deposits that are similar to drusen, extracellular deposits that build up beneath the basement membrane of the RPE within BrM. These changes occur prior to the formation of both focal choroidal atrophy and RPE attenuation, which progress to large areas of RPE loss. The abnormalities are age dependent and increase in severity over time.

Choroidal change impact on RPE

The RPE–choroid interactions do not merely function to serve the choroid; changes in the choroid have also been shown to impact the RPE. Choroidal ischemia has been reported to lead to opaque RPE lesions and subsequent serous retinal detachment. Furthermore, very early studies on the effects of choroidal congestion on RPE–retina function revealed that increased choroidal pressure may

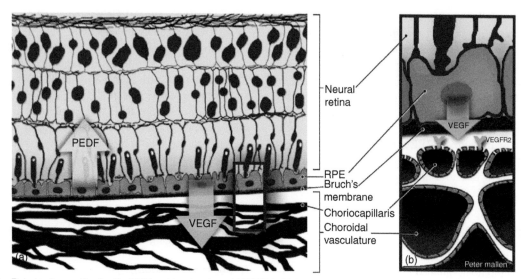

Figure 4 Polarized secretion by the RPE and localization of VEGFR2. (a) The RPE secretes both VEGF and PEDF in polarized fashion in opposing directions. PEDF is secreted to the apical surface of the RPE to support the neurons and photoreceptors of the retina, whereas VEGF is primarily secreted to the basolateral surface where it acts on the choroidal endothelium. RPE-derived VEGF is essential for both survival and maintenance of the underlying choriocapillaris endothelium, and PEDF functions not only as a neuroprotective factor, but also as an antiangiogenic factor to maintain the avascularity of the outer nuclear layer of the retina. (b) Higher magnification of selected area in (a) showing fenestrations in choriocapillaris and localization of the VEGFR2 receptor. In parallel with the VEGF secreted by the RPE, the choroid expresses the corresponding VEGF receptor, VEGFR2. VEGFR2 is observed primarily in the choriocapillaris adjacent to the RPE, with substantially less VEGFR2 expression observed in the major vessels of the choroid.

cause RPE–retina dysfunction by altering normal fluid movement across the RPE, modifying the integrity of the subretinal space.

RPE–Choroid Changes with Age

Various changes occur in the RPE–choroid complex with aging. In light of the close association among the RPE, BrM, and choroid, it is not surprising that an alteration in a single component of this complex compromises the normal RPE–choroid interaction and ultimately leads to disease. Studies have shown that the density and diameter of the choriocapillaris and medium-sized choroidal vessels substantially decline with age, resulting in decreased choroidal blood volume and blood flow. The aging RPE exhibit changes such as a reduction in cell density and a loss of RPE melanin. Melanin pigmentation is believed to play a protective role, acting to protect cells against oxidative stress. Oxidative changes in RPE melanin may be attributed to complexing of melanin with lipofuscin, pigment granules composed of lipid-containing residues of lysosomal digestion, which generate reactive oxygen species upon excitation with blue light, thereby making the aged RPE more susceptible to oxidative damage.

BrM Changes

Changes in BrM include increased thickness, accumulation of lipids, and subsequent alterations of the BrM

permeability. Collectively, these changes can limit the diffusion of water-soluble proteins, leading to a range of problems. Furthermore, drusen are known to accumulate in the aging eye (**Figure 5**). There is a strong correlation between the presence of drusen and ocular pathology, such as AMD. It has been speculated that drusen lead to a gradual reduction in the diffusion of RPE-derived factors, such as VEGF, which may be causal in the atrophy of segments of the RPE and underlying choriocapillaris, known as geographic AMD. Drusen accumulation and coalescence may also lead to breaks in BrM that are believed to be the initiating event in the formation of choroidal neovascularization (CNV), in which new blood vessels sprout from preexisting choroidal vessels and invade the overlying RPE and retina. The development of CNV associated with wet AMD leads to visual loss due to the fact that the neovessels are leaky and cause damage to the surrounding tissues.

Gene Expression

Gene expression of the RPE–choroid also changes with age. There is considerable upregulation of the expression of genes and proteins involved in leukocyte extravasation, and the accumulation of leukocytes at the RPE–BrM interface suggests that leukocytes are possibly recruited to aid in the removal of cellular waste. Furthermore, there have been reports of increased macrophages in the aged RPE–choroid; mice harboring mutations that render them

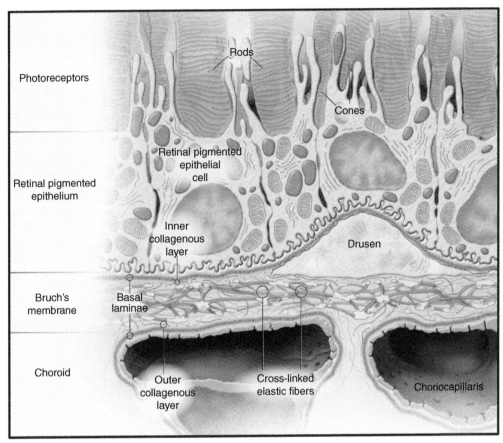

Figure 5 The accumulation of drusen beneath the RPE within BrM. In the aging eye, it is common for drusen to accumulate at the interface between the inner collagenous layer of BrM and the basal lamina of the RPE. Drusen are extracellular deposits that are strongly correlated with ocular pathology, such as age-related macular degeneration (AMD). It is proposed that drusen deposits between the RPE and the BrM as well as within BrM, and causes a gradual reduction in the diffusion of RPE-derived factors, such as VEGF, which may lead to the atrophy of segments of the RPE and underlying choriocapillaris, also known as geographic AMD. Drusen accumulation and coalescence may also cause breaks in BrM that are believed to be the initiating event in the formation of choroidal neovascularization (CNV), in which new blood vessels sprout from preexisting choroidal vessels and invade the overlying RPE and retina. Adapted from Johnson, L. V. and Anderson, D. H. Age-related macular degeneration and the extracellular matrix. *New England Journal of Medicine* 351(4): 320–322. Copyright © 2004 Massachusetts Medical Society. All rights reserved.

deficient in macrophage recruitment display hallmarks of AMD. It is also thought that the aged RPE–choroid synthesizes proteins that not only attract leukocytes, but that also activate the complement pathway, which is a part of the immune response and can lead to inflammation. Recent associations between polymorphisms in a member of the complement pathway reinforce the role of inflammation in the development of AMD.

Conclusions

Despite the fact that the RPE and choroid are separated by BrM, there is a great deal of interaction between the tissues. The presence of an intact, fully differentiated RPE is not only required for proper choroidal development, but is also essential for survival and maintenance of adult choriocapillaris endothelium specializations

(fenestrations) and integrity. The RPE secretes a variety of factors, one of the most notable being VEGF. The secretion of VEGF by the RPE is somewhat of a double-edged sword. Although VEGF is vital to choroidal homeostasis during both development and in adult, breakdown of the RPE barrier upon direct contact with choroidal endothelial cells is thought to involve a VEGF-mediated mechanism, for VEGF has been shown to mediate the vessel growth and permeability associated with wet AMD. Thus, maintenance of proper RPE–choroid interaction is vital to normal function, and any perturbation to this system can ultimately lead to disease.

See also: Breakdown of the RPE Blood–Retinal Barrier; Choroidal Neovascularization; Developmental Anatomy of the Retinal and Choroidal Vasculature; Immunobiology of Age-Related Macular Degeneration; RPE Barrier.

Further Reading

Gogat, K., Le Gat, L., Van Den Berghe, L., et al. (2004). VEGF and KDR gene expression during human embryonic and fetal eye development. *Investigative Ophthalmology and Visual Science* 45(1): 7–14.

Hartnett, M. E., Lappas, A., Darland, D., et al. (2003). Retinal pigment epithelium and endothelial cell interaction causes retinal pigment epithelial barrier dysfunction via a soluble VEGF-dependent mechanism. *Experimental Eye Research* 77(5): 593–599.

Ivert, L., Kong, J., and Gouras, P. (2003). Changes in the choroidal circulation of rabbit following RPE removal. *Graefe's Archive for Clinical and Experimental Ophthalmology* 241(8): 656–666.

Korte, G. E., Reppucci, V., and Henkind, P. (1984). RPE destruction causes choriocapillary atrophy. *Investigative Ophthalmology and Visual Science* 25(10): 1135–1145.

Mancini, M. A., Frank, R. N., Keirn, R. J., Kennedy, A., and Khoury, J. K. (1986). Does the retinal pigment epithelium polarize the choriocapillaris? *Investigative Ophthalmology and Visual Science* 27(3): 336–345.

Marneros, A. G., Fan, J., Yokoyama, Y., et al. (2005). Vascular endothelial growth factor expression in the retinal pigment epithelium is essential for choriocapillaris development and visual function. *American Journal of Pathology* 167(5): 1451–1459.

Ramrattan, R. S., van der Schaft, T. L., Mooy, C. M., et al. (1994). Morphometric analysis of Bruch's membrane, the choriocapillaris, and the choroid in aging. *Investigative Ophthalmology and Visual Science* 35(6): 2857–2864.

Rousseau, B., Larrieu-Lahargue, F., Bikfalvi, A., and Javerzat, S. (2003). Involvement of fibroblast growth factors in choroidal angiogenesis and retinal vascularization. *Experimental Eye Research* 77(2): 147–156.

Saint-Geniez, M. and D'Amore, P. A. (2004). Development and pathology of the hyaloid, choroidal and retinal vasculature. *International Journal of Developmental Biology* 48: 1045–1058.

Saint-Geniez, M., Maldonado, A. E., and D'Amore, P. A. (2006). VEGF expression and receptor activation in the choroid during development and in the adult. *Investigative Ophthalmology and Visual Science* 47(7): 3135–3142.

Saint-Geniez, M., Maharaj, A. S., Walshe, T. E., et al. (2008). Endogenous VEGF is required for visual function: Evidence for a survival role on Müller cells and photoreceptors. *PLoS ONE* 3(11): e3554.

Steinberg, R. H. (1985). Interactions between the retinal pigment epithelium and the neural retina. *Documenta Ophthalmologica* 60: 327–346.

Strauss, O. (2005). The retinal pigment epithelium in visual function. *Physiological Reviews* 85(3): 845–881.

Zhao, S. and Overbeek, P. A. (2001). Regulation of choroid development by the retinal pigment epithelium. *Molecular Vision* 2(7): 277–282.

Retinal Pigment Epithelium: Cytokine Modulation of Epithelial Physiology

S S Miller, A Maminishkis, R Li, and J Adijanto, National Eye Institute, Bethesda, MD, USA

Published by Elsevier Ltd., 2010.

Glossary

Cytokines – A large and diverse family of polypeptide regulators that are used in cell regulation.

ELISA – The enzyme-linked immunosorbent assay is a biochemical technique that uses an enzyme-linked antibody to detect its corresponding ligand within a sample. It can be used to assay for chemokines and cytokines.

Subretinal space (SRS) – The extracellular space between the retinal photoreceptors and the apical surface of the retinal pigment epithelium.

Tight junctions – A structure of specialized proteins that form a seal between neighboring cells and regulate ion and small molecule selectivity and resistance in the paracellular pathway between epithelial cells. Tight junction proteins include occluding, claudins, and junction adhesion molecules (JAMs).

Introduction

In the back of the vertebrate eye, the apical membrane of the retinal pigment epithelium (RPE) and the photoreceptor outer segments form a very tight anatomical relationship (**Figure 1**). This structural feature supports a whole host of mechanical, electrical, and metabolic interactions that maintain the health and integrity of the neural retina throughout the life of the organism. Like all epithelia, the RPE plasma membrane contains a wide variety of proteins, enzymes, and small molecules that are specifically segregated to the apical or basolateral sides of the epithelium, which face the neural retina and choroidal blood supply, respectively (**Figure 2**).

The asymmetrical distribution of these functionally distinct molecules is maintained by junctional complexes that surround each cell and by the continuous synthesis and regulated traffic of these molecules to each membrane. Epithelial polarity is defined by the steady-state maintenance of this asymmetric distribution and is critical for the ongoing vectorial transport of ions, metabolites, fluid, and waste products across the RPE. Epithelial polarity is also fundamentally important for controlling changes in the volume and chemical compositions of the extracellular spaces on either side of the RPE, following transitions between light and dark. In the distal retina, the extracellular or subretinal space (SRS) separates the photoreceptor outer segments and the RPE apical processes. The chemical composition of this space is tightly buffered by the cells which surround it (Müller cells, photoreceptors, and RPE). On the opposite side of the RPE, an extracellular space is formed between its basolateral membrane and Bruch's membrane, which is adjacent to the choriocapillaris. The physiological and pathophysiological states of the RPE/distal retina complex are significantly affected by changes in the chemical composition of these extracellular spaces as evidenced in disease processes such as age-related macular degeneration (AMD) or uveitis. AMD develops within the RPE/distal retina complex and eventually leads to RPE impairment and loss of photoreceptor function. The RPE's ability to control and respond to varying levels of oxidative insult from light quanta, outer segment phagocytosis, vitamin A uptake and delivery, and oxygen consumption diminishes with age. These changes significantly affect the chemical composition of the surrounding extracellular spaces, SRS and choroid, and are a major factor in disease pathogenesis. In recent years, significant advances have been made in identifying the role of the immune system in neurodegenerative disease in general and in AMD, in particular (summarized by Hageman and colleagues and by Nussenblatt and Ferris). This article summarizes recent experiments from our lab and others, which show that inflammation induced changes in the environment surrounding human RPE can significantly alter intracellular signaling and physiology. This study provides a basis for understanding disease progression and regression.

This article is divided into three main parts. We begin with a description of our development of a robust and well-defined primary cell culture model of human fetal retinal pigment epithelium (hf RPE). We use this model to analyze how metabolic waste products, produced in the retina following light/dark transitions, can be disposed of by CO_2/HCO_3 and lactate transporters located in the apical and basolateral cell membranes. In the second part, we use this cell culture model to analyze RPE antioxidant mechanisms that are protective against disease processes, such as AMD or uveitis. In the third part, we describe a series of experiments that use this model to define the impact of cytokines on human RPE function. Finally, we briefly focus on the role of interferon gamma (INFγ) in controlling RPE physiology.

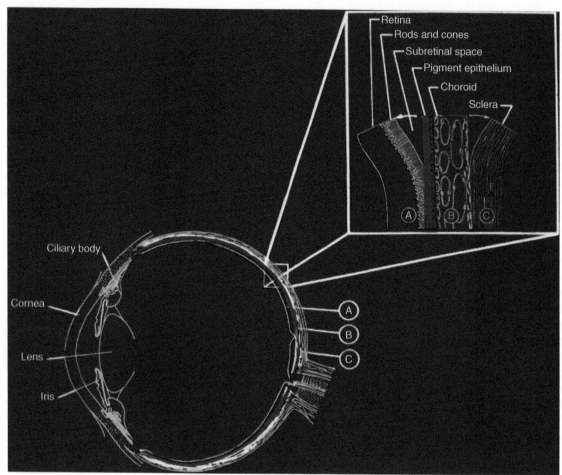

Retina
Rods and cones
Subretinal space
Pigment epithelium
Choroid
Sclera

Ciliary body

Cornea

Lens

Iris

Figure 1 Schematic diagram of the eye. The retinal pigment epithelium is located in the back of the eye between neural retina and choroid.

Human RPE: Morphology, Polarity, and Function

The availability of native human tissue, fetal or adult, is limited and extant models of cultured human RPE have been, in varying degrees, less than adequately characterized or understood, not reproducible or available in large quantities, and lacking expression of melanin pigment and key functional proteins such as bestrophin (Best1), and RPE-specific protein 65 (RPE65). Therefore, we developed a set of standard procedures for producing confluent monolayers of hfRPE cells and demonstrated that they have the morphology, polarity, and function of the native tissue from which they were derived (**Figure 3**). Light and electron microscopy (EM) studies confirmed the presence of apical processes (microvilli) and basal infoldings that increased the elaboration of the apical and basolateral membranes, respective, in cultured hfRPE cells. We carried out immunoblot and immunofluorescence experiments for a variety of proteins to help define the polarity of this model. In the course of these experiments, we discovered that human RPE tight junctions contain a variety of

membrane proteins (claudins) that are important for regulating the selectivity and conductance of the paracellular pathway, and we also confirmed the presence of several visual cycle and cytoskeleton proteins. Intracellular recordings confirmed many membrane physiological properties and demonstrated the polarity of purinergic and adrenergic receptors at the apical membrane, which serve to regulate cell calcium and transepithelial fluid absorption (**Figure 2**). By enzyme-linked immunosorbent assay (ELISA), we showed that these monolayers constitutively secrete pigment epithelium-derived factor (PEDF) to the apical bath and vascular endothelial growth factor (VEGF) to the basal bath; the former provides neuroprotection for the retina and the latter would allow active regulation of endothelial cell fenestration, a structural feature critical for choroidal circulation.

pH_i – Induced Changes in Fluid Absorption

In previous experiments we showed that acetazolamide, a carbonic anhydrase (CA) inhibitor, reduced net ^{36}Cl flux across frog RPE. Subsequent animal models and clinical

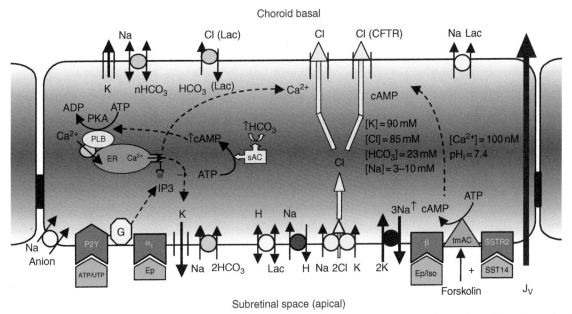

Figure 2 Schematic diagram of retinal pigment epithelium (RPE) summarizing some membrane proteins, channels, and receptors that are responsible for a variety of RPE functions such as fluid transport, pH maintenance, or cell-volume regulation. The arrows highlighted in yellow indicate a main pathway for solute-driven fluid transport across the RPE, consisting of a sodium, potassium, 2 chloride co-transporter (Na/K/2Cl co-transporter) at the apical membrane, and a cyclic AMP-activated chloride channel, CFTR (cystic fibrosis transmembrane conductance regulator), and Ca^{2+}-activated Cl-channels at the basolateral membrane. The apical membrane contains a variety of receptors, for example, purinergic (P2Y), adrenergic (α-1 and β), and somatostatin (SSTR2), which when bound to their specific ligands, activate Ca^{2+} and cAMP second-messenger signaling systems. The plasma membrane localization of previously described ion transporters such as the $Na/2HCO_3$ co-transporter, Na/H exchanger, 3Na/2K ATPase, H/Lac co-transporter, Cl/HCO_3 exchanger, and potassium channels are also shown. From Maminishkis, A., et al. (2002). *Investigative Ophthalmology and Visual Science* 43(11): 3555–3566. © Association for Research in Vision and Ophthalmology.

trials have utilized CA inhibitors to reduce disease-induced abnormal accumulation of retinal fluid. CAs catalyze the reversible hydration of CO_2 to HCO_3 and protons, which are transported across the plasma membrane (e.g., Na/HCO_3 or H/lactate co-transporters) to regulate cell pH. As a first step, we have identified and localized several highly expressed CAs in human RPE and begun study of their physiology. A total of 16 CAs have been identified in human tissues. In human fetal RPE cell cultures, 14 of the 16 known isozymes have been confirmed by quantitative real-time polymerase chain reaction (qRT-PCR). Immunocytochemical studies indicate that CA II is localized intracellularly, as in many other cell types. CA IV, XII, and XIV are localized to the apical surface, while CA IX, the most abundantly expressed isozyme in hfRPE cultures, is expressed apically and laterally (**Figure 4**). However, it should be noted that CA IX messenger RNA (mRNA) and proteins are not expressed in native adult or fetal human RPE. CA inhibitors have had limited success in alleviating the effects of retinal disease, partly because of systemic side effects, but mainly because of their lack of specificity. The positive clinical outcomes for some patients with retinal edema suggest that nonspecific CA inhibitors, such as acetazolamide, may be affecting multiple CAs or other transport-related mechanisms that can either increase or decrease

net fluid absorption across the RPE in varying degrees in different patients. This is supported by *in vivo* animal studies, in which intravenous administration of acetazolamide to rabbits increased fluid clearance from the SRS. The regulatory role of CAs is potentially important in human RPE, which critically depends on HCO_3 transport to maintain fluid absorption (J_V) out of the SRS (**Figure 5**).

Modulation of SRS Metabolic Load and Chemical Composition

In the intact eye (cat/monkey), the transition from light to dark causes significant alterations in SRS pH, Ca^{2+}, and K. In addition, the transition from light to dark increases photoreceptor O_2 consumption by ≈ 2-fold as measured *in situ* in cat and nonhuman primate retina. The rates of retinal O_2 consumption in light and in dark were used by Linsenmeier and Winkler and their colleagues to estimate the associated changes in glucose metabolism that leads to the concomitant release of carbon dioxide, lactic acid, and water from the photoreceptor inner segments into the SRS. This metabolic acid load is potentially damaging to all of the cells that surround the SRS (i.e., photoreceptors, Müller cells, and RPE). It raises the

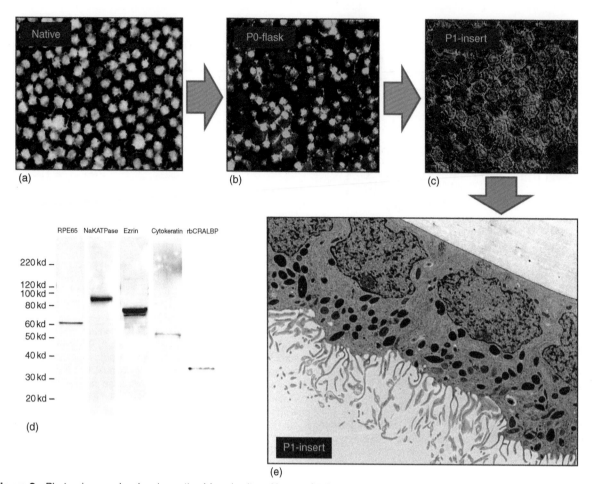

Figure 3 Photomicrographs showing native (a) and cultured human fetal retinal pigment epithelium (hfRPE) ((b) P_O cultured on flask; (c) P_1 cultured on insert). (d) Westerns blots for five hfRPE specific proteins. (e) Transmission-electron micrograph of hfRPE cells grown on inserts. From Maminishkis, A., et al. (2006). *Investigative Ophthalmology and Visual Science* 47 (8): 3612–3624. © Association for Research in Vision and Ophthalmology.

question of how the RPE could help prevent this accumulation of metabolic acid and water in the SRS.

In *in vivo* studies of rabbit eye, it was estimated that $\approx 70\%$ of fluid absorption across the RPE is linked to metabolite transport to the choroidal blood supply. In addition, *in vitro* studies of frog RPE showed that steady-state fluid absorption decreased by $\approx 70\%$, following the removal of HCO_3 from both bathing solutions, implicating HCO_3 transport in a regulatory role on fluid transport. The RPE functionally expresses several different HCO_3 transport proteins at the apical and basolateral membranes as illustrated in **Figure 5**. In an earlier study, a 4,4'-diisothiocyano-2,2'-stillbene-disulfonic acid (DIDS)-sensitive electrogenic Na/2HCO$_3$ co-transporter was localized to the apical membrane of frog and bovine RPE; DIDS is a bicarbonate transport inhibitor. At the basolateral membrane, HCO_3 is transported out of the RPE through a pH-sensitive Cl/HCO$_3$ exchanger with a possible contribution from a Na/HCO$_3$ co-transporter. These HCO_3 transporters in the RPE are linked to Na and Cl transport, which are major driving forces for

fluid transport. Recently, the identities of some of these HCO_3 transporters have been characterized in our laboratory and by other groups. NBC1 (Na/2HCO$_3$ co-transporter) and NBC3 (NBCn1; electroneutral Na/HCO$_3$ co-transporter) were localized to the apical membrane. AE2 (Cl/HCO$_3$ exchanger) mRNA transcripts were detected, but protein expression in the RPE remains to be determined. The identity of the basolateral membrane Na/nHCO$_3$ co-transporter (NBC) is still unknown.

In vitro, we mimic the increased retinal CO_2 production, following the transition from light to dark by increasing apical bath CO_2 level from 5% to 13%. This maneuver increased NaCl uptake at the apical membrane and can enhance CA-mediated Na/HCO$_3$ co-transport across the RPE, thus increasing net NaHCO$_3$ absorption. This increase in solute transport would drive additional fluid across the RPE as observed in *in vitro* experiments. The transport of metabolic waste products from the SRS to the choroidal blood supply by the RPE helps maintain ionic and pH homeostasis of the SRS. The RPE handles the increased metabolic load by transporting

Carbonic anhydrase mediated HCO₃-transport

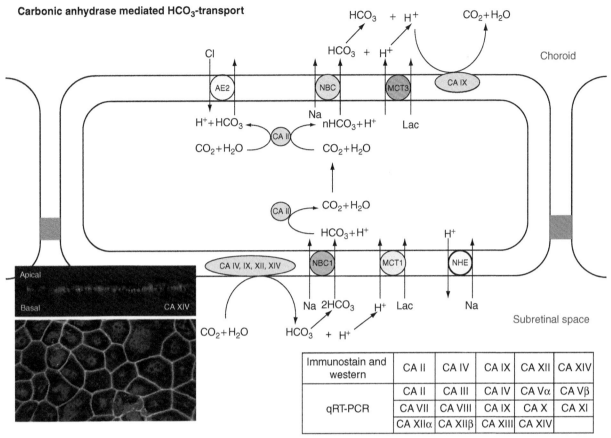

Figure 4 Carbonic anhydrase (CA)-mediated CO_2/HCO_3 transport in retinal pigment epithelium (RPE). Membrane-bound CA IV, XII, and XIV are expressed exclusively at the apical membrane (immunostaining of CA XIV shown in insert on lower left). CA IX is expressed at both the apical and basolateral membranes of cultured human fetal RPE. CA II is expressed in the cytosol and can be recruited to the inner leaflet of the membrane. Experimental data (immunostaining, Western blots, and RT-PCR) that support the localization of the various CAs are listed in the table (lower right). Membrane-bound CAs hydrate CO_2 into HCO_3 and H^+, which are substrates for NBC1 (sodium bicarbonate co-transporter) and MCT1 (proton lactate co-transporter) at the apical membrane. Cytosolic CA II regulates CO_2 and HCO_3 equilibrium in the RPE. The anion exchanger isoform 2 (AE2) mediates Cl/HCO_3 exchange at the basolateral membrane, while a sodium proton exchanger (NHE) at the apical membrane helps regulate intracellular pH.

CO_2 across the RPE in the form of HCO_3 through HCO_3-transporters, and this process is mediated by the catalytic activity of CAs. This increase in Na and HCO_3 absorption provides the driving force for increased net fluid absorption across the RPE, which dehydrates the SRS and creates retinal adhesion, thus allowing the RPE to maintain proper anatomical relationship with the photoreceptors.

In the retina, ≈95% of glucose consumption is metabolized through glycolysis into lactic acid, which is subsequently deposited into the SRS. In addition to the high glycolytic activity of the retina, several other mechanisms cause additional lactic acid to be released by the retina following light–dark transition: (1) increased glucose metabolism at the outer retina; (2) reduced retinal oxygen level in the dark-adapted eye, leading to an increased anaerobic lactate production; and (3) glutamate-induced lactate release from Müller cells. The RPE disposes of this metabolic load by transporting lactic acid to the choroid

through monocarboxylate transporters (MCTs) of the MCT family. We previously demonstrated that the RPE is extremely resistant to pH change compared to other epithelia and that part of this regulation comes from H/Lac co-transporters at the apical and basolateral membranes.

In human RPE, Philp and colleagues showed that monocarboxylate transporter 1 (MCT1), a H/Lac co-transporter, is immunolabeled at the apical membrane (**Figure 5**). They also showed that MCT3, a H/Lac co-transporter expressed exclusively in the RPE and choroid plexus basolateral membranes, mediates lactate efflux from the RPE into the choroidal blood supply (**Figure 5**). In addition, a Cl/Lac exchanger, possibly anion exchanger 2 (AE2), has been shown to transport lactate at the basolateral membrane. The importance of lactate transport in the mammalian eye has also been demonstrated in mice lacking MCT1, MCT3, and MCT4 expression – the mutant mice gradually lose photoreceptor function and

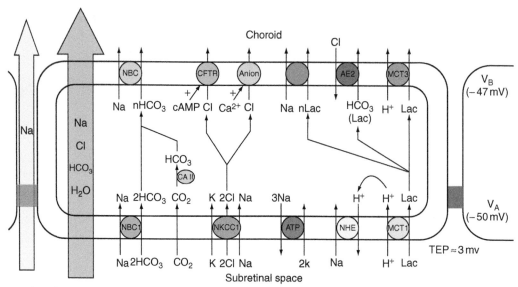

Figure 5   Na, Cl, HCO₃, and lactate transport mechanisms in retinal pigment epithelium (RPE). CO_2 enters the apical membrane via diffusion and HCO_3 is transported into the cell by NBC1 (sodium bicarbonate co-transporter). As CO_2 enters the cell, it can be hydrated into HCO_3 in a reversible reaction catalyzed by carbonic anhydrase II. Cl enters the apical membrane via NKCC1 (sodium potassium chloride co-transporter – Na/K/2Cl) and exits the basolateral membrane via CFTR or Ca^{2+}-activated Cl channels. Lactic acid is transported across the apical membrane by MCT1, and out of the basolateral membrane through MCT3 (H/Lac co-transporter) and AE2 (Cl/Lac exchanger). These ion-transport mechanisms at the apical and basolateral membranes mediate net solute transport across the RPE, which drives fluid absorption. The apical and basolateral membrane potentials are given by V_A and V_B, respectively, and their difference is the transepithelial potential (TEP).

were completely blind after 41 weeks. Further, altered visual function in MCT3-null mice demonstrates the importance of lactate transport specifically in the RPE.

Oxidative Stress

The RPE encounters significant levels of oxidative stress on a daily basis and this onslaught promotes mitochondrial (mt) damage and decreases in mt potential and respiration, which may contribute to inflammation and the onset of age-related diseases such as AMD (summarized by Jarrett and colleagues). In opposition to these oxidative stresses, there exist mt protective mechanisms that provide direct antioxidant protection and those that enhance glutathione (GSH) production; furthermore, there is evidence that all of these protective mechanisms weaken with age. Different cell types can exert different levels of protection; for example, it has been shown that hfRPE monolayers are significantly more resistant to oxidative stress than ARPE-19 cells. Voloboueva and colleagues used the hfRPE primary cultures to examine mt and other pathways that are putative targets for therapeutic intervention against oxidative stress. In one set of experiments they studied the protective effects of α-lipoic acid (R-form), a potent intracellular antioxidant that has been shown in other systems, to induce all three cellular protective mechanisms. The R form of lipoic acid is a coenzyme in mt that has

been shown to reverse the age-related decrease in mt function. Measurements of cell viability, mt potential, cell death, oxidative stress, apoptosis, and GSH/GSSH show that lipoic acid can protect hfRPE cells in three ways: (1) directly scavenge reactive oxidative species; (2) repair and protect mt enzymes; and (3) activate antioxidant defenses through phase 2 enzymes. Our results suggest that (R)-α-lipoic acid can be used as an all-purpose therapeutic intervention against the slow accumulation of oxidative damage that can occur in AMD.

Cigarette smoke is an important risk factor for AMD and causes significant oxidative damage in RPE that also can be mitigated by (R)-α-lipoic acid. RPE mitochondria are themselves a main generation site of oxidants and a critical and sensitive target of specific cigarette-smoke components. Acrolein is present in the gas phase of cigarettes (25–140 μg per cigarette), and it is estimated that the gas phase of one cigarette reaches a concentration of 80 μM in the airway surface fluid. Acrolein has a high hazard risk in cigarette smoke and causes oxidative stress in cells by reacting with sulfhydryl groups. In the physiological range (0.1–100 μM), it causes significant mt damage in hfRPE that can be ameliorated by (R)-α-lipoic acid, for example, by inducing GSH and other phase-2 antioxidant protective enzymes. Pretreatment by (R)-alpha-lipoid acid has a protective effect against peroxide induced mt oxidative stress in several ways: (1) by lowering cell calcium; (2) by increasing mt electron chain complexes I, II, and III activity levels; (3) by increasing mt membrane potential; and (4) by increasing total antioxidant power as well as GSH peroxidase/GSH/superoxide dismutase levels. Collectively, these data indicate that lipoic acid may be an effective therapeutic strategy against age-related, oxidant-induced RPE degeneration.

As summarized recently by Dunaief, AMD patients have elevated levels of iron within the RPE that also can lead to oxidative damage to mitochondria. Based on that observation, Voloboueva and colleagues showed that ferric ammonium citrate increased intracellular iron and oxidant production and decreased GSH and mt complex IV activity in human fetal cultured RPE. They also showed that N-*tert*-butyl hydroxylamine (Nt-BHA), a known mt antioxidant, reduced oxidative stress, mt damage, and age-related iron accumulation. These data show that the application of Nt-BHA may be an effective therapeutic strategy against AMD.

RPE–Immune System Interactions in and around the SRS

The integrated effect of proinflammatory molecules on RPE function depends on the polarized location of the cognate receptors and the access of their ligands (cytokines and chemokines) to the apical and basolateral membranes, and the interactions of downstream signaling pathways. For the experiments summarized in **Table 1**, we used a mixture of three proinflammatory cytokines, interleukin 1 beta (IL-1β), interferon gamma (IFNγ), and tumor necrosis factor-alpha (TNF-α) to stimulate confluent monolayers of hfRPE. These proinflammatory cytokines are elevated in patients with uveitis and are detected in the vitreous and blood of patients with proliferative diabetic retinopathy (PDR) and AMD with choroidal neovascularization (CNV).

As a first step in understanding how the RPE *in vivo* can actively control the inflammatory environment in the SRS and choroid, Shi and colleagues used confluent monolayers of human fetal RPE primary cultures to (1) measure the constitutive and polarized secretion of angiogenic/angiostatic cytokines by the RPE; (2) determine how this pattern of polarized secretion changes in the inflammatory state; and (3) demonstrate that the inflammatory state alters RPE physiology. Constitutively, the human RPE secretes massive amounts of monocyte chemoattractant protein 1 (MCP-1) to the SRS and lesser amounts of IL-6 and IL-8 (**Table 1**), all of which contribute to the ongoing downregulation of the immune environment of the retina. RPE activation was achieved using a cocktail of IL-1β, TNF-α, and INFγ with similar concentrations as that detected in the diseased eye. We showed that IL-1β receptors are mainly localized to the apical membrane and TNF-α and INFγ (subunit 1)

receptors are mainly localized at the basolateral membrane. This cocktail significantly increased the secretion of various cytokines/chemokines to both baths, but significantly more to the apical bath. The increase in angiogenic cytokine secretion exceeds the increase in angiostatic cytokine secretion. However, two chemokines generally thought to be angiostatic, interferon-inducible T-cell α-chemoattractant (I-Tac) and monokine induced by γ interferon (MIG), were secreted to the apical bath in significant quantities. The mechanisms by which these chemokines exert their effects and their role in eye physiology are not yet known. Similarly intriguing and not understood are the secretions into the apical bath of interferon-inducible protein 10 (IP-10), monocyte chemoattractant protein 3 (MCP-3), and the Rantes chemokine. In animal model experiments from Charlotte Reme's group, blue-light-induced oxidative damage induces the invasion of blood-borne monocytes and activation of retinal microglia, thus stimulating the secretion of cytokines to induce an inflammatory response. Our experiments strongly suggest that the RPE is a significant source of cytokines and chemokines. Thus both retinal microglia and RPE can contribute to the inflammatory response in a diseased eye. Our further demonstration that basolateral addition of the cocktail acutely increases fluid absorption across the RPE (**Figure 6**), from the apical to basal baths (retina to choroidal side of tissue), and significantly decreases transepithelial resistance after a 24-h treatment is important because of the possibility that with age or accumulated oxidative stress these changes can alter chemokine/cytokine gradients across the RPE. These gradients regulate the attraction of monocytes to the RPE basement membrane and, thus, play a role in the accumulation of drusen with age. We believe that this concept is important for understanding early events that underlie chronic disease processes, such as AMD, a notion revisited below.

Modulation of RPE Proliferation and Migration by Cytokines and Growth Factors

Breakdown of the inner or outer blood–retinal barrier can lead to significant alterations in the chemical composition of the SRS, including cytokines and growth factors, which trigger the activation of normally quiescent RPE cells. In proliferative vitreoretinopathy (PVR), RPE cells proliferate and migrate to the vitreous cavity along with other types of cells (e.g., glial cells, fibroblasts, and macrophages) and form fibrocellular membranes on the retinal surface or in the vitreous. These newly formed membranes, if left untreated, eventually contract, resulting in retinal detachment and eventual vision loss.

Several isoforms of platelet-derived growth factor (PDGF) are present in retinal membrane from patients

Table 1   Inflammatory cytokine mixture alters polarized secretion of chemokines and cytokines by cultured human fetal retinal pigment epithelium (hfRPE)

(a)

Stimuli	Secretion	Angiogenic										Angiostatic				
		MCP-1 pg/ml	IL-8 pg/ml	GRO-α pg/ml	IL-6 pg/ml	MCP-2 pg/ml	MDC pg/ml	MIP-1α pg/ml	MIP-1β pg/ml	12p40 pg/ml	12p70 pg/ml	MIG pg/ml	ITAC pg/ml	MCP-3 pg/ml	RANTES pg/ml	IP-10 pg/ml
None	Apical	9,495	647	70.2	55.1	9.6	3.0	7.9	18.6	9.7	1.1	8.8	8.2	14.4	0.6	11.1
	SD	5,285	979	45.9	28.2	6.9	1.4	3.1	17.0	12.9	0.5	6.0	8.6	5.8	0.8	4.0
	Basal	992	95.2	24.4	3.5	3.6	1.4	6.2	9.2	3.7	0.6	3.4	3.7	4.0	0.4	5.0
	SD	1,343	238	29.7	3.1	2.0	1.5	3.6	9.3	7.8	0.3	5.4	2.2	3.3	0.0	4.1
ICM (Ap)	Apical	212,408	110,865	72,097	125,167	34,185	21.9	143	39.5	63.3	40.1	8,518	89,126	22,431	21,067	239,839
	SD	52,434	33,309	1,518	6,565	2,575	2.8	115	13.6	7.2	13.5	3,550	3,409	2,381	2,237	37,493
	Basal	75,019	4,415	2,328	3,221	2,900	10.8	36.8	26.2	51.3	19.9	132	1,335	1,452	4,915	11,700
	SD	8,694	213	191	330	501	0.5	8.7	5.6	4.7	3.8	28.8	250	428	707	1,291
ICM (Ba)	Apical	179,247	22,567	8,375	7,108	6,613	12.9	44.2	24.2	66.2	36.3	202	3,623	1,741	1,088	17,626
	SD	36,462	11,316	2,710	1,457	1,844	1.7	7.5	3.6	6.5	5.8	61.6	1,390	431	244	5,605
	Basal	48,470	2,053	1,227	1,550	1,527	9.0	31.9	21.1	42.4	26.0	94.6	838	622	2,497	5,549
	SD	16,646	309	297	281	492	1.2	1.8	3.8	6.9	5.6	13.1	34.4	263	635	1,661
ICM (Ap & Ba)	Apical	314,290	112,416	39,406	52,479	21,120	11.7	249	81.8	29.2	58.6	6,668	66,863	8,415	9,277	132,548
	SD	99,378	20,528	19,909	26,200	8,537	7.6	104.9	30.3	27.2	55.9	3,754	22,992	4,342	6,159	62,068
	Basal	73,521	4,599	3,363	17,929	2,900	5.0	69.9	51.3	15.1	16.9	138	1,395	623	1,984	7,546
	SD	15,277	2,402	2,039	24,194	1,469	1.9	16.1	15.1	21.5	6.0	37.9	714	252	2,004	4,079

(b)

Stimuli	Secretion	Angiogenic										Angiostatic				
		MCP-1 pg/ml	IL-8 pg/ml	GRO-α pg/ml	IL-6 pg/ml	MCP-2 pg/ml	MDC pg/ml	MIP-1α pg/ml	MIP-1β pg/ml	12p40 pg/ml	12p70 pg/ml	MIG pg/ml	ITAC pg/ml	MCP-3 pg/ml	RANTES pg/ml	IP-10 pg/ml
IL-1β(Ap)	Apical	107,865	51,298	31,578	7,572	373	14.5	22.4	30.2	50.7	9.5	74.6	27.4	706	580	236
	SD	18,455	14,956	12,362	3,219	309	3.1	7.4	7.2	8.7	1.5	25.3	8.2	470	419	111
	Basal	19,902	1,420	805	318	13.4	4.4	5.8	7.0	23.8	2.6	50.9	30.1	17.5	25.4	184
	SD	9,520	971	634	15.7	3.2	0.5	2.5	2.1	10.3	0.4	53.6	32.4	2.0	16.8	270
[TNF-α, IFNγ](Ba)	Apical	73,113	2,358	514	729	14.0	6.5	3.1	10.2	29.7	4.3	153	184	30.3	40.8	3,155
	SD	1,294	1,422	504	30.1	5.0	1.5	0.0	2.1	8.7	1.3	57.0	49.9	7.7	20.2	472
	Basal	28,098	437	98.8	283	11.1	4.1	3.1	5.3	16.1	2.5	161	64.5	15.3	72.8	678
	SD	6,962	98.1	3.9	33.3	1.5	0.2	0.0	0.5	2.8	0.2	72.9	2.6	1.5	31.2	167
IL-1β(Ap) & [TNF-α, IFN]	Apical	169,963	66,722	42,890	9,935	586	16.5	18.4	28.6	42.1	8.9	282	453	584	1,005	5,524
	SD	35,717	2,751	27,066	4,532	170	4.4	6.4	11.6	5.5	2.4	33.1	544	479	638	1,954
	Basal	59,938	7,959	4,134	2,636	310	6.8	6.3	13.9	42.9	10.3	174	88.6	72.9	207	1,283
	SD	8,945	2,286	904	498	25.2	1.5	2.8	4.7	10.4	6.2	1.4	38.4	7.1	67.9	273

P< 0.05 | 0.01 | 0.001   P>0.05

Source: From Shi, G., et al. (2008). Investigative Opthalmology and Visual Science 49: 4620–4623. Copyright Association for Research in Vision and Opthalmology.
The table summarizes the polarized secretions of simultaneously measured chemokines (12) and cytokines (3) using multiplex sandwich ELISAs. The left-hand column of this table defines the stimuli and the site of addition of pro-inflammatory mediators. These data are reported as actual concentrations and therefore corrected for transwell volume differences.
Each Table entry is the mean ± SD for three experiments and the levels of significance are color coded for ease of comparison.

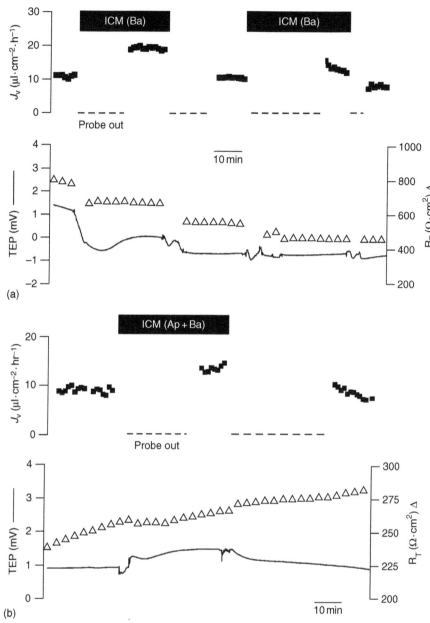

Figure 6 Inflammatory cytokine mixture (ICM) induced changes in hfRPE fluid transport (J_V). In all panels, the top trace is J_V, which is plotted as a function of time; net fluid absorption is indicated by positive values and TEP and total tissue resistance (R_T) are plotted in the lower traces. (a) Addition of ICM to the basal bath increased J_V by ≈ 13 $\mu l \cdot cm^{-2} \cdot hr^{-1}$ with no significant changes in TEP and R_T. (b) Concomitant addition of ICM to apical and basal baths increased J_V by ≈ 10 $\mu l \cdot cm^{-2} \cdot hr^{-1}$ with no change in TEP and a slight increase in resistance that is not statistically significant. From Shi, G., et al. (2008). *Investigative Ophthalmology and Visual Science* 49(10): 4620–4630. © Association for Research in Vision and Ophthalmology.

with PVR and PDR and are elevated in the vitreous of PVR eyes. Recently we showed that PDGF-C, -D are highly expressed in human fetal and adult RPE and that the mRNA levels of these two isoforms are up to 100-fold higher than PDGF-A and -B. PDGF-C and -D have been implicated in PVR and lens epithelial cell proliferation, but relatively little is yet known about their function in RPE. In other systems, they play an important role in angiogenesis and wound healing.

PDGFR-α and PDGFR-β, the receptors for PDGF-C and -D, respectively, are mainly localized to the apical membrane of human fetal RPE as shown in **Figure 7**. PDGF-CC, -DD, and -BB significantly stimulated hfRPE cell proliferation, while PDGF-DD, -BB, and -AB significantly stimulated cell migration. Furthermore, the stimulatory effects of PDGF were abrogated by a proinflammatory cytokine cocktail composed of TNF-α, IL-1β, and IFNγ. Comparison of the component effects

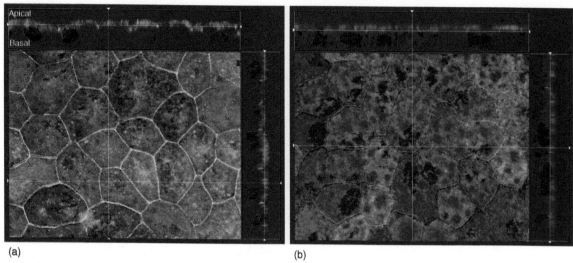

(a)   (b)

Figure 7 Immunofluorescence localization of PDGF receptors on hfRPE. The main part of each panel is an *en face* view of a cell culture monolayer shown as a maximum intensity projection through the *z*-axis. The top and right side of each panel is a cross section through the Z-plane of multiple optical slices obtained using the Apotome. In all experiments shown, the nucleus was stained with DAPI (blue), and the tight junction protein (ZO-1) was immunolabeled in red. The platelet-derived growth factor receptor PDGFR-α (green) is shown in (a), and while the beta subunit, PDGFR-β (green) is shown in (b). Both were detected on the apical membrane of hfRPE. From Li, R., et al. (2007). *Investigative Ophthalmology and Visual Science* 48(12): 5722–5732. © Association for Research in Vision and Ophthalmology.

showed that IFNγ was more effective in suppression than the entire cocktail or any subset of the cocktail, indicating that the downstream cytokine signaling pathways are interactive. Identifying the elements of this putative network and the specific nature of these interactions could provide targets for therapeutic intervention. For example, the proinflammatory cocktail may activate PDGF secretion by the RPE. In preliminary experiments, we showed that IFNγ increased the polarized secretion of PDGF-AA to the apical bath, providing a possible autocrine signal mediating RPE proliferation/migration. We have shown that the cytokine cocktail induces cell apoptosis, alters cytoskeleton distribution, and significantly decreases transepithelial resistance, which can help mediate leukocyte traffic to the SRS. The cytokine cocktail-induced in hibition of RPE proliferation/migration indicates a potential therapeutic role against proliferative responses at the retina/RPE/choroid interfaces.

Since IFNγ has a strong inhibitory effect on RPE proliferation and migration, it is natural to ask how this signaling pathway might provide the basis for inhibition. Native human adult RPE and hfRPE cells constitutively express two transcription factors, interferon regulatory factors 1 and 2 (IRF-1 and IRF-2), which are well-characterized members of the IFN regulatory family and key factors in the regulation of cell growth through their effects on cell cycle. In hfRPE, we showed that stimulation by IFNγ significantly increased IRF-1 protein levels with no effect on IRF-2. If these two transcription factors are mutually antagonistic, as shown in other systems, this may explain the strong inhibitory effect of IFNγ on RPE proliferation

and migration, which we hypothesize is caused by an increase in the ratio of IRF-1/IRF-2.

IFNγ Regulation of RPE Fluid Transport

Immunoblots, immunofluorescence, intracellular recordings, pharmacology, and fluid transport data indicate a basolateral location of cystic fibrosis transmembrane conductance regulator (CFTR), a chloride channel, in native adult and fetal human RPE. As a first step in unraveling the network of cytokine interactions, we focused on INFγ since it is a main determinant of several key effects produced by the inflammatory cocktail. IFNγ has been implicated in the pathogenesis of a number of inflammatory diseases of infectious or presumed autoimmune origin and it has been detected in vitreous aspirates of patients with uveitis, PVR, and other inflammatory ocular diseases. In human RPE, IFNγ activates several intracellular signaling pathways, including the canonical janus-activated kinase and signal transducers and activators of transcription protein (JAK/STAT) pathway and P38 mitogen-activated protein kinase (MAPK), leading to the elevation of cyclic adenosine monophosphate (cAMP) and the subsequent activation of protein kinase A-dependent chloride channels – CFTRs. This results in a significant increase in net fluid absorption across the epithelium (**Figure 8**). These data and the data summarized below provide a possible basis for the etiology of chronic inflammatory diseases, such as posterior uveitis and AMD. In the diseased eye, the IFNγ-induced dehydration of the SRS

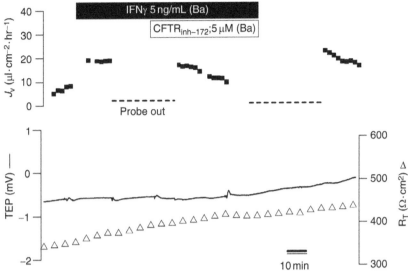

Figure 8 IFNγ-stimulated fluid transport (J_V) increase is inhibited by 5 μM CFTR$_{inh}$-172, an inhibitor of the cystic fibrosis transmembrane conductance regulator (CFTR), added to basal bath. J_V is plotted as a function of time in the top trace and net fluid absorption (apical to basal bath) is indicated by positive values; TEP (−) and R_T (△) are plotted as function of time in the two lower traces.

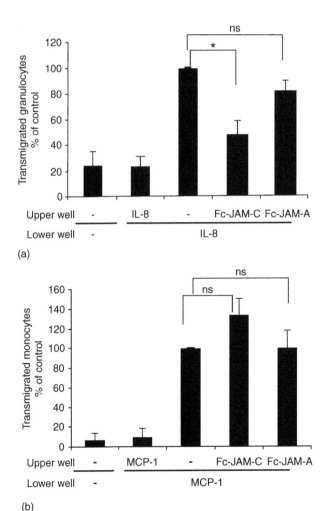

(a)

(b)

Figure 9 IL-8 and MCP-1 regulated transmigration of leukocytes. (a) Basolateral to apical transepithelial migration of

could increase the concentration of the already-accumulating chemokines and thereby help draw monocytes and neutrophils to the RPE basement membrane or across the RPE to the SRS. This helps control the continuing accumulation of debris from incompletely digested photoreceptors, oxidative stress, and accumulation of drusen that normally occur in and around the RPE over the first five decades of life. Based on a variety of risk factors that activate the immune system (e.g., monocytes), these protective gradients may dissipate with age, aided perhaps by the loss of RPE barrier function and the steady buildup of an immunologically hostile environment.

Leukocyte Migration across the RPE: A Model of Disease Progression

The integrity of the RPE monolayer depends on the inter-epithelial junctions that include tight and adherens junctions and desmosomes. The main constituents of tight junctions are three families of transmembrane proteins: occludins, claudins, and junctional adhesion molecules

granulocytes. The number of transmigrating granulocytes is shown as % of control. Exogenous addition of modified junctional adhesion molecule Fc-JAM-C competes with JAM-C/JAM-C interactions between adjacent cells, significantly reducing IL-8 induced transmigration of granulocytes. (b) In contrast, Fc-JAM-C had no significant effect on the basolateral to apical transepithelial migration of monocytes toward MCP-1. Addition of another RPE JAM isoform Fc-JAM-A, did not alter transmigration of granulocytes or monocytes. From Economopoulou, M., et al. (2009). *Investigative Ophthalmology and Visual Science* 50(3): 1454–1463. © Association for Research in Vision and Ophthalmology.

(JAMs). The third member of the JAM family, JAM-C, has been identified in various cell types and implicated in inflammatory processes and shown to participate in the transmigration of leukocytes through endothelial and gut epithelial cells. Economopoulou and colleagues found that JAM-C is localized at the tight junctions of intact monolayers of adult and fetal human RPE, it is found at the initial cell–cell contacts of newly forming junctions, and that it helps initiate hfRPE junction formation and polarization. JAM-C also promotes the transepithelial migration of granulocytes through intact monolayers of cultured hfRPE driven by physiological gradients of interleukin 8 (IL-8). Thus, in the intact eye, JAM-C may be an important determinant of RPE initial junction formation, cell polarization, and immune-system-mediated pathophysiology at the retina–RPE interface (**Figure 9**). Recent animal model studies have implicated monocyte chemoattractant protein 1 (MCP-1) and fractalkine receptor in retinal microglia as critical regulators of drusen accumulation, local inflammation, and the development of AMD. As demonstrated by Shi and colleagues, the RPE secretes significant amounts of MCP-1 and IL-8 to the apical side in a polarized manner (**Table 1**). Both chemokines could form gradients across the RPE that coordinate monocyte and neutrophil movement to the RPE basement membrane. This could provide local surveillance/protection against the accumulation of immunologically active debris (drusen). The transformation over time of monocytes into macrophages would slowly degrade the RPE's ability to maintain protective chemokine gradients for the removal of immunologically active debris and eventually lead to degeneration/disease.

See also: Injury and Repair: Light Damage; Phototransduction: The Visual Cycle; Secondary Photoreceptor Degenerations: Age-Related Macular Degeneration.

Further Reading

Adijanto, J., Banzon, T., Jalickee, S., and Miller, S. S. (2009). CO_2-induced ion and fluid transport in human retinal pigment epithelium. *Journal of General Physiology* 133(6): 603–622.

Blaug, S., Quinn, R., Quong, J., Jalickee, S., and Miller, S. S. (2003). Retinal pigment epithelial function: A role for CFTR? *Documenta Ophthalmologica* 106: 43–50.

Bryant, D. M. and Mostov, K. E. (2008). From cells to organs: Building polarized tissue. *Nature Reviews. Molecular Cell Biology* 9(11): 887–901.

Daniele, L. L., Sauer, B., Gallagher, S. M., Pugh, E. N., Jr., and Philp, N. J. (2008). Altered visual function in monocarboxylate transporter 3 (Slc16a8) knockout mice. *American Journal of Physiology Cell Physiology* 295: C451–C457.

Donoso, L. A., Kim, D., Frost, A., Callahan, A., and Hageman, G. (2006). The role of inflammation in the pathogenesis of age-related macular degeneration. *Surveys of Ophthalmology* 51: 137–152.

Dunaief, J. L. (2006). Iron induced oxidative damage as a potential factor in age-related macular degeneration: The Cogan lecture. *Investigative Ophthalmology and Visual Science* 47: 4660–4664.

Economopoulou, M., Hammer, J., Wang, F., et al. (2009). Expression, localization, and function of junctional adhesion molecule-C (JAM-C) in human retinal pigment epithelium. *Investigative Ophthalmology and Visual Science.* 50: 1454–1463.

Fisher, S. K., Lewis, G. P., Linberg, K. A., and Verardo, M. R. (2005). Cellular remodeling in mammalian retina: Results from studies of experimental retinal detachment. *Progress in Retinal and Eye Research* 24: 395–431.

Gehrs, K. M., Anderson, D. H., Johnson, L. V., and Hageman, G. S. (2006). Age-related macular degeneration – emerging pathogenetic and therapeutic concepts. *Annals of Medicine* 38: 450–471.

Illek, B., Fu, Z., Schwarzer, C., et al. (2008). Flagellin activates inflammatory response and cystic fibrosis transmembrane conductance regulator-dependent Cl secretion: Role for p38. *American Journal of Physiology Lung Cell Molecular Physiology.* 295: L531–L542.

Jarrett, S. G., Lin, H., Godley, B. F., and Boulton, M. E. (2008). Mitochondrial DNA damage and its potential role in retinal degeneration. *Progress Retinal Eye Research* 6: 596–607.

Jia, L., Liu, Z., Sun, L., et al. (2007). Acrolein, a toxicant in cigarette smoke, causes oxidative damage and mitochondrial dysfunction in RPE cells: Protection by (R)-alpha-lipoic acid. *Investigative Ophthalmology and Visual Science* 48: 339–348.

Li, R., Maminishkis, A., Wang, F. E., and Miller, S. S. (2007). PDGF-C and -D induced proliferation/migration of human RPE is abolished by inflammatory cytokines. *Investigative Ophthalmology and Visual Science* 48: 5722–5732.

Maminishkis, A., Chen, S., Jalickee, S., et al. (2006). Confluent monolayers of cultured human fetal retinal pigment epithelium exhibit morphology and physiology of native tissue. *Investigative Ophthalmology and Visual Science* 47: 3612–3624.

Nussenblatt, R. B. and Ferris, F., 3rd (2007). Age-related macular degeneration and the immune response: Implications for therapy. *American Journal of Ophthalmology* 144: 618–626.

Philp, N. J., Ochrietor, J. D., Rudoy, C., et al. (2003). Loss of MCT1, MCT3, and MCT4 expression in the retinal pigment epithelium and neural retina of the 5A11/basigin-null mouse. *Investigative Ophthalmology and Visual Science* 44: 1305–1311.

Shi, G., Maminishkis, A., Banzon, T., et al. (2008). Control of chemokine gradients by the retinal pigment epithelium. *Investigative Ophthalmology and Visual Science* 49: 4620–4630.

Strauss, O. (2005). The retinal pigment epithelium in visual function. *Physiological Reviews* 85: 845–881.

Voloboueva, L. A., Killilea, D. W., Atamna, H., and Ames, B. N. (2007). N-tert-butyl hydroxylamine, a mitochondrial antioxidant, protects human retinal pigment epithelial cells from iron overload: Relevance to macular degeneration. *FASEB Journal* 21: 4077–4086.

Voloboueva, L. A., Liu, J., Suh, J. H., Ames, B. N., and Miller, S. S. (2005). (R)-alpha-lipoic acid protects retinal pigment epithelial cells from oxidative damage. *Investigative Ophthalmology and Visual Science* 46: 4302–4310.

Wangsa-Wirawan, N. D. and Linsenmeier, R. A. (2003). Retinal oxygen: Fundamental and clinical aspects. *Archives of Ophthalmology* 121: 547–557.

Winkler, B. S., Starnes, C. A., Twardy, B. S., Brault, D., and Taylor, R. C. (2008). Nuclear magnetic resonance and biochemical measurements of glucose utilization in the cone-dominant ground squirrel retina. *Investigative Ophthalmology and Visual Science* 49: 4613–4619.

RPE Barrier

L J Rizzolo, Yale University School of Medicine, New Haven, CT, USA

Glossary

Adherens junctions – A component of the apical junctional complex that provides mechanical strength to cell–cell adhesions and, together with the tight junction, regulates cell size, shape, and proliferation.

Apical junctional complex – An assembly of tight, adherens, and gap junctions that join neighboring cells of an epithelial monolayer together. The junctions form a belt that completely encircles each cell at the apical end of the lateral membranes.

Apical membrane – The region of the plasma membrane that interdigitates with the photoreceptors of the neural retina. It is separated from the basolateral membranes by the apical junctional complex.

Basolateral membrane – The region of the plasma membrane that rests on Bruch's membrane and faces the choroid. It is separated from the apical membrane by the apical junctional complex.

Claudins – A family of proteins that forms tight junctional strands and determines the selectivity and permeability of the tight junctions.

Paracellular space – The space between the neighboring cells of an epithelial monolayer.

Subretinal space – The thin space that lies between the apical membrane of the retinal pigment epithelium and the photoreceptors. It becomes a wide space with retinal edema and detachment.

Tight junctions – A component of the apical junctional complex that regulates transepithelial diffusion through the paracellular space, retards diffusion of lipids and membrane proteins between the apical and basolateral membranes, and, together with the adherens junction, regulates cell size, shape, and proliferation.

Transepithelial electrical resistance (TER) – An amalgam of the electrical resistances of the apical membrane, basolateral membrane, and paracellular space. It is commonly used as a reflection of the electrical resistance of tight junctions. When the sum of the membrane resistances greatly exceeds the paracellular (shunt) resistance, the TER approximates the electrical resistance of the tight junctions.

Introduction

Blood–tissue barriers were first revealed by the inability of injected proteins, or protein-bound dyes, to move from the blood into certain tissues. Only the brain, testes, and placenta shared this property. The cells that formed the barrier exhibited reduced transcytosis and were bound together by seemingly impermeable tight junctions. Transcytosis is one mechanism to move serum solutes across the cells of the barrier, whereas tight junctions partially occlude the paracellular spaces to reduce transepithelial diffusion between the cells. This initial conception has since been expanded to include all mechanisms of transcellular transport and the metabolic and catabolic pathways that alter solutes during transport.

The blood–retinal barrier has two divisions. The inner layers of the retina are supplied by a vascular bed, whose endothelia form the inner blood–retinal barrier. This endothelial barrier typifies most of the blood–brain barrier. This article focuses on the outer blood–retinal barrier, which is more similar to the choroid plexus and testes blood–tissue barriers. These barriers are a collaboration of a fenestrated capillary bed with an epithelium. In the outer retina, the collaboration is between the choriocapillaris and the retina pigment epithelium (RPE). As fenestrae make the capillaries porous, the RPE forms the barrier to serum components (**Figure 1**). Certainly, the Bruch's membrane that separates the capillaries and RPE serves as a filter; however, this aspect of the outer blood–retinal barrier is discussed elsewhere in the encyclopedia. A good example of a metabolic pathway that participates in barrier function is the visual cycle. Vitamin A, transported by the serum, is endocytosed and transformed into *cis*-retinal as it is transported across the cell and exported to the photoreceptors. Ultimately, the paracellular and transcellular transport of ions and small organic solutes should be considered as a unit. Because explorations of how these pathways interact remain in their infancy, the two topics are discussed separately in this encyclopedia. This article focuses on the paracellular pathway, but includes the transcellular pathway whenever a connection between the two can be made. We discuss the assembly of RPE tight junctions, their retina-specific properties, and how the RPE and its tight junctions are regulated by the surrounding tissues.

Structure and Function

Tissue Level

The distinctive structure of the outer blood–retinal barrier leads to unique functions. Unlike other epithelia and endothelia, the RPE separates two solid tissues – the choroid and the neural retina. Early in embryogenesis, the apical surface of the RPE borders a fluid-filled lumen, the lumen of the optic vesicle. As development proceeds, this lumen is reduced to a potential space known as the subretinal space (**Figure 2**). In this space, the microvilli of the RPE's apical pole interdigitate with the outer segments of the photoreceptor cells. The intimate contact of the RPE and photoreceptors allows retinoids of the visual cycle to readily shuttle back and forth across the subretinal space, and allows disk membranes shed by the photoreceptors to be phagocytosed

by the RPE. The ionic composition of the subretinal space is carefully regulated to support the functions of the photoreceptors and the RPE. To this end, the RPE absorbs water. Water continuously enters the retina from the inner vascular bed and vitreous and is transported by the RPE into the choroid for removal by the choroidal circulation. Failure of this process results in retinal edema and even retinal detachment. Contrast this with another region of the blood–brain barrier, the epithelium of the choroid plexus. The epithelium of the choroid plexus is also derived from the neuroepithelium, but in this case, the lumen of the neural tube expands to form the ventricular system. Rather than absorb fluid, the choroid plexus secretes copious volumes of cerebral spinal fluid. To understand the functional differences between the RPE and the epithelium of the choroid plexus, we need to look more closely at the structure of the barrier.

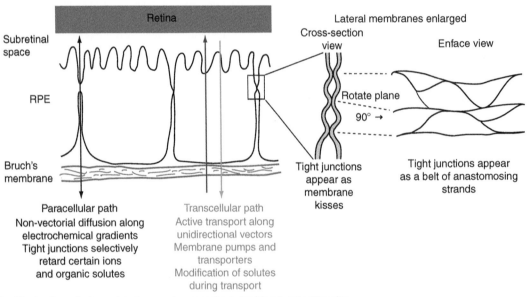

Figure 1   Mechanisms that regulate transport across the outer blood–retinal barrier.

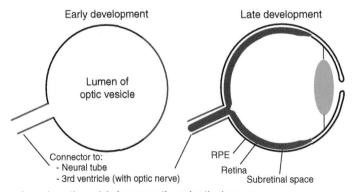

Figure 2   The lumen of the embryonic optic vesicle becomes the subretinal space.

Cellular Level

The RPE is a monolayer of cells that are joined by a complex of junctions known as the apical junctional complex. The apical junctional complex encircles each cell to bind the monolayer together much like the plastic rings that hold together a six-pack of canned beverages. The complex consists of three junctions (tight, adherens, and gap) whose functions are intertwined. Adherens junctions bind neighboring cells together. Tight junctions form a partially occluding seal that semiselectively retards diffusion through the paracellular spaces of the monolayer (**Figure 1**). Both junctions regulate proliferation, cell size, and the polarized distribution of plasma membrane proteins. The junctions of the complex work in concert, but we focus on how tight junctions contribute to barrier function.

Tight junctions form a barrier between the apical and basolateral poles of the cell. They work in conjunction with intracellular trafficking pathways to create and maintain an apicobasal polarity that is essential for the blood–retinal barrier to function. Unlike most epithelia, the Na,K-ATPase (ATP, adenosine triphosphate) is enriched in the apical membrane rather than localized to the basolateral membrane. Although this initial discovery suggested that the RPE is an upside-down epithelium, it is now known that only a few RPE proteins have an atypical distribution. What is crucial for RPE function is the distribution of membrane channels and transporters. The RPE and the epithelium of the choroid plexus both have an apical Na,K-ATPase that provides the energy for vectorial transport. It is the distribution of the various ion channels and transporters that determines whether the RPE absorbs water or the epithelium of the choroid plexus secretes it. Briefly, the polarized distribution of transporters in the RPE results in the active transport of chloride from the apical to basal side of the cell. As a result, the apical side of the monolayer has a positive charge relative to the basal side. Sodium and potassium are transported down this electrical gradient to balance the chloride transport. The osmotic gradient that results pulls water in the apical to basal direction. The polarized distribution of the various channels and transporters differs in the epithelium of the choroid plexus to support the opposite, basal to apical, transport of water. There are general housekeeping mechanisms that recognize the targeting signals encoded in a protein's structure to deliver it to the correct membrane. In some cases, different tissue-specific isoforms encode different targeting signals; in others, tissue-specific variations in the targeting machinery give each epithelium its unique character.

The vectorial transport mechanism outlined above would have little effect, if transepithelial ion gradients were dissipated by the paracellular pathway. Early microscopists believed that a zonular band of junctions occluded the paracellular space by completely encircling each cell. Their name for this junction, zonula occludens or tight junction, is misleading because the junction is selectively leaky. The degree of leakiness, and selectivity for certain ions, not only varies among epithelia, but is essential for epithelial function as well. The permeability and semiselectivity of the junctions are matched to transcellular transport mechanism. For example, RPE tight junctions need to be leakier to sodium than to chloride, because RPE pumps chloride across the cell and needs a leak for cations to passively follow the chloride flux. The inadequate capacity for sodium and potassium to cross the cell is ameliorated by the sodium-selective leak through the tight junction.

A compelling example of how transcellular transport is matched to tight junction selectivity is provided by the kidney. The diuretic hormone, aldosterone, changes the flux of water by acting simultaneously on a membrane sodium channel and a tight junction protein that regulates sodium selectivity. In the retina, the volume and composition of the subretinal space vary with the day/night cycle. It remains to be investigated how the properties of tight junctions and membrane transporters are coordinated to manage the subretinal space during this cycle.

Molecular Level

Proteins of the apical junctional complex fall into transmembrane, adaptor, and signaling/regulatory categories. Adaptors link the transmembrane proteins to signaling proteins and a cortical band of actin filaments. Besides known members of the tight junction (**Figure 3**), proteomics suggest that over 912 proteins are associated with the tight junction. The adaptor proteins express multiple copies of PDZ domains. This protein-binding domain of approximately 100 amino acids takes its name from the three proteins that defined this class of protein-binding domain: postsynaptic density protein 95 (PSD-95), disks large, and zonula occludins 1 (ZO-1). PDZs are the largest family of protein-binding domains and form the basis of many protein complexes. Each adaptor protein expresses

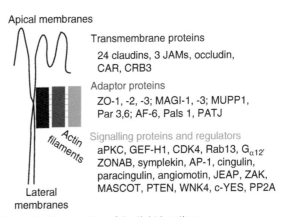

Figure 3 Composition of the tight junctions.

multiple homologs of PDZ domains that have distinct, but sometimes overlapping binding specificities. Together with other protein-binding motifs, for example, SRC homology 3 domain, guanylate kinase, bi-tryptophan domain that binds proline-rich peptides, Dilute domain, and Phox and Bem1P domain, the known adaptor proteins have the capacity to bind the large number of regulatory proteins that proteomics suggests. A junction of such complexity would be inconsistent with the original view of the tight junction as a static barrier.

The tight junction is a highly dynamic structure with the potential to rapidly respond to environmental stimuli. Photobleaching studies demonstrate that ZO-1 and occludin, a regulator of permeability, rapidly associate and dissociate from the junction. Occludin has a very short half-life and its degradation is regulated by endocytic and ubiquitin pathways. Therefore, the cell exerts a fine control over the properties of the tight junction that can rapidly respond to changes in the environment. This flexibility extends to the claudin family of transmembrane proteins. Membrane proteins typically have a half-life on the order of days, but the claudins that have been studied have a half-life of 4–12 h. Claudins form the anastomosing network of strands that are observed by freeze-fracture electron microscopy (**Figure 4**) and schematized in **Figure 1**. There are 24 or more claudins. Each epithelium expresses a subset of claudins. The subset of claudins, expressed and localized to the tight junction, determines the selectivity and permeability of the junction. In the kidney example given above, aldosterone decreases the expression of claudin 4 to increase the sodium leak through the tight junctions.

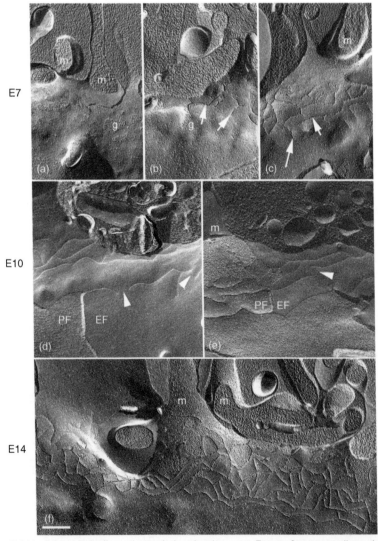

Figure 4 Strands of the tight junction gradually coalesce during development. Freeze-fracture replicas show how sparse, disconnected strands on E7 become a necklace of strands with discontinuities by E10 and a continuous, uninterrupted network by E14. Microvilli (m) at the top of each panel indicate the apical end of the lateral membrane. Arrows, tight junctional strands; Arrowheads, discontinuities. EF, E-face; PF, P-face; Bar = 0.25 μm.

Although the basic structure of the outer blood–retinal barrier is conserved among species, there are species-specific variations in the composition of the subretinal space, the properties of transmembrane transport, and the composition of the tight junctions. In human RPE, the principal claudins appear to be claudins 3, 10, and 19. Claudin 19 has been linked to kidney disease and visual impairment. By contrast, chick RPE expresses primarily claudins 1 and 20 with lesser amounts of claudins 2, 4L2, 5, and 12. Claudins 1 and 3 are fairly ubiquitous with most epithelia expressing one or the other. Claudin 2 increases sodium permeability in some contexts. However, the study of how claudins affect permeability is in its infancy particularly in regard to the effects of cellular context on function.

Regulation of RPE Tight Junctions

Clues from Embryonic Maturation

What is a mature, differentiated RPE monolayer? Which markers and how many should we use to render this judgment? Might enhancing the expression of some markers in a culture experiment lessen the expression of others? If cellular pathways form an integrated web, would over- or underexpression of a protein have deleterious effects? A default path for human embryonic stem cells appears to be an RPE-like cell, but those cells lack some RPE proteins and express non-RPE proteins. The RPE is the first retinal cell to form during development, but despite its undeniable RPE character, the early RPE cell is only partially differentiated. It will undergo many transformations, as the neural retina and choroid differentiate on either side of it. The RPE is very plastic. Depending upon pathology or culture conditions, one can observe many partially differentiated states, or even transform RPE to other phenotypes. This may be the reason why RPE transplants fail when retinal degeneration is advanced. Dedifferentiation would be a normal response of healthy RPE to this abnormal environment. In support of this hypothesis, RPE transplants were most successful when the RPE and neural retina were co-transplanted. In some ways, culture is like a disease state where a degenerate neural retina and choroid no longer send RPE the signals that maintain key functions. In the chick culture model described, we found that retinal secretions promote RPE differentiation over the course of days rather than the months required without the retina. More work will be needed to determine whether these findings apply to mammalian eyes.

A study of chick embryonic development illustrates the maturation process. We can describe early, intermediate, and late phases of development. The early phase extends from the time that the RPE forms on embryonic day 3 (E3) until the inner segments of photoreceptors protrude the outer limiting membrane on E9. The intermediate phase extends from E9 to E15, when photoreceptors begin to elaborate outer segments. The late phase extends from E15 until hatching on E21. During the intermediate phase, the layers of Bruch's membrane gradually form. Fenestrations in the walls of the choroidal capillaries begin to form in the intermediate phase, but are not fully elaborated until the middle of the late phase. In parallel with the formation of fenestrations in the capillaries, infoldings of the basolateral membrane begin to form in the intermediate phase, but are not completely elaborated until the middle of the late phase. As the basolateral membranes elaborate infoldings, apical membrane microvilli elongate in coordination with the elongation of photoreceptor inner and outer segments. Some plasma membrane proteins have a distribution that is polarized between the apical and basolateral membranes as early as E7. Nevertheless, some proteins become polarized later in development in parallel with the morphological changes of the apical and basolateral membranes. Examples include basigin, monocarboxylate transporters, and the Na,K-ATPase. This coordination of the development of photoreceptors, RPE plasma membranes, Bruch's membrane, and the choriocapillaris is conserved between chickens and mammals. It is in the context of this maturing environment that the RPE completes its differentiation.

The RPE is the first retinal layer to overtly differentiate. Despite an epithelial morphology and the expression of RPE markers, early embryonic RPE is still immature. Between E7 and E18 of chick development, 40% of the transcriptome changes with substantial effects on the extracellular matrix, junctional complexes, cell surface receptors, signal transduction pathways, cytoskeleton, regulators of gene expression, and transmembrane transport proteins. Some genes turned on and others off, but most changed their level of expression relative to one another. These data suggest why RPE cultures that express the same tissue-specific markers can function so differently. Without direction from the neural retina or choroid to establish balanced gene expression, cultured RPE can adopt the range of behaviors it displays during development *in vivo*.

Assembly of Tight Junctions during Differentiation

Most studies of the assembly of the apical junctional complex were performed in mouse blastocysts or in culture using kidney or intestinal cell lines. The rapid kinetics of assembly in these models makes it difficult to parse the role of the putative assembly proteins. The consensus is that a primordial adherens junction forms first, followed by the segregation of tight junction components into a nascent tight junction. Nectin, E-cadherin, and junctional adhesion molecule-A (JAM-A) trigger the formation of the primordial adherens junction. These transmembrane proteins form homodimers with their counterparts on the neighboring cell. Their cytoplasmic domains crystallize a complex of proteins by binding an adaptor protein. For example, JAM-A binds the adaptors, AF-6 (Afadin), PAR3 (Partitioning defective-3 homologue), and ZO-1. Their multiple protein-binding domains enable these adaptors to assemble a complex. For example, PAR-3 binds PAR-6 (Partitioning defective-6 homologue) and the atypical protein kinase C, which contributes to cell polarity. JAM-A also localizes ZO-1 to the apical side of the complex to initiate the formation of a tight junction. The tight junction transmembrane proteins, occludin and claudin, bind this nascent complex. Anti-JAM-A antibodies block the formation of tight junctions, as evidenced by the mislocalization of occludin and a low transepithelial electrical resistance (TER). Nevertheless, the adherens junction did form, as evidenced by the localization of E-cadherin and ZO-1. Taken together, these data suggest that JAM-A may play a role in assembling adherens junctions, but plays a more critical role in assembling tight junctions.

As the apical junctional complex assembles slowly during normal RPE development, the process may be studied in greater detail. In chick, primordial adherens junctions are already present in the neuroepithelium that forms the RPE on E3, days before rudimentary tight junctions begin to form on E7. Many of the proteins described above in the assembly of the apical junctional complex are present at this time, but the adherens junction will remodel throughout development both morphologically and molecularly. In each phase of development, different cadherins will appear and disappear. The early phase includes many tight-junctional proteins: ZO-1, occludin, and the assembly proteins AF-6, JAM-A, PAR3, and PAR6. Nonetheless, tight junctions are absent until claudins expression begins on E7 and short, sparse tight-junctional strands begin to appear (**Figure 4**). During the intermediate phase, tight-junctional strands grow in number and length to gradually coalesce into a complete network that encircles each cell. The tight junction first becomes functional between E10 and E12 (defined by the ability to block the transepithelial diffusion of horseradish peroxidase). During the late phase, structural modifications of the tight junctions continue. Like the adherens junctions, these morphological changes are accompanied by molecular changes. Some claudin messenger RNAs (mRNAs) appeared early, but others appeared during the intermediate or late phases. The expression of some of the early-appearing claudins decreased during the late phase. During the intermediate phase, there was a switch in the expression of ZO-1 isoforms, an event also observed during tight-junction formation in pre-implantation embryos. ZO-3 did not appear until the late phase of development. Although changes in protein expression parallel gene expression to some extent, it appears that the claudins and ZO proteins are also regulated by effects on protein stability and subcellular localization.

The molecular and morphological changes in the apical junctional complex imply the function of the outer blood–retinal barrier changes during this long maturation process. The changes in claudin expression imply changes in the selectivity and permeability of the tight junctions. It would be reasonable to expect that this would be coupled with changes in transepithelial transport. Among the changes in the transcriptome, many involve membrane transporters. Several should be mentioned, because there is also physiological and cell biological data to corroborate the changes in gene expression. Changes in the expression and polarized distribution of the monocarboxylate transporters and Na,K-ATPase have already been mentioned. Early in development, several facilitated glucose transporters are expressed, but more are expressed later in development, including a sodium-coupled glucose transporter. The appearance of the latter transporters corresponds to the time that the tight junctions become relatively impermeable to glucose. These changes are essential because the retina has a high demand for glucose. It appears that housekeeping transporters that are sufficient for the RPE's individual needs are replaced by a transcellular, active transport mechanism at the time the blood–retinal barrier forms. These conclusions are based on studies of a primary cell culture model of RPE maturation.

Culture Models to Study Regulation of the Outer Blood–Retinal Barrier

RPE can be cultured on matrix-coated filters that are suspended in a culture dish (**Figure 5**). This architecture separates the media compartment into apical and basal chambers. The cultures spontaneously polarize with the basal membrane against the filter substrate and the apical microvilli projecting into the upper chamber. The culture architecture allows the cells to feed from the basolateral membranes, as *in vivo*. By contrast, plastic-grown cells need to feed from the apical membranes. Another advantage is that it is easy to measure barrier function by placing tracers in one medium chamber and measuring their flux

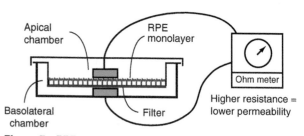

Figure 5 RPE cell culture.

across the monolayer into the opposite chamber. Further, electrodes may be placed to measure a TER. This flexibility is important because selectivity and permeability of tight junctions are regulated semi-independently. The TER is commonly used to assess tight junctions, but TER is an amalgam of transcellular and paracellular resistances to a current that is carried by all the ions of the extracellular space. The TER often approximates the resistance of the tight junctions, because the transjunctional current is often much greater than the transcellular current. Junctions of epithelia with a similar TER can differ in their ion selectivity. Therefore, it is valuable to measure ion fluxes directly. Further, the permeation of mannitol (a small organic tracer) can be modulated independently of TER, and vice versa. Accordingly, multiple assays of barrier function are required for a full assessment.

Although it is relatively easy to culture RPE from many species, including human adults, it is difficult to establish cultures that form an effective barrier. A good example of the problems encountered is the human-derived ARPE19 cell line. This spontaneously transformed cell line has many RPE-specific properties, but its tight junctions are immature. Many claudins were expressed that are undetected in native tissue, and some of the claudins expressed *in vivo* were undetected in ARPE19. By modifying culture conditions, barrier function, morphology, and melanin expression could be improved. Although claudin expression was affected somewhat, large differences between native RPE and ARPE19 remained. Like the chick studies described above, genomic analyses of native and cultured human RPE have become or are becoming available, which will allow a molecular definition that can compare native to cultured RPE.

Several culture systems have been devised that allow RPE to form a barrier that resembles native RPE. The most highly differentiated are primary or secondary cultures that were isolated from the intermediate phase of RPE development. This stage corresponds to E14 in the chick, postnatal day 5 in rat, and 18–22 weeks gestation in the human. Notably, human fetal RPE appears to have a broader window, as cultures isolated from 13-week fetuses also have excellent properties. Each culture relies on highly specialized medium that includes low amounts, or no, serum. In some cases, the medium appears to remove the need for choroidal or retinal stimulation, but these cultures require 1–2 months in culture to fully mature. Rat and chick cultures that were maintained in serum-free medium were very sensitive to the addition of serum to the apical medium chamber, as might be seen in pathology. For each species, apical serum decreased barrier function, as measured by the TER.

In contrast to the other models, the chick model was designed to study tissue interactions. It used primary cultures that formed incomplete tight junctions with a low TER in a serum-free medium. However, the cultures were very sensitive to retinal secretions. A medium conditioned by the organ culture of neural retinas induced the formation of complete tight junctions with a TER that was similar to native RPE. Reconstitution experiments showed that contact with the neural retina was required for the proper polarized distribution of the Na,K-ATPase and certain integrins. A central finding was that retinal interactions promoted differentiation on a timescale of days rather than months. Besides cellular junctions, the retinal conditioned medium affected the expression of genes related to the visual cycle, phagocytosis, cytoskeleton, and transmembrane transport. Besides gene expression, retinal conditioned medium affects the half-life and subcellular localization of claudins and ZO proteins. By regulating membrane transporters and tight junctions, the retina and RPE appear to collaborate in regulating the subretinal space. This is an area of research that needs to be explored in greater detail.

RPE in the Larger Context of Ocular Biology and Disease

Many diseases are coming to be viewed as a low-grade inflammatory process, including age-related macular degeneration. Investigations of inflammatory diseases of the intestine (Crohn's) and central nervous system (multiple sclerosis) demonstrate that disease can affect the tight junctions in part through the action of inflammatory cytokines. Certainly, disease also affects the membrane transporters of epithelia and endothelia. In the renal field, there has been progress in understanding the interrelationships of tight junctions and membrane transporters and how they are coordinately regulated. The retina field lags behind, but the availability of good culture models and the advances in genomics and systems biology hold great promise for the future.

See also: Breakdown of the RPE Blood–Retinal Barrier; Phototransduction: The Visual Cycle; Retinal Pigment Epithelium: Cytokine Modulation of Epithelial Physiology.

Further Reading

Bradbury, M. W. B. (1979). *The Concept of a Blood–Brain Barrier.* New York: Wiley.
Burke, J. M. (2008). Epithelial phenotype and the RPE: Is the answer blowing in the Wnt? *Progress in Retinal and Eye Research* 27(6): 579–595.
Cereijido, M. and Anderson, J. M. (eds.) (2001). *Tight Junctions.* Boca Raton, FL: CRC Press.
Cereijido, M., Contreras, R. G., Shoshani, L., Flores-Benitez, D., and Larre, I. (2008). Tight junction and polarity interaction in the transporting epithelial phenotype. *Biochimica et Biophysica Acta* 1778: 770–793.

Grunwald, G. B. (1996). Cadherin cell adhesion molecules in retinal development and Pathology. *Progress in Retinal Eye Research* 15: 363–392.

Guillemot, L., Paschoud, S., Pulimeno, P., Foglia, A., and Citi, S. (2008). The cytoplasmic plaque of tight junctions: A scaffolding and signalling center. *Biochimica et Biophysica Acta* 1778: 601–613.

Klimanskaya, I., Hipp, J., Rezai, K. A., et al. (2004). Derivation and comparative assessment of retinal pigment epithelium from human embryonic stem cells using transcriptomics. *Cloning and Stem Cells* 6: 217–245.

Le Moellic, C., Boulkroun, S., Gonzalez-Nunez, D., et al. (2005). Aldosterone and tight junctions: Modulation of claudin-4 phosphorylation in renal collecting duct cells. *American Journal of Physiology – Cell Physiology* 289: C1513–C1521.

Radtke, N. D., Aramant, R. B., Petry, H. M., et al. (2008). Vision improvement in retinal degeneration patients by implantation of retina together with retinal pigment epithelium. *American Journal of Ophthalmology* 146: 172–182.

Rajasekaran, S. A., Beyenbach, K. W., and Rajasekaran, A. K. (2008). Interactions of tight junctions with membrane channels and transporters. *Biochimica et Biophysica Acta* 1778: 757–769.

Rizzolo, L. J. (2007). Development and role of tight junctions in the retinal pigment epithelium. *International Review of Cytology* 258: 195–234.

Strauss, O. (2005). The retinal pigment epithelium in visual function. *Physiological Reviews* 85: 845–881.

Van Itallie, C. M. and Anderson, J. M. (2006). Claudins and epithelial paracellular transport. *Annual Review of Physiology* 68: 403–429.

Wilt, S. D. and Rizzolo, L. J. (2001). Unique aspects of the blood–brain barrier. In: Anderson, J. M. and Cereijido, M. (eds.) *Tight Junctions*, pp. 415–443. Boca Raton, FL: CRC Press.

Retinal Vasculopathies: Diabetic Retinopathy

N C Steinle and J Ambati, University of Kentucky, Lexington, KY, USA

Glossary

Cotton-wool spots (CWSs) – Also known as soft exudates, these may be found in nonproliferative diabetic retinopathy (NPDR). They are composed of accumulations of neuronal debris within the retinal nerve fiber layer, and result from disruption and stasis of axoplasmic flow.

Diabetes control and complications trial (DCCT) – A study that contributed to our understanding that intensive glycemic control is associated with a reduced risk of newly diagnosed retinopathy and a reduced progression of existing retinopathy in people with diabetes.

Diabetes mellitus (DM) – A metabolic disorder characterized by sustained hyperglycemia secondary to lack or diminished efficacy of endogenous insulin.

Diabetic retinopathy (DR) – A retinal disease consequent to development of DM.

Insulin-dependent diabetes mellitus (IDDM) – A term sometimes used to refer to type I diabetes.

Intraretinal microvascular abnormalities (IRMAs) – The tortuous, hypercellular micro vessels that develop in NPDR.

New vessels elsewhere (NVE) – The neovascularization of the retina found greater than one disk diameter from the optic nerve head.

New vessels on disk (NVD) – The neovascularization on or within one disk diameter of the optic nerve head.

Non-insulin-dependent diabetes mellitus (NIDDM) – An older term for type II diabetes or adult-onset diabetes.

Nonproliferative diabetic retinopathy (NPDR) – DR characterized by intraretinal microvascular changes which precede the proliferative phase.

Optical coherence tomography (OCT) – A noninvasive imaging technique that permits analysis of retinal structure in the living eye.

Proliferative diabetic retinopathy (PDR) – DR characterized by the presence of retinal neovascularization.

The early treatment diabetic retinopathy study (ETDRS) – A large study of progression and treatment of DR.

United Kingdom Prospective Diabetes Study (UKPDS) – A study that contributed to our understanding that intensive glycemic control is associated with a reduced risk of newly diagnosed retinopathy and a reduced progression of existing retinopathy in people with diabetes.

Vascular endothelial growth factor (VEGF) – A growth factor that promotes development of endothelial cells.

Background

Diabetes mellitus (DM) is a metabolic disorder characterized by sustained hyperglycemia secondary to lack or diminished efficacy of endogenous insulin. The terminology used for classification of different types of diabetes is evolving and can be a source of confusion. Traditionally, there have been two types of DM: type I diabetes and type II diabetes (these terminologies are used throughout the remainder of this article). Immune-mediated diabetes is the latest, and perhaps most descriptive, terminology applied to type I diabetes. Autoimmune destruction of insulin-producing pancreatic islet cells is postulated as instrumental in the pathogenesis of type I diabetes. Previous terminology used for type I diabetes included insulin-dependent diabetes mellitus (IDDM), and juvenile-onset diabetes. Type II diabetes is characterized by relative deficiencies of insulin and/or peripheral insulin resistance. It was previously known as non-insulin-dependent diabetes mellitus (NIDDM) or adult-onset diabetes.

DM is a common medical problem and is a major global source of morbidity and mortality. The incidence of DM is thought to be increasing throughout the world in part due to an increasing incidence of obesity and sedentary lifestyles. In the United States, it is estimated that 7.8% of the total population has DM, and it causes a vast array of long-term systemic complications which have a significant impact on both quality and quantity of life. Patients with DM have heart disease, death rates, and stroke rates that are 2–4 times higher than adults without diabetes, and DM is the leading cause of end-stage renal disease in the United States. Further, people with diabetes are more susceptible to many other illnesses and, once acquired, often have worse prognoses (e.g., pneumonia).

From an ophthalmic standpoint, DM causes numerous complications. Chief among these complications are diabetic retinopathy (DR), unstable refractions, accelerated cataracts, rubeosis iridis, which can lead to neovascular glaucoma, cranial nerve palsies, reduced corneal sensitivity, papillopathy, and poor wound healing. The incidence of blindness is 25 times higher in patients with diabetes than in the general population. Furthermore, DR is the most common cause of blindness in patients aged 20–74 years, accounting for 12 000–24 000 new cases of blindness in the United States each year.

Risk Factors for DR

The prevalence of DR in the diabetic population increases with the duration of diabetes and patient age. Studies have shown that after 20 years of diabetes, nearly 99% of patients with type I DM and approximately 60% of patients with type II DM have some degree of DR (**Figure 1**). DR rarely develops in children younger than 10 years of age, regardless of the duration of diabetes. The risk of DR increases after puberty. Approximately 5% of type II diabetics have DR at presentation; this observation is a reflection of the typically insidious onset of hyperglycemia in type II diabetes many years before the diagnosis is firmly established.

In addition to duration of DM, other risk factors for the development of DR include poor glycemic control, the type of diabetes (type 1 more than type 2), and the presence or absence of associated conditions such as hypertension, smoking, dyslipidemia, nephropathy, and pregnancy. The Diabetes Control and Complications Trial (DCCT) and

the United Kingdom Prospective Diabetes Study (UKPDS) demonstrated that intensive glycemic control is associated with a reduced risk of newly diagnosed retinopathy and a reduced progression of existing retinopathy in people with DM (type I in DCCT (**Figure 2**) and type II in UKPDS). According to the DCCT, intensive insulin therapy reduced the incidence of new cases of DR by as much as 76% compared with conventional therapy. The UKPDS found similar results in type II diabetics; each 1% point reduction in glycosylated hemoglobin was associated with a 37% reduction in development of retinopathy. Further, the UKPDS showed that control of hypertension was also beneficial in reducing progression of DR. Pregnancy is occasionally associated with rapid progression of DR; thus, women with diabetes who become pregnant require more frequent evaluation of the retina. Pregnant women without any DR are at a 10% risk of developing nonproliferative diabetic retinopathy (NPDR) during their pregnancy. Of those with preexisting NPDR, 4% progress to proliferative retinopathy (**Figure 3**).

Pathogenesis

The pathogenesis of DR is the current subject of intense research. It is theorized that exposure to chronic hyperglycemia results in a number of biochemical and physiologic alterations that ultimately produce retinal vascular changes and subsequent retinal injury and ischemia. The list of hematologic and biochemical abnormalities theorized to play a role in the development of DR

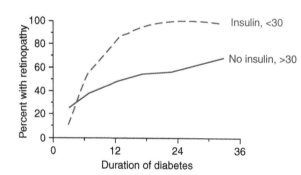

Figure 1 Incidence of diabetic retinopathy (DR) increases over time. Duration of DM is directly associated with an increased prevalence of DR in people with both type I and type II DM. The figure represents the percent of diabetic patients with retinopathy according to duration of disease in patients under the age of 30 years who were treated with insulin (primarily type I diabetics) and patients over the age of 30 years who were not treated with insulin (primarily type II diabetics). Retinopathy increased over time in both groups, affecting virtually all patients with type I diabetes by 20 years. The increased incidence in type II diabetes at 3 years is likely secondary to the difficulty in determining the exact time of onset of type II DM. Data from Klein, R., Klein, B. E., Moss, S. E., Davis, M. D. and DeMets, D. L. (1984). The Wisconsin Epidemiologic Study of Diabetic Retinopathy: III. Prevalence and risk of diabetic retinopathy when age at diagnosis is 30 or more years. *Archives of Ophthalmology* 102: 527.

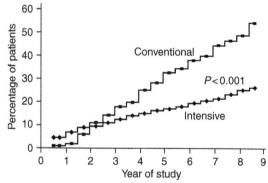

Figure 2 Intensive glycemic control slows progression of retinopathy. Cumulative incidence of progressive retinopathy in patients with type 1 diabetes and early nonproliferative retinopathy who were treated with either conventional or intensive insulin therapy for 9 years. Intensive glycemic control reduced the risk of DR progression over time by 54%, although intensive therapy was associated with transient worsening in the first year ($p < 0.001$). Data from Diabetes Control and Complications Trial Research Group (1993). The effect of intensive treatment of diabetes on the development and progression of long-term complications in insulin-dependent diabetes mellitus. *The New England Journal of Medicine* 329: 977.

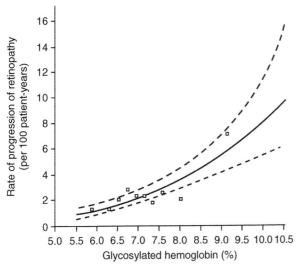

Figure 3 Progression of DR in relation to glycemic control. The figure shows the rate of progression of retinopathy in patients with type 1 diabetes according to mean glycosylated hemoglobin values (solid line). Better glycemic control was associated with slower rates of DR progression. The dashed lines represent the 95% confidence intervals. Data from Diabetes Control and Complications Trial Research Group (1993). The effect of intensive treatment of diabetes on the development and progression of long-term complications in insulin-dependent diabetes mellitus. *The New England Journal of Medicine* 329: 977.

includes the following: impairment of retinal blood vessel autoregulation, the occurrence of retinal microthrombosis and subsequent ischemia, accumulation of advanced glycosylation end products, and damage caused by reactive oxygen species. The role that growth factors (e.g., vascular endothelial growth factor (VEGF)) play in the formation of DR is discussed later. There have also been recent considerations of categorizing DR as an inflammatory disease. Trials investigating anti-inflammatory agents for prevention or treatment of DR in humans are ongoing.

Specific retinal vascular changes theorized to be instrumental in DR include the loss of pericytes, basement membrane thickening, and impaired endothelial cell function. The walls of retinal capillaries consist of endothelial cells and pericytes and are devoid of smooth muscle and elastic tissue. Endothelial cells form a single layer on a basement membrane and are linked by tight junctions that form the inner blood–retinal barrier. Pericytes are found external to the endothelial cells and have pseudopodial processes that envelop the capillary. It is believed that pericytes have contractile properties and are thought to participate in autoregulation of the microvascular capillary circulation (analogous in function to the smooth muscle found in larger arteries). The classic histologic finding of early DR in the human retina is the loss of microvascular pericytes; however, the exact mechanism by which pericytes are preferentially lost early in DR is

unknown. Thickening of the retinal capillary basement membrane is another well-known lesion found in DR. In addition to basement membrane thickening, patients with DR are also found to have vacuolization and deposition of fibrillar collagen in their basement membranes. Similar to the loss of pericytes, the exact biochemical events that lead to basement membrane alterations in DR are not fully evident. Several studies implicate the sorbitol pathway in this process. The sorbitol pathway is the name given to the sequence of reactions that convert glucose to fructose involving the enzymes aldose reductase and sorbitol dehydrogenase. In this pathway, glucose is reduced first to sorbitol, which is then oxidized to fructose. However, since the latter reaction occurs slowly in many cells, sorbitol may build to high, and possibly, toxic concentrations. The toxicity of sorbitol is theorized to perhaps lead to basement membrane alterations. The final vascular change that appears to be instrumental in DR is the loss of endothelial cell function and the subsequent breakdown of the blood–retinal barrier. One possible cause of the blood–retinal barrier breakdown is opening of the tight junctions (zonulae occludentes) between adjacent microvascular endothelial cell processes. Several proteins are known to be involved with tight junction function, namely ZO-1, occluding and claudin. Studies have shown that high glucose levels appear to inhibit ZO-1 expression. Further experiments have shown reduced expression and anatomical distribution of occludin in experimental diabetes. Finally, studies have shown that intravitreal injections of the growth factor VEGF in rats increased production of nitric oxide (NO) and increased phosphorylation of ZO-1 and occludin, changes that result in increased breakdown of the blood–retinal barrier.

Advanced stages of DR are marked by the proliferation of new blood vessels. Retinal neovascularization is a devastating process that can lead to blindness in DR. Neovascularization develops through angiogenesis, in which capillaries develop from preexisting blood vessels. In DR, the regulatory mechanisms of angiogenesis can become compromised, which leads to uncontrolled endothelial cell division. On a molecular level, angiogenesis is a complicated pathway involving interplay between a number of angiogenic messengers, proteolytic enzymes, and the preexisting vessels themselves. The common final product in angiogenesis is activation of vascular endothelial cells. Several endothelial cell mitogens have been isolated and studied, including VEGF, platelet-derived growth factor (PDGF), insulin-like growth factor (IGF), basic fibroblast growth factor (bFGF), protein kinase C (PKC), antiopoietins, integrins, and ephrins. VEGF is commonly considered the most potent angiogenic factor, and some of the other molecules may act indirectly through VEGF. As the activated endothelial cells proliferate, they secrete proteolytic enzymes that degrade the parent vessel's basement membrane as well as the extracellular matrix. Once the endothelial cells gain access to the extravascular space, they migrate and form new capillary sprouts. Mesenchymal cells are then recruited to form smooth muscle cells in arterioles and a new basement membrane is deposited to complete the process of angiogenesis. The resulting abnormal vessels lack structural integrity and are prone to leak fluid, leading to retinal edema. Further, the abnormal vessels often are associated with a fibrovascular membrane. This membrane can become adherent to both the retina and the posterior hyaloid face. As the vitreous contracts, the fibrovascular membrane can cause tractional forces on the retina, leading to retina edema, vitreous hemorrhages, retinal heterotropia, retinal tears, and tractional retinal detachments. Fortunately, the majority of DR follows a fairly predictable course; thus, screening examinations and prophylactic interventions can be implemented to reduce the devastating and potentially blinding consequences of advanced DR.

Classifications

The classification of DR is classically based on the severity of intraretinal microvascular changes and the presence or absence of retinal neovascularization. Thus, DR is divided into two main forms: nonproliferative and proliferative. NPDR is characterized by intraretinal microvascular changes which precede the proliferative phase. Proliferative diabetic retinopathy (PDR) is characterized by the presence of retinal neovascularization. These two classifications of DR (NPDR and PDR) have been useful for the analysis of treatment efficacy in the literature and serve as general indicators for treatment strategies. However, it should be noted that an individualized approach to the treatment of DR is prudent, as every patient with DR has a unique combination of findings, symptoms, and rate of progression.

Nonproliferative Diabetic Retinopathy

The retinal microvascular changes found in NPDR are, by definition, limited to the confines of the retina and do not extend beyond the innermost retinal layer – the internal limiting membrane. Characteristic findings in NPDR include microaneurysms, retinal hemorrhages, retinal edema, hard exudates, cotton-wool spots (CWSs), areas of capillary nonperfusion, intraretinal microvascular abnormalities (IRMAs), and venous beading. NPDR primarily causes visual decline through either capillary nonperfusion leading to macular ischemia, or through increased vascular permeability, resulting in macular edema.

The first visible sign of NPDR is the retinal capillary microaneurysm. Clinically, microaneurysms are identified as red dots from 15 to 60 μm in diameter (**Figure 4**).

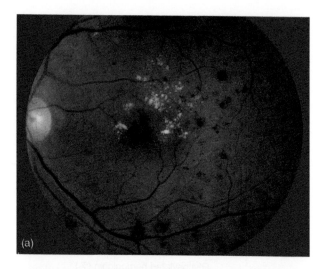

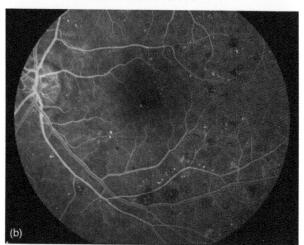

Figure 4 (a) Severe non-proliferative diabetic retinopathy. Color fundus photograph and a red-free fundus photograph of the left eye of a patient with severe NPDR. The photographs demonstrate numerous diffusely scattered microaneurysms, dot-blot hemorrhages, and hard exudates. (b) Microaneurysms and retinal hemorrhages. Arteriovenous phase fluorescein angiogram image of the same left eye (a). The microaneurysms demonstrate marked early hyperfluorescence, whereas the retinal hemorrhages block fluorescein and thus appear hypofluorescent.

On histologic examination, microaneurysms are hypercellular saccular outpouchings of the capillary wall. They are often found in relation to areas of capillary nonperfusion. Postulated mechanisms behind microaneurysm formation include the release of vasoproliferative factors (e.g., VEGF) with endothelial cell proliferation, weakness of the capillary wall secondary to the loss of pericytes, abnormalities of the adjacent retina, and increased intracapillary pressure. Microaneurysms can be differentiated from punctate retinal hemorrhages, which are also seen in DR, through fluorescein angiography. Microaneurysms will demonstrate marked early hyperfluorescence against the darker choroidal background; whereas, retinal hemorrhages will block fluorescein and thus appear hypofluorescent (**Figure 4(b)**). Late fluorescein angiography frames often demonstrate leakage emanating from microaneurysms as a result of the breakdown in the blood–retinal barrier. Individual microaneurysms typically appear and disappear over time. Microaneurysms often foreshadow progression of DR as an increase in microaneurysms often is associated with progression of DR.

The retinal hemorrhages most commonly seen in NPDR include both dot-blot hemorrhages and retinal nerve fiber layer hemorrhages. Dot-blot hemorrhages are punctate, intraretinal hemorrhages that arise from the venous end of retinal capillaries and are located within the compact middle layers of the retina. These compact, vertically aligned, middle retinal layers confer upon the retinal hemorrhages their characteristic red, dot-blot appearance (**Figure 4**). Retinal nerve fiber layer hemorrhages arise from the more superficially located precapillary arterioles. The horizontal alignment of the retinal nerve fiber layer gives these hemorrhages their classic flame shape.

As discussed previously, another consequence of DR is excessive vascular permeability, which can result in retinal edema, usually in the macular region. Retinal edema is often accompanied by macular hard exudates, which are lipid deposits that accumulate in association with lipoprotein leakage from decompensated endothelial tight junctions. On clinical examination, hard exudates are yellowish intraretinal deposits often found at the border of edematous and nonedematous retinal tissue (**Figure 4**). Macular edema initially accumulates between the outer plexiform and inner nuclear layers. With chronic edema, the entire thickness of the retina becomes edematous and can assume a cystoid appearance. Retinal thickening secondary to macular edema is best detected by indirect slit-lamp biomicroscopy; in addition, optical coherence tomography (OCT), can be used to detect thickening and may be used to assess response to therapy (**Figure 5**).

CWSs, also known as soft exudates, may often be found in NPDR. CWSs are composed of accumulations of neuronal debris within the retinal nerve fiber layer. These result from disruption and stasis of axoplasmic flow. As CWSs heal, debris is removed from the nerve fiber layer by autolysis and phagocytosis. Clinically, CWSs are seen as yellowish, fluffy superficial lesions which obscure the underlying blood vessels. Interestingly, CWSs are only found in the postequatorial retina where the nerve fiber layer is of sufficient thickness to allow visualization of the CWSs.

As NPDR progresses, it can lead to the obliteration of retinal capillaries. These areas of capillary nonprofusion are seen on fluorescein angiography as patches of hypofluorescence. Adjacent to areas of nonperfusion, tortuous, hypercellular vessels often develop. It is difficult to determine whether these vessels are actually dilated preexisting capillaries or whether they represent new vessels forming within the retina. These vessels have been

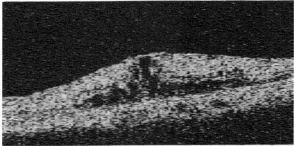
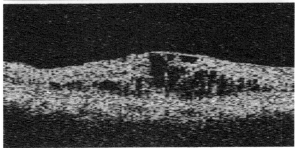

Figure 5 Macular edema secondary to diabetic retinopathy. OCT (two images) demonstrating retinal thickening and cystoid intraretinal spaces created by extensive capillary leakage and secondary macular edema in a patient with severe NPDR.

referred to as IRMAs, a term which encompasses both possibilities. The main distinguishing features of IRMAs are their intraretinal location, failure to cross major retinal blood vessels, and absence of leakage on flouroscein angiography. As areas of capillary nonperfusion become extensive, it is common to see an increase in intraretinal hemorrhages or dilated segments of retinal veins (referred to as venous beading). The degree of retinal capillary nonperfusion is directly associated with the severity of IRMAs, intraretinal hemorrhages, and venous beading.

NPDR is further categorized into four levels of severity: mild, moderate, severe, and very severe (**Table 1**). The clinical extent of microaneurysms, retinal hemorrhages, venous beading, and IRMA determine the level of severity of nonproliferative disease. Mild and moderate NPDR are characterized by relatively few microaneurysms and intraretinal hemorrhages and only minimal venous changes or IRMA. Severe NPDR is characterized by diffuse intraretinal hemorrhages, two quadrants of venous beading, or moderate IRMA in at least one quadrant. If any two of these features are present, the retinopathy is considered to be very severe NPDR. The Early Treatment Diabetic Retinopathy Study (ETDRS) found that severe NPDR had a 15% chance of progression to high-risk PDR within 1 year. Very severe NPDR had a 45% chance of progression to high-risk PDR within 1 year.

Macular Edema

Macular edema is the most common cause of visual impairment in patients with NPDR. Due to the

Table 1 Classification of diabetic retinopathy

Nonproliferative diabetic retinopathy (NPDR)

Mild NPDR:
At least one microaneurysm
Criteria not met for other levels of DR

Moderate NPDR:
Hemorrhage/microaneurysm ≥standard photograph #2A
or
Soft exudates (cotton-wool spots), venous beading, and
    intraretinal microvascular abnormalities definitely present
Criteria not met for severe NPDR, very severe NPDR, or PDR

Severe NPDR:
Hemorrhage/microaneurysm ≥standard photograph #2A in all
    four quadrants
or
Venous beading in at least two quadrants
or
Intraretinal microvascular abnormalities ≥standard photograph
    #8A in at least one quadrant

Very servere NPDR:
Any two or more of criteria for severe NPDR
Criteria not met for PDR

Proliferative diabetic retinopathy (PDR)

Early PDR:
New vessels
Criteria not met for high-risk PDR

High-risk PDR:
Neovascularization of the disk ≥1/4 to 1/3 disk area
or
Neovascularization of the disk and vitreous or preretinal
    hemorrhage
or
Neovascularization elsewhere ≥1/2 disk area and vitreous or
    preretinal hemorrhage

Advanced PDR:
Posterior fundus obscured by preretinal or vitreous hemorrhage
or
Center of macula detached

---

breakdown of the blood–retinal barrier, leakage of fluid and plasma constituents leads to retinal edema (**Figure 6**). If the retinal edema threatens the center of the fovea, there is a higher risk of visual loss. In the ETDRS, the 3-year risk of moderate visual loss was 32% (moderate visual loss was defined as a doubling of the initial visual angle or a decrease of three lines or more on a logarithmic visual acuity chart). The ETDRS investigators classified macular edema by its severity. More specifically, macular edema was defined as clinically significant macular edema (CSME) if any of the following features were present: (1) thickening of the retina at or within 500 μm of the center of the macula; (2) hard exudates at or within 500 μm of the center of the macula, if associated with thickening of the adjacent retina; or (3) a zone of thickening larger than one disk area if located within one disk diameter of the center of the macula (**Table 2**). Many of the current treatment paradigms for the management of diabetic

macular edema are derived from the ETDRS. The ETDRS demonstrated that eyes with CSME benefited from focal argon laser photocoagulation treatment when compared to untreated eyes in a control group. Furthermore, focal argon laser photocoagulation treatment for CSME reduced the risk of moderate visual loss, increased the chance of visual improvement, and was associated with only minor losses of visual field. Specifically, in patients with CSME involving the center of the macula, focal treatment reduced moderate visual loss by 60% after 3 years of follow-up. In patients with less than CSME, little difference was noted between the untreated and treated groups during the first 2 years of follow-up, after which there was a trend toward less frequent visual loss in the treated group.

Treatment patterns regarding CSME continue to evolve. In patients with refractory CSME, intravitreal administrations of corticosteroids have been shown to be beneficial. Intravitreal anti-VEGF agents have also been shown to improve CSME. Currently, several trials investigating corticosteroid use as well as anti-VEGF agents in the treatment of CSME are underway. Pars plana vitrectomy and detachment of the posterior hyaloid may also be useful for treating CSME. Surgical intervention may prove particularly beneficial when there is evidence of posterior hyaloidal traction and diffuse macular edema.

Proliferative Diabetic Retinopathy

As the degree of retinal ischemia increases, extensive hypofluorescent areas representing retinal nonperfusion are seen on fluorescein angiography. Eventually neovascularization may develop in an attempt to revascularize hypoxic retinal tissue. This neovascularization is the hallmark of PDR. It has been estimated that over one-quarter of the retina has to be nonperfused before PDR develops. PDR affects 5–10% of the diabetic population (**Figure 7**). Type I diabetics are at particular risk for PDR with an incidence of about 60% after 30 years. Although new vessels may arise from anywhere in the retina, neovascularization is particularly common on the optic disk itself (this location is termed new vessels on disk, or NVD). NVD is defined as neovascularization on or within one disk diameter of the optic nerve head. It can be differentiated from normal vessels by utilizing fluorescein angiography, which demonstrates profuse leakage in NVD but not in normal vasculature (**Figure 8**). Neovascularization of the retina found greater than one disk diameter from the optic nerve head is termed new vessels elsewhere, or NVE. NVE typically is found along the course of the major retinal vessels. The rate of growth of NVD or NVE is extremely variable. In some patients, neovascularization may show little change over many months, while in others definite changes in neovascularization may be seen in as little as 1–2 weeks. As neovascularization

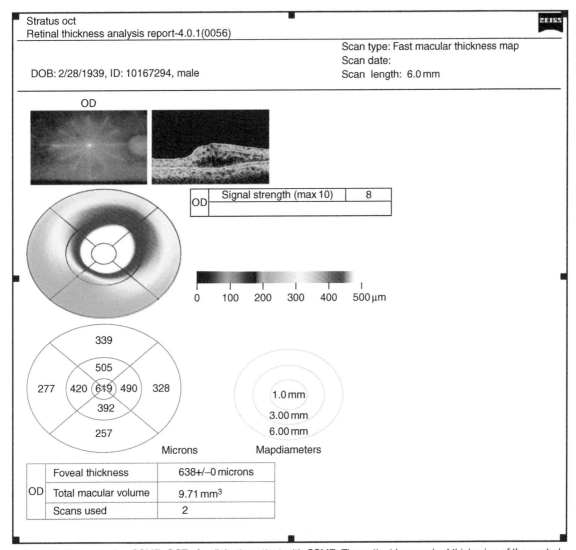

Figure 6 OCT demonstrating CSME. OCT of a diabetic patient with CSME. The patient has marked thickening of the central macula secondary to CSME. The OCT retinal cross section reveals cystic intraretinal and subretinal changes.

Table 2 Clinically significant macular edema as defined by ETDRS

Thickening of the retina ≤500μm from the center of the macula
or
Hard exudates and adjacent retinal thickening ≤500 μm from
   macular center
or
Zone of retinal thickening at least 1 disk area in size located
   <1 disk diameter from the center of the macula

progresses, a white fibrous membrane composed of fibrocytes and glial cells accompanies the growth of the new vessels. As mentioned previously, this membrane can become adherent to both the retina and the posterior hyaloid face. As the vitreous contracts, the fibrovascular membrane can cause tractional forces on the retina, leading to retina edema, vitreous hemorrhages, retinal heterotropia, retinal tears, and tractional retinal detachments.

The main therapy used to prevent the devastating complications of PDR is the application of thermal laser photocoagulation in a panretinal pattern in order to induce neovascular regression. Numerous panretinal photocoagulation (PRP) application protocols exist.

The classification of PDR is determined by the location and size of the neovascularization, along with the presence or absence of vitreous hemorrhage (**Table 1**). The Diabetic Retinopathy Study (DRS) defined high-risk PDR as any one of the following: (1) mild NVD with vitreous or preretinal hemorrhage, (2) NVD ≥ 1/4 to 1/3 disk area with or without vitreous hemorrhage, (3) NVE ≥ 1/2 disk area with vitreous or preretinal hemorrhage. The DRS was designed to evaluate the effectiveness of photocoagulation for treating diabetic retinopathy. As complications from PDR can result in severe vision loss (SVL), the primary outcome measured in the DRS was SVL, defined as visual acuity of less than 5/200. The DRS found that PRP

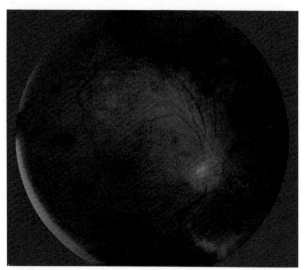

Figure 7 High-risk PDR. Color fundus photo of the right eye of a patient with high-risk PDR. Patient has significant NVD and NVE. This patient also has several areas of preretinal hemorrhages. CWSs can be seen inferior to the optic disk.

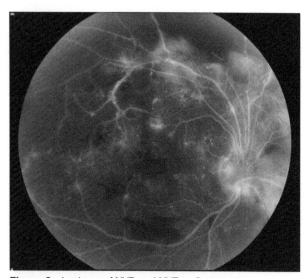

Figure 8 Leakage of NVD and NVE on fluorescein angiography. Fluorescein angiogram of the same patient seen in **Figure 7**. This patient has developed severe NVD and NVE causing significant hyperfluorescences secondary to profuse vascular leakage.

produced a 50% reduction in the rates of SVL in eyes treated with PRP compared to untreated control eyes during a follow-up of over 5 years. Treated eyes with high-risk PDR achieved the greatest benefit; thus, the DRS recommended prompt PRP treatment of eyes with high-risk PDR because this group had the highest risk of SVL.

Surgical management of PDR can be employed in patients with severe, persistent vitreous hemorrhages, progressive tractional retinal detachments, combined tractional and rhegmatogenous retinal detachments, and dense macular preretinal hemorrhages. Investigation into the use of anti-VEGF agents in the treatment of PDR is ongoing.

Screening for Diabetic Retinopathy

Proper screening for DR is critical as diabetic patients often do not experience visual disturbances until late in the disease course. Screening examinations allow for the prompt diagnosis and treatment of DR before the development of potentially blinding complications. In patients with type I diabetes, initiating screening examinations 3–5 years after diagnosis is recommended. In patients with type II diabetes, an initial examination upon diagnosis is recommended. Pregnant women with preexisting diabetes should undergo screening early in the first trimester. More frequent retinal evaluations are subsequently required during pregnancy and in the early postpartum period.

See also: Pathological Retinal Angiogenesis.

Further Reading

Aiello, L. M. (2003). Perspectives on diabetic retinopathy. *American Journal of Ophthalmology* 136: 122–135.

Centers for Disease Control and Prevention (2007). *National Diabetes Fact Sheet.* http://www.cdc.gov/diabetes/pubs/pdf/ndfs_2007.pdf (accessed June 2009).

Diabetes Control and Complications Trial Research Group (1993). The effect of intensive treatment of diabetes on the development and progression of long-term complications in insulin-dependent diabetes mellitus. *The New England Journal of Medicine* 329: 977–986.

Diabetes Control and Complications Trial Research Group (1995). Progression of retinopathy with intensive versus conventional treatment in the Diabetes Control and Complications Trial. *Ophthalmology* 102: 647–661.

Diabetic Retinopathy Study Research Group (1981). Photocoagulation treatment of proliferative diabetic retinopathy: clinical application of Diabetic Retinopathy Study (DRS) findings. DRS report 8. *Ophthalmology* 88: 583–600.

Early Treatment Diabetic Retinopathy Study Research Group (1987). Treatment techniques and clinical guidelines for photocoagulation of diabetic macular edema. ETDRS report 2. *Ophthalmology* 94: 761–774.

Early Treatment Diabetic Retinopathy Study Research Group (1991). Early photocoagulation for diabetic retinopathy. ETDRS report 9. *Ophthalmology* 98: 766–785.

Early Treatment Diabetic Retinopathy Study Research Group (1995). Focal photocoagulation treatment of diabetic macular edema. Relationship of treatment effect to fluorescein angiographic and other retinal characteristics at baseline. ETDRS Report 19. *Archives of Ophthalmology* 113: 1144–1155.

Kanski, J. J. (2007). *Clinical Ophthalmology*, 6th edn. Philadelphia, PA: Butterworth Heinemann-Elsevier.

Klein, R., Klein, B. E., Moss, S. E., Davis, M. D., and DeMets, D. L. (1984a). The Wisconsin Epidemiologic Study of Diabetic Retinopathy: II. Prevalence and risk of diabetic retinopathy when age at diagnosis is less than 30 years. *Archives of Ophthalmology* 102: 520–526.

Klein, R., Klein, B. E., Moss, S. E., Davis, M. D., and DeMets, D. L. (1984b). The Wisconsin Epidemiologic Study of Diabetic Retinopathy: III. Prevalence and risk of diabetic retinopathy when age at diagnosis is 30 or more years. *Archives of Ophthalmology* 102: 527–532.

Preferred Practice Patters Committee (2003). Retina Panel. *Diabetic Retinopathy*. San Francisco: American Academy of Ophthalmology; November 2003.

Ryan, S. J., Hinton, D. R., Schachat, A. P., and Wilkinson, C. P. (2006). Retina. 4th edn. Philadelphia, PA: Mosby-Elsevier.

United Kingdom Prospective Diabetes Study Group (1998a). Intensive blood-glucose control with sulphonylureas or insulin compared with conventional treatment and risk of complications in patients with type 2 diabetes. UKPDS 33. *Lancet* 352: 837–853.

United Kingdom Prospective Diabetes Study Group (1998b). Tight blood pressure control and risk of macrovascular and microvascular complications in type 2 diabetes. UKPDS 38. *British Medical Journal* 317: 703–713.

UpToDate (2008). *Prevention and treatment of diabetic retinopathy*. http://www.uptodate.com/online/content/topic.do?topicKey=diabetes/12336&selectedTitle=1~90&source=search_result (accessed June 2009).

Retinopathy of Prematurity

M E Hartnett, Moran Eye Center, University of Utah, Salt Lake City, UT, USA

Glossary

Aggressive posterior retinopathy of prematurity (APROP) – Severe ROP manifesting at young ages in zone 1 with flat neovascularization. Outcomes may be poor with conventional management.

Avascular retina – Retina that lacks blood vessel growth in the inner capillary plexus.

Intravitreous neovascularization – Endothelial budding or blood vessels that grow above the inner limiting membrane of the neurosensory retina into the vitreous.

Oxygen-induced retinopathy – Includes a number of models of recently born animals of species that complete retinal vascular development after birth. Exposure to oxygen stresses varies depending on the model but shares the features of first avascular retina followed by intravitreous neovascularization.

Peripheral severe ROP (PSROP) – It refers to zone II, stage 2–3 ROP, plus disease, in this article.

Plus disease – A feature of severe ROP, it refers to the dilation and tortuosity of retinal arterioles and veins.

Postgestational age – The age measured in weeks that is the sum of the gestational age (age of preterm infant from conception or last menstrual period) and the chronologic age (age since the birth of the infant). For example, a 24-week gestational age infant who was born 16 weeks ago would have a postgestational age of 40 weeks. Similar terms used include postmenstrual age, postconceptual age, or corrected age.

Retinal detachment – A fluid develops between the photoreceptor outer segments and the retinal pigment epithelium. A traction retinal detachment occurs by vitreous tractional forces, a rhegmatogenous retinal detachment because of a break in the retina, and a serous retinal detachment from exudation often as a result of inflammation or leaky vessels, and is sometimes seen after treatment for severe ROP.

Clinical Background

Epidemiology

The Institute of Medicine reported that preterm births were up 30% from 1981 and now account for 12.5% of all births in the US. Furthermore, developing countries are witnessing an increase in retinopathy of prematurity (ROP); therefore, ROP has now become a leading cause of childhood blindness worldwide. A report from a national registry of children in the US (Babies Count) found that ROP was the earliest cause of visual impairment and one of the three most prevalent conditions to cause visual impairment along with cortical visual impairment and optic nerve hypoplasia. ROP of any stage affects approximately 16 000 infants yearly in the US. Most early stages of ROP resolve, but about 1100 infants require treatment. Even with treatment, blindness occurs in 550 infants per year. In the US, ROP is seen more commonly in Caucasians than African-Americans, but once severe ROP occurs, the outcomes appear similar. Asians also have an increased risk of ROP. ROP has been reported in preterm infants of larger birth weight and older gestational ages in developing nations compared to those in the US, possibly because of variations in ethnic groups, regulation and monitoring of oxygen delivery, and the availability of prenatal care.

Clinical Classification of ROP

Based on the International Classification of ROP (ICROP), it is characterized by several parameters: zone, stage, extent of stage, and the presence of plus disease.

The zone of ROP is the retinal area supplied by the retinal vasculature and is an indicator of the extent of retinal vascular development (**Figure 1(a)**). Zone I is the smallest, having the largest area of avascular retina. Zone II has a greater area of retinal vascularization than zone I, and less than zone III. When vascularization completely extends to the ora serrata, it is termed complete vascularization. The risk of a poor visual outcome is greatest when severe ROP occurs in zone I (**Figure 1(b)**) and is still substantial in zone II. The risks of severe ROP and poor vision are rare when retinal vascularization extends into zone III. There are five stages of ROP. Stages 1 through 3 are the acute forms of ROP and are named for the appearance of the retina at the junction of vascular and avascular retina. A line (stage 1) or ridge (stage 2, **Figure 2(a)**) can often regress and vascularization of the previously avascular retina can occur. Stage 3 ROP has intravitreous neovascularization, a feature of severe ROP (**Figure 2(b)**). Stages 4 (**Figure 2(c)**; also see **Figures 6(a)**) and 5 (**Figures 2(d)** and **2(e)**) define partial or complete retinal detachment, respectively, and are associated with vitreous and fibrovascular changes. The extent of ROP indicates the number of clock hours of a stage. Plus disease is the presence of dilated and tortuous retinal vessels in two or more quadrants around the optic nerve and is a feature of severe ROP (**Figure 3**).

There is a useful distinction to note between retinal drawings and images of dissected retinal flat mounts. The retina covers the inner sphere of the eyeball and, when dissected, must be cut with relaxing incisions in order to flatten it onto a microscope slide. The result is a clover-leaf appearance (**Figure 4(a)**). However, clinicians and surgeons represent the retina as a round clock face and use clock hours to describe the location of pathologic features on the retina determined in clinical examinations (**Figure 4(b)**).

Management of ROP

Based on the American Academy of Pediatrics and American Academy of Ophthalmology, infants born at or younger than 30 weeks gestational age or less than 1500 g birth weight are screened for retinal vascular development

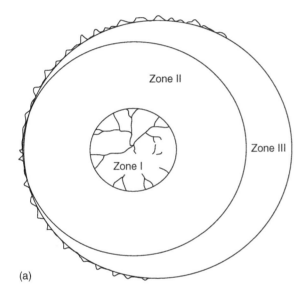

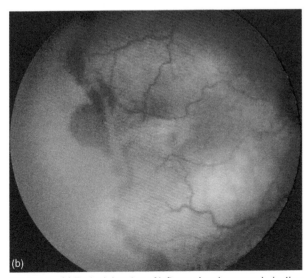

Figure 1 (a) Retinal drawing of left eye showing vascularization into zone I and the areas of the retina that would encompass zone II or zone III. (b) Image taken with wide-angle viewing system (Retcam, Clarity) of the right eye of an infant with zone I ROP (optic nerve barely visible at right of image); Courtesy Sarah Moyer, CRA, OCT-C.

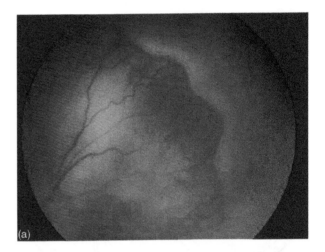

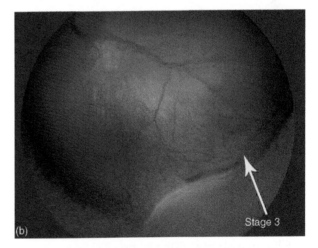

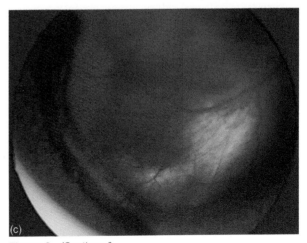

Figure 2 (Continued)

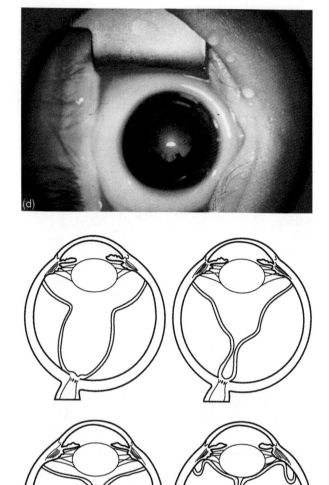

Figure 2 Images taken with wide-angle viewing system (Retcam, Clarity) of infant left eyes with (a) stage 2 ROP with early ridge and (b) stage 3 ROP with areas of intravitreous neovascularization (arrow) and hemorrhage adjacent to avascular retina (3–5 o'clock in image). Images of infant right eye with (c) early stage 4A ROP and (d) stage 5 ROP showing total retinal detachment with white pupil. (e) Diagram showing cross section of possible retinal appearances in stage 5 ROP; Figures 2(a), 2(b), and 2(c) courtesy Sarah Moyer, CRA, OCT-C. **Figure 2(e)** from Schepens' Retinal Detachment and Allied Diseases, 2nd edn (eds. Schepens, Hartnett, and Hirose) copyright 2000. Butterworth-Heinemann, Boston, MA: Figure 26-19, p. 534.

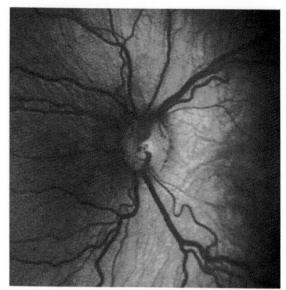

Figure 3 Plus disease (dilated and tortuous vessels in all four quadrants around optic nerve) of right eye in infant with severe ROP; Courtesy Sarah Moyer, CRA, OCT-C.

and the presence of ROP. The screening examination is performed 4–6 weeks after birth (chronologic age) or at 31 weeks postgestational age, whichever is older. (The postgestational age is the gestational age + chronologic age from birth in weeks and is similar in meaning to postmenstrual, postconceptual, or corrected ages). At any examination, the risk of a bad outcome depends on the presence of plus disease or stage 3 ROP (both features of severe ROP). Severe ROP develops at about 35–37 weeks postgestational age, regardless of the gestational age or birth weight of the infant.

At the screening examination, the extent of retinal vascular development is determined (i.e., the zone of ROP), along with the presence and stage of ROP and presence of pre-plus or plus disease. (Pre-plus is less severe than plus disease and alerts the clinician to follow the infant closely.) If ROP is not severe, follow-up examinations are performed until full vascularization of the retina occurs or until severe ROP develops, at which time the treatment for acute neovascular ROP is performed.

Treatment of ROP

Acute neovascular stages

Based on the Cryotherapy for Retinopathy of Prematurity (CRYO-ROP) and Early Treatment for Retinopathy of Prematurity (ETROP) studies, the treatment for ROP is strongly considered for infants with type 1 prethreshold ROP and almost always performed for infants with threshold ROP (**Table 1**). Laser is preferred to cryotherapy because it causes less inflammation, is less destructive, and is less often associated with myopia. The laser is applied to the peripheral avascular retina (**Figure 5**). Infants are then followed up weekly for regression of severe ROP or for the development of progressive stage 4 ROP.

Fibrovascular stages/retinal detachment

Stage 4 ROP refers to partial retinal detachment and develops as a result of fibrovascular changes and

Table 1 Definitions of threshold and type 1 prethreshold retinopathy of prematurity

Threshold ROP (CRYO-ROP) (risk of unfavorable outcome approaches 50%)	Type 1 prethreshold (risk of unfavorable outcome is ≥15% – ETROP)
Zone I or II, stage 3 (5 contiguous or 8 total clock hours with plus disease*)	Zone I, any stage with plus disease*
	Zone I, stage 3 without plus disease*
	Zone II, stage 2 or 3 with plus disease*

CRYO-ROP – Cryotherapy for Retinopathy of Prematurity Study.
ETROP – Early Treatment for Retinopathy of Prematurity.
*The ETROP recognized plus disease as two quadrants of dilated and tortuous vessels whereas CRYO-ROP defined it as four quadrants.

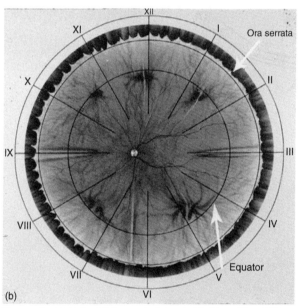

Figure 4 (a) Retinal flat mount stained with Alexa Fluor 568 conjugated isolectin B4 lectin (lectin-B4) to demonstrate vasculature from postnatal day 14 rat pup from the rat 50/10 oxygen-induced retinopathy model. (b) Artist's drawing of retina with clock hour delineations, often used clinically to localize regions of pathology on the retina. Innermost circle at vortex vein ampullae indicates equator of the globe; middle circle indicates ora serrata; and outermost circle indicates pars plana region in adult.

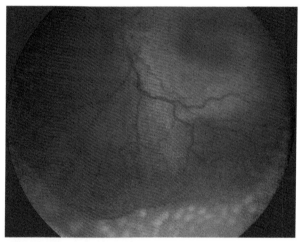

Figure 5 Type 1 prethreshold ROP (zone II, stage 2 severe ROP) after laser; it shows white spots from recently delivered laser to avascular retina and skipped areas between laser spots that will need to be filled in with laser in the left eye.

vitreoretinal traction that particularly occur at the junction of vascular and avascular retina and optic nerve to lead to retinal detachment (**Figure 6(a)**). The timing of the development of stage 4 ROP is often between 37 and 44 weeks postgestational age and represents a change

from neovascular to fibrovascular changes. Once progressive stage 4 ROP is diagnosed, surgery, preferably with a lens-sparing vitrectomy (**Figure 6(b)**), is performed to release vitreous tractional forces that detach the retina. The main forces addressed are those around the optic nerve, between the ridge and anterior aspect of the eye and lens, between the ridge and optic nerve, between the ridge and ora serrata, and from ridge to ridge (**Figure 6(b)**). The retina then reattaches in the postoperative period (**Figure 6(c)**), although the ridge and pulled-up retina often persist. Occasionally, a scleral buckle is performed, often when a break, that is, rhegmatogenous component, is present.

The decision to operate must be carefully considered in the preterm infant eye because there are surgical difficulties that make operating on an infant eye different from operating on an adult one. The infant eye is about two-thirds the diameter of the adult eye and the region of safe

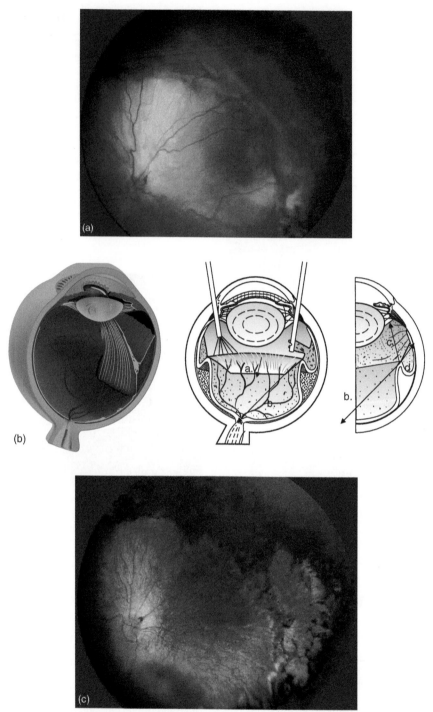

Figure 6   Image taken with wide-angle viewing system (Retcam, Clarity) of infant left eyes with (A) stage 4B ROP showing superior and temporal region of ridge with incorporated retinal detachment and retinal detachment extending posteriorly toward the optic nerve. Pigmented laser spots are in the peripheral region of avascular retina. Focus is on the retinal detachment posterior to the ridge; therefore ridge, which is more anterior in the eye, and optic nerve, which is posterior in the eye, are out of focus. Detachment extends from about 11 o'clock until 4 o'clock and involves the macular region; Courtesy of Sarah Moyer, CRA, OCT-C. (B) An artist's representation of traction retinal detachment and of lens-sparing vitrectomy on left. On the right, the vitrectomy is done to address the vitreous between the (a) ridge and the optic nerve, (b) around the optic nerve, (c) ridge to the anterior portion of the eye, and (d) ridge to the ora serrata. Care is taken not to cut the elevated ridge, which can have retina drawn into it, and risk causing a retinal break. At the end of the case, an air bubble is placed into the vitreous cavity to maintain the globe form while the sclerotomies are sewn closed. (C) 2-week postoperative image taken with wide-angle camera (Retcam, Clarity), showing the resolution of retinal detachment posterior to the ridge and elevated ridge/retinal detachment in the region of the pigmented avascular retina. In addition, note the reduction in plus disease. Residual vitreous hemorrhage is present along the inferior retinal arcade. Lens remains clear; Courtesy Sarah Moyer, CRA, OCT-C.

entry without damage to the retina is less than 1 mm in width compared to about 6 mm in the adult eye. Unlike in the adult, a retinal break can lead to an inoperable retinal detachment and blindness in an infant. The ridge/junction region often has retinal detachment incorporated into it. Therefore, in order to reduce the risk of causing an iatrogenic retinal break, the surgical strategy is to release vitreous traction rather than to remove the ridge or dissect preretinal tissue from it. In addition, injury to the lens can lead to cataract, amblyopia, and poor vision. Therefore, the timing of surgery is prior to that when the retina and ridge are pulled anteriorly to contact the posterior lens capsule.

Visual rehabilitation in preterm infants

Infants with ROP are more likely to be myopic and develop strabismus in childhood than full-term infants. The more severe the ROP, the greater the risk of developing high myopia. Infants who develop stage 5 ROP or total retinal detachment have very poor vision, and are usually legally blind (bilateral visual acuity <20/200) even after successful surgery to reattach the retina. Therefore, the goal in managing ROP is to prevent stage 5 ROP. Surgery at early stage 4 ROP often permit the retention of the lens which is important in visual development. Infants who have their lenses removed have compromised visual development and reduced visual acuity from aphakic amblyopia, even with optical means to correct aphakia.

Genetics Related to ROP

Based on retrospective analysis of monozygotic and dizygotic twins, a 70% variance in the susceptibility of ROP was found to be from genetic factors, suggesting that genetics plays a strong role in ROP. However, studies of candidate genes are not in agreement and suggest that a more complex situation exists, involving genetics and environmental factors such as nutrition, oxygen, and the health of the infant.

The Norrie disease gene produces the gene product, norrin, which is also a downstream ligand for receptors in the Wnt signaling pathway. Norrie disease is usually x-linked and causes visual and hearing loss. The Wnt pathway and norrin are important in retinal and vascular development. Genetic mutations in the Norrie disease gene [Xp11.2-11.3] were reported to account for 3% of the cases of advanced ROP, but were not found in a study in infants with severe ROP compared to control preterm infants with no or minimal ROP within a racially diverse population. Another study reported that mutations within the cysteine knot configuration of the Norrie disease gene

were associated with severe retinal dysplasia, whereas other polymorphisms within the gene had less severe vitreoretinopathies.

Severe ROP has been reported in infants with certain polymorphisms in the gene of vascular endothelial growth factor (VEGF), but the same polymorphisms have not been confirmed in other studies. Despite the finding that low serum insulin-like growth factor-1 (IGF-1) was associated with more severe ROP, one study failed to show a relationship between a prevalent polymorphism in the IGF-1 receptor and the presence of ROP. The role of genetics requires greater study and will continue to be elucidated along with the effect of environmental factors on gene function.

Pathophysiology of ROP

To understand ROP, it is helpful to understand the known processes in human retinal vascular development. It is believed that vasculogenesis or *de novo* development of the central vasculature around the optic nerve occurs from angioblasts, or endothelial cell (EC) precursors. Angioblasts lack markers commonly thought to be present on ECs such as CD31, CD34, and von Willebrand's factor, but do express CD39 and CXCR4. Retinal vascular development is believed to be completed mainly through a process of angiogenesis, but the role of circulating endothelial precursors is being appreciated more. During angiogenesis, a front of migrating cells, astrocytes in cat or angioblasts in dog, sense physiologic hypoxia and express VEGF. The ensuing ECs are attracted to VEGF and migrate to create blood vessels. The VEGF signaling pathway has been found to regulate and integrate several cell processes important during sprouting angiogenesis. Whereas VEGF concentration is thought to regulate EC division rate, the presentation of VEGF, as in a gradient, may regulate filopodia formation of endothelial tip cells at the migrating front and direct the growth of ECs. The delta-like ligand 4/Notch1 (Dll4/Notch1) signaling pathway regulates VEGF-induced endothelial tip/stalk cells at the junction of vascular and avascular retina and permits ordered angiogenesis.

Role of Oxygen in Retinal Development

The oxygen level has been long recognized as important in the development of ROP. High unregulated oxygen at birth likely accounted for many of the early cases of ROP first described in the 1940s and 1950s. With improved oxygen monitoring and avoidance of hyperoxia, ROP virtually disappeared. However, as infants of younger gestational ages and lower birth weights survived, ROP re-emerged. Currently, it is recognized that fluctuations in oxygen, as well as high oxygen at birth, are important risks for severe ROP. Although a multicenter clinical study (Supplemental Therapeutic Oxygen for Prethreshold Retinopathy of Prematurity) to test the effects of supplemental oxygen to prevent severe ROP found no adverse effect from supplemental oxygen and, perhaps, a benefit in a subgroup, some reports indicate that severe ROP is associated more commonly with infants who have had high oxygen saturations during their courses.

High oxygen early in development causes loss of perfused central capillaries and is the basis for the development of several animal models of oxygen-induced retinopathy (OIR) in cats, mice, beagles, and rats. One mechanism proposed is that hyperoxia inhibits basic fibroblast growth factor (bFGF)-induced angioblast differentiation into ECs *in vitro*. Although most models of OIR recapitulate this high constant hyperoxia, it is not relevant to what preterm infants experience in neonatal intensive care units (NICUs) in which oxygen is well regulated, in whom minute to minute fluctuations in oxygen have been measured rather than constant oxygen. Furthermore, almost all studies of the mechanisms of oxygen stress on retinal vessels have been performed using models that subject animals to high constant oxygen.

Animal Models of Severe ROP

Models of severe ROP take advantage of the fact that several species undergo retinal vascular development after birth. Furthermore, the newly developed capillaries are susceptible to oxygen stresses such that high oxygen will cause loss of capillaries. No model uses premature animals. In addition, no model develops stage 4 or 5 ROP (retinal detachment). The beagle OIR model comes closest to stage 4 ROP with the development of tractional retinal folds, but does not develop stage 5 ROP.

Mouse OIR (model of aggressive posterior ROP)

The mouse OIR model also uses high constant oxygen and is probably the most useful to study mechanisms of extreme oxygen insult by using genetically modified animals (**Table 2** and **Figure 7**). This model may mimic aggressive posterior ROP (APROP, **Figure 8**), a less-common form of severe ROP, in which there are broad areas of central avascular retina. APROP may share the retinal oxygenation pattern as does the avascular hypoxic retina during relative hypoxia in the mouse OIR model. Fluorescein angiograms of infants with APROP demonstrated extensive capillary loss centrally that appeared similar to that seen in perfused retinas from the mouse OIR, and several studies have shown avascular retina in OIR models to be hypoxic.

Rat 50/10 OIR (model of peripheral severe ROP)

The rat 50/10 OIR model mimics the more common form of severe ROP, peripheral severe ROP (PSROP). Here,

Table 2   Mouse and 50/10 rat models of OIR

Constant high hyperoxia: Mouse OIR model of human APROP

p12 Low retinal VEGF
    Central avascular retina
p17 Intravitreous neovascular budding

Fluctuations in oxygen: Rat 50/10 OIR model of more common PSROP

p14 Rat 50/10 OIR; ~32 weeks premature infant
    High retinal VEGF

p18 Rat 50/10 OIR; ~35-37weeks premature infant
    High VEGF164 and VEGFR2 signaling
    Intravitreous neovascularization

~p30 Rat 50/10 OIR; 40 weeks preterm infant
    Intraretinal vascularization and regression of IVNV

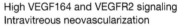

newborn rat pups are placed into an oxygen environment that cycles concentrations between 50% and 10% oxygen every 24 h for 14 days. The oxygen extremes in the rat 50/10 OIR model cause rat arterial oxygen levels to be similar to the transcutaneous oxygen levels measured in preterm infants who developed severe ROP. In addition, rather than constant oxygen used in other models, the 50/10 OIR model exposes pups to repeated fluctuations in oxygen, a risk factor for severe ROP. As in PSROP, in which there is reduced peripheral retinal vascularization and perfusion with minimal central nonperfusion, the rat 50/10 OIR model has mainly peripheral avascular retina and minimal central capillary loss (**Figure 4(a)**). The 50/10 OIR model reproducibly and consistently first develops avascular retina (analogous to human zone II ROP) and, subsequently, vessel tortuosity (analogous to human plus disease) and intravitreous neovascularization (analogous to human stage 3 ROP). All OIR models have regression of tortuosity and intravitreous neovascularization with later intraretinal vascularization; however, the beagle model takes the longest to regress.

Role of Avascular Retina

Although the size of the avascular retina does not always correlate with the presence of intravitreous neovascularization in animal models, multicenter clinical trials have shown that eyes with the largest peripheral retinal avascular zones have the worst outcomes from ROP. Furthermore, the avascular, hypoxic retina is believed to be a source of angiogenic

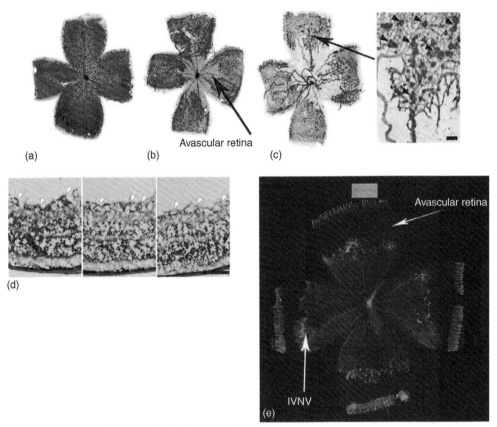

Figure 7 Retinal flat mounts from (a) room air raised mouse at postnatal day (p)12; (b) mouse raised in oxygen-induced retinopathy (75% constant oxygen for 5 days) with central capillary loss (arrow) at p12. ((c), left) After 5 days in room air (relative hypoxia) with capillary budding into the vitreous at p17, possibly mimicking aggressive posterior ROP (see **Table 2**). ((c), right) Enlargement of area centered at arrow; arrowheads indicate capillary budding; and (d) cross section of (c) showing endothelial budding into the vitreous (white arrowheads). (e) Lectin B4-stained retinal flat mount from rat 50/10 OIR model following oxygen fluctuations between 50% and 10% oxygen until p14, followed by 4 days of room air. At p18, intravitreous neovascularization (IVNV) appears at the junction of vascularized and peripheral avascular retina similar in appearance to peripheral severe ROP (see also **Table 2**).

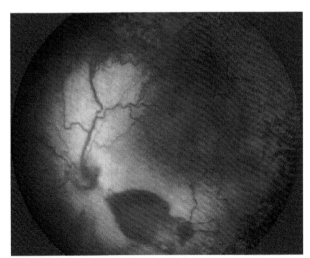

Figure 8 Preterm infant left eye with APROP after laser treatment showing pigmentation of laser to the right of image, plus disease, and vitreous hemorrhage inferior to optic nerve; Courtesy Sarah Moyer, CRA, OCT-C.

factors that cause pathologic intravitreous neovascularization in some models of ROP. What causes avascular retina remains largely unknown. Delayed retinal vascular development in B-cell lymphoma protein 2 ($bcl2^{-/-}$)-deficient mice that have a defect in protection against apoptosis supports the thinking that increased apoptosis of ECs or their precursors may contribute to avascular retina. Protection of newly formed capillaries from hyperoxia-induced endothelial death occurs by giving growth factors or nutritional supplements prior to the hyperoxic insult, and these protective agents largely prevent intravitreous neovascularization that would occur in the hypoxic phase of the mouse OIR model. The activation of nicotinamide adenine dinucleotide phosphate [NADPH] oxidase from repeated oxygen fluctuations in the rat 50/10 OIR model contributed to the avascular retina through apoptosis. Thus, the area of avascular retina appears to be one factor involved in the severity of human ROP, and apoptosis of ECs or their precursors may contribute to its size.

Role of Growth Factors

Molecular mechanisms of OIR have been identified mainly from the mouse OIR model, which mimics APROP, but does not mimic most cases of severe ROP.

Vascular Endothelial Growth Factor

VEGF is an important factor in retinal vascular development and is also neuroprotective. However, it is also one of the most important angiogenic factors involved in pathologic retinal and choroidal vascular diseases. Mice in OIR had reduced VEGF expression in association with capillary loss centrally when exposed to high constant hyperoxia. Subsequently, when placed into relative hypoxia that occurred in room air, VEGF messenger ribonucleic acid (mRNA) was overexpressed in association with the development of endothelial budding above the internal limiting membrane. If VEGF was given during hyperoxia, capillary loss could be reduced and, if agents to inhibit VEGF were given during relative hypoxia, endothelial budding into the vitreous was also reduced.

In the rat 50/10 OIR model, a relevant model of most cases of severe ROP in the US currently, neutralizing VEGF with an antibody made against $VEGF_{164}$ reduced tortuosity (analogous to plus disease) and intravitreous neovascularization (analogous to stage 3 ROP). Too low a dose appeared to lead to a rebound in intravitreous neovascularization and persistence of the avascular retina.

Clinical trials are underway, testing intravitreous injections of antibodies to VEGF in severe ROP. There are concerns regarding the effect of dose based on animal studies described above, and the possible adverse effect of inhibiting VEGF in the developing preterm infant. VEGF is neuroprotective and besides the possible local effect on the retina, there is the potential adverse effect systemically from the absorption of the antibody into the bloodstream. Compared to the adult, an intravitreous drug in the newborn can achieve a higher concentration in the bloodstream and affect measurable outcomes, such as body weight gain in animal models. However, in some forms of severe ROP, such as APROP, there are few other options to prevent retinal detachment and permit the development of vision in these infants. Therefore, clinical trials are necessary and the data obtained will be important in developing improved treatments for severe ROP.

IGF-1 – IGF 1BP3

IGF-1 is important in the physical growth of the infant. However, IGF-1 levels that occur *in utero* are not maintained upon birth in preterm infants. Low serum IGF-1 was found to correlate with greater avascular retinal area in human preterm infants. Furthermore, transgenic mice expressing a growth hormone antagonist gene, or wild-type mice treated with an inhibitor to growth hormone, had reduced intravitreous neovascularization in the mouse OIR model. IGF-1 was also found to be important for signaling through the mitogen-activated protein (MAP) kinase pathway, which is important in cell proliferation. In addition, VEGF and IGF-1 synergistically triggered the serine-threonine kinase, Akt, which is important in cell survival. Based on these findings, it is theorized that IGF-1, which is low in the preterm infant, is necessary for early retinal vascular survival and growth, but can result in later intravitreous neovascularization in ROP. However, the timing and dose of IGF-1 appear to be critical.

A hypoxia-regulated binding protein of IGF-1, IGF-1BP3, was shown to be important in reducing hyperoxia-induced capillary loss and in promoting vascular regrowth into the retina in the mouse OIR model. IGF-BP3 was shown to promote differentiation of endothelial precursor cells into ECs and in promoting angiogenic processes, such as cell migration and tube formation.

Erythropoietin

Erythropoietin is angiogenic, erythropoietic, and neuroprotective. It is upregulated after the stabilization of hypoxia-inducible factor (HIF)-1α in response to hypoxia. In the mouse OIR model, hyperoxia reduces the expression of erythropoietin. The administration of exogenous erythropoietin prior to hyperoxia reduced capillary loss, whereas giving erythropoietin during relative hypoxia in room air enhanced pathologic neovascularization. Clinical studies have reported an association between the number of administrations of erythropoietin for anemia of prematurity and the prevalence of severe ROP and have found that recombinant erythropoietin is an independent risk factor for severe ROP.

HIF 1α

The stabilization of HIF1α occurs under hypoxic conditions or secondary to reactive oxygen species (ROS) generated from NADPH oxidase, nitric oxide, mitochondria, and other enzymes. HIF1α binds to the hypoxia response elements to cause transcription of several genes including angiogenic factors, VEGF, and erythropoietin. A knock-out to HIF 1α is lethal, but a knock-out to the HIF-1α-like factor (HLF)/HIF-2α provided evidence that erythropoietin was a major gene involved in intravitreous neovascularization after relative hypoxia from hyperoxia-induced capillary loss in the mouse OIR model.

Role of Oxidative Stress

Oxidative stress has been proposed to be important in the development of ROP because the retina is susceptible to oxidative damage given its high metabolic rate and rapid rate of oxygen consumption. In addition, the premature infant has a reduced ability to scavenge ROS, increasing its vulnerability to oxidative stress. End products of ROS, lipid hydroperoxides, were increased in the 50/10 OIR model at time points corresponding to intravitreous neovascularization. When injected into the vitreous, these compounds caused intravitreous neovascularization in the rabbit. In addition, ROS can trigger signaling pathways relevant to apoptosis or angiogenesis, both important in the pathogenesis of ROP.

The treatment of pups in the 50/10 OIR model or humans with ROP using a broad antioxidant, N-acetylcysteine, failed to show a reduction in clock hours of intravitreous neovascularization, or the avascular retinal area. In a clinical trial in preterm infants, there was no difference in the incidence in ROP between those receiving N-acetylcysteine or control. However, reduction in ROS with preparations of vitamin E or liposomes containing the antioxidant enzyme, manganese superoxide dismutase, reduced OIR severity. Also, a meta-analysis of human preterm infants treated with vitamin E showed a significant reduction in severity of ROP. Reducing the activation of NADPH oxidase, an enzyme that produces ROS, can also reduce the size of the avascular areas and subsequent intravitreous neovascularization in certain OIR models.

Light was proposed to be important in ROP development through photooxidation of polyunsaturated fatty acids within photoreceptor outer segments. On the other hand, during the dark, photoreceptors are more metabolically active. A clinical trial testing the effect of light or shade on the development of ROP showed no significant difference.

Future Treatment Considerations

There are difficulties in studying treatments for ROP in developing infants. Preterm infants with severe ROP often have other developmental and health problems, making it difficult to assess long-term complications of a drug in clinical trials. Inhibiting the bioactivity of VEGF has been reported to be beneficial in adult diseases, such as proliferative diabetic retinopathy, but this strategy has been reported to interfere with neuronal and endothelial survival in some animal models, and these are issues of concern in the developing preterm infant. Inhibition of ROS may be detrimental to the preterm infant, whose abilities to combat infection are limited. A reduction in inspired oxygen concentration may also be detrimental to the developing preterm infant brain and long-term effects of such reductions are unknown. The timing and dose of neuroprotective agents such as erythropoietin and growth factors are critical but vary among individual infants, so that an agent may worsen rather than lessen ROP severity. Currently, the optimal time or dose of an agent cannot be

safely determined for an individual infant. However, the current standard of care laser treatment for severe ROP does not address vascular development that is ongoing in the developing human preterm infant. Better treatments for severe ROP are needed.

See also: Anatomy and Regulation of the Optic Nerve Blood Flow; Choroidal Neovascularization; Evolution of Opsins; Genetic Dissection of Invertebrate Phototransduction; Microvillar and Ciliary Photoreceptors in Molluskan Eyes.

Further Reading

Capone, A., Jr., Hartnett, M. E., and Trese, M. T. (2005). Treatment of retinopathy of prematurity: Peripheral retinal ablation and vitreoretinal surgery. In: Hartnett, M. E., Trese, M. T., Capone, A., Keats, B., and Steidl, S. (eds.) *Pediatric Retina*, pp. 417–424. Philadelphia, PA: Lippincott Williams and Wilkins.

Chen, J., Connor, K. M., Aderman, C. M., and Smith, L. E. (2008). Erythropoietin deficiency decreases vascular stability in mice. *Journal of Clinical Investigation* 118: 526–533.

Coats, D. K. (2005). Retinopathy of prematurity: Involution, factors predisposing to retinal detachment, and expected utility of preemptive surgical reintervention. *Transactions of the American Ophthalmological Society* 103: 281–312.

Geisen, P., Peterson, L. J., Martiniuk, D., et al. (2008). Neutralizing antibody to VEGF reduces intravitreous neovascularization and does not interfere with vascularization of avascular retina in an ROP model. *Molecular Vision* 14: 345–357.

Hartnett, M. E. (2003). Examination and diagnosis in the pediatric patient. In: Steidl, S. M. and Hartnett, M. E. (eds.) *Clinical Pathways in Vitreoretinal Disease*, pp. 341–373. New York: Thieme Medical Publishers.

Hartnett, M. E. and Toth C. A. (2010). Retinopathy of prematurity. In: Levin L. and Albert, D. (eds.) *Ocular Disease: Mechanisms and Management*. London: Elsevier.

Hartnett, M. E., Trese, M. T., Capone, A., Keats, B., and Steidl, S. (eds.) (2005) *Pediatric Retina*. Philadelphia, PA: Lippincott Williams and Wilkins.

Hartnett, M. E., Martiniuk, D., Byfield, G., Zeng, G., and Bautch, V. (2008). Neutralizing VEGF decreases tortuosity and alters endothelial cell division orientation in arterioles and veins in rat model of ROP: Relevance to plus disease. *Investigative Ophthalmology and Visual Science* 49: 3107–3117.

Hasegawa, T., McLeod, D. S., Prow, T., et al. (2008). Vascular precursors in developing human retina. *Investigative Ophthalmology and Visual Science* 49: 2178–2192.

McColm, J. R. and Hartnett, M. E. (2005). Retinopathy of prematurity: Current understanding based on clinical trials and animal models. In: Hartnett, M. E., Trese, M. T., Capone, A., Keats, B., and Steidl, S. (eds.) *Pediatric Retina*, pp. 387–410. Philadelphia, PA: Lippincott Williams and Wilkins.

McLeod, D. S., Hasegawa, T., Prow, T., Merges, C., and Lutty, G. (2006). The initial fetal human retinal vasculature develops by vasculogenesis. *Developmental Dynamics* 235: 3336–3347.

Penn, J. S., Henry, M. M., and Tolman, B. L. (1994). Exposure to alternating hypoxia and hyperoxia causes severe proliferative retinopathy in the newborn rat. *Pediatric Research* 36: 724–731.

Penn, J. S., Madan, A., Caldwell, R. B., et al. (2008). Vascular endothelial growth factor in eye disease. *Progress in Retina and Eye Research* 27: 331–371.

Saito, Y., Uppal, A., Byfield, G., Budd, S., and Hartnett, M. E. (2008). Activated NADPH oxidase from supplemental oxygen induces neovascularization independent of VEGF in retinopathy of prematurity model. *Investigative Ophthalmology and Visual Science* 49: 1591–1598.

Smith, L. E. H., Wesolowshi, E., McLellan, A., et al. (1994). Oxygen induced retinopathy in the mouse. *Investigative Ophthalmology and Visual Science* 35: 101–111.

Stone, J., Itin, A., Alon, T., et al. (1995). Development of retinal vasculature is mediated by hypoxia-induced vascular endothelial growth factor (VEGF) expression by neuroglia. *Journal of Neuroscience* 15: 4738–4747.

Relevant Websites

http://www.iom.edu – Institute of Medicine Statement on Prematurity.
http://www.nei.nih.gov – National Eye Institute statement on ROP.

Rhegmatogenous Retinal Detachment

S C Wong, Moorfields Eye Hospital, London, UK
Y D Ramkissoon, Royal Hallamshire Hospital, Sheffield, UK
D G Charteris, Moorfields Eye Hospital, London, UK

Glossary

Retinopexy – A Laser- or cryotherapy-induced chorioretinal adhesion surrounding a retinal break as prophylaxis for or during surgical treatment of rhegmatogenous retinal detachment.
***Rhegma* (Greek)** – A rupture, rent, or fracture.
Rhegmatogenous retinal detachment – A separation of the neurosensory retina from the retinal pigment epithelium associated with a full-thickness hole, break, or tear in the retina. The detachment occurs secondary to the passage of vitreous fluid through the break.
Tamponade – The plugging on a retinal break so that fluid can no longer pass through it (e.g., with a gas bubble).
Vitreous – Pertaining to the vitreous body of the eye located in the posterior chamber of the eye.

Rhegmatogenous retinal detachment (RRD) is defined as a separation of the neurosensory retina from the underlying retinal pigment epithelium (RPE) due to accumulation of subretinal fluid (SRF) via one or more full-thickness retinal breaks. With an incidence of 1 in 10 000 among the general population, RRD is relatively uncommon but remains an important cause of vision loss. In 1918, Jules Gonin recognized the importance of retinal breaks in the etiology of RRD and transformed the prognosis of this previously untreatable, blinding condition. Using thermocautery to seal retinal breaks, he demonstrated successful retinal reattachment in 30–40% of his cases. Modern vitreoretinal surgery can now achieve a final retinal reattachment rate of >95%.

Pathophysiology

Under physiological conditions, retinal attachment is principally maintained by the following mechanical and metabolic factors: (1) fluid or intraocular pressure (IOP) differentials (hydrostatic and osmotic forces); (2) glue-like interphotoreceptor matrix at the photoreceptor outer segment and RPE interface; and (3) RPE pump (NA^+–K^+ adenosine triphosphate (ATP)ase metabolic pump transporting ions and fluid across the subretinal space).

For RRD to occur, the following factors need to be present to overcome the attaching forces: (1) full-thickness retinal break in the neurosensory retina and (2) vitreous liquefaction and vitreoretinal traction. Vitreoretinal traction may coexist at the site of the retinal break, and varies depending on the type of break.

Full-thickness retinal breaks are classified as tractional tears, round holes, or dialyses.

Retinal tractional tears (also known as U, horseshoe, or flap tears) develop following posterior vitreous detachment (PVD) and are the most common causes of RRD. They occur at sites of enhanced vitreoretinal adhesion, such as in peripheral retinal lattice degeneration, chorioretinal scars, or at the posterior edge of the vitreous base. Persistent vitreous traction exerted on the flap of a tear promotes rapid, continuous recruitment of liquefied vitreous into the subretinal space and progression to retinal detachment, which usually occurs more rapidly than with a round hole or dialysis (where there is no posterior vitreous separation).

Retinal hole formation is not usually associated with overt vitreous traction. This is caused by localized retinal atrophy and occurs more commonly in myopic patients. The pathophysiology of progression from retinal hole to RRD is not entirely clear, as 5–10% of postmortem eyes have full-thickness retinal holes without RRD.

Retinal dialyses are circumferential retinal breaks occurring along the ora serrata with concurrent avulsion of the overlying vitreous base. There is often a history of ocular trauma, although spontaneous cases are not uncommon.

Clinical Features

Symptoms

Many patients present acutely with classic symptoms of flashing lights (photopsia) and visual floaters in the affected eye, usually resulting from PVD. Photopsia lasts for seconds, is more noticeable in dim light conditions, may be precipitated by ocular movements, and commonly occurs in the temporal field but is not localizing. Floaters in the visual field vary from a solitary floater (e.g., Weiss ring formation) to innumerable minute dark spots (pigment or hemorrhage in vitreous). As an RRD develops and progresses, patients may notice peripheral visual-field loss, which is often described as a black curtain, shadow, or half-moon. Central vision is lost if SRF spreads to involve the macula and detaches the fovea.

In a minority of patients, RRD may be asymptomatic. Typically, this occurs in young myopic patients with atrophic round holes who often present late only when the detachment encroaches the macula.

Signs

Nonspecific signs of RRD may include reduced visual acuity, relative afferent pupillary defect, mild anterior uveitis, and low IOP compared to the fellow eye. Up to 2% of patients may present with raised IOP or Schwartz–Matsuo syndrome. This typically occurs in young patients with chronic RRD, and is due to rod photoreceptor outer segments passing into the anterior chamber and blocking trabecular outflow. A positive Shaffer's sign, defined as the presence of tobacco dust or pigment granules in the anterior vitreous, is highly predictive of a retinal break particularly when associated with any of the above symptoms.

Stereoscopic fundal examination reveals an elevated neurosensory retina in the area of the detachment, with reduced visibility of underlying choroidal markings. The detached retina has a convex configuration and may have a corrugated surface due to intraretinal edema. This is in contrast to tractional non-RRD, which has a concave configuration.

In addition, one or more full-thickness retinal breaks may be visible as evidenced by discontinuity of the retinal surface. These typically occur anterior to the equator. Visualization of the type, number, and distribution of all retinal breaks is important for surgical planning. This is best achieved with the aid of indirect ophthalmoscopy and scleral indentation. However, this may be difficult in the presence of media opacities. Identification of break type in RRD is crucial, as it determines the choice of appropriate surgical technique for repair.

Break type – identification

Retinal tears

1. Horseshoe or U-shaped tear (may be irregular; **Figure 1**).
2. Persistent point of vitreous traction present at anterior edge of flap.
3. Giant retinal tear (GRT) is a circumferential retinal tear of 3 or more clock hours.
4. Avulsion of the anterior flap of a retinal tear may result in a round break with overlying retinal operculum. A tear with an operculum that is fully separated from the retina is known as a fully operculated round hole (FORH).
5. Retina is visible anterior to break.

Retinal round holes

1. Round atrophic hole without overlying retinal operculum (**Figure 2**).
2. May be associated with retinal degeneration (e.g., peripheral lattice degeneration).
3. Retina is visible anterior to break.

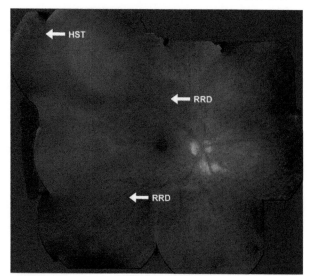

Figure 1   Right eye. Retinal tear in the superotemporal retinal periphery causing an acute macular-sparing retinal detachment. HST, horseshoe tear; RRD, rhegmatogenous retinal detachment.

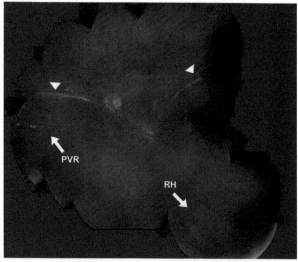

Figure 2   Left eye. Retinal round hole in the inferotemporal retinal periphery causing a chronic macular-sparing retinal detachment. A demarcation line (arrowhead) indicates several months of stasis and nonprogression of the retinal detachment. Proliferative vitreoretinopathy in the form of subretinal band formation is consistent with the chronicity of the detachment. RH, round hole; PVR, proliferative vitreoretinopathy.

Retinal dialyses

1. Circumferential retinal break at the ora serrata without a PVD (**Figure 3**).
2. No visible retina anterior to retinal break.

Lincoff first described the principles of predicting retinal break location, taking into account the effect of gravity on the spread of SRF from the primary break and the resultant topography of an RRD. This is influenced by anatomical limits such as the disc, the ora serrata, and any chorioretinal adhesions that might be present.

Break localization – principles based on topography of RRD

Superotemporal and superonasal

1. SRF descends on the same side of break toward the disc, then revolves around the inferior pole of the disc and rises on the opposite side.
2. SRF may rise as high on the opposite side of the disc as the level of the primary retinal break, but never as high as the fluid level on the primary side.
3. The primary break will be found within 1.5 clock hours of the highest border of the detachment in 98% cases.

Midline (12 o'clock meridian) and total detachments

1. Detachments that cross the 12 o'clock meridian originate from breaks at or near 12 o'clock.
2. These detachments can become total.
3. In subtotal detachments, if the break is slightly to one side of 12 o'clock, the fluid front will be more extensive on the side coincidental with the break.
4. The more posterior the break, the more it can deviate from 12 o'clock position and still cause a detachment that will cross the vertical meridian.

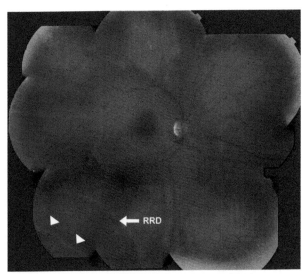

Figure 3 Right eye. Retinal dialysis (arrowhead) in the inferotemporal retinal periphery causing a macular-sparing retinal detachment.

Therefore, contour edges of the detachment are of less localizing value.

Inferior

1. The SRF that arises from breaks below the level of the optic disc develops first around the break and then advances toward the disc and macula, rising higher on that side of the disc where the break lies.
2. The break need only be 1 or 2 mm from the 6 o'clock position for it to cause a difference in fluid levels.
3. When the levels are equal, the hole is at the 6 o'clock meridian.
4. Such detachments are never bullous.

Chronic RRD

The distinction between an acute and chronic RRD is important, as it may determine both the urgency of treatment and prognosis. Signs of chronicity are:

1. Retinal atrophy – retina may be thinned and atrophic.
2. Intraretinal cysts – these can develop in long-standing RRD, typically those over 12 months duration. At the time of surgery, they can be left undisturbed as they do not usually interfere with retinal reattachment and can subsequently collapse.
3. Subretinal demarcation lines – also known as high watermarks or tidemarks, these develop at the advancing front of the RRD as a result of proliferation of RPE cells. Their presence simply indicates a period of stasis of the RRD at the point of the demarcation line. However, it does not confer the strength of a chorioretinal scar that is generated by retinopexy; thus, RRD progression may still occur.
4. Proliferative vitreoretinopathy (PVR) (**Figure 4**) – This is a process of cellular proliferation and fibrocellular membrane contraction that complicates between 5% and 12% of all RRD, and is the most common reason for final failure of RRD surgery. PVR has a higher incidence in RRD secondary to GRTs (16–41%) and in eyes sustaining penetrating trauma (10–45%). Once PVR has developed, visual recovery is usually limited despite improving anatomical surgical success rates. The location and extent of epiretinal or subretinal proliferation have been classified by the Retina Society Terminology Committee in 1991 (**Table 1**).

Differential Diagnoses

RRD must be distinguished from other causes of retinal or choroidal elevation. Differential diagnoses include PVD, exudative RD, tractional RD, retinoschisis, choroidal lesions, and artifacts (**Table 2**).

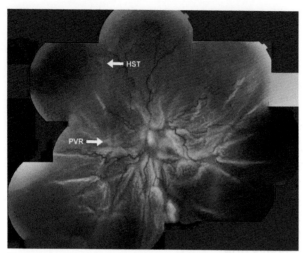

Figure 4   Right eye. Total rhegmatogenous retinal detachment secondary to a small superior retinal tear. Proliferative vitreoretinopathy is evidenced by posterior starfolds involving approximately 11 clock hours (PVR CP11). HST, horseshoe tear; PVR, proliferative vitreoretinopathy. Reproduced with kind permission from Paul M. Sullivan.

Table 1   Proliferative vitreoretinopathy classification by the Retina Society Terminology Committee

Grade (stage)	Characteristics
A	Vitreous haze; vitreous pigment clumps; and pigment clusters on inferior retina
B	Wrinkling of the inner retinal surface; rolled and irregular edge of retinal break; retinal stiffness; vessel tortuosity; and decreased vitreous mobility
C P 1 – 12 (clock hours)	Posterior to equator: focal, diffuse, or circumferential full-thickness retinal folds; and subretinal strands
C A 1 – 12 (clock hours)	Anterior to the equator: focal, diffuse, or circumferential full-thickness retinal folds; subretinal strands; anterior displacement; and condensed vitreous with strands

From Machemer, R., Aaberg, T. M., Freeman, M., et al. (1991). An updated classification of retinal detachment with proliferative vitreoretinopathy. *American Journal of Ophthalmology* 112: 159–165, with kind permission from Elsevier.

Management

The primary aim in the management of RRD is to prevent complete loss of central and peripheral vision, that is, progression to no perception of light. In patients with symptomatic RRD, the secondary aim of management is restoration/improvement of peripheral visual field and central vision. Depending on the type of retinal break, size, and rate of progression of the RRD and PVD status, treatment may be either conservative or active. Conservative management involves observation for progression of RRD without intervention. Active management includes either barrier laser demarcation to wall off the detachment

and prevent further progression, or surgery to reattach the retina. The various treatment options are discussed in turn, categorized by type of causative retinal break.

Conservative

Conservative management consists of either self-monitoring by the patient for development or progression of RRD symptoms, and/or observation by regular clinical examinations charting any signs of an enlarging RRD. This option is usually reserved for patients with chronic RRD without clinical signs of recent progression who are often asymptomatic and discovered to have RRD as an incidental finding. Another group of patients, where conservative management may be appropriate, comprises those with such significant systemic comorbidity that active treatment cannot be safely administered. When considering conservative management regardless of the specific indication, the ophthalmologist should discuss in detail its risks and benefits versus active therapies with the patients, allowing them to make informed decisions regarding their care. In addition, it is important to ensure that the patient understands the symptoms of RRD progression, and are able to reliably monitor their vision monocularly to detect changes before macula detachment occurs.

RRD due to retinal tear

RRD due to retinal tears occurs following PVD and tends to be rapidly progressive in nature due to persistent vitreoretinal traction on the anterior edge of the tear. It is therefore unusual for this to be asymptomatic or present as chronic RRD, and so much less likely to be managed conservatively.

RRD due to retinal hole or dialysis

Both retinal holes and retinal dialyses are usually associated with an attached vitreous, that is, without a PVD. This is thought to account for the relatively slower progression of RRD compared to those caused by retinal tears. As such, these patients often present with signs of chronicity and nonprogression, and many are asymptomatic until the macula becomes involved. Conservative management may be considered with appropriate patient selection.

Laser Demarcation

Laser photocoagulation results in the formation of a chorioretinal scar, conferring enhanced adhesion between the neurosensory retina and RPE. Laser demarcation for RRD is applied to the area of attached retina immediately adjacent to the most posterior edge of the RRD, walling off the area of detachment and preventing further

Table 2   Differential diagnosis of rhegmatogenous retinal detachment

Differential diagnosis	Features
Posterior vitreous detachment	– Weiss ring, vitreous syneresis
	– Attached retina
Exudative RD	– RD configuration convex, but smooth without corrugated surface
	– Shifting SRF – configuration of RD responds to changes in posture and its gravitational effects
	– No full-thickness retinal breaks present
	– May be associated with uveitis, scleritis, tumor, vascular, and other disorders
Tractional RD	– Associated with proliferative diabetic retinopathy or penetrating ocular trauma
	– RD has concave configuration
	– No full-thickness retinal breaks present, unless combined tractional and rhegmatogenous
	– Limited retinal mobility; no shifting fluid
	– Retina less elevated, and associated with sites of vitreoretinal traction
Retinoschisis	– Age-related or congenital X-linked
	– Occurs secondary to intraretinal splitting (age-related: splitting at outer plexiform or less commonly inner nuclear layer. X-linked: splitting at nerve fiber layer)
	– Often bilateral, and more common in hypermetropes
	– Retinal elevation has concave configuration
	– No full-thickness retinal breaks present
	– Partial thickness retinal breaks may be visible involving the inner and/or outer leaf
	– Does not produce demarcation line
	– Produces absolute scotoma, unlike RD which produces relative scotoma
	– Laser reaction test – this can help differentiate retinoschisis from a retinal detachment. Low-energy argon green laser is applied to the area of retinoschisis. Intensity of gray burn reaction should be equal to a control shot in peripheral flat retina. No reaction is seen in RRD.
Choroidal lesions	– Choroidal detachments or choroidal masses (tumors or inflammatory lesions)
	– No full-thickness retinal breaks present
	– May be associated with exudative RD
Artifacts	
– Vitreous hemorrhage	– Hemorrhage in vitreous cavity
	– Absence of true retinal vessel configuration
– Lens opacity – cataract or pseudophakic capsule opacification	– No relative afferent pupillary defect
	– Visible lens opacity
	– No vitreous haze or pigment granules
	– No mobility of opacity
	– Attached retina on ultrasound B scan

progression. Two to three contiguous rows of laser are required, extending anteriorly up to the ora serrata on either side of the RRD. This is usually achieved using indirect laser ophthalmoscopy with scleral indentation. Increased adhesion at the site of laser photocoagulation is evident within 24 h in histological studies of animal and human eyes, and is twice normal by 2–3 weeks. Therefore, laser demarcation is only appropriate for RRD with signs of slow progression or recent stasis. Cryotherapy is not appropriate for RRD demarcation as a large area of retinopexy is required, causing more extensive tissue destruction and is less precise than laser.

RRD due to retinal tear

As discussed previously, acute RRD due to retinal tears tends to progress rapidly and is therefore not usually suitable for laser demarcation. Possible exceptions are localized RRD from a single small tear, particularly if within inferior retina.

RRD due to retinal hole or dialysis

Patients with limited or no symptoms from RRD secondary to retinal hole or dialysis, with associated signs of chronicity, may be suitable candidates for laser demarcation. Exceptions are patients with significant symptoms, especially visual field loss, signs of recent rapid progression, or when the SRF has progressed to the major vascular arcades. Laser demarcation limits rather than alleviates RRD symptoms. Laser-induced chorioretinal scars in the posterior pole can expand up to 13% per year, potentially causing a symptomatic central scotoma, if this is close to the macula.

Pneumatic Retinopexy

Pneumatic retinopexy for RRD was introduced by Hilton and Grizzard in 1986 as a two-step outpatient procedure without conjunctival incision. This involves an intravitreal gas injection to temporarily close the retinal break preventing further recruitment of SRF from liquefied vitreous, followed by retinopexy (laser or cryotherapy) to permanently seal the break. Postoperative posturing for a minimum of 5 days is required in all patients. Careful case selection is required as not all patients are suitable for pneumatic retinopexy.

This procedure is not suitable in the following cases:

1. multiple breaks spanning over 3 clock hours;
2. giant retinal breaks;
3. inferior breaks involving inferior 4 clock hours;
4. presence of PVR grade C; and
5. inability to maintain postoperative posturing.

RRD due to retinal tear

Pneumatic retinopexy is most suitable for RRD with a single superior tear situated in the superior 8 clock hours. Patients must be able to posture, and attend clinic regularly in the early postoperative period to monitor the outcome of treatment. Additional gas injection or further surgery with scleral buckling or vitrectomy may be required in the event of treatment failure.

RRD due to retinal hole or dialysis

Pneumatic retinopexy is not suited to either of the following pathologies: RRD due to retinal hole or dialysis. The presence of an attached vitreous in both retinal holes and dialyses increases the risk of complications of the procedure, including new retinal break and formation of multiple small gas bubbles (see the section titled 'Complications').

Scleral Buckling

The concept of scleral buckling was introduced by Custodis in 1949. Prior to the advent of noncontact wide-angle viewing systems for vitrectomy, scleral buckling was the most commonly performed operation for RRD, and remains so in certain countries due to high primary success rates of over 90%. It is typically performed under general anesthesia, although local anesthesia may be used.

The basic principle of scleral bucking (**Figure 5**) is to close retinal breaks by indentation of the sclera, preventing further recruitment of fluid into the subretinal space. Scleral indentation is achieved by using a variety of explant materials sutured externally to the scleral surface. Encircling explants afford a permanent 360° indent, while segmental explants provide a localized indent, which supports the break for a period of several months until the indent fades. Segmental explants are preferred due to lower associated ocular comorbidity. Permanent break closure must be ensured by application of retinopexy, typically cryotherapy. Intraoperative drainage of SRF

may be performed either to aid more rapid resolution of the RRD, or to create space within the vitreous cavity for the scleral indent by substituting SRF volume for the volume of indent, thus preventing excessive IOP elevation. Drainage of SRF involves surgical penetration of the sclera and choroid to the subretinal space. Indications for drainage remain controversial as this converts scleral buckling surgery from an external to an internal procedure, with associated increased risk of intraoperative complications (see the section titled 'Complications').

RRD due to retinal tear

RRDs due to a single tear, or a cluster of tears spanning 2–3 clock hours at the same anteroposterior position, are suitable for scleral buckling. Scleral buckling should be considered in patients with inferior tears, particularly as these are more difficult to effectively tamponade with a vitrectomy and gas procedure. Relatively young phakic patients are ideal candidates for scleral buckling as accommodation is preserved and the risk of developing cataract is small. Scleral buckling is not suitable for patients with significant media opacity, large tears, GRTs, or posterior breaks.

RRD due to retinal hole or dialysis

Patients with retinal holes or dialyses respond well to scleral buckling, typically with segmental explants. Primary reattachment rates of up to 100% have been reported.

Vitrectomy

Machemer performed the first pars plana vitrectomy in 1971. Since then, significant progress has been made in our understanding of vitreoretinal pathology alongside improvements in microsurgical instrumentation and techniques. In the United Kingdom, vitrectomy is now the most commonly performed procedure for both simple and complex RRD.

Vitrectomy (**Figure 6**) is performed via three pars plana sclerostomy ports. Standard 20-gauge (0.9 mm) instruments are most widely used. The technique involves removal of vitreous with relief of vitreous traction at the site of retinal breaks, followed by intraocular drainage of SRF, application of retinopexy (cryotherapy or laser), and intravitreal injection of gas or silicone oil tamponade. Noncontact wide-angle viewing systems allow for superior depth of field, visualization, and break localization compared to scleral buckling. This is particularly useful in pseudophakic patients and those with media opacities.

RRD due to retinal tear

In both the UK and USA, vitrectomy has become the treatment of choice for RRD due to retinal tears as there is usually a preexisting PVD. This simplifies

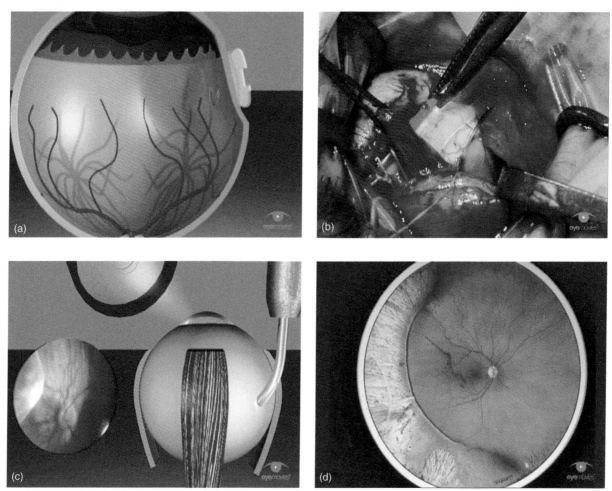

Figure 5 Scleral buckling surgery. (a) Schematic of explant placement on scleral surface. An underlying retinal tear is visible. (b) Silicone explant being secured to scleral surface using a 5/0 Ethibond suture. Part of the explant is positioned under a rectus muscle as shown. (c) Schematic of cryotherapy application to a retinal break viewing through an indirect ophthalmoscope. Inset photo demonstrates corresponding retinal whitening during cryotherapy application. (d) Schematic of postoperative scleral indent. Chorioretinal scars from previous cryotherapy are visible over the indent with successful retinal reattachment. Reproduced from Aylward, G. W., Sullivan, P. M., and Vote, B. (2007). *Vitreoretinal. Volume 1: Basic Techniques (DVD)*. Surrey: Eye Movies, with kind permission from Eye Movies.

vitrectomy surgery as induction of a PVD is not required and direct relief of vitreoretinal traction at the tear can be achieved. A PVD is present in approximately 75% of people aged over 65. Therefore, the majority of patients with retinal tears is already presbyopic, minimizing the consequences of postoperative cataract formation.

RRD due to retinal hole or dialysis

Vitrectomy is rarely indicated in RRD secondary to retinal holes or dialyses, which usually have an attached vitreous. Surgically inducing a PVD in these typically young patients is often difficult, adding an unnecessary layer of complexity with increased risk of complications, for example, iatrogenic break formation. In addition, in the case of a dialysis, PVD induction would convert the break into a GRT.

Outcomes

Pneumatic Retinopexy

Pneumatic retinopexy is less invasive than scleral buckling or vitrectomy, with less local tissue damage and inflammation. Primary success rate in a selected patient group has been shown to be noninferior to scleral buckling in a multicenter, randomized, controlled trial by Tornambe and co-workers. A recent comprehensive review of 4128 eyes reported in the literature over a 21-year period (1986–2007) showed that pneumatic retinopexy has an overall primary retinal attachment rate of 74.4%, lower than scleral buckling or vitrectomy, with a final reattachment rate of up to 96.1%. Aphakic or pseudophakic patients appear to do less well, with primary retinal reattachment rates of 41–67%.

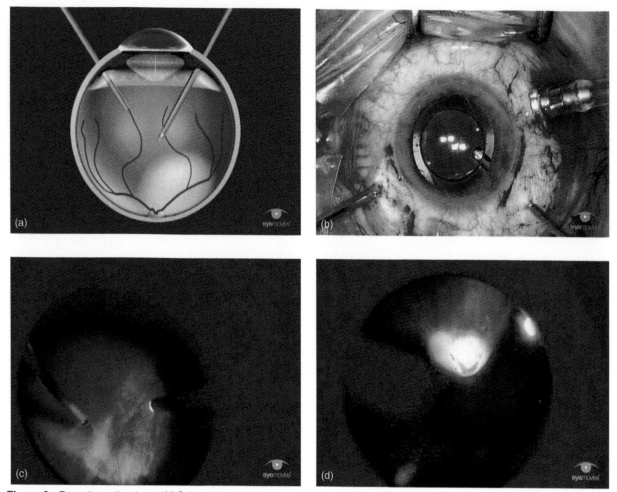

Figure 6 Pars plana vitrectomy. (a) Schematic of vitrectomy. Instruments are inserted 3.5–4mm posterior to the corneal limbus, entering the posterior segment through the pars plana. (b) Microscope view of standard instrumentation. Three scleral ports are usually required for vitrectomy. Instrumentation consists of a fluid infusion line, endoillumination, and vitrector (cutter). (c) Intraoperative view of vitrectomy for rhegmatogenous retinal detachment due to a retinal tear. Vitrectomy is being performed close to a retinal tear. (d) Intraoperative view of cryotherapy application to a retinal tear. Reproduced from Aylward, G. W., Sullivan, P. M., and Vote, B. (2007). *Vitreoretinal. Volume 1: Basic Techniques (DVD)*. Surrey: Eye Movies, with kind permission from Eye Movies.

Scleral Buckling

Primary retinal reattachment occurs in over 90% of cases with scleral buckling alone, although up to 100% has been reported in small series. In patients with RRD and an attached macula, 90% achieve visual acuity of 20/30 or better at 6 months, although reduction in two Snellen lines of visual acuity occur in 10% despite anatomical success. Causes include macular epiretinal membrane and cystoid macular edema. Patients, who have their macular detached preoperatively, achieve a visual acuity of 20/50 or better in 38–71% of cases. A recent study involving foveal optical coherence tomography (OCT) imaging, by one of the authors (DGC), showed that postoperative resolution of foveal SRF can be delayed by over 6 weeks in 55% of patients, and complete resolution can take over 12 months. These patients have worse visual outcomes.

Pars Plana Vitrectomy

Primary retinal reattachment rates with vitrectomy are similar to scleral buckling, ranging from 65% to 100% with a mean of 85%. Final reattachment rate is 96–99%. Median visual acuity is 20/30 in macula-attached RRD, and 20/40 in macula-detached RRD. In an OCT imaging study by the same author (DGC), delayed resorption of foveal SRF was less common following vitrectomy, occurring in 15% of patients.

Complications

Intraoperative and postoperative complications for each procedure are discussed below.

Pneumatic Retinopexy

The main intraoperative complication of pneumatic retinopexy is formation of multiple small intravitreal gas bubbles or fish eggs during injection of gas, rather than a single large bubble as desired. This can result in subretinal gas migration, inadequate break tamponade, and treatment failure. This is principally managed by postoperative posturing to allow the gas bubbles to coalesce.

Postoperatively, new retinal break formation occurs in up to 20% of patients, and is a significant cause of surgical failure. PVR occurs in approximately 5% of patients.

Scleral Buckling

Intraoperative complications include inadvertent scleral perforation during suture placement, or SRF-drainage-related problems. Consequences include hypotony, choroidal or subretinal hemorrhage (with or without subfoveal tracking of blood), and retinal incarceration.

Postoperative complications include glaucoma, anterior segment ischemia (following encirclement), diplopia secondary to ocular dysmotility, change in refractive error and astigmatism, explant extrusion/intrusion or infection, and PVR. Contact lens users should be forewarned about possible lens-wear intolerance due to an irregular tear film and ocular surface following surgery, particularly as many patients who undergo scleral buckling are young myopes.

Pars Plana Vitrectomy

Intraoperative complications include iatrogenic retinal breaks, vitreous and retinal incarceration into the sclerostomy ports, lens trauma, and suprachoroidal hemorrhage.

Postoperatively, cataract formation is common and occurs in 81% of patients by 6 months. Raised IOP in the early postoperative period is common, and there appears to be an associated risk of late secondary glaucoma. PVR complicates between 5% and 12% of RRDs. Endophthalmitis following 20-gauge vitrectomy is rare.

Treatment Failure

In approximately 1 in 100 patients, it is not possible to reattach the retina despite multiple surgical attempts resulting in loss of vision. Consequences include chronic uveitis, rubeosis iridis, glaucoma, hypotony, band keratopathy, leukocoria, phthisis bulbi, and chronic ocular discomfort which may be severe enough to necessitate evisceration or enucleation of the eye.

New Developments

Smaller 25-gauge vitrectomy instruments were first introduced in 1990. Since then, techniques have evolved to enable sutureless transconjunctival surgery, with the benefits of reduced surgical time, less postoperative inflammation, more rapid wound healing, and improved patient comfort. As yet, small-gauge sutureless vitrectomy has not replaced 20-gauge systems, principally due to concerns regarding higher rates of endophthalmitis. A large retrospective series of 8601 patients from the Wills Eye Hospital demonstrated a 12-fold higher incidence of endophthalmitis with 25-gauge compared to 20-gauge vitrectomy. A smaller difference has been supported by other studies. Until this concern is adequately addressed, 20-gauge systems with conjunctival peritomy and suturing of ports will remain the gold standard.

The need for gas tamponade and postoperative posturing following vitrectomy for RRD has recently been challenged. A small prospective study of 60 patients by Martínez-Castillo and co-workers demonstrated a 98.3% primary reattachment rate without tamponade following complete drainage of SRF and laser retinopexy. This has not yet been supported by larger studies.

PVR remains to be the most important cause of failure of RRD treatment. Novel prophylactic treatment with adjunctive intraoperative 5-fluorouracil and low-molecular-weight heparin showed promise with decreased incidence of PVR in high-risk cases. However, this has not been reflected in unselected RRD undergoing primary vitrectomy, and may in fact be detrimental to visual acuity outcomes in patients with macula-attached RRD. Reducing the risk of PVR formation remains an important area of research.

See also: Proliferative Vitreoretinopathy.

Further Reading

Aylward, G. W., Sullivan, P. M., and Vote, B. (2007). *Vitreoretinal. Volume 1: Basic Techniques (DVD)*. Surrey: Eye Movies.
Chan, C. K., Lin, S. G., Nuthi, A. S., and Salib, D. M. (2008). Pneumatic retinopexy for the repair of retinal detachments: A comprehensive review (1986–2007). *Survey of Ophthalmology* 53: 443–478.
Charteris, D. G. and Wong, D. (2007). The role of combined adjunctive 5-fluorouracil and low molecular weight heparin in proliferative vitreoretinopathy prevention. In: Kirchhof, B. and Wong, D. (eds.) *Vitreo-Retinal Surgery*, 1st edn., pp. 33–37. New York: Springer.
Kunimoto, D. Y. and Kaiser, R. S. (2007). Wills eye retina service. Incidence of endophthalmitis after 20- and 25-gauge vitrectomy. *Ophthalmology* 114: 2133–2137.
Lincoff, H. and Gieser, R. (1971). Finding the retinal hole. *Archives of Ophthalmology* 85: 565–569.
Machemer, R., Aaberg, T. M., Freeman, M., et al. (1991). An updated classification of retinal detachment with proliferative

vitreoretinopathy. *American Journal of Ophthalmology* 112: 159–165.

Martínez-Castillo, V., Zapata, M. A., Boixadera, A., Fonollosa, A., and García-Arumí, J. (2007). Pars plana vitrectomy, laser retinopexy, and aqueous tamponade for pseudophakic rhegmatogenous retinal detachment. *Ophthalmology* 114: 297–302.

Peyman, G. A., Meffert, S. A., and Conway, M. (eds.) (2007). *Vitreretinal Surgical Techniques,* 2nd edn. London: Informa UK.

Ryan, S. (ed.) (2006). *Retina, Vol. III: Surgical Retina,* 4th edn., 3 vols. Philadelphia, PA: Elsevier.

Wilkinson, C. P. and Rice, T. A. (eds.) (1997). *Michels Retinal Detachment,* 2nd edn. St. Louis, MO: Mosby.

Rod and Cone Photoreceptor Cells: Inner and Outer Segments

D H Anderson, University of California, Santa Barbara, CA, USA
D S Williams, UCLA School of Medicine, Los Angeles, CA, USA

Glossary

Cilium – An organelle that projects from a cell and contains a defined array of microtubules.

Electron microscope – A microscope that uses a particle beam of electrons to illuminate a specimen and create a highly magnified image. An electron microscope has much greater resolving power than a light microscope because the wavelength of an electron is much smaller than that of visible light.

Glycoconjugate – A class of carbohydrates covalently linked to proteins, lipids, and other types of molecules.

Mitochondrion – A cellular organelle that provides most of the chemical energy, in the form of adenosine triphosphate, for metabolism in eukaryotic cells.

Plasma membrane – A bilayer of lipid molecules that physically separates the interior of cells (i.e., cytoplasm) from the extracellular environment.

In this article, we focus on the organization of the photoreceptor inner and outer segments, with an emphasis on the structural similarities and differences between rods and cones. The process by which the photoreceptor outer segments replace their disk membranes; soma and synapse; and cilim, phototransduction, soluble protein dynamics, and the visual cycle are described elsewhere in this encyclopedia.

In the late nineteenth century, Schultze proposed an organizational framework for vertebrate photoreceptors that, today, has become a cornerstone of visual science. According to this duplicity theory, photoreceptors may be anatomically and functionally divided into two main groups: rods and cones. These terms emerged from early anatomical observations showing that the distal portions of photoreceptors, the so-called outer segments, are either cylindrical or conical in shape. However, it soon became clear that there were exceptions to the outer segment shape criterion, most notably in the foveas of some species where cone outer segments possess a distinct rod-like configuration. Therefore, with the passage of time, the definitions of rods and cones were relaxed somewhat to include the combined shapes of the outer and inner segments (**Figure 1**). More recently, light-sensitive ganglion cells without obvious outer segments have been identified.

Duplicity theory also stipulated that rods and cones may be distinguished functionally by the light levels to which they are tuned. Rods were regarded primarily as a sensitivity mechanism; whereas cones were considered to be an acuity mechanism, and a prerequisite for color vision. The predominance of rods in nocturnal species and of cones in diurnal species provided compelling evidence in support of this generalization.

With the advent of electron microscopy in the latter half of the twentieth century, it became apparent that there were also ultrastructural differences between rods and cones. In longitudinal sections, rod outer segments are visualized as a stack of hundreds of free-floating disks that resemble a stack of coins. The disk stack in rods is enclosed by, and separated from, the cell's plasma membrane, except at the base where new disks are formed (**Figure 2**, left). In cones, however, many and perhaps all of the disk membranes appear to be in continuity with the plasma membrane (**Figure 2**, right; **Figure 4**). This ultrastructural difference between rod and cone outer segments has been confirmed in a wide variety of vertebrates, and still remains a distinction with no known exceptions. In cross section, single-rod disks have a scalloped margin all of which are in alignment; in contrast, cone disks possess a single incisure that extends from the margin to the center of the disk.

The photoreceptor cell is the epitome of a specialized neuronal cell. It is dedicated to the absorption of light energy (photons) and transduction of that energy into an electrochemical signal that is transmitted throughout the retina and, ultimately, to the brain. The packing density of the inner and outer segments defines the spatial resolution of the retina. Where visual acuity is at a premium, the segments form a fine array, the density of which is limited only by the wave nature of light. Hence, in many retinas, the inner and outer segments are long, narrow structures that may be as small as 1 μm in diameter.

The photoreceptors form the outermost layer of the neural retina (**Figure 3**). Rod and cone outer segments are enveloped by microvilli (mv) that emanate from the apical surface of the retinal pigment epithelium (RPE). In rods, the outer segments abut the apical surface; but in many species, including humans, cone outer segments are slightly recessed from the apical surface (see **Figure 1**). A highly organized array of mv known as the cone sheath extends down from the apical RPE surface and ensheaths the cone outer segments. Cone outer and inner segments are also ensheathed by an extracellular matrix rich in glycoconjugates known as the cone matrix sheath (**Figure 4**). The inner segments lie between the outer segments and the photoreceptor cell nuclei. Broadly speaking, the inner segment provides the metabolic support for the outer segment.

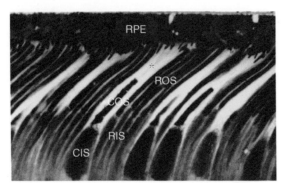

Figure 1 Light micrograph showing the photoreceptor cell layer in a monkey retina. RPE, retinal pigment epithelium; ROS, rod outer segment; RIS, rod inner segment; COS, cone outer segment; CIS, cone inner segment; *, cone sheath.

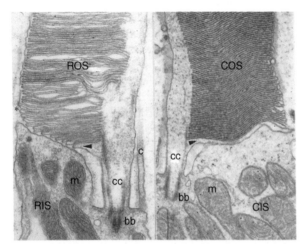

Figure 2 Electron micrographs illustrating the organization of disk membranes at the base of mammalian rod and cone outer segments. The only cytoplasmic link between inner and outer segments is through a connecting cilium. Arrowheads indicate growth points for the evagination of new disk membranes from the plasma membrane adjacent to the centric face of the connecting cilium. In cones, most disks retain a connection to the enclosing plasma membrane; whereas, in rods, only the most basal disks (below the asterisk) retain a plasma membrane connection. bb, basal body; c, ciliary process; cc, connecting cilium; CIS, cone inner segment; m, mitochondrion; RIS, rod inner segment.

The distal region of the inner segment, known as the ellipsoid, contains a high concentration of mitochondria that are excluded from the outer segment. The only cytoplasmic link between inner and outer segments is a connecting cilium (**Figure 2**). The proximal region of the inner segment, the myoid, is the primary site for the synthesis of proteins destined for the outer segment (**Figure 3**). Projecting from the lateral margin of the ellipsoid is an array of calycal processes containing longitudinally oriented, filamentous actin and myosin. In monkey cones, each ellipsoid contains an array of actin-containing cables that originate in the myoid region, converge as they course through the

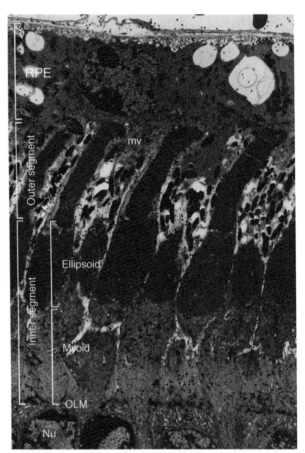

Figure 3 Low power electron microscopic autoradiogram of squirrel photoreceptors, illustrating the ellipsoid and myoid regions, the outer segments, and their relationship to the RPE. The black dots over the tissue are developed silver grains that signify sites of incorporation of a sugar residue (fucose) that had been labeled with a radioactive isotope (tritium). mv, microvilli; Nu, nucleus; OLM, outer limiting membrane; RPE, retinal pigment epithelium.

ellipsoid, and terminate within the calycal processes that form a circumferential ring at the base of the outer segments (**Figure 5**).

Just proximal to the myoid is the outer limiting membrane (OLM) (**Figure 3**). In longitudinal sections, the OLM appears as a demarcation line between the inner and outer retina. In reality, it is not a membrane but a series of aligned adherens junctions between photoreceptor cells, and between photoreceptor cells and adjacent radial glial cells (i.e., Mueller cells), that are linked to the cells' actin cytoskeleton. In lower vertebrates, the myoid region of the inner segment is quite labile, and is capable of expansion and contraction in response to changes in ambient lighting. These so-called retinomotor movements are considered to be a light-adaptive mechanism in many species of lower vertebrates.

Approximately 90% of the membrane protein in the rod disk and plasma membranes consists of the visual pigment opsin, a light-sensitive guanine nucleotide-binding protein

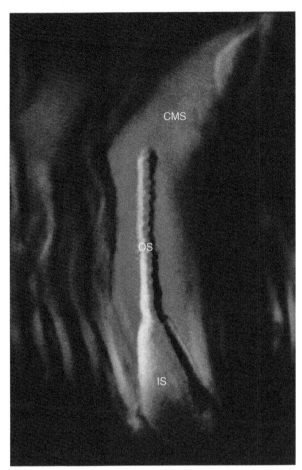

Figure 4 Longitudinal section of a human cone photoreceptor inner segment (IS) and outer segment (OS) labeled with a plant lectin (peanut agglutinin) conjugated to a fluorescent probe. The cone matrix sheath (CMS) (shown in yellow) is a distinct domain of the retinal interphotoreceptor matrix that is rich in glycoconjugates and envelops cone outer and inner segments. Modified from Hageman, G. S. and Johnson, L. V. (1991). Progress in Retina Research. In: Osborne, N. and Chader, J. (eds.) *Structure, Composition, and Function of the Retinal Interphotoreceptor Matrix*, Vol 10, Ch.9, p. 226.

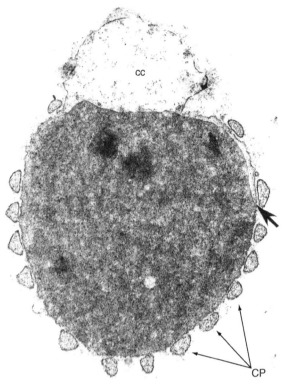

Figure 5 Electron micrograph of a tangential section through a monkey cone outer segment. The calycal processes (small arrows) that project from the ellipsoid form a basket at the base of the cone outer segment. At one point, the rim of the cone disk and the outer plasma membrane appear to be in continuity (large arrow), signaling a potential growth point. cc, connecting cilium; CP, calycal processes.

(G-protein)-coupled receptor. Cone outer segments contain homologous visual pigment proteins, also known as photopsins or iodopsins. Numerous other proteins that participate in the phototransduction cascade are also present in the outer segments. Rod outer segments, especially from bovine and rodent retinas, have been used widely in biochemical studies of structure and function. The ciliary connection between the inner and outer segments is quite fragile, such that outer segments are readily broken off in solution by shaking retinas that have been detached from the adjacent RPE. Following purification over a density gradient, relatively pure biochemical preparations of outer segments can be obtained. With additional steps, the disk and plasma membranes of rod outer segments can be separated from each other, and have thus been shown to contain some notable differences in protein composition.

For example, the light-dependent cyclic nucleotide-gated channel and Na^+, K^+, Ca^{2+} exchanger proteins are restricted to the plasma membrane, while the retinal degeneration slow (peripherin/rds), rod outer segment membrane protein 1 (rom-1), and the ATP-binding cassette family A 4 (ABCA4) transporter proteins are present only in disk membranes. Opsin is present in both membrane domains. The disk membranes also contain two distinctive domains. Opsin is a ubiquitous component of rod disk membranes, except at their rims where a complex of peripherin/rds and rom-1 enables the formation of a loop that gives the disk its characteristic bilamellar structure.

In development, the outer segments form by repeated outgrowths or evaginations of the distal plasma membrane of the connecting cilium. In a mature photoreceptor cell, new rod and cone disks are also formed from evaginations originating from the centric face of the connecting cilium (**Figures 2** and **6**). The mature outer segment remains connected to the inner segment by what corresponds to the transition zone of the cilium, and what has been referred to historically as the ciliary stalk or connecting cilium. In recent years, it has been recognized that, while the photoreceptor cilium is highly unusual with its attached

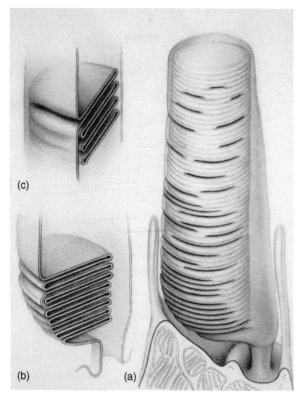

Figure 6 Schematic of the cone photoreceptor outer segment, inner segment, and connecting cilium (a). (b) Base of the cone outer segment and its relationship to the plasma membrane of the connecting cilium. (c) Illustration of the continuity between cone disks and the outer segment plasma membrane. From Anderson, D. H., Fisher, S. K., and Steinberg, R. H. (1978). Mammalian cones: Disc shedding, phagocytosis, and renewal. *Investigative Ophthalmology and Visual Science* 17(2): 130.

and outer segments, the connecting link between them is an organelle that is highly conserved. This point has been made most poignantly by demonstrations in a variety of syndromic disorders, such as Senior Loken and Bardet Biedel syndromes. These disorders are characterized, in part, by retinal degeneration, as well as kidney and other disorders, caused by mutations in genes which encode ciliary proteins that subserve shared functions.

See also: Circadian Photoreception; Cone Photoreceptor Cells: Soma and Synapse; Fish Retinomotor Movements; Light-Driven Translocation of Signaling Proteins in Vertebrate Photoreceptors; The Photoreceptor Outer Segment as a Sensory Cilium; Phototransduction: Adaptation in Rods; Phototransduction: Inactivation in Cones; Phototransduction: Inactivation in Rods; Phototransduction: Phototransduction in Cones; Phototransduction: Phototransduction in Rods; Phototransduction: The Visual Cycle; Rod and Cone Photoreceptor Cells: Outer Segment Membrane Renewal; Rod Photoreceptor Cells: Soma and Synapse.

Further Reading

Anderson, D. H., Fisher, S. K., and Steinberg, R. H. (1978). Mammalian cones: Disc shedding, phagocytosis, and renewal. *Investigative Ophthalmology and Visual Science* 17(2): 117–133.

Besharse, J. C. and Horst, C. J. (1990). The photoreceptor connecting cilium: A model for the transition zone. In: Bloodgood, R. A. (ed.) *Ciliary and Flagellar Membranes*, pp. 409–431. New York: Plenum Press.

Cohen, A. I. (1970). Rods and cones. In: Fjuortes (ed.) *Handbook of Sensory Physiology*, vol. 7, part 1B, pp. 63–110. Berlin: Springer.

Molday, R. S. (2004). Molecular organization of rod outer segments. In: Williams, D. S. (ed.) *Photoreceptor Cell Biology and Inherited Retinal Degenerations*. Singapore: World Scientific Publishing.

Schultze, M. (1866). Anatomie und Physiologie der Nezhaut. *Archiv für mikroskopische Anatomie* 2: 175–286.

Spitznas, M. and Hogan, M. J. (1970). Outer segments of photoreceptors and the retinal pigment epithelium. *Archives of Ophthalmology* 84: 810–819.

Young, R. W. (1970). Visual cells. *Scientific American* 223: 80–91.

network of outer segment disks, it is homologous to other primary cilia located elsewhere in the body. For example, like other primary cilia, it possesses microtubule motor proteins that may function in the transport of outer segment components. There are numerous other proteins shared by primary cilia and photoreceptor cilia. Thus, despite the extraordinary specializations of the inner

Rod and Cone Photoreceptor Cells: Outer Segment Membrane Renewal

D S Williams, UCLA School of Medicine, Los Angeles, CA, USA
D H Anderson, University of California, Santa Barbara, CA, USA

Glossary

Autoradiography – A technique used to localize radioactivity emitted by cells, tissues, or organisms that have been treated or injected with a radioactive isotope.

Electron microscope – A microscope that uses a particle beam of electrons to illuminate a specimen and create a highly magnified image. An electron microscope has much greater resolving power than a light microscope because the wavelength of an electron is much smaller than that of visible light.

Lysosome – An organelle in a cell that contains digestive enzymes.

Phagocytosis – The engulfing and internalization of particles by a cell.

Retinal pigment epithelium – A single layer of pigmented epithelial cells that borders the back of the sensory retina. The outer segments of the photoreceptor cells interdigitate with the apical processes of the retinal pigment epithelium cells.

Based upon their observations of photoreceptor disk membrane-like inclusions in the cytoplasm of rodent retinal pigmented epithelial (RPE) cells, A. Bairati and N. Orzalesi proposed in 1963 that the disk membranes of photoreceptor outer segments were in a dynamic state of turnover. A few years later, in a classic series of autoradiographic studies, Richard Young and his colleagues provided the first direct evidence that vertebrate rod photoreceptors continually replace their disk membranes. Following injection of radiolabeled amino acids into a series of frogs, a transverse band of radioactive protein became apparent at the base of rod outer segments within 24 h following injection (**Figure 1**). We now know that the main proteinaceous component of the phototransductive disk membranes is the visual receptor, opsin. Over the ensuing days, the band became progressively displaced toward the tip of the outer segment, and eventually became evident in inclusions within the cytoplasm of the RPE cells. Electron microscopic observations later revealed that packets of disks derived from the tips of rod outer segments were engulfed by microvillous processes on the apical surface of the RPE cells and internalized into the RPE cytoplasm. These membrane-bound inclusions, called phagosomes, were then degraded by the RPE. Young proposed the term outer-segment renewal to refer to the overall turnover process. It encompasses a number of stages, including protein synthesis, transport to the outer segment, disk formation, disk displacement, shedding, phagocytosis and degradation, all of which result in the vectorial flow of membrane from the photoreceptor inner segment to the RPE (**Figure 2**).

Many cell types have short lives and are replaced on a regular basis. However in a terminally differentiated cell, such as a photoreceptor cell, the components must be turned over to prevent the formation and build up of macromolecular byproducts that can interfere with cellular function. Accordingly, it is widely believed that the turnover of phototransductive membrane is part of a normal preemptive process to replace macromolecules before they become dysfunctional. Because phototransductive membrane is typically an extremely amplified membrane system, its turnover involves an extraordinarily high rate of synthesis and degradation of membrane proteins. Some nocturnal arthropods provide the most extreme examples, with the turnover of nearly all of their phototransductive membrane each day. Even at the more pedestrian rate of 10% per day, as found in rodent and primate rods, the turnover of disk membranes represents a major metabolic challenge for the photoreceptor and RPE cells. In each human retina, which contains approximately 100 million photoreceptor cells, an average of 9 billion opsin molecules turn over every second. Not surprisingly, therefore, the processes involved in disk membrane turnover are critical for photoreceptor cell viability, as shown with some of the first-studied rodent models of retinal degeneration. For example, in the RCS rat, phagocytosis by the RPE is defective, and in the retinal degeneration slow (rds) mouse disk membrane morphogenesis is blocked; photoreceptor cell death and blindness ensues in both these models.

Whereas the experiments using radiolabeled amino acids convincingly demonstrated the renewal of the rod outer-segment membranes, and the involvement of the RPE in the degradation phase, the case for cone outer-segment renewal was less compelling. In contrast to rods, no discrete band of radioactivity was detected at the base of frog cone outer segments; instead, it appeared to be distributed diffusely throughout the cone outer segment. However, in the cones of diurnal rodents, quantitative electron microscopic autoradiographic studies showed that a

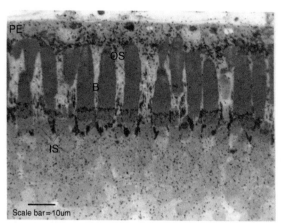

Figure 1 Light micrograph autoradiograph of a retina from a frog injected with radiolabeled amino acids, illustrating the band of radiolabeled protein (B) near the base of the outer segments (OS). IS, inner segments. Scale bar = 10 μm. From Hall MO, Bok D, and Bacharach ADE (1968) Visual pigment renewal in the mature frog retina. *Science* 161: 787–789.

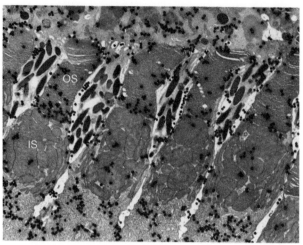

Figure 3 Electron micrograph autoradiograph of a retina from a ground squirrel, shortly after injection with radiolabeled fucose, illustrating the concentration of radiolabel near the base of the cone outer segments. OS, outer segment; IS, inner segment.

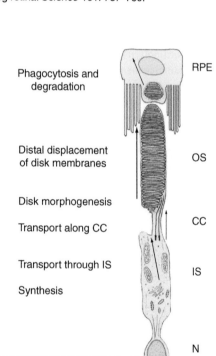

Figure 2 Illustration of the stages involved in the turnover of phototransductive membrane in vertebrate photoreceptor cells. Modified from Williams DS (2002) Transport to the photoreceptor outer segment by myosin VIIa and kinesin II. *Vision Research* 42: 455–462.

concentration of radioactive glycoconjugates could be detected in the proximal portion of cone outer segments within a few hours after injection of radiolabled fucose (**Figure 3**). In retrospect, it is now appreciated that the banding pattern in rods is due to the confinement of newly synthesized membrane proteins to rod disks, which

mature into discrete units. In contrast, cone outer-segment disks remain interconnected, so that radiolabeled protein is free to diffuse longitudinally throughout the disk stack. Studies demonstrating the presence of disk shedding in cones, and the presence of phagosomes in RPE cells overlying regions of high cone density, such as in the human fovea and in all-cone retinas, also helped to demonstrate that the unidirectional flow of disk membrane proteins, from inner segment to outer segment, and then to the RPE, occurred in cones as well as rods.

The *de novo* synthesis of opsin-containing membranes occurs mainly in the proximal region of the inner segment (i.e., the myoid). Membranes containing newly synthesized protein are then transported from the *trans*-Golgi network, through the ellipsoid, to the base of the connecting cilium (**Figure 2**). Opsin contains motifs that appear to be important for its targeting to the outer segment. The best characterized are its C-terminal amino acids. Missense mutations that affect this region result in mistargeting of opsin and underlie forms of inherited retinal degeneration in humans. It seems likely that a molecular motor, like a dynein, which travels toward the minus ends of microtubules, is involved in this vectorial transport. In support of this notion is the finding that opsin binds to tctex-1, a light chain of dynein, *in vitro*. Further studies have reported roles for a number of other proteins that appear to be important for the targeting and fusion of membrane near the base of the cilium. These proteins include small GTPases, such as ADP ribosylation factor 4 (ARF4) and the RAS oncogene-related protein RAB8. It is likely that RAB8 promotes membrane fusion near the base of the cilium, and may be recruited to the post-Golgi membranes by a complex of Bardet–Biedl syndrome (BBS) proteins; mutations in BBS proteins are responsible for BBS, which includes photoreceptor degeneration along with defects in cilia throughout the body. In photoreceptors that have a high rate of delivery of disk precursor membrane to the outer segment, the site of membrane fusion appears as a periciliary ridge complex, where the surface area of the plasma membrane has been greatly amplified by forming a series of ridges and grooves around the base of the cilium.

Transport along the cilium to the site of disk membrane morphogenesis involves molecular motors and associated proteins. Myosin VIIa, an actin-based motor, participates in this transport, although there is a clearer requirement for the kinesin-2 family of microtubule motors. In the absence of functional heterotrimeric kinesin-2, opsin delivery to the outer segment fails, and the ectopic build up of opsin causes rapid photoreceptor cell death. A homodimeric kinesin-2, KIF17, may also function in ciliary transport of outer-segment proteins. A group of intraflagellar transport proteins appear to function in concert with kinesin-2 motors. These proteins, which function in anterograde transport in primary cilia in other cells, are present in the photoreceptor axoneme, and IFT88 is required for the assembly and turnover of the outer-segment disk membranes. The specific cargoes of the different motor systems along the photoreceptor cilium are not known, but it is likely that there are a variety of different routes along the cilium. For example, evidence indicates that opsin and the rim-specific protein, peripherin-rds, are transported by different mechanisms.

Historically, the process of disk formation in rods was thought to result from an invagination of the outer plasma membrane toward the centric face of the connecting cilium. However, in 1980, an alternative evagination model of disk membrane morphogenesis was proposed that could account for disk morphogenesis in both rods and cones. Over the years, experimental evidence has been generated in support of this model. According to this model, as successive evaginations of new disk membrane occur, a second membrane growth phase forms the nascent rims around the newly formed evaginations. In rod cells, rim formation occurs relatively quickly, so that only a small number of disks at the base of the stack remain continuous with the outer plasma membrane. The vast majority of rod disks are discrete units that form a stack, completely surrounded by the plasma membrane. In cone cells, however, the process of rim formation remains incomplete and a connection(s) between the disk and the outer plasma membrane is retained in many, if not all cone disks.

Although there have been some challenges to the evagination model of disk membrane morphogenesis, some key observations provide strong support. First, in contrast to mature rod disks, the evaginating membranes can be labeled with tracer molecules, showing that they are open to the extracellular milieu. Second, the nascent disks contain specific membrane proteins that are not found in mature disks, and they lack the proteins added at the stage of rim formation. This latter observation indicates that there must be a mechanism(s) to sort different groups

of outer-segment proteins prior to disk membrane morphogenesis.

Most cells are responsible for the complete turnover of their lipids and proteins. However, vertebrate photoreceptor cells are unusual in that they have recruited another cell type, the RPE cell, for the catabolic phase of disk membrane turnover. The disposal of the distal outer-segment disks requires an interaction between the photoreceptor and RPE cells. When the photoreceptors are detached from the RPE, the distal disks are not shed, indicating that disk membrane shedding and phagocytosis by the RPE are not independent events (**Figure 4**). In 1976, Matthew LaVail reported that rod disk shedding in rodents followed a daily rhythm, and that a peak of shedding was apparent near the

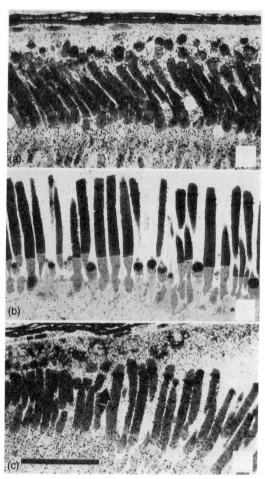

Figure 4 Light micrograph autoradiographs of retinas from frogs injected with radiolabeled amino acids. The band of radiolabel had migrated to the distal disks, and the eyecups were removed and placed in culture for the next shedding phase. Where the retina has remained attached to the RPE (a, and to the left of arrow in (c)), radiolabeled phagosomes are evident in the RPE. Where the retina was detached from the RPE (b, and to the right of arrow in (c)), the band of radiolabel is still evident at the distal end of each outer segment, indicating that the distal disks have not been shed. Scale bar = 50 μm. From Williams DS and Fisher SK (1987) Prevention of rod disk shedding by detachment from the retinal pigment epithelium. *Investigative Ophthalmology and Visual Science* 28: 184–187.

time of light onset. Subsequent studies have shown that the shedding of phototransductive membrane occurs at dawn in other vertebrates, as well as invertebrates. However, cone photoreceptors of most vertebrate species provide an exception to this rule; the peak of cone disk shedding typically occurs shortly after dusk (or light-offset). In crepuscular and nocturnal arthropods, the morphogenesis and shedding phases occur at dusk and dawn, respectively, resulting in different-sized light-absorbing structures (known as rhabdoms) between day and night. In this manner, the turnover of phototransductive membrane is coordinated to increase the efficiency of photon absorbance by individual photoreceptors at night.

The signaling mechanisms involved in the phagocytosis of disk membranes are not well understood; however, the requirement for a number of molecules has been determined. The c-mer proto-oncogene (*Mertk*) gene encodes a receptor tyrosine kinase, which is localized to the apical membrane of the RPE. In the absence of a functional form of this receptor, the ingestion of disk membranes does not occur, as is found in the retina of the RCS rat which carries a mutation in the *Mertk* gene. A role for the $\alpha v \beta 5$ integrin receptor and its ligand, milk fat globule EGF factor 8 (MFG-E8), has also been detected. This receptor appears to be required for the diurnal rhythm of shedding and phagocytosis since, in $\beta 5$ integrin knockout mice, disk membranes are still phagocytosed, but there is no peak of disk shedding at light onset.

Following internalization, the phagosomes are transported out of the apical region of the RPE where they subsequently fuse with lysosomes, and are degraded enzymatically. In mice lacking myosin VIIa, phagosome transport in the cytoplasm is impaired, thus indicating a role for this actin-based motor in the transport process. Microtubule motors also appear to be required in facilitating phagosome–lysosome fusion. A major enzyme in the degradation of opsin is cathepsin D, which is highly concentrated in the RPE lysosomes. Lysosomal enzymes have also been detected in RPE melanosomes, which may have a minor degradative role. The degradation of disk membrane proteins and lipids represents the end of the catabolic phase of the turnover process.

See also: Injury and Repair: Light Damage; Photoreceptor Development: Early Steps/Fate; The Photoreceptor Outer Segment as a Sensory Cilium; Phototransduction: Phototransduction in Cones; Phototransduction: Phototransduction in Rods; Phototransduction: Rhodopsin; Primary Photoreceptor Degenerations: Retinitis Pigmentosa; Rod and Cone Photoreceptor Cells: Inner and Outer Segments; Secondary Photoreceptor Degenerations: Age-Related Macular Degeneration.

Further Reading

Besharse, J. C. (1986). Photosensitive membrane turnover: Differentiated membrane domains and cell–cell interaction. In: Adler, R. and Farber, D. B. (eds.) *The Retina*, Part I, pp. 297–352. New York: Academic Press.

Deretic, D. (2006). A role for rhodopsin in a signal transduction cascade that regulates membrane trafficking and photoreceptor polarity. *Vision Research* 46: 4427–4433.

Insinna, C. and Besharse, J. C. (2008). Intraflagellar transport and the sensory outer segment of vertebrate photoreceptors. *Developmental Dynamics* 237: 1982–1992.

LaVail, M. M. (1976). Rod outer segment disk shedding in rat retina: Relationship to cyclic lighting. *Science* 194: 1071–1074.

Papermaster, D. S., Schneider, B. G., and Besharse, J. C. (1985). Vesicular transport of newly synthesized opsin from the Golgi apparatus toward the rod outer segment. Ultrastructural immunocytochemical and autoradiographic evidence in Xenopus retinas. *Investigative Ophthalmology and Visual Science* 26: 1386–1404.

Steinberg, R. H., Fisher, S. K., and Anderson, D. H. (1980). Disc morphogenesis in vertebrate photoreceptors. *Journal of Comparative Neurology* 190: 501–508.

Williams, D. S. (ed.) (2004). *Photoreceptor Cell Biology and Inherited Retinal Degenerations*, chs. 1, 3–6, and 13–15. Singapore: World Scientific Publishing.

Young, R. W. (1967). The renewal of photoreceptor cell outer segments. *Journal of Cell Biology* 33: 61–72.

Young, R. W. and Bok, D. (1969). Participation of the retinal pigment epithelium in the rod outer segment renewal process. *Journal of Cell Biology* 42: 392–403.

Rod Photoreceptor Cells: Soma and Synapse

R G Smith, University of Pennsylvania, Philadelphia, PA, USA

Glossary

Endocytosis – Active incorporation of external membrane into the cell, used to recover vesicular membrane that has fused with the external membrane at a chemical synapse.

Invagination – A permanent infolding of a cell's external membrane, associated in photoreceptors with their synaptic ribbons, and containing fine dendritic processes of bipolar and horizontal cells.

Mesopic – The 3-log unit range of background illuminance in which rod and cone signals temporally sum in cones and the cone bipolar pathway.

Poikilotherm – A species such as fish, turtles, and frogs whose body temperature varies with the environment.

Refractory period – A period immediately after an event, for example, the release of a vesicle of neurotransmitter, during which the next event cannot occur.

Rhodopsin – The rod pigment molecule that absorbs photons and starts the visual transduction cascade.

Ribbon – A presynaptic structure that collects vesicles of neurotransmitter for release.

Scotopic – The range of background illuminance from starlight to moonlight in which rods absorb one photon or less per integration time (~200 ms).

Spherule – The rod terminal, normally in the shape of a sphere, that contains synaptic ribbons.

Telodendria – Fine axonal processes extending laterally from the base of the cone terminal, which contact neighboring rod and cone terminals.

Introduction

Photoreceptors are the vertebrate retina's primary site for transduction of light into a neural signal. The rod photoreceptor is responsible for vision at night. Rods are essential for vision over the scotopic range (starlight through moonlight) into the mesopic range (twilight) where their signals mix with cone signals. In starlight they must transduce single-photon signals, but in twilight they function more like sensitive cones to temporally integrate photon signals and to adapt before finally saturating in bright daylight. The rod's exquisite combination of signal-processing mechanisms makes it a marvel of signal processing for a dedicated purpose.

Structure

Morphology and Topology

The rod is a specialized neuron consisting of an outer segment, inner segment, soma, axon, and axon terminal (**Figure 1(a)**). The biochemical pathways responsible for transducing light into an electrical signal are contained in the outer segment. The electrical signal passes to the inner segment where it is transformed by voltage-gated channels. The inner segment of the vertebrate rod lies just above the external limiting membrane. The rod soma lies in a variable location in the outer nuclear layer, connected to the inner segment above and the terminal below by the axon. For rods of most species, the soma is larger in diameter (~3–5 μm) than the outer segment. It contains the nucleus which holds the cell's DNA, necessary for development and to maintain the cell's biochemical machinery. Many mammalian species have ~20-fold more rods than cones, so the outer nuclear layer consists mainly of several layers of rod somas. Rods of most species fill the space between the cones, and are closely spaced (~100 000–200 000 mm^{-2}) to capture as many photons as possible.

Axon and Terminal

The rod axon, which carries the electrical signal from the soma to the axon terminal, varies in length depending on the species and the eccentricity (distance from central retina) of the rod. In para-foveal rods near the center of the primate eye, rod axons extend along with cone axons up to 200–400 μm laterally to allow the cone outer segments to be packed tightly together in the fovea and the axon terminals to be given adequate space outside the fovea to make their synaptic connections. In other mammals, the rod axon is shorter; in a cat it is vertical and typically extends ~50 μm through 6–10 layers of rod somas in the outer nuclear layer, and in guinea pig the axon is shorter, typically 10–25 μm, depending on the number of layers of rod somas. Typically, rod axons are ~0.5 μm in diameter, interspersed between rod somas and cone axons of 1–2 μm diameter (**Figure 1(a)**). The mammalian rod axon terminal is typically 2–3 μm in diameter, and is spherical; hence, it is

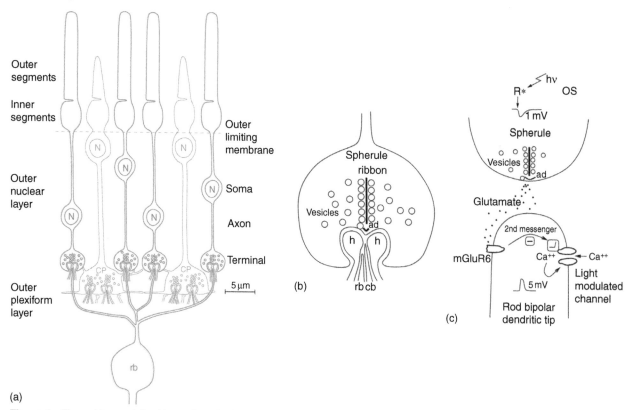

Figure 1 The rod is specialized to capture single photons in starlight but also functions in twilight. (a) Rods (blue-green) in most mammals and in distal primate retina extend vertically interspersed between cones (light red). The rod spherules are electrically coupled to cones at gap junctions on basal processes that emanate from cone terminals. At the bottom, a rod bipolar (green) collects signals from 20 to 100 rods, depending on the species. (b) The rod ribbon synapse sits at the apex of an invagination into the basal surface of the rod spherule. Attached to the ribbon are two rows of vesicles that contain glutamate. The vesicles are released at the active zone near the arciform density (ad). For each ribbon, two horizontal cell processes (h) and one invaginating rod bipolar cell dendrite (rb) extend into the spherule to receive glutamate from vesicles released at the ribbon. One type of OFF-cone bipolar (cb) also invaginates the spherule. (c) A photon absorption ($hv \rightarrow$ R*) in the rod outer segment (OS) generates a 1 mV hyperpolarizing signal, which travels down the axon to the rod spherule, where it slows glutamate release by the ribbon synapse. The glutamate level falls via transporters in the rod spherule's membrane and by diffusion out of the invagination. Glutamate unbinds from the mGluR6 receptors on the rod bipolar dendrite, which through an inverting second-messenger cascade opens the ion channel. The single-photon R* signal passes through two nonlinearities: the high-gain release of glutamate and the threshold nonlinearity in the second-messenger cascade. Calcium entering the rod bipolar's ion channel closes the channel after a delay, generating a ~5 mV response more transient than the rod's signal.

commonly called a spherule. The spherules of adjacent rods are located at different heights, allowing them to fit just above the cone terminals where they can be reached by cone telodendria and fine dendritic processes of bipolar cells.

Synapse

The rod makes chemical synaptic contacts onto one type of rod bipolar cell, one or more types of OFF-cone bipolar cell, and one type of horizontal cell at its axon terminal (**Figure 1(b)**). The synapse releases glutamate via small packets of membrane called synaptic vesicles, which are small (~30 nm) organelles, created by endocytosis from the cell's external membrane and filled with glutamate by transporter proteins. The vesicles diffuse

freely around the cytoplasm while they are being filled. The presynaptic machinery in the terminal contains one or two dense structures called ribbons, because in cross section they are thin and extend vertically away from the cell's membrane, easily seen in electron micrographs or by confocal visualization of specific presynaptic proteins (ribeye, kinesin). The ribbon is thought to be a specialization to allow high release rates, for it collects several rows of vesicles and tethers them for release. The mechanism for vesicle release is complex and contains several dozen proteins, and its details are not yet understood, but it is known to be initiated by calcium ions binding to a receptor protein which causes the vesicle to fuse with the external membrane and thus release its contents into the extracellular space. The ribbon is thought to gather vesicles ready to be docked to provide a larger readily releasable pool. As the rod is depolarized for long periods at night, its ribbon mechanism is specialized to release vesicles at a high rate continuously.

Invagination

The rod's ribbon synapse is located in a way similar to the cone's ribbon, in an extension of extracellular space into the bottom surface of the spherule called an invagination, where very fine processes of horizontal and bipolar cells extend to form postsynaptic specializations. Each invagination contains two ribbons, each of which is presynaptic to two horizontal cell dendrites from different horizontal cells. In addition, each ribbon is presynaptic to one rod bipolar cell, for a divergence from a rod to two rod bipolars. In addition, the rod terminal in some species contacts one or more types of OFF bipolar cells. The function of the invagination is unknown, but it has been suggested to limit diffusion of the neurotransmitter released by the cone or by horizontal cells for negative feedback.

Biophysical Properties

The inner segment and axon terminal of rods are similar to those of cones in that they contain several membrane-bound ion channels, including K_v, K_{Cn}, BK, and L-type Ca^{2+}. The K^+ channels (K_v, K_{Cn}, and BK) of the inner segment and soma are activated by depolarization, providing an outward current to balance the inward dark current through the light-modulated cyclic guanosine monophosphate (cGMP)-gated channels of the outer segment. These channels provide an adaptational influence, opposing the dark signal, and indirectly opposing the light signal by deactivating with hyperpolarization. The K_{Cn} channels in the soma, axon, and terminal underlying the hyperpolarization-activated current (I_h) provide a delayed depolarization when activated by hyperpolarization. The rod terminal's L-type Ca^{2+} channels ($Ca_v 1.4$) uniquely include α 1F subunits and may have special gating properties in conjunction with CaBP4, a calmodulin-like binding protein that shifts the $Ca_v 1.4$ gating activation curve to provide higher gain. In addition, the terminal contains a calcium-sensitive chloride current ($I_{Cl(Ca)}$) which provides signal enhancement because the chloride gradient is depolarizing in rods. However, the resulting chloride efflux is thought to downmodulate the calcium current. Further, the rod terminal contains a calcium-sensitive potassium current (BK) which limits depolarization and calcium entry, and calcium-induced calcium release (CICR) from internal stores which amplifies the internal calcium signal. The rod terminal contains a high-affinity calcium system which includes buffers, plasma membrane calcium ATPase pumps (PMCAs), ryanodine receptors, and inositol triphosphate (IP_3) receptors to modulate calcium release from internal stores. One possible reason for these specializations

is that the calcium channels that trigger vesicle release must be modulated by a tiny single-photon signal and thus the system must have high gain. The rod terminal contains several transporters with special functions. Some of them transport glutamate, including at least two isoforms in the membrane of vesicles to load them with glutamate. A glutamate transporter sitting in the external membrane of the rod terminal is important for uptake of glutamate from the extracellular space.

Gap Junctions

The rod spherules in most vertebrates are electrically coupled by gap junctions to the neighboring cones. The gap junctions are made between the rod spherule and cone telodendria which are basal processes emanating from the cone terminals. This coupling allows rod signals to enter cones at twilight and to be carried through cone pathways to ganglion cells. In addition, rods in some, possibly all, mammals are directly coupled by small gap junctions which may reduce voltage noise in the rod terminal that originates in transduction and membrane ion channels.

Function

The Single-Photon Signal and Noise

The rod is faced with a difficult task. Vertebrate species active in night, such as most mammals, can see in starlight backgrounds when photons are rare and a rod receives a photon only every 20 min. Over the ~3-log unit range of scotopic backgrounds, a rod receives one photon or less per integration time (200 ms) so that its signal is binary. To generate a continuous visual image at such low backgrounds requires that signals from many rods be summed. A typical mammalian ganglion cell sums signals from several thousand rods, and can detect the signal from just one photon in that huge field. This extraordinary feat is a challenge because the single-photon signal is tiny, about 1 mV in amplitude. To detect such a tiny signal in a huge field of rods would be straightforward if all the rods were silent in the dark; but the mechanisms of transduction and synaptic transmission are noisy. The noise from the rods converging to a ganglion cell, if linearly summed with the single-photon signal, would completely mask it, preventing its transmission through the visual pathway. Therefore, the rod pathway has evolved mechanisms to remove the dark continuous noise before it is summed with photon signals.

Sources of Noise

The mammalian rod generates dark continuous noise of thermal origin in its transduction cascade with an amplitude of 10–30% of the peak single-photon signal.

In addition, the rod generates thermal isomerizations of its pigment molecule rhodopsin, called dark thermal events, at a rate equivalent to starlight on a cloudy night. The signals generated by dark thermal events are identical to the single-photon signal, so they appear as real photons from the visual environment and are sometimes referred to as dark light. Another source of noise is variability in the single-photon signal's amplitude. Further, the rod synapse generates robust noise from fluctuation in vesicle release, which masks single-photon events in a way similar to the dark continuous noise. These sources of noise would not be a problem for detecting a single-photon signal from one rod, but if the signals of more than 10 rods were linearly summed, the noise would mask the single-photon signal. To mitigate the problem, synaptic convergence in the rod pathway is accomplished in several stages. The first stage of convergence, from rods to the rod bipolar, is limited to between 20:1 and 100:1 depending on the species, but even this limited amount of convergence would mask the rod signal without special synaptic mechanisms.

Synaptic Transfer Function: High Gain and Temporal Filtering

The rod ribbon synapse is specialized to transmit the rod signal at scotopic backgrounds. It is thought to transmit a binary single-photon signal because, over the range of starlight to moonlight, single-photon events predominate, and when measured with brighter flashes, the variance of the rod bipolar response saturates at the single-photon level. The rod's presynaptic release function (vesicle release as a function of voltage) has very high gain, possibly as high as ~ 1 mV e^{-1}-fold change, allowing the tiny single-photon signal to modulate a large fraction of the dark release rate. Indeed, the rod's calcium channels, high-affinity buffering, internal calcium stores, and PMCA are thought to participate in the high-gain release, to maximize modulation of vesicle release by the 1 mV hyperpolarization of the single-photon signal. In addition, the postsynaptic half of the rod synapse in the rod bipolar dendrite includes several mechanisms specialized for its unique single-photon function (**Figure 1(c)**). It contains a second-messenger cascade that inverts the signal. A photon signal causes a drop in the rod's glutamate release, which reduces glutamate binding to the postsynaptic mGluR6 receptor to deactivate the second-messenger cascade, opening the postsynaptic ion channels and depolarizing the rod bipolar. The second-messenger cascade includes several steps of temporal filtering, which remove the high-frequency components of the dark continuous noise. In some species (e.g., salamander), the synaptic filter in the rod synapse is $\sim$10-fold slower than the corresponding cone synapse. This matches the slower rod transduction response and allows more complete removal of the noise. In mammals, the light-modulated

ion channel in the rod bipolar dendrite is blocked after a short delay by calcium entering the channel. The effect is to generate a transient that limits the duration of the light response in the rod bipolar.

Rate of Vesicle Release

The random release of vesicles by the rod ribbon is a major source of noise for the rod pathway. Although the details of the ribbon's release mechanism are not known, it is stochastic (noisy), similar to a modulated Poisson distribution, for which the standard deviation is equal to the square root of the mean. The rod's ribbon synapse is similar to other ribbon synapses in the cone and bipolar cell, but its challenge is simpler and yet more extreme. The rod's binary signal simplifies the requirements for the synapse; it must only transmit two discriminable levels, signifying the presence or absence of a photon. However, to transmit a discriminable level, the rod must drop its calcium level, in response to a 1 mV hyperpolarization, by a fraction adequate to modulate its random release by more than one standard deviation over the single-photon signal's rise time (100 ms). The rod's rate of vesicle release is thought to be ~ 100 s^{-1}, equivalent to 10 vesicles in 100 ms, with a standard deviation of $\sim$3 vesicles, which would be adequate if the release were modulated by more than 30%, but the 1 mV single-photon signal is thought to modulate the rod's calcium channels by only 20%. This implies either (1) a higher release rate, for example, 250 ves/s, (2) a higher voltage gain for calcium-channel gating, (3) an unknown mechanism to remove vesicle fluctuation noise, or (4) an unknown mechanism for amplifying the single-photon signal so that it can modulate a greater fraction of release. There is some evidence of a refractory period that could generate more regular release. Protons released along with a cone photoreceptor's vesicles bind to the local calcium channels which may generate a short refractory period, allowing release to be more regular. If release in this case could be more regular than a Poisson distribution, it raises the question of why synaptic release is typically found to be Poisson-like.

Synaptic Transfer Function: Nonlinear Threshold

The second-messenger cascade in the rod bipolar dendritic tip contains a nonlinear threshold to process the single-photon signal embodied in the binding of glutamate to the mGluR6 receptor. The rod bipolar's dendritic ion channels do not open unless the rod signal rises above the nonlinear threshold. This mechanism removes much of the dark continuous noise remaining after the cascade's temporal filter. When the peak amplitude of the single-photon signal falls below the nonlinear threshold, the photon signal is lost, resulting in a false-negative, reducing the quantum efficiency of vision. When the peak amplitude of the continuous dark noise rises above the nonlinear threshold, a false-positive event is generated, which may mask individual real photon events, confounding their detection. A high nonlinear threshold will reduce the false-positive rate, but cannot reduce the rate of dark thermal events because they are identical to real photon events. Therefore, the dark thermal rate will mask a low false-positive rate, obviating the need to reduce the false-positive rate to zero. The optimal level for the nonlinear threshold thus is a compromise between false-negative and false-positive rates, and depends on the level of continuous noise, the variability of the single-photon signal, and the thermal event rate. The optimal level for the threshold is suggested by some studies to be in the range of 0.4–0.8 times the peak amplitude of the single-photon signal, which will cause a false-negative rate of ~30–50%. In some studies, the optimal level was suggested to be 1.3 times the peak amplitude of the single-photon signal, thus losing most of the photon signals. The discrepancy between these studies emphasizes that not all the details of the rod synapse are understood. However, it is clear that when signals from 20 rods are summed in a rod bipolar cell, the nonlinear threshold reduces the noise level enough to allow a single-photon signal to be detected downstream in the visual pathway.

Electrical Coupling in Starlight

The electrical coupling between rods not only reduces their voltage noise by averaging their signals, but also reduces the amplitude of the single-photon signal. The amplitude of the single-photon signal is reduced proportionate to the equivalent number of rods, but the noise amplitude is reduced only by the square root of the equivalent number of rods coupled, which produces a net reduction in signal-to-noise ratio at each rod synapse. This type of lateral electrical coupling would not affect performance if the rod bipolar were to sum rod signals linearly, for example, without the nonlinear threshold, but in that case the noise would swamp single-photon signals as described above. The nonlinear synaptic threshold of the rod synapse depends, for its noise-reduction effect, on the amplitude of the single-photon signal, implying that electrical coupling between rods reduces the signal-to-noise ratio of the single-photon signal transmitted to the rod bipolar cell. This paradoxical result suggests that the gap junctions may be modulated by light or by the circadian clock. Alternately, the distribution of each single-photon signal to several nearby rod terminals by electrical coupling may allow vesicle fluctuation noise from rod ribbons to be averaged downstream. The amount of improvement in this case would depend on the gain and degree of nonlinearity and the amount of noise in rod vesicle release.

Electrical Coupling in Twilight

At mesopic backgrounds, rods receive ~1–1000 photons per integration time, so they temporally integrate photon signals and function more like cones except that their response gain is 30-to100-fold higher than the cone signal. As the rod synapse is specialized for single-photon signals and the rod bipolar pathway is specialized for high-gain spatial summation, the robust signals in rods at twilight need an alternate pathway. The electrical coupling between rods and cones allows twilight signals to pass from rods into neighboring cones and thence into cone pathways, where, as the background level increases, the rod component drops as rods saturate or adapt and the rod-cone coupling is downmodulated. In most mammals, the rod–cone coupling is not selective for cone type, so different cone spectral types have a similar rod contribution. This is convenient because opponent color signals computed by subtraction in retinal circuits can then remove the rod contribution. The mixing of the cone's signal with the higher-gain rod signal represents a form of adaptation because it extends by 2–3 log units the background range over which the cone ribbon synapse can transmit a signal. Rods of some species (frog, turtle, and fish) are much larger in diameter than mammalian rods and therefore, when warm, generate more dark thermal events and are more likely to receive multiple photon events at scotopic backgrounds; so in this case rod–rod coupling imitates mesopic cone–cone coupling and is advantageous. However, at low temperatures these poikilotherms (cold-blooded species) have a low dark thermal event rate and can respond to stimuli that evoke single-photon signals.

Negative Feedback

The rod ribbon synapse is similar to the cone ribbon synapse in that two horizontal cell dendritic processes are postsynaptic to each ribbon. The horizontal cell is the type B axon terminal (HBat; H1 in primate) which contacts only rods and is known to carry exclusively rod signals. The HBat is therefore considered to be electrically isolated from its cone-driven soma and dendritic tree. The function of this horizontal cell appendage is thought to be feedback to rods, but little evidence exists for negative feedback at the rod terminal. In salamander, the buffer 4-(2-hydroxyethyl)-1-piperazineethanesulfonic acid (HEPES) when applied in the perfusion bath *in vitro* blocks the shift in the calcium current originating in horizontal cells, suggesting that rod horizontal cell feedback functions in a way similar to cone horizontal cell feedback. The HBat collects from several hundred to several thousand rods, making it a good candidate to collect an average rod signal in starlight to control, through negative feedback, the rods' release of neurotransmitter. One likely possibility is that feedback to rods helps to closely regulate the synaptic threshold for neurotransmitter release, to provide maximal gain for transmitting the tiny single-photon signal. Corroborating evidence for this role exists in recordings from ganglion cells at scotopic backgrounds where the antagonistic receptive field surround has been reported as hidden (summing nonlinearly with the receptive field center) and is only evident when a light stimulus is also applied to the center. This would be expected in the case where negative feedback from the HBat to the rod must pass through the rod synapse's nonlinear threshold to be sensed by the rod bipolar pathway.

Conclusion

The rod functions both at night and in twilight, but is specialized to transduce signals over the scotopic range of illuminance from starlight to moonlight where photons are rare. The single-photon signal is tiny and would be swamped by the dark continuous noise present in a large array of rods, but the rod ribbon synapse contains several mechanisms to reduce the noise and amplify the single-photon signal. Horizontal cells are thought to provide negative feedback at the rod synapse, and may provide a means to regulate the presynaptic release of neurotransmitter to optimize the single-photon signal. The balance between signal and noise in the rod's signal processing mechanisms is delicate. However, the rod's striking ability to capture and transmit single-photon signals suggests that each of its components has a specific function, and that biology has found for each a nearly optimal solution.

Acknowledgment

This work was supported by NEI grant EY016607.

See also: The Physiology of Photoreceptor Synapses and Other Ribbon Synapses; Rod and Cone Photoreceptor Cells: Inner and Outer Segments.

Further Reading

Berntson, A., Smith, R. G., and Taylor, W. R. (2004). Postsynaptic calcium feedback between rods and rod bipolar cells in the mouse retina. *Visual Neuroscience* 21: 913–924.

Berntson, A., Smith, R. G., and Taylor, W. R. (2004). Transmission of single photon signals through a binary synapse in the mammalian retina. *Visual Neuroscience* 21: 693–702.

Field, G. D., Sampath, A. P., and Rieke, F. (2005). Retinal processing near absolute threshold: From behavior to mechanism. *Annual Review of Physiology* 67: 491–514.

Hagins, W. A., Penn, R. D., and Yoshikami, S. (1970). Dark current and photocurrent in retinal rods. *Biophysical Journal* 10: 380–412.

Heidelberger, R., Thoreson, W. B., and Witkovsky, P. (2005). Synaptic transmission at retinal ribbon synapses. *Progress in Retinal and Eye Research* 24: 682–720.

Hsu, A., Tsukamoto, Y., Smith, R. G., and Sterling, P. (1998). Functional architecture of primate cone and rod axons. *Vision Research* 38: 2539–2549.

MacLeish, P. R. and Nurse, C. A. (2007). Ion channel compartments in photoreceptors: Evidence from salamander rods with intact and ablated terminals. *Journal of Neurophysiology* 98: 86–95.

Migdale, K., Herr, S., Klug, K., et al. (2003). Two ribbon synaptic units in rod photoreceptors of macaque, human, and cat. *Journal of Comparative Neurology* 455: 100–112.

Okawa, H. and Sampath, A. P. (2007). Optimization of single-photon response transmission at the rod-to-rod bipolar synapse. *Physiology (Bethesda)* 22: 279–286.

Rodieck, R. W. (1998). *The First Steps in Seeing.* Sunderland, MA: Sinauer Associates.

Smith, R. G., Freed, M. A., and Sterling, P. (1986). Microcircuitry of the dark-adapted cat retina: Functional architecture of the rod–cone network. *Journal of Neuroscience* 6: 3505–3517.

Sterling, P. and Matthews, G. (2005). Structure and function of ribbon synapses. *Trends in Neuroscience* 28: 20–29.

Taylor, W. R. and Smith, R. G. (2004). Transmission of scotopic signals from the rod to rod bipolar cell in the mammalian retina. *Vision Research* 44: 3269–3276.

Thoreson, W. B. (2007). Kinetics of synaptic transmission at ribbon synapses of rods and cones. *Molecular Neurobiology* 36: 205–223.

van Rossum, M. C. W. and Smith, R. G. (1998). Noise removal at the rod synapse. *Visual Neuroscience* 15: 809–821.

The Role of the Vitreous in Macular Hole Formation

W E Smiddy, Bascom Palmer Eye Institute, Miami, FL, USA

Glossary

Henle's nerve fiber layer – The inner layer of cones at the fovea lacking overlying nerve fiber or inner nuclear layer cells that have the most attenuated internal limiting membrane covering; could be the most susceptible location for a breach in that layer that might initiate macular hole formation.

Internal limiting membrane – The confluence of inner footplates of Müller cells of retina that forms a membrane-like structure delimiting inner retinal surface where it contacts the posterior hyaloidal elements.

Macula – The retina that subserves central vision, populated by the most dense distribution of cone photoreceptors to offer the best visual resolution; this term is somewhat imprecisely used among clinicians (broader, perhaps 2 disk diameter zone) and histologists (more narrowly used to central point).

Posterior hyaloid – The most posterior elements of the vitreous body; this structure is more complex than previously thought, frequently exists as a multilaminated structure, and composed primarily of type IV collagen, but may be impregnated with cellular components.

Vitreoschisis – A phenomenon that is usually not visible, but could be important pathogenically whereby the vitreous body develops a splitting of some of its layers, most notably just anterior to the posterior hyaloidal attachment to the macula.

The pathogenesis of macular holes has undergone at least one full cycle of thought. Since macular holes were first identified following trauma, it was natural for early clinicians to deduce that a coup–countrecoup force transmitted through the vitreous caused macular hole formation. However, as it became apparent that trauma was only infrequently associated with the finding of a macular hole, other mechanisms were proposed as first summarized by Aaberg. Three principal, possibly coexisting, enhanced mechanisms have been proposed including unroofing or dissolution of the attenuated inner retinal layer of cystoid edema due to a range of etiologies, atrophy or degeneration of the retina, and vitreomacular traction. Advances in imaging have allowed unprecedented resolution which has ratified vitreomacular traction as the most prominent component, if not the basic underlying mechanism, of macular hole formation.

Undeniably, the vitreoretinal interface is the battlefront of macular hole formation and therapeutics. Some observations do not fully conform to the model of vitreomacular traction as the sole and universal cause of macular hole formation, but these may be consequences of vitreomacular traction or factors contributing to or potentiating vitreous traction. Thus, it is possible that the vitreous may play a prime-mover role, an innocent bystander role, or a collaborative role in the formation of macular holes. This article reviews the evidence and perspective of the role of the vitreous in macular hole formation.

Vitreous Anatomy and Biochemistry

The vitreous body has a much more complex structure and physiology than might initially be apparent from its transparent, paucicellular, predominantly aqueous appearance. It is a nonhomogeneous biochemical complex of collagen and hyaluronan. The physical structure of the vitreous is an extensive cisternal system of cavitated, more liquefied portions lined by extenuated and thickened walls. This seems to be accentuated with age (syneresis) or in certain disease states. Most specifically, the concept of a posterior precortical vitreous pocket and, consequently, a cortical vitreous layer has been defined and to a large degree imaged which, by its proximity to the fovea, mostly likely plays some role in the genesis of vitreous forces at the fovea. Improved optical visualization of the fine structure of the vitreous has allowed clinical detection of this pocket, and its size has been hypothesized to be proportional to the amount of tangential traction generated at the fovea.

The vitreous dual senescent processes of synchesis and syneresis render a posterior vitreous detachment (PVD) to be a much more complex process than the casual observer might suspect.

Vitreous Traction

Peripheral retinal breaks form at the time of a PVD. Thus, it was a natural step to conceive of the PVD as playing a formative role in causing macular holes. Gass' seminal grading scheme standardized observations of the stages involved in macular hole formation and it remains generally accepted. Its clinical relevance was immediately seized upon and preventative surgery aimed at inducing the posterior vitreous separation before a macular break could occur led to several surgical investigations that seemed to offer promising results initially, but a randomized trial failed to establish the efficacy of this preemptive strategy. The surgical strategy employed at the time was to remove the cortical vitreous, which itself may reflect an incomplete or overly simplified understanding of the pre-hole dynamics.

Various clinical studies identified a lack of PVD to be associated with a higher risk of macular hole formation in the fellow eyes of patients with macular holes compared to fellow eyes with incident posterior vitreous separations. Histopathologic series not only established that posterior vitreous separation was more common among patients with full-thickness macular holes, but also made the observation that persistent attachments of the vitreous and associated cystic changes may imply a more complex mechanism than simple posterior vitreous separation.

The earliest imaging studies included ultrasound tests which also seem to establish the strong association of vitreous separation with macular hole formation. Subsequently, imaging with optical coherence tomography (OCT) depicted a compelling role for vitreous traction.

However, other observations have suggested that while traction at the time of posterior vitreous separation may well be important, it may be part of a more complex scenario. The fellow eyes of patients develop macular holes much more frequently than in the general populations, suggesting a constitutional susceptibility to macular hole formation, presumably at the time of posterior vitreous separation.

Vitreoretinal Interface

The second iteration of the vitreous separation at the fovea has focused on the nature and actions that occur at the vitreofoveal interface. Formerly, a vitreous separation was simplistically thought of as a clean, total separation of the hyaloid from the internal limiting membrane (ILM). However, the concept of an anomalous posterior vitreous separation at the fovea is what is probably most consistent with both the clinical behavior and the imaging appearances. Vitreous cortex remnants have been identified at the fovea and perifoveal area after apparent spontaneous posterior vitreous separation. In this way, the vitreous may be a necessary substrate for action at the fovea. Histopathologic studies have also suggested or have identified numerous other possible substrates that might mediate traction, but more of a delayed or tangential nature than the expected anterior-to-posterior traction of the simplified posterior vitreous separation. Anatomic features at the fovea are unique and may also account for what are

incidental vitreoretinal forces elsewhere. For example, the fovea is thin centrally and the ILM as well as the basal lamina of the vitreous are extremely thin at the fovea. At the fovea, the outer plexiform layer and the photoreceptor layer (Henle's nerve fiber layer) are more exposed and the ILM is even more attenuated in this area. The tangential orientation of the fibers, their delicate nature, and the thin overlying ILM may make this layer more susceptible to incidental trauma at the time of vitreous separation at the fovea.

These anatomic and theoretic features seem to be consistent with other clinical observations, including the apparent predecessor lesion as a cyst. Thus, the question of the pathogenic role of the vitreous becomes open to its possible chronic role in inducing an inner cystic change, or its subsequent action on the weakened inner cystic layers to precipitate the full-thickness hole. Careful biomicroscopic study of the vitreous at the fovea can demonstrate vitreous separation or, more importantly, a partial vitreous separation at the fovea. Other imaging modalities besides OCT, including laser biomicroscopy kinetic ultrasound and retinal thickness analysis, also demonstrate these findings. The most convincing evidence for this role comes from the more newly advanced and available optical coherence tomography images which have corroborated the apparent impact of the vitreous at the fovea (**Figure 1**). This concept is demonstrated by cases with spontaneous closure of a macular hole after release of the persistent vitreofoveal attachment, whether in traumatically induced or in idiopathic cases. Applying this observation therapeutically, others have performed vitrectomy relieving the traction with peeling of the posterior hyaloid, vitrectomy with removal of the ILM, or even vitreolysis of its persistent attachment without a larger vitrectomy with resolution of a macular hole as well. However, the indication for intervention is not well established, and occasionally, the apparent vitreofoveal traction will spontaneously release (**Figure 2**).

Numerous case reports of macular hole formation associated with seemingly unrelated diseases share themes of vitreous traction in the setting of intravitreal injection, fungal endophthalmitis, and laser-assisted *in situ* keratomileusis (LASIK) associated with formation of full-thickness macular holes. A corollary of the role of the vitreous attachment at the fovea is that other case reports have suggested a sort of accomplice role that the vitreous may play in transmitting tractional effects from seemingly unassociated conditions as they burst as hemorrhage from a retinal microaneurysm, LASIK, transpupillary thermal therapy, yttrium aluminum garnet (YAG) laser injury, and subhyaloid hemorrhage. The possible contractile-potent quality of tissue in the vitreous is demonstrated by the ultrastructural analyses of removed tissue specimen from patients with impending macular holes which

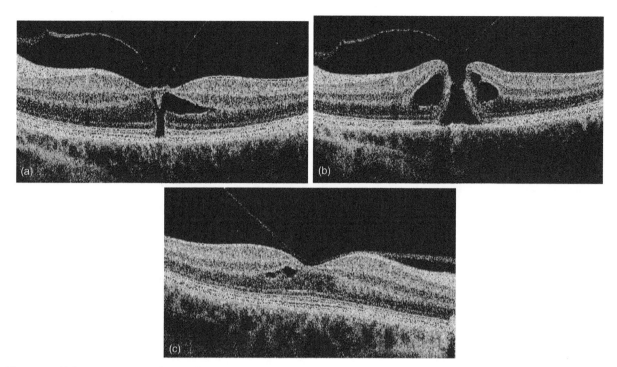

Figure 1 (a) Spectral domain OCT of right eye of a 60-year-old female with a visual acuity of 20/40 and a 2-week history of central visual disturbance. There is vitreous adherence at the fovea with associated cystic changes. (b) Spectral domain OCT 2 weeks later; the visual acuity has decreased to 20/70 with completion of a macular hole. (c) Fellow eye showing normal foveal contour with mild cystic change and with vitreous adherence at the fovea.

demonstrate glial-type tissue that might cause such a secondary tractional effect.

Confounding Observations

However, some macular holes have been observed under circumstances that do not seem to follow the mold of a direct or secondary, tractionally induced macular hole formation via active vitreous traction. These include macular hole cases that seemed to have been observed distantly after a documented short-term or long-term PVD, following the repair of retinal detachment repair or even after a vitrectomy for seemingly unrelated disorders. In addition, macular holes have been found to reopen clearly in the absence of any vitreous which would have been systematically removed at the time of previous surgery.

These observations have caused some to consider other possible mechanisms that may cause macular hole formation or may influence the vitreous role in macular hole formation. A leading mechanism besides vitreous traction as a primary element is the possible role of cystoid foveal changes. The finding of cystoid changes may represent a confounding of previously, undetected vitreous traction that might have been released, but may play a prominent role in the circumstances apparently leading to the macular hole by compromising the integrity of the already-delicate foveal elements. Unroofing of cystoid changes has been implicated in a variety of again, seemingly unrelated associations of macular hole formation with conditions including idiopathic juxtafoveal telangiectasis following uncomplicated phacoemulsification surgery and the presumed pseudophakic cystoid edema, macular edema for a branch vein occlusion, eyes with retinitis pigmentosa, and eyes as part of the Alport syndrome. The reopening of macular holes has been demonstrated to be preempted by some cystoid changes that seem to have no relationship to the vitreous since the vitreous was previously removed. Another case report of a formation of a macular hole after an OCT documented release of the posterior cortical remnant had been demonstrated a year and a half previously further brings into question the tractional induced role. The cystic changes observed might possibly be secondary to small breaks in the ILM or Henle's nerve fiber layer, which subsequently lead to hydration with secondary stiffening of the retina and extension of a separation into the deeper layers of the retina.

There are numerous elements that have been delineated above that may persist in the foveal area that could have contractile potential and lead to the separation of the foveal layers.

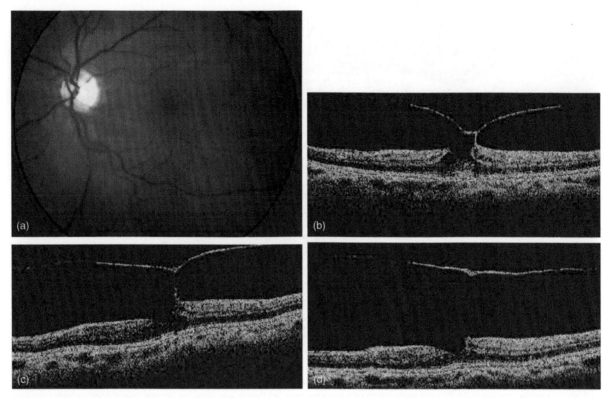

Figure 2   (a) Fundus appearance of left eye of a 70-year-old female with a visual acuity of 20/50 and a few months' history of an apparently nonprogressive central visual disturbance. (b) Time domain OCT shows vitreofoveal traction and mild cystic changes. (c) The clinical appearance and symptoms are unchanged 6 months later, but the OCT shows a more elongated vitreous attachment at the fovea. (d) There was a mild clinical improvement 3 months later, with visual acuity at 20/40, and the OCT appearance depicts spontaneous release of the vitreofoveal attachment.

Summary

The pathogenesis of macular holes has been widely reviewed. Undoubtedly, the vitreous plays a pivotal role in the formation of most, if not all, macular holes. Its actual role could vary from being the prime initiator and executor of traction at the fovea to being a sort of accomplice by transmitting traction generated elsewhere to being an innocent bystander to some other anatomic circumstances.

Cystoid changes seem to occur at least after a macular hole has occurred and very possibly before a macular hole is actually apparent. This could be because of hydration from a discontinuity in the Henle's nerve fiber layer or ILM break, or it could be a reflection of some other biochemical process that is occurring, possibly as a response to persistent vitreous traction. The formation of a macular hole in at least many cases is probably not simply from the vitreous separating, but rather from some abnormal features that are consummated by otherwise inconsequential persistent vitreous traction.

See also: Acuity; Adaptive Optics; Histogenesis: Cell Fate: Signaling Factors.

Further Reading

Aaberg, T. M. (1970). Macular holes. A review. *Survey of Ophthalmology* 15: 139–162.

de Bustros, S. (1994). Vitrectomy for prevention of macular holes. Results of a randomized multicenter clinical trial. Vitrectomy for Prevention of Macular Hole Study Group. *Ophthalmology* 101: 1055–1059 discussion 1060.

Ezra, E., Wells, J. A., Gray, R. H., et al. (1998). Incidence of idiopathic full-thickness macular holes in fellow eyes: A 5-year prospective natural history study. *Ophthalmology* 105: 353–359.

Gass, J. D. M. (1988). Idiopathic senile macular hole: Its early stages and pathogenesis. *Archives of Ophthalmology* 106: 629–639.

Gass, J. D. M. (1995). Reappraisal of biomicroscopic classification of stages of development of a macular hole. *American Journal of Ophthalmology* 119: 752–759.

Green, W. R. (2006). The macular hole. Histopathologic studies. *Archives of Ophthalmology* 124: 317–321.

Hee, M. R., Puliafito, C. A., Wong, C., et al. (1995). Optical coherence tomography of macular holes. *Ophthalmology* 102: 748–756.

Kiry, J., Ogura, Y., Shahidi, M., et al. (1993). Enhanced visualization of vitreoretinal interface by laser biomicroscopy. *Ophthalmology* 100: 1040–1043.

Kishi, S., Demaria, C., and Shimizu, K. (1986). Vitreous cortex remnants at the fovea after spontaneous vitreous detachment. *International Ophthalmology* 9: 253–260.

Lipham, W. J. and Smiddy, W. E. (1997). Idiopathic macular hole following vitrectomy: Implications for pathogenesis. *Ophthalmic Surgery and Lasers* 28: 633–639.

McDonnell, P. J., Fine, S. L., and Hillis, A. I. (1982). Clinical features of idiopathic macular cysts and holes. *American Journal of Ophthalmology* 93: 777–786.

Sebag, J. (1998). Macromolecular structure of vitreous. *Progress in Polymer Science* 23: 415–446.

Sebag, J. (2004). Anomalous posterior vitreous detachment: A unifying concept in vitreo-retinal disease. *Graefe's Archive for Clinical and Experimental Ophthalmology* 242: 690–698.

Smiddy, W. E. and Flynn, H. W., Jr. (2004). Pathogenesis of macular holes and therapeutic implications. *American Journal of Ophthalmology* 137: 525–537.

Smiddy, W. E., Michels, R. G., Glaser, B. M., and deBustros, S. (1988). Vitrectomy for impending idiopathic macular holes. *American Journal of Ophthalmology* 105: 371–376.

Worst, J. G. F. (1977). Cisternal systems of the fully developed vitreous body in the young adult. *Transactions of the Ophthalmological Societies of the United Kingdom* 97: 550–554.

Secondary Photoreceptor Degenerations: Age-Related Macular Degeneration

L V Johnson, University of California, Santa Barbara, CA, USA

Glossary

Drusen – The abnormal deposits containing cellular debris and inflammatory molecules that form beneath the retinal pigmented epithelium (RPE) and are associated with the development of age-related macular degeneration.

Macula – The portion of the retina that is responsible for fine acuity vision in the central visual field. It is a small (~6 mm diameter) area located at the posterior pole of the eye near the optic nerve that is structurally unique and contains the highest concentration of cone photoreceptor cells in the retina.

Photoreceptor – The cells of the retina responsible for the absorption of light and its conversion to electrical signals that are transmitted to the brain. In humans, there are two types of photoreceptor cells: rods and cones. Rod photoreceptor cells are very sensitive in dim light situations but do not provide high resolution or color vision. Cone photoreceptor cells are less sensitive than rods but in bright light distinguish colors and provide for high-acuity vision owing to their concentration in the macula.

Retina – The light-sensitive part of the eye that lines its inner surface. It is a multilayered tissue that collects light, processes electronic signals produced by the light, and transmits those signals to the brain.

Retinal pigmented epithelium(RPE) – A pigmented monolayer of cells lying directly adjacent to the retina that provides support for the photoreceptor cells. Its functions include absorption of excess light; processing of molecules required for transduction of light into electrical signals; removal of cellular debris shed by photoreceptor cells; and transport of nutrients, waste products, and ions.

Introduction

Age-related macular degeneration (AMD) is a disease of the eye that causes loss of vision in the center of the visual field. Central vision is mediated by a region of the retina, known as the macula that is specialized for fine acuity vision such as that used for reading and other tasks that require fine focus and high resolution. While the macula comprises only 4% of the retina (it is only 6 mm in diameter), it is responsible for essentially all scotopic (bright light) vision and almost 10% of our visual field. In AMD, the light-sensitive photoreceptor cells in this critical region of the retina become dysfunctional and die resulting in symptoms ranging from slight visual distortions to complete loss of central vision.

Incidence

As the name implies, AMD is most common in the elderly, its incidence being highest in those over 50 years of age. It is the most common cause of irreversible blindness in elderly individuals worldwide and its clinical symptoms are recognized in more than one-third of persons over the age of 75 in industrialized societies. In 2004, it was estimated that over 9 million persons were affected by some form of AMD in the United States alone. It is expected that this number will reach almost 15 million by the year 2020. There are other diseases that affect the macula in younger individuals, but the frequencies of these are significantly less than for AMD and many of these are known to be monogenic diseases, that is, caused by single gene mutations (e.g., vitelliform macular dystrophy, Sorsby's fundus dystrophy, and malattia leventinese).

The Macula

It is not yet clear why the macula is preferentially affected in AMD. The term macula is derived from the Latin *macula lutea* (yellow spot) that describes the appearance of the macula in life due to the accumulation of two carotenoid pigments, lutein and zeaxanthin, in the macular retina. Lutein and zeaxanthin are thought to function as antioxidants and filters of high-energy blue light, and thus protect the macula from light-induced oxidative damage. The fovea is the centermost 2 mm of the macula and is the region of the retina where cone photoreceptor density is highest (up to 200 000 cells mm^2) and visual acuity is maximal. One factor that may contribute to macular sensitivity is its relative lack of vasculature compared to the rest of the retina. Despite the fact that photoreceptor cell density is highest in this region of the retina and photoreceptor cells consume oxygen at a higher rate than any other cells in the body, the inner retina here is thinned and avascular. As a consequence, macular photoreceptors appear to rely on the choroidal capillary bed, the choriocapillaris, as their primary source for oxygen and nutrients. Age-related decreases in choroidal vascular volume and flow could thus compromise macular photoreceptor cells. Age-related changes in Bruch's membrane that are more pronounced in the macular region than elsewhere in the eye may also contribute to the macula's particular susceptibility to pathogenic changes leading to AMD.

Clinical Symptoms

AMD is typically classified into early and late forms. Early AMD (dry or atrophic AMD) is characterized by the appearance of abnormal extracellular deposits known as drusen that form below the retinal pigmented epithelium (RPE) and by focal areas of increased or decreased pigmentation in the RPE and choroid. Drusen are classified as hard (small, distinct, and hemispherical) or soft (large, diffuse, and amorphous) based on their fundoscopic and histologic appearance. Late AMD has two forms: neovascular AMD (wet or exudative AMD) and

geographic atrophy, the end stage of dry AMD when neovascularization does not occur (**Figure 1**).

Neovascular AMD is typified by the growth of blood vessels from the choroid into the RPE and the subretinal space (choroidal neovascularization or CNV) and subretinal hemorrhage that cause severe damage to the retina, leading to precipitous vision loss. Physicians often use a procedure known as fluorescein angiography to assess the extent of CNV in patients suspected of having neovascular AMD. If untreated, neovascular AMD can lead to the development of a disk-shaped fibrovascular scar underlying the macula and severe loss of central vision (central scotoma). Individuals who develop neovascular AMD in one eye are likely to have it develop in the other (contralateral) eye and thus the risk of bilateral vision loss is high in neovascular AMD. Geographic atrophy is characterized by widespread areas of depigmentation that represent areas of degeneration of the RPE and adjacent photoreceptor cells of the retina. Atrophy of the capillary bed (choriocapillaris) of the choroid is also commonly associated with geographic atrophy. Ultimately, these areas coalesce to form a large distinct area of degeneration that encompasses the macula and causes loss of central vision.

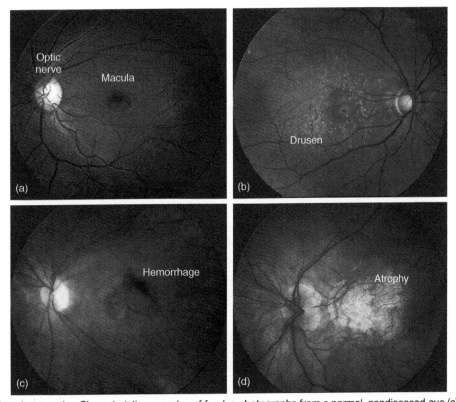

Figure 1 Fundus photographs. Characteristic examples of fundus photographs from a normal, nondiseased eye (a) and eyes with dry AMD (b), wet AMD (c), and geographic atrophy (d). In (b) the hard drusen (numerous small spots) that are characteristic of dry AMD can be seen surrounding the macula. In (c) an area of bleeding (hemorrhage) is present near the macula of an eye with wet AMD. In (d) a large area of atrophy encompassing the entire macular region can be seen in an eye with geographic atrophy. Photographs: courtesy of Dr. Robert Avery.

In early, dry AMD, vision loss is gradual and progresses over many years. As such, individuals with early AMD are often unaware of their visual dysfunction. More rapid and severe vision loss is seen in individuals where early AMD progresses to geographic atrophy or neovascular AMD; this occurs in 10–15% of individuals in each case.

Histopathology

Numerous studies show that the earliest pathologic changes associated with AMD occur in the RPE and the adjacent extracellular matrix complex known as Bruch's membrane, which undergoes a substantial number of changes in association with aging and AMD. These include thickening, accumulation of lipids and cholesterol, decreased permeability, and development of deposits of extracellular debris between Bruch's membrane and the RPE. The most prominent of these deposits are known as drusen (German for geodes, because of their glittering appearance when first described) (**Figure 2**).

Large numbers and extensive areas of drusen deposits in the macula (especially soft drusen) are potent risk factors for the development of AMD. It is thought that drusen diminish access to vital nutrients and oxygen diffusing from the choroidal vessels, leading to the compromise of overlying RPE cells and secondarily to the dysfunction and death of the photoreceptor cells in the adjacent neural retina.

The process of drusen formation is poorly characterized. Drusen are comprised of protein and lipid molecules, many of which are known to be involved in inflammatory processes. These include activators of the complement system, activated complement components, and complement regulatory molecules. As such, it has been proposed that drusen formation is the byproduct of chronic local inflammatory processes. Many drusen components are normal constituents of circulating plasma

and it has been suggested that reduced choroidal blood flow associated with AMD leads to increased hydrostatic pressure in the choroidal capillaries forcing plasma proteins extravascularly into the sub-RPE space. However, some drusen-associated molecules are known to be biosynthetic products of RPE cells and cellular debris, likely of RPE origin, is frequently observed in drusen. It is thus likely that RPE-derived molecules also contribute to drusen biogenesis; however, causative conditions and precise mechanisms remain to be identified. It has also been noted that some molecular constituents of drusen have been oxidatively modified, consistent with the idea that oxidative damage may be an important component of drusen formation and AMD pathogenesis.

Risk Factors

The two most significant risk factors for AMD are age and heredity. As noted above, the disease is most commonly diagnosed in individuals over 50 years of age. There is a significant increase in AMD risk if close family members are afflicted with the disease. Individuals who have a first-degree relative (parent or sibling) with the disease are more than twice as likely to be afflicted as those without afflicted relatives. They also tend to be affected at an earlier age. These results indicate that specific genetic attributes are likely to be involved in influencing an individual's susceptibility to AMD (see the section titled 'Genetics').

Of environmental risk factors, smoking is the most significant modifiable risk factor for AMD. One study indicates that approximately 25% of AMD cases in individuals over 70 may be attributable to risk conferred by smoking. The preponderance of epidemiological data shows that smokers have a two- to threefold increased risk of having late AMD, neovascular AMD, or geographic atrophy, compared to individuals who have never smoked. There is a dose–response effect in that the risk of

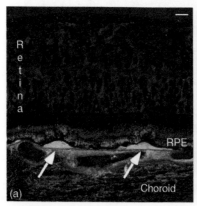
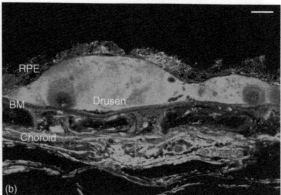

Figure 2 Drusen histology. (a) Two large drusen (green, arrows) are seen lying between the retinal pigmented epithelium (RPE, blue) and the choroid. The adjacent retina is stained red. Scale = 10 μm. (b) At higher magnification, two adjacent drusen (green) are seen displacing the RPE (red) from Bruch's membrane (BM). Scale = 5 μm.

developing AMD increases in relation to the duration and intensity of smoking history. It also appears that the average age of onset of AMD symptoms is earlier in smokers compared to nonsmokers, as is the likelihood of disease in both eyes. Stopping smoking, however, can have beneficial effects. While ex-smokers still have an increased AMD risk compared to never-smokers, it is significantly lower than for current smokers. It has been suggested that the increased AMD risk conferred by smoking is related to oxidative damage in the retina, induction of proinflammatory mediators, reduced choroidal blood flow, and decreased levels of protective carotenoid pigments (lutein, zeaxanthin) in the macular retina.

Obesity (or elevated body mass index) has also been linked to an increased risk of progression to late-stage AMD. It has been proposed that this relationship is due to increased oxidative stress, reduced antioxidant defense mechanisms, and/or increased incidence of chronic low-grade inflammation in obese individuals. Low-grade systemic inflammation may contribute to AMD pathogenesis directly through inflammatory damage at the level of the RPE and Bruch's membrane, or secondarily by reducing systemic availability of the macular pigments, lutein and zeaxanthin (increased body fat is associated with decreased serum and macular levels of carotenoids).

Reduced progression of early AMD and reduced risk for neovascular AMD is associated with dietary supplementation with antioxidant vitamins and minerals (the Age-Related Eye Disease Study (AREDS) supplement: 500 mg vitamin C, 50 mg vitamin E, 15 mg beta carotene, 80 mg zinc, and 2 mg copper daily). Diets high in foods that are good sources of the macular carotenoid pigments (e.g., certain green leafy vegetables) and omega-3 fatty acids (e.g., fatty fish and flaxseed) are also associated with decreased incidence of AMD. Interestingly, some epidemiological studies have noted a decreased risk of AMD in individuals who regularly use antacids or anti-inflammatory medications.

Genetics

Genes in two chromosomal regions, one on chromosome 1 (1q32) and one on chromosome 10 (10q26), have now been linked to a significant proportion of AMD cases. In both cases, certain single-base differences in DNA sequence (single-nucleotide polymorphisms or SNPs) in specific genes have been shown to be present at higher frequency in individuals with AMD than in those not afflicted with the disease. Together, genetic variants in these two chromosomal loci confer significant risk for AMD and can account for over half of AMD cases. The implicated gene on chromosome 1 is that for complement factor H (*CFH*), which lies within an area of the chromosome known as the regulator of complement activation (RCA) locus that is known to be involved in modulating activity of the complement system. The involvement of the complement system in AMD that is implied by these observations is consistent with pathological evidence suggesting that chronic inflammation has a primary role in the AMD disease process (see the section titled 'Histopathology'). The AMD-associated genes on chromosome 10 are tightly linked and include *PLEKHA1, ARMS2 (LOC387715)*, and *HTRA1*. The functions of the proteins encoded by these genes and their potential relationship to AMD pathogenesis are less well defined than for *CFH*. *HTRA1* encodes a serine protease and *PLEKHA1* a plekstrin-homology-domain-containing protein; the product of the *ARMS2* gene has not been completely characterized but may be a mitochondrial protein. Polymorphisms in the genes in these two chromosomal loci, especially *CFH* and *ARMS2/HTRA1* (see below), can account for a substantial number of AMD cases. Several additional genes, including those encoding complement component C3, complement factor B, the adenosine triphosphate (ATP)-binding cassette transporter (ABCA4), fibulin-5, apolipoprotein E, human leukocyte antigens, and others have been associated with smaller percentages of AMD cases.

CFH

A SNP (rs1061170) in the gene for *CFH* has been shown in numerous studies to be associated with increased risk for both forms of late AMD. The *CFH* gene is located on human chromosome 1 (1q32) in an area often linked to AMD in familial and population-based studies. The risk-conferring polymorphism in *CFH* results in a thymidine (T) to cytosine (C) substitution at nucleotide position 1277 in exon 9 of the gene. Individuals carrying one copy of the C allele (heterozygotes) are 2–5 times more likely to be afflicted with AMD and carriers of two copies (homozygotes) are 3–7 times more likely to be afflicted. The T → C nucleotide change leads to a tyrosine to histidine shift at amino acid position 402 (Y402H) of the CFH protein. At the time of writing, the functional implications of this amino acid change have not been fully characterized but the affected site in the protein is one that is known to contribute to a number of CFH functions, including the binding of heparin and C-reactive protein. *CFH* is an important negative regulator of the alternative pathway of complement activation and it is hypothesized that the 402H variant of the molecule may have reduced inhibitory activity. Compromised regulation of the complement system could lead to inflammation and complement-mediated damage to the RPE, and secondary damage to the retina. Combined analyses of polymorphisms in *CFH* and genes encoding some other members of the complement system (complement factor

B, complement component 2) indicate that their AMD risk-conferring variants can together account for nearly 75% of AMD cases. There is additional evidence that environmental risk factors, such as smoking and obesity, can combine with genetic risk to increase AMD susceptibility even further. For example, the risk of AMD is 8–10 times higher in individuals who are both homozygous for the risk-conferring C allele in the *CFH* gene and smoke. Additional variants in the *CFH* and related genes have also been linked to AMD risk. For example, deletion of two *CFH*-related genes (*CFHR1* and *CFHR3*) has been shown to have protective effects and decrease the likelihood of AMD.

ARMS2/HTRA1

Both the *ARMS2* and *HTRA1* genes contain polymorphisms that have been associated with increased risk for AMD and they exhibit a high degree of linkage disequilibrium (i.e., they are essentially always transmitted together). Because of this latter property it has been difficult to determine which of the polymorphisms is responsible for the associated AMD risk. In addition, biological evidence supporting a role for both has been presented. One *ARMS2* polymorphism (rs10490924) leads to a change in a region of the gene that is predicted to result in an amino acid change from alanine to serine at position 69 (A69S) of the ARMS2 protein. Some data suggest that the ARMS2 protein is located in mitochondria and that it is involved in modulating oxidative stress and associated apoptosis, but there is no direct evidence yet linking ARMS2 protein dysfunction and the AMD disease process. An additional *ARMS2* polymorphism that involves both deletion and insertion of genetic material has been shown to lead to messenger RNA (mRNA) instability and marked reduction in the levels of *ARMS2* transcripts, and is significantly associated with AMD risk. ARMS2 protein is highly expressed in the retina and it has been proposed that dysfunctional ARMS2 and/or reduced levels of ARMS2 in mitochondria-rich retinal photoreceptor cells may enhance susceptibility to age-associated mitochondrial dysfunction that leads to photoreceptor compromise and visual loss.

The AMD-linked polymorphism in the *HTRA1* gene (rs11200638) lies in the gene's promoter region and some data have suggested that the risk-conferring allele leads to altered binding of transcription factors, increased transcription of the gene, and elevated HTRA1 protein levels, but other studies have not confirmed this observation.

Treatments

Most therapeutic approaches for AMD have historically focused on the neovascular form of the disease, as it is the one which is responsible for the most severe and precipitous vision loss. Therapies for neovascular AMD have included thermal laser photocoagulation of leaky neovessels and photodynamic therapy that utilizes laser activation of an intravenously administered compound that damages the endothelial cells of neovessels, leading to thrombosis. These approaches were applicable to less than half the patients with neovascular AMD and thus, to less than 5% of all AMD patients. More recently, molecular inhibitors of vascular growth factors (primarily vascular endothelial growth factor, VEGF) have been exploited to limit neovascularization in a variety of ocular neovascular processes. Such VEGF inhibitors are generally administered in multiple intravitreal injections and act by inhibiting the interaction of VEGF with its receptor on the surfaces of vascular endothelial cells. The binding of VEGF to its receptor promotes endothelial cell proliferation, vessel growth, and increased vascular permeability, leading to the development of CNV in affected eyes. The most striking results to date have been provided by antibody-based compounds (Lucentis and Avastin) that are directed against VEGF and inhibit its function. These compounds have provided unprecedented benefits in terms of visual stabilization in most patients with CNV and visual improvement in many. It is anticipated that advances in the specificity, efficacy, durability, and deliverability of anti-VEGF compounds, as well as compounds directed against additional vascular growth factors, will continue to advance the physician's ability to combat neovascular AMD. However, similar therapeutic advances have not been made for the dry or atrophic forms of AMD that represent 85–90% of disease cases.

See also: Developmental Anatomy of the Retinal and Choroidal Vasculature; Physiological Anatomy of the Retinal Vasculature; Primary Photoreceptor Degenerations: Retinitis Pigmentosa; Primary Photoreceptor Degenerations: Terminology; Retinal Pigment Epithelium: Cytokine Modulation of Epithelial Physiology; Secondary Photoreceptor Degenerations.

Further Reading

Baird, P. N., Robman, L. D., Richardson, A. J., et al. (2008). Gene–environment interactions in progression of AMD – the CFH gene, smoking and exposure to chronic infection. *Human Molecular Genetics* 17: 1299–1305.

Berger, J. W., Fine, S. L., and Maguire, M. G. (1999). *Age-Related Macular Degeneration*. St. Louis, MS: Mosby.

Chong, E. W., Kreis, A. J., Wong, T. Y., Simpson, J. A., and Guymer, R. H. (2008). Dietary omega-3 fatty acid and fish intake in the primary prevention of age-related macular degeneration: A systematic review and meta-analysis. *Archives of Ophthalmology* 126: 826–833.

Cong, R., Zhou, B., Sun, Q., Gu, H., Tang, N., and Wang, B. (2008). Smoking and the risk of age-related macular degeneration: A meta-analysis. *Annals of Epidemiology* 18: 647–656.

Gehrs, K. M., Anderson, D. H., Johnson, L. V., and Hageman, G. S. (2006). Age-related macular degeneration – emerging pathogenic and therapeutic concepts. *Annals of Medicine* 38: 450–471.

Grisanti, S. and Tatar, O. (2008). The role of vascular endothelial growth factor and other endogenous interplayers in age-related macular degeneration. *Progress in Retinal and Eye Research* 27: 372–390.

Jager, R. D., Mieler, W. F., and Miller, J. W. (2008). Age-related macular degeneration. *The New England Journal of Medicine* 358: 2606–2617.

Johnson, E. J. (2005). Obesity, lutein metabolism and age-related macular degeneration: A web of connections. *Nutrition Reviews* 63: 9–15.

Klein, R. (2007). Overview of progress in the epidemiology of age-related macular degeneration. *Ophthalmic Epidemiology* 14: 184–187.

Lotery, A. and Trump, D. (2007). Progress in defining the molecular biology of age related macular degeneration. *Human Genetics* 122: 219–236.

Montezuma, S. R., Sobrin, L., and Seddon, J. M. (2007). Review of genetics in age related macular degeneration. *Seminars in Ophthalmology* 22: 229–240.

Pieramici, D. J. and Rabena, M. D. (2008). Anti-VEGF therapy: Comparison of current and future agents. *Eye* 20: 1330–1336.

Provis, J. M., Penfold, P. L., Cornish, E. E., Sandercoe, T. M., and Madigan, M. C. (2005). Anatomy and development of the macula: Specialization and the vulnerability to macular degeneration. *Clinical and Experimental Optometry* 88: 269–281.

Rattner, A. and Nathans, J. (2006). Macular degeneration: Recent advances and therapeutic opportunities. *Nature Reviews Neuroscience* 7: 860–872.

Scholl, H. P. N., Fleckenstein, M., Issa, P. C., et al. (2007). An update on the genetics of age-related macular degeneration. *Molecular Vision* 13: 196–205.

Sunness, J. S. (1999). The natural history of geographic atrophy, the advanced atrophic form of age-related macular degeneration. *Molecular Vision* 5: 25.

Relevant Websites

http://www.ahaf.org – American Health Assistance Foundation.

http://www.macular.org – American Macular Degeneration Foundation (AMDF).

http://www.eyesight.org – Macular Degeneration Foundation (MDF).

http://www.mayoclinic.com – Mayo Clinic.

http://www.nei.nih.gov – National Eye Institute (NEI).

http://www.nlm.nih.gov – National Library of Medicine, National Institutes of Health.

Secondary Photoreceptor Degenerations*

M B Gorin, Jules Stein Eye Institute, Los Angeles, CA, USA

Glossary

Epigenetic – The heritable modifications of gene expression that are not the result of changes in the DNA sequence. This can include methylation of DNA that results in inactivation of gene transcription or factors that modify how RNA transcripts are spliced to form the final transcripts that are translated into peptide sequences.

Extrinsic factor – An agent external to the organism that contributes to or is causative of a disease state. This can include drugs, foods, normal nutrients (excess or deficiency), toxins, inhaled chemicals, infectious agents, and exposures to radiation such as light, sound, and high-energy particles.

Intrinsic factor – An agent that is inherent to the organism that contributes to or is causative of a disease state. While commonly these factors are genetic variants in the organism's DNA that may predispose (or be protective) of specific conditions, other intrinsic factors include epigenetic changes, aging changes, and the effects of the biology of symbiotic bacteria in the skin or gut. Another intrinsic agent is an organism's immunologic response behavior and memory (though obviously the immunologic memory is heavily affected by the exposure to extrinsic agents, such as viral infections).

Microbiomics – The genetic information expressed by the microbes that are indigenous to a host organism (e.g., bacteria colonized to the skin or intestinal tract).

Introduction

Secondary retinal degeneration occurs when cells in the retina die by a process triggered by factors not inherent to retina. Secondary retinal degeneration can be caused by trauma, infection, inflammation, toxins, anti-retinal antibodies, or as an adverse effect of medications.

* All of the genes that are mentioned in this article are described in Table 2 of Chapter 210 (for the retinal degenerations), RetNet (www.sph.uth.tmc.edu/retnet/), and/or Online Mendelian Inheritance of Man (OMIM) (www.ncbi.nlm.nih.gov/Omim/).

In the past, clinicians have tended to view genetic and nongenetic etiologies of retinal degeneration as easily separated categories. The molecular studies of hereditary retinal degenerations have shown that, while some retinal conditions are caused by mutations in genes with photoreceptor-specific expression, many retinal conditions are the results of mutations in genes that are widely expressed in the body as well as from the secondary effects of metabolic changes caused by the expression of mutated genes in ocular cell types other than photoreceptors as well as from other organs and tissues distant from the eye. Based on our understanding of complex genetic disorders, we now realize that there can be interplay of genetic and nongenetic factors that run the entire spectrum of possibilities. For example, rhegmatogenous retinal detachments, which can lead to secondary photoreceptor degeneration, may be influenced or caused by genetic variants (e.g., COL11A1, VCAN, COL9A1, and COL2A1) that are expressed in nonretinal cells, and whose expression may be limited to a particular period in ocular development. Thus, we have to consider this continuum of causality as we attempt to make useful classifications that can guide diagnostics and therapy. In light of these complexities, we offer the following operational distinctions among primary and secondary photoreceptor and retinal degenerations that may be relevant to therapeutic approaches.

- If the genetic defect is such that it would require actual alteration of the gene expression in the photoreceptors to correct the abnormality and arrest the degeneration, then this can be considered a primary photoreceptor degeneration. The genetic alteration is necessary and sufficient to cause photoreceptor degeneration. The gene that is mutated may (e.g., opsin, peripherin/rds, cone transducin, AIPL1, and GUCY2D) or may not (e.g., splicing factors PRPF8, PRPF3, and PRPF31, IMPDH1, and CA4) be photoreceptor specific. For a primary photoreceptor degeneration, one would expect that the correction of the genetic alteration outside of the photoreceptors would not be sufficient to prevent photoreceptor degeneration. However, a secondary photoreceptor degeneration that results from loss of expression or expression of a mutated protein in either other retinal cells or the retinal pigment epithelium (RPE) (e.g., RPE-65, RGR, and LRAT) might be corrected by gene therapy to the key nonphotoreceptor cells in the retina or RPE.

- If one reviews the genes attributed to primary photoreceptor degenerations, it is clear that many of these causative genes are not limited to photoreceptor-specific expression. Mutations in these genes have been attributed to nonsyndromic primary photoreceptor degenerations (such as retinitis pigmentosa (RP), Leber congenital amaurosis (LCA), cone dystrophy, and cone–rod dystrophy) as well as syndromic forms (e.g., Usher syndrome, Bardet–Biedl syndrome, Alstrom disease, and Cohen syndrome). Most of these conditions are further described and discussed elsewhere in the encyclopedia. In some instances, the mechanisms of action of these genes may not be solely mediated through their direct effects on the photoreceptors, thus raising the possibility that, in some cases, the photoreceptor degeneration is mediated through a mixed primary and secondary photoreceptor degeneration model (see below). Two unique examples of this potential ambiguity are the ABCA4 (Stargardt disease, cone–rod dystrophy) and RS1 (X-linked retinoschisis) genes. Both genes are specifically expressed in the photoreceptors, but their mechanism of action appears to be mediated through other retinal/RPE cells that lead to a secondary photoreceptor degeneration (see below).

- If the photoreceptors degenerate as the result of an alteration in a gene whose expression is primarily in other retinal or RPE cells, then this would be a primary retinal degeneration with secondary photoreceptor degeneration. Correction of the genetic defect would require modification of the effects of those retinal/RPE cells. A primary retinal degeneration without photoreceptor degeneration can occur such as with optic neuropathies that lead to retinal ganglion cell loss without significant loss of photoreceptors (**Table 1**).

- A mixed model of primary and secondary photoreceptor degeneration can be considered in two different modes. One is when a genetic alteration within the photoreceptors themselves would not be sufficient to cause photoreceptor degeneration by itself but would

predispose to degeneration in the presence of an extrinsic or intrinsic agent. The genetic alteration within the photoreceptors could be necessary, conditional, or probabilistic but not sufficient. An intrinsic agent could be a genetic alteration in nonphotoreceptor retinal cells or due to expression elsewhere in the body. As noted above, a number of genes that are expressed in photoreceptors and for whom there are mutations that are known to be responsible for photoreceptor degeneration also have expression in other retinal cells as well as in other tissues. In some of these cases, it is not always clear if the expression in the photoreceptors alone is sufficient to cause cell death or whether or not there is a component of photoreceptor degeneration that is secondary to the effects on other cells and tissues. The only way to distinguish a secondary effect from a primary one would be to create animal models in which the genetic alteration is limited to specific cell populations and to determine if the photoreceptors are spared when their gene expression is normal. This is especially true for the forms of RP that are associated with mutations in genes that affect metabolic processes throughout the body. Examples of these conditions include gyrate atrophy, Bietti crystalline retinopathy, abetalipoproteinemia, and Refsum disease. At this time, we simply cannot establish if the effects of these genetic mutations are mediated by a primary effect on the photoreceptors or by secondary mechanisms. In the case of gyrate atrophy and Refsum disease, there is evidence that nutritional therapy can ameliorate the progression of the condition, which suggests an interplay of a person's intrinsic genetic makeup and diet (an extrinsic agent), but we still do not know if the effect is due to the systemic reduction of toxic metabolites or a photoreceptor-specific mechanism is also involved. Similarly, with Bietti crystalline retinopathy, the defect in CYP4V2 has multitissue consequences but it is not known if a systemic correction of the metabolic defect would be sufficient to overcome the enzyme deficiency in photoreceptor cells. Only future studies will be sufficient to distinguish if these conditions are representative of a mixed model of photoreceptor degeneration or secondary photoreceptor degenerations of an intrinsic type (see below).

An extrinsic agent can be a drug or environmental exposure (including something in the diet). There are relatively few established human examples of this model for retinal degenerations, though retinal degeneration-B (rdgB) mutants in *Drosophila* show light-dependent photoreceptor degeneration. This mixed model could possibly account for some of the cases of photoreceptor degenerations with incomplete penetrance (individuals who have the disease-causing mutation but show no clinical evidence of retinal degeneration).

- If one could prevent the photoreceptor degeneration by preventing an individual's exposure to an extrinsic agent

Table 1 Secondary photoreceptor degenerations associated with primary retinal/RPE degeneration/dystrophy

Gene involved, site of cell/tissue expression related to retinal degeneration (RPE-retinal pigment epithelium, RVE-retinal vascular endothelium, RVP-retinal vascular pericytes, MGC-Muller glial cells), and phenotype (LCA-Leber congenital amaurosis, RP-retinitis pigmentosa) (from RetNet)

RPE65	RPE	LCA and RP
MERTK	RPE	RP
CRALBP	RPE, MGC	Bothnian dystrophy
LRAT	RPE, liver	RP
RGR	RPE	RP and dominant choroidal sclerosis
TIMP3	RPE, RVP	Macular dystrophy
C1QTNF5	RPE	Macular dystrophy
ABCC6	REV	Macular dystrophy
AMD-related genes	RPE, liver	Macular dystrophy
BEST1	RPE	Macular dystrophy

or condition (e.g., toxin, drug, infectious agent, light, and trauma), then this is secondary photoreceptor degeneration of the extrinsic type (even if the body converts that agent to a toxic form as part of a normal metabolic pathway – such as methanol to formaldehyde). Clearly, the primary method of management is to avoid exposure to the extrinsic conditions that would induce the degeneration. This form of degeneration can be due to exposure to an external agent as well as deprivation of a mandatory nutrient (such as vitamin A). The deficiency can be the result of a lack of intake or synthesis of the key nutrient (vitamin-A- or zinc-deficient diet) or due to the inability to process or use such a metabolite/nutrient. Examples would be malabsorption of vitamin A and zinc due to intestinal disorders or drugs which block utilization, such as fenretinide or accutane (**Table 2**).

- A second mode of a combined primary and secondary photoreceptor degeneration is when one group of photoreceptors, such as the rod photoreceptors, undergoes a primary degenerative process due to a mutation in a gene that is expressed in those photoreceptors that precipitates apoptosis. At the same time, there is a second group of photoreceptors, the cone photoreceptors, which undergoes a secondary degenerative process due to alterations in the cellular environment induced by the death of neighboring cells. This situation is actually very common among patients with retinal dystrophies such as rod–cone (e.g., RP) or cone–rod forms. Recent studies of several mouse models of RP due to rod-photoreceptor specific genes have showed that the nonautonomous death of the cone photoreceptors is influenced by activation of the rapamycin pathway that can be modified by exogenous

insulin, suggesting a possible intrinsic mechanism that could be influenced by a systemic therapeutic approach. The importance of this mechanism cannot be overemphasized since preservation of cone photoreceptor cells and function in a patient with RP would have a dramatic impact on maintaining useful visual function and it does not necessarily require the correction of the primary photoreceptor degeneration mechanism in the rod photoreceptors.

- If one can prevent photoreceptor degeneration by correcting or reversing a systemic or ocular metabolic or immune process, then it is a secondary photoreceptor degeneration of the intrinsic type. A number of these conditions are driven or influenced by genetic etiologies (necessary and sufficient in the case of metabolic syndromes, but often conditional or probabilistic in immune-related conditions), but the retinal degeneration is still secondary. Intrinsic causes are not exclusively genetic, one may have to consider epigenetic factors as well as immunologic memory and the microbiomics of the natural flora. Clearly, one would primarily direct therapy to correcting the primary metabolic or immune disturbance rather than focusing on modifying the behavior of the photoreceptors. Therapy might be directed specifically to the affected eye(s), (such as periocular or intraocular steroid therapy) rather than systemically, but it would be intended to primarily modify effector cells in the tissue, rather than the photoreceptors themselves (**Table 3**).

- A mixed intrinsic and extrinsic etiology for a secondary photoreceptor degeneration would be when a person has a genetic variant that creates a toxic metabolite in the presence of an extrinsic molecule that would normally

Table 2   Retinotoxic drugs and agents, nutrient deficiencies, infectious agents, light injury, and trauma

Drugs
Ethambutol, aminoglycosides, epinephrine, desferroximine, antimalarials (hydroxychloroquine, chloroquine, quinine), vigabatrin, phenothiazines (e.g., fluphenazine, mellaril, and stellazine).
Nutrient deficiencies
Zinc, vitamin A, omega-3 fatty acids.
Infectious
Toxoplasmosis, cytomegalovirus, herpes simplex, varicella zoster, HIV, DUSN (nematode), rubella, syphilis, prion, corona virus, others.
Toxins
Cadmium, iron (siderosis), lead, mercury (suspected), copper (intraocular chalcosis), cobalt, iodoacetic acid (IAA), methanol.
Light
Solar, laser chronic exposure.
Trauma
Commotio, retinal detachment.
Vascular
Occlusive disease, embolic, inflammatory, retinopathy of prematurity (ROP), Coats disease.

Table 3   Intrinsic factors: genes, phenotypes (e.g., RP nonsyndromic, RP syndromic, and macular degeneration), mechanism (e.g. metabolic, immune, inflammatory (inflamm))

OAT	Gyrate atrophy	Metabolic
CYP4V2	Bietti crystalline retinopathy	Metabolic
PEX1, PEX2	Zellweger Syndrome	Metabolic
PEX7, PHYH	Refsum disease (adult)	Metabolic
MTP	Abetalipoproteinemia	Metabolic
PANK2	Hypoprebetalipoproteinemia	Metabolic
	Niemann–Pick	Metabolic
	neuronal ceroid lipofuscinosis	Metabolic
CTNS	Cystinosis	Metabolic
CA4 (carbonic anyhydrase 4)	RP	Metabolic
LRP5	FEVR	Metabolic
HLA-B27, A29, B7	Ankylosing spondylitis	Immune
	Birdshot choroidopathy	Immune
	Bechet's disease	Immune
Unknown, retinal antigens, cancer	Cancer-assoc. retinopathy	Immune
	Autoimmune retinopathy	Immune
CFH	Hemolytic uremia – mac deg	Inflamm

not be encountered. A normal person would not experience a retinal degeneration under the same exposure conditions. This set of conditions has overlap with the purely extrinsic and intrinsic etiologies if the genetic variation simply shifts the dose–response characteristics of the host. For example, a person is genetically predisposed to react to an extrinsic molecule at levels in the normal environment, while another person would experience similar photoreceptor degeneration only when the exposure is at levels that would exceed normal exposures. Reduction of the extrinsic exposure below the normal levels could be beneficial for these individuals (such as Refsum disease or gyrate atrophy). Alternatively, correction of the genetic variant would allow the person to cope with normal exposure levels. An animal model of the mixed intrinsic and extrinsic secondary photoreceptor degeneration would be the RPE65-MET450 mutants (intrinsic) that have varying reduced sensitivity to light-induced (extrinsic) photoreceptor degeneration as compared to animals that have the LEU450 variant in the RP65 gene.

• We are only beginning to understand these types of situations, although it is likely that many of the idiosyncratic reactions that some patients experience to certain situations or medications are the result of genetic variations that affect drug bioavailability, mechanism of action, and elimination. One such example would be the patient who develops cystoid macular edema (CME) after uncomplicated surgery. The surgical intervention would be considered an extrinsic agent. While CME is common in cases of complicated surgery and postsurgical inflammation, it is relatively uncommon (but not rare) in individuals whose surgery and postoperative care are uneventful and have no predisposing clinical conditions. Yet, this is most likely due to intrinsic (nonphotoreceptor-specific) genetic factors that govern inflammation. Persistent CME can lead to secondary photoreceptor degeneration.

If the extrinsic exposure cannot be manipulated, then essentially, one is forced to treat the mixed etiology as a purely intrinsic issue. For example, if a person had a genetic condition from an intrinsic metabolic defect that is light sensitizing such that normal ambient light would trigger photoreceptor degeneration, the distinction between an intrinsic etiology and a mixed intrinsic/extrinsic etiology becomes almost meaningless, since having a person avoid all light exposure to prevent photoreceptor degeneration is neither feasible nor desirable. However, reduction of the light exposure might alter the rate of disease progression, but therapy directed toward the intrinsic factor(s) would be essential to preserve vision under normal exposure circumstances. This situation is comparable to the mixed primary and secondary photoreceptor degeneration category (such as a mutation in an photoreceptor-specific gene that is

responsible for light-dependent degeneration) except that, instead of the intrinsic etiology being disconnected from the photoreceptors themselves, the photoreceptors are directly affected by a genetic variant that renders the photoreceptors vulnerable to the extrinsic factor (e.g., light). While the reduction of the extrinsic exposure would be desirable, it may not be realistic and thus therapy would also have to be directed to the photoreceptors themselves.

The combination of extrinsic and intrinsic factors that affect photoreceptor degeneration is comparable to the genetic and environmental interactions that are often discussed in the context of complex genetic diseases such as age-related macular degeneration. At this time, our understanding of these interactions is very limited, but there are some examples for simpler retinal conditions. Mice that are heterozygous for a deletion in the PDE6B subunit (the rd mouse), a dose of sildenafil citrate (Viagra) that would normally have no effect in the normal mouse, will show a major change in the electroretinogram. It is likely that some of the individuals who experience visual side effects from this medication may have a genetic variant that reduces the overall level of phosphodiesterase activity in their retinas, thus conferring sensitivity. Another mixed etiology of secondary photoreceptor dysfunction (which can ultimately lead to degeneration) can be seen in an individual with a normally adequate intake of vitamin A, who becomes vitamin A deficient due to an acquired or hereditary malabsorption syndrome (including postsurgical bowel resection or remodeling). The etiology may be intrinsic or iatrogenic, but the treatment is directed toward the extrinsic agent by increasing the dose or mode of absorption of the vitamin A.

Mechanisms of Secondary Photoreceptor Death

Photoreceptors can die from several mechanisms, including physical lysis, destruction by thermal denaturation (such as by laser), or by triggering the apoptotic pathways. Apoptosis can be triggered by a number of disruptions, including loss of key trophic factors such as vascular endothelial growth factor (VEGF), energy depletion through mitochondrial failure, oxidative damage of proteins and lipids, release of calcium by shifts in membrane permeability, which can be caused by deregulation of ionic channels, or from the fixation of complement to the membrane surface. Light levels below those that cause thermal denaturation can lead to direct activation of caspases, calpain 2, and cathepsin D. In addition, mitochondrial-dependent apoptotic pathways also appear to be activated.

In a number of cases, signaling of the apoptotic pathway appears to be governed by at least two pathways: the Wnt pathway and the Jak–STAT pathway. A number of research groups are attempting to identify nonspecific therapies that can block or inhibit the pathways that result in photoreceptor death. The use of ciliary neurotrophic factor (CNTF) as a trophic factor to inhibit activation of the apoptosis pathway is currently in clinical trials to treat primary and secondary photoreceptor degenerations.

As one considers these multiple mechanisms of photoreceptor death, it becomes clear that a major value of experimental animal models for these conditions is to specifically determine the extent to which photoreceptor death is a primary event and if genetic defects within the photoreceptors themselves are necessary and sufficient to initiate apoptosis. At the same time, this does not negate the importance of understanding the intrinsic (both genetic and nongenetic) factors and extrinsic factors that either trigger or modify cell death that may be amenable to therapeutic intervention at a systemic level. Finally, even in the presence of a combination of factors that lead to photoreceptor death, there is the possibility of interrupting or inhibiting the common apoptosis signaling pathways within the photoreceptor cells and other retinal neurons in order to preserve function and vision.

See also: Primary Photoreceptor Degenerations: Retinitis Pigmentosa; Primary Photoreceptor Degenerations: Terminology; Retinal Ganglion Cell Apoptosis and Neuroprotection.

Further Reading

Glazer, L. C. and Dryja, T. P. (2002). Understanding the etiology of Stargardt's disease. *Ophthalmology Clinics of North America* 15(1): 93–100, viii.

Hackam, A. S. (2005). The Wnt signaling pathway in retinal degenerations. *IUBMB Life* 57(6): 381–388.

Ling, C. P. and Pavesio, C. (2003). Paraneoplastic syndromes associated with visual loss. *Current Opinion in Ophthalmology* 14(6): 426–432.

Poll-The, B. T., Maillette de Buy Wenniger-Prick, L. J., Barth, P. G., and Duran, M. (2003). The eye as a window to inborn errors of metabolism. *Journal of Inherited Metabolic Disease* 26(2–3): 229–244.

Punzo, C., Kornacker, K., and Cepko, C. L. (2009). Stimulation of the insulin/mTOR pathway delays cone death in a mouse model of retinitis pigmentosa. *Nature Neuroscience* 12(1): 44–52.

Rattner, A. and Nathans, J. (2006). An evolutionary perspective on the photoreceptor damage response. *American Journal of Ophthalmology* 141(3): 558–562.

Samardzija, M., Wenzel, A., Aufenberg, S., et al. (2006). Differential role of Jak–STAT signaling in retinal degenerations. *FASEB Journal* 20(13): 2411–2413.

Siu, T. L., Morley, J. W., and Coroneo, M. T. (2008). Toxicology of the retina: Advances in understanding the defence mechanisms and pathogenesis of drug- and light-induced retinopathy. *Clinical and Experimental Ophthalmology* 36(2): 176–185.

Stone, J., Maslim, K., Valter-Kosci, K., et al. (1999). Mechanisms of photoreceptor death and survival in mammalian retina. *Progress in Retinal and Eye Research* 18(6): 689–735.

Wenzel, A., Grimm, C., Samardzija, M., and Remé, C. E. (2005). Molecular mechanisms of light-induced photoreceptor apoptosis and neuroprotection for retinal degeneration. *Progress in Retinal and Eye Research* 24(2): 275–306.

Wu, J., Seregard, S., and Algvere, P. V. (2006). Photochemical damage of the retina. *Survey of Ophthalmology* 51(5): 461–481.

Yang, L. P., Zhu, X. A., and Tso, M. O. (2007). A possible mechanism of microglia–photoreceptor crosstalk. *Molecular Vision* 13: 2048–2057.

Relevant Website

http://www.sph.uth.tmc.edu – Retinal information network.

Unique Specializations – Functional: Dynamic Range of Vision Systems

A C Arman and A P Sampath, University of Southern California, Los Angeles, CA, USA

Glossary

Dynamic range – The full range of light levels over which retinal cells or pathways can dynamically modify their function to encode.

Mesopic vision – Intermediate light level vision mediated by both rod and cone photoreceptors under conditions where both are responsive, allowing a seamless transition from rod to cone vision.

Photopic vision – High light level vision mediated by cone photoreceptors.

Ribbon synapse – Synapses where a ribbon-like structure is present that optimizes the continuous release of neurotransmitter.

Rod spherule – The specialized synaptic terminal of the rod photoreceptor.

Scotopic vision – Low light vision mediated by rod photoreceptors.

Introduction

Nearly all sensory systems must find a way to represent a wide range of input signals and translate them into meaningful neural responses. The human visual system is able to operate effectively from starlight to bright sunlight, a range that spans about 12 orders of magnitude of light intensity. The pupil serves as the first stage of sensitivity control by changing the amount of the light reaching the photoreceptors. The diameter of the pupil can change by a factor of 4, allowing light intensity to change by a factor of 16. However, this alone cannot account for the complete range of light to which the mammalian retina is sensitive. To reliably transmit changes in light stimuli over this range, the mammalian retina has evolved several specializations to report changes in the light environment that include: (1) the evolution of two photoreceptor types, the rods and cones, that operate at different light levels; (2) several neural pathways with which to encode the output of these photoreceptors; and (3) adaptive mechanisms at all levels of retinal processing to modulate light sensitivity based on light history

Rods versus Cones

The evolution of the duplex retina in vertebrates with two classes of photoreceptors, the rods and cones, marks a departure from the single type of photoreceptor present in many invertebrates. The use of two photoreceptor subtypes with different sensitivities to light allows the vertebrate retina to respond over a greater range of light intensities. By switching between rods and cones, the vertebrate retina is thus able to maximize visual sensitivity depending on the ambient light level. It is now believed that such an arrangement allows the vertebrate retina to reduce energy consumption in daylight, when the rods are not responsive.

Rods mediate vision when photons are scarce; their design and cytoarchitecture are optimized for maximal sensitivity to incoming photons and are capable of generating a reproducible response to a single absorbed photon, which is critical for setting the sensitivity of scotopic vision when the retinal circuitry pools thousands of rods (see the section titled 'The pathways concept' below). As the mean background light level increases, rods themselves are able to adapt, which allows them to signal light intensities up to ~ 1000 R* or more per second.

Cone photoreceptors are ~ 100-fold less sensitive than rods, and are critical for our daytime vision under conditions when the exquisitely sensitive rod photoreceptors are saturated. The reduced sensitivity of cones arises from reduced amplification within cone phototransduction, and mechanisms designed to shut off phototransduction more quickly than rods. Furthermore, even in the brightest light the cone photocurrent does not remain saturated, which leaves open the ability to adapt and signal changes in light intensity even under conditions where a majority of the photopigment is bleached. The optimizations of cones to function in bright light with virtually little overlap with rod light levels allows for a smooth transition from rod to cone vision, in what is referred to as the mesopic range.

The Pathways Concept

The adaptive features of the rod and cone light response allow these photoreceptors to remain responsive over a larger range of light intensities by preventing response saturation. However, the dynamic range of the rods and cones themselves cannot account for the ~ 12 orders of magnitude in light intensity we experience. Another strategy used by the mammalian retina to extend further the dynamic range of vision is to utilize multiple neural pathways to carry light-evoked signals. The functional properties of these pathways (e.g., convergence, gain, and adaptation) can then be adjusted to maximize the visual system's ability to remain responsive over the largest range of light intensities. To date, the most-studied retinal pathways in mammals are those that carry signals from rod photoreceptors to ganglion cells. These include the circuits that carry light information near scotopic threshold (rod bipolar pathway or classical rod pathway), and those that operate at higher rod light levels that may provide a seamless mesopic transition (rod–cone and rod–OFF pathways) to photopic vision. Rather less is known about the functional properties of the cone pathways that maximize dynamic range.

Rod Bipolar Pathway

Psychophysical studies have demonstrated at the limits of scotopic vision that the human visual system is capable of detecting the absorption of a few photons of light. This remarkable sensitivity arises from the fact that individual rods can reliably signal the absorption of a single photon, and that a specialized retinal circuitry referred to as the rod bipolar pathway can combine these signals such that a ganglion cell projecting centrally can pool thousands of rod signals. The discovery of a dedicated depolarizing bipolar cell and an amacrine cell that carries rod signals in the rabbit retina leads to the finding that this circuit appears conserved across all mammalian species. A hallmark of this pathway is the convergence of rod signals at many stages of processing that is critical for our scotopic sensitivity. For instance, as many as 20–100 rods converge on a rod (ON) bipolar cell, and 20–30 rod bipolar cells converge on an AII amacrine cell. Thus, a ganglion cell that sums the output of many AII amacine cells can pool upward of 10 000 rods. Ultimately, by pooling rod signals and eliminating rod noise to preserve best the single-photon response from individual rods, the rod bipolar pathway can extend the dynamic range of vision down to light levels where a small fraction of the rods absorb a photon.

Signal transfer from rods to rod bipolar cells

At scotopic threshold, vision relies on a sparse number of photons at the retina that produce few photon absorptions per thousands of rods within the 0.2-s integration time of the rod photoresponse. Under these conditions, the transmission of a small, graded hyperpolarization upon photon absorption requires that rod synapse is appropriately optimized. The transmission of small, graded single-photon responses at the rod synaptic terminal is aided by two specializations. First, the resting dark membrane potential, or voltage, sits at approximately $-40\,mV$, near the steepest point in the relationship between voltage and L-type Ca^{2+} channel opening (**Figure 2**). Thus, small changes in membrane potential produce substantial changes in the number of open channels, thereby altering

glutamate release. Second, if the rod bipolar cell is sensing reductions in glutamate release due to photon absorption, then statistical lapses of glutamate release in darkness would mimic light absorption. Thus, the high rate of glutamate release generated in darkness by the specialized synaptic ribbon in the rod spherule reduces the probability of these lapses. Together, these synaptic properties allow the small, light-evoked signals from rods to be reproducibly transferred to downstream neurons.

Despite the rod synaptic specializations for the transmission of single-photon absorptions, the depolarization in darkness due to open cyclic guanosine monophosphate (cGMP)-gated channels is also a complicating factor in the detection of these sparse signals. Open cGMP-gated channels in turn will report internal fluctuations in cGMP, produced by the phototransduction mechanism, which are commonly referred to as dark noise. Since rods generate a small, graded hyperpolarization upon photon absorption, the downstream convergence of thousands of rod signals would cause the light-evoked response from a single rod to be overwhelmed by the dark noise of the majority. Given the magnitude of dark noise in individual rods, it has been proposed that some type of nonlinear combination of rod signals would be required to increase the detection of the single-photon responses in downstream cells. Since rod photoreceptors are relatively depolarized in darkness, the steady release of glutamate from the synapse provides some insights into potential mechanisms. Postsynaptic saturation at the rod-to-rod bipolar synapse would allow noise generated by open cGMP-gated channels in the rod outer segment to be eliminated. It was proposed that the saturation of postsynaptic glutamate receptors would provide a nonlinear way to eliminate the rod noise, since the synapse would not be able to relay small changes in membrane potential that reflect rod noise. Later work suggested that such thresholding is critical for maximizing the detection of the single-photon response in retinal neurons downstream of the rods (**Figure 1**). In particular, the extent of nonlinear signaling appears to be set to separate optimally the rod single-photon response from rod noise, allowing scotopic vision to reach the highest possible sensitivity.

The mechanism that underlies the nonlinear threshold at the rod synapse has been studied to some extent, but is hindered by a lack of identification of the components of the signaling pathway. Light-evoked signaling between rod photoreceptors and rod (ON) bipolar cells results in a membrane depolarization, effectively inverting the sign of the rod's hyperpolarizing light response. The postsynaptic mechanism underlying this sign inversion is a G-protein signaling pathway initiated by the metabotropic glutamate receptor, mGluR6. mGluR6, in turn, activates a guanine nucleotide-binding protein, $G_o\alpha$, which leads to a series of unidentified events that close a cationic transduction channel of unknown identity. Thus, upon

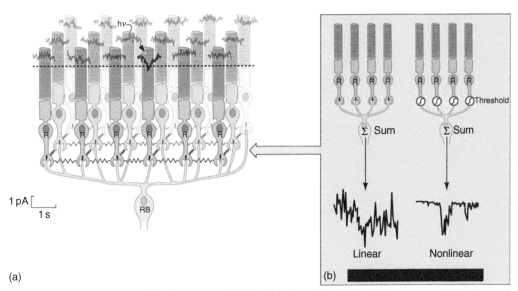

Figure 1 Convergence at the rod-to-rod bipolar synapse. (a) A rod bipolar cell pools inputs from many rods, but near absolute visual threshold only one rod may absorb a photon (red), while the remaining rods are generating electrical noise (blue). A nonlinear threshold (dashed) may improve photon detection at this synapse by retaining responses in rods absorbing a photon and discarding responses of the remaining rods. (b) Nonlinear signal processing can improve the fidelity of rod signals. If rod outputs from (a) are simply summed, the resulting trace is noisy, but when summed after applying a threshold for each rod in (a), the response is more detectable. From Okawa, H. and Sampath, A. P. (2007). Optimization of single-photon response transmission at the rod-to-rod bipolar synapse. *Physiology* 22: 279–286. With kind permission from The American Physiological Society.

light-absorption, glutamate release from rods is reduced, thereby reducing the activity of the mGluR6 signaling pathway and allowing transduction channels to open and depolarize the cell (**Figure 2**).

In the context of the mGluR6 signaling cascade, it now appears that the nonlinear threshold that eliminates rod noise is due to saturation within the signaling cascade, and not at the level of the glutamate receptors. Furthermore, evidence from axotomized rod bipolar cells indicates that nonlinear signal transfer does not arise due to feedback in the inner plexiform layer. Saturation of the mGluR6 signaling cascade allows the elimination of noise by making the rod bipolar cell insensitive to small fluctuations in glutamate, driven by noise in the rod photoreceptor. Only when the rod's membrane potential is hyperpolarized sufficiently does the glutamate concentration in the synaptic cleft reduce enough to relieve the synapse from saturation. Such an operation thus allows larger hyperpolarizations due to light absorption to cross the rod synapse, while masking smaller fluctuations that are more likely due to noise in the rod photocurrent or synaptic transmission. Near absolute visual threshold, such synaptic processing is necessary to maximize the detectability of rod signals.

Signal transfer from rod bipolar cells to AII amacrine cells

The convergence of the rod bipolar pathway moving from rods to rod bipolar cells and, finally, AII amacrine cells, requires the further accentuation of the single-photon response. Two main specializations between these cells appear well tuned to further improve the detection of the single-photon response, and thus push the dynamic range of vision to lower light intensities. First, a specialized ribbon synapse between the rod bipolar cell and the AII amacrine cells allows the coordinated release of multiple vesicles upon stimulation. Such multivesicular release increases the amplitude of the AII amacrine cell response, allowing it to be distinguished from vesicular release due to noise in the rod bipolar cell. Second, the electrical coupling of AII amacrine cells by connexin 36 appears to reduce noise in the network, allowing an improved signal-to-noise ratio.

Ganglion cell sensitivity

Retinal ganglion cells can also relay single-photon responses to higher visual centers, a requirement for the high sensitivity of rod vision. Through the process of neural convergence, as well as the mechanisms described above, ganglion cells and AII amacrine cells have about the same flash sensitivity. Recordings of many groups from dark-adapted cat retinal ganglion cells indicate bursts of ~2–3 action potentials occurred with a frequency consistent with an upstream origin that may be the rods. The frequency of these bursts increased with background light, suggesting that the bursts were related to photon absorption. Similar conclusions have been drawn using a cross-correlation analysis from paired ganglion cell recordings in the presence and absence of background lights.

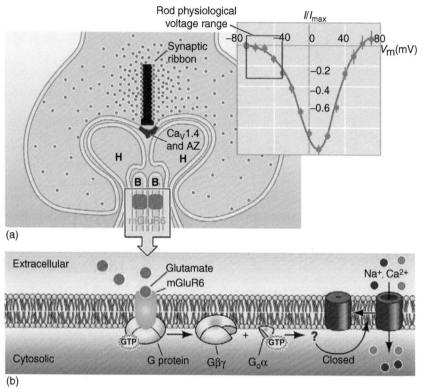

Figure 2 Structure and signal transfer at the rod-to-rod bipolar synapse. (a) The rod spherule is a specialized invaginating structure where the dendrites of horizontal (H) and rod bipolar cells (RB) are apposed to a glutamate release site controlled by a ribbon. Ca^{2+} channels ($Ca_V1.4$) are located near the active zone (AZ) and allow the continuous release of glutamatergic vesicles in darkness. In ON rod bipolar cells, glutamate is sensed by mGluR6 receptors located near the mouth of the invagination. (Inset) Release of glutamate is dependent on Ca^{2+} influx through Ca^{2+} channels, which is graded by voltage over the physiological range. (b) The signaling cascade in rod bipolar cell dendrites is poorly understood. mGluR6 activation leads to the activation of $G_o\alpha$, which through unknown mechanisms leads to the closure of a nonselective cation channel whose identity is also unknown. The light-evoked reduction in glutamate release relieves activity in this cascade and opens cation channels leading to depolarization. From Okawa, H. and Sampath, A. P. (2007). Optimization of single-photon response transmission at the rod-to-rod bipolar synapse. *Physiology* 22: 279–286. With kind permission from The American Physiological Society.

Rod–Cone Pathway

A consequence of the high sensitivity of the rod bipolar pathway is that it saturates at modest light levels where the rods themselves are not saturated. To capture light-evoked signals from rods until they themselves saturate additional pathways are required. Studies of the ultrastructure of the outer plexifom layer reveal that gap junctions exist between rods and rods, and rods and cones. The nature of the rod-to-rod gap junction is unclear, but has been proposed to dissipate small rod signals into the network allowing signal averaging at the cost of a twofold elevation in visual threshold. The rod-to-cone gap junctions have been studied in more detail and are found in a majority of rods. Evidence points to the expression of the gap junction subunit connexin 36 in cones, but not in rods. Thus, the electrical gap junctions between rods and cones must be heterologous. Given the lower input impedance of cones, signal transfer will preferably travel from rods to cones, which would allow these signals to be relayed to ganglion cells through the cone circuitry.

Physiological recordings have indicated robust rod signals in the cone photoreceptors of several species. In particular, measurements from dark-adapted retina indicate that rod-to-cone coupling must exist, going against the idea that, to maximize visual sensitivity, the rods and cones should be uncoupled in darkness. Recordings from mouse retinal whole mounts suggest that the signals through this pathway are 10-fold less sensitive than for the rod bipolar pathway. These physiological data correspond well with psychophysical experiments indicating a secondary rod pathway with a 10-fold lower sensitivity than the primary pathway.

Rod–OFF Pathway

A third rod pathway has been identified in some species of mammals. Under conditions where the rod–bipolar and rod–cone pathways are blocked or eliminated, OFF signals with rod sensitivity have been shown to persist in the mouse retina. Direct synaptic contacts have been identified anatomically in mice, rats, and rabbits. However, direct contacts between rods and OFF bipolar cells have not yet been reported in higher mammals. It is possible that an OFF rod pathway evolved as a specialization in nocturnal rodents to sense dark objects against a light sky.

The rod–OFF pathway is believed to be active at 5–10-fold higher light intensities than the rod bipolar pathway. Its prominence as a signaling circuit in rodents appears to vary based on the percentage of the OFF cone bipolar cell population that sees direct signals from rods. It is now believed that between 5% and >25% of cone OFF bipolar cells may see direct contacts from rods, the upper end suggesting that this pathway may play an integral role in visual processing.

Despite attempts by various groups to characterize the physiological response properties of the rod pathways, the exact sensitivities and operating ranges of each pathway remain unclear. The lack of physiological evidence elucidating the role of these pathways in rod vision arises from common signaling mechanisms used by each. Thus, genetic or pharmacological manipulation of each rod-signaling pathway will also influence other pathways, making it impossible to unambiguously identify the properties of each. Nevertheless, the evidence now points to the rod bipolar pathway as the primary carrier of single-photon responses near absolute visual threshold, with an ~5–10-fold reduction in the sensitivity of the rod–cone and rod–OFF pathways that may carry signals at higher light levels where cone function is merged.

Cone Pathways

Our understanding of the cone pathways is far less complete than our understanding of rod pathways in the mammalian retina. The physiological properties of the cones and cone pathways remain as one of the frontiers in retinal neurobiology, especially as cone function dominates our visual experience. To date, as many as nine types of cone bipolar cells have been identified in mammalian retinas, which may connect to as many as 10–15 types of retinal ganglion cells. Other than the specific pathway for S-cones, there is presently little understanding of specific circuits that carry cone responses to ganglion cells.

Adaptation to Mean Background Light

A common feature of all sensory systems is the ability to adapt to increases in the mean level of a stimulus by reducing the gain of the system. Such sensitivity adjustments allow the sensory system to remain maximally responsive as the stimulus intensity changes. Both rods and cones in the retina, as well as their circuitry,

exhibit adaptive mechanisms that are designed to increase the dynamic range of the receptor. However, in the context of the dynamic range of the visual system, the influences of adaptation on the lower limit of scotopic and upper limit of photopic vision are opposite. To retain maximal sensitivity near absolute visual threshold, the retina must maximize its gain for the single-photon response and any adaptive mechanism would allow the rod circuitry to aid in the transition from scotopic to mesopic vision. Conversely, the upper light limits of our visual experience are ultimately defined by adaptation in the cone photoreceptors, which continue to operate even when a majority of the photopigment is bleached.

Adaptation: Rod Pathways

Adaptation within a neural circuit with considerable convergence will begin centrally and move peripherally. Under these circumstances, downstream cells will be the first to detect sufficient signal to adapt for the weakest stimuli, and the rod pathways are no exception. In the rod bipolar pathway, weak background light begins to reduce the gain of ganglion cells and AII amacrine cells before adaptation is detectable in rod bipolar cells or rod photoreceptors. Some mechanisms that provide this gain reduction have been identified, particularly at the rod-to-rod bipolar, and the rod bipolar-to-AII amacrine synapses. For instance, at the rod-to-rod bipolar synapse, the influx of Ca^{2+} through mGluR6 transduction channels reduces the bipolar cell gain for subsequent stimulation. In addition, at the rod bipolar-to-AII amacrine synapse depression mediated by depletion of the available pool of vesicles can be evoked at individual synapses by single-photon responses. These adaptive mechanisms allow the rod bipolar pathway to extend its range to higher light levels where they merge with the other rod pathways. However, the extension of the dynamic range of vision to lower light levels requires that the retina remains maximally responsive to single photons, and thus these adaptive mechanisms would impair absolute threshold.

Adaptation: Cone Pathways

It has been well documented that adaptation of ganglion cell responses in the cone pathways occurs at lower light levels than where adaptive features of cone pathways have been documented. Cone and horizontal cell recordings have demonstrated that cone adaptation is observed at light levels that exceed those required for the adaptation of cone-driven signals in ganglion cells. This adaptation of the postsynaptic cone circuitry, in turn, prevents these pathways from saturating, thereby allowing the extension of the cone operating range through mesopic to photopic vision. Adaptation, both in the cones and in the postreceptoral circuitry, have been found to be mutually exclusive. Ultimately, the upper limits of cone vision are directly imposed by receptoral adaptation of the cones themselves and not by the postsynaptic circuitry.

Conclusions

The vertebrate retina has developed many strategies to maximize the dynamic range of vision. At the initial stages of light detection the evolution of two photoreceptor types, the rods and cones, allows the visual system to signal a wide range of light intensities. By dividing the output of these receptors across many retinal pathways, each of which is subject to its own optimization and adaptation, the human eye is capable of providing the brain with information that extends the range of vision to encompass approximately 12 orders of magnitude of light intensity.

See also: Anatomically Separate Rod and Cone Signaling Pathways; Information Processing: Bipolar Cells; Information Processing: Ganglion Cells; Morphology of Interneurons: Bipolar Cells; Phototransduction: Adaptation in Cones; Phototransduction: Adaptation in Rods; Phototransduction: Phototransduction in Cones; Phototransduction: Phototransduction in Rods; Rod Photoreceptor Cells: Soma and Synapse.

Further Reading

Barlow, H. B. (1956). Retinal noise and absolute threshold. *Journal of the Optical Society of America* 46: 634–639.

Barlow, H. B., Levick, W. R., and Yoon, M. (1971). Responses to single quanta of light in the retinal ganglion cells of the cat. *Vision Research Supplement* 3: 87–101.

Baylor, D. A., Lamb, T. D., and Yau, K. W. (1979). Responses of retinal rods to single photons. *Journal of Physiology* 288: 613–634.

Dacheux, R. F. and Raviola, E. (1986). The rod pathway in the rabbit retina: A depolarizing bipolar and amacrine cell. *Journal of Neuroscience* 6: 331–345.

Dunn, F. A. and Rieke, F. (2008). Single photon absorptions evoke synaptic depression in the retina to extend the operational range of rod vision. *Neuron* 57: 894–904.

Dunn, F. A., Lankheet, M. J., and Rieke, F. (2007). Light adaptation in cone vision involves switching between receptor and post-receptor sites. *Nature* 449: 603–606.

Field, G. D. and Rieke, F. (2002). Nonlinear signal transfer from mouse rods to bipolar cells and implications for visual sensitivity. *Neuron* 34: 773–785.

Hecht, S., Schlaer, S., and Pirenne, M. H. (1942). Energy, quanta, and vision. *Journal of General Physiology* 25: 819–840.

Hornstein, E. P., Verweij, J., Li, P. H., and Schnapf, J. L. (2005). Gap-junctional coupling and absolute sensitivity of photoreceptors in macaque retina. *Journal of Neuroscience* 25: 11201–11209.

Okawa, H. and Sampath, A. P. (2007). Optimization of single-photon response transmission at the rod-to-rod bipolar synapse. *Physiology* 22: 279–286.

Sampath, A. P. and Rieke, F. (2004). Selective transmission of single photon responses by saturation at the rod-to-rod bipolar synapse. *Neuron* 41: 431–443.

Singer, J. H., Lassova, L., Vardi, N., and Diamond, J. S. (2004). Coordinated multivesicular release at a mammalian rod synapse. *Nature Neuroscience* 7: 826–833.

Smith, R. G., Freed, M. A., and Sterling, P. (1986). Microcircuitry of the dark-adapted cat retina: Functional architecture of the rod–cone network. *Journal of Neuroscience* 6: 3505–3517.

Sterling, P., Freed, M. A., and Smith, R. G. (1988). Architecture of rod and cone circuits to the on-beta ganglion cell. *Journal of Neuroscience* 8: 623–642.

van Rossum, M. C. and Smith, R. G. (1998). Noise removal at the rod synapse of mammalian retina. *Visual Neuroscience* 15: 809–821.

Xenopus laevis as a Model for Understanding Retinal Diseases

O L Moritz and D C Lee, University of British Columbia, Vancouver, BC, Canada

Glossary

AP20187 – A small molecule modeled on a dimer of the immunosuppressive drug FK506. One molecule of AP20187 can bind with high affinity to two FK506-binding protein (Fv) domains. Thus, Fv domains can be used to produce fusion proteins that will dimerize in the presence of AP20187. This in turn can be used to control the activity of proteins or enzymes whose activities are influenced by dimerization, such as caspases.

Bardet–Biedl syndrome – An autosomal recessive disorder characterized by obesity, retinal degeneration, polydactyly, hypogonadism, developmental delay, and mental retardation. Genes implicated in this syndrome are involved in ciliary transport processes.

Caspase-9 – An initiator caspase in the apoptotic cascade. Caspases are a family of cysteine proteases, which play essential roles in programmed cell death. Once activated, caspase-9 triggers activation of other caspases, precipitating the apoptotic process.

Optical coherence tomography (OCT) – A technique that permits three-dimensional imaging within tissues that scatter light. The technique is frequently used in ophthalmology to noninvasively image the cell layers of the retina.

Peripherin/RDS – A transmembrane glycoprotein found in the outer segment of both rod and cone photoreceptor cells. It is thought to be a structural protein important for disk morphogenesis. Mutations in the gene encoding peripherin/RDS are associated with a variety of autosomal dominant retinal dystrophies; also referred to in publications as rds, peripherin, or peripherin-2.

Rab proteins and Arf4 – The members of the Ras superfamily of monomeric guanine-nucleotide-binding proteins (G proteins), which are involved in the regulation of membrane trafficking.

Retinal degeneration – A phenotype associated with many different retinal disorders, involving progressive death of retinal cells, usually of a specific cell type.

Retinitis pigmentosa – A hereditary retinal dystrophy characterized by defective dark adaptation, progressive loss of peripheral vision that may eventually extend to loss of central vision, and the appearance of black pigment in the fundus.

RNA helicase Ddx39 – A member of the DEAD box protein family of putative RNA helicases. Family members are characterized by the conserved motif Asp-Glu-Ala-Asp (D-E-A-D), and are involved in RNA metabolism.

Stargardt's disease – An autosomal recessive juvenile-onset form of macular dystrophy arising from mutations in the *ABCA4* gene. The *ABCA4* gene product is expressed in photoreceptor cells and is thought to be an ATP-dependent transporter for N-retinylidene-PE.

Introduction

The amphibian retina has many unique properties that have intrigued visual scientists for decades. Many studies have been conducted on the retina of *Xenopus laevis*, a frog species commonly used as a laboratory animal. These frogs were introduced to the research community in the 1930s and soon became widely available, after the discovery that they can be induced to lay eggs by injection of human chorionic gonadotropin (known as the Hogben test for pregnancy). As a research subject, *X. laevis* have advantages over other amphibians in that they have low maintenance requirements (they are entirely aquatic and do not require live food), and they can easily be induced to lay eggs. This made them a desirable model organism for developmental biologists studying fertilization and embryonic development; however, they have also been adopted by vision researchers as a model amphibian retina. The large size of the principle rod photoreceptors makes them highly amenable to biochemical, morphological, and electrophysiological studies (**Figure 1**). Early studies on the properties of the *X. laevis* visual system laid the groundwork for the use of this animal as a model organism for the study of visual disorders, while more recently developed techniques for genetic manipulation have resulted in *X. laevis* models of inherited retinal disease that are particularly amenable to certain forms of analysis.

Early Work on *X. laevis* – Biochemistry, Electrophysiology, and Microscopy

Modern research on the amphibian retina dates back to the extraordinary anatomical studies of Cajal and

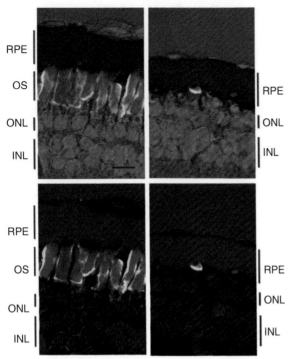

Figure 1 Expression of a bovine rhodopsin P23H mutant transgene causes retinal degeneration in *Xenopus laevis*. Combined confocal/DIC (upper panels) or confocal micrographs (lower panels) of developmental stage 47/48 wild-type retinas (left panels) or transgenic retinas expressing bovine-rhoP23H (right panels). Animals were raised in bright cyclic light for 2 weeks, beginning at fertilization. The 12-μm retinal cyrosections were stained with wheat germ agglutinin (red), B630N anti-rhodopsin (green), and Hoechst 33342 nuclear stain (blue). The central retinas of animals expressing bovine P23H rhodopsin (right panels) have almost no remaining rods, with those that remain having short and irregular outer segments. Note the relatively large diameter (7 μm) of the rod photoreceptors relative to those of the mammalian retina. Due to the relatively large diameter of the photoreceptors, only a single row of nuclei is required in the ONL. RPE, retinal pigment epithelium; OS, outer segment; INL, inner nuclear layer; ONL, outer nuclear layer. Scale = 10 μm.

others in the 1800s that defined many of the principal cell types of the retina. Most early studies dealt with *Rana* and *Bufo* species. However, *X. laevis* have been used as a subject of retinal research dating back more than 50 years. In the 1950s, Wald and co-workers used *X. laevis* for investigations of visual pigment biochemistry. From the most abundant rod photoreceptors, it was found that the principal pigment exhibited maximal absorbance at 523 nm.

The amphibian retina is also of interest to electrophysiologists. Although the first electroretinogram (ERG) was recorded by Holmgren in the 1860s from a frog eye, *X. laevis* frogs were not used extensively for electrophysiological studies until the 1970s, with initial studies of *X. laevis* retinal physiology performed by Ripps and co-workers, correlating visual pigment content and photoreceptor threshold.

In the 1950s and 1960s, detailed analysis of the ultrastructure of photoreceptors was obtained by electron microscopy, leading to the current understanding of photoreceptor disk membrane structure, and the mechanisms of disk membrane synthesis and renewal. Many of these studies utilized amphibian retina, again typically *Rana* species, although Lanzavecchia examined the ultrastructure of *X. laevis* rods and cones in 1960. Further studies by Kinney and Fisher further characterized the morphogenesis and ultrastructure of the principal rod photoreceptor in *X. laevis*.

In the 1940s and 1950s, work by Sperry and co-workers demonstrated regeneration of the optic nerve after transection in various amphibians. Beginning in the late 1960s, Jacobson and co-workers conducted investigations of retinal development in *X. laevis*, with particular emphasis on the development of retinotectal projections. These studies employed embryonic surgery (e.g., inversion of the eye) to identify the origin of signals for optic nerve axon guidance, and demonstrated that the location of synapses of ganglion cell axons in the optic tectum are specified by the location (dorsal, ventral, nasal, or temporal) of the cell bodies in the retina.

These early investigations established the *X. laevis* retina as a viable subject for retinal research that is still in use, including ongoing studies of *X. laevis* disk shedding and renewal, electrophysiology, biochemistry, and retinal development, that are further improving our understanding of retinal function. Additionally, several studies discussed below have directly modeled human retinal disorders in *X. laevis*, in order to better understand retinal dysfunction.

X. laevis as a Model for Vitamin A Deprivation

One of the first instances of modeling retinal disease in *X. laevis* was a study by Witkovsky and co-workers on vitamin A deprivation in tadpoles. This model reproduced features of night blindness (i.e., decreased rod sensitivity) seen in vitamin-A-deprived patients. Among other novel findings, the authors demonstrated that bleaching of photopigment causes a greater reduction in photoreceptor sensitivity than can be accounted for by reduction in pigment quantity alone. This was accounted for in later studies, which demonstrated that opsin has a greater tendency to activate the visual transduction cascade than rhodopsin. More recently, studies on vitamin A deprivation in *X. laevis* have been continued by Solessio and co-workers, who have also pioneered the use of psychophysical measurements of *X. laevis* visual sensitivity.

X. laevis as a Model for Glaucoma

Early studies by Jacobson and Keating involving optic nerve transection in *X. laevis* are similar to paradigms currently used in research of optic nerve injury or glaucoma, although more typically the research subject is a mammal. However, unlike a mammalian optic nerve, a severed *X. laevis* optic nerve regenerates (a property that would be highly desirable in glaucoma patients). Research focused on *X. laevis* models of retinal regeneration of this type is currently being continued by Belecky-Adams and co-workers, who recently identified a possible role of the RNA helicase Ddx39 in the regulation of stem cells in the retina.

X. laevis and Studies of the Transport of Rhodopsin

The first investigations of the mechanism of outer segment renewal utilized amphibians injected with radioactive amino acids in a pulse–chase paradigm that allowed newly synthesized membranes to be visualized by autoradiography. The original experiments by Young and co-workers demonstrated that new disks were formed at the base of the rod outer segment, and subsequently became discontinuous with the outer segment plasma membrane. These studies were further extended by Besharse and Hollyfield, who examined the influence of light on disk synthesis in *X. laevis* photoreceptors, demonstrating diurnal regulation of disk membrane synthesis. Collectively, these studies demonstrated the extremely high rate of rod photoreceptor disk membrane synthesis in *X. laevis* retina, estimated at roughly 50-fold higher than in mammalian photoreceptors.

These findings suggested that, due to the enormous rate of outer segment membrane synthesis, the amphibian retina would be an excellent choice for studying the biosynthesis of rhodopsin. This work was pioneered by Papermaster and Dreyer, and subsequently continued by Papermaster and co-workers. These studies largely used *Rana* and *Xenopus* species, and resulted in identification of molecular and ultrastructural details of the rhodopsin transport pathway. Components of the biosynthetic machinery and transport mechanisms first identified in amphibia included associated features of the connecting cilium (the pericilliary ridge complex), and vesicles transporting rhodopsin (RTCs or rhodopsin transport carriers). Deretic was able to reconstitute rhodopsin transport in amphibian retinal extracts, and using this assay demonstrated that the rhodopsin transport signal was located in the cytoplasmic C-terminal domain. This *in vitro* assay and associated methodologies subsequently led to identification of a number of molecules associated with rhodopsin transport, including the small G proteins, rab6, rab8, rab11, and arf4.

However, due to the lack of an effective system for culturing photoreceptors, it was not possible to incorporate molecular biology approaches into the study of

amphibian rhodopsin transport pathways until the 1990s, when techniques for the production of transgenic *X. laevis* developed by Kroll and Amaya, and cloning of promoters suitable for driving expression in *X. laevis* rods, allowed the study of rhodopsin transport in a genetically manipulated amphibian retina. Tam and colleagues found that rhodopsin-green fluorescent protein (GFP) fusion proteins expressed in *X. laevis* retina were transported correctly to rod outer segments. This allowed further identification and *in-vivo* demonstration of the function of the outer segment localization signal QVAPA, located at the extreme C-terminus of rhodopsin. Several mutations affecting this region cause retinitis pigmentosa (RP). These studies also demonstrated the extraordinary utility of transgenic *X. laevis* for conducting comparisons between transgenic animals, as numerous primary transgenic animals carrying different transgenes can be generated in a relatively short amount of time. For example, the rhodopsin outer segment localization signal was identified and refined using 15 distinct transgene constructs.

Using the same transgenic *X. laevis* system, the function of rab8 in rhodopsin transport was explored using dominant-negative and constitutively active mutants. Expression of these mutant rab8-GFP fusion proteins in *X. laevis* rods generated the first *X. laevis* models with an inherited retinal degeneration (RD) phenotype, as these fusion proteins proved to be quite toxic to retinal rods, and the phenotype was passed to F1 offspring. The expression of dominant-negative GFP-rab8T22N caused a particularly rapid death of photoreceptors associated with accumulation of rhodopsin-containing intracellular vesicles in the vicinity of the base of the connecting cilium. Although it was not clear at the time, subsequent studies indicate that the resulting phenotype may be closely related to the RD associated with Bardet–Biedel syndrome (BBS). Similar investigations of the small G-protein Arf4, which binds the rhodopsin outer segment localization signal, are ongoing.

The RD observed in this study demonstrated a unique central-to-peripheral distribution subsequently seen in all other *X. laevis* models of RD. This is associated with the rapid growth of the eye in young *X. laevis*, which results in continuous addition of new photoreceptors to the peripheral retina. Thus, a single cryosection can demonstrate all stages of photoreceptor degeneration moving from central retina to periphery.

Modeling RP in Transgenic *X. laevis*

Subsequently, genetically modified transgenic *X. laevis* was used in a number of studies that directly examined mutations associated with the human disorder RP, an inherited form of RD. The pioneering study involved transgenic expression of mutant forms of peripherin/rds, a protein found at the periphery of rod outer segment disks. Peripherin/rds and its mutants were expressed as GFP fusion proteins using the rod opsin promoter to drive expression in retinal rods, at levels sufficiently high to cause RD in some cases. Confocal microscopy of the intrinsic GFP fluorescence showed that several fusion proteins had unique localization patterns distinct from wild type that respectively suggested either specific disruptions in normal function, or misfolding and endoplasmic reticulum (ER) retention. Furthermore, electron microscopy revealed unique abnormalities in disk organization, possibly associated with disruption of the normal functions of the peripherin/rds C-terminus.

Previous attempts to use rhodopsin-GFP fusion proteins to develop similar models of RP were unsuccessful, most likely due to low expression levels. In order to adapt the system for the study of RD induced by rhodopsin mutants, a system was devised for detection of nonfluorescent transgene products based on nonconserved rhodopsin antibody epitopes, such that epitope tags involving minimal (or no) sequence changes could be introduced. Initially, this system was applied to the study of an RP-causing mutation (Q348ter) that disrupts the previously identified rhodopsin outer segment localization signal. In these studies, the power of the *X. laevis* system for drawing comparisons between transgenes was further expanded. Rhodopsin mutants defective in signal transduction properties were combined with rhodopsin mutants responsible for RP to dissect the role of rhodopsin signal transduction in rod cell death pathways. The results demonstrated rhodopsin mislocalization was associated with axonal sprouting and cell death, regardless of whether rhodopsin signal transduction properties were inhibited.

The same system was subsequently applied to the study of the rhodopsin mutation P23H, the most common cause of autosomal dominant RP in North America. Despite numerous studies of this rhodopsin mutant in cultured cells and transgenic rodents, there was no clear consensus as to the effects of this mutation on rhodopsin function; in cultured cells, it was classified as a mutant defective in folding and ER exit, while transgenic animal studies suggested it was transported correctly to rod outer segments, where it caused RD that was exacerbated by light.

Studies in transgenic *X. laevis* compared several different forms of P23H rhodopsin, including P23H rhodopsins based on different species, and P23H rhodopsins defective in signal transduction and chromophore binding. Interestingly, all forms of P23H rhodopsin caused RD, but varied in terms of ER retention, expression level, and light sensitivity. P23H rhodopsins that exhibited dramatic ER retention (*X. laevis* P23H rhodopsin) caused RD under all circumstances, while P23H rhodopsins that were

transported in small quantities to the OS (bovine P23H rhodopsin) caused RD only on light exposure (**Figure 1**). Furthermore, for bovine P23H rhodopsin, disruption of the chromophore-binding site was associated with reduced expression levels and RD, regardless of light exposure. This result reconciles the differences seen between previous studies, suggesting that in some forms of P23H-induced RD, chromophore binding promotes ER exit of newly synthesized rhodopsin. The dramatic sensitivity of these phenotypes to the underlying rhodopsin sequence was confirmed by Zhang and colleagues, who demonstrated light-sensitive RD in an *X. laevis* rhodopsin that differed from that used previously only in the sequences of the epitope tags. In addition to providing insight into the mechanisms underlying RD, these studies dramatically emphasize the difficulties in extrapolating results reported from a single disease model to human disease states.

A unique finding in this system was the presence of considerable quantities of truncated P23H rhodopsin, in which a significant portion of the N-terminal domain (including the mutated H23 residue) was removed; in fact, this was the dominant species observed in retinas expressing bovine P23H rhodopsin. This truncated species was also previously observed in cultured cells, although in smaller quantities. Identification of this species in other transgenic models would be difficult due to lack of a suitable reagent for detection, but was readily achieved in *X. laevis* due to the availability of both N- and C-terminal specific antibodies that did not cross-react with endogenous rhodopsin.

Subsequent studies of the same *X. laevis* models of P23H-rhodopsin-induced RD probed the association of chromophore binding and ER exit. In order to address the question of whether the causative factor in light-induced RD was a reduction in the supply of free 11-*cis* retinal, or isomerization of 11-*cis* retinal bound to P23H rhodopsin as chromophore, the sensitivity of RD to different wavelengths of light was examined. It was determined that the profile of light sensitivity was consistent with photoisomerization of rhodopsin (which maximally absorbs green light) rather than free chromophore (with maximal absorbance in the UV). This also brings to mind similar studies of the constitutively active rhodopsin mutant K296E, classified as misfolding by some studies in cultured cells, suggesting that the active conformation of rhodopsin and/or loss of chromophore can be associated with altered kinetics of ER exit.

Studies of additional RP-causing rhodopsin mutations in transgenic *X. laevis* are ongoing, and have been reported at international meetings, including K296E rhodopsin and the glycosylation-defective mutants T4K and T17M. Glycosylation-defective rhodopsin mutants are also reported to be associated with light-exacerbated RD in *X. laevis*.

Inducible RD

In an alternate approach to modeling RP in the transgenic *X. laevis* retina, Hamm and co-workers designed a drug-inducible model of RD driven by a modified form of caspase-9. Dimerization and subsequent activation of this caspase-9 transgene is driven by AP20187, a small molecule based on a dimer of FK506. Administration of AP20187 to these transgenic *X. laevis* induces rapid RD that is not dependent on any particular environmental condition (such as special lighting). The system is designed to examine the effects of rod degeneration on other cell types, including cone photoreceptors and cells of the inner nuclear layer, and to examine the capacity for regeneration of rods in the *X. laevis* retina. This study was also the first to provide functional (i.e., electrophysiological) data for an *X. laevis* model of RD.

Interestingly, this model demonstrated a dramatic reduction in electrophysiological responses to stimuli designed to isolate cone function (e.g., a rapid flicker stimulus), despite the fact that no associated cone death was detected. This reduction in cone sensitivity may be associated with the cones themselves, or with other cells associated with the cone pathway and the ERG B-wave (e.g., cone bipolar cells and/or Müller glia). The restoration of responses to flicker stimuli was associated with a thickening of the inner nuclear layer, and a similar thickening can be observed by optical coherence tomography (OCT) in human RP patients.

Other Transgenic *X. laevis* Models of Retinal Disease

In addition to rhodopsin and peripherin/rds mutations responsible for RP, gene products related to Stargardt's disease have been expressed in *X. laevis* retina as GFP fusion proteins in order to examine their localization properties, although this has not yet resulted in a replication of a RD phenotype.

In studies by Kefalov and colleagues, cone opsins were expressed in *X. laevis* rods. As cone photoreceptors are considerably noisier than rod photoreceptors, this allowed a determination of the proportion of cone dark noise (activation of transduction in the absence of photons) that is purely due to the cone pigment sequence, and not other aspects of the transduction cascade or photoreceptor environment. Although generating a model of disease was not a goal of this study, the resulting animals could be considered a model for congenital stationary night blindness, which is due to abnormally high activity of the visual transduction cascade in the absence of photons.

X. laevis as a Model for Eye Development/Developmental Disorders

The rapid development of the *X. laevis* embryos makes these animals of particular interest to developmental biologists, including those concerned with eye development. The first studies of the development of the *X. laevis* eye were conducted by Hollyfield through radioactive monitoring of the growth of the developing retina. Chung and colleagues histologically monitored the structural changes in the developing larval *X. laevis* retina and correlated these changes with electrophysiological changes, notably that while the receptive field of a ganglion cell remains constant in the developing larvae through metamorphosis, the inhibitory peripheral region expands to the entire retina of the adult *X. laevis*.

The development of the *X. laevis* eye has been manipulated by both the overexpression and by the knockdown of transcription factors. El-Hodiri and colleagues have extensively studied the transcriptional regulation of photoreceptor development. More recently, they identified a retinal homeobox gene family member, Rx-L, which regulates photoreceptor-specific gene expression. Expressed in developing embryos, the knockdown of Rx-L expression adversely affected photoreceptor development, causing subtle phenotypes of altered photoreceptor morphology.

Using a similar embryonic transfection paradigm, Knox and colleagues found that overexpression of the transcription factors, *Nrl* and *Nr2e3*, in *X. laevis* retina resulted in an increase in numbers of rods, with concomitant reduction in cone photoreceptors, indicative of the roles of these factors in determining the developing photoreceptor cell fate. This system may prove extremely useful in modeling developmental disorders of the retina with similar underlying mechanisms.

X. laevis Models of Retinal Regeneration

Some recent studies have investigated the fascinating capacity of the *X. laevis* retina to repair itself after severe traumatic injury. In these studies, the entire retina is excised from an *X. laevis* tadpole eye. The retina subsequently demonstrates a dramatic capacity to completely regenerate by transdifferentiation of the remaining cells of the retinal pigment epithelium (RPE). Certain aspects of this transdifferentiation can be reproduced in culture, and it appears to be dependent on diffusible factors (possibly fibroblast growth factor 2 (FGF2)) released from the choroid. These results could have implications for traumatic eye injuries such as retinal detachment, retinal degenerative disorders, and glaucoma.

Summary

As an unconventional system for modeling retinal disease, *X. laevis* presents a number of advantages. As it is quite easy to generate transgenic *X. laevis*, they are an excellent system for comparing the effects of multiple transgenes. Other advantages include the relative ease of microscopic and electrophysiological studies due to the large size of the photoreceptor cells, regenerative capacity of the retina, and non-cross-reactivity of mammalian antibodies. However, there are also significant disadvantages, such as the current lack of knock-out or gene-replacement capabilities, long generation time (1 year), pseudotetraploid genome, and relatively small eyes, such that it is clearly not an ideal system appropriate for all experiments. Rather, *X. laevis* models of retinal disease are a very useful addition to the library of systems and models available to vision researchers.

See also: The Photoreceptor Outer Segment as a Sensory Cilium; Primary Photoreceptor Degenerations: Retinitis Pigmentosa; Primary Photoreceptor Degenerations: Terminology; Retinal Degeneration through the Eye of the Fly; Secondary Photoreceptor Degenerations: Age-Related Macular Degeneration; Secondary Photoreceptor Degenerations; Zebra Fish as a Model for Understanding Retinal Diseases; Zebra Fish–Retinal Development and Regeneration.

Further Reading

Araki, M. (2007). Regeneration of the amphibian retina: Role of tissue interaction and related signaling molecules on RPE transdifferentiation. *Development, Growth and Differentiation* 49: 109–120.

Besharse, J. C., Hollyfield, J. G., and Rayborn, M. E. (1977). Turnover of rod photoreceptor outer segments. II. Membrane addition and loss in relationship to light. *Journal of Cell Biology* 75: 507–527.

Deretic, D., Williams, A. H., Ransom, N., et al. (2005). Rhodopsin C terminus, the site of mutations causing retinal disease, regulates trafficking by binding to ADP-ribosylation factor 4 (ARF4). *Proceedings of the National Academy of Sciences of the United States of America* 102: 3301–3306.

Hamm, L. M., Tam, B. M., and Moritz, O. L. (2009). Controlled rod cell ablation in transgenic *Xenopus laevis*. *Investigative Ophthalmology and Visual Science* 50(2): 885–892.

Hollyfield, J. G. (1971). Differential growth of the neural retina in *Xenopus laevis* larvae. *Developmental Biology* 24: 264–286.

Pan, Y., Nekkalapudi, S., Kelly, L. E., and El-Hodiri, H. M. (2006). The Rx-like homeobox gene (Rx-L) is necessary for normal photoreceptor development. *Investigative Ophthalmology and Visual Science* 47: 4245–4253.

Papermaster, D. S., Schneider, B. G., Zorn, M. A., and Kraehenbuhl, J. P. (1978). Immunocytochemical localization of opsin in outer segments and Golgi zones of frog photoreceptor cells. An electron microscope analysis of cross-linked albumin-embedded retinas. *Journal of Cell Biology* 77: 196–210.

Sperry, R. W. (1944). Optic nerve regeneration with return of vision in Anurans. *Journal of Neurophysiology* 7: 57–69.

Tam, B. M. and Moritz, O. L. (2007). Dark rearing rescues P23H rhodopsin-induced retinal degeneration in a transgenic *Xenopus laevis* model of retinitis pigmentosa: A chromophore-dependent mechanism characterized by production of N-terminally truncated mutant rhodopsin. *Journal of Neuroscience* 27: 9043–9053.

Tam, B. M., Moritz, O. L., Hurd, L. B., and Papermaster, D. S. (2000). Identification of an outer segment targeting signal in the COOH terminus of rhodopsin using transgenic *Xenopus laevis*. *Journal of Cell Biology* 151: 1369–1380.

Tam, B. M., Xie, G., Oprian, D. D., and Moritz, O. L. (2006). Mislocalized rhodopsin does not require activation to cause retinal degeneration and neurite outgrowth in *Xenopus laevis*. *Journal of Neuroscience* 26: 203–209.

Witkovsky, P., Gallin, E., Hollyfield, J. G., Ripps, H., and Bridges, C. D. (1976). Photoreceptor thresholds and visual pigment levels in normal and vitamin A-deprived *Xenopus* tadpoles. *Journal of Neurophysiology* 39: 1272–1287.

Young, R. W. and Droz, B. (1968). The renewal of protein in retinal rods and cones. *Journal of Cell Biology* 39: 169–184.

Zhang, R., Oglesby, E., and Marsh-Armstrong, N. (2008). *Xenopus laevis* P23H rhodopsin transgene causes rod photoreceptor degeneration that is more severe in the ventral retina and is modulated by light. *Experimental Eye Research* 86: 612–621.

Zebra Fish as a Model for Understanding Retinal Diseases

A A Lewis, C C Heikaus, and S E Brockerhoff, University of Washington, Seattle, WA, USA

Glossary

Achromatopsia – A disease characterized by defects in the cone photoreceptors resulting in extreme light sensitivity and color blindness or rod monochromacy.

Apoptosis or programmed cell death – A form of cell death characterized by a series of biochemical and morphological changes resulting in the formation of apoptotic bodies and removal by the immune system.

Bystander effect – This describes the transmission of death from mutant or injured cells to healthy neighboring cells.

Cyclic guanosine monophosphate (cGMP) – A cyclic nucleotide derived from guanosine triphosphate (GTP). cGMP acts as a regulator of the cyclic-nucleotide-gated ion channels in photoreceptors.

Cyclic nucleotide phosphodiesterases (Pde) – A family of enzymes that hydrolyze the phosphodiester bond in the second-messenger molecules cAMP and cGMP. They regulate the localization, duration, and amplitude of cyclic nucleotide signaling.

Electroretinography (ERG) – A method for evaluating visual response. Electrodes are placed against the cornea and used to measure the electrical responses of various cell types in the retina to a light flash of varying intensity.

GAF domains – A large group of protein domains that bind small molecules; in the Pde proteins these domains bind cyclic nucleotides. The GAF acronym comes from the names of the first three different classes of proteins identified to contain them: cGMP-specific and-regulated cyclic nucleotide phosphodiesterase, adenylyl cyclase, and *E. coli* transcription factor FhlA.

Gap junctions – The intercellular connections that occur between some types of cells. These junctions allow the movement of various molecules and ions between cells.

Optokinetic response (OKR) – A method for evaluating zebrafish vision. Fish are placed in a small dish in the center of a rotating drum decorated with vertical stripes. If the fish can see, its eyes will follow the rotating stripes with regular reflexive saccades.

Retinitus pigmentosa (RP) – A group of diseases characterized by defects in the rod photoreceptors resulting in night blindness. Progressive RP often results in cone loss and tunnel vision in some cases progressing to total blindness.

Scotopic vision – The low light vision that is produced exclusively by rod function.

Zebrafish or *Danio rerio* – A tropical freshwater fish of the minnow family that has gained prominence as a scientific animal model. For further information see the Zebrafish Information Network (ZFIN), an online database of zebrafish genetic, genomic, and developmental information.

Introduction

Inherited photoreceptor degenerations are a major cause of incurable blindness. Degenerations can affect rods, causing night blindness, cones, causing color and daylight blindness, or both cell types, leading to complete blindness. Although there are many models of retinal degeneration caused by variety of mutations in different genes, it is still not possible to completely describe the molecular cascade causing cell death in any of these disorders and therefore it is equally difficult to prevent the degenerative process. Fundamental new information about the biochemistry of photoreceptor cell death is required to enhance our understanding of retinal degeneration and to develop new successful therapies. Zebrafish have gained prominence as a model organism for studies of retinal development, disease, and vision because they offer some distinct advantages over other genetically tractable systems. Zebrafish develop rapidly *ex utero* and can be maintained transparent allowing cells in the retina to be visualized in live larvae in real time using confocal and multiphoton imaging techniques. This provides the opportunity to visualize morphological and biochemical changes occurring in diseased photoreceptors with cellular and subcellular resolution. Here, we describe two mutations in the zebrafish cone phosphodiesterase (*pde6c*) gene that result in retinal degeneration. These mutants provide a unique opportunity to learn more about the biochemical triggers and inhibitors of cell death within the retina.

Retinal Disease

Photoreceptors are the primary sensory cells within the visual system. There are two main types of photoreceptors within the vertebrate eye: rods, which are monochromatic and respond in low light levels, and cones, which respond to higher light levels and specific wavelengths within the visual spectrum. The system of light absorption, ion fluctuation, and neuronal transmission are processes that require significant energy and, thus, the retina is one of the highest-energy-consuming tissues in the body. The high level of oxygen consumption by photoreceptors makes them particularly susceptible to injury and perturbation often resulting in cell death. Retinal degeneration is a leading cause of blindness in the developed world. The most common form of degeneration is age-related macular degeneration, which first affects the cones within the central retina (macula) and then progresses to the periphery.

Retinal degeneration can also occur as a result of genetic mutations. Achromatopsia and Retinitis pigmentosa (RP) together define a large class of heritable diseases that affect vision in humans. Both of these diseases are caused by a wide variety of mutations that disrupt visual transduction and photoreceptor maintenance. Achromatopsia is characterized by defects in the cone photoreceptors while rods remain functional, resulting in extreme light sensitivity and color blindness or rod monochromacy. Achromatopsia symptoms are generally seen at birth and the vision loss is only rarely progressive. RP, in contrast, develops during childhood and in later stages of life starting with degeneration of the rod photoreceptors, and progressing in some cases to total blindness. RP is estimated to affect 1 in 10 000 people. Due to the diversity of genes associated with these disease states, most therapeutics have focused on the general prevention of cell death as a method for limiting progressive vision loss.

Animal Models of Vision

Many animal models are used to study the visual system, and each possesses various strengths and weaknesses. By using a variety of systems scientists are able to capitalize on the strengths of all of them. The two most common systems for retinal modeling have been the fruit fly, *Drosophila melanogaster*, and mice, with a variety of work occurring in related species such as rabbit or ferret.

Drosophilae have mainly been used to study ocular development and patterning. Surprisingly, despite the obvious structural differences between the insect compound eye and our own, many of the same signaling cascades are used to establish the structure and patterning of the *Drosophila* eye. *Drosophilae* have an extremely short

gestation period and well-established methods for rapid genetic manipulation, and these features have been used to understand the roles of multiple genes in eye development. However, there are still many limitations in the use of this organism, particularly in the modeling of retinal disease. Several differences exist between the insect and mammalian phototransduction cascade, and the structural differences of the compound eye also limit the applicability of this organism for disease studies.

The most commonly used mammalian animal model of the retina is the mouse. Mice have a number of advantages as a model system. There is an extensive literature on a variety of mutants that have been studied for many years. There are also multiple techniques for genetic manipulation, including the ability to modify genes and genomic loci and several techniques for retinal explantation for *in vivo* imaging. However, mice are nocturnal creatures and depend primarily on their olfactory system for foraging and predator identification. As a result, unlike the human retina, the mouse retina is dominated by rod photoreceptors and contains only 3% cone photoreceptors. While this makes mice ideal for studying rod photoreceptor disease and function, the study of cone physiology and function is less straightforward in this system. Additionally, mouse eye development occurs *in utero*, making it difficult to image and understand the early stages of eye development and retinal perturbation.

The Advantages and Techniques of the Zebrafish Model System

The zebrafish system has a number of benefits for studying visual development and disease. First, zebrafish vision, similar to humans, is cone dominated. They rely upon their vision for food acquisition and have four different cone types: red, green, blue, and ultraviolet. In the zebrafish, ocular development occurs externally over the first 5 days postfertilization (dpf). During zebrafish development, the cone photoreceptors mature first between 3–5 dpf followed by the rods, which mature between 15–20 dpf. Thus, early zebrafish vision is dependent solely on cone-mediated vision. The rapid development of the visual system and early cone dominance have facilitated several genetic screens using young embryos to select specifically for cone related defects (see below).

Zebrafish were originally developed as a model system because they are inexpensive to keep, and take only 3 months to reach sexual maturity after which a mating pair will produce 200–300 eggs per week for at least a year. Further, development of many different genetic tools has added greatly to the versatility of this model for scientific study. In particular, mutagenesis protocols were optimized in the mid-1990s to introduce high frequencies of mutations. This made it possible to conduct large-scale forward genetic screens. Hundreds of different mutants affecting many aspects of vertebrate development and function have been identified. The constantly improving annotation of the nearly completed genome makes identifying mutated genes straightforward, and an impressive collection of vision mutants is available.

Another advantage is that cellular behavior can be observed in the intact, living organism. During the first 2 weeks of development, zebrafish larvae can be maintained in a translucent state and visualized live using either exogenous or genetically encoded fluorescent markers (see examples in **Figure 1**). Zebrafish can be maintained alive and healthy in agar on the microscope stage for days. Thus, cells can be imaged over the course of development or degeneration without perturbing the extracellular environment. At this point, transgenic lines containing fluorescent markers in various cell types are simple to generate and maintain. Transient injection of plasmid DNA at the one-cell stage will also produce a mosaic expression of genes throughout the fish allowing for the study of isolated cells within the intact animal. Multiple promoters have been developed to drive expression in various types of cells within the eye.

While no procedures exist to manipulate specific genes within the genome, several methods have been developed which allow scientists to circumvent this limitation. Morpholinos are synthetic RNA-like oligonucleotides that can be injected at the one-cell stage or electroporated into the fish at later stages. These morpholinos bind to corresponding RNA sequences within the cell, resulting in the degradation of the RNA and loss of protein expression. The loss-of-protein function is not as complete as is seen in knockouts and is only transient, but it has allowed

researchers to study the effects of specific knockdowns during development. Additionally, several labs have developed a method known as targeting-induced local lesions in genomes (TILLING) in which high-throughput polymerase chain reaction (PCR) methods are used to identify specific gene mutations from libraries of randomly mutagenized fish.

Very recently, another method has been established in which nucleases are used to produce double-strand breaks at specific loci within the genome. These double-strand breaks are repaired by nonhomologous end joining, often resulting in deletions or insertions at the break site that can lead to frame-shift mutations in the target gene. The targeting specificity of these double-strand breaks is established by fusion to an array of zinc finger domains that bind to specific DNA sequences. Each zinc finger recognizes a 3-bp sequence and three zinc fingers are fused together to create a 9-bp recognition domain. Further, these zinc finger nuclease arrays must dimerize to activate the nuclease such that a total 18-bp recognition sequence is required. Currently, the technology for generating the zinc finger arrays is cumbersome, but soon this will be a rapid and convenient way to generate targeted zebrafish mutants.

Methods have also been developed to create chimeric fish allowing for the rapid evaluation of the cell autonomous nature of genetic phenotypes. To produce these fish, eggs are grown to the blastula stage and then cells from one egg are removed and inserted into another egg. This procedure does not affect fish development and wild-type chimeras grow normally. The production of chimeric fish can be used to determine the critical cell population for the phenotypic changes associated with a mutation. It can also be useful for the evaluation of neighboring and surrounding effects.

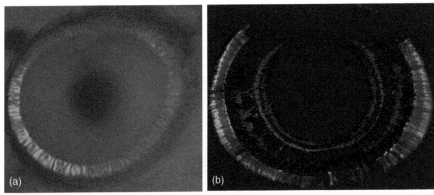

Figure 1 The zebrafish eye. These images illustrate the translucence of the eye and the ability to image the eye *in vivo*. For these experiments, zebrafish are anesthetized and embedded in a 0.5% agar solution during imaging. The zebrafish can survive under these conditions for up to 2 days. (a) A fluorescent image of the eye with the cone photoreceptors expressing the transgene for green fluorescent protein (GFP) under the control of a cone-specific promoter (TαCP), shown over the differential interference contrast (DIC) image of the eye. (b) An eye showing the photoreceptors in blue, expressing the transgene for cyan fluorescent protein under a cone-specific promoter (Tg(TαCP:MmCFP)) and the secondary layer of neurons or bipolar cells in red, expressing yellow fluorescent protein under a bipolar-specific promoter, (Tg(nyc:mYFP)).

Zebrafish have one other major advantage for the study of disease, which is the aqueous environment in which they reside. Methods are being developed for rapid high-throughput drug screens using zebrafish by adding compounds to their water and evaluating the effects. Most of the current studies with this method use fluorescent cell markers to indicate the presence or absence of various cells. This technique has been used with fluorescent hair cells to identify compounds that prevent hair cell death in the presence of the ototoxic agent neomycin. As zebrafish are small and easy to maintain, they are the best vertebrate model for this type of shotgun approach to drug development.

Evaluating zebrafish vision

There are several ways of evaluating zebrafish vision. The simplest involves a manipulation of the fish instinct to maintain its position within a moving stream. In this test, groups of fish can be placed in a long dish with a series of moving bars along its side. As the bars move, the fish respond to the apparent current by swimming to maintain their position with the bars. Thus, shoals of fish can be tested simultaneously for their ability to see and respond to the moving lines. This is known as the optomotor response. A related test called the optokinetic response (OKR) is done with a single fish placed in a small dish in the center of a rotating drum decorated with vertical stripes. In this test, the fish's eye will follow the stripes with periodic involuntary saccades. This is a very sensitive measure of an individual fish's ability to perceive its visual environment. Several labs have identified blind fish in mutagenesis screens using the optomotor response and/or the OKR. These screens have yielded a variety of mutants that can be used as models for retinal disease.

Another method that has been used to measure fish vision is the electroretinogram (ERG), which is a stimulated measure of the electrical response of the eye to a flash of light. In these measurements, the fish are dark adapted and a small electrode is placed on the cornea. The eye is then stimulated with a light flash and the electrical response is recorded. This technique records both the primary photoreceptor response, which appears as a negative spike at the beginning of the recording known as the a-wave, and the response of the secondary neurons, which is a large secondary positive response following the a-wave, known as the b-wave.

The Visual System: Phosphodiesterase and Phototransduction

Cyclic GMP (cGMP) phosphodiesterase (Pde) is an important enzyme in the process of phototransduction. During phototransduction the 11-*cis* retinal absorbs a single photon of light and alters the conformation of the opsin protein to activate the associated heterotrimeric guanosine triphosphate (GTP)-binding protein, transducin. The activated transducin removes the inhibitory gamma subunit from phosphodiesterase, which then degrades cGMP in the outer segments (for a schematic see **Figure 2**). The lowering of cGMP levels stimulates the closure of cyclic-nucleotide-gated ion channels in the plasma membrane, causing hyperpolarization and a decrease in neurotransmitter release at the synapse, initiating light signaling to downstream neurons. Channel closure also interrupts Ca^{2+} influx leading to a decrease in intracellular $[Ca^{2+}]_i$. The drop in intracellular $[Ca^{2+}]_i$ activates a Ca^{2+}-sensitive guanylyl cyclase, restoring cGMP to presignaling levels. Both types of photoreceptors have a similar phototransduction cascade, but use different

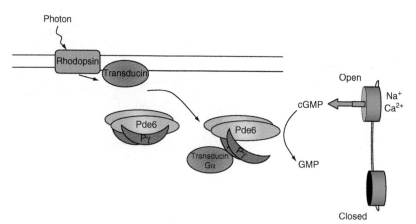

Figure 2 A schematic of the initial steps of phototransduction. The absorbance of a photon activates the 11-*cis* retinal of rhodopsin, which in turn activates the heterotrimeric GTP-binding protein, transducin. The Gα subunit of transducin binds to the inhibitory Pγ subunit of the Pde6 holoenzyme and relieves the inhibition of the catalytic domain. Pde6 then cleaves cGMP, resulting in a drop in cGMP levels that causes the closure of cGMP-gated ion channels in the plasma membrane.

genetically encoded enzymes to accomplish each task. Therefore, a mutation in the cone phosphodiesterase will not affect rod phototransduction.

The photoreceptor Pde (Pde6) is a multisubunit enzyme that differs slightly in rods and cones. In both cells the catalytic domain is a dimer that is bound and inhibited by the regulatory gamma subunit called Pγ. During phototransduction, activated transducin binds to Pγ and exposes the catalytic cGMP-binding site allowing the catalytic domain to cleave cGMP. In rod photoreceptors, two related but different proteins, Pα and Pβ, form the Pde6 catalytic domain. In cones, the catalytic domain consists of a dimer of one protein Pα', also known as Pde6c.

Retinal Degeneration in *pde6* Mutants: Primary Degeneration

Mutations in the Pde gene (*pde6*) are found in several families with RP and similar mutations have been used for many years as a model for RP in mice. The oldest and most commonly used mouse model of RP is the retinal degeneration 1 (*rd1*) mouse that contains a mutation in the Pβ rod-specific subunit of Pde6, the *pde6b* gene. In this model, rod degeneration is apparent by postnatal day 8 and nearly all of the rods are lost by 3 weeks of age. Despite many years of study, the process of primary photoreceptor degeneration in the *rd1* mouse is not well understood. Initial work has implicated programmed cell death pathways. The death of rods is associated with the extreme DNA cleavage that accompanies apoptosis, and this DNA cleavage can be detected with the terminal deoxynucleotidyl transferase biotin-dUTP nick end labeling (TUNEL) assay. However, several studies suggest that the standard apoptotic cascades, including a group of cysteine proteases known as the caspases, and other typical apoptotic effectors, are not involved. Instead, recent work implicates intracellular calcium concentration ($[Ca^{2+}]_i$) and the calpains, a calcium-activated set of proteases found in the mitochondria, as the relevant initiators of photoreceptor programmed cell death.

In *rd1* mice, the absence of Pde6 results in elevated levels of cGMP even under dark conditions, and it has been hypothesized that this results in the greater open probability of the cyclic-nucleotide-gated ion channels leading to increased $[Ca^{2+}]_i$ levels and apoptosis. However, little is known about the distribution of $[Ca^{2+}]_i$ during the death of cones. An initial report using the Ca^{2+} channel blocker, D-*cis*-diltiazem, to inhibit $[Ca^{2+}]_i$ accumulation showed a decrease in apoptosis. However, other labs using the same and other Ca^{2+} channel blockers have been unable to repeat these results. Due to the difficulty of imaging a mouse retina *in vivo* over long time periods, none of these studies has examined the levels of $[Ca^{2+}]_i$ within the cells *in vivo*. The translucence

of the zebrafish retina enables *in vivo* imaging of both the morphological changes of degeneration as well as changes in levels of signaling molecules, such as Ca^{2+} or cGMP.

Retinal Degeneration in *pde6* Mutants: Secondary Retinal Degeneration, the Bystander Effect – Models and Mechanisms

In humans, RP is characterized by a lack of night vision, but often progresses over time to tunnel vision and, in some cases, complete blindness. This progression of the disease is due a gradual death of cones by apoptosis. Cell death begins in the peripheral retina where the number of rods is highest and moves toward the central retina. The cone photoreceptors are fully functional and do not use the mutated genes for phototransduction. However, the death of the mutant rods causes healthy cones to die apoptotically. This transmission of death to healthy neighboring cells is a process known as the bystander effect. The death of these cones represents the most debilitating part of this disease, and has significant potential for therapeutic intervention.

In addition to retinal degeneration, the bystander effect has been seen in a variety of diseases, including cancerous tumors, and can be either beneficial or detrimental to therapeutic efforts. One of the first descriptions of the bystander effect occurred during studies of gene therapy for cancer in which malignant tumors were injected with viruses containing suicide genes, which convert a prodrug into a lethal compound inside cells. Researchers found that, although only a small population within the tumor expressed the detrimental genes, significant portions of the tumor mass still died, suggesting that virally transfected cells were able to induce death in untransfected neighboring cells. Recently, researchers have found a similar occurrence in cells exposed to radiation therapy. In this case, cells that have not been irradiated show the genetic instability associated with radiation exposure.

In general, apoptosis does not affect the health of neighboring cells and it is unclear why in some instances there is a spread of death across a population. There are currently several hypotheses for how healthy cells are induced to die. One possibility is that live cells release a trophic factor that stimulates the growth and differentiation of their neighbors and is required for their proper maintenance. For instance, it has been suggested that rods release a factor that stimulates the growth of cones. Another possibility is that dying cells release toxic factors that kill neighboring cells. A potential corollary to both of these hypotheses is that either toxic or trophic factors are released to neighboring cells through gap junctions. It is also possible that the immune response triggered by the

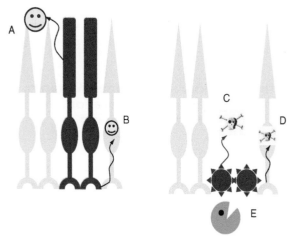

Figure 3 Schematic of some of the possible sources of the bystander effect. (a, b) Live cells release trophic factors (happy faces) that help neighboring cells and are lost when the mutant cells (in red) die. These factors could be released exogenously (a) or through gap junctions between cells (b). (c, d) Dying cells release toxic factors (skulls) that kill neighboring cells, again either exogenously (c) or through gap junctions (d). (e) The immune response to the presence of dying cells could have a deleterious effect on the remaining healthy cells.

removal of apoptotic cells has a deleterious effect on the neighboring cells. See **Figure 3** for a schematic of these possibilities. Understanding the source of the bystander effect will suggest other methods by which it might be prevented.

Levels of the Bystander Effect in Photoreceptor Degeneration

The extent of bystander cell death is not the same for all cases of retinal degeneration. The death of cones in rod–cone dystrophies has been extensively studied in human patients and in a variety of animal models, including the *rd1* mouse. Mutations in rod phototransduction, resulting in rod death, almost always lead to some cone degeneration. However, the number of cells that die within the central fovea varies significantly from patient to patient. Mutations in genes required for cone phototransduction and their effects on rods are less well studied. Cone–rod dystrophies are conditions in which degeneration of the cones causes a decrease in scotopic vision, low light vision that is produced exclusively by rod function. Little is known about the state of rods in these patients, but it is thought that there is some degeneration associated with the visual loss. In contrast, some patients with cone dystrophies have cone death without affecting the rods or scotopic vision.

Differences in the connections between cells in a population might account for the variable levels of bystander death. In particular, it is thought that gap junctions

between rods and cones in mammals may be predisposed to allow the flow of materials from the rods to the cones, but not from the cones to the rods. Thus, it may be possible that the flow of information would be asymmetrical between cell types and this could lead to differences in the effects on neighboring cells.

In order to better understand the bystander effect, an important first step is to determine how death progresses throughout the population and which populations of cells are capable of propagating apoptotic signals to healthy neighbors. For this type of analysis, zebrafish provide an ideal system. Not only are mosaic animals easy to generate, but the transparency of the embryo make it possible to analyze the transmission of death *in vivo*. This feature combined with the other tools described above provides a novel and powerful approach to examining the bystander effect within the photoreceptor population.

Zebrafish Models of Retinal Degeneration: Mutations in *pde6c*

Mutations in the zebrafish cone phosphodiesterase have been identified in two separate genetic screens for fish lacking OKR at 5 dpf. One of the mutations is a null mutant (*w59*), while the other is a missense mutation in a conserved amino acid (*els*). Recently, recessive mutations in *pde6c* were also identified in mice and in three human families with achromatopsia. Preliminary results indicate that in humans this form of achromatopsia may be progressive, suggesting the possibility of a bystander-associated death of rods.

pde6c^*w59*^

pde6c^*w59*^ is a mutation in the splice site between exons 11 and 12 in the *pde6c* gene. The abnormal splicing caused by this mutation introduces a premature stop codon generating a null mutant. Zebrafish, homozygous for the *pde6c*^*w59*^ mutation, are viable and form normal swim bladders, a marker of general fish health, but must be raised with higher-than-normal food concentrations, as their visual defects impair acquisition of food (**Figure 4(a)**). In these fish, cone photoreceptors initially develop normally. However, at 4 dpf, when fish first respond to light, the cone photoreceptors in the central retina begin to degenerate. The *pde6c*^*w59*^ mutants exhibit a flat ERG at 5 dpf, indicating a total lack of photoreceptor response to light and no signaling to downstream neurons. At this stage in development, zebrafish vision relies on cones, so a flat ERG is consistent with the rapid cone degeneration observed in these fish. As the cone photoreceptors die, they retract their synaptic connections and outer segments and become spherical. Time-lapse images of this process are shown in **Figure 4(b)**. In the final stages of cell death, only the

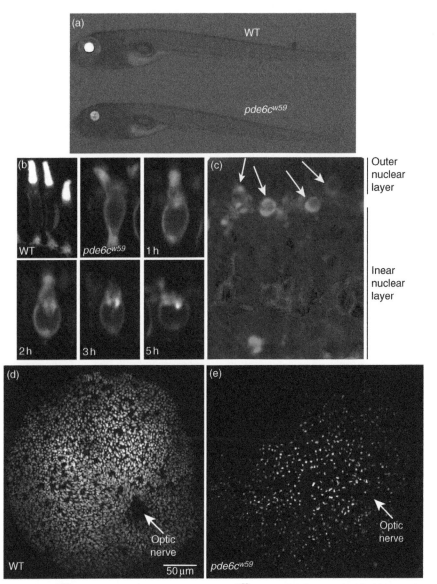

Figure 4 Images of the *pde6c^{w59}* mutant fish. (a) Homozygous *pde6c^{w59}* mutant fish have swim bladders, and are generally healthy. These fish contain the transgene *Tg(TαCP:MmCFP)*, which causes expression of membrane-tagged cyan fluorescent protein (MmCFP) specifically in cone photoreceptors. The level of fluorescence in the mutant eye is decreased due to the degeneration of cone photoreceptors. (b) A time-lapse sequence of a single mutant cone photoreceptor undergoing apoptosis. The initial picture is an image of wild-type cones. The subsequent pictures are a single *pde6c^{w59}* mutant cell over a 5-h time course. As the cells die, they retract their synaptic connections and round up to form apoptotic bodies which are eventually removed by the immune system. (c) Apoptotic photoreceptors (green) viewed using the membrane label boron-dipyrromethene (BODIPY), which labels all cellular membranes in the fish. Apoptotic bodies are clearly visible in the outer nuclear layer (arrows). Bipolar cells express yellow fluorescent protein (*Tg(nyc: mYFP)*) and are shown here in red. (d, e) The dramatic loss of fluorescently labeled photoreceptors in the whole eye of *Tg(TαCP:MmCFP)* *pde6c^{w59}* mutant. Eyes were removed from euthanized and fixed 6-day-old *Tg(TαCP:MmCFP)* WT (d) and *pde6c^{w59}* (e) mutant animals. Fluorescently labeled cone photoreceptors were viewed with confocal microscopy. From Lewis, Wong and Brockerhoff, in preparation.

rounded apoptotic bodies remain (**Figure 4(c)**, arrows). This cell debris travels out of the outer nuclear layer and is disposed of by the macrophages of the immune system.

Within the central retina of the *pde6c^{w59}* mutant, a majority of cones die by 5 dpf. **Figures 4(d)** and **4(e)** show the central retina of a wild-type (d) and mutant (e) eye, expressing membrane-tagged cyan fluorescent

protein (MmCFP) specifically in the cone photoreceptors. The MmCFP accumulates in the outer segments of the cones. In **Figures 4(d)** and **4(e)**, the eye is visualized by removing it from a day-6 zebrafish larva and inverting it onto a microscope slide. In the wild-type eye, the central retina is densely populated with cones that show a regular mosaic pattern (**Figure 4(d)**). In the *pde6c^{w59}* eye, most of the cone photoreceptors have died and been removed from the retina, and many of the remaining outer segments appear dystrophic (**Figure 4(e)**).

Unlike humans or mice, fish continue to produce photoreceptor cells throughout their life. At the periphery of the eye is a region of cells known as the circumferential marginal zone, in which new photoreceptors are generated by multipotent stem cells. In this zone, young cones are constantly differentiating throughout the life of the animal. Even in adult $pde6c^{w59}$ mutants, there are always cones in this region of the eye, indicating that there is no defect in cone morphogenesis or differentiation, but that cones die as they mature. The circumferential marginal zone is not visible in **Figures 4(d)** and **4(e)** but can be seen in **Figure 4(a)** as a faint ring of fluorescence around the periphery of the $pde6c^{w59}$ eye.

The rods in the central retina also deteriorate during early development in $pde6c^{w59}$ mutants but later recover. At 7 dpf, the rods appear normal and are slightly more clustered than in wild type, but not reduced in number. At 8–9 dpf, the number of rods begins to decrease in the central retina. The outer segments of the remaining rods in the central retina appear dystrophic. The deterioration of the rods continues through at least 6 weeks postfertilization. These data were the first evidence that zebrafish can undergo a bystander effect in the retina (i.e., mutant cones kill neighboring healthy rods).

The multipotent stem cells in the circumferential marginal zones of the eye can differentiate into rods throughout the life of the animal. Additionally, in cases of damage or injury, the Müller glia can also enter a mitotic state and produce stem cells capable of forming all types of retinal neurons. In the $pde6c^{w59}$ mutants, this continuous regeneration eventually replenishes the small number of rods that have died. Thus, by 3 months postfertilization, the retina is completely populated with rods. Interestingly the $pde6c^{w59}$ mutants never develop a scotopic ERG or OKR, indicating that even the remaining rods are unable to form proper connections with the downstream neurons in the eye. Unlike human and mouse eyes, zebrafish do not have separate rod and cone bipolar cells. The bipolar cells that connect to rods also connect to cone photoreceptors. The lack of scotopic vision in the $pde6c^{w59}$ mutant suggests that the cones are necessary for the proper connections of the rods with their bipolar targets. In support of this theory, the bipolar cells of the $pde6c^{w59}$ embryos often show an altered morphology with axons that extend into the ganglion cell layer and dendritic branches that send filopodia into the photoreceptor layer.

The $pde6c^{w59}$ mutant allows for exceptional visibility of photoreceptor degeneration. Due to the rapid degeneration of the cones and rods in the central retina, it is possible to monitor many aspects of cell death as they are occurring. However, in the human diseases of RP and Achromatopsia, retinal degeneration can be a slow process occurring over several years. Thus, although it is easier to visualize cell death and develop drugs that will combat the rapid loss of photoreceptors in the $pde6c^{w59}$ mutation, it would also be useful to have a mutant where degeneration occurred over a more protracted period of time. Recently the *els* mutant was identified as a mutation in *pde6c* that results in slow cone degeneration.

els

The *els* mutation produces a single amino acid change; methionine (M) 175 is mutated to an arginine (R) (M175R) in the first GAF domain of Pde6c. At 5 dpf, when zebrafish vision is cone dependent, fish that are homozygous for the *els* mutation have a flat ERG and no OKR, although initially the cones appear morphologically normal. This indicates that phototransduction is disrupted but surprisingly cell death has not been triggered. This finding suggests that the *els* allele is not null for the Pde6c protein, but that secondary defects within the *els* cones may be disrupting phototransduction, resulting in a flat ERG. One finding in support of this idea is that, although all four types of cones are initially present in the *els* mutant, the localization of various opsins within the photoreceptors is abnormal. Generally, opsins are found only in the outer segments, but in the *els* mutant, opsin proteins are found throughout the cell. At this stage, there is also a small but significant increase in the number of apoptotic cells in the *els* retinas. By 3 weeks postfertilization, the cones and rods in the central retina have begun to show an altered morphology, and the number of cone cells has decreased. However, despite their slightly altered morphology, rod maturation occurs in these fish and by 3 weeks postfertilization the fish respond to OKR under scotopic conditions.

As the *els* fish matures, the cones continue to deteriorate and by 6 months postfertilization the retina no longer contains cones, but consists entirely of rods. The rods are morphologically normal and, unlike those found in $pde6c^{w59}$, functional. This suggests that the significantly slower death of cones in *els* compared to $pde6c^{w59}$ fish allows rods to form proper synaptic connections with bipolar cells. Interestingly, the total number of cells in the inner retina is decreased in this mutant; however, the ratio of inner retinal cells to rods has increased over wild type, suggesting that the rods are forming more connections in the mutant than they do in the wild-type retina. The number of mitotically dividing cells in the retina is also increased in the mutant, suggesting that there is an increase in cellular proliferation probably resulting from the death of the cone photoreceptors.

Pde6 Structure

Recent work on the structure and function of Pde6c has shed some light on the potential effects of the *els* mutant for the Pde6c holoenzyme. The catalytic subunit of the Pde6 protein consists of two domains: a regulatory domain with two GAF domains, one of which (GAF A) binds cGMP, and a catalytic domain that has phosphodiesterase activity. The inhibitory subunit Pγ binds to the GAF A domain though its C terminus and the catalytic domain through interactions with its N terminus. Recent work with the rod Pde6 has suggested that the binding of cGMP to the GAF A domain may increase the binding affinity of the N terminus of Pγ for the catalytic domain. Thus, the binding of cGMP to the GAF A domain may help to regulate the activity of Pde6c *in vivo*. Met175 is located within the allosteric cGMP-binding site of the GAF A domain of Pde6c (**Figure 5(a)**). A recently determined crystal structure of the GAF A domain from chicken Pde6c reveals that the side chain of Met175 (equivalent to Met179 in chicken Pde6c) is in close proximity to the cyclic phosphate group of cGMP, although it does not directly interact with the cyclic nucleotide. As in other cyclic-nucleotide-binding Pde GAF domains, the phosphate group of the ligand is stabilized through the positive dipole of helix α3 (**Figure 5(b)**). The mutation M175R (equivalent to M179R in chicken Pde6c) introduces a larger and positively charged side chain into the binding pocket and thereby changes the cGMP-binding environment and, potentially, the binding affinity for the allosteric regulator cGMP (**Figure 5(c)**). A model in which a methionine side chain is substituted for an

arginine side chain suggests three potential consequences of the M175R mutation. First, the R175 side chain clashes with helix α3, which may disrupt the dipole interaction between the helix and the phosphate group. Second, the R175 side chain clashes directly with the phosphate group of cGMP. Third, R175 might form a salt bridge with the phosphate group. In the former two scenarios, R175 would disrupt cGMP binding and lower the binding affinity of the GAF domain for cGMP, whereas in the latter scenario, R175 may cause a higher affinity for cGMP.

In either model, it is likely that the M175R mutation interferes significantly with cGMP binding and the sensitive regulatory mechanism of Pde6 through its GAF A domain. This, in turn, may lead to impaired Pde6 function and cause over- or underexpression of Pde6 protein, thereby disrupting the cone photoreceptor function. No data are available on the stability or functionality of the *els* mutant protein. However, the phenotypic comparison with the *w59* mutant suggest that the *els* mutant is not a null mutation, but creates a more subtle effect on the levels or activity of the Pde6c protein. The proximity of the mutation to the cGMP-binding site in the GAF A domain implicates a possible disruption of the intramolecular allosteric regulation of catalytic activity ascribed to this portion of Pde6c.

The *els* mutant represents a unique opportunity to understand more about the enzymatic functions of Pde6 and its role in retinal degeneration. The slower degeneration

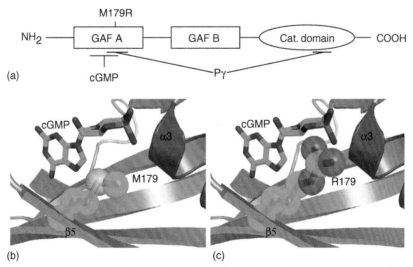

Figure 5 Structural consequences of M179R mutant in Pde6c GAF A. (a) Domain organization of Pde6c. The C-terminal catalytic domain is regulated by allosteric noncatalytic binding of cGMP to the N-terminal GAF A domain and binding of the inhibitory Pγ-subunit. The methionine (M)179 to arginine (R) mutation in chicken Pde6c, for which there is a crystal structure, is equivalent to the *els* Met175R mutation in zebrafish. (b) Experimentally determined structure of wild-type cGMP-binding pocket of chicken Pde6c GAF A (pdb-code: 3dba). α-Helices are shown in red, β-strands are shown in blue. cGMP and Met179 are shown in sticks with carbon atoms in cyan. Met179 is also highlighted through spheres. The figure was prepared with PyMOL, a molecular modeling program. (c) Model of M179R mutant cGMP-binding pocket of Pde6c GAF A. M179 was mutated to R179 through the mutation-function of PyMOL. In the absence of some structural rearrangements in the binding pocket, the longer side chain of R179 clashes with the phosphate group of cGMP and helix α3. α-Helices are shown in red; β-strands are shown in blue. cGMP and R179 are shown in sticks with carbon atoms in *cyan*. R179 is also highlighted through spheres.

of this mutation compared to the *w59* allele more accurately reflects slower human forms of degeneration. These two zebrafish *pde6c* mutants will provide complementary tools for studying Achromatopsia and apoptosis due to phosphodiesterase deficiency.

Conclusion

Zebrafish have gained prominence as a model for retinal disease. Cone-based vision, visual translucence, inexpensive maintenance, and rapid external embryologic development help make this model particularly exciting for retinal studies. Several genetic screens for blind fish have identified a variety of mutants that mimic human retinal disease. Among these, the *pde6c* mutants are a particularly good example of a retinal model that has been studied for many years in mice, and will benefit from the types of study available in the zebrafish system. In particular, the potential for live imaging of cells *in vivo* in an intact animal presents a novel opportunity to visualize and understand the source of photoreceptor degeneration.

One of the largest differences between the eyes of zebrafish and humans is the continuous growth of the zebrafish eye, and its regenerative ability in response to damage. Although this regenerative potential can complicate the evaluation of zebrafish as a model organism, it also presents a novel possibility to understand and imitate a natural system of retinal stem cell regeneration. By studying the differences between the zebrafish and mammalian systems it may be possible to stimulate our own potential for retinal regeneration.

The zebrafish model also offers an unprecedented potential for high-throughput drug screening. Using fluorescent cell markers and fluorescent plate readers, it will soon be possible to do large-scale screening of drugs that affect the levels of retinal degeneration. This method can also be used to test permeability and the toxicity of drugs. This is the first vertebrate animal model that provides a method for this type of rapid drug development.

See also: Color Blindness: Inherited; Phototransduction: Phototransduction in Cones; Phototransduction: Phototransduction in Rods; Phototransduction: Rhodopsin; Primary Photoreceptor Degenerations: Terminology; Secondary Photoreceptor Degenerations; Zebra Fish–Retinal Development and Regeneration.

Further Reading

Brockerhoff, S. E., Hurley, J. B., Janssen-Bienhold, U., et al. (1995). A behavioral screen for isolating zebrafish mutants with visual system defects. *Proceedings of the National Academy of Sciences of the United States of America* 92(23): 10545–10549.

Cote, R. H. (2007). Photoreceptor phosphodiesterase (PDE6): A G-protein-activated PDE regulating visual excitation in rod and cone photoreceptor cells. In: Beavo, J. A., Francis, S. H., and Houslay, M. D. (eds.) *Cyclic Nucleotide Phosphodiesterases in Health and Disease*, pp. 165–193. Boca Raton, FL: CRC Press/Taylor and Francis.

Doyon, Y., McCammon, J. M., Miller, J. C., et al. (2008). Heritable targeted gene disruption in zebrafish using designed zinc-finger nucleases. *Nature Biotechnology* 26(6): 702–708.

Goldsmith, P. and Harris, W. A. (2003). The zebrafish as a tool for understanding the biology of visual disorders. *Seminars in Cell and Developmental Biology* 14(1): 11–18.

Hamada, N., Matsumoto, H., Hara, T., and Kobayashi, Y. (2007). Intercellular and intracellular signaling pathways mediating ionizing radiation-induced bystander effects. *Journal of Radiation Research (Tokyo)* 48(2): 87–95.

Martinez, S. E., Heikaus, C. C., Klevit, R. E., and Beavo, J. A. (2008). The structure of the GAF A domain from phosphodiesterase 6C reveals determinants of cGMP binding, a conserved binding surface, and a large cGMP-dependent conformational change. *Journal of Biological Chemistry* 283(38): 25913–25919.

Morris, A. C., Scholz, T. L., Brockerhoff, S. E., and Fadool, J. M. (2008). Genetic dissection reveals two separate pathways for rod and cone regeneration in the teleost retina. *Developmental Neurobiology* 68(5): 605–619.

Muto, A., Orger, M. B., Wehman, A. M., et al. (2005). Forward genetic analysis of visual behavior in zebrafish. *PLoS Genetics* 1(5): e66.

Nishiwaki, Y., Komori, A., Sagara, H., et al. (2008). Mutation of cGMP phosphodiesterase 6alpha'-subunit gene causes progressive degeneration of cone photoreceptors in zebrafish. *Mechanisms of Development* 125(11–12): 932–946.

Paquet-Durand, F., Johnson, L., and Ekstrom, P. (2007). Calpain activity in retinal degeneration. *Journal of Neuroscience Research* 85(4): 693–702.

Ripps, H. (2002). Cell death in retinitis pigmentosa: Gap junctions and the 'bystander' effect. *Experimental Eye Research* 74(3): 327–336.

Sancho-Pelluz, J., Arango-Gonzalez, B., Kustermann, S., et al. (2008). Photoreceptor cell death mechanisms in inherited retinal degeneration. *Molecular Neurobiology* 38(3): 253–269.

Stearns, G., Evangelista, M., Fadool, J., and Brockerhoff, S. E. (2007). A mutation in the cone specific pde6 gene causes rapid cone photoreceptor degeneration in zebrafish. *Journal of Neuroscience* 27(50): 13866–13874.

Wissinger, B., Chang, B., Dangel, S., et al. (2007). Cone phosphodiesterase defects in the murine cpfl1 mutant and human achromatopsia patients. *Investigative Ophthalmology and Visual Science* 48(5): 4521.

Zhang, X. J., Cahill, K. B., Elfenbein, A., Arshavsky, V. Y., and Cote, R. H. (2008). Direct allosteric regulation between the GAF domain and catalytic domain of photoreceptor phosphodiesterase PDE6. *Journal of Biological Chemistry* 283(44): 29699–29705.

Zebra Fish–Retinal Development and Regeneration

T J Bailey and D R Hyde, University of Notre Dame, Notre Dame, IN, USA

Glossary

BrdU (bromodeoxyuridine) labeling – The synthetic nucleoside which is incorporated into nascent DNA during replication and is detectable by antibodies to indicate what cells have divided since exposure to BrdU.

CMZ (circumferential marginal zone) – The region of the retina distal to the optic stalk and proximal to the ciliary margin and lens. Here, retinal cells are continually born throughout the life of the zebrafish. CMZ cells express genes found in the neuroretina of the developing zebrafish embryo.

Homeobox transcription factors – The proteins that pattern tissue in that they regulate gene transcription by binding to a specific DNA sequence, the homeobox, in the target gene promoter.

Morpholino – Similar to RNA, this polymerized oligomer with a morpholino (rather than ribose) backbone, can base pair with ribonucleic acid (RNA) molecules, and persists in the cell as it is not easily degraded by RNases. Morpholinos interfere either with ribosomal processing of the messenger RNA into protein or spliceosome processing of premessenger RNA into mRNA, thus depleting the amount of protein produced.

Notch signaling – The plasma membrane-bound receptor that regulates the cell fate choice of individual neurons. Intracellular cleavage product can act as a transcription factor and regulate the expression of pro-neural genes such as basic helix–loop–helix (bHLH) transcription factors.

Pcna (proliferating cell nuclear antigen) – A marker for DNA replication in that it is a protein that functions as a trimer to promote DNA polymerase δ processivity.

Shh (sonic hedgehog) signaling – The Shh family members act as morphogens to pattern tissue. Signaling pathway proteins and pathways are reutilized in more specific cell fate specification as tissues are patterned.

TUNEL (terminal deoxynucleotidyl transferase dUTP nick end labeling) – A hallmark of apoptotic cells is genomic DNA fragmentation by specific nucleases that cleave DNA between histones. This process results in semi-uniform lengths of DNA with overhanging hydroxyl groups. Terminal transferase efficiently polymerizes labeled nucleotides to DNA hydroxyl groups without the need of a template. TUNEL is used to detect the genomic fragmentation of an apoptotic cell.

Introduction

Zebrafish has rapidly become a leading model system to study a variety of developmental processes, due to its large clutch size, external development of transparent embryos, and the rapid development of the embryo. Large and small genetic screens identified hundreds of mutants that affect the development of various tissues, including the retina. The ease in generating transgenic zebrafish lines has permitted a detailed cellular analysis of organ development without harm to the embryo. For example, transgenes that label specific cells are used to follow cell fates in the transparent embryo during tissue development. Alternatively, transgenes that express molecules that lead to cell death have been used to ablate cells to study either the disruption of development or organization of a tissue. Furthermore, the nearly complete sequence of the zebrafish genome has allowed comparative analyses with other vertebrate genomes to predict the presence of orthologous genes and developmental processes. Combined with the ability to direct the transient reduction in expression of desired proteins, it is possible to functionally analyze the potential role of different signaling pathways in the development of various tissues. Our understanding of zebrafish retinal development, from a sheet of neuroepithelial cells to a laminated and functional neural tissue, has benefited significantly from all of these approaches.

In addition to being an excellent model system to study early development, zebrafish has quickly become the premier model system to study tissue regeneration. In addition to exhibiting rapid and functional regeneration of the fin, liver, and heart, zebrafish also regenerate neuronal tissues, including the spinal cord, brain, and retina. Using mutants and transgenic lines that are readily available in zebrafish, a detailed comparison of the genes, molecules, and process that are required for retinal development and retinal regeneration is starting to be generated. While it may initially seem reasonable that regeneration would recapitulate the mechanisms that are involved in retinal development, recent studies revealed that regeneration may utilize the same genes and proteins as development, but in a different context. This becomes most obvious when one compares the development of a laminated retina from an unpatterned neuroepithelium to the regeneration of single neuronal cell type from an existing laminated retina. To fully appreciate the differences between these two processes and the mechanisms involved in regeneration, it is important to contrast our understanding of the general events involved in retinal development with our recently acquired knowledge of the processes underlying retinal regeneration.

Embryonic Eye Patterning

Zebrafish Eyes Form from a Single Field in the Anterior Neural Plate

There is a large body of knowledge regarding the genes and genetic pathways that are required for the proper formation of the vertebrate eye. Around 10 h postfertilization (hpf), several secreted proteins induce the most anterior region of the anterior ectoderm to become the neural plate. Several signaling pathways (wingless (Wnt), fibroblast growth factor, and insulin-like growth factor) then further subdivide the anterior neural plate to produce the presumptive eye field. Some anterior-most cells then begin expressing several homeobox transcription factors, such as the retinal progenitor genes visual system homeobox 2 (*vsx2*), paired box gene 6 (*pax6*), and retinal homeobox (*rx*), which further restrict their fates to be retinal progenitor cells. The midline of the underlying head mesoderm then expresses several secreted, signaling molecules (such as the sonic hedgehog-related protein, sonic-you), which induces the eye field in the anterior neural plate to split into two distinct regions. This yields the first morphological sign of the developing visual system, the bilateral evagination of a single-cell thick epithelium from the anterior end of the neural keel, which will develop into the optic lobe. The optic lobes are morphologically visible by 12 hpf (**Figure 1(a)**). Each optic lobe expresses diffusible signals that induce the overlying naïve epithelium to commit to form the lens. As the lens placode starts to thicken (18 hpf, **Figure 1(b)**), it produces soluble signaling molecules that promote the underlying optic lobe to proliferate and invaginate, which results in the formation of a concave neuroepithelium – the optic cup. Expression of the transcription factor genes microphthalmia-associated transcription factor (*mitf*) in the ventral optic cup and *vsx2* in the dorsal optic cup commits those cells to develop into the retinal pigmented epithelium and neural retina, respectively. The region that lies at the junction of the *mitf* and *vsx2* expressing cells later becomes one region of persistent retinal neurogenesis in the adult retina, the circumferential marginal zone (CMZ, discussed below).

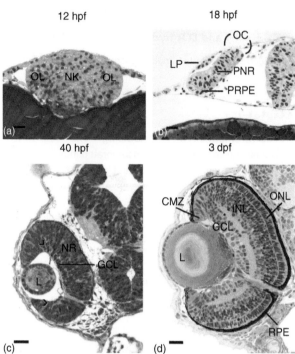

Figure 1 Zebrafish retinal development. (a) Cells evaginate bilaterally from the anterior neural keel (NK) to form the optic lobes (OL), which are morphologically distinguishable by 10–12 h postfertilization (hpf). (b) By 18 hpf, the proliferating cells of the presumptive neuroretina (PNR) have induced the overlying ectoderm to thicken into the lens placode (LP). The retina then invaginates to form the OC, with the thicker dorsal cells, or presumptive neuroretina, expressing *vsx2* and the thinner ventral cells, or presumptive retinal pigmented epithelium (PRPE), expressing *mitf*. (c) The cells in the neural retina continue to proliferate until a wave of sonic hedgehog (Shh) signaling induces the first retinal ganglion cells (RGCs) in the most basal nuclear layer, the ganglion cell layer (GCL: between the arrowheads), to differentiate around 40 hpf in the center of the retina, whereas cells in the retinal margin remain in an uncommitted state. (d) Additional waves of Shh signaling produce three nuclear layers, the ganglion cell layer (GCL), inner nuclear layer (INL), and outer nuclear layer (ONL). The ONL is immediately below the retinal pigmented epithelium (RPE), which is now darkened with pigment granules by 3 days postfertilization (dpf). Scale bars represent 20 µm. OC, optic cup; NK, neural keel; NR, neural retina; RPE, retinal pigmented epithelium; CMZ, circumferential marginal zone; GCL, ganglion cell layer; INL, inner nuclear layer; L, lens; LP, lens placode; OL, optic lobe; ONL, outer nuclear layer; PNR, presumptive neuroretina; PRPE, presumptive retinal pigmented epithelium.

The Laminar Structure of the Retina Forms as Cells Exit the Cell Cycle and Differentiate

The number of cells in the NR increases through cell proliferation until the diffusible signaling protein sonic hedgehog (Shh) is produced in a ventronasal patch of the NR. The Shh protein initiates a wave of expression of the basic helix–loop–helix (bHLH) transcription factor atonal homolog 7 (Atoh7) that sweeps radially toward the dorsal retina that induces the differentiation of the retinal ganglion cells (RGCs) (**Figure 1(c)**). Shh is then secreted from the newly specified RGCs to sequentially induce the apically located naïve mitotic cells to exit the cell cycle and commit to the other retinal neuronal identities. Later, Shh signaling from the RPE is required for proper photoreceptor differentiation. This process results in the formation of the terminally laminated retina, which is composed of three nuclear layers (ganglion cell layer, inner nuclear layer (INL), and outer nuclear layer) and two synaptic layers (inner plexiform layer and outer plexiform layer) by 72 hpf (**Figure 1(d)**).

Intrinsic commitment and specification of the different retinal cell types also requires the expression of homeobox transcription factors and pro-neural genes of the Notch signaling pathway. Experiments aimed at determining when the different retinal cell types are committed (neuronal birthdating) established a bias of early committing cells to the RGC, amacrine cell, cone, and horizontal cell classes, followed by the bipolar and rod photoreceptor cells, and lastly the Müller glial cells. The reproducible timing observed with these different cell types suggests that a molecular clock modulates their commitment and differentiation. This model proposes that the commitment of a retinal progenitor cell to a particular neuronal cell type corresponds to when the progenitor cell exits the cell cycle, with early committed cells localizing more basally in the retina. Thus, retinal progenitor cells divide and some daughter cells become ganglion cells, while the remaining daughter cells continue to divide. Some of these retinal progenitors exit the cell cycle and differentiate as amacrine cells and others continue as retinal progenitors. These progenitors continue their asymmetrical cell division to produce some retinal neurons with each round of cell division, until the final retinal progenitors are committed to become Müller glial cells. This suggests that the Müller glia is the retinal cell type that is most recently differentiated from the retinal progenitor cell. This model also allows for the presence of external signals to influence the commitment of the cell, which also changes over time. These mechanisms appear to be conserved across species.

Addition of Retinal Cells Throughout the Life of a Zebrafish

Unlike mammals, the zebrafish eye continuously grows throughout the lifetime of the fish. This growth requires the continual generation of new retinal neurons in a process called persistent neurogenesis. These additional retinal cells are produced from two adult stem cell niches, the CMZ (**Figure 1(d)** and **Figure 2(a)**) and an INL stem cell niche (**Figure 2(a)**). The stem cells within the CMZ continue to express the cell cycle genes and the embryonic retinal progenitor genes, such as orthodenticle homolog 2 *otx2*, *pax6*, and *rx*, throughout the life of the fish. These stem cells proliferate to yield daughter cells that ultimately

differentiate into ganglion cells, amacrine cells, horizontal cells, cone photoreceptors, bipolar neurons, and Müller glial cells, but not rod photoreceptors. Rods arise from the INL stem cell niche as described below. The addition of these newly differentiated cells to the region adjacent to the CMZ (**Figure 2(c)** and **2(d)**) results in the radial growth of the adult retina.

The INL stem cells have recently been demonstrated to correspond to the Müller glial cells. Unlike the CMZ stem cells, however, the asymmetric division of the Müller glia ultimately produce only rod photoreceptors during persistent neurogenesis (**Figure 2**). Relatively few Müller glia are actively dividing at any given moment. The asymmetric division of the Müller glial cell produces neuronal progenitor cells (**Figure 2(a)**), which continue to proliferate as they migrate to the outer nuclear layer, where they are called rod precursor cells and continue to undergo cell division (**Figure 2(b)**). Unlike the pluripotent CMZ stem cells, these Müller glial-derived rod precursor cells are committed to differentiate into only rod photoreceptors during persistent neurogenesis (**Figure 2(c)**). As the adult eye enlarges, the distance between the originally differentiated rod photoreceptors increases. The Müller glial-derived rods fill in this space. Thus, persistent neurogenesis in the zebrafish retina encompasses both the radial growth of the retina by addition of all retinal cell types by the CMZ and the slow production of additional rod photoreceptors by the Müller glia (**Figure 2(d)**).

Regeneration in the Zebrafish Retina

Zebrafish regenerate all retinal neurons

Persistent neurogenesis involves the continual generation of new neurons in the adult retina, without any prior loss of retinal neurons that must be replaced. It should be noted that the scientific literature often uses retinal regeneration to define the reprojection of axons from viable neuronal soma to replace damaged axons, such as the reprojection of axons from RGCs subsequent to an optic nerve crush or severing. For this discussion, retinal regeneration refers to the replacement of entire neuronal cells that were lost through retinal insult or genetic causes. Zebrafish respond to the loss of retinal neurons by significantly increasing both the number of Müller glia that reenter the cell cycle and the rate of proliferation in the neuronal progenitor cells, relative to that observed during persistent neurogenesis. This amplified proliferation response appears to be proportional to the amount of damage suffered. In contrast to persistent neurogenesis, these Müller-glial-derived neuronal progenitors are not committed to become only rod photoreceptors. These neuronal progenitors proliferate, migrate to the retinal layer that contains the missing neurons, and differentiate specifically into the lost neurons. Thus, the damage response alters the persistent neurogenesis program to

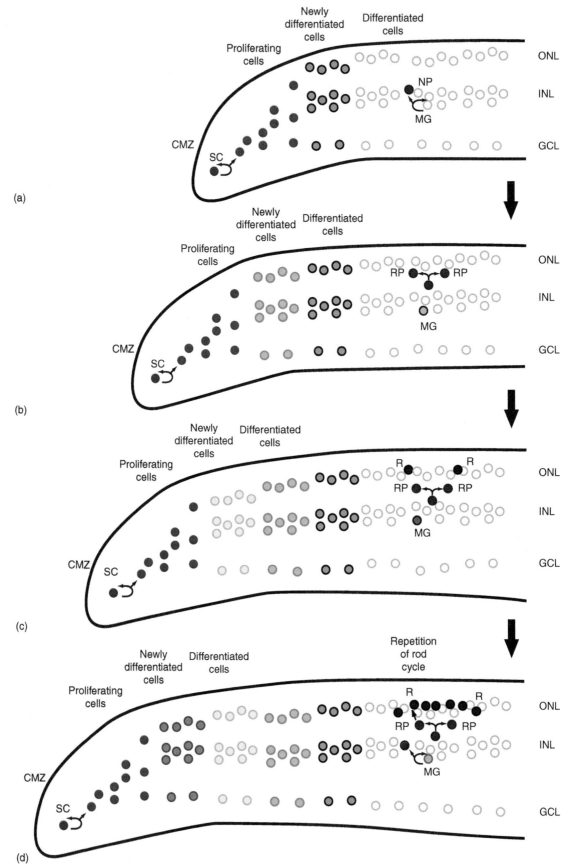

Figure 2 Continued

significantly increase the number of neuronal progenitors and allow them a greater breadth of cell differentiation potential.

A variety of damage paradigms have been studied in the zebrafish retina. For example, simple surgical lesion induces cells proximal to the cut site to die, which usually results in the death of all the retinal cell types in a small area of the retina. Similarly, heat caused by either a high temperature probe or laser ablation, or high concentrations of ouabain (Na, K-ATPase inhibitor), often causes loss of cells from all three retinal layers. In contrast, other damage models exhibit more restricted cell-type loss. For example, intravitreal injection of low concentrations of ouabain causes the loss of ganglion cells and INL neurons, without the loss of significant numbers of rod and cone photoreceptors. Constant intense light, in contrast, causes apoptosis of only the rod and cone photoreceptors, primarily in the dorsal and central retina, with no detectable cell death in the INL or GCL layers. An advantage of the light damage model is that only two neuronal classes are lost, rod and cone photoreceptors, which limits the complexity of the regeneration response. Furthermore, the light damage extends across a large region of the retina, which induces the participation of a very large number of Müller glia and neuronal progenitor cells. This results in an amplification of the signals and processes that are required for regeneration, which should increase the likelihood of their identification.

Constant Intense Light Kills Photoreceptors, Which Are then Regenerated by Müller Glia

Zebrafish rods and cones die by apoptosis upon exposure to prolonged high-intensity light. Apoptosis is a genetically programmed cell death mechanism, in which the dying cell fragments its DNA and the cell body (blebs) to produce easily phagocytosed cell corpses. Terminal deoxynucleotidyl transferase-mediated dUTP nick-end labeling (TUNEL), is a standard method to detect this fragmented genomic DNA. Fragmentation of the DNA produces large

numbers of free 3'-hydroxyl ends that can be used by terminal deoxynucleotidyl transferase to add dUTP nucleotides that are covalently modified to a detectable molecule, such as biotin or fluorescein (**Figure 3**).

The light-induced photoreceptor cell death is rapid, with TUNEL-positive cells first detected in the ONL as early as 3 h after initiating the light treatment. By 16 h of constant light, the photoreceptor cell death is evident based on the reduction in the number of ONL nuclei relative to undamaged retina and is striking by 3 days of constant light (**Figure 4(a)** and **4(b)**). In addition to the strong TUNEL-positive signal that is detected in the ONL, a weaker TUNEL signal is observed throughout the Müller glial cells in the INL. Strikingly, this INL TUNEL signal is neither restricted to, nor predominantly localized in, the Müller glia nuclei, which demonstrates that the Müller glial cells are not apoptotic. Rather, the colocalization of some proteins derived from apoptotic rods with the TUNEL signal in the Müller glia suggests that the Müller glia selectively engulf dying rod photoreceptors. These data are consistent with results in the degeneration model of the *Tg(Xops:mCFP)* transgenic line, which expresses membrane-bound cyan fluorescent protein (mCFP) from the *Xenopus rod opsin* promoter (*Xops*) in only zebrafish rods. The *Tg(Xops:mCFP)* retina exhibits a persistent loss of only rod photoreceptors, even in the absence of any retinal insult, such as constant bright light. Furthermore, the *Tg(Xops:mCFP)* retina revealed TUNEL labeling in the INL in a morphology similar to Müller glia. Thus, the engulfment of apoptotic rod photoreceptors by Müller glia is a common early feature of the damaged zebrafish retina and may be required for regenerative Müller glial cell stimulation. TUNEL labeling in the Müller glia can be detected early, within the first 12 h of light treatment, and represents one of the earliest signs of Müller glial response to apoptotic photoreceptor damage. Robust regeneration of the light-damaged zebrafish retina follows and a normal complement of rod and cone photoreceptors reform within 28 days after terminating the constant light treatment (**Figure 4(c)**).

Figure 2 Schematic of persistent neurogenesis in the adult zebrafish retina. (a) In the adult retina, stem cells (SC) in the circumferential marginal zone (CMZ) continue to proliferate through asymmetric cell division to produce neuronal progenitor cells (pink circles), which then exit the cell cycle to become newly differentiated cells (blue circles). These CMZ-derived cells differentiate into cone, horizontal, bipolar, amacrine, and ganglion neurons, as well as Müller glia (not labeled), but not rod cells. Throughout the remainder of the retina, a limited number of Müller glial cells (MG) divide asymmetrically to produce a Müller glial cell and a neuronal progenitor cell (NP). (b) As the CMZ stem cells continue to produce neuronal progenitors (pink circles), there is a radial growth of the adult retina, with the most recently differentiated neurons (green circles) located closer to the CMZ than the older neurons (blue circles). In the central region of differentiated retina, the new NP cell continues to divide and migrates to the ONL. Once these NP cells reach the ONL, they are termed rod precursor cells (RP). (c) The CMZ continues producing new retinal cells (yellow circles), which are located closest to the CMZ. The RP cells in the ONL continue to proliferate, with some of the daughter cells differentiating into rod photoreceptors (purple circles). The newly differentiated rod cell intercalates between differentiated cone cells to maintain the density of rod photoreceptors during the radial expansion of the retina. (d) Persistent neurogenesis continues as the CMZ-based retinal stem cells and the Müller-glial-derived neuronal progenitors continue to produce new neurons at the margin and central retina, respectively. CMZ, circumferential marginal zone; GCL, ganglion cell layer; INL, inner nuclear layer; MG, Müller glial cell; ONL, outer nuclear layer; NP, neuronal progenitor; RP, rod precursor cell; SC, stem cell.

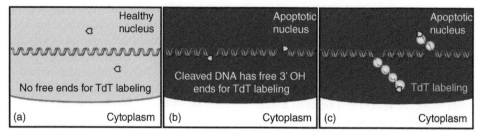

Figure 3   TUNEL assay detects the fragmented DNA in apoptotic nuclei. (a) A healthy nucleus contains intact DNA, which contains only a single 3′ hydroxyl group at the end of each chromosome. The terminal deoxynucleotidyl transferase enzyme (TdT), which can add polymerized stretches of deoxyuridine triphosphate (dUTP), without the need for a complementary strand of DNA with which to pair nucleotides (template-independent). TdT binds to both ends of the linear chromosomal DNA and adds only a few dUTPs per healthy nucleus. (b) Apoptosis results in the fragmentation of the DNA, which generates a very large number of free 3′ hydroxyl ends. (c) TdT adds dUTPs to each of the now many generated 3′ hydroxyl ends. The large number of dUTP molecules can either directly fluoresce because they are tagged with a fluorochrome or, in the case of biotinylated-dUTP, are detected by tagged Streptavidin molecule to generate a strong fluorescent signal in the apoptotic nucleus. Because of the absence of the genomic DNA in the cytoplasm, little, if any, fluorescent signal is present in the cytoplasm of an apoptotic cell.

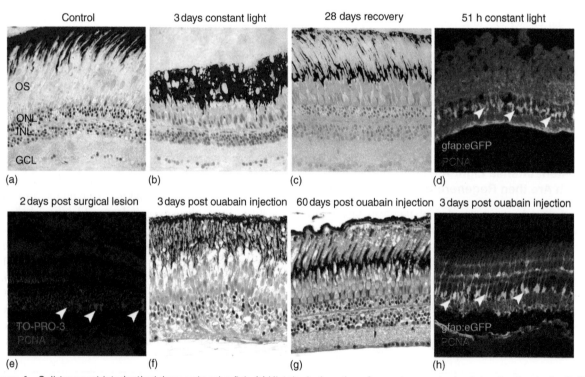

Figure 4   Cell-type restricted retinal damage in zebrafish. (a) Histological section of an undamaged zebrafish retina (control) with the following labeled: outer segments (OS), outer nuclear layer (ONL), inner nuclear layer (INL), and ganglion cell layer (GCL). (b) After 3 days of constant light, the loss of photoreceptor nuclei in the ONL and reduced outer segment integrity relative to the control are evident. (c) Photoreceptors are regenerated only 28 days post light treatment based on the nuclear density and thickness of photoreceptor nuclei in the ONL and the restoration of the photoreceptor outer segments. (d) A transgenic zebrafish line, *Tg(gfap: eGFP)*, expresses EGFP in the Müller glia from the *gfap* promoter. Light-induced damage of this transgenic line induces proliferating cell nuclear antigen (Pcna-red) in a subset (arrowheads) of the Müller glial cells (green) after only 51 h of constant intense light. (e) Proliferating cell nuclear antigen (Pcna-red) immunolabeling Müller glial cells (arrowheads) at 2 days after a surgical lesion (TO-PRO-3 labeling of nuclei-blue) is very similar to the response observed in the light-damaged retina (d). (f) Three days after intraocular injection of ouabain, the damaged retina revealed lost retinal ganglion cells and INL neurons, without significant loss of ONL photoreceptors. (g) Sixty days after ouabain injection, the regenerated retina revealed a GCL and INL that is nearly indistinguishable from wild type (a). (h) The *Tg(gfap:eGFP)* line revealed that Pcna was detected in a subset of Müller glial cells (arrowheads) at 3 days after intraocular injection of ouabain.

In light-damaged zebrafish retinas, the Müller glia exhibit increased expression of proliferating cell nuclear antigen (Pcna), which is a component of the DNA replication machinery. The expression of this protein in the Müller glia indicates their reentry into the cell cycle (**Figure 4(d)**). To confirm that the Müller glia were proliferating, light-damaged *Tg(gfap:EGFP)* retinas, which express enhanced green fluorescent protein (EGFP) specifically in the Müller glial cells from the zebrafish glial fibrillary acidic protein (*gfap*) promoter, were coimmunolabeled for the expression of the proliferation marker Pcna. The number of actively dividing Müller glia, coexpressing EGFP and the red fluorescing anti-Pcna antibody, increases through 51 h of constant intense light treatment (**Figure 4 (d)**), at which point the Müller glial-derived INL neuronal progenitor cells continue to proliferate to produce clusters of 8–12 progenitor cells associated with a single Müller glial cell. While this suggests that all the neuronal progenitor cells in a cluster are derived from a single Müller glial cell and remain associated with that glial cell, this has not been formally demonstrated. This Müller glial cell-based regeneration response is conserved throughout a number of different damage models. For example, either surgical lesion or ouabain injection results in the death of many different neuronal cell types in the retina, and they both exhibit increased proliferation of the Müller glial cells (**Figure 4(e)**–**4(h)**).

Curiously, not all of the Müller glial cells proliferate in response to constant intense light treatment. Earlier reports suggested that a threshold of rod cell death was required to induce the Müller glial proliferation response. However, intravitreal injection of a low concentration of ouabain into the zebrafish retina resulted in massive death of the neurons in both the ganglion cell and INLs, with minimal cell death of photoreceptors (**Figure 4(f)**). Regeneration of the ouabain-damaged retina takes longer than the light-damaged retina, but still produces a relatively normal retina by 60 days post ouabain injection (**Figure 4(g)**). This demonstrates that significant rod or cone cell death is not required to induce retinal regeneration from the Müller glia (**Figure 4(h)**). Furthermore, this suggested that the number of apoptotic neurons, rather than the type of neuron, was critical for inducing this regeneration response.

To address if only rod photoreceptor cell death was sufficient to induce a Müller glial-derived regeneration response, the *Tg(Xops:mCFP)* transgenic line and the phosphodiesterase 6c (*pde6c*) mutant line were analyzed. The *pdge6* mutant line fails to maintain cone cells, in contrast to the rod photoreceptor cell death in the *Tg(Xops:mCFP)* fish. While no detectable Müller glial proliferation response was observed in the *Tg(Xops:mCFP)* line, a small, but significant, Müller glial proliferation response was detected in the *pdge6* mutant line. This suggested that loss of rods requires only increased proliferation of the ONL rod precursor cells

that were derived from the neuronal progenitor cells during persistent neurogenesis, while loss of cones requires the increased proliferation of the pluripotent Müller glia.

To further test this hypothesis, a *Tg(zop:ntr)* transgenic line was generated that expresses the bacterial *nitroreductase b* (*ntr*) gene from the zebrafish *rhodopsin promoter* (*zop*). The NTR enzyme converts the prodrug, metronidazole, into a cell-autonomous toxin within NTR producing cells. Exposing a *Tg(zop:ntr)* transgenic line, expressing NTR in all rod photoreceptors, to metronidazole results in the death of only rod photoreceptors and the induction of a Müller glial proliferation response. Addition of metronidazole to a similar transgenic line that expresses NTR in only a subset of rods, however, failed to induce the Müller glial response. These data suggest that it is the magnitude of the cell death that determines if the Müller glia exhibit a robust proliferation response. The failure of the *Tg(Xops:mCFP)* transgenic line to induce the Müller glial proliferation response may be due to the rod cell death being small and chronic relative to the massive and acute cell death observed in the metronidazole-treated *Tg(zop:ntr)* transgenic retinas that express NTR protein in all the rod photoreceptors.

Discovery and Analysis of Candidate Genes Involved in Retinal Regeneration

Several groups have performed microarray analyses of mRNA expression patterns in different retinal damage models to determine what genes might change their expression during specific points of regeneration. In both the surgical-lesioned and light-damaged models, signal transducer and activator of transcription 3 (*stat3*) and its negative regulator suppressor of cytokine signaling 3 (*socs3*) exhibited increased expression shortly after the retinal insult (e.g., within 16 h of starting the constant intense light treatment). The increased *stat3* expression suggests that Gp130 receptor signaling is involved in the early damage response. Gp130 is a promiscuous receptor that binds a number of different extracellular signaling molecules, such as cytokines, to activate the Stat3 transcription factor in neuronal niches to promote cell proliferation in the adult vertebrate brain. The *socs3* gene, which is transcriptionally activated by the Stat3 protein, encodes a protein that binds the activated receptor to prevent further Stat3 activation. Increased *stat3 expression* is also known to occur in RGCs following optic nerve crush, further supporting Stat3's role in the damage response. Microarrays also revealed that both the achaete-scute complex-like 1a (*ascl1a*) pro-neural gene and the notch pathway genes were upregulated in the surgical-lesioned and light-damaged retinal models. By contrast, retinal progenitor genes, such as *rx* and *vsx2*, whose expression are maintained in the CMZ throughout life, are not significantly increased in expression in the damaged retina.

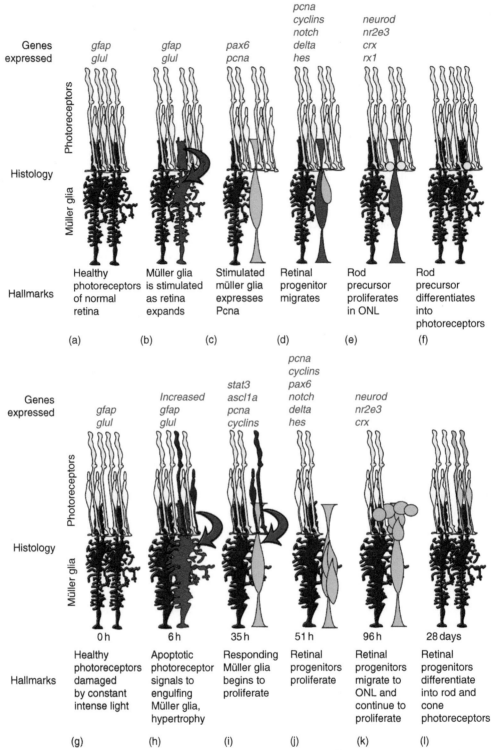

Genes expressed — gfap glul | gfap glul | pax6 pcna | pcna cyclins notch delta hes | neurod nr2e3 crx rx1 |

Histology — Photoreceptors / Müller glia

Hallmarks:

(a) Healthy photoreceptors of normal retina

(b) Müller glia is stimulated as retina expands

(c) Stimulated müller glia expresses Pcna

(d) Retinal progenitor migrates

(e) Rod precursor proliferates in ONL

(f) Rod precursor differentiates into photoreceptors

Genes expressed — gfap glul | Increased gfap glul | stat3 ascl1a pcna cyclins | pcna cyclins pax6 notch delta hes | neurod nr2e3 crx |

Histology — Photoreceptors / Müller glia

0 h | 6 h | 35 h | 51 h | 96 h | 28 days

Hallmarks:

(g) Healthy photoreceptors damaged by constant intense light

(h) Apoptotic photoreceptor signals to engulfing Müller glia, hypertrophy

(i) Responding Müller glia begins to proliferate

(j) Retinal progenitors proliferate

(k) Retinal progenitors migrate to ONL and continue to proliferate

(l) Retinal progenitors differentiate into rod and cone photoreceptors

Figure 5 Schematic comparison of zebrafish retinal development and regeneration. (a) The healthy retina is made up of rod (tall dark blue) and cone (short light blue) photoreceptors interdigitating with the processes of Müller glial cells (purple). Other retinal neurons are neglected for simplicity. Müller glial cells express genes indicative of their differentiated state, glial fibrillary acidic protein (gfap) and glutamine synthetase (glul). (b) As the retina expands, a Müller glial cell becomes stimulated to proliferate (brown), possibly due only to slow or sporadic progression through the cell cycle. (c) The stimulated Müller glial cell (green) expresses *pcna* and *pax6* as it begins to divide. (d) The cell division in panel C produces a daughter Müller glial cell (brown) and a neuronal progenitor cell (green), which expresses *pcna, cyclins, notch, delta,* and *hes* genes as it migrates to the photoreceptor layer. (e) This neuronal progenitor cell reaches the ONL and is now called a rod precursor cell (yellow), as it expresses *neurod* and the rod specification genes *nr2e3, crx,* and *rx1*. (f) The rod precursor cell divides with one daughter cell differentiating into a rod photoreceptor (green). (g) The undamaged retina,

This suggests that Müller glia exhibit a state of competency that is downstream of the CMZ stem cells, but upstream of retinal neuron differentiation.

The ability to test the function of these various candidate genes in the regenerating retina has been recently advanced by the development of a method to electroporate morpholinos into the adult retina. Morpholinos are modified oligonucleotides that contain a morpholine ring, rather than deoxyribose. The morpholinos, which are complementary to a specific mRNA sequence, can base pair with and transiently block the efficient translation of the target mRNA. Because some protein could still be correctly translated in the presence of a morpholino, these loss-of-function experiments and organisms are termed knockdowns and morphants, respectively. Morpholinos have traditionally been introduced into zebrafish embryos by direct injection into either the relatively large cells of the early embryo (1–32 cell stage) or the yolk. The morpholinos then diffuse into the daughter cells as they divide during embryonic development. This approach has tested the function of numerous proteins in zebrafish development. Recently, morpholinos that are covalently attached to a positively charged fluorochrome, lissamine, have been injected into the vitreous and then electroporated into the adult retina. These morpholinos can disrupt retinal regeneration if they are electroporated into the retina to knockdown the expression of a target protein prior to its role in regeneration.

Proof-of-principle experiments to test the effectiveness of this method were shown for the requirement of Pcna in retinal regeneration. Morpholino-induced knockdown of Pcna expression resulted in the Müller glia failing to proliferate in the light-damaged retina, which led to the premature death of the stimulated Müller glial cells due to their inability to proceed through the S phase of the cell cycle. Retinas that were injected and electroporated with anti-Pcna morpholinos also failed to upregulate expression of the retinal progenitor cell marker *pax6* in the Müller glial-derived neuronal progenitor cells. Similar morpholino knockdown studies revealed that Stat3 and Pax6 are also required at different steps in the regeneration process. Microarray analyses of mRNA expression at different time points during regeneration of the surgical-lesioned and light-damaged retinas

have revealed numerous candidate genes for functional study. The electroporation of morpholinos will permit a relatively rapid loss-of-function analysis to elucidate the genes that are required for regeneration and the key steps and processes underlying retinal regeneration.

Events Underlying Regeneration of the Light-Damaged Retina

Adult Müller glial cells are characterized by the expression of cell-specific markers (**Figure 5(a)**), such as glutamine synthetase (Glul) and glial fibrillary acidic protein (Gfap) among others. During persistent retinogenesis, rods are added to the established repeating mosaic of cone and rod photoreceptors from the Müller glial cell population. It is not clear if there is a signal (arrow) to stimulate Müller glia proliferation or if a small subset of Müller glia remain in a slow cell cycle (**Figure 5(b)**). This limited number of proliferating Müller glia can be detected by Pcna and Pax6 expression (**Figure 5(c)**). The small number of dividing Müller glial cells produce a daughter Müller glial cell (**Figure 5(d)**, brown) and a neuronal progenitor cell (green), which continues to express genes important in the cell cycle (Pcna and cyclins) and genes required for cell specification signaling (*notch, delta,* and *hes*). The neuronal progenitor continues to proliferate and migrates to the ONL. As the neuronal progenitor reaches the ONL (where it is now called a rod precursor), it continues dividing or begins to differentiate, expressing rod specification genes (neurogenic differentiation (*neurod*), nuclear receptor subfamily 2e3 *nr2e3, crx,* and *rx1*) and giving rise only to rod photoreceptors (**Figure 5(f)**, green).

Upon light-induced retinal damage (or other forms of retinal insult), Müller glia hypertrophy and transiently increase their expression of Glul and Gfap (**Figure 5(h)**). If the number of dying rods and cones (**Figure 5(h)**, red) is sufficiently large, the Müller glial cells increase their expression of the *ascl1a* and *signal transducer* (*stat3* genes (**Figure 5(i)**)) through unclear mechanisms. Electroporation of morpholinos into the retina prior to inducing retinal damage revealed that both *stat3* and *ascl1a* are independently required for the Müller glial proliferation response. The reentry of these Müller glia into the cell cycle is accompanied by the increased expression of the

---

identical as in 4A is repeated here for comparison. (h) Photoreceptors (light blue) begin to undergo apoptosis (red) within 6 h of entering constant intense light treatment (rod precursors have been ignored for simplicity). The damaged photoreceptors signal (arrow) a subset of the Müller glial cells (brown) to hypertrophy, increase expression of *gfap,* and *glul,* and phagocytose the apoptotic rod cell bodies. (i) At 35 h, responding Müller glial cells (green) increase expression of the early response genes, such as *stat3* and *ascl1a,* and the cell cycle regulatory genes, the *cyclins*. Loss of Pcna expression by morpholino-induced knockdown, results in the failure of responding Müller glial cells to proliferate and regenerate the lost rods and cones. Similarly, morpholino-induced knockdown of Stat3 expression significantly reduces the number of proliferating Müller glial cells. (j) Pcna-positive neuronal progenitor cells are clustered around a Müller glial cell and exhibit increased expression of retinal progenitor genes, such as *pax6,* and genes involved in intracellular signaling pathways, such as Notch. (k) Pcna-positive neuronal progenitors display several neuronal and photoreceptor differentiation markers, including *neurod, nr2e3,* and *crx,* as they migrate to the damaged ONL. (l) One month after exiting the constant intense light treatment, regenerated photoreceptors (green) have differentiated and are indistinguishable from the undamaged photoreceptors.

cell cycle regulatory proteins, the cyclins. The Müller glial cell divisions produce neuronal progenitor cells, which continue to proliferate (expressing Pcna and the cyclins) and begin to express the retinal progenitor gene *pax6* and genes in the Notch and Delta neuronal signaling pathways (**Figure 5(j)**). As the neuronal progenitor cells migrate toward the ONL, they begin to express genes that are required for the commitment and differentiation of rod and cone photoreceptors (*neuroD, nr2e3,* and *crx*; **Figure 5(k)**). Within 28 days of ending the constant intense light treatment, the rod and cone photoreceptors have regenerated (**Figure 5(l)**).

Vestigial Retinal Regeneration Activity in Mammals

The finding that the Müller glial cell acts as an adult neuronal stem cell in the regeneration of the damaged zebrafish retina suggests that it potentially could produce a similar regeneration response in the damaged mammalian retina. Recent studies support the hypothesis that mammalian Müller glia possess some of the features of adult neuronal stem cells. Transplanted rodent Müller glia into a damaged retina will produce rhodopsin-positive cells. As expected, intraocular injection of neurotoxic doses of *N*-methyl-D-aspartic acid (NMDA), which is a synthetic amino acid that binds a glutamate receptor, results in the death of many retinal cell types. However, NMDA damage in the rat retina is followed by a small number of Müller glia proceeding through one cell division to yield a limited number of cells that differentiate into photoreceptors and bipolar cells. This Müller glial cell proliferation and regeneration response can be slightly enhanced with the addition of various growth factors. Injection of the toxin *N*-methyl-*N*-nitrosourea (NMU) into the adult rat eye, which specifically kills photoreceptors, induces Müller glial hypertrophy and increased expression of GFAP and the neural stem cell marker, nestin, followed by the proliferation of some Müller glia. While BrdU labeling confirmed that the Müller glia actively divided in response to the NMU damage, only 42% of the BrdU-labeled cells remained 2 weeks after the NMU injection. Of these surviving BrdU-labeled cells, which were all located in the INL, 58% expressed glutamine synthetase (suggesting they corresponded to Müller glial cells that initially divided) and only 8% expressed rhodopsin. Thus, there were very few rhodopsin-positive cells produced by this Müller cell division and the ones that were generated, failed to properly migrate to the ONL. Expression or addition of various growth factors increased the number of proliferating Müller glia only to a small extent. Activation of various signaling pathways, such as Notch, similarly resulted in only a slight increase in the number of Müller glial-derived neuronal progenitors in the damaged rodent retina. While retinal damage induced some mammalian Müller glia to proliferate and various growth factors or

signaling pathways slightly increased that number, there remained an insufficient number of proliferating Müller glial cells and neuronal progenitor cells to properly regenerate all the neurons lost from the retinal damage. Analyzing the mechanisms underlying the robust Müller glial proliferation response in the light-damaged zebrafish retina will provide important clues as to why the regenerative capacity of the mammalian retina is so limited.

Similarities and Differences of Development with Retinal Ontogeny/Genesis

As discussed above, Müller glia and the neuronal progenitor cells respond to retinal damage by increasing the expression of many of the retinal progenitor genes, such as *pax6*, that are expressed during embryogenesis and near the CMZ throughout the life of the fish. The inability of the mammalian Müller glial cells to mount a sufficiently robust proliferative and regenerative response may be due to differences in the signals that stimulate the Müller glia to respond. All attempts to stimulate sufficient levels of Müller glial proliferation in the mammalian retina have met with very limited success. Conversely, the failure to robustly proliferate may be due to intrinsic differences between the mammalian and zebrafish Müller glial cells. For example, transplanted mammalian Müller glial cells into a damaged retina will differentiate into rhodopsin-positive cells that are predominantly restricted to the INL, rather than repopulating the ONL. Thus, a further understanding of the signals that stimulate the zebrafish Müller glial and neuronal progenitor cells to proliferate and migrate may reveal insights into how to better manipulate the mammalian retina, and by extension, potentiate human retinal regeneration.

See also: Color Blindness: Inherited; Injury and Repair: Light Damage; Retinal Histogenesis.

Further Reading

Bailey, T. J., El-Hodiri, H., Zhang, L., et al. (2004). Regulation of vertebrate eye development by Rx genes. *International Journal of Developmental Biology* 48: 761–770.

Bernardos, R. L., Barthel, L. K., Meyers, J. R., and Raymond, P. A. (2007). Late-stage neuronal progenitors in the retina are radial Muller glia that function as retinal stem cells. *Journal of Neuroscience* 27: 7028–7040.

Fausett, B. V. and Goldman, D. (2006). A role for alpha1 tubulin-expressing Muller glia in regeneration of the injured zebrafish retina. *Journal of Neuroscience* 26: 6303–6313.

Fimbel, S. M., Montgomery, J. E., Burket, C. T., and Hyde, D. R. (2007). Regeneration of inner retinal neurons after intravitreal injection of ouabain in zebrafish. *Journal of Neurosciences* 27: 1712–1724.

Kassen, S. C., Ramanan, V., Montgomery, J. E., et al. (2007). Time course analysis of gene expression during light-induced photoreceptor cell death and regeneration in albino zebrafish. *Developmental Neurobiology* 67: 1009–1031.

Malicki, J. (2000). Harnessing the power of forward genetics – analysis of neuronal diversity and patterning in the zebrafish retina. *Trends in Neurosciences* 23: 531–541.

Morris, A. C., Scholz, T. L., Brockerhoff, S. E., and Fadool, J. M. (2008). Genetic dissection reveals two separate pathways for rod and cone regeneration in the teleost retina. *Developmental Neurobiology* 68: 605–619.

Morris, A. C., Schroeter, E. H., Bilotta, J., Wong, R. O., and Fadool, J. M. (2005). Cone survival despite rod degeneration in XOPS-mCFP transgenic zebrafish. *Investigative Ophthalmology and Visual Science* 46: 4762–4771.

Otteson, D. C. and Hitchcock, P. F. (2003). Stem cells in the teleost retina: Persistent neurogenesis and injury-induced regeneration. *Vision Research* 43: 927–936.

Raymond, P. A., Barthel, L. K., Bernardos, R. L., and Perkowski, J. J. (2006). Molecular characterization of retinal stem cells and their niches in adult zebrafish. *BMC Developmental Biology* 6: 36.

Raymond, P. A. and Hitchcock, P. F. (2000). How the neural retina regenerates. *Results and Problems in Cell Differentiation* 31: 197–218.

Thummel, R., Kassen, S. C., Enright, J. M., et al. (2008). Characterization of neuronal progenitors in zebrafish adult retinal regeneration. *Experimental Eye Research* 87: 433–444.

Thummel, R., Kassen, S. C., Montgomery, J. E., Enright, J. M., and Hyde, D. R. (2007). Inhibition of Muller glial cell division blocks regeneration of the light-damaged zebrafish retina. *Developmental Neurobiology* 68: 392–408.

Vihtelic, T. S. and Hyde, D. R. (2000). Light-induced rod and cone cell death and regeneration in the adult albino zebrafish (*Danio rerio*) retina. *Journal of Neurobiology* 44: 289–307.

Vihtelic, T. S., Soverly, J. E., Kassen, S. C., and Hyde, D. R. (2006). Retinal regional differences in photoreceptor cell death and regeneration in light-lesioned albino zebrafish. *Experimental Eye Research* 82: 558–575.

Yurco, P. and Cameron, D. A. (2005). Responses of Muller glia to retinal injury in adult zebrafish. *Vision Research* 45: 991–1002.

Subject Index

Note: Page entries followed by an 'f' refer to Figures; those followed by 't' refer to Tables.

N

Printed and bound by CPI Group (UK) Ltd, Croydon, CR0 4YY

08/05/2025

01865034-0003